11th EDITION

THE
WORLD
GUIDE

GLOBAL REFERENCE
COUNTRY BY COUNTRY

The World Guide 11th Edition

New Internationalist Publications Ltd has an exclusive license to publish and distribute the English language print and CD-ROM editions of The World Guide throughout the world. Whilst every care has been taken in preparing this edition, the publisher and distributors make no representation, express or implied, with regard to the accuracy of the information contained in this book and cannot accept any legal responsibility or liability for any errors or omissions, or for any action taken or not taken as result of its content.

Third World Guide (English language edition) 1986, 1988, 1990, 1992.

The World Guide (English language edition) 1995, 1997, 1999, 2001, 2003, 2005, 2007.

Copyright © 2007 by Instituto del Tercer Mundo

A catalogue record for this book is available from the British Library.
A catalogue record for this book is available from the Library of Congress.

New Internationalist™ Publications Ltd,

Registered Office: 55 Rectory Road,

Oxford OX4 1BW United Kingdom.

www.newint.org

Printed on recycled paper by C&C Offset Printing Co. Ltd., China, who hold the environmental accreditation ISO 14001

General index
World Guide 11th edition

How to read The World Guide

SECTION I: Global issues

Theme

Article Summary

Article Text

Table
Statistics
Figures

SECTION II: Countries

History

A range of sources has been used to update the texts on each country. In many cases this meant checking with local sources and contributors, documentation centers linked to electronic mail networks and grass-roots organizations throughout the world. This input went into our database in Montevideo, where the final editing was done.

Efforts were made to avoid the most frequent bias of Western reference books, such as to appearing to make history start with the arrival of the Europeans (particularly in the case of African and Latin American countries) or ignoring the role of women.

The last overall updating of the database before printing was done in June 2006, but in several cases events as late as October were included. ▪

Profile

ENVIRONMENT

Estimates of area are based on UN official estimates according to internationally recognized borders; these includes inland waters but not territorial ocean waters. Except where otherwise specified, territories claimed by certain countries but not under their effective jurisdiction are not taken into account, though this implies no judgment as to the validity of the claims.

SOCIETY

Peoples, Languages and Religions: Very few countries keep official ethnic and religious data, and UN-related institutions definitely do not do so. The ethnic and religious make-up of a society is a historical-cultural factor which is in constant change. Certain forces promote tribal or ethnic divisions to favor their own plans of domination, while others favor whatever makes for integration and national unity. In many countries the political and social problems cannot be grasped without reference to these factors.

Main Political Parties and Social Organizations: It is practically impossible to make a complete listing of all parties and other movements of any country, since they are permanently changing and in most cases would add up to several hundred names. Thus only major organizations are mentioned, even though 'major' is subjective when votes, parliamentary representation or number of members cannot be verified, where parties are outlawed or the right to associate is restricted.

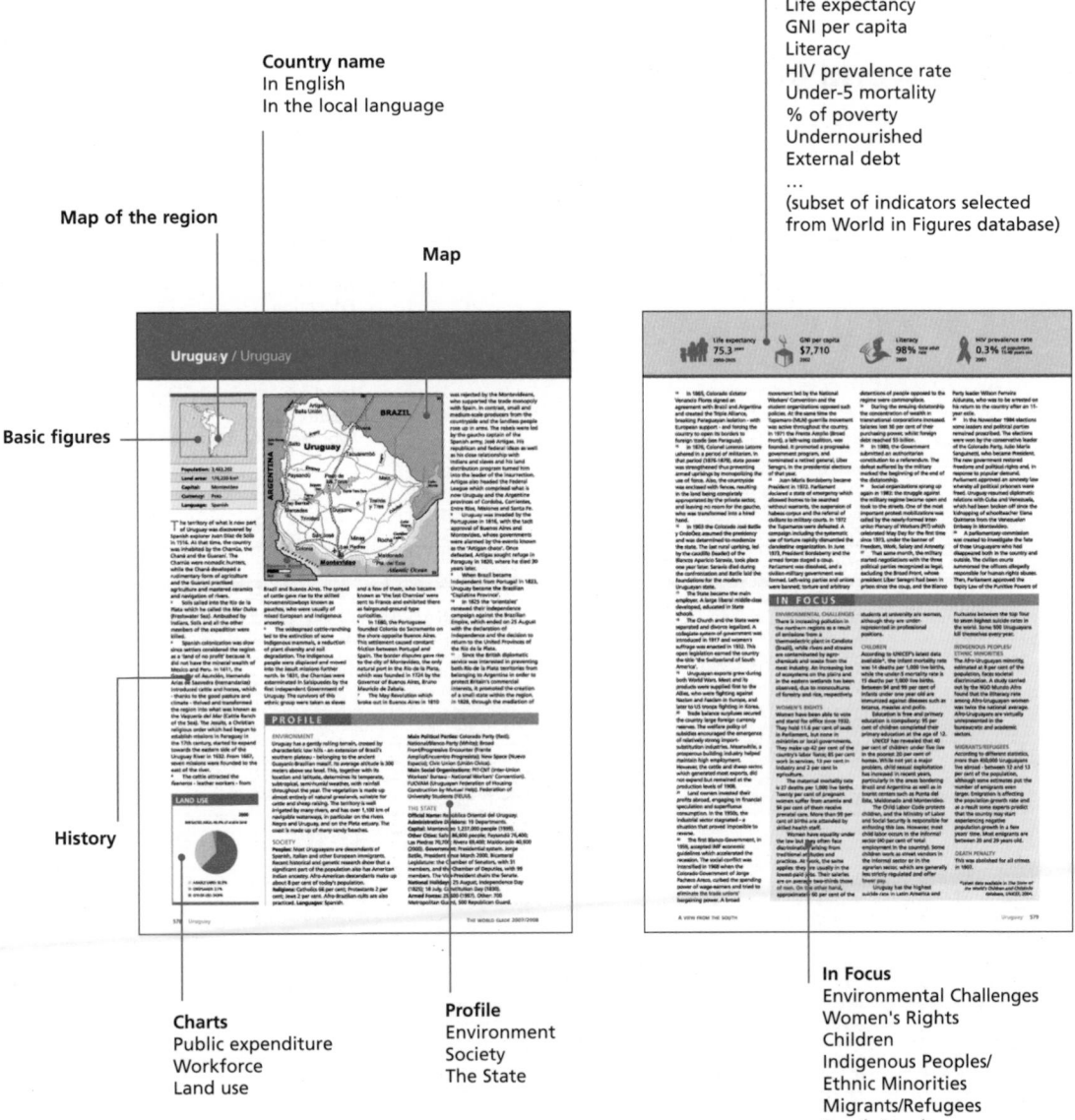

Map of the region

Country name
In English
In the local language

Facts
Life expectancy
GNI per capita
Literacy
HIV prevalence rate
Under-5 mortality
% of poverty
Undernourished
External debt
...

(subset of indicators selected
from World in Figures database)

Map

Basic figures

History

Charts
Public expenditure
Workforce
Land use

Profile
Environment
Society
The State

In Focus
Environmental Challenges
Women's Rights
Children
Indigenous Peoples/
Ethnic Minorities
Migrants/Refugees
Death Penalty

THE STATE

Official Name: Complete name of the state in the official language.

Administrative divisions, the capital and other cities. Population figures are for the most recent available year. The legal limits of a city frequently do not match its real borders. Thus the information may refer only to the core area, which is a minor part of the whole city. **Government:** Names of the major authorities and institutional bodies as of June 2006.
National Holiday: When more than one holiday is commemorated, Independence Day (if it is a holiday) is indicated.
Armed Forces and Others: The total number of personnel for the year indicated. 'Other' forces are those trained and equipped beyond the level of a Police Force (though some fulfill this role) and whose constitution and control means they can be used as regular troops.

In focus

ENVIRONMENTAL CHALLENGES

The main issues compromising sustainable development of the country's land resources.

WOMEN'S RIGHTS

Respect for women's rights; gender discrimination; political and economic involvement; access to the labor market; education and health services.

CHILDREN

Their health and education. Statistics and data on infant mortality, domestic violence, HIV/AIDS, child labor, exploitation and trafficking.

INDIGENOUS PEOPLES / ETHNIC MINORITIES

The current situation of indigenous peoples and/or of minorities who are displaced, discriminated against or at risk.

MIGRANTS/REFUGEES

Recent emigration from and immigration to the country. The situation of refugees and asylum seekers in the country - and of local citizens who have sought refuge in other nations.

DEATH PENALTY

Information on whether or not capital punishment applies and under what circumstances. The date of the last execution.

SECTION III: The world in figures

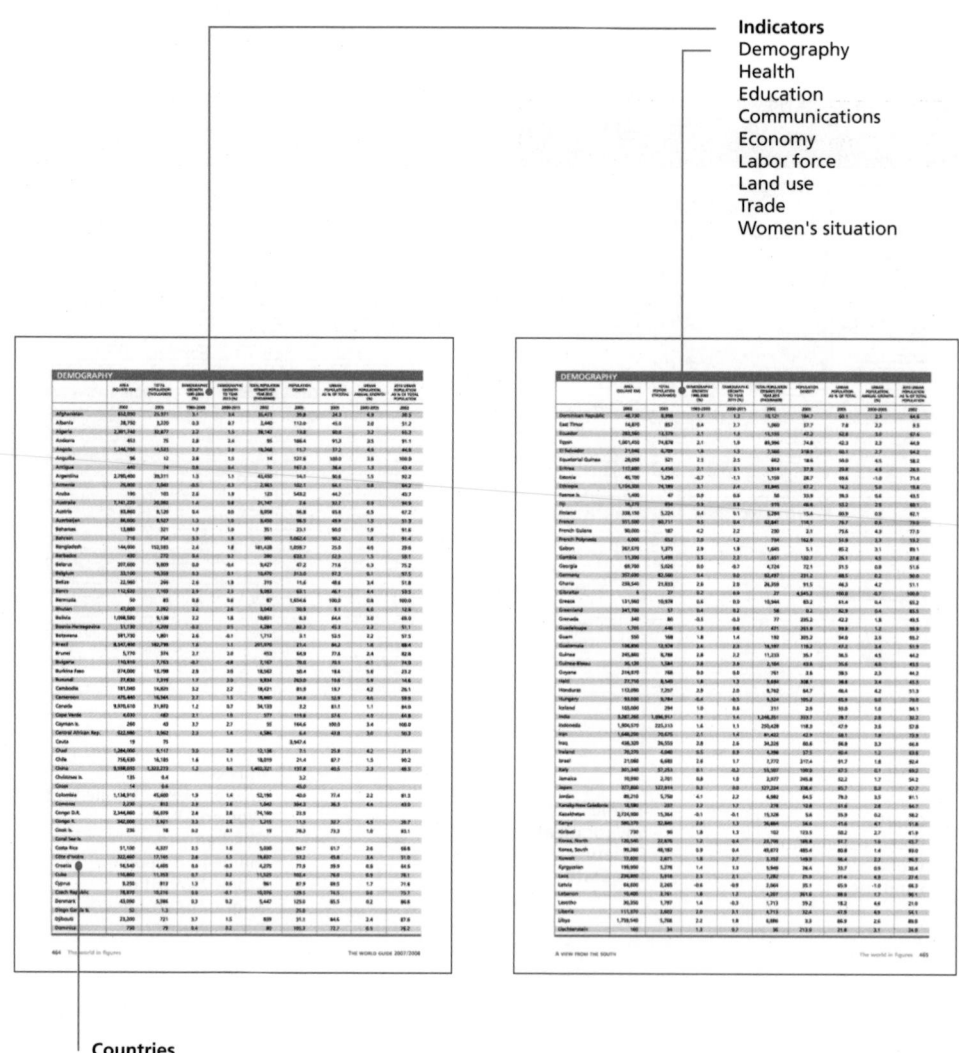

Indicators
Demography
Health
Education
Communications
Economy
Labor force
Land use
Trade
Women's situation

Countries

Statistics database

HEALTH

Life expectancy at birth
Indicates the number of years a newborn infant, male or female, would live if prevailing patterns of mortality at the time of birth were to stay the same throughout child's life (2000-2005). SOURCE: 2

Total fertility rate (children per woman)
Represents the number of children that would be born per woman were she to live to the end of her child-bearing years and bear children at each age in accordance with prevailing age-specific fertility rates (2002). SOURCE: 3

Crude birth rate
Indicates the number of live births occurring during the year, per 1,000 population estimated at midyear (2000-2005). SOURCE: 2

Crude death rate
Indicates the number of deaths occurring during the year, per 1,000 population estimated at midyear (2000-2005). SOURCE: 2

Contraceptive use
Contraceptive prevalence rate is the percentage of women who are practicing, or whose sexual partners are practicing, any form of contraception. It is usually measured for married women ages 15-49 only (1995-2002). SOURCE: 3

Maternal mortality (per 100,000 live births)
Maternal mortality ratio refers to the number of female deaths that occur during pregnancy and childbirth per 100,000 live births. Due to changes in the model of estimation, 1995 and 2000 data are not comparable. The data are official estimates from administrative records, survey-based indirect estimates, or estimates derived from a demographic model developed by the World Health Organisation (WHO) and the United Nations Children's Fund (UNICEF) (2000). SOURCE: 3

Births attended by trained health personnel
Percentage of deliveries attended by personnel trained to give the necessary supervision, care, and advice to women during pregnancy, labor, and the postpartum period, to conduct deliveries on their own, and to care for the newborns (1995-2002). SOURCE: 3

Infant mortality rate
Number of infants dying before reaching one year of age, per 1,000 live births in a given year (2002). SOURCE: 3

Under-5 mortality rate
Number of infants dying before reaching five years of age, per 1,000 live births in a given year (2002). SOURCE: 3

Low weight at birth
Newborns weighing less than 2,500 grams, with measurement taken within the first hours of life, before significant postnatal weight loss has occurred (1998-2002). SOURCE: 3

Child malnutrition
Prevalence of child malnutrition (weight for age) is the % of children under five whose weight for age is less than minus two standard deviations from the median for the international reference population ages 0 to 59 months (1995-2002). SOURCE: 3

Udernourished people
Undernourishment is the result of food intake that is insufficient to meet dietary energy requirements continuously. The World Health Organization recommended that the average person needs to take a minimum of 2,300 Kcal per day to maintain body functions, heath and normal activity. This global minimum requirement of calories is broken down into country-specific differentials that are a function of the age-specific structure and body mass of the population (1998-2000). SOURCE: 5

Breastfeeding
Exclusive Breastfeeding Rate (% of under-6-months children) (1995-2002). SOURCE: 3

Calorie consumption
Daily per capita calorie supply. Shown as a national average, though a country's income distribution may create a wide gap between the average, the highest and the lowest strata. Minimum calorie requirements vary in different countries, depending on climate and nature of the main activities (2001). SOURCE: 7

Doctors per 100,000 people
Traditional medicine and community health care practiced by not officially recognized health personnel are not included in these statistics, although they may be the only health service available for the majority of the population in many countries (1990-2002). SOURCE: 4

Nurses per 100,000 people
All persons who have completed a program of basic nursing education and are qualified an registered or authorized by the country's authorities to provide responsible and competent service for the promotion of health, prevention of illness, care of the sick and rehabilitation (1997). SOURCE: 5

Access to improved water sources
The United Nations includes the percentage of population with 'reasonable' access to safe water sources. They include treated surface waters and untreated but uncontaminated water from springs, wells and protected boreholes in the 'reasonably safe water' category (2000). SOURCE: 3

Access to sanitation services
Percentage of the population with at least adequate excreta disposal facilities (private or shared, but not public) (2000). SOURCE: 3

EDUCATION

Literacy
Indicates the estimated percentage of people, male or female, over the age of fifteen who can read and write (2000). SOURCE: 3

School net enrolment ratio
Primary / secondary net school enrolment ratio. Number of people, male or female, (of the age group officially corresponding to primary /secondary school level) enrolled in primary / secondary school level, divided by the population of the age group officially corresponding to that level (2000). SOURCE: 1

Tertiary gross enrolment ratio
Ratio of total enrolment, regardless of age, to the population of the age group that officially corresponds to the level of education shown. Tertiary education, whether or not to an advanced research qualification, normally requires, as a minimum condition of admission, successful completion of education at secondary level (1997). SOURCE: 1

Primary pupil to teacher ratio
Primary school teachers / students ratio (students per teacher) (2000). SOURCE: 1

COMMUNICATIONS

Mass media
Estimation of the print run of daily newspapers (1997/2001), he number of working radio receivers (1997/2001) and TV sets (2001) per 1,000 people. The latter figure may rely on the number of licences granted or the number of declared receivers. SOURCE: 1

Telephones
Telephone mainlines per 1,000 people (2001). SOURCE: 1

Computers
Personal computers per 1,000 people (2001). SOURCE: 1

ECONOMY

International Poverty Line
Percentage of the population living on less than $1.08 a day at 1993 international prices (equivalent to $1 in 1985 prices, adjusted for purchasing power parity).

GNI per capita
Gross national income, converted to US dollars using the World Bank Atlas method, divided by the midyear population. GNI is the sum of value added by all resident producers plus any product taxes (less subsidies) not included in the valuation of output plus net receipts of primary income (compensation of employees and property income) from abroad (2002). SOURCE: 1

GDP per capita
GDP (PPP, current $) is the value of the total production of goods and services of a country's economy within the national territory. GNP is GDP plus the income received from abroad by residents in the country (such as remittances from migrant workers and income from investments abroad), minus income obtained in the domestic economy which go into the hands of persons abroad (such as profit remittances of foreign companies) (2002). SOURCE: 1

GDP annual growth
Annual growth rate (%) of GDP per capita based on constant local currency (2002). SOURCE: 1

Annual inflation rate
Annual inflation rate as GDP implicit deflator (average annual $ growth), measures the average annual rate of price change in the economy (2002). SOURCE: 1

Consumer price index
Reflects changes in the cost to the average consumer of acquiring a fixed basket of goods and services (1995 = 100) (2002). SOURCE: 1

External debt
Public state guaranteed and private foreign debt (million $) accumulated by 2001. Per capita debt ($) was calculated from total external debt and total population (2001). SOURCE: 1

Debt service
Debt service as % of exports of goods and services. The service of a foreign debt is the sum of interest payments and repayment of principal (capital loaned, regardless of

yield). The relation between debt service and exports of goods and services is a practical measurement commonly used to evaluate capacity to pay the debt or obtain new credits. These coefficients do not include private foreign debts without state guarantees - a considerable amount in some countries (2001). SOURCE: **1**

Official development assitance
ODA consists of money flows from official governmental or international institutions for the purpose of promoting economic development or social welfare in developing countries. These funds are supplied in the form of grants or 'soft' loans, i.e. long-maturity loans at interest rates lower than those prevailing on the international market. We included the total and per capita net ODA received in dollars and as % of receptor country GDP; and total net ODA disbursed by the countries in dollars and as % of donors GDP (2001). SOURCE: **4**

Energy use consumption (oil equivalent) per capita (kg)
These statistics refer only to commercial energy, and do not include, for example, that which rural people in poor countries produce by their own means (mainly firewood). Energy consumption of the country is measured in kilograms of 'oil-equivalent' per capita (1.000 kWh electrics = 0.222 million tep -centrales thermiques classiques) Energy imports are given as a percentage of energy consumption. Figures are negative in those countries that are net exporters of energy products (2000). SOURCE: **1**

Public expenditure
Defense expenditure (2001), health services (2000) and education services (2001) as % of GDP (2001). SOURCE: **1**

TRADE
Imports and exports
Annual value in US dollars f.o.b. (free on board) for *exports* and c.i.f. (costs, insurance and freight) for *imports* (2002). SOURCE: **1**

Cereal imports
The cereals are wheat, flour, rice, unprocessed grains and the cereal components of combined foods. This figure includes cereals donated by other countries, and those distributed by international agencies (2002). SOURCE: **7**

Food production and imports
Food production per cápita index (1981-91=100).
Imported food in relation to the food available for internal distribution. This means the total of food production, plus food imports, minus food exports (2001). SOURCE: **1**

Weapons imports and exports
Imports and exports of conventional weapons ($ million, 1990 prices) (2002). SOURCE: **4**

LABOR FORCE
Labor force as % of total population
Total labor force comprises people who meet the International Labour Organization definition of the economically active population: all people who supply labor for the production of goods and services during a specified period. It includes both the employed and the unemployed. While national practices vary in the treatment of such groups as the armed forces and seasonal or part-time workers, in general the labor force includes the armed forces, the unemployed, and first-time job-seekers, but excludes homemakers and other unpaid caregivers and workers in the informal sector (2002). SOURCE: **1**

Unemployment rate
Unemployment refers to the share of the labor force that is without work but available for and seeking employment. Definitions of labor force and unemployment differ by country (2002). SOURCE: **6**

Employment and unemployment
The male and female labour force percentage (2002), excluding homemakers and other unpaid workers; Employment rate by sex and activity sector (agriculture, industry and services, years). Unemployment rate (1995/2002). SOURCE: **1, 4**

LAND USE
Land use
Forest and woodland as percentage of land area (2000). Arable land as percentage of land area (2000). Irrigated area as percentage of arable land area (2000). Fertilizer use (kgs per ha) (2000). SOURCE: **1**

Fertilizer use
Refers to purchases of nitrate, potassium and phosphate based fertilizers used on arable land (1999). (kgs per ha) (2000). SOURCE: **1**

WOMEN'S SITUATION
% of women professionals and technicians; % of women legislators, senior officials and managers; Earned income shared (% to women), % of ministerial posts occupied by women, % of parliamentary seats occupied by women. SOURCE: **4**

Sources

1   World Development Indicators 2006, World Bank

2   World Population Prospects - The 2004 Revision. United Nations

3   The State of the World's Children 2006, UNICEF

4   Human Development Report 2005, UNDP

5   WHOSIS - WHO Statistical Information System, Web Site WHO 2006

6   LABORSTA database, ILO Web Site

7   FAOSTAT - Statistical Database - FAO Web site 2006

SECTION I: Global issues

I. ONE PLANET, TWO WORLDS

NORTH AND SOUTH SEEM TO SPEAK DIFFERENT LANGUAGES. THERE IS A COMMUNICATION GAP, RATHER THAN AN INFORMATION GAP. THE MORE WE HEAR ABOUT THE GLOBALIZED WORLD – AN EMPORIUM OF COMMON INTERESTS BOOSTED BY INFORMATION AND COMMUNICATION TECHNOLOGIES – THE MORE WE PERCEIVE A FUNDAMENTAL DISCONNECTION BETWEEN REAL NEEDS AND CURRENT PRIORITIES.

19

Who's aided by aid?

Aid is more likely to serve the needs of wealthy countries than those of the poorer nations receiving it. Some critics maintain that development co-operation as a whole, far from reducing the gap between rich and poor, has tended to increase it.

(Box) **Genetically modified aid.** 19
(Box) **Aid for oil companies.** 20
(Table) **Debt in figures.** 21

22

North and South in the Information Society.

According to the World Summit on the Information Society declaration of principles, the Information Society has to be people-centered, inclusive and development-oriented. Nevertheless, effective solutions have not yet been implemented to deal with the main problem areas identified at the Summit: internet governance, the digital divide and intellectual property rights.

(Box) **Internet: predominantly for men.** 22
(Table) **Communications in figures.** 23

II. STATE SECURITY VS. HUMAN SECURITY

A SENSE OF INSECURITY AND FEAR IS GAINING GROUND. SECURITY, A GOAL SHARED BY STATES AND INDIVIDUALS, HAS BECOME THE SOURCE OF POLICIES THAT EITHER FEED OUR FEAR OR THROW MILLIONS OF PEOPLE INTO TOTAL INSECURITY.

25

State of insecurity.

'Human security' should be ensured by states. However, states often face difficulties when it comes to deciding to what extent they offer their guarantees. Terrorist attacks and natural disasters challenge the state-individual relationship and show that 'human security' is one thing while 'state security' is something very different.

26

Resorting to 'just war'.

Those involved in a war tend to regard their own fight as legitimate. Whether they are defending themselves from attack, or launching an attack on another, they uphold the idea that justice endorses their war. The 'just war' doctrine, among other things, underlies 'pre-emptive' or other types of wars launched by the US Government, which tend to threaten or affect Islamic countries.

28

The rule of fear.

The world is increasingly perceived as an insecure, dangerous and menacing place. The global ideology of fear seems to have become the basis for all analysis.

(Table) **Sexual minorities and the law: a world survey.** 29

III. LATIN AMERICA, A NEW TWIST

NEW ACTORS EMERGE ON THE LATIN-AMERICAN POLITICAL STAGE. SELF-PROCLAIMED LEFTIST PARTIES HAVE COME TO POWER BY ELECTORAL MEANS IN SOME COUNTRIES. INDIGENOUS LEADERS REACH POSITIONS OF POWER AT NATIONAL LEVEL. AT FIRST SIGHT, CHANGES ARE APPARENT. BUT ARE THEY FOR REAL?

IV. ENERGY SCENARIOS

SO FAR, NO EFFICIENT AND PROFITABLE ALTERNATIVES HAVE BEEN DEVELOPED TO REPLACE NON-RENEWABLE ENERGY SOURCES. WITHIN THE CURRENT MODEL, POSSIBILITIES RANGE FROM FURTHER DEVELOPMENT OF TECHNOLOGIES THAT OBTAIN ENERGY FROM OTHER SOURCES - SUCH AS NUCLEAR FISSION OR WIND - TO A NEVER-ENDING SERIES OF WARS AND CONFLICTS OVER THE CONTROL OF EXISTING RESOURCES. BUT THE REAL SOLUTION COULD LIE IN A PARADIGM SHIFT.

V. GENDER

ALL OVER THE PLANET, WHATEVER THE LEVEL OF THEIR COUNTRY'S DEVELOPMENT, WOMEN ARE AT A DISADVANTAGE. THEY ARE WORSE PAID - IF THEY ARE PAID AT ALL - THAN MEN FOR DOING THE SAME JOB AND THEY ARE LESS REPRESENTED IN DECISION-MAKING POSITIONS AND PROCESSES. IN SOME REGIONS, JUST BECAUSE OF THEIR GENDER, THEIR VERY LIVES ARE AT RISK.

41

Missing women.

Women and girls are not as highly valued as men in many societies. UN estimates from 2006 say that between 113 and 200 million women are 'missing'. Every year between 1.5 and three million of them lose their lives as a result of gender-based violence or neglect. There are several explanations: in some countries (China is an example), boys are preferred to girls. This can lead to female infanticide or selective abortion. Those that survive may suffer the neglect of their parents, because food and healthcare are prioritized for the male members of the family.

(Box) The murdered women of Latin America. 42

42

The GEI and the gender gap.

No country treats its women and men equally; no country has reached gender equity. However, in the last two decades there has been a change, since many countries have acknowledged gender inequity as a problem. This new reality calls for the creation of new tools to measure this multi-faceted concept. One of them is the Gender Equity Index developed by Social Watch.

(Table) Table 1. GEI 2006. 43
(Table) Table 2. Income gap (women/men) by geographic region. 43
(Map) Gender equity index. 44

44

Female genital mutilation and Islam.

Nearly 130 million women in about 30 countries have undergone female genital mutilation or cutting, performed for various 'traditional' or 'cultural' reasons. Each year, some two million girls are subjected to these practices. The custom is often associated with Islam, yet, in 2006, representatives of 50 Muslim states issued a declaration condemning female genital mutilation and emphasizing that it is contrary to Islamic principles.

VI. ECONOMY: HEADS AND TAILS

IF MAINTAINED, CURRENT TRENDS IN ECONOMIC POWER AND HUMAN DEVELOPMENT WILL RESULT
IN AN EVER-WIDENING GAP BETWEEN THEM. THE SHIFT OF CAPITAL TOWARDS SOME COUNTRIES IN THE
SOUTH (INDIA AND BRAZIL AMONG THEM) WILL NOT TRANSLATE INTO BETTER INCOME DISTRIBUTION OR
LIVING CONDITIONS, NOR INTO THE SUSTAINABLE EXPLOITATION OF NATURAL RESOURCES. ALTERNATIVES
LIKE THE GREEN ECONOMY SEEM TO ARISE IN THE GAP BETWEEN CAPITAL AND THE REAL WORLD.

46
China, India and the rotating axis of capital.
Economic forecasts suggest a different world order in the making, which in the next half-century will see emerging economies at the top. Only the US and Japan will remain among the six major powers, which will be led by China and India. The US will no longer be 'the world's locomotive'; the new economic train will be pulled by the 'poor giants'.

50
Sukuk and Islamic finance.
In recent years there has been a significant increase in the use of Islamic financial procedures compatible with the requirements of Sharia, a legal code based on the Qur'an. The idea of finances compliant with Sharia - and parallel to the rules of capitalism - emerged in Egypt at the beginning of the 1960s and later spread to Saudi Arabia and the United Arab Emirates. Sukuk certificates have provided a way of operating within the dual framework of Islam and international financial capital. Sukuk market was launched in 2002 and by 2006 it had a turnover of around $500,000 million and was growing at an annual rate of 15%. Among other financial prescriptions, Sharia forbids the payment or charging of interest (*riba*, which more precisely means excess or addition) and requires stated fixed value and a benefit commensurate with risk (*gharar*) and liquidity.

52
Prisons: Workforce behind bars.
Prisons are institutions associated with the development of capitalism, particularly in its industrial phase. Throughout history, governments have not ceased in their efforts to make prisoners produce wealth, either through forced or voluntary labor or, as in recent times, through the privatization of prison services.

53
Green economics: a new model.
A green economy would not only consider financial matters but would encompass social, environmental and even spiritual concerns and would prioritize qualitative rather than quantitative growth. It would seek to evaluate the real worth of products and services rather than simply giving them an exchange or monetary value. It would finally make visible many of the important tasks performed outside the formal economy while also factoring in any damage done to the earth's natural systems.

55
Ecological debt: who owes whom?
The concept of ecological debt encompasses the social and environmental impacts suffered mainly by Southern countries as well as the intensive exploitation of natural resources to support production and consumption patterns in the rich world.

58
Petrodollars versus Petroeuros.
The US dollar became the universal currency 70 years ago. The selling pressure on the dollar decisively faded in 1971, when the US suspended the free convertibility of notes into gold at fixed rates of exchange. Since then, the dollar has become linked to the oil trade. With a strong currency in Europe and the decision by some countries to price their oil in euros, US hegemony is at serious risk.

SECTION II: Countries

Index to Special Boxes

SECTION III: The world in figures

Preface

The cauldron

Nasreddin Hodja, a good humored teacher who lived in the Middle East almost a thousand years ago, had borrowed a cauldron from his neighbour. When he didn't return it for a long time the neighbour came knocking on the door.

— `Hodja Effendi, if you are finished with the cauldron could I take it back? The wife needs it today.´

When Hodja came back with the cauldron, the neighbour noticed that there was a small pot in it:

— `What is this?´

— `Well, neighbour, congratulations, your cauldron gave birth to a baby pot.´ said Hodja.

The neighbour, delighted, thanked Hodja, took his cauldron and the new little pot, and went home.

A few weeks after, Hodja borrowed the cauldron again. And once more it was taking him forever to return it back. The neighbour had no choice but to go asking for it again.

— `Ahh…´ bemoaned Hodja, `I am afraid your cauldron is dead.´

— `Hodja Effendi, that's not possible, a cauldron cannot die!´ exclaimed the disbelieving neighbour.

— `My dear fellow, if you accepted it can give birth, why can't you believe that it can also die?´

Aren't we all a little like Hodja's neighbor? We are bombarded every day by messages from demagogues and advertisers keen to sell us a line that we may anyhow want to believe. Yet they know better than Hodja and never confront us with the sour logical consequence of our own credulity. In the modern media, to please the public, cauldrons always give birth but they never die.

At the World Guide we respect our readers, the more so because so many of them are students. They are not ready to believe in dying cauldrons, but neither do they want to hear about them giving birth to start with. This is a reference book, and its main purpose is to provide facts and details. But we choose to call it a "Guide", because we want to offer more than a collection of numbers, dates and names - we want to present them in a way that stimulates critical thinking about our complex contemporary world.

Since its beginnings in the mid-1970s, the Guide has departed from ready made explanations, in an explicit attempt to highlight the point of view of what some call the "third world" the and others "Global South": the excluded majorities, the victims of globalization, the unheard voices of the poor, of women, of indigenous peoples.

The editorial team of the Third World Institute led by Amir Hamed has worked with dedication and pride over the last two years to keep this tradition and at the same time renew the product in order to make of this new edition of the World Guide not just an update of the previous one, but a completely reformulated tool, with new sections, more graphs, layout that is easier to read and the highest standards of accuracy.

After all, as my grandmother used to teach, you never return your neighbor an empty cauldron… and in that spirit we offer the reader food for thought with the best of our local flavours.

Roberto Bissio
(Montevideo, 2006)

THE WORLD GUIDE

INSTITUTO DEL TERCER MUNDO

Executive Board

President:
Clara Píriz

Secretary:
Luis Álvarez

Director:
Roberto Bissio

THE WORLD GUIDE

Editor:
Amir Hamed

Editorial Co-ordinator:
Ana Zeballos

Research and Editing:
Gustavo Alzugaray
Joaquín Olivera
Andrea Tutté

Collaborators:
Gustavo Espinosa
Sandra López Desivo
Carlos Rehermann
Lucy Gray Donald

Statistics:
Social Watch Social Sciences
Research Team
Karina Batthyány (Coordinator)
Mariana Sol Cabrera
Graciela Dede
Daniel Macadar
Ignacio Pardo

Translators:
Álvaro Queiruga
Patricia Draper
Liliana Battipede
David Reed
Valentina Vidal
Lori Nordstrom
Ana Mesa

Proofreading:
Susana Ibarburu
María Laura Mazza

Design and layout:
MONOCROMO
Graphic design: Valentina Ordoqui,
Pablo Uribe
Layout: Magdalena Sayagués,
Myriam Bustos
Phone: +598(2) 4001685
E-mail: info@monocromo.com.uy

Administration:
María Lucía Rivero
Maika Flores

**NEW INTERNATIONALIST
PUBLICATIONS LTD**
(English language edition)

Editors:
Chris Brazier and Troth Wells

Production:
Fran Harvey and Alan Hughes

Marketing and Distribution:
Jo Lateu and Dan Raymond-Barker

This book is the result of documentation, research, writing, editing and design work done by the Instituto del Tercer Mundo (Third World Institute), a nonprofit institution devoted to information, communication and education, based in Montevideo, Uruguay.

The World Guide 2007/2008 is a fully updated, corrected and expanded edition of *The World Guide 2005/2006*. This reference book was published for the first time in Mexico in 1979, on the initiative of Neiva Moreira, to complement the dissemination work of *The Third World* magazine he had begun publishing. In 1980 the first Portuguese version of the *Guide* was published. The first English version was published in 1984. New Internationalist Publications Ltd, in Oxford, England, has been in charge of the English version and its distribution in English-speaking countries since 1996. In 1999, SERMIS-EMI (Editrice Missionaria Italiana), from Bologna, launched the Italian version of *The World Guide*. Since 1992, *The World Guide* has been published on CD-ROM. Since 1997, through the Internet and together with the updating service, the Spanish version of *The World Guide* has been available in Web Pages.

The World Guide is a cumulative work, building on the efforts of former staff members. We would like to mention the contributions of: Carlos Abín, Mohiuddin Ahmad, Carlos Afonso, Andrés Alsina, Claude Alvares, Iván Alves, Elizabeth Ardans, Juan José Argeriz, Maria Teresa Armas, Marcos Arruda, Iqbal Asaria, Gonzalo Abella, Susana de Avila, Víctor Bacchetta, Edouard Balby, Artur Baptista, Roberto Bardini, Luis Barrios, Cedric Belfrage, Max van den Berg, Alicia Bidegaray, Beatriz Bissio, Ricardo de Bittencourt, Samuel Blixen, Gerardo Bocco, Inés Bortagaray, Lucila Bortagaray, José Bottaro, Dirk-Jan Broertjes, Alberto Brusa, José Cabral, Luis Caldera, Juan Cammá, Altair Campos, Paulo Cannabrava Filho, Carmen Canoura, Cristina Canoura, Gerónimo Cardozo, Diana Cariboni, Gustavo Carrier, Virgilio Caturra, María Fernanda Cortinas, Macário Costa, Susan Day, Anne-Pieter van Dijk, Carlos María Domínguez, Rodrigo Egaña, Roberto Elissalde, Wáshington Estellano, Marta Etcheverrigaray, Carlos Fabião, Marcelo Falca, Helena Falcão, Wilson Fernández, Cecilia Ferraría, Alejandro Flores, Lidia Freitas, Sonia Freitas, Marc Fried, Héctor García, Leo García, José Carlos Gondim, Leo van Grunsven, Goran Hammer, Mario Handler, David Hathaway, Ann Heidenreich, Bill Hinchberger, Etevaldo Hipólito, Mohammed Idris, Leticia Jorge, Dina Yael Kaganovicius, Sytse Kujik, Heraclio Labandera, Peter Lenny, Arne Lindquist, Cecilia Lombardo, Linda Llosa, Geoffrey Lloyd Gilbert, Fernando López, Paulina López Rivero , Hakan Lundgren, Carlos Mañosa, Virginia Martínez, Pablo Mazzini, Daniel Mazzone, Susana Medina, María Eugenia Méndez, Carol Milk, Fernando Molina, Fabián Muro, João Murteira, Abdul Naffey, Claudia Neiva, Ruben Olivera, Hattie Ortega, Pablo Piacentini, Ana Pérez, Christopher Peterson, Virginia Piera, Carlos Pinto Santos, Paola Pintos , Sieni C. Platino, Hans Poelgrom, Artur José Poerner, Graciela Pujol, Roberto Raposo, Felisberto Reigado, Sofi Richero, Ana Ridao, Carmen Ridao, Malva Rodríguez, Maria da Gloria Rodrigues, Julio Rosiello, Ash Narain Roy, Ana Sadetzky, María José Santacreu, Alexandru Savulescu, John Sayer, John Schlanger, Dieter Schonebohm, Gregorio Selser, Irene Selser, Eunice H. Senna, Anthony Shaw, Baptista da Silva, Herbet de Souza, Yesse Jane V. de Souza, José Steinleger, Firiel Suijker, Sjef Theunis, Victoria Swarbrick, Fernanda Trías, Carolina Trujillo, Elena Vasilisky, Pedro Velozo, Horacio Verbitsky, Amelia Villaverde, Anatoli Voronov, Germán Wettstein and Asa Zatz.

The editorial staff acknowledges the valuable contributions made by the libraries of the following institutions: UNDE, UNESCO, CLASH, FE, the UN Information Center in Montevideo, Iepala in Madrid and the UNCLAD in Geneva.

The Third World Network, Penang, Malaysia provided ideas, documents and links with many contributors.

To all of them we offer our thanks. This fully expanded *World Guide 2007/2008* has been benefited with the contributions from Jens Martens and Jonna Schürkes, from the World Economy, Ecology & Development Association (WEED), Germany, and from Fernanda Carvalho and Mauricio Santoro, from the Instituto Brasilero de Análises Sociais e Econômicas (IBASE).

Thanks also to the hundreds of readers who keep volunteering their comments, suggestions and detailed information. The judgements and values contained in this book are the exclusive responsibility of the editors and do not represent the opinion of any of the people or institutions listed, except for the Third World Institute.

INSTITUTO DEL TERCER MUNDO
Juan D. Jackson 1136, Montevideo 11200, Uruguay.
Phone: 598 2 419 6192; Fax: 598 2 411 9222.
E-mail: item@item.org.uy/
Website: http://www.item.org.uy

NEW INTERNATIONALIST™ PUBLICATIONS LTD
55 Rectory Road, Oxford OX4 1BW, United Kingdom
Phone: (44) 1865 811400; Fax: (44) 1865 793152
E-mail: ni@newint.org
Website: http://www.newint.org/

Global issues

Hopes and fears

I. ONE PLANET, TWO WORLDS

North and South seem to speak different languages. There is a communication gap, rather than an information gap. The more we hear about the globalized world – an emporium of common interests boosted by Information and Communication technologies – the more we perceive a fundamental disconnection between real needs and current priorities.

Who's aided by aid?

AID IS MORE LIKELY TO SERVE THE NEEDS OF WEALTHY COUNTRIES THAN THOSE OF THE POORER NATIONS RECEIVING IT. SOME CRITICS MAINTAIN THAT DEVELOPMENT CO-OPERATION AS A WHOLE, FAR FROM REDUCING THE GAP BETWEEN RICH AND POOR, HAS TENDED TO INCREASE IT.

The growing number of armed conflicts and the worsening of natural disasters since the 1990s has made for a considerable increase in humanitarian action - though still almost always insufficient. Until 1994, the concept of security held by the United Nations was directly related to conflicts between nation-states. That year, however, the term 'human security' was for the first time included in the *Human Development Report* published by the United Nations Development Programme (UNDP). This was described by many as the equivalent of a small Copernican revolution. Human beings, and no longer the State, were now at the core of security. For the first time in history there was an awareness that threats to security not only came from wars and criminal violence, but also from hunger, the spread of new pandemics (HIV/AIDS and Ebola, for example), environmental disasters, terrorism, religious and ethnic fundamentalisms, etc.

The traditional response to this is to argue that aid to poor countries should be increased. Yet aid's effectiveness is coming ever more into question.

Official Development Assistance (ODA) is the name given to donations or soft loans granted by rich countries belonging to the Development Assistance Committee (DAC) of the Organization for Economic Co-operation and Development (OECD). A loan granted at a lower interest rate than the market's is considered to be assistance, or aid, even though the receiving country will have to repay the whole amount plus interest.

In addition, such assistance is usually accompanied by a list of conditionalities

that must be enforced. The most common are reducing government deficits, privatizing State services, reducing protectionist measures and ending controls on the flow of capital. These 'conditionalities' are defined by industrialized nations, usually through their controlling role in the World Bank and the IMF.

Critics say that ODA is an inefficient way of allocating funds, full of political interference and vested interests. 'It should disappear and give way to market mechanisms,' proposes Eric

Toussaint, President of the Brussels-based Committee for Cancellation of the Third World Debt. Such critics contend that since aid is subject to the political agenda of Northern nations it has endorsed a development model that has served to increase the North-South, Rich-Poor divide. According to US writer Paul Theroux, Africa does not need indiscriminate economic aid, but should be allowed to take care of itself. In recent decades, Africa has received huge amounts of humanitarian aid - but with far from adequate results. Some analysts argue that the West seeks to maintain

Genetically modified aid

THE UNITED STATES AGENCY FOR International Development (USAID) is Washington's main vehicle lending economic and humanitarian aid to developing countries. However, many activists believe that promoting the interests of US foreign policy is the central goal of that assistance.

In recent years there has been particular concern about US agricultural aid that has spread genetically modified crops throughout the world - to the advantage of US-based transnational corporations.

In 1991, USAID launched the Agricultural Biotechnology Support Project (ABSP). Headed by the University of Michigan, this involved a group of private companies and state think tanks drawing up strategies for the production of transgenic crops. Although ABSP was not commercially successful, it led many scientists to co-

operate with US companies and trained those researchers in the production of transgenic crops. It also helped launch a political campaign on intellectual copyright, among others.

A new initiative took shape in 1998 splitting the original ABSP in two. The Collaborative Agricultural Biotechnology Initiative (CABIO) was made up of two programs, the ABSP II and PBS (Program for Biosecurity Systems). The former was focused on the creation of marketing projects for genetically modified products, while the latter was meant to establish certain 'system' countries that would introduce transgenic crops in the market. The ultimate goal was to harmonize legislation among the regions, creating regional markets for transgenic crops, with uniform regulation processes. ∎

the continent's dependence. Others say it is Africa that is unconcerned with its development, preferring to play the role of a pauper eternally demanding aid and invoking past wrongs.

To many, however, the main problem with aid is rather that there is not enough of it. The UNDP *Human Development Report 2005* maintains, for example, that aid is justified both by social justice and by an interest in collective well-being and security. The report stresses that, without more efficient aid, a large number of countries will lack the financial resources to build the needed social and economic infrastructure they will need if they are to meet the Millennium Development Goals. According to this view: 'well-guided aid accelerates human development'.

Even the UNDP warns, however, that additional aid, by itself, is not enough - the way in which countries are granted funds must be changed. Southern countries need predictable aid, which minimizes transaction costs and maximizes money value. In contrast, most countries receive unpredictable, uncoordinated aid, with conditionalities that do not reflect the dynamics of domestic reform, and tied to purchases in the donor countries. According to the UNDP, with aid distributed as it is at present, developing countries have a hard time planning expenditure on key elements such as teachers' wages or building new infrastructure. Less than 40 per cent of aid to poor countries gets to where it is needed.

'Tied aid remains one of the most egregious abuses of poverty-focused development assistance. By linking development assistance to the provision of supplies and services provided by the donor country, instead of allowing aid recipients to use the open market, aid tying reduces value for money,' according to the *2005 Human Development Report*. 'We conservatively estimate the costs of tied aid for low-income countries at between $5 billion and $7 billion dollars.'

Since the early 1970s, OECD and DAC member countries have pledged to devote 0.7 per cent of their national income to ODA, yet the only countries that have reached and exceeded that target are Denmark, Luxembourg, Norway, Holland and Sweden. Increasing aid to the 0.7 per cent level would have a minimal effect on donor country finances but it would make a major difference to recipient countries.

Aid for oil companies

THE WAR AGAINST TERRORISM, launched and led by Washington after the attacks of 11 September 2001, has exacerbated the trend for ODA to be 'securitized', turning it into another foreign policy tool to reduce the terrorist threat. This has changed aid's priorities. In the late 1990s, for example, only a quarter of the foreign aid supplied by the US was related to security issues. In 2004 that proportion had risen to more than half.

The official DAC line is that development co-operation can be useful in the anti-terrorist fight, although it must aim at prevention and try not to abandon development goals. It must, for example, direct funds towards building stronger and more stable political structures, create employment programs for young people at risk of being absorbed by terrorist groups, and help to improve countries' education systems.

The invasion of Iraq has certainly led to a highly questionable use of aid by both the US and Britain. US aid to the Iraqi people has in effect been assistance to the expansion of US oil and construction companies. In addition, some argue that the cost of the destruction has been effectively transferred to Iraq, with deductions from Iraqi oil income to come once the country has been reconstructed. ∎

A 2005 plan designed by UK finance minister Gordon Brown proposed increasing aid by an additional $25 billion a year until 2010 and by $50 billion more per year between 2010 and 2015. The plan was based on the 'poverty trap' theory. Devised in the 1950s by Polish economist Paul Rosenstein-Rodan, this says that since poor economies suffer many problems at the same time, addressing one without mending the rest will not solve anything; what is needed is a 'big push' in a short period of time.

The Brown plan also seeks to forgive the debt African countries have incurred to multilateral agencies (mainly to the IMF and World Bank) and urges the elimination of agricultural tariffs and subsidies in order to let African products have equal access to Western markets. As with every other aid plan driven by Northern governments, the recipient nations are to be given duties to perform. In this case, African governments have to begin by fighting against corruption and bureaucratic inefficiency, the main culprits, according to some economic analysts, for the meager productive investment in the continent.

Certainly debt remains one of the main forces keeping poor countries poor. Interest payments on debt mean there is a huge and constant capital transfer to rich nations and an unbearable burden for Southern countries. As just two examples, Zambia spends more on debt service than on education while Malawi sends a third of its government budget directly to wealthy countries, which is twice as much as it spends on health.

But another major factor entrenching poverty is the current structure of international trade. Trade, far from being a growth engine, has had the effect of 'impoverishing' Southern countries.

On this issue, the World Trade Organization (WTO) plays a major role. The WTO forces poor countries to open their markets yet does little or nothing to require powerful rich countries to reduce their tariff barriers - it has been estimated that these tariff barriers cost Southern farmers as much as $100 billion a year. While farmers from the South are 'thrown to the wolves', Western agriculture receives more than a billion dollars per day in subsidies. The US, for example, grants its 25,000 cotton growers four billion dollars per year, an average of $160,000 dollars each, way above the value of their crop.

In comparison with the vast sums involved in the structural problems of trade and debt, aid is certainly minimal in its scope. Its value to rich countries in public-relations terms - in persuading the public that 'something is being done' about world poverty - may be greater than any concrete benefit to the developing world. ∎

Sources: Eric Toussaint, *La insignia*.
Human Development Report 2005, UNDP
Fundación Seminario de Investigación Para la Paz
USAID
rebelion.org

DEBT IN FIGURES

A= Total external debt (million $), 2004. **Source:** World Development Indicators 2006, World Bank.

B= Per capita external debt ($), 2004. **Source:** World Development Indicators 2006, World Bank.

C= External debt service (% of exports of goods and services), 2004. **Source:** World Development Indicators 2006, World Bank.

	A	B	C		A	B	C
Albania	1,549	490	2.6 2003	Liberia	2,706	784	
Algeria	21,987	649		Lithuania	9,475	2,784	14.3 2004
Angola	9,521	564	14.8 2004	Macedonia, TFYR	2,044	1,002	10.5 2004
Argentina	169,247	4,281	28.5 2004	Madagascar	3,462	177	6.0 2003
Armenia	1,224	408	8.0 2004	Malawi	3,418	254	7.6 2002
Azerbaijan	1,986	233	5.2 2004	Malaysia	52,145	1,987	7.9 2003
Bangladesh	20,344	138	5.2 2004	Maldives	345	997	4.6 2004
Barbados	702	2,593	5.2 2004	Mali	3,316	232	5.8 2003
Belarus	3,717	385	2.1 2004	Mauritania	2,297	707	
Belize	959	3,419	62.5 2004	Mauritius	2,294	1,811	7.4 2004
Benin	1,916	214	7.6 2003	Mexico	138,689	1,265	22.9 2004
Bhutan	593	263		Moldova	1,868	446	12.1 2004
Bolivia	6,096	640	18.6 2004	Mongolia	1,517	559	2.9 2004
Bosnia-Herzegovina	3,202	817	3.7 2004	Morocco	17,672	545	14.0 2004
Botswana	524	299	1.2 2003	Mozambique	4,651	227	4.5 2004
Brazil	222,026	1,160	46.8 2004	Myanmar-Burma	7,239	141	3.8 2004
Bulgaria	15,661	2,056	17.1 2004	Nauru			5.5 2004
Burkina Faso	1,967	140	11.9 2001	Nepal	3,354	119	
Burundi	1,385	170	66.0 2003	Nicaragua	5,145	900	5.8 2004
Cambodia	3,377	231	0.8 2004	Niger	1,950	131	7.5 2003
Cameroon	9,496	563	2004	Nigeria	35,890	262	8.2 2004
Cape Verde	517	975	5.3 2003	Oman	3,872	1,451	6.9 2004
Central African Rep.	1,078	260		Pakistan	35,687	217	21.2 2004
Chad	1,701	165		Panama	9,469	2,832	14.3 2004
Chile	44,058	2,649	24.2 2004	Papua New Guinea	2,149	351	12.7 2001
China	248,934	187	3.5 2004	Paraguay	3,433	533	13.5 2004
Colombia	37,732	804	33.0 2004	Peru	31,296	1,087	17.1 2004
Comoros	306	364	2004	Philippines	60,550	705	20.9 2004
Congo DR	11,841	2,794		Poland	99,190	2,579	34.6 2004
Congo R.	5,829	95	4.0 2004	Romania	30,034	1,394	17.2 2004
Costa Rica	5,700	1,276	7.3 2004	Russia	197,335	1,391	9.8 2004
Côte d'Ivoire	11,739	625	6.9 2004	Rwanda	1,656	175	11.2 2004
Croatia	31,548	6,926	27.2 2004	Samoa	562	3,006	
Czech Republic	45,561	4,468	10.5 2004	São Tomé and Príncipe	362	2,213	25.6 2002
Djibouti	429	522		Senegal	3,938	322	2004
Dominica	226	2,808	9.7 2002	Serbia and Montenegro *	15,882	1,514	
Dominican Republic	6,965	761	6.4 2004	Seychelles	615	7,490	8.1 2004
Ecuador	16,868	1,239	36.0 2004	Sierra Leone	1,723	297	10.9 2004
Egypt	30,291	394	7.6 2004	Slovakia	22,068	4,086	13.8 2003
El Salvador	7,250	1,019	8.8 2004	Solomon Is.	176	351	
Equatorial Guinea	291	552		Somalia	2,849	325	
Eritrea	681	145	3.1 2000	South Africa	28,500	597	6.4 2004
Estonia	10,008	7,579	15.7 2004	Sri Lanka	10,887	516	8.5 2004
Ethiopia	6,574	81	5.3 2004	St Kitts-Nevis	316	7,248	24.5 2002
Fiji	202	234		St Lucia	413	2,529	7.9 2002
Gabon	4,150	2,904	10.9 2003	St Pierre and Miquelon			7.3 2002
Gambia	674	423		St Vincent	257	2,136	
Georgia	2,082	474	11.2 2004	Sudan	19,332	512	6.0 2004
Ghana	7,035	306	6.6 2004	Swaziland	470	459	1.7 2004
Grenada	433	4,108	15.9 2002	Syria	21,521	1,077	3.5 2004
Guatemala	5,532	418	7.4 2004	Tajikistan	896	134	6.8 2004
Guinea	3,538	361	19.9 2004	Tanzania	7,799	196	5.3 2004
Guinea-Bissau	765	455	16.1 2003	Thailand	51,307	786	10.6 2004
Guyana	1,331	1,769	5.8 2004	Togo	1,812	280	2.0 2003
Haiti	1,225	140	4.0 2003	Tonga	81	788	2.5 2002
Honduras	6,332	842	7.8 2004	Trinidad and Tobago	2,926	2,228	3.8 2003
Hungary	63,159	6,288	25.2 2004	Tunisia	18,700	1,812	13.7 2004
India	122,723	108	18.9 2003	Turkey	161,595	2,150	35.9 2004
Indonesia	140,649	617	22.1 2004	Uganda	4,822	156	6.9 2004
Iran	13,622	191	10.7 2000	Ukraine	21,652	476	10.7 2004
Jamaica	6,399	2,394	14.8 2004	Uruguay	12,376	3,526	34.9 2004
Jordan	8,175	1,370	8.2 2004	Uzbekistan	5,007	183	
Kazakhstan	32,310	2,183	38.0 2004	Vanuatu	118	539	1.4 2003
Kenya	6,826	190	8.6 2004	Venezuela	35,570	1,285	16.0 2004
Kyrgyzstan	2,100	390	14.2 2004	Vietnam	17,825	206	6.0 2002
Laos	2,056	332	9.0 2001	Yemen	5,488	246	3.5 2004
Latvia	12,661	5,544	21.1 2004	Zambia	7,279	604	20.2 2000
Lebanon	22,177	6,070		Zimbabwe	4,797	364	
Lesotho	764	428	4.5 2004				

* All data available are previous to the political separation of Serbia and Montenegro.

North and South in the Information Society

ACCORDING TO THE WORLD SUMMIT ON THE INFORMATION SOCIETY DECLARATION OF PRINCIPLES, THE INFORMATION SOCIETY HAS TO BE PEOPLE-CENTERED, INCLUSIVE AND DEVELOPMENT-ORIENTED. NEVERTHELESS, EFFECTIVE SOLUTIONS HAVE NOT YET BEEN IMPLEMENTED TO DEAL WITH THE MAIN PROBLEM AREAS IDENTIFIED AT THE SUMMIT: INTERNET GOVERNANCE, THE DIGITAL DIVIDE AND INTELLECTUAL PROPERTY RIGHTS.

There is no universally accepted definition of 'The Information Society'. Most analysts agree that around 1970 the way that societies function began to change. The factor underlying this ongoing change is the gradual transfer of wealth generation from industrial to service sectors. In other words, in today's society, the tendency is for fewer jobs to be associated with the manufacture of tangible products and more with the generation, storage and processing of all types of information. As a result sectors related to information and communication technologies (ICTs) play a predominant role in the new order.

The concept of an 'Information Society' was introduced in the 1960s by Austrian economist Fritz Machlup in his book *The production and distribution of knowledge in the United States*. He attempted to outline the economic consequences of the development of information and communication activities and concluded that there were more jobs based on information use and management than those related to some type of physical effort.

In spite of its lack of definition the Information Society concept has enthusiastic supporters and detractors. A perception that the Information Society fosters technological dependence is the main reason why a significant number of critics claim that it is simply a revised and updated version of cultural imperialism imposed by rich countries on poor ones. On the other side are those who claim that the universal introduction of ICTs into production processes would help developing countries make their way in the fiercely competitive global market.

What did the World Summit achieve?

The 'digital revolution' arising from the development of ICTs has changed the way people across the planet act, think, communicate and work. Realizing that this new dynamic demanded discussion at a global level, in 1998 the International Telecommunication Union decided to hold a World Summit on the Information Society (WSIS) under the auspices of the United Nations. It took place in two stages, the first in December 2003 in Geneva and the second in November 2005 in Tunis.

The objective of the WSIS first phase was to create and promote a clear declaration of political will to instigate effective measures that would create the foundations of an inclusive 'information society', taking into account all the different interest groups involved.

Almost 50 heads of state, heads of government or vice-presidents, 82 ministers, 26 deputy ministers and other government delegates from 175 countries, along with representatives of international organizations, the private sector and civil society, attended this first phase and gave political support to the Geneva Declaration of Principles and the Geneva Plan of Action, approved in December 2003.

The Declaration contains 11 key principles and the Plan of Action 147 proposals. Several issues raised in discussions on the information age became the subject of WSIS documents, such as the right to information, equity, access, the right to cultural identity and diversity, local content and capacity building for the marginalized poor. However some analysts, particularly Malaysian economist Martin Khor, took the view that several fundamental issues, such as the impact of intellectual property rights on access to information, were hardly addressed.

Internet governance was one of the issues that generated major controversy during the first phase of the Summit. It was the subject of much discussion between representatives of countries from the South and the North. Developing countries argued that domain name administration and other aspects of global internet management should be the responsibility of an intergovernmental body and not of a private organization as is currently the case.

The second issue that caused controversy was how to finance the Plan of Action's proposals for bridging the digital divide. The proposal presented by countries of the South called for the creation of a Digital Solidarity Agenda with the objective of mobilizing 'human, financial and technological resources for the inclusion of all men and women in the emerging Information Society'. However, this proposal was rejected by the majority of industrial countries on the grounds that

Internet: predominantly for men

WITH THE GROWTH OF THE INTERNET a new type of poverty has arisen that denies developing countries access to information, again dividing the rich from the poor, the young from the old, urban inhabitants from rural and women from men.

The International Labor Organization (ILO), in its *World Employment Report,* reveals that globally the typical internet user is a 36-year-old male with university education and a high income who lives in an urban zone and speaks English. According to ILO, women are in a minority amongst internet users both in developing and developed countries. For example, women make up only 38 per cent of internet users in Latin America, 25 per cent in the European Union, 19 per cent in Russia, 18 per cent in Japan and a mere 4 per cent in the Middle East.

Even at an early age there is a tendency for information and technology to be more accessible to males than females. In the US five times more boys than girls use computers at home and parents spend twice as much on technological products for their sons than for their daughters.

The information divide however, does not only relate to gender: in the web world high income and knowledge of English facilitate access. In Latin America 90 per cent of internet users belong to high-income groups while more than 60 per cent of internet sites are available exclusively in English although this tendency is diminishing. ∎

Sources: www.itu.int/wsis/index-es.html
Choike
rebelion.org
Wikipedia
ILO

A = Telephones mainlines (per 1,000 people), 2004. **Source:** *World Development Indicators 2006*, World Bank.

	A	B		A	B		A	B
Afghanistan	1.7		France	560.9	487.1	Myanmar-Burma	8.5	6.5
Albania	90.0	11.7	French Polynesia	215.3	308.7	Namibia	63.7	109.5
2002			Gabon	28.4	29.4	Nepal	15.1	4.4
Algeria	70.7	9.0	Gambia	27.4	15.6	Netherlands	482.8	682.4
Angola	6.2	3.2	Georgia	151.2	42.5	Netherlands Antilles	460.8	
Antigua	474.5		Germany	661.1	561.1	New Zealand-Aotearoa	443.4	473.8
Argentina	226.7	96.4	Ghana	14.5	5.2	Nicaragua	39.9	37.2
Armenia	192.5	66.1	Greece	466.5	89.2	Niger	1.8	0.7
Australia	540.6	682.2	1995			Nigeria	8.0	6.7
Austria	460.4	418.4	Greenland	447.5	107.5	Norway	668.7	572.8
Azerbaijan	118.4	17.9	Grenada	309.3	151.3	Oman	94.9	46.6
Bahamas	439.0		Guam	506.5		Pakistan	29.6	4.9
Bahrain	267.6	169.0	Guatemala	92.1	18.8	2003		
Bangladesh	5.9	11.9	Guinea	2.9	4.8	Palestine	101.9	48.2
Barbados	504.8	126.4	Guinea-Bissau	7.1		Panama	118.4	40.9
Belarus	328.8		Guyana	136.9	36.0	Papua New Guinea	12.1	63.6
Belgium	456.4	348.0	Haiti	16.7		Paraguay	50.4	59.2
Belize	119.5	132.0	Honduras	52.7	15.6	Peru	74.4	97.6
2002			Hungary	353.9	146.0	Philippines	42.1	45.1
Benin	8.9	3.7	Iceland	652.1	472.4	Poland	321.8	192.8
Bermuda	870.9	528.8	India	40.7	12.1	Portugal	403.5	133.5
2002			Indonesia	45.9	13.9	Puerto Rico	285.5	
Bhutan	33.0	12.3	Iran	219.5	109.6	Qatar	245.7	171.2
Bolivia	69.4	35.5	Iraq	36.9	7.5	Romania	202.4	113.0
Bosnia-Herzegovina	239.4	44.6	2002			Russia	255.8	132.2
2002			Ireland	496.3	494.3	Rwanda	2.6	
Botswana	77.1	45.2	Israel	441.3	741.0	Samoa	72.9	6.6
Brazil	230.4	105.2	Italy	450.9	315.3	2002		
Brunei	251.7	84.8	Jamaica	189.1	62.8	San Marino	738.9	857.1
Bulgaria	356.9	59.4	Japan	460.1	541.6	São Tomé and Príncipe	46.6	
Burkina Faso	6.3	2.2	Jordan	113.5	55.1	Saudi Arabia	154.3	353.9
Burundi	3.4	4.7	Kanaky-New Caledonia	231.7		Senegal	20.6	21.3
Cambodia	2.7	2.8	Kazakhstan	166.7		Serbia and Montenegro*	329.6	47.7
Cameroon	6.9	10.0	Kenya	8.9	13.2	Seychelles	253.4	179.3
Canada	634.5	700.2	Kiribati	47.2	10.2	Sierra Leone	4.9	
Cape Verde	148.3	96.9	Korea, North	44.0		Singapore	439.6	763.2
Central African Rep.	2.5	2.8	Korea, South	541.9	544.9	2003		
Chad	1.4	1.6	Kuwait	202.1	183.0	Slovakia	232.3	296.0
Chile	205.8	132.6	Kyrgyzstan	78.6	17.1	Slovenia	407.0	352.5
China	241.1	40.9	Laos	9.7	3.8	Solomon Is.	13.7	42.9
Colombia	195.2	66.7	Latvia	272.8	216.6	Somalia	25.1	6.3
Comoros	23.0	8.5	Lebanon	178.0	113.0	South Africa	105.2	82.2
Congo D.R.	0.2		Lesotho	20.7		Spain	415.8	256.7
Congo R.	3.6	4.4	Liberia	2.2		Sri Lanka	51.0	27.3
Costa Rica	315.8	238.4	Libya	133.2	23.6	St Kitts-Nevis	532.1	234.1
Côte d'Ivoire	12.6	14.7	2002			St Lucia	321.2	158.9
Croatia	424.9	189.5	Liechtenstein	587.7		St Vincent	160.6	135.1
Cuba	68.3	26.7	Lithuania	238.7	155.1	Sudan	29.0	17.1
Cyprus	506.5	301.5	Luxembourg	800.2	653.0	Suriname	182.1	45.7
Czech Republic	337.7	239.8	Macedonia, TFYR	308.2	68.9	2001		
Denmark	643.0	655.6	Madagascar	3.4	5.0	Swaziland	41.8	32.1
Djibouti	14.3	27.0	Malawi	7.4	1.6	Sweden	708.1	763.0
Dominica	293.4	125.9	Malaysia	178.6	196.8	Switzerland	710.5	826.2
Dominican Republic	106.8	0.5	Maldives	98.1	112.1	Syria	143.1	32.3
2002			Mali	5.7	3.2	Tajikistan	38.6	
Ecuador	123.6	55.5	Malta	522.0	314.0	Tanzania	4.0	7.4
Egypt	130.3	31.7	Marshall Is.	75.6	81.7	Thailand	106.7	58.3
El Salvador	131.3	43.9	Mauritania	13.2	14.1	Togo	10.4	28.6
Equatorial Guinea	20.0	14.2	Mauritius	286.7	278.7	Tonga	110.7	49.0
Eritrea	9.3	3.5	Mexico	174.1	108.0	Trinidad and Tobago	246.9	105.3
Estonia	329.2	920.7	Micronesia	109.4		Tunisia	121.2	47.5
Ethiopia	6.3	3.2	Moldova	204.7	26.6	Turkey	266.6	51.6
Faeroe Is.	419.0		Mongolia	55.7	124.1	Turkmenistan	80.1	
Fiji	122.4	52.3	Morocco	43.9	20.8	Uganda	2.6	4.3
Finland	452.9	481.1	Mozambique	4.1	5.8	Ukraine	255.9	28.0

* All data available are previous to the political separation of Serbia and Montenegro.

current levels of assistance are sufficient and that developing countries should themselves provide the necessary funds.

The third and last unresolved source of discord was the way present and future intellectual property regulations increase information and communication costs for consumers and hinder public access to information and use of ICTs, particularly in the case of people with few economic resources.

Internet Governance Forum

The second phase in Tunis was supposed to find effective mechanisms for internet governance and for at least making progress towards bridging the digital divide (a modern version of the technological divide), but many analysts and civil society representatives claim that the Summit's concrete results were limited and did not address actual problems. The Summit's final document is interesting as a declaration but in reality contributes few practical solutions.

There were no effective solutions, for example, for the main issues involved in internet governance such as address management and assistance for poor countries to develop their technologies. The US maintained its stance as controller and sole administrator of the internet system and rich countries argued that investment in the development of new technologies is a matter for each country to deal with and finance on its own. Michael Gallagher, White House representative at the Summit, said: 'Global digital communication is so important from a strategic point of view that the US cannot allow intervention in its administration or technical organization

by any country and not even by the United Nations.'

The countries of the South were demanding UN participation in internet governance policy-making but between their stance and Washington's position, an alternative arose: the Internet Governance Forum. The Forum's function is to facilitate dialogue; it is not a decision-making body. Through it governments will be able to monitor governance processes for transparency and inclusiveness, make recommendations and propose that matters be reconsidered. Although the Forum does not enable governments directly to supervise or control the administration of IP names and numbers in the internet domain system and other related issues, it at least represents a change of stance and some flexibility on the part of the US.

The digital divide and development

The digital divide results from differing degrees of access to new information technologies (telecommunications and computing). For many analysts it began with the analog telephone, was emphasized by the computer and now yawns even wider as a result of internet and the communications revolution.

Gradually, access to the internet is becoming the main indicator of progress. It may not be an exaggeration to say that soon the critical differentiation will not be made in terms of rich and poor but in terms of the literate, those who are connected to internet, and the illiterate, those who are not.

Today the exchange value of goods and services is not determined by the

value of the material they are made of but by the knowledge associated with them. As Eduardo Samán, director of the Autonomous Intellectual Property Service, points out: 'The value of a mobile phone is not determined by the amount of silicon, plastic or copper in it but rather by the technology and knowledge associated with it.' As a result knowledge has acquired a high value and the concept of intangible assets has arisen.

The connection between the economic development of nations and the technological and digital divide is already evident. Countries that promote technological development projects are creating a better future and of course it is the more developed nations that are channeling greater resources into technological and digital infrastructure.

This becomes all the more evident if you consider the diverging experience of the world's regions. Guillermo Perry, the World Bank's Regional Chief Economist for Latin America and the Caribbean, has pointed out that most countries of the region suffer from a large and growing divide in education, technology and income. While in Latin America incomes doubled during the second half of the 20th century, in industrialized nations they tripled and in 'Asian tiger' countries they quadrupled. In 1960 these Asian countries had educational levels similar to those in Latin America whereas now it is normal for students to finish secondary school there (while in Latin America only 53 per cent do) and the average worker has had two years more education. In sub-Saharan Africa, meanwhile, even the levels of education and access to technology achieved in Latin America remain a distant dream. ∎

II. STATE SECURITY VS. HUMAN SECURITY

A sense of insecurity and fear is gaining ground. Security, a goal shared by states and individuals, has become the source of policies that either feed our fear or throw millions of people into total insecurity.

State of insecurity

'HUMAN SECURITY' SHOULD BE ENSURED BY STATES. HOWEVER, STATES OFTEN FACE DIFFICULTIES WHEN IT COMES TO DECIDING TO WHAT EXTENT THEY OFFER THEIR GUARANTEES. TERRORIST ATTACKS AND NATURAL DISASTERS CHALLENGE THE STATE-INDIVIDUAL RELATIONSHIP AND SHOW THAT 'HUMAN SECURITY' IS ONE THING WHILE 'STATE SECURITY' IS SOMETHING VERY DIFFERENT.

The concept of 'human security' emerged in the mid-1980s as a substitute for 'national security', which had been the term used until then. The simplest way to think about this is to associate it with 'non-violence' and think of it as a state of daily life in which individuals are not threatened with some type of physical aggression that puts their lives at risk. War, civil conflict, crime, terrorism and state repression all have an impact on the degree of security a society may offer its members.

However, there is a more comprehensive concept of 'human security' which is not exclusively focused on violence. The very etymology of the word 'security' goes beyond the mere idea of 'peace' since it stems from the Latin words 'securitas' and the latter 'securus' (unconcerned). It is clear that people's concerns in today's world are not (and should not be) exclusively about being alive and safe.

Thus, according to experts on the subject, there are two major ways of defining 'human security'. Those who opt for the first, restricted definition determine the degree of human security simply by measuring the amount of violence that threatens individuals in a community. This threat can be external (war), internal (social insecurity, armed conflicts) or involve the state (repression, harassment) - and in the last 100 years many more people have died at the hands of their own government than as a result of international armed conflict.

A more comprehensive definition adds factors such as hunger, diseases or natural disasters to the threat of violence since these kill more people than wars, genocides and terrorism put together. The United Nations Development Programme (UNDP), in its *Human Development Report 1994*, for example, follows this line of analysis.

All analysts agree that states - according to Max Weber, the only political entities authorized to exert legitimate violence[1] - have the absolute responsibility as guarantors of the security of individuals. For many theorists, the institutions of the state originally had no other purpose.

It is the state's role to do the optimum to make society more or less vulnerable (unsafe) - something that does not necessarily relate to its level of development or wealth. Every now and then terrorism, war, internal violence or natural disasters - such as the terrorist attacks in London and Madrid or Hurricane Katrina in the US - prove that human security is not guaranteed by the level of development and that state responses, even in the most powerful of countries, can be late, deficient and chaotic.

Security and development
In the poorest countries, the weaknesses are self-evident. Data on infant and maternal mortality, malnutrition, unemployment, illiteracy or low life expectancy reveal situations of extreme insecurity. Many countries in sub-Saharan Africa and South Asia - as well as some in Latin America, such as Haiti, Nicaragua and Guatemala - perform poorly in almost all the indicators that contribute to the security of their people. Certainly any areas affected by armed conflict threaten their inhabitants - including displaced people, refugees and orphans - with the crudest forms of human insecurity.

The developing world, on the other hand, experiences another kind of vulnerability. Even countries with medium and medium-high development indicators - such as Uruguay, Argentina and Brazil - show extremely impoverished social sectors, where human security is highly deficient.

When war, poverty, debt and hunger shape the whole reality of a country, little is to be expected in terms of security from a dismantled state, which will limit itself to managing the extremely scarce resources and fighting for its own survival in the storm. A weak state is unable to protect anybody.

In other less critical situations, the state has to focus on managing different options and choosing the paths that lead to the development of the country. This management is basically reduced to distributing the somewhat less limited resources available and fulfilling its international commitments. Such states are usually more stable but have limited room for maneuver. An overwhelmed state can offer limited protection.

---

1. Max Weber, 'Politics as a vocation'.

Finally, there is the case of countries with strong states, which manage plentiful resources and have countless possibilities for action. However, as we have already seen, the security of their people cannot always be assured. In part this is due to the fact that the state spends too much on its own security. A state that places itself as top priority is unable to protect its citizens in an effective way.

In view of the terrorist attacks carried out since 2001 in the US, London and Madrid, the idea of 'security' should be re-analyzed. These hammer blows turned out to be successful in spite of the huge sums of money that modern industrialized powers (particularly, the US and UK) allocate to 'national security'. Even so, none of the attacks, no matter how successful, posed a threat to state security, although they all were dreadful blows against 'human security'.

Dream of invulnerability shattered
On 11 September 2001, people in the US had their dream of invulnerability - which had led their government to waste billions in a quest for a 'Star Wars' style missile shield - turn into a nightmare, as a few suicide attackers, using their own education, technology and fuel, could bring a little of the 'outer world' to their home territory.

Four years later, a long-announced hurricane - among the many that sweep along the coast of the Caribbean and Southeast Asia - much photographed and analyzed by satellites, went even further. The 'outer world' met the US then, like a wedge embedded in the south, and produced TV pictures showing floating wrecks, crying children and destroyed families, seeming no different from those in Thailand, Indonesia or Sri Lanka. The images of armed civilians trying to survive the chaos, looking for food or looting stores, brought New Orleans closer to Puerto Principe, Mogadishu or Sarajevo.

The idea of 'state security' vaguely revealed its logic when, in the middle of the crisis and with paralyzed rescue systems, the authorities explained that no officers could be sent to the disaster area because it would not be safe for them.

The state machinery capable of plowing the seas and skies all over the world in the name of invasion and occupation did not feel safe to rescue its own stricken people.

Again, notwithstanding the magnitude of the crisis, and unlike cases such as Haiti, Somalia or Sarajevo, the US State was never in danger, which shows that 'human security' and 'state security' are far from going together.

These terrorist attacks and natural disasters have changed the relationship between the state and civil society in developed countries by revealing a crack in the foundations.

The situation is not equivalent to that of other countries in conflict. The American, Spanish and British people are not under daily threat of violence. Madrid is not Baghdad with car bombs in every street, nor are starving people lurking in the corners of London or New York to steal food. Yet people living in those cities have learned to suspect that neither the state's wealth nor its 'legitimate violence' can ensure their security. ∎

Resorting to 'just war'

THOSE INVOLVED IN A WAR TEND TO REGARD THEIR OWN FIGHT AS LEGITIMATE. WHETHER THEY ARE DEFENDING THEMSELVES FROM ATTACK, OR LAUNCHING AN ATTACK ON ANOTHER, THEY UPHOLD THE IDEA THAT JUSTICE ENDORSES THEIR WAR. THE 'JUST WAR' DOCTRINE, AMONG OTHER THINGS, UNDERLIES 'PRE-EMPTIVE' OR OTHER TYPES OF WARS LAUNCHED BY THE US GOVERNMENT, WHICH TEND TO THREATEN OR AFFECT ISLAMIC COUNTRIES.

Classical Antiquity ignored the concept of a just war. In Ancient Greece the concept of supremacy prevailed, which legitimated interventions against barbarians, regarded as inferiors. Talking about justice made no sense when the consequences of war, even for the conquered enemy, were positive: they benefited from Greek civilizing action.

Rome had a collective security concept. What legitimated the wars of the Republic, and later on those of the Empire, was the need to secure an area of international stability. But Rome was different from Greece: it admitted diversity, rapidly absorbed foreigners and incorporated many of their customs.

Christianity, which supplied the powers ruling the Western world until the fall of the Roman Empire with important ethical foundations, included commands as radical as love thine enemy, forgive those who offend you or pray for those who do you wrong.

Augustine of Hippo (354-430), better known as St Augustine, was one of the first theologians to attempt to reconcile the teachings of Jesus with the defense of an empire that was beginning to accept Christianity and tried to survive the onslaught of barbarians. The Augustinian synthesis admitted private pacifism but accepted the legitimacy of the Empire's military defense.

The scholastic doctrine of just war, widely treated by Thomas Aquinas (St Thomas, 1225-1274), was articulated around three axes: legitimacy of self-defense, commensurate response and possibility of success.

The Spanish Dominican Francisco de Vitoria (1483-1546), the father of international law, had a foot in scholastic doctrine but with a humanist emphasis. He only admitted defensive wars as just and stated that, in a case of serious aggression, a response may be admitted, as long as it is proportional to the aggression received. Among conditions for a just war were declaration of war by the governing authority (usually a prince), the inevitability of conflict in order to protect security, and a commensurate use of victory.

Vitoria did not regard as just wars waged over religious differences, wars of conquest or for glory. He condemned the cruelty of the Spanish conquistadors in America and the slaughter of innocents and prisoners. He proposed what today is known as 'conscientious objection': 'If a subject is convinced of the injustice of a war, he may not serve in it, even though his sovereign commands'.

Holy war

For French philosopher René Guénon (1886-1951) holy war, referred to by multiple holy texts (especially Eastern), should be understood as a symbol of restoring a spiritual order, more than inciting to the destruction of an external enemy. Often the Arab term *jihad* is interpreted as 'holy war', although it actually means 'effort'. It refers to the struggle the faithful must carry out to overcome adversity, often represented by the opposition of others.

In the Middle Ages, in the 12th century, when the Crusades began - a typical holy war initiated by Christianity - St Bernard said: 'War must be waged as little as possible, analyzing each case... Among Christians it is only just when the unity of the Church is in danger; against Jews, heretics, pagans violence shall be avoided, since truth is not imposed through force. Christians must convince, and only a defensive war is justified'.

Neither Christian nor Islamic texts encourage a war of aggression against the different, although both religions endorse active conversion of infidels. Reality showed that both faiths could co-exist when Al-Andalus, the Iberian land ruled by Muslim princes, became a place of fruitful exchange between Muslims and Christians. Religious intolerance only arose as a justification of political enmity.

Contemporary thinkers Michael Hardt and Antonio Negri say that renewed interest in the just war doctrine is a renaissance of the idea of 'empire'. According to them, the development of US military hegemony has turned all world conflicts into 'domestic' ones. The ease with which the US Government defines its enemies as rogues (including them in an 'axis of evil') makes sense within a conception of globalization as an imperial sphere.

Pre-emptive war

The current US Government maintains that it has the right to use military force against any alleged enemy, even before being attacked, to prevent a possible aggression. This peculiar way of legitimizing belligerent action is based on the idea that some governments have information about aggressive intentions from other governments or terrorist groups. Only military secret services have the power to identify and define potential risks to US safety and interests. According to this point of view, any country may be subject to a legitimate attack from the US due to assumptions or potential dangers.

When in early 2003 the US attacked Iraq based - so it said - on secret service reports regarding the existence of weapons of mass destruction, the chain of justification was sustained on extremely fragile links. First, the mere existence of weapons becomes the means to justify an attack. There is never an attempt to show that weapons will be used. Second, those who apparently decide to launch the war are not the rulers, but certain agencies supplying information. In the case of the war on Iraq, both the US and UK governments eventually admitted that the information from their intelligence services was mistaken.

The structure of both points of departure - the enemy's alleged intention as reason for war, and the information supplied by secret services - should suffice to invalidate the pre-emptive war doctrine. On the one hand, war is waged to avoid war, which is obviously contradictory; on the other hand, prevention is forced to take into account information not wholly confirmed, and therefore, even without considering the existence of illegitimate interests, it is subject to error.

The just war doctrine was used by the US to explain its interventions in Korea and Vietnam. It was used to justify its invasion of Panama in order to carry out the detention order by a US judge against an alleged drug dealer (General Manuel Noriega, the ruler), thereby extending Washington's jurisdiction throughout Central America and ignoring the principle of national sovereignty. Just war doctrine also lay behind the use of force to enforce UN resolutions (the first Gulf War). Later on, 'humanitarian intervention', a variation of just war, replaced international law in Kosovo.

The invasion of Iraq is the first pre-emptive war carried out by the US, but its warnings to Iran and North Korea lead many to believe it is only the beginning of a war cycle in which pre-emption will replace justice as a source of legitimacy.

Pre-emptive holy war

While the just war doctrine has advocates among theologians (like Augustine) and philosophers (like Thomas Aquinas or the American John Rawls [1921-2002]), pre-emptive war is advocated by US President George W Bush's administration and not by the lucubration of any thinker.

In philosophical terms pre-emption relates most closely to Manichean conceptions of good and evil (Augustine was, before he converted to Christianity, a believer in the dualist doctrine of Mani, a Persian sage who lived in the third century in what is now Iraqi territory).

During the annual State of the Union address by the US President to Congress in January 2002, Bush referred to an 'axis of evil' in which he placed Iran, North Korea and Iraq. He defined the Iraqi threat in these terms: 'This is a regime that has something to hide from the civilized world.' With this vague definition of the obscure character of its enemy, the US launched a war of aggression that cannot be defended by just war doctrine.

Although the US Government does not officially identify Islam as the enemy, there is a clear trend in most of the Western mass media to identify Islamic communities with forces opposed to 'civilization', a term used to define the culture of Western capitalism in developed countries. The use of terms such as 'axis of evil' and 'civilization' recalls the feudal princes in medieval Europe rallying people to holy war.

Like the ancient Greeks, the US Government seems convinced that its actions are beneficial, even toward its enemies; like the Romans, its wars of conquest are complemented by a practice of cultural assimilation and multi-ethnic immigration. But in modern terms the US notion of war is certainly breaking new ground. ∎

The rule of fear

THE WORLD IS INCREASINGLY PERCEIVED AS AN INSECURE, DANGEROUS AND MENACING PLACE. THE GLOBAL IDEOLOGY OF FEAR SEEMS TO HAVE BECOME THE BASIS FOR ALL ANALYSIS.

Reports from organizations monitoring armed conflicts and the plight of refugees and displaced people worldwide seem to indicate that between 1992 and 2004 there were fewer wars.

The *Human Security Report 2005 (HSR 2005)* [1] offers concrete data reflecting this trend: between 1992 and 2004, the number of armed conflicts worldwide fell from 44 to 32 (a reduction of over 40 per cent). Wars between countries only amounted to five per cent of armed conflicts in that year. Military coups d'état or attempted coups (which totalled 25 in 1963) amounted to ten in 2004, all of which failed.

Wars are also reported to be currently less deadly than they were 50 years ago: in the 1950s the average number of casualties was 38,000 a year, whereas they amounted to 600 in 2002.

Meanwhile *Alerta 2005* [2], a report on conflict and human rights published by the School of Peace Culture at the Autonomous University of Barcelona, documents 25 armed conflicts at the end of 2005 and 21 during the first quarter of 2006. [3]

As regards terrorism, the *HSR 2005* states that this started to decrease as from 1980 and, although an increase has been registered as of 2001, terrorist activities still produce a negligible fraction of the mortality associated with wars.

According to the most recent report by UNHCR [4], the current number of refugees in the world (a fact closely linked to armed conflicts) amounted to 9.2 million at the beginning of 2006, this being the lowest figure registered over the past 25 years, though there has been an increase in internal displacement.

Despite these data, the general perception prevails that we are living in an increasingly insecure world, with more wars and a greater number of casualties as a result of violence. Reality and the perception of that reality not only differ, they are utterly divergent. The *HSR 2005* suggests that the feeling of insecurity is brought about by a series of myths, 'some of them originating in the media; others propagated or restated by international organizations and NGOs'.

Swiss writer Tariq Ramadan argues that global terrorism and the 'war on terror' promoted by the US after the attacks of 11 September 2001, both equally foster the 'global ideology of fear'. According to Ramadan, in the North as well as in the South, particularly in those communities where the population is mainly Muslim, 'fear is omnipresent, deeply rooted everywhere'. He adds that such fear has an unmistakable impact on human beings' perception of the world. This fear, in a natural and often subconscious way, engenders mistrust and fosters potential conflict with the 'Other'. 'We observe facts,' he says, 'condemn their consequences, reject individuals together with their motivations and actions, but every principle of causation seems to have vanished from the analytic horizon'.

The global 'war on terror' seems to have changed the landscape: we are living in a land of suppostion, not of analysis. The US and its allies maintain that it was the terrorist attack on the Twin Towers that changed the world irreversibly. The logic of fear, reconverted into state security, completely disregards Human Rights and security. Thus, UNHCR reports that 'since the 11 September 2001 attacks in the United States, state security concerns have come to dominate the migration debate, at times overshadowing the legitimate protection needs of individuals.' It is the fear itself that, paradoxically, leads to defenselessness. ∎

1. Human Security Centre, *Human Security Report 2005, War and peace in the 21st Century*, Oxford University Press, January 2006.

2. Escola de Cultura de Pau (ECP). *Alerta 2005. Informe sobre conflictos, Derechos Humanos y construcción de paz*, Universitat Autònoma de Barcelona, Icaria, 2005.

3. Each organization uses different criteria for including an event as an example of 'armed conflict' and thus their overall numbers differ. However, the trend is similar in both cases.

4. UNHCR, *The state of the world's refugees. Human displacement in the new millennium*. 2006, www.unhcr.org

SEXUAL MINORITIES AND THE LAW: A WORLD SURVEY

Today the issues of homosexuality and transgender are being hotly debated in parts of the world where they had been just a hushed whisper. It has become harder for political and religious leaders to maintain the 'homosexuality is not part of our culture' as home-grown lesbian, gay, bisexual or transgender groups have sprung up in Africa, Asia and Latin America, to fight unjust laws and demand freedom from discrimination and persecution. This 'internationalization' of sexual minority rights has coincided with the rise in internet activism and support from international human rights organizations. However, the situation for many lesbian, gay, bisexual or transgender (LGBT) people in the world remains dire. In nine countries homosexuality incurs the death penalty, while in some 80 states it is illegal, sometimes incurring long prison sentences. Currently, many governments are making constitutional and legal changes to combat centuries of discrimination. Pioneers in this area have included countries like the Netherlands or Denmark, with established liberal traditions, but also South Africa, Brazil and, more recently, the Philippines.

F= female. M= male. LGBT: Lesbian, gay, bisexual and transgender.

AFGHANISTAN
Homosexuality: Illegal. F/M. Imprisonable for up to 15 years.
Transgender: No data or legal situation unclear.

ALBANIA
Homosexuality: Legal. Age of consent equal. LGBT citizens have been granted asylum by other countries.
Transgender: Gender reassignment ('sex change') is illegal.

ALGERIA
Homosexuality: Illegal. F/M. Imprisonable for up to 3 years. LGBT citizens have been granted asylum by other countries.
Transgender: No data or legal situation unclear.

ANDORRA
Homosexuality: Legal.
Transgender: Gender reassignment ('sex change') is illegal.

ANGOLA
Homosexuality: Illegal. F/M.
Transgender: No data or legal situation unclear.

ANTIGUA AND BARBUDA
Homosexuality: Legal.
Transgender: No data or legal situation unclear.

AOTEAROA / NEW ZEALAND
Homosexuality: Legal. Age of consent equal. Legal protection for sexual orientation under the Human Rights Act. Same-sex civil unions recognized. Prepared to grant asylum to LGBT refugees.
Transgender: Gender reassignment ('sex change') legal or openly performed without prosecution. All official documents may be reissued to reflect change.

ARGENTINA
Homosexuality: Legal. Civil unions recognized in some regions. Anti-discrimination laws apply. Social intolerance can be extreme and LGBT citizens have been granted asylum by other countries.
Transgender: Gender reassignment ('sex change') legal or openly performed without prosecution.

ARMENIA
Homosexuality: Legal. Some residence rights for bi-national gay couples. LGBT citizens have been granted asylum by other countries, but now some support for LGBT refugees from elsewhere.
Transgender: No data or legal situation unclear.

ARUBA
Homosexuality: Legal. Protection for sexual orientation under Dutch law but challenged by Arubian authorities.
Transgender: No data or legal situation unclear.

AUSTRALIA
Homosexuality: Legal. Age of consent higher for gay men in some states. Anti-discrimination laws apply. Domestic partnership recognition in some states but federal ban on same-sex marriage. Same-sex adoption and donor insemination services available in some states. Prepared to grant asylum to LGBT refugees.
Transgender: Gender reassignment legal in some states. Specific protection from discrimination exists for transgendered people.

AUSTRIA
Homosexuality: Legal. Anti-discrimination laws apply. Prepared to grant asylum to LGBT refugees.
Transgender: Gender reassignment legal or openly performed without prosecution. All personal documents may be reissued following change.

AZERBAIJAN
Homosexuality: Legal. Age of consent equal.
Transgender: No data or legal situation unclear.

BAHAMAS
Homosexuality: Legal. Age of consent higher for lesbians and gay men.
Transgender: No data or legal situation unclear.

BAHRAIN
Homosexuality: Illegal. F/M. Imprisonable for up to 10 years; deportation for 20 years.
Transgender: Gender reassignment illegal.

BANGLADESH
Homosexuality: Illegal. F/M. Imprisonable for life. LGBT citizens have been granted asylum by other countries
Transgender: No data or legal situation unclear.

BARBADOS
Homosexuality: Illegal. F/M. Imprisonable but rarely enforced against private behaviour. Laws currently under review.
Transgender: No data or legal situation unclear.

BELARUS
Homosexuality: Legal. Equal age of consent. But severe discrimination persists.
Transgender: Gender reassignment legal or openly performed without prosecution. No data on reissue of documents.

BELGIUM
Homosexuality: Legal. Age of consent equal. Same-sex couples can marry and adopt. Anti-discrimination laws apply. Prepared to grant asylum to LGBT refugees.
Transgender: Gender reassignment legal or openly performed without prosecution. All personal documents may be reissued following change.

BELIZE
Homosexuality: Illegal. F/M. Imprisonable for 10 years.
Transgender: No data or legal situation unclear.

BENIN
Homosexuality: Illegal. F/M.
Transgender: No data or legal situation unclear.

BHUTAN
Homosexuality: Illegal. F/M. Imprisonable for life.
Transgender: No data or legal situation unclear.

BOLIVIA
Homosexuality: Legal. No anti-discrimination laws.
Transgender: No data or legal situation unclear.

BOSNIA & HERZEGOVINA
Homosexuality: Legal. Equal age of consent. Anti-discrimination laws apply. No legal recognition of same-sex partnerships.
Transgender: No data or legal situation unclear.

BOTSWANA
Homosexuality: Illegal. M (F not mentioned in law). Imprisonable for 5 years.
Transgender: No data or legal situation unclear.

BRAZIL
Homosexuality: Legal. Equal age of consent. Anti discrimination and anti-vilification laws exist in several states; civil unions recognized in some. But high levels of homophobic violence; LGBT citizens have been granted asylum by other countries.
Transgender: Gender reassignment ('sex change') legal or openly performed without prosecution.

BRUNEI
Homosexuality: Illegal. F/M. Imprisonable for 10 years.
Transgender: No data or legal situation unclear.

BULGARIA
Homosexuality: Legal. Age of consent equal. Anti-discrimination laws apply.
Transgender: Gender reassignment ('sex change') legal or openly performed without prosecution.

BURKINA FASO
Homosexuality: Legal. Age of consent equal.
Transgender: No data or legal situation unclear.

BURUNDI
Homosexuality: Illegal. M (F not known). Punishable as an 'immoral act'.
Transgender: No data or legal situation unclear.

CAMBODIA
Homosexuality: Legal. Age of consent equal. Former King Sihanouk has called for legalization of gay marriage.
Transgender: No data or legal situation unclear.

CAMEROON
Homosexuality: Illegal. F/M. Imprisonable for 5 years.
Transgender: No data or legal situation unclear.

CANADA
Homosexuality: Legal. Age of consent higher for anal sex (18). Legal recognition of same-sex partnerships, marriage and adoption rights. Constitutional laws against discrimination apply. Prepared to grant asylum to LGBT refugees.
Transgender: Gender reassignment legal in some states and provinces.

CAPE VERDE
Homosexuality: Illegal. F/M. 'Repeat offenders' may be imprisoned.
Transgender: No data or legal situation unclear.

CAYMAN ISLANDS
Homosexuality: Legal. British laws apply.
Transgender: No data or legal situation unclear.

CENTRAL AFRICAN REPUBLIC
Homosexuality: Legal. Age of consent equal.
Transgender: No data or legal situation unclear.

CHAD
Homosexuality: Legal. Age of consent higher (18).
Transgender: No data or legal situation unclear.

CHILE
Homosexuality: Legal. Age of consent higher (18). No anti-discrimination laws. LGBT citizens have been granted asylum by other countries.
Transgender: No data or legal situation unclear.

CHINA
Homosexuality: Not illegal but considered 'unacceptable'. Legal in Hong Kong with equal age of consent. LGBT citizens from mainland China have been granted asylum by other countries.
Transgender: Gender reassignment ('sex change') legal or openly performed without prosecution.

COLOMBIA
Homosexuality: Legal. Age of consent equal. Social intolerance can be extreme and LGBT citizens have been granted asylum by other countries.
Transgender: First country to restrict genital mutilation of intersex children without their, or before age of, consent.

COMOROS
Homosexuality: Legal.
Transgender: No data or legal situation unclear.

CONGO
Homosexuality: Legal. Age of consent equal.
Transgender: No data or legal situation unclear.

CONGO DEM REP
Homosexuality: Illegal. F/M. Imprisonable for 5 years under 'crimes against the family' law.
Transgender: No data or legal situation unclear.

COOK ISLANDS
Homosexuality: Illegal. M (F not mentioned in law). Imprisonable for 7 years.
Transgender: No data or legal situation unclear.

COSTA RICA
Homosexuality: Legal, though 'scandalous' homosexuality illegal. Age of consent equal. Laws against discrimination apply but same-sex marriage is banned.
Transgender: No data or legal situation unclear.

CÔTE D'IVOIRE
Homosexuality: Legal.
Transgender: No data or legal situation unclear.

CROATIA
Homosexuality: Legal. Age of consent higher for lesbians and gay men (18). Registered partnerships. LGBT citizens have been granted asylum by other countries.
Transgender: No data or legal situation unclear.

CUBA
Homosexuality: Legal but LGBT associations are banned. No laws against discrimination. LGBT citizens have been granted asylum by other countries.
Transgender: No data or legal situation unclear.

CYPRUS
Homosexuality: Legal. Age of consent equal.
Transgender: No data or legal situation unclear.

CZECH REPUBLIC
Homosexuality: Legal. Age of consent equal. Registered partnerships. Some anti-discrimination protection.
Transgender: Gender reassignment ('sex change') legal or openly performed without prosecution. Some personal documents may be reissued.

DENMARK
Homosexuality: Legal. Age of consent equal. Legal recognition of same sex partnerships (also applies in Greenland). Legal recognition of non-biological parents. Anti-discrimination laws apply. Prepared to grant asylum to LGBT refugees.
Transgender: Gender reassignment ('sex change') legal or openly performed without prosecution. All personal documents may be reissued following change.

DJIBOUTI
Homosexuality: Illegal. F/M.
Transgender: No data or legal situation unclear.

DOMINICAN REP.
Homosexuality: Legal. Age of consent equal. No laws against discrimination.
Transgender: No data or legal situation unclear. Cultural acceptance of transgendered guevedoche or 'pseudo- hermaphrodites'.

ECUADOR
Homosexuality: Legal. Anti-discrimination written into the Constitution. Custody rights for lesbians.
Transgender: No data or legal situation unclear.

EGYPT
Homosexuality: Technically legal but effectively illegal. A variety of laws are applied.
Transgender: Gender reassignment ('sex change') legal or openly performed without prosecution. Civil Law provisions exist for reissue of personal documents to reflect change.

EL SALVADOR
Homosexuality: Legal. No laws against discrimination. LGBT citizens have been granted asylum by other countries.
Transgender: No data or legal situation unclear.

EQUATORIAL GUINEA
Homosexuality: Illegal. F/M. Imprisonable for 3 years.
Transgender: No data or legal situation unclear.

ERITREA
Homosexuality: Legal. M (F situation unclear) Age of consent 18.
Transgender: No data or legal situation unclear.

ESTONIA
Homosexuality: Legal. Age of consent equal.
Transgender: Gender reassignment ('sex change') legal or openly performed without prosecution. Documents cannot be reissued.

ETHIOPIA
Homosexuality: Illegal. F/M. Imprisonable for 3 years.
Transgender: No data or legal situation unclear.

FIJI
Homosexuality: Illegal. M (F not mentioned in law). Imprisonable for 14 years. Law has been ruled unconstitutional as Constitution protects against sexual orientation discrimination.
Transgender: No data or legal situation unclear.

FINLAND
Homosexuality: Legal. Age of consent equal. Registered partnerships. Anti-discrimination laws apply. Legal recognition of non-biological parents; access to state donor insemination services. Prepared to grant asylum to LGBT refugees.
Transgender: Gender reassignment ('sex change') legal or openly performed without prosecution. All personal documents may be reissued following change.

FRANCE
Homosexuality: Legal. Age of consent equal. Civil unions available to all regardless of sexual orientation. Anti-discrimination laws apply. Prepared to grant asylum to LGBT refugees.
Transgender: Gender reassignment ('sex change') legal or openly performed without prosecution. Some personal documents may be reissued after change.

FRENCH GUYANA
Homosexuality: Legal. French laws apply.
Transgender: No data or legal situation unclear.

GABON
Homosexuality: Legal. Age of consent higher for lesbians and gay men.
Transgender: No data or legal situation unclear.

GAMBIA
Homosexuality: Illegal. M (F situation unclear). Imprisonable for 14 years.
Transgender: No data or legal situation unclear.

GEORGIA
Homosexuality: Legal. Age of consent equal. Some anti-discrimination protection.
Transgender: Gender reassignment ('sex change') legal or openly performed without prosecution.

GERMANY
Homosexuality: Legal. Age of consent equal. Registered partnerships and adoption rights. Regional anti-discrimination laws apply. Prepared to grant asylum to LGBT refugees.
Transgender: Gender reassignment ('sex change') legal or openly performed without prosecution. All personal documents may be reissued after change.

GHANA
Homosexuality: Illegal. M. Punishable under 'unnatural carnal knowledge' laws. LGBT citizens have been granted asylum by other countries.
Trangender: Gender reassignment ('sex change') is illegal.

GREECE
Homosexuality: Legal. Anti-discrimination in employment law applies. Access to state donor insemination services. Prepared to grant asylum to LGBT refugees.
Transgender: Gender reassignment ('sex change') legal or openly performed without prosecution. All personal documents may be reissued after change.

GRENADA
Homosexuality: Illegal. M (F not mentioned in law). Imprisonable for 10 years.
Transgender: No data or legal situation unclear.

GUAM
Homosexuality: Legal. No laws against discrimination.
Transgender: No data or legal situation unclear.

GUATEMALA
Homosexuality: Legal. Anti-discrimination laws apply.
Transgender: No data or legal situation unclear.

GUINEA
Homosexuality: Illegal. F/M. Imprisonable for 3 years.
Transgender: No data or legal situation unclear.

GUINEA-BISSAU
Homosexuality: Legal.
Transgender: No data or legal situation unclear.

GUYANA
Homosexuality: Illegal. M (F situation unclear). Imprisonable for life.
Transgender: No data or legal situation unclear.

HAITI
Homosexuality: Legal. No laws against discrimination.
Transgender: No data or legal situation unclear.

HONDURAS
Homosexuality: Legal. Same-sex marriage and adoption by same-sex couples banned. LGBT citizens have been granted asylum by other countries.
Transgender: No data or legal situation unclear.

HUNGARY
Homosexuality: Legal. Legal recognition of same-sex partnerships. Anti-discrimination laws apply.
Transgender: Gender reassignment ('sex change') legal or openly performed without prosecution. Legal situation unclear on document issue.

ICELAND
Homosexuality: Legal. Age of consent equal. Some legal protection for sexual orientation. Legal recognition of same-sex partnerships. Parenting: Legal recognition of non-biological parents.
Transgender: Gender reassignment ('sex change') legal or openly performed without prosecution. Legal situation unclear on document issue.

INDIA
Homosexuality: Illegal. M (but law has been used against women and transsexuals too). Imprisonable for life, but provisions rarely applied. Judicial review of the law being sought.
Transgender: Hijari (eunuchs) can have ID cards changed to reflect female status.

INDONESIA
Homosexuality: Legal. The Justice Ministry recently attempted to criminalize homosexuality.
Transgender: Gender reassignment ('sex change') legal or openly performed without prosecution. Transgender people have an accepted place in society.

IRAN
Homosexuality: Illegal. F/M. Death penalty applies. In July 2005 two gay teenagers were executed. Some 4,000 LGBT people are reported to have been executed since 1979.
Transgender: Gender reassignment ('sex change') is legal and government support available.

IRAQ
Homosexuality: Situation unclear. Current Government has issued a decree allowing Sharia laws (death penalty for homosexuals) to be enforced. LGBT Iraqis are now targeted for prosecution and execution.
Transgender: No data or legal situation unclear.

IRELAND
Homosexuality: Legal. Age of consent equal. Anti-discrimination laws apply. Prepared to grant asylum to LGBT refugees.
Transgender: Gender reassignment ('sex change') legal or openly performed without prosecution. It is illegal to change birth certificate or marry after gender reassignment.

ISRAEL
Homosexuality: Legal. Age of consent equal. Some legal protection for sexual orientation, including within the armed forces.
Transgender: Gender reassignment ('sex change') legal or openly performed without prosecution.

ITALY
Homosexuality: Legal. Age of consent equal. Civil unions in some regions.
Transgender: Gender reassignment ('sex change') legal or openly performed without prosecution. Some personal documents may be reissued after change.

JAMAICA
Homosexuality: Illegal. M (F not mentioned in law). Imprisonable for 10 years (can be accompanied by hard labor). High level of social intolerance and violence towards LGBT people.
Transgender: No data or legal situation unclear.

JAPAN
Homosexuality: Legal.
Transgender: Gender reassignment ('sex change') legal or openly performed without prosecution. All personal documents may be reissued after change.

JORDAN
Homosexuality: Legal. But LGBT citizens at risk of vigilante 'honour' killing and have been granted asylum by other countries.
Transgender: No data or legal situation unclear.

KAZAKHSTAN
Homosexuality: Legal. No anti-discrimination laws.
Transgender: No data or legal situation unclear.

KENYA
Homosexuality: Illegal. M (F not mentioned in law). Imprisonable for 14 years.
Transgender: No data or legal situation unclear. Traditionally transgendered people had an accepted place in society.

KIRIBATI
Homosexuality: Illegal. M (F not mentioned in law). Imprisonable for 14 years.
Transgender: No data or legal situation unclear.

KOREA, N
Homosexuality: Legal.
Transgender: No data or legal situation unclear.

KOREA, S
Homosexuality: Legal. Anti-discrimination law exists but Government continues to discriminate.
Transgender: No data or legal situation unclear.

KUWAIT
Homosexuality: Illegal. F/M. Imprisonable for 7 years. Laws denying freedom of expression and/or association also apply.
Transgender: No data or legal situation unclear.

KYRGYZSTAN
Homosexuality: Legal. No anti-discrimination laws.
Transgender: No data or legal situation unclear.

LAOS
Homosexuality: Unclear as to whether acts in private between consenting adults are legal or not.
Transgender: No data or legal situation unclear.

LATVIA
Homosexuality: Legal. Same-sex marriages banned. No anti-discrimination laws. Prepared to grant asylum to LGBT refugees.
Transgender: Gender reassignment ('sex change') legal or openly performed without prosecution. Some personal documents may be reissued after change.

LEBANON
Homosexuality: Illegal. F/M. Imprisonable for 1 year. Laws denying freedom of expression and association apply. LGBT citizens have been granted asylum by other countries.
Transgender: No data or the situation unclear.

LESOTHO
Homosexuality: Legal. Homosexuality not mentioned in law.
Transgender: No data or legal situation unclear.

LIBERIA
Homosexuality: Illegal. F/M.
Transgender: No data or legal situation unclear.

LIBYA
Homosexuality: Illegal. F/M. Imprisonable for 5 years.
Transgender: No data or legal situation unclear.

LIECHTENSTEIN
Homosexuality: Legal. Age of consent equal. Laws against discrimination apply.
Transgender: No data or legal situation unclear.

LITHUANIA
Homosexuality: Legal. Age of consent equal. Some legal protection against discrimination.
Transgender: Gender reassignment illegal.

LUXEMBOURG
Homosexuality: Legal. Age of consent equal. Registered same-sex partnerships. Laws against discrimination apply.
Transgender: Gender reassignment ('sex change') legal or openly performed without prosecution. Some personal documents may be reissued after change.

MACEDONIA (TFYR)
Homosexuality: Legal. No laws against discrimination. Gays barred from entering legal profession.
Transgender: Gender reassignment ('sex change') is illegal.

MADAGASCAR
Homosexuality: Legal.
Transgender: No data or legal situation unclear. Transgendered people have an accepted place in society.

MALAWI
Homosexuality: Illegal. F/M.
Transgender: No data or legal situation unclear.

MALAYSIA
Homosexuality: Illegal. M (situation unclear for F). Imprisonable for 20 years. LGBT citizens have been granted asylum by other countries.
Transgender: Gender reassignment is recognized but right to marry denied.

MALDIVES
Homosexuality: Illegal. M (F not mentioned in law). Imprisonable for life.
Trangender rights: No data or legal situation unclear.

MALI
Homosexuality: No laws against, but LGBT may be prosecuted under 'Public Morals' law.
Transgender: No data or legal situation unclear.

MALTA
Homosexuality: Legal. Age of consent equal.
Transgender: No data or legal situation unclear.

MARSHALL ISLANDS
Homosexuality: Illegal. M (F not mentioned in law). Imprisonable for 10 years.
Transgender: No data or legal situation unclear.

MARTINIQUE
Homosexuality: Legal. French laws apply.
Transgender: No data or legal situation unclear.

MAURITANIA
Homosexuality: Illegal. F/M. Death penalty applies. LGBT citizens have been granted asylum by other countries.
Transgender: No data or legal situation unclear.

MAURITIUS
Homosexuality: Illegal. F/M. Imprisonable for 3 years.
Transgender: No data or legal situation unclear.

MEXICO
Homosexuality: Legal. Laws against discrimination apply. Social intolerance can be extreme and LGBT citizens have been granted asylum by other countries.
Transgender: No data or legal situation unclear.

MICRONESIA
Homosexuality: Not mentioned in law.
Transgender: No data or legal situation unclear.

MOLDOVA
Homosexuality: Legal.
Transgender: Gender reassignment ('sex change') legal or openly performed without prosecution. All personal documents may be reissued following change.

MONACO
Homosexuality: Legal. Age of consent equal.
Transgender: Gender reassignment ('sex change') legal or openly performed without prosecution. All personal documents may be reissued following change.

MONGOLIA
Homosexuality: Not mentioned in law but penal code prohibiting 'immoral gratification of sexual desires' is used against gay people.
Transgender: No data or legal situation unclear.

MOROCCO
Homosexuality: Illegal. F/M. Imprisonable for 3 years. LGBT citizens have been granted asylum by other countries.
Transgender: Legal situation unclear.

MOZAMBIQUE
Homosexuality: Illegal. F/M. Imprisonable for 3 years with hard labour.
Transgender: No data or legal situation unclear.

MYANMAR/ BURMA
Homosexuality: Illegal. M (F not mentioned in law). Imprisonable for life.
Transgender: No data or legal situation unclear. Traditionally transgendered people have accepted place in society.

BERMUDA
Homosexuality: Legal. Higher age of consent.
Transgender: No data or legal situation unclear.

NAMIBIA
Homosexuality: Illegal. M (situation unclear for F). Legal precedent gave immigration rights to same-sex lesbian partnership. Some anti-discrimination provision applies, in spite of illegality.
Transgender: Gender reassignment recognized. Documents can be changed on social acceptance.

NAURU
Homosexuality: Legal situation unclear.
Transgender: No data or legal situation unclear.

NEPAL
Homosexuality: Illegal. F/M. Imprisonable for life. Law being challenged in Supreme Court.
Transgender: No data or legal situation unclear.

NETHERLANDS
Homosexuality: Legal. Age of consent equal. Same-sex civil unions and marriages performed. Anti-discrimination laws apply. Legal recognition of non-biological parents and same-sex adoption rights. Access to state donor insemination services. Prepared to grant asylum to LGBT refugees.
Transgender: Gender reassignment ('sex change') legal or openly performed without prosecution. Some personal documents may be reissued after change.

NETHERLANDS ANTILLES
Homosexuality: Legal. Dutch law applies but while same-sex civil unions and marriages are recognized they cannot be performed.
Transgender: No data or legal situation unclear.

NICARAGUA
Homosexuality: Illegal. F/M. Imprisonable for 3 years. Law is being challenged in the High Court. LGBT citizens have been granted asylum by other countries.
Transgender: No data or legal situation unclear.

NIGER
Homosexuality: Legal. Age of consent 21.

Same-sex marriages are allowed.
Transgender: No data or legal situation unclear.

NIGERIA
Homosexuality: Illegal. M (F not mentioned in law). Imprisonable for 14 years.
In Northern provinces death penalty under Sharia law applies. A new law forbids same-sex marriage and prohibits gays from assembling and petitioning the government. It also allows prosecution of newspapers that publish information about same-sex relationships and religious groups that allow same-sex unions. Those who violate the law can be sentenced to 5 years in prison.
Transgender: No data or legal situation unclear.

NIUE
Homosexuality: Illegal. M (F not mentioned in law). Imprisonable for 10 years.
Transgender: No data or legal situation unclear.

NORWAY
Homosexuality: Legal. Age of consent equal. Legal recognition of same-sex partnerships. Legal recognition of non-biological parents. Prepared to grant asylum to LGBT refugees.
Transgender: Gender reassignment ('sex change') legal or openly performed without prosecution. Some personal documents may be reissued to reflect change.

OMAN
Homosexuality: Illegal. (F/M). Imprisonable for 3 years.
Transgender: No data or legal situation unclear. Transgendered people have an accepted place.

PALESTINE
Homosexuality: Not illegal but LGBT people targeted. Many leave for Israel.
Transgender: No data or legal situation unclear

PAKISTAN
Homosexuality: Illegal. F/M. Death penalty applies under Sharia law. LGBT citizens have been granted asylum by other countries.
Transgender: Some official recognition of gender reassignment and right to marry subsequently.

PANAMA
Homosexuality: Legal.
Transgender: Gender reassignment ('sex change') legal or openly performed without prosecution. Some personal documents may be reissued to reflect change.

PAPUA NEW GUINEA
Homosexuality: Illegal. M (F not mentioned in law). Imprisonable for 14 years.
Transgender: No data or legal situation unclear.

PARAGUAY
Homosexuality: Legal. Age of consent equal. No anti-discrimination laws.
Transgender: No data or legal situation unclear.

PERU
Homosexuality: Legal. Some anti-discrimination protection applies. LGBT citizens have been granted asylum by other countries.
Transgender: No data or legal situation unclear.

PHILIPPINES
Homosexuality: Legal. Age of consent equal. Same-sex marriage banned.
Transgender: Gender reassignment ('sex change') legal or openly performed without prosecution.

POLAND
Homosexuality: Legal. Age of consent equal. High level of social intolerance. LGBT citizens have been granted asylum by other countries.
Transgender: Gender reassignment ('sex change') legal or openly performed without prosecution. Some personal documents may be reissued to reflect change.

PORTUGAL
Homosexuality: Legal. Age of consent equal.

Same-sex civil unions allowed but not same-sex adoption.
Transgender: Gender reassignment ('sex change') legal or openly performed without prosecution. Personal documents may be reissued to reflect change.

QATAR
Homosexuality: Illegal. F/M. Imprisonable for up to 5 years.
Transgender: No data or legal situation unclear.

ROMANIA
Homosexuality: Legal. Some anti-discrimination protection.
Transgender: Gender reassignment ('sex change') legal or openly performed without prosecution. Some personal documents may be reissued to reflect change.

RUSSIA
Homosexuality: Legal in all states except Chechyna (where the death penalty applies under Sharia law). Age of consent equal. Three attempts to recriminalize homosexuality in Russia between 2002 and 2004 failed. LGBT citizens have been granted asylum in other countries.
Transgender: Gender reassignment ('sex change') legal or openly performed without prosecution. All personal documents may be reissued to reflect change. Traditionally, transgender people had an accepted place in Siberian society.

RWANDA
Homosexuality: Legal. Age of consent higher.
Transgender: No data or legal situation unclear.

SAINT KITTS AND NEVIS
Homosexuality: Illegal. M (F not mentioned in law). Imprisonable for 10 years.
Transgender: No data or legal situation unclear.

SAINT LUCIA
Homosexuality: Illegal. M. (F not mentioned in law). Imprisonable for 10 years.
Transgender: No data or legal situation unclear.

SAINT VINCENT/GRENADINES
Homosexuality: Illegal. F/M. Imprisonable for 10 years.
Transgender: No data or legal situation unclear.

SÃO TOMÉ AND PRÍNCIPE
Homosexuality: Illegal. F/M.
Transgender: No data or legal situation unclear.

SAUDI ARABIA
Homosexuality: Illegal. F/M. Death penalty, imprisonment and flogging under Sharia law. Executions have taken place during past 10 years.
Transgender: No data or legal situation unclear.

SENEGAL
Homosexuality: Illegal. F/M.
Transgender: No data or legal situation unclear.

SERBIA
Homosexuality: Legal. Age of consent higher for male homosexual anal sex (18).
Transgender: No data or legal situation unclear.

SEYCHELLES
Homosexuality: Illegal. M (F not mentioned in law).
Transgender: No data or legal situation unclear.

SIERRA LEONE
Homosexuality: Illegal. M (situation unclear for F).
Transgender: No data or legal situation unclear.

SINGAPORE
Homosexuality: Illegal. M (F not mentioned in law). Imprisonable for life but law rarely enforced. LGBT citizens have been granted asylum by other countries.

Continues on next page...

Transgender: Gender reassignment ('sex change') legal or openly performed without prosecution.

SLOVAKIA
Homosexuality: Legal. Same-sex registered partnerships. Anti-discrimination laws apply.
Transgender: Gender reassignment ('sex change') legal or openly performed without prosecution. Some personal documents may be reissued to reflect change.

SLOVENIA
Homosexuality: Legal. Same-sex registered partnerships. Anti-discrimination laws apply.
Trangender: No data or legal situation unclear.

SOLOMON ISLANDS
Homosexuality: Illegal. F/M. Imprisonable for up to 14 years.
Transgender: No data or legal situation unclear.

SOMALIA
Homosexuality: Illegal. F/M. Death penalty in some areas ruled by Sharia law and applies to both women and men. Imprisonable for up to 3 years.
Transgender: No data or legal situation unclear.

SOUTH AFRICA
Homosexuality: Legal. First country in the world to include equality and sexual orientation protection in its Constitution. Age of consent higher (19). Same-sex marriage and adoption of children allowed. Prepared to grant asylum to LGBT refugees.
Transgender: Gender reassignment ('sex change') legal or openly performed without prosecution. Personal documents may be reissued to reflect change.

SPAIN
Homosexuality: Legal. Age of consent equal. Same-sex marriage and adoption of children allowed. Anti-discrimination laws apply. Access to state donor insemination services.
Transgender: Gender reassignment ('sex change') legal or openly performed without prosecution. All personal documents may be reissued to reflect change.

SRI LANKA
Homosexuality: Illegal. M (F not mentioned in law). Imprisonable for 10 years.
Transgender: No data or legal situation unclear.

SUDAN
Homosexuality: Illegal. F/M. Death penalty applies under sharia law or 5 years imprisonment.
Transgender: No data or legal situation unclear.

SURINAME
Homosexuality: Legal but severe discrimination in criminal law. Age of consent higher (18).
Transgender: No data or legal situation unclear.

SWAZILAND
Homosexuality: Illegal. F/M. Imprisonment and /or fine.
Transgender: No data or legal situation unclear.

SWEDEN
Homosexuality: Legal. Age of consent equal. Anti-discrimination laws apply. Legal recognition of same-sex partnerships and adoption of children allowed. Access to state

donor insemination services. Prepared to grant asylum to LGBT refugees.
Transgender: Gender reassignment ('sex change') legal or openly performed without prosecution. All personal documents may be reissued to reflect change.

SWITZERLAND
Homosexuality: Legal. Age of consent equal. Registered same-sex partnerships. Anti-discrimination on the basis of 'lifestyle' clause in the Constitution. But no access to state donor insemination for lesbians.
Transgender: Gender reassignment ('sex change') legal or openly performed without prosecution. All personal documents may be reissued to reflect change.

SYRIA
Homosexuality: Illegal. F/M. Imprisonable for up to 3 years. LGBT citizens have been granted asylum by other countries
Transgender: No data or legal situation unclear.

TAIWAN
Homosexuality: Legal. Age of consent equal. Pending law allows same-sex marriage.
Transgender: Gender reassignment ('sex change') legal or openly performed without prosecution.

TAJIKISTAN
Homosexuality: Illegal. M (F not mentioned in law).
Transgender: No data or legal situation unclear.

TANZANIA
Homosexuality: Illegal. F/M. Imprisonable for life New laws have crimminalized lesbianism and made same-sex marriage illegal and imprisonable for 7 years. LGBT citizens have applied for asylum in other countries.
Transgender: No data or legal situation unclear.

THAILAND
Homosexuality: Legal. Age of consent equal.
Transgender: Gender reassignment ('sex change') legal or openly performed without prosecution.

TOGO
Homosexuality: Illegal. F/M. Imprisonable for up to 3 years.
Transgender: No data or legal situation unclear.

TONGA
Homosexuality: Illegal. M (F not mentioned in law). Imprisonable for up to 10 years.
Transgender: No data or legal situation unclear.

TOKELAU
Homosexuality: Illegal. M (F not mentioned in law). Imprisonable for 10 years.
Transgender: No data or legal situation unclear.

TRINIDAD AND TOBAGO
Homosexuality: Illegal. F/M. Imprisonable for 10 years.
Transgender: No data or legal situation unclear.

TUNISIA
Homosexuality: Illegal. F/M. Imprisonable for up to 3 years. LGBT citizens have been granted asylum by other countries.
Transgender: Rights only for born hermaphrodites.

TURKEY
Homosexuality: Legal. Age of consent equal. Lesbians and gay men are banned from the armed forces. LGBT citizens have been granted asylum by other countries.
Transgender: Gender reassignment ('sex change') legal or openly performed without prosecution. All personal documents may be reissued to reflect change.

TURKMENISTAN
Homosexuality: Illegal. M (F not mentioned in law). Imprisonable for 3 years.
Transgender: No data or legal situation unclear.

TURKS AND CAICOS
Homosexuality: Legal.
Transgender: No data or legal situation unclear.

TUVALU
Homosexuality: Illegal. M (F not mentioned in law). Imprisonable for 14 years.
Transgender: No data or legal situation unclear.

UGANDA
Homosexuality: Illegal. M (F not mentioned in law, but has been used against women). Imprisonable for life. In 2005 same-sex marriage was criminalized. LGBT citizens have been granted asylum in other countries.
Transgender: No data or legal situation unclear.

UKRAINE
Homosexuality: Legal. Age of consent equal. No anti-discrimination laws.
Transgender: Gender reassignment ('sex change') legal or openly performed without prosecution. All personal documents may be reissued to reflect change.

UNITED ARAB EMIRATES
Homosexuality: Illegal. M (situation for F unclear). Death sentence applies. In 2006 11 men were imprisoned for 5 years for attending a 'gay wedding'.
Transgender: No data or legal situation unclear.

UNITED KINGDOM
Homosexuality: Legal. Age of consent equal. Anti-discrimination laws apply. Same-sex civil unions recognized in 2005. Same-sex couples allowed to adopt in some areas. Prepared to grant asylum to LGBT refugees.
Transgender: Gender reassignment ('sex change') legal or openly performed without prosecution. All personal documents may be reissued to reflect change.

UNITED STATES
Homosexuality: Legal. In 2003 a US Supreme Court ruling overturned anti-sodomy laws which applied in 20 states. Progressive anti-discrimination laws and legal recognition of same-sex partnerships apply in some states and municipalities. Same-sex couples can adopt in some states. Prepared to grant asylum to LGBT refugees.
Transgender: Gender reassignment ('sex change') legal or openly performed without prosecution in some states. Personal documents may be reisssued to reflect change in most states. Traditional Native American acceptance of transgender.

URUGUAY
Homosexuality: Legal. Anti-discrimination laws apply.
Transgender: No data or legal situation unclear.

UZBEKISTAN
Homosexuality: Illegal. M (F not mentioned in law). Imprisonable for three years. LGBT citizens have been granted asylum by other countries.
Transgender: No data or legal situation unclear.

VANUATU
Homosexuality: Legal.
Transgender: No data or legal situation unclear.

VATICAN/HOLY SEE
Homosexuality: Technically legal but *de facto* forbidden.
Transgender: Condemned as 'repugnant'.

VENEZUELA
Homosexuality: Legal. Anti-discrimination laws apply. Social intolerance can be extreme and LGBT citizens have been granted asylum by other countries.
Transgender: No data or legal situation unclear.

VIETNAM
Homosexuality: Legal. Age of consent equal. Same-sex marriage banned since 1998.
Transgender: No data or legal situation unclear.

AMERICAN SAMOA
Homosexuality: Illegal. F/M. Imprisonable for 7 years.
Transgender: No data or legal situation unclear. Some traditional acceptance of transgender people.

YEMEN
Homosexuality: Illegal. F/M. Death penalty applies.
Transgender: No data or legal situation unclear.

ZAMBIA
Homosexuality: Illegal. M (F not mentioned in law). Imprisonable for 14 years.
Transgender: No data or legal situation unclear.

ZIMBABWE
Homosexuality: Illegal. M. (F not mentioned in law). Pro-LGBT clergy risk imprisonment. LGBT citizens have been granted asylum in other countries.
Transgender: No data or legal situation unclear.

Sources: International Gay and Lesbian Human Rights Commission (**IGLHRC**) 2006
Web: www. iglhrc.org; International Lesbian and Gay Association *World Survey* (**ILGA**) 2006
Web: www.ilga.org; *Legal Wrap Up on the Laws over the World Affecting LGBT Persons*, research by Daniel Ottosson, Public Law at Sodertorn University, Stockholm, Sweden, 2006, accessible at http://www.ilga.org/Statehomophobia/ WorldLegalWrapUpSurvey_Daniel_Ottoson.pdf **Sodomy Laws** www.sodomylaws.org
Behind the Mask 2006 web: www.mask.org.za Amnesty International AI-LGBT website www.ai-lgbt.org/status_worldwide.htm **Sexo, Amor y Homofobia** by Vanessa Baird, Amnesty Internacional/ Editorial EGALES, 2006. Press for Change Web: www.pfc.org. uk *Integrating Transexual and Transgendered People:A Comparative Study of European, Commonweatlh and International law*, last modified 2004 available on http://www.pfc.org. uk/liga/liba-4a6.htm

Notes:
1. Legality and social or cultural acceptance are not necessarily related.
2. Where there are no specific laws against homosexuality, other laws may be used against LGBT people.
3. Laws against male homosexuality may also be used against lesbians in some countries.
4. In the countries of the European Union cases of discrimination against sexual minorities can be challenged by referring to the Human Rights Act of 2000 providing another right has also been breached.

Laws concerning the status of LGBT people are not always clear and subject to change. Any updates, corrections or amendments will be gratefully received by Vanessa Baird on vanessab@newint.org

III. LATIN AMERICA, A NEW TWIST

New actors emerge on the Latin-American political stage. Self-proclaimed leftist parties have come to power by electoral means in some countries. Indigenous leaders reach positions of power at national level. At first sight, changes are apparent. But are they for real?

Left turn

LEFT-LEANING GOVERNMENTS HAVE BEEN TAKING POWER IN MANY LATIN AMERICAN COUNTRIES. BUT WHAT KIND OF LEFT DO THEY BELONG TO? AND WHY IS THE INTER-AMERICAN DEVELOPMENT BANK SO HAPPY TO WORK WITH THEM?

In recent years, a diverse cast of political figures have achieved power in several Latin American countries, all of them belonging to what might be called 'the Left'. This has been interpreted and advertised in each case as a ground-breaking event, announcing the possibility of a future essentially different from the past. The winning politicians proclaim that their victory marks the advent of those who have been historically excluded - whether because of their politics, culture, class or gender.

This happened in Brazil, for example, when former labor leader Luiz Inácio 'Lula' da Silva, candidate of the Workers Party (PT), won the 2002 elections, marking the first time that a member of the working class had reached the presidency. In October 2004 the Socialist Tabaré Vázquez, leader of the Broad Front coalition (ranging from Communists and former guerrillas to traditional party dissenters) won the elections in Uruguay. For the first time in Uruguay's 174-year history as an independent country the hegemony of the Colorado and Blanco parties was broken. A year later, Aymara leader Evo Morales, leading the Movement Towards Socialism (MAS) won the elections in Bolivia. The electoral campaign claimed that he would be the first indigenous president in the Americas and, during the celebrations following his taking of office, Uruguayan writer Eduardo Galeano said that Morales' victory had initiated 'a new history'. In 2006, Michelle Bachelet, from the Socialist Party, a former political prisoner and the daughter of a military officer murdered by Augusto Pinochet's dictatorship, became the first woman to reach the presidency in Chile.

In the case of Argentina there was no such heady expectation. After a financial, economic and institutional crisis which some called the worst in the nation's history, Néstor Kirchner, belonging to the Justicialista (Peronist) Party, and linked in the 1970s to his country's radical left, reached the presidency with scant political support and surrounded by gloomy predictions. He soon consolidated his political position, however, thanks to firm decisions regarding the Army and his stance on the endemic corruption in the justice and social security systems, as well as his initial stand on the question of Argentina's foreign debt.

Meanwhile, Venezuela's Hugo Chávez Frías has become one of the most conspicuous players on the international scene. This has been achieved not only by his exuberant and charismatic speeches (for example, on his radio and TV show *Aló, presidente - Hello, President* - presented as if he is in direct phone communication with Venezuelans) and his foreign policy - described by some as hyperactive and unusually aggressive for a Latin American president - but also by the radical reforms he has introduced at home.

Chávez has also gained great visibility due to Venezuela's privileged situation in relation to other Latin American nations. It is the world's fourth-largest oil producer and second-largest among the members of the Organization of the Petroleum Exporting Countries (OPEC). Chávez's interventions in OPEC have aimed to turn it into a strategic weapon for developing countries. He attempted unsuccessfully to redirect the Andean Community of Nations (CAN) towards an anti-imperialist stand (until Venezuela left the bloc in

2006). He has supplied oil to Cuba and other Latin American nations under soft financial conditions or in exchange for other goods and services. He has mounted firm opposition to the US-driven Free Trade Area of the Americas and has consistently called for anti-imperialist action in Latin American summits, from Brasilia 2000 to Mar del Plata 2005.

At home there have also been radical interventions, including the expropriation of idle lands for thousands of rural families and the pouring of oil profits into social missions, including the Misión Robinson mass literacy drive and new primary healthcare posts in the *barrios*.

Most commentators, from both ends of the political spectrum, tend to detach Chávez from the other new leftist leaders. Leftist US political scientist James Petras, for example, excludes Chávez when he talks of: 'a series of new *popular* elected presidents (Lula in Brazil, Gutiérrez in Ecuador, Vázquez in Uruguay and Kirchner in Argentina) who have respected privatized companies, rigorously pay foreign debt, who apply the IMF's fiscal policies and send military forces to Haiti to endorse the puppet government placed by the US and repress the struggle of the poor to restore the democratically elected Aristide administration.'

A fervent Chávez fan such as Cuban Enrique Ubieta Gómez, referring to assessments of Latin American political leaders made by the right-wing press, comments with irony: 'Lagos (Bachelet's predecessor) is idolized - something continuously stressed as a reminder to Bachelet; Lula and Tabaré are naughty democrats, but are people they can talk to; Kirchner is a contradiction, bothersome because it was unexpected; and of course Chávez and Fidel are two despicable extremists.'

From the other end of the ideological scenario, Rafael Rojas says: 'The region's more self-controlled experts, wishing to calm things down, insist that left-wing diversity makes it virtually impossible to create a sub-continental bloc against US hegemony, and more so against representative democracy and market economy... They think if politicians such as Lula in Brazil, Bachelet in Chile, and López Obrador in Mexico, align with a moderate left, willing to preserve democratic institutions and the market and respectfully negotiate cohabitation with the US, they would tilt other governments towards that current, such as Kirchner in Argentina, Vázquez in Uruguay, or Torrijos in Panama, and would contain the more radical and destabilizing pole.'

These speculations, even though they come from opposing points of view, are part of the distinction between a 'premodern' and 'modern' left. The latter distances itself from the 'reform or revolution' dichotomy and is based on a socialist and democratic political theory that critically questions the former left and its revolutionary goals. From some radical viewpoints, like that of the Chilean Helio Gallardo (now based in Costa Rica), this modern left - manifested in the progressive governments of Argentina, Bolivia, Brazil, Chile and Uruguay - 'carries out a practical and relatively comfortable neoliberalism which does not pitch it against the right

even though it may be complemented by a social concern rhetoric and targeted (and disintegrated) care mechanisms towards the poorer sectors.'

Other readings, such as that of the Uruguayan Yamandú Acosta, blame both lefts: 'If the left's theoretical-revolutionary perspective in the 1960s was responsible for the counter-revolutionary transformations coming from the authoritarian regressions of the 1970s, the theoretical-progressive perspective of the two lefts at the turn of the century is also responsible for the new dynamic of capital accumulation and its destructive impact on societies and the environment'.

This progressive or modern institutional Left derives its identity from the period when it was either underground or in exile during the dictatorships of the 1970s. In those circumstances it primarily sought the recovery of political democracy (putting social and economic transformation on the backburner), which had previously been despised and depicted as 'bourgeois democracy'. Once representative democracy has been reinstated, however - in a global context of neoliberal hegemony - it becomes a factor to control, restrict and deactivate every radical impulse.

The progressive movement has reached government at just the point when the Inter-American Development Bank (IDB) is proposing a new role for the State in Latin America. This 'rectification' is a consequence of the Bank's having recognized the inefficiency (the ultimate

sin in the eyes of neoliberals) of policies implemented in the 1990s. These policies had caused, amongst other things, exclusion, corruption, *latifundios* (large landholdings), explosive social situations, low growth, burgeoning foreign debt and inefficient privatization of State companies. As a result, then IDB president Enrique Iglesias proposed in 2004 that the State should break with the inefficiency and corruption of former governments, acting both as a regulator and as a compensator. This 'new state' should guide the market, launching programs to leave crises behind, and should redistribute social spending through third parties such as social activists, faith-based and philanthropic organizations, etc.

It seems that the new left-wing governments in Latin America are seen as appropriate vehicles for this kind of 'rectification'. ■

Sources:

Acosta, Yamandú: La Izquierda en América Latina: Reflexiones desde/sobre dos visiones. La Gaceta. Nº 39, Montevideo, December 2005; Evo Morales y Bolivia: gestos populistas y fondo neoliberal, www. rebelión.org

Gallardo, Helio: 'Siglo XXI. Militar en la izquierda', Editorial Arlekín, San José, Costa Rica, 2005.

Lesgart, Cecilia: "Usos de la transición a la democracia", Homo Sapiens, Santa Fe, Argentina, 2003.

Petras, James: Chile's New President, www. rebelión.org

Rojas, Rafael: 'Las nuevas izquierdas, España y Cuba', in El País, Madrid, 6 February 2006.

Ubieta Gómez: 'Enrique: Izquierda y Democracia',

Indigenous cultures

THE ELECTION VICTORY OF EVO MORALES IN BOLIVIA HAS BEEN ACCLAIMED AS AN EPOCHAL MOMENT FOR INDIGENOUS PEOPLES IN LATIN AMERICA. BUT ONE LEADER CANNOT CHANGE EVERYTHING. IN THE CONTINENT AS A WHOLE, MULTICULTURALISM HAS BEEN GAINING GROUND OVER THE OLD APPROACHES OF INTEGRATION OR DOWNRIGHT OPPRESSION - BUT DOES IT GO FAR ENOUGH?

In December 2005, Evo Morales (an Aymara), leader of the Movement Towards Socialism (MAS), won elections in Bolivia by an absolute majority - the largest electoral victory in the country for 50 years. On the eve of the swearing-in ceremony in January 2006, there was a colorful celebration with people from native communities. Amid the festivities the indigenous people 'anointed' Morales as their moral authority. His electoral campaign made much of the fact that he is among the first indigenous presidents in the Americas[1] - and when he was sworn

in he stated that: 'We have struggled for five centuries, we shall hold power for five centuries.'

The rise of Morales as a leader of indigenous Americans can be set against a series of transformations that range from academic and publishing events - such as the ascendancy of intercultural philosophy in the 1980s and the testimonial book *I, Rigoberta Menchú* in 1982 - to political events such as the fame accorded to Menchú when she received the Nobel Peace Prize in 1992. It is timely to look

again at the interaction between the State and indigenous cultures in the region. Is multiculturalism a new model within which these relations can develop? If so, is this the best one to guide us in critiquing and repairing past situations of injustice and oppression?

In the late 15th century when Europeans arrived in the Americas they found (but did not always acknowledge) great cultural diversity: societies with complex economic structures and socio-political organizations; hunting and gathering communities;

---

1. Benito Juarez (1806-1872), a Zapotec indian, was President of Mexico.

peaceable groups along with warriors; oppressors and oppressed; coastal and continental cultures, nomadic and settled groups, and some settlements larger and more splendid than any European city of the time.

Many of these institutions, ideas and images used by the native peoples - either invented by them, adopted from other cultures or inherited - were, in varying degrees, destroyed, weakened or dismissed by Europeans, especially the Spanish and Portuguese. Although Pope Paul III acknowledged that every American aborigine had a soul and banned their enslavement in the 16th century, the process of emerging European modernism had devastating effects. Conquest and colonization, and their legacy even after independence, had genocidal tendencies. War, weapons, disease and brutal labor systems decimated the indigenous population.

The diverse and rich cultures that survived those catastrophes are commonly lumped together, seen as uniform, and represented by the word 'Indian'.

This simplification does not reflect the genetic, linguistic and anthropological realities, and so Latin American states are having to face up to the problem of cultural diversity. In the past this had been seen by some state institutions as an insurmountable obstacle for which solutions had been proposed, ranging from massacres[2] to the enactment of special laws, whose paternalistic origins could not always be shaken off.

During the 19th century some of the young American states granted the Catholic Church the power to manage native communities, establishing a continuity with the missionaries' work during colonization. The Argentine Constitution of 1853 stated that: 'the Government shall maintain peaceful relations with the Indians and do the utmost to convert them to Catholicism'. In Colombia, an 1890 law regarding the forest people stated: '... the general legislation of the Republic cannot be applied to the natives the missions are attempting to civilize. The Government, in accordance with religious authorities, shall establish the way in which these populations should be governed.'

Later, even though state and church power became relatively weaker, laws with similar attributes continued to appear. In 1895 Nicaraguan Congress placed Indians 'under protection of the Republic'. In 1897 the Ecuadorian Constitution established the need to 'defend and protect Indians'. The Brazilian Civil Code of 1916 considered Indians to be legal minors. The 1942 Code continued this infantilization by allowing the protection status over an indigenous person to be gradually removed according to how his or her behavior evolved. In Chile, a law from 1931 barred Indians from selling their goods without court consent, unless the seller had some education.

According to some economic interpretations, the subjugation of native cultures by Europeans was an inevitable outcome of the clash between two incompatible modes of production. Subsistence economies, that generally do not accumulate capital, succumbed to an economic structure that already had the typical features of capitalism - production for a global market, accumulation, financial capital, commercial activity, wages and so on, while retaining some feudal elements (*latifundio* - large landholding - and serfdom). This weakness of native peoples seems to have been noticed by the labor legislation of several Latin American states. In Peru the 1933 Civil Code treats the Indian worker differently from any other of the waged categories. In Ecuador the 1964 Constitution also regulated the work of Indians differently.

During the first half of the 20th century several nations of the continent instituted areas of indigenous ownership where the inhabitants' traditions were to rule: Argentina (1912), Colombia (1927), Chile (1931), Panama (1946) and Venezuela (1947). However, in some countries this type of solution would have meant isolating most of the population in *resguardos* (reservations) or *reducciones* (centralized villages), as they were called in Colombia and Chile respectively. So, in the first half of the past century, countries such as Peru, Ecuador and Bolivia acknowledged the legitimacy of indigenous communities, reinstating some of their rights.

Likewise, several agencies were created to monitor the compliance of 'indigenous laws', such as the Indian Protection Service (Brazil, 1910), the General Board of Indigenous Affairs (Mexico, 1935), the Indian Protection Board (Argentina, 1946), and the Indigenous Affairs Board (Panama, 1952).

Indigenous peoples and the Constitution

THE CONSTITUTIONS of 15 Latin American states include regulations that recognize the rights of indigenous peoples. Here are some examples:

Article 75 of the Constitution of Argentina grants Congress the responsibility of '...acknowledging the ethnic and cultural pre-existence of Argentine indigenous peoples. Guaranteeing respect for their identity and the right to a bilingual and intercultural education. Recognizing the legal status of its communities and the communal possession and ownership of the lands they have traditionally occupied...'

The Paraguayan Constitution, Article 62, states: 'This Constitution recognizes the existence of indigenous peoples, defined as cultural groups existing before the formation and organization of the State...'; and in its Article 140: 'Paraguay is a pluricultural and bilingual country. Its official languages are Spanish and Guarani. The law shall dispose the ways in which one or the other will be used.'

Meanwhile, the Bolivian Constitution (Article 171) states: 'The social, economic and cultural rights of the indigenous peoples that inhabit national territory are recognized, respected and protected within the law, especially with regards to their original communal lands, guaranteeing the use and sustainable exploitation of natural resources, their identities, values, languages, customs and institutions...' ▪

---

2. On 11 April 1831 the Uruguayan army, led by the country's first constitutional President, Fructuoso Rivera, exterminated the Indian population in an episode known as the 'Salsipuedes massacre'. Between 1878 and 1879 the Argentine Minister of War, General Julio Argentino Roca, heading an army of 6,000 men, developed the so-called 'Desert campaign', which seized large areas of Indian territory.

Towards a new paradigm

Since the 1980s, there have been changes in integration policies - or at least in the documents defining indigenous communities. According to certain interpretations, many of these policies go beyond territorial claims - or beyond the classical focus of claims based on the modern concept of 'property', regulated by civil legislation - and betoken a change of model. These policies have been influenced by academic ideas such as Intercultural Philosophy, which has gained particular currency in Latin America. This takes issue with the idea of 'the other', and takes a new approach to arrive at a deeper understanding between various cultures.

In 1982 a Mayan Quiché woman born in Guatemala received acclaim for her book *I, Rigoberta Menchú*, and the issues it tackles. The book became what many consider a classic text among US multiculturalists, placed Menchú on the road to her Nobel Prize, and fueled an academic debate.

More recently, in the political arena, the *Draft Declaration on the Rights of Indigenous Peoples*[3] states: 'In the last two decades, multiculturalism as a new concept of unity-in-diversity has had growing acceptance as a political and constitutional principle in Latin America. With different approaches and content (multiethnic and pluricultural nations; intercultural education and public services) multiculturalism has developed as the dominant paradigm. This has happened not only in countries with proportionately large indigenous populations (eg Ecuador, Bolivia, Guatemala, Mexico) but also in Brazil, Argentina and Colombia, where these groups make up a minority of the national population.'

This assessment is based on more recent constitutional texts, such as that of Bolivia (modified by a 2004 law), which starts by defining the country as 'free, independent, sovereign, multiethnic and pluricultural.' The Constitution of Ecuador (1998) expands the concept of 'territory' linking it to the environment and respect for production and exchange systems originated in Quechua traditions. Section 119 of the Venezuelan Constitution (1999) legitimizes ancestral ownership of the land and states the obligation to guarantee the lifestyles of native communities. Likewise the Constitution of Brazil (1988) institutes the concept of indigenous communities' 'habitat', including among the conditions defining it 'the areas necessary for cultural reproduction and survival as a community'.

Integration, reparation and criticism

All these policies contain an element of criticism of - and reparation for - not only the policies of past empires but also the current actions of nation-states. As modernized states they introduced education. However, much of the teaching operated in a cultural vacuum which tended to delegitimize the collective indigenous memory through the concept of *tabula rasa* - 'clean slate' - suggesting that indigenous people's minds were a kind of blank page open to receive the imprint that 'civilization' brought to them.

Another way to assimilate cultural identities is through the idea of a 'culture of poverty'[4] - meaning the way of life of people whose traditional social and economic structures have been usurped or destroyed. However, peasant or village communities with simple technologies and subsistence economies that manage to preserve their 'habitat' may be poor financially but do not necessarily develop the typical traits of a culture of poverty.

Some, as we have seen, believe that multiculturalism is the way to right the wrongs. Others are not so sure. Writer and critic Fernando Aínsa thinks: 'It is important to stress that multicultural presence in a society does not mean equal opportunities for all. Indeed, increased options emphasize even more the lack of equal opportunities...'[5] Some go further and suggest that multiculturalism actually perpetuates poverty and injustice. ∎

New era

THE ANDEAN CALENDAR divides solar cycles (4,000 years) into two parts. The first 2,000 years of each new cycle are ascending, while the latter fall. Each of these stages is also divided into 500-year microcycles, in which abundance alternates with hardship. According to this concept, 1992 would have been the end of a dark 500-year period which started in 1492, and the end of that cycle. Thus, we are witnessing the rise of a new solar cycle in whose first 500 years would be included the promise of five centuries of indigenous rule announced by Bolivian President Evo Morales. ∎

3. Osvaldo Kreimer: Report of ohe Rapporteur Meeting of the Working Group on the Fifth Section of the Draft Declaration with special emphasis on Traditional Forms of Ownership and Cultural Survival, Right to Land and Territories. Washington, DC, November 7-8, 2002.

4. Lewis, Oscar: *Anthropology of Poverty*, FCE, Mexico, 1961.

5. Ainsa, Fernando: 'El desafío de la identidad múltiple en la sociedad globalizada' in Cuadernos Americanos, 63, UNAM, Mexico, 1997.

IV. ENERGY SCENARIOS

So far, no efficient and profitable alternatives have been developed to replace non-renewable energy sources. Within the current model, possibilities range from further development of technologies that obtain energy from other sources - such as nuclear fission or wind - to a never-ending series of wars and conflicts over the control of existing resources. But the real solution could lie in a paradigm shift.

In search of a new cultural paradigm

AS OIL RESERVES BECOME INCREASINGLY DEPLETED, GOVERNMENTS OF INDUSTRIALIZED COUNTRIES ARE SIZING UP THEIR OPTIONS FOR THE FUTURE. THINKING THAT CLEAN RENEWABLE ENERGY COULD NOT MEET CURRENT DEMAND, MANY GOVERNMENTS ARE TURNING AGAIN TO NUCLEAR FISSION, WHICH PRODUCES LARGE AMOUNTS OF RADIOACTIVE WASTE - AND THE QUOTA MARKET IN CARBON EMISSIONS IS ALSO LEADING TO PRESSURE FOR THE NUCLEAR SOLUTION. EVERYTHING SEEMS TO INDICATE THAT A TRANSITION TOWARDS CLEAN AND RENEWABLE POWER SYSTEMS WILL BE IMPOSSIBLE WITHOUT A PROFOUND CULTURAL CHANGE. THE PROBLEM LIES NOT IN TECHNOLOGICAL SOLUTIONS, BUT IN SOCIAL PRIORITIES.

Renewable energy sources are those that regenerate themselves through natural cycles (for instance, the energy of tides or rivers), through artificial cycles (wood from forests, biodiesel or biogas), or that are permanent or virtually permanent (such as solar and wind energy).

Twenty per cent of the electricity used in the world comes from renewable sources. Hydroelectric generation amounts to 90 per cent of renewable energies used today; other kinds have minimal shares - biomass 5.5 per cent, geothermic 1.5 per cent, wind 0.5 per cent and solar 0.05 per cent.

Some 80 per cent of energy needs in Western industrial societies come from the heating and air-conditioning of buildings and transportation - cars, trains, planes. However, most renewable energies are applied in the production of electricity.

Renewable energies contrast with sources such as oil, coal or natural gas, which will be depleted within a few decades according to even the most optimistic estimates. Nuclear fission is also classed as a non-renewable energy since it requires the mining of uranium, but reserves of uranium would be sufficient to last for thousands of years of energy production.

A problem typical of renewable energies is their scattered nature and relatively low intensity. We are used to talking of power-producing 'centers', be they nuclear, hydroelectric or geothermic. But some energy systems cannot be centralized, such as those producing heat energy from the sun (sun collectors and ovens) or electric power from light (photovoltaic panels).

In order to obtain a megawatt hour of electricity (the annual per capita consumption among industrialized countries), the dweller of a home located in a cloudy region of the Northern hemisphere must install at least eight square meters of photovoltaic panels. Four square meters of solar thermic collectors are enough to cover hot water needs.

Clean/polluting energies

Both renewable and non-renewable energies can be divided into two categories, clean and polluting.

Amongst the clean ones are the sun, wind, freshwater bodies (hydroelectric energy), seas and oceans (tidal power), and heat from the Earth (geothermal energy).

The polluting energy sources cause two kinds of damage. Those based on burning fuels (whether oil, coal or biomass) produce greenhouse gases that upset the planet's thermal balance and eventually issue poisonous substances. Nuclear power based on fission produces highly radioactive waste, which is extremely dangerous to all life-forms if released in the environment.

Energy from polluting renewable sources has the same problem as that produced by fossil fuels, when they burn they issue carbon dioxide, a greenhouse gas. They are often more contaminating than fissile fuels, since they issue more soot and solid particles than natural gas and other oil derivatives.

It has recently been proposed that even hydrogen-burning energies should be classified among those that pollute, since the product of burning hydrogen is water (instead of carbon oxides) - and although water is a harmless substance in itself, excessive water vapor in the atmosphere contributes to global warming.

Clean energies such as hydroelectric power, which releases no emissions into the environment, may have negative consequences for regional climates if it increases water tables, causing changes in local rains and disturbing fisheries, upsetting the ecosystem.

Radioactive waste: the most lasting human product

The energy produced by the controlled fission of heavy atoms - the proccess by which the nucleus of an atom splits, releasing substantial amounts of energy - is used in many countries. Some 99 per cent of electric power in France is produced by nuclear plants; Italy, meanwhile, does not produce electric power, but buys it from French nuclear plants. Now other countries are looking to replace carbon or oil-based generators with nuclear reactors, in order to comply with the Kyoto Protocol.

The process of producing nuclear power itself is clean. The problem caused by the use of radioactive materials in reactors is that once a year the fuel has to be changed - and once used it is highly radioactive. There are systems that separate its components, recover and stabilize the residues, but a significant amount of the spent fuel will remain for tens of thousands of years issuing radiation that is hazardous to living beings.

Another much-discussed nuclear alternative is fusion - the proccess by which multiple nuclei join together to form a heavier nucleus - which would leave no contaminating waste, or at least would leave waste that some believe could be rendered harmless within a century. However, humanity has as yet only been able to produce hydrogen bombs through this process, and is still some way from understanding how to keep that huge energy potential under control. Iter, a multinational project of an experimental reactor whose goal is to keep fusion under control for 400 seconds, will start to operate in France towards 2016. The first industrial fusion reactor could not be operational until the last third of the 21st century.

European countries such as Germany and Belgium that have developed alternative, sustainable energies - especially wind power - have decided to end their nuclear programs and have set dates for the closure of their fission power plants. But others, who have signed the Kyoto Protocol, such as Sweden, believe that nuclear energy will help them to comply with the carbon-emission goals.

The problem of spent radioactive fuels has, however, no solution. There is no human institution that has lasted as long as radiation from the spent fuel of a reactor. This means that, even if in the short term we are able to store the waste safely, we are creating a serious problem for the future of humanity.

Until ways are found to completely recycle and prevent radioactive waste, nuclear energy will not be a responsible alternative, however efficiently it might reduce carbon-dioxide emissions. ∎

Solar power

IN ANCIENT TIMES, both Greeks and Romans learned too late that forests grow slower than the needs of their stoves. After turning their forests into wastelands and marshes, both cultures developed solar architecture strategies that utilized solar energy through the correct orientation of their buildings.

Romans enhanced the technique by introducing glass in windows, and their juridical culture even had rules for shade. Ulpian, a lawyer in the second century, successfully defended owners who sued their neighbors for erecting buildings that cast a shadow over their homes.

Roman experience did not, however, prove very useful to the rest of Europe, which almost wiped out all its forests over several centuries during the Middle Ages.

Cultures with low population growth have been protected from these energy catastrophes, and peoples that lived in cold climates - like the northern Chinese - developed important cultural strategies to exploit solar architecture. The rules of Feng Shui, for example, form a bio-climatic system that assures the good use of available solar radiation. Another strategy uses soil as a natural accumulator of solar energy. Today in China some 10 million people are living in homes dug up from sedimentary soils in order to use the natural temperature regulation of the land.

The easy availability of fossil fuels over the last two centuries has led humanity to forget the vital principles of solar design. ∎

The geopolitical battle

UNCERTAINTY ABOUT THE REAL SIZE OF OIL DEPOSITS RAISES FEARS THAT A GLOBAL ENERGY COLLAPSE COULD TAKE PLACE WHEN DEMAND FOR OIL EXCEEDS SUPPLY. THE IMPORTANCE OF OIL IS SO GREAT THAT IN SPITE OF THE GLOBAL ECONOMIC TREND TOWARDS PRIVATIZATION, SOME OF THE BIGGEST OIL COMPANIES IN THE WORLD ARE STATE-OWNED AND HAVE CONTROL OVER 70 PER CENT OF WORLDWIDE RESERVES. THE US IS AT THE CENTER OF ALL OIL-RELATED GEOPOLITICAL TENSIONS. OTHER ENERGY SOURCES SUCH AS NATURAL GAS ARE STARTING TO BECOME ENORMOUSLY SIGNIFICANT IN THE GEOPOLITICAL ENERGY EQUATION.

Over the last 40 years serious errors have been made in estimating the size of oil reserves. Some analysts in the early 1970s confidently forecasted that oil reserves would be exhausted by the 1990s, and these errors have resulted in a general discrediting of such predictions. One factor contributing to the continued flow of oil is that technological advances have made it possible to extract oil from deposits that it was previously considered unfeasible to exploit.

Nevertheless there will come a point in the near future when demand for oil will exceed supply. This point, called 'peak oil' by the experts, is considered by some to be very close, even before 2010, while others predict that it will arrive around 2020. Whenever it happens, peak oil will trigger an irreversible process of increasing prices for oil and its derivatives. However, peak oil will not mean the end of the oil age. The US Geological Service estimates that oil exhaustion will not occur before 2060.

More pessimistic scientists predict that by 2030 oil extraction will have fallen to a level so low that a global energy collapse will ensue. According to a 2004 British Petroleum report reserves will be exhausted in the 2040s.

One view, not often considered, has it that oil reserves will never be exhausted because high prices will give impetus to the development of alternative technologies (for the transformation of wind, tide and solar energy amongst others). According to this view, just as coal has continued to be used as a fuel during the oil derivatives age, so oil will continue to be one source of energy alongside many others for the near future, although it will lose its present hegemony.

Alarms and readjustments

Natural gas has become a key element in world energy dynamics. In the European winter of 2005-2006 the distribution company Gaz de France reported a 30-per-cent decrease in the amount of gas received from Russia via Ukraine.

This decrease - caused by Ukraine claiming considerable amounts of gas for itself - produced panic in several countries as the European energy matrix is significantly dependent on Russian gas. Ukraine gets 20 per cent of the proceeds from Russia's gas sales as rent for gas pipelines and tolls.

Towards the end of the year the announcement that a gas pipeline is to be constructed connecting Russia and Germany through the international waters of the Baltic Sea caused relief in European governments. Germany will become a distributor of Russian gas and several countries of the ex Soviet Union, which hitherto were important players in economic relations between Russia and Europe, will lose the benefits derived from providing transit services. Up to now alliances between the US and the former Soviet republics have enabled Washington to intervene in European energy policies through restraints imposed on the transport of gas and oil.

China is a very important consumer and net importer of oil and the increase in oil prices during 2005 has been attributed to its consumption growth. Chinese

South American energy ring

IN SOUTH AMERICA Venezuela is at the center of a conflict in which the US is also a protagonist. South American fossil-fuel deposits are concentrated in Venezuela and Colombia in the form of oil and in Bolivia and Argentina in the form of natural gas.

In the 1960s Argentina created a national gas pipeline network. When Brazil began to consume Bolivian gas in the 1970s Argentina rapidly constructed trans-Andean pipelines to supply Chile in an effective attempt to pre-empt any deal between Chile and Bolivia that could further strengthen the latter's position.

There are several factors that place Bolivia at the center of important geopolitical changes related to energy and put at risk Argentina's central role in gas distribution for the Southern Cone: the growth of consumption in Argentina; an adjustment in the estimate of Argentina's reserves suggesting that they will be exhausted in little more than a decade; the discovery of medium-sized deposits in Peru; and the confirmation that Bolivia's reserves are particularly large.

In 2005 the term 'energy ring' began to be used to describe a system of interconnected gas pipelines between Argentina, Brazil, Chile and Uruguay. The gas to feed this energy ring would be produced by Bolivia with the possibility of eventual contributions from the newly discovered Peruvian deposits, ensuring supplies for at least the next 30-40 years.

Venezuela has shown interest in joining this ring, which would significantly boost regional integration and energy independence but would inevitably generate serious conflict with the US. A substantial part of Venezuelan oil is currently exported to the US. Bolivia also faces disagreement with Washington over the proposed destination of its natural gas.

Up to now problems connected with energy have not led to armed conflicts in South America. However a glance at the world map shows that the main areas of conflict are associated with oil (Iraq, Iran and Sudan, for example). It is not inconceivable that if Washington succeeds in strengthening its position in the Middle East it will then focus its attention on South America and its incipient energy independence. ■

Alternative fossil fuels

THE EXHAUSTION of oil reserves will provoke the development of alternative energy sources. Some scientists consider that as humanity's energy consumption is so great and its growth so accelerated renewable energies (wind, tide, solar, bioclimatic) will be insufficient to fulfill demand.

They look instead to the exploitation of new fossil fuels. In the case of natural gas, this is already happening. Some decades ago natural gas that was unavoidably extracted because of its presence in oilfields was burned on site as its exploitation was not considered profitable. Today gas is an essential element in the world energy equation.

Extra heavy oils, oil sands and shale, low permeability sandstones, methane between coal layers and coal gasification all seem possible fuel sources that would not require large-scale industrial restructuring. But even if it proves possible to exploit these new types of fossil fuel, the technological developments necessary may not take place before the oil runs out. ▪

companies invest heavily in oilfields located outside the US zone of influence and in coming years this could result in a situation similar to that during the Cold War period, only this time with China as one of the poles of conflict. Its investments in Iran and Sudan, important producers of gas and oil, endanger what for the US are very sensitive balances. The US has repeatedly indicated that it will not tolerate an independent production capacity in Iran. Its warnings to Tehran about activities relating to the development of nuclear technology are very similar to, and have the same rhythms as, the threats formerly made against Iraq.

Washington's show of force
At the same time China is becoming a threat in economic terms to a degree unimaginable until recently. In the middle of 2005 the Chinese oil giant Chinese National Offshore Oil Corporation (CNOOC) was unable to buy the US company Unocal (known before as The Union Oil Company of California) due to an explicit prohibition by the United States Congress. Had it taken place, the purchase would have represented the largest transaction ever made by a Chinese company in the US; it was able to offer a higher purchase price than the powerful North American corporation Chevron.

This offer provoked intense resistance in Washington as oil plays a very significant strategic role in the US economy. Unocal owns substantial reserves in Asia and, fearing that China would gain control over valuable sources of oil and gas, Congress prohibited the acquisition of Unocal by CNOOC on the grounds of national security.

The US has demonstrated that it will not be deterred by any legal obstacle from maintaining control over energy sources deemed necessary for the functioning of its industry. The war against Iraq, undertaken in spite of UN opposition, is a demonstration of force that China, the European Union and Russia must take into serious consideration. US hegemony in economic and military spheres is not to be questioned, especially where oil is concerned. ▪

V. GENDER

All over the planet, whatever the level of their country's development, women are at a disadvantage. They are worse paid - if they are paid at all - than men for doing the same job and they are less represented in decision-making positions and processes. In some regions, just because of their gender, their very lives are at risk.

Missing women

WOMEN AND GIRLS ARE NOT AS HIGHLY VALUED AS MEN IN MANY SOCIETIES. UN ESTIMATES FROM 2006 SAY THAT BETWEEN 113 AND 200 MILLION WOMEN ARE 'MISSING'. EVERY YEAR BETWEEN 1.5 AND THREE MILLION OF THEM LOSE THEIR LIVES AS A RESULT OF GENDER-BASED VIOLENCE OR NEGLECT. THERE ARE SEVERAL EXPLANATIONS: IN SOME COUNTRIES (CHINA IS AN EXAMPLE), BOYS ARE PREFERRED TO GIRLS. THIS CAN LEAD TO FEMALE INFANTICIDE OR SELECTIVE ABORTION. THOSE THAT SURVIVE MAY SUFFER THE NEGLECT OF THEIR PARENTS, BECAUSE FOOD AND HEALTHCARE ARE PRIORITIZED FOR THE MALE MEMBERS OF THE FAMILY.

In societies that value males more highly than females, the killing of women - 'femicide' - is all too common; women's lives are considered cheap. But some femicide is dressed up as 'honor' killings. In countries where women are considered the property of men, their menfolk can murder them for choosing their own sexual partners (thereby supposedly besmirching family 'honor'). Young brides may be killed if their parents did not pay sufficient money to the husband - these are called 'dowry' deaths. The international sex trade kills many women, as do wars. However, the main cause of women's death worldwide is domestic violence.

The term femicide is used to denote women's murders that result from their gender. The word was coined in 1992, by Jill Radford and Diana Russell in the book *Femicide: the Politics of Woman Killing*, and refers to the misogynist killing of women - killed just because they are female. Another similar term is *gendercide* coined by Mary Anne Warren in *Gendercide: the Implications of Sex Selection* (1985).

According to Jill Radford, femicide is the killing of women that is condoned or even promoted by states and/or religious institutions. It serves as a means of control over women as a sexual class and to maintain the patriarchal *status quo*. The absence and inadequacy of legal and policing institutions, and also their male profile, reinforces patriarchy. Radford claims that femicide, as seen in court and in the media, is surrounded by the myth of 'woman-blaming'.

Women's behavior is scrutinized and measured against men's idealized views of how women should behave. The message to women is: 'Step out of line and it may cost you your life', while for men it is: 'You can kill her and get away with it'. These messages for example 'advise' women not to live alone; not to go out unaccompanied (meaning without a man); not to go to certain areas. Such 'advice' controls women by constraining their public activities. A women's place, according to such patriarchies, is in the home. But the worst and most ironic fact is that even there women are not safe: home can be the deadliest place - women's abuse is out of sight.

More about femicide

The concept of femicide has differing interpretations. For example Jacqueline Campbell and Carol Runyan (1998) used the term femicide to refer to all killings of women. Desmond Ellis and Walter De Keseredy (1996) took into account intention in such crimes, so that femicide is crimes that involve premeditation. For Russell and Radford, the key is in the relationship between the murdered woman and her killer and the motives of the killing.

In countries where femicide has been investigated, it occurs most frequently within the family, usually by the woman's lover, husband or other male relative. This has led to the notion of 'intimate femicide'. Killings that happen outside the domestic or familial sphere are called non-intimate or sexual femicide, according to the type of relationship in which they occur, or the conditions of rape for example. Murders of sex workers are commonly perpetrated by clients. In war or conflicts, meanwhile, the victims are on the 'enemy' side: they are raped and murdered to emphasize the defeat of one side or the other.

Other studies have noted ritualistic female murders based on beliefs associated with women's sexual organs, for example in Zambia, Zimbabwe and South Africa.

Sharon Hom (2001) suggests reconceptualizing female infanticide in China as 'social femicide' because those crimes are caused by a social order that devalues women's life. Russell suggests the classification of 'massive femicide' for all the killings of women and girls caused by male conducts of power and domination.

Femicide is currently outlawed under the Statute of the International Criminal Court. However, agreements and treaties remain to be ratified, both at international and state levels, to enforce laws and guarantee non-violence towards women. ■

The murdered women of Latin America

IN RECENT YEARS, activists in Latin America have often used the term femicide to denounce an alarming phenomenon in the region: the killing of women just because they are women. This trend has grown since 1993 when a series of killings started in the Mexican city of Ciudad Juárez. Activists have called it femicide because they want to draw attention to the frequent deaths of women due to their gender, and their abuse through torture, mutilation, sexual cruelty and/or violence. States frequently downplay the systematic nature of such abuse and deaths.

In 1994, around the time the term came into being, the Inter-American Convention on the Prevention, Punishment and Eradication of Violence against Women was held in Belem do Para, Brazil, and was later ratified by all Latin American and Caribbean nations. Article 1 states that: 'Violence against women shall be understood as any act or conduct based on gender which causes death, or physical, sexual or psychological harm or suffering to women, whether in the public or in the private sphere'. Article 7b describes the state's duty to 'apply due diligence to

prevent, investigate and impose penalties for violence against women'.

In Latin America, the gender-based violence obviously varies in specifics from country to country. But, however it happens, it is an abuse of Human Rights: the right to life, personal integrity, liberty, security and justice. In addition, it hinders development and peace within societies. States that do not act to prevent femicide, and then investigate, legislate and punish, are failing in their duties to half their population.

Because in many countries corruption, ineffectiveness and patriarchy have allowed femicide to flourish, it is only through the activism of non-governmental women's groups and human rights organizations that we have learned of the epidemic scale of the crimes. As a result, the Inter-American Commission of Human Rights (IACHR) prepared its *Report on the Situation of the Rights of Women in Ciudad Juárez, Mexico: the Right to be Free from Violence and Discrimination* to raise the profile of the problem. It is estimated that in that city between 1993 and 2005

almost 300 women were killed and 4,500 remain missing, according to the National Human Rights Commission.

Many governments do not gather the data that could provide an accurate picture of the scale of the femicide problem. Where official data is given, the figures are always lower than those gathered by NGOs and similar organizations. Some information systems in Latin American countries do not specify sex, age or ethnic background, nor information about the relationship between the victim and the perpetrator, thus making it difficult to carry out comparative studies within each country or the region.

In this regard, Amnesty International's 2005 annual report noted that in Latin America, gender-based violence against women is still 'a problem of epidemic proportions both in the home and in the community', and that the governments of the region continue to ignore the Belem do Para Convention provisions. ∎

The GEI and the gender gap

NO COUNTRY TREATS ITS WOMEN AND MEN EQUALLY; NO COUNTRY HAS REACHED GENDER EQUITY. HOWEVER, IN THE LAST TWO DECADES THERE HAS BEEN A CHANGE, SINCE MANY COUNTRIES HAVE ACKNOWLEDGED GENDER INEQUITY AS A PROBLEM. THIS NEW REALITY CALLS FOR THE CREATION OF NEW TOOLS TO MEASURE THIS MULTI-FACETED CONCEPT. ONE OF THEM IS THE GENDER EQUITY INDEX DEVELOPED BY SOCIAL WATCH.

Since 1979, when the United Nations General Assembly adopted the Convention on the Elimination of All Forms of Discrimination Against Women, gender inequity has been a central issue in the global development agenda. After the World Summit for Social Development in 1994 and the Fourth World Conference on Women (Beijing) in 1995, the international community focused two of the Millennium Development Goals, to be met by 2015, on the situation of women. Number three seeks to promote gender equality and empower women, with particular attention to equitable representation in decision-making processes; and number five seeks to reduce maternal mortality by three-quarters.

In spite of these gestures, acting on this consensus remains an uphill task, since

there are 47 UN member countries that have still neither signed nor ratified the Convention, and 43 others have done so with reservations. In the meantime, the global gender divide remains disturbingly wide. Out of the 1,300 million poor in the world, 70 per cent are women; more than two-thirds of the 860 million illiterate people are women; globally, female income reaches between 50 per cent and 80 per cent that of male income. Each day, complications during pregnancy and birth kill 1,600 women.

Social justice is not possible unless this situation is redressed. Former UN Secretary General Kofi Annan has stressed that: 'There is no tool for development more effective than the empowerment of women. No other policy is as likely to raise economic

productivity, or to reduce infant and maternal mortality. No other policy is as sure to improve nutrition and promote health - including the prevention of HIV/AIDS. No other policy is as powerful in increasing the chances of education for the next generation. And I would also venture that no policy is more important in preventing conflict, or in achieving reconciliation after a conflict has ended.'

The converse is also true: unless current trends change fundamentally, the Millennium Development Goals for 2015 will not be met.

The Gender Equity Index
Gender equity is a complex, multi-faceted concept that can be difficult to measure. In order to help monitor

TABLE 1. GENDER EQUITY INDEX (GEI) - 2006

Countries ranked according to their performance on a range of indicators of gender equity.
A score of 100 would indicate complete gender equity.

Country	Score	Country	Score	Country	Score	Country	Score	Country	Score
Sweden	89	Switzerland	74	Jamaica	65	Malta	58	Mali	46
Finland	86	Hong Kong	73	Kazakhstan	65	Mozambique	57	Niger	46
Norway	86	Hungary	73	Sri Lanka	65	Tajikistan	57	Turkey	46
Denmark	81	Israel	73	Suriname	65	Uzbekistan	57	Bahrain	45
Aotearoa/New Zealand	81	Portugal	73	Viet Nam	65	Albania	56	Bangladesh	45
Bahamas	80	Slovenia	73	El Salvador	64	Ghana	56	Egypt	45
Iceland	80	Ukraine	73	France	64	Korea, Rep.	56	Eritrea	45
Australia	79	Austria	72	Azerbaijan	63	Cape Verde	55	Guinea-Bissau	45
Barbados	79	Czech Republic	72	Chile	63	Lesotho	55	Kuwait	45
Latvia	79	Panama	72	Dominican Republic	63	Mauritius	55	Algeria	44
Lithuania	79	Argentina	71	Italy	63	Nicaragua	55	Equatorial Guinea	44
Canada	78	Romania	71	Belize	62	Laos	54	Morocco	44
Moldova	78	Thailand	71	Kenya	62	Madagascar	54	Oman	44
United States of America	78	Ireland	70	Armenia	61	Senegal	53	Syrian Arab Republic	44
Colombia	77	Macedonia	70	Cambodia	61	Solomon Islands	53	Congo, Rep.	43
Estonia	77	Trinidad and Tobago	70	Ecuador	61	Zambia	53	Nigeria	43
United Kingdom	77	Uruguay	70	Japan	61	Guatemala	52	Saudi Arabia	43
Netherlands	76	Belarus	69	Malaysia	61	Indonesia	52	United Arab Emirates	43
Philippines	76	Georgia	69	Maldives	61	Tunisia	51	Sudan	42
Spain	76	Brazil	68	Mexico	61	West Bank and Gaza	51	Nepal	41
Croatia	75	South Africa	68	Swaziland	61	Angola	50	Burkina Faso	40
Namibia	75	St Lucia	68	Uganda	61	Zimbabwe	50	Togo	40
Russia	75	Venezuela	68	Fiji	60	Iran	48	India	39
Rwanda	75	Costa Rica	67	Kyrgyzstan	60	Gambia	47	Central African Republic	38
Slovakia	75	Honduras	67	Peru	60	Guinea	47	Pakistan	38
Belgium	74	Tanzania	67	Bolivia	59	Jordan	47	Sierra Leone	37
Botswana	74	Cuba	66	Burundi	58	Benin	46	Chad	36
Bulgaria	74	Cyprus	66	China	58	Ethiopia	46	Côte d'Ivoire	36
Mongolia	74	Paraguay	66	Guyana	58	Lebanon	46	Yemen	26
Poland	74	Greece	65	Luxembourg	58	Malawi	46		

women's position worldwide, the Social Watch coalition of citizen organizations has developed a Gender Equity Index (GEI). The GEI positions and classifies countries according to a selection of relevant indicators of gender inequity, based on internationally available and comparable information on economic activity, empowerment and education. In its 2006 issue, the GEI classifies 149 countries and confirms that no country currently offers women the same opportunities as men. The Index shows that, although the situation of women has improved to some extent in recent years, it is clear that their economic and political opportunities are still limited.

Disaggregating the GEI: Different regions

Results obtained for the GEI 2006 indicate that Sweden, Norway, Finland and Denmark are the highest-ranking countries.

The income gap

The economic participation dimension of the GEI seeks to reflect inequity via two indicators - the percentage of women in paid work in the non-agricultural sector, and the ratio of estimated income between women and men.

Women worldwide have less access to the job market, and are subject to labor discrimination, seen in lower wages. In global average terms women earn little more than half what men do; in other words, for the same job women receive 53 per cent of the income earned by men. This situation varies regionally. The smallest gap is found in the North America (0.63) and Central Asia (0.62) regions, and the largest in the Middle East and North Africa (0.32) and Latin America and the Caribbean (0.43).

Disparity: less and more

There is a smaller disparity in education than in income or empowerment, though the gap is below 0.5 (meaning girls are half as likely as boys to be in school) in Chad, Central African Republic, Guinea-Bissau, Guinea, Sierra Leone, Benin and Yemen.

On the other hand, gender inequity is at its greatest when it comes to analyzing the percentage of women who are professionals and managers, who hold parliamentary seats and ministerial office. Despite being more than half of the world's population, women only hold six per cent of cabinet

INCOME GAP (WOMEN'S EARNINGS AS % OF MEN'S) BY GEOGRAPHIC REGION

Region	Average
East Asia and the Pacific	0.59
Europe	0.58
Central Asia	0.62
Latin America and the Caribbean	0.43
Middle East and North Africa	0.32
South Asia	0.46
Sub-Saharan Africa	0.56
North America	0.63
Total	0.53

posts in national governments. A few years ago, only Norway, Sweden and Finland had rates above 40 per cent. In 1995, Sweden had the world's first cabinet with 50 per cent of women. The global average is for parliaments to be composed of 15 per cent of women. Their general absence from government structures results in priorities being set without their contribution or opinion, although their life experience and subjectivity could mean major differences in the perception of community needs, concerns and priorities.

Since 2004 there has been an improvement in the number of women taking part in decision-making processes; the 2006 GEI shows 17 countries from both South and North to be above 30 per cent: Rwanda, Sweden, Norway, Finland, Denmark, Netherlands, Argentina, Cuba, Spain, Costa Rica, Mozambique, Belgium, Austria, Iceland, South Africa, New Zealand/Aotearoa and Germany. Other countries have now followed Sweden's example and appointed cabinets in which at least 50 per cent are female: Spain in 2004 and Chile in 2006.

Gender inequity by regions and country income

Except for Australia, all of the highest-ranked countries in the GEI are European. Gender inequity tends to be at its worst in countries from the Middle East, the northern half of Africa and South Asia. ∎

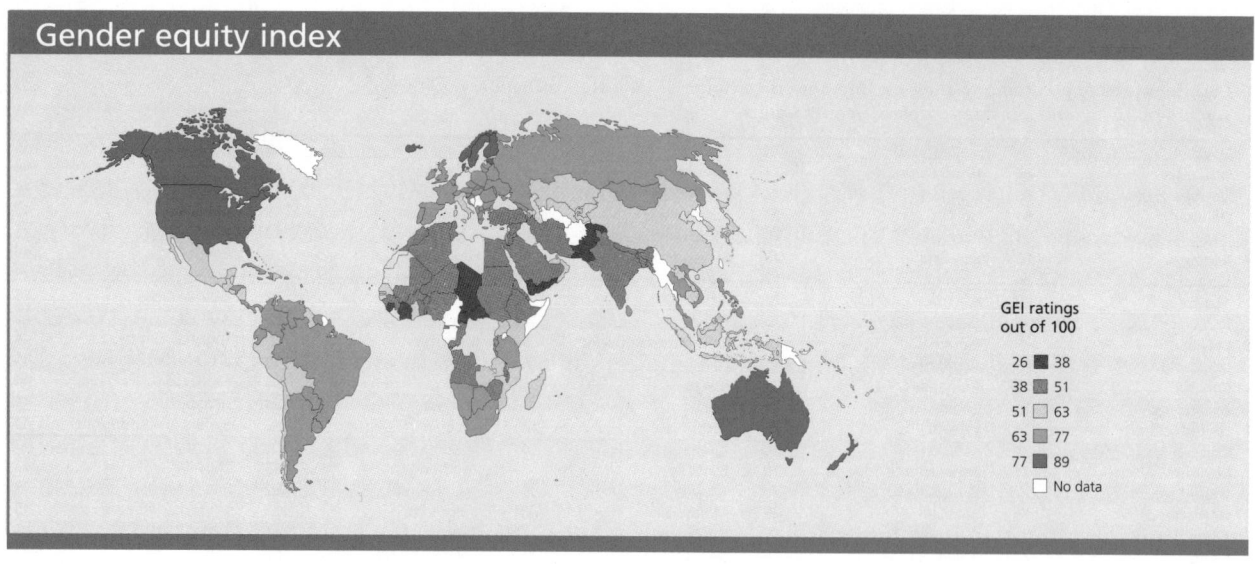

GEI ratings
out of 100

26 ■ 38
38 ■ 51
51 □ 63
63 ■ 77
77 ■ 89
□ No data

Female genital mutilation and Islam

NEARLY 130 MILLION WOMEN IN ABOUT 30 COUNTRIES HAVE UNDERGONE FEMALE GENITAL
MUTILATION OR CUTTING, PERFORMED FOR VARIOUS 'TRADITIONAL' OR 'CULTURAL' REASONS. EACH
YEAR, SOME TWO MILLION GIRLS ARE SUBJECTED TO THESE PRACTICES. THE CUSTOM IS OFTEN
ASSOCIATED WITH ISLAM, YET, IN 2006, REPRESENTATIVES OF 50 MUSLIM STATES ISSUED A DECLARATION
CONDEMNING FEMALE GENITAL MUTILATION AND EMPHASIZING THAT IT IS CONTRARY TO ISLAMIC
PRINCIPLES.

Female genital mutilation or cutting (FGM/C) is mainly practised in parts of sub-Saharan Africa, Northeast Africa, India and Sri Lanka. It is also found in countries that have received immigrants from such places. FGM/C has different characteristics according to the community in which they are practised; but all pose a significant threat to the health of millions of women. On top of this, they are a brutal form of gender discrimination, since - in addition to the 'cultural' underpinnings that endorse the practices - they are also seen, and felt, as instruments of a woman's subordination by controlling her sexuality.

The World Health Organization (WHO) distinguishes the following types of FGM/C:

Type I: Clitoridectomy - Excision of the prepuce with or without removal of part or all of the clitoris.

Type II: Excision - Excision of the prepuce and clitoris together with total or partial excision of the labia minora.

Type III: Infibulation - Excision of part or all of the external genitalia and stitching/narrowing of the vaginal opening.

Type IV: No classification - All other procedures involving total or partial removal of female genitalia and/or injuries to the female genital organs whether for cultural or any other non-therapeutic reasons.

In general, FGM/C is performed with no anesthetic, which makes it extremely painful. It is often done by older women using knives or broken bottles, which may be dirty, as cutting instruments - with the attendant risk of septicemia. Infections are commonly transmitted, including HIV.

With infibulation - a common practice in Mali, Sudan, Somalia, part of Ethiopia and Nigeria - a girl's external genitalia are removed and her vagina sewn up. After marriage the vagina is surgically opened for sex, and further widened before childbirth. This scarred tissue does not stretch in childbirth, so more cutting may be required. After the baby is born, the mother is stitched up again - all of this is without anesthetic and in unsterile conditions.

The procedures not only pose risks and complications to mother and child at birth but can lead to chronic difficulties in passing urine and menstrual blood, cause infections and turn sexual intercourse into a painfully traumatic experience.

Today, around 130 million women are reckoned to have undergone FGM/C and two million girls potentially face the knife (or broken bottle) each year. Although circumcision is usually performed to mark the passage from childhood into adulthood, some procedures are commonly performed on girls aged between four and twelve. At times excisions are performed on babies a few days old, and in other cases immediately before marriage.

Obscure origins
This custom is old but its origins are unclear. Some people claim that it began in ancient Egypt, although there is no archeological evidence, in mummies for instance. It seems that the practice is long-standing in those countries where it is still performed.

One theory is that the Romans performed mutilation for cosmetic reasons or as an indication of slavery or subordination. In the eighth century, the poet Al-Farazdak mentions an Azd community in the Arabian Peninsula where uncircumcised women were considered inferior. Later, in the 16th century, Cardinal Pietro Bembo describes these sorts of rituals performed in regions around the Red Sea.

The reasons given for maintaining this tradition vary according to the different communities. In some places it is thought that a woman who has not undergone mutilation is likely to have stillborn babies (when in fact, infant mortality rate is higher among mutilated women's babies). In other regions it is believed that if the clitoris is not excised, it will grow to the size of a penis. Others justify FGM/C on hygiene grounds, saying that female secretions are dirty. The most common justification is to preserve chastity and fidelity, that is, to control a woman's sexuality.

After mutilation, a girl may be showered with gifts. In a reinforcement of the role ahead of her, she may be taught how to cook, to use herbs and to produce handicrafts as well as being allowed to wear attractive female clothing.

Among some groups like the Maasai in East Africa, a woman who endures mutilation without showing pain is regarded as brave and dignified, capable of enduring childbirth pain in the future. In certain regions (for example, Nigeria), a woman's cutting is associated with her fertility and obedience. Women who have not undergone the ritual are often excluded from community activities, and - no matter what their age - are perceived not to have attained adulthood, thus proscribing their participation.

From a human rights perspective, FGM/C is obviously a violation in terms of gender equity. In a typical irony, although FGM/C

is regarded as an essential practice for 'womanhood' in some societies, men (according to studies in Sudan) prefer to have sex with non-mutilated women - and a high percentage of mutilated women get divorced or are abandoned.

On the other hand, among those who encourage the circumcision of girls are those who have suffered the practice themselves. Often they have internalized male/patriarchal values and/or fear the consequences of non-conformity.

No god demands mutilation

The fact that FGM/C is a widespread in Muslim communities in sub-Saharan and northeast Africa links this practice with Islam; many Muslims believe it is a religious obligation. However, it is not mentioned in the Qur'an, nor does it exist in most Muslim countries. In addition, in all these regions, FGM/C is practised among Christians, Muslims, Ethiopian Jews and followers of other indigenous African religions. It is likely that following its expansion throughout these territories, Muslims could have assimilated the rituals from a pre-existing practice.

During the First Islamic Ministerial Conference on the Child, organized by the Islamic Educational, Scientific and Cultural Organization, held in Rabat, Morocco, in 2006, 50 politicians, ministers and religious leaders issued a declaration calling for the need 'to take the necessary measures to eliminate all forms of discrimination against girls and all harmful traditional or customary practices, such as child marriage and female genital mutilation'. The declaration calls upon Muslim countries where FGM/C is practiced to act strongly against this non-Islamic tradition.

Institutional condemnation of FGM/C by the international community has become increasingly strong in recent decades. In

1979, the WHO published a document denouncing it; since then, different African women's organizations have been created aiming at fighting these practices. In 1984, a Committee Against Traditional Practices Affecting the Health of Women was established in Dakar (Senegal) and 10 years later, the International Conference on Population and Development, held in Cairo, approved the first international document condemning FGM/C. At the same time, although it has been punished by legislation in some countries (Kenya, Egypt and Burkina Faso), this has failed to achieve the desired results. Indeed in some cases it has made things worse by driving the practice underground.

Organizations working on this issue recommend a careful, holistic and culturally sensitive approach in order to eradicate the cruel practice. One alternative solution proposed by a non-governmental organization in Kenya is a rite known as *Ntanira na Mugambo* (circumcision by words) which has started to prove effective in replacing the bloodletting ceremonial being practised in rural areas of that country.

In June 2006, a World Health Organization (WHO) report on a study involving 30,000 African women warned that women who have had the procedure are more likely to need cesareans, and the death rate among their babies is up to 50 per cent higher. The WHO described FGM/C as a form of 'torture' that must be stamped out, even if performed by trained medical personnel and/or in clinics.

'By medicalizing it, we will be endorsing this practice, this violation of a child's body and a basic human right of an individual and I think that's the worst thing we can possibly do,' said Joy Phumaphi, WHO assistant director-general for family and community health. ∎

VI. ECONOMY: HEADS AND TAILS

If maintained, current trends in economic power and human development will result in an ever-widening gap between them. The shift of capital towards some countries in the South (India and Brazil among them) will not translate into better income distribution or living conditions, nor into the sustainable exploitation of natural resources. Alternatives like the Green Economy seem to arise from the divorce between capital and the real world.

China, India and the rotating axis of capital

ECONOMIC FORECASTS SUGGEST A DIFFERENT WORLD ORDER IN THE MAKING, WHICH IN THE NEXT HALF-CENTURY WILL SEE EMERGING ECONOMIES AT THE TOP. ONLY THE US AND JAPAN WILL REMAIN AMONG THE SIX MAJOR POWERS, WHICH WILL BE LED BY CHINA AND INDIA. THE US WILL NO LONGER BE 'THE WORLD'S LOCOMOTIVE'; THE NEW ECONOMIC TRAIN WILL BE PULLED BY THE 'POOR GIANTS'.

In 2004, the International Monetary Fund (IMF) warned that global economic growth would cool off in 2005 due to the imbalance caused by the US deficit and surpluses in Asian countries. This interdependence between the world's largest economy and the most promising economic blocs of the 21st century has emerged following the huge changes that have taken place in the world over the last quarter of a century.

The US trade imbalance - the country imports far more than it exports - arises from the high purchasing power of its consumers, who are increasingly demanding foreign goods. At the end of 2004, US imports exceeded exports to the tune of $617.7 billion - a figure equivalent to 5.5 per cent of its gross domestic product (GDP). The economic law of supply and demand would normally do its own work: the abundance of dollars in the market would imply a depreciation in its value.

On the other hand, the fiscal deficit (the amount by which government spending exceeds state revenues) resulting from tax cuts and the high costs of war has forced the country deeper into debt and brought about increased interest rates.

Dollar depreciation has the paradoxical effect of raising the price of goods from other countries - including those from Europe and Asia - and rendering US goods more competitive.

The Chinese box
At the end of the 1970s, China began to open up its economy to the outside world. However, unlike other world regions where this opening up was marked

by an euphoric irruption of capitalism, channeling investments according to its own rules, in China the process followed a predetermined plan.

In 1980, five special economic zones were opened up in the provinces of Guangdong (3), Fujian and Hainan. These were followed by another 14 cities and 7 regions, between 1984 and 1985 (which made up the so-called open coastal belt). In 1990, the Pudong area was opened up, including Shanghai and other cities on the Changjiang river. Since 1992, more border cities have been opened as well as all capital cities of provinces and autonomous regions. In addition, 15 free trade zones, 32 technological and industrial development zones and 53 high-tech zones were established in some large and medium-sized cities.

Through these zones, the Chinese Government aimed at attracting foreign investments by offering an adequate infrastructure, banking services, communications, transportation and housing. In exchange, it intended to increase the level of employment and technology transfer, as well as acquire foreign currencies. Besides, the intention was to use the zones as a testing ground for the projected economic and administrative reforms.

The great initial dynamism upon the arrival of foreign companies somewhat declined during the 1980-1991 period but, once the Pudong zone - which has undergone abrupt and constant growth - was established in 1992, the impulse was revitalized. This zone has been increasingly focusing on high computing technology and biotechnology.

Major progress is being made in the development of integrated circuits, to the point that Pudong is about to become a world center for semiconductor manufacturing. The Semiconductors Manufacturing International Corporation and the Shanghai Grace Semiconductors Manufacturing Corporation are two huge factories that have attracted a large number of related companies in order to set up an important industrial chain.

In the area of biotechnology, the zone aims at becoming Asia's main biopharmaceutical center. There are more than ten pharmaceutical research and development centers, such as the Institute of Materia Medica of the Chinese Academy of Sciences and the National Human Genome Center.

Chinese investment has borne fruit to the extent that between 1999 and 2002 each yuan invested by the State attracted 14 yuans of Chinese private capital and 12 dollars of foreign capital. In 2004, 100 innovative companies and 16 research and development organizations - both national and international - arrived in Pudong, among them IBM, General Electric and Intel.

In the period 1986-2006 the country's average economic growth was nine per cent a year. According to the Organization for Economic Co-operation and Development (OECD), China will replace the US and Germany as the world's leading exporter by 2010. By then, again according to the OECD, Chinese goods and services will have grown from six per cent of world trade to at least ten per cent.

India's response

In 2000, India decided to imitate its neighbor and passed a law allowing for the creation of its own special economic zones. Since the initial purpose was to boost exports, already existing export-processing zones were reconverted.

The first of these was set up in Gujarat and included private capital as well as associations between the private sector and the State. The Government manages the setting up and development of the zone, while the private sector is in charge of infrastructural investments.

The huge industrial boost experienced by the country in recent years has resulted in a higher than expected growth. In the second quarter of 2005, the Indian economy registered an 8.1 per cent increase, after growing 7 per cent in the first quarter. This accelerated growth is explained by industrial exports and higher domestic demand.

According to the World Trade Organization, India's share in the world textile market will grow from three per cent in 2004 to fifteen per cent in 2010. The elimination of quotas - which regulated textile exports for 30 years - in 2005 has been key to this boom.

Poor giants

A 2003 report by Goldman Sachs[1], one of the world's leading investment banks, asserts that by 2050, Brazil, Russia, India and China - the 'BRIC economies' - will become the largest force in the global economy. Economists state that of the current superpowers gathered in the G7 (the US, Japan, Italy, France, Canada, Britain and Germany) only the two first will be among the six largest economies in 2050.

1. Goldman Sachs, www.gs.com/insight/research/ reports/docs/G8_change112.pdf

CORRUPTION PERCEPTIONS INDEX

The Transparency International Corruption Perceptions Index 2005 (CPI) ranks countries in terms of the degree to which corruption is perceived to exist among public officials and politicians. The CPI focuses on corruption in the public sector and defines corruption as the abuse of public office for private gain. The sources do not distinguish between administrative and political corruption or between petty and grand corruption. The highest ranking countries are those which are least corrupt.

RANK SCORE	COUNTRY	2005 CPI SCORE*	CONFIDENCE RANGE**	SURVEYS USED***	RANK SCORE	COUNTRY	2005 CPI SCORE*	CONFIDENCE RANGE**	SURVEYS USED***
1	Iceland	9.7	9.5 - 9.7	8		Bulgaria	4.0	3.4 - 4.6	8
2	Finland	9.6	9.5 - 9.7	9	55	Colombia	4.0	3.6 - 4.4	9
	New Zealand-Aotearoa	9.6	9.5 - 9.7	9		Fiji	4.0	3.4 - 4.6	3
4	Denmark	9.5	9.3 - 9.6	10		Seychelles	4.0	3.5 - 4.2	3
5	Singapore	9.4	9.3 - 9.5	12		Cuba	3.8	2.3 - 4.7	4
6	Sweden	9.2	9.0 - 9.3	10	59	Thailand	3.8	3.5 - 4.1	13
7	Switzerland	9.1	8.9 - 9.2	9		Trinidad and Tobago	3.8	3.3 - 4.5	6
8	Norway	8.9	8.5 - 9.1	9		Belize	3.7	3.4 - 4.1	3
9	Australia	8.8	8.4 - 9.1	13	62	Brazil	3.7	3.5 - 3.9	10
10	Austria	8.7	8.4 - 9.0	9		Jamaica	3.6	3.4 - 3.8	6
11	Netherlands	8.6	8.3 - 8.9	9	64	Ghana	3.5	3.2 - 4.0	8
	United Kingdom	8.6	8.3 - 8.8	11	65	Mexico	3.5	3.3 - 3.7	10
13	Luxembourg	8.5	8.1 - 8.9	8		Panama	3.5	3.1 - 4.1	7
14	Canada	8.4	7.9 - 8.8	11		Peru	3.5	3.1 - 3.8	7
15	Hong Kong	8.3	7.7 - 8.7	12		Turkey	3.5	3.1 - 4.0	11
16	Germany	8.2	7.9 - 8.5	10		Burkina Faso	3.4	2.7 - 3.9	3
17	United States	7.6	7.0 - 8.0	12	70	Croatia	3.4	3.2 - 3.7	7
18	France	7.5	7.0 - 7.8	11		Egypt	3.4	3.0 - 3.9	9
19	Belgium	7.4	6.9 - 7.9	9		Lesotho	3.4	2.6 - 3.9	3
	Ireland	7.4	6.9 - 7.9	10		Poland	3.4	3.0 - 3.9	11
21	Chile	7.3	6.8 - 7.7	10		Saudi Arabia	3.4	2.7 - 4.1	5
	Japan	7.3	6.7 - 7.8	14		Syria	3.4	2.8 - 4.2	5
23	Spain	7.0	6.6 - 7.4	10		Laos	3.3	2.1 - 4.4	3
24	Barbados	6.9	5.7 - 7.3	3	77	China	3.2	2.9 - 3.5	14
25	Malta	6.6	5.4 - 7.7	5	78	Morocco	3.2	2.8 - 3.6	8
26	Portugal	6.5	5.9 - 7.1	9		Senegal	3.2	2.8 - 3.6	6
27	Estonia	6.4	6.0 - 7.0	11		Sri Lanka	3.2	2.7 - 3.6	7
28	Israel	6.3	5.7 - 6.9	10		Suriname	3.2	2.2 - 3.6	3
	Oman	6.3	5.2 - 7.3	5	83	Lebanon	3.1	2.7 - 3.3	4
30	United Arab Emirates	6.2	5.3 - 7.1	6		Rwanda	3.1	2.1 - 4.1	3
31	Slovenia	6.1	5.7 - 6.8	11	85	Dominican Republic	3.0	2.5 - 3.6	6
32	Botswana	5.9	5.1 - 6.7	8		Mongolia	3.0	2.4 - 3.6	4
	Qatar	5.9	5.6 - 6.4	5		Romania	3.0	2.6 - 3.5	11
	Taiwan	5.9	5.4 - 6.3	14	88	Armenia	2.9	2.5 - 3.2	4
	Uruguay	5.9	5.6 - 6.4	6		Benin	2.9	2.1 - 4.0	5
36	Bahrain	5.8	5.3 - 6.3	6		Bosnia-Herzegovina	2.9	2.7 - 3.1	6
37	Cyprus	5.7	5.3 - 6.0	5		Gabon	2.9	2.1 - 3.6	4
	Jordan	5.7	5.1 - 6.1	10		India	2.9	2.7 - 3.1	14
39	Malaysia	5.1	4.6 - 5.6	14		Iran	2.9	2.3 - 3.3	5
40	Hungary	5.0	4.7 - 5.2	11		Mali	2.9	2.3 - 3.6	8
	Italy	5.0	4.6 - 5.4	9		Moldova	2.9	2.3 - 3.7	5
	Korea, South	5.0	4.6 - 5.3	12		Tanzania	2.9	2.6 - 3.1	8
43	Tunisia	4.9	4.4 - 5.6	7	97	Algeria	2.8	2.5 - 3.3	7
44	Lithuania	4.8	4.5 - 5.1	8		Argentina	2.8	2.5 - 3.1	10
45	Kuwait	4.7	4.0 - 5.2	6		Madagascar	2.8	1.9 - 3.5	5
46	South Africa	4.5	4.2 - 4.8	11		Malawi	2.8	2.3 - 3.4	7
47	Czech Republic	4.3	3.7 - 5.1	10		Mozambique	2.8	2.4 - 3.1	8
	Greece	4.3	3.9 - 4.7	9		Serbia and Montenegro****	2.8	2.5 - 3.3	7
	Namibia	4.3	3.8 - 4.9	8	103	Gambia	2.7	2.3 - 3.1	7
	Slovakia	4.3	3.8 - 4.8	10		Macedonia, TFYR	2.7	2.4 - 3.2	7
51	Costa Rica	4.2	3.7 - 4.7	7		Swaziland	2.7	2.0 - 3.1	3
	El Salvador	4.2	3.5 - 4.8	6	107	Yemen	2.7	2.4 - 3.2	5
	Latvia	4.2	3.8 - 4.6	7		Belarus	2.6	1.9 - 3.8	5
	Mauritius	4.2	3.4 - 5.0	6		Eritrea	2.6	1.7 - 3.5	3

Table continues over page...

CORRUPTION PERCEPTIONS INDEX

RANK SCORE	COUNTRY	2005 CPI SCORE*	CONFIDENCE RANGE**	SURVEYS USED***	RANK SCORE	COUNTRY	2005 CPI SCORE*	CONFIDENCE RANGE**	SURVEYS USED***
	Honduras	2.6	2.2 - 3.0	7		Papua New Guinea	2.3	1.9 - 2.6	4
	Kazakhstan	2.6	2.2 - 3.2	6		Venezuela	2.3	2.2 - 2.4	10
	Nicaragua	2.6	2.4 - 2.8	7	137	Azerbaijan	2.2	1.9 - 2.5	6
	Palestine	2.6	2.1 - 2.8	3		Cameroon	2.2	2.0 - 2.5	6
	Ukraine	2.6	2.4 - 2.8	8		Ethiopia	2.2	2.0 - 2.5	8
	Vietnam	2.6	2.3 - 2.9	10		Indonesia	2.2	2.1 - 2.5	13
117	Zambia	2.6	2.3 - 2.9	7		Iraq	2.2	1.5 - 2.9	4
	Zimbabwe	2.6	2.1 - 3.0	7		Liberia	2.2	2.1 - 2.3	3
	Afghanistan	2.5	1.6 - 3.2	3		Uzbekistan	2.2	2.1 - 2.4	5
	Bolivia	2.5	2.3 - 2.9	6		Congo DR	2.1	1.8 - 2.3	4
	Ecuador	2.5	2.2 - 2.9	6	144	Kenya	2.1	1.8 - 2.4	8
	Guatemala	2.5	2.1 - 2.8	7		Pakistan	2.1	1.7 - 2.6	7
	Guyana	2.5	2.0 - 2.7	3		Paraguay	2.1	1.9 - 2.3	7
	Libya	2.5	2.0 - 3.0	4		Somalia	2.1	1.6 - 2.2	3
	Nepal	2.5	1.9 - 3.0	4		Sudan	2.1	1.9 - 2.2	5
	Philippines	2.5	2.3 - 2.8	13		Tajikistan	2.1	1.9 - 2.4	5
126	Uganda	2.5	2.2 - 2.8	8		Angola	2.0	1.8 - 2.1	5
	Albania	2.4	2.1 - 2.7	3	151	Côte d'Ivoire	1.9	1.7 - 2.1	4
	Niger	2.4	2.2 - 2.6	4	152	Equatorial Guinea	1.9	1.6 - 2.1	3
	Russia	2.4	2.3 - 2.6	12		Nigeria	1.9	1.7 - 2.0	9
130	Sierra Leone	2.4	2.1 - 2.7	3		Haiti	1.8	1.5 - 2.1	4
	Burundi	2.3	2.1 - 2.5	3	155	Myanmar-Burma	1.8	1.7 - 2.0	4
	Cambodia	2.3	1.9 - 2.5	4		Turkmenistan	1.8	1.7 - 2.0	4
	Congo R	2.3	2.1 - 2.6	4		Bangladesh	1.7	1.4 - 2.0	7
	Georgia	2.3	2.0 - 2.6	6	158	Chad	1.7	1.3 - 2.1	6
	Kyrgyzstan	2.3	2.1 - 2.5	5					

* CPI Score relates to perceptions of the degree of corruption as seen by business people and country analysts and ranges between 10 (highly clean) and 0 (highly corrupt).

** Confidence range provides a range of possible values of the CPI score. This reflects how a country's score may vary, depending on measurement precision. Nominally, with 5 percent probability the score is above this range and with another 5 percent it is below. However, particularly when only few sources (n) are available an unbiased estimate of the mean coverage probability is lower than the nominal value of 90%.

*** Surveys used refers to the number of surveys that assessed a country's performance. 16 surveys and expert assessments were used and at least 3 were required for a country to be included in the CPI.

**** All data available are previous to the political separation of Serbia and Montenegro.

According to prospects drawn up by Jim O'Neill - one of the economists in charge of the Goldman Sachs report - China will overtake Germany in 2007, Japan in 2015 and the US by 2035. India will be ranked third, following the US and China, in 30 years' time. The country experiencing the most rapid growth (more than five per cent a year) over the next 30-50 years will be India.

The same study makes it clear that the Chinese and Indian populations - due to their huge size - will continue to be poorer than that of the existing G7, notwithstanding the fact that they will have become economic superpowers. This explains why GDP - which measures the size of the economy and determines the position of a country in the world economy - is one thing, and per-capita income (which helps measure the level of development) is something very different.

Competition and growth

Since the setting up of special economic zones, both China and India have experienced a kind of symbiosis based on India's technological capacities and China's resource wealth. The World Bank ranks India first in terms of the cost-quality ratio in software exports (80 per cent of US companies prefer India when it comes to importing specialized software).

The Shanghai Pudong Software Park has received a large influx of Indian experts and capital in the area of programming. In early 2002, three Indian software giants became established there - Satyam, Infosys and TCS - giving a key boost to the region in the development of this type of product.

In O'Neill's opinion, competition for economic leadership will be, in the coming decades, the engine of the global economy. Both countries will compete for the position of best market so as to attract foreign capital. This race will also involve factors such as education levels, commercial legislation, levels of state corruption and social tensions (directly associated with the distribution of wealth, equity and social policies in general).

The United Nations Conference on Trade and Development (UNCTAD) reported that, in 2004, foreign resources flowing into the emerging economies reached $233 billion - 40 per cent up on 2003. In the developed world, meanwhile, global investment declined by 14 per cent. According to UNCTAD, most resources allocated to investment and development arrived in China, India, Thailand and Singapore.

Fear and distrust that move the world

As the Asian bloc's spectacular growth becomes increasingly evident, the world starts to look at it with a mixture of surprise, admiration and some fear. Wary of the damage strong competition like that envisaged in China and India could imply, some countries have requested protective measures. Among those is Mexico, traditionally a major exporter to the US, which sees the future of its textile industry threatened by the invasion of Asian products.

In other sectors, China has been accused of distorting the global monetary market. From the mid-1990s onwards, the yuan was artificially tied to the dollar, with an exchange rate of 8.28 yuans per dollar. To maintain this parity, the Chinese State spent an annual 12 per cent of its GDP and had its currency depreciated (undervalued by 40 per cent according to some analysts) since it had to come out into the market to buy dollars, which were then reinvested in US debt (bonds and assets).

A low yuan makes Chinese goods more competitive. That was the reason for complaints by the US, the WTO and the IMF to Beijing.

As of 2005, however, China moved to a more flexible exchange rate policy, with a band system that fluctuates 0.15 per cent above or below the reference price against a basket of currencies including the dollar, the euro and the yen, among others.

Economic agents gave a warm welcome to the announcement of this measure. China, which seems to be sacrificing itself for the sake of the economic system's health, has managed - with a more expensive yuan - to reduce imports of its raw materials by billions of dollars, while pressure on its competitors slackens and international institutions applaud. These are market tools the giant is learning to respect, but also to use. ∎

TRADE IN FIGURES

A= Imports of goods and services (current million $), 2004. **Source:** *World Development Indicators 2006*, World Bank.

B= Exports of goods and services (current million $), 2004. **Source:** *World Development Indicators 2006*, World Bank.

Country	A	A year	B	B year
Albania	2,586	2003	1,167	2003
Angola	10,635		13,798	
Antigua	510	2002	437	2002
Argentina	28,152		39,702	
Armenia	1,514		985	
Aruba	3,782		3,955	
Australia	131,417		112,514	
Austria	155,304		161,062	
Azerbaijan	6,312		4,235	
Bahamas	3,033		2,714	
Bahrain	7,069		9,179	
Bangladesh	13,089		9,234	
Barbados	1,820		1,517	
Belarus	17,019		15,666	
Belgium	284,718		297,953	
Belize	626		506	
Benin	1,073	2003	713	2003
Bolivia	2,319		2,546	
Bosnia-Herzegovina	7,111		2,914	
Botswana	2,780	2003	3,689	2003
Brazil	80,069		109,059	
Bulgaria	16,465		13,975	
Burkina Faso	650	2001	260	2001
Burundi	175	2003	43	2003
Cambodia	3,663		3,243	
Cameroon	1,608	1995		
Canada	336,733		377,646	
Cape Verde	547	2003	277	2003
Central African Rep.	244	1994		
Chad	411	1994		
Chile	29,542		37,981	
China	606,543		655,827	
Colombia	19,929		19,496	
Comoros	103	1995		
Congo R.	995	2003	1,546	2003
Costa Rica	9,140		8,610	
Côte d'Ivoire	6,181		7,650	
Croatia	20,180		17,828	
Cyprus	7,853		7,376	
Czech Republic	76,966		76,569	
Denmark	98,925		111,355	
Djibouti	292	1995		
Dominica	156	2002	123	2002
Dominican Republic	9,049		9,283	
Ecuador	9,306		8,734	
Egypt	26,915		26,516	
El Salvador	7,029		4,301	
Equatorial Guinea	477	1996		
Eritrea	500	2000	98	2000
Estonia	9,674		8,794	
Ethiopia	3,778		1,684	
Fiji	1,043	1999		
Finland	60,636		71,099	
France	526,635		531,488	
Gabon	1,882	2003	3,351	2003
Gambia	282	1997		
Georgia	2,491		1,631	
Germany	912,587		1,051,303	
Ghana	5,356		3,487	
Greece	61,380		48,824	
Grenada	270	2002	175	2002
Guatemala	8,483		4,608	
Guinea	964		811	
Guinea-Bissau	102	2003	71	2003
Guyana	782		748	
Haiti	1,375	2003	469	2003
Honduras	4,430		3,066	
Hungary	69,425		66,351	
Iceland	5,250		4,517	
India	93,918	2003	82,735	2003
Indonesia	79,116		89,789	
Iran	17,503	2000	29,727	2000
Ireland	124,724		152,172	
Israel	52,040		51,445	
Italy	423,241		435,871	
Jamaica	5,272		3,899	
Japan	542,380		636,611	
Jordan	9,407		5,983	
Kazakhstan	18,800		22,602	
Kenya	5,115		4,202	
Kiribati	45	1994		
Korea, South	269,782		299,174	
Kuwait	18,510		33,543	
Kyrgyzstan	1,135		942	
Laos			708	
Latvia	8,180		6,001	
Lesotho	1,398		771	
Libya	10,532		17,862	
Lithuania	13,321		11,751	
Luxembourg	37,859		46,853	
Macedonia, TFYR	3,247		2,080	
Madagascar	1,654	2003	1,126	2003
Malawi	795	2002	472	2002
Malaysia	96,820	2003	118,577	2003
Maldives	725		688	
Mali	1,471	2003	1,152	2003
Malta	4,428		4,022	
Mauritania	471	1998		
Mauritius	3,603		3,460	
Mexico	216,589		202,003	
Moldova	2,122		1,331	
Mongolia	1,405		1,211	
Morocco	19,860		16,632	
Mozambique	2,381		1,759	
Myanmar-Burma	2,458		3,181	
Namibia	2,495		2,310	
Nepal	2,186		1,224	
Netherlands	341,622		388,899	
Netherlands Antilles	2,803		2,655	
New Zealand-Aotearoa	28,791		28,305	
Nicaragua	2,851		1,653	
Niger	681	2003	415	2003
Nigeria	16,064		26,993	
Norway	73,557		109,104	
Oman	10,613		14,175	
Pakistan	22,057		16,079	
Panama	9,172		8,859	
Papua New Guinea	1,594	2001	2,098	2001
Paraguay	3,540		3,397	
Peru	12,581		14,530	
Philippines	50,492		42,829	
Poland	99,935		95,333	
Portugal	65,411		51,899	
Romania	34,029		27,099	
Russia	130,144		203,741	
Rwanda	493		201	
Samoa	140	1999		
São Tomé and Príncipe	41	2002	19	2002
Saudi Arabia	66,746		131,849	
Senegal	2,657	2003	1,826	2003
Seychelles	629		625	
Sierra Leone	342		215	
Singapore	206,796		238,522	
Slovakia	25,649	2003	25,241	2003
Slovenia	19,927		19,519	
Solomon Is.	198	1999		
South Africa	57,888		56,734	
Spain	307,365		269,030	
Sri Lanka	9,108		7,284	
St Kitts-Nevis	257	2002	155	2002
St Lucia	402	2002	328	2002
St Vincent	217	2002	177	2002
Sudan	4,651		3,822	
Suriname	1,011		924	
Swaziland	2,448		2,438	
Sweden	134,855		163,934	
Switzerland	146,291		181,568	
Syria	7,915		8,175	
Tajikistan	1,445		1,220	
Tanzania	3,196		2,179	
Thailand	107,512		114,019	
Togo	959	2003	693	2003
Tonga	105	2002	41	2002
Trinidad and Tobago	4,283	2003	5,890	2003
Tunisia	14,099		13,308	
Turkey	102,199		91,048	
Turkmenistan	1,680	1997		
Uganda	2,154		1,153	
Ukraine	34,846		39,719	
United Kingdom	604,562		533,167	
United States	1,769,031		1,151,448	
Uruguay	3,673		4,008	
Vanuatu	146	2003	122	2003
Vatican				
Venezuela	22,042		39,846	
Vietnam	21,458	2002	19,654	2002
Yemen	4,918		5,045	
Zambia	1,318	2000	871	2000
Zimbabwe	2,515	1994		

Sukuk and Islamic finance

IN RECENT YEARS THERE HAS BEEN A SIGNIFICANT INCREASE IN THE USE OF ISLAMIC FINANCIAL PROCEDURES COMPATIBLE WITH THE REQUIREMENTS OF *SHARIA*, A LEGAL CODE BASED ON THE QUR'AN. THE IDEA OF FINANCES COMPLIANT WITH SHARIA - AND PARALLEL TO THE RULES OF CAPITALISM - EMERGED IN EGYPT AT THE BEGINNING OF THE 1960s AND LATER SPREAD TO SAUDI ARABIA AND THE UNITED ARAB EMIRATES. *SUKUK* CERTIFICATES HAVE PROVIDED A WAY OF OPERATING WITHIN THE DUAL FRAMEWORK OF ISLAM AND INTERNATIONAL FINANCIAL CAPITAL. THE *SUKUK* MARKET WAS LAUNCHED IN 2002 AND BY 2006 IT HAD A TURNOVER OF AROUND $500,000 MILLION AND WAS GROWING AT AN ANNUAL RATE OF 15 PER CENT. AMONG OTHER FINANCIAL PRESCRIPTIONS, SHARIA FORBIDS THE PAYMENT OR CHARGING OF INTEREST (*RIBA*, WHICH MORE PRECISELY MEANS EXCESS OR ADDITION) AND REQUIRES STATED FIXED VALUE AND A BENEFIT COMMENSURATE WITH RISK (*GHARAR*) AND LIQUIDITY.

Sharia, literally 'way to the watering place', is not so much a system of criminal justice as a daily way of life adopted to a greater or lesser degree by Muslims as a question of personal conscience. It stems from permanent discussion on how to adapt the Qur'an to the order, harmony and fundamental law governing nature and the cosmos. In addition to being a matter of individual conscience it has been incorporated into legislation in some Muslim states.

Sharia is rigid: no remunerated checking accounts, no bank loans or mortgages that charge interest, no life insurance or other types of insurance. Profit gained without physical work is considered sinful or immoral and is clearly apparent in the charging of interest and usury where the beneficiary gets richer just by lending money with the sole aim of making a profit. Another of the Islamic legal code's basic rules for financial management prohibits *gharar*, the taking on of excessive risk. It also forbids the making of profit through speculation.

Riba, gharar, haram and halal

Riba and *gharar* prohibition are grounded in four basic principles: a) risk-sharing: risk must be shared between the parties to a contract; b) materiality: a financial transaction needs to have a 'material finality', in other words it must be directly or indirectly linked to a real economic transaction; c) no exploitation: a financial transaction must not lead to the exploitation of any party to that transaction; and d) no financing of *haram* (literally 'that which excludes') or sinful activities: *haram* applies, among other things, to alcoholic drinks, harmful or poisonous substances, intoxicating

plants or drinks, animals killed by suffocation, strangulation or beating and products derived from prohibited animals or animals not slaughtered in accordance with *halal*, the Islamic slaughter rite.

Islam and money

Money is considered to be exclusively a means for exchanging assets and not an asset in itself. It may only be multiplied by trading assets. However, at some stage of their life Muslims can need finance for buying their house, insurance or to make a will and bequeath legacies. And there are powerful Muslim investors who want to obtain a return from the huge amounts of money, derived mostly from the oil business, that have been accumulating in banks of the principal tax havens such as Switzerland, Andorra, Monaco, Panama, Liechtenstein, etc, or that have been invested in the world's main stock-markets under established free market rules where most 'traditional' instruments involve some form of yield or return. To meet these needs, a wide range of financial instruments that fulfill and respect Islamic dictates, *Sukuks*, has been developed in the Middle East over more than three decades.

What is Sukuk?

Sukuk is the generic name given to different types of financial instruments that embody the idea of risk being shared between the user and supplier of funds. The basic principle underlying these instruments is that the owner has an undivided interest in a certain asset and as a consequence has the right to a financial return generated by that asset. There are two broad categories of Sukuk:

negotiable certificates (*mudarabah* and *musharakah*, which are basically shares in companies or businesses) and non-negotiable or zero coupon bonds (the word 'coupon' replaces 'interest').

The former, unlike shares in Western limited liability companies, represent the ownership of a proportional but undivided part of an asset or company, which bestows the right to periodical fixed returns from that asset or company. Zero coupon bonds are almost the opposite of a common bond: instead of buying debt and earning interest on it, the client buys a bond issued at a discount and receives a pre-determined payment at maturity. Even though the concept of Sukuk is simple, in practice it can become complex as some banks combine two or more types of instruments in adaptation to their client's needs (see box).

The classic Sukuk structure involves the acquisition of an asset by a Special Purpose Company[1] set up in a neutral fiscal jurisdiction. The greatest disadvantage of Sukuk instruments is their lack of liquidity due to the absence of an inter-bank market, which in the West is used to adjust assets and liabilities. This is solved by establishing secondary markets[2] that exclude classic Western certificates (for example hedging certificates); there, already outstanding stocks or securities are traded and the degree of liquidity depends on their quality. Their advantage is that they yield a periodic return and are revalued in the same way as other financial assets.

In every case, before being issued on the market, any type of Sukuk will need to be vetted by scholars to confirm that the

---

1. A Special Purpose Company (SPC) is created exclusively to provide a legal basis for a project and its cashflow. Investor risk is limited to the quality of credits owned by the SPC.

2. The primary or issue market is the market where stocks (shares, securities, etc) issued by a company, or a private or public institution, are sold for the first time, while the secondary market is where stocks previously sold in the primary market are traded. So the secondary market is a 'second hand' or resale market. The stock exchange is the most important and more organized part of the secondary or trade market; the rest of the secondary market is known as the unlisted market. See: www.ucm.es/info/assuarez/art1.pdf

The most common Islamic finance certificates

DEBT CERTIFICATES: *Mudarabah*, a purchase and resale contract by which the bank buys tangible assets from a supplier at the request of the client and the resale price is fixed on the basis of cost plus a profit margin; *Salam*, a contract for the purchase of goods with deferred delivery (the opposite of *Mudarabah*), used mainly for financing agriculture; *Istisna*, a financing and leasing certificate with advance delivery used to finance long-term projects; and *Qard al-Hasan* (benevolent loan) a non-interest credit, usually secured.

Quasi-debt certificates: *Ijara*, a leasing contract by which one party leases an asset for a specified period and fee. The asset's owner (the bank) takes on all the risk associated with ownership. The asset may be sold at an agreed market price incorporating the sale of the *Ijara* contract. This contract can be structured as a leasing and purchase contract under which each lease payment includes a part-payment of the agreed purchase price and it can cover a period equivalent to the anticipated life of the asset.

Profit and loss sharing certificates: *Musharakah*, a contract for the joint provision of capital where both bank and client contribute to the financing of a project. Ownership is proportional to the share in financing; *Mudarabah*, a financing contract under which one party contributes capital for the project and the other contributes work. The parties agree on a share of profit and the finance provider bears any loss except in cases of undue behavior, negligence or violation of agreed conditions.

Non-negotiable certificates or zero coupon bonds: these bonds are sold at a discount, that is, the purchase price is less than their nominal value. At maturity the actual nominal value is received. They are very useful for investors who prefer a single payment on a fixed date instead of a continuous flow of payments through a certain period. ▪

Certain conditions are essential to foster further development of the Sukuk market:

a) standardization of Sharia (Oman's differs from Indonesia's, Saudi Arabia's differs from Iran or Iraq's);
b) standardized operational systems;
c) credit risk classification;
d) standardized legal documents and contracts;
e) establishment of payment terms, custody agents and specialized investment funds;
f) regional co-ordination, a free flow of information and the fostering of capital availability and efficient markets.

The huge potential of this market, still in its infancy, has engendered great interest amongst bankers, fund administrators, business consultants, accountants and lawyers in Asia, Europe and the US. In recent times finance institutions such as Citigroup and HSBC have created subsidiaries that work exclusively with the Islamic market.

Up to now the most popular assets have been those related to real estate, in which rents provide a cash return to property owners and property repurchase terms ensure repayment on pre-determined maturation dates. Other kinds of assets currently available are airlines, gas pipelines, large-scale air conditioning units and car fleets.

Today these markets occupy a small niche in global terms but their projected growth is immense and although it has arisen from a religious perspective Sukuk, it seems, speaks the global language of profitability. ▪

Sources: **Mohammed El Qorchi,**
Las finanzas islámicas en expansión,
(Islamic Finance in Expansion)
www.imf.org/external/pubs/ft/fandd/
spa/2005/12/pdf/qorchi.pdf
www.ahorrando.org/print.cfm/6/117

transaction complies with the dictates of Sharia.

Islamic finance market

In 2004 there were 265 banks in the Islamic finance market with assets and investments valued at more than $262 billion and $400 billion respectively, according to a report of the International Organization of Securities Commissions.[3] The report points out that although currently representing only a modest part of the global finance, Islamic capital markets, banking and investment grew by between ten per cent and twenty per cent annually during the last decade and concludes that by 2015 almost half of the money saved by Muslims worldwide will be deposited in Islamic banks.

According to bankers and analysts an ever-increasing number of Muslims are seeking opportunities to invest their capital in Islamic businesses and as a result the variety and sophistication of options offered is growing. This expansion of investment is due to factors such as record oil prices and the subsequent flood of money into the Middle East, an increased availability of Sharia-compliant instruments and a greater awareness of religious obligations on the part of Muslim investors.

---

3. The International Organization of Securities Commissions (IOSCO) is a forum whose main objectives are: to maintain justice and efficiency in the markets; to promote the development of domestic markets through an exchange of information and experience; to establish standards for the efficient monitoring of international transactions in stocks and derivative instruments; and to ensure the integrity of markets within a framework of mutual assistance.
See: www.cnbv.gob.mx/noticia.asp?noticia_liga=no&com_id=0&sec_id=380&it_id=56

Prisons: workforce behind bars

PRISONS ARE INSTITUTIONS ASSOCIATED WITH THE DEVELOPMENT OF CAPITALISM, PARTICULARLY IN ITS INDUSTRIAL PHASE. THROUGHOUT HISTORY, GOVERNMENTS HAVE NOT CEASED IN THEIR EFFORTS TO MAKE PRISONERS PRODUCE WEALTH, EITHER THROUGH FORCED OR VOLUNTARY LABOR OR, AS IN RECENT TIMES, THROUGH THE PRIVATIZATION OF PRISON SERVICES.

There are at present nine million prisoners worldwide. One in every four of them is a US citizen, which turns that country into the world's biggest jailer (nearly 800 every 100,000 US citizens are in prison, a rate five times higher than the average for the world's eight richest countries). Half of the world's prison population is concentrated in three countries: the US, China and the Russian Federation, which account for 25 per cent of the world population.

Prisons in history

In the majority of cultures, prisons have their origin in the lodgings allocated for slave accommodation in ancient times. The possibility of committing a crime presupposes the existence of a culture with a code of laws capable of defining crimes, and therefore the prison institution is characteristic of complex civilizations, with well-organized strong repressive forces and a solid government.

The gradation of punishment imposed by law - called by French thinker Michel Foucault 'principle of modulation of penalties' - is characteristic of modern times. Up until the final institutionalization of prisons both the petty thief and the brutal murderer were punished with equal harshness.

Being sold as a slave, forced labor, shameful public exhibition, physical torment (including different kinds of mutilation), deportation and the death penalty; all of these were punishment alternatives that have effectively been replaced now by compulsory confinement. An important function of prison facilities was, and still is, to serve as lodging for the political enemies of powerful people - and this hostage-keeping may have originally been their only purpose. Thieves and murderers were simply not valuable enough for someone to be willing to keep them alive. Nations on the Mediterranean coast often sent offenders to work as galley rowers. The French monarch Louis XIV (1638-1715), who wanted a powerful fleet at his disposal, requested that judges sentenced convicted criminals to the galleys instead of sentencing so many of them to death.

Thus dungeons were simply used as a transitory accommodation for condemned slaves, tortured criminals, prisoners under death sentence or hostages of the political power.

Isolated and productive colonies

Colonial powers used to encourage infringers to become enlisted in the armies instead of being locked up in prison. In peaceful times, meanwhile, deportation to 'uninhabited' colonies was a frequent solution, which at the same time contributed to the colonization of new territories and kept undesirable people far away from the metropolis.

Australia received some 800 British prisoners as its first European inhabitants on a date which is currently celebrated as its national day. Throughout the following decades, tens of thousands of convicted criminals were sent to inhabit the forced labor and detention centers of the colony's different correctional facilities. France created penal colonies in its South American possession, French Guiana, which became famous for the brutal treatment inflicted on prisoners until they were closed down towards the middle of the 20th century.

There seems to be quite a close relationship between the development of capitalism as from the 16th century onwards and the growth in prison populations, as the authorities looked to turn the masses of unemployed people who were migrating from rural to urban areas into a profitable force. Even Columbus had convicted criminals among the crew of his first ships. It could certainly be said that the unemployment brought about by the surplus workforce resulting from industrialization was the major promoter of prison institutions in

Europe and its colonies. Even now a very large percentage of prisoners (nearly 50 per cent) are unemployed at the time of their arrest. Capitalism finds a way to punish the unemployed and, once they are placed into a disciplinary system, it finds a way to make them productive.

Until very recent times, some countries offered prisoners the choice to fight for their homeland in order to commute their sentences, as Britain did during World War Two.

Many penal colonies were planned for production purposes; during the 19th and 20th centuries they became characteristic facilities of authoritarian governments which brutally exploited their prisoners. Nazi Germany, the Soviet Union and China during the Cultural Revolution all had numerous important colonies where prisoners were forced to perform slave or semi-slave labor.

Disciplinary architecture and surveillance

Until the 20th century, almost no building was specifically built to be a prison; instead, the thick walls and stable guards of solid fortresses, barracks and castles were put to good use.

It was during the Industrial Revolution, when the large concentration of population in urban centers resulted in an increased number of prisoners that authorities started to feel the need to build special facilities. At the same time, certain humanist ideas

Prison population boom

THE GROWTH IN PRISON population all over the world has resulted in widespread overpopulation in prisons. There are very few countries in which prison facilities are not overcrowded.

Institutional weakness and authoritarian regimes are associated with overpopulation, which in some poor countries - including Kenya, Barbados and Zambia - exceeds 300 per cent. But even in developed countries such as the US, France, Belgium, Italy and Spain, occupancy rates are in excess of 100 per cent.

One of the current problems at world level is the percentage of inmates imprisoned

without sentence. In some countries, this figure accounts for over 90 per cent of inmates, as it does in Paraguay. This can apply in countries with both high and low levels of human development: Morocco, Belgium, Malaysia, Italy, Côte d' Ivoire and Switzerland all have 40 per cent of their inmates imprisoned without sentence.

Another problem is the number of children who live with their mothers in correctional institutions. In general, there are far fewer female inmates than male, but women suffer the same kind of rights deprivation: lack of space, lack of opportunity for personal development and imprisonment without sentence. ■

associated with the dissemination of utopian socialism - seeing criminals as victims of the social order who needed education or rehabilitation - paradoxically resulted in increased prison terms. Many philosophers recommended that governments increase detention sentences so that the corrective effects of imprisonment could have an impact.

At the end of the 18th century, when German anatomist Franz Gall postulated phrenology - the study of personality through the shape of the skull - the British philosophers John Howard and Jeremy Bentham defended the idea that prisons should be institutions for correction rather than punishment. Bentham took the idea for his famous *panopticon* - a building where all prisoners could be simultaneously watched from only one surveillance post - from a workers' monitoring system that his brother, in the service of Prince Potemkin, was using in large public works in Russia.

Prisoners knew they were being observed although they could not see the watcher. This earthly duplication of the heavenly role - God's watchful eye - was considered to be beneficial to develop a feeling of repentance and thereby the purification of prisoners, just like the monitoring of workers implied an improvement of their work performance. The paternalism of jailers was based on the phrenologic certainty that some features of criminals were not their own fault but the result of their physically and morally ill constitution.

At present, the ubiquity of electronic surveillance has made the panopticon an architectural irrelevance. On the other hand, invisible surveillance is not limited to prisons but has invaded all community spaces. Phrenology no longer justifies the differences: all citizens deserve surveillance by powerful institutions.

Prisons for profit

A form of exploitation of prisoners currently being put into practice is becoming disseminated at a fast pace. Inmates are offered work with much lower wages than those paid at factories since, unlike free workers, they do not have to cover transportation, food or housing expenses in prison. Prisons can therefore sign very profitable work contracts with large corporations for the assembly of their products. Inmates usually accept this work, which allows them to save small sums and at the same time keeps them away from the pernicious inactivity of cells.

Those who are responsible for these practices defend their legitimacy based on their non-compulsory character. However, even if prisoners refuse to work they turn out to be profitable for some companies. Prison services have started to be privatized in many countries, alleging that the costs of private administration are lower than those of public administration. Yet a recent study by the Department of Justice of the US - where four per cent of prisoners are detained in private institutions - reveals that state-run prison costs are lower, notwithstanding the fact that employees are paid much higher wages than those earned by their peers in the private system. Those who are against privatization believe that the logic of the free market cannot be applied to prisons, given that the self-regulation involved in providing goods for mass consumption cannot take place within this extremely limited market, where the only clients are state governors or national presidents.

The figures involved are huge: in the US it is estimated that more than $35 billion a year are spent on the maintenance of prisoners. As a result of the large amounts of money involved and the limited number of clients, privatization - according to its detractors - favors corruption.

According to some people in charge of large private security firms (Wackenhut Corrections and the Correctional Services Corporation, among others) private institutions should cover 50 per cent of the 'inmate market'. This would turn it into a $20 billion a year business for security companies in the US alone. The so-called PIC (Penal Industrial Complex, similar to the Military Industrial Complex defined by Eisenhower) is now quoted on the stock exchange, and in the prisons of one private contractor - the Corrections Corporation of America - profit indicators are shown at the gates. ∎

Green economics: a new model

A GREEN ECONOMY WOULD NOT ONLY CONSIDER FINANCIAL MATTERS BUT WOULD ENCOMPASS SOCIAL, ENVIRONMENTAL AND EVEN SPIRITUAL CONCERNS AND WOULD PRIORITIZE QUALITATIVE RATHER THAN QUANTITATIVE GROWTH. IT WOULD SEEK TO EVALUATE THE REAL WORTH OF PRODUCTS AND SERVICES RATHER THAN SIMPLY GIVING THEM AN EXCHANGE OR MONETARY VALUE. IT WOULD FINALLY MAKE VISIBLE MANY OF THE IMPORTANT TASKS PERFORMED OUTSIDE THE FORMAL ECONOMY WHILE ALSO FACTORING IN ANY DAMAGE DONE TO THE EARTH'S NATURAL SYSTEMS.

Green economics, also known as ecological economics, is not limited to environmental concerns as its name might imply. It encompasses social, environmental and spiritual concerns, all of which have been historically overlooked in the study of economics, and proposes that we design a new model for the economy. This model, according to Brian Milani, the author of *Designing the Green Economy* [1], must establish direct democracy, meet everyone's needs and harmonize human activity with nature.

Many aspects of our daily lives are excluded from mainstream economics, which measures industrial production and the exchange of money. In general terms, producers and consumers only take into consideration their own direct costs and benefits when making decisions rather than the costs and benefits to society as a whole. Examples of these externalities (costs or benefits that are passed along to society at large) can be positive or negative. A common example of a negative externality is pollution which is produced by one manufacturer but affects many others. This pollution harms factory neighbors and workers, but as it stands at present does

1. Brian Milani, *Designing the Green Economy*, Rowman & Littlefield 2000.

not directly affect the bottom line of the manufacturer - which means that the manufacturer does not consider the effects of that pollution. An example of a positive externality is something like a vaccine which not only benefits the recipient but those in society at large by reducing the spread of the disease to others. Current economic measures were not designed to consider externalities and leave the job of managing these to governments, which must discourage or encourage them as they see fit.

Green economics versus environmental economics
Environmental economics offers ways of measuring externalities and therefore gives lawmakers better tools to work with. Ecological or green economics goes beyond this to look at how to redesign the economy so as to discourage negative effects and encourage positive ones from the very start. Green economics assumes that humanity can regenerate community and ecosystems and that there can be positive qualitative change, while environmental economics continues to focus on quantitative control. Whereas environmental economics has inserted itself snugly into the current economic paradigm without making any fundamental change, green economics is about designing a new economic system that takes more than materials and money into account. Green economics may use the tools of environmental economics to build social and environmental costs into prices but recognizes that these actions in the long run will not bring about the changes that are needed. While environmental economics asks how the industrial economy can do less damage to the environment and people, green economics asks why the economy needs to be destructive in the first place.

Some aspects of green economics
1) The problem with GDP
Only activities involving the exchange of money are considered by mainstream economists to contribute to economic progress. This is particularly problematic since measures of economic progress are often used as indicators of social progress. In addition, any transaction that involves the exchange of money is measured, whether or not it is of real social value. As a result, environmental disasters that require a lot of cleaning up, such as oil spills, increase the gross domestic product (GDP) if the economic activity created by the clean-up is greater than would have been the case if the spill had not occurred. Since GDP is conventionally used as a measure of progress, when it increases due to this kind of disaster it sends an unclear signal about an event that is devastating to ecosystems and to human health. To

give an example of something excluded by the GDP, we need look no further than care for children and the elderly, work that is generally unpaid and carried out by women. Although no-one would deny the value of this work, it is ignored by GDP and its contribution to society goes unmeasured. Instead, economic indicators measure and encourage material production in itself without considering whether this material accumulation is useful or serves a social purpose.

2) Valuing the unvalued
The green economy, in contrast, aims to value what truly has worth by focusing on the 'use value' of transactions and products rather than the exchange value. This would make many invisible activities finally visible, such as work done in the home, which is the base and support for all other work in society, or volunteer work in the community, which is equally ignored by the monetary economy. One way of doing this is through time-use surveys which look at how people spend their time, instead of only considering what time they spend getting paid for work. In 1995 the Human Development Report (UNDP) estimated that $16 trillion of global output was invisible, with $11 trillion of that output being performed by women. This compared with an estimated GDP of $23 trillion that year. Another study[2] categorized 17 different ecosystems and assigned them value in terms of their ability to provide support to human needs. The total services of these ecosystems were estimated at $33 trillion dollars, which is larger than global GDP.

3) A real service economy
By focusing on use values it is possible to move towards a real service economy. Green economics encourages us to forget the mainstream definition of a service economy, in which all manufacturing jobs move from industrialized nations to developing ones, while the labor force in industrialized nations becomes concentrated in 'service jobs' such as catering and retail. Services in the green economy are those provided to meet a need. This means talking about mobility instead of the number of cars manufactured and focusing on ways of producing warmth and light other than fossil fuel-powered plants. A concrete example is Interface Flooring, an American company that has tried to do this. Once a carpet manufacturer, Interface Flooring changed its focus to flooring and providing its customers with this service. This has meant a different way of carrying out business. Instead of selling their customers physical flooring, they lease their customers the service, transferring the material costs onto themselves. Now instead of replacing hundreds of square metres of carpet every time some portion of the carpet becomes worn or dirty, they use their carpet tile design to replace only the portion of the carpet that needs replacing - thereby reducing their own costs as well as the burden on landfills and the environment.

4) Working within natural systems
Moving away from reliance on the production and consumption of materials and energy will also allow us to begin working within natural systems to follow natural flows. One way to do this is to close the loop in material and energy use by using the extra or waste materials and energy as inputs in other processes, either in the facility or in the community. This type of 'industrial ecology', which recognizes that one organism's waste is another's food, is just one example of measures being taken to reduce material flows. Working at an appropriate scale for the ecosystem and recognizing bioregions will necessarily involve more decentralization and may provide the basis for more direct democracy. ∎

2. R Constanza, R d'Arge, R de Groot, S Farber, M Grasso, B Hannon, K Limburg, S Naeem, RV O'Neill, J Paruelo, RG Raskin, P Sutton, M van den Belt, 'The value of the world's ecosystem services and natural capital', in *Nature*, 387 (6230):255, 1997.

Ecological debt: who owes whom?

THE CONCEPT OF ECOLOGICAL DEBT ENCOMPASSES THE SOCIAL AND ENVIRONMENTAL IMPACTS SUFFERED MAINLY BY SOUTHERN COUNTRIES AS WELL AS THE INTENSIVE EXPLOITATION OF NATURAL RESOURCES TO SUPPORT PRODUCTION AND CONSUMPTION PATTERNS IN THE RICH WORLD.

Although as a concept, 'ecological debt' is relatively new, the phenomenon apparently began in colonial times - in the 16th century - and acted as engine of the Industrial Revolution. According to the international environmental organizations that are the main proponents of the notion of ecological debt, this is an historical obligation that Northern countries have towards Southern ones, due to the looting and exploitation of their natural goods. Countries have become rich as a result of the continued usurpation of poor countries' natural resources, of unfair trade and of using the global environment as garbage dump. One of the clearest examples of this is climate change - a phenomenon that has been almost exclusively created by industrialized countries but the consequences of which are suffered by all the world's inhabitants. But there is also deforestation to feed demand for timber or meat products in the North. Southern governments generally lack the capacity to impose minimum environmental or social restrictions, since they are in such dire need of investments to alleviate their poverty.

The South as creditor

Ecological debt has four components: the carbon debt, owed by industrialized countries as a result of their disproportionate pollution of the atmosphere through greenhouse gas emissions; biopiracy (the intellectual appropriation of local and indigenous knowledge for trade purposes by laboratories from the rich world - banned under the Cartagena Protocol); damage to the natural environment caused by the activities of transnational corporations in the developing world; and the dumping

Carbon Debt

INEQUALITIES in terms of carbon emissions between Northern and Southern countries are huge: the average US citizen produces seven tons of carbon gas a year, while an Indian citizen barely reaches 0.5 tons.

Energy consumption in Northern countries is almost exclusively based on the burning of fossil fuels, resulting in large emissions of carbon dioxide (CO_2) - the gas mainly responsible for the greenhouse effect and, therefore, for climate change. Such pollution has global consequences, and is considered responsible for the increased strength and frequency of extreme natural events such as floods and long droughts. These natural disasters mainly affect those countries situated in the tropics and with poor infrastructure, despite their limited contribution to the overall carbon problem.

Extreme natural events entail the loss of human lives and agricultural crops and the destruction of road infrastructure and housing. In order to prevent and repair the damage caused, Southern states often have to resort to new foreign loans. This is the reason why it is said that the North has a debt towards the South - though it is also necessary to consider all the greenhouse gases currently absorbed by the forests and ocean waters of Southern countries. ■

Sources:
www.portaldelmedioambiente.com
Acción ecológica

of Northern toxic waste in Southern countries (banned the Basel Convention, signed by all rich countries but the US).

In addition, however, ecological debt acts as a form of vindication for Southern countries, and aims at counteracting the effects of the unsustainable and unpayable external debt. From the point of view of activists, both kinds of debt should be weighed against each other in order to conclude who owes more to whom and what is the real value of each.

Early in the 21st century, annual debt-servicing by poor countries amounted to more than two trillion dollars, which amounts to one-seventh of the estimated carbon debt of $14.5 trillion. This is one reason why many experts state that the South continues to finance the development of the North. According to Joan Martínez Alier, Professor of Economics and Economic History at Barcelona University, global environmental sustainability requires that structural adjustment plans in the South become environmental adjustment plans in the North. ■

Sources:
www.portaldelmedioambiente.com
Acción ecológica
Unidad de Economía Ambiental,
Universidad del País Vasco
www.gloobal.info
Juan Carlos Galindo, Agencia de Información
Solidaria

WORKERS IN FIGURES

A= Labor force (% of total population), 2004. **Source:** Calculated from *World Development Indicators 2006*, World Bank.

B= Female employment in agriculture (% of female labor force), 1995-2002. **Source:** World Development Indicators 2006, World Bank and World Population Prospects 2004 .

C=Female employment in industry (% of female labor force) 1995-2002. **Source:** Human Development Report 2005 UNDP

D=Female employment in services (% of female labor force) 1995-2002. **Source:** Human Development Report 2005 UNDP

E=male employment in agriculture (% of male labor force) 1995-2002. **Source:** Human Development Report 2005 UNDP

F=male employment in industry (% of male labor force) 1995-2002. **Source:** Human Development Report 2005 UNDP

G=male employment in services (% of male labor force) 1995-2002. **Source:** Human Development Report 2005 UNDP

H=Unemployment (% of labor force), 2004. **Source:** World Development Indicators 2006, World Bank

	A	B	C	D	E	F	G	H			A	B	C	D	E	F	G	H
Albania	42.7							15 2003		Finland	50.1	4	14	82	7	40	53	9
Algeria	38.2							27 2001		France	44.2	1	13	86	2	34	64	10
Angola	40.5									French Polynesia	41.3							
Argentina	45.4		12	87	1	30	69	16 2003		Gabon	40.8							
Armenia	42.6							36 1997		Gambia	40.1							
Aruba								7 1997		Georgia	51.9	53	6	41	53	12	35	12 2003
Australia	49.5	3	10	87	6	30	64	5		Germany	49.3	2	18	80	3	44	52	10
Austria	48.0	6	14	80	5	43	52	5		Ghana	41.6							8 2000
Azerbaijan	47.1	43	7	50	37	14	49											
Bahamas	46.4	1	5	93	6	24	69	10		Greece	45.5	18	12	70	15	30	56	10
Bahrain	44.2									Grenada		10	12	77	17	32	46	15 1998
Bangladesh	42.4	77	9	12	53	11	30	3 2000		Guam	44.0							
Barbados	57.0	4	10	63	5	29	49	11 2003		Guatemala	29.9	18	23	56	50	18	27	3 2003
Belarus	49.5									Guinea	44.1							
Belgium	42.7	1	10	82	3	36	58	7		Guinea-Bissau	36.8							
Belize	39.5	6	12	81	37	19	44	10 2002		Guyana	43.4							9 2001
Benin	35.5									Haiti	40.7	37	6	57	63	15	23	7 1999
Bolivia	42.3	3	14	82	6	39	55	6 2002		Honduras	39.9	9	25	67	50	21	30	5 2003
Bosnia-Herzegovina	52.1									Hungary	41.9	4	26	71	9	42	49	6
Botswana	35.3	17	14	67	22	26	51	19 2001		Iceland	57.9	3	10	85	12	33	54	3
Brazil	47.0	16	10	74	24	27	49	10 2003		India	37.6							4 2000
Brunei	41.1									Indonesia	46.1	43	16	41	43	19	38	10
Bulgaria	41.3							14 2003		Iran	36.8							12 2003
Burkina Faso	40.1									Iraq								28 2003
Burundi	44.9									Ireland	47.0	2	14	83	11	39	50	4
Cambodia	45.2							2 2001		Israel	38.4	1	12	86	3	34	62	11 2003
Cameroon	36.6							8 2001		Italy	41.2	5	20	75	6	39	55	8
Canada	52.9	2	11	87	4	33	64	7		Jamaica	43.6	10	9	81	30	26	45	11
Cape Verde	30.2									Japan	52.2	5	21	73	5	37	57	5
Cayman Is.								4 1997		Jordan	30.3							13 2000
Central African Rep.	43.4									Kanaky-New Caledonia	39.0							19 1996
Chad	34.6									Kazakhstan	53.7							9 2003
Chile	38.7	5	13	83	18	29	53	7 2003		Kenya	42.0	16	10	75	20	23	57	
China	57.7							4 2002		Korea, North	46.8							
Colombia		7	17	76	33	19	48	14 2003		Korea, South	50.1	12	19	70	9	34	57	4
Comoros	29.2									Kuwait	46.7							
Congo D.R.	36.4									Kyrgyzstan	41.2	53	8	38	52	14	34	10 2003
Congo R.	34.6									Laos	36.9							
Costa Rica	42.3	4	15	80	22	27	51	7 2003		Latvia	47.9	12	16	72	18	35	47	11 2003
Côte d'Ivoire	35.5									Lebanon	37.6							9 1997
Croatia	43.0	15	21	63	16	37	47	14 2003		Lesotho	35.7							39 1997
Cuba	47.2							3 2002		Liberia	34.3							
Cyprus	47.5	4	13	83	5	31	58	4 2003		Libya	37.0							
Czech Republic	50.8	3	28	68	6	50	44	8		Lithuania	47.8	12	21	67	20	34	45	12 2003
Denmark	52.1	2	14	85	5	36	59	5		Luxembourg	41.8							5
Djibouti	37.5									Macedonia, TFYR	42.1							37 2003
Dominica		14	10	72	31	24	40	23 1997		Madagascar	42.5							5 2002
Dominican Republic	41.0	2	17	81	21	26	53	16 2001		Malawi	43.2							1 1998
East Timor	32.9									Malaysia	40.9	14	29	57	21	34	45	4
Ecuador	45.3	4	16	79	10	30	60	11 2003		Maldives	32.3	5	24	39	18	16	55	2 2000
Egypt	29.0	39	7	54	27	25	48	11 2003		Mali	37.1							
El Salvador	38.1	4	22	74	34	25	42	7 2003		Malta	41.1	1	21	78	3	36	61	8 2003
Equatorial Guinea	36.6									Marshall Is.								31 1999
Eritrea	36.5									Mauritania	36.0							
Estonia	50.2	4	23	73	10	42	48	10 2003		Mauritius	44.4	13	43	45	15	39	46	10 2003
Ethiopia	38.1							8 1999		Mexico	38.7	6	22	72	24	28	48	3
Fiji	44.1							5 1995		Moldova	51.1	50	10	40	52	18	31	8 2003

WORKERS IN FIGURES

	A	B	C	D	E	F	G	H
Mongolia	43.1							14 2003
Morocco	33.7	6	40	54	6	32	63	11
Mozambique	44.5							
Myanmar-Burma	52.3							
Namibia	30.9	29	7	63	33	17	49	31 2001
Nepal	36.3							1 1999
Netherlands	52.1	2	9	86	4	31	64	4 2003
Netherlands Antilles	44.2							14 2000
New Zealand-Aotearoa	51.7	6	12	82	12	32	56	4
Nicaragua	34.4							8 2003
Niger	38.5							
Nigeria	34.0	2	11	87	4	30	67	17 1995
Norway	53.6	2	9	88	6	33	58	4
Oman	35.3							
Pakistan	33.1	73	9	18	44	20	36	8 2002
Palestine	18.7	26	11	62	9	32	58	26 2003
Panama	42.7	6	10	85	29	20	51	14 2003
Papua New Guinea	40.9							3 2000
Paraguay	43.1	20	10	69	39	21	40	8 2001
Peru	45.1	6	10	84	11	24	65	10 2003
Philippines	41.8	25	12	63	45	18	37	10 2001
Poland	45.1	19	18	63	19	40	40	19
Portugal	52.0	14	23	63	12	44	44	7
Puerto Rico	36.2							12 2003
Qatar	52.0							4 2001
Romania	48.2	45	22	33	40	30	30	7 2003
Russia	51.5	8	23	69	15	36	49	9 2002
Rwanda	43.2							1 1996
Samoa	34.6							
San Marino								3 2003
São Tomé and Príncipe	29.1							
Saudi Arabia	29.7							5 2002
Senegal	36.6							
Serbia and Montenegro*	37.4							15 2003
Sierra Leone	39.3							
Singapore	48.7		18	81		31	69	5 2003
Slovakia	49.2	4	26	71	8	48	44	18

	A	B	C	D	E	F	G	H
Slovenia	52.1	10	29	61	10	46	43	7 2003
Solomon Is.	37.6							
Somalia	38.9							
South Africa	40.0	9	14	75	12	33	50	28 2003
Spain	46.7	5	15	81	8	42	51	11
Sri Lanka	39.3	49	22	27	38	23	37	9 2003
St Lucia	46.9	16	14	71	27	24	49	25 2003
St Vincent	46.1							
Sudan	27.2							
Suriname	33.1	2	1	97	8	22	64	14 1999
Swaziland	32.4							25 1997
Sweden	51.2	1	11	88	3	36	61	7
Switzerland	57.2	3	13	84	5	36	59	4
Syria	36.4							12 2002
Tajikistan	31.6							
Tanzania	47.6							5 2001
Thailand	54.0	48	17	35	50	20	30	2
Togo	36.4							
Tonga	38.3							
Trinidad and Tobago	47.2	3	13	84	11	36	53	10 2002
Tunisia	36.3							14 2003
Turkey	35.3	56	15	29	24	28	48	10
Turkmenistan	43.1							
Uganda	37.1							3 2003
Ukraine	49.3	17	22	55	22	39	33	9
United Arab Emirates	54.0		14	86	9	36	55	2 2000
United Kingdom	50.6	1	11	88	2	36	62	5
United States	50.6	1	12	87	3	32	65	6
Uruguay	49.2	2	14	85	6	32	62	17 2003
Uzbekistan	40.6							
Vanuatu	47.2							
Venezuela	44.8	2	12	86	15	28	57	17 2003
Vietnam	49.9						2	
Virgin Is. (Am.)	47.3							
Yemen	25.5	88	3	9	43	14	43	12 1999
Zambia	40.3							12 1998
Zimbabwe	43.2							8 2002

* All data available are previous to the political separation of Serbia and Montenegro.

The commoditization of the Kyoto Protocol: the deal of the 21st century

ON 16 FEBRUARY 2005 the Kyoto Protocol came into force. Since then, quotas of greenhouse gas emissions that cannot be exceeded have been allocated to every country that has ratified it. The Protocol's first target is to reduce emissions by 5.2 per cent in relation to 1990 levels within the 2008-2012 period.

Each ratifying country had to draw up a National Allocation Plan to distribute emission allowances. When a country or company exceeds the allocated emission cap, other mechanisms are put into motion, such as the Clean Development Mechanism (CDM) or Emissions Trading.

However, negotiations left room for two ways of avoiding the emission caps set by the Protocol. The first one consists of helping a developing country to reduce emissions thanks to investments in clean, renewable or less contaminating energies - that is to say, through the application of the CDM. In theory it is as if the reduction had been carried out in a Northern country but with the advantage that in the South it is much cheaper.

On the other hand, if a country exceeds the pollution limits permitted, it may purchase emission allowances from countries whose emissions fall below the limits. In other words, the Protocol's requirements can be fulfilled without any reduction in pollution.

Setting this agreement into motion has generated nearly 30 billion euros aimed at 'carbon financing' and at clean energy markets and technologies in developing countries. It is estimated that by 2010 investments will have reached the amount of 200 billion euros.

The Protocol has been turned into a business deal. There is a European futures and options market for trading in emissions of toxic gases, especially carbon dioxide. This implies that, since the Protocol came into force, any citizen, company or institution is able to invest in pollution. It did not take long for consultants, investors and banks to discover 'the deal of the 21st century' and 'CO$_2$ experts' or 'climate change advisors' have become frequent consultants to the world of finance. ■

Sources:
www.portaldelmedioambiente.com
Acción ecológica
Juan Carlos Galindo, Agencia de Información Solidaria

Petrodollars versus Petroeuros

THE US DOLLAR BECAME THE UNIVERSAL CURRENCY 70 YEARS AGO. THE SELLING PRESSURE ON THE DOLLAR DECISIVELY FADED IN 1971, WHEN THE US SUSPENDED THE FREE CONVERTIBILITY OF NOTES INTO GOLD AT FIXED RATES OF EXCHANGE. SINCE THEN, THE DOLLAR HAS BECOME LINKED TO THE OIL TRADE. WITH A STRONG CURRENCY IN EUROPE AND THE DECISION BY SOME COUNTRIES TO PRICE THEIR OIL IN EUROS, US FINANCIAL HEGEMONY IS AT SERIOUS RISK.

World War Two allowed the US to become a dominant power as well as a reliable source of reference when it came to establishing trade agreements between countries. The dollar started to be accepted as an international currency under the White Plan, proposed by a Treasury official (Harry White, who was apparently a KGB agent effectively infiltrated into the US Government, according to Soviet documents recently made public). According to this plan, the US dollar was based on gold and all other currencies could only be converted into dollars. Based on the fact that the US supplied half the world's gross product, international trade started to use the dollar as the official currency.

There was no need for Washington to be too careful when it came to print dollar notes, which started to be in high demand all over the world. Many countries accumulated dollars from the sale of their products and soon started to claim gold, which according to the letter of the law, was equivalent to every dollar they had. It was not until 1971, under the Richard Nixon administration, that the obligation to deliver gold in return for dollars was declared void. By then, the US had gained control over the world oil and gas market, and the dollar, which no longer operated on the 'gold standard' was tied to a kind of 'oil standard'. At the end of any chain of transactions involving debts in dollars something objectively valuable could therefore still be obtained: oil instead of gold.

Oil price hike

The first sudden jump in oil prices took place precisely when Nixon departed from the old gold standard. The fact that this dollar crisis was labeled an 'oil crisis' is quite meaningful: all governments felt that energy prices were skyrocketing, when actually what was happening was an effective devaluation of currency.

At present 70 per cent of the world's currency reserves are held in dollars and oil is held captive by dollar values as if it were a US commodity. But what would happen if someone decided to use another currency?

From petrodollars to petroeuros

The worst nightmare of the Federal Reserve (the Central Bank of the US) is on the verge of becoming true: Organization of the Petroleum Exporting Countries (OPEC) or at least some of its members are considering abandoning the dollar and starting to use the euro as currency for natural gas and crude oil transactions.

According to experts, such a dollar-for-euro switch could result in a sudden devaluation within a 20-40 per cent range, thus unleashing a crisis similar to that of 1929.

In 2000 Saddam Hussein, then Iraq's head of government, decided to price his oil in euros. Few people believe that the real reasons for the US invasion of Iraq were unconnected with that decision to abandon the dollar as trading currency. The only serious US ally in the war against Iraq is the UK, which is part of the European Union (EU) but has not yet adopted the euro. The International Petroleum Exchange (IPE) - one of the two large international stock exchanges for oil - operates in Britain and prices crude oil and natural gas in US dollars. Other European countries, particularly the main actors in the eurozone - France and Germany - were opposed to the Iraq war from the beginning and are warning now about the ill-effects of a possible invasion of Iran.

Venezuela under Hugo Chávez, with the proposal to unite South America by means of a huge gas pipeline network, has mentioned the idea that the euro could become its oil transaction currency - and Venezuela is the third largest crude oil supplier to the US.

The real headache for Washington, however, is Iran. Not only did Iran switch to euros from 2003 onwards but it has announced a plan to open an euro-based international oil market. This would no longer simply involve bilateral purchase and sale agreements but would be an oil bourse where the free euro-based trading mechanism could compete against the two leading oil exchanges: New York's NYMEX (New York Mercantile Exchange) and London's IPE.

Both the value of US dollars and oil prices are two parts of the same phenomenon of international control exerted by the US Government. The existence of a new market out of its control would have ominous effects on its economy.

The China syndrome

China is a major exporter to the US. It invests $600 billion in US treasury bonds, which turns it into the second largest holder of US currency reserves (Japan is the largest holder with $800 billion). The situation is delicate for the US, which has allowed hundreds of domestic companies to sell out to Japanese capital but now places political obstacles in the way of similar Chinese purchases in its territory.

On the other hand, China aims at maintaining the US as a market for its goods, and thereby has kept its currency (the yuan) undervalued compared to the dollar. The Chinese Government has taken measures towards de-linking the yuan from the dollar, which could itself result in an immediate fall in the dollar's value.

China is also, however, dependent on Iran for oil, which supplies about 13 per cent of its energy needs. In October 2004, Iran and China signed an oil and gas trade agreement valued at $70-100 billion. US military action against Iran would therefore also threaten China.

Even if the only result of conflict was to halt Iranian oil exports to China, US interests could end up severely damaged. The Asian giant might then decide to buy oil from Russia, which would end up strengthening the standing of the euro - Russia is well disposed towards pricing its hydrocarbons in euros, which certainly makes sense given the fact that its major client is the EU.

In addition, experts warn that if the flow of Iranian crude oil were halted it would unfailingly result in prices in all markets climbing. Despite the desire of Washington élites to enforce petrodollar hegemony by military force, therefore, the geopolitical risks of an attack on Iran would be huge.

The wheel of fate

Little would change for the poorest countries if petrodollars were replaced by petroeuros. The US would probably lose its dominant role but would maintain a strong industrial and economic leadership in many areas, in the same way as Europe - despite the dominance of petrodollars - has been among the wealthiest regions in the world in recent decades.

The situation would, however, be different from that of half a century ago: natural reserves are closer to exhaustion now than when the US started to build its petrodollar empire. Will a 'universal commodity' equivalent to oil be found, or will there be a new multiplicity of currencies, with exchange-based bilateral trading systems, once fossil fuels are depleted?

The true possibility of a switch to petroeuros still seems remote. Oil-exporting countries have the greater part of their currency reserves placed in dollar accounts, so in order to switch to euros they would have to get rid of them. And who would accept dollars if the whole world was looking to dispose of them? The suspicion alone that such a massive offer of dollars was on the cards would be enough to cause the US currency to plummet, and everybody is so fearful of that possibility that no drastic moves are made.

Nevertheless, the EU and Russia could make it happen - and could even be able to cover any losses suffered by many poor countries as a consequence of the move towards the petroeuro system. The future of the US is now, much more than during Cold War days, in the hands of Russia. An international order based on Russia's alliances with the EU and China could yet emerge which would force Washingon from its position as the world's exclusive leading actor. ∎

The world according to...

...The World Bank ranked by income (Gross National Income, GNI, per capita).

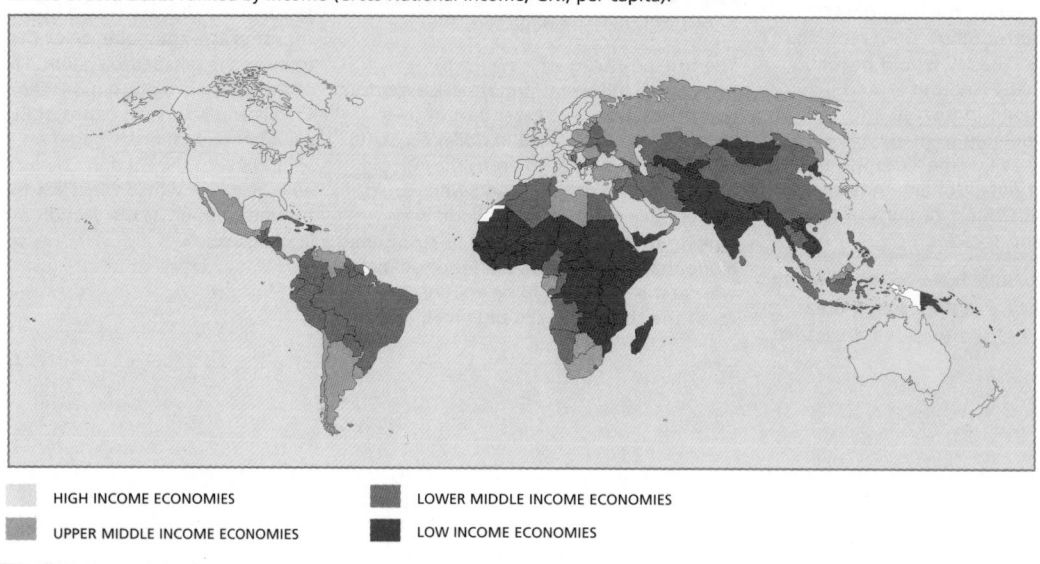

- ▢ HIGH INCOME ECONOMIES
- ▨ UPPER MIDDLE INCOME ECONOMIES
- ▨ LOWER MIDDLE INCOME ECONOMIES
- ■ LOW INCOME ECONOMIES

...UNDP (UN Development Program) ranked by Human Development Index.

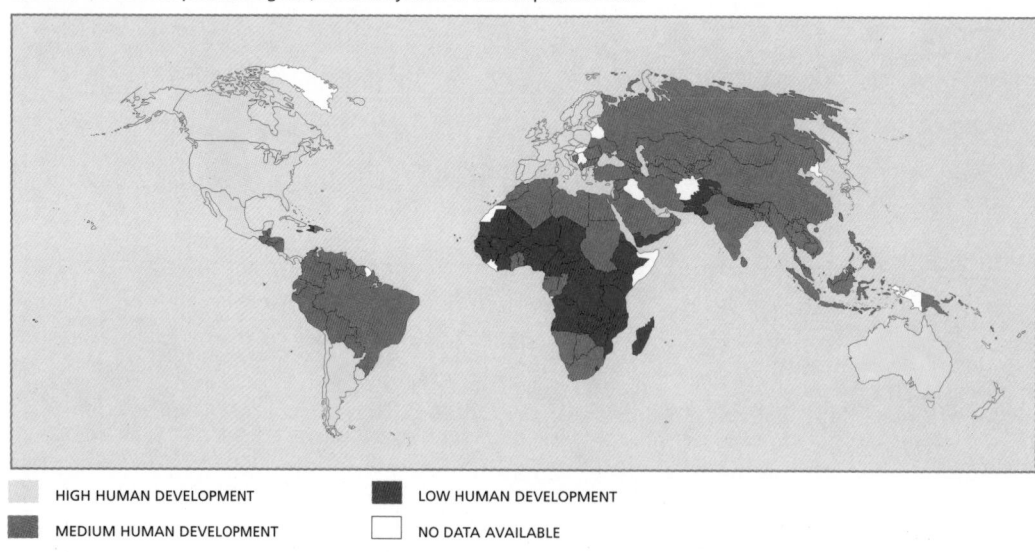

- ▢ HIGH HUMAN DEVELOPMENT
- ■ MEDIUM HUMAN DEVELOPMENT
- ■ LOW HUMAN DEVELOPMENT
- ▢ NO DATA AVAILABLE

...UNICEF (UN Children's Fund) ranked by under-five mortality rate (per 1,000 live births).

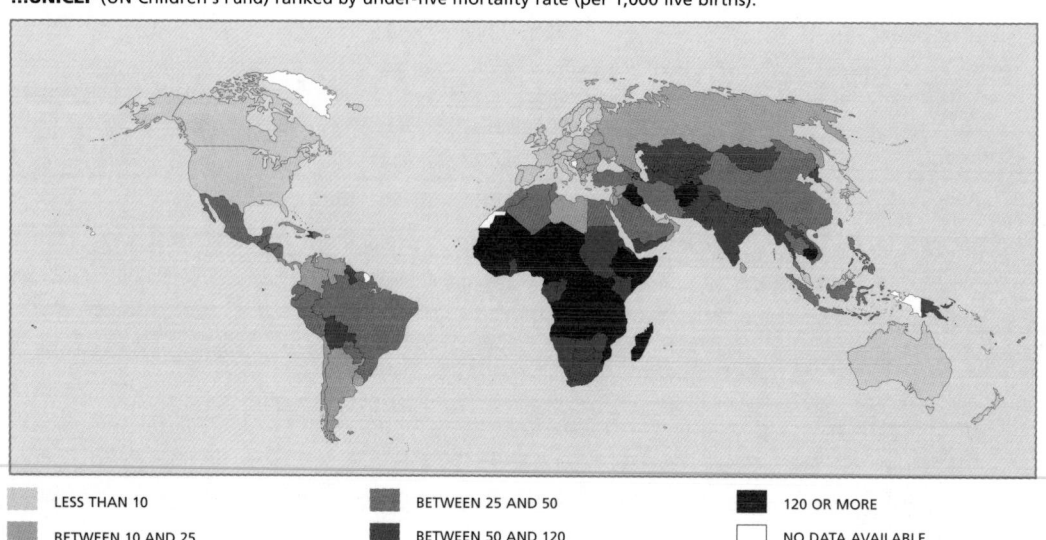

- ▢ LESS THAN 10
- ▨ BETWEEN 10 AND 25
- ▨ BETWEEN 25 AND 50
- ■ BETWEEN 50 AND 120
- ■ 120 OR MORE
- ▢ NO DATA AVAILABLE

NON-GOVERNMENTAL ORGANIZATIONS/CAMPAIGNS

ACORD (Agency for Co-operation and Research in Development)
Development House
56-64 Leonard Street
London EC2A 4JX
UK
www.acord.org.uk
An international consortium of NGOs working under the trusteeship of its member agencies. There are 11 member agencies, each in a different country. Its main role is to help establish or strengthen local non-governmental structures with a view to promoting self-reliant, participatory development. It also acts in emergency situations which seem likely to give rise to new development needs.

Amazon Basin Indigenous Organizations Co-ordinating Committee
www.coica.org
Founded in 1984, its aim is to strengthen local organizations and ties among indigenous peoples in their struggle to defend their rights and their native lands.

Amnesty International (AI)
The Human Rights Action Centre
17-25 Newe Inn Yard
London EC2A 3EA
UK
www.amnesty.org
Founded in 1961, AI is a worldwide campaigning movement (active in over 140 countries) that works to promote all the human rights enshrined in the Universal Declaration of Human Rights and other internationally recognized human-rights agreements. In particular, AI campaigns to free all prisoners of conscience; ensure fair and prompt trials for political prisoners; abolish the death penalty, torture and other cruel treatment of prisoners; and end political killings and 'disappearances'.

AMREF (African Medical Research Foundation)
www.amref.org
AMREF is Africa's largest indigenous health charity. It was founded in 1957 in Kenya and works with local communities and governments in sub-Saharan Africa to research and alleviate the region's health problems, with a special emphasis on primary healthcare. It also incorporates the Flying Doctor Service. It is supported by funds raised by 11 offices in Europe and North America.

Anti-Slavery International
Thomas Clarkson House
The Stableyard
Broomgrove Road
London SW9 9TL
www.antislavery.org
Founded in 1839, Anti-Slavery International is the world's oldest human-rights organization and has consultative status with the UN Economic and Social Council. It exposes current cases of slavery and campaigns for their eradication, pressing for more effective implementation of international laws against slavery. It focuses on all forms of slavery, including exploitative child labor, bonded labor, forced labor, trafficking of women and children, forced marriage and 'chattel' slavery.

ATTAC (Association for the Taxation of Financial Transactions for the Aid of Citizens)
www.attac.org
The international ATTAC movement was created at a meeting in Paris in December 1998. It describes itself as 'the international movement for democratic control of financial markets and their institutions'. It is a network with neither hierarchical structures nor a geographical center. The network now covers 33 countries and 15 languages.

Care International
www.care.org
An international network of 12 national organizations aiming to serve individuals and families in the poorest communities in the world. It works in the following fields: water; agriculture; emergency relief; health; urban priorities; micro-finance; and education.

Caritas Internationalis
www.caritas.org
A worldwide confederation of 162 Catholic relief and development organizations.
Palazzo San Calisto
00120 Vatican City

CEE Bankwatch Network
www.bankwatch.org
An international network with member organizations in 12 countries of the CEE/CIS region. The basic aim of the network is to monitor activities of International Financial Institutions (IFIs) in the region, and to propose constructive alternatives to their policies and projects, with particular attention to the environment, energy and transport.

EarthAction
www.earthaction.org
Created in 1992, its goal is to mobilize growing numbers of people around the world to press their governments (or sometimes corporations) for stronger action to solve global problems: environmental degradation, poverty, war and the abuse of human rights. Has international offices in Chile, the UK and the US and partner organizations in 162 countries.

Earth Rights International (ERI)
1612 K St NW, Suite 401
Washington, DC 20006
US
www.earthrights.org
ERI is a nonprofit group of activists, organizers and lawyers with expertise in human rights, the environment, and corporate and government accountability. It also has an office in Southeast Asia.

Focus on the Global South
c/o CUSRI, Chulalongkorn University,
Wisit Prachuabmoh Building,
Bangkok-10330
Thailand
www.focusweb.org
Focus on the Global South is a program of development policy research, analysis and action. It was founded in 1995, the same year the World Trade Organization came into existence, and reflects the priorities of a people's movement grappling with the impact of corporate-driven globalization on the daily lives and struggles of the poor and marginalized people in the South.

Friends of the Earth International
PO Box 19199
1000 GD Amsterdam
The Netherlands
Tel: +31 20 622 1369
Fax: +31 20 639 2181
info@foei.org
www.foei.org
Friends of the Earth International is a federation of autonomous environmental organizations from all over the world. Its members, in 68 countries, campaign on the most urgent environmental and social issues of the day, while simultaneously catalyzing a shift toward sustainable societies.

Greenpeace International
Ottho Heldringstraat 5
1066 AZ Amsterdam
The Netherlands
www.greenpeace.org
Greenpeace focuses on the most crucial threats to our planet's biodiversity and environment. It campaigns to: stop climate change, protect ancient forests, save the oceans, stop whaling, say no to genetic engineering, stop the nuclear threat, eliminate toxic chemicals and encourage sustainable trade. Formed in 1971, it now has a presence in 40 countries across Europe, the Americas, Asia and the Pacific. It uses research, lobbying and quiet diplomacy to pursue its goals, as well as high-profile, non-violent conflict to raise the level and quality of public debate.

HelpAge International
www.helpage.org
Founded in 1983, this is a global network of not-for-profit organizations with a mission to work with and for disadvantaged older people worldwide to achieve a lasting improvement in the quality of their lives.

Human Rights Watch
www.hrw.org
Originally a US organization, it now also has offices in London, Brussels, Sarajevo, Moscow, Tbilisi, Tashkent, Dushanbe and Rio de Janeiro. It is dedicated to protecting the human rights of people around the world, investigating and exposing human-rights violations and holding abusers accountable. It also challenges governments to end abusive practices and respect international human-rights law.

Institute for War and Peace Reporting
48 Grays Inn Road
London WC1X 8LT
UK
www.iwpr.net
Also has offices in Armenia, Azerbaijan, Georgia, Kazakhstan, Kosovo, Kyrgyzstan and Serbia. Works to inform the international debate on conflict and support the independent media in regions in transition. Publishes the magazine WarReport.

International Alert
346 Clapham Road
London SW9 9AP
UK
www.international-alert.org
A charity committed to the just and peaceful transformation of violent conflicts. International Alert seeks to advance individual and collective human rights by helping to identify and address the root causes of violence.

International Baby Food Action Network (IBFAN)
www.ibfan.org
IBFAN consists of public interest groups working around the world to reduce infant and young child morbidity and mortality. It co-ordinates international campaigning against and boycotts of companies breaching the International Code of Marketing of Breastmilk Substitutes, which continue to cause the deaths of babies and young children in the developing world.

International Campaign to Ban Landmines
!CBL, Chemin Balexert 7
1219 Geneva
Switzerland
www.icbl.org
An international network of more than 1,400 NGOs in 90 countries working for a global ban on landmines. The organization won the 1997 Nobel Peace Prize.

International Community of Women Living with HIV/AIDS
Unit 6, Building 1
Cannonbury Yard
190a New North Road,
London N1 7BJ
UK
www.icw.org
Formed at the Internationa AIDS Conference in Amsterdam in 1992 in response to the desperate lack of support and information available to HIV-positive women worldwide.

International HIV/AIDS Alliance
Queensberry House
104–106 Queens Road
Brighton BN1 3XF
UK
www.aidsalliance.org
Established in 1993 as an international NGO supporting community action on HIV and AIDS in developing countries.

International Planned Parenthood Federation (IPPF)
4 Newhams Road
London SE1 3UZ
UK
www.ippf.org
Founded in Mumbai, India, in 1952, IPPF is the world's leading voluntary organization in this field. It exists to support sexual and reproductive health programmes – including family planning – through more than 150 national family planning associations in over 100 countries.

International Red Cross and Red Crescent Movement
International Committee of the Red Cross
www.icrc.org
International Federation of Red Cross and Red Crescent Societies
www.ifrc.org
The International Red Cross and Red Crescent Movement is the largest independent humanitarian organization in the world. It was inspired by a Swiss entrepreneur who, appalled by the suffering of thousands left to die after the Battle of Solferino in 1859, proposed setting up national relief societies of volunteers that would provide neutral and impartial help in times of war. The movement's key principles are: humanity; impartiality; neutrality; independence; voluntary service; unity; and universality.
The Federation is the focal point for the national societies and is primarily concerned with the victims of natural disasters; the ICRC is concerned with the victims of war. The Movement enforces the Geneva Conventions.

Médecins Sans Frontières (MSF)
www.msf.org
Formed in 1971, MSF provides emergency medical assistance to populations in danger in more than 80 countries. MSF works in rehabilitation of hospitals and dispensaries, vaccination programmes and water and sanitation projects. MSF also works in remote healthcare centres, slum areas and provides training of local personnel. All this is done with the objective of rebuilding health structures to acceptable levels.

Minority Rights Group (MRG)
54 Commercial Street
London E1 6LT
UK
www.minorityrights.org
MRG works to secure rights for ethnic, religious and linguistic minorities worldwide, and to promote co-operation and understanding between communities. To this end it engages in advocacy, training, publishing, facilitating and outreach.

One World International
2nd Floor
River House
143-145 Farringdon Road
London EC1R 3AB
UK
www.oneworld.org
Originally launched as One World Online in 1995, it became a global organization in 1999 and now has centres in Austria, Costa Rica, Finland, India, Italy, the Netherlands, the UK, the US and Zambia. Partner organization form a global network sharing a common aim of using the internet to promote human rights and sustainable development.

Oxfam International
www.oxfam.org
Originally a British aid organization formed in 1942, Oxfam International is now a confederation of 12 organizations working together with over 3,000 partners in more than 100 countries to find lasting solutions to poverty, suffering and injustice. It seeks increased worldwide public understanding that economic and social justice are crucial to sustainable development.

Peoples' Global Action (PGA) c/o Canadian Union of Postal Workers (CUPW),
377 Bank Street
Ottawa, Ontario, Canada
www.agp.org
Formed in 1998 to co-ordinate worldwide resistance to the global market, the alliance's full title is Peoples' Global Action against 'Free' Trade and the World Trade Organization. Its major activity has been coordinating decentralized Global Action Days around the world to highlight the global resistance of popular movements to capitalist globalization.

Pesticide Action Network (PAN)
www.pan-international.org
Pesticide Action Network (PAN) is a network of over 600 participating NGOs, institutions and individuals in over 90 countries working to replace the use of hazardous pesticides with ecologically sound alternatives. Its projects and campaigns are co-ordinated by five autonomous regional centers in North America, Latin America, Europe, Africa and Asia-Pacific.

Save the Children
International Save the Children Alliance
Second Floor, Cambridge House
100 Cambridge Grove
London W6 0LE
UK
www.savethechildren.net
27 Save the Children organizations make up the International Save the Children Alliance, which works in over 115 countries. The organization fights for children's rights while aiming to deliver immediate and lasting improvements to children's lives worldwide.

Third World Network
131 Jalam Macalister
10400 Penang
Malaysia
www.twnside.org.sg
An independent non-profit international network of organizations and individuals involved in issues relating to development, the Third World and North-South issues. Its objectives are to conduct research on economic, social and environmental issues pertaining to the South; to publish books and magazines; to organize and participate in seminars; and to provide a platform representing broadly Southern interests and perspectives at international fora such as the UN conferences and processes. It also has offices in Delhi, India; Montevideo, Uruguay; Geneva; and Accra, Ghana.

Transparency International (TI)
Alt Maabit 96
10559 Berlin
Germany
www.transparency.org
Founded in 1993 as a global coalition against corruption, TI has more than 85 independent national chapters and annually publishes the Global Corruption Report.

World Rainforest Movement (WRM)
International Secretariat
Maldonado 1858
11200 Montevideo
Uruguay
www.wrm.org.uy
Established in 1986, the WRM is an international network of citizens' groups of North and South involved in efforts to defend the world's rainforests. It works to secure the lands and livelihoods of forest peoples and supports their efforts to defend the forests from commercial logging, dams, mining, plantations, shrimp farms, colonization, settlement and other projects that threaten them.

Worldwide Fund for Nature (WWF)
www.panda.org
World Wildlife Fund was founded in 1961 to preserve wildlife and natural habitats. In 1989 it changed its name to the Worldwide Fund for Nature and its key aims are now expanded to: conserving the world's biological diversity, ensuring that the use of renewable natural resources is sustainable and promoting the reduction of pollution and wasteful consumption.

World Social Forum
www.worldsocialforum.org
The World Social Forum, which takes place annually in January, is an open meeting place where groups and movements of civil society opposed to neo-liberalism and a world dominated by capital or by any form of imperialism, but engaged in building a planetary society centred on the human person, come together to debate ideas, to formulate proposals, to share experiences freely and network for effective action. Regional and local social forums are also increasingly active and important.

Countries of the world

Afghanistan / Afghanestan

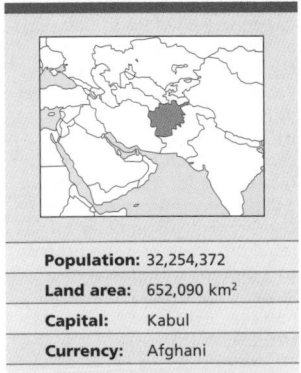

Population:	32,254,372
Land area:	652,090 km²
Capital:	Kabul
Currency:	Afghani
Language:	Pashtu, Dari

The territory now known as Afghanistan is believed to have been inhabited in the Neolithic era, 100,000 years ago. In a cave at Darra-i-Kur, in Badakhshan, fragments were found of a Neanderthal skull. During the Bronze Age, in the third and second millennia BC, with the rise in commerce with Mesopotamia and Egypt, and the export of lapis lazuli extracted from the Badakhshan mines, the first urban centers were founded: Mundigak and Deh Murasi Ghundai. While the mesa region of Persia, the Central Asian steppes and the Indo valley saw important population growth, the region became a migration route, and the Khyber Pass turned into a gateway to northern India.

[2] Throughout its history, the territory has been known primarily by three names: Ariana, when the Arian tribes settled there around 2000 BC, Khurasan in the medieval period, and Afghanistan in the modern era. It is thought that Kabul was founded during the Arian settlement and that the *Rig-Veda*, one of the founding texts of Hinduism, was written in Ariana.

[3] The region was incorporated into the Persian Empire of Cyrus the Great in the 6th century BC,

and there is speculation that around this time the Zoroastrian religion was introduced in what is known as Bactria. Three centuries later, Alexander of Macedonia (who founded Alexandropolis, present-day Kandahar) pushed out the Persians and incorporated the territory into the Alexandrian empire. Upon his death, in 323 BC, the eastern Satraps came under the dominance of the Seleucid dynasty, which governed from Babylon. In 250 BC, Diodoth, a local Greco-Bactrian governor, declared the Amu River plains independent. The Greco-Bactrians expanded southwards and, in 180 BC, established their dominion in Kabul and Punjab. The Parthians of eastern Iran broke off from the Seleucids, taking over Seistan and Kandahar.

[4] A confederation of five Central Asian nomad tribes, known as the Yüeh-chih, deposed the Greco-Bactrian kingdom and, united under the Kusana tribe, conquered the rest of the territory, establishing a kingdom of the same name, which turned into a trade center between Rome, India and China, opening the way for the 'silk route'. In the 2nd century AD, the empire of King Kaniska stretched from Mathura, in north and central India, to the Chinese borders in Central Asia.

[5] The Sassanian Persian Empire annexed part of the Kusana empire in the 3rd century and in the following

century a new wave of Central Asian nomads, known as Heftali, took control of the territory. In 550 AD, the Persians re-conquered what is Afghanistan today, though they faced repeated uprisings by the Afghan tribes.

[6] Islam was introduced when Muslim forces defeated the Sassanians in 642 in Nahavand (near present-day Hamadan, in Iran) and spread to Afghan territory. The 9th and 10th centuries saw the rise of various local Islamic dynasties, one of the first being that of the Tahirids, established in Khorasan, and whose dominion included Balkh and Herat. That dynasty was succeeded by the Saffarids, natives of Seistan. The northern areas soon became feudatories of the powerful Samanids, who ruled from Bujara, making their grandeur known in Samarkand, Balkh and Herat.

[7] The Mongols, under the command of Genghis Khan, invaded the eastern portion of Sultan Ala ad-Din's empire in 1219, and conquered the territory in 1221, making it part of their own vast empire. With the breakup of the empire following Genghis' death in 1227, some local leaders were able to maintain autonomous fiefdoms, while others swore servitude to the Mongol princes. In 1360, they fell under the reign of Timur Lenk (Tamerlane),

the Turkish conqueror of Islamic faith, whose descendants governed Khorasan until the beginning of the 16th century.

[8] With the rise of the third Persian Shi'a empire (1502) and the empire of the Great Mogul in India (1526), the region became the scene of constant battles between the Mongols who dominated Kabul, the Saffavid Persians who controlled the southern region, and the Uzbek descendants of Tamerlane who ruled the northwest. The battles and political upheaval gave way to unification in 1747 when an assembly of local chieftains elected Shah Ahmad Durrani, a military commander who had previously served Persian sovereigns. Ruling by military might, the new shah consolidated the national borders, which would be threatened by the expansionism of Czarist Russia and the interests of Britain, which controlled India.

[9] The first British-Afghan War (1839-1842), which the British Empire lost, reinforced Dost Muhammad Shah's slightly pro-Russian sympathies. He sought to increase his influence in northern India by encouraging anti-British movements. His son Sher Ali Shah continued this policy, which led British forces to invade the country once again.

[10] As a result of the second British-Afghan War (1878-1880), the Durrani dynasty

LAND USE

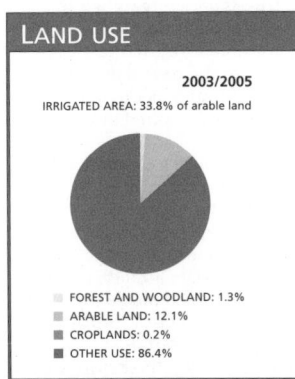

2003/2005

IRRIGATED AREA: 33.8% of arable land

- FOREST AND WOODLAND: 1.3%
- ARABLE LAND: 12.1%
- CROPLANDS: 0.2%
- OTHER USE: 86.4%

was overthrown. Afghanistan lost its territories south of the Khyber and became a buffer state between Czarist Russia and India. The country thus lost control over its foreign relations. In 1893, the Durand Line, which was not intended as a political border, delineated the zones of responsibility for maintaining law and order between British India and Emir Abdor Rahaman Khan, who governed from Kabul.

[11]   In 1919, after a third British-Afghan war, lasting four months, the country was freed from British protection. Independence leader Emir Amanullah Khan (the heir and grandson of the British-imposed ruler) came to power and modernized the country. He enacted a relatively liberal Constitution and became the first head of state in the world to establish diplomatic ties with the Soviet Union, marking the beginning of special relations that would last seven decades.

[12]   Amanullah was overthrown in 1929 by the Mohammadzai clan, which crowned Muhammad Nadir Shah. A new Constitution in 1931 recognized the autonomy of local leaders. The new Shah was assassinated in 1933 and the crown went to his son Zahir, who for the first 20 years of his reign attempted to consolidate the country, expanding foreign relations and promoting internal development. Once Pakistan became independent in 1947, the old Durand Line left the country with the problem of determining the political status of the Pashtuns who lived on the Pakistani side.

[13]   Lt-Gen. Muhammad Daud Khan, the Shah's cousin and brother-in-law, became Prime Minister in 1953 and launched a new modernization process: he nationalized utilities, built roads, irrigation systems, schools and hydroelectric facilities (with US funding); he reorganized the armed forces (with Soviet assistance); and maintained neutrality throughout the Cold War. Furthermore, he abolished the obligatory use of the *chador* (veil) by women, and *purdah*, or the prohibition of women in the public sphere.

[14]   Finding itself in the middle of the zone of conflict, Afghanistan attempted early in the Cold War to remain equidistant between the US and the USSR. But it grew increasingly dependent upon the USSR due to the ongoing US support of Pakistan. As of 1955, thousands of Afghans were regularly being sent to study in the Soviet Union, and particularly to receive military training.

[15]   The Pashtuns' demands for independence prompted Daud to take repressive measures against them, leading Pakistan to close its border with Afghanistan in 1961. Soviet influence became evident in some Marxist leanings in the press and the government, which displeased those allied with the king. In March 1963, King Zahir 'accepted Daud's resignation' and, two months later, Pakistan reopened the border.

[16]   Muhammad Yusuf was appointed prime minister. He proposed a Cabinet of technocrats and intellectuals, and pushed for a new Constitution based on the principle of individual freedom that at the same time upheld the values of Islam and the monarchy. Enacted in 1964, the Constitution for the first time allowed the creation of political parties and the holding of elections, but it indirectly banned the participation of Marxist parties.

[17]   The People's Democratic Party of Afghanistan (PDPA), founded as an underground organization, staged its first anti-monarchy demonstrations in 1965. Shortly thereafter, the PDPA was split between the Khalk (consisting of the ethnic Takjik or Afghan-Persians), which advocated revolution through a worker-peasant alliance, and the *Parcham*, or Banner (of the Pashtun), which sought to establish a broad-based front involving intellectuals, the national bourgeoisie, the urban middle class and the military.

[18]   Workers and students began to organize actively in the country's industrial regions. Demonstrations and criticisms of the king grew more frequent. Moscow, which had not fully accepted Daud's replacement, supported the naming of Daud as president in 1973 while King Zahir was abroad. With the support of the PDPA, he proclaimed Afghanistan a republic and annulled the 1964 Constitution.

[19]   Daud designed a platform based on democracy and socialism, nearly identical to the one published four years earlier in the first edition of *Parcham* newspaper, particularly in the areas of agrarian reform, bank nationalization, industrial development and social justice. The new single-party Constitution, based on the models of Algeria and Nasser's Egypt, was approved in April 1977. Daud, who had thrown out the communist ministers and lost Moscow's backing, was elected President for a 10-year term.

[20]   Daud travelled to Kuwait, Saudi Arabia and Egypt, in an attempt to re-establish ties with the Islamic world. In a desperate bid, he tried to reconcile with the Shah of Iran in 1978, which only accelerated his demise. The military organized by the Parcham assassinated Daud and his entire family, and replaced him with Nur Muhammad Taraki, who was also named Secretary-General of the PDPA. Hafizulah Amin, leader of a rival communist faction, and Babrak Karmal, leader of the Parcham, were appointed vice-premiers. Conflicts between the two were resolved in April 1979, when Amin was appointed Prime Minister. In September, Amin overthrew and assassinated his one-time ally Taraki.

[21]   Amin introduced changes such as a literacy campaign based on secular values, equality for women, agrarian reform and the abolition of the dowry system, which shook up the country's traditional standards. Though he asserted that Afghanistan considered itself a non-aligned nation, the peasants, familiar with the radio broadcasts from Moscow, assumed that the new government was Marxist, pro-Soviet and, therefore, atheist. In February 1979, the US ambassador to Kabul was kidnapped and murdered. The US withdrew economic assistance and increased its hostilities towards what it considered a pro-Soviet government.

[22]   Amin was assassinated in a coup, which was backed by the Soviet troops that had entered the country in December 1979 for strategic reasons. Babrak Karmal was installed as Prime Minister, president of the Revolutionary Council and PDPA Secretary-General. In several areas of the country, resistance against the Soviet invaders began to grow and *mujahedin*, or Islamic guerrillas, were organized. Islamic fundamentalists traveled to Afghan territory in volunteer expeditions financed by Saudi Arabia. Meanwhile, millions of Afghan peasants sought refuge in neighboring Pakistan and Iran.

[23]   The mujahedin guerrillas, divided among various factions backed by different countries (Iran, Pakistan, Saudi Arabia and the US), coincided with the deepening divisions in Kabul. In May 1986, Mohammed Najibullah, a young doctor of Pashtun origin, replaced Karmal as PDPA Secretary-General. He announced a unilateral ceasefire in January 1987, with guarantees for guerrilla leaders willing to negotiate with the Government, an amnesty for rebel prisoners, and the promise of a prompt withdrawal of Soviet troops. The mujahedin, however, continued their armed struggle.

[24]   After six years of negotiations, an Afghan-Pakistani accord was signed in Geneva, guaranteed by both the US and USSR. The agreement ensured the voluntary return of refugees, who by then numbered more than four million. Another document, signed by Afghanistan and the USSR, provided for the withdrawal of Soviet troops. The PDPA was renamed Watan Party, or Party of the Homeland.

[25]   In September 1991, the US and USSR agreed to stop sending arms to the Afghan Government and guerrillas, sparking confrontation between Saudi Arabia and Iran, and the Afghan mujahedin groups they each financed. The Kabul regime was left without foreign support following the break-up of the USSR. After Najibullah sought refuge at the UN headquarters in Kabul, in April 1992 (marking the collapse of the communist regime), the Government was left in the hands of the four vice-presidents.

[26]   The Government announced its willingness to negotiate with the rebel groups, but its meeting with commander Ahmed Shah Massud, of the Jamiat-i-Islami, triggered demonstrations among the mujahedin belonging to the Pashtun majority in the country's south and east. From Pakistan, Gulbuddin Hekhmatyar, the head of fundamentalist group Hezb-i-Islami, threatened to start bombing the capital if the Government refused to step down. In the days that followed, the forces of Massud and Hekhmatyar clashed in Kabul itself.

[27]   The alliance of 'moderate' Muslim groups, headed by Massud (the new defense minister), gained control of the capital, expelling the Islamic fundamentalists led by Gulbuddin Hekhmatyar. In May, the Interim Council formally dissolved the Watan Party (former PDPA). The KHAD, or secret police, and the National

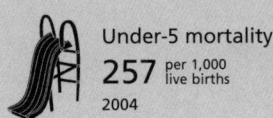

Under-5 mortality
257 per 1,000 live births
2004

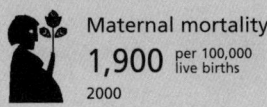

Maternal mortality
1,900 per 100,000 live births
2000

Assembly were also dismantled.
[28] Some of the Government's measures reimposed Islamic law: the sale of alcohol was outlawed, and new rules required women to cover their heads and wear traditional Islamic dress. Hekhmatyar continued fighting Kabul, demanding the withdrawal of Massud and of the militia loyal to Abdul Rashid Dostam, who had been a member of the communist government, but had defected to join the Muslim guerrillas that took power.
[29] By this time, the economy had come to a standstill and 60 per cent of its productive structure had been destroyed. Afghanistan had become the world's largest producer of opium. The Pakistani Government decided to put a stop to the arms and food contraband across its border with Afghanistan in order to weaken Hekhmatyar, whom it held responsible for the deterioration in relations between the two countries.
[30] Beginning in 1993, the president in Kabul and head of the Jamiat-i-Islami, Buranuddin Rabbani, Hekhmatyar and Dostam were the main leaders of the conflict, marked by pacts and betrayals. The emergence in the south in 1995 of the armed

Taliban (Persian for 'students of the Qur'an'), changed the course of the war. These guerrillas, trained in Pakistan, aimed to create a united Islamic government in Afghanistan. The Taliban proclaimed that the three aforementioned leaders were a fundamentalist-communist alliance that offended Islam.
[31] Pakistan, Saudi Arabia and the US supported the Taliban army, which took control of Kabul in September 1996, while the Government fled to the north. Mohammed Omar Akhunzada (Mullah Omar) was elected in April 1996 as 'Commander of the believers' (*amir ol momumin*) in the Taliban territories. In June 1997 the United National Islamic Front for the Salvation of Afghanistan, better known as Northern Alliance (NA) or United Front, made up mainly by Tajik, Uzbek and Hazara factions, was formed to fight the Taliban regime.
[32] Once in control of Kabul, and following its principle to rule according to its own interpretation of the *Qur'an*, the Taliban banished women from the public sphere, excluding them from education and reviving purdah. At the same time, it outlawed music and

songs (except religious hymns), cinema, theater and alcohol, declaring them 'non Islamic'. By late 2000, the Taliban army controlled more than 95 per cent of Afghan territory.
[33] On 3 September 2001, the NA leader Massud was assassinated - supposedly under orders from Mullah Omar - which would have been a mortal blow for the opposition's aspirations had it not been for the terrorist attacks of September 11 against New York and Washington. These unleashed US wrath in the 'war on terrorism' against terrorist organization al-Qaeda, led by Saudi Osama bin Laden, a former mujahed who lived in Afghanistan along with thousands of his men, sheltered by the Taliban.
[34] In September 2001, the Council of Elders, meeting in Kabul, asked the Taliban regime to persuade Bin Laden to leave the country on his own free will. The council also resolved to call for a *jihad* (holy war) in case the US attacked Afghanistan.
[35] Air raids by the US-led coalition, with involvement of the UK, Australia and Canada and the support of the EU and NATO (including Turkey), China, Russia, Israel, India, Saudi

Arabia and Pakistan, began on 7 October. Iran and Iraq condemned the attacks.
[36] As the war evolved, Rabbani's leadership weakened and he eventually was replaced by a triumvirate formed by Foreign Minister Abdullah Abdullah - main spokesperson for the Front - Home Minister Yunus Qanuni (both from a more secular perspective), and Defense Minister Mohammad Qasem Fahim.
[37] On 13 November 2001, the NA took Kabul and was welcomed by part of the population, which wanted the end of the Taliban regime and the bombings.
[38] In the inter-Afghan Conference held in Bonn, an Interim Administration of 30 members was created, with monarchist Pashtun Hamid Karzai as chair. An 18-month schedule to prepare for general elections was agreed on. Before the voting there would be an emergency *Loya Jirga* (assembly), a Transitional Authority and a Constitutional *Loya Jirga*, assisted by the UN. The NA got 18 out of the 29 Ministries. Abdullah, Qanuni and Fahim were confirmed in their roles, while Zahir was appointed to open the *Loya Jirga*. Karzai took office on 22 December.
[39] In February 2002 trade relations with the US - suspended since the Soviet occupation - were resumed. In spite of the declared end of the war, different factions kept fighting, and there was still no sign of Osama bin Laden or Mullah Omar.
[40] In July 2002 Vice President Haji Abdul Qadir - a Pashtun and former NA commander - and his driver were killed. In September, Karzai survived an assassination attempt by al-Qaeda in Kandahar. Most of the country was under control of powerful warlords supported by the US. President Karzai's authority was mostly limited to Kabul.
[41] In August 2003, NATO launched a peace mission in the country, its first outside Europe. NATO was to be in charge of planning, supervising, commanding and controlling the International Assistance Security Force in Afghanistan, under the auspices of the UN.
[42] In November, a draft Constitution was sent by the Constitutional Revision Commission to President Karzai and the UN special envoy, Lakhdar Brahimi. The draft provided for the creation of an Islamic republic with equal rights

PROFILE

ENVIRONMENT
The country consists of a system of highland plains and plateaus, separated by east-west mountain ranges (principally the Hindu Kush) which converge on the Himalayan Pamir. The main cities are located in the eastern valleys. The country is dry and rocky though there are many fertile lowlands and valleys where cotton, fruit and grain are grown. Coal, natural gas and iron ore are the main mineral resources.

SOCIETY
Peoples: Pashtun 44 per cent, Tajik 25 per cent, Hazara 10 per cent and Uzbek 8 per cent. The rest is composed of peoples of Turkic origin and nomads of Mongolian origin.
Religions: 99 per cent of the population is Muslim (74 per cent Sunni, 15 per cent Shi'a, and 10 per cent others).
Languages: Pashtu and Persian (Dari) are the official languages. There are also many languages, mainly of Persian or Turkic root: Hazaragi, Turkmen, Uzbek, Aimaq, among others.
Main Political Parties: Islamic United Party of Afghanistan (Hezb Wahdat Islami Afghanistan); Islamic Party Jamiat of Afghanistan (Hezbe Jamiate Islami Afghanistan); Islamic Party of Afghanistan (Hezb-i-Islami Afghanistan); National Congress Party of Afghanistan (Hezb-e-Congra-e-Mili Afghanistan); National Movement of Afghanistan (Hezb-e-Nuhzhat-e-Mili Afghanistan).
Main Social Organizations: Trade unions are

very weak because industrial activity is not significant. Other social organizations operate especially among Afghans living abroad, such as: Revolutionary Afghan Women's Association (RAWA); Humanitarian Assistance for the Women and Children of Afghanistan (HAWCA), Co-ordinating Council for National Unity and Conciliation in Afghanistan.

THE STATE
Official Name: Jomhuri-ye Eslami-ye Afghanestan (Islamic Republic of Afghanistan).
Administrative divisions: 34 provinces.
Capital: Kabul 2,967,000 people (2004).
Other cities: Kandahar 381,200 people; Herat 267,500; Mazar-e-Sharif 292,000 (2004).
Government: Hamid Karzai was elected president 5 October 2004. The bicameral National Assembly consists of the *Wolesi Jirga* or House of People and the *Meshrano Jirga* or House of Elders. Parliamentary and provincial council elections were held on 18 September 2005, with results finally announced on 12 November 2005. These were the first parliamentary elections for over 30 years. Approximately 12 million people were eligible to vote for the 249-seat Wolesi Jirga and 34 provincial councils. The 102 members of the Meshrano Jirga are indirectly elected by the provincial councils.
National holiday: 19 August, Independence Day (1919).

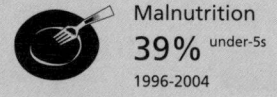

Malnutrition
39% under-5s
1996-2004

Water source
13% of population using improved drinking water sources
2002

IN FOCUS

ENVIRONMENTAL CHALLENGES
The increasing rate of deforestation, excessive use of the land and soil erosion are severe threats. Safe water resources are very scarce. Huge areas of land and buildings were devastated by the war.

WOMEN'S RIGHTS
Women have voted and run for office since 1963. In 2005, three years after the fall of the Taliban regime, the country remained one of the world's least developed. Women, condemned to a life of malnutrition, exclusion from public life, violence, rape and forced marriage, are the worst off. Among other restrictions they still face inadequate access to healthcare and nutrition. Afghanistan has one of the highest maternal mortality ratios, together with low life expectancy and severe malnutrition.

In 2003, girls made up 33 per cent of all primary school students, a huge increase compared to the 7 per cent in 2000. However, in southern and eastern Afghanistan female enrolment remained very low.*

CHILDREN
Although the situation of children has improved greatly, the data up to 2004 show that 1 in 9 children will die before their first year, while 1 in 6 will die before the age of 5. Some 45 per cent of boys and girls are not in school; 30 per cent of under 5s have diarrhea-related diseases, related to the fact that 60 per cent of homes lack drinking water. About 39 per cent of children under 5 are moderately or seriously underweight, while 54 per cent are moderately or seriously stunted.

As of December 2004, some 4,000 children between 14 and 17 were demobilized and disarmed.

INDIGENOUS PEOPLES/ETHNIC MINORITIES
The Hazara speak Farsi, are Shi'a Muslims and live in the central hills (Hazarajata) of Afghanistan. They subsist on their crops, sheep raising and trade with other nomad peoples such as the Pashtun, Afghanistan's largest ethnic group. Their political, economic and cultural status is vulnerable although they amount to 9 per cent of the country's population. Between 1998 and 2000 the Taliban - a Pashtun-related cultural group with strict religious discipline - carried out massacres of Hazara civilians.

The Uzbek, Sunni Muslims, are Turkic from an ethnic and linguistic point of view. They live and work in the northern agricultural region of the country, bordering their relatives in Uzbekistan. Uzbek women are renowned as carpet weavers, and this has historically supplied their community with extra income. This economic advantage has given way to political ones, and Uzbek men have held political office in several Afghan governments.

MIGRANTS/REFUGEES
According to the treaties between Pakistan and Afghanistan, 2,740,000 refugees were repatriated by UNHCR since 2002. In the same period, 1,400,000 returned from Iran. In 2006 it is expected some 400,000 will return to their homes. The repatriates receive between $4 and $7 for the trip (depending on the distance) and another $12 to cover expenses upon arrival. New agreements are being sought for after 2006.

DEATH PENALTY
It is currently applied as a punishment, even for common crimes.

* Latest data available in The State of the World's Children and Childinfo database, UNICEF, 2006.

---

for all its citizens. It made no reference to the *sharia* or Islamic law, but it was assumed no laws would contradict Islam.

[43] The Constitution, which established a strong presidency, was passed by the Loya Jirga in January 2004. Details of the abuse committed by US forces against detainees in detention centers were issued between March and May.

[44] In late March, presidential and parliamentary elections announced in the Bonn schedule for June 2004 were postponed until September due to security problems and delays in the registration of voters. During a trip to Berlin that month, Karzai requested economic aid to rebuild his country and warned that, unless things improved, Afghanistan was at risk of falling once again into chaos and war.

[45] In April, the UK, US, Japan, Germany and the UN pledged $8.2 million in aid for the following three years. That month the country signed cooperation agreements to fight drug traffic with China, Pakistan, Iran, Turkmenistan, Uzbekistan and Tajikistan.

[46] In July 2004, elections were postponed once again, due to the growth of violence in several regions of the country. The date set was October 9 and Karzai was one of the more than 20 candidates. Among his stronger rivals was Abdul Rashid Dostum, a military advisor for the interim government and one of the most powerful warlords. Karzai warned he was a larger threat to the country than the Taliban, while for observers his candidacy was a sign that legitimized the new democratic regime.

[47] On 3 November, Karzai won with 55 per cent of the vote and took office 30 days later, as stated by the Constitution. UN observers dismissed fraud reports by the opposition.

[48] In January 2005 the US army freed 80 suspects who allegedly were Taliban supporters. In September, the first parliamentary and provincial elections in more than 30 years were held, and three months later the new Parliament met for the first time.

[49] More than 30 people died in January 2006 in a series of suicide attacks in the southern province of Kandahar. Government troops and suspected Taliban fighters clashed in February in the southern province of Helmand.

[50] That month, donors meeting in London pledged to contribute to the country's reconstruction more than $10 billion in the following five years.

[51] In late 2006, NATO took over the southern region (including the provinces of Day Kundi, Helmand, Kandahar, Nimroz, Uruzgan and Zabul), traditionally dominated by the Taliban and drug traffickers. The handing over of power - from the US to an organization made up of 37 countries, which increased its presence from 3,000 to 9,000 troops in the territory - took place amid growing violence in the area, where occasionally there are sometimes more daily attacks than in Iraq (see history of Iraq). Apart from the attacks on government officials and the international forces, violence also affected schools. The education system has been specially targeted for destabilization by the Taliban because of its opposition to the schooling of girls.

[52] Lieutenant General David Richards - commander of NATO forces in Afghanistan - said in August 2006 that the British forces posted there had been fighting the UK's longest and bloodiest conflict in the last 50 years. Meanwhile, shadow UK foreign secretary William Hague stated that the rest of Britain's allies in NATO should also contribute with troops so 'not only the UK suffered the brunt' of the conflict.

[53] October 2006 marked the fifth anniversary of the ousting of the Taliban Government, yet Taliban-led violence continued to increase, reaching its highest peak since 2001. More than 3,000 people were already thought to have died during 2006 to date, compared with 1,600 deaths during 2005.

Over 150 foreign soldiers from the international force had died during 2006 – more than three times the number of deaths in 2003 and 2004.

[54] The spiralling insurgency of Islamist militants is by no means the only threat to stability in Afghanistan. The Government has little power outside Kabul, and vast areas of the country are controlled by warlords, many of whom are involved in drug trafficking. Afghanistan still produces 90 per cent of the world's opium, which is said to account for a third of the country's economy, and poppy cultivation jumped by almost 60 per cent during 2006. According to the Ministry of Counter Narcotics, the area under poppy cultivation has risen from 8,000 hectares in 2001 to 165,000 hectares in 2006.

[55] In October 2006 NATO took over command of military operations in eastern Afghanistan from the US-led coalition. Dozens of civilians were killed during NATO air strikes in the insurgency-ravaged southern province of Kandahar as the insurgency continued to take hold. In the same month, the UN-assisted Afghan repatriation programme came to an end. As if the ongoing conflict were not trouble enough, some 2.5 million Afghans were hit by drought and, having lost their crops, faced food shortages. ∎

Albania / Shqipëria

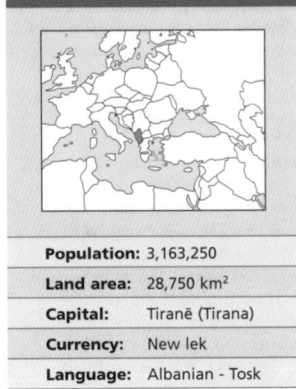

Population:	3,163,250
Land area:	28,750 km²
Capital:	Tiranë (Tirana)
Currency:	New lek
Language:	Albanian - Tosk

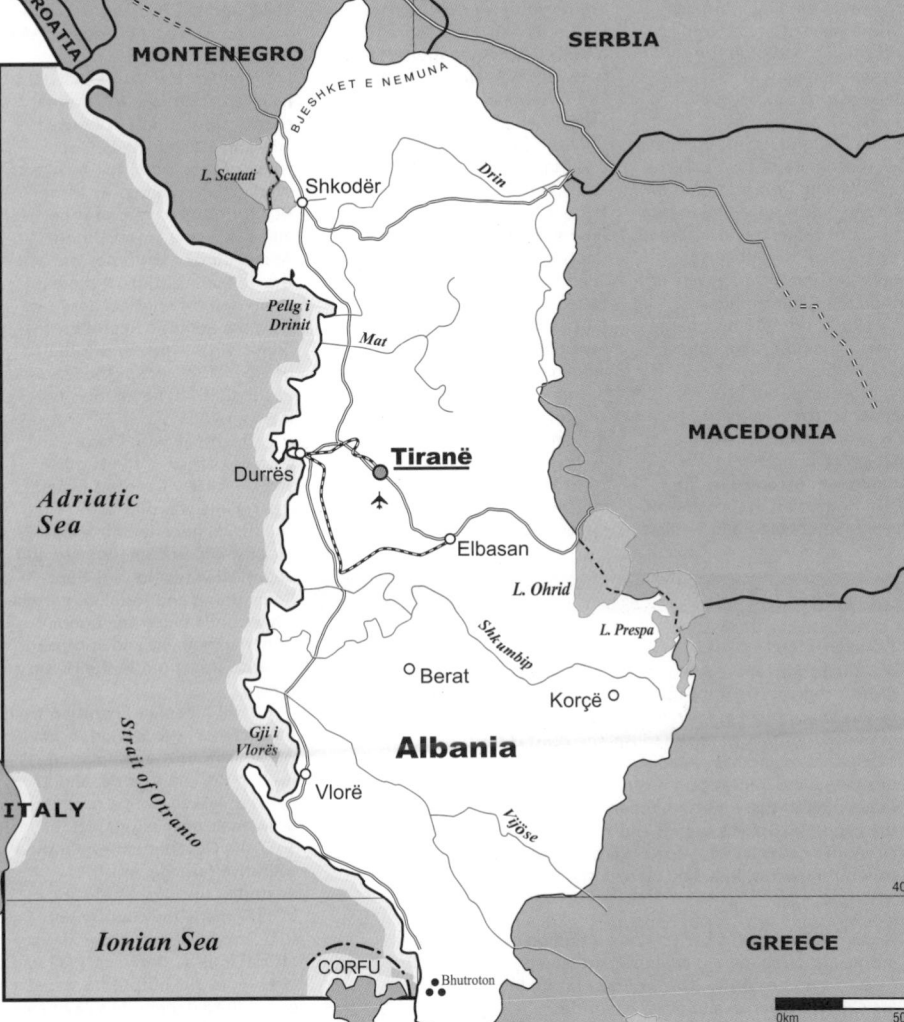

Albanians are descended from the ancient Illyrians, an Indo-European people who had migrated southward from Central Europe to the north of Greece by the beginning of the Iron Age. The southern Illyrians were much in contact with Greek colonies, while the northern Albanian tribes were mostly united under local kings. The most important of them was Argon, whose kingdom (latter half of the 3rd century BC) expanded from Dalmatia in the north to the Vijose river in the south. In the year 168 BC, the Romans conquered all of Illyria and then the Albanians became part of the prosperous Roman province of Illyricum. After 395 AD, with the decline of the Roman Empire, the area remained connected with Constantinople (now Istanbul) for administration purposes. From 491 to 565 AD three Byzantine emperors were of Illyrian origin: Anastasius I, Justin I and the most acclaimed emperor of all, Justinian I.

2   Despite the Hun incursions during the 3rd to 5th centuries, and the Slavic invasions during the 6th and 7th centuries, the Illyrians kept their own language and customs. With the passing of centuries and under the impact of the Roman, Byzantine and Slavic cultures, the old Illyrian population turned into Albanian. Between the 8th and 11th

centuries, the name of Illyria started to make way for that of Albania. When the Christian Church split in 1054, southern Albania maintained links with Constantinople, while the north fell under Roman jurisdiction, marking the first significant religious fragmentation of the country.

3   When the Turks invaded in 1431, the Albanians put up stiff resistance, but succumbed 47 years later. In the late 16th century, the Ottoman Turks imposed Islam upon the country, and they kept up the Islamic pressure over the following century. By the early 18th century, two-thirds of the Albanian population had converted to Islam. More than 25 of the great Viziers of Turkey were of Albanian origin.

4   The ideologists of the nationalist movement of the 19th century, in order to overcome the religious divisions and encourage

national unity, adopted the slogan 'The religion of the Albanians is Albania'. The Albanian League - which had both political and cultural aims - was founded in 1878 in the Kosovan town of Pritzen. The League tried without success to bring the Albanian territories of Kosovo, Monastir, Shkodër and Jänina (which were then divided into four provinces) together as a single state within the Ottoman Empire. On the cultural front, language, literature and an educational system which would foster nationalism were promoted.

5   The Empire suppressed the League in 1881. In 1908, Albanian leaders met in Monastir (part of present-day Macedonia) and adopted a national alphabet mainly based on Latin, which supplanted the previous Arabic and Greek versions. Also that year, the Young Turks took power in Istanbul and ignored their promises of bringing in democratic

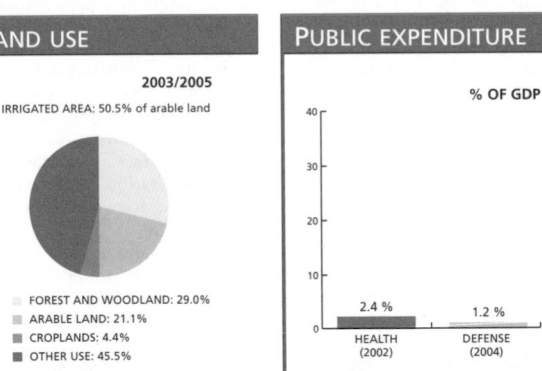

LAND USE

2003/2005

IRRIGATED AREA: 50.5% of arable land

- FOREST AND WOODLAND: 29.0%
- ARABLE LAND: 21.1%
- CROPLANDS: 4.4%
- OTHER USE: 45.5%

PUBLIC EXPENDITURE

% OF GDP

2.4 %	1.2 %
HEALTH (2002)	DEFENSE (2004)

WORKERS

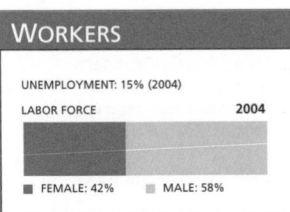

UNEMPLOYMENT: 15% (2004)

LABOR FORCE   2004

- FEMALE: 42%   ■ MALE: 58%

Life expectancy
74 years
2005-2010

GNI per capita
$2,120
2004

Literacy
99% total adult rate
2000-2004

reforms and guaranteeing Albanian autonomy, which led to three years of armed conflict. Turkey finally accepted the Albanian demands in 1912, and independence was declared.

[6] Due to pressure from Albania's neighbors, the European powers assigned Kosovo to Serbia when drawing up the borders of the new country, granting Greece the bulk of Sameria and a portion of Epirus. Half the Albanian population and territory were left outside the Albanian borders.

[7] In 1914, the international powers appointed German prince Wilhelm zu Wield as King of Albania, but he quickly abandoned the country due to his lack of familiarity with the Albanians. During World War I Albania was occupied by the Austro-Hungarian, French, Italian, Greek, Montenegran and Serbian armies. US President Woodrow Wilson vetoed a plan by the French, British and Italians that aimed to divide up the territory between its neighbors.

[8] A national congress, established in January 1920, set the basis for a new government and that same year Albania was admitted to the League of Nations. In 1927, Ahmed Zogu, president since 1925, signed a treaty with Italian fascist leader Benito Mussolini, turning the country into a virtual Italian protectorate. In 1928 Zogu proclaimed the country a monarchy.

[9] In April 1939, Italy occupied and formally annexed Albania to the kingdom of Victor Emmanuel III. The Communists, led by Enver Hoxha, organized guerrilla resistance. The anti-fascist front forced the withdrawal of the occupying forces on 29 November 1944, and the People's Republic was declared on 11 January 1945.

[10] Despite its desire for independence, Albania matched its internal options to external events. Upon the split between Tito and Stalin in 1948 (see former Macedonia) the Albanian Workers' Party sided with the KOMINFORM (communist parties allied with the Soviet Union). Albania broke with Moscow after the de-Stalinization of the early 1960s and established close ties with the People's Republic of China, with which it eventually broke in 1981, when the Maoists of the Cultural Revolution fell from power.

[11] The break with China became official at the Eighth Workers' Party Congress, where a party line was put forth against US imperialism, Soviet socialist-imperialism, Chinese and Yugoslavian revisionism, Euro-Communism and social democracy, while condemning the policies of non-alignment and European detente, as set out in the Helsinki accords.

[12] Until Albania's liberation from Italian fascist occupation in 1944, 85 per cent of the population lived in the countryside, and 53 per cent lacked even a patch to grow their own vegetables. In 1967 the Government established collective farming which, according to official statistics, in 1977 made the country self-sufficient in wheat. According to official figures, between 1939 and 1992 industrial output increased over 120-fold, materials 262 times over, and electricity more than 300-fold.

[13] In 1989, Ramiz Alia - head of state since Hoxha's death in 1985 - initiated a process of restructuring. Border immigration procedures were simplified to encourage tourism; religious worship was authorized; home ownership was permitted; capital punishment was abolished for women and the number of crimes punishable by death was reduced from 34 to 11. One year later, Alia authorized the activity of independent political parties and called for general elections after 46 years of communist rule. Diplomatic relationships were resumed with the USSR in July 1990, and with the US in February 1991.

[14] In March 1991, one thousand candidates from 11 political parties stood for election.. Amidst accusations of fraud the Communists won 156 of the 250 parliamentary seats. In May, more than 300,000 workers went on strike demanding the resignation of the Government, and a 50 per cent salary increase. Prime Minister Fatos Nano dissolved his cabinet, seeking an alliance with the opposition.

[15] At the height of the economic collapse, with an 80 per cent abstention rate, the parliamentary elections of March 1992 gave the Democratic Party (DP), a spectacular triumph over the Socialist Party (SP). Sali Berisha, leader of the DP, replaced Ramiz Alia as president, thus becoming the first non-Marxist president since the end of World War II.

[16] In 1993, the Government put the main political figures of the previous regime on trial. Naxhmija Hoxha, widow of Enver, ex-President Alia and ex-Premier Nano, among others, were sent to prison convicted of misuse of public funds.

[17] The parliamentary elections of 1996, with the main opposition leaders proscribed, gave an overwhelming victory to the DP, which was accused of fraud by the US and several European countries interested in investing in Albania.

[18] In 1997, the collapse of a series of pyramid investment funds led to a bloody social and political uprising. One in six Albanians was left penniless and the biggest cities, including the capital, Tirana, rose in arms. Eighty per cent of weapons fell into civilian hands, after people sacked forts and barracks abandoned by the army and the police, who had joined the rebels. Armed confrontations caused 1,500 deaths. President Berisha was forced to bring forward parliamentary elections to June 1997. The SP, led by Nano (now out of jail) won the elections.

[19] Nano, back as Prime Minister, had regained control of the country by the end of the year and was able to reduce inflation. The socialist government pledged to carry out reforms that included the revitalization of the economy through an extensive privatization plan.

[20] In March 1998, the IMF praised Nano's management and promised funding for three years to support Albania's economic reform. But, in September that year an internal quarrel within the SP led the Prime Minister to stand down. The administration was left in the charge of Pandeli Majko, who in turn had to hand over to Ilir Meta, also from the Socialist Party, in 1999.

[21] As civil war was well advanced in the former Yugoslavia (now Serbia and Montenegro), Albania called for NATO intervention to protect Kosovan Albanians and both countries broke off

PROFILE

ENVIRONMENT

A Balkan state on the Adriatic Sea. Albania comprises two distinct regions: from the border with Serbia and Montenegro to the Bay of Vlöre are alluvial plains, which become swampy in the winter; further to the south, the coast is surrounded by mountains and has a Mediterranean climate. The soil of the mountainous region is very poor and cattle-raising predominates there. Cotton, tobacco and corn are grown on the plains; rice, olives, grapes and wheat are produced in the valleys. The country has large areas of forest and is rich in mineral resources, including oil deposits.

SOCIETY

Peoples: Albanians (95 per cent) are a homogeneous ethnic group, although there is an important division between the Gegs - from the north - and the Tosks - from the south. There are Greek, Bulgarian and other minorities.
Religions: Freedom of worship was authorized in 1989, having been banned since 1967. Islam (70 per cent): Sunni majority and a Shi'a-bektasi minority; Christians: Orthodox (20 per cent) and Catholic (10 per cent).
Languages: Albanian-Tosk (official) and other dialects; Greek; Macedonian, Romanian and Romani.
Main Political Parties: Democratic Party of Albania (*Partia Demokratike e Shqipërisë*) and the socialist (post-communist) Socialist Party of Albania (*Partia Socialiste e Shqipërisë*). Other parties include ALDM-Republican Party of Albania (*Partia Republikane e Shqipërisë*); ALDM-New Democratic Party (*Partia Demokrate e Re*); ALDM-Christian Democratic Party of Albania (*Partia Demokristiane e Shqipërisë*).
Social organizations: The Union of Independent Trade Unions of Albania (BSPSH) and the Confederation of the Trade Unions of Albania (KSSH) are the main union federations.

THE STATE

Official Name: Repúblika e Shqipërísë.
Administrative divisions: 36 districts.
Capital: Tirana (Tiranë) 367,000 people (2003).
Other cities: Elbasan 101,300 people; Durrës 98,400;Shkoder 84,300; Vlöre 81,000 (2000).
Government: Parliamentary republic. Alfred Moisiu, President of the Popular Assembly since July 2002. Sali Berisha, Prime Minister, since September 2005. Single-chamber legislature: People's Assembly, made up of 140 deputies elected by universal suffrage every four years.
National holiday: 28 November, Independence from the Turks (1912).
Armed Forces: 27,000 (2001).

	Under-5 mortality			Poverty			Debt service			Maternal mortality
	19 per 1,000 live births 2004			**2%** of population living on less than $1 per day 2002			**2.6%** exports of goods and services 2003			**55** per 100,000 live births 2000

IN FOCUS

ENVIRONMENTAL CHALLENGES

Farming in wooded areas is causing deforestation and threatens fauna. Grazing, floods and droughts increase soil erosion. Earthquakes and tsunamis occur in the southwest.

Oil and mineral extraction have polluted air, soil and underground water, especially in the central region. Air pollution brought about by metallurgic, chemical and oil plants is particularly serious, while water resources are affected by non-treated domestic and industrial waste.

WOMEN'S RIGHTS

Women have been able to vote and run for office since 1920. Between 1995 and 2005, the percentage of seats held by women in parliament fell from 11 to 7.5 per cent, while the percentage of ministerial and equivalent positions grew from 0 to 11 per cent. Illiteracy among women was 2 per cent; among men it was 1 per cent *.

After the Kosovo War, the number of cases of neonatal tetanus increased as more

mothers transmitted the disease. Some women refused to be vaccinated because of the rumor that Serbian doctors used the vaccination to sterilize Albanian women to limit population growth.

Women make up 41 per cent of the workforce. Some 27 per cent work in agriculture, 45 per cent in the industrial sector and 28 per cent in services. The female unemployment rate rose from 10.9 per cent in 1990 to 19.1 per cent in 2003, while male unemployment rose from 8.4 to 13.6 per cent in that period.

CHILDREN

Basic health and education services, freely available during the Communist era, are severely diminished. Child and maternal mortality rates are very high by European standards, due to malnutrition (19 per cent of children under 5 and 17 per cent of children under 1*) and poor access to basic health services, especially in rural areas. An outbreak of polio in 1996 led to international co-operation to increase vaccination coverage for common childhood diseases.

Sixty per cent of children experience violence in their family. Institutionalized children are subject to various sorts of violence, a reality common in several Central European countries. It was estimated that 15,000 children were trafficked via Italy or Greece in 2002. Hundreds of children are locked up by their parents, deprived of schooling and social life, for fear of the *Kanun* - a traditional code of honor in Albania and Kosovo, applied in poorer regions, which among other things allows bloody *vendettas*.

INDIGENOUS PEOPLES/ ETHNIC MINORITIES

This is the European country with the greatest ethnic homogeneity, with Albanians making up more than 95 per cent of the population.

MIGRANTS/REFUGEES

It is estimated that there are about 7 million Albanians throughout the world, most of them in bordering countries. Kosovo has an Albanian majority but there are also communities in Italy, Greece, Bulgaria and Rumania, and since 1970, the diaspora has been

extended to the rest of Europe and the US.

In 2002, almost 9,900 Albanians sought asylum in other countries: more than 5,000 in the US and the rest in the UK. Surveys carried out in 2003 show that 44 per cent of young people intend to emigrate when they come of age. Notwithstanding, most of those who emigrated during the 2001 conflict with Macedonia returned voluntarily towards the end of that year.

Unlike 1998 and 1999, a period of massive migratory inflow, refugee situations were scarce in the country during 2002, when the National Commission for Refugees revoked a temporary protected status that had been agreed with Kosovars.

DEATH PENALTY

This still applies, though there have been no executions since 1995.

* Latest data available in *The State of the World's Children* and *Childinfo* database, UNICEF, 2006.

---

diplomatic relationships. Local mafias supplied weapons to the Kosovo Liberation Army (KLA), while Albania accepted tens of thousands of refugees and served as a base for the Alliance troops. Diplomatic relationships were re-established in 2001.

[22] Although the SP won a second term in the 2001 elections, new internal disputes paralyzed the Government, which was denounced by Nano as corrupt and incompetent, leading his followers to block the appointment of new ministers. With both the Government and the economy crippled, a power shortage was compounded by a particularly harsh winter. Without funds for imports, Meta declared a state of emergency. Some areas spent up to 20 hours a day without electricity.

[23] In August 2002, after new political disputes and a series of resignations, Nano became head of government for the third time and also chair of the SP. By now the pretender to the Albanian throne, Leka Zogu, son of former King Ahmed Zogu, returned to Albania with his family after 63 years of exile.

[24] A gun culture prevails in Albania. Hundreds of thousands of weapons stolen from the police in 1997 are still circulating - many

of them are trafficked in different countries around the world - and any Albanian can get hold of Russian guns, Chinese machine-guns or anti-aircraft weapons.

[25] In March 2003, as a party to the Chemical Weapons Convention, Albania declared it had such weapons. Their destruction began one month later, although the Organization for the Prohibition of Chemical Weapons did not receive any detailed information about the number or type of weapons.

[26] That year Albania took the first steps to join the European Union. Illegal immigration from Albania to neighboring countries, mainly to Italy and Greece, resulted in frequent incidents, as it did in January 2004, when 21 Albanians died due to the sinking of their boat.

[27] In February 2004, thousands of Albanians took to the streets under the slogan 'Nano out', accusing Prime Minister Fatos Nano of leading the country into corruption and organized crime. Nano was also accused of electoral fraud. More than 2,000 police controlled the march, held 13 years after 100,000 demonstrators toppled communist leader Hoxha's statue, marking the start of the current transition. The Prime Minister was under pressure due

to price rises in public utilities and increased unemployment, among other factors.

[28] The opposition Democratic Party, headed by Berisha, won parliamentary elections in July 2005, with 55 seats. Nano and the SP got 41 seats. Observers said the elections were mostly fair.

[29] In April 2006, in an attempt to end a serious energy crisis, Berisha reached an agreement with a Swiss, Italian and US consortium for the construction of gas and electricity plants in the south. By 2009, the company plans to build a 1,200 megawatt thermoelectric generator and a plant to extract 10 billion cubic meters of gas annually - of which 80 per cent is to be exported through an underwater pipeline linking Albania and Italy.

[30] That month the parliament banned speedboats and other small private vessels from coastal waters, to tackle drugs and people smuggling. The three-year ban would affect around 2,000 speedboat owners. A report published in 2005 had identified the country as playing a key role in what was considered to be an 'alarming increase' in human trafficking from Eastern Europe to the EU. According to the report, human traffickers were attracted to Albania by lax border

controls, a weak judiciary and corrupt police. Tiranë had been told to curb human trafficking if it wanted to ease the visa regime for Albanians seeking entry to the EU and Prime Minister Berisha, justifying the ban, stated that speedboats were a 'bad symbol' for Albania since the country became a transit point for other illegal immigrants wanting to cross into the EU. According to Berisha, this was a price that had to be paid, since the law affected between 1,000 and 2,000 Albanians who possessed speedboats but gave 'the possibility for thousands of others to be issued visas and go abroad'.

[31] In June 2006, Tiranë signed a Stabilization and Association Agreement with the EU. The agreement, considered a milestone on Albania's path to EU membership, outlined a set of political and economic criteria that authorities in Tiranë were expected to meet and confirmed the good progress of transition reforms in Albania. According to national NGOs some of the remaining challenges in the process of integration were to substantially increase social spending, and to link that social spending to its impacts so as to achieve tangible results in terms of social justice. ∎

Algeria / Al Jaza'ir

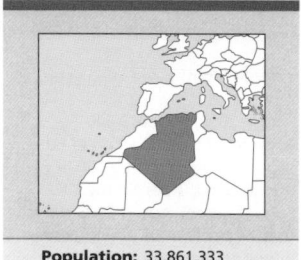

Population:	33,861,333
Land area:	2,381,740 km²
Capital:	Algiers (Alger)
Currency:	Dinar
Language:	Arabic, Tamazight

The kingdoms that arose in Algeria dating back to antiquity were linked to the region's two power centers: Tunisia, from the Carthage era (10th century BC), and Morocco, following the Arab conquest of the Iberian Peninsula (711 AD). That intermediate position facilitated the concentration there of dissent against discrimination practised between born Muslims and recent converts. This dissent later led people to the Jariyite sect that promoted egalitarian principles and the belief that caliphs did not have to be descendants of Muhammad or his family: thus, any Muslim could ascend to the caliphate regardless of race, color or social status. This sect appealed especially to the Berbers (the name comes from the Romans' name for them: 'barbarians'). They had resisted the Arab invasion in the 7th century AD but eventually converted to Islam and played a role in the Arab conquest of the Iberian Peninsula. Berbers remained subordinate to the Arabs who were politically dominant, though fewer in number. In the 12th century, invading Bedouin Arabs destroyed the Berbers' peasant economy in coastal North Africa; as a result, many became nomadic.

2   With the downfall of the Almohad Empire in 1212, Yaglimorossen ibn Ziane founded a new state on the Algerian coast.

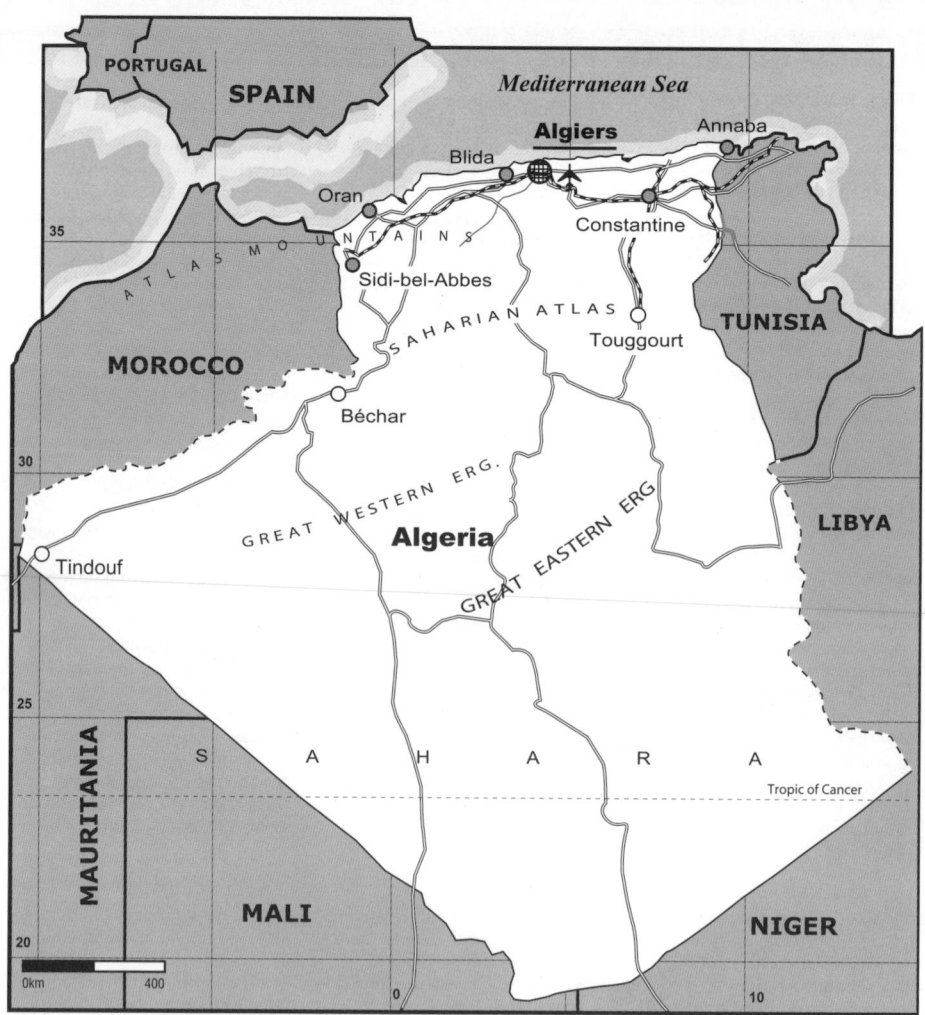

LAND USE

2003/2005

IRRIGATED AREA: 6.9% of arable land

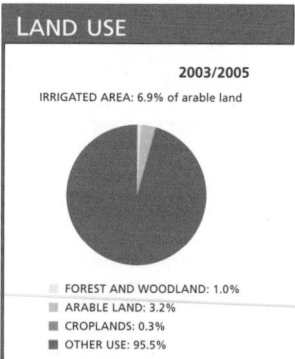

- FOREST AND WOODLAND: 1.0%
- ARABLE LAND: 3.2%
- CROPLANDS: 0.3%
- OTHER USE: 95.5%

WORKERS

UNEMPLOYMENT: 27% (2004)

LABOR FORCE	2004

- FEMALE: 30%   - MALE: 70%

Its borders were consolidated as economic prosperity and cultural development led nomadic peoples to settle. Ziane and his successors governed the country between 1235 and 1518. After the Christians had put an end to seven centuries of Muslim domination, in 1492 the Zianids were confronted with a series of Spanish military incursions in which various strategic sites, like Oran, were taken.

3   Algeria and Tunisia both became part of the Ottoman (Turkish) Empire in the 16th century. The Arroudj and Kheireddine brothers drove the Spanish from the Algerian coast and expanded the state's authority over a sizeable territory. The Empire's mighty fleet won respect for the nation, and its sovereignty was acknowledged in a series of treaties (with the Low Countries in 1663, France in 1670, Britain in 1681 and the US in 1815).

4   Wheat production gradually increased until it became an export crop once again, for the first time since the Hilalian invasion (see Mauritania). Wheat exports became the indirect cause of European intervention. At the end of the 18th century, the French revolutionary government bought large amounts of wheat from Algeria but failed to pay. Napoleon, and later the Restoration monarchy, delayed payment until the Dey of Algiers demanded that the debt be paid. Reacting to further excuses and delays, he slapped a perplexed French official in the face - a show of temper that would cost the Turkish Pasha dearly. 36,000 French soldiers disembarked to avenge the offense, a pretext used by the French to carry out a long-standing project: to re-establish a colony on the African coast opposite their own shores. However, the French encountered heavy resistance, and were defeated.

5   In 1840, disembarking with 115,000 troops, the French set out once again to conquer Algeria. Successive rebellions were launched against the invaders. In the South the nomadic groups remained virtually independent and fought the French until well into the 20th century (see Western Sahara). In 1873, France decided to expropriate land for French settlers, or *pieds-noirs* as they were known, who wished to remain in the colonies (they numbered 500,000 by 1900, and over a million after World War Two). As a result, the French *pieds-noirs* came to monopolize the fertile land, and the country's economy was restructured to meet French interests.

6   Nationalist resistance grew stronger from the 1920s until in 1945 it exploded when the celebration of the victory over Nazi-Fascism turned into a popular rebellion. The French forces tried to put down the rebellion and, according to official French reports, 45,000 Algerians and 108 Europeans were killed in the

Life expectancy
72 years
2005-2010

GNI per capita
$2,270
2004

Literacy
70% total adult rate
2000-2004

HIV prevalence rate
0.1% of population 15-49 years old
2003

ensuing massacre.

7   Shortly thereafter, the Algerian People's Party, founded in 1937, was restructured as the Movement for the Triumph of Democratic Liberties (MTLD) which participated in the 1948 and 1951 elections called by the colonialists.

8   Convinced of the futility of elections under colonial control, nine leaders of the OS (Special Organization, the military wing of the MTLD) founded the Revolutionary Committee for Unity and Action (CRUA). In November 1954, this committee became the National Liberation Front (FLN) that led the armed rebellion. Frantz Fanon, a doctor from Martinique, who had fought in the liberation of France during World War Two, joined the FLN and came to exert great intellectual influence not only in Algeria but also throughout the sub-Saharan region.

9   In order to maintain 'French Algeria' and the *pieds-noirs*, the French colonial regime destroyed 8,000 villages, killing more than a million civilians, made systematic use of torture and deployed more than 500,000 troops. Right-wing French residents of Algeria formed the Secret Army Organization (OAS), a terrorist group which blended Neo-Fascism with the demands of the French colonials, who resented the growing power of the Algerians. Finally, on 18 March 1962, French President De Gaulle signed the Evian Agreement, agreeing to a cease-fire and a plebiscite on the proposal for self-determination.

10   Independence was declared on 5 July 1962, and a Constituent Assembly was elected later that year. Ahmed Ben Bella was named Prime Minister. Almost all foreign companies were nationalized; 600,000 French nationals abandoned the country taking everything they could with them, and 500,000 Algerians returned, to share the lot of 150,000 landless and hungry peasants. A system of local agricultural and industrial operations was introduced by the new government.

11   However, Ben Bella's self-management program came up against the reality of a government with little administrative expertise. In June 1965 a revolutionary council headed by Houari Boumedienne took power and jailed Ben Bella. A new emphasis on centralization and state power began to prevail over Ben Bella's self-management notions. Under Boumedienne, there was further nationalization and a program of industrialization based on oil and liquid natural gas exports. The country entered a period of economic expansion, which was not reflected in the countryside. Population grew more

rapidly than agricultural production, and Algeria went from exporting to importing food. Strikes broke out in several cities.

12   Houari Boumedienne died in December 1978, after a long illness, just as the country's political institutions were beginning to consolidate. In 1976, a new National Charter was approved and, in 1977, the new members of the National People's Assembly were elected, with Colonel Chadli Ben Jedid appointed president.

13   The new President initiated a policy of reconciliation by releasing Ben Bella, who had been imprisoned for 14 years. Restrictions on travel abroad were lifted, taxes reduced and prohibitions removed on private housing. The restructuring of inefficient public enterprises gave an impetus to private companies.

14   Ben Jedid was re-elected in January 1984. In October 1988, a wave of protests broke out in several cities due to the lack of water and basic consumer goods; the legitimacy of the FLN and the military was called into question. Among the main groups participating in these mass protests were militant Islamists. Some mosques - especially in poorer neighborhoods - became sites of political demonstrations.

15   Some sectors of the most radical forms of Islam, influenced by Iran, began sending volunteers to fight in Afghanistan, to carry out the *jihad* or 'holy war'; against the Soviet-backed Kabul regime. In mid-1989, against a background of protest and upheaval, Ben Jedid presented a new constitution

which introduced a modified multiparty system, breaking the monopoly which the FLN had held.

16   More than 20 opposition groups - including Muslims - openly expressed their views. The most significant were the Islamic Salvation Front (FIS), the Dawa Islamic League, the communist-socialist Avant Garde Party (PAGS) and the mainly Kabbyle (an ethnic minority of Berber origin) Rally for Culture and Democracy (RCD). Mouloud Hamrouche, a leading reformer, was appointed Prime Minister.

17   Hamrouche and his cabinet resigned in June 1991, against a background of social agitation from the mosques. In early June, a state of siege was declared throughout the country in the face of massive protests by FIS agitators who demanded that presidential elections be held ahead of schedule and that an Islamic state be proclaimed. Sid Ahmed Ghozali, an oil technician who had been Foreign Minister during the previous administration, was appointed Prime Minister. Legislative and presidential elections were scheduled for later in the year, and the FIS suspended its campaign of social and political agitation.

18   The country requested loans from the IMF in order to alleviate fluctuations in the price of oil. Ghozali proposed parliamentary reforms in order to ensure the transparency of the electoral system, but the proposals, which included the possibility of withdrawing a man's right to vote for his wife, were boycotted by the FLN's parliamentary majority.

19   In the December 1991 elections, 40 per cent of the 13 million registered voters abstained. The first round, for 430 seats in Parliament, gave the victory to the FIS. The anti-fundamentalists - alarmed by the FIS victory in the first round and headed by the Workers' Center, UGTA and the FFS - raised a demonstration of 100,000 in central Algiers. Women's, professional and intellectual movements also participated.

20   President Chadli Ben Jedid resigned under strong pressure from the military and politicians fearful of a FIS victory. A Security Council made up of three military leaders and the Prime Minister was put in power. Shortly after, a High Council of State was appointed, headed by Mohamed Boudiaf, a dissident FLN leader. The arrest of FIS leaders followed immediately and the election was annulled. In February, the High Council of State proclaimed a nationwide state of emergency for a year.

21   In March 1992, the FIS was outlawed. The Government dissolved nearly 400 local councils controlled by FIS members, and the Supreme Court ratified the illegality of the FIS. In June, Boudiaf was assassinated by one of his bodyguards whilst making a public speech. The Government of Prime Minister Belaid Abdelsalam decreed a series of 'anti-terrorist' measures, including extending the death penalty for various crimes.

22   In February 1993, the High Council of State extended the state of emergency indefinitely, imposed a curfew on Algiers and in five provinces, and dissolved

PROFILE

ENVIRONMENT

South of the fertile lands on the Mediterranean coast lie the Tellian and Saharan Atlas mountain ranges, with a plateau extending between them. Further south is the Sahara desert, rich in oil, natural gas and iron deposits. Different altitudes and climates in the north make for agricultural diversity, with Mediterranean-type crops (vines, citrus fruits, olives and so on) predominating.

SOCIETY

Peoples: Algerians are mostly Arab (80 per cent) and Berber (20 per cent), although current Arab population results from the mixture of ancient populations (Amazigh/Berber) with various invading peoples from the Middle East, southern Europe and sub-Saharan Africa. Arab invasions in the 8th and 11th centuries provided a very limited number of new people but resulted in the Arabization and Islamization of most of the indigenous population. Nomadic groups linked to the Tuareg of Nigeria and Mali live in the south. Nearly a million Algerians live in France.
Religion: Islam
Languages: Arabic and Tamazight (Berber) are the official languages. Many people speak French,

but Arabic has gradually been replacing it in education and public administration.
Main Political Parties: National Liberation Front, Movement for National Reform, National Rally for Democracy, Movement of Society for Peace, Workers' Party.
Social Organizations: General Union of Algerian Workers (UGTA), National Union of Algerian Peasants, National Union of Algerian Women, National Youth Union.

THE STATE

Official Name: Al-Jumjuriya al-Jazairia ash-Shaabiya.
Capital: Algiers (Alger) 3,060,000 people (2003).
Other cities: Oran 712,300 people; Constantine 501,900; Annaba 350.000.
Government: Abdel-Aziz Bouteflika, President since 1999, re-elected in April 2004. Abdelaziz Belkhamed, Prime Minister since May 2006.
National Holiday: 1 November, Anniversary of the Revolution (1954).
Armed Forces: 121,700 (65,000 conscripts).
Other: 180,000 (Gendarmerie, National Security Forces, Republican Guard).

Under-5 mortality
40 per 1,000 live births
2004

Poverty
2% of population living on less than $1 per day
1995

Maternal mortality
140 per 100,000 live births
2000

all associations linked to the FIS. Following a long series of failed negotiations, the Government named Defense Minister Lamine Zeroual president of the country for three years.

[23] In a climate of increasing factionalism among all political groupings, the Islamic guerrillas split into the Armed Islamic Group (GIA) and the Armed Islamic Movement. In one of their most spectacular acts, the fundamentalists allowed 1,000 prisoners to escape from the Tazoult high security prison.

[24] In early 1995, following a meeting in Rome, the FIS, FLN, FFS and some moderates from the Islamic group Hamas proposed an end to violence, the release of the political prisoners and the formation of a national unity government to organize elections. Despite international support, the proposal was not accepted by Zeroual, who responded by setting elections for November.

[25] The FIS, FLN and FFS boycotted the elections, which Zeroual won with 61 per cent of the vote against the moderate Islamic Mahfoud Nahnah's 25 per cent. Despite the presence of international observers, there were still strong doubts over whether the elections had truly been free and fair.

[26] In early 1996, Zeroual's government, apparently supported by the new leaders of the FLN, gained important military victories and followed an IMF-recommended structural adjustment plan which increased poverty across a large section of the middle class and the most deprived sections of the population.

[27] As the June 1997 elections approached, violence increased. The elections gave a relative majority to the ruling party, which took 155 of the 380 seats. The FIS, which had called for a boycott on the elections, declared itself satisfied with an abstention rate of 34 per cent.

[28] In August 1997, the recently freed FIS leader Abasi Mandana confirmed his movement's desire to end violence through dialogue with the Government. However, the massacre of some 300 people in a village south of Algiers once again reduced the possibility of bringing an end to the conflict.

Witnesses said members of the Algerian army could have prevented the event, but had preferred not to intervene.

[29] The President stood down and called new elections. In April 1999 Abdelaziz Bouteflika took over, immediately calling a referendum on a law of reconciliation. The response was hugely in favor (98.6 per cent), with overwhelming FIS participation and support. The President announced a general amnesty for those giving up their weapons and joining the legal political battle, which was rejected by the GIA and the Salafist Group for Preaching and Combat (GSPC).

[30] Abdelkader Hachani, the number three in FIS, was murdered in a fundamentalist neighborhood of Algiers in November 1999. The murder was seen as an attack on the peace process by radical Muslims opposed to Hachani's line of dialogue with the Government.

[31] The army's behavior in the conflict was condemned by public opinion and part of the international community after it was exposed in the book *The Dirty War*, by retired colonel Habib Souaidia, which stated that Algerian troops disguised as rebels took part in civilian massacres in the 1990s, and that the Army tortured Islamic militants to death. The Bouteflika Government refused to carry out a thorough and public investigation, as human rights activists demanded.

[32] In early 2001, while Bouteflika was unable to put a stop to the GIA's increasingly bold and bloody attacks, a series of violent clashes between Berber demonstrators and security forces left 60 civilians dead. The Rally for Culture and Democracy (RCD) left the Government.

[33] In August, tens of thousands of Berbers carried out an 'official' ceremony in the Soummam valley, in the heart of the Kabbyle region.

[34] In June 2002 the FLN won elections marked by violence and a Berber boycott.

[35] An earthquake in May 2003 killed more than 2,000 people and injured 7,000. That month, Ahmed Ouyahia, from the National Rally for Democracy (RND), was appointed Prime Minister.

[36] The Government launched an offensive against the GSPC in September in Setif province. In November, GIA leader Rachid Abou Tourab was arrested.

[37] Bouteflika was re-elected in April 2004 with 84 per cent of the vote. Former Prime Minister Ali Benflis, the FLN candidate, claimed electoral fraud. However, international observers said the results had been fair.

[38] Two journalists who had reported human rights violations were detained and jailed in June 2004. In July, the correspondent for Qatari news channel al-Jazeera was suspended by the Government when the network broadcast a report that heavily criticized the administration.

[39] In March 2005 official reports revealed that the security forces had been responsible for the disappearance of more than 6,000 civilians during the conflict in the 1990s.

[40] In the November local elections, opposition parties kept their lead in the Kabila regions, where there is a Berber majority. In December, Bouteflika returned from France after surgery.

[41] High international oil prices emboldened Algeria during 2006 to reverse its earlier decision to liberalize the oil sector. The Government introduced a windfall tax on excessive profits but also limited the role of foreign investors in oil production. The new provisions require that the state-owned company Sonatrach, Africa's largest company by revenue, takes a mandatory minimum 51 per cent stake in all exploration and production ventures. ∎

IN FOCUS

ENVIRONMENTAL CHALLENGES
Several species of mammals, reptiles and birds run a serious risk of extinction. Desertification affects regions bordering the Sahara desert. Twelve million hectares are severely eroded. Oil industry waste and the dumping of untreated domestic and industrial run-off have caused significant pollution of waterways and the Mediterranean coast.

WOMEN'S RIGHTS
Algerian women have been able to vote and run for office since 1962. The number of Parliamentary seats held by women fell from 7 to 6.2 per cent between 1995 and 2002, while ministerial or similar positions rose from zero in 2000 to 11 per cent in 2003. Women made up 30 per cent of the workforce in 2003. More than 49 per cent lived in rural areas and worked in agriculture, though not always on a paid basis. Access to primary education was 96 per cent for men and 94 per cent for women. In 2000, illiteracy among women stood at 40 per cent. Nine per cent of childbirths are unattended by qualified staff *. In major cities, 96 per cent of pregnant women receive prenatal care; 75 per cent in smaller towns and only 46 per cent in rural areas*.

CHILDREN
Although security has improved lately in part because of disarmament achieved by the Government, isolated terrorist acts continue to affect the population, especially women and children.

In 2004, the annual death rate for children under five was 40 per 1,000 live births.

The percentage of underweight children under five fell from 13 per cent in 1995, to 6 per cent in 2000, to 3 per cent in 2004*. Only 13 per cent of children under 6 months are fully breast-fed.

In 2004 the Government fully funded a basic immunization program. There is considerable mother-to-child transmission of HIV.

INDIGENOUS PEOPLES/ ETHNIC MINORITIES
The Berbers who still speak their ancestral language (Tamazight) represent roughly 20 per cent of the population. (Even if almost every Algerian is of Berber descent, Arab invasions in the 8th and 11th centuries resulted in the Islamization and Arabization of most of the indigenous population). Algeria has the largest Berber community in North Africa. The Berber Cultural Movement, closely linked to the two Berber political parties, succeeded in making Tamazight an official language in 2002. Between 100,000 and 300,000 Tuaregs live in Algeria, Libya, Mali, Nigeria and Burkina Faso. Before they were colonized, they had a system of nomad confederations to administer their huge territory. After 30 years of Tuareg resistance, the colonial powers split the confederation into several smaller groupings.

MIGRANTS/REFUGEES
In late 2004 there were 169,000 refugees, more than half under 18. The authorities withhold information and limit investigations by international humanitarian agencies. In February 2006, heavy rains caused flooding in three Sahrawi refugee camps (Aoserd, Smara and L'ayoun) destroying half the homes and damaging many more.

DEATH PENALTY
Currently there are more than 600 people under sentence of death. However, there have been no executions since 1993.

* Latest data available in *The State of the World's Children* and *Childinfo* database, UNICEF, 2006.

Andorra / Andorra

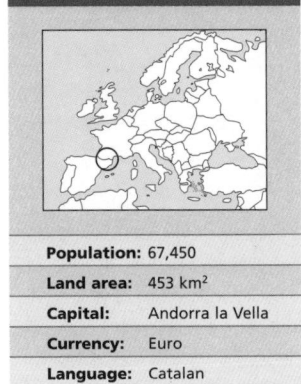

Population:	67,450
Land area:	453 km²
Capital:	Andorra la Vella
Currency:	Euro
Language:	Catalan

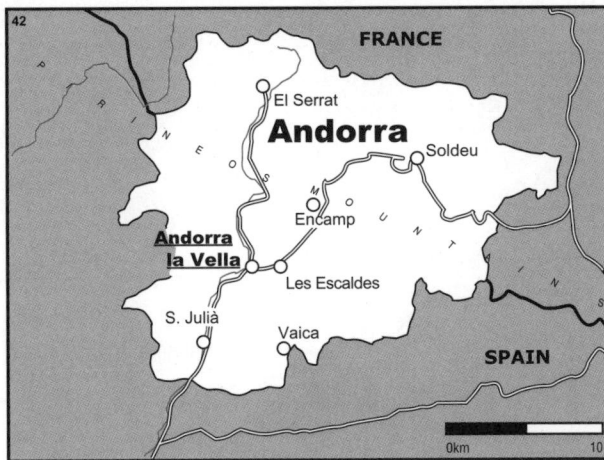

Evidence from cave paintings and funerary offerings shows that the territory of present day Andorra was inhabited in Neolithic times. Some historians believe that the original settlers of the Andorra valley are related to the Basque peoples of northern Spain, and that the name is therefore derived from the Basque language (Euskera). Other pre-Roman settlers, mentioned by Greek historian Polybius in his first century BC description of the Punic Wars, included some migratory Celtic and Southern Iberian peoples, along with a small group called Andosinos (the ancestors, some believe, of Andorrans).

2 After the fall of the Roman Empire, Andorra once again became a thoroughfare for barbarians from the northern Gaulish Roman provinces to the Iberian peninsula. Several of these left their mark, including the Alans, Visigoths and Vandals.

3 When the Muslims arrived, having made their way to northern Spain from Africa, the people in Andorra were predominantly Christian with some pagan settlements in more remote areas.

4 The Holy Roman Emperor Charlemagne is said to have liberated the territory from Muslim occupation in 803 AD, but the first document citing the name Andorra is an order from Charles the Bald, Charlemagne's grandson. This document, dated 843 AD, cedes the valleys of Andorra to Sunifred, Count of Urgell, from the nearby Spanish city of La Seu d'Urgell. The Act of the Consecration of La Seu cathedral, dated 860, mentions this Pyrenean parish as a dominion of the Counts of the nearby Segre valley.

5 During the Middle Ages, in-fighting between the small kingdoms and constant Arab invasions forced the Spanish Bishop of Urgell to seek help from a French noble family in the area (the Caboets, later known as the Foix) to protect Andorran settlements. Disputes over control of the territory continued between the Bishop of Urgell and the Count of Foix until 1278, when a landmark agreement was concluded sharing the sovereignty of Andorra between the two parties.

6 In 1479 the Count of Foix became King of Navarre and the French side of co-sovereignty passed to that royal house. In 1589 the King of Navarre became King Henri IV of France and the joint rights over Andorra passed to the French monarchy.

7 The French Revolution (1789) exterminated the monarchy in France and left the Principality orphaned from secular rule, and at the mercy of the Spanish, until the arrival of Napoleon in 1806. He re-established the co-principality at the request of the Andorrans. Since that day, the Head of State has been the 'co-prince,' a post held by all presidents of the French Republic.

8 As time passed, Andorra modified its structures, but maintained its sovereignty. In 1866 the New Reform was announced, bringing in the General Council of the Valleys - a Parliament. In 1933 all Andorran men aged over 25 were given the right to vote; previously only the head of the family had had the vote. In 1970 women were granted suffrage.

9 In 1982 the first Government of the Principality was created, separating the Legislative Power from the Executive Power. In 1988 the Universal Declaration of Human Rights (adopted by the United Nations in 1948) became law.

10 In December 1989 the first elections were held under the new constitution. A majority of moderate reformers, led by Oscar Ribas Reig, took over the General Council of the Valleys. Reig had already run the country from 1982 to 1984, until he was ousted for being too slow with reforms.

11 In 1993, the General Council approved a new constitution. The first article defines Andorra as an independent state in law, democracy and society, with a political regime described as 'institutionalized parliamentary co-principality' as befitting tradition, with the co-princes - the Bishop of Urgell and the President of the French Republic - as Heads of State, in a joint and indivisible manner.

12 The principality joined the UN in 1993 and in 1994 it became the 33rd member of the Council of Europe.

13 In the first legislative elections, held in 1993, Oscar Ribas Reig was re-elected, but he left the post that same year after losing the support of his allies over disagreements about the budget.

14 In 1997 the Andorran Government approved the Law of Linguistic Ordering, aiming to guarantee the use of the official language, Catalan, in all areas of public life, the media, cultural activity and sport.

15 Legislative elections were held in 2001. Marc Forné Molné, President of the current government and of the Liberal Party (LP) of Andorra, gained an absolute majority in the General Council.

16 In 2002, the Organization for Economic Cooperation and Development (OECD) included Andorra in a list of seven tax havens that had not met its standards on financial transparency. This could have led to sanctions and the resulting reduction of investments.

17 In July 2004, the Forné administration said it was willing to compromise with the EU and to gradually impose taxes on savings. Taxes would be in line with a formula agreed on by Austria, Belgium and Luxembourg. However, Forné stated that Andorra would not lift bank secrecy.

18 In April 2005 the LP won the elections and Albert Pintat became Prime Minister.

19 In August 2006, a US investment broker charged with evading millions of dollars in tax detailed how he had set up multiple sham trusts in Andorra as one of his strategies for financial concealment, again focusing international attention on the Principality's standards of financial transparency. ∎

PROFILE

ENVIRONMENT
Located in the eastern Pyrenées, the Principality of Andorra is made up of deep ravines and narrow valleys surrounded by mountains between 1,800 and 3,000 meters high. The Valira de Ordino and Valira de Carrillo rivers join in Andorran territory under the name of the Valira river. Wheat is grown in the valleys, but livestock (especially sheep) has given way to tourism as the primary economic activity.

SOCIETY
Peoples: Spanish 44.4 per cent; Andorran 20.2 per cent; Portuguese 10.7 per cent; French 6.8 per cent; other 6.6 per cent.
Religions: Catholic 92 per cent; Protestant 0.5 per cent; Jewish 0.4 per cent; other 7.1 per cent. **Languages:** Catalan (official), French and Spanish. **Main Political Parties:** Liberal Party of Andorra (ruling party); Social Democratic Party (main opposition).
Social Organizations: There are no organized trade unions. Many workers have joined French trade unions.

THE STATE
Official Name: Principat d'Andorra. **Administrative divisions:** 18 provinces. **Capital:** Andorra la Vella 21,000 people (2003). **Other cities:** Les Escaldes 16,000 people; Encamp 10,800 (2000). **Government:** Parliamentary republic. The Bishop of Urgell (Spanish jurisdiction) and the President of France, respectively represented by Joan Enric Vives i Sicilia and Philippe Massoni, are 'co-princes' of the territory. Albert Pintat Santolària, member of the Liberal Party, is the executive council president (head of government) since 27 May 2005. Single-chamber Legislature (General Council), with 28 members, elected by direct vote every four years. At some point in the future, total independence from France and the Spanish bishopric is foreseen. **National holiday:** September 8, Our Lady of Meritxell Day (1278).

Angola / Angola

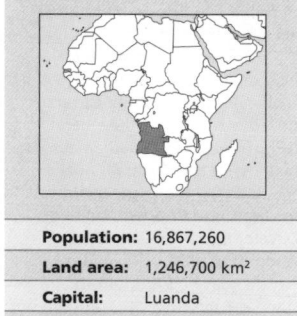

Population:	16,867,260
Land area:	1,246,700 km²
Capital:	Luanda
Currency:	New kwanza
Language:	Portuguese

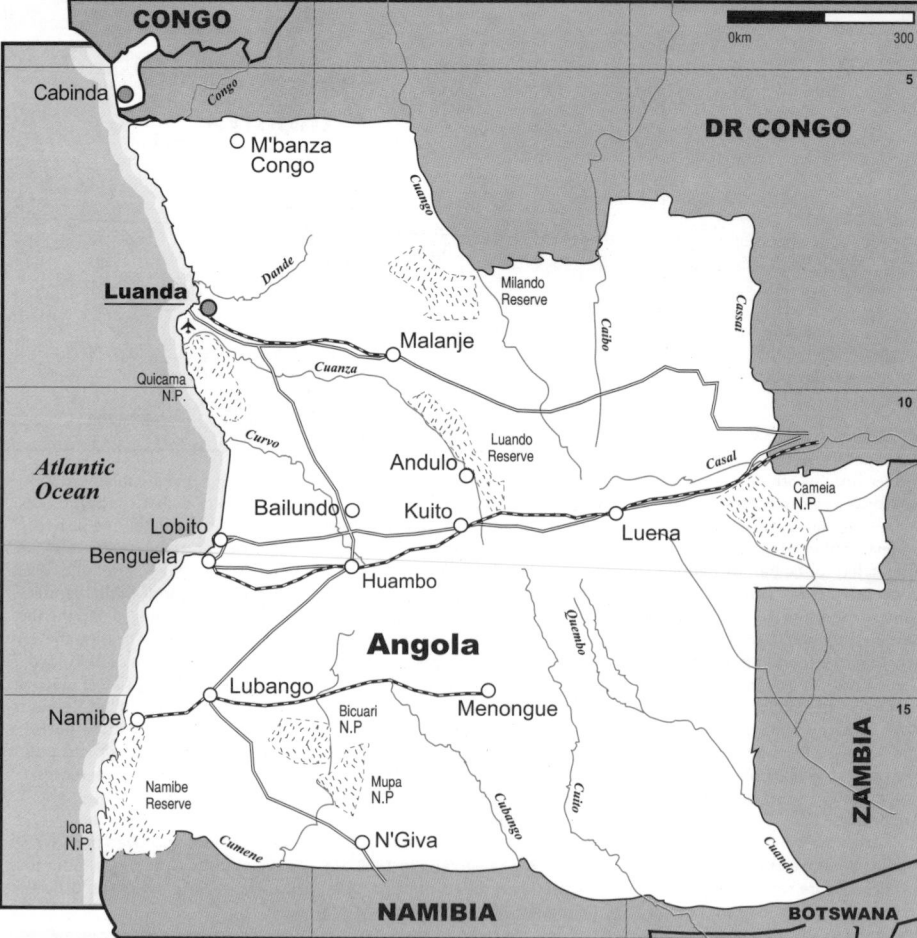

The original inhabitants of the current territory of Angola were Khoisan-speaking hunter-gatherers. The large-scale migrations of Bantu-speaking peoples in the first millennium AD made them dominant in the Khoisan area. The Khoisan - named Bushmen by the Europeans - still live in small groups in some zones of southern Angola.

[2] The Bantu-speakers were farming, hunting and gathering people who probably began their migration from the rainforest in what is the present-day border of Nigeria and Cameroon. Their expansion took place as small groups relocated in response to political and economic circumstances. Between the 14th and 17th centuries, the Bantu set up a series of kingdoms. In Angola the most important one was the Congo, covering the strip that today forms the frontier between Angola and Zaire, and reaching its apogee from the mid-13th to the 14th centuries.

[3] In 1482, a Portuguese fleet commanded by Diogo Cao entered the Congo river mouth. He made the first contact with the Angolans of the old kingdom of Congo and began the colonization process. This process was started by missionaries and traders, later giving way to military expeditions against the peoples of the Angolan interior.

[4] Various kingdoms within the country opposed foreign occupation until the mid-18th century. Wars and slavery reduced the Angolan population from 18 million in 1450, to barely eight million in 1850. Even so, the Angolan people never gave up their opposition to Portuguese colonization, with figures like Ngola Kiluange, Nzinga Mbandi, Ngola Kanini and Mandume leading the resistance.

[5] Portugal intensified its military incursions following the 1884 Berlin Conference that divided Africa among the European colonial powers. Nonetheless, it took 30 years of military campaigns (1890-1921) to 'pacify' the colony.

[6] From then on, Portuguese settlers arrived in ever-increasing numbers. In 1900, there were an estimated 10,000, in 1950, 80,000 and in 1974, less than a year before independence, 350,000. Only one per cent lived on farms inland. The colonial economy was parasitic, built upon the exploitation of mineral and agricultural wealth (diamonds and coffee), with the bulk of profits going to Portuguese merchants.

[7] In 1956, several small nationalist groups joined together to form the Popular Movement for the Liberation of Angola (MPLA). Their aim was to pressure the Portuguese Government into recognizing the Angolan people's right to self-determination and independence. When Britain and France began to withdraw from their overseas colonies in the 1960s, Portugal did not follow suit and frustrated all Angolan attempts to win independence by peaceful means.

[8] In 1961, a group of MPLA militants from the most underprivileged classes stormed Luanda's prisons and other strategic points in the capital. This spurred resistance in other Portuguese colonies. Their agenda was clear: they were fighting not just colonialism but also the international power system that sustained it. In addition, they were fighting racism and ethnic chauvinism.

[9] In the years that followed, other independence movements with different regional origins sprang up: the National Front for the Liberation of Angola (FNLA) led by Holden Roberto; the Cabinda Liberation Front (FLEC); and the Union for the Total Independence of Angola (UNITA)

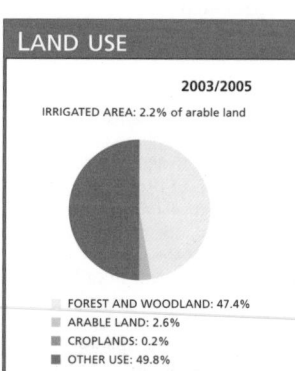

LAND USE

2003/2005

IRRIGATED AREA: 2.2% of arable land

- FOREST AND WOODLAND: 47.4%
- ARABLE LAND: 2.6%
- CROPLANDS: 0.2%
- OTHER USE: 49.8%

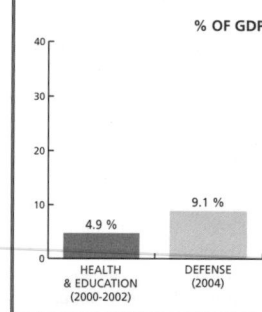

PUBLIC EXPENDITURE

% OF GDP

- HEALTH & EDUCATION (2000-2002): 4.9 %
- DEFENSE (2004): 9.1 %

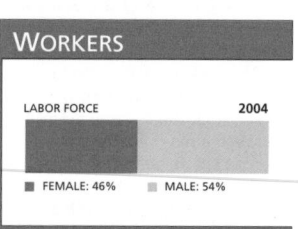

WORKERS

LABOR FORCE   **2004**

- FEMALE: 46%
- MALE: 54%

Life expectancy
42 years
2005-2010

GNI per capita
$930
2004

Literacy
67% total adult rate
2000-2004

HIV prevalence rate
3.9% of population 15-49 years old
2003

led by Jonas Savimbi. Under the direction of Agostinho Neto, the MPLA militants held a conference in January 1964 to discuss and define their strategy of a prolonged people's war.

[10] Portugal's domestic problems, coupled with military setbacks in Angola, Mozambique and Guinea-Bissau and repeated shows of international solidarity with the independence fighters, dashed Portuguese army hopes of a military solution. An uprising led by the Armed Forces Movement (MFA) overthrew the Portuguese regime of Oliveira Salazar and Marcelo Caetano in1974. The MFA expressly recognized the African colonies' right to self-determination and independence.

[11] The MFA immediately invited the MPLA, FNLA and UNITA to participate with Portugal in a transitional government for Angola in the interim period, the mechanisms of which were established in the Alvor Accords, signed in January 1975. By this time, political and ideological divergences among the three groups had become irreconcilable; the FNLA was directly assisted by US intelligence services and received military aid from Zaire (now DR Congo). UNITA received overt backing from South Africa and Portuguese settlers while the MPLA was aligned ideologically with the socialist countries. The accords were never implemented.

[12] The FNLA and UNITA unleashed a series of attacks on MPLA strongholds in Luanda, and a bloody battle for control of the capital ensued. Between September and October 1975, Angola was attacked on all sides: Zaire invaded from the north while South Africa, with the complicity of UNITA, attacked from the south to prevent a Marxist Government.

[13] On 11 November 1975, the date agreed for the end of colonial rule, the MPLA unilaterally declared independence in Luanda, pre-empting the formal transfer of sovereignty. Some 15,000 Cuban troops aided the new government in fighting off the South African invasion. In 1976, the United Nations recognized the MPLA Government as the legitimate representative of Angola. However, the South African attacks supporting UNITA continued from Namibia.

[14] The Angolan economy was severely debilitated. The war had paralyzed production in the far north and south of the country. The Europeans had emigrated in mass, taking all that they could with them and effectively destroying the productive capacity.

[15] Under these circumstances, the Angolan Government began to restore the chief production centers, and to train the largely unskilled and illiterate workforce. In this way, a large public sector emerged which was to become the economy's driving force, and banking and strategic activities were nationalized.

[16] In 1977, Nito Alves led a faction of the MPLA committed to 'active revolt', in a coup attempt. Six leading MPLA members were killed but the conspiracy was successfully put down within hours. Seven months later, at its first congress, the MPLA declared itself Marxist-Leninist and adopted the name of MPLA-Workers' Party. In 1978, closer political and economic links were established with the countries of the socialist Council for Mutual Economic Assistance.

[17] The country's first president, Agostinho Neto, died of cancer in Moscow in 1979 and was succeeded by Planning Minister José Eduardo dos Santos.

[18] In August 1981, South Africa launched 'Operation Smokeshell', in which 15,000 soldiers with tanks and air support advanced 200 kilometers into Cunene province. Pretoria justified this aggression as an operation against guerrilla bases belonging to the Namibian liberation movement, the South West African People's Organization (SWAPO). But their real aim seemed to be to establish a 'liberated zone' where UNITA could install a parallel government inside Angolan territory capable of obtaining some degree of international recognition.

[19] This incursion and successive attacks in the years that followed were contained by effective Angolan and Cuban military resistance. The cost of the war, plus international pressure and the mounting anti-apartheid campaign at home, obliged South Africans to resume diplomatic discussions with the MPLA. In December 1988, Angola, South Africa and Cuba signed a Tripartite Accord in New York, which put an end to the war between Luanda and Pretoria, and provided for the independence of Namibia and withdrawal of South African and Cuban troops from Angola.

[20] In Lisbon, 1990, the Angolan authorities announced they would resume conversations with the UNITA, with the intention of achieving a final ceasefire. That year Jonas Savimbi officially acknowledged José Eduardo dos Santos as Chief of State. The MPLA introduced some reforms and, in May 1991, a law on political parties was passed, which brought one-party rule to an end. That same month, the law banned political participation by active members of the armed forces, the police or the judiciary. Political amnesty was granted, and the last Cubans left Angola.

[21] After 16 years of civil war, a peace settlement was signed on 31 May by the Angolan Government and UNITA, in Estoril, Portugal. This agreement included an immediate cease-fire, as well as a promise to hold democratic elections in 1992 and the creation of a Joint Politico-Military Commission (CCPM), charged with establishing a national army made up of soldiers from both opposition groups. Portugal, the US and the Soviet Union were involved in the discussion and drawing up of the agreement, as was the United Nations, which was put in charge of supervising compliance with the terms of the peace agreement.

[22] Holden Roberto, leader of the FNLA, and Jonas Savimbi, president of UNITA, returned to Luanda in August and September 1991 respectively - after 15 years of exile - to

PROFILE

ENVIRONMENT

The 150 kilometer-wide strip of coastal plain is fertile and dry. The extensive inland plateaus, higher to the west, are covered by tropical rainforests in the north, grasslands at the center and dry plains in the south. In the more densely populated areas (the north and central west) diversified subsistence farming is practiced. Coffee, the main export crop, is grown in the north; sisal is cultivated on the Benguela and Huambo plateaus; sugarcane and oil palm along the coast. The country has abundant mineral reserves: diamonds in Luanda, petroleum in Cabinda and Luanda, iron ore in Cassinga and Cassala. The port of Lobito is linked by railway to the mining centers of Zaire and Zambia.

SOCIETY

Peoples: As a consequence of centuries of slave trade the population density is still very low. To maintain control over the country, the Portuguese colonizers fostered local divisions between the various ethnic groups: Bakongo (13 per cent), Kimbundu (25 per cent), Ovimbundu (37 per cent), others (22 per cent). There are also small minorities of Europeans (1 per cent) and Afro-Europeans (2 per cent).
Religions: The majority practise traditional African religions; 38 per cent are Catholic, and 15 per cent Protestant. However, there are forms of syncretism which make it impossible to establish clear boundaries between one religion and another.
Languages: Portuguese (official) and 41 African languages including Ovidumbo, Kimbundu and Kikongo.
Main political parties: The People's Movement for the Liberation of Angola (MPLA), founded by Agostinho Neto on December 10 1956 dominates the Government of Unity and National Reconciliation. The main opposition party is the National Union for the Total Independence of Angola (UNITA).
Social Organizations: National Union of Angolan Workers (UNTA); Organization of Angolan Women (OMA).

THE STATE

Official Name: República Popular de Angola.
Administrative divisions: 18 provinces.
Capital: Luanda 2,623,000 people (2003).
Other cities: Huambo (Nova Lisboa) 165,700 people; Lobito 133,100; Benguela 129,800 (2000).
Government: José Eduardo dos Santos, President since September 1979, re-elected in 1992; Fernando da Piedade Dias dos Santos, Prime Minister since December 2002. Unicameral legislature. 220-member National Assembly, elected by direct popular vote.
National holiday: 11 November, Independence Day (1975).
Armed Forces: 120,000 (2001) Other: 20,000 Internal Security Police.

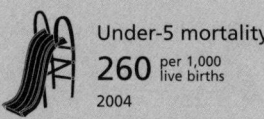
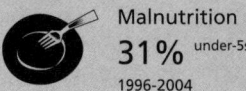

IN FOCUS

ENVIRONMENTAL CHALLENGES
In recent years there has been an extensive use of pasture land and further erosion of the soil, due to the overpopulation of certain areas and the deforestation of the tropical rainforests for the export of wood and fuel. Though Angola is the second-biggest oil exporter in Africa, only in the capital Luanda is it used as an alternative to firewood. Deforestation is particularly serious in the central hill region, where there is not only loss of biodiversity but also water pollution and sedimentation in rivers and dikes. More topsoil is washed away every year during the rainy season.

WOMEN'S RIGHTS
Women have been able to vote and be elected since 1975. In 2003 their parliamentary representation amounted to 16 per cent; their places in ministerial or equivalent positions fell from 14 per cent in 2000 to 6 per cent in 2003. Women made up 46 per cent of the workforce between 1990 and 2003.

Although the situation has improved in comparison to previous years, only 66 per cent of Angolan women receive prenatal care, and only 45 per cent of births are attended by qualified staff *. Data available in 2000 showed that 1,700 mothers died for every 100,000 live births*.

CHILDREN
Sixty per cent of Angolans are children, who grow in the context of a 40-year war history. In spite of the peace achieved in 2002, while the health system barely starts to recover, the death rate for children under 1 and 5 years old in 2004 was the same as in 1990 (154 and 260 for every 1,000 children born, respectively). Two generations of children - many of them soldiers - disappeared in the last decade of the civil war. Some 100,000 minors were separated from their families and 4.5 million were not registered upon birth.

Almost 50 per cent of children do not attend school and 45 per cent suffer from chronic malnutrition*.

INDIGENOUS PEOPLES/ETHNIC MINORITIES
The ethnic groups at risk during the war were the Bakongo (13 per cent), the Kimbundu (25 per cent), the Cabinda and the Himba. In 2002 the sacred places of the Himba community were affected by international tourism promoted by the Government after the end of the war. Likewise, in their search for oil, transnationals like Chevron and Texaco have altered the way fisherpeople and other tribal groups that inhabit the coastal areas of the south of the country earn their living.

MIGRANTS /REFUGEES
In 2002-2005 more than four million displaced people within the territory returned to their homes and around 350,000 Angolan refugees returned from Zambia, DR Congo, Congo, Namibia, and Botswana. In 2005, the Government, with the support of UNHCR, launched the Sustainable Reintegration Initiative, which set places in the provinces of Moxico, Uige and Zaire to receive the repatriated people.

DEATH PENALTY
Abolished in 1992.

Latest data available in The State of the World's Children and Childinfo database, UNICEF, 2006.

launch their election campaigns. The US continued to support UNITA, and as a result tensions increased as the 1992 elections drew near.
[23] Plagued by a foreign debt of more than $6 billion, the Government appealed to the international community for economic aid. The US refused to suspend the economic and diplomatic blockade, alleging that Angola was a Marxist nation and announcing it would not grant diplomatic recognition until after the 1992 elections. Consequently, the US companies in Angola were unable to get loans from banks in their own country.
[24] Following intense negotiations between the Government and UNITA, elections were set for September 1992. The ruling MPLA obtained nearly 10 per cent more votes than UNITA. Savimbi refused to recognize defeat and hostilities resumed. In their advance, the UNITA troops occupied the diamond mines of the interior, leaving the Government with oil as the only stable source of income (between $1.6 and $1.7 billion per year).
[25] In November 1993, peace talks were re-started in Lusaka, the capital of Zambia. A year later in November 1993 a peace agreement was signed. The main points (a ceasefire and constitutional changes so that Savimbi could become Vice-President) were not put into practice until late 1995 and the fighting continued.
[26] Some progress was seen during 1996. In May, an amnesty law was approved and UNITA soldiers began to be absorbed by the armed forces. Savimbi withdrew most of his troops to barracks and handed over some of the weapons. The civil war caused the most serious social and economic crisis in Angolan history. The adoption of IMF and World Bank economic liberalization measures did not improve matters for ordinary people.
[27] In 1997, hard negotiating ended with UNITA accepting to join the Government at executive, legislative and military levels. Even though its position in the capital was weak, UNITA's troops still controlled 40 per cent of the territory. The fall of Mobutu Sese Seko in Zaire (now DR Congo) in May weakened UNITA even further, and they were forced to abandon areas of the northern frontier. The Angolans wanted to avoid the infiltration of Mobutu's troops - formerly allied to Savimbi - into their territory. Mobutu's soldiers were fleeing from the government of the new DR Congo leader Laurent Kabila, a former ally of dos Santos.
[28] In 1998, thousands of demobilized soldiers mostly with little or no education encountered serious difficulties in returning to their home villages, given the scarce employment possibilities there and the slow arrival of economic aid. Since 1994, only 300,000 of the 4.5 million people uprooted by the civil war had been able to resettle, according to United Nations figures. The presence of large numbers of land mines across the country increased the insecurity of the population.
[29] In April 1999, the Government announced the formation of a self-defense front with Zimbabwe, Namibia and DR Congo. This reflected the interconnected nature of regional conflicts, which transcended frontiers established by former colonial powers. In late 1999, after winning back Andulo and Bailundo - the main cities under opposition control - and following a run of military victories the Government felt confident in announcing that the end of the war was in sight.
[30] Fighting resumed in 2000 and the UN withdrew its peace mission, which had been in the country since 1995.
[31] Savimbi was killed in combat in February 2002, in the province of Moxico. His right-hand man, Antonio Dembo, took over the organization's leadership while the Government called for peace. In order to ensure security before calling general elections, President dos Santos began to make contact with Dembo.
[32] In April a formal ceasefire was signed. Four months later UNITA finally disbanded its armed wing and the Defense Minister announced the end of the 27-year war, the longest in Africa.
[33] In agreement with UNITA, mineral exploration - which had been tainted by the illegal trafficking of 'blood' diamonds exchanged for weapons - was expanded. A UN report in October 2002 stated that six months after the ceasefire UNITA still kept illegal diamonds hidden. Before its independence, Angola was the world's fourth-largest diamond exporter.
[34] Transformed into a political party, UNITA elected Isaias Samakuva as its new leader in June 2003.
[35] A Human Rights Watch report in January 2004 stated that corruption and mismanagement had cost the country $4 billion in lost oil production in the last five years.
[36] The Government's drive to put an end to illegal mining led to the expulsion of tens of thousands of foreign miners and some 300,000 diamond traffickers.
[37] In February 2005 thousands of demonstrators demanded autonomy for the province of Cabinda, an Angolan enclave in DR Congo. The rebel group Frente de Libertação do Enclave de Cabinda fought for the independence of the oil-rich province.
[38] In early 2006, a cholera epidemic affected several provinces and left 1,000 dead in four months, 167 of them in Luanda.
[39] Abundant mineral resources have led to an oil boom in Angola, which may soon overtake Nigeria as Africa's biggest oil exporter. The country's national budget has recently almost doubled, from $13 billion to $25 billion and in June 2006 Chinese prime minister Wen Jiabiao visited, offering $9 billion in loans and credits targeted on infrastructure projects such as the building of roads and schools. ∎

Anguilla / Anguilla

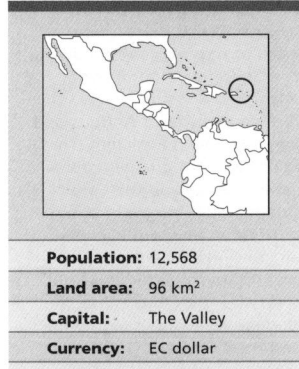

Population:	12,568
Land area:	96 km²
Capital:	The Valley
Currency:	EC dollar
Language:	English

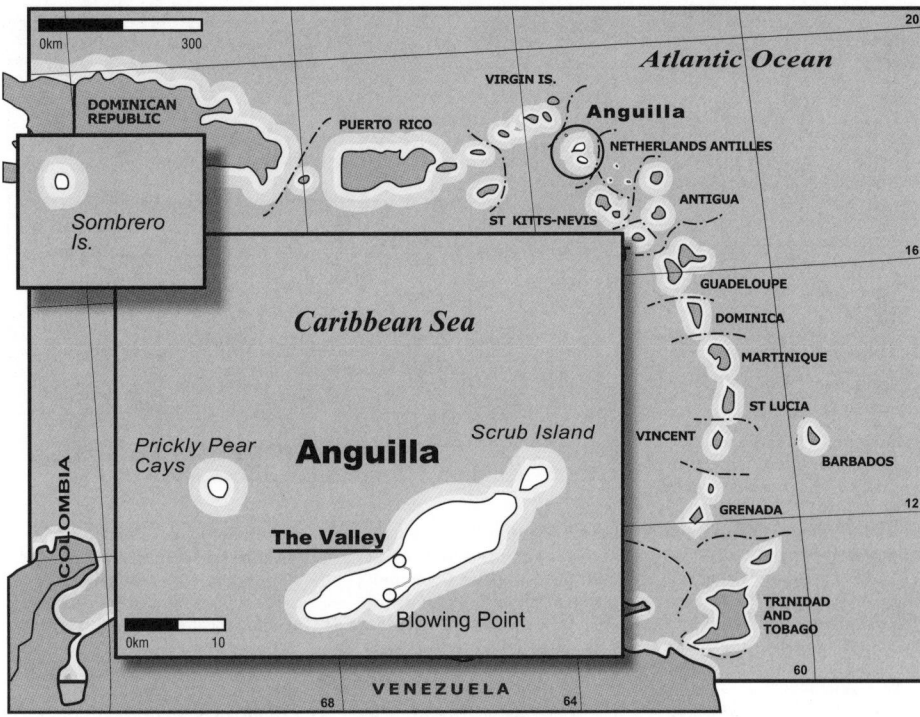

Anguilla is the most northerly of the Leeward Islands. Called Malliouhana by the Arawaks, who inhabited it until their extermination by the Caribs, its present name was first mentioned by explorer Pierre Laudonnaire, after it was sighted by a French expedition in 1556. In 1650 it was colonized by the British Empire but in 1656 the Caribs attacked the island, to be followed by several failed attempts of domination from Ireland (1698) and France (1745 and 1756). Its small size and its agriculturally unfit land made the island unattractive to the British Empire, factors that determined the failure of slavery on the island, although abolition was formalized as late as 1838.

[2] From 1816 to 1871 Anguilla, St Kitts-Nevis and the Virgin Islands were administered together as one colony. The Virgin Islands were split off in 1871, leaving the others as a colonial unit ruled from St Kitts.

[3] Colonialism lasted until the St. Kitts-Nevis-Anguilla colony became one of the five Caribbean 'States in Association with the United Kingdom'. Anguillans opposed the agreement and in 1967 rebelled against the St Kitts Government, deposing the authorities. Under the leadership of local entrepreneur Ronald Webster, Anguilla demanded a separate constitution through a referendum and headed its own administration until 1969, when British troops regained control of the island.

[4] In 1976 a new constitution was approved by Britain, establishing a parliamentary system of government under the patronage of the British Commissioner. But it was not until 1980 that Anguilla was able formally to withdraw from the Associated State arrangement with St Kitts-Nevis, gaining the status of 'British Dependent Territory'.

[5] The 1976 constitution provided for a governor appointed by the British Crown, responsible for defense, foreign relations, internal security (including the police), utilities, justice and the public audit. The Governor presides over the Executive Council.

[6] The first general elections, held in March 1976, voted in the People's Progressive Party (PPP) leader, Ronald Webster, as Chief Minister. In the following changes of government, Webster's party (which dissolved the PPP in 1981 and created the Anguilla People's Party) and the opposition Anguilla National Alliance (ANA), headed by Emile Gumbs, alternated in government until 1994, when alliances among the parties were necessary in order to rule.

[7] The Constitution was changed in 1982 to modify the number of ministers. This led to a series of constitutional discussions and eventually to the creation, in 1990, of the Constitutional and Electoral Reform Consultative Forum of Anguilla.

[8] Through the 1980s construction for the tourism industry reduced unemployment from 26 per cent to 1 per cent, and income generated from livestock, salt production, lobster fishing, and remittances from émigrés were gradually overtaken by the construction, tourism and international financial service sectors.

[9] In the 1994 elections, Hubert Hughes was elected Chief Minister after an alliance between the Anguilla United Party and the Anguilla Democratic Party (ADP).

[10] In 1999 Anguilla joined the Caribbean Community Common Market (CARICOM) as an associate member.

[11] The 2000 parliamentary elections were won by the United Front, a coalition between the ANA and the ADP. Osbourne Fleming became Chief Minister.

[12] The Organization for Economic Co-operation and Development (OECD) listed Anguilla as a tax haven in November. According to the OECD, the countries on this list were supposed to have adopted a program to eliminate practices allowing tax evasion by 31 December 2001.

[13] The ANA and ADP, in the United Front, were the major winners of the February 2005 parliamentary elections, getting four out of the seven seats. The ANA got 34.1 per cent of the vote (3 seats) and the ADP got 10.8 per cent. The Anguilla United Movement (AUM) was second, with 12.1 per cent of the vote and two seats. Fleming was re-elected.

[14] Anguilla opened its first golf course, part of an 'ultra-deluxe' tourist complex, in November 2006. ■

PROFILE

ENVIRONMENT
Anguilla covers 91 sq km and its dependency, the island of Sombrero, is 5 sq km. They are part of the Leeward Islands of the Lesser Antilles. The climate is tropical with heavy rainfall. There are considerable salt deposits on the island. In some regions there are problems with the distribution of drinking water.

SOCIETY
Peoples: Most are descendants of African slaves integrated with European settlers. There is a British minority. Many Anguillans live permanently in the Virgin Islands or other US possessions.
Religions: Anglican 40 per cent; Methodist 33 per cent; Adventist 7 per cent; Baptist 5 per cent; Roman Catholic 3 per cent; other, 12 per cent (1996).
Languages: English (official).
Main Political Parties: Anguilla National Front (Anguilla National Alliance and Anguilla Democratic Party), Anguilla United Movement, Anguilla Strategic Alliance, Anguilla Progressive Party.

THE STATE
Official Name: Anguilla.
Capital: The Valley, 1,000 people (2003).
Government: Queen Elizabeth II, Chief of State, since February 1952. Andrew George since July 2006. Osbourne Fleming, Chief Minister since 2000. The single chamber Legislative Assembly has 11 members, 7 elected by direct vote.
Status: Overseas territory of the United Kingdom.

Antigua and Barbuda / Antigua and Barbuda

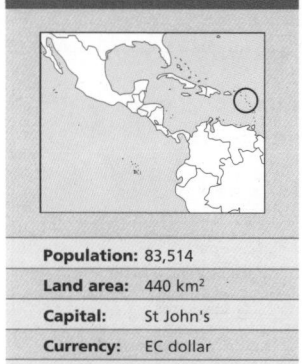

Population:	83,514
Land area:	440 km²
Capital:	St John's
Currency:	EC dollar
Language:	English

Barbuda

17°40'

Codrington

0km 10

Atlantic Ocean

VIRGIN IS.
ANGUILLA
RICO
NETHERLANDS ANTILLES
ST KITTS-NEVIS
Antigua & Barbuda
MONTSERRAT
GUADELOUPE
DOMINICA
MARTINIQUE
ST LUCIA
BARBADOS
GRENADA
TRINIDAD AND TOBAGO

Caribbean Sea

62
Saint John's
Cedar Grove
All Saints
Parham
Bolands
Willikies
Liberta
Sweets
Harbour Town
Antigua
Guadeloupe Passage
0km 10

COLOMBIA
ARUBA
VENEZUELA
0km 300
68
64
60

The Caribs inhabited most of the islands in the sea which took their name, but abandoned many of them, including Antigua, in the 16th century due to the lack of fresh water.

2   In 1493 the name Antigua was given to one of the Antilles by Christopher Columbus in honor of a church in Seville. The Spanish attempted to settle in 1520 and the French in 1629, but left again because of the scarcity of water. However, a few English were able to settle by using appropriate techniques to store rainwater. By 1640, the number of English families on Antigua had increased to 30. The few Indians who had stayed were eventually wiped out by the settlers, who imported African slaves to work in tobacco plantations and later sugar plantations.

3   In 1666, the French governor of Martinique invaded the island, kidnapping the African slaves. When England regained control in 1676, a rich colonist from Barbuda, Colonel Codrington, acquired large landholdings and brought in new African slaves, reinstating sugar production.

4   The neighboring island of Barbuda was colonized in 1678. The Crown granted the island to the Codrington family in 1685.

5   Slavery was abolished in the British colonies in 1838, but Antigua continued with it until labor unions were formed in the 20th century.

6   The first trade union, led by Vere Bird, was founded on 16 January 1939, and gave rise to the Antigua Labor Party (ALP).

7   In the elections of April 1960, Bird's party won and he became Prime Minister. In 1966, a new constitution introduced self-government with a Parliament elected by Antiguans and Barbudans. Britain remained responsible for defense and foreign relations. Bird was again successful in the 1967 elections.

8   On 1 November 1981, Antigua and Barbuda became independent as a sovereign state and was admitted into the UN and the Caribbean Community (CARICOM). By the end of the year, IMF and World Bank loans had increased its foreign debt to half the GDP.

9   An alliance with the US was consolidated in 1983, when Antigua participated in the US invasion of Grenada. This paved the way for Bird's re-election in April 1984, despite accusations of corruption during the campaign. There have been bases on the island ever since.

10   Lester Bird, Vere's son, won the elections in 1994 and 1999 as head of the ALP. His government was affected by a series of scandals, including money laundering, arms trafficking and a case brought against his brother Ivor for cocaine smuggling. In June 1999 Vere Bird died.

11   In 2002 Antigua and Barbuda joined other Caribbean states in a program by the Organization for Economic Co-operation and Development (OECD) to improve governmental transparency and prevention of tax fraud. This action meant the country avoided inclusion on the OECD blacklist, which would have entailed severe sanctions.

12   In 2003, the Caribbean Financial Action Task Force (CFATF), an organization formed by 29 States of the Caribbean Basin to fight money laundering and financing of terrorists, appointed Ronald Sanders - then foreign minister of Antigua and Barbuda - as its President. That same year, Antigua joined the Latin America and the Caribbean Organization for the Prohibition of Nuclear Weapons (OPANAL), an intergovernmental organization created by the Treaty for the Proscription of Nuclear Weapons in Latin America and the Caribbean (Tlatelolco Treaty).

13   In March 2004, Baldwin Spencer - leader of the United Progressive Party - became Prime Minister when he defeated Lester Bird in the general elections, ending the political dominion of the Bird family.

14   An anti-corruption law with fines and jail sentences for poor ministerial performance was passed by Parliament in October 2004.

15   A personal income tax, eliminated in 1975, was reinstated by the Government in April 2005 in an attempt to reduce the deficit inherited from previous administrations.

16   A seminar against money laundering and financing of terrorists took place in St John's in March 2006, organized by the CFATF, along with the government of Antigua and Barbuda, the IMF and the US Treasury. ■

PROFILE

ENVIRONMENT
The islands - Antigua with 280 sq km, and its dependencies Barbuda with 160 sq km and Redonda with 2 sq km - belong to the Leeward group of the Lesser Antilles. Antigua is endowed with beautiful coral reefs and large dunes. Its wide bays distinguish it from the rest of the Caribbean islands because they provide safe havens. Barbuda is a coral island with a large lagoon on the west side. It consists of a small volcano joined to a calcareous plain. Redonda is a small uninhabited rocky island, and is now a flora and fauna reserve. Sugar cane and cotton are grown along with tropical fruits; seafood is exported. The reduction of habitats due to the reforestation of native forests with imported species is the main environmental problem of most of the Caribbean islands.

SOCIETY
Peoples: The majority of Antiguans and Barbudans are of African origin (91.3 per cent); Europeans; Mestizo; Syrian-Lebanese; Indo-Pakistanis.
Religions: Protestants 73.7 per cent (Anglicans 32.1 per cent, Moravians 12 per cent; Methodists 9.1 per cent; Seventh Day Adventists 8.8 per cent; and others); Catholics 10.8 per cent; Jehovah Witnesses 1.2 per cent; Rastafarians 0.8 per cent.
Languages: English is the official language, but in daily life a local Patois dialect is spoken.
Main Political Parties: Antigua Labor Party (ALP), Barbuda People's Movement (BPM), National Democratic Congress, United Progressive Party (UPP, a coalition of three opposition parties - Antigua Caribbean Liberation Movement - ACLM, Progressive Labor Movement - PLM, United National Democratic Party - UNDP).
Main Social Organizations: Antigua Workers' Union, linked to the UNDP; Antigua Trades and Labor Union, with ALP leadership; People's Democratic Movement.

THE STATE
Official Name: Associated State of Antigua and Barbuda.
Capital: St John's 28,000 people (2003).
Other cities: Parham 1,400 people; Liberta 1,400 (2000).
Government: James Carlisle, Governor-General since June 1993, representative of Queen Elizabeth II (Head of State). Baldwin Spencer, Prime Minister since March 2004. Bicameral Legislature: House of Representatives, with 19 members, and the Senate, with 17 appointed members.
National holiday: 1 November, Independence Day (1981).
Armed Forces: 90.

nas

Palau

Palestine

Palmira & Kingman

Panama

Papua-New Guinea

Paraguay

Peru

Philippines

Poland

Portugal

Puerto Rico

Puntland

Q

Qatar

R

Réunion

Romania

Russia

Rwanda

S

St Helena

St Kitts-Nevis

St Lucia

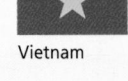

St Pierre and Miquelon

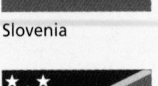

St Vincent and the Grenadines

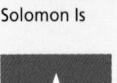

Samoa

Samoa (American)

San Marino

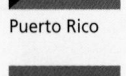

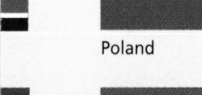

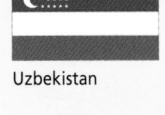

São Tomé and Príncipe

Saudi Arabia

Senegal

Serbia

Seychelles

Sierra Leone

Singapore

Slovakia

Slovenia

Solomon Is

Somalia

Somaliland

South Africa

South Korea

Spain

Sri Lanka

Sudan

Suriname

Swaziland

Sweden

Switzerland

Syria

T

Taiwan

Tajikistan

Tanzania

Thailand

Timor-Leste

Togo

Tokelau

Tonga

Trinidad and Tobago

Tunisia

Turkey

Turkmenistan

Turks and Caicos

Tuvalu

U

Uganda

Ukraine

United Arab Emirates

United Kingdom

United States

Uruguay

Uzbekistan

V

Vanuatu

Vatican City

Venezuela

Vietnam

Virgin Is (US)

Virgin Is (Br.)

W

Wake

Wallis and Futuna

Western Sahara

West Papua

Y

Yemen

Z

Zambia

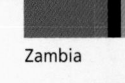

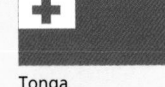

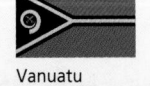

Zimbabwe

80

Arctic Circle
Lena

US
Bering
Sea

60

Okhotsk
Sea

Sakhalin

50

Aleutian Is.

0 500 1000 1500 2000 km
0 500 1000 1500 miles
Scale: 1:100.000.000

Amur

Novosibirsk

Ulaanbaatar

MONGOLIA

Vladivostok
Sea
of
Japan

Hokkaido

40

CHINA

Beijing

N.KOREA
Pyongyang
Seoul
S.KOREA

Hosnu

JAPAN
Tokyo

Kuril'skje
Ostrova

Bishkek
KGYZ.

Huanghe

Nanjing

Kyushu

Ryukyu Is.

30

Pacific

KASHMIR
nabad

Tibet

Chanjiang

Chongqing

Osawara
Gunto

Tropic of Cancer

Wake
(US)

New
Delhi

NEPAL
Kathmandu

BHUTAN

Dacca

Taipei

TAIWAN

20

INDIA

Ganges

BANGL.

MACAU

Xianggang (Hong Kong)

20

umbai

MYANMAR/
BURMA

LAOS
Vientiane

Hanoi

Hainan

Luzon

PHILIPPINES

NORTHERN
MARIANAS IS.

Hyderabad

Bay
of Bengal

Rangoon

THAILAND
Bangkok

VIETNAM

GUAM

Ocean

Andaman
(INDIA)

Phnom Penh

CAMBODIA

Manila

10

Laccadive Is..
(INDIA)

SRI LANKA
Colombo

Nicobar Is.
(INDIA)

Mindanao

PALAU

MARSHALL
IS.

10

MALDIVES

MALAYSIA

BRUNEI

MICRONESIA

Howla
(US)

Kuala Lumpur

Borneo

180

Baker
(US)

70

80

90

SINGAPORE

100

120

130

140

150

160

Equator

170

180

INDONESIA

New Guinea

NAURU

KIRIBATI

0

Diego Garcia
(MAUR.)

Jakarta

Java

WEST PAPUA

PAPUA NEW GUINEA

SAM
ian Ocean

TIMOR-LESTE

Port Moresby

SOLOMON IS.

TUVALU

10

Wallis a
Futuna

Timor Sea

Coral
Sea

VANUATU

FIJI

FIJI

Cocos Is.
(AUSTL)

Christmas Is.
(AUSTL)

20

20

Tropic of Capricorn

KANAKY/
NEW CALEDONIA

AUSTRALIA

Norfolk Is.
(AUSTL)

30

USTRIA

Graz

HUNGARY

20

ROMANIA

Darling

Sydney

Lord Howe Is.
(AUSTL)

30

Maribor

Szeged

Arad

SLOVENIA

Drava

Pécs

Timisoara

Canberra

Ljubljana

Zagreb

CROATIA

Melbourne

Tasman
Sea

Wellington

40

Rijeka

Save

Osijek

Vojvodina

Novi Sad

Craiova

45

Tasmania

NEW ZEALAND/
AOTEAROA

Dugi Is.

BOSNIA-
HERZEGOVINA

Belgrade

Danube

ncona

Split

Sarajevo

SERBIA

Stewart Is.

Auckland Is.

50

Hvar Is.

Nis

MONTENEGRO

scara

KOSOVO

Sofia

BULGARIA

Macquarie Is.
(AUSTL)

Campbell Is.

60

Foggia

ITALY

Dubrovnik

Tirana

Skopje

MACEDONIA

ALBANIA

GREECE

Bari

0 300 km
0 200 miles

Balleny Is.

70

Antarctic Circle

80

80

The World Guide

Argentina / Argentina

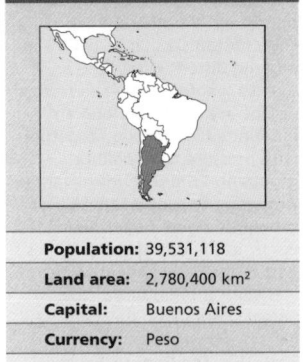

Population:	39,531,118
Land area:	2,780,400 km²
Capital:	Buenos Aires
Currency:	Peso
Language:	Spanish

A round 300,000 indigenous people inhabited the territory currently occupied by Argentina when the first explorers of the Spanish Crown arrived searching for gold towards the beginning of the 16th century. The Pampean Patagonian peoples in the south, such as the Tehuelch and the Pehuelch, were nomadic hunters and gatherers, while the tribes of the Chaco in the north-east the Mataco, the Guaycurúe and the Guaraníes had begun to settle down. The Andean peoples, such as the Diaguita and the Huarpes, through contact with the Incas, perfected their agricultural system, introducing terracing and artificial irrigation. This gave rise to commercial trade with the northeastern and western regions. They also made handicrafts and raised llamas. Little is known about the Omahuacas, the Patamas, the Capayanes, the Comechingones and the Algarrobero peoples.

[2] In 1526, Sebastian Cabot founded a fort on the banks of the Carcarañá river which was the first European settlement of present-day Argentina.

[3] To check the Portuguese advance, Spain sent Pedro de Mendoza to the region on a contract granting the conquistador established political and economic privileges. In 1536, de Mendoza founded Santa Maria del Buen Ayre (Buenos Aires), a small town which was abandoned in 1541 after being besieged by the indigenous peoples.

[4] Using Asunción (in today's Paraguay) as a focus for colonization, the Spanish founded several cities in what is now Argentine territory (Santiago del Estero, Córdoba, Santa Fe), until they came to the second founding of Buenos Aires in 1580. Under colonial administration, the region was in principle subject to the Vice-Royalty of Peru. Three cities succeeded each other in predominance. Tucumán, linked with gold exploitation in Upper Peru, was the center during the 16th century. Córdoba, where the first university of the region was established in 1613, was the intellectual center during the 17th and 18th centuries. And after the new administrative divisions in 1776, the port city of Buenos Aires became capital of the new Viceroyalty of the River Plate, which covered what are now Bolivia, Paraguay, Argentina and Uruguay.

[5] In the 17th and 18th centuries, the Spanish conquest pushed the Mapuche - known as Araucans by the Europeans - from Chile into the center and south-east of present-day Argentina resulting in the 'Araucanization' of the local inhabitants. The abundance of cattle, the main product the viceroyalty sent to Spain, caused great ethnic and cultural changes. Outside the cities, vast plains of good pasture were ranged by horsemen, some of whom were indigenous people (who modified their diet to feed on beef). The rest were gauchos - mixed-race cowboys who lived by working the cattle.

[6] The strong bourgeoisie of the port area who favored free trade initiated the revolutionary movement of 1810, which created the United Provinces of the River Plate and overthrew the Viceroy, accusing him of a lack of loyalty to Spain, which was at that time occupied by Napoleon's troops. Gauchos and Indians swelled the ranks of armies organized in Buenos Aires to fight the Spanish crown beyond the frontiers of the Viceroyalty. General José de San Martín led the armies which defeated the royalists and contributed decisively to the independence of Chile and Peru.

[7] Even though the Spanish were quickly expelled, discrepancies between Buenos Aires and the rest of the United Provinces of the River Plate (including the present-day Republic of Uruguay) kept the region in a permanent state of war. The Unionists defended the centralism of Buenos Aires, which threatened the economy of the interior, while the Federalists pursued a more equitable agreement for all the provinces.

[8] In 1829 Juan Manuel de Rosas, a land-owner with federal roots (opposed to Buenos Aires' centralism), took over as governor of Buenos Aires. Rosas, who had himself proclaimed Restorer of the Laws, gained fame in the 'Desert Campaign' in which the indigenous peoples were expelled from the areas surrounding Buenos Aires. He used political means and force to pacify the interior, bringing most of the governors together and deploying troops from Buenos Aires as far as the frontiers with Bolivia and Chile.

[9] Since Buenos Aires was the gateway to the Paraná and Uruguay rivers - main arteries for French and British trade with the provinces of the interior and Paraguay - Rosas brought in a customs law which restricted access to foreign vessels and products. The Rosas Government was attacked by armies from both powers, which blockaded the port of Buenos Aires intending to crush the Government and support Rosas' more liberal rivals. During the campaign to put pressure on Rosas in 1833 Britain occupied the Malvinas (Falkland Islands).

[10] The 'Great War' (waged

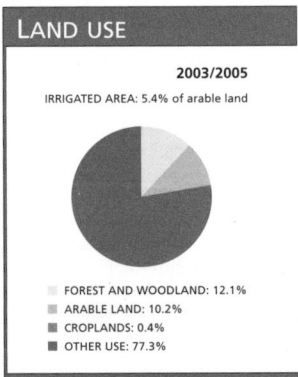

LAND USE

2003/2005

IRRIGATED AREA: 5.4% of arable land

- FOREST AND WOODLAND: 12.1%
- ARABLE LAND: 10.2%
- CROPLANDS: 0.4%
- OTHER USE: 77.3%

PUBLIC EXPENDITURE

% OF GDP

8.5 % — HEALTH & EDUCATION (2000-2002)

1.0 % — DEFENSE (2004)

WORKERS

UNEMPLOYMENT: 16% (2004)

LABOR FORCE — 2004

- FEMALE: 42%
- MALE: 58%

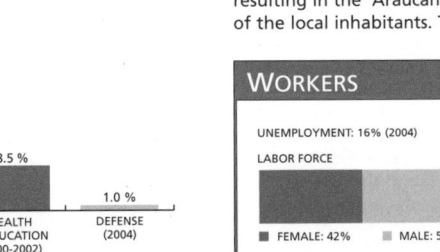

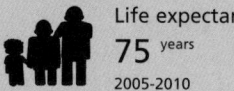

Life expectancy
75 years
2005-2010

GNI per capita
$3,580
2004

Literacy
97% total adult rate
2000-2004

HIV prevalence rate
0.7% of population 15-49 years old
2003

in Argentina and Uruguay between 1839 and 1852) involved Argentina, Uruguay, Paraguay and Brazil, and saw direct intervention from Britain and France, bringing an end to Rosas' 20 years in power.

[11]   Justo José de Urquiza, Governor of Entre Rios, then presided over the Argentine Confederation with its capital in Paraná in accordance with the federal constitution of 1853. However, there was resistance from Buenos Aires, which proclaimed independence and declared itself a separate state, even having diplomatic representation abroad. After 10 years of fighting, in 1861 Buenos Aires got its own way.

[12]   Bartolomé Mitre (President from 1862 to 1868) made an alliance with Pedro II, Emperor of Brazil, and Venancio Flores, President of Uruguay, to wage the 'War of the Triple Alliance' on Paraguay. The fighting started in 1865 and ended in 1870 with the death of Paraguayan President Francisco Solano López and most of the Paraguayan population. By the time 'victory' came, Domingo Faustino Sarmiento had taken over from Mitre.

[13]   Once the war was over, the rest of what is now Argentina was occupied through successive incursions against the indigenous peoples of Patagonia, the Chaco and the Andean regions of Río Negro. In 1879, General Julio Roca (President from 1880-86 and 1898-1904) ended Argentina's 60-year military campaign to exterminate the indigenous peoples. The railway, symbol of modernization, ensured the Buenos Aires Government's control over the territory. The mass immigration of European workers, above all Spaniards and Italians, changed the face of the nation which underwent unprecedented industrial, agricultural and commercial growth. The population grew from less than a million people in 1869 to almost eight million in 1914.

[14]   Up until the electoral law of President Roque Sáenz Peña in 1912, which guaranteed universal, secret and obligatory suffrage for adult men, political power had never been a matter of democratic elections, rather the product of permanent fraud. This was the cause of the 1874, 1890, 1893 and 1905 rebellions led by Mitre, Leandro Alem, Aristóbulo del Valle and Hipólito Irigoyen, respectively.

[15]   The Sáenz Peña Law helped the Radical Party, led by Irigoyen, into power in 1916. During World War I, in which Argentina was neutral, the country sold food to Europe and experienced industrial growth. One result of this was that the unions became far stronger.

[16]   Despite being re-elected by a margin of two to one in 1928, Irigoyen was unable to ride the consequences of the world economic crisis of 1929, which drastically undermined the agro-exporter model. Irigoyen was ousted a year later in a military coup led by General José Félix Uriburu. The coup marked the end of a period of constitutional continuity which had lasted 68 years, as well as a long period of economic expansion based on the export of raw materials which had doubled between 1913 and 1928.

[17]   During World War II, President Ramon S. Castillo maintained Argentina's neutrality, provoking opposition. Another military coup ousted Castillo in 1943. The ensuing dissent within the army - which wanted neither to restore democracy nor to prolong an indefinite dictatorship - together with pressure from political groups and the US (which wanted Argentina to support the Allies) brought Colonel Juan Domingo Perón to the presidency from his post in the Employment Ministry.

[18]   With the support of the unions, Perón won the 1946 elections by a narrow margin. Backed by the General Labor Confederation (CGT), he declared a state of civil war, which allowed him to get round the opposition and extend his authoritarian power. He nationalized foreign trade, the banks, the railway, gas and telephones; extended the fleet and created the air force; increased the workers' share of the national income to 50 per cent and framed advanced social legislation. Furthermore, he organized workers and bosses into national confederations with whom he negotiated economic and social policy. His wife Eva ('Evita') formed an exceptionally charismatic nexus between Perón and the workers, whom she dubbed 'the Shirtless Ones'. It was Eva Perón who was largely responsible for Argentina giving women the vote in 1947. In foreign affairs Perón took the 'Third Position' between the two superpowers of the Cold War, which brought him into conflict with the US.

[19]   Re-elected in 1951, Perón lost army support by confronting the Church. He was ousted in a coup in 1955 and forced into exile. A dictatorship headed by Pedro Aramburu replaced him, embracing the National Security Doctrine which passed responsibility for defending the region against 'the enemies of democracy' to the US. The dictatorship imposed a regressive wealth distribution system and opponents from both army and civilian ranks were executed in the 1956 'Operation Massacre', a precursor of decades of political violence.

[20]   'Peronism' was outlawed, but it was the Peronist vote which brought the pro-development government of Arturo Frondizi to power in 1958, opening the country to oil and automobile transnationals, and implanting a model of growth and wealth concentration which fed serious social confrontations. Frondizi legalized 'Peronism', and in the 1962 parliamentary elections the Peronists won in 10 provinces. This provoked another coup from the army which toppled

PROFILE

PROFILE

ENVIRONMENT

There are four major geographical regions. The Andes mountain range marks the country's western limits. The sub-Andean region consists of a series of irrigated enclaves where sugarcane, citrus fruits (in the north) and grapes (central) are grown. A system of plains extends east of the Andes: in the north, the Chaco plain with sub-tropical vegetation and cotton farms; in the center, the Pampa with deep, fertile soil and a mild climate where cattle and sheep are raised, and wheat, corn, forage and soybeans are grown, and to the south stretches Patagonia, a low, arid, cold plateau with steppe vegetation where sheep are extensively raised and oil is extracted. Argentina claims sovereignty over the Malvinas (Falkland) Islands and a 1,250,000 sq km portion of Antarctica.

SOCIETY

Peoples: Most Argentineans are descendants of European immigrants (mostly Spaniards and Italians) who arrived in large migrations between 1870 and 1950. Among them is the largest Jewish community in Latin America. According to unofficial figures, the indigenous population of 447,300 is made up of 15 indigenous and 3 mestizo peoples mainly in the north and southeast of the country, and in the marginal settlements around the major cities. The Mapuche, Kolla and Toba constitute the largest ethnic groups. The indigenous peoples in the east, center and southernmost tip are in decline.
Religions: Catholic (92 per cent, official); Protestant, Evangelical, Jewish and Islamic minorities.
Languages: Spanish. Minor groups maintain their languages: Quechua, Guarani and others.
Main Political Parties: 'Justicialist' or 'Peronist' Party (*Partido Justicialista*, PJ), currently in power, Front for Victory (*Frente para la Victoria*), Radical Civic Union (*Unión Cívica Radical*), Alternative for a Republic of Equals (*Alternativa por una República de Iguales*), Republican Initiative Alliance (*Alianza Propuesta Republicana*).
Main Social Organizations: The General Labor Confederation (CGT), 'Peronist' in orientation, founded in 1930. In reaction to the Government's economic and labor policies in the 1990s, the confederation has split into three factions. Argentine Workers' Central; Mothers of Plaza de Mayo (different groupings); Argentine Agrarian Federation; Argentine University Federation; Ecumenical Movement for Human Rights; Indigenous Peoples.

THE STATE

Official Name: República Argentina.
Administrative divisions: 4 Regions with 23 Provinces and the Federal Capital of Buenos Aires.
Capital: Buenos Aires 3,047,000 people; Greater Buenos Aires 13,047,000 people (2003).
Other cities: Córdoba 1,521,700 people; Rosario 1,339,100; Mendoza 957,400; La Plata 813,800 (2000).
Government: Presidential system. Néstor Kirchner has been president since 2003. The National Congress (Legislature) has two chambers: Chamber of Deputies of the Nation, with 257 members, and the Senate of the Nation, with 72 members. Each province and the Federal District have three seats in the Senate.
National Holidays: 25 May, Revolution (1810); 9 July, Independence Day (1816).
Armed Forces: 67,300: army 60 per cent, navy 26.8 per cent, air force 13.2 per cent.

Under-5 mortality
18 per 1,000 live births
2004

Poverty
7% of population living on less than $1 per day
2003

Debt service
28.5% exports of goods and services
2004

Maternal mortality
82 per 100,000 live births
2000

Frondizi. Following two serious confrontations between separate factions of the army, General Juan Carlos Onganía emerged as the new strongman.

21   With Peronism banned once again, the Radical Party's Arturo Illia was elected in 1963 in the first administration for 40 years not to apply a State of Emergency nor other special measures for repression or cultural censorship. His term in office was marked by friction with the Peronist unions who organized strikes, demonstrations and occupations of factories.

22   In June 1966, Illia was overthrown by the 'Argentinean Revolution' of General Juan Carlos Onganía, who brought in a new authoritarian model, politically clerical and corporatist, economically liberal and defender of the 'ideological borders' in foreign policy. Onganía consecrated the country to the Sacred Heart, banned political parties, intervened in the universities and the CGT and denationalized the economy: bankrupted companies were bought up extremely cheaply by US, British and German consortia.

23   A succession of social uprisings, like the 1969 'Cordobazo', and the emergence of a guerrilla force, threatened to split the army. General Agustín Lanusse took over the presidency in 1971 and in order to preserve

the military institution he announced elections, although he banned Perón, exiled in Madrid, from taking part.

24   Héctor Cámpora, the 'Justicialista' Liberation Front candidate - a Peronist electoral coalition - received 49 per cent of the vote in the March 1973 elections, taking office in May 1973, and resigning two months later to allow new elections after Perón's return to the country. On 23 September, Perón was re-elected in a new contest with 62 per cent of the vote, and the vice-presidency went to María Estela 'Isabelita' Martínez de Perón, his third wife.

25   The Peronists resumed diplomatic relations with Cuba, proposed a reorganization of the Organization of American States (OAS) to serve Latin America's interests, promoted Argentina's participation in the Non-Aligned Movement and increased trade with socialist countries. After the Ezeiza massacre on 20 June, the day Perón returned to Argentina, friction grew among various Peronist factions and open warfare broke out between the old-time labor leaders and the 'special squads', as Perón used to call the guerrillas.

26   When Perón died in 1974, his widow Isabelita took office. Her Social Welfare Minister (Perón's former secretary), José López Rega, set up the 'Triple

A' (Argentine Anticommunist Alliance), a paramilitary hit squad that murdered Marxist opponents and left-wing Peronists.

27   On 24 March 1976, a military coup put an end to Isabelita's inefficient and corrupt administration. A military junta led by General Jorge Videla suspended all civil liberties and set in motion a cycle of kidnapping, torture, betrayal and murder. The term 'missing person' became ominously commonplace, and government priorities were dictated by the newly adopted 'National Security' doctrine. Human rights organizations drew up a list of over 25,000 missing people, who had been arrested by the police in front of witnesses. Their fate has never been determined with certainty.

28   The Junta encouraged imports to the point of liquidating a third of the country's productive capacity. Fifty years of labor gains were wiped out, real wages lost half their purchasing power, and regional economies were choked by high interest rates. The country's cattle herds decreased by 10 million head and foreign debt climbed to $60 billion, a quarter of which had been spent on arms. It was the era of what Argentines called the *patria financiera* (financial fatherland), when governmental economic policies encouraged most of the country's productive sector to turn

to speculation. In 1980, a wave of bankruptcies hit banks and financial institutions.

29   In 1981, Videla was replaced by Roberto Viola who was in turn replaced by General Leopoldo Galtieri. Galtieri thought he had US President Reagan's unconditional support and decided to ward off the domestic crisis by recovering the Malvinas Islands, in British hands since 1833. His troops landed there on 2 April1982. Galtieri's error soon became evident. Britain deployed a powerful fleet that included nuclear submarines, and the US backed its North Atlantic ally. After 45 days of fighting Argentina surrendered on 15 June; 730 Argentineans and 250 British soldiers had been killed during the conflict. Two days later Galtieri was forced to resign from both his military and presidential posts.

30   The junta set elections for 30 October 1983. The Radical Civic Union's new leader, Raúl Alfonsín, won the election with 52 per cent of the vote.

31   The new government started well but ended with the complete discrediting of Alfonsín. On the political front, the Government wanted to try the military leaders who had taken part in the 'Dirty War', responsible for the disappearance of more than 30,000 people. On the economic front, it aimed to tackle inflation head on, as this had reached

IN FOCUS

ENVIRONMENTAL CHALLENGES
Deforestation, soil degradation, desertification and pollution of the air and water are significant problems. Untreated sewage has polluted several rivers, particularly the Matanza-Riachuelo river in Buenos Aires.

Soil erosion is growing, mainly in the north of the humid Pampa. In compliance with the 1999 Convention on Climate Change Argentina has taken significant steps to increase use of compressed natural gas in the transportation system.

WOMEN'S RIGHTS
Argentinean women have been able to vote and run for office since 1947. In 2005, 41.7 per cent of senators and 35 per cent of representatives were women. In 2003 female unemployment was 19 per cent. The participation of women in the workforce rose from 28 per cent in 1990, to 33 per cent in 2000 and 35 per cent in 2003. That year, 12 per cent of female workers were in industry and 87 per cent in services. In 2003

there were 54 births per 1,000 women aged between 15 and 19. Some 98 per cent of deliveries are attended by qualified staff*. There is universal enrolment in primary education but more girls than boys stay in education. Some 60 per cent of students in tertiary education are female.

CHILDREN
The mortality rates of children under 1 and under 5 fell considerably in the 1990-2004 period: the former from 26 to 16, and the latter from 29 to 18 per 1,000 births. In 2006, 95 per cent of children reached fifth grade and more than 90 per cent enrolled in secondary school*.

However, in the period 1995-2000, between 5 and 19 per cent of the child population suffered from hunger, depending on the province. The economic and financial crisis that took shape towards the end of 2001 led to sharp increases in child mortality and hunger, especially in the northern provinces, and affected children's access to education,

nutrition and health services.

Soup kitchens set up for school children were the means of keeping many of the poorest children at school. In 2003 it was decided that more than 1,500 schools should remain open during the holidays in three provinces, providing 330,000 children with one meal a day and basic care.

INDIGENOUS PEOPLES/ ETHNIC MINORITIES
Indigenous organizations estimate that there are between 800,000 and 2,000,000 native inhabitants in the country. Some provinces' populations are 17-25 per-cent indigenous. Indigenous people can only file complaints at the National Office for Native Affairs, an agency that, in practice, does not function.

In 2006, the mobilization of the Mapuche people led to the amendment of the Neuquen province Constitution granting them economic, cultural and political rights over their ancestral lands, as well as certain autonomy from central Government.

MIGRANTS/REFUGEES
A challenge for countries in the region is to find homes in places with high unemployment rates for the 10,000 or so immigrants, most of whom come from Colombia. The Government works with Uruguay in joint resettlement programs similar to those in Chile and Brazil.

The emigration of Argentineans owing to the economic crisis peaked in 2001-2, with most of them bound for Spain, the US and Italy. Almost 155,000 people left the country in those years, many never to return. In 2003 this population drain was reduced, based on hopes raised by the new Government's proposed changes.

DEATH PENALTY
In 1984 it was abolished for ordinary crimes.

* Latest data available in *The State of the World's Children* and *Childinfo* database, UNICEF, 2006.

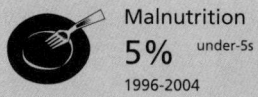

Malnutrition

5% under-5s

1996-2004

Doctors

301 per 100,000 people

1990-2004

Primary school

100% net enrolment rate

2004

astronomical proportions - 688 per cent by the end of 1984 - by reducing public spending and launching the Austral Plan. This froze prices and fees for services as well as establishing a new currency, the Austral, initially worth more than the dollar.

[32] On the basis of revelations made in studies by the National Commission of Missing Persons, nine Commanders in Chief of the dictatorship were put on public trial accused of having ordered the crimes of that period. The sentences meted out to several high-ranking army leaders - including former President Videla - and the later extension of the trials to lower ranking officers prompted a strong reaction from the military. Between 1987 and 1989, Alfonsín had to put down four military uprisings. In 1987, the President sent Congress the 'Due Obedience' bill, which was approved, exempting most military personnel accused of human rights violations, claiming they were simply obeying orders from above.

[33] The failure of the Austral Plan led to thousands of jobs being lost between December 1983 and April 1989, salaries were drastically reduced and some 10 million inhabitants - almost 30 per cent of the population - were virtually pushed out of the consumer market. In this period, the CGT organized 14 general strikes and shops were raided in various areas of the capital and some cities of the interior.

[34] The May 1989 presidential elections brought the Peronist Carlos Saúl Menem to power. The magnitude of the economic troubles resulted in the new president being asked to take office a couple of months early. Menem, who had been governor of La Rioja, established a program of privatizations based on the August 1989 State Reform Act. In his first year in power, the privatization of the state oil company was promoted, as well as that of several mass media and communications companies and the state airline. The economic liberalization policy caused a schism in the CGT between the sectors supporting the Government and those opposing them.

[35] Menem re-established relations with the United Kingdom (leaving the key issue of sovereignty over the Malvinas under an ambiguous 'protective umbrella') and in two stages pardoned all the army officers responsible for the 'Dirty War'.

[36] Despite constant scandals and accusations of corruption, Menem maintained his image

thanks to the economic stability achieved by the Convertibility Plan, which drastically reduced public spending and established parity between the new monetary unit - the peso - and the US dollar. Inflation fell and reached a historical low: in 1993 it was 7.4 per cent. There was strong economic growth but the unequal wealth distribution across the various regions worsened.

[37] The big surprise of the 1994 constituent elections was the performance of the leftist Frente Grande (Broad Front), which became the third strongest political force in the nation and triumphed in the Federal Capital with 37.6 per cent of the vote. It also won in the southern province of Neuquén. In Buenos Aires it became the second political force with 16.4 per cent of the vote. However, Peronists and Radicals obtained the majority necessary to ensure the constitutional reform which allowed Menem to stand for re-election.

[38] Shortly before Menem won the elections with 50 per cent of the vote in May 1995, trouble struck at the heart of his family, when his son, Carlos Menem Junior, was killed in a dubious helicopter crash.

[39] In the first quarter of 1996, Menem sacked Economy Minister Domingo Cavallo for having reported facts which linked the President with the murder of José Luis Cabezas, a photographer for a magazine critical of the government. The new Economy Minister, Roque Fernández, kept the economy going along the same lines as his predecessor had.

[40] In the 1999 presidential elections, the candidate for the Alliance (a coalition formed by the UCR and the Frepaso, derived from the Frente Grande), the radical Fernando de la Rúa, was victorious with almost 50 per cent of the vote in the first round.

[41] Economic recession deepened in 2000 and the new administration, which had promised to fight corruption, was involved in a scandal. In September, Vice President Carlos 'Chacho' Álvarez reported that the Government had bribed several members of Parliament in April. Since De la Rúa refused to remove the officials involved, the Vice President resigned in October.

[42] In late 2000, the IMF granted Argentina $40 billion in aid to cover social needs, but the funds disappeared fast and the country did not attract foreign private investment.

[43] After failing to revitalize the economy, Economy Minister José Luis Machinea resigned in March 2001. The rest of the Cabinet fell

with him. De la Rúa appointed Cavallo as Economy Minister once again, and the Senate passed a law with a 13 per cent cut in state wages and pensions.

[44] In December, the IMF denied Argentina a new loan, arguing that the economic policy was unsustainable, and demanded more budget cuts. By then, Argentina owed international organizations more than $140 billion (54 per cent of its GDP) and had lost approximately $19 billion in investments.

[45] The massive withdrawal of deposits forced the Government to apply a series of temporary restrictions on accounts. The measure, announced on 1 December for an initial period of 90 days, came to be known as the *corralito* - little corral. A nationwide strike demanding employment ended with Cavallo's resignation.

[46] De la Rúa resigned on 20 December, amidst massive street protests. The demonstrations, which included looting of supermarkets and other businesses, were severely repressed by security forces, which opened fire on the demonstrators, killing six and injuring dozens.

[47] After De la Rúa resigned, Senate chairman Ramón Puerta, a Justicialist, took office as President. Shortly after, the Legislative Assembly named another Justicialist, Adolfo Rodríguez Saá, who resigned five days later after defaulting on Argentina's private foreign debt, which amounted to $81.8 billion. Since Puerta declined to take office, Eduardo Camaño, Chamber of Deputies chairman and also from the JP, was sworn in as interim President and summoned the Legislative Assembly. Finally, Eduardo Duhalde - Menem's main rival inside the JP - was chosen President. Duhalde was allowed to preside over the Government until September 2003 and elections scheduled for 3 March were cancelled.

[48] The Government announced in January 2002 that the *corralito* would be enforced until 2003 and people would only recover their savings in installments. The Central Bank was forced to intervene in the currency market to prevent the peso from crashing. Protests continued, causing destruction in banks and automated teller machines. Duhalde broke his promise that people who had savings in dollars would receive dollars in return, and announced the blocked deposits would be returned only in devalued pesos.

[49] The population's discontent increased when, on 26 January, judges freed former President Menem, who had been detained

for arms smuggling to Croatia and Ecuador.

[50] In February the Government allowed the dollar to float freely, which led to a sharp rise in its value. Without political backing from Congress to approve the Bonex plan - to turn dollar deposits into pesos - Economy Minister Jorge Remes Lenicov resigned in April. Duhalde replaced him with Roberto Lavagna, who sought to stop the dollar's uncontrollable rise and started difficult negotiations with the IMF.

[51] A moderate agreement reached with the IMF in January 2003 temporarily postponed Argentina's repayments.

[52] Duhalde brought the elections forward to May and backed Néstor Kirchner, a lawyer who was governor of Santa Cruz province and who had had a secondary role in the political sphere. In the first round, Kirchner came second, with 22 per cent of the vote, behind Menem (24 per cent). The backing of the other candidates for Kirchner led Menem to withdraw from the race.

[53] Although Menem tried to leave the new President isolated by withdrawing from the second round, within a a few months Kirchner's popularity rating was around 80 per cent.

[54] Kirchner appointed ministers who were mainly of his own generation and went for a bold style of leadership. He immediately retired dozens of military officers involved in repression in the past dictatorship, challenged the immunity laws and pushed research on the 'missing' people issue. In August he 'screened' the Federal Police, removing officers involved in corruption. The same happened with Supreme Court judges and former President Fernando De la Rúa, who were all banned from leaving the country.

[55] On the economic front, he increased wages and returned the money frozen in the *corralito* accounts to small creditors.

[56] Foreign debt climbed towards $178 billion. Before the UN General Assembly in September, Kirchner demanded a rescheduling of the payment plan to the IMF, saying that 'multilateral organisms that encouraged indebtedness will now have to take on the responsibility.' In November, he stressed that 2004 budget funds were not going to pay the foreign debt, but would be used on social spending.

[57] In spite of Government measures, the unemployed held constant protests by blocking streets. This, plus a wave of kidnappings, became a significant

destabilizing factor after 2003. That year, the economy grew 6 per cent and the dollar remained stable at around 3 pesos.

[58] Kirchner and Brazilian president Luiz Ignacio (Lula) da Silva started a plan in July to enlarge and strengthen MERCOSUR, aiming to improve the negotiating position with Europe and the US and to prevent the advance of the Free Trade Area of the Americas advocated by Washington.

[59] In April 2004 a court issued an international arrest warrant for former President Menem, accused of fraud.

[60] Hundreds of demonstrators tried to storm Parliament in July to protest against the reform of the urban cohabitation code, which included harsh sentences for street protests, prostitution and street vendors.

[61] Menem returned in December from self-imposed exile in Chile, after fraud charges were dropped.

[62] In January 2005 Argentina offered its private creditors a deal to pay its $100 billion debt with a 75 per cent discount. That month, former army officer Adolfo Scilingo, accused of ejecting political prisoners into the sea from military planes, was taken to Spain for trial. In April, he was sentenced to 640 years in prison.

[63] The October elections reflected the President's popularity. His party, Front for Victory, was well supported. Menem was elected Senator. Late that month, Kirchner appointed Felisa Miceli in Lavagna's place.

[64] The arrival of US President George W Bush to attend the Summit of the Americas in November 2005 led to massive and violent protests.

[65] In January 2006 Argentina paid-off its multi-million debt with the IMF. In early May, a conflict with Uruguay over the construction of two cellulose paste plants on the Uruguay river brought the case before The Hague International Court of Justice. In July, the Court ruled out the immediate suspension of the construction of the plants, while the trial carried on.

[66] In September 2006, a senior judge lifted the pardon given in 1990 to former dictator Jorge Videla, ruling that the pardon was unconstitutional. Videla led the military government from 1976 to 1983. Along with two former ministers, Jose Alfredo Martinez de Hoz and Albano Harguindeguy, who also had their pardons overturned, he faced new charges related to the kidnapping of two entrepreneurs between November 1976 and April 1977. ∎

Falkland Islands / Islas Malvinas

Population: 3,063

Land area: 11,410 km²

Capital: Port Stanley (Puerto Argentino)

Currency: Pound sterling

Language: English

When the islands were sighted for the first time in 1520 by a Spanish ship they were uninhabited. In the 18th century, they were baptized the 'Malouines' in honor of Saint Malo, port of origin of the French fisherpeople and seal hunters who settled there. In 1764, Louis Antoine Bougainville founded Port Louis on Soledad Island. This move brought Spanish protests, and France recognized Spain's prior claim. That same year the British, who since 1690 had called the islands 'Falkland', founded Port Egmont, renamed Puerto Soledad when the islands were returned to Spain in exchange for £24,000.

[2] In 1820, shortly after independence, Argentina appointed Daniel Jewit as first Governor of the Malvinas. In 1831, Governor Vernet impounded two US ships on charges of illegal fishing. A US fleet that was visiting South America avenged this act of 'piracy' by destroying houses and military facilities at Port Soledad. On 3 January 1833, the English corvette Clio landed a contingent of settlers, which the small local force was unable to repel.

[3] After World War II, the United Nations Decolonization Committee included the Malvinas and their dependencies on the list of 'non-autonomous' territories and established that, as the inhabitants were British, the principle of self-determination was not applicable. The only legally valid solution to the problem was to recognize Argentinean sovereignty.

[4] On 2 April 1982, Argentinean forces occupied the Malvinas. British Prime Minister Margaret Thatcher, whose popularity rating had been rock bottom before the conflict, responded by going to war. Two months later, at the cost of more than 1,000 lives, the Union Jack was once again hoisted over the islands.

[5] At the beginning of his administration, Menem renewed relations with London, but the intractable issue of the Malvinas' sovereignty was left unresolved.

[6] In March 1994, the Argentinean Minister of Defense reported that Argentinean soldiers had died at the hands of British forces, in circumstances that violated the Geneva Convention on the treatment of prisoners of war.

[7] An understanding reached between President Carlos Menem and Prime Minister John Major at the United Nations, in September 1995, established that Argentina and the United Kingdom would jointly exploit the west of the islands, along the border with Argentina.

[8] In October 1999, as a result of an agreement between the Argentinean and British Governments, a group of Argentineans flew to the island, the first such visit since 1982. The ex-combatants and journalists who made up the party paid tribute to those who died in the war.

[9] In June 2004, the UN Committee on Decolonization demanded that Argentina and the UK negotiate the situation in the Malvinas in order to put an end to the 171-year sovereignty dispute.

[10] A book with the first British official history of the conflict between Argentina and the UK was published in June 2005. According to different reviews, the Argentinean demand was 'very close to being legitimate', which sparked great controversy on the question of British over sovereignty of the islands.

[11] In May 2006, in an act commemorating the 24th anniversary of the conflict and paying tribute to the war dead of the National Gendarmerie, the Interior Minister, Aníbal Fernández, predicted that 'at some point' Argentina would recover the Malvinas. ∎

PROFILE

ENVIRONMENT

An archipelago with nearly 100 islands located in the South Atlantic. It includes South Georgia, South Sandwich and South Shetland Islands. There are two main islands - Soledad (East Malvina) and Gran Malvina (West Malvina) - separated by the San Carlos Channel. The coast is rough and mountainous. More than half of the population lives in the capital on Soledad. The main economic activity is sheep rearing. The islands' territorial waters are believed to contain oil reserves. There is also hope that krill (a microscopic crustacean rich in protein) can be marketed. Its proximity to Antarctica gives the archipelago strategic importance.

SOCIETY

Peoples: 3,000 descendants of English colonists (est. 2005).
Language: English.
Religion: Mostly Anglican; Catholics, other Protestant churches.

THE STATE

Capital: Port Stanley (Puerto Argentino) 2,000 people (2003).
Government: Howard Pearce, Governor since 2002, appointed by Britain. The Legislative Assembly has 10 members, 8 members elected for a four year term and 2 members ex officio. In November 2005 elections, 8 non-partisans were elected.
Armed Forces: 4,000 UK soldiers since June 1982.

Armenia / Hayastan

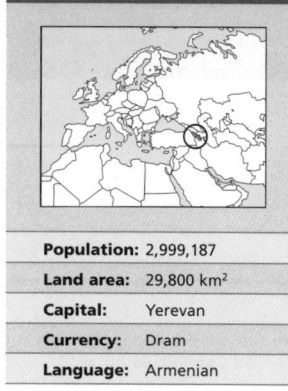

Population:	2,999,187
Land area:	29,800 km²
Capital:	Yerevan
Currency:	Dram
Language:	Armenian

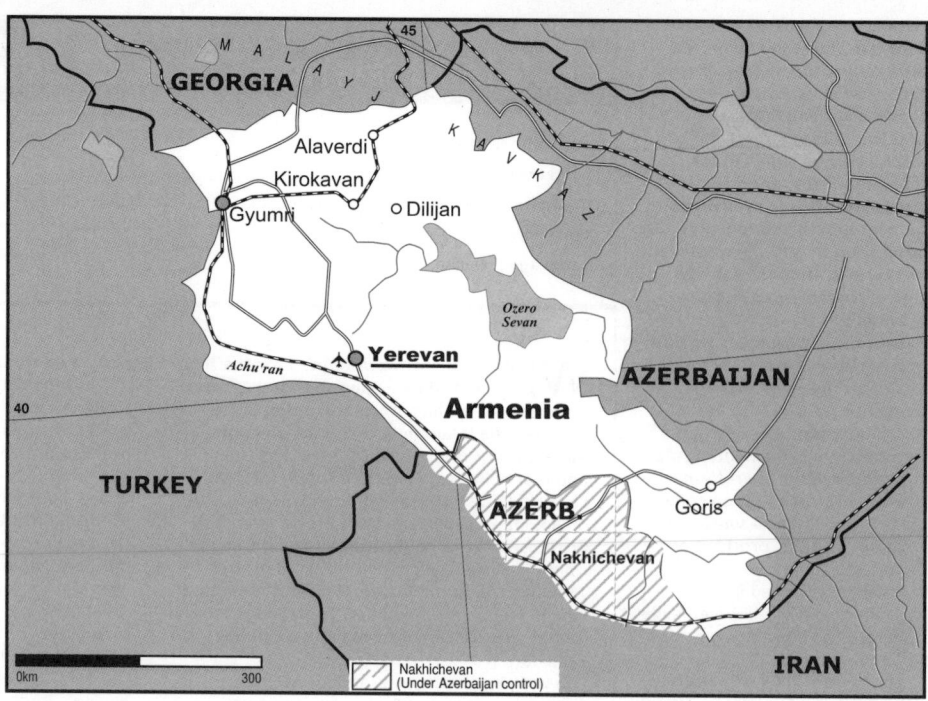

Nakhichevan
(Under Azerbaijan control)

The first historical reference to the country 'Armina' (Armenia) was made in the cuneiform writings from the era of King Darius I of Persia (6th-5th centuries BC). But the name Hayk, as the Armenians are called, comes from the name of the country, Hayasa, mentioned in the Hittite writings from the 12th century BC. The Urartians, direct ancestors of the Armenians, founded a powerful state in the 9th to 6th centuries BC; its capital was the city of Tushpa (today Van, in Turkey). In the year 782 BC, they founded the fortress of Erebuni, in the north of the country (today Yerevan, capital of Armenia).

2   With the collapse of Ur state, the ancient Kingdom of Armenia emerged in its territory. The first rulers were the *satraps* (viceroys) of the shahs of Persia. This period was recorded in the works of Xenophon and Herodotus. In *Anabasis*, Xenophon described how the Armenians turned back 10,000 Greek mercenaries between 401- 400 BC. His writings also describe Armenia's prosperous production, and its wealth in wheat, fruit and delicious wines.

3   After the expeditions of Alexander the Great and the rise of the Seleucid Empire, Armenia came under extensive Greek influence, which gave a boost to the country's cultural life. The Seleucid State fell into the hands of the Romans in 190 BC, and Armenia became independent. The local government named Artashes (Artaxias) King of Greater Armenia.

4   Armenia reached the pinnacle of its prosperity during the reign of Tigranes the Great (95-55 BC), an age known as the Golden Age. King Tigranes united all Armenian-speaking regions and annexed several neighboring areas. Its borders extended as far as the Mediterranean to the south, the Black Sea to the north and the Caspian Sea to the east.

Tigranes' empire soon fell into the hands of the Romans and the Parthians, and Armenia was proclaimed 'friend and ally of the Roman people', a euphemism for the vassals of Rome.

5   In 301 AD, Armenia became the first country in the world officially to adopt Christianity as a state religion. At that time, St Gregory the Illuminator, the first pontiff, founded the monastery at Echmiadzin, still extant as the headquarters for the patriarchs of the Armenian Church. The country lost its state integrity and disappeared in the year 428, when it was divided between the Roman Empire and the new Persian kingdom. Garni, a Greek temple dating from the 1st Century near Yerevan, is one of the monuments that reflect this period. The Church identified with Armenian national feeling, allowing people to remain united even after they lost their organization as a state.

6   In 405, the monk Mesrop Mashtots devised an alphabet that formed the basis of the Armenian writing system. The characters of this alphabet have remained unchanged, achieving a continuity which spans the centuries and links ancient, medieval and modern cultures. The 5th century was the golden age of religious and secular literature and of Armenian historiography; the natural sciences also developed during a later period. In the 7th century, Ananias Shirakatsi wrote that the world was round and formulated the hypothesis that there were several worlds

inhabited by beings endowed with some form of intelligence.

7   In the 5th and 6th centuries, Armenia was divided between Byzantium and Persia. The Persians tried to stamp out all traces of Christianity in the eastern Armenian regions triggering a massive rebellion. Prince Vartan Mamikonian, commander of the Armenian army, assumed the leadership of the rebellion. In the year 452, he led an army of 60,000 troops into battle against a vastly superior Persian force, in the Avaraev valley. The Armenians were defeated and Prince Vartan was killed, but the Persians also suffered heavy losses, and subsequently gave up their attempts to convert the Armenians to Islam. All those who lost their lives in this battle were later canonized by the Armenian Church.

8   In the 7th century, Arab forces invaded Persia, bringing about the collapse of the Persian Empire. The new Muslim leaders also established control over the Armenian regions. The people resisted, fighting for their independence until the late 9th century, when Prince Ashot Bagratuni was named King of Armenia, and established an independent government.

9   The prosperity of the Bagratids' reign was short-lived, for in the 11th century the Byzantines and Seleucids began bearing down on the Transcaucasian region from Central Asia. Many Armenian princes ceded their lands to the

Byzantine Emperor, in exchange for lands in Cilicia (in modern-day Turkey). The inhabitants of other Armenian regions began flocking to Cilicia, fleeing the Turkish raids.

10   In the 11th century, the Rubenid dynasty founded a new Armenian state in Cilicia, which lasted 300 years. Cilicia had close ties to the western European states; Armenian troops took part in the Crusades, and intermarriage with other ruling dynasties introduced the Rubenids to the circle of European rulers. In 1375, Armenian Cilicia fell to the Mamelukes of Egypt, who retained Cilician science, culture and literature. In the meantime, the region that had originally been Armenia was devastated by invasions and wars.

11   In the 13th century the Ottoman Turks replaced the Seleucids and began their conquest of Asia Minor. In 1453, they took Constantinople and marched eastwards, invading Persia. Armenia was the scene of numerous wars between Turkey and Persia, until the 17th century, when the country was divided between the two Islamic empires.

12   In 1722, Russian troops sent an expedition to Transcaucasia, occupying the city of Baku and other territories belonging to Persia. Armenian princes in Nagorno-Karabakh and other neighboring areas seized the opportunity to join forces with the Russians, and organized a revolution against the Persians. The uprising was led by the Armenian national hero David

Life expectancy
72 years
2005-2010

GNI per capita
$1,060
2004

Literacy
99% total adult rate
2000-2004

HIV prevalence rate
0.1% of population 15-49 years old
2003

Bek. However, Armenian hopes were dashed when the Russian Czar, Peter the Great, died. He had promised to support the Armenians, but on his death Russia signed a peace treaty with Persia. Another war between Russia and Persia, one hundred years later, ended in 1813 with the Treaty of Gulistan. According to the terms of this, Karabakh and other territories that had historically belonged to Armenia became part of the Russian Empire.

[13] Russia was at war with either Turkey or Persia during most of the 19th century; with each war, Russia annexed more and more Armenian territory. Finally, almost the entire eastern part - home to more than 2 million Armenians - was swallowed up. However, the major part of Armenia's historic lands - with a population of more than 4 million - belonged to Turkey.

Protected by Russia against wars and invasions, Eastern Armenia prospered, while within the Ottoman Empire the Armenians were the objects of abuse and persecution. There were frequent disturbances and riots, which were cruelly suppressed by the Turks in 1915. During World War I, citing the Armenians' pro-Russian sympathies, the 'Young Turk' Government perpetrated the genocide that killed almost two million Armenians. While the men were executed in the villages, the women and children were sent to the Syrian deserts, where they starved to death. Survivors of these atrocities sought refuge in Armenian expatriate communities.

[14] At the fall of the Russian Empire, Armenia's independence was proclaimed in Yerevan. Turkey attacked Armenia in 1918 and again in 1920. In spite of some resounding victories on the part of the Armenian troops, the young republic's economy suffered and it also lost a significant part of its territory. In late 1920, a coalition of communists and nationalists proclaimed the Soviet Republic of Armenia. The nationalists were eased out of power and in February 1921 the Communist Government was brought down. However, with the help of the Red Army - which came into Armenia from Azerbaijan - the Communists were back in power after three months of fighting.

[15] In 1922, Armenia, Georgia and Azerbaijan formed the Transcaucasian Soviet Federated Socialist Republic, which became a part of the USSR. In order to avoid friction between Christian Armenians and Muslim Azeris, the Soviet regime adopted the policy of separating nationalities into different political/administrative entities, which involved relocating large segments of the population. In 1923, the Nakhichevan (Nachicevan) Autonomous Soviet Socialist Republic was created as a dependency of Azerbaijan, from which the entire Armenian population had been removed. Azerbaijan was also given Upper Nagorno-Karabakh, a region which had historically been Armenian and which Azerbaijan had previously relinquished in 1920.

[16] In 1936, the Transcaucasian Federation was dissolved, and the republics joined the Soviet Union as separate constituent republics.

[17] In 1965, Armenians around the world commemorated the 1915 genocide for the first time. In the Armenian capital, demonstrators clamored for the return of their lands, referring to the region of Upper Karabakh. The first petition for the reunification of Nagorno-Karabakh and Armenia - signed by 2,500 inhabitants of the former - was submitted to Nikita Krushchev, president of the USSR, in May 1963. Since that time, there have been two diametrically opposed positions: Armenia, in favor of reunification, and Azerbaijan, against it. In 1968, fighting broke out between Armenians and Azeris in Stepanakert, the capital of Nagorno-Karabakh (see Azerbaijan).

[18] In 1988, encouraged by the political opening of the USSR (known as *glasnost*), Nagorno-Karabakh Armenians (80 per cent of the local population) began a campaign to join Armenia. Karabakh's Regional *Soviet* (Parliament) approved the resolution and in Armenia, the Karabakh petition for reunification was received enthusiastically. Moscow reacted violently and sent in troops to crush the demonstrations in Yerevan and Stepanakert.

[19] In the 1991 referendum, 99.3 per cent of the electorate voted in favor of secession from the USSR. The Armenian Soviet proclaimed independence and in October Levon Ter-Petrosian was elected President with 83 per cent of the vote.

In October 1991, Nagorno-Karabakh also declared independence after 99 per cent of the electorate approved separation. Azerbaijan responded with an economic and military blockade, sparking a war between the two republics. In December 1991 Armenia joined the Commonwealth of Independent States (CIS) and in 1992 was admitted to the UN.

[20] In 1993, pro-Armenian forces achieved important victories in Nagorno-Karabakh, but Yerevan withdrew its unconditional support - at least officially. In 1994 - when, according to Azerbaijan, the Armenian forces had taken over 12,000 square km of disputed territory - Russian pressure made a ceasefire possible. Some 20,000 people had been killed in the war and one million displaced from their homes.

[21] After the disintegration of the USSR, Soviet investment in and support for industry disappeared. The closure of the borders with Azerbaijan and Turkey devastated the economy, which was dependent on oil and raw materials from abroad.

[22] In 1995, the President was granted more powers and the Government declared the liberalization of prices and a series of privatizations. Ter-Petrosian started a second term as President after winning the September 1996 elections, although fraud was suspected. The social situation - 20 per cent unemployment and public protests against the policies on Nagorno-Karabakh - forced him to resign in 1998. In the elections for the remaining term, Robert Kocharian, a native of Nagorno-Karabakh, defeated Karen Demirchian, leader of the Communist Party during the Soviet era.

[23] Demirchian, now in the Miasnutiun (Unity) Alliance, was avenged in the June 1999 parliamentary elections, when he was elected president of the legislature. But in October an armed group with no ties to political organizations entered Parliament and killed Prime Minister Vazgen Sarkisian, Demirchian and six legislators. As a result of negotiations with Kocharian, the attackers stepped down in exchange for guarantees of their personal safety and Aram Sarkisian (brother of their dead leader) was appointed Prime Minister. In 2000, the opposition accused the President of obstructing the investigations into the attack on Parliament. To control the crisis, Kocharian removed Aram from office and appointed Andranik Markarian in his place.

[24] In 2001, Armenia joined the Council of Europe as a full member. The European Parliament

PROFILE

ENVIRONMENT

Armenia is a mountainous country, bounded to the north by Georgia, in the east by Azerbaijan and in the south by Turkey and Iran. With an average altitude of 1,800 meters, high Caucasian peaks - like Mount Aragats (4,095 meters) - alternate with volcanic plateaus and deep river valleys. The most important river is the Aras, a tributary of the Kura, which forms a natural boundary with Turkey and Iran. The climate is dry and continental; the summers are long and hot and the winters extremely cold. On the plains, wheat, cotton, tobacco and sugar beet are grown. There are also vineyards, from which good quality wine is made. Cattle-raising is generally limited to the mountains. There are important copper, aluminum and molybdenum deposits.

SOCIETY

Peoples: 93.3 per cent Armenian; 2.6 per cent Azeri; 2.3 per cent Russian; 1.7 per cent Kurd.
Religions: Most belong to the Armenian Church (a branch of the Christian Orthodox Church).
Languages: Armenian (official), Russian, Azerbaijani and Kurdish.
Main Political Parties: Republican Party of Armenia (*Hayastani Hanrapetakan Kusaktsutyun*), Justice (*Ardartyun*), Rule of Law (*Orinants Erkir*), Armenian Revolutionary Federation (*Hai Heghapokhakan Dashnaktustyune*), National Unity (*Azgajin Miabanutiun*).

THE STATE

Official Name: Hayastani Hanrapetut'yun (Republic of Armenia). **Administrative divisions:** 10 provinces.
Capital: Yerevan (Erevan) 1,079,000 people (2003).
Other cities: Gyumri 130,400 people; Alaverdi 30,800; Dilijan 27,800; Goris 27,900 (2000).
Government: Robert Kocharian, President since February 1998; Andranik Markaryan, Prime Minister since May 2000. Single chamber National Assembly with 131 members.
National holiday: 28 May, Independence (1918).
Armed Forces: 57,400 (1996).

Under-5 mortality	Poverty	Debt service	Maternal mortality
32 per 1,000 live births 2004	**2%** of population living on less than $1 per day 2003	**8%** exports of goods and services 2004	**55** per 100,000 live births 2000

unanimously upheld a 1987 resolution establishing that Turkey could only enter the EU as a full member once it publicly recognized the 1915 Armenian genocide.

[25] Pope John Paul II visited the country for the first time in 2001 and resumed contact with the Armenian Apostolic Church, which had broken off relations with the Vatican in the 6th Century.

[26] That same year, Russian President Vladimir Putin made the first visit by a Russian leader to independent Armenia. Moscow and Yerevan signed a treaty on economic co-operation, which also authorized the Russian army to defend Armenia's Turkish and Iranian borders. This treaty was the basis for the creation of the Collective Security Committee, consisting of Armenia, Russia, Belarus, Kazakhstan, Kyrgyzstan and Tajikistan, which was strengthened a year later in Yerevan to include military cooperation against Islamic extremism.

[27] Robert Kocharian won the 2003 elections with almost 50 per cent of the vote. His main opponent, Stepan Demirchian (son of Karen Demirchian, who had been murdered), member of the Popular Party and born in Yerevan, obtained almost 30 per cent of the vote. The elections were held even though fraud was suspected,

and while some partisans of Demirchian were being placed under arrest, both in the capital city and in the provinces.

[28] The Constitutional Court of Armenia determined that the arrests broke the European Convention on Human Rights, while the Council of Europe repeated its call to review the Code of Administrative Crimes, which dated back to the Soviet era.

[29] Up to Armenia's independence, the economy had been based on the chemical, machinery, electronic goods, processed foods, and synthetic rubber industries, and depended mostly on foreign resources. Soviet industrial investments and aid disappeared after the fall of the USSR. The closing of the borders with Azerbaijan and Turkey devastated the economy, which was dependent on oil and commodities from abroad. The country's little potential in coal, gas and oil has not been developed.

[30] Since 1995, the traditional productive sectors changed toward the processing of gemstones, jewellery, communications technology and tourism, achieving a strong growth of the economy. This progress opened the way to loans from the IMF and the World Bank, as well as from foreign

countries. To continue to grow in the 21st century, Armenia will have to reduce its budget deficit, stabilize the currency, favor the development of agriculture, food processing, and transportation, and strengthen its health and education sectors.

[31] Parliament abolished the death penalty in September 2003 by 92 votes to 1. President Kocharian commuted the death sentences of 42 prisoners to life imprisonment.

[32] In December 2003, six people were sentenced to life imprisonment for their participation in the 1999 Parliament shootings in which the prime minister and six others were killed.

[33] At the end of May 2004, the Foreign Affairs Ministers of Armenia and Azerbaijan met in Stirin (Czech Republic) to try and find a peaceful solution to the long-running Nagorno-Karabakh conflict. The meeting had been called by the Russian, French and US members of the Minsk Group of the OECD.

[34] In June 2004, the EU froze more than $100 million in aid because Armenia refused to set a date to close an old Russian nuclear power plant. The Metsamor plant, 40 km west of Yerevan - closed down in 1988 and re-opened in 1995 - was built

in one of the world's most active seismic zones.

[35] In April 2005, hundreds of thousands of Armenians took to the streets of Yerevan on the 90th anniversary of the 1915 genocide, to pay their respects to the victims of the Ottoman Empire and to demand international recognition for what most Armenians consider the first genocide of the 20th century. President Kocharian was leading an effort for such recognition, but Turkey's view was that the dead were merely casualties of war. France, Russia, Poland and Germany were among the 15 nations that put pressure on Ankara when Turkey was about to enter the European Union.

[36] In January 2006, the gas supply to the country was severely disrupted after explosions in Russia damaged the pipeline to Armenia, via Georgia. By April, the price of Russian gas had doubled.

[37] In April 2006, the Russian energy company Gazprom announced that it was to take control of Armenian pipelines and a power station in exchange for setting gas prices at half European levels until 2009. The move provoked protest in Armenia, not least because local gas prices were then a quarter of the European level and the deal thus involved an immediate doubling in price. ■

IN FOCUS

ENVIRONMENTAL CHALLENGES
The energy crisis in 1990 caused massive deforestation, since firewood became an alternative source of energy. The pollution of the Hrazdan and Aras rivers, and the desiccation of Lake Sevan, which were also a consequence of the extensive use of water as a source of energy, compromised water supplies. Another serious challenge Armenia faces is the reopening of the Metsamor nuclear power plant, which is located in an area of seismic activity. Part of the soil is contaminated with highly toxic chemicals, such as DDT.

WOMEN'S RIGHTS
Armenian women have been able to vote and stand for election since 1921.

In 2003, in spite of a slight increase over the previous period, the number of seats held by women was just 5.3 per cent; they had no representation in ministerial or equivalent posts.

In 2003 women made up 49 per cent of the labor force. In 2003, female unemployment was 13.1 per cent, twice the

rate for men.

Almost 8 per cent of women receive no prenatal care, but 97 per cent of births are attended by trained health staff*.

In 2003, among women over 15, illiteracy was 0.8 per cent*. The same year, enrollment in higher education was 31 per cent for women and 24 per cent for men*.

CHILDREN
Within the period 1990-2004, under-1 and under-5 mortality rates were nearly halved. The former figure fell from 52 to 29 per 1,000 live births and the latter from 60 to 32 deaths per 1,000 live births. In 2004, 13 per cent of children under five years old had moderate or severely stunted growth.

In 2002, more than 55 per cent of the population lived in poverty and 8.5 per cent lived in extreme poverty. Families with children under 5 years old comprised 60 per cent of poor people. Low income and high unemployment have made it extremely difficult to support households and children are

being increasingly abandoned by their parents.

In 2000, UNICEF ranked Armenia among the seven countries with the highest levels of child sexual exploitation and trafficking. Malaria reappeared after 30 years, and 2,000 cases were reported in 1999.

INDIGENOUS PEOPLES/ ETHNIC MINORITIES
In Armenia ethnic minorities only represent around 3 per cent of the population. In 2003, there were 20 ethnic minority groups, among them 45,000 Yezidis, 8,000 Assyrians, 6,000 Greeks, 4,000 Ukrainians, more than 1,000 Kurds, some Georgians, Germans and Polish, besides the Armenian gypsies and Armenian 'Tats'. The Armenian subgroups are distinguished by their religious beliefs, their customs, and especially by their language, in the case of the gypsies a sort of Armenian slang that is on its way to extinction and has been corrupted and, in the case of the Tats, Farsi, which nowadays is only spoken by people over 50 years old.

MIGRANTS/REFUGEES
In 2003 and 2004, the efforts of the Armenian Government were directed at finding a lasting solution for integrating refugees in the country into the economy.

At the end of 2002, around 256,000 ethnic Armenians from Azerbaijan lived 'as refugees', although they were socially integrated.

In the early 1990s there were around 50,000 internally displaced Armenians, especially farmers uprooted from near the Azerbaijan border.

In the period 2003-2005, nearly 20,000 Armenians were exiled into different industrialized countries.

DEATH PENALTY
Parliament abolished the death penalty in September 2003 and ratified Protocol 6 of the European Convention of Human Rights.

* Latest data available in *The State of the World's Children* and *Childinfo* database, UNICEF, 2006.

Aruba / Aruba

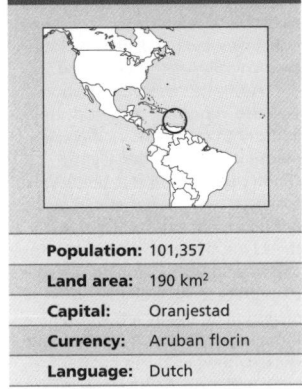

Population:	101,357
Land area:	190 km²
Capital:	Oranjestad
Currency:	Aruban florin
Language:	Dutch

The Caiquetios, an Arawak nation originating in the Orinoco River basin, settled on the island now known as Aruba some 2,000 years ago. Their culture and livelihood, based on intensive cultivation of manioc/cassava, sweet potato and maize, as well as hunting and gathering, underwent a dramatic change in 1499 when the island caught the attention of Spaniard Alonso de Ojeda. Within two decades, most of the population had been decimated by the diseases and slavery introduced by the Europeans.

[2] Since the Spanish Crown was not interested in the island then called 'La Española', in the 16th century some convicts were sent from South America to inhabit the territory.

[3] In 1633, the Dutch, who were engaged in the 80 Years War with Spain, took the island. By the Treaty of Westphalia (1648), the Netherlands ruled over Aruba, Curaçao and Bonaire.

[4] Throughout the 18th century, the colonial authorities used Aruba as a feed-lot for their horses. At the beginning of the 19th century, land began to be sold to the colonists.

[5] In 1800, with the burial of the last of the Caiquetios the culture and language of the first islanders disappeared.

[6] The Netherlands retained control over the island except for a period of UK domination (1805-1815). Aruba required little labor to support its horse and cattle-raising industry. For this reason there was little need for slaves, and only 12 per cent of the population was of African origin by the time slavery was abolished (mid-19th century).

[7] Gold was found in 1825. However, by 1913 the gold mines were so depleted that mining was phased out.

[8] The discovery of petroleum and the installation of large oil refineries on Aruba at the beginning of the 20th century generated a wave of migrant skilled workers, particularly from the US. Aruba's economic growth created problems with Curaçao, which was by then the colonial capital city.

[9] In 1954, the Dutch Government gave Aruba and the other five islands of the colony (Bonaire, Curaçao, St Maarten, St Eustacio and Saba) their autonomy. This gave rise to the Netherlands Antilles Confederation. The political parties in power were the Aruba People's Party (AVP) and the Aruban Nationalist Union (UNA).

[10] Juancho Irausquin, a former AVP member, founded the Aruban Patriotic Party (PPA) in 1971 and was its leader through the following decades. Gilberto 'Betico' Cores was the opposition leader and founder of the People's Electoral Movement (MEP). This supported each island's right to its own Constitution and autonomy within the Dutch commonwealth.

[11] In 1979 the Antiyas Nobo Movement (MAN), won a significant electoral victory in Curaçao. With the MEP and the Bonaire Patriotic Union (UPB), it established the first left-of-center coalition Government. But while MAN favored a federation with broad autonomy for each island, the MEP insisted on the secession of Aruba. These differences led to the disintegration of the governmental alliance in 1981.

[12] The MEP demanded separation from the Curaçao administration. In 1985, the Dutch Government granted Aruba separate status, and on 1 January 1986, Aruba became a separate entity within the Dutch commonwealth, then made up of The Netherlands, Aruba and the Netherlands Antilles.

[13] Aruba's economy continued to be underwritten by Holland. The bulk (98.9 per cent) of its economic resources came from refining Venezuelan oil. In 1985, the Exxon refinery withdrew from the country. Since then, the Government has turned to tourism, which accounted for most of the island's GDP.

[14] Henry Eman, of the AVP, was elected Prime Minister in 1994 and re-elected in 1997. That year, the Government of Aruba, along with those of the Netherlands and Netherlands Antilles, decided to postpone indefinitely transition to complete independence.

[15] The People's Electoral Movement won the 28 September 2001 elections, and for the first time since 1980 a single party had overall control of Parliament. The MEP obtained 12 of the 21 seats and its leader Nelson Oduber was named Prime Minister.

[16] At the beginning of 2002, the OECD announced that Aruba, among other countries, was no longer on their 'black list' of tax havens - ie places where 'harmful tax practices' such as tax evasion are permitted.

[17] In 2003, the Government of Aruba decided that it would accept euros, the main European currency, as a means of encouraging more visitors from that region.

[18] In an effort to encourage tourism, the Government has used the symbol of the *Divi Divi* tree as an emblem of the island in postcards, cards and ads.

[19] In April 2005, the Ministry of Tourism and Transportation announced investments of around $274 million in all branches of the tourism industry.

[20] The Central Bank of Aruba reported in September 2006 that tourist numbers were down by 10 per cent over the first seven months of the year. ∎

PROFILE

ENVIRONMENT
Located off the coast of Venezuela, the island of Aruba was until January 1986 one of the 'ABC Islands', otherwise known as the Netherlands Antilles, together with Curaçao and Bonaire. The climate is tropical, moderated by ocean currents. Tourism is the main economic activity.

SOCIETY
Peoples: Predominantly of European and Carib origin, its inhabitants intermingled with Latin American and North American immigrants. **Religions:** Mainly Catholic (82 per cent). There is also a Protestant (8 per cent) minority and small Jewish, Muslim and Hindu communities.
Languages: Dutch (official). The most widely spoken language, as on Curaçao and Bonaire, is Papiamento, a local dialect based on Spanish with elements of Dutch, Portuguese (spoken by the Jewish community), English and some African languages.

Main Political Parties: People's Electoral Movement (MEP), Aruban People's Party, Aruban Patriotic Movement, Network (*RED*), Real Democracy.
Main Social Organizations: The Aruba Workers' Federation.

THE STATE
Official Name: Aruba.
Capital: Oranjestad 29,000 people (2003).
Other cities: St Nicolaas 17,400 people (2000).
Government: Fredis Refunjol, Governor appointed by Holland, since May 2004.
Nelson Oduber, Prime Minister and Minister of General Affairs since 2001. Holland continues to be in charge of defense and foreign relations. Single-chamber legislature; Parliament made up of 21 members elected for a 4-year period.
National holiday: 18 March, Flag Day (1976).

Australia / Australia

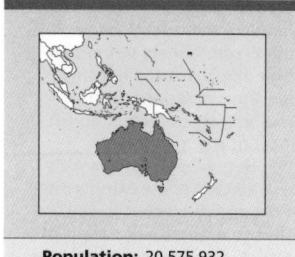

Population:	20,575,932
Land area:	7,741,220 km²
Capital:	Canberra
Currency:	Australian dollar
Language:	English

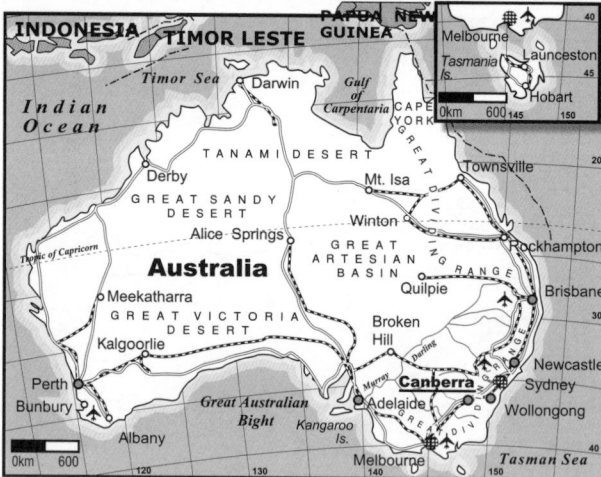

The first inhabitants of Australia are thought to have come from Southeast Asia between 60,000 and 120,000 years ago, but whether they were an homogeneous ethnic group or not is unknown.

[2] These peoples, named 'Aborigines' by the Europeans, spoke over 260 different languages that embodied distinctive cultures. However, they had common traits: they were all semi-nomadic hunters and gatherers. The clan, the basic economic and social unit, administered the use of the land and resolved ecological and social differences. Their principal social unit was the family, which was self-sufficient for food and supplies within the clan.

[3] In the 12th century AD Portuguese cartographers and navigators debated the existence of an unknown *terra australis*. In the 17th century, Dutch navigators sighted the northwest and southeast parts of the Australian coastline but took no territorial interest. It was not until the 18th century that the voyages of the British navigator James Cook began the colonization of Australia.

[4] In 1788, Britain established a penal settlement on the east coast of Australia at Botany Bay, to which it first sent 736 convicted criminals: it had recently lost its American colonies, which had previously absorbed some of its convict population. The new arrivals divested the natives of their richest cultivating and grazing lands, even though international and British legislation recognized the Aborigines' right to their territory until such time as it was abolished by mutual agreement. British officers, then free settlers, farmed the land and from 1797 built successful industries (such as sheep farming and textile trades) that were to become the backbone of the Australian economy for more than a century. To do this, they utilized slave-like labor, since they used the convicts to build bridges and roads; clear and farm the land; and act as servants to an emerging landed gentry.

[5] From 1787 to 1857, Britain transported 160,000 of its convicts (petty thieves, deserters from the Royal Navy and members of Irish opposition groups) to Australia. In the various colonies across the country, penal administration and local government remained in the hands of high-ranking military officers, appointed by the Crown until well into the 19th century. Their functions were to oversee the prison population and take charge of defense against possible attacks by other European powers.

[6] The battle to gain control of the land claimed the lives of 80 per cent of the Aborigines. As resistance was broken, Aborigines were progressively relegated to Australia's more inhospitable areas. If genocide was not intended, in some parts of the country this was the effect. By 1871, records reveal only 879 Aborigines in Victoria, while in Tasmania less than 10 were on record as having survived. Exposure to European diseases claimed many. Entire families were poisoned, forcibly removed from their lands or confined to British reservations. Many of those that survived, completely defeated, were forced to sign with a mark 'work contracts' written in English - a language they did not understand - committing them to work in slave-like conditions as household servants or farmhands

subject to severe disciplinary measures. In the process, Aborigines' rights over land and, little by little, their culture were undermined.

[7] Australia's strategic location was also important for British trade, in order to control its worldwide maritime network. Initially the country also served as a safety valve for social tensions generated by Britain's rapid industrialization. By the 1830s, the poor classes in Britain and Ireland were being encouraged to migrate to Australia by both their government and by letters from those who had already settled and were by then enjoying plentiful work and meat on their tables.

[8] The development of the livestock industry and the subsequent discovery of gold and other precious metals boosted the economy between 1830 and 1860. Prospectors in search of gold from Britain, America, Poland, Germany, Italy and China rushed into the colonies of New South Wales and Victoria during the early 1850s. In November and December 1854, gold miners at Ballarat, Victoria, revolted against exorbitant license fees and those that imposed them. Under the Southern Cross flag, 10,000 of them met at Bakery Hill and demanded the right to vote in elections. Thirty of them later died defending the Eureka Stockade, which the miners had built to resist the aggression of the colony's soldiers. The Southern Cross flag remains a popular symbol of

PROFILE

ENVIRONMENT
Australia occupies the continental part of Oceania and the island of Tasmania, and has a predominantly flat terrain. The Great Dividing Range runs along the eastern coast. Inland lies the Central Basin, a desert plateau surrounded by plains and savannas. The desert region runs west to the huge Western Plateau. Rainfall is greatest in the north where the climate is temperate, with dense rainforests. Some 75 per cent of the population is concentrated around the southeastern coastline, which has a temperate climate and year-round rainfall. Oats, rye, sugar cane and wheat are grown there; Australia is one of the world's largest producers of the latter. Australia has the largest number of sheep of any country in the world and is also the world's largest exporter of wool. It also exports meat and dairy products and is one of the world's largest producers of minerals: iron ore, bauxite, coal, lead, zinc, copper, nickel and uranium. It is self-sufficient in oil and has a large industrial zone concentrated in the southeast.

SOCIETY
Peoples: When the British 'discovered' Australia in 1788, there were 250,000 Aborigines in approximately 500 different tribes. In 1901, only 66,000 of their descendants were still alive. Today, there are 460,000, representing just 2.2 per cent of the national population. Descendants of British immigrants make up two-thirds of the population.

The rest are immigrants from Asia, Europe and Latin America.
Religions: Christians 74 per cent (Catholic 27 per cent, Anglican 24 per cent, Methodist 8 per cent). Buddhist, Muslim, Confucian and other, 13 per cent.
Languages: English.
Main Political Parties: Liberal Party, a right-of-center party, National Party (NP), which represents the interests of farmers, Australian Labor Party (ALP) founded in 1901, Australian Greens (progressive ecologist).
Main Social Organizations: The Australian Council of Trade Unions (ACTU) is the largest labor confederation, with 133 union affiliates.

THE STATE
Official Name: Commonwealth of Australia.
Administrative divisions: 6 states and 2 territories.
Capital: Canberra 373,000 people (2003).
Other cities: Sydney 3,985,800 people; Melbourne 3,317,300; Brisbane 1,535,300; Perth 1,365,600; Adelaide 1,115,900 (2000).
Government: Parliamentary monarchy. Michael Jeffery, Governor General, appointed by the British Queen in August 2003. John Howard, Prime Minister since March 1996, re-elected in 1998, 2001 and 2004. Bicameral Legislature: the House of Representatives, with 150 members, and the Senate, with 76 members.
National holiday: 26 January, Australia Day.
Armed Forces: 59,000 (7,500 women included).
Dependencies: Cocos Islands, Coral Sea Islands, Christmas Island, Norfolk Island.

 **Life expectancy**
81 years
2005-2010

 GNI per capita
$27,070
2004

HIV prevalence rate
0.1% of population
15-49 years old
2003

resistance to authoritarian rule to this day.

9 Large tracts of land leased by British administrators meant that by the end of the Gold Rush all the good land in the colonies was held by wealthy farmers ('squatters'). Following broad support for 'unlocking the land', the colonies passed laws to enable the squatters' land to be subdivided and made available to 'selectors' - small farmers who would settle in the land. Although the success of these laws was limited, the stranglehold of the squatters was broken in parts of South Australia, Queensland and Victoria.

10 In the meantime, a labor movement began to appear in the cities, and it soon had a significant following. In the 19th century, the Australian unions gained important victories and concessions which Europe's working classes were still a long way from obtaining (such as the eight-hour day obtained by stonemasons in New South Wales and Victoria in 1856). Urbanization went hand-in-hand with industrial development. Sydney and Melbourne turned into large urban centers. A demand for Australia's products on the world market and the low cost of land encouraged massive waves of immigrants, mainly British. A middle class developed alongside a wealthy industrial bourgeoisie, utterly transforming Australian life. Liberal governments dominated the country's political scene between 1860 and 1890.

11 With major strikes by miners, wharfies and shearers during the 1890s, a political power base emerged to unify and represent the working classes. In 1891 labor candidates contested 45 seats in the New South Wales Legislative Assembly and won 36 to hold the balance of power. By 1908 an Australian Labor Party (ALP) politician had become prime minister.

12 Confederation of the separate Australian colonies arrived after a constitution, drafted in 1897-1898, was approved by the British parliament. In 1901 the six colonies (New South Wales, Victoria, South Australia, Western Australia, Queensland and Tasmania) were federated in the 'Commonwealth of Australia', governed by a federal parliament. The Commonwealth took over the administration of the Northern Territory and the federal capital in 1911. One of the first laws passed by the new nation's parliament authorized a dictation test to screen out 'undesirable' immigrants of particular nationalities from entering Australia, effectively implementing the 'White Australia' policy. This legislation was not repealed until 1958.

IN FOCUS

ENVIRONMENTAL CHALLENGES
Unique species, animal and vegetable, are in danger of extinction due to the destruction of their habitat. The erosion of the land due to extensive grazing, industrial development, unplanned urbanization and inadequate farming are of great concern; there is also an increase in salinity. The largest coral reef in the world, near the northeast coast, is threatened by shipping and by its increasing popularity as a tourist site. As the country contains some of the driest places on Earth, natural water resources for human use are limited.

WOMEN'S RIGHTS
White Australian women have been able to vote and run for office since 1902. However, the right to political participation was not extended to indigenous women until 1967, and even then, their registration on the electoral roll was not compulsory. In 2004 women held 24.7 per cent of parliamentary seats; in ministerial or equivalent posts, the number of women increased from 14 to 20 per cent from 2000 to 2003.

In 2003 women made up 44 per cent of the labor force. Of these, 3 per cent worked in agriculture, 10 per cent in industry and 87 per cent in services. Women belonging to ethnic minorities face discrimination due to gender and to race. In 2003 strong measures were taken against trafficking of women for prostitution, and arrests were made in accordance with the 1999 sexual slavery legislation.

CHILDREN
Australian children enjoy maximum access to all basic elements which will ensure their wellbeing (education, nutrition, health, clothing, shelter, social participation). In theory this also covers the indigenous population. But the life expectancy of Australian Aborigines is 20 years lower than that of Australians of European origin (see Indigenous Peoples).

INDIGENOUS PEOPLES/ETHNIC MINORITIES
In 2002, around 458,500 Aborigines lived in Australia (nearly 2.4 per cent of its total population). Even though discrimination is considered illegal in the country since the 1975 Law was passed, and despite the government's investments in programes to improve Aborigines' economic and social status, they are still victims of discrimination throughout the country. The Australia Reconciliation organization was created in 2001 seeking, with Government support, to integrate Aboriginal peoples into the rest of society. In spite of that, Aborigines comprise over 20 per cent of the population in adult prisons and over 40 per cent of the youths arrested since 1997: Aboriginal youths are 21 times more likely to be arrested than white youths. The suicide rate among Aborigines is 6 times higher than the average for the whole population. Since Federation, only two Aboriginal people (both men) have sat as members of the Federal Parliament.

MIGRANTS/REFUGEES
Towards the end of 2002, Australia was hosting around 25,000 refugees and people of diverse nationalities seeking asylum, among them 4,000 Kosovo-Albanians and 1,800 Timorese. The majority of Kosovars had returned home by 1999, and some of the Timorese temporary refugees were repatriated in 1999 and 2000 - in some cases under pressure from government action, such as the interruption of basic services.

Since September 2001, the country has been in the international spotlight for its zero-tolerance policy on the unauthorized arrival of boats with people seeking asylum. Through the so-called 'Pacific Solution', which remained in effect during 2002, the Government denied the boats access and, in most cases, sent the refugees to other countries in the Pacific, such as Nauru and Papua New Guinea. Towards the end of 2002, over 500 Afghans and Iraqis remained in those two places outside Australian territory. Some 300 Afghans obtained economic assistance to return to their own country.

DEATH PENALTY
The death penalty was abolished in 1985, the last execution having taken place in 1967.

13 A prolonged period of economic prosperity financed a series of social reforms, impelling the country towards an open society. South Australia granted women the right to vote in 1893; in 1902, the Commonwealth of Australia became one of the first countries in the world to grant women the right to vote when it gave the vote to all British subjects of six months' residence and over 21 years of age - except for Aborigines, Asians and Africans.

14 World War II, in which 30,000 Australians died and 65,000 were wounded, loosened the ties between Britain and Australia, and the US guaranteed security in the region.

15 The Korean War (1950-1953) triggered a sharp increase in the price of wool, and consolidated Australia's economy, helping to diminish the gap between well-being in urban and in rural areas.

16 During the Cold War, the ANZUS military assistance treaty was signed in 1951 by Australia, New Zealand/Aotearoa and the US. The aim of the treaty was to guarantee the security of US and allied interests in the region. This military alliance also committed the Australians to fight in the Vietnam War (as well as in other conflicts), which damaged the treaty's image internationally, and triggered an important anti-war movement.

17 Japan became the main market for Australian minerals, financing coal exploration in Australia in the 1960s. The large coal deposits discovered have supplied local industry ever since, while also meeting Japan's huge industrial power needs.

18 A plebiscite held in 1967 granted Australian Aborigines full citizenship rights (including the right to vote) and placed the Aboriginal issue under the jurisdiction of the Federal Government.

19 In 1972, a group of Aborigines set up a tent embassy outside the Federal Parliament, swearing to stay until land rights were achieved. In the following five years both Liberal and Labor governments began to put in place policies and laws for the return of some indigenous lands. But in 1983, the commitment to pass a law in defense of the Aborigines' territorial claims was shelved, when the enterprises developing mineral resources (gold, uranium, bauxite and iron) argued, against public opinion, that the Aborigines' defense of their territorial claims could compromise the country's economic growth.

20 Aborigines in Australia by now represented 2.4 per cent of the country's total population, many retaining their original languages.

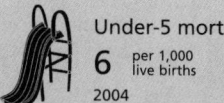

Under-5 mortality
6 per 1,000
live births
2004

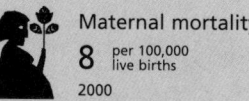

Maternal mortality
8 per 100,000
live births
2000

Two thirds of them no longer live in tribal groups. A substantial minority continues to live in areas which Europeans consider to be inhospitable, like the central desert, where they have managed to keep their own religious and social traditions alive.

²¹ After slowly dismantling the White Australia policy, leaders on both sides of politics in the 1970s proclaimed Australia to be multicultural. Closer economic and political ties with Asia were formed. In 1989 the Asian Pacific Economic Co-operation Organization (APEC) was formed, led by Australia. The project promoted the formation of a common market in the region, and hoped to become the spokesperson for food-exporting countries, promoting the 'Cairns Group' in the Uruguay Round of the GATT.

²² In 1991, unemployment figures reached more than a million, putting constant pressure on Bob Hawke's Labor Government. The economic liberalization driven by Hawke in the 1980s continued under the government of his successor, Paul Keating. The following year, unemployment reached 11.1 per cent. The Government approved legislation aimed at increasing employment and reducing immigration by 29 per cent.

²³ In the 1992 Mabo judgment, the Australian High Court recognized for the first time that those indigenous people who could demonstrate that they and their ancestors had occupied and worked Crown land, possessed native title and could claim the land. In the

Cocos Islands

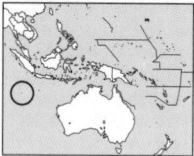

Population: 630
Land area: 14 km²
Capital: West Island
Currency: Australian dollar
Language: English

The Cocos were discovered in 1609 by captain William Keeling of the East Indies Company, but it was not until 1826 that a settlement was established by Alexander Hare from Britain.
² The Clunies-Ross company, founded by Scottish sailor John Clunies-Ross in the early 1800s, established itself on the islands and became their real owner, in spite of their formal status as a British colony. Queen Victoria ceded the islands to the Clunies-Ross family in 1886, in exchange for the right to use the land for public purposes. Ross brought in Malayan workers - in conditions that were close to slavery - to work the coconut groves.
³ In 1978, after many years of negotiations, Australia bought the islands from the Clunies-Ross company. However, the company kept a monopoly on the production and commercialization of copra. Australia, on the other hand, established a military base on West Island, purchased in 1951.
⁴ In a referendum held in 1984, the population voted in favor of having Australian nationality and for a full annexation of the islands to Australia. In December of the same year, the UN General Assembly validated the results of the referendum and Australia was freed from the obligation of reporting to the Decolonization Committee.
⁵ Australia purchased the remaining Clunies-Ross properties on the islands in 1993.
⁶ In June 2000 an earthquake measuring 7.5 on the Richter scale hit the Indian Ocean, near the Cocos Islands. Strong tremors were felt on the islands but there was no structural damage.
⁷ In December 2001, the population of West Island doubled as a result of a wave of illegal immigrants - mostly from Sri Lanka. Tourist operators warned that this could end tourism on the islands.
⁸ The diplomat Evan Williams was appointed by the Australian Government as administrator of Cocos in October 2003. In January 2006, Williams was replaced by Neil Lucas, who was also administrator of Christmas Island. ■

ENVIRONMENT
A group of coral atolls in the Indian Ocean, southwest of Java, Indonesia. Of the 27 islets, only two are inhabited, West and Home. The climate is tropical and rainy, and the land is flat and covered with the coconut groves that give the islands their name. Direction Island has a military base.

SOCIETY
Peoples: The inhabitants of Home Island are descended from Malayan workers. Australians and Sri Lankans predominate on West Island. **Religions:** Sunni Muslims (57 per cent). Christians (22 per cent). Other: 21 per cent. **Languages:** English and Malay.

THE STATE
Capital: West Island 250 people (1999).
Government: Neil Lucas, administrator appointed by the Australian Government since January 2006. There is an unicameral Cocos Shire Council with 7 seats. The territory has a five-person police force.

Christmas Island

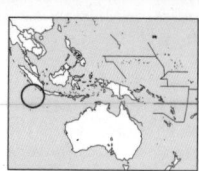

Population: 433
Land area: 135 km²
Capital: The Settlement
Currency: Australian dollar
Language: English

The island, formerly a dependency of the British colony of Singapore, was transferred to Australia on 1 October 1958 and its residents were accepted as Australians as of 1981. In 1984, social benefits and political rights were granted, and in 1985 income tax was imposed.
² Phosphate mining, the core economic activity and sole source of employment, was shut down in 1987 and reopened by private companies in 1990 under strict environmental conservation measures. Two-thirds of the island was declared a national park, and since 1991 the Government has attempted to promote tourism as an economic alternative. In 1993 a private casino was opened with state support and an investment of $34 million. Also under consideration is the possibility of building a space rocket launch site.
³ In August 2001, some 460 illegal migrants, mostly Afghans, though also Sinhalese and Thais, were rescued by a Norwegian freighter when their boat foundered. The ship tried to place the migrants on Christmas Island, but Australia refused them entry. A new Australian law excluded Christmas Island and other Australian dependencies from its migration zone. At the same time Canberra asked other Pacific islands - like Nauru, Palau, Fiji and Tuvalu - to install temporary camps for migrants seeking asylum, who were rejected by Australia.
⁴ In January 2003, an Iraqi girl, who had been detained on the island, was taken to a hospital in west Australia and died. The extremely high levels of cadmium registered on the island, which prevent vegetation from thriving, turn it into an unsuitable place for the hosting of refugees.
⁵ The Indonesian Government criticized the Australian decision to grant 42 protection visas to Papuan separatists, who were hosted on Christmas Island. ■

ENVIRONMENT
An island in the Indian Ocean, 2,500 km northeast of Perth, Australia, and 380 km south of Java, Indonesia. Mountainous and arid, it has a dry climate.

SOCIETY
Peoples: Nearly two-thirds of the population are of Chinese origin. There are some Malays and a minority of Australians. There are no native inhabitants. **Religions:** Christianity (Protestants), Confucianism and Taoism. **Languages:** English, Malay, Mandarin and Cantonese.

THE STATE
Capital: The Settlement. **Government:** Neil Lucas, Administrator appointed by the Australian Government, since January 2006. The assembly of 9 members is renewed every two years. There are no political parties.

Water source
100% of population using improved drinking water sources
2002

Doctors
249 per 100,000 people
1990-2004

Primary school
95% net enrolment rate
2004

1993 general elections, Keating won by a narrow margin. One of Keating's campaign promises was to give the Aboriginal issue a high priority. However, his government presented a package of measures that limited the effect of the Mabo judgment by protecting mineral exploitation and livestock herding on land that could be claimed by the Aborigines. Keating's package divided Aboriginal leaders, with many opposing the measures and accusing Keating of going back on his electoral promises.

[24] The resumption of French nuclear testing on Mururoa atoll provoked massive public protests throughout Australia, causing the Keating Government to remonstrate with France. The British Government's refusal to confront France strengthened the argument for Australia to sever its ties with Britain and become a republic.

[25] The Liberal Party (LP) under John Howard won the general elections held in 1996. In a further ground-breaking judgment for Aboriginal people that year, the High Court upheld a claim by indigenous Australians to land leased by governments to pastoralists. The Howard Government responded immediately, limiting the effect of the decision. In 1997, the start of new uranium mining activities in the north caused strong resentment among Aboriginal groups. Their leaders claimed the land would be contaminated. Since then, Aboriginal groups have filed suits for the recognition of their property over vast coast and sea areas, though few of these have been successful.

[26] The coalition between the LP and the National Party (NP) won the 1998 elections and Howard was re-elected. The 1999 referendum on replacing the monarchy with a Republic was won by the monarchists, with 54 per cent of the vote. Britain's Queen Elizabeth II remained head of State.

[27] In 2000 a UN report criticized Australia for its treatment of Aboriginal people. The Government did not apologize but stated its 'regret' for constant mistreatment of Aboriginals in the past.

[28] Howard obtained his third mandate in the 2001 general elections. It appeared that many voters supported his decision to turn away boats carrying Afghan and Iraqi refugees.

[29] In spite of pressures from the UNHCR, Australian policy towards refugees in 2002 - especially towards Afghans fleeing after the fall of the Taliban - continued to be wary of granting asylum. That year, after the terrorist attacks in Bali that killed 94 Australian tourists, Howard promised justice for the victims and reaffirmed his support for the US in the war against terrorism.

[30] The deployment of Australian troops to the Persian Gulf in support of the imminent war headed by the US provoked strong public protests and the first vote of no confidence in the history of the Senate against a sitting prime minister.

[31] Governor General Peter Hollingworth resigned in May 2003 after admitting that in the 1990s, as an Anglican archbishop, he had allowed a known pedophile priest to continue in the ministry.

[32] A parliamentary investigative committee said in March 2004 that intelligence services had failed both in the Bali bombings case and in finding the alleged weapons of mass destruction in Iraq. However, it absolved the Government of lying or deliberately manipulating the information. In April, the main adviser of the Defense Science and Technology Organization of the Defense Department, Jane Errey, who had refused to state that Iraq possessed weapons of mass destruction, was fired.

[33] A change in immigration policy in July enabled thousands of asylum seekers to obtain permanent residence. Immigration minister Amanda Vaston said: 'Those who make a significant contribution to the Australian community may remain here'.

[34] In August 2004, Australia announced the start of a multi-million dollar program to make Cruise missiles. In September, a bomb went off in the Australian embassy in Jakarta (Indonesia) and killed nine people. The attack hindered the normalization of the relationship between both countries, which had deteriorated as a result of Australia's support for the Iraq war and its attempt to extend the war against terrorism to the Asia-Pacific region.

[35] Howard's LP, allied with the NP, won its fourth consecutive Government term in October 2004.

[36] In November, while Parliament debated a new anti-terrorist law, Australian police announced it had thwarted a major-scale attack plan against the country.

[37] The signing, in January 2006, of an agreement between Australia and Timor-Leste to share the anticipated earnings from the exploitation of oil and gas reserves in the Timor Sea led to the postponement of the debate on maritime borders between the countries.

[38] In September 2006, the country was rocked by a scandal involving hundreds of millions of dollars paid in bribes by the Australian Wheat Board (AWB) to officials in Saddam Hussein's former regime in Iraq in order to secure wheat sales.

[39] 2006 saw Australia suffer its worst drought in a century, raising public concern about the impact of climate change. ■

Norfolk Island

Population: 1,853
Land area: 36 km²
Capital: Kingston
Currency: Australian dollar
Language: English

There are no records of the existence of an indigenous population before the arrival of the Europeans. British sailor Captain Cook arrived in 1774, and the island was used as a prison site between 1825 and 1855. It was transferred to Australia as an overseas territory in 1913.

[2] In November 1976, two-thirds of the Norfolk Island electorate opposed annexation to Australia. Since 1979, the island has had autonomy. In December 1991, the local population rejected a proposal to become part of Australia's federal electorate.

[3] A steady increase in tourism, the main economic activity, has brought Norfolk a level of prosperity unusual among inhabitants of the Pacific islands. On the other hand, the island has become self-sufficient in the production of beef, poultry and eggs.

ENVIRONMENT
The island is located in southern Melanesia, northwest of the North Island of New Zealand/Aotearoa. The subtropical climate is tempered by sea winds.

SOCIETY
Peoples: 1,853 inhabitants in 2003. A large part of the population descends from the mutineers of the British vessel HMS Bounty, who came from Pitcairn Island in 1856. **Religions:** Protestant. **Languages:** English (official). **Main Political Parties:** There are no political parties.

THE STATE
Official Name: Norfolk Island. **Capital:** Kingston 1,000 (1999).
Government: Grant Tambling, Administrator appointed by the Governor-General of Australia in November 2003; Geoffrey Robert Gardner, Chief Minister since December 2001. Legislative Assembly with 9 members. ■

Coral Sea Islands

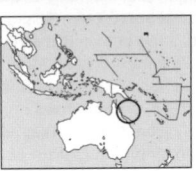

Created in 1969 as a separate administrative entity, the territory consists of several islets located east of Queensland (eastern Australia). The major ones are Cato and Chilcott in the Coringa group, and the Willis archipelago. With the exception of a weather station on one of the Willis islands and lighthouses in several islands, the rest of the islands are uninhabited.

[2] The Constitutional Act by which the territory was created did not provide for Australian administration of the islands, but only for control over foreign visitors by the Canberra Government. However the discovery of oil fields and the expanding fishing industry may change this situation.

[3] In 1997, the charter was amended in order to extend its boundaries and include the reefs of Elizabeth and Middletown, 160 km north of Lord Howe island. At present, these reefs are considered part of the Australian continental shelf.

[4] In 2004, Australia maintained meteorological stations on the islands and claimed a 200-km exclusive fishing zone.

[5] There is no economic activity on the islands and they are uninhabited except for a small meteorological staff of six people in 2006. ■

Austria / Österreich

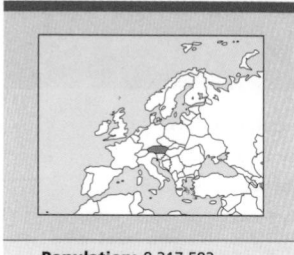

Population:	8,217,583
Land area:	83,860 km²
Capital:	Vienna (Wien)
Currency:	Euro
Language:	German

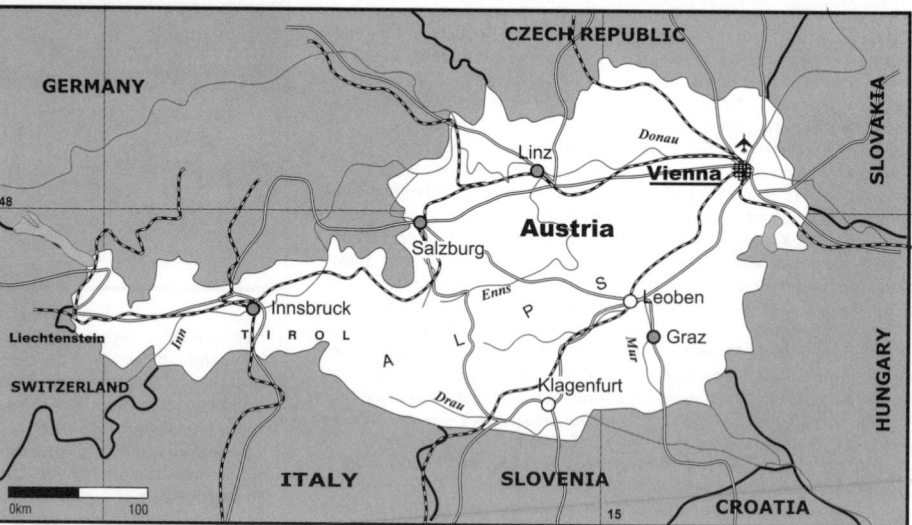

The first traces of human settlement in the land that is now Austria date back to the Early Paleolithic age. From then on, the territory was occupied by various ethnic groups. The northern region of Hallstatt gave its name to the main culture in the Iron Age, from 800 to 450 BC. The origin of the Hallstatt culture cannot be ascribed to only one group; it was developed by the Illyrians and Venetians (among other peoples), but the main growth was achieved by the Celts, owing to their great knowledge of iron production techniques. This metal allowed the construction of vehicles which facilitated the crossing of the Alps, thus increasing commercial exchange. Archaeologists found 2,000 tombs within the region, containing the corpses of salt miners. Cattle breeding, suitable for mountain areas, replaced agriculture; salt was used to preserve meat and meat consumption increased.

2   Celtic tribes moved into the Eastern Alps in the year 400 BC, and founded the kingdom of Noricum; in the west, the ancient race of Raetians was able to maintain its lands. Later, the Romans also settled here, attracted by the iron ore in the area and by its strategic military importance: Roman troops conquered the whole of the country around the year 15 BC. Raetia, Noricum and Pannonia became Roman provinces which were subdivided into municipalities. The Empire's dominions were extended as far as the Danube through a vast network of roads. The Pax Romana ended with the arrival of Germanic tribes between 166 and 180 AD. Although Marcus Aurelius repulsed this invasion, the region did not recover its prosperity. Between the 4th and 6th centuries, the Huns and the Germans raided the area, putting an end to the Empire on the Danube.

3   In the 5th century, according to written records from the era, the Germanic tribes of the Rugii, Goths, Heruli and the *Langobardi* (Lombards), successively invaded

the territory. In 488 a part of the population of the devastated province of Noricum was forced to emigrate to Italy. Following the departure of the Lombards in 568, after successive battles with Slavic tribes, the sway of the Bavarians - under the political influence of the Franks - reached as far as the Avar frontiers in the 6th century. When the Frankish king Dagobert I died, the Bavarian dukes were left virtually independent. Christianity survived through the Roman missionaires who remained in the west region and thanks to the support of the dukes. Under the protection of the Christian churches of Salzburg and Passau, led by the Slav apostles Cyril and Methodius, the Bavarians expanded both militarily and economically during the 8th century.

4   Charlemagne, king of the Franks, deposed Duke Tassilo III of Bavaria, and between the years 791 and 796 annexed to his kingdom the Avar lands in the south. The surviving Avars (mounted nomads possibly from Central Asia, who built an empire in Eastern Europe between the 6th and 9th centuries) were forced to settle in the western part of Low Austria, between the Fischa and Leitha rivers, and soon disappeared from the historical record, probably mixing in with the native population. Charlemagne took over as Holy Roman Emperor in the year 800, becoming the model of a Christian king and emperor. Even though the Empire disintegrated after his death, the German medieval monarchies - just like the French - derived their constitutional traditions from the Carolingian Empire.

5   At the end of the 9th century, the Magyar invaders took control of the low lands to the Rivers Enns and Styria to Koralpe. However, the Germans and Slavs continued to settle and after the German

king Otto started expelling the Magyars in 955, the territory was predominantly German again.

6   Between the 10th and 13th centuries, during the period of Babenberg control, Austria was contested by the Pope and the Holy Roman Empire on several occasions in the battle for control of the German church. Meanwhile, the reformists gained ground, founding the monasteries of Gottweig, Lambach, and Admont in Styria. The Babenbergs maintained the duchies of Austria and Styria, expanding them north and south. New settlements were made by clearing the forests and moving into mountain areas. The colonization process changed the distribution of the German-speaking population and except for some Alpine regions, the Slavs were gradually assimilated, as were the Roman populations in Salzburg and the northern Tyrol.

7   The expansion of the German language was also encouraged by the attraction the Babenberg court held for the leading German poets. Decorated texts proliferated in monasteries, and in the early 13th century, the saga of the Niebelungs was composed by an unknown Austrian poet. In this era Austria also saw the flowering of the best romanesque and early gothic architecture.

8   Following the death of Frederick II, the Babenbergs' dominions were coveted by their neighbors. Premysl Otakar II of Bohemia became king in 1253, facing the opposition of the Austrian nobility after appointing foreigners in official positions, destroying fortresses built without his consent and dissolving his marriage. Rudolf IV of Habsburg came to the German throne in 1273 and pushed Otakar out with the help of the Hungarians.

9   Even though they were initially rejected by the local nobility and

their neighbors, the Habsburgs managed to maintain control over their dominions. In 1322, several defeats by the Swiss, and in particular that of Frederick I at the hands of Louis IV of Bavaria, threatened their dominion over the area of southern Rhine and Lake Constance. By passing the last years of his life on Austrian territory, and being buried in the Carthusian monastery of Mauerbach in 1330, Frederick was the first of his dynasty to consecrate Austria as a home for the Habsburgs. The Habsburg government and territories were known as *dominium austriae*, a term which would later be replaced by the concept of the House of Austria. Its consolidation was achieved through inheritance and marriage alliances. On the death of Frederick III, Maximilian I inherited the House of Austria and the German Empire. His son Philip I, married in 1496 to the Infanta Juana, gained the throne of Spain. A famous saying of the time ran: 'Let others make wars: you, fortunate Austria, get married'.

10   The desire to expand Lutheranism, supported by the noble families, involved the Holy Empire in armed conflicts. In 1521, Protestant pamphlets were printed in Vienna, and bans on their dissemination had no practical effect in 1523. There were peasant revolts in Tyrol, Salzburg and Innerösterreich. Although the Anabaptists (opposed to the christening of children and rebaptized as adults) were joined by many peasants, they had no support from the powerful, which meant they suffered greater persecution. In 1528, in Vienna, Balthasar Hubmaier, reformist leader in the Danube and southern Moravia, was burned at the stake. In 1536, in Innsbruck, Jakob Hutter, a Tyrolian, was sentenced to the same fate after he had led his followers into Moravia.

11   With the death of King Jagiellon of Bohemia and Hungary,

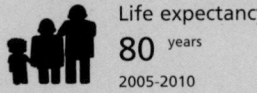
Life expectancy
80 years
2005-2010

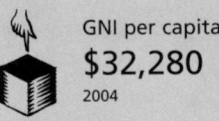
GNI per capita
$32,280
2004

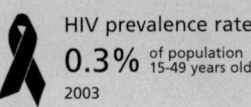
HIV prevalence rate
0.3% of population
15-49 years old
2003

Vienna took the chance to extend the power of the Habsburgs, who held the union of Austria, Bohemia and Hungary as their driving policy. Ferdinand I was proclaimed King of Bohemia in 1526 but his troops were repelled with the help of the Turks when he tried to impose himself on the Hungarians. The 1562 Constantinople Peace Treaty divided Hungary into three possessions: the north and west which went to the Habsburgs; the center held by the Turks; and Transylvania, with its neighboring territories, to Hungary's Janos Zapolya and his successors.

[12] In Austria, the Counter-Reformation started with the Jesuits, strong in Vienna, Graz and Innsbruck, and with the impetus of Melchior Klesl, apostolic administrator in Vienna, then Bishop and Cardinal as well as a key figure in Austrian politics. Maximilian II - successor of Ferdinand I as Emperor of Bohemia, a part of Hungary and the Austrian Danube - who had Protestant inclinations, promised his father he would keep the Catholic faith. His successor, Rudolf II, educated in Spain as a strict Catholic, expelled the Protestants from the court and put Klesl in charge of the conversion of the cities and markets. The Counter-Reformation caused mass emigration, including members of the nobility, to Protestant states and the imperial cities in southern Germany.

[13] The Catholic-Protestant controversy suffered ups and downs which made war inevitable. Upon the death of Emperor Matthias in 1619, Ferdinand II, who had been recognized a year earlier as king of Bohemia and Hungary, succeeded him as Head of the House of Habsburg, and tried to impose Catholicism on his subjects. Lower Austria claimed the resignation of Ferdinand to Bohemia by means of a peace treaty and proposed religious concessions. The Bohemians were forced to withdraw and the territory was occupied by imperial troops.

[14] During the same year, the Diet (Legislature) - predominantly Protestant - unilaterally deposed Ferdinand, choosing to replace him with Frederick V on the Bohemian throne. Two days later, Ferdinand II was named Holy German Roman Emperor; who as a secular branch of the church, committed himself to continue imposing Catholicism. The conflict over the crown ran beyond the borders of the Empire and led to a series of conflicts known as the Thirty Years War.

[15] Bavaria and Saxony joined Ferdinand II, as did Spain - then at war with the Low Countries - to sustain Catholicism. After five years, the Bohemian army was defeated, and an imperial edict put down the Diet. Catholicism was imposed by force. Protestants emigrated en masse to Germany, which was invaded in 1630 by imperial troops of Adolf II, who won over many German princes to his anti-Catholic and anti-Roman cause. Although Germany was from then on the Gordian knot of the war, no throne in Europe was free of the conflict which drew in France, Poland and Denmark.

[16] In 1648, the Peace of Westfalia brought an end to the Thirty Years War and marked a new order in Europe. Holland became an independent republic, and the member states of the Holy Roman Empire were granted complete sovereignty. The old notion of a Catholic empire of Europe, led spiritually by the Pope and secularly by the Emperor, was abandoned for good. The modern structure of a community of sovereign states was established.

[17] Ferdinand II's heir Leopold was threatened by Hungarian rebels and by the Ottoman Turk Empire (over frontier disputes), which led him to form an alliance with Poland. In 1683, Vienna was besieged by the Turks, but the Austrians were rescued by Bavarian, Saxon, Frank and Polish forces, under the leadership of Polish king John III. In 1685, the Emperor signed a pact with Poland and the Republic of Venice, establishing the Holy League.

[18] Between the 17th and 19th centuries, the Habsburgs were involved in all the European conflicts, several of them due to dynastic disputes, but the nature of these disputes was changed by the French Revolution. The Napoleonic Wars virtually dismantled the Austrian Empire, and it was only after Napoleon's abdication in 1814 that the House of Austria recovered most of its territory. In order to prevent a revolutionary uprising, the Austrian Chancellor Clemens Metternich created the Holy Alliance of the European powers at the Congress of Vienna in 1815. This upheld the principles of Christian authoritarianism and foreign intervention against liberal movements.

[19] In 1848, the repercussions of the Paris Commune reached Austria and a revolt broke out in Vienna, led by crowds demanding the liberalization of the regime. Metternich's resignation, rather than bringing peace, unleashed a revolution throughout the Empire. At the same time, in Hungary, the liberal government demanded independence. In Germany the revolution installed a National Assembly in Frankfurt, which incorporated Austro-German liberals and conservatives interested in separation from the Habsburg Empire. The Emperor accepted the Budapest petitions, except on two key points: budget and military autonomy. The Hungarian Parliament declared the power of the Habsburgs null and void, and proclaimed a republic in 1849 but, soon after, the revolution was crushed.

[20] From the 18th century, known as the Age of Enlightenment, until the 20th century, Austria was cradle and shelter of some of the most important personalities of European art and intellectual thought, including musicians like Joseph Haydn, Wolfgang Amadeus Mozart and Franz Schubert; thinkers like Sigmund Freud (founder of psychoanalysis); and the philosopher Ludwig Wittgenstein.

[21] Counter-revolution annulled the Frankfurt Assembly, but the Austro-Prussian dispute persisted. The Habsburg Empire weakened; it lost, ceded and decentralized its dominions until it disappeared in 1918, following its defeat in World War I. In that same year, a national assembly declared German Austria an independent state and, following the abdication of the Emperor, the Austro-German republic was proclaimed a component of the Republic of Germany. The Socialist Karl Renner headed the first republican government, a coalition in which his deputy was the Social Democrat Otto Bauer.

[22] The economic chaos and hunger inherited from the war forced the new government to confront - without consulting the old regime - the social unease and communist activism, that had been inspired by the Russian revolution in 1917 and by the Hungarian revolution of 1919. The personal prestige of Renner and Bauer helped them survive two attempted coups led by the Communists. The Social Democrats, who had support among the peasants and the Conservatives, had a majority in Vienna, where one-third of the population lived, while German nationalism fed on the urban middle classes.

[23] The League of Nations supported Austrian postwar economic recovery, on condition that the country remained independent and did not join Germany. In 1922, the Government stabilized the country's finances by means of a loan until the great depression of 1929 which brought the Austrian economy to the verge of collapse. A customs agreement with Germany was fiercely opposed by the rest of Europe. Together with the rise of Nazism, German nationalism in Austria was showing signs of strengthening. In 1932, the Social Christian government of Engelbert Dollfuss attempted to take an authoritarian stance against the Social Democrats and the Nazis simultaneously. The Social Democrats rebelled and were declared illegal and in 1934 the Nazis murdered Dollfuss in a failed coup.

[24] Taking advantage of the

PROFILE

ENVIRONMENT
Austria is a landlocked country in central Europe. The Alps stretch over most of its territory. The western mountain provinces of Vorarlberg and Tyrol represent a major tourist attraction. Rye and potatoes are farmed in the lower woodlands of the Eastern Alps. The rich agricultural basins of Klagenfurt and Styria, in the eastern Alps, produce corn, wheat, fruit, cattle and poultry. Austria's iron ore and coal mines feed its large iron and steel industry. The Danube river valley runs across the northern foothills of the Alps and is a major natural route for river transportation.

SOCIETY
Peoples: Mostly of Germanic origin. There are Slav, Polish, Hungarian and Roma minorities. **Religions:** Christian (Catholics, 84.3 per cent; Protestants, 6 per cent; Lutheran Evangelists, 5 per cent; **Muslims**, 4.2 per cent. **Languages:** German (and various local dialects); Serbo-Croatian; Romani; Slovene and Hungarian. **Main Political Parties:** The Social Democratic Party of Austria; Austrian People's Party. **Main Social Movements:** The Austrian Trade Union Federation, with more than 1.5 million affiliates; Österreichische Berbauernvereinigung (a farmer organization).

THE STATE
Official Name: Republik Österreich. **Administrative Division:** nine states. **Capital:** Vienna (Wien) 2,179,000 people (2003). **Other cities:** Graz 237,000 people; Linz 188,200; Salzburg 143,300 (2000). **Government:** Heinz Fischer, President since July 2004. Wolfgang Schüssel, Chancellor (Prime Minister) since February 2000. Bicameral Legislature: National Council, with 183 members elected by direct vote every four years; Federal Council, with 64 members elected by the provincial parliament. **National holiday:** 26 October, National Day (1955). **Armed Forces:** 55,750; from 20,000 to 30,000 of whom are conscripts.

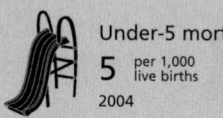
Under-5 mortality
5 per 1,000 live births
2004

Maternal mortality
4 per 100,000 live births
2000

internal crisis and the Government's weakness, German troops invaded Austria in 1938, unchallenged by the European powers. A plebiscite carried out that same year in greater Germany recorded a vote of more than 99 per cent in favor of Hitler. In 1945 after his defeat in World War II, Austria was divided into 4 zones, occupied by US, French, British and Soviet troops.

²⁵ In the first post-War election, the Conservatives obtained 85 seats in the National Council and the Social Democrats 76. Between 1945 and 1952, Austria fought for survival, since after being liberated from the Nazis, it suffered a severe economic collapse which was only overcome with the aid of the United Nations and the US, under the Marshall Plan. Heavy industry and banking were nationalized in 1946, and inflation was controlled by price and salary agreements. The interference of military groups with political and economic matters, within the Soviet zone of occupation, caused a considerable migration of capital and industry from Vienna and Lower Austria, to the agricultural areas of the

western states. In the long run, this migration brought about a very important change in the social and economic structure of the country.

²⁶ Conservatives and Socialists shared the government of the Second Austrian Republic, which only recovered full independence in 1955, with the Treaty of State and the withdrawal of the allied troops The country became a member of the UN in 1955 and of the Council of Europe in 1956. Since then, the key issues in foreign policy have been the dispute with Italy over Sudtirol (Bolzano), resolved in 1969, and its association with the European Economic Community.

²⁷ The coalition ended in 1966, when the People's Party was elected. In the postwar period, Austria did not take part in any military alliance and during the Cold War, the country was liberal in its acceptance of political refugees from Poland; it was also a transit station for Soviet Jewish émigrés. The Socialist Party (SPÖ) won a narrow victory in 1970 forming a minority government led by Bruno Kreisky, an agnostic Jew, born in Vienna. Between 1971 and 1975, the SPÖ monopolized

government, supported by great economic stability and a policy of moderate social reforms. Kreisky resigned when the SPÖ lost its majority in 1983. In coalition with the Liberal Party (FPÖ), the SPÖ maintained its social welfare policy and active neutrality in the international sphere.

²⁸ In the last decade of the 20th century, the ultra-nationalist FPÖ led by Joerg Haider became the second strongest political force in Vienna. Haider was removed from his post as governor of Carinthia in 1991 for praising the full employment policy of the Hitler's Third Reich. He also accused foreigners resident in Austria of 'stealing' jobs from Austrians. In a widely circulated book, Hans Henning Scharsach highlighted the similarities between Haider and Hitler. The rise of Haider coincided with a debate on Austria's role in World War II. In 1992, after repeated attacks on foreign residents, the Government passed a law which punished neo-Nazi activities. In the same year, Thomas Klestil of the ÖVP was elected President.

²⁹ A referendum in 1994 decided that Austria would join the

European Union (EU). In theory this integration did not affect the country's neutrality. Almost 1,600 companies went bust in 1996, which was associated with an increase in economic competitiveness due to joining the EU. The Freedom Alliance (headed by the FPÖ) gained the same number of seats as the SPÖ in the European Parliament, overcoming the ÖVP. The Chancellor (head of government) Franz Vranitsky resigned in 1997 and was replaced by Viktor Klima, who masterminded an austerity economic plan. In 1998, Klestil was re-elected President.

³⁰ Haider was re-elected in Carinthia and reinforced the FPÖ's position as the second political force in 1999. The Greens obtained 13 MPs, not enough for an alliance with the SPÖ. Klima accepted a conservative-liberal alliance. Even though Haider was not in the cabinet, the 14 remaining EU members decided in 2000 to curtail diplomatic contact and abstain from supporting Austrian candidates in the EU or other international bodies.

³¹ In January 2001, Austria agreed to compensate Jewish victims of the Nazis during World War II. In November, after long disputes with the Czech Republic, Chancellor Schüssel and the Czech Government reached an agreement to monitor Temelin nuclear station, 60 km away from the Austrian border.

³² In the same year, 15 per cent of the Austrian electorate signed a petition presented by the FPÖ, demanding that the Czech Republic be refused entry to the EU and that the Temelin nuclear power station be closed down.

³³ In September 2002, Schüssel called for early elections after Vice-Chancellor (and FPÖ leader) Suzanne Riess-Passer, Finance minister Karl-Heinz Grasser and two other members of the cabinet left the coalition due to a political dispute with Haider.

³⁴ In February 2003, a new coalition of Conservatives with the FPÖ was formed, promoted by Schüssel's negotiations with the Social Democrats and the Green Party. In October, the Government passed a law on political asylum which was considered among the most restrictive in Europe.

³⁵ Heinz Fischer (SPÖ) was elected President in April 2004 with 52.4 per cent of the vote. Foreign Minister Benita Ferrero-Waldner received 47.6 per cent.

³⁶ In April 2005 Haider announced the formation of a new political party, the Alliance for the Future of Austria (AFA), after divisions within the Freedom Party jeopardized the Government coalition. All the FP ministers joined the AFA. In May, Parliament ratified the EU Constitution. ■

IN FOCUS

ENVIRONMENTAL CHALLENGES
Emissions from industry and traffic cause air pollution. The use of agrochemicals causes soil erosion and forest degradation.

WOMEN'S RIGHTS
Women have been able to vote and stand for election since 1918. This is one of the countries where women's political representation has increased between 1987 and 2003. In 2003, 34 per cent of parliamentary seats and 35 per cent of ministerial or equivalent positions were held by women. Women made up 42 per cent of the country's workforce. Of these, 80 per cent worked in services, 14 per cent in industry and 6 per cent in agriculture.

Since 1990, more than 17 NGOs on reproductive health, HIV/AIDS, sex education, counseling for teenagers on contraception and support and information for sex workers have been created. Other organizations provide assistance to victims of domestic violence.

Since 1998, measures have been taken against human trafficking: the 'humanitarian visa' has been introduced and an intervention center has been set up for victims of trafficking and exploitation.

CHILDREN
In 2004 the mortality rate for infants under one year old was 5 per 1,000 live births, the same rate as for children under five. Enrolment in primary education was 90 per cent.

Austria is the home of the SOS Children's Villages, an organization founded in 1949 to care for thousands of children after World War II.

INDIGENOUS PEOPLES/ ETHNIC MINORITIES
In 1999, there were between 40,000 and 60,000 Slovenes in the country. Under the post-World War II Treaty, which founded the Second Republic of Austria, the Slovenes are provided with minority rights in the areas of organization, education and administration, and all activities hostile to minorities are prohibited. However, pressure from nationalists has prevented full compliance with the Treaty. Within the Styria region, the existence of the Slovene minority is not recognized. In contrast, the ethnic minority in Carinthia, supported by Slovenia, operates with two central organizations and has developed autonomous financial, scientific and cultural activities.

The Roma (Gypsy) population includes the descendants of those who have lived in Austria for generations (many of Turkic or

Indian origin), the immigrants or descendants of immigrants who came to the country in recent decades, and refugees and asylum-seekers from Central and Eastern Europe. Only people from the first group are part of the *Roma-Gypsy Volksgruppe*, which entitles them to special rights (state financial support for cultural projects, bilingual schooling). In general, the other gypsy groups face serious social disadvantages and prejudice in the areas of employment, housing and public spaces. Despite the small size of the Jewish community in Austria (there were about 7,000 in 2000), antisemitism is a serious problem, manifesting itself in harassment, in the circulation of antisemitic material and graffiti, and in the desecration of cemeteries.

MIGRANTS/REFUGEES
In 2004 Austria received 24,600 new asylum applications; it hosted some 18,000 refugees and 38,000 asylum seekers. There are 38,300 cases pending. Asylum applications were 20 per cent up on the previous year, most of them from Yugoslavs, Iraqis, Afghans, Turks and Indians.

DEATH PENALTY
The death penalty was abolished in 1968; the last execution was in 1950.

Azerbaijan / Azärbaycan

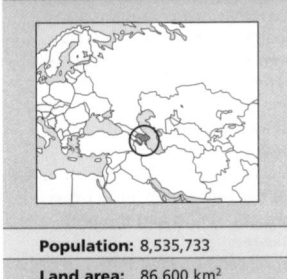

Population:	8,535,733
Land area:	86,600 km²
Capital:	Baku
Currency:	Manat
Language:	Azeri

Map legend:
- Nakhichevan (Under Azerbaijan control)
- Nagorno-Karabakh (Under Armenian control)

Azeris came from the mix of ancient peoples of eastern Caucasus. Archeological expeditions confirm that the territory has been inhabited by human beings since the Stone Age. A considerable number of primitive settlements - belonging to different periods of the Stone Age - refer in cave paintings to the existence of ancestors of Azeri people. The most famous cave is Azykh, located in the southern part of the Karabakh region. Items like utensils, stone tools and fireplaces were found there. Fire characterized the ancestors of Azeri people to the point that Azerbaijan was identified as 'The land of fire' by the Persians. Drawings on rocks at Gobustan (a city located near Baku), date back to the 8th century BC.

2   In the 9th century BC, the states of Mana, Media, Caucasian Albania and Atropatene emerged in the area of present-day Azerbaijan. General Atropates proclaimed the independence of this province in the year 328 BC, when Persia was conquered by Alexander the Great. The above-mentioned states were incorporated to the Persian Arsacid and Sassanid kingdoms. In the year 642, the Arab caliphate conquered Azerbaijan, which was still inhabited by several ethnic groups. The Arabs united the country under Shi'a Islam, despite some resistance. Between 816 and 837, an anti-Arab revolt was led by Babek.

3   Between the 7th and 10th centuries, the territory provided an important trade route which united the Near East with Eastern Europe. From the 11th to the 14th centuries the Seleucid Turks occupied Transcaucasia and the north of Persia. The peoples in the region adopted the Turkish language, and the Azeris' ethnic identity was consolidated. In the 15th and 16th centuries, the region of Sirvan (north of Azerbaijan), became an independent State.

4   Between the 15th and 16th centuries, the Setevid State emerged. Shah Ishmael I, founder of the dynasty, was supported by the nomadic Azeris who became the main power behind the State. The Azeri nobility transferred its support to the Iranians and between the 16th and 17th centuries East Transcaucasia was the scene of Iranian-Turkish rivalry. In the 18th century, Azerbaijan was disputed by the Russian Empire, which resulted in the emergence of more than 15 Azerbaijani Khanates dependent upon Iran. After several Russian wars against Turkey and Persia, the peace treaties of Gulistan (1813) and Turkmenchai (1828) were signed, granting Russia northern Azerbaijan (Baku and Yelisavetpol, corresponding to modern Gyanja).

5   The peasant reform of 1870 in Russia accelerated the development of Azerbaijan, which was supported by the abundance of oil in the region. With the Russian Revolution of 1905, the bourgeois nationalist Musavit (Equality) Party, which had a pan-Turkish, and pan-Islamic platform, was founded in Baku in 1911. After the triumph of the Bolsheviks in October 1917, the Commune of Baku established Soviet power in Azerbaijan. In 1918, Turkish-British intervention ousted the Commune and brought the Musavatists to power. Azerbaijan was declared an independent state but Baku remained under the Communist Government, aided by the local Armenian army. In 1920, the Red Army re-established Soviet power throughout the country, proclaiming the Soviet Socialist Republic (SSR) of Azerbaijan.

6   In 1920, within the framework of Soviet policy on inter-ethnic problems, Moscow incorporated the regions of Upper Nagorno-Karabakh and Nakhichevan into Azerbaijan. These had previously belonged to Armenia as the ancient khanates of Karabakh and Nakhichevan. In 1922 Azerbaijan became part of the Transcaucasian Federation of Soviet Socialist Republics, along with Armenia and Georgia, which experienced considerable economic, urban and industrial development. Although Azeris held powerful positions in education, the republic was controlled by Moscow, especially during Stalin's period. The country was divided between traditional rural areas and the cosmopolitan city of Baku.

7   In 1923, the Autonomous Region of Nagorno-Karabakh was founded, and in 1924, the Autonomous Region of Nakhichevan; both had formerly belonged to the SSR of Azerbaijan. In 1929 an attempt was made to substitute Azeri script (Arabic characters) for the Latin alphabet. In 1936, the Transcaucasian Federation was dissolved, and the SSR of Azerbaijan joined the USSR on its own. In 1940, the Cyrillic alphabet was introduced.

8   During 1980, socio-economic, political and ethnic problems exacerbated the feeling of discontent among Azeris. Between 1969 and 1982, the main leadership of the Communist Party of Azerbaijan was in the hands of Heydar Aliyev, who was known and trusted by the Secretary of the Soviet Communist Party, Leonid Brezhnev. In 1986, the new Soviet leader Mikhail Gorbachev initiated economic reforms (perestroika) and openness (glasnost) in the administration of the country, which encouraged popular discontent within the Union.

PROFILE

ENVIRONMENT
Located in the eastern part of Transcaucasia, Azerbaijan is bordered by Iran to the south, Armenia to the west, Georgia to the northwest, Dagestan (an autonomous region of the Russian Federation) to the north, and the Caspian Sea to the east. The mountains of the Caucasus Range occupy half of its territory; the Kura-Araks river valley lies in the center of the country; and the Lenkoran Valley in the southeast. The climate is moderate and subtropical, dry in the mountains and humid on the plains. The vegetation ranges from arid steppes and semi-deserts to Alpine-like meadows. The mountains are covered with forests. The country has important deposits of oil, gas, copper and iron and foreign corporations moved in to exploit the oil reserves during the 1990s.

SOCIETY
Peoples: Azeris, 90 per cent; Dagestanis, 3.2 per cent; Russians, 2.5 per cent; Lezghi 2.2 per cent; Armenians, 2.0 per cent. There are also Ukrainians, Tatars, Kurds and Talysh Georgian.
Religions: Muslim (majority Shía, 93.4 per cent), orthodox Russian (2.5 per cent), orthodox Armenian (2.3 per cent). **Languages:** Azeri (official) and Russian. Armenian, Kurmanji, Talysh, etc.
Main Political Parties: New Azerbaijan Party; Popular Front of Azerbaijan; Musavat Party; Azerbaijan National Independence Party; Citizens' Solidarity Party; Communist Party.
Main Social Organizations: Sadval, movement of the Lezgian people; movement of the Talysh people. There is also the self-proclaimed Nagorno-Karabakh Republic of Armenian People. There are some independent trade unions, such as the journalists' union.

THE STATE
Official Name: Azärbaycanärbaycan Respublikasi.
Capital: Baku 1,816,000 people (2003).
Other cities: Gyanja (formerly Kirovabad) 299,300 people; Sumgait 277,300; Mingechaur 98,900; Nakhichevan 67,100 (2000).
Government: Ilham Aliyev, president since 2003. Artur Tahir Rasizade, prime minister since 2003. The legislative body has 125 members.
National holiday: 28 May, Independence (1918).
Armed Forces: 70,700 (1996). Other: Militia (Ministry of Internal Affairs) 20,000; Popular Front (Karabakh People's Defense) 12,000 (est).

Life expectancy
67 years
2005-2010

GNI per capita
$940
2004

Literacy
99% total adult rate
2000-2004

HIV prevalence rate
<0.1% of population 15-49 years old
2003

9   There was a wave of strikes, political rallies and demonstrations. New political movements came into being, like the leading People's Front of Azerbaijan (PFA) with a platform stressing civil rights, free elections and political and economic independence for the country, but opposing the long-standing aspiration of the Armenians of Nagorno-Karabakh to rejoin the Armenian republic. In 1989, Azerbaijan was proclaimed a sovereign state within the USSR. The ethnic conflicts between Azeris and Armenians led to the formation of extremist groups and Armenians were massacred in Sumgait and Baku. After the events in Baku, the Communist Government decreed a state of emergency and called in troops from the USSR to re-establish order; which led to the killing of more than 100 people.

10   The Nagorno-Karabakh issue increased friction with Armenia and the Soviet (Council) of the Autonomous Region of Karabakh proclaimed its independence from Azerbaijan. In 1989, Armenia's Soviet approved the reunification, which Azerbaijan denounced as interference in its internal affairs. Karabakh, besieged and bombed by Azerbaijan forces, voted 99.3 per cent in favor of its independence in a plebiscite; the Azeri minority population refrained from voting. In 1991, the Republic of Nagorno-Karabakh declared independence from both Azerbaijan and Armenia. The status of Autonomous Region was annulled and the troops of the Commonwealth of Independent States (CIS) were withdrawn, though later on fighting between Azeri and Armenian guerrillas intensified. The People's Front of Azerbaijan (PFA) demanded the creation of an Azeri national army. Baku accused CIS troops of facilitating the union of Armenia and Nagorno-Karabakh.

11   In 1991 Moscow approved Azerbaijan's independence. The state of emergency was lifted in Baku and the first presidential elections were held, the Communist Party of Azerbaijan having been dissolved. Former Azeri Communist leader, Ayaz Mutalibov, who had supported the aborted coup against Gorbachev, was elected. The PFA called the elections 'undemocratic', and withdrew their candidate. Azerbaijan joined the CIS in 1991 and in 1992 it was admitted to the UN.

12   In 1992, Mutalibov resigned, accused of being responsible for a massacre in Khojala (a territory of Karabakh), and Yuri Mamedov assumed presidential powers. The fighting between Armenians and Azeris extended to Nakhichevan. Azerbaijan's Parliament, controlled by former communists, reinstated the deposed Mutalibov, who suspended planned elections and imposed a curfew. The PFA leaders, with the support of the national militia, responded by seizing the Parliament building in Baku, declaring the reinstatement of Mutalibov illegal. Despite fierce clashes in the capital, Mutalibov was ratified as president. The PFA assumed the directorship of the security services and of the official media.

13   Later that year, due to the defeat of Azeri troops and the siege of Parliament by PFA forces, Mutalibov was forced yet again to resign and Abulfaz Elchibey of the PFA was elected. This prompted the return of Aliyev (former Soviet Communist Party and KGB) to a senior post. Over the following weeks, the Armenians launched a strong offensive and as a result there was a coup. Rebels led by Colonel Guseinov took control of five regions of the country. Elchibey fled when the offensive against Baku started, leaving Aliyev as interim President. Armenian attacks intensified, and in May 1993 an offensive was launched on Stepanakert, the administrative center of Nagorno-Karabakh. During that year, Aliyev won the elections and successfully attacked Karabakh.

14   In 1994, under Moscow's pressure, a ceasefire was declared and negotiations between the warring factions began. Parliamentary elections were held, without the participation of the powerful Musavit party, or communist or Islamic groups. Aliyev's party, New Azerbaijan (NA), won the elections. Two years later, NA won the majority of parliamentary seats in the country's first presidential and legislative multi-party elections. A new Constitution was approved by referendum.

15   In 1998, opposition leaders, among them Abulfez Elchibey, boycotted the presidential elections in which Aliyev obtained 76 per cent of the vote. Observers, the opposition and the press alleged the elections had been unfair and pressed for them to be declared null and void.

16   In 1997 Azerbaijan opened its oil fields in the Caspian Sea. Azerbaijan, Kazakhstan and Turkmenistan together possess the third largest oil reserves in the world.

17   Despite reports of irregularities in the 2000 elections, President Aliyev's NA won by an overwhelming majority. Azerbaijan became a full member of the European Council that year.

18   Azerbaijan supported the US 'war on terror' after the 2001 attacks in New York and Washington; in exchange the US lifted the sanctions imposed on Azerbaijan in 1992 when Baku blocked a railroad to Armenia.

19   In 2002 Aliyev collapsed during a television address and was hospitalized in Turkey; he appointed his son Ilham as Prime Minister.

20   In October 2003, Ilham Aliyev won the presidential elections. Street demonstrations were repressed by police, who arrested hundreds of protesters.

21   Pakistan's President Musharraf visited Baku in July 2004, to strengthen relations and encourage bilateral agreements. In spite of the differences in population size and their historically low trade exchange, in 2003 Azerbaijan and Pakistan signed a cooperation defense treaty which enabled Pakistan to deploy Azeri troops in its territory.

22   In 2005 several demonstrations were severely repressed in Baku. In March, journalist Elmar Huseynov was shot. In May, just before the opening of a gas pipeline, police action again left hundreds of demonstrators injured. In September and October, as parliamentary elections neared, more protests were also repressed by the police.

23   In November 2005, the election victory of the NA was followed by a new wave of opposition protests and police repression.

24   US mediator Steven Mann announced in February 2006 that his attempts to broker an agreement between Azerbaijan and Armenia on the Nagorno-Karabakh conflict had failed. ∎

IN FOCUS

ENVIRONMENTAL CHALLENGES
There is soil contamination from pesticides such as DDT. Highly toxic defoliants have been used extensively on cotton crops. Water pollution is a serious problem; approximately half the population lacks sewage facilities and only a quarter of all water is treated. The Caspian Sea, especially near the Abseron Peninsula, is the most ecologically degraded area in the world.

WOMEN'S RIGHTS
Women have been able to vote and stand for election since 1921. In 2005 they held 12.3 per cent of Parliamentary seats and in 2003 had a 15 per cent representation in ministerial or equivalent positions. In 2003, they made up 43 per cent of the workforce. In 2004, 66 per cent of pregnancies had prenatal care, while 84 per cent of births were attended by skilled health staff.

There are no family planning advice services, so abortion remains the main method of birth control. Maternal mortality stood at 95 per 100,000 live births, but the UN reports much higher figures.

CHILDREN
Infant and under-five mortality rates fell slightly in the 1990-2004 period, but continue to be high. The former fell from 84 to 75 per 1,000 live births and the latter from 105 to 90. The principal causes of under-five mortality are respiratory conditions and parasitic infections. Vaccination has improved, reaching levels higher than 90 per cent. Some 13 per cent of children under five are moderately or seriously stunted. In 2003, malaria became a new problem affecting mainly children.

INDIGENOUS PEOPLES/ ETHNIC MINORITIES
The Lezgins, a Sunni Muslim people whose lands are divided by the border between Russia and Azerbaijan, make up 2.5 per cent of the country's population. Today, most Lezgins speak Azeri as a second language, and are fairly well-integrated into the society of Azerbaijan. The most frequent complaints by Lezgins include the lack of Lezgin language education and media in Baku.

MIGRANTS/REFUGEES
In 2003, Armenians from the Azeri region of Nagorno-Karabakh declared the independence of 20 per cent of the territory, leading to the displacement of one million Azeris. The decade-long conflict in that area has caused the largest inflow and outflow of exiles along the country's borders. The Government reports being host to 11,400 refugees and asylum seekers, mainly from Chechnya and Afghanistan.

DEATH PENALTY
Was abolished in 1998.

Bahamas / Bahamas

Population:	331,643
Land area:	13,880 km²
Capital:	Nassau
Currency:	Bahamian dollar
Language:	English

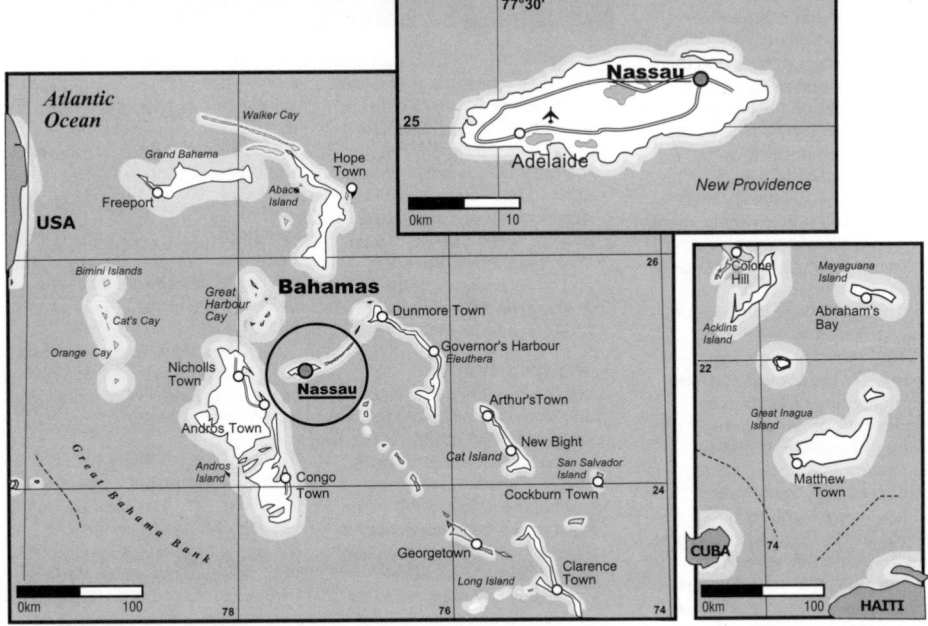

1 The Bahamian archipelago was one of the few areas of the Caribbean from which the Arawak Indians, who lived mainly on fishing and the harvesting of seafood and molluscs, were not displaced by the Caribs. The few remaining traces of their culture are pots, potsherds and petroglyphs.

2 Columbus probably first trod American soil on the Bahamian island of Guanahani (or San Salvador or Watling).

3 The Spanish historian Francisco de Gomara pointed out, 'over a period of 25 years, the Spaniards enslaved 40,000 Indians who were sent to work in mines on the other islands', such as Santo Domingo.

4 The Spaniards did not colonize the islands that lacked mineral resources. Instead, British privateers and pirates sought refuge in these islands after seizing the gold extracted by the Spaniards in other American territories.

5 From 1640 the British began to settle the Bahamas. Sugarcane and other tropical crops were grown in plantations worked by African slaves whose descendants today make up most of the local population. In 1873, the Treaty of Madrid settled the dispute over control of the Bahamas in favor of the British.

6 The British refused to accept the independence of this

strategic archipelago, and it was not until 1973 that the Bahamas proclaimed their independence within the British Commonwealth. This change actually meant little to the islanders because in the meantime the country had become increasingly dependent on the US.

7 In fact, most of the three million tourists who visit the Bahamas come from the US, drawn to the beaches and casinos. The transnationals which use the Bahamas as their formal headquarters are also North American, taking advantage of the exemptions that make the country a 'tax-haven'. Also, US citizens are the main buyers of lottery tickets, a source of fiscal revenues that contributes heavily to the State budget. The country's second economic activity is banking: in 2006 there were approximately 350 banks. In 1942, the US installed a naval base at Freeport, which helped control traffic from the Gulf of Mexico to the Atlantic, via the Florida Strait.

8 While other Caribbean nations sought ways to bring about regional integration, the Bahamas never joined any regional organization.

9 In 1956 the Assembly passed an anti-discriminatory resolution aimed at promoting ethnic equality. Thus, the Afro-Caribbean population was given access to places where they had never before been admitted.

10 The Progressive Liberal Party (PLP), whose leaders were black, won the 1967 election, putting an end to white supremacy. Lynden Pindling became Prime Minister.

11 In 1977, when the economic and social crisis started to bite,

the Government decided to give even greater incentives to foreign capital. In an attempt to diminish the massive unemployment which threatened to create social tensions and changes in the archipelago, the Government opened an

industrial estate of 1,200 hectares near a deep water port in Grand Bahama. It was meant to be used as a storage point for merchandise which was later to be re-exported after minimal local processing.

12 During the 1977 electoral

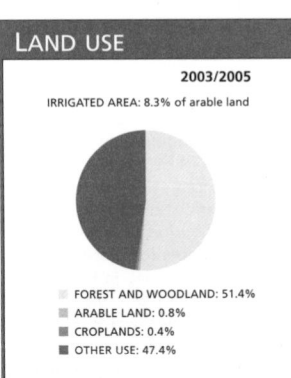
PROFILE

ENVIRONMENT
The territory comprises over 750 islands, only 30 of them inhabited. The most important are: New Providence (where the capital is located), Grand Bahama and Andros. These islands, made of limestone and coral reefs, built up over a long period of time from the ocean floor. Despite the subtropical climate, the lack of rivers has prevented the favorable climatic conditions being fully exploited for agriculture. Farming is limited to small crops of cotton and sisal. The main economic activity is tourism, centered on New Providence.

SOCIETY
Peoples: Descendants of African slaves, 86 per cent, plus North Americans, Canadians and British (12 per cent); Asian and Hispanic people (2 per cent).
Religions: Predominantly Christian. Baptists 32 per cent; Anglicans 20 per cent; Roman Catholics 19 per cent and Methodists 6 per cent.
Languages: English (official) and Creole.
Main Political Parties: Progressive Liberal Party (PLP) founded in 1953; Free National Movement (FNM).
Main Social Organizations: Trade Union Congress; trade unions of the different sectors: hotel caterers and related activities, teachers-professors, civil servants, airport employees, cab drivers, musicians and theatre workers.

THE STATE
Official Name: Commonwealth of the Bahamas.
Capital: Nassau 222,000 people (2003), on New Providence Island.
Other Islands: Freeport/Great Bahama 42,400 people; Eleuthera 3,300; Andros 2,800; Long Island 1,800 (2000).
Government: Queen Elizabeth II is Head of State, represented by Governor General Arthur D. Hanna since February 2006. Prime Minister Perry Christie since May 2002. There is a bicameral Legislature, with a 16-member Senate and a 40-member Assembly.
National Holiday: 10 July, Independence (1973).
Armed Forces: 2,550: Police (1,700); Defense Force (850).

Life expectancy

72 years

2005-2010

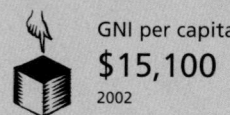

GNI per capita

$15,100

2002

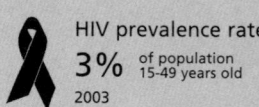

HIV prevalence rate

3% of population
15-49 years old

2003

campaign the opposition parties - the Free National Movement, the Democratic Party of Bahamas, and the Vanguard Party - accused the Government of corruption and squandering public funds. Both left and right criticized the Government's policy towards transnationals. However, Pindling again won by a landslide and promised to lower unemployment. He opened the country's coasts to the transnationals, which resumed oil prospecting in 1979.

[13] In 1984, the political scene was further upset when the US NBC Network News directly charged Prime Minister Pindling with receiving large sums of money for authorizing drug traffic through Bahamian territory. An investigation immediately confirmed that Government officials were involved in smuggling but cleared Pindling of any responsibility in the affair.

[14] In 1987, the unemployment rate was estimated to be above 18 per cent, with that of young people under 25 as high as 35 per cent. Although they had strongly supported the Government in the past, labor unions now criticized the authorities for the lack of constructive, long-term programs aimed at solving the grave unemployment problem.

[15] Lynden Pindling won his sixth consecutive election on 19 June 1987. The PLP obtained 31 out of 49 seats. The opposition Free National Movement (FNM), led by Cecil Wallace Whitfield repeatedly asked Pindling to step down, charging him with fraud, corruption, and being 'soft' towards drug dealers.

[16] In 1991, there was a sharp decrease in the number of tourists, estimated at three million per annum. This was attributed to the fact that there had been a rise in crime, compounded by the Bahamas' extremely high cocaine consumption rate. Although there had been a slowdown in inflation and a slight fiscal surplus, this did not prevent people from taking to the streets.

[17] Tourism had generated over 65 per cent of the country's GDP, but in 1992 this fell by 10 per cent in relation to 1990 figures. Meanwhile the banking system lost customers to its competitors in the Cayman Islands. In addition, the Government was unable to increase agricultural production, making it necessary for the country to import 80 per cent of its food.

[18] Pindling's 25 years in office ended in 1992 when Hubert Ingraham, a former protégé of Pindling's and leader of the National Free Movement, won the parliamentary elections with 55 per cent of the vote. The new

Government aimed to reduce unemployment through the liberalization of foreign investment laws and the re-establishment of the Bahamas as a major tourist destination.

[19] In February 1994, one of the prominent figures from Pindling's time, lawyer Nigel Bowe, was incarcerated in Miami for drug trafficking. That year, the Government appointed a commission to investigate Pindling, who was accused of having used the Hotel Corporation's funds to increase his wealth. Pindling denied the charges and sought protection in bank secrecy, which limited the Commission's progress.

[20] In 1995, foreign investors carried out the privatization of several hotels belonging to the Hotel Corporation. The most significant problems for the Government were unemployment and the presence of several thousands of Haitian and Cuban refugees. A repatriation agreement with Haiti was achieved, but the Cuban refugees refused to be sent back, and demanded to be transferred to the United States.

[21] In January the following year the conflict was resolved when the Bahamas signed an agreement with Cuba whereby Cubans living in Bahamian detention camps were returned to their home country.

During 1996 250 Cubans from the camps, and a further 70 living illegally, were returned to Cuba. The Cuban Government agreed not to take retaliatory measures against the deportees.

[22] The death penalty was applied in March 1997 for the first time in 12 years when two prisoners were hanged. The National Free Movement won the Parliamentary elections held that month with 35 seats. The Progressive Liberal Party won five seats. Hubert Ingraham was reelected as Prime Minister.

[23] The Bahamas were ravaged in September 1999 by Hurricane Floyd, the worst storm in decades. Tens of thousands of people were evacuated, amid winds reaching speeds of 240 km per hour. They swept through whole villages, destroying boats and severing communications in Nassau.

[24] In spite of pleas for pardon from different human rights groups, a man was hanged in January 2000 for the murder of two German tourists. Amnesty International asked Bahamas Governor Orville Alton Turnquest to abide by international treaties, stop the execution and consider abolishing the death penalty.

[25] Late that year the Government enacted a series of laws to restructure the financial services sector and comply with certain

demands of the Financial Action Task Force (FATF), the OECD and others that had included Bahamas on their 'blacklists'. A package of nine laws came into effect which covered, among other things, a complete revision of the Central Bank and the Private Banks and Trust Funds Acts, revenues from illegal acts, reports of financial transactions and the creation of a Financial Intelligence Unit.

[26] In November 2001, for the first time in the country's history, a woman, Ivy Dumont, took office as Governor General of Bahamas.

[27] The Bahamian oil tanker 'Prestige' was wrecked near the Galician coast of Spain in November 2002, spilling 60,000 tons of oil.

[28] In September 2005, Bahamas and several Caribbean countries signed the PetroCaribe agreement with Venezuela, which provided them with oil at concessionary prices (they should only pay market prices for part of this oil; the rest is to be financed at an interest rate of 1 per cent over 25 years).

[29] The abolition of mandatory death sentences for those convicted of murder, in March 2006, marked a slight progress in terms of human rights. This resulted in the review of at least 30 cases of people who had been sentenced to death. ■

Bahrain / Al Bahrayn

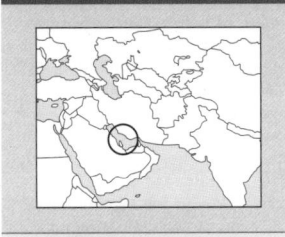

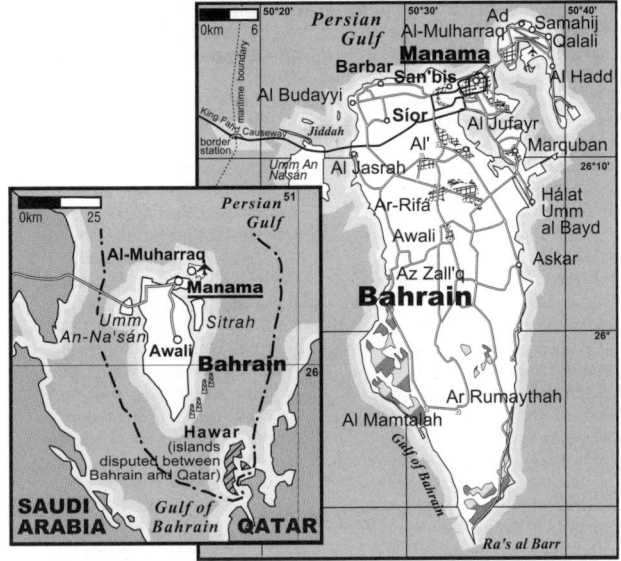

Population:	751,320
Land area:	710 km²
Capital:	Manama
Currency:	Dinar
Language:	Arabic

From the time of the Sumerians, the area now called Bahrain was central to the intense maritime trade between Mesopotamia and India. This trade became particularly prosperous between the 11th and 15th centuries, as Islamic civilization expanded over all the territory from the Atlantic Ocean to the South Pacific.

2 Portuguese sailors occupied the island in 1507 and stayed there for a century until the Persians expelled them. Iran's claim to sovereignty over this part of the Persian or Arab Gulf dates from this period. Sheikh al-Khalifah took power in 1782, displacing the Persians the following year; his descendants are still in power. Independence lasted until 1861, when another Khalifah, afraid of Persian annexation, agreed to declare a 'protectorate' under the British.

3 During the two World Wars Bahrain was an important British military base. In 1932 the first oil wells in the Gulf were opened on the island. Nationalist movements demanding labor rights, democracy and independence grew during the 1950s, as they did in other parts of the Arab world. In December 1954, a strike broke out in the oil fields. British troops quashed the rebellion and, slowly, some reforms were carried out and local participation in public administration increased.

4 Finally, beginning in the early 1970s, the British decided to withdraw from their last colonies 'east of Suez', though they maintained their economic and strategic interests in Bahrain. Bahrain and Qatar refused to join the United Arab Emirates, so on 14 August 1971 the country became independent under Sheikh Isa ibn-Sulman al-Khalifah.

5 The new nation authorized the US to set up naval bases in its ports. These were dismantled in 1973 following the Arab-Israeli conflict. Local elections were held the same year and the National Assembly came under the control of progressive candidates calling for freedom to organize political parties and greater electoral representation. The British felt their interests were being threatened, and, in August 1975, they backed al-Khalifah's decision to dissolve Parliament.

6 In the 1970s, Iran managed to eclipse Saudi control of the Emirates, forcing a virtual protectorate over them. Meanwhile, heavy migration into Bahrain threatened to create a large and active Iranian minority. The Emir responded by cracking down even harder on both Iranian or Shi'a immigrants and on all progressive movements. At the same time he drew closer to the other Arab governments, rejecting the Camp David agreement and signing mutual defense treaties with Kuwait and Saudi Arabia.

7 In 1981, the country joined the Gulf Cooperation Council (GCC). This was set up with US help to guarantee military and political control over the area, to stop the influence of the Iranian Islamic revolution, and to keep an eye on opposition groups in the member states.

8 One of the most closely watched movements was the Bahrain National Liberation Front, based in the UK and supported mainly by oil workers, students and professionals. This group was a member of the Gulf Liberation Front until 1981, along with other regional movements.

9 In November 1986, a super-highway was opened between Saudi Arabia and Bahrain, making Bahrain no longer truly an island. During the first year alone, the highway was used by more than a million vehicles.

10 In March 1991, after the Iraqi defeat in the Gulf War, the foreign ministers of Egypt, Syria and the six Arab member States of the GCC signed an agreement with the US in Riyadh in order to 'preserve regional security'. After Kuwait, Bahrain was the emirate most affected by the conflict.

11 In December 1994, Shi'a leader Sheikh al-Jamri was arrested, after signing a claim for the restoration of the Constitution and Parliament, dissolved in 1975. His arrest provoked anti-government demonstrations, in which two students and a police officer died. In April 1995, Emir Isa ibn-Sulman al-Khalifah met with 20 opposition leaders in an attempt to put an end to growing violence. In August, both parties reached an agreement that concluded with the release of 1,000 political prisoners.

12 In 1996, the demonstrations spread across the country, and some ended in violent confrontations with the police. The Government decided to use the death penalty to punish those 'responsible', a measure endorsed by the Courts. The UN Working Group on Arbitrary Detentions issued three declarations on the situation of the inmates of Bahrain's prisons.

13 On 6 March 1999, after ruling Bahrain for 37 years, Emir Isa

---

PROFILE

ENVIRONMENT
This flat, sandy archipelago consists of 33 islands in the Persian Gulf between Saudi Arabia and the Qatar Peninsula. The largest island, also called Bahrain, is 48 km long and 15 km wide. The climate is warm, moderately humid in summer and slightly dry in winter. Manama, the capital and main trade center, is located on the island of Bahrain. Bahrain has the same environmental problems that are characteristic of all the Gulf countries. Industrialization led to the occupation of what few fertile lands there were, in the northern part of the main island. Bahrain is developing its petroleum industry; the refineries, the large storage tanks and transport pipelines have a major impact on the environment.

SOCIETY
Peoples: Bahrainis are Arab people. The petroleum industry has attracted a number of Iranian, Indian and Pakistani immigrants.
Religions: Islamic (85 per cent), predominantly of the Sunni sect in the urban areas and Shi'a in the rural areas. Christians 8.5 per cent. There are also Jews, Hindus, Baha'i and others.
Languages: Arabic, English, Farsi/Persian, Urdu.
Main Political Parties: Islamic National Accord Association (INAA); Liberation Front of Bahrain; Ba'ath Arab Socialist Party; Arab Nationalist Movement.
Main Social Organizations: General Union of Bahrain Workers, Bahrain Society for Human Rights, Bahrain's Women Union.

THE STATE
Official Name: Mamlakat al-Bahrayn.
Administrative divisions: 12 municipalities.
Capital: Manama (Al-Manamah) 139,000 people (2003).
Other Cities: Al-Muharraq 81,800 people; Ar-Rifa' 82,000; Madinat'Isa 61,600. (2000).
Government: Sheikh Hamad ibn Isa al-Khalifah, Emir since March 1999 and King since February 2002; Khalifah ibn Salman al-Khalifah, Prime Minister since January 1970, assisted by an 11-member cabinet. The National Assembly, partially elected by popular vote, was dissolved in August 1975. In October 2002 after parliamentary elections the two chamber National Assembly (comprised of the Shura Council and the Chamber of Deputies, each with 40 seats) was restored.
National Holidays: 15 August, Independence (1971); 16 December, National Day (1971).
Armed Forces: 10,700. Other: (Ministry of Interior): Coast Guard: 400; Police: 9,000.

Life expectancy
75 years
2005-2010

GNI per capita
$14,370
2004

Literacy
88% total adult rate
2000-2004

HIV prevalence rate
0.2% of population 15-49 years old
2003

Ibn-Salman Al-Khalifah died. He was succeeded by his son, Hamad Ibn Isa Al-Khalifah, who pursued conciliatory policies with the Shi'a - who represent almost 70 per cent of the population but have traditionally been ruled by the Sunni minority - to which the royal family belongs. During May and June he freed more than 300 Shi'a political prisoners, but another 1,000 remained in jail awaiting trial.

[14] In December 1999, in his Independence Day address, the new Emir spoke about democratic openness and promised to reinstate municipal councils. The following year, in addition to publicly promising to reinstate Parliament, the Emir for the first time named as members of the Consultative Council - a body created in 1992, made up of 40 people who monitor most of the government's policies - non-Muslim men and women, including a Jewish entrepreneur and four women, one a Christian.

[15] In a February 2001 referendum, the Bahrainis overwhelmingly supported the political reforms proposed by the royal family. The reforms entailed converting the State into a constitutional monarchy and separating the branches of government.

[16] One month later, after lengthy conflicts between Bahrain and Qatar over the Hawar islands - which contain important reserves of natural gas - the International Court of Justice at The Hague delivered its verdict, in favor of Bahrain.

[17] In December, the Ministry of Information tried journalist Hafez al-Shaikh for publishing articles critical of the Shi'a community. Al-Shaikh was accused of violating the country's press laws. He argued that the persecution was really due to his negative comments, published in a Lebanese daily, about Bahrain's collaboration with US incursions in Afghanistan.

[18] Throughout 2001 several political associations and non governmental organizations were officially recognized. In June, the General Committee of Workers was officially registered as the General Union of Bahrain Workers, and in September a trade union law was enacted. The Bahrain Women's Union, an advocate of women's rights, was recognized in November.

[19] Advancing the dates set the previous year, on 14 February 2002, the State of Bahrain declared itself a constitutional monarchy. The King (formerly the Emir) called for early elections (the first in 27 years) to be held on 24 October of that same year.

[20] In May 2002 local elections were held in Bahrain and for the first time women were allowed to vote and stand as candidates. This was also the case for October's parliamentary elections. Despite the calls for a boycott from the Islamic National Accord Association (INAA), the main political party representing the majority Shi'a, there was a 50 per cent voter turnout. The Shi'a INAA opposition - which was the party of choice in the elections - had considered the electoral process undemocratic, since the legislative power would be divided between the elected chamber and an advisory council appointed by the King, who is Sunni.

[21] In March the authorities blocked access to several websites, including that of the Bahrain Freedom Movement, which had strongly criticized the constitutional reforms approved by the King. In May, the Arab television network Al-Jazeera was banned within the country. A royal decree enacted in October a new Press and Publications Act, whose article 68 punished with up to 5 years in prison the publication of articles offensive to the State religion, critical of the king or inciting people to depose or change the government.

[22] Also in May, a colonel from the Security and Intelligence Service, Adel Jassem Fleifel, fled to Australia when the authorities began to investigate him on corruption charges. Opposition groups had long been accusing him of torturing and ordering the torture of detainees and political prisoners. That month, Amnesty International urged the Government to investigate all human rights violations supposedly committed by Fleifel and the Security and Intelligence Service. Fleifel returned to Bahrain in November and was arrested. In October, the King issued Decree 56, which clarified the contents of the general amnesty of February 2001 (Decree 10), effectively banning any judicial action against civilians or military officers who had committed or been implied in human rights violations before February 2001.

[23] The UN High Commissioner for Human Rights visited Bahrain in March and stated the need to investigate human rights violations committed in the past and take those responsible to court. Shortly after, Bahrain ratified the UN Convention on the Elimination of All Forms of Discrimination against Women, with reserves on articles 2, 9, 15, 16 and 29.

[24] The 2003 report issued by the Bahrain Human Rights Society warned about different forms of discrimination: among others, against women and Shi'a citizens in their access to public administration, and about the nepotism exercised by the royal family.

[25] In April 2004, King Hamad bin Issa al-Khalifa appointed Nada Haffadh as Health Minister. She was the first woman to head a government ministry in the country.

[26] Following the confrontations that took place in Manama in May 2004, when the police clashed with more than 5,000 demonstrators that were protesting against the US military presence in Iraq, the King sacked the Interior Minister, Sheik Mohammed bin Khalifa al Khalifa. According to the Government, events could have been prevented so as not to make repression necessary. The King stated that people's right to express anger and protest 'against the excess and oppression suffered by our brothers in Palestine and Israel, and against the violation of human rights and holy cities such as Najaf and Karbala in Iraq' was legitimate. He added that his government would share those sentiments with regards to injustices committed. In July, more than 500 US citizens residing in Bahrain had to be evacuated due to the threat of terrorist attacks in the country.

[27] The following year, some 50 people were beaten and arrested in Manama when the police dispersed protesters who demonstrated against unemployment. Among those beaten was Abdulhadi al-Khawaja, head of the Bahrain Center for Human Rights, which had been banned by the Government the previous year. The detainees were soon released.

[28] In the November 2006 parliamentary and municipal elections, the United Nations Development Program (UNDP) provided $8,000 in support for each female candidate. The UNDP co-ordinator in Bahrain said that the basis for the aid was 'contained in Article 4 of the Convention on the Elimination of All Forms of Discrimination Against Women.' There were 18 women among the 220 candidates in the parliamentary election and 5 women out of 171 in the municipal poll. ■

IN FOCUS

ENVIRONMENTAL CHALLENGES
The process of desertification, resulting from the degradation of arable lands during periods of drought and dust storms, continues to advance. The different stages of the oil refining process have a major impact on the environment which is mainly reflected in coastal degradation. Oil extraction in the region produces about 4.7 per cent of the oil industry's contribution to world pollution.

WOMEN'S RIGHTS
Women have been able to vote and run for office since 1973, but interruptions in the country's democracy mean that they have enjoyed full citizen rights only since 2002. In 2005, women held 7.5 per cent of parliamentary seats; their representation in ministerial or equivalent posts stood at 10 per cent.
In 2003, women made up 23 per cent of the workforce.

The illiteracy rate among women amounted to 83 per cent in 2005.

CHILDREN
The economic boom of the 1970s and early 1980s established the basic infrastructure of socio-economic development, such as education, health, drinking water, environmental health and electricity. Apart from improving overall living standards, the Gulf countries have implemented extensive networks of free or subsidized services for their citizens, which have contributed to huge improvements in the survival, development and protection of women and children. Improvements are still needed in educational materials, school infrastructure and legislation.

INDIGENOUS PEOPLES/ETHNIC MINORITIES
The Shi'a - members of the branch of Islam that regards Ali and his descendants as the legitimate successors to the Prophet Muhammad and reject the first three caliphs - face political and economic discrimination. Most Shi'a belong to the middle and lower economic and social classes. The new Constitution, adopted in the year 2000, paved the way for democracy in the country, significantly reduced tensions between the Sunni and Shi'a and opened dialogue. Opposition is mostly moderate and violence between the Sunni and Shi'a has decreased in recent years.

MIGRANTS/REFUGEES
Human Rights Watch estimated the immigrant population at 720,000, of whom 290,000 were non-Bahrainis (40 per cent in all). In 1991 foreigners made up 36 per cent of the population. The total number of workers in 2001 was 332,521, of whom 213,007 (or 64 per cent) were foreigners.

DEATH PENALTY
Although it still applies even for minor crimes, there have been no executions since 1996.

Bangladesh / Bangladesh

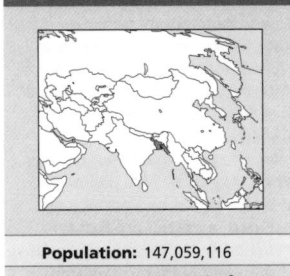

Population:	147,059,116
Land area:	144,000 km²
Capital:	Dhaka (Dacca)
Currency:	Taka
Language:	Bangla/Bengali

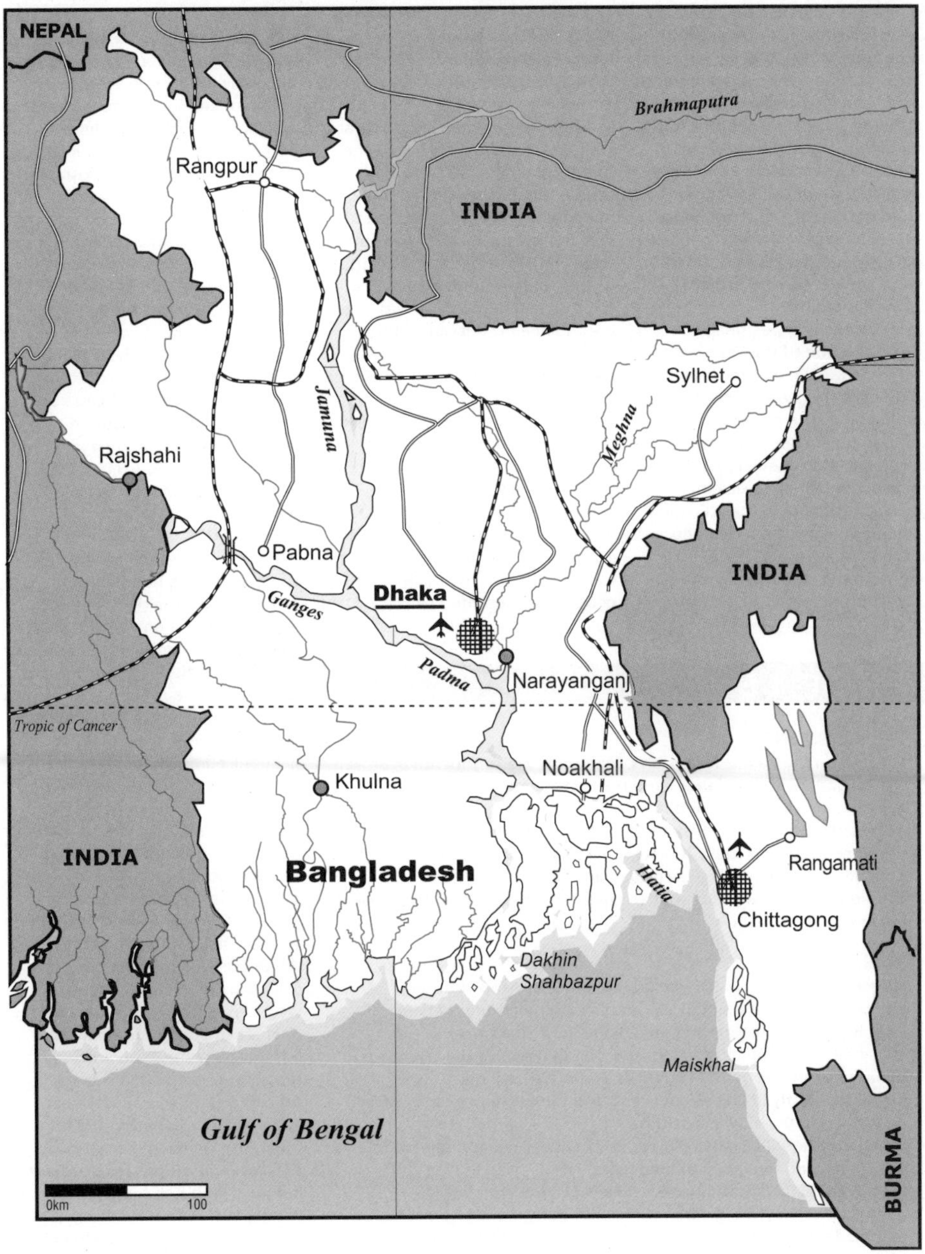

The area comprising present-day Bangladesh and the Indian state of West Bengal was settled in about 1000 BC by Dravidian peoples who were later known as the Bang. The first Dravidian empire to spread over most of present-day India, Pakistan, and Bangladesh was the Mauryan Empire between the 3rd and 1st century BC. During the Mauryan Empire, a Buddhist chief named Gopala came to Bengal and took power, becoming the first ruler of the Pala Dynasty, which lasted from AD 750-1150. He and his successors provided stable government, security, and prosperity while Buddhism was spread throughout the State and into neighboring territories.

2   The Senas, militant Hindus, replaced the Buddhist Palas as rulers of a united Bengal until the Turkish conquest in 1202. Opposed to the Brahmanic Hinduism of the Senas with its rigid caste system, vast numbers of Bengalis, especially those from the lower castes, would later convert to Islam.

3   Turks ruled Bengal for several decades before the conquest of Dhaka by forces of the Mughal emperor Akbar the Great in 1576. The British East India Company, a private company formed in 1600 during the reign of Akbar and operating under a charter granted by Queen Elizabeth I, established a factory on the Hooghly River in

Bengal in 1650 and founded the city of Calcutta in 1690.

4   Although the initial aim of the East India Company (EIC) was to seek trade under concessions obtained from local governors, the steady collapse of the Mughal Empire (1526-1858) enticed the Company to take a more direct involvement in the politics and military activities of the subcontinent. Siraj ud Daulah, governor of Bengal, provoked a military confrontation with the British at Plassey in 1757. He was defeated by Robert Clive, a young official of the Company. By 1815 the supremacy of the EIC was unchallengeable, and by the 1850s British control and influence had extended into territories that by 1947 would be the independent states of India and Pakistan.

5   In 1905 the British governor general, Lord George Curzon, divided Bengal into eastern and western sectors in order to improve administrative control of the huge and populous province. Thus, two new provinces were established: East Bengal, which had its capital at Dhaka, and West Bengal (the present-day state of West Bengal in India), with its capital at Calcutta, which also was the capital of British India. Many Bengali Muslims viewed the partition as recognition of their cultural and political differences from the majority Hindu population - the All-India Muslim League (ML) supported the partition, but Curzon's decision was ardently challenged by the Indian National Congress (INC), a political organization dominated by Hindus,

LAND USE

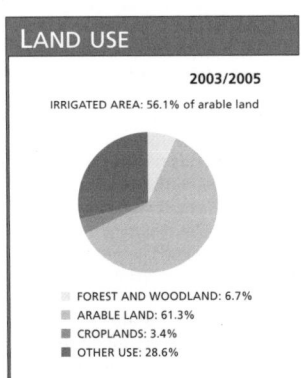

2003/2005

IRRIGATED AREA: 56.1% of arable land

- FOREST AND WOODLAND: 6.7%
- ARABLE LAND: 61.3%
- CROPLANDS: 3.4%
- OTHER USE: 28.6%

PUBLIC EXPENDITURE

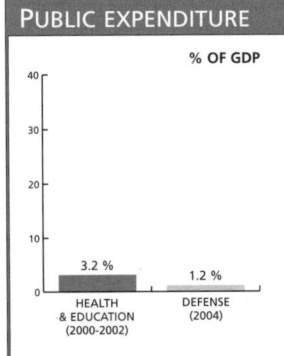

% OF GDP

3.2 %	1.2 %
HEALTH & EDUCATION (2000-2002)	DEFENSE (2004)

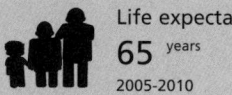
Life expectancy
65 years
2005-2010

GNI per capita
$440
2004

Literacy
41% total adult rate
2000-2004

founded in 1885 and supported by Calcutta's elite. In 1912 the British voided the partition of Bengal, reconstituting the reunited province as a presidency, while moving the capital of India from Calcutta to the less conflictive area of New Delhi.

6   During World War I, the INC supported Great Britain in the hope that the British Crown would reward Indian loyalty with political concessions after the war. The ML was more ambivalent, perhaps because the dismemberment of the Ottoman Empire presaged the destruction of the last great Islamic power. In 1920 the Khilafat Movement (KM) was launched, combining Indian nationalism and pan-Islamic sentiment with strong anti-British overtones. KM leaders and Mahatma Gandhi, the leading figure in the INC, came to an agreement resulting in the joint advocacy of self-rule for India, agitation for the protection of Islamic holy places

and the restoration of the caliph of Istanbul. In 1922 the Hindu-Muslim accord suffered a double blow when their noncooperation movement miscarried and the KM lost its purpose when the postwar Turkish nationalists abolished the sultanate, proclaimed Turkey a secular republic, abolished the religious office of the caliph, and sent the last of the Ottoman ruling family into exile. After the eclipse of the Hindu-Muslim accord, the spirit of communal unity was never reestablished in the subcontinent.

7   In August 1947, the British divided India, transforming part of the territory into Pakistan. Bengal was also divided. The mostly Muslim regions, known as East Bengal, became part of Pakistan, and the mostly Hindu regions became part of India. In 1956, the new Constitution of Pakistan changed the name of East Bengal to East Pakistan. The Bangladeshi people - discontented with the

vast transfer of resources from the region to the rest of the country and with the military-bureaucratic oligarchy installed in West Pakistan - demanded regional autonomy, with the goal of establishing an autonomous government. Reacting to these attempts, the Pakistani military took strong measures against civilians. The people of East Pakistan - which adopted the name of Bangladesh - declared their independence, began a movement of armed resistence and formed a government in exile, in India, with Sheik Mujibur Rahman as president. In December 1971 the occupation forces were expelled, and a Constitution adopting parliamentary democracy was adopted in November 1972. Democracy, secularism, socialism, and nationalism were declared basic pillars of the State, and industries, banks and insurance companies were nationalized.

8   However, the challenge to improve an economy ravaged by war and other problems turned out insurmountable for the governing party - the Awami League - and its inexperienced leaders. Nationalist fervor was brief, giving way to generalized discontent and the appearance of armed political movements. In December 1974 the Government declared a state of national emergency, suspended fundamental civil rights, banned political parties and trade union activity. The only party remaining was Bakshal, formed mostly by Awami League members and pro-Soviet communists. Newspapers were closed and a new press act banned all opposition opinion.

9   In this context, an army rebellion of active and retired officers murdered the president, Sheik Mujibur Rahman, and his family, proclaiming martial law. After several coups, General Ziaur Rahman, who had founded the Bangladesh Nationalist Party (BNP), was assassinated in a failed coup d'état. In March 1982 the army overthrew the Government and placed General Hossein Mohammed Ershad as President. Ershad dissolved Parliament after the Jatiya Party (JP) won the 1986 national elections, amidst electoral fraud claims that led to massive popular uprisings. Confidence in political integrity had dwindled so that the 1988 elections were boycotted by the main parties, followed by a large percentage of voters.

10   The gradual islamization of politics led in 1989, through a constitutional amendment, to the declaration of Islam as official religion. According to the Constitution, a woman shall inherit only half of what her brother may inherit. In practice, this fraction ends up in the hands

of her husband or kept for her dowry. Bangladeshi feminists state that women are treated as objects and not as individuals, since they belong to their parents in childhood, to their husbands in marriage (most marry at 13 years of age) and to their children when they grow old. Their work at home and as harvesters is not included in official production statistics, and divorce (a prerogative of men, under Islamic law) can be easily obtained if women's productivity falls. However, after Ershad was overthrown in 1991 and elections were called, both the Awami League (AL) and the Bangladesh Nationalist Party (BNP) slated women as their main candidates, both widows of former political leaders. Begum Khaleda Zia, from the BNP, was elected in March and declared her support for the establishment of a parliamentary regime. Five months after her victory at the polls, and with the unanimous approval of legislators of both parties, the Congress of Bangladesh replaced the presidential system with a parliamentary one.

11   The identity of Bangladesh as a Muslim country was reaffirmed with the Gulf War (1990-1991). This sentiment was intensified in 1992 with the repatriation of Muslim refugees. Early that year, Bangladesh received some 250,000 Bihari Muslims that had supported Pakistan in 1971. In June, some 270,000 Rohingya Muslims arrived, escaping from persecution in Myanmar, a mostly Buddhist country. Repatriation agreements signed by both countries in 1992 did not stem the flow of refugees.

12   Bangladesh has always been a predominantly rural society. Agriculture makes up approximately half the GDP, while only 10 per cent results from manufacturing. In the 1970s and 1980s, more than 7,000 varieties of rice were planted in the country. In recent times only one is extensively cropped.

13   In a country where the average annual income did not exceed $170 per capita and 50 per cent of the population lived in mud houses without sanitation facilities - one bathroom was shared among 50 families - there was a rapid migration to urban areas, thus enlarging deprived neighborhoods. Within that framework, 95 per cent of the country's budget was allocated to external debt payments, since by 1992 Bangladesh was strongly dependent on international aid. Almost 95 per cent of development programs were financed by the US, Japan, the Asian Development Bank and the World Bank. In 1990 Polli Sree,

PROFILE

ENVIRONMENT

Located on the Padma River Delta, formed by the confluence of the Meghna with the Ganges and the Brahmaputra, Bangladesh is a fertile, alluvial plain where rice, tea and jute are grown. There are vast rainforests and swamps. A tropical monsoon climate predominates, with heavy summer rains from June to September generally accompanied by hurricanes and floods of catastrophic consequences. Low-quality coal and natural gas are the only mineral resources.

SOCIETY

Peoples: The people of Bangladesh are ethnically and culturally homogeneous, as a result of 25 centuries of integration between the local Bengali population and immigrants from Central Asia. There are small Urdu and Indian minorities. The Chittagong Hill Tracts are home to 11 ethnic groups known as a whole as Jumma.
Religions: Mostly Islamic (83 per cent) and Hindu (16 per cent), with Buddhist and Christian minorities.
Languages: Bangla/Bengali (98 per cent); others: dialects related to the Tibeto-Burman group of languages.
Main Political Parties: Bangladesh Awami League (AL), in favor of a mixed economy; Bangladesh Nationalist Party, right of center (Bangladesh Jatiyatabadi Dal, BNP); National Party (coalition), an Islamic-inspired alliance of five nationalist parties; Islamic Conference Bangladesh; Islamic Unity Front; National Socialist Party-Rob and other minor parties.
Main Social Organizations: Jana Sanghati Samiti, fighting for the rights of the tribal people; Women for Women.

THE STATE

Official Name: Gana Prajatantri Bangladesh.
Administrative divisions: 4 districts.
Capital: Dhaka (Dacca) 11,560,000 people (2003).
Other Cities: Chittagong 2,500,900 people; Khulna 1,168,800; Rajshahi 687,300 (2000).
Government: Parliamentary republic. Iajuddin Ahmed, President and Head of State since September 2002. Khaleda Zia, Prime Minister and Head of Government since October 2001. Single-chamber legislature: Parliament made up of 330 members (300 elected by direct vote and 30 reserved for women, nominated by Parliament, for 5-year term).
National Holidays: 26 March, Independence Day (1971); 16 December, Victory Day (1971).
Armed Forces: 115,500 (1995). Other: Bangladesh Rifles: 30,000 (border guard); Ansars (Security Guards): 20,000.

 Under-5 mortality **77** per 1,000 live births 2004

 Poverty **36%** of population living on less than $1 per day 2000

 Debt service **5.2%** exports of goods and services 2004

Maternal mortality **380** per 100,000 live births 2000

a women-led non-governmental organization, organized community groups seeking to improve the economic conditions, health and education of Bangladeshis. Programs developed by Polli Sree are aimed at raising awareness about gender issues, as well as empowering poor women and girls, by means of increasing their knowledge and skills and granting them better access to financial aid.

[14] Although the 1996 elections - carried out under army control - were considered fraudulent, Khaleda Zia remained in power. The Awami League called for national strikes which paralyzed the country, followed by clashes between police and opposition activists. Violence did not stop with Zia's fall, and the Government was transferred by former Supreme Court of Justice chief, Mohammed Habibur Rahman, to Sheik Hasina Wajed, chosen prime minister in the second elections within four months. Social conflict continued throughout 1997, increasing in December with several strikes called by opposition parties protesting against the agreement signed by the government to put an end to armed resistance in the southeast. For several days, both supporters and opposers of these strikes marched in various cities - Dhaka, Chittagong, Barisal, Sylhet and Rajsani, among others - and in many small towns, leaving the country in a state of semi-paralysis.

[15] In order to keep exports competitive, the Central Bank of Bangladesh decided in November 1999 on a further devaluation of the taka by 3 per cent. The main opposition party, the Bangladesh Nationalist Party (BNP), argued that the measure advised by the International Monetary Fund (IMF) - which warned that the currency was overvalued - was against the people, among other things because it was the 16th devaluation carried out by the Government since 1996.

[16] In the context of the US war on terrorism, in September 2001 Dhaka granted Washington's request to use its air space, ports and airports in case of an invasion of Afghanistan. In November, head of state Badruddoza Chowdhury travelled to Washington to meet with Secretary of State Colin Powell. While the US sought to confirm the support of Bangladesh - one of the largest Muslim countries - in its campaign against Afghanistan, Chowdhury sought economic benefits for his country in exchange for its support. After three days of talks, high officials from Bangladesh and India announced, in March 2002, new measures to reduce the tension throughout the 4,000 kilometers of

IN FOCUS

ENVIRONMENTAL CHALLENGES
A large part of the population is landless and lives and farms on sites prone to flooding. Poor sanitation and the tainting of groundwater by arsenic limit people's access to drinking water. On the other hand, there is large water pollution of fishing areas resulting especially from the use of pesticides and agrochemicals. The indiscriminate felling of trees is accelerating the process of soil erosion and degradation.

WOMEN'S RIGHTS
In 2005, women held 2 per cent of parliamentary seats, while in ministerial or equivalent positions the rate stood at 8 per cent. In 2003, they made up 43 per cent of the country's workforce, comprising 71 million people. Seventy-seven per cent of women worked in agriculture; 12 per cent in the area of services; and 9 per cent in industry.

More than 20,000 women die each year due to pregnancy-related complications. Only 40 per cent of pregnant women receive prenatal care; while barely 14 per cent of births are attended by trained staff.

Discrimination and different forms of violence against women are common (see history).

CHILDREN
Although the death rate among children under five has diminished almost 50 per cent between 1990 and 2004, more than 30 per cent of newborns are underweight, which may lead to serious problems in their later development*. Cases of child prostitution and child trafficking have been registered as well as cases of children used for domestic labor, minors living in irregular settlements either in urban or rural areas, and minors belonging to tribal groups that have been abandoned in orphan homes or jailed.

A gradual reduction in population growth (due to decreased fertility) has been noticed in recent years. There has also been an improvement in the quality of life of women and children, particularly in terms of health and education, as well as in life expectancy.

MIGRANTS/REFUGEES
Of the more than 250,000 people that have arrived in the country from Myanmar since 1993 (about half of them are Rohingya people), barely 10 per cent have been aided by UNHCR and acknowledged by Dhaka; the rest are considered illegal immigrants. Some 300,000 Biharis - who arrived in what was then East Pakistan in 1947 from the Indian state of Bihar - still live as refugees.

Refugees belonging to the Rohingya people, from Myanmar, seldom have access to the legal proceedings that might grant them the legal status of refugees. At the same time, there are no government policies on refugees.

INDIGENOUS PEOPLES/ ETHNIC MINORITIES
From the late 1990s, thousands of local inhabitants, intimidated by the army, have been displaced from the Chittagong Hill Tracts, declared as 'reserved forest' by the Environment and Forest Ministry. Their lands were occupied by transmigration of other ethnic groups.

Neglected by the Government, the Chakma people had to emigrate to India due to the flooding of their lands. Transnational oil companies, such as Shell and Halliburton, and institutions such as the World Bank, are putting pressure on the Government to export oil and natural gas, threatening lands belonging to the Jumma, Bawm and Khumi with indiscriminate digging and deforestation.

DEATH PENALTY
The death penalty still applies; in 2004 sentences were passed and executions carried out, although the exact number remains unknown.

* Latest data available in *The State of the World's Children* and *Childinfo* database, UNICEF, 2006.

their common border. Among other things, they decided joint patrols of the area, and regular meetings between their commanders in charge. In June 2002, Chowdhury resigned the presidency and in September 2002, Iajuddin Ahmed was sworn in as president.

[17] A dramatic rise in violence against women led, in March 2002, to the adoption of laws punishing attacks with sulphuric acid with the death sentence. While the use of acid rose 50 per cent between 2000 and 2002, according to police records, in 2001 there were 13,339 cases of domestic violence, six times more than those registered in 1995 (2,048). In Bangladesh - and Myanmar, Cambodia, Pakistan, among others - sulphuric acid, cheap and easily obtained, is used by men to disfigure, and sometimes kill, women and girls; the reasons for these attacks are refusal to accept marriage proposals, domestic fights and conflicts about the property of goods. In addition, throughout 2002, some 3,189 cases of rape and death by torture of women and children - 49 more than

the previous year - were reported. In view of this, Khaleda Zia, who was reelected prime minister in 2001, introduced two additional laws to stem the tide. Throughout the year, 2,343 people were arrested for domestic violence. One year later, not one had been punished for their actions.

[18] In May 2004 the AL accused the ruling Bangladesh National Party of seeking to increase its majority when Parliament amended its constitution to reserve 45 seats for women. Female MPs would be selected in proportion to each party's support in the latest election.

[19] In July, Bangladesh was lashed by heavy rains; severe floods affected two-thirds of the country and left more than 20 million people isolated or homeless. Forty per cent of the capital, Dhaka, was under water due to the overflowing of rivers. More than 200 people died across the country and in some areas, water reached record levels.

[20] In December a tsunami devastated South Asia. In Bangladesh a strong earthquake brought about by the catastrophe

left two people dead.

[21] In 2005, Transparency International released a report ranking Bangladesh among the world's two most corrupt countries, together with Haiti. Meanwhile, another report by Reporters Without Borders pointed out that Bangladesh, for the third year running, was the country with the largest number of journalists being attacked or threatened with death. During the year, four journalists were murdered and ten more were arrested.

[22] In April 2006, the AL staged demonstrations to demand electoral reforms and the resignation of Khaleda Zia, whom they accused of corruption. During the protests there were numerous confrontations between demonstrators and security forces, and several activists were injured or arrested.

[23] At the end of October 2006, the BNP Government came to the end of its term of office amid a furore of strikes and protests by the opposition over the organization of the next national election, due to take place in late January 2007. ∎

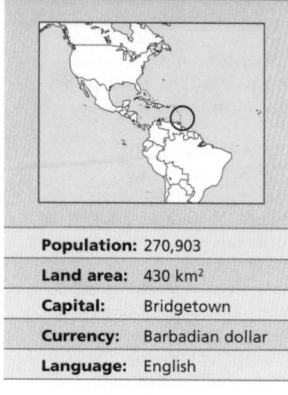

Population:	270,903
Land area:	430 km²
Capital:	Bridgetown
Currency:	Barbadian dollar
Language:	English

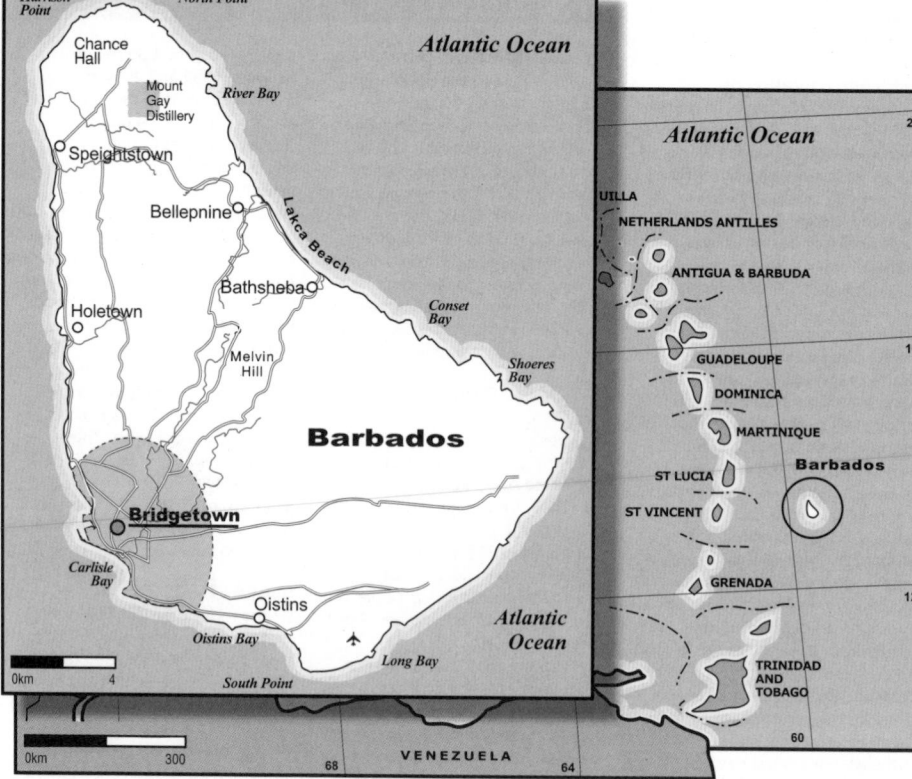

The peaceful and nomadic Arawak people expanded throughout the Caribbean region, and although they were dislodged from many islands by the Caribs, they remained on others - as in the case of Barbados.

[2] The Spanish landed on the island early in the 16th century and christened it the 'island of the bearded fig tree'. Satisfied that there was no natural wealth they withdrew, but not before massacring the native population, taking a few survivors with them to amuse the Spanish court. In 1625 the British arrived and found a fertile, uninhabited territory.

[3] Around 1640 the island had close to 30,000 inhabitants. The majority were farmers and their families, and some were political and religious dissidents from Britain. The settlers grew tobacco, cotton, pepper and fruit on small plots, raising cattle, pigs and poultry.

[4] Sugarcane was introduced, with the support of the British, and caused extensive social change. Plantation owners purchased large plots to improve profitability, and small landowners - most of them in debt - sold off their land to the plantation owners. The importation of slaves from Africa to work the sugar plantations began during this period.

[5] In 1667, 12,000 farmers emigrated to other Caribbean islands or to the 13 colonies of North America. Nonetheless, the island had a commercial fleet of 600 vessels and, as recorded by a French traveler in 1696, was 'the most powerful island colony in America'.

[6] Towards the end of the 18th century, the island was one huge sugar-producing complex which included 745 plantations and over 80,000 African slaves. By that time there was no woodland left on an island described in a 16th century account as 'entirely covered with trees'. Its ecological balance seriously impaired, the island fell victim to drought in the early 19th century and parts of it suffered soil exhaustion.

[7] The pursuit of increased profitability and an economy oriented toward foreign trade resulted in underdevelopment. A different approach might have led to development along the lines of the other North American colonies.

[8] Slavery was abolished in 1834 but the plantation economy still dominated the island's fiscal system. European landowners controlled local politics until well into the 20th century.

[9] In 1938, following the gradual extension of political rights, the Barbados Labor Party (BLP) led by

PROFILE

ENVIRONMENT
Of volcanic origin, Barbados is the easternmost island of the Lesser Antilles. The fertile soil and rainy tropical climate favor intensive farming of sugarcane, rotated with cotton and corn.

SOCIETY
Peoples: Most are of African origin (92.5 per cent), with a minority of Europeans (3.2 per cent) and 2.8 per cent of mixed descent. Barbados is one of the most densely populated countries in the world, with an average of 616 people per square kilometer.
Religions: 33 per cent are Anglican, 29.8 per cent belong to other Protestant faiths; 4.4 per cent are Catholic.
Languages: English (official), Creole (Bajan).
Main Political Parties: Barbados Labor Party (BLP); Democratic Labor Party (DLP). **Main Social Organizations:** Barbados Workers Union; National Organization of Women; Barbados Gays and Lesbians against Discrimination; Mothers' Union of Barbados.

THE STATE
Official Name: Barbados. **Administrative divisions:** 11 parishes.
Capital: Bridgetown 140,000 people (2003).
Other Cities: Speightstown 820 people; Holetown 720; Bathsheba 720 (2000). **Government:** Owen Arthur, Prime Minister since September 1994, re-elected in 1999 and ratified in 2003; Sir Clifford Husbands, Governor-General, appointed by Queen Elizabeth II in 1996. Parliament is bicameral: the Senate, with 21 members, and the Legislative Assembly, with 30 members.
National Holiday: 30 November, Independence Day (1966).
Armed Forces: 610 army.

LAND USE

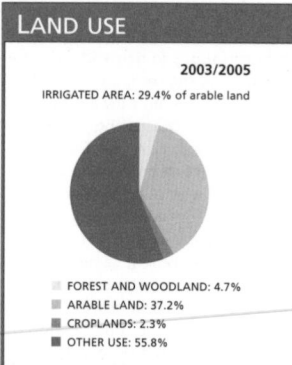

2003/2005

IRRIGATED AREA: 29.4% of arable land

- FOREST AND WOODLAND: 4.7%
- ARABLE LAND: 37.2%
- CROPLANDS: 2.3%
- OTHER USE: 55.8%

PUBLIC EXPENDITURE

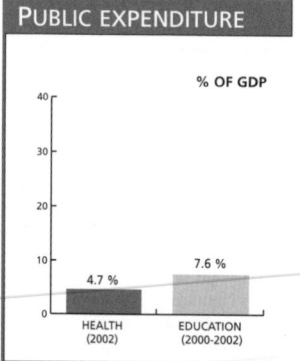

% OF GDP

4.7 % HEALTH (2002)
7.6 % EDUCATION (2000-2002)

Life expectancy
76 years
2005-2010

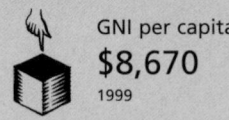

GNI per capita
$8,670
1999

Literacy
100% total adult rate
2000-2004

HIV prevalence rate
1.5% of population 15-49 years old
2003

Grantley Adams developed from within the existing labor unions. Universal suffrage was declared in 1951, and Adams became leader of the local government.

10 Internal autonomy was granted in 1961, and in 1966 independence was proclaimed within the British Commonwealth. Errol Barrow was elected Prime Minister. Unlike the rest of the West Indies, Barbados never severed its links with the colonial capital in spite of its political independence.

11 After 1966, Errol Barrow's Democratic Labor Party (DLP) contributed to the creation of the Caribbean Free Trade Association (CARIFTA) which became CARICOM in 1973, involving 12 islands of the region. Barrow showed great interest in the Non-Aligned Movement. In 1996 Barbados became a UN member-country, and in 1997 it joined the Organization of American States (OAS).

12 Free education and new electoral laws were not followed by any significant change in DLP policies towards the owners of sugar refining plants. The rise in unemployment diminished DLP support and resulted in the Party's electoral defeat.

13 In 1970, the country became a member of the International Monetary Fund (IMF).

14 The BLP won 17 of the 24 available seats in 1976. Tom Adams, Grantley Adams' son, was elected Prime Minister. He promised to fight corruption and described himself as a social democrat (the BLP became a member of the Socialist International in 1978). However, the Government protected the interest of investors in sugar and tourism transnationals, while encouraging foreign investment.

15 Adams pressured Washington to withdraw from its naval base on nearby St Lucia, which it did in 1979. In 1981, Adams was re-elected and consolidated relations with Washington. Barbados supported the US invasion of Grenada.

16 In order to attract foreign capital, the Government passed new tax exemption laws and liberalized ship registration. In 1986, an agreement between the US and Barbados led to 650 new companies registering in the off-shore sector. Unemployment and inflation within Barbados continued to grow.

17 The 1986 election was won by the DLP. Prime Minister Errol Barrow committed the Government to changing the policy toward the US that had been followed by the BLP Government in the preceding 10 years. In 1987, Barrow died of a heart attack and was succeeded as Prime Minister by Erskine Sandiford.

Two years later, as a result of a split within the DLP, a new opposition group was founded, the right-wing National Democratic Party (NDP), under the leadership of Richard Heynes.

18 In 1991, the DLP was re-elected with 49 per cent of the vote although it lost 2 of its 20 seats. In November, riots and protests due to an 8 per cent cutback on civil servants' wages led to a general strike. Despite this challenge, Sandiford remained in office.

19 In 1992 the Government obtained an IMF loan amounting to $64.9 million and the Prime Minister announced further wage reductions in the public sector, a rise in interest rates and cutbacks in the social system, as well as the privatization of oil and cement production and the tourism industry. After two successive finance ministers resigned, Sandiford himself assumed the post.

20 In June 1994, the BLP withdrew its support for the Prime Minister and won the early elections held in September. Owen Arthur, the new Prime Minister, canceled planned wage reductions, and this became law in February 1995.

21 The legislative assembly voted in February 1996 for civil servants salaries to be brought back up over a two-year period. The measure, promoted by the BLP, was a response to the salary reduction policy applied in 1994 by the DLP.

22 The January 1999 elections saw a broad victory for the BLP, which won 26 out of the 28 seats, while the other two remained in DLP hands. Owen Arthur was confirmed in his post.

23 A Financial Intelligence Unit (FIU) was set up to fight money laundering. The new body was advised by members of the UN's International Drug Control Program (UNIDCP).

24 Barbados enacted a law on international companies in 2001 in order to attract foreign investment. The law created a 2.5 per cent maximum income tax, exemptions on every other tax and on tariffs on the imports of materials for production, the exemption of foreign exchange controls, and a deduction of 150 per cent on the cost of research and development activities on exports.

25 In June 2001, Barbados hosted the preparatory Regional Round Table for the World Summit on Sustainable Development, held in Johannesburg, South Africa, in September 2002. The General Assembly of the OAS was held in June 2002 in Bridgetown, the capital of Barbados.

26 In the May 2003 elections, the BLP won again, and Owen Arthur

IN FOCUS

ENVIRONMENTAL CHALLENGES
Untreated domestic waste is seriously polluting rivers and the sea. There is an excessive exploitation of maritime resources and a marked erosion of the soil and coastal areas.

WOMEN'S RIGHTS
Women have been able to vote and run for office since 1950. In 2005, women held 17.6 per cent of seats in Parliament; their representation at ministerial or equivalent posts stood at 45 per cent. In 2003 women made up 46 per cent of the workforce. That year, 12 per cent of the female workforce was unemployed, while the rate of unemployed men was below 9 per cent.

CHILDREN
Immunization rates are in excess of 90 per cent. The incidence of HIV/AIDS emerged as a big challenge in 2002. The number of children abandoned in orphanages has increased in recent years mainly due to two causes: the high rate of teenage pregnancies and the high mortality caused by HIV/AIDS*.

Domestic violence against women and children is also a rapidly growing problem, as reported by various NGOs.

MIGRANTS/REFUGEES
In February 2003, UNHCR announced that, although states have the right to adopt their own policies on migration, they should also take into consideration the jurisprudence of international organisms that supervise human rights. It encouraged Barbados to strengthen its legal framework for the protection of asylum-seekers, potential refugees and stateless people, since the country did not adopt the 1951 Convention relating to the Status of Refugees, or later instruments. Less than 20,000 people emigrated between 1995 and 2000.

DEATH PENALTY
Barbados applies the death penalty for ordinary crimes to those aged 16 and over.

* Latest data available in *The State of the World's Children* and *Childinfo* database, UNICEF, 2006.

was ratified as Premier. The BLP won 21 seats and the DLP only 7 out of the 30 at stake. Both parties focused their electoral campaign on reducing taxes - Barbados has one of the highest tax rates in the Caribbean - and job creation. In 2002 the Government presented a plan to reduce income tax from 25 per cent to 22.5 per cent, announcing it would fall to 20 per cent in 2004. The Opposition emphasized the Government's mismanagement of public funds and, especially, the millions of dollars spent on paying off hotel industry debts.

27 An alarming proportion of the Barbados population lives with HIV/AIDS, according to UN reports issued in 2003. It was estimated that reported cases amounted to only one-fifth of the total population living with HIV in the country. In 2001, the Barbados HIV/AIDS Prevention Program, previously managed by the Ministry of Health, was placed under the National HIV/AIDS Commission. The latter managed to agree on the design of common policies and advice with other Caribbean Community and Common Market (CARICOM) countries and with the Joint United Nations Programme on HIV/AIDS (UNAIDS). According to data published in 2005 by the Pan American Health Organization

(PAHO), a reduction in the number of new cases was registered for the first time.

28 In May 2004, Washington offered its official apologies to Bridgetown. The US had published wrong information on the internet about the existence of violent crimes and an inept police force on the island. The US admitted that the information - which had stayed on the world wide web for four months - was extremely inaccurate and made it clear that it had been removed and replaced by information portraying Barbados' real situation.

29 The only prison in the country had to be evacuated in April 2005 after being seriously damaged in a fire, followed by a two-day uprising among inmates. Two prisoners died and 25 people were injured, among them six guards. Seventeen inmates, who said riots had been sparked when authorities of the correctional facility dismissed sexual assault complaints from an inmate, were accused on arson charges.

30 After more than 20 years of disputes between Barbados and Trinidad and Tobago over fishing rights, a court of arbitration in The Hague ruled, in April 2006, that the boundary between both fishing areas should run halfway between the two countries. ∎

Belarus / Belarus

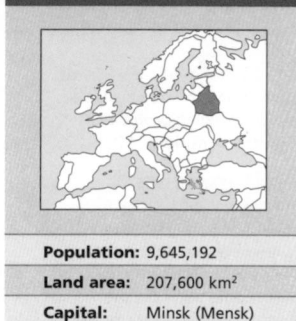

Population:	9,645,192
Land area:	207,600 km²
Capital:	Minsk (Mensk)
Currency:	Rouble
Language:	Belarusian

According to archeological evidence, the territory of present-day Belarus has hosted Upper Paleolithic and Neolithic cultures. Between the 6th and 8th centuries, the territory was inhabited by the Slav tribes: Krivichi, Dregovichi, and Radimichi. In the 9th century, the eastern Slav peoples formed the Kiev Rus, the ancient Russian State (see history of Russia) which gave rise to modern Russia, Ukraine and Belarus.

2   In the 14th century, Lithuanian principalities annexed themselves to the Western Rus. Between the 14th and 16th centuries, the Belarusian culture began to differentiate itself from that of the Russians and the Ukrainians, and to develop its own language, different from the old Russian language. Gueorgui Skorina became the first Belarusian printer.

3   With the first partition of Poland in 1772, Russia kept the eastern part of Belarus. Between 1793 and 1795, the rest of Belarus became part of the Russian Empire.

4   In 1861, the Russian Czar put an end to the serf system and the peasants' feudal bondage to the landed nobility, but this failed to solve the land problem. Only 35 per cent of the land was handed over to Belarusians, which triggered a number of uprisings, the most important in 1863 led by K. Kalinovski.

LAND USE

2003/2005

IRRIGATED AREA: 2.3% of arable land

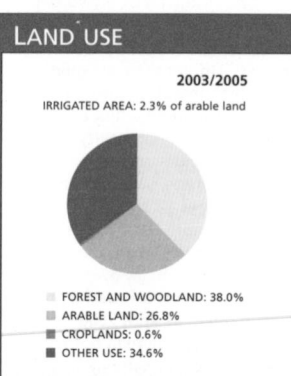

- FOREST AND WOODLAND: 38.0%
- ARABLE LAND: 26.8%
- CROPLANDS: 0.6%
- OTHER USE: 34.6%

5   In March 1899, the 1st Congress of the Social Democratic Workers' Party of Russia (SDWPR) was held secretly in Minsk. This group was inspired by Marxist socialism and was determined to bring down the czarist regime.

6   During World War I, part of Belarus was occupied by Germany. After the Russian Revolution of February 1917, councils (*soviets*) made up of workers' representatives were formed in Minsk, Gomel, Vitebsk, Bobruisk and Orsha.

7   Soviet power began to be established at the end of 1917. In February 1918 the large landholdings were nationalized; soon after, land began to be distributed among the peasants. At the insistence of the Bolsheviks (socialist revolutionaries of the SDWPR), the first collective farms (*koljoses*) were set up, while the land gradually went into State control.

8   The 6th Party Conference of the Russian Communists (Bolsheviks) - formerly the SDWPR - held in Smolensk, approved the decision to found the Soviet Socialist Republic of Belarus (SSRB). In January 1918, the 1st Congress of the SSRB soviets decided to join Lithuania. Lithuania approved this motion. On 28 February 1919, in the city of Vilno (today's Vilnius, the capital of Lithuania) the Government of the Soviet Socialist Republic of Lithuania and Belarus was elected, with Mickevicius-Kapsukas as head of state.

9   After the occupation of a considerable part of Belarus in February 1919, and according to the Treaty of Riga, signed between Soviet Russia and Poland (1921), Western Belarus became part of Poland. On 1 August 1920 the Assembly of representatives of the Lithuanian and Belarusian Communist Parties and of labor organizations in Minsk and its surrounding areas approved the foundation of the independent Belarusian republic.

10   On 30 December 1922, Belarus joined the Union of Soviet Socialist Republics (USSR) as one of its founders, along with the Russian Federation, Ukraine and the Transcaucasian Federation (Armenia, Georgia and Azerbaijan).

11   The industrialization and collectivization of agriculture began in the second half of the 1920s. On 19 February 1937 the 12th Congress of the Soviets of Belarus approved the new constitution. In November 1939, as a result of the Molotov-Ribbentrop Pact, Western Belarus was reincorporated into the Soviet Union.

12   In June 1941 the state became the first of the Soviet republics to suffer Hitler's aggression. The fortress at Brest offered fierce resistance. Guerrilla warfare extended throughout the country and by the end of World War II, more than two million Belarusians had lost their lives.

13   After Germany's defeat in 1945, Belarus' current borders were established. It became a charter member of the UN that same year, with a delegation independent of the USSR.

14   Between the 1920s and the 1980s, Belarus ceased to be a rural country (with 90 per cent of the population living from agriculture and livestock production) to become an urban, industrialized one. It took several decades for the economy to be rebuilt. Belarus became an important producer of heavy trucks, electrical appliances, radios and television sets.

15   Because of the policies of *glasnost* and *perestroika* initiated by President Mikhail Gorbachev, there was no strong pressure within Belarus to secede immediately from the USSR, although there were movements

PROFILE

ENVIRONMENT

Belarus is located between the Dnepr (Dnieper), Western Dvina, Niemen and Western Bug rivers. It is bounded in the west by Poland, in the northwest by Latvia and Lithuania, in the northeast by Russia and in the south by Ukraine. It is a flat country, with many swamps and lakes, and forests covering a third of its territory. Its climate is continental and cool, with an average summer temperature of 17-19°, and 4-7° below zero in winter.

SOCIETY

Peoples: Belarusians, 81.2 per cent; Russians, 11.4 per cent; Poles, Ukrainians and others 7.4%. **Religions:** Christian Orthodox in the east and Catholic in the west. **Languages:** Belarusian (official); Russian (second language for most of the population); Polish; Ukrainian; Yiddish and Tatar. **Main Political Parties:** Communist Party; Agrarian Party; Liberal Democratic Party. **Main Social Organizations:** Labor Union Federation of Belarus; Free Union, formed by the new strike committees; Belarusian Assembly of Democratic NGO's; Charter '97, democratic movement for civil rights and civil freedom; Women's Christian Association; Zubr, young people's movement in favour of democracy.

THE STATE

Official Name: Respublika Belarus. **Administrative divisions:** Six regions (Brest, Gomel, Grodno, Mensk, Moguiliov and Vitebsk). **Capital:** Minsk (Mensk) 1,75,000 people (1999). **Other Cities:** Gomel 503,400 people; Mogilev 372,700; Vitebsk 356,000; Grodno 311,500; Brest 304,200 (2000). **Government:** Aleksander Lukashenka, President since July 1994, re-elected in March 2006; Sergei Sidorsky, Prime Minister since July 2003. The National Assembly has two chambers: Representatives with 110 members and the Soviet with 64 seats. **National Holiday:** 25 August, Independence (1991). **Armed Forces:** 86,654 (2003). Other: Border Guards (Ministry of Interior): 8,000.

Life expectancy
69 years
2005-2010

GNI per capita
$2,140
2004

Literacy
100% total adult rate
2000-2004

favoring a multiparty political system, as well as protest demonstrations against high food prices.

[16] The catastrophe at the Chernobyl nuclear plant, Ukraine, in 1986, brought about serious consequences in Belarus, the most affected neighboring country. According to foreign researchers, in the following years there was a rapid increase in cases of cancer, leukemia and birth defects.

[17] In June 1991, Belarus declared independence and in October it signed an economic integration agreement with Kazakhstan and Uzbekistan. On 8 December that same year, the presidents of the Russian Federation and Ukraine, Boris Yeltsin and Leonid Kravchuk, and of the Belarus Parliament, Stanislav Sushkevich, signed an agreement that put an end to the USSR and founded an association of sovereign states. On 21 June, in Alma-Ata (now Almaty, in Kazakhstan), 11 republics signed an agreement creating the Commonwealth of Independent States (CIS), whose members requested admission to the UN as separate countries.

[18] In March 1994, once the new Constitution was approved, the country became a presidential republic with a 260-seat Parliament. Aleksandr Lukashenko became President after having obtained 80 per cent of the vote.

[19] In spite of having criticized in his electoral campaign his predecessors' policy of rapprochement with Moscow, in April 1996 Lukashenko signed a political and economic integration agreement with Russia. By the end of the year, a constitutional reform had granted more powers to Lukashenko. The opposition denounced the establishment of a dictatorship. In April 1999 Senate approval was given for integration with Russia, which caused some protest in the south of the country but was broadly supported in the north, where Russian influence is stronger. Among other things, it was agreed that citizens of both countries would enjoy identical rights on either side of the border.

[20] Belarus entered the 21st century with over 70 per cent of the population living below the poverty line. In the 2001 presidential election, the opposition, several of whose candidates had been banned, called for a boycott of the election. However, Lukashenko was re-elected to a second five year term amid controversy over the election's validity.

[21] In September 2002, the Parliamentary Assembly of the Council of Europe expressed

IN FOCUS

ENVIRONMENTAL CHALLENGES
Although there are large intact areas of wetlands, considered to be unique in the world, vast tracts have also been drained and there are even larger areas of lands and marshes contaminated by radioactive rain resulting from the Chernobyl catasprophe in 1986.

WOMEN'S RIGHTS
Belarusian women have been able to vote and stand for election since 1919. In 2005 women held 30.1 per cent of the seats in Parliament, and almost 10 per cent of ministerial or equivalent positions. In 2003, women represented 49 per cent of the country's total workforce. Women, children and young people of Belarus continue to bear the costs of the lengthy economic and social transition from communism to capitalism. Despite modest economic growth poverty remains widespread.

CHILDREN
The increasing number of orphans is one of the most worrying manifestations of the social and economic crisis. There is a clear problem of child abandonment and abuse. Since the Chernobyl disaster in 1986, there has been a significant increase in the number of child cancer cases and other diseases. Under-five mortality in 2003 stood at 17 deaths per 1,000 live births.

INDIGENOUS PEOPLES/ ETHNIC MINORITIES
The Polish minority lives near the border with Poland. Poles have only recently experienced significant discrimination or disadvantage. Their situation started to change after the election of Aleksandr Lukashenko as President in 1994. In 1995, the State Committee for Religious and Ethnic Affairs suggested that changes should be made to the Law on Public Associations, in order to debar national and cultural associations from political activity. This was directly addressed to the Association of Belarusian Poles (ABP), blamed for 'straining relationships with state agencies and taking part in opposition activities'. The proposal violated the Polish-Belarusian treaty which guaranteed the right of setting up organizations that would represent national minorities. There are also signs of religious discrimination against Poles, manifested in the strong official support to the Belarusian branch of the Russian Orthodox Church, and by the restriction of activities of other religious groups.

MIGRANTS/REFUGEES
At the end of 2002, about 3,600 asylum seekers and refugees were living in Belarus. Among these, 656 were recognized as refugees by the Government and 2,010 (from Russia and Afghanistan) had been rejected by Minsk but remained under the protection of the UNHCR. Belarus lacks a formal policy on providing humanitarian protection to refugees fleeing generalized violence that do not meet the criteria for asylum under the UN Refugee Convention.

Since 1997 about 600 people have been granted refugee status, mostly Afghans. In 2002 refugee status was granted to half of 106 cases (Afghans and Georgians). During 2002, more than 4,400 persons from Belarus sought asylum abroad, most of them in Western Europe. Another 16,900 stateless persons of former Soviet origin were living in Belarus in refugee circumstances. In 2001, 10,500 stateless persons were granted citizenship.

DEATH PENALTY
The death penalty still applies.

concern about the violation of human rights and fundamental freedoms. Yet abuses continued: in October 2003, the president of the Auto and Agricultural Machine-Building Workers' Union, was jailed for staging a protest action in the city center.

[22] In February 2004, Minsk reacted strongly to Moscow's decision to cut gas supplies to Belarus. Lukashenko accused Russia of 'large-scale terrorism', since the gas cut upset life in Belarus, a country which endures temperatures below zero. The Russian attitude was termed by the Government as 'blackmail' to control the gas pipeline that crosses Belarus and sends fuel to western European markets. Moscow declared that the measure was taken because Belarus had failed to pay Russia the due price for the gas. Following Lukashenko's angry reaction, Moscow and Minsk signed a 'temporary' agreement on the resumption of gas supplies.

[23] In April, a report by the Council of Europe accused the Government of involvement in the disappearance of four men, among them a Belarusian high-profile personality and some opposition activists. The report charged several government officials as well as Lukashenko himself with being suspects of involvement in the disappearances. The Government denied the charges, saying it had done everything possible to clarify the situation.

[24] Meanwhile, Mikhail Marinich, opposition candidate in the 2001 elections, former minister, ambassador and mayor of Belarus, was arrested on charges of stealing documents and trafficking with firearms. The opposition as a whole released a declaration accusing the Government of repressing civil society and democratic forces. According to the opposition, Marinich's arrest marked another wave of open political repression in the country.

[25] In October 2004, following a referendum, the Constitution was changed to allow Lukashenko to run for a third presidential term.

[26] In March 2006, Lukashenko won the presidential election with 82.6 per cent of the vote. The result sparked demonstrations in Minsk by thousands who claimed the election had been rigged; hundreds of opposition supporters were arrested.

[27] According to international observers in the country, the March 2006 elections were held in an 'irregular' climate. The Organization for Security and Co-operation in Europe labeled the electoral process as 'neither free, nor fair or democratic'.

[28] In May 2006, the EU froze any assets held in its territories by Lukashenko and 30 top aides and ministers, as well as imposing a visa ban to stop them traveling to EU countries. The list included chief of staff Gennady Nevyglas, information minister Vladimir Rusakevich, justice minister Viktor Golovanov and KGB chief Stepan Sukhorenko.

[29] Also in May, Belarusian journalists accused the Government of carrying out a campaign against press freedom in the period following the controversial elections. Alexander Starikiévich, head of the independent newspaper *Salidarnast*, accused Lukashenko of imposing an 'opinion monopoly', making sure that any newspaper or journalist expressing disagreement or criticism was immediately punished. ∎

Belgium / België - Belgique

Population: 10,453,494
Land area: 33,100 km²
Capital: Brussels (Brussel - Bruxelles)
Currency: Euro
Language: Dutch, French

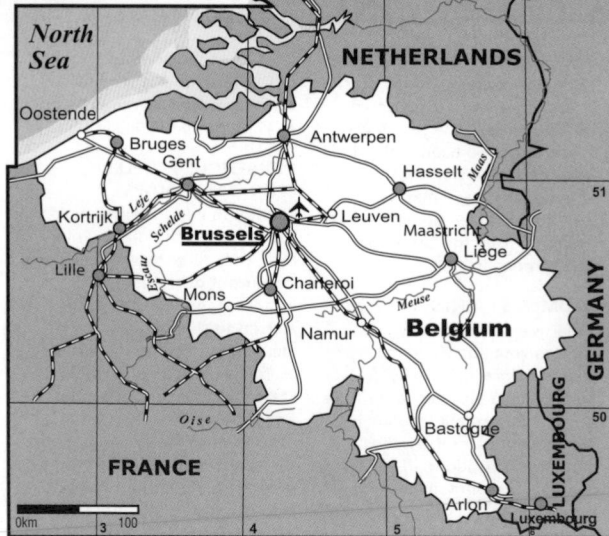

Belgium, Holland, Luxembourg and a part of northern France make up the Low Countries, which had a common history until 1579 (see Netherlands). The linguistic separation between the Roman and Germanic languages coincided with the borders of the Holy Roman Empire, which divided the Low Countries in two.

2 Between 1519 and 1814, the southern provinces were successively ruled by Spain (1519), Austria (1713) and France (1794). After the fall of Napoleon in 1814, the European powers enforced reunification with the north. The southern provinces had already forged an identity of their own, and were unwilling to accept Dutch authority.

3 The region's economy was based on the production of linen and textiles. Industrialization was facilitated by the fact that manufacturers and landowners were often one and the same, and that textile mills were concentrated in the hands of a few owners.

4 In 1830, Brussels' bourgeoisie took up arms against the Dutch authorities. When the conflict spread and proved substantial, European powers recognized the independence of the southern provinces, which were henceforth known as Belgium. Congress adopted a parliamentary monarchy, which has survived until today.

5 By the end of the 19th century, workers forced the government to pass laws aimed at providing housing for working-class people, and improving conditions in the workplace, especially for women and children. At the same time, Parliament changed the constitution and in 1893, established limited male suffrage. General male suffrage, without limitations, was introduced in 1919.

6 Between 1880 and 1885, Leopold II financed international expeditions to the Congo, making it effectively his 'private' colony. Substantial deficits and protests from several European nations against the severe repression and exploitation occurring in the Congo, forced the Belgian Government to take it as a colony in 1908. With this, the worst excesses ended, but the Belgian Government, businesses and the Church increased their influence over the following years.

7 In the 20th century, the Belgian democratic system was threatened twice by a major world war, but each time democracy prevailed. In 1914, Belgium refused to give way to the Germans and was drawn into war. The Treaty of Versailles granted the territories of Eupen and Malmedy to Belgium. In Africa, Belgium colonized the former German colonies of Rwanda and Burundi after the League of Nations granted them a mandate. In 1920 Belgium signed a military assistance treaty with France, and the following year, it formed an economic alliance with Luxembourg.

8 Belgium was occupied by Germany between 1940 and 1944. The return of the King from imprisonment by the Germans unleashed a major controversy. In a plebiscite, 57 per cent of the population voted in favor of his return to the throne, but continuing tension in Wallonia forced Leopold III to abdicate in favor of his son Baudouin (1950).

9 In 1947, Belgium, the Netherlands and Luxembourg formed an economic association known as the Benelux within the European Economic Community. Belgium also became a member of NATO in 1949. In 1960 the Belgian

LAND USE

2003/2005

IRRIGATED AREA: 4.5% of arable land

- FOREST AND WOODLAND: 20.3%
- ARABLE LAND: 26.6%
- CROPLANDS: 0.6%
- OTHER USE: 52.5%

PROFILE

ENVIRONMENT
Northwestern Belgium is a lowland, the Plains of Flanders, composed of sand and clay deposited by its rivers. In southern Belgium the southern highlands rise to 700 m on the Ardennes Plateau. One of the most densely populated European countries, its prosperity rests on trade, helped by its geography and by the transport network covering the northern plains, converging at the port of Antwerpen. Belgium has highly intensive agriculture, and a major industrial center. Heavy industry was located near the coal fields of the Sambre-Meuse valley, and textiles were traditionally concentrated in Flanders.

SOCIETY
Peoples: The country's two major language-based groups are the Flemish (55 per cent) and the Walloons (44 per cent). There is also a German minority (0.7 per cent). Over 7 per cent of the economically active population (about 250,000 people) are immigrants (Italian, Moroccan and, in lesser numbers, Turkish and African).
Religions: Mainly Catholic. There are Protestant, Muslim and Jewish minorities.
Languages: Dutch (60 per cent) and French (40 per cent) are the official languages. French is the main language spoken in the south and east, and Flemish in the north and west. German is spoken by about 0.6 per cent of the population.
Main Political Parties: Flemish Liberals and Democrats; Socialist Party (Flemish social democratic party); Spirit; Christian Democratic and Flemish; Socialist Party; Flemish Interest.
Main Social Organizations: Confederation of Christian Labor Unions of Belgium (ACV/CSC) 1,300,000 members; the General Labor Federation of Belgium (ABVV/FGTB) 1,100,000 members.

THE STATE
Official Name: Koninkrijk België/Royaume de Belgique.
Administrative divisions: 10 provinces.
Capital: Brussels (Brussel/Bruxelles) 998,000 people (2003). Brussels is also the capital of the European Union.
Other Cities: Antwerpen 945,800 people; Liège 620,900; Gent 223,000 (2000); Charleroi 201,700.
Government: Federal parliamentary state since 1993, under a constitutional monarch. Belgium has in total 6 regions and communities, each of them having a parliament, a government and an administration: Flemish region and community, Walloon Region, French speaking community, German speaking community and Brussels Capital. There are three official language communities: Flemish, French and German. King Albert II, Head of State since August 1993. Guy Verhofstadt, Prime Minister since July 1999, reelected on 2003. Bicameral Legislature: Chamber of People's Representatives, with 150 members; Senate, with 71 members.
National and Regional Holiday: 21 July, National Day (the day King Leopold I came to the throne in 1831). 11 July (Flemish official holiday). 27 September (Walloon official holiday).
Armed Forces: 53,000 (including women). Conscription was abolished in 1994. Belgium now has only a professional army.

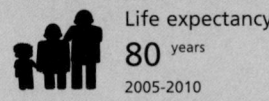

	Life expectancy		GNI per capita		HIV prevalence rate
	80 years		**$31,280**		**0.2%** of population 15-49 years old
	2005-2010		2004		2003

Congo became independent, but Belgium and the Western powers continued to intervene in the former colony (which became Zaire, and is now the Democratic Republic of Congo). In 1962 Rwanda and Burundi became independent.

[10] In 1970, linguistic communities were granted autonomy in cultural matters. In 1980, a new federal structure was approved by Parliament, made up of Flanders, Wallonia, and Brussels as the bilingual capital district. This marked the beginning of a process of decentralization that would continue for the next 30 years and is still ongoing.

[11] In 1975, Belgian women conquered the right to equal pay.

[12] In 1983, the installation of NATO nuclear missiles on Belgian soil unleashed a nationwide controversy. In 1984 and 1985 the facilities were attacked. The missiles were withdrawn in 1988, after an arms reduction agreement was signed between the US and the USSR. The mass peace movement disappeared along with the missiles.

[13] On 31 July 1993, King Baudouin died after a reign of 42 years and was succeeded by his brother Albert.

[14] Meanwhile, Parliament had agreed on major constitutional changes that turned the country into a federal state. In May 1995, voters not only elected their members of parliament, but also the 75 members of the three new regional assemblies of Brussels, Flanders and Wallonia. Premier Jean-Luc Dehaene, in office since 1992, had a clear victory in the national elections.

[15] The Government closed its 1997 budget with a national deficit of 2.7 per cent of GDP. In spite of this enormous public debt, which clearly exceeded the criteria of the Maastricht Treaty, Belgium was allowed to join the European Monetary Union.

[16] In 2001, the Lambermont agreements continued the reform of the state through decentralization. Federal powers concerning agriculture and foreign trade were transferred to Flanders and Wallonia. Food security however remained a federal issue due to the different food scandals that had threatened the country for years. Swine fever, mad cow disease and foot-and-mouth disease brought Dehaene down and allowed into the political arena a coalition headed by Guy Verhofstadt, leader of the Flemish Liberal Party (VLD). Verhofstadt formed a coalition with Socialists and Greens from both communities, leaving the Christian Democrats out of office for the first time in 41 years.

[17] On the basis of the genocide

IN FOCUS

ENVIRONMENTAL CHALLENGES
Water protection laws, passed in 1971, did not prevent the River Mosa's pollution by steelworks. This river provides water resources to 5 million people. The increase in the concentration of nitrates and the proliferation of algae in many rivers are due to contamination by manure and the intensive use of fertilizers in agriculture.

WOMEN'S RIGHTS
The first women able to stand for election were war widows, in 1921. Since 1948 there have been no restrictions on women's political participation. In 2003 women held 35 per cent of parliamentary seats, and 21 per cent of ministerial or equivalent positions. Women represent 41 per cent of the four million workers that make up the labor force. Of them, 1 per cent work in agriculture, 10 per cent in industry, 82 per cent in services and the rest in various other areas).

law of 1993, crimes against humanity, committed anywhere in the world, can be prosecuted in Belgium. This happened in 2001, when four Rwandan citizens were sentenced in Brussels for taking part in the 1994 genocide. In the same year, the existence of the genocide law prompted the Israeli premier Ariel Sharon to cancel his visit to Belgium as result of denunciations made by several Palestinians, survivors of the 1982 Sabra and Shatila massacre. Following diplomatic pressure, Louis Michel considerably weakened the Genocide law.

[18] In 1996 Marc Dutroux was arrested for the abduction, rape and murder of several children. Public opinion was shocked that this could occur and lost confidence in the police and justice apparatus. The demission of a highly appreciated inquiry judge during the inquiry procedure unleashed a mass protest against the judicial changes (White March). This led to reforms of the police and the criminal justice system.

[19] In 2001 Belgium resumed diplomatic links with DR Congo. Premier Verhofstadt and Minister of Foreign Affairs Michel visited the former colony to further the peace negotiations that should end the civil war in the eastern part of the country.

[20] The flow of refugees from Third World and former communist states became a political issue in Europe. Belgium followed the

The country is one of the main destinations for trafficked African women, particularly from Ghana and Nigeria.

CHILDREN
School is compulsory up to the age of 18. Those between 15 and 18 years old are able to combine work and school.

Belgium is a country of transit and also a destination for child trafficking, mainly from African countries. Besides child prostitution, in 1996 a group involved in violent pedophile activities and child pornography was discovered. The Government passed laws and trained police staff to fight both crimes, among other measures.

INDIGENOUS PEOPLES/ ETHNIC MINORITIES
The country has a multicultural and multi-linguistic population. In addition, ethnic minorities have been integrating into Belgian society as immigrants, refugees or people in need of assistance from non-industrialized countries.

European trend, abolishing all financial support to refugees and introducing a fast deportation procedure.

[21] In July 2001, Belgium took over the chair of the European Union. It put the Tobin Tax, the taxation of international financial transactions, on the European agenda. Since July 1, 2004 a law on the Tobin Tax has been voted in the Belgian Parliament. The law will only come into force when all other EU member states also vote a Tobin Tax law. On 1 January 2002, the Euro replaced the Belgian franc as the national currency.

[22] In August 2002 the Government agreed to sell military equipment to Nepal. According to the opposition, this transaction was against the law prohibiting arms trading with countries in a state of civil war. The governing coalition survived until the federal elections held in June 2003, where two of the coalition parties got most of the votes.

[23] In September 2002 Belgium became the second country in the world, after the Netherlands, to legalize euthanasia.

[24] A report by a committee of renowned historians stated that 10 million Congolese had been killed by the private army of king Leopold II. Brussels assumed moral responsibility and apologized for the death in 1961 of Patrick Lumumba, Prime Minister of the Democratic Republic of Congo, after a coup d'etat led by the head of the

Some initiatives were launched to fight racism and to promote intercultural dialogue. Belgium has had an anti-racism law since 1981. In 1993, a Centre for equal Opportunities and against Racism was established. Its main task is to act against racism, and to follow up Belgian integration policy.

MIGRANTS/REFUGEES
The number of asylum requests filed in 2004 fell to 15,357 almost half the number of the previous year. Although there are no official figures for the precise number of immigrants and refugees living in Belgium illegally, it is estimated there are between 50,000 and 75,000.

In 2005, Belgium granted a record number of 3,059 residence permits to political refugees, most of them coming from Russia, Rwanda, Congo, Serbia-Montenegro, Iran and Iraq.

DEATH PENALTY
Abolished in 1996. The last execution took place in 1950.

armed forces, Mobutu Sese Seko.

[25] In July 2004, 15 people died and more than 120 were seriously burned by an explosion at an industrial plant in Ghislenghien at Ath, a town located 30 km southwest of Brussels. Several of the dead had been investigating a gas leak in a pipeline taking the fuel from Belgian coast towards France. The Belgian Government considered the explosion 'one of the worst disasters of recent times'. France sent helicopters and emergency medical aid.

[26] After months of failed negotiations regarding linguistic rights in a district including parts of Flanders and Brussels the Government won a confidence vote in Parliament in May 2005. The final decision on the issue was postponed for two years and cast doubt on the future of Belgium as a unified country. The dispute, concerning political parties from Flanders and Wallonia, was complex. Many French-speaking residents had moved to Flemish suburbs in recent years and the political parties were not willing to lose their control over those voters.

[27] In April 2006 Parliament passed a law by a one-vote margin making Belgium the fourth country in the world - after Spain, Netherlands and Canada - to allow the adoption of children by gay couples. In 2003 it had been the second country in the world to allow same-sex marriage. ■

Belize / Belize

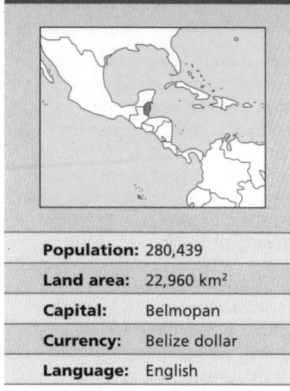

Population:	280,439
Land area:	22,960 km²
Capital:	Belmopan
Currency:	Belize dollar
Language:	English

Native American people known as the Itzae were the original occupants of what is now Belize (formerly British Honduras). Belize, together with Guatemala and southern Mexico, was part of the Mayan Empire. In Belize, the Mayas built the cities of Lubaatún, Pusilhá, and a third which archaeologists call San José.

[2] In 1504 Columbus sailed into the bay naming it the Gulf of Honduras. Spain was nominally the colonial power in the region, but never pushed further into Belize, because of tough resistance from the local people. According to the terms of the Treaty of Paris (1763) Spain allowed the British to start exploiting timber in the area. This authorization was later confirmed in the Treaty of Versailles (1783). In 1798 the British gained control of the colony, although Spain retained sovereignty until it became a British colony in the 1840s, and its name changed to British Honduras.

[3] The first settlers in Belize were British Puritans, attracted by the cedar, campeche wood and timber. They began to establish themselves in the coastal areas, importing African slaves to work their estates. Shortly afterwards, slaves outnumbered Europeans, and in 1784 only 10 per cent of the population was of European extraction.

[4] The ethnic base became more heterogeneous in the early 19th century. By that time, Garifuna migrants had settled on the Southern coast of Belize. The War of Castes in Yucatan between 1847 and 1901 displaced thousands of Spanish-speaking inhabitants to the northern coast of Belize, while several Maya communities re-settled in the northern and western areas of the country. These immigrants introduced changes in agricultural techniques that set the pattern for subsistence farming and for sugar, banana and citrus production. During the 1860s and 1870s, sugarcane plantation owners sponsored the migration of several thousand workers from China and India. By the end of that century, Maya and Kekchi indigenous peoples, who had escaped oppression in Guatemala, established self-sufficient communities in the south and west of Belize.

[5] By the early 20th century, the economy was stagnant and the British colonial administration prevented any democratic participation. In 1931, a hurricane destroyed a large part of Belize City. That same year, a series of strikes and demonstrations by workers and the unemployed gave rise to unions and increasing demands for democracy. Eligibility for voting was legally introduced in 1936, but it was heavily restricted by level of literacy, land title and gender.

[6] In 1949 when the Governor devalued the national currency, union leaders and the local middle class joined in a People's Committee that demanded constitutional changes. As a result, in 1950 the People's United Party (PUP) was founded, led by George Price. First organized as a 'people's committee' to fight against arbitrary treatment by the colonial administration, the PUP won its first elections by a landslide majority. In 1954 the direct election of the legislative representatives was approved. In 1961 a ministerial system of government was established, and in 1964 the country was granted internal autonomy, with Price becoming Prime Minister. On 1 June 1973 the country changed its name to Belize.

[7] Guatemala claimed to have inherited sovereignty over Belize from Spain, and did not recognize the Guatemala-Belize border. In March 1981, Guatemala and Britain signed a 16-point agreement. Britain assured the future independence of Belize in exchange for some concessions to the Guatemalan regime, such as free and permanent access to the Atlantic, joint exploitation of the marine resources, the building of a pipeline, and an 'anti-terrorist' agreement.

[8] Price and the PUP were accused of partiality towards the 'revolutionary' Cuba and Nicaragua, an issue exploited by the right-wing opposition, the United Democratic Party (UDP), which won the December 1984 elections.

[9] The new Prime Minister, Manuel Esquivel, a US-educated physics professor, adopted a liberal economic policy and supported the

LAND USE

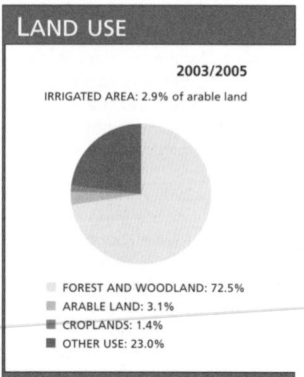

2003/2005

IRRIGATED AREA: 2.9% of arable land

- FOREST AND WOODLAND: 72.5%
- ARABLE LAND: 3.1%
- CROPLANDS: 1.4%
- OTHER USE: 23.0%

PROFILE

ENVIRONMENT
Belize covers the southeastern tip of the Yucatan Peninsula. The land is low, and the climate warm and rainy in the north. In southern Belize, the hillsides sustain a variety of crops. The northern coastline is marshy and flanked by low islands. In the south there are excellent natural harbors between reefs. Significant oil and gas deposits are believed to exist off the coast.

SOCIETY
Peoples: Spanish-Indian descendants 43.6 per cent; Creole (predominantly black) 29.8 per cent; Mayan Indian 11.0 per cent; Garifuna (African-Carib Indian) 6.7 per cent; white 3.9 per cent; East Indian 3.5 per cent; other or not stated 1.5 per cent.
Religions: 60 per cent are Catholic, most of the rest are Protestant (Anglican, Methodist, Seventh-Day Adventist, Pentecostal, Jehovah's Witness).
Languages: English (official); common language is Creole. Spanish, Quiché, Yucatan Mayan and Garifuna are also spoken.
Main Political Parties: The People's United Party (PUP), founded in 1950; The United Democratic Party (UDP).

Main Social Organizations: The General Workers' Union, the Christian Workers' Union (CWU), the United General Federation, the General Federation of Workers and the Public Service Union of Belize.

THE STATE
Official Name: Belize.
Administrative divisions: 6 districts.
Capital: Belmopan 9,000 people (2003).
Other Cities: Belize City 50,200 people; Orange Walk 13,800; San Ignacio 13,600; Dangriga (formerly known as Stann Creek) 9,000; Corozal 8,100 (2000).
Government: Parliamentary monarchy, Queen Elizabeth II of England is Head of State. Colville Young, Governor-General since November 1993. Said Wilbert Musa, Prime Minister since August 1998, re-elected in 2003. Legislative power lies with the House of Representatives, which has 29 members and is elected by universal suffrage, and a 8-member Senate.
National Holidays: 10 September, National Day; 21 September, Independence Day (1981).
Armed Forces: 1,065 (1995).

Life expectancy
72 years
2005-2010

GNI per capita
$3,940
2004

Literacy
77% total adult rate
2000-2004

HIV prevalence rate
2.4% of population 15-49 years old
2003

private import-export sector, which was in the hands of inexperienced family business ventures. He also developed favorable policies to encourage foreign investment and attracted US, Jamaican and Mexican investments in tourism, energy, and agriculture.

[10] Sugarcane generated 50 per cent of the country's revenue, but a fall in international prices badly affected the economy. The industry only survived because of import quotas guaranteed by the US and EEC markets, which took 60 per cent of the sugar output, with the remainder being sold at a loss.

[11] In March 1986, Prime Minister Esquivel proposed a plan for the sale of Belizean citizenship, aimed primarily at Hong Kong business people. Anyone investing $25,000 in government bonds - really worth only $12,500 - would be granted instant citizenship of Belize.

[12] During the 1980s, the country received approximately 40,000 Salvadoran, Guatemalan, Honduran and Nicaraguan refugees. In spite of official tolerance, some officials blamed immigrants for the rise in marijuana trafficking and crime.

[13] When Vinicio Cerezo became President of Guatemala in 1986, relations between the countries changed substantially. In December 1986, Cerezo's government re-established diplomatic relations with Britain, broken off two decades earlier over Guatemalan claims to Belizean territory. A Permanent Joint Commission was formed with Belizean, Guatemalan and British representatives to find a peaceful solution to the issue.

[14] Meanwhile, the number of US embassy personnel grew sixfold after independence and the number of Peace Corps volunteers was ten times higher. Dean Barrow, Minister of Foreign Affairs and Economic Development, admitted that he was aware of the country's dependence upon the US. But he also stated that Belize could defend its territorial integrity through the Non-Aligned Movement.

[15] Drug-trafficking - or the sale of 'Belize breeze' as marijuana is known - showed spectacular growth. According to some foreign economists, it became the country's main export, and it was estimated that some 700 tons - worth at least $100 million - had been taken into the US, its market price there being 10 times this sum. The area under cultivation increased by at least 20 per cent in spite of a US-sponsored herbicide-spraying campaign.

[16] George Price was returned to power when the PUP won the September 1989 elections. In 1991, Guatemalan President Jorge Serrano Elias finally recognized Belize's sovereignty and right to self-determination. The

Government of Belize in return gave Guatemala free access to the Gulf of Honduras.

[17] At the end of the 1980s and early 1990s a large number of Haitians went to Belize as agricultural laborers, attracted by higher wages than at home. Lower inflation rates and more equal income distribution were incentives for immigration, and they also had access to health and education services. The social services consequently became overloaded, triggering a backlash from Belizeans who demanded that the immigrants be deported.

[18] Banana, sugar and citrus production increased, employing 40 per cent of the working population and representing four-fifths of the country's exports. Tourism became the sector with the greatest potential. Mayan archeological finds attracted many visitors.

[19] In the June 1993 elections the PUP was defeated by the UDP, led by Esquivel and dominated by Spanish-Indian descendants. The Government formed a group of economic advisors to look into establishing facilities for the capital and tourist markets, and the development of more free zones.

The fear of new territorial demands from Guatemala and the withdrawal of British troops led to an increase in defense spending and new recruitment for the armed forces.

[20] Victory in the August 1998 general elections went to the PUP and Said Wilbert Musa became the new Prime Minister. The UDP was reduced to just three seats.

[21] On 24 February 2000 four members of the Belize security force were apprehended in Guatemala for allegedly entering that country's territory illegally. The incident intensified the dispute between Belize and its neighbor. Almost immediately the Caribbean Community (CARICOM) accused the Guatemalan Government of invading territory and kidnapping Belizeans.

[22] In 2001 Hurricane Iris buried entire towns and left more than 13,000 homeless. The hurricane flattened forests, destroyed banana plantations and claimed around 20 lives.

[23] On 7 February 2003 Belize, Guatemala and Honduras, under the supervision of the Organization of American States (OAS), signed an agreement to give Guatemala access to the Caribbean Sea. At the same

time, a tripartite commission was set up to oversee fishing in the Gulf of Honduras.

[24] On 15 April 2005 strikes and protests broke out against budget cuts and corruption in the Musa administration, leaving 37 injured (among them 10 police agents) and the country without telecommunications. Musa, who had said the telecommunications system had been 'toppled' in an act of 'vandalism and sabotage' by employees of the privatized main telephone company (which most of the population wanted returned to State hands) blamed the opposition. The demonstrations, started by secondary students in the capital, were peaceful at first, but later less peaceful demonstrators blocked streets and railroads and looted businesses. More than 100 demonstrators were arrested.

[25] In February 2006, negotiations to reach an agreement in the territorial dispute between Belize and Guatemala over jurisdictional waters were resumed after three years. The final agreement will be subject to a referendum in both countries. If no agreement is reached, an OAS arbitration court will intervene. ∎

IN FOCUS

ENVIRONMENTAL CHALLENGES
Sea pollution, soil erosion and deforestation are problems shared with other Caribbean countries. The Belize Barrier Reef is a UNESCO World Heritage Site; there are signs of damage to the reef from agricultural effluent, global warming and tourism. Close to the cities, the surface of the water is covered with residual wastes and by-products of sugarcane production. Industrial waste, dumped in the water, has generated public health problems and killed fish stocks. Final disposal of household waste and sewage water treatment are unsatisfactorily resolved issues.

WOMEN'S RIGHTS
Women have been able to vote and run for office since 1954. In 2003 women held 11 per cent of Parliament seats, while their representation in ministerial or equivalent positions is 6 per cent.

In 2003 women made up 25 per cent of the workforce. Of them, 72 per cent were wage workers, 25 per cent were independent workers and 3.5 per cent were underemployed. Female unemployment amounted to 20.3 per cent.

Some 96 per cent of women receive pre-natal care, but only 83 per cent of deliveries are assisted

by qualified staff.* The total birth rate is 3.1 per woman*.

CHILDREN
Over 48 per cent of the population is under 18 years old. High rates of crime and violence persist among youngsters. Between 1990 and 2000, there has been an increase in the number of institutions devoted to childhood and adolescence. There are 3,600 confirmed cases of HIV/AIDS. In 2001 and 2002 there was a 150 per cent increase in the AIDS infection rate amongst infants under one year old. Most mother-to-child-transmissions during delivery were in HIV-positive single mothers. The percentage of pre-school national health coverage is low: a total of 27.5 per cent, with higher figures in Belize City (60.4 per cent) than in other districts, particularly Toledo (2.7 per cent). The infant mortality rate is 32 per 1,000 live births, while the under-five mortality rate is 39 per 1,000. Both rates have improved significantly in the last 15 years.

INDIGENOUS PEOPLES/ETHNIC MINORITIES
Garifuna (African-Carib Indian) which are 7 per cent of the population, were recognized by the Government as a people in 2001. The Government announced its intention to work with UNESCO in protecting their culture. However their land ownership claims

are still questioned.

Maya culture (11 per cent of the population) is threatened by the plans to build a dam by Belize's electricity company and Fortis (a Canadian transnational), which could flood an 1,100 hectare region. Experts say it could threaten the environment and also ancient Maya sites and tourism.

MIGRANTS/REFUGEES
Belize is one of the countries with the highest immigrant rate compared to its native population. Immigration's impact is felt throughout the land and at every ethnic, cultural, social and economic level. Immigrants come mainly from Central America, especially from El Salvador, although also from Asia and North America. Between 14 per cent and 20 per cent of the population are Central American immigrants (35,000 to 50,000). Of these, 35 per cent are refugees, 25 per cent are legal immigrants and 40 per cent illegal.

DEATH PENALTY
It is currently applied.

* Latest data available in *The State of the World's Children* and *Childinfo* database, UNICEF, 2006.

Benin / Benin

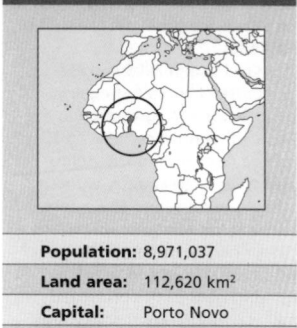

Population:	8,971,037
Land area:	112,620 km²
Capital:	Porto Novo
Currency:	CFA franc
Language:	French

Benin (known as Dahomey until 1975) is among the poorest countries in the world. It lies in the region of the Yoruba culture, which developed at the ancient city of Ife. It was here that the Ewe peoples, who came from the same linguistic family, evolved into two distinct kingdoms during the 17th century: Hogbonu (today known as Porto Novo) and Abomey, further inland. These states developed around the booming slave trade, serving as intermediaries.

² The traditional rulers of Abomey, the Fon, built a centralized state that extended east and west beyond Benin's present-day frontiers. A well-disciplined army, with European rifles and a large contingent of female soldiers enabled them to end the patronage of the Alafin of Oyo (Nigeria) and capture various Yoruba cities. After the 17th century, Ouidah became the main port for British, French and Portuguese slave traders receiving their human cargo.

³ The ruling group of Abomey suffered a setback in 1818 when Britain banned the slave trade, although Ghezo who ruled between 1818 and 1856 maintained a thriving clandestine traffic to Brazil and Cuba. He also promoted the development of agriculture and established a strict state monopoly on foreign trade.

⁴ In 1889, Ghezo's grandson Benhanzin inherited a prosperous State, although already threatened by colonialism. In 1891, Fon troops resisted the French invasion only to be defeated a year later. The King and his army retreated to the forests, where they held out until 1894. Benhanzin, who became a symbol of anti-colonial resistance, died in exile in Martinique in 1906.

⁵ The colonists destroyed the centralized political structure of the ancient Fon state. Traditional Fon society was dismantled and replaced with a system based on the exploitation of farm labor. The French also declared a monopoly on the palm-oil trade and ruined families that had resisted foreign penetration for nearly a century.

⁶ By the beginning of the 20th century, the colony of Dahomey (as the French called it), was no longer self-sufficient. When it gained independence in August 1960, oil-seed exports stood at 1850 levels, while the population had tripled.

⁷ Independence came as a direct consequence both of France's weakness at the end of World War II and the activities of European-educated nationalists, led by Louis Hunkanrin, who waged a stubborn 20-year struggle against the compulsory labor imposed by the French. All forms of political organization were banned and in retaliation, Hunkanrin created the Human Rights League. A period of ruthless repression followed: hundreds of villages were burned down, nearly 5,000 people were killed, and Hunkanrin took refuge in Mauritania.

⁸ By 1960, Dahomey had become an unbearable economic burden and France agreed to independence. The new Government inherited a bankrupt economy and a corrupt infrastructure. A series of 12 military and civilian governments marked a 16-year period of instability.

⁹ The neocolonial élite collapsed in 1972 when then-Major Mathieu Kérékou headed a coup by a group of young officers opposed to political corruption and official despotism. Two years later, a Marxist-Leninist State with a communitarian political and economic system was proclaimed and its name was changed to Benin,. All foreign property was nationalized, and a single-party system was introduced with the creation of the People's Revolutionary Party.

¹⁰ The revolutionary government became the target of several conspiracies plotted abroad. There was an unsuccessful invasion in January 1977 with the participation of French mercenaries and the support of Gabon and Morocco.

¹¹ In 1980, a new Revolutionary Assembly was elected through direct vote. The Government switched to a more pragmatic foreign policy and diplomatic relations with France were resumed. Although palm-oil production continued to fall as the trees grew older and were not replaced, cotton and sugar sales rose. High unemployment continued, but in 1982 offshore oil was discovered thus guaranteeing energy

LAND USE

2003/2005

IRRIGATED AREA: 0.4% of arable land

- FOREST AND WOODLAND: 21.3%
- ARABLE LAND: 24.0%
- CROPLANDS: 2.4%
- OTHER USE: 52.3%

PUBLIC EXPENDITURE

% OF GDP

HEALTH (2002)	EDUCATION (2000-2002)
2.1%	3.3 %

PROFILE

ENVIRONMENT
Benin is a narrow strip of land that extends north from the Gulf of Guinea. Its 120-km sandy coast lacks natural ports. Several physical regions cut across the country from south to north: the coastal belt where oil palms are cultivated; the tropical wooded lowlands; and the plateau which rises gradually towards the headwaters of the Queme, Mekrou, Alibori and Pendjari rivers, in a region of tropical hills.

SOCIETY
Peoples: Benin's people stem from 60 ethnic groups. The Fon (47 per cent), Adja, Yoruba and Bariba groups are the most numerous and before French colonization they had already developed stable political institutions. There is a European minority.
Religion: Around 70 per cent practice traditional African religions, 15 per cent are Muslim and 15 per cent Catholic.
Languages: French (official). Other widely-spoken languages are Fon, Fulani, Mine, Yoruba and Massi.
Main Political Parties: Democratic Renewal Party; Social Democratic Party; Benin Rebirth Party or Renaissance Party of Benin; African Movement for Development and Progress; Movement for the People's Alternative.
Main Social Organizations: The Benin Workers' National Trade Union (UNSTB) is the only union.

THE STATE
Official Name: République Populaire du Bénin.
Administrative divisions: 6 provinces.
Capital: Porto Novo 238,000 people (2003).
Other Cities: Cotonou 704,900 people; Djougou 177,300; Parakou 141,100; Abomey-Calavi 86,900 (2000).
Government: Presidential republic with a strong Head of State. Yayi Boni, President since April 2006. Since 1998 he has also taken the role of Prime Minister. Single chamber parliament of 83 members.
National Holiday: 1 July, Independence (1960); 30 November, Revolution Day (1974).
Armed Forces: 4,800. Other: Gendarmerie and People's Militia: 4,000.

Life expectancy
56 years
2005-2010

GNI per capita
$450
2004

Literacy
34% total adult rate
2000-2004

HIV prevalence rate
1.9% of population 15-49 years old
2003

IN FOCUS

ENVIRONMENTAL CHALLENGES
Deforestation and desertification are among the main environmental problems, made worse in recent years by a significant decrease in rainfall. Only 12 per cent of rural people have access to adequate sanitation.

WOMEN'S RIGHTS
Women have been able to vote and stand for election since 1956. In 2003 women held 7 per cent of Parliament seats; representation in ministerial or equivalent positions amounted to 19 per cent.

In 2005, women represented 48 per cent of the total workforce (65 per cent of them in agriculture, 4 per cent in industry and 30 per cent in services). The fertility rate stood at 5.2 children per woman*. Meanwhile, 81 per cent of women receive perinatal assistance; while 34 per cent of deliveries are not assisted by qualified staff.* For every 100,000 live births, 850 women die*. The frequency of HIV/AIDS

in women aged 15 to 20 was 2.2 per cent (compared with 0.9 per cent for men of the same age); for women aged 20 to 24 the prevalence rate increased to 4.8 per cent.

CHILDREN
Poverty, illiteracy and illnesses are some of the factors hindering progress towards the achievement of child rights in the country. Only 38 per cent of children under 6 months old were exclusively fed on breast milk*.

In 2006, 23 per cent of children under 3 years old suffered from chronic malnutrition - an improvement of 2 percentage points on the 1996 figure of 25 per cent - and 8 per cent of children suffered from acute malnutrition - an improvement on the 1996 figure of 14 per cent. Chronic malnutrition is almost equally distributed among girls and boys, while acute malnutrition is more prevalent among boys.

In 2003, there were 34,000 children orphaned by HIV/AIDS,

and 340,000 children orphaned for other reasons*. By the end of that same year, 12,000 children between 0 and 12 years old were carrying the virus*. Forty per cent of children under 5 years old who had fever as a symptom of malaria were not receiving medication against the disease.

INDIGENOUS PEOPLES/ ETHNIC MINORITIES
Benin is part of French-speaking Africa but there are still 51 ancestral languages alive in the country, among them Yoruba, Bariba and Awuna. Some of these are also still spoken in parts of Latin America, having been carried there by slaves transported from the region of present-day Benin. In addition, Benin and Nigeria are the cradles of *Vodou*, a form of ancestral cult which was also carried to Latin America and the Caribbean by slaves and was syncretized with other religious practices to create voodoo. One of the current dangers faced by indigenous peoples such as the Ogoni and the Ijaw in Benin

and neighboring countries is the threat to their environment by transnational oil companies. This is part of an agreement with the Government of Nigeria for the construction of a gas pipeline designed to traverse 600 miles (1,000 km) from Benin up to Togo and Ghana.

MIGRANTS/REFUGEES
In 2005, after the May elections in Togo, some 25,000 Togolese emigrated toward Benin seeking asylum. Many were taken to refugee camps set up by UNHCR, and also to the homes of relatives living in the country. In February 2006 a series of violent clashes between local residents and refugees led some 9,000 refugees to flee from a camp in Lokossa, 18 km from the border with Togo.

DEATH PENALTY
It is still applied, though there have been no executions since 1987.

> * Latest data available in *The State of the World's Children* and *Childinfo* database, UNICEF, 2006.

self-sufficiency. In addition large phosphate deposits were discovered in the northern Mekrou region. However, hopes of recovery dimmed as a serious drought reached the northern provinces.
[12] The economic crisis forced the Government to accept the terms of the International Monetary Fund (IMF), including a 10 per cent income tax and a 50 per cent reduction in non-wage social benefits.
[13] On 8 December 1989, disappointed with results and besieged by street demonstrations, President Kérékou announced that he was abandoning his Marxist-Leninism. A new Constitution was drawn up, providing for a series of political and economic reforms, especially the promotion of free enterprise.
[14] On 24 March 1991 Prime Minister Nicéphore Soglo defeated President Kérékou with 68 per cent of the votes in the country's first presidential election in 30 years. In 1992, former president Kérékou, who had been prosecuted for his activities following the 1972 coup, was granted an amnesty, and political prisoners were released.
[15] Soglo continued the economic liberalization and privatization policy initiated by Kérékou in 1986. Debt

servicing still represented a high percentage of the resources annually obtained (in 1992 debt service amounted to 27 per cent of the country's income).
[16] The 100 per cent devaluation of the CFA franc decreed by France in January 1994 had contradictory effects on Benin's economy. GDP continued to grow at a 4 per cent annual rate and cotton exports increased. However, public expenditure cuts brought about drastic reductions in social spending schemes.
[17] In the general elections held in March 1996, Soglo suffered a narrow defeat by Kérékou, the former leader of the Marxist regime.
[18] Between August and October 1997, disease killed around 60,000 pigs - 10 per cent of the country's stock of swine.
[19] The country's five unions called for a strike in February 1998 against 'the antisocial budget for 1998, the dictates of the IMF, the World Bank and the European Union, as well as widespread corruption encouraged by Kérékou's government'.
[20] The March 1998 parliamentary elections saw the presidential faction win over the other 55 parties. Prime Minister Adrien Houngdedji stood down and his role was taken on by Kérékou.

[21] In January 2000, the President denounced a conspiracy against his government by some of the military, particularly former members of regional peacekeeping forces. A group like this had led a coup in Côte d'Ivoire in December 1999 and a similar attempt in Mali had been narrowly avoided when another group tried to claim outstanding pay.
[22] In July 2000, Benin qualified for the Broad Initiative for Heavily Indebted Poor Countries with a $460 million reduction in its foreign debt.
[23] A boat carrying dozens of children made Benin the center of an international search in April 2001, highlighting slave trafficking in that country and the region. In October 2003, the Presidents of Nigeria and Benin decided to undertake a joint mission for the repatriation of slave children, with the support of UNICEF and the NGO Terre des Hommes.
[24] In January of the same year, a law was passed banning all female genital mutilation, setting fines and penalties ranging from six months to three years in prison for practitioners. The penalties increase up to five years if the woman is a minor, and could be 10 years if she dies as a result of mutilation.
[25] In the parliamentary

elections of March 2003, the Kérékou faction again won a parliamentary majority with 55.8 per cent of votes.
[26] A UNICEF report published in May 2005 revealed that some 50,000 children were trafficked every year from the country. Traffickers assured parents their children would earn enough money to send home large sums and would receive a good education. Most are trafficked to Nigeria to work in quarries and others to Europe, to serve as domestic slaves. Most only got to see their parents years later, if ever. A way to frighten the trafficked children was to force them to take a vow in voodoo ceremonies, and warn them they would die if they broke it.
[27] Boni Yayi, a former banker and newcomer to politics, won the second round of the March 2006 presidential elections with almost 75 per cent of the vote. Thus came to an end the era of Kérékou, who had held power for more than 30 years. In his 70s, he had exceeded the constitutional age limit for presidential office.
[28] In June 2006 Parliament took the decision that capital punishment would not be abolished, on the grounds that abolition might make the country a refuge for international criminals. ∎

Bermuda / Bermuda

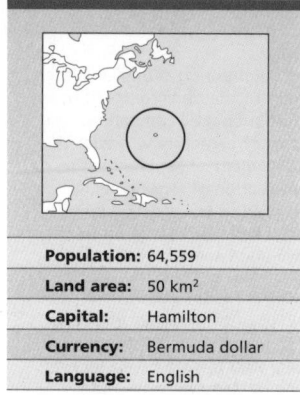

Population:	64,559
Land area:	50 km²
Capital:	Hamilton
Currency:	Bermuda dollar
Language:	English

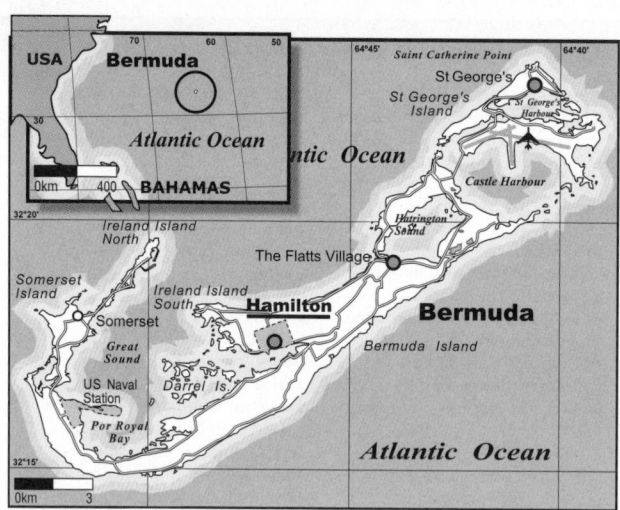

The Bermuda archipelago was the first colony of the British Empire. Sighted by the Spanish navigator Juan Bermúdez in 1503 (the place was named after him), it was settled in 1609 when British emigrants on their way to America were shipwrecked nearby.

2 From 1612, the colony welcomed religious and political dissidents. In 1684, it began to be administrated by the British Crown, and the first parliament was installed. Even though African slaves formed the majority, only plantation owners could elect representatives.

3 Agriculture almost disappeared in the 20th century, being replaced by tourism, gambling and transnational corporations lured by numerous tax exemptions. Bermuda is one of the most densely populated places in the world (more than 1,000 people per square kilometers). The residents of this tourist and tax haven do not receive much of the huge wealth that circulates there. In the last century, during the Prohibition (1919-1933) traffickers living in Bermuda smuggled rum into the United States. From 1941, Washington set up air and naval bases on the islands.

4 After the creation of the Bermuda Industrial Union (BIU), in 1963, workers founded the Progressive Labor Party (PLP), which favored the country's total independence and the introduction of an income tax. The following year, the Right organized the United Bermuda Party (UBP).

5 Britain granted the island greater administrative autonomy in 1968, and the majority party was given the right to appoint the prime minister. The period leading up to the elections was marked by racial and political violence. The assassination of the governor led to intervention by British troops. When the election was held, the UBP won by a large margin.

6 The UBP won again in the 1976 election, though the PLP increased its number of seats. From its position in opposition, the PLP continued to demand greater autonomy for the island.

7 In 1977, two members of the 'Black Cadre' were sentenced to death, for participating in armed anti-colonial activities. Their execution unleashed a wave of protests, with British troops intervening once again. The Minister of Communal (Race) Relations was sacked, and his duties delegated to the Bermuda Regiment, responsible for putting down protest and discontent.

8 The 1979 Bermuda Constitutional Convention failed to reach a consensus on representation and the minimum voting age. However, a reduction in the number of non-Bermudan voters went into effect that December.

9 The US recession continued to affect tourism in the first few months of 1992. London rejected plans for independence.

10 That year, the economic crisis caused the French naval base to close as well as cutbacks in personnel at the US air base.

11 In August 1995 the question of independence was put to a referendum. The result was against independence, so Bermuda's status as a British colony was maintained.

12 The PLP won the 1998 elections and Jennifer Smith became Prime Minister, putting an end to the 30-year-UPB government. However, despite this fact, Smith's leadership was questioned by vast factions of the PLP and she was forced to resign. Alex Scott took her place.

13 Although there had been no executions in Bermuda since 1977, capital punishment had remained on the statute book. However, the death penalty and corporal punishment were finally abolished in December 1999.

14 In June 2000 Bermuda was excluded from the Organization for Economic Co-operation and Development (OECD) report on tax havens, as a result of the Government's commitment to reform their tax system before 31 December 2005.

15 In March 2003, a constitutional reform reduced the members of the Assembly from 40 to 36 and in December, the Education Law was reformed, establishing that parents are responsible for their children's behavior at school, and imposing a fine on those who break the law. The Black Alliance, an organization for the protection of African descendants' rights, claimed that the Law was discriminatory, and that it tended to criminalize the poor. The law is only applied to public education, which is mostly used by the working class and Afro-descendants.

16 In September, the worst hurricane in four decades swept across Bermuda. 'Fabian' reached winds of up to 200 kilometers per hour and caused waves nearly 10 meters high, thus forcing the evacuation of hundreds of people from low-lying areas.

17 The Pan American Health Organization released a report in December 2005, saying that despite the high HIV/AIDS rates in Caribbean countries, some places see a 'light of hope'. That includes Bermuda where, for the first time since the virus became known, a reduction in new cases has been registered.

18 In October 2006, Ewart Brown, formerly minister of tourism and transport, took over as premier following a vote at the PLP delegates conference. ∎

PROFILE

ENVIRONMENT
This Atlantic archipelago is made up of 360 small coral islands, characterized by chalky, permeable soil. Bermuda includes 150 of these islands, of which 20 are uninhabited. The warm Gulf Stream current produces a mild climate which attracts tourists, mainly from the US. Pollution, especially from the US, caused by former military bases, is one of the country's main environmental problems.

SOCIETY
Peoples: Approximately 60 per cent are of African descent; there are also descendants of Portuguese from Madeira and the Azores; mixed European and Indian descendants and a minority of European origin.
Religions: Anglican majority (28 per cent), in addition to Methodists (12 per cent), Adventists (6 per cent), Catholics (15 per cent) and members of other religions.
Languages: English (official): Portuguese.
Main Political Parties: Progressive Labor Party (PLP); The United Bermuda Party (UBP).
Main Social Organizations: Bermuda Industrial Union, Bermuda Public Service Association, Audubon Society of Bermuda, the Bermuda Human Rights Alliance and the Black Alliance.

THE STATE
Official Name: Bermuda. Administrative division: 9 counties.
Capital: Hamilton 1,000 people (2003).
Other Cities: St George's 1,800 people (2000).
Government: British dependency. John Vereker, Governor since April 2002; nominated by Queen Elizabeth II. Ewart Brown, Prime Minister since October 2006. The Prime Minister appoints the cabinet on approval by the Governor. Bicameral parliament: Senate has 11 members, 5 elected by the Prime Minister, 3 by the opposition leader and 3 by the Governor. Assembly has 36 members elected by direct voting for a 5-year term.
National Holiday: 24 May, Bermuda Day.

Bhutan / Druk Yul

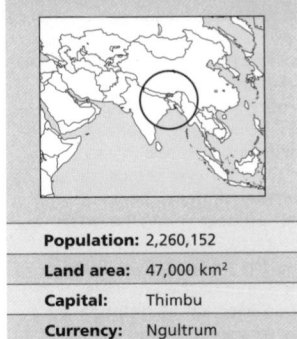

Population:	2,260,152
Land area:	47,000 km²
Capital:	Thimbu
Currency:	Ngultrum
Language:	Dzongkha

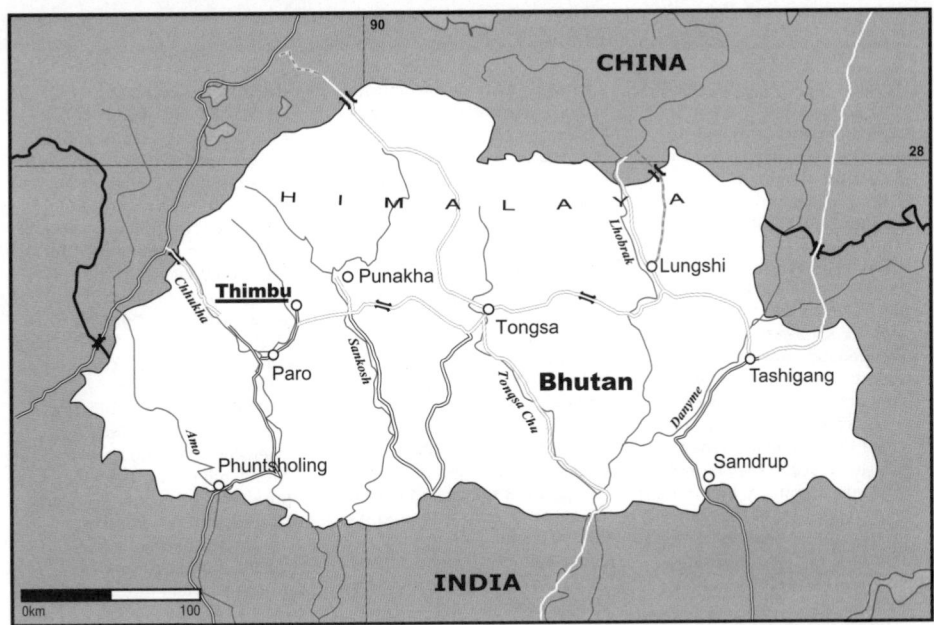

B hutan lies in the heart of the vast Himalayan mountains. Early explorers and envoys of the British colonial Government called it Bootan, land of the Booteas, or sometimes Bhotan. Wedged between giant neighbors China and India and cut off by some of the world's highest peaks, it was little known to the rest of the world. Even today, the origin of its name remains unknown. Perhaps it came from the Sanskrit. To the people of Bhutan it is Druk or Druk Yul, land of the Thunder Dragon.

[2] From the 12th century, the Drukpa Kargyud tradition became dominant. After a long period of rivalry among various groups, the country was united in the 17th century by a Drukpa Kargyud Lama named Ngawang Namgyal. Druk, the country's endogenous name, derives from the Kargyud sect of the Mahayana Buddhism (*drukpa*), which is currently the official religion. Namgyal, popularly known as Shabdrung ('at whose feet one submits'), was both the country's spiritual and secular ruler. Factionalism gradually eroded the power of the subsequent Shabdrungs. On 17 December 1907 Ugen Wangchuk reunified the country and established Bhutan's first hereditary monarchy.

[3] The British colonial administration in India signed important treaties with Bhutan in 1774 and 1865. The 1910 Treaty of Punakha stipulated that the British would not interfere in Bhutan's internal affairs, but made the country a British protectorate in terms of external relations. Similar provisions were included in the 1949 treaty signed between Bhutan and independent India.

[4] Bhutan emerged from its isolation in the 1960s and joined the Colombo Plan for Co-operative, Economic, and Social Development in Asia and the Pacific and the UN in 1962. Bhutan also joined the Non-Aligned Movement.

[5] Bhutan is still one of the poorest countries in Asia, although the emergent industrialization and recent development make exploitation of its natural resources feasible.

[6] President-monarch Jigme Singye Wangchuck was crowned in 1972 and at that point the

LAND USE

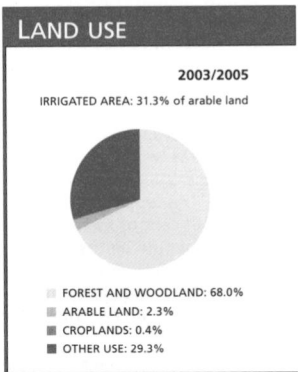

2003/2005

IRRIGATED AREA: 31.3% of arable land

- FOREST AND WOODLAND: 68.0%
- ARABLE LAND: 2.3%
- CROPLANDS: 0.4%
- OTHER USE: 29.3%

PROFILE

ENVIRONMENT
This Himalayan country is made up of three distinct climatic and geographical regions. The Duar plain in the south, humid and tropical, is densely wooded and ranges in height from 300 to 2,000 meters. At the center lies a temperate region with mountains of up to 3,000 meters. Finally, there are the great northern heights, with year-round snow and peaks up to 8,000 meters. Forests are the country's mainstay. Rivers and waterfalls have been exploited and they supply energy to the country and neighboring areas. There are graphite, marble, granite and limestone deposits. Approximately half of the arable land lies on steep slopes; of this nearly 15 per cent is merely topsoil.

SOCIETY
Peoples: The main ethnic groups are the Bhutias, Drukpas and the Tibetans (50 per cent). The ethnic minorities are geographically divided: the Ngalong live in the west area, the Scharchops live in the east and the Nepalis (35 per cent of the population; known as Lhotsampas) live in the south. There are also Lepchas, indigenous people, and Santal, descended from Indian immigrants. 95 per cent of the teachers and 55 per cent of public servants are of Indian origin.
Religions: Buddhist 69.6 per cent; Hindu 24.6 per cent; Muslim 5.0 per cent; other 0.8 per cent (1980).
Languages: Dzongkha (official). Nepali and other dialects are also spoken.
Main Political Parties: there are no legal parties. Political groups from the Nepali minority operate from the exile: the Bhutan People's Party (BPP), founded in Nepal in 1990 represents the Nepali minority; the National Bhutanese Congress (DNC) founded in 1992, in Nepal and India; the United People's Liberation Front (UPLF), also founded in 1990.
Social Organizations: the National Front for Democracy (NFD); the National Women's Association.

THE STATE
Official Name: Druk Yul.
Administrative divisions: 18 Districts.
Capital: Thimbu 35,000 people (2003).
Other Cities: Phuntsholing 54,300 people; Punakha 20,700; Samdrup 13,200 (2000).
Government: Jigme Singye Wangchuk, King since July 1972; hereditary absolute monarchy. The monarch is assisted by a council of 9 members, of whom 5 are elected by the people, 2 are appointed by the monarch and 2 by the Buddhist religious dignitaries whose 6,000 lamas (monks) are headed by the Je Khempo. Khandu Wangchuk, Prime Minister since September 2006. There is also a consultative assembly (Tsogdu) of 150 members 105 of whom are elected, 35 appointed and 10 representatives of Buddhist groups. The ruler also dispenses justice.
National Holidays: 8 August, Independence Day (1949, from India); 17 December, National Day (1907); 2 August, Buddhist Lent; 30 October, end of Buddhist Lent.
Armed Forces: 7,000 royal troops.

Life expectancy
65 years
2005-2010

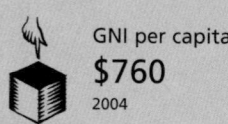
GNI per capita
$760
2004

IN FOCUS

ENVIRONMENTAL CHALLENGES
Most arable soil is thin and is located in steep areas, which makes the territory very susceptible to erosion. Access to drinking water is becoming increasingly difficult.

WOMEN'S RIGHTS
Women, who have been able to vote and stand for election since 1953, have been excluded, discriminated against and exploited. They have never been appointed to ministerial or equivalent positions. In 2003, women held nine per cent of Parliament seats. They have no economic autonomy, and more than 90 per cent of them are illiterate. Their health situation is poor, with high maternal mortality and anemia during pregnancies.

Sexual exploitation of women is common. A discriminatory law has been in force since 1988, which regulates mixed marriages between men from dominant ethnic groups (Drukpa or Ngalong) and Lhotsampa women (Lhotsampas are Bhutanese Hindus of Nepali origin), or between Lhotsampa men and non-Bhutanese women. The Bhutanese men who marry Lhotsampa women - and their children - automatically lose their civil, economic and social rights. In addition, foreign wives of

Lhotsampa men are discriminated against - particularly Nepalese or Indian women - and the couple is deprived of civic rights: this does not happen when the Drukpas marry foreigners. This legislation is retroactive, which means that it applies to marriages solemnized before 1988. More than 10,000 Lhotsampas' wives are deprived of their nationality.

CHILDREN
In 1990, 65 per cent of the population was under 30 years old, and 40 per cent was under 15. In 1999, the percentage of working minors aged 10 to 14 years old was the highest in Asia (55 per cent), despite the fact that in 1990 Bhutan had signed the Convention on Children's Rights. This is the main reason for the low school enrolment rate - despite the Kingdom's claim to have universal primary education. Only 47 per cent of girls go to school, 11 percentage points below the rate registered for boys.

The under-five mortality rate was 80 per 1,000 live births.

Malnutrition and anemia are particularly high. In 2003, 40 per cent of children suffered from arrested development. By the end of the 1990s, 5 per cent of the armed forces were under 18 years old. Roughly 90,000 children under 17 are orphans.

INDIGENOUS PEOPLES/ ETHNIC MINORITIES
The Lhotsampas - Bhutanese Hindus of mainly Nepali origin, whose language derives from Sanskrit - have very different traditions from those of the Drukpas or the Ngalongs. They have mainly settled in the warmer southern areas of the country, while the Ngalong - Buddhists who speak the Tibetan language - live in the colder north of the country.

The 1985 Citizenship Act deprived the Lhotsampas of citizenship; claiming they were illegal immigrants, the Bhutanese authorities arrested, tortured and murdered many of them, confiscated their properties and documents and forced them into exile. In 1997, the Bhutanese National Assembly decreed that the 'Nepali nationals' - the Lhotsampas - could not work in Bhutan, and allowed the Drukpas or the Ngalongs to settle on the land of those Lhotsampas who had fled to neighboring countries. In 2001, Bhutan and Nepal began jointly to 'verify' the potential candidates for repatriation, but by the end of 2002, disagreements about the verification process - together with the continuing occupation of Lhotsampas' ancestral

land by Bhutanese people - prevented the repatriation process from succeeding. In October 2003, after the verification process in one of the 7 refugee camps was completed, only 3 per cent of the refugees were allowed to return to Bhutan.

The six NGOs that visited the Khundunabari refugee camp rejected this 'solution' to the refugee issue because of the low repatriation rate and the irregularities found in the verification process: the UN High Commissioner for Refugees was not allowed to monitor it and was denied access to the places of return.

MIGRANTS/REFUGEES
In 2006, some 105,000 Bhutanese lived as refugees in Nepal; 49 per cent of them were women and 40 per cent children. Almost all were Lhotsampas or Bhutanese of Nepali or Indian origin, who lived in the southern plains of Bhutan. Most of them fled to Nepal and India during the first years of the 1990s to escape the hostility, the harassment and expulsion by the Bhutanese authorities.

DEATH PENALTY
Abolished in 2004. The last execution was in 1964.

kingdom began a slow process of opening up to the world. Alongside this, a policy was initiated for the Bhutanization of the country. Following the Decree of Citizenship in 1985, many southerners were declared illegal immigrants. After the 1988 census, the monarch imposed the use of national dress and the Dzongkha language in public places. Education in Nepali was banned and work permits for foreigners were stopped. The political crisis arising from these measures led to the displacement of 100,000 Nepali Bhutanese to Nepal, where they lived in refugee camps. This event constitutes one of the largest ethnic expulsions in human history. Since 1990, Bhutan has considered the refugees, known as Lhotsampas, as people of no nationality.

[7] The seventh five-year plan started in 1992, with the objective of increasing exports, environmental conservation, regional balance and the institutional promotion of women. The Government encouraged

foreign investment and began a cautious privatization program. In the 1990s, Bhutan signed bilateral agreements with the Netherlands, Norway, Japan and Switzerland (among others), which provided funding for development programs.
[8] In 1998, the King, who was also head of state, government and head of the Court of Appeals, appointed a prime minister and a cabinet, whilst also granting the Consultative Assembly the right to initiate ideas or veto his decisions, and even call for his abdication. The opposition accused him of planting his followers in key positions in the Assembly.
[9] Singye Wangchuk set out his own philosophy of development, which states that economic growth and material progress are not the only way to achieve personal success, as equal emphasis must be placed on emotional and spiritual security. As State ideology, this implies the aim to be achieved is Gross National Happiness (GNH) above Gross Domestic Product. GNH formed the framework for

an agreement signed in 2001 by Bhutan and the Asian Development Bank. The agreement prioritizes poverty reduction in the country, and the Bank is supposed also to take into account the non-material elements of well-being.
[10] Bhutan is one of the countries which signed bilateral agreements that grant US citizens charged with genocide, crimes against humanity and war crimes, immunity from prosecution by the International Criminal Court.
[11] In September 2003, Bhutanese and Indian authorities agreed to try to find a solution which allowed the armed groups who fought for the independence of the Indian state of Assam to leave Bhutan.
[12] In December, the Government launched a new offensive against Indian separatist groups and transferred hundreds of soldiers to the western region of the country in order to eliminate rebels hiding in the forests after their bases were destroyed in September. Thimbu was in state of alert; the capital was a potential target for rebel attacks.

[13] In July 2005 the National Front for Democracy (NFD), a coalition made up by three political parties in exile, strongly criticized the Government's draft Constitution. The NFD said the draft aimed to distract international attention from the country's refugee crisis. The Constitution, if approved by referendum, would establish a two-party system, a 'limited democracy' in the words of the NFD.
[14] Lyonpo Sangay Ngedup, former health minister, was named Prime Minister in September 2005.
[15] A top Government official said Bhutan could become in 2006 the largest hydroelectric power producer in Asia. From 2005 to 2015 it would export to India some 5,000 megawatts per month. This would represent almost 45 per cent of its total exports, and would be the turning point for the kingdom's economic development. 'A major step in the goal of becoming an economically independent nation in the future', according to the official. ∎

Bolivia / Bolivia

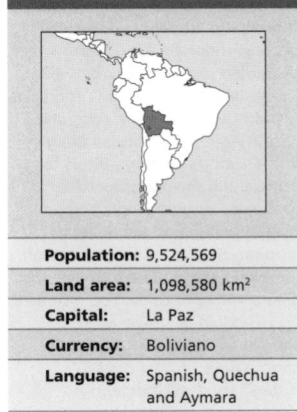

Population:	9,524,569
Land area:	1,098,580 km²
Capital:	La Paz
Currency:	Boliviano
Language:	Spanish, Quechua and Aymara

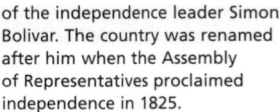

In 2000 BC the region of modern-day Bolivia was inhabited by farmers in the Andes, and forest hunter-gatherers in the east. Their terrain ranged from high mountain areas, *puna*, to hot valleys and forests. They raised livestock and grew potatoes, cotton, maize, and coca; they also fished and mined. This region, rich in natural resources, sustained several kingdoms and fiefdoms around Lake Titicaca, with the Tiahuanaco-Huari (600 BC-1000 AD) civilization at their center.

[2] The basic social unit was the *ayllu* kinship group, in which there was no private land ownership. The society was stratified into farmers, artisans and the ruling ayllu, of priests and warriors, who appointed the *malku* (chief).

[3] By 800 AD, Tiahuanaco formed the first Pan-Andean empire. By 1100 the Incas, from the Cuzco Valley in Peru, had colonized the other Andean peoples and formed a confederation of states called the Tahuantinsuyo. Also known as the Inca Empire, it adopted elements of Tiahuanaco culture, technology, religion and economics, particularly the ayllu social unit.

[4] Through the *mita*, each worker would render service to the centralized state. This system was later cruelly exploited by the Spaniards. Social organization was based on self-sufficient, communal production.

[5] When the Spanish arrived at the beginning of the 16th century, the Tahuantinsuyo extended from southern Colombia, through Ecuador and Peru, to northern Chile, and from Lake Titicaca and the *altiplano* highlands to northern Argentina, embracing the mountain valleys and the eastern plains. One million people were estimated to have been living within the area of the present Bolivia, and from 12 to 13 million in the Tahuantinsuyo as a whole, making it the most densely populated area of South America. This society included a number of ethnic groups, predominantly Aymara (around Lake Titicaca) and Quechua. The eastern plains were inhabited by dispersed groups of Tupi and Guaraní, with no central nucleus. To this day, Aymara and Quechua are the most widespread languages in Bolivia.

[6] In 1545 the Spanish discovered silver at Potosí. They extracted immense quantities of the metal, which contributed to the capital accumulation of several European powers. Millions of Native Americans died there,

cruelly exploited to the point of exhaustion. Potosí was one of the three largest cities of the 17th century, growing up at the foot of the hill. It became the economic nerve center for vast regions of Chile and Argentina, and nurtured a rich mining bourgeoisie, guilty of corruption, ostentation and wastefulness.

[7] Decades of popular struggle against the Spanish reached their peak with the successive rebellions of Tupac Katari (1780-82) and in the Protective Board of La Paz (1809). The pro-independence movement was subsequently taken up by the *criollos* (Spaniards born in America), who distorted it by advocating social, economic and political systems based on the models of emerging European capitalist powers. A British blockade interrupted the supply of mercury, essential for treating the silver, and the Bolivian mining industry went into decline. The Buenos Aires based trading bourgeoisie soon lost interest in Upper Peru (Bolivia) and offered little resistance when it fell under the influence

of the independence leader Simon Bolivar. The country was renamed after him when the Assembly of Representatives proclaimed independence in 1825.

[8] Peru exerted great influence over the independent Bolivia until 1841. Bolivian president Marshall Andrés de Santa Cruz tried to modernize Bolivia, founding universities and the Supreme Court of Justice, and compiling law codes.

[9] A mine-owning oligarchy including Patiño, Aramayo and Hochschild worked with the politicians and generals, who were their associates, treating the Bolivian republic as if it were part of the tin business. British imperialist interests, initially in the saltpeter at Antofagasta and later in Bolivia's southern oil reserves, triggered two wars in South America: the Pacific War (Chile against Bolivia and Peru from 1879-83), and the Chaco War (Paraguay and Bolivia, from 1932-35). As a result of these conflicts, Bolivia lost its coast and three-quarters of its territory in the Chaco region. The ceding of Amazonian Acre to Brazil, in 1904, completed the country's dismemberment.

[10] On 21 July 1946 Bolivian president Gualberto Villarroel - accused by left and right parties as 'fascist'- was overthrown, assassinated, and his body hung from a lamp-post in downtown La Paz. Villarroel who had overthrown elected-president General Enrique Peñaranda, confronted the owners of the mines imposing taxes and organizing the miners' trade unions. He also mobilized peasants for the first time, gathered in the Indigenous Congress.

[11] Nationwide frustration at these humiliations gave way to a powerful current of reformism and anti-imperialism. The MNR grew up alongside progressive labor and peasant movements. After several uprisings, and a 1951 electoral victory that was not honored, the MNR led a popular insurrection in 1952. Civilians defeated the oligarchy army in the streets, and they carried first Victor Paz Estenssoro, and then Hernan Siles Zuazo, to the presidency. The Bolivian revolution nationalized the tin mines, carried out agrarian reform and proclaimed universal suffrage. Workers' and peasants' militias were organized and together with the Bolivian Workers' Confederation (COB) formed a coalition with the MNR.

[12] Troubled by internal divisions, the MNR gradually lost its drive and was defeated in November 1964 by a military junta led by René Barrientos. Ernesto 'Che' Guevara tried to establish a guerrilla nucleus in the Andes to spread revolutionary war throughout

LAND USE

2003/2005

IRRIGATED AREA: 4.1% of arable land

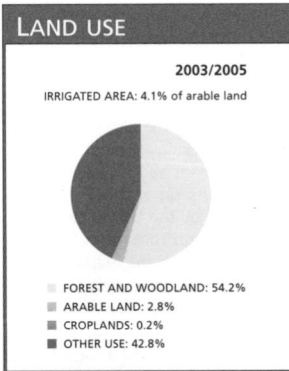

- FOREST AND WOODLAND: 54.2%
- ARABLE LAND: 2.8%
- CROPLANDS: 0.2%
- OTHER USE: 42.8%

PUBLIC EXPENDITURE

% OF GDP

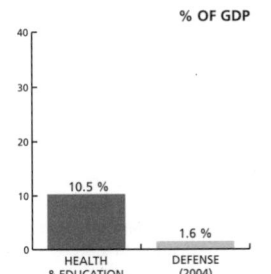

HEALTH & EDUCATION (2000-2002)	10.5 %
DEFENSE (2004)	1.6 %

WORKERS

UNEMPLOYMENT: 6% (2004)

LABOR FORCE **2004**

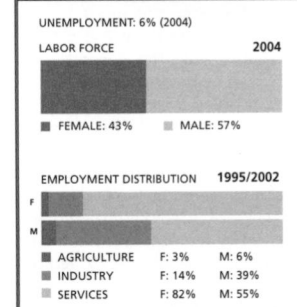

- FEMALE: 43%   - MALE: 57%

EMPLOYMENT DISTRIBUTION **1995/2002**

F
M

AGRICULTURE	F: 3%	M: 6%
INDUSTRY	F: 14%	M: 39%
SERVICES	F: 82%	M: 55%

Life expectancy
66 years
2005-2010

GNI per capita
$960
2004

Literacy
87% total adult rate
2000-2004

HIV prevalence rate
0.1% of population 15-49 years old
2003

PROFILE

ENVIRONMENT

A landlocked country with three geographic regions. 70 per cent of the population live in the cold, dry climate of the altiplano, an Andean highland plateau with an average altitude of 4,000 meters. This region holds the country's mineral resources: tin (second largest producer in the world), silver, zinc, lead and copper. The subtropical valleys (*yungas*) of the eastern slopes of the Andes form the country's main farming area, where coffee, citrics, cocoa, sugarcane, coca and bananas are grown. The tropical plains of the East and North, a region of jungles and grasslands, produce cattle, rice, corn, and sugarcane. The area is also rich in oil. Bolivia is made up of three drainage basins: Lake Titicaca (closed basin), the Amazon in the north and the Río de la Plata in the south.

SOCIETY

Peoples: 57 per cent of Bolivians are Quechua and Aymara. Mestizos and Cholos account for 25 per cent of the population. A minority of European descent has ruled the country since the Spanish conquest. The Tupí and Guaraní peoples live in the eastern forests.
Religions: Mainly Catholic (95 per cent). Syncretism in indigenous communities. Protestant and Jewish minorities. Freedom of religion.
Languages: Spanish, Quechua and Aymara (all official). More than half the population speaks native languages (including Guaraní); there are 39 ethnic-linguistic groups.
Main Political Parties: Movement Toward Socialism; Democratic and Social Power; National Unity Front; Revolutionary Nationalist Movement; Indigenous Pachakuti Movement.
Main Social Organizations: The Bolivian Workers' Confederation (COB) and the Sole Labor Union Confederation of Farm Workers of Bolivia (CSUTCB), mainly indigenous. Indigenous Confederation of the Eastern region, Chaco and Bolivian Amazon Area (CIDOB); the Guaraní People's Assembly (APG); the Aymara People's Parliament (PPA); the Federation of Campesino Women; the Federation of Neighborhood Commissions; and the Bolivian Forum on Environment and Development (FOBOMADE).

THE STATE

Official Name: República de Bolivia.
Administrative divisions: 9 Departments.
Capital: Sucre (212,000 people) is the constitutional capital, and seat of the judiciary. La Paz 1,477,000 people (2003) - including El Alto (766,100 people) which became a separate city in 1988 - functions as the seat of government.
Other Cities: Santa Cruz de la Sierra 1,089,400 people; Cochabamba 558,500 (2000).
Government: Evo Morales, President and Head of the Government since January 2006. Bicameral legislature: Chamber of Deputies, made up of 130 members; Senate with 27 members.
National Holiday: 6 August, Independence Day (1825).
Armed Forces: 32,000 (2003). Other: 23,000 Police.

South America, but he was caught by US-trained counter-insurgency troops, and assassinated on 8 October 1967.

[13] Division within the army coupled with pressure from the grassroots level led to an anti-imperialist faction taking over government in 1969, with General Juan José Torres as the leader. During his short time in office, there was a significant increase in the number of grassroots organizations. The People's Assembly was formed, with links to the COB and the parties of the Left. In August 1971 he was ousted by Colonel Hugo Bánzer Suárez, who formed a government with MNR support. The civilian-military coalition government remained in power until July 1978, with an

authoritarian but development-oriented administration encouraging agribusiness and stressing infrastructure projects, bolstered by the high price of oil and other minerals.

[14] Military uprisings and disregard for election results occurred repeatedly between 1978 and 1980. On 29 June 1980, the elections were won by the Democratic Popular Union (UDP), a center-left coalition whose candidate, Hernan Siles Zuazo, was prevented from taking office by another bloody coup engineered by General Luis Garcia Meza. According to reports received by Amnesty International, thousands of people were killed or tortured.

[15] By 1982 internal dissent and the erosion of the regime's

international standing because of its connections with drug trafficking, plus the dogged popular resistance led by the Bolivian Workers' Confederation (COB), resulted in the fall of the military regime. On 10 October Hernan Siles Zuazo took office after 18 years of military regimes.

[16] Siles Zuazo, with a populist/nationalistic outlook, handed administration of the state-owned mines to the labor unions. He also announced the non-payment of Bolivia's foreign debt. The labor movement and the peasants exerted pressure on the government through demonstrations, and several laws passed by his administration allowed these groups to participate in the economic policy decisions of large businesses, and in local committees dealing with food, health and education issues. In response to these measures, creditor banks, the IMF and the World Bank blocked credits to Bolivia placing an embargo on its international trade, provoking a fiscal crisis and uncontrolled hyperinflation.

[17] Under heavy pressure from all social sectors, the Government cut short its term and called elections for July 1985. As neither candidate secured over 50 per cent of the vote, the decision was put to Congress, who elected Victor Paz Estenssoro (MNR) as constitutional president, though Hugo Banzer (ADN) had marginally more votes.

[18] The Estenssoro Government decreed a program of neo-liberal measures to end subsidies and close down state enterprises. It eliminated price controls and the official listing of the dollar against the local currency. Mines were closed down, and others rented out, leaving thousands of miners jobless, while investment ground to a standstill.

[19] In the 1989 national elections, Jaime Paz Zamora's Revolutionary Movement of the Left (MIR) proved itself a potent new political force, coming third with 19 per cent of the votes (double that of previous elections). The MNR candidate, Gonzalo Sánchez de Losada, obtained 23 per cent and Banzer's Nationalist Democratic Action (ADN), 22.6 per cent. As no party had achieved a majority, an agreement known as the Patriotic Accord (AP) between the MIR and the ADN made it possible for Paz Zamora to be nominated president by the National Congress. The Patriotic Accord continued Paz Estenssoro's neo-liberal economic policy.

[20] The Government embarked upon privatization of state enterprises, except those considered strategic. Congress passed a law permitting the state to sell off 22 of

the existing 64 public enterprises, though the Supreme Court ruled it unconstitutional. The Government promoted joint ventures between the state-owned Mining Corporation (COMIBOL) and private companies.

[21] In April 1991, Bolivia's Congress authorized US military officers to come and train local personnel in the war against drugs. Despite military action and the policy of crop substitution - a program called 'development in place of coca' - the area under coca cultivation increased.

[22] The loss of power in the workers' movement was compensated for by new organizations of indigenous peoples and communities. Several congresses of the Confederation of Indigenous Peoples of the Eastern Region, the Chaco and the Bolivian Amazon Region (CIDOB) and of the Guaraní People's Association, among other organizations, were held. Their demands included assignation of lands, habitat preservation and the use of their native languages in the educational system. The population of Bolivia's eastern region included 10 linguistic groups and 35 ethnic groups.

[23] In September 1990, a group of peoples from this region carried out a 750 km march from the east to the capital with the slogan 'Land and Dignity'. Paz Zamora's government approved a National Plan for the Defense and Development of the Indigenous Peoples, and in August 1991, recognized the Santa Ana de Horachi.

[24] In January 1992, Presidents Paz Zamora and Alberto Fujimori (of Peru) signed an agreement whereby Peru ceded an area of 327 hectares to Bolivia for it to develop a free zone at the port of Ilo, so Bolivia gained a free port for its international trade.

[25] Gonzalo Sánchez de Lozada, of the MNR, won the national elections of June 1993, with 36 per cent of the vote.

[26] In its first year the Government established the right to education in the Indian languages (Aymara, Quechua, Guaraní). The Capitalization Bill aimed to privatize 50 per cent of the main public industries (telecommunications, electricity, oil, gas, railways, airlines) on the basis of transferring half the shares to the Bolivian citizens as pension funds. The aim was to attract foreign investment, reduce unemployment and increase GDP.

[27] The burning of coca plantations by the US led to continuous confrontations between the peasants and the military.

[28] The Capitalization Bill, unpopular amongst workers, led to a series of strikes in 1995. The Government declared a state of emergency on two occasions,

Under-5 mortality	Poverty	Debt service	Maternal mortality
69 per 1,000 live births 2004	**23.2%** of population living on less than $1 per day 2002	**18.6%** exports of goods and services 2004	**420** per 100,000 live births 2000

granting the police special powers and imposing a curfew.

29 In the national elections in June, the ADN won with 22 per cent of the vote, followed by the MNR, MIR, UCS and CONDEPA. Hugo Banzer became president after complicated negotiations.

30 The 100 coca syndicates agreed voluntarily to reduce production to comply with US requirements, as the US had promised to give $40 million to help in the war on drugs. In January 1999 Washington stated the plan had been a success and that Bolivian production had been reduced by 50 per cent.

31 Jorge Quiroga replaced Banzer as president in August 2001, because the latter was suffering from cancer.

32 In December, poor farmers rejected a government offer of $900 per head per year to stop planting coca. The same month, farmers' leader Casimiro Huanca was shot dead by police.

33 In Cochabamba, the country's third-largest city, local citizens rebelled in 2000 against a World Bank-imposed privatization of the local public water system, leased off to the US transnational, Bechtel. Angry over steep increases in water prices, Cochabambinos shut down their city for a week and forced Bechtel out of the country.

34 Two candidates topped the poll in the June 2002 national election: Sánchez de Lozada, with 22.46 per cent of the vote and Evo Morales, the Indian leader of the coca growers and candidate for Movement towards Socialism (MAS), who received 20.94 per cent. Since neither won an outright majority, Congress had to vote and Sánchez de Lozada was elected president in August.

35 In January 2003 10 peasants and 2 soldiers were killed in a crude act of aggression that occurred after the blockade of the country's main road by coca planters. Sanchez de Lozada started a dialogue with peasants, who wanted an increase in the authorized coca plantation quota and a radical change in the Government's policies.

36 On 1 September 2003, a group of peasants and workers marched from Caracollo town to La Paz, opposing the export of gas, through Chile to the US, and demanding the resource be reserved in the first instance for domestic use and internal development. The protest grew for a month, with pickets, marches and road blockades.

37 On 10 October, demonstrators virtually laid siege to La Paz. Unrest spread throughout the country after the army opened fire on the crowd, leaving 26 dead and many injured in El Alto on 12 October.

38 The biggest march in Bolivia's history took place on 16 October 2003. Human rights activists, intellectuals and middle-class people on hunger strike demanded that the President step down in favor of Vice-President Carlos Mesa. On the 17th, Sanchez de Lozada resigned, fleeing to Miami with his family and some of his ministers. That night Mesa was declared President in a special session of the Congress.

39 The new President promised to organize a referendum on gas policy, responding to the main claims of the people and opposition parties.

40 The so-called 'gas war' had left 74 dead and hundreds injured, mostly indigenous people.

41 Finally, in July 2004, the referendum was carried out. The government-backed option won with 75 per cent of the vote. The Government's strategy was to use a route via Peru for natural gas exports. In August, Bolivia and Peru signed a deal to allow Bolivia to export gas through a Peruvian port. By means of this agreement, Bolivia was granted access to the sea for the first time in 125 years.

42 In October, thousands of peasants marched towards La Paz, demanding that former president Sánchez de Lozada was put on trial, blaming him for the death of 58 people during the 'gas war'. At the same time, they demanded a bill was passed for the nationalization of hydrocarbons.

43 Overwhelmed by the massive social protests demanding the nationalization of oil and by regional demands for autonomy, Mesa resigned. Street protests had intensified to the extent that the staff at the Quemado Presidential Palace had to be evacuated.

44 Evo Morales became Bolivia's first indigenous President, winning an absolute majority in December 2005 elections. Upon taking office in January 2006, he stated that nobody would be marginalized by his government and that he would 'respect the voice of the people'. Likewise, in order to make it clear that there would be no room under his government for those who used to profit at the expense of the State and people's poverty, he quoted an Inca proverb: 'do not lie, do not steal and do not be lazy'.

45 On 1 May 2006, speaking at the San Alberto oil field, Morales announced the nationalization of hydrocarbons. The announcement had large repercussions, both at the political and the social level - it had been the centerpiece of Morales' political campaign and was celebrated by thousands of people during events held on Workers' Day. As expected, the political opposition, producer regions, businesspeople and oil companies all expressed their concern.

44 In October 2006, Bolivia agreed energy deals with 10 foreign gas and oil firms, just before a deadline for foreign firms to agree new contracts or leave the country. ∎

IN FOCUS

ENVIRONMENTAL CHALLENGES
Intensive agricultural production using poor methods of cultivation and uncontrolled logging are among the biggest problems. These have to increases in soil erosion, desertification and biodiversity loss, and have threatened forest and fauna diversity and water resources. Air contamination has increased in La Paz, as a consequence of industrial emissions and the proliferation of vehicles. Drinking and irrigation water resources are polluted by untreated industrial waste waters.

WOMEN'S RIGHTS
All Bolivian women have had the vote and have been able to stand as candidates since 1952 (between 1938 and 1952 a limited number of women were allowed to participate in politics). In 2005, parliamentary seats held by women amounted to 14 per cent, while their representation in ministerial or equivalent positions reached almost 10 per cent.
According to the the most recent figures, dating from 2002, women made up 38 per cent of the total workforce. Twenty per cent of women over 15 were illiterate, while for men in the same age group this rate was at least 8 percentage points lower. Almost 79 per cent of pregnant women receive prenatal care and 67 per cent of births are attended by skilled health staff*. In 2003, it was estimated that 1,300 women aged between 15 and 49 were living with HIV/AIDS*.

CHILDREN
Although under-five mortality has declined as a result of government policies aimed at strengthening the Basic Health Insurance to cover pregnant women and children in this age group, anemia and chronic infant malnutrition persist among children under 5 years old. Eight per cent of these children are underweight*.
Immunization against common illnesses has improved, as has school enrolment, but both are worse in the rural areas. Approximately 800,000 people under 18 years old worked in mines, harvesting sugarcane or in prostitution*. The spread of sexually transmitted diseases, including HIV/AIDS, among children and teenagers is still on the increase.

INDIGENOUS PEOPLES/ ETHNIC MINORITIES
There are several indigenous peoples' associations to defend their rights and work against the discrimination they have suffered for five centuries. After the December 2005 presidential elections, Evo Morales (an indigenous Aymara-Quechua), of the Movement towards Socialism (MAS) became the first indigenous president in the history of Bolivia. Also for the first time in history, 14 out of the 16 ministerial positions were to be held by indigenous leaders. For the different indigenous groups, Morales' victory marked the beginning of a time of hope that their rights would finally be recognized.

MIGRANTS/REFUGEES
Those recognized as refugees can get a permit to live in Bolivia indefinitely or temporarily, with travel and identity documents. However, even recognized refugees still find it difficult to get hold of such documents.
In the 1990s, over 200,000 Bolivian emigrants chose South American countries, especially Brazil and Argentina, as their destination, seeking informal sector jobs; they stayed as illegal immigrants for long periods of time. Many of them returned to Bolivia during the Argentinean economic crisis of 2001. Since 2003 Bolivians are included among those people from the Americas for whom there will be no restrictions on mobility, residence and work within the MERCOSUR countries. Apart from possible new emigrants, this benefits many Bolivians who have been living illegally in these countries for decades.

DEATH PENALTY
The death penalty was abolished in 1997. The last execution took place in 1974.

* Latest data available in The State of the World's Children and Childinfo database, UNICEF, 2006.

Bosnia-Herzegovina / Bosna i Hercegovina

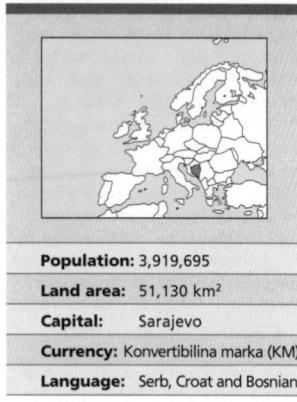

Population:	3,919,695
Land area:	51,130 km²
Capital:	Sarajevo
Currency:	Konvertibilina marka (KM)
Language:	Serb, Croat and Bosnian

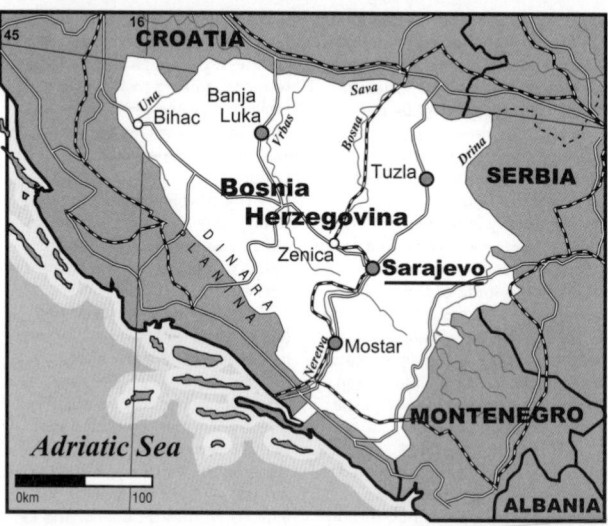

The earliest inhabitants of what is now Bosnia-Herzegovina were Illyrians and Celts. The Roman Empire crossed the Adriatic Sea in the mid-2nd century BC, and created the province of Illyria, where the border between East and West was drawn according to the division of the Roman Empire. Slavs settled in the area in the 7th century AD, coming from present-day Poland and Ukraine, gradually absorbing Illyrians and Celts.

[2] In the mid-12th century, the region came under the jurisdiction of the Hungarian archbishop of Kalocsa. The combined efforts of the papacy and of the Hungarians to impose their religious authority, gave rise to strong national resistance. Bosnia was a bastion of the Bogomils (or Cathari), one of southern Europe's main heretical movements. Neighboring Christians - both Orthodox Serbs and Catholic Croats - organized several crusades against them.

[3] Ban Prijezda founded the Kotromanic dynasty (1254-1395) under which Bosnia conquered the province of Hum (Herzegovina, took its name from the Duke [Herceg] Stejpan Vukcic, who ruled the southern area of the present republic until the arrival of the Turks). In 1377, Tvrtko crowned himself King of Serbia, Bosnia and the coastlands. The Turkish invasion of 1386 defeated the Serbs in Kosovo (1389), but Tvrtko carried out further conquests in the west, and in 1390 was crowned King of Rashka, Bosnia, Dalmatia, Croatia and the coastlands.

[4] The Ottoman Empire conquered Constantinople in 1453 and occupied Serbia in 1459. Bosnia became a province of the Ottoman Empire in 1463. Hum resisted, but in 1482 the port of Novi (now Herzegnovi) fell, and Herzegovina too became an Ottoman province. Bosnian Bogomils converted to Islam. In addition to Catholic Slavs and Christian Orthodox Slavs, there were now Muslim Slavs. Muslims were the élite and Christians the *raia* (poor); relations between the three communities were strained, and religion became the decisive social factor.

[5] The Turkish governor (Pasha) had his headquarters in Banja Luka, but later transferred them to Sarajevo. In 1580, Bosnia was divided into 8 *sanjaks* (sub-regions), under the jurisdiction of 48 hereditary Kapetans who exercised a feudal power over their territories where the manufacture of wrought metals and weapons was developed. In the 16th and 17th centuries, Bosnia played an important role in the Turkish wars against Austria and Venice. In 1697, Prince Eugene of Savoy captured Sarajevo. By the Treaty of Karlowitz (1699) the Sava river (Bosnia's northern border), also became the northern boundary of the Ottoman Empire. Herzegovina and the part of Bosnia east of the Una river were ceded to Austria in 1718, and returned to Turkey in 1739.

[6] In the 19th century, Bosnia's nobility resisted Turkish interference. In 1837, Herzegovina's regent declared independence. Uprisings became chronic, bringing Christians and Muslims together, despite their differences, against the bureaucracy and corruption of the Empire. In 1875, a local Herzegovinian conflict unleashed a rebellion, which spilled over into Bosnia. Austria, Russia and Germany tried unsuccessfully to mediate between Turkey and the rebels. The Sultan's promise to reduce taxes, grant religious freedom and install a provincial assembly was rejected.

[7] By a secret agreement in 1877, Russia authorized Austria-Hungary to occupy Bosnia-Herzegovina, in exchange for its neutrality in Russia's upcoming war against Turkey. After the Russo-Turkish War (1877-1878), the Congress of Berlin disregarded Serbian wishes, and assigned Bosnia and Herzegovina to the Austro-Hungarian Empire (although nominally they continued to be under Turkish control). In 1878, Vienna put down armed resistance from Bosnia-Herzegovina with an army of 200,000 soldiers.

[8] The revolution launched by the Young Turks in 1908 brought on a crisis within the Ottoman Empire. The Turkish Government asked Bosnia-Herzegovina to participate in the new parliament at Istanbul, which strengthened the nationalist feeling. Austria-Hungary ended that process by annexing the two provinces in 1908, with Russian consent. Vienna established a provincial assembly (*Sabor*), without representation in Vienna or Budapest. The 1910 Constitution was promoted by the Empire to consolidate social and religious differences by establishing three electoral colleges - Orthodox, Catholic and Muslim - each with a fixed number of seats in the Sabor.

[9] The influence of the *Mlada Bosna* (Young Bosnia) movement and other revolutionary groups led the Empire's authorities to close Bosnia's Sabor, and dissolve several Serbian political groups. In 1914, the Archduke Franz Ferdinand (heir to Austria's crown) and his wife the Duchess of Hohenberg were assassinated in Sarajevo by a Bosnian Serb student. Austria declared war on Serbia and thus triggered World War I.

[10] In 1915, emigrants in Peru founded the Yugoslav Committee (Yugoslav means 'southern Slavs'), which began an intense campaign in favor of independence and the unification of the 'Yugoslavs'. The kingdom of the Serbs, Croats and Slovenians was proclaimed on December 1 that year, and included Bosnia-Herzegovina. In 1919 the Yugoslav Communist Party was founded, and won 14 per cent of the parliamentary seats. It was banned in 1920. The country was renamed the Kingdom of Yugoslavia in 1929, following an authoritarian coup that persecuted communists, trade unionists, and opponents of Serb dominance.

[11] The Nazis occupied Yugoslavia in 1941. Bosnia-Herzegovina was subjected to the puppet administration of Croatia. In the two provinces, Croat Ustashes (Fascists) massacred the Serbs. The continuing rivalry between Muslim, Serb and Croat degenerated into deep hostility. The Communists, led by Tito, organized a guerrilla resistance movement with the support of the allies. At the end of World War II the country remained a federation of republics, one of which was Bosnia-Herzegovina. The slogan of the Yugoslav federated socialists was 'Brotherhood and Unity', but ethnic confrontation was visible in the arts, and literature.

[12] The federal system and Tito's leadership achieved a half-century of domestic peace. Development plans favored the poorer regions and diverse communities were successfully integrated. Following Tito's death in 1980, a collegial executive was established with representation from all the republics and a yearly rotation of the presidency among them. But instead of pacifying the rivalries between the federated entities, this mechanism seemed to exacerbate them.

[13] After the fall of the Berlin Wall in 1990, the Yugoslav Communist League eliminated the monopoly over the political system. Political demagogues incited discontent, stirring up local and ethnic demands. That year, in the post-war period's first multi-party

LAND USE

2003/2005

IRRIGATED AREA: 0.3% of arable land

- FOREST AND WOODLAND: 42.7%
- ARABLE LAND: 19.6%
- CROPLANDS: 1.9%
- OTHER USE: 35.8%

WORKERS

LABOR FORCE	2004

- FEMALE: 48%
- MALE: 52%

legislative elections, the Bosnian electorate chose candidates who were ethnic standard-bearers. The nationalist parties elected 73 Serbs and 44 Croats, while the candidates of the Democratic Reform Party (ex-Communist) and the liberal technocrats lost political ground.

[14] The Muslims were represented by the Democratic Action Party (DAP). Their leader, Alija Izetbegovic, who held a doctorate in theology, was elected president of the republic. The Bosnian Croat and Muslim leadership sought to follow the example of Slovenia and Croatia in seceding from Yugoslavia, encouraged by Western Europe and fearful of the advance of Serbian nationalism. The Bosnian Serbs favored remaining within the Yugoslav federation. In 1991, Bosnia-Herzegovina's Sabor approved a declaration of independence and in 1992 called for a plebiscite on the issue of separation. Izetbegovic, in order to maintain the unity and integrity of the republic, promised that Bosnia-Herzegovina would not become a Muslim state, and guaranteed the rights of all nationalities. During that year, conflict broke out when independence was ratified in the referendum by 99.4 per cent of the Muslims and Croats.

[15] In 1992, the European Union (EU) and the US recognized the independence of Bosnia-Herzegovina. The Bosnian Republic was accepted as a member state in the Conference of Security and Co-operation in Europe, and joined the UN. At the same time, the Serb community proclaimed the independence of the 'Serb Republic of Bosnia-Herzegovina' in the areas under Serb control (Bosnian Krajina, with its center at Banja Luka). The conflict quickly extended throughout the entire region. Local Croat forces also controlled certain areas of the Republic; there were sporadic confrontations with Bosnian government troops. Finally, that year, Croatia and Bosnia signed a mutual recognition pact.

[16] In 1993, Serb troops killed Bosnian deputy prime minister, Hakija Turajlic, in Sarajevo. The UN decreed a cease-fire in that city: at that point, there were numerous reports on the existence of Serb concentration camps, as well as an 'ethnic cleansing' campaign (forced and violent expulsion and sometimes murder of members of rival ethnic groups). According to Amnesty International, thousands of civilians, as well as soldiers that had been captured or wounded, were deliberately executed and prisoners were submitted to torture. According to UN figures, a total of 40,000 women had been raped. Although atrocities were

committed by all sides, the Serbs bore the major responsibility, while the Muslims were the main victims. The UN Protection Forces (UNPROFOR) sent in some 20,000 peacekeeping troops. The US refused to send troops to Bosnia, despite pressure from the UN and European countries. Several security zones and successive ceasefires were not observed.

[17] In 1993, Serb occupation of Bosanki Brod opened up a corridor between Serbia and Bosnian Krajina. Serbs controlled 70 per cent of the territory, due to their superior artillery and armored vehicles, as well as their control of the bridges over the Drina river - border between Serbia and Bosnia - which allowed them illegally to receive arms and other supplies from the Yugoslav Federation. On account of this support, the UN called for an economic blockade on the Federation, and an arms embargo aimed at Bosnians and Croats. The Muslims found themselves cornered in Sarajevo and a few other minor sites, receiving what little financial and moral support they could from a few Islamic countries and sporadic UN humanitarian aid flights, subject to the authorization of Serbia-Bosnians besiegers.

[18] During the same year, Serb President Slobodan Milosevic and his Croatian counterpart, Franjo Tudjman, announced the partition of Bosnia into three ethnic entities (Serb, Croatian and Muslim), within the framework of a federal state, coinciding with a peace proposal by the UN and the EU that the territory should be divided into three semi-autonomous provinces, controlled by each ethnic group. The Croats, faced with the partition of Bosnia, sought to gain the upper hand to negotiate from a position of strength and launched an offensive against Mostar, capital of Herzegovina. The UN Human Rights Commission reported that 10,000 Muslims had been held in Croatian concentration camps, suffering torture and summary executions.

[19] In the meantime, conditions in Sarajevo grew worse: there were epidemics, and no electricity, water or food. The 300,000 inhabitants managed to survive on minimal rations. In 1994, the UN and EU put forward a proposal which included the partition of Bosnia into three ethnically homogeneous republics: Serbs would receive 52 per cent, Muslims 30 per cent and Croatians 18 per cent. The Bosnian Government rejected the proposal since it called for the transfer of people from one sector to another and implied the legitimacy of 'ethnic cleansing'.

[20] In 1994, Croatians and Muslims

approved a federal agreement between the two communities: 51 per cent of the territory would remain among Bosnians and Croatians while Serbs would be in control of 49 per cent, without the need to divide Bosnia into three ethnically distinct states. The agreement was supported by the EU, Washington and Moscow, but the Serbs rejected it. Negotiations were hampered because Milosevic (who represented the Serbs

diplomatically) stated he had no authority over the self-proclaimed Republic of Srpska.

[21] In 1995, the Bosnian-Serbs held several UN troops hostage and took Bihac. The situation was radically changed by NATO's bombing of Bosnian-Serb positions in the siege of Sarajevo. Almost at the same time, Croatia expelled Serb-Croat forces from the eastern side of the country forcing their delegates to negotiate. Under US

PROFILE

ENVIRONMENT

Bosnia-Herzegovina has a 20-kilometer coastline on the Adriatic Sea. In the north and west it is bordered by Croatia; in the southeast by Montenegro and in the east by Serbia. The major part of the country lies in the Dinaric Alps, with elevations of around 4,265 meters, making overland communication difficult. The country is drained by the Sava and Neretva rivers and their tributaries. The territory takes its name from the Bosna River, a tributary of the Sava. Half of the country's area is covered by forests (on account of which there is an important timber industry), and another part consists of arable lands, mainly in the Sava and Drina river valleys. The main crops are grains, vegetables and grapes; there is also livestock rearing. There is a wealth of mineral resources, including coal, iron, copper and manganese. Because of air pollution, respiratory ailments are very common in urban areas. Barely half of the region's water supplies are considered safe, the Sava River being the most polluted of all. There are multiple environmental problems due to the conflict of 1992-1996.

SOCIETY

Peoples: From a common Slav origin, ethnic differences are historo-religious: for example Muslim Slavs (Bosnian), 49.2 per cent; Orthodox Serbs, 31.3; Catholic Croats, 17.3. Serbs are a majority in north-east Bosnia, with the center in Banja Luka; and the Croats in Herzegovina in the west, centered on Mostar. Ethnic distinctions are less clear in other regions. In the capital, Sarajevo, there are Muslims (majority), Croats and Serbs. Until 1992 there was a 1,200 strong Jewish community. There is also a large Roma (gypsy) minority.
Religions: The majority is Muslim. Other: Christian Orthodox and Roman Catholic.
Languages: Serb, Croat and Bosnian (all official) (very similar and previously known as Serbo-Croat).
Main Political Parties: Alliance of Independent Social Democrats (SNSD); Bosnian Party (BOSS); Civic Democratic Party (GDS); Croat Christian Democratic Union of Bosnia and Herzegovina (HKDU).
Main Social Organizations: Unions are currently in the process of being reorganized as are other organizations; some environmental groups operated even during the conflict.

THE STATE

Official Name: Republika Bosna i Hercegovina.
Administrative divisions: 50 Districts.
Capital: Sarajevo 579,000 people (2003), reduced to less than 50,000 in September 1995.
Other Cities: Banja Luka 175,700 people; Tuzla 111,900; Mostar 72,000 (2000).
Government: There are two entities, each with their own government and national assembly: the Bosnia-Herzegovina Federation (the Croat and Muslim zones) and the Srpska Republic (mostly Serbians) although the city of Brcko is autonomous. At state level, Bosnia i Herzegovina (BiH) has a 2 chamber parliament. The current BiH President is Sulejman Tihi (since February 2006), sharing this role on a rotating basis with Ivo Miro Jovi and Borislav Paravac (all three from different ethnic groups). Prime Minister: Adnan Terzic, since December 2002.
National Holiday: 1 March, Independence Day (1992).
Armed Forces: 25,000 (2003) Due to the conflict the United Nations Protection Force (UNPROFOR) deployed several thousand troops. As part of the peace agreements two separate armies are kept in the nation, one for the Federation and one for the Serb Republic.

Under-5 mortality
15 per 1,000
live births
2004

Malnutrition
4% under-5s
1996-2004

Debt service
3.7% exports
of goods
and services
2004

Maternal mortality
31 per 100,000
live births
2000

military pressure, the peace process launched in Dayton (Ohio, US) stipulated elections to be held in 1996, with the aim of promoting more tolerant leaders among each of the nationalities in conflict. The presence of American troops forced a peaceful settlement that was signed in Paris (1995) and froze the political situation.

[22] The Dayton Accords acknowledged two ethnically based mini-states (the Bosnian-Serb Republic - Srpska - and the Croatian-Muslim Federation) that resulted from the physical elimination or expulsion of ethnic minorities. The International Criminal Tribunal at the Hague convicted Radovan Karadzic, leader of the Srpska Republic and his military commander Ratko Mladic, of genocide. The Dayton peace accords banned the electoral participation of people accused of war crimes. In spite of being convicted and banned, neither was incarcerated, and they retained considerable influence in the republic's political life.

[23] Seventy-three per cent of voters took part in the general elections of 1996. Izetbegovic's DAP-Muslim Party won the majority of votes with 19 of the 42 parliamentary seats. Momcilo Krajisnik's Democratic Serbian Party (DSP) obtained nine seats with 24 per cent of votes, and Kresimir Zubak's Croat Democratic Union (CDU), gained 14 per cent of votes and eight seats. In 1997, the Croatian and Bosnian presidents met in Split to relaunch the Muslim-Croatian federation, pledging once again to facilitate the return of the refugees. Radovan Karadzic questioned the legislative elections, accusing Western representatives of having fixed the results in favor of the Muslim and Croatian parties.

[24] High Commissioner for the republic, Carlos Westendorp, imposed unification measures, like the creation of a common flag and national symbols for Bosnia-Herzegovina, which brought new confrontation. Each group maintained their own armed forces and the federation between Croats and Muslims was shaped as a combination of both groups and not a unified identity.

[25] The new currency, the mark, introduced in 1998, was widely accepted in the domestic market. It was not until 2001 that the privatization of companies and banks was accelerated, having been initiated two years earlier with a loan from the World Bank. In 1991, all the communist era payment offices had been closed. The country had depended on humanitarian aid from the international community for its reconstruction.

[26] In 2000 the international High Commissioner, Wolfgang Petritsch, sacked the Croat president in the rotating presidency, Ante Jelavic, accusing him of diminishing the Dayton Accords by supporting the creation of a Croat mini state. Finally, that year, a new, non-nationalist government took power.

[27] At the end of the war there had been 60,000 international peace-keepers, but by 2001 barely 18,000 were left. The US proposed the withdrawal of its 3,000 blue helmets. According to Petritsch, a further three to four years were needed before the country could function independently, without an international administrator.

[28] The 2002 elections meant the return to power of the nationalist parties, which came to occupy the tripartite presidency. This return of the nationalists to key positions was mainly due to the poor performance of the pro-reform parties.

[29] In 2003, Mirko Sarovic, Serb member of the presidency, resigned after being accused by a Western intelligence service of participating in the illegal sale of weapons to Iraq. There were also accusations of espionage against international officers. Borislav Paravac, of the DSP, became president in his place. During that year, the international High Representative, Paddy Ashdown (in office since 2002) abolished the Republika Srpska's Supreme Defense Council and removed some provisions of both the Federation of Bosnia-Herzegovina and Srpska that made any references to the entities having state power.

[30] In April 2003, Commander Naser Oric was arrested and charged at the Tribunal at The Hague with 'violations of the laws and customs of war', including murder, persecution, wanton destruction, and plunder against Serbs in Srebrenica between 1992-1995. The Tribunal decided not to make his indictment public.

[31] In June 2004 some 60 Government officials, among them two political leaders, were dismissed due to the Government's failure to arrest Radovan Karadzic, an alleged war criminal. Parliamentary speaker Dragan Kalinic and Home Minister Zoran Djeric were removed from their positions. High Commissioner Ashdown explained that he wanted to punish the so-called 'little group of corrupt politicians' which complicated Bosnia's future membership in NATO and the EU.

[32] In July, Milosevic faced three indictments before an international court for the former Yugoslavia, accused of genocide in Bosnia from 1992 to 1995 and serious violations of Human Rights in Croatia (1991-1992) and Kosovo (1999).

[33] Hundreds of corpses were found in August in a tomb within a coal mine in Miljevina, a town near Foca, some 70 km southeast of Sarajevo. Previously, in that same location, another 100 corpses had been found. Foca was one of the first towns seized by the Bosnian Serbs during the war in 1992.

[34] For the first time, in January 2005, days after the US cancelled the payment of $10 million in aid due to the lack of arrests, Bosnian-Serb authorities sent an alleged war criminal to the Hague court.

[35] In February 2006, the Council of Europe, an agency that monitors human rights in the region, published a list of five countries - including Bosnia-Herzegovina - accused of withholding information on a series of secret flights carried out by the US Central Intelligence Agency (CIA), which had allegedly transported terrorism suspects over European territory. ∎

IN FOCUS

ENVIRONMENTAL CHALLENGES
Only half of the region's water supplies are considered safe; the Sava River being the most polluted of all. In addition, air pollution is caused by iron and steel factories in urban areas; other environmental problems result from the widespread destruction of infrastructures during the 1992-1996 war. There are insufficient facilities for the final disposal of solid waste.

WOMEN'S RIGHTS
Women have been able to vote and be elected since 1949. In 2005, women held 12 per cent of seats in Parliament; in 2003, their representation in ministerial or equivalent positions was 11 per cent of the total.

That year, women were 38 per cent of the labor force.

Almost all pregnant women receive perinatal care*.

Lured by the promise of lucrative jobs in western Europe, women and girls from Moldova, Ukraine and Romania have instead found themselves trapped and sold as servants or forced into prostitution in Bosnia. This problem grows daily but the response is inadequate, since the authorities do not recognize its magnitude.* However, some progress is noted regarding the indictment of individuals accused of forcing women, girls and boys into prostitution.

CHILDREN
Four per cent of children suffered from low birth weight*. Also, 10 per cent of children under five were moderately or severely stunted*. Immunization rates for children have risen, but have not reached pre-war levels yet.

All children and young people cope with some traumatic experience relating to the war and post-war period. The one million landmines still scattered throughout the country in 2001 posed a serious danger for both children and adults, and especially for displaced people wanting to return to their homes.

According to UNICEF reports, the birth registration system is inadequate: approximately two per cent of all newborns go unregistered. Access to schooling is a problem for Roma children. Early childhood development is not taught as part of medical academic courses, with the result that medical staff often lack the skills to deal with the needs of small children.* It is estimated that HIV/AIDS cases will increase among young people in the near future, given the trend towards high-risk behavior (injecting drug use, sex trade and risky sexual practices).

MIGRANTS/REFUGEES
Between 1995 and 2005 more than one million people displaced during the war (see history) returned to Bosnia-Herzegovina. Of these, 440,000 had left the country, mainly for Serbia-Montenegro, Germany, Croatia and Switzerland, while some 56,000 had been internally displaced. Serbia-Montenegro and Croatia still held some 100,000 Bosnian refugees, 50,000 were living throughout Europe, and another 300,000 were displaced within Bosnia-Herzegovina. According to UNHCR, some 500,000 may have definitely settled in other countries.

In spite of the huge number of returnees, labor discrimination towards ethnic minority workers is one of the most serious obstacles for the repatriation of refugees and the internally displaced.

DEATH PENALTY
Still applies for exceptional crimes; it was abolished for common offenses in 1997.

* Latest data available in *The State of the World's Children* and *Childinfo* database, UNICEF, 2006.

Botswana / Botswana

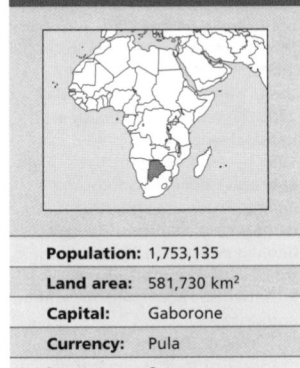

Population:	1,753,135
Land area:	581,730 km²
Capital:	Gaborone
Currency:	Pula
Language:	Setswana

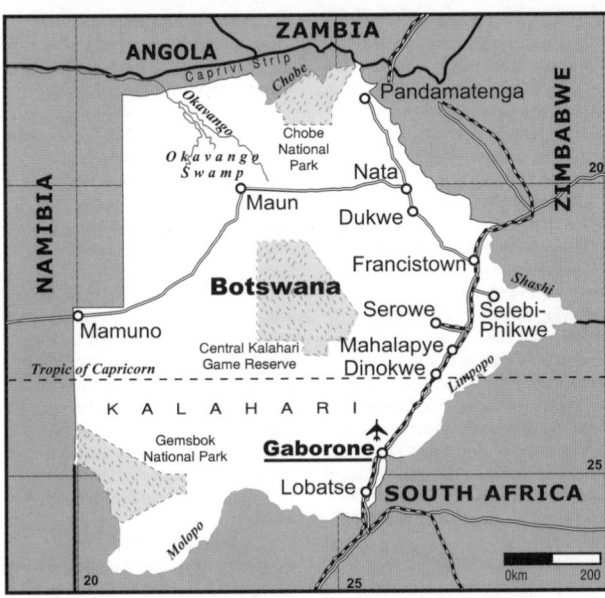

The first inhabitants of what is now Botswana were probably the ancestors of the San (also known as Bushmen), hunters and gatherers who today inhabit the semi-arid plains of southwestern Botswana, and the Khoikhoi from the north. Bantu-speaking populations reached the region in the first century BC. The ancestors of the Tswana, currently the country's majority group, settled between the 11th and 12th centuries in the plains of the Vaal River (in what is now South Africa). The Tswanas were divided among eight powerful clans. Clan rivalry stood in the way of establishing a kingdom like those of other nations in southern Africa.

[2] The history of Botswana - the 'fatal crossroads' located in the heart of southern Africa - is the history of the Kalahari Desert, a passage between the northeastern inhabited savanna and the southeastern plains. The pre-colonial movement facilitated settlement for British, Dutch and Portuguese colonists since the 18th century. The British tried to unite the continent from south to north (from South Africa to Egypt), taking the 'missionaries' road'. The Portuguese wanted to unite the colonies of Angola and Mozambique. The region became a real focal point for the different colonial strategic interests, and for the clash between these interests and the Tswana who had inhabited those areas since the 17th century.

[3] In 1840, Boer colonists of Dutch origin (also known as Afrikaners), fleeing from Cape Town to escape the British, were settled in eastern Botswana. The Boers (farmers) fought with the Tswanas for the scant fertile lands, thus provoking clashes between Tswanas and the Zulus who had been driven out from southern Africa by settlers. In 1895, three Tswana kings traveled to London, seeking support against the Boers and against the German expansion in Southwest Africa (that became Namibia). Later, Botswana became a British protectorate, known as Bechuanaland. The kings had to allow, in exchange for

protection, the construction of a railway between their lands and Zimbabwe (Southern Rhodesia) by the British South Africa Company. British trusteeship prevented political absorption by South Africa but paved the way for the Afrikaners' economic supremacy.

[4] In spite of its large semi-arid area, Botswana became one of southern Africa's major cattle and meat exporters. At the beginning of the 20th century, 97 per cent of the population lived in rural areas, every family owned at least a couple of cows, and the richest had oxen to plough their fields. In 1966, when Botswana gained independence, the urban population reached 15 per cent and almost 40 per cent of the rural population had no cattle. Due

to the economic concentration, the Afrikaners dominated agriculture and controlled 60 per cent of meat exports.

[5] The struggle for independence had become entangled with the wedding of Seretse Khama, a leader of the major Bamagwato chiefdom who went to England to study law. He married Ruth Williams, a British woman. This upset both the British and the Afrikaners who prevented Seretse from returning home. He withstood every pressure including offers of money from the British and, firmly supported by his people, he retained leadership of the country's main ethnic group. He did not come back until 1956. Nine years later, his Botswana Democratic Party (BDP) polled 80 per cent of the vote

in general elections.

[6] At independence, Seretse was elected as first president of the country. In 1967, he was knighted by the British. The BDP carried out a conciliatory policy towards Europeans, who managed 80 per cent of the economy. Botswana was one of the 'frontline nations' fighting apartheid (see South Africa), and a member of the Southern African Development Community (SADC), a grouping of the nine southern African countries seeking to end the economic dependence on South Africa.

[7] Seretse Khama died of cancer in 1980 and was succeeded by the Vice-President, Quett Masire, who had studied economics at Oxford. Strong pressure was exerted on Masire by revolutionary socialist groups to limit the concentration of arable land in European hands and to increase the area allotted to cooperatives. The rural poor accused large landowners of overgrazing, causing a rapid deterioration in the quality of the land. A movement also arose demanding the nationalization of rich diamond, iron, copper and nickel deposits exploited by South African companies.

[8] Between 1978 and 1988, Botswana became the third largest diamond producer in the world, behind Australia and the DR Congo (then Zaire). The national economy grew at a record rate of 12 per cent a year. Despite this, three-fifths of the population lived on subsistence crops or 'non-institutionalized' activities, that is, off the statistical record, beyond fiscal control and the commercial circuit.

[9] In 1985 there were repeated flare-ups along the frontier with South Africa due to the Botswana

Life expectancy
34 years
2005-2010

GNI per capita
$4,360
2004

Literacy
79% total adult rate
2000-2004

HIV prevalence rate
37.3% of population 15-49 years old
2003

Government's support for the African National Congress (ANC) anti-apartheid campaign. In 1987, South Africa applied pressure blocking the roads to Botswana's capital, Gaborone.

¹⁰ In 1989, Masire was re-elected, and the Government had to confront successive economic and political problems, fundamentally due to a fall in the international demand for diamonds. Government corruption came into the open, forcing several ministers to resign. In 1991, three of the seven opposition parties created the People's Progressive Front (PPF) in opposition to the ruling BDP. That same year, the country suffered the biggest strikes since independence.

¹¹ In 1992, unemployment reached 25 per cent. In an attempt to increase employment and lift the flagging fortunes of the BDP, the Government initiated an incentive policy for non-mining industries. Severe drought forced the authorities to declare a state of emergency. Public spending was drastically reduced. Despite the economic and social problems, the BDP kept its majority in the 1994 parliamentary elections.

¹² The country - today the second most important diamond exporter after Russia - has always depended on the export of its minerals. Partly as a result of its diamond wealth, the economy flourished and, according to the World Bank (WB), it had the

fastest-growing economy in the world during the period 1965-1996, when per capita income grew at 9.2 per cent. During those years, tourism became the second largest income earner. The rapid growth led to increasing disparities between rich and poor people.

¹³ Rates of HIV infection exploded in the 1990s, leaving Botswana with one of the world's highest HIV prevalence rates, and crippling the economy and social services. Following action taken by South Africa and Kenya to import generic - cheaper - versions of HIV/AIDS drugs, corporations agreed to reduce prices on medicines. Debswana Diamond Company, a joint venture between the Government and South African corporation De Beers, began subsidizing medication against the disease for its employees and their spouses.

¹⁴ The civil war in Namibia in the Caprivi Strip, a corridor 460 kilometers by 30 kilometers, affected Botswana's relations with its neighbor. In 1999, almost 2,000 residents of this area - many of them separatists - fled to Botswana. The decision to grant them asylum worsened relations with Windhoek. Both countries were also involved in a border dispute over an island in the Chobe river. During that same year, the BDP was again the winner in the elections, taking 33 of the 40 seats open to direct vote.

Its candidate, Festus Gontebanye Mogae, was reconfirmed in the presidency he had held since Masire stood down in 1998.

¹⁵ In 2002, the Government relocated the last 2,200 San people into settlements, having first deprived them of water and food, removing them from lands they had occupied for 30,000 years. In 2003, there was conflict between Gaborone and the organization Survival International (SI), which was opposed to the mistreatment and relocation of the San and Basarwa people. The Kalahari zone, the San ancestral lands, is rich in diamonds. Under Botswana's legislation, indigenous communities may not engage in mining activities nor possess minerals, despite the fact that they reside in mining areas.

¹⁶ In September 2003, the Government started building a fence along the border with Zimbabwe to stop the inflow of illegal immigrants from that country, who were fleeing from political and economic problems. The Government estimated that more than one million Zimbabweans were living illegally in Botswana.

¹⁷ Some 1,000 workers from the Debswana Diamond Company were fired after a strike - declared illegal by the Government - in August 2004. The Botswana Mining Workers Union (BMWU) demanded a 16 per

cent wage rise, and a 24 per cent bonus for 2004 and 2005.

¹⁸ Citing its involvement in the 'global fight against terrorism' the Government rejected in May 2005 the requests to abolish a national security law, passed in 1986 in response to the then aggressive policy of South Africa against Botswana and other neighboring countries. The law banned any person from publishing official information and any information about the army without authorization.

¹⁹ In May 2006 there was an outbreak of foot and mouth disease in the southeast, the area with the highest production of beef in the country. The losses from the lack of exports and the closing of meat packers amounted to several million dollars and threatened the survival of the beef industry.

²⁰ Also in May 2006 came the conclusion of Botswana's longest-running court case, brought by the San/Bushmen against the Botswana Government. Over 10 per cent of the 243 applicants had died in government resettlement camps since the case was first filed in April 2002, following evictions from the Central Kalahari Game Reserve in February that year.

²¹ The landmark ruling on whether or not the Bushmen have the right to return was due in December 2006. ∎

IN FOCUS

ENVIRONMENTAL CHALLENGES
The spread of cattle-raising has meant a reduction in the areas originally occupied by wild species. Intensive cattle-raising rapidly depletes the soil, bringing about desertification and erosion. Water resources available for human consumption are limited.

WOMEN'S RIGHTS
Women have been able to vote and stand for office since 1966. In 2004, women held 11 per cent of seats in Parliament and their representation in ministerial or equivalent positions increased from 14 per cent in 2000 to 27 per cent in 2004.

In 2003, women made up 44 per cent of the country's workforce. Illiteracy among women over 15 was 18 per cent.

Maternal mortality is estimated at 100 deaths for every 100,000 live births. More than 90 per cent of pregnant women receive prenatal care and 98 per cent of births are assisted by qualified staff.*

An average 39 per cent of pregnant women receiving prenatal care were HIV-positive.*

CHILDREN
Since independence, there has been better access to basic services (water, primary education, health) and wider vaccination coverage against common infant diseases. However, Botswana has the highest HIV/AIDS infection rate in the world (39 per cent of adults are HIV- positive). Among children, the greatest impact is seen in girls, who are at least 4 times more susceptible to early infection than boys.* In 2003 there were approximately 120,000 orphans, aged 17 or less, due to HIV/AIDS. Traditional systems of support and care for orphaned children based on the extended family or the local community are breaking down under the strain; 9,500 babies become infected by their mothers each year.*

Some 35 per cent of babies admitted in urban hospitals had HIV/AIDS-related problems, and almost 70 per cent of infant deaths in hospitals were due to that cause.*

Although access to primary education has improved, only 9 per cent of pre-school children have access to educational services, since such facilities are few in number,

mainly in urban areas and have high fees.

INDIGENOUS PEOPLES/ ETHNIC MINORITIES
Some of the San - Bushmen - are among the world's last nomadic hunter-gatherers. They speak a variety of Khoisan languages and are not related to the majority Tswanas. San people are concentrated in the central region of Kalahari Desert. Since 1997, the Government has been resettling them in reserves, where the basic sanitation, water and health services are minimal. This Government plan has been repeatedly accused of mistreatment and repression against the San. They face restrictions in social services, as well as social exclusion. As a result, alcoholism is widespread and there is a high rate of arrests; sentences are harsher than for other ethnic groups. Their traditional chiefs are denied recognition. Since 1989, they have begun to organize and publicize their main grievances: lack of access to traditional lands, recognition for their groups and leaders, more say in government,

protection and promotion of their culture and language. The recent discovery of diamond reserves in their territories brings to light a potential conflict between Government economic development policies and the San's claims over the land.

MIGRANTS/REFUGEES
Some 4,000 refugees and asylum seekers, most of them from Angola and Namibia, from where they had fled during the 1998 insurrection in the Caprivi Strip, reached Botswana in 2002. Some Barakwena refugees had difficulty adapting to sedentary life in camps, since they were nomadic. With help from UNHCR, some of them returned to Namibia in 2002 when violence diminished.

DEATH PENALTY
It is currently applied, even for common crimes.

* Latest data available in The State of the World's Children *and* Childinfo *database, UNICEF, 2006.*

Population:	191,341,355
Land area:	8,547,400 km²
Capital:	Brasilia
Currency:	Real
Language:	Portuguese

T he huge territories which were to become today's Brazil were initially inhabited by small bands belonging mainly to the Tupi Guarani, Carib and Arawak linguistic groups. The indigenous peoples from the Amazon basin fished and farmed while the inhabitants of the dry savanna lived by hunting and gathering. When the first Portuguese ships arrived at the coast of what today is Bahia de Todos os Santos in 1500, more than two million people inhabited those lands.

2   Under the Tordesillas Treaty (1494) which divided the non-European world between Spain and Portugal, the latter was assured its rights over these lands. Sailing towards India, Pedro Alvarez Cabral landed at Bahia de Todos os Santos in 1500 and christened the area the Island of Vera Cruz. Between 1501 and 1502 a naval expedition headed by Gaspar de Lemos charted the region between Rio Grande and the Río de la Plata; Cabral's error (he thought he had reached an island) was thus corrected and the region came to be known by the Europeans as Santa Cruz Land.

3   For the Portuguese, these lands were far less lucrative than Africa and India. The strip of coast did not reveal major deposits of precious metals and was inhabited by semi-settled indigenous peoples, the Tupi, related to the Guaraní, whom the Spanish would later find

in Paraguay. At first the Portuguese crown only valued those lands as places to trade slaves or barter metals and trifles with local people in exchange for brazil wood. The pulp from this tree, used to make a fire-colored dye, would give the country its final name - 'brasa' in Portuguese means embers.

4   Wood industry did not lead to the creation of major cities or other signs of European development in the region, in spite of being quite significant. Although the Indians were used to felling trees in order to clear the forests, they lacked a commercial tradition in wood and could not cut trees on a large scale.

The Portuguese supplied axes and saws, and trading agents had the timber ready for shipment. Trading posts were often established on islands in the Atlantic ocean and, shortly after, the first Portuguese settlements were also established on these islands. Only a few Portuguese exiles inhabited the continent by that time, together with the indigenous communities. On several occasions, these exiles helped other mainland Portuguese make fruitful alliances with the natives.

5   Around 1530, the Portuguese were forced to increase their involvement with Brazil. Other European traders, particularly the French, started to arrive. Commerce with India had slumped and the achievements of Spanish conquistadors in other parts of the continent represented both an incentive and a threat. In order to expel the French and establish its authority, the Portuguese crown sent an expedition along with some settlers. In 1532, the first official Portuguese settlement, São Vicente, was established on an island close to São Paulo.

6   The Spanish expanded their empire through the conquest of lands under the strict control of the Crown. The Portuguese however, due to their maritime commercial

tradition, divided the Brazilian coast into captaincies which were awarded to *donatarios*, prominent individuals who supposedly had the wealth required to occupy and exploit the lands. The captaincies were hereditary, with extensive judicial and administrative powers, although several were never occupied and others survived for a brief time. Four of them became permanent settlements and two (Pernambuco to the north and São Vicente to the south) would turn out to be viable and lucrative.

7   Just like in the Spanish colonial outposts, the first Portuguese settlements had to be fortified to defend them from Indian attacks. Procurement of supplies was difficult and for a while the Portuguese obtained most of their food by trading with the indigenous peoples and began to eat cassava instead of wheat, which was hard to grow in most of the region. Two agricultural systems developed, the *rozas* or farms and the large *fazendas*, given over to exports, mainly of sugarcane. In spite of favorable conditions, the fazendas took a long time to prosper, due to the lack of capital and labor. Besides, agriculture and the discipline imposed on the plantations were foreign to the local peoples, whom

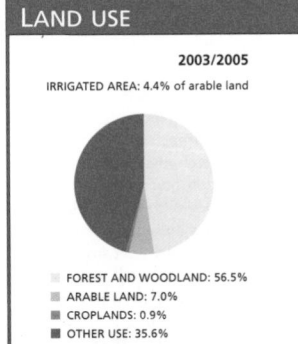

LAND USE

2003/2005

IRRIGATED AREA: 4.4% of arable land

- FOREST AND WOODLAND: 56.5%
- ARABLE LAND: 7.0%
- CROPLANDS: 0.9%
- OTHER USE: 35.6%

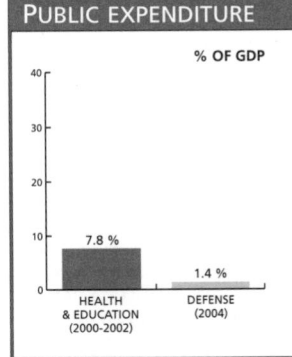

PUBLIC EXPENDITURE

% OF GDP

7.8 %
HEALTH & EDUCATION (2000-2002)

1.4 %
DEFENSE (2004)

Life expectancy
72 years
2005-2010

GNI per capita
$3,000
2004

Literacy
88% total adult rate
2000-2004

HIV prevalence rate
0.7% of population 15-49 years old
2003

the Portuguese forced to toil in exchange for European products. The settlers decided to obtain slave workers either through expeditions which directly hunted for Indians or through other Indians acting as intermediaries.

8   During the second half of the 16th century, Indians had already been decimated by European illnesses (influenza, smallpox, measles) or had fled to other areas. African slaves were then imported to grow sugar and the trade grew to the point that, between the 16th and 19th century, it is estimated that three to four million Africans arrived in Brazil. In 1548, as a result of pressures similar to those of 1530, the Crown opted for a direct representation in Brazil and appointed a general governor, who with 1,000 men established the capital of the country in Bahia (northeastern coast). A bishopric was created in 1551. Only 50 years after the first contact, Brazil had reached the same degree of European institutionalization which characterized the Spanish territories.

9   The Jesuits had started to arrive by that time, and soon became the strongest branch of the Catholic Church, unlike in Spanish America, where they arrived much later than other orders. The Jesuits learnt the Guaraní language to convert Indians to Catholicism and established villages very similar to the missions in the Spanish areas. The main contacts between Indians and Europeans (war, trade, slavery and missions) were the same as in the Spanish areas. Due to these contacts, Guaraní became the language used in the 16th century for all kinds of exchanges.

10   Brazil carried out major expansions towards the west of the Tordesillas line, whose meridian was drawn up 370 leagues to the west of Cape Verde. The expansion reached the slopes of the Andes mountain range and, from north to south, from the Amazon to the Río de la Plata. In the north, the movement led by the Jesuits established several missions along the Amazon. In the northeast, cattle farmers from the sugar areas of Pernambuco and Bahia ventured to the heart of the continent, looking for new grazing lands, and reached today's Piauí, Goiás and Maranhao regions.

11   The westward march was led by the *paulistas* (São Paulo settlers) who in the search for Indian slaves, gold and precious stones, set up major expeditions to the interior, known as *bandeiras*. The incorporation of Portugal into the Spanish kingdom in 1580 facilitated the paulistas' incursions, since internal borders were broken and the Tordesillas division became ineffective. The bandeiras took the paulistas to Peru's mining regions and even to Bogotá (Colombia), also exploring the region of Mato Grosso. To the south, they attacked local *reducciones* (missions), particularly those of Guaira, where the Guaraní people were already relatively immune to disease and were accustomed to collective agricultural work. The Indians and Jesuits who were protecting them, offered resistance, but human hunts were so devastating that the missions were forced to move further and further south until they were finally located in the 'Seven Towns' (today's Rio Grande state).

12   Apart from the paulistas who penetrated the dense forests, thousands of Africans sought refuge there, fleeing from the coastal plantations. Africans, indigenous Americans and their mixed descendants joined forces in constant war against the colonial military expeditions, founding villages known by the African terms *quilombo* or *mokambo*. The most famous of these were at Palmares (1630-1695), in northeastern Brazil, where the legendary Zumbi led the struggle. The Brazilian anti-racist movement still commemorates the date of Zumbi's death in battle, 20 November, as Black Consciousness Day.

13   Brazil found itself involved in a war of independence between the Netherlands and Spain. Flanders and the Netherlands had been 'inherited' by the Spanish Crown. Between 1630 and 1654, the Dutch reasserted their control over Pernambuco after a failed attempt to conquer Bahia, which was repulsed by the joint efforts of native Americans, Africans and Portuguese. The separation of Spain and Portugal that followed could not reinstate the Tordesillas Treaty since events had surpassed the shifting of colonial borders. In 1696, a *bandeirantes* expedition struck the first gold in what is now Minas Gerais. Gold mining peaked in the 18th century. The viceregal capital was transferred from Salvador to Rio de Janeiro in 1763 mostly due to the impact of these mines on the Brazilian economy.

14   The local ruling class began to benefit from an expanding export economy and soon expressed a growing desire to dispense with the Portuguese role as intermediaries in their trade with Europe. The first moves towards independence came in the late 18th century and were rapidly crushed by the colonial power. Brazil's main freedom figure, Tiradentes, was executed in 1792 for his leading role in the Minas Conspiracy of 1792.

15   When Napoleon invaded the Iberian peninsula in 1808, Portuguese King Dom João VI had to transfer his court to Brazil, making the country semi-independent. Portugal ceased to be an intermediary and Brazil then dealt directly with its main customer, Britain. The Brazilian merchants prospered at the expense of other sectors linked to the Portuguese monopoly. The 1821 Oporto revolution in Portugal was an attempt to reinstate the old colonial system based on monopoly. When the King returned to Lisbon, the Brazilian merchants - determined to secure their gains - declared independence with British blessing. Brazil became an empire and Pedro I, the prince-regent, became its emperor.

16   The following decade was one of the most turbulent in Brazilian history. From 1831 to 1835, a regency unsuccessfully tried to

PROFILE

ENVIRONMENT

There are five major regions in Brazil. The Amazon Basin, in the North, is the largest tropical rainforest in the world. It consists of lowlands covered with rainforest and rivers. The Carajás mountain range contains one of the world's largest mineral reserves, rich in iron, manganese, copper, nickel and bauxite. The economy is mainly extractive. The northeastern sertao consists of rocky plateaus with a semi-arid climate and scrub vegetation. Raising cattle is the main economic activity. The more humid coastal strip, situated on the 'Serra do mar' (Coastal Sierra), has numerous sugarcane and cocoa plantations. In the southeast, the terrain consists of huge plateaus bordered in the east by the Serra do Mar mountain range. The main crops are coffee, cotton, corn and sugarcane. The southern plateau, with a sub-tropical climate, is the country's main agricultural region, where coffee, soybeans, corn and wheat are grown. In the far south, on the Rio Grande do Sul plains, cattle-raising is the main economic activity. Finally, the mid-west region is made up of vast plains where cattle-raising predominates.

SOCIETY

Peoples: Brazilians come from the ethnic and cultural integration of indigenous people (mainly Guaraní), African slaves and European (mostly Portuguese) immigrants. Arab and Japanese minorities have also settled in the Rio-São Paulo area. There are also many Indian peoples. Contrary to what is commonly admitted, racial discrimination does exist although the 1988 Constitution includes racism as a crime.
Religion: Most are baptized Catholic; but there is considerable merging into syncretic Afro-Brazilian cults (macumba, candomblé and umbanda).
Language: Portuguese is the official and predominant language. Many Indian languages are spoken (i.e. Bariwa and Guajajára).
Main Political Parties: Workers' Party; Party of the Liberal Front; Party of the Brazilian Democratic Movement; Party of Brazilian Social Democracy.
Main Social Organizations: Workers are grouped together primarily in the Consolidated Union of Workers (CUT), the General Confederation of Workers (CGT) and the Labor Union Force. Many labor unions do not belong to any of these, preferring to remain independent. Landless Movement of Brazil (MST), an association of workers without land whose agenda is agrarian reform in rural areas, and land for the construction of housing in urban areas. National Union of Indigenous Peoples (UNI), an association of Brazil's different indigenous groups. Pastoral Commission of the Earth (CPT) and Indigenous Missionary Council (CIMI), pastoral groups of the Catholic Church involved in social action in these areas. Defense Network of the Human Race (REDEH), an eco-feminist organization. 'Torture No More', state groups committed to the defense of human rights.

THE STATE

Official Name: República Federativa do Brasil.
Administrative divisions: 26 States, 1 Federal District.
Capital: Brasilia 3,099,000 people (2003).
Other Cities: São Paulo 17,800,000 people; Rio de Janeiro 10,600,000; Belo Horizonte 4,310,000; Porto Alegre 3,576,500; Recife 3,377,600.
Government: Luis Inacio (Lula) da Silva, President since 2003, re-elected in 2006. Bicameral legislature: The National Congress has two chambers: the Chamber of Deputies has 513 members, and the Federal Senate has 81 members.
National Holiday: 7 September, Independence Day (1822).
Armed Forces: 303,000 troops (2003). Other: 243,000 Public Security Forces.

Under-5 mortality
34 per 1,000 live births
2004

Poverty
7.5% of population living on less than $1 per day
2003

Debt service
46.8% exports of goods and services
2004

Maternal mortality
260 per 100,000 live births
2000

put an end to the civil war in the provinces and to the army's insubordination. In 1834, the Constitution was amended to decentralize the government, with the creation of provincial assemblies with considerable local powers, and to choose a regent for a period of four years. In 1835, Diego Antonio Feijóo, a priest, was elected as regent and fought for two years against the break-up in the revolts in Rio Grande do Sul, known as *Guerra dos Farrapos*, 1835-1845 (War of the Ragged Ones). Feijóo was forced to resign in 1837, being replaced by Pedro Araújo Lima. Impatient with the regency, Brazilians hoped to find in an Emperor the figurehead that would unite them. In 1840, Pedro de Alcántara was declared of age and enthroned with the name of Pedro II.

[17] During the Empire that lasted from 1822 to 1889, Brazil consolidated its national unity and extended its borders over the areas settled by bandeirantes (17th and 18th centuries). The territorial expansion included: the annexation of the Cisplatine Province (now Uruguay); the war of the Triple Alliance against Paraguay, in which Brazil annexed 90,000 square kilometres of Paraguayan territory; and, towards the end of the century, the annexation of Bolivia's Acre territory.

[18] Under the rule of Pedro II, the population grew from 4 to 14 million, but the economy continued to depend on large estates and the export of tropical agricultural products, mainly coffee. Slavery was not abolished until 1888, which accelerated the fall of the monarchy, but brought little change in the social and political conditions of blacks who, as illiterate people, were denied the right to vote and therefore political freedom.

[19] By the end of the 19th century, the gap between urban and rural areas was widening. The urban middle class, the military and coffee growers pressed for the country's modernization. In 1889, the army backed a conspiracy of modernists. Pedro II abdicated and was exiled in Europe. The abolition of slavery, together with the fall of the Empire, brought about social, political and economic changes which accelerated modernization, as well as causing of political, social and religious disorder.

[20] The establishment of republican institutions was difficult. Prudente de Morais was elected first civilian President in 1894. The Federal Republican Party was founded in 1893, but the Navy rebelled both that year and the following. From 1893 to 1895, Brazil suffered a federalist rebellion in Rio Grande do Sul. From 1896 to 1897, the

Canudos war, in which a religious community was wiped out by the republic's troops, took place in the northeastern provinces. This war was due to a clash between the interior - poor, illiterate and desperate - and the coast - literate, with a developed economy and seeking modernization.

[21] Until 1920, uprisings, outbreaks of authoritarianism and fights among the regional oligarchies were frequent. Neither electoral justice nor the secret ballot was yet instituted and dissatisfaction with election results was widespread. Elections were rigged since the electoral rolls often included deceased voters. The period from the declaration of the Republic up to 1930 was known as the Old Republic.

[22] Coffee was consolidated as the major export. Between 1914 and 1918, World War I brought an economic boom to Brazil, since the country was one of the main suppliers of commodities to the powers at war. However, between 1920 and 1930, coffee prices tumbled as international competition reduced sales and, in 1929, due to that year's world slump, 29 million bags of coffee were left unsold. In 1930 a coup resulted in the appointment of Getulio Vargas as president. The Revolution of 1930 marked the end of the landowners' predominance, already weakened by the crisis that had destroyed the coffee industry. Vargas introduced the 'import substitution' model and gave priority to national industrial development, especially in iron and steel during World War II. Vargas held power from 1937 to 1945 as dictator of the New State.

[23] Fearing Vargas would attempt to retain power, the military forced him to resign in 1945. General Eurico Gaspar Dutra won the presidential election in that year and Vargas was elected to a seat in the Senate. In reaction against Vargas, the constitution promulgated in 1946 sought to prevent the rise of another all-powerful president. The three branches of government were separated, ensuring Congress of its independence and freedom in the election of its members. The Constitution set restrictions to prevent abusive federal intervention in the internal affairs of the states.

[24] In 1950 Vargas was elected again as constitutional president. Nationalism and reformist defense of workers' rights, *trabalhismo* (laborism), the name of his movement, were the two outstanding features of his government. In 1953 the state oil monopoly was established with the creation of Petrobras and social security laws were passed. Vargas committed suicide in 1954 leaving a letter accusing 'dark forces'

(referring to imperialism and its local accomplices) of blocking his efforts to govern according to popular and national aspirations.

[25] By means of a development policy, Juscelino Kubitschek's administration (1956-61) admitted transnationals to the Brazilian market, granting them exceptional privileges. Brasilia city was built during his term in office, to mark a new era in the country's economic development. In 1960, the federal capital, previously Rio de Janeiro, was transferred there.

[26] In 1961, Vice-President João Goulart, the Labor Party leader and Getulio Vargas' political heir, assumed the presidency. High-ranking military officers opposed his appointment but he was backed by a civilian/military movement which wanted a legal government, led by Leonel Brizola, governor of Rio Grande do Sul. A compromise parliamentary solution was adopted, with Tancredo Neves becoming Prime Minister. In 1963 a national referendum reinstated the presidential system. Goulart attempted to introduce measures like agrarian reform and legislation to regulate profit transfers abroad by foreign companies. A US-backed military coup deposed him in 1964.

[27] The new government passed Institutional Act No. 1, which repealed the 1946 liberal constitution, allowing the revocation of parliamentary mandates and the suspension of political rights. A string of arrests all over the country forced major political leaders such as João Goulart, Leonel Brizola, Miguel Arraes and later even Juscelino Kubitschek into exile or to fight underground. The military junta appointed General Humberto de Alencar Castello Branco head of government for the remainder of the constitutional period, but his mandate was later extended until 1967. In some of the 1965 state elections, opposition candidates won in Rio de Janeiro and Minas Gerais. The military retaliated with Institutional Act No. 2, stating that the President would henceforth be appointed by an electoral college and banning existing political parties. A two-party system was created, formed by the majority and pro-government National Renewal Alliance (ARENA), and the Party of the Brazilian Democratic Movement (PMDB), from the opposition but without the possibility of reaching power.

[28] During 1967, a new Constitution came into effect and General Arthur da Costa e Silva became President. In 1968, to confront growing electoral support for the popular opposition, Institutional Act No. 5 was passed, granting full autocratic powers to the military regime. In 1969, Costa

e Silva was replaced by a military junta that remained in power for a month, when another army general, Emilio Garrastazú Médici, former head of the National Information Services (SNI), became President. His administration was marked by extreme repression of both the legal and illegal opposition, and by an economic policy which fed middle class consumerism.

[29] In 1974, General Ernesto Geisel was appointed President. He put an end to the state oil monopoly, signed a controversial nuclear agreement with West Germany and granted further prerogatives to foreign investors. Its arms industry placed Brazil fifth among the world's main arms exporters. Under the Geisel administration there was a gradual relaxation of political controls, which allowed the democratic process to evolve. Between 1974 and 1978, despite media censorship, the PMDB achieved significant victories at the polls. At the end of his mandate, Geisel delivered the reins of government to General João Baptista Figueiredo (former SNI head). Figueiredo came to power in 1979, announcing that he intended to complete the softening of political restrictions. A month later, a strike by 180,000 metalworkers in Sao Paulo led by Luis Inácio 'Lula' da Silva was settled without violence, as a result of negotiations between the Labor Ministry and the Union. By the end of that year, the Congress passed a bill granting a far broader amnesty to political opponents than the Executive had originally intended; political prisoners were released and exiles began to return.

[30] In the economic-financial arena, the after-effects of the monetary policies applied by successive military governments were felt during the Figueiredo administration. Foreign debt spiraled and in the early 1980s, Brazil became an early exporter rather than importer of capital in its efforts to find funds to pay off the interest on its $100 billion debt. In 1985, according to official data, there were six million unemployed and 13 million under-employed, out of a population of more than 130 million, of which over 50 per cent lived below the poverty line and outside the formal economy, in cities alone. Ministry of Labor officials stated that 'not even seven per cent growth per year for 20 years would be enough to improve these peoples' living conditions'.

[31] The opposition's electoral victory in 1983 reflected enormous popular discontent. The central government held only 12 states while the opposition won 10, including the economically decisive states of São Paulo, Rio de Janeiro and Minas Gerais which accounted for 59 per cent of the population

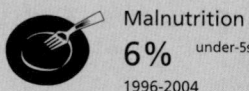
Malnutrition
6% under-5s
1996-2004

Water source
89% of population using improved drinking water sources
2002

Doctors
206 per 100,000 people
1990-2004

Primary school
97% net enrolment rate
2004

and 75 per cent of the country's GDP. Tancredo Neves, governor of Minas Gerais, was the chief co-ordinator of the opposition front. Neves was elected President and José Sarney, who had been formerly president of the Government party, became Vice President. Neves announced plans for a new social order: the New Republic.

32 The day before his inauguration, Neves was hospitalized and rushed into surgery. Sarney was sworn in as interim president and came to power after Neves died in 1985, legalizing the Communist Party and left-wing organizations, which had been banned more than 20 years before. Democratization was strengthened: direct election for President of the Republic and mayors of state capitals were approved, a national assembly was convened for 1987 to draft a new constitution and illiterate people were granted the right to vote.

33 In 1986 Sarney declared a moratorium on the foreign debt and launched the Cruzado Plan that aimed at fighting inflation. The plan produced impressive short-term results: a boom in consumption and economic growth. This startling prosperity coincided with the 1987 parliamentary elections which the PMDB won by a landslide.

34 The Cruzado Plan could not be maintained without fighting the excessive speculation and halting pressure from the financial sector. After the Parliamentary election, the price freeze came to an end and inflation leapt to a monthly rate in double figures. The planned agrarian reform was gradually reduced.

35 In 1988, landowners from the Acre region murdered Chico Mendes, leader of the *seringueiro* (rubber tappers) movement and the Amazonian indigenous peoples. Mendes had organized an original *empates* struggle (linking arms to stop trees being felled) to prevent the clearing of the forests, and proposed the creation of reserves, to guarantee their right to live and work in the forest without destroying it.

36 The first direct presidential elections in 29 years were held in 1989. Nearly 80 million people voted and the final round was fought between conservative candidate Fernando Collor de Mello and Workers' Party (PT) leader, 'Lula'. Collor de Mello, a young politician who had begun his career under the military regime, won the second round with 42.75 per cent to his opponent's 37.86 per cent of the vote. Collor adopted the neo-liberal economic model, with the privatization of state enterprises and the reduction of tariff barriers for foreign products, but he failed to control inflation, and neither held off recession nor unemployment.

37 Besides this complex economic setting, the Government faced a critical social situation and the escalation of violence. In 1991, more than 350 street children were murdered in Rio de Janeiro. The parliamentary commission that investigated these murders estimated that over 5,000 children had been killed in this way within three years. The same commission reported that the persecution of homeless children - 7 million according to estimates of the Brazilian Center for Childhood and Adolescence - was carried out by paramilitary groups financed by shop owners.

38 The accelerated destruction of the tropical forest in order to exploit its mining and timber potential and to turn it into grassland or mining areas, kept devastating indigenous peoples. Apart from suffering epidemics, the degradation or loss of natural resources, pollution and a systematic fall in their standard of living, they were being murdered and subject to violence by miners and the police.

39 In 1991, thousands of people from the Landless Movement of Brazil (MST) staged a march in the state of Rio Grande do Sul. The protesters demanded settlements to work and that the 4,700 million cruzeiros earmarked for agrarian reform should be spent, of which only 800 million had actually been used. Figures provided by the Pastoral Commission of the Earth in 1992 showed that there were 15,042 rural slaves - triple the number recorded the previous year. According to data gathered by the Federal Bureau of Statistics, nearly four million people living in rural areas worked under conditions of virtual slavery.

40 In 1992, a Parliamentary Investigating Commission studied government corruption, which took the form of influence-peddling in exchange for deposits made to the President's personal accounts. Demonstrations against corruption and the discovery of evidence implicating other government figures in these schemes led all the political parties to vote for the President's impeachment. Congress voted to relieve Collor of his duties, so that he could be put on trial. Vice-President Itamar Franco became President. In 1992, Collor was found guilty by the Senate of 'criminal responsibility'. His presidential mandate was removed and his political rights suspended until the year 2000. Franco officially assumed the presidency.

41 In 1993, Economy Minister Cardoso presented the Plan Real, an economic stabilization project which ended index-linking and created a new currency, the Real, in 1994. The success of this anti-

IN FOCUS

ENVIRONMENTAL CHALLENGES
The Amazon region is being devastated by the unrestrained felling of trees, which has destroyed the habitat of many plant and animal species, as well as that of indigenous peoples. There are serious pollution problems in cities like Rio de Janeiro and São Paulo. The main causes for the pollution of important aquiferous sources are mining activities, degradation of wetlands and oil leakages.

WOMEN'S RIGHTS
In 2006, women held 9.1 per cent of Parliament seats and 11 per cent of ministerial and similar positions. In 2003*, 97 per cent of women aged 15 to 24 were literate (two percentage points higher than men in the same age bracket), as well as 88.6 per cent of women over 15.
Maternal mortality continues to be a problem, although 96 per cent* of deliveries are assisted by qualified staff and 86 per cent of women receive prenatal care. There is unequal access to health services for different segments of society and regions of the country.* In 2003, women made up 35 per cent of the 82-million work force.

CHILDREN
The infant mortality rate has fallen to 29 per 1,000 live births within the last decade*, but remains disproportional to national production capacity and available technology. Brazil is, together with Mexico, the country of the Americas with the most street children. There are many young women - especially from the poor Northeast - who migrate to the big southern cities looking for domestic work, but in countless cases they end up in prostitution or pornography.
Brazil has the highest child prostitution rate in Latin America, even taking into account the lack of precise estimates, which range from 500,000 to 2,000,000.

INDIGENOUS PEOPLES/ ETHNIC MINORITIES
The 200 indigenous groups comprise about 350,000-500,000 people (from 0.2 per cent to 0.3 per cent of the country's population). They are mostly located in the Amazon and central regions (small communities, missions, national parks - there are four - and reserves appointed by the Government). There are groups of farmers, hunter-gatherers and others which are semi-nomadic. The Amazon indigenous groups are highly dependent on the land and river to develop their way of life, but they are threatened by the gold mining, agriculture and timber and oil industries. The exploitation of these natural riches has caused the decline of the indigenous population, partly as a result of confrontations with non-indigenous people, some of whom consider them 'less than complete people', and also because of the introduction of new diseases in their habitat. During Fernando Henrique Cardoso's administration, the privatization of part of their lands was encouraged, threatening their way of life. The organization of the different groups is mainly local, because of the remoteness and distances between them.

MIGRANTS/REFUGEES
It is estimated that approximately two million Brazilians live in foreign countries, including Germany, Switzerland and Italy. Although Brazil is one of the world's largest economies, the income distribution is far from equitable which makes many people seek better opportunities in other countries.
The policy on refugees in Brazil has changed frequently but since the early 1990s, the country has specific legislation. In 2002, practical programs arising from the Refugee Act began to be implemented, giving priority initially to Afghan asylum-seekers by settling them in Brazilian cities, organized in cooperation between the National Commission for Refugees and the UNHCR. In 2005 the country registered more than 3,000 refugees.

DEATH PENALTY
The death penalty has not been used in Brazil since the proclamation of the Republic in 1889. The Federal Constitution of 1988 expressly prohibits the death penalty except in times of declared war.

* Latest data available in The State of the World's Children *and* Childinfo *database, UNICEF, 2006*

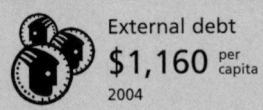

External debt	**Imports (millions)**	**Exports (millions)**	**Received aid**
$1,160 per capita	$80,069 goods and services	$109,059 goods and services	$2 per capita
2004	2005	2004	2003

inflationary policy rapidly made Cardoso the most popular candidate for the October elections. In the first round he beat the previous favorite Luis Ignacio 'Lula' da Silva, of the PT. Cardoso began the privatization of state companies, including part of Petrobras and the telecommunications sector, but economic recession led to the rise of unemployment, urban labor conflicts, crime and land seizures by poor peasants.

⁴² That year, Parliament passed a constitutional reform that permitted presidential re-election, and the Government issued a decree which guaranteed indigenous peoples the exclusive use of 23 plots of land covering 10 per cent of Brazilian territory. The marches and seizures by the MST were supported by the Vatican's Pontifical Council of Justice and Peace in 1998, in the document 'For better land distribution: the land reform challenge'.

⁴³ In 1998, Cardoso was re-elected President. In 1999, former president Itamar Franco, Governor of Minas Gerais, declared a moratorium on debt to the Federal Government. The Government freed the real against the dollar (a measure that was denied until the day before its implementation) resulting in a 10 per cent devaluation of the real. The president of the Central Bank resigned 'so that investors can regain confidence' but the real continued its slide with devaluation reaching 50 per cent in a month. The real crisis hit consumption, industry (in need of imported resources) and Brazil's relations with its Mercosur partners, but benefited the country's export sector.

⁴⁴ In 2000, to 'counter-commemorate' the fifth centennial of the Europeans' arrival in Brazil, 2,000 Indians met at a beach in northeast Bahia, carrying protest banners. Of the five million inhabitants there when the Portuguese arrived, only 350,000 remained. The 1,000 groups that existed at the time of colonial contact were reduced to 210 (of which 50 have stayed beyond contact).

⁴⁵ In order to reduce the inequality in land ownership, the Government suspended the property titles of some 1,900 landowners, since they were not able to justify the origin or legality of their property titles. Some 62 million hectares of land (roughly the size of Central America) was unlawfully held. The acquisition of lands with fake documents is a centuries-old practice in Brazil. It is estimated that a third of all landowners have built their estates with false papers. Nowadays, 90 per cent of all arable lands are in the hands of 20 per cent of the population, while 40 per cent of the poor population holds

barely one per cent of land fit for agriculture. This makes Brazil one of the least equitable countries in the world.

⁴⁶ In 2001 Cardoso closed two government development agencies, which were accused of corruption involving more than $1 billion. Federal police discovered that projects promoted by these agencies, to develop the Amazon and poor areas in the northeast, were no more than façades to embezzle money. Both the Senate chairman Jader Barbalho and his wife were investigated.

⁴⁷ In 2001, José Nilson Pereira da Silva and Juliano Filipini Sabino were sentenced to 21 years in prison for the death of Edison Neris da Silva, a gay man who was beaten to death in early 2000 by a group of skinheads in São Paulo. The trial and sentence were considered landmark rulings by human rights and gay groups, since it was the first time a sexual discrimination crime was sentenced in Brazil. According to Bahia's Grupo Gay, 299 homosexuals were murdered in Brazil between 1999 and 2000; 30 per cent of the victims were transvestites and three per cent were lesbians. Two thirds of the murders were committed in northeastern Brazil.

⁴⁸ During that same year, the Government, as part of its AIDS prevention program, hired a producer of pornographic films, Sexy Videos, to show movies in which the actors wore condoms. A special message about AIDS prevention which was shown at the beginning of the films became compulsory for all 'adult' films produced within the country. The AIDS program includes lessons in school where teenagers are taught how to put on a condom using a clay model. According to the organization Médecins sans Frontières (MSF), AIDS-related deaths in Brazil fell more than 60 per cent. The HIV and AIDS program in Brazil is renowned worldwide for its success.

⁴⁹ Free access to the AIDS drugs-cocktail treatment is the main basis of the program. A 1996 law states that, if foreign firms do not make their drugs locally, they will lose their patent in Brazil after a three-year period. The law states that in situations of 'public interest', the local industry can break the patents. Currently, Brazil manufactures 8 out of the 12 drugs that are part of the drug-cocktail received freely by patients at state clinics. Those opposing the program argued that poorer patients might not be able to follow the cocktail's complicated instructions. The Government solved the problem by putting labels with suns, moons and drawings of food on each medication.

⁵⁰ In 2002, with the title 'Towards a Globalization of the World Social Forum (WSF)', the second meeting

of the WSF was held in Porto Alegre, thus institutionalizing this global assembly which had been launched a year earlier in that same city. The WSF, which is held around the same time as the World Economic Forum, was attended by 5,000 organizations to analyze, share ideas, debate and define parameters and alternatives in the anti-globalization struggle.

⁵¹ That year, at his fourth attempt to win the presidency, Lula da Silva triumphed over José Serra, the candidate of the party in power, and led Brazil's left to government. That victory, with an overwhelming majority in the second round of elections, was helped by the alliance of the Workers' Party with conservative elements from center and right, in an attempt to allay the market's fears.

⁵² The Workers' Party committed its administration to complying with the repayment schedules agreed with multilateral credit organizations. In spite of the many economic and political conditions, Lula came into government backed by 52 million votes and promised to strengthen Brazil's economic independence, supporting the common strategies of the Southern Cone Common Market (Mercosur) and applying a gradual economic redistribution policy that would reduce the extreme inequality that has prevailed for decades.

⁵³ In May 2003 Brazil won the global health award Gates 2003, for its national program to fight HIV/AIDS. The program was considered a model in the battle against the pandemic, since it combined free access to anti-retroviral treatments with aggressive AIDS-prevention publicity campaigns.

⁵⁴ In October, Asma Jahangir, a UN expert in summary executions, condemned the country's human rights situation. Jahangir had visited Brazil to investigate allegations of torture and murders committed by the Brazilian police. She said that, in spite of the Government's efforts to reduce violence and prevent impunity, it was indispensable to introduce reforms that guaranteed the observance of human rights in the country.

⁵⁵ Brazil's southern coast was hit by a cyclone in March 2004. Government sources said that during the unusually strong storm two people died and several went missing. Beaches were evacuated, more than 500 homes were destroyed, a large number of buildings were damaged and the power supply was interrupted in major areas. The subtropical cyclone, formed in the Atlantic Ocean, some 440 km from Brazil, affected mostly the coastal municipalities in the states of Santa Catarina and Rio Grande do Sul. The US National Hurricane Centre said that

there were no precedents for a meteorological phenomenon of this type in the South Atlantic Ocean, near the Brazilian coast.

⁵⁶ A public survey, published that month, showed the President's popularity had fallen from 39.9 per cent in February to 34.6 per cent in March. Another survey indicated that the Government's approval rate had fallen from 66 per cent in December to 54 per cent. The Government was strongly criticized by the opposition and its political allies (including sectors of the WP) for its conservative economic policy and the lack of results in social policies. On top of all this, Cabinet chief José Dirceu was involved in a corruption scandal when a videotape showed him requesting money for himself and to finance the WP's campaigns in 2002.

⁵⁷ In August 2004, Brazil, through its state firm Petrobras, signed an agreement with Ecuador to build drilling facilities in one of the largest environmental reserves in the Amazon, Ecuador's Yasuní National Park. The Park has been declared a biosphere reserve by UNESCO. It covers 982 hectares of jungle and is a haven to 90 species of amphibians and more than 500 types of birds.

⁵⁸ In October, a delegation from the International Atomic Energy Agency (IAEA) visited a new nuclear facility with uranium-enriching capacity, the Resende plant, located in the state of Rio de Janeiro. The goal of IAEA experts was to confirm that uranium-enriching techniques were not being used for military purposes. The Government insisted that its nuclear program pursued energy generation goals, and announced it would only allow a visual inspection limited to the plant's centrifuges in order to protect its technology.

⁵⁹ In December 2005, Brazil paid off all its debts to the IMF. The $15.5 billion payment was financed with international reserve funds from the Central Bank of Brazil, which amounted to $67 billion.

⁶⁰ In May 2006, in São Paulo, the transfer of more than 700 prisoners from the criminal group First Capital Commando (FCC) to a top-security prison unleashed a wave of violence which caused panic and chaos and left the city virtually paralyzed. Hundreds of attacks against police and civilian targets took place between the afternoon of 12 May and the evening of 15 May, as well as riots in 36 jails. The riots left at least 115 dead, most of them security forces and members of the FCC, and some 50 injured.

⁶⁰ In October 2006, Lula da Silva won a second four-year term as Brazil's President, in a resounding victory over his challenger Geraldo Alckmin. ■

Brunei / Brunei

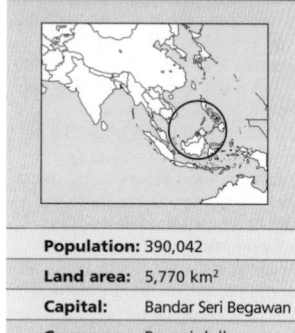

Population:	390,042
Land area:	5,770 km²
Capital:	Bandar Seri Begawan
Currency:	Brunei dollar
Language:	Behasa Malay

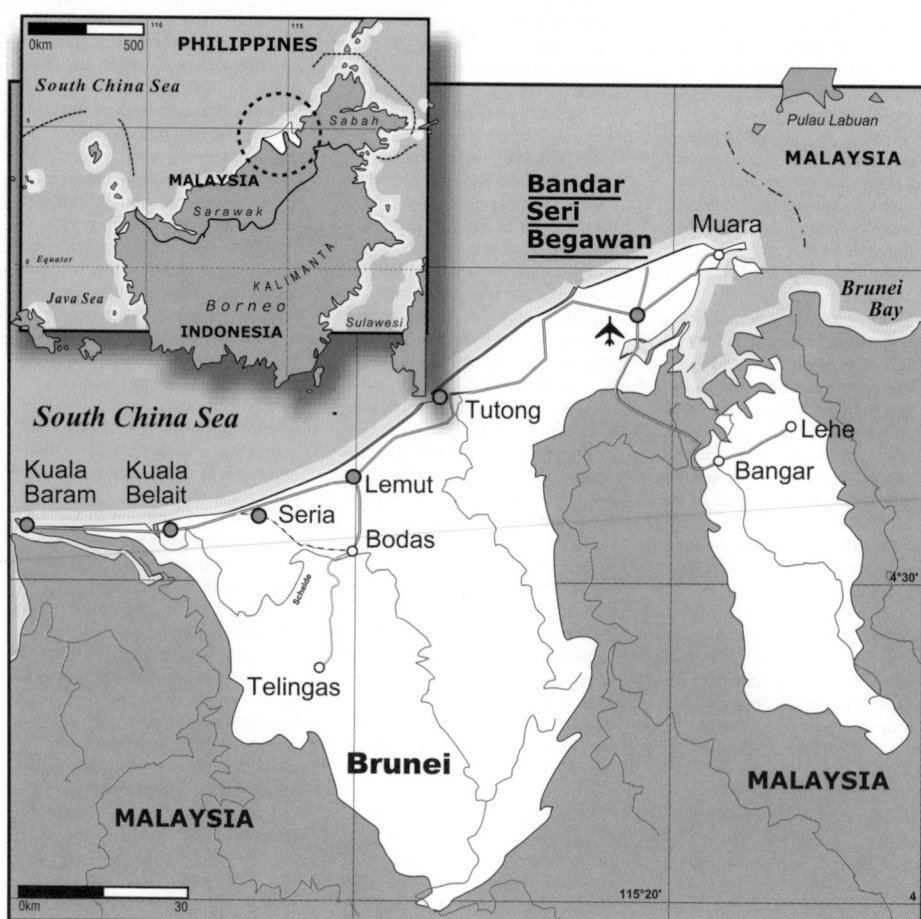

The ancient history of Brunei is not well documented, although there is information that its inhabitants were trading with and paying tribute to China in the 6th century AD. Later, it came under Hindu influence through allegiance to the Majapahit kingdom in Java. From the 13th century, Brunei was under the Islamic empire that covered most of the island of Borneo, from which it derives its name. Although the European presence - Portuguese, Spanish and Dutch - became constant in the region from the 16th century onwards, it was only in the early 19th century that the colonial powers decided to occupy the large island. The Dutch advanced from the south, while Sultan Bolkiah (1473-1521), turned to the British in an attempt to preserve independence.

[2] In 1841, as payment for help in quelling the 1839-1840 rebellion, the Sultan had to turn the province of Sarawak over to James Brooke, who became a European Rajah over a Malayan state. In 1846 the British annexed the strategic island of Labuan, and in following years paved the way for the secession of the province of Sabah. In 1888, the British consolidated their position and established separate protectorates over Brunei, Sarawak and Sabah.

[3] Following World War II, despite Britain's efforts, the island began the decolonization process. An agreement signed with the incumbent Rajah Brooke in 1946 made Sarawak and Sabah into British colonies while Kalimantan (former Dutch Borneo) gained independence in 1954 as part of Indonesia.

[4] All that was left of the British protectorate was the Sultanate of Brunei, reduced to a tiny enclave between two Malaysian provinces, scarcely 40 km from the border with Indonesia. In 1929, the transnational Shell discovered oil deposits in the area. In the following decades, drilling for oil and natural gas began, reaching production rates of 175,000 barrels a day.

[5] In 1962, Sultan Omar Ali Saiffudin accepted a proposal from Malaysian Premier Abdul Rahman to join the Federation of Malaysia, which at the time joined Sabah and Sarawak to Singapore and the provinces of the Malayan peninsula. The Brunei People's Party, (Rakyat) which held 16 seats in the 33-member Legislative Council, opposed the move and pressed for the creation of a unified state comprising Northern Borneo, Sarawak and Sabah, but excluding peninsular Malaya.

[6] During that year, a mass uprising broke out, staged by the Rakyat, backed by the Barisan Sosialis (Socialist Party) of Singapore, with support from the anticolonialist Sukarno regime in Indonesia. The rebels opposed integration to the Federation of Malaysia, demanding participation in administration and the end of the autocratic regime. The rebellion was rapidly stifled, the People's Party outlawed and the leaders arrested or forced into exile. Finally, in spite of ethnic, historical, and cultural ties with Malaya, Sultan Omar decided to keep out of the Federation. He was not satisfied with arrangements for power-sharing with the other Malayan rulers and least of all with Federation hopes of a share in his territory's oil resources.

PROFILE

ENVIRONMENT

Brunei comprises two tracts of land located on the northwestern coast of Borneo, in the Indonesian Archipelago. It has a tropical, rainy climate slightly tempered by the sea. Rubber is tapped in the dense forests. There are major petroleum deposits along the coast. The country is one of the world's main exporters of liquid gas.

SOCIETY

Peoples: Malay 67 per cent; Chinese 15 per cent; indigenous 6.0 per cent; Indian, European and other.
Religions: Islam is the official religion. Muslims, 67 per cent; Buddhists 12.8 per cent; and Christians 10 per cent.
Languages: Bahasa Malay (official), Chinese, English, local languages.
Main Political Parties: were banned in 1988. There are a few illegal parties: Brunei National Solidarity Party (PPKB), Brunei People's Awareness Party (PAKAR); National Development Party (NDP).

THE STATE

Official Name: Negara Brunei Darussalam.
Administrative divisions: 4 districts.
Capital: Bandar Seri Begawan 61,000 people (2003).
Other Cities: Kuala Belait 26,600 people; Seria 23,000; Tutong 16,300 (2000).
Government: Hassanal Bolkiah Muizzaddin Waddaulah, Sultan since 1967. He is also Prime Minister, Minister of Finance and the Interior, assisted by five Councils: religious, private, ministerial, legislative and of succession. The Legislative Council is made up of 20 members and has only an advisory role.
National Holiday: 1 January, Independence Day (1984); 15 July, the Sultan's birthday.
Armed Forces: 4,900 (1995). Others: Gurkha Reserve Unit: 2,300; Royal Brunei Police: 1,750.

Life expectancy
77 years
2005-2010

Literacy
93% total adult rate
2000-2004

HIV prevalence rate
<0.1% of population 15-49 years old
2003

IN FOCUS

ENVIRONMENTAL CHALLENGES

Brunei is one of the world's main exporters of liquid gas, production of which threatens the environment in various ways. In drought periods, fires in the Indonesian forests result in heavy smoke or haze cover.

WOMEN'S RIGHTS

The participation of women (and men) in politics is restricted by the Sultan's or his counselor's decisions. Whatever participation there is takes place within the framework of an interim regime which is renewed monthly.

Women make up 47 per cent of the total population. In 2003, they represented 36 per cent of the workforce. Since 2000, there has been a rise in women's employment, accompanied by an increasing variety in their jobs (traditionally, they only did domestic commerce). Ninety per cent of women over 15 are literate*. Since 1999, there has also been an increase in the number of women attending university; women now make up almost two-thirds of university registration. In July 1999, a Married Women's Law came into effect, improving the rights of non-Muslim women over maintenance, property and domestic violence. In the same year, the Islamic Family Law was also amended to improve Muslim women's rights in marriage and divorce.

CHILDREN

In 2004, there were 40,000 children under 5 years old and the under-5 mortality rate was 8 per 1,000 live births. During the three previous decades, all indicators relating to child welfare (including under-5 mortality, education, health and maternal care) had improved, mirroring the country's growth in prosperity derived from its energy industry.

Women married to foreigners or bearing children by foreign fathers cannot pass on citizenship to their children, even when such children are born in Brunei. This has resulted in a population of 5,000 or so stateless children who can live in Brunei and have travel documents, but who may not enjoy the full privileges of citizenship, including the right to own land.

INDIGENOUS PEOPLES/ ETHNIC MINORITIES

Most of the population is Malay and includes Kedayan, Tutong, Belait, Bisaya, Dusun and Murut - groups that in 2000 made up 67 per cent of the population. Other indigenous groups like Iban, Dayak and Kelabit represented 5.9 per cent (19,600 people); 11.6 per cent of the population belongs to other non-specified racial groups.

MIGRANTS/REFUGEES

The economy is dependent on the oil industry and historically has used migrant labor from Southeast Asia. It is estimated that migrant labor constitutes more than 35 per cent of the total workforce.

DEATH PENALTY

The death penalty is in force for ordinary crimes but the last execution was in 1957.

---

[7] In 1976, with Malaysian prompting and UN support, the renegotiation of the anachronistic colonial statute was taken into consideration, when the newly-elected Malaysian Prime Minister Datuk Hussain Onn promised to respect Brunei's independence. Full independence from the UK became effective on 1 January 1984. Power was formally transferred on that date, but celebrations were postponed until 23 February 1985, so that foreign guests could attend.

[8] One month after independence, Hassanal Bokiah, son of Sultan Omar, who had abdicated in his favor in 1967, dissolved the Legislative Council and went on to govern by decree. From that time, the main sources of tension were the power struggle within the ruling family and the presence of foreigners in all the key positions of public office, the economy and the army.

[9] Brunei obtained its independence in particularly favorable conditions for a Third World country. It had a relatively small population, an annual per capita income of $20,000, low unemployment, a generous social security system and considerable foreign exchange reserves ($14 billion in 1984). However, 20 per cent of the population was living below the poverty line and 90 per cent of consumer goods, including food, were imported, which made the cost of living extremely high.

[10] The Sultan, however, was aware that the country depended on a non-renewable resource, and food imports. Consequently, with a view to achieving self-sufficiency in food production, he attempted to diversify the economy and promote a new land-owning class. Only 10 per cent of the arable land was cultivated, and small farmers, especially rubber-tree growers, tended to emigrate to the city.

[11] Economically, Brunei depended upon the complex interplay of transnational interests. The Government's partnership in exploiting natural gas reserves with Brunei Shell Petroleum Co. and Mitsubishi, a shipping contract with Royal Dutch Shell and oil field concessions to Woods Petroleum and Sunray Borneo, introduced powerful new parties into the process of national decision-making. In 1985, the Government created an Energy Control Board to supervise the activities of the Brunei Shell Petroleum Company, a company funded equally by the Government and Shell.

[12] In 1987, it was reported that a request from Colonel Oliver North, of President Reagan's administration in the US, for 'non-lethal' aid to the Nicaraguan contras resulted in the Sultan of Brunei to deposit a $10 million donation in a Swiss bank account.

[13] In 1991, Sultan Hassanal Bolkiah freed six political prisoners who had been detained after the failed 1962 revolt. The release was ascribed to political pressure by the British Government. This same year Brunei signed a contract for almost $150 million with the United Kingdom in order to modernize its army.

[14] In 1992, Brunei joined the Non-Aligned Movement, together with Vietnam and India. Together with other members of ASEAN - Indonesia, Singapore, Thailand, Malaysia and the Philippines - it signed an agreement to create the first integrated market in Asia in 2007. This project stipulates the creation of 'growth triangles' - association between some ASEAN members to deregulate trade in certain economic sectors, allowing the overall liberalization planned for 2007.

[15] In 1994, Brunei created - with the Philippines, Malaysia and Indonesia - a sub-regional market to intensify trade in tourism, fishing, and transport by sea and air. In 1995, the country joined the World Bank (WB) and the IMF. Negotiations with southeast Asian leaders were intensified following the regional stock market crisis in 1997, in order to co-ordinate policies to stabilize the regional economy and establish a recovery strategy. In the negotiations, the possibility was discussed of using local currencies instead of the US dollar for trade transactions within the area.

[16] In 1998, the Sultan named his eldest son, Al-Muhtadee Billah, as successor to the oldest Islamic dynasty in the region - in existence for over six centuries. The announcement was made in the midst of the worst economic crisis since independence. Construction and the export sectors began a slow recovery during 1999. The Government announced plans to train a large part of the workforce over five years, so as to diversify the economy and develop sectors other than tourism.

[17] In 2000, the authorities indicted the Sultan's younger brother, Prince Jefri, for embezzling public funds during his terms as minister of Finance and president of the State Investment Agency, where he served until 1998. Jefri had been removed and forced to declare bankruptcy of his private company, to which he had transferred large sums of government funds during his ten years in office. A year after the trial, an auction was held in which some 10,000 of the prince's belongings were sold.

[18] During the ASEAN summit in Brunei in 2001, the leaders agreed to cooperate in the US-led war against terrorism. They also postponed the launch date for the Asian Free Trade Area (AFTA), to be set for sometime between 2006 and 2010.

[19] In February 2003, Brunei and Zambia established diplomatic relations. In March, the National Petroleum Company signed an agreement with a consortium of three foreign companies, TotalFinaElf Deep Offshore Borneo BV, BHP Billiton, and Amerada Hess.

[20] Two former members of the country's security forces and a transport entrepreneur were detained by the Government in March 2004. They were accused of treason, subversion and incitement to hate Sultan Bolkiah. As dictated by internal security laws, they were jailed for an indefinite time and without the possibility of being put to trial. Authorities claimed that they jeopardized the country's security and stability.

[21] In September the Sultan reinstated Parliament, which had been suspended for 20 years. In May 2005 Bolkiah dismissed four of his Cabinet members, the most dramatic political reform since independence. For the first time, ministers with private sector experience and a non-Muslim were included. The Sultan's older son was given a position in the Prime Minister's office, which was seen as a sign of a future succession.

[22] In April 2006 Brunei became a member of the Asian Development Bank (ADB).

[23] In July 2006, the Sultan celebrated his 60th birthday with a lavish banquet, and announced the first pay rise in 20 years for the kingdom's civil servants. ∎

Bulgaria / Balgarija

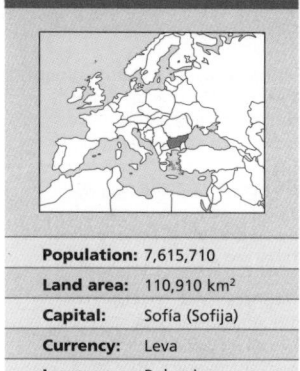

Population:	7,615,710
Land area:	110,910 km²
Capital:	Sofía (Sofija)
Currency:	Leva
Language:	Bulgarian

Evidence of human habitation in the Bulgarian area dates from the Middle Paleolithic period (100,000 to 40,000 BC). The first agricultural communities appeared in the Neolithic Period (7,000 to 3,000 BC). Thracian peoples - ethnically Turks - arrived from Central Asia around 3,500 BC, settled there and developed their culture, the traces of which remain in their monuments devoted to horse-worship and acting. Thanks to their aptitude for combat, the Thracians resisted several Macedonian and Persian attacks as well as the Roman Empire for 150 years, which finally managed to subdue them in the first years of our era. In Roman times Bulgaria was divided between the provinces of Moesia and Thrace and was crossed by the main land route from the west to the Middle East.

[2] The Bulgarians came to the region towards the end of the 5th century, among the peoples that followed the devastating Mongolian invasion, led by Attila. They were fierce warriors who lived by warfare and plunder. They first settled temporarily on the steppes north of the Black Sea and northeast of the Danube.

[3] Towards the end of the 6th century, the region was occupied by Slav immigrants coming from the east, who spread across the area between the Danube and the Aegean Sea. That land was deserted after the Goths (Germanic), the Huns (barbaric people who came from Upper Asia during the 5th century; they built a powerful state in the Danube area, survived the Roman Empire and disappeared after it fell) and the Avars (nomadic warriors from Central Asia) passed by. During the 7th century, the Avars forced the Bulgarians across the Danube and subdued the Slavs.

[4] In the year 681, after the Bulgarians won several battles for territory - Moesia was under the control of the Byzantine Empire - Emperor Constantine IV formally recognized the State of Bulgaria.

[5] When the Patriarch of Byzantium recognized the independence of the Bulgarian Church in the year 864, Bulgaria's political autonomy was finally consolidated. This achievement favored the cultural assimilation of Slavs and Bulgarians, who developed an original identity, as seen in the creation of the Slav alphabet and the establishment of the official language in 863. Literature and arts flourished during that period and several schools and universities were opened.

[6] Under Simeon (893-927), the Bulgarian State extended its domain as far as the Adriatic, subduing the Serbs, and becoming the most powerful kingdom of Eastern Europe.

[7] Bulgaria's power declined after Simeon's death. Internal disputes among the nobility, the opposition of the peasants and renewed attacks from abroad led to its downfall. In 1014, Bulgaria lost all its territory to the Byzantine Empire, which kept control for more than 150 years. After a large uprising in 1185, the northern part of Bulgaria recovered its independence.

[8] During the reign of Ivan Asen II (1218-41) Bulgaria regained power. However, none of Ivan's successors managed to impose a central authority over these diverse areas where feudalism was the norm. By 1393 the whole of Bulgaria had fallen under Turkish rule.

[9] In the 17th and 18th centuries, after the wars with Austria and the unsuccessful siege of Vienna, the Ottoman Empire began to decline, though it still retained much of its territory. The former Bulgarian State was twice invaded by Russia, in 1810 and 1828.

[10] Throughout the invasion period the Bulgarians maintained their cultural identity; keeping their language, music and folklore alive. Under Turkish domination, the Greek Orthodox Church assumed religious leadership, suppressing the independent patriarch. In this way Bulgarian monks were among the precursors of the national liberation movement.

[11] The Bulgarian Church fought for 40 years to recover its independence. In 1870, the Sultan gave permission for the Church to create an *exarchate* (sub-Patriarchy). The first *exarch* (deputy patriarch) and his successors were declared schismic and were excommunicated by the Greek patriarch.

[12] One of Moscow's conditions for the Treaty of St Stephen (1878) was the creation of a Bulgarian state, but the European powers feared the creation of a Russian satellite in the Balkans. That year, in the Congress of Berlin, the 'autonomous province' of Rumelia was created in the south, and the State of Bulgaria in the north. Macedonia was to continue as part of the Ottoman Empire. An Assembly of Notables was in charge of drawing up a law and appointing the ruler of the new State.

[13] The assembly approved a liberal Constitution, establishing a constitutional monarchy. Prince Alexander of Battenberg, grandson of Alexander II of Russia, was elected and assumed the Bulgarian throne in July 1878, swearing to uphold the Constitution. He suspended it two years later.

[14] The Prince set up a dictatorship, headed by Russian general Leonid Sobolev and other conservatives. The Russian Emperor's death modified Alexander's behavior, making him more attuned to Bulgarian issues. In 1885, he supported the liberal rebellion in Rumelia, the governor there was replaced and union with Bulgaria was proclaimed.

[15] In 1886, several treaties were signed that recognized Prince Alexander as the ruler of Rumelia and Bulgaria. However, he was subsequently taken to Russia against his will and forced to abdicate. Looking for someone who would be acceptable to Russia, as well as the rest of Europe, the Bulgarians finally appointed Ferdinand of Saxe-Coburg-Gotha as his replacement.

[16] Although initially distrusted, Ferdinand was able to gain the support of Vienna, London, Rome and Russia. He then concentrated on the reunification of the Bulgarians. Prince Ferdinand proclaimed Bulgaria's independence in 1908.

[17] In 1912, Bulgaria encouraged the formation of the Balkan League, together with Greece, Serbia and Montenegro to fight Turkey in the first Balkan War that began in October. In May, Turkey ceded its European dominions on the Black Sea.

[18] The allies did not agree with the distribution; Bulgaria confronted Greece, Serbia and Romania. The Second Balkan War quickly ended in Bulgarian defeat. In Bucharest, in August 1913, Macedonia was divided between Greece and Serbia, and Romania gained an area of northern Bulgaria, rich in natural resources.

[19] That year, the Bulgarian Government abandoned its traditionally pro-Russian stance, seeking closer ties with Germany. When World War I broke out the Bulgarian people and the army disapproved of the official policy, even though Serbia was beaten. Ferdinand surrendered to the Allies in 1918 and abdicated in favor of his son Boris.

[20] Having lost much territory, Bulgaria was disarmed and forced to pay extensive war damages. With the restoration of the 1878 Constitution, elections were held in 1920. The anti-war reaction gave the Agrarian Party a wide margin. Working on a Soviet model, the Government started radical agrarian reforms. The Government was not pro-Soviet, however, and local Communists were persecuted.

[21] Bulgaria joined the League of Nations and followed a conciliatory line of diplomacy for some time.

Life expectancy
73 years
2005-2010

GNI per capita
$2,750
2004

Literacy
98% total adult rate
2000-2004

HIV prevalence rate
<0.1% of population 15-49 years old
2003

However, its territorial losses and the pressure exerted by expatriates abroad soon led to new tensions with its neighbors. Aleksandur Stamboliyski, the leader of the Agrarian Party and head of the Government, was ousted and assassinated by a conspiracy of Macedonians and opposition figures in 1923.

22   Aleksandur Tsankov assumed control of the Government, heading a multiparty alliance which excluded the Liberal, Communist and Agrarian parties. Uprisings and armed activity by the opposition led to hundreds of executions and assassinations. The Government declared martial law and reinforced the army in order to avoid outright rebellion.

23   In 1934, fearing the effects of the worldwide economic depression and taking a cue from his neighbors, King Boris III set up a dictatorship. All political parties were proscribed, there was censorship of the press, the universities were closed and an ultra-right youth movement was established.

24   Tension with Turkey eased and in 1937, a peace and friendship treaty was signed with Yugoslavia. The following year, Bulgaria signed a non-aggression pact with the Balkan alliance, in exchange for the rearmament of the Bulgarian army. While the King was once again seeking rapprochement with Germany, the Bulgarian dream of re-establishing its former borders was gathering strength.

25   In 1940, Germany made Romania return the parts of Bulgaria which it had won in the second Balkan War. Bulgaria signed the Anti-Komintern pact and German troops set up bases aimed at Greece and Yugoslavia on Bulgarian soil. In exchange, Bulgarian troops were allowed to occupy the part of Thrace belonging to Greece, and the part of Macedonia belonging to Yugoslavia, and part of Serbia.

26   When Bulgaria refused to declare war on the Soviet Union, King Boris was assassinated, and a new pro-German Government was formed. Growing anti-Nazi resistance, led by the Communists, contributed to the formation of the Patriotic Front in 1942. The Republicans, left-wing Agrarians, Democrats and independents all subsequently joined.

27   In May 1944, paralyzed by the civil war, the pro-German Boshilov resigned, and was replaced by Bagrianov. While Soviet troops advanced toward the Danube, Bagrianov sought an agreement with the Allies. In August, Bulgaria proclaimed its neutrality, and ordered the disarmament of the German troops on its soil.

28   The USSR and the Red Army entered Bulgarian territory on 5 September. The resistance fanned local insurrection. On 8 September, the Red Army took the capital,

and the Patriotic Front formed a Government headed by the Republican Kimon Georgiev.

29   Sofia signed a treaty with the allies in October 1944. Bulgarian troops, under Soviet command, collaborated in the defeat of the German forces in Hungary, Yugoslavia and Austria.

30   In a referendum held in September 1946, 92 per cent of the electorate approved the creation of the Republic of Bulgaria. The Patriotic Front won the elections, and Communist leader Giorgi Dimitrov became Prime Minister.

31   In 1947, Britain and the US recognized the government. The National Assembly ratified the peace treaty with the Allies, the new Constitution went into effect and, at the end of the year, the Soviet troops withdrew from the country. After joining the opposition, some of the Patriotic Front's former leaders were arrested and sentenced to death for conspiracy.

32   Under the leadership of the Communist Party (BCP), the Bulgarian State adopted the Soviet socio-economic model. A process of accelerated industrialization was set in motion, without taking into account the lack of raw materials or the technical preparation of the labor force. This industrialization led to urbanization that determined the demographic expansion of Sofia, whose population increased fourfold in two generations. In the countryside, collective farming was enforced.

33   Dimitrov resigned from the administration in 1949. Vulko Chervenkov succeeded him. In 1954, Todor Zhivkov was named First Secretary of the BCP, becoming Prime Minister in 1962. Bulgaria was the USSR's closest ally among the Warsaw Pact countries, and in 1968 Bulgarian troops accompanied Soviet troops in the invasion of Czechoslovakia.

34   In 1988, Bulgaria and Turkey signed a protocol governing bilateral economic relations. The dialogue was interrupted in the following year, when it was revealed that the Bulgarian militia had used violence to put down a protest by 30,000 Turks, who had been demonstrating against the Government's policy of assimilation.

35   The largest protest since the War was held in November in front of the National Assembly. A group known as 'Eco-glasnost' demonstrated against a proposed nuclear power plant on an island in the Danube and against the construction of a reservoir in one of the country's largest nature reserves.

36   The Central Committee of the BCP replaced Zhivkov as Secretary General, a post he had held for 35 years and as president of the State Council. He was succeeded by Petur Mladenov, who was considered to be a proponent of liberalization of

the regime.

37   Political demonstrations demanded reforms and elections. Social pressure forced the Government to amend the Constitution to introduce an electoral law that allowed elections to take place.

38   In March 1990, the Union of Democratic Forces (UDF), made up of 16 opposition parties, and the BCP agreed to the election of a Constitutional Assembly. The July election was won by the Bulgarian Socialist Party, BSP (formerly BCP). In October, the BSP was forced to enter a new coalition, led by Yelio Yelev, a dissident during the 1970s and leader of the Social Democratic wing of the UDF.

39   The new coalition government adopted a program of economic reforms in consultation with the IMF and the World Bank. They reached an agreement with the labor unions for a 200-day 'social peace' until the reforms could go into effect.

40   In July 1991, the new Constitution was approved, establishing a parliamentary system, allowing personal property, and freedom of expression. After the October election, Parliament named

Filip Dimitrov as Prime Minister. Virtually alienated from the Social Democrats and the 'Greens' - co-founders of the UDF - Dimitrov was chosen because of the support he had from the right-wing of the opposition coalition and the Movement for Freedom and Human Rights (of the Turkish minority).

41   In May 1992 Bulgaria joined the Council of Europe. Also in that year, former Communist leader Todor Zhivkov, three former prime ministers and another former member of government prior to 1991, were arrested and charged with corruption in the exercise of their duties. The economic situation led the Movement for Rights and Freedoms (MRF) to withdraw its support of Dimitrov, which caused the downfall of his cabinet.

42   The new Prime Minister Liuben Berov stated he was willing to restore the lands confiscated by the Communists to the Turkish minority.

43   Transition from a centralized planned economy to a free market economy continued to be difficult and caused paradoxical situations. When the former Soviet Union stopped buying two-thirds of Bulgaria's exports, foreign trade

PROFILE

ENVIRONMENT

Located in the Balkans, Bulgaria comprises four different natural regions. The fertile Danubian plains in the north are wheat and corn-producing areas. South of these lie the wooded Balkan Mountains, where cereals and potatoes are cultivated, cattle and sheep are raised, and the country's major mineral resources - iron ore, zinc and copper, are found. Cattle are also raised on the Rhodope Mountains in southern Bulgaria. South of the Balkan mountains there is a region of grassland, crossed by the Maritza River, where tobacco, cotton, rice, flowers and grapes are cultivated.

SOCIETY

People: The majority are Bulgars of Slav origin (85 per cent). There are also Turks (9.4) and Roma (gypsies, 3.6), plus immigrant communities of Macedonians, Armenians, Tatar, Gagauz, Circassians, Russians. **Religions:** Orthodox Christian Church of Bulgaria, 83 per cent, Muslim 13, Catholic 1.5, Jewish 0.8 and Protestants, Gregorian Armenians. **Languages:** Bulgarian (official and predominant); also spoken: Romani, Turkish, Macedonian, Gagauz and Armenian. **Main Political Parties:** There are many small parties, hence the formation of alliances like the ruling Coalition for Bulgaria (eight parties); National Movement Simeon the Second; Movement for Rights and Freedoms; National Union Attack (nationalist coalition); United Democratic Forces (eight parties). Ethnic parties are banned. **Main Social Organizations:** The Confederation of Independent Unions of Bulgaria was founded in 1990 and is the biggest (1.6 million members). Since January 1990 the Confederation of Labor Podkrepa has been legal, becoming a new union axis, headed by Constantin Trenchev. Importance of the agrarian movement led by the National Agrarian Union of Bulgaria and the national 'Nikola Petkov' union. There are many regional, ethnic and national groups with varied agendas, one of the most active is the Internal Macedonian Revolutionary Organization.

THE STATE

Official Name: Narodna Republika Balgarija. **Capital:** Sofia (Sofija) 1,076,000 people (2003). **Other Cities:** Plovdiv 344,500 people; Varna 293,600; Burgas 192,900 (2000). **Government:** Georgi Parvanov, President since January 2002; Sergey Dmitrievich Stanishev, Prime Minister since August 2005. National Assembly has 240 members. **National Holiday:** 3 March, National Liberation Day - Independence (1878). **Armed Forces:** 68,450. Other: Border Guards (Ministry of Interior): 12,000; Security Police: 4,000; Railway and Construction Troops: 18,000.

 Under-5 mortality
15 per 1,000
live births
2004

 Poverty
2% of population
living on less
than $1 per
day
2003

 Debt service
17.1% exports
of goods
and services
2004

Maternal mortality
32 per 100,000
live births
2000

was significantly reduced. UN sanctions imposed on neighboring former Yugoslavia caused losses of $1.5 million to Bulgaria. In 1993, Berov went on with the transition to a market economy at a rate considered excessively slow by the IMF, which caused a certain amount of tension between Sofia and the international organization.

[44] In June 1994, Berov was able to pass the privatization law. His economic policies and the rights of Bulgarians of Turkish origin were being sharply questioned and so three months later he resigned. President Yelev dissolved Parliament and called new elections. The BSP won an absolute majority in the National Assembly.

[45] In January 1995, socialist leader Zhan Videnov formed a new Government which included members from the BSP, the Bulgarian Agrarian National Union and the Eco-glasnost Political Club. His cabinet was the first in the history of post-Communist Bulgaria with an absolute majority at the National Assembly.

[46] Petar Stoyanov's Union of Democratic Forces won the presidential elections in 1996. That year Bulgaria became a member of the World Trade Organization.

[47] Harrassed by his political enemies, Prime Minister Zhan Videnov resigned. The conservative groups that backed Stoyanov called for the dismissal of the Government that was supported by the parliamentary majority of the socialists and their allies. A series of public demonstrations in the capital and other cities demanded new elections. In January 1997, Stoyanov became President.

[48] The legislative elections were won by the Union of Democratic Forces (UDF) in April 1997. The new government implemented a purely neo-liberal economic policy, following IMF guidelines, planning the privatization of state companies considered to be loss-making and the elimination of 60,000 jobs in the public sector. Services were liberalized with the aim of reducing inflation. In May 1998 agricultural subsidies were abolished and the State telecommunications company, several banks and the Bulgarian airline were privatized. The Government declared its objective would be fiscal balance. A few weeks later the IMF agreed an $800 million loan.

[49] To be accepted as a member of the European Union in 2007, Bulgaria was forced to shut down two of its oldest nuclear reactors. Of the remaining four, another two were closed down in 2006.

[50] Former king Simeon II - crowned at the age of six and reigning for three years from 1943 - launched the Simeon II National Movement in 2001. Simeon Saxe-Cobourg won the parliamentary elections and became Prime Minister in July, thus becoming the first former monarch of Eastern Europe to return to power. His campaign agenda did not include the restoration of the monarchy, but promised to end poverty, unemployment and other ills that plagued Bulgaria after the fall of Communism. Bulgaria is currently the major entry point into Western Europe for heroin and cocaine smugglers from Asia.

[51] However 100 days after the election, thousands thronged the streets of the capital protesting the promises had not been fulfilled.

That same month the Socialist Party leader, Georgi Parvanov, was elected president. The electoral turnout was the lowest since the fall of communism (41 per cent).

[52] In October 2003 local elections were held. The BSP won with 33 per cent of the votes, followed by the UDF with 21 per cent, while Simeon's party got 10 per cent. Voter participation was below 40 per cent.

[53] A report about the trafficking of people in southeast Europe by the United Nations, the Organization for Security and Cooperation in Europe (OSCE) and the Stability Pact for Southeastern Europe, published in December 2003, stated that in spite of legislative reform and police cooperation by European governments, human traffic continued to rise. Most of the trafficked people were women and children.

[54] In April 2004, Bulgaria (along with Estonia, Lithuania, Latvia, Romania, Slovakia and Slovenia) became a member of NATO, in the largest expansion of the organization since it was founded in 1949; the number of member countries reached 26. Bulgarian public opinion favored the move. During the ceremony in Brussels, Jaap de Hoop Scheffer, NATO's secretary general, said the expansion would be 'honest and productive'.

[55] In June, two Bulgarians were kidnapped in Iraq. Truckers Georgi Lazov and Ivaylo Kepov were captured near Mosul (city located 400 km to the north of Baghdad) by rebel Iraqi militias. The kidnappers demanded the freedom of all Iraqi prisoners held by US forces.

[56] Two months later, the Government confirmed that both corpses had been found on the river Tigris, which crosses Iraq, Syria and Turkey. In spite of the hostages' death, Bulgaria kept its 500 troops in Iraq.

[57] In April 2005 Bulgaria signed a treaty to become a member of the EU. To join the union on 1 January 2007 the country had to implement reforms, mainly in relation to the fight against corruption, though it also had to ensure that all citizens were given a fair trial.

[58] US Secretary of State, Condoleezza Rice, and Bulgarian Foreign Affairs Minister, Ivailo Kalfin, signed an agreement on April 2006 to share military facilities in Bulgaria. The 10-year agreement allowed the joint use of Bulgarian training facilities and would strengthen the ability of both countries to carry out joint military operations. In that period, up to 2,500 US soldiers would be deployed in Bulgaria on a rotating basis.

[59] In October 2006 Parvanov won the presidential election – the first Bulgarian President to win a second term. ∎

IN FOCUS

ENVIRONMENTAL CHALLENGES
Heavy metals, nitrates, petroleum derivatives and cleansing agents damage the medium and lower courses of the most important rivers that flow into the Black Sea, which is highly contaminated. One quarter of the forests suffer the effects of air pollution, such as acid rain, which is a consequence of metallurgical and industrial plants' emissions.

WOMEN'S RIGHTS
Bulgarian women have been able to vote since 1937 and to be elected since 1944. In 2005 women held 22 per cent of Parliament seats; also, 24 per cent of ministerial or equivalent posts were held by women. In 2003, women represented 48 per cent of the labor force. In 2000, 10,000 Bulgarian women were forced to work as prostitutes in industrialized countries and in 2003, 13 Bulgarian citizens were arrested in Italy for being part of the chain 'distributing' women, for fees ranging from $3,000 to $8,000, depending on the woman's 'quality'.

CHILDREN
The indicators of progress regarding commitment to child welfare have improved between 1970 and 2000 (mortality rate of children under five and mothers decreased, while the percentages of vaccination and access to health care and education of children increased). Under-five mortality among the Roma are higher than the national average, while their enrolment and school completion rates are lower. Very few Roma children finish secondary school, and even fewer attend university.

Education is compulsory up to the age of 14. The enrolment rate of school-age children was 98 per cent in 2004*. Approximately 45,000 children drop out of school every year. The highest dropout rates (32 per cent) are among children from families of Roma origin, followed by Bulgarian girls (8 per cent) and Turkish girls (6 per cent).

INDIGENOUS PEOPLES/ ETHNIC MINORITIES
Most Turks live in two main areas where they form the majority of the inhabitants; one in the northeast (Silistra-Varna) and another in the southeastern corner (Haskovo-Kurdzali). The forced 'Bulgarization' of the Turk minority in the 1980s is daily becoming less common. In 2000 a law that allowed broadcasting in Turkish was approved, as well as a new law on minorities by which everybody could communicate in his/her mother tongue. In political terms, the Turks have representatives in the legislature and the executive as well as at the local level in areas where they form the majority. Their main grievances have to do with the economic situation and with needing more governmental support for their language and cultural traditions.

Roma are discriminated against in terms of access to housing, public services, education and health care. They remain a focus for police violence. The Government took some measures towards the implementation of a 'Framework Program for Equal Integration of Roma in Bulgarian Society', signed in April 1999. Among other things, this promised to improve Roma settlements and to grant them funds for the diffusion of their culture - but neither had been implemented by the end of 2005.

MIGRANTS/REFUGEES
Most asylum requests come from Iraq, Afghanistan, Armenia and Nigeria. In 2002, the country sheltered around 3,000 refugees and people seeking asylum. Some of those denied asylum were granted residence permits for humanitarian reasons for different periods of time. Although the precise number is unknown, in the last five years some 10,000 Bulgarians, apparently of Roma origin, requested asylum in other European countries.

DEATH PENALTY
It was abolished in 1998.

* Latest data available in *The State of the World's Children* and *Childinfo* database, UNICEF, 2006.

Burkina Faso / Burkina Faso

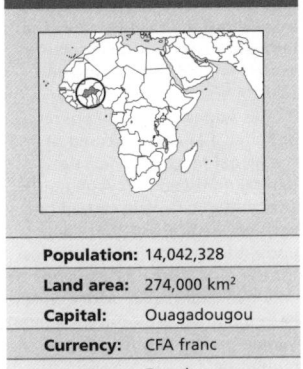

Population:	14,042,328
Land area:	274,000 km²
Capital:	Ouagadougou
Currency:	CFA franc
Language:	French

During the 11th century, indigenous peoples, ancestors of the Lobi and Bobo, settled in the western region of the Mouhoun River (or Volta Noire/Black Volta). Though they never developed a centralized political system, the populations that migrated from the Dagomba region (present-day Ghana) founded the Mossi dynasty in the north. The Mossi kings imposed their military aristocracy over the region of the Volta Rivers (Volta Noire, Volta Blanche and Volta Rouge - black, white and red).

[2] Several well organized kingdoms were created in the following two centuries. TheYatenga and Uagadugu were the most important kingdoms. The monarchs of the Uagadugu kingdom were chosen from the members of the royal family by four ministers who had to maintain the balance between the Mossi aristocracy and the Mande people. This election system was to be maintained until the 20th century.

[3] The Mossi and Mande resisted attempts at annexation by the Mali and Songhai empires (see Mali and Guinea) and remained independent throughout the Fulah invasions of the 18th and 19th centuries.

[4] In a series of military incursions between 1895 and 1904, the French laid waste the central plains, burning houses, and slaughtering people and animals. The ensuing reign of terror finally sparked off an insurrection in 1916, which met with such violent repression that millions were forced to emigrate, mostly to Ghana.

[5] Called Upper Volta by the French, the country was administered as part of the colony of Haute Sénégal-Niger, when it became an independent protectorate. In 1932 it was split between Côte d'Ivoire, Niger and Sudan, and reconstituted as a separate colony in 1947. Finally, in line with French neo-colonial strategy, the country was declared independent in 1960.

[6] Elections held in 1960 were won by the Voltanese Democratic Union (UDV), a party backed by landowners and private enterprise. Maurice Yameogo was elected President and was re-elected in 1965 amidst intense trade-union agitation, which occurred in response to administrative chaos and austerity measures. Yameogo was overthrown a year later in a military coup led by army chief-of-staff, General Sangoulé Lamizana.

[7] The 1970s witnessed a succession of elections, military coups, and more or less fraudulent re-elections orchestrated by General Lamizana. Starvation was widespread, herds were dwindling and an estimated quarter of the population had emigrated to neighboring states.

[8] In 1980 Colonel Saye Zerbo led a coup d'état, but was in turn ousted in 1982 by Major Doctor Jean-Baptiste Ouedraogo. Ouedraogo was overthrown by a young officer, Thomas Sankara, who was popular among soldiers and the rural poor as he brought in an anti-corruption campaign and organized brigades to assist victims of the prolonged drought and ensuing deforestation.

[9] Under Sankara's government Upper Volta was renamed Burkina Faso - 'land of the incorruptible'; the national anthem was sung in African languages; land reform was carried out and popular courts were set up to dispense justice.

[10] Sankara, leading a National Revolutionary Council, set a target of two meals and 10 liters of water per person per day. The implementation of such measures presented a considerable challenge in a country where 82 per cent of adults were illiterate and absolute poverty was the norm. Sankara defined his revolutionary ideology as anti-imperialist in a speech in October 1983, and stated that he was to fight corruption and hunger, and promote reforestation. Education and health were his government's priorities. He was an inspirational figure, whose ideas and example resonated all over Africa.

[11] On 15 October 1987, Sankara was overthrown, prosecuted and executed along with 12 of his supporters in a coup led by his second-in-command, Blaise Compaoré, who pledged to continue the Popular Front and 'rectify' the regime. In spite of initial tensions with Washington over the strong links between Sankara and Libyan leader Muammar al-Qadhafi, the US soon replaced France as Burkina Faso's main aid donor.

[12] Contrasting with Sankara's principled austerity, Compaoré built a government palace and purchased a presidential plane. His economic strategy included encouraging private enterprise and foreign capital, and consideration of denationalization, deregulation and agreements with international lending institutions.

[13] From November 1987

PROFILE

ENVIRONMENT
A landlocked country, Burkina Faso is one of the most densely populated areas of the African Sahel (the semi-arid southern rim of the Sahara). The Mossi plateau slopes gently southwards and is traversed by the valleys of the three Volta Rivers (Black Volta, White Volta and Red Volta). Export crops such as millet, peanuts and cotton are produced, mainly in the southwest.

SOCIETY
Peoples: Over half of the population are Mossi. Peulh herders and the Tamajek clans with their vassals, the Bellah, amount to around 20 per cent. Djula peasants and traders are indigenous minorities. The language of these three population groups is the linguistic bridge between the various regions. Senufo and Bobo-Fing cultures inhabit the western plains where the savanna merges into the forest. In the south, the predominant Lebi, Bobo-Ule, Gurunsi and Bissa cultures extend into several states. The savanna to the east holds the great Gurmantehé civilization, while the Sampo, Rurumba and Marko cultures live in the desert regions to the north and northeast.
Religions: 50 per cent are Muslims; traditional African religions, 40 per cent; Christian (mostly Catholic), 10 per cent. **Languages:** French (official); 71 languages from the Sudanic family, spoken by 90 per cent of the population (most common are Mossi, Bobo, Bissa and Gurma). **Main Political Parties:** Congress for Democracy and Progress (CDP - Compaoré's ruling party); African Democratic Rally-Alliance for Democracy and Federation (RDA-ADF); Confederation for Federation and Democracy (CFD); Tolerance and Progress Movement (MTP); Party for African Independence (PAI).
Main Social Organizations: Burkina General Work Confederation (CGTB); Burkina Human Rights Movement (MBDHP); February 14 Group; Burkina Workers National Confederation (CNTB); Free Syndicates National Organization (ONSL); surveillance groups on political action in different organizations and communities.

THE STATE
Official Name: République de Burkina Faso.
Administrative Division: 45 provinces, 300 departments and 7,200 villages.
Capital: Ouagadougou 821,000 people (2003).
Other Cities: Bobo-Dioulasso 474,300 people; Koudougou 124,400; Ouahigouya 74,000 (2000).
Government: Republican parliamentary system, with a powerful head of state. Blaise Compaoré, President and Head of State, in power since October 1987 after a coup d'état; elected in 1991 and re-elected in 1998 and 2005. Ernest Paramanga Yonli, Prime Minister and Head of Government since November 2000. Bicameral Parliament: the National Assembly, with 111 members, and the House of Representatives, with 178 members.
National Holiday: 5 August, Independence (1960).
Armed Forces: 8,700 (including Gendarmerie). Others: 1,750.

 **Life expectancy**
49 years
2005-2010

 **GNI per capita**
$350
2004

 **Literacy**
13% total adult rate
2000-2004

 **HIV prevalence rate**
4.2% of population 15-49 years old
2003

onwards, political groups opposed to the rectification process were organized, including the Popular and Democratic Union, which demanded trade union freedom, amnesty for political prisoners and free elections, and the Revolutionary Burkina's Workers Party (PRTB), which maintained that the cause of Sankara's overthrow was his anti-corruption campaign.

[14] In December 1992 and March 1993 there were two general strikes against the implementation of the structural-adjustment plan recommended by the IMF. Devaluation of the CFA franc in January 1994 led to further popular dissatisfaction. In July, in spite of protests from the opposition, Parliament passed a law that allowed the Government to privatize 19 national companies.

[15] The Government continued its policy of economic liberalization and officially took Burkina Faso into the World Trade Organization in June 1995.

[16] In June 1996, the IMF approved a new $57 million loan to support the implementation of a structural-adjustment program in the subsequent three years. In 1997, a drought decimated food production and left the country in a critical situation. The deficit in food production, which added up to 32 per cent of the GDP and employed 90 per cent of

the population, was estimated at 156,000 tonnes for the 1997-1998 period. In December 1997, Ouagadogou had to call on the international community for 67,000 tonnes of grain to tackle the crisis.

[17] The falling prices of traditional cash crops encouraged marijuana plantations to grow swiftly in Burkina Faso, as well as in other countries of the region. The devaluation of the CFA franc caused a reduction in the import of medicines, which further complicated the fight against the two major diseases that ravage the population: malaria and HIV/AIDS. In 1998, there was just one doctor for every 25,000 people in the country.

[18] The 1998 elections, boycotted by the opposition, were won by Compaoré. A series of protests led to the resignation of Ouagadogou cabinet, but the President confirmed most of the ministers in their posts in January 1999.

[19] The murder of journalist Norbert Zongo, in December 1998, brought the Government under suspicion. Zongo had made important revelations about the supposed responsibility of François Compaoré, the President's brother, for the murder of his chauffeur, following torture inflicted by members of the presidential security staff. A year after Zongo's death, despite Compaoré's attempts to silence the protests, 30,000 people demonstrated in Ouagadougou,

clamoring for justice. Four of the six guards accused of Zongo's murder were finally condemned, but the President's brother testified only once.

[20] In February 2000, the Commission for National Reconciliation, formed by Compaoré as a way of easing the tension caused by the Zongo case, requested the rehabilitation of former president Thomas Sankara, assassinated in Compaoré's coup.

[21] A report by the UN stated in 2000 that Compaoré acted as intermediary in arms smuggling from the Soviet bloc to rebel groups such as UNITA in Angola and the RUF in Sierra Leone in violation of an international embargo on arms to these groups. His services were paid for with diamonds, whose trade for this purpose is also forbidden. In December, the Government finally accepted the presence of a United Nations' supervisory committee to monitor the import of weapons.

[22] The opposition took part in the legislative elections of May 2002, having boycotted the Presidential elections in 1998 and the local elections in 2000.

[23] In October 2003, 12 members of the presidential guard and the armed forces, accused of planning a coup against Compaoré, were imprisoned. In April 2004, Luther Ouali, a Navy captain, was sentenced to 10 years in jail after

admitting his involvement. Ouali said before the military court that his main reason for taking part in the attempted coup was to end social inequity in the country.

[24] In August 2004, the police detained 14 people accused of the illegal practice of female genital mutilation on young girls. The mutilations were carried out without medical assistance and under life-threatening conditions for the girls. Burkina Faso is one of only two countires in Africa which have mounted prosecutions against people performing female genital mutilation.

[25] In November 2005, after 18 years in power, Compaoré won the presidential election with more than 80 per cent of the vote, extending his mandate for five more years. Opposition candidates pointed out the vast sums of money available to the President's campaign, which distributed free T-shirts and caps all over the country and put billboards on almost every street. Compaoré has taken up residence in a new presidential palace in the 'millionaire quartier' of Ouaga 2000, on the edge of the capital.

[26] Over 300,000 people in the country depending on cotton for their living were hit by the falling international price of cotton during 2006. Farmers blamed the decline in price firmly upon the US Government's subsidies to its own cotton farmers. ∎

IN FOCUS

ENVIRONMENTAL CHALLENGES
Burkina Faso suffers from desertification, caused by serious droughts and the intensive production of export crops such as millet, peanuts and cotton. One of the harmful consequences of desertification is the lack of wood for domestic use. In the main cities, air pollution (due to usage of petroleum derivatives as vehicle fuel and also to industrial emissions) is the major environmental problem.

WOMEN'S RIGHTS
Since 1958 women have been able to vote and stand as candidates. In 2002, women held 11.7 per cent of Parliament seats and 15 per cent of ministerial or similar positions. In 2003, women comprised 48 per cent of the country's work force. Moreover poverty in general disproportionately affects women. Most of the polygamous households (58 per cent), with three or more women per man, were poorer in 2001 than monogamous or

single person households.

Some 73 per cent of pregnancies benefit from antenatal care but only 31 per cent of deliveries are attended by qualified personnel*. Approximately 2.3 per cent of pregnant women between 15 and 19 years old are HIV-positive. It is estimated that 150,000 women between 15 and 49 years old live with HIV/AIDS.

Since 1996, female genital mutilation has been forbidden, and Burkina Faso is one of only three African countries that has actually prosecuted people, but the practice is still widespread. Only 8 per cent of adult women are literate - the lowest figure in the world - compared to 19 per cent of men.

CHILDREN
Approximately 56 per cent of the population is under 18.* In 2003 there were 260,000 orphans under 17 years of age due to HIV/AIDS. By the end of that year it was estimated that 31,000 children between 0 and 14 years old were HIV-positive. Before the year

2000, the child mortality rate was decreasing (6 per cent between 1990 and 2001). Now, however, malaria, respiratory and intestinal infections and undernourishment, due to generalized poverty, are worsening the health of both women and children. Some 38 per cent of children under five years old are classified as underweight; 14 per cent are critically so*. Some 19 per cent of the children in this age range are weakened by undernourishment, and 39 per cent suffer from moderate or severe growth delay*. Only 25 per cent of the population have access to safe drinking water.

INDIGENOUS PEOPLES / ETHNIC MINORITIES
There are more than 60 ethnic groups in the country, each one with its particular social and cultural characteristics, even though all of them are of Burkinabe origin. The major group is the Mossi, descendants of the Moro-Naba dynasty. Tuaregs face discrimination and, in some cases, danger.

MIGRANTS/REFUGEES
In 2003 the country had almost a thousand refugees and people requesting asylum, mainly from Chad, Rwanda, Congo and Burundi. Since 1998 Burkina Faso has been one of the few countries which has agreed to give permanent residency to those refugees who could no longer stay in other African countries.

Thousands of Burkinabes, who had emigrated in the 1990s to Côte d'Ivoire searching for jobs, were obliged to return in the year 2000 due to anti-foreigner pressure in that country. Some of them, descendants of Burkinabes but born in Côte d'Ivoire, thus had little connection with Burkina Faso.

DEATH PENALTY
The country retains the death penalty for murder. No person has been executed during the past 10 years.

* Latest data available in The State of the World's Children and Childinfo database, UNICEF, 2006.

Burundi / Burundi

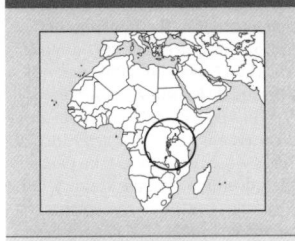

Population:	8,140,849
Land area:	27,830 km²
Capital:	Bujumbura
Currency:	Burundi franc
Language:	Rundi, Kirundi and French

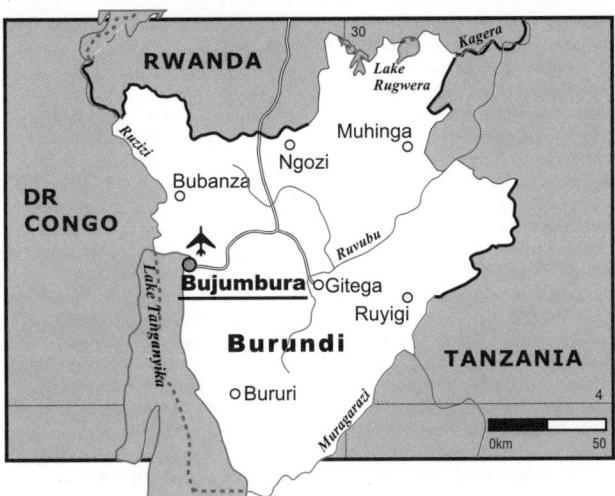

The scarce pieces of metal found in what is now Burundi are the only archeological evidence of what took place in that territory between the 7th century BC and the 15th century AD. It is believed to have been a period of progressive land occupation and agricultural development using natural irrigation. There was also herding and some hunting and gathering.

2 In the 19th century, anthropologists characterized the population of the area now known as Burundi as having three ethnic groups: the Twa (Pygmies indigenous to the area), Hutus (a Bantu group that came from Chad and Nigeria) and Tutsis (a Hamitic group from the Nile valley area).

3 The progressive organization of Burundian society, between the 15th and 16th centuries, led to the formation of a monarchy ruled by a Mwami (King). The State was structured as a feudal system (feudal and semi-autonomous organization and division of power held by *ganwas*, the equivalent of feudal lords in the Middle Ages) which assimilated political and religious functions and assured the pacific coexistence among ethnic groups. No ethnic conflict was registered until the second phase of the colonial period (20th century).

4 The first Catholic mission was set up in Burundi in 1879. In 1885, the Congress of Berlin placed Burundi in the German sphere of influence and it became part of German East Africa. In 1897, Germany founded its general headquarters in Usumbura, within a district that included Rwanda and Burundi, which became known as Ruanda-Urundi. On 6 June 1903, King Mwezi Gizabo, after resisting the Germans, accepted the Treaty of Kiganda and acknowledged the protectorate. The Germans established an indirect administration policy and maintained local customs and authorities. Large quantities of ivory were exported, coffee-growing was started, a railway was built and the first German civil settlements were set up in 1906.

5 After World War I, in 1919

Belgium was granted a mandate over Ruanda-Urundi by the Allied Supreme Council. In 1922 the League of Nations gave Belgium the mandate for the region, which Belgium took up in 1924. Belgians weakened the local monarchy to the point that it lost its traditional symbolic content. Between 1926 and 1933, they carried on significant administrative restructuring, favoring the ganwa aristocracy to the detriment of most Tutsis and especially of Hutus, terminating their own ritualistic aristocracy.

6 By breaking up the traditional political-administrative ties, Belgium destroyed the political organization that kept the balance within Burundian society. The notion of 'chief' was radically changed from defender and regulator of social relationships to a bureaucratic official, who made excessive use of force. Privileges were accorded to the minority Tutsis, which caused severe disruption and violence among the ethnic groups.

7 A law passed in 1925 established the administrative union of Ruanda-Urundi within the Belgian Congo (now DR Congo). This forced Burundi to provide cheap labor to mining centers in Katanga. As regards the economy, the Belgians put a paternalistic policy into practice which among other things led to a system of barter, with little money in circulation.

8 In 1946, after the end of World War II, Burundi became a UN trust territory under Belgium, with the condition that its economic development was assured and the country was led towards independence. However, the Belgian control on local authorities was intensified: the Mwami was personally appointed by the Vice-Governor in charge of the region. This administrative change, which ended the traditional aristocracy, together with the extension of the Belgian Congo's control over Ruanda-Urundi, brought about the loss of national identity. The Belgian administration

was supported by the Catholic Church, which expanded further in Burundi than in any other African country.

9 On 18 September 1961, Burundi's first multiparty elections were held, although the franchise was limited. The Union for National Progress (UPRONA) won. Its founder, Tutsi Prince Louis Rwagasore, was appointed Prime Minister. His pressure for independence led to his assassination by the Belgians on 30 October. He was replaced by another

member of the Tutsi élite, a puppet ruler who changed the country's policy by giving more support to the Hutus.

10 In the midst of this political turmoil, Belgium granted Burundi's independence on 1 July 1962. From that point, disputes over the control of political power fueled ethnic conflicts between Tutsis and Hutus. In 1972 a series of massacres between the groups became known as 'the 1972 crisis'. During the first four years of autonomous rule, there were five different Prime Ministers.

11 In November 1966, Prime Minister Captain Michael Micombero staged a coup and proclaimed the Republic of Burundi. The new President carried out a purge of Hutu government officials. In 1971, 350,000 Hutus were killed and an additional 70,000 went into exile.

12 In 1976, Lt-Col Jean Baptiste Bagaza seized power in a coup and proclaimed the Second Republic. Bagaza broadened UPRONA, put a land reform program in operation, in defiance of the Tutsi élite and their foreign capitalist allies, and also rehabilitated labor unions. Until 1980, violent confrontations were prevented through a policy of silence on the topic of ethnicity which was considered a taboo subject, and religious freedom was restricted. In

PROFILE

ENVIRONMENT

Most of the land is made up of flat plateaus and relatively low hills covered with natural pastures. Located in the Great Lakes region (Lakes Tanganyika, Victoria), the Ruvubu River valley stretches through the country from north to south. Tropical forests are found in the low, western regions. Most inhabitants engage in subsistence agriculture (corn, cassava, sorghum, beans). Coffee is the main export. Internal communications are hampered by natural barriers and the nearest sea outlet is 1,400 km beyond the border, a fact which makes foreign trade difficult.

SOCIETY

Peoples: Most Burundians (86 per cent) belong to the Hutu ethnic group, an agricultural people of Bantu origin. They were traditionally dominated by the Tutsi or Watusi (13 per cent), pastoralists of Hamitic descent. There is a small minority (1 per cent) of Twa pygmies. **Religions:** Christians, 67 per cent; 32 per cent follow traditional African religions and 1 per cent are Muslim. **Languages:** Rundi, Kirundi and French, official, with Swahili the business language. **Main Political Parties:** National Council for the Defense of Democracy–Forces for the Defense of Democracy (NCDD-FDD), was formerly the main rebel group; Front for Democracy in Burundi (FRODEBU); Union for National Progress (UPRONA); The Movement for the Rehabilitation of Citizens-Rurenzangemero (MRC-Rurenzangemero); Party for National Recovery (PARENA). **Main Social Organizations:** As a result of civil war, organizations are restructuring but there are many foreign non-governmental organizations (NGOs) and local NGOs like the Association for the Economic Promotion of Women.

THE STATE

Official Name: Republika y'u Burundi. **Administrative divisions:** 15 Provinces. **Capital:** Bujumbura 378,000 people (2003). **Other Cities:** Gitega 23,500 people (1999). **Government:** Pierre Nkurunziza, a Hutu, president since August 2005. Martin Nduwimana, a Tutsi, First Vice-President since August 2005. Bicameral Parliament: The Senate may have between 37 and 54 members who serve five-year terms. After elections in July 2005 it has 49 members. The National Assembly (Lower House) has 100 directly elected deputies and between 18 to 21 co-opted members who serve five-year terms. (After the July 2005 elections it has 118 members). **National Holiday:** 1 July, Independence (1962). **Armed Forces:** 52,000 (2003).

Life expectancy
46 years
2005-2010

GNI per capita
$90
2004

Literacy
59% total adult rate
2000-2004

HIV prevalence rate
6% of population 15-49 years old
2003

foreign policy, the new government drew closer to Tanzania, and China sent aid to develop Burundi's mineral resources.

[13] A new Constitution became effective in 1981. It prevented the exploitation of the Hutu majority by the Tutsi minority, prompted the modernization of the political structure, adopted a socialist stance and gave men and women equal rights. The new constitutional reforms provoked strong feelings between the Government and the Catholic Church, leading to the confiscation of properties and deportation of 63 missionaries.

[14] Elections held in 1982 (the first by universal suffrage), upheld Bagaza's policies. Despite Burundi's established political independence, the economy was in tatters. Being a landlocked country raised the price of both imports and exports (coffee was the leading export). The vast majority of Burundi's population lives in the north where overuse of the land has caused soil erosion, affecting its fertility.

[15] In September 1987, Bagaza was overthrown by army major Pierre Buyoya (UPRONA) in another coup. In August 1989, strife between Hutus and Tutsis broke out in the north; several thousands of people, mostly Hutus, were massacred by the military, which was dominated by Tutsis. About 60,000 Hutus sought refuge in Rwanda. The Government responded by appointing a Hutu Prime Minister, Adrien Sibomana, and a new cabinet with equal representation of Hutus and Tutsis.

[16] In 1992, Buyoya enacted a multiparty constitution and called elections for October 1993, in which he was defeated by Melchior

Ndadaye, from the opposition Front for Democracy in Burundi (FRODEBU) composed mainly of Hutus.

[17] Three months after he was elected, Ndadaye was assassinated during an attempted military coup. Although the coup failed, Ndadaye's assassination led to one of the worst massacres in Burundi's history. Supporters of the former president attacked UPRONA members - Tutsis or Hutus - causing the death of tens of thousands of people and the flight of some 700,000. The so-called 'extremist armed militias' - hostile to living alongside the other ethnic groups - were consolidated by this time, as were the 'Undefeated' Tutsis and the *Intagohekas* ('those who never sleep') Hutus. Violence spread.

[18] On 6 April 1994, Ntaryamira died along with Rwandan president Juvenal Habyarimana when their plane was shot down. Another Hutu, Sylvestre Ntibantunganya, replaced the assassinated president. Violence intensified, especially between militias backing Hutu power and the Tutsi-controlled army. In February 1995, UPRONA left the Government in order to force Prime Minister Anatole Kanyenkiko to resign. The resignation paved the way for nomination of Tutsi Antoine Nduwayo and UPRONA's return to the coalition government alongside FRODEBU.

[19] Fearing that the Burundian conflict would extend to neighboring countries, the UN and the Organization for African Unity decided to intervene. Pierre Buyoya, claiming the need to prevent an intervention by inter-African forces in the country, staged a new and successful coup in July 1996 and became the new president of Burundi. After this action, Ethiopia,

Kenya, Rwanda, Tanzania, Uganda and Zaire (now DR Congo) imposed an embargo on the country.

[20] In late 1997, the UN declared that international sanctions had had devastating effects, significantly degrading the living conditions of poor Burundians and questioned the embargo's usefulness. In September, Buyoya accused Tanzania of protecting more than 200,000 Hutu rebels and of intending to 'annex' Burundi.

[21] On 16 July 1998, the Transitional National Assembly was created, with the admission of 40 new representatives from the two political parties and the public. Until then, FRODEBU held 65 seats and UPRONA 16. At the end of 1998, the number of victims of the civil war since 1993 was estimated at more than 200,000.

[22] Domitien Ndayizeye, the country's fourth Hutu President, took office in April 2003 as President of the transitional Government, which had been agreed in Arusha in 2001; however some felt that these negotiations were running into the sand.

[23] In November, Ndayizeye and Pierre Nkurunziza, leader of the FDD (Forces for the Defense of Democracy), signed a peace treaty in Arusha (in the north of Tanzania) in the context of the African leaders' summit. Nkurunziza was appointed minister for Good Governance in December 2003.

[24] In May 2004, Nkurunziza left his ministry, although he emphasized that there was no reason for the war to recommence. The FDD accused the Government of not taking the necessary steps for achieving permanent peace, pointing out that, during the six-month ceasefire, no FDD representation

had been obtained in the Ndayizeye administration. The incorporation of FDD rebels to the country's armed forces (in November 2003) was not affected by Nkurunziza's resignation.

[25] In August 2004, 20 parties signed an agreement under which Hutus and Tutsis would share and take part in the government of the country. Other 11 parties frustrated the agreement by not signing it.

[26] The new Constitution, passed in February 2005 by 90 per cent of the votes, marked the end of a system which had left power in Tutsi hands almost without interruption since independence in 1962. All Hutu or pro-Hutu parties (85 per cent of the population) voted in favor; Tutsi parties (14 per cent of the population) voted against. The new Constitution, which established proportional ethnic representation, offered the Hutu majority the possibility of coming to power and winning the democratic victory that had been denied to them in 1993 with Ndadaye's assassination. On the other hand, a restructure of the army and the police ensured that both Hutus and Tutsis were equally represented within their ranks. In June, Hutu parties won the parliamentary elections, gaining control over the National Assembly and appointing Nkurunziza as President.

[27] In April 2006, Burundian authorities revoked the 13-year curfew, after the Government stated that: 'security has highly improved; 95 per cent of the territory has been pacified.' Of the rebel groups, only the Hutu Forces for Liberation remained active, although they were expected to participate in peace talks before long. That month, the UN Security Council authorized the transfer of over 5,000 blue helmets to Burundi. ∎

IN FOCUS

ENVIRONMENTAL CHALLENGES
Deforestation, due to the indiscriminate felling of trees for firewood, is one of the country's worst problems, aggravated by the growing incidence of farming on unsuitable lands. By 2000, 22 per cent of population had no access to drinking water.

WOMEN'S RIGHTS
Burundian women have been able to vote and stand for office since 1961. In 2000, women held 6 per cent of seats in Parliament and their representation in ministerial positions amounted to 8 per cent. In 2005, women held 32 per cent of seats in Parliament, a 500 per cent increase compared to 2000. In 2003, female representation in ministerial and similar positions amounted to 11 per cent.
Burundi has some of the worst health indicators in the

world and continues to register worsening infant and maternal mortality rates.
Only 25 per cent of births are assisted by qualified medical staff, while 78 per cent of pregnant women receive prenatal care*.
The fertility rate remained constant in 1960, 1990 and 2001 at 6.8 children per women*.

CHILDREN
In 2000, severe malnutrition, epidemic levels of malaria and smallpox and a rise in cholera and dysentery hit the country. That year, the decline in the rate of exclusive breastfeeding to 89 per cent of babies under four months old signalled deteriorating child-care practices. Under-five chronic malnutrition rates rose from 48 per cent in 1987 to 56.8 per cent in 2000*. Some 16 per cent of newborn babies are underweight

and 57 per cent of under-fives are undersized. The number of HIV/AIDS cases continues to rise dramatically, particularly in rural areas. In 2001, the total estimated number of HIV-positive children under 14 years old was 55,000. Infection rates in girls aged 15 to 19 were four times higher than boys of the same age. Over 200,000 children have been orphaned by HIV/AIDS in the country*.

MIGRANTS/REFUGEES
At the end of 2002, more than 400,000 Burundians were living as refugees in other countries, most of them in Tanzania, DR Congo, Malawi, Rwanda and South Africa. In addition, some 47,000 Burundians were living in western Tanzanian settlements without official refugee status. Approximately 400,000 or more Burundians were internally displaced by the year's end; out of the 1 million or so people that had

been uprooted during the year, only 300,000 were in refugee camps.
Nearly 50,000 Burundians returned to some provinces in the north and east of the country, primarily from Tanzania since 2002. Unable to return to their villages because of poor security, many Burundians crowded into camps for internally displaced people.
During late 2005 and early 2006, over 4,000 Burundians crossed the border into Tanzania. UNHCR currently attends to 195,000 Burundian refugees in Tanzania.

DEATH PENALTY
It is applied to all types of crimes.

* Latest data available in The State of the World's Children and Childinfo database, UNICEF, 2006.

Cambodia / Kâmpuchéa

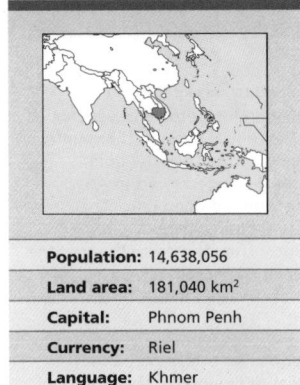

Population:	14,638,056
Land area:	181,040 km²
Capital:	Phnom Penh
Currency:	Riel
Language:	Khmer

There is still much debate about how long people have lived in what is now Cambodia, where they came from, or what languages they spoke before the introduction of writing in 300 BC. Carbon-dating evidence shows that there was pottery-making in the region and therefore human habitation from as early as 4000 BC. Ongoing studies indicate that those early inhabitants built their houses on piles of wood and lived on fish, pork and buffalo.

2   The land of Cambodia takes its name from a *rishi* (sage), for the first Cambodian kings proclaimed themselves descendants from the great sage Cambu Svayambhuva. The first kingdom of Cambodia was known in Chinese texts under the name of Funan and was founded by the Mon and Khmer peoples who migrated from the north in the first century AD. During these first centuries, a strong Indian influence should be registered since Indian and Chinese pilgrims used to visit these coasts and exchanged silk and metals for spices, aromatic wood and gold.

3   For centuries Khmer people have practised the cult of Devaraja or god-king, who demanded absolute obedience in exchange for prosperity. Among the legends of this community is the 'Churning of the Ocean Milk', according to which the God Vishnu becomes

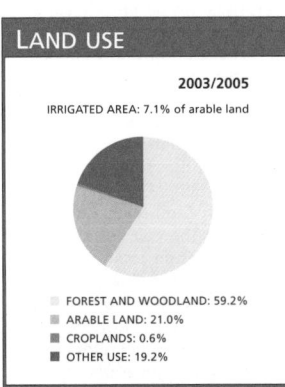
a turtle and offers its shell to support a mountain. Gods and demons coil a rope around it and start to swivel it back and forth, beating the surrounding ocean of milk in order to create the ambrosia that will bring happiness and wealth. The most beautiful and impressive architectural works in the country date from the Angkor Empire period (the Khmers dominated the region from the 9th to the 15th century AD). The most outstanding are Angkor Wat, the 'capital that is a temple', and Angkor Thom, the 'great capital'.

4   Chinese sources reveal the scope of Indian influence by the example of the sage Kaundimnya, who married princess Soma, of Nagi origin (that is, semi-divine) and transformed the Khmer institutions along Indian lines. One of the modifications was probably the introduction of large-scale irrigation, which allowed the production of up to three rice harvests a year. Another was the worship of the Indian god Shiva, who became seen as a guardian ancestor and the spirit of land by the Khmer. A third modification ascribed to Kaundimnya was the coexistence of Buddhism and Hinduism, which characterized Cambodia for more than 1,000 years.

5   According to Chinese writings, the Funan created a centralized state machinery, headed by an absolute ruler, who ran agricultural labor and used farming surpluses to maintain his lifestyle and that of a caste of priests, and also to build fortresses, palaces and temples.

6   The first Sanskrit writings date from the 6th century and the first ones found in the Khmer language are from the 7th century. The developments between the decline of Funan and the foundation, three centuries later,

of a new centralized state in the Cambodian north-eastern region (called Chenla in Chinese texts of that period) have not been clarified as yet. Chinese sources cite the existence of at least two Cambodian kingdoms, which - under the sovereignty of the Java Kingdom - requested recognition from China. Sanskrit and Khmer sources reveal that, during the 7th, 8th and 9th century, multiple kingdoms paid homage and fealty to the Java Kingdom.

7   In 790, a Khmer prince who had been raised in the Javanese court of the Sailendra dynasty was installed as vassal ruler of the area. In 802, however, he declared the Khmer territory independent of Java and was crowned King Jayavarman II. During his reign he established the devaraja cult as the official state religion and reunified the old kingdom of Chenla which was expanded and formed into the Khmer Empire. He died in 850 and was succeeded by his son Jayavarman II.

8   In 887, Indravarman I usurped the throne that belonged to his cousin Jayavarman III. During his reign, a large dam surrounding the capital city Roluos was built, which was the first of a vast system of dams, channels and irrigation canals that made these lands productive and allowed the Khmer to maintain a densely populated and centralized state in the area, which otherwise would have remained infertile. In 889, Yasovarman I became king of Khmer and built Angkor city (then called Yasodharapura) on the same site, which became capital of the Khmer Empire. In 1002, Suryavarman I usurped the throne and extended the Kingdom of Angkor to parts of today's territories of Thailand and Laos. In 1080, Angkor being conquered by the Kingdom of Champa, a Khmer

ruler of a northern province proclaimed himself king under the name of Javayarman VI, ruling from his home province rather than from Angkor.

9   In 1177 Angkor was conquered again by Champa forces. Jayavarman VII became king in 1181 and conquered Vijaya, capital of Champa (in present-day Vietnam). Under his rule, the Khmer Kingdom reached its height, including virtually the whole of today's Thailand and Laos and even reaching Myanmar/ Burma, Malaysia and Vietnam. Jayavarman VII converted from Hinduism to Buddhism, which became the national religion.

10   In 1200, work began on building the new capital city of Angkor Thom. This undertaking depleted the royal coffers and the Kingdom experienced economic problems in the following decades. The decline of Angkor coincided with the rise of the Thai Kingdom to the west and the Vietnamese one to the east. Turned into a small buffer state between both, the Khmer Kingdom alternately depended on the power of the Thais or the Vietnamese, since in order to be free from one conqueror, it needed the aid of the other. In 1432, when the Thais conquered Angkor once again, the Khmer people abandoned the city to the jungle.

11   During more than 400 years, the Kingdom was alternately conquered by the Thai and Vietnamese forces, until 1864, when Cambodian King Norodom accepted the status of French 'protectorate' for his country, in the hope that the French would protect them from Siam (Thailand) and Vietnam. The French were not able to prevent the Siamese (Thais) from temporarily annexing some west Cambodian areas, including the city of Battambang. However, by recognizing French authority, King Norodom managed to prevent the country from being divided and distributed between Vietnam and Siam (Thailand). During the previous centuries, Vietnam had taken control of large tracts of Cambodia. The Mekong Delta area was inhabited by Cambodians until the 18th century.

12   In 1884, with the consent of King Norodom, Cambodia became a French colony. France's political influence increased and, together with Vietnam and Laos, Cambodia became a part of the Indo-Chinese Union. Later on, the colonial empire installed a European administrative system in Cambodia and developed the country's infrastructure. The economic development of the French Indo-Chinese Union did not reach the

levels attained by Burma or India under British rule.

13 After France was invaded by Germany, Japanese forces occupied Indochina in 1940, with virtually no resistance. This enabled the Japanese to use the military facilities in exchange for allowing France to maintain administrative control. As a result Cambodia suffered less damage during World War II than the South Pacific islands, which were fiercely disputed by opposing forces.

14 In 1941, French authorities appointed the 18-year-old Prince Norodom Sihanouk as King of Cambodia, in the hope of controlling the politics of an inexperienced youngster. However, when the Japanese invaders withdrew, in 1945, Norodom Sihanouk declared independence. But the French occupied Indochina again. While Vietnam began an anti-colonial war, in Cambodia the King managed to gain greater autonomy from France.

15 In 1947, a new constitution kept King Sihanouk on the throne, his power being limited by parliament. In 1949, Cambodia's legal system was re-negotiated, with Paris retaining control over military and foreign affairs and the remaining functions carried out by the 'protected' local Government. In 1953, the Vietnamese offensive forced French troops to withdraw from Cambodia, leaving the country fully independent. In 1955, Sihanouk abdicated in favor of his father, in order to participate in political life (constitutionally barred to the King). In 1960, the former King, now Prince Sihanouk, became head of state.

16 During the first years of US aggression against Vietnam, Cambodia sought to maintain political and territorial neutrality. In 1965, however, Cambodia was bombed by South Vietnamese forces, under US orders, and Sihanouk's government moved closer towards the Chinese/Russian camp. In 1968 the US began air strikes on the so-called Ho Chi Minh Trail which came through Cambodian territory and was the supply-line from North Vietnam for the Viet Cong resistance in the South.

17 While taking diplomatic action abroad to protect his country's sovereignty, Sihanouk was ousted in a CIA-backed coup in 1970. He was replaced by Lon Nol who, during his 5-year rule, received $1.6 billion in aid from Washington. During that period, the US invaded Cambodia to bomb the Communist Khmer Rouge (KR) guerrillas. As a result of this bombing, 100,000

Cambodians were killed. Sihanouk went into exile in Peking (now Beijing), and from there, supported by the Khmer Rouge, he organized the National United Front of Kâmpuchéa (NUFK) using the name of the country in Khmer language.

18 In 1975, before the Viet Cong took Saigon, the Khmer Rouge entered Phnom Penh and proclaimed the Democratic Republic of Kâmpuchéa. In 1976, together with a new Constitution, the People's Congress confirmed Sihanouk and Khieu Samphan as head of state and government, respectively. However, on his return from exile, Sihanouk was forced to resign and kept under house arrest, while Pol Pot rose to the fore as the regime's new leading force. Cambodia closed its borders even to diplomats of friendly countries. Under Pol Pot's regime, money was eliminated and large numbers of the urban population were transferred to the countryside, enforcing the return to an agricultural lifestyle. Mass purges and executions, hunger and illness, left at least a million people dead.

19 The crimes committed by the Khmer Rouge from 1975 - which they styled Year Zero - were in 1979 described by the UN Human Rights Commission as the worst to have occurred anywhere in the world isnce Nazism. By 1985 the UN Special Rapporteur on Genocide ruled that what the Khmer Rouge had done was genocide even under the most restricted definition.

20 The Pol Pot regime strengthened ties with China and broke off relations with Vietnam. In 1978, a pro-Vietnamese faction of the KR created the United Front for the Salvation of Kâmpuchéa (KNUFNS), under the presidency of General Heng Samrin. In 1979, the Vietnamese forces and those of the United Front entered Phnom Penh and proclaimed the People's Republic of Kâmpuchéa (PRK), establishing a People's Revolutionary Council to rebuild the country.

21 While Pol Pot was tried in absentia for war crimes, his forces turned to guerrilla warfare. With the support of China and the US, the KR maintained UN recognition as the country's legitimate Government. In 1982, Sihanouk and the Khmer Serei, led by Son Sann, joined the Government of Democratic Kâmpuchéa in exile, together with the Khmer Rouge. The Government of Phnom Penh aimed at reactivating the industrial, agricultural and transportation sectors. The Soviet Union and other socialist countries gave aid which was focused on

the reconstruction of power sources and the formation of a very basic industrial infrastructure.

22 Municipal and legislative elections were held in 1981, with many candidates for each post. The Non-Aligned Movement summit meeting in New Delhi, decided to leave Cambodia's seat vacant, as they could not decide which of the two civil war protagonists was the legitimate representative. In 1985, the Vietnamese announced the withdrawal of 150,000 troops over a period of five years. In spite of the dissolution of the Communist Party in 1981, the Khmer Rouge's entry into the political arena posed the main obstacle for a negotiated settlement.

23 Difficulties stemming from the war affected the Cambodian Government, especially the international isolation since it was only officially recognized by some 30 countries. Still, it managed to bring about a steady increase in the production of rice, cattle, pigs and poultry.

24 In 1986 the Phnom Penh Government made overtures to

Sihanouk within the context of a general agreement, offering him the position of head of state in a government from which the KR would be excluded. Between 1987 and 1989, Prime Minister Hun Sen met with Sihanouk on six occasions. The main disagreement each time was whether the KR would be admitted to a new provisional government or not. In 1989, a curfew in effect since 1979 was lifted and private ownership of some 'non-strategic' enterprises was permitted. Land and transportation services were privatized. During that year, to complete the transformation and diminish international isolation, the country changed its name to the State of Cambodia.

25 In Paris, that same year, the International Conference for Peace in Cambodia ended in failure as the factions of the armed opposition could not come to any agreement. The two main points of disagreement were UN monitoring of the Vietnamese withdrawal and Khmer Rouge participation in government. Vietnamese troops withdrew

Under-5 mortality	Poverty	Debt service	Maternal mortality
141 per 1,000 live births 2004	**34.1%** of population living on less than $1 per day 1997	**0.8%** exports of goods and services 2004	**450** per 100,000 live births 2000

completely that year. In 1990, a Phnom Penh military offensive eliminated rebel bases and opposition forces retreated to the Thai border. Bangkok sponsored a new Sihanouk-Hun Sen meeting and both leaders agreed to the installation of a supernational agency which would symbolize the sovereignty and national unity of Cambodia, under 'adequate' UN supervision. In 1990, the US announced they would withdraw recognition to Democratic Kâmpuchéa and would initiate negotiations with Vietnam over a peace settlement.

26   In 1991, a peace treaty was signed. The Supreme National Council was created, with representatives of the Phnom Penh Government and part of the opposition, chaired by Sihanouk who was to govern the country until elections in 1993. In 1992, the UN sent a peacekeeping force to enforce the cease-fire and organize the elections.

27   In the constituent elections of 1993, boycotted by the KR, the supporters of FUNCINPEC, led by Sihanouk's son Norodom Ranariddh, won the majority of seats. In the new Government, Ranariddh and Hun Sen shared the post of Prime Minister. That year, a new Constitution was enacted which turned the National Assembly into a Parliament and established a parliamentary monarchy. Sihanouk was appointed King, 'independent' of all political parties. The King continued to advocate a 'national reconciliation' Government, with the inclusion of the KR.

28   In 1996, many defections suggested that the Government policy aimed at splitting the KR was proving successful. Accused of treason and of planning a civil war alongside the KR, Prince Ranariddh was ousted in 1997 by Cambodian army officials loyal to the 'second Prime Minister' Hun Sen. From China, King Sihanouk approved Hun Sen's proposal to appoint foreign minister Ung Huot (FUNCINPEC), as first Prime Minister. In 1998 the monarch pardoned his son - sentenced to a 30-year prison term - in an attempt to delay the political chaos which followed Ranariddh's expulsion and as a means to secure international financial aid which had been interrupted since the coup.

29   After the death of Pol Pot in Thailand in 1998, Ranariddh returned to Cambodia to compete in the election held that year, which Hun Sen won by a narrow margin. Ranariddh's FUNCINPEC and the opposition party's leader, Sam Rainsy (SRP), denounced irregularities in the elections but, according to the 600 observers, there were no defects in the counting of votes.

30   In 1999, Cambodia joined ASEAN as its tenth member. Spien Kizuna, the first bridge over the Cambodian stretch of the Mekong river, was opened. The Government hoped that the bridge, which was built with Japanese money and connects the country by land from east to west, would stimulate domestic trade.

31   In 2002, the first local elections in many years were held in a violent atmosphere, with more than 20 opposition candidates killed during the campaign. Hun Sen's Cambodian Pracheachon Party won 1,597 of the 1,620 communes in the whole country, while the opposition only won 23 (10 for FUNCINPEC and 13 for Sam Rainsy's SRP).

32   That year, the Government imposed a three month deadline on UN attempts to restart negotiations on the trial of KR leaders. (In 1999, Cambodia had agreed with the UN for a tribunal to try former KR leaders for crimes against humanity committed under the Pol Pot regime). According to the Cambodian Government, low-ranking officials will not be judged and the trial will be focused on leaders with direct responsibility for genocide. The UN withdrew its support for the process in 2002, stating that the independence and impartiality of the court could not be guaranteed, since national law would take precedence over the agreement with the UN.

33   Hun Sen's Pracheachon Party again won the July 2003 elections, although it did not obtain an absolute majority.

34   After intense negotiations during the Cancún Ministerial Conference in September 2003, the World Trade Organization (WTO) approved Cambodia's membership. This would allow Cambodia to achieve greater regional and global economic integration. Australia and Cambodia signed a bilateral agreement which covered current and prospective export interests in goods and services of both countries.

35   Union leader Che Vichea, also linked to FUNCINPEC, was murdered in Phnom Penh. Sam Rainsy accused the Government of targeting Vichea due to his struggle against corruption and human rights abuses.

36   A deal between the Pracheachon and FUNCIPEC parties, allowed Hun Sen to become re-elected in July 2004. One month later, Parliament ratified Cambodia's entry into the WTO.

37   In October 2004 King Sihanouk, aged 81, abdicated the throne and was succeeded by his son Norodom Sihamoni. In the same month, an agreement signed by Vietnam's Prime Minister over a controversial border issue, unleashed strong criticism from opposition groups.

38   Sam Rainsy was sentenced to nine months in prison in December 2005, having been accused of defamation of the Prime Minister. Rainsy had been exiled in Paris since February of that year.

39   In February 2006, Rainsy returned to Cambodia, having been granted a royal pardon, promising to be less confrontational in his dealings with Hun Sen.

40   In September 2006, human-rights activists warned that forced evictions in Phnom Penh were spiraling out of control. Thousands of slum residents were being forced to leave so that property developers could move in to build luxury apartments and shopping centers. ■

IN FOCUS

ENVIRONMENTAL CHALLENGES
Deforestation has been caused by defoliants and mine and bomb damage during the war (1965-1975), in which the country lost approximately three-quarters of its fauna. The illegal exploitation of forest products throughout the country and the plundering of precious stones from mines on the border with Thailand have weakened biodiversity. The destruction of the mangrove swamps, in particular, threatens fish stocks which have already been affected by overexploitation and illegal fishing.

WOMEN'S RIGHTS
Cambodian women have had the right to vote since 1995. In 2003, women held 9.8 per cent of seats in the lower House of Parliament and 13.6 per cent in the upper House, while they accounted for 7 per cent of posts at ministerial level. In 2004, the literacy rate among adult women was 64 per cent, while male literacy reached 85 per cent.

At the end of 2003, it was estimated that 51,000 women were living with HIV/AIDS. In 2003, women made up 52 per cent of the country's workforce. Female unemployment amounted to 2.2 per cent.

CHILDREN
The poor situation of Cambodian children is a direct effect of many years of armed conflicts which only very recently were resolved. Cambodia is still one of the poorest countries in Asia and its infrastructure is only beginning to be rebuilt. Almost 50 per cent of children suffer from malnutrition and one in eight dies before the age of five (mostly due to preventable causes). Cambodia has one of the highest HIV/AIDS prevalence rates in Asia and morbidity and mortality in newborn infants still pose a serious threat. The infant mortality rate rose from 80 per 1,000 live births in 1990 to 97 in 2004. Yet some progress has been made: nearly 90 per cent of children have entered primary school, with girls completing the course at the same rate as boys. There are community action programs reaching more than one million people as well as interventions supported by UNICEF in terms of health, nutrition and immunization aimed at breaking the cycle of poverty-hunger-ignorance.

INDIGENOUS PEOPLES/ ETHNIC MINORITIES
Vietnamese people are widely dispersed across the territory. Most of them moved there in the post-1945 era. Another group migrated to Cambodia in 1979, after the Vietnamese invasion. They speak the Khmer language, and most of them practise Buddhism. The status of the Vietnamese living in Cambodia has been affected by the relations between Hanoi and Phnom Penh, and by disputes between the different ideological factions battling for control of the Government.

MIGRANTS/REFUGEES
In 2001, more than 1,000 Montagnards (mountain people from Vietnam's central region) settled in the northern provinces of Mondolkiri and Ratanakiri, escaping ethnic conflicts. Around 250 of them returned to Vietnam that same year. Although at first some members of this group were deported - at Hanoi's request - as a result of international pressure, Phnom Penh allowed UNHCR to negotiate with both countries and open a refugee camp within Cambodian territory. At the end of 2005, refugee groups which at first opposed both a forced return to Vietnam and relocation to a third country, were made to choose one of the two options and started to return to their homes.

DEATH PENALTY
It was abolished for all crimes in 1989.

* Latest data available in The State of the World's Children and Childinfo database, UNICEF, 2006.

Cameroon / Cameroun

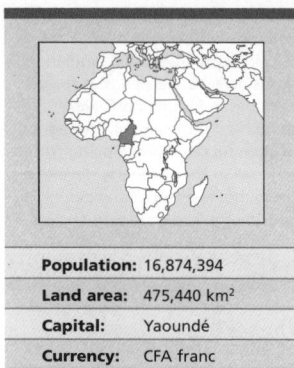

Population:	16,874,394
Land area:	475,440 km²
Capital:	Yaoundé
Currency:	CFA franc
Language:	French and English

Archeological evidence shows that humans lived in the region 50,000 years ago, in large kingdoms and states, and also in more recent times, for example the kingdom of Sao near Lake Chad, around the 5th century AD. The indigenous inhabitants of Cameroon, the Bakas (pygmies), live today in the forests in the south and east of the region. Bantu-speaking people, originally from the highlands, were the first group to migrate east and south after 200 BC, spreading new crop varieties and methods for working iron.

2 The migration of Fulani herders from the western Sahel region - they were among the first Muslims to spread out across West Africa - helped transform the local economies into part of a regional nexus which later gave rise to the Emirate of Adamaua in the north central region, dominating and displacing non-Muslim inhabitants. The nomadic Fulanis lived in portable huts and rarely killed cattle for meat. The urban Fulanis were strong Muslims, whereas the herders were less inclined to religion.

3 In 1472, Portuguese explorer Fernando Po named the river Wouri *Cameroes* (Portuguese for shrimp) because of the great number of crustaceans in it. The name eventually became Cameroon. The Portuguese arrived in 1500, working with sugarcane and the slave trade. Malaria and other diseases hindered the proliferation of European colonies and the conquest of the hinterland until 1800, when quinine (medicine for malaria) became available.

4 German penetration began in 1884 when envoy Gustav Nachtigal agreed with the Doualas, a coastal people, to make the region a protectorate. In 1885 the Berlin Conference awarded Cameroon to Germany. In 1894 Adamaua, which the British wanted, was formally included. The Protectorate was marked by conflict: the Doualas dominated the trade that the Germans wanted to control between the coast and Yaoundé, the trade

center between Adamaua and the south. In 1897, the Doualas began a bloody 4-year war against the Germans, who appropriated the most fertile lands, while the Africans died of hunger by the thousands.

5 In 1918 France and Britain invaded Cameroon: the French took three-quarters of the territory, and the British took the rest. The cause of the independence organizations was facilitated by tensions between the colonial powers. In 1945, the People's Union of Cameroon (UPC) was founded, headed by Rubem Um Nyobé. With great popular support it

launched a series of legal campaigns between 1948 and 1956, when it was outlawed. Nationalist leaders fled to the British-held western sector and organized a guerrilla movement. The UPC established liberated zones in the southern forests, setting up autonomous governments - a first for sub-Saharan Africa. The efficiency of the guerrillas made it possible for them to resist constant French attacks, until 1960 (Nyobé died in 1958, but the revolution continued).

6 UPC resistance forced the French to adopt a new strategy, adding political maneuvering to their repressive tactics. Paris created the National Union of Cameroon (UNC), merging two conservative, predominantly Islamic, northern-based parties. In 1960 UNC leader Alhaji Ahmadou Ahidjo became president after the French Cameroon gained full independence as the Republic of Cameroon. In 1961 it became the Federal Republic of Cameroon, after unification with the southern part of British Cameroon. UPC nationalists were unable to help shape their long-held dream of an independent Cameroon as they were mostly underground or in exile. Ahidjo developed one of sub-Saharan Africa's most efficiently repressive systems. European human rights groups revealed the existence of thousands of political prisoners in the country.

7 In 1982 Ahidjo suddenly resigned and was succeeded by his former Prime Minister, Paul Biya. Politicians loyal to Adhidjo supported an attempted coup by a group of military officers. Biya attempted to reinforce his control by calling early elections in 1984. Political parties

ENVIRONMENT

The country is divided into three regions: the plains of the Lake Chad basin in the northern region (savanna where cattle are raised and corn and cotton are grown); the central part, made up of humid grasslands; and the southern part, an area with rich volcanic soil where the main cash crops (coffee, bananas, cocoa and palm oil) are grown. It is in the latter that most of the population lives.

SOCIETY

Peoples: There are some 200 ethnic groups, the main ones being the Doualas, Bamilekes, Tikars and Bamauns in the south; the Euondos and Fulbes in the west, and the Fulanis in the north. In the southeast live the Baka Pygmies, who live by hunting and fishing.
Religions: Half the population practice traditional African religions. Christians are a majority in the south while Muslims predominate in the north.
Languages: French and English are the official languages. There are nearly 200 African languages (Beti and Bulu being the most widely spoken). German is also spoken.
Main Political Parties: Cameroon People's Democratic Movement, Social Democratic Front, National Union for Democracy And Progress,

Alliance for Democracy and Development, Union of the Peoples of Cameroon.
Main Social Organizations: In 1971, the Government banned the Workers' Union of Cameroon, heir to the labor movement from the previous century. The Government-run National Workers' Union of Cameroon was created. Trade Union Confederation of Cameroon Workers. There is a strong separatist movement in the English-speaking area of Cameroon. People's Conference of 'Southern Cameroons' (SCPC) unites several parties and organizations.

THE STATE

Official Name: République du Cameroun.
Capital: Yaoundé 1,616,000 people (2003).
Other Cities: Douala 1,409,200 people; Garoua Boulai 180,900; Ncongsamba 106,800 (2000).
Government: Parliamentary republic. Paul Biya, President since 1982, re-elected in 1992, 1997 and 2004. Ephraim Inoni, Prime Minister since December 2004. Unicameral Legislature: National Assembly, with 180 members.
National Holidays: 1 January, Independence Day (1960); 1 October, Reunification (1961); 20 May, proclamation of the Republic (1972).
Armed Forces: 14,600.

	Life expectancy		GNI per capita		Literacy		HIV prevalence rate
	46 years 2005-2010		**$810** 2004		**68%** total adult rate 2000-2004		**6.9%** of population 15-49 years old 2003

were banned, and he was re-elected. Nevertheless, overall instability led to a new coup attempt, followed by a series of bloody incidents. In those years, the UNC changed its name to 'Democratic Group of the People of Cameroon' (RDPC), but its political line remained the same.

[8] Biya created new northern provinces to reduce the economic and political power of the Muslim north. Disagreements over oil revenues aggravated inter-ethnic and inter-regional friction. French companies controlled almost 44 per cent of the export market. The Government implemented a structural adjustment program and sought to stabilize its finances through World Bank and IMF support and the renegotiation of the country's debts with the Paris Club.

[9] The economic crisis, aggravated by the devaluation of the franc, affected Cameroonian society. Police corruption and an inefficient judicial system led to the creation of groups within the militia or police force which committed flagrant abuses of power.

[10] In 1990, social and political organizations denounced the repression. The Government authorized the creation of political parties and the following year, legislative elections were set for 1992. The opposition demanded, in vain, a modification of the constitution and electoral law prior to the elections. In 1992, in the country's first multiparty elections, seven parties fielded candidates: among them, the Social Democratic Front (FSD) led by John Fru Ndi, the main opposition party representing the English-speaking community; and the Democratic Union of Cameroon, led by Adamu Ndam Njoya. Amid widespread accusations of fraud, the ruling RDPC's victory (39.98 per cent against the FSD's 35.97 per cent) provoked incidents in the English-speaking Northwest Province, Fru Ndi's territory. International observers confirmed the fraud, but the Supreme Court refused to annul the election. Fru Ndi proclaimed himself President, and the Government decreed a state of emergency in the Northwest. Fru Ndi and his supporters were immediately placed under house arrest.

[11] In late 1992, as a result of international pressure, Biya ended the state of emergency and released Fru Ndi. Biya was finally re-elected; press censorship was intensified and the position of Prime Minister was eliminated.

[12] In 1997, the RDPC again won most seats in Parliament. Biya was re-elected President in a second round that was boycotted by the majority of opposition parties. Just three million of the six million registered voters were called on to

IN FOCUS

ENVIRONMENTAL CHALLENGES
Forest biodiversity is seriously endangered. Cameroon's mountain range is isolated and so its inhabitants depend on it for their basic resources. Deforestation has caused many streams and much wildlife to disappear. There is a high incidence of water-borne diseases. Soil is exhausted as a result of extensive cattle rearing and desertification. Indiscriminate hunting and fishing are ongoing practices.

WOMEN'S RIGHTS
In 2003, the proportion of women in Parliament stood at 9 per cent and their representation at ministerial level amounted to 11 per cent. The female literacy rate in 2004 was 60 per cent compared to male literacy of 77 per cent. In terms of primary school enrolment, rates stood at similar levels: 73 per cent for women and 76 per cent for men. Only 62 per cent of births were attended by skilled health staff. At the end of 2003, 290,000 women aged 15-49 were living with HIV/AIDS.

CHILDREN
In 2004, 11 per cent of newborn infants were underweight. The infant mortality rate rose from 85 deaths per 1,000 live births in 1990 to 87 per 1,000 in 2004. The situation is even worse for children under five years old, whose mortality rate increased from 139 to 149 deaths per 1,000 live births over the same period.

Human trafficking, mainly for cheap labor in the informal sectors of the economy, still poses a major threat to Cameroonian children. At the end of 2003, 43,000 children under 15 were orphaned by HIV/AIDS.

INDIGENOUS PEOPLES/ETHNIC MINORITIES
During the 17th century the Fulani - an Islamic nomadic group - entered the territory, and 100 years later they built the Adamaua Emirate in the north-central region. The Bamileke (27 per cent) and a group of Bantu peoples dominate the economic and cultural life of western Cameroon; the westerners (around 20 per cent) are Christian English-speaking minorities; the Fulani (8 per cent), whose power base is in the north, are made up of 21 chieftaincies. The Baka, the Gyeli and the Tikar, nomadic peoples also called Pygmies, live in the southeast and southwest forests. The Government and the Catholic Church have tried to make them settle in 'pilot' villages. In a September 2005 report, Amnesty International accused the Governments of Chad and Cameroon, as well as the companies in charge of the construction of the oil pipeline between both countries, of violating the rights of indigenous peoples living in the affected region. In some cases they are denied access to their own lands or to drinking water sources (sometimes the sole source for many kilometers). The oil pipeline also affects fisherfolk who work off Cameroon's coast.

MIGRANTS/REFUGEES
During the 1970s and 1980s, large groups of Chadians arrived in Cameroon, fleeing from Chad's civil war and the insurrection. Although 7,000 people have returned, helped by the voluntary repatriation operation launched in 1999 by UNHCR, by the end of 2002 there were still around 30,000 refugees in Cameroon. More than 20,000 Nigerians - mostly Fulani who had entered the country with hundreds of thousands of livestock after escaping ethnic persecution - found refuge during 2002, joining others who had already settled there. Around 8,000 Nigerians returned home that year, but some 15,000 remained in Cameroon. In April 2005, an agreement between UNHCR and the governments of Cameroon and Nigeria allowed for the return of 7,500 Nigerians to their country. The operation was due to be completed in 2006, with the return of the rest of the refugees. As of 2006, Cameroon hosted nearly 40,000 refugees and had 6,800 pending asylum applications from Central and West African countries (including Nigerian people still to be repatriated).

DEATH PENALTY
The death penalty is in force, even for ordinary crimes.

* Latest data available in The State of the World's Children and Childinfo database, UNICEF, 2006.

cast ballots. The rest were excluded by the Government, which insisted they were 'foreigners'.

[13] In the wake of the 1999 publication of a Transparency International report that ranked Cameroon as the second most corrupt country in the world, Biya announced in 2000 a campaign to fight corruption as well as continued economic reforms. He promised to implement the constitutional changes that had been approved by Parliament in 1996, which included the creation of a Senate, regional councils and a constitutional council.

[14] In 2002 the International Court at The Hague ruled in favor of Cameroon over its border dispute with Nigeria. The case, which began in 1994 with a claim by Cameroon over sovereignty of the Bakassi peninsula - which is rich in fishing and oil fields - turned into a question of defining the border between the two countries, and also affected Equatorial Guinea. Nigeria rejected the ruling, keeping its occupation forces in the area.

[15] In 2003, after talks in Cameroon, Nigeria decided not to leave the Bakassi peninsula, at least for a further three years. Likewise, in December, Nigeria handed over 32 villages under its control to Cameroon.

[16] An agreement for joint patrolling of the border was signed in January 2004. The meeting of Biya, Olusegun Obasanjo (Nigeria) and UN Secretary-General Kofi Annan in Geneva proved also useful for discussing future joint control over Bakassi's oil fields.

[17] In August 2004, both countries reached a final agreement on Bakassi. Authorities from both countries held a series of meetings and agreed that Bakassi should remain under Cameroonian jurisdiction until 15 September. Minister Amadou Ali, of Cameroon, and Prince Bola Ajibola, of Nigeria, met with Ahmedou Ould-Abdalla, United Nations representative for West Africa, to ensure a peaceful transition in the peninsula, according to the International Court of Justice ruling.

[18] In December 2004, Biya appointed a new cabinet with Ephraim Inoni as Prime Minister in charge of decisively fighting corruption. The previous year, Transparency International had pointed out that bribes were 'omnipresent' in the country.

[19] With a stagnated economy as a result of the economic crisis, debt liabilities and a succession of natural disasters, in October 2005 Cameroon faced an increasingly widespread famine in the North Province, where more than one million people were in an extreme emergency situation.

[20] About 27 per cent of Cameroon's debt to international institutions (including the IMF and World Bank) was cancelled. The amount of $4.9 billion was written off when Cameroon became the 19th state to complete the Heavily Indebted Poor Countries (HIPC) initiative.

[21] In August 2006, a ceremony was held in the disputed oil-rich Bakassi peninsula to mark its transfer from Nigeria to Cameroon. ∎

Canada / Canada

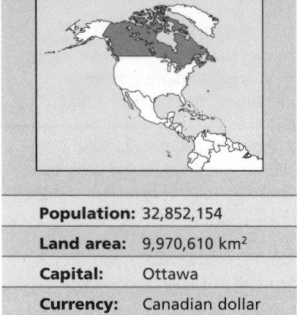

Population:	32,852,154
Land area:	9,970,610 km²
Capital:	Ottawa
Currency:	Canadian dollar
Language:	English and French

S ome time between 20,000 and 35,000 years ago, during the last Ice Age, the first humans to make their homes in North America migrated from Asia. Unknown numbers of people moved southward along the western edge of the North American ice cap. Archaeological evidence and oral traditions help rebuild the period that preceded contact with Europeans. There were 12 major language groups among the peoples living in what is now Canada: Algonquin, Iroquois, Siouan, Athabascan, Kootenaian, Salishan, Wakashan, Tsimshian, Haidan, Tlinglit, Inuktitut, and Beothukan. Within each language group there were usually political and cultural divisions. These subgroups were also divided; at the time of contact the Iroquois had organized themselves into a confederacy, the Iroquois League, consisting of the Mohawk, Oneida, Onondaga, Cayuga, and Seneca. A sixth group, the Tuscarora, joined later. The Inuit (Eskimo) who live in Canada's Arctic regions today were the last of the aboriginal peoples to reach Canada. The European colonists who came to North America in the 16th century estimated the indigenous population of the continent to be between 10 and 12 million.

[2] The peoples of the Eastern Woodlands (Huron, Iroquois, Petun, Neutral, Ottawa, and Algonquin),

created a mixed subsistence economy of hunting and agriculture supplemented by trade. The Huron and the Iroquois formed political and religious confederacies and both created extensive trade systems and political alliances with other groups. Peoples living in the far north did not appear to have formed larger political communities, while those of the West Coast and the Eastern Woodlands formed sophisticated political, social, and cultural institutions. All the groups were self-governing and politically independent.

[3] In 985 AD Norse (Viking) sailors sailing from Iceland to Greenland were blown far westward off their course and sighted the coast of Labrador. In 1000 Leif Ericson became the first European to land in North America. A colony was established in what the Vikings described as Vinland, which died out during the 14th and 15th

centuries.

[4] Italian navigator Giovanni Caboto (Cabot) sailed from Bristol in 1497 under a commission from the English king to search for a short route to Asia (what became known as the Northwest Passage). As Cabot and his sons explored the coasts of Labrador, Newfoundland, and possibly Nova Scotia they discovered that the cold northwest Atlantic waters were teeming with fish. Soon Portuguese, Spanish, and French fishing crews braved the Atlantic crossing to fish in the waters of the Grand Banks. Some landed on the coast of Newfoundland to dry their catch before returning to Europe. The English paid little heed to the Atlantic fishery until 1583, when Sir Humphrey Gilbert laid claim to the lands around present-day St John's in Newfoundland. The French also claimed parts of Newfoundland, primarily on the north and west coasts of the island. The initial period of contact between the Indians and the Europeans was based on the fishery. Although each was deeply suspicious of the other, the fishing crews and the Indians carried on a sporadic trade. Chiefly

as a sideline of the fishing industry, there continued an unorganized traffic in furs.

[5] From 1756 until 1763 a fierce war was fought between France and Britain over colonized territories in this area. In spite of some early French victories, the British managed to achieve the surrender of Quebec settlements and advanced on Montreal. In 1763, the Treaty of Paris granted Britain all French colonies to the east of the Mississippi river. French language and Catholic religion were recognized in Quebec in 1774. In 1791, this colony was split into Upper Canada (present-day Quebec) and Lower Canada (present-day Ontario). Colonization increased thanks to the profitable fur trade, with the local population growing to almost half a million by the end of the 19th century. The British North America Act of 1867 determined that the Canadian constitution would be similar to Britain's, with executive power vested in the King and delegated to a Governor General and Council. The legislative function would be carried out by a Parliament composed of a Senate and a House of Commons.

[6] In 1931, the Statute of Westminster released Britain's dominions from the colonial laws under which they had been governed, giving Canada legislative autonomy. That same year, Norway recognized Canadian sovereignty over the Arctic regions to the north of the main part of its territory. In 1981, the

LAND USE

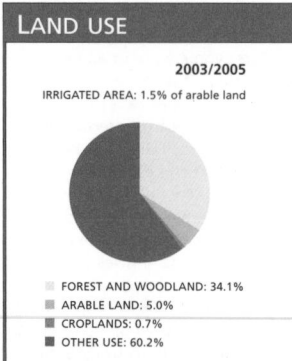

2003/2005

IRRIGATED AREA: 1.5% of arable land

- FOREST AND WOODLAND: 34.1%
- ARABLE LAND: 5.0%
- CROPLANDS: 0.7%
- OTHER USE: 60.2%

PUBLIC EXPENDITURE

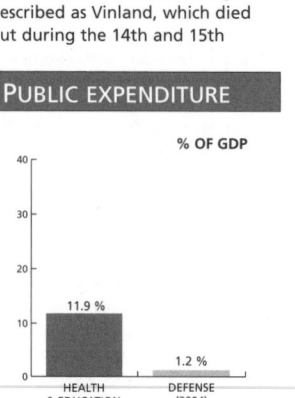

% OF GDP

11.9 % — HEALTH & EDUCATION (2000-2002)

1.2 % — DEFENSE (2004)

WORKERS

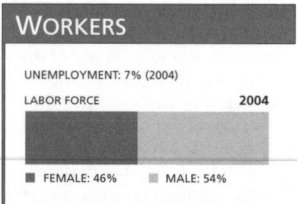

UNEMPLOYMENT: 7% (2004)

LABOR FORCE — 2004

- FEMALE: 46%
- MALE: 54%

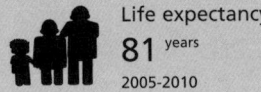
Life expectancy
81 years
2005-2010

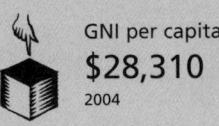
GNI per capita
$28,310
2004

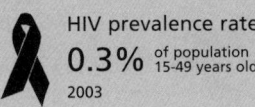
HIV prevalence rate
0.3% of population
15-49 years old
2003

Canadian Government reached an agreement with the British Parliament over constitutional transition. The following year, the 1867 Constitution was replaced by the Act of Canada, which granted Canada the autonomy to reform its own constitution. The Constitution Act of 1982 included a Charter of Rights and Freedoms, which recognized the country's pluralistic heritage and the rights of its indigenous peoples. It set forth the principle of equal benefits among the country's ten provinces, and the sovereignty of each province over its own natural resources. The largely French-speaking province of Quebec did not sign the agreement.

7   The religious sentiment and political nationalism evoked in Quebec's French population when the territory was occupied by English Protestants in 1760 led Quebec to take on a special role as guardian of the Catholic faith, the French language and the French heritage in North America. It was a role that it performed well. From a French population of 6,000 in 1769, the number of Québecois had increased to 6 million by 1960. In Quebec, four-fifths of the population speak French as a first language and preserve their cultural identity. Provincial autonomy was, and remains, a delicate issue. In 1977, the ruling separatist Parti Québecois (PQ), led by René Lévèsque, adopted French as the official language of education, business and local public administration. Lévèsque discarded unilateral separation, proposing instead a concept of 'sovereignty-association' with a monetary and customs union, but voters rejected this proposal by 59.5 to 40.5 per cent in a plebiscite held in 1980. In 1990, a poll revealed that 62 per cent of Québecois were in favor of their province's secession.

8   Antiquated British law governed gender issues in North America for a long time, and it was only in 1929 that Canadian women obtained full legal rights. True sexual equality was finally established with the 1982 Constitution. Each province has an equal rights law, guaranteeing access to housing, jobs, services and other facilities, without discrimination on the basis of race, religion, age, nationality or gender. The marriage of gay and lesbian people is now legal in Canada - one of the few places where this is the case.

9   Liberal governments, led by Pierre Trudeau, were elected in 1968, 1972, 1974 and again in 1980, after a brief Conservative interlude. Trudeau loosened Canada's traditional ties with Western Europe and the US, and strengthened those with the Far

East, Africa and Latin America. In addition, he refused to participate in the economic blockade against Cuba. Economic difficulties stemming from the worldwide recession triggered a sharp drop in the Liberal Party's (LP) popularity, in favor of the Conservative Party (CP). In 1983, Conservative leader Brian Mulroney, a labor relations lawyer and entrepreneur from Quebec, became Prime Minister, replacing John Turner, who had succeeded Trudeau as head of the LP. Mulroney re-established a 'special relationship' between Canada and the US, starting negotiations for a free trade agreement in 1985. This agreement, which went into effect in January 1989, provoked criticism from the Liberals and other members of the opposition, who claimed that the terms of the agreement were overly favorable to the US. Nevertheless, it received majority support from the voters in the 1988 election.

10   The Conservative victory was made possible by the Québecois votes (reflecting the growing economic importance of French-speaking voters) and by the impact of the Meech Lake Accord initiated by Mulroney and signed in 1987. In it the Federal Government ceded important powers to the provinces, granting Quebec for the first time recognition of its unique cultural status. To be implemented, the Meech Lake Accord had to be ratified by the unanimous consent of each of the provinces but Newfoundland and Manitoba blocked it. In Manitoba the one vote against came from a Native Canadian who did not accept the 'distinct society' clause for Quebec.

11   Canada's first trading partner is the US, and vice versa, but while the volume of Canada's sales to the US accounts for one-fifth of its total production, American exports to Canada represent less than 3 per cent. No other major Western economy maintains a trade imbalance of such magnitude. It is comparable only to the dependence of Southern countries in relation to the industrialized countries. With the 1989 free trade agreement Canada's integration with the American economy was accentuated. Through trade, the extension of credit, and investments in Canada, the US secured ever greater control over Canadian natural resources and majority control over shares in some Canadian industries. This degree of dependence has been labeled 'colonial', even though Canada is the world's eighth largest industrial power, with a standard of living ranked tenth in the world (according to OECD statistics).

12   The Canadian Armed Forces (CF) are charged with protecting

PROFILE

ENVIRONMENT

Canada is the second largest country in the world in land area, divided into five natural regions. The Maritime Provinces along the Atlantic coast are a mixture of rich agricultural land and forests. The Canadian Shield is a rocky region covered with woods and is rich in minerals. To the south, along the shores of the Great Lakes and the St Lawrence River, there is a large plain with fertile farmlands, where over 60 per cent of the population is concentrated, and the major urban centers are located. Farming (wheat, oats and rye) is the mainstay of the central prairie provinces. The Pacific Coast is a mountainous region with vast forests. The north is almost uninhabited, with very cold climate and tundra. There are ten provinces, and three territories; the Yukon, Northwest and Nunavut (the latter created in 1999 as land for Inuit people). Canada has immense mineral resources; it is the world's largest producer of asbestos, nickel, zinc and silver and the second largest of uranium. There are also major lead, copper, gold, iron ore, gas and oil deposits.

SOCIETY

Peoples: There are about 800,000 First Nations People or Native Americans, Métis (mixed race) and Inuit (Eskimos) ranging from highly acculturated city-dwellers to traditional hunters and trappers living in isolated northern communities. There are six distinct culture areas and ten language families; many native languages such as Cree and Ojibwa are still widely spoken. About 326,000 native people are classified as such: that is, they belong to one of 577 registered groups and can live on a federally protected reserve (though only about 70 per cent actually do so). Métis and those who do not have official status as 'native people' have historically enjoyed no separate legal recognition, but attempts are now being made to secure them special rights under the law. About 45 per cent of the Canadian inhabitants descend from British settlers, while 29 per cent are French. Today Canada is multicultural with many people from Asia, the Caribbean and Europe (Germans 3.4 per cent; Italians 2.8 per cent; Chinese 2.2 per cent; Ukrainian 1.5 per cent, Dutch 1.3 per cent; also Portuguese, Polish, Pakistani, Filipino).
Religions: Roman Catholic 45.7 per cent; Protestant 36.3 per cent. There are also Jews, Muslims, Buddhists, Hindus; 12.4 per cent of the population considers itself non-religious.
Languages: English and French, both official. 13 per cent of the population is bilingual, 67 per cent speak only English, 18 per cent only French, and 2 per cent speak other languages (Italian, German and Ukrainian and indigenous languages).
Main Federal Political Parties: Conservative Party; Liberal Party; Bloc Québecois; New Democratic Party; Green Party.
Main Social Organizations: Canadian Labour Congress, with more than 2.3 million members; Confédération des syndicats nationaux; Fédération de travailleurs et travailleuses du Québec; Canadian Federation of Students. Non-union: Assembly of First Nations; Council of Canadians.

THE STATE

Official Name: Canada.
Administrative Divisions: 10 Provinces and 3 Territories.
Capital: Ottawa 1,093,000 people (2003).
Other Cities: Toronto 5,411,300 people; Montreal 3,490,600; Vancouver 1,922,000 (2000).
Government: Stephen Harper, Prime Minister and Head of Government since February 2006. Canada is a federation of ten provinces and a member of the British Commonwealth, with a parliamentary system of government. Senate has 105 members and the House of Commons 308. Governor-General Michaelle Jean (2005) represents Queen Elizabeth II.
National Holiday: 1 July, Canada Day (1867).
Armed Forces: 52,000 (2003). Others: 6,400 (Coast Guard).

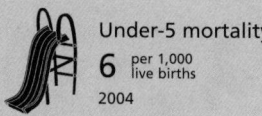

Under-5 mortality
6 per 1,000 live births
2004

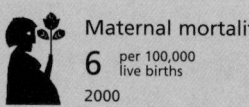

Maternal mortality
6 per 100,000 live births
2000

national interests both inside and outside the country, with defending North America in cooperation with the US, and complying with its NATO commitments; in addition, it participates in UN Peacekeeping. Relations between Canada and the US became tense in 1985, when a US Coast Guard vessel passed through the Northwest Passage without Canadian authorization. The US recognizes Canadian sovereignty over the Arctic islands but not over its waters. A similar dispute exists with relation to the waters surrounding the French islands of St Pierre and Miquelon. Canada made little headway in getting the US to control industrial pollutants, which drift over Canadian territory, bringing with them acid rain. The Canadian Government made a commitment to reduce its industrial emissions by half, anticipating a reduction of 20 per cent by the year 2005.

[13]   In 1991, the second phase of a huge hydroelectric project at James Bay - which borders Quebec and Ontario and into which hundreds of rivers flow - was launched. The first phase, called '*La Grande*' ('the big

one') would generate more than 10,000 megawatts. Almost 11,000 Cree and 7,000 Inuit have been hunting and fishing in this area for 5,000 years. The project's second phase will complete the damming of the rivers and flood a further 10,000 sq km.

[14]   In the same year, the International Work Group for Indigenous Affairs (IWGIA) stated that the Canadian Government was not respecting the religious rights of the Mohawk population and reported several acts of aggression towards them, including allowing the building of golf courses on their sacred sites. Also that year, delegates from 22 American countries attended a conference organized by the Canadian '500 Indigenous Women Committee'. They rejected the male domination and discrimination and vowed to regain some of the leadership women held in indigenous society before the arrival of the Europeans.

[15]   Canada's original inhabitants became increasingly well organized and between the 1960s and 1970s this resulted in the creation of the National Indian

Brotherhood (NIB) to represent them before public opinion and the federal government. The NIB was subsequently replaced by the Assembly of First Nations (AFN). Indigenous peoples fight for government respect of the treaties that affirm their rights to their land and resources. Those who still live from hunting and fishing - Cree, Dene, Innu, Haida and Iroquoians - are committed to the self-government of indigenous communities.

[16]   In 1992 the Tungavik Federation of Nunavut (TFN), the federal Government and the Government of the Northwest Territories (NT) signed in Iqalut the Nunavut Political Accord, confirming the splitting of these lands and the creation of a new government in Canada: the Government of the Nunavut territory. That same year, among intense debate over Quebec's growing demands for autonomy, Canada, the US and Mexico signed the North American Free Trade Agreement (NAFTA).

[17]   In 1993 Mulroney - amid record-low popularity ratings -

resigned from the CP and therefore from the government. Kim Campbell, former Defense Minister, became Canada's first woman Prime Minister. In the 1993 general elections, the LP regained power with a major victory over the CP, whose representation in the House of Commons dropped from 155 seats to just 2. Liberal Jean Chrétien took office in November and the following month Kim Campbell resigned from her post as leader of the CP.

[18]   The reduction of public deficit and Federal Government spending, the unemployment rate and separatist tendencies in Quebec were the main Government concerns during 1994 and 1995. Chrétien's popularity had increased 13 per cent since the elections, but the PQ triumphed in Quebec. The provincial Premier, Jacques Parizeau, promised to do everything possible to make Quebec a sovereign state.

[19]   In 1995, separation was rejected by 50.6 per cent of the electorate. The Quebec secession threat faded in 1996, despite the high number of votes received

Water: plentiful and badly managed

CANADA HAS 9 PER CENT of the world's renewable fresh water. It is also one of the countries that uses it least efficiently. Most of this is underground water and its volume is estimated to be 37 times greater than that of the surface water in the country's lakes and rivers. Underground water is known to supply 22 per cent of Lake Erie and 42 per cent of Lakes Huron and Ontario. More than one quarter of Canadians are supplied with underground water for domestic use. In spite of having so much drinking water available, the population only has access to 40 per cent of it. This is due to the fact that in many regions, subterranean water is not replenished in proportion to its use. Canada is suffering from pollution in some areas from petrochemical industries, pesticides, sewage, nitrates, chemical waste and bacteria. Contaminated aquifers can kill people, especially children and vulnerable older people. The daily

consumption of water for household use is on average 343 liters per capita.

Some 20,000 Canadian lakes are affected by acid rain. It is estimated that in the 21st century, water supply will be a greater problem than that of food and energy resources. According to Terence Corcoran, editor of the Toronto-based *Financial Post*, water will be 'the oil of the 21st century'. Canadians, including the Council of Canadians, are concerned that a water-hungry and profligate US will look north as a way of solving its water needs.

Some economists are interested in creating an international water market to avoid possible conflicts over water. In this context, the US demands that Canada sells its water resources in accordance with NAFTA.

However, in late 2005 both nations

signed a treaty banning the channeling of water from the Great Lakes toward other regions. But the treaty allows the extraction of water, which according to analysts, could trigger desiccation in the Great Lakes.

WATER-HUNGRY NATIONS
Canada and the US have the world's highest per capita consumption of fresh water. The largest demand comes from agriculture and power generation, which jointly account for 80 per cent of fresh water used in both countries.

Most irrigation water comes from underground aquifers which take thousands of years to regenerate. For example, one of the world's largest aquifers is that of Ogallala, located under the Great Plains region in the US. This runs in a great north-south belt from Wyoming through to northwest Texas but has been depleted due to over-extraction. ∎

Water source
100% of population
using improved
drinking water
sources
2002

Doctors
209 per 100,000
people
1990-2004

Primary school
100% net
enrolment
rate
2004

by the separatist option (49 per
cent). The Federal Government
transferred some powers to the
provinces in an attempt to appease
the nationalist movement. Lucien
Bouchard, a separatist leader,
replaced Parizeau following
the defeat in the referendum.
Unemployment continued to rise
in 1996, despite the economic
growth. Disagreements continued
throughout the year with the US
over Canadian companies dealing
with Cuba.

[20]   In 1997, Chrétien's party again
won the parliamentary elections,
winning 155 of the 301 seats,
followed by the Bloc Québecois,
the New Democratic Party (NDP)
and the Conservatives. That year,
the Supreme Court ruled that
Quebec could not secede without
the consent of Federal Government,
which in turn promised to
negotiate the secession, if the
Québecois majority agreed.

[21]   After 50 years of federal rule,
in 1999 the Inuit finally achieved
the devolution of one-fifth of
Canadian territory, in Nunavut.
Given the lack of infrastructure to
tap Nunavut's mineral resources,
the 25,000 Inuit that live there
remained dependent on Canadian
Government subsidies and
welfare.

[22]   In response to a wave of
criticism towards his Government,
Chrétien - against the opinion of
members of his party - brought
forward the elections to 2000. He
won an overall majority in the
House of Commons, and got 17
more seats than in 1997. Also, his
LP won seats in Quebec, which
weakened the separatists. The
performance of the Canadian
economy encouraged voters to
support the Prime Minister and
his team. The country ranked top
in the UN's Human Development
Index.

[23]   Following secret negotiations
between the Quebécois
Government and leaders of
the Cree community in 2001,
it was agreed that the Cree
would receive Can $2.2 billion in
exchange for an environmental
survey that would enable the
construction of a hydroelectric
dam. For many of the 12,000
Cree, who had struggled for
decades against the project, the
agreement amounted to a sell-out
by their leaders and went against
their conviction that land is sacred
and should not be harmed.

[24]   After the September 11 attacks
on the US, Canadian troops took
part in US President Bush's war on
terrorism. In 2002, Chrétien stated
in a TV interview that the West
should learn from the 9/11 attacks
on New York and Washington that
'you cannot exert your powers to
the point of humiliating others',

IN FOCUS

ENVIRONMENTAL CHALLENGES
Lakes and forests have been
seriously damaged by air
pollution and acid rain caused
by industrial and vehicle
emissions. In addition to this,
toxic waste from industrial,
mining and agricultural
activities pollutes the sea.

WOMEN'S RIGHTS
During World War I, nurses
serving in Europe were allowed a
postal vote, and became in 1917
the first Canadian women to
vote. However, it was not until
1920 that they were allowed to
stand for office. In 2003, women
held 23 per cent of ministerial
positions. In 2006, 21 per cent
of seats in the Lower House and
35 per cent of seats in the Upper
House of Parliament were held
by women.
    Women made up 46 per
cent of the labor force in 2003.
Female unemployment stood at
7 per cent; one percentage lower
than male unemployment.

CHILDREN
The infant mortality rate is 5
deaths for every 1,000 live births,
while the under-5 mortality rate
is 6 deaths for every 1,000 live
births*. Six per cent of newborn
babies are underweight*.
    Schooling is free and
compulsory until the age of

16. Legislation protects children
against sexual abuse, child labor
and discrimination, imposing
severe penalties on those who
infringe these laws. In spite of
existing controls, trafficking in
young women mainly from Asia
and Eastern Europe to work
in the sex industry still poses a
problem for which no lasting
solution has been found.

INDIGENOUS PEOPLES/
ETHNIC MINORITIES
Two per cent of the population
are First Nations peoples, some
living a traditional lifestyle in
isolated areas, particularly in
the far north, some living on
protected reserves, and some
living in cities. They mostly speak
their native languages - the most
common are Inuit, Cree, Ojibwa,
Inukitut - and one of the official
languages, French or English.
Life expectancy for indigenous
peoples is lower than the
Canadian average and they have
higher rates of alcoholism, suicide
and drug abuse. They also have
higher birth and unemployment
rates and are more likely to suffer
from poor housing conditions.
Only 25 per cent of aboriginal
students complete secondary
education and young people
under 25 are at greater risk of
being imprisoned. Although
discrimination is banned by law,
indigenous peoples generally

suffer a lower standard of
living.

MIGRANTS/REFUGEES
In 2002, in response to the
imminent enforcement of the
'third safe country' agreement
signed that year by Washington
and Ottawa (which allowed
Canadian immigration
authorities to send back would-
be migrants entering from the
US), several asylum-seekers
entered from the Middle East,
South Asian and North African
countries. The number of
asylum permits granted in the
period 2003-2005 has declined
steadily compared with
previous years. In 2003, 32,000
permits were granted while
in 2004 the number stood at
25,500 and until August 2005,
at 14,000.

DEATH PENALTY
On 14 July 1976 Parliament
passed Bill C-84 abolishing
capital punishment from the
Canadian Criminal Code,
replacing it with a mandatory
life sentence without possibility
of parole for 25 years for
first-degree murder. The last
execution was in 1962.

** Latest data available in The State
of the World's Children and Childinfo
database, UNICEF, 2006.*

and that 'the West is becoming
excessively rich in relation to poor
countries'. At the same time,
Canada refused to support any
kind of military action against Iraq
without UN backing. This earned
Chrétien much fierce criticism
from conservative opponents, who
demanded he apologize to the US
Government.

[25]   In 2003, the LP won Quebec's
local elections, ending the PQ's
9-year rule. Jean Charest became
the Province's Premier, succeeding
Bernard Landry (Bouchard's
successor). The LP won 75 of
Quebec's 125 National Assembly
seats. Paul Martin (Finance Minister)
became Prime Minister after the
national elections held that year;
Chrétien retired after ten years in
office.

[26]   In December 2003, the
Progressive Conservatives and the
right-wing Canadian Alliance Party
merged to form the Conservative
Party of Canada. Stephen Harper
was elected leader of the new party
in March 2004.

[27]   A financial scandal placed

Prime Minister Martin in an
extremely compromising situation
in February 2004. Canada's auditor-
general issued a report on the
misuse of several millions of dollars
allegedly spent by the Government
in political advertising and support
programs. During the federal
investigation, the judge decided
that the LP could only attend
the commission charged with the
investigation but would not be
allowed to question witnesses.

[28]   In June 2004, the LP won
the elections once again but lost
30 seats; for the first time in 25
years a minority government had
to be formed. The polls gave 135
of the 208 seats in the House
of Commons to the LP, 99 to
Conservatives, 54 to the separatist
Bloc Québecois, 19 to the NDP and
one to independents.

[29]   The corruption scandal that
had erupted in 2004 eventually led
to the ousting of Martin's minority
government as a result of a no-
confidence vote in Parliament.
Conservative Stephen Harper took
office in January 2006, putting an

end to 12 years of Liberal rule.

[30]   In June, after an intense
investigation that involved more
than 400 agents, the Toronto
police detained 22 people planning
attacks with explosives in the
country. Seventeen were indicted.
Canadian authorities said that the
three tons of ammonium nitrate
- a substance commonly used in
fertilizers that can also be used to
make bombs - found in possession
of the detained were three times
more than the amount used in the
1995 attack in Oklahoma, US, which
killed 168 people.

[31]   In May the cabinet had voted
to extend Canadian military activity
in Afghanistan until 2009. In early
August US forces gave over control
of southern Afghanistan to NATO
troops comprising Australian, British
and Canadian soldiers (see history
of Afghanistan). Growing violence
in several southern Afghanistan
provinces - traditionally under
Taliban domain - kept NATO forces
on permanent alert. Very often
there were more daily attacks than
in Iraq (see history of Iraq). ■

Cape Verde / Cabo Verde

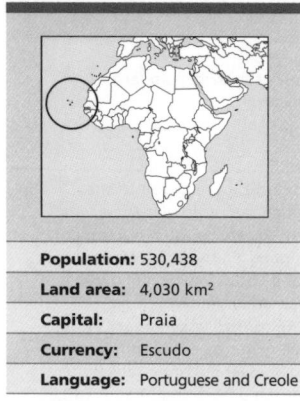

Population:	530,438
Land area:	4,030 km²
Capital:	Praia
Currency:	Escudo
Language:	Portuguese and Creole

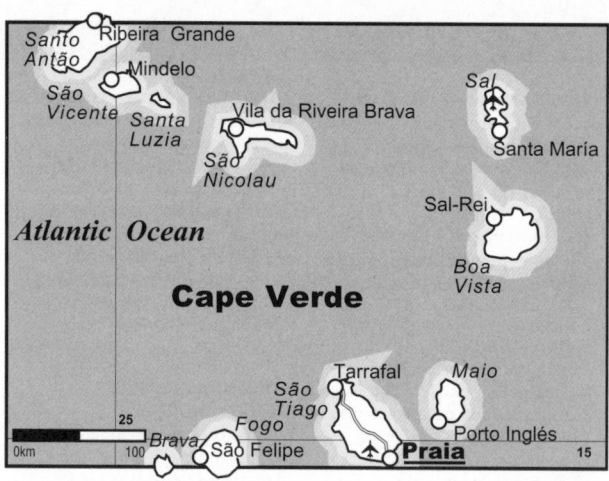

When the Portuguese settled on the Cape Verde archipelago in the 15th century, the islands were deserving of their name (Green Cape). They were covered by lush tropical vegetation that stood out against the black volcanic rock and the blue sea. There is no evidence of the islands having been inhabited prior to the arrival of settlers, although it is believed that the Moors had visited the Island of Sal to collect salt supplies in previous centuries. In 1462, the first settlers landed on what today is known as São Tiago and founded the oldest European city in the tropics (now called Cidade Velha). The Portuguese introduced the cultivation of sugarcane which was not successful because of the dry climate. Cape Verde's prosperity came as a result of the slave trade (mainly from the West African coast). After 400 years of Portuguese colonization the islands were transformed into a 'floating desert', with most of the population emigrating to escape starvation.

2   For Portugal, the islands' position between Africa, South America and Europe represented an important strategic interest. In the 16th century, Cape Verde was port of call for ships carrying slaves to America (slavery was abolished in 1876). The frequent incursions by French, British and Dutch pirates made Portugal bring

farmers from its Alentejo region (in southeastern Portugal and known as the 'barn' of that country) to the islands. Their style of agriculture eroded the fertile soil and periodic droughts have struck the country ever since. Between the 17th and 19th centuries, cotton was the prized export, but later falling farm production caused huge out-migration of Cape Verdeans. Most went to Guinea-Bissau, another former Portuguese colony with close ties to the archipelago. There was further migration to Angola, Mozambique, Senegal, Brazil and in particular the US. In 1800, the prosperity of the islands was slowly disappearing due to poverty, hunger, bad administration and the colonial government's corruption.

3   In 1951, Cape Verde's status changed to overseas province. During the independence struggle, the islands developed closer ties to Guinea-Bissau (see Guinea-Bissau). In 1956, the African Party for the Independence of Guinea and Cape Verde (PAIGC) was created, with supporters in both colonies. Amilcar Cabral, founder and ideologist, thought that they could fight together for freedom and development, on the basis that they were two economically similar countries. In 1961, guerrilla war broke out on the mainland, and hundreds of Cape Verdeans joined the fight. Cabral was assassinated in 1973. The following year, the colonial regime was overthrown and after a transition government in 1975, independence was proclaimed. For the first time ever one political party, the PAIGC, took

office in two different countries at the same time. Aristides Pereira became President of the Republic of Cape Verde and commander Pedro Pires its Prime Minister. The PAIGC took the first steps toward s a federation between Cape Verde and Guinea-Bissau: the national assemblies of the two countries sat together as the Council of the Union.

4   From 1975 onwards, the forested areas of Cape Verde increased from 3,000 to 45,000 hectares. The Government

predicted that in the next 10 years a further 75,000 hectares would be planted making the islands self-sufficient in firewood. In rainy seasons, men and women left their homes and offices to spend a week planting trees. The land reform program was implemented, giving priority to food production to meet the needs of the population (local production amounted barely to 5 per cent) instead of favoring the export crops of the colonial period. In spite of these actions, agricultural production declined due to severe droughts which led the Government to invest in fisheries.

5   Cape Verde supported Angola during its 'second liberation war' (see Angola) by allowing Cuban planes to land on the archipelago, helping to defeat the invasion of Angola by former Zaire and South Africa and adopted a policy of non-alignment, declaring that no foreign military bases would be allowed in their territory.

6   In 1981, while the PAIGC was discussing a new constitution for Guinea-Bissau and Cape Verde, Guinea-Bissau's President Luiz Cabral was overthrown. João Bernardo Vieira took office in his place, and was hostile to integration with Cape Verde. That

LAND USE

2003/2005

IRRIGATED AREA: 6.1% of arable land

- ☐ FOREST AND WOODLAND: 20.8%
- ☐ ARABLE LAND: 11.4%
- ☐ CROPLANDS: 0.7%
- ■ OTHER USE: 67.1%

WORKERS

LABOR FORCE	2004

- ■ FEMALE: 34%   ■ MALE: 66%

PROFILE

ENVIRONMENT

An archipelago of volcanic origin, composed of Windward Islands Santo Antao, São Vicente, São Nicolau, Santa Luzia, Sal, Boa Vista, Branco and Raso; and Leeward Islands Fogo, Santiago, Maio, Rombo and Brava. The islands are mountainous (heights of up to 2,800 meters) without permanent rivers. The climate is arid, influenced by the cold Canary Islands' current. Agriculture is poor; nonetheless it employs most of the population.

SOCIETY

Peoples: Cape Verdeans are descendant from Africans, mainly from Bantu-speaking peoples, and from Europeans. Today the population is 71 per cent of mixed descent; 28 per cent African and 1 per cent European.
Religions: Mainly Roman Catholic 93.2 per cent (with influence of local practices); Protestant 6.8 per cent (mainly from the Nazareth Church).
Languages: Portuguese is the official language, but the national language is Creole, based on old Portuguese with West African vocabulary and structures.
Main Political Parties: African Party for the Independence of Cape Verde (PAICV); Movement for Democracy (MPD); Democratic and Independent Cape Verdean Union (UCID); Democratic Renewal Party (PRD); Social Democratic Party (PSD).
Main Social Organizations: National Cape Verde Workers' Union-Central Trade Union Committee (UNTC-CS); Cape Verde Confederation of Free Trade Unions.

THE STATE

Official Name: República do Cabo Verde.
Administrative Divisions: 9 islands and 14 counties.
Capital: Praia 107,000 people (2003).
Other Cities: Mindelo 64,000; Santa Maria 13,400 (2000).
Government: Parliamentary republic. Pedro Pires, President since March 2001, re-elected in 2006. José Maria Neves, Prime Minister, since February 2001. National Assembly with 72 members.
National Holiday: 5 July, Independence Day (1975).
Armed Forces: 1,100.

Life expectancy
72 years
2005-2010

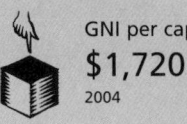

GNI per capita
$1,720
2004

Literacy
76% total adult rate
2000-2004

same year, the PAIGC held an emergency meeting in Cape Verde to discuss political developments in Guinea-Bissau. After the ratification of the principles of Cabral, the party changed its name to 'the African Party for the Independence of Cape Verde' (PAICV) stressing their independence from the party of Guinea. Relations between the two governments became tense, until mediation efforts by Angola and Mozambique in 1982 paid off and Mozambique's President Machel brought Pereira (re-elected in 1981) and Vieira together in Maputo. At the Conference of Former Portuguese Colonies in Africa (1982), held in Praia, Cape Verde, Vieira met with colleagues from Angola, Mozambique, Cape Verde and São Tomé. Diplomatic relations returned to normal, although the party was not reunited and plans for reunification were abandoned.

[7] In 1984, after severe drought crop yields fell 25 per cent below those of the previous 5 years, trade deficit stood at $70 million and foreign debt reached $98 million. The food distribution system and efficient state management prevented famine. Being poor in natural resources, with just 10 per cent of arable land, Cape Verde is highly dependent on imported food, mainly from aid agencies. Dependency on foreign aid also limited the scope of the First Development Plan. In 1986, the Second Development Plan prioritized private enterprise especially in the informal sector, and the fight against desertification. Up to 1990, the plan was to recover more than 500 sq km, and establish a centralized system of administering water reserves for the whole country. During the first phase, more than 15,000 dams were built to store rainwater, and 231 sq km were forested. Despite droughts, farm production gradually increased to make the country practically self-sufficient in meat and vegetables, without resorting to imports.

[8] In 1991, Antonio Mascarenhas Monteiro (President of the Supreme Court during the previous decade) was elected President in the first free and multiparty elections. The transition to a free-market economy with the privatization of insurance companies, fishing and banking began, in line with the requirements of international agencies. Foreign aid accounted for 46 per cent of the GDP, while remittances from the 700,000 Cape Verdeans residing abroad, constituted another 15 per cent of it. The MPD government (center-right), faced with a 25 per cent unemployment rate, announced

the restructuring of the state. In 1993, a 50 per cent reduction of the 12,000 civil servants began, together with a gradual deregulation of prices. The 1994 budget, despite making cuts in public expenditure, increased public investment (in transport, telecommunications and rural development) from $80 million in 1993 to $138 million in 1994.

[9] In 1995, Prime Minister Carlos Veiga introduced changes to facilitate the country's transition to a free-market economy and merged the Ministries of Finance, Economic Coordination and Tourism, and Industry and Commerce into a single Ministry of Economic Coordination. In 1997, the African Development Bank granted a $4.9 million loan for road building. Cape Verde also received economic support from China and created an association with Angola to invest in health and social welfare.

[10] Reports of police brutality towards prisoners recurred throughout 1998 and 1999. The number of prisoners exceeded

prison capacity. Prisons were way below minimum reasonable conditions. Self-censorship is frequent in the media.

[11] In 2001, presidential elections tainted by fraud allegations resulted in too slight a margin (50.05 per cent against 49.95 per cent), and new elections were held. The final result was decided by the Supreme Court after complaints of irregularities. Pedro Pires, from the PAICV, won by 17 votes and succeeded Monteiro as the third President since independence. Jose Maria Pereira Neves was elected Prime Minister.

[12] Pires' Government began to step up efforts to decentralize and privatize the public sector, and for that purpose it signed a 610-million-euro cooperation agreement with France.

[13] Following privatizations, the cost of basic services increased and access to clean water, especially outside the capital, was poor. The Government planned for all schools to have access to at least one computer within five years. Premier Neves announced the

start of a development plan called 'Operation Hope' and emphasized that his office was 'a guarantee to the future of Cape Verdean children'.

[14] Finance Minister João Pinto Serra promised in an official letter sent to the International Monetary Fund (IMF) in September 2004 that he would accelerate structural reforms in his administration in order to speed up privatizations. Reforms would affect the energy, water, telecommunications, transport, fishing and navigation sectors.

[15] In May 2005, Prime Minister Neves said the country could try to join the North Atlantic Treaty Organization (NATO). A month earlier NATO had chosen Cape Verde to test, for the first time in Africa, its Reaction Force. In June, the opposition Movement for Democracy called for urgent discussion of the special relationship between Cape Verde and the European Union.

[16] The PAICV won the January 2006 parliamentary elections and Pires was elected President. ∎

IN FOCUS

ENVIRONMENTAL CHALLENGES
Although an archipelago, Cape Verde lies within the Sahel belt, which is undergoing increased desertification, with periodic droughts and a high demand for wood - used as fuel - that generates serious deforestation. These phenomena are made worse by factors such as the topography of rolling hills, high winds and the islands' small size. The environmental damage threatens different species of birds and reptiles. Illegal beach-sand extraction and indiscriminate fishing have been recorded.

WOMEN'S RIGHTS
Cape Verdean women have been able to vote and stand for election since 1975. In January 2006 women held only 15 per cent of seats in Parliament. Adult illiteracy - accompanied by a wide gender gap - was still high in 2004: it affected 15 per cent of men and 32 per cent of women.
Poverty affects a significant portion of the population, especially women and children. Population growth exacerbates the situation. Food production only provides 10 per cent of the country's needs, the rest coming from multilateral aid agencies and remittances sent by emigrants.

CHILDREN
A high percentage of children show signs of chronic malnutrition. In 2004, the infant mortality rate was 27 per 1,000 live births. Girls frequently drop out of school as a result of sexual abuse and teenage pregnancy, which becomes intertwined with sexually transmitted infections and the potential spread of HIV/AIDS. Adolescents face numerous risks, among them sexual exploitation, alcohol, tobacco and drug abuse and delinquency. Child prostitution is a particular threat to girls, but also affects boys, as on the island of Sal. The Government has taken measures in education to fight discrimination towards girls and to raise awareness of sexually transmitted diseases, especially HIV/AIDS.

INDIGENOUS PEOPLES/ ETHNIC MINORITIES
According to Marian Aguiar, from Boston University, the bi-racial origins of Cape Verdeans are a legacy of Portuguese settlement and the African slave trade. The small and elitist white population, former slaveholders, determined social position according to origins as well as race. In addition, as the islands' mestizo (or mixed, of indigenous and European descent) population grew, racial lines became increasingly difficult to draw. From almost the moment of settlement, the Portuguese brought slaves to or through the islands. The Africans

who left their mark on Cape Verde were called pretos (blacks). More than 50 per cent of the resident slaves were female. Often 'whiteness' or 'blackness' was as much a signifier of class position as it was of 'blood'. Today the connotation of other markers, such as class or education, makes it possible for an individual to become 'more white' by moving up in society.

MIGRANTS/REFUGEES
The emigration rate is among the highest in the world; the diaspora outnumbers the resident population and every family has at least one emigrant among its members. According to the 2000 census, half of all emigrants during the period 1995-2000 chose Portugal as their destination, followed by the US, France and the Netherlands. In proportion to the total financial flow, remittances to Cape Verde have increased by 35 per cent within the last 20 years. In spite of the country's situation, Cape Verde hosts refugees from Sierra Leone, Mauritania, Liberia, Rwanda and Côte d'Ivoire.

DEATH PENALTY
It was abolished in 1981. The last execution was in 1835.

* Latest data available in *The State of the World's Children* and *Childinfo* database, UNICEF, 2006.

Cayman Islands / Cayman Islands

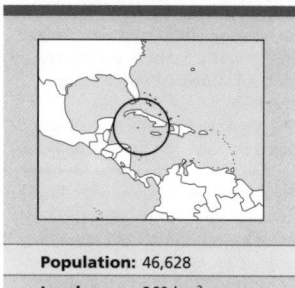

Population:	46,628
Land area:	260 km²
Capital:	George Town
Currency:	Cayman dollar
Language:	English

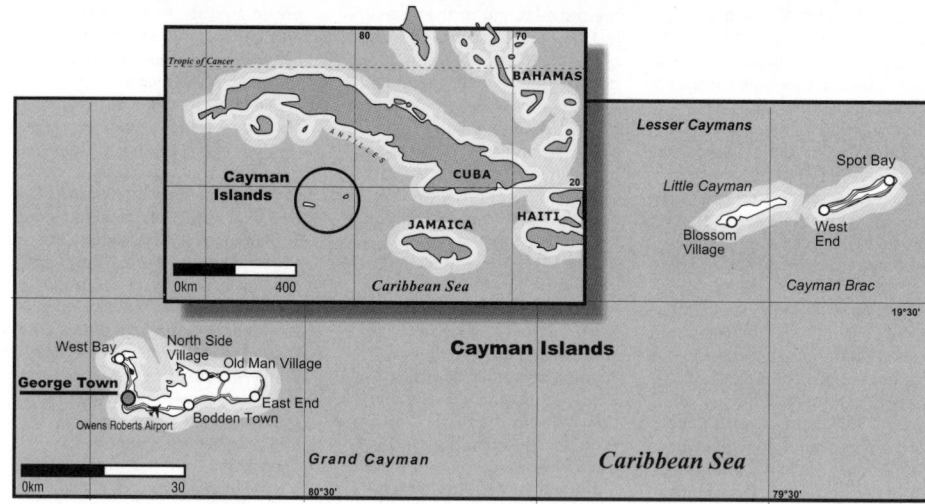

In his last voyage to the West Indies in 1503 Christopher Columbus sighted the Cayman Islands (Grand Cayman, Cayman Brac and Little Cayman) and named them the 'Turtles' because of the vast numbers of turtles there (in the late 18th century; over-fishing virtually killed off this resource). Shortly afterwards, buccaneers and privateers took over the islands which were loosely under French control. For a good part of the 17th century, the Caymans were the pirate headquarters of the Antilles.

2 Conflicts among European powers affected the Antilles and particularly the Cayman Islands. When Jamaica was ceded to Britain under the Treaty of Madrid (1670), the Cayman Islands became a Jamaican dependency. The first permanent colonial settlement was established on Grand Cayman and most of the settlers were British sailors, castaways from buccaneer ships, African slaves and Jamaican landowners. When slaves were freed (1835), the islands' society became homogeneous. In 1959, with Jamaican independence, they were given the status of British dependent territory.

3 In 1972, a new constitution granted some local autonomy on domestic matters. The Governor, appointed by the Crown, became responsible for defense, foreign affairs, internal security, and some social services. There was also an Executive Council and a Legislative Assembly.

4 Their tax-exempt status has encouraged the establishment of a strong financial center: in 1987 there were 515 banks in the Cayman Islands. By the end of the decade, a quarter of the economically active population worked in tourism (the islands were visited that year by more than 600,000 tourists). There are many hotels and the influx of visitors has produced considerable immigration, which in 1988 accounted for 35 per cent of the population.

5 With the exception of turtle farming (which replaced fishing), local industry and agriculture only meet domestic needs. The Caymanese are renowned boat-builders and sailors. One of the country's main sources of revenue is the money sent home by sailors. Fishing declined seriously in the 1980s and has not recovered.

6 In 1991, McKeeva Bush, a member of the Assembly, founded the first organized political party in the Caymans. The DPP (Democratic Progressive Party) aimed to change the legal status of the islands from being British dependencies. The DPP campaigned for constitutional reform, seeking the creation of a party system, more members of the Executive Council, the creation of the post of Prime Minister, and an increase in the Legislative Assembly to 15 members. The Governor would become the President of the extended Executive Council.

7 In 1996, the capture of a drug trafficker from Mexico - the third main country supplying cocaine to the US - revealed the role of the Caymans' banking system in a money-laundering circuit running from Texas to Switzerland. In 1998 the Government denied entry to 910 gay tourists, mostly US citizens, alleging that they did not fulfill the 'appropriate behavior requirements'. The British Government distanced itself from this 'discriminatory' policy.

8 In 2000 concern about illegal financial practices led international monitors to categorize the Cayman Islands as 'reluctant' to collaborate in fighting money-laundering. The country was removed from the blacklist in June 2001, after enacting regulations that banking authorities said were already being followed. Even so, anyone engaged in the finance industry, from banks to lawyers, must now verify the identity and source of income of each client. In 2002 it was discovered that US company Enron used some 700 societies registered in the archipelago to evade federal taxes. The islands signed an agreement with the US Government to share tax information to uncover offenders.

9 In 2003 the Caymans were implicated in the Italian company Parmalat's fraudulent accounting - almost 4 billion euros were missing - leading to the largest corporate scandal of the year. The Bank of America (third largest US bank) stated that a Parmalat certificate for $3,950 million in cash and bonds was false. It was held by Caymans-registered Bonlat Financing Corporation.

10 Hurricane Ivan, which lashed the Islands in September 2004, caused the postponement of parliamentary elections. These were finally held in May 2005 and the opposition People's Progressive Movement (PPM) obtained 9 of the 15 seats, thus displacing the United Democratic Party (UDP) from power. The modernization of the Constitution started by the UDP before the election therefore never reached completion. ∎

Central African Republic / République Centrafricaine

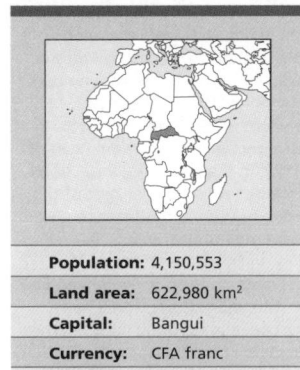

Population:	4,150,553
Land area:	622,980 km²
Capital:	Bangui
Currency:	CFA franc
Language:	French

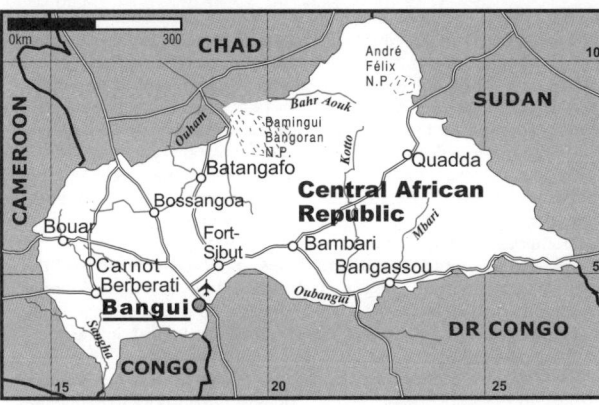

Diamonds and carved quartz that are at least 8,000 years old and a megalith set up about 500 BC attest to the existence of humans here before 800 BC. The first inhabitants of the present-day Central African Republic (CAR) were Pygmies: hunters and gatherers whose descendants still live in the forest (to the southwest).

[2] The first state structure was the Kanem Kingdom, founded in the 9th century; it became Muslim in the 11th century, spreading from Chad. It is presumed that peoples related to the Nubians (that came from Egypt and Sudan) settled two Kingdoms during the 15th and 16th centuries in other Central African regions.

[3] In the 1750s, the Sudan sultanates deported many people who lived by the Oubangui river (Sango-speaking traders). From 1780, the Portuguese removed many in the slave trade. These were mainly Gbaya-Mandja and peoples that lived in the central and northern savanna who, at that time, numbered half the country's current population.

[4] Most of today's Central African people are descended from different ethnic groups that were settling by the end of the 18th century and the formal abolition of slavery. Images made with butterfly wings, and some ebony and ivory sculptures that have been found, attest to the art that was disrupted by European colonization. Today several of these small diverse groups (ranging from hunters to farmers), still practise some of the same animist rituals, singing and drumming.

[5] During the last decades of the 19th century, the Belgians, British and Germans competed for control over the territory in search of raw materials. In 1887, France and Belgium divided their domain along both sides of the Oubangui river. In 1888 in this region, the French founded an administrative entity known as French Congo, with an outpost at Bangui, capital of today's CAR. In 1891, the French-occupied

territory was called Oubangui-Chari; colonial status was conferred in 1903 so that it could be annexed to French Equatorial Africa in 1910.

[6] During World War I, hundreds of people from the region were sent to fight for the French against German forces in Africa in a bid to conquer Germany's territories. From the outset, the French military occupation of today's CAR, was brutal, with the system of *corvée*, or forced labor, and constant expeditions against Sultan Rabah (in Sudan). About 40 Europeans controlled the exploitation of rubber, ivory and diamonds as well as the cotton and coffee plantations. Railways were built to make transport possible.

[7] There were further rebellions in the 1930s, provoked by the forced labor system. During World War II, cotton and diamond exports reached record levels and France was reluctant to relinquish this source of wealth. In the end, in 1946, France granted the status of 'overseas territories' to Oubangui-Chari and other colonies. This new statute gave the right of representation in the French Legislative Assembly. The first Deputy from the region was the Catholic priest Barthélemy Boganda (1910-1959). In spite of these reforms, France retained control of foreign trade, defense and tax collection.

[8] From 1949, after founding the Movement for the Social Evolution of Black Africa (MESAN), Boganda led the fight for independence. The French carried out campaigns to discredit him, and in 1956 they managed to bribe his main advisers. His nephew and associate David Dacko took over but MESAN degenerated into a French controlled agency. Abel Goumba broke away to form the Movement for the Democratic Evolution of Central Africa (MEDAC) which was banned by the French in 1960 and its leaders arrested.

[9] In 1958, the country became the Central African Republic as part of the Franco-African Community

(CFA). Boganda became Prime Minister and clearly had vision for his country and the region. However, in 1959, he died in a mysterious plane crash.

[10] On 13 August 1960, independence was declared. Dacko became the President and in 1964 promoted a constitutional reform which granted him full powers. He purged the more progressive elements within the MESAN and drew increasingly closer to the French. He made French the official language and Sango and national language and granted US companies concessions for mining the uranium and cobalt reserves.

[11] Dacko rapidly lost the support he had inherited from Boganda, as a result of his corrupt administration. The country plunged into a deep crisis. In 1965, Dacko was ousted in a coup led by one of his cousins, General Jean-Bedel Bokassa (1921-1996). Bokassa was pro-France, having served in the French Army for 22 years. In 1972, he proclaimed himself President-for-life.

[12] In 1977, Bokassa decided to crown himself Emperor of the country (renaming it as Central African Empire). The coronation ceremony cost $28 million and was financed by France, Israel and South Africa. The Emperor gave away diamond-rich lands to foreigners and hired notorious international arms dealers as his military advisers.

[13] The Emperor and the then President of Zaire (now DR Congo), Mobutu Sese Seko, established close military collaboration to repress the continuing popular rebellions in both countries. In 1979, a massacre of students shook the world's public opinion.

[14] France decided to depose the Emperor, attempting to erase the negative image it had created by supporting his coronation. On 20 September 1979, he was overthrown and former President David Dacko came back to power, protected by 1,000 French troops.

PROFILE

ENVIRONMENT

This is a landlocked country in the heart of Africa. It is located on a plateau irrigated by tributaries of the Congo River, like the Oubangui, the main export route, and of Lake Chad. The southwestern part of the country is covered by a dense tropical forest. Cotton, coffee and tobacco are the basic cash crops. Diamond mining is a major source of revenue.

SOCIETY

Peoples: Most Central Africans belong to the ethnic group Baya (Gbaya), 33 per cent; followed by Banda, 27 per cent; Mandjia, 13 per cent; Sara, 10 per cent; Mboum, 7 per cent; M'baka, 4 per cent; Yakoma, 4 per cent; Ba'aka, 1.3 per cent and other 2 per cent.
Religions: There is no official religion. 24 per cent practise traditional African religions; 25 per cent are Protestant; 25 per cent Roman Catholic; and 15 per cent Muslim.
Languages: French (official); Sango is the national language, used for communication between the various ethnic groups.
Main Political Parties: National Convergence (Kwa Na Kwa); Movement for the Liberation of the Central African People (MLPC); Central African Democratic Rally (RDC); Social Democratic Party (PSD); Patriotic Front for Progress (FPP).

THE STATE

Official Name: République Centrafricaine / Ködrö tî Bê-Afrika.
Administrative Divisions: 16 Prefectures, 52 Sub-prefectures.
Capital: Bangui 698,000 people (2003).
Other Cities: Berberati 61,400 people; Bouar 53,800; Carnot 51,900; Bambari 49,900 (2000).
Government: General François Bozize, President and Head of State since 15 March 2003 by a coup, officially elected in May 2005. Prime Minister and Head of Government: Élie Doté, since June 2005, appointed by the President.
Parliament: National Assembly with 105 members, elected for a five-year term.
National Holiday: 1 December, Independence Day (1960).
Armed Forces: 1,400 (1996). Other: 2,700 Gendarmes.

Life expectancy
40 years
2005-2010

GNI per capita
$310
2004

Literacy
49% total adult rate
2000-2004

HIV prevalence rate
13.5% of population 15-49 years old
2003

15   Dacko dissolved the Empire and reinstated a republic, granting France a 10-year lease on the big Bouar (west) air base. Dacko's return amounted to no more than a change of name as corruption and repression continued. Political opposition was ruthlessly repressed and almost all opposition leaders were sent to prison or exiled.

16   Amidst countless conspiracies, Dacko was overthrown by another military coup in September 1981, which brought General Kolingba to power. Kolingba asked the French to pay the salaries of 24,000 civil servants for the period of one year. In an attempt to draw closer to the US, he granted them new economic concessions, allowing the exploitation of uranium resources.

17   The planned return to democracy, initially set for 1982, became rather an election barred to the opposition on 21 November 1986. Kolingba was elected President with his CAR Party of Democratic Recovery (PCRD). A constitution was passed establishing a one-party system, and in July 1987, the members of the General Assembly were chosen from PCRD members.

18   During that period, an IMF structural adjustment plan was launched, which did not improve the catastrophic situation shown by socio-economic indicators. The intensification of malnutrition,

infant mortality and the spread of diseases like HIV/AIDS (which reached 12 per cent of the population in 2003), among others, are clear signs of the suffering of Central Africans after the adoption of successive economic policies.

19   In 1986 Bokassa returned from exile in France. Being accused of murder, cannibalism and misappropriation of public funds, he had been sentenced to death in absentia. Upon his return, he was arrested, tried again and sentenced to life imprisonment.

20   Municipal elections were held in May 1988, with universal suffrage. In 1991 a plan was approved for constitutional reform and for the adoption of a multiparty system. Elections were scheduled for October 1992, but shortly after the voting started Kolingba annulled the process claiming there had been irregularities. A Provisional National Council of the Republic was formed, made up of the five presidential candidates.

21   The first round of the presidential elections took place in August 1993. Kolingba, who came fourth, annulled the result by decree once again. However France threatened to suspend its military and financial aid, forcing him to allow the second round to go ahead and to free political prisoners, including former emperor Bokassa.

22   On 19 September, Ange-Felix Patassé, Bokassa's former Prime Minister, was returned as President with 52.47 per cent of the vote. In 1996, Patassé requested French military intervention to crush rebel soldiers. After the intervention, Bangui saw some violent expression of hostility towards the presence of French soldiers.

23   In June, the President announced a new Government of national unity and named former ambassador in France, Jean-Paul Ngoupande, Prime Minister.

24   In spite of a truce, in early 1997 France launched an offensive against rebel troops in Bangui in retaliation for the death of two French soldiers. This led Patassé and the rebel leader Anicet Saulet to agree the replacement of French troops by a guard from African countries, although funded by Paris.

25   In February, Patassé formed a new Government, including some opposition members, and most of the rebels returned to the barracks. Some months later, the President demanded French withdrawal from the military bases in the country. At the same time, he tried to strengthen links with the US, which had a growing influence over the region.

26   In 1998, the UN Security Council authorized the deployment of a UN Mission in the Central African Republic.

27   In elections held that year, the ruling National Liberation Movement (NLM), led by Patassé, won 49 of the 109 seats. However, the nomination of the new cabinet sparked violent street demonstrations in Bangui, in early 1999. In the presidential elections Patassé recived 51 per cent of the vote, but the opposition accused the Government of fraud.

28   In December 2000, civil servants went on strike demanding outstanding wages; some had not been paid for 30 months. The leaders of the Economic Community of Central African States (ECCAS), meeting in Cameroon in 2001, called for dialogue. However, a presidential spokesperson declared that there was no conciliation possible with the leader of the uprising, François Bozize, former Chief of Staff of the Army.

29   In February 2002, the population of Bangui demonstrated against the arrival of Libyan troops, ostensibly there to protect the President. Opposition leader Bozize, who had taken refuge in Chad and France, returned in October. At the same time, rebels loyal to him crossed the border from Chad, attacking many towns and taking over a third of Bangui, demanding that Patassé resume the dialogue or else resign.

30   In March 2003, Bozize seized power, proclaimed himself president and dissolved Parliament. Four years after withdrawing 'definitively' from its military bases in the Central African Republic, France sent troops, along with Chad and the Republic of Congo, to secure the coup. The intervention forces agreed with the new President that democratic elections would be held in 2004. From exile in Togo, Patassé had declared in May that he was prepared to negotiate to achieve a 'consensual transition'.

31   In August 2004 it was announced that an internationally financed election would be held in 2005. An information campaign - involving debates and meetings with representatives of women, civilian groups, political parties and religious institutions - encouraged citizens from all over the country to register to vote. In December a new Constitution was approved by referendum.

32   Bozize won the presidential elections in May 2005 on a second round with 64.6 per cent of the vote, against the 35.4 per cent of former Prime Minister Martin Ziguele. In June the President was sworn in.

33   A plane with 50 armed men landed in the north of the country in April 2006. The Government claimed that they were back-up forces for Chadian rebels, coming from Sudan. ∎

IN FOCUS

ENVIRONMENTAL CHALLENGES
Bad land management has increased soil erosion and reduced fertility. The scarcity of water and pollution of rivers pose serious problems. Illegal fishing and poaching have damaged the country's reputation as one of the world's largest nature reserves.

WOMEN'S RIGHTS
Women have had the right to vote and to stand for election since 1986. In 2005, 10.5 per cent of members of Parliament were women.

According to data gathered in 2004, female illiteracy was 67 per cent, while male illiteracy was 35 per cent. The primary school attendance rate was 39 per cent for women and 47 per cent for men. Only 62 per cent of pregnant women received health care before delivery and just 44 per cent of the births were assisted by a specialist.

CHILDREN
In 2004, 149,000 births were recorded. The under-five mortality rate has worsened in recent years, rising from 168 per

1,000 live births in 1990 to 193 in 2004. The main causes of disease and mortality in children are HIV/AIDS and malnutrition. In 2003, it was estimated that 110,000 children under 18 years old were orphans due to HIV/AIDS.

INDIGENOUS PEOPLES/ ETHNIC MINORITIES
Mbum, Mbororo, Ba'aka and Hausa people are the main minorities. In the 20th century, the Mbum (4 per cent of the population) fled the mountainous region of the country escaping Mbororo raids. Today they are marginalized, and live in extreme poverty.

The Mbororos (of the Fulani ethnic group) descend from semi-nomadic herders of the western grasslands. Because of their relative cattle wealth, they have been the subject of attacks and extortion.

The Muslim Hausa are less than one per cent of the total population, yet constitute three-quarters of the small traders.

The Ba'aka people (also known as pygmies, a group of some 20,000 members) are mainly hunter-gatherers and inhabit the southern tropical forest of the country. They are particularly threatened by the

political instability of the region (especially in DR Congo).

MIGRANTS/REFUGEES
Between June and August 2005, 12,000 people from Central African Republic (CAR) fled to Chad to escape insecurity, increasing the total number of CAR refugees in that country to 42,000. At the same time, 1,300 Chadians were repatriated from Boubou after the local economy - highly dependent on cotton crops - collapsed. In 2004, 3,000 Congolese people returned to their country (DR Congo) and another 800 expressed their wish to be repatriated in 2005.

It is estimated that 44,000 people from CAR live in Chad and Cameroon, including those exiled in 2005. During 2006, 12,000 displaced people are expected to return to CAR.

DEATH PENALTY
Although it still applies, there have been no executions since 1981.

* Latest data available in *The State of the World's Children* and *Childinfo* database, UNICEF, 2006.

Chad / Tashad - Tchad

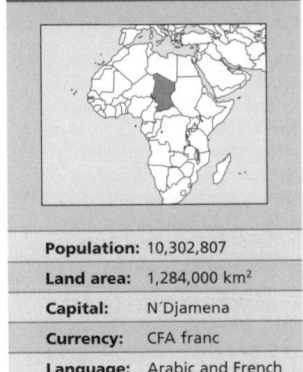

Population:	10,302,807
Land area:	1,284,000 km²
Capital:	N'Djamena
Currency:	CFA franc
Language:	Arabic and French

The Sahel region, of which the Republic of Chad is part, has been inhabited from time immemorial and could be the place where the first hominids lived, according to the July 2002 discovery of a 7-million-year-old fossil skull of a specimen named *Sahelanthropus tchadensis*.

[2] By the end of the 18th century, European missionaries had converted some of the southern peoples to Christianity and given them a European-style education. These converts sided with the Europeans against the native peoples of the north.

[3] France 'acquired' Chad in 1885 after the Berlin Conference, but did not establish in the territory until a 1920 invasion by the notorious French Foreign Legion which defeated the northern Muslim groups.

[4] The colonizers introduced cotton farming in 1930, and small parcels of land were distributed to peasants to grow this crop, while the French monopolized its trade. The result was a cotton surplus and a food shortage, accompanied by famine.

[5] In August 1960 when France granted Chad its independence, southern leaders who had been negotiating with the colonizers since 1956 assumed power. However Chad's first president, François Tombalbaye, leader of Chad Progressive Party, was unable to unite a country where frontiers still reflect the arbitrary colonial divisions.

[6] The Chad Liberation Front (FROLINAT), founded in 1966, was crushed by the French troops. The Front had become fully active about the time that southern peasants rebelled against the French company Cotonfran which controlled the cotton industry.

[7] In 1970, the Front controlled two-thirds of the national territory, and by 1972 FROLINAT guerrillas were within range of N'Djamena, the capital. In 1975 Tombalbaye was ousted and killed in a French-staged coup that put General Felix

Malloum in power.

[8] Paris supported Hissene Habré's faction - the Armed Forces of the North (AFN) - because he opposed Goukouni Oueddei, FROLINAT's president and leader of the People's Armed Forces (PAF),

who was receiving support from Muammar Qadhafi's Government in Libya.

[9] In 1979 Chad's 11 major political groups formed a Provisional Government of National Unity (PGNU). Habré was named

Minister of Defense in Malloum's Government. However, the French were displeased with the make-up of the Cabinet because of their strategic interest in Chad - linked to the Maghreb and to the uranium and oil discoveries in the 1960s. In March 1980 Habré resigned, broke the alliance, and unleashed civil war.

[10] The PGNU split into three factions. In May 1980, Oueddei requested military aid from Libya, and Qadhafi sent 2,000 troops.

[11] That year, over 100,000 refugees fled the country, after the bombing of N'Djamena by Habré's forces. In October, Libyan troops reached the capital, mediation efforts by the Organization for African Unity (OAU) failed, and the defeated Habré fled to Cameroon in December 1980.

[12] France led an international campaign against Libyan expansionism in Africa, with support from the United States, Egypt, Sudan, and other African countries, fearful that Qadhafi's revolutionary drive would eventually 'infect' poor Islamic populations of the Sahel region, south of the Sahara.

[13] Habré was accused of being opportunistic and corrupt; however, Libyan support made it possible for France to divide Oueddei's allies. In April 1981, backed by France, Habré reorganized his followers in Sudan. In July, in Nairobi, the OAU decided to send a peacekeeping force to Chad, with the help of the

PROFILE

ENVIRONMENT

Forty per cent of the territory is part of the Sahara Desert, with the great Tibesti volcanic highlands. The central region, the Sahel, which stretches to the banks of Lake Chad, is a transition plain where nomadic pastoralism is common. The lake, only half of which lies within Chad's borders, is shallow and mostly covered with swamps. Thought to be the remnant of an ancient inland sea, its waters are fed by the rivers Logane and Chari. The banks of these rivers, fertilized by flooding, contain the country's richest agricultural lands and are the most densely populated areas. In colonial times, economic activity was concentrated there. Cotton is the main export product but subsistence agriculture, though hampered by droughts, still predominates. In recent years, mineral reserves of uranium, tungsten and oil have attracted the attention of the transnationals.

SOCIETY

Peoples: Northern Chadians, mostly nomadic shepherds of Berber and Tuareg (Toubou, Ouaddai, Kotoko, Maba) origins, have traditionally opposed the majority population of the south, where the farming Sara, Massa, Mundani and Hakka peoples predominate. Due to droughts in the Sahel region, there is a constant process of internal migration, from the north to the fertile regions of the south, which heightens conflicts.

Religions: An estimated 50 per cent of the population is Muslim, 27 per cent practice traditional African religions, 23 per cent Catholic.

Muslims are mostly concentrated in the north, non-Muslims in the south.

Languages: Arabic and French are the official languages. There are more than 100 local languages, the most widely spoken being Sara (in the southern region).

Main Political Parties: Patriotic Salvation Movement (MPS) led by President Idriss Déby; Rally for Democracy and Progress (RDP); Front of Action Forces for the Republic (FAR); National Rally for Development and Progress (RNDP); National Union for Democracy and Renewal (UNRD).

Main Social Organizations: Chad Federation of Labor Unions, Chadian Human Rights League, Chad Non-violence, Chadian Association for the Promotion and Defense of Human Rights.

THE STATE

Official Name: République du Tchad.
Administrative Divisions: 14 Prefectures.
Capital: N'Djamena 797,000 people (2003).
Other Cities: Moundou 111,200; Sarh 84,400 people; Abéché 61,100 (2000).
Government: Idriss Déby, President since December 1990, re-elected in 2001. Pascal Yoadimnadji, Prime Minister since February 2005. Council of State with 31 members, appointed by the President. National Assembly with 155 members elected every four years.
National Holiday: 11 August, Independence Day (1960).
Armed Forces: 25,200. Other: 4,500 gendarmes.

Life expectancy
44 years
2005-2010

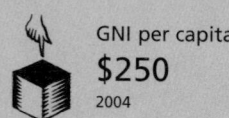

GNI per capita
$250
2004

Literacy
26% total adult rate
2000-2004

HIV prevalence rate
4.8% of population 15-49 years old
2003

French and troops from six African countries. Oueddei, yielding to foreign pressure and tensions within the PGNU, requested the withdrawal of Qadhafi's troops in November.

14 After his defeat by Habré's forces in June 1982, the exiled Oueddei set up a Provisional National Salvation Government in October. The new civil war split the country in two: northern Chad, under the control of a recently formed National Liberation Council with Libyan support; and southern Chad, with the Habré Government dependent on French troops.

15 When Habré seized power, bringing this phase of the civil war to an end, the country was in ruins. The population of N'Djamena had dropped to 40,000, and half of its businesses and small enterprises had closed. Outside the cities, 2,000 wells and all the water towers had been destroyed, and the health and educational infrastructures were practically non-existent.

16 In 1987, the southern forces supported by France took Fada, Faya Largeau and the frontier strip of Aozou, claimed by Libya. In 1989, Chad and Libya signed an agreement on this 114,000 sq km territory, which included the return of prisoners and the presentation of a territorial lawsuit before the International Court in The Hague.

17 In 1990, Idriss Déby (leader of the Patriotic Salvation Movement, supported by France) ousted Habré, who fled to Senegal. During the deposed president's term, some 40,000 people had been executed or 'disappeared'.

18 Déby inaugurated a national conference in 1993 to 'democratize' Chad, with the participation of some 40 opposition parties, another 20 organizations and six armed rebel groups. Fidele Moungar was appointed interim Prime Minister during the transition period.

19 In February 1994, the International Court of The Hague ruled that the Aozou strip belonged to Chad. In April the Transition High Council postponed the elections for one more year. In May, Libya officially returned the Aozou strip to N'Djamena.

20 In March 1995, the Transition High Council postponed the elections for another 12 months.

21 In the Presidential elections finally held in June and July 1996, Déby was elected President with 69 per cent of the vote. Several rebel groups signed peace agreements with the Government and in August, Déby signed an agreement with the armed forces in the south to establish a federal republic, bringing the fighting to a close.

22 Southern Chad was at the heart of an international controversy in December 1997 when the Campaign for the Reform of the World Bank,

IN FOCUS

ENVIRONMENTAL CHALLENGES
Desertification and drought are endemic to the region and affect all aspects of daily life. Only 34 per cent have access to improved drinking water sources. Ineffective waste disposal contributes to soil and water pollution.

WOMEN'S RIGHTS
Women have had the vote and been able to stand for office since 1958. The percentage of parliamentary seats held by women rose from 2 per cent in 2000 to 6.5 per cent in 2002. While in 2000 there were no women in ministerial positions, in 2003 women occupied 12 per cent of those positions.

Women make up 45 per cent of the country's workforce. In 2004 the adult literacy rate for women was 13 per cent, and for men 41 per cent. The net primary school attendance rate was 51 per cent for girls and 75 per cent for boys.

CHILDREN
Between 1990 and 2004, the rates of infant and under-five mortality remained practically unchanged at 117 and 200 per

1,000 live births*. The percentage of children born underweight was 10 per cent while 28 per cent of children under five suffered from severe and moderate low weight. By the end of 2003, 18 per cent of children under 15 lived with HIV/AIDS and 96,000 people under 18 had been orphaned by the disease.

INDIGENOUS PEOPLES/ ETHNIC MINORITIES
As a result of contacts with Sudan and Egypt, the southern and eastern regions are largely populated by Arabs, while southerners integrated into European culture under French colonial rule. Southerners - Sara, Massa, Moundang and Hakka - comprise 46 per cent of the country's total population. They are mainly Christians or Animists, living by agriculture in up to 10 per cent of the country's total area.

Other ethnic groups include Toubou, Teda and Daza in the northern prefectures of Bourkou, Ennedi and Tibesti (known as the BET region - one third of the country, with 6 per cent of the population). Hadjerai people live in the mountainous central region; Zagawa or Bidaye (one of the Ouaddian groups) live in the north

and east straddling the Sudan-Chad border; and the Buduma, fishing people, around Lake Chad. The Arabs comprise 25 to 30 per cent of the total population; most of them are nomads or semi-nomads living in the Salamat area in the southeast or settled in the central regions. They are predominantly Muslim.

MIGRANTS/REFUGEES
UNHCR aimed in 2006 to resettle 12,500 of the total 42,000 Central African Republic refugees in the south of the country. Roughly 20,000 Sudanese lived in border areas, waiting to be repatriated when conditions turned favorable, while another 200,000 remained hosted in 12 camps. In 2006, between 3,000 and 5,000 Chadians refugees living in Sudan are expected to return voluntarily to their country.

DEATH PENALTY
The death penalty still applies for ordinary crimes.

* Latest data available in The State of the World's Children and Childinfo database, UNICEF, 2006.

supported by dozens of NGOs, opposed a mega-project planned by the Bank in this zone. The project involved digging and transporting oil to the Atlantic Ocean through Cameroon. Despite the opposition, the project - which had been supported by the Bank itself - went ahead.

23 From 1998 onwards the most active guerrilla group, fighting in the country's desert north, was the Movement for Democracy and Justice in Chad (MDJT), led by Youssouf Togoimi, former Minister of Defense in Déby's Government. The armed group increased its activity in 2000.

24 In the presidential elections of 20 May 2001, Déby was re-elected with 67.4 per cent of the vote. The six defeated candidates claimed electoral fraud and called for the poll to be annulled, even though international monitors stated they were mostly satisfied with the electoral process.

25 In February 2002, the Government and the rebels of the MDJT signed a peace treaty, ending three years of civil war. The parliamentary elections of 2002 gave Déby 110 seats in the 155-member legislature. However, in March the clashes between the MDJT and the Government forces recurred.

26 In January 2003 a ceasefire with the National Resistance Army (operating in the southeast) was

signed in Gabon. Chad was forced to reimburse the IMF a $7.5 million loan in June 2003 since the Government had given false accounting information. The Chad-Cameroon Oil Development and Pipeline Project was officially inaugurated in October 2003. The 1,070 km-long pipeline was financed by Exxon Mobil, Chevron Texaco, Petronas and the World Bank, despite strong opposition from several NGOs. The Government promised to devote 80 per cent of the oil revenues to education, health and the environment, and to provide access to safe drinking water. This promise came after officials admitted that, in 2000, $4 million had been spent in arms.

27 Towards the end of 2003, the Senegalese President agreed to hold Habré until an international court sought his extradition, even though the Government of Senegal had declared in 2001 that its courts had no jurisdiction over crimes committed outside the country. The case was presented to the Belgian judge who has been investigating the numerous accusations of human rights violations against the former dictator made by relatives of the victims of the Habré regime. A new peace agreement was signed between the Government and the MDTJ in Burkina Faso.

28 In January 2004 the UN

started resettling more than 95,000 Sudanese refugees who fled to Chad because of the armed conflict in Darfur, western Sudan. There were clashes between the Chad military forces and Sudan militia crossing the border in Darfur.

29 In order to allow President Déby to seek a third mandate in the 2006 elections, an amendment to the Constitution was approved with 66 per cent of the vote in a referendum in June 2005. In November, former president Habré was arrested in Senegal, accused of crimes against humanity. Rebel forces attacked the village of Adre, near the border with Sudan in December; N'djamena accused Khartoum of being behind these attacks.

30 Déby backed a law to reduce the allocation of oil revenues to development in January 2006. As a result, the World Bank suspended its loans and froze the Government's accounts. In March, official sources claimed to have thwarted a coup attempt. In April, rebel forces fought against Government troops outside the capital in an attempt to oust Déby. Chad broke off diplomatic relations with Sudan, accusing the country of backing the rebels.

31 The presidential election was held in May, with the country still divided. The main opposition groups boycotted it. ■

Chile / Chile

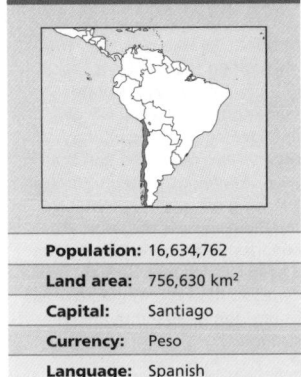

Population:	16,634,762
Land area:	756,630 km²
Capital:	Santiago
Currency:	Peso
Language:	Spanish

At the beginning of the 16th century, the north of Chile formed the southernmost part of the Inca empire (see: Peru, Bolivia and Ecuador). The area between Copiapo to the north and Puerto Montt to the south was populated by the Mapuche, later called Araucanians by the Europeans. Further south lived the fishing peoples: the Yamana and the Alacalufe.

2   Diego de Almagro set off from Peru in 1536 to begin the conquest of Chile. With an expeditionary force of Spanish soldiers and enslaved native Americans he had covered nearly 2,500 km, when a mutiny by Spanish soldiers in Lima forced him to turn back.

3   Between 1540 and 1558, Pedro de Valdivia settled in what is now the port of Valparaiso and founded several cities including Santiago. The Mapuche, led by chief Lautaro, a capable military strategist, beat the invaders on several occasions, adapting their military tactics to changing conditions in the region. Valdivia died in one of these battles. His successor, Francisco de Villagra, defeated and killed Lautaro in 1557. In a case unique in colonial America, the Mapuche maintained an independent territory on the Bio-bio river for more than 300 years, officially recognized by Spain as Araucaria. *Creole* -Spanish dominion only spread to all the territory in the

second half of the 19th century.

4   In 1810 Santiago's town council became an autonomous ruling Junta. In 1811, led by José Miguel Carrera the Junta instigated the independence process. General Bernardo O'Higgins, the son of a former viceroy in Peru, joined this movement. War broke out between the independence army and the royalist forces, with their strongholds in Valdivia and Concepción. Helped by the army of José de San Martín, which crossed the Andes to fight the royalists, the independence army finally defeated the Europeans on 5 April 1818 in the Battle of Maipú.

5   In 1817, O'Higgins was designated Supreme Head of State, while the royalist troops still maintained pockets of resistance. He laid the political foundations of the country, which were reinforced in the 1833 Constitution, during the Presidential term of Diego Portales.

This 'aristocratic republic' denied all forms of political expression to the new urban sectors, the middle class and the rising proletariat. English companies, in alliance with the creole oligarchy, organized an export economy based on the rich saltpeter deposits of the north along the maritime coast of then non-landlocked Bolivia to Peru. Soon the English controlled 49 per cent of Chile's foreign trade. Chilean and British capital also owned 33 per cent of Peru's saltpeter, but they wanted total control.

6   The 'Nitrate War' or Pacific War of 1879-1884 was caused by this Chilean-British alliance. Chilean territory increased by a third and left Bolivia in its present landlocked state. The victory brought about the rapid growth of the saltpeter industry and its labor force.

7   Jose Manuel Balmaceda was elected President in 1886 and tried to break the oligarchic order. Nationalism fostered by war, economic growth, social diversification and the education of the wealthy helped to gain him support. He encouraged protectionism to develop national industry. The oligarchy reacted violently, supported by the English. The army defeated the President's partisans, and Balmaceda committed suicide in the Argentine embassy in 1891.

8   In 1900, the first union was founded in Iquique. In 1904, 15 unions with 20,000 members joined to form a federation called the National Convention. That year the unions clashed with the military in Valparaíso and three years later automatic weapons supplied by the US were used to massacre 2,500 workers and their families in a school in Iquique.

9   Housing, railroads and new mines continued to expand, stimulating trade, new services and public administration. In 1920, populist politician Arturo Alessandri became the leader of the new social factions that sought to subvert the oligarchic order and achieve representation in politics. Alessandri's Government promoted constitutional reform and welfare legislation with electoral rights for literate men over 21, direct presidential elections, an 8-hour working day, social security, and labor regulations.

10   Chile's economy, based on farm and mineral exports, was severely affected by the 1929-30 depression. Recovery did not come until the bourgeoisie was able to impose an industrialization program to produce previously imported goods. Their proposal served as a base for the 1936 Popular Front, which marshaled support from Communists and Socialists. The

front was led by Pedro Aguirre Cerda. The armed forces were purged and withdrew from the political scene for almost 40 years. Although the oligarchy was weakened, it made a pact with the Government on agrarian policies and Aguirre Cerda never allowed land reform or the formation of rural workers' unions.

11   The alliance worked out in the 1930s broke down during González Videla's term of office, 1946-1952. The climate created by the Cold War was used to legitimize the Law for the Permanent Defense of Democracy, which banned the Communist Party and deprived its members of the vote. On 9 January 1949, to compensate for these measures, the Government brought in female suffrage. The Radical Party's repressive and de-nationalizing policies, and the disabling of the Left enabled the populist Carlos Ibáñez to win the 1952 election.

12   Economic deterioration rapidly eroded the strength of the populist Government. In 1957, offshoots from the National Falange (populist) and from the old Conservative Party (oligarchic), founded the Christian Democratic Party (PDC). The Communist Party was legalized and the Left rebuilt its alliances, forming the Popular Action Front. The people wanted change, but, sensitive to an aggressive anti-communist campaign, voted for Eduardo Frei's 'revolution in freedom', which initiated agrarian reform in 1964.

13   The UP (Popular Unity) coalition led by Salvador Allende won the 1970 election, obtaining 35 per cent of the vote, while the rest of the electorate was split between the Christian Democrats and the conservative parties. The UP coalition included the Socialist Party, the Communist Party, the United Popular Action Movement (MAPU) and the Christian Left. The following year the UP won almost 50 per cent of the vote in the local elections, leading the Right to fear a definitive loss of its majority.

14   Allende nationalized copper and other strategic sectors, together with private banks and foreign trade. He increased land reform, promoted collective production and created a 'social sector' in the economy, managed by workers.

15   The traditional élite, now out of power, conspired with the Pentagon, the CIA and transnational corporations, particularly ITT, to topple the Government. The Christian Democrats were indecisive, but finally supported the coup. Inflation, a shortage of goods and internal differences within Popular Unity contributed to the climate of

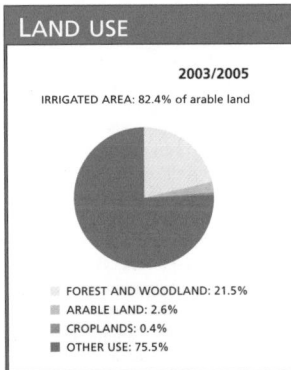

LAND USE

2003/2005

IRRIGATED AREA: 82.4% of arable land

- FOREST AND WOODLAND: 21.5%
- ARABLE LAND: 2.6%
- CROPLANDS: 0.4%
- OTHER USE: 75.5%

WORKERS

UNEMPLOYMENT: 7% (2004)

LABOR FORCE   2004

- FEMALE: 35%   MALE: 65%

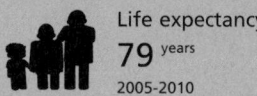
Life expectancy
79 years
2005-2010

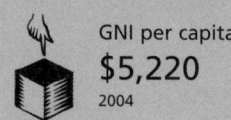
GNI per capita
$5,220
2004

Literacy
96% total adult rate
2000-2004

HIV prevalence rate
0.3% of population 15-49 years old
2003

instability.

[16] On 11 September 1973 General Augusto Pinochet led a coup. The Presidential Palace at La Moneda was bombed by the air force, and President Allende died, possibly committing suicide, during the fighting. Violent repression ensued: people were shot without trial, sent to concentration camps, tortured, or simply 'disappeared'.

[17] The Chilean military dictatorship was one of many that ravaged South America during the 1970s, inspired by the National Security Doctrine. This was supported by the Chilean oligarchy and the middle classes, as well as by transnational corporations who recovered the companies that had escaped their control.

[18] After the coup, Chile's economic policy started to be based on neo-liberal doctrines. Inflation dropped below ten per cent per year, unemployment practically disappeared and imported manufactured goods flooded the market. Alongside this ran a loss of earning power in workers' salaries and an overall impoverishment of the poorer classes.

[19] The detrimental influence of these neo-liberal economic policies became evident in 1983, two years after a constitution containing a mandate for 'continuity' was approved in a referendum by 60 per cent of the voters. Unemployment reached 30 per cent, real salaries had been reduced by 22 per cent in just two years and 55 per cent of all families were living below the poverty line. This situation provided the background for the violent popular uprising which took place in November 1983, led by the National Labor Co-ordination Board and the National Workers' Command.

[20] In 1984 the Church started talks. To participate in these, the opposition formed the Democratic Alliance, led by the PDC. The talks with Home Minister Sergio Onofre Jarpa failed and the break with the Church became evident. After this, the Vicarate for Solidarity of the Archbishopric of Santiago played a key role in defending human rights.

[21] The political left joined together under the Popular Democratic Movement (MDP), vindicating all forms of struggle against the dictatorship. The MDP and the PDC attempted to join in 1985, but the controversy on armed struggle kept them apart. One faction formed the Manuel Rodriguez Patriotic Movement (FPMR), an armed group which carried out numerous attacks, the most important being an attempted assassination of Pinochet on 7 September 1986.

[22] The international isolation of the Chilean Government during the US Carter administration eased a little with the election of Ronald Reagan and Margaret Thatcher in the UK. In August 1985 Chile authorized US space shuttle landings on Easter Island. At the beginning of 1986 a US delegation proposed that Chile be condemned before the UN Human Rights Commission, thereby avoiding having to take more drastic action against the country.

[23] On 5 October 1988, an 8-year extension of Pinochet's rule was submitted to a referendum. Widespread opposition ended his Government, bringing elections the following year.

[24] Facing up to the defeat, Pinochet negotiated constitutional reform. Proposed changes included further restrictions of the power of future governments, an increase in the number of senators, shortening the Presidential term from 8 years to 4 and a liberalization of the proscription of left-wing parties. The reform was approved by referendum on 30 July 1989.

[25] Elections were held on 14 December 1989. Patricio Aylwin, leader of the Christian Democrats, gained 55.2 per cent of the vote, taking office on 11 March, 1990.

[26] In April, Aylwin appointed a Truth and Reconciliation Commission to investigate the issue of missing people. The Commission confirmed that there were at least 2,229 missing people, who were assumed dead. It also made a detailed study of repression during the dictatorship. When the facts were made public in March 1991 the President asked the nation for forgiveness in the name of the State. He announced that judicial procedures would follow and he requested the co-operation of the armed forces in these proceedings

[27] The armed forces and the Supreme Court justified their conduct during the Pinochet dictatorship, and denied the validity of the Government report - thus discrediting the President.

[28] On 2 April 1991, Senator Jaime Guzmán, former adviser and ideologue of the military regime, was murdered and the assassination attributed to radical left wing groups. This enabled the Right to raise the issue of terrorism once more. Political life was slowly brought back to normal. On 23 April, the FPMR announced its decision to abandon armed struggle.

[29] The Chilean economy maintained a 10-year expansion, with annual growth rates of over six per cent, mainly due to high levels of investment (especially fixed capital) and the expansion of the external sector.

[30] During Aylwin's presidency, social indicators improved. In 1993, real salaries increased by five per cent, the unemployment rate fell to 4.5 per cent, social spending increased 14 per cent in two years and inflation stabilized around 12 per cent.

[31] In August 1993, the Special Commission on Indigenous Peoples (a government agency) proposed introducing indigenous language instruction in Mapuche, Aymara and Rapa Nui at primary schools in communities where Spanish is not their mother tongue. This was considered vital to reduce the loss of cultural identity among indigenous children.

[32] In 1993, Eduardo Frei, candidate for the Christian Democratic Party and the *Concertación* (agreement) coalition, won the presidency with 58 per cent of the vote. However, he did not achieve the parliamentary majority required to do away completely with the old authoritarianism because of eight 'designated' seats in the Senate which were a legacy from the Pinochet regime.

[33] Before handing over the presidency, President Aylwin pardoned four FPMR activists, sentenced to death for the assassination attempt against Pinochet in 1986.

[34] The Frei Government announced a plan to reduce the poverty which affected nearly a quarter of the population. In May 1995, the minimum salary was increased by 13 per cent. Taxes were introduced on the sale of cigarettes and motor vehicles to fund a 10 per cent increase in the lowest pensions and a 5 per cent increase in the education budget. In June, Chile requested associate membership of the Mercosur market and negotiated entry into the North American Free Trade Association (NAFTA).

PROFILE

ENVIRONMENT

Flanked by the Andes in the east and the Pacific in the west, the country is a thin strip of land 4,200 km long and never wider than 360 km. Its length explains its variety of climates and regions. Due to the cold ocean currents, the northern territory is a desert. The central region has a mild climate which makes it good for agriculture. The southern part of the country is colder and heavily wooded. The major salt and copper mines are located in northern Chile. 65 per cent of the population live in the central valleys. Chile exercises sovereignty over Easter Island/Isla de Pascua (Rapa Nui).

SOCIETY

Peoples: Chileans are descended from the native American population and European immigrants. 300,000 Mapuche Indians live mainly in southern Chile.
Religions: Mainly Catholic (77 per cent), Protestant 13 per cent.
Languages: Spanish, Mapudungun (Mapuche language), Rapa nui and other minority languages.
Main Political Parties: Concertation of Parties for Democracy (CPD), a coalition of the Christian Democratic Party (PDC), the Socialist Party (PS), the Democratic Party (PPD) and the Social Democratic Radical Party (PRSD); Alliance for Chile, an alliance between the National Renovation Party (RN) and the Independent Democratic Union (UDI); Together We Can Do More, an alliance between the Communist Party of Chile (PCCh) and the Humanist Party; Regionalist Action Party of Chile.
Main Social Organizations: The Central Workers' Union (CUT), was the main labor organization until 1973, when it was forbidden. Legalized in 1990, it became the most important union again in the early 90s; the Copper Workers' Confederation (CTC); the United Workers' Front (FUT); the National Labor Co-ordination Board (CNS). The indigenous movement has independent organizations such as the Action Group for the Bío Bío (GABB), the Mapuche Inter-regional Council (CIM) and the Mapuche Ad-Mapu Organization.

THE STATE

Official Name: República de Chile.
Administrative Divisions: 12 numbered Regions and the unnumbered Metropolitan Region of Santiago.
Capital: Santiago 5,478,000 people (2003).
Other Cities: Viña del Mar 356,800 people; Concepción 963,800; Valparaíso 888,300; Temuco 280,200 (2000).
Government: Michelle Bachelet Jeria, President since March 2006. Bicameral Legislature: the National Congress is formed by the Chamber of Deputies, with 120 members, and the Senate, with 38 members.
National Holiday: 18 September, Independence Day (1810).
Armed Forces: 78,000 troops (2003); 31,000 police; 50,000 reservists.

Under-5 mortality	Poverty	Debt service	Maternal mortality
8 per 1,000 live births 2004	**2%** of population living on less than $1 per day 2000	**24.2%** exports of goods and services 2004	**31** per 100,000 live births 2000

IN FOCUS

ENVIRONMENTAL CHALLENGES

Pollution caused by industrial and vehicle emissions reaches very high levels in Santiago. Native forests in the south were severely damaged by indiscriminate logging. The mining sector also poses a threat to the environment. Water pollution is caused by industrial waste.

WOMEN'S RIGHTS

In 1931 women over 25 were allowed to vote in municipal elections, but they were not given the franchise to vote in national elections until 1949. In 2005, only 15 per cent of deputies and 4.2 per cent of senators were women. On the other hand, female ministers make up 50 per cent of President Bachelet's cabinet. In 2003 women represented 35 per cent of the country's workforce. Female unemployment was 8.5 per cent, a percentage point higher than male unemployment.

CHILDREN

According to UNICEF, basic social indicators compare favorably to other countries in the region. In 2002, the Government launched 'Chile Solidario', a social protection system which focuses on assistance to the poorest households (56,055 in 2002 and 60,318 in 2003). The program aims to increase the time children spend in the classroom through measures such as expanding pre-school education, lengthening the school day and increasing the years of compulsory education from 9 to 12.

It is estimated that approximately 65,000 children between 12 and 17 years of age work (many of them as farm workers) or are seeking work.

INDIGENOUS PEOPLES/ ETHNIC MINORITIES

Indigenous groups comprise 4.2 per cent of the Chilean population. The Mapuche community, which includes the Pehuenche and the Huilliche, makes up 90 per cent of the indigenous population and is the most organized. About a third of them are scattered in urban areas and in reservations first established by the Government in the 17th century. However their claims for ethnic and cultural recognition have often set them on a collision course with government policies.

For 350 years the Mapuche resisted first Spanish settlers and then successive governments and the army. In 1553, indigenous people led by Lautaro were very close to invading Santiago while Spanish expeditionary forces and Governor Pedro de Valdivia were founding the southern city of Concepción. When Chile became independent at the beginning of the 19th century, the Government and the wealthy thought that the 'Mapuche problem' had to be solved. In 1881 after the genocide of these people by the army, the 'Araucanian territory' was recovered. In 1883, after a series of pacts, Mapuche people were confined to 300,000 hectares (they had owned 5,000,000 hectares) and during the Pinochet era lost a further 200,000 hectares, which have since been exploited by forestry companies and hydroelectric developments by private companies such as Forestal Mininco and Endesa.

The Mapuche communities on the south bank of Bío-Bío river have for many years been reporting the predation of their lands, destruction of their crops and the impunity granted to timber and electric energy corporations in the 200,000 hectares taken away during the military dictatorship (1973-1989).

Since the 1980s Mapuche communities have lost 60 per cent of their territory. In 1997 the Mapuche Community Union was created - in 1999 they marched on Santiago - to demand that the State give back their lands and prevent the companies' expansion on their territory.

MIGRANTS/REFUGEES

Before 1973, Chile sheltered thousand of refugees and political exiles fleeing from authoritarian governments or who wanted to give support to what they considered a unique socialist experiment. In the three years following the coup, foreign refugees who were unable to stay in Chile for political reasons were resettled in other countries by UNHCR. In September 1973 UNHCR opened an office in Santiago and a National Committee for Refugee Aid (CONAR) was created. Refugees from Chile's dictatorship dispersed and were granted asylum in about 110 countries.

There are no accurate figures about the number of exiled people during the military dictatorship; the Intergovernamental Committee for European Migrations allowed 20,000 people to leave for Europe in 1980. Other estimates put the total number of people exiled by the regime, either voluntarily or expelled, at no less 200,000 people.

DEATH PENALTY

It was abolished for common crimes in 2001, but still applies in exceptional cases.

---

[35] Brigadier Pedro Espinoza and retired General Manuel Contreras were sentenced to imprisonment for their parts in the murder of former foreign minister Orlando Letelier in Washington in 1976. Pinochet reiterated his support for the sentenced officers, but then called for respect for the civil authorities. Then, the Government suspended investigations into corruption charges against the former dictator's son. In a later hearing, during February 1998, Contreras stated that the true leader of the DINA (the political police during the dictatorship) was Pinochet himself.

[36] The Chamber of Deputies approved the trade agreement with Mercosur member nations by 76 votes to 26. Two right-wing parties the UDI and RN, came out against the agreement. The signing of the free trade agreement took place on 25 June 1996, establishing the 'four plus one' association between Argentina, Brazil, Paraguay, Uruguay and Chile. This formula was to remain in place until Chile became a full member of the agreement. The treaty with Mercosur came into operation on 1 October 1996.

[37] A report from the National Society of Farmers (SNA) said the agreement would mean annual losses of $460 million for the Chilean agricultural sector, but, balancing this, the country would become integrated into a market of more than 200 million people.

[38] Various studies in 1996 suggested the average economic growth of six per cent in the last 11 years had led to a reduction in 'extreme poverty' - but not in social inequality. Thus 20 per cent of the population still controlled 57 per cent of the national wealth, while the poorest 20 per cent handled only 3.9 per cent. During the first half of 1996, military and civil tribunals closed 21 cases of disappearances and extra-judicial killings involving 56 victims, without finding anyone responsible.

[39] Concertación retained the majority in the 11 December 1997 legislative elections, but lost ground to the Right. The Government coalition took 50.6 per cent of the vote (compared with 56.1 per cent in the 1996 municipal elections), while the right took 39 per cent (35 per cent in 1996).

[40] After passing on the post of Commander-in-Chief of the armed forces to General Ricardo Izurieta, Pinochet entered the Senate on 11 March 1998, amidst general indignation. After the Senate vetoed President Frei's proposal for a referendum on whether or not to eliminate the post of 'senator for life', (Pinochet's position), Frei called for a new popular consultation. At the same time 21 parliamentarians stated that the constitutional clause which gave former presidents a Senate seat for life should not apply to Pinochet, since he had not been elected.

[41] Pinochet's arrest in Britain in October 1998, following an extradition bid by Spanish judge Baltasar Garzón, deeply shook the Chilean political process. The Government and opposition united in calls for the General to be returned to Chile. But while the Right accused the Government of only making luke-warm efforts, Concertación suffered internally from having to defend the former dictator under the guise of defending national sovereignty. The legal comings and goings in London were accompanied by often violent demonstrations in Santiago both for and against the arrest.

[42] Confrontations between Mapuche and forestry planters in southern Chile led to the mobilization of an armed group of police officers that surrounded the village of Temucuicui in October 1999. The Government announced a plan to give the Mapuche economic, technical and educational assistance. However, Mapuche leaders saw this as a collection of old unfulfilled promises disguised as new. The Chilean Constitution offers no special treatment to ancestral communities.

[43] Ricardo Lagos, the socialist Concertación coalition candidate, won a slim victory in the January 2000 elections, which left the new Government very little room for maneuver. The right, which had received 49 per cent of the vote, promised an attitude of 'vigilant collaboration' with the President.

[44] Pinochet returned to Chile in March, after the British justice system considered that 'his bad physical and mental health' exempted him from being tried in Spain. Chilean Congress approved a constitutional amendment guaranteeing immunity to former presidents. In May, Government representatives (including the army) and the civil society established a Human Rights Committee to find the remains of the 'disappeared'. In August, the Supreme Court ruled that Pinochet had lost his immunity.

[45] In the first months of 2001, judge Juan Guzmán confirmed the charges of homicide and kidnapping against the former dictator and

 Malnutrition
1% under-5s
1996-2004

 Water source
95% of population using improved drinking water sources
2002

 Doctors
109 per 100,000 people
1990-2004

 Primary school
86% net enrolment rate
2004

confined him to house arrest for his involvement with the 'Death Caravan', a military squad that traveled through Chile by helicopter, executing 75 political prisoners at the beginning of the dictatorship.

[46] In July 2001, a Chilean court suspended the charges against Pinochet, claiming he was not able to stand trial since he was not 'in a state of mental capacity' enabling him to exercise 'efficiently the rights of due process'. In September of the same year, a US court began a trial against former secretary of state Henry Kissinger for his involvement in the plan leading to the murder of the Chilean general René Schneider in 1970. The Supreme Court of Chile approved a request from judge Guzmán to interrogate Kissinger about the death of the American journalist Charles Horman.

[47] On 21 February 2002, members of the US Congress requested attorney Roscoe Howard to present formal charges against Pinochet for terrorism and for his alleged responsibility in the murder of former Chilean Foreign Minister Letelier. The arrest, in the same month, of the former Minister

of Public Works, Carlos Cruz - an associate of President Lagos - together with the removal of five members of Parliament accused of corruption, further tarnished Chile's image as the least corrupt country in Latin America.

[48] Between August and October 2003 Chile signed a free trade agreement (FTA) with the US that came into force on 1 January 2004. This agreement removed tariffs from 85 per cent of Chilean export goods to the US. The rest of the tariffs were to be gradually phased out by 2014.

[49] In December 2003 retired general Manuel Contreras and another two chiefs from the DINA were convicted by judge Guzmán for their involvement in the disappearance of nine people during Operation Condor, the South American military regimes' program of repression during the 1970s. Within this framework, a new appeal to remove Pinochet's privileges was taken to the Appeal Court. It was based on an interview given by Pinochet in which he appeared to 'be both lucid and having a good memory'.

[50] In May 2004, Contreras was

sentenced to 15 years in prison for the disappearance and death of the journalist Daiana Aaron, in 1974. In July, the US Congress declared that the US bank Riggs had helped Pinochet to hide between four and eight million dollars while the former dictator was in London. Soon after, Pinochet faced charges of fiscal fraud, embezzlement and bribery.

[51] Declarations of Alejandro Toledo (Peruvian President) about the possibility of revising the maritime limits treaty with Chile (signed in the 1950s) caused uneasiness in Santiago. The Chilean army carried out military exercises (which according to the Government had been planned beforehand) based on the possibility of armed conflict with bordering nations. In August, a group of Parliament members traveled to the US to investigate Pinochet's accounts.

[52] Also in August 2004, photographs of a Peruvian military base in the Andean city of Arequipa were published in the press and contributed to heightened tension between Santiago and Lima. Defense Minister Michelle Bachelet stated that the location of the base was

widely known in military circles and did not pose any threat to national security. On 26 August, the Supreme Court of Chile removed Pinochet's privileges, thus allowing the prosecution of the former dictator.

[53] Pinochet was confined to house arrest in January 2005, as the court found him in good health to face the charges for murder and kidnapping. Former Defense Minister, Bachelet, won the first round of the elections in December 2005 but fell short of the majority she needed. In the second round she had to compete against the multimillionaire former senator Sebastián Pinera.

[54] In January 2006, Bachelet confirmed her victory in the second round and became the first woman to be elected president of Chile and the fourth consecutive head of State from a left-center coalition. The President took office in March and appointed women to half the posts in her Cabinet.

[55] In June 2006 hundreds of thousands of school students returned to school after a strike claiming that their action had won extra funding for education. ∎

Operation Condor versus the truth

IN THE 1970S AND 1980S, the military dictatorships that ruled the Southern Cone countries (Argentina, Bolivia, Brazil, Chile, Paraguay and Uruguay) implemented and applied a US-backed cross-border intelligence and repression plan called 'Operation Condor' or the 'Condor Plan'. This scheme aimed to destabilize the internal opposition movements who were fighting those regimes.

Lawsuits and demands for justice and for clarification of the fate of the disappeared have intensified in recent years, partly due to the shift to the left in those countries' governments.

In 2003, the Argentinian Government declared null and void the Due Obedience and Full Stop laws enacted in the 1980s and repealed a 2001 decree which prevented the extradition of human rights violators. Argentina heads the list of countries that committed crimes against humanity in the region, with 30,000 disappeared people according to human rights organizations. On the other hand, it is also the country which has made the most progress towards bringing former repressors to justice. Of the six countries that took

part in Operation Condor, Brazil and Uruguay are the only ones that have not brought any former head of a repressive dictatorship to trial until the present time.

In Brazil, the amnesty protected both opposition insurgents and the dictatorship's repressors who committed abuses, murders and tortures; most of the perpetrators having been identified. The Brazilian justice system attempted to investigate the deaths of members of the Communist Party, who were killed near the Araguaia river in northern Brazil between 1972 and 1974. However, the Government refused to declassify the relevant military documents.

In Uruguay, a 1986 law (*Ley de Caducidad* or Amnesty Act), ratified by a referendum in 1989, put an end to prosecutions of military and police officers accused of Human Rights abuses during the 1973-1985 dictatorship. The administration of Tabaré Vázquez (see Uruguay's History) resumed the search for the disappeared people in order to find out what had happened to them. In December 2005, the remains of two Communist Party activists who had disappeared during the dictatorship were

found. That year, for the first time, the courts called top army officers, including former dictator Gregorio Álvarez, to clarify their links with the disappearances and tortures that took place during the dictatorship. Up to then, no military officer had ever been tried, because of the Amnesty Act. However, Vázquez stated this law did not apply to former commanders of the Armed Forces.

The Chilean Government failed to repeal the amnesty law decreed by the Pinochet dictatorship covering the crimes committed from March 1973 until 1978. It was precisely during that period that most of the 3,000 disappearances and political murders were committed. However, when democracy was reinstated judges ruled that the amnesty law did not apply to cases of disappearances, on the grounds that these were ongoing crimes.

In April 2006 a Chilean judge prosecuted 18 people, including senior ranks from the secret police (DINA), for abuses committed during the dictatorship in *Colonia Dignidad*, an underground torture center and graveyard for opposition victims of Pinochet's regime. ∎

China / Zhonghua Renmin Gongheguo

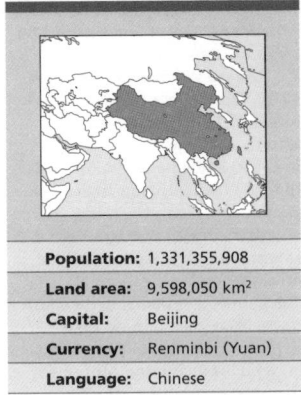

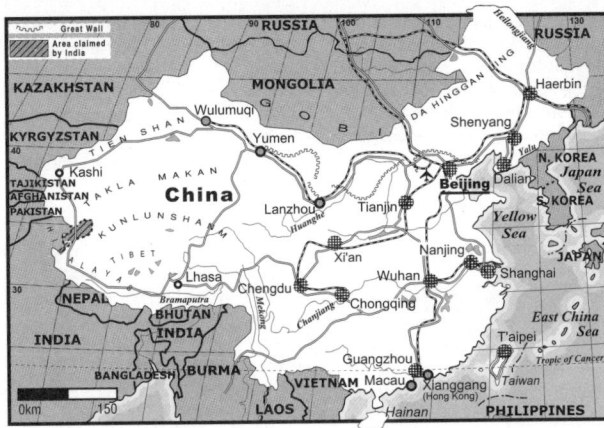

Population:	1,331,355,908
Land area:	9,598,050 km²
Capital:	Beijing
Currency:	Renminbi (Yuan)
Language:	Chinese

Chinese civilization is one of the oldest in the world, having reached unparalleled refinement long before Western cultures appeared. Its birth can be traced back to the 29th century BC, to the times of the mythical emperor Fu Hsi, in the Huanghe river basin, at the centre of present-day China.

2   The transition from a tribal to a feudal way of life took place between the 18th and 12th centuries BC, during the Shang dynasty. In that period iron replaced bronze, and new irrigation techniques were invented for agriculture.

3   By then, a cosmology that integrated opposites (life and death, micro and macro, material and immaterial, known and unknown, chaos and order, feminine and masculine, etc) as basic forces of vital cycles was deeply ingrained in Chinese culture. These notions were included in the I Ching (Book of Changes, 4th century BC) through the Yin ('the dark side', related to Earth) and Yang ('the light side', related to heaven) concepts.

4   The belief in the meaning of human existence as not being above that of any other living being, and as a product of an incomprehensible cosmic order, enabled the flourishing of a celestial mythology up to the beginning of the warring states period, between the 6th and 3rd centuries BC.

5   The Tao Te Ching, written by Lao Tzu in the 6th century BC, paid attention to human beings and their role as custodians of the laws of nature on Earth. Thus, it defined the precepts of an ethics of 'virtue'. A detailed analysis of human anatomy lay at the heart of its approach: the body's flow of blood would correspond to that of rivers, its 369 bone joints to the days of the ritual calendar, its five openings and senses to the basic elements of nature (water, fire, earth, wood and metal; air would be the essential source of energy). Its studies on the human body, biology, chemistry

and physics laid the foundations for the development of a medicine frequently, even today, more efficient and economic than that of the West.

6   Concepts within the Tao Te Ching ('The Way to Virtue') were interpreted and adapted several times throughout Chinese history. Power struggles between the kingdoms, typical of the warring states period, led learned man Kung Fu-Tzu (Confucius, 551-479 BC) to grant a social and political dimension to the ethics of individual submission. The stress Confucius put on respecting hierarchy and the preservation of order in every realm of social life (from the family to the highest political spheres), provided the ideological basis for the establishment, in the 3rd century, of a centralized and bureaucratic administration. The positive essence of Confucianism is present nowadays in popular beliefs, although relatively emptied of its sacred nature. In spite of several uprisings and foreign invasions, society was disciplined and means of domination were increasingly rationalized starting with the Ch'in dynasty (221-206 BC).

7   What would later become China was mostly unified for the first time in 221 BC, when the king of Qin (or Ch'in) state adopted the title of Shi Huangdi (First Emperor), a formulation previously reserved to deities or mythological emperors from sagas. The submission of the other six main states was mostly carried out through academic and legal advisors; centralization focused on the uniformity of legal codes, bureaucratic procedures, writing and coining systems, patterns of thought and study. A single system of writing through ideograms was created, and construction of the 5,000-km Great Wall was started to repel 'barbarians' from the north (it was reconstructed in the Sui, Jing and Ming periods). Many dissident Confucian scholars were banned or

executed, their books confiscated and burned.

8   Taocracy (defined as neo-Taoism because of Confucius' doctrinary reforms and because of Buddhist influence) expanded to all of modern China throughout the 2nd and 3rd centuries, when light paper was invented under the Han dynasty (206 BC-220 AD). The first reference to Buddhism in China is found in writings from 65 AD, which revealed the interest shown by the imperial family in meditation techniques from India. Cloth and porcelain were painted to represent the coexistence of the 'three branches of the Chinese tree': Confucius (the scholar of society), Buddha (the monk) and Lao Tzu (the ascetic).

9   Far from living in harmony, these three doctrinary branches and their extensions became the excuse for various separatisms and territorial claims. Buddhist monasteries did not pay taxes and became autonomous centres of economic and military power, mainly in Tibet. These privileges caused protests in dominated regions burdened by taxes.

10   Between the 3rd and 13th centuries, territorial break-up and penetration by Mongols and Turks was avoided (partially in the case of the latter) through successive divisions of the territory among the aristocrats. With sporadic reunifications, six dynasties came to coexist, with conspiracies that led the eunuchs (castrated servants and counsellors) to attempt a coup d'état in 835. Foreign empires were the subject of satire and called barbarian by several literary works of the period in which Chinese culture had its first golden age (between the 8th and 10th centuries - T'ang dynasty), with the expansion of paper currency.

11   The Turks dominated Central Asia between the 6th and 13th centuries and by paying taxes occupied lands in China's far west. They traded some Chinese goods (such as silk and tea) abroad and

introduced Islam to the region.

12   During the Song dynasty (1127-1279), in the south, the 'second golden age' of Chinese culture took place, when the printing press, the compass and gunpowder were invented. In 1206, Genghis Khan, advised by Tangut scholar Yeh-lu, invaded China and gradually imposed Mongolian domination until 1368. He created the Yuan dynasty and a system of dual government (Mongolian military administration allied with traditional Chinese bureaucracy and cultured aristocracy). China was finally conquered by his grandson, Kublai Khan.

13   Instead of fighting Chinese culture, Mongolians were overwhelmed by it and absorbed its doctrines (giving way to neo-Confucianism) and knowledge. In this period the Chinese made discoveries in astronomy, mathematics, navigation and weaponry, and had a first look at Christianity.

14   Tibetans maintained their autonomy under Mongolian power.

15   In 1368, amidst popular uprisings and a partially reassembled army, the Chinese overcame the Mongolians within their land and the Ming dynasty came to power until 1644. In order to control the economy the free movement of goods and individuals was banned, while Mongolians, Turks, Japanese and European sea expeditionaries, accused of piracy, were held back. However, in 1557 the Chinese granted the port of Macao to the Portuguese and allowed Europeans the use of some ports, although under strict conditions.

16   The northern Manchu took Beijing in 1644, taking advantage of the Ming dynasty's weakness, after several uprisings against large landowners, internal purges and its military defeat against Japan in Korea (1612). They founded the last Chinese dynasty, which fell in 1911.

17   The Manchu dynasty sought to consolidate its dominion in the continent's hinterland. In 1696 it recovered the protectorates of Tibet and Mongolia. This facilitated European penetration, interested in obtaining goods sought by metropolitan markets. In 1682, the Portuguese occupied Taiwan (Formosa), while the English and French attempted, at first, a peaceful approach.

18   French Jesuits reached the highest spheres of the Manchu dynasty and had an important role in the so-called 'third golden age' (1736-1796). Traditional academies were closed and the study of European Enlightenment ideas was imposed, especially with mathematics and astronomy that contradicted orthodox

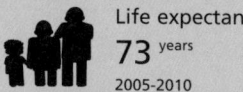
Life expectancy
73 years
2005-2010

GNI per capita
$1,500
2004

Literacy
91% total adult rate
2000-2004

HIV prevalence rate
0.1% of population 15-49 years old
2003

neo-Confucianism. In its effort to secularize culture, the Government promoted the proliferation of bookshops in the cities. But Chinese authorities continued to refuse trade with Europe.

19   In the latter half of the 18th century, British traders started to import opium from India, and were paid with silver, which was sent to London through the port of Canton. This enabled entrepreneurs to buy silk, tea and porcelain to be sold mostly in India, which allowed for the appearance of mafia groups in Canton. The widespread use of opium brought down the economy.

20   When China outlawed the opium trade in 1839, the British Crown declared war, sending 16 ships to attack Canton in 1840. London obtained the island of Hong Kong and five ports on the China Sea through the Nanking treaty (1841 - end of the First Opium War). An anti-European and anti-Japanese movement was formed in Canton, which led to a Second Opium War (1854-1860) that ended with Beijing (1860) falling to the English and French.

21   The Ch'ing Dynasty also suffered the Taiping political and religious rebellion (1851-1864), which was the largest civil war in Chinese history, devastating 17 provinces and killing 20 million people. Under the spiritual leadership of Hung Hsiu-ch'uan (God of Heavens), the Taiping practised a synchretic monotheism with influences from Protestant Christianity (individual relationship with God), neo-Confucianism (intolerance), and Taoism (extreme egalitarianism). They attacked the Manchu foreigners that ruled the country, formed an army of one million men with hungry peasants, workers and miners. Under the slogan a 'general treasure' (common property of land) they controlled part of southern China to carry out land reform and practised a type of primitive Communism of Taoistic inspiration. They were crushed with the help of Western troops.

22   The Chinese army was rapidly defeated in 1895 when the Japanese invaded Taiwan and part of the Korean peninsula.

23   An anti-Western rebellion broke out in 1898, headed by a secret society of Harmonious Fists or Boxers (with the unofficial support of Chinese authorities), but was crushed by a joint British, Russian, German, French, Japanese and US expedition. The victorious armies divided the country into regions of influence and demanded that China pay huge war reparations, as well as 'concessions' (of land) to put up factories near the Shanghai port.

24   Japanese victory over Russia in the 1904-1905 war awoke in China a call for constitutionalism, which the Imperial Court eventually supported. Sun Yat-sen, a non-aristocrat ignorant of Chinese culture, obtained a pragmatic education in Japan and the US. He used army differences to get economic support and proclaim the republic in 1912, heading a disorganized revolutionary movement headquartered in Tokyo.

25   The first Parliamentary elections (1913) were won by the Kuomintang (KTM; anti-Manchu) nationalist movement, which hindered negotiations between Yuan, the chief of Government (committed to the Manchu) and foreign banks to get loans for China's modernization. Chinese society, shaken by the civil wars, awaited the elaboration of a final Constitution. Yuan closed the Parliament in 1914 and established a dictatorship.

26   During World War One (1914-1918), Japan supported the Allied forces. In 1915, the Japanese secretly presented Yuan with the so-called 21 Demands that aimed to turn China into a Japanese dependency. Yuan could not uphold his initial refusal, due to a lack of external support. Japan obtained privileges and concessions in Manchuria, and demanded exclusive trade. Yuan died in June 1916.

27   Vice President Li Yang-hung became President and continued the pro-Japanese dictatorship until 1927.

28   In 1915, Chen Duxiu, a young man who had studied in Japan and France, founded 'New Youth' magazine. This was the main propaganda source against Yuan. After his death it spread the iconoclastic ideas of Beijing University's avant-garde. In May 1919, students marched against the Chinese Government's acquiescence to the decision of the Versailles Peace Conference, that year, to transfer the former German concessions; they organized strikes and boycotts against Japanese assets.

29   Many of these university students, inspired by Karl Marx (1818-1883) and the Russian Revolution (1917), belonged to the Socialist Youth League (with 3,000 members). In 1921, some of them founded the Chinese Communist Party (CCP), in a congress with the presence of a Kremlin representative. In the two ensuing years, the Communists carried out a propaganda campaign favoring a nationalist revolution among train and factory workers, and sent members to Moscow to study the Soviet revolutionary process.

30   The CCP had 300 members in 1923, while the KMT, reorganized under the leadership of Sung, its founder, was in process of consolidation and writing a constitution based on the Soviet model. That year, during the third congress, the CCP made an alliance with the KMT to join forces against the Government, which had the joint military support of the United States, Britain, France and Japan.

31   Moscow chose Chang Kai Shek (1887-1975), a KMT member, to organize an armed section of the Chinese revolutionary movement. Trained in the Soviet army, he was named commander of the Chinese revolutionary army in 1926. In 1927 his troops took the main cities. In Shanghai, conservative nationalist leaders and Chinese entrepreneurs convinced Chang Kai Shek to crush the General Workers' Union (controlled by the CCP, jointly with 700 other unions) and to expel the

PROFILE

ENVIRONMENT

The terrain of the country is divided into three main areas. Central Asian China, comprising Lower Mongolia, Sinkiang and Tibet, is made up of high plateaus, snow-covered in winter but with steppe and prairie vegetation in summer. North China holds the vast plains of Manchuria and Hoang-Ho, with extensive plantations of wheat, barley, sorghum, soybean and cotton (though China no longer has a surplus in farm products and has become a net importer). The North also has coal and iron ore deposits (Manchuria is the country's leading mining region). South China is a hilly region crossed by the Yangtse Kian and Si-kiang Rivers and has a warm, humid, monsoon climate. The country holds great mineral wealth: coal, petroleum, iron and non-ferrous metals.

SOCIETY

Peoples: There are 56 official recognized nationalities. Han (91.96 per cent), Chuang (1.37 per cent), Manchu (0.87 per cent), Hui (0.76 per cent), Miao (0.65 per cent), Uighur (0.64 per cent), Yi (0.58 per cent), Tuchia (0.50 per cent), Mongol (0.42 per cent), Tibetan (0.41 per cent), Puyi (0.23 per cent), Tung (0.22 per cent), Yao (0.18 per cent), Korean (0.17 per cent), Pai (0.14 per cent), Hani (0.11 per cent), Kazak (0.1 per cent), Tai (0.09 per cent), Li (0.09 per cent), other (0.51 per cent). **Religions:** No religion 59.2 per cent. Confucianism (a moral code, not a religion) combined with mystical elements from Taoism and Buddhism are what could be called the predominant 'beliefs'. Buddhist 6.0 per cent; Muslim 2.4 per cent; Christian 0.2 per cent; other 0.1 per cent.
Languages: Chinese (official) is a modernized version of northern Mandarin. Variants of this (many of which are mutually unintelligible) can be found in the rest of the country, the most widespread being Cantonese, in the south. Ethno-cultural diversity is reflected by the 205 registered languages.
Main Political Parties: The Chinese Constitution states that the Communist Party is the 'leading nucleus of all the Chinese people'. There are eight minor parties that participate in political life, including the Democratic Party of Workers and Peasants, the Revolutionary Committee of the Kuomintang, the China Democratic League and the League for the self-governance of democratic Taiwan. Hong Kong and Macau also have their own parties.
Main Social Organizations: Chinese Federation of Unions is the largest. Opposition groups of unknown size, based on ethnicity and religion, extending to the diaspora: movements in Tibet, Inner Mongolia, Xinjiang, as well as the Falun Gong group.

THE STATE

Official Name: Zhonghua Renmin Gongheguo.
Administrative Divisions: 23 provinces (including Taiwan), 5 autonomous regions, 4 municipalities and 2 special administrative regions (Macau and Hong Kong). **Capital:** Beijing 10,848,000 (2003). **Other Cities:** Shanghai 12,900,000 people; Tianjin 9,200,000; Xianggang (Hong Kong) 8,087,700; Shenyang 6,326,000 (2000). **Government:** Socialist Republic. Hu Jintao, General Secretary of the Communist Party, President since March 2003. Wen Jiabao, Prime Minister since March 2003. The 2,979-member National People's Congress (NPC) sits for about two weeks a year to ratify laws. NPC delegates are drawn from various geographical areas and social sectors (army, minorities, women, religious groups).
National Holiday: 1 and 2 October, proclamation of the People's Republic of China. **Armed Forces:** 2,555,000 (2003). Other: 1,200,000 Armed Peoples' Police, Defense Department.

Under-5 mortality
31 per 1,000 live births
2004

Poverty
16.6% of population living on less than $1 per day
2001

IN FOCUS

ENVIRONMENTAL CHALLENGES

Water pollution (mainly in the northern areas) is caused by industrial waste. Erosion and desertification, together with urban and industrial development, have caused arable lands to reduce. Less than 13 per cent of the land is still forested. The widespread use of coal as the main source of power causes acid rain, pollution and high carbon dioxide emissions (China is second only to the US in terms of CO_2 emissions). The Three Gorges dam project (the world's largest) seriously threatens the ecology in the affected areas.

WOMEN'S RIGHTS

Chinese women have been able to vote and run for office since 1949. In 2003, 20.3 per cent of members of Parliament were women. Women make up 45 per cent of the workforce, which totals 773 million people. In 2004, the adult literacy rate was 87 per cent for women and 95 per cent for men, while the primary school attendance rate reached 99 per cent for both sexes*.

The fertility rate in 2004 was 1.7 children per woman*. Between 1990 and 2005, maternal mortality dropped from 89 to 53 per 100,000 births. In 1979, due to the extraordinary growth rate of its population, which was nearing 1,000 million, the Government established a harsh family planning policy called 'one family, one child'. It fined parents who had more than one child, increased the legal age for marriage and demanded the use of contraceptive methods such as IUDs, abortion and sterilization. This policy was effective in reducing population growth (1.6 per cent between 1970 and 1990, 0.9 per cent between 1990 and 2004), but it also promoted the hiding and killing of millions of female children before or after they were born.

Preference for boys is traditional in China. Some studies suggest that a reduction in the number of women was reported in the chronicles of Imperial China and the republican era. Due to the present limitation of one child, many couples only have one legal chance to have a son and resort to ultrasound machines to identify the unborn and abort the girls. In rural areas - where approximately two-thirds of the people live - the program allows for a second child, but if the firstborn is a girl, frequently the second girl is aborted or abandoned once born in order to have a boy.

CHILDREN

Opening up to global market forces could increase inequalities and put pressure on the poor. Meanwhile, it is difficult to measure the impact of recent disasters such as the SARS epidemic and natural catastrophes - floods, blizzards, earthquakes - to which children are always most vulnerable.

The main causes of death for children under 5 are pneumonia, diarrhea and injuries. Child malnutrition occurs mostly in rural areas: severe malnutrition is three times worse in rural than urban areas; in the cities, obesity is a problem.

There are major regional differences in economic development. The country has 30 million poor, mostly in western China. Child and maternal death rates are notoriously higher in the west than on the coast.

It is estimated that currently there are one million people living with HIV/AIDS in the country*. The Academy of Preventive Medicine of China estimates there will be 7 million people with HIV in 2005, and between 10 and 15 million in 2010. The epidemic is transmitted mostly through heterosexual contact and blood transfusions. It is estimated that 150,000 children live and work on the streets.

INDIGENOUS PEOPLES/ ETHNIC MINORITIES

Hui Muslims - of mixed Arab, Persian and Chinese descent, tracing back to the middle of the 8th century, when Arab and Persian merchants came to China - make up 0.7 per cent of the population (8.7 million people) and live mostly in Ningxia and the provinces of Gansu, Qinghai, Henan, Hebei, Shandong, Yunnan and Xinjiang. The Chinese Government continues to allow the Hui to bury their dead in Muslim cemeteries - while all Han must be cremated - and exempts them from many of the restrictions imposed by the birth-control program.

Government interest in the development of western provinces could contribute to improving the group's status. They lack any political representation. In the past there have been conflicts between different Sufi sects among the Hui, between the group and the Chinese State and between Hui and Han groups. The Hui do not seek autonomy: their major concern lies in the restriction of religious practices, economic backwardness and environmental deterioration in Ningxia. More than 5,000 Chinese make an annual pilgrimage to Mecca; the first Islamic university started to operate in the country, and Xian has a system of Islamic education for children over 4 years old.

Tibetans - who have lived in the region since127 BC and whose empire peaked between the 7th and 10th centuries and collapsed between the years 824 and 1247 - make up 0.44 per cent of the population (6 million people). Half live in the Autonomous Region of Tibet and the rest in the neighboring provinces of Qinghai, Gansu, Sicuani and Yunnan. They are highly organized and integrated as a group, with strong transnational support. In spite of Tibet's economic development the potential for rebellion persists, while China keeps limiting their cultural and religious activity, refuses any direct talks with their leader, the Dalai Lama, and insists on ignoring the Dalai Lama's request for regional autonomy.

Turkmen (Uigurs and Kazakhs; Turkic minorities that arrived in China in the 8th century) make up 0.7 per cent of the population (8.658 million people). Uigurs (who are Muslim) live mostly in the Xinjiang region, which they call East Turkestan, and suffer religious and economic discrimination, in addition to receiving in their territory a constant influx of Han groups. In the 1930s and 1940s, the Uigur moved to neighboring countries, but when the Soviet Union collapsed, they returned to Xinjiang (there are exiled communities in Afghanistan, Kyrgyzstan and Turkey, from which they receive help).

Throughout 2002, China kept repressing the Uigurs. An unknown number of Uigurs moved to Kyrgyzstan, Kazakhstan and other Central Asian countries, although most failed to find security. The United Nations Security Council added the Islamic Movement of East Turkestan - one of numerous separatist groups - to a list of terrorist groups, responding to a request from Beijing, with US support. However, Amnesty International has stated that China uses anti-terrorism as an excuse to suppress the Uigurs.

MIGRANTS/REFUGEES

Between 1999 and 2001 China was among the top 10 countries of destination for refugees. In 2002 it held more than 396,000 refugees and asylum seekers; most (296,000) came from Vietnam (mostly ethnic Chinese), having left that country in 1979 during the China-Vietnam war and settled in the southern provinces, while some 100,000 were from North Korea (although some NGOs estimate this figure to be as high as 300,000). Tens of thousands of North Koreans were repatriated during 2002, but thousands more kept coming to China. An unknown number of Kachin refugees from Myanmar were in the province of Yunnan.

Beijing kept committing human rights violations in Tibet during 2002 - repressing political dissidents and religious activity - while 2,000 Tibetans entered Nepal, where the UNHCR helped move them to India. An unknown number of Chinese asylum seekers, mostly from the province of Fujian, fled by ship to Canada, Australia, Japan and the United States. Several asylum seekers paid organized smugglers to take them, frequently in unsafe ships. To get to Europe they paid between $10,000 and $15,000 per person; to get to the US they paid some $30,000. Although some receiving countries regard them as economic migrants, most asylum seekers say they were persecuted, because of the one-child policy or for being members of the Falun Gong spiritual group. In the last decade there has been intense Chinese emigration that has added to the Chinese Diaspora, amounting to between 30 and 50 million people. In 2006, the Government was still trying to find a permanent solution for these refugees. It is estimated that there are approximately between 100,000 and 300,000 Koreans living in China.

DEATH PENALTY

It applies to all kinds of offences. Official reports say that 726 people were executed in 2003, but the real number is estimated to be much higher.

*Latest data available in *The State of the World's Children* and *Childinfo* database, UNICEF, 2006.

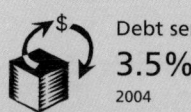

Debt service

3.5% exports of goods and services
2004

Maternal mortality

56 per 100,000 live births
2000

CCP from the Central Committee.

[32] Between 1927 and 1937, the Chang Kai Shek Government had not been able to control the country nor organize its economy. Communists formed 15 agrarian bases in central China, defending them with their armed section (the Red Army, made up of peasants). In 1934 they advanced westward, taking land along the way (the Long March). In 1931 the Japanese occupied Manchuria; uprisings took place in several regions.

[33] In 1935 the CCP Central Committee chose Mao Zedong (1893-1976), one of the party's founders, as Secretary General, while promoting the struggle against the Japanese as a first step to power. Thus, it concluded secret agreements with the KMT.

[34] In 1937 both parties formed a United Front against the Japanese invasion in July. Never before had the Chinese army been so large (1,700,000). Agreements between the KMT and the CCP were broken several times until the Japanese withdrew in 1945, defeated in World War Two (1939-1945).

[35] The fight against the Japanese Empire enabled growth in the CCP, which was excluded from the Government, while Chang Kai Shek received financial aid from the US in 1946 to counter high inflation. By then, the KMT armed forces were larger than the Red Army's, but two years later, Mao Zedong had recruited more than 500,000 peasants willing to defend the land reform initiated by the CCP in several regions.

[36] On 1 October 1949, after defeating the KMT, the communist leadership proclaimed the People's Republic of China and established a one-party system, similar to the Soviet Union's. The remnants of the KMT Government and army moved to the island of Taiwan. There, Chang Kai Shek claimed to be the sole legitimate authority of China and, with US support, laid plans to reconquer the mainland.

[37] As leader of China Mao faced the triple challenge of ensuring a transition from stagnation to economic growth, from social disintegration to discipline and from military to civilian rule. In December 1949 Mao made an official visit to Moscow and signed a treaty of friendship, alliance and bilateral assistance with Joseph Stalin (1879-1953).

[38] The Government started by instituting land reform to the rest of China and nationalizing foreign property. Massive health and education plans were implemented. Most of the population was illiterate, partly because of the complex ideograms, so writing was simplified. Female prostitution was banned.

[39] In 1950, the year after the CCP came to power, the Red Army faced the US in Korea. Although this gave Mao the support of the Chinese population, it forced him to pay less attention to domestic needs.

[40] The first 5-year plan adopted by the CCP Central Committee, with technical and financial support from the Soviet Union, was launched in 1953. The process of land collectivization and the socialization of industry was begun, while business activity was regulated. The State gave peasants, which made up 80 per cent of the population, small lots of land and their means of production. This allowed them to trade their surpluses, while industrial production was left to the State.

[41] Although China kept receiving financial aid from the USSR (allowing it, among other things, to build the atomic bomb in 1958), after Stalin's death relations between the countries deteriorated. Mao deemed it convenient for China to adopt a more neutral role in the Cold War (1950-1991) and to create a third pole of hegemonic power in world geopolitics.

[42] In the 1950s, Nikita Khruschev (1894-1971) started the USSR's 'destalinization'. In 1958 Mao implemented an opposite domestic policy, implementing terror to restructure the economy and increase production, whose growth rate was below that of the population. These procedures caused 20 million deaths and were carried out until 1961 in what Mao called the Great Leap Forward.

[43] Demonstrations for civil rights in Tibet in 1959 were met with severe repression by Chinese authorities and forced the Dalai Lama (the Tibetan Buddhist leader) to flee to India. This caused a row between both countries over Tibet's sovereignty.

[44] In 1962 Mao delivered a public self-criticism of his economic policy mistakes and was replaced by Lao Shaoqi as chief of State, but kept on as head of the Party and had the support of the People's Liberation Army. In spite of this self-criticism, Mao became increasingly radical and accused Lao and his minister Deng Xiaoping (1904-1977) of 'revisionism' (reformism), for favoring a fledgling free trade in the cities.

[45] In 1963 Beijing accused the Soviet Union, jointly with the United States and Britain, of conspiring against it, and broke off relations with Moscow. In 1965, the Red Army installed troops in Cambodia.

[46] Throughout 1965 orthodox Maoists became radicalized in opposition to 'revisionist' sectors. In 1966, the army and young students of the Red Guards followed Mao

in what they called the Great Proletarian Cultural Revolution (1966-1969). Radicalized by Mao and his *Little Red Book*, they launched ideological persecution in all levels of society. Lao Shaoqi was murdered and a civil war was unleashed, with millions of people killed or confined to 'rehabilitation' camps.

[47] Mao feared that a Soviet invasion would interrupt the Cultural Revolution. His foreign policy focused on stopping the USSR from extending its area of influence to countries that were starting revolutionary processes. This led him to finance and oversee pro-Chinese parties and governments in Asia, Africa and Latin America. The covert control he had over Albania and the founding of the Non-Aligned Movement (1961) were the pillars of his 'third way' policy.

[48] In 1968, the Red Guard, used by Mao in his increasingly intense confrontations with the CCP's reformist sectors, withdrew its support. In 1969, Chou Enlai was named prime minister.

[49] In 1971, the UN admitted the Communist Government to replace Taiwan as representative of China. This was possible due to US abstention in the vote. Chou Enlai started talks with Washington, stating his intention of beginning a gradual process of industrial modernization using foreign capital. In 1971 Deng Xiaoping (restored to the Central Comittee) planned the future opening of the so-called Special Economic Regions near Macao and the world financial hub of Hong Kong.

[50] US President Richard Nixon travelled to Beijing in 1972 to attend a ceremony marking the establishment of diplomatic relations with China.

[51] In 1975, after Mao's death, reformist bureaucrats took control of the CCP, and the so-called Gang of Four (which included Mao's widow), were arrested and charged with conspiracy. They became scapegoats for the failures and excesses of the Cultural Revolution. In 1981 they were put on public trial, televised and broadcast throughout the world.

[52] The first Special Economic Regions opened in 1978. Their lands were rented at low cost and their activities were tax-free.

[53] Between 1978 and 1979 the Government implemented a contract system of employment replacing the lifetime assignment to a production unit. The State appropriated lands given by Mao to peasants, who became tenants. Production quotas were replaced by taxes, and peasants were permitted to sell their surpluses for cash.

[54] Small businesses were allowed and price subsidies of consumer

goods were gradually removed. Likewise, more decision-making power was passed to plant managers, in matters such as hiring or dismissing workers. The social security system was gradually phased out and control on labor activity was regulated, breaking international agreements.

[55] The Deng Xiaoping Government rehabilitated victims of the Cultural Revolution, greater expression was tolerated, and censorship on music, dress and other cultural goods was lifted. The education system was redirected toward technical and scientific specialization, according to world standards. A Constitution was passed in 1982 that entails, in the long term, the institution of the right to property.

[56] Between 1982 and 1985, the policy called by Deng Xiaoping 'one country, two systems' caused inflation and encouraged corruption within the CCP. In 1984 the Government authorized 14 more Special Economic Regions. In 1985 it implemented a 'rectification' campaign within the CCP, excluding corrupt officials, others who opposed economic reforms and still others who wanted more civil and political rights.

[57] In that period the Government launched a birth-control plan called 'one family, one child', which included the widespread practice of abortion and sterilization. Spouses had to have a minimum age and request their bosses' permission to get married.

[58] The reduction of purchasing power, linked to price hikes, caused popular discontent. The people could not protest through organized labor, therefore expressed themselves through numerous student marches between 1986 and 1989. Hu Yaobang, Secretary General of the CCP and in favor of political reforms, became a symbol of the struggle for democracy, after his expulsion from the Party in 1986 during the 'rectification' campaign.

[59] The death of Hu Yaobang in April 1989 served as a pretext for massive protests of students and workers in a dozen cities, which did not stop until June, when army tanks were sent into Tiananmen Square (Gate of Heavenly Peace square) in Beijing. The image of tanks crushing demonstrators outraged Western public opinion. In the three days that ensued repression was unbridled, leaving hundreds dead, injured or jailed and subject to torture.

[60] That year, police in Tibet opened fire on demonstrators protesting against cultural and religious persecution and demanding greater political rights. Widespread rioting resulted in

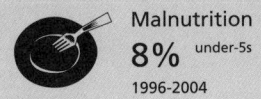
Malnutrition
8% under-5s
1996-2004

Water source
77% of population using improved drinking water sources
2002

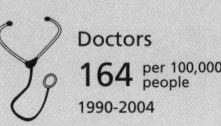
Doctors
164 per 100,000 people
1990-2004

Tibet

Little is known about the origin of the Tibetan people, the descendants of warring nomadic tribes known as Qiang (Chiang). The first known religion in the region was Bon, which combined a belief in gods, demons and ancestral spirits who were responsive to priests or shamans. Chinese Buddhism was introduced in ancient times -the first scriptures date back to the 3rd century AD - but mainstream Buddhist teachings came to Tibet from India in the 7th century. The blend of both produced the particular Lamaist Buddhism of the region and its many different sects.

[2] When the Yarlung Empire spread throughout Central Asia, during the 8th century, Tibet controlled the Silk Road and received taxes from the Tang empire in China. In the 13th and 14th centuries, the Mongol empire conquered China and accepted Tibet's submission, without invading it, before conquering the majority of Eurasia. The Tibetans developed a phonetic alphabet about AD 600 and, after centuries of rivalry, a theocratic kind of feudal state was established in the early 10th century. Both political and religious powers were conferred on the lamas - Tibetan priests divided into sects with a complex hierarchy - who, as the ruling class, controlled the peasants and the produce of the land.

[3] In 1247, Köden, the younger brother of Güyük Khan, symbolically invested the Sa-skya lama with temporal authority over Tibet. Kublai Khan, grandson of Genghis Khan, appointed the lama Phags-pa as his 'Imperial preceptor'. The politico-religious relationship between Tibet and the Mongol Empire was characterized as a personal bond between the Emperor as patron and the lama as priest. For one century, many Sa-skya lamas, living at the Mongol court, became viceroys of Tibet on behalf of the Mongol emperors.

[4] During the Chinese Ming dynasty (1368-1644), Tibet was ruled independently by the Tibetan Pagmodru, Rinpung and Tsangpa dynasties. In the 17th century, China was reconquered by the Manchu, while Tibet was ruled between 1642 and 1682 by the fifth Dalai Lama. The country was gradually demilitarized and, to avoid having to keep an army, a protective alliance was negotiated with the Manchu Emperor around 1650.

[5] The assassination of two Chinese high commissioners in 1751 brought an immediate and bloody response from the Manchu dynasty: the Emperor sent a military expedition to Lhasa. From that time onward, the Dalai Lama's relationship with China became more difficult than that of the Panchens - the other leaders of the religious hierarchy. Worldly competition between the two heads of Lamaist Buddhism often fostered division and sectarianism.

[6] In 1910 Chinese troops entered Lhasa. The Dalai Lama appealed to the British to help expel them - and was refused assistance. But the Chinese empire was in its death throes and fell to the Nationalists in 1911. Tibetans seized the chance to expel the invaders and in June 1912 the Dalai Lama proclaimed Tibetan independence. About ten years later, disagreements between the Dalai Lama and the Panchen Lama ended in the flight of the latter to Beijing. A boy born of Tibetan parents about 1938 in Tsinghai province, China, Bskal-bzang Tshe-brtan, was recognized as his successor by the Chinese Government and brought to Tibet in 1952. Eventually, he entered Lhasa under Communist military escort and was enthroned as head monk of the Tashilhunpo Monastery.

[7] China invaded Tibet in 1950, the year after the Communist revolution led by Mao Zedong. Most of the Tibetan artistic, literary and architectural treasures were destroyed by the occupiers. Before the invasion there were some 6,250 Buddhist monasteries in Tibet; in 1979 only 13 remained. Monks and nuns were detained in concentration camps, murdered or simply forced to leave monastic life.

[8] The Dge-lugs-pa (Yellow hat sect) ruled political and religious life from the 17th century until 1959, when they launched a failed uprising against the Chinese Communists, who had abolished the feudal system, created the first 'Communist' communes and fought against the Tibetan religious system.

[9] When the 1959 uprising failed, the Panchen Lama remained in Tibet while the Dalai Lama went into exile in India. From there he lobbied against the occupation and advocated a return to traditional society. Since the Panchen Lama refused to denounce the Dalai Lama as a traitor, the Chinese Government jailed him in 1964. He was freed in the 1970s and died in 1989.

[10] In 1994 Chinese and Tibetan religious authorities declared a five-year-old Tibetan child Panchen Lama, but the Dalai Lama did not recognize the nomination. In late 1999, the 14-year-old Buddhist leader Karmapa Lama managed to flee Tibet to India, meeting with the Dalai Lama. This heightened hopes abroad for independence in Tibet, and India was concerned over possible confrontation with Beijing.

[11] Tibet - presently one of the five Chinese autonomous regions - had a mostly Tibetan population until a large Han migration was encouraged by Beijing. The Han form the Chinese majority. Tibetans are no longer in a majority in the capital Lhasa. Furthermore, the region is settled by the Hui (Muslim Chinese), the Hu and the Monba, amongst others.

[12] It is estimated that 1.2 million Tibetans have fled the region since Chinese occupation.

[13] In November 2000, a delegation from the European Union approached the Chinese leader Li Peng with the proposal that the Dalai Lama be appointed governor of Tibet. This idea was rejected on the grounds that he would have to renounce his Tibetan citizenship and become Chinese before this could happen.

[14] The UN General Assembly has not implemented the resolutions adopted on Tibet in 1959, 1961 and 1965. The Tibetan Youth Congress, one of the most radical groups fighting for freedom in Tibet, started a hunger strike in front of the UN headquarters in New York, on 2 April 2004, demanding that the UN stop the execution by the Chinese authorities of a Tibetan religious leader and investigate the situation of the Panchen Lama, the second most important figure in Tibetan Buddhism. He is detained by the Chinese authorities in an undisclosed location.

[15] China has implemented a surveillance system in Tibet's internet to control the activity of the occupied territory's activists on the web, according to the International Campaign for Tibet (ICT).

[16] The first railroad line between the Chinese city of Qinghai and Lhasa was opened in October 2005.

[17] In January 2006 the National Audience of Spain - based on a decision by Spain's Constitutional Court allowing that nation's courts to try crimes against humanity committed in other countries - declared itself capable of trying an alleged case of genocide in Tibet. The case involved several Chinese leaders, including former president Jiang Zemin. Since the invasion, Tibetan activists have accused the Chinese Government of causing the deaths of thousands of people and the destruction of the region's religious legacy. ■

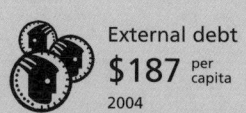

External debt
$187 per capita
2004

Imports **(millions)**
$606,543 goods and services
2004

martial law, lifted only in April 1990. Tibetan exiles reported detentions and several executions. Tibet had been invaded and annexed to China in 1950 and converted into an autonomous region in 1965.

[61] The Tiananmen events caused political changes. Li Peng became Prime Minister and Beijing was briefly estranged from the West. In September 1991 Britain was the first to send representatives to the capital to sign an agreement on the construction of a new airport in Hong Kong, as part of the negotiations to return the British enclave to Chinese sovereignty in 1997.

[62] That year, after the breakup of the USSR, China re-established diplomatic relations with Moscow and Hanoi.

[63] In late November, the Government freed student leaders from the Tiananmen demonstrations. The following year, Amnesty International reported there were 20,000 political prisoners.

[64] At the same time, Deng Xiaoping, aged 87, was called to lead international financial negotiations in an attempt to counter the deficit left by the estrangement of foreign investors following the Tiananmen episode.

[65] In 1992 the People's National Congress decided to maintain Jiang Zemin as Secretary General of the CCP and appoint him both President of the Republic and commander of the Armed Forces, becoming the first person since Mao to combine those functions. Prime Minister Li Peng was confirmed in his post.

[66] That year it was announced that the giant Three Gorges dam would be constructed with foreign funds. The construction would be finished in 2009 and would flood 10 cities and more than 800 towns. Hydrologists, seismologists, economists, human rights defenders and activists against economic globalization opposed the project, saying it would destroy the habitats of endangered species and leave millions of people exposed to earthquakes, landslides and floods.

[67] In September, the Government said pro-independence action in Tibet would be 'implacably repressed'. An austerity plan was launched and taxes were increased on the rural population. A series of protests and demonstrations forced the Government to lift the measures a few months later.

[68] In 1995 the Government maintained subsidies for state enterprises to stem unemployment which had affected 30 million workers in the sector without unemployment benefits.

[69] That year, the CCP first

secretary in Beijing, Cheng Xitong, was forced to leave his post for misappropriation of funds, along with important local leaders. In April, the deputy-mayor of Beijing, Wang Baosen, committed suicide, accused of having embezzled $37 million of Government funds.

[70] In 1996 Amnesty International condemned the Chinese repression of Buddhist monks in Tibet. According to AI, 80 monks were injured for refusing to respect a ban on the public exhibition of pictures of the Dalai Lama. That year, two student leaders of the 1989 uprising were sent to prison for 11 and three years, accused of promoting the overthrow of the Government.

[71] In 1997 China recovered its sovereignty over Hong Kong and named it a Special Administrative Region. Several foreign companies were authorized to convert local money into dollars or yen. A few months earlier, Jiang had made the first-ever visit of a Chinese President to South Korea.

[72] Following the death of Deng Xiaoping in February 1997, the 15th CCP Congress confirmed Jiang Zemin as leader, reaffirming the political system. In March 1998 the People's National Congress ratified the changes decided by the CCP, by re-electing Jiang Zemin as Head of State and commander of the armed forces, with 98 per cent of the vote.

[73] Hu Jintao, mentioned as a possible successor to Zemin, was elected Vice-President, while outgoing Prime Minister, Li Peng, became Head of Parliament. The Constitution prevented him from serving a third term as Head of Government. Zhu Rongji, former deputy Prime Minister in charge of the economy, was elected Prime Minister. The new cabinet, made up mostly of economic experts, faced the preparation of the 370,000 State companies for free market rules.

[74] In June 1999, the Government issued an arrest warrant on militants of the 'exercise and meditation movement' Falun Gong and asked Interpol to arrest Li Hongzhi, the man who founded the sect in 1992 before emigrating to the United States. At the same time, the authorities destroyed more than one and a half million books on the group's beliefs. This confrontation with Falun Gong - seen by the Chinese leadership as the greatest threat to the Government since the demonstrations in 1989 - started after the group staged a silent protest against Government hostility in April when 25,000 of its supporters demonstrated opposite Jiang Zemin's residence.

[75] China, which had signed in

1992 the Non-Proliferation Nuclear Treaty, added in November 1999 100 missiles to its southeastern coast facing Taiwan. In 2003 China had 300 nuclear warheads, half the number held that year by the European Union.

[76] In December 1999, Macao was reunified with China, and became a Special Administrative Region, at least for the next 50 years.

[77] From 1996 to 2001, the CCP added 7.5 million members under 45 years of age to its ranks. Although Chinese leaders asserted that these figures were a reaffirmation of Marxist and revolutionary ideologies among young people, other commentators think that these new members have been motivated by the possibility of making connections in the business world.

[78] Jiang Zemin and the Russian President, Vladimir Putin signed a friendship treaty between their countries in July 2001. The treaty stated that neither country had territorial claims on the other. Voices of concern were raised in Russia about the potential for Chinese immigrant incursions in Russia's semi-desert areas of the far east and in the Siberian regions, which China claimed from Moscow during the Cold War. In December 2001, China was admitted to the World Trade Organization (WTO).

[79] Beijing, which adopted anti-terrorist measures demanded by Washington after September 2001, abstained, in the UN Security Council, from declaring itself in favor of the invasion of Iraq in 2003.

[80] In March 2003, the central Committee of the CCP chose Hu Yintao as President and Wen Jiabao as Vice President. In a closed-door agreement, India and the new authorities of China agreed in June on a new status for Tibet, which the press of both countries hailed as a 'triumph', although it was actually a meeting to make all kind of agreements, such as increasing trade and building a highway between the two countries. According to Indian authorities, China recognized Indian sovereignty over Sikkim; according to the Chinese, India recognized Chinese sovereignty over Tibet.

[81] In June 2003, the floodgates of the Three Gorges (see In Focus) dam were closed to fill its reservoir. That same month the Government announced that the two largest state banks (Bank of China and Construction Bank of China) would be turned into corporations in 2004, and that private insurance companies would be allowed to operate.

[82] In 2003, Chinese GDP grew twice as much as that of the US. Based on these figures, US investment bank Goldman Sachs

estimated that, if that growth rate were to be maintained, China would become the world's strongest economy in 2040.

[83] In December, the explosion of a mixture of natural gas and hydrogen sulphide in a *China National Petroleum Corporation* plant near the southwest city of Chongqin killed more than 250 people and caused the evacuation of over 60,000.

[84] Following a policy of gradual reform for Hong Kong, the Government announced in April 2004, that Hong Kong representatives for the 2007 elections would not be directly elected. In September, Jiang Zemin announced his resignation ('for the party's sake', he declared) to his post as commander of the armed forces. President Hu Jintao took his place.

[85] In March 2005, China's National People's Congress passed an anti-secession law which directly affected Taiwan's plans for rewriting its Constitution and attaining international recognition. That same month the president of Hong Kong, Tung Cheehwa, announced his resignation.

[86] In April, thousands of Chinese protesters demonstrated against the publication of Japanese textbooks which whitewashed the crimes committed by the Japanese army during World War II. The tension was increased by Japan's campaign for a seat in the UN Security Council and exploitation permits for the East China Sea. Also in April, the head of Taiwan's nationalist Party traveled to China for the first meeting between nationalist and communist leaders since 1949.

[87] China and Russia carried out their first joint military exercises in August 2005. In November, an explosion in a chemical plant poisoned the waters of the river Songhua and shut off the water supply for millions of people.

[88] In May 2006, the Chinese Government demanded again that the US turn over five of their Muslim citizens, who had been held in the miltary prison at Guantanamo Bay for five years. According to the US, the Chinese citizens held at Guantanamo were terrorist suspects; however, Washington refused to turn them over, citing fears that the prisoners might 'face prosecution' in China. Finally the five men were taken to Albania as refugees, redirecting Beijing's criticism towards Tirana's Government.

[89] In September 2006, Chen Liangyu, the top political figure in China's richest city, Shanghai, was sacked following his involvement in a multi-million-dollar corruption scandal. ■

Exports **(millions)**
$655,827 goods and services
2004

Received aid
$1 per capita
2003

Hong Kong (Xianggang)

Hong Kong (Xianggang) island was ceded to Britain 'in perpetuity' in 1842, when the British attacked China in the first Opium War. Eighteen years later, the British gained the rights over Kowloon, the mainland peninsula facing the island. In 1898, the British forced the Chinese to give them a 99-year lease on the rural zone north of Kowloon, known as the 'New Territories'.

[2] Hong Kong was initially used as a trade center, constituting a point of entry to China. But in the 1950s, following the Communist victory in China, the United States and Britain imposed a trade embargo. Hong Kong was forced to import all its basic goods from overseas, therefore having to develop exports, rapidly transforming itself into a light industry center, exporting textiles, clothing, plastic and electronic goods. Just as in Taiwan and South Korea, this development was generously supported by Western powers interested in these 'bastions' of the Cold War.

[3] Growth of trade and the export industry converted Hong Kong into a financial, communications and transport center. Government policy also contributed, setting low taxes and minimal customs tariffs, offering trustworthiness and freedom in the movement of capital.

[4] In the late 1970s, China announced a stage of liberalization toward foreign trade and investment, and Hong Kong - armed with one of the best natural ports in the world, sophisticated investment and trade systems, as well as large terminals for containers - was in a position to take advantage of the situation.

[5] Almost all the population was of Chinese origin, having arrived in successive migration waves, particularly after the establishment of the Republic of China in 1912.

[6] London and Beijing began negotiations on the future of the colony in the early 1980s, as the 99-year lease was due to end in 1997. The people of Hong Kong were not represented in these discussions.

[7] Agreement was reached in 1984, giving China sovereignty over the entire territory, but allowing Hong Kong 'a high degree of independence' as a Special Administrative Area.

[8] In 1991, the people of Hong Kong elected the members of the Legislative Council for the first time in 150 years.

The United Democrats of Hong Kong (UDHK) candidates - critical of the colonial government of Hong Kong and of China and calling for a strengthening of democracy - won most of the seats.

[9] British Governor, Chris Patten, reformed the electoral system, totally separating the executive and legislative powers. China said this reform contradicted the principles of the Basic Law.

[10] In 1994, Patten proposed a plan to increase the number of voters in the 1995 elections, causing further confrontations with Beijing. In September 1995 the Democrat Party, opposed to the official Chinese interpretation of the Basic Law, triumphed in the Legislative Council elections.

[11] In July 1997, China recovered control over Hong Kong. Tung Chee Hwa was appointed head of the new Executive branch with the support of a Legislative Council.

[12] According to the new law, Hong Kong would retain its rights and liberties, its legal independence and nature as an international financial and trade center, along with its way of life, for 50 years. Beijing would handle defense and foreign relations.

[13] Reunification put China in a 'one country, two systems' situation, combining the free market economy of Hong Kong with rigid political control over the rest of the country.

[14] The transfer took place against a backdrop of spectacular growth in the Chinese economy, which showed no signs of slowing down. Hong Kong's economy, meanwhile, was considered the third most powerful in the world, and the former enclave was also the third most important international financial center behind New York and London.

[15] Hong Kong was much more severely hit by the Asian financial crisis than the rest of China. Exports declined, and the economy shrank 7 per cent in the third quarter of 1997. In the meantime, unemployment rose to its highest level since 1982.

[16] In June 1999, Beijing was forced to intervene in Hong Kong's judicial system to avoid a wave of massive immigration. The local Supreme Court had extended the right of citizenship to all Chinese children born to Hong Kong citizens. The Court decision put Hong Kong at risk of being 'invaded' by almost two million people (equivalent to one third of its population).

[17] In the May 1998 elections, the Democratic Party of Hong Kong won the majority of votes, but due to the inclusion of special constituencies (based on occupational groupings or corporations), the pro-China sectors gained control of the Legislative Council. In spite of having obtained 60 per cent of the votes, the so-called 'democratic' parties (pro-West), only won 33 per cent of the seats.

[18] Between 2001 and 2002, the economy was affected by a deep recession. Unemployment peaked at 6.7 per cent in 2002, the highest level for 20 years.

[19] In July 2003, Tung Chee Hwa announced that voting on an anti-subversion bill would be postponed. This followed public protests against the bill on National Security, which would have granted life imprisonment for any person convicted of subversive acts, sedition, or treason against China, while granting greater powers to the police. As a result of protests, the bill was filed away.

[20] In April 2004, China's key legislative body decreed that Hong Kong could reform its electoral laws as of 2005, but first had to request Beijing's permission. Pro-democracy leaders stated that the measure would undermine the autonomy of the territory.

[21] In early July 2004, more than 250,000 people took to the streets of Hong Kong to demand greater democracy, and to criticize the government for its mismanagement of the Severe Acute Respiratory Syndrome (SARS) epidemic, which caused the death of 299 people between March and June 2003. They were also protesting at the high level of unemployment and, mainly, to express resentment at Beijing's meddling in elections.

[22] Donald Tsang was appointed in June 2005 by an Electoral Committee as successor for Tung Chee Hwa - who had resigned for health reasons - until the end of his term in office in 2007.

[23] In February 2006, the Chinese Government warned Hong Kong's Archbishop Joseph Zen - a well-known critic of Beijing's religious policy - not to express his opinion on political issues, soon after he was appointed Cardinal by Pope Benedict XVI. In May, ignoring this warning, the prelate declared that the Vatican should suspend talks with China on the restoration of diplomatic ties. ∎

Taiwan / T'aiwan

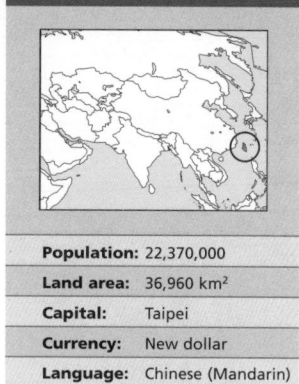

Population:	22,370,000
Land area:	36,960 km²
Capital:	Taipei
Currency:	New dollar
Language:	Chinese (Mandarin)

The island of Taiwan was originally inhabited by Austronesian peoples, who also spread to the Polynesian archipelago and New Zealand/ Aotearoa, and are still found among a dozen ethnic groups that have not been assimilated into the Chinese culture.

2 In 1590, the Portuguese landed on the island, renaming it Ilha Formosa. After a brief occupation by the Dutch in the early 17th century, the island was incorporated into the Chinese Empire during the Qing (Manchu) dynasty in 1683 and proclaimed a province of China in 1887. During this time, the political and administrative systems of mainland China were extended to Taiwan, and many people migrated there from the continent.

3 After China's defeat in the 1895 Sino-Japanese war, Taiwan became a Japanese colony, but was returned to China after the defeat of Japan at the end of World War II.

4 At first, the people of Taiwan rejoiced at the end of Japanese colonialism, but they soon discovered that life under the authoritarian Kuomintang (KMT) party led by Chiang Kai-Shek resembled colonialism.

5 On 28 February 1947, there was a major demonstration against the KMT authorities. The KMT reacted at first by lifting martial law and inviting the opposition to form a Settlement Committee of politicians, trade unionists and student groups to discuss possible political reforms. Meanwhile, they drafted in 13,000 additional troops and when the opposition came forward, the KMT massacred large numbers of them, imprisoning others.

6 In 1949, the entire KMT government, the remnants of its armies, and their relatives and supporters, fled to Taiwan after losing the mainland civil war to the Communist armies. From its refuge, backed by the United States, the KMT declared Taiwan to be the temporary base of the Republic of China pending recovery of the mainland.

7 When the Korean War erupted, with China supporting the North Koreans, the US redoubled its military and economic commitment to Taiwan, protecting it as a front-line state in the battle to defend the 'free world'. During the Cold War Taiwan became a champion of anti-communism, and forged close relations with right-wing dictatorships, including Chile, Paraguay, Uruguay and South Africa.

8 Democracy disappeared. Human rights were violated, demonstrations, strikes and political parties banned, and martial law imposed - all in the name of the battle to reconquer the mainland. The KMT set up a governmental system which claimed to represent the whole of China, with legislators representing each mainland province.

9 Taiwan's remarkable industrialization began in the 1970s, when World Bank and US technocrats helped the Government apply an export-oriented development strategy. The US granted financial, trade and aid advantages to bolster an authoritarian political regime which, in turn, hectored Taiwan's disenfranchised and politically disorganized workforce.

10 Output grew at an annual average of 8.6 per cent between 1953 and 1985, and Taiwan became one of the four newly industrialized 'tigers' of East Asia, based on highly developed plastics, chemicals, ship-building, clothing and electronics. Growth was entirely export-oriented, and the island developed the world's second-largest trade surplus, along with the US, after Japan.

11 In 1971, the US decided to seek closer ties with China, no longer vetoing the latter's admission to the UN. Taiwan therefore lost its representation in the world body. A few years later, in 1979, the US officially broke off diplomatic relations with Taiwan.

12 The country found itself at an economic crossroads. It was over-dependent on a few export markets and labor-intensive industries, importing nations were pressuring for more balanced trade and Taiwan's labor was no longer as cheap as that of many of its Asian neighbors.

13 Technocrats insisted that Taiwan needed a more open political system in order to compete. The KMT also faced an internal succession crisis as aging politicians-clung precariously to power.

14 A significant portion of the opposition began to reject both the authoritarian policies of the KMT and Deng Xiaoping's unification proposal of 'one nation, two systems' with Taiwan becoming a Chinese dependency but maintaining its socioeconomic

PROFILE

ENVIRONMENT

Located 160 km southeast of continental China, Taiwan is part of a chain of volcanic islands in the West Pacific which also includes Japan. A mountain range stretches from north to south along the center of the country. A narrow plain along the island's western coast constitutes its main agricultural area where rice, sugar cane, bananas and tobacco are cultivated. More than two-thirds of the island's area is densely wooded. Taiwan has considerable mineral resources: coal, natural gas, marble, limestone and minor deposits of copper, gold, and oil.

SOCIETY

Peoples: Most are Chinese who have migrated from the mainland since the 17th century and are known as 'Taiwanese'. Hundreds of thousands of Kuomintang Chinese fled to Taiwan during 1949-50. The island's indigenous inhabitants are of Malayo-Polynesian origin (recognized as a sub-group of the Austronesian-speaking family) and make up about 1.7 per cent of the population. They are concentrated on the east coast, where they comprise 25 per cent of the population.
Religions: More than half are Chinese Buddhist. There are also Muslim and Christian minorities. Indigenous religions in some areas.
Languages: Chinese (Mandarin), official. Taiwanese, a derivative of the Chinese dialect from Fujian province, is the language of the majority. Hakka is the second Chinese dialect in Taiwan. There are several indigenous languages, such as Amis.
Main Political Parties: There are 99 political parties registered, but only four of them are significant: Democratic Progressive Party (DPP), formed in 1986 with a very broad platform to restore political democracy in Taiwan; Kuomintang (Chinese Nationalist Party/KMT) founded in China in 1919,

took political monopoly control of Taiwan when it fled from the continent in 1949 and ruled under martial law up to 1987; People's First Party (PFP).
Main Social Organizations: Every trade union has to be a member of the Chinese Federation of Labor, controlled by the Kuomintang, but in 1987 a few independent trade unions appeared. These are organized in several federations, including the National Federation of Independent Unions and Tao-Chu-Miao Brotherhood. There are several diaspora-based organizations; and groups campaigning for the rights of Taiwan's indigenous peoples.

THE STATE

Official Name: Republic of China.
Administrative divisions: 7 municipalities and 16 counties.
Capital: Taipei (T'aipei) 7,423,000 people (1999).
Other Cities: Kaohsiung 2,478,600 people; Taichung 2,070,300; Tainan 770,100 (2000).
Government: Chen Shui-bian, President since May 2000, re-elected in 2004. Su Tseng-chang, Premier since January 2006. Bicameral Legislature: Yuan (225 members serve three-year terms); National Assembly (300 members). As a result of constitutional amendments approved by the National Assembly in June 2005, the number of members of the Yuan will be reduced from 225 to 113 beginning with election in 2007; the amendments will also eliminate the National Assembly, thus giving Taiwan a unicameral legislature.
National Holiday: 1 January, Day of the Republic; 25 February, Constitution Day.
Armed Forces: 376,000 troops. Other: Military police 25,000.

model, in similar fashion to Hong Kong.

[15] Martial law was formally lifted on 15 July 1987. A huge upsurge in union activity took place and strong independent sections were formed in the trade union system. Its leaders founded parallel political structures in the Labor Party and Workers' Party.

[16] While the KMT was resolutely opposed to independence, some DPP members formed the New Wave group proposing self-rule. Other groups advocated a referendum for self-determination.

[17] From 1994, voices rose demanding an independent way for Taiwan, leaving behind the assumption of being a government representative of all Chinese. However, due to Beijing's opposition to any measure which could further lead the island toward independence, Taipei's efforts to be accepted into the UN were in vain.

[18] The first multi-party municipal elections were held in December. Most votes were either for the governing Kuomintang Party or the recently formed Democratic Progressive Party (DPP).

[19] In 1995 economic relations with Beijing intensified. Taiwan became the second 'foreign' investor in the People's Republic of China after Hong Kong. However political relations between both countries deteriorated after a private visit by Taiwanese President Lee Teng to the US in June. In July and August, Beijing carried out a series of missile tests in the sea, 140 kilometers from Taiwan.

[20] Lee Teng, of the KMT, triumphed on 20 March 1996 in the first presidential elections with direct suffrage in the history of the island.

[21] In July China made a show of military force off the Taiwanese coast, coinciding with the transfer of power in Hong Kong. A month later Prime Minister Lien Chan resigned and was replaced by Vincent Siew.

[22] In April 1998, China and Taiwan decided to reopen a direct dialogue, having broken off relations in 1995.

[23] In the weeks running up to the presidential elections of 2000, Beijing announced it could resort to force if Taipei refused unification. The announcement was made as a response to the DPP election campaign in which candidate Chen Shui-bian had stated he would hold a referendum to decide the future status of the island were he to win.

[24] This external pressure did not benefit the KMT, which lost power on the island for the first time in its history in the March elections. The DPP won, with the independent candidate James Soong - formerly a member of the KMT hierarchy but now leader of a new faction - coming second.

IN FOCUS

ENVIRONMENTAL CHALLENGES
The country suffers from the negative impact of its enormous industrial expansion. Low-level radioactive waste disposal increases the already high levels of water, air and soil contamination. Trading in endangered species exists in the country.

WOMEN'S RIGHTS
Women have been able to vote and stand for election since 1947, when the Constitution was passed. In 2000, for the first time a woman was elected Vice-President.

There are frequent cases of domestic violence, sexual harassment and rape against women. The level of discrimination is such that when a single woman dies her body can only be put in the family vault if she is posthumously married to a living man. In this way the soul of the dead woman is 'saved' and avoids being condemned to total oblivion. At the beginning of 2006 the trafficking of women from continental China into Taiwan - where they worked as prostitutes - was still going on and there was no solution in sight. Traffickers made a profit of up to $15,000 for each woman brought into the country.

CHILDREN
In 2006 infant mortality was slightly over 6 per 1,000 live births. There was a significant difference between the figure for boys (almost 7 per 1,000) and for girls (5.5 per 1,000). Education is free and compulsory between the ages of 6 and 15, and almost 100 per cent of children are registered at birth. The state provides free health care, immunization and medicines. There is concern about the high number of cases of sexual abuse against minors: in 2002 alone some 5,000 incidents of abuse were reported.

INDIGENOUS PEOPLES/ ETHNIC MINORITIES
The indigenous population (some 438,700 people, comprising 1.7 per cent of the total population in 2003) descends from indigenous peoples of the Pacific islands. Their ancestors have inhabited Taiwan since 3,000 BC, long before the creation of the Chinese state and current Taiwan. Today they live mostly in the central mountains and the east of the country, while a quarter live in cities such as Taipei and Kaushiong. They are divided into lowland and highland communities which number ten and nine, respectively. There are regional and communal divisions, but no political differences among indigenous peoples. Some

groups have sought better health care, social services, education and economic status. However, they lack a political organization to represent them. Several indigenous peoples of Taiwan have become 'tourist attractions' because the Government has promoted visits to 'model' villages.

MIGRANTS/REFUGEES
In spite of the announcement in 2003 of proposed legislation on refugees, in 2004 their situation remained unchanged. Illegal immigrants, particularly from the People's Republic of China (PRC), were indefinitely held in detention centers or repatriated, even those who had been subjected to people trafficking (see 'Women's rights'). During that year the authorities deported 1,440 illegal immigrants from the PRC.

DEATH PENALTY
Although the number of executions had been decreasing and there was a clear intention to abolish it, the death penalty was still being applied at the beginning of 2006 and 'terrorism' had been added to the list of crimes that could receive such a punishment.

[25] In late 2000, Taipei allowed direct trade contacts between the Taiwanese islands of Qinmen and Matsu and the People's Republic. Until then, business ties had been indirect and, like the rest of Chinese-Taiwanese trade, had gone through Hong Kong or Macau. In early 2001, after more than 50 years of isolation, Taiwan and China inaugurated the first direct links between their territories.

[26] In the legislative elections of December 2001 the Kuomintang lost its parliamentary majority. This result put an end to the political hegemony of the party that had ruled the island with an iron fist since 1949 and was responsible for the 'Taiwanese miracle'.

[27] In late 2003 President Chen proposed holding a 'defense referendum' on how to maintain the status quo in Taiwan with the objective of helping to consolidate defense in terms of 'people's psychological attitude' as well as 'drawing global attention to the military threat to Taiwan posed by mainland China'.

[28] Chen survived an assassination attempt in March 2004 and was re-elected. In his re-election campaign he employed once again his slogan:

'a country on each side'.

[29] In May 2004 Chen announced constitutional reforms and promised that these would not affect delicate matters such as sovereignty and independence. In the parliamentary elections of December 2004 the DPP won 35.7 per cent of the vote. The indecisive nature of the result limited Chen's ability to radically reform his relations with Beijing. Chen had threatened China with rewriting the Constitution, removing the name 'China' from its overseas diplomatic offices and seeking membership of international bodies.

[30] In March 2005 Beijing responded to Taipei's moves. The People's National Congress approved an anti-secession law directly applicable to Taiwan that provided a legal framework for Beijing's repeated threats to attack the island. A month later, and for the first time since 1949, nationalist and communist leaders met when Chen traveled to China. The way forward 'lies in seeking what we have in common in spite of our differences and generating good will' said Chen during his visit, while emphasizing that both countries wanted dialogue and reconciliation, not confrontation.

[31] Taipei Mayor Ma Ying Jeou was elected as the new KMT leader in July 2005 with the sole declared objective of preparing the party to win the 2008 elections. In December, the KMT won municipal elections with what was seen as a protest vote against the Chen Government.

[32] In February 2006 Taiwan decided to dissolve the National Unification Council - a body that sought to implement reunification with China. Beijing labeled this move 'disastrous' and Chen as 'problematic for Taiwan and the entire Asia Pacific region'.

[33] After a corruption scandal involving some of his relatives, Chen announced in June 2006 that he was delegating some of his governmental powers to the Prime Minister, although he would keep control over foreign policy, defense and relations with China.

[34] In November 2006, President Chen's wife was indicted on corruption charges and it was announced that the President could face similar charges in future, though he was then protected by presidential immunity. Many critics saw the announcement as a vindication of the independence of Taiwan's judicial system. ■

Colombia / Colombia

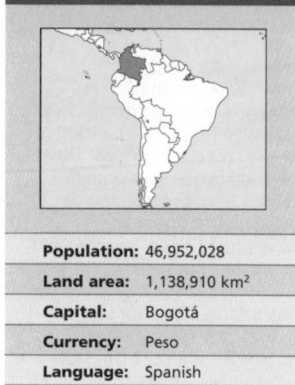

Population:	46,952,028
Land area:	1,138,910 km²
Capital:	Bogotá
Currency:	Peso
Language:	Spanish

The best known of Colombia's indigenous cultures is that of the Chibcha or 'Muisca' peoples. Living in the north of Colombia, in present-day Panama, they were farmers and miners and developed a stratified society of feudal lords and vassals, with a matrilineal succession of power and patrilineal inheritance of property. The legend of *El Dorado* is attributed to a Chibcha ceremony in which a newly appointed chief was covered with gold dust and then bathed in a holy lake. The Chibcha culture and language remained isolated and, in spite of Spanish dominance, survive today in some areas of northern Colombia.

2   Spain conquered Colombia between 1536 and 1539. Gonzalo Giménez de Quesada's forces decimated the Chibchas and founded the city of Santa Fé de Bogotá which became the center of the Viceroyalty of New Granada in 1718. The population was subjected to thinly disguised forms of slavery, and by the 19th century, a large part of the indigenous population had been killed.

3   Extensive, export-oriented agriculture (coffee, bananas, cotton and tobacco) replaced traditional crops (potatoes, cassava, corn, wood and medicinal plants). African slaves were imported to work in the plantations.

4   The Revolt of the Comuneros (1781) started the process leading to the declaration of independence in Cundinamarca, in 1813. The road to independence was marked by constant struggles between the advocates of centralized Government and the federalists, headed by Camilo Torres. Antonio Nariño (who had drafted the declaration of independence) represented the urban bourgeoisie, linked to European interests, while Torres presided over the Congress of the United Provinces, representing the less privileged.

5   In 1816, Pablo Morillo reconquered this territory, executing Torres. Three years later, Simon Bolívar counter-attacked from Venezuela, liberated Colombia and founded the Republic of Greater Colombia, including Venezuela, Ecuador and the province of Panama. After the secession of Venezuela and Ecuador in 1829-30 the Republic of New Granada was proclaimed, and in 1886 adopted the name of Colombia.

6   From 1830 to the early 20th century, the country went through nine civil wars, 14 local wars and two with Ecuador, three military uprisings and 11 constitutions. The Liberal and Conservative parties were created in 1849.

7   Between 1861 and 1885, a Liberal Party administration separated the Church from the State and created nine state companies.

8   Between 1921 and 1957 US firms made a profit of $1,137 million tapping Colombia's oil reserves, which led to their disappearance. These companies controlled the mining industry, 80-90 per cent of the banana trade and 98 per cent of electricity and gas.

9   In 1948, the 'Bogotazo' popular uprising killed leader Jorge Eliécer Gaitán. From that year - when a Liberal mayor organized the first guerrilla group - until 1957, the civil wars caused between 250 and 300 thousand deaths.

10   The Revolutionary Armed Forces of Colombia (FARC), led by Manuel 'Tiro Fijo' (Sure Shot) Marulanda and Jacobo Arenas, appeared on the scene in 1964. Guerrilla strategists included Camilo Torres Restrepo, a priest and co-founder of the National Liberation Army (ELN), who was killed in combat in 1965.

11   Large landowners organized, armed and paid 'self-defense' groups to fight these rural-based guerrilla movements. These were supported by members of the army and, in some cases, by foreign mercenaries. Closed out of official circles, the army also created paramilitary groups, later condemned by Amnesty International.

12   In 1974, President Alfonso López Michelsen, a Liberal, tried to give greater attention to popular demands, but vested economic interests led to the failure of this policy. Figures for 1978 reveal that only 30 per cent of industrial workers and 11 per cent of the rural workforce had social security benefits. Colombia was dependent on international coffee prices on the US and German markets for its foreign exchange, as these countries consumed 56 per cent of the Colombian product at that time.

13   Guerrilla movements, particularly FARC and the April 19 Revolutionary Movement (M-19), continued their activities into the late 1970s. Military repression grew during the Government of President Julio C Turbay Ayala (1978-82).

14   In 1980, M-19 guerrilla leader Jaime Bateman proposed a high-level meeting in Panama. Bateman subsequently died in an airline 'accident' and talks were suspended. The FARC and the Government reached an agreement which led to a ceasefire between them and to the adoption of political, social and economic reforms.

15   In 1982, a divided Liberal Party nominated two candidates, thus handing the victory to the Conservative Party's Belisario Betancur, a journalist, poet and humanist, who had actively participated in the peace process in Central America. Betancur proposed that Colombia join the Non-Aligned Movement and reaffirmed the right of debtor nations to negotiate collectively with creditor banks. Also, in 1983 he entered peace talks with leaders of the M-19.

16   Large landowners fiercely opposed talks between the Government and the guerrillas. The rural oligarchy, holding 67 per cent of the country's productive land, denounced the peace process as 'a concession to subversion' and proposed the creation of private armies. Paramilitary action started up again; subsequent investigations revealed the hand of the *Muerte a los Secuestradores* (MAS- 'death to the kidnappers'), which had opposed the withdrawal of the army from guerrilla-controlled areas. A one-year truce came into effect but M-19 withdrew five months later, claiming that the army had violated the ceasefire.

17   In January 1985, the Government passed a series of unpopular economic measures which worsened the recession: price hikes, and a currency devaluation. The aim was to increase exports and reduce the $2 billion fiscal deficit by 30 per cent. A commission of 14 banks, presided over by the Chemical Bank, stated that the Government would have to reach a formal agreement with the IMF.

18   According to the Human Rights Commission, 80 prisoners had disappeared in one year, political detainees had been tortured and 300 clandestine executions were confirmed. The number of political activists who had disappeared rose to 325.

19   On 6 November 1985, 35 guerrillas from the M-19 took over the Palace of Justice in Bogotá. The army attacked, causing a massacre. All the guerrillas were killed, along with 53 civilians.

20   Over 2,000 left-wing activists were killed by terrorists, and two presidential candidates were assassinated: Jaime Pardo Leal, a member of the Patriotic Union, in 1987, and Liberal senator Luis Carlos Galán, in 1989. They had promised to dismantle the paramilitary groups and fight drugs. War broke out between the Government and the drug mafia. In March 1990, Bernardo Jaramillo, the Patriotic Union's presidential candidate, was assassinated, with Carlos Pizarro (replacing Jaramillo) also killed 20 days later.

Life expectancy
73 years
2005-2010

GNI per capita
$2,020
2004

Literacy
94% total adult rate
2000-2004

HIV prevalence rate
0.7% of population 15-49 years old
2003

21 More than 140 paramilitary groups were present in the country, some of them financed by the drug mafia. Meanwhile, the US Drug Enforcement Agency allegedly bombarded coca plantations with chemical herbicides.

22 The presidential elections of 27 May 1990 were won by Liberal Party candidate Cesar Gaviria, who received 48 per cent of the vote in an election where the abstention rate was 58 per cent. The following year, in elections for the constituent assembly, abstentions were at 65 per cent, and the ADM-19 (Alianza Democrática M-19/Democratic Alliance M-19) won 19 seats.

23 In June 1991, members of Gaviria's Government met in Caracas with representatives of the Revolutionary Armed Forces of Colombia (FARC), the National Liberation Army (ELN) and the People's Liberation Army (EPL) - members of the Simón Bolívar Guerrilla Coordinating Committee, which controlled 35 per cent of the country. Talks dealt with the demobilization of guerrillas, the subordination of the armed forces to civilian authority, the dismantling of paramilitary groups and the reintegration of guerrilla fighters into areas where they could exert political influence.

24 The new 1991 Constitution created the office of vice-president; abolished presidential re-election; legalized civil divorce for Catholic marriages; promulgated direct election of local authorities; guaranteed democratic rights for indigenous peoples; added the democratic tools of referendum and grassroots legislative initiatives; guaranteed equal opportunities for women, and banned the extradition of Colombians for trial overseas. The Constitution met criticism from the Left for failing to give civilian courts jurisdiction over members of the military accused of committing crimes against civilians and for granting judicial powers to State security agencies.

25 In October an agreement between the President and the three major political forces participating in the constituent assembly reduced the number of representatives. After the dissolution of Congress, parliamentary elections were held. The Liberal Party, split into several groups, obtained 60 per cent of the vote and support for the ADM-19 dropped to 10 per cent. In the March 1992 municipal elections, marked by a 70 per cent abstention rate, this trend was continued.

26 The peace process reached a low point in 1992. After talks had been discontinued, the Government promoted its so-called 'Integral War', which authorized intervention in civilian organizations with suspected links to rebel groups.

27 The Simón Bolívar Coordinating Committee resisted the army offensive, continuing its campaign. In response, paramilitary groups resumed their activities. Violence caused major population displacement from conflict areas to the interior, for fear of the children being recruited into the conflict.

28 In November 1992, the Government decreed a state of emergency after Pablo Escobar Gaviria (head of the Medellín Cartel, a powerful drug-trafficking ring) had escaped from prison earlier in the year, stepping up the Cartel's violence. In January 1993 a group appeared known as 'PEPES' - People Persecuted by Pablo Escobar. Within a two-month period, they killed 30 cartel members, destroyed several of Escobar's properties and harassed members of his family. The confrontation reached serious proportions, with car-bombs causing dozens of deaths. On 2 December, Escobar was killed in a shoot-out with police forces in Medellín. His death was a serious blow to the political and social power of the Medellín Cartel.

29 The Supreme Court of Justice decriminalized the use of cocaine, marijuana and other drugs, with the opposition of several political and religious circles, headed by President Gaviria.

30 The coffee market crisis and the 1993 drought, as well as the reduction of banana quotas to the European Union, affected exports. However, with $2 billion per year from drug-trafficking and the discovery of oil in Casanare province, the country achieved a sustained growth of 2.8 per cent per capita. However, 45 per cent of the population were still living in deep poverty.

31 President Gaviria was elected secretary-general of the Organization of American States (OAS), with the support of the US, which welcomed the victory of Ernesto Samper, his party's candidate in the 1994 elections. Samper defeated Conservative Andrés Pastrana by 50 per cent to 48.6 per cent of the vote. Support for the ADM-19 dropped to four per cent of the vote, while abstentions slightly decreased to 65 per cent.

32 The Samper Government stroke a series of successful blows to drug-trafficking, but in September 1995 political scandal broke out when the Cali Cartel revealed details of that organization's contributions to both the Samper and Pastrana campaigns. Defense Minister Fernando Botero, Samper's former campaign director, was sent to prison for embezzling.

33 In August 1996, Samper decreed a state of emergency to curb a wave of violence and kidnappings, a move considered an attempt to protect himself from drug-linked scandals. However, the murders of several opposition leaders and actions by FARC and the ELN, which attacked high-tension power lines, oil pipes, police and military facilities, continued. With fighting in almost 100 places, these two groups controlled growing areas in the economically powerful coffee region, the Caribbean and even in the vicinity of Bogotá and Medellín.

34 Efforts to eradicate coca and opium poppy plantations continued, as well as armed operations against the Cartel's bases. Some of the Cali Cartel's main leaders, responsible for 70 per cent of cocaine traffic worldwide, gave themselves up. In March 1996 the US took Colombia off its list of countries which co-operate in the war on drugs. This measure ended bilateral aid to Colombia and blocked its access to foreign financial sources. Washington denied an entry visa to Samper, in an attempt to corner him diplomatically.

35 Some 1,900 candidates decided not to run for the 26 October local elections because 49 mayors and city councilors had been killed and more than 180 kidnapped since the beginning of the year. In spite of the traditionally low turnout, over 5 million people enclosed symbolic 'vote for peace' slogans in their ballot papers.

36 In November the Human Rights prosecutor revealed that since August 1995 his office had ordered disciplinary measures, including 50 dismissals, against 126 military and police officers for human rights abuses. In the same period, more than 600 cases were investigated against security forces members, related with 1,338 victims of murder, torture or disappearance. Some 500 kidnappings mainly by FARC and ELN were reported in the period.

37 Several organizations estimated that since early 1997 one million Colombians had been displaced from their homes in conflict areas, mainly due to the activity of paramilitary groups. According to the Government, guerrilla groups obtained an annual net income of $750 million, substantially more than that earned by coffee. The only sectors with higher earnings were

PROFILE

ENVIRONMENT

The Andes cross the country from north to south, in three ranges: the western range on the Pacific coast, and further inland, the central and eastern ranges, separated by the large valleys of the Cauca and Magdalena Rivers. North of the Andes, the swampy delta of the Magdalena River opens up, leaving flat coastal lowlands to the west - along the Pacific coast - and to the east, plains covered by jungle and savannas extend downward to the Orinoco and Amazon Rivers. This diversity results in great climatic variety, from perpetual snows on Andean peaks to tropical Amazon rain forests. The country's Andean region houses most of the population. Coffee is the main export item, followed by bananas. Abundant mineral resources include petroleum, coal, gold, platinum, silver and emeralds. Coca plantations in Colombia increased by almost 25 per cent in 2001, some 33,600 hectares out of a total of 169,800.

SOCIETY

Peoples: Colombians are descended from native Americans, Africans and Europeans. **Religions:** 93 per cent are Catholic. Although this is the country's official religion, there is religious freedom. **Language:** Spanish (official); there are dozens of Indian languages including wayuu, camsá and cuaiquer. **Main Political Parties:** Primero Colombia (created by Alvaro Uribe in 2002), Liberal Party; Conservative Party; Independent Democratic Pole; Social National Unity Party. **Main Social Organizations:** There are four major labor organizations: the Confederation of Colombian Workers; the Confederation of Workers' Unions, Workers' Union and the General Labor Confederation. Eighty per cent of all salaried workers affiliated to trade unions are members of the Unitary Workers Federation (CUT), founded in 1986. Regional Indigenous Council of Tolima (CRIT). Colombian Association of Peasant and Indigenous Women.

THE STATE

Official Name: República de Colombia. **Administrative Divisions:** 32 departments and the capital district. **Capital:** Bogotá 7,290,000 people (2003). **Other Cities:** Cali 1,718,900 people; Medellín 1,621,400; Barranquilla 1,064,300; Cartagena 745,700 (1995). **Government:** Álvaro Uribe Vélez, President since August 2002. The Congress (Legislature) has two chambers: the Chamber of Representatives, with 161 members, and the Senate of the Republic, with 102 members. **National Holiday:** 20 July, Independence Day (1810). **Armed Forces:** 207,000 troops (2003).

Under-5 mortality	Poverty	Debt service	Maternal mortality
21 per 1,000 live births 2004	**7%** of population living on less than $1 per day 2003	**33.0%** exports of goods and services 2004	 **130** per 100,000 live births 2000

the drug cartels of Medellín and Cali.

[38] In February 1998, US President Clinton decided, for 'national interest' reasons, to include Bogotá as a cooperating state in the war on drugs. According to the World Bank, high murder rates reduced GDP growth by 2 per cent each year.

[39] That year, in June, Pastrana, the leader of the conservative New Democratic Force, was elected President. He obtained 50.4 per cent of the vote, ending 12 years of liberal rule.

[40] In August 2000, Pastrana announced the Plan Colombia, which sought to eradicate 60,000 hectares of coca crops. The plan encompassed the formation of three anti-narcotics battalions, trained and equipped by the US, backed by 60 helicopters. The objective was to weaken the guerrillas' and drug barons' finances, rather than confronting them in the battlefield. After the 11 September 2001 attacks in the US, Colombia was included by Washington among the targets in its 'war on terror'.

[41] Peace talks between the Government and the FARC, which had just started, broke down on 20 February 2002, after the guerrillas kidnapped several politicians in order to influence electoral results. In partial elections held in March, the abstention rate reached 55 per cent and sabotage attempts by the guerrillas caused several deaths among rebels and paramilitary groups. Right-wing candidate Alvaro Uribe Vélez, took the lead with 53 per cent of the vote. Uribe Vélez had been linked to paramilitary forces, the United Self-Defense Groups of Colombia (AUC) and also to drug traffickers. This led many to believe that Uribe's pre-electoral promise 'security for all Colombians' was a declaration of war aimed at the FARC.

[42] On 4 May during a battle between FARC and AUC forces, a mortar-bomb hit a church where the population of Bojaya was taking refuge, killing 117 people, at least 40 of them children. That month, Washington granted the country $2.6 billion - twice the amount of Plan Colombia. At the same time, Pastrana held peace talks in Cuba with the ELN.

[43] In January 2003, Uribe requested Washington's direct intervention in the fight against guerrilla and - surprisingly - also against paramilitary groups, calling them terrorists. US special forces were deployed in the province of Arauca, becoming the first military troops from that country to get directly involved in Colombia's civil war.

[44] In October 2003, the election of Luis Eduardo Garzón, from the center-left party Independent Democratic Pole (PDI), as mayor of Bogotá (the most important office in the country after the presidency) meant a historical shift in Colombian politics and the consolidation at the national level of a left-wing movement. In November, the disarmament of AUC members began.

[45] In May 2004, Ricardo Palmera, a FARC commander and the highest-ranking guerrilla leader ever captured, was sentenced to 35 years in prison. In June, the ELN agreed to Mexican mediation in the Colombian civil conflict and seemed inclined to become reincorporated into the country's civil life. A month later, within the framework of the peace talks between the Government and the AUC, some leaders of the paramilitary group addressed Congress. That month, more than 200 British members of Parliament - mainly from the Labour Party - requested that Prime Minister Tony Blair suspend military assistance to Colombia, due to the close link between Bogotá's Government and paramilitary groups.

[46] In August 2004, several leaders of the indigenous community Paéz Nasa - a widely recognized project promoting sustainable development and education and fighting poverty in the south-west area of Cauca, which had been awarded the Nobel Peace Prize in 2000 - were kidnapped by the FARC. The guerrillas accused the Nasa of collaborating with paramilitary groups and supporting Uribe's 'security' policy. The following month, the leaders were released after talks between the FARC and a commission comprising more than 250 members of the Indigenous Guard of the Nasa.

[47] In June 2005, a new law, which established the legal framework for the demobilization of the AUC, granted sentence reductions and protection against extradition requests to the paramilitaries who agreed to turn in their arms. The schedule anticipated the demobilization of about 10,000 irregular fighters during the year. Pro-human rights groups warned the Government about the leniency of the law.

[48] In December, new exploratory talks seeking a long-term peace agreement with the ELN started in Cuba.

[49] Colombia and the US signed a free trade agreement in February 2006. The document awaits ratification by both countries' parliaments. In March, Uribe's party won an overwhelming victory in the parliamentary elections. ∎

IN FOCUS

ENVIRONMENTAL CHALLENGES

Pesticide abuse, intensive crops and mining have degraded water and soil quality. Deforestation is significant. Two-thirds of bird species are at risk. Air pollution, especially in Bogotá, is caused by vehicle carbon emissions.

WOMEN'S RIGHTS

Colombian women have been able to vote and run for office since 1954. In 2002, 12.1 per cent of members of Parliament were women.

In 2003, 87 per cent of girls enrolled in primary school and only 58 per cent in secondary school. That year, women made up 40 per cent of the workforce: of them, 17 per cent worked in industry, 76 per cent in services and the rest in the farming sector. The percentage of unemployed women was 20 per cent.

Illiteracy among women over 15 was halved between 1980 and 2000, from 16.8 to 8.4 per cent*. In 2000, 19 per cent of adult women were victims of physical violence. In 2005 the rate of maternal deaths remained high (130 per 100,000 live births in 2000). One of the reasons for this high level is the poor access to prenatal and childbirth care*.

CHILDREN

Ninety-two per cent of the population has received primary education, but the average number of school years completed is less than four. In 2000, 47 per cent of children had access to pre-school education.

Indigenous and Afro-Colombian children from rural areas have little access to education *.

More than one million children between 5 and 17 work. Sexual exploitation and international trafficking have increased. More than one million children have been displaced in the last 15 years. Some 7,000 fighters between 10 and 18, recruited by guerrillas and paramilitary groups, make up 20 per cent of these forces. A high percentage of landmines victims are children.

In 2002, some 300 children were kidnapped.

INDIGENOUS PEOPLES/ ETHNIC MINORITIES

Blacks, descendants of slaves brought from Africa in the 18th century, make up four per cent of the population. The abolition of slavery after 1850 coincided with the movement of black workers to cities such as Medellín and Bogotá and the influx of whites in search of jobs in the mining, business, and wood industry sectors. In the cities, even today, blacks work mostly as domestic employees and in non-qualified areas. They make up the majority of the workforce in the Antioquia coffee plantations and in the Choco mines.

Indigenous peoples make up one per cent of the population and live in 27 of the 32 departments, with large concentrations in the Amazon, Orinoquia, Pacific coast, Sierra Nevada de Santa Marta, Perija mountains, Guajira Peninsula and the *cordillera*. There are between 80 and 90 groups, identified by the *cabildo* or department they live in, their languages (more than 64) and dialects (300), and by the collective claims established by their local organizations.

The efforts of indigenous organizations to implement land reform or claim ancestral lands lost during colonial and post-colonial periods have historically encountered legislative resistance and violence from military and paramilitary groups. Their main demands are: regional autonomy with limited power, greater political rights over their regions, equality of civil rights, equal distribution of funds and public services, protection of lands and resources from foreign oil companies, protection from attack and occupation by military groups.

MIGRANTS/REFUGEES

In 2005, due to the long and complex internal armed conflict in Colombia, between two million and three million people were internally displaced - though some NGOs say that the real number was 3.5 million - and there were approximately 50,000 refugees in neighboring countries. In the course of the year, the armed conflict forced 19,000 indigenous men, women and children to leave their lands (14,000 of them belonged to the Nasa group living in the west of the country).

DEATH PENALTY

Abolished in 1910.

* Latest data available in *The State of the World's Children* and *Childinfo* database, UNICEF, 2006.

Comoros / Komori - Comores

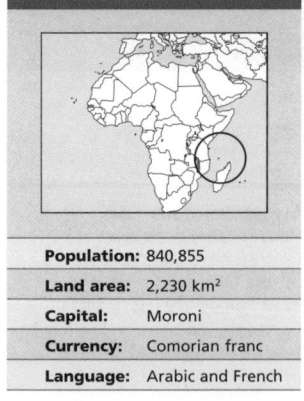

Population:	840,855
Land area:	2,230 km²
Capital:	Moroni
Currency:	Comorian franc
Language:	Arabic and French

Comoros was inhabited around the 5th century by one of the last Indonesian migrations (see Madagascar). The Comoros Islands remained isolated from the continent until the 12th century when Muslim traders from Kilwa settled on the archipelago, founded ports and reproduced the civilization of the eastern African coast. In the 16th century Comoros had a prosperous economy but the Portuguese seized the islands and destroyed their active trade. When the Sultan of Oman finally drove the Portuguese from the region, Comoros came under the influence of Zanzibar and as a result of the slave trade the Bantu-speaking population increased considerably.

[2] In the 19th century, after Zanzibar split from the sultanate of Oman, European pressure became stronger and finally France occupied Mayotte in 1843. Colonial domination eventually spread to the entire archipelago, motivated by spice production and the islands' strategic position on the Cape route.

[3] The Comoros National Liberation Movement (MOLINACO) joined forces with the local Socialist Party (PASOCO) to form the United National Front (FNU) which pressured the French Government into holding a referendum in 1974. A large majority, 154,182 people, voted in favor of the islands' independence, and only 8,854 voted against.

[4] Most of those who wished to remain under French rule were Mayotte residents (63 per cent of its voters). France had air and naval bases on Mayotte and the economy was controlled by a few dozen Catholic families, sympathetic to France and politically represented by the Mahorés People's Movement (MPM), led by Marcel Henry.

[5] Ahmed Abdallah, the archipelago's leading rice exporter and Prime Minister of the semi-autonomous local Government, proclaimed the independence of Comoros in July 1975, before the French announced the result of the referendum. Abdallah was afraid

that his *Udzima* (Unity) Party would lose out to the FNU in a future assembly to draft a new constitution. The MPM took advantage of the situation to declare that Mayotte would continue under French rule. Paris supported the secession in order to maintain its military presence in the Indian Ocean, violating its previous commitment to respect the result of the referendum. France did not oppose the islands' membership in the UN but vetoed specific Security Council resolutions to reincorporate Mayotte into the archipelago.

[6] Less than a month after the declaration of independence, a small group of FNU youths seized the national palace in Moroni and appointed their leader, the socialist Ali Soilih, as president, in place of Abdallah. France reacted by sending a task force of three warships and 10,000 soldiers to Mayotte; one soldier for every three inhabitants.

[7] In May 1978, a mercenary force, under the command of Ahmed Abdallah, in exile in Paris, landed on Grand Comoro (now Njazidja) overthrowing and assassinating Ali Soilih. At the head of the operation was the notorious French mercenary Bob Denard, who had been tried in 1977 for mercenary acts of war against the Government of Benin. His presence in Comoros triggered international protests, to the point of the Comoros Islands' delegation being expelled from the Organization of African Unity. Madagascar tightened its diplomatic relationships with the islands and the UN threatened to apply economic sanctions to the regime.

[8] From then on, Denard became a key figure in Comoros' politics, and Abdallah's control of the Government came to depend on support from Denard and his 650 troops. On 26 November 1989, a coup led by Denard ousted Abdallah, who was killed in the fighting.

[9] The French Government suspended all economic aid to the islands and initiated negotiations designed to oust Denard and his mercenaries. They surrendered to

the French troops on 15 December and left for South Africa, where they were imprisoned.

[10] Restrictions on the formation of political parties ceased after Abdallah's death. A number of opposition groups returned from exile.

[11] In August 1991, the Supreme Court of the Islamic Federal Republic of the Comoros found President Said Mohammed Djohar unfit to govern and guilty of serious negligence. After a failed attempt to oust the President, the political parties and Djohar signed a national reconciliation pact. A new Government was formed, headed

by Mohammed Taki, leader of the National Union for Democracy in Comoros (UNDC), who was dismissed in July by President Djohar, accusing him of appointing a former French mercenary to his cabinet.

[12] In October 1993, when Djohar, aged about 80, was outside the country, Prime Minister Caabi el Yachourtu Mohamed proclaimed himself 'interim President', refusing to give up power on Djohar's return.

[13] Djohar recovered a 'symbolic' Presidency in January 1996, while Taki, leader of the UNDC, won the elections held in March.

[14] The October 1996 Constitution replaced the 1992 one and included, among other innovations, the formation of a Council of Ulemas (Islamic teachers) to monitor the compliance of the laws with their interpretation of Islamic law. The death penalty, suspended in 1975, was reinstated.

[15] In August 1997, a separatist movement headed by Abdallah Ibrahim demanded the independence of Anjouan Island, where half the Comoran population lives. In March 1998, more than 99 per cent of Anjouan citizens voted for secession in a referendum, while Mohéli island also called for its independence. On 23 April 1999, in Madagascar, an agreement was proposed to grant more autonomy

PROFILE

ENVIRONMENT
The Comoros Islands are located at the entrance of the strategic Mozambique Channel, on the oil tanker route between the Arab Gulf and western consumer nations. The four major islands of this volcanic archipelago are: Njazidja, formerly Grand Comoro; Nzwani, formerly Anjouan; Mwali, formerly Moheli; and Mahore, also known by its former name of Mayotte. Njazidja has an active volcano, Karthala, 2,500 meters in altitude. The mountainous island is covered by tropical forests. Only 37 per cent of the cultivated land (100,000 hectares) is used to grow cash crops - vanilla and other spices; the rest is devoted to subsistence farming which is carried out without permanent rivers. The rainwater, which is stored in reservoirs, is easily polluted.

SOCIETY
Peoples: The original Malay-Polynesian inhabitants were absorbed by waves of Bantu and Arab migrations. Today, the latter groups predominate, co-existing with minority Indian and Malagasy communities. **Religions:** Islam (official 98 per cent), Catholic (2 per cent). **Languages:** Arabic and French are official. Most people speak Comoran, a Swahili dialect, and some groups speak Malagasy. **Main Political Parties:** Convention for the Renewal of the Comoros, Camp of the Autonomous Islands Forces for Republican Action (FAR); National Front for Justice (Islamic); National Union for Development (conservative alliance). **Main Social Organizations:** The failure of the political parties has led to the foundation of several social organizations, the main one being the Comoran Workers' Union. Citizen's Initiative has played a key role in the peace negotiations.

THE STATE
Official Name: Union des Comores. **Capital:** Moroni 53,000 people (2003). **Other Cities:** Mutsamudu 25,400 people; Domoni 14,000; Fomboni 12,200 (2000). **Government:** Parliamentary federal Islamic republic. Ahmed Abdallah Sambi, Federal President of the Comoro Union since May 2006. Abdoul Soule Elbak is President of Njazidja, Mohamed Bacar of Nzwani and Said Fazul of Mwali. Legislature: General Assembly of 33 members (15 elected by local assemblies and 18 by universal suffrage). **National Holiday:** 6 July, Independence Day (1975).

Life expectancy
65 years
2005-2010

GNI per capita
$560
2004

Literacy
56% total adult rate
2000-2004

to Anjouan and Mohéli and to establish an interim Government and a rotating Presidency among the three islands. Anjouan delegates refused to sign it, alleging they had to consult their people first; this led to violent clashes both in Njazidja and Anjouan, which led to a bloody military coup on 30 April. Colonel Azzali Assoumani became President, and promised to hold elections within 10 months.

[16] While Anjouan's political leaders were in Mohéli scrutinizing a constitutional draft on redefining the islands' relationships in August 2001, a military committee led by Major Mohamed Bacar took over Anjouan and granted the island 'regional autonomy from Comoros, but not total separation'.

[17] Two separatist coups within the military élite failed in Anjouan, and control was left in the hands of Bacar. Finally, a referendum in Njazidja supported the new constitution (already approved in the other two islands) that held the Federation, but granted more autonomy to each part. The Presidency would rotate among the three islands. The first four-year term went to Njazidja.

[18] After an election rife with protests of fraud, Assoumani was confirmed as federal President of Comoros by a new ad hoc electoral body, the Commission of Ratification, in May. The Commission declared Mohamed Fazul had won the elections in Mohéli, which had also been denounced as fraudulent. Bacar had won Anjouan's presidential elections and Abdou Soule Elbak was elected President of Njazidja (Grand Comoro).

[19] In elections to the federal Parliament held in April 2004, the national parties of the three autonomous islands won the majority of seats. The parties were held together only by their opposition to President Assoumani, who would have to govern with an opposition-dominated federal Parliament. The Parliament started to hold sessions in June.

[20] French journalist Morad Aït-Habbouche was arrested in September 2004, accused of 'plotting a coup d'état', shortly after his arrival in Grand Comoro. Opposition party leader Said Larifor was also arrested as an alleged accomplice of Aït-Habbouche.

[21] Assoumani's visit to Paris, in January 2005, was the first one paid by a Comoran leader to France in 30 years.

[22] Religious leader Ahmed Abdallah Sambi was elected president in May 2006 with 58 per cent of the vote, after defeating another two candidates from Anjouan - the island which would take over the rotating federal presidency. ■

Mayotte

Population:	163,366	**Currency:**	Euro
Land area:	400 km²	**Language:**	French, Arabic,
Capital:	Mamoutzou	Swahili and Comoran	

Known as the Island of Perfumes, Mayotte lies in the Indian Ocean, about halfway between Madagascar and the coast of Mozambique, 1,500 kilometers away from Réunion. Mahore (Mayotte in Creole) comprises two main islands, Petite Terre and Grand Terre separated by two kilometers of sea and surrounded by a coral reef of 1,000 square kilometers.

[2] As a consequence of successive invasions since the 10th century, Mayotte is a combination of Arabic, Malgasy and French influence. After the treaty of 25 April 1841, Mayotte became a French colony. In 1974, after pressure from independence movements, France agreed to a referendum in the Comoros islands, of which Mayotte was part. Nationalist forces won and in 1975 the independence of the Comoros was proclaimed. However, those who wished to remain under French rule gathered in Mayotte and, supported by Paris, encouraged secession from the Comoros. Taking advantage of a movement that overthrew Ahmed Abdallah - prime minister of the semi-autonomous local government - 10,000 French troops occupied Mayotte.

[3] The policy statement of the Eighth Conference of Non-Aligned Countries (1986), declared that 'the Comoran island of Mayotte, still under French occupation, is an integral part of the sovereign territory of the Federal Islamic Republic of the Comoros'. In 1991, the UN General Assembly reaffirmed the sovereignty of the Comoros over Mayotte by an overwhelming majority.

[4] France maintains an important air-naval base on the island which is still under a French protectorate. In 1997, the islands of Nzwani and Mwali declared their independence from the Comoros, stating their desire to return to French rule. In 1999 a peace agreement was signed and the islands remained within the federation, but enjoying greater autonomy. Voters on Nzwani rejected the peace agreement and endorsed independence in January 2000. The Organization of African Unity imposed economic sanctions on the island.

[5] In July 2000, the people of Mayotte accepted in a referendum an offer by Paris to increase the island's autonomy. The agreement granted Mayotte a status similar to that of a department, instead of an overseas territory.

[6] First originating in Tanzania, the Chikungunya virus, a non-lethal disease mainly affecting joints and transmitted by the same mosquito that causes dengue, had devastating effects on Reunion Island and the Comoros archipelago in the early months of 2006. A virus outbreak of unprecedented magnitude in the region, also affected Mayotte, which registered about 6,000 cases during the early part of 2006. ■

ENVIRONMENT

A mountainous island with tropical climate and heavy rainfall throughout the year. The island is of volcanic origin, has dense vegetation and is located at the entrance to the Mozambique Channel. Cyclones (hurricanes) are quite frequent.

SOCIETY

Peoples: Mayotte inhabitants are of the same origin as Comorans. There is an influential community of French origin that controls the island's commerce. Religions: 98 per cent Muslim. **Languages:** Arabic, Swahili, Comoran, and French (official). **Main Political Parties:** Mahoré Departementalist Movement; Union for a Popular Movement; Socialist Party; Citizen and Republican Movement.

THE STATE

In October 1991, the UN General Assembly reaffirmed the sovereignty of the Comoros over Mayotte by an overwhelming majority. **Official Name:** Collectivité Territoriale de Mayotte. **Capital:** Dzaoudzi 13,500 people (1999). **Other cities:** Mamoudzou 40,900 people (2000).- **Government:** Jean-Paul Kihl, Prefect since January 2005, appointed by the French Government. Unicameral Legislature: General Council, with 19 members. **National Holiday:** All French holidays. **Armed Forces:** Defense is the responsibility of France; a small contingent of French forces is stationed on the island.

IN FOCUS

ENVIRONMENTAL CHALLENGES
Soil degradation and erosion result from crop cultivation on slopes without proper terracing. Furthermore, there is deforestation and cyclones occur during the rainy season.

WOMEN'S RIGHTS
Women have been able to vote and run for office since 1956. In 2003, women made up 43 per cent of the workforce.

In 2004, they only held 3 per cent of total seats in Parliament. That year, the literacy rate for adult women was 49 per cent, whereas for men it was 63 per cent*.

CHILDREN
Both rape and physical abuse of boys and girls are disturbingly common. These generally occur within the family circle and are never taken to court but settled between the offender and the victim's parents over a sum of money.

INDIGENOUS PEOPLES/ ETHNIC MINORITIES
There are Arabs, descendants of Shirazi settlers, who arrived in the 15th century; the Cafres, an African group that settled on the islands before the Shirazi; a second African group, the Makoa, descendants of slaves brought by Arabs from the East African coast; and three groups of Malayo-Indonesian peoples - the Oimatsaha, the Antalotes, and the Sakalava, the latter having settled mainly in Mahoré.

Creoles, descendants of French settlers who married indigenous people, form a tiny (less than 100 people) but influential group in Mahoré. They are Roman Catholic and own small plantations.

MIGRANTS/REFUGEES
An estimated 80,000-100,000 Comorans live abroad, mostly in Tanzania, Madagascar, and other parts of East Africa. Approximately 40,000 Comorans live in France.

DEATH PENALTY
The death penalty applies for ordinary crimes.

Congo Democratic Republic / République Démocratique du Congo

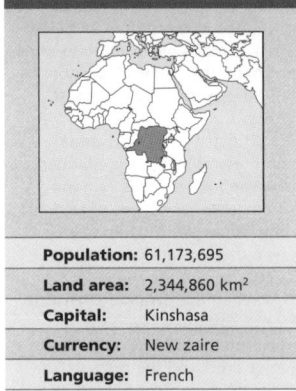

Population:	61,173,695
Land area:	2,344,860 km²
Capital:	Kinshasa
Currency:	New zaire
Language:	French

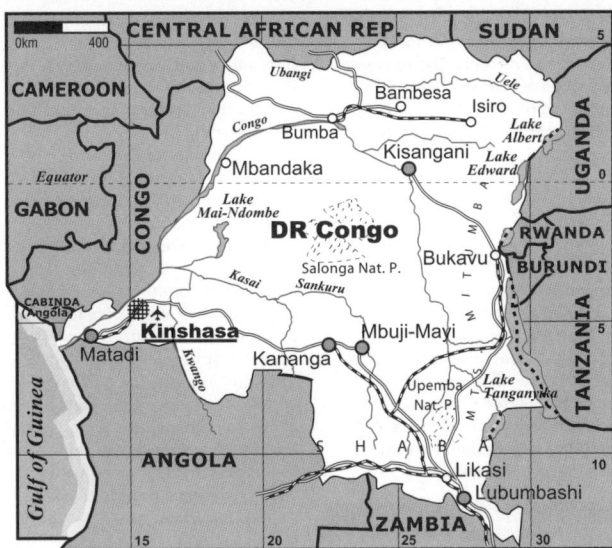

The first known state to emerge in what is now DR Congo was the Luba kingdom - Katanga (Shaba) region - created when a warrior named Kongolo subdued the small chiefdoms in the area and established a highly centralized state. To the northwest was the Kuba, a federation of numerous chiefdoms that reached its peak in the 18th century. Dr David Livingstone brought the region to the notice of the Western world through his explorations between 1840 and 1870. He met up with Henry Stanley, a journalist and adventurer, in Ujiji in 1871. In 1876, King Leopold II of Belgium founded the International Association for the Exploration and Civilization of Congo, a private organization that financed Stanley's expeditions. Stanley succeeded in signing more than 400 trade and/or protectorate agreements with local leaders along the Congo River. These treaties, and the Belgian trading posts established at the mouth of the river, devised a system for the economic exploitation of the Congo. The Berlin Conference (1884-1885) decided that the 'Free State of Congo'; was the Belgian King's personal property. Consequently, Leopold II stopped British colonialist expansion, and made a fortune using slave labor to exploit rubber and ivory. It is estimated that 10,000,000 Congolese died from forced labor, starvation and systematic extermination during the period of Leopold II's kingdom.

2   The harsh working conditions did not change when the region formally became a Belgian colony in 1908. Military force was systematically employed to suppress anticolonial opposition and to protect the flourishing copper mining industry in Katanga (now Shaba).

3   In 1957, liberalizing measures permitted the formation of African political parties. This led countless tribal-based movements to enter the political arena, all trying to benefit from the general discontent. Only the National Congolese Movement led by Patrice Lumumba had a national outlook, opposing secessionist tendencies and supporting independence claims.

4   In 1959, the police suppressed a peaceful political rally triggering a series of bloody confrontations. King Baudouin of Belgium tried to appease the demonstrators by promising independence in the near future, but European residents of the Congo reacted with more oppressive measures. Independence was finally achieved in 1960, with Joseph Kasavubu as President and Lumumba as Prime Minister. A few days later, Moise Tshombe, then Premier of the Province of Katanga, initiated a secessionist movement.

5   Belgium sent in paratroopers and the United Nations, acting under US influence, intervened with a 'peacekeeping force'. Kasavubu staged a coup and arrested Lumumba, delivering him to Belgian mercenaries in Katanga who killed him. The civil war continued until 1963. Secessionist activity ceased when Tshombe, who represented the neo-colonial interests, was appointed Prime Minister. With the help of mercenaries, Belgian troops and US logistical support, he defeated the revolutionary forces. In 1965, he was forced to resign by Kasavubu who was in turn overthrown by army commander Joseph Desiré Mobutu. For the transnationals, Mobutu was the only person who could restore the conditions required for them to continue operating there.

6   Under the doctrine of 'African authenticity', Mobutu changed the name of the country to Zaire and his own to Mobutu Sese Seko. However, his nationalism went little further than this, and the 'Zairization' of copper, which he declared in 1975, only benefited an already wealthy economic elite and the state bureaucracy.

7   Although these measures caused some discomfort among US diplomats, Mobutu offered Washington his services in the region.

8   Zaire sheltered and actively supported the so-called National Front for the Liberation of Angola (FNLA). Mobutu encouraged secessionist groups in the oil-rich Angolan province of Cabinda, and Zaire's troops effectively cooperated with the South African racist forces in their war against the Angolan nationalists.

9   Meanwhile, in Zaire guerrillas continued the struggle in the interior. In 1978 and 1979, the major offensive launched by the Congolese Liberation Front was checked with the aid of French and Belgian paratroopers and Moroccan and Egyptian troops, again with US logistical support.

10   At the end of 1977, international pressure led to parliamentary elections being held for an institution which had been given limited legislative functions. This helped to divert international attention from human rights violations against students and intellectuals in the cities, the establishment of concentration camps for Mobutu's opponents and the brutal reception given to refugees who returned under an 'amnesty' decreed in 1979.

11   At this time Zaire was the world's largest cobalt exporter, the fourth biggest diamond exporter and among the top ten world producers of uranium, copper, manganese and tin. But corruption was rampant throughout the country's administration, worsening the already unstable economic situation.

12   During 1980 and 1981, the major Western powers decided to seize direct control of the strategic mineral reserves in the country. The International Monetary Fund (IMF) took special interest in Zaire's economy, facilitating the renegotiation of its foreign debt, and imposing drastic measures against corruption. Zaire's economy was under direct IMF control, and the Fund's representatives in Kinshasa began to supervise the country's accounts personally.

13   In April 1981, Prime Minister Nguza Karl I-Bond sought political asylum in Belgium; he presented himself to the Western powers as a 'decent alternative' to the official corruption in Zaire.

14   The elections of June 1984 officially gave Mobutu 99.16 per cent of the vote. In February 1985 Zaire signed a security pact with Angola to improve their relations.

15   The Angolan Government denounced the fact that $15,000,000 of covert aid for the FNLA was channeled by the Reagan administration through Zaire, and that Zaire effectively acted as an arsenal for the Front.

16   In April 1990, anticipating the process of democratization which he considered imminent, Mobutu decreed the end of the one-party system, opened up the labor movement and promised to hold free elections within a year. A rapid process of political organization began. Hundreds of associations and political groups demanded legal recognition from the Government. The extent of the popular reaction frightened the authorities and in May, Mobutu issued a statement saying that no party had yet been legalized and that it would be necessary to modify the Constitution before holding elections, because the head of state wished to 'preserve his authority without exposing himself to criticism'.

17   The students at the University of Lubumbashi, in Shaba province, began calling for Mobutu's resignation; he reacted by sending in his presidential guard to silence the protests.

18   The troops stormed the university campus at dawn on 11 May. More than 100 students were killed and the terrified survivors fled to other provinces and Zambia, from where they condemned the massacre.

19   President Mobutu partially stifled the massacre's outcome, but the European Community called for an international investigation, while Belgium stopped its economic aid. The political reform plan lost momentum, at least temporarily.

20   The massacre at Lubumbashi University generated an anti-Mobutu wave which led to a series

Life expectancy
45 years
2005-2010

GNI per capita
$110
2004

Literacy
65% total adult rate
2000-2004

HIV prevalence rate
4.2% of population 15-49 years old
2003

of strikes, like that at state-owned Gecamina, the country's most important mining company.

[21] In October 1990, under growing internal and external pressure, Mobutu decided to carry out a new political 'democratization' process and he authorized the unrestricted creation of new political parties. In December, the opposition - grouped together under the Holy Union, a front made up of the largest parties - demanded Mobutu's resignation and called for a National Conference to decide on the political future of Zaire without presidential intervention.

[22] Mobutu faced another national uprising in September 1991, provoked by a general price hike and the failure of a conference called in August to introduce democratic reforms. The uprising led to dozens of deaths and the intervention of France and Belgium, which sent hundreds of troops to evacuate their citizens.

[23] In November 1991, the Holy Union formed a shadow government, and appealed to the armed forces to depose Mobutu. The same month, the President appointed Nguza Karl I-Bond as his new Prime Minister, his fifth in that year. Nguza, a former opposition leader who had been Mobutu's head of government 10 years earlier, took office amidst a worsening economic crisis and growing international pressures, especially from the US.

[24] Early in 1992, the National Conference was set up. The opposition had long awaited this opportunity to press for constitutional reform and transition to democracy.

[25] In February of the same year, Prime Minister Nguza Karl-I Bond suspended the Conference, causing a faction of the army to rebel, taking over a state-run radio station and demanding President Mobutu's resignation. Some hours later the rebels were defeated and the army clashed with thousands of demonstrators, leaving many dead and injured. The European Community suspended financial aid to Zaire immediately, until the National Conference was reinstated.

[26] In March 1992, after meetings with Conference president Archbishop Monsegwo Pasinya, Mobutu announced the reopening of the National Conference. Etienne Tshisekedi, leader of the Holy Union, was made Prime Minister, to replace Nguza Karl-I Bond.

[27] Inter-ethnic strife erupted again in 1992. In Shaba there were outbreaks of violence after Karl-I Bond's dismissal. Lunda people, Karl-I Bond's group, attacked members of the Luba community, Tshisekedi's people. Some 2,000 people were killed, and thousands of Luba left Shaba as their homes had been destroyed.

[28] The 12 commercial banks operating in Zaire closed indefinitely in 1992, due to a lack of funds. Inflation reached 16,500 per cent. According to a report by the Washington-based Population Crisis Committee, in 1992 Zaire was among the ten poorest countries in the world.

[29] In December, galloping inflation prompted the Prime Minister to put a new currency in circulation. Mobutu nonetheless ordered that troops receive their back pay in the old currency. In early 1993, battles broke out between the soldiers - furious at having been paid in worthless bills - and Mobutu's personal guard. This confrontation resulted in over 1,000 deaths in Kinshasa.

[30] On 24 February, Mobutu's soldiers surrounded the building housing the High Council of the Republic, a transitional body formed by the National Conference. They demanded that the legislators approve the old currency which Mobutu had put back into circulation.

[31] With the worsening situation, the US, Belgium and France sent a letter to Mobutu demanding that he share power with a provisional government headed by Tshisekedi. Mobutu responded by dismissing Prime Minister Tshisekedi.

[32] The US Department of State suggested Belgium and France should block Mobutu's wealth - calculated in more than $4 billion - as a punitive measure that would not harm the country's economy nor European and US business interests.

[33] The genocide in Rwanda and the arrival of large numbers of Rwandan refugees - amongst whom were thousands of soldiers responsible for the massacre - created great tension in eastern Zaire.

[34] When the Rwandan Patriotic Front (FPR) guerrillas came to power in Rwanda, several Western nations reduced the pressure on Mobutu - newly considered as a potential ally following the victory of 'the English-speaking Tutsis' in the neighboring country. This reinforced the President's power, facilitating the nomination of Léon Kengo Wa Dondo as Prime Minister.

[35] The tension between Zaire and Rwanda increased in 1996, after Rwandan and Zaire militias began ethnic cleansing in the Masisi region, ousting and killing the *Banyamulenge*, Tutsis who had lived in this part of eastern Zaire for generations.

[36] At the end of 1996, the conflict between armed Tutsi groups and the remnants of the Rwandan army (mostly Hutu), assumed civil war proportions. The confrontation spread with government forces participating to put a stop to the advance of the Tutsi rebels, who took several cities in the east of the country. The Mobutu regime became seriously threatened when various opposition forces formed an alliance led by the veteran guerrilla leader Laurent Kabila.

[37] In the first few months of 1997, opposition forces easily took over practically the whole country, and a group of states - South Africa, the US, France and Belgium included - attempted to mediate and seek a solution. Nelson Mandela arranged a meeting between Mobutu and Kabila which failed when Kabila demanded that Mobutu stand down.

[38] In March, Parliament sacked Prime Minister Kengo Wa Dondo; Tshisekedi was appointed for his third term in office, with Mobutu's approval. Tshisekedi offered Kabila six ministries, including Defense and Foreign Affairs, but Kabila rejected this and Tshisekedi was dismissed. Foreseeing the final outcome, the powerful mining companies began to negotiate with Kabila to protect their interests in Zaire.

[39] On 16 May, Mobutu fled to Morocco and opposition troops entered Kinshasa the following day. Kabila declared himself President. Mobutu died in Rabat on 7 September and the new

PROFILE

ENVIRONMENT

Located in the heart of the African continent, DR Congo's territory covers most of the Congo River basin and has a narrow outlet into the Atlantic. The center and northern regions are covered with rainforests and are sparsely populated. Small plots of subsistence farming are found here. In the southeast, a plateau climbs to 1,000 meters in the Shaba region (former Katanga). Most of the nation's resources and population are located in the southern grasslands, where cotton, peanuts, coffee and sugarcane are grown. There are also large palm-oil and rubber plantations. The country's mineral wealth is concentrated in Shaba: copper, zinc, tin, gold, cobalt and uranium. Local industry is found in the mining areas too. The small eastern region, Ituri, is home to the world's richest goldfield, the Kilo Motu, and is already the locus of significance for oil exploration. Eastern Congo also produces the majority of the world's supply of coltan, used in chips in cell phones and computers, and which has at times rivaled gold in price per ounce.

SOCIETY

Peoples: The people of Congo are made up of several major African ethnic groups. West African people live in the northwest, while Nilo-Hamitic descendants live in the northeast and a large pygmy minority lives in the eastern center. There are more than 200 ethnic groups, mainly: Luba (18 per cent), Mongo (13.5 per cent), Azande (6.1 per cent), Bangi and Ngale (5.8 per cent), Rundi (3.8 per cent), Teke (2.7 per cent), Boa (2.3 per cent), Chokwe (1.8 per cent), Lugbara (1.6 per cent), Banda (1.4 per cent), other (16.6 per cent).
Religions: It is difficult to pinpoint the religion of peoples in the DRC due to high levels of religious syncretism. There is a large Christian population, above all Catholic (41 to 50 per cent), Protestant (32 per cent), a Muslim minority (1.2 to 10 per cent) and many traditional religions.
Languages: French (official). The most widely spoken local languages are Swahili, Shiluba, Kikongo and Lingala (the army's official language).
Main Political Parties: Congolese Rally for Democracy; Movement for the Liberation of Congo; Alliance of Democratic Forces for the Liberation of Congo-Zaire (AFDL); Congolese National Movement-Lumumba; Democratic Social Christian Party.

THE STATE

Official Name: République Démocratique du Congo.
Administrative Divisions: 10 provinces and a capital.
Capital: Kinshasa 5,277,000 people (2003).
Other Cities: Lubumbashi 1,044,200 people; Mbuji-Mayi 1,018,100; Kisangani 792,400; Kananga 521,900 (2000).
Government: Joseph Kabila, President since January 2001. There is a provisional Senate with 120 appointed members and an also provisional 500-member National Assembly.
National Holiday: 30 June, Independence Day (1960).
Armed Forces: Between 20,000 and 40,000 of Kabila's troops constitute the new Armed Forces of the country.
Other: Gendarmerie: 21,000; Civil Guard: 19,000.

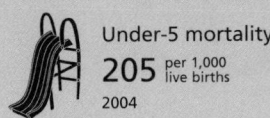 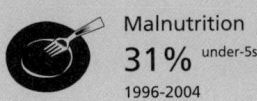 

	Under-5 mortality	Malnutrition	Maternal mortality
	205 per 1,000 live births 2004	**31%** under-5s 1996-2004	**990** per 100,000 live births 2000

IN FOCUS

ENVIRONMENTAL CHALLENGES
Water pollution - especially untreated waste that flows into rivers - is a major source of disease. Some species of fauna run the risk of becoming extinct, due to poaching: among them, elephants and rhinoceros whose horns are sold to produce aphrodisiac powders. Wooded areas have deforestation problems.

Mining is the cause of serious ecological damage. The extraction of coltan (a rare natural superconductor resistant to temperature changes and coveted by the aerospace industry, and for computer, videogames, mobile telephony and military purposes, among others) is particularly harmful. Some 80 per cent of the coltan found in the world is from DR Congo.

WOMEN'S RIGHTS
Women have been able to vote since 1967.

Their life expectancy is 42.8, nine years lower than in 1980. Illiteracy among women over 15 decreased between 1980 and 2000*, from 79.2 to 50 per cent (for men it was 27 per cent in 2000). That year, women made up 43 per cent of the workforce.

In 2003, 12.7 per cent of representatives and 2.5 per cent of senators were women.

The maternal death rate in childbirth is 990 per 100,000 births. In 2003, 48 per cent of women over 15 were illiterate.

CHILDREN
In less than five years of civil war - one of the bloodiest conflicts since World War II - some 3.3 million people were killed, mostly civilians. Children were the most vulnerable: hundreds died in confrontations, and others from malnutrition and other preventable or curable diseases.

Thousands of children were recruited, one third of them by force; many are barely 10 years old. In the province of Ituri hundreds of women and children were raped, mutilated and murdered in 2002 and 2003.

Twelve per cent of children are born underweight and 31 per cent with moderate low weight*. In 2004, some 572,000 children under 5 years old died (205 per 1,000). For children under 1, the rate was 129 per 1,000 live births.

In late 2001 it was estimated that some 170,000 children between 0 and 14 were HIV-positive and some 930,000 had been orphaned by HIV/AIDS.

INDIGENOUS PEOPLES / ETHNIC MINORITIES
There are more than 200 ethnic groups, mostly Bantu-speakers; the four largest groups - Mongo, Luba, Kongo and Mangbetu-Azande - make up 45 per cent of the population.

From the 18th onwards Hutus were treated as slaves or serfs in kingdoms ruled by Tutsis, who formed militias loyal to the royal authority.

The Twa, a hunting and gathering people of the tropical forests, face a dismal future. Deprived of rights, compensation or justice, and exposed to discrimination from other sectors of society, the Twa are suffering an alarming rise in malnutrition and disease. Together with the Twa, the Hutus and Tutsis were one nationality, called *banyarwanda*. This collapsed with the genocide in Rwanda in 1994, when Hutu extremists murdered more than one million Tutsis and moderate Hutu. After the genocide, over a million Hutu refugees fled to DR Congo, fearing reprisals from the Tutsi, who formed the Rwandan Government and had recently carried out sporadic attacks against them.

The Tutsi minority (2 per cent of the population) are dominant politically and militarily, with sporadic outbreaks of violence against Hutus (3 per cent). These live in the eastern province of Kivu, densely populated and of difficult access due to its mountainous terrain.

MIGRANTS/REFUGEES
More than 2.4 million people were displaced in 2002, including 2,000,000 internally displaced and almost 410,000 refugees and asylum-seekers. Around 140,000 people from DRC took refuge in Tanzania, 80,000 in the Republic of Congo, and tens of thousands in other African countries.

At the beginning of 2006, DRC sheltered nearly 200,000 refugees: 90,000 from Angola, 50,000 from Rwanda, 13,500 from Sudan and 40,000 from other countries. Also, 63,500 Congolese had returned to the country; according to UNHCR their number would reach 123,000 by the end of the year.

DEATH PENALTY
Still applies, even for ordinary crimes.

* Latest data available in The State of the World's Children and Childinfo database, UNICEF, 2006.

---

strongman took over as president that month with full military, legislative and administrative powers. The new government changed the name of Zaire to the Democratic Republic of Congo (DR Congo/DRC) and announced a series of economic recovery measures.
[40] A study by Médecins Sans Frontières (MSF) in late 1997 exposed the massacres committed by Kabila's forces and the Rwandan troops that supported him in ousting Mobutu. Members of the Union for the Republic and Democracy (URD, former members of the AFDL founded by Kabila himself) accused Kabila of returning to tribalism, and under the leadership of Ernest Wamba dia Wamba - an intellectual - armed groups of Tutsi refugees and demobilized Congolese soldiers took half the country using weaponry and officers provided by the Rwandan and Ugandan governments. Kabila increased resistance efforts.
[41] The conflict rapidly raised international concern and, in April 1999, in response to the attack, Kabila and the Angolan, Zimbabwean and Namibian leaders announced an alliance to provide a joint response if any of the members were attacked. Hence Angola, Zimbabwe and Namibia supplied Kabila with troops and weapons, while Uganda and Rwanda stepped up support of the rebels.
[42] Two years into the civil war which brought Kabila to power, whole provinces were under Ugandan and Rwandan control. The Congolese Parliament was dissolved in 2000, and the President replaced it with a 300 member assembly.
[43] Kabila was assassinated by one of his bodyguards in January 2001 while the Franco-African summit was meeting in the presidential palace. His son Joseph immediately replaced him as President, with support from the former president's allies.
[44] In July 2002, Kabila and Paul Kagame, President of Rwanda, signed a peace treaty ending four years of civil war in the DRC. The conflict, called the 'World War of Africa', involved the armies of six nations, divided the DRC into regions controlled by rebel forces and by the Government and caused the death of almost 3,000,000 people, mostly from illness and starvation.
[45] In December, the Kabila Government, the main guerrilla groups - the Congolese Rally for Democracy, backed by Rwanda, and the Movement for the Liberation of Congo, backed by Uganda - and opposition political parties signed an agreement ending the war. The agreement included the formation of a bicameral Parliament and the distribution of ministries between the Government, the guerrillas and opposition parties.
[46] In July 2003 a transition Government was installed. This was a military junta presided over by Kabila and four vice-presidents: two leaders from the largest rebel groups, one from the opposition and one from the Government. It undertook to lead the country toward its first elections in 40 years.
[47] Kabila neutralized a coup attempt led by the head of his Presidential Guard in June 2004. In December, there was fighting in the east of the country between the army and soldiers from a former pro-Rwanda rebel group.
[48] In May 2005, parliament gathered to adopt a constitution that would replace transitional laws. In the referendum held in December, the constitution was backed by 84.31 per cent of voters, thus paving the way for elections in 2006.
[49] A new national flag was adopted in February 2006 as the new constitution entered into force.
[50] Finally, in July, Congo held the first free elections in more than 40 years of independent existence.

Kabila won 45 per cent of the votes and came out as winner in the western provinces of the country; former rebel leader Jean-Pierre Bemba won 20 per cent of support, with a clear victory in the eastern provinces, where Swahili is the dominant language; a former Deputy Prime Minister, Antoine Gizenga, gained 13 per cent, while other candidates totalled 9 per cent.
[51] The result, which prevented Kabila from being elected in the first round, meant that the new government would not be revealed until November. Bemba alleged fraud and in August, in view of the mounting tension, some 400 UN troops were deployed in the capital to prevent a widespread outbreak of violence. That same month, 14 foreign ambassadors, who had arrived to promote a meeting among political opponents, had to be rescued by peacekeeping forces on the banks of the Congo river.
[52] Following the run-off in the presidential election at the end of October, Joseph Kabila was confirmed as President in November, winning 58 per cent of the vote. But his opponent Bemba contested the result and UN peacekeepers were put on alert for trouble in the capital, Kinshasa, a Bemba stronghold. ∎

Coltan: the deadly semiconductor

THE DEMOCRATIC REPUBLIC of Congo (DRC), one the world's poorest and most war-torn countries within the world's poorest and most war-torn continent, is home to 80 per cent of the world's known coltan reserves. Coltan, also found in Brazil, Thailand, Canada and Australia, is nowadays the most sought-after mineral. And, once again, a natural resource stirs up greed, increases poverty, destroys nations and finances wars - as with diamonds, gold or oil in the past and maybe water in the future.

This mineral ore, whose name is a combination of the words columbite and tantalite, was discovered in the 19th century and initially used to make filaments for the first electric light bulbs - now made with tungsten. It is extracted in a very rudimentary way, by digging holes with shovels and sloshing around the mud in washtubs, where the mineral ore is then allowed to settle on the bottom. When refined, coltan turns into a metallic powder - tantalum - whose unique electrical properties make it an ideal semiconductor[1]. It can be found in all kinds of high-tech electronic components and different industries such as those of video games[2], cell phones, satellites, missiles, magnetic resonance equipment, organ implants or computers.

These recently discovered many uses for coltan aroused the interest of several transnational companies, while the possibility of having a new source of wealth seemed to give renewed hope to a region depressed by the devaluation of agricultural products and desertification, among other things. The illusion was short-lived. From the mid-1990s onwards, companies such as Nokia, Ericsson, Siemens, Sony, Bayer, Intel, Hitachi and IBM, among others, joined a myriad of shell companies that entered into partnership with local governments, military forces or guerrilla groups to extract - often illegally - the valuable ore. Since then, large capitals have been struggling for control over coltan deposits with help from their native allies, while small ones try to profit by all possible means.

These disputes contributed to complicate and deepen the already existing conflicts in the region, where inter-ethnic confrontations - including ethnic cleansing and widespread genocide in Rwanda - had played a leading role. The military intervention in DRC by Rwandan president Paul Kagame, the ousting of Mobutu Sese Seko in 1997, and the appointment and later assassination of DRC President Laurent Kabila[3] unleashed what former US Secretary of State Madeleine Allbright called 'Africa's first world war'.[4]

In 1999, the Lusaka Agreement outlined the new borders of DR Congo. The treaty allowed the joint mining exploitation between different joint ventures, from which international capitals could not stay away. The Great Lakes Mining Company (SOMIGL) - formed by Belgian company Africom, Rwandan Promeco and South African Cogecom - is the largest among them. With Rwanda's military support, it has the monopoly of coltan trade, thus avoiding intermediaries. This civic and military partnership allows a two-way flow of goods, since the same trucks, helicopters and planes that transport the mineral - using the airports of Entebbe in Uganda and Kigali in Rwanda, among others - are also used to transport weapons and equipment for the different militia groups that continue to fight over control of the mines.

The main buyers of coltan are the United States, Germany, Belgium and Kazakhstan. Starck - a Bayer subsidiary - produces half the world's tantalum powder, while the Kigali-based Bank of Commerce, Development and Industry - Citibank's subsidiary in the region - manages funds resulting from the trade in coltan, gold and diamonds. About 20,000 people are engaged in the extraction of coltan, 'protected' by militia groups that supervise each mine. They are 'granted' a small part of the mineral extracted on a daily basis - usually a spoonful per person per day, which brings several thousands of dollars a month to soldiers.[5] Every worker earns $10 per kilo of coltan, which is later sold in London for $1,000.

In addition to the problem of violence and work under extreme conditions, the uncontrolled exploitation of coltan is causing irreparable damage to the region's environment. Soldiers are indiscriminately killing elephants and gorillas to trade in leather and ivory, but also to feed miners - who have been settled in large national parks such as Kahusi Biega. According to reports by several civil organizations, elephant and gorilla numbers have undergone a sharp and dangerous decline. ∎

---

1. A semiconductor is a material which shows a double behavior, depending on the corresponding electric field. Sometimes it behaves as an insulator and on other occasions it has an appreciable electrical conductivity. At the same time, unlike typical conductors - usually metals - which show a stable conductivity, that is, hardly variable with regards to temperature, the conductivity of these substances can be regulated - either by increasing or decreasing the energy supplied.

2. Sony was forced to delay the market launch of its Playstation 2 due to a shortage of coltan.

3. See Rwanda and Democratic Republic of Congo.

4. In four years, three million people were killed in this war, in which at least eight African nations and several militias and guerrilla groups were directly involved. It also indirectly involved the participation of several powers - mainly the US. One of the factions comprised Rwanda, Uganda and Burundi, together with last-minute guerrilla groups - such as the Movement for the Liberation of Congo or the Congolese Rally for Democracy - supported by the US. The other faction comprised the DRC, Angola, Namibia, Zimbabwe and Chad.

5. In 2003, it was estimated that the Rwandan army had obtained $250 million by way of coltan sales on the black market.

Congo / République du Congo

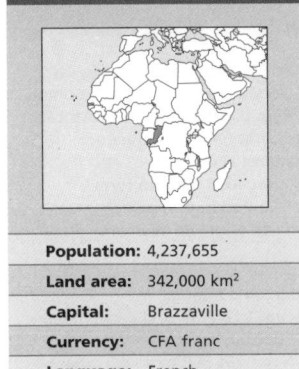

Population:	4,237,655
Land area:	342,000 km²
Capital:	Brazzaville
Currency:	CFA franc
Language:	French

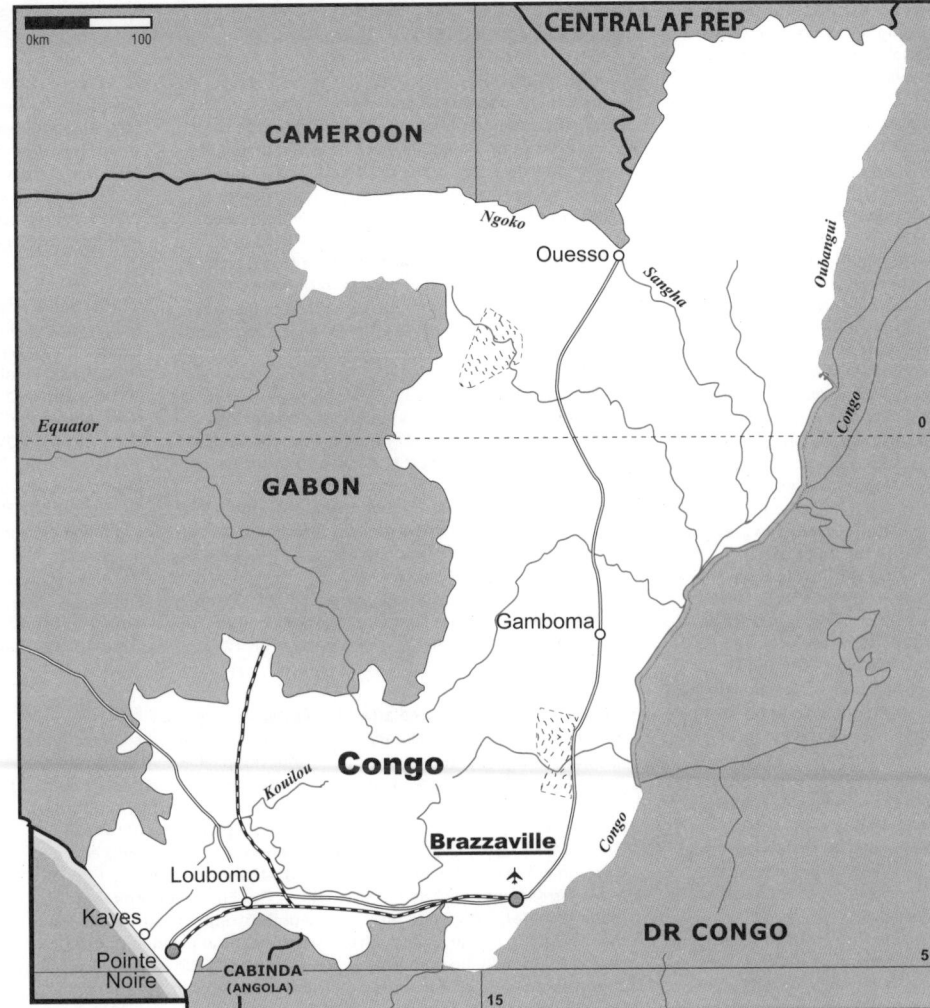

Today's Congo was originally populated by Pygmies and Bushmen (San). By the 16th century it comprised the Bantu states of Luango and Kacongo, for many years ruled by the Manicongo. These nations managed to stave off early Portuguese attempts at colonization. Instead, for three centuries, under the Batekes of Anzico, they acted as suppliers and intermediaries of British and French slave dealers.

[2] Towards the end of the 19th century, trade in rubber and palm oil replaced slave traffic and brought with it French colonization.

[3] In 1880 French troops led by Savorgnan de Brazza began to colonize the Congo by force, and by the 1920s, two-thirds of the local population had been killed, which resulted in the emergence of independence movements with strong religious beliefs, under the leadership of Matswa.

[4] At the start of the decolonization process, the French sponsored Friar Fulbert Youlou who led the Democratic Union for the Defense of African Interests and was the first president of independent Congo in 1960. The growing mass participation never accepted Youlu's neo-colonialist policies and a wave of demonstrations against corruption and the banning of trade unions ended in a popular uprising during the 'three glorious days' (13-15 August) of 1963.

[5] Youlou resigned and the self-proclaimed socialist Alphonse Massemba-Debat, took office and forced the French troops stationed in the country to withdraw. He then founded the National Movement of the Revolution (MNR) as the country's sole party.

[6] This force could not coexist with the neo-colonial army, equipped and trained by the French, and the consequent crisis led Massemba-Debat to resign on 1 January 1969. He was replaced by Major Marien N'Gouabi, backed by left-wing army officers.

[7] Political life was re-organized and a Marxist-Leninist party called the Congolese Workers' Party (PCT) was created. In 1973, a new constitution was approved, proclaiming the Congo a People's Republic.

[8] In December 1975, N'Gouabi made a public self-appraisal, issued a call to 'extend the revolution', and launched a general review of party structures, the state apparatus and mass organizations. N'Gouabi began educational reforms and modified the structure of Government.

[9] On 18 March 1977, N'Gouabi was assassinated by followers of ex-president Massemba-Debat. The conspirators failed in their efforts to seize power, however, and Massemba-Debat was executed.

[10] The new president, Colonel Joachim Yombi Opango, did not maintain the previous austere style and was forced to resign on 6 February 1979, charged with corruption and abuse of power. Denis Sassou N'Guesso replaced him.

[11] N'Guesso launched an anti-corruption campaign within the public sector, administrative and ministerial reforms and a pragmatic foreign policy, maintaining economic relations with Eastern Europe, US and France. He played a leading role in negotiations between Angola, South Africa and Cuba. This culminated in the signing of a peace settlement for the region in 1988, in Brazzaville, paving the way for Namibia's independence.

[12] The demise of the Soviet Union precipitated political and economic changes. In December 1990, the country adopted a multiparty system. In July 1991, André Milongo was appointed Prime Minister.

[13] This situation caused confrontation in the streets and the formation of a transitional Government with the participation of the Army. In the following elections, in August 1992, Pascal Lissouba succeeded Sassou N'Guesso.

[14] In the legislative elections of May 1993, the ruling party won 62 seats, and the opposition coalition got 49. The opposition

LAND USE

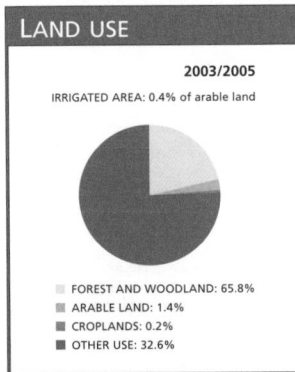

2003/2005

IRRIGATED AREA: 0.4% of arable land

- FOREST AND WOODLAND: 65.8%
- ARABLE LAND: 1.4%
- CROPLANDS: 0.2%
- OTHER USE: 32.6%

PUBLIC EXPENDITURE

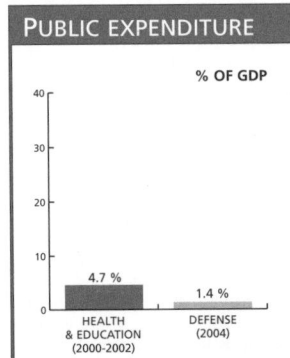

% OF GDP

4.7 % HEALTH & EDUCATION (2000-2002)

1.4 % DEFENSE (2004)

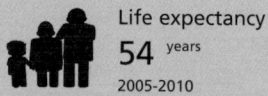
Life expectancy
54 years
2005-2010

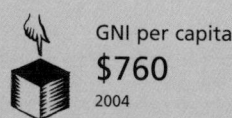
GNI per capita
$760
2004

Literacy
83% total adult rate
2000-2004

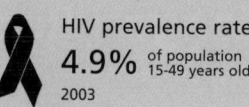
HIV prevalence rate
4.9% of population 15-49 years old
2003

accused the Government of fraud, leading to further clashes between demonstrators and the military. President Lissouba appointed Jacques Yhombi-Opango as Prime Minister, which led the opposition to form a parallel government led by Bernard Kolelas.

[15] An agreement between the opposition and the Government, in mid-March 1994, marked the beginning of a ceasefire. The election of Kolelas as mayor of Brazzaville further calmed the situation, allowing a national ceremony of public reconciliation.

[16] That year, Lissouba accepted the IMF structural adjustment program. Competition between various oil multinationals challenged the dominance of the French Elf in Congo, leading the company to increase the proportion of profits reinvested in the country from 17 to 31 per cent.

[17] Civil war began when the Government sought to disarm and arrest Denis Sassou N'Guesso in June 1997. In November N'Guesso won, with the aid of Angolan troops. He formed a new Government and sought to wipe out 'ninja' militias led by Kolelas.

[18] A ceasefire was signed in November 1999. In compliance with the agreement, N'Guesso freed prisoners and thousands of rebels gave up their weapons. In February 2000, Kolelas ratified his recognition of N'Guesso as President.

[19] The first Presidential elections after the end of the civil war were held in February 2002. N'Guesso won with 74.7 per cent of the vote. Although the elections were peaceful, the main opposition candidates were either in exile or withdrew their candidacy at the last minute. Soon after the elections,

PROFILE

ENVIRONMENT
The country comprises four distinct regions: coastal plains; a central plateau (separated from the coast by a range of mountains rising to 800 meters); the Congo River basin in the northeast; and a large area of marshland. The central region is covered with dense rain forests and is sparsely populated. Two-thirds of the population live in the south, along the Brazzaville-Pointe Noire railroad. Lumber and agriculture employ more than one third of the population. The country has considerable mineral resources including oil, lead, gold, zinc, copper and diamonds.

SOCIETY
Peoples: The Congolese are of Bantu origin; the Bakongo prevail in the South, the Teke (or Bateke) at the center; the Sanga and Vilil in the North. There is also a Pygmy minority.
Religions: Approximately half of the population are Christians and the others practise traditional African religions, with some fusion of the two. There is a Muslim minority (2 per cent).
Languages: French (official), Kongo, Lingala, Téké.
Main Political Parties: Congolese Workers Party (PCT), founded in 1969, the only party until 1990.

the Government denounced attacks by 'ninja' rebels against military positions in the Pool region. By April, the conflict had spread and in June attacks were launched on Brazzaville.

[20] In March 2003, the Government reached a new peace agreement and 2,300 rebels voluntarily surrendered their weapons. On 30 August, the National Assembly unanimously approved an amnesty for rebels. However, the NGO Congolese Human Rights Observatory called the amnesty 'selective', since it did not include all protagonists of the conflict, especially Lissouba, Opango and Kolelas, who

remained in exile.

[21] In 2003-2004, the UN continued to develop extensive aid programs, focused on the emergency needs of people displaced by the conflict, medical assistance and control of the HIV/AIDS epidemic and reduction of poverty.

[22] The Energy Information Administration (EIA), an US agency, announced in mid 2004 that the oil crisis caused by a fall in investments in the country - which had begun in 2000 due to the 1990's civil wars and the ongoing political chaos - was coming to an end. An industrial boom was expected and the plants of Libondo, Tchibeli, Litanzi and

Since then, more than 30 parties and movements have been formed. United Democratic Forces; Union for Democracy and Republic (Mwinda); Pan-African Union for Social Democracy; Party of the Poor.
Main Social Organizations: The four existing labor organizations merged in 1964 to form the Congolese Labor Confederation (CSC), Revolutionary Union of Congolese Women.

THE STATE
Official Name: République du Congo.
Administrative Divisions: 9 regions and 6 communes.
Capital: Brazzaville 1,080,000 people (2003).
Other Cities: Pointe Noire 765,300 people; Loubomo 84,500 people (2000).
Government: Parliamentary republic. Denis Sassou N'Guesso, President since October 1997, re-elected in March 2002. Prime Minister: Isidore Mvouba, since January 2005. Two-chamber Parliament: the National Assembly made up of 153 members and the Senate made up of 66 members.
National Holiday: 15 August, Independence Day (1960).
Armed Forces: 10,000 (2003). Other: 6,100: Gendarmerie (1,400); People's Militia (4,700).

Yanga-Sud were refurbished. Huge damages caused by the conflicts on the rest of the country's industries increased oil dependency.

[23] The Government stated in April 2005 that a group of Army officers, arrested in January for theft of weapons, had planned a coup d'état. In October, Kolelas returned to the country after eight years in exile, to attend his wife's funeral. The former prime minister had been sentenced to death accused of war crimes, but was granted an amnesty in November.

[24] In January 2006 Congo took over the presidency of the African Union. ∎

IN FOCUS

ENVIRONMENTAL CHALLENGES
Problems arise from haphazard urban development, waste and lack of sewage facilities and drinking water, contagious diseases, pollution, deforestation and disappearance of fauna.

WOMEN'S RIGHTS
Women have been able to vote and be elected since 1963. In 2002 they held 8.5 per cent of seats in the Chamber of Representatives and 12.3 per cent in the Senate. In late 2003, 45,000 women aged between 15 and 49 lived with HIV/AIDS. Female illiteracy was 33 per cent in 2003. UNICEF data from 2000 show that maternal mortality

amounted to 510 deaths per 100,000 births.*

CHILDREN
Armed conflict and economic crisis have worsened the human rights situation of women and children, who are most vulnerable to bad sanitary conditions, diseases and death. The main causes of death among children are malaria and diarrhea. In 2004 the infant mortality rate amounted to 81 per 1,000 live births; for children under 5 it was 108 per 1,000 live births. The births of more than one third of children go unregistered.

INDIGENOUS PEOPLES/ ETHNIC MINORITIES
There are several ethnic groups in different regions. The Lari

live in the south, along with the Bakongo, Vilil, Yombe and Bembe peoples. Historically, the Lari - who make up 20 per cent of the population - have been political rivals of the M'boshi, from the north, along with the Teke (17 per cent of the population). The Lari are politically discriminated against. The M'boshi, President N'Guesso's people, are favored by the Government.

MIGRANTS/REFUGEES
As of early 2006 the Congo Republic hosted 34,800 refugees from DR Congo, who lived off the land, fished and engaged in small-scale trade, but benefited from no health or education services. There were also approximately 5,200 Rwandans, which the authorities aimed to have

voluntarily repatriated. Until September 2005, 2,200 Angolans continued to live in the country. It was estimated that some 15,000 Congolese refugees are in Gabon, DR Congo and Benin, and that some 2,000 would return voluntarily to the country during 2006.

DEATH PENALTY
Although it is still in force, there have been no executions since 1982.

*Latest data available in *The State of the World's Children* and *Childinfo* database, UNICEF, 2006.

Costa Rica / Costa Rica

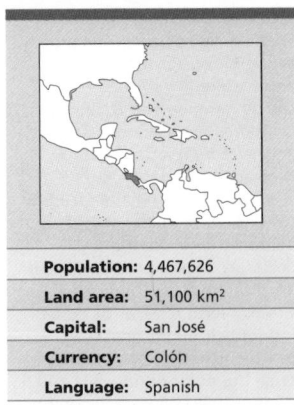

Population:	4,467,626
Land area:	51,100 km²
Capital:	San José
Currency:	Colón
Language:	Spanish

When Christopher Columbus reached the coast of Costa Rica - 'the rich coast' - on his last voyage in 1502, the territory of present-day Costa Rica was populated by small groups of the Chorotega, Cobici, Carib and Boruca nations. Although the first contact was friendly, Europeans did not easily dominate the local people, and it took nearly 60 years for a permanent settlement to be established in the region.

2   Gaspar de Espinosa, Hernán Ponce de León and Juan de Castañeda traveled along the coast of the territory between 1514 and 1516. Between 1560 and 1564, they were followed by Juan de Cavallón, Juan de Estrada Rabago and Juan Vásquez de Coronado. The conquest of the territory was consolidated in the second half of the 16th century.

3   The Spanish established the first permanent settlement, the town of Cartago, on the central plain in 1564. This was assigned to the political jurisdiction of the Captaincy-General of Guatemala, while remaining under the spiritual guidance of the bishop of Nicaragua.

4   The Indians' resistance kept the colonizers isolated. They were unable to establish a system of *encomiendas* - the virtual enslavement of the Indian work force. Thus, a patriarchal society of small landowners was formed, with no powerful land-owning oligarchy as in the neighboring countries. This might explain how, instead of becoming a nation scourged by civil wars and military dictatorships, modern Costa Rica maintained greater democratic stability and has not established a regular army like the other countries in the region.

5   When Mexico declared independence from Spain in 1821, Costa Rica, along with other former Spanish colonies in Central America, formed part of the Mexican empire. In 1823, Costa Rica helped create the United Provinces of Central America and remained a part of this federation until its dissolution in 1840. A persistent opponent to the 'Balkanization' brought about by British imperialism, Costa Rica's territory was used as a base for operations by Francisco Morazán - an advocate of Central American unity - until its independence in1848.

6   In the mid-19th century, William Walker - a US national who had taken control of Nicaragua - tried to extend his dominion over the Central American isthmus. Forces commanded by the Costa Rican President Juan Rafael Mora defeated him. Material progress reached Costa Rica during the regime of General Tomás Guardia, who ruled the nation from 1870 to 1882. While his administration took away some liberties and increased foreign debt, coffee and sugar production rose and more schools were built. The Constitution adopted in 1871 remained in place until 1949.

7   The last decades of the 19th century were marked by a gradual reduction of Church influence in secular affairs. Cemeteries were secularized and the Jesuits expelled from the country for a few years. In 1886, primary education became obligatory and free; schools were founded, along with a museum and a national library. In 1890, José Joaquín Rodríguez was elected President in what were considered the first free and fair elections in Central America.

8   In 1916, Nicaragua gave the US permission to use the San Juan river, which forms the frontier with Costa Rica. The San José Government protested that its rights had been overlooked and the complaint was taken to the Central American Court of Justice, which ruled in favor of Costa Rica. Nicaragua rejected the verdict and withdrew from the Court. This was one of the main reasons for the dissolution of the Court a year later.

9   Costa Ricans had their first elections with direct voting in 1913. There was no outright winner and the Legislative Assembly designated Alfredo González Flores as President. General Federico Tinoco Granados, unhappy with the

LAND USE

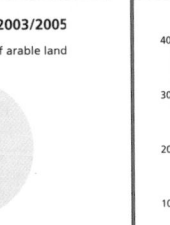

2003/2005

IRRIGATED AREA: 20.6% of arable land

- FOREST AND WOODLAND: 46.8%
- ARABLE LAND: 4.4%
- CROPLANDS: 5.9%
- OTHER USE: 42.9%

PUBLIC EXPENDITURE

% OF GDP

HEALTH (2002)	EDUCATION (2000-2002)
6.1%	5.1 %

WORKERS

UNEMPLOYMENT: 7% (2004)

LABOR FORCE **2004**

- FEMALE: 34%
- MALE: 56%

EMPLOYMENT DISTRIBUTION **1995/2002**

F

M

AGRICULTURE	F: 4%	M: 22%
INDUSTRY	F: 15%	M: 27%
SERVICES	F: 80%	M: 51%

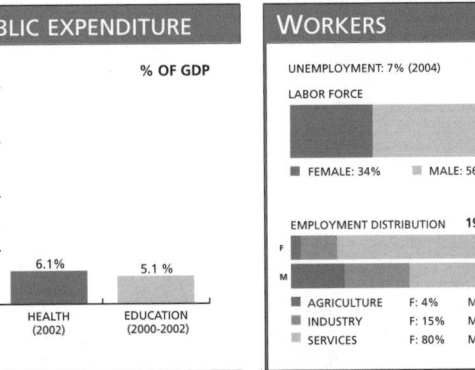

Life expectancy
79 years
2005-2010

GNI per capita
$4,470
2004

Literacy
96% total adult rate
2000-2004

HIV prevalence rate
0.6% of population 15-49 years old
2003

PROFILE

ENVIRONMENT

A mountain range with major volcanic peaks stretches across the country from northwest to southeast. Costa Rica has the highest rural population density in Latin America, with small and medium sized farmers who use modern agricultural techniques. Coffee is the main export crop. The lowlands along the Pacific and the Caribbean have different climate and vegetation. Along the Caribbean coast there is dense, tropical rainforest vegetation. Cocoa is grown in that region. The Pacific side is drier; extensive cattle raising is practiced along with artificially irrigated sugarcane and rice plantations.

SOCIETY

Peoples: Costa Ricans, usually called *ticos* in Central America, are descended from the integration between native Americans and European migrants, mainly Spanish. African descendants, who were brought in from Jamaica, make up three per cent of the population and are concentrated along the eastern coast. Indigenous peoples, one per cent.
Religions: 76.3 per cent of the population are Catholic; Evangelical Protestant 13.7 per cent, Jehovah's Witnesses 1.3 per cent.
Languages: Spanish is the official language, spoken by the majority. Mekaiteliu, a language derived from English, is spoken in the province of Limón (east coast). Several indigenous languages.
Main Political Parties: National Liberation Party, Citizen's Action Party, Libertarian Movement Party, Social Christian Unity Party, Union For Change Party.
Main Social Organizations: The Unitary Workers' Union (CUT); the 50,000-member CUT groups both workers and peasants from the National Federation of Civil Servants, the National Peasant Federation; the Federation of Industrial Workers, regional federations and smaller unions. 375 cooperatives.

THE STATE

Official Name: República de Costa Rica.
Administrative Divisions: 7 Provinces: Alajuela, Cartago, Guanacaste, Heredia, Limon, Puntarenas, San José.
Capital: San José 1,085,000 people (2003).
Other Cities: Alajuela 716,286 people; Cartago 432,395; Limón 339,295 (2000).
Government: Oscar Arias, President since May 2006. Unicameral Legislature: the Legislative Assembly, made up of 57 members, elected by proportional representation in each province for a four-year term.
National Holiday: 15 September, Independence Day (1821).
Armed Forces: Abolished in 1949.
Other: 7,500: Civil Guard (4,300), Rural Guard (3,200), (1995).

reforms proposed by González, led one of the few coups experienced by the nation, in 1917. But the lack of US recognition for his Government and the threat of intervention forced Tinoco to resign in 1919.
[10] Between 1940 and 1948 the Government was backed by coffee plantation owners and bankers. However in 1948 opposition leader Otilio Ulate, nominated by the National Unity Party, won a presidential election which was annulled by Congress. This unleashed civil war which ended with a junta seizing power, presided over by José Figueres Ferrer. The junta issued a call for the election of new representatives, who in turn confirmed Ulate's victory. A year later, the new Constitution was promulgated.
[11] The new Constitution of 1949 not only gave women and African descendants the right to vote but it also abolished the army (replaced with a civil guard). This decision subsequently marked the country internationally as being pacifist and a promoter of demilitarization. Costa Rica stood apart from the other Central American countries because it did not suffer military coups and because it earmarked greater resources to areas key to the country's development, such as education and health.
[12] This heralded the so-called 'Welfare State Capitalism'. Figueres was elected President in 1954 and during his administration Costa Rica became a strongly anti-communist welfare state. In 1958, the conservatives defeated Figueres and imposed a US-backed import-substitution development model.
[13] The traditional antagonism between liberals and conservatives gave way to new tensions between the National Liberation Party (PLN), led by Figueres, and a heterogeneous group consisting of various small parties. In 1966 the United National Opposition elected José Joaquín Trejos to the presidency.
[14] After the 1970 election, the PLN returned to power with Figueres, who remained in office until 1974, when Daniel Odúber Quirós, co-founder of the Party in 1950, was elected President.
[15] Odúber tried to restore unity to the Central American Common Market, left in a critical situation after the 1969 war between El Salvador and Honduras. However, his obvious pro-democracy stance did not meet with the approval of Somoza's regime in Nicaragua. Costa Rica was constantly harassed by its neighbor and became a safe haven for thousands of political refugees.
[16] 1975 saw a rise in wages due to favorable conditions resulting from the nationalization of transnational oil companies and the rise in coffee prices on the world market.
[17] In 1978 presidential elections were won by a conservative coalition which had been critical of Figueres's administration. The leftist bloc, grouped in the United People's Coalition, made considerable gains.
[18] The new President, Rodrigo Carazo Odio, imposed an unpopular economic policy prescribed by the International Monetary Fund (IMF), which resulted in growing confrontation with labor and left groups. In 1979 however, encouraged by popular sympathy towards the Sandinista rebels, and under threat of invasion by neighboring dictator Anastasio Somoza, the Costa Rican Government actively supported the Nicaraguan Sandinistas.
[19] A radically different attitude was taken in 1980 with regard to El Salvador's insurgency. In spite of human rights violations in that country, the San José Government supported the Salvadoran military junta. In 1981 President Carazo broke off diplomatic relations with Cuba.
[20] In January 1982, the US-backed Central American Democratic Community was set up in San José with the purpose of isolating revolutionary Nicaragua.
[21] Luis Alberto Monge, a right-wing PLN candidate, became President in February 1982. He proclaimed his alignment with Western democracies, announced economic austerity measures and fostered closer ties with the governments of El Salvador, Guatemala and Honduras, thus aggravating relations with Nicaragua.
[22] Costa Rica's hostile attitude towards its neighbor was clearly demonstrated when the US declared a commercial embargo on Nicaragua's revolutionary Government. A series of border incidents brought relations between the two countries to breaking point during July and August 1985. However, prompt action taken by the Contadora Group (organization of Latin American countries - Mexico, Panama, Colombia and Venezuela - formed in 1983, at a meeting on Panama's Contadora island, to achieve a peace agreement for Central America) prevented mounting tension in the area. Both governments agreed to place neutral observers along the common frontier to arbitrate any further border clashes.
[23] The winner of the February 1986 presidential election was Social Democrat Oscar Arias, who won a tight victory with 52 per cent of the vote. Arias devoted himself to the task of designing a policy that would break both the logic of war and the escalating tension within the region.
[24] In August 1987 he presented a peace plan at a summit meeting held in Esquipulas, Guatemala. This was accepted and signed by the Presidents of El Salvador, Nicaragua, Guatemala and Honduras. The focal points of the plan were: a simultaneous cease-fire in Nicaragua and El Salvador, an immediate end to American aid to the Nicaraguan contras, a democratization time-table for Nicaragua which included holding free elections and putting an end to the use of foreign territory as supply or attack bases.
[25] The signing of this peace plan, known as 'Esquipulas II', earned Costa Rica a special place in international relations, and constituted a personal triumph for President Arias, who received the Nobel Peace Prize in October 1987.
[26] During his term, Arias instituted two structural adjustment programs (PAE I and PAE II), with World Bank support. Their objective was the transformation of industry through technological modernization, increased efficiency and greater productivity. Neo-liberal formulas were applied, following the dictates of international financial organizations. However, according

Under-5 mortality
13 per 1,000 live births
2004

Poverty
2.2% of population living on less than $1 per day
2001

Debt service
7.3% exports of goods and services
2004

Maternal mortality
43 per 100,000 live births
2000

to the labor unions, these 'prescriptions' were formulated without looking at their social effects.

27   In July 1989, a parliamentary commission on drug trafficking produced a report. It stated that both the main political parties (the PLN and the PUSC) were guilty of receiving drug money during the 1986 electoral campaign. At the same time, another scandal broke out over the financing of electoral campaigns, with accusations that both parties - and Oscar Arias individually - had received money from Panamanian General Noriega in 1986.

28   In 1990, women officially made up 29.9 per cent of the economically active population, but their number in the informal sector was reckoned to be around 41 per cent. Teenage prostitution increased through organized crime rings, which have become multi-million-dollar businesses for their owners.

29   Under the slogan of 'change' and focusing specifically on low-income groups with lower educational levels, the Social Christian candidate Rafael Angel Calderón won the election held in February 1990 and obtained an absolute majority in the legislature.

30   The application of a severe economic adjustment program led to a reduction in the State apparatus as well as in the fiscal deficit, which had reached 3.3 per cent of GDP. As a result of these cuts, unemployment rose and popular discontent increased.

31   In the 1994 elections, social democrat candidate José María Figueres Olsen (José Figueres' son) defeated Government candidate Miguel Rodriguez by a narrow margin, after a campaign with few differences in their political positions but with notably harsh speeches from the rival parties.

32   A free trade agreement was signed with Mexico in January 1995. However, the deterioration of the economy, with rising inflation and a fiscal deficit led the Government to increase taxes in order to balance the budget. The World Bank rejected the economic plan and refused to finance the structural adjustment schedule. In April, the PLN accepted the liberalization of the banking system and the privatization of the insurance, petroleum and telecommunications state enterprises proposed by the Christian opposition in exchange for the tax package approval. Unions organized strikes against these measures, particularly the dismissal of several civil servants.

33   In 1996, the governing PLN agreed a budget plan with the opposition which limited the Government deficit to one per cent of the GDP. In late July, Hurricane Caesar struck Costa Rica, especially the south, causing some 30 deaths. The damage was estimated at around $100 million.

34   In February 1998, Miguel Angel Rodríguez, the Social Christian Unity Party (PUSC) candidate was elected president, with the PUSC also gaining the majority of seats in the parliamentary elections.

35   In April 2002, the PUSC candidate Abel Pacheco was elected in a second round, with 58 per cent of the vote, defeating PLN candidate Rolando Araya. The prevailing two-party system was challenged by an independent party, Citizen Action Party (PAC), which captured 26 per cent of the vote.

36   In May 2003 electricity and telecommunication workers,

opposed to the Government's privatization plans, together with teachers demanding a wage rise, organized a strike which led to the resignation of three ministers.

37   In December 2003 Costa Rica's representatives walked out of negotiations with the US on a Central America Free Trade Agreement (CAFTA). Separate negotiations set for 2004 were motivated by the threat to state insurance and telecommunications monopolies.

38   Pacheco lost part of his cabinet again in September 2004, with resignations including the Foreign Trade, Transport and Finance ministers. According to observers, disagreements over public expenditure and trade union protests prompted the resignations.

39   In October 2004, former presidents José María Figueres, Ángel Rodríguez, and Rafael Calderón were investigated for

payments allegedly made to their private accounts by foreign companies with public contracts in the country. Shortly after being elected, Rodríguez stepped down from his post as Secretary General of the OAS when he was investigated by the prosecutor's office.

40   A state of national emergency was declared by the Government in January 2005, when serious floods affected the Caribbean coast.

41   Ottón Solís, of the PAC, was narrowly defeated in the February 2006 elections. Oscar Arias, who had been awarded the Nobel Peace Prize in 1987, became president for a second time. This time Arias faced a country with divided opinions regarding CAFTA. Costa Ricans received him with a mixture of expectation and skepticism. In his acceptance speech he promised to reduce the high cost of living and to create more jobs and security. ∎

IN FOCUS

ENVIRONMENTAL CHALLENGES
Deforestation (largely due to the expansion of farming) has been partially responsible for soil erosion. Coastal marine pollution and air contamination are observed. Land reduced fertility, particularly in the Region Huetar Norte, is caused by the expansion of wood production forests. Reduced soil fertility, particularly in the Huetar Norte Region, is caused by the expansion of timber plantations. The rich biodiversity of the forests (there are about 150 different tree species in the country) actively needs safeguarding.

WOMEN'S RIGHTS
Women have been able to vote and run for office since 1949. In 2006, 35 per cent of deputies were women. In 2003, they made up 33 per cent of the work force, and 8 per cent were unemployed. In 2004, the illiteracy rate for both women and men was 4 per cent*.

CHILDREN
In 2004, 11 of every 1,000 children died before their first birthday, and 13 died before their fifth birthday. Sexual abuse, exploitation of minors and child labor are frequent.
Although the primary school enrollment rate was 91 per cent in 2000*, 3 out of 10 drop out; 4 out of 10 drop out from secondary school.
According to the

Responsible Paternity Law, adopted in 2001, when a mother registers the birth of an illegitimate child she must declare both paternity and maternity. If the father is absent, the mother is authorized to declare and register paternity (assuming she knows who the father is). The man may attempt to prove he is not the father, using DNA evidence. If the alleged paternity is proven the father is obliged to pay child support. 30 per cent of the children lack official acknowledgement of paternity. There are still institutional and cultural difficulties in applying this law, but it is nonetheless an international milestone in this field.

INDIGENOUS PEOPLES/ ETHNIC MINORITIES
The black community, descendant from immigrants who have come mainly from Jamaica since 1870, is divided into an élite that seeks to revitalize the culture of their ancestors and the poor who live in dire conditions.

MIGRANTS/REFUGEES
By the end of 2002, there were 12,750 refugees living in Costa Rica: 7,600 Colombians, 2,700 Nicaraguans, 1,100 Cubans and 1,350 people from another 28 countries.
Another 20,000 to 50,000 Colombians were living in Costa Rica in refugee-like circumstances. 58 per cent of their asylum applications had been approved. Costa Rica grants equal rights to its citizens and refugees, who can

apply for permanent residence after living in the country for two years. The UNHCR provides legal, economic, social and psychological assistance to refugees through a non-governmental local agency.
In 2005, more than 8,000 Colombians (out of a total of 10,400 refugees) lived in Costa Rica, which hosts the second-largest Colombian refugee group in Latin America. Although the number has dropped in recent years, asylum applications average 1,440 per year. In general, Costa Rica grants equal rights to citizens and refugees, who can apply for permanent residence after living in the country for two years. The UNHCR provides legal, economic and psychological aid to refugees through a local non-governmental agency.
Refugees in Costa Rica are largely urban, middle-class and live in the biggest cities of the country. Though they are allowed to work, their relative unemployment rate is high, since many employers are not aware that it is legal to employ refugees. Some 300,000 Nicaraguans have migrated to Costa Rica over the last 20 years.

DEATH PENALTY
The death penalty was abolished in 1877.

*Latest data available in *The State of the World's Children* and *Childinfo* database, UNICEF, 2006.

Côte d'Ivoire / Côte d'Ivoire

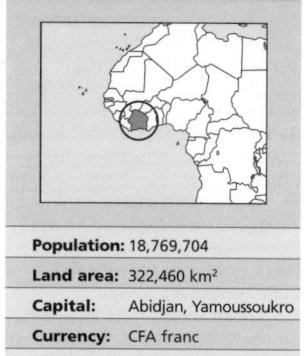

Population:	18,769,704
Land area:	322,460 km²
Capital:	Abidjan, Yamoussoukro
Currency:	CFA franc
Language:	French

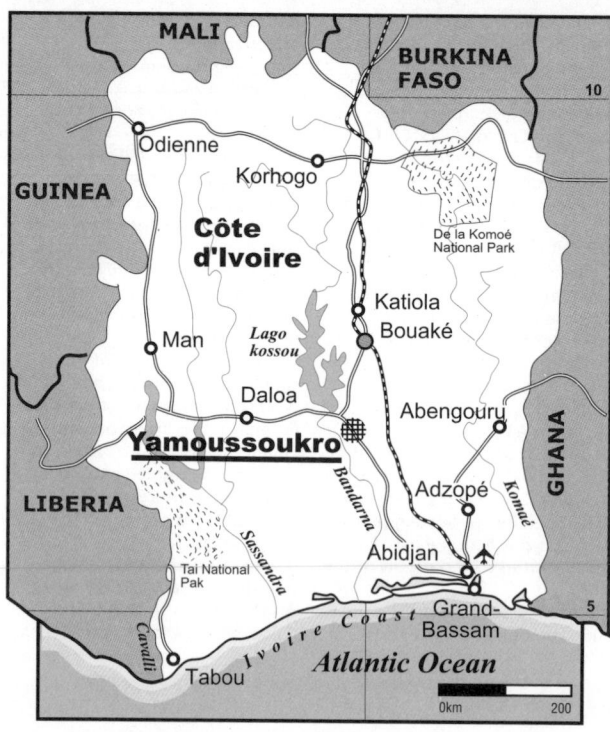

According to Baulé tradition, in 1730 Queen Aura Poka emigrated westwards with her people and founded a new state in the center of a territory known since the 15th century as the Côte d'Ivoire (or Ivory Coast) because of the active trade in elephant tusks.

2   This Ashanti state soon grew and became a threat to the small states of Aigini, on the coast, and Atokpora, inland. In 1843, these states requested French protection, thus enabling France to obtain exclusive rights over the coastal trade and, later, the opportunity to annex Côte d'Ivoire to its other territories, now called Guinea, Mali and Senegal.

3   However, the French met with stiff resistance from Samori Touré. He had set up a state in the heart of the region that the Europeans were planning to unify (See Guinea).

4   In 1898 Touré was defeated and the leaders of the dominant groups signed colonial agreements with the French.

5   Th e area became known as French West Africa, comprising Senegal, French Sudan (now Mali), Guinea and Côte d'Ivoire. Later, what are now Chad, Burkina Faso and Mauritania were also annexed. The French hoped to balance the poorer regions of Chad and Burkina Faso with the better off Senegal and Côte d'Ivoire.

6   Modern political activities began in 1946 with the creation of the African Democratic Union (ADU), a political party with branches in Senegal, Mali and Guinea. The ADU promoted independence and unity for French colonies in the area.

7   Felix Houphouët-Boigny, a physician and well-to-do farmer, was appointed party president, based on his experience leading a farmers' association which had fought against colonial policies.

8   Tactically allied to the French Communist Party, the ADU staged strikes and boycotts of European businesses. The nationalist cause was savagely suppressed, leaving many dead and thousands imprisoned, giving Houphouët-Boigny grounds for ending his alliance with the

French communists in 1950. He soon reached a new agreement with Mitterrand, then Minister for Overseas Territories. This move undermined the ADU's standing, and Houphouët-Boigny was only just able to maintain his prestige at home.

9   Between 1958 and 1960 all of French West Africa became independent and the new states joined the UN. Aware of their meager economic prospects, the political leaders proposed to form a federation. However Houphouët-

Boigny, confident in the relative prosperity of his country and his privileged neo-colonial relations with France, opposed the idea.

10   As a major producer of cocoa, coffee, rubber and diamonds, Côte d'Ivoire was able to attract transnational investors, offering them the political stability produced by Houphouët-Boigny's paternalistic, authoritarian rule and cheap labor, mostly supplied by neighboring countries.

11   The economic growth rate, which remained between eight and ten per cent a year from 1966 to 1976, declined as the West went into a recession after 1979. Agricultural exports fell from $4 billion to barely $1 billion between 1980 and 1983. Half the industries set up between 1966 and 1976 closed down, pushing unemployment figures up to 45 per cent. The foreign debt in 1985 was five times greater than in 1981.

12   In 1985 the 8th congress of the Côte d'Ivoire Democratic Party (PDCI) proposed Houphouët-Boigny for a sixth presidential term. The appointment was confirmed by apparently 99 per cent of the vote in that year's election.

13   The Government built the largest basilica in Africa in a country where only 12 per cent of the population is Christian.

14   Côte d'Ivoire suffered a blow to its main export, cocoa. Between July and October 1987, cocoa prices plummeted 50 per cent on the international market.

15   After the death of the President in December 1993, Henri Konan-Bédié, leader of the National Assembly, assumed the post. He consolidated his power within the ruling Democratic Party in spite

WORKERS

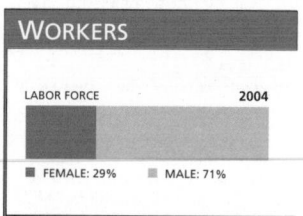

LABOR FORCE   2004

■ FEMALE: 29%   ■ MALE: 71%

PROFILE

ENVIRONMENT

Located on the Gulf of Guinea, the country is divided into two major natural regions: the South, with heavy rainfall and lush rainforests, where foreign investors have large plantations of cash crops like coffee, cocoa and bananas; and the North, a granite plain characterized by its savannas, where small landowners raise sorghum, corn and peanuts.

SOCIETY

Peoples: The population includes five major ethnic groups, the Kru, Akan, Voltaic speakers, Mande and Malinke, some from the savannas and others from the rainforests, sub-divided into approximately 80 smaller groups. There are many 'non-Ivorians' living in the country, mostly from the neighboring countries.

Religions: It is difficult to quantify the religions and mixture of beliefs of 'non-Ivorians'. They are divided more or less equally among traditional beliefs, Islam and Christianity.

Languages: French (official). There are as many languages as ethnic groups; the most widely spoken are Diula, in the North; Baule, in center and West;

and Bete in the Southeast.

Main Political Parties: Democratic Party of Côte d'Ivoire; Ivorian Popular Front; Ivorian Workers' Party; Rally of the Republicans; Union of Democrats of Côte d'Ivoire.

Main Social Organizations: The General Workers' Union (UGT) is the only recognized union. Several opposition unions work underground.

THE STATE

Official Name: République de Côte d'Ivoire.
Administrative Divisions: 49 departments.
Capital: Abidjan (economic center) 3,337,000 people; Yamoussoukro (political center since 1983) 416,000 (2003).
Other Cities: Bouaké 741,100 people; Daloa 184,300; Korhogo 164,400 (2000).
Government: Parliamentary Republic. Laurent Gbagbo, President since October 2000. Charles Konan Banny, Prime Minister since December 2005. Unicameral Legislature: National Assembly, with 225 members.
National Holiday: 7 August, Independence (1960).
Armed Forces: 8,400. Other: 7,800.

Life expectancy
46 years
2005-2010

GNI per capita
$760
2004

Literacy
48% total adult rate
2000-2004

HIV prevalence rate
7% of population 15-49 years old
2003

of opposition from former prime minister Alassane Ouattara.

[16]   In 1994 the Government faced a strong labor movement demanding compensation after the 100 per cent devaluation of the CFA franc in January. In what was seen as a reward for having accepted the devaluation - imposed by France and the IMF - half of Côte d'Ivoire's debt to the Paris Club was cancelled. However, the country continued to have the world's highest per capita foreign debt.

[17]   In 1995, Bédié won a presidential election that the opposition boycotted. For the first time since the country's independence, residing foreigners were not allowed to vote, nor were Côte d'Ivoire citizens who had foreign parents. This permitted Bédié to dispose of Ouattara, whose father was from Burkina Faso.

[18]   A military coup led by General Robert Guéi overthrew Bédié, who fled the country. The US and the European Union urged the military junta to return to democracy. In January 2000, after declaring that the state coffers were empty, Guéi took over as interim president and announced a referendum that modified the Constitution and set elections for October.

[19]   The constitutional changes gave the right to vote to those aged 18 and over (which Bédié had originally opposed, fearing that the younger segment of the population - affected by his social policies - would tend to vote against him). Last minute amendments stipulated that both parents of presidential candidates must be 'of Ivorian origin', a change that once again excluded Ouattara.

[20]   The abstention rate was over 60 per cent in the October 2000 elections. Two days later, when the re-count gave the victory to the socialist candidate of the Ivorian People's Front (FPI), Laurent Gbagbo, with 51 per cent of the vote, Guéi declared himself President. The presidential guard took over the Electoral Commission and the director said errors had been made in the re-count and denounced fraud committed by Gbagbo's party. The streets filled with protesters, but neither the army nor the police intervened. The Electoral Commission reappeared, but this time with the true results: Gbagbo 60 per cent; Guéi 32 per cent. Gbagbo took office while Guéi was in Benin.

[21]   Initially the three majority parties were united against Guéi, but after the official announcement of Gbagbo's victory, Ouattara's supporters demanded new elections. The dispute triggered an anti-North uprising. In the South, mosques were burned and and FPI

IN FOCUS

ENVIRONMENTAL CHALLENGES
Côte d'Ivoire currently has one of the fastest rates of deforestation in the world. Its forests, once the largest in West Africa, have been almost completely cleared. Industrial and agricultural wastes contaminate water sources.

WOMEN'S RIGHTS
Women have been able to vote and be elected since 1952. In 2000, women held 8 per cent of seats in the Lower House of Parliament.

In 2003, they made up 33 per cent of the workforce. Their life expectancy had declined from 52 years in 1980 to 46 in 2003.Illiteracy among adult women reached 62 per cent in 2004. By the end of 2003,300,000 women aged between 15 and 49 were living with HIV/AIDS.

CHILDREN
Côte d'Ivoire is one of Africa's richest countries, but internal conflicts have disrupted access to basic health and education services. Hundreds of thousands of primary school children had no regular lessons as schools

closed in the conflict-affected areas (north and west). Many health centers have been forced to cut services as staff fled and provisions of essential medical supplies dwindled.

In 2004, under-one and under-five mortality rates reached 117 and 194 per 1,000 live births, respectively. Seventeen per cent of newborns were underweight and 21 per cent of children under five years old had severely or moderately stunted growth. It was estimated that some 310,000 children under 17 years old were orphaned by HIV/AIDS.

INDIGENOUS PEOPLES/ETHNIC MINORITIES
The population of Côte d'Ivoire is diverse, with over 60 ethnic groups. The main ones include the Akan-speaking peoples of the south-east, Mande peoples in the north, the Voltaic groups of the north-east, and the Kru of the south-west. About 12 per cent of the Ivorian inhabitants are Baoulé, an Akan-speaking people and the largest group in the country. Meanwhile, immigrants from Burkina-Faso, Mali and Ghana form 30 per cent

of the country's population. About one quarter of the population is Muslim, while one in eight people is Christian; the remainder follow traditional beliefs.

MIGRANTS/REFUGEES
In early 2006, there were 34,400 Liberian refugees (the vast majority of them living in special areas in the west of the country); only about 400 lived in Abidjan. There were also some 2,600 urban refugees (over 50 per cent women) mainly students and professionals from DR Congo, Republic of Congo, Rwanda, Burundi and the Central African Republic. It is also estimated that after the civil war in 2002, 500,000 Ivorians remained internally displaced and 25,000 went into exile.

DEATH PENALTY
The death penalty was abolished in 2000.

Latest data available in The State of the World's Children and Childinfo database, UNICEF, 2006.

militants supported by the army and the police massacred some 500 Muslim northerners.

[22]   The new President refused to re-run the elections. The Gbagbo administration's first actions included investigating and punishing the perpetrators of the deaths of young Muslims linked to the Rally of the Republicans (RDR); and also creating a national reconciliation committee.

[23]   In April 2002, the European Union and the Paris Club cancelled $911 million of Côte d'Ivoire's debt. In exchange, the Government promised to implement broad structural reforms, which would include privatizing state companies, reducing subsidies and liberalizing trade.

[24]   In September, armed violence broke out in Abidjan and other cities of the country. The rebellion - led by the so-called Côte d'Ivoire Patriotic Movement - was backed by the majority of the northern population, mostly Muslim. The Minister of Interior, Emile Boga Doudou, and the former president Guéi were murdered in Abidjan during the uprising.

[25]   In October, on the arrival of a group of Ministers from Economic Community of West African States (ECOWAS), the rebels agreed to a ceasefire. However a new offensive by Government forces broke this truce and rebels captured the city of Daloa. After

the Government troops expelled them, a new ceasefire was signed in the northern bastion of Bouaké, backed by several neighboring countries which supported the creation of an intermediary regional force.

[26]   In November 2003 the UN Security Council extended its mission in Côte D'Ivoire, after a new peace agreement was signed in France between the Ivorian army and the rebel *Forces Nouvelles*. The blue helmets would focus on protecting civil rights and media security in the run-up to the 2005 elections.

[27]   Clashes during the repression of an opposition rally against Gbagbo's administration, in March 2004, left 120 people dead and led to the withdrawal of rebel leaders from the coalition. In May, the UN reported summary executions and torture by security forces. Gbagbo replaced FN ministers with FPI members.

[28]   In November 2004, the air force bombed rebel-held towns in the north. After the deaths of French soldiers, French President Jacques Chirac ordered the destruction of most Ivorian warplanes and helicopters, sparking riots in Abidjan and Yamoussoukro. French helicopters swept in to rescue Europeans stranded in Abidjan and troops killed some demonstrators. Meanwhile, the Government urged people to rise

up against the old colonial power. The UN Security Council called an emergency meeting to discuss imposing an arms embargo, while Ivorian officials were banned from travelling.

[29]   In spite of the violence in early 2005, a ceasefire agreement was reached in April between rebels and the Government. The agreement also provided for presidential elections to be held that same year.

[30]   Plans for elections were shelved in October, when President Gbagbo invoked a law which he said allowed him to stay in power. The UN extended its mandate to stay in the country.

[31]   In December, Economist Charles Konan Banny was appointed Prime Minister by mediators, in the expectation that he would disarm militias and call for elections.

[32]   In January 2006, Gbagbo's supporters took to the streets to protest against what they regarded as excessive UN interference in the country's internal affairs. The next month, presidential elections were announced for October.

[33]   In October the planned presidential election could not take place and a summit of West African leaders in Lagos failed to agree on a way forward for the divided country. President Gbagbo's term of office was set to expire at the end of October. ∎

Croatia / Hrvatska

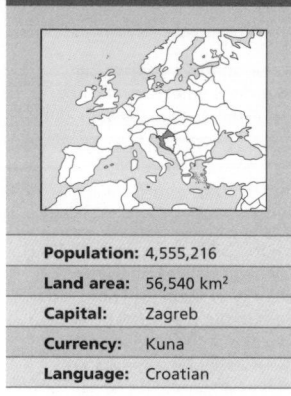

Population:	4,555,216
Land area:	56,540 km²
Capital:	Zagreb
Currency:	Kuna
Language:	Croatian

The Croats, a Slav people, emigrated in the 6th century AD from White Croatia, a region today in the Ukraine. They moved on toward the Adriatic Sea, where they conquered the Roman stronghold of Salona, in 614. Once established in Pannonia and Dalmatia, the Croats freed themselves from the Avars and began developing independently. Even though the territory was under the Byzantine empire, the Croats accepted the Roman Catholic Church, whilst preserving the Slav liturgy.

2   In the 8th century, Croats set up the dukedoms of Pannonia and Dalmatia. The first one was under the French Empire's rule and the second under Byzantium, as a result of the 812 peace treaty between the two empires. Both dukedoms broke free by the mid-9th century and joined to develop the first independent Croatian kingdom.

3   The independent Croat State developed in Dalmatia. During the reign of Tomislav (910-928) this achieved great military development. Tomislav and his heirs defended themselves from the Bulgar Empire in Pannonia and the Venetian expansion along the Dalmatian coast. The Byzantine Empire helped King Stjepan Drzislav (969-997) to defend himself from the Venetians, and re-established its influence in

the Adriatic. King Peter Kresimir (1058-1074) broke with Byzantium and strengthened links with the papacy. In that period, Croatia reached the peak of its power and territorial expansion.

4   During Kresimir's rule, the country split in two, with one group favoring the king, and an opposition group backed by popular support. When Dimitrije Zvonimir tried to involve the Kingdom in a war against the Seljuk Turks, the opposition accused him of being the Pope's

vassal and assassinated him in 1089. The civil war that was then unleashed marked the beginning of the decline of the Croatian Kingdom.

5   The Byzantines recovered Dalmatia. Lazlo I of Hungary conquered Pannonia in 1091, laying claim to the Croatian crown. He founded a bishopric at Zagreb in 1094, which became the center of the Church's power in the region. The Dalmatians crowned Petar Svacic - the last king of Croat blood - but the Pope considered him a rebel, and turned to King

Kalman of Hungary who invaded the country and unseated Svacic in 1097.

6   After an extended war Kalman signed a treaty, the *Pacta Conventa*, with the Croat representatives. Only Bosnia, then a part of the Croatian kingdom, refused to submit to a foreign monarch. For the next eight centuries, Croatia was linked to Hungary. In the 14th century, Dalmatia became a part of Venice, which ruled over it for 400 years.

7   After the defeat of the Croatian and Hungarian forces in the battles of Krbavsko Polje (1493) and Mohacs (1526), most of Pannonia and Hungary fell into Turkish hands.

8   Turkish domination altered the ethnic composition of Pannonia, as many Croats migrated northwards, some even going into Austria. In the meantime, the Turks brought in German and Hungarian settlers, and gave incentives for Serbians fleeing the Balkans to settle in the Vojna Krajina.

9   When the Turks were driven back in the 17th century, Austria tried to limit Croatia and Hungary's state rights, to make them mere provinces of the Austrian Empire. The Croatian and Hungarian nobility conspired together to organize an independence movement, which failed. The Croatian leaders were executed and their lands were distributed among foreign nobles.

10   After the annexation of Rijeka (Fiume) in the 1770s, Hungary tried to impose its language, but this triggered a nationalist reaction

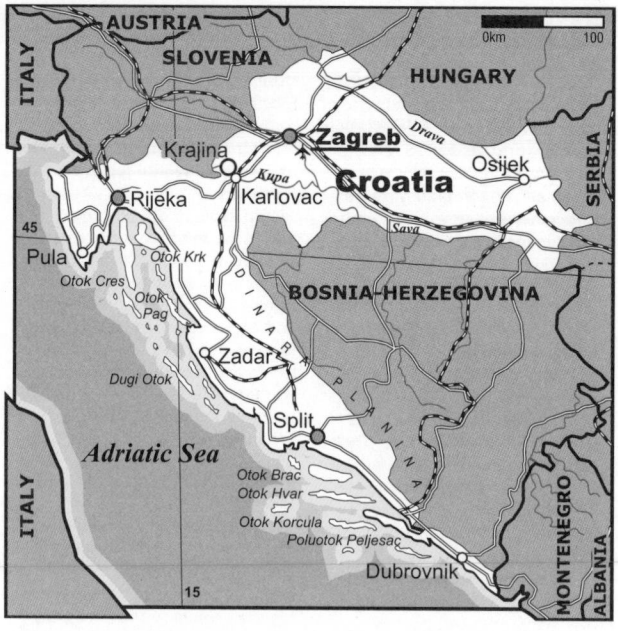

LAND USE

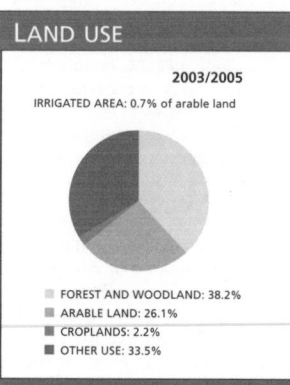

2003/2005

IRRIGATED AREA: 0.7% of arable land

- ■ FOREST AND WOODLAND: 38.2%
- ■ ARABLE LAND: 26.1%
- ■ CROPLANDS: 2.2%
- ■ OTHER USE: 33.5%

PROFILE

ENVIRONMENT
Croatia is bounded to the north by Slovenia and Hungary, and to the east by Serbia. The Dalmatian coast - a 1,778-kilometer coastline on the Adriatic, with numerous ports and seaside resorts, as well as a thousand islands - lies in southern and western Croatia. The territory is made up of three regions with different landscapes: rolling hills in the north, around Zagreb; rocky mountains along the Adriatic coast and the inland valleys of the Pannonian Basin. The coast mountains are the Dinaric Alps and the Valebit and Velika Kapela mountain systems, between 700 and 2,200 meters high. The inland valleys are washed by the mid and upper streams of River Sava, which runs through the country from the northwest to the southeast, and it is part of the borderline shared with Bosnia-Herzegovina. The River Drava forms part of the border between Serbia and Croatia.

SOCIETY
Peoples: 78.1 per cent Croats; 12.1 Serbs; Slav-Muslims 1.0; Hungarians 0.5; Slovenians 0.5. Czechs, Italians, Roma.
Languages: Croatian (official) 96 per cent. Istrian, Romani, Czech, Slovac, Italian.
Religions: Roman Catholic 76 per cent; Orthodox

11 per cent; Muslims one per cent.
Main Political Parties: Croatian Bloc (HB); Croatian Christian Democratic Union (HKDU); Croatian Democratic Union (HDZ); Croatian Party of Rights (HSP); Croatian Peasant Party (HSS).
Main Social Organizations: Association of the Independent Trade Unions of Croatia, Union of Autonomous Trade Unions of Croatia, Railworkers' Union of Croatia, Metalworkers' Trade Union of Croatia, Croatian Journalists' Association.

THE STATE
Official Name: Republika Hrvatska.
Administrative Divisions: 102 Districts.
Capital: Zagreb 688,000 people (2003).
Other Cities: Split 200,800 people; Rijeka 180,000 (2000). **Government:** Stjepan Mesi, President since February 2000, re-elected in January 2005. Ivo Sanader, Prime Minister since December 2003. The unicameral Assembly (Sabor) is one of Europe's oldest: it has 152 members elected for four-year terms. (One seat was added in the November 2003 parliamentary elections).
National Holiday: June 25, Independence (1991).
Armed Forces: 105,000 (1995). Other: 24,000 Police (1,000 deployed in Bosnia).

Life expectancy
76 years
2005-2010

GNI per capita
$6,820
2004

Literacy
98% total adult rate
2000-2004

HIV prevalence rate
<0.1% of population 15-49 years old
2003

among the Croatians. The French Revolution and the Napoleonic Wars which had incorporated Dalmatia, Pannonia and the area south of the Sava river to the French Empire, further stimulated Croatian nationalism. Upon the fall of Napoleon, relations between Hungary and Croatia rapidly deteriorated.

[11] In April 1848, the Hungarian Parliament adopted a series of measures limiting Croatian autonomy. The Croatian Diet (Parliament), dissolved in 1865, declared its separation from Hungary, abolishing serfdom and approving equal rights for all its citizens. Hungary's troops were weakened by this conflict, making it easier for the Habsburgs to put down the Hungarian Revolt and regain power later that year.

[12] With the division of the crown, Germany and Hungary became the major nations of the Austro-Hungarian Empire. In 1868, Hungary accepted the union of Croatia, Slavonia and Dalmatia as a separate political entity, though Austria refused to relinquish its claim to Dalmatia.

[13] In the early 20th century, Croatian nationalists intensified their activity. An alliance of Croatian and Serbian leaders adopted the 'Rijeka resolution', a plan of action which enabled them to win the 1906 elections. The Croatian Peasant Party began political activity among the peasants. The Crown responded by increasing repression.

[14] In 1915, Croatian, Serbian and Slovenian leaders organized the Yugoslav Committee in Paris to push for separation from the empire and union with an independent Serbia. Austria-Hungary's defeat in World War I accelerated the creation of the Yugoslav kingdom in 1918.

[15] The Serbs implemented an unifying policy that was in conflict with Croatian desires for independence and autonomy. Croatia demanded the creation of a Yugoslav federation. From 1920, the Peasant Party led by Stjepan Radic headed the Croatian opposition, until the leader's assassination in 1928.

[16] Yugoslavia was divided by Hitler in 1941 during World War II. The German army launched a racist campaign against Serbs, Jews, Gypsies/Roma and Croats.

[17] Local committees were created during the communist-led resistance. After the anti-Nazi guerrillas occupied Zagreb in May 1945, the Anti-Fascist Council of National Liberation of Croatia took over the Government. By the end of the year, Croatia joined the new People's Federated Republic of Yugoslavia.

ENVIRONMENTAL CHALLENGES
Air pollution (from metals industry) and resulting acid rain is damaging the forests. Coastal pollution from industrial and domestic waste is observed. The number of active landmines - buried during the 1990s conflict - has declined significantly. At the end of 2001, anyway, some 50,000 still active landmines remained in an area of 1,700 square kilometers.

WOMEN'S RIGHTS
Croatian women have been able to vote and run for office since 1945. In 2003, they held 22 per cent of seats in Parliament, as well as 33 per cent of ministerial positions. That year, they made up 45 per cent of the labor force. Female unemployment stood at 17 per cent; of those who worked, 63 per cent were in services, 21 per cent in industry and 15 per cent in agriculture.

CHILDREN
As in other former Eastern bloc countries, some children in state orphanages and institutions suffer ill-treatment and abuse. In spite of being illegal, domestic violence persists. Underweight, child mortality and malnutrition rates are low; immunization is close to 100 per cent in all cases and primary education reaches 90 per cent of children.

INDIGENOUS PEOPLES/ ETHNIC MINORITIES
The number of Roma (Gypsies) is estimated at between 200,000 and 300,000. Discrimination against the Roma in Croatia has been persistent over the last decade. Some 219,000 Serbs lived in Croatia in 2002. Although it is expected that the country's expressed intention to join the EU will result in a better treatment of its minorities, the long history of animosity between Serbs and Croats does not create favorable conditions for the Serb minority. Apart from being a small minority, Serbs are excluded from political and economic decisions and cannot count on Serb military support.

MIGRANTS/REFUGEES
In December 2005, Croatia hosted 2,545 refugees from Bosnia-Herzegovina, 376 from Serbia and Montenegro and 4,800 internally displaced people.

DEATH PENALTY
It was abolished in 1990.

* Latest data available in *The State of the World's Children* and *Childinfo* database, UNICEF, 2006.

[18] At the end of 1980, the Yugoslav League of Communists (YLC) repealed the monopoly and leading role it was granted by the Constitution, and in April 1990, the first multiparty elections since World War II were held (see Serbia and Montenegro).

[19] Communists were defeated by the Democratic Croat Union (DCU), led by Franjo Tudjman. In 1991, Croatia declared independence but the European Community and the US withheld recognition. The eastern Serbo-Croats took over one-third of Croat territory and expelled the Croats with help from the Yugoslav army. Croatia supported the Bosnian Croats, first against the Bosnian Serbs, then against the Muslims, during the war in Bosnia-Herzegovina (1992-95). Serbs proclaimed the republic of Krajina as a new member of the Yugoslav federation.

[20] In 1992 a peace plan for Serbia and Croatia was brokered by the European Community. In May, the UN accepted membership of former Yugoslavian republics Croatia, Slovenia and Bosnia-Herzegovina. By the end of 1992, the Bosnian Serb leader Radovan Karadzic announced the formation of the Serb Republic of Bosnia-Herzegovina and the Serb Republic of Krajina, leading to renewed hostilities. The war in Croatia caused thousands of civilian deaths. In late 1993, Tudjman and President Alija Izetbegovic of Bosnia-Herzegovina signed a ceasefire agreement.

[21] Tudjman, re-elected President in 1992, restored diplomatic relations with Serbia in 1996. Even though the Croat re-occupation of Krajina was considered one of the biggest ethnic cleansing operations of the war in the former Yugoslavia, by May 1996 only eight Croats had been tried *in absentia* by The Hague War Crimes Tribunal.

[22] Tudjman died in 1999 and in January 2000, the DCU was defeated by a coalition led by the Social Democrats and the Social-Liberals. In February, when Stjepan Mesi - of the Croat Peoples Party - became President, he announced Croatia's intention of joining NATO and the European Union (EU).

[23] In February 2001, hundreds of thousands of people, led by war veterans, protested against attempts to arrest General Mirko Norac for war crimes. The Right accused the Government of treason, with the intention of bringing down Mesi and Prime Minister Ivica Racan. The latter was criticized when he decided to comply with the Hague Tribunal request for the extradition of generals Ademi and Gotovina, accused of war crimes. A parliamentary vote allowed the premier to stay in office.

[24] The DCU won the November 2003 elections, defeating the center-left coalition led by former Prime Minister Racan. The new leader, Ivo Sanader, undertook to keep Croatia's international commitments, including the co-operation with the UN War Crimes Tribunal. Other campaign pledges were to join NATO in 2006 and the EU in 2007.

[25] In June 2004, Serb leader Milan Babic was sentenced to 13 years in prison by The Hague tribunal for his part in war crimes against non-Serbs in the self-proclaimed Krajina Serb republic in the early 1990s. Babic was accused of having participated in repeated 'ethnic cleansing' operations. According to the judge presiding the tribunal, Alphons Orie, Babic was responsible for the death of more than 200 civilians, including women and children, and for confining and jailing hundreds of civilians in inhuman conditions.

[26] In December 2004, Croatia started talks with the EU with the goal of joining the bloc in March 2005.

[27] Mesi won wide support in the January 2005 presidential elections. He narrowly missed an outright victory in the first round, when he won 49 per cent of the votes. In the second round, he defeated Deputy Prime Minister Jadranka Kosor by 66 to 34 per cent. Although the election platforms of both Mesi and Kosor were almost identical (improve the economy; develop friendly relationships with other countries; join the EU), it was Mesi who managed to win the support of center candidates.

[28] The failure to arrest General Ante Gotovina, wanted by international courts for war crimes, delayed Croatia's accession to the EU. In October, the green light was given for EU accession talks to go ahead and finally, in December, Ante Gotovina was arrested in Spain.

[29] Early in 2006, there were demonstrations in support of Gotovina. The general was considered a war hero since 1995, when he recovered the Krajina region, which had fallen under Serb control during the conflict. ∎

Cuba / Cuba

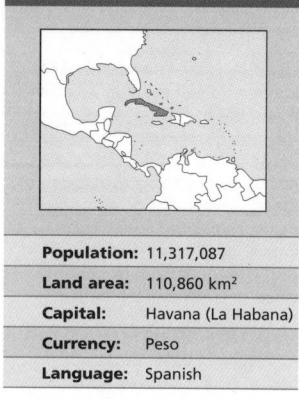

Population:	11,317,087
Land area:	110,860 km²
Capital:	Havana (La Habana)
Currency:	Peso
Language:	Spanish

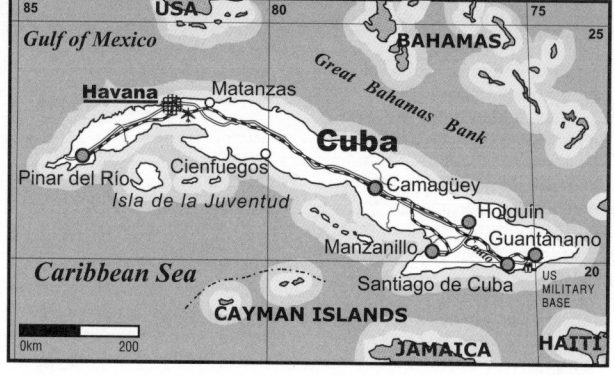

Until the 16th century the island of Cuba was inhabited by several ethnic groups, mainly the Taino or Arawak and Ciboney. Christopher Columbus arrived there on 27 October 1492 but it was not until 1509 that a voyage was made around the coast proving that Cuba was an island. The conquest which was begun in 1511 by Diego Velázquez de Cuéllar, ended in 1514 when the Spanish were defeated by the local forces led by chieftains Hatuey and Guama.

2   The Spanish expeditions which would subsequently conquer a large part of the Caribbean, Mexico and Central America, departed from Cuba. It was described in a mid-16th century document as 'high and mountainous' with small rivers 'rich in gold and fish'. In 1511, colonists from Santo Domingo started mining Cuban gold. It was a short-lived economic cycle, probably because the Indian population was rapidly exterminated. As the number of African enslaved workers on the island was insufficient, economic life soon declined and did not recover until the end of the 16th century, with the advent of sugar production.

3   As early as the 17th century, economic diversification was achieved through shipbuilding and the developing leather and copper industries. The economic center of the island gradually moved from Santiago, on the southern coast, to Havana, in the north, a port of great importance in the mid-17th century.

4   Towards 1840, the slave labor in the sugar plantations represented 77 per cent of the total Cuban workforce. There is evidence that there were *palenques*, settlements of escaped slaves, similar to the so-called *quilombos* in other parts of America (see Brazil) on the island, and there was an abortive slave revolt known as 'La Escalera' in 1843.

5   In addition to the British efforts to end slavery, actions taken by the slaves themselves during the Ten Years War contributed

decisively to the abolition of slavery in 1886.

6   As in the other Spanish colonies in the Americas, the struggle for independence began in the first few decades of the 19th century. Spain reinforced its military presence and, in 1818, a liberalization of trade policies took effect, allowing the export of sugar to the US. The second war of independence began in 1895, led by José Martí, Antonio Maceo and Máximo Gómez. In 1898, aware that the victory of the Cuban patriots was inevitable, the US declared war on Spain and landed at Guantánamo.

7   US occupation forces ruled the country from 1899 to 1902, imposing a constitution including the so-called 'Platt Amendment'. This secured the US rights to intervene in Cuba and to retain Guantánamo, where they set up a powerful military base, still in operation. The 'right' to intervene has been exercised on various occasions, with US Marines remaining on Cuban soil for extended periods of time.

8   In 1933, a popular uprising overthrew Machado from power. Grau San Martín tried to implement several popular and anti-imperialist measures but he was forced to resign by US pressure. Fulgencio Batista emerged as a key-figure during a turbulent period characterized by corruption and gangsterism under the auspices of the US. On 10 March 1952, Batista engineered yet another coup, establishing a dictatorial regime which was responsible for the death of 20,000 Cubans.

9   On 26 July 1953, Fidel Castro and a group of revolutionaries attacked the Moncada Army Base in Santiago de Cuba. Although the attack itself failed, it marked the beginning of the revolution. Castro's revolutionary program had been defined during his trial after the failed initial uprising, ending with his well-known words: 'History will absolve me'. After a period of imprisonment and subsequent exile

in Mexico, he landed in Cuba with a small force in December 1956.

10   At the end of 1958, Batista fled from Cuba as guerrilla forces led by Ernesto 'Che' Guevara and Camilo Cienfuegos, vanguard of the Rebel Army, closed on Havana. In just over two years, the guerrillas of the '26th of July' Movement broke the morale of Batista's corrupt army.

11   In 1961, counter-revolutionaries backed by the US disembarked at Playa Giron (Giron Beach), in the Bay of Pigs, in an attempt to bring down the regime which had carried out agrarian reforms and expropriated various American enterprises. They had counted on a popular uprising against the revolutionary Government, but this did not materialize. After 72 hours of fierce fighting, the Bay of Pigs invasion ended in the defeat of the invading forces. Two days before the invasion, on 15 April, while the victims of the Havana Airport bombing were being buried, Castro proclaimed the socialist nature of the revolution and its political alignment with the Soviet bloc.

12   Also in that year, all the pro-government organizations joined together in a common structure. This was initially known as the Integrated Revolutionary Organizations (ORI), and later became the United Party of the Socialist Cuban Revolution (PURSC).

13   In 1962, the US had Cuba excluded from the Organization of American States (OAS). It also put pressure on other countries to sever diplomatic relations, engineering an economic blockade of the island under the pretext of Cuban support to revolutionary movements in Latin America. In October, the setting up of Soviet nuclear missile launching sites on the island made the possibility of a war between the US and the Soviet Union (USSR) into a near-probability. The end of the crisis was negotiated between Washington and Moscow. Cuba was denuclearized and the US pledged not to invade it, but disregarded the Cuban Government's demands for an end to the blockade, the

withdrawal of US troops from Guantánamo and an end to US-managed terrorist activities.

14   The literacy campaign during these years soon bore fruit, and by 1964, Cuba was free of illiteracy. Improvements in health were also one of the Government's priorities. In October 1965, the PURSC was turned into the Cuban Communist Party (PCC). In 1967 Che Guevara was killed in Bolivia.

15   The economic and political ties with the USSR were strengthened in the following years. Cuba began lending technical assistance to like-minded peoples and governments o`f the Third World. In 1975, troops were sent to countries like Ethiopia and Angola, who requested help to resist invading forces.

16   The revolution began to be institutionalized after the first PCC congress in 1975. A new constitution was approved in 1976 and there were subsequent elections of representatives for the governing bodies at municipal, provincial and national levels. In 1979, Castro and the leaders of the PCC launched a campaign of revolutionary requirements to correct weaknesses in the administrative and political management areas of the revolutionary process.

17   Tension between Cuba and the US increased when Ronald Reagan came to the White House in 1979. In the mid-1980s, some 120,000 people left Port Mariel bound for Florida. During this period, Washington's two-sided policy towards Cuban immigration was intensified. On the one hand, the illegal departure from the island was supported and propagandized by the US, but on the other hand entry applications were restricted.

18   Cuba's relations with many Latin American countries improved after the 1982 Malvinas/Falklands war. During 1985, several meetings were held in Havana on the debt issue, which also drew Cuba closer to the rest of the region.

19   After the Third PCC Congress in 1986, a 'process of the rectification of errors and negative tendencies' was initiated. This coincided with the changes which were taking place in the USSR, but the Cubans avoided adopting the eastern European model.

20   In June 1989, a high-ranking group of Army officers and officials of the Ministry of the Interior were brought to trial and four of them were executed for being involved in drug trafficking. Among them was Arnaldo Ochoa, the main military chief after Fidel's brother (and potential successor), Raúl.

21   In early 1990, the George Bush (senior) administration increased US pressure on Cuba, with important military maneuvers at Guantánamo Base and in the Caribbean. The

Life expectancy
79 years
2005-2010

Literacy
100% total adult rate
2000-2004

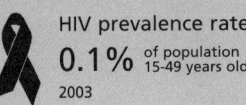

HIV prevalence rate
0.1% of population 15-49 years old
2003

US also violated Cuban television airspace with transmissions by 'Televisión Martí', which was supported by 'The voice of America'. However, this broadcast was only picked up on the island for one day, after which it was jammed.

22　At about this time, the last of Cuba's troops returned home from DR Congo, Ethiopia and Angola. More than 300,000 Cubans had served in Angola, with the loss of 2,016 lives.

23　The 4th PCC Congress took place in October 1991. Together with changes in the top leadership it was decided to reform the Constitution so that members of the National Assembly could be elected directly, the single-party system was ratified, religious freedom was extended and the Government recognized the need for joint ventures, especially with Latin American investors.

24　With the changes in Eastern Europe, and the disappearance of the Council for Mutual Economic Assistance allies, some of Cuba's basic supplies dropped to critical levels. To ease this crisis, the Government strengthened its ties with China, Vietnam and North Korea, and looked for ways to capitalize on its recent technological advances, particularly in biotechnology and medicine (like the meningitis vaccine).

25　With the collapse of the Soviet Union - Cuba's main trade partner - the Cuban economy went into free fall. From 1989 to 1993, its gross domestic product (GDP) fell by half, from $19.3 to $10 billion. When Moscow scrapped its oil-for-sugar deal and cut oil shipments by 25 per cent between1989 and 1991 the country lost most of its petroleum supplies. Imports fell by a huge 75 per cent - much of that was food, spare parts, agro-chemicals and industrial equipment.

26　The years since 1989 are known as the 'special period', a time when the Cuban people had to figure out a way of coping with this massive economic dislocation. The State moved dramatically to restructure the economy. In essence it adopted its own, self-imposed structural-adjustment program.

27　The loss of Cuba's main trading partners and the re-enforcement of the US blockade led to the creation of a Special Plan, aimed at distributing scarce resources equitably. In 1990, bread was rationed at 100 grams per person per day, three newspapers ceased publication, and the official newspaper, *Granma*, was forced to cut its circulation by over half. The tourist industry was strengthened. A further initiative was joint ventures with tens of investors, specially from Spain.

28　In early 1993, National Assembly (parliament) elections were held, and deputies were elected directly for the first time. Ricardo Alarcón, foreign minister, was designated president of the National Assembly, and Roberto Robaina, secretary-general of the Communist Youth organization, became foreign minister.

29　On 26 July 1993, the 40th anniversary of the attack on Moncada barracks, Castro announced that it would henceforth be legal for Cubans to possess and use foreign currency, and to be self-employed. Later that year, the National Assembly organized a debate on the financial crises in workplaces. These 'Workers' Parliaments', held in 1994, provided opinions about the workplaces and their economic management. The sugar harvest (then Cuba's top foreign-exchange earner) had plummeted from 8.4 million tons in 1990 to 4.2 million tons in 1993. In the same year, at the height of the crisis, Cuba was spending 60 per cent of its import bill on food and oil.

30　In April 1994, a meeting entitled 'The Nation and Emigration' was held in Havana on the initiative of Foreign Minister Robaina, and it was attended by some groups of Cubans living abroad. Shortly afterwards, there was a series of incidents in Havana involving people who wanted to leave the country illegally in frail boats. When Cuba announced it would not stop the 'boat people' from leaving, the US started official negotiations to regulate the illegal departure of the immigrants.

31　In July, Cuba entered the Association of Caribbean States (ACS) as a full member. Its participation in the ACS, a group emerging as a new economic bloc, encouraged greater integration of the Cuban economy in the region, offering tariff benefits and trade facilities.

32　In 1995, the fiscal deficit fell for the third year running, due to the reduction of public services and cutbacks in subsidies. A system of convertibility of the peso with the dollar was introduced and holding US currency was legalized. The Cuban parliament approved a new investment law, allowing for totally foreign-owned companies to be established, including by Cuban residents abroad. US Congress approved the Helms-Burton law which penalized companies dealing with Cuba through third-party countries. The international community, especially the EU, harshly criticized this measure for violating the WTO and GATT agreements on free trade.

33　In February 1996, the Cuban air force shot down two light aircraft flown by a group of Cuban exiles in Miami known as 'Brothers to the Rescue'. According to Havana, they had violated Cuban air space to drop anti-governmental leaflets on the island. Amnesty International denounced the imprisonment of several people linked to the Cuban Council. According to Amnesty, the Council included 140 groups of opponent journalists, professionals and union activists, while the Cuban Government linked it with US intervention.

34　The Helms-Burton law barred Cuban access to loans from international institutions like the World Bank or the IMF. However foreign companies continued to invest in Cuba by using various subterfuges such as pseudonyms, and the economic reforms continued to be implemented. In mid-1996, productivity was up 8 per cent on 1995 figures, while the

PROFILE

ENVIRONMENT

The Cuban archipelago includes the island of Cuba, the Isle of Youth (formerly the Isle of Pines) and about 1,600 nearby keys and islets. Cuba, the largest island of the Antilles, has a rainy, tropical climate. With the exception of the southeastern Sierra Maestra highlands, wide and fertile plains predominate in the country. Sugarcane farming takes up over 60 per cent of the cultivated land, particularly in the northern plains. Nickel is the main export mineral resource. Caribbean beaches are exploited as a resource for the tourism industry.

SOCIETY

Peoples: Cubans call themselves 'Afro-Latin Americans', because of the European and African influence in their heritage. Population of mixed descent (mulattos) 51 per cent; whites 37 per cent, blacks 11 per cent, Chinese and others 1 per cent.
Religions: Catholic 39.6 per cent; atheist 6.4 per cent; Protestant 3.3 per cent; and Afro-Cuban syncretists.
Languages: Spanish (official). Some Lucumi words are used in Santería (voodoo) rituals.
Main Political Parties: The Cuban Communist Party (PCC) is defined by the constitution as the 'supreme leading force of society and the state'. It was founded in October 1965, stemming from the United Party of the Socialist Cuban Revolution which had been created in 1962 from the Integrated Revolutionary Organizations. This political front was formed in 1961 by the merger of the July 26th Revolutionary Movement, the People's Socialist Party, and the March 13th Revolutionary Directory. Illegal parties: Cuban Liberal Union; Christian Democrat Party; Democratic Solidarity Party, Social Democratic Coordinator, Democratic Party November 30 'Frank País'.
Main Social Organizations: Cuban Workers' Union (CTC). The CTC has nearly 3,000,000 members, and represents 80 per cent of the Cuban active labor force. The National Association of Small Farmers, 200,000 members with more than 3,500 grassroots organizations (member of the farmers' international network Via Campesina). The Federation of Cuban Women (FMC), with more than 2,000,000 members. The University Student Federation (FEU) and the Federation of Secondary School Students (FEEM), 500,000 student members. The José Martí Pioneers' Union, 2,000,000 children and young people; the Revolution Defense Committees that are organized in neighborhoods.

THE STATE

Official Name: República de Cuba.
Administrative Divisions: 14 Provinces, 169 Municipalities including the special municipality of Isla de Pinos.
Capital: Havana (La Habana) 2,189,000 people (2003).
Other Cities: Santiago de Cuba 534,600 people; Camagüey 342,900; Holguín 305,000; Guantánamo 264,100; Pinar del Río 172,300 (2000).
Government: Fidel Castro Ruiz, President of the Council of State and the Council of Ministers since December 1976, elected by the National Assembly of People's Power (with 609 members). The 1976 constitution states that the power is exercised through the Assemblies of People's Power. The local Assemblies delegate power to successively more encompassing representative bodies until a pyramid is formed which peaks in the National Assembly. Representatives are subject to recall by the voters.
National Holidays: 1 January, Liberation Day (1959); 26 July, Assault on the Moncada barracks (1953).
Armed Forces: 105,000 (1995). Other: 1,369,000 Civil Defense Force, Territorial Militia, State Security, Border Guard.

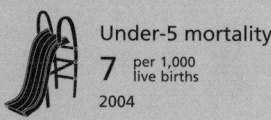
Under-5 mortality
7 per 1,000 live births
2004

Malnutrition
4% under-5s
1996-2004

Maternal mortality
33 per 100,000 live births
2000

IN FOCUS

ENVIRONMENTAL CHALLENGES

There is evidence of air and water pollution (especially noticeable in Havana Bay). The environmental pollution caused by nickel plants in Moa, Holguin Province, has damaged the health of residents, who have complained to the local authorities (but apparently still with no response). Biodiversity is weakened as a result of pollution, and overhunting threatens wildlife.

WOMEN'S RIGHTS

Women have been able to vote and stand for office since1934. In 2003, they held 36 per cent of seats in Parliament and 16 per cent of ministerial positions. Also in 2003, they made up 40 per cent of the labor force, comprising 6 million people.

Iron deficiency is the main cause of anemia, which affects 27 per cent of pregnant women and 25 to 35 per cent of women of child-bearing age. The literacy rate was 100 per cent for both men and women in 2004. Both men and women have an average of 12 years of schooling. In 2004, 100 per cent of births were attended by skilled health staff*.

CHILDREN

The under-one and under-five mortality rates were nearly halved between 1990 and 2004. The first one fell from 13 to 7 per 1,000 live births, while the latter was reduced from 11 to 6 per 1,000 live births. In 2000*, the net school registration and attendance rate was 99 per cent. HIV/AIDS is perceived by the Cuban authorities as an epidemic that can be contained, partly because for example blood transfusions are regulated. However the prevalence rate in the 15-34 age group went up from 9.5 per 100,000 inhabitants in 2000 to12.1 in 2001. If the present trend continues, the number of people living with HIV/AIDS will at least triple within the next 10 years.

INDIGENOUS PEOPLES/ETHNIC MINORITIES

Afro-Cubans make up between 34 and 62 per cent of the population; the variance is because statistics are based on people's self-perception. The Cuban mulatto community is highly integrated within itself: there is a process of ethnic mix. The Afro-Cubans live mainly in the eastern part of the island and Havana neighborhoods.

MIGRANTS/REFUGEES

In 2002, about 34,200 Cubans were seeking refuge in foreign countries, mainly in the US. Approximately 25,300 arrived there by boat or through the Mexican border. They were admitted with conditions, but were free to seek permanent residence, under the Cuban Adjustment Act. The US Committee for Refugees defines them as people that require international protection. The US admitted 1,900 Cubans as refugees directly from Havana. There were about 1,200 Cuban refugees or asylum-seekers in Spain, 1,100 in Costa Rica, 900 in Peru and hundreds scattered worldwide. Throughout 2002, the US refused entry to 700 Cubans coming by boat and repatriated them.

DEATH PENALTY

It is still in force for all types of crimes.

* Latest data available in The State of the World's Children and Childinfo database, UNICEF, 2006.

GDP had increased 9.6 per cent.

[35] Cuban exiles in Miami suffered two setbacks in late 1997. The first was the death of their leader Mas Canosa, and the second came with Castro's political coup with Pope John Paul II's visit to the island. Following the PCC meeting in 1991 and constitutional amendments of 1992, relations between Cuba and the Vatican improved considerably. As a run-up to the visit, the Pope condemned the US embargo on the island and Castro declared Christmas 1997 a national holiday.

[36] In April 1998, Grenada and Cuba re-established diplomatic relations which had been suspended since the US invasion of the former in 1983. Grenada's prime minister Keith Mitchell made an official visit to Cuba where agreements on economic co-operation were signed.

[37] In February, the UN Historical Verification Commission blamed the US for backing the Guatemalan army in massacres of the population, whilst pointing out that Cuba's Government had supported the left-wing guerrillas here, even sending supplies of weapons.

[38] The tensions between Castro and successive US administrations were aggravated in November 1999 when a child, Elián González, was the only survivor of a Cuban boat that sank on its way to Miami. Cuba demanded the return of Elián while US political groups used the boy's situation as a banner for the 2000 elections. Ignoring the US justice system, which ruled that he should be returned to Cuba, they requested that Congress made the boy a US citizen. In Cuba, Elián also became a symbol of the struggle against the US and its 40-year blockade. Finally, in June 2000, the boy returned to the island.

[39] In November 2001, Hurricane Michelle hit the island, killing 20 people and destroying homes and plots. 481,300 people were evacuated. The US, for the first time in 40 years, exported food to the island to help overcome the disaster's effects.

[40] Hundreds of Afghan prisoners were confined in the US military base at Guantanamo Bay, in January 2002, awaiting interrogation as suspected members of the terrorist Al-Qaeda network. The US, without granting them prisoner of war status, called them 'illegal warriors', a legal term unknown in international law.

[41] The Russian military base, Lourdes, located 20 km away from Havana, which had been set up in 1964 to monitor US movements and communications, was closed in January 2002. Although Moscow claimed the decision was taken due to its high costs, Cuban authorities accused Russian President Vladimir Putin of closing it as a gesture to the US, part of a rapprochement policy adopted after the September 11 terrorist attacks in New York and Washington, and by US threats of suspending financial aid to Russia.

[42] On 20 April 2002, the UN High Commissioner for Human Rights (UNHCR) passed a resolution against Cuba, urging the Government to grant more individual liberties and political rights, and to allow the visit of a UN envoy to monitor progress in this area. The initiative was begun by Uruguay (see Uruguay). Granma reacted angrily against the resolution, denouncing that Washington planned more trickery for Cuba. The resolution was supported by 23 countries and rejected by 21, while 9 abstained. Of the Latin American countries, Brazil and Ecuador abstained and Venezuela and Cuba voted against.

[43] That month, Castro released the tape of a phone call with Mexican President Vicente Fox to confirm that the Mexican Government had asked him to cancel or cut short his stay at the March 2002 UN summit held in Monterrey (Mexico). He had also been told not to criticize the US in his speech.

[44] In May 2002, former US President Jimmy Carter visited Cuba. He was the first US President (either in office or not) to do so since Castro came into power. After visiting a laboratory that Washington alleged was manufacturing biological weapons, Carter declared that this was not the case. According to previous accusations from Washington, Cuba was part of the 'axis of evil '.

[45] In June 2003, after the imprisonment of 75 dissidents and 29 journalists, the EU imposed diplomatic sanctions on Cuba, on the grounds of ongoing violations of Human Rights.

[46] Almost one year later, in April 2004, the UN Human Rights Commission passed a motion - by a vote of 22 for and 21 against with 10 abstentions - censuring Cuba for its human rights policies.

[47] EU officials dealing with Latin American issues met in Brussels in October to decide whether the bloc should soften its diplomatic stance toward Havana, upon evidence that sanctions had not improved respect for democracy or Human Rights in the country. That same month, for the 13th consecutive year, the UN voted - by 179 votes to 4, with 1 abstention - against the US blockade of the island.

[48] In May 2005, around 200 dissidents gathered in Havana, in the first opposition meeting since the 1959 revolution. The event, organized by the Assembly for the Promotion of Civil Society in Cuba, was aimed at promoting democracy in the country. Cuban authorities did not intervene but had earlier prevented the participation of several European politicians who planned to attend.

[49] In January 2006, various messages to the Cuban people were projected onto a big public screen on the US diplomatic mission in Havana, among them quotes from the Universal Declaration of Human Rights.

[50] In February, Castro unveiled a monument which blocked the view of those messages. It comprised 138 huge black flags each with a white star symbolising the people who had died - more than 3,400 - as a result of violent US acts against the island since the Revolution. In July 2006 Fidel Castro temporarily ceded power to his brother, Raul, for the first time in 47 years while he underwent intestina surgery. Rumours that Fidel had intestinal cancer and would be unable to resume power were denied by Havana. ∎

Cyprus / Kipros

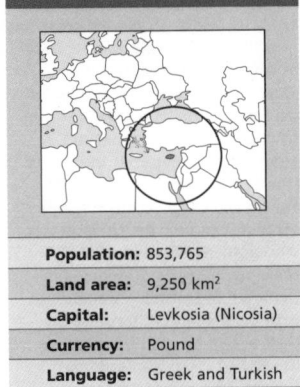

Population:	853,765
Land area:	9,250 km²
Capital:	Levkosia (Nicosia)
Currency:	Pound
Language:	Greek and Turkish

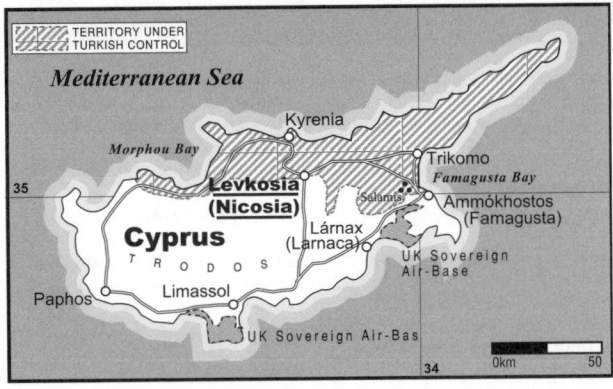

There is early evidence of human presence in Cyprus, tools and instruments dating back around 10,000 years. The first known settlement was in Khirokitia (near the south coast), where some 2,000 inhabitants built their houses of stone. The discovery of small quantities of obsidian - a type of volcanic rock not found on the island - is the only sign of contact with other cultures. Khirokitia and other smaller associated settlements disappeared after a few centuries, leaving the island deserted for about 2,000 years. Hittites, Phoenicians, Greeks, Assyrians, Persians, Egyptians, Romans, Arabs and Turks trooped through its valleys and over its hills until 1878, when the British Empire - in need of a base for eastward expansion - negotiated its occupation with Turkey. The Ottomans, after over 300 years of domination, ceded Cyprus to England, in exchange for British protection against Czarist Russia.

² In 1931, various movements appeared favoring *enosis* (annexation) of Cyprus by Greece (in view of the example of the incorporation of Crete to Greece in 1913). Enosis was promoted by the Greek Orthodox Church, the religion of the Greek Cypriots, who make up the majority of the island's population but Turkey - Greece's rival - feared being surrounded by a hostile neighbor on its Mediterranean coast. The British exiled various Greek Cypriot priests but, after World War II, the country's most important political figure was Archbishop Vaneziz Makarios (original name was Mikhail Khristodolou Mouskos), who had been exiled in the Seychelles. Makarios headed the Cypriot pro-independence movement from exile.

³ In 1959, representatives of the Greek and Turkish communities, of Makarios' Democratic Party and of British colonial interests, reached an agreement for the creation of the Republic of Cyprus, with constitutional guarantees for

the Turkish minority, and British sovereignty over the island's military bases. Independence was proclaimed on 16 August, 1960, and Makarios (considered the 'Mediterranean Fidel Castro' by the US Government) took office as President. An active supporter of anti-colonialism, the President played an important role in the Movement of Non-Aligned Countries (he was re-elected in 1968 and 1973). Tensions between Greece and Turkey persisted and had frequent repercussions in Cyprus where conflicts erupted between the Greek and Turkish communities.

⁴ In 1963 there was an unsuccessful coup attempt by members of the radical right (who supported enosis). In 1974, the Cypriot National Guard, under the command of Greek army officers, ousted Makarios (who fled to Britain) and Nikos Sampson, who favored annexation by Greece, was appointed President. Turkey immediately invaded Northern Cyprus, bombed Nicosia, and drove 200,000 Greek Cypriots southward, under the pretext of protecting the Turkish minority. That same year, Sampson turned the presidency over to Glafcos Klerides (President of the House of Representatives) and facing the prospect of a war with Turkey - added to domestic opposition and world-wide repudiation - the Greek military junta (in power since 1967) also stepped down.

⁵ Towards the end of 1974, Makarios returned to Cyprus and held the Presidential office until his death in 1977. Spyros Kyprianou succeeded Makarios and followed the same policy (refusing to recognize the division of Cyprus and retaining membership of the Non-Aligned Movement). The Turkish forces, who were occupying 40 per cent of the island, refused to return to the situation which had prevailed prior to the coup, and remained in the country: a Turkish Cypriot Federal State was proclaimed in the northern part of the island, under the presidency of Rauf Denktash (1975).

⁶ Denktash and Makarios had set four basic conditions for a negotiated peace settlement: a) the establishment of a binational, non-aligned and independent federal republic; b) an exact delimitation of the territories that each community would administrate; c) the discussion of - among other issues - internal restrictions on traveling, ownership rights and a federal system with equal rights for both communities; d)

sufficient federal power to ensure Cyprus' unity. The Turkish refusal to withdraw the troops, an essential condition for the Greek Cypriots, meant that no substantial progress was made: in 1983, the Turkish Republic of North Cyprus (TRNC) was proclaimed, but only Turkey acknowledged the new state.

⁷ Since 1983, Cyprus has enjoyed a period of economic prosperity brought by tourism, foreign aid, and international business that made the island an international financial center, (replacing Beirut, whose money markets had been paralyzed by the Lebanese civil war). The main beneficiary of this was the Greek Cypriot bourgeoisie. In the northern part of the island there was a large influx of Turkish immigrants (more than 40,000 people) which, when added to almost 35,000 Turkish soldiers and the migration of some 20,000 Turkish Cypriots, created a change of profile in the population. In the 1990s there was one continental Turk for every Turkish Cypriot.

⁸ In 1985, a Constitution for the TRNC was submitted to

PROFILE

ENVIRONMENT
Once part of continental Europe, the island of Cyprus is located in the eastern Mediterranean, close to Turkey. Two mountain ranges - the Troodos in the southwestern region, and the Kyrenia in north Cyprus - enclose a fertile central plain. The temperate Mediterranean climate, with hot, dry summers and mild, rainy winters, is good for agriculture.

SOCIETY
Peoples: Cypriots are divided into Greek (85 per cent) and Turkish (12 per cent) communities; they have separate political, cultural and religious organizations.
Religions: Greek Orthodox and Islam.
Languages: Greek and Turkish (official); English.
Main Political Parties: Progressive Party of Working People (Anorthotikon Komma Ergazomenou Laou, socialist); Democratic Rally (Dimokratikos Sinagermos); Democratic Party (Dimokratikon Komma); Movement for Social Democracy-United democratic Union of Centre (Kinima Sosialdimokraton-Eniaia Dimokratiki Enosi Kentrou); New Horizons (Neoi Orizontes).
Main Social Organizations: There are two major organizations representing approximately 30 unions, the Pancypriot Federation of Labor, and the Cyprus Worker's Confederation.

THE STATE
Official Name: Kypriaki Dimokratia-Kibris Cumhuriyeti.
Capital: Levkosia (Nicosia) 205,000 people (2003).
Other Cities: Larnaca 44,600 people; Limassol 26,700; Paphos 16,300 (2000).
Government: Tassos Papadopoulos, Head of State and of the Government since February 2003. Single-chamber legislature with 71 members. Since April 2005, Ferdi Sabit Soyer is president of the Turkish Republic of North Cyprus, since April 2005. This Government, based upon Turkish occupation and recognized only by Turkey, exercises an independent administration, including its own Judicial System and a 50-member Legislative Assembly, elected every five years.
National Holiday: 1 October.
Independence: 16 August 1960. The Turkish Republic of North Cyprus (declared in 1983) celebrates Independence on 16 November.
Armed Forces: 10,000 (including 400 women). Other: Armed Police: 3,700.

Life expectancy
79 years
2005-2010

GNI per capita
$16,510
2004

Literacy
97% total adult rate
2000-2004

IN FOCUS

ENVIRONMENTAL CHALLENGES
Although the atmospheric pollution index has risen on account of traffic and industry, it is at an acceptable level. Soil deterioration is mainly due to the excessive use of agrochemicals. The coastal ecosystem has borne the adverse effects of tourism, particularly the construction of large hotels and high-rise apartment buildings. There are water scarcity and salinity problems, especially in the North, due to the seepage of sea water into the aquifers.

WOMEN'S RIGHTS
Cypriot women have been able to vote and stand for office since 1960. In 2003, 11 per cent of Parliament seats were held by women; the percentage of women in ministerial positions was nil. That year they made up 39 per cent of the work force. Women's unemployment rate is 4.2 per cent compared with the island's total unemployment rate of 3.3 per cent.

The illiteracy rate was 4.9 per cent among women over 15, while it was 1.4 per cent among men.

CHILDREN
In 2003, 100 per cent of elementary school students reached 5th grade.
The net rate of elementary

school enrollment and attendance was 95 per cent. The death rate for under-fives was 6 per cent in 2002.

INDIGENOUS PEOPLES/ ETHNIC MINORITIES
Even though historically Turkish and Greek settlers had lived together peacefully, their relations deteriorated after independence from British rule in 1960, and ended in the division of the island by the the Turkish army in 1974 with Greeks in the South, Turks in the North, and thousands of people murdered and displaced. The main conflict was for control and administration of the territory: the violent confrontations between communities at independence revealed the lack of a Cypriot identity and the fact of two peoples closely linked to their mainlands.

In 1983, the Turkish minority (12 per cent of the island's population) established the Turkish Republic of North Cyprus. Although it has only been recognized by Ankara since its formation, poor relations between Turks and the Greek Cypriot majority (85 per cent) have ground to a halt. In the North, Turkish is spoken and Islam is the religion.

At present, there is tension on both sides which prevents reaching a settlement that would allow their entry to the EU. The history of tension suggests that violence

could continue between the communities, but international attention would inhibit Turkish military activities - backed by Ankara - and reduce the political isolation.

Other minorities also reside in Cyprus: since 1960, some 8,000 Maronites, Armenians and Italians have emigrated to the Greek sector.

MIGRANTS/REFUGEES
At the end of 2002, Cyprus accommodated over 1,800 refugees plus around 950 asylum-seekers - most of these from Iran (400) and the Gaza Strip (170). The total includes over 1,700 asylum-seekers awaiting decisions and 90 whose applications had been granted during the year (of these, 62 were Iranians).

Around 265,000 people were still internally displaced in 2002: 200,000 Greek Cypriots in the south and about 65,000 Turkish Cypriots in the north.

The Cypriot Government has been processing asylum applications since 1 January 2002. Prior to that, UNHCR processed them in Cyprus (it still does in the Turkish sector and had 5 applications in 2002).

DEATH PENALTY
It was abolished in 2002; the last execution was in 1962.

to meet Denktash. The imminent entry of Cyprus to the EU breathed new life to the dialogue. However, as Ankara had accepted the TRNC's entry conditional on it its own admission, the TRNC refused to negotiate EU membership alongside the Republic of Cyprus (the Greek sector).
[15] In 2002, the EU again seemed set to accept only the Greek sector if no agreement was reached. However, that year Klerides and Denktash re-started negotiations with UN mediation, focusing on their aim of EU membership. Toward the end of 2002, UN Secretary General Kofi Annan presented a peace plan that proposed a federation of the two sectors, governed alternately. That year, the EU invited Cyprus to the 2004 Copenhagen summit, prefigured in the UN plan in the event of an agreement being reached in 2003.
[16] Tassos Papadopoulos was elected in the 2003 elections in the Greek sector. A few weeks later the UN deadline for agreement on the island's future passed without a reunification settlement. Annan admitted his plan had failed. That same year, for the first time in three decades, both the Turkish and Greek Cypriots crossed the 'green line' dividing the country.
[17] In twin referendums in April 2004 the Greek Cypriots rejected a UN plan for reunification of the island while the Turkish Cypriots backed it. On 1 May Cyprus became, along with nine other countries, a full EU member, but only the Greek part gained the benefits of membership.
[18] Ankara stated in December that it would recognize Cyprus as a EU member before the start of the negotiations, planned for October 2005, for its own entry into the Union.
[19] Mehmet Ali Talat was elected Turkish Cypriot president in April 2005. In May, Greek Cypriot authorities and UN officials started talking about the possibility of a new peace agreement.
[20] The Cypriot Parliament passed the European Constitution draft in June.
[21] In March 2006, in spite of its pledge, Ankara had still not recognized Cyprus as a EU member and kept troops in the TRNC. Although stalled, negotiations about the Turkish entry into the EU continued, while blue helmets patrolled the 'green line'. In mid May, 470,000 Greek Cypriots and 1,000 Turkish Cypriots were to vote to renew the Chamber of Representatives.
[22] In July 2006, the President Papadopoulos met the Turkish Cypriot leader Mehmet Ali Talat for the first time in two years. ∎

referendum: 65 per cent voted in favor. However, 30 per cent of the registered voters abstained, and this was seen as a reflection of distrust in the State's legitimacy.
[9] Giorgis Vassiliu was elected President of Cyprus in 1988. He re-established the negotiations with Denktash that had been stalled since 1985: leaders and representatives of the 350 thousand Cypriots in exile called for 'constructive flexibility' on the part of the two leaders. The Greek community - buoyed by their economic prosperity and the fact that they were numerically the majority - wanted independence to be guaranteed by the UN. They also sought freedom of movement and property rights across the island. The Turkish Cypriots - based on the superior strength of the Turkish army - demanded a binational federation under Turkey's protection. Talks were again broken off in 1989.
[10] In 1990, Denktash was re-elected President of the TRNC. With the support of US President George Bush in 1991 Turkey proposed a summit

among representatives from Ankara, Athens and both Cypriot communities. Greece and the Greek Cypriot authorities considered Washington's support of the initiative as compensation for the aid given by Turkey during the Gulf War. In 1992, the UN declared Cyprus a bi-communal and bi-regional country, with equal political rights for both communities. In 1993, Glafkos Klerides defeated Vassiliu in the presidential elections.
[11] In 1994 the European Court of Justice declared that direct trade between the TRNC and the EU was illegal. In 1996 violence and tension increased in the north of the island. Klerides and Denktash met throughout 1997, with UN mediation: their aim was reunification, ahead of admission to the EU. However, direct negotiations between the EU and Greek Cypriot authorities stalled dialogue within Cyprus and the drive for reunification broke down. The EU stated that, if the island were not reunited, the Greek area could be admitted as if it were an independent state, leaving out the

Turkish area. Denktash considered that the political legitimacy of the TRNC had to be acknowledged and the whole of Cyprus should be admitted to the Union.
[12] In 1998, the EU agreed to initiate talks for the incorporation of the Greek sector. In 1999, Greece lifted its veto on Turkey's entry to the EU, and provided aid to the Turkish Government after a devastating earthquake (see Turkey), marking a change in the political climate with regard to Cyprus.
[13] In the year 2000, negotiations were renewed between Klerides and Denktash within a UN framework, but ended with no progress. TheGreek Cypriots advocated a reunified bi-communal federation; the Turkish Cypriots wanted one based on equal sovereignty. Once the Geneva rounds ended, Denktash threatened not to negotiate as long as the TRNC did not have international recognition.
[14] In 2001, Turkey withdrew its veto, conditionally, on the agreement between the EU and NATO. Shortly after, Klerides crossed the 'green line' in Nicosia

Czech Republic / Ceska Republika

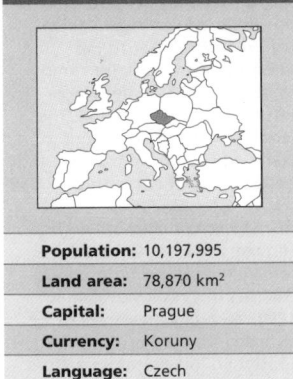

Population:	10,197,995
Land area:	78,870 km²
Capital:	Prague
Currency:	Koruny
Language:	Czech

The exact origin of the first inhabitants of the Mid-Danube region remains uncertain, although there are signs left by the Boii, a Celtic people whose name gave rise to the Latin name 'Bohemia'. The Celts were displaced without major conflict by Germanic peoples and later, in the 6th century, by the Slavs, while the Germans continued their migration southward.

2   The inhabitants of mountain and forest areas had the natural protection of these areas. However, the lowlands were repeatedly invaded by the Avars. The Slavs were able to repel these invasions when they had leaders strong enough to unite the tribes, such as the Frankish merchant Samo. In the 8th century, peace was restored to Bohemia with the defeat of the Avars at the hands of Charlemagne.

3   In the early 9th century, three potential political centers emerged: the plains of Nitra, the Lower Morava Basin and Central Bohemia. The Slavs of Bohemia, called the Czechs, gained the upper hand in most of the region. The first Czech Prince was Mojmir I, who extended his realm as far as Nitra. His successor, Rostislav I, institutionalized the State and consolidated political relations with the Eastern

Frankish Empire, as a way of maintaining his own sovereignty.

4   The Franks organized the first missions at Nitra and in Bohemia, but Rostislav would not allow Latin to be taught, and asked the Byzantine emperor to send preachers who spoke the Slavic language. Constantine (Cyril) and Methodius arrived in 863, leading a group of Greek missionaries; they developed the first Slavic alphabet and translated religious texts.

5   Methodius won recognition from Rome for his work in Moravia and in Panonia, which became an ecclesiastical province linked to the Archbishop of Sirmium. The Franks came to consider Methodius an enemy; he was captured and kept prisoner until 873, when he returned to Moravia.

6   Rostislav was the founder of Great Moravia, uniting the territories inhabited by the Slavs of the region and Slovakia, bordered by the northern ring of the Carpathian Mountains and by the Morava River.

7   After the death of

Methodius in 885 the Frankish Bishop Wiching displaced Methodius' disciples and the new Pope outlawed Slavic liturgy.

8   King Arnulf sent a military expedition to Moravia in 892, allying himself later with the Magyars to defeat the principality. Between 905 and 908, Great Moravia went through several foreign occupations, until an agreement was reached between Mojmir II and Arnulf.

9   In the 10th century, between the strengthening of Germania and the restoration of the Holy Roman Empire, Bohemia lost the major part of its possessions. When Bfetislav I ascended the throne in 1034, the principality recovered part of Moravia and invaded Poland in 1039. However, King Henry III of Germania forced a retreat, and the Hungarian Crown kept Slovakia.

10   In order to maintain its independence, Bohemia found it necessary to be actively involved in the campaigns of the Holy Roman Empire. Thus, the situation arose whereby the Bohemian princes were being crowned king by both the rival powers.

11   In the early 13th century the Church separated from the State, and the feudal lords began demanding greater political participation. At the same time, Germanic immigration increased the population, building new urban centers and exploiting mineral resources, which gave rise to a new class of traders and entrepreneurs.

12   Under the Przemysl dynasty, which lasted until 1306, Bohemia controlled part of Austria and the Alps; at one point a single king ruled over Bohemia and Poland.

This dynasty was succeeded by the Luxembourgs in 1310. With the coronation of Emperor Charles I in 1455, Bohemia and the Holy Roman Empire were joined together. As the capital of the Kingdom and the Empire, this was Prague's greatest moment.

13   The religious reform movement strengthened in the 14th century and became even more radical under the influence of Father John Huss. Excommunicated by the Pope, Huss was later tried for heresy and sedition by the Council of Constance, and was burned alive in 1415 after refusing to recant.

14   The anger that followed Huss' execution marked the birth of the Hussite movement in Bohemia and Moravia. The Germanic peoples remained faithful to Rome, however, and in addition to these religious differences, the ethnic issue remained, triggering political conflict between them. The Holy Roman Empire, allied with the German princes, launched several military campaigns in Bohemia, but they were repulsed by the Hussites.

15   Religious differences prevented political union between Bohemia and its former possessions for many years. Vladislav II reigned over Bohemia from 1471 but Moravia, Silesia and Lusacia were ruled by Mathias, of Hungary. Only after Mathias's death in 1490 was Vladislav II elected king of Hungary and the territories reunited.

16   The death of Louis II in 1526 paved the way for the rise of the Habsburgs. Ferdinand I, Louis' brother-in-law, became king by currying favor with

LAND USE

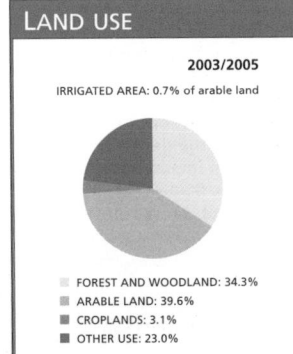

2003/2005

IRRIGATED AREA: 0.7% of arable land

- FOREST AND WOODLAND: 34.3%
- ARABLE LAND: 39.6%
- CROPLANDS: 3.1%
- OTHER USE: 23.0%

PUBLIC EXPENDITURE

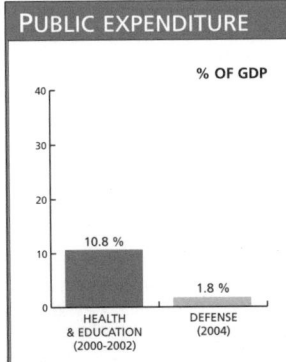

% OF GDP

10.8 %

1.8 %

HEALTH & EDUCATION (2000-2002)   DEFENSE (2004)

Life expectancy
76 years
2005-2010

GNI per capita
$9,130
2004

HIV prevalence rate
0.1% of population
15-49 years old
2003

the nobility. Austria's victory over the Protestant Society of Schmalkaldica in 1547 permitted Ferdinand to impose the right of hereditary succession to the throne upon Bohemia and its states.

17   The Habsburgs strengthened the Counter-reformation throughout the region. Slovakia remained within its realm because the Habsburgs had retained it when Hungary was invaded by the Ottoman Empire, in 1526.

18   Rudolf II (1576-1612) transferred the seat of the empire to Prague, making it once again one of the continent's most important political and cultural centers. Many key positions of the kingdom were filled by Catholics during Rudolf's reign, as he himself was Catholic. However, this triggered a rebellion by the non-Catholic (Reformed church) majority, and the king was deposed in 1611.

19   After a stormy succession, Ferdinand II of Styria, with the support of Maximilian I of Bavaria, defeated the Protestants and ruled with a strong hand. The Germanic language was added to the traditional use of Czech, and only the Catholic religion was authorized.

20   Unlike Bohemia, Moravia did not become involved in the fight against the Habsburgs and therefore did not suffer the effects of civil and religious strife as severely. In Moravia there was religious tolerance allowing the growth of Protestantism in the state, which remained separate from the Austrian Crown until 1848.

21   Despite the hegemony of the Germans, the Czechs conserved their ethnic identity, their language and their culture. Something similar had also occurred in the Hungarian counties inhabited by Slovaks. This set the scene for a resurgence of nationalism in the early 19th century, which strengthened the traditional ties between these two peoples.

22   Czechs and Slovaks, together with the inhabitants of the German republics, helped put a stop to the absolutist doctrine, amidst a revolutionary wave which swept Europe in 1848. In 1867, the empire split in two: Austria, where ethnic Germans outnumbered the Czechs, Poles and other nationalities; and Hungary, where the Magyars subdued the Slovaks.

23   With World War I the fall of the Austro-Hungarian Empire finally brought about the recognition of the Republic of Czechoslovakia. The new state's borders established in 1919 by the victorious powers included parts of Poland, Hungary and the Sudetenland, where there lived around three million Germans, source of potential conflict.

24   Czech and Slovak leaders charged the National Assembly with drawing up a Constitution. The Assembly opted for a strict parliamentary system, in which the President and his cabinet would be responsible to two legislative chambers. Women's right to vote or to be elected was obtained for the first time.

25   The 1930s worldwide depression affected the Sudetenland intensely, as it was a highly industrialized region. It also accentuated nationalist feeling among the German people there, who developed a separatist movement alongside Hitler's rise to power in 1933. Britain, France and Italy negotiated the ceding of the Sudetenland to Germany in 1938, thus paving the way for the German occupation of Czechoslovakia in 1939.

26   After Hitler's defeat in World War II, and under the occupation of the Soviet army, Czechoslovakia recovered its 1919 borders, while the German population was almost entirely expelled from the country.

27   The Communist Party (CKC) obtained 38 per cent of the vote in the 1946 election, increasing to 51 per cent in 1948. In June, a People's Republic was proclaimed, and the CKC applied the economic model in effect at the time in the USSR. Czechoslovakia joined the Council for Mutual Economic Assistance (CMEA) and the Warsaw Pact.

28   In 1960, the People's Republic of Czechoslovakia added 'socialist' to its name. In political terms, this decade was a turning point. Slovak leaders, expelled from the party in the 1950s, were rehabilitated. The Slovak struggle for autonomy (which had been even further restricted by the new socialist constitution) together with the 1967 student strikes brought an end to Antony Novotny's leadership of the CKC.

29   In early 1968, the election of Alexander Dubcek as Secretary of the CKC, and of Ludwik Svoboda as the country's president, led to the implementation of a program to decentralize the economy, and affirm national sovereignty, against a background of broad popular support.

30   The USSR and other members of the Warsaw Pact viewed the possibility of Czechoslovakian reforms as a threat to the integrity of the socialist camp; in August 1968, Soviet forces intervened in the country. The leaders of the 'Prague Spring' were expelled from the CKC and political alignment with the USSR was re-established.

31   Soviet Communist Party Secretary Mikhail Gorbachev's reform process brought about changes in Czechoslovakia. In 1989, despite violent repression, anti-Government protests continued, precipitating a crisis within the regime.

32   The Government was forced to negotiate with the Civic Forum, an alliance of several opposition groups. Among other reforms, Parliament approved the elimination of the CKC's leadership role. In late 1989, a provisional Government was formed, with a non-communist majority.

33   In December 1989, the Civic Forum declared that the CKC had redistributed cabinet positions in such a way as to keep its own people in the key positions. 200,000 people gathered in Prague to demand a greater opposition representation in the cabinet. Gustav Husak resigned the presidency of the federation and was replaced by Vaclav Havel, who immediately granted amnesty to all political prisoners and called elections for June 1990. Havel was confirmed in the presidency and the Czech and Slovak Federal Republic was proclaimed.

34   Following the elections, the Forum divided into the Civic Democratic Party, the self-proclaimed 'right with a conservative program', and the Civic Movement. The Slovakian Party 'Public against Violence' split into two opposition groups. The Movement for a Democratic Slovakia - the most important one - summoned the people to fight against the right wing profile the country was acquiring.

35   In the June legislative elections, the Czech Civic Democratic Party and the Slovak group, Democratic Slovakia, won in their respective republics. When negotiations over the statutes of the new federation came to an impasse, Czech and Slovak leaders admitted that separation was inevitable. Czechoslovakia disappeared from the world map, replaced by the Czech Republic, with its capital Prague, and the Republic of Slovakia, with its capital Bratislava.

36   Vaclav Havel, the former

PROFILE

ENVIRONMENT

The Bohemian massif occupies the western region, bordered by the Moravian plains to the southeast. Cereals and sugar beet are cultivated in the lowlands, where cattle and pigs are also raised. Rye and potatoes are grown in the Bohemian valleys. The region has rich mineral deposits: coal, lignite, graphite and uranium, while Moravia is rich in coal.

SOCIETY

Peoples: Czechs, 81.2 per cent; Moravians, 13.2 per cent; Slovaks, 3 per cent.
Religions: Catholic (39 per cent); Protestant (4.3 per cent); Orthodox (3 per cent). 40 per cent of the population are atheist.
Languages: Czech (official). Romani is spoken by the Roma (Gypsy) community (200,000).
Political parties: Czech Social Democratic Party (eská strana sociáln demokratická); Civic Democratic Party (Ob anská demokratická strana); Communist Party of Bohemia and Moravia (Komunistická strana ech a Moravy); Christian Democratic Union-Czechoslovakian People's Party (K es ansko-demokratická unie - eskoslovenská strana lidová, continued to exist after the country's separation).
Social Organizations: Czech and Moravian Confederation of Trade Unions; Autonomous Democracy Movement of Moravia and Silesia; groups of activists against economic globalization (INPEG).

THE STATE

Official Name: Ceska Republika.
Administrative Divisions: 8 regions, 73 districts and 4 municipalities.
Capital: Prague 1,170,000 people (2003).
Other Cities: Brno 382,800 people; Ostrava 320,900; Olomouc 102,800 (2000). **Government:** Parliamentary republic. Vaclav Klaus, President since March 2000, re-elected in 2003. Nirek Topolánek, Prime Minister since September 2006. The Parliament of the Czech Republic (Legislature) has two chambers: the Chamber of Representatives, with 200 members, and the Senate, with 81 members. National holiday: 28 October, Czech Founding Day (1928); 1 January, Independence (1993, from Czechoslovakia).
Armed Forces: 58,000 (2000).

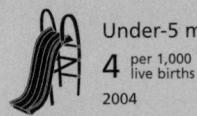 **Under-5 mortality**
4 per 1,000 live births
2004

 Poverty
2% of population living on less than $1 per day
1996

 **Debt service**
10.5% exports of goods and services
2004

Maternal mortality
9 per 100,000 live births
2000

IN FOCUS

ENVIRONMENTAL CHALLENGES

Air and water pollution in the north and northwest pose a health risk. Sulfur dioxide emissions produced by electricity generation are very high, causing acid rain. Large forested areas have been destroyed or damaged. Approximately three-quarters of the country's trees show some degree of defoliation. Waste from industry, mining and intensive farming threatens water purity, both above and below ground.

WOMEN'S RIGHTS

Women have been able to vote and run for office since 1920. Seventeen per cent of seats in the lower house and 12 per cent of seats in the upper house of Parliament were held by women in 2004.

The female labor force was 47 per cent of the total in 2003. Female unemployment reached 10.6 per cent.

Some 87 per cent of girls were enrolled in primary school in 2004.

CHILDREN

In spite of good indicators (mortality, access to education and health, among others), children face several threats. The Czech Republic, Romania, Poland and Slovakia have criminal networks that exploit children between 9 and 13 years, using them to steal and act as drug couriers.

In addition, there is child trafficking from the Czech Republic into Germany, Netherlands, Belgium and the UK.

INDIGENOUS PEOPLES/ETHNIC MINORITIES

In August 2003, the UN Committee on the Elimination of Racial Discrimination advised the Government to approve legislation outlawing discrimination against national minorities. The Committee praised 'the numerous measures, programs and strategies adopted in order to improve the situation of Roma/Gypsies and other marginalized groups, including refugees'. However, the Committee was concerned about the acts of racial violence, the persistence of racial hatred and intolerance and the de facto segregation particularly of Roma (Gypsies).

Discrimination against Roma in employment, housing and education was marked in 2002. This group of 267,000 people - 2 per cent of the population, most of them very poor - was the victim of hate crimes and violence, including murders by neo-Nazis who were punished lightly. The Government announced plans to stop Roma seeking asylum in EU countries and to limit public assistance for Roma who returned after unsuccessfully seeking asylum abroad. The authorities also proposed creating a special police force to fight money-lending in Roma communities. Fear of money-lenders' retribution when they default on debt is apparently a major reason why some Roma migrate.

The Slovak population represents another minority group (309,000 people), mainly in Prague, Brno, Karvina, Olomouc, Tabor and Kladno. Unlike the Roma, this minority group has improved living conditions over the last years. They do not face political or cultural restrictions (although there is lack of Slovak-language education) and there are several organizations representing their interests. They have no representation in Parliament, mainly because Czech Slovaks tend to vote for the broader political parties.

MIGRANTS/REFUGEES

In the first half of 2005 the region of Central Europe and the Baltic countries received some 12,500 asylum requests. This figure is 17 per cent lower than in the same period of 2004. The 10 new members of the EU received 11,800 applications. Three countries, among them Slovakia, Czech Republic and Poland, registered significant reductions in requests in comparison to the first half of 2004: 78 per cent, 48 per cent and 25 per cent, respectively. In very few countries (including Cyprus and Slovenia) the figure increased in 2005.

DEATH PENALTY

It was abolished in 1990.

* Latest data available in The State of the World's Children and Childinfo database, UNICEF, 2006.

president of Czechoslovakia who had resigned on 17 July 1992 after the National Assembly declared the independence of Slovakia, was elected President of the Czech Republic in 1992. Vaclav Klaus became the first Prime Minister of the new independent State.

[37] Klaus launched a rapid privatization plan. In 1995, the Czech Republic became a member of the Organization for Economic Co-operation and Development (OECD).

[38] In December 1996, the Czech Republic and Germany signed a reconciliation document, in which Germany apologized for the behavior of the Nazi regime during World War II, while the Czech Government apologized for the expulsion of three million Germans from the Sudetenland after the War.

[39] During 1997, the devaluation of the national currency, the koruny, brought political chaos.

Havel called for the resignation of Klaus, leader of the Civic Democratic Party (CDP), accused of receiving bribes from financial firms. Klaus was replaced in November by Josef Tosovsky.

[40] In April 1998, the Czech Republic was admitted to NATO. The Social Democratic Party (SDP) won the June elections with 32.3 per cent of the vote. The CDP agreed to give the SDP control over the Government, in exchange for the appointment of its leader, Klaus, as parliamentary speaker. The Government aimed to wipe out corruption, raise the national minimum wage and halt the 'devolution' of properties to the Catholic Church. However, by the end of the year unemployment had reached record levels and inflation exceeded 11.5 per cent.

[41] In November 1999, the Government ordered that a wall in the north of Ústí nad Labem - built to separate Roma houses

from those of other residents - be pulled down.

[42] In late September 2000, the World Bank and the International Monetary Fund - meeting for the first time in Prague - attempted to showcase the Czech Republic as their first economic success in a former Iron Curtain country. But widespread acts of civil disobedience and anti-globalization protests forced the meeting to end a day earlier than planned.

[43] In October 2000 the first reactor at Temelin nuclear plant was activated, causing conflict with Austria, which tried to block the Czech Republic's entry to the European Union (EU). In November 2001, the Government settled the dispute with Austria by agreeing to implement safety measures and to monitor the environmental impact.

[44] In 2002 Parliament unanimously rejected the neighboring countries' request

to annul the 'Benes decrees' - the post-war decrees under which 2.5 million ethnic Germans and Hungarians were expelled from Czechoslovakia. After forming a coalition with the center alliance between the Christian Democratic Party and the Free Union, the SDP, led by Vladimir Spidla, won 70 of the 200 parliamentary seats. In December, at the Copenhagen summit, the Republic was invited to join the EU.

[45] Former Prime Minister Vaclav Klaus was appointed President by the Parliament in 2003. In June, Czechs decided in a referendum to continue negotiations to join the EU.

[46] The Czech Republic formally joined the EU on 1 May 2004. Prime Minister Vladimir Spidla resigned in June. He was replaced by Stanislav Gross, after a new Government coalition was formed.

[47] The new premier faced a political crisis in March 2005, when he was accused of financial irregularities. He was forced to resign in April, and the coalition that had supported him named Jiri Paroubek, a Social Democrat, as Prime Minister.

[48] A disagreement between Paroubek and President Klaus arose in May 2006, when the latter said that, two years after joining the EU, there were no benefits to the country. Paroubek said Klaus ignored statistics which clearly indicated the contrary. That month, Iraq paid off a $200 million debt - 80 per cent of the total - to the Czech Republic.

[49] The country was left in political stalemate following the general election of June 2006. The right-wing Civic Democrats, having campaigned for a flat 15-per-cent income tax and a flat basic income to replace welfare benefits, emerged as the largest party with 35 per cent of the vote. The ruling Social Democrats, however, recovered from what had looked like inevitable defeat to achieve 32 per cent, their own largest-ever vote.

[50] The wrangling over possible coalitions involving the Greens, Christian Democrats and the Communists continued through the summer, until Civic Democrat leader Miroslav Topolánek formed a minority administration on 4 September involving 'non-political technocrats'. This Government was toppled in a vote of confidence on 4 October, though Topolánek continued in office while President Klaus tried to persuade the warring parties to accept a technocratic cabinet that would see the country through to an early election. ∎

Denmark / Danmark

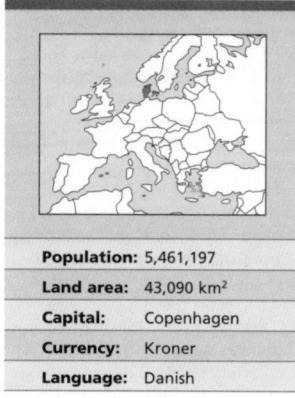

Population:	5,461,197
Land area:	43,090 km²
Capital:	Copenhagen
Currency:	Kroner
Language:	Danish

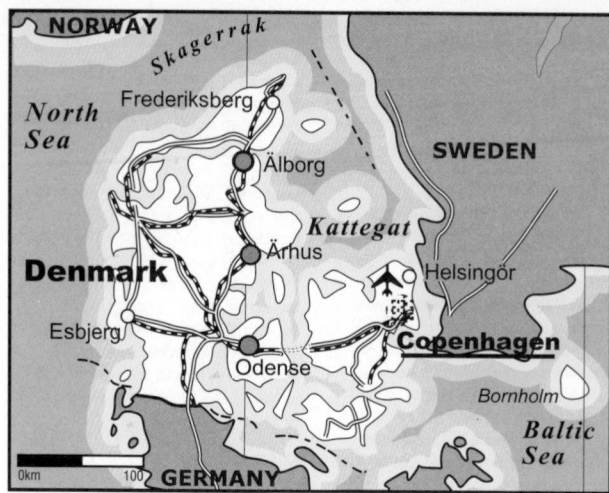

The first hunting peoples established themselves in Denmark in about 10,000 BC as the Neolithic period drew to a close. This was followed by a flourishing Bronze Age civilization, about 1000 BC. Around the year 500 AD, northern Germanic peoples began settling on the islands as fishers and navigators. Certain place names bear witness to the worship of Scandinavian gods, such as Odin, Thor and Frey.

2 The first evidence of hierarchical society in Denmark comes from the Viking age, mostly from cemeteries and settlement sites. The Vikings were Scandinavian farmers, navigators, merchants and above all raiders who ruled the northern seas between the 8th and 10th centuries AD.

3 Archeological remains indicate that Roskilde on the island of Sealand, Hedeby, south of Jutland, and Jelling in the north, were the most intensely populated areas. After the Danish victory over the Germans the Eider River became the final southern border. A huge wall was built to the south and west of Hedeby.

4 In the 10th century, after continuous conflict with rival kingdoms, the center of the kingdom's power was transferred to Jelling, where Gorm became king of Jutland. His son Harald Bluetooth (Blatand) is credited with uniting Denmark and conquering parts of Norway.

5 Subsequent Viking reigns extended Danish possessions as far as modern-day England and Sweden. In 1397 Queen Margrethe managed to unite Denmark, Norway, Iceland, Greenland, Sweden and Finland in the 'Union of Kalmar'.

6 The introduction and spread of Christianity and strengthening of the Hanseatic League went hand-in-hand with the weakening of Denmark's military power.

7 The Danish kings were involved in successive wars between themselves, campaigns which were interspersed with peasant rebellions and bourgeois revolts; a large and powerful middle class had developed as a result of growing mercantile activity. These conflicts ceased in the 17th century, when a weakened nobility gave the King power as absolute sovereign. He was then able to create laws to be imposed throughout the land.

8 During the 18th century, Denmark colonized the Virgin Islands. The Danish colonists organized local production using African slave labor and in 1917 the islands and their population were sold to the US. (See US Virgin Islands)

9 Peace existed in Denmark and Norway from 1720 until the Napoleonic Wars. After Napoleon's defeat Sweden attacked Denmark and, under the Kiel Peace Treaty, annexed Norway in 1814.

10 The loss of Norway, combined with British trading impositions brought on an economic collapse which worsened as a result of low wheat prices. The ensuing agricultural crisis forced land reform to a standstill. The situation later improved when agricultural prices stabilized, trade increased and industrialization began. In 1814 an educational reform made schooling obligatory.

11 After the European revolutions of 1848, King Frederick VII called an assembly which established parliamentary monarchy and abolished absolutism. The 1849 Constitution guaranteed freedom of the press, of religion and of association, as well as the right to hold public meetings.

12 A dispute with Germany over the Duchies of Schleswig and Holstein reinforced nationalistic sentiment. In 1864, when Denmark was defeated by Prussia and Austria, it lost its claim to these lands and the national-liberal government was brought down.

13 The 1866 Constitution maintained the monarchy. In 1871, Louis Pio, a former military officer, attempted to form a socialist party. A series of strikes and demonstrations organized by the socialists was put down by the army, and Pio was deported to the US. The Social Democratic Party, mainly supported by intellectuals and workers, was formed in 1876.

14 The peasants and emerging middle class weakened the monarchy on three fronts: the farm co-operative movement, a liberal bourgeois party (popularly referred to as 'leftist') and the social democrat party. In 1901 the United Left (Liberal) came to power establishing a new government. The emergence of the UL and the Social Democrats as a leading force at the turn of the century was the result of agrarian reform, industrialization and the development of railroads. The growth of urbanization and overseas trade (accelerating the formation of labor unions throughout the country and the rise of co-operatives in the countryside) was the main reason for these changes.

15 After the 1870-71 Franco-

PROFILE

ENVIRONMENT

With an average altitude of only 35 meters above sea level, Denmark is one of the lowest countries in Europe. The land is divided into the mainland - the Jutland peninsula - and the islands (including the islands of Sjaeland and Jelling) which represent one third of the territory. The continental summers are relatively warm and rainy. The land is intensely cultivated, in spite of cold winters. Denmark is a supplier of livestock products and has important maritime activity.

SOCIETY

Peoples: 95.5 per cent of the population is of Danish origin. A German minority lives in Jutland. There are immigrant communities from Turkey (1.5 per cent), Asia (1.6 per cent), Africa (0.3 per cent), former Yugoslavia (0.5 per cent) and other Scandinavian countries (0.4 per cent). There are 250,000 immigrants from different countries.
Religions: 96 per cent of the population belong to the Lutheran Church; minorities of Catholics and Jews. **Languages:** Danish (official). English is spoken as a second language.
Main Political Parties: Liberal (Venstre); Social Democrats; Danish People's Party; Conservative People's Party. **Main Social Organizations:** The Danish Confederation of Trade Unions (LO) has 1,500,000 members (49 per cent women) from 40 different unions; Confederation of Salaried Employees and Civil Servants in Denmark (FTF), 330,000 members; Danish Confederation of Professional Associations (AC), 100,000 members; and the Women Workers' Union in Denmark (KAD).

THE STATE

Official Name: Kongeriget Danmark. **Administrative Divisions:** 14 departments (Amtskommuner), Frederiksberg, Copenhague. **Capital:** Copenhagen 1,066,000 people (2003). **Other Cities:** Århus 645,300 people; Odense 145,200; Ålborg 119,800; Frederiksberg 91,200 (2000).**Government:** A constitutional parliamentary monarchy. Queen Margrethe II, since January 1972. Prime Minister, Anders Fogh Rasmussen, since November 2001. Unicameral Parliament: People's Diet, with 179 members, including 2 from Greenland and 2 from the Faeroe Islands. **National Holiday:** 5 June, Constitution Day (1953); 16 April, Queen's birthday (1940).**Armed Forces:** 33,100 (1995).

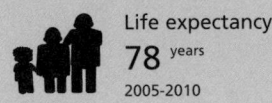

Life expectancy
78 years
2005-2010

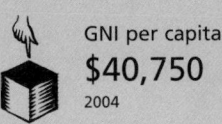

GNI per capita
$40,750
2004

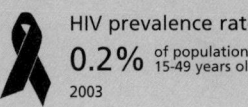

HIV prevalence rate
0.2% of population
15-49 years old
2003

IN FOCUS

ENVIRONMENTAL CHALLENGES

The North Sea is polluted with nitrates and phosphates from industrial and power plant emissions. Groundwater has been polluted by animal and factory wastes and agrochemicals. In 2003, Denmark banned the spraying of glyphosate because the pollution it caused was five times higher than the permissible level for drinking water.

WOMEN'S RIGHTS

Danish women have been able to vote and stand for office in all elections since 1915. In the 1980s women held 23 per cent of all parliamentary seats and in 1982 women headed the ministries of Labour and Religious Affairs.

In 2005, women held 38 per cent of parliamentary seats and 33 per cent of ministerial or equivalent positions.

Women make up 47 per cent of the workforce: 2 per cent work in agriculture, 83 per cent in services and 15 per cent in the industrial sector.

An unknown number of women and children are trafficked every year from Southeast Asia and eastern Europe for sex, but although this is a growing problem, the Government has done little.

CHILDREN

Low birthweight affects 5 per cent of all newborns*. Approximately 96 per cent of children were immunized against polio, measles and tetanus*.

In recent years, the under-5 mortality rate has stabilized at five per 1,000 live births. Although there is not much information on child prostitution, reports from several NGOs state that a significant number of children are involved.

INDIGENOUS PEOPLES/ ETHNIC MINORITIES

The Inuit live in Greenland, with a population of 44,000. The Danish Government expelled them in 1950 from Thule and the US installed a military base there. A Danish court confirmed in December 2003 that

the expulsion had been illegal. The Inuit demand the closing of the base and the return of their land; however they lost their case in the High Court in January 2004.

The Faeroese, from the Faeroe Islands, have lived in the area for the last 12,000 years.

There are small groups of Turks, and a larger number of Germans. The Germans' ancestors came to what is now Denmark 11,000 years ago. Today they are seeking greater recognition, access to public and political positions. They are Protestants, and although their official language is Danish, most of them speak German.

MIGRANTS/REFUGEES

In 2004 Denmark received 3,200 refugees and asylum-seekers, mostly from Afghanistan and Iraq. Iraqis, Afghans and Somalis have the highest acceptance rates for asylum applications.

Denmark, like other European countries, has experienced a growing rate of xenophobia in

the last two decades. When the Government decided in August 2002 to repatriate the Somali refugees who insisted on staying, the populist extreme right proposed dropping the refugees by parachute over Somalia.

Immigration has exceeded emigration in the last decades. There are some 300,000 immigrants living in Denmark.

The proportion of foreigners amounted to 4.8 per cent of the population in 1999, compared to 2 per cent in 1984. More than half of all foreign citizens live in metropolitan Copenhagen, and more than one quarter of them come from Nordic or EU countries.

DEATH PENALTY

Abolished in 1978.

*Latest data available in *The State of the World's Children* and *Childinfo* database, UNICEF, 2006

German War, Denmark adopted a neutral international stance. World War I gave Copenhagen trading opportunities with the warring nations, but also affected its supplies. During World War II Denmark was invaded by Germany, although it officially maintained its independence until 1943.
[16] When Hitler attacked the USSR, Denmark created a volunteer army and outlawed communist activity. The 1943 election was an anti-Nazi plebiscite, with the electorate throwing its support behind the democratic parties. Resistance to the Nazi regime, strikes, and the Government's refusal to enforce Nazi rule, led the German occupation forces to declare a state of emergency dissolving Denmark's police and armed forces.
[17] In September 1943, the Danish Freedom Council was created to co-ordinate the anti-Nazi opposition. When Germany finally surrendered, a transition government was formed, with representatives from the Council and the traditional parties. The 1945 election was won by the liberals.
[18] The Faeroe Islands, under Danish control since 1380, were occupied by Britain during the war and subsequently returned to Denmark. The 1948 Constitution gave the islands greater autonomy, though the Danish Parliament retained control of defense and

foreign affairs.
[19] Denmark was responsible for foreign policy and justice of Greenland, a territory controlled by the Danes since 1380. In 1979, the island obtained the right to have its own legislative assembly (*Landsting*), which has power over internal affairs. Other areas are dealt with by the Danish Parliament which has two representatives from the island.
[20] Denmark joined NATO in 1949, increasing its military power with the help of the US. A US proposal for setting up air bases on Danish soil was turned down.
[21] The Constitution was amended in 1953, reducing the Legislature to a single chamber (*Folketing*). In 1954 the post of 'ombudsman' was created to ensure that municipal and national government complied with the law, and to protect 'ordinary' citizens from the misuse of power on the part of government officials or agencies.
[22] Denmark became the founding member of the European Free Trade Association (EFTA) in 1959. In 1972, 63.7 per cent of the electorate voted yes in a referendum to join the European Economic Community.
[23] In 1986, the Parliament passed a strict environmental protection law which entailed significant costs for industry and agriculture, in a country without nuclear plants.
[24] Marriage between people

of the same sex was authorized in 1989. An amendment to the social security system in 1990 enabled parents to take up to 52 weeks' leave in the case of serious illness of a child under 14 years old.
[25] In 1993 the Danes finally approved the Maastricht Treaty, the formation of the European Union (EU), after rejecting it the previous year, on condition that the country would not join the Economic and Monetary Union. That year, the liberal-conservative government fell after Justice minister Erik Ninn Hansen was found guilty of preventing the family reunion of Tamil refugees from Sri Lanka.
[26] Social Democratic Prime Minister Poul Nyrup Rasmussen, elected in 1993, was re-elected in 1994.
[27] Danish cinema, flourishing since 1910, became internationally famous in 1995 with movies produced according to the 'Dogma manifesto'. Established by directors Lars Von Trier and Thomas Vinterberg, the Dogma stated that genre movies were not acceptable; that cameras should be hand-held; that sound is incidental and that the final work should not be credited to the director.
[28] Pia Kjaersgaard, leader of the Danish People's Party (DPP), received more than 6 per cent of the vote in the local Copenhagen elections in November 1997.

She based her campaign on the 'danger' posed to her country by Third World immigrants. In early 2000, surveys showed the DPP had risen to third place in Danish politics. The party slogan stated: 'Muslims are just as good as us, but they are a problem for a Christian country'.
[29] After 7,000 years of geographic separation, a 16-kilometer-long bridge and tunnel connected Copenhagen and Malmo, Sweden, in 15 minutes. The work was opened by King Carl Gustav of Sweden and Queen Margrethe II of Denmark in July 2000.
[30] In a referendum in September 2000, Denmark rejected replacing its kroner with the euro.
[31] In January 2001, Denmark reacted against the independence plan put forward by the Faeroe Islands, which included extending Danish subsidies until 2012. Oil prospects in the islands' sea platform had increased calls for independence.
[32] In Jutland, the livestock region, Danish farmers complained that the arrival of Dutch people - wealthier than the Danes - had increased the price of land, making it difficult for many to compete (ten per cent of milk in Denmark was produced in Dutch-owned farms and 60 per cent of farms on sale were bought by Dutch farmers).
[33] The Liberal party won the

Under-5 mortality
5 per 1,000 live births
2004

Maternal mortality
5 per 100,000 live birth
2000

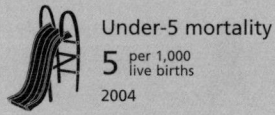
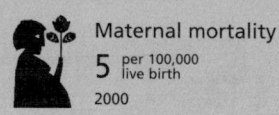

Faeroe Islands

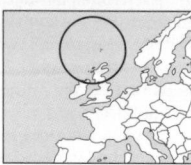

Population: 47,603
Land area: 1,400 km²
Capital: Torshvn
Currency: Danish kroner
Language: Faeroese, Danish

ENVIRONMENT

The climate is generally rainy and cloudy, with mild temperatures both in summer and winter. Only six per cent of the land is cultivated. Agricultural production mainly consists of vegetables as the land is not suitable for grain. Sheep are raised throughout the islands and there is mining in Suderoy. 20 per cent of the population are employed in handcraft production and 21 per cent in fisheries, which provide 90 per cent of the islands' exports. The Faeroe Islands are one of 10 largest salmon breeders in the world. To protect the fishing industry, a 200-mile zone was established in 1977. Modernization of the island's fishing fleet and methods was financed through a series of foreign loans, which totaled $839 million in 1990 - $18,000 per capita. Maritime oil prospecting was initiated.

SOCIETY

Peoples: The population is of Scandinavian origin.
Religions: Lutheran. There are a large number of Baptists and a small Catholic community.
Languages: Faeroese, Danish.
Main Political Parties: Social Democratic and Union (liberal) parties, which favor ties with Denmark; the Republican 'Left', Popular and Progressive parties, favor independence.

THE STATE

Capital: Torshavn 18,000 people (2003).
Other Cities: Klaksvik 4,800 people; Runavík 2,500 (2000).
Government: The parliament (Løgtinget) has 32 members elected based on proportional representation. Parliament names a cabinet (Landstyret) of six ministers. The Prime Minister is Jóannes Eidesgaard (2003). Birgit Kleis (since 2001) is the High Commissioner (Rigsombudsman) who represents the Crown. The islands send two representatives to the Danish Parliament. Diplomacy: In January 1974 the Løgtinget resolved not to join the EEC and the aspiration for independence continues.
National Holiday: 29 July, Olaifest (local holiday).
Armed Forces: Defense is the responsibility of Denmark; no organized native military forces; only a small Police Force and Coast Guard are maintained.

Greenland

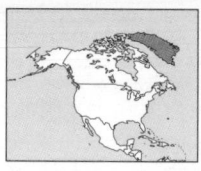

Population: 57,217
Land area: 341,700 km²
Capital: Nuuk (e-Godhaab)
Currency: Danish kroner
Language: Greenlandic, Danish and Inuit

ENVIRONMENT

Located in the Arctic Ocean, the island is the second largest tract of frozen land on the planet. Nearly four fifths of its surface is covered by an ice cap. In the month of June, soon after the rapid thaw, moss and lichen vegetation appear on certain parts of the coast. Most of the population is concentrated in the western region, where the climate is less severe. The country has lead, zinc, and tungsten deposits. Cryolite from the large reserves in Ivigtut is also exported. Fishing forms the basis of the economy, although fish stocks are approaching exhaustion in national waters, salted and frozen fish is exported, as well as whale oil. The island is encouraging tourism as an alternative source of income, with 35,000 visitors each year. There are highly protective policies on the nordic environment. Its ice covered coastlines are melting at a rate of more than a meter each year.

SOCIETY

Peoples: 80 per cent are Inuit. The remaining 20 per cent are Danish or other short-term European residents.
Religions: Lutheran. The Greenlandic Church comes under the jurisdiction of the Bishop of Copenhagen and the minister of ecclesiastical affairs.
Languages: Greenlandic, Inuktitut (Inuit/Aleut language) and Danish (official). There are three linguistic groups amongst the Inuit population; Kalaallit (west coast), Inughuit (north) and Lit (east coast).
Main Political Parties: Siumut (Forward), Democrats; Inuit Atgatigiit (Eskimo Community); Atssut (Feeling of Community).

THE STATE

Official Name: Kalaallit Nunaat.
Administrative Division: 3 districts: Avannaa, Tunu y Kitaa.
Capital: Nuuk-14,000 people (2003).
Other Cities: Sisimiut 5,200 people; Ilulissat 4,100 (2000).
Government: Queen Margrethe II of Denmark, Head of State since 1972; Srøen Hald Møller, High Commissioner since 2005; Hans Enoksen, Prime Minister since 2002. Unicameral Legislature: The *Diet*, with 31 members elected for a four-year term. Greenland elects two representatives to the Danish *Folketing*, and has one representative on the Nordic Council.
National Holiday: 21 June, Longest Day; 5 June, Danish Constitution (1953).
Armed Forces: Defense is the responsibility of Denmark.

general elections in November 2001, after nine years of Social-Democrat Government. The EU-Russia summit was moved from the Danish capital to Brussels in November 2002, in the midst of a diplomatic row between Copenhagen and Moscow regarding Chechen exiles. Russian president Putin had threatened to boycott the summit in Copenhagen when Denmark did not extradite Akhmed Zakayev, a Chechen guerrilla leader.
³⁴ When US troops detained captured Iraqi president Saddam Hussein (see Iraq) in December 2003, and after President Bush stated that Saddam deserved the death penalty once he was brought to trial in Baghdad,

Prime Minister Fogh Rasmussen - who had supported the US invasion of Iraq - stressed his opposition, shared widely among Danes, to the capital punishment.
³⁵ In August 2004, Annemette Hommel, a secret service agent, was accused of abusing Iraqi prisoners when she was captain of an allied battalion. Hommel returned to Denmark along with other Danish soldiers involved in the case. It had a big impact on Danish public opinion, which had previously supported Copenhagen's alliance with Washington.
³⁶ On 4 May 2005, the 60th anniversary of the end of German occupation was celebrated amid controversy regarding collaboration with Nazi Germany.

³⁷ Cartoons of the prophet Muhammad - first published by a Danish newspaper in September 2005 - appeared in several Arab and European papers in January 2006 and provoked violent protests from Muslims throughout the world. In February, a group of Syrian demonstrators in Damascus stormed the Danish, Chilean and Swedish embassies, setting them alight.
³⁸ In May 2006 Denmark announced it would reduce its troops in Iraq from 530 to 80. However, it offered a Hercules plane and 70 troops to the UN for the country's reconstruction.

Foreign Minister Per Stig Moeller said 'Iraqis are so well-trained that we do not need to send our soldiers to teach them, so we decided to reorganize our contingent'.
³⁹ That month, Canada and Denmark launched a joint expedition to map beneath the Arctic Sea, to help them resolve sovereignty claims over areas with potential oil and gas reserves.
⁴⁰ In July 2006, a British academic produced a survey rating Denmark as the happiest country on earth. ■

Djibouti / Djibouti

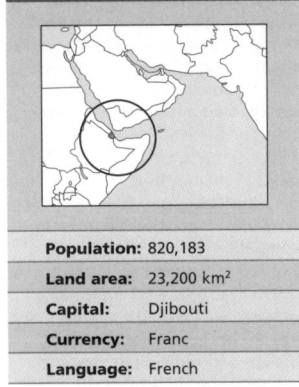

Population:	820,183
Land area:	23,200 km²
Capital:	Djibouti
Currency:	Franc
Language:	French

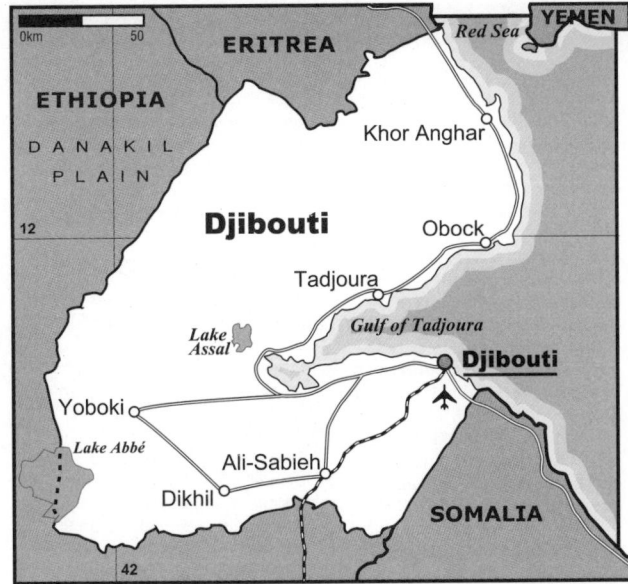

A round the 3rd century BC Ablé immigrants came from Arabia and settled in the north and parts of the south. The Afars, or Danakil, are descendants of these peoples. Later the Somali Issas pushed the Afars out of the south and settled in the coastal regions. In AD 825 Islam was brought to the area by missionaries. Arabs controlled the trade in this region until the 16th century, when the Portuguese competed for it. In 1862, Tadjoura, one of the Sultanates on the Somalian coast (see Somalia), sold the port of Obock and adjoining lands to the French for 52,000 francs and in 1888 French Somaliland (Côte Français des Somalis) was established.

2   Djibouti became the official capital of this French territory in 1892. A treaty with Ethiopia in 1897 reduced the territory in size. A railway was built to connect Djibouti with the Ethiopian hinterland, reaching Dire Dawa in 1903 and Addis Ababa in 1917. The interior of the area was effectively opened up between1924 and 1934 by the construction of roads and administrative posts. After World War II Djibouti port lost trade to the Ethiopian port of Asseb (now in Eritrea). In 1946 French Somaliland acquired the status of an overseas territory (from 1967 called the French Territory of the Afars and Issas), and in 1958 it voted to become an overseas territorial

member of the French Community under the Fifth Republic.

3   Independence and the reunification of neighboring Somalia stimulated the emergence of anti-colonialist movements such as the Somaliland Liberation Front and the African League for Independence, both of which used legal and armed branches.

4   During the 1970s, renewed resistance forced acting governor Ali Aref to resign. France called a plebiscite on 8 May 1977, and 85 per cent of the population voted for independence. Hassan Gouled Aptidon, main leader of the African League for Independence, became President of the fledgling Republic.

5   In an attempt to overcome old ethnic divisions, Gouled granted

governmental participation to various groups. He even appointed several Afar ministers. Though French remained the official language, Djibouti was admitted into the Arab League.

6   The new state, created for strategic reasons by colonialism, today transports much of Ethiopia's foreign trade through its port, earning revenue.

7   Ethiopia and Somalia, its two neighbors, both had territorial designs on Djibouti. Ethiopia's interest was geopolitical, because of Djibouti's strategic location as a route to the Red Sea. When Eritrea gained independence, Ethiopia became a landlocked territory. If an agreement could not be reached for the use of

Eritrean ports, Djibouti would be its only available port. Somalia's interest was in unifying the Somali nation.

8   In mid-1979, President Hassan Gouled resumed relations with Ethiopia and Somalia, signing trade and transportation agreements with them. The participation of Afars in the Government and in the newly-formed army was encouraged as a way of securing national unity. Foreign aid was basically used for irrigation works and to improve the situation of refugees from the Ogaden war.

9   After 8 years as an independent country, Djibouti was determined to follow its own path, contrary to earlier expectations that it would be annexed by Ethiopia or Somalia.

10   In October 1981, President Gouled amended the Constitution to introduce a single-party system. The official Popular Association for Progress (RPP) became the sole legal party and the other groups were banned. It was argued that they had racial or religious aims and in consequence were potentially harmful to national unity.

11   During 1983, the first steps were taken in a project to radically alter Djibouti's economy and transform it into the Middle East's Hong Kong. The plan included the creation of a financial hub and a free trade port. Six foreign banks opened offices in Djibouti, mainly attracted by a solid national currency backed by dollar deposits in the US.

12   By the end of 1984, the first results were not encouraging: the number of passengers and goods in transit to Ethiopia and Somalia

PUBLIC EXPENDITURE

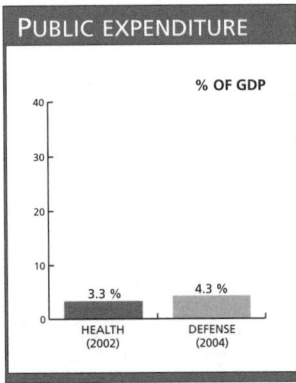

% OF GDP

3.3 % HEALTH (2002)

4.3 % DEFENSE (2004)

PROFILE

ENVIRONMENT
Located in the Afar triangle, facing Yemen, Djibouti is one of the hottest countries in the world (average annual temperature 30º C). The land is mostly desert; its only green area is found in the basalt ranges of the northern region. Extensive cattle-raising is practised by nomads. Economic activity is concentrated around the port.

SOCIETY
Peoples: Djiboutians are divided into two major ethnic groups of equal size: the Afars, scattered throughout the country, and the Issas, of Somalian origin, who populate most of the southern territory and predominate in the capital. There are also French, Yemeni, Ethiopian, Italian, Greek and Pakistani minorities, among others.
Religions: Sunni Muslims. There is a small Christian minority (5 per cent).
Languages: Afar and Issa (Somali), French (official) and Arabic (religious).
Main Political Parties: Union for Presidential Majority (People's Rally for Progress, Front for Restoration of Unity and Democracy); Union for

a Democratic Change (Republican Alliance for Democracy; Movement for Democratic Renewal and Development).
Main Social Organizations: Union of Djiboutian Workers (UDT), a national labor federation; Union of Construction and Public Works Workers (SB-BTP).

THE STATE
Official Name: République de Djibouti.
Administrative Divisions: 5 districts.
Capital: Djibouti 502,000 people (2003).
Other Cities: Ali-Sabieh 12,200 people; Tadjoura 11,500; Dikhil 9,900 (2000).
Government: Ismail Omar Guelleh, President since May 1999. Dileita Muhammad Dileita, Prime Minister since March 2001. The Prime Minister is traditionally Afar to counterbalance the power of Gouled, who is Issa. The National Assembly has 65 members.
National Holiday: 27 June, Independence Day (1977).
Armed Forces: 9,000 (1997) Other: Gendarmerie (Ministry of Defense): 600; National Security Force (Ministry of Interior): 3,000.

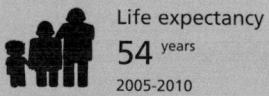
Life expectancy
54 years
2005-2010

GNI per capita
$950
2004

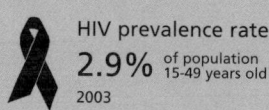
HIV prevalence rate
2.9% of population
15-49 years old
2003

IN FOCUS

ENVIRONMENTAL CHALLENGES

Most of the land is desertified and not suitable for farming. Agriculture, concentrated in oases and some coastal areas, produces only about a quarter of local needs. Drinking water is scarce. A great variety of animal and plant species are in danger due to the encroachment of the desert.

WOMEN'S RIGHTS

Women have had the vote since 1946 and been able to stand for office since 1986. However, the first time a woman was elected to Parliament was in 2003. In 2005, women held 11 per cent of parliamentary seats. In 2004* only 32 per cent of girls attended school. Anemia affects 40 per cent of pregnant women*. Some 67 per cent receive prenatal care and 61 per cent of births are assisted by qualified staff*. The maternal mortality rate is 730 for every 100,000births*. Genital mutilation is common.

At least 5 per cent of women between 15 and 45 live with HIV/AIDS.

CHILDREN

Although it fell more than 20 per cent in the last 15 years, Djibouti's under-five mortality rate is one of the highest in the world: 126 deaths per 1,000 live births*. One of the main causes of infant mortality is polio. Poor access to medicines and lack of skilled birth attendants increase the number of deaths. Some 18 per cent of children under 5 suffer malnutrition or low weight, and 26 per cent suffer some type of stunting.*

Child prostitution is increasing. According to UNICEF, 73 per cent of streetchildren aged 8-17 are prostitutes. Most are girls who do not speak French, trafficked from Ethiopia.-

The presence of military bases, whose number has increased because of the war in Iraq, has increased the demand for prostitutes.

INDIGENOUSPEOPLES/ ETHNIC MINORITIES

The Afar in Djibouti and Ethiopia are a homogenous group. They did not support the 1977 division of their traditional lands between Ethiopia, Djibouti and Eritrea. Their situation has improved over the previous decade. However, in spite of some political reforms, the Somali Issas who belong to the ruling party are dominant. This has caused friction between the two groups.

MIGRANTS/REFUGEES

In 2004 there were 23,200 refugees in Djibouti, most of them from Somalia and Ethiopia. Most of the Somali refugees come from the north (Somaliland) and are nomads that arrived between 1988 and 1990, fleeing from civil war.More than 40,000 Ethiopians arrived during Ethiopia's civil war. Once it ended,in 1996, more than 90 per cent returned, but a small number decided to remain in Djibouti.

DEATH PENALTY

Abolished in 1995; the last execution took place in 1977.

*Latest data available in *The State of the World's Children* and *Childinfo* database, UNICEF, 2006.

---

had dropped considerably and, therefore, customs revenues and bank activities had decreased. The continuing conflict in the area has been seen as the major cause for the withdrawal of European capital.
[13] With the support of the UN High Commissioner, Gouled's Government resumed the voluntary repatriation of more than a hundred thousand refugees, a process which had been interrupted in 1983. At the same time, bilateral agreements were signed with Ethiopia to combat contraband and promote peace in the border areas. These had been closed in 1977, at the outset of the conflict with Somalia over the Ogaden region.
[14] In August 1987, foreign military presence in Djibouti grew as French bases were also used by US and British forces participating in maneuvers in the Persian Gulf.
[15] France wrote off Djibouti's debt in 1990 by granting it $40,000,000 as public development aid. In 1991, confrontations between the Government and the Front for the Restoration of Unity and Democracy (FRUD) guerrillas were renewed. In November, Amnesty International accused the Government of the torture of 300 prisoners.
[16] In the May 1993 presidential elections, Gouled was again re-elected, with over 60 per cent of the vote. However, encouraged by the FRUD, half of the electorate

abstained from voting and the opposition considered the elections a 'fraud'.
[17] Armed confrontations between government troops and guerrillas escalated in the weeks following the elections, which led thousands to seek refuge in Ethiopia. Acting as mediator, the French Government sought a cease-fire and negotiations began. Meanwhile, Gouled believed the rebellion was part of a plan orchestrated by Ethiopia.
[18] In June 1994, Gouled and the FRUD jointly decided to end the two-and-a-half year war. That month, a demonstration of mostly Afar residents from the Arhiba district, opposed to the demolition of their homes for 'security reasons', was quelled by the police. The intervention left four dead, 20 injured and 300 arrests, including that of Muhammad Ahmad Issa, United Opposition Front president.
[19] The movement's leaders in October banned Ahmad Dini Ahmad and Muhammad Adoyta Yussuf, another breakaway leader, from holding any 'activity or responsibility' in the FRUD. In 1995, after the Constitution was revised, a section of the FRUD formed an alliance with the ruling party. Dini Ahmad said this alliance amounted to 'treason'.
[20] In mid-1995 under pressure from the IMF, the Government reduced public spending and took measures to increase fiscal income.

The following year, the IMF granted Djibouti a $6.7 million credit.
[21] In Addis Ababa, Ethiopia, 17 opponents of the Gouled regime were abducted on 26 September and made prisoners of the Djibouti Government.
[22] On 18 February 1998, despite Djibouti's having signed the African Human Rights Charter, guaranteeing freedom of speech and information, the editor of the bimonthly *Al Wahda*, Ahmed Abdi Farah, and journalist Kamil Hassan Ali were arrested for an article critical of the Government published a year earlier.
[23] April 1999 marked the first time since independence for citizens to vote for a president, with Ismail Omar Guelleh elected by a wide margin as the second President in the history of the country. Guelleh's campaign centered on promises to alleviate poverty. His political rivals accused him of having assassinated members of the opposition during two decades when he was a key advisor to and director of special police corps.
[24] General Yacin Yabeh Galab, chief of police, led a failed coup attempt in December 2000. The Government and the most radical faction of the FRUD signed a peace treaty in May 2001 that ended the civil war. Ahmad Dini returned to Djibouti after 9 years in exile.
[25] As a first gesture toward the FRUD, Guelleh restructured the

Government. However, in spite of previous speculation, no FRUD representative was named among the nine new Cabinet members.
[26] In January 2002, the US installed an anti-terrorist military base for the Horn of Africa in Djibouti, for which it promised to pay $31 million that year.
[27] The 1992 law allowing only three opposition parties to compete with the ruling party expired in September 2002, opening the way for multiparty politics. However, the Government has shown little tolerance for groups and individuals that question the political status quo.
[28] Also in September, Djibouti stressed it would not be used as a base for attacking other countries in the region. Nine hundred US troops were stationed there, supporting the war against terrorism.
[29] In January 2003 the coalition supporting President Omar Guelleh - the Union for Presidential Majority - won the first multiparty elections since independence in 1977.
[30] In May 2003, France agreed to increase its annual payment from $20-million to $34 million for its military base of 2,700 troops.
[31] At least 50 people died and 4,000 fled their homes in the capital after heavy rains burst the banks of the Ambouli river in April 2004, causing severe floods and forcing banks, schools and businesses to close.
[32] Guelleh was re-elected with 100 per cent of the vote in the April 2005 elections, which had a 79 per cent turnout. The last opposition candidate had withdrawn from the race in March, after the opposition decided to boycott the elections, anticipating fraud.
[33] In November 2005, all union leaders and workers detained after taking part in a general strike in the port were freed.-But 36 of them - including 11 leaders of the port workers' union who had been under surveillance and then detained - were fired.
[34] A report published in February 2006 by the International Confederation of Free Trade Unions (ICFTU) denounced the serious labor rights violations that have affected union members. According to the document, the country's laws did not comply with the International Labor Organization's agreements on forced labor, freedom of association and equal pay for men and women.
[35] In April 2006, an Amnesty International report claimed that Djibouti was one of the countries where prisoners allegedly abducted and mistreated by the US were held. ∎

Dominica / Dominica

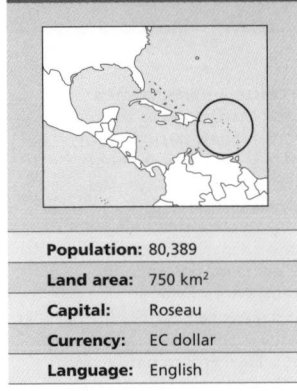

Population:	80,389
Land area:	750 km²
Capital:	Roseau
Currency:	EC dollar
Language:	English

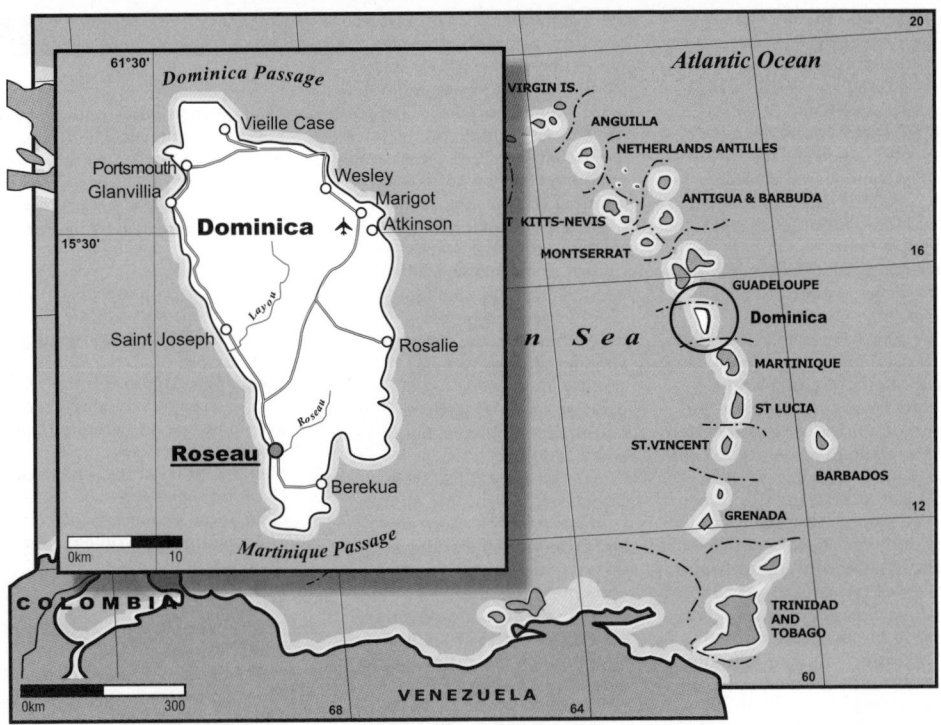

At first the island was inhabited by Arawak tribes. In the 14th century, with the arrival of the Carib, the Arawak were practically exterminated and the few survivors were forced to leave the island.

2 On Sunday 3 November 1493, the Genoese pilot Christopher Columbus, during his second voyage to America, arrived on an island that he named Dominica (referring to the day of his arrival) and turned it into a Spanish possession. During 1500, Spanish ships made frequent stops on the island to try to settle in it, but the strong Carib resistance drove Spain to abandon its colonization attempt.

3 In 1627 the British monarchy launched its attempts to take control of the island, but also failed. As it happened in other Caribbean islands, the indigenous population was gradually exterminated. In spite of resistance to European colonization, reliable sources point out that in 1632 only some 1,000 Carib had survived on the island.

4 Finally, in 1635, the island became a dominion of France, which introduced coffee and cotton and started to operate with sugar cane. The forests were wiped out to make place for plantations, toiled by

LAND USE

2003/2005

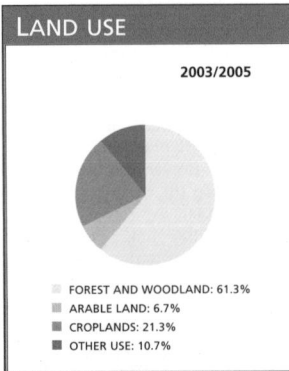

- FOREST AND WOODLAND: 61.3%
- ARABLE LAND: 6.7%
- CROPLANDS: 21.3%
- OTHER USE: 10.7%

thousands of slaves brought over from Africa.

5 France and Britain fought for the control of the island until 1763, when France gave up its possession to the United Kingdom, which turned it into a colony in 1805.

6 After five centuries of colonization, the Dominicans inherited a rudimentary agrarian economy based on the mono-cultivation of bananas for export. These are the only significant export item, because most of the land went over to banana production when sugarcane became unprofitable at the end of the 19th century.

7 The political system adopted on achieving internal autonomy in 1967 was a copy of the British model. The Constitution established the 'free association' of the Associated States of the West Indies with Britain. Britain retained responsibility for defense and foreign relations and each member island elected its own state government. The

WORKERS

UNEMPLOYMENT: 23% (2004)

EMPLOYMENT DISTRIBUTION **1995/2002**

F
M

AGRICULTURE	F: 14%	M: 31%
INDUSTRY	F: 10%	M: 24%
SERVICES	F: 72%	M: 40%

seat of the federal government was in Barbados. The Legislative Council was replaced by the Assembly Chamber; the administrator by the Governor, and the Chief Minister was renamed Prime Minister.

8 After the emancipation of African slaves in 1834, relations with the British empire were interrupted. In 1838 Dominica became the first and only British colony in the Caribbean to have a Parliament with a black majority. In 1896, the British regained control of the island once again, which became a colony of the Crown. Fifty years later it became a province of the Federation of West Indies (1958-1962).

9 In May 1979, the police fired at demonstrators protesting against two decrees limiting union activities and press freedom. At the same time, secret agreements between John and the South African regime were revealed. The Dominican premier had helped to plan a mercenary attack on Barbados and intended to supply Pretoria with oil refined in Dominica. Patrick John was forced to resign. His successor, Oliver Seraphine, from the progressive wing of the Labor Party, called elections and strengthened Dominica's ties with progressive neighboring governments.

10 In the July 1980 election, Mary Eugenia Charles won a

landslide victory and became the first woman to head a Caribbean government. In the 1985 election, Charles' Dominica Freedom Party (DFP) obtained 15 of the 21 Parliamentary seats, although the opposition demanded an investigation into a $100,000 donation made by the CIA for alleged government collaboration in the US invasion of Grenada.

11 The third election since independence was won by the ruling party, which held on to 11 of the 21 seats. The United Workers' Party (UWP), led by Edison James, won six seats, while Douglas' Labor Party took the four remaining seats.

12 In 1990, Prime Minister Charles signed an agreement with her counterparts James Mitchell of St Vincent, John Compton of St Lucia and Nicholas Braithwaite of Grenada, for the four islands to form a new state. In the same year the Regional Constitutional Assembly of the Eastern Caribbean was set up. It comprised government officials, religious authorities and representatives from social organizations.

13 This integration project was in line with the Chaguaramas Treaty, which aimed at the unification of the entire English-speaking Caribbean community, although the three islands of St Kitts, Montserrat and Antigua

refused to commit themselves to the process. The opposition parties of Grenada and St Lucia also stated they were against it. The seven mini-States of the Eastern Caribbean have a common Central Bank, which coins their shared currency.

[14] In 1991, the Government decided on a series of measures to stimulate the national economy, centering its efforts on the development of agriculture and communications.

[15] The Prime Minister narrowly escaped censure in April through a motion presented by the United Workers' Party. At the end of 1991, Charles attempted to enact legislation making it illegal for civil servants to protest against the Government.

[16] In April 1994, the Government decision to increase the number of transport vehicle licenses caused protests and public disorder in Roseau. The Prime Minister accused the opposition of trying to bring forward the elections, and the crisis continued until an agreement was signed on May 6.

[17] That month, Dominica voted against the creation of a whale sanctuary in the South Atlantic. The country was threatened with a tourism boycott from the International Wildlife Coalition. The measure would have seriously affected the island's economy - dependent on tourism - but was not implemented.

[18] In June 1995 the UWP won 11 of the 21 Parliament seats. The new Prime Minister, Edison James, supported the banana industry and privatized state companies to invest in social infrastructure.

[19] In August and September 1995, a succession of hurricanes and tropical storms destroyed the plantations and crops for export. Houses, bridges, roads and hotels were also destroyed and had to be rebuilt, at great cost.

[20] In October 1998 the Assembly appointed Vernon Lorden Shaw, a retired civil servant, seventh president of Dominica. Shaw promised to boost Dominica's sources of income which had been badly hit by climatic disasters and the global economic crisis. That year, the Government announced plans to make the country into the main supplier of offshore financial services, 'not only in the Caribbean, but in the whole world'.

[21] In February 2000, Rosie Douglas became Prime Minister. That same month, Dominica and Cuba signed a bilateral co-ordination and consultation agreement, to strengthen cultural, social and political co-operation. Cuban-Dominican co-operation on education would extend to tourism.

[22] Douglas died unexpectedly on 1 October, aged 58, and was succeeded by Pierre Charles, leader of the Labor Party. Charles had drawn attention to himself after criticizing the US invasion of Grenada in 1983 and, more recently, the US embargo on Cuba and its intervention in Afghanistan.

[23] In February 2002, Parliament passed a law against money-laundering aimed at allowing foreign investigators to scrutinize offshore bank accounts. In May Charles announced a tight budget to help the economy battered by economic and financial crises, low exports and fewer tourists.

[24] In the October 2003 elections, Nicholas Liverpool was elected President.

[25] In January 2004, Pierre Charles died at the age of 49. The Labor Party chose former teacher Roosevelt Skerrit as his successor.

[26] Dominica broke off its 20-year diplomatic relations with Taiwan, in favor of ties with China. Beijing agreed to give $122 million in aid over five years. Eugene Chien, Foreign Minister of Taiwan, criticized what he called China's 'dollar diplomacy.'

[27] In April 2004, WRB Enterprises Ltd bought 72 per cent of Dominica Electricity Services, Domlec. The World Bank disapproved of the purchase, arguing that it threatened the Government's plans to downsize its budget and address the country's chronic fiscal deficit.

[28] In the May 2005 parliamentary elections, the ruling Labor Party won 12 out of the 21 seats, while the Workers Party won 8. One seat went to an independent candidate.

[29] Skerrit won the May 2006 elections. The Government's top priority was to eliminate the restrictions preventing the country from joining the Caribbean Single Market (CSM).

[30] In October 2006, Skerrit and his Cabinet met with Dominicans residing overseas at a function at the State House, and urged them not to put the country down, especially on the internet. ∎

PROFILE

ENVIRONMENT
The largest Windward Island of the Lesser Antilles, located between Guadeloupe to the north and Martinique to the south, Dominica is a volcanic island. Its highest peak is Morne Diablotin, at 2,000 meters. The climate is tropical with heavy summer rains. The volcanic soil allows for some agricultural activity, especially banana and cocoa plantations.

SOCIETY
Peoples: Descendants of Africans 89 per cent; descendants of Europeans and Africans 7.2 per cent; Caribs living in reservations 2.4 per cent; Europeans 0.4 per cent; Other 0.7 per cent.
Languages: English (official). Creole, a local French patois, with African elements, is widely spoken.
Religions: Roman Catholic 77 per cent; six largest Protestant groups 17.2 per cent.
Main Political Parties: Dominica Labour Party; United Workers' Party; Dominica Freedom Party; Dominica Progressive Party.
Main Social Organizations: National Workers' Union; Officials of Dominica Labor Union; Association for the Conservation of Dominica (environmentalist).

THE STATE
Official Name: Commonwealth of Dominica.
Administrative Divisions: 10 parishes.
Capital: Roseau 27,000 people (2003).
Other Cities: Portsmouth 3,600 people; Marigot 2,900; Atkinson 2,500 (2000).
Government: Parliamentary system. Dr. Nicholas Liverpool, President since October 2003; Roosevelt Skerrit, prime minister since January 2004, re-elected in 2006.
National Holiday: 3 November, Independence Day (1978) and Discovery by Christopher Columbus (1493).
Armed Forces: Commonwealth of Dominica Police Force (includes Special Service Unit, Coast Guard).

IN FOCUS

ENVIRONMENTAL CHALLENGES
Dominica s known as the 'Caribbean Paradise' due to its wonderful scenery and great animal and plant diversity. It has an extended system of protected natural parks, including Boiling Lake, one of the largest cauldrons of volcanic thermal waters in the world.

WOMEN'S RIGHTS
Women have been able to vote since 1951. In 2005, they held 13 per cent of Parliamentary seats. Women made up 43 per cent of the workforce in 2000.-

Just about all women receive medical care before, during and after birth.

Between 45 and 48 per cent of primary and secondary students are girls. Only 9 per cent of women reach tertiary education.

Between 25 and 45 per cent of households headed by women are below the poverty line.-

In the 15-19 age bracket, five times more girls than boys are living with HIV/AIDS*.

CHILDREN
Ten per cent of children have low birth weight*.

Malnutrition affects 5 per cent of children under 5, one of the lowest rates in the Caribbean*. Some 99 per cent of children are immunized against measles, polio and tetanus. Juvenile delinquency and boys and girls in conflict with the law are growing social problems*.

INDIGENOUS PEOPLES/ ETHNIC MINORITIES
The indigenous population, the Caribs, were virtually exterminated during the Spanish colonization. Currently there are some 2,000 Carib descendants living in reservations.

DEATH PENALTY
It applies to several offenses.

*Latest data available in *The State of the World's Children* and *Childinfo* database, UNICEF, 2006

Dominican Republic / República Dominicana

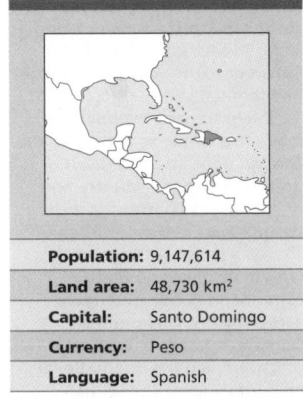

Population:	9,147,614
Land area:	48,730 km²
Capital:	Santo Domingo
Currency:	Peso
Language:	Spanish

The island of Quisqueya is now two countries: Haiti and the Dominican Republic. The first inhabitants belonged to several ethnic groups: the Lucayo, the Ciguayo, the Taino and the Carib. They were fishers and gatherers who practiced basic agriculture. There was always a great deal of contact between the Caribbean islands and trade between the groups.

[2] In December 1492 Christopher Columbus reached the island of Quisqueya, which he renamed Hispaniola. With the wood from one of his vessels he built a fort, initiating the European colonization of America. Within a few years, the Europeans had appropriated the whole island, subduing the Carib population. The terrible living and working conditions imposed by the Spaniards nearly exterminated the indians. Faced with this shameful situation, Bishop Bartolomé de Las Casas proposed that they replace local slave labor with Africans, millions of whom were distributed in the centuries that followed throughout the American continent.

[3] Dominican historical records show that in 1523 a group of rebel African slave workers founded the first *quilombo* (former slave settlement) on the island. Subsequent rebel groups, in 1537 and 1548, set up their own quilombos. The replacement of indigenous labor with African slaves accompanied a change from panning for gold to sugar plantations and extensive cattle raising. As historian Pierre Vilar pointed out, the gold cycle in Hispaniola was destructive, not of raw materials, but rather, of the labor force. During the colonial period the extraordinary economic potential of the Dominican Republic was comparable only with that of Brazil. The island was successively the greatest gold producer in the Antilles, one of the largest producers of sugar in the New World between 1570 and 1630, and finally, such an important cattle producer that there were 40 cattle per person on the island.

[4] 'Santo Domingo is like a microcosm of all American history', said one contemporary historian. 'Its history not only anticipates but also highlights evolutions that in other places occur less noticeably'.

[5] As a major sugar producer with a key position on the trade route from Mexico and Peru to Spain, Hispaniola was coveted by the other colonial powers. In 1586, the English buccaneer Francis Drake raided the capital and in1697 the French occupied the island's western half. When they were given official ownership under the Treaty of Ryswick, they renamed it Haiti. Later, the whole island fell under French rule but was partially recovered by Spain in1809, after the first Afro-American republic had been established (in Haiti).

[6] Haiti's government regained control over the whole island in 1822. The Spanish descendants' (*criollo*) resistance came to a head after an uprising in Santo Domingo. The independence of the Dominican Republic was proclaimed, but in 1861 the Government asked Spain to reinstate colonial status, in an attempt to gain support for the criollos, whose dominance was threatened by the black and mulatto majorities.

[7] However, Spain did not defend its colony effectively and the Dominican Republic became independent again in 1865 after a mulatto uprising. The economic system remained unchanged.

[8] By that time the US, fully recovered from its civil war, began to gain influence in the West Indies. In 1907 the US imposed an economic and political treaty on the Dominican Republic, prefiguring dollar diplomacy, which laid the groundwork for its 1916 inavasion. It imposed a protectorate that lasted until 1924.

[9] In 1930, when the country was autonomous again, Rafael Leónidas Trujillo seized power. He was Chief of Staff of the National Guard, elected and trained by the American occupation forces. He set up a dictatorial regime, with US backing, without nominally occupying the presidency. His crimes were so numerous and so apparent that he finally became too embarrassing even for the US, and the CIA planned his assassination which was carried out in May 1961.Trujillo owned 71 per cent of the country's arable land and 90 per cent of its industry.

[10] In 1963, following a popular rebellion, the first democratic elections were held and writer

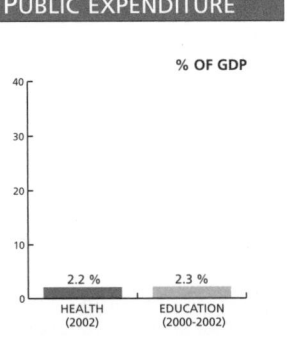

LAND USE

2003/2005

IRRIGATED AREA: 17.2% of arable land

- FOREST AND WOODLAND: 28.4%
- ARABLE LAND: 22.7%
- CROPLANDS: 10.3%
- OTHER USE: 38.6%

PUBLIC EXPENDITURE

% OF GDP

2.2 % HEALTH (2002)

2.3 % EDUCATION (2000-2002)

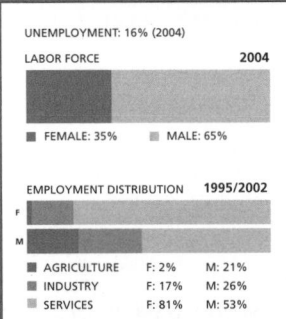

WORKERS

UNEMPLOYMENT: 16% (2004)

LABOR FORCE **2004**

- FEMALE: 35%
- MALE: 65%

EMPLOYMENT DISTRIBUTION **1995/2002**

F

M

	F	M
AGRICULTURE	2%	21%
INDUSTRY	17%	26%
SERVICES	81%	53%

Life expectancy
69 years
2005-2010

GNI per capita
$2,100
2004

Literacy
88% total adult rate
2000-2004

HIV prevalence rate
1.7% of population 15-49 years old
2003

Juan Bosch was elected president. Seven months later, he was overthrown by military officers from the Trujillo regime. In April 1965, Colonel Francisco Caamaño Deñó led a constitutional armed rebellion. Accusing the nationalists of having pro-Castro (communist) sympathies, the US intervened once again, sending in 35,000 Marines who suppressed the insurgency.

[11] Before leaving the country, the Marines paved the way for a Trujillo supporter, Joaquín Balaguer, to rise to power. In return, he opened the country to transnationals. The sugar industry fell under Gulf and Western control. The corporation also bought shares in local banking, agro-industry, hotels and the cattle industry, becoming very influential.

[12] The nationalist opposition kept up its resistance and in 1973 Francisco Caamaño was killed while leading a guerrilla group. The Dominican Revolutionary Party (PRD), originally led by Juan Bosch, split. The right wing (led by landowner Antonio Guzmán) eliminated the main reformist measures from its program. This action made it acceptable to the State Department and in1978, when the PRD won the elections, the US used its influence - in the name of human rights - to ensure that the results would be respected.

[13] The PRD program re-established democratic freedoms and popular organizations took advantage of the new situation to reorganize their weakened structures after decades of harsh repression.

[14] New presidential elections were held on 16 May 1981, and Salvador Jorge Blanco became president in the PRD's second successive victory. José Francisco Peña Gómez, one of the Latin American leaders in the Socialist International, was elected mayor of Santo Domingo. On July 4, departing president Antonio Guzmán killed himself, generating political tension which ended with the announcement of the electoral result.

[15] Blanco tried to tackle the situation by applying IMF-tailored austerity measures. But during 1983, the international price of sugar fell 50per cent below the cost of production, and sugar accounted for 44 per cent of Dominican exports. In 1984, the Government withdrew the subsidies on several products, and imposed a 200 per cent price increase on staple and medical goods. These measures brought about protest rallies led by leftist organizations and labor unions. In return, the Union headquarters were occupied by soldiers leaving 100 dead, 400 injured and over 5,000 imprisoned.

[16] In 1985, the US reduced its Dominican Republic sugar quota again, causing another decrease in exports. Unemployment rose abruptly. The Government continued to toe the IMF line, harshly repressing all the strikes and protests against its policies. The Dominican Republic Association of Economists noted that since 1980, poverty - which had been limited to the low-income groups - had spread to the middle classes.

[17] For the poorer people, the situation was untenable. Committees for Popular Struggle began to appear and grass roots movements organized to resist price rises on basic goods and services. In April, dockworkers found 28 young Dominican Republic women asphyxiated in a ship's container. They could not find work and had expected to find a way of making a living on another island. It was disclosed that every two weeks a cargo of young women from the Dominican Republic left for the Franco-Dutch island of St Martin, where a brothel manager sold them to other Caribbean islands for between 800 and 1,000 German marks.

[18] Chaotic national elections were held on16 May 1986, with three presidential candidates: Jacobo Majluta for the Dominican Revolutionary Party (PRD), Joaquín Balaguer for the Social Christian Reformist Party (PRSC) and Juan Bosch for the Marxist-inspired Dominican Liberation Party (PLD). Balaguer was elected President by a slight margin, while the opposition claimed there had been fraud.

[19] Clearly a conservative, without a parliamentary majority and faced with highly organized social opposition, Balaguer did not even have the necessary power to impose a more restrictive economic program to get new loans from the IMF.

[20] Balaguer was re-elected in the May 16 1990 elections. His rival, Juan Bosch, leader of the PLD, accused him of an electoral fraud.

[21] In 1990 and 1991 the already precarious situation worsened for the thousands of Haitian immigrants working as cane-cutters in the Dominican Republic sugar industry. In June 1991, Balaguer expelled them as illegal immigrants.

[22] Negotiations by Balaguer made it possible to refinance the foreign debt. In 1993 tourism grew, making the country one of the most important tourist destinations in the Caribbean.

[23] As a result of the political crisis in neighboring Haiti, contraband goods were taken across the border from the Dominican Republic into Haiti. This practice foiled the international embargo against the Haitian regime. Oil was the main product smuggled into the country.

[24] Economic problems caused hundreds of Dominican Republicans to leave the country each year with forged visas and documents. Many sailed in unseaworthy vessels, bound for Puerto Rico, normally as a stopover on their way to the US.

[25] Although he had announced his retirement, Balaguer, 87, sought re-election in 1994. His perennial opponent Juan Bosch also ran. To avoid another case of electoral fraud, four of the five participating parties signed a civility pact, with the Catholic Church acting as guarantor. In spite of this, the electoral campaign turned violent, with hundreds of people injured and 12 people killed.

[26] International observers were called in to supervise the elections. The PRD condemned the voting stating that some 200,000 voters were unable to vote as a result of official party manipulation. According to official figures, Balaguer obtained 43 per cent of the vote, leading Peña Gómez by 1.5 per cent.

[27] Peña Gómez and Balaguer agreed to hold national elections on 16 November 1995, and to amend the Constitution to ban presidential re-election. Meanwhile, Balaguer was sworn in as President.

[28] On 30 June 1996, Leonel Fernández Reyna of the PLDP won the elections after forming an alliance with the conservative PRSC. On August 18 he succeeded Joaquín Balaguer, who had served seven terms as president.

[29] Increasing prices, unemployment of more than 30 per cent, and the poverty that affected 70 percent of the population led to an increase in social tension. There were demonstrations in the streets, some violent. Despite this, the overall economy saw exceptional growth of 6.9 per cent in the first half of the year.

PROFILE

ENVIRONMENT

The Dominican Republic comprises the eastern part of the island of Hispaniola, the second largest of the Antilles group. The Cordillera Central, the central mountain range, crosses the territory from northwest to southeast. Between the central and northern ranges lies the fertile region of the Cibao. Sea winds and ocean currents contribute to the tropical, rainy climate. Between 1962 and 1990, the country lost a significant portion of its woodlands. Coral reefs are suffering the effects of pollution, which has harmed marine habitats and reduced fish populations. Hurricanes cause serious damage.

SOCIETY

Peoples: Most are of Spanish and African descent, with a small native American component.
Languages: Spanish.
Religions: Roman Catholic 91.3 per cent; other 8.7 per cent.
Main Political Parties: Dominican Liberation Party; Dominican Revolutionary Party; Social Christian Reformist Party.
Main Social Organizations: Most workers are represented by the General Workers' Union (CGT) and the Unity Workers' Union (CUT). In March 1991, four major labor groups, 57 federations and 366 labor unions merged within the CUT.

THE STATE

Official Name: República Dominicana.
Administrative Divisions: 26 Provinces, 1 National District.
Capital: Santo Domingo 1,865,000 people (2003).
Other Cities: Santiago de Los Caballeros 446,800 people; La Vega 416,300: San Pedro de Marcorís 257,700 (2000).
Government: Leonel Antonio Fernández Reyna, President since August 2004. Presidential system with a bicameral legislature. Congress of the Republic: a 30-member House of Senators and a 149-member House of Representatives, all elected by popular vote.
National Holiday: February 27, Independence Day (1844).
Armed Forces: 24,500 (1995). Other: National Police, 15,000.

	Under-5 mortality		Poverty		Debt service		Maternal mortality
	32 per 1,000 live births 2004		**2.5%** of population living on less than $1 per day 2003		**6.4%** exports of goods and services 2004		**150** per 100,000 live births 2000

30   A law approved in June allowed private capital to invest in state companies, including the sugar and electricity sectors. The Government aimed to thus balance the state accounts.

31   Towards the end of 1997, the Presidents of the Dominican Republic and Haiti agreed to stop the large-scale repatriation of Haitians and to respect human rights. Haitians often worked in the worst conditions.

32   On 16 April 1998, the Dominican Republic and Cuba restored diplomatic relations. The Government sent a consular representative to Havana, later followed by a delegation of ministers that formally inaugurated the diplomatic headquarters. The US protested against the measure, deeming it inappropriate.

33   New disturbances shook Santo Domingo in January 1999 following the outcome of elections for the new president of the municipal league, a body which handled a $100 million budget to aid local governments.

34   The May 2000 elections gave the Presidency to Hipólito Mejía, candidate of the PRD, who won by a big enough margin to avoid a second round.

35   The new President took office in August, promising to create jobs and fight corruption and poverty. In November, a march held by Fernández Reyna to denounce the imprisonment of four members of his government was suppressed by the police. The former President, who was hospitalized as a result of police tear gas, claimed the prison sentences aimed to incriminate him in corruption cases.

36   In October 2000, the Government refused to extradite seven Haitian police accused of taking part in a plot to destabilize the Haitian Government. Hugo Tolentino Dipp, Dominican Republic's Minister of Foreign Affairs, stated the seven would be held in his government's custody.

37   The following year, in May, the Court of Appeal annulled a trial against former President Salvador Jorge Blanco on corruption charges.

38   Since a wave of murders and kidnappings jeopardized the arrival of tourists - one of the country's main sources of income - Mejía sent the army to patrol the streets, seeking to contain crime.

39   In November 2001 an American Airlines plane from New York to Santo Domingo crashed in Queens, US, killing 225 people including many from the Dominican Republic. At first it was thought it might be linked to the

IN FOCUS

ENVIRONMENTAL CHALLENGES

The country lost a significant amount of its forested areas between 1962 and 1980. Coral reefs suffer from pollution and degradation, which significantly reduce the fish population. Eighty per cent of water tables have less water as a result of deforestation and soil erosion; productivity of the land has severely declined.

The country has plenty of surface water, in excess of demand, but there is inadequate infrastructure to deliver enough drinking water; sanitation is also a problem.

WOMEN'S RIGHTS

Since 1942 women have had the vote and been eligible to stand for office. Women held 4 per cent of ministerial or equivalent positions in 2003, while their representation in Parliament stood at 17 per cent.

Of the 4 million people that make up the country's total workforce, 32 per cent are women (of them, 81 per cent work in services, 17 per cent in industries and 2 per cent in agriculture).

Some 99 per cent of pregnant women receive prenatal medical care, and 98 per cent of births are attended by skilled health staff; there are 180 maternal deaths per 100,000 births*.-

It is estimated that there are over 100,000 sex workers.

CHILDREN

Thirty-two children under 5 die for every 1,000 live births, while 11 per cent suffer from low birth weight*.

About 2.2 per cent of people are living with HIV/AIDS; children under 5 account for more than 5,000 cases.

The country has a large sex industry. Over 25,000 children and teenagers between 12 and 17 are involved in prostitution. Many youngsters are taken to Europe as prostitutes.

Some 436,000 children aged between 5 and 17 are exploited as laborers in the country. Over 18 per cent of them perform adult labor in plantations, which affects both their health and future growth.

INDIGENOUS PEOPLES/ ETHNIC MINORITIES

Approximately 11 per cent of the population descends from Europeans; the rest are of Euro-African origin. There is racial discrimination against blacks.

Haitian immigrants are the poorest social group, earning 60 percent less than the average. They are under-nourished, do not receive adequate medical care and do not participate in the country's political, social or cultural life. In most cases they

work in sugar plantations, under slave-like conditions. They work for $1.5 per day, and in some cases receive no pay. In 1937, on the orders of dictator Rafael Leonidas Trujillo, the army murdered more than 25,000 Haitians.

MIGRANTS/REFUGEES

There are 700,000 people from the Dominican Republic in the US, most of them living in New York.

In 2004, about 300 Haitians were granted refugee status in the Dominican Republic. But in spite of the political and social turmoil in Haiti, none of the 400-plus that requested asylum that year were granted asylum status.

Dominican Republic authorities estimate that more than 1 million Haitians live in the country, but often some are deported. The few that are granted asylum have to overcome myriad official procedures. The Government regards most Haitians as 'in transit', therefore denying them citizenship. The children of Haitians born in the Dominican Republic are considered citizens.

DEATH PENALTY

Abolished in 1966.

*Latest data available in *The State of the World's Children* and *Childinfo* database, UNICEF, 2006.

9/11 attacks by al-Qaeda on the US, but this was later dismissed.

40   In July 2002, Balaguer died at 95.

41   After months of demonstrations, a general strike took place in November 2003. The police arrested hundreds of demonstrators, and six people were killed. Mejía declared there would be 'no mercy' in the response to public order disturbances.

42   In December 2003, tropical storm Odette forced the evacuation of 10,000 people. Civil defense forces warned the population to leave their homes, but without giving them alternative shelters.

43   After the Spanish Government announced that it was pulling its troops out of Iraq, Mejía ordered the withdrawal of Dominican troops in April 2004. Dominican troops were part of the Plus Ultra Brigade, under Spanish command, which also included troops from Honduras, El Salvador and Nicaragua.

44   On 17 May 2004 Mejía lost his

bid for re-election, to Fernández Reyna. Given the final result - 56 per cent of the votes were for Fernández - a second round was not necessary. According to analysts, Fernández's victory was due to a punishment vote against Mejía for the economic crisis suffered by the country since 2003, unleashed after the collapse of one of the main Dominican private banks.

45   That month, intense rains caused severe floods (also in Haiti), leaving more than 500 dead and 13,000 missing. Maize crops were destroyed and thousands of Dominicans were left homeless. Jimaní, in the southwest of the country, was the area worst affected.

46   In March 2006, Marino Vinicio Castillo, government advisor on drug trafficking, released a series of statements that sparked controversy. Castillo pointed out at a police event that political parties, the military and almost all state institutions were directly or indirectly

involved in drug-trafficking. At the same time, he added that this problem had become worse during Mejía's administration.

47   The religious organization Solidaridad Fronteriza reported in April that Dominican military officials were taking bribes to allow the smuggling of goods and Haitian immigrants across the border. The organization gave photographic evidence to the army. Army chief Major General José Ricardo Estrella Fernández, undertook to investigate the claims.

48   In October 2006, the Government announced that ten Dominican agricultural products will be protected for at least 10 years once the DR-CAFTA free trade agreement with the United States and Central America comes into effect. Rice, milk, pork and chicken, beans, corn, onions and garlic are all on the list. The announcement was made in the hope of calming fears about the impact of the free trade agreement on local agriculture. ∎

Ecuador / Ecuador

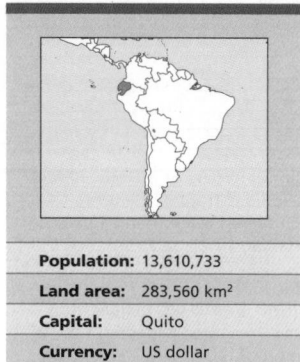

Population:	13,610,733
Land area:	283,560 km²
Capital:	Quito
Currency:	US dollar
Language:	Spanish

The territory now known as Ecuador has been inhabited at least since 2500 BC. The region was largely a border zone influenced by the Nazca, Tiahuanaco-Huari, Chibcha and also Mexica civilizations, among others. It has been suggested that there was contact with peoples of the Pacific, Japanese or Polynesians, though there is still much debate on that point. In the early 15th century AD, the Cara nation, led by the Shiri dynasty, expanded to the north and central Andean foothills. The Cara settled in the Quito kingdom, which was the largest unit of a confederation that left no historical records. In the same era, both the Chimu nation, originating in Peru's northern coastal zone, and the Inca empire began to exert pressure on the Cara and other peoples settled in the region.

[2] In 1478, Inca Topa Yupanqui united the Ecuadorean agricultural peoples. Within a few years the northern region of the Tahuantinsuyu acquired great economic importance and Quito became its commercial center. But the rivalry for succession between Atahualpa (from Quito) and Huascar (from Cuzco) weakened the power of the Empire (see History of Peru).

[3] The Spanish conquistadors, under the command of Sebastián de Benalcázar, took advantage of the situation to usurp the kingdom of Quito in 1534. In the first era of the colonial period, the territory formed part of the Viceroyalty of Peru and was known as the Real Audiencia of Quito. Crude textiles were the only industry of the Royal District of Quito at the time.

[4] With the reorganization carried out by the Bourbons in 1717, Quito was transferred to the Viceroyalty of Nueva Granada, which comprised present-day territores of Ecuador, Colombia, Panama and Venezuela. In 1809 there was an uprising against the authority of the Crown. In 1822, the armies of Simón Bolívar and Antonio José de Sucre invaded from Colombia, in support of the patriot rebels. On 24 May of that year, in Pichincha, near Quito, Sucre defeated the Spanish and won the emancipation of Ecuador, which was thus incorporated into Bolívar's Greater Colombia scheme.

[5] In 1830, the Real Audiencia of Quito seceded from Greater Colombia and adopted the name Republic of Ecuador.

[6] In 1895, the Liberal Revolution led by Eloy Alfaro raised the hopes of the majority of peasants for a solution to the agrarian issue. Church property was nationalized, but the large landowners were not affected. Alfaro wa s assassinated in 1912.

[7] In 1914 Ecuador ceded to Colombia the territory between river Caquetá and river Putumayo. In 1916, the elected president - the liberal Alfredo Baquerizo Moreno - promulgated the eight-hour working day.

[8] A military coup led by young officers paved the way for a new reformist period in 1925, but the regime did not survive the world economic crisis of 1929. A period of instability began and saw 23 different presidents from 1925 to 1948.

[9] In 1941, after a brief war with Peru, Ecuador was forced to renounce its claims to sovereignty over an extensive area of the Amazon, losing the province of El Oro. The Peace Protocol signed in 1942 in Rio de Janeiro, with Argentina, Brazil, Chile and the United States as guarantors, established the border between the two countries, but much of it was not demarcated within the territory.

[10] A popular uprising in 1944 ousted President Carlos Arroyo, ushering in a populist government led by José María Velazco Ibarra and consisting of conservatives, communists and socialists, under the name of Democratic Alliance. The Cold War made it impossible for this alliance to continue and soon the left became the target of persecution. In 1962, under pressure from the United States, the government of Carlos Arosemena broke off diplomatic relations with Cuba.

[11] In 1972, Ecuador began to export petroleum, which became the leading sector of the economy, replacing bananas, coffee and cocoa. That same year, the political situation changed, as veteran populist leader Velazco Ibarra was ousted for the fourth time by the armed forces. Under the government of General Guillermo Rodríguez Lara, the country joined OPEC, the state purchased 25 per cent of Texaco-Gulf shares and asserted its rights over the 200 miles of territorial waters when faced with US pressure for fishing interests, leading to what was dubbed the 'tuna war'.

[12] Jaime Roldós, nominated by the Convergence of People's Forces (CPF) and the People's Democratic Party, became president in August 1979. Ecuador renewed diplomatic relations with Cuba, China and Albania. The Government also initiated a program aimed at integrating marginalized rural and

PROFILE

ENVIRONMENT

The country is divided into three natural regions: the coast, the mountains and the rainforest. More than half of the population lives along the coast, where cash crops of bananas, cocoa, rice and coffee are grown. In the highlands, extending between two separate ranges of the Andes, subsistence crops dominate. In the eastern Amazon region, the exploitation of oil fields generates profits at the cost of environmental destruction. The Colón archipelago or Galápagos Islands belongs to Ecuador.

SOCIETY

Peoples: 65 per cent mestizos, 25 per cent indigenous, three per cent Afro-Americans, seven per cent Indo-European. There are several indigenous nationalities: Huaorani, Shuar, Achar, Siona-Secoya, Cofan, Quechua, Tsachila and Chachi. There are over 1.5 million Quechua living in the inter-Andean valley.

Religions: Mainly Catholic (95 per cent).

Languages: Spanish (official), although 40 per cent of the population speaks Quechua and other indigenous languages.

Main Political Parties: Ecuadorian Roldosist Party; Democratic Left; Institutional Renewal Party of National Action.

Main Social Organizations: The Ecuadorean Central Organization of Class Unions (CEDOC), and the Central Organization of Ecuadorean Workers (CTE) are coordinated with the United Workers' Front (FUT); Confederation of Indigenous Nationalities of Ecuador (CONAIE); University Students Federation of Ecuador (FEUE).

THE STATE

Official Name: República del Ecuador.

Administrative Divisions: 22 Provinces.

Capital: Quito 1,451,000 people (2003).

Other Cities: Guayaquil 2,627,900 people; Cuenca 271,400; Machala 211,300 (2000).

Government: Alfredo Palacio, President appointed by the National Congress on January 2005. One-chamber parliament: National Congress with 100 members.

National Holidays: 24 May, Independence (1822); 10 August, First Shout of Independence of Quito (1809).

Armed forces: 57,100 troops (conscripts). 100,000 reserves (1996). Other: 200 Coast Guard and six coastal patrol units (1993).

 Life expectancy
75 years
2005-2010

 GNI per capita
$2,210
2004

Literacy
91% total adult rate
2000-2004

HIV prevalence rate
0.3% of population 15-49 years old
2003

urban populations. However, it encountered a hostile Congress, as well as US opposition to its human rights policies and to its antagonism to the dictatorships that were ruling South America.

[13] Toward the end of January 1981, the Five Day War broke out between Ecuador and Peru, with skirmishes along borders which had not been clearly delineated by the 1942 Protocol.

[14] Roldós died in a suspicious plane accident in 1981, and was replaced by vice-president Osvaldo Hurtado. In 1982 a deep social crisis was caused by the implementation of IMF policies and the increasingly evident intent to match the military might of Peru's armed forces.

[15] In the following decade, the Durán Ballén government promoted privatization of state companies and rigid structural adjustment; however, the economy did not improve, a factor which aided the populist Abdala Bucaram into power. He was later deposed following massive price rises on essential services (and after being declared 'insane' by parliament) and Jamil Mahuad took office in his place.

[16] In 1990 the indigenous people's movement erupted onto the political stage for the first time. The Confederation of Indigenous Nationalities of Ecuador (CONAIE) campaigned for the 1997 Constitution to recognize the pluri-cultural and multi-ethnic nature of the State (and, therefore, the right of indigenous people and Afro-Ecuadorians to participate in equality), and there were seven large demonstrations.

[17] Amidst nation-wide demonstrations, Mahuad announced the economy would be pegged to the dollar, precipitating an indigenous uprising. Parliament and Government buildings were invaded, with the support of military groups. A government council formed by representatives of the army, judicial power, indigenous people and unions, was well received by the population. However, pressure from the US led to the arrest of the insurrectionist soldiers - among whom was the current president Lucio Gutiérrez. Mahuad was replaced by Vice-President Gustavo Noboa, an Opus Dei (right-wing Catholic organization) militant, who implemented the dollarization of the economy.

[18] In February 2002, indigenous groups from Sucumbíos and Orellana, two northeastern provinces, demanded that oil production in the country be stopped. They began a strike in protest against a new pipeline for crude oil built by the consortium OCP Ecuador Inc. The provinces demanded that the company gave $10 million to be used in social programs as compensation for the damage caused by their works. After a fortnight the strike ended with four people dead and almost $3 million in losses as a result of the interruption in oil production.

[19] Lucio Gutiérrez - a retired colonel involved in the insurrection against Mahuad - and banana tycoon Alvaro Noboa (unrelated to former President Gustavo Noboa) made it to the second round of the 24 November presidential elections. Gutiérrez, head of the Patriotic Society Party (PSP), won the presidency with 54.3 per cent of the vote.

[20] Former President Noboa exiled himself in the Dominican Republic in August 2003, after corruption charges were brought against him. In October began the trial of a Chevron-Texaco subsidiary accused of destroying large areas of rainforest and contaminating land and rivers in Nueva Loja province.

[21] In November, President Gutiérrez was accused of accepting money for his political campaign from people linked to drug-trafficking. This prompted the resignation of his entire Cabinet.

[22] After 10 days of violence in jails, in April 2004 the police regained control over detention centers with 11,000 inmates. The protests, in demand of better confinement conditions, led to bloody confrontations between rival bands.

[23] After a significant defeat of the PSP in the October local elections, the opposition demanded Gutiérrez' resignation, but he refused.

[24] In December, Congress removed most of the Supreme Court judges, accused by Gutiérrez of being partial to the opposition.

[25] Tension mounted when the Government faced growing protests as a result of corruption charges brought against former Presidents Noboa and Bucaram by the Supreme Court in August 2005. The Government decreed martial law but lifted it 20 hours later, after a general state of civil disobedience led Congress to remove Gutiérrez. Vice-president Alfredo Palacio took his place while Gutiérrez sought refuge in the Brazilian embassy. In October, Gutiérrez was arrested, accused of jeopardizing national security.

[26] He was freed in March 2006 after the judges withdrew the charges against him. That month, there was a nationwide protest against a potential free trade agreement with the US.

[27] In May 2006, the Government seized the assets of a US oil company, Occidental, following a long-running dispute. The Government said the dispute was particular to Occidental and that other foreign companies had nothing to fear but the US responded by cancelling free-trade negotiations with Ecuador. ∎

IN FOCUS

ENVIRONMENTAL CHALLENGES
In the coastal region, 95 per cent of the woodlands have been felled. Soil depletion and consequent desertification has increased by 30 per cent over the last years. Oil companies cause serious pollution, especially in fragile areas like the Galápagos Islands.

WOMEN'S RIGHTS
Although Ecuador was the first Latin American country to grant the vote to literate women (1929), suffrage did not become universal until 1967.

In 2000, women held 16 per cent of parliamentary seats and 14 per cent of ministerial positions.

In 2003, women made up 29 per cent of the labor force: 4 per cent in agriculture, 16 per cent in industry and 80 per cent in services.

The illiteracy rate for women over 15 is 10 per cent, half the 1980 rate*.

Maternal mortality is higher in indigenous areas, where only 20 per cent of births take place in health centers. Approximately 69 per cent of births are assisted by trained staff, and 69 per cent of pregnant women receive prenatal care.

CHILDREN
Nearly 70 per cent of children live in poverty. More than 250,000 children work. Malnutrition affects 12 per cent of children under five years old and childhood development programs reach only 8.4 per cent of children*.

Boys and girls have equal access to education*, but children of indigenous and African descent do not: 90 per cent of these live in poverty and only 39 per cent complete primary school.

Seven out of 10 children under one year old are anemic*. HIV/AIDS has not reached pandemic levels, but its incidence has increased seven-fold since 1990. About 50 per cent of infant deaths are considered avoidable and the mortality rate for indigenous and children of African descent is higher than the rest.*

INDIGENOUS PEOPLES/ ETHNIC MINORITIES
Afro-descendants (10 per cent of the population) live on the northern coast, mainly in the province of Esmeraldas. They first arrived in the country as slaves; in 1851 slavery was abolished and they started to control the political and economic life of the regions they inhabited. Nowadays, they are poor and suffer social discrimination.

Ecuadorean Coastal and Sierra Indians represent 28.5 per cent of the population, mostly mestizos (55 per cent). While Sierra Indians produce food for domestic markets, Coastal Indians mainly work in export agriculture.

Although there is still political and economic discrimination against indigenous peoples, the mobilization of Ecuador's indigenous groups is remarkable. These communities are supported by international organizations.

MIGRANTS/REFUGEES
Since 2000, more than 25,000 Colombians have requested asylum and refuge in the country. Of that number, only 6,080 had been recognized in late 2004 as refugees by the Ecuadorean Government. Most come fleeing from the civil war. An article published in 2002 by *The New York Times* estimated that more than 200,000 Colombians had entered Ecuador in the previous years and remained there.

The Ecuadorean Government and people have welcomed Colombians with fewer restrictions than Venezuela and Panama: they enjoy all rights accorded to any foreigner and they can even apply for citizenship. Undocumented Colombians usually live in towns like Ibarra, Santo Domingo de los Colorados and Lago Agrio.

Ecuador's population is, at the same time, traditionally emigrant. Ecuadoreans represent the second largest South American group in the US, after Colombians. Throughout the 1970s, emigration to the US grew at an annual average 8.5 per cent, increasing even more during the 1980s and 1990s. It is estimated that in the last decade more than 500,000 Ecuadoreans left the country and did not return. In 2005, Ecuador received $1,800 million by way of remittances, representing some 8 per cent of its GDP.

DEATH PENALTY
It was abolished in 1906.

*Latest data available in *The State of the World's Children* and *Childinfo* database, UNICEF, 2006.

Discrimination in Latin America

LATIN AMERICA'S Afro-descendant and indigenous populations have endured centuries of exclusion, and even today the vast majority of them live in poverty. Taking indigenous, black and mixed-race people together, they constitute a clear majority of the population of Latin America, around two-thirds - yet have been effectively treated as minorities and marginalized throughout much of the region's history. There are around 60 million indigenous peoples from more than 400 ethnic groups, who live in all the region's countries except in Uruguay, where they were exterminated in the 19th century.

Although it is a remarkable fact that Bolivia elected an indigenous person as President in 2005 (see History) for the first time in its history, the reality is that, compared to the white population, African descendants and indigenous peoples have very poor access to health, education, employment, justice and political participation. In many cases, they have lost their main livelihood, their land and natural resources. This is the reason why, for several decades, they have been moving to the cities in the hope of a better life. But often all they find are precarious and badly paid jobs which are dangerous and even life-threatening.

According to the Economic Commission for Latin America and the Caribbean (ECLAC), the 'progressive loss of lands and the rupture of community economies' are two of the main factors that contribute to poverty in these countries.

The health of the minority groups is significantly worse than that of the general population, because of inadequate diets and lack of basic medical services. In many of the region's countries, big development projects often have negative effects on indigenous populations. The indiscriminate clearing of natural forests, oil extraction and the construction of dams and reservoirs have brought devastation to many communities.

Education is another area where minority people lose out. In Ecuador, for example, slightly more than 50 per cent of the indigenous population has access to primary education, less than 20 per cent reaches secondary school and barely one per cent goes to university.

AFRO-DESCENDANTS
Afro-descendants and mestizos are not much better off than indigenous peoples in this region. The Inter-American Development Bank (IDB/IADB) considers Afro-descendants to be the 'most invisible among the invisible': they are absent from political, economic and educational leadership indicators. Also, many Latin American countries do not take into account the African origin of a part of their population in the national census; Colombia, Brazil, Bolivia, Ecuador and Costa Rica are the only ones that offer the possibility of classifying their citizens as Afro-Latin.

In spite of the difficulties in obtaining reliable data, according to different NGOs the high rates of poverty among Afro-descendants and their limited access to basic services are evident. More than 80 per cent of the Afro-Colombian population lives in poverty and their average per capita income stands at $600, while national income is over $1,500.

Brazil offers a similar picture. The problems of unemployment, low salaries and lack of access to key political positions impact more deeply on the Afro-descendant population.

THE AMAZON AND EXTINCTION
The bleak truth is that even if governments and laws were to guarantee the existence of the isolated indigenous groups still living in the Brazilian, Ecuadorian and Peruvian Amazon forest and in Paraguay's Chaco region, it may already be too late. Their path to extinction appears to be mapped out already, since these isolated communities are facing a cultural genocide.

When the Europeans arrived in the Americas, the Amazon region was home to some 2,000 indigenous groups that amounted to about 7 million people. More than five centuries later, after slavery and exploitation, persecution, and European diseases to which local people had no immunity, less than 400 groups and only about 2 million individuals remain. Of these, some 5,000 still shun contact with 'civilization'. ∎

Egypt / Misr

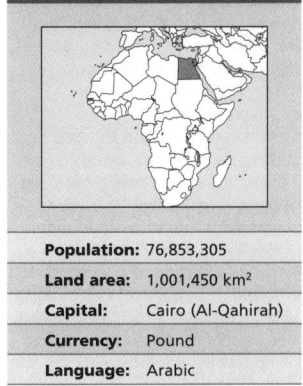

Population:	76,853,305
Land area:	1,001,450 km²
Capital:	Cairo (Al-Qahirah)
Currency:	Pound
Language:	Arabic

S ix thousand years ago, a civilization began to form in the Nile valley (Nahr-an-Nil) and developed into a centralized state. Around 6000 BC, Pharaoh Menes, from Upper Egypt, united the two kingdoms - Upper (south) and Lower (north) Egypt - that coexisted for some time.

2   A powerful and centralized empire arose in the third millennium BC (the Ancient Empire); during those times Egypt carried out several successful military campaigns against the Nubians and the Libyans, beginning a prosperous sea trade. The Ancient Empire saw its demise with the collapse of the central authority during the Sixth Dynasty, due in large part to widespread famine.

3   The Third Dynasty developed the practice of constructing monumental tombs for the monarchs, the Pharaohs. These monuments or pyramids, also built for members of the élite, and the funeral ceremonies of the poor, were closely linked to the belief in life after death. The pyramids and the decorated walls of the pharaonic tombs, the creation of a complex writing system and broad knowledge of medicine and agriculture are just a few examples of the level of civilization achieved by Egyptian society.

4   Egypt reached its height of power, wealth and territorial dominance around 1500BC.

Reconverted into a military state, Egypt launched a series of campaigns to win control of Palestine, Syria and the north of the Euphrates. Territorial expansion led to the development of a complex diplomatic system of alliances and treaties.

5   In the last millennium BC, the decline of this remarkable civilization opened the way to foreign Pharaohs (Libyan and Sudanese dynasties). Toward the end of the 20th dynasty, the weakening of the Pharaohs led to renewed division between Upper and Lower Egypt and, after the Persian Cambises deposed the Pharaoh, the territory formed part of other empires (Persian, Greek and Roman).

6   During the period of Greco-Roman domination, Alexandria (Al-Iskandariyah) was one of the most influential cultural centers in the classical world, and its famous library was the largest until it was burned down in Julius Caesar's time. It brought together the most outstanding philosophers, scientists and scholars of the era. In 642 AD, when the Arabs conquered the

country, little remained of its highly developed past and the Egyptians adopted Islam and the Arabic language.

7   Three centuries later, under the government of the Fatimid Caliphs, the new capital, Cairo (Al-Qahirah), became one of the major intellectual centers in the Islamic world and scholars, particularly African Muslims, were attracted by its university.

8   Between the 10th and 15th centuries, Egypt benefited from its geographic location, becoming the trade center between Asia and the Mediterranean. The Venetians and Genoese came to trade here, and even the constant warfare provoked by the European Crusades in Palestine, in the 11th to 13th centuries, did not stop active trade.

9   Once the Crusaders were driven out, it seemed that Egypt would naturally become the center of the ancient Arab empire. However the Sultanate of the Ottoman Turks was the rising power in the Islamic world at the beginning of the 16th century, and it soon conquered Egypt. The opening of the sea route between Europe and the Far East had already put an end to Egypt's previous trade monopoly, reliant on its dominion over the Red Sea, and Egypt had begun its economic decline.

10   Until the 19th century, Turkish domination was little more than nominal and the real power lay in the hands of Mameluke leaders. In 1805, Muhammad Ali, an Albanian military leader, took power. He forcibly eliminated local Mameluke leaders, established a centralized regime, reorganized the army, declared a state monopoly on foreign trade of sugarcane and cotton and achieved increasing

autonomy from the Sultan of Istanbul, laying the foundations of a modern economy.

11   The economy was poorly managed by Muhammad Ali's successors, the crisis deepened and dependency on Europe increased. In 1874, Egypt was forced to sell all its shares in the Suez Canal, built as a joint Egyptian-French project between 1860 and 1870, to pay its debts to the British.

12   The situation continued to deteriorate, loans piled up and in 1879 the creditors imposed a Bureau of Public Debt formed by three ministers, one English, one French and one Egyptian. This bureau assumed the management of the country's finances.

13   This degree of interference awakened an intense nationalistic reaction, supported by the army. That year, the military overthrew Muhammad Ali's successor Khedive Ismail, forcing his son, Tawfiq, to expel the foreign ministers and appoint a nationalist cabinet. The colonial power reacted promptly: in 1882 English troops landed in Alexandria, seizing military control of the country.

14   The occupation was 'legalized' in 1914, when Egypt was formally declared a protectorate. The British installed King Fuad I on the throne in 1917. This situation continued until 1922, when an Egyptian committee in London negotiated their independence. But as before, despite the nominal independence, the British retained control.

15   In 1948 the state of Israel was created in Palestine and Egypt and other Arab nations launched an unsuccessful war against the new state. The frustration of defeat brought about massive demonstrations against the royal government. Against a background of widespread government corruption, a nationalist group known as the 'Free Officials' was formed within the Egyptian Army, led by General Mohammed Naguib and Colonel Gamal Abdel Nasser.

16   On 23 July 1952 this group ousted King Faruk and, in June 1953, proclaimed a Republic. Three years later, Nasser became President.

17   The new regime declared itself nationalist and socialist, deciding to improvethe living conditions of the *fellahin*, the country's impoverished peasants. Land reform was started, limiting the landowners' monopoly of the majority of the land.

18   The Government's reform program gave priority to the construction of the Aswan dam, one of the world's largest dam projects. The construction was carried out by the Soviet Union, after the Western powers had refused to take it on. The dam, which had been hailed as the key to the country's industrialization and 'development',

LAND USE

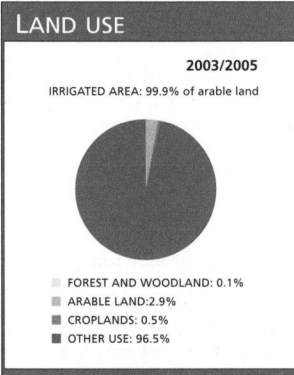

2003/2005

IRRIGATED AREA: 99.9% of arable land

- FOREST AND WOODLAND: 0.1%
- ARABLE LAND:2.9%
- CROPLANDS: 0.5%
- OTHER USE: 96.5%

PUBLIC EXPENDITURE

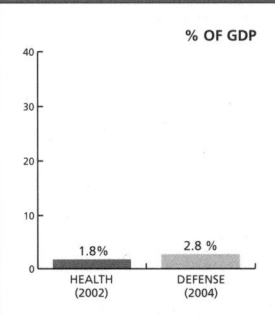

% OF GDP

HEALTH (2002)	DEFENSE (2004)
1.8%	2.8 %

Life expectancy
71 years
2005-2010

GNI per capita
$1,250
2004

Literacy
56% total adult rate
2000-2004

HIV prevalence rate
<0.1% of population 15-49 years old
2003

in fact caused serious environmental disruption.

[19] In 1955, Egypt was one of the leading organizers of the Bandung Conference, a forerunner neutralist Afro-Asian movement which preceded the Movement of Non-Aligned Countries. Twenty-nine Afro-Asian countries condemned colonialism, racial discrimination and nuclear armament.

[20] In October 1956, after the nationalization of the Suez Canal, French, British and Israeli troops invaded Egypt. The Government responded by distributing weapons to civilians. A diplomatic battle was also launched; as a result of UN intervention and joint US-Soviet disapproval, France, Britain and Israel were forced to withdraw and the Canal finally came under Egyptian control.

[21] After Nasser's re-election in 1965, Egypt gave high priority to the conflict with Israel. However, its attempt to economically paralyze Israel by blockading the Gulf of Aqaba failed during the Arab-Israeli conflict, the 'Six-Day War' of June 1967. This ended in another defeat of the Arab countries - Egypt, Syria and Jordan - when Israeli forces occupied the Sinai Peninsula, the Gaza strip, the West Bank, and the Syrian Golan Heights. The cost of the war aggravated Egypt's financial problems.

[22] Gamal Abdel Nasser died in 1970, and Vice-President Anwar Sadat, a member of the right wing of Nasser's Arab Socialist Party, took his place. Sadat put the *infitah* into practice. This was a government plan that meant an opening to Western influence, the de-nationalization of the Egyptian economy and the end of the one-party system. Furthermore, the new Government broke off relations with the Soviet Union, and US economic and military aid flowed into Egypt.

[23] In 1973, Egyptian troops crossed the Suez Canal to bring to an end Israeli occupation of Sinai, beginning the fourth Arab-Israeli war. The brief war led OPEC to substantially increase oil prices. This move did not produce the desired effect, and Israel retained the rest of the occupied territories.

[24] Substantial price rises and unemployment worsened the living conditions of workers and resulted in massive anti-government demonstrations in 1976 and1977. Peasants rebelled against the land redistribution of 1952, and the Islamic parties began to conspire openly against Sadat, accusing him of paving the road for a new period of foreign domination.

[25] Sadat's visit to Jerusalem in November 1977 raised a wave of protest in the Arab world. The process of rapprochement with Israel reached its apex in March1979, with the signing of the Camp David Agreement, wherein the US negotiated the return of Sinai to Egypt. From then on, Egypt became the main beneficiary of US military aid, aimed at turning the country into the new US watchdog in the Arab World, as Shah Pahlevi of Iran had recently been deposed.

[26] In October 1981, Sadat was killed in a conspiracy organized by sectors of the military opposed to infitah and the repression of fundamentalist Islamic movements. Vice-President Hosni Mubarak became President on 14 October.

[27] Mubarak ordered an inquiry into the wealth accumulated by the Sadat family in an attempt to neutralize the general discontent. He also extended further concessions to foreign companies.

[28] There were some improvements in Egyptian foreign affairs during 1984. Egyptian diplomacy managed to overcome the most adverse reactions to the Camp David agreements. Their new position on the Palestinian question argued that any fair settlement of the Middle East crisis had to contemplate the rights of the Palestinian people and that Arab solidarity was 'the only way to recover the usurped rights'. Between1980 and 1986 the role of foreign capital in the national economy increased dramatically.

[29] In August 1990 Iraqi troops invaded Kuwait; Egypt was among the first Arab countries to condemn the action, sending troops to the Gulf immediately. When the land offensive started in January 1991, the US announced the cancellation of the Egyptian military debt, which amounted to $7 billion.

[30] In 1991, foreign minister and deputy minister Esmat Abdel Meguid was named the new secretary-general of the Arab League. Coming upon the heels of the return of Arab League headquarters to Cairo (from Tunis) this appointment meant Egypt's recovery of its leadership role within the Arab world.

[31] That year, Islamic fundamentalists began to rebel, seeking the country's conversion into a theocratic state. These groups had been fighting against Mubarak's government for ten years. In spite of the fact that the Government extended the state of emergency and executed 15 people in 1993, by early 1998 an estimated 1,251 people had fallen victim to attacks and political killings, whilst the estimated number of political prisoners was between 10,000 and 30,000, according to different sources.

[32] In 1995, Mubarak still could not find a solution to the confrontation with the Islamic fundamentalists. In January, Interior Minister al-Alfi met the Ministers of Internal Affairs of the Arab countries in an attempt to coordinate the fight against violent Islamic movements.

[33] In November, the ruling National Democratic Party won the 1995 parliamentary elections amidst violence and allegations of fraud.

[34] In July 1996 the Health Minister banned female circumcision - the removal of the clitoris or part of it and/or the sewing together of the vagina/labia - a practice common in some regions of the country.

[35] Attacks by armed Islamic groups continued throughout 1996 and 1997, along with government repression of all such groups, including those opposed to the use of violence, like the Muslim Brotherhood.

[36] There was heated parliamentary debate in March 1999 over decrees on female circumcision. The rulings continued to provoke resistance in traditionalist sectors.

[37] The Parliamentary discussion over the status of women was resumed in January 2000. The Government had put forward an amendment to family law that authorized women to file for divorce and allowed them to leave the country without their husband's consent. However, when filing for divorce, women have to return all money, property and gifts received during the marriage and forgo alimony. The proposal was considered as 'non-Islamic', while the groups in favor of women's rights considered the measures too 'restricted'.

[38] The August 2000 Parliamentary elections were not free and fair, since opposition supporters were not allowed to vote.

[39] Due to domestic and regional pressure, Egypt was forced to resume relations with Iraq, disrupted since the1991 war. Anti-American and anti-Israeli sentiment had grown among the population

PROFILE

ENVIRONMENT

Ninety-nine per cent of the population live in the Nile valley and the delta although this constitutes only 30 per cent of the land. The remaining land is covered by desert except for a few isolated oases. The floods of the Nile set a pattern for the country's economic life thousands of years ago. Although controversial, and causing many people to be displaced, the construction of dams, especially the Aswan in the south, has benefitted agriculture, particularly cash crops, and the economy. In addition to the traditional crops, wheat, rice and corn, cotton and sugarcane are also grown. The hydroelectric power supply, together with the northeastern oil wells, in the Sinai Peninsula, favored industrial development.

SOCIETY

Peoples: Egyptian society is ethnically diverse. While most people are Semitic-Hamitic in origin, there is a nomadic Bedouin minority in the desert east of the country. The other major group are the Nubians, African people settled for thousands of years in the Upper Nile region. The ethnic legacy of peoples that have passed through the territory is apparent, from the Romans and Greeks, to the Turks and Circasians, and more recently, to the English and French.
Religions: Islamic, mostly Sunni. Orthodox Copts (under 10 per cent) and other Christian churches.
Languages: Arab (official); English and French are used by an educated élite; Nobiin/Nubian; Berber; Coptic (for religious matters).
Main Political Parties: National Democratic Party; Tomorrow Party; New Wafd Party.
Main Social Organizations: The Egyptian Labor Federation is the only central labor organization; Union of Egyptian Students; The Muslim Brotherhood. Several NGOs work in defense of human rights.

THE STATE

Official Name: Jumhuriyah Misr al-'Arabiyah.
Administrative Divisions: 26 provinces.
Capital: Cairo (Al-Qahirah) 10,834,000 people (2003).
Other Cities: Alexandria (Al-Iskandariyah) 3,723,000 people; El-Giza 2,485,000; Subra al-Haymah 974,000 (2000).
Government: Presidentialist Republic. Hosni Mubarak, President since October 1981, re-elected in September 2005 for fourth time in a row. Ahmed Nazif, Prime Minister since July 2004. Bicameral system: the Majlis al-Sha'b (People's Assembly) has 454 members, of which 10 are appointed by the President; and Majlis al-Shura (Advisory Council) which functions only in a consultative role, has 264 members, of which 88 are appointed by the President.
National Holiday: 23 July, Revolution Day (1952).
Armed Forces: 440,000 (incl. 270,000 conscripts). Other: Coast Guard, National Guard and Border Guards, 74,000.

|  Under-5 mortality **36** per 1,000 live births 2004 | Poverty **3.1%** of population living on less than $1 per day 2000 | Debt service **7.6%** exports of goods and services 2004 | Maternal mortality **84** per 100,000 live births 2000 |

after Israel responded violently to the second *intifada* (see history of Palestine). Likewise, Egypt withdrew its ambassador in Israel in protest against the escalation of violence against Palestinians.

[40] In December, Egypt, Lebanon and Syria signed an agreement to build a pipeline that would carry Egyptian natural gas under the Mediterranean to the Lebanese port of Tripoli, and also to Syria. Another pipeline would take gas to Turkey and the European markets.

[41] In 1997, the rebel organization al-Gamaa al-Islamiya claimed responsibility for several attacks that had taken place since 1981. Among others, they admitted to the murder of 58 foreign tourists in Luxor.

[42] Tourism - which was one of the country's main sources of income - suffered a strong fall after the terrorist attacks of September 2001 against the US. Egypt had to resort to loans from international organizations to cover its losses.

[43] In April 2002 - a month that commemorates Israeli withdrawal from Egyptian territories in Sinai - Mubarak criticized Israel and accused it of resorting to state terrorism to crush Palestinians and abuse human rights. Likewise, it claimed Israel erased all traces of its crimes, for example in the Jenin refugee camp, and suggested that some international powers - referring to the US - did not honor their international responsibilities, losing all credibility. Cairo cut back its diplomatic relations with Israel to just those contacts that could help the Palestinians.

[44] In September 2003, the Egyptian authorities set free the radical leader who had ordered Sadat's murder in October 1981. Karam Zohdy was one of the leaders of the Islamic group al-Gamaa al-Islamiya, the larger rebel group in the country.

[45] Early in January 2004, Iranian Vice-President Mohammad Ali Abtahi confirmed that his country and Egypt would re-establish diplomatic relations. Tehran had cut diplomatic ties with Egypt in 1980, one year after Cairo signed a peace treaty with Israel and granted asylum to ousted Reza Pahlevi.

[46] The Egyptian Supreme Council for Human Rights, a state-backed institution whose members are appointed by the Government, issued its first annual report in April 2005. This alleged that Egyptian security forces commonly used to arrest anyone around the scene of a crime and torture them to obtain information. For many years, such allegations had been made by independent human rights groups, but this was the first time they had been supported by a Government-backed organization.

[47] In May 2005 Parliament

IN FOCUS

ENVIRONMENTAL CHALLENGES
The unchecked growth of cities has swallowed up fertile lands. Oil pollution from industry affects beaches and destroys marine habitats and coral reefs. The Nile's waters are seriously polluted by agrochemicals, untreated sewage and industrial waste.

WOMEN'S RIGHTS
Egyptian women have been able to vote and be elected since 1956. In 2005, they held 2 per cent of seats in Parliament and 6 per cent of ministerial positions.

Women made up 31 per cent of the total labor force in 2003. That year, female unemployment reached 22.7 per cent, while the male unemployment rate was 5.1 per cent. Thirty-nine per cent of women worked in agriculture, 7 per cent in the industrial sector and 54 percent in services.

In 2003, 69 per cent of pregnant women received prenatal care and the same percentage of births were attended by skilled health staff.

Ninety-seven per cent* of Egyptian women have been subjected to genital mutilation, which has been banned since 1996.

CHILDREN
In the last decade there have been achievements in child rights. In 2004 the infant mortality rate and under-5 mortality rate were more than halved, reaching 26 and 36 for each 1,000 live births, respectively.

Immunization reaches an average of 97.5 per cent of

children*.

The net primary school enrolment ratio which stood at 86 per cent in early 1990, has seen a steady increase, exceeding 96 per cent in 2002*. Child labor is still a high impact problem, while the number of boys and girls who work and do not attend school in Upper Egypt is disproportionately higher than elsewhere in the country.

INDIGENOUS PEOPLES/ ETHNIC MINORITIES
Although not ethnically distinct from other Egyptians, the Copts -the Egyptian Christians (9 per cent of the population) - are politically weak and have suffered persecution. They live mainly in Alexandria, Cairo and urban areas in Upper Egypt. Most belong to the Coptic Orthodox Church. Although they are an economically advantaged group, they receive little public spending and feel discriminated against, for example in getting university places, and in marriage, divorce and inheritance laws (based on Islamic practice). During the1990s they were victims of attacks by Muslims.

The Nubians inhabit the Upper Nile valley. They are considered descendants of the ancient Kush kingdom and the Kushite dynasty. In the 1960s, several Nubian homelands were flooded when the Aswan dam was built and about 100,000 Nubians had to leave. They emigrated to the north of Aswan, Sudan, Uganda and Kenya.

The Bedouin people (1,300,000 or so, who speak Badawi) migrated to the Sinai peninsula, while others now live along the northern edge of the Sahara Desert.

The 5,800 Berbers, whose language is Tamazight, are

dispersed throughout the country.

MIGRANTS/REFUGEES
By January 2006, Egypt hosted some 30,000 refugees and asylum-seekers, mostly-Sudanese, but also Eritrean, Somali and Ethiopian. In addition, there were about 70,000 Palestinian refugees who had been displaced from the West Bank and Gaza in the 1967 Arab-Israeli war.

As of 2002, over 2,000 Egyptians were seeking asylum in Western countries like the US, Canada and Australia.

Egypt is not a signatory to the UN Refugee Convention and has no domestic asylum laws. In 2002, the Government allowed UNHCR to determine the refugee status of asylum-seekers. UNHCR granted refugee status to 5,000 people during that year, and 20,000 cases are still outstanding (most of them Sudanese applications).

More than 3,000,000 Sudanese lived in Egypt in 2002. Until the late 1980s, Egyptian law made migration from Sudan to Egypt easy, but as a result of the large registered migration throughout the following decade, the authorities limited their numbers.

DEATH PENALTY
It is in force for all types of crimes. People were sentenced to death during 2005, but no-one was actually executed.

*Latest data available in *The State of the World's Children* and *Childinfo* database, UNICEF, 2006.

passed a constitutional amendment proposed by Mubarak that would allow multiple candidates to stand in the September presidential elections. Opposition parties had denounced that the new legislation favored Mubarak's re-election since independent candidates faced serious constraints, such as the requirement to present the endorsement of at least 65 members of Parliament in order to register as presidential candidates. For many people, the reform represented a superficial change in Egyptian politics. For others, however, the amendment meant a change in the political attitude of the ruling class, opening possibilities that could not be controlled by Mubarak. The reform was approved by 83 per cent of Egyptians.

[48] Presidential elections - marred

by fraud and intimidation of opposition supporters – were won by Mubarak, who was re-elected for a fifth consecutive term with 88.6 per cent of the votes. The Tomorrow Party led by Ayman Nour, Mubarak's main challenger, was left out of the competition, and in the parliamentary elections - which lasted months - the opposition was headed by the banned Muslim Brotherhood, which registered its candidates as independent.

[49] In February 2006, a ferry carrying about 1,300 people from Saudi Arabia to Egypt sank in the Red Sea in stormy conditions; at least 200 people died.

[50] That same month, Parliament put back local council elections for two years. According to the Government, this measure was needed to draft new constitutional

amendments that would grant increased power to municipal authorities. The opposition criticized the decision alleging that 'elections were postponed to halt the influence of the Muslim Brotherhood'.

[51] In April 2006, Egyptian authorities decided to release some 900 members of the radical group al-Gamaa al-Islamiya, considered for quite a long time as the largest organization in the country. However, human rights groups warned that some 15,000 detainees who had never been brought to trial remained in Egyptian jails.

[52] The local elections due in April 2006 were postponed for two years. Analysts saw this as a setback for democratic reform, saying the Government feared that the Muslim Brotherhood would repeat its success the previous year. ∎

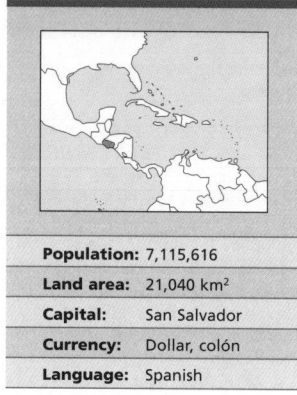

Population:	7,115,616
Land area:	21,040 km²
Capital:	San Salvador
Currency:	Dollar, colón
Language:	Spanish

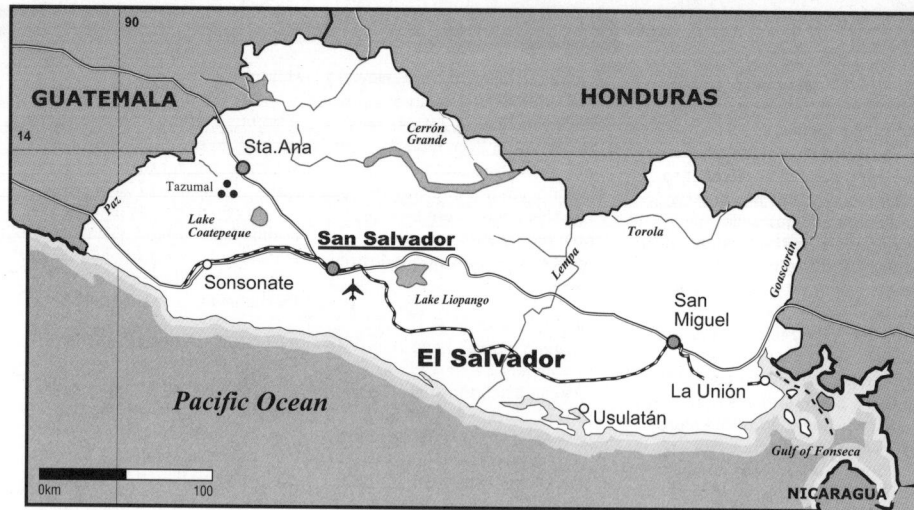

The region of El Salvador was inhabited from early times by Chibcha people (Muisca: see Colombia), most of whom were Pipile and Lenca. The Maya also lived within the region.

[2] The Spaniards subdued the Aztecs in Mexico, and subsequently began the conquest of Central America under the leadership of Pedro de Alvarado. In 1525, Alvarado founded the city of El Salvador de Cuscatlan. The territory formed part of the Captaincy-General of Guatemala, a dependency of the Viceroyalty of Mexico. Central America became independent from Spain in 1821 and organized itself into a federation.

[3] In 1827, internal rivalries between 'imperialists' and 'republicans' led to civil war. In 1839, General Francisco Morazán, president of the Central American Republic, tried to prevent the break-up of the federation. From El Salvador, Morazán struggled to preserve the union with the support of some Nonulco natives, headed by Anastasio Aquino.

[4] When the federation was dissolved, Britain took advantage of the situation dominating the isthmus. In 1848, President Doroteo Vasconcelos refused to bow under British pressure, and the British blockaded Salvadoran ports.

[5] At the end of the century, the invention of artificial coloring destroyed the demand for indigo - El Salvador's principal export product - and its price fell to rock bottom. Indigo was replaced with coffee. Coffee required larger and more extensive farming areas, and the Liberal Revolution of 1880 drove thousands of peasants from their communal lands, forming a rural working class and a countryside full of anger and conflict. The coffee plantation owners became the dominant oligarchy and the ruling class of El Salvador.

[6] The 1929 financial crash caused the coffee market to collapse, crops were left unharvested, and thousands of sharecroppers and poor peasants starved. This led to a mass uprising on 22 January 1932, headed by the Communist Party of El Salvador and Farabundo Martí, a former secretary of Augusto Sandino in his campaign against the US invasion of Nicaragua.

[7] The rebellion was ruthlessly crushed by the troops of General Maximiliano Hernández Martínez, who had taken power in 1931. This started a series of military regimes that lasted half a century. Twelve thousand people died as a consequence of the repression.

[8] In 1960, the Alliance for Progress sponsored an industrialization program, within the Central American Common Market. High economic growth rates were attained without reducing the rampant unemployment that had caused 300,000 landless peasants to emigrate to neighboring Honduras. Population growth and competition between the local industrial interests, led to war between El Salvador and Honduras in June 1969. The regional common market collapsed after the 100-hour conflict, severely damaging Salvadoran industry.

[9] In the early 1970s unions and other civilian movements took on new life. Guerrilla fighters appeared in El Salvador and the legal opposition parties joined into a national front; the UNO, formed by the Christian Democrats (PDC), the Communists (UDN) and the Social Democrats (MNR). Colonel Arturo Molina, presidential candidate for the official National Conciliation Party, defeated the UNO's candidate Napoleón Duarte in a fixed election in 1972.

[10] In 1977, another fraudulent election made General Carlos Humberto Romero president. Mass riots broke out in protest but social unrest was suppressed, leaving 7,000 dead.

[11] A civilian/military junta seized power on 15 October 1979. It included representatives of the social democrats and the Christian democrats. The Junta lacked real power and had no control over the ruthless repression campaigns carried out by police and military forces. Civilian members resigned and were replaced by right-wing Christian democrats from Duarte's party.

[12] On 24 March 1980, the Archbishop of San Salvador, Monsignor Oscar A. Romero, was assassinated while performing mass in a clear reprisal for his constant defense of human rights.

[13] In October 1980, the five anti-regime political-military organizations agreed to form the Farabundo Martí Front for National Liberation (FMLN). On 10 January 1981 the FMLN launched a 'general offensive' and increased their actions throughout most of the country.

[14] In August 1981, the Mexican and French governments signed a joint agreement recognizing the FMLN and the Democratic Revolutionary Front (FDR) as 'a representative political force'.

[15] The US administration, led by President Ronald Reagan, saw the situation in El Salvador as a national security issue. The US became directly involved in the political and social conflict, and was the military and economic mainstay of the 'counter-insurgency' war, which the Salvadoran Army was unsuccessfully waging.

[16] On 28 March 1982, as instructed by Washington, the regime held an election for a Constituent Assembly. In response, the rebels launched an offensive ending in a one-week siege of Usulatán, a provincial capital.

[17] After continuous internal tussling for power, the presidency of the constitutional convention went to Roberto D'Aubuisson, the main leader of the ultra-right Nationalist Republican Alliance (ARENA) and the power behind the assassination of Archbishop Romero.

LAND USE

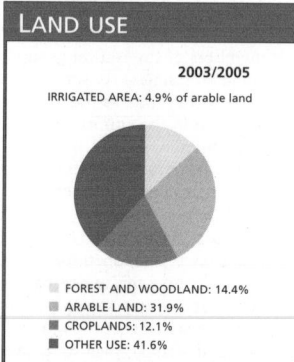

2003/2005

IRRIGATED AREA: 4.9% of arable land

- FOREST AND WOODLAND: 14.4%
- ARABLE LAND: 31.9%
- CROPLANDS: 12.1%
- OTHER USE: 41.6%

PUBLIC EXPENDITURE

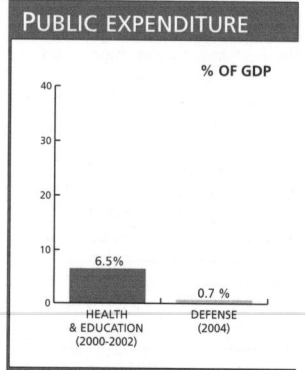

% OF GDP

HEALTH & EDUCATION (2000-2002): 6.5%
DEFENSE (2004): 0.7%

WORKERS

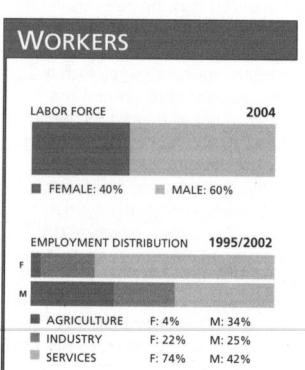

LABOR FORCE   **2004**

- FEMALE: 40%
- MALE: 60%

EMPLOYMENT DISTRIBUTION   **1995/2002**

- AGRICULTURE   F: 4%   M: 34%
- INDUSTRY   F: 22%   M: 25%
- SERVICES   F: 74%   M: 42%

	Life expectancy		GNI per capita		Literacy		HIV prevalence rate
	72 years		$2,320		80% total adult rate		0.7% of population 15-49 years old
	2005-2010		2004		2000-2004		2003

18 Against a background of an upsurge in fighting, general elections were held in March 1984. These were boycotted by the FDR-FMLN; the abstention rate by voters was 51 per cent. Ostensibly supported by the US, the PDC - led by Napoleon Duarte - obtained 43 per cent of the vote, against the 30 per cent obtained by ultra-right candidate Major Roberto D'Aubuisson.

19 In October 1986, a strong earthquake brought about a virtual cease-fire, which eventually led to the renewal of negotiations in October 1987. These talks took place within the new framework of regional peace making. The Central American governments had signed the Esquipulas agreements in August 1987, agreeing to strive for peace.

20 Elections were held in October 1989; these were boycotted by some of the guerrillas, but civilian sectors of the FDR (members of the social democratic and social Christian parties) participated, with Guillermo Ungo as their presidential candidate. Alfredo Cristiani, the ARENA party's candidate, won the election.

21 In November 1989, the FMLN launched an offensive occupying several areas of the capital and surrounding regions. The Government responded by bombing several densely populated areas of the capital. Six Jesuits, including the rector of the University of Central America, Ignacio Ellacuria, were brutally killed by heavily armed soldiers. This provoked a world-wide outcry, especially from the Catholic Church, and American economic aid was threatened.

22 On 10 March 1991, the legislative and local elections reflected a new spirit of negotiation. For the first time in ten years the FMLN did not call for the boycott of the elections, instead they decreed a three-day unilateral truce. Abstention was still above 50 per cent, and there were acts of paramilitary violence immediately prior to the polls. On 12 March, the confrontations started over again.

23 In Mexico on 4 April 1991 delegates of the Cristiani Government and the FMLN started negotiations for a cease-fire agreement. On 19 April, two weeks before the Congress' mandate and the term granted by the FMLN for the peace-making came to an end, 10,000 demonstrators, from 70 social organizations, gathered in the Permanent Committee for National Debate (CPDN), demanded that the Constitution be reformed.

24 On 27 April, after several attempts, representatives of the Government and the Farabundo Martí Front signed the 'Mexico Agreements' restricting the function of the armed forces to the defense of national sovereignty and territorial integrity. The formation of paramilitary groups was banned, and it was agreed to reform article 83 of the Constitution to say that sovereignty 'resides in the people, and that it is from the people that public power emerges'.

25 On 16 November, new talks began in the UN headquarters. This time, the FMLN declared an indefinite unilateral truce until a new, definite, cease-fire was signed. Meanwhile, a Spanish parliamentary delegation visited El Salvador and wrote a report on the murders of the six Spanish Jesuits from the Central American University (UCA). The report, submitted to the Spanish, European, Salvadoran, and US parliaments, accused the Salvadoran Government and the army of concealing evidence.

26 On 1 January 1992, after 21 weeks of negotiation and 12 years of civil war, both parties met in New York to sign agreements and covenants establishing peace in El Salvador. The war left 75,000 people dead, 8,000 missing, and nearly one million in exile. The period 1 February-3 October 1992 was designated as the time to cease all armed confrontation and to create an appropriate environment to apply the agreements and negotiations, which continued under the supervision of the UN and the OAS.

27 The final agreements were signed in the Mexican city of Chapultepec on 16 January 1992. They included substantial modifications to the Constitution and to the armed forces. They guaranteed to change rural land tenure and to alter the terms of employee participation in the privatization of State companies; they established the creation of human-rights organizations, and guaranteed the legal status of the FMLN.

28 According to the peace accords the Government was to reduce its troops by half by 1994, bringing the number down to 30,000; in addition, it was to disband its intelligence service. As of 3 March, a new civilian police was to be created, made up in part by former members of the FMLN. The FMLN, a legal political party since 30 April 1991, held its first public meeting on 1 February 1992 and called for the unification of all opposition forces for the 1994 election. After years of being underground, it was presided over by guerrilla commanders Shafick Handal, Joaquín Villalobos, Fernán Cienfuegos, Francisco Jovel and Leonel González.

29 In January 1992, the Law of National Reconciliation granted amnesty to all political prisoners. In addition, the Government pledged to turn over lands to the combatants and provide assistance to campesinos belonging to both bands.

30 On 15 February 1993, the last 1,700 armed rebels turned over

ENVIRONMENT

It is the smallest and most densely populated country in Central America and the only one with no Caribbean coastline. A chain of volcanoes runs across the country from east to west and the altitude makes the climate mild. Coffee is the main cash crop in the highlands. Subsistence crops such as corn, beans and rice are also grown. Along the Pacific Coast, where the weather is warmer, there are sugarcane plantations.

SOCIETY

Peoples: 89 per cent of the Salvadoran population are mixed descendants of American natives and Spanish colonizers, 10 per cent are indigenous peoples, and one per cent are European. **Religions:** Mainly Catholic. **Languages:** Spanish is the official and predominant language. Indigenous minority groups speak Nahuatl and Kekchi. **Main Political Parties:** The Nationalist Republican Alliance (ARENA) currently in power; The Farabundo Martí National Liberation Front (FMLN), founded in October 1980, comprises five political-military organizations; United Democratic Centre-Christian Democratic Party; Party of National Conciliation. **Main Social Organizations:** National Union of Salvadoran Workers (UNTS); MUSYGES and the National Coordinator against Hunger and Repression (CNHR), formed in 1988. Coordinating Indigenous Salvadoran Council (CCNIS); Peasant Democratic Movement (ADC).

THE STATE

Official Name: República de El Salvador. **Administrative Divisions:** 14 Departments. **Capital:** San Salvador 1,424,000 people (2003). **Other Cities:** Santa Ana 538,800 people; San Miguel 473,300. **Government:** Antonio Elías 'Tony' Saca, President since June 2004. Parliament: Legislative Assembly with 84 members. **National Holiday:** 15 September, Independence Day (1821). **Armed forces:** 30,500 troops (1995). Other: National Civilian Police, made up of former guerrillas, soldiers and police.

their weapons in a ceremony which was attended by several Central American heads of state and by UN Secretary-General Boutros Ghali. The National Civil Police was created, as well as a Human Rights Defense Commission and a Supreme Electoral Court.

31 The result of the investigation of human-rights violations, carried out by the Truth Commission created by the UN, led to the resignation of Defense Minister General René Emilio Ponce, as being the one who ordered the assassination of six Jesuits in 1989. According to the Commission's final document, the military, the death squads linked to these and the State were responsible for 85 per cent of the civil rights violations committed during the war.

32 The Truth Commission recommended the dismissal of 102 military leaders and that some former guerrilla leaders be deprived of their political rights. President Cristiani proposed a general amnesty for cases where excesses had been committed; this proposal was approved on 20 March 1993, only five days after the Truth Commission document had been made public. With this measure, the most serious crimes committed during the war met with total impunity.

33 A year later on 20 March 1994, the first elections since the civil war were held. The candidate of the left coalition, Democratic Convergence - made up of the FMLN and other groups - won 25.5 per cent in the first round, against 49.2 per cent for the right-wing candidate, Armando Calderón Sol, from the ARENA party. Although the Left considered the elections fraudulent, the UN observers confirmed that the elections had been fair.

34 The long-promised award of land to demobilized fighters was slow and inefficient. By mid-1994, only one-third of the potential beneficiaries - 12,000 of a total 37,000 former members of the army or the guerrillas - had obtained plots. The rest remained inactive, living in substandard temporary housing and often drifting into organized crime.

35 An agreement concluded on May 1995 between ARENA and the Democratic Party - split from the FMLN - enabled a three per cent rise of the valued added tax from ten to 13 per cent. This raise was explained by the need to collect funds to finance land reform, infrastructure works and reconstruction of the country's electoral and judicial apparatus.

36 In August 1996 demonstrators affected by the slowness of the process occupied streets and government buildings in downtown San Salvador. In May, the Democratic Party withdrew from its deal with ARENA, which left the Government without a parliamentary majority. In March 1997, the opposition's FMLN

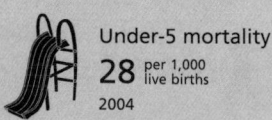 **Under-5 mortality**
28 per 1,000 live births
2004

 **Poverty**
19.0% of population living on less than $1 per day
2002

 Debt service
8.8% exports of goods and services
2004

Maternal mortality
150 per 100,000 live birth
2000

IN FOCUS

ENVIRONMENTAL CHALLENGES

The country has some of the worst deforestation problems in Latin America, with erosion, water and soil pollution from toxic waste. A study in the 1990s stated that 90 per cent of all rivers were polluted, while two-thirds of all land suffered serious erosion.

WOMAN'S RIGHTS

In 1939 the Legislative Assembly extended the franchise to educated women over 21. Since the 1950 Constitution, all women have been able to vote and run for office. In 2003, nearly 11 per cent of parliamentary seats and 35 per cent of ministerial or equivalent positions were held by women.

In 2003, women made up 39 per cent of the total workforce; of them, 4 per cent worked in agriculture, 22 per cent in the industrial sector and 74 per cent in services. Eighty-six per cent of pregnant women receive prenatal care and 92 per cent of births are attended by skilled health staff*.

CHILDREN

Although the Constitution bans all child labor, there is a significant number of working children. Most are boys and work in agriculture or fishing, while others work as street vendors, vulnerable to sexual exploitation. This is more acute in poor communities and when the parents have little schooling.

The under-five mortality rate has more than halved in the last 15 years, falling from 60 deaths per 1,000 live births in 1990 to 28 per 1,000 in 2004*.

The net primary school enrolment rate amounts to 90 per cent*. Seven per cent of children are born underweight*, while 19 per cent of children under five years old suffer from stunting*.

INDIGENOUS PEOPLES/ ETHNIC MINORITIES

Indigenous people account for 5 per cent of the population (310,000 people). They live mostly in the southwestern region (Sonsonate, Ahuachapan, La Libertad and Santa Ana - a community better known as Panchimalco). Most speak Spanish as a first language,

and few are familiar with Nahuatl, their own language. Although the Government implemented policies in 1992 to improve their situation and acknowledged past human rights abuses, they face economic discrimination. They are descended from the Pipils, a nomadic group of the Nahoa people in central Mexico. Since the beginning of the Spanish conquest, indigenous people and Europeans have coexisted in the same regions.

In the 16th century intermarriage was common. In 1932, during an indigenous peoples' protest against government policies, 35 *ladinos* (non-indigenous inhabitants) were murdered. In retaliation, between 35,000 and 50,000 native Americans were massacred, in an act called 'La Matanza' (The Massacre). During the 1980-92 civil war, thousands of indigenous people were victims of death squads. The present Constitution makes no specific provision for the rights of these peoples or their participation in decisions over the use of the land, their culture or the exploitation of natural resources.

MIGRANTS/REFUGEES

The civil war in the 1980s caused wide-scale emigration, mostly to the US. At the present time, a large number of Salvadorans emigrate to First World countries seeking better living conditions. In 1996 there were 40,000 Salvadorans living in Canada. Meanwhile, it is estimated that there are nearly 900,000 Salvadorans, either registered or not, living in the US - the country that is chosen by the vast majority of migrants. The US deported 3,743 people to El Salvador in 1997. In 2004 El Salvador received more than $2.5 billion in remittances, which account for 16 per cent of GDP.

DEATH PENALTY

Abolished for ordinary offenses in 1983. The last execution took place in 1973.

*Latest data available in *The State of the World's Children* and *Childinfo* database, UNICEF, 2006.

---

obtained an important victory in municipal elections, after winning in the capital and dozens of provincial cities, although ARENA still kept the majority of the votes. The turnout was 40 per cent.

[37] The March 1999 elections ratified ARENA's dominance. At the age of 39 the elected candidate Francisco Flores became the youngest president in South America, providing an image of rejuvenation. Flores' government program, 'The New Alliance', had four priorities: work, social investment, citizen security and sustainable development.

[38] Dollarization of the economy was enforced as of 1 January 2001, aiming to revitalize the economy and attract investors. Although the US currency was the new unit of the banking sector, citizens could carry out their transactions in the local currency, *colones*.

[39] The earthquakes of 13 January and 13 February 2001 left 1,159 dead, 8,122 injured and 1.5 million homeless (25 per cent of the population). It is estimated the earthquakes aggravated environmental deterioration and caused 225,000 new poor.

[40] The World Food Program began giving out food in August 2001, as a result of one of the worst droughts in recent years. Approximately two tons of corn, beans and cooking oil were distributed among 20,000 families

who had lost all of their yearly harvest.

[41] US President George W Bush visited San Salvador in March 2002, on the same day as the 22nd anniversary of the murder of Archbishop Romero. Bush promised to support free trade policies, as the only solution to the region's problems. Even though US intervention during the civil war years had been widely proven, US officials claimed the purpose of the visit was to celebrate a story with a 'happy ending', which had involved George Bush Sr., George W's father, as one of the signatories of the peace in 1992.

[42] In July 2002, a Florida (US) Court found Carlos Eugenio Vides and José Guillermo García, two Salvadoran retired generals, guilty of committing torture and other human rights violations in their country during the 1980s. The generals were ordered by the US jury to pay $54.6 million in compensatory damages to the three plaintiffs: Juan Romagoza, Neris González and Carlos Mauricio.

[43] In March 2004 general elections, ARENA's candidate, Anthony 'Tony' Saca, was elected president with 57 per cent of the vote, defeating FMLN's Shafick Handal, who only obtained 35.6 per cent. Saca, whose election meant the fourth consecutive victory for the ARENA, had promised during his campaign to crack down on

the so-called *maras* (armed gangs which have acts of murder as a basic requirement for membership), and to strive for transparent government.

[44] In August, riots at La Esperanza prison left 31 people dead and dozens injured. Violence was unleashed when members of the 'Mara 18' gang and other convicts clashed, using grenades and knives against each other. Given the fact that La Esperanza was designed for 800 inmates but held more than 3,000, riots were commonplace within the facility. The Government and prison authorities promised to transfer members of the 'Mara 18' to another prison in the country.

[45] Hurricane Stan, which devastated Guatemala in October 2005, also hit El Salvador; though weakened, it killed at least 25 people and caused damage which amounted to millions of dollars. A national emergency was declared and a relief network was organized both within the country and overseas to aid the victims.

[46] Former guerrilla commander Schafik Handal, one of the country's most famous leaders, died at 75 in January 2006.

[47] On 1 March 2006 a Free Trade Agreement (FTA) between the US and El Salvador came into force. The FTA's implementation was made possible after El Salvador promoted a series of changes to its legislation and regulatory system.

The previous day, several protests against the FTA had taken place in the streets of the capital, involving sellers from the informal sector, students and union members. According to union leaders, the FTA will only benefit large companies and business people in El Salvador.

[48] In the local and parliamentary elections held that month, Violeta Menjívar, an FMLN candidate,-was elected mayor of San Salvador - the fourth time the FMLN had won the mayoralty. Meanwhile, parliamentary elections ended in a virtual draw: ARENA got 34 seats and the FMLN 32; the remaining 22 were divided among other parties.

[49] Shortly after the elections, Venezuela signed an agreement to provide cheap fuel to a group of Salvadoran municipalities headed by left-wing mayors (of the FMLN). According to the agreement, PDV Caribe - a branch of the Venezuelan state oil firm PDVSA - and the Inter-municipal Energy Association for El Salvador (ENEPASA) would set up a joint venture. 'If the Government of El Salvador had offered us the opportunity to cooperate with Salvadoran people, we would have started to do so a long time ago, but this has not been possible since there are governments in these lands that still believe in the American dream,' stated Hugo Chávez, Venezuelan President, upon signing the agreement. ∎

Equatorial Guinea / Guinea Ecuatorial

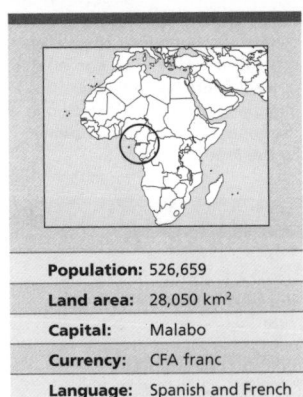

Population:	526,659
Land area:	28,050 km²
Capital:	Malabo
Currency:	CFA franc
Language:	Spanish and French

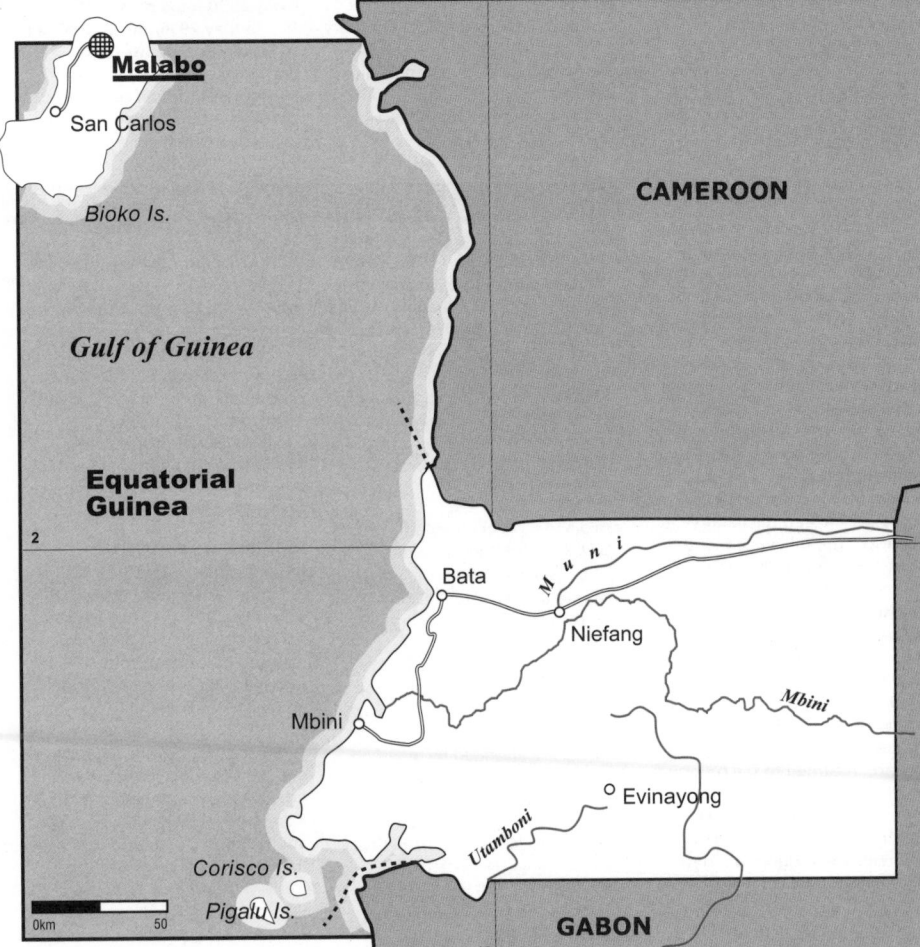

From the 13th to the 15th century, Fang and Ndowe people settled in the River Muni area, north of Gabon, subduing the Bayele pygmy population, who now only exist in a few isolated groups. From the continental coast, these nations expanded onto the nearby islands that were described as 'densely populated' in the 15th century.

2   In the colonial division of Africa, the Rio Muni area and the islands received the name of Equatorial Guinea. The kings of Portugal, proclaiming themselves the lords of Guinea, ceded the entire District of Biafra to Spain under the treaties of San Ildefonso and Pardo in 1777 and 1778, in exchange for Spanish territory in southern Brazil. In 1778, an expedition sailed out from Montevideo to occupy the islands. They lost their commanding officer, Argelejos, in a battle against the Annoboneuses, and the survivors, under their new leader, Lieutenant Primo de Rivera, turned back. The French and British gradually took over sections of the territory, with the British finally occupying it, founding the first settlements.

3   Between 1843 and 1858, the District was militarily re-conquered by Spain, re-establishing their rights over the area. Bioko Spanish settlers supported Franco during the Spanish civil war in 1936 and then obtained control over the archipelago.

4   From 1963, the colony had a regime of internal autonomy which allowed for the legal existence of several political parties. Meanwhile, international pressure became so strong that Franco's Spain had to recognize the independence of Equatorial Guinea, officially proclaimed on 12 October 1968.

5   Francisco Macías Nguema took over the presidency a year later on the pretext of an alleged coup attempt. He instigated violent repression of the opposition, leaving a trail of thousands of political prisoners, killings, disappearances and 160,000 exiles.

6   In August 1979, a coup led by Lieutenant Colonel Teodoro Obiang Mba Nzago, the President's nephew, brought an end to Macías's power, and he was executed for crimes against humanity. Obiang set up the Supreme Military Council.

7   The new regime continued the repression of its forerunner, and the groups which were benefited by the regime remained untouched, all now linked to the Democratic Party of Equatorial Guinea (PDGE), the only legal party. The National Alliance for the Recovery of Democracy formed in exile. Its attempts to negotiate with Obiang were fruitless.

8   Initial rapprochement with the Soviet Bloc was undone when Obiang took the nation into the Customs and Economic Union of Central Africa, linking the national currency - the *ekwele* - to central banks of the region. This measure led to a drastic reduction in facilities offered by the Soviets and East Germans.

9   Following his re-election in 1989, Obiang visited France and Equatorial Guinea later asked to join the French-speaking African countries using the franc (CFA) as their currency.

10   Ten political parties were legalized in early 1993. In late March, the freeing of all political prisoners was agreed with the Joint Opposition Platform (a 10-party coalition formed in November1992 from legalized parties). In August the first multi-party elections were held but the PDGE boycotted them and took 68 of the 80 seats.

11   Amidst accusations, arbitrary detentions and torture, the Government set presidential elections for early 1996. In February of that year, shortly before election day, the Government dissolved the Opposition Platform and detained several of its members. Obiang duly romped home with 99 per cent of the vote in an election described as a farce by the opposition. There was an 80 per cent abstention rate.

12   That year, huge oil deposits

LAND USE

2003/2005

- FOREST AND WOODLAND: 58.2%
- ARABLE LAND: 4.6%
- CROPLANDS: 3.6%
- OTHER USE: 33.6%

PUBLIC EXPENDITURE

% OF GDP

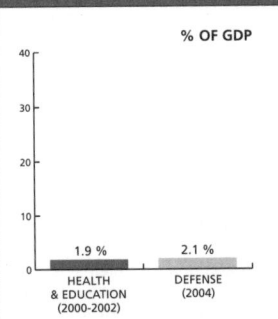

1.9 %	2.1 %
HEALTH & EDUCATION (2000-2002)	DEFENSE (2004)

Life expectancy
42 years
2005-2010

GNI per capita
$710
2001

Literacy
84% total adult rate
2000-2004

were found. Oil exports brought about an extraordinary growth of the economy. In 1997 GDP grew by 71.2 per cent, by 22 per cent in 1998 and 15 per cent in 1999, doubling the size of the economy in less than three years.

[13] Although the per capita income grew from $370 to $2,000 between 1995 and 2000, this disguised the continuing deeply unequal distribution of wealth. However, the economy's spectacular growth did allow for some positive improvements in health and education.

[14] In February 2002, Cándido Muatetema Rivas was appointed Prime Minister, after Teodoro Obiang Nguema accused Ángel Serafín Seriche Dougan of causing a constitutional crisis and ousting him for 'not respecting the opinion of the people and the nation's interests'. The following month, eight opposition parties in exile formed a coalition in Spain to monitor the internal politics of the nation, stating that democracy under the Obiang government was only a headline.

[15] In July, the exiled Florentino Ecomo Nsogo, leader of the Party for Reconstruction and Social Welfare, returned to the country.

[16] The discovery of natural gas deposits in 2001 caused a new leap in growth of over 50 per cent. There was significant new investment in the construction sector that generated-growth in other production areas, thus bringing some of the newly created wealth to more disadvantaged sectors.

[17] In June 2002, 68 people - including the main opposition leader Plácido Micó Abogo - were arrested. They were accused of plotting to overthrow Obiang.

[18] In the December 2002 elections the President was re-elected with 100 per cent of the vote, according to official figures. Opposition leaders said the election was rigged.

[19] In January 2004, Silvestre Siale Bileka, president of the Supreme Court, resigned. In his resignation letter to Obiang, Siale said that despite his best efforts in the Justice Department, there had been no major improvements.

[20] In March, 15 mercenaries accused of attempting to overthrow Obiang's government were arrested. Links were suspected between these mercenaries and a group of detainees in Zimbabwe.

[21] In the parliamentary and municipal elections of April, Obiang's PDGE and its allies won 98 of the 100 parliamentary seats and almost all of the 244 municipal seats. International observers were critical of both the way in which the elections were run and their

results.

[22] A cholera epidemic in February 2005 became the main cause of death in the island region of Bioko, where the capital is. About four bodies a day were buried, mainly in Malabo, but this number rose to as many as 30 some days.

[23] In May 2006, the Committee to Protect Journalists (CPJ) published a report showing that among African countries, Equatorial Guinea had the most repressive media censorship. The report claimed that 'In these countries the media is completely controlled by the élite who hold power.' There was only one 'independent' radio-TV-station in the country - and this was owned by the President's son. ∎

IN FOCUS

ENVIRONMENTAL CHALLENGES
A major environmental concern is the use of Pigalu Island as a dump for industrial and toxic radioactive waste, resulting in pollution and diseases from contamination. Deforestation is in progress. Only 2 per cent of the population take advantage of 80 per cent of the country's natural resources.

WOMEN'S RIGHTS
Women have been able to vote and stand for election since 1963. In 2004, 18 per cent of parliamentary seats and 5 per cent of ministerial positions were held by women.

Life expectancy for women is 52 years*. Eighty-six per cent of pregnant women receive prenatal care and only 65 per cent of all births are attended by qualified staff.* Maternal mortality is 880 per 100,000 births*.

In 2003 the female labor force was 36 per cent of the total.

CHILDREN
Only 62 per cent of children attend primary school and some 18.5 per cent receive secondary education*.

At the end of 2001, 420,000 children under 15 years old were HIV-positive or living with AIDS. Only four per cent of young people aged 15 to 24 had sufficient information to protect themselves against the disease*.

Thirteen per cent of babies are born underweight and 39 per cent have moderate or severe growth problems*.

UNICEF has backed the development of national health and educational policies aimed at eradicating polio.

INDIGENOUS PEOPLES/ ETHNIC MINORITIES
The most important groups are formed by Bubis (15,000 people, mainly Catholic), Fernandinos - people descended from intermarriage when the British administered Fernando Po in the 19th century; from Sierra Leone, or expelled from Cuba - Pagalos (Annoboneses), Ndowes, Combes, Bengas, Bujebas and Fangs or Pámues (including Okaks and Ntumus). According to oral tradition, Fangs originally came from what are now Gabon and Cameroon.

The Fang people's main social institution, the 'house of word', is the place where neighbors get together to explain and solve their disputes openly in front of the whole community.

There are around 35,000 Ibo and Hausa immigrants from Nigeria.

MIGRANTS/REFUGEES
Since 2000, this country with its rich mineral resources such as petroleum and minerals, has become a main destination for skilled and non-skilled workers. From 1970 to 1980 many agricultural workers from Angola, Cameroon, the Central African Republic and Chad, started working on the cocoa, coffee and sugar plantations.

During the Macías regime (1968-1979) 160,000 people left the country. Then, during the 20 years after the Obiang Nguema coup (1979), a few refugees returned; some were imprisoned and beaten. On release, they went into exile again.

DEATH PENALTY
It is applicable to all kind of crimes. In 2004 death sentences were handed down but no executions were carried out.

*Latest data available in *The State of the World's Children* and *Childinfo* database, UNICEF, 2006.

PROFILE

ENVIRONMENT
The country consists of mainland territory on the Gulf of Guinea (Río Muni, 26,017 sq km) and the islands of Bioko (formerly Fernando Po, and Macías Nguema) and Pigalu (formerly Annobon, Corisco, Greater Elobey and Lesser Elobey). The islands are of volcanic origin and extremely fertile; Río Muni is a coastal plain covered with tropical rainforests but without natural harbors. It is one of the most humid and rainy countries of the world, a characteristic that limits the variety of possible crops. The main exports are cocoa, wood and oil.

SOCIETY
Peoples: The population is mostly of Bantu-speaking origin. In the islands there are also Igbo and Efik peoples who migrated from Nigeria, subduing the local Bubi population. In Río Muni the inhabitants are mainly Fang and Ndowe. Nearly all of the Europeans and a third of the local population emigrated during the Macías regime.
Religions: Mainly Christian on the islands; traditional African beliefs in Rio Muni.
Languages: Spanish is the official and predominant language. French is also official. In Rio Muni, Fang is also spoken, and on the islands, Bubi, Ibo and English.
Main Political Parties: Democratic Party of Equatorial Guinea (PDGE); Convergence for Social Democracy.
Main Social Organizations: National Alliance for the Restoration of Democracy and other associations in exile (mainly in Spain) like the Platform for Peace and Human Rights in EG, Movement for the Self Determination of Bioko Island, Union of Workers of Equatorial Guinea.

THE STATE
Official Name: República de Guinea Ecuatorial.
Administrative Divisions: Four continental and three island regions.
Capital: Malabo 95,000 people (2003).
Other Cities: Bata 43,000 people.
Government: Teodoro Obiang Nguema Mbasogo, President since August 1979, re-elected in December 2002. Ricardo Mangué Obama Nfubea, Prime Minister since August 2006. Unicameral Parliament: Chamber of People's Representatives, with 100 members.
National Holiday: 12 October, Independence Day (1968).
Armed forces: 1,300.

Eritrea / Ertra

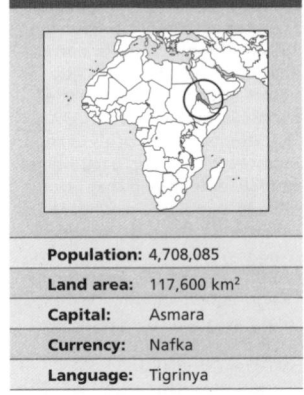

Population:	4,708,085
Land area:	117,600 km²
Capital:	Asmara
Currency:	Nafka
Language:	Tigrinya

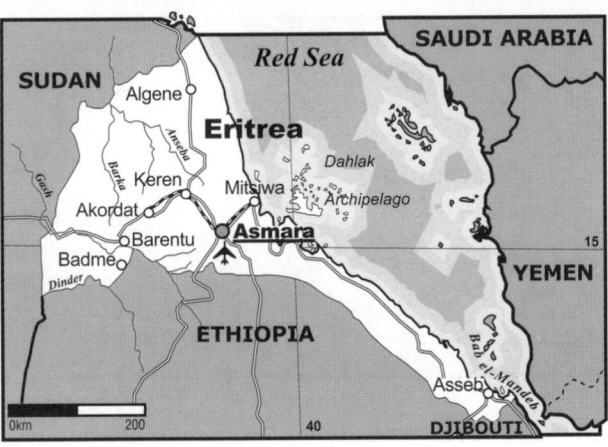

About 300 BC, Aksumite herders, who emigrated from Arabia to Mesopotamia, reached the shores of the Red Sea in the region now known as Eritrea. This area was linked to the beginnings of the Ethiopian kingdom, but it retained much of its independence until it fell under Ottoman rule in the 16th century. From the 17th to the 19th century control over the territory was disputed among Ethiopia, the Ottomans, the kingdom of Tigray, Egypt and Italy. In 1890, the Treaty of Wichale between Italy and Menilek II of Ethiopia recognized Italian possessions on the Red Sea, and the colony, created on 1 January 1890, was named by the Italians for the *Mare Erythraeum* ('Red Sea') of the Romans.

[2] Eritrea was used as the main base for the Italian invasions of Ethiopia in 1896 and 1935-36. Italian rule continued until 1941 when the area came under British administration.

[3] On 2 December 1950, the United Nations declared that Eritrea should become a federated state within the Ethiopian Empire. The resolution rejected Ethiopian demands for outright incorporation, but also left the process of Eritrean self-determination undefined.

[4] In Eritrea, a national assembly was elected which enjoyed some autonomy until 1962, when Ethiopian leader Haile Selassie forced a group of Eritrean politicians to vote for its complete incorporation into Ethiopia. The decision was contested by nationalist groups, sparking a rebellion.

[5] The Eritrean Liberation Front (ELF) founded in 1958, in Cairo, by journalist and union leader Idris Mohamed Adem, began guerrilla activities in September 1961. In 1966, a split produced the Eritrean Popular Liberation Front (EPLF). In 1974, with Sudanese mediation, the two groups agreed to coordinate their actions and in the next few years, the EPLF imposed its leadership upon the rebel movement.

[6] During the pro-Soviet Mengistu government in Ethiopia, the Eritreans felt that the changes in Addis Ababa did not bring their self-determination closer, so they had no reason to stop fighting.

[7] In February 1990, the EPLF captured Asmara and the ports of Massawa and Asseband almost all the Eritrean territory. The road between Asseb and Addis Ababa was the only way supplies could reach the Ethiopian capital by land.

[8] In 1991, Asmara and Addis Ababa started to relate as separate States. The Red Sea ports were opened again to enable the arrival of international aid.

[9] In the April 1993 referendum, 99.8 per cent of voters opted for independence. The EPLF formed a provisional government, led by Isaias Afwerki, which turned into a political party the following year, the People's Front for Democracy and Justice (PFDJ). After becoming a member of the UN in 1993, Eritrea joined the IMF in February 1994.

[10] Fighting with Ethiopia resumed in February 1999. The war sucked in Nigeria, which supplied weapons to Eritreans, and Kenya, which mobilized forces along its border with Ethiopia.

[11] In June, Afwerki and his Prime Minister Meles Zenawi accepted a proposal presented by the Organization of African Unity (OAU) for an immediate ceasefire and withdrawal of troops from the area in dispute. In September 1999, Eritrea accepted the peace plan proposed by the UN, while Ethiopia maintained minor differences.

[12] Eleven ruling party officials were arrested in September 2001, charged with 'treason' for criticizing the Government and demanding democratic reform.

[13] In April 2002, the Permanent Court of Arbitration at The Hague decided on the border dispute between Eritrea and Ethiopia. The 1,000 km border was set by a court of five international specialists. The border cities of Zalembessa, Alitena and Badawere given over to Ethiopia, while Badme - the flashpoint of the war in 1997 -was given to Eritrea.

[14] In 2003 the major droughts that hit Africa triggered one of the worst food crisis since Eritrea's independence.

[15] Information minister Ali Abdu Ahmed denounced in October 2004 that Sudan was intensifying its attempts to sabotage peace and stability in Eritrea and the region. The country seemed bound towards complete isolation with the closure of its borders.

[16] After rising tensions along their common border, in March 2006 Eritrea and Ethiopia reached an agreement to renew the border demarcation process based on The Hague's ruling, which Addis Ababa had ignored until then. A report from the UN Mission for Ethiopia and Eritrea (UNMEE) stated: 'a lasting peace will not be possible between Eritrea and Ethiopia or in the region unless the border is fully demarcated'.

[17] Meanwhile, the UN Security Council considered the possibility of turning UNMEE into an observation and assistance mission for both countries in the border demarcation. ∎

LAND USE

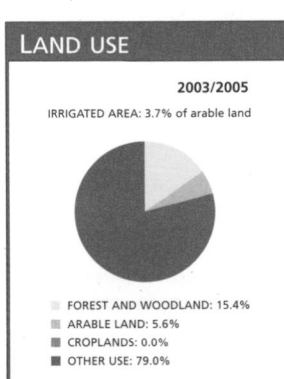

2003/2005

IRRIGATED AREA: 3.7% of arable land

- FOREST AND WOODLAND: 15.4%
- ARABLE LAND: 5.6%
- CROPLANDS: 0.0%
- OTHER USE: 79.0%

PROFILE

ENVIRONMENT

Eritrea is in the Horn of Africa. The 1,000 km northeast coast is on the Red Sea; to the northeast lies Sudan; to the south, Ethiopia and to the southeast, Djibouti. The dry plains and extremely hot desert are inhabited by pastoralist herders.

SOCIETY

Peoples: The nine ethnic groups are the Tigrinya, Tigre, Bilen, Afar, Saho, Kunama, Nara, Hidareb and Rashaida. The majority are pastoralists or farmers; 20 per cent are urban workers. Half a million Eritrean refugees live in Sudan, 40,000 in Europe and 14,000 in the US.

Religions: Almost half of all Eritreans are Coptic Christians; most of the rest are Muslim, although there are Catholic and Protestant minorities.

Languages: Tigrinya; Afar, Beni-Amir, Tigré, Saho, Kunama, Arab and other local languages.

Main Political Parties: People's Front for Democracy and Justice (PFDJ) is the only party recognized by the Government. Formerly the Eritrean People's Liberation Front (EPLF), it adopted its new name in 1994.

Main Social Organizations: Eritrean Liberation Front, split into several factions; Eritrean Islamic Jihad.

THE STATE

Official Name: Hagere Ertra.
Administrative Divisions: 8 provinces.
Capital: Asmara 556,000 people (2003).
Other Cities: Asseb 53,600 people; Keren 36,600; Mitsiwa 28,800 (2000).
Government: Parliamentary Republic. Isaias Afwerki, President since May 1993. Unicameral National Assembly.
National Holiday: 24 May, Independence (1993).
Armed Forces: 35,000 (1997).

Estonia / Eesti

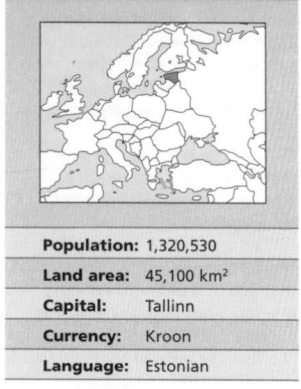

Population:	1,320,530
Land area:	45,100 km²
Capital:	Tallinn
Currency:	Kroon
Language:	Estonian

The region was settled some 6,000 years ago. Estonians, a branch of the Finno-Ugric nations, have greater cultural and linguistic ties with the Finns to the north than with the Indo-European Balts to the south.

[2] Around the year 400 AD, hunting and fishing activity began to be replaced by agriculture and cattle raising. At the same time, navigation and trade with neighboring countries along the Baltic Sea intensified. In the 11th and 12th centuries, combined Estonian forces successfully repelled Russia's attempts to invade the territory.

[3] The Germans, Russians and Danes, invading Estonia in the 13th century, found a federation of states with a high level of social development and a strong sense of independence, keeping them united against foreign conquerors.

[4] In the 13th century, the Knights of the Sword, a Germanic order which was created during the Crusades, conquered the southern part of Estonia and the north of Latvia, creating the kingdom of Livonia and converting the inhabitants to Christianity. German traders and landowners brought the Protestant Reformation to Estonia in the first half of the 16th century.

[5] Between 1558 and 1583 Livonia was repeatedly attacked by Russia before being broken up in 1561. Poland conquered Livonia in1569; a hundred years later, it ceded the major part of the kingdom to Sweden. In the Nordic Wars (1700-1721), Russia took Livonia from Sweden, and kept these lands under the Treaty of Nystad.

[6] Russia received the Polish part of Livonia in 1772, with the first partition of Poland. The former kingdom of Livonia became a Russian province in 1783. Power was shared between the Czar of Russia and local German nobles, who owned most of the lands and the peasants lived in serfdom.

[7] The abolition of serfdom in Russia and peasant land ownership rights (1804), strengthened Estonian nationalism.

[8] In 1904, Estonian nationalists seized control of Tallinn, ousting the German-Baltic rulers. After the fall of the Czar in February 1917,a demonstration by 40,000 Estonians in Petrograd forced the Provisional Government to grant them autonomy.

[9] In November 1917, with the election of a constituent assembly, the Estonian Bolsheviks obtained 35.5 per cent of the votes. On 24 February 1918, Estonia declared its independence from the Soviet Union and set up a provisional government. The following day, German troops occupied Tallinn and the Estonian Government was forced to go into exile.

[10] After World War I, the Estonians successfully fought both the Red Army and the Germans. On 2 February 1920, with the Treaty of Tartu the Soviet Union recognized Estonia's independence. That year Estonia started the construction of what would become the world's first shale-oil distillery.

[11] Estonia passed legislation guaranteeing the rights of minorities, and ensuring that all ethnic groups had access to schools in their own languages. The economic depression of the 1930s led the country to become a virtual dictatorship in 1933, before adopting a presidential-parliamentary system in 1937.

[12] The secret protocols of the Molotov-Ribbentrop pact, signed in 1939, determined that Estonia - like its two Baltic neighbors, Latvia and Lithuania - would remain within the Soviet sphere of influence. At the same time, Tallinn signed a mutual assistance treaty with Moscow giving the USSR the right to install naval bases on Estonian soil.

[13] In June 1940, after demanding the right for his troops to enter Estonian territory under the pretext of searching for missing soldiers, Stalin deposed the Tallinn government and replaced it with members of the local Communist Party (CP). Elections were held during the Soviet occupation, after which the CP seized power.

[14] Following the examples of Latvia and Lithuania, the new government adopted the name 'Soviet Socialist Republic of Estonia', joining the USSR. More than 60,000 Estonians were

PROFILE

ENVIRONMENT

Located on the northeastern coast of the Baltic Sea, Estonia is bounded by the Gulf of Finland in the north, Russia in the east and Latvia in the south. The Estonian landscape was formed by glaciers; there are numerous rivers and more than 1,500 lakes, the largest of which are Lake Peipsi (Europe's fourth largest) and Lake Vorts. Forests make up 38 per cent of Estonia's territory; the highest elevation is Mount Suur Muna Magi (317 m). The climate is temperate, with average temperatures of +28° C in summer and around -5°C in winter. The Baltic coast is 1,240 km long and includes a number of fjords. Some 1,500 islands close to the coast make up around one-tenth of Estonia's territory. The Gulf of Finland has numerous ice-free bays, of which Tallinn is the largest. Estonia's most important mineral resources are shale-oil (which meets most of Estonia's energy needs), and phosphates.

SOCIETY

Peoples: Estonians 64.2 per cent; Russians 28.7 per cent; Ukrainians 2.6 per cent; Belarusians 1.5 per cent Finnish and others 3.3 per cent (1998).
Religions: Lutheran (majority), Orthodox, Baptist.
Languages: Estonian (official); Russian and others.
Main Political Parties: Estonian Center Party; Union for the Republic; Estonian Reform Party; Estonian Peoples Union.
Main Social Organizations: Confederation of Estonian Trade Unions (EAKL)

THE STATE

Official Name: Eesti Vabariik.
Administrative Division: 15 counties.
Capital: Tallinn 391,000 people (2003).
Other Cities: Tartu 98,400 people; Narva 72,100; Kohtla-Järve 45,200; Pärnu 42,800 (2000).
Government: Parliamentary republic. Arnold Rüütel, President since October 2001. Andrus Ansip, Prime Minister since April 2005. Unicameral Legislature: State Council (Riigikogu), with 101 members elected every 4 years.
National Holiday: 24 February, Independence (1918).
Armed Forces: 3,800 (2006). Other: 2,000 (Coast Guard).

LAND USE

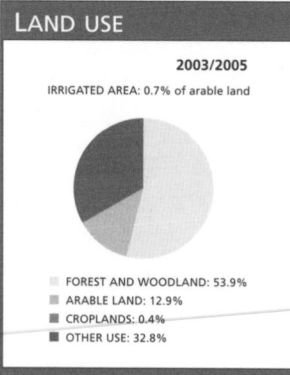

2003/2005

IRRIGATED AREA: 0.7% of arable land

- FOREST AND WOODLAND: 53.9%
- ARABLE LAND: 12.9%
- CROPLANDS: 0.4%
- OTHER USE: 32.8%

PUBLIC EXPENDITURE

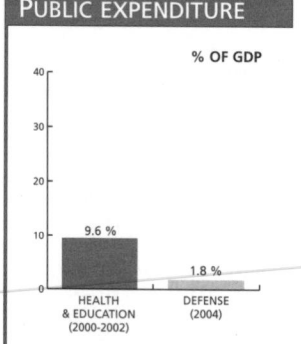

% OF GDP

9.6 %
HEALTH & EDUCATION (2000-2002)

1.8 %
DEFENSE (2004)

Life expectancy
73 years
2005-2010

GNI per capita
$7,080
2004

Literacy
100% total adult rate
2000-2004

HIV prevalence rate
1.1% of population 15-49 years old
2003

deported to Siberia.

15 When the German offensive against the USSR began in 1941, Nazi troops invaded Estonia, establishing a reign of terror. The USSR recovered the Baltic States in 1944.

16 The Soviet regime imposed forced industrialization and collectivization of the countryside. Some 80,000 Estonians emigrated to the West, while Russian colonization gradually altered the traditional ethnic composition of the population.

17 Around 20,000 Estonians were deported between 1945 and 1946. The third wave of mass deportations took place in 1949, when another 40,000 Estonians were sent to Siberia, most of them farmers who refused to accept collectivization of the land.

18 The reforms set in motion in 1985 by Soviet President Mikhail Gorbachev stimulated social and political activity within Estonia. In August 1987, a demonstration in Tallinn demanded the publication of the Molotov-Ribbentrop pact. Latvians and Lithuanians also asked that the contents of the protocols be revealed.

19 In January 1988, former Estonian political prisoners founded the Estonian Independence Party, defending the country's right to self-determination. In addition, this group called for the establishment of multiparty democracy, and the restoration of Estonian as the country's official language. Another group, the Estonian Heritage Society, began trying to locate and restore the country's historical monuments.

20 The first congress by the Popular Front of Estonia (FPE) - a coalition of nationalists and communists - organized in October 1988, reaffirmed Estonia's demand for autonomy, and asked Moscow to admit that the country had been occupied against its will in 1940. The following month, Estonia's Parliament declared the country's sovereignty, and affirmed its right to veto laws imposed by Moscow without consent.

21 In August 1989, some two million Estonians, Latvians and Lithuanians formed a 560-km human chain from Tallinn to Vilnius to demand the independence of the Baltic States. In February 1990, a convention of Estonian representatives approved the Declaration of Independence, based on the 1920 Peace Treaty of Tartu.

22 In the May 1990 elections, the FPE and other nationalist groups won an ample majority within parliament. Moderate nationalist leader Edgar Savisaar was named as the leader of the

IN FOCUS

ENVIRONMENTAL CHALLENGES
Estonia has an important fishing industry, although its fleet now has to fish further from its coastal waters because of pollution in the Baltic Sea, which contains toxic waste dumped by industries in several countries with Baltic coastlines. Neither Tallinn nor other cities have satisfactory sewage systems. The northeast is polluted by power plant emissions, but the level of emissions has fallen 80 per cent since 1980.

WOMEN'S RIGHTS
In 2003, 19 per cent of parliamentary seats and 15 per cent of ministerial positions were held by women. Female school enrolment stands at 94 per cent*. In 2004 adult illiteracy remained at 0.2 per cent, both for women and men*.

Some 2,600 women aged between 15 and 49 years old were living with HIV/AIDS in 2003.

Women made up 49 per cent of the labor force in 2003. Of them, 4 per cent worked in agriculture, 23 per cent in industry and 73 per cent in services. Female unemployment reached 9.7 per cent in 2003.

CHILDREN
Four per cent of all children are underweight at birth*.

Some children and youngsters from the Tallinn and Ida-Virumaa regions have drug problems. Young addicts may spend between 10,000 (approx $800) and 30,000 Estonian krooni per month on drugs. In 2002 the minimum wage was 1,600 Estonian krooni, then worth about $90; the average wage was 4,500 Estonian krooni, or $25). These adolescents have a hard time finding jobs, so they often end up in illegal activities such as drug-dealing, smuggling and prostitution.

A report by the ILO and the International Program on Elimination of Child Labor (IPECL) stated that over the last years the number of 15 and 16 year-olds who have taken hard drugs has doubled. Fifty-three per cent of those seeking medical assistance started taking drugs before they were 18. The main first hard drug is heroin. In 2000 there were 1,581 drug-related crimes.

INDIGENOUS PEOPLES/ETHNIC MINORITIES
The Russian community is made up of more than 400,000 people living mainly in Tallinn and the border cities of Narva and Sillamae. This minority suffers political, economical and social discrimination and does not have any support for their demands from the Russian Government. The new citizenship act that promotes voting and political representation of Russians should lead to improvements. Nevertheless, Russians are better-off in Estonia than those living in other former Soviet countries.

Ukrainians are another minority group, making up 2.5 per cent of the total population.

MIGRANTS/REFUGEES
The Government sends asylum-seekers to reception centers, like the one located in Illika, bordering Russia. Those who can financially support themselves are exempt.

Estonia is among the 10 countries with the highest percentage of immigrants in its total population (26.2 per cent). Most are Russian immigrants who settled there after independence.

DEATH PENALTY
Abolished in 1998.

*Latest data available in *The State of the World's Children* and *Childinfo* database, UNICEF, 2006.

first elected government since 1940. In August, parliament proclaimed the independence of Estonia, but Moscow did not consider it to be valid. In September 1991, the USSR recognized the independence of the three Baltic States.

23 In January 1992, Savisaar and his government resigned in the face of growing criticism over his economic policy. Parliament named former transportation minister Tiit Vahi to head the new government. Estonia had to ration food and fuel when the Russian Federation began restricting and raising the price of its products.

24 On 20 June 1992, the new Constitution (based on the 1938 Constitution) was ratified by referendum. In September, the *Riigikogu* (Parliament) was elected. Lennart Meri, of the National Country Coalition Party (NCCP), was elected President on 5 October.

25 In June 1993, an overtly nationalistic law was approved, targeting foreigners - especially those of Russian origin, who made up 30 per cent of the total population. The law obliged foreigners to apply for a

discretionary residence permit.

26 The March 1995 elections led to the defeat of the coalition which had ruled Estonia since the former Soviet republic broke away from the USSR. Prime Minister Tiit Vahi caused a controversy when he named a 'disproportionate' number of former communist ministers in his government. In October his cabinet was forced to resign due to corruption charges against the Minister of the Interior. The new government was formed with the inclusion of Reform Party (RP) members.

27 In February 1998, Estonia, Latvia and Lithuania signed a Letter of Association with the US. In it, Washington committed itself to supporting the three states' integration into NATO.

28 Mart Laar was appointed Prime Minister in March 1999.

29 Laar resigned in January 2002, citing the 'betrayal' of the Reform Party, which had 'given up' the city of Tallinn to the opposition. During his term, Estonia, as well as becoming the strongest economy of the former Soviet republics, had started talks on European Union (EU) membership.

30 In a referendum held in 2003, 67 per cent of voters agreed to join the EU in May 2004.

31 In January 2004, the Presidents of Estonia and Cyprus signed two cooperation agreements, one on education and culture and the other to fight organized crime.

32 Estonia and nine other countries joined the EU on 1 May 2004, taking the total number of member states to 25.

33 Prime Minister Parts resigned in March 2005, after Justice Minister Ken-Marti Vaher received a no-confidence vote due to his handling of the anti-corruption program. Andrus Ansip, from the centre-right Reform Party, was appointed Prime Minister.

34 On 9 May 2006, Europe Day, the Estonian Parliament ratified the European Constitutional Treaty by 73 votes to 1, so becoming the 15th member state to ratify the European Constitution.

35 A week later, Estonia withdrew voluntarily from its aim of joining the euro zone in 2007. Inflation in the country had exceeded the required level, so the Government made a new plan to reduce it and join in 2008. ∎

Ethiopia / Ityop'iya

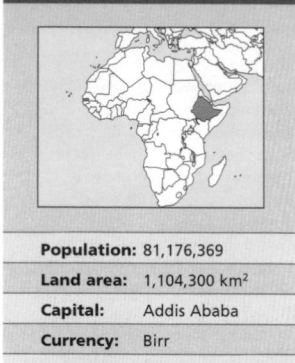

Population:	81,176,369
Land area:	1,104,300 km²
Capital:	Addis Ababa
Currency:	Birr
Language:	Amharic

The first hominid, called Australopithecus or southern ape, appeared in the Ohma river valley in Ethiopia, two million years ago. The Australopithecus was characterized by walking upright; it made tools, fed on meat and lived in groups. The appearance of the Australopithecus marked the beginning of the cultural evolution of humankind.

[2] Axum, in the north of present-day Ethiopia, was the center of trade between the Upper Nile valley and the Red Sea ports which traded with Arabia and India; it reached its height in the first centuries AD. Ethiopia was a rich and prosperous state, which was able to subdue present-day Yemen, but which went through a crisis in the 7th century. Trade routes moved as Arab unification and expansion dominated the area, conquering Egypt.

[3] The Ethiopian ruling élite had converted to Christianity in the 4th century, further contributing to their isolation. Expansion towards the south, excessive growth of the clergy, and declining trade led to a process of social and economic stratification similar to that in feudal western Europe. By the 16th century, one third of the land belonged to the 'king of kings'; another third belonged to the monasteries and the rest was

divided among the nobility and the rest of the population.

[4] The Muslim population that had developed a powerful trade economy on the coast of the Red Sea, instigated an insurrection, leading Ethiopia to resume its relations with Europe to request assistance. The aid took almost a century to arrive, but the Portuguese fleet, when it finally arrived in 1541, was decisive in destroying the Sultanate of Adal (See Somalia).

[5] For 150 years, Ethiopian emperors focused their efforts on the coast, giving Oromo (a people akin to the Hausa) a chance to gradually penetrate from the west until they became a majority. Their influence grew so great that an Oromo became emperor, between 1755 and 1769; though

the Amhara ruling elite took great pains to oust him.

[6] This state of affairs continued until 1889 when Menelik II came to power. Designated heir to the throne in1869, he spent the next 20 years training an army (with British and Italian assistance) and organizing the administration of his own territory, the state of Shoa. In 1895 his former allies, the Italians, invaded the country claiming that previous commitments had not been honored. The final battle was fought in Adua in 1896 where 4,000 of the 10,000 Italian soldiers were killed. It was the most devastating defeat suffered by European troops on African soil until the Algerian War.

[7] In the diplomatic negotiations that followed their defeat, the Italians succeeded in obtaining two territories that Ethiopia did not really control: Eritrea and the southern Somalian coast. In 1906, the world powers recognized the independence and territorial integrity of what

was then known as Abyssinia, in exchange for certain economic privileges.

[8] This arrangement saved Ethiopia from direct colonization until 1936, when Italian Fascist dictator Benito Mussolini invaded the country and overthrew Haile Selassie, Menelik's heir.

[9] During the five-year occupation, several industries as well as coffee plantations were started, and a system of racial discrimination was installed, similar to that of apartheid in South Africa.

[10] In 1948, Ethiopians won their autonomy back from Britain, which had taken over the country after Mussolini's defeat. When Selassie took the throne, his country was floundering in unprecedented crisis: foreign occupation had disrupted production; nationalist political movements had strengthened in the struggle for autonomy and rejected a return to feudalism; and poverty in the interior had grown considerably.

[11] Selassie denounced colonialism, favored non-alignment and supported the creation of the Organization for African Unity, which finally set up headquarters in Addis Ababa. He also maintained close links with Israel.

[12] In 1974, after a series of strikes, student rallies and widespread protests against

LAND USE

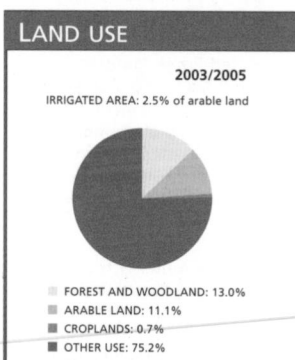

2003/2005

IRRIGATED AREA: 2.5% of arable land

- FOREST AND WOODLAND: 13.0%
- ARABLE LAND: 11.1%
- CROPLANDS: 0.7%
- OTHER USE: 75.2%

PUBLIC EXPENDITURE

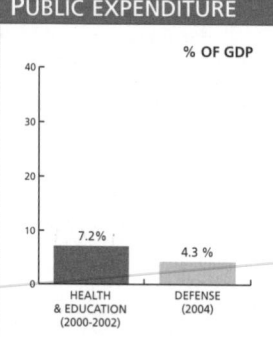

% OF GDP

- 7.2% HEALTH & EDUCATION (2000-2002)
- 4.3 % DEFENSE (2004)

WORKERS

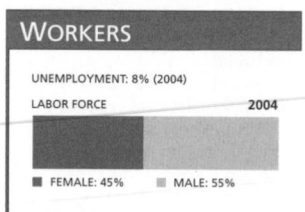

UNEMPLOYMENT: 8% (2004)

LABOR FORCE   2004

- FEMALE: 45%
- MALE: 55%

Life expectancy	GNI per capita	Literacy	HIV prevalence rate
49 years	**$110**	**42%** total adult rate	**4.4%** of population 15-49 years old
2005-2010	2004	2000-2004	2003

absolutism and food shortages, Selassie was overthrown.

[13] An Armed Forces Coordination Committee, the *Dergue* ('committee' in Amharic), headed by General Aman Andom abolished the monarchy and proclaimed a republic, suspending the Constitution and dissolving Parliament.

[14] After subsequent internal crisis, Colonel Mengistu Haile Mariam rose to power in December 1977. He managed to hold the Dergue together and put an end to the military's internal struggles.

[15] The military government nationalized foreign banks, insurance companies and heavy industry. US military bases were closed down. The key to the 'National Democratic Revolution' was the State's takeover of land, which put an end to the power of landowners. 'Scientific Socialism' was adopted as the official ideology in June 1976. The opposition was put down by the 'red terror' between 1977 and 1978. Thousands of people were executed during this period.

[16] After settling this crisis, the Government confronted two separatist movements which had been gaining strength since 1977, in Eritrea and the Ogaden desert.

[17] Soviet and Cuban support was decisive in the defeat of Somali troops for the Ogaden. The Eritrean separatists were forced to retreat after a major offensive in 1979. Meanwhile, a peasant-based guerrilla warfare had broken out in the Tigray region.

[18] In 1984, the country was struggling under the effects of a drought which had begun in 1982, causing thousands of deaths from starvation. The drought affected 12 provinces, threatened five million lives and killed over half a million people.

[19] That year, the Ethiopian Workers' Party (PWE) held their founding congress, approving a program to transform the country into a socialist state.

[20] The newly elected Assembly (the Shengo, or parliament) proclaimed the People's Democratic Republic of Ethiopia on 12 September, ratifying Mengistu Haile Mariam as head of state. Separatists extended operations in Eritrea and Tigray, as well as in Wollo, Gondar and Oromo in the south.

[21] The new constitution provided for the creation of five autonomous regions and 25 administrative regions. Eritrea was able to legislate on all matters except defense, national security, foreign relations and its legal status in relation to central government. The separatists rejected the proposal, calling it 'colonial'.

[22] The rebel military activity took a heavy toll on the Ethiopian army and in 1989 12 divisions (150,000 troops) stationed in the front line attempted a coup. Mengistu returned hastily from West Germany and put down the coup.

[23] In September 1989, the last Cuban soldiers withdrew from Ethiopia. The Government had signed a peace agreement in April 1988, and no longer needed their services.

[24] In 1990, within the framework of political changes in the Soviet bloc, the Central Committee of the Ethiopian Workers' Party decided to restructure the party and change its name to the Ethiopian Democratic Union Party (EDUP). While excluding the possibility of a multi party system, the changes sought to lay the basis of 'a party of all Ethiopians', open to 'opposition groups'. The Marxist-Leninist tag was dropped. The Government established a mixed economy, including state enterprises, cooperatives and private businesses.

[25] In May 1991, after overwhelming guerrilla victories in the north, Mengistu unexpectedly fled the country. The Government was left in the hands of Vice-President Tesfaye Gabre Kidane, considered a moderate, who initiated his transitional government by negotiating a cease-fire with the Eritrean rebels.

[26] Kidane's government took part in peace talks, in London, presided over by the US, with the participation of the most important rebel groups. They aimed at reaching an agreement which would stave off civil war. Kidane resigned in late May, when the US advised the forces of the Ethiopian People's Revolutionary Democratic Front (EPRDF) to take control of Addis Ababa.

[27] Ata Meles Zenawi, leader of the EPRDF, became interim president, until a multi-party conference could be held. He promised to bring the civil war to an end, re-establish democracy and put an end to hunger. Three months later, upon reopening Parliament and passing a new Constitution, Zenawi pledged to hold elections within a year.

[28] In March 1992, the new regional Councils were elected. However the Oromo Liberation Front announced its withdrawal from the Council of Representatives, made up of 87 members, which legitimized the transitional government.

[29] In May 1994, the Council of Representatives approved a draft Constitution which created the Federal Democratic Republic of

PROFILE

ENVIRONMENT

A mountainous country with altitudes of over 4,000 meters. Ethiopia is isolated from neighboring regions by its geography. In the mountains and plateaus, the vegetation varies with altitude. The Dega are cool, rainy highlands, above 2,500 meters, where grain is grown and cattle raised. The deep valleys which traverse the highlands are warm and rainy with tropical vegetation, known as the Kolla (up to 1,500 m). The drier, cooler, medium-range plateaus where coffee and cotton are grown (1,500 to 2,500 m) are the most densely populated parts of the country. To the East lies the Ogaden, a semi-desert plateau inhabited by nomadic shepherds of Somali origin.

SOCIETY

Peoples: There are more than 90 ethnic groups of which only seven have more than one million people. One-third of the population are Oromo, approximately a quarter are Amhara and a tenth are Tigrai. There are also Gurage, Somali, Sidama and Wolaita. At present there are 22 recognized minorities.
Religions: The Amhara and Tigrai are mostly Christians. The Somali, Afar and Aderi are mostly Muslims. African traditional religions are also practised.
Languages: There are four major language families: Semitic (Amhara), Cushitic (Oromo, Somali, Afar), Omotic and Nilo-Saharan. Amharic is the official language, among other 80 registered languages. Tigrai speak Tigrigna.
Main Political Parties: Ethiopian People's Revolutionary Democratic Front (EPRDF) is made up of several members: Tigray People's Liberation Front; Oromo People's Democratic Organization; Amhara National Democratic Movement; Southern Ethiopian People's Democratic Union. Coalition for Unity and Democracy is made up of several members: Ethiopian Democratic League; All Ethiopian Unity Party.

THE STATE

Official Name: Federal Democratic Republic of Ethiopia.
Administrative Divisions: There are nine ethnically based states and 2 self-governing administrations (Addis Ababa and Dire Dawa).
Capital: Addis Ababa (Adis Abeba) 2,723,000 people (2003).
Other Cities: Dire Dawa 202,700 people; Harar 93,900 (2000).
Government: Federal Republic. Girma Wolde-Giorgis, President since October 2001; Meles Zenawi, Prime Minister since August 1995, re-elected in 2000 and 2005. Bicameral Legislature: the Federal Parliamentary Assembly is formed by the Council of People's Representatives, with 547 members, and Council of the Federation, with 117 members.
National Holiday: 28 May, Overthrow of the Dergue (1991).
Armed Forces: 120,000 (1995)

Ethiopia. This draft was based on the 'ethnic federalism' doctrine, which put an end to the previous official unified vision of the nation. According to the approved document, the 'sovereignty resides in the nations, nationalities and peoples of Ethiopia' and not the people as a whole.

[30] In June, elections were held for a Constituent Assembly, but were boycotted by main opposition parties, like the Oromo Liberation Front and the Ogaden National Liberation Front.

[31] In May and June 1995, parliamentary elections were held, also boycotted by most of the opposition parties. The new federal republic was officially established in August, when Negasso Gidada, a Christian Oromo from the Welega region in the west of Ethiopia, took over the presidency. The outgoing president, Meles Zenawi, became Prime Minister and the 17 members of government were carefully selected to reflect 'the ethnic balance' of the country.

[32] The Government went ahead with the privatization of state companies - 144 in 1995 - and the annual grain deficit stood at around a million tons.

[33] In early 1998, food shortages threatened millions of Ethiopians. Due to the price increases set by the Government (13 per cent between August and December 1997) the satisfaction of basic needs became increasingly difficult for the poorer people. The Ethiopian Disaster Prevention and Preparedness Commission (DPPC)

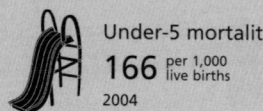

formally called on international organizations for help to avoid another famine.

34 Fighting against Eritrea began again in February 1999, following the short war that in May 1998 killed more than a thousand people. The UN Security Council called for an immediate cease-fire and a weapons and communications warfare embargo.

35 In March 1999, US President Bill Clinton proposed the cancellation of $70 billion of foreign aid paid to 46 African countries in a meeting of ministers in Addis. At this time, Clinton was under pressure from Congress to approve its 'Trade initiative with Africa'.

36 Heavily armed troops from Ethiopia entered Eritrea in May 2000, capturing 300 prisoners. As in 1998, the UN Security Council issued a three-day ultimatum to stop fighting, and the US proposed an arms embargo on both countries which was rejected by Russia. Arms were allegedly being sold by the former Soviet republics of Eastern Europe. The ultimatum expired and the conflict continued.

37 Thousands of people celebrated the decision of the Permanent Court of Arbitration at The Hague, in April 2002, to mark the boundaries of Ethiopia's 1,000 kilometer border with Eritrea. The government in Addis considered it a victory for its demands, although its claim over the port of Asseb was not taken into account by the commission.

38 In late April, Ethiopia decided to close its border with Eritrea to UN officials, accusing them of having taken journalists from Eritrea to the area without Ethiopian visas.

39 In April and May 2002, demonstrations over the results of local elections spiraled into bloodshed between Ethiopian security forces and ethnic Sheko and Mezehenger populations in the south western Ethiopian town of Tepi. The violence killed more than 150 civilians, uprooted nearly 5,000 others, and destroyed some 1,000 homes.

40 In December 2002, the Swiss transnational Nestlé demanded $6 million in compensation for the nationalization carried out in 1975 by the Tafari Benti (1974-1977) Government. Nearly 40,000 people sent e-mails calling on Nestlé to withdraw its claim. In January, the transnational agreed to reduce its claim to $1.5 million which would be reinvested in national feeding programs.

41 In September 2003, Eritrea accused Ethiopia of violating international law with its refusal to accept a ruling on the boundary drawn up by an independent commission in 2002. Eritrea said

IN FOCUS

ENVIRONMENTAL CHALLENGES
Many regions once rich in vegetation are now rocky, desert areas. Desertification and erosion have increased within the last years as a result of deforestation, intensive grazing and inappropriate use of water in agriculture. Lack of water affects more than four million people.

WOMEN'S RIGHTS
Ethiopian women have been able to vote and be elected since 1955. In 2005, they held 21.5 per cent of seats in Parliament and 6 per cent of ministerial positions.

Women's life expectancy is 42 years old*. Twenty-seven per cent of pregnant women receive prenatal care*, while only 6 per cent of births are attended by trained health staff*. Maternal mortality amounts to 850 deaths in every 100,000 live births.*

Although there are no official statistics, it is estimated that 80 per cent of women have suffered genital mutilation.

In some regions, abduction is used to take a girl as wife. The girl is taken by a group of men and then raped by the prospective husband. The elders from the man's village apologize to the family of the victim and ask them to agree to the marriage. The family often consents because a girl who has lost her virginity will not be able to marry. Articles 558 and 599 of the 1957 Penal Code allow rapists to escape punishment by marrying the woman they rape.

CHILDREN
Forty-three per cent of all children aged between 5 and 14 work.* Ethiopia is one of the countries worst affected

by HIV/AIDS: 4.4 per cent of the adult population carries the virus. In 2003, there were 120,000 children under 14 living with HIV/AIDS*. More than 720,000 children were orphaned by the disease.* Droughts, almost constant in the region, may accelerate HIV transmission, as people are forced to move and may become involved in sex work as a survival strategy.

Forty-seven per cent of children under 5 are severely underweight; while 52 per cent of these have physical development problems.

INDIGENOUS PEOPLES/ ETHNIC MINORITIES
The Afar (2 per cent of the population) live in the east, practise both Islam and Christianity and are mostly nomadic pastoralists. They have inhabited these lands for over 2,800 years. They were among the first victims of colonialism in Africa, which decided the break-up of their territorial unit into the three states of Djibouti, Eritrea and Ethiopia.

Amhara people (27 per cent of the population) reside mainly in the north and in Addis Ababa. In 1,500 BC, together with the current Tigray peoples, they created the Axum Empire. In spite of being a minority within the population, their traditional leaders took part in the bureaucratic hierarchy of the empire.

The Oromo (30 per cent of the population) are the largest ethnic group in the country, mainly living in the south. They farm and raise cattle; practising Islam or Christianity, except for about 15 per cent of their rural population that follow traditional religions. Together with the Amhara and the Tigrayans they dominated the Government and military classes of

the Ethiopian Empire.

Somalis (39 per cent of the population) practise Islam and are pastoralists. On account of their opposition to the Ethiopian State, they did not take part in the Government until 1995. They inhabit two semi-autonomous regions: Dire Dawaand Ogaden, in eastern Ethiopia.

Tigrayans (5 per cent of the population) are Christians and mostly inhabit the northern Tigray province, where they work in agriculture. They are descendants of Semites who arrived in the region about 3,000 years ago. After a long exclusion from political power, they led resistance to the Mengistu regime. The Tigrayan People's Liberation Front is part of the governing coalition and its leader Meles Zenawi is Prime Minister.

MIGRANTS/REFUGEES
In early 2006, Ethiopia hosted more than 100,000 refugees and asylum seekers, mostly from Sudan, Somalia and Eritrea. In February 2006, Ethiopia and Sudan signed an agreement to repatriate 73,000 Sudanese.

Between 1991 and 2002, about 800,000 Ethiopian refugees came back after having fled their country during the 1974-1991 dictatorship. Some 90,000 Ethiopians were internally displaced at year's end.

DEATH PENALTY
In force for all crimes. After a period of time without executions, the practice was restored in 1998.

*Latest data available in The State of the World's Children and Childinfo database, UNICEF, 2006.

Ethiopia was still claiming the village of Badme, the starting point of the war, despite the fact that it had been allotted to Eritrea by the commission.

42 In July 2004 the UN World Food Programme (WFP) called for international solidarity with Ethiopia's drought victims. The authorities claimed seven million people needed assistance that year. Instead of food, the WFP asked for cash to buy cereals from local farmers with excess produce.

43 That month, some 40 African leaders met in Addis Ababa for the opening of a heads of sate summit of the African Union (AU).

44 Elections held in May 2005 were won by Prime Minister Zenawi, who said the large

turnout - 90 per cent - meant a victory for his democratic policies.

45 In June, police in Addis Ababa opened fire against demonstrators protesting against alleged electoral fraud. More than 20 people died and several businesses were destroyed.

46 In March 2006 Ethiopia and Eritrea agreed to resume the process to draw up their common border (see Eritrea).

47 In November 2005, after post-election street protests, freedom of the press deteriorated seriously. The number of newspapers critical of the Government declined and 14 journalists faced charges punishable by death.

48 On 29 March 2006, the World Bank, the International

Monetary Fund (IMF) and the African Development Bank (ADB) cancelled the debt of 13 African countries, including Ethiopia. The cancellation would take effect from July that year.

49 The Somali region of Ethiopia, known as Ogaden by most people of Somali origin, was at the center of increasing tension at the end of 2006. The year had seen heightened military activity as tens of thousands of Ethiopian troops were sent into the region. The troops were dispatched not only to fight the rebel Ogaden National Liberation Front, which has fought for independence for 20 years, but also to secure the border and counter the perceived threat from the Union of Islamic Courts in Somalia. ∎

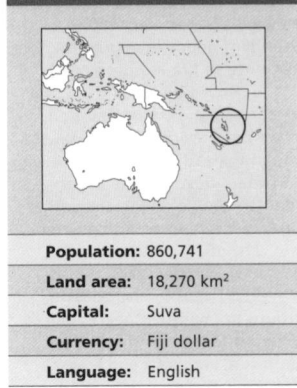

Population:	860,741
Land area:	18,270 km²
Capital:	Suva
Currency:	Fiji dollar
Language:	English

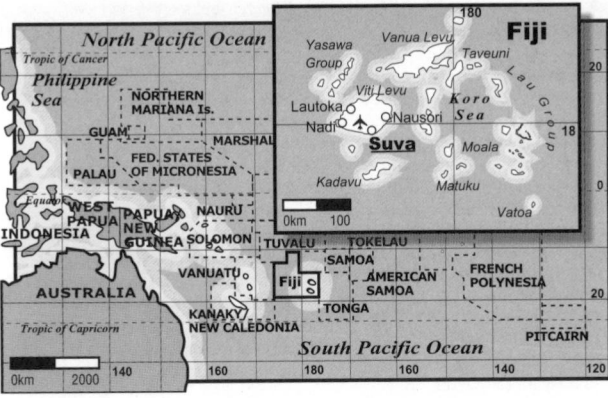

F our thousand years ago the Fijian archipelago was already populated. Melanesian migrations first reached the islands in the 6th century BC, and the Fijians had one of the leading Pacific cultures. In 1789, British Captain William Bligh (Mutiny of the Bounty fame) visited the islands, writing the first detailed account of Fijian life.

[2] Social life on the islands was organized in families and clans which gradually formed larger communities. One of these, ruled by traditional leader Na Ulivau, extended its influence from Ngau over the rest of the islands, achieving unification.

[3] In 1830, the first Christian missionaries arrived from Tonga and 24 years later achieved a major victory by christening ruler Thakombau, son of Na Ulivau. This 'king' (formerly reported to be a 'cannibal') became such an enthusiastic admirer of the Western world that he offered to annex Fiji to the US. The White House, caught up in the turmoil of the Secession War, missed the opportunity to tack another star on the flag. The British, faced with a shortage of sugar, 'discovered' the archipelago's potential for growing cane and officially annexed the islands on 10 October 1874.

[4] Fijian officials raised no objections to the purchase or expropriation of large tracts of land for the new crops, though the peasants were unwilling to leave their communal estates to work on the plantations. Consequently, there was an influx of bonded laborers, first from the Solomon Islands, and then from India.

[5] At the end of their contracts, Indian workers brought their families to Fiji and became small shop owners, craftspeople or bureaucrats in the colonial administration, and retained their language, religion and caste system.

[6] Initial moves toward local autonomy resulted in a complicated electoral system securing political control by the 'natives' who were in the minority. Prime Minister Sir Kamisese Mara initiated an elaborate scheme of representation, which proved useful in preventing major disruptions during the transition to independence in 1970. His Alliance Party claimed to be multiracial, and marshaled support from some Indians and other minor groups, in addition to the Fijian electorate.

[7] In 1976, the Government turned down a constitutional reform project which would have required multi-group cooperation in the country's administration.

[8] The 1970s economic difficulties forced thousands of Fijians to emigrate to New Zealand/Aotearoa in search of jobs. Individualism and private land ownership were encouraged, seeking to bolster the economic situation of 'natives' over that of the Indians. However, under these conditions, the rural workers gained few improvements; urban income increased by 3.5 per cent in 1978, but rural income increased only by 0.3 per cent.

[9] The difficult economic situation swelled the number of 'racist' groups such as the Nationalist Party, militant against the Indian majority, though Indians had been on the islands for five generations.

[10] Former Prime Minister Kamisese Mara's conservative administration was deeply marked by racial and ideological trends. His government was the only one in the Commonwealth to maintain relations with the racist Rhodesian and South African regimes and to welcome Chilean dictator Augusto Pinochet on an official visit in early 1980.

[11] In the 1982 election, the incumbent Alliance Party retained office with a small margin over the opposition National Federation Party (NFP), which had united its two rival factions despite growing conflict within the party.

[12] In the 11 April 1987 election, the Indian majority won and Timoci Bavadra became premier, ending 16 years of government by Melanesians. A few days after the election, Bavadra was overthrown by a military coup led by Colonel Sitiveni Rabuka, which justified himself as 'attempting to solve the ethnic problem'. However, his real objective seemed to be the removal of the Indian Government, who believed in an independent foreign policy, and planned to join the treaty of Rarotonga. The treaty promoted regional denuclearization; endorsed by Australia and New Zealand, but criticized by Britain and the US.

[13] The main Indian and Melanesian political parties reached an agreement, with added pressure from the Commonwealth, which appeared to appease the military. However, on 6 October

LAND USE

2003/2005

IRRIGATED AREA: 1.1% of arable land

- FOREST AND WOODLAND: 54.7%
- ARABLE LAND: 10.9%
- CROPLANDS: 4.7%
- OTHER USE: 29.7%

PROFILE

ENVIRONMENT

Fiji consists of nine large islands and 300 volcanic and coral islets and atolls, of which only 100 are inhabited. The group is located in Melanesia, in the Koro Sea, between Vanuatu (formerly New Hebrides) to the west and Tonga to the east, five degrees north of the Tropic of Capricorn. The largest islands are Viti Levu, where the capital is located, Vanua Levu, Taveuni, Lau, Kandavu, Asua, Karo, Ngau and Ovalau. The terrain is mainly mountainous. Fertile soils in flatland zones plus a tropical rainy climate, mildly tempered by sea winds, make the islands suitable for plantation crops, sugarcane and copra.

SOCIETY

Peoples: Half of Fijians are of Melanesian origin with some Polynesian influence, the other are descendants of Indian workers who came to the archipelago early in the 20th century; plus those of European and Chinese origin. Banabans (see Kiribati) have bought the island of Rambi, between Vanua Levu and Taveuni, with the purpose of settling there since their own island was left uninhabitable by phosphate mining.

Religions: 53 per cent of Fijians are Christian (mainly Methodist and other Protestant sects) 38 per cent are Hindu and eight per cent Muslim.

Languages: English (official), Urdu, Hindi, Fijian, Chinese. Rotuma and Kiribati (minority languages).

Main Political Parties: Fiji Labour Party; United Fiji Party (*Soqosoqo Duavata ni Lewenivanua*); National Federation Party; Conservative Alliance (*Matanitu Vanua*).

Main Social Organizations: The Association of Fijian Young People and Students, Fijian Trade Unions Congress (FTUC).

THE STATE

Official Name: Republic of the Fiji Islands.

Administrative Divisions: 5 Regions divided into 15 Provinces.

Capital: Suva 210,000 people (2003).

Other Cities: Lautoka 45,000 people; Nadi 32,100; Nausori 22,500 (2000).

Government: Ratu Josefa Iloilo, President since July 2000. Prime Minister, Laisenia Qarase, since March 2001; elected September 2001. Parliament: House of Representatives (Vale) and the Senate (Seniti).

National Holiday: 10 October, Independence Day (1970).

Armed Forces: 3,900 (1995).

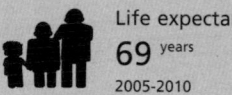
Life expectancy
69 years
2005-2010

GNI per capita
$2,720
2004

Literacy
93% total adult rate
2000-2004

HIV prevalence rate
0.1% of population 15-49 years old
2003

IN FOCUS

ENVIRONMENTAL CHALLENGES
Overfishing and pollution of the coastal waters threaten the marine environment. Deforestation and soil erosion on the islands are caused by intensive agriculture.

WOMEN'S RIGHTS
Fijian women have been able to vote and stand for office since 1963. Since 2001, 8.5 per cent of Lower Chamber seats and 12.5 per cent of Upper Chamber seats have been held by women. In 2003 women held 9 per cent of ministerial positions. That year, they constituted 33 per cent of the country's workforce.

The fertility rate fell from 3.5 to 2.6 children per woman during the period 1980-2003. In 2004, 99 per cent of births were attended by qualified staff and maternal mortality was 38 per 100,000 births.

CHILDREN
In 2004 under-one and under-five mortality rates were 16 and 20 per 1,000 live births respectively*. Ten per cent of babies were underweight at birth and eight per cent of children under five suffered from moderate or severe emaciation*.

Fiji is a popular destination for child sex tourists (tourism generates almost one third of the nation's income). Child abuse within the family is one of the main reasons children go into prostitution. Immigrant children (especially from China and the Philippines) are particularly at risk.

INDIGENOUS PEOPLES/ ETHNIC MINORITIES
Ethnic Fijians (a mixture of Melanesians and Polynesians who arrived many centuries ago) constitute 51 per cent of Fiji's population while descendants of Indian immigrants brought to work on sugar plantations during the last 20 years of the 19th century and the first 15 years of the 20th century comprise 44 per cent. While 83 per cent of arable land belongs to ethnic Fijians, 75 per cent of Indian Fijians work on sugar plantations.

The majority of the ethnic Fijians have settled in urban areas and the southern islands. They speak Fijian and most of them are Methodist Christians. When the British colonized Fiji in 1874, Fijians (cooperating with them) maintained their political and economic advantages over other ethnic groups in the country.

Indian Fijians form the majority population in the east, where sugar plantations are located. They speak Fijian Hindi (a local version of Hindi) and are mostly Hindus or Muslims. During British colonization they lost most of their rights. At present they are struggling to eliminate the political, religious and cultural restrictions that they experience.

MIGRANTS/REFUGEES
The number of Fijians seeking asylum abroad (mostly in Australia, New Zealand and Japan) diminished during the period 2003-2005. In 2003 there were around 170 applications, just under 100 in 2004 and less than 60 in the first nine months of 2005.

DEATH PENALTY
The Death penalty for ordinary crimes was abolished in 1979. There have been no executions since 1964.

*Latest data available in *The State of the World's Children* and *Childinfo* database, UNICEF, 2006.

1987, Rabuka retaliated by proclaiming a Republic in a move intended to disavow the authority of the head of state, the British-appointed governor.

[14] In December, Rabuka resigned as head of state in an attempt to create an image of a joint civilian-military government, aimed at improving its foreign image. Penaia Ganilau was named President and Kamisese Mara, Prime Minister - a regime never subjected to the approval of the electorate.

[15] In July 1990 a new constitutional decree based on apartheid went into effect, assuring the indigenous people 37 of the Chamber of Representative's 70 seats and 24 of the Senate's 34 seats. A constitutional referendum announced for 1992 was finally cancelled. The following year, the UN's General Assembly, as well as Mauritius and India denounced the apartheid model.

[16] Rabuka founded the Fijian Political Party (FPP/SVT) and in 1992, amid growing political and social strain, a military officer was named Prime Minister.

[17] Fiji was not able to overcome its chronic balance of payments deficit: most of its fuel and manufactured products were imported and its main sources of income - sugar exports and tourism - were not enough to balance the budget.

[18] In November 1993 six FPP/SVT members voted together with the opposition against the budget, forcing general elections. Rabuka retained power with 31 of the 37 Fijian seats and support from independent and General Vote Party members. Dissidents formed a Fijian Association obtaining only five seats.

[19] In November 1994, the Government began a timid revision of the racist Constitution. In 1995, Rabuka had to reorganize his Cabinet several times due to internal divisions in the coalition.

[20] In September 1996, a Commission completed a report on the new Constitution, which created a 'multiracial council' and reserved a certain number of seats in Parliament for certain ethnic groups. The council eventually came into operation in 1998.

[21] In September 1997, Fiji was readmitted into the British Commonwealth, 10 years after the coup which took it out.

[22] The first truly democratic elections after the coup, held in May 1999, were won by Mahendra Chaudhry of the Labor Party (LP). The Fijian Political Party (FPP/SVT) headed by Rabuka, won just six of the 71 parliamentary seats.

[23] An armed group led by Fijian entrepreneur George Speight entered Parliament on 19 May 2000 and abducted Prime Minister Mahendra and another 30 people, in an attempt to force a Constitutional reform to prevent the Indian minority from having access to power.

[24] President Kamisese dismissed the abducted Prime Minister. Commodore Frank Bainarama led a new coup on 29 May and overthrew the President, saying he had become weak and ineffective.

[25] The rebels were granted amnesty after releasing the hostages. In July, Ratu Josefa Iloilo was appointed acting President, while Laisenia Qarase became acting Prime Minister. The army arrested Speight together with over 350 rebels.

[26] The coup intensified racial divisions. In February 2001 Fiji's High Court ruled that the existing military-backed Government was illegal and the Government announced that elections would be held in August of that year.

[27] Speight, the Conservative Alliance (MV) leader, was elected Prime Minister in the general elections of August 2001while he was in prison awaiting trial for treason. The ousted Mahendra Chaudhry won a seat but his party lost by a narrow margin. Qarase excluded Chaudry's party from the Cabinet but the Supreme Court ordered him to include Indians in it.

[28] In December, Speight was removed from office, accused of treason and sentenced to death. However, in February 2002, President Iloilo commuted his sentence to life imprisonment.

[29] In November 2002 the sugar sector began a transformation, after the Government announced the privatization and restructuring of state industries. The sugar industry employs 200,000 workers and its revenues comprise seven per cent of Fiji's GDP.

[30] In November 2003, the US Government asked Fiji to send troops to 'maintain peace' in Iraq, offering to pay for the soldiers' uniforms, weapons and transport. The Fijian Government said it could not afford the costs of such a mission. About 100 Fijian soldiers took part in the invasion of Iraq as part of the British Armed Forces.

[31] The Asian Development Bank (ADB) granted Fiji $40 million a year in loans for the period2004-2006. The programs included infrastructure projects to rebuild airports, upgrade roads and promote urban development.

[32] In April 2004 Kamisese Mara, considered the father of Fijian independence, died aged 83. In August, vice-president Ratu Jope Seniloli was found guilty of treason for taking part in the attempted coup of May 2000, but served only a few months of his four-year sentence.

[33] In November the Labor Party rejected an offer to join the cabinet so as not to compromise its opposition role.

[34] In July 2005 Army Commander Frank Bainimarama declared that he would not hesitate to overthrow the Government if its proposal of amnesty for those involved in the coup of 2000 went ahead.

[35] In March 2006 the Great Council of Chiefs elected President Iloilo to serve a second term of five years. In May, former Prime Minister Rabuka was accused of organizing the unsuccessful coup.

[36] The Soqosoqo Duavata Ni Lewenivanua Party, led by Qarase, won 30 out of 59 seats in the general elections of May 2006. Chaudhry's Labor Party obtained 25 seats.

[37] No violence was reported in the racially charged election campaign but Fiji faced the prospect of political instability as its top military officer, Commodore Frank Bainimarama expressed his unhappiness at Qarase's election victory.

[38] The rift between the Government and the military widened in the ensuing months, with Bainimarama threatening violent opposition if a law was introduced offering an amnesty to plotters in the 2000 coup. In November 2006, Qarase, having failed to dismiss the military leader from his post, withdrew the proposed law and offered direct talks. ∎

Finland / Süomi

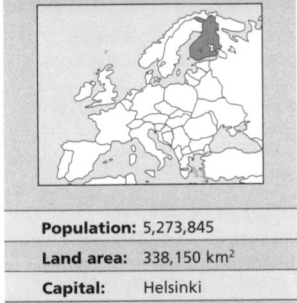

Population:	5,273,845
Land area:	338,150 km²
Capital:	Helsinki
Currency:	Euro
Language:	Finnish, Swedish

The ancestors of today's Sami people lived in nomadic groups, 7,500 years BC, in what is now Finland. According to archeological evidence, there were two Stone Ages, before the people were pushed to the north around 4,000 years BC, after the arrival of the Finno-Ugrics from the Urals area. It was then that Samis began practicing agriculture and domesticating animals.

[2] The Finno-Ugric immigration was made up of two groups: the ancestors of the Tavastlanders who came from the Gulf of Finland, and the Carelians who arrived from the southeast. At that period, Scandinavian peoples (mostly Swedes) occupied the west coast, the archipelagos and Ahvenanmaa Island. The Scandinavians prevailed as an ethnic group in the subsequent evolution of the Finnish population.

[3] In the 1st century AD, groups of Finns that had emigrated from the Volga regions of Russia, settled and fought against the Tavastlanders and Carelians who, simultaneously, were fighting between themselves over land.

[4] In the Viking Age (between the 8th and 9th centuries), the Finns did not take part in the expeditions. They mainly settled in the north, where they sold furs along the trade routes to Russia. Towards the end of this period, Finland's territory was a route for German and Russian goods.

[5] Finland was claimed by both the Russian and Swedish empires from the 12th century until their respective demises. In 1172, the Pope advised the Swedes to control the Finns, to avoid their being proselytized by the Russian Orthodox Church. The Church of Sweden joined the Protestant Reformation in the 16th century. This unleashed a major peasant revolt. Peasants were dissatisfied with their poor living conditions and the foreign policy imposed by Sweden (Club War 1596-97). The translation of the New Testament in 1548 was the first book printed in Finnish.

[6] Between 1634 and 1721, the territory of Finland was incorporated into the Kingdom of Sweden. This period of political stability and the spread of Protestantism helped unite the Finns.

[7] The Kingdom of Sweden began to weaken after the Great Northern War (1721). Alexander I of Russia succeeded in occupying Finland in 1808 and in 1809 it became a grand duchy of Imperial Russia. He granted Finland certain autonomy, allowing the *Diet* (parliament) to operate, as well as the army and the local judicial system.

[8] Once they became subjects of the Russian Empire, the Finns began to develop a national movement. They defended their religious identity by actively supporting the Lutheran Church. They also collected medieval pagan legends and myths about the early Finno-Ugric groups, as well as epic tales like the *Kalevala*, which was written in Finnish

[9] Until the mid-19th century, Swedish - which was only spoken by a minority - was the only language allowed in schools and universities and there were almost no publications in Finnish. In that period, the campaign for Finnish, the language of the majority, to be the main language became a key issue. The climax came in 1850 when the authorities banned the printing of all books in Finnish, except prayer books and bibles.

[10] The Russian Empire allowed the opening of the first Finnish Grammar School in 1858 and promised to make Finnish the official language 25 years later. But, on the unification of Germany(1861-70), Czar Nicholas II strengthened the Russian military and civil presence in Finland and restricted the relative autonomy granted by his predecessors.

[11] The main opposition came from the Labor Party, founded in 1899, which four years later changed its name to the Social Democratic Party (SDP), after joining the Second Socialist International. The SDP formed a single front, together with the Constitutionalists who had been expelled from the Diet.

[12] After its defeat by Japan, Russia's weakness facilitated the complete reform of the parliamentary and voting systems in 1905. In 1906, a single-chamber Parliament was created, and universal and equal suffrage was established. Parliament was dissolved several times by the Russian Emperor until it was finally closed in 1910.

[13] During World War I (1914-1918), Germany gave financial support and military training to the Finnish Liberation Movement.

[14] After the triumph of the Bolshevik Revolution in Russia, Finland declared independence in December 1917. In January 1918, the SDP took Helsinki, with the support of Soviet troops, and the major industrial centers. The Government counterattacked, backed by Czarist and German troops under the command of General Mannerheim. In May, the civil war came to an end after causing the death of 30,000 people. The Socialist leaders that could not escape to the USSR were sent to prison camps.

[15] The country became a monarchy, and German Prince Frederick Charles of Hessen was

PROFILE

ENVIRONMENT

Finland is a flat country (the average altitude is 150 meters above sea level) with vast marine clay plains, low plateaus, numerous hills and lakes formed by glaciers. The population is concentrated mainly on the coastal plains, the country's main farming area. The economy is based on forest (coniferous) products and mobile phones. Main export products are from energy, high technology in electronics and communications, metallurgy, wood (paper paste and paper) and chemicals sectors.

SOCIETY

Peoples: 92.1 per cent of the population is Finnish and 7.5 per cent Swedish. There are Roma and Sami minorities. **Religions:** official churches: Evangelic Lutheran Church (more than 94 per cent of the population), Finnish Orthodox Church (2 per cent). **Languages:** Finnish (official and spoken by 93.2 per cent). Swedish (official and spoken by 6 per cent). Sami and Russian are spoken by minorities.

Main Political Parties: Social Democratic Party of Finland (SDP); Finnish Centre; National Coalition; Swedish Peoples Party. **Main Social Organizations:** Central Organization of Finnish Unions, about 1,000,000 members and 28 member unions. Finnish Confederation of Salaried Employees.

THE STATE

Official Name: Suomen Tasavalta. **Administrative Divisions:** 6 provinces. **Capital:** Helsinki 1,075,000 people (2003). **Other Cities:** Espoo 219,400 people; Tampere 197,200; Vantaa 182,700; Turku (Abo) 175,100 (2000). **Government:** Tarja Halonen, President since March 2000 (re-elected in January 2006). Matti Vanhanen, Prime Minister since June 2003. Unicameral Legislature: the Diet, with 200 members. Since 1996 there is a Parliament for the Sami minority (the samediggi) with limited autonomy on cultural matters. **National Holiday:** 6 December, Independence (1917, from Russia). **Armed Forces:** 31,200 (1994). Other: Border guards: 4,400.

Life expectancy
79 years
2005-2010

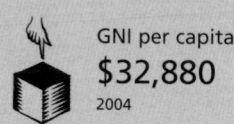

GNI per capita
$32,880
2004

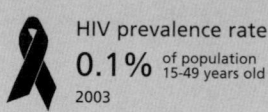

HIV prevalence rate
0.1% of population 15-49 years old
2003

elected king. After Germany's defeat in World War I, the monarchy collapsed and General Mannerheim was appointed regent, on condition that he would set up a republican system.

[16] In the 1919 elections, in order to assure Mannerheim's defeat, the SDP supported Kaarlo Juho Stahlbergal, presidential candidate of the National Progress Party (center), who was elected.

[17] The same year a Constitution was approved, establishing a parliamentary system with a strong presidential figure. In addition, it made the Prime Minister the head of the government, and the President head of state. Both Finnish and Swedish were recognized as official languages. Nationalist groups later demanded that Finnish be given priority which brought about a controversy that led to the creation of the Swedish People's Party.

[18] The Constitution ratified the principle - established in 1868 and still in operation - of the relationship between the Lutheran Church and the Finnish State by which the latter promised to finance the Church, but at the same time it reserved its right to appoint bishops.

[19] The country was ruled almost uninterruptedly after independence by alliances between the Social Democrats and the Centrists. From 1920 to 1930, these two political forces backed programs for the modernization of agricultural production and the timber industry, as well as a progressive legislation for the workers' rights. Agriculture employed 70 per cent of the active population until the 1960s. In the early 20th century, the Rural Party was founded, which later became the Center Party of Finland.

[20] In 1926, the Marxist wing left the SDP and founded the Communist Party (FCP). The subsequent growth of the FCP gave rise to the ultra-right-wing Lapua (Lappo) Movement, which gained the support of conservative groups and some Rural Party sectors and banned the FCP in 1930. In 1931 the Lappo movement began to carry out attacks against the Social Democrats and the following year it attempted a coup which was put down by the Government.

[21] After the German invasion of Poland in 1939, the USSR seized Karelia province, a naval base on the Hanko peninsula and some islands in the Gulf of Finland. After a year of conflict, Finland had to formally cede them on signing the Treaty of Moscow in 1940.

[22] At the beginning of World War II (1939-1945) Finland declared its neutrality. However, it entered into co-operation with Germany, aiming to recover the territories ceded to

IN FOCUS

ENVIRONMENTAL CHALLENGES
Atmospheric pollution caused by sulfur dioxide emissions from factories and electric power plants contributes to acid rain. Contamination of the Baltic Sea with agrochemicals and industrial waste causes serious problems in the environment and has resulted in habitat loss and a threat to the survival of flora and fauna species.

WOMEN'S RIGHTS
Finland was the first country in the world to grant women full political rights. They have been able to vote and stand for office since 1906. In 2003, women held 37.5 per cent of seats in Parliament and 47 per cent of ministerial or equivalent positions. In the same year women made up 48 per cent of the labor force and unemployment amongst them rose to 9 per cent. There is discrimination at work; although women have access to important positions they are paid lower wages than men holding the same positions and doing the same tasks.

In 2004 perinatal care (before, during and after birth) was available for 100 per cent of pregnant women. Maternal mortality was 6 per 100,000 births, one of the lowest in the world*.

Many women are victims of actual or threatened violence

or sexual abuse from the age of 15 (frequently at the hands of a family member). Immigrant women and those belonging to minority groups face double discrimination based on ethnic origin and gender.

CHILDREN
In 2004, child mortality for under-ones and under-fives (3 and 4 per 1,000 live births respectively) was one of the lowest in the world.

ECPAT International reported Government concern about the growing sex trade carried out by phone and online, particularly because of the risk to children having access to these services.

INDIGENOUS PEOPLES/ ETHNIC MINORITIES
The Sami are one among several other indigenous groups in Europe. Their total population is 75,000 and they inhabit lands in Norway, Sweden, Russia and Finland. More than half of them speak Sami or some related dialect. In Finland there are 6,500 Sami, 4,000 of whom live in a 35,000 sq km reserve that includes a special region for the Skolt Sami who were located in this area after the end of World War II.

They are a majority in the municipality of Utsjoki, but a minority elsewhere. In recent years, both the conditions of their lives and their participation in the social activities of the country have improved.

In 1994, the Government passed a new law that guaranteed Sami people the right to freely use *skolt* (their native language, belonging to a sub-family of Uralic languages) and practice their culture.

MIGRANTS/REFUGEES
Between 2001 and 2005 more than 15,000 people applied for asylum in Finland. In 2005 the number of refugees hosted by Finland fell from 321 in January to 242 in December.

Finland grants temporary asylum to a significant number of foreigners who are unable to furnish conclusive proof of persecution but come from countries in conflict or show signs of torture. This type of asylum is granted for a three-year period and cannot be renewed. The holders of temporary asylum have the same rights as those who are granted full asylum. After residing in Finland for three months asylum-seekers receive permission to work, although this is limited to specific job opportunities: those that citizens are unwilling to accept.

DEATH PENALTY
This was abolished in 1972.

*Latest data available in *TheState of the World's Children* and *Childinfo* database, UNICEF, 2006.

Moscow. The 1944 Soviet counter-offensive and re-occupation of Karelia triggered the resignation of President Ryti. He was succeeded by General Mannerheim who, in 1944, signed an armistice with Moscow - recognizing the 1940 Treaty - and organized the fight against Germany.

[23] After the war, Finland was disarmed and forced to pay economic reparations to the USSR for six years. The country developed its heavy industry and operated in markets on both sides of the Berlin Wall. After joining the UN in 1955 Finland was the only member of the Nordic Council that maintained relations with the EC and COMECON (the Communist bloc) during the Cold War (1950-1991).

[24] With the disintegration in 1991 of the USSR, which had been purchasing over 25 per cent of Finland's exports, the economy stagnated and unemployment reached 20 per cent. Centrist Prime Minister Esko Aho launched a drastic readjustment plan which affected social security programs.

[25] Finland joined the European Union (EU) in 1995. The internal impact of this was lessened by

special subsidies for the farming sector. The euro was adopted in 1999. Finland systematically criticized NATO actions.

[26] Social-Democrat Tarja Halonen won the January 1999 elections, becoming the country's first woman head of state. However, due to the parliamentary system established by a constitutional reform in 2000, she only carried out representative functions.

[27] In March 2003, Anneli Jaatteenmaki, of the Center Party (CP), was elected Prime Minister by a margin of just 6,600 votes. The CP had to form a coalition with the SDP and the Swedish Peoples' Party. Two months later Jaatteenmaki resigned, citing death threats against her.

[28] In June, Jaatteenmaki was succeeded by Matti Vanhannen (also of the CP) who expressed reservations about EU defense policy. He also opposed the construction of a fifth nuclear reactor and voiced concern about unemployment and increasing racial tensions resulting from the arrival of immigrants from Africa and Southeast Europe.

[29] In March 2004, former Prime

Minister Jaatteenmaki faced charges of illegally obtaining documents relating to the Iraq war while she was leader of the opposition. The Court declared her innocent on the grounds of insufficient evidence.

[30] After six weeks of strikes and seven months of disputes, management and workers in the paper-making industry signed a new labor agreement at the end of June 2005. The dispute involved the country's principal timber companies, such as Stora Enso and PM-Kymmene, and threatened the regular supply of paper to Europe.

[31] In January 2006 Tarja Halonen was re-elected President for a second term. During the campaign she presented herself as a 'president for all the people', promising to preserve equality and social welfare.

[32] In May 2006, Parliament approved the European Constitution. European leaders were endeavoring to reinvigorate the initiative, which had suffered serious setbacks with its rejection by France and Holland in 2005. ∎

France / France

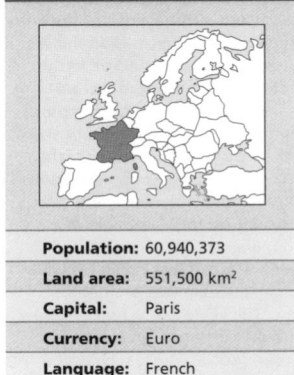

Population:	60,940,373
Land area:	551,500 km²
Capital:	Paris
Currency:	Euro
Language:	French

The two regions occupied by the Celts were known to the Romans as Gaul. The region between the Alps and Rome was Cisalpine Gaul, and beyond the Alps was Transalpine Gaul. With natural borders on all sides - the Alps, the Pyrénées, the Atlantic Ocean and the Rhine - Gaul covered not only what is now France, but also Belgium, Switzerland and the western banks of the Rhine.

2 Gaelic society was essentially agricultural, with almost no urban life. The few cities were used as fortresses, where the peasants sought refuge when under attack. Society was divided into the nobles (who were also warriors), the people and the Druids, keepers of Celtic wisdom and religious traditions.

3 The Romans came to Gaul in 125 BC. They conquered the area along the Mediterranean, the Rhone valley and Languedoc, calling the combined area 'Provincia'. Caesar divided Gaul into two regions; Provincia and Free Gaul. Free Gaul was subdivided into Belgian Gaul in the north, between the Rhine and the Seine; Celtic Gaul in the center, between the Seine, the Garonne and the lower Rhine; and Aquitaine, in the southwest.

4 In 27 BC, Augustus Caesar set up administrative centers in Gaul, to manage Rome's affairs encouraging urbanization. Bridges and an extensive road network

were built throughout the region, facilitating an increase in trade. Wheat production was increased and vineyards were planted, with wine replacing beer as the traditional beverage. After a series of invasions by the Visigoths in the south and the Burgundians along the Sane and the Rhone, the northern Gauls conquered the rest of Gaul under the leadership of Clovis, adopting the name, 'Franks'.

5 Between the 5th and the 9th centuries France emerged as the Merovingian and Carolingian dynasties brought the entire region under the influence of Christianity. With the Islamic expansion and the fall of the Roman Empire, trade ceased, urban civilization was almost completely wiped out, the population decreased and the culture became decadent.

6 By the 9th century, feudalism had become firmly established. Centralized authority practically disappeared, as local people were unable to repel the Scandinavians, Hungarians, Saracens, and other invaders of this era. By the end of the century, the previously united land was a conglomerate of more than 300 independent counties.

7 From the 10th century onward, the royal dynasties slowly recovered their power. They established hereditary succession to the throne,

they shared power with the Church and became the main feudal landowners.

8 In the 13th century, an increase in commercial activity led to remarkable rebirth of the cities, and agricultural techniques were improved, as the population increased. The Crusades led to greater circulation of people and goods, and the gradual disappearance of serfdom gave rise to greater social mobility. This was the 'Golden Age' of the French Middle Ages, when France had great power over, and influence upon Western civilization.

9 Paris was one of Europe's most important cities, and the prestige of its University was linked to its cultural pre-eminence. The University trained lawyers in Roman law, and their influence helped form a new concept of the State where the king was no longer a feudal lord, but rather the embodiment of the law. Over a period of time, nationalistic feelings began to develop.

10 Louis XIV, the 'Sun King', personified the concept of absolute monarchy. He came to the throne in 1661, and established the 'Divine Right of Kings'. He consolidated the unity of France, giving rise to the concept of the modern State. During his reign, French cultural influence reached its apogee.

11 The 1789 Revolution opened up a new era in the history of France. The National Assembly, convened in July of that year, replaced the absolute monarch with a constitutional monarchy. The fall of the Bastille on 14 July and the Declaration of the Rights of Man, on 27 August, brought the old regime to an end, thus

paving the way for the rise of the bourgeoisie - the prevailing class of the towns - whose reforms came into direct conflict with the Church and the King. Finally, the Assembly overthrew the monarchy and proclaimed the First French Republic.

12 The rest of Europe joined forces against revolutionary France. Danton and Robespierre declared the nation 'to be in peril' and formed a citizen army. This Committee of Public Salvation was able to forestall foreign invasion but internal confrontations resulted in the 'Reign of Terror'. Robespierre and his companions were overthrown and executed in July 1794.

13 For the next five years, the revolutionaries tried to regain control of the country, which had fallen victim to corruption, internal strife and instability. Napoleon Bonaparte's coup (1799) brought an end to the dying regime. Seizing power, he had himself named Consul for Life in 1802 and then Emperor in 1804.

14 Although Napoleon's reign represented a return to absolutism, it preserved the main achievements of the Revolution. Legal, administrative, religious, financial and educational reorganization changed the country irrevocably. Napoleon strove hard to bring the rest of Europe under his control and his armies occupied the whole of the continent from Madrid to the outskirts of Moscow. Finally, exhausted by war, France was defeated at Waterloo, in 1815.

15 Rebellions in 1830, 1848 and 1871 rocked the country. In spite of this, the Industrial Revolution brought factories, railroads, large companies and credit institutions to France. The Third Republic, beginning in 1870, was to be France's longest-lasting regime in almost a century and a half.

16 With the establishment of universal male suffrage in 1848, the peasants and the urban middle class had the greatest electoral power. The Government managed to win their support by protectionism and the establishment of free, secular and mandatory primary education, which raised aspirations of greater social mobility.

17 France started a period of colonial expansion with the conquest of Algeria in 1830, and continuing with other territories in Africa and the Far East. A large empire was built, with colonies in the Caribbean, Africa, the Middle East, the Indo-Chinese peninsula and the Pacific.

18 World War I enabled France to recover Alsace and Lorraine, territories annexed by Germany in 1870. The War left France devastated. More than 1.5 million

LAND USE

2003/2005

IRRIGATED AREA: 13.3% of arable land

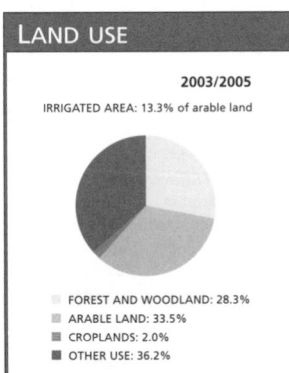

- FOREST AND WOODLAND: 28.3%
- ARABLE LAND: 33.5%
- CROPLANDS: 2.0%
- OTHER USE: 36.2%

WORKERS

UNEMPLOYMENT: 10% (2004)

LABOR FORCE   **2004**

- FEMALE: 46%
- MALE: 54%

Life expectancy
80 years
2005-2010

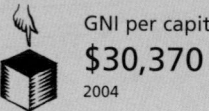
GNI per capita
$30,370
2004

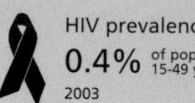
HIV prevalence rate
0.4% of population
15-49 years old
2003

people had been killed. Property damage coupled with the internal and foreign debt added up to more than 150 billion gold francs, and the nation's currency lost its traditional stability.

¹⁹ The world-wide recession that started in the US in 1929 reached France in 1931. In 1936, the parties of the left joined together to form the Popular Front, won the legislative elections and carried out important social reforms, such as paid vacations and the 40-hour working week, but they were unable to hold back unemployment or the looming economic crisis.

²⁰ Germany's invasion of Poland led France and England to declare war on Germany in 1939. Germany went on to occupy almost one-third of France. In 1940, Marshall Pétain signed an armistice proclaiming the 'national revolution' transforming non-occupied France into a satellite of Berlin. The resistance fighters (or *maquis*) in the south did not militarily affect the occupation.

²¹ Once the German Army had been defeated on the Russian front, the Allied landing in Normandy led to France's liberation in 1944 and to its participation in the invasion of Germany. In October 1946, the Fourth Republic was proclaimed, led by General de Gaulle. In the 1946 elections the French Communist Party (PCF) obtained one of the highest votes in its history.

²² Funds from the US Marshal Plan led to the economic and social reconstruction of the country. Production reached a six per cent annual growth rate. Per capita income increased 47 per cent between 1949 and 1959; women were granted the right to vote; the banks were nationalized and a social security program came into effect.

²³ After 1945, France was unable to re-establish its pre-war control over its colonies. This was partly a result of the growing sense of democracy and human rights. Having seen the anti-Fascist alliance and the birth of the League of Nations, many colonies (not just French ones) began to breathe the air of independence. French colonialism was based on the concepts of 'unity of the Republic' and 'cultural assimilation', which resulted in a centralized administration, with no autonomy for local governments. Thus, decolonization took place through fierce independence movements, with little room for negotiation.

²⁴ In 1945, Syria and Libya were the first French colonies to become independent, followed by Morocco, Tunisia and Madagascar. Vietnam, Laos and Cambodia became independent only in 1954, after a long and bloody war. In May1958, four years after the Algerian revolution began, the *pieds-noirs* - French people residing in the colony - dealt a mortal blow to the Fourth Republic, and the Government called in General de Gaulle to deal with the crisis.

²⁵ The establishment of the Fifth Republic in 1958 and the 1962 decision to elect the President by direct universal suffrage laid the foundation for a regime with strong presidential powers. After the independence of Algeria in 1962 and the last remaining African colonies, France sought to achieve greater stability, based on strengthening the currency, growth of vanguard industries, scientific research, and development of a national independence strategy. France made use of 'deterrent' atomic power and in 1966 withdrew from the military structure of the North Atlantic Treaty Organization (NATO), but kept its membership in case of a 'surprise attack'.

²⁶ France maintained enormous influence over its former African colonies south of the Sahara. Diplomatic relations with Algeria, Vietnam, and other countries that had fought bloody wars for independence were not restored until 1982.

²⁷ In May 1968, the greatest social and political crisis of the Fifth Republic took place. The regime's growing authoritarianism in the educational and social sectors gave rise to huge student protests and labor strikes throughout the country. For a whole month, the Government seemed to be seriously threatened. However, there were no political forces capable of toppling the Government, and the general strike was called off when a salary increase was promised.

²⁸ The years which followed saw the birth of other groups based around social issues, such as the feminist, ecological and antinuclear movements. In 1972 the Socialist Party and the Communist Party created the Union of the Left. François Mitterand, the Socialist candidate, was elected President in 1981. His was the first left-wing cabinet since 1958.

²⁹ The new government nationalized industrial and banking groups, granted new labor rights - the 35-hour working week, an increase in social benefits, retirement at the age of 60 - and decentralized power. However, unemployment, the economic crisis and an increase in imports led the Government to enforce a harsh economic policy and to carry out restructuring of the industrial sector, which made communist ministers resign. In 1981, a Ministry of Women's rights was created.

³⁰ In March 1986, a right-wing coalition led by neo-Gaullist Jacques Chirac, the Mayor of Paris (1977-1995) defeated the Left in the legislative elections that saw the progress of Jean Marie Le Pen's ultra-right National Front. Chirac formed a new government and for two years the country experienced its first 'cohabitation' between a left-wing president (Mitterand) and a conservative council of ministers.

³¹ Chirac's government wiped out some of the 1981 and 1982 reforms with the privatization of several companies nationalized by the left but kept most of the social gains. In the field of individual liberties, the strong hand regarding legislation concerning foreigners living in France was criticized by several humanitarian organizations.

³² In 1988, Mitterand defeated Chirac in presidential elections. Without 'cohabitation' this time, socialist Michel Rocard was appointed Prime Minister.

³³ The Socialist economic policy did not differ substantially from that of the right and unemployment kept on the rise. In 1991 Edith Cresson became France's first female Prime Minister.

³⁴ In 1993, the Left was defeated in legislative elections once again and Mitterrand named conservative

PROFILE

ENVIRONMENT

In the north is the Paris Basin, which spreads out into fields and plains. The Massif Central, in the center of France, is made up of vast plateaus. The Alps rise in the south-east. The southern region includes the Mediterranean coast, with mountain ranges, such as the Pyrénées, and plains. Grain farming is the main agricultural activity; wheat is grown all over the country, especially in the north. Grapes are grown in the Mediterranean region for wine exports. The main mineral resources are coal, iron ore and bauxite. France has the highest nuclear energy production per capita in the world. It runs second to the US in its nuclear power capacity; 77 per cent of the country's electricity comes from its 58 nuclear reactors.

SOCIETY

Peoples: Most of the population stem from the integration of three basic European groups: Nordic, Alpine and Mediterranean. Approximately 7 per cent of the population is of foreign descent, mainly from Northern Africa (Algeria, Morocco, Tunisia), from the former French colonies in sub- Saharan Africa and from Europe (Spain, Italy and Portugal).
Religions: Mainly Catholic (81.4 per cent); Islam (6.8 per cent) practiced mostly by North African and West African immigrants; Protestants (2 per cent) and Jews (1 per cent).
Languages: French is the official and predominant language. There are also regional languages: Breton in Brittany, a German dialect in Alsace and Lorraine, Flemish in the northeast, Catalan and Basque in the Southwest, Provençal in the south-east, Occitan in the central south, Corsican on the island of Corsica. Immigrants speak their own languages, mainly Portuguese, Arab, Berber, Spanish, Italian and African languages.
Main Political Parties: Rally for the Republic; Nacional Front; Socialist Party; Union for the French Democracy.
Main Social Organizations: General Labor Confederation (CGT), communist; French Democratic Labor Confederation (CFDT), socialist; Workers' Force (FO); French Confederation of Christian Workers (CFTC). France has the lowest level of unionization in the European Community (about 10 per cent).

THE STATE

Official Name: République Française.
Administrative Divisions: 22 Regions with 96 Departments in France; 4 overseas departments (French Guiana, Guadeloupe, Martinique, Réunion); 4 overseas territories (French Polynesia, New Caledonia, Wallis and Futuna and the Southern and Antarctica French Lands) and 2 overseas territorial collectivities (Mayotte, St Pierre and Miquelon).
Capital: Paris 9,794,000 people (2003).
Other Cities: Lyon 2,800,000 people; Marseille 2,800,000; Toulouse 800,000; Nice 933,080 (2000).
Government: Jacques Chirac, President since May 1995, re-elected in 2002. Dominique de Villepin, Prime Minister since May 2005. Bicameral Parliament: the National Assembly, with 577 members, and the Senate, with 321 members.
National Holiday: 14 July, Bastille Day (1789).
Armed Forces: 416,000 (2001). Other: 100,000 Gendarmes.

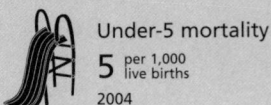
Under-5 mortality
5 per 1,000
live births
2004

Maternal mortality
17 per 100,000
live births
2000

IN FOCUS

ENVIRONMENTAL CHALLENGES

Dependence on nuclear energy poses a serious problem. Nuclear reactors operating in the country generate three-quarters of the national consumption of electricity, making France the second largest producer of nuclear energy, after the US. There is also a nuclear-reprocessing plant (Cap de la Hague), which generates plutonium, with serious sanitary and environmental risks. Acid rain causes serious forest damage. Industrial and vehicle emissions contribute to air pollution. Rivers are polluted by urban and industrial waste as well as by agricultural runoff.

WOMEN'S RIGHTS

Women have been able to vote and run for office since 1944. In 2004, 12 per cent of Lower House seats and 17 per cent of the Senate were held by women. In 2003, 18 per cent of ministerial or equivalent positions in government were held by women. That year, women comprised 46 per cent

of the country's workforce. In 2000-2004, the maternal mortality rate was 17 per 100,000 live births*.

CHILDREN

France, like other Western European countries, is a major destination for child-traffickers. There are 12,000 to 15,000 street sex workers, of which 7,000 are based in Paris, but it is unclear how many of these are under 18. Some organizations place the total number of child victims of prostitution between 2,000 and 3,000, with hundreds in Paris. According to ECPAT, every year 2,000 to 3,000 new sex workers populate French streets and often their age is not clear because passports are confiscated or the victims do not have birth certificates from their home countries.

INDIGENOUS PEOPLES/ ETHNIC MINORITIES

Most Muslims came from North Africa after World War II to meet the demand for low-paying labor. They mainly speak French and several Arabic dialects; most of them are Sunni Muslims. They suffer from social, economic, racial and cultural discrimination; they are often

victims of attacks by ultra-right-wing groups.

Corsicans - from Corsica, a French dependency since 1768 when it was sold by Genoa - speak Corsican, a mixture of French and Italian. Many support autonomy for their island. The National Liberation Front of Corsica (FLNC) has been fighting for this since 1976. The Front has been behind several terrorist attacks in France.

The Roma-Gypsies (called *gitanes* in French) began to arrive in France in 1400. They spread out, searching for better economic opportunities, fleeing discrimination; they speak different languages and have varied religious beliefs. They are largely excluded from French society; many do not pay taxes and do not have access to health and education.

The Basques comprise a very small proportion of the country's population, and live in southwestern France. In contrast with Basques from Euskadi, many suffer the effects of economic marginalization and very few speak Euskera, their original language.

MIGRANTS/REFUGEES

In spite of a 15 per cent fall in asylum applications in 2005 over

the previous year, France took the largest number of refugees (50,000), followed by the US (48,800), the UK (30,500), Germany (28,900), and Austria (22,500). However, taking into account the proportion of refugees to the receiving countries' population, France, the UK and Germany rank lower than Austria, Sweden, Cyprus or Norway, among other European countries. During the 2001-2005 period, France received the third highest amount of asylum applications (281,600), after the US (379,500) and the UK (325,800). In 2005 most of the refugees in the country came from Haiti, Serbia and Montenegro, Turkey, DR Congo, Russian Federation, China, Bosnia-Herzegovina, Moldova, Algeria and Sri Lanka.

DEATH PENALTY

The death penalty was abolished in 1981.

*Latest data available in *The State of the World's Children* and *Childinfo* database, UNICEF, 2006.

Edouard Balladur as Prime Minister. The corruption scandals, whose main targets had been socialist leaders, affected this time renowned right-wing politicians. Three of Balladur's ministers resigned in 1994.

[35] In the April 1995 presidential elections Chirac defeated socialist Lionel Jospin and appointed conservative Alain Juppé as Prime Minister.

[36] In December, the largest civil servants' strike since 1968 paralyzed the country for over three weeks. The social situation remained tense in 1996, not only because of growing unemployment, but also due to unfair income distribution.

[37] The Juppé administration continued its austerity policy that caused confrontation with labor unions in 1996. Over 150,000 people took part in a protest against a new law passed in 1997 to restrict the entrance and residence of immigrants.

[38] Unexpectedly, Chirac called for early legislative elections. In the second round, on 25 May 1997, the leftist opposition obtained an important victory. The Greens took seven seats. Lionel Jospin became the new Prime Minister.

[39] Scandal broke out in the political system when an illegal scheme of funding for political parties was revealed in August

2000. According to the public prosecutor, during Chirac's presidency (1990-1995) business people had made 'gifts' to the parties. A 1995 law banned companies from making donations to political parties, which are funded by public money.

[40] In April 2002, in a presidential election with the lowest turnout in the 44 years of the Fifth Republic, Le Pen unexpectedly ended in second place with 17.02 per cent of the vote. Jospin, left out of the second round, decided to quit politics, leaving the Socialist Party without a leader. He called on his followers to vote in the second round for Chirac, who had won the first round with 19.67 per cent of the vote.

[41] In the second round, in May, Chirac obtained 82 per cent of the vote, while Le Pen maintained his 17 per cent. Chirac greeted 'those French people committed to solidarity and freedom, open to Europe and the world'. World political leaders expressed their 'relief' at the results, but Le Pen's electoral advance was a clear triumph for the European right.

[42] Jean-Pierre Raffarin, Chirac's Prime Minister, reduced taxes by five per cent, cut back the state infrastructure and privatized gas and electricity.

[43] In late 2002, together with

Germany and Russia, France vigorously opposed a UN resolution authorizing the use of force to disarm Iraq, threatening to use its veto power if the resolution was approved in the Security Council.

[44] In March 2003, amendments to the Constitution granting more autonomy on economic, tourism, cultural and educational issues and the power to call for local referenda were approved.

[45] Chirac promised to end chronic unemployment in France by 2004. He trimmed unemployment benefits - as well as pension contributions and support to young job-seekers - and reduced taxes on higher incomes. Criticism came from social and workers' organizations, the opposition and his own party.

[46] In December 2003 a Government commission recommended passing a law banning conspicuous religious symbols in school, for example the Islamic veil (*hijab*), the Jewish *kippa* and large Christian crosses. Most of the French supported this idea, on the basis that the state education's secular tradition should be defended from religious excess, particularly from Islamic radicalization.

[47] Some Muslim, Christian and Jewish leaders felt that the law would only bring more religious

discrimination. In January 2004 40,000 Muslims demonstrated against the measure in Paris, Marseilles, Lille and other cities. There were also protests in London, Berlin, Brussels, Cairo and Bethlehem. There are some five million Muslims in France.

[48] Chirac's Union for a Popular Movement (UPM) was defeated in the second round of the regional elections held in March, getting 37 percent of the vote, against more than 50 per cent obtained by the left. This led to the resignation of Raffarin, but Chirac appointed him premier once again. In November, Nicolas Sarkozy resigned as Finance minister and took up the leadership of the UPM.

[49] A labor legislation reform passed by Parliament in March2005 left in place the 35-hour work week and authorized a maximum 48-hour workweek, the longest in the EU.

[50] An error of judgment, stemming from his confidence in previous surveys, led Chirac to opt for the ratification of the European Constitution through a referendum instead of taking it to Parliament. In the May 2005 referendum the Constitution was rejected by 54.87 per cent of the vote, and the result destabilized its approval process in the rest of Europe. A few days later, Holland also rejected it.

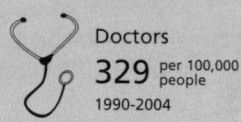

Doctors
329 per 100,000 people
1990-2004

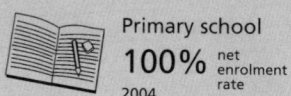

Primary school
100% net enrolment rate
2004

[51] After an accident in which two immigrant teenagers died in an electricity sub-station near Paris, the country was struck by a wave of riots. Violent protests - mostly by young immigrants and children of immigrants from the Maghreb and sub-Saharan Africa - spread rapidly and worsened with police repression.

[52] Paris, Marseilles, Rennes, Nantes and Lille were some of the cities with the most significant clashes. On 5-6 November, 1,400 cars were burnt, as well as dozens of supermarkets and public buildings, including schools and sports centers.

[53] Home Minister Nicolas Sarkozy said it was necessary to 'the clean the filth from the suburbs', which inflamed the situation. Although Chirac called for calm, dismissed far right-winger Jacques Bompard's demand to take the army to the streets and rejected Sarkozy's words, the Home Minister insisted that the riots had been planned by criminal bands and religious extremists. When calm returned to the streets in November, more than 10,000 cars had been burnt.

[54] A new law on youth employment mobilized the French once again between March and April 2006. Massive demonstrations in Paris and other major cities drove the Government to discard the law. ∎

St Pierre and Miquelon

Population: 5,883
Land area: 242 km²
Capital: St Pierre
Currency: Euro
Language: French

ENVIRONMENT
Archipelago formed by two main islands: St Pierre (26 sq km) and Miquelon (which, with Langlade, amounts to 216 sq km) and some ten smaller islands off the Canadian coast in the North Atlantic. Economy almost exclusively depends on fishing.

SOCIETY
Peoples: the majority are descendants of French settlers.
Religion: Catholic.
Language: French.
Main Political Parties: Rally for Construction - Saint Pierre and Miquelon 2000; Miquelon Objectives; Road to Future; Archipelago Tomorrow.

THE STATE
Official Name: Departement de Saint Pierre et Miquelon (Territorial Collectivity of Saint Pierre and Miquelon - conventional name).
Capital: St Pierre 6,000 people (2003).
Other Towns: Miquelon 1,100 people (2000).
Government: A French overseas department. The Head of State is Jacques Chirac (French President). Administered by a Prefect, Yves Fauqueur, since August 2006, who represents the French state. Unicameral Legislature: the General Council, with 19 members elected for a six-year term in single-seat constituencies. The territory has one deputy and one senator in the French Parliament and has representation in the European Parliament.

OVERSEAS DEPARTMENTS AND TERRITORIES

DEPENDENCIES:
Guadeloupe; Martinique; French Guiana; Réunion; St Pierre and Miquelon (see entries).

TERRITORIES:
Mayotte; New Caledonia (see Kanaky); Wallis and Futuna (see entries); French Polynesia (see entry).

SOUTHERN AND ANTARCTIC TERRITORIES:
Comprising two archipelagos: Kerguelen (7,000 sq km, with 80 people in Port-Aux-Français) and Crozet (500 sq km, with 20 people); two islands: New Amsterdam (60 sq km, with 35 inhabitants) and St Paul (7 sq km, uninhabited), located in the southern Indian Ocean; the Land of Adélie (500,000 sq km with 27 people at the Dumont Durville Base), in Antarctica. Administered from Paris by Administrateur Superieur Francois Garde (since 24 May 2000), assisted by Secretary General Jean-Yves Hermoso (since NA). Technical personnel at weather stations are the only inhabitants.

Wallis and Futuna

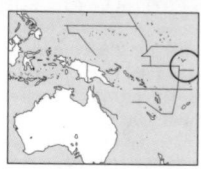

Population: 14,888
Land area: 274 km²
Capital: Mata-Utu
Currency: Cfp franc
Language: French

Wallis was named after Samuel Wallis, a navigator who 'discovered' it in 1767. Marist missionaries arrived in the archipelago in 1837 and converted the inhabitants to Catholicism. It became a French protectorate in 1888, and in December 1959, after a referendum, the country adopted the status of French Overseas Territory.

[2] Unlike other French dependencies in the Pacific, there are no independence movements on the islands. In 1983, the two kingdoms of Futuna achieved separation from Wallis, but maintained their relationship with France. The islands' economic prospects are poor: in addition to the devastating effect of cyclones that periodically hit them, the only bank on the islands was closed. Wallis and Futuna received 55 million francs in aid from France in 1987 and still relies heavily on grants. Approximately 50 per cent of the economically active population has had to emigrate to other parts of Polynesia in search of work. Their remittances, together with public works projects, constitute the main source of income for the islands.

[3] In the 1992 elections for the Territorial Assembly, the Left managed to defeat the neo-Gaullist Rassemblement pour la République (RPR), a party of the right, for the first time in 20 years. In 1997, the neo-Gaullist candidate Victor Brial gained the seat of deputy for Wallis in the French National Assembly.

[4] The population growth rate is extremely weak. The steady immigration flow of citizens from Wallis and Futuna to other Polynesian islands - particularly to Kanaky/New Caledonia, where they made up 10 per cent of the population - mostly comprises people of reproductive age who do not return to their home islands. ∎

ENVIRONMENT
The territory consists of the Wallis archipelago (159 sq km), formed by Uvea Island - where the capital is located - and 22 islets, plus the Futuna (64 sq km) and Alofi Islands (51 sq km). This group is located in western Polynesia, surrounded by Tuvalu to the north, Fiji to the south and the Samoa archipelago to the east. With a rainy and tropical climate, the major commercial activities are copra and fishing.

SOCIETY
Peoples: Of Polynesian origin. Approximately two thirds of the population live on Wallis and the rest on Futuna. Nearly 12,000 inhabitants live in Kanaky and Vanuatu.
Religions: Catholic.
Languages: French (official) and Polynesian languages.

THE STATE
Official name: Territoire des îles de Wallis et Futuna.
Administrative divisions: There are no defined administrative divisions, but there are three kingdoms: Wallis, Sigave and Alo.
Capital: Mata-Utu (located on Uvea) 1,000 people (2003).
Other cities: Utuofa 820 people; Vailala 800 (2000).
Government: Overseas territory administered by a French-appointed Chief Administrator, Xavier de Furst, since January 2005, assisted by a 20-member Territorial Assembly elected for a 5-year term. The kingdoms of Wallis and Futuna (in Sigave and Alo) from which the country was formed, have very limited powers. They send one deputy to the French National Assembly, and another to the Senate.
Armed forces: The defense is the responsibility of France.

French Guiana / Guyane Française

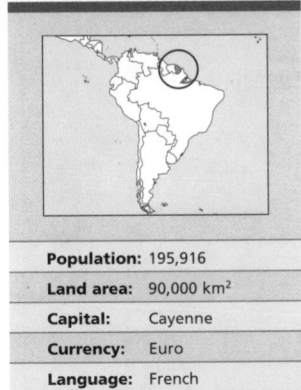

Population:	195,916
Land area:	90,000 km²
Capital:	Cayenne
Currency:	Euro
Language:	French

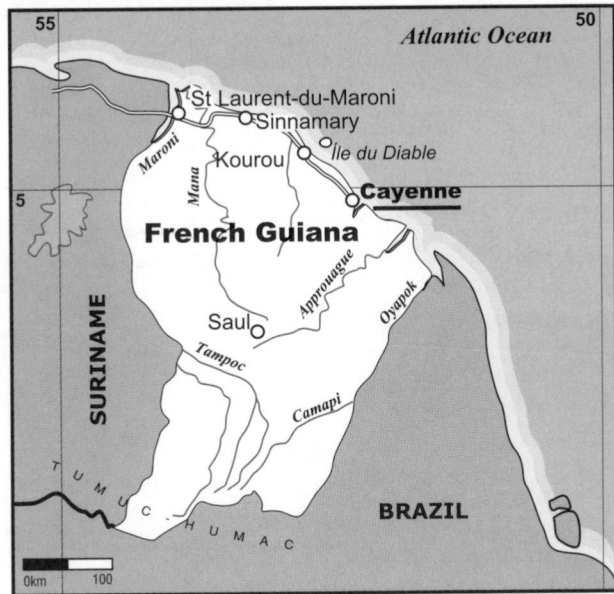

The Arawaks were copper-skinned people with straight, black hair. They grew corn, cotton, yams and sweet potatoes. They built round huts with thatched, cone-shaped roofs, and slept in hammocks - an Arawak word which survived the culture which gave rise to it. They lived in balance with their surroundings. The Caribs displaced them from the area and later resisted the Spaniards who began to arrive toward the beginning of the 16th century.

[2] In 1604 the French occupied Guiana, despite Carib resistance. The colony passed successively into Dutch, English and Portuguesehands, until the beginning of French domination in 1676. Towards the end of the18th century, France sent more than 3,000 colonists to settle the interior. Few survived the tropical diseases, but those who did sought refuge in a group of islands off the coast, which they named Health Islands. The most famous among them is known as the Devil's Island, which was turned into a prison in the 19th century.-

[3] In 1946, Guiana became a French 'Overseas Department'. The country is heavily reliant on French funding, which provided 70 per cent of the GNP in 1989.

[4] In 1967, the National Center for Space Studies was established. Over 1,300 foreign technicians work there, earning First World salaries, as well as 1,500 French Guyanan nationals. Thirty Ariane voyages have been launched from the base at Kourou, with all the satellites sent into space financed by European consortia.

[5] During the 1970s, the autonomist Socialist Party of Guiana (PSG) became the majority party at the local level. Early in the 1980s, armed groups attacked 'colonialist' targets, but tensions were defused with the victory in France of the French Socialist Party in 1981.

[6] At the 'First Conference of the Last French Colonies' held in Guadeloupe in 1985, there was severe criticism of the availability of French visas and citizenship for east Asians, while Haitians, Brazilians and Guianans, with closer cultural ties, underwent persecution and discrimination.

[7] In 1986, Guiana's representation in the French National Assembly increased to two members. In the 1989 municipal elections, Cayenne and 12 other districts, out of the 19 at stake, were won by the Left. Georges Othily, a PSG dissident was elected for the French Senate.

[8] In 1992, a week-long general strike was called by unions andbusiness leaders. As a result, Paris agreed to finance a plan to improve infrastructure and education. In 1994, French Guiana joined the Association of Caribbean States as an associate member.

[9] The French Government, after acknowledging the critical economic situation of its possession, announced a new 'development plan' for French Guiana in November 1997.

[10] The March 2000 visit by Jean-Jacques Queyranne, under-secretary of Overseas Territories, triggered protests that ended in clashes between the police and the demonstrators who demanded independence. The crowds looted businesses and set cars on fire, and several people were injured. The pro-independence Workers' Union of Guiana led a general strike.

[11] In 2001, French and Spanish authorities seized nearly seven tons of cocaine from two ships in territorial waters.

[12] In December, many communities filed complaints with the French Government for its inaction in enforcing the environmental laws in the mining sector. Mercury, used in extracting gold from the ore, contaminates rivers and kills fish, the main source of food for the Wayana and Emerillon peoples.

[13] Europe's largest and most expensive satellite was launched from the Kourou base in French Guiana in 2002, to monitor the health of the planet from outer space. Until that year, more than 125 commercial satellites had been launched from French Guiana.

[14] During 2003, European scientists continued to launch more space probes. The 2004 European Agenda proposed strengthening the space research program, which would increase the flow of scientists and specialists from all over the world to Kourou, with benefits for the economy.

[15] The French Government approved in March 2006 an agreement with Brazil to build a bridge over the river Oiapoque - the natural boundary between French Guiana and Brazil. The idea, first mooted in 1997, had been raised by Brazilian President Lula da Silva during a visit to Paris in July 2005. ■

WORKERS

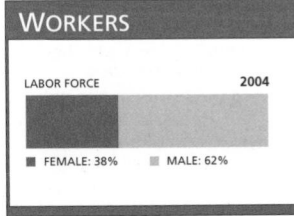

LABOR FORCE | 2004

■ FEMALE: 38% ■ MALE: 62%

PROFILE

ENVIRONMENT
French Guiana, the smallest and least populated country in South America, is located slightly north of the Equator. Due to its hot and rainy climate, only the coastal alluvial lowlands are suitable for agriculture (cocoa, bananas, sugar cane, rice and corn). The hinterland mountains are covered with rainforests. The country has large reserves of bauxite and gold. Mining activities have polluted the Moroni River, the principal source of food for the indigenous peoples of the area

SOCIETY
Peoples: Mostly of mixed European and African slave roots. There are 4,000 indigenous peoples, including the Wayana, Oyami and Emerillon in the interior jungles, and the Arawak, Galibi and Palikur along the coast. There are minority groups from China, India and France. Also in the jungles there are Cimarrones, descendants of rebel or fugitive African slaves.
Religions: Mainly Roman Catholic, also Hindu and Islam.
Languages: French (official), local Creole. Some groups continue to speak their own languages.

Main Political Parties: Socialist Party of Guiana (PSG), Union for the Republic (RPR), Union for French Democracy (UDF), De-Colonization and Emancipation of Guiana Movement.
Main Social Organizations: Workers' Union of Guiana.

THE STATE
Official Name: Département d'Outre-Mer de la Guyane française.
Administrative Divisions: Two Districts.
Capital: Cayenne 56,000 people (2003).
Other Cities: Kourou 21,000 people; St Laurent-du-Maroni 21,000 (2000).
Government: Jacques Chirac, Chief of State since May, 1995; re-elected May 2002. Jean Pierre Laflaquière, Prefect appointed by France in 2006. Regional parliament (consultative body): General Council, made up of 19 members, and Regional Council, with 31 members. The department has two representatives in the National Assembly and two in the French Senate.
National Holiday: All French holidays.
Armed Forces: 8,400 French troops.

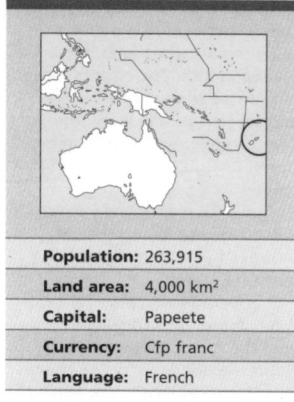

Population:	263,915
Land area:	4,000 km²
Capital:	Papeete
Currency:	Cfp franc
Language:	French

There are various theories on the origins of Polynesia's first inhabitants. Some say they came from Latin America, others say it was from Indonesia; but the evidence is inconclusive.

² In 1840 France occupied the islands and in 1880, despite native resistance, they were officially made a colony under the name of 'the French Establishments of Oceania'. In 1958, they became an Overseas Territory.

³ Except for some concessions on the domestic front, France maintains its control over the islands, with a hard-line policy because of their strategic location and the atomic tests that have taken place on the Mururoa atoll since 1966. In 1975 France also carried out tests on the Fangataufa atoll, despite strong opposition from the population and countries such as New Zealand/Aotearoa.

⁴ The 'nuclearization' of the area rapidly resulted in the destruction of French Polynesia's traditional economic base, which depends upon the French military budget. In a few years, the economy went from self-sufficiency to dependence upon imported goods. By the end of 1980, 80 per cent of basic foodstuffs were imported.

⁵ Since 1984, French Polynesia is ruled by an autonomy statute (ratified in 1996), which stated power would be shared between France and the Territory.

⁶ In 1992, French President François Mitterrand decreed a temporary suspension of nuclear tests and Paris began negotiations with Papeete for an economic plan to be carried out after the permanent closure of the experiment centers in Mururoa and Fangataufa. In January 1996 the permanent suspension of nuclear tests was announced.

⁷ Gaston Flosse, the President of Polynesia and a member of the French Senate, was prosecuted and found guilty of corruption by a French court of law. He received a prison sentence for having accepted tens of thousands of dollars in exchange for illegal gambling permits in Tahiti. Flosse refused to resign.

⁸ In May 2001, The President's party, the Tahoeraa Huiraatira/ Rally for the Republic (RPR), won Parliamentary elections obtaining 28 seats, compared to 13 obtained by the Tavini Huiraatira (TH). Flosse was re-elected President on 18 May.

⁹ In early 2002, official statistics showed that, after the terrorist attacks of September 2001 in the US, tourism had fallen by 9.7 percent in early 2002. French Polynesia is the second most popular tourist destination in the region, after Fiji.

¹⁰ In July 2003 President Chirac visited French Polynesia for the first time. While Flosse welcomed Chirac, pro-independence leader Oscar Temaru boycotted the reception. He organized a demonstration to remind the French President that France 'has the duty to lead the peoples under its responsibility toward independence and sovereignty'.

¹¹ Chirac gave assurances that, according to International Atomic Energy Agency (IAEA), people's health would not be affected by the French nuclear tests in the area. However, Chirac announced that, as a precaution, France would monitor the atolls where the tests took place.

¹² In October 2004, Temaru, chair of the cabinet, received a no-confidence vote in Parliament and was replaced by Flosse. Temaru, who called for a general strike in protest for his removal and for a new vote, accused France of mischief-making against him; Paris denied the accusations. Demonstrators took to the streets of Papeete to protest Temaru's removal.

¹³ In February 2005, the Assembly passed a no confidence vote against the Government. The election of Temaru as president put an end to two decades of rule by Flosse. In July, Anne Boquet was appointed High Commissioner in place of Michel Mathieu, who would take that office in Kanaky.

¹⁴ The Government Council of the French Polynesia refused to accompany the French nuclear energy representative, Marcel Jurien de la Graviere, on his visit to the Mururoa atoll, in May 2006. Their spokespeople said a four-hour visit to one of the places where the French nuclear tests had been carried out was of interest to no-one. ∎

PROFILE

ENVIRONMENT
The territory is located in the southeast portion of Polynesia. Most of the French Polynesian islands are of volcanic origin but they also have major coral formations. The islands are largely mountainous, with a tropical climate and heavy rainfall. Relatively fertile soils favor agricultural development. The Windward and Leeward Islands form the Society Islands archipelago. Nuclear tests on atolls such as Mururoa and Fangataufa carried out by France over a 26-year period have caused damage to the environment and people that is difficult to evaluate.

SOCIETY
Peoples: Polynesian 78 per cent, Chinese 12 per cent, French descendants 6 per cent. **Religions:** Mostly Christian, 54 per cent Protestant, 30 per cent Catholic. **Languages:** French (official), Tahitian (national). **Main Political Parties:** Tahoeraa Huiraatira (Popular Rally); Union for the Democracy (is made up of several members: Tavini Huiraatira; Aia Api; Tapura Amui No Te Faatereraa Manahune - Tuhaa Pae; Tapura Amui No Raromatai); Alliance for a New Democracy (includes Fetia Api; No Oe E Te Nunaa). **Main Social Organizations:** The Pacific Christian Workers' Central (CTCP), the Federation of French Polynesian Unions (FSPF) and the Territorial Union of the General Labor Confederation 'Workers' Force' (UTSCGT).

THE STATE
Official Name: Territoire d'Outre-Mer de la Polynésie française. **Administrative Divisions:** Windward Islands, which include Tahiti, Moorea, Maio; Papeete constitutes the center of the district. Leeward Islands, with the capital in Utoroa on Ralatea island. It also includes the islands of Huahine, Tahaa, Bora-Bora and Maupiti. Tuamotu and Gambier Archipelagos, the Austral Islands and the Marquesas Islands. **Capital:** Papeete 126,000 people (2003). **Other Cities:** Punaauia 22,700 people; Pirae 16,200 (2000). **Government:** The French Government is represented by a High Commissioner, who controls defense, foreign relations and justice, Anne Boquet since October September 2005. President: Oscar Temaru, President since March 2005. A local 57-member Territorial Assembly is elected for a five-year term by proportional representation. **National Holiday:** All French holidays. **Armed Forces:** Defense is the responsibility of the French Government. French Forces (includes Army, Navy, Air Force), Gendarmerie.

Gabon / Gabon

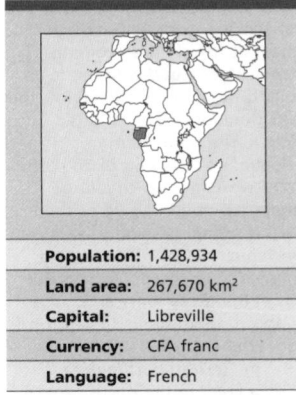

Population:	1,428,934
Land area:	267,670 km²
Capital:	Libreville
Currency:	CFA franc
Language:	French

Tools found in the Gabon forests indicate that this region has been inhabited since the Palaeolithic age. In the 16th century, the same migration that triggered the crisis in the ancient state of Congo brought the Myene and, in the 17th century, the Fang to Gabon. Thereafter they monopolized the slave and ivory trade together with the Europeans.

2   The Portuguese arrived in 1472. Around the middle of the 19th century, the French, Dutch and British established a permanent trade of ivory, precious woods and slaves. In 1849, Libreville was founded and established as a settlement for freed slaves from other French colonies. The territory was of little economic interest to the French, who used it as a base for expeditions into the heart of the continent.

3   The quest for independence was relatively uneventful because the two local parties (the Joint Mixed Gabonese Movement of Leon M'Ba and the Democratic and Social Union of Jean-Hillaire Aubame) were willing to accept neo-colonialism. In 1960, independence was declared and a military treaty was signed between Libreville and Paris.

4   Gabon has abundant resources: iron ore, uranium, manganese, timber and oil. Until

a few years after independence, timber was the only local industry of any size. The discovery of large oil reserves prompted the transnational companies Shell, Elf, Amoco and Braspetro to exploit them and to initiate a process of industrialization when they realized that Gabon could serve as a spearhead to penetrate the markets of Central Africa.

5   This 'development', relying on foreign capital has only exacerbated social conflicts and

the promise of jobs in the cities has encouraged urban migration. Gabon's social structure is changing as independent producers become suppliers of cheap labor for transnational industries.

6   When M'Ba died in 1967, he was succeeded by Omar Bongo, who faithfully followed his predecessor's style. Applying the US thesis of 'sub-imperialisms' to French interests, Bongo became its watchdog in central Africa, with Gabon a base for aggression

against neighboring progressive regimes. In January 1977, Gabon provided the planes and arms used by a mercenary group in an unsuccessful attack on Benin.

7   Bongo's foreign policy maintained good relations with several states in the region, including Angola, without altering the country's privileged relationship with France. Like Senegal, Ivory Coast, Chad, and the Central African Republic, Gabon also had French troops on its soil.

8   In 1979 and in 1986, Bongo was re-elected with 99 per cent of the vote, in presidential elections in which he was the only candidate.

9   His squandering of the country's income led to violent protests in the early 1980s. The revolt spread to the police who in 1982 organized a demonstration, demanding an increase in wages and the withdrawal of French advisors. The protests were brutally repressed.

10   The Government repressed the National Reorientation Movement (MORENA), formed by intellectuals, workers, students and nationalist politicians. The movement was accused of having expropriated 30 tons of weapons, in 1982, when the Bongo family and French military installations were the target of armed attacks. At least 28 MORENA leaders were sentenced to 15 years' imprisonment.

11   With the advent of political changes in Eastern Europe, in late 1989 there were signs of

LAND USE

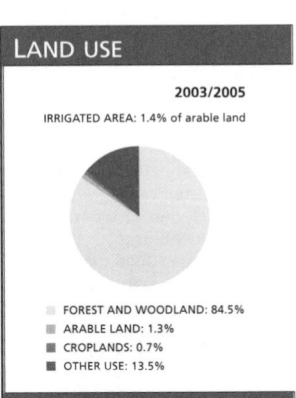

2003/2005

IRRIGATED AREA: 1.4% of arable land

- FOREST AND WOODLAND: 84.5%
- ARABLE LAND: 1.3%
- CROPLANDS: 0.7%
- OTHER USE: 13.5%

PROFILE

ENVIRONMENT

Irrigated by the Ogooué River basin, the country has an equatorial climate with year-round rainfall. The land is covered by dense rainforest. The timber industry - especially that of okoumé wood (used in plywood) - employs a large part of the population, together with the extractive industries (manganese, petroleum, uranium and iron), which make up 80 per cent of the country's exports.

SOCIETY

Peoples: Gabon was populated by Babinga (pygmies). In the 16th century it was invaded by Myene and other groups. Today over half of the population are Bantu-speaking peoples divided into more than 40 different groups including Galoa, Nkomi, and Irungu, among others. One third of the population are Fang and Kwele, in the northern part of Gabon, while there are Punu and Nzabi minorities in the south; plus Baka and Babongo pygmy people.
Religions: Mainly Christian. More than one third practise traditional African religions and there is a small Muslim minority.
Languages: French (official). There are as many languages as ethnic groups. Bantu languages are predominant, divided into 10 main linguistic groups; pygmy languages are also spoken.

Main Political Parties: Gabonese Democratic Party (PDG), pro-government; Union of the Gabonese People (UPG); Gabonese Socialist Party (PSG); Rally of Democrats (RDD); National Woodcutters' Rally.
Main Social Organizations: Gabonese Confederation of Free Trade Unions (CGSL).

THE STATE

Official Name: République Gabonaise.
Administrative Divisions: 9 provinces and 37 prefectures.
Capital: Libreville 611,000 people (2003).
Other Cities: Port Gentil 99,300 people; Masuku 39,100; Oyem 28,100 (2000).
Government: Parliamentary republic with strong head of state. El Hadj Omar Bongo, President and head of State, since November 1967 and re-elected in 2005 for a seven-year term. Jean Eyeghe Ndong, Prime Minister since January 2006. Bicameral Legislature: National Assembly, with 120 members elected every five years, and the Senate, with 91 members.
National Holiday: 17 August, Independence Day (1960).
Armed Forces: 4,750.
Other: Coast Guard: 2,800; Gendarmerie: 2,000.

Life expectancy
53 years
2005-2010

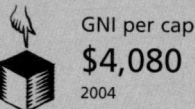
GNI per capita
$4,080
2004

HIV prevalence rate
8.1% of population
15-49 years old
2003

IN FOCUS

ENVIRONMENTAL CHALLENGES
Deforestation is one of the most serious environmental problems, together with the depletion of the country's wildlife.

WOMEN'S RIGHTS
Women have been able to vote and be elected since 1956. By 2003 women held 9.2 per cent of seats in the Lower Chamber, 15.4 per cent in the Upper Chamber and 12 per cent of ministerial positions.

In 2003 women made up 44 per cent of the one-million-strong labor force.

In spite of 86 per cent of births being attended by skilled health staff and 94 per cent of pregnant women receiving prenatal care, in 2004 the maternal mortality rate was 420 per 100,000 births*.

CHILDREN
In 2004 child mortality for under-ones and under-fives was 60 and 91 per 1,000 live births respectively.

That year, 12 per cent of children under five were underweight and 21 per cent measured moderately or severely below standard height.*

Immunization coverage was still very poor: 55 per cent of babies were not vaccinated against tetanus at birth; only 55 per cent of children under five had been vaccinated against measles, 31 per cent against polio, 38 per cent with the DPT triple vaccination and 11 per cent had still not received a tuberculosis vaccination. In 2004 almost half of the children under five were treated for acute respiratory infections.

INDIGENOUS PEOPLES/ ETHNIC MINORITIES
Baka people are one of the Pygmy groups of Africa (the name derives from the Greek word *pyme* which means 'a cubit in height'). They were traditionally hunters and gatherers and were probably the first inhabitants of Equatorial Africa. Today, the different groups inhabit Rwanda, Burundi, Uganda, DR Congo, Gabon, Central African Republic and Cameroon, totaling 200,000 people.

There are also Bantu-speaking groups in the region, that have deeper historical roots in the territory: the Babongo or Obongo (in southern and central Gabon), the Barimba and Bagama (southwest). The Babongo group (formerly nomadic and hunter-

gatherer communities) is made up of some 2,000 people settled in Gabon and Congo.

MIGRANTS/REFUGEES
In 2005 almost 14,000 refugees (the great majority of them from DR Congo) were living in Gabon. In spite of the endeavors of organizations such as UNHCR, only approximately 300 refugees were waiting to be repatriated at the end of that year.

A majority of refugees has received some assistance, including food, special rations for malnourished children, tools and seeds for farming, and education benefits.

The mining sector attracted many migrants from the Africa's central region, who arrived in Gabon during the 1990s. In 2000, 20.3 per cent of the population were immigrants.

In Europe, the greater part of the Gabonese diaspora has settled in France (3,000), Germany (238) and Italy (194).

DEATH PENALTY
It is applied to all types of crimes.

*Latest data available in *The State of the World's Children* and *Childinfo* database, UNICEF, 2006.

West African countries, child trafficking exists in Gabon, with most of the children coming from Nigeria and working in plantations, domestic service, the sex trade or on the streets.
[22] In the parliamentary elections at the end of 2001 the PDG won again with a large majority, obtaining 84 of the 120 National Assembly seats. In January 2002 Abessole and his party accepted an invitation from Bongo to leave the opposition and join a coalition government.
[23] In 2002 an outbreak of Ebola fever in the north caused approximately 50 deaths. The Mekambo region and the border with the Republic of Congo were the most affected areas.
[24] In January 2003, after the PDG election victory, Bongo pledged that his new administration would be an open one. In May, Health Minister Faustin Boukoubi announced the outbreak of Ebola fever was under control.
[25] In July the Constitution was amended to remove limits on the number of times that the president can stand for re-election. Opposition leader Pierre Mamboundou demanded intervention by the international community, claiming that the amendment was designed to keep Bongo in power. Since Bongo took over the presidency in 1967, the Constitution had been amended 16 times.
[26] In February 2004 the French oil company Total Gabon signed an agreement with the Chinese company Unipec to export oil to China. Co-operation between Libreville and Beijing had begun in 1974, during a visit by Bongo to China. From September 2004, the country's oil industry was greatly invigorated by the discovery of new deposits and production increased.
[27] In the November 2005 presidential elections Bongo won again by a wide margin, with nearly 80 per cent of the vote and at 69 became the longest-serving African head of State. The main opposition candidates, Mamboundou and former minister Zacharie Myboto, claimed that the election was fraudulent.
[28] In January 2006 the new Prime Minister, Jean Eyeghé Ndong, included 12 women in his 49-member cabinet.
[29] In February 2006 the governments of Gabon and Equatorial Guinea agreed to launch negotiations on the sovereignty of several small islands - potentially rich in oil reserves - in the Gulf of Guinea. ∎

democratic participation for Gabon's opposition forces. After violent confrontations in the streets, the President had the constitution amended, introducing a multiparty system and lifting censorship of the press. In the meantime, he invited opposition leaders to join the cabinet.
[12] In May 1990 Joseph Redjambe, president of the Gabonese Progressive Party, was murdered. His death provoked strong reactions against the Government. The resulting incidents left six dead and a hundred wounded. A state of rebellion through the Port Gentil region lasted ten days and the French Government evacuated 5,000 French residents, charging the presidential guard with re-establishing order.
[13] In June, all political and social groups met in a National Conference and an agreement was reached whereby free presidential elections would be held. Although this was a major victory for the opposition, who extracted promises of multiparty elections from the President, Bongo advanced the election date from 1992 to late 1990, contrary to what the opposition wanted. Bongo used the power base he

had built up over the past 20 years in office to carry out his campaign, while the opposition - repressed for decades - had no time to organize itself adequately and faced the election with internal divisions. In September 1990, President Bongo's Gabonese Democratic Party (PDG) obtained a majority in the National Assembly. The new constitution, approved in March 1991, formally established a multiparty system.
[14] In 1991, the country was rocked by a new outbreak of political and social violence. At the same time, the economic crisis continued to worsen. Stabilization and adjustment plans put forward by the IMF and World Bank failed to generate much hope among social and political groups.
[15] The 1993 presidential elections, which returned Bongo to power, were questioned by the opposition. In February 1994, the protests continued and repression left 30 dead. A coalition government ruled from September to November 1994, when free elections were held.
[16] In July 1995, the President gained the support of 96 per cent of the voters in a referendum for constitutional reform in order

for presidential and legislative elections to be held. The National Assembly elections in December 1996 gave the Democratic Party (PDG), led by Bongo, 47 of the 55 seats. Meanwhile, the opposition leader Paul Mba-Abessole became Mayor of the capital, Libreville.
[17] Despite the fact Gabon had per capita income of $3,490 per year due to the money earned from oil, in 1997 life expectancy was below 55 and the infant mortality rate stood at 87 per 1,000 live births.
[18] Early in 1998, Mayor Mba-Abessole requested the UN to supervise presidential elections that were to be held later that year in order to avoid the 'fraudulent re-election of Omar Bongo'. The elections were once again won by Bongo, who defeated Mbú Abessole and Pierre-André Kombila.
[19] The authorities sought the UN's humanitarian assistance due to the constant influx of refugees from the Republic of Congo. As of October 1999, the country was hosting around 10,000 refugees.
[20] In July 2000 two freighters, one Greek and one French, collided in Gabonese waters, causing a 400-ton oil spill.
[21] As in other Central and

Gambia / Gambia

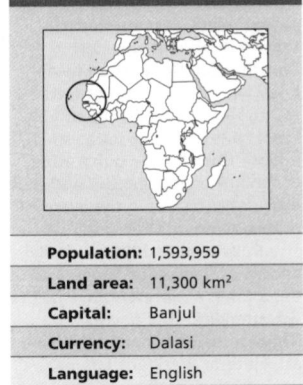

Population:	1,593,959
Land area:	11,300 km²
Capital:	Banjul
Currency:	Dalasi
Language:	English

The earliest settlers of the Gambia river valley came from what is now Senegal. Attracted by Gambia's coast, which lent itself to trade and navigation, they settled along the river, carrying out subsistence farming.

² In the 15th century, the region was colonized by the Mandingo who, together with the Mali Empire, founded several kingdoms in the Gambia valley, which controlled coastal trade and enabled them to develop economically and culturally.

³ With the arrival of Portuguese sailors in 1455, most internal trade was transferred to the Atlantic coast. For the Portuguese, Gambia became the point of departure for their precious metals and a prosperous enclave on their route to the Orient.

⁴ However, in 1618 the Portuguese Crown sold its commercial and territorial rights to the British Empire, which at the height of its naval prowess was trying to assert its dominance as a colonial power by acquiring a foothold in Africa.

⁵ At that time, a war began between Britain and France (controlling all of what is now Senegal) which was to last for over 200 years. As of 1644, the British used this coastal area as a source of slaves: British merchants set up alliances with princes inland to obtain slaves for Britain's colonies and for its slave trade with other colonial powers.

⁶ The British therefore limited themselves to establishing a rudimentary trading post in the area, founded in 1660. Border disputes between the British and French increased during the 18th century.

⁷ Throughout the 19th century a series of religious wars resulted in the complete Islamization of the country, with an increase in Muslim immigration from other parts of Africa.

⁸ The enclave lost all economic significance with the abolition of the slave trade. It did however gain strategic importance by being a British enclave inserted into the heart of Senegal, a region instrumental in France's designs on sub-Saharan Africa. Slavery actually continued within the colony until the 20th century, not being abolished until 1906.

⁹ In 1889, France and Britain reached an agreement as to the boundaries of their respective colonies, ensuring peace in the region and the formal recognition of British sovereignty over Gambia by other European powers.

¹⁰ Gambia's status as a British colony remained unchanged throughout the first half of the 20th century. In 1963, it received partial administrative autonomy under the decolonization process begun after World War II.

¹¹ In 1965, Gambia obtained full independence and joined the British Commonwealth. At the time of its independence, some felt that Gambia did not constitute a nation as such because of its ethnic, cultural and economic complexity.

¹² Following independence, the territory's social and economic structures remained unchanged. Peanut exports continued to be vital to the economy. Traditional social structures remained so powerful that they were finally legitimized by the 1970 Constitution, which guaranteed legislature seats to five regional leaders.

¹³ Dawda Jawara, a veterinary doctor and the founder of the People's Progressive Party (PPP), dominated Gambian politics as of the 1960s. He won the 1962 elections, but failed to assume office as the opposition forced a vote of no confidence. However, he again won in 1970 when the country was proclaimed a republic and embraced the presidential system.

¹⁴ The success of Alex Haley's 1976 book Roots placed Gambia in the limelight. In the late 1970s it became a popular destination for tourists. Prostitution and drug-trafficking increased.

¹⁵ The lack of border controls in Gambia led to it becoming a paradise for West African smuggling and a large part of Senegal's agricultural produce was illegally shipped through the port of Banjul. This close economic relationship between the countries led the government of Dawda Jawara to accept a project for union with Senegal in 1973.

¹⁶ In July 1981, Muslim dissidents attempted to overthrow Jawara, aiming to end official corruption through the establishment of a revolutionary Islamic regime. The revolt was crushed by Senegalese troops who entered Gambia at the request of President Dawda, who was in London at the time.

¹⁷ The proposed union with Senegal had been planned for 1982, but the coup attempt against Gambia's government accelerated plans for creating the Senegambian confederacy.

¹⁸ Senegambia officially existed from February 1982 to the end of 1989 with Abdou Diouf of Senegal as its first president, assisted by a confederate council of ministers and a bi-national parliament. The treaty ensured that Dawda Jawara gained protection against internal rebellions and Senegal gained greater control over the leak of export tax revenues through smuggling. Both countries retained their individuality and internal organization.

¹⁹ In the mid-1980s, Jawara became reluctant to consolidate his ties with Senegal and failed to comply with military aid agreements. In late 1989 Gambia signed a mutual defense pact with Nigeria which in practical terms meant the dissolution of Senegambia.

²⁰ Cordial relations were

 **Life expectancy**
58 years
2005-2010

 **GNI per capita**
$280
2004

HIV prevalence rate
1.2% of population
15-49 years old
2003

reinstated with a treaty of friendship and co-operation in 1991; however, the confederacy was not restored.

[21] In 1993, agriculture and tourism were hit by the consequences of the European economic crisis. Gambia's trade with Senegal was damaged when the Central Bank of the Western Africa States decided to stop financing trade based upon the African franc (CFA) outside the area comprised by countries using this monetary system. That same year, the Government took measures to initiate a national reconciliation process, including an amnesty granted to rebel movements fighting to oust the regime.

[22] In July 1994, a military coup headed by Yahya Jammeh overthrew President Dawda Jawara, who sought asylum in Senegal after taking refuge on a US warship that was visiting the country. The presence of this ship in Banjul suggested complicity between the US and the military.

[23] Two members of the Provisional Armed Forces Council were arrested in January 1995 after trying to hand the Government over to civilians. In March, Jammeh also arrested the former Justice minister and general prosecutor for promoting the return of civilian rule. In November, the military junta expanded the powers of the security forces.

[24] In August 1996, following a referendum, a new Constitution was approved and Jammeh, until then chief of the Armed Forces Government Junta, became Gambia's second elected President. Arrests of Islamic leaders were frequent in 1998.

[25] In May 1999 Ousainou Darboe, of the opposition UDP, accused the Government of arresting the party's followers and maintaining a democracy with 'illegitimate' laws.

[26] In September, Jammeh criticized the UN's lack of responsibility and inaction regarding the conflicts faced by Africa. A month later, the Press Union denounced governmental measures against freedom of the press (raids on their buildings and the power of the minister of information to confiscate their records). The security forces prevented an attempted coup in January 2000 and arrested the two officers allegedly responsible for the uprising.

[27] Gambia, with another 44 countries - mostly African - lost its right to vote in the UN General Assembly on 2 February 2000, for 'falling into arrears with its debt repayments'.

[28] The Community of Sahelian-Saharan States (COMESSA), accepted Gambia, Senegal and Djibouti as new members. The 11 members agreed not to interfere in members' domestic affairs and not to aid hostile forces of any of the countries.

[29] The President decided to lift the ban on political parties for the coming elections. In October 2001, Jammeh won the presidential elections, receiving the approval of international observers. However, the opposition argued strongly that the elections had been fraudulent.

[30] In May 2002, Parliament passed a law on mass media that went against the 1997 Constitution. Under this law a body was created with powers to register all journalists, compel them to reveal their sources, impose fines for the publication of 'unauthorized' articles and shut down newspapers in the event of orders being disobeyed.

[31] In December 2003, a scandal known as 'Babagate' broke out. Baba Jobe, a representative of the Government APRC party in the National Assembly and long-time associate of President Jammeh, was arrested on charges of money-laundering and fraud.

[32] In January 2004, five senior officials of the Central Bank were accused of multiple financial crimes against the State, related to the embezzlement of nearly nine million Swiss francs intended for stabilizing the Gambian currency, which ended up financing four private companies through illegal contracts. Baba Jobe was the main shareholder of two of these companies.

[33] In February, Jammeh announced the discovery of 'large scale' oil deposits, quoting a comprehensive study carried out in Gambian coastal waters and on land. He added that exploration would begin in relevant areas, based on the findings of this study. Gambia has no oil industry or experience in this field.

[34] In December, journalist Deida Hydara, co-owner of *The Point* journal and correspondent for a foreign agency, was shot dead. In the preceding months there had been several attacks on journalists and media infrastructure that were not thoroughly investigated. The organization Reporters Without Frontiers suggested that the attacks seemed to be part of a Government campaign against the independent press.

[35] During Jammeh's visit to Brasilia in February 2005, Gambia and Brazil signed several co-operation agreements including one for the Brazilian national oil company Petrobras to help Gambia with the exploitation of the oil discovered in 2004.

[36] In March 2006 several military and civilian personnel were arrested under suspicion of taking part in an attempted coup headed by Army Chief of Staff, Lieutenant Colonel Mbure Cham, who according to the Government escaped to Senegal. Amongst those arrested were ex-Intelligence Chief Abdulaye Kujaby, and ex-Treasury Director Alieu Jobe.

[37] In September 2006, President Jammeh was re-elected for a third term with 67.3 per cent of the vote.∎

IN FOCUS

ENVIRONMENTAL CHALLENGES
Large areas of forest have disappeared as the land was taken over for growing export crops. Wood is also used as a major source of fuel. As a result, desertification has increased. The low-lying capital, Banjul - one meter above sea-level - risks being submerged in the next few decades by rising sea-levels, a consequence of global warming.

WOMEN'S RIGHTS
Women have been able to vote and be elected since 1960. In 2004 women held 13 per cent of parliamentary seats and 20 per cent of ministerial positions.

Female life expectancy increased from 42 to 55 years between 1980 and 2003.

That year, women comprised 45 per cent of the labor force; the total workforce was 1 million.

Maternal mortality was 540 per 100,000 live births. Only 55 per cent of births were attended by qualified personnel, while 91 per cent of pregnant women received prenatal care.*

An estimated 60 to 90 per cent of women have been circumcised (female genital mutilation). By the end of 2003 there were 3,600 Gambian women aged 15 to 49 living with HIV/AIDS.

CHILDREN
In 2004, 47 per cent of the population was under 18. In spite of significant decreases during the period 1990-2004, mortality rates for children one and under five years old continued to be high: 89 and 122 per 1,000 live births respectively in 2004. Some 53 per cent of children received elementary schooling. At birth, 17 per cent of babies were underweight and the same percentage of children under five were moderately or severely underweight with 19 per cent measuring moderately or severely less than normal height.*

In 2004 approximately 500 children under 14 were living with HIV/AIDS and there were around 2,000 orphans as a result of this disease.

INDIGENOUS PEOPLE/ ETHNIC MINORITIES
The Mandingo groups (42 per cent) along with other Mande-speaking groups, originally were part of the 13th-century Malian Empire. The Wolof (16 per cent) mainly live south of the Gambia river, where they work in agriculture or trade. The Jola people (10 per cent) are one of the oldest groups in the region; during the 18th century, they had to pay tribute to the Mandingo. The Serahuli (9 per cent), who live in what was the Wuli Kingdom, are a mix of Berbers, Mandingos and Fulanis. The Fulani (18 per cent) are pastoralists; they dominated during the Tekrur Kingdom until the 11th century. The Aku peoples (3.5 per cent) are mostly descendants of Yorubas who were freed there before they could be shipped as slaves to the Americas. They settled in the country during the 1820s and 1830s. By the late 19th century and during the 20th century, some had positions in government; they adopted Western customs and excelled as traders.

MIGRANTS/REFUGEES
At the end of 2004 the number of foreign refugees living in Gambia was uncertain. UNHCR estimated the figure at around 12,000 while other organizations spoke of up to 30,000 spread over the whole country. Most were from Sierra Leone and Senegal but other countries of origin were Liberia, Somalia, Ethiopia, Rwanda, Iraq and Eritrea.

DEATH PENALTY
It is still applicable even for ordinary crimes, but the last recorded execution took place in 1981.

*Latest data available in *The State of the World's Children* and *Childinfo* database, UNICEF, 2006.

Georgia / Sak'art'velo

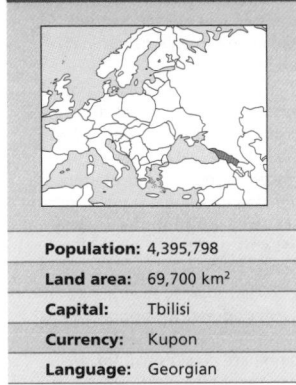

Population:	4,395,798
Land area:	69,700 km²
Capital:	Tbilisi
Currency:	Kupon
Language:	Georgian

The cultivation of grain in what is now Georgia dates back to the last period of the Stone Age, some 12,000 years ago. The Caucasian peoples are thought to be the inventors of metallurgy, with their use of bronze, in the 3rd millennium BC.

2   Within the 1st millennium BC, Georgia was inhabited by groups of Assyrians (from Babylon), Armenians and Cimmerians (of Thracian origin, Turkey). The fusion of the latter with indigenous peoples would have formed the group called Kolkhida.

3   The kingdom of Kolkhida in the 6th century BC, situated in the eastern end of the Black Sea, was conquered by Greece at the end of the 5th century BC, and then by Rome in the year 66 AD. Also invaded was the kingdom of Iberia, created in what is now southeastern Georgia during the 2nd century BC. As a result of Pompey's (106 BC - 48 BC) campaigns, Roman hegemony was established over the whole territory of present-day Georgia.

4   Just like the other dependencies of the Roman Empire, Georgia was converted to Christianity by Constantine in 337. During the three following centuries, the country was drawn into-the conflict between Constantinople and Persia. The ancient kingdom of Kolkhida fell under the control of Constantinople

and the previous territory of Iberia came under Iranian control.

5   Apart from the local authority exercised by the faithful magnates of each province, after 654, Arab caliphs established an emirate at Tbilisi (the capital of Georgia previously known as Tiflis). At the end of the 8th century, Bagratid Ashot I profited from the weakness of the Byzantine emperors and the Arab caliphs to set himself up as hereditary Prince in Iberia.

6   King Bagrat III (975-1014) united eastern and western Georgia into one state. Tbilisi was not recovered from the Muslims until 1122. Georgia reached an apex under Queen Tamara, who formed a pan-Caucasian empire between 1184 and 1213 marked by the flourishing of architecture and religious arts.

7   The Mongol invasions of 1220 brought Georgia's golden age to an end. The massacres of Turk conqueror Tamerlane (1336-1405) displaced Mongols and destroyed Georgia's economy and culture. The introduction of Christianity brought about the emergence of the alphabet and literature, enhanced by contributions from Arabic culture.

8   The last king of united Georgia was Alexander I, between 1412 and 1443, under whose sons the realm was divided and subsequently disintegrated.

9   The fall of Constantinople to the Ottoman Turks, in 1453, left it isolated from the rest of the Christian world. Within the three following centuries, Georgia became the object of territorial disputes between Turkey and Iran.

10   During that period of time, several revolts took place and hundreds of Christians were deported to Iran. During a respite registered between 1638 and 1723, Iran introduced printing to Georgia and appointed a commission of scholars to edit the Georgian annals.

11   After centuries of attempts, the-Russians managed to enter

Georgia in 1789, and occupied the whole territory after defeating both Turks and Persians in 1878.

12   The local language was eliminated from administrative documents and the use of Russian language-imposed. The Georgian Church started to be ruled by Russian Orthodox bishops.

13   Following the liberation of the Russian serfs (1861), the Georgian peasants also received freedom in 1864. The affluence of

entrepreneurs from Western Europe (after the Napoleonic invasions) and the construction of connecting railroads, fostered the emergence of nationalist and freedom movements. In 1893, the Social Democratic Party was secretly founded and five years later, the Georgian Joseph Stalin became one of its members.

14   After frequent rebellions, the Georgian population confronted the Cossacks during the 1905 Russian Revolution.

15   When World War I broke out (1914-1918), the Caucasus saw fighting between Russia and Turkey. Georgians formed a legion to fight alongside the Turks and in February 1917 they brought down the czarist regime.

16   In November, after the Bolshevik triumph in Petrograd, power in Transcaucasia (Georgia and adjacent territories) fell into the hands of the Mensheviks. Stalin's group abandoned the Caucasus to join the Bolsheviks led by Lenin. In May 1918, in Tbilisi, the United Government of Transcaucasia announced its separation from Soviet Russia - requesting German protection.

17   Between 1918 and 1920,

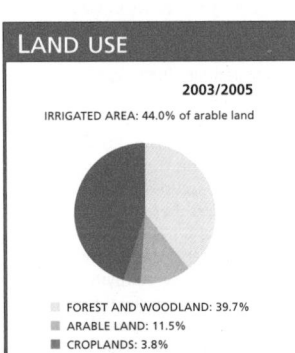
PROFILE

ENVIRONMENT
Located in the central western part of Transcaucasia, Georgia is bordered in the north by Russia, in the east by Azerbaijan and in the south by Armenia and Turkey, with the Black Sea to the west. Most of its territory is occupied by mountains. Between the Little and the Great Caucasus lies the Kolkhida lowland and the Kartalinian Plain; the Alazan Valley lies to the east. Subtropical climate in the west, moderate in the east; there is heavy rainfall in the western area, along the shores of the Black Sea. Principal rivers are the Kura and Rioni. fourty per cent of the republic is covered by forests. Georgia has important carbon and manganese deposits and is famous for its wine.

SOCIETY
Peoples: Georgians, 70 per cent; Armenians, 8.1 per cent; Russians, 6.3 per cent; Azeris, 5.7 per cent; Ossetians, Abkhazians and Adzharians.
Religions: Georgian Orthodox (65 per cent), Muslims (11 per cent), Russian Orthodox (10 per cent), and Armenian Orthodox (8 per cent).
Languages: Georgian (official), Russian, Abkhazian, Armenian, Azeri, Greek, Kurmanji, Turkish and various Caucasian languages.
Main Political Parties: United National Movement (*Ertiani Natsionaluri Modzraoba*); New Rights Party (*Axali Memarjveneebi*); Industry will save Georgia (*Mretsveloba Gadaarchens Sak'art'velos*); Democratic Union for Revival (*Demokratiuli Aghordzinebis Pavshiri*); The Georgian Labour Party (*Sakartvelos Leoboristuli Partia*).
Main Social Organizations: the traditional trade unions have 2.6 million members. Currently, the Georgian Trade Union Association is the leading organization.

THE STATE
Official Name: Sak'art'velos Respublika.
Capital: Tbilisi 1,064,000 people (2003).
Other Cities: Batumi 143,800 people; Sukhumi 63,800; Kutaisi 264,600; Rustavi 178,900 (2000).
Government: Mikhail Saakashvili, President since 2004. Zurab Noghaideli, Prime Minister since February 2005. Single-chamber parliament with 235 members.
National Holiday: 26 May, Independence (1991).
Armed Forces: 13,000.

Life expectancy

71 years

2005-2010

GNI per capita

$1,060

2004

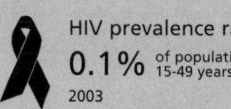

HIV prevalence rate

0.1% of population 15-49 years old

2003

IN FOCUS

ENVIRONMENTAL CHALLENGES

Bacterial pollution of 70 per cent of the Black Sea constitutes a serious problem. Only 18 per cent of the residual waters in the main port of Batumi undergo adequate treatment. Air pollution is marked, particularly in Rust'avi. Overuse of pesticides has resulted in soil deterioration.

WOMEN'S RIGHTS

Although women's right to vote was declared in 1918, it was not until 1921 that Georgian women were able to vote and stand for office freely. In 2004, they held 9.4 per cent of seats in Parliament and 22 per cent of ministerial positions.

In 2003, women made up 47 per cent of the labor force. In 2004, 96 per cent of births were attended by skilled health staff and 95 per cent of pregnant women received prenatal care.

CHILDREN

In 2004, the under-one and under-five mortality rates stood at 41 and 45 per 1,000 live births, respectively. This rates have remained almost unchanged since 1990. The primary school enrolment rate in 2004 was 89 per cent.

In 2004, 99 per cent of children under five years old had been treated for acute respiratory infections.

INDIGENOUS PEOPLES/ ETHNIC MINORITIES

There are some 450,000 Armenians, 310,000 Azeris, 120,000 Pontics, 100,000 Abkhazis, 100,000 Greeks, 52,000 Ukrainians, 33,000 Kurds, 3,650 Lezgis, 3,000 Tartars and 3,000 Turks. Georgians make up the natural majority of the population (70 per cent).

MIGRANTS/REFUGEES

Within the 2004-2005 period, nearly 16,000 Georgians were registered as refugees in industrialized countries; 98 per cent in European countries (mostly in Austria and France).

In 2006, the UNHCR registered more than 200,000 internal displaced people in the provinces of Abkhazia and South Ossetia, while the situation of Chechen refugees in Pankisi was still being closely monitored.

DEATH PENALTY

It was abolished in 1997.

Latest data available in The State of the World's Children and Childinfo database, UNICEF, 2006.

German, Turkish and British troops entered Georgia. In early 1921, the Red Army occupied Georgia and the Soviet Socialist Republic of Georgia was proclaimed. In March 1922, Georgia, Azerbaijan and Armenia were reorganized as the Transcaucasian Federation. In 1936, the federation was dissolved and Georgia became one of the 15 republics of the Soviet Union (USSR).

[18] During Stalin's despotic rule (1928-1953), Georgia suffered from repression of all expressions of nationalism, the forced collectivization of agriculture and also witnessed many purges.

[19] In the Soviet period, Georgia was industrialized. The Kremlin promoted the creation of Georgian cultural and administrative elites destined to take part in the central government. After Stalin's death, a freewheeling 'second economy' was developed, which supplied goods and services.

[20] In 1953, Moscow made Eduard Shevardnadze, a Communist Youth leader until then, chief of police. When Shevardnadze was appointed First Secretary of the Communist Party in 1972, a new nationalist feeling began to emerge with declarations in defense of the Georgian language and violent acts of sabotage against the Soviet administration.

[21] In 1978, the new constitution of the USSR triggered a wave of protest, as it made Russian the official language. However, Shevardnadze managed to have the measure abolished and allowed an anti-Stalin movie, 'Repent', to be screened. This marked the development of *glasnost* (openness) and *perestroika* (restructuring) process in Georgia.

[22] In 1985, Shevardnadze was made Foreign Minister of the Soviet Union. After the political changes made by Mikhail Gorbachev in the late 1980s, several independent movements and parties emerged in Georgia - frequently repressed by the Red Army - among them, a Green Party that denounced industrial pollution in the Black Sea and deforestation problems.

[23] In the 1990 elections for the Soviet (parliament), the 'Free Georgia Round Table' coalition won, led by Zviad Gamsakhurdia, a well-known opponent to the Soviet regime. After the disintegration of the USSR, Georgia declared independence in April 1991 and Gamsakhurdia was elected President. But Gamsakhurdia's authoritarian behavior soon drove many of his supporters into opposition.

[24] Georgia's economy, dependent on the USSR for its energy supply, was the most affected of all former Soviet Republics after the rupture.

[25] A civil war broke out and in January 1992 the President was deposed by a Military Council, which was subsequently replaced in March by a State Council, headed by Shevardnadze. Tbilisi's authorities fought against Gamsakhurdia's supporters and Ossetian separatists. In June, a cease-fire was called in South Ossetia, supervised by a peacekeeping force of Russians, Georgians and Ossetians.

[26] The following month, the Abkhazian authorities (in the northeast) resolved to limit the jurisdiction of Georgia's central government and in August government troops occupied Sukhumi (Abkhazia's capital). In July 2003, a Russian-mediated armistice came into effect. Three months later, a strong Abkhazian offensive took the Georgians by surprise and occupied Sukhumi.

[27] In November, supporters of former president Gamsakhurdia launched a broad offensive, but were defeated with the help of Russian troops. In early 1994, Gamsakhurdia committed suicide according to official reports. In February, Georgia signed a friendship treaty with Russia, and in April a peace treaty with the Abkhazian rebels.

[28] In August 1995, Parliament approved a new Constitution making-Georgia-a presidential republic. Shevardnadze won the first presidential elections in November and his party, the Union of Citizens of Georgia won an absolute majority in Parliament. In 1997, the country became the first former Soviet republic to abolish the death penalty.

[29] Between 1998 and 1999, Shevardnadze narrowly escaped two assassination attempts. In both cases there were rumors of Moscow's involvement. Russian President Vladimir Putin accused Georgia's Government of harboring Chechen rebels on their territory.

[30] The Abkhaz parliament declared independence in October 1999 but Tbilisi did not recognize the new state. In March 2001, Georgia and Abkhazia signed an accord renouncing the use of force between them, but in October, new confrontations broke out.

[31] In October 2001, following the closing of the private Rustavi-2 TV station, which was critical of the failure to fight corruption, massive popular protests demanded the resignation of the Shevardnadze government. His administration was blamed for corruption, civil war and impoverishment. Shevardnadze answered by sacking the entire administration. In November 2001,

Nino Burdzhanadze, then Foreign Minister, was elected head of Parliament in an extraordinary session.

[32] The arrival in Georgia of US military advisors, in February 2002, prompted Putin's threat to Tbilisi. Washington's stated objective was to support the local army in fighting an Islamic terrorist group allegedly linked to the al-Qaeda network and based on the Chechen border.

[33] That month, several opposition parties agreed to support Mikhail Saakashvili as presidential candidate. Saakashvili (of the National Movement party) was a US-educated lawyer, who led the 'Rose Revolution', a popular revolt that ousted Shevardnadze. Even though he was among the Ministers expelled in 2001, Saakashvili gained popularity by exposing government corruption and promising to fight poverty. Meanwhile, Nino Burdzhanadze was appointed as interim President.

[34] In January 2004, Saakashvili won the presidential election with an overwhelming majority (more than 96 per cent of the votes). The new President announced that 'special measures' would be taken in order to 'eradicate organized crime and corruption'.

[35] In March, Tbilisi imposed a partial economic blockade on the semi-autonomous region of Ajaria (southeast coast of the Black Sea) after his leader, Aslan Abashidze, failed to recognise the authority of Saakashvili.

[36] That month, Saakashvili won the majority of seats in Parliament.

[37] South Ossetia held parliamentary elections in May, which were not recognized by Tbilisi. In August 2004, tensions between the region and the central government erupted and several military clashes were reported.

[38] In February, following the death of Prime Minister Zurab Zhvania, Economy Minister Zurab Nogaideli was appointed as his successor.

[39] Tbilisi and Moscow reached an agreement in May 2005 to close four Soviet-era bases in Georgia. The closures were to be completed by 2008.

[40] In January 2006, explosions damaged a gas pipeline and an electricity transmission line, cutting energy supplies from Russia to Georgia during freezing weather. Saakashvili accused Moscow of being behind these explosions to put pressure on Tbilisi.

[41] In May, and in a context of increasingly impaired relationships with Russia, Saakashvili threatened to leave the Commonwealth of Independent States, made up of the former Soviet Republics, and to seek closer ties with the European Union. ∎

Germany / Deutschland

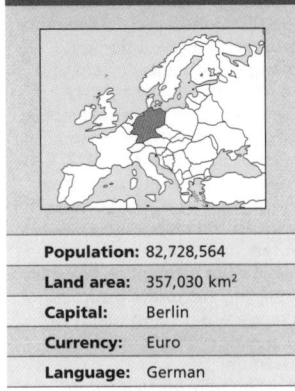

Population:	82,728,564
Land area:	357,030 km²
Capital:	Berlin
Currency:	Euro
Language:	German

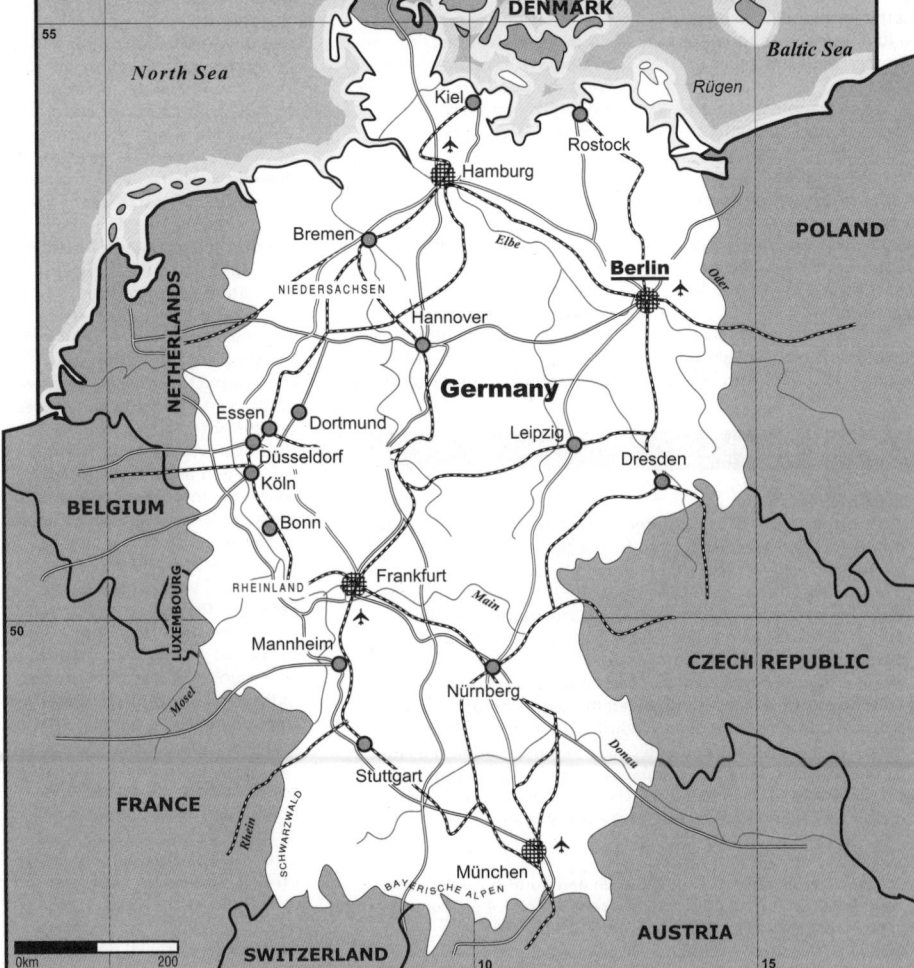

Germanic peoples have lived in the area of what is now Germany since around 50 BC. The Gauls lived west of the Rhine. Greek and Roman writers such as Julius Caesar, who invaded part of Gaul around 50 BC, repeatedly mentioned the Germans. Between 260 and 276 AD, two German peoples, the Alemanni and the Franks, attacked Gaul.

[2] When the Roman Empire fell in 476, the German tribes west of the Rhine were politically divided. Among the largest tribes were the Merovingians, Carolingians and Ottomans. Around 800 Charlemagne built a great empire that occupied the present Germany and France. After his death in 814 came a period of great social unrest. However, Otto I (936- 973) consolidated the empire once again.

[3] During the 12th and 13th centuries Germany experienced a rapid growth in population, and continuous expansion.

[4] The instability of royal dynasties strengthened secular and religious principalities. Princes were free to build fortresses, exploit natural resources and administer justice in their dominions.

[5] In 1356, the authority of the king in relation to the papacy was consolidated in the Golden Bull of Charles IV (1346-1378) establishing the right to appoint the King without the Pope's approval, and strengthened the hand of the principalities, on whose support the king depended.

[6] During the 15th and 16th centuries internal instability persisted. In 1517 Martin Luther led the Protestant Reformation, which, coupled with political rivalries, frustrated reunification efforts.

[7] The Reformation provided a forum for criticisms of secularization and corruption in the German Church, which had become an increasingly prosperous economic and financial institution. It owned a third of the land in some districts and profited from selling indulgences - one of the contributory factors to its downfall. After a succession of internal wars, including peasant uprisings, the Peace of Augsburg was reached in 1555, giving equal rights to both Catholics and Lutherans.

[8] Political and religious factions combined their strengths; in 1608 the Protestant Union was created, and the Catholic League a year later. The Bohemian rebellion triggered the Thirty Year War (1618-1648), which spread to the entire continent, reduced the population of central Europe by around 30 per cent, and ended in 1648 with the Peace of Westphalia.

[9] In the 18th century, the Kingdom of Prussia emerged as an economic and political unit, responsible for creating tensions among the German states. Napoleon's victories over Prussia in 1806 and the formation of the Confederation of the Rhine put an end to the Holy Roman Empire.

[10] In Central Europe during the 18th century, culture formed an outlet for intellectual energies, which could not be expressed through politics given that autocratic rulers dominated political life. Philosophers like

LAND USE

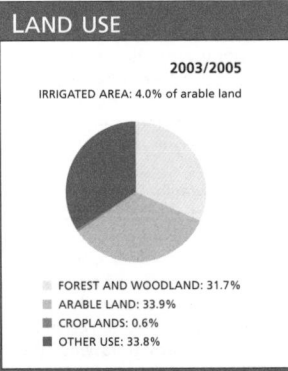

2003/2005

IRRIGATED AREA: 4.0% of arable land

- FOREST AND WOODLAND: 31.7%
- ARABLE LAND: 33.9%
- CROPLANDS: 0.6%
- OTHER USE: 33.8%

PUBLIC EXPENDITURE

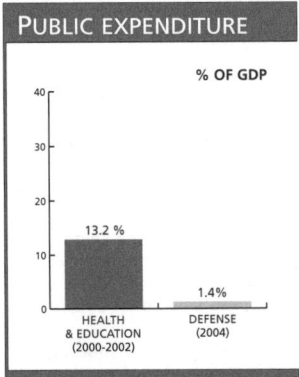

% OF GDP

13.2 % HEALTH & EDUCATION (2000-2002)

1.4% DEFENSE (2004)

WORKERS

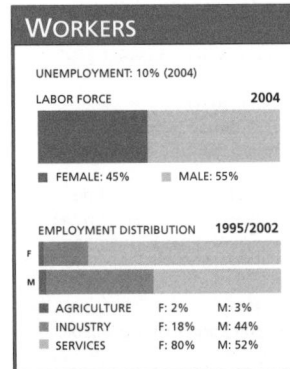

UNEMPLOYMENT: 10% (2004)

LABOR FORCE **2004**

- FEMALE: 45%
- MALE: 55%

EMPLOYMENT DISTRIBUTION **1995/2002**

F
M

- AGRICULTURE   F: 2%   M: 3%
- INDUSTRY   F: 18%   M: 44%
- SERVICES   F: 80%   M: 52%

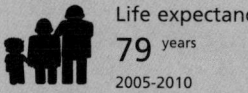
Life expectancy
79 years
2005-2010

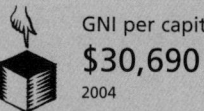
GNI per capita
$30,690
2004

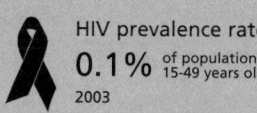
HIV prevalence rate
0.1 % of population 15-49 years old
2003

Kant and Herder, and writers like Goethe and Schiller, expressed the idealism and spiritualism that characterized German art and literature of this time.

[11] When Napoleon fell, the German princes created a confederation of 39 states, which were independent except for foreign policy. Austrian and Prussian opposition to broader forms of representation increased popular unrest, leading to the 1830 revolts and increased repression.

[12] In 1834, Prussia threw its growing economic weight into the political realm, with the establishment of the German Customs Union, from which Austria was excluded. The main consequences of the Union were the duplication of trade among its members over the next ten years, as well as the formation of industrial centres and the emergence of a working class. Due to the rapid growth of the urban population, the supply of labor greatly exceeded the demand. The resulting impoverishment of industrial workers and artisans served as a breeding ground for the rebellions of subsequent years, culminating in the revolutionary wave of 1848-49.

[13] A National Assembly first met in Frankfurt on 18 May 1848; its representatives belonged mostly to liberal democratic sectors. They campaigned for German unity, and guarantees of political freedom. However, internal divisions facilitated the regrouping of the forces of the former regime, leading finally to the dissolution of Parliament in June 1849 and the repression of opposition organizations.

[14] With the revolutionary tendencies crushed, Austria and Prussia were free to dispute their respective roles in German unification. The issue was settled in 1866 with Prussia's victory in the Seven Weeks' War. The union was forged around the North German Confederation, a creation of the Prussian chancellor, Otto von Bismarck, which aimed to halt liberalism. The Parliament (*Reichstag*) was inaugurated in February 1867.

[15] Three years later, war broke out with France. Prussia's victory in 1871 was the last step in Bismarck's scheme to unite Germany under a single monarch, and Prussian domination.

[16] The Empire had to deal with opposing internal forces - the Church and social democracy. Bismarck passed the May Laws, secularizing education and some other activities. He later reversed this, securing the Church as an ally against socialism. Alarmed at the growth of social democracy, the regime used repression and social reform to neutralize the latter's potential.

[17] Bismarck's government introduced commercial protectionism to increase domestic income and foster national industry and the German economy grew substantially, especially in heavy industry, chemicals, the electro-technical area and production. The creation of the Triple Alliance with Austria and Italy, and the acquisition of colonies in Africa and Asia after 1884 made the German Empire a leading world power.

[18] German rivalry with France and Britain in the west, and Russia and Serbia in the east, triggered World War I. The capitulation of the Austro-Hungarian Empire and Turkey in November 1918 led to Germany's final defeat. The crisis was aggravated by an internal revolution that led to the abdication of the Emperor. Government was handed over to the socialist Friedrich Ebert, who was to call a National Constituent Assembly.

[19] German social democracy split into a moderate tendency favoring a gradual evolution to socialism, and a radical tendency promoting a revolutionary change. The radical group, the Spartacists, headed by Karl Liebknecht and Rosa Luxemburg, identified with the Russian revolution of October 1917 and wanted to set up a system similar to that of the Soviets. The Spartacist leaders were executed in January 1919 following a failed coup attempt. A few days later, the electorate returned a moderate socialist majority to power in the Constituent Assembly.

[20] Formally proclaimed in August of that year, the Weimar Constitution was welcomed as the most democratic of its time. The president elect had power to nominate the Chancellor, whose government needed approval of the lower chamber of parliament, or the Reichstag. It also provided for the constitution of an upper house or Länder, formed by delegates designate by the governments of the liberal states.

[21] The Weimar Republic had a short and hazardous life. Despite the virtues attributed to the Constitution, various factors came together to undermine it. One of the main destabilizing elements for the Republic was the conditions imposed on the country in Versailles by the victorious powers. These affected not only the economy but also the morale of the population, which refused to consider Germany 'guilty' of having caused the war in 1914 and was unable to accept the stipulations of the treaty that judged any German, from the Kaiser down, as a war criminal.

[22] Even though the Republican Government managed to weather the serious economic crises, as in 1920-23 and events following the Wall Street Crash in 1929, the destabilizing influence of the Communists and the National Socialist Party, led by Adolf Hitler, were set to topple it. The National Socialist or Nazi Party had attempted a coup in 1923. Although this failed, the Party saw a sustained increase in its popularity throughout the decade, with rocketing support after the crisis in 1929. With barely 170,000 members in 1929, the number grew to 1,378,000 by 1932.

[23] Field-Marshal Paul von Hindenburg, elected President in 1925, dissolved parliament in 1930. In the elections later that year, Communists and Nazis both saw a great increase in votes. The Nazis' promises to rebuild Greater Germany following the humiliation of the post-war

PROFILE

ENVIRONMENT

The northern part of the country is a vast plain. The Baltic coast is jagged, with deep, narrow gulfs. The center of the country is made up of very old mountain ranges, plateaus and sedimentary river basins. Of the ancient massifs, the most important are the Black Forest region and the Rhineland. The southern region begins in the Danube Valley, and is made up of plateaus (the Bavarian Plateau), bordered to the south by the Bavarian Alps. There are large deposits of coal and lignite along the banks of the Ruhr and Ens rivers, which provided the backbone of Germany's industrial development. Heavy industry is concentrated in the Ruhr Valley, mid-Rhineland and Lower Saxony. The south of the former German Democratic Republic is rich in coal, lignite, lead, tin, silver and uranium deposits. The chemical, electrochemical, metallurgical and steel industries are concentrated there.

SOCIETY

Peoples: German 91.1 per cent; Turkish and Kurdish 2.3 per cent; Yugoslav 0.7 per cent; Italian 0.7 per cent; Greek 0.4 per cent; Polish 0.4 per cent; Spanish 0.2 per cent; other 2.0 per cent.
Religions: Christian; Protestants (32 per cent) in the North and East; Catholics (33 per cent) were a majority in West Germany before reunification. Jewish and Muslim minorities (6 per cent).
Languages: German (official) and local dialects which, in spite of restrictions, are regaining popularity. Turkish, Kurdish.
Main Political Parties: Social Democratic Party of Germany (SPD); Christian-Democratic Union (CDU); Christian Social Union in Bavaria (CSU); Alliance-90/The Greens; Free Democratic Party (FDP).
Main Social Organizations: Workers' Federation (DGB); Social Watch; Heinrich Böll Foundation, linked to the Green Party; Association of German Development NGOs (Venro); Human Rights Forum.

THE STATE

Official Name: Bundesrepublik Deutschland.
Administrative Divisions: Federal parliamentary state made up of 15 *Länder* (federated states), as of 3 October 1990; 11 Länder made up what was formerly West Germany (Schleswig-Holstein, Hamburg, Bremen, Niedersachsen, Nordrhein-Westfalen, Hessen, Rheinland-Pfalz, Saarland, Baden-Württemberg, Bavaria and Berlin) while the former German Democratic Republic was divided into five Länder (Mecklenburg, Brandenburg, Sachsen-Anhalt, Sachsen and Thueringen).
Capital: Berlin (since 1990) 3,327,000 people (2003).
Other Cities: Hamburg 3,258,500 people; München (Munich) 2,342,500; Dresden 1,031,100; Köln (Cologne) 966,500; Frankfurt 645,500 (2002).
Government: Horst Köhler, President since July 2004; Angela Merkel, Chancellor since 2005. Parliament: The Bundestag (Federal Diet) has 598 members and the Bundesrat (Federal Council) has 69 members. Both have been in Berlin since 1999, while the Länder Chamber is still in Bonn.
National Holiday: 3 October, Unity Day (1990).
Armed Forces: 271,923 (2002). Other: Federal Border Guard 24,800; Coast Guard 535.

	Under-5 mortality		Maternal mortality
	5 per 1,000 live births 2004		8 per 100,000 live births 2000

treaties, and their campaign to blame the Jews and Communists for the economic crisis, gained ground when the Nazi Party doubled its share of the vote in the 1932 elections (37 per cent of the total).

[24] The rise of Hitler and the National Socialist Party proved unstoppable. Despite the fact that Hitler's demands to become Chancellor had been refused, in January 1933 - as unemployment topped six million - Hindenburg was forced to hand him the reins of office. Hitler dissolved Parliament and called elections which the Nazis won.

[25] In Potsdam in March 1933, the new parliament granted Hitler the power to issue decrees outside the Constitution without the approval of the legislative body or the President for four years. He had the power to set the budget, request loans and sign agreements with other countries, reorganize both his cabinet and the supreme ranks of the armed forces, and proclaim martial law. In July, Hitler abolished the German Federation and instituted absolute central power. He outlawed trade unions, strikes and all parties except his own. Germany also withdrew from the Disarmament Conference and the League of Nations. The Nazis dubbed their government the Third *Reich* (Third Empire).

[26] After the death of President Hindenburg in August 1934, the cabinet was forced to swear personal allegiance to the Chancellor, Adolf Hitler. In 1935, in open violation of the Treaty of Versailles, he began to re-arm Germany. The European powers protested, but were unable to stop him. With the 1935 'Nuremberg Laws', the regime provided a legal framework for its racist ideology, laying the foundation for its subsequent policies of ethnic and religious persecution.

[27] Germany and Italy signed a co-operation agreement in October 1936 which included support for General Franco in the Spanish Civil War. The following month Germany and Japan (the Axis powers) agreed to set up a military exchange, and in November 1937 Germany, Italy and Japan signed the Anti-communist Pact in Rome.

[28] In March 1938, German troops invaded Austria, and Hitler annexed it. That same year, the pressure of Hitler and German nationalism on the Sudetenland forced the European powers to cede this Czechoslovakian region to Germany. On 'Kristallnacht', 9-10 November, the Government

ENVIRONMENTAL CHALLENGES
There are hazardous waste dumps, though mechanisms have been established to stop using nuclear energy by 2015. In the Eastern region the air is polluted by chemical, electrochemical and metallurgical industries, as well as by coal-based energy sources. Sulfur dioxide emissions contribute to acid rain. Industrial waste mixed with heavy metals has affected eastern rivers, some of them tributaries of the Baltic Sea. The aim is to identify nature preservation areas in line with the EU's Flora, Fauna, and Habitat directive.

WOMEN'S RIGHTS
Women have been able to vote and be elected since 1918. From 1995 to 2003, the number of parliamentary seats held by women rose from 26 to 31.8 per cent; they hold 36 per cent of managerial positions.

In 2003 they made up 43 per cent of the workforce (of these, 2 per cent worked in agriculture, 18 per cent in industry and 80 per cent in services). Almost 92 per cent were waged workers, 6 per cent were independent and 2 per cent were underemployed. Female unemployment that year amounted to 8.3 per cent, 0.5 percentage points lower than among males.

Between 150,000 and 500,000 women enter the country illegally each year, mostly from Central or Eastern Europe.

CHILDREN
The estimated number of children under 18 and adults under 50 living with HIV/AIDS rose from 8,400 in 2001 to 43,000 at the end of 2003.

Education and health indicators are good for children born in German families, but there are no clear figures for immigrants or refugees.

Germany was a pioneer of pre-school education in the Western world. It was begun by Fröbel in 1840 and focused on the pedagogical care and supervision of 3- to 6-year-old children.

INDIGENOUS PEOPLES/ ETHNIC MINORITIES
The Turks are an historical ethnic minority in Germany. The inflow of Turkish immigrants has been more or less constant since the 1950s. Now their birth rate is higher than the German population's; that could change their present marginalization in the not-so-distant future. Immigrant Turks are politically marginalized and still face barriers that prevent them from becoming citizens.

All immigrant workers face legal problems. The State may restrict their freedom of association, movement or choice of occupation. Workers from other EU countries receive residence permits, automatically renewed. Non-EU workers, including Turks, become residents only after 8 years of constant work and residence. For those in the second and third generations, who no longer speak Turkish, integration to Turkish society would be as difficult as for any other German.

MIGRANTS/REFUGEES
By the end of 2004, Germany had the third largest number of exiles (876,622) in the world, after Iran's (1,046,000) and Pakistan's (960,000). At the beginning of 2004 the number stood at 960,400, so there was an 8.7 per cent decrease throughout the year. The country was also ranked third in terms of asylum requests (35,600), following France and the UK. The main countries of origin of exiles were: Afghanistan (46,975), Bosnia and Herzegovina (38,688), Iraq (73,489), Iran (47,131), Lebanon (18,410), Poland (13,796), Russia (45,568), Serbia and Montenegro (168,980), Sri Lanka (15,121), Syria (15,745), Turkey (140,702), Ukraine (57,309), Vietnam (25,357).

DEATH PENALTY
Abolished in West Germany in 1949, and in East Germany in 1987. The date of the last execution before abolition remains unknown.

carried out a systematic destruction of Jewish commercial property and religious and cultural institutions.

[29] In 1939, taking advantage of the disagreements between Czechs and Slovaks, German troops advanced into Prague; Bohemia, Moravia, and Slovakia became protectorates.

[30] Britain assured Poland, Romania, Greece and Turkey that it would protect their independence, and Britain and France attempted to establish an alliance with the Soviet Union. In August 1939, however, the USSR signed a non-aggression treaty with Germany. On 1 September, German troops invaded Poland. Britain and France issued Germany an ultimatum which was disregarded. World War II broke out.

[31] By 1940, Germany had invaded Norway, Denmark, Belgium, the Netherlands, Luxembourg and France. In 1941 Hitler started his offensive against the USSR, but German troops were halted a few miles outside Moscow. They were finally defeated after the siege of Stalingrad (Volgograd) in 1943.

[32] From the outset, German aggression against its neighbours was accompanied by the systematic extermination of the Jewish population in concentration camps, primarily in Poland. Over six million Jews, and a million other people were killed.

[33] The advance of the Red Army - culminating in the capture of Berlin - and the 1944 Allied landing in Normandy forced Germany to surrender in May 1945.

[34] Four million Germans from neighboring countries, and from the territories annexed by Poland and the USSR, were forced to move to the four zones that Germany was now divided into. The country remained occupied by the United States, France, Britain and the USSR. In 1949, discord between the former Allies over the future governance of Germany led to the creation of the Federal Republic of Germany (FRG), in the West, and the German Democratic Republic (GDR), in the East. The issue of the two Germanies became a bone of contention in post-war relations between the USSR and the US.

[35] In 1955, the sovereignty of both Germanies was recognized by their occupying forces. During the Cold War, West Germany became a member of NATO while East Germany joined the Warsaw Pact. Foreign forces continued to be based in their territories and the two republics were still subject to limitations on their armed forces and a ban on nuclear weapons.

[36] The United Socialist Party (SED), formed from the union of Communists and Social Democrats in 1946, took over the government of East Germany. The USSR partly compensated them for war losses with money, equipment and cattle, and a social system similar to that of the USSR was set up. In 1953, the political and economic situation of East Germany led to a series

Water source

100% of population using improved drinking water sources

2002

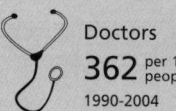

Doctors

362 per 100,000 people

1990-2004

of protests, which were put down by Soviet troops. In the meantime, emigration to the Federal Republic of Germany (FRG) increased.

[37] Between the state's formation and 1961, when the East German Government forbade all emigration to the West, some three million East Germans emigrated to West Germany. To enforce their resolution, the East closed its borders and built the Berlin Wall between the eastern and the western sections of the city. In 1971, Erich Honecker took over the leadership of the SED party, and later the GDR Government.

[38] Between 1949 and 1963, Chancellor Konrad Adenauer, a conservative Christian Democrat, oversaw the reconstruction of West Germany/FRG, under the slogan '(establishing) a social market economy'. With US support (the Marshall Plan) and huge amounts of foreign capital, the FRG became one of the most developed capitalist economies, playing a key role in the founding of the European Community (EC).

[39] With the victory of the Social Democratic Party (SPD) in the 1969 elections, the government of Chancellor Willy Brandt launched a policy of rapprochement toward Eastern Europe and the German Democratic Republic. In 1970 the first formal talks between the FRG and the GDR began, and in 1971 the occupying powers agreed to free access of FRG citizens to the GDR. A Basic Treaty of bilateral relations was signed by both Germanies in 1973; in September they were admitted to the United Nations.

[40] In the 1970s, the number of power stations producing nuclear energy increased. In response to this, a strong environmental movement was formed, with a network of hundreds of grassroots groups throughout the country.

[41] In 1974, after the discovery that his private secretary was an East German spy, Brandt resigned as Chancellor and was succeeded by Helmut Schmidt. The modernization of Soviet medium-range missiles in East Germany, and the December 1979 NATO decision to do the same with its arsenal in West Germany, gave a strong impetus to the anti-nuclear movement in both states, with massive protests in the FRG.

[42] In 1982, the Social Democratic Party (SPD) had to leave power when the Liberal Party withdrew from the government alliance after 13 years. It was succeeded by the liberal-conservative alliance (CDU/CSU, FDP) led by Chancellor Helmut Kohl of the Christian Democratic Union (CDU).

[43] In mid-1989, Hungary liberalized regulations regarding transit across the border with Austria. Within a few weeks, some 350,000 Germans from the GDR had emigrated to the FRG. Street demonstrations demanding change brought on a crisis in the GDR. In August, Honecker resigned and was replaced by Egon Krenz. On 9 November the GDR opened up its borders and the Berlin Wall fell. Kohl immediately proposed the creation of a confederation of East and West Germany.

[44] In February 1990, the East German Government agreed to German unification and the withdrawal of all foreign troops from its territory. The fusion of the two states was officially recognized in August 1990, as the Federal Republic of Germany. Political union was made possible when the USSR accepted the East German entry into NATO. The only important difference between the two Germanies over the next two years was the East's preservation of a more liberal abortion law.

[45] In the first parliamentary elections of the new Federal Republic, in December 1990, the governing coalition of Christian Democrats and Liberals obtained 54 per cent of the vote and stayed in power. From 1991, the extreme right made important gains in places such as Bremen, where they increased their share of the vote by seven per cent. During 1992, there were 2,280 attacks on foreigners and Jewish monuments, which left 17 people dead. After one incident which caused the death of a Turkish woman and two children, the Government outlawed three neo-Nazi organizations.

[46] Throughout the following year, the closure of most of the industry in the east of the country and economic recession - the worst since 1945 - caused a constant increase in unemployment. In May 1994, the conservative, Roman Herzog, backed by Kohl, was designated president of Germany by a special electoral assembly, after defeating the Social Democrat Johannes Rau. Kohl triumphed again in the October general elections, although his majority fell to only 10 seats of the 672 in play.

[47] In 1995, the constant weakening of the Liberal Party (FDP) in various local elections prompted the resignation of Foreign Minister Klaus Kinkel. On the social front, Parliament adopted a new law which, despite again permitting abortion during the first 12 weeks of pregnancy, was more restrictive than the one annulled in 1993 for being unconstitutional.

[48] The discontent of many foreigners resident in Germany led to the formation of the Democratic Party of Germany, which stood for greater access to the electoral roll and citizenship - often restricted to people of German origin - for the children of immigrants.

[49] In 1996, five years after unification and three years after the frontiers fell for workers of the European Union, unemployment stood at around 10.6 per cent on a national level, and up to 16 per cent in the states of the former East Germany. Five million people were out of work.

[50] Racism and antisemitism continued to be issues for the Kohl administration. In October 1997, a television broadcast showed an army battalion giving Nazi salutes and shouting anti-Semitic and anti-US slogans. Amnesty International indicated that police abuse of foreigners could not be seen as isolated cases, due to the systematic repetition of the offences.

[51] In early 1998, the Deutsche Bank paid back Jewish organizations the money made from the sale of gold suspected to have been robbed from the Jews by the Nazis. At the same time, a Swiss foundation began to provide economic compensation - although with 'token' sums of money - to Roma (gypsy) residents in Germany who had survived the Nazi holocaust.

[52] The Social Democrats won the 1998 elections and Gerhard Schröder was named Federal Chancellor. The SPD leader Oskar Lafontaine headed the Finance ministry. The former communists of former East Germany won 36 seats in the Federal Parliament.

[53] A series of defeats in the 1999 local elections and Lafontaine's resignation put the SPD in a difficult position. It seemed that public support had returned to the Christian Democrats. But in November former Chancellor Kohl was accused of authorizing the sale of armored vehicles to countries at war without informing Parliament, and soon after, of accepting illegal contributions from private party donors. Kohl, who refused to hand over the list of donors, had to stand down from honorary presidency of the CDU and was found guilty of illegal handling of funds.

[54] In March 2002 the lower chamber initiated the ratification process for the Kyoto Protocol, unanimously passing it into law. The country committed itself to reducing its carbon dioxide emissions by 20 per cent by the year 2012.

[55] During the electoral campaign of that year, Chancellor Schröder made public his opposition to any pre-emptive US attack on Iraq. This boosted his popularity with voters, but worsened his relationship with Washington.

[56] In 2003, Germany maintained its opposition to war. However, it allowed the US Navy to cross its territorial waters to supply airplanes bombing Iraq.

[57] The Government made cuts in public spending and public health budget in 2003; among the major changes were the reduction of unemployment and welfare benefits. Industrial workers were required, according to this policy, to increase their working week from 35 to 40 hours and delay retirement by five years (until they were 65).

[58] In May 2004, Horst Koehler - former IMF head and representative of the conservative and liberal opposition - was elected German President with 604 of the 1,204 votes of the assembly.

[59] Tens of thousands of demonstrators took to the streets in August to protest against the labor policy reforms, which were cutting many social benefits.

[60] In March 2005, in spite of the long legal battle by former Chancellor Schröder to prevent it, former East German secret police files including compromising information were made public.

[61] Parliament ratified the EU constitution in May. In July, the president called early elections, after Chancellor Schröder deliberately lost a confidence vote in Parliament in order to trigger the poll.

[62] The September elections gave a very tight victory to the CDU. In November, Angela Merkel took office.

[63] In May 2006, Chancellor Merkel - who had harshly criticized the US foreign policy of recent years - visited Washington seeking to improve the tense relationship between Germany and the US.

[64] In August 2006, Chancellor Merkel and Interior Minister Wolfgang Schaeuble laid out plans for new anti-terror measures following a failed bomb attack on two trains in July, including increased video surveillance and a national database of terror suspects. ∎

Ghana / Ghana

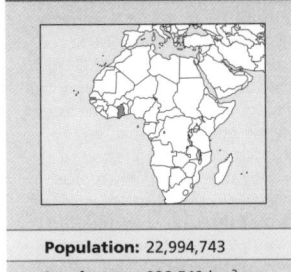

Population:	22,994,743
Land area:	238,540 km²
Capital:	Accra
Currency:	Cedi
Language:	English

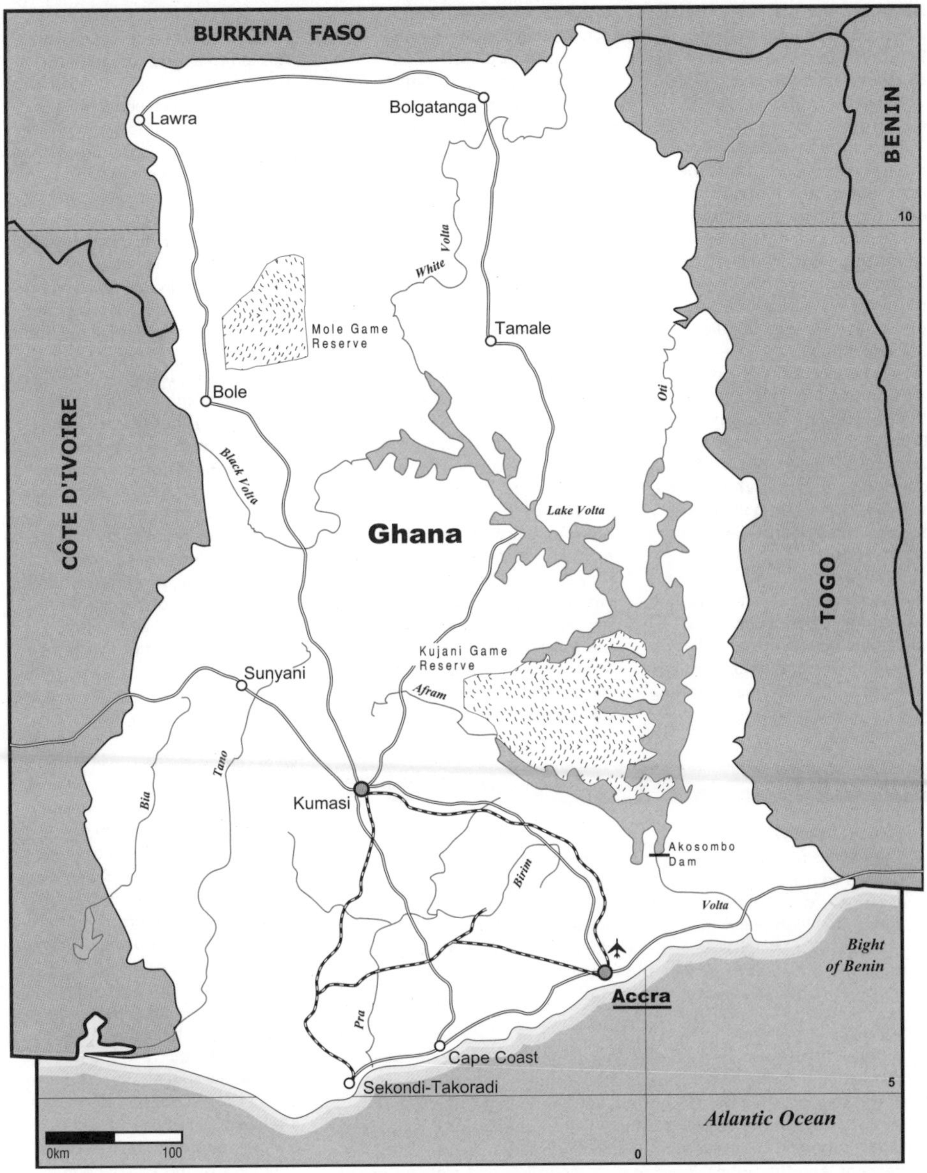

In about 1300, the Akans or Ashantis moved into Ghana from the north. The Fanti State of Denkyira was already established on the coast so the Akans settled in the inland jungles where they founded a series of small states.

2   Around the 15th century the Ashantis began to trade in the markets of Sudan, on the border of present-day Côte d'Ivoire. They traded slaves and gold for fabrics and goods from other places (an ornate pitcher that had once belonged to Richard II, King of England between 1367 and 1400, was discovered in the treasure of a Kumasi ruler).

3   In the 17th century, the migration of peoples threatened the existence of the small states. The Akans united to confront and defeat the Doma invaders.

4   With the decline of the Songhai, the Moroccan incursions and the beginning of the Fulah expansion, the North African trade network collapsed with serious repercussions for the inland economy. The Ashanti were deprived of access to the coastal trading posts as these were in Denkyira Territory. So, they declared war on the Denkyira, defeated them and then organized a centralized state, ruled by the 'Ashantihene' (leader of the Ashanti nation) endowed with a powerful army. By 1700, the Ashanti had control of the slave

LAND USE

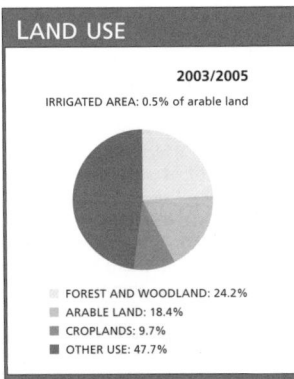

2003/2005

IRRIGATED AREA: 0.5% of arable land

- FOREST AND WOODLAND: 24.2%
- ARABLE LAND: 18.4%
- CROPLANDS: 9.7%
- OTHER USE: 47.7%

trade and the flow of European goods to the interior.

5   The British stopped trading in slaves; meanwhile, the Ashanti were attempting to seize the shoreline from the Fanti, who had retained a significant share of the coastal trade. English backing of the Fanti led to the Anglo-Ashanti wars, 1806-1816, 1825-1828 and 1874. The British turned the Fanti territory into a colony, and in 1895 they proclaimed a protectorate over the northern territories.

6   While the coastal and northern regions were under British rule, the central region belonged to the autonomous Ashanti nation. In 1896, another Anglo-Ashanti war broke out. The capital, Kumasi, was razed by cannon fire, and the ruler was deposed and exiled. In 1900, an attempt to collect the debt owed by the Ashanti (50,000

ounces of gold in compensation for 'war damages') coupled with the British Governor's desire to sit on the gold throne, led to a general rebellion in which thousands of people were killed. In 1902 the Ashanti state was formally annexed to the Gold Coast Colony.

7   Early in the 20th century, a strong nationalist current developed in spite of the ethnic and religious differences within Ghana. There was economic opposition between the north, where the traditional structures had remained intact, and the south, where a Westernized middle class and a relatively significant working class had developed.

8   Popular pressure on the colonial administration led to political concessions. In 1946, London admitted a few Africans into the colonial administration

and in 1949 Kwame Nkrumah formed the Convention People's Party (CPP) to campaign for greater reform.

9   Nkrumah was one of the founding fathers of Pan-Africanism and African nationalism. He established a solid rural and urban party structure and, in1952, he became Prime Minister of the colony. He proclaimed himself a 'socialist, Marxist and Christian', promising to fight against imperialism.

10   Nkrumah represented Ghana at the Bandung Conference in 1955 which marked the birth of the Movement of Non-Aligned Countries, together with Tito, Nasser, Nehru and Sukarno. In 1957 he achieved a major victory and Ghana became the first country in West Africa to gain independence. Known as *Osagyefo* (redeemer),

Life expectancy
58 years
2005-2010

GNI per capita
$380
2004

Literacy
54% total adult rate
2000-2004

HIV prevalence rate
3.1% of population 15-49 years old
2003

Nkrumah enthusiastically embraced African anti-colonialism. He initiated a series of internal changes based on industrialization, agrarian reform and socialist education.

[11] Traditional and neo-colonial interests conspired against the Government and Nkrumah was overthrown. The pro-British coup leaders drew up a parliamentary constitution and held elections for a civilian government, where Nkrumah was not allowed to participate. He died in exile in Bucharest in 1972.

[12] That year, Colonel Ignatius Acheampong led a coup replacing the Government of Dr Kofi Busia. Acheampong dropped the ambitious industrialization and development plans, substituting an essentially agrarian policy which favored the owners of large cocoa plantations.

[13] Acheampong's Government survived eight coup attempts, but his economic policy was not as successful. Ghana had an inflation rate of 36 percent, an oppressive foreign debt, a devalued currency and hundreds of imprisoned

intellectuals and students, condemned for questioning government policy.

[14] In 1977, the so-called revolt of the middle class occurred. A period of social unrest began which ended with Acheampong's resignation in 1978. A new military regime was established, led by General William Frederick Akuffo. The opposition claimed this was simply a continuation of the previous government under a different figurehead. In June 1979, a coup led by Lt Jerry Rawlings overthrew Akuffo and called fresh elections. The Convention People's Party, which included Nkrumah's followers, won a large majority.

[15] In October 1979, with the acquiescence of the Revolutionary Council of the Armed Forces, Hilla Limann, a PNP leader, took over the presidency. Limann abandoned many of Nkrumah's nationalistic economic policies, replacing them with International Monetary Fund policies, in an attempt to reduce the fiscal deficit. To attract foreign investors and overcome the sharp fall in cocoa export revenues, the Government reduced all imports

and, as a result, the purchasing power of workers fell drastically. This led to a series of strikes in 1980 and 1981.

[16] Inflation exceeded 140 per cent and the unemployment rate was more than 25 per cent. This created an unstable situation which culminated in a further coup, led by Rawlings on 31 December 1981. In spite of this, Rawlings still retained his popularity.

[17] A campaign against corruption in the public sector was launched, promising a social justice revolution in the country. Within a few months, more taxes had been paid, and cocoa-smuggling to neighboring countries was reduced. People's Courts were set up to pass judgement on irregularities committed by the authorities of the previous government.

[18] It became necessary to develop policy, and look for sources of foreign capital. Rawlings, who was willing to carry out an austerity program dictated by the IMF, strengthened relations with them in 1983. Ghana was granted generous loans by the IMF and the World Bank as a result.

[19] Taxes were increased, subsidies were removed, the Government salaries budget was reduced and financing for inefficient private enterprises was eliminated. These measures helped to reduce inflation from 200 to 25 per cent, the banking system improved and better prices were obtained for cocoa producers. However the costs soon became apparent.

[20] New loans for $500 million were approved for Ghana. The loans were to be disbursed in two installments, the first one due in 1994, at an annual interest rate of five per cent and the second with a ten-year grace period and 0.5 per cent interest rate. The major drawback of this scheme was that Ghana's finances were under rigid IMF control. The foreign debt approached $4 billion and the servicing of the debt used up two-thirds of the country's export earnings.

[21] The social cost of the structural adjustment program was high. Consumer prices rose by 30 per cent between 1983 and 1987; 45,000 public employees lost their jobs and there was a general loss of earning power leading to an increase in hunger, infant mortality and illiteracy.

[22] In the early 20th century there were eight urban centers in the country, and by 1984 this number had grown to 180. A strong migration to the largest cities resulted in the creation of shanty towns without access to safe drinking water or sanitation. In order to alleviate this situation, the Government launched a program

to transfer 12,000 people a year to rural areas. In 1987, a Program of Action to Mitigate the Social Costs of Adjustment (PAMSCAD) was implemented, to analyze the impact of the economic recession on the poorest people, and to prioritize the action required.

[23] By the 1980s, the tropical forest - that used to cover 34 per cent of the land - was reduced to seven per cent. Meanwhile, 42 per cent of the area which is officially considered 'forest', is in fact covered with timber plantations, secondary vegetation, or immature trees. For the people who live in the country, and for most of those who live in the cities, forest plants constitute the basis of their traditional medicine. Seventy-five per cent of the population rely on bush meat for their basic source of protein and forests also provide firewood, a basic household necessity.

[24] These activities are ignored by official plans, which see the forest merely as a source of timber for export. Around 70,000 people are employed in the timber industry.

[25] After the loss of millions of dollars because of the fall in export prices, Ghana was granted further loans of $900,000,000 in 1989. But, the hopes of growth collapsed, the current account deficit increased and earnings for exports were reduced. Between 1988 and 1989, around 120 industries closed down because their products could not compete with substitutes imported from China, South Korea and Taiwan

[26] The crisis led Rawlings to initiate a democratization program and a devolution plan, favoring local administrations. In 1991, an advisory assembly of 260 members was elected to draft a new constitution. The National Council for Women and Development, a body with ministerial status, won ten seats.

[27] In December 1991, Amnesty International denounced the Ghanaian policy of silencing or intimidating its opponents. The country's first human rights organization, the Committee on Human and People's Rights, was created in 1992.

[28] That year, a new constitution was approved and elections were held. Rawlings took 58.3 per cent of the vote in November and the three parties which supported him took 197 of the 200 seats in December. The opposition, which accused Rawlings of fraud and intimidation, boycotted the legislative elections which registered a voter turnout of only 29 per cent.

[29] In 1993, representatives from 80 countries and international organizations attended Rawlings'

PROFILE

ENVIRONMENT

The southern region is covered with dense rainforest, partially cleared to plant cocoa, coffee, banana and oil palm trees. Wide savannas extend to the north. The rest of the territory is low-lying, with a few high points near the border with Togo. The Volta, Ghana's main river, has an artificial lake formed by the Akosombo dam. The climate is tropical, with summer rains. The subsoil is rich in gold, diamonds, manganese and bauxite.

SOCIETY

Peoples: Ghanaians come from seven main ethnic groups: the Akan (Ashanti and Fanti), 44 per cent, located in the mid southern part of the country; the Ewe, 13 per cent, and Ga-Adangbe, 8 per cent, on both sides of the Volta in southern and southeastern Ghana; the Mossi-Dagomba (16 per cent) in the northern savannas; the Guan, 4 per cent, and the Gurma, 3 per cent, in the valleys and plateaus of the northeastern territory.
Religions: 50 per cent Christian, 32 per cent traditional religions, 13 per cent Muslim.
Languages: English (official); Ga is the main local language, Fanti, Hausa, Fantéewe, Gaadanhe, Akan, Dagbandim and Mamprussi are also spoken.
Main Political Parties: New Patriotic Party (NPP); National Democratic Congress (NDC); Grand Coalition (Peoples National Convention - PNC); Convention Peoples Party.
Main Social Organizations: Ghana Trade Union Congress, of nearly 50 unions. The 31st December Women's Movement, headed by Nna Konadu Agyeman Rawlings, wife of the former president.

THE STATE

Official Name: Republic of Ghana.
Administrative Divisions: 10 regions, subdivided into 110 districts.
Capital: Accra 1,847,000 people (2003).
Other Cities: Kumasi 906,400 people; Tamale 259,200; Sekondi-Takoradi 164,400 (2000).
Government: John Agyekum Kufuor, President, Head of State and Government since January 2001, re-elected in 2004; Alhaji Aliu Mahama, Vice-President. Unicameral Legislature, with 230 members elected for a four-year term.
National Holiday: 6 March, Independence Day (1957).
Armed Forces: 5,000. Other: People's Militia 5,000.

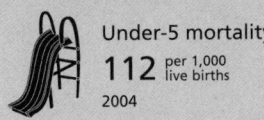
Under-5 mortality
112 per 1,000 live births
2004

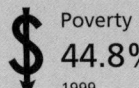
Poverty
44.8% of population living on less than $1 per day
1999

Debt service
6.6% exports of goods and services
2004

Maternal mortality
540 per 100,000 live births
2000

IN FOCUS

ENVIRONMENTAL CHALLENGES

Desertification, affecting the northwest, deforestation, soil deterioration caused by intensive grazing; hunting and habitat loss that threaten wildlife; water pollution and a lack of drinking water are major environmental issues.

WOMEN'S RIGHTS

Women have been able to vote and stand for office since 1954. In 2004, 11 per cent of parliamentary seats were held by women.

In 2003, women made up 50 per cent of the total workforce. Female literacy rate in 2004 was 46 per cent, while male literacy rate reached 63 per cent. Although 92 per cent of pregnant women had access to prenatal care, only 47 per cent of the births received specialized assistance. Maternal mortality was 540 per 100,000 live births. It is estimated that five per cent of women are subject to genital mutilation.

CHILDREN

Mortality rates for children under one and five years old have shown little decline between 1990 and 2004: the former dropped from122 to 112 per 1,000 live births and the latter from 75 to 68 per 1,000 live births. In 2004, 16 per cent of children were born with severe or moderate underweight, while 22 per cent of children under five years suffered from severe or moderate underweight. The stunted growth rate was 30 per cent*.

The biggest causes of child mortality are malaria, respiratory infections, diarrhea and measles.

It is estimated that in 2003, 24,000 children under 14 years old were HIV-positive and 170,000 children under 17 years were AIDS orphans. There are 800,000 working children in Ghana; it is estimated that some 20,000 of them live and work in the streets of greater Accra. Some are caught up in ritual servitude *(trokosi)*. Trokosi means in the Ewe language 'slaves of the gods'.

INDIGENOUS PEOPLES/ ETHNIC MINORITIES

Under Kwame Nkrumah (1947-1966), most people identified themselves as Ghanaians since his Convention Peoples Party (CPP) opened its membership to everyone, regardless of ethnic origin.

The Akan Ashanti people and Ewes have vied for power and influence. The Ewes(13 per cent) were favored while Jerry Rawlings ruled. They do not face political restriction or discrimination, but they are not as economically advanced as the Ashanti (28 per cent), which is the largest ethnic group, concentrated in the inland region of the country. In 2000, John Agyekum Kufuor came into power with Ashanti support.

The Mossi-Dagomba (16 per cent) live in the northern part and form a cohesive group as a result of their relative isolation. In 1740, the Dagomba were dominated by the Ashanti, and by 1874, their kingdom had fallen apart.

In comparison with the other main groups, they have less access to health and education. They inhabit an area with little mineral wealth, poor climate and soil conditions and they have to compete with the Ashanti for usable farm land.

MIGRANTS/REFUGEES

In early 2006, Ghana hosted 62,500 refugees and asylum-seekers from various countries, including 39,200 Liberians and 16,200 Togolese. The number of asylum applications was 6,800 (half of them made by Togolese people).

DEATH PENALTY

Although the death penalty still applies, there have been no executions since 1993.

* Latest data available in *The State of the World's Children* and *Childinfo* database, UNICEF, 2006.

investiture. This foreign support was attributed to the rigorous application of the IMF imposed structural adjustment plans and the prompt payment of debt service.
30   In 1994, confrontations over land ownership caused the death of more than a thousand people and the migration of 150,000. A state of emergency was declared and in June an agreement was reached to bring an end to the violence.
31   Despite the New Patriotic Party and the People's Convention trying to settle their differences and form an alliance to present a common electoral platform, Rawlings won the December 1996 elections with 57.2 per cent of the vote. Amidst a severe energy crisis - produced by a fall in the water level of the Akosombo dam - Rawlings' Government sought for alternative sources which proved not to be feasible. In 1998, the Ministry of Science and Technology said Ghana could not pay the initial investment needed to develop a possible nuclear energy-based project. The crisis was partly resolved with help from Côte d'Ivoire, which increased the amount of electricity supplied to Ghana from 20 to 35 megawatts per day.
32   In August 1999, the police harshly put down a student demonstration against an increase in fees and costs passed by the Government. The authorities decided to close the University of Ghana temporarily until the demonstrators could be pacified.

33   In December 2000 Rawlings stepped down. The next month John Kufuor from the NPP put an end to 20 years of NDC Government. He beat John Atta Mills (who had been backed by Rawlings). The NPP also won the legislative elections, taking 97 of the 200 seats at stake, while the government party won only 86 seats. Kufuor assumed the Presidency - the first occasion in Ghana's history that a change of power had taken place through democratic means.
34   In late 2001, violent confrontations took place between Mamprusis and Kusasis, leaving 50 dead. New clashes took place in March 2002, causing the death of KngYa-Na Yakubu Andani II, an Andani, and of 27 other people. The Government declared a state of emergency and deployed its troops in order to calm both groups. Historically, Mamprusi people tended to favor the NPP, while the Kusasis tend to support the NDC.
35   Kufuor approved the creation of the National Reconciliation Commission, with the aim of investigating cases of human rights abuses that took place during the 22 years of military dictatorships. This Commission would grant immunity to those who testify and would try to solve the 200 cases of disappeared persons, during Rawlings' military regime.
36   In May 2003, an arrangement under the Poverty Reduction and Growth Facility was approved by

the IMF, granting Ghana $258 million for the Government's economic reform programs for the period 2003-2005. Despite tight public expenditure control, an annual growth of 4.9 per cent was expected together with a reduction in inflation.
37   In June of that year, the Government of Ghana and the African Development Fund signed a loan agreement to finance a health service rehabilitation project. The mission of the so-called Health Project III would be to control HIV/AIDS, malaria and blood transfusions. It also aimed at reducing maternal and infant mortality rates.
38   In July, Parliament renewed the state of emergency in the north, in Dagbon, to keep the conflict there under control. However, this measure badly affected trade with neighboring Burkina Faso.
39   In September 2003, a group of 173 Ghanaian children, who had been sold by their parents to local fishermen for $180, were reunited with their families as part of an International Organization for Migration (OIM) program. The children had been forced to do hard labor and some did not survive. In addition, they were poorly fed and ill-treated. The OIM offered the fishermen counseling and equipment, and also gave micro-credit to the children's parents so that they could set up small businesses to try and generate more income.

40   A thaumatin production facility was started in late 2003. Thaumatin is a low-calorie, natural sweetener, derived from the *katemfe* bush, which is 2,500 times sweeter than sugar. This enterprise could turn into a multi-billion industry, although the patents on the genes that contain the sweetening agent of the plant have been registered in the US, thus threatening the Ghanaian industry prospects.
41   In February 2004, former president Jerry Rawlings testified before a commission set up to investigate Human Rights abuses committed under his rule. Rawlings was questioned about the disappearance of over 200 people, as well as extra-judicial execution orders, among other crimes.
42   Kufuor was re-elected for a second four-year term with 52.75 per cent of the vote in the elections held on 10 December 2004. His main rival, opposition leader and former vice-president John Atta Mills, received 44.32 per cent. International observers said 'the election was peaceful and well organized'.
43   Between April and May 2005, thousands of Togolese refugees arrived in Ghana, fleeing from the political violence in their country.
44   In April 2006, former first lady Nana Konadu Agyemang, Rawlings's wife, appeared before a court, accused of defrauding the state. The charges related to the privatization of a state-owned company in the 1990s. ∎

Greece / Ellás

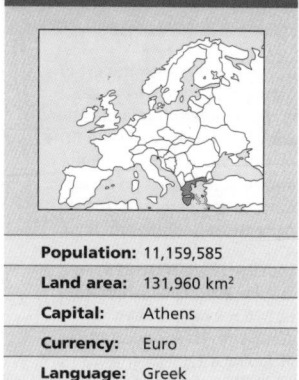

Population:	11,159,585
Land area:	131,960 km²
Capital:	Athens
Currency:	Euro
Language:	Greek

Although the oldest records of Greece's official history date from 800 BC, agricultural tools were found that belonged to a period earlier than 6000 BC. The first Greeks, engaged in different cultivation activities and devoted to the cult of a Mediterranean goddess of fertility, survived more than 3,000 years without fearing invasions, within a period evoked by the Greek legend as 'the golden age'.

2   During the third millennium BC, the Minoic civilization emerged in Crete. Owing to its intense maritime trade, urban development and unique artistic talent, the Minoic was regarded as the first European civilization.

3   In 1600 BC, the Achaeans (a fusion of Asia Minor and Indo-European peoples, in the third millennium BC) from Thessaly, in central Greece, invaded the Minoic, who originally came from the Middle East. The mixture of cultures, one peaceful and female-oriented (fertility), and the other one warlike, having Dyaus (Indo-European god of the sky, predecessor of Zeus) as their main god, gave rise to the Mycenian civilization.

4   Mycenae dominated Athens, Thebes, Pylos and Tiryns, among other cities. According to the Greek epic, Agamemnon, King of Mycenae, conquered the city of Troy (now Hisarlik in Asia Minor, where the Greek alphabet was created) in the 12th century BC, after a long war that would have led to the end of Mycenian civilization.

5   The decline of Mycenae began during that century with the Dorian invasions. The Dorians, also of Indo-European origin, destroyed all that they came across in Crete and its dependencies (Thessaly, the Peloponnese and the Cyclades Islands) and subdued the population with iron weapons, which were unknown in Greece.

6   The three centuries that followed the Dorian invasion were known as the Dark Age of Greek culture, characterized by the militarization of society. This period of terror resulted in the emergence of autonomous and fortified city-states in Athens, Sparta, Thebes, Corinth and Argos.

7   Dorians and Mycenians developed a common religion and language. Homer's epic poems, written during the 9th century BC, incorporated deities from different locations into a coherent story (unified by Zeus) focused on the summit of Mount Olympus.

8   Gods and heroes of the past (mortals gifted with god-like powers) took a leading part in those episodes that explained the cosmic laws that governed life on Earth. Likewise, the Greeks recognized obscure powers, hidden in an underworld, which were activated by guilt.

9   The cult of each god - whose rituals included human sacrifice - took place both in private and in multitudinous festivals, where passages of the *Iliad*, the *Odyssey* and other Homeric poems were staged. The first event registered in the history of Greece were the Games organized at Olympia in 776 BC, to honor Zeus and the goddess Hera, which also marked the beginning of the Archaic Age.

10   In the 8th century BC and the two following centuries, a demographic explosion spurred the Greeks to sail to the north of Africa, Spain, France, Italy and Asia Minor in search of land.

11   Greek art, influenced by Asia Minor and Egypt, recovered the use of curves which had been replaced by the rectilinear geometrical forms imposed by Dorians since the Mycenian period. Athens, with its Attic style, displaced the Phoenicians (see Lebanon) from the pottery and painting market.

12   By the end of the 7th century BC, commerce was radically changed by the use of coins which the Greeks had learnt from the Lydians. This accelerated the rise of merchants to political power. They joined the peasants to displace the nobility. Governmental structures underwent democratization in the different cities, except for Sparta, which remained an oligarchic state.

13   In the 6th century BC, philosophers of Troy started to develop a rational critique of religious practices and beliefs. The Sophists continued this line of skeptical thought, as did dramatists like Euripides and Aristophanes. The lyric poetry of Lesbos (7th and 6th century BC) shocked the Greeks; some mathematics works by Pythagoras that date from the 6th century BC were never surpassed.

14   The beginning of the Classical period (from 499 BC to 323 BC) was marked by the devastation of two Persian invasions as well as by the victory and consolidation of Athenian culture. The triumph of the confederated cities (in the Medic Wars, 499-479 BC) over a more powerful invader (the Persian Empire) gave Athenians - whose military intervention had been of the utmost importance - the glorious enthusiasm necessary to undertake the re-building of their city and the expansion of their own empire.

15   After seizing the treasure of the Delos sanctuary, in 454 BC - a symbol of the union of the Greek city-states against the Persian enemy - Athens imposed the payment of tributes on all confederated cities of the League of Delos (476 BC).

16   Pericles, governor of Athens between 443 BC and 429 BC, improved the governmental system - established in 594 BC - which would be later referred to as a 'slave-holding democracy'. This was due to the exclusion of slaves and women from citizen status. The first definitions of citizenship and political pluralism involving rights and duties were set down in writing by Solon (640-558 BC).

17   The works and ideas of scholars were exposed and debated in open spaces. Socrates (470-399 BC) developed the notion of dialectic thought and applied it to introspection; Plato (427-348 BC) went deep into the ethical and methodological principles of Socrates and founded his own school, the Academy; Herodotus (484-420 BC) used, for the first time, an objective method (examination of evidence) for historical research.

18   Greek painters and sculptors engaged in the study of the human body, which turned into the main subject of their works. Sculpture was the most widely-known art and the practice of sports represented a sign of spiritual grandeur. Man, who was considered as 'the measure of all things', was examined in all his dimensions.

19   The city of Sparta was a completely militarized state founded in the 9th century BC. It disputed with Athens the hegemony over Greece in continuous struggles between 431-404 BC (Peloponnesian War), which resulted in the fall of the Athenian empire (404 BC).

20   The supremacy of Sparta (404-371 BC) was followed by that of Thebes (372-362 BC). These constant struggles weakened the Greeks and enabled Macedonians to conquer this domain under the rule of Philip II, in 338 BC, who aimed at annexing this territory to the Persian Empire, at a later stage.

21   King Philip II had Demosthenes (384-322 BC), an Athenian master of rhetoric, as his main advisor. Likewise, he chose Aristotle (384-322 BC), founder of formal logic and author of treatises on politics, metaphysics and physics, as tutor of his son Alexander.

22   After the death of Philip II, in 340 BC, Alexander continued the plan set out by his father. He spread the area of influence of Greek culture by conquering territories in the east of the Mediterranean, the Arabian Peninsula, Mesopotamia and India. The exchange of knowledge and customs between the Greek culture and that of the conquered peoples, gave rise to what is known as 'Hellenism'

23   The political and cultural centers of the Hellenic world were Alexandria (Egypt) and Babylon (Mesopotamia). Upon the death

Life expectancy
79 years
2005-2010

Literacy
91% total adult rate
2000-2004

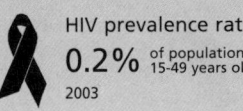
HIV prevalence rate
0.2% of population 15-49 years old
2003

of Alexander the Great in 323 BC, the empire was divided and Greece remained under the rule of the kings of Macedonia.

[24] Rome formally established its rule over Greece in 146 BC, after defeating the Macedonian army and the Achaean League (of Greek confederated cities) in a war that lasted for 50 years. Although they subdued Greece from the economic and political point of view, in order to consolidate and spread their empire, the Romans (Italics, of Indo-European origin, who arrived in the 2nd millennium BC) had to assimilate the Greek culture. On account of this, Athens was granted special benefits and became the headquarters of the main university of the Roman Empire.

[25] In Greco-Roman paganism, the field of ethics belonged to philosophers. Opposed doctrines, such as the Stoic (austere) or the Epicurean (sensual), oriented the life of learned citizens, without any conception of life after death. When Christianity became the official religion (3rd and 4th centuries), some of those doctrines that had emerged in Greece, acted as counter-cultures until the 6th century. At the same time, other doctrines like Neo-Platonicism contributed to strengthen what would turn into the new Western paradigm.

[26] Christian sermons were well received among Greek slaves. Being mostly farmers, they hoped some authority would minimize the difference between the extreme wealth of urban centers and the desolation of rural areas. From the 4th century, monasteries would partially fulfill this role.

[27] Constantine I (280-337), crowned as Emperor in 313, transferred the capital of the Roman Empire to Byzantium (present-day Istanbul), which had been colonized by the Greeks in the 7th century BC. It was there that he founded Constantinople between 324 and 330, in order to avoid a Persian invasion and to control the Danube border, through which barbaric invaders entered.

[28] Between the 3rd and 11th centuries, Greece witnessed devastating incursions of Germanic (Goths, Heruls and Vandals), Asiatic (Huns and Avars), and Norman (Scandinavians) peoples, as well as of Bulgars and Arabs.

[29] During the 4th century, schools in Athens declined while paganism abandoned mythology. In 394, the prohibition of the Olympic Games was declared. Likewise, in the 6th century, the Emperor Justinian prevented pagan philosophers from teaching and limited access to texts by the great Greek thinkers to the elite groups in Constantinople.

[30] Although some merchants prospered under the rule of

Byzantium, the Greek economy was weakened. Byzantium was capital of the Eastern Roman Empire (394-1054), of the Greco-Oriental Empire (after the Schism of the Roman Church in 1054 when Greek Christians pledged obedience to the Orthodox Church), and of the Latin Eastern Empire, under the Fourth Crusade occupation between 1204 and 1261.

[31] The crusaders were displaced by Venetians, who gained control over wide land and maritime areas of the ancient Latin Empire, even after the taking of Constantinople by the Turkish-Ottoman Empire in 1454. It was not until 1718, when the Peace of Passarowitz was signed, that Venice promised to withdraw from Greece, and did not actually leave until 1797.

[32] During the late Byzantine period (1204-1453), the Greek territories were occupied by Serbs, Catalans, Sicilians, French and mercenary companies, besides Venetians and Ottoman-Turks. Byzantium's political weakness brought about a revival of Hellenism in Greece, which was reinstated as a cultural center.

[33] Within the first three centuries of Turkish-Ottoman occupation, Greece put up little resistance. The first sultans allowed Greeks to engage in trade activities and practice their own language and religion.

[34] After the religious uprising of 1770, a partial victory over the Turks, together with the news about the US (1776) and French (1789) Revolutions, stirred a desire for freedom in Greeks which they were not able to define by common criteria. Their lack of national unity and strength to fight the Turkish-Egyptian forces enabled the direct intervention of Russia, Britain and France, in the War of Greek Independence (1821-1832).

[35] The anti-Turkish revolution comprised a first stage of local uprisings (from 1821 until 1825), which was followed, from 1826, by armed confrontations between the Turkish-Egyptians and the three European powers. The latter signed the Treaty of London in 1827, which declared Greece an independent state under their protection.

[36] In 1832, the 17-year-old Otto of Bavaria was appointed king by the London Convention. Otto's permissive behaviour towards the outbursts of a Bavarian rebel group in Greece, called forth several insurrections between 1833 and 1843. That year, the rebel forces surrounded the royal palace, demanding Otto's resignation and forcing him to grant a constitution, which was promulgated in 1844.

[37] In 1863, after being discredited by both the rivalries between the European powers and Turkish

threats, Otto was replaced by William of Denmark. This was accomplished by a protocol signed in London - similar to the one signed in 1832 - which also stipulated the withdrawal of Britain from the Islands of Troy.

[38] William named himself George I. He reigned in Greece between 1863 and 1913, set up an elective government and defined the king's role as a passive instrument of the people's will within a system defined as 'monarchic democracy'. Eleutherios Venizelos gained popularity during the anti-Turkish struggles in Crete and managed to obtain its re-annexation to Greece in 1908. He also amended the Constitution in order to be elected Prime Minister in 1910, with 80 per cent of the vote.

[39] Once a new Constitution was approved in 1911, Venizelos created the Balkan League. In November 1912, war was declared on Turkey, with Serbia and Bulgaria as allies, in order to recover the territory of Salonica, which had been lost after the Treaty of London of 1827. One month later, a second Balkan War broke out when Bulgars launched attacks on their two allies and Romania. Both wars, which ended

in 1913, allowed Greece to spread its territory.

[40] World War I (1914-1918) unleashed a confrontation between King Constantine and Venizelos. As a result, the Greek population was divided. The King proclaimed his neutrality but Venizelos denounced his alignment with Germany, on account of which he was expelled from government in 1915. In 1916, after establishing a rival republican government in Salonica, Venizelos helped the allied troops.

[41] The interwar period was shaken by a warlike conflict with Turkey (1922-1923) - which caused over one million deaths - and another border conflict with Albania, which prompted Italy to launch an air raid (1923). Both conflicts prompted the intervention of the League of Nations. Venizelos was removed from the political arena in 1935.

[42] Greece remained under Nazi occupation between 1941 and 1944. In 1949, the US launched its direct intervention to eliminate communist guerrillas, and from then on, it took control of the Greek economy. Greek soldiers were recruited to fight in the Korean War (1950-1953). Under

PROFILE

ENVIRONMENT

Located at the southeastern end of the Balkan peninsula, in the eastern Mediterranean, the country consists of a continental territory between the Aegean Sea and the Ionian Sea and numerous islands, including the island of Rhodes. Greece is a mountainous country, it has a Mediterranean climate with hot, dry summers. It is an essentially agricultural country (producing wine, olives, tobacco, wheat, barley); sheep and goats are raised in the mountains, where the soil is poor. Development of salt-water fish farming is under way. Greece produces lignite, bauxite and nickel. The traditional manufacturing industries include food, beverages, apparel, leather and paper. Other major industries are cement, chemicals, petrochemicals and mining.

SOCIETY

Peoples: Most people are of Greek origin and there is a small Turkish minority (1 per cent), plus Albanian, Macedonian and Roma.
Religions: Greek Orthodox. There is a Muslim minority.
Languages: Greek.
Main Political Parties: New Democracy, conservative; Pan-Hellenic Socialist Movement (PASOK); Communist Party of Greece; Coalition of the Radical Left.
Main Social Organizations: The General Confederation of Greek Workers.

THE STATE

Official Name: Hellenikē Demokratía.
Administrative Divisions: 10 regions divided into 51 administrative units.
Capital: Athens/Piraeus 3,215,000 people (2003).
Other Cities: Patras 172,100; Thessaloniki 793,900 (2000).
Government: Head of State, Karolos Papoulias, President since March 2005. The President is elected by Parliament every five years. Kostantinos (Kostas) Karamanlis, Prime Minister since March 2004. The Prime Minister is appointed by the President. Unicameral Legislature: Greek Parliament (Vouli ton Ellinon), with 300 members (Parliamentary Republic).
National Holiday: 25 March, Independence Day (1821).
Armed Forces: 168,300 (1996). Other: Gendarmerie: 26,500; Coast Guard and customs: 4,000.

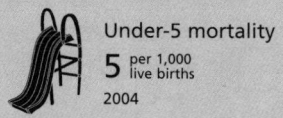
Under-5 mortality
5 per 1,000
live births
2004

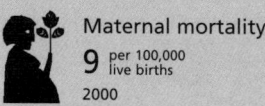
Maternal mortality
9 per 100,000
live births
2000

IN FOCUS

ENVIRONMENTAL CHALLENGES

Increasing urbanization over the last five decades has put pressure on coastal areas and caused sea pollution levels to rise. Solid and toxic waste is poorly managed. Arid areas suffer from a lack of drinking water and some desertification.

WOMEN'S RIGHTS

Greek women have been able to vote since 1952. In 2004, 13 per cent of parliamentary seats were held by women.

In 2003, they made up 38 per cent of the 5-million labor force. That year, 14.6 per cent of women were unemployed.

In 2004, the net primary school enrolment rate for girls reached 99 per cent*.

CHILDREN

In the 1990-2004 period, mortality rates for children under 1 and 5 years old dropped by more than 50 per cent. The former went from 10 to 4 per 1,000 live births and the latter from 11 to 5 per 1,000 live births.-

Eight per cent of children suffered from low birth weight*. The enrolment rate in primary school reached 99 per cent of all children.

INDIGENOUS PEOPLES/ ETHNIC MINORITIES

Turkish Muslims (1.2 per cent) are concentrated in the area of Thrace and have a higher birth rate than Greeks. There are few teachers in Turkish schools and students are not allowed to learn Greek, which hurts them economically. Likewise, there are allegations of discrimination against non-Greek speakers seeking to enter university. The Government makes it easy to obtain land from Turks mainly through providing low interest loans. Meanwhile, there are restrictions on the purchase and sale of property by Turks.

The Roma make up 1.7 per cent of the population. When confronted with social and cultural repression, they have traditionally preferred to move to a new location. Due to their nomadic lifestyle, the Roma as a group are not cohesive and organized. They are easily identifiable due to their physical appearance and are considered lazy, dirty and prone to crime. Also, their refusal to assimilate is not well received in Greece's nationalistic society. They have higher birth rates and poorer health conditions than the rest of the population.

MIGRANTS/REFUGEES

In the 2004-2005 period, most of the countries that have traditionally received refugees - such as Slovakia, Poland, Switzerland, UK, Sweden, Canada, Germany and France - reported significant falls in asylum requests. However, in Greece and the Netherlands there was a rise in asylum requests during the period. In Greece, the total number increased by 102 per cent. Something similar had happened in the 2002-2003 period, although the increase then had been of 45 per cent. In 2001-2005, Greece received almost 33,000 refugees. In 2004, refugees came mostly from Iraq, Afghanistan, Nigeria, Georgia, Pakistan, Iran and Bangladesh.

DEATH PENALTY

In 1993, it was abolished for ordinary crimes and the last execution took place in 1972.

* Latest data available in *The State of the World's Children* and *Childinfo* database, UNICEF, 2006.

US influence, Greece became a member of the Council of Europe (1949) and of NATO (1951).

[43] In 1952, a new Constitution was approved. In the 1956 presidential elections women voted for the first time. Elections were won by Konstantinos Karamanlis, of the National Radical Union, who was forced to resign in 1963, after being accused of fraud.

[44] In April 1967, upon the threat of a possible victory of the left in the upcoming elections, a group of colonels staged a coup. Martial law was enforced, the Constitution was suspended and democratic movements were harshly repressed.

[45] The Colonels' regime was supported by the US and by tycoons such as Onassis. Attempts were made to mask the dictatorship behind a unicameral Parliament in 1968, but in reality the military junta ruled by decree.

[46] Between 1973 and 1974 the military government grew weaker. In July 1974, the junta promoted a coup in Cyprus by collaborating with the Cypriot National Guard. The coup succeeded in deposing president Archbishop Vaneziz Makarios; and a minister in favor of Greek annexation was appointed. The Turkish army invaded Cyprus, allegedly defending the Turkish minority in the country. The Greek military government became even more discredited and internationally condemned, and they relinquished power immediately at the prospect of war with Turkey.

[47] Karamanlis returned from exile and took over the government. In the 1974 elections his party won a parliamentary majority and a later referendum abolished the monarchy. In 1975, Parliament adopted a new constitution and Konstantinos Tsatsos, a Karamanlis partisan, was elected first President of the Republic.

[48] In the 1981 parliamentary elections, PASOK (Pan-Hellenic Socialist Movement) led by Andreas Papandreu gained an absolute majority. That year, Greece joined the EEC. Papandreou's Government recognized the PLO and led a worldwide campaign in favour of handing back works of art that had been stolen during colonial domination to their countries of origin.

[49] In 1984 PASOK won the elections again. In 1986, successive austerity plans and salary freezes, applied since 1983, fuelled new protests and strikes. In June 1989, the Greek Left and the Communist Party formed the Left Alliance.

[50] In the 1989 elections PASOK lost its majority, and the conservative New Democracy party received a large part of the vote. As there was no parliamentary majority and no agreement to form a government, the presidency went to the leader of the Left Alliance that formed a temporary government with New Democracy, aiming to investigate financial scandals. In November, a coalition government was formed.

[51] Between 1983 and 1989, Greece and the US signed various agreements including the maintenance of four US military bases in the country, in return for economic and military assistance as well as US diplomatic support for Greece in its disputes with Turkey. However, in 1990, Washington and Athens announced a new agreement to close two military bases.

[52] In March 1990, a law was passed that established free negotiations between workers and bosses, putting an end to 50 years of State intervention, and including norms for the organization of company and union committees.

[53] Following Karamanlis' triumph in the 1990 presidential elections, a new government was formed headed by the conservative Constantinos Mitsotakis, who promoted public spending cuts, price liberalization and privatizations. The social cost of these measures contributed to the defeat of the conservative government in the 1993 legislative elections, in favour of Papandreou.

[54] The public debt and pressures from the European Union (EU) regarding economic matters complicated Papandreou's administration.

[55] In 1995, at the end of Karamanlis' second term as President, Papandreou supported Kostis Stefanopulos, candidate of the small Political Spring party, who enjoyed a strong political reputation. He was elected in 1995 and re-elected in 2000, when he defeated the conservatives led by Karamanlis by a narrow margin.

[56] Ill and increasingly criticized, Papandreou resigned in January 1996 and died five months later. His former Industry Minister, Konstantinos Simitis, replaced him. In September the Socialist Party, led by the new Prime Minister, won the legislative elections.

[57] Backing its pro-regional integration position, in January 1998, Athens abolished a discriminatory constitutional article allowing 'non-ethnic Greeks' seeking to leave the country to be stripped of citizenship, which dated from the so-called 'dictatorship of colonels'. The article had been used against the Muslim minority of Turkish origin and led 60,000 people to lose their Greek nationality.

[58] In 2000, Simitis was appointed Prime Minister for the second time and stated that he would continue to adopt measures to reduce public spending, which still reached 50 per cent of the Greek GDP.

[59] In 2002, Greece and Turkey signed an agreement to build a gas pipeline for Turkey to supply natural gas to Greece.

[60] In March 2004, Kostas Karamanlis (nephew of the former president) became Prime Minister after his party, ND, won parliamentary elections that month.

[61] Parliament ratified the EU Constitution in April 2005.

[62] In May 2006, in one of the largest scandals in the history of the Greek secret service, two high officials were formally accused of kidnapping and interrogating two Pakistanis detained in the country in 2005, suspected of being involved in the subway attacks in London in July that year. The prosecutor said charges could not be brought against a British intelligence agent that had supposedly taken part in the interrogations, because he had diplomatic immunity. ■

Grenada / Grenada

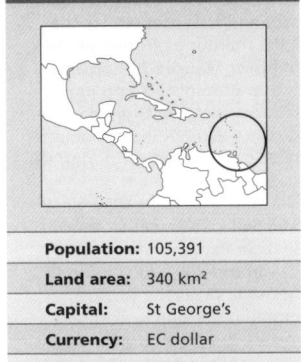

Population:	105,391
Land area:	340 km²
Capital:	St George's
Currency:	EC dollar
Language:	English

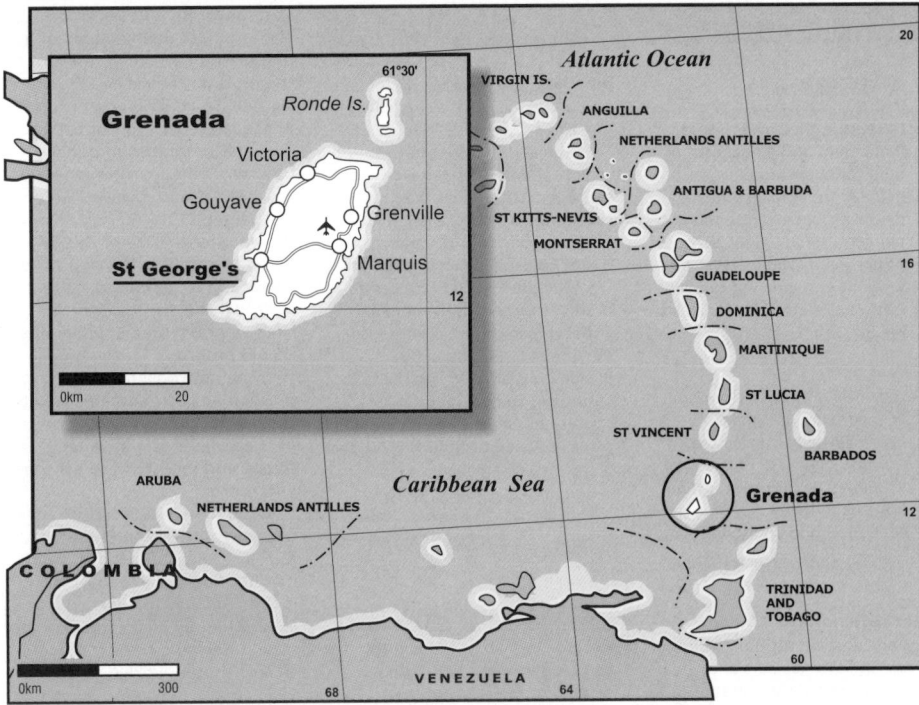

The Carib Indians inhabited Grenada when Christopher Columbus arrived, around 1498, and named the island Concepción. This European visit did not disrupt the island's peace but two centuries later in 1650 the governor of the French possession of Martinique, Du Parquet, decided to occupy the island. By 1674, France had established control over Grenada, despite fierce resistance from the Caribs.

² In 1753, French settlers from Martinique had around 100 sugar mills and 12,000 slaves on Grenada. The indigenous population had been exterminated.

³ The British took control of the island towards the end of the 18th century and cultivated cocoa, cotton and nutmeg, using slave labor. In 1788 there were 24,000 slaves, a number which remained stable until the abolition of slavery in the following century.

⁴ The severe living conditions of workers resulted in the creation of the first union in the mid-20th century, the Grenada Manual and Metal Workers' Union. In 1951, a strike broke out and the labor struggle won considerable wage increases. Eric Matthew Gairy, a young adventurer who had lived away from the island most of his life,

formed the first local political party, the Grenada United Labor Party (GULP) which favored independence. In 1951, GULP won a legislative election and Gairy became leader of the assembly.

⁵ In 1958 Grenada joined the Federation of the British West Indies, which was dissolved in 1962. The country became part of the Associated State of the British Antilles in 1967. That year, Gairy was appointed Prime Minister, and his main objective was total independence from Britain.

⁶ GULP soon obtained semi-independence, which gradually led to full independence. By that time, left-wing groups such as the New Jewel Movement (NJM), led by Maurice Bishop), had appeared on the island, opposing separation from Britain. Though apparently paradoxical, many Grenadians considered that Gairy was seeking independence for his personal benefit, manipulating a politically unprepared population.

⁷ In January 1974, an anti-independence strike broke out to prevent Gairy from seizing power. After some weeks of

total paralysis of the country, the Mongoose Squad - similar to the Haitian *Tonton-Macoutes* - appeared. Its paramilitaries were on the Prime Minister's payroll and brutal repression was used to end the strike. Independence was proclaimed the following week.

⁸ As feared, Gairy exploited power for his own personal benefit. He distributed government jobs among the members of his party, and promoted the paramilitary squad to the level of a 'Defense Force', making it the only military body

LAND USE

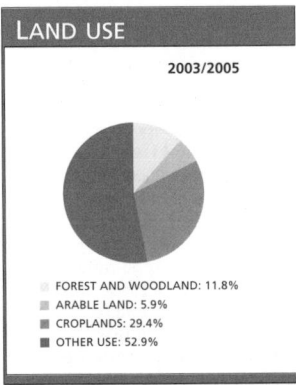

2003/2005

- ○ FOREST AND WOODLAND: 11.8%
- ▨ ARABLE LAND: 5.9%
- ▤ CROPLANDS: 29.4%
- ■ OTHER USE: 52.9%

PROFILE

ENVIRONMENT

Grenada is the southernmost Windward island of the Lesser Antilles. The island is almost entirely volcanic. Lake Grand Etang and Lake Antoine are extinct volcanic craters. The rainy, tropical, climate, tempered by sea winds, is fit for agriculture, which constitutes the country's major source of income. It is famous for its spices, and is known as 'the Spice Island of the West'. The territory includes the islands of Carriacou (34 sq km) and Petite Martinique (2 sq km), which belong to the Grenadines.

SOCIETY

Peoples: Most are descended from African slaves (85 per cent); most of the others are of mixed descent (13 per cent) and European (0.7 per cent) and Indo-Pakistani (3 per cent) minorities.

Religions: Mainly Catholic (53.1 per cent); Protestant (38.1 per cent); other (7.4 per cent).

Languages: English (official and predominant). A patois dialect derived from French, and another from English are also spoken.

Main Political Parties: New National Party, conservative (NNP); National Democratic Congress, liberal (NDC); Grenada United Labor Party (GULP), People's Labor Movement (PLM).

Main Social Organizations: Grenada Trade Union Council, which includes 8 organizations including those for public sector workers, teachers, seamen and dockers; Rural Farmers'Association; National Council of Students of Grenada.

THE STATE

Official Name: Grenada.

Administrative Divisions: six parishes and one dependency (Carricou and Petite Martinique).

Capital: St George's 33,000 people (2003).

Other Cities: Gouyave 3,200 people; Grenville 2,300 (2000).

Government: Queen Elizabeth II, Head of State; Daniel Williams has been the representative of the British Crown, (Governor-General) since August 1996. Keith Mitchell, Prime Minister since June 1995, re-elected in 1999 and 2003. Bicameral Legislature: a 15-member House of Representatives elected by direct popular vote and a 13-member Senate with ten of them appointed by the Governor-General and three by the leader of the opposition.

National Holiday: 7 February, Independence (1974).

Armed Forces: Royal Grenada Police Force (includes Special Service Unit), Coastguards.

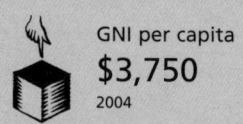

GNI per capita
$3,750
2004

IN FOCUS

WOMEN'S RIGHTS
In 2003, women held 27 per cent of seats in the Lower House of Parliament and 40 per cent of ministerial positions.
In 2004, the female primary school enrolment rate was 80 per cent. Ninety-eight per cent of pregnant women received prenatal care and 100 per cent of births were attended by skilled health staff*.

CHILDREN
HIV/AIDS is a major challenge in the island. It is mainly heterosexually transmitted. Currently the rate of infection is estimated at two per cent, but appears to be increasing steadily. Teenagers and youngsters -

especially girls between 15 and 19 - are particularly vulnerable, due to early sexual initiation.
In many countries in the region, HIV-positive female teenagers outnumber their male counterparts 5 to 1, and 4.5 to one for reported AIDS cases. The risk of infant death is highest for the estimated 3,000 born to HIV-positive mothers. Governments in the region together with the Caribbean Community (CARICOM), UNAIDS and the Caribbean Epidemiological Center have developed national plans of action to fight HIV/AIDS.

*Latest data available in *The State of the World's Children* and *Childinfo* database, UNICEF, 2006.

on the island. The Mongoose Squad received military training from Chilean advisers and recruited former inmates of St George's Prison.
9   In the December 1976 election, the opposition People's Alliance, made up of the NJM, the National Party of Grenada and the United Popular Party, increased from one to six representatives out of a total of 15. On 12 March 1979, while Gairy was out of the country, the opposition seized power in a coup. Widespread popular support enabled them to establish the People's Revolutionary Government (PRG), led by Maurice Bishop.
10   In four years, the PRG stimulated the formation of grass-roots organizations, and they created a mixed economy, expanding the public sector through agro-industries and state farms. However, Bishop allowed some private enterprise to continue.
11   The PRG was constantly harassed by the US and some conservative neighboring countries because of its socialist policies. The US started a coordinated campaign to stifle Grenada economically, alleging that the modern Grenadian Point Saline airport could be used by Cuba to transport its troops to Africa.
12   The Government based its foreign policy on the principles of anti-imperialism and non-alignment. Cuba agreed to collaborate in the construction of an international airport, conceived as a way to stimulate tourism, which employed 25 per cent of the country's workforce.
13   Prime Minister Bishop faced

constant pressure from the NJM's extreme left wing. Strife within the Party led to tragic consequences when Bishop was overthrown, possibly by his Minister of Finance backed by the military, in October 1983. Bishop was placed under house arrest, while General Hudson Austin, head of the army, seized power. Bishop was freed by a crowd of sympathizers, only to be shot dead by troops, as were his wife Jacqueline Creft - Minister of Education, the Foreign and Housing Ministers, two union leaders and 13 members of the crowd.
14   The US used the upheaval as a pretext to invade - something that had been planned for over a year. On 25 October, 5,000 marines and Green Berets landed on the island. They were followed several hours later by a symbolic contingent of 300 police from six Caribbean countries: Antigua, Barbados, Dominica, Jamaica, St Lucia and St Vincent, who joined the farce of a 'multinational intervention for humanitarian reasons'.
15   Resistance from the Grenadian militia and some Cuban technicians and workers meant that the operation lasted much longer than expected. The US suffered combat casualties, and the press was barred from entering Grenada until all resistance had been eliminated. This made it impossible to verify how many civilians had been killed in attacks on a psychiatric hospital and other non-military targets.
16   While strict US military control continued, Sir Paul Scoon, official British crown representative in Grenada, assumed the leadership of an interim government with the task of organizing an election.

Voting was held in December 1984 to elect the members of a unicameral parliament. In turn, the representatives appointed Herbert Blaize Prime Minister. Blaize led a coalition of parties that was presented to public opinion as the New National Party (NNP) and received support from the US.
17   Neither NATO nor the OAS dared to condone the aggression. Twenty days later, Barbados was rewarded for its 'co-operation' in the invasion with a $18.5-million US aid program.
18   The new government reached a classic agreement with the IMF including a reduction of the civil service, a wage-freeze and incentives to private enterprise.
19   The Regional Security System (RSS) permitting the Prime Minister to call on troops from neighboring Caribbean islands if Grenada was threatened was established by Blaize in December 1986, under the pretext that the trial of those involved in the 1983 coup was coming to an end. Bernard Coard, his wife Phyllis, and former army commander Hudson Austin were sentenced to death, along with 11 soldiers. Three others were tried, receiving prison sentences of 30 to 45 years.
20   During his first years in government, Blaize achieved considerable economic growth, at between five and six per cent per year, basically from tourism. However, youth unemployment continued to increase, along with crime and drug addiction.
21   In 1989, after Blaize's death, Ben Jones of the National Party took office, backed by big business and the landowners. On 12 March 1990 - the anniversary of the 1979 coup that overthrew Eric Gairy - the general elections were won by Nicholas Braithwaite, interim head of government after the invasion, and warmly regarded by the US.
22   In 1994 unemployment reached 30 per cent and there was a clear pattern of emigration. The fall in international prices of bananas, coconuts, wood and nutmeg influenced this trend.
23   In the June 1995 general elections, the governing NDC was pushed out by Keith Mitchell, former mathematics professor at the Howard University in Washington and leader of the New National Party (NNP). His victory was attributed to his promise to remove income tax, which had been revoked in 1986 and re-imposed by the NDC in 1994.
24   In March 1997, the Government rejected a request from the Grenada Council of

Churches to free the two men serving life sentences for the 1983 murder of former Prime Minister Maurice Bishop and a group of ministers and union leaders. In April, Grenada and Cuba re-established diplomatic relations, broken off during the US invasion.
25   In the January 1999 elections, Mitchell's NNP won all the 15 seats in Parliament.
26   In the 2001-2002 period, Grenada's was the fastest-growing Caribbean economy, mainly due to a boom in tourism and foreign investment, as well as greater farming production. However, unemployment rates remained high, mainly among young people.
27   In October 2003, on the 20th anniversary of Bishop's murder and the US invasion, Amnesty International published a report classifying the Grenada 17 detainees as the 'last Cold War prisoners', and requesting the Grenadian authorities to set up an independent review of the case. Among the prisoners there was Bernard Coard, Bishop's Minister of Finance, convicted of instigation of the coup that toppled Bishop. The report denounced, among other shortcomings, irregularities in jury selection, the lack of legal representation of the accused, questionable evidence and confessions obtained under torture.
28   In September 2004, the island was hit by hurricane Ivan, which left 34 people dead and some 5,000 families homeless. Only one building in 10 survived unscathed and even some of the designated hurricane shelters were severely damaged. The leaders of the Caribbean Community (CARICOM) called for a moratorium on Grenada's external debt payments in order to contribute to its recovery after the disaster. However, during 2005, reconstruction efforts proved ineffective and the agency NERO - in charge of the island's reconstruction - was dubbed 'ZERO' by angry Grenadians.
29   In July 2004, hurricane Emily swept over Grenada, killing one person, destroying crops and damaging homes. It was estimated that damage caused amounted to approximately $110 million.
30   In October 2006, the Acting Chief Justice of the Eastern Caribbean Supreme Court, Brian Alleyne, dismissed media reports that his appointment was being blocked by the Mitchell Government in Grenada, which had accused him of bias during his stint as a high court judge on the island. ■

Guadeloupe / Guadeloupe

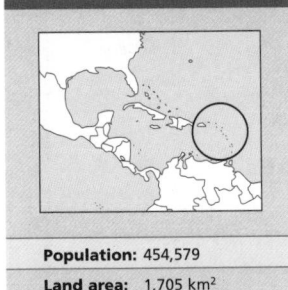

Population:	454,579
Land area:	1,705 km²
Capital:	Basse-Terre
Currency:	Euro
Language:	French

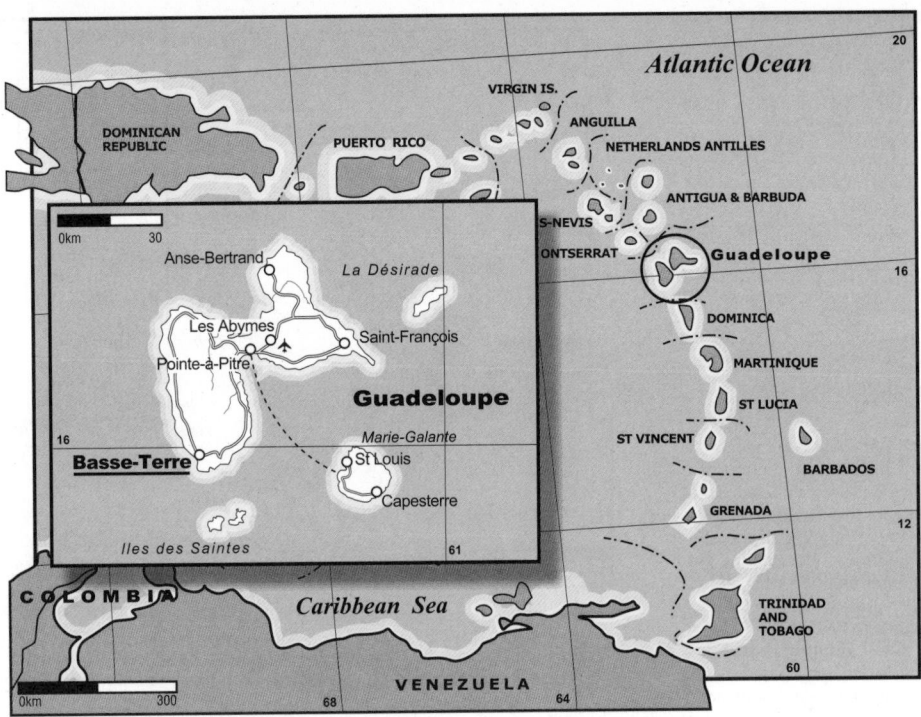

The entire archipelago of present-day Guadeloupe was inhabited by the Caribs. Originally from South America, they dispersed throughout the islands after overpowering the Arawak people. They resisted the Spanish invasion in 1493 but were defeated by the French two centuries later.

2   The French colonizers built the first sugar mill on the island in 1633 and began importing African slave laborers. By the end of the 17th century, Guadeloupe had become one of the world's main sugar producers. The colonists killed the last of the archipelago's surviving Caribs in the early 18th century.

3   With the abolition of the slave trade in 1815, France restructured its formal links with the Caribbean islands, giving them the status of colonies. In 1946, with the initiation of the new French constitution, Guadeloupe achieved greater political autonomy, as an Overseas Department.

4   French Government subsidies increased per capita income and the consumption of imported goods, but ruined the local economy.

5   Several independence movements swept through Guadeloupe after World War II. However, French President Charles de Gaulle visited the islands in 1956, 1960, and 1964 and managed to undermine separatists and to persuade most of the population that the islands should remain a part of France. As of the 1970s, separatist groups became more active, launching bombings in Paris during that decade and the following one. Notwithstanding this, the support for the status quo continued to be strong among the population until the end of the century.

6   The traditional sugarcane production has been slowly replaced by other crops, such as banana (which represents about 50 per cent of export earnings), eggplant and flowers. By the end of the century, an important reduction in productive activities deepened the trade balance deficit. Exports,

which amounted to barely 9.9 per cent of imports in 1995, decreased to 5.2 percent in 1996. Due to declining foreign demand and the competition from US-multinationals that managed plantations in Central America, the banana sector exported in 1996 less than half than in 1993.

7   Banana workers began a massive strike in late 1997. Union representatives accused the plantation owners of creating private militias and threatening to kill the workers to make them lift the strike. In February 1998, in a country with 40 per cent of the working population unemployed

and with growing social inequity, the strike had extended to other sectors.

8   French president Jacques Chirac visited the island in March 2000 to negotiate a special trade agreement between the EU and France's Overseas Departments in the Caribbean. Chirac's visit to the island coincided with a peak in demands for greater autonomy in French dependencies in the region.

9   In August 2001, Haitian immigrants complained to the public prosecutor about a TV show host for inciting hate and xenophobic racial violence against Haitian, Dominican and other

residents. The accused, Ibo Simon, was also a Councillor on the Regional Council of Guadeloupe.

10   In August 2003, several people were injured in a hold-up at a fast-food restaurant in Abymes. Sylvère Selbonne, a photo-journalist for the daily *France-Antilles Guadeloupe*, was arrested and roughed up by the police after taking pictures in the restaurant.

11   In recent years, the lack of economic improvements and the endemic unemployment - particularly among young people - has continued to encourage pro-independence groups. ■

PROFILE

ENVIRONMENT
Includes the dependencies of Marie Galante, La Désirade, Les Saintes, Petite-Terre, St Barthélemy and the French section of St Martin forming part of the Windward Islands of the Lesser Antilles. The tropical, rainy climate is tempered by sea winds, and sugarcane is grown.

SOCIETY
Peoples: 90 per cent of African descent. There is a small European minority. There are immigrants from Lebanon, China and India (five per cent).
Religions: mostly Catholic. There are also African, Hindu and several Protestant religions.
Languages: French (official).
Main Political Parties: Rally for the Republic (RPR), and Union for French Democracy (UDF), are the local branches of the French parties; the Communist Party of Guadeloupe (PCG); the Popular Union for the Liberation of Guadeloupe (UPLG).

Main Social Organizations: The Guadeloupe General Labor Confederation; the Department Organization of Trade Unions (CGT-FO), General Union of Guadeloupe Workers (UGTG).

THE STATE
Official Name: Département d'Outre-Mer de la Guadeloupe.
Administrative Divisions: three Arondissements, 36 Cantons.
Capital: Basse-Terre 12,700 people (2003).
Other Cities: Pointe-à-Pitre 21,400 people; Les Abymes 64,400 (2000).
Government: Jacques Chirac, Chief of State, represented by Jean-Jacques Brot, since August 2004. Legislative power: a 42-member General Council (chaired by Jacques Gillot since March 2001), and 41-member regional Council (chaired by Lucette Michaux Chevry since 1992). Guadeloupe has four deputies and two senators in the French parliament.
National Holiday: 14 July, Bastille Day (1789).

Guam / Guam

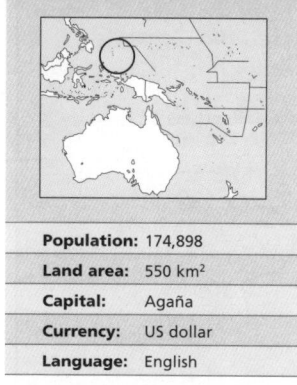

Population:	174,898
Land area:	550 km²
Capital:	Agaña
Currency:	US dollar
Language:	English

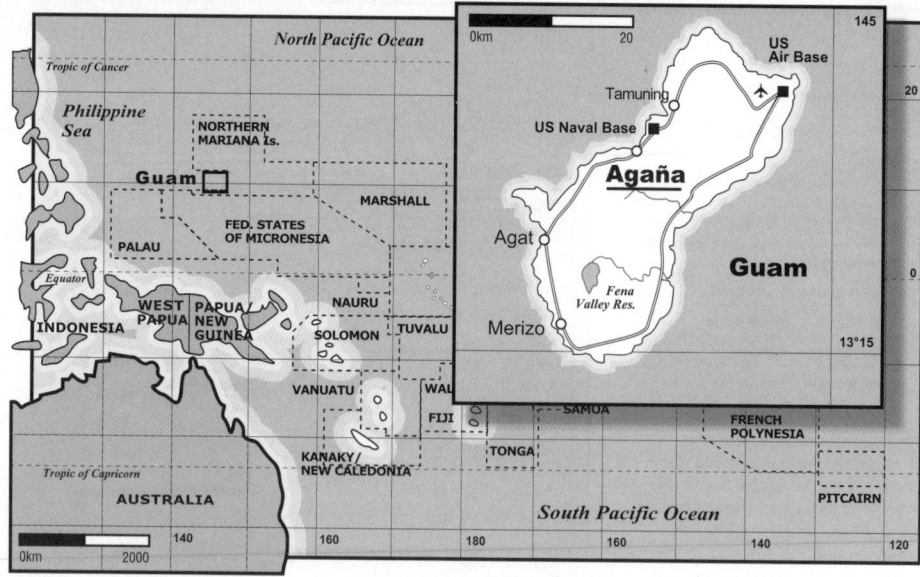

G uam shares a common history with the rest of the Micronesian archipelago (see Micronesia). The population, which settled on the island thousands of years ago, became the victim of extermination campaigns at the hands of Spanish colonizers, between 1668 and 1695.

2   As a result of armed aggression and epidemics (the people lacked immunity against European illnesses) the population declined from 100,000 at the beginning of the 17th century to fewer than 5,000 in 1741.

3   The few survivors intermarried with Spanish and Filipino immigrants, producing the Chamorro people who presently populate the island.

4   For three centuries, Guam was a port of call on the Spanish galleon route between the Philippines and Acapulco (Mexico), a major depot on the trade route to Spain.

5   Under the terms of the Treaty of Paris of 1898, Guam changed hands from Spain to the US, together with the Philippines. The island continued to serve as a stop over until it was invaded by the Japanese in 1941. Recovered in 1944, it became a US military base.

6   Since 1973, the United Nations has unsuccessfully urged Washington to permit the islanders to exercise their right to self-determination. In January

1982, a plebiscite was held on the issue of US federated status but the vote failed to reach the absolute majority required.

7   In a plebiscite on self-determination, 75 per cent of voters supported a political system in association with the US.

8   The UN General Assembly held in December 1984 recommended that the US implement Guam's decolonization. The UN also reiterated its conviction that military bases were a major obstacle to ensure the full exercise of people's right to self-determination.

9   In February 1987, former governor Ricardo Borballo, who had been elected in 1984, was found guilty of bribery, extortion and conspiracy.

10   Negotiations with the UN and US on the right to political self-determination and the creation of a Free Associated State were renewed in 1996. A landowners' organization demanded that the Government include the territory occupied by the US military bases in the talks. Washington considered Guam a key geo-strategic enclave.

11   The US Anderson airbase in Guam is also used for Marine exercises, military maneuvers and war exercises.

12   World War II hit Guam

particularly hard. In 2003 the US Federal Advisory Committee Act established a Review Commission which heard testimony from some 70 survivors. The Commission sought to determine whether the US had treated Guam in the same way as other territories (such as the Philippines and Micronesia) when it paid the

island compensations linked to Japanese occupation during the war.

13   Since 1975, Guam's rate of soil erosion has doubled. This has been mainly due to road construction and forest destruction by fire. According to estimates, Guam's erosion rate stands at 243 tons per acre per year. ∎

LAND USE

2003/2005

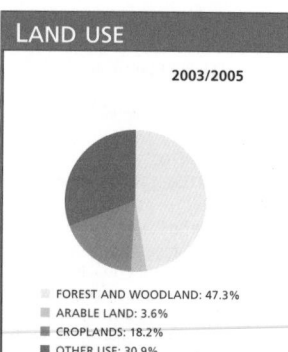

- FOREST AND WOODLAND: 47.3%
- ARABLE LAND: 3.6%
- CROPLANDS: 18.2%
- OTHER USE: 30.9%

WORKERS

LABOR FORCE	2004

- FEMALE: 39%   ■ MALE: 61%

PROFILE

ENVIRONMENT

Guam is the southernmost island of the Marianas archipelago, located east of the Philippines and south of Japan. Of volcanic origin, its contours are mountainous except for the coastal plain in the northern region. The climate is tropical, rainy from June to November (over 300 mm a month) and drier and colder from December to May. It has rainforest vegetation. One-third of the island is occupied by military installations.

SOCIETY

Peoples: Chamorro indigenous people account for approximately 47 per cent of the population; Filipino, 25 per cent. US troops and dependents, 10 per cent. Japanese, Chinese, Korean and other, 18 per cent.
Religions: 98 per cent of all Guamanians are Catholic.
Languages: English (official), Chamorro (a dialect derived from Indonesian), and Japanese.
Main Political Parties: Republican Party and Democratic Party, as in the US.

THE STATE

Official Name: Territory of Guam.
Capital: Agaña 140,000 people (1999).
Other Cities: Tamuning 11,800 people.
Government: Felix P.P. Camacho, Governor since January 2003. His office reports to the US Interior Department and has less autonomy than the local military commander, as one-third of the island is under control of the US Navy and Air force. Guamanians formally possess US citizenship and elects one nonvoting delegate to the US House of Representatives. Unicameral Parliament: The Guam Legislature has 15 members, elected by popular vote to serve two-year terms.
National Holiday: First Monday in March, Discovery Day (1521).
Armed Forces: Defense is the responsibility of the US.

Guatemala / Guatemala

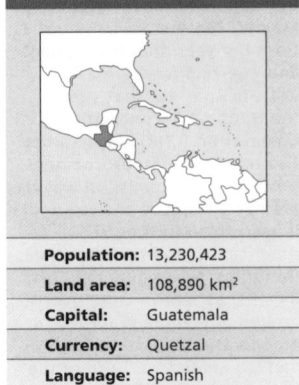

Population:	13,230,423
Land area:	108,890 km²
Capital:	Guatemala
Currency:	Quetzal
Language:	Spanish

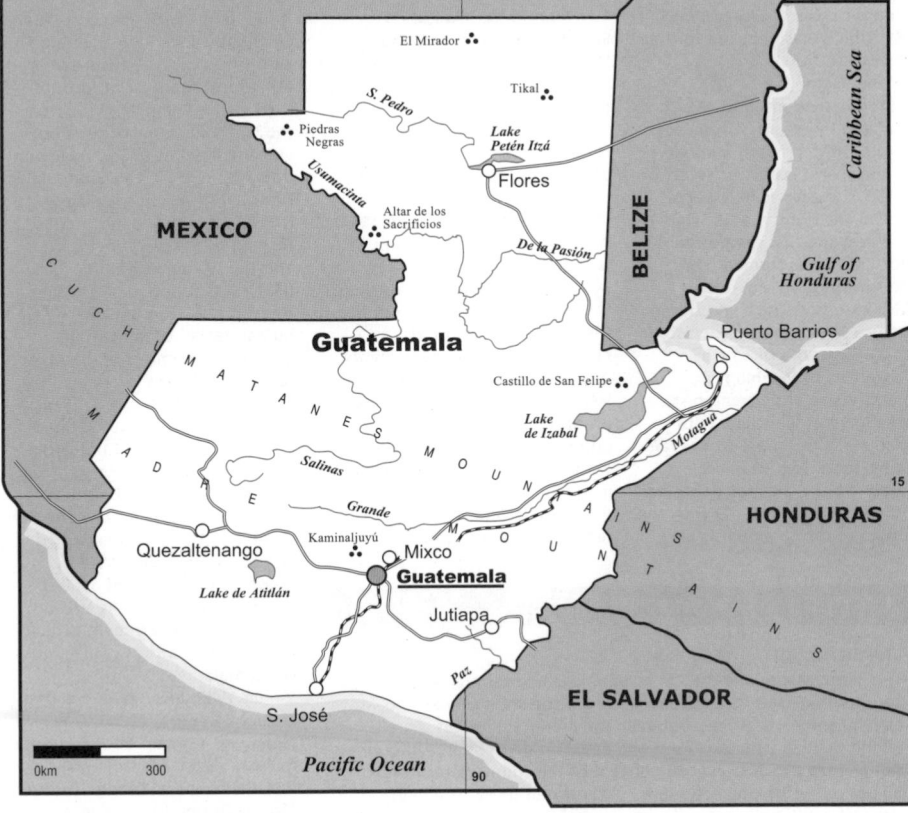

It is thought that the Maya culture appeared in 9,000 BC when hunters-gatherers lived on the *puuc* (hilltops) and along the Caribbean coast. An early complex social structure developed around farming. A new pre-classical era, from around 2,000 BC to 300 AC, has been determined after the discovery of pre-Maya figurines. The Maya civilization flourished during the first ten centuries AC in what is now Guatemala and parts of Mexico, Honduras, Belize and El Salvador.

2 Historical research suggests that a religious caste ruled the Maya civilization, controlling government and religious affairs. Important events such as deaths, births, royal marriages, victories and defeats were carved on stone. Routes (*sacbes*), some longer than 100 km, linked important towns. These settlements were abandoned by the end of the classical period, probably because the resources could not support the population, but also perhaps because they were conquered by invaders from Mexico.

3 Spanish troops, under the command of Pedro de Alvarado, entered the country in 1524 and founded the city of Guatemala, gaining total control over the country two years later. This process was facilitated by the fact that the country was undergoing a gradual transition and readjustment among its various ethnic groups - the K'iche', Kaqchi', Mam, Q'eqchi, Poqomchi', Q'anjob'al, Tz'utujiil and others - all of which stemmed from a common Mayan ancestry. Though the situation favored the invaders, there were nevertheless frequent incidents in which they faced stiff resistance.

4 In the 18th century, the invention of synthetic dyes in Europe brought about a severe economic crisis in Guatemala since its most important export was vegetable dyes. After that, coffee became the prominent crop, grown on large plantations. On 15 September 1821, large landowners and local business interests joined forces with colonial officials, to peacefully proclaim the independence of the Vice-royalty of New Spain, including the five countries of Central America. In 1823, after independence from Mexico, Guatemala became the administrative and political center of the United Provinces of Central America (UPCA).

5 In 1831, under great debt pressure, the Government yielded large portions of territory to Britain for timber. This area later became British Honduras, now the independent nation of Belize. The UPCA collapsed in 1838 for two main reasons. One was the military coup by Rafael Carrera (in power until he died in1865); the other was that British policy was to divide the American nations. In 1847 the State of Guatemala was formally created. In the 1871 Liberal Reform, indigenous people were deprived of their communal land, which was annexed to the coffee plantations. During the late 19th century, Guatemalan politics were dominated by the antagonism between liberals and conservatives.

6 Toward the end of the 19th century, Manuel Estrada Cabrera rose to power and governed Guatemala until 1920. He initiated an 'open door policy' (*cabrerismo*) for US transnationals, which eventually owned the railroads, ports, hydroelectric plants, shipping and international mailing services, in addition to the enormous banana plantations of the United Fruit Company (Unifruco).

7 General Jorge Ubico Castañeda, (the last of a generation of military leaders that went back to 1871), was elected President in 1931. He was at first a 'patron' for indigenous people because of his paternalism towards them. However his was a harsh regime that banned unions and introduced a Vagrancy Law, forcing people into work whatever the conditions.

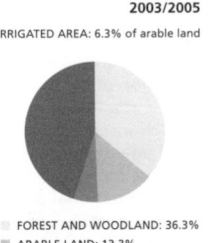
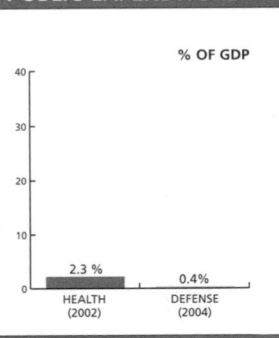
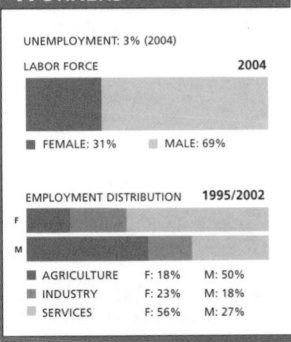

Life expectancy
68 years
2005-2010

GNI per capita
$2,190
2004

Literacy
69% total adult rate
2000-2004

HIV prevalence rate
1.1% of population 15-49 years old
2003

Discontent grew with the economic crisis of World War II in 1941. Ubico was deposed in the 1944 Revolution.

[8] The 'October Revolution' resulted in new elections which were won by reformist Juan Jose Arévalo. He promoted democracy and social and economic reforms. In 1945, literate women were granted the right to vote. That same year the first campesino labor union was formed. The land reform program, under which extensive tracts of unused Unifruco land were expropriated, was considered a threat to US interests by Washington. An aggressive anti-communist campaign was launched, with the sole aim of harassing Arévalo and his successor, President Jacobo Arbenz Guzman (who was elected President in 1950 with the support of the communists and continued the reforms)

[9] John Foster Dulles, US Secretary of State, but also a United Fruit Company shareholder and company lawyer, pressured the Organization of American States (OAS) to condemn Arbenz's reforms. Allen Dulles, director of the CIA and also a Unifruco shareholder, organized an invasion from Honduras in 1954. Arbenz was overthrown in this US-backed coup and was replaced by Colonel Carlos Castillo Armas. He gave the land back to Unifruco (renamed United Brands), reinstating foreign investments. Armas rooted out any communist influence, ended the land reform program and crushed labor unions. His ruthlessness was met with violence: he was assassinated in 1957.

[10] Two decades of military regimes followed. Elections in 1970, 1974, 1978 and 1982, were fraudulent - with the top military candidates invariably elected. This kind of political atmosphere bred armed insurgency with groups such as the Rebel Armed Forces, the Guerrilla Army of the Poor and the Revolutionary Movement (MR-13). In 1982, the Guatemala National Revolutionary Unity (URNG), a guerrilla movement formed by various rebel groups and the Guatemala Labor Party (PGT), was founded. According to estimates from several humanitarian organizations, government repression had taken some 80,000 lives between 1954 and 1982.

[11] On 31 January 1980, the Embassy of Spain in Guatemala was assaulted by military forces, under the orders of President Fernando Lucas García. A fire started by the assaulters killed 36 people, including Vicente Menchú, the father of activist Rigoberta Menchú-. As a result, the Spanish Government broke diplomatic relations with Guatemala.

[12] In 1982, after elections that brought General Anibal Guevara to power, a group of discontented military seized control and installed General Efraín Ríos Montt. During his first year more than 15,000 people were murdered, 70,000 were forced into exile (mostly to Mexico) and about 500,000 fled to the mountains to escape the army. Hundreds of rural towns were razed. The number of 'model hamlets' increased systematically. Peasants were taken by force to these hamlets, where they were required to produce cash crops for export, rather than growing subsistence crops. Social and economic conditions grew worse due to the governmental inefficacy and brutality, which led to a higher level of political violence.

[13] In August 1983, another coup staged by the CIA deposed Ríos Montt and General Oscar Mejía Víctores came into power, promising a quick return to a democratic system. In 1984, a Constituent Assembly was set up to draft a new constitution (to replace the 1965 one), to include new constitutional guarantees, *habeas corpus* and electoral regulation. The Constituent Assembly approved the right to strike for civil servants, authorized the return from exile of leaders of the Socialist Democratic Party and called for elections in November 1985. These were won by Marco Vinicio Cerezo; some democratic progress followed.

[14] In October 1987, representatives of the URNG and Vinicio Cerezo's government met in Madrid, the first direct negotiations between the Government and guerrilla forces in 27 years of conflict. That year, the National Reconciliation Commission (CNR) played a decisive role in the rapprochement process. The commission was created as a result of the Esquipulas II peace plan for Central America, (signed by Guatemala, Honduras, Nicaragua and Costa Rica). That year in Oslo, guerrillas and government agreed on an operational pattern for CNR and UN mediators. Despite the persistence of political persecution and assassination, a basic agreement was signed in Madrid by the National Commission for Reconciliation, political parties and the URNG.

[15] During the last few months of 1990, negotiations came to a standstill and a high degree of skepticism developed among voters, which led to a 70 per cent abstention rate in the 11 November 1990 presidential elections. During the second round of the elections, in 1991, Jorge Serrano Elías, of the Solidarity Action Movement (MAS), was elected President. The Serrano Government and the URNG decided to take up peace negotiations in Cuernavaca, Mexico, after three decades of violence, committing themselves to reach a lasting agreement. The agenda included topics such as: democratization, human rights, the strengthening of civil groups, rights of the indigenous peoples, constitutional reforms, the resettlement of the landless, the incorporation of the URNG into legal political life.

[16] Human rights organizations reported that in the first nine months of Serrano's rule there had been more than 1,700 human rights violations, including 650 summary executions and the murder of street children.

[17] In July, the US Senate suspended military aid to Guatemala. The URNG demanded that human rights - violations cease immediately. Serrano recognized the sovereignty and self-determination of Belize, the former British colony which proclaimed its independence in 1981. The announcement caused the resignation of chancellor Alvaro Arzú, the leader of the National Advancement Party (PAN), and one of the ruling party's main allies.

[18] In 1992, a national debate began on the existence of government armed civilian groups, such as the Civilian Self-Defense Patrols (PACs). The Catholic Church criticized

PROFILE

ENVIRONMENT

The Sierra Madre and the Cuchumatanes Mountains cross the country from east to west, and these are the areas of volcanic activity and earthquakes. Between the mountain ranges there is a high plateau with sandy soil and easily eroded slopes. Although the plateau occupies only 26 per cent of the country's territory, 53 per cent of the population is concentrated there. The long Atlantic coastline is covered with forests and is less populated. In the valleys along the Caribbean coast and in the Pacific lowlands there are banana and sugar plantations.

SOCIETY

Peoples: Approximately 90 per cent are of Mayan descent. Amid the country's great cultural and linguistic diversity, four major peoples can be distinguished: the Ladino (descendants of Amerindians and Spaniards), the Maya, the Garifuna (of the Caribbean region) and the Xinca.
Religions: Mainly Catholic. In recent years a number of Protestant groups have appeared. The Mayan religion has also survived.
Languages: Spanish is official but most of the population speak one of the 22 Maya dialects.
Main Political Parties: The Guatemalan Republican Front (FRG); Great National Alliance (GANA); National Advancement Party (PAN); National Unity for Hope (UNE); New Nation Alliance (ANN); the Guatemalan National Revolutionary Unity (URNG).
Main Social Organizations: Union of Labor and Popular Associations; National Labor Union Alliance; National Workers' Coordinating Committee; Labor Union of Guatemalan Workers; Altiplano Farmers' Committee; Campesino Unity Committee (CUC). The indigenous people's movement has grown stronger since the 1996 peace accords. It is organized in: National Coordinating Committee of Indigenous Campesinos (CONIC); National Committee for Mayan education (CNEM); Office of Human Rights of the Archbishopric; Centre of Studies of Mayan Culture (CECMA); Communities of People's Resistance.

THE STATE

Official Name: República de Guatemala.
Administrative Divisions: 22 Departments.
Capital: Guatemala City 951,000 people (2003).
Other Cities: Mixco 268,300 people; Villa Nueva 129,600; Quetzaltenango 115,900 (2000).
Government: Oscar Berger Perdomo, President since January 2004. Unicameral Congress: National Assembly with 158 members elected by popular vote for four-year terms.
National Holiday: 15 September, Independence Day (1821).
Armed Forces: 44,200 troops (1994). Other: 10,000 National Police, 2,500 Hacienda Guard; 500,000 Territorial Militia.

Under-5 mortality	Poverty	Debt service	Maternal mortality
45 per 1,000 live births 2004	**13.5%** of population living on less than $1 per day 2002	**7.4%** exports of goods and services 2004	**240** per 100,000 live births 2000

the Government's economic policy and spoke out in favor of agrarian reform. In the meantime, organizations representing the indigenous peoples demanded the ratification of ILO Agreement 169, dealing with indigenous and tribal peoples. The Government created the *Hunapú* force, made up of the Army, the National Police and the hacienda Guard. In April, members of the Hunapú provoked an incident during a student demonstration demanding improvements in the education policies. One student was killed and some injured. The World Bank, the US Government and the European Parliament urged the Guatemala Government to end political violence. That year, at the time of the quincentenary of Columbus' arrival in America, Rigoberta Menchú Tum, a leader from the Quiché people, won the Nobel Peace Prize.

[19] In 1993, President Serrano, backed by a group of military officers, carried out a coup, revoking several articles of the constitution and dissolving Congress and the Supreme Court. After national and international, as well as US, pressure, Serrano was ousted. Former human rights attorney Ramiro De Leon Carpio took over and dismissed Serrano's military supporters. Shortly afterwards, Jorge Carpio Nicolle, the President's cousin, was assassinated. The 1994-95 Government Plan, presented in August, reaffirmed the structural adjustment program prioritizing the end of state intervention in the economy, along with fiscal reform and the privatization of state companies.

[20] Despite the intense campaign against the Civilian Self-Defense Patrols (PACs) and compulsory military service, President De Leon Carpio stated that he would maintain both institutions as long as the situation of armed conflict persisted. The Government said that the records kept on citizens considered a 'danger' to State security had disappeared - thus eliminating evidence against those responsible for human-rights violations. De Leon Carpio's stated goal was to fight corruption in the public sector. In 1994, the President called for the resignation of legislative deputies and members of the Supreme Court, causing a confrontation between the President and Congress. This led to a clash of economic and political interests, culminating in the Executive and Congress agreeing on constitutional reform proposals.

IN FOCUS

ENVIRONMENTAL CHALLENGES
The area of forests, which covered 41.9 per cent of all the land in 1980, had been reduced to 33.8 per cent by 1990, putting at risk the diversity of the ecosystem. Logging is particularly heavy in the Petén tropical forest. There is water pollution.

WOMEN'S RIGHTS
Women have been able to vote and stand for election since 1946. In 2003, 8 per cent of parliamentary seats were held by women. They made up 31 per cent of the labor force and female unemployment reached 2.3 per cent.

Although 84 per cent of pregnant women received prenatal care in 2004, only 41 per cent of births were attended by skilled health staff*. Maternal mortality was 240 per 100,000 live births*.

A high number of women die as a result of domestic violence.

CHILDREN
Although the infant mortality rate was reduced during the 1990-2004 period, it is still high. Among under-five children, it stands at 45 per 1,000 live births (in 1990 it amounted to 82); among children under one year old it is 33 per 1,000 (in 1990 it reached 60).

In 2004, 12 per cent of children were born underweight, 23 per cent of those under five years old suffered from moderate and severe low weight and 49 per cent were moderately and severely stunted (these figures are considerably higher among the indigenous population). The primary school enrolment rate amounts to nearly 80 per cent, but of those who attend, only 65 per cent reach 5th grade. The secondary school enrollment rate stands at 23 per cent.

Sexual abuse and incest affect a large number of children.

INDIGENOUS PEOPLES/ ETHNIC MINORITIES
Indigenous people number about 5,000,000 (42 per cent of the total population). The majority are Maya and are scattered mainly in the rural districts in the north and in the west of Guatemala. The most common languages from the 26 spoken in the country are Quiché, Cakchiquel, Maya, Tzutujil, Achi and Pokomán.

They have only poor health services, and poor nutrition. Indigenous people make up 73 per cent of households living in poverty and 93 per cent of those living in extreme poverty. They have been internally displaced and excluded both socially and politically. In spite of agreements signed and the Government's commitment to recognize the economic, cultural and political rights of these peoples, by late 2003 no significant changes had been registered. The investigation of crimes committed by the army and paramilitary groups against indigenous peoples (almost all of them between 1978 and 1984 and in Maya territory) is still pending.

MIGRANTS/REFUGEES
Guatemala is used as a transit country by traffickers of undocumented migrants seeking to enter the US illegally via Mexico. In early 2006, over 100 Ecuadorians were captured while attempting to reach the country and their boat was nearly wrecked. Some months earlier, a boat heading towards the same destination had capsized near Colombian coast and only 12 passengers (all of them Ecuadorian) managed to survive.

In 2003, 5,800 Guatemalan refugees living in Mexico were voluntarily repatriated to Guatemala and another 1,100 were applying for Mexican citizenship. In 2004, 1,500 Guatemalans sought refuge in the US. In 2005, this figure stood at 1,600. Ninety-five per cent of Guatemalan migrants in the US live in Los Angeles, New York and Miami.

Remittances sent by Guatemalans living abroad account for 5 per cent of the country's GDP and are equivalent to 30 per cent of exports.

DEATH PENALTY
It still applies although President Alfonso Portillo postponed the execution of condemned people in 2002.

* Latest data available in The State of the World's Children and Childinfo database, UNICEF, 2006.

[21] An OAS office was occupied by members of the Committee for Campesino Unity and the National Commission of Widows of Guatemala and some 5,000 members of indigenous groups carried out a march demanding the dissolution of the PACs. In 1994, the Government and guerrillas signed agreements for the resettlement of the population displaced by the armed conflict, without a mediated ceasefire. As a result of the accords 800 people were able to settle in the zones of Chalkily, Newton and Huehuetenango, but the majority of the resettlement areas remained still under army control. That year the Minister of Foreign Affairs recognized Belize as an independent State, but upheld Guatemala's territorial claim, meaning no frontier could be established. The Government and URNG agreed to disband the PACs and to involve the UN in human rights issues. Shortly after, the president of the Constitutional Court - Epaminondas González Dubón - was murdered.

[22] The URNG and the Government signed a draft agreement for the resettlement of the 'People Uprooted by the Armed Confrontation', in Oslo, Norway, in 1994. The Communities of Peoples were recognized as non-fighting civilians and the vital importance of land for these uprooted populations was explicitly stated. The second agreement in Oslo enshrined the principle of not individualizing responsibility for human rights violations as a way of neutralizing the action of those opposed to a negotiated outcome. The UN stated that impunity for the perpetrators was the main obstacle to justice in the case of human rights' violations such as illegal detention, torture and execution.

[23] In the 1995 elections, Alvaro Arzú (National Progress Party) won over Alfonso Portillo Cabrera of the FAG. Abstentions reached a record 63 percent. In 1996, Arzú and the URNG signed a series of peace agreements which put an end - after 36 years - to a civil war which had cost more than 200,000 lives. The cease-fire was respected. A round 80 per cent of the population were living below the poverty threshold at this time.

[24] In 1998, Hurricane Mitch caused more than $5,000,000 worth of damage and left 24,000 dead in the region - 256 from Guatemala. More than 100,000 people were made homeless.

[25] Investigation of human-rights violations during the war produced an avalanche of

Malnutrition
23% under-5s
1996-2004

Water source
95% of population using improved drinking water sources
2002

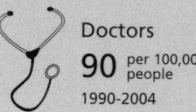
Doctors
90 per 100,000 people
1990-2004

Primary school
93% net enrolment rate
2004

threats against investigators and members of the judiciary. After presenting a report condemning military officers for several massacres, Bishop Juan Gerardi was murdered in 1998. The public prosecutor working on the case fled to the US where he requested asylum, claiming he had been under heavy pressure and had received several death threats. That year, the Government ordered hundreds of bodies to be dug up in the grounds of an elite police unit in the capital. A UN report estimated that 96 per cent of the deaths during the war were the responsibility of the army and armed forces. In the second round of the elections held in late December 1999, Alfonso Portillo of the Guatamalan Republican Front (FRG) defeated Oscar Berger of the Great National Alliance (GANA). Abstention reached 59 per cent.

[26] The drought in 2002 brought famine, killing 41 people. President Portillo declared a 'state of national disaster', and requested international aid. A UN report estimates that 80 per cent of Guatemalans live in extreme poverty.

[27] Due to threats to their lives two public prosecutors and a judge investigating the murder of Bishop Gerardi exiled themselves: before leaving, Judge Yassim Barrios sent three soldiers and a bishop to prison for the crime. According to the Supreme Court, in 2001, 23 judges were menaced. Param Cumaraswamy, member of UN Special Envoy for Judicial Independence, said impunity still reigned in Guatemala and the Government had given no signs of political will to end that situation.

[28] In 2002, the Guatemala Constitutional Court removed parliamentary immunity from Ríos Montt, bringing him closer to trial for changing tax laws to favor an alcoholic drinks company. The Rigoberta Menchú Foundation (FRM) took the case to the Spanish Courts in 2001 to find an alternative route to bring Ríos Montt and other officer to trial for the disappearances, torture and deaths of 200,000 people. That year, Guillermo Ovalle de León of the FRM was murdered. Many human rights activists, journalists and other people seeking the truth about the military regime's crimes were also killed.

[29] In 2002, Guatemala and Belize agreed on a referendum to end the long dispute over borders. President Portillo was accused of diverting public funds to personal accounts abroad. Journalists Rodolfo Flares (*Siglo*

XXI, Guatemala) and Roland Rodríguez (*La Prensa*, Panamá) revealed that public officials, friends and relatives of Portillo had opened local and off-shore bank accounts in Panama. Portillo accused the press of making political capital by exaggerating the allegations of corruption.

[30] In 2003 the Spanish courts accepted the FRM's case for the trial of six military officials, including Ríos Montt, and two citizens on charges of the murder of four priests and three Spanish diplomats. In Guatemala the Center for Human Rights Legal Action (CALDH) laid charges against Ríos Montt for the genocide of indigenous people. Human Rights organizations hope that the Spanish courts will be able to extradite the accused.

[31] That year, the conservative Oscar Berger (GANA), defeated center-left Alvaro Colom, from the National Unity for Hope (UNE). Ríos Montt (FRG) also ran, but was roundly defeated. UN observers reported that activists and opposition leaders had been intimidated, persecuted and murdered during the elections. Former officials from the Civilian Self-Defense Patrols (PACs) contributed to the violence, according to Comité Campesino del Altiplano/CCDA (Altiplano Farmers' Committee).

[32] Berger came to power in a country deep in poverty and insecurity. He promised to strengthen institutions, and to invest in public safety, health, education and technology as well to fight against corruption. He also promised to rebuild the infrastructure - roads, ports and airports; to modernize the National Police, prosecute drug-dealers and organized crime rings.

[33] President Berger repeatedly invited Rigoberta Menchú and Helen Mack Chang (a business administrator who set up many social projects, and whose sister was assassinated) to work together with the Government. Finally, both accepted. Menchú agreed to help oversee the application of the 1996 peace accords. That year, an agreement was signed between the UN and the Government to set up a Commission for the Investigation of Illegal Bodies and Clandestine Security Apparatus,-in order to investigate the people responsible for political violence.

[34] In November, following a national day of protest, Congress passed a law, as a matter of 'national emergency',-to compensate former paramilitary group members. About 10, 000 demonstrators closed highways,

airports and seaports to demand the immediate payment of approved resources - over $100 million. Days before, the Constitutional Court had rejected a similar law in response to a request submitted by Human Rights organizations which stated that former paramilitary members should not be compensated given the fact that they had committed serious abuses during the civil war between 1960 and1996.

[35] The 2005 *Human Development Report* indicated that five out of every ten people continued to be poor and among these, one in five was extremely poor, thus showing that indicators remained unchanged since 2002. The country's worst affected area was the north-western - which hosted the largest number of indigenous people - where 1.2 million people were poor. It was followed by the northern area, which registered 809,000 people living in poverty conditions. In general, 60 per cent of Guatemalan people were living in poverty, while 20 per cent of them lived in extreme poverty.

[36] In May 2005,the representative of the International Program on the Elimination of Child Labor of the International Labor Organization (ILO) denounced the fact that 436,000 children aged between 5 and 17 were exploited as laborers in Guatemala. The worst conditions were registered in rural areas where 18.4 per cent of children performed adult labor. Many children started to work in agriculture at five or six years old. At this age, work affects both their health and their growth.

[37] Tropical Storm Stan hit the country in October 2005, causing deaths and destruction and sinking the country even further into poverty and violence. The death toll stood at over 1,000 and about 250,000 people - most of them indigenous people and minors - lost their homes and belongings. The UN declared that the hurricane could plunge even more of the population of Guatemala into extreme poverty and called on the international community to pull together at least $22 million in aid. President Berger requested urgent assistance to address the immediate needs for water, food and health services.

[38] By May 2006, it had been established that the number of women killed daily in Guatemala continued to rise (see section 'Themes'), and reached two per day. However, not a single murderer had been convicted.

[39] Violence against children

and teenagers became something usual in recent times. On 23 March 2006, a group of children on their way to school in Tegucigalpa was shot at by alleged band members. Four days later, an 11-year old was kidnapped and murdered. Organizations working on the issue claim the government's inaction led to the murder of two children a day, in average.

[40] In the first semester of 2006 the country received 2,051 million in remittances, which implied a 21% growth regarding the same period in 2005. Remittances sent by Guatemalans living abroad, mostly in the US, were the second largest source of foreign currency income for the country, only surpassed by exports.

[41] In late July, Berger announced he would report the United States to the World Trade Organization (WTO) for unfair trade practices, or dumping. According to the president, the US was selling poultry parts at a price lower than that set by the WTO.

[42] On 29 May the former military ruler of Guatemala, Fernando Lucas García, died at 81 in Puerto la Cruz, Venezuela, where he lived since 1982. Garcia's extradition had been requested by a Spanish judge in 2005 to be tried for the killing of seven Spanish citizens during his government. The request was rejected by the Venezuelan courts, alleging that the necessary papers had not been presented on time.

[43] An ancient type of punishment the Maya applied on criminals was adopted in several communities for people accused of selling children. The punishment consisted in flogging the men and shaving the women's hair and keeping them on their knees over pebbles. The accused were also expelled from their communities. Community leader Antonio Cotí said it was 'not right they had children only to sell them, that is why they were subjected to the punishment and expelled from the community, so they do not involve any more people in this type of actions'.

[43] In October 2006, Guatemala put itself forward as a candidate for one of the seats on the UN Security Council allocated to Latin America. The country's candidacy was backed by the US but firmly opposed by Venezuela, which was also keen to stand. Guatemala consistently outpolled Venezuela in voting but failed to attain the required two-thirds majority. Ultimately Panama was elected as a compromise candidate. ∎

Guinea / Guinée

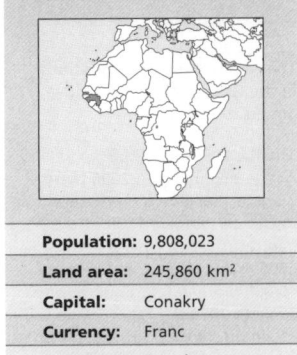

Population:	9,808,023
Land area:	245,860 km²
Capital:	Conakry
Currency:	Franc
Language:	French

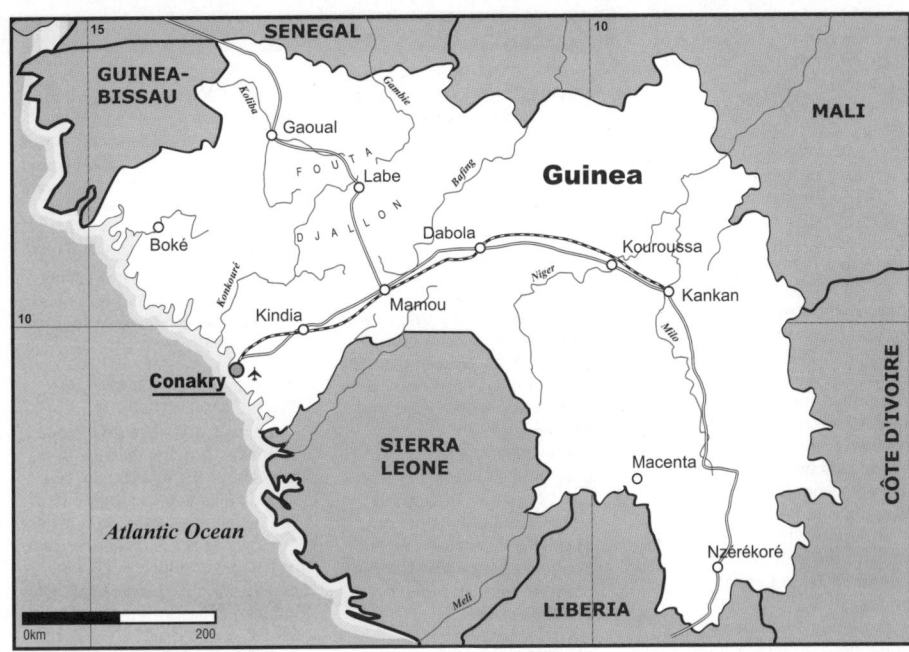

Hunters and gatherers were living in Guinea around 30,000 years ago. Late in the first millennium, the towns of Baga, Koniagi (Coniagui) and Nalu (Nalou) were sacked by Mandinkas (Malinkes) and Sussus. The settlements in upper Guinea had been annexed to Mali since the mid-13th century. From the 16th to the 19th century, Fulbe (Fulani, see Cameroon history), controlled Fouta Djallon, where Bambuk goldmines supported the Mediterranean economy for centuries.

2 Samori Touré (1840-1900, Mandinka leader, reformist and warlord), founded a powerful empire of Muslim states which included regions from the present Mali and Côte d'Ivoire. In 1886, Samori confronted French troops which came from Senegal. The *Almani* (the religious and political title he had adopted from 1879) fought against colonization until 1898; after that he was expelled from the country to Gabon where he died. His putative grandson, Ahmed Sekou Touré (brought up in a poor and illiterate family) founded de Democratic Party of Guinea (DPG).

3 With the loss of Indochina in 1954, Tunisia and Morocco in 1956 and the Algerian revolution in 1954, French colonialism collapsed. In its place, French President Charles de Gaulle created the French Community in 1958 to protect French interests in sub-Saharan Africa. His approach was neocolonialist, to try and guarantee that the large monopolies remained under the French influence. In the 1958 referendum on the constitution for the French Fifth Republic Guinea voted against membership in the French Community. On 2 October 1958, independence was proclaimed and Sekou Touré became the first President of Guinea. In response, Paris withdrew its qualified technical personnel, thereby paralyzing the infant industries, and blocked trade with Guinea.

4 In 1959 the State took control of the economy and created its own currency, freeing the country from the French franc. Industry and agriculture were diversified in an attempt to become self-sufficient; bauxite production exceeded one million tons per year. French aggression continued, including blocking bank accounts in Paris.

5 In 1970, Portuguese mercenaries invaded in an attempt to overthrow the Government. They also wanted to destroy the African Party for the Independence of Guinea and Cape Verde (PAIGC) which was fighting for the independence of Portuguese Guinea. Local revolutionary committees were set up to keep control; to strengthen the political campaigns and to stamp out corruption and theft.

6 In 1978, Congress changed the official name to Revolutionary People's Republic of Guinea and normalized relations with France. Sekou Touré, after a long isolation period, visited African and Arab capitals to diversify mining and attract investments to help endure foreign debt. French companies developed iron deposits in Mount Nimba, oil and bauxite. Guinea became the world's second largest producer of bauxite.

7 In 1984, Sekou Touré died in a US hospital. Immediately, Colonel Lansana Conté led a coup d'état and overthrew Louis Beauvoguii. Conté abolished the 'party-state', the Constitution, labor unions and the Assembly; he illegalized the PDG and erased Revolutionary People's from the country's name. Conté supported the private

sector, eliminated para-state companies and requested the help of France, the United States and African states to revitalize the economy. With an 800-million dollar debt, the national currency - the syli - was 100% devalued, and public spending was reduced, a pre-requisite for joining the CFA franc bloc.

8 That year, Conté reduced the cabinet and took for himself the positions of Head of State, Prime Minister and Minister of Defense. He supported rice crops and other farming, commercial, and industrial companies, within an economic recovery context. Even so, the country did not achieve food self-sufficiency. He also reduced public spending, through the massive retrenchment of civil servants. Military discontent with low wages grew, while consumer prices tripled and several demonstrations forced the government to lower the price of basic goods and rents.

9 In 1992, 650,000 refugees crossed the border between Sierra Leone and Liberia and set up precarious camps. The government increased the number of peace forces from the Economic Community of West African States along the borders to prevent the arrival of new refugees.

10 That year a multi-party system was approved. Leader Alpha Condé returned from exile and formed the National Democratic Forum, along with 30 opposition groups. Political tensions and persecutions continued. Lansana Conté was re-elected in the 1993 presidential elections, with at least 51 per cent of the vote. Condé

LAND USE

2003/2005

IRRIGATED AREA: 5.4% of arable land

- FOREST AND WOODLAND: 27.4%
- ARABLE LAND: 4.5%
- CROPLANDS: 2.6%
- OTHER USE: 65.5%

PUBLIC EXPENDITURE

% OF GDP

- HEALTH & EDUCATION (2000-2002): 2.7%
- DEFENSE (2004): 2.9%

WORKERS

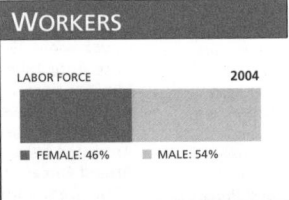

LABOR FORCE — 2004

- FEMALE: 46%
- MALE: 54%

Life expectancy
54 years
2005-2010

GNI per capita
$410
2004

HIV prevalence rate
3.2% of population
15-49 years old
2003

accused him of being responsible for a coup d'état and there were serious confrontations between the police and opposition followers. On 1994, dozens of people died in riots in the border with Liberia. In 1995, pro-government parties won 76 of the 114 seats in parliamentary elections. The Paris Club wrote off $85 million of Guinea's debt, due to the improvement in its GDP.

[11] In 1996, army elements unhappy with low wages attempted to overthrow the government. The uprising – which took over downtown Conakry and bombed the presidential palace – was crushed. Government officials accused of supporting the rebellion were sent to prison. That year, economist Sidia Touré was appointed Prime Minister and announced his priority was to reactivate the economy.

[12] Ahmed Tejan Kabbah, president of Sierra Leone overthrown in 1997, fled to Guinea; political instability in that country drove thousands of people to take refuge in Guinea. In 1998, humanitarian organizations sent food to the northwest to alleviate a long drought that had killed dozens of people. That year, Sierra Leone and Liberia signed a non-aggression pact in Conakry, with the mediation of American Jesse Jackson, sent by President Bill Clinton. Each country had accused the other of sheltering and aiding rebel groups. In the 1998 presidential elections, Conté was re-elected and the opposition reported serious irregularities. Condé was sent to prison.

[13] Economic growth helped advance the government goals, stated in the *Guinea vision 2010* document from 1999, which aimed to reduce poverty and the development of education and health. A meeting of

the Occidental African States Economic Community (OASEC) took place in the Ghanaian capital Accra in 2001: Ghana, Guinea, Nigeria, Sierra Leone, Gambia and Liberia, agreed to move towards a common currency in 2003.

[14] In 2002, Guinea, Sierra Leone and Liberia agreed frontier security measures in case of future rebellions. The region has been ravaged by internal conflicts in the last two decades, with a disastrous impact on human development.

[15] New presidential elections were held on 21 December 2003, under repressive conditions. The Government closed the airport and

frontiers in the days running up to the election and only allowed cars with diplomatic or special licences on the roads. The opposition boycotted the elections, but some of its leaders were still arrested in the lead-up to the election, along with some army officers. President Lansana Conté was duly elected for the third time.

[16] In February 2004 Lounseny Fall was appointed Prime Minister, but he resigned in April during a visit to the US, on the grounds that Conté had not allowed him sufficient room for maneuver to revive the country's economy. In November, the Government signed

an agreement with a Japanese company to build an aluminum refinery in the mining city of Sangaredi, in north-east Guinea. At an estimated cost of $2,000 million it would be the world's biggest bauxite complex and the largest project in West Africa.

[17] In January 2005 there was an armed attack against a convoy carrying President Conté, but the assassination attempt failed. Investigations did not reveal the attackers' identity.

[18] In July 2005, thousands of Alpha Condé's supporters welcomed him on his return from exile in France.

[19] In March 2006 President Conté traveled to Switzerland for medical treatment. In his absence secretary-general at the presidency Fode Bangoura took day-to-day control. Bangoura then sacked prime minister Cellou Dalein Diallo in early April - but had to rescind the decree when soldiers stormed the national radio station in protest.

[20] The succession to Conté remained the key political issue - with the influence of the country's richest man, Al Hajj Mamadou Sylla, looming large and ethnic tensions between the dominant Sussu minority, the Malinke and the Fulani providing a vital backdrop to the personal intrigue. Guinea's West African neighbors viewed the potential for conflict in the post-Conté era with alarm. ■

PROFILE

ENVIRONMENT
The central massif of Futa-Dyalon, where cattle are raised, separates a humid and densely populated coastal plain, where rice, bananas and coconuts are grown, from a dryer northeastern region, where corn and manioc/cassava are cultivated. Rainfall reaches 3,000-4,000 mm per year along the coast. The country has extensive iron and bauxite deposits.

SOCIETY
Peoples: Guineans comprise 16 ethnic groups, of which Fulani, Mandingo, Malinke and Sussu are the most numerous.
Religions: 65 per cent are Muslim, 33 per cent practice traditional religions, and 2 per cent are Christian and other minor groups.
Languages: French (official). The most widely spoken local languages are Malinke and Sussu.
Main Political Parties: Party for Unity and Progress; Rally of the Guinean People; Party for Renewal and

Progress; Union for the New Republic; National Union for Prosperity of Guinea.
Main Social Organizations: National Confederation of Guinean Workers.

THE STATE
Official Name: République de Guinée.
Administrative Divisions: 33 regions.
Capital: Conakry 1,366,000 people (2003).
Other Cities: Kankan 124,200 people; Labe 90,200; Nzérékoré 77,400 (2000).
Government: General Lansana Conté, President since April 1984. Cellou Dalein Diallo, Prime Minister since 2004. Parliament: National Assembly with 114 members, elected for a four year term.
National Holidays: 2 October, Republic Day (1958); 3 April, The Second Republic Anniversary (1984).
Armed Forces: 9,700 (1996). Other: People's Militia: 7,000; Gendarmerie: 1,000; Republican Guard: 1,600.

Guinea-Bissau / Guiné-Bissau

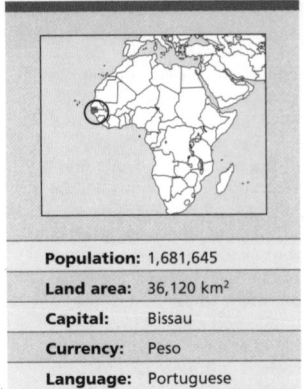

Population:	1,681,645
Land area:	36,120 km²
Capital:	Bissau
Currency:	Peso
Language:	Portuguese

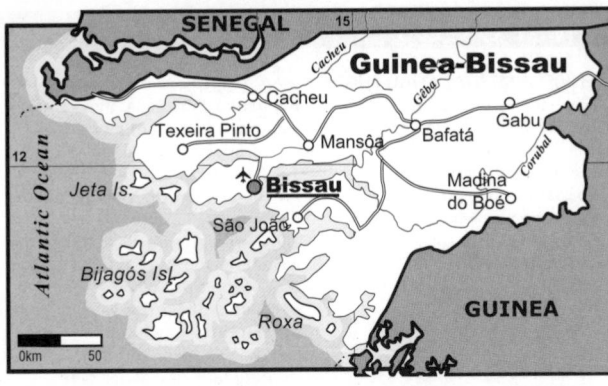

G uinea-Bissau was the first Portuguese colony to gain independence in Africa. This was achieved even before the fall of the Portuguese dictatorship in Lisbon, in a successful political and military struggle led by Amilcar Cabral's African Party for the Independence of Guinea and Cape Verde (PAIGC).

2 After belonging to the Mali and Songhai empires, the peoples in the Geba river valley became independent. This independence was soon threatened by the Portuguese, who had settled on the coast from the end of the 15th century, and by the Fulani, coming from the interior in the 16th century. Inland, the state of Gabu remained autonomous until the 19th century while the coastal population suffered the consequences of the slave trade and forced displacement to the Cape Verde Islands.

3 Resistance against the European colonists began in 1500 when the Portuguese arrived in Guinea. At that time, the country was inhabited by several different groups, immigrants from the state of Mali along with the Fulani and Mandingo groups, who lived in organized autocratic societies in the savannas. During the 17th century Guineans made their first contact with the inhabitants of the Cape Verde islands, a mandatory stopover for the ships carrying slaves to Brazil.

4 The Portuguese colonization of Guinea was harsh. As the country was small and poor, the monopoly on trade and agriculture was dealt with by a private company, the Unio Fabril. Guineans were forced to cultivate export crops while massively reducing the acreage available for subsistence farming. In the 1950s, infant mortality reached the remarkable rate of 600 deaths per 1,000 births. There were only 11 doctors in the country and only one per cent of the rural population was literate. In the early 1960s, only 11 Guineans had completed secondary education.

5 It was against this setting of misery and exploitation that Amilcar Cabral founded the Athletics and Recreational Association in 1954. This organization developed into the African Party for the Independence of Guinea and Cape Verde (PAIGC) two years later. The Party called on all Guineans and inhabitants of Cape Verde to unite in anti-colonial resistance, regardless of color, race or religion. In September 1959, after trying fruitlessly to engage the Portuguese in negotiations for three years, the PAIGC embarked upon guerrilla warfare. The fighting spread quickly and by 1968 the Portuguese were confined to the capital, Bissau, and a few coastal strongholds. A Popular National Assembly was elected and on 24 September 1973, the 'democratic, anti-imperialist and anti-colonialist republic of Guinea' was proclaimed. Two months later the UN General Assembly recognized the independent state.

6 Amilcar Cabral was assassinated in Conakry, Guinea, in February 1973, by Portuguese agents. He left many books and studies on the struggles for freedom in the African colonies. His successor, Luis Cabral, set up the Government Council in the heart of the liberated area.

7 The impact of the unilateral independence of Guinea-Bissau and its immediate recognition by the UN shook the infrastructure of Portuguese colonialism. General Spinola, commander of the 55,000 colonial soldiers, demanded that political changes in Portugal were needed. The Captains' Movement was born in Bissau, and later became the Armed Forces' Movement, the group which was responsible for the coup that overthrew the dictatorial regime in Portugal on 25 April 1974. Four months after the coup, Portugal recognized the independence of Guinea-Bissau.

8 The PAIGC Government diversified agriculture giving priority to feeding the population.

Foreign companies were nationalized, agrarian reform was implemented together with a mass literacy campaign. In foreign relations, the new government opted for non-alignment and unconditional support for the struggle against apartheid and colonialism in Africa. Top priority was given to economic integration with the archipelago of Cape Verde, with a view towards uniting the two countries.

9 In 1980, João Bernardino (Ninho) Vieira, a former guerrilla commander, staged a coup and installed a Revolutionary Council.

10 Talks with Cape Verde were cut short while the two countries were discussing a united constitution (See Cape Verde). The new government in Bissau was

immediately recognized by the neighboring Republic of Guinea, which had been in dispute with former president Cabral over offshore oil rights in an area presumed to be rich in petroleum deposits.

11 The first development plan of 1983-86 proposed an initial investment of $118.6 million, of which 75 per cent would be financed by international funds. In 1984, the construction of five ports was started, and the construction of the Bisalanca Airport was completed. The Government started a campaign against corruption and inefficiency in public administration, and as a result in 1984 Vice-President Victor Saude Maria was asked to resign. The Popular Assembly eliminated the position of prime minister, and the Revolutionary Council became the Council of State.

12 The 1984 stabilization plan failed, causing further deterioration of the economic and financial situation. Sixty per cent of the country's income came from the export of peanuts and dates whose price had fallen sharply.

13 The Government adopted a 'corrective' policy, including freezing salaries and reducing public investment. This was an attempt to bring it into line with IMF conditions for refinancing the servicing of the foreign debt. The economy was subsequently opened

PROFILE

ENVIRONMENT
The land is flat with slight elevations in the southeast and abundant irrigation from rivers and canals. The coastal area is swampy, suitable for rice. Rice, peanuts, palm oil and cattle are produced in the drier eastern region.

SOCIETY
Peoples: Balante 27.2 per cent; Fulani 22.9 per cent; Malinke 12.2 per cent; Mandyako 10.6 per cent; Pepel 10.0 per cent; other 17.1 per cent.
Religions: Two-thirds profess traditional African religions; nearly one-third are Muslim and there is a small Catholic minority.
Languages: Portuguese (official). The *crioulo* dialect, a mixture of Portuguese and African languages, is used as the lingua franca. The most widely spoken native languages are Mande and Fulah.
Main Political Parties: African Independence Party of Guinea and Cape Verde; Party for Social Renewal; United Social Democratic Party; Workers' Party.
Main Social Organizations: National Workers' Union of Guinea-Bissau (UNTG); Confederation of Independent Unions (CSI).

THE STATE
Official Name: República da Guiné-Bissau.
Administrative Divisions: 8 Regions and 1 Autonomous Sector.
Capital: Bissau 336,000 people (2003).
Other Cities: Bafatá 19,400 people; Gabu 12,200 (2000).
Government: João Bernardo 'Nino' Vieira, President since October 2005; Aristides Gomes, Prime Minister since November 2005. Single-chamber legislature: People's National Assembly, with 102 members elected among the members of regional councils.
National Holiday: 24 September, Independence declared unilaterally (1973).
Armed Forces: 9,250 (2001). Other: 2,000 Gendarmes.

Life expectancy
45 years
2005-2010

GNI per capita
$160
2004

up to foreign capital in the hope of attracting resources from Portugal and France, particularly in the area of telecommunications.

[14] In February 1991, the PAIGC approved a political reform which anticipated elections for 1992. Because of the economic instability and political tensions in 1992, the Government postponed the elections. However, in 1993, they were put back once more after the murder of a high military commander.

[15] Finally, in 1994 João Bernardo Vieira defeated Kumba Ialá of the Party for Social Renewal (PSR). During the campaign, Ialá accused him of supporting tribalism and racism. In the parliamentary elections, Vieira's PAIGC took 64 of the 100 seats at stake. Believing that the ruling party had 'bought' votes, Ialá refused to take part in a national unity government. That same year, the West African Economic Monetary Union (WAEMU) was founded, with eight member countries and a market of 72 million consumers, including Guinea-Bissau.

[16] In 1995, the IMF granted a new $14 million loan in support of economic reforms. In June, the visit of Senegalese President Abdou Diouf led to a rapprochement with Dakar. Both countries agreed to exploit joint energy and mineral resources.

[17] In late 1995, Guinea-Bissau ratified the border agreement signed with Senegal in 1993, resetting its maritime frontiers and stipulating the joint exploitation of an area which was supposedly rich in oil.

[18] Guinea-Bissau continued to house pro-independence rebels from the Senegalese Movement of Democratic Forces of Casamance (MFDC). However, a Guinea-Bissau's military attack on a refugee camp on the frontier with Senegal, in January 1998, fed rumors of rapprochement between Bissau and Dakar.

[19] In November 1998, a peace agreement was signed by both the rebels and the Government, in Abuja, Nigeria. In January 1999, fight was resumed in the capital, between the forces of General Ansumana Mané, who had been accused of supplying weapons to Senegalese rebels, and Vieira's Government troops. After four days of bloody battles, from which many fled, a ceasefire was agreed.

[20] On 4 May, the UN asked 'donor' countries to help Guinea-Bissau recover following eight months of civil war. Three days later, however, General Mané rose up again and Vieira was defeated; he sought political asylum in Portugal. The army accused Vieira of corruption and treason. France

IN FOCUS

ENVIRONMENTAL CHALLENGES

The need to increase exports led to over-cultivation of the soil; in addition, rice plantations are replacing part of the coastal forest. Slash-and-burn cultivation, as well as numerous forest fires, have contributed to deforestation.

WOMEN'S RIGHTS

Women have been able to vote and stand for office since 1977. In 2004, 14 per cent of parliamentary seats and 38 per cent of ministerial positions were held by women.

In 2003, women made up 41 per cent of the workforce.

In 2004, the net female primary school enrollment rate reached 37 per cent*.

Only 62 per cent had access to prenatal care and 35 per cent of births were assisted by skilled staff*. It is estimated that 50 per cent of women have undergone genital mutilation (between 70 and 80 per cent in areas inhabited by the Fulani and Mandinka and between 20 and 30 per cent in urban areas).

CHILDREN

Although the mortality rates for children under one and five

years of age have decreased in the period 1990-2004, they remain very high. The infant mortality rate dropped from 153 to 126 per 1,000 live births, while mortality for children under five years old dropped from 253 to 203 per 1,000 live births. In 2004, low birth weight newborns comprised 22 per cent of the total, while 25 per cent of children under five years of age suffered from moderate or severe low weight. Severe or moderate stunting rate was 30 per cent*.

UNICEF is carrying out a campaign to eradicate polio and to create an environment conducive to ensuring immunization services, vitamin A supplements, malaria control, HIV/AIDS prevention and breastfeeding promotion. At the end of 2001, 1,500 children under 14 were HIV-positive and 4,300 children were orphaned by this disease.

INDIGENOUS PEOPLES/ ETHNIC MINORITIES

The Balante (30 per cent of the population) inhabit the central and northern coastal region. Their political organization is based on the council of family authorities. In 1984, Ntombikte, regarded as a prophet, created the Kiyang-yang movement, whose leaders were

arrested by the Government a year later, banning their religious and curative activities.

The Fulani (22 per cent of the population) live in the central region. Meanwhile, the Mandyako peoples (14 per cent of the population) are located in the western and southern area of the Gambia River. The Pepel represent 10 per cent of the population and live on Bissau Island and the southern coastal regions. It is estimated that the Pepel have lived in part of their current territory since the 12th century. Historically, this group had a strong presence in urban areas and in the army.

MIGRANTS/REFUGEES

In early 2006, Guinea-Bissau and Gambia hosted almost 8,000 refugees from Senegal. There were about 900 Liberians in Guinea-Bissau, Senegal and Mali, waiting to be repatriated to their country.

DEATH PENALTY

This was abolished in 1993.

Latest data available in The State of the World's Children and Childinfo database, UNICEF, 2006.

condemned the coup and Mané was accused of violating the Abuja and Lomé agreements, signed three months earlier.

[21] Five months after the coup, a mass grave was found containing 18 bodies in the town of Portogole, including former vice president Correira. Meanwhile, the military junta presented evidence to Portugal of the 'crimes' committed by Vieira in order to get him repatriated. On 17 November, two weeks before the national elections, General Mané stated that 'any president who is elected and does not fulfil his promises will be immediately deposed'.

[22] In the second round of presidential elections on 16 January 2000, Kumba Ialá of the PSR was elected President with 72 per cent of the vote.

[23] In November, Mané proclaimed himself leader of the army and attempted a coup, but he was killed together with eight followers after cross-fire with government forces in Quinhamel, 30 kilometers from Bissau.

[24] The Guinea-Bissau Resistance-Ba Fata Movement (RGB), with the second largest representation in the National Assembly, abandoned the government coalition in

January 2001. In May, both the IMF and the World Bank stopped their monetary help due to the 'loss' of millions of dollars of aid for development. In September, Ialá removed the President of the Supreme Court and three Judges from office. In November, he sacked his Minister of Foreign Affairs for criticizing him. In December, accused of attempting a coup, Prime Minister Faustino Imbali was also dismissed.

[25] In November 2002, Ialá dissolved Parliament and promised anticipated elections. In February 2003, after the arrest of several opposition members, the elections were postponed until October.

[26] In September, after the elections had been postponed for the fourth time, Army Chief Verissimo Correia Seabre led a bloodless coup, which was supported by the majority of the population. A military junta appointed interim authorities to lead a transitional government until the parliamentary elections set for March 2004 and the presidential elections set for March 2005. Henrique Rosa, an economist who in 1994 had headed the Electoral Commission, was appointed President. Antonio Arthur Sanha, leader of the Party

for Social Renewal and outspoken critic of the deposed president, became Prime Minister.

[27] In January 2004, Maria do Ceu Silva Monteiro was appointed President of the Supreme Court. She was in charge of validating the results of the March elections, won by PAIGC.

[28] In October 2004, a military revolt to get six months of outstanding pay - earned during peacekeeping missions in Liberia - and also to get better living conditions, resulted in the deaths of two senior officers, former Defense Chief Verissimo Correia Seabre - who had led the coup - and Domingo Barros.

[29] In April 2005, former president João Bernardo 'Nino' Vieira returned from six years' exile in Portugal. Three months later he won the presidential election in the run-off with the 52.35 per cent of the vote. The PAIGC candidate Malam Bacai Sanhá had 47.65 per cent.

[30] Former Minister of the Interior Marcelino Lopes Cabral was arrested in April 2006, accused of giving support to Senegalese rebels who had been fighting Guinea-Bissau troops along the southern border of the country since March. ∎

Guyana / Guyana

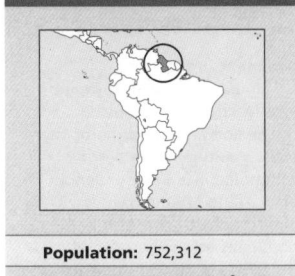

Population:	752,312
Land area:	214,970 km²
Capital:	Georgetown
Currency:	Guyana dollar
Language:	English

The original inhabitants of what is now Guyana, the Arawaks, were displaced from the area by the Caribs, warriors who dominated the region before moving on to the nearby islands which were later called after them.

2 Both the Arawaks and the Caribs were nomads. Organized into families of 15 to 20 people, they lived by fishing and hunting. There are thought to have been half a million inhabitants at the time of the arrival of Europeans in Guyana. There are around 45,000 Indians, divided into nine ethnic groups, of which seven maintain their cultural identity and traditions.

3 Led on by the legend of *El Dorado*, in 1616 the Dutch built the first fort. Guyana was made up of three colonies: Demerara, Berbice and Essequibo. But in 1796, the Dutch colony was taken over by the British, who had already begun a wide-scale introduction of slaves. A slave, Cuffy, led a rebellion in 1763 which was brutally put down. To this day, Cuffy is considered a national hero.

4 Those slaves that escaped from the plantations went into the forests to live with the indigenous peoples, giving rise to the 'bush blacks'. The English brought in Chinese, Javanese and Indian workers as cheap labor. Since 1950,

LAND USE

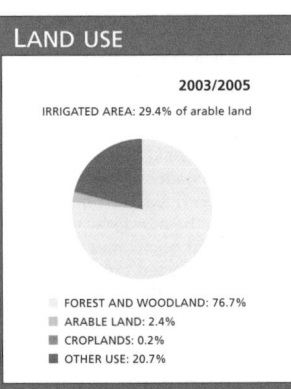

2003/2005

IRRIGATED AREA: 29.4% of arable land

- FOREST AND WOODLAND: 76.7%
- ARABLE LAND: 2.4%
- CROPLANDS: 0.2%
- OTHER USE: 20.7%

Guyana's population managed to channel independence ideas into a single movement, the People's Progressive Party (PPP), with policies of national independence and social improvements, and long-term aims for a socialist country. Cheddi Jagan, the first Prime Minister of the colony, was in power for three successive terms.

5 After years of violence, Britain recognized Guyana's independence within the Commonwealth on 26 May 1966. By that time, the PPP had split; most of the Afro-Guyanese population joined the People's National Congress (PNC), while indigenous people remained loyal to Jagan. Forbes Burnham, leader of the PNC, took office, supported by other ethnic minorities.

6 This process was influenced by ethnic conflict and by foreign interests, particularly from the US, which felt its hegemony in the Caribbean threatened by Jagan's socialism.

7 Even though Burnham came to power with Washington's blessing, he kept his distance. He declared himself in favor of non-alignment and proclaimed a Cooperative Republic in 1970. The bauxite, timber and sugar industries were nationalized in the first half of the 1970s, and by 1976 the State controlled 75 per cent of the country's economy. At the same time, regional integration was implemented through CARICOM (Caribbean Community), the Latin American Economic System (SELA), and the Caribbean Merchant Fleet.

8 In 1976, Cheddi Jagan stated the need to 'achieve national anti-imperialist unity', when disputes broke out with Brazil over the border. The PPP representatives returned to Parliament, from which they had withdrawn three years earlier to protest over electoral corruption. Afterwards, Burnham announced the creation of a Popular Militia.

9 Elections were postponed in order to hold a constitutional referendum, with Parliament drawing up a new constitution. This led to the PPP's withdrawal from legislative activity for the second time. In 1980 Burnham was elected President. According to international observers, the election had been plagued by fraud. Burnham granted authorization for transnational corporations to carry out oil and uranium operations and he turned to the IMF to obtain credit.

10 In June of the same year,

PROFILE

ENVIRONMENT
Ninety per cent of the population and most of the country's agriculture are concentrated on the coastal plain which ranges between 15 and 90 km in width. Rice and sugar cane are the main crops. As most of the shore is below sea level, dams and canals have been built to prevent flooding. The inner land consists of a 150 km-wide rainforest where the country's mineral resources are concentrated (bauxite, gold, and diamonds). To the west and south, the rest of the country is occupied by an ancient geological formation, the Guyana mountain range. Guyana is a native word meaning 'land of waters'. There are many rivers, as a result of the tropical climate and year-round rains.

SOCIETY
Peoples: Half the population are descended from Indian indentured workers, one-third from African natives and the rest are native Americans, mixed European and Indian descendants, Chinese and Europeans.
Religions: Protestant 34 per cent; Catholic 18 per cent; Hindu 34 per cent; Muslim 9 per cent. Traditional religions are practiced by American Indian groups.
Languages: English (official); Creole. Several indigenous languages are spoken, mainly belonging to Carib and Arawak linguistic groups. Hindi and Urdu are used in religious ceremonies.
Main Political Parties: People's Progressive Party (PPP), socialist; People's National Congress(PNC), socialist; coalition between: The Guyana Action Party and the Working People's Alliance; The United Force, conservative.
Main Social Organizations: Trade Union Congress (TUC) with 22 member unions. Youth National Council; Amerindian Peoples Association of Guyana.

THE STATE
Official Name: Cooperative Republic of Guyana.
Administrative Divisions: Ten Regions.
Capital: Georgetown 231,000 people (2003).
Other Cities: Linden 43,800 people; New Amsterdam 31,500 (2000).
Government: Bharrat Jagdeo, President since August 1999, re-elected in 2001. Samuel Hinds, Prime Minister since August 1999. Unicameral Legislature: National Assembly, with 65 members, of which 12 are regional representatives and 53 are elected through direct vote and proportional representation.
National Holiday: 23 February, Proclamation of the Republic (1970).
Armed Forces: 1,600 (1995). Other: 4,500 People's Militia, national service.

Life expectancy
65 years
2005-2010

GNI per capita
$1,020
2004

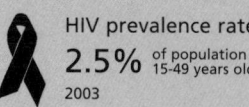
HIV prevalence rate
2.5% of population
15-49 years old
2003

Walter Rodney, the famous Guyanese intellectual and founder of the opposition Working People's Alliance (WPA) was killed by a car bomb. The culprits were never found.

[11] In the post-election period, border disputes escalated. Venezuela claimed the Essequibo region, approximately 159,000 sq km (three quarters) of Guyanese territory, arguing that British imperialism illegally deprived Venezuela of that area in the 19th century. In 1983, both countries turned to the UN. In 1985 direct negotiations started again to settle the dispute. Negotiations were focused on an outlet to the Atlantic Ocean for Venezuela.

[12] While financial difficulties increased during 1984 and the Government faced a new crisis in its relations with labor unions, Burnham resumed contacts with the IMF to obtain a $150 million loan, but considered the conditions on the loan 'unacceptable'. The US invasion of Grenada - and Guyana's criticism of this action - led to a deterioration of the relations between the two countries. Guyana made overtures to the socialist countries.

[13] Burnham died in 1985 and was replaced by Desmond Hoyte. The PNC won the general elections that year, but the opposition complained of alleged fraud. In 1986, five of the six opposition parties formed the Patriotic Coalition for Democracy, whereby all the seats went to the PNC. Hoyte announced in January 1987 that his Government would return to 'Co-operative Socialism'.

[14] Parliament met in December 1991, after the Government declared a state of emergency in order to postpone the elections planned for that month. The state of emergency was extended until June 1992. In October general elections, Cheddi Jagan defeated President Desmond Hoyte (54 per cent to 41 per cent).

[15] In 1993 President Jagan allowed US troops to train in the country's forests. He also accepted US military collaboration to combat drug-trafficking and to help bring clean water and sanitation to the interior. The President also intended to modify the adjustment plan launched by Hoyte, in agreement with the IMF. He proposed a market economy strategy to solve the problem of the poverty affecting 80 per cent of the population, whose emigration rate exceeded the demographic growth index. The celebration of the first anniversary of Jagan's

IN FOCUS

ENVIRONMENTAL CHALLENGES
Compared with deforestation elsewhere in the world, Guyana has suffered little and until 1990, only a small fraction of the extensive forests had been felled. However, foreign companies are pressing for the intensification of timber exploitation. In some areas, there has been no reforestation after logging, and this has led to soil erosion. There is water pollution from industrial and agricultural chemicals.

WOMEN'S RIGHTS
Women have been able to stand for election since 1945 and to vote since 1953. In 2003, women held 20 per cent of parliamentary seats and 22 per cent of ministerial positions.

That year, women made up 36 per cent of the country's workforce.

In 2004, female primary school attendance rate was 98 per cent. Prenatal care reached 81 per cent of women and 86 per cent of births received specialised assistance.

CHILDREN
Although they are still high, mortality rates for children under one and five years old dropped between 1990 and 2004. The former decreased

from 64 to 48 per 1,000 live births and the latter from 88 to 64 per 1,000 live births. The percentage of low birthweight was 12 per cent while 14 per cent of children under five years suffered from moderate or severe low weight. The same percentage of children suffered from moderate or severe stunting*.

INDIGENOUS PEOPLES/ ETHNIC MINORITIES
Ethnicity has been a key factor in Guyana's political history, even before independence. While the descendents of Africans were assimilated into the dominant European culture, the Indian community maintained its culture. Half of today's population is of Indian descent, 39 per cent are Afro-Guyanan and the rest are of European, Asian or indigenous origin.

African slaves were brought by the Dutch to work on sugar plantations in the 17th century. The British later took over the colony, and after the British Empire abolished slavery in 1838, a labor shortage on the plantations resulted as Afro-Guyanans moved to the cities or established collective farms. To replace slave labor, the British hired workers from China and India. The Indian population eventually became one of the country's

dominant groups.

Indian and Afro-Guyanan intermarriage, religious conversion and adoption of each other's cultures have become common. Since independence, each ruling regime has identified with one of the dominant ethnic groups. When the PPP is in power, the Afro-Guyanan minority complains of discrimination; and when the PNC rules, the Indo-Guyanan majority makes the complaints.

Although ethnic tensions seemed to decline between the late 1980s and early 1990s, tension rose during Janet Jagan's months in office, mainly sparked by allegations of electoral fraud; there was also resentment over the fact that she was born outside Guyana, in the US.

MIGRANTS/REFUGEES
Between January 2003 and September 2005, approximately 1,600 Guyanans sought refuge in the US and Canada.

DEATH PENALTY
The death penalty still applies for ordinary crimes.

* Latest data available in *The State of the World's Children* and *Childinfo* database, UNICEF, 2006.

Government was tarnished by a strike in the national electricity company, due to the Government's failure to implement its promise to increase public workers' pay by 300 per cent. In 1996, the country had $500 million - nearly a quarter - of its foreign debt pardoned.

[16] Following Jagan's death in 1997, his wife Janet Jagan took over as interim Prime Minister. In the December elections she was elected President with 55.3 per cent of the vote against the 40.6 per cent for Hoyte's PNC. Sam Hinds was appointed Prime Minister. After serving for 20 months, the 79-year-old President resigned for health reasons. Her Economy minister, Bharrat Jagdeo, took her place. The National People's Party criticized this transition of power.

[17] In March 2000, Venezuelan President Hugo Chávez reiterated his country's claims to the Essequibo region, but affirmed that Venezuela would submit to UN arbitration. Chávez also criticized a US company planning to build a rocket-launching site in the area in question.

[18] The permit granted by

Guyana to a Canadian oil company for the exploitation of territorial waters caused conflict with Suriname. After a round of talks mediated by Jamaican Prime Minister Percival James Patterson, agreement was reached to hold future meetings between both parties to decide the future of the region.

[19] In March 2001, the general election planned for 1997 was held, with international observers who confirmed that virtually everything was in good order, in spite of some irregularities due to the absence of some voters' names in the register. Finally, Jagdeo was re-elected.

[20] According to a World Trade Organization (WTO) report published in October 2003, Guyana's economy is now largely dependent on natural resources such as sugar, gold, bauxite and rice. Production of these resources has been growing very slowly over the last 15 years despite the liberalization of Guyana's trade and investment policies.

[21] In January 2004 the Paris Club of creditor countries agreed to reduce Guyana's debt by $95 million, under the enhanced

Heavily Indebted Poor Countries (HIPC) Initiative, created by the IMF and the World Bank. Most creditors also committed to grant additional debt relief to Guyana so that the debt will be reduced by a further $33 million.

[22] In face of the pressure from opposition leaders to investigate a possible ministerial involvement in death squads - accused of killing hundreds of suspected criminals - Home Affairs minister Ronald Gajraj resigned in May 2004 to allow an inquiry to proceed. In June, the UN set up a tribunal to settle the long-standing Guyana-Suriname maritime boundary dispute.

[23] In April 2005, Gajraj was re-appointed as Home Affairs minister after inquiries cleared him of direct involvement with death squads.

[24] In April 2006, Agriculture minister Satyadeow Sawh was shot together with two members of his family. The murder was part of a series of gun crimes with previously selected targets that had resulted in the deaths of 50 people since January. The police linked these crimes with groups involved in illegal arms trade and drug-trafficking. ■

Haiti / Haïti

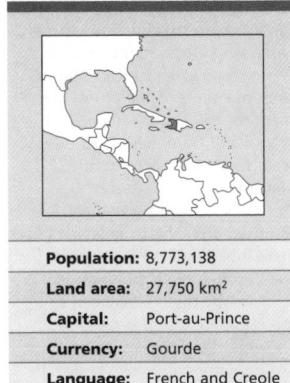

Population:	8,773,138
Land area:	27,750 km²
Capital:	Port-au-Prince
Currency:	Gourde
Language:	French and Creole

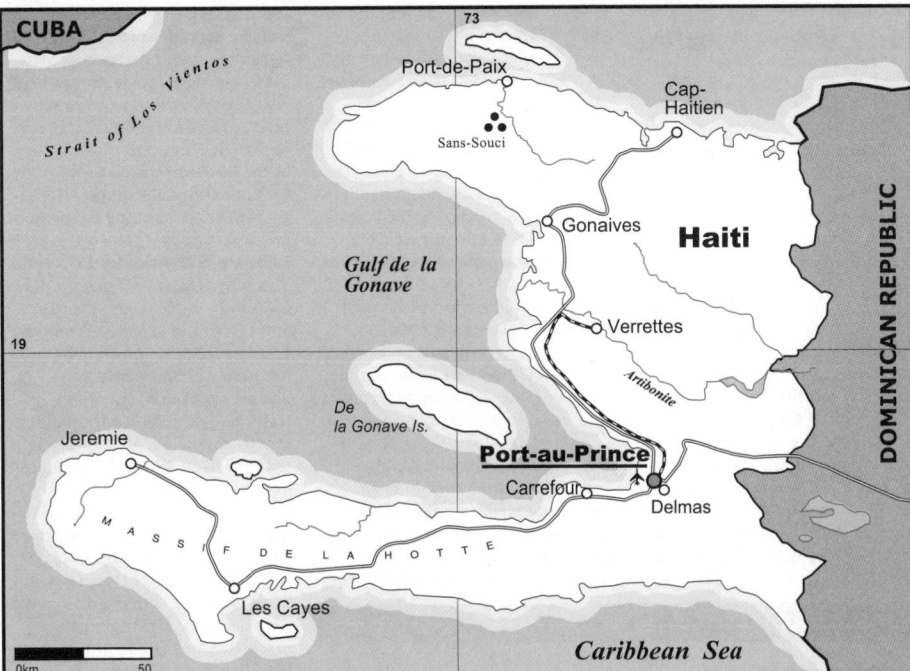

The island of Hispaniola or Quisqueya, its local name, is currently divided into two republics: Haiti and the Dominican Republic (see Dominican Republic). When it was 'discovered' by Christopher Columbus in 1492, the island was inhabited by numerous Arawak peoples, who almost entirely disappeared over the next few decades. Spanish colonists, supported by the Dominican missionaries, called the islands after their patron saint, St Dominic (Santo Domingo). The island was later colonized by the French and other European settlers who were attracted by the sugar plantations. Disputes arose among the Europeans and in 1697 Spain ceded the west of the island to France, under the Treaty of Ryswick.

[2] After gaining control of the island, France began to exploit it, introducing about 20,000 African slaves per year, leading to rapid racial mixing. Sugar soon became the principal export product of the region, and during the 18th century Haiti became the most important French possession in the Americas.

[3] This prosperity was based on African slave labor. By 1789, the number of African slaves in the colony had reached 480,000. There were 60,000 mulattos and free 'colored' people, while the rich land-owning Europeans constituted a minority of no more than 20,000. The Haitians were influenced by

the revolutionary movement that had started in the colonial capitals and they waged a revolutionary war, led by former slave Toussaint L'Ouverture. The war lasted from 1791 to 1803 and ended with the proclamation of the first black republic in the world.

[4] It was L'Ouverture who gave the *marrons* or *quilombolas* (rebel slaves) their direction, rallying them to the call of 'general freedom for all', transforming the different groups into a disciplined army. On 4 February 1794, taking advantage of the splits in the French colonial system, he succeeded in getting the French National Convention to ratify a decree abolishing slavery in Santo Domingo and appointing himself as a general. After the coup of the 18 Brumaire (1799), Napoleon Bonaparte sent a large military expedition to reconquer the colony and re-establish slavery. L'Ouverture responded with a general uprising, but he was

imprisoned and died in exile in France in 1803.

[5] Jean-Jacques Dessalines took over leadership of the war of independence, aided by Henri Christophe and Alexandre Pétion, who together radicalized L'Ouvertures's legacy. They succeeded in uniting the Africans and mulattos and after a series of heroic campaigns, they forced the French troops to capitulate. Independence was proclaimed on 28 November 1803, and Haiti became the first independent state in Latin America.

[6] In 1818, JP Boyer was elected president instead of Pétion. Boyer recovered the north of the country in 1820, putting an end to Christophe's monarchic experiment. Two years later he conquered Santo Domingo in the east of the island, thus achieving a fragile reunification, which lasted for a quarter of a century. In 1843, a revolution led by the Santo Domingan Creoles divided the island into two definite, independent States: the Dominican Republic in the east, and the Republic of Haiti in the west.

[7] From 1867, a bloody civil war launched a period of political instability and economic crisis which lasted until 1915, when the country was occupied by US marines for non-compliance with 'its commitments'. A year later the US invaded the Dominican Republic, gaining control of the whole island.

[8] The invasion of Haiti was heroically resisted by Charlemagne Péralte's 'Revolutionary Army'. Péralte was treacherously murdered

in 1919. The US troops finally defeated the resistance and controlled the country until 1934, turning it into a virtual colony. That year, President Vincent succeeded in getting the US troops to withdraw from the island, but he could not eliminate US influence from the country's domestic affairs.

[9] The national army, or Garde d'Haiti, took a central role in national politics, staging coups against presidents Lescot, 1941-1946; Estime, 1946-1950; and Magloire, 1950-1957. In 1957, François Duvalier, a middle-class doctor, seized power supported by the army and the US.

[10] The army, the commercial bourgeoisie, the ecclesiastic authorities, the state bureaucracy, and the US State Department used Duvalier to control the country for over 30 years. In 1964, Duvalier or 'Papa Doc' proclaimed himself President-for-life, passing on the title to his son Jean-Claude or 'Baby Doc' on his death in 1971.

[11] Assailed on the international level by continual condemnations of human rights violations, and on the domestic level by active opposition, Jean-Claude Duvalier's government called elections in 1984. Sixty-one per cent of the population abstained. The opposition grew, organizing itself into parties and trade unions, while the regime was becoming a burden to the US.

[12] Repression grew, and by 1985 it was estimated that Baby Doc's regime had been responsible for 40,000 murders. The country was enveloped by a growing wave of protests and strikes. Duvalier fled

LAND USE

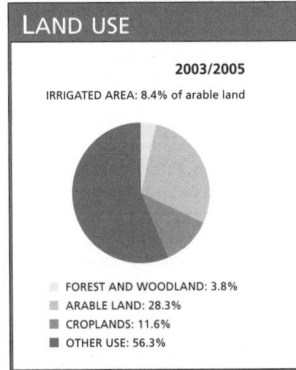

2003/2005

IRRIGATED AREA: 8.4% of arable land

- FOREST AND WOODLAND: 3.8%
- ARABLE LAND: 28.3%
- CROPLANDS: 11.6%
- OTHER USE: 56.3%

WORKERS

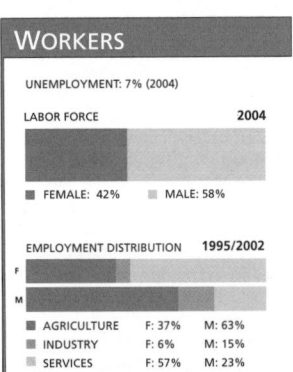

UNEMPLOYMENT: 7% (2004)

LABOR FORCE	2004

- FEMALE: 42%   ■ MALE: 58%

EMPLOYMENT DISTRIBUTION	1995/2002

F
M

■ AGRICULTURE	F: 37%	M: 63%
■ INDUSTRY	F: 6%	M: 15%
■ SERVICES	F: 57%	M: 23%

the country in a US airforce plane and received temporary asylum in France.

[13] A National Governing Council (CNG) led by General Henri Namphu assumed control of the government, promising 'free and direct' elections by the end of 1987.

[14] The dictator's flight did not end the mobilization of the people. Mass lynching by Tontons-Macoutes forced the National Governing Council to dissolve this repressive force.

[15] In October 1986, the CNG called elections to elect a Constituent Assembly to draw up a new constitution. Less than 10 per cent of the three million Haitians participated in the election and, in March 1987, a referendum approved the new constitution with 99.8 per cent of the vote. The new constitution established a Parliamentary system, limited the presidential term to five years, and

divided the power with a prime minister chosen by Parliament.

[16] The elections were to be held in November 1987, but a few hours after the polling stations were opened, they were sabotaged by factions of the armed forces and by former Tontons-Macoutes, and the elections were suspended, finally taking place in January 1988. In a very volatile atmosphere, Leslie Manigat, the 'official' candidate, was elected, only to be deposed in June, in a coup led by General Namphu. In September 1988, a movement of sergeants and soldiers deposed General Namphu, putting Prosper Avril, the éminence grise of the Duvalier period, into power.

[17] In March 1990, General Avril was ousted by General Abraham, who relinquished control to a provisional civilian government headed by Judge Ertha Pascal-Trouillot, the first woman to occupy the presidency in Haiti. The

provisional government created suitable conditions to put the Constitution into practice, and called elections for December 1990.

[18] These elections were won by a priest, Jean-Bertrand Aristide, who obtained 67 per cent of the vote as the leader of the National Front for Change and Democracy (FNCD). He was voted in mostly by the poor urban sectors, and took office on 7 February 1991. Aristide, a Liberation Theology activist, had been censured in 1988 by the church authorities and expelled from the Salesian order. His governmental program was based on a war against corruption and drug trafficking, including a thorough literacy campaign, and a project to move from 'extreme poverty to poverty with dignity'.

[19] On 30 September, General Raoul Cédras staged a bloody coup. In protest the Organization of American States (OAS) declared a trade embargo, starting diplomatic negotiations in the region and in the UN. Meanwhile, the rebels tried to avoid international isolation by officially recognizing the sovereignty and operation of parliament.

[20] In February 1992, OAS representatives, Haitian members of parliament, and the deposed Aristide signed an agreement in Washington to re-establish democracy and reinstate the former president.

[21] In January, the de facto government held legislative elections which were designed only to partially replace Parliament. Less than 3 per cent of those registered to vote took part. Months later, Marc Bazin resigned as Prime Minister.

[22] On 27 June, indirect talks began in New York between General Cédras and ousted president Aristide. Meanwhile, the UN Security Council imposed a financial, oil and arms embargo on Haiti. In July, Aristide and Cédras signed an agreement that guaranteed the return of the President as well as an amnesty for all military leaders involved in the coup. In accordance with the agreement, Aristide named Robert Malval Prime Minister.

[23] However, a new wave of violence soon broke out in Haiti, to prevent the agreement from going into effect. In October, a US warship patrolled the coast near the capital. An armed mob threatened the American troops, so US President Clinton ordered the ship back to US Guantanamo military base in Cuba. In the meantime, the Security Council reinstated the naval embargo.

[24] Malval's Justice Minister, Guy Malary, was assassinated in 1993. The attack was carried out almost

in the same place where Antoine Izmery, a pro-Aristide businessman, had been killed a month earlier. Those responsible for the killings were members of the pro-military Front for the Advancement and Progress of Haiti, who used the same tactics as the Tontons-Macoutes.

[25] On 15 October 1994, Aristide returned after the coup leaders had gone into exile and the country was occupied by a multinational force led by the US. While in exile, the President had promised to implement a structural adjustment program prescribed by the IMF. In December, troops were demobilized in order to create a new national police force.

[26] In spite of meaning to punish those responsible for human rights abuses, Aristide - under sustained pressure from Washington - was forced to offer merely symbolic gestures in most cases, such as a gravestone in memory of the victims of death squads.

[27] In 1995, a contingent of UN troops replaced the multinational forces. In November, René Préval, an Aristide supporter, won the elections with 88 per cent of the vote. The new President took office on 7 February 1996, in a country where 80 per cent of the population was living below the poverty line. Préval requested the UN peace forces to remain for an additional period of time due to the numerous conflicts and acts of violence which involved national police. Kidnappings and murders - including those of eight policemen out of duty - did not stop, while popular uprisings were violently repressed.

[28] In July 1996, Claude Raymond, a general and minister in Duvalier's cabinet, was arrested for terrorist activities. Four days later, André Armand, former Army sergeant and leader of the retired soldiers lobby group, was killed by unknown assailants after publicly stating that retired army members were plotting to assassinate Préval and former president Aristide. The resignation of Prime Minister Rosny Smarth in 1997 and the President's decision to dissolve Parliament and govern by decree heightened political tensions and confrontation with the opposition. Jacques-Edouard Alexis, of the Lavalas political organization, was named in March 1999 to take over as Prime Minister.

[29] Elections slated for March of that year were postponed until May in order to resolve voter registration problems. One month after the elections and under pressure from violent and strong protests that paralyzed the island, the Electoral Council announced the results which gave Lavalas 16 out of the 17 Senate seats. The US, UN and OAS

PROFILE

ENVIRONMENT

Haiti occupies the western third of the island of Hispaniola, the second largest of the Greater Antilles. Two main mountain ranges run from east to west, extending along the country's northern and southern peninsulas, contributing to their shape. The hills and river basins in between form the center of Haiti. The flatlands that open up to the sea in the west, are protected from the humid trade winds by the mountains to the north and east. Coffee is the main export product.

SOCIETY

Peoples: Nearly 95 per cent of Haitians are descendants of African slaves. There are also minorities of European and Asian origin, and integration has produced a small mestizo group. Thousands of Haitians have emigrated in recent years, especially to Colombia, Venezuela and the United States.
Religions: Voodoo, a mixture of Christianity and various African beliefs. Catholics (80 per cent); Protestants (16 per cent).
Languages: French and Creole (both official). French is spoken by less than 20 per cent of the population. Most people speak creole, a combination of Spanish, English, French and African languages.
Main Political Parties: Front for Hope (Fwon Lespwal/Front de l'Espoir); Rally of Progressive National Democrats (Rassemblement des Démocrats Nationaux Progressistes); Respect (Respè); Christian National Union for the Reconstruction of Haiti (Union Nationale Chrétienne pour la Reconstruction d'Haiti); Christian Movement for a New Haiti (Mouvement Chrétien pour Batir une Nouvelle Haiti)
Main Social Organizations: The 'Grassroots' Church groups, of Catholic origin; Creole Language Movement; Solidarity of the Women of Haiti (SOFA); Confederation of Haitian Workers; Federation of Workers Trade Unions; Movement of Support for Victims of Violence (MAP VIV); Platform of Haitian Human Rights Organizations (POHDH).

THE STATE

Official Name: Republik Dayti.
Administrative Divisions: 9 departments.
Capital: Port-au-Prince 1,961,000 people (2003).
Other Cities: Carrefour 356,400 people; Delmas 301,200; Cap-Haitien 119,400 (2000).
Government: René Préval, President since 2006. Jacques-Édouard Alexis, Prime Minister since 2006. The National Assembly (Legislature) has two chambers: the Chamber of Deputies, with 83 members, and the Senate, with 27 members.
National Holiday: 1 January, Independence Day (1804).
Armed Forces: 1,500.

|  Under-5 mortality **117** per 1,000 live births 2004 |  Poverty **53.9%** of population living on less than $1 per day 2001 |  Debt service **4.0%** exports of goods and services 2003 | Maternal mortality **680** per 100,000 live births 2000 |

IN FOCUS

ENVIRONMENTAL CHALLENGES

Copper extraction ceased in 1976 and bauxite deposits are almost exhausted. The northern coastline has the heaviest rainfall and contains the country's most developed area, though the land there is suffering from serious erosion. The felling of trees in order to use land for agriculture and wood as fuel, has accelerated the process of erosion. Forests make up less than 2 per cent of the land area.

WOMEN'S RIGHTS

Women have been able to vote and stand for office since 1950. In the context of many years of extreme economic and political instability, the vulnerability of children and women rose significantly.

Skilled health staff attends only 24 per cent of births and the maternal mortality rate is 680 per 100,000 live births*. Chronic malnutrition is widely spread among pregnant women. The fertility rate is four children per woman*.

Many women enter the labor market at an early age; about 10 per cent of girls aged between five and nine and 33 per cent aged between ten and 14 can be considered economically active.

Although there are no recent studies available, domestic violence is a phenomenon that grows on a daily basis. According to UNICEF, the country's context of instability has favored the increase in domestic violence.

CHILDREN

The school enrolment rate is 60 per cent; boys attend school for just four years on average, and girls for half of that time. Meanwhile, only 20 per cent of the population has access to secondary education*.

The infant mortality rate is 74 deaths per 1,000 live births*. Roughly 20 per cent of children dying under the age of five in Latin America and the Caribbean are Haitian. The percentage of children immunized against measles before their first birthday is less than half the percentage registered in Sub-Saharan Africa, the world's poorest region*. Only 40 per cent of boys and girls have access to the most basic health services. The main causes of infant mortality are diarrhea, acute respiratory infections and malnutrition.

More than 200,000 boys and girls have lost one or both parents to HIV/AIDS; yet, only 20 per cent of young people know how to protect themselves from the virus*.

INDIGENOUS PEOPLES/ ETHNIC MINORITIES

When Christopher Columbus arrived, in 1492, numerous Arawak people were living on the island. Over the next few decades they almost entirely disappeared as a result of death and disease.

MIGRANTS/REFUGEES

There has been a considerable (though not measurable) number of internally displaced Haitians since the last wave of political and social conflicts started in 2004.

Some 2,000 Haitians fled the country in February of that year to seek asylum in the US. At least 90 per cent of them were stopped by the US Cost Guard and repatriated to Haiti.

In 2005, the number of Haitians fleeing their homeland to seek asylum in other countries rose to 27 per cent. Of them, 51 per cent requested asylum in the US, 46 per cent in Guadeloupe and the rest in Dominican Republic, Jamaica and Cuba.

DEATH PENALTY

It was abolished in 1987.

* Latest data available in The State of the World's Children and Childinfo database, UNICEF, 2006.

---

questioned the recount.

[30] In April 2000, the country's most noted journalist, Jean Dominique, owner of Radio Haiti Inter, was assassinated on arrival at work. The case, which had not been solved two years later, became the best-financed and organized crime investigation in the country's history. A senator loyal to Aristide used his legislative immunity to avoid testimony, evidence for the case disappeared, and one of the witnesses was killed and his body vanished.

[31] The opposition boycotted the February 2001 presidential elections, forcing their leading candidates not to participate. They appointed an 'alternative president'. Despite the boycott and the lack of international observers, the elections took place and gave an overwhelming victory to Aristide, who thus took office as President for a third time. Talks between Aristide's governing party and the opposition alliance failed in an attempt to establish a new Electoral Council to review the ballot. The motive of the dispute was the number of parliamentary seats belonging to each party. International economic aid had been suspended since the 2000 elections.

[32] In 2002, most of Haiti's resources were in the hands of 15 per cent of the population (the one per cent of the population of European origin owned half of the country's wealth). Seventy-three per cent of Haitians were living in extreme poverty and lacked access to sanitation, water, and electricity. Two-thirds of Haitians did not eat a proper meal a day; the remaining one-third ate only once a day, usually a meal lacking in basic nutrients.

[33] During 2002, at least 30 journalists were attacked or threatened with death by alleged government supporters, and several reporters and their relatives decided to leave Haiti. The Inter-American Commission for Human Rights (ICHR) published a report expressing concern about the weakness of the state of law in Haiti and about the threats to several journalists.

[34] On 1 January 2004, Haiti celebrated the 200th anniversary of independence with large demonstrations throughout the country. The opposition used it as an opportunity to launch a general strike. The opposition - the 'Group 184', which comprised that number of political parties and civil society organizations - took advantage of the anniversary to draw the international community's attention to the possibility of rigged legislative elections in 2004 and of Aristide's attempt to be re-elected in 2005. The President of the Caribean Community (CARICOM) and Jamaica's Prime Minister, Percival Patterson, sent Aristide a letter, pointing out that the Caribbean 'is worried about reports on the increase of political instability in Haiti'.

[35] The OAS continued to put pressure both on the Government and the opposition in order to negotiate the setting up of the Electoral Council that would monitor and validate the legitimacy of the 2004 elections. The opposition requested the creation of a transitional government, headed by a representative of the Supreme Court and a 9-member council. In February 2004, the uprising against the government intensified. That same month, Washington and Paris called for Aristide's resignation. Washington sent troops for a multinational army composed mainly of French and Canadian soldiers and contracted a plane to carry Aristide and his small party to the Central African Republic. Once on African soil, Aristide announced he had been kidnapped. In March, CARICOM representatives reiterated a call for an investigation under the auspices of the UN. France and the US announced they would veto any UN investigation.

[36] In July, at a donors conference held in Washington, different countries and multilateral institutions pledged to provide more than $1 billion, in loans and grants, for Haiti's reconstruction over the two following years. Resources would be targeted at improving the security forces and health and energy infrastructure, in order to strengthen social and political areas. The funds aimed to create 31,000 public sector jobs and re-open 1,500 schools.

[37] In September, tropical storm Jeanne left more than 600 people dead and over 80,000 injured in the north-west of the country. In the northern region of Artibonite, only in the city of Gonaives, 500 people were killed, most of them being children. The main roads were transformed into fast-flowing rivers. Half of the city of Gonaives remained under water. The almost complete deforestation of the country contributed to the devastating effect of floods.

[38] One year after the coup, the interim government - which had received unprecedented support from Washington - was a complete failure. In spite of the presence of UN blue helmets, violence was widespread and security non-existent. Dead bodies were found in the streets and armed bands made up of former soldiers moved with absolute freedom. Several members of the Lavalas party had been murdered, schools and hospitals were being closed and roads were completely destroyed. By May 2005, some 10,000 Haitians had been killed and more than 1,000 were in jail, most of them Aristide followers.

[39] The Electoral Council declared René Préval winner of the February 2006 general elections, after the parties reached a deal to change the way blank ballot papers were allocated, thus avoiding a second round of voting.

[40] In June 2006, a new democratic government was finally sworn in under Prime Minister Jacques-Edouard Alexis, bringing two years of uncertainty to a close. In July, major donors agreed to an aid package of $750 million to help economic recovery. ∎

Honduras / Honduras

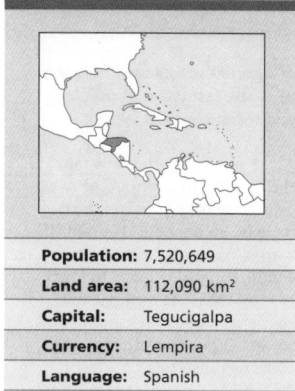

Population:	7,520,649
Land area:	112,090 km²
Capital:	Tegucigalpa
Currency:	Lempira
Language:	Spanish

E vidence shows the Copan valley, in the western part of present-day Honduras, has been occupied for about 4,000 years by Maya people, although the city of Copan was inhabited for only about 500 years. Copan reached its peak around 500-800 AD during the Classic Period, when about 15,000 people lived in the area, but like other Maya city-states it was abandoned mysteriously around 900 AD. A variety of other peoples lived in the rest of the territory, including Chibcha (see Colombia) and Lenca.

2   Gold stimulated Spanish conquest of the area early in the 16th century. Despite strong resistance from the Native Americans led by Lempira, Pedro de Alvarado - who was in charge of the final Spanish conquest - joined the territory to the Captaincy-General of Guatemala. The initial mining centers were located near the Guatemalan border. The demand for labor led to revolts and accelerated the decimation of the native population. As a result, African slaves were brought in to Honduras, and by 1545 the captaincy may have had as many as 2,000 slaves.

3   For long periods the Spanish utilized a soft defense against the Caribbean pirate attacks; thus, the British came to control the Mosquito region on the Caribbean coast. In the 18th century, however, the Spanish Bourbon kings made a sustained effort to recover these areas, and their success in the Gulf of Honduras was made evident with the completion of a fort at Omoa by 1779.

4   In 1821, Honduras gained independence from Spain. Together with the other Central American provinces it joined the short-lived Mexican Empire of Iturbide, which collapsed two years later. Francisco Morazán and other Honduran leaders sought in vain to set up an independent Central American federation. Their efforts were no match for Britain's 'Balkanization' tactics.

5   With the liberal reform of 1880, mining became the backbone of the economy. To encourage the development of this sector the country was opened to foreign investment and technology. Toward the end of the 19th century, the United Fruit Company (Unifruco) was established in the country; the US company took over vast tracts of land. It produced almost the entire fruit output of the country, ruled railroads, ships and ports, and dictated many key political decisions.

6   US Marines invaded Honduras in 1924, imposing a formal democracy and allowing Unifruco to establish a monopoly in banana production by buying out its main competitor, the Cuyamel Fruit Company. Washington eventually handed power over to Tiburcio Carías Andino, who governed Honduras from 1933 to 1949.

7   Border disputes with Guatemala led to US arbitration in 1930. In 1969, friction resulting from Salvadoran peasants emigrating to Honduras led to a further war, which was finally ended with mediation by the Organization of American States (see El Salvador).

8   In 1971, nationalists and liberals signed the Unity Pact. General Osvaldo López Arellano, in power since 1963, permitted elections and Ramón Ernesto Cruz of the National Party was elected President.

9   In 1972 López Arellano overthrew the Cruz administration. He demonstrated his sensitivity to peasant demands for land reform and began to impose controls upon United Brands (as Unifruco was now called). The affected parties reacted, and López Arellano was replaced by Colonel Juan A Melgar Castro.

10   The army commander-in-chief, General Policarpo Paz García, took power in 1978. This regime became closely allied to that of the dictator Anastasio Somoza in neighboring Nicaragua. Nicaragua's Sandinista Revolution hastened the election of a constituent assembly in Honduras which promptly ratified Paz García as President.

11   When Roberto Suazo Córdova, assumed the presidency in January 1982, he authorized increases in the price of consumer goods and enacted an 'anti-terrorist' law forbidding strikes as 'intrinsically subversive'. Death squads acted with impunity and opposition political figures 'disappeared' daily.

12   Honduras tolerated the presence of US troops and Nicaraguan counter-revolutionary bases in its territory. It is estimated that, in 1983, the Pentagon had 1,200 soldiers in Honduras. Apart from intervening directly in armed operations, they gave military instruction, logistic support and built infrastructure works. The Nicaraguan 'contras' had some 15,000 fighters, together with some 30,000 Nicaraguan refugees.

13   Rafael Callejas, candidate for the National Party, had an easy victory in 1989 in elections that were considered fraudulent. Backed by the US and business circles, Callejas began a complete liberalization of the economy.

14   In 1990, the Government decreed an amnesty for political prisoners and outlaws. The anti-terrorist law was abolished and a climate of complete political consensus was created. The murder of opposition members

PROFILE

ENVIRONMENT
Mountains and rainforests cover 80 per cent of the land. Both population and economic activity are concentrated along the Caribbean coast and in the southern highlands, close to the border with El Salvador. The coastal plains have the largest banana plantations in Central America. Coffee, tobacco and corn are grown in the southern part of the country. Electricity is widely available. Approval is underway for a hydroelectric project in Piedras Amarillas, which will be one of the largest in the country, capable of generating 100 mW.

SOCIETY
Peoples: Most Hondurans are of mixed Mayan and European descent. There are 10 per cent of Native Americans and 2 per cent of African descent. The Garifunas, descendants of fugitive slaves and Native Americans, live along the Caribbean coast and on the nearby islands, maintaining their traditional lifestyles.
Religions: Roman Catholic 85 per cent; Protestant 10 per cent.
Languages: Spanish (official), Garifuna, various indigenous languages (such as Lenca and Miskito); a small number of people speak English.
Main Political Parties: Liberal Party of Honduras (LPH); National Party (NP); Democratic Unification Party (DUP); Christian Democratic Party of Honduras (CDPH); Party for Innovation and Unity-Social-Democracy (PIU).
Main Social Organizations: The Confederation of Honduran Workers (CTH) founded in 1964 and affiliated to the ORIT (Regional Interamerican Labor Organization). Also: The General Workers' Central Union (CGT), (Social-Christian), the United Federation of Workers (FUT), the Federation of Honduran Workers' Unions (FESITRAH), the Independent Workers' Federation and United National Peasants' Front of Honduras (FUNACAMPH).

THE STATE
Official Name: República de Honduras.
Administrative Divisions: 18 departments.
Capital: Tegucigalpa 1,007,000 people (2003).
Other Cities: San Pedro Sula 616,500 people; Ceiba 108,900; El Progreso 106,500; Choluteca 93,100 (2000).
Government: Manuel Zelaya, President since January 2006. Unicameral Legislature: National Congress, with 128 members.
National Holiday: 15 September, Independence Day (1821).
Armed Forces: 18,800, including 13,200 conscripts (1995). Other: 10,000 members of the Public Security Force.

 **Life expectancy**
69 years
2005-2010

 **GNI per capita**
$1,040
2004

 **Literacy**
80% total adult rate
2000-2004

HIV prevalence rate
1.8% of population 15-49 years old
2003

and other abuses committed by the military were denounced by the Honduran Committee in Defense of Human Rights.

[15] Opposition candidate Carlos Roberto Reina triumphed in the 1993 elections. One of the Government's first resolutions was the dissolution of the National Board of Investigations, accused of torturing prisoners.

[16] While the armed forces kept policing the cities, the Legislative Assembly began the process of constitutional reform which resulted in granting control of public security forces to the civilian power. The Unit of Criminal Investigation, led by civilians, began operating in January 1995, replacing the secret police dissolved the previous year. The new body, formed initially by 1,500 agents, was trained by Israeli police and the US Federal Bureau of Investigation (FBI). At the time, more than 50 people were murdered each day in Honduras.

[17] Senior government officials were jailed in 1995 for their involvement in the sale of official passports. The Supreme Court of Justice revoked former president Callejas' immunity so he could testify regarding the forged documents and misappropriation of public funds. President Reina was also investigated on the use of state funds for private purposes.

[18] On taking office in January 1998, President Carlos Flores was willing to form a national unity government and to reach a truce with the opposition linked to the military. With 80 per cent of the population living in extreme poverty, 228 landowners held more than 75 per cent of the country's lands.

[19] In October that year, the destruction wreaked by Hurricane Mitch amounted to over $5.36 billion. It killed 24,000 people throughout Central America, 14,000 of whom were Hondurans. Likewise, two million Hondurans lost their homes. One year later floods killed 35 people and covered 14,000 hectares of arable land, causing $20 million in losses. The authorities stated that, had they received the aid promised by developed countries after Mitch had struck, they could have dredged the rivers and avoided the new catastrophe.

[20] The conflict between residents in Olancho and the Energisa corporation intensified in 2001. The company was planning to install a hydroelectric project in a protected area of the Sierra de Agalta National Park, known as the 'Meso-American biological

IN FOCUS

ENVIRONMENTAL CHALLENGES
Deforestation and uncontrolled development are contributing to soil deterioration. Mining activities are causing water pollution, especially in Yojoa Lake, the country's largest source of fresh water.

WOMEN'S RIGHTS
Women have been able to vote and stand for office since 1955. In 2005, women held about 24 per cent of seats in Parliament, while in ministerial or equivalent positions their representation stood at 14 per cent.

Women made up 35 per cent of the total labor force in 2003 (of them, 9 per cent worked in agriculture, 25 per cent in industry and 67 per cent in services). Thirty-five per cent of pregnant women are anemic*. The overall fertility rate is estimated at 4 children per woman*. Health skilled staff attend 56 per cent of births, while 83 per cent of women receive prenatal care*.

CHILDREN
Sixty-eight per cent of Honduran families live in poverty (75 per cent in rural areas and 57 per cent in cities)*. High unemployment rates, which oscillate between 15-20 per cent, have been feeding the informal economy since the early 1990s. Approximately 400,000

children and adolescents aged 5-18 work*. Of these, 73.6 per cent are boys and 69.2 per cent come from rural areas; almost half are under 15. Only 39 per cent of working children are paid. Pre-school coverage (among children of the corresponding age) is 35 per cent and primary school enrolment rate reaches 86 per cent*.

Thirty-five per cent of children under 6 months old are exclusively breast-fed*; 14 per cent of newborns are underweight and 29 per cent of children under five years old have moderately or severely stunted growth*. Almost 4,000 children under 14 are living with HIV/AIDS*. During 2001 and 2002 there were many extra-judicial executions of youth gang members, and in total there were over 1,500 deaths of children and young people. In many cases, there was official acquiescence in the security forces' participation.

Honduras has achieved 13 of the 27 World Summit for Children goals and has enacted the Children and Adolescents Plan of Action (PANNA). Among the achievements, child mortality rates have fallen significantly and there has been an increase in coverage and access to primary school*.

INDIGENOUS PEOPLE/ETHNIC MINORITIES
There are eight main groups: the Lenca, Pech, Garifuna, Chortis, Tawahkas, Tolupanes, Xicaques

and Miskito - who comprise the majority within the indigenous population - and the English-speaking Afro-American population. The country's indigenous population is estimated at less than 500,000 people, which accounts for nearly 8 per cent of the total. Their main claim, and the main reason for their persecution by those with power, is their demand for 14,000 hectares of land in the west; an issue that is likely to lead to conflict. The Miskito are the poorest sector of society, and they live isolated from the other indigenous peoples in the rural southeast and on the Caribbean coast. This contributes to their lack of access to educational and health services.

MIGRANTS/REFUGEES
Internal migration has shown that women mainly migrate toward cities and men toward agricultural areas. Between 1995 and 1989, migration shows a net negative balance: more people left than came into the country.

DEATH PENALTY
The death penalty was abolished in 1956.

* Latest data available in The State of the World's Children and Childinfo database, UNICEF, 2006.

corridor', located in central Honduras. According to residents, the project had been authorized without having performed the due environmental impact assessments. Four activists were murdered and demonstrations were brutally repressed.

[21] In January 2002 the National Party candidate, Ricardo Maduro, became the country's new President after receiving 52.2 per cent of the votes. Shortly after his coming to office, Maduro - who had promised he would put an end to the growing violence and delinquency - formed a force of 10,000 soldiers to fight crime.

[22] The year 2004 started with heated public debate on the so-called 'anti-maras' law ('maras' is a term used to refer to youth gangs) which had been promulgated in 2003. Some jurists considered its provisions unconstitutional, discriminatory and brutally repressive. There was pressure on various fronts to rescind it, but President Maduro declared that in case the derogation was effected,

Congress would pass an equally severe law to replace it.

[23] In April 2004, Maduro called for an early withdrawal of Honduran troops deployed in Iraq (which should be returning in July), after the Spanish Government announced the immediate withdrawal of its troops.

[24] In October, Congress voted unanimously in favor of a Constitutional reform to ban same-sex marriage and adoption by same-sex couples, as a result of pressure exerted by the Catholic and Evangelical Churches. The amendment was due to a decision made by Maduro's Government to grant legal status, for the first time in the country, to gay rights groups that had been waiting 15 years for such decision.

[25] By mid-2005, the Committee of Relatives of Prisoners-Missing Persons in Honduras (COFADEH) reported threats, arbitrary arrests, house searches and harassment against several

activists who continued to oppose the-hydroelectric complex in Sierra de Agalta.

[26] Manuel Zelaya, of the opposition Liberal Party, won the presidential elections held in November. Upon taking office in January he was in charge of the second poorest country in Central America and the third in Latin America, following Haiti and Nicaragua. Six out of 10 Honduran people were living in poverty and, among them, four were living in extreme poverty. Although the country had been granted the partial cancellation of its foreign debt - which was reduced from $5.08 to $2.21 billion - on condition that the remaining funds were allocated to poverty reduction, Honduran people were still to see the results.

[27] In March 2006, in spite of protests from different sectors of the population, Honduras became the second country in Central America, following El Salvador, to ratify a free trade agreement with the US. ∎

Hungary / Magyarország

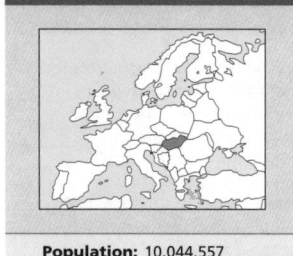

Population:	10,044,557
Land area:	93,030 km²
Capital:	Budapest
Currency:	Forint
Language:	Hungarian

The Roman Empire extended into the western part of present-day Hungary: the area up to the Danube formed the province of Pannonia. By the end of the 4th century, however, Rome had lost Pannonia, which had been occupied by German and Slavic peoples. The central plains were inhabited in Roman times by the Sarmatians and later by Huns, Bulgars and Avars, nomadic peoples from the steppes north of the Black Sea. The Avars ruled over the Danube basin during the seventh and eighth centuries, until they were conquered by Charlemagne.

[2] Charlemagne's successors set up a number of duchies in the western and northern parts of the basin, while the southern and eastern parts fell under the sphere of influence of the Byzantine Empire and Bulgaria. The Duchy of Croatia became independent in 869 and Moravia put up stiff resistance to the Carolingians until the appearance of the Magyars. The latter had organized a federation of tribes west of the lower Don. These were made up by several clans led by a hereditary chieftain. The federation was called On-Ogur (Ten Arrows); the term 'Hungarian' is a derivation of this word in the Slavic language. In 892, the Carolingian Emperor Rudolf sought the support of the Magyars to break Moravian resistance.

[3] Led by Arpad, the prince elected by the chiefs of the seven

LAND USE

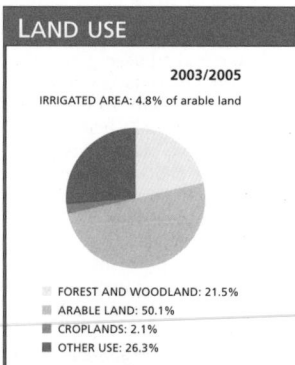

2003/2005

IRRIGATED AREA: 4.8% of arable land

- FOREST AND WOODLAND: 21.5%
- ARABLE LAND: 50.1%
- CROPLANDS: 2.1%
- OTHER USE: 26.3%

groups of Hungarians, the Magyars crossed the Carpathian mountains in 895 and conquered the inhabitants of the central plateau. By 907 the Magyars were threatened by a European army charged by the Holy Roman Emperor Louis III with the task of eliminating the last remaining pagan people in Europe who refused to convert to Christianity. The invading army was defeated at the Battle of Pozsony, thereby preserving Hungarian language and culture, though Arpad was killed. The Hungarians then expanded northwards and carried out numerous raids into the rest of Europe. The German Emperor Otto I halted Magyar expansion in 955. Arpad's heirs reunified the groups and adopted Western Christianity.

[4] Stephen I was crowned by Rome on Christmas Day in the year 1000, subsequently putting in place the foundations of the Hungarian State; after his death Stephen became a saint and his crown became a symbol of divinity, more important than the king or queen who wore it thereafter. The fights for succession upon Stephen's death triggered two centuries of instability, but Hungary consolidated its dominions as far as the Carpathian mountains and Transylvania in the north, and the region between the Sava and Drava rivers in the south. In addition, it ruled over Croatia, Bosnia, and Northern Dalmatia (although the latter remained a separate state).

[5] After the Mongol invasion of the 13th century - in which Hungary lost half of its population - the kingdom opened its doors to new settlers. However, it was forced to make various concessions to the Cuman overlords and immigrants, who further weakened it. Finally, the country found stability when Charles Robert of Anjou (1307-1342), the royal candidate who was

favored by the Pope, was appointed to the throne. As the struggles between the Holy Roman Empire and the papacy did not involve Hungary, the 14th century was the country's golden age. The kingdom established friendly relations with Austria, Bohemia and Poland, and strong ties with Bosnia, although it seized Dalmatia from Venice, and other territories from Serbia.

[6] Due to the long absences and arbitrary rule of Sigismund of Luxembourg (1387-1437), who was also a German and Czech king, the Hungarian Diet (parliament), which was made up of nobles, was enabled to pass laws. Taxes were continually being exacted from the peasants, who staged revolts in the north and in Transylvania. After another controversial succession, Matthias Corvinus became king of Hungary in 1458. He ruled his country with an iron fist. With the help of the Black Army, made up of mercenaries, Matthias subdued his enemies within the country and expanded his dominion over Bosnia, Serbia, Walachia, and Moldova, engaging in campaigns against Bohemia and Austria. Upon Matthias' death in 1490, the richest nobles appointed Vladislav II, who was King of Bohemia and known for his weak character. The Black Army was disbanded, but the oppressed peasants rebelled again in 1514, and the rebellion was ruthlessly put down.

[7] Much of Hungary was conquered by the Ottoman Empire at the Battle of Mohács in 1526 but Austria still claimed the throne. A Hungarian noble, János Zapolya, gained support for the idea that no foreigner should sit on the Hungarian throne and was supported by the Ottoman Sultan, Suleiman the Magnificent, until Zapolya's death in 1540 - at which point the Turks occupied

Budapest and annexed a large portion to the south and center of the country. Croatia and the western and northern strip of the country remained under the rule of Ferdinand of Habsburg, who had to pay tribute to the Turkish Empire. In 1686 the Ottoman Turks were finally pushed out of Hungary but the nation was still under the thrall of an Austrian empire whose only interest in the country lay in the tributes it paid. Conflict intensified when the majority of the population embraced the Reformation, and Vienna attempted to re-establish Catholicism. A new awareness emerged, even among the nobles, against absolutism, the poverty and oppression suffered by the peasants, and the country's stagnation.

[8] Inspired by the 1848 revolution in Paris, the Hungarian Diet passed the so-called April Laws, which introduced changes in agriculture by transferring ownership to those who worked the land, and in fiscal matters by broadening the tax base. In addition, parliament was reorganized on a more representative basis. The reunification of the country and the creation of a separate administration in Budapest were proposed. The reform was met with distrust on the part of large landowners and the Serbian, Romanian, and Croatian minorities. When the revolution was defeated, Austria, aided by Russia, annulled the reforms and regained control over Hungary. When Austria was defeated by Prussia in 1866, Vienna subdivided its empire and accepted the April Laws. A Nationalities Law guaranteed respect for the rights of minorities, giving way to the establishment of the Austro-Hungarian Empire in 1867.

[9] With the collapse of the Habsburg Empire during World War I, a provisional government

Life expectancy
74 years
2005-2010

GNI per capita
$8,370
2004

Literacy
99% total adult rate
2000-2004

HIV prevalence rate
0.1% of population 15-49 years old
2003

took power and proclaimed the Republic of Hungary. But Serbians, Czechs and Romanians seized two-thirds of the country and the central government was paralyzed. In 1919, a communist rebellion led to the formation of a Soviet Republic. Bela Kun's Bolsheviks were forced to flee when the Romanian troops took over the capital. The European powers pushed Romania into withdrawing and installed a provisional government. The 1920 parliament restored the monarchy and appointed Admiral Miklós Horthy as provisional ruler.

10   In the Trianon Treaty, the victorious allies recognized Hungarian independence, but Yugoslavia, Romania, and Czechoslovakia kept control of most of the country's territory and 60 per cent of its population. Austria, Poland, and Italy also benefited from this partition. Hungary was forced to pay heavy reparations. Unemployment rose to unprecedented levels and nearly 400,000 refugees arrived from the territories it had lost. Middle-class Hungarians and refugees joined together to set up right-wing armed groups, blaming the left for their ruin.

11   Funds granted by the League of Nations, followed by private investments, alleviated domestic tensions, but the depression of the 1930s had a severe impact. Horthy formed an extreme-right government which sided with Germany, while not consenting to the deportation of Hungarian Jews to extermination camps. The alliance with Berlin enabled Budapest to recover part of Slovakia, Ruthenia, and northern Transylvania. Hungary co-operated in the German attacks on Romania, Yugoslavia and the USSR, but nothing could stop the Red Army counter-offensive. In the Treaty of Paris, Hungary was forced to retreat to the borders determined by the Treaty of Trianon, pay reparations and reduce its army, under the supervision of the Russian occupation.

12   In 1944 a provisional assembly formed a coalition government. Its program included the expropriation of large estates, the nationalization of the banking system and heavy industry, guarantees for small landowners and private initiative, democratic rights and liberties. The communists, at that time represented by the Workers' Party, assumed control of the government. A 1945 election which resulted in an 83-per-cent opposition vote was ignored by the communists who, backed by the Russian Army, took complete control. The constitution of the People's Republic of Hungary was promulgated in 1946. In 1948, agriculture was forcibly collectivized and a series of development plans

which put a priority on heavy industry were implemented. In 1953, Matyas Rakosi was replaced as head of government by Imre Nagy, who promised political changes, generating expectations among the population. In 1955, Nagy was deposed and expelled from the Workers' Party. He was replaced by András Hegedus; Erno Gero remained as first secretary of the party.

13   The resolutions adopted by the 20th Congress of the Soviet Union Communist Party prompted students to organize a demonstration in Budapest in October 1956 that drew protesters from many other sectors of society. Gero reacted harshly. The police were instructed to open fire on the crowd, the demonstration turned into a popular revolt, and Nagy returned as head of government. He announced Hungary's withdrawal from the Warsaw Pact, and asked the UN to recognize the country's neutrality. Hungary was, however, seen by the West as being within the Soviet Union's sphere of influence and in 1956 Soviet forces reinstated the communist government, led by Janos Kadar, who closely followed the Soviet line.

14   Central planning became less strict around 1968. Living standards rose, but bureaucracy and corruption grew. Discrimination against Hungarian women continued, even though they were largely incorporated into the labor market. In 1981, women made up 45 per cent of all workers, although they were paid less than men and their job opportunities were restricted to specific fields. Between 1986 and 1988, Hungarian and Austrian environmentalists protested over the construction of a dam on the Danube, a Hungarian-Czechoslovakian project supported by Austria. In the end, the government in Budapest was forced to shelve the project.

15   Kadar was elected first secretary of the Hungarian Workers' Socialist Party (WSP) and head of government. In 1988, he faced a demonstration in Budapest in which the protesters demanded reforms. The Government subsequently relaxed press censorship and permitted the formation of trade unions and independent political groups such as the Hungarian Democratic Forum (HDF). As perestroika (restructuring) developed in the Soviet Union, in 1989 Hungary's parliament passed a law legalizing strikes, public demonstrations and political associations. In May 1989 the Government's decision to open its border to Austria was a major contributor to the fall of the Berlin Wall a few months later. The WSP also approved the abolition of the single party system and agreed to

PROFILE

ENVIRONMENT

The country is a vast plain with a maximum altitude of 1,000 m partially ringed by the Carpathian Mountains. The mountainous region has abundant mineral resources (manganese, bauxite, coal). Between the Danube and its tributary, the Tisza, lies a highly fertile plain, the site of most of the country's farming activity. Cattle are raised on the grasslands east of the Tisza.

SOCIETY

Peoples: Hungarian 92 per cent; and minority groups: Croatian; German; Roma/Gypsies; Romanian; Jewish; Serbian; Slovak and Slovene. **Religions:** Roman Catholic 57.8 per cent, Protestant 21.6 per cent; no religion 18.5 per cent; other 1.9 per cent. **Languages:** Hungarian (official), 98 per cent; German; Roma; and Slovak. **Main Political Parties:** Hungarian Civic Party (Fidesz-MPP); Hungarian Socialist Party (MSzP); Alliance of Free Democrats (SzDSz); Hungarian Justice and Life Party (MIEP); Centrum Party (CP). **Main Social Organizations:** Autonomous Union Confederation; Democratic Union Confederation; National Confederation of Hungarian Trade Unions; Hungarian Feminist Network.

THE STATE

Official Name: Magyar Koztarsasag (Hungarian Republic). **Administrative Divisions:** 19 counties and the capital. **Capital:** Budapest 1,708,000 people (2003). **Other Cities:** Debrecen 209,600 people; Miskolc 181,900; Szeged 170,600 (2000). **Government:** László Sólyom, President since August 2005. Ferenc Gyurcsany, Prime Minister since September 2004. Both are elected by the single-chamber National Assembly (legislature, 386 representatives elected for four-year terms). The Assembly is the supreme authority in the Republic. Eight seats in the Assembly are reserved for each of the country's minorities. **National Holiday:** 15 March, Independence Day (1848); 20 August, Constitution Day; 23 October, Declaration of the Hungarian Republic (1989). **Armed Forces:** 64,300 (1996). Other: Border Guard, 15,900. Civil Defense Troops, 2,000. Internal Security Troops, 2,500.

celebrate Independence Day on 15 March, the date of the 1848 revolt against Austria. An austerity plan which reduced subsidies and devalued the currency was adopted; as a consequence, unemployment and inflation climbed. About 100,000 people demonstrated in Budapest, demanding elections and the withdrawal of Soviet troops. The opposition candidates won the provincial elections held that year, and two million workers went on strike, protesting against price increases.

16   After an agreement between the WSP and the opposition, the Republic of Hungary was proclaimed on 23 October 1989, and the single-party system was abolished. Hungary soon broke its Cold War alliances. Within a short time, Budapest had established relations with Israel, South Korea, and South Africa. In 1989, the WSP had become the Hungarian Socialist Party (MSzP), with one faction deciding to keep its old name. In the 1990 elections, the HDF won 43 per cent of the vote, forming a coalition government with two smaller parties; the MSzP and the WSP took 10.3 and 3.5 per cent of the vote, respectively. Agricultural and industrial output

shrank 10 per cent due to IMF policies which stood in the way of domestic capital accumulation and favored foreign investment. In 1990, inflation climbed to 30 per cent, with the average family spending 75 per cent of its income on essential goods. By 1991, the spending had increased to 90 per cent of income. Of the country's 10 million inhabitants, two million were living below the poverty line.

17   Women's representation in politics declined: in the 1990 elections, women won 7.5 per cent of the seats, down from 21 per cent in 1985. Women made up 46 per cent of the active workforce of 4.8 million, but this began to change with the newly expanding view that the 'natural' order should be restored. However, the idea that women should stay at home clashed with the new economic reality of the country, where two salaries were needed to cover even the most basic needs of a nuclear family.

18   In 1992, the Government decided to reduce the public deficit, which exceeded $900 million in 1991, by slashing public spending. Growing popular discontent forced center-right Prime Minister Joszef Antall to back down on planned

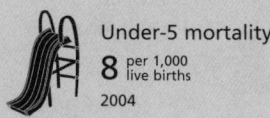
Under-5 mortality
8 per 1,000 live births
2004

Poverty
<2% of population living on less than $1 per day
2002

Debt service
25.2% exports of goods and services
2004

Maternal mortality
16 per 100,000 live births
2000

IN FOCUS

ENVIRONMENTAL CHALLENGES
Large investments will be necessary in order to meet EU standards on waste management, energy efficiency and pollution. Oil and natural gas deposits have been found in the Szegia and Zala river basins, and exploitation regulations need to be established. Forty-one per cent of the population is exposed to sulfur dioxide and nitrogen dioxide emissions. The sulfurous emissions are greater than in most western European countries.

WOMEN'S RIGHTS
Women have been able to vote and stand for office since 1918. In 2002, women held 9 per cent of total seats in Parliament, while in ministerial or equivalent positions their representation amounted to12 per cent.

In 2003, women made up 45 per cent of the labor force(4 per cent in agriculture, 25 per cent in industry, and 71 per cent in services), while female unemployment was 5.4 percent.

According to data from different NGOs, almost once a week a woman in Hungary is killed by her partner. In spite of the existence of laws for the protection of women against domestic violence, these are not applied and the number of victims is getting increasingly higher. Since 1990, Hungary has been identified as being involved

in the largely forced trafficking of women for sex trade, as both a country of origin and a destination for victims.

CHILDREN
The basic indicators of child well-being are similar to those of industrialized countries*. However, the Roma minority children are often segregated in several aspects, particularly in access to education. Roma children comprised a majority in special needs schools or classes for mentally disabled children.

INDIGENOUS PEOPLES/ ETHNIC MINORITIES
Geographically dispersed throughout the country, the Roma people are the most disadvantaged minority, suffering the worst discrimination. Culturally, racially and linguistically distinct from the majority population, they have been a frequent target of discrimination and prejudice by the authorities and society as a whole. In 2002, systematic racial discrimination against them was documented.-Also noted was the failure of the Hungarian legislature and authorities to protect them against violence and address their needs for access to education, housing and public services. Although their situation is alarming- and in many cases has worsened since the communist era - the Government has adopted policies, including new systems of self-rule, as well as measures to

root out discriminatory behavior in the police force. However, these policies do not go far enough in dealing with the depth of general discrimination suffered by the Roma, who not only face violent attacks by right-wing groups, but also lack of protection and prejudice from the police and courts.

MIGRANTS/REFUGEES
Hungary was one of the countries that hosted more refugees during World War II. At the present time it grants asylum to victims of the civil war in the former Yugoslavia and is considered by Romanian refugees in the country as a nation that offers work and economic prosperity.

According to UNHCR, at the end of 2005 the country hosted about 377 refugees and asylum-seekers. Many people who come to the country as refugees or asylum-seekers regard Hungary as an entrance door to the EU; once they are granted asylum or refugee status, they emigrate to the countries with the strongest economies.-

DEATH PENALTY
Capital punishment was abolished in 1990.

* Latest data available in *The State of the World's Children* and *Childinfo* database, UNICEF, 2006.

oil price hikes. The economic crisis fuelled nationalistic and xenophobic demonstrations.

[19] In the 1994 elections, the MSzP led by Gyula Horn won 209 of the 386 seats, and he was sworn in as prime minister. In 1995, Horn's honeymoon with the electorate came to an end when his unpopular package of economic measures was approved. The education budget and unemployment and maternity benefits were cut, in order to reduce the fiscal deficit. Horn continued to forge closer ties with the West. He also requested neighboring countries for more help to Hungarian minorities living within their borders, with a view to preserving their cultural identity. In 1996, Hungary signed an agreement with Slovakia for the protection of ethnic minorities and, soon after, reached a similar accord with Romania.

[20] In 1998, Hungary agreed to the installation of nuclear weapons and NATO troops in Hungarian territory as a pre-condition to joining the military alliance. That year, admission to NATO was approved

by 85 per cent of those who voted in a referendum (in which around 4 million people abstained). Bilateral relations with Romania, home to a large Hungarian minority, improved following former communist Ion Iliescu's defeat and the removal of the ultra-nationalist Romanian parties, which were hostile to the Hungarian and Roma/Gypsy minorities. In 1998, Hungary continued to focus its diplomatic efforts on admission into the EU. In 1999, Parliament voted overwhelmingly in favor of joining NATO. That year, Hungary, Poland and the Czech Republic were admitted into NATO, just before the Alliance bombing of Yugoslavia began.

[21] In 2000, a 100,000 cubic meter cyanide spill - at a mining site in Romania - in the Tisza river valley led to the region's worst environmental catastrophe since the 1986 explosion at Chernobyl nuclear reactor in Ukraine. The river, which flows into the Danube, contaminated drinking water in Yugoslavia, Romania and Hungary.

Two smaller acid spills also occurred that year, prompting authorities to take legal action against the Australian mining company, which refused to pay compensation for financial damage in excess of $28 billion.

[22] In 2000, the National Assembly elected the independent Ferenc Mádl as president, the second democratically elected leader of the country. In 2001, parliament passed a controversial Status Law to authorize Hungarian descendants living in Romania, Ukraine, Croatia and Slovenia to carry a special identity card that would grant them temporary work, education, healthcare and travel benefits while they were in Hungary. That year, the Hungarian economy grew 4.2 per cent, with a significant increase in GDP and a steady decline in unemployment. In 2001 - the 1,000-year anniversary of Hungary's consolidation as a nation - a monument originally built in 1934 was re-erected to commemorate four major national tragedies: the defeats by the Tatars,

Turks and Habsburgs and the Peace of Trianon. During the communist period, the monument had been demolished.

[23] In the 2002 elections - the most hard-fought since the return to democracy - a coalition of Socialists and Liberal Democrats won a majority in parliament, and Peter Medgyessy became prime minister. Former premier Viktor Orbán was criticized by the new government for his decision to sell more than 500,000 hectares of state-owned arable land. Orbán said he had intended to grant land to small farmers before the new government sold the property to large companies. That year, Medgyessy admitted he had worked as a secret service officer from 1970 to1982, but denied having collaborated with the KGB. He said his job had been to steer Hungary towards IMF membership without Moscow's knowledge.

[24] In 2003, Parliament amended the 2001 Status Law in order to remove several key aspects, including a reference to a 'unified Hungarian nation spanning borders'. Romania and Slovakia, both home to large Hungarian minorities, complained that the law interfered with their sovereignty and discriminated against other ethnic groups. But according to media reports, the law amendments responded to EU guidelines.

[25] On 1 May 2004, Hungary joined the EU as full member, raising the total membership of the bloc to 25.

[26] In August 2004, Medgyessy resigned and was eventually replaced as Prime Minister by Ferenc Gyurcsány. In November, Gyurcsány announced at a military ceremony in Budapest that Hungary would withdraw all of its 300 troops deployed in Iraq by the end of March 2005. The Premier pointed out that the country was obliged to keep its troops until after Iraqi elections in January. There had been intense pressure from the Hungarian public opinion and opposition groups to pull them out.

[27] In April 2006, Gyurcsány's Socialist-Liberal coalition won the legislative elections and the Prime Minister announced a period of important reforms, which he defined as the 'most intensive' since the post-communist transition.

[28] In September 2006 rioting broke out in Budapest, in the first major unrest since the fall of communism. Protesters demanded the resignation of Gyurcsany as Prime Minister, after he admitted that he lied to voters during the election campaign. The protests continued through October - over the 50th anniversary of the 1956 uprising - but at the end of the month Gyurcsany won a vote of confidence in Parliament. ■

Iceland / Island

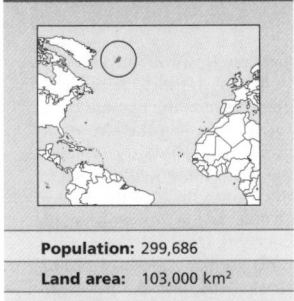

Population:	299,686
Land area:	103,000 km²
Capital:	Reykjavik
Currency:	Kronur
Language:	Icelandic

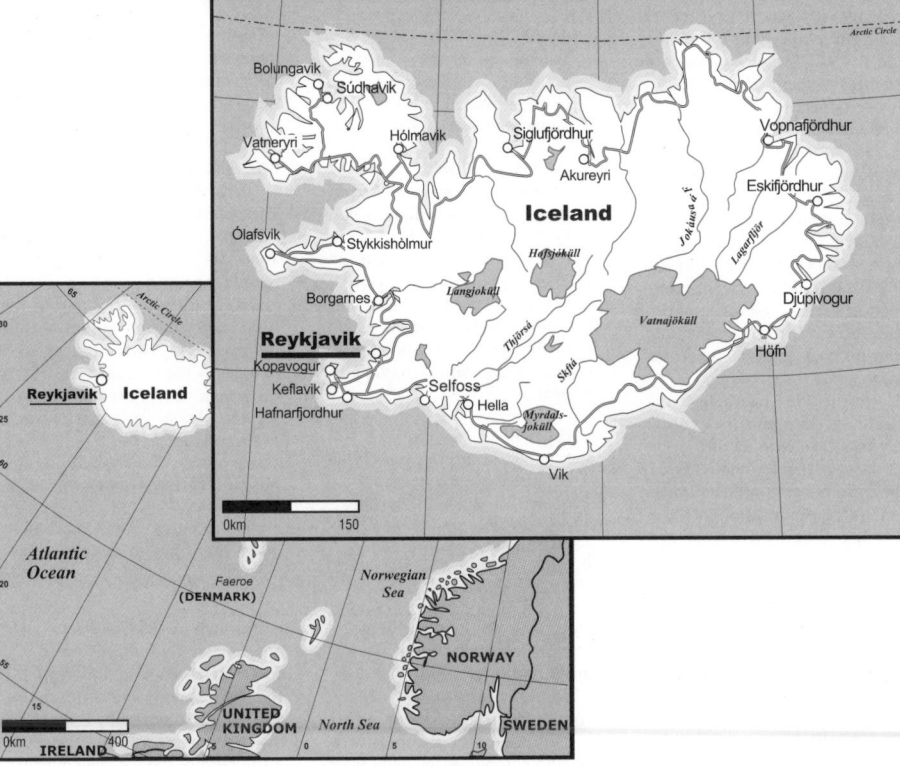

According to accounts written after the territory was discovered in the Viking era, Scandinavian peoples settled in Iceland. The Book of Icelanders (*Íslendingabók*), written around 1130 AD, places Scandinavian colonization between 870-930 AD. Another source of information from the 12th century, The Book of Settlers (*Landnámabók*), states that the first Scandinavian settler, Ingólfr Arnarson, arrived in Reykjavik in 874 AD. In 930 AD, a Constitution was adopted and an Assembly (*Althingi*), the world's first bicameral parliament, was created. Most of Iceland's population is originally of Norwegian, Scottish and Irish descent.

[2] Iceland was an independent country until 1262, but under the 'Old Treaty' of 1263 it became part of the kingdom of Norway. In the 14th century Iceland and Norway were conquered by Denmark. Norway separated from the Danish Crown in 1814, but Iceland remained under its dominion. By the late 19th century Iceland had no proper roads or bridges.

[3] In 1915 Danish women were allowed to vote and stand for office; for Icelandic women this happened in 1920. In 1918, Iceland became an associate state of Denmark, until it recovered its independence and a Republic was proclaimed in June 1944.

[4] After World War I, agriculture was revitalized through new legislation and the introduction of modern equipment. From 1920 onwards, the fishing industry grew steadily, and communications improved through the building of a network of inland roads.

[5] In 1949 Iceland joined the Council of Europe and NATO. Since the country had no army or navy, the US provided it with defense forces, within NATO's strategic framework. The island has armed coast guard ships and helicopters to prevent illegal fishing in its territorial waters. In 1952, Iceland, Sweden, Norway, Finland and Denmark founded the Nordic Council in Copenhagen. This was a consultative organization on legislative matters, whose mission was to study and boost Scandinavian regional cooperation.

[6] Due to the fishing industry's importance for Iceland - it is a major source of employment - and the fear of over-fishing by foreign fleets, Reykjavik extended its territorial waters to 12 nautical miles in 1964, and to 50 miles i n 1972. This triggered two serious conflicts known as the 'cod wars' with the United Kingdom, which were settled by a treaty signed in 1973. In 1975, Iceland extended its territorial waters to 200 miles, citing the need to protect the environment and its economic interests. The failure to reach a new agreement led to the third and most serious 'cod war'.

[7] NATO membership became controversial in the 1970s. The pro-NATO People's Alliance (PA) won the 1978 elections and formed a ruling coalition with the moderate Social Democratic Party (SDP). In 1980, Vigdis Finnbogadottir, an independent candidate opposed to US military bases on the island, won the presidential election with the support of the Left, taking 34 per cent of the vote. She was the country's first female head of state. However the post did not confer on her the power to alter government policy towards NATO membership and the US bases. In 1984 she was re-elected unopposed.

[8] In 1988, the US claimed that Iceland's decision to catch 80 adult and 40 young rorqual whales violated the moratorium imposed by the International Whaling Commission (IWC). Iceland answered that the whales were captured as part of a scientific program and refused to modify its decision. The US threatened to boycott Icelandic seafood products and two Icelandic whaleboats were sunk by environmental activists in Reykjavik Bay. In 1987, Iceland announced the reduction of its catch to 20 whales. The threats of sanctions continued, and the following year it further reduced the quota.

[9] In 1988, Finnbogadottir was re-elected for a third term, with the backing of the main parties and over 90 per cent

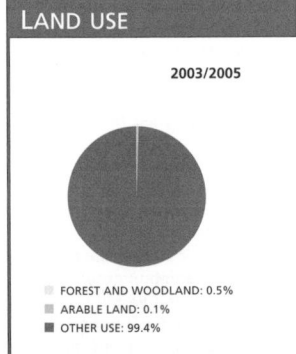

LAND USE

2003/2005

- FOREST AND WOODLAND: 0.5%
- ARABLE LAND: 0.1%
- OTHER USE: 99.4%

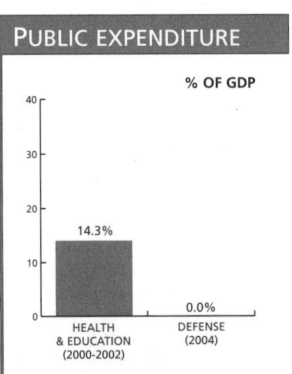

PUBLIC EXPENDITURE

% OF GDP

14.3% — HEALTH & EDUCATION (2000-2002)

0.0% — DEFENSE (2004)

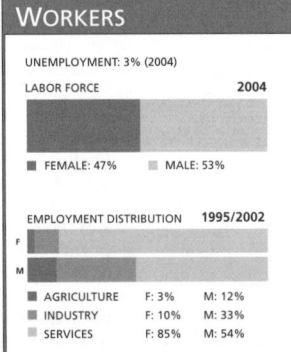

WORKERS

UNEMPLOYMENT: 3% (2004)

LABOR FORCE **2004**

- FEMALE: 47%
- MALE: 53%

EMPLOYMENT DISTRIBUTION **1995/2002**

F
M

AGRICULTURE	F: 3%	M: 12%
INDUSTRY	F: 10%	M: 33%
SERVICES	F: 85%	M: 54%

Life expectancy
81 years
2005-2010

GNI per capita
$37,920
2004

HIV prevalence rate
0.2% of population 15-49 years old
2003

of the electorate. Steingrimur Hermannsson became prime minister, as head of a center-left coalition made up of the SDP and the PA. The new government committed itself to an economic austerity program, designed to curb high inflation and recession.

[10] In 1989, as a result of growing international pressure, including a boycott of Icelandic products organized by Greenpeace, Reykjavik announced a two-year suspension of its whale hunting. The fishing industry went into crisis as certain species, especially cod, suffered from over-fishing. Foreign ship access to territorial waters was restricted.

[11] In the 1991 parliamentary elections, Hermannsson's center-left coalition won 32 of the 63 seats in the Althingi. David Oddsson became prime minister. Iceland's incorporation into the EU was approved by the Althingi in 1993.

[12] In the 1995 parliamentary elections, the Independence Party won 25 of the 63 seats and gained a majority in the Althingi in alliance with the Progressive Party, which won 15 seats. The economy enjoyed a slight recovery: GDP grew by three per cent and inflation was below two per cent. Unemployment - aggravated by layoffs at the NATO base - rose to five per cent.

[13] After ruling the country for 16 years, Finnbogadottir did not run in the 1996 elections. Olafur Ragnar Grimsson (former finance minister and member of the PA) won with 41.4 per cent of the vote. That year, fishing quotas were increased after several years of strict limits aimed at

IN FOCUS

ENVIRONMENTAL CHALLENGES
Several aquifers have been polluted by fertilizer runoff. Sewage treatment is not adequate in all cases.

WOMEN'S RIGHTS
Women have been able to vote and stand for election since 1915. In 2003, they held 33 per cent of parliamentary seats and 27 per cent of ministerial or equivalent positions. That year, women made up 45 per cent of the workforce, while female unemployment reached 3 per cent.

CHILDREN
Iceland is a sparsely populated country with low population growth. In 2004, the fertility rate was low (there were 4,000 births in the country), as well as

infant mortality (2 in every 1,000 live births) and under-5 mortality (3 in every 1,000)*. Child welfare is generally good, although more so in urban than rural areas.

INDIGENOUS PEOPLES/ ETHNIC MINORITIES
The population is of Norwegian, Finnish, Scottish and Irish descent. There are no organized indigenous peoples groups or ethnic minorities. However, Iceland forms part of the Arctic Council, which represents the interests of regional indigenous peoples in their respective governments (for example, the Sami in Norway - see that country's *In Focus* box)

MIGRANTS/REFUGEES
The country does not receive many asylum requests - rarely more than 100 a year - due to its geographic

location. However, according to official statistics, there has been a rise in the number of foreigners going there to work in the fishing industry. While in the 1990s some 5,000 arrived in the country, in 2004 there were 10,000, mostly from Poland (17 per cent), Denmark (8.4 per cent), former Yugoslavia (6.3 per cent), Philippines (6.1 per cent), US (4.8 per cent), Thailand (4.6 per cent), Lithuania (4 per cent), Portugal (3.4 per cent) and UK (3.2 per cent).

DEATH PENALTY
The death penalty was abolished in 1928.

*Latest data available from *The State of the World's Children* and *Childinfo* database, UNICEF, 2006.

preserving the species, fuelling economic growth. The economy continued to grow in 1997 due to domestic consumption. The bulk of investments were aimed at aluminum production. Several foreign companies - mostly from Switzerland, the US and Norway - intensified exploitation of this mineral resource.

[14] In 2000, as a member of FATF (Financial Action Task Force on Money-Laundering, a 29-nation group founded in 1989), Iceland launched a review of anti-money-laundering measures, which continued through 2001 and 2002.

[15] In 2001, despite rejoining the IWC (from which it had

withdrawn in 1993), Iceland announced it would return to commercial whaling, ignoring the Commission's moratorium. In 2002 the IWC voted by a slight margin to reincorporate Iceland as a full member, since the country planned to restrict the capture of whales for scientific purposes in the near future and to limit commercial whaling after 2006.

[16] In the 2003 general elections, Oddsson was re-elected Prime Minister. In August, Iceland resumed whale hunting, suspended since 1989. The Government argued that their overpopulation due to the moratorium on the whale trade

endangered the stock of other species, crucial for the country's fishing industry.

[17] Grimsson was re-elected President in June 2004. In September, Oddsson turned over his position as Prime Minister to former Foreign Affairs minister Halldór Ásgrímsson, from the Progressive Party.

[18] In October, 30 years after an historic one-day strike that had paralyzed the country in 1975, Iceland's women demanded equal pay with men. On 24 October 1975 some 25,000 women had attracted world attention by leaving their homes and workplaces to go on strike. For two hours they occupied Reykjavik's downtown area, in what was possibly the largest political act in Iceland's history. In October 2005 there were 50,000 demonstrators in the capital and another 10,000 in other locations - almost 16 per cent of the total population.

[19] In early 2006, geologists developed an ambitious project to drill directly into the heart of a hot volcano. Some 90 per cent of homes in Iceland are already heated by geothermal energy but the new technology could yield 10 times more power.

[20] In June 2006, Prime Minister Ásgrímsson resigned following the Progressive Party's poor performance in municipal elections. He was succeeded by Geir Hilmar Haarde, chair of the Independence Party.

[21] In October 2006, Iceland angered conservation groups by announcing that it would resume commercial whaling, hunting 9 finwhales (an endangered species) and 30 minke whales each year. ∎

PROFILE

ENVIRONMENT
Located between the North Atlantic and the Arctic Ocean, Iceland is an enormous plateau with an average altitude of 500 meters. A mountain range crosses the country from east to west passing through an extensive ice-covered region, the source of the main rivers. The coastline falls sharply from the plateau forming fjords. Most of the population lives in the coastal area in the southern and western parts of the country, where ocean currents temper the climate. Reykjavik, the capital and economic center, is located in a fertile plain where the largest cities are found. The northern coast is much colder as a result of Arctic Ocean currents. Geysers and active volcanoes are used as a source of energy. Agricultural output is poor; fishing accounts for 80 per cent of all exports.

SOCIETY
Peoples: 96 per cent of Icelanders are descendants of Norwegian, Scottish and Irish immigrants.
Religions: Protestants (95.8 per cent), mainly evangelical Lutheran (91.5 per cent), who practice the official religion. Catholics (0.9 per cent). No religion (1.5 per cent). Other (1.8 per cent)

Languages: Icelandic.
Main Political Parties: Independence Party (conservative); People's Alliance (socialist); Progressive Party (center-left); Left -Green Movement.
Main Social Organizations: Icelandic Workers Federation; Confederation of State and Municipal Employees of Iceland.

THE STATE
Official Name: Lydhveldid Island.
Administrative Divisions: 26 districts and 105 municipalities.
Capital: Reykjavik 184,000 people (2003).
Other Cities: Kopavogur 24,400 people; Hafnarfjordhur 19,800 (2000).
Government: Olafur Ragnar Grimsson, President and head of state since August 1996. Geir Hilmar Haarde, Prime Minister and head of government since June 2006. Unicameral parliament composed of 63 members, elected by popular vote for a four-year term.
National Holiday: 17 June, Independence Day (1944).
Armed Forces: 120 coastguards (1996). Other: 2,200 NATO troops (1996).

India / Bharat

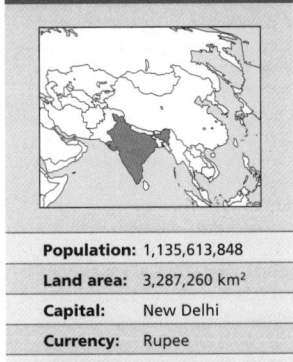

Population:	1,135,613,848
Land area:	3,287,260 km²
Capital:	New Delhi
Currency:	Rupee
Language:	Hindi

The roots of Indian civilization stretch back to prehistoric times. The earliest human activity in the subcontinent can be traced back to the Early, Middle and Late Stone Ages (400,000-200,000 BC). These peoples were semi-nomadic hunters and gatherers for many millennia. The first evidence of agricultural settlements on the western plains of the Indus is contemporaneous with similar developments in Egypt, Mesopotamia and Persia.

2   The earliest known civilization in India, the starting point in its history, dates back to about 3000 BC. It was a highly developed urban civilization, and two of its towns, Mohenjodaro and Harappa, located in the Indus Valley in present-day Pakistan, represent the high watermark of the settlements. They erected huge temples, engaged in irrigated agriculture and maintained active trade with peoples from the Persian Gulf and Sumeria (Iraq).

3   In the 16th century BC, the Aryans (Indo-Europeans) entered India and subdued the local population. They brought with them horses, iron armor and the Sanskrit language, which is the basis of the majority of Indian languages. Cavalry warfare facilitated the rapid spread of Aryan culture across north India, and allowed the emergence of large empires. The Aryans did not have a written language, but they developed a rich tradition, composing the hymns of the Vedas, the great philosophic poems that are at the heart of Hindu thought. The civilization they created, later called Vedic, was based on a rigid caste system in which the conquerors constituted the dominant nobility.

4   The 6th century BC was a time of social and intellectual ferment in India. It was then that Mahavira Jain and Gautama Buddha started to preach. The two great religions, Jainism and Buddhism, became the cornerstones of the Indian culture. Later, Buddhist monks were to spread their religion to what is now China, Japan, North and South Korea, Sri Lanka and Southeast Asia.

5   By the end of the 3rd century BC, Chandragupta Maurya unified north India and formed the first great Indian empire. Its greatest emperor was Ashoka (286-231 BC).

6   A golden age of Indian culture began with the Gupta empire (300-500 AD). Art and literature flourished; treatises were written on mathematics, astronomy and medicine. It was during this era that the Kamasutra appeared, the famous work on the art of love.

7   The invasions of the Huns signalled the end of the Gupta empire; north India broke up into a number of separate kingdoms and was not unified again until the arrival of the Muslims.

8   Great rival dynasties that rose in the south were the Cholas, Pandyas and Pallavas; the latter created the baroque-style Dravidian architecture.

9   The Muslim invasions that began in 700 had great impact on Indian culture, including language, dress, architecture and social values. In 1192, Muslim power arrived in India on a permanent basis. The most important Islamic empire was that of the Mughals, a Central Asian dynasty founded by Babur early in the 16th century. During the reign of Shahjehan, the capital was moved to Delhi and the Taj Mahal was built (around 1650).

10   In 1296 Ala-ud-din Khalji proclaimed himself Sultan of Delhi and by 1311 the whole of India was under the Sultanate. In 1336 the Vijayanagara Empire, the kingdom of Hindu alliance, was founded with its capital at Hampi to counter the Muslim power. Over time uprisings divided the Empire and the Muslim Sultanates formed a new alliance. In 1565 the Sultanate coalition defeated the Vijayanagar army. As a result, the power in the region passed to Muslim rulers and later their kingdoms were annexed to the Mughal Empire (1529-1857).

11   The arrival of the Europeans marked the next major phase in India's history. In 1687, the British East India Company settled in Bombay and throughout the 18th century, its private army waged war against the French, emerging victorious in 1784. From 1798, Company troops led by Richard Wellesley methodically conquered Indian territory in various campaigns.

12   The British ruled over India through the East India Company. India later became the Jewel in the Crown of the British Empire, giving an enormous boost to the nascent Industrial Revolution by providing cheap raw materials, capital and a large captive market for British industry. The Indian economy was dismantled. Exports of Indian handmade, high-quality cloth were stopped because they impaired the growth of the British textile industry. The ruin of this industry brought widespread impoverishment to the countryside. The land was reorganized under the harsh Zamindari (landlord) system to facilitate the collection of taxes to enrich British coffers. Farmers were forced to switch from subsistence farming to commercial crops (indigo, jute, coffee and tea). This resulted in severe famines.

13   By 1820, the English were in control of almost all of India, except for the Punjab, Kashmir and Peshawar, which were governed by their Sikh ally, Ranjit Singh. After his death in 1849, the British annexed these territories. The loyal allies retained nominal autonomy and were allowed to keep their courts, great palaces and immoderate luxury, much to the satisfaction of European visitors.

14   Divide and rule was a motto of British domination. Mercenaries recruited in one region were used to subdue others. Such was the case with Nepalese Gurkhas and Punjabi Sikhs. Religious strife was also fomented; an electoral reform at the beginning of the 20th century stated that Muslims, Hindus and Buddhists could each vote only for candidates of their own faiths. Throughout the colonial period manipulation led to innumerable social uprisings.

15   The most serious of these were the 1857-1858 rebellions by *sepoys*, Indian soldiers in the British army. These began as a barracks movement, eventually incorporating a range of grievances and growing into a nationwide revolt. Hindus and Muslims joined forces and even proposed the restoration of the ancient Mughal Empire. By the end of the rebellion, the East India Company was dissolved and the country became a British domain ruled by a Viceroy. Queen Victoria was proclaimed Empress of India.

16   The educational system was conceived to train Indians for colonial administration; however, it did not exactly fulfill this purpose. What it did was to create an intellectual élite fully conversant with European culture and thinking. Years later, it was that intelligentsia that formed the Indian National Congress (1885) which included British liberals and, for a long time, limited itself to proposing superficial reforms to improve British administration.

17   When Mohandas K Gandhi, a lawyer educated in England, returned to India in 1915, the independence cause won wide popular support. Gandhi had taken part in the struggle against apartheid in South Africa, where he had developed a technique of non-violent action which he called *satyagraha* (moral domination). He was a devout Hindu and espoused a philosophy of tolerance, brotherhood of all religions and non-violence (*ahimsa*). His ties with the Indian National Congress, where young Jawaharlal Nehru was an activist, strengthened the movement's most radical wing. In 1919, the Amritsar massacre occurred; a demonstration was savagely repressed leaving, according to British sources, 380 dead and 1, 200 wounded.

18   Under Gandhi's leadership, the Congress launched the Non-

Life expectancy
65 years
2005-2010

GNI per capita
$620
2004

Literacy
61% total adult rate
2000-2004

HIV prevalence rate
0.9% of population 15-49 years old
2003

Cooperation Movement (1920-1922) and the Civil Disobedience Movement (1930). One of the most important mass movements was the Salt March, in which Gandhi led a band of followers on a trip to the remote village of Dandi on the west coast to collect salt. This act implied a symbolic violation of British law, which had imposed a monopoly on the collection and selling of the vital mineral.

[19]   The campaign showed the effectiveness of peaceful civilian opposition. The movement spread nationwide at all levels and included non-participation in elections or administrative bodies, non-attendance at British schools, non-violence, refusal to consume British products, and passive acceptance of the ensuing legal consequences. For the first time, the British saw women flocking to demonstrations. Jails overflowed with prisoners who did not resist arrest, posing an immense problem for the colonial authorities.

[20]   Gandhi came to be called Mahatma (Great Soul) in recognition of his leadership. He became the movement's representative in contacts with the British, who after World War II were left with no option but to negotiate independence. India became independent in August 1947.

[21]   The Indian Union brought together a great diversity of ethnic, linguistic, and cultural groups, and there was conflict from the start. The massacres that followed were only partially mitigated by Gandhi's hunger strikes, which he continued until his assassination by a fanatical Hindu in 1948. The subcontinent was finally divided into two states: the Indian Union and Pakistan, which was created to concentrate the Muslim population in one area (see Pakistan and Bangladesh). The Sikhs (2 per cent of the population), who had a long list of martyrs in the struggle for independence, also demanded an independent state in the Punjab, which they did not achieve.

[22]   The division of Pakistan and India led to 562 principalities having to choose which state they would belong to in 1947. The local government in Kashmir - with a largely Muslim population - tried to avoid the question, but an invasion by Pakistani fighters led it to opt for India in return for military aid. After the Indo-Pakistan war in 1948-49, Kashmir was divided into two parts: Azad (Free) Kashmir that remained in Pakistani hands and Jammu and Kashmir state, of Muslim majority, that joined India.

[23]   After independence, Prime Minister Jawaharlal Nehru, along with Sukarno of Indonesia, Gamal Abdel Nasser of Egypt and Tito of Yugoslavia, advanced the concept of political non-alignment for newly decolonized countries. In India, he applied development policies based on the notion that the industrialization of society would bring prosperity.

[24]   In a few decades India made rapid technological progress, which enabled it to place satellites in orbit and, in 1974, to detonate an atom bomb, making India the first nuclear power in the non-aligned movement. However, it did not manage to solve the country's food problems.

[25]   Indian economy was severely hit by the oil crisis of the early 1970s, as it was dependent on oil imports. The export-oriented industrial growth was not sufficient to make up for the rising prices of imports and the growing demand for food by a population which was expanding at a rate of 15 million people a year. In 1975, the economic crisis, and popular resistance to the government's mass sterilization campaigns, led Indira Gandhi (Nehru's daughter, who had taken over as Prime Minister when her father died in 1966) to declare a state of emergency and impose press censorship.

[26]   Abandoning the Congress Party's traditional populist policies, Indira Gandhi followed World Bank economic guidelines, losing mass support for the government, without winning wholehearted backing from business sectors (particularly those linked to foreign capital), which demanded even greater concessions. The Government was forced to call parliamentary elections in March 1977. The Congress Party was roundly defeated by the Janata Party, a heterogeneous coalition formed by a splinter group of rightwing Congress Party members, the Socialist Party headed by trade union leader George Fernandes, and the Congress for Democracy, led by Jagjivan Ram, a former minister in Indira Gandhi's cabinet.

[27]   India's foreign policy of non-alignment remained basically unchanged under ageing Prime Minister Morarji Desai, who was unable to fulfill his promises of full employment and economic improvement.

[28]   Indira Gandhi returned to power in January 1980. Her administration was marked by a growing concentration of power, and accusations of excessive bureaucracy and corruption in the government, which gradually tarnished her image. In the Punjab, the Government faced increasingly strident demands from Sikh separatists. Small groups of militant Sikhs harassed Hindus, to drive them out of the Punjab and create an absolute Sikh majority in the province. After that, the next step would be secession and the formation of independent Khalistan. Indira accused forces from abroad (namely Pakistan and the US) of destabilizing the country.

[29]   After Indira Gandhi's assassination by militant Sikhs in 1984, thousand s of Sikhs fell victim to indiscriminate retaliation by Hindu paramilitary groups. Bypassing party and institutional formalities, Indira's son Rajiv was rapidly promoted to the office of prime minister and leader of the Congress Party. In December 1984 a gas explosion at US company Union Carbide factory at Bhopal killed and injured thousands of people.

[30]   Elections in January 1985 gave Rajiv overwhelming support. Despite the landslide victory, Gandhi lost in Karnataka, Andhra Pradesh and in Sikkim, a Himalayan kingdom annexed to India in the 1970s, where the separatist Sikkim Sangram Parishad party won.

[31]   The new premier appointed a conciliatory figure as governor of the Punjab, released political prisoners and ordered that militants within his own party who had participated in the anti-Sikh violence should be tried and punished. These measures paved the way

PROFILE

ENVIRONMENT

The nation is divided into three major geographic regions: the Himalayan mountains, along the northern border; the fertile, densely populated Ganges plain immediately to the south, and the Deccan plateau in the center and south. The Himalayas shelter the country from the cold north winds. The climate is subject to the influence of the monsoons; hot and dry for eight months of the year, and heavy rains from June to September. Rice cultivation is widespread. Coal and iron ore are the main mineral resources. There is a long-standing territorial dispute with Pakistan over Kashmir in the northwest, where important oil deposits are found.

SOCIETY

Peoples: The population of India is a multitude of racial, cultural and ethnic groups. Most are descendants of the Aryan peoples who developed the Vedic civilization. The north still bears the influence of invasions by the Arabs (between the 7th and 13th centuries) and Mongols (12th century). Peoples of Dravidian origin still predominate on the Deccan plateau in the center and south of India.

Religions: 83 per cent Hindu, 11 per cent Muslim, 2.5 per cent Sikh, 2 per cent Christian, 1 per cent Buddhist and 0.5 per cent other.

Languages: 400 registered languages, of which 18 are officially recognized, including Hindi, Bengali, Tamil or Urdu. English is a lingua franca, widely used for administrative purposes. There are 16 official regional languages and an infinity of local variants.

Main Political Parties: The Indian National Congress (INC), also known as the Congress Party, founded in 1885, fought for independence from Britain under the leadership of Mahatma Gandhi; The Bharatiya Janata Party (BJP) is a rightwing nationalist party with roots in the militant Rashtriya Swayam Sevak Sangh (RSS); the Communist Party of India-Marxist (CPI-M).

Main Social Organizations: INTUC (Indian National Trade Union Congress); the Bharatiya Mazdoor Sangh; the All-India Trade Union Congress (AITUC); Foundation of Investigation in Science, Technology and Policies of Natural Resources, Chipko Movement, an eco-feminist movement that fights to protect the country's forests. Separatist movements in Kashmir, Punjab and Assam, and the Tamils who support the Tamil Tigers (see Sri Lanka).

THE STATE

Official Name: Bharat (Hindi).

Administrative Divisions: 25 states and 7 union territories.

Capital: New Delhi 14,146,000 people (2003).

Other Cities: Greater Mumbai (Bombay) 18.1 million; Kolkata (Calcutta) 12. 9 million; Hyderabad 6. 3 million; Bangalore 6.2 million (2000); Chennai (formerly known as Madras) 7,6 million.

Government: Federal Republic. Abdul Kalam, President since July 2002. Manmohan Singh, Prime Minister since May 2004. Bicameral Parliament: Upper House (250 members) and Lower House (545 members).

National Holidays: 15 August, Independence Day (1947); 26 January, Day of the Republic (1950).

Armed Forces: 1,145,000 troops (1996). Other: 1,421,800: Army, Navy (including naval air force), Air Force, various security or paramilitary forces (Border Security Force, Assam Rifles, Rashtriya Rifles, and National Security Guards).

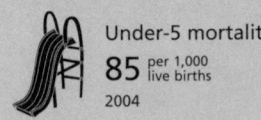
Under-5 mortality
85 per 1,000 live births
2004

Poverty
34.7% of population living on less than $1 per day
2000

Debt service
18.9% exports of goods and services
2003

Maternal mortality
540 per 100,000 live births
2000

for dialogue with Akali Dal, the regional majority party of Sikhs and other dissident groups. The Punjab autonomists proposed that the Indian central government should maintain control over defense and foreign affairs, the emission of currency, mail, highways and telecommunications. Meanwhile, the local government would enjoy greater autonomy than India's other states.

[32] In 1987, India intervened in the Sri Lankan conflict. It sent troops to press for a ceasefire agreement between the Tamils and Sinhalese. Three years later, the Indian peacekeeping force was quietly withdrawn after suffering heavy casualties.

[33] India's foreign policy remained loyal to non-alignment but a few changes were announced on the domestic front. Rajiv promised the private sector he would lift restrictions on imports and on the purchase of foreign technology.

[34] In March 1990, tensions between India and Pakistan mounted as Pakistan increased its support for the Kashmir independence movements. In November, clashes between Hindus and Muslims escalated amidst a general worsening of the economic crisis. Prime Minister Singh was replaced by Chandra Shekhar, also from the Janata Dal.

[35] An election campaign during which more than 280 people were killed gave way to parliamentary elections in May 1991. But the elections were adjourned following the assassination of Rajiv Gandhi, who was killed in an attack by a member of the Tamil liberation movement. A week later, Narasimha Rao was appointed Gandhi's successor as head of the Congress Party. The elections - the most violent in India's history as an independent state - resumed in June. The Congress Party obtained a majority of seats.

[36] The new government announced a drastic shift toward liberalism that would change the economic policy in force since independence. Prime Minister Rao opened up India's market to foreign investment, reduced the role of the state, allowed the rupee to float against the dollar, and removed import controls.

[37] The process of economic liberalization launched in 1992 has been intensified and has severely damaged the economic, social and cultural rights of the population. The State reduced its role in providing healthcare, education and electricity, and encouraged private companies' participation. The situation is particularly aggravated in the case of water, a resource that was traditionally seen as a public good in a country where

two-thirds of the territory is prone to drought.

[38] Numerous acts of violence took place in 1992 by Hindu fundamentalists against the Islamic population in the cities of Bombay and Ayodhya. The clashes between communities, prompted by the demolition of Babur's mosque in Ayodhya, left approximately 1,300 dead and extended to neighboring countries such as Pakistan and Bangladesh.

[39] Economic reforms triggered protests from several sectors, especially agriculture. Strong resistance was put up to transnational fertilizer and seed companies. As part of the green revolution and capital-intensive agriculture, the World Bank had granted loans for the purchase of genetically engineered seeds while the Government granted subsidies to farmers. Following instructions from the World Bank, the government decided to eliminate those subsidies. The Farmers' Association of the State of Karnataka (KKRS) - with a membership of ten million farmers - headed rural protests which were responsible for a number of direct attacks against representatives of transnational corporations since 1991.

[40] At the International Conference on Rights of Third World Farmers held in Bangalore in October 1993, farmers declared that the seeds, plants, biological material and wealth of the Third World form part of the Collective Intellectual Property of the peoples of the Third World. They pledged to develop these rights in the face of the private patenting system which encourages the spread of monoculture and threatens biodiversity.

[41] In 1994, India signed two agreements with China to reduce the number of troops stationed along the 4,000 km of common borders and to encourage trade relations. When former Pakistani Prime Minister Nawaz Sharif claimed his country had nuclear weapons, relations between New Delhi and Islamabad were strained. Pakistan closed its consulate in Bombay.

[42] During 1995, Prime Minister Narasimha Rao changed his cabinet three times. Elections in several states revealed an increasingly weakened Congress Party. Economic stability and assistance schemes announced by Rao, which included a school dinners plan for 110 million children and the construction of ten million rural homes, were not enough to stop his popularity from plummeting. Rao resigned in May 1996, following his party's defeat in the general elections.

[43] The BJP (Bharatiya Janata Party, Hindu nationalist) did not win a large enough majority in parliament to allow it to govern. A new political crisis brought about the appointment and resignation of two other prime ministers (Atal Bihari Vajpayee, HD Deve Gowda). Nearly a month after the fall of Gowda, Inder Kumar Gujral, also of the United Front, was appointed prime minister. In July, KR Narayanan was elected president.

[44] In November, Gujral was forced to resign when an official commission revealed alleged links between a governing coalition party, the Dravida Munnetra Kazagham (DMK), and Sri Lanka's Tamil guerrillas, a group involved in the murder of former Prime Minister Rajiv Gandhi.

[45] The BJP triumphed in the February 1998 parliamentary elections and Atal Bihari Vajpayee was designated prime minister. In May, a series of nuclear tests in India heightened tensions with Pakistan, and gave the neighboring country the pretext to carry out similar tests that same month.

[46] A bill to reserve one third of parliamentary seats for women was boycotted in parliament throughout the month of July. The bill's opponents, led by two socialist parties, protested outside parliament, interrupting proceedings. According to them, the proposal was unacceptable because it did not include quotas for women from the lower castes. The bill's principal defenders were the ruling party and the major opposition force, the Indian National Congress Party, in agreement for the first time.

[47] The government coalition collapsed in April 1999. The Tamil AIA-DMK party forced two of its ministers to resign after Prime Minister Vajpayee rejected AIA-DMK demands that Defense Minister George Fernandes be removed from office and investigated for having sacked India's Navy Commander, Admiral Vishnu Bhagwat.

[48] New confrontations with Pakistan erupted in June after Pakistani forces crossed the border delineated by the UN. Some 1,000 people died in the conflicts. The international organization Human Rights Watch reported serious violations on either side of the border, attributed to officials and agents from both governments. India's security forces were accused of carrying out summary executions, rape and torture.

[49] In October, after five electoral rounds, the BJP once again led the administration. Despite the coalition's victory, this election also marked a BJP decline that benefited leftist and regional parties. More

than half of parliamentary seats remained outside the control of the two largest national parties (the BJP and the Congress Party, which suffered the worst defeat in its history).

[50] More than a million people were left homeless in an earthquake in Gujarat State in western India in February 2001. At least 30,000 were killed and 55,000 injured in a region that had been devastated by a cyclone in 1998.

[51] The battle for control of world food supplies gave a legal victory to the less well-off in a ruling on rights over Basmati rice - which had been produced on the Indian subcontinent for centuries. The rice battle broke out in 1997, when RiceTech in the US patented Kasmati, a type of Basmati rice, and ended in May 2001 when the US Patent Office rejected RiceTech's application. A ruling the other way would have meant an end to Indian rice exports to the US, and would have forced Asian farmers to pay intellectual property rights on one of their traditional crops.

[52] In 2001, a report on AIDS funded by UNICEF but drawn up by the Human Rights Commission of Madhya Pradesh was criticized for suggesting that prostitution is a way of life for the Bedia caste. The view was rejected by non-governmental organizations (NGOs), UNICEF and the Bedia community, who demanded that the report be withdrawn on grounds that it promoted caste discrimination. Brinda Karat, of the Joint Action Council and All India Democratic Women's Association (AIDWA), pointed out that the very concept of a caste-based survey is repugnant to Human Rights and democratic thinking as it is premised on a belief that there is something intrinsic to the caste which makes women prostitutes and men pimps. A year earlier, the founders of the NGO Sahyog, Abhijeet and Yashodhara Das barely escaped lynching in the Himalayan area of Almora for producing a leaflet entitled 'AIDS and Us', which indicated that incest was widespread in the region. India is, after South Africa, the country with the greatest number of people living with HIV/AIDS.

[53] In July 2001, Vajpayee met Pakistani President Pervez Musharraf in the first summit between the neighboring countries in more than two years. The meeting ended with no resolution to the situation in Kashmir. Chancellor Singh accused Pakistan of standing in the way of an agreement by making the Kashmir issue the main subject of the talks instead of advancing on other bilateral issues. Meanwhile, the Pakistani delegation responded

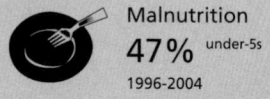
Malnutrition
47% under-5s
1996-2004

Water source
86% of population using improved drinking water sources
2002

Doctors
51 per 100,000 people
1990-2004

Primary school
87% net enrolment rate
2004

IN FOCUS

ENVIRONMENTAL CHALLENGES

The use of fertilizers and pesticides introduced with the Green Revolution has produced lower soil fertility and yields, while the selection of high-yielding seeds has resulted in the disappearance of traditional crop varieties.

Extensive hunting by the British and Indian rajahs, large-scale clearing of forests for agriculture, poaching, potent pesticides and the ever-increasing population have had a devastating effect on the environment in India. The entire country suffers from a shortage of clean drinking water.

WOMEN'S RIGHTS

Women have been able to vote and stand for office since 1950. In 2004, 8.3 per cent of the lower house seats and 11.6 per cent of the Senate were held by women. In 2003, they made up 33 per cent of the 473-million-strong workforce.

In 2004, 60 per cent of pregnant women received prenatal care and 43 per cent of births were attended by skilled health staff*. The maternal mortality rate was 540 deaths in every 100,000 live births.* More than 50 per cent of women are anemic. In 2004, the percentage of women who could read and write stood at 66 per cent of the male level.*

CHILDREN

Although mortality rates for infants and children under 5 have fallen between 1990 and 2004, they remain high. The former went from 84 to 62 for every 1,000 live births, and the latter from 123 to 85 for every 1,000. With 26 million children born each year, this means more than 2 million deaths per year. About 30 per cent of children had low birth weight; 47 per cent of children under 5 had low weight (in 18 per cent of cases they were seriously underweight); 16 per cent were moderately and seriously undernourished, and 46 per cent were moderately and seriously stunted.

The primary enrollment rate was 77 per cent.

It is estimated that between 2.2 million and 7.6 million people live with HIV/AIDS.

INDIGENOUS PEOPLES/ ETHNIC MINORITIES

There are around 500 groups of indigenous peoples or *adivasis* including the Kanikar, Muthuvan, Urali and Mala Arayan who have born the brunt of caste oppression. There are also non-*adivasi* minorities such as Kashmiris and Sikhs, who are involved in disputes particularly over territory on the borders with Pakistan. Muslims are another minority, some emphasizing their religious differences to distinguish themselves from the Hindu majority.

MIGRANTS/REFUGEES

In November 2003, Human Rights Watch denounced police abuses during a demonstration by Burmese refugees whose status had been recognized by UNHCR.

Some 600,000 Indians were internally displaced and nearly 17,000 Kashmiris from the Indian-controlled area of Kashmir remained in Pakistan at year's end.

India was one of the 16 countries of origin with a reduction of at least 25 per cent in their number of refugees in 2005. That year, the number of Indians leaving the country - mostly toward industrialized nations - fell by 38 per cent. In early 2006 the authorities were considering granting citizenship to 8,500 Sikhs and repatriating 1,200 Afghans. Sri Lankan Tamils who were refugees in camps in the south were being repatriated. Some 850 of them had returned to Sri Lanka.

DEATH PENALTY

India is among the countries in which the death penalty is still applied to common crimes.

* Latest data available in The State of the World's Children and Childinfo database, UNICEF, 2006.

that lead to a person's death. In November, Parliament and the central government stated that they favored the extension of the death penalty to the crime of rape.

[62] In 2004, 20 years after the 1984 Bhopal tragedy - when 8,000 people died and 150,000 were left chronically ill after inhaling poisonous gas from the Union Carbide plant - a coalition led by the survivors of the disaster, stepped up pressure on Dow Chemicals, the new owners of Union Carbide, to face criminal charges.

[63] After winning the 2004 elections, Congress Party leader Sonia Gandhi - Rajiv's widow - declined taking office as prime minister. Former Finance Minister Manmohan Singh, a Sikh, became India's first non-Hindu prime minister.

[64] In September, along with Brazil, Germany and Japan, India formed a group pressing the UN to become permanent members of its Security Council. In November India started a partial withdrawal of its troops in Kashmir.

[65] The tsunami that hit Southern Asia at the end of 2004 swept along more than 2,000 km of coast in the states of Tamil Nadu, Andhra Pradesh and Kerala. India was the most affected country after Indonesia. Some 7,000 people died and more than 130,000 were left homeless. Massive funerals were held in the south; many of the victims were children.

[66] In March 2005 the Government passed a patent act, in compliance with the World Trade Organization, regulating for the first time India's pharmaceutical production, which had become one of the world's largest. In October, thousands of people died in an earthquake in Kashmir, an event that revealed major shortcomings in the aid and rescue services and a heavy reliance on the army.

[67] In an attempt to address widespread poverty, exacerbated by economic liberalization, in February 2006 India launched-the largest rural employment project in its history, aiming to reach some 60 million peasant families.

[68] During the trip US President George W Bush made in March 2006, both countries signed a nuclear treaty which would give India access to US technology and improve the security of its nuclear programs.

[69] In September and October 2006 there were major strikes and protest demonstrations during which at least four people died in the capital, Delhi, over plans to crack down on illegal or informal trading. Municipal authorities had begun sealing up shops in residential areas following a ruling by the Supreme Court. ∎

that the main problem was that Indian officials insisted on changing the text of an agreement already approved by the rulers of both countries.

[54] In September, in the wake of the attacks on New York and Washington, the US lifted sanctions it had imposed on India and Pakistan following the nuclear tests staged by the two nations three years previously. This was a reward for the support they offered Washington in its global war on terrorism.

[55] In October, Kashmir was once more the center of conflict when Indian troops fired on Pakistani military posts. Suicide squads attacked parliament in New Delhi, killing several police officers. Both countries made widescale deployments of troops on the frontier, fearing a new war. In early 2002, India ran successful trials of the Agni, a missile with nuclear capacity.

[56] In February 2002, a wave of violence against the Muslim community spread in Gujarat state. More than 2,000 people were killed and women were

specially targeted. They were gang-raped before being burned alive. Rebels burned and looted shops, homes, and mosques. Some 15,000 Muslims were driven from their homes. According to the report issued by Amnesty International, the government of Gujarat and the state police took insufficient action to protect civilians, instead colluding with the attackers. Twenty-one suspects accused of the murder of 14 people burned to death in Baroda were acquitted. After the trial, several witnesses said they lied in court because they had received death threats. The National Human Rights Commission carried out a new investigation and asked the Supreme Court to provide witness protection and to ensure a retrial outside Gujarat to guarantee fair proceedings. In January 2003, the Gujarat government, which had turned a blind eye to the massacre, easily won the elections.

[57] In June 2002, although Defense Minister George Fernandes said Indian troops would remain along the border with Pakistan as long as necessary, there was a marked

drop in incursions in Kashmir due to intervention by the US, Russia and other countries.

[58] The new National Water Policy adopted in 2002 laid great emphasis on encouraging private participation in the sector.

[59] Kashmiri separatist groups called for a boycott of the October 2002 state assembly elections. For security reasons, the elections were held over four phases, with an average turnout of 44 per cent. After a 30-year rule of the-Jammu and Kashmir National Conference, the voters gave the victory to the People's Democratic Party (PDP) and its ally, the Congress Party. PDP leader Mufti Mohammed Sayeed became Primer Minister.

[60] In January, Gujarat's ruling BJP, which had covered up the February 2002 massacre, won the elections once again by an ample margin.

[61] In 2003, some 29 people were estimated to have been executed (the Government does not release these figures). In March, the promulgation of POTA (Prevention of Terrorism Act) extended capital punishment to acts of terrorism

Indonesia / Indonesia

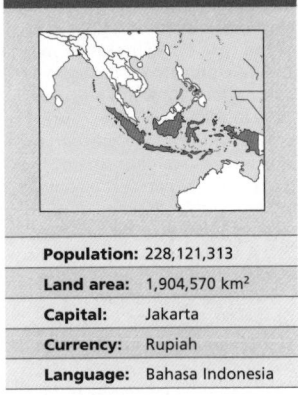

Population:	228,121,313
Land area:	1,904,570 km²
Capital:	Jakarta
Currency:	Rupiah
Language:	Bahasa Indonesia

Indonesia has some of the world's earliest *Homo Sapiens* remains. The ancestors of the present-day population were Malay immigrants who arrived on the islands of Java and Sumatra around 400 BC. They brought with them the cultural and religious influences of India. Indonesian civilization reached its height in the 15th century when the kingdom of Majapahit extended far east - beyond Java, Bali, Sumatra, and Borneo - becoming commercially and culturally linked with China.

[2] At the end of the 13th century Islam was introduced to the archipelago. It was not imposed through Arab conversion campaigns, but was freely adopted because it represented a simple, egalitarian faith suited to local conditions. When Muslim traders sent Indonesian spices to Europe, the Europeans' interest was aroused, and they set their colonial sights on the region.

[3] In 1511, the Portuguese arrived in Melaka; in 1521, the Spanish reached the Moluccas and in 1595, private Dutch merchants organized their first expedition. The Netherlands had just won its independence from Spain and wanted to secure its own supply of spices. In 1602, various Dutch trading groups founded the Dutch East India Company, obtaining a trade monopoly, with a colonial

mandate from the Governor of the region.

[4] During the 17th and 18th centuries, Indonesia was fought over by Spain, Portugal, the Netherlands and Britain, the latter creating another private company. Cash crops, coffee and sugar were introduced, yielding excellent profits but seriously upsetting the local socio-economic organization - intact until then - with the subsequent outbreak of anti-colonial revolts. Toward the end of the 19th century, rubber, palm oil and tin became the main export products, though industry only began to develop during World War II, when the Netherlands was unable to meet its production needs at home.

[5] In December 1916, nationalist pressure caused the formation of the People's Council or *Volksraad*, a body designed to defend the rights of the local population. Although its proposals were largely ignored, the Council encouraged political participation among the local population.

[6] In 1939, eight nationalist organizations formed a coalition called the GAPI (*Gabungan Politik Indonesia*), demanding democracy, autonomy and national unity within the framework of the anti-Fascist struggle. GAPI adopted the red and white flag and *Bahasa* Indonesia as the national

language.

[7] After the outbreak of World War II, the Netherlands were invaded by Germany, and Indonesia by Japan in 1942. The Japanese, who claimed to be the 'Asiatic brothers' of the Indonesians, freed nationalist leaders like Sukarno and Muhammad Hatta who had been imprisoned under Dutch rule. On 11 August 1945 (four days prior to the Japanese surrender) they invested Sukarno and Hatta with full powers to establish a local government.

[8] Indonesia proclaimed its independence on 17 August and its intention to be an 'independent, united, sovereign, just and prosperous' republic. The Dutch tried to recover the archipelago, forcing the Indonesians to fight back. The Arab and Indian communities actively supported the pro-independence guerrillas, while Britain backed the Netherlands. The US pressed for a negotiated settlement.

[9] Unable to recover military control, the Netherlands accepted a partial transfer of sovereignty in 1949 under a Dutch-Indonesian confederation.

[10] In 1954, the Dutch-Indonesian Union, which was never fully implemented, was denounced by the Sukarno Government

and the archipelago became fully independent. Sukarno's government soon began its own colonial policies in the region. In 1963, reacting to The Hague's refusal to withdraw from the island of New Guinea, Indonesia occupied West Irian, the former Dutch colony whose territory took up half of the island. The Indonesian independence process, together with the Indian and Pakistani independence, the Cuban Revolution, the nationalization of the Suez Canal, and the French defeats in Vietnam and Algeria, heralded the arrival of the Third World onto the international political scene. Sukarno was one of the main leaders of the movement for Third World solidarity and in 1955 the Indonesian city of Bandung played host to the first meeting of the main Third World leaders.

[11] The Communist Party - which, with three million members, was the second most powerful in Asia after the Chinese - supported Sukarno. He launched nationalist development programs, aimed at raising the living standards of a population with one of the world's lowest per capita incomes. Petroleum provided a strong foundation on which to base economic development; Sukarno created a state petroleum company, PERTAMINA, to break the domination of the Anglo-Dutch transnational, Royal Dutch Shell.

[12] In 1965, Indonesian oil deposits were nationalized. In October of that year, a small force of soldiers led by General Suharto seized power under the pretext of stemming the 'communist penetration'. This bloody coup left nearly 700,000 dead, and some 200,000 political activists were imprisoned.

[13] Though deprived of any real power, Sukarno remained the nominal president until 1967 - he died in 1970 - when Suharto was officially named Head of State.

LAND USE

2003/2005

IRRIGATED AREA: 13.1% of arable land

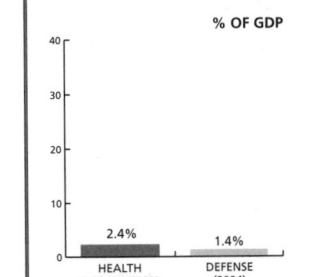

- FOREST AND WOODLAND: 48.8%
- ARABLE LAND: 11.6%
- CROPLANDS: 7.4%
- OTHER USE: 32.2%

PUBLIC EXPENDITURE

% OF GDP

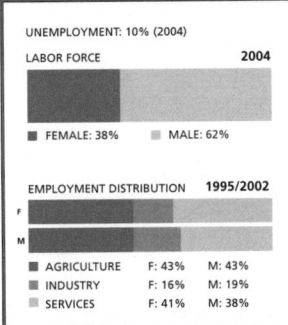

2.4%	1.4%
HEALTH & EDUCATION (2000-2002)	DEFENSE (2004)

WORKERS

UNEMPLOYMENT: 10% (2004)

LABOR FORCE   **2004**

- FEMALE: 38%   MALE: 62%

EMPLOYMENT DISTRIBUTION   **1995/2002**

F

M

AGRICULTURE	F: 43%	M: 43%
INDUSTRY	F: 16%	M: 19%
SERVICES	F: 41%	M: 38%

Life expectancy
69 years
2005-2010

GNI per capita
$1,140
2004

Literacy
88% total adult rate
2000-2004

HIV prevalence rate
0.1% of population 15-49 years old
2003

Suharto opened the doors to foreign oil companies seeking drilling rights. But with the rise in oil prices, the influx of capital, and a liberal economic policy, not only did national income rise; the gap between high and low incomes also widened. Many of the millions of rural dwellers had to leave their lands, enlarging the shantytowns in the large cities.

[14] In 1971, defying repression, students took to the streets in protest at the alliance between corrupt generals, Chinese merchants, and Japanese investors. In an attempt to neutralize local dissent under the flag of national unity, in 1975 Suharto invaded East Timor, shortly after its independence from Portugal. However, the people of Timor did not see the Indonesians as liberators but as new colonists. The unyielding resistance on the island only deepened Indonesia's internal problems.

[15] In the elections of May 1977, discontent surfaced once again. Despite repression, the banning of leftist parties and press censorship, the official Golongan Karya (Golkar) Party lost in Jakarta to a Muslim coalition which had campaigned against the rampant corruption. The governing party also lost ground in rural areas.

[16] To ensure victory in the election five years later, the regime clamped down on political activity and returned to an electoral system dependent on the Ministry of Home Affairs. In March 1983, in spite of growing opposition, the People's Consultative Assembly unanimously re-elected Suharto for a fourth five year presidential term.

[17] Indonesia adopted a birth control policy that led to a reduction in the population growth rate: while in 1984 it amounted to 2.3 per cent, the average rate between 1980 and 1990 was 1.8 per cent. Even so, demographic pressure, particularly on the island of Java, together with the radical reorientation of the country's economy toward the world market, as well as rapid industrialization, have all led to a deterioration in the quality of the environment and the depletion of agricultural lands. From 1979, the Government reacted to this situation with a population transfer project known as 'Transmigrasi', which involved moving 2.5 million Javanese to other less populated islands.

[18] In the 1980s over 300 ethnic groups saw their standards of living drop sharply. The most energetic protests were those of the inhabitants of West Papua (Irian Jaya) who demanded self-determination and freedom of movement to and from the neighboring territory of Papua New Guinea, with which they had a high degree of cultural and historical affinity.

[19] In 1991 the fighting between the army and the Aceh liberation movements in Sumatra became more acute when the Commander of the army called for the annihilation of the insurgents. In March 1992, an armed offensive of several separatist groups was started in West Papua. At the same time the US granted $2,300,000 for the training of the security forces of Indonesia.

[20] In March 1993 the People's Consultative Assembly chose Suharto - who had also been elected in 1988 - as President for the sixth time.

[21] In 1996 the pre-election debates reached a new low when complaints about the illegal enrichment of the Suharto family and its circle grew. The military saw Islamic groups as well as Sukarno's daughter, Megawati Sukarnoputri as the main threats to Suharto's power.

[22] In early 1997, the Indonesian population reached 200 million. The Government announced it would continue its program to transfer the people from 'over-populated' regions to less populated areas.

[23] The ruling Golkar party won the May 1997 parliamentary elections, obtaining 74 per cent of the vote (325 out of 400 seats). The new parliament included 12 of President Suharto's relatives - six children, two wives, two brothers-in-law, one brother and a cousin - as well as many of the leader's trade partners or cronies. In March 1998, Parliament re-elected Suharto.

[24] A severe stock market crisis increased inflation and caused a risk of hyperinflation. The currency had lost 50 per cent of its value since mid-1997. The price rise affected mostly basic goods and the economic situation deteriorated. Two million workers lost their jobs between October 1997 and March 1998. Following widespread social unrest and severe repression - there were hundreds of dead - Suharto stepped down in May 1998 to be succeeded by Bacharuddin Jusuf Habibie.

[25] In October, violent student protests to demand democracy and the removal of army chief Wiranto overwhelmed the capital. Five students were killed in confrontations with anti-insurrection forces. The October 1999 elections were won by Abdurraman Wahid, until then leader of the Nahdlatul Ulama Muslim organization. One of Wahid's first actions as President was to offer Aceh broad autonomy and an increase in economic support if the province were to remain a part of Indonesia. He set a seven-month deadline to carry out a referendum, similar to the one held in August 1999 in Timor Leste (formerly East Timor, see Timor Leste).

[26] Wahid was soon implicated in financial scandals, which in August 2000 led Parliament to launch an unprecedented investigation. Procedures were started to try Wahid and remove him from office, but these were abandoned in February 2001, after thousands of people took to the streets in support of the President and demanded that the opposition Golkar party be dismantled.

[27] In Kalimantan, the following month, combatants of the Dayak peoples took over parts of the province in the worst outbreak of violence in the region since 1997. Within one week, at least 1,000 Madurese refugees were killed and tens of thousands were forced from their homes.

[28] In May, the Parliament - the country's sole legislative body - voted 365 to 4 to begin impeachment proceedings against Wahid, who refused to resign and declared a state of emergency,

PROFILE

ENVIRONMENT

Indonesia is the largest archipelago-state in the world, made up of approximately 13,700 islands. The most important are Borneo (Kalimantan), Sumatra, Java, Sulawesi, Bali, the Moluccas, West Papua and Timor. Lying either side of the equator, the island group has a tropical, rainy climate and dense rainforest vegetation. The population, the fourth largest in the world, is unevenly distributed: Java has one of the highest population densities in the world (640 people per sq km), while Borneo has fewer than 10 people per sq km. Cash crops - mainly coffee, tea, rubber and palm oil - are cultivated, along with subsistence items, especially rice. Indonesia is the tenth largest oil producer and the third largest tin producer in the world.

SOCIETY

Peoples: Malay, Javanese, Sundanese, Madurese, Balinese, Ambon, Alfur, Toraja, Dayak, Batak, Minahasa and Papuan. There are also Chinese, Acehnese and Indian minorities.

Religions: 86 per cent of the population is Muslim, nearly 10 per cent Christian, 2 per cent Hindu (mostly in Bali), 1 per cent Buddhist, and small minorities that practice traditional beliefs. The State recognizes Islam, Protestant and Catholic Christianity, Hinduism and Buddhism.

Languages: Bahasa Indonesia (official), very similar to Bahasa Malaysia, official language of Malaysia. The governments of both countries have agreed to gradual unification, based on Melayu, the shared mother tongue. Javanese, language of 60 million inhabitants. English, language of business. There are hundreds of regional/local languages - more than 200 are concentrated in the province of West Papua (Irian Jaya).

Main Political Parties: Democratic Party; Indonesian Democratic Party-Struggle; Golkar; National Mandate Party.

Main Social Organizations: All-Indonesia Workers' Union (SPSI), since 1985, founded in 1973 under the name Labor Federation of Indonesia (FBSI); National Federation of Indonesian Unions; Indonesian Federation of Peasant Unions; Women's Solidarity Movement; Indonesian Environmental Forum.

THE STATE

Official Name: Republik Indonesia.

Administrative Divisions: 26 provinces.

Capital: Jakarta 12,296,000 people (2003).

Other Cities: Surabaya 3,683,200 people; Bandung 3,834,300; Medan 2,977,000 (2000).

Government: Susilo Bambang Yudhoyono, President since October 2004. Elections in July 2004 were by direct vote, in accordance with constitutional changes. Legislature, House of Representatives or Dewan Perwakilan Rakyat, single-chamber, made up of 550 members.

National Holiday: 17 August, Independence Day (1945).

Armed Forces: 297,000 (2001). Other: Police, 215,000; Auxiliary Police, 1.5 million.

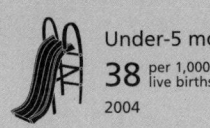

	Under-5 mortality			Poverty			Debt service			Maternal mortality
	38 per 1,000 live births 2004			**7.5%** of population living on less than $1 per day 2002			**22.1%** exports of goods and services 2004			**230** per 100,000 live births 2000

which was not heeded by the police or the army.

29 Nearly two years after taking office, and after the Supreme Court ruled the decreed state of emergency unconstitutional, a vote by the MPR (Consultative Assembly) of 591 members present (out of a total of 700) removed Wahid in July 2001. Vice-President Megawati Sukarnoputri, Sukarno's daughter, assumed the presidency.

30 In August, the President issued an apology to the provinces of Aceh and West Papua - which Jakarta had exploited over decades for their natural resources: petroleum and natural gas in Aceh, minerals in West Papua; however, she asserted that these provinces would never be allowed to secede as East Timor had done two years earlier.

31 In January 2002, Jakarta inaugurated the Human Rights Court to try the army for the atrocities committed in East Timor. Three generals, including the commander at the time of the massacres, Adam Damiri, appeared before the tribunal which, according to its creators, 'is better than an International Court'. East Timor assumed its complete independence in May.

32 In March, Tommy Suharto, son of the former president, was accused of assassinating a Supreme Court judge who had sentenced him to prison on corruption charges as a tough test of the country's legal and judicial system, still considered vulnerable to corruption.

33 Over 200 people, most of them tourists, died in October as a result of a bomb attack on a night club in Bali. That same day another bomb exploded near the US Consulate in Sanur, but this time there were no victims. The Jemmah Islamiya (JI) group, allegedly the local branch of the Islamist network al-Qaeda was blamed for the attacks. The Government provided the police with powers to prosecute alleged terrorists. Abu Bakar Ba'asyir, the spiritual leader of the JI, was arrested that same month, accused of having ordered the church bombings and of plotting the assassination of President Sukarnoputri. This led to confrontations between Bakar Ba'asyir's followers and the police.

34 In December 2002 the Jakarta Government and the separatist Aceh Liberation Movement (GAM) signed a peace treaty in Geneva attempting to put an end to 26 years of violence. In May 2003 the negotiations failed, the Government launched a military attack against rebels and put the province under martial law.

35 In August 2003 an explosion

IN FOCUS

ENVIRONMENTAL CHALLENGES
Deforestation caused by the expansion of the paper industry and timber exports affects certain groups such as indigenous peoples in West Papua (Irian Jaya). There are significant air and water pollution problems, especially in urban areas.

WOMEN'S RIGHTS
Women have been able to vote and run for office since 1945. In 2004, 11 per cent of Parliament seats were held by women. In 2003 they comprised 42 per cent of the workforce (107 million people). That year, illiteracy affected 83 per cent of women. In 2004, 92 per cent of pregnant women received prenatal care, and 72 per cent of births were attended by skilled staff*. In spite of that, 230 women died per 100,000 live births.

CHILDREN
The tsunami that devastated the country's coasts in late 2004 destroyed almost 1,000 schools in Aceh province. Apart from the loss of equipment and buildings, 2,500 teachers died and another 3,000 lost their homes. In 2006, a permanent care center for child survivors provides education and psychological care. In late 2005 there were 21 similar centers in Aceh and North Sumatra, caring for 17,000 children. A year after

the disaster there were children living in camps for internally displaced people; many of them, as a result of the trauma, could not remember their address, describe their parents or recognize their neighborhood, making it difficult to trace relatives.

INDIGENOUS PEOPLES/ ETHNIC MINORITIES
The Acehnese, the Chinese and the West Papuans (1.7 per cent, 4 per cent and 0.5 per cent of the population, respectively) are the groups facing the greatest risks. The Acehnese, orthodox Muslims, bear the brunt of the Government's political, economic and religious repression. They have been seeking independence for Aceh province since 1976. Even though Jakarta and the GAM - the group leading the pro-independence movement - signed a peace treaty in 2000, the number of Acehnese suffering discrimination had not diminished. In 2003 the Government put the province under martial law.

The Chinese and their descendants, mainly Christians, have traditionally enjoyed high economic status. Despite this they have faced political, social and cultural discrimination and repression. Their mother tongue has been outlawed for everyday use and in school since 1966. In 2002 over 20 laws, as well as numerous local and military regulations, were clear examples of discrimination.

Conflict between West Papuans and the Government have been escalating since Indonesia took control in the late 1960s. Between 10,000 and 30,000 Papuans have been killed. Political abuse and economic repression have been systematic since the occupation. At the beginning of this century, Papuans were still not allowed to form groups or make political statements, and were executed, tortured and abused by the military and the police.

MIGRANTS/REFUGEES
In early 2005, hundreds of thousands of Indonesians were in camps for internally displaced people, living on international aid and trying to recover after the tsunami. Most of them spent months after the disaster looking for surviving relatives. The authorities registered some 977,000 internally displaced, more than half of them in Aceh. In August that year there was an attempt to repatriate of at least 20,000 people affected by the 28 March earthquake in Aceh.

DEATH PENALTY
Indonesia is one of the countries where the death penalty is still enforced for ordinary crimes.

* Latest data available from The State of the World's Children and Childinfo database, UNICEF 2006.

in front of a hotel in Jakarta killed 14 people. Jemmah Islamiya was blamed for the attack.

36 In September two members of JI were sentenced to death for the Bali attack. Abu Bakar Ba'asyir was sentenced to 5 years in prison for other crimes; his links to the attacks were not proven, which led many to question Indonesia's commitment to the fight against terrorism. JI came into being in 1970 when Suharto requested Islamic extremists' assistance to fight the 'communist threat'.

37 In September 2004 a car-bomb exploded at the entrance to the Australian embassy in Jakarta. Nine people died and 180 were injured. The Jemaah Islamiya claimed responsibility for the attack.

38 That month, retired general Bambank Yudhoyono won the second round of the country's first direct elections. The former Security Minister pledged to carry out an ambitious set of reforms to eradicate nepotism, terrorism and corruption.

39 A massive tsunami devastated

South Asia in December 2004. In Indonesia there were 220,000 dead and more than 130,000 people missing. The western island of Sumatra was the closest one to the epicenter. Dozens of buildings were destroyed in the quake previous to the tsunami. The wave hit the provinces of Aceh and Northern Sumatra.

40 In March 2005, Ba'asyir was found guilty of conspiracy in the Bali attacks of 2002. The court sentenced the Muslim clergyman to 30 months in prison. An earthquake in Sumatra killed at least 1,000 people, most of them from the island of Nias.

41 In mid-August, a peace agreement between the Government and the Aceh Liberation Movement led to the release of 1,500 prisoners linked to the armed separatist movement.

42 In October, three suicide attackers carrying bombs killed 23 people in Bali.

43 The meeting between Presidents Xanana Gusmao (Timor Leste) and Yudhoyono

enabled the establishment of diplomatic relations between their countries in February 2006, after a UN report about Indonesia's 47-year occupation of Timor Leste accused Indonesian forces of abetting the murder of some 180,000 Timorese.

44 A powerful earthquake hit Java in May 2006, leaving more than 6,000 dead and 200,000 homeless, overstretching emergency services. The renewed activity of the Mount Merapi volcano near Yogyakarta triggered fears of a new humanitarian catastrophe.

45 In July 2006, the Indonesian Parliament unanimously passed a new law giving more autonomy to the country's northernmost province of Aceh. The law followed the peace agreement in 2005 between the Government and the former separatist group, the Free Aceh Movement, which ended nearly 30 years of fighting in the province. The new legislations gives Aceh more autonomy than any other province in Indonesia. ∎

Iran / Iran

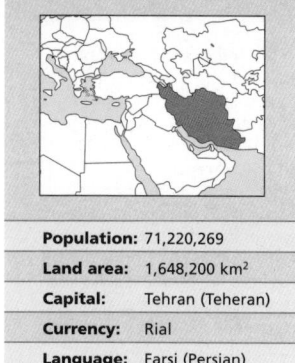

Population:	71,220,269
Land area:	1,648,200 km²
Capital:	Tehran (Teheran)
Currency:	Rial
Language:	Farsi (Persian)

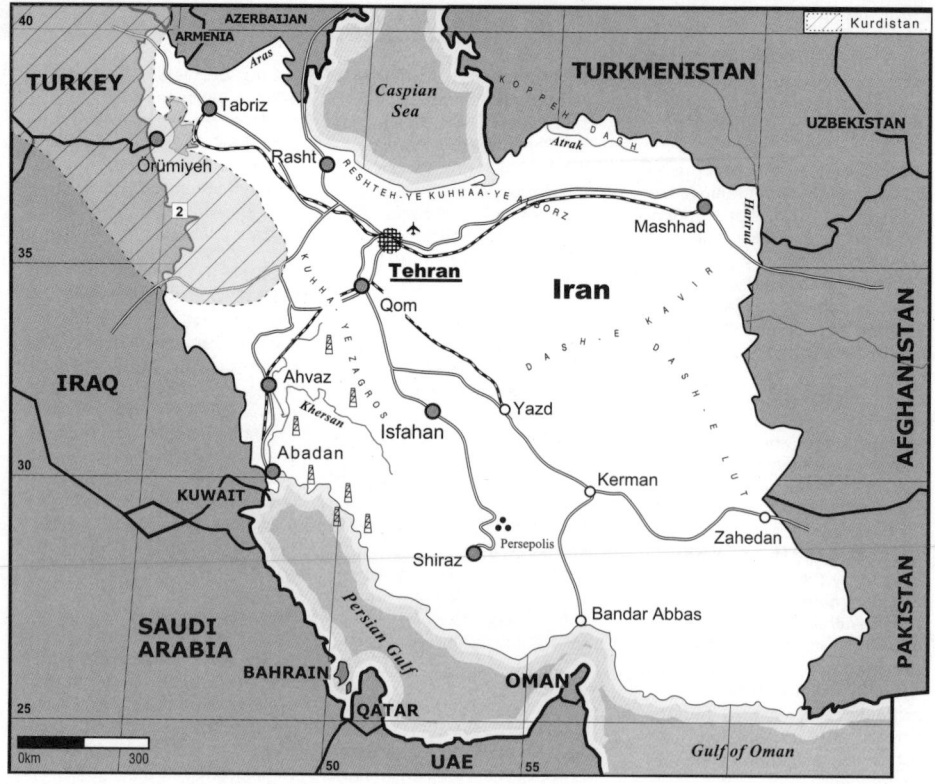

Shortly before the 18th century BC, Indo-European peoples reached the plains of Iran, subduing the shepherds who inhabited the region. Indo-Europeans continued to arrive up until the 10th century, contributing to the Mesopotamian cultural mix. They later became known by different names: Medes, from the name of the ruling group; Iranians, the name they adopted in Persia and India (from Sanskrit *ayriana* meaning nobles); Persians (a Greco-Latin term alluding to Perseus, the mythological ancestor that the Greeks foisted on Iranians), with its corrupted forms: Parsis, Farsis, Fars, or Parthians (according to the time and source). Whatever their names, they first won control of the mountainous region, then conquered the Mesopotamian plains under the reign of Ciaxares. During the rule of Cyrus the Great (559-529 BC), this wave of expansion reached as far west as Asia Minor, and as far east as present-day Afghanistan. These borders were later extended as far as Greece, Egypt, Turkestan and part of India.

[2] Towards the end of the 4th century BC, this vast empire fell into the hands of Alexander of Macedonia. Alexander's successors, the Seleucids and Romans (see Syria), lost their hold on the eastern part of the empire to the Persians, who recovered their independence with the Arsacid dynasty (2nd century BC to 3rd AD). They remained independent under the Sassanids until the 7th century, though constantly at war with the Romans and Byzantines.

[3] After the Arab conquest reached the region in 641 (see Saudi Arabia), Islamic thought and practices became dominant. Unlike the people in most other provinces of the Arab Empire, they retained their own language and distinctive styles in arts and literature. With the fall of the Caliphate of Baghdad, Persia attained virtual independence, first under the descendants of Tahir, the last Arab viceroy, and later under the Seleucid Turks and the Persian dynasties. Despite political restlessness, the period was remarkably rich in cultural and scientific progress, personified by the poet, mathematician, philosopher and astronomer, Ummar al-Khayyam.

[4] In 1501 Shi'ism became the state religion in Iran. The Qajar and Pahlavi dynasties would maintain Islam as the state religion. Iran was a Shi'a stronghold in the region, but not being an Arab country and not being on good terms with the Sunni representatives - Arabs at first, Ottomans later - its influence was limited.

[5] The Mongol invasion led by Hulagu Khan that began in 1258 was an altogether different matter. Three centuries of Mongol domination brought dynastic strife between the descendants of Timur Lenk (Tamburlaine) and the Ottomans. The dispute paved the way for the Persian Ismail Shah whose grandson Abbas I (1587-1629) succeeded in uniting the country. He expelled the Turks from the west, the Portuguese from the Ormuz region, and also conquered part of Afghanistan.

For a short time Iran ruled a region extending from India to Syria.

[6] A 1909 treaty divided the country into two areas - one of Russian and the other of British economic influence. A British firm was given the opportunity to exploit Iranian oil fields. Military occupation by the two powers during World War I, in addition to government corruption and inefficiency, led to the 1921 revolution headed by journalist Sayyid Tabatabai, and Reza Khan, commander of the national guard. Reza, the revolution's war minister, became prime minister in 1923. Two years later, the National Assembly dismissed Tabatabai, and Reza ousted the Shah, occupying the throne himself.

[7] Reza repealed all treaties granting extra-territorial rights to foreign powers, abolished the obligatory use of the veil by women, reformed the education and health systems, and cancelled oil concessions that favored the British. His attempt to establish a militarily strong, neutral modern state met with fierce resistance. However, he insisted on remaining neutral and did not allow the passage of allied arms to the Soviet Union through Iranian territory. The country was invaded and occupied in 1941 by British and Soviet forces. The Shah was overthrown and sent into exile. He was replaced by his son, Muhammad Reza Pahlavi, more

LAND USE

2003/2005

IRRIGATED AREA: 41.9% of arable land

- FOREST AND WOODLAND: 6.8%
- ARABLE LAND: 9.9%
- CROPLANDS: 1.3%
- OTHER USE: 82.0%

PUBLIC EXPENDITURE

% OF GDP

HEALTH & EDUCATION (2000-2002)	7.8%
DEFENSE (2004)	3.4%

WORKERS

UNEMPLOYMENT: 12% (2004)

LABOR FORCE · 2004

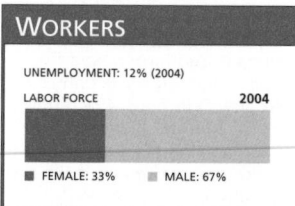

- FEMALE: 33%
- MALE: 67%

Life expectancy
72 years
2005-2010

GNI per capita
$2,320
2004

Literacy
77% total adult rate
2000-2004

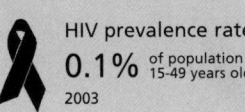
HIV prevalence rate
0.1% of population 15-49 years old
2003

amenable to European interests, who ruled under Anglo-Soviet tutelage until the end of the war.

8   A 1949 constitution curtailed the Shah's authority, and progressive, nationalist forces won seats in Parliament. With their support, Prime Minister Muhammad Mossadegh attempted to nationalize oil reserves and expropriate the Anglo-Iranian Oil Company.

9   In 1953, Mossadegh's audacity was met by an economic embargo and a coup backed by the US Central Intelligence Agency (CIA), which restored the Shah's almost absolute power. Nationalist and left-wing leaders were massacred and thousands were imprisoned. The Shah encouraged multinational corporations into Iran - citing 'modernization' which consisted of promoting Western consumerism. These measures were resisted by Islamic clergy and several social sectors, notably small farmers and the urban poor. By the end of the 1960s, due to the growing power of foreign corporations and rapidly changing consumption patterns, the Shah had lost the backing of the powerful commercial élites.

10   Opposition groups included the National Front founded by Mossadegh, the Tudeh Communist Party, Fedayin (Marxist) and Mujahedin (Islamic) guerrillas and the exiled cleric Ayatollah Khomeini. The recordings of Khomeini's preaching in Paris became familiar and encouraged the masses to organize. Demonstrations began in secondary schools in 1977 and were widespread by 1978. The Shah was forced to flee in January 1979 and Khomeini made a triumphant return. On 11 February crowds invaded the imperial palace; the Shah's prime minister resigned, and the army accepted the new situation. The Islamic Revolution was introduced as an alternative to Western models and was welcomed enthusiastically throughout the Muslim world.

11   Prime Minister Mehdi Bazargan of the National Front sought to reconcile Muslim traditions with a mixed economy model. However he did not find the required support. Muslim fundamentalists, bolstered by the 'revolutionary guards' and Khomeini's popularity, excluded their former allies from government.

12   In early November 1979, a student group stormed the US embassy in Tehran, taking the staff hostage and showing documentary evidence of CIA meddling in Iranian politics. A US attempt to rescue the hostages was unsuccessful. In 1980, war with Iraq

broke out; it continued until 1988.

13   In 1981, the Islamic Revolutionary Party (IRP) won the presidential elections with more than 90 per cent of the vote. A bombing of the party's headquarters killed 72 political leaders, among them the President and the Prime Minister. Ali Khamenei, former secretary-general of the IRP, was elected president. A theocratic regime was imposed, which suppressed opposition. Many people were jailed for political reasons, others went into exile, and still others were executed - between 500 and 1,500 people were sentenced to death in 1989, mostly for drug trafficking.

14   In 1985, Iran posted a huge trade surplus, despite the war with its neighbor. Oil made up almost all the exports.

15   Iran's international image deteriorated in early 1989 when Khomeini issued a *fatwa* - death sentence - against writer Salman Rushdie, citing his *Satanic Verses* book as blasphemous. Rushdie went into hiding, but in 1990 returned to public life and was reconciled with the main Islamic authorities, although the *fatwa* was not formally lifted.

16   Contrary to Western expectations, the death of Ayatollah Khomeini on 3 June 1989 did not lead to widespread instability. As more than eight million mourners gathered to bury him, outgoing President Ali Sayed Khamenei was appointed *faghih* - Iran's spiritual leader - by a vote of the Assembly of Experts.

17   In August, Ali Akbar Hashemi Rafsanjani won a landslide victory in the presidential elections. The new constitution granted the president, until then a largely ceremonial position, real powers. Rafsanjani's election as president meant a strengthening of the 'pro-Western' wing in the regime. Under the Iranian constitution, religious and lay authorities share power.

18   In 1990, Iran condemned the Iraqi invasion of Kuwait, and remained neutral when the Gulf War broke out in 1991. Eventually Iraq was expelled from Kuwait by US military forces in coalition with many other countries. Iran's neutrality was designed to give it an advantage over Baghdad, by gaining acceptance in regional and international diplomacy. Diplomatic relations with Britain were resumed in 1990 and with Saudi Arabia in 1991.This diplomatic initiative was hindered by domestic violence - with multiple political attacks - and by Tehran's links with groups involved in regional conflicts, such as the Lebanon-

PROFILE

ENVIRONMENT

Central Iran is a steppe-like plateau with a harsh climate, surrounded by deserts and mountains (the Zagros on the western border and the Elburz to the north). Underground water irrigates the oases where several varieties of grain and fruit trees are cultivated. The shores of the Caspian Sea are suited for tropical and subtropical crops (cotton, sugar cane and rice). Modern industries (petrochemicals, textiles, building) have grown, but the industry of hand-made rugs and other textiles remains significant.

SOCIETY

Peoples: Persian 51 per cent; Azeris 24 per cent; Kurds 7 per cent; Gilaki and Mazandarani 8 per cent; Lur 2 per cent; Arab 3 per cent; Baloch 2 per cent; Turkmen 2 per cent.
Religions: Shi'a Muslims 89 per cent; Sunni Muslims 10 per cent; Bahai, Christians, Zoroastrians and Jews 1 per cent.
Languages: Persian/Farsi (official) and minority languages (Kurd 7 per cent).
Main Political Parties: Reformist Coalition, whose main party is the Front of Islamic Participation; Islamic conservative parties: Assembly of the Followers of the Imam's Line, Islamic Solidarity of Iran Party, Association of Militant Clergy, and others. Out of Parliament: National Resistance Council; Kurdish Democratic Party; Communist Party; Party of the Masses (Tudeh).
Main Social Organizations: People's Mojahedin is the largest and most active armed militant group, whose philosophy is a mix of Marxism and Islam. With it are the Iranian National Liberation Army and the Union of Islamic Student Societies. Student groups have a significant degree of involvement.

THE STATE

Official Name: Dshumhurije Islâmije Irân. **Administrative Divisions:** 28 provinces, 472 towns and 499 municipalities.
Capital: Tehran 7,190,000 people (2003).
Other Cities: Mashhad 2,020,400 people; Isfahan 2,535,000; Tabriz 1,207,600 (2000).
Government: Presidential republic. Mahmoud Ahmadinejad, President since August 2005. Religious power is exercised by Sayed Ali Khamenei since June 1989. Unicameral Legislature: Islamic Consultative Assembly with 290 members elected by direct popular vote every four years. The members of parliament are elected by secret, universal ballot. **National Holiday:** 1 April, Revolution Day (1979).
Armed Forces: In 2003 there were 12,000,000 personnel available for military service. Other: BASIJ 'Popular Mobilization Army' volunteers, mostly youth; called up in wartime; strength has been as high as 1 million; Gendarmerie: (45,000 border guards).

based Hizbullah.

19   To attract foreign investment, Rafsanjani privatized enterprises nationalized by the Islamic Revolution in 1979. With the disintegration of the Soviet Union, Iran set its sights on a new area of influence: the Islamic republics of the Caucasus and Central Asia, signing agreements and opening new channels of communication with its neighbors.

20   In the April 1992 legislative elections, the 'moderates' who supported President Rafsanjani won a clear victory over the 'radicals'. In July, Iran's spiritual leader Khamenei launched a campaign to 'eradicate Western influence', clashing with Rafsanjani and his more moderate vision of Islam. However, in June 1993, Rafsanjani was confirmed as president by 63 per cent of the vote. In February 1994 he survived an attempt on his life.

21   Muhammad Khatami, regarded as the most pro-reform

candidate, won the May 1997 elections. The new government took over in August and Khatami announced the country would open up to the West. The reformists, headed by Khatami, also won the February 2000 elections, taking 226 of the 290 parliamentary seats. In June 2000, Khatami was re-elected.

22   In January 2002, US President George W Bush described Iraq, Iran and North Korea as terrorist-supporting countries and part of the 'axis of evil'.

23   The following year, after the US invaded and occupied Iraq, Ayatollah Khamenei predicted a violent future for the Bush administration if it did not leave the region.

24   In September 2003, the construction of Iran's first nuclear reactor raised alarm in several Western nations, especially the US. After complicated diplomatic efforts, Iran accepted UN inspectors, who concluded that the

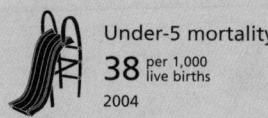
Under-5 mortality
38 per 1,000 live births
2004

Poverty
2% of population living on less than $1 per day
1998

Debt service
10.7% exports of goods and services
2000

Maternal mortality
76 per 100,000 live births
2000

IN FOCUS

ENVIRONMENTAL CHALLENGES
There is air pollution from industrial emissions - mainly from refineries - and cars. There is deforestation and desertification. Persian Gulf waters are polluted with oil. The soil is degraded and salinized. Untreated waste water from cities causes pollution. There are problems with the drinking water supply.

WOMEN'S RIGHTS
Women have been able to vote and stand for office since 1963. In 2001, only 4 per cent of Parliament seats were held by women. In the last term of the Khatami administration several executive positions were held by women. In 2003, Shirin Ebadi was awarded the Nobel Peace Prize. Ebadi was the first woman to preside over a courtroom in Tehran, but after the 1979 Islamic Revolution she was forced to resign. She taught and practised as a lawyer. In early 2006 a series of organizations that fight - among other things - for equal rights for women and against the application of *sharia* (Islamic law) met in Stockholm to remember Fadime Shahindal, a Swedish-Kurdish student victim of an 'honor' killing. The organizations asked Tehran to ban religious schools and the veil for young girls, and demanded the end of financial aid for religious organizations and institutions, help for secular organizations and others who fight for women

and children's rights, more help and support for women who flee their homes and support for the movement that fights against gender *apartheid* in Iran.

INDIGENOUS PEOPLES/ ETHNIC MINORITIES
There are three main minority groups: other nationalities (Arab, Azeri, Turkmen, Armenian); religious minorities (Sunni, Baha'i, Christian); and ethnic minorities (Kurd, Bakhtiar, Baloch). Discrimination - sometimes oppression - of these groups is based mainly on religious and linguistic differences.

CHILDREN
Children's health has improved in the last two decades. Preventive health services have been extended, reducing maternal and child mortality rates. About 90 per cent are immunized, and polio has been almost totally eradicated*. Ninety per cent of births are attended by qualified medical personnel; 11 per cent of children are moderately to severely underweight; and 15 per cent are weak or stunted*. In February 2005, an earthquake measuring 6.4 on the Richter scale hit the southwestern city of Zarand. The destruction of several towns affected more than 30,000 people, including many children. Aid agencies working with children whose schools were destroyed aim to give them temporary facilities and portable libraries. A 7-year drought has forced many rural people to sell

their herds and seek casual seasonal jobs. This meant that older children stayed at home to help, leaving school. The large distances that separate secondary schools from some rural homes lead many children to leave the education system after primary school.

MIGRANTS/REFUGEES
In June 2005 the governments of Iran and Afghanistan, together with UNHCR, extended until March 2006 the agreement regulating the voluntary repatriation of Afghan refugees in Iran. In the first eight months of 2005, more than 160,000 had returned to Afghanistan, in addition to the 350,000 that had done so in 2004. Another 28,000 living in the Zabol district were to be repatriated or moved, for national security reasons, according to the Tehran Government. The February 2005 earthquake killed at least 600 people and injured 1,400 more, destroying 8,000 homes and displacing 12,000 people.

DEATH PENALTY
Applicable to a wide range of crimes; according to Amnesty International, 108 people were executed in 2003. The methods used are stoning and hanging.

* Latest data available in The State of the World's Children and Childinfo database, UNICEF, 2006.

Islamic nation'. He added that Iranians had 'deeply humiliated' the US with their 'democratic transparency'.

[33] Despite international pressure, Tehran announced it would resume uranium enrichment in its Isfahan plant, which had been closed by the IAEA, and reiterated that its nuclear plans were for peaceful ends. The Natanz plant, also closed by the IAEA, was re-opened in December.

[34] The Natanz plant started to operate once again in February 2006 and the IAEA decided to report the Iranian case to the UN's Security Council. In March, Tehran announced its uranium enrichment program at Natanz had been successful.

[35] The offensive launched by Israel against Hizbullah in Lebanese territory (see histories of Israel and Lebanon) in July-August 2006 had, for many analysts, the goal of sending a clear message to Iran and Syria – accused by Tel Aviv of giving military support to the Lebanese guerrillas – as well as destroying the firepower of the rebel group.

[36] Iran saw the conflict as an excuse used by Israel to play the role of US police in the Middle East at the expense of innocent lives and the Lebanese economy, seriously affected by the destruction of its infrastructure. An investigation published by *The New Yorker* stated that Washington helped to plan the Israeli attack. According to the document, this would be a test for a future US attack on Iran before which Hizbullah's response capacity had to be wiped out first. The investigation said President Bush was not willing to leave the White House without dealing with what he called the 'Iranian nuclear threat'.

[37] After the ceasefire imposed by the UN Security Council in mid-August, Tehran stated that 'the Lebanese resistance and Hizbullah had achieved complete victory', and had ended the 'myth of the Zionist regime's invincibility.'

[38] In October 2006 Ayatollah Khamenei asserted that the country would continue developing nuclear technology at the same time as Iranian scientists reported that they had installed a second centrifuge cascade for uranium enrichment. The announcements came at a time of rising international tension over Iran's nuclear program. US Secretary of State Condoleezza Rice urged the UN Security Council to adopt a resolution imposing sanctions on Iran over its nuclear program. She argued that Iran must be held to account if the international community's credibility was to be saved. ∎

Iranian nuclear program was not for military purposes.

[25] That year, conservatives were upset when a woman, Judge Shirin Ebadi, won the Nobel Peace Prize. This was taken as an international gesture of support for Khatami's reformers.

[26] In December, an earthquake rocked the city of Bam, declared a UNESCO World Heritage Site. The quake left some 40,000 dead. Global solidarity included an easing of US economic sanctions, although tension remained high between Tehran and Washington.

[27] In the October 2004 legislative elections the Council of Guardians - a body controlled by Islamic conservative clerics with the power of veto over parliament - vetoed more than 2,500 candidates, mostly reformers and Khatami followers. Although half the registered voters did not show up at the polls, conservatives won an absolute majority in Parliament.

[28] In June 2004, the International Atomic Energy Agency (IAEA) sharply condemned Tehran for failing to cooperate with the investigation on Iran's nuclear activity. Although the IAEA warned

Tehran it was essential to deal with these issues, it neither threatened to report this case to the UN Security Council for possible sanctions nor gave a deadline for Iran to fulfill its obligations.

[29] Tehran had accused the authors of the condemnation report (Britain, France and Germany) of being on the side of Washington, which had previously charged the Iranian Government with secretly developing a nuclear weapons program.

[30] Two years after Iran's nuclear program first came to light there was no convincing explanation for the uranium contamination found in and around the uranium enrichment facilities. Washington maintained that Iran was carrying out experiments to develop a facility to create weapons-grade nuclear fuel. The IAEA board urged Iran to take 'all necessary measures to clarify remaining questions'. In November 2004 the Government decided to suspend most of its uranium enrichment programs, following the EU's recommendation.

[31] In January 2005 the Government allowed a UN

inspection of a military facility in Parchin, which Washington claimed was involved in the development of nuclear weapons. Tehran insisted its nuclear program was only for nuclear power. In April, Parliament voted to ease abortion-related legislation. Pregnancies could be terminated within the first four months if the fetus suffered physical or mental disabilities, or if the mother's life was at risk. Both parents had to consent to the abortion, and three doctors had to confirm the diagnosis.

[32] Mahmoud Ahmadinejad, a radical conservative and former agent of the Revolutionary Guard (religious police, created in 1979 to protect Khomeini's revolutionary principles, which during the war against Iraq became part of the regular army) won the June 2005 elections by a wide and unexpected margin. Analysts believe this marked the end of the reform period led by Khatami. Ahmadinejad, who won with 61.8 per cent of the vote, called on the people to leave their differences aside and 'create a powerful, advanced and exemplary

Iraq / Al Iraq

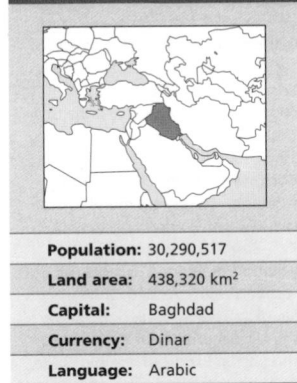

Population:	30,290,517
Land area:	438,320 km²
Capital:	Baghdad
Currency:	Dinar
Language:	Arabic

Iraq occupies the territory that was the site of one of the oldest Western civilizations. The Sumerian culture flourished in Mesopotamia around 5000 BC. In 2371 BC, King Sargon of Akkad gained control over the region and established the first Assyrian dynasty. The Assyrian empire expanded its dominion to include modern Turkey, Iran, Syria and Israel. The empire collapsed with the fall of its capital Nineveh (modern-day Mosul) in 612 BC, and was replaced by the Babylonian civilization. King Hammurabi (reigning circa 1792-50 BC), made Babylon the capital and established the first Code of Laws. King Nebuchadnezzar II (c. 605-562 BC), a splendid builder, developed hanging gardens that helped make Babylon one of the greatest cities of the ancient world.

[2] The Babylonian era came to an end when the Persians, ruled by Cyrus the Great, invaded in 539 BC and dominated the region until Alexander the Great's conquests in 331 BC. His successors, the Seleucids, ruled for 175 years, until the new Persian invasions under the Parthians, who constructed extensive irrigation systems and canals. Later on, the Sassanids established a new capital at Ctesiphon near the Tigris.

[3] After the Arab conquests of the 7th century, Mesopotamia became the center of an enormous empire (see Saudi Arabia). A century later, the new Abbas dynasty moved the capital east from Damascus. Caliph al-Mansur built the new capital, Baghdad, on the banks of the Tigris and for three centuries the city of 'A Thousand and One Nights' was the center of a new culture.

[4] This culture led to the greatest flourishing of the arts and sciences in the Mediterranean region since the days of the Greeks. However, the empire's enormous size led to its collapse after the death of Harun al-Raschid. The African provinces were lost and the region north and east of Persia won independence under the Tahiris (the Kingdom of

Khorasan). The caliphs were forced to depend increasingly on armies of slaves or mercenaries (Sudanese or Turks) to retain their grip on an ever-shrinking empire. When the Mongols assassinated the last caliph in Baghdad in 1258, the title had already lost its political meaning.

[5] The conquests of Genghis Khan devastated the region's agricultural economy, and the region was subsequently ruled in whole or part by Seleucids or Ottomans, Turks, Mongols, Turkomans, Tartars, and Kurds. The movement of steppe peoples (see Afghanistan) brought great instability to the fertile crescent, which finally achieved unification under the Ottoman Turks in the 16th century, having repelled an attack by Timur Lenk (Tamburlaine) in the 14th century.

[6] In the early 16th century, Sunnism held power in Iraq, under Ottoman rule. But the Shi'as from southern Iraq, who identified with the Irani regime, continued to enjoy considerable prestige, which limited Ottoman authority. Efforts were aimed at keeping open the trade routes that ran through the territory, joining East and West to the Mediterranean, as an alternative to the sea routes around Africa. To achieve this, it was necessary to confront the intractable Arab and Kurdish groups as well as the continuous encroachments by Iran. Süleyman imposed strict and direct rule over Iraq with that aim.

[7] In the early 17th century, the authority of local leaders within Iraq had grown significantly. Around that time, Bakr Su Bashi, military chief of a garrison in Baghdad, joined the Safavid Shah Abbas I, who managed to gain control over central Iraq, while Mosul and Shahrizor remained under Ottoman control. The central area was under Safavid rule from

1623-1638.

[8] The Treaty of Qasr-i Shirin (also known as the Treaty of Zuhab) of 1639 brought an end to the conflict, restoring Ottoman control over Baghdad. Aside from unrest between various groups, Iraq maintained a certain level of stability. Control was finally gained over southern Iraq in 1668, and the problems that followed reflected the state of affairs in Istanbul, center of the Ottoman Empire.

[9] The 18th century brought important changes in the region. The reign of Sultan Ahmed III in Istanbul was marked by political stability and reforms influenced by European models.

[10] In Baghdad, Hasan Pasha (1704-1724), of Georgian origin, was succeeded by his son, Ahmed Pasha (1724-1747), who introduced the Mamluks from Georgia. The Mamluks were mostly Christian slaves from the Caucasus, who were trained to perform military and administrative duties. When Ahmed died, the Mamluks assumed power, appointing his son-in-law, Süleyman Abu Layla, as the first Mamluk pasha of Iraq.

[11] From the second half of the 18th century, the Mamluk regime alternated between periods of prosperity and order and periods plagued with internal strife and corruption.

[12] In the early 20th century, Arab Renaissance movements were active in Iraq, paving the way for the rebellion that rocked the Turkish realm during World War I (see Saudi Arabia, Jordan and Syria). The British were keen to expand their influence in the region. With the defeat of the Turks, Iraq entertained hopes of independence. These, however, were dashed when the secret Sykes-Picot treaty of 1916 became known, whereby France and Britain divided the Arab

territories between themselves. Faisal, son of sharif Hussein, was expelled from Syria by the French. In 1920, Britain was awarded a mandate over Mesopotamia by the League of Nations, triggering a pro-independence rebellion.

[13] In 1921, Emir Faisal ibn Hussain was appointed King of Iraq in compensation. In 1930, General Nuri as-Said, who had taken office as Prime Minister, signed a treaty with the British, under which the country became nominally independent on 3 October 1932.

[14] That year, the Baghdad Pact was signed, making Iraq part of a military alliance with Turkey, Pakistan, Iran, Britain and the US. The pact was resisted by Iraqi nationalists. In July 1958, anti-imperialist agitation resulted in a military coup led by Abdul Karim Kassim, which led to the execution of the royal family.

[15] In 1959, the new regime banned all political parties and proclaimed Iraq's annexation of Kuwait. The Arab League, dominated by Egypt, authorized the deployment of British troops to protect the oil-rich enclave.

[16] Close ties with the Soviet Union and China fomented predictions that Iraq could become 'a new Cuba'. Steps towards economic change were taken, the power of landowners was weakened by agrarian reform, and a greater share of the Iraq Petroleum Company's revenues went to the state. In 1963, Kassim was deposed by pan-Arabian sectors within the army. Several unstable governments followed, until July 1968, when a military coup placed the Ba'ath party in power.

[17] Founded in 1947, the Arab Ba'ath Socialist Party (ba'ath meaning 'renaissance' in Arabic) was inspired by the ideal of Pan-Arabism, which regarded the Arab World as an indivisible political and economic unit where no country can be self-sufficient. The Ba'athists proclaimed that 'socialism is a need which emerges from the very core of Arab nationalism. It is organized on a national (Arab) level, with several regional leaders in each country.'

[18] Iraq nationalized foreign-owned companies and defended the use of oil as a 'political weapon in the struggle against imperialism and Zionism'. It advocated protected prices and the consolidation of OPEC. A land reform program was launched, and ambitious development plans encouraged the reinvestment of oil money into national industrialization.

[19] In 1970, the Baghdad Government gave the Kurdish language official status, and granted Kurdistan domestic autonomy. However, encouraged by

Life expectancy
61 years
2005-2010

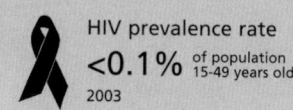
HIV prevalence rate
<0.1% of population 15-49 years old
2003

Iran, the traditional regional leaders rose up in arms. In March 1975, the Iran-Iraq border agreement deprived the Kurds of their main foreign support, and the rebels were defeated. The Baghdad Government decreed the teaching of Kurdish in local schools, greater investment in the region, and the appointment of Kurds to key administrative positions.

20   In July 1979, President Ahmed Hassan al-Bakr resigned and was replaced by Vice-President Saddam Hussein, who attempted to establish Iraq as a leader of the Arab world.

21   Hussein rejected the Camp David accord between Egypt, Israel and the US, while Iraqi relations with other Arab countries also worsened.

22   Iraqi forces staged a pre-emptive strike against Iran in September 1980, unleashing a war that lasted for eight years. The West backed Iraq against the fundamentalist regime of Ayatollah Khomeini in Iran.

23   On 17 June 1981, under the pretext that Iraq was producing nuclear weapons, Israeli planes destroyed the Tamuz nuclear plant.

24   During the war, the Saudis and Kuwaitis - in an attempt to stop Iranian fundamentalism - granted Baghdad many loans which were used both in the conflict and for strengthening the country's infrastructure. An oil pipeline was built through Turkey as an alternative to the one which ran

across Syria to the Mediterranean; Syria had closed the pipeline down in solidarity with Iran. The roads to Jordan were also upgraded.

25   In November 1984, 17 years after they had broken off diplomatic relations, official ties with the US were re-established. In spite of Washington's declaration of neutrality in the Iran-Iraq conflict, the Iran Contra scandal (see Nicaragua) revealed the superpower's double-dealing.

26   Through the 1988 armistice, Iraq retained 2,600 square kilometers of Iranian territory and a powerful and skilful army that soon found another pretext to take action again.

27   The Iran-Iraq war had been in part the outcome of massive arms-buying by both countries during the 1970s. The oil price rises in the early 1970s had swollen both countries' coffers and the West was keen to recover this money through arms sales. By 1975, Iran had become the single largest purchaser of US arms. Equally, without imported weapons and technology, Saddam Hussein would not have been able to invade Kuwait in 1990, and the Gulf Wars would not have taken place.

28   Neighboring Kuwait was extracting more oil than allowed from deposits along the border, and refused to establish export quotas. As the US hinted that it would remain neutral in the event of conflict, on 2 August, Iraq invaded Kuwait and took thousands of

foreign hostages.

29   Four days later the UN decided on a total economic and military embargo until Iraq retreated unconditionally from the occupied territory. Withdrawal was rejected, but a proposal for an international conference to discuss the problems of the Middle East issue was submitted. When Iraq started to release the hostages and to make new attempts at negotiating, the US refused to talk and demanded an unconditional surrender.

30   On 17 January 1991, an alliance of 32 countries led by the US launched an attack on Iraq. When the land offensive began in March, Saddam Hussein had already announced his unconditional withdrawal. The Iraqi army did not fight back, and merely attempted to stage an organized retreat, yet it suffered great losses. The war ended early in March, with the total defeat of the Iraqis.

31   Towards the end of the offensive, the US encouraged an internal revolt against Hussein by the southern Shi'a and the northern Kurds. However, the political differences between the two groups stood in the way of an alliance, and the rebels were crushed by the still-powerful Iraqi army. Over one million Kurds sought refuge in Iran and Turkey to escape the troops from Baghdad, and thousands starved or froze to death with the onset of winter.

32   Between 150,000 and 200,000 people, mostly civilians, died in the war. An estimated 70,000 Iraqis died as a result of the subsequent embargo, including 20,000 children. At the end of 1991, both the Turkish and Iraqi armies were continuing to harass the Kurds in the border area.

33   The conditions for lifting the embargo became very strict due to the US determination to bring about the downfall of Hussein. In addition, according to *The New York Times* and *The Sunday Telegraph*, the US introduced huge amounts of counterfeit dinars, smuggled across the Jordanian, Saudi Arabian, Turkish and Iranian borders. Baghdad established the death penalty for anyone participating in these operations.

34   Toward the end of 1991, the Iraqi Government authorized UN inspections of military establishments. In 1992, Iraq was found to be engaged in a uranium enrichment project, which had been developed using German technology. UN inspection teams destroyed 460 x 122 mm warheads armed with sarin, a poisonous gas. They also dismantled the nuclear complex at al-Athir, the uranium enrichment installations at Ash-Sharqat and Tarmiuah, and the chemical weapons plant at

Muthana.

35   In 1994 a border crossing was opened with Turkey to allow certain UN authorized foodstuffs and medicines to enter the country - the only exceptions to the trade embargo. But in March 1995, Turkish troops invaded Iraqi Kurdistan - under allied military tutelage - to repress members of the Kurdish Workers Party (PKK).

36   Baghdad's international isolation deepened in 1996, when Jordan's relations with Kuwait and Saudi Arabia improved. However, the UN Security Council voted for the partial lifting of the blockade, to allow restricted sales of crude oil, in order to buy food and medicines for the Iraqi population.

37   A UN report revealed in April 1997 that the number of dead due to hunger or lack of medicines arising from the embargo exceeded one million, of which 570,000 were children. UNICEF reported that 25 per cent of children under 5 were suffering from severe clinical malnutrition.

38   In October, the Security Council threatened to enforce new sanctions if a new inspection was not allowed, with the aim of verifying that the Iraqi administration lacked the capacity to develop chemical and biological weapons. Iraq rejected the presence of US inspectors, which hardened US President Clinton's stance. Clinton, backed by British Prime Minister Tony Blair, launched a missile attack on several Iraqi cities, starting on 16 December. The attack, 'Operation Desert Fox', killed hundreds of Iraqi civilians and troops.

39   In December 1999, the Security Council approved the resumption of weapons inspections in Iraq and the suspension of economic sanctions in case Baghdad decided to cooperate. Russia, France, China and Malaysia abstained from voting. Iraq, alleging it was an attempt by the US to impose its 'evil' will on the Security Council, refused to co-operate and demanded that all sanctions be lifted.

40   When George W Bush took office as US President in January 2001, he announced a hard-line stance and stiffened sanctions against Iraq. After the attacks on New York and Washington DC in September that year, the US focused on Baghdad. However, the allies, including Britain, withheld their support. Meanwhile, Saddam regained popularity in the Arab world by supporting the second Palestinian *intifada* and proposing that Muslim countries pursue their common interests through controlling oil prices.

41   In January 2002, in his State of the Union address, Bush dubbed Iraq, Iran and North Korea the 'axis of evil'. He advocated the 'need'

PROFILE

ENVIRONMENT

The Mesopotamian region, between the rivers Tigris (Dijlah) and Euphrates (Al-Furat) in the center of the country, is suitable for agriculture, and contains most of the population of Iraq. There are significant oil deposits in the mountainous areas in the north, in Kurdistan. In Lower Mesopotamia, on the Shatt-al-Arab channel, where the Tigris and the Euphrates merge, palm trees produce 80 per cent of the dates sold worldwide.

SOCIETY

Peoples: Three-fourths of the population is Arab. In the north there is a significant Kurdish minority (20 per cent) and the rest are small minorities of Syrians, Armenians and others. **Religions:** Mainly Muslim. About 62 per cent of the population is Shi'a, concentrated in the south. Sunnis are around 35 per cent. Most of the political élite is Sunni. Northern Kurds blend Sunnism with their traditional religion, Yezidi. There is a Christian minority. **Languages:** Arabic (official and predominant); in Kurdistan it is taught as a second language, after Kurdish. **Main Political Parties:** United Iraqi Alliance; Democratic Patriotic Alliance of Kurdistan; Iraqi Accord Front; Iraqi National List. **Main Social Organizations:** As in the area of politics, the newly emergent order is still formative. The situation regarding trade unions and social organisations had not yet been clearly defined.

THE STATE

Official Name: Al-Jumhuriyah al-'Iraqiyah. **Administrative Divisions:** 15 provinces and 3 autonomous regions. **Capital:** Baghdad 5,620,000 people (2003). **Other Cities:** Mosul 1,099,700 people; al-Basrah 1,004,800; Irbil 692,100; Karkuk (Kirkuk) 688,500 (2000). **Government:** Jalal Talabani, President since April 2005. Nouri al-Maliki, Prime Minister since May 2006. **National Holiday:** 14 July, Proclamation of the Republic (1958).-**Armed Forces:** Undergoing reorganization (as is the police).

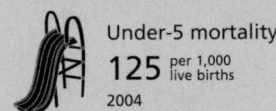
Under-5 mortality
125 per 1,000 live births
2004

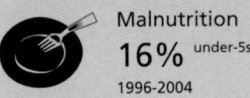
Malnutrition
16% under-5s
1996-2004

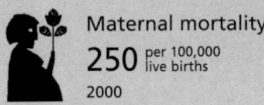
Maternal mortality
250 per 100,000 live births
2000

IN FOCUS

ENVIRONMENTAL CHALLENGES
The armed conflict launched in March 2003 has affected the environment, as well as other aspects of daily life. Most of the country's infrastructure continues to be devastated by the recent wars - since 1980 Iraq has been involved in 3 wars, lasting 12 years in total - and now the occupation led by the US. Tank and troop movements, and bombing raids over insurgent areas have destroyed the ground and soil, especially on the border with Saudi Arabia. There are problems with access to drinking water and soil erosion, while salinization and desertification are major concerns.

WOMEN'S RIGHTS
Women have been able to vote and stand for office since 1980. In 2005, 25 per cent of seats in parliament were held by women. When the 2003 war started, women made up 22 per cent of the 7-million workforce. In 2004 the maternal mortality rate stood at 250 deaths per 100,000 deliveries. Current accurate data are not available as a result of the chaotic situation in the country. Women's rights organizations talk of sexual apartheid when referring to issues such as death by stoning; gender segregation in the workplace; the permits that fathers or husbands have to grant women to allow them to work, study or travel; and the fact that divorced women are not allowed to see their children.

CHILDREN
Every Iraqi's daily experience is affected by the war and occupation, and these particularly impact on children's current and future opportunities. Since the educational and sanitation infrastructure has collapsed, the country has to try to educate and care for millions of children (in 2004 half the population was under 18) who, before the conflict, already had high levels of undernourishment and growth problems. Thousands of schools were destroyed, and half those remaining lack sewage or drinking water. International aid and the teachers' efforts keep schools operational. Infant and under-five mortality grew 150 per cent in the 1990-2004 period. The former went from 40 to 102 per 1,000 live births and the latter from 50 to 125 per 1,000. Fifteen per cent of children have low birth weight, 16 per cent of children under 5 are moderately or seriously underweight and 22 per cent have moderately or seriously stunted growth.

INDIGENOUS PEOPLES/ ETHNIC MINORITIES
Kurds were an ethnic minority persecuted by the Saddam Hussein regime. After his fall, the Shi'a majority - which had been dominated by the Sunni minority that supported the President - were voted into government and have the control. Now Sunnis feel persecuted and marginalized and are intensifying their fight against the new administration. The degree of violence in early 2006 meant that the conflict had really become a civil war which is just being contained by the US occupation forces as they continue to serve as the new regime's political police.

MIGRANTS/REFUGEES
In early 2006 it was estimated about one million Iraqis were refugees in neighboring countries. There were also 1.2 million internally displaced due to the war, some 250,000 who had returned and 46,000 refugees from abroad (of these, 34,000 were Palestinians arrived in 1948, 1967 and 1991, and the rest were Turkish, Iranians and Syrians). There were also between one and two million people living in the country without any official status. In 2005 asylum requests rose by more than 27 per cent.

DEATH PENALTY
The death penalty still applies for common crimes.

* Latest data available in The State of the World's Children and Childinfo database, UNICEF, 2006.

to attack Iraq, spuriously linking it with the al-Qaeda network and emphasizing that the danger of Saddam Hussein's regime lay in its 'potential' to develop weapons of mass destruction (WMD).
[42] In August, Blair talked Bush into taking the US case for attacking Iraq to the UN. Meanwhile, Saddam invited the UN's chief weapons inspector to negotiate.
[43] In September, during the 57th UN General Assembly, Bush asked a skeptical audience of world leaders to confront the 'serious and growing threat to peace' posed by the Iraqi regime, or else allow the US to act. The following month, Baghdad let the UN visit dozens of 'sensitive' locations. However the UK and US rejected this as they wanted the Security Council to approve a new resolution which would authorize military attacks if Iraq did not comply with the demands.
[44] Backed by a new UN resolution, which was more in line with US and UK wishes, the UN weapons inspectors returned to Iraq in November. The January 2003 report found no evidence of the existence of WMD.
[45] Even without that evidence or a new resolution by the Security Council explicitly authorizing the use of force, the US, UK and coalition forces launched an attack on Iraq in March 2003, entering the southern part of the country.
[46] In April, US troops entered Baghdad and continued to advance towards northern Iraq, meeting strong resistance only in the main cities like Kirkuk and Mosul. Looting became widespread while the allied forces searched for Saddam Hussein and 54 other 'most-wanted' leaders.
[47] In May 2003, the UN Security Council lifted economic sanctions against Iraq. The occupation forces destroyed the Ba'ath Party institutions. The US announced the end of major combat operations.
[48] On 14 December, it was announced that Saddam Hussein had been captured in an underground refuge. Images of the former Iraqi leader were broadcast around the world.
[49] In February 2004, while Shi'as in al-Basrah continued to demand direct elections, Kofi Annan - in line with Washington's position and in opposition to the Shi'a majority - announced that the best solution for Iraq would be an interim government (to be installed in June 2004). For the first time since World War II, Japanese troops were deployed overseas in a conflict zone, helping with Iraq's reconstruction.
[50] In late February, UN Secretary-General Kofi Annan and Japanese Prime Minister Koizumi announced that elections in Iraq could be held in late 2004 or early 2005.
[51] In March 2004 the Iraqi Council agreed - after tough negotiations - on a draft of the interim Constitution which would operate until the permanent Constitution could be written by an assembly elected by the Iraqi people.
[52] Photographs showing US soldiers' physical abuse of prisoners of war were revealed in April 2004. More evidence of torture and mistreatment emerged later.
[53] In June 2004, after the self-dissolution of the Governing Council, the new interim government took office. Ghazi Yawer, a US-educated civil engineer and leader from the northern town of Mosul was appointed as interim President. Iyad Allawi, known for his close ties to the CIA, took office as Prime Minister. The insurgency - through suicide bombings, kidnappings and the execution of foreigners - intensified and was focused in Fallujah.
[54] Some 8 million people voted in the January 2005 elections for the National Transition Assembly. The United Iraqi-Shi'a Alliance won, followed by Kurdish parties.
[55] In early April, after nine weeks of negotiations, Parliament elected a three-person presidency. Kurdish leader Khalal Talabani would be the interim leader of the Presidential Council, along with Ghazi Yawer, a Sunni, and then Finance minister Adel Abdul Mahdi, a Shi'a. Another Shi'a, Ibrahim Jaafari, one of the country's most popular politicians, was appointed Prime Minister.
[56] In June 2005 Massoud Barzani, the leader of the Kurdistan Democratic Party, was sworn in as President of Iraqi Kurdistan. In July, an Iraqi NGO estimated that approximately 25,000 civilians had died since the 2003 invasion.
[57] A draft Constitution was passed in August by Kurdish and Shi'a representatives, but not by Sunnis. In October, Saddam Hussein was charged with crimes against humanity. That month a new Constitution was passed which turned Iraq into an Islamic federal democracy.
[58] In the December 2005 general elections Iraqis chose their first non-interim government since the invasion. The degree of violence had not abated and many Sunni leaders reported irregularities during the elections. In January 2006 the victory of the United Iraqi-Shi'a Alliance was declared. It had also won a majority (but not an absolute one) in the Transition Assembly.
[59] Growing instability threatened in August to turn into civil war. Although this was not admitted by the US administration other voices - including that of the British ambassador in Iraq, William Paty - saw the division of the country according to ethnic groups as a growing possibility.
[60] In November 2006, Saddam Hussein was sentenced to death by hanging for the murder of 148 people in the Shi'a town of Dujai. In the same month, the defeat of US President Bush's Republicans in mid-term elections led to the dismissal of Defense Secretary Rumsfeld and speculation about a modified Washington approach to Iraq. ∎

Ireland / Êire

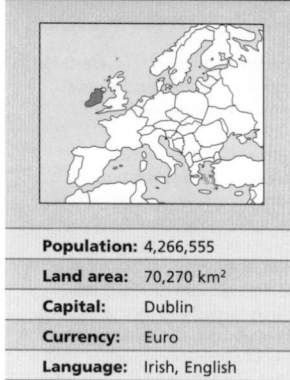

Population:	4,266,555
Land area:	70,270 km²
Capital:	Dublin
Currency:	Euro
Language:	Irish, English

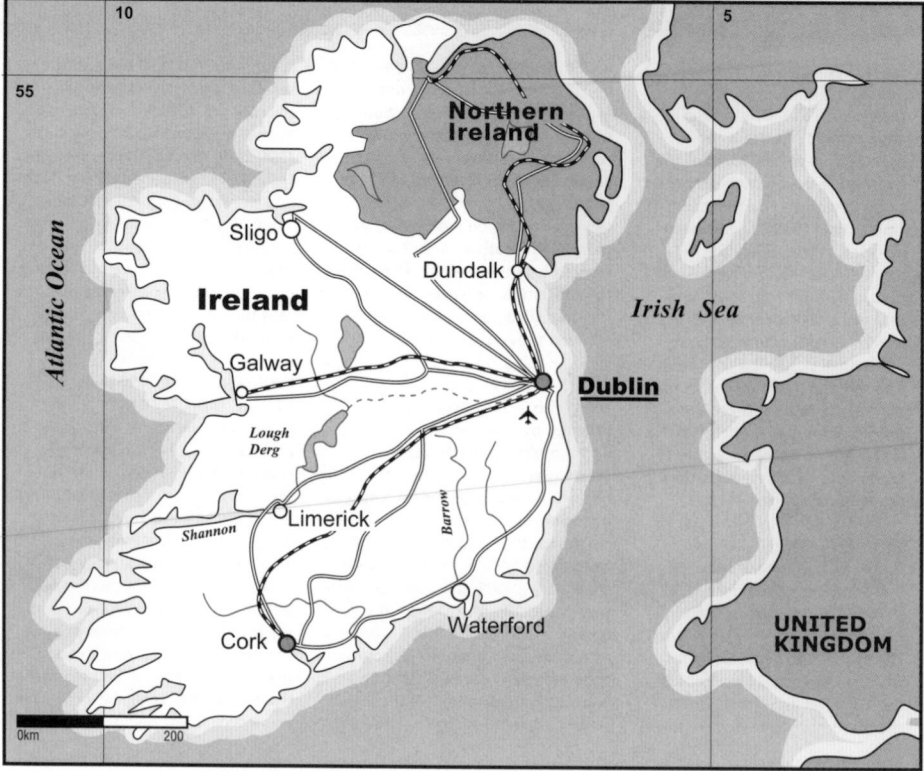

Ireland's original inhabitants were Mesolithic hunter-gatherers who used stone implements. Around 3000 BC, they evolved into Bronze Age people who cultivated crops, raised domestic animals and made weapons, tools and jewelry out of bronze. Starting about 2000 BC, they built the massive stone sanctuaries and tombs (megaliths) that still dot the Irish countryside.

[2] The first Celts arrived in Ireland about 1600 BC. Before their arrival, the basic units of Irish society were the *tuatha*, or petty kingdoms; perhaps 150 tuatha for a population of less than 500,000. This societal structure suited the Celts, who were predisposed towards relatively small and autonomous units. At the beginning of the Christian era, the Celts divided Ireland politically into five provinces: Leinster, Munster, Meath, Ulster and Connaught.

[3] A bishop and missionary coming from England, St Patrick (c.389-461), arrived in Ireland to convert the inhabitants to Christianity. He was able to make important converts among the royal families and, through the monastery schools, introduced the written word in Latin. By the death of St. Patrick the Irish elite recorded their history in writing. Ireland became almost exclusively Christian, as well as a center of scholarship and culture, but most of this legacy was destroyed in the Viking raids of

the 9th and 10th centuries. Pagan customs were incorporated into Christian practice.

[4] By the end of the 10th century Brian Boru, the king of a small state called Dal Cais, conquered neighboring Munster and became the strongest king in the southern half of Ireland. But Mael Morda, King of Leinster, began to plot against him and made an alliance with Sitric, the Viking king of Dublin, who got help from the Vikings of the Orkney Islands and the Isle of Man. The battle of Clontarf near Dublin in 1014 ended in victory for Boru's army, but Boru himself was killed in his tent by Vikings fleeing from the battle.

[5] In 1170, a party of Normans coming from England landed near Waterford, which fell into their hands along with Dublin. By 1300, the Normans controlled most of the country but they did not succeed in conquering Ireland because there was no central government that they could take control of. From about 1350, the Irish chieftains - who had acquired many of the

weapons used by the Normans and had learnt some of their tactics - began to recover their territories.

[6] Queen Mary I was the first English monarch to attempt to subdue Ireland by confiscating land and giving it to English settlers. Her half-sister Elizabeth I continued the policy and sent armed expeditions aimed at subduing rebellion, winning a major battle at Kinsale in 1601.

[7] Under King James I from 1608, Ulster was settled by Scottish and English Protestants in a conscious attempt to spread the religion. The leading colonists, who paid rent to the King, were required to clear their estates completely of native Irish inhabitants. Native resentment at the Plantation of Ulster, a planned process of colonization, led to a major rebellion in 1641. In 1649, following his execution of King Charles I, Oliver Cromwell led a suppressing army and saw atrocities committed in the 1641 rebellion as justification for massacring 4,600 people at Drogheda and Wexford. The power of Protestant landowners was reinforced.

[8] When the Catholic King James II was deposed in 1688 he raised an army in Ireland and quickly took control of all but the cities of Derry and Enniskillen. The siege of Protestant Derry became a vital battleground for the whole of Europe: its people held out for eight months before they were

relieved, ultimately enabling the Protestant King William III to confirm his own power and that of Protestants in Ireland at the Battle of the Boyne in 1690.

[9] By virtue of the 1 January 1801 Act of Union, Ireland was incorporated into the United Kingdom. During the 19th century most of the population outside the Protestant-dominated northeast supported independence, which led to the formation of a strong nationalist movement.

[10] The failure of the potato crop due to blight from 1845 to 1849 resulted in the Great Famine - the diet of the poor was heavily dependent on the tuber. Over 1.1 million Irish - mostly impoverished rural people - died due to undernourishment, typhus and other famine-related diseases, while at least a million others emigrated, mainly to the US. Ireland continued to export food throughout this period - neglect by absentee British landlords and the laissez-faire attitude of the British Government exacerbated the famine.

[11] An organized labour movement developed late in Ireland: there was little industry outside the northeast and the political priority was nationalism. The Dublin Lock-out of 1913 marked a sea change: recently unionized workers refused to relinquish union membership and prompted a wave of sympathetic

LAND USE

2003/2005

- FOREST AND WOODLAND: 9.7%
- ARABLE LAND: 17.2%
- CROPLANDS: 0.0%
- OTHER USE: 73.1%

WORKERS

UNEMPLOYMENT: 4% (2004)

LABOR FORCE	2004

- FEMALE: 42%
- MALE: 58%

strikes.

[12] In 1916, the republican Easter Rising in Dublin was crushed by the occupying forces but it marked the foundation of the Irish Republican Army (IRA) and the final stage of the long struggle for freedom. Although the Rising did not have widespread support, the subsequent execution of many of its leaders and other oppressive measures by the British galvanized support for the republican party Sinn Fein, which won the majority of seats in the 1918 general election. The IRA's campaign forced the British in 1921 to grant independence to the 26 counties with Catholic majorities. Southern Ireland became a self-governing region within the UK. The remaining six northeastern counties became Northern Ireland, with a devolved government in Belfast and representation in the British parliament in Westminster.

[13] Controversy over this settlement caused a bitter civil war in which at least 4,000 died. Though pro-Treaty forces prevailed, relations between the Free State and the British Government remained strained until after World War II, setting back economic development by decades. The 1937 Irish Constitution considered Ireland to be a single country, where all the inhabitants - North and South - have citizenship rights. In 1948 Ireland became a republic, and broke away from the British Commonwealth.

[14] In Northern Ireland, meanwhile, the Protestant majority - which tended to be Unionist, seeing Ulster as part of Britain - had exclusive control, led by the provincial prime minister and a governor acting as a representative of the British Crown. The Catholic minority in Northern Ireland - which tended to be nationalist - was thus excluded from domestic political affairs and faced discrimination. This led to the creation of an active civil rights movement in the 1960s. Although it was non-violent, the civil rights movement was considered by Unionist extremists to be a threat to the region's status and their dominant position and they reacted to it with violence. This in turn led to the re-emergence in 1969 of armed republican resistance led by the Provisional IRA.

[15] In April 1969, amidst growing disturbances, the Northern Ireland Government requested British troops to protect the region's strategic installations. In August, Belfast and London agreed that all the Province's security forces should come under British command.

[16] As a counterpart to the Provisional IRA, the 'loyalists' (loyal to the British Crown) formed a number of paramilitary organizations, including the Ulster Volunteer Force and the Ulster Defence Association. Between 1969 and the middle of 1994, more than 3,100 people died at the hands of the Protestant and Catholic paramilitaries, the British army and the Ulster police force, the Royal Ulster Constabulary (RUC).

[17] On 30 January 1972 (Bloody Sunday) 13 Catholics were murdered by British troops while demonstrating peacefully in the city of Derry. In the same year, political status was given to paramilitary prisoners, but this amendment was abolished in 1976.

[18] The growing violence provoked London to take over full responsibility for law and order in Northern Ireland. The Government in Belfast was abolished and a system of 'direct rule' from Westminster installed in 1972. In a plebiscite held in 1973, 60 per cent of the population of Northern Ireland voted in favor of union with Britain.

[19] At the end of 1973, a Northern Irish Assembly and Executive was created in Belfast in which Protestant and Catholic representatives were supposed to share power. In December, the London and Dublin governments agreed on the establishment of an Irish Council. Both this agreement and the new power-sharing Assembly were bitterly opposed by Unionists. In 1974 a general strike was declared, which resulted in the resignation of the Executive and London took over direct rule once more.

[20] In the 1973 election in the Irish Republic, Fianna Fáil (FF) - a party which had been in power for 35 of the previous 41 years - was defeated. A coalition between the conservative Fine Gael (FG) and the Labour Party (LP) took office. It was committed to power-sharing between the two communities in Ulster, but it rejected the immediate withdrawal of British troops from the region. Ireland joined the European Community in 1973, a decisive shift away from economic dependence on Britain.

[21] In 1976, after the IRA murdered the British ambassador in Dublin, and two years after bombings in Dublin and Monaghan killed 33 and injured hundreds more, the Irish Government took stricter anti-terrorist measures. Fianna Fáil was elected in 1977 and it maintained the friendly relations with London established by the previous administration. The new *Taoiseach* (head of the Irish Government) Jack Lynch supported the creation of a devolved parliament in the North, instead of total unification.

[22] In August 1979, Dublin agreed to increase border security after the murders, on the same day, of Lord Mountbatten (a prominent British public figure related to the Royal Family) in the Irish Republic, and 18 British soldiers in Warrenpoint, Northern Ireland. In December, Lynch resigned and was replaced by Charles Haughey, who went back to the old idea of reunification with some form of autonomy in Ulster.

[23] A referendum held in 1983 approved a constitutional amendment affirming the ban on abortion. In 1986, a government proposal to remove the constitutional prohibition on divorce was defeated in a referendum.

[24] Talks between the Irish and British heads of Government from 1980 onwards led to the signing of the Anglo-Irish Agreement in 1985. With it, the Dublin Government would have a say in political, judicial, security and border issues in Northern Ireland. Most Unionists in Northern Ireland were strongly opposed, but the Agreement guaranteed that no constitutional change could be made without the consent of the North's population.

[25] In the 1980s, the Irish economy suffered high unemployment levels (an average of 16.4 per cent between 1983 and 1988) and emigration levels, added to high inflation rates and industrial recession. Strict austerity measures, applied from 1987, allowed for growth in the second half of the decade.

[26] In November 1990, Mary Robinson, the Labour Party candidate and a lawyer who stood up for the rights of gays, women and the legal recognition of illegitimate children, became the first woman president of Ireland.

[27] Multi-party negotiations began in Belfast in April 1991, in another attempt to define Northern Ireland's political future, with representatives from Ulster's constitutional parties and the British Government. Sinn Féin, the political arm of the Provisional IRA, which supported the immediate withdrawal of British troops, the disarmament of the Royal Ulster Constabulary (RUC) and Ireland's reunification, was excluded from the talks for having refused to condemn the IRA's acts of violence.

[28] Meanwhile, the IRA launched one of its largest military offensives, both in Ulster and England. In early 1992, in Ulster, independent Republicans were responsible for several acts of arson in businesses and shops.

PROFILE

ENVIRONMENT

The country comprises most of the island of Ireland. The south is made up of rocky hills, none over 1,000 meters in height. The central plain extends from east to west and is irrigated by many rivers and lakes. With a humid ocean climate and poor soil, much of the country is covered with grazing land. Farming is concentrated mainly along the eastern slopes of the hills and the Shannon valleys. The main agricultural products are wheat, barley, oats, potatoes and beet. Meat and milk processing are among the major local industrial activities.

SOCIETY

Peoples: The Irish comprise 94 per cent of the population, with a small English minority. The great famines which struck the island over the last two centuries led to the emigration of about 4 million Irish people, especially to the US.

Religions: 91.6 per cent Catholic, 2.3 per cent Anglican, 0.4 per cent Presbyterian (1991).

Languages: Irish, English and Shelta (language of the Travelers/ Gypsies/Roma people).

Main Political Parties: Fianna Fáil (FF), conservative; Fine Gael (FG), Christian democratic linked to the rural population; Labour Party; Progressive Democrats.

Main Social Organizations: The Irish Congress of Trade Unions, with 750,000 members and 64 affiliated unions.

THE STATE

Official Name: Poblacht na h'Eireann.

Administrative Divisions: 26 Counties.

Capital: Dublin 1,015,000 people (2003).

Other Cities: Cork 123,100 people; Limerick 54,000; Galway 65,800; Waterford 44,600 (2002).

Government: Mary McAleese, President since November 1997. Bartholomew (Bertie) Ahern, Prime Minister since June 1997. Bicameral Parliament: the Senate with 60 members and the House of Representatives, with 166 members.

National Holiday: 17 March (St Patrick's Day).

Armed Forces: 10,559 (2002).

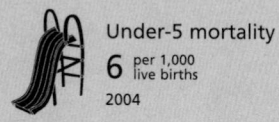
Under-5 mortality
6 per 1,000 live births
2004

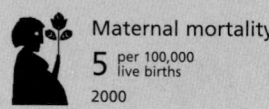
Maternal mortality
5 per 100,000 live births
2000

IN FOCUS

ENVIRONMENTAL CHALLENGES
Water pollution is the main environmental challenge, particularly in lakes due to agricultural runoff.

WOMEN'S RIGHTS
There was a partial franchise for women in 1918, and this was extended to them all in 1928. In 2003, women held 13 per cent of seats in Parliament and 21 per cent of ministerial positions. There were 15 births per 1,000 women aged between 15 and 19. Women made up 36 per cent of the labor force, comprising a total of 2 million people. Within the female labor force, 2 per cent were in agriculture, 14 per cent in industry and 83 per cent in services. Female unemployment reached nearly 4 per cent. Ireland is the only EU country where abortion is illegal.

CHILDREN
There is access to education, good nutrition, health, housing and social services. However, revelations of abuse by Catholic priests has shocked the country; in the archdiocese of Dublin, over 100 priests were suspected of sexually abusing at least 350 children between 1962 and 2002.

INDIGENOUS PEOPLES/ETHNIC MINORITIES
Some sectors of the population speak Gaelic and maintain some traditional ways of life, but Ireland's ethnic diversity today comes more from late 20th-century immigration (70 per cent of immigrants are from European countries) than from its original inhabitants.

Many Catholics in Northern Ireland, while generally supporting the independence movement which has for decades confronted the Unionists (the majority Protestant population), still feel they are socially discriminated against by being linked to the IRA, even though it announced a ceasefire in July 2005.

MIGRANTS/REFUGEES
According to official figures, between 2000 and 2005 Ireland hosted more than 6,800 refugees. In 2005 it received 4,323 new asylum applications: 1,278 came from Nigeria, 385 from Romania, 367 from Somalia, 203 from Sudan and 202 from Iran. Most of the refugees that arrived in the country in 2004 (a total of 4,766) came from these same five countries.

DEATH PENALTY
It was abolished in 1990. The last execution was in 1954.

From late 1991, Protestant paramilitary groups had reinforced their attacks on Catholics. These groups said they were willing to reply with 'an eye for an eye' to every IRA action.

[29] During January 1993, Albert Reynolds (FF) was confirmed as head of the FF/Labour Party Government in the Republic. At the same time, a referendum on abortion led two-thirds of voters to support the right to information on birth control and abortion and the right to travel abroad to carry out an abortion.

[30] In January 1994, the British Government lifted all restrictions on TV and radio broadcasts of interviews with Sinn Féin members. In August, the IRA declared it was ceasing all military operations.

[31] The Reynolds Administration fell in November, after the Labour Party withdrew its support due to the controversy on the delay of extradition to Northern Ireland of a priest accused of paedophilia. The new Prime Minister was John Bruton, who led a coalition of his FG Party with the Labour Party and the Democratic Left.

[32] Widespread sexual and physical abuse of children by priests and religious orders was exposed during the 1990s. Inquiries and compensation tribunals began considering the cases of up to 10,000 victims of abuse over the previous 50 years. The state has underwritten the compensation fund - while the religious orders are contributing cash and property transfers, the taxpayer will bear the brunt of the cost.

[33] The Roman Catholic Church, pre-eminent in social and political affairs since before independence, suffered a major loss in confidence, with mass attendance falling rapidly.

[34] In a referendum in 1995, the Irish approved (by 50 to 49 per cent of the vote) a constitutional reform authorizing divorce for couples separated for more than four years.

[35] In the June 1996 round of talks held in Belfast, Irish leader Bruton supported the British proposal to exclude Sinn Féin from the negotiating table, since both governments regarded the IRA's attitude as un-constructive. In February the IRA ended its unilateral ceasefire by setting off a bomb in London that killed two people and hurt several others.

[36] In April 1997, several attacks by the IRA on streets, train stations and airports paralyzed the British transport system during the election campaign. Sinn Féin asked the IRA to declare a new ceasefire, which was decided in July. Negotiations were resumed in September, with the inclusion of Sinn Féin.

[37] Fianna Fáil won the June 1997 Irish general election with 77 out of 166 seats, forming a minority coalition with the Progressive Democrats and the support of independents. Bertie Ahern became Prime Minister. Mary McAleese became President.

[38] After tough negotiations, on 10 April 1998 a peace agreement on Northern Ireland was reached in Belfast (the Good Friday Agreement). The text, negotiated by eight political parties, mediated by London, Dublin and Washington, included limited autonomy for Northern Ireland, with the creation of a power-sharing executive (cabinet), legislative assembly and cross-border co-operation bodies. The plan also called for the gradual release of political prisoners and for the disarmament of paramilitary organizations whose parties took part in the negotiations and the reconstitution of the police force to encourage nationalists to join.

[39] Most political forces, including Sinn Féin and the political wings of loyalist paramilitaries, initially favored the Agreement, which overcame its first hurdle when Ulster Unionist leader David Trimble received the support of his party, the Ulster Unionist Party (UUP). Only two Unionist parties were opposed: the Democratic Unionist Party (DUP), of Dr Ian Paisley, and Robert McCartney's United Kingdom Unionist Party (UKUP).

[40] In late May a referendum ratified the Agreement with 94 per cent of the vote in the Republic and 72 per cent in the North. A month later the first elections were held to form the Assembly as of February 1999. The new agreement paved the way for the people of Ulster to decide the Province's future, through a vote, to reunite with Ireland or to remain within the UK.

[41] In August, a bomb in Omagh, Northern Ireland, killed 30 people and endangered the fragile peace process. But the uproar caused by the attack on both sides of the border led the republican group responsible for the bomb (the Real IRA, a splinter group from the Provisional IRA) to declare a unilateral ceasefire.

[42] The establishment in December 1999 of the first Government of Northern Ireland in 25 years, although 'cross-border', confirmed the 'devolution' of sovereignty to the Province by the British Parliament. This new ruling body (Executive Council) was formed by the Prime Minister, David Trimble, from the UUP, with the deputy Prime Minister, Seamus Mallon, of the Social Democratic and Labour Party (SDLP) and ten ministers, three each belonging to the pro-British UUP and the moderate Catholic SDLP, and two each from Sinn Féin and the DUP.

[43] Coinciding with the inauguration of the new Government, the Republic of Ireland withdrew its constitutional claim over Northern Ireland, which it had maintained since Ireland's independence from the UK. The commitment was signed while the cross-border bodies were being set up in which Northern and Southern Irish would take part.

[44] The Nice Treaty, which paved the way for 12 Eastern Europe countries to join the EU, was rejected by the Irish in a referendum held in June 2001 but accepted in a second referendum in October 2002. Unlike the UK, Ireland adopted the euro as its currency in January 2002.

[45] The FF/PD coalition was returned to power with an overall majority in the 2002 general election.

[46] In October 2002, Sinn Féin members were arrested in Belfast, charged with espionage and attempting to obtain official information for terrorist purposes. London suspended Ulster's Executive Council and the Legislative Assembly and resumed control of Northern Ireland.

[47] The outcome of November 2003 Northern Ireland Legislative Assembly elections showed gains of ten seats for the DUP and six seats for Sinn Féin, strengthening the two parties at opposite ends of the local political spectrum. London has not yet restored the authority of the Assembly nor of the Executive.

[48] Economic growth in the Republic, which was eight per cent for the 1995-2002 period, went down to just 2.7 per cent in 2003.

[49] In June 2005, Irish was recognised by the EU as its 21st official language. Previously it had been granted the status of a treaty language. Irish Foreign Minister Dermot Ahern expressed his satisfaction and said that this 'affirmed at European level the dignity and status of the Irish people's first official language'.

[50] In January 2006, the Irish Government continued to complain at the UN about London's lack of response to the problem of polluting emissions from the British nuclear reprocessing plant at Sellafield on the Irish Sea coast. ∎

Israel / Yisra 'el

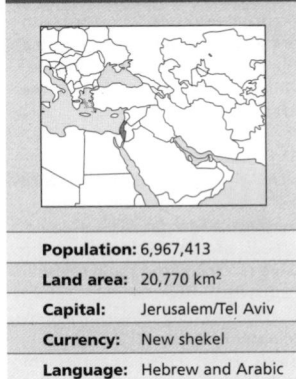

Population:	6,967,413
Land area:	20,770 km²
Capital:	Jerusalem/Tel Aviv
Currency:	New shekel
Language:	Hebrew and Arabic

Toward the 11th century a group of Hebrew speaking tribes occupied part of what is now Palestine. After a series of alliances and wars against the Canaanite and Philistine peoples settled there (see history of Palestine) a kingdom was formed which in 926 split into Israel, to the north, with Sichem as its capital, and Judah, to the south, with Jerusalem as its capital. These kingdoms were successively conquered by: Assyria, Babylon, Persia, Macedonia, the Seleucid Empire, the Roman Empire and, later on, by Byzantium. The presence of Jews diminished constantly due to massive expulsions, mainly those launched by the Romans between 66 and 73 AD, after the First Jewish Revolt was defeated. The Byzantines were expelled by the Arabs in 639. From that point, several Muslim states dominated the region until 1517, when the Ottoman Empire took control of the land.

2 Around 1800, five per cent of the population in Palestine was Jewish. The proportion grew in the 1880s when antisemitism in Europe led to a migratory inflow of European Jews. In 1896, Viennese journalist Theodore Herzi published the book *The Jewish State*, which gave birth to the Zionist movement - a word derived from 'Zion', one of the Biblical names for Jerusalem - whose goal was to find a place for the Jewish nation-state to settle. Uganda and Madagascar were some

LAND USE

2003/2005

IRRIGATED AREA: 45.3% of arable land

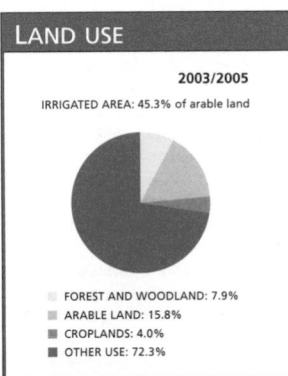

- FOREST AND WOODLAND: 7.9%
- ARABLE LAND: 15.8%
- CROPLANDS: 4.0%
- OTHER USE: 72.3%

Mediterranean Sea

LEBANON
Kiryat Shemona
SYRIA
Akko
Haifa
Wadi al Arab
Nazareth
Netanya
Nahr az-Zarqa'
Tel Aviv - Jaffa
Nablus **Palestine**
Holon
Ramallah
Jericho
Rehovot
Wadi shu'ayb
Jerusalem
Hebron
(Al Khalil)
Gaza
Qiryat Gat
Al Bahr al Mayyit
Beersheba
Israel
H A N E G E V
EGYPT
AL - KARAK
JORDAN
30
SHIBH JAZIRAT SINAI
Elat
0km   100
35

OCCUPIED BY ISRAEL
1 Syria - Golan Heights - occupied in 1967
2 Southern Lebanon - occupied 1983-2000
3 West Bank - occupied in 1967
4 Gaza Strip - occupied 1967-1994
Recovered by Palestine from 1994:
Gaza Strip, Jericho, Nablus, Ramallah

of the places considered - before Palestine - for the buying of lands. Between 1904 and 1914 the second *Aliya* (migration) took place, which took some 40,000 Jews toward Palestine, as part of the Zionist program.

3 After World War I, England and France shared the remains of the Ottoman Empire in the Middle East. The British Minister of Foreign Affairs, Arthur Balfour, declared in 1917 his support for the establishment of a Jewish National Home in Palestine (although, in 1905, as Prime Minister, he opposed Jewish immigration in Great Britain). The declaration stated that 'nothing shall be done which may prejudice the civil and religious rights of existing non-Jewish communities in Palestine', which were 90 per cent of the population at that time.

4 The Zionist goal of forming a State had a staunch opposition in the local Arab leaders, who saw in the Ottoman defeat the chance to create their own State or join a larger Arab entity and re-establish the Arab Empire of Islam.

5 At the end of World War II, England retained its control over

Palestine, arguing it was bound by the Balfour Declaration. Jewish militia groups grew. The largest and most important, Haganah (Defense) was an armed branch of the Jewish Agency, in charge of attracting Jews to Israel. These armed organizations were affiliated to right and left-wing Zionist political groups. The extremist group Irgun Zvai Leumi and its even more radical fraction Lehi (also known as Stern Gang) were affiliated to the ultraconservative Revisionist Party, founded by Vladimir Zeev Jabotinsky. Although Palmaj was actually an elite fraction of Haganah, it recruited many of its members among Socialist-leaning *kibbutzim* (agricultural colonies). Members of these militias included Yigal Alon, Moshe Dayan and Yitzhak Rabin.

6 In the early 20th century there were half a million Arabs and 50,000 Jews living in Palestine. During the 1930s the number of Jews amounted to 300,000, above the legal 'quotas' imposed by Palestine, due mostly to the fleeing of Jews from Nazi persecution in Germany.

7 The British saw their hegemony at risk in Palestine because Arabs also grew - due to high birth rates and immigration - from 440,000 to around one million in 1940.

8 In 1939 London stated its goal was to establish an independent Palestine state 'with both peoples sharing the government'. Fugitives fleeing from Europe under Hitler's rule were rejected from Palestine ports. Zionists organized sabotage and terrorist acts to get England to keep its promise.

9 Using donations from Jews all over the world, the Zionists purchased Palestinian lands from Arab owners living in Beirut or Paris, who cared little about the fate of their tenants, the Palestinian *fellahin* (peasants). The Jews then arrived, deeds in hand, to expel peasant families that had lived there for generations. They set up co-operative farms, *kibbutzim*, defending themselves from a now-hostile climate through armed militia.

10 In November 1947, in view of intensified anti-British attacks, London submitted the Palestinian problem to the United Nations. A special committee recommended partition of the territory into two independent states; one Arab, the other Jewish. Jerusalem would remain under international administration.

11 In that Cold War era, the Soviet Union preferred a Jewish state to a British military base, and its support was decisive in the creation of Israel. At the same time, London and Washington considered the partition unfeasible.

12 The UN General Assembly finally approved the partition plan in a 33 to 13 vote (Arab countries and India), with ten abstentions. Hardline Zionist militias began to expel Palestinians *en masse* from major cities and towns, alleging an imminent Arab attack. This policy culminated in a massacre at the village of Deir Yasin in April 1948, when the *Irgun*, an extremist group led by Menachem Begin, murdered its entire population.

13 On 14 May 1948, the British High Commissioner withdrew from Palestine, and David Ben Gurion proclaimed the State of Israel. The Jordanian, Egyptian, Syrian, Iraqi and Lebanese armies attacked immediately.

14 The war ended in January 1949 and Israel obtained 40 per cent more territory than it had been due under the partition plan. Although weapons and aircraft purchased from the Soviet Union proved decisive in the Israeli victory, the Government of Israel supported the West in the Cold War and formed a strategic alliance with the US. This relationship, lasting to this day, experienced difficult moments when

Life expectancy
81 years
2005-2010

GNI per capita
$17,360
2004

Literacy
97% total adult rate
2000-2004

HIV prevalence rate
0.1% of population 15-49 years old
2003

in 1956, Israeli troops backed by France and Britain invaded Egypt in response to the nationalization of the Suez Canal. The US and the USSR firmly opposed the action, forcing the invaders to withdraw with agreement from Egypt that it would stop sending fighters into Israeli territory.

15 Egyptian President Gamal Abdel Nasser, deprived of US credit and arms, turned to the USSR, which led to even closer relations between Israel and the US. Egypt and Syria, who were backed by the Soviet Union, and Jordan, supported by Britain, maintained a constant pressure of guerilla raids on Israeli civilians.

16 Several years later, the border with Syria remained a scene of constant conflict, particularly as both sides squabbled over access to fresh water from the Sea of Galilee. It was later revealed that the Soviet Union had intentionally escalated the situation in the Middle East by sending false messages to the various Arab states that the Israelis were massing their forces at the border with Syria.

17 On 17 May 1967, Nasser demanded that the UN Emergency Force (UNEF) - a UN peacekeeping force set up after the 1956 Suez war - leave Egypt, which the then UN Secretary-General U Thant complied with. Nasser immediately began remilitarizing the Sinai Peninsula, increasing tensions with Israel. On 23 May, the Egyptian Navy closed off the Straits of Tiran blockading Israeli shipping and the port of Eilat at the north of the Gulf of Aqaba. Egypt and Jordan had also signed a mutual defense treaty on 30 May (treaties with Syria already existed). Israel launched an attack on Egypt on 5 June and thus the Six-Day War began.

18 The war resulted in the Israeli seizure of the Jordanian-controlled West Bank of the Jordan River and East Jerusalem, Syria's Golan Heights, the Egyptian-controlled Gaza Strip and the Sinai Peninsula. Soon after the end of the Six-Day War, Prime Minister Eshkol's cabinet voted in favor of withdrawing from all the occupied territories (except East Jerusalem) in exchange for a comprehensive peace treaty.

19 In response, in September 1967, eight Arab leaders agreed the Khartoum Declaration which condemned Israel and guided Arab states' policy towards Israel until the mid-1970s. Specifically, the Declaration called for 'no peace with Israel, no recognition of Israel, no negotiations with it, and insistence on the rights of the Palestinian people in their own country'.

20 Resolution 242 of the UN Security Council (22 November 1967), calls for peace and recognition of the 'right of every nation to live free from threat within secure and recognized boundaries', in return for Israel's withdrawal from the occupied territories.

21 The armed conflict that broke out in 1973, known as the Yom Kippur war as it took place on one of Judaism's holiest days, began when Egyptian troops crossed the Suez Canal and Syrian forces attempted to regain the Golan Heights. They were supported militarily and financially by many Arab states including Saudi Arabia, Kuwait, Iraq, Algeria, Libya, Tunisia, Sudan and Morocco. The war ended the myth of Israeli military invincibility, but it did not lead to a significant modification of the border lines.

22 In 1977, Menachem Begin was elected Prime Minister, breaking the historical continuity of Labor Party rule.

23 Begin refused to negotiate with the Palestine Liberation Organization (PLO) and expressed his intention to annex the West Bank. Meanwhile in 1977, the US persuaded Egyptian President Anwar Sadat to sign the Camp David Agreement, leading to peace between Cairo and Tel Aviv and the return of Sinai to Egypt.

24 In June 1982, Israel launched 'Operation Peace for Galilee', invading Lebanon and devastating Beirut, under the pretext of stopping infiltration by Palestinian guerrilla groups. Arafat's forces withdrew from Lebanon in exchange for the deployment of a joint force of Italians, French, and North Americans to guarantee the security of Palestinian civilians.

25 Despite the agreement, in September 1982 hundreds of Palestinian refugees were murdered by right-wing militia in the Sabra and Shatila refugee camps, within Israeli-controlled areas. Dissent in Israel led to a 400,000-strong demonstration sponsored by the group Peace Now. Begin was forced to appoint a commission of inquiry, which found then Defense Minister Ariel Sharon and other military leaders 'indirectly responsible' for the murders.

26 In December 1987, the funerals of several young Palestinians killed in a clash with Israeli military patrols led to further confrontations, general strikes and civil protests. This marked the beginning of the first *intifada* (uprising - literally, 'shaking off') and Middle Eastern politics were shaken from the least expected quarter - the unarmed grassroots.

27 On 17 January 1991, in answer to the start of the first Gulf War, Iraq launched several Scud-missile attacks on Israel, aimed at provoking it to war. However, Israel did not do so, leaving its defense to Patriot anti-missile units operated by US troops.

28 When the war ended in March 1991, the US presented diplomatic circles with a 'land for peace' proposal. Two months later in Damascus, Syria and Lebanon signed a 'brotherhood, co-operation and co-ordination' treaty. To Israel, this treaty constituted a Syrian threat to a region in the north rich in water sources.

29 That year, Israel asked the US to approve loans of $10 billion to ease its economic difficulties, caused by the resettling of between 250,000 and 400,000 Soviet Jews from 1989 to 1991.

30 The Government created new settlements for immigrants on the West Bank where in 1991, despite efforts to create new jobs, unemployment reached 11 per cent. Attempting to promote peace negotiations in the region, Washington imposed the condition that loans granted would not be invested in settlements in occupied territories.

31 The settlements became a double-edged sword for Yitzhak Shamir's Government. The US loans and new settlements were both vitally important, while Palestinians and other Arabs demanded an end to new settlements, so the peace talks could continue.

32 On 30 October 1991, a Middle East Peace Conference was held in Madrid, sponsored by the US and the USSR. Shortly afterwards, hundreds of thousands of Israelis held demonstrations, calling on their government to maintain a dialogue with the Palestinians and Israel's Arab neighbors. Delegations from Jordan, Lebanon, Syria and Israel attended the conference; the Palestinians formed part of the Jordanian delegation, as Israel refused to negotiate directly with them.

33 In June 1992, Labor won a decisive victory in the general elections and Yitzhak Rabin became prime minister. Construction of housing in the occupied territories came to an abrupt standstill, and the US lifted the embargo on loans for Israel.

34 After months of secret negotiations in Oslo, Norway, Israeli authorities and PLO leaders signed the Declaration of the Principles of the Interim Self-Government Arrangements, in Washington in September1993. The Oslo Accords foresaw the installation of a limited autonomy system for Palestinians in the Gaza Strip and the city of Jericho for a five-year period, which would then be extended to include all of the West Bank.

35 The agreement was questioned, due to the opposition of *Hamas* and the Iranian-backed *Hizbullah*, two radical Islamist movements.

Meanwhile, Jewish settlers in the occupied territories - supplied with arms by the Government - rejected the agreement (as it stipulated their withdrawal, along with that of the Israeli army).

36 This situation worsened on 25 February 1994 when Baruch Goldstein, a member of Rabbi Meir Kahane's ultra right-wing Jewish movement *Kach* (*lit.* 'Only Thus'), massacred several Palestinians who were praying at the Tomb of the Patriarchs in Hebron.

37 In early May 1994, Israeli Prime Minister Rabin and PLO leader Yasser Arafat signed an agreement in Cairo granting autonomy to Gaza and Jericho. Late that month, the Israeli army withdrew from Gaza, ending 27 years of occupation (see Palestine).

38 1995 was marked by a growing division of Israeli society regarding the peace process with the Palestinians, leading to Rabin's assassination by a young far-right Israeli. Shimon Peres, also from the Labor Party, took Rabin's place but was defeated in the general elections of May 1996 by right-wing leader Binyamin Netanyahu.

39 The return to power of the conservatives hindered negotiations with Palestinians and heightened tensions, putting the country on the brink of a new war. In September the Government authorized the opening of a tunnel under the Temple Mount, the most sacred site to Jews, but also the home of the Al-Aqsa mosque, the third most sacred place of Islam. This provoked a Palestinian reaction with subsequent disturbances and deaths.

40 Netanyahu, harshly critical of Rabin for having held talks with Arafat and the PLO, met with the Palestinian leader on numerous occasions in late 1996 and early 1997 to negotiate the complete withdrawal of Israeli troops from the city of Hebron in the West Bank.

41 In March 1997, the Government announced its plan to build a new settlement in the Har Homa hills, on the Palestinian outskirts of Jerusalem. The Palestine Authority and the US rejected the project (which ran counter to the spirit of the Oslo Accords) and the negotiations stalled.

42 Ehud Barak, a retired general, had replaced Peres as leader of the Labor Party and became leader of the opposition in 1996. The Labor Party wanted an agreement with the conservative Likud Party to form a unity government that would allow the resumption of the suspended peace process.

43 After strong international pressure, Netanyahu accepted the US proposal to seek an accord to restart the peace process. He met US President Clinton and Arafat at Wye River in the US. It was agreed

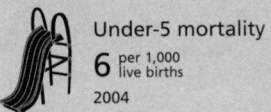
Under-5 mortality
6 per 1,000 live births
2004

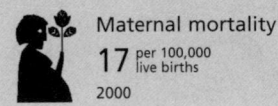
Maternal mortality
17 per 100,000 live births
2000

that Israel would return Palestinian territory in exchange for Arafat's acceptance of supervision by US intelligence organizations in his fight against terrorism by Islamic groups. The accord, which weakened Arafat in the eyes of his internal enemies, also led to the fall of the Likud Government. Labor backed the signing of the accords, but the governing coalition dissolved and several Likud legislators abandoned the party. The elections of 17 May 1999 were won by Barak's Labor Party. Breaking with a 50-year tradition, in November Israel lifted the state of emergency, which had been in effect since 1948.

[44] In late May 2000, Prime Minister Barak ordered the total withdrawal of troops from South Lebanon, occupied since 1982.

[45] Another summit was held at Camp David in the US from 11-25 July, but failed in part because agreements on Israeli withdrawal did not materialize and the Palestinian territories were blockaded following a terrorist attack committed by extremist groups.

[46] On 28 September 2000, Likud leader Ariel Sharon's visit to the Temple Mount (*Al-Haram Ash-Sharif* to Muslims) - sacred to both Muslims and Jews - triggered a new Palestinian uprising, now commonly referred to as the Second Intifada or the Al-Aqsa Intifada. Hundreds of people, mostly Arabs, died in clashes in the following months. This period was also marked by the increased frequency of suicide bombings against Israeli civilian targets by groups such as Islamic Jihad, Hamas, Hizbullah and the Al-Aqsa Martyrs' Brigade. Ariel Sharon also stepped up his policy of 'targeted assassinations' against Palestinian leaders.

[47] Barak resigned in December and Sharon won the elections. Sharon's mandate turned into one of effective rupture with the Oslo Accords. Arafat stepped up repression against Islamic extremist organizations, but the critical social situation breathed new life into the Intifada. In the following months the armed clashes multiplied and all attempts to reach a truce were frustrated.

[48] After the 11 September 2001 terrorist attacks on Washington and New York, Sharon intensified the offensive against the Palestinian uprising. The effects of the war began to take a toll on the Israeli economy, which was left not only without cheap labor, but also without consumers, as the Palestinian territories had been Israel's main external market.

[49] Israeli forces searched for the murderers of Tourism Minister Rehavim Ze'evi in the Palestinian areas, and in October 2001 they

besieged Arafat's compound in Ramallah. Sharon demanded that Arafat hand over the alleged murderers. Any efforts at negotiation were cut off. Protests against the Israeli reprisals spread around the world.

[50] In January 2002, the Israeli army withdrew from the Palestinian territories, but continued the siege of Arafat's headquarters, shortly before the arrival of US envoy Anthony Zinni.

[51] On 23 February 2002 Israeli helicopters launched missiles into Arafat's compound. The Palestinian leader was not hurt. The attack came hours after Palestinian gunmen killed six Israeli soldiers at a West Bank checkpoint. In May, Likud voted against the creation of a Palestinian State as requested by Netanyahu, Sharon's main rival within the party. Netanyahu argued that a Palestinian State on the West Bank of the River Jordan would pose a 'mortal threat' to Israel, while the top Palestinian negotiator, Saeb Erekat, responded that the vote undermined any attempt to reach a peace agreement. The Likud vote was a heavy blow to Sharon, who had publicly supported the creation of a Palestinian State in some areas currently occupied by Israel.

[52] In June and July 2002, terrorist attacks, murders and arrests of Palestinian leaders stepped up. Israel started building a separation wall on the West Bank border, saying this would help prevent suicide bombers from entering Israel. As well as protest about the construction of the wall, there was criticism of its route, seen as an attempt to take more Palestinian land since it does not follow the UN 'Green Line' 1948 borders but cuts into areas of the West Bank.

[53] Tension with Lebanon was re-ignited in September 2002, after Beirut's project to divert 10 per cent of the water in the Wazzani river. The river rises in Lebanon and supplies Israel with potable water. Sharon said the project could lead to war, and filed a complaint with the UN. Lebanon, on the other hand, argued that the volume of water they intended to extract - less than 10 million cubic metres of the 50 million annual flow - was in line with international law.

[54] Likud won the January 2003 elections, taking 37 of 120 seats. The Labor Party's defeat, and reduction in seats from 25 to 19, constituted another problem for Sharon, who was unable to create a coalition with control of parliament. Likud was then again forced to consider an alliance with ultra right-wing and orthodox sectors, which could damage Israel's relationship with Washington. The US supported the creation of a Palestinian State, but the conservative Israeli political

parties did not.

[55] 2003 was marked by the war on Iraq. In Israel, fears that it could be hit by Iraqi missiles compounded concerns over continued Palestinian suicide attacks, while the Israeli army continued its incursions in the Gaza Strip and the West Bank.

[56] At the end of April 2003, after the end of 'major combat operations' in Iraq, the US, Russia, the EU and the UN released their 'road map' peace plan for a Middle East settlement, which envisaged a Palestinian State by 2005.

[57] In November, with the creation of a new Palestinian Government with Ahmed Qorei as Prime Minister, possibilities for dialogue were renewed. On 27 November, Israeli and Palestinian representatives met in London to launch a new peace plan backed by the US. That day, in statements to the press, Sharon announced 'positive unilateral steps' aimed at easing the tension with

PROFILE

ENVIRONMENT

Israel, with 20,770 sq km within the pre-1967 borders (see Palestine), is bordered to the west by the Mediterranean Sea. In the south it has a small outlet to the Gulf of Aqaba on the Red Sea. The following territories are currently occupied, both militarily and with settlements of Jewish colonists: Golan Heights (Syria) 1,150 sq km, West Bank 5,879 sq km, Greater Jerusalem 70 sq km. The land comprises four natural regions: the coastal plains, with a Mediterranean climate, the country's agricultural center; a central hilly and mountainous region, stretching from Galilee to Judea; the western lowlands, bound on the north by the Jordan River, which flows into the Dead Sea; and the Negev Desert, to the south, which covers half of the total territory. The main agricultural products are citrus fruits for export, grapes, vegetables, cotton, beets, potatoes and wheat. There is considerable livestock production. Natural resources include timber, potash, copper ore, natural gas, phosphate rock, magnesium bromide, and clays. Industrial production, particularly in the high-tech sector, is growing rapidly and there are serious difficulties with water and pollution. The country has over 2,000 sq km of irrigated land - part of the Negev has been reclaimed through irrigation projects.

SOCIETY

Peoples: Jews 80.1 per cent (Europe/America-born 32.1 per cent, Israel-born 20.8 per cent, Africa-born 14.6 per cent, Asia-born 12.6 per cent); Arabs, Druze, Circassians, Armenians, and others 19.9 per cent.
Religions: Judaism (official). Arabs are mostly Muslim with nearly 10 per cent Christian; also Druze (*Mowahhidoon*) and Bahá'í.
Languages: Hebrew and Arabic (official), several languages of immigrants' countries of origin. Also Ladino (Judeo-Spanish), Circassian, and Amharic.
Main Political Parties: Kadima (Forward); Ha Avoda (Labor Party and Meimad); Shas (Sephardic Religious Party); Likud (conservative); Yisrael Beytenu (*Our Home Israel*).
Main Social Organizations: The Histadrut Haoudim Haleumit (National Labor Federation) is the main trade union. Gush Emunim, nationalists demanding Jewish settlements in the Gaza Strip and West Bank; Peace Now supports territorial concessions in the West Bank. *Kibbutzim* - co-operative communities - played a significant role in the development of the country, but are now in decline.

THE STATE

Official Name: Medinat Yisra'el (Hebrew); Daulat Isra'il (Arabic).
Administrative Divisions: 6 districts, 31 municipalities, 115 local councils and 49 regional councils.
Capital: In 1980, Jerusalem (population 686,000 in 2003) was proclaimed the 'sole and indivisible' capital of Israel. The UN condemned the decision. Many diplomatic missions are in Tel Aviv-Jaffa (360,400 people).
Other Cities: Haifa 270,800 people; Rishon Le Ziyyon 211,600; Ashdod 187,500.
Government: Moshe Katzav, President since August 2000. Ehud Olmert, Prime Minister since April 2006, but was Acting PM since January. Parliamentary system, with 120 member legislature (Knesset). There is no constitution but some of the functions of a constitution are filled by the Declaration of Establishment (1948), the Basic Laws of the Knesset, and the Israeli citizenship law.
National Holiday: 14 May, Independence Day (1948).
Armed Forces: 175,000 (2004; 430,000 reservists). Other: Border Police (Magav): 6,000.

Water source
100% of population using improved drinking water sources
2002

Doctors
391 per 100,000 people
1990-2004

Primary school
99% net enrolment rate
2004

IN FOCUS

ENVIRONMENTAL CHALLENGES
Limited arable land and natural fresh water sources pose serious constraints. There is significant air pollution from the burning of fossil fuels in urban and industrial areas. However, solar energy is used for many functions, and research on new energy sources is ongoing. Drip irrigation technology and other water-saving measures are widely used.

WOMEN'S RIGHTS
Israel is a Jewish state, although not a theocracy. The status of Jewish women in politics is related to their position within Judaism. For Orthodox Jews, women may not become rabbis, and so may not hold religious and some public posts.

Outside of religion, gender equality is guaranteed by law. In 2003, women held 15 per cent of seats in Parliament. Equal rights and duties entail compulsory military service for both men and women - three years for men and two for women. Women with strong religious convictions can choose to do their military service in health or education institutions, among other options. The status of Arab women relies to some degree on the political

temperature between Palestine and Israel. In July 2005, Parliament passed an immigration law preventing 'family reunions' for Israeli men married to Palestinian women under 25 or Israeli women married to Palestinian men under 35.

CHILDREN
In education, discrimination is widely reported, especially in terms of resource allocation by the Government. Spending on Jewish children is notoriously higher than spending on Arab children.

Infant and under 5 mortality was halved between 1990 and 2004, falling from 10 to 5 per 1,000 live births and from 12 to 6 per 1,000, respectively. Israeli (and Arab) children have to learn young about potential invasions and terrorist, chemical or nuclear attacks, due to the region's conflictual status. War and violence are part of their daily life.

INDIGENOUS PEOPLES/ETHNIC MINORITIES
Almost one million of Israel's inhabitants are Arabs, Druze, Bedouins and Circassians. The situation of Arab citizens who stayed in Israel after the wars of 1948 and 1967, remains complex (see Palestine).

The Druze (*Mowahhidoon*)

broke off from Islam in the 10th century, and have always lived in the Middle East. They are loyal to the State they live in, and in Israel they serve in the army. There are also Druze minorities in Syria and Lebanon.

The Government is trying to encourage its Bedouins to adopt a settled way of life, contrary to their traditions.

MIGRANTS/REFUGEES
All Jews are eligible to immigrate and become citizens under Israel's Law of Return. In late 2002, 18,000 Ethiopian Jews immigrated to Israel. Several thousand Argentinians and Uruguayans took refuge in Israel after Argentina's 2001-2 economic collapse. But in December 2003, over 1,000 of them returned home, mostly because of fear of terrorism, according to Israel's *Ma'ariv* newspaper. Between January 2003 and June 2005 more than 1,000 Israelis sought asylum in Canada and the US.

DEATH PENALTY
The death penalty is applicable only in the case of war crimes. Adolf Eichmann was put to death on 31 March 1962 - the only execution carried out since the creation of the Israeli State in 1948.

the Zionist regime to install itself', which in his view would settle the 'problem'.

[70] Sharon suffered a stroke in December, and was temporarily replaced by Deputy Prime Minister Ehud Olmert. Sharon had formed a new centrist party, Kadima, under which he planned to run for a third term in office.

[71] When Hamas won the Palestinian elections in January 2006, Israel declared it would not negotiate unless the group renounced the use of arms.

[72] Olmert won the elections held in March and committed himself to go ahead with plans to draw Israel's final borders. His party, Kadima, won 28 out of the 120 seats in parliament. The Likud party came in fifth place with just five seats.

[73] In May 2006, Hamas leader, Ibrahim Hamad, was captured by Israeli troops in a raid in Ramallah.

[74] In July, under the pretext of rescuing two Israeli soldiers that had been kidnapped by Lebanese militants of Hizbullah, Israel launched air and maritime attacks on southern Lebanon. The stated purpose was to destroy a group that, according to Tel Aviv, was ruling southern Lebanon with the support of Syria and Iran, and to wipe out its capacity to launch missiles against the northern cities of Israel.

[75] With the unconditional support of the US - considered as co-author of the attack by many analysts - and in view of Hizbullah's unexpectedly strong response - even managing to hit a warship with its missiles - thousands of Israeli troops entered Lebanon in August. Meanwhile, the bombing of cities continued and there was increased international pressure for an immediate ceasefire.

[76] By mid-August, a very precarious ceasefire was achieved. By then, over 1,000 Lebanese people - mostly civilians - and some 150 Israelis - mostly soldiers - had been killed. The damage caused to the Lebanese infrastructure and economy was beyond calculation; Israel's losses were estimated at $5 billion. None of the initial goals were fulfilled and, while hundreds of thousands of Lebanese displaced people returned to their destroyed homes and Hizbullah celebrated what it regarded as a military victory, there was growing criticism of Olmert, Defense Minister, Amir Peretz and armed forces commander, Dan Halutz.

[77] At the end of October 2006 Israel renewed its onslaught on Gaza, targeting Hamas and Islamic Jihad militants. A tank attack killed 18 civilians in the town of Beit Hanoun, prompting Palestinian outrage and widespread international condemnation. Ehud Olmert said the strike on a civilian area was the result of a 'technical failure'. ■

Palestinians, and recognized that Israel would have to make 'painful concessions'.

[58] Meanwhile, unofficial negotiations were taking place on a Draft Permanent Status Agreement, more commonly known as the Geneva Accord. An informal agreement was reached by prominent leaders from peace camps on both sides of the conflict, among them Yossi Beilin, leader of *Yachad* (*lit.* 'Together') who was also instrumental in bringing about the Oslo Accords, and former Palestinan Authority Minister Yasser Abed Rabbo. As of December 2003, the vast majority (78 per cent) of Palestinians knew little or nothing of the Geneva Accord, published in both the *al-Ayyam* and *al-Quds* newspapers. Of these, less than 10 per cent had read it. Among those who had, a majority disagreed with its central concepts (withdrawal, statehood, Jerusalem, refugees, and ending the conflict).

[59] Among Israelis there was much greater awareness of the Accord's content and it was debated in the Israeli press. Public support for the agreement was around 30 per cent, according to polls.

[60] Two and a half years after the

beginning of the Second Intifada, reports estimated that there were around 5,000 dead and 43,000 injured people on both sides. The Israeli army counted 900 dead, 600 of whom were killed in attacks by suicide bombings. 60,678 people had been arrested by Israel.

[61] In February 2004, Prime Minister Sharon unveiled his plan, which included dismantling 17 of the 21 Jewish settlements in Gaza with 5,000 settlers and some others in the West Bank.

[62] In the meantime, giving credence to Palestinian complaints that Sharon was attempting to encroach on West Bank territory, construction of the wall continued, with some changes to the original route and the dismantling of some sections (such as the one that isolated Baka al-Sharqia for a year, before it returned to the West Bank). The Palestinian Authority denounced plans for a second wall along the Jordan river valley.

[63] Reactions for and against the barrier were widespread. Meanwhile, new Palestinian suicide bombings took place in Jerusalem, unleashing Israeli raids on Ramallah.

[64] In October, Israeli forces demolished the homes of hundreds

of Palestinians, killing more than 70 people in the worst attack in the Gaza Strip in years.

[65] That month, Arafat was taken to Paris to be treated for an undisclosed illness. He died on 11 November.

[66] After the January 2005 elections, Mahmoud Abbas became President of the Palestinian Authority. It was hoped that his appointment would open the way for new negotiations between Palestinians and Israelis.

[67] Immediately after he was sworn in, several Palestinian attacks against Israeli targets seemed to put an end to all dialogue. However, in February, Abbas convinced Hamas and Islamic Jihad to agree on an unofficial ceasefire. Abbas and Sharon announced they would meet in Egypt to start negotiations.

[68] In spite of protests from Israeli settlers, the planned withdrawal from Gaza - carried out by the Israeli army, which continued to dismantle the settlements - continued throughout September and October.

[69] In late 2005, Iranian President Mahmoud Ahmadinejad said Israel should be 'wiped off the map' and that 'Germany and Austria could supply two or three provinces for

Italy / Italia

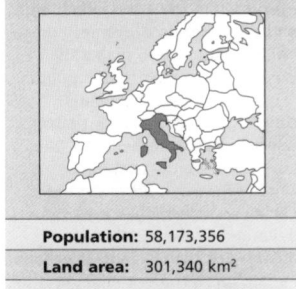

Population:	58,173,356
Land area:	301,340 km²
Capital:	Rome (Roma)
Currency:	Euro
Language:	Italian

From 2000 to 1000 BC, the Italian peninsula received Indo-European peoples from Central Europe. In that period two homogenous cultural areas developed: one in the north, characterized by the construction of lake dwellings and the cremation of their dead; and the other, in the south, which was influenced by Mediterranean civilizations. At the end of the second millennium BC strong migratory currents weakened the northern and fragmented the southern cultures. Numerous regional cultures arose (Latin, Ligur, Veneta, Villanovan and Iliric, among others). The foundation of Greek colonies beginning in the 8th century BC was culturally significant. The cities on Sicily and the city of Cagliari, in Sardinia, were founded by the Phoenicians.

2   With the fall of the Hittite empire around 900 BC, the Etruscans established themselves to the north of the Tiber River. Their influence extended throughout the Po valley until the end of the 6th century, when the Celts bore down on them, destroying their territorial unity.

3   According to legend, Romulus founded the city of Rome upon the Palatine hill in the year 753 BC. During the following century, this settlement was united with those on the Quirinal, Capitoline and Esquiline hills. The first form

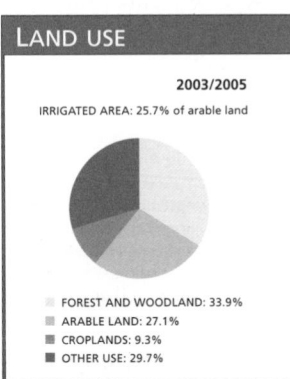

LAND USE

2003/2005

IRRIGATED AREA: 25.7% of arable land

- FOREST AND WOODLAND: 33.9%
- ARABLE LAND: 27.1%
- CROPLANDS: 9.3%
- OTHER USE: 29.7%

of government was an elective monarchy. Its powers were limited by a senate and a people's assembly of clans which held the power of imperium or mandate to govern.

4   There were two social classes: the patricians, who could belong to the Senate, and the plebeians, who had to band together to protect themselves from the abuses of the large landowners.

5   Under King Tarquinius Priscus (616-578), Rome entered the Latin League. The poverty of the plebeians and the system of debt-induced slavery led to the expulsion of the kings in 509. In the 5th century, the traditional laws were written down. This Law of the Twelve Tables extended to the plebeians, who after a lengthy struggle had managed to win some rights.

6   The Punic Wars against Carthage in the 3rd century allowed Rome to expand its possessions once again; in the early 2nd century, after displacing the Macedonians, Greece became a protectorate. Within a few years, Asia Minor, the northeast of Gaul, Spain, Macedonia and Carthage (including the northern part of Africa) had fallen into Roman hands.

7   Toward the end of the 2nd century, the Gracchi brothers, Tiberius and Gaius - both Roman representatives - were assassinated by the nobles, along with 3,000 followers, for supporting the plebeians.

8   The Roman Empire controlled the land from the Rhine in Germany to the north of Africa, and also included the entire Iberian Peninsula, France, Britain, Central Europe and the Middle East as far as Armenia. The 2nd century brought internal disputes which plunged Rome into chaos.

9   In 330, the Emperor Constantine transferred the capital of the Empire to Byzantium - called New Rome - and converted to Christianity. In 364, the Empire split into two parts: the Western and Eastern Roman Empires.

10   The end of the 5th century was marked by the invasions of the Mongols and other northern tribes, and by the attempts of the Byzantine Empire to recover its lost territories. In the mid-6th century, Italy became a province once again, but the Lombards conquered the northern part of the peninsula.

11   When the capital of the empire had been transferred to Byzantium, the bishops of Rome had presented themselves as an alternative to Byzantine power with a separate power base in Rome. When the Lombard kings began taking up arms in defense of Christianity against Rome's enemies, the bishops broke the alliance, in order to maintain their temporal power.

12   In 754, Pope Steven II asked for help from Pepin the Short, and in exchange, crowned him King of the Franks. After the defeat of the Lombards, Pepin turned over the center of the peninsula to the Pope. Charlemagne, Pepin's son, was crowned king and emperor of Rome in 800, but the Muslim invasions which took place mid-

century once again left the region without government.

13   Between the 9th and 10th centuries, the Church formed Pontifical States in the central region, including Rome itself. In the 12th century, self-government arose in some cities because of the lack of a centralized power.

14   In the 14th century, when the struggle intensified between the Guelfs (those who favored the Pope) and the Ghibellines (the defenders of the German Empire), the Holy See was transferred to Avignon, where it remained for the next seven papacies. Two centuries later, the prosperity and stability of cities like Venice, Genoa, Florence and Milan produced the intellectual and artistic flowering of the Renaissance.

15   In the early 16th century, the peninsula was attacked by the French, the Spanish and the Austrians, who all craved control of Italy. In 1794, Napoleon Bonaparte entered the country expelling the Austrians. Four years later, he occupied Rome and created the Roman Republic and the Parthenopean Republic, in Naples. Only the two Italian states of Sicily and Sardinia were not under Napoleon's control, as they were governed by Victor Emmanuel I. The French Emperor rescinded the temporal power of the popes and deported Pius VII to Savona.

16   Before the fall of Napoleon in 1815, Victor Emmanuel II named Camillo Benso di Cavour president of the council of ministers. Cavour was to be the architect of Italian unification, forging a single kingdom of Italy from those of Sardinia and Piedmont, with only Rome and Venice remaining outside the realm. In 1870, the Italians invaded Rome and, given Pope Pius IX's refusal to renounce his temporal power, they confined him to the Vatican, where his successors would remain until 1929. In 1878, the King, Humberto I, brought Italy into the Triple Alliance with Austria-Hungary and Germany. Italy's colonial conquest of Eritrea, Ethiopia and Somalia, in eastern Africa, also began.

17   In 1872, influenced by the events of the Paris Commune, Italy's first socialist organization was formed, giving rise in 1892 to the Socialist Party (PSI). The encyclical Rerum Novarum (1891) oriented Catholics towards militant unionization and the union movement expanded rapidly. The Italian invasion of Ethiopia in 1896 ended in defeat for Italy.

18   When World War I broke out, Italy proclaimed its neutrality; however, in the face of growing pressure from nationalist groups on the Left, it ended up declaring war against its former allies of the Triple Alliance.

Life expectancy
81 years
2005-2010

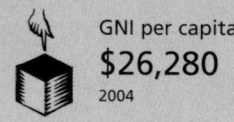

GNI per capita
$26,280
2004

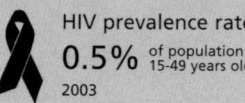

HIV prevalence rate
0.5% of population
15-49 years old
2003

[19] Benito Mussolini, who had been expelled from the PSI for supporting Italy's entry into the war, was able to manipulate resentment over the poor outcome through a blend of nationalism and pragmatism.

[20] In 1921, a group headed by Amadeo Bordiga and Antonio Gramsci split off from the PSI to form the Communist Party (PCI), leaving the PSI without its radical wing.

[21] Having confronted one government crisis after another, and following Mussolini's impressive march on Rome, Victor Emmanuel III turned over the government to Mussolini. An electoral reform, giving Mussolini's Fascist Party a majority, was denounced by socialist leader Giaccomo Matteotti, who was subsequently assassinated by followers of *Il*

Duce (Mussolini) in 1924. A new constitution established censorship of the press; in 1929 the Pact of Letran was signed with the Vatican, re-establishing the temporal power of the popes, and thereby gaining Catholic support for the Government.

[22] Mussolini's foreign policy was directed almost exclusively toward the acquisition of colonies. In 1936, Italy invaded Ethiopia, and a year later the Italian East African Empire was formed. During the Spanish Civil War, closer ties developed with Hitler's Germany, forming the basis for what was to become the Rome-Berlin axis. In April 1939, Italian troops took Albania.

[23] In June 1940 Italy declared war on France and Britain, invading Greece in October. Military defeats in North Africa and Greece brought Germany to its aid. Allied forces

invaded Sicily in July 1943. A few days later, the Fascist Grand Council asked the king to reassume his powers. Humberto I dismissed and jailed Mussolini, naming Pietro Badoglio Prime Minister. When Badoglio negotiated an armistice with Allied forces, Germany invaded Italy and rescued Mussolini. *Il Duce* then founded the Italian Social Republic (at Salo), where he became a puppet for Hitler until he was captured and executed in April 1945 by the Communist resistance. The resistance, known as partisans, also made up of Christian Democrats, Socialists, Republicans, Radicals and Liberals, played an essential role in the fall of Fascism, especially between 1943 and the end of the war.

[24] At the end of the war, 444,000 Italians were dead - including more than 280,000 civilians - and all of Italy's colonies had been lost. The king handed power over to his son Humberto II. A June 1946 referendum decided on the formation of a republic. Under the leadership of Alcide de Gasperi, the Christian Democrats (DC) managed to form a minority government. These first elections marked the beginning of the Christian Democrats' hold on power. In May 1948, Luigi Einaudi, also of the DC, was elected Italy's first president.

[25] The 1946 International Conference authorized Italy to continue administering Somalia, which it continued to do until 1960. During the 1950s, Italy participated in the reconstruction of Europe. In 1957, it became one of the charter members of the Common Market.

[26] Between 1952 and 1962, the average income of Italians doubled, as a result of the development of industry, which had come to employ 38 per cent of the national workforce. At the same time, agricultural employment dropped by 11 per cent, triggering migration from the countryside to the cities, and from the south to the north. The industrial triangle of Milan, Turin and Genoa attracted a concentration of millions of people, living in overcrowded conditions inferior to those of the rest of Europe.

[27] After successive victories at the polls, in 1961 the DC began opening up to the Left, seeking alliances with the Socialists and Social Democrats. The powerful Communist Party, in spite of its strong electoral presence, was permanently excluded from the cabinet. The economic and institutional crises which took place during that decade led radical groups from the right and left alike to turn to violence as a means of bringing about change: the far right through bombings, while the Red Brigades of the Left

used political kidnapping as their main tool. In 1978, the kidnapping and assassination of former Prime Minister Aldo Moro sealed their isolation from mainstream politics.

[28] According to the 1948 constitution, the president must select a prime minister who will have parliamentary support. Until 1978, when socialist Sandro Pertini was elected, all presidents belonged to the DC. Charges against Christian Democrat Arnaldo Forlani's government - linking it to an organization called Propaganda Due - brought the Government down in May 1981. Francesco Cossiga, elected in 1985, returned Italy to the tradition of Christian Democrat presidents.

[29] Government cabinets that took lengthy negotiations to form generally only managed to last a few months, with the exception of Bettino Craxi's Socialist administration (1983-1987).

[30] In February 1991, the PCI became the Democratic Party of the Left (PDS), which sought admission into the Socialist International. In December 1991, dissenters from the official party line decided to create the Refounded Communist Party (PRC).

[31] In the April 1992 elections, the DC did not achieve a parliamentary majority, the first defeat of a Christian Democratic government since 1946. Days afterwards, Prime Minister Giulio Andreotti announced the dissolution of his government and President Cossiga resigned. The national turmoil caused by the late May assassination of Judge Giovanni Falcone - the Mafia's number one enemy - in Sicily influenced the outcome of the elections, in which DC candidate Oscar Scalfaro, the former president of the chamber of deputies, won a landslide victory.

[32] Two months after Judge Falcone's assassination, the Mafia killed Paolo Borsellino, who on Falcone's death had taken over the investigation against organized crime.

[33] In 1993 an investigation revealed a complex corruption network that involved politicians of all tendencies, members of the business community and the Mafia. More than a thousand political and business leaders were tried under Operation Clean Hands, including former Prime Ministers Bettino Craxi and Giulio Andreotti.

[34] Between 1980 and 1992, corruption deprived State coffers of some $20 billion. Because of illegal payments to officials and politicians, Italian public spending was 25 per cent more expensive than in the rest of the European Community.

[35] In 1986, spurred by the opening of Italy's first McDonald's fast-food restaurant in Rome, leftist

PROFILE

ENVIRONMENT

The northern region of the country consists of the Po River plains which extend as far as the Alps. It is the center of the country's economic activity, having the main concentration of industry and farming. Cattle are raised throughout the peninsula; important crops include olives and grapes, with vineyards extending along the southern coastal strip. The country includes not only the peninsula - which is divided by the Apennines - but also the islands of Sicily and Sardinia.

SOCIETY

Peoples: Italians 94 per cent. Others, particularly Sardinians and Germans in Alto Adige and immigrants from Africa.
Religions: Predominantly Catholic (more than 90 per cent); Catholic and Jewish communities are deep-rooted. Growing Muslim community through immigration.
Languages: Italian (official). Several regional languages, like Neapolitan and Sicilian, are widely spoken. French is spoken in Val d'Aosta and German in Alto Adige. Immigrants speak their own languages, particularly African languages.
Main Political Parties: The Union (which includes: Democrats of the Left; Daisy-Democracy is Freedom; Communist Refoundation Party); supporting the Union: Federation of the Greens; Party of Italian Communists; United Consumers. House of Freedom (which includes: Forza Italia; National Alliance; Union of Christian and Centre Democrats).
Main Social Organizations: three central unions: CGIL, a Left Democrat affiliate; CISL, centrist; and the UIL, of Social Democratic tendency, represent a combined total of 11 million workers (2004). The three unions signed an agreement with the government and corporations in 1993, to which smaller unions did not adhere, to to hire workers collectively. Organizations representing workers in industry and commerce (Confindustria; Confcommercio), farmers associations (Confcoltivatori, Confagricoltura), Slow Food Movement.

THE STATE

Official Name: Repubblica Italiana.
Administrative Divisions: 20 Regions divided into 95 Provinces.
Capital: Rome (Roma) 2,665,000 people (2003).
Other Cities: Milan (Milano) 4,047,500 people; Naples (Napoli) 3,620,300; Turin (Torino) 1,619,400; Palermo 947,300 (2000).
Government: Giorgio Napolitano, President since May 2006. Romano Prodi, Prime Minister since May 2006. Bicameral parliamentary system: the Chamber of Deputies has 630 members and the Senate of the Republic with 316 members.
National Holiday: 2 June, Anniversary of the Republic (1946).
Armed Forces: 216,800 (2001). Other: 111,800. The phasing-out of military service by 2006 was approved in June 2000. The system has been replaced by a professional army in which women are also able to serve.

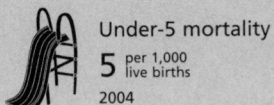
Under-5 mortality
5 per 1,000 live births
2004

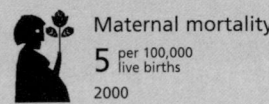
Maternal mortality
5 per 100,000 live births
2000

journalist Carlo Petrini founded the International Slow Food Movement to protest the homogenization of food around the world, and to preserve local foods.

[36] In April 1993, former Central Bank president Carlo Azeglio Ciampi was appointed prime minister.

[37] For the March 1994 legislative elections, within the space of just a few months, media magnate Silvio Berlusconi created the Forza Italia party which, allied with Umberto Bossi's federalist Northern League and Gianfranco Fini's neo-fascist National Alliance (NA), won an absolute majority in parliament.

[38] Berlusconi was appointed prime minister and consolidated his popularity in June when Forza Italia triumphed in the European elections. However, increasingly tense relations with the Northern League, which had persistently criticized Berlusconi, and the NA fascists, began to complicate government action.

[39] In October, the unions opposed retirement pension reforms proposed by the prime minister as, in their view, the reforms limited benefits, increased contributions and privatized part of the system. In December, Bossi - who despite a poor showing in the election had nearly a fifth of the deputies behind him - withdrew his support for the Government and Berlusconi resigned.

[40] In April 1996, the center-left Olive Tree Coalition, led by former Christian Democrat Romano Prodi and supported by the PDS, won the elections. Appointed prime minister, Prodi formed a government with prominent PDS leaders and leading conservative personalities like Ciampi and Fini, which also included former communists in the cabinet and support from the Refounded Communists.

[41] The Communist Party threatened to withdraw support for Prodi in October 1997 if the Government would not accept a reduction of the working week to 35 hours. In October 1998 the conflict flared up again, and Prodi resigned. In order to avoid passing power to the right, the center-left alliance proposed Massimo D'Alema, former communist and leader of the PDS, as prime minister. D'Alema was confirmed in his post by President Scalfaro on 21 October.

[42] In July 1998 a Milan court sentenced Silvio Berlusconi and seven other people to 33 months in prison for bribing the financial police. However, none of them served time because the 'Sinconi Act', which had taken effect two weeks earlier, only stipulated imprisonment for those who received sentences of over 3 years in jail.

[43] Former premier Andreotti was acquitted in October 1999 of having used political influence to help the Mafia. Carlo Azeglio Ciampi, a 78-year-old banker and former prime minister, won the presidency with broad support from both right and left in May 1999.

[44] In June 2001, Berlusconi became prime minister once again, leading a center-right coalition, the House of Freedom, in the general elections. The new premier formed a coalition government identical to his previous one (with Bossi and Fini).

[45] In January 2002, the euro replaced the lira, but Italy was the only country which did not celebrate the change of currency. In response to the unenthusiastic statements by right-wing colleagues in the cabinet, Foreign Minister Renato Ruggiero resigned and Berlusconi himself took over that role in addition to that of Prime Minister.

[46] In November 2002 former President Andreotti (84) was sentenced to 24 years in prison for ordering the Mafia to murder a journalist in 1979. But due to his advanced age, he was not sent to prison.

[47] Despite mass opposition to the war in Iraq, in March 2003 Berlusconi committed troops to the US-led coalition forces.

[48] In May, Berlusconi testified before a Milan court which was trying him for allegedly bribing judges in Rome in connection with the sale of state-controlled SME food company in 1985. In June, Parliament passed an immunity law for high government officials, to save Berlusconi from 'dishonor', since Italy was to take over the European Union (EU) presidency in July. In November, 19 Italian soldiers were killed by a car bomb in Nasiriyah, Iraq. It was the biggest Italian military loss since World War II.

[49] In December Italy was at the center of an international scandal when it emerged that Parmalat, the country's leading food corporation, had committed fraud amounting to 14.3 billion euros. The Government intervened to 'protect jobs, not the shareholders or the board of directors' of the empire made up of 197 factories and 36,000 jobs worldwide, including 4,000 in Italy.

[50] In December 2003 more than a million people gathered in Rome to protest the pensions system proposed by Berlusconi.

[51] In January 2004, the Constitutional Court ruled that the law which granted immunity to Berlusconi and other top government officials during their term in office was unconstitutional.

[52] When US President George W Bush visited in June, thousands of demonstrators protested in Rome

IN FOCUS

ENVIRONMENTAL CHALLENGES
Air pollution caused by industry emissions (particularly sulfur dioxide), coastal areas and land polluted by industrial effluents and agriculture. Some industrial and domestic waste disposal installations do not provide adequate treatment for waste. Acid rain has damaged lake ecosystems.

WOMEN'S RIGHTS
Women have been able to vote and stand for office since 1945. In 2003, female representation in Parliament stood at 12 per cent, while in ministerial or equivalent positions it was 8 per cent. Women made up 39 per cent of the 25-million strong workforce. Some women are trafficked from poorer countries (mainly Albania, Ukraine, Moldova and Latin American nations) and exploited as sex workers.

CHILDREN
In Italy, as in other industrialized countries, generally the situation is good. However, there are differences between the south and the north. As in other industrialized countries, drugs are one of the major problems for young people.

INDIGENOUS PEOPLES/ ETHNIC MINORITIES
The historical ethnic minorities are the Sards, Tyrolese and Roma (Gypsies). Other minority groups are recent immigrants, a product of economic crisis and military conflicts in less industrialized countries, mostly from the Balkans, Africa and Latin America. Of the last group, many are descendants of Italians who had emigrated to the Americas during World War I and World War II, and have now obtained residency permits due to their Italian ancestry.

MIGRANTS/REFUGEES
In 2004 there were 15,000 refugees and almost 10,000 asylum seekers in Italy. These were low figures compared to other EU countries: Germany, UK and France received considerably higher numbers that year. In 2002-2004, refugees came mostly from Iraq, Liberia, Sri Lanka, Serbia and Montenegro, Pakistan, Somalia, Eritrea, Romania, Nigeria and Sudan.

DEATH PENALTY
Capital punishment was abolished in 1994, but the country's last execution had taken place in 1947, when the death penalty was abolished for common offenses.

* Latest data available in The State of the World's Children and Childinfo database, UNICEF, 2006.

against the Iraq war, shortly after Pope John Paul II repeated his criticism of US intervention in Iraq.

[53] Berlusconi was acquitted of corruption charges. However, analysts said the four-year trial was a tough blow to his political image.

[54] The Government coalition suffered an overwhelming defeat in the April 2005 regional elections. Berlusconi resigned and was immediately reinstated by President Carlo Ciampi. That month, Parliament ratified the European Constitution, which was signed in October by the EU's 25 heads of state in Rome - in the same room where the group was created in 1957. The outgoing chair of the European Commission, Romano Prodi, warned that 'signing the European Constitution does not mean we have crossed the finish line'.

[55] Both President Ciampi and Prime Minister Berlusconi announced, in April 2006 - at the ceremony in Rome for three Italian soldiers killed in Iraq - that the 2,500 Italian troops remaining in that country would return home by the end of the year. That month, Prodi won the general elections by a narrow margin, as the head of a centre-left coalition. Berlusconi refused to resign at first, but then accepted the results after the international community recognized Prodi as the new Prime Minister. Also in April, Italy's most wanted man - Bernardo Provenzano, believed to be the head of the Sicilian Mafia - was captured on a farm in Sicily. The *tratturi* - or tractor, as he was known in his circles - had been sentenced *in absentia* to life in prison for the murder of two judges in 1992.

[56] Former Communist Giorgio Napolitano was elected President in May 2006, with 543 out of 1009 votes in Parliament. When Prodi was sworn in as premier he said there was a 'moral crisis in Italy'. He planned to withdraw Italian troops from what he said was the 'grave mistake of the occupation' of Iraq.

[57] In October 2006, Italy's prime minister said that troops could be sent to Naples to deal with a rash of Mafia-linked crimes. ∎

Jamaica / Jamaica

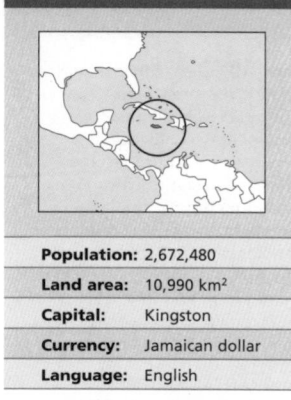

Population:	2,672,480
Land area:	10,990 km²
Capital:	Kingston
Currency:	Jamaican dollar
Language:	English

The name Jamaica comes from the Arawak Indian name for the island - Xamayca ('land of springs', 'land of wood and water') a reference to the abundant waters of its luxuriant forests. The Arawaks had pushed out the Guanahatabey, the original inhabitants, who had come from North America. They were skillful sailors, fishers and craftspeople, who carved and polished seashells. Singing and dancing were important activities in their lives. They worshipped idols (*zemis*), but believed in only one God, creator of nature.

[2] Columbus reached Jamaica on his second voyage to the New World in May 1494, but it was his son, Diego Columbus, who conquered the island in 1509. From then on, the number of Arawaks decreased dramatically. Around 1545, Spanish historian Francisco López de Gomara wrote that 'Jamaica resembles Haiti in all respects - here the Indians have also been wiped out'. It is estimated that the number of Arawaks stood at about 60,000 prior to the Spanish Conquest.

[3] The Spanish, now absolute rulers of the island, began to plant sugarcane and cotton, and to raise cattle. There were incursions by the British in 1596 and 1636 and in 1655, when 6,500 British soldiers under the command of William Penn

dislodged the 1,500 Spaniards and Portuguese. Jamaica rapidly became a haven for pirates who ravaged Spanish trade in the Caribbean. The last important enemies that the English had to face on the island were the enclaves of rebel slaves or *quilombos*, hidden in remote areas like the Blue Mountains. In 1760 a general rebellion in the colony was put down, and in 1795 a further revolution shook the island.

[4] By the late 19th century there were approximately 800 sugar mills and more than 1,000 cattle ranches in Jamaica. The economy was built on the labor of 200,000 African slaves. Anti-slavery and anti-colonial rebellions of the 18th and 19th centuries were followed by labor union struggles in the first few decades of the 20th century. The two main contemporary political parties, the Jamaica Labour Party (LP) and the People's National Party (PNP), grew out of workers' organizations. Independence was proclaimed in 1962, but successive LP governments failed to rescue the economy from foreign hands.

[5] In 1942 rich deposits of bauxite were discovered, and the

aluminum transnationals ALCOA, ALCAN, Reynolds and Kaiser quickly established themselves on the island, virtually replacing the sugar industry.

[6] The transnationals exploited Jamaica's bauxite by shipping the raw metal out of the country, making all the decisions on production, and paying minimal customs duties. After independence a few plants were

built to transform bauxite into aluminum, but the bulk of the mineral extracted continued to be shipped unprocessed to the US.

[7] The PNP, with a clearly progressive program, won the 1972 elections. The new Prime Minister, Michael Manley, raised the bauxite export tax and began negotiations with the foreign companies to exert greater control over their activities. The

PROFILE

ENVIRONMENT

Jamaica is the third largest of the Greater Antilles (10,991 sq km). A mountain range, occupying two-thirds of the land area, runs across the island from east to west. A limestone plateau covered with tropical vegetation extends to the west. The plains are good for farming and the subsoil is rich in bauxite. The climate is rainy, tropical at sea level and temperate in the eastern highlands.

SOCIETY

Peoples: Most Jamaicans are of African descent. There are small Chinese, Indian, Arab and European minorities.
Religions: Protestants 56 per cent; Catholics 5 per cent, Rastafarians 5 per cent.
Languages: English (official). A dialect based on English called 'patois English' or 'Creole' is also spoken.
Main Political Parties: People's National Party (PNP), founded in 1938 by Norman Manley, a member of the Socialist International since 1975. The Jamaica Labour Party, founded in 1943 by Alexander Bustamante.
Main Social Organizations: National Workers' Union of Jamaica (NWUJ); Bustamante Industrial Trade Union (BITU). Jamaican Student's Union. Rastafarians groups.

THE STATE

Official Name: Jamaica.
Administrative Divisions: 14 parishes.
Capital: Kingston 575.000 people (2003).
Other Cities: Spanish Town 127,300 people; Montego Bay 90,500 (2000).
Government: Parliamentary monarchy. Head of State: Elizabeth II of Britain. Kenneth Hall, Governor General, representing Britain since February 2006. Prime Minister, Portia Simpson Miller, since March 2006. Bicameral Legislature: House of Representatives, 60 members elected by direct popular vote every 5 years and the Senate, with 21 members appointed by the Governor General.
National Holiday: 6 August, Independence Day (1962).
Armed Forces: 523,550 available troops for the army, the navy and the air force (2002).

LAND USE

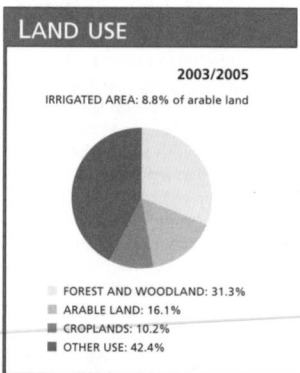

2003/2005

IRRIGATED AREA: 8.8% of arable land

- FOREST AND WOODLAND: 31.3%
- ARABLE LAND: 16.1%
- CROPLANDS: 10.2%
- OTHER USE: 42.4%

WORKERS

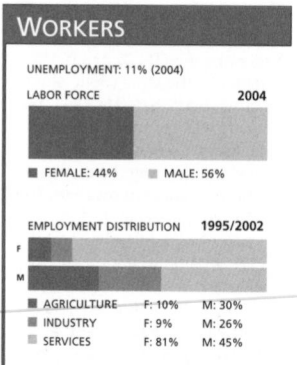

UNEMPLOYMENT: 11% (2004)

LABOR FORCE — **2004**

- FEMALE: 44%
- MALE: 56%

EMPLOYMENT DISTRIBUTION — **1995/2002**

	Female	Male
AGRICULTURE	F: 10%	M: 30%
INDUSTRY	F: 9%	M: 26%
SERVICES	F: 81%	M: 45%

Life expectancy
71 years
2005-2010

GNI per capita
$3,300
2004

Literacy
88% total adult rate
2000-2004

HIV prevalence rate
1.2% of population 15-49 years old
2003

PNP Government also supported Caribbean integration. A bi-national bauxite marketing company was created with Venezuela. Jamaica became a member of the Caribbean Multinational Merchant Fleet.

[8] In December 1976 the PNP won again, with an increased majority. Manley advocated socialism within the existing constitutional structure. Jamaica took active part in the Movement of Non-Aligned Countries and supported African anti-colonial movements. This stand strained relations with the US and the transnational mining companies reduced their production, transferring operations to other countries. In consequence, export revenues fell, and with them, funding for government social programs.

[9] In 1979 Manley sat down to negotiate with the IMF, but in 1980 suspended the talks saying the financial institution's conditions implied a drastic fall in the standard of living of the population. Manley called for early elections in 1980. The elections were marked by the instability caused by the right-wing opposition, which was to achieve a major triumph. The new LP government, headed by Edward Seaga, expelled the Cuban ambassador, and imposed policies that opened the country to unconditional foreign investment. This resulted in an increase in unemployment and the doubling of foreign debt between 1981 and 1983.

[10] In 1983, Jamaica was part of the small group of Caribbean countries that gave diplomatic support and symbolic military assistance to the US invasion of Grenada. A month later, taking advantage of the favorable political atmosphere, Seaga decided to call early parliamentary elections. The PNP boycotted the elections; only the governing party nominated candidates, thereby winning all 60 seats.

[11] In 1989, the PNP came to power again, strengthened by its triumph in the 1986 municipal elections. Manley presented a very different program from that of 1976, based on free enterprise and good relations with the US. Manley re-established relations with Cuba and stated the agreements with the IMF would be respected, although he made it clear he would not accept conditions which would worsen social inequalities. His aim was to maintain economic growth, but with better wealth distribution.

[12] In April 1992, Percival Patterson replaced Manley as

IN FOCUS

ENVIRONMENTAL CHALLENGES

Soil loss from deforestation and erosion is put at 80 million tons per year. In some metropolitan areas, the lack of sewerage and the dumping of industrial wastes have polluted drinking water supplies and coastal areas, threatening the urban population and destroying large extensions of coral reefs.

WOMEN'S RIGHTS

In 2003, women held nearly 12 per cent of parliamentary seats and 18 per cent of ministerial positions.

Illiteracy rate among women over 15 years is 9 per cent*. About 98 per cent of pregnant women receive prenatal care and 79 per cent of births are attended by health skilled staff*. In 2003, women made up 48 per cent of the total workforce of one million, most of them working in services (81 per cent), while the rest were engaged in agriculture (10 per cent) and industry (9 per cent). That year, the female unemployment rate amounted to almost 22 per cent.

CHILDREN

Of the 2.7 million people living in the country, 39 per cent are children. They account for 43 per cent of all people living below the poverty line*.

In 2000*, primary school

enrolment reached 95 per cent for both boys and girls. However, the quality and efficacy of learning and teaching are a problem. Almost 30 per cent of students, mostly boys, are functionally illiterate at the end of primary level and only 3.6 per cent in the 0 to 3 year age group are enrolled in pre-schools or playgroups.

Jamaica has a high HIV/AIDS prevalence; nearly 8 per cent of those infected are children under 10 years of age*. Of these, 80 per cent live in poor households and one out of four will be abandoned by her or his parents*. Among adolescents, infection rates have doubled every year since 1995, and adolescent girls are three times as likely as boys to become infected. Sexual initiation can occur as early as 10 years old; only 50 per cent of adolescents use condoms on a regular basis*.

About 22,000 children work and almost 2,500 - mostly boys - live on the streets. Sexual exploitation is an emerging problem. Child abuse has increased*. There is violent discrimination against gay males.

INDIGENOUS PEOPLES/ETHNIC MINORITIES

Afro-descendants account for 90.9 per cent of the population. There are also people from the Indian subcontinent, Chinese and Europeans. Arawaks or *taínos* were

Jamaica's first inhabitants (the name Jamaica comes from the Arawak word *Xamayca*). in 700 BC they reached the Antilles and Jamaica, coming by raft from Guyana.

In the 15th century, by the time Columbus arrived, the Arawak indigenous population had been drastically reduced as a result of a bloody war over land with invading Caribs.

MIGRANTS/REFUGEES

Jamaica is the largest receiver of remittances within CARICOM. In 2004, nearly $1.5 billion entered the country, amounting to 17 per cent of GDP. Jamaicans are the third largest Caribbean group living in the US and Canada. Emigration is an old issue: during the 1940s Jamaican farmers were recruited to work in the US. Until the 1990s, between 10,000 and 12,000 people had gone to the US to work on sugar plantations in Florida or in orchards on the East Coast. Today it is estimated that some 450,000 Jamaicans are living in the US. There are also many in the UK.

DEATH PENALTY

Applicable even to common crimes.

* Latest data available in The State of the World's Children and Childinfo database, UNICEF, 2006.

Prime Minister, as the latter resigned following a long period of illness. Promising labor guarantees, in 1992 the Government started the privatization of around 300 state companies and public services. This package included the entire sugar industry.

[13] In March 1993 Patterson was re-elected. The LP - led once again by Seaga - refused to participate in the partial elections in 1994 because of its disagreement with the electoral system. Defeat hastened the splitting up of the JLP, and the National Democratic Movement was founded. Patterson's policies continued as before, adopting measures favored by multilateral credit organizations such as the IMF and the World Bank.

[14] During the campaign for the December 1997 parliamentary elections, violence was such as to cause the massive resignation of candidates.

[15] Jamaica withdrew from the Organization of American States (OAS) Inter-American Commission on Human Rights in 1998, since the latter opposed the death

penalty wich was in force in Jamaica and other Caribbean countries.

[16] In 1999 the army took to the streets to control riots triggered by price hikes. For example, fuel prices had increased by 30 per cent.

[17] In 2002, while celebrating the 40th anniversary of Jamaica's independence, Patterson was elected for the third consecutive time as Prime Minister. This time the elections were relatively peaceful. Patterson continued his policy of economic liberalism. This gave an initial boost to the financial sector, but soon it faded. On foreign policy, Patterson moved for greater self-determination and in 2003 he proposed a constitutional amendment to make Jamaica a republic, and sever the ties with its colonial past.

[18] In September 2004, when Hurricane Ivan hit the south of the island with winds of 248 kilometers per hour, several areas suffered from floods and mud slides. At least one person was killed by the cyclone and roads remained blocked by landslides.

Several residential areas were left without basic services such as electricity, water and phone lines. Thousands of people sought refuge in other parts of the island and the Government declared a state of emergency to prevent any public disorder.

[19] The following year, in spite of not touching land, Hurricane Emily killed four people when sweeping along the coasts of Jamaica.

[20] On 30 March 2006, Portia Simpson Miller, of the People's National Party, became the first Jamaican woman to hold the office of Prime Minister. She replaced Patterson after winning an internal party vote. Soon after she took office, Jamaica Labour Party leader Bruce Golding submitted a series of proposals to address the issues of human rights, the economy and constitutional reform.

[21] Amnesty International's 2006 Report highlighted Jamaica's many human rights violations and especially the use of excessive force by the police. The Jamaican police force murder rate is one of the highest in the world. ■

Japan / Nihon

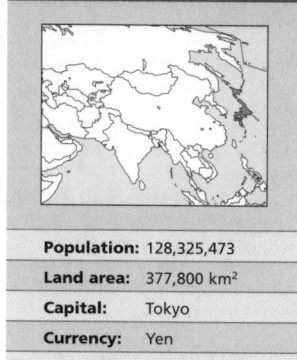

Population:	128,325,473
Land area:	377,800 km²
Capital:	Tokyo
Currency:	Yen
Language:	Japanese

The Paleolithic Age in Japan dates back 10,000 to 30,000 years ago. This was followed by the Jomon Neolithic culture which lasted until 200-300 BC and spread throughout the Japanese archipelago. They were hunters and gatherers, creators of fine pottery and stone and bone utensils.

[2] The arrival of the Yayoi, probably from the continent by land bridges across the straits of Korea, Tsushima, Soya and Tsugaru, introduced rice cultivation, horses and cows, the potter's wheel, weaving and iron tools into Japanese culture.

[3] According to Chinese chronicles, at the beginning of the Christian era the Wo region (Wa in Japanese) was divided into more than 100 states. During Himito's reign, some 30 of them were grouped together. The Wa people were divided into social classes and paid taxes. Buildings were technically advanced and there were large markets.

[4] Isolation from the end of the civil war in 266 until Yamato's consolidation as Emperor in 413 led to the unification of the nation, a prerequisite for further expansion. In the year 369, Yamato subdued the Korean kingdoms of Paekche, Kaya and Sila, from where he controlled the region. The Yamato Empire suffered a rapid decline, partly as a result of resistance from its Korean subjects

LAND USE

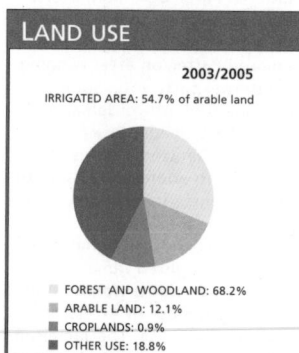

2003/2005

IRRIGATED AREA: 54.7% of arable land

- FOREST AND WOODLAND: 68.2%
- ARABLE LAND: 12.1%
- CROPLANDS: 0.9%
- OTHER USE: 18.8%

but also because of internal fighting within the court.

[5] The introduction of Buddhism between the years 538 and 552 initially stirred curiosity and admiration because of its majestic temples and supposed magical powers.

[6] Buddhism (modified through contact with Central Asia, China and Korea) plus the grid system of land division, still visible today, and the characters used in writing are the most important traces of Chinese presence in Japan, which dates back 1,500 years.

[7] This cultural heritage underwent successive adaptations to the local conditions, language and habits, especially in the 17th and 18th centuries. So-called 'Japanization' was particularly apparent in architecture and language, which shows evidence of very different local linguistic influences.

[8] In the 9th century, chieftains were replaced by a hereditary court. The aristocracy made Buddhism a controlling force which served to reinforce the power of the State. Japan's first permanent capital was Nara, established in the year 710. After several conflicts, Kammu (781-806) re-established the Empire's independence and transferred the capital to Heian (Kyoto).

[9] In the new capital, the Fujiwara family consolidated their power and established the Regent as ruling figure, above the Emperor. With imperial approval, the Tendai and Shingon Buddhist sects (closer to Japanese culture) developed in Heian, bringing an end to Nara's religious hegemony.

[10] The imperial land tenure system fell increasingly into private hands. Aristocrats and religious institutions took over large tracts of tax-free land (*shoen*) and organized private armies for themselves, leading to the creation of a new rural warrior class - the *samurai*.

[11] The predominant Taira and Minamoto clans vied for power in a number of military confrontations. The Taira were in power from 1156 until their defeat in the Gempei War (1180-85). The *shogun* (general), Minamoto Yoritomo, founded the Kamakura shogunate, the first of a series of military regimes which ruled Japan until 1868.

[12] The Kamakura defeated the Mongol invasions of 1274 and 1281 aided by providential storms, called *kamikaze* (divine winds). During this period, several new Buddhist sects emerged, such as Pure Land Buddhism, True Pure Land and Lotus.

[13] In the early 14th century, the Kamakura shogunate was destroyed by Emperor Go-Daigo's Kemmu Restoration. Shortly thereafter military clans expelled Go-Daigo from Tokyo and replaced him with a puppet emperor. Go-Daigo established his court in Yoshino, and for 56 years there were two parallel imperial courts.

[14] The Onin War (1467-77), over succession within the Ashikaga shogunate, became a civil war which lasted almost 100 years. New military chiefs, independent of imperial or shogun authority, established their vassals within fortified cities, leaving the

surrounding villages to run themselves and to pay tribute.

[15] In the cities, trade and manufacturing ushered in a new way of life. Portugal began trading with Japan in 1545, and the missionary Francis Xavier introduced Catholicism in 1549. Christianity led to conflict with feudal loyalties, so it was proscribed in 1639. At the same time, all Europeans were banished from Japan, except the Dutch.

[16] In the late 16th century, the warlords became increasingly isolated from the rest of society because of their use of firearms (initially supplied by Europeans), their fortresses, the disarming of peasants, and their tightening of control over the land. This situation helped to pacify and unite the country around a single national authority.

[17] During the 17th century the Tokugawa clan gained supremacy, ruling from the city-fortress of Edo (Tokyo) until 1867. A careful distribution of the land among their relatives and local chieftains guaranteed them the control of the largest cities - Kyoto, Osaka and Nagasaki - as well as of the most important mines.

[18] Local chieftains were compelled to spend half their time on the shogun's affairs while their families remained behind, as hostages. Transformed into military bureaucrats, the samurai were the highest level of a four-class system, followed by the peasants, artisans and traders (who, although despised, were essential in urban life).

[19] A national market arose for textiles, food, handcrafts, books and other products, mainly as a result of the almost total isolation from the outside world effected by the Tokugawas since 1639. Nagasaki was the only exception; here, the Chinese and the Dutch were allowed to open trading posts, although the latter were restricted to a nearby island.

[20] In the 19th century, the old economic and social order went into a state of collapse. Peasant revolts were more and more frequent, and the samurai and local chieftains found themselves heavily in debt with the traders. In 1840, the government tried to carry out a series of reforms, but these failed and weakness allowed the US to force the opening of japanese ports.

[21] Japan was forced out of its isolation by the arrival of US warships under the command of Commodore Matthew Perry. He successfully negotiated the opening up of Japanese markets in 1854. The signing of unfavorable trade agreements with the US and several European countries deepened

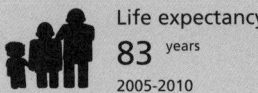
Life expectancy
83 years
2005-2010

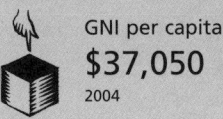
GNI per capita
$37,050
2004

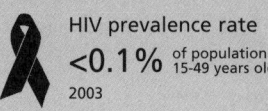
HIV prevalence rate
<0.1% of population 15-49 years old
2003

the crisis. The samurai carried out several attacks against the foreigners and then turned against the shogun, forcing him to resign in 1867.

[22] Imperial authority was restored with the young Meiji emperor, in 1868. During the Meiji Restoration, Japan's modernization process began, following the Western model. The US, England, France and Germany exerted influence in education, the sciences, communication and on Japanese culture.

[23] In less than 50 years, the closed, feudal Japan was transformed into an industrialized world power. Western advisers and technology were brought in for education, trade and industry. An army based on the draft replaced the military authority of the samurai, defeated when they tried to rebel in 1877.

[24] In 1889, succumbing to internal political pressure, the emperor approved a constitution which turned Japan into a constitutional monarchy, with a bicameral legislature (Diet). However, only one per cent of the population was eligible for office, and the prime minister and his cabinet were responsible to the emperor (seen as a divine figure).

[25] In the late 19th century and early 20th century, Japan won two major wars: against China (1894-95), which enabled Japan to maintain its control over Korea (annexed in 1910), and in the Russo-Japanese War (1904-05), after which it annexed the Sajalin Peninsula. Japan entered World War I as a British ally, since a treaty had been signed to that effect in 1902.

[26] The war allowed Japan to gain control of several German possessions in East Asia, including the Chinese territory of Kiaochow. In 1915, China was forced to accept Japanese influence over Manchuria and Inner Mongolia. In 1918, Hara Takashi became the head of the first government to have a parliamentary majority.

[27] Economic difficulties caused by the international depression of the 1930s gave militarists the excuse they were seeking to attack the Government. They proposed that the country's problems could only be solved by expanding its military power and conquering new markets for Japanese products and as sources of raw materials.

[28] When Japanese officers occupied Manchuria without official backing in 1931, the Government, unable to deter the military, accepted the creation of the puppet state of Manchukuo in February 1932. Three months later, the Government was turned over

to the militarists, who retained it until 1945.

[29] In 1940, Japan invaded Indochina hoping to open up a route through to Southeast Asia. The US and Britain imposed a total embargo on Japanese merchandise. The Japanese attack on Pearl Harbor, in Hawaii, and on the Philippines, Hong Kong and Malaysia, sparked the war with the US and opened up a new phase of World War II.

[30] Japan surrendered on 15 August 1945, after the US had dropped two atomic bombs on Hiroshima and Nagasaki on August 6 and 9 respectively. US troops occupied Japan and imposed a government of the Supreme Command of the Allied Powers (SCAP) under the leadership of General Douglas MacArthur, between 1945 and 1952. SCAP forced Japan to abandon the Meiji institutions, to renounce the Emperor's claim to divinity, and transfer the power to a Parliament, which was charged with electing the Prime Minister, and establishing an independent judiciary.

[31] Although imposed upon the Japanese from the outside, the principles laid down in the 1947 constitution were accepted by all sectors of society and in 1952 the country recovered its independence. Japanese sovereignty was restored over the Tokara archipelago in 1951, over the Amami islands in 1953, over the Bonin islands in 1968, and over the rest of the Ryukyu, including Okinawa, in 1972.

[32] SCAP also took other measures to weaken the hierarchical model of the Meiji family-state. These ranged from giving tenants the right to purchase the land they lived on (until then they had to pay taxes to their lords), and laws aimed at strengthening free trade and preventing the return of monopolies. However, the Japanese financial system remained intact and provided the basis for economic recovery at the end of the occupation.

[33] The 1947 Constitution restricted the development of Japanese military power. Japan bowed to US strategy for the region and formed alliances with Taiwan and South Korea. In 1956 it joined the UN and re-established relations with the USSR.

[34] In 1955, opposing the country's conservative and nationalist sectors, which had supported the war policy, the center-right Liberal Democratic Party (LDP) was formed.

[35] The return to independence found the Japanese economy in a state of growth and change. Although agriculture suffered from small-scale production and urban migration, industrialization and full

employment triggered the need for rapid technological innovation.

[36] During the 1960s, Japan specialized in the production of high technology products, establishing trade relations with more industrialized countries instead of its previous Asian partners. The oil crisis of 1973 did not halt the growth of the Japanese industry, which led the world in steel, ship building, electronics, and automobile manufacturing.

[37] Although Prime Minister Kakuei Tanaka's visit to Beijing in 1972 signalled Japanese recognition of the People's Republic of China, it damaged the country's relations with Taiwan. The scandal following Tanaka's bribing by the Marubeni Corporation (a representative of the US Lockheed Aircraft Corporation), adversely affected the LDP's popularity and in 1976, for the first time in its history,

it lost its absolute majority in parliament.

[38] During the 1960s and 1970s, Japan had a large trade surplus in its trade with the US. Japan began to rank first or second with all its trading partners. With direct investment and the establishment of subsidiaries, it expanded worldwide.

[39] The impressive development of the Japanese economy is based mostly on a policy of foreign investment in projects which quickly deplete non-renewable natural resources. This policy has caused irreversible damage to rainforests and serious alterations to the Third World ecosystem.

[40] Emperor Hirohito's death in January 1989 ended the Showa era, begun in 1926. The coronation of Akihito, in the traditional Japanese style, launched the Heisei era (achievement of universal peace). The coronation was attended by

PROFILE

ENVIRONMENT

The country is an archipelago made up of 3,400 islands, the most important being Hokkaido, Honshu and Kyushu. The terrain is mountainous, dominated by the so-called Japanese Alps, which are of volcanic origin. Since 85 per cent of the land is taken up by high, uninhabitable mountains, 40 per cent of the population lives on only 1 per cent of the land area, in the narrow Pacific coastal plains, where demographic density exceeds 1,000 inhabitants per sq km. The climate is sub-tropical in the south, temperate in the center and cold in the north. Located where cold and warm ocean currents converge, Japanese waters have excellent fishing, and this activity is important to the economy. Japan's intensive and highly mechanized farming is concentrated along the coastal plains (rice, soybeans and vegetables). There are few mineral resources. Highly industrialized, the economy revolves around foreign trade, exporting manufactured products and importing raw materials.

SOCIETY

Peoples: The Japanese are culturally and ethnically homogeneous, having their origin in the migration of peoples from the Asian continent. There are Korean, Chinese, Ainu and Brazilian minorities.
Religions: Buddhism and Shintoism 84 per cent.
Languages: Japanese.
Main Political Parties: Liberal Democratic Party (LDP); Social Democratic Party of Japan (SPDJ); New Komeito Party and the Japanese Communist Party.
Main Social Organizations: The General Council of Japanese Trade Unions has 4,500,000 members.

THE STATE

Official Name: Nihon- Koku
Capital: Tokyo 34,997,000 people (2003).
Other Cities: Yokohama 3,518,000 people; Osaka 2,641,000; Nagoya 2,243,400; Kyoto 1,488,800 (2000).
Government: Parliamentary constitutional monarchy. Emperor Akihito has been Head of State since January 1989, although his official coronation did not take place until 12 November 1990. Shintzo Abe, Prime Minister and head of the Government since September 2006. Legislature (The Diet) is bicameral: House of Representatives, made up of 480 members; House of Counsellors, with 252 members, elected by direct popular vote every four and six years, respectively.
National Holidays: 23 December, Emperor's Birthday (1933).11 February, Founding of the Country (1889).
Armed Forces: 239,500 (including 8,000 women). Other: 12,000 (non-combat Coast Guard, under the jurisdiction of the Ministry of Transport).

Under-5 mortality
4 per 1,000
live births
2004

Maternal mortality
10 per 100,000
live births
2000

more heads of state than had ever gathered together for any such event. The expense incurred generated internal protests.

[41] At the end of the Cold War, Japan emerged as one of the three main world economic powers, together with the US and what is now the European Union (EU).

[42] In November 1991, Prime Minister Toshiki Kaifu lost the support of the Takeshita clan - the most influential of the official party's five factions - and resigned. Elected in August 1989, he had never been involved in a corruption scandal. He was replaced by 72-year-old Kiichi Miyazawa (also with Takeshita support), who had been elected president of the LDP nine days before.

[43] In his inaugural speech, Miyazawa outlined his administration's objectives: expand relations with, and aid to China; negotiate with the US and normalize relations with the former Soviet Union. He announced his willingness to liberalize the rice market, making concessions similar to those already made by the EC and the US, to ward off a failure of the Uruguay Round of the trade talks, GATT.

[44] In mid-1992, after heated debate, a law was passed authorizing troops to be sent abroad for the first time since World War II. In September, some troops joined the UN peacekeeping forces in Cambodia.

[45] In early 1993, Miyazawa promised to broaden the scope of Japanese forces, both in terms of funding and personnel, in the UN peacekeeping missions.

[46] Also in 1993 the Prime Minister was censured by the Diet for failing to carry out electoral reforms needed to end the endemic corruption in Japanese political life. The official party split, losing 47 seats, and elections were moved up to July.

[47] The two dissident groups, led by Tsutomu Hata and Masayoshi Takemura, formed new parties; the Reformation Party and the Pioneer Party.

[48] Diplomatic relations with the Russian Government improved slightly when Japan announced that, while not relinquishing its claim (dating from 1950) to the Kuril Islands it would not put conditions on its economic aid to Russia.

[49] In March 1993, Shin Kanemaru, the LDP's historic leader, was arrested. He had received bribes in 1992, part of a wider corruption scandal in the Government. Shortly after his arrest he was granted an amnesty.

[50] On 18 July, general elections drew the lowest rate of voter participation since the war (67.3 per cent). The LDP, in power since 1955, lost its majority in the *Kokkai* (Diet), This result altered the balance of power in effect since World War II. Miyazawa resigned the LDP's presidency, and assumed responsibility for the defeat. Yohei Kono was named party President.

[51] The 'new majority' of the Socialist, Reformation, Komeito (Buddhist), Democratic Socialist, Unified Democratic Socialist and Pioneer parties agreed to form a new Government based upon a limited platform which would not attempt any major changes.

[52] Hosokawa, former governor of the province of Kunamoto, was elected Prime Minister in August. On taking office, he announced far-reaching political reforms aimed at fighting corruption and recession, and modernizing the pension and health systems. In his first speech to the Diet, he spoke of the aggression with which Japan had treated its Asian neighbors, from the 1930s until the end of World War II. On 15 August, the anniversary of Japan's surrender, he offered condolences and apologies to the victims of Japanese colonialism.

[53] When Hosokawa succeeded Miyazawa, the Confederation of Industry announced it would discontinue its contributions to the LDP. According to the local press, these had amounted to $1 billion per year.

[54] In August 1993, 165 political leaders and 445 business people were arrested for irregularities committed during the election campaign. 5,500 cases of electoral fraud were brought to court. In October, Shinji Kiyoyama - president of Kajima, the country's second largest construction firm - was arrested for paying a bribe in exchange for a building permit.

[55] In December the Japanese Government opened up the rice market, importing four per cent of its ten million tons requirement for domestic consumption.

[56] Hosokawa's 'romance' with public opinion lasted until early 1994 when the press became more critical, accusing him of having received money in exchange for favors.

[57] A summit between Japan and the US ended in failure in March when Hosokawa rejected mandatory quotas to open up the Japanese automobile, telecommunications, pharmaceutical and insurance markets. Japanese businessmen ignored Hosokawa and felt inclined to wage a commercial war with the US before yielding to its pressures. At that time the automobile industry employed 11 per cent of Japan's workforce and contributed 30 per cent of the GDP.

[58] LDP member of parliament Nakamura was arrested for accepting bribes from Kajima and other construction firms. Hosokawa, unable to shake off accusations from the opposition about his own involvement in illicit business deals, resigned in April and asked 'sincerely for forgiveness from the people of Japan'.

[59] Tsutomu Hata was nominated Prime Minister and headed Japan's first minority Government in four decades. The socialists had walked out of the ruling coalition leaving the Government with only 182 of the 512 seats in the lower house.

[60] On visits to Europe aimed at establishing closer trade links with the EU, Hata admitted that Japan's trade surplus had caused the trade crisis with the US. After an initial agreement was reached between the two countries, he launched a scheme to promote economic deregulation.

[61] Socialist Tomiichi Murayama was elected Prime Minister in June 1994 and took office on 18 July. His party, the Social Democratic Party of Japan (SDPJ) had no parliamentary majority but formed an alliance with its traditional rival, the LDP, and with a new party, the Sakigake.

[62] The opening of Kansai airport, located on an artificial island, led to a blossoming of island-city projects in a bid to ease congestion in densely populated cities.

[63] On 17 January 1995 an earthquake hit the area of Hanshin. Over 6,000 died, 100,000 buildings were destroyed in the city of Kobe and over 300,000 were left homeless.

[64] In March, a series of attacks with poisonous Sarin gas killed 12 people and intoxicated 5,500 others in Tokyo's subways. A similar attack had taken seven lives in Matsumoto in June 1994. Shoko Asahara, the leader of a religious sect called Aum Shinriyko (Supreme Truth), was charged with the attack and arrested along with 16 other leaders from the movement.

[65] In January 1996 Ryutaro Hashimoto (LDP president) replaced Murayama as Prime Minister and called for general elections in October, winning a small majority in parliament. In November, Hashimoto formed a cabinet with only LDP members.

[66] In a referendum held in late 1997, slightly more than half of the Nago population, on Okinawa island, decided against the construction of a US heliport on the island (held by the US since the Japanese surrender in 1945). Both governments claimed the construction was a step towards the dismantling of a sophisticated air base on the island.

[67] The financial and economic crisis which struck Southeast Asia in 1997 also affected Japan, the second world power and first creditor of the world. The greatest challenges facing the new Government of Keizo Obuchi were to cut taxes in order to stimulate consumption and to take action against the bad loans that paralyzed the Japanese banking system.

[68] A major radioactive uranium leak at a JCO company reprocessing plant, near the Tokoaimura nuclear plant north of Tokyo, caused radiation levels to rise to 15,000 times higher than normal. The environmental organization Greenpeace denounced the accident as a symptom of the problems plaguing Japan's nuclear safety system.

[69] In August, right-wing groups hailed the re-adoption after 50 years of the old imperial flag, while dozens of teachers - who refused to honor imperial symbols from militarist times - were sacked. The growth in nationalism was also reflected in the armed forces, which saw recruitment reach the highest levels for several decades.

[70] Following yet another nuclear accident, this time at the Tokaimura plant in September 1999, Obuchi ordered the inspection of all installations using nuclear fuel. The JCO firm, which recycled nuclear material, admitted that for years it had been using procedures that did not comply with the minimum safety requirements established by the Government. Greenpeace claimed that the plant continued to emit radiation five times higher than recommended safety levels, while Government officials from the nuclear safety area were criticized for their slowness in measuring contamination. *Yomiuri* newspaper reported that the three operators charged with negligence in the handling of radioactive materials did not even know what a nuclear chain reaction meant.

[71] Obuchi, who had achieved his country's economic revival, went into a coma in April 2000. He was replaced by Yoshiro Mori. In May, days before Obuchi's death, the new Prime Minister spoke at a meeting of Shintoist followers (who during World War II worshipped Emperor Hirohito as a living deity). Mori shocked national and international public opinion with his description of Japan as 'a divine nation that has the Emperor at its center'.

[72] Responding to demands for his resignation from the press, the opposition and factions within his own party, Mori appointed a new cabinet and made administrative reductions. To prop up his eroded power base, Mori gave key

Water source

100% of population using improved drinking water sources

2002

Doctors

201 per 100,000 people

1990-2004

Primary school

100% net enrolment rate

2004

positions to former prime ministers Kiichi Miyazawa and Ryutaro Hashimoto.

⁷³ In February 2001, Mori continued to play golf after hearing that a Japanese fishing vessel had collided with a US nuclear submarine, raising press calls for his resignation due to his 'lack of sensitivity.'

⁷⁴ Finally, in April, Mori admitted he had lost public confidence and resigned. He was replaced by Junichiro Koizumi who, as well as promising to revitalize the economy and clean up the government image, included five women in his cabinet, including Makiko Tanaka - daughter of the former premier - who became the first female Foreign Minister. This government, with a record female contingent, was classed as a 'Hollywood' cabinet aiming chiefly for popularity. When he came to power, Koizumi had 90 per cent support.

⁷⁵ That month, China and South Korea condemned a history book approved by the Japanese authorities. Seoul and Beijing said the text 'glossed over' atrocities by the Japanese army during World War II. The book was written by a group of nationalist historians who stated that Japanese action during the War benefited the South East Asian nations because it prepared them for independence, and that the Nanjing massacre of 1937 - where 300,000 civilians were killed - was 'very far from being a holocaust'. Following the protests, the Japanese Education Minister said 137 changes had been made to the text.

⁷⁶ Further friction with China and South Korea was caused by Koizumi's visit to the Shinto altar of Yasukuni. The 2.4 million soldiers honored there include some figures considered war criminals - for example, the executed Prime Minister Hideki Tojo, who led Japan during World War II.

⁷⁷ In October, Koizumi visited Seoul and offered apologies for the suffering of that country under the Japanese colonial government.

⁷⁸ Koizumi's approval rating was hit by a series of scandals, and fell to 40 per cent in April 2002. In January he had sacked Chancellor Tanaka, whom he accused of lying in an argument with his bureaucrats. This was followed by resignations from allies and ministers, for various reasons, and in April, Yutaka Inoue, speaker of the ruling party in the upper house, also resigned.

⁷⁹ In September 2002 Koizumi became the first Japanese leader to visit North Korea. North Korean leader Kim Jong Il apologized for the abduction of Japanese citizens in the 1970s and 1980s,

IN FOCUS

ENVIRONMENTAL CHALLENGES

Air pollution - mainly in large urban areas (Tokyo, Osaka and Yokohama) - plus pollution in several coastal areas and acid rain are among the challenges. Acidification in lakes and reservoirs has reduced water quality and threatened habitats. Japan is one of the largest consumers of fish and tropical forest woods, contributing to the depletion of these in Asia and other regions of the world.

WOMEN'S RIGHTS

Japanese women have been able to vote since 1947. In 2000, 11 per cent of Parliament seats were held by women, and female representation at ministerial level was seven per cent. In 2003 women made up 42 per cent of the 68 million workforce, while female unemployment stood at 5.1 per cent. In 2000, 73 per cent of the female labor force worked in services, 21 per cent in industry and five per cent in agriculture. Since 1990*, 100 per cent of births have been attended by trained staff.

CHILDREN

There are six million children under five years old. Infant mortality rates are among the lowest in the world. At least since 1995*, primary school enrolment and attendance has been 100 per cent.

INDIGENOUS PEOPLES/ ETHNIC MINORITIES

There are more than 700,000 Koreans in Japan, who make up 85 per cent of the foreign population. The first Koreans arrived in Japan in the early 20th century. During the 1930s and 1940s when Korea was under Japanese rule, several were recruited by Japan. Koreans in Japan receive political support from the governments of South Korea and, to a lesser degree, North Korea. The Korean community is around 710,000 people (85 per cent of the foreign population). They mainly live in Osaka and other urban areas. Koreans face cultural and political restrictions: their education and financial opportunities are limited, and they suffer discrimination, not just from the Government - although they are not allowed to vote - but also from individual and corporate practices. They are mainly concentrated in small communities and do not present a cohesive group. As a result of no employment or low-paid work,

their lower income affects nutrition and health. Although some have become integrated into Japanese culture (adopting name, language and religion), a significant portion resist assimilation and live in Korean neighborhoods, attend Korean schools and take an active part in North and South Korean political relations with Japan.

MIGRANTS/REFUGEES

In 2004 there were almost 2,000 refugees and 500 asylum seekers. In 1990 the Japanese Government decreed that all illegal workers were to be deported and their employers sanctioned, with fines ranging from $18,000 to 3 years in jail. Currently there are one million foreigners living in Japan. The total population level is falling, and it is estimated that in 2010 the country will be short of 1.87 million workers.

DEATH PENALTY

Japan maintains the death penalty for ordinary offenses.

* Latest data available in The State of the World's Children and Childinfo database, UNICEF, 2006.

and confirmed that eight had been killed. A month later, five of the kidnapped were returned to their relatives.

⁸⁰ During 2003 Japan had a long trade dispute with the US. The Bush administration set tariffs on steel imports, some as high as 30 per cent, harming Japanese exports. Japan's Trade minister, Soichi Nakagawa, threatened sanctions against US products unless Washington obeyed WTO rules by November.

⁸¹ On 4 December 2003, after joint pressure through the WTO from the EU, China, Brazil and other countries, the US lifted the controversial tariffs, putting an end to the dispute.

⁸² That month, the Japanese Government announced it would install a 'purely defensive' US missile-shield. This caused outcry from China, which had already protested when Prime Minister Koizumi appeared in May before the Special Commission on Emergency Legislation of the upper chamber and stated that the Self-Defense Forces were actually 'Japan's army'.

⁸³ In February 2004, the Government ordered the deployment of 'non-combat' soldiers to Iraq, raising charges

of unconstitutionality from the opposition. The order was based on a 1992 law and the precedent of peacekeeping forces in Cambodia. This was the first involvement of Japanese troops in a combat zone since World War II.

⁸⁴ On 9 August, four people died and seven were injured in a nuclear accident in the Mihama nuclear plant, 350 km west of Tokyo. One of the plant's turbines had a vapor leak, although the authorities concluded there had been no radioactive escape.

⁸⁵ The new incident put into question Japan's reliance on nuclear power. Surveys showed that half of those asked were in favor of reducing the number (52) of nuclear plants in the country.

⁸⁶ In October, typhoon Tokage - the worst to hit Japan in a decade - left 48 dead and 200 injured, with dozens missing. It reached 229 km/h winds, which led to the evacuation of thousands of people. The south of the country was paralyzed; schools were closed and public transport was suspended.

⁸⁷ Tension with China reached its highest level in decades after Japan published school books that downplayed atrocities committed by its troops during the first half of the 20th century. In March, thousands

of demonstrators took to the streets across China to protest against Japan. Adding to the tension were Japan's aspirations to a permanent seat in the UN Security Council and exploitation rights in the East China Sea.

⁸⁸ In the September 2005 parliamentary elections, Koizumi's Democratic Party won 296 of the 480 seats at stake.

⁸⁹ In October, Parliament approved the privatization of the postal service; this was one of the largest privatizations in history since the service not only delivers mail, but is a bank. In fact it is the world's largest bank, with 260,000 employees and roughly $3 trillion in assets.

⁹⁰ The country's second-largest nuclear reactor was shut down in March 2006, after a court found in favor of citizens living near the plant. They argued that the plant would not withstand earthquakes.

⁹¹ Following the by-laws of the ruling Liberal Democratic Party (LDP), Koizumi was forced to leave the office of Prime Minister in the midst of his second mandate, as well as the LDP's presidency. After internal elections, the Liberal Democrats chose Shinzo Abe, considered more conservative than the outgoing Koizumi. Abe took office in September 2006. ■

Imports **(millions)**
$542,380 goods and services
2005

Exports **(millions)**
$636,611 goods and services
2004

Radioactive homeless people

MORE THAN 70,000 people work in nuclear power plants and reactors scattered across Japan. Although nuclear power stations have their own employees in technical positions, more than 80 per cent of the non-technical staff is made up by untrained workers who accept short-term contracts. Homeless people are recruited to perform the most dangerous tasks such as cleaning reactors and decontaminating facilities. The *yakuza* (Japanese mafia) finds, selects and illegally hires homeless workers for companies which are totally reliant on this labor force to keep their operations going. The most probable fate for these hidden workers is death from bone cancer, caused by the considerable levels of radioactivity in their bodies, higher than those allowed in most countries.

They are called 'nuclear gypsies' on account of the nomadic life they lead, going from one power station to another, until they become ill and then die. To hire these rootless poor people is only possible through the connivance of the Government. The Japanese authorities have stipulated that the annual amount of radioactivity a person can be exposed to is 50 mSv (milli-sieverts). However, the European Union (EU) set 100 mSv as the maximum dosage a worker in a nuclear reactor can be exposed to in five years, while one mSv is the annual amount allowed for the general public. According to Spanish daily *El Mundo*, the 'companies operating nuclear stations hire homeless people until they have been exposed to the maximum radiation levels and then they fire them *for the sake of their health*, sending them onto the street again'. Then, within days or months, those same workers are hired again under different names.

MAFIA MOBS
'The *yakuza* (mafia) acts as an intermediary. Companies pay 30,000 yen (216 euros) per working day, but the hired worker only gets 20,000 yen (144 euros). The yakuza get to keep the difference' says Kenji Higuchi, a Japanese journalist who has been investigating this issue for 30 years. The daily wage for working at power plants - in jobs commonly performed by robots in other countries - is approximately twice the amount paid in the construction industry. Recruitment of homeless people for such work has been carried out in Japan since the 1970s. According to Fujita, at least half of the 5,000 temporary workers employed by nuclear plants are homeless people.

Up to 17 per 10,000 workers in the power plants have 100 per cent chances of dying from cancer, and a larger number have a 'high likelihood'. It is estimated that over the last 30 years more than 300,000 temporary staff have been recruited into the Japanese nuclear plants.

GENERATING WORK FOR HOMELESS PEOPLE
Panasonic, Toshiba and Hitachi are among the transnational companies that subcontract homeless people. The soaring demand for electricity in high-tech Japan, with its more than 120 million inhabitants, has fueled the need for nuclear energy. ∎

Jordan / Al Urdunn

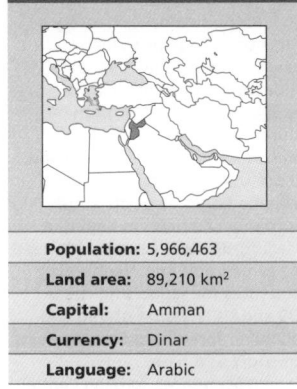

Population:	5,966,463
Land area:	89,210 km²
Capital:	Amman
Currency:	Dinar
Language:	Arabic

OCCUPIED BY ISRAEL
1. Southern Lebanon - occupied 1983-2000
2. Syria - Golan Heights - occupied in 1967
3. West Bank - occupied in 1967 (some individual towns now have autonomy)

0km 50

B y 2000 BC, groups of Semitic nomads known as Amorites had entered the region that would come to be called Canaan (present-day Jordan). By the middle of the second millennium they settled in the Jordan Valley, which became a Semitic language area. During the 15th to 13th centuries BC, small tribal kingdoms that are mentioned in the Old Testament: Edom, Moab, Bashan, Gilead, dominated Canaan until its complete conquest by the Israelites between 1220 and 1190 BC. In 722 BC Israel fell to the Assyrians, who divided the Jordan region into provinces.

2 Between the 3rd and 1st century BC, three peoples coexisted in Jordan: Jews, Greeks and Nabataeans (nomadic Arabs who arrived in Edom in the 7th century BC). The Greeks were mainly veterans of Alexander's military campaigns who fought one another for regional hegemony. By the 1st century BC, Roman legions removed the last Seleucids from Syria, converting the area into a full Roman province.

3 In order to check Muslim expansion, in the year 636 the Byzantine emperor Heraclius fought the Arabs at the Yarmuk river. Victory gave the Muslims access to the Fertile Crescent (see Saudi Arabia). During the first stage of the Crusades, the western part of the Jordanian territory

was used in warfare against the European strongholds.

4 Between the 16th century and the start of World War I, the territory was ruled by the Ottoman Turks, and formed part of the district of Damascus.

5 Jordanian peoples joined the widespread Arab rebellion against the Turks, and their participation was decisive in the defeat of the Ottomans. Under the Sykes-

Picot Agreement of 1916, Britain controlled Iraq and Palestine, in which present-day Jordan was included. The British had promised Sherif Hussein of Mecca that they would form a single Arab kingdom with these territories and the Arab Peninsula.

6 The situation of Faisal - son of the Sherif - in Syria (under French dominion) and his expulsion in 1920, led Prince Abdullah

- another of Hussein's sons - to support his brother with Bedouin forces. The British convinced him to accept rule by the Emirate of Transjordan.

7 This Emirate remained under British mandate until 1928, when the frontiers with Palestine were set and a law was passed giving Abdullah and his heirs power over the State. Foreign affairs and the military remained under British control until 1946, when the Emirate became the 'Hashemite Kingdom of Transjordan'. After the Arab-Israeli war in 1948, King Abdullah annexed the West Bank of Jordan. The country became known as Jordan, but had new problems regarding the situation of Palestinian refugees, the legal status of Jerusalem, and the doubling in length of its frontier with Israel.

8 In 1951 Abdullah was assassinated and succeeded by his son Talal. He was anti-British and promised a progressive government, but was deposed a year later. In 1953 his son Hussein took over the throne at the age of 17.

9 In 1967, the Six Day War - in which Israel on one side fought Egypt, Syria and Jordan on the other - brought serious consequences for Hussein's kingdom. In addition to suffering great military losses, he lost one-third of the most fertile lands and the cities of Bethlehem, Hebron,

PROFILE

ENVIRONMENT
Seventy-five per cent of the country is a desert plateau, between 600 and 900 meters in altitude, which forms part of the Arabian Desert. The western part of this plateau has a series of cleavages at the beginning of the great Rift Valley fault, which crosses the Red Sea and stretches into East Africa. In the past, these fissures widened the Jordan River valley and formed the steep depression that is now the Dead Sea. This region is suitable for agriculture. The country has a a rainy winter and very dry summer. As most of the country is made up of dry plains, farming is limited to cereals (wheat and rye) and citrus fruits. Sheep and goats are also bred.

SOCIETY
Peoples: Most of the population is made up of Palestinians who immigrated following the wars with Israel in 1948 and 1967. Native Jordanians are from 20 large Bedouin ethnic groups of which about one third are still semi-nomadic. There is a Circassian minority (of around 100,000 people) who came from the Caucasus in the 19th century, and now play a major role in trade and administration. Armenians, Kurds, Azeris.
Religions: Muslim (mainly Sunni Muslim but the Shi'a numbers are increasing rapidly) 92 per cent; Christian 8 per cent (mainly orthodox).
Languages: Arabic (official), English.
Main Political Parties: In 1991 political parties were legalized. The most important are: National Constitutional Party (a nine-party pro-monarchy

coalition). The only real party of opposition is the Islamic Action Front (*Al-Jabhat al-Amal al-Islami*).
Main Social Organizations: The most important union body is the General Federation of Unions of Jordan. The Union of Jordanian Women has participated in the democratization process and in defense of the political rights of women. Others: National Union of Students of Jordan and the Muslim Brotherhood.

THE STATE
Official Name: al-Mamlakah al-Urdunniya al-Hashimiyah.
Administrative Divisions: 12 provinces.
Capital: Amman 1,237,000 people (2003).
Other Cities: Irbid 537,600 people; Az-Zarqa 471,200; ar-Rusayfah 184,300; as-Salt 64,000 (2000).
Government: Abdullah II, King since February 1999. Marouf Al-Bakhit, Prime Minister (since 27 November 2005). Legislative branch: National Assembly (bicameral), with a 40 member Senate appointed by the King, and an 80-member Chamber of Deputies elected by direct popular vote; the latter can be dissolved by the King.
National Holiday: 25 May, Independence Day (1946).
Armed Forces: 98,650 personnel (1996). Other: 6,000 soldiers under the authority of the Department of Public Security; 200,000 militia in the 'People's Army'; 3,000 Palestinians in the Palestinian Liberation Army, under the supervision of the Jordanian Army.

LAND USE

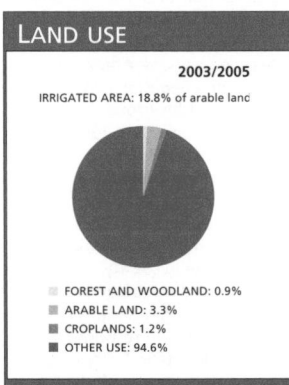

2003/2005

IRRIGATED AREA: 18.8% of arable land

- FOREST AND WOODLAND: 0.9%
- ARABLE LAND: 3.3%
- CROPLANDS: 1.2%
- OTHER USE: 94.6%

Life expectancy
72 years
2005-2010

GNI per capita
$2,190
2004

Literacy
90% total adult rate
2000-2004

HIV prevalence rate
<0.1% of population 15-49 years old
2003

IN FOCUS

ENVIRONMENTAL CHALLENGES

Water scarcity is the most pressing environmental problem. Desertification and urban expansion are leading to the loss of large portions of arable land, especially near the Jordan River.

WOMEN'S RIGHTS

Women have been able to vote and run for office since 1974.

King Abdullah appointed six women, all liberal professionals, to the 55-seat Senate. As a result, since 2003 female representation stands at 11 per cent, the highest proportion in the Arab world, and women hold 11 per cent of ministerial positions. However, the illiteracy rate among women over 15 was 15.7 per cent in 2000, three times higher than the male rate and considerably lower than the 44.6 per cent rate recorded in 1980*.

In 2003, women comprised 26 per cent of the total labor force. At the same time, female unemployment amounted to 21 per cent, whereas the male unemployment rate was 12 per cent. Ninety-nine per cent of pregnant women receive prenatal care and 100 per cent of births are attended by skilled health staff, which represents a significant progress if compared to the previous five-year period. The fertility rate is 3.4 children per woman.

CHILDREN

Five per cent of children under five were underweight and 9 per cent had stunted growth*.

Under-five mortality amounts to 27 deaths per 1,000 live births, with an annual average decline of 2.8 per cent since 1990*.

Education is compulsory until the age of 16; however, there are no penalties for parents who fail to comply. Since 2000, Iraqi children are not allowed to attend state schools unless they are either legal residents or refugees recognized as such by the UNHCR.

The State subsidizes food and transport for large or very poor families. There are free vaccination programs and free access to healthcare for minors in state clinics.

Even though it is difficult to determine, sexual abuse within families is more prevalent than reported. Although there are laws prohibiting child labor, it is common to see children hawking their wares on Amman streets. When the police return them to their homes, they are soon back on the streets.

INDIGENOUS PEOPLES/ ETHNIC MINORITIES

The more than two million Palestinians living in Jordan (about 40 per cent of the total population) are mainly Sunni Muslim (85 per cent) and have come in different waves. They first arrived in 1946 during the incorporation of Palestine into the recently founded Transjordan, during the War of Independence from Israel in 1948, and during and after the Six Day War in 1967, when Israel occupied Palestinian territories. Palestinians in Jordan are dispersed and their status is varied, ranging from prominent individuals completely

integrated into Jordanian culture to refugees living in deep poverty.

MIGRANTS/REFUGEES

In 2004, Jordan hosted more than 150,000 Palestinian refugees from the Gaza Strip. That year, 1,100 new refugee applications and 12,500 asylum applications were filed, most of them by Iraqi citizens. An estimated 300,000 Iraqis live in Jordan, although it is not clear how many of them are refugees. Large contingents have arrived since 2002, fleeing the US invasion, economic sanctions and persecution.

There are thought to be 800,000 Palestinians, displaced from the West Bank after the Arab-Israeli War in 1967. They live as Jordanian citizens, unlike the Gaza Palestinians who do not have access to citizenship and remain refugees.

As a result of the global war against terrorism unleashed in 2001, the country began to accept fewer refugees. The Jordanian Government restricted entry of Palestinians with Jordanian documents living on the West Bank. On the other hand, more than 500 Jordanians requested asylum in Canada and Sweden during 2004.

DEATH PENALTY

The death penalty still applies even for common crimes.

*Last available fact in *The State of the World's Children* and *Childinfo* database, UNICEF, 2006.*

shrinking Dead Sea.
[19] In the first parliamentary elections of Abdullah II's reign, held in June 2003, the independent candidates who supported the King won two-thirds of the seats.
[20] In August 2003, 11 died and more than 50 were wounded when a bomb went off in the Jordanian Embassy in Baghdad, Iraq. Jordan was strongly criticized by some Iraqis and neighboring countries for supporting the US and UK invasion of Iraq and allowing foreign soldiers to use its territory as a base.
[21] In September, Jordan's Central Bank went back on its decision to freeze accounts belonging to leaders of the Hamas Islamic movement (see history of Palestine).
[22] In April 2004, a Jordanian court sentenced eight Islamic militants to death for killing US diplomat Laurence Foley in October 2002. Among those sentenced, although in absentia, was Abu Musab al-Zarqawi, a Jordanian-born al-Qaeda suspected member. Zarqawi was accused by the Jordanian government of masterminding the attack.
[23] On 11 November 2005, simultaneous attacks on three US-based hotels in Amman - Grand Hyatt, Days Inn and Radisson SAS - killed 67 people and left 300 wounded. Most victims were Jordanian nationals although there were also many Palestinians. An organization operating in Iraq and apparently led by Zarqawi claimed responsibility for the attacks in a statement posted on the internet.
[24] Jordan and Spain agreed in April 2006 to coordinate and strengthen efforts to 'close their doors to intolerance and fanaticism' of all kinds, underlying 'global threats and global security', particularly terrorism.
[25] That month, Jordan accused the Islamic Resistance Movement (Hamas), which now headed the Palestinian Government, of plotting attacks against Amman. According to the Jordanians, the attacks were prevented thanks to the detention of a 'considerable number' of suspects.
[26] Meanwhile, Hamas spokesperson Sami Abu Zuhri denied the allegations and asserted that his group did not have armed groups in Jordan or any other country.
[27] In October 2006, King Abdullah, on a visit to the Netherlands, said that Muslims living in Europe should take an active role in society, obey local laws and not be afraid of losing their religious identity. His comments were in response to the rising debate in Europe over Muslim integration and identity. ■

Jericho, Nablus, Ramallah and Jerusalem.
[10] With the Israeli occupation of Transjordan, the Kingdom received many Palestinians, expelled by the Israelis from their lands initially in 1948 and then from their refugee camps in 1967. In 1970, a series of guerrilla actions, hijackings and Hussein's desire to replace the Palestine Liberation Organization (PLO) as the Palestinians' representative body were among the reasons leading to the Jordanian army massacre of Palestinians known as 'Black September'.
[11] After the Arab-Israeli war in 1973, the King re-established relations with the PLO, and in 1979 recognized it as the only legitimate representative of Palestine people.
[12] In 1985, King Hussein and Palestinian leader Yasser Arafat were reconciled; in July 1988

the King renounced any claim to the West Bank and gave the PLO responsibility for the Israeli-occupied territory.
[13] When Iraq invaded Kuwait in August 1990, Jordan found itself in a difficult position, militarily flanked by Israel and Iraq, with a majority of subjects who supported Iraq, and depending on Saudi Arabia for finance and on Baghdad for oil. Although it joined the trade embargo against Baghdad, it opposed the use of military force to enforce the UN Security Council resolutions.
[14] Despite its dependence on Iraqi oil, Jordan cut back on trade with Baghdad in 1996. Exports to Iraq were reduced 50 per cent, and Washington was authorized to use an air base within the territory.
[15] King Hussein died on 7 February 1999. He had designated a new heir, his 37-year-old son Abdullah ibn al-Hussein, instead of

his brother Hassan.
[16] When Abdullah II took the throne, the Arab-Israeli peace process had reached a stalemate, and Iraq was under attack by the US.
[17] In February 2002, Abdullah II backed Washington's labelling of Iran, Iraq and North Korea as 'the axis of evil'. The Government was pressured into taking a less cordial position towards Israel. Chancellor Marwan Muasher called the Israeli Ambassador to a meeting in March and threatened to take measures to protest the attack on the Palestine Authority headquarters. Similarly, he asked the UN Security Council for immediate deployment of troops to the Palestinian territories.
[18] In September 2002, in their largest joint initiative to date, Jordan and Israel agreed on an $800 million plan to transport water from the Red Sea to the

Kanaky - New Caledonia / Kanaky

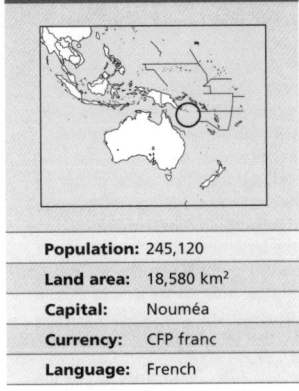

Population:	245,120
Land area:	18,580 km²
Capital:	Nouméa
Currency:	CFP franc
Language:	French

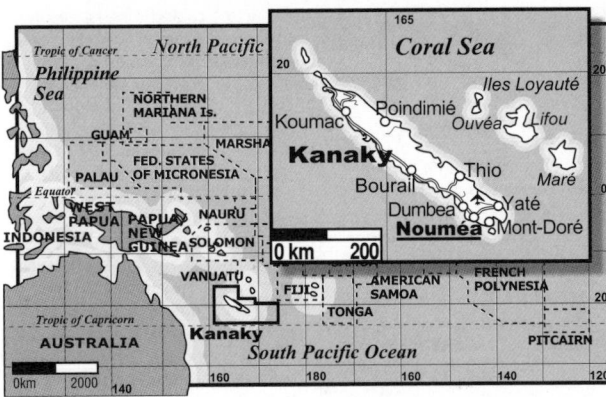

Kanaky was populated by Melanesians (Kanaks) 3,000 years ago. The islands were named New Caledonia by Captain Cook in 1774, as the tree-covered hills reminded him of the Scottish - Caledonian - landscape. In 1853, the main island was occupied by the French Navy which organized a local guard to suppress frequent indigenous uprisings. Nickel and chrome mining attracted thousands of French settlers. The colonizers pushed out the original inhabitants, and traditional religions, crafts and social organizations were obliterated. Many landless natives were confined to 'reservations', and the system of terraced fields were trodden over by cattle. The last armed rebellion, stifled in 1917, only accelerated European land appropriation.

² After Algerian independence, in July 1962, colonization increased with the arrival of *pieds-noirs*, the former French colonists in Algeria. By 1946, New Caledonia had become a French Overseas Territory, but the resulting political autonomy did not favor the Kanaks, reduced to a minority group in relation to the *caldoches* (descendants of Europeans who had settled a century ago).

³ In the 1970s, discontent with the economic situation produced by colonial domination generated strikes, land invasions, experiments in co-operative work, and a powerful campaign to restore traditional lands to the local groups. These had been totally occupied by settlers and were used mostly as grazing land for cattle. The rescue of *coutume* (cultural traditions) and the Kanak identity became a priority, and the proposed luxury tourist Club Méditerranée camps were firmly rejected.

⁴ In September 1981, pro-independence leader Pierre Declercq, a Catholic of European origin, was murdered at his home by right-wing extremists, changing the malaise to a full-blown political crisis.

⁵ A key reason why France was hesitant to grant Kanaky independence is that it has the world's second largest nickel deposits, and extensive reserves of other minerals including chrome, iron, cobalt, manganese, and polymetallic nodules, discovered recently on the ocean floor within territorial waters.

⁶ Furthermore, the islands' strategic position is of great military value. Its ports, facilities and bases house 6,000 troops and a small war fleet (including a nuclear submarine), considered by the military command as a 'vital point of support' for the French nuclear-testing site on Mururoa atoll.

⁷ The election of President Francois Mitterrand in 1981 rekindled the hopes of the pro-independence parties. The French socialist leader was supported by most Kanaks, who saw independence as a way to end the unfair income distribution on the island. This stood at $7,000 per capita (the highest in the Pacific except for Nauru) but the vast majority of the money was concentrated in the hands of European - mostly French - business people, the *métros*, who enjoyed fiscal benefits, and the caldoches, who held top public posts.

⁸ In July 1984, the French National Assembly passed special bills concerning the colony's autonomy, though it rejected amendments submitted by pro-independence parties, confirming Kanak fears that the socialist government of France had no intention of granting independence. In November, the main opposition force, the Socialist Kanak National Liberation Front (FLNKS) called for a boycott of local Territorial Assembly elections, which were sure to endorse the French government plan of postponing Kanak independence indefinitely.

⁹ In December 1984, local government became fully controlled by the caldoches with no indigenous Kanak representation, and the FLNKS unilaterally declared New Caledonia independent, proclaiming it a Kanak state.

The resulting election was boycotted by 80 per cent of the Kanak population, forcing the Government to call it off, and prepare for negotiations.

¹⁰ In December 1986, the United Nations General Assembly proclaimed the right of the Kanak people to self-determination and independence, proposing that the FLNKS be recognized as their legitimate representative.

¹¹ One year later a referendum was held to determine whether or not ties with France should be maintained. Voting was open to all residents of the island, even Europeans and immigrants who had arrived within the previous three years and, for this reason, the FLNKS boycotted the referendum. According to the opposition and the Australian and New Zealand governments, the high abstention rate of around 41.5 per cent invalidated any claim to legitimacy for continued colonial domination.

¹² When all attempts at negotiation failed for the Kanaks, the French attacked the island of Ouvéa, and 19 people were killed, most of them apparently executed rather than killed in combat.

¹³ In June 1988, FLNKS leader Jean-Marie Tjibaou, and Jacques Lafleur (leader of the Caledonian Popular Assembly for the Republic and strongly opposed to independence) signed Section 1 of the Matignon Accord, supported by French Prime Minister Michel Rocard, in Paris. From July that year direct government over Kanaky was re-established from Paris. Section II of the Accord stipulated the adoption of preparatory measures for voting on self-determination in 1998 and the freezing of the electoral register,

PROFILE

ENVIRONMENT
The territory consists of the island of Kanaky/New Caledonia (16,700 sq km), the Loyauté/Loyalty Islands (Ouvéa, Lifou, Maré and Walpole), the archipelagos of Chesterfield, Avon, Huon, Belep, and the island of Nouméa. The whole group is located in southern Melanesia, between the New Hebrides (Vanuatu) to the east and Australia to the west. Of volcanic origin, the islands are mountainous with coastal reefs. The climate is rainy, tropical, and suitable for agriculture. The vegetation is dense and the subsoil is rich in nickel deposits.

SOCIETY
Peoples: Indigenous Kanaks/New Caledonians are of Melanesian origin (the Kanaka group), 42.5 per cent; there are French and descendants of French (known as caldoches), 37.1 per cent; as well as Wallisian, 8.6 per cent, Vietnamese, Indonesian, Chinese and Polynesian minorities.
Religions: Roughly 60 per cent Catholic, 16 per cent Protestant and around 5 per cent Muslim.
Languages: French (official) and more than 30 Melanesian and Polynesian dialects.
Main Political Parties: Rally for Caledonia within the Republic (RPCR), anti-separatist; Kanak and Socialist National Liberation Front (FLNKS); Federation of Committees for the Co-ordination of Independentists, separatist; National Front, nationalist; Socialist Kanak Liberation/Kanaky Future, extreme left separatist.
Main Social Organizations: The Caledonian Workers' Confederation (CTC); the Federation of New Caledonian Miners' Unions (FSMNC); the New Caledonian Federation of Laborers' and Employees' Unions (USOENC); and the Union of Exploited Kanak Workers (USTKE).

THE STATE
Official Name: Nouvelle-Calédonie.
Administrative Divisions: Three provinces: Loyauté, Nord and Sud.
Capital: Nouméa 140,000 people (2003).
Other Cities: Mont-Doré 22,700 people; Dumbéa 15,200; Poindimié 4,700 (2000).
Government: Head of State, French President Jacques Chirac. High Commissioner named by France, Michel Mathieu, since September 2005. Head of Government, Marie-Noëlle Thémereau since June 2004. Legislature: 54-member Territorial Assembly.
National Holiday: 14 July, Bastille Day (1789).
Armed Forces: French troops 3,700 (1993).

Life expectancy
76 years
2005-2010

GNI per capita
$14,020
2000

IN FOCUS

ENVIRONMENTAL CHALLENGES
Approximately 80 per cent of the territory has been deprived of its original forest and plant cover to serve the mining industry and the production of agricultural exports like rice, pineapples and oranges, which has led to a process of accelerated soil erosion.

WOMEN'S RIGHTS
Some 97 per cent of pregnant women receive prenatal healthcare*, and 98 per cent of births are attended by qualified medical personnel*. Maternal mortality stands at 10 deaths per 100,000 live births*. The fertility rate is 2.5 children per woman*.

CHILDREN
The child mortality rate is seven deaths per 1,000 live births*, while the average under-five mortality rate is 10 deaths per 1,000 live births*.

DEATH PENALTY
Capital punishment was abolished by France in 1981.

* Latest data available in The State of the World's Children *and* Childinfo *database, UNICEF, 2006.*

to prevent France from increasing the number of voters by sending new colonists.

[14] The territory was divided into three regions, two with a majority of Kanak voters. One of the aims of this division was to create a Melanesian (Kanak) political and financial 'élite', taking over power from the pro-independence groups in most of the territory.

[15] In a first referendum that same year, the agreements were ratified. In May 1989, Tjibaou and another independence leader who supported the Matignon agreements were assassinated in Ouvéa.

[16] In 1991, the balance of trade was affected by a drop in the international prices of nickel and fish. A new generation of leaders emerged in the provinces controlled by the pro-independence groups, but most Melanesians saw their living standards decline even further. The imbalance of income became pronounced amongst the Kanaks and greater access to material goods distanced many Melanesians from their community structures and traditions.

[17] In the caldoche areas, mainly covering the capital Nouméa, social inequalities also increased, partly due to the arrival of Melanesian farmers who built shanty-towns on the outskirts of the city, but also due to the impoverishment of some caldoches. In a context of increasing social tension, street disturbances became more common.

[18] The political repercussions of the social gaps were reflected in the 1995 provincial elections. The Palika, one of the members of FLNKS, presented separate lists criticizing the Front representatives' administration of both provinces controlled by the pro-independence groups.

[19] The negotiations for Kanaky independence changed course in April 1998. The FLNKS and Paris established the basis for a general agreement, known as the Nouméa Accord. The coexistence of two different systems - one that follows Kanak traditions and the other imposed by France - proved to be the most difficult issue to resolve. The Kanaks wanted respect for their culture and their traditional civil society organizations. The Nouméa Accord allowed for the transference of powers that would assure a 'nearly sovereign' territory within 15 to 20 years. The Kanaks and the Caldoches agreed to share a common 'citizenship', while France acknowledged the 'blotches' remaining from the colonial period.

[20] In November, a referendum was held to ratify the Nouméa agreements. The Yes-voters won with 69.14 per cent of the vote. In December, the text of the law defined the application of the Noumea Accord. On 23 December, the National Assembly voted on the legal foundations of the new country, which covered the creation of new institutions as well as a 'gradual' transfer of state powers.

[21] In late 2002, a demonstration was held to demand the repeal of a permit for a nickel mine in Prony, granted by French authorities to the Goro Nickel Company, that belonged to the Canadian mining company INCO. The work sites were blocked, and Goro Nickel withdrew its workers from the islands and cancelled the mining operations.

[22] The victory in 2004 of the anti-independence *Avenir ensemble* (Future Together, made up of Caucasians and Polynesians), put an end to the predominance of the RPCR (Caledonian People's Assembly for the Republic), which was also against independence but was seen as the voice of white Caledonians. According to the Nouméa agreement, Kanaky would have to decide its final status through a referendum held before 2014. ∎

Melanesians and Polynesians: surviving cultures

MICRONESIA, Melanesia and Polynesia were originally inhabited by Melanesians and Polynesians. Melanesians live in a group of South Pacific islands which include New Guinea, Kanaky (New Caledonia), Vanuatu, Solomon Islands, and Fiji as well as on the Bismarck and Louisiade archipelagos. They are a homogeneous group, although recent studies link them to the Papuans and even to Aboriginal communities in Australia. Polynesians come from the same ethnic origins as Melanesians, and while they are physically distinct, there are some common traits in language and-appearance although their societies have taken different directions.

The first Melanesians arrived 40,000 years ago, probably from the south of the Asiatic continent. About 9,000 years ago, they began to domesticate indigenous root crops and organize their social life around agriculture. Later on, they also specialized in trade and maritime technology as well as fishing. They generally moved in small groups, with a stay in any one place limited by the duration of crop cycles. Polynesians, by comparison, probably descended from Austronesians, ancient seafarers who arrived on the Pacific archipelagos from South Asia around 4,000 BC, populating the Samoan Islands, French Polynesia, Tonga, Tahiti, Hawaii and other smaller islands where they still live.

The range of languages spoken by Melanesians and Polynesians - although decreasing in number - clearly displays the exceptional linguistic variation of Oceania, where one quarter of all world languages are spoken. Melanesian and Polynesian languages belong to the eastern branch of the Malayo-Polynesian family, which includes over 800 dialects spoken by approximately five million people. Melanesians speak more than 400 of the dialects in this group. In Fiji, the main dialect is spoken by nearly half the population - some 334,000 people - and it is used in official journals and publications. Other dialects include Motu, Roviana, Bambatana, Tolai and Yabem.

Christianity has been gaining ground and progressively replacing traditional forms of religion. Few communities have managed to maintain old beliefs faced with the thrust of Christianity.

This great cultural, social and linguistic diversity was devastated first by the arrival of the Europeans and then by the globalization drive. Western culture - in the form of new foodstuffs, clothing, music and dances - has reached-practically every corner of the world. Native Melanesian and Polynesian youth have rapidly adapted to this new way of life, leaving ancestral practices aside. ∎

Kazakhstan / Qazaqstan

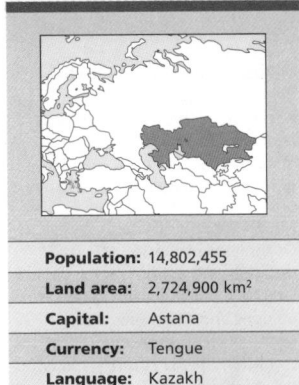

Population:	14,802,455
Land area:	2,724,900 km²
Capital:	Astana
Currency:	Tengue
Language:	Kazakh

In the Bronze Age (about 2000 BC), the territory of Kazakhstan was inhabited by people who lived by farming and raising livestock. In the Iron Age, around 500 BC, an alliance was formed among the Saka peoples, who had by then developed the ability to write. In the 3rd century BC, the Usune and Kangli peoples - who lived near the Uighur, Chechen and Alan - subdued the other groups in the area. Later, Attila's Huns occupied the region until they were expelled by the Turks.

² In the mid-4th century, the Turkish Kaganate (*Khanate* or kingdom) was formed and later divided into eastern and western parts, the latter inhabited by the Turkish-speaking Usune, Kangli, Turguesh and Karluk. Turkish conquerors built mosques and tried to impose Islam upon the local population. These were times of famous scientists such as Farabi (870-950), Biruni (973-1048) and Makhmud of Kashgar, author of the *Dictionary of Turkish Dialects*.

³ Between the 9th and 12th centuries, the region was occupied by the Oghuz, Kimak, Kipchak and Karajanid groups. The Kipchaks never achieved political unity and remained outside the realm of Islamic influence, which was concentrated in the cities along the Caspian Sea. Until the 13th century, successive waves of Seleucid, Kidan and Tatar invasions swept across the great steppes. Kipchak chiefs and Muscovite princes joined to resist foreign domination, but did not achieve independence until the fall of the Mongols.

⁴ Most of these peoples were nomadic shepherds but gradually settled groups of farmers and artisans formed. The Silk Road, uniting Byzantium, Iran and China, passed through Kazakhstan. Trade relations extended as far as Western Europe, Asia Minor and the Far East.

⁵ By the late 15th century, the Khanate of Kazakh had been formed, composed of Elder, Middle and Lesser *zhuzes* (hordes), an alliance of like-minded peoples. By the 16th century, an ethnic

identity had been forged among the Kazakhs, whose predecessors were Usune, Kangli, Kipchaks and other groups. The khans of the Kazakh *zhuzes* passed on their power to their heirs, who claimed to be descendants of Juchi, the eldest son of Tenguis-Khan (Genghis Khan).

⁶ In the 17th century, the Khanate of Dzhungar occupied and looted in successive raids the Kazakh region. Russian colonial expansion from the north began in the 18th century. The Russians built a line of forts and then began working their way southward, creating a line of defense against the Dzhungars. The Middle and Lesser *zhuzes* fell under Russian protection and lost their autonomy in the 1820s because of frequent rebellions. Defeat of the Elder *zhuze* in 1848 completed Kazakhstan's annexation by the Russian Empire.

⁷ Russia installed its government institutions, collected taxes, established areas close to the Kazakhs and built new cities, declaring the entire territory property of the State. The conquest of the 'steppe territory', as Kazakhstan was known, was a long process of

wars against local tribes, which were completely defeated in 1880.

⁸ In the late 19th and early 20th centuries, Russia built huge railroads across the region, uniting it with distant urban centers and facilitating the exploitation of Kazakhstan's fabulous mineral wealth. The country held a third of Russia's coal reserves, half its copper, lead and zinc reserves, strategic metals like tungsten and molybdenum, iron, and oil in the west and under the Caspian sea.

⁹ At the beginning of the 20th century, a small nationalist movement emerged in Kazakhstan, and after the Russian Revolution of 1905 the Kazakhs had their own representatives to the first and second *Duma* (parliament) convened by the Czar. In 1916, when the czarist regime ordered the mobilization of all men between the age of 19 and 43 for auxiliary military service, the Kazakhs rebelled, led by Abdulghaffar and Amangeldy Imanov. The revolt was brutally crushed, but in November 1917, after the triumph of the Soviet revolution in Petrograd, the Kazakh nationalists demanded total autonomy for their country. A

nationalist government was installed in Alma-Ata (now Almaty) in 1918, but the country soon became a battleground.

¹⁰ Fighting between the Red Army and the White Russians - the latter defending the overthrown regime - lasted until 1920, when the counter-revolution was defeated. In 1936, Kazakhstan became one of the 15 republics of the USSR and in 1937 the local Communist Party was founded. Kazakhstan received significant migratory inflows of Ukrainians, Belorussians, Germans, Bulgarians, Poles, Jews and Tatars, many of them deported by Joseph Stalin's regime.

¹¹ In addition to developing its industrial potential, the Soviet regime increased the amount of land under cultivation, which came to constitute 15 per cent of all agricultural land in the USSR. Production included wheat, tobacco, mustard, fruit and cattle (particularly cows). Bringing virgin territory under cultivation was an achievement associated with Leonid Brezhnev, during his period as head of the Communist Party (CP) of Kazakhstan.

¹² Until 1985, the most powerful person in Kazakhstan was Dinmujamed Kunaev, a member of the Politburo of the Soviet Communist Party Central Committee, and first secretary of the CP of Kazakhstan. In 1989, the forced resignation of Kunaev triggered student riots, which were crushed by the army.

¹³ After the transformations set in motion in the USSR by President Mikhail Gorbachev, Kazakhstan declared its independence. During this period two social movements arose - Birlik and Zheltoksan - as well as the anti-nuclear movement Semipalatinsk-Nevada.

PROFILE

ENVIRONMENT

Kazakhstan is bordered to the southeast by China; to the south by Kyrgyzstan; and to the north by the Russian Federation. In the western part of the country lie the Caspian and Turan plains; in the center, the Kazakh plateau; and in the eastern and southeastern regions, the Altai, Tarbagatay, Dzhungarian Alatau and Tien Shan mountains. The country has a continental climate, with average January temperatures of -18 ºC in the north, and -3 ºC in the south. In July, the temperature varies from 19 ºC in the north to 28 ºC in the south. Major rivers include the Ural, Irtysh, Syr Dar'ya, Chu and Ili. There is also Lake Balkhash and the Caspian and Aral Seas. The vegetation is characteristic of the steppes, but vast areas have come under cultivation (wheat, tobacco, etc) or are used for cattle-raising. The region's abundant mineral wealth includes coal, copper, semi-precious stones and gold.

SOCIETY

Peoples: Kazakhs, 46.5 per cent; Russians, 35 per cent; Ukrainians 5 per cent; Uzbeks 2 per cent; Tatars 2 per cent. **Religion:** Islam and Christian Orthodox. **Languages:** Kazakh (official), Russian, German, Ugric,

Korean, Tatar. **Main Political Parties:** Republican Party of the Country (OTAN); Coalition for a Just Kazakhstan; People's Communist Party of Kazakhstan; Democratic Party of Kazakhstan Bright Path. **Main Social Organizations:** Birlik Movement, Zheltokso and Semipalatinsk-Nevada, an anti-nuclear movement.

THE STATE:

Official Name: Qazaqstan Respublikasy. **Administrative Divisions:** 14 regions and 3 cities. **Capital:** Astana (formerly called Aqmola, inaugurated as the new capital in 1997), 332,000 people (2003). **Other Cities:** Almaty (formerly called Alma-Ata) 1,250,000 people (1995); Karaganda 420,500; Pavlodar 300,000; Kokchetav 123,000 (2000). **Government:** Nursultan Nazarbayev, President since December 1991, re-elected in 1999 and 2005. Daniyal Akhmetov, Prime Minister since June 2003. Bicameral legislature made up of the 77-member Mazhilis (Assembly) and 47-member Senate. **National Holidays:** 16 December, Independence (1991); 25 October, Republic Day (1991). **Armed Forces:** 41,000 army, 19,000 air force.

Life expectancy
64 years
2005-2010

GNI per capita
$2,250
2004

Literacy
100% total adult
rate
2000-2004

HIV prevalence rate
0.2% of population
15-49 years old
2003

IN FOCUS

ENVIRONMENTAL CHALLENGES
Toxic and radioactive waste - mainly from the armaments industry - have caused serious health problems. A reduction in the water volume of the Aral Sea as a result of the intensive cultivation of cotton on its riverbanks has increased desertification in the area and given rise to a number of environmental problems. Lake Balkhash is suffering similar effects due to the growing industrialization of northern China and the intensive use of water from the Ili river for irrigation. To that is added the continuing thaw of the glaciers in the Tien Shan mountains caused by global warming, leading to a glacial retreat of two cubic kilometers per year, which has affected the volume of water in nearby rivers.

WOMEN'S RIGHTS
Women have been able to vote and stand for election since 1993. The percentage of women in Parliament decreased from 10 per cent in 2000 to 8 per cent in 2004. In the same period,
the percentage of women holding ministerial or equivalent positions rose from 5 to 18 per cent.

In 2004, women comprised 45 per cent of the total labor force of eight million; of them, 58 per cent worked in the service sector, 21 per cent in agriculture, and 21 per cent in industry. Women earn 61.7 per cent of men's wages for the same job.

The maternal mortality rate stood at 210 for every 100,000 births. Some 91 per cent of pregnant women received pre-natal care, while 99 per cent of births were attended by qualified medical personnel. Only 32 per cent of women between 15 and 24 years old had used a condom in their latest high-risk sexual relationship.

An estimated 4,000 to 7,000 women are trafficked from Kazakhstan to other Asian countries and Western Europe for prostitution. More than 2,000 Kazakh women are working in the South Korean sexual market.

CHILDREN
Kazakhstan is one of the few countries in Asia that has signed up for the International Program
for the Eradication of Child Labor. Thousands of children in Kazakhstan work in servitude and prostitution in hazardous conditions and occupations. Many of them have run away from home, often escaping abuse or parents in the grip of alcoholism or drug addiction, turning to the streets in search of a better life. Authorities are concerned because of the increasing number of children falling into sexual exploitation at an early age - sometimes as young as eight. Under-five mortality stands at 73 per 1,000 live births, a rise from the 67 per 1,000 live births in 1990.

INDIGENOUS PEOPLES / ETHNIC MINORITIES
German and Russian minorities are among the largest of the groups living in Kazakhstan. Most of the Germans are farmers or skilled workers. In the 1990s, more than 350,000 emigrated to Germany, and 70 per cent of those still living in the country would like to emigrate due to unemployment and low salaries, according to opinion polls.

Most of the Russians in Kazakhstan live in the city of Kustana. In the mid-1990s, groups
of Russians staged uprisings to demand independence from Kazakhstan. The Government has addressed most of the Russian minority's demands, for example granting them dual citizenship.

MIGRANTS/REFUGEES
Of the more than 15,000 refugees and asylum seekers in late 2004, most were Afghans, Iranians and Chechens. Due to the difficult economic conditions in the country, many refugees lack food and housing. International organizations such as the International Red Cross and UNHCR assist 12,000 refugees and asylum seekers. These organizations give them medical care and basic services (such as education, health and counseling), and also act as a link to the Government to strengthen the State's support for such people.

DEATH PENALTY
Capital punishment is applicable to a wide range of crimes, although a moratorium on executions was again adopted in 2004.

[14] In September 1989, Kazakhstan presented a seven-point plan for the creation of a new union treaty, which was approved by Gorbachev and ten republics. On 1 December 1991, Nazarbayev was elected as the first president of the Republic. The Communist Party became the Socialist Party of Kazakhstan.

[15] Several political groups were formed in 1992, such as the Socialist Party, the People's Congress Party and the pro-government People's Unity Party. Nazarbayev's party won the first multi-party legislative elections, in March 1994.

[16] In April the Government launched a vast privatization plan, which included 3,500 state companies, 70 per cent of all state firms. The opening up of the economy and the country's natural resources attracted many foreign investors during 1995.

[17] A customs union between Kazakhstan, Kyrgyzstan, Belarus and the Russian Federation was formed in March 1996, to promote a common market in goods, capital and workers.

[18] Nazarbayev won the January 1999 Presidential elections with 78 per cent of the vote. Although their fairness was questioned, the elections had an 80 per cent turnout.

[19] Prime Minister Nurlan Balgimbayev resigned in October 1999, after having been censored by Parliament when he sought
approval of a very limited budget. Kasymzhomart Tokayev was appointed as the new premier.

[20] The economy had better prospects of growth with the inauguration, in May 2001, of its first large oil pipeline - built by the Kazakh, Russian and American governments - which would carry 20 million tons of crude oil from the Caspian to the Black Sea, and thence to world markets.

[21] The country still suffers the environmental and health impact of the more than 500 nuclear tests carried out between 1949 and 1989 by Russia in Semipalatinsk. These caused the discharge of toxic waste and fuel into the sea by rockets launched from Baykonur.

[22] The Kazakh Parliament elected Imangali Tasmagambetov as new Prime Minister in February 2002, after Tokayev's resignation. The former Prime Minister had no choice but to leave, because of constant internal divisions in his cabinet.

[23] Kazakhstan and other countries supported US-led military actions against Afghanistan and Iraq by sending troops to help deactivate landmines, and by giving the US access to an airport for re-fueling and emergency landings. For its part, the US pledged $5 million in military equipment and training to bolster security for pipelines and oil installations on the shores of the Caspian Sea.

[24] In June 2003, Daniyal Akhmetov
was appointed as new premier by Parliament after Tasmagambetov was forced to resign due to a dispute over a law governing private land ownership.

[25] In December 2003, Parliament began to discuss a bill to regulate the press. The law established new requirements for obtaining permission to practise journalism, and allowed the Government to intervene in cases in which the media were used for 'propaganda', 'incitement to riot' or 'divulging confidential State information'. Media corporations found to commit such 'crimes' would be subject to temporary or permanent closure. In addition, the law introduced a ban on the 'exhibition of products designed to excite sexual interest'. Although the law was approved, organizations campaigning for press freedom successfully pressed for modifications. However, several key articles remained unchanged.

[26] In May 2004 Kazakhstan and China signed an agreement to build an oil pipeline that would cross the Caspian Sea toward China's western border.

[27] Opposition leader Galymzhan Zhakiyanov was freed in August, after serving two years of his seven-year prison sentence. He was sent to internal exile a month before parliamentary elections were held.

[28] Nazarbayev and his party, OTAN, managed to retain control of the lower house of Parliament after
the elections held in September and October, considered 'flawed' by international observers.

[29] At a summit on foreign investment held in Almaty in June 2005, Nazarbayev warned that a hasty 'import' of Western-style democracies into Central Asia would be destabilizing. He was alluding to popular uprisings that leaders of the countries concerned felt were organized by the US. Nazarbayev, who a month before had brutally repressed demonstrators in the capital, said that although his country welcomed growing foreign investments in the Caspian Sea oil industry, Western partners had to refrain from introducing their political principles into Kazakhstan. Democracy, according to Nazarbayev, is a culture that must be assimilated by societies in due time. He added that the formula for prosperity was the union of the Caspian republics.

[30] In spite of the continuing corruption allegations against the Government, related to foreign investments, observers say the country's economic conditions are improving.

[31] Nazarbayev won another 7-year presidential term with more than 90 per cent of the vote in the December elections. There were irregularities in the polling stations, the opposition was intimidated and the media were biased towards the Government. ∎

Kenya / Kenya

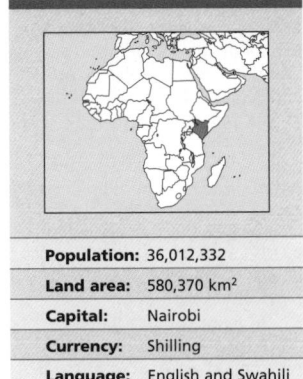

Population:	36,012,332
Land area:	580,370 km²
Capital:	Nairobi
Currency:	Shilling
Language:	English and Swahili

B antu-speaking populations
migrated toward the coast
from the forests of central Africa
3,000 years ago, settling along the
shores of Lake Victoria (in the east
of present-day Kenya). Some 2,000
years later Nilotic and Cushitic
peoples (ancestors of the Oromo
and Somali) entered from Sudan and
Ethiopia and spread throughout the
Kenyan territory.

² In the 16th century, attracted
by Kenya's ivory and metals and
precious stones from neighboring
areas, Arab merchants founded port
cities on the shores of the Indian
Ocean. They engaged in heavy
maritime trade with the Arab world,
Persia and India. In coastal cities
such as Malindi and Mombasa, the
admixture of the merchants and
the Bantu people, whose steady
expansion had reached the center
and the eastern limits of the country,
gave birth to the Swahili culture.

³ Malindi and Mombasa were
at their height from 1100 to 1500,
when Portuguese squadrons
attacked both cities. After
establishing a network of trade in
slaves and ivory, the Portuguese
were finally expelled in 1698 by
Oman's Sultans, who also took over
Zanzibar (a small island off the
coast, which later became part of
Tanzania) and took advantage of
the Portuguese infrastructure and
markets.

⁴ Despite violent attacks by the
Masai (a Nilotic shepherd people,

like the Luo), Arabs and Swahilis
penetrated the country's hinterland
in caravans, reaching Lake Victoria
and the extreme southern regions.
The country's inner regions,
inhabited mostly by Kikuyu and
Kamba (Bantu-speaking peoples)
never developed a culture or
organization comparable to those
of the coastal area.

⁵ With the abolition of the
slave trade in 1873, and incursions
into Kenyan territory by German
expeditionaries, Arabs and (less
numerous) Indians began to leave
the cities in greater numbers.

⁶ In 1890 Germany and Britain
agreed to divide up power over
East Africa. Kenya went to the
British, who already dominated the
territory of present-day Uganda.
The regime of the protectorate
from 1890 - Kenya later became
a British Crown colony in 1920
- created the conditions for the
development of the British East
Africa Company.

⁷ Railways linking Mombasa with
Uganda and Lake Victoria were laid
in 1896 and about 5,000 European

families settled on land destined for
the plantation of coffee and tea,
using African labor. These families
appropriated four-fifths of Kenya's
fertile lands, while the remaining
one-fifth was divided up among
one million Kikuyu.

⁸ During World War II,
contingents of Kenyans were
forced to fight neighboring Italian
territories, such as Ethiopia and
Somalia, while the British admitted
just one local representative to the
Legislative Council.

⁹ In 1944 the Kenya African
Union (KAU) was formed. The KAU
was a multi-ethnic anti-colonialist
organization, which focused on
Kikuyu demands. Led by a Kikuyu
named Jomo Kenyatta, the KAU
organized strikes and peasant
marches on the cities.

¹⁰ In 1952 massive uprisings were
unleashed under the command
of a secret society linked to the
KAU, called Mau Mau. A state of
emergency was declared in 1953
by the colonial administration
in response to the increasing
attacks on the lives and property
of colonists. Political parties were
banned, their main leaders arrested
- Kenyatta among them - and
thousands of Kikuyu were interned
in concentration camps.

¹¹ Kenyatta remained in prison
until 1959, when he was transferred
to house arrest. In 1960 the KAU
was legalized under a new name,
the Kenyan African National Union
(KANU). Kenyatta assumed the
presidency of KANU once he was
released in 1961.

¹² Kenyatta became a member of
the Legislative Council in 1962, as
part of the British plan for African
access to political participation.
KANU - supported mainly by the

Kikuyu and the Luo, from urban
as well as rural areas - won the
seat on the Legislative Council
in elections in which it faced off
with the KADU (Kenyan African
Democratic Union), which was
supported by several pro-colonialist
ethnic groups. The KADU fell apart
in 1964.

¹³ From 1961 on, the British
administration set in motion a land
sales scheme involving two million
hectares. Thus, the colonists - who
had been living on Kenyan land
- earned $55 million from selling
these lands to the people of Kenya.
British financial companies opened
up credit lines to support the sale
of land.

¹⁴ General elections were held
in May 1963 and Kenyatta became
prime minister, while a new
constitution, which anticipated
the country's autonomy, was
approved. After a year of lengthy
discussions with London, Kenya
achieved formal independence in
December. A year later, the republic
was proclaimed and Kenyatta was
elected president, while Oginga
Odinga - a Luo - became vice-
president.

¹⁵ Shortly after Kenyatta took
office he was accused of favoring
his own ethnic group (Kikuyu)
and of neglecting the rest. He
encouraged private enterprise and
the installation of subsidiaries by
transnational corporations. Farmers
who had won back their lands lost
them again due to their heavy debt
burden. A black bourgeoisie with
close ties to Kenyatta replaced the
white colonists in managing the
country's economic affairs.

¹⁶ Odinga left KANU in 1966,
complaining that it compromised
too much, and founded the KPU
(Kenyan Popular Union). The KPU
was banned after the murder of
Tom Mboya, a Luo who had served
as a government minister.

¹⁷ Throughout the 1970s, the
repression of ethnic movements
intensified under the direction
of Charles Njonjo. The largest
organization of ethnic groups,
the GEMA - an association of
Kikuyu, Embu and Meru, led by the
wealthy Karume Njegay - became
a powerful pressure group linking
tribal leaders who grew rich doing
business with interests from the US
and the UK.

¹⁸ Kenyatta died at the age of
85 in September 1978. He was
succeeded by his vice-president,
Daniel Arap Moi, a member of
the smaller Kalenjin ethnic group.
The climate of distrust continued,
aggravated by the tough economic
situation: economic activities
of transnational companies
generated structural imbalances,
which aggravated the crisis and
social tensions. The ruling party,
KANU, was the only political force

LAND USE

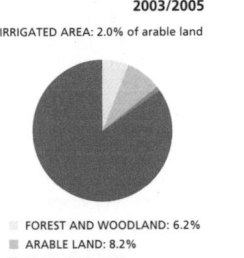

2003/2005

IRRIGATED AREA: 2.0% of arable land

- FOREST AND WOODLAND: 6.2%
- ARABLE LAND: 8.2%
- CROPLANDS: 1.0%
- OTHER USE: 84.6%

PUBLIC EXPENDITURE

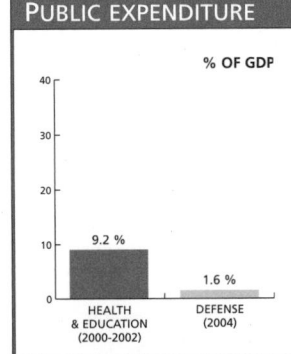

% OF GDP

9.2 % — HEALTH & EDUCATION (2000-2002)

1.6 % — DEFENSE (2004)

Life expectancy
50 years
2005-2010

GNI per capita
$480
2004

Literacy
74% total adult rate
2000-2004

HIV prevalence rate
6.7% of population 15-49 years old
2003

allowed to present candidates in the November general elections, in which Moi was confirmed as president.

[19] In early 1979, President Moi declared an amnesty for political prisoners and launched a campaign against corruption. In 1980 Njonjo was appointed minister of Home Affairs.

[20] At the beginning of Moi's term of office, foreign banks provided credit for planting export crops. Many Kenyan farmers planted sugarcane, coffee, tea and flowers for export under contracts from large US and European firms. As a result, the Government was forced to import enormous quantities of corn and wheat from the US and South Africa.

[21] Difficulties caused by a large deficit in public finances prompted President Moi to seek reconciliation with old political rivals who had been driven out of politics. One of the main beneficiaries of the regime's new openness was former vice-president Oginga Odinga, who returned to Parliament in 1981 after 11 years of ostracism.

[22] In August 1982, the Air Force organized a plot that culminated in massive demonstrations and widespread rioting and looting in Nairobi. Moi obtained support from the Army, which suppressed the rebellion and disbanded all Air Force units. The university was closed down indefinitely and Odinga was placed under house arrest.

[23] The aborted coup sowed distrust among the different factions in KANU. In May 1983, President Moi denounced an attempted coup by minister Njonjo, who according to Moi had the support of Israel and South Africa. In the midst of the confusion, President Moi called early elections, and won a landslide victory over Njonjo and his supporters.

[24] President Moi reopened the border with Tanzania in November 1983, after a summit meeting in Arusha. The summit was meant to be a starting-point for the gradual renewal of economic co-operation between Kenya, Tanzania and Uganda after the failure of the East African Economic Community in 1977.

[25] After the 1983 victory, Moi expanded presidential powers, to the detriment of Parliament. He made it compulsory for civil servants to join KANU, and in the internal party elections, he replaced the secret ballot with a public vote. The moderate women's opposition group, Mandeleo Ya Wanawake, was taken over by the Government in 1986.

[26] In late 1987 Muslim demonstrations in Mombasa triggered a further wave of repression in which Nairobi University was once again closed. Amnesty International denounced human rights violations in which the Government was implicated, including the torture and the murder of opponents of the Government, especially members of the Mwakenya group.

[27] Foreign Affairs Minister Robert Ouko, who had denounced corruption within the cabinet, was assassinated in February 1990. An inquest carried out by Scotland Yard disclosed that close advisers of the President had taken part in the murder, which triggered a new wave of anti-government protests.

[28] Ouko's murder led to an outcry by the international diplomatic community which, added to the numerous reports of human rights violations, prompted Norway to break off diplomatic ties with Kenya in 1991. However, Moi's alignment with the multinational force during the 1991 Gulf War enabled the country to receive economic aid from Britain and military support from Washington. Starting in 1986, Kenya and Uganda engaged in military clashes. Furthermore, the Kenyan Government accused Sudan of protecting groups hostile to Nairobi, and accused Ethiopia of concealing the smuggling of wild animals captured in Kenya.

[29] In mid-1991, KANU convened the party council to discuss the introduction of democratic reforms. Pressure groups such as the Forum for the Restoration of Democracy (FORD), led by Odinga, and the Moral Alliance for Peace (MAP), were legally recognized as political parties. However, the setting of a timetable for holding general elections was postponed until late 1992.

[30] Early in 1992, lawyer James Orengo and environmentalist Wangari Maathai were arrested and accused of 'spreading malicious rumors' against the Government. They had stated that President Moi had a plan aimed at cutting short the democratization process that got under way in 1991. That year, a minority opposition group created the Democratic Party (PD), while women's groups demanded access to decision-making positions in the different organizations. Women constituted 52 per cent of voters and 80 per cent of the agricultural work force.

[31] The December 1992 general elections were preceded by a march in Nairobi organized by FORD. The Forum, despite its recent break-up into three factions - Kenya, Asili and People - rallied over 100,000 people demanding a definite date for elections and the end of repression and press censorship. Although the six opposition parties took 60 per cent of the vote, they won only 88 seats, compared to KANU's 95 seats. Meanwhile, 2,000 died in western Kenya in conflicts between ethnic groups.

[32] In January 1993 Moi took office for a fourth consecutive term amidst accusations of fraud and corruption. In February the Government engaged in negotiations with the IMF and the World Bank. The agreements led the World Bank to grant the country a $350 million credit. That year, the local currency depreciated 23 per cent and in 1994 Nairobi eliminated exchange controls.

[33] In the general elections of December 1997, Moi was re-elected with 40 per cent of the vote, after the dissolution of Parliament the previous month. The widespread questioning of the legitimacy of his presidency forced Moi to appoint both Raila Odinga (son of Oginga Odinga) and Mwai Kibaki to the office of vice-president.

[34] That year, environmental groups and fishing authorities warned that the environment in the region of Lake Victoria - one of the world's largest fishing reserves - would be seriously damaged if the governments of Kenya, Uganda and Tanzania did not curb the use of toxic chemicals by fishing people, which was poisoning fish stocks and polluting water sources.

[35] In August 1998 a bomb killed 248 people at the US Embassy in Nairobi. Three men, allegedly linked to Osama bin Laden, were held responsible for the attack.

[36] In the year 2000 Kenya suffered the worst drought in a century. The subsequent loss of crops led to food shortages, water rationing and power cuts, in households as well as industry.

[37] That year, US President George W Bush cut his government's funding for family planning programs, arguing that some of the money intended to promote contraception was being used for abortions. In 2002 one out of three deaths of women in Kenya were caused by unsafe illegal abortions, according to the Feminist Majority Foundation. At the same time, the Moi Government ordered the purchase of condoms from a German firm, using World Bank funds, to help curb the spread of HIV. HIV/AIDS patients occupy half of the hospital beds in the country.

[38] In 2001 deaths resulting from police brutality amounted to 90

PROFILE

ENVIRONMENT

Kenya is located on the east coast of central Africa. There are four main regions, from east to west: the coastal plains with regular rainfall and tropical vegetation; a sparsely populated inland strip with little rainfall which extends towards the north and northwest; a mountainous zone linked to the eastern end of the Rift valley, with a climate tempered by altitude, and volcanic soil fit for agriculture (most of the population and the main economic activities are concentrated here); and the west which is covered by an arid plateau, part of which benefits from the moderating influence of Lake Victoria.

SOCIETY

Peoples: Kenyans are descended from the main African ethnic groups: Bantu, Nilo-Hamitic, Sudanese and Cushitic. Numerically and culturally, the most significant groups are the Kikuyu, the Luyia and the Luo. Others include the Kamba, Meru, Gusii and Embu. There are Indian and Arab minorities. **Religions:** 66 per cent of the population are Christian, 6 per cent are Muslim and 20 per cent practice traditional religions. **Languages:** English and Swahili are the official languages. There are more than 50 languages spoken, such as Kikuyu and Kamba. **Main Political Parties:** Kenya African National Union (KANU) founded in 1943 by Jomo Kenyatta; National Rainbow Coalition (NARC) led by Mwai Kibaki; Forum for the Restoration of Democracy-Asili (FORD-Asili); FORD-People; FORD-Kenya; Democratic Party (DP). **Main Social Organizations:** Central Organization of Trade Unions (COTU), founded in 1965, is the only labor federation. Nairobi University Students Organization. Green Belt Movement, environmentalist; several human rights NGOs.

THE STATE

Official Name: Jamhuri ya Kenya. **Administrative Division:** 7 provinces and 1 area. **Capital:** Nairobi 2,575,000 people (2003). **Other Cities:** Mombasa 685,000 people; Kisumu 266,300; Nakuru 319,200; Machakos 173,700 (2000). **Government:** Mwai Kibaki, President since 2002 (Head of State and Government); Moody Awori, Vice-President since September 2003. Unicameral Legislature: 224-member National Assembly. **National Holiday:** 12 December, Independence Day (1963). **Armed Forces:** 24,200 troops. Other: 5,000.

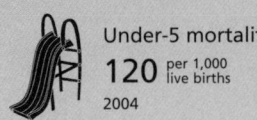
Under-5 mortality
120 per 1,000 live births
2004

Poverty
22.8% of population living on less than $1 per day
1997

Debt service
8.6% exports of goods and services
2004

Maternal mortality
1,000 per 100,000 live births
2000

IN FOCUS

ENVIRONMENTAL CHALLENGES
The main environmental problems include soil degradation, erosion and desertification, deforestation and pollution of fresh water sources, mostly near large cities like Nairobi and Mombasa. Meanwhile, thousands of fish in Lake Victoria die each year due to the increased use of pesticides and fertilizers.

WOMEN'S RIGHTS
Women have been able to vote and run for office since 1963, although it was not until 1990 that two women won seats in Parliament. In 2003, there were 16 female members of Parliament, holding 7 per cent of all seats; there was a 10 per cent female representation in ministerial or equivalent posts.

Maternal mortality stands at 1,000 for every 100,000 live births*. One of the main factors contributing to that high rate is that only 42 per cent of births are attended by qualified medical personnel*.

Over 50 per cent of Kenyan women have undergone genital mutilation, a proportion that rises to 80 and 90 per cent in some western districts and in the provinces of Nyanza and Rift Valley. UNICEF estimates that more than 720,000 women between 15 and 49 are living with HIV/AIDS.

CHILDREN
Ten per cent of newborns are underweight and 35 per cent are undersize, which tends to lead to anemia, frailty and rickets*. Between 80 and 85 per cent of children under the age of one are immunized against the most common childhood diseases, like measles, polio, and tetanus*. The under-five mortality rate rose from 97 to 120 deaths per 1,000 live births between 1990 and 2004*. Most of the cases of death and illness among small children are caused by diarrhea, respiratory infections, malnutrition, malaria and HIV/AIDS. Of the 1.2 million people living with HIV/AIDS in Kenya in 2003, more than 100,000 were under 14, and there were over 650,000 children orphaned by AIDS.

Life expectancy has been reduced from 57.7 years in 1985 to 45.6 years in 2005, as a consequence of the growing HIV/AIDS pandemic.

The harsh drought affecting the Horn of Africa (northern Kenya) could double in a short time the mortality of children, as well as their malnutrition, diarrhea, malaria and other illnesses related to the lack of food.*

INDIGENOUS PEOPLES/ ETHNIC MINORITIES
Kenya is home to many minority ethnic groups, the largest of which are the Kalenjin, Kisiis, Luhya, Luo, Masai and Somalians.

Kikuyu are the largest ethnic group, making up 22 per cent of the population. Most of them live in Nairobi. They arrived in Kenya in small groups in the 19th century.

The Masai differ greatly from other ethnic groups in Kenya, with different customs and religious rituals, and they suffer discrimination from the rest of Kenyan society.

The Luo also stand out from the rest of the ethnic groups because of their social customs, race and religious beliefs.

MIGRANTS/REFUGEES
In 2005, Kenya hosted approximately 200,000 refugees and asylum-seekers, including more than 154,000 from Somalia, nearly 70,000 from Sudan, and around 10,000 from other countries in the region, such as Ethiopia, Eritrea, Republic of Congo, Tanzania, Burundi and Uganda.

Most Somali refugees abandoned their home country during the early 1990s, fleeing civil war and famine. The Kenyan Government has repeatedly blamed them for the deterioration of the country's economic situation, which has led to increasing hostility towards them by the local population.

Sudanese, Ethiopians and refugees from other countries depend almost entirely on international aid to survive. Less than 10 per cent of them have jobs, earning wages of less than a dollar a day.

Political confrontations, land disputes, and ethnic tensions led to the internal displacement of 400,000 Kenyans, mainly farmers, in the last decade.

DEATH PENALTY
The death penalty is applicable to common crimes.

* Latest data available in *The State of the World's Children* and *Childinfo* database, UNICEF, 2006.

---

per cent of all killings. That year, conflicts broke out among ethnic groups over the right to the land located along the Tana river, in the southern part of the country as well as in Nairobi - between Luo and Nubian peoples from Kibera district. Displacement of the population occurred after villages were burnt down.

[39] A law banning genital mutilation of girls under 18 and stipulating prison sentences for those found guilty of practising it was passed in February 2002. Genital mutilation, practised in half of the country's rural districts, was traditionally seen as a way to reduce promiscuity among women.

[40] In November 2002 a terrorist group crashed a car-bomb into an Israeli-owned hotel near Mombasa, killing ten Kenyans and three Israelis. In a simultaneous attack, an Israeli airliner came under missile fire as it took off from Mombasa airport, but the plane was not hit and landed safely in Tel Aviv, Israel. The al-Qaeda network claimed responsibility for the attacks and promised there would be further 'lethal' assaults against Israel and the US. Vice-President Musalia Mudavadi said the country had become a battleground for other people's wars.

[41] On 27 December 2002, Mwai Kibaki, a 71-year-old former vice-president and finance minister - the candidate of the National Rainbow Coalition - won the presidential elections with 63 per cent of the vote. KANU candidate Uhuru Kenyatta - son of Kenya's first president Jomo Kenyatta, and Moi's hand-picked successor - conceded defeat. This marked the end of Moi's 24-year rule and of KANU's 40 years in power.

[42] The new President, known for his moderate and conciliatory character, pledged to conduct his duties without fear, favoritism or malice, and to fight corruption. In January 2003, he created an anti-corruption commission that filed legal charges in June against former president Moi for embezzlement, in a case that became a major bank scandal. However, in December the Government granted Moi immunity from prosecution.

[43] The price of coffee, Kenya's main export, plunged to the lowest level in history. The crisis ruined most of Kenya's small farmers.

[44] In March 2004, a new Constitution was drafted, as Kibaki had promised. The document, which requires Parliament approval, restricts presidential powers and creates the post of Prime Minister.

[45] In July, the statute to approve the new Constitution went missing. A few days later, Kibaki announced the Constitution's approval would be delayed, causing protests in Nairobi and Kisumu that were repressed by the police. This caused conflict within the Government and several clashes in the capital. More than 100 people were arrested in the riots.

[46] In October, environmental activist and human rights advocate Wangari Maathai won the Nobel Peace Prize. Norway's Nobel Committee said Maathai was chosen 'for her contribution to sustainable development, democracy and peace'. Maathai - by now a minister for the environment in the Kibaki Government - was the first African woman to receive such a distinction.

[47] In December 2004, a tsunami devastated South Asia. The huge waves originating in the Indian Ocean reached Kenya's coast. Mombasa and several locations in the north and south were seriously affected.

[48] The Government announced in August 2005 the construction - with Uganda - of an oil pipeline that would run from Eldoret in the west to Kampala, Uganda's capital. The construction would begin in August 2006 and it would be operational in 2007. Both governments would cover 49 per cent of the costs, while the private sector would contribute with 51 per cent.

[49] A report issued by Oxfam International in March 2006 warned that nations affected by the drought in the Horn of Africa would take 15 years to recover unless they received urgent assistance. In some areas, pastoralists had lost up to 95 of their animals, while more than 400,000 people were helped by international organizations.

[50] In April, a plane carrying 18 people - among them Government officials and politicians traveling to a meeting on security in the Marsabit region - crashed, causing the death of 14 passengers.

[51] In May, the Government revoked an agreement that would have given Shell Petroleum control over 50 per cent of oil operations in the country. It was the first time in Kibaki's administration that the agency that monitors monopolies and prices acted to stop a purchase by an oil company. Analysts pointed out that this prevented Shell's domination of Kenya's oil industry.

[51] In January 2007, the World Social Forum was held in Nairobi. This was the first time that the Forum had been held on the African continent. ∎

Kiribati / Kiribati

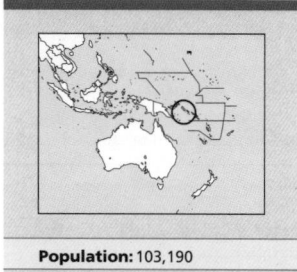

Population:	103,190
Land area:	730 km²
Capital:	Bairiki
Currency:	Australian dollar
Language:	English and Kiribati

Originally inhabited by Micronesian people, the islands that make up the Republic of Kiribati (Makin, Butaritari, Marakei, Abaiang, Tarawa, Maiana, Abemama, Kuria, Aranuka, Nonouti, Tabiteuea, Beru, Nikunau), witnessed the arrival of the British in the 18th century. In 1765 John Byron landed at Nikunau. In 1788 Thomas Gilbert arrived at Tarawa, and John Marshall at Aranuka. Around 1820 the islands were named the Gilbert Islands.

[2] Missionaries arrived in 1857, and three years later, trade in palm oil and copper began. In 1892 the islands became a British protectorate. In 1915 the islands were joined with the neighboring Ellice archipelago (now Tuvalu), to form the colony of the Gilbert and Ellice Islands.

[3] In 1916 Banaba Island became part of the colony. The island had large deposits of guano (bird droppings rich in phosphate). The fertilizer, exploited by the UK since 1920, was exported to Australia and New Zealand.

[4] In 1943, during World War II, the islands were occupied by the Japanese. The intensity of the fighting led to the evacuation of the Banabans.

[5] Since the open-cast mining of guano had made the island uninhabitable, the evacuees were unable to return. Afterwards, they obtained an indemnity of £19 million from the British Government.

[6] In 1957 Britain, as part of its nuclear armament program, detonated three hydrogen bombs near Christmas Island.

[7] The Ellice Islands broke off in 1975 due to their ethnic, historical and cultural differences with the Melanesian majority of the Gilbert Islands.

[8] The inhabitants of the Gilbert Islands proclaimed independence on 12 July 1979 adopting the name: Republic of Kiribati (the equivalent of 'Gilbert' in their 13-letter alphabet). Ieremia Tabai

assumed the presidency.

[9] In 1986 Kiribati was recognized by the UN as one of the world's poorest countries, and began negotiations with the IMF. Since the country has the highest-grade manganese deposits in the world, it hoped to exploit its offshore mineral deposits.

[10] A 1989 UN report on global warming and the possible rise in sea level said that unless urgent measures were taken, Kiribati could disappear under the rising sea.

[11] In the 1991 elections, Teatao Teannaki won with 46 per cent of the vote, leading to the first change of president since independence.

[12] Teannaki, accused of misuse of state funds, resigned in May 1994. In July, the opposition coalition Maneaba Te Mauri won a parliamentary majority and in September Teburoro Tito was elected president.

[13] In 1995, due to the continuation of French nuclear tests in Mururoa (French Polynesia), Kiribati broke off ties with France.

[14] In 1996, Kiribati signed a trade agreement with China. The following year Japan contributed $40 million towards the construction of Betio port, one of Kiribati's main ports.

[15] The country joined the UN in 1999 and the International Labour Organization the following year.

[16] In March 2002, together with Tuvalu and Maldives, Kiribati announced that it would take measures against the US in the face of its refusal to sign the Kyoto Protocol on greenhouse gas emissions - the cause of global warming and rising seawaters.

[17] In July 2003, after having defeated his brother Harry in the elections, Anote Tong was elected president.

[18] China broke off diplomatic relations with Bairiki after Kiribati established ties with Taiwan in November 2003.

[19] According to UNICEF's 2006 Report, although many countries in the region had reached high levels

of development, Kiribati had some of the worst statistics regarding infant mortality and preventable childhood diseases, such as respiratory infections and diarrhea.

[20] In March 2006, the UN announced that one of the world's largest marine parks will be created in Kiribati so as to protect an extraordinary untouched coral ecosystem. ■

PROFILE

ENVIRONMENT

Kiribati consists of 33 islands and coral atolls, with a total land area of 810 sq km, scattered over 5,000 sq km in Micronesia in the Pacific. The islands include Banaba (Ocean Island), the Phoenix and the Line Islands, except for Jarvis Island, which is a US possession. The sandy soil is made of coral rock, and is only suitable for palm trees. The climate is tropical and rainy, tempered by the effect of sea winds. The country's large phosphate deposits are now virtually exhausted. Fishing (carried out in agreement with Japanese, Taiwanese, US and Korean fleets), and copper are its main export industries. Underwater mineral deposits offer great economic potential.

SOCIETY

Peoples: The population is mostly of Micronesian descent. Kiribati 97.4 per cent; mixed (Kiribati and other) 1.5 per cent; Tuvaluan 0.5 per cent; European 0.2 per cent; other 0.4 per cent.
Religion: Catholics (53.4 per cent); Protestants (39.2 per cent), including Adventists (1.9 per cent) and Mormons (1.6 per cent); Baha'is (2.4 per cent); others (1.5 per cent).
Languages: English and Kiribati (Gilbertese).
Main Political Parties: Boutokanto Koaava; Maneaban Te Mauri; and Maurin Kiribati Pati.
Main Social Organizations: there are several labor unions, but the main one since 1979 has been Kiribati General Labor Confederation; Kiribati Student Association.

THE STATE

Official Name: Republic of Kiribati.
Capital: Bairiki 42,000 people; (2003).
Other Cities: Bikenibeu 7,000; Abaiang 5,300 (2000).
Government: Anote Tong, president since July 2003. Legislature: House of Assembly, made up of 42 members, one of whom represents the Banabans, elected by direct popular vote.
National Holiday: 12 July, Independence Day (1979).

Korea / Choson

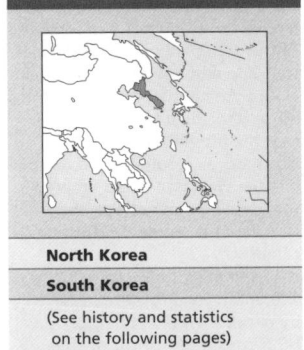

North Korea

South Korea

(See history and statistics on the following pages)

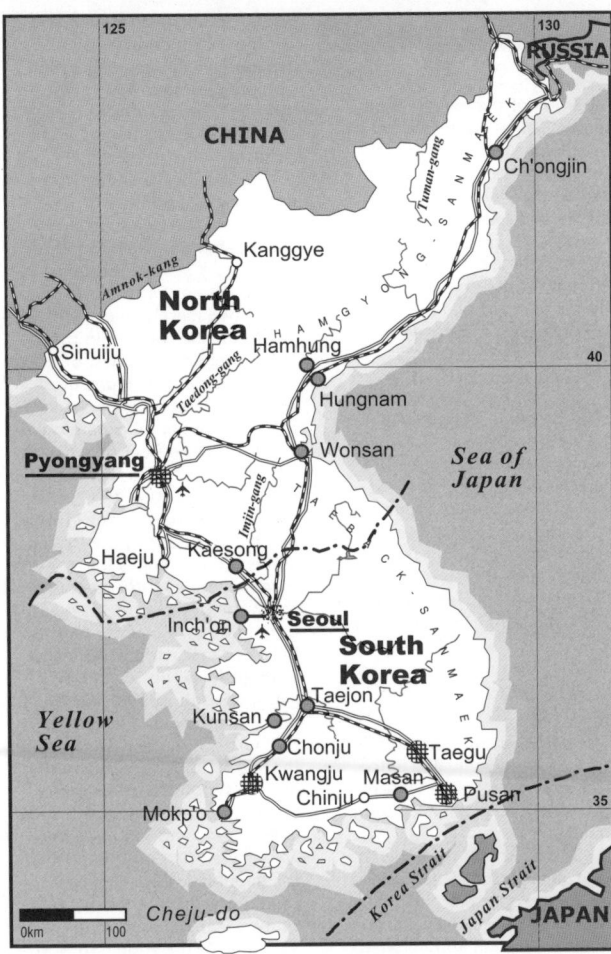

The Korean peninsula lies between China and Japan, and this position has shaped the nation's history and the character of its people. The territory has frequently been the arena of struggles between armies from China, Mongolia and Japan. The peninsula was first inhabited by tribes speaking the Tungu language who emigrated to Siberia. Between the 10th and 8th centuries BC several tribal states were established, of which the most complex was the one known as Old Choson, in the Taedong river basin. Towards the 4th century BC, Choson developed into a league of tribes grouped between the basins of the Liao and Taedong rivers. At this time, Choson inhabitants used weapons made of iron, harnesses for horses and war carriages. In 108 BC, the Chinese empire defeated the kingdom of Choson and replaced it with four Chinese colonies. During the height of Old Choson, other tribal states flourished in the peninsula: Puyo, in the Manchurian region of the Sungari river basin, and Chin, which arose during the 2nd century BC to the south of the Han river and was later divided into three tribal states (Mahan, Chinhan and Pyonhan).

2   The different leagues expanded through the peninsula which, as of the 1st century BC, was divided into the rival kingdoms of Koguryo, Paekche and Silla. Koguryo (which in the 6th century covered an area similar to present-day North Korea) was founded by Chu-mong in 37 BC. Paekche was founded by Onjo in 18 BC, and Silla by Pak Hyokkose in 57 BC. These three states were consolidated by King Taejo (53-146) in Koguryo, King Koi (204-286) in Paekche and King Naemul (356-402) in Silla. The three kingdoms developed into states through successive wars which led to the organization of centralized military and administrative systems.

3   With the support of the Chinese T'ang dynasty, Silla conquered Paekche in the year 660 and Koguryo in 668. However, the survivors of the defeated Koguryo, led by General Tae Cho-yang, established the Kingdom of Perhae to the north of present-day Manchuria and soon, impeded by China from integrating with the other Korean kingdom, entered into direct confrontation with Silla. The peninsula was divided into two states, one in the north and one in the south, both paying taxes to China. Perhae became a very sophisticated state, called 'the prosperous state of the East' by the Chinese, but after it fell into the hands of the Khitan, a nomadic people from the north, its territory no longer participated in the history of Korea.

4   United Silla became an absolute monarchy, which minimized the influence of the aristocracy. The Chinese variant of Avatamsaka Buddhism offered the ideological base for the monarchy and the aristocracy, while the underprivileged were attracted to Pure Land Buddhism, which promised redemption in the next world. In the capital

(now Kwangju, in South Korea), the monarchs built giant palaces and royal tombs, while the State's administration was divided, following China's example, into provinces, prefectures and counties.

5   The decline of Silla began towards the end of the 8th century, as a result of peasant uprisings and conflicts with the aristocracy, which had abolished royal despotism. A new system was established which increased the power of some landlords. Two provincial leaders, Konhwon and Kungye, established the kingdoms of Late Koguryo and Late Paechke respectively. During this period, known as the Three Late Kingdoms Period, Zen Buddhism was the most popular religion, with emphasis on individual fulfillment through contemplation. This doctrine was much less hierarchical than that of Avatamsaka Buddhism, and favored individual autonomy.

6   In 918, Wang Kong founded Songak (today's Kaesong, in North Korea) and in 936 he united the peninsula once again, incorporating the survivors of Late Paechke, devastated by the

Khitan. Wang Kong proclaimed himself the legitimate successor of Koguryo and repeatedly confronted the Khitan, expanding Koguryo's territory up to the Yalu river. Wang Kong was the founder of the Koryo dynasty, from which the western name Korea derives. The ruling class of Koryo was made up mainly of provincial lords, owners of castles, and by Silla's old aristocracy. Koryo was ruled by a Supreme State Council, formed by aristocrats, who adopted Buddhism as a religion to achieve spiritual goals and personal happiness, while they practised the political principles and ethical values of Confucianism.

7   A coup broke out in 1170 and, taking advantage of the subsequent chaos, general Ch'oe Ch'ung-hon established a military regime from 1197 to 1258. In the 13th century, Koryo was repeatedly invaded by the Mongols who came to have great influence in the court. In 1392, Confucian master Yi Song-gye overthrew the tottering dynasty and founded the Choson dynasty (also known as the Yi dynasty, after its founder), which lasted until 1910. In 1394, Yi Song-gye founded Seoul and turned it into the capital of the Kingdom.

8   Choson was dominated by a hereditary aristocratic class, called *yangban*, which devoted itself to the study of neo-Confucian doctrines. During the reign of Sejo, the seventh monarch, a government structure emerged, led by the *yangban* ideology. The country was divided into eight provinces and the central government appointed the administrative chiefs of the state. Legal codes were written and the State Council took charge of the administration. During the 15th century many teachers were recruited to serve the government. They criticized the bureaucracy and recommended several radical measures to implement the ideas of Confucius. However, these teachers had to leave the administration due to strong pressure.

9   In 1597, Toyotomi Hideyoshi, a Japanese military leader who had just reunited his country, invaded Shohon under the pretext of invading China. The national crisis made people from all spheres of life - including Buddhist monks - join in the struggle against the invaders. In 1598, with China's help, the Koreans forced the aggressors to retreat, but most of Shohon was devastated. Many palaces, public buildings and homes were burnt, numerous national treasures destroyed and a large number of craftspeople and experts were kidnapped and taken

PROFILE

ENVIRONMENT

North Korea comprises the northern portion of the peninsula of Korea, located east of China, between the Sea of Japan and the Yellow Sea. The territory is mountainous with wooded ranges to the east, along the coast of the Sea of Japan. Rice, the country's main agricultural product, is cultivated on the plains, 90 per cent of the land being worked under a co-operative system. There are abundant mineral resources (coal, iron, zinc, copper, lead and manganese).

South Korea is located in the southern part of the Korean Peninsula, east of China, between the sea of Japan and the Yellow Sea. The terrain is more level than North Korea's and the arable land area, mainly used for rice farming, is larger.

SOCIETY

Peoples: Both North and South Koreans, probable descendants of the Tungu people, were influenced by their Chinese and Mongolian conquerors over many centuries. Homogeneous ethnic and cultural features in Korea stand out in sharp contrast from those of most other Asian countries. There are no distinct minorities.

Religions: Religious practices are frowned upon in the North. Buddhism, Confucianism (a code of ethics rather than a religion), Chondokio (which combines Buddhist and Christian elements) are practised throughout the country, while traditional Shamanic cults prevail in the interior.

Languages: Korean (official).

to Japan. At the beginning of the 17th century, nomad Manchurians invaded Shohon, took over the northern part of the territory, and captured Seoul in 1636, demanding the unconditional surrender of the king. In 1640 the Manchurians overthrew the Ming dynasty in China and replaced it with the Ch'ing dynasty. The taxes Korea paid to the Ming went to the Ch'ing.

¹⁰ During the second half of the 17th century and throughout the 18th century, Korean society underwent great changes. Rice became widespread and irrigation systems were improved. Agricultural production was increased and the peasants' standard of living improved. Tobacco and ginseng crops encouraged domestic and foreign trade, thus intensifying contacts with European traders and Catholic priests. Meanwhile, radical ideological changes were taking place in Korea, since many teachers no longer concerned themselves only with theoretical speculation but also with matters of practical importance. This gave birth to *silhak* or education based on pragmatism, which urged the Government to undergo changes. Also, a school of silhak devoted itself to the study of the Korean language and history and facilitated, jointly with the development of popular art, the access of the people to written texts. By the end of the 18th century some teachers of silhak had converted to Catholicism, followed by members of the aristocracy. Large sectors of the population, encouraged by the hope of finding equality before God after death, were seduced by the new religion which spread rapidly.

¹¹ The decline of the Yi dynasty was characterized by economic and religious factors, plus outside pressures. The yangban or aristocracy had appropriated public lands and did not pay taxes, which led to an increase of taxes on the poor, who could not pay and lost their lands. Also, the State banned Christianity, because of its incompatibility with Confucianism. During the persecutions of 1801, 1839 and 1866, the converted teachers faced death or apostasy, while foreign missionaries were beheaded. At the same time, Japan pressed Korea to open its foreign trade and China increased its interference in the peninsula to counterbalance the Japanese influence. In 1860, the scholar Ch'oe-u founded a popular religion called Tonhak (oriental teachings) which combined elements of Confucianism, Christianity, Shamanism and Buddhism. Soon, these new teachings, in the name of resistance against foreigners and corruption, gained a large following among peasants and by 1893 had turned into a political movement. In May 1893 the Tonhak followers took the city of Chonju, in the southwest, and the two intervening powers, China and Japan, sent troops to crush them.

¹² To justify their military presence in the peninsula, Japan proposed that China carry out a joint reform in Korea, but China's negative reply led to a military conflict which ended with a Japanese victory in 1895. Japan occupied Korea in 1905 and in 1910 the country was formally annexed, putting an end to the Yi dynasty.

¹³ Under the Japanese, Korea was used as a supplier of foodstuffs and as a source of cheap labor. Japanese landlords and factory owners settled in Korea, with an infrastructure developed merely to extract the wealth of the country. In the 1930s, the northern part of Korea saw industrial development of war materials, to supply Japan's goal of continental expansion.

¹⁴ With the defeat of Japan in World War II, Korea was occupied on each side of the 38th parallel: the northern part by Soviet troops and the southern by the United States. Korean hopes for a united, independent country seemed on the verge of being realized but a complex struggle of interests between the two powers made it impossible. The Soviet Union maintained its influence and did not allow general elections in the north of the peninsula. In the south, under the supervision of a United Nations Temporary Commission, elections were held in May 1948 and Syngman Rhee was elected first president of the Republic of Korea, whose capital was established in Seoul. Meanwhile, the Supreme Assembly of the People of North Korea wrote a new Constitution, which came into effect in August 1948. Kim Il-Sung was appointed Prime Minister and on 9 September the People's Republic of Korea was proclaimed, with its capital in Pyongyang.

¹⁵ On 12 October the Soviet Union recognized this state as the only legitimate government of Korea. In December, the General Assembly of the United Nations recognized the exclusive sovereignty of the southern Republic of Korea. Most foreign troops left both countries the following year but US troops remained in the south.

¹⁶ In June 1950, North Korea launched a carefully planned offensive against South Korea. The United Nations convened its members, to put a stop to the invasion. In the meantime US President Truman called his army to assist South Korea, without asking Congress to declare war.

¹⁷ Likewise, Truman failed to seek UN permission before sending the US Fleet to the Strait of Formosa to protect one of the US Army's flanks and to assist Chiang Kai Shek's anti-Communist Chinese regime. The disastrous military situation of the South Koreans was saved by General Douglas MacArthur, who landed some 160 kilometres south of the 38th parallel and managed to divide and defeat North Korean troops. China, concerned over the advancing allied troops, warned that the presence of the US in North Korea would force it to join the war. MacArthur ignored the warning and in November launched his 'Home by Christmas' offensive.

¹⁸ However, China sent 180,000 troops to Korea and by mid-December it had driven US troops back, south of the 38th parallel. On 31 December 1951, China launched a second offensive against South Korea, subsequently taking up positions along the former border.

¹⁹ After differences of opinion on military strategy, MacArthur was relieved of his command by Truman. It was later revealed that MacArthur had outlined plans to use nuclear weapons against Chinese cities, advocating full-scale war with China.

²⁰ In 1953, an armistice was signed and Korea was officially divided in two by the 38th parallel. The conflict had lasted 17 months and left approximately four million dead.

²¹ On 8 August 1990, the UN Security Council unanimously approved the admission of both Koreas. On 13 December 1991, Prime Ministers Yon Kyong Muk (North Korea) and Chong Won Shik (South Korea) signed a 'Reconciliation, Non-Aggression, Exchange, and Cooperation Agreement', regarded as an important step towards reunification.

²² In late 2002, North Korea reinitiated nuclear activity and, in January 2003, withdrew from the Treaty of Non Proliferation of Nuclear Weapons, straining relations with the US, whose government included Pyongyang among the 'axis of evil' countries. South Korea offered to mediate between both nations, and this was accepted by North Korea.

²³ In February 2004, the second round of negotiations between China, the United States, Russia, Japan and the two Koreas took place in Beijing. Pyongyang stated it would cease its nuclear program if Washington assured it that no retaliation would be sought.

²⁴ In October 2006 it was announced that a diplomatic breakthrough at an informal meeting in Beijing between North Korea, China and the US meant that six-party talks on North Korea's nuclear programme were likely to resume soon. The other countries involved in the talks would be South Korea, Russia and Japan. ∎

North Korea / Choson

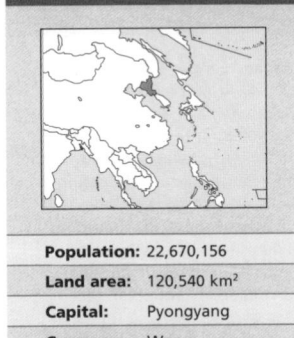

Population:	22,670,156
Land area:	120,540 km²
Capital:	Pyongyang
Currency:	Won
Language:	Korean

The Kingdom of Koguryo, the seventh-century state now known as North Korea, had a special way of treating foreign enemies: it avoided direct confrontation and maintained an impressive military capacity within its cities (see history of Korea).

[2] The region was invaded by several peoples, among them the Japanese (1592-1598) and the Manchus (1627-1636). Japan annexed the region in 1910, and occupied it until the end of World War II. After the end of this conflict, the USSR occupied Korea north of the 38th parallel, while the US occupied the southern section.

[3] On 3 September 1948 the People's Republic of Korea was proclaimed in Pyongyang and Kim Il Sung, leader of the Korean Workers' Party (KWP), was elected Prime Minister. North Korea followed a similar strategy to that of the former kingdom, limiting its contact with the outside world and concentrating on building up its military strength. Despite its geographical proximity to the People's Republic of China, North Korea tried to remain neutral in the Sino-Soviet conflict.

[4] The KWP enjoyed uninterrupted power for five decades. The party's philosophy is called *Juche*, a blend of self-reliance, nationalism and centralized control of the economy.

[5] Under the socialist regime, agrarian reform collectivized part of the country's agriculture. Industrialization had begun during the Japanese occupation, and large textile, chemical and hydroelectric plants were established.

[6] After the devastation of the Korean War (1950-53, see Korea), North Korea started the reconstruction of the country. When Syngman Rhee was ousted in South Korea (1960), North Korea attempted rapprochement with its neighbor, a move interrupted when the military took power in Seoul.

[7] Korea adopted a centrally-planned socialist economy, with 90 per cent of industry in the hands of the State and the rest organized in co-operatives.

[8] Between 1954 and 1961 North Korea signed military assistance treaties with China and the USSR.

In 1972, the new constitution made Kim Il Sung President as well as Prime Minister.

[9] Kim Jong Il, Kim Il Sung's son, was named head of the Government in 1980. In 1984 North Korea provided relief to flood victims in South

CHINA

RUSSIA

North Korea

Pyongyang

Sea of Japan

Yellow Sea

SOUTH KOREA

Korea Strait

Japan Strait

JAPAN

Korea. That same year mixed enterprises were authorized in the construction, technology and tourist sectors.

[10] In 1988 North Korea began 'ideological rectification' - contrary to the liberalization process introduced by Gorbachev in the USSR.

[11] North Korea was internationally isolated, dependent on oil imports and financially tied to Japan, while receiving goods and remittances amounting to $1,000 million a year from 200,000 of the 700,000 North Korean exiles living in Japan.

[12] Kim Il Sung's death at 82, in July 1994, complicated talks with the US and delayed the summit planned between the two Koreas. Kim Jong Il succeeded his father without having the same authority, which led to a power struggle among leading cadres.

[13] In 1995 widespread floods affecting some five million people, with the loss of almost two million tons of crops, led the North Korean Government to make an unprecedented appeal for foreign aid. Japan donated 300,000 tons of rice and South Korea 150,000 tons.

[14] In October 1997 Kim Jong Il, the de facto leader after his father's death, was officially appointed head of the Workers' Party. But due to internal rivalries he was not formally appointed President.

[15] In 1998 it was estimated that some 100,000 people had died of hunger, cold and lack of medical attention since 1995. The daily ration of rice per person was just 100 grams.

[16] In June 2000 an historic summit between the two Koreas was held in Pyongyang. Kim Dae-Jung, South Korea's President,

LAND USE

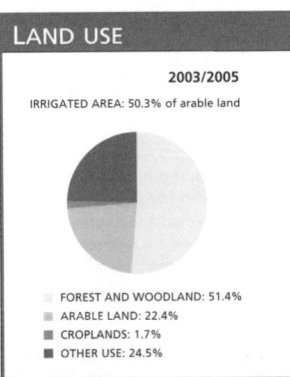

2003/2005

IRRIGATED AREA: 50.3% of arable land

- FOREST AND WOODLAND: 51.4%
- ARABLE LAND: 22.4%
- CROPLANDS: 1.7%
- OTHER USE: 24.5%

WORKERS

LABOR FORCE   2004

- FEMALE: 39%
- MALE: 61%

PROFILE

SOCIETY

Main Political Parties: Democratic Front for the Reunification of the Fatherland, which includes: Workers Party of Korea; Korean Social Democratic Party; Chondoist Chongu Party.

Main Social Organization: The General Federation of Trade Unions is the only workers' organization, while the Union of Agricultural Workers is a peasants' association.

THE STATE

Official Name: Choson Mintschutschui Inmin Konghwaguk.
Administrative Divisions: 9 Provinces, 1 District.
Capital: Pyongyang 3,228,000 people (2003).
Other Cities: Hamhung 808,300 people; Chongjin 663,400; Sinuiju 371,300; Kaesong 195,300 (2000).
Government: The presidency is vacant since Kim Sung Il died in July 1994. His son, Kim Jong Il, is de facto President of the Republic. Kim Yong Nam, Chair of the Presidium of the Supreme People's Assembly since September 1998. Pak Pong Ju, Premier since September 2003. The State's highest body is the Supreme People's Assembly, with 687 members.
National Holiday: 9 September, Republic Day (1948).
Armed Forces: 1,054,000 (1994). Other: 3,800,000 Peasant Red Guard, 115,000 Security Troops, of the Ministry of Public Security.

Life expectancy
64 years
2005-2010

HIV prevalence rate
<0.1% of population
15-49 years old
2003

IN FOCUS

ENVIRONMENTAL CHALLENGES
Pollution of rivers, as well
as inadequate supplies of
safe drinking water, result in
frequent water-borne diseases,
such as diarrhea and cholera.
Uncontrolled deforestation leads
to soil erosion and degradation
and loss of fertile land. Air
contamination started becoming
a problem in the 1970s, as did
pollution of rivers and seas (the
country's west coast on the
Yellow Sea is severely polluted).

WOMEN'S RIGHTS
Women have been able to vote
and be elected for office since
1946.
In 2003 women held 20
per cent of seats in Parliament,
although they occupied no
ministerial posts.
Qualified medical personnel
attend 97 per cent of births and
almost all pregnant women
receive prenatal healthcare*.
Jailed women are forced to
work from morning till night,
in unhealthy conditions, even
those who are pregnant or old.
Prisons lack the basic facilities to

cover their needs.
Reports have condemned the
trafficking of women from North
Korea to China, on the promise of
jobs. Once they reach China their
identification documents are taken
from them and they are often
forced into prostitution.

CHILDREN
Twenty-three per cent of children
under five are moderately or
severely underweight *. Chronic
food shortage, along with natural
disasters (such as frequent floods)
have direct consequences for the
population's nutrition. A report by
the Government and the World Food
Programme showed that 33 per cent
of children under 7 are moderately
or seriously undernourished.*
Education is compulsory and
universal between the ages of 6
and 16, and there is a 100 per cent
enrollment rate.
Many children are subjected
to intense political indoctrination.
There have been reports of young
children being submitted to
military training and compulsory
indoctrination at school.
According to the Government,

sexual exploitation of children is not
a problem in North Korea. However,
the US State Department reports
that young women and girls are sold
as wives in China.

**INDIGENOUS PEOPLES/
ETHNIC MINORITIES**
See South Korea.

MIGRANTS/REFUGEES
At least 100,000 North Korean
refugees are living in China, of
whom 75 per cent are women.
Although there are no exact figures,
there are also significant numbers
of North Koreans in Russia and
elsewhere in Asia, while over 5,000
have emigrated to South Korea and
adopted that nationality.
Poverty and the food crisis,
which started in the mid-1990s,
are the main causes of emigration.
Almost 3.5 million North Koreans
(nearly 18 per cent of the
population) have died since 1994
from hunger or famine-related
diseases.
Several international NGOs
have suspended operations in North
Korea, citing the Government's
failure to provide a transparent food

management and distribution
system. NGOs accuse the
Government of dividing the
population into categories based
on their loyalty and utility to
the regime, and of channeling
food aid accordingly. Under the
country's Penal Code, desertion
is punished with a minimum of
seven years in prison and the
death penalty applies in the
case of deserters who establish
contact with South Koreans,
Christians or foreigners once they
reach China.
The number of internally
displaced people is estimated
at over 100,000. Despite
the Government's policy of
controlling internal migration,
they move from one place to
another to avoid being caught.

DEATH PENALTY
The death penalty applies even
for ordinary crimes.

* Latest data available in *The State
of the World's Children* and *Childinfo*
database, UNICEF, 2006.

---

met with Kim Jong Il to discuss
security issues. They signed an
agreement to work together
toward the reunification of the
Korean peninsula.
[17] A delegation from the
European Union, led by Swedish
Prime Minister Göran Persson,
visited the two Koreas in May
2001 to support the process
of reconciliation between the
countries.
[18] US President George W
Bush stated in January 2002
that North Korea was part of
the 'axis of evil' dedicated to
terrorism, along with Iran and
Iraq. Pyongyang described Bush's
statement as a virtual declaration
of war. Military maneuvers carried
out jointly by South Korea and
the US in March 2002 led the
North Korean Government to
threaten to break the nuclear
agreement in force since 1994,
under which Pyongyang had
accepted abandoning its own
nuclear program in exchange for
US assistance with energy and
building nuclear power stations.
[19] In April 2003, delegations
from North Korea, the US and
China held the first talks on
the nuclear crisis, in Beijing.
In October the Government
announced that 8,000 spent
fuel rods had been reprocessed,
producing enough material for
eight atomic bombs.
[20] Five young men and women

born in North Korea of Japanese
parents who had been kidnapped
by Pyongyang agents 25 years
earlier, arrived in Tokyo in May
2004. Until two years before,
these young men and women had
not known about their origins,
and had grown up under the
North Korean regime. In 2002 Kim
Jong Il had offered an apology to
Japanese Prime Minister Junichiro
Koizumi, during his visit to North
Korea, for the abductions of
Japanese citizens in the 1980s
and 1990s. Japan had supplied
North Korea with 250,000 tons of
rice and $10 million of medical
supplies in return for the release
of the young 'abductees'.
[21] North Korea's
undernourishment levels
continued to be the highest in
the world. The UN's World Food
Programme (WFP) had been
feeding up to six million of the
poorest and most vulnerable
people there, ensuring that
the supplies reached them.
Meanwhile, China and South
Korea delivered cereals to North
Korea, without any restrictions.
In September 2005, Pyongyang
asked the UN to end its aid by
the end of that year. Deputy
Foreign Minister Choe Su-Hon
accused the US of politicizing
food aid by linking it to human
rights. The WFP warned that
millions of North Koreans would
be left in the verge of famine.

[22] According to most analysts,
multilateral dialogue was the
best way to try to resolve the
crisis arising from the country's
nuclear ambitions. In November
2005, the fifth round of talks
between Pyongyang, Russia,
South Korea, the US, China and
Japan ended with a statement
that only reaffirmed the parties'
commitment to implement a
preliminary agreement drafted
during the previous round in
September.
[23] After six months of talks, in
May 2006 Pyongyang allowed
new shipments of international
food aid. The WFP would now
aid only 1.9 million of the
'neediest' people in the country.
[24] In July, North Korea test-
fired seven missiles, including
a long-range Taepodong
2, believed to be capable
of reaching Alaska. The
international community was
divided over how to respond.
Japan presented in the UN
Security Council a draft
resolution - backed by the US,
Britain, France, and five other
members - calling North Korea
a 'threat to international peace
and security' and invoking
Chapter Seven of the UN charter.
Resolutions made under Chapter
Seven can lead to sanctions or
even military action. China and
Russia threatened to veto it, and
insisted on the need to find a

diplomatic solution.
[25] Finally, on 15 July, the
Security Council unanimously
passed a resolution demanding
Pyongyang end the missile
launching. Also, the resolution
urged UN member countries to
avoid importing or exporting
materials and technology from
or to North Korea that could be
used in the making of weapons
of mass destruction.
[26] The strong rains that fell
that month on the central region
of the country caused serious
floods. These left 10,000 dead
and missing, and some 60,000
displaced persons. The floods
caused the loss of 100,000 tons of
food, which severely affected the
country's neediest population.
[27] In October 2006, North Korea
conducted a controlled explosion of
a nuclear device. The explosion was
relatively small – less than a tenth
the size of the bomb dropped on
Hiroshima in 1945 – but confirmed
that Pyongyang had joined 'the
nuclear club' in defiance of the
international community. There
was widespread international
condemnation of the development,
including from China, which has
historically been supportive of
Pyongyang. International demands
for sanctions against North
Korea were defused by
North Korea's agreement to rejoin
six-party talks over its nuclear
program. The talks had stalled in
2005. ∎

South Korea / Han'guk

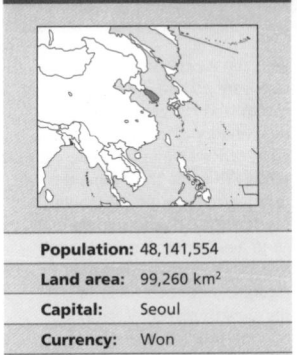

Population:	48,141,554
Land area:	99,260 km²
Capital:	Seoul
Currency:	Won
Language:	Korean

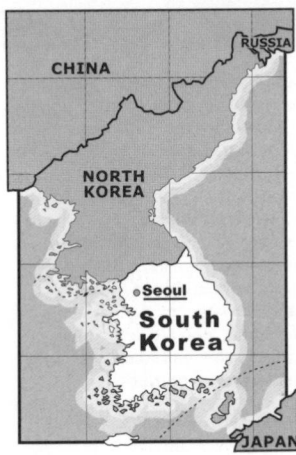

The Republic of South Korea was created on 15 August 1948. Its first President was Syngman Rhee, following elections held with US troops in the country. For 14 years Rhee ruled the country as a dictator, and imposed a Constitution to perpetuate his rule.

2   The Rhee administration was an unconditional ally of US policy. Americans were very much aware of events in China, where the nationalist Kuomintang had been defeated by the communists who promised rural peoples 'land to the tiller'. They then urged South Korea to carry out a land redistribution scheme, compensating landowners and limited to three hectares per person.

3   A draconian National Security Law in 1958 enabled Syngman Rhee to imprison political dissidents of all types. Rhee was re-elected in 1952, 1956 and 1960, amid charges of electoral fraud. There were protests in Seoul and the threat of revolution forced Rhee to resign on 27 April 1960.

4   New elections were held, and Po Sun Yun, a member of the Democratic Party, was elected President. The head of government, John Chang, attempted to lead the country toward effective economic development and put an end to corruption.

5   In May 1961 a military coup ousted Chang. In July, General Chung Hee Park took command of the junta and proceeded to suspend all democratic freedoms and imprison all members of the previous regime.

6   The new Government initiated a National Reconstruction policy, including a planned strategy against communism and corruption, and promised free elections upon completion of these 'revolutionary tasks'.

7   In 1963, Park held elections and won by just 1.4 per cent. He declared martial law to quell the protests that followed, and suppressed all political and labor freedoms.

8   The military regime established centralized economic planning and, with the help of western technocrats, South Korea became an exporting country. In 1965 the Government signed a treaty with Japan renouncing their claim to war reparations in exchange for economic aid. After the agreement, South Korea started to receive Japanese funds.

9   The country developed into an industrial economy dominated by large, Korean-owned transnational corporations producing steel, ships, cars and electronic goods. Low grain prices impoverished the peasantry, who were forced into cities. Some of the world's lowest wages, longest hours and most unsafe working conditions were the norm for workers here.

10   Eighteen years after taking power, having won four fraudulent elections, Park was killed in October 1979, shot by the director of his Intelligence Agency in unclear circumstances.

11   On 17 May 1980, the military established martial law once again, arresting opposition members. The following day, factory workers and students took control of Kwangju city, in a historic uprising. Repression was brutal: the army killed thousands. Kim Dae Jung, a prominent opposition leader, was sentenced to life imprisonment, charged with instigating the protests.

12   In the same manner as his predecessor, military leader Chun Doo Hwan held elections in an attempt to legitimize and civilianize his rule; he won the elections in 1981.

13   In October 1983, several members of the South Korean cabinet were killed by a bomb at the Martyrs' Mausoleum in Burma during a state visit. The Burmese, claiming proof of North Korean involvement, broke off diplomatic relations with Pyongyang.

14   Opposition to Chun's regime continued to grow as repression escalated, reminiscent of Park's worst excesses. The US withdrew their support of the Marcos regime in the Philippines, after claims of election fraud and human rights abuses, and a worried Korean regime instituted some reforms. Censorship was lifted somewhat and Kim Dae Jung's imprisonment was changed to house arrest.

15   During 1987, hundreds of thousands of Korean workers joined in strikes and factory occupations in an unprecedented wave of protests. They demanded the right to form democratic unions independent of the government-run Federation of Korean Trade Unions, higher wages, an end to forced overtime and a larger share of the benefits of the nation's spectacular growth.

16   In July 1987, Chun appointed Roh Tae Woo both as his successor and as president of the official Democratic Justice Party. Demonstrations followed amid protests that Roh would continue the dictatorship once in power. Demands that Chun face trial for his part in the Kwangju massacre were also voiced.

17   Faced with the possibility of larger street demonstrations and concern over its international image (South Korea was to host the Olympic Games in 1988) political restrictions were eased during the 1987 election campaign.

18   In the elections, the newly freed opposition together polled a majority, but failed to unite the two factions, led by Kim Yong Sam of the Reunification Democratic Party and Kim Dae Jung of the Party for Peace and Democracy. The split enabled the incumbent government to win the elections.

19   In January 1990, opposition groupings formed a merger with the official Democratic Justice Party, and became the Democratic Liberal Party, controlling 220 seats in the 298-member National Assembly. The Party for Peace and Democracy remained the only real parliamentary opposition.

20   In April 1990, in a new offensive against independent trade unions, police stormed the Hyundai shipyards and arrested over 600 union activists, ending a 72-hour worker occupation protesting the arrest of union leaders. A few days later, 400 striking workers occupying the Korean Broadcasting System's headquarters were also arrested. The resulting nationwide protests precipitated the biggest drop in the history of the country's stock market.

21   In September 1991, US President George Bush made the decision to withdraw tactical nuclear weapons from South Korea, and in November this was accomplished. This significant step met one of North Korea's requirements before allowing nuclear inspections in its territory.

22   In December 1991, Seoul and Pyongyang signed a Reconciliation, Non-aggression, Exchange and Cooperation Accord, improving bilateral relations (see Korea).

23   In May 1992, President Roh Tae Woo named Kim Young Sam, who had obtained 41.4 per cent in the December presidential election, as his successor. Kim's election coincided with a weakening of the opposition, worsened by the resignation in February 1993 of Chung Ju-Yung, leader of the United People's Party, charged with having accepted illegal contributions from a major corporation during the electoral campaign.

24   In 1995, two former presidents, Chun Doo Hwan (1979-1988) and Roh Tae Woo (1988-1993), were arrested for their role in the coup that put Chun in power in December 1979. They were accused of treason and embezzlement.

25   In December 1996 a Seoul court sentenced former Defense Minister Lee Yanh-ho to four years in prison for accepting illegal commissions from Daewoo, the country's third largest car manufacturer.

26   In 1997 the country was rocked by the Asian financial crisis. The IMF

PROFILE

SOCIETY
Main Political Parties: Millennium Democratic Party; Grand National Party; Democratic Labour Party.
Main Social Organizations: Legally, all unions must belong to the Government-controlled Federation of Korean Trade Unions (FKTU).

THE STATE
Official Name: Taehaen-min guk (Republic of Korea).
Administrative Divisions: 9 Provinces.
Capital: Seoul 9,714,000 people (2003).
Other Cities: Pusan 4,266,100 people, Taegu 2,947,400; Incheon 2,403,700 (2000).
Government: Roh Moo Hyun, President since 2003. Han Myeong Sook, Prime Minister since 2006. Single-chamber Legislature: National Assembly, with 299 members, elected every 4 years.
National Holiday: 13 August, Liberation Day (1945).
Armed Forces: 680,000 (1995). Other: 3,500,000 Civil Defense Corps, 4,500 Coast Guard.

Life expectancy
78 years
2005-2010

GNI per capita
$14,000
2004

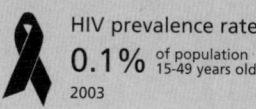

HIV prevalence rate
0.1% of population 15-49 years old
2003

IN FOCUS

ENVIRONMENTAL CHALLENGES
There is significant air pollution, especially in the big cities. Acid rain affects most of the country. Rivers and part of the seas are polluted by discharge of waste and industrial waters; the west coast on the Yellow Sea is badly affected. Trawling is destroying marine habitats and also threatening species with extinction.

WOMEN'S RIGHTS
Women have been able to vote and stand for election since 1948. In 2006, for the first time in the country's history, a woman took office as Prime Minister. In 2004, female representation in Parliament stood at 13.4 per cent. The previous year, women made up 41 per cent of the country's total labor force: 68 per cent were engaged in services, 19 per cent in industry and 13 per cent in agriculture.

Maternal mortality rate is one of the world's lowest at 20 deaths for every 100,000 live births*. All births are attended by qualified medical personnel*.

The rate of female-headed households living in extreme poverty was doubled after the financial crisis in 1997, increasing from 8.4 per cent to 16.9 per cent. Sixty per cent of poor adults are women. In 2001, legislation was strengthened as special laws were passed to prevent domestic violence and discrimination against women. However, these legal provisions are still barely applied.

CHILDREN
Between 97 and 99 per cent of all children were immunized against diseases such as polio, measles and tetanus. Under-five mortality rate was one of the lowest in Asia and in the world, at six deaths per 1,000 live births*.

Public education is compulsory until the age of 15, with universal primary school enrollment rates. Almost 100 per cent of children have access to high quality healthcare.

The 1999 Youth Protection Act provides for up to 10 year prison sentences and $7,750 fines for those who employ youngsters under 19. A sexual protection law was passed in 2000, setting 20 years

as the minimum age for practicing prostitution, and also establishing penalties for those who hire prostitutes under 20.

In spite of Government efforts, the sex trade grows daily in South Korea. It is estimated that 500,000 of the country's sex workers are underage. Human trafficking from neighboring countries has also increased, mostly of people under 18, who are destined for prostitution in brothels, massage parlors and nightclubs.

INDIGENOUS PEOPLE / ETHNIC MINORITIES
In contrast with most of the rest of the world, Korea is ethnically, culturally and linguistically homogeneous. However, it has ancestral regional tensions dating back to the unification of the three Korean Kingdoms in the 7th century. But economic and political differences came to surface after World War II, influenced by the Soviet bloc in the North and the US bloc in the south.

MIGRANTS/REFUGEES
In 2004, South Korea received nearly 300 requests for asylum and refuge;

with the exception of six, all the rest were accepted. Refugees recognized by the Government receive annual temporary visas, renewable up to three times. After the third time, the refugee can request permanent residence, work and sign up in state health programs, although healthcare does not reach all refugees since according to the Government there are not enough funds to cover both citizens and refugees.

Since 2001 the Government grants a temporary status to those people who, for reasons such as guerrilla insurgency in their own countries, are refused refugee status. Those granted this status do not have the right to work and annual visas are renewed provided the Government deems it necessary.

DEATH PENALTY
The death penalty is applicable to all types of offenses.

*Latest data available in *The State of the World's Children* and *Childinfo* database, UNICEF, 2006.

intervened with a $67 billion loan, but demanded more flexible labor conditions and the privatization of the *chaebols* (industrial conglomerates such as Samsung, Hyundai, Daewoo, of which most South Korean political and military leaders are shareholders).

[27] Kim Dae Jung, of the opposition Democratic Party (DP), won the December presidential elections. Upon taking office he announced an amnesty for political prisoners and a national unity government.

[28] A large crowd received Kim Dae Jung upon his return in June from a trip to Pyongyang for an historic summit between the two Koreas. He discussed with the North Korean leader Kim Jong Il issues of security, including the 37,000 US soldiers stationed in South Korea, and missile programs.

[29] In October 2000, Kim Dae Jung received the Nobel Peace Prize for his 'work for democracy, for Human Rights in South Korea and East Asia and for peace and reconciliation with North Korea'.

[30] In November, the Daewoo car manufacturer declared itself bankrupt. In the previous months, billions of dollars from public coffers had gone toward bailing it out. However, company chairman, Kim Woo Joong, had disappeared with a large portion of those funds in

1999. After the announcement that more employees would be laid off, in February 2001, Daewoo workers went on strike. This was put down violently by the police, leaving hundreds injured and imprisoned.

[31] The publication of a Japanese history textbook whitewashing the country's colonial past and the atrocities committed by its army in Korea caused protests. A South Korean social movement demanded that Japan withdraw the book.

[32] Relations between the two Koreas were tested in June 2002 by a naval collision that killed five South Korean sailors and several North Koreans. In July, Pyongyang expressed regret for the incident. Afterwards there were two summits and a detente process began. In May, North and South Korean families were reunited. In an unprecedented gesture, North Korean athletes competed in the Asian Games in September, in Pusan (South Korea). That month, both states agreed on the construction of railway and highway links, and on the de-mining of two points of the border that has separated them since 1953.

[33] In February 2003, Roh Mooh Hyun, a human rights lawyer, took office as President. He stated his desire to continue the rapprochement with Pyongyang and to achieve greater

independence from US foreign policy.

[34] In April, parliament supported sending non-combat troops to Iraq (300 engineering and medical military personnel) to cooperate with the US in that country's reconstruction. Later on, the opposition and the press accused close presidential aides of corruption.

[35] The crisis between the US and North Korea (labeled by US president George W Bush as one of the 'axis of evil' countries) affected South Koreans because of the impact on them of any possible confrontation. Demonstrations for peace and against the US policy took place. Pyongyang accepted Seoul's offer to mediate in the conflict.

[36] In September 2004, the Government admitted that South Korean scientists had secretly carried out an experiment to enrich uranium, without official support or knowledge. The Government denied any intention to develop a nuclear weapons program and assured it had been a one-off event.

[37] The main opposition force, the Grand National Party (GNP), scored a landslide victory in parliamentary by-elections held in October 2005, winning all four seats at stake in Daegu, Ulsan, Bucheon and Gwangju, which left it with 127

seats in Parliament. The ruling party suffered a huge defeat, failing to win a single seat.

[38] In February 2006, Seoul and Washington held talks on a free trade agreement to be signed by the end of the following year. Once signed, it would be the largest free trade deal involving the US in Asia. Seoul regarded the negotiations as the most important event since the Korean War, when a military alliance was signed with Washington.

[39] In June 2006, in a joint celebration held in the city of Kwangju to mark the sixth anniversary of the landmark inter-Korean summit, former president Kim Dae Jung pointed out that reunification was the final goal.

[40] That month, after a long period marked by tension, South Korea and Japan began talks to solve a dispute over the Dokdo islands. They hoped to reach a reasonable deal for both nations, drawing the maritime boundaries for their exclusive economic zones based on international law.

[41] Following North Korea's nuclear test explosion in October 2006, South Korea announced in retaliation that some Northern officials would be banned from the South under new travel rules and that Seoul would tightly control inter-Korean trade. ∎

Kuwait / Al Kuwayt

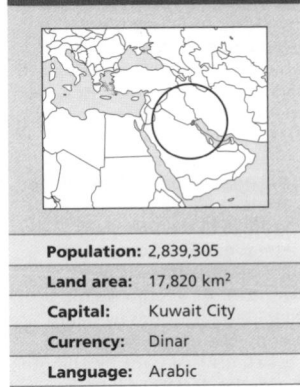

Population:	2,839,305
Land area:	17,820 km²
Capital:	Kuwait City
Currency:	Dinar
Language:	Arabic

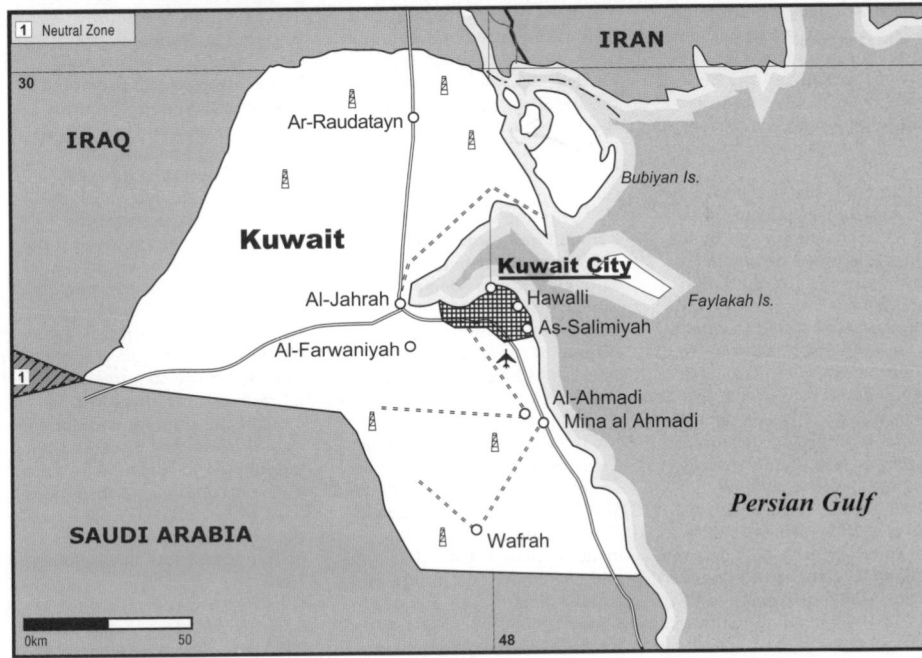

The history of the territory currently known as Kuwait was always intimately linked to that of the Mesopotamian civilizations (see Iraq) but, beginning in the 13th century, after the Mongol invasion caused the collapse of the Caliphate, the region entered a long period of isolation.

[2] In the 16th century several clans from the Al Aniza tribe migrated toward the Northern shore of the Persian Gulf from Najd, their homeland in Central Arabia, seriously affected by hunger. The clans settled in the current region of Qatar for more than 60 years. Later on they resettled in various areas, including what is now Kuwait.

[3] In the 18th century, although nominally subject to the Ottoman Empire, the local Arab groups were virtually independent. They decided to elect a *shaij* (sheikh) to conduct sporadic negotiations with the Turks. In 1756, the leader of the Anaiza tribe, Abdul Rahim al-Sabah, founder of the current reigning dynasty, was appointed to this task.

[4] At that time, the place, formerly known as the Qurain (horn), began to be called Kuwait, a diminutive for al-Kout, which is a local Arabic term to describe the fortified houses on the coast.

[5] The Ottoman Empire, in alliance with the Sabah family, ruled the country until the 19th century. But when the Ottomans threatened to annex Kuwait in 1899, the Sabahs requested protectorate status from Britain. In exchange, Britain guaranteed Kuwait's territorial integrity. This treaty and the subsequent British military presence in the region frustrated Turkish attempts to expand the Berlin-Baghdad railway to reach the Gulf through Kuwaiti territory.

[6] At the end of World War I, France and Britain divided up the remains of the Ottoman Empire. Kuwait was now considered a British protectorate, separate from the newly created kingdom of Iraq, which claimed it as a province, invoking the historical domination of the region by the Government of Baghdad.

[7] In 1938, oil began to flow in Kuwait. After World War II, Emir Ahmad Jabet al-Sabah granted the concession to the Kuwait Oil Company, owned by British BP and US Gulf, and oil was first exported in 1946.

[8] In 1961, independence was negotiated. Emir Sabah proclaimed himself Emir and took up full powers. Iraq refused to recognize the new state, claiming that it was an artificial creation of the British to maintain access to oil. Consequently, the British troops remained to defend the emirate until they were replaced by the troops of the Arab League.

[9] In 1962 a Constitution was adopted creating a National Assembly of 50 members, elected individually by male citizens over the age of 21, whose fathers or grandfathers had resided in Kuwait before 1920.

[10] Oil entirely changed the country. The Bedouins replaced their camels with luxurious air-conditioned cars. Pearl fishing - the main economic activity at the time - disappeared. The entire population settled in brand new cities, where stylized mosque towers stood side-by-side with shopping centers which replaced the old *souks* (markets). The population's educational standards and life expectancy rose. All manual labor and work in the oil industry was done by immigrant workers, who by 1985 outnumbered the Kuwaiti population by almost two-to-one.

[11] Kuwaiti rulers became concerned that so much prosperity in such a poor region could put its legitimacy into question. In 1961 the Arab Fund for Economic Development was created, in order to channel 'soft' loans and donations to Third World countries. When the Organization of Petroleum Exporting Countries (OPEC) succeeded in raising prices in 1973, Kuwait increased its revenue immensely.

[12] However, in contrast with other Persian Gulf countries, Kuwait was generous with its wealth. In the late 1980s it was the country with the largest official development aid in proportion to its grosss domestica product.

[13] Most of the Third World countries supported OPEC. They hoped to receive help to establish a 'New International Economic Order', in which they would obtain better prices for the other raw materials that they supply to the countries of the North. However, instead of investing their oil revenues in their own countries or in other Third World nations, the Gulf monarchs placed their fortunes in transnational banks. This added to the excess liquidity in these banks, which started to grant loans to the Third World quite indiscriminately. This situation was one of the main factors that provoked the 'debt crisis' in 1982.

[14] When the Iran-Iraq war broke out in 1979, Kuwait

LAND USE

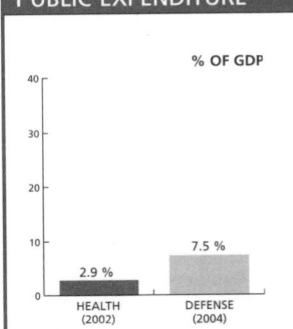

2003/2005

IRRIGATED AREA: 72.2% of arable land

- FOREST AND WOODLAND: 0.3%
- ARABLE LAND: 0.8%
- CROPLANDS: 0.2%
- OTHER USE: 98.7%

PUBLIC EXPENDITURE

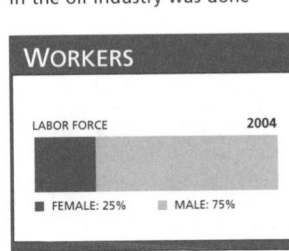

% OF GDP

2.9 % HEALTH (2002)

7.5 % DEFENSE (2004)

WORKERS

LABOR FORCE   2004

- FEMALE: 25%   - MALE: 75%

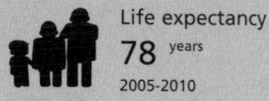
Life expectancy
78 years
2005-2010

GNI per capita
$22,470
2004

Literacy
83% total adult rate
2000-2004

officially remained neutral, but in fact supported Iraq with large donations and loans. Like the rest of the Gulf Security Council monarchies, Kuwait considered Iraq to be a 'first line of defense' against the Iranian Islamic revolution.

[15] In 1981, the new Prince, Jaber Al-Sabah, called for national elections. Forty of the 50 seats at stake belonged to candidates loyal to the ruling family. Only 6.4 per cent of the population qualified to vote.

[16] In the late 1980s, the Kuwait Investments Office (KIO) had capital assets outside the country estimated at $1 billion, which included hotels, art galleries, European and US real estate, and major shares in large transnational corporations: 10 per cent of British Petroleum, 23 per cent of Hoechst, 14 per cent of Daimler-Benz, and 11 per cent of the Midland Bank.

[17] In 1987, alleging that Iraq was using the port of Kuwait to export oil and import weapons, the Iranian navy attacked Kuwaiti merchant ships. In response, Kuwait requested and obtained permission from the major powers - US, France, Britain, and the USSR - to use their flags for the Kuwait merchant navy. The US and Britain sent their navies to protect Kuwaiti ships in the Gulf.

[18] Once the war between Iran and Iraq came to an end in 1988, tension with Iraq started to mount. Kuwait demanded the payment of $15 billion on account of war loans, which Iraq refused to repay, alleging that those sums had been used to protect Kuwait.

[19] Iraq accused Kuwait of 'stealing' the country's oil from the common deposits which lie under the border, and demanded $2.4 billion in compensation.

[20] On 2 August 1990, Iraq invaded Kuwait. Emir al-Sabah and his family took refuge in Saudi Arabia. Nearly 300,000 Kuwaitis fled the country. A pro-Iraqi provisional government, led by al-Hussein Ali, sought the merging of Iraq and Kuwait.

Ten days later, the Emirate was declared an Iraqi province.

[21] The US reacted strongly to the invasion and promoted a series of tough UN Security Council measures. On 6 August, the trading, financial, and military boycott of Iraq was voted through, and on 29 November, the use of force against Iraq was approved if the country refused to withdraw from Kuwait before 15 January 1991. Only Cuba and Yemen abstained from voting.

[22] The ensuing Gulf War devastated the country, not so much in terms of lives lost, for most of the fighting was done in Iraqi territory, but because the bombings and forced withdrawal of the occupation troops left most of Kuwait's oil wells burning and the country was unable to produce any oil until 1992.

[23] The Kuwait Institute for Scientific Research stated that 900 km of desert were affected by the traffic of military vehicles, which caused the land to shift, and frequent sandstorms.

[24] It was estimated that the clean-up and reconstruction cost between $150 and 200 billion. Kuwait became a debtor nation, owing $22 billion to the Allied countries.

[25] After the War was over, more than 1,300 people were killed by mines laid during the conflict.

[26] In March 1991 Amnesty International denounced extrajudicial killings, arbitrary detentions and torture of Palestinians, Jordanians and Iraqis living in Kuwait by Kuwaiti soldiers and civilians. Meanwhile, during the visit to the county by US Secretary James Baker, the Government promised to lead the Emirate towards democracy.

[27] In June 1991, Emir Jabir al-Sabah called a National Council to discuss the elections, and female and foreign suffrage. The opposition was still demanding the re-establishment of the 1962 constitution and the formation of a democratic parliament, though no opposition leaders were allowed on the Council.

[28] When the war ended, the Emir launched the slogan of 're-Kuwaitizing' the country to drastically reduce the number of foreigners. Over 300,000 Palestinians were expelled.

[29] In mid-December 1991, Saudi Arabia, the United Arab Emirates, Oman, Qatar, Bahrain, and Kuwait held a summit aimed at creating a collective security mechanism, and establishing a new defensive framework for the region which holds 40 per cent of the world's oil reserves.

[30] After resuming oil production, Kuwait initially accepted the OPEC quota of two million barrels per day. However in February 1993 it confronted OPEC's other members and unilaterally decided to raise daily production to 2.16 million barrels.

[31] In 1993 the UN finally determined the frontier with Iraq, despite protests by Baghdad. As a response to supposed incursions by Iraqi troops, the US bombed Iraq repeatedly (see Iraq). Washington decided to install missiles in Kuwait and it began building a wall 130 kms long equipped with mines along the new frontier.

[32] In August the ex-chief of the Temporary Free Government imposed by Baghdad, al-Hussein Ali, was condemned to death, as well as five Kuwaitis and 10 Jordanians accused of collaborating with the Iraqi occupation.

[33] The following year, the Government moved towards privatizing State companies and carried out major arms investments.

[34] In February 1996, Amnesty International again denounced summary executions, torture and deportations without trial.

[35] After several legislators threatened to throw out the Minister of Islamic Affairs for errors printed in 120,000 copies of the Qur'an, the Emir dissolved Parliament in May 1999 and called legislative elections in July.

[36] The burning of an oil refinery in early 2002 led to harsh criticisms of the new Petroleum Minister, who was forced to resign. In Parliament, a series of debates on the matter ended in denunciations of corruption, incompetence and nepotism, which for politicians and analysts alike underscored the Government's operational problems. Given the context, the candidates proposed for the ministerial seat refused the offer, paralyzing decision-making at the Petroleum Ministry. The Information Minister, Sheikh Ahmad Fahd al-Sabah, eventually took over the Ministry until the 2003 elections.

[37] After the 11 September 2001 attacks against the US, the Emir expressed his support for the US led anti-terrorism coalition. It was later revealed that the spokesperson for al-Qaeda, Suleiman Abu Ghaith, was a Kuwaiti citizen, and that other Kuwaitis were held at the US military base in Guantanamo, Cuba. In January 2002, US Treasury Secretary Paul

---

PROFILE

ENVIRONMENT

Nearly all the land is flat, except for a few ranges of dunes. The inland is desert with only one oasis, the al-Jahrah. The coast is low and uniform. The city and port of Kuwait is located in the only deep-water harbor. The hot climate is tempered by ocean currents but the temperature is high in summer. Winters are warm with frequent dust and sand storms. Petroleum is the main economic resource, with three refineries in Shuaiba, Mina al-Ahmadi and Mina Abdulla.

SOCIETY

Peoples: Kuwaitis, of Arab descent, account for less than half of the population, 60 per cent of which is made up of immigrant workers from Palestine, Egypt, Iran, Pakistan, India, Bangladesh, the Philippines and other countries.
Religions: Muslim 85 per cent, of which Sunni account for 70 per cent and Shi'a 30 per cent; other (mostly Christian and Hindu) 15 per cent.
Languages: Arabic (official).
Main Political Parties: There are no legal parties. In the National Assembly are represented: the Islamic Constitutional Movement, a moderate Sunni group; the Kuwaiti Democratic Forum, liberal; the Salafeen, a fundamentalist Sunni group, and the National Islamic Alliance (Shi'a).
Main Social Organizations: Federation of Unions, founded in 1967 with 12,000 members. The Kuwait Students' National Union. Kuwaiti Women's Cultural and Social Society.

THE STATE

Official Name: Dawlat al-Kuwayt (State of Kuwait). **Administrative Divisions:** 5 governances.
Capital: Kuwait City (Al-Kuwayt) 1,222,000 people (2003).
Other Cities: As-Salimiyah 142,700 people; Hawalli 90,100; Al-Farwaniyah 58,100 Al-Jahra 16,000 (2000).
Government: Constitutional Monarchy. Sabah al-Ahmad al-Jabir al-Sabah, Emir since January 2006. Nasser Muhammad al-Ahmad al-Sabah, Prime Minister appointed by the Emir in February 2006. The 65-seat National Assembly has 50 members elected for a four-year term, and 15 seats for appointed cabinet members.
National Holiday: 25 February, National Day (1950).
Armed Forces: 15,500.
Others: The National Guard has 6,600 troops (paramilitary).

Under-5 mortality
12 per 1,000 live births
2004

Malnutrition
10% under-5s
1996-2004

Maternal mortality
5 per 100,000 live births
2000 ■

O'Neill visited Kuwait to ask the Government to block all sources of financing for al-Qaeda.

[38] The US, UK and coalition forces invaded Iraq in March 2003, amid strong international controversy. The Arab League unanimously condemned the invasion, with the exception of the Kuwaiti delegation.

[39] According to the Amnesty International 2003 annual report, Kuwait sentenced to death four men, one of them found guilty of murdering a journalist in March 2001. The other three, a Kuwaiti and two Saudis, were convicted of raping and killing a 6-year-old *bidun* (stateless) girl in May. The three testified at the beginning of the trial that they had confessed under torture. In spite of this, the court accepted the confessions. The four were hanged in public and the bodies left on display for some minutes.

[40] In the National Assembly elections of July 2003, Islamic groups won 21 seats, government supporters 14, independents 13 and Liberals 3.

[41] In January 2004, the Government charged former Iraqi leader Saddam Hussein and his close aides of having committed more than 200 war crimes during the 1990-1991 occupation. Kuwait also announced it was willing to negotiate a significant reduction of Iraqi debts.

[42] In February, Japanese soldiers arrived in Kuwait to help with humanitarian activities in Iraq: the first Japanese troops to enter a conflict area sinceπ World War II.

[43] After meeting US envoy, former Secretary of State James Baker, Kuwait announced it was prepared to waive a 'significant proportion' of the Iraqi debt. Iraq's debt amounted to $16 billion and, with this gesture, Kuwait aligned with other Gulf states that were also Baghdad creditors such as the United Arab Emirates and Qatar, which had cancelled most part of their debt.

[44] In May 2004, religious authorities issued a *fatwa* (religious edict) banning concerts with female singers as 'un-Islamic'. The ban followed an outcry over a Lebanese TV show, *Star Show*, which is based on a hit French TV show of the same name in which male and female teenagers from different Arab countries live together before competing in a talent contest.

[45] In the July 2003 elections to the National Assembly, Islamic groups won 21 seats, Government loyalists won 14, Independents 13 and Liberals 3.

[46] A scandal broke out in October

IN FOCUS

ENVIRONMENTAL CHALLENGES
The worst environmental damage was caused by the burning of oil wells during the Gulf War in 1991. Air and water pollution are problems, as well as a shortage of drinking water, and rapidly advancing soil erosion leading to desertification.

WOMEN'S RIGHTS
Women over 21 have been able to vote and be elected since 2005. In 2003, they made up 24 per cent of the labor force of 1 million people. Maternal mortality stands at 5 per 100,000 live births. Forty per cent of pregnant women are anemic, 95 per cent receive prenatal care and 98 per cent of births are attended by qualified staff.*

CHILDREN
Kuwait has made progress regarding the situation of children and women. The economic boom drawn from oil enabled the setting of the social and economic infrastructure needed for development, among which are the distribution and access to basic services such as education, health and energy. Kuwait, like other Gulf countries, has thus improved its social indicators relating to life expectancy, development, child care, etc., equalling in many cases those from industrialized Northern countries. Mortality rates among children are low: 10 per 1,000 live births for infants and 12 per 1,000 in children

under-5*. Ten per cent of under-5 children are underweight and 24 per cent have stunted growth. Between 94 and 99 per cent of children were immunized in 2002* against common diseases (polio, measles, diphtheria, and tetanus). More than 80 per cent of all children have access to primary education and 77 per cent to secondary education*. The authorities are planning to universalize preschool education, which at present covers only a small fraction of the population.

There are no reliable sources on domestic violence, abuse, or help for these problems.

MIGRANTS/REFUGEES
Immigrants can request 5-year residence permits, which grant them some advantages over illegal residents. In order to obtain Kuwaiti citizenship they must present a passport and other documentation from their countries of origin, which is difficult for those who have fled their homelands. The country does not recognize the presence of refugees, but tolerates that of foreign laborers.

Kuwait is not a member of the Office of the United Nations High Commissioner for Refugees (UNHCR), and lacks legislation and proceedings to handle requests or applications from refugees. In spite of this, in 1996 the Kuwaiti Government and UNHCR signed an agreement allowing the UN agency to provide aid to refugees, on its own or in conjunction with other organizations.

Kuwait received thousands

of Iraqi refugees during the Gulf War in 1991. Once the conflict was over they were deported to Iraq, accused of collaborating with the Iraqi regime during the occupation of Kuwait.

According to UNHCR estimates, in 2004 there were more than 100,000 refugees, mostly from Palestine, Iraq, Afghanistan and Somalia, unrecognized by the State.

INDIGENOUS PEOPLES/ ETHNIC MINORITIES
The Government denies the *Bidun* (see above) the rights and guarantees granted to Kuwaiti citizens. Mostly Bedouin, they come from the northern Arab provinces and arrived in Kuwait in 1920, when the present borders in the region were demarcated.

However, in May 2000, Parliament passed a law to allow some 2,000 *Biduns* the possibility of becoming citizens. The rest will have to obtain passports from their countries of origin to be able to reside in Kuwait.

A significant number of Arab and Iranian nomads living in coastal regions along the Gulf since the 1940s also do not have Kuwaiti citizenship.

DEATH PENALTY
Applicable to a wide range of offenses.

* Latest data available in *The State of the World's Children* and *Childinfo* database, UNICEF, 2006.

2005, when an MP - after deeming the Government 'chaotic' - asked for the resignation of Crown Prince sheik Saad al-Abdullah al-Sabah, the Emir's cousin, who had been hospitalized. Salem al-Sabah, Head of the National Guard, called for the creation of a committee to 'support the Emir's leadership'. Meanwhile the Emir, who had suffered a heart attack in 2001 and rarely appeared in public, stated his total confidence on Prime Minister Sheikh Sabah al-Sabah.

[47] In 2005, migrant workers - who make up most of the country's labor force - protested on several occasions for the non payment of their wages, arbitrary cutbacks in their salaires, illtreatment, inadequate living conditions, and the non renewal of their residence permits.

[48] In January 2006, Emir Jaber al-Sabah died at 80. Saad al-Abdalah

al-Sabah was named Emir amid controversy - some believed that, at 75, he was not fit to rule - that led him to resign a week before taking office. Finally, the emirate went to his younger brother, Sabah al-Ahmad al-Sabah.

[49] After the Foreign minister of Iran, Manuchehr Mottaki, visited Kuwait in May 2006, the Emir said - about-the conflict between Tehran and the US due to the former's nuclear program - that 'Iran has been and is the major central axis in the region and always handles matters with wisdom'.

[50] After rejecting several initiatives on the issue during the previous years, Parliament passed a bill recognizing the political rights of women, allowing them to vote and stand for office.

[51] The Emir closed Parliament in 21 May and called for

early elections due to a major disagreement between legislators and the Government over an electoral reform bill.

[52] In the parliamentary elections held in June, reformist candidates won 33 out of 50 elected seats in the National Assembly. In spite of the opposition's victory, the Emir appointed a new cabinet dominated by his own family. The election was the first in which women were allowed to take part, although no women were elected.

[53] Analysts considered that the reformist tilt of the voting could largely be accounted for by the participation of women, who made up 57 per cent of the electorate because military personnel were denied the vote. The woman candidate who came closest to being elected, Rula Dashti, came fourth in Kuwait City's tenth district. ■

Kyrgyzstan / Kyrgyzstan

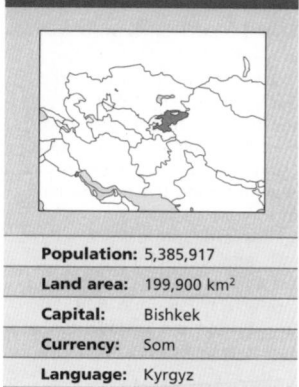

Population:	5,385,917
Land area:	199,900 km²
Capital:	Bishkek
Currency:	Som
Language:	Kyrgyz

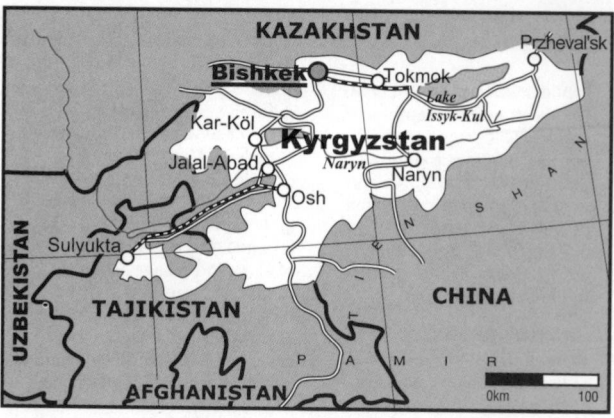

B eginning in the 3rd century BC or earlier, groups of nomadic and shamanistic Kyrgyz herding people settled on the steppes in eastern Siberia (to the north of present-day Mongolia).

² As they began to move towards the west, starting in the 1st century BC, the Kyrgyz merged with the Tashtika - a mix of Asian and European descendants - a process that continued through the 5th century. Between the 6th and 8th centuries, the Kyrgyz populated the land where the Yanisey emerged, in central Siberia.

³ Between the 6th and 10th centuries, the Kyrgyzstan region remained under the control of different Turkish groups. In 751, the area was the scene of a battle in which Turks, Arabs and Tibetans joined forces to expel the vast army of the Chinese Tang Dynasty (618-907) from Central Asia.

⁴ In the 9th century, the Kyrgyz set out on a journey that lasted four centuries. They occupied Tula territory (in present-day Russia) temporarily and returned to Yanisey in the 12th century. After that, they began slowly to descend into the Yanisey valleys, and reached the territory of modern-day Kyrgyzstan in the 15th century.

⁵ In 1207, the Kyrgyz surrendered to Mongol Jochi (Genghis Khan's son) to avoid falling under the dominion of Islamic (Arab and Turkish) peoples

spreading from southwestern Asia. Until then they had managed to escape the Mongol conquest, thanks to their mobility and ability to adapt to the cold, treeless steppes.

⁶ In the 16th and 17th centuries, the Kyrgyz occupied both sides of the Tien Shan mountains (the area of the present-day frontier between Kyrgyzstan, China and Kazakhstan), provoking constant attacks from Zungaros (Mongols from the northeast) starting in 1703. Many Kyrgyz fled to the Kyrgyzstan region. The clashes ended in 1757 when the Manchu Empire under the Ching Dynasty (1644-1911/12) gained formal, but not effective, control over the Tien Shan mountains.

⁷ Between 1835 and 1830, the Kyrgyz people were subdued by the Khan of Kokand (from Uzbekistan, to the west of Kyrgyzstan) who founded the city of Bishkek (the present-day capital). As a

consequence, Islam began to be adopted, contributing to the emergence of a kind of national Kyrgyz culture.

⁸ Two groups from Tien Shan became entangled in a fratricidal war that lasted from 1835 to 1858. Both groups were supported alternately by the Khanate of Kokand Khan and the Russians. In 1855, the Russian army intervened.

⁹ The mass immigration of Russian serfs after the abolition of serfdom in 1861 was the main cause of the displacement, between 1862 and 1872, of one-third of the Kyrgyz to the Tien Shan mountains. Russian peasants escaping famine set fire to Kyrgyz towns, appropriated their pasture land and applied their own agricultural techniques. In 1875, the Russian Empire annexed the entire territory and granted the colonists legal title to the land.

¹⁰ Tension between the Russians and Kyrgyz over land ownership

and the imposition of compulsory military service triggered a revolt that began in 1916 and continued for years after the October Revolution of 1917. Before that, many Kyrgyz sought refuge in China.

¹¹ In December 1917, the Bolshevik government opened a regional assembly in Bishkek (which was renamed Frunze in honor of a Red Army general). In 1926, the present borders were drawn up, and Kyrgyzstan was granted the status of Autonomous Soviet Socialist Republic, as part of the USSR.

¹² In 1921, the USSR began to collectivize land and the industrial production of the region's raw materials such as cotton, wool, leather, tobacco, fish, wood, water and metal. Factories were built for processing antimony and other metals. Planned production, tight social control and persecution put an abrupt end to the nomadic Kyrgyz way of life.

¹³ From 1926 to 1959, the Kyrgyz proportion of the population shrank from 60 to 40 per cent. Russians, Ukrainians and Germans who had been captured by the Red Army in 1941 immigrated to Kyrgyzstan.

¹⁴ Industry underwent further expansion, with the production of machinery, electricity and construction materials, although the modernization and improvement of the economy failed to ease the tensions.

¹⁵ In the early 1990s, the Kyrgyz Communist Party (PCK) opposed the legalization of non-Communist parties promoted by Moscow. In August 1991, during the attempted

LAND USE

2003/2005

IRRIGATED AREA: 78.5% of arable land

[pie chart]

- FOREST AND WOODLAND: 4.5%
- ARABLE LAND: 6.8%
- CROPLANDS: 0.3%
- OTHER USE: 88.4%

PROFILE

ENVIRONMENT
Located in the northeastern part of Central Asia, Kyrgyzstan lies in the heart of the Tian Shan mountain range. It is bounded by China and Tajikistan in the south, Kazakhstan in the north, Uzbekistan in the west, the Pamir and Altai mountain ranges in the southwest and the Tian Shan range in the northeast. It has plateaus and valleys: in the north, the Chu and Talas valley; in the south, the Alai Valley; and in the southwest, the Fergana Valley. The climate is continental, with sharp contrasts between day and night temperatures. The eastern part of Tian Shan is dry, while the southwestern slopes of the Fergana range are rainy. The main rivers are the Narym and the Kara-Suu. Lake Issyk-Kul is the most important of the country's lakes. In the mountains there are forests and meadows, while desert and semi-desert vegetation abounds at lower altitudes. Metal deposits include lead and zinc; in addition, there are large coal reserves and some oil and natural gas deposits.

SOCIETY
Peoples: Kyrgyz, 52.4 per cent; Russians, 21.5 per cent; Uzbeks, 12.9 per cent; Ukrainians, 2.5 per cent, Germans 2.4 per cent.
Religions: 75 per cent Muslim (Sunni); 6 per cent

Christian (Russian Orthodox Church) and others.
Languages: Kyrgyz (official); Russian (co-official); Uzbek; Dungan; Ukranian and others.
Main Political Parties: Union of Democratic Forces; Party of Communists of Kyrgyzstan; Ate-Meken Socialist Party.
Main Social Organizations: Council of Free Trade Unions, Kyrgyz Committee on Human Rights, Movement for the National Democratic Union, Business Union.

THE STATE
Official Name: Kyrgyz Respublikasy (Kyrgyz Republic)
Administrative Division: 6 provinces and 1 city (Bishkek).
Capital: Bishkek 806,000 people (2003).
Other Cities: Osh 217,000 people; Jalal-Abad 73,200; Kara-Köl 66,900; Tokmok 61,800; Naryn 41,700 (2000).
Government: Chief of State: Kurmambek Bakiyev, President since July 2005. Feliks Kulov, Prime Minister since September 2005. There is a two-chamber Supreme Council: the Legislative Assembly (60 members) and the People's Representatives Assembly (45 members).
National Holiday: 31 August, Independence Day (1991).
Armed Forces: 10,900 troops.

coup against Gorbachev staged by the conservative wing of the Communist Party, Askar Akayev, a prominent liberal member of the Kremlin, supported Gorbachev (as did Boris Yeltsin and others). Akayev was considered one of the brains behind the political reforms undertaken by the USSR in the 1980s. A few days later, Akayev was appointed President of the Kyrgyzstan Federal Republic.

[16] In August 1991, Kyrgyzstan was proclaimed an independent, democratic and republican State by its Parliament (or *Soviet*) from which the PCK was excluded.

[17] The PCK was banned until after the 1991 presidential election in which Akayev's presidency was confirmed unopposed. In December, Kyrgyzstan and ten other former USSR republics signed the founding charter for the Commonwealth of Independent States (CIS), which established, among other conditions, the stationing of joint military forces in each republic.

[18] In 1992, after pledging to preserve democracy - the ban on the PCK was lifted that year - and to modernize the economy, still dependent on Russia through the CIS, Akayev pushed for membership of the UN, IMF and the Organization for Security and Co-operation in Europe. In July, a package of stringent economic measures was adopted in agreement with the IMF, which entailed the privatization of industry and land ownership, as well as the liberalization of the financial and banking system, to develop a fully functioning free market economy.

[19] Exacerbated by the lack of support from the CIS, Akayev's economic policy led to a significant loss of jobs and gave rise to disputes in parliament. In 1993, Parliament approved the first Kyrgyz Constitution, which established that legislative elections would be held in 1995.

[20] In 1995, the President attempted to extend his term through a referendum, but Parliament forced him to run in elections in December. Akayev was re-elected, taking more than 60 per cent of the vote. In another referendum in 1996, a broad majority of voters approved constitutional amendments that gave him wider powers. As a result, opposition members and journalists critical of the Government were hounded, jailed for slander and even tortured.

[21] In December 1999, the entire cabinet resigned after Akayev and the National Security Council held it responsible for rising inflation, fiscal deficit and currency devaluation.

[22] In February and March 2000, parliamentary elections were

IN FOCUS

ENVIRONMENTAL CHALLENGES
Water pollution caused by mining industry waste is a major concern since one-third of the population is supplied directly from rivers, streams or wells. In the meantime, the salinization of soil is increasing due to the use of salt-water for irrigation.

WOMEN'S RIGHTS
Women have been able to vote and stand for election since 1918. In 2005, they had no representation in Parliament. In 2003, women made up 47 per cent of the country's workforce of two million. The majority (53 per cent) worked in agriculture, 38 per cent in services and the remaining 9 per cent in industry.

Ninety-seven per cent of pregnant women receive prenatal care and 98 per cent of births are attended by qualified medical personnel*. Violence against women, including domestic violence, is a serious problem.

According to Interior Ministry statistics, 300 women are murdered each year, although unofficial figures are much higher. Most of the crimes against women go unreported due largely to cultural traditions and psychological pressure.

After Kyrgyzstan gained independence from the Soviet Union in 1991, 'bride abduction' became a frequent practice, especially in the southern part of the country. Each year, between 10 and 30 women are estimated to be kidnapped and forced into marriage.

CHILDREN
The infant mortality rate stands at 58 deaths per 1,000 live births*. Seven per cent of newborn babies are underweight and 25 per cent of children under-five years old have stunted growth*. Approximately half of the deaths of babies under two are due to tuberculosis.

In 2004*, the literacy rate stood at 99 per cent*; primary education is obligatory and parents are fined or even sentenced to a year of forced labor if they fail to send their children to school.

From 1999 to 2003, 36 people were prosecuted for involvement in child sex exploitation and pornography, and ten were convicted of trafficking in children.

INDIGENOUS PEOPLES/ ETHNIC MINORITIES
Russians came to the region as representatives of the Russian Empire in 1861 to supervise the colonization of Central Asia. Russians complain of discrimination. Since the break-up of the Soviet Union, thousands of Russians have left Kyrgyzstan, mainly for Russia and other former Soviet Republics.

Uzbeks are concentrated in the southern regions of Osh, Batken and Jalal-Abad. In 1990, there were armed clashes between Uzbeks and Kyrgyz in the region of Osh. Since 2002, the Kyrgyz Government has been working to grant Uzbeks more participation in politics and to ease tension among the country's three main ethnic groups (Kyrgyz, Russians and Uzbeks).

MIGRANTS/REFUGEES
In 2005, 250 Kyrgyz had fled the country and sought refuge in Sweden (190) and the Czech Republic (60). That year, Kyrgyzstan received 4,200 refugees and asylum-seekers coming from Tajikistan, Afghanistan, Chechnya and China. Those from China were Uighurs, a Muslim Turkish people.

Refugees recognized as such by the Government are allowed to live in the country indefinitely, enjoy labor rights, and are eligible for Kyrgyz identity documents. After the 11 September 2001 terrorist attacks on the US, a number of Afghans were arrested in Kyrgyzstan, including a few who had already been granted official refugee status. Under bilateral arrangements with the Chinese Government, Kyrgyzstan does not grant asylum to refugees from China.

DEATH PENALTY
Applicable even to common crimes.

* Latest data available in The State of the World's Children and Childinfo database, UNICEF, 2006.

held in which the PCK won the majority of votes, followed by the Union of Democratic Forces. International observers and members of the Kyrgyz Human Rights Committee were forced to leave Kyrgyzstan after suffering reprisals for protesting the arrest of parliamentarian Feliks Kulov. He had been arrested to keep him from participating in the second round of the elections.

[23] In September 2001, the Government allowed the installation of a US air base at Manas airport in Bishkek, saying it was co-operating in the war on Islamic terrorism. In September 2003, Akayev agreed with Moscow to open a rapid reaction military base in Kant, 30 km from Manas, for the same purpose.

[24] In 2002, the imprisonment of opposition leader Azimbek Beknazarov, the 10-year sentence handed down to Kulov, the increase in deaths during crackdowns on street demonstrations and international organizations' constant criticism of human rights violations, created a situation that led to Akayev's isolation.

[25] In May 2004, Colonel Chynybek Aliyev, head of the anti-corruption department, was murdered. Aliyev led the division in charge of fighting organized crime and had carried out several inquiries into contract killings.

[26] Thousands of people blocked streets and roads and took over official buildings in February 2005 in support of two candidates who had been barred from standing in the elections. Numerous candidates had been disqualified for having allegedly violated election campaign rules.

[27] In Bishkek, protesters broke into the presidential palace while Akayev fled the country for Russia. Opposition leader Kurmanbek Bakiyev was appointed as interim president.

[28] In the July 2005 elections, Bakiev was finally elected President. The economy posed one of the main challenges, with 40 per cent of the population living below the poverty line. Another challenge was to set up an administration that would reflect the different ethnic and regional groups.

[29] At the end of April 2006, opposition members convened between 10,000 and 15,000 people outside the government buildings in Bishkek to demand constitutional and judicial reforms, press freedom and punishment for corrupt officials. Other claims were focused on the resignation of the President's chief of staff, the State Secretary and the General Procurator of the Republic.

[30] On 2 May, 13 of the 15 Kyrgyz ministers handed in their resignations - with the exception of the Culture and Transport ministers - after a parliamentary resolution calling the Government's performance unsatisfactory. However, Bakiyev refused to accept their resignations and stressed his full support for cabinet members.

[31] In November 2006, after eight days of mass protests in Bishkek's main square, Bakiyev agreed to give up some of his power and give more authority to parliament.■

Laos / Lao

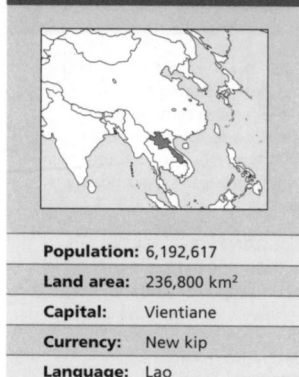

Population:	6,192,617
Land area:	236,800 km²
Capital:	Vientiane
Currency:	New kip
Language:	Lao

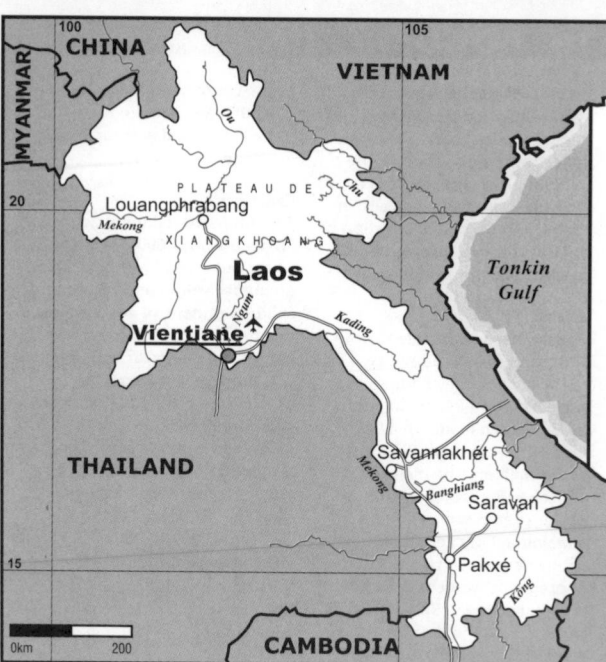

The earliest inhabitants of Laos were migrants from southern China. From the 11th century onward, parts of Laos fell under the Khmer Empire (in the region now known as Cambodia), and later under Siamese influence from the Sukhothai dynasty. With the fall of Sukhothai in 1345, the first kingdom of Laos emerged under Fa Ngum, a Burmese prince brought up in the court of Angkor Wat. As the Khmer Empire crumbled, Fa Ngum welded together a new empire, founding the flourishing state of Lang Xang, the 'Land of a Million Elephants'. In the late 18th century, the country split into three regions: Champassac, Vientiane and Luang Prabang. In the early 19th century, the Thais had established dominion over those territories. Tiao Anuvong, Prince of Vientiane, led an ill-fated nationalist rebellion in 1827. In 1892, the French invaded and by 1893 had established a protectorate over Luang Prabang, with the rest of the country becoming part of French Indochina.

[2] During World War II, Japan occupied Laos and a pro-independence movement arose in Vientiane, led by three princes, Phetsarat, leader of the Free Laos government-in-exile, Suvana Fuma, leader of the National Progressive Party, and Tiao Sufanuvong, head of the Neo Lao Issara (Laos National United Front). In September 1945, they set up a provisional government and declared the independence of Laos.

[3] In 1946, the country was again under French occupation. The Pathet Lao provisional government sought refuge in Bangkok, where the Pathet leaders organized the anti-colonialist struggle through the Lao Issara or 'free Laos' movement.

[4] On 19 July 1949, the Franco-Laotian Convention recognized Laotian independence 'as part of the French Union'. The Pathet Lao leaders saw this as mere formal independence. Suvana Fuma opted for negotiation, but the opposition coalition decided on active resistance; their military victories led to a new treaty in 1953.

[5] Differences between the Lao Issara, Suvana Fuma government, and the Pathet Lao were reconciled in November 1957. An agreement was reached, whereby the Pathet Lao would participate in the political life of the country under the name 'Neo Lao Haksat' (Laotian Patriotic Front), led by Tiao Sufanuvong.

[6] In 1958, Suvana Fuma's ruling party joined forces with the Independent Party to form the so-called Laotian People's Demonstration, together achieving a small majority in Parliament.

[7] The 1958 elections were won by the left. A center-left coalition government was formed, led by 'neutralist' Suvana Fuma, with Sufanuvong as planning minister.

[8] Opposition from the US and its threat to cut off economic aid destabilized the Government, and in August the leaders of the Committee for the Defense of National Interests took over. The new Government, supported by the US, launched a military offensive against the Pathet Lao, forcing Sufanuvonghim to return to armed struggle.

[9] In late 1959, the army seized power, while the Pathet Lao resistance forces controlled the strategic northern provinces and the central Plain of Jars.

[10] General Fumi Nosavang and his troops seized Vientiane on 13 December 1960, driving out Pathet Lao troops, which had taken over the capital for two days.

[11] Thailand and the US supported the Revolutionary Anti-Communist Committee, headed by Fumi Nosavang and Prince Bun Um, which declared itself the legitimate government. On 20 December 1960, however, Princes Suvana Fuma and Sufanuvong signed a declaration in favour of forming a government of national unity.

[12] The 'neutralists' joined forces with the Pathet Lao. By the end of 1960, half of the country's territory was under the control of the Pathet Lao and a similar area had been taken over by 'neutralist' forces.

[13] In Geneva in 1961, Britain and the Soviet Union began negotiations for a peaceful solution, and in January 1962 a final agreement was signed to form a government of national unity.

[14] Growing US intervention resulted in the internationalization of the Vietnam War, which led to bombings in Laos. In nine years Laos was pulverized by more bombs than was the whole of Europe during World War II.

[15] The Pathet Lao declared an armistice in 1973, and the Vientiane Government formed a new cabinet, including Pathet Lao members, with a Council of Ministers headed by Suvana Fuma. The US defeat in Vietnam deprived the right-wing groups of the only support they had received.

[16] In 1975, the national unity cabinet gave way to a majority of Neo Lao Haksat ministers, and in

PROFILE

ENVIRONMENT

Laos is the only landlocked country in Indochina. The territory is mountainous and covered with rainforests. The Mekong river valley, stretching down the country from north to south, is suited to agriculture, basically rice. It is estimated that 40 per cent of the arable land was left barren as a result of the 25-year war. The climate is tropical and the lowlands are prone to disasters such as the 1978 flood and the 1988 drought.

SOCIETY

Peoples: Three-fifths of Laotians are descendants of the Lao ethnic groups who inhabit the western valleys. The inhabitants of the mountains account for more than one-third of the total population, while five per cent are of Chinese and Vietnamese origin.
Religions: Buddhist 57.8 per cent; traditional religions 33.6 per cent; Christian 1.8 per cent; Muslim 1.0 per cent; atheist/no religion 4.8 per cent; Chinese folk-religions 0.9 per cent.
Languages: Lao (official); French; English and many minor ethnic group languages.
Main Political Parties: The Lao People's Revolutionary Party (PPRL), in power with 97 of the 99 National Assembly's seats.

Main Social Organizations: Union of Lao People's Revolutionary Youth and the Association of Patriotic Women are mass organizations of the PPRL; Lao Students Movement for Democracy; the Chao Fah, formed by members of the Meo minority - or Hmong - maintains an insurgent group.

THE STATE

Official Name: Sathalanalat Paxathipatai Paxaxon Lao (Lao People's Democratic Republic).
Administrative Divisions: 16 provinces, one special zone (Xaisomboun) and one municipality (Viangchan).
Capital: Vientiane (Viangchan) 716,000 people (2003).
Other Cities: Savannakhét 154,900 people; Louangphrabang 116,000 (2000).
Government: Choummaly Sayasone, President since March 2006, elected by the National Assembly. Bousone Bouphavanh, Prime Minister since June 2006, appointed by the President. Unicameral Legislature: 99-member National Assembly.
National Holiday: 2 December, Proclamation of the Republic (1975).
Armed Forces: 29,100 troops (2002). Other: 100,000 members of the Self-defense Militia Forces.

Life expectancy
56 years
2005-2010

Literacy
69% total adult rate
2000-2004

HIV prevalence rate
0.1% of population 15-49 years old
2003

December a peaceful movement put an end to the monarchy. A People's Democratic Republic was proclaimed with Prince Sufanuvong as President, led by the renamed Lao People's Revolutionary Party (PPRL). Real power was wielded by Kaysone Phomvihan, the General Secretary of the Party. Entrepreneurs and state bureaucrats left the country en masse, ruining the economy and crippling the public administration.

[17] Within its two first years in power, the PPRL launched different policies, which included the collectivization of agriculture. More than 40,000 people were sent to 're-education' camps and nearly 30,000 were incarcerated for political reasons.

[18] The Government nationalized the banks and reorganized the public sector. Rice production rose from 700,000 tons in 1976 to 1.2 million in 1981, when self-sufficiency in grain was achieved for the first time. To gain access to the sea and to reduce dependence on Thailand, a road was constructed to the Vietnamese port of Danang, and an oil pipeline to Vietnam's refineries. The 1980s were characterized by an 'economic opening'.

[19] In October 1982, General Phoumi Nosavan, a 'conservative' who had been living in exile since 1965, formed an anti-communist government which was named the 'Royal Lao Democratic Government'. Soon after, several exiles and members of the resistance movement who belonged to the United Lao National Liberation Front (ULNLF, created in September 1980), withdrew from the Government and settled in southern Laos.

[20] In 1988 diplomatic relations were renewed with China, and in early 1989 the first co-operation agreements were signed with the US to combat the cultivation and trafficking of opium. By the late 1980s practically all political prisoners had been released and 're-education' camps had been closed.

[21] Upon the disintegration of the Soviet Union, all economic aid was suspended and bilateral trade was reduced by 50 per cent.

[22] In 1991, the economic crisis was exacerbated by floods and pest infestations in a quarter of the country's farmlands.

[23] In that context, Laos established closer ties with Thailand. Also in 1991, the two governments signed a Co-operation and Security Treaty. Thai investments were mainly concentrated in banking and trade.

[24] In December 1992, legislative elections were called; only the PPRL and a few independent government-authorized candidates took part. Scores of government opponents were imprisoned.

[25] Deforestation became a serious environmental problem. The timber felled expanded from 6,000 cubic meters in 1964 to more than 600,000 in 1993. That year, the Government restricted lumber exports. In March 1994 the World Bank granted Laos a loan for reforestation. International environmental groups criticized the project because it gave the funds directly to the Government, with little input from local communities.

[26] In April 1994, the 'Friendship Bridge' over the Mekong River was opened, uniting Laos and Thailand. This reconciliation represented a distancing from Vietnam and an integration with Thailand, a more prosperous nation.

[27] In early 1995, Thailand-based capital dominated investment in Laos. A law passed in March eliminated the last vestiges of the planned economy system. Another law updated labor legislation. However, the Government was determined to maintain its communist identity.

[28] In December 1997, the PPRL retained its dominance, winning 99 seats in the National Assembly. Just four of the 159 candidates were from outside the country's only party.

[29] In February 1998, the National Assembly elected Prime Minister Siphandon as the new President. Sisavat Keobounphan became the new Prime Minister.

[30] Burma/Myanmar, Thailand and Laos agreed in April 1999 to co-ordinate their fight against the production and export of drugs, especially opium. These three countries are known as the 'Golden Triangle' - the area is the world's leading producer of opium (which can be refined into heroin).

[31] A wave of bomb attacks shook the country in 2000-2001. According to the Government they were the work of the guerrilla Chao Fa group, or of anti-communist groups based abroad. However, several analysts suspected that the violence could be linked to friction within the regime itself.

[32] In his opening address before the PPRL congress in March 2001, President Siphandon said the party aimed to triple per capita income in Laos by the year 2020, though it recognized the failures in managing the country's fragile economy. A statute passed by the congress reaffirmed the Party's support for socialism and 'opposition to multi-party systems and political pluralism'.

[33] In September 2001, Japan granted a loan for the building of the second 'Friendship Bridge' over the Mekong, which started in early 2002 and would be finished in 2005.

[34] In the February 2002 elections, Khamtay Siphandone was re-elected Head of State. The ruling party confirmed its hold on power, since only one out of the 166 candidates was not a member of the LPRP.

[35] In November 2004, the World Bank approved loans for the construction of two hydroelectric dams in Nam Theun - in central and southern Laos - which would produce electricity for export.

[36] Two European journalists and an American translator were expelled from the country for 'interfering in Laos' internal affairs and obstructing Government policies' in June 2005. The deported people were members of the California-based Fact Finding Commission (FFC), who according to the Ministry of Foreign Affairs had distorted and prevented the implementation of policies in Laos, and also disseminated anti-government material. In practice, they had prevented the application of resettlement measures aimed at ethnic groups, encouraging people living in Hmong, in the northern province of Xieng Khouang, to cause public disorder. According to the FFC, the Lao Citizens Movement for Democracy (LCMD) had started a revolution in 11 provinces, but the Government dismissed it.

[37] The construction of Nam Theun's hydroelectric dams started in November 2005.

[38] In March 2006, President Khamtay Siphandone, aged 82, stepped down as leader of the PPRL - which has held power for 30 years. He was replaced by Vice-President Choummaly Sayasone. The move was announced at the end of the Party congress in Vientiane.

[39] In elections in April 2006, the PPRL won 113 of the 115 seats. It is the only legal political party; the other two seats went to independents. ∎

IN FOCUS

ENVIRONMENTAL CHALLENGES
The most pressing environmental problems are the heavy logging and the resultant dwindling of water supplies and loss of 70 per cent of natural habitats. Less than five per cent of the land is suitable for agriculture, though farming generates 80 per cent of employment. There are many unexploded mines in the countryside.

WOMEN'S RIGHTS
Laotian women have been able to vote and stand for office since 1958. In 2003, 23 per cent of seats in Parliament were held by women, but they held no ministerial positions.

That year, they made up 43 per cent of the workforce; of these 81 per cent worked in agriculture, 14 per cent in services and 5 per cent in the industrial sector.

The literacy rate among women over 15 years old reached 61 per cent, and among those aged between 15 and 24, it amounted to 75 per cent. Female net primary school enrollment rate was 78 per cent*. In 2004, only 27 per cent of pregnant women received prenatal care and just 19 per cent of births were attended by skilled health staff. Maternal mortality stood at 650 per 100,000 live births.

CHILDREN
In 2004, infant and under-5 mortality rates continued to be high although they had been slightly reduced since 1990. The former fell from 120 to 65 deaths per 1,000 live births while the latter was reduced from 163 to 83 per 1,000 live births. Fourteen per cent of newborns were underweight. Among children under five years old, 40 per cent were moderately or severely underweight and 42 per cent were moderately or severely stunted.

INDIGENOUS PEOPLES/ ETHNIC MINORITIES
The Hmong, also referred to as the Meo, constitute a population of 193,000 people, equivalent to four per cent of the total population. They come from the mountains and live in and around the Plain of Jars. Having been recruited and trained by the CIA for a secret war in the officially neutral Laos, the Hmong aided the US during the Vietnam War. The Hmong protected important US radar installations and also infiltrated North Vietnamese troops. According to UNHCR, 20,000 Hmong soldiers died in combat during the conflict, 50,000 civilians were wounded and 120,000 were displaced from their homes.

MIGRANTS/REFUGEES
Thailand is the destination of many Laotians who emigrate in search of work. Since 1975, over 200,000 Hmong have fled Laos as refugees. In August 2005, over 15,000 - who had spent years in Thailand - were resettled in the US. In August, nearly 6,500 were deported.

DEATH PENALTY
The death penalty is applicable even for common crimes.

*Latest data available in *The State of the World's Children* and *Childinfo* database, UNICEF, 2006.

Latvia / Latvija

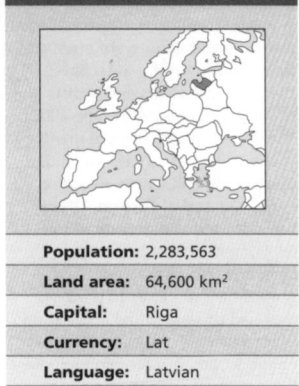

Population:	2,283,563
Land area:	64,600 km²
Capital:	Riga
Currency:	Lat
Language:	Latvian

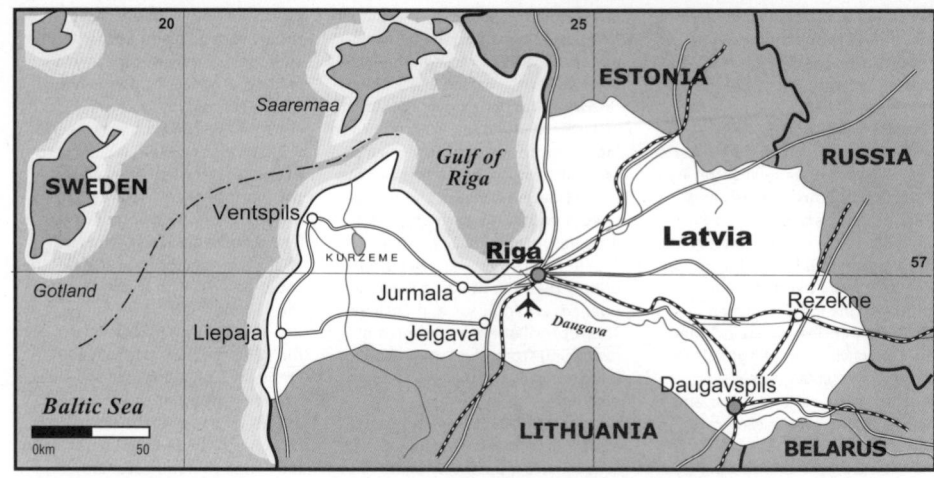

The first inhabitants of present-day Latvia were nomadic hunters, fishers and gatherers who migrated to the forests along the Baltic coast, after the last glaciers had retreated. Around 2,000 BC, these groups were replaced by the Baltic peoples, Indo-Europeans who began farming and established permanent settlements in Latvia, Lithuania and eastern Prussia.

[2] The ancient Baltic peoples had come into contact with the Roman Empire through the amber trade. This activity, which reached its peak during the first two centuries of the Christian era, was brought to a halt by Slav expansion toward central and eastern Europe.

[3] The Swedes and the Russians both claimed these lands during the 10th and 11th centuries and in the 12th century, German warriors and missionaries came to the Latvian coast. As it was inhabited at the time by the Livs, the Germans called it Livonia. In 1202, the bishop of the region, under authorization from Rome, established the Order of the Knights of the Sword (see Estonia).

[4] Before becoming the Knights of the Teutonic Order, in 1237, the Germans had subdued and converted the tribal groups of Latvia and Estonia to Christianity. The Teutonic Knights created the so-called Livonian Confederation, consisting of areas controlled by the Church, free cities and regions governed by knights.

LAND USE

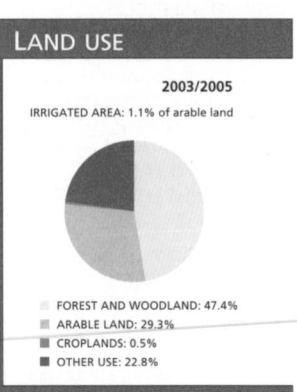

2003/2005

IRRIGATED AREA: 1.1% of arable land

- FOREST AND WOODLAND: 47.4%
- ARABLE LAND: 29.3%
- CROPLANDS: 0.5%
- OTHER USE: 22.8%

[5] When Russia invaded the region in 1558, to halt Polish-Lithuanian expansion, the Order fell apart and Livonia was partitioned. At the end of the Livonian War in 1583 Lithuania annexed the area north of the Dvina river; the south remained in Polish hands and Sweden kept the north of Estonia. In 1621, Sweden occupied Riga and Jelgava; Estonia and the northern part of Latvia were subsequently ceded to Sweden by the Truce of Altmark (1629).

[6] The region west of Riga, on the Baltic Sea, was organized into the Duchy of Courland, becoming a semi-independent vassal of Poland. In the mid-17th century, Courland became known as a major naval and trade center for northern Europe, and even had colonial aspirations.

[7] Sweden kept these territories until the Great Northern War, when it was forced to cede them to Russia under the Peace of Nystad. In 1795,

after the three partitions of Poland, Livonia was finally subdivided into three regions within Russia: Estonia (the northern part of Estonia); Livonia (the southern portion of Estonia and northern portion of Latvia) and Courland. The Russian Revolution of 1905 gave rise to the first expressions of Latvian national sentiment.

[8] The peasants revolted against their German feudal lords, and the Russian rulers. Although the rebellion was put down by czarist troops, it set the stage for the war of independence 13 years later. After the Russian Revolution of 1917, the Latvian People's Council proclaimed the country's independence on 18 November 1918. A government led by the leader of the Farmers' Union, Karlis Ulmanis, was formed.

[9] Far from having its desire for independence and sovereignty

respected, Latvia was attacked by German troops and by the Red Army. Only in 1920 was Latvia able to sign a peace treaty with the USSR, in which the latter renounced its territorial ambitions. In 1922, a constituent assembly established a parliamentary republic. The international economic crisis of the 1930s, and the polarization of socialists and Nazi sympathizers led to the collapse of the Latvian Government. In 1934, Prime Minister Ulmanis suspended parliament and governed under a state of emergency until 1938.

[10] With the outbreak of World War II, according to the secret Russo-German pact, Latvia remained within the USSR's sphere of influence. In 1939, Latvia was forced to sign a treaty permitting the Soviets to install troops and bases on its soil. In 1940, it was invaded by the Red Army, and a

PROFILE

ENVIRONMENT

Latvia's terrain is characterized by softly rolling hills (highest point: Gaizins, 310 m), and by the number of forests, lakes and rivers, which empty into the Baltic Sea and the Gulf of Riga, in the northeastern part of the country. Latvia has 494 kms of coastline. The country's most fertile lands lie in the Zemgale Plain, which is known as the country's breadbasket. The plain is located in the south, extending as far as the Lithuanian border. The highlands, which make up 40 per cent of the land, lie in the western and northern parts of the country, crossing over into Estonia. The climate is humid and cold, due to cold air masses coming from the Atlantic. Summers are short and rainy, with an average temperature of 17° C; winters last from December to March, with temperatures below zero, sometimes as low as -40° C. Two-thirds of all arable land is used for grain production, and the rest is pasture land. The main industries are metallurgical engineering (ships, automobiles, railway carriages and agricultural machinery), followed by the production of motorcycles, home appliances and scientific research equipment.

SOCIETY

Peoples: Latvians and Lithuanians constitute the two main branches of the Baltic Indo-European peoples,

with distinct languages and cultures that set them apart from the Germans and Slavs. Ethnic Latvians make up 57 per cent of the country's population, followed by Russians, 30 per cent; Poles, Belarussians, Ukrainians, Lithuanians and Estonians account for the remaining 13 per cent. **Religions:** The majority is Protestant (Lutheran), followed by Catholics. **Languages:** Latvian (official); Russian, Lithuanian and Polish.
Main Political Parties: New Era (JL), For Human Rights in United Latvia, People's Party (TP), Latvia First Party (LPP), Union of Greens and Farmers.
Main Social Organizations: Confederation of Free Trade Unions.

THE STATE

Official Name: Latvijas Respublika.
Capital: Riga 733,000 people (2003).
Other Cities: Daugavspils 113,600 people; Liepaja 94,100; Jelgava 70,700; Jurmala 58,900 (2000).
Government: Parliamentary republic. Vaira Vike-Freiberga, President, elected in July 1999. Aigars Kalvitis (TP), Prime Minister since December 2004. Single-chamber parliament (Saeima), made up of 201 members elected by direct vote.
National Holiday: 18 November, Independence Day (1918). **Armed Forces:** 6,950 troops (1996).

Life expectancy
73 years
2005-2010

GNI per capita
$5,580
2004

Literacy
100% total adult rate
2000-2004

HIV prevalence rate
0.6% of population 15-49 years old
2003

new government was formed, which subsequently requested that the republic be admitted to the USSR.

[11] During the German offensive against the USSR, between 1940 and 1944, Latvia was annexed to the German province of Ostland, and its Jewish population was practically exterminated. The liberation of Latvia by the Red Army meant the re-establishment of Soviet government. Before the Soviet forces arrived, 65,000 Latvians fled to Western Europe.

[12] In 1945 and 1946, about 105,000 Latvians were deported to Russia, and the far northeastern corner of Latvia - with its predominantly Russian population - was taken away from Latvia to form part of the USSR. In 1949, forced collectivization of agriculture triggered a mass deportation of Latvians, with about 70,000 being sent to Russia and Siberia. In 1959 the President of Latvia's Supreme Soviet, Karlis Ozolins, was dismissed because of his nationalist tendencies.

[13] Armed Latvian resistance to the Soviet regime was finally put down in 1952. Russian became the official language, and massive immigration of Russians and other nationalities began, in an attempt by Moscow to diminish the influence of the indigenous population.

[14] Until the 1980s, Latvian resistance was expressed in isolated actions by political and religious dissidents, which were systematically repressed by the regime; in addition, some nationalist campaigns were carried out by exiles. From 1987, the policy of *glasnost* (openness) initiated by Mikhail Gorbachev in the USSR, gave hope to Latvian aspirations, permitting public political demonstrations, and the reinstatement of the national symbols.

[15] In October 1988, close to 150,000 people gathered to celebrate the founding of the Popular Front of Latvia (LTF), which brought together all of Latvia's recently formed social and political groups, as well as militant communists. A month later, for the first time since Soviet occupation, hundreds of thousands of Latvians commemorated the anniversary of the 1918 declaration of independence. The LTF began to have influence with the local government, and with the Moscow authorities.

[16] A year later, the LTF Congress endorsed the country's political and economic independence from the USSR. Despite Moscow's resistance to Latvia's secession, LTF's policy of carrying out peaceful changes by means of public demonstrations, free elections and parliamentary decisions received widespread popular support. Latvia's 1938 Constitution went into effect once again, for the first time since the

IN FOCUS

ENVIRONMENTAL CHALLENGES
The most pressing challenges are improving the quality of potable water and the sewage disposal system. Toxic waste disposal and curbing air pollution are major concerns. The Government has committed itself to implementing all EU environmental directives by 2010.

WOMEN'S RIGHTS.
Women have been able to vote and stand for election since 1917. In 2003, women held 21 per cent of seats in Parliament and 24 per cent of ministerial positions. They made up 51 per cent of the country's one-million-strong workforce. Female unemployment reached 11 per cent. Since 1990, some women have been trafficked into prostitution via Lithuania en route to Western Europe. Sexual harassment in the workplace is reportedly common, even though it is illegal. Cultural factors tend to dissuade women from reporting the problem. Although labor and wage discrimination is banned, it is common.

CHILDREN
On joining the EU in 2004, Latvia's under-one and under-five mortality rates were higher than the bloc's average. Under-one mortality was 10 per 1,000 live births and under-five mortality was 12 per 1,000. Domestic violence and sexual abuse continued to be major challenges, in spite of Government/UNICEF programs. Some young people are recruited by criminal organizations for robberies or to be used as dealers.

INDIGENOUS PEOPLES/ ETHNIC MINORITIES
The population comprises the following ethnic groups: 57.4 per cent Latvian, nearly 30 per cent Russian, 4 per cent Belorussian, 3 per cent Ukrainian and 2 per cent Polish. Since 1990, discrimination against Russians has taken the following forms: a 'quota' for the naturalization of minorities, which hinders their access to political life; different rules on property ownership, which puts them at an economic disadvantage; and the restriction on receiving education in their own language. As a result, Russians have come together in their own non-governmental organizations and political parties

since 1994, to fight political discrimination and push for the right of their children to be educated in the Russian language.

MIGRANTS / REFUGEES
The difficulty of obtaining refugee status has been questioned by the UNHCR since 2002, when the Government decided to modify selection procedures and transferred the responsibility for reaching a decision on applications for asylum to the Interior Ministry. In the view of the UNHCR, the Latvian police lacks the training and experience for the task. After joining the EU, no significant increase was registered in terms of applications received. During 2005, only 18 were filed (11 adults and 7 minors).

DEATH PENALTY
In May 2003, the country took the first step toward abolishing the death penalty, signing protocol 13 of the European Convention on Human Rights. The last execution was carried out in 1996.

1940 Soviet occupation.

[17] On 4 May 1990 Latvia marked the Declaration of the Re-establishment of Independence, as well as the reinstatement of the 1922 Constitution. In September 1991, the new Council of State of the USSR, in its inaugural session, formally recognized the independence of the Baltic republics.

[18] A new Parliament elected in June 1993 appointed Guntis Ulmanis as President. The beginning of an economic liberalization process led to a sharp rise in unemployment. In 1994, the economy was still dependent on Russia, its main supplier of fuel and its chief market for exports.

[19] The September 1995 legislative elections did not reveal any clear winner since nine parties obtained between five per cent and 16 per cent of the vote. An agreement between the conservative National Block and two left-wing parties led to Andris Skele being appointed Prime Minister in December.

[20] The People's Party, with 21.2 per cent of the vote, beat the Union Latvia's Way, which garnered 18.1 per cent in the October elections.

[21] The New Era party, a new center-right party led by Einars Repse, won the October 2002 parliamentary elections and immediately began talks to

form a coalition with the other conservative forces.

[22] On 20 September 2002, Latvia's accession to the EU was submitted to referendum. Sixty-seven per cent voted in favor with 32.3 per cent voted against.

[23] In February 2004, Prime Minister Repse was forced to resign after the September 2003 break-up of the ruling coalition. Following the referendum approving Latvia's entry into the EU, three of the four parties in the coalition accused Repse of using 'extortion, threats and lies' to rule the country. President Vaira Vike-Freiberga appointed Indulis Emsis, of the Peasants Union Alliance-Green Party, as the new Prime Minister.

[24] Latvia was admitted to the North Atlantic Treaty Organization (NATO) and on 1 May joined the EU.

[25] In October, the new minority coalition government, led by Indulis Emsis, had its draft budget for 2005 rejected by Parliament. After Parliament voted against the budget, Emsis was forced to resign. By the end of November, Vike-Freiberga appointed Aigars Kalvitis as the new Prime Minister.

[26] US President, George W Bush, visited Riga in May 2005. That month, President Vike-Freiberga was the sole Baltic Head of State to attend Moscow's Victory Day celebrations.

[27] In June 2005, Parliament ratified the proposed EU constitution by 71 votes to five. Foreign Affairs Minister Artis Pabriks insisted that the EU Constitution was not dead in spite of the rejection by France and the Netherlands - two founding members. In October, the Interior Minister resigned after disagreements with the Government over the budget allocation for police and emergency services. In December, amidst criminal investigations being launched into his business dealings, Defense Minister and former Prime Minister Repse resigned.

[28] Latvia is now trying to re-establish the national language and identity after the years of Soviet occupation when Russian was compulsory.

[29] In June 2006, the Latvian Parliament defied the European Union by refusing to introduce a law banning discrimination at work on the grounds of sexual orientation. This was despite the law on employment discrimination being a condition of Latvia's acceptance as a member of the European Union in 2004. During the parliamentary debate, homosexuality was described as a sin. Latvia has changed its constitution so as to prevent same-sex marriages. ∎

Lebanon / Lubnan

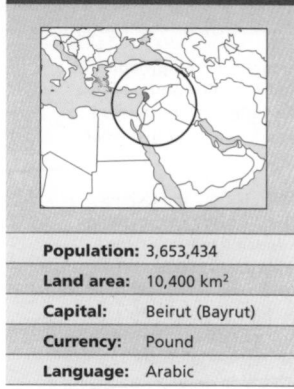

Population:	3,653,434
Land area:	10,400 km²
Capital:	Beirut (Bayrut)
Currency:	Pound
Language:	Arabic

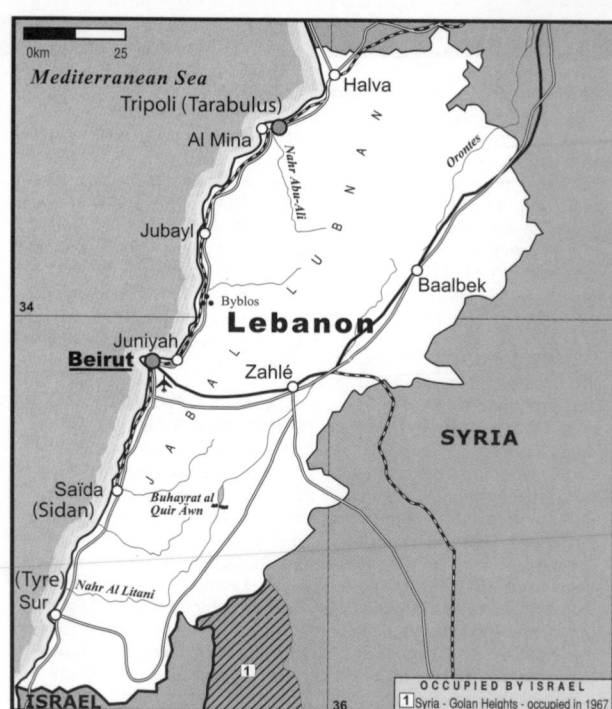

In 3000 BC, the Hellenes stated that the coastal areas they called Phoenicia were occupied by groups coming from the Persian Gulf. Byblos, the first Phoenician city, was founded between 3050 BC and 2850 BC. Its inhabitants established commercial and religious connections with Egypt as of the 25th century BC.

2 The total destruction of Byblos by fire in 2150 BC, was a consequence of an invasion by the Amorites (a Semitic people), who rebuilt the city and intensified ties with Egypt. In the 18th century BC, new invaders - called Hyksons (another Semitic people) - destroyed the Byblos Amorite government and the Egyptian Middle Kingdom (in 1720 BC).

3 After ousting the Hyksons in 1567 BC, Egypt began its imperial expansion, which created the conditions for the development of Phoenician commerce on a large scale. Under Egyptian guard, Phoenician merchants regularly distributed papyrus, ivory, gems, timber, horses, silk and other goods between the Orient and the Mediterranean Sea.

4 The Egyptian domination of Phoenicia declined during the reign of Ramses III (1187-1156 BC), upon the invasion of Syria by groups from Asia Minor and Europe. Between the withdrawal of Egypt and the advance of the Assyrians (10th century BC), the history of Phoenicia is primarily the history of the city-state of Tyre, that founded colonies on islands and on the African and European coasts of the Mediterranean Sea. The Phoenicians formed connections with the Greeks, to whom they transmitted their arts and their alphabet in the 8th century BC.

5 In 538 BC, the Phoenicians backed the conquest of the Babylonian territories by the Persians, who allowed them to trade there in their own currency. In 332 BC Tyre capitulated to the army of Alexander the Great after resisting for eight months. The surviving local inhabitants were sold into slavery.

6 In 64 BC Phoenicia was incorporated into the Roman province of Syria. During the Roman period, the Phoenician language died out, and Aramaic, spoken by Semitic peoples who arrived from the Orient, was adopted as the local language. During that period, Lebanon produced numerous writers in Greek, among whom the Neo-Platonist Porphyry (3rd century BC) stands out. The Beirut Law School, which flourished in the 6th century, made essential contributions to Roman jurisprudence.

7 Between 608 and 630, Persians and Byzantines disputed supremacy over Lebanon and Syria, whose people accepted the Muslim Arab conquest in 630.

8 The Muslim occupation facilitated the settlement of Arab peoples in southern Lebanon. In the meantime, Monothelitic Christian groups fleeing persecution in Syria settled in the north after being declared heretics in 681. They adopted the Arab language and, together with the indigenous peasants, founded the Maronite Church. Even though they rebelled against the Muslims on various occasions, the Maronites enjoyed their protection against constant attacks from Constantinople until the early 11th century.

9 During the 11th century, the southern Arab settlers, dissidents from the Ismaelian Shi'a Islam (followers of Caliph Ali, 656-661), founded a religion only accessible to the initiated that they called the Druze faith. In the coastal towns the population became mainly Sunni Muslim (orthodox). A great number of Christians grouped in diverse sects, both in the cities and in the country, where they spoke Arabic as did the Maronites.

10 For a 100-year period (from the late 11th century to 1187) the Muslims lost control of Lebanon to the first Papal crusaders. After it was reconquered, thanks to support from Egypt, the Muslims managed to repel Mongolian attacks. At the end of the 13th century, it became part of the Mamluk state (military oligarchy) of Syria and Egypt, and enjoyed certain autonomy. Trade was promoted, leading to prosperity for the city of Tripoli.

11 After defeating the Mamluks in 1516-17, the Ottoman Turk Empire gained control over Lebanon.

12 Between the 16th and 18th centuries, the Shi'as became secure in the south under the control of Damascus (capital of Syria), as were most of the Druze and some Maronites. Christians and Druze settled along the length of Mount Lebanon, enjoying a semi-autonomous status and finding a common interest in consolidating their power against the coastal Sunni, under the leadership of bureaucrats from Istanbul. The Ottoman Tripoli governed the northern territory of Lebanon.

13 In 1697, the Mount Lebanon notables elected a prince belonging to the Sunni Shihab family, who governed in close cooperation with the Druze until 1842. Throughout this period, European influence over Lebanese politics grew, while the Ottomans suffered the wear of continuous attacks from Egypt. French merchants, settled in Lebanese ports, influenced the Maronites, who joined the Roman Catholic Church in 1736.

14 As well as sustained economic growth, the 19th century brought social changes and political crises. The Ottoman Turks ended the local rule of the Shihab Dynasty in 1842, irreversibly exacerbating already poor relations between the Maronites (supported by the French) and the Druze (aided by the British). These relations reached a low ebb with the Druze massacre of Maronites in 1860.

15 In 1861, the French, together with the Ottoman authorities, imposed a basic set of laws - which prevailed until World War I (1914-18) - in which the Ottomans established their direct control over Mount Lebanon.

16 In 1923, the League of Nations awarded the administration of Syria and Lebanon to France which, in fact, had never ceased to exercise control over the areas. The first 20 years of French administration were favorable to the Maronites, who made up half of the population of Lebanon. In 1926, the French agreed that the President would be a Maronite, the Prime Minister a Sunni Muslim and the Head of the Senate a Shi'a Muslim.

17 During their first years of management in Lebanon, the French developed production, communications and Jesuit-based education. After the stock market crash and the start of the global depression at the end of the 1920s, both friction and nationalism grew between religious groups in Lebanon. The total withdrawal of the French army took place at the end of 1946. Lebanon immediately became a member of the UN and the Arab League.

18 The Nationalist Maronite President Bishara al-Khuri, elected in 1943 - the year of Lebanon's declaration of independence - was forced to resign in 1952. The escalation of violence, supported by the Syrian Ba'ath Arab Socialist Party (pan-Arabist) since 1949, was the product of the favoritism and corrupt dealings of the Khuri Government (allied with the Sunni) and a controversial constitutional amendment that allowed the President to run for a second term.

19 The presidency of Maronite Camille Chamoun, elected by parliament in place of Khuri, coincided with the presidency of Gamal Abdel Nasser, Egyptian anti-colonialist pan-Arabist leader. When Nasser attempted to seize the Suez Canal from

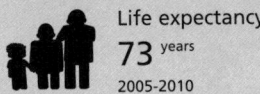
Life expectancy
73 years
2005-2010

GNI per capita
$6,010
2004

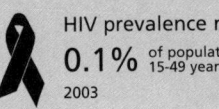
HIV prevalence rate
0.1% of population 15-49 years old
2003

the British in 1956, Chamoun denied Nasser's request to cut off diplomatic ties with Europe. The 1957 parliamentary elections were marked by confrontations between those favoring Lebanon's integration in the United Arab Republic and pro-Western supporters. Parliament was manipulated to ensure the re-election of Chamoun.

[20] By the following year the rioting had escalated into an all-out civ il war. In July, Chamoun, who was no longer obeyed by the Muslim members of the army, allowed 10,000 US Marines to disembark. The foreign troops remained in Lebanon until October of that year, when parliament appointed General Chehab as president.

[21] Between 1958 and 1969, the governments of Chehab and subsequently of Helou (both of Maronite extraction) disavowed the traditional system of political-religious representation, and the army was called out to clamp down on the violent protests by civilians.

[22] During that period, due to the influx of peasants - who made up half of the economically active population at the time - Beirut became home to 40 per cent of the country's population. The peasants were forced to abandon their lands which, due to the deterioration resulting from decades of intensive farming, only produced one per cent of GDP. In Beirut, each neighborhood identified with a religious affiliation whose sectarianism was on the rise.

[23] Even though it had received numerous Palestinians evicted from their land by Israel, Lebanon's failure to intervene in the 1967 Arab-Israeli War revived the antagonism among Lebanese people concerning their country's role in the Arab world. The Muslims, most of whom were united under the Muslim Lebanese Nationalist Movement - which backed the Palestine Liberation Organization (PLO) - called for Lebanon's annexation to Syria, as had been demanded prior to the French occupation.

[24] Following the 1973 Arab-Israeli War, Lebanon granted refuge to 300,000 Palestinians who were allowed to settle in the southern part of its territory. Concentrated in camps, between 1973 and 1975 the refugees were the object of segregation and violent attacks by Maronite paramilitary groups. They were also bombed by the Israeli army on several occasions.

[25] The agreement signed by Israel and Egypt in September 1975 raised the possibility that the Palestinians and Lebanon's Muslims would be left on their own by the Arab nations.

[26] In April 1975, a civil war again erupted throughout Lebanese territory. At the beginning of 1976, as a result of the daily armed clashes since the start of the civil war, the central government was dissolved and the Maronites were forced to accept defeat.

[27] According to the Syrian authorities, the prospect of a pro-Palestinian left-wing government in Lebanon in 1976, in addition to the possibility of a division of the Lebanese territory, could trigger an Israeli invasion. As a result, Syrian President Hafez al-Assad backed the restoration of the Maronite Government and refrained from interfering in the attacks on Palestinian refugee camps launched by Maronite militia between July and September.

[28] In September and October 1976, Lebanon was divided by a 'green line' that split Beirut into eastern and western zones, as well as the rest of the country, following the route to Damascus. The northern region remained under the control of the Maronite Government, whose president was Elias Sarkis. The southern region was placed under the administration of left-wing Kamal Jumblat (who was assassinated in March 1977) with the participation of Druzes, Muslims and Palestinians and the intervention of a 30,000-soldier Arab-League peace force.

[29] In 1978 Syria once again gave formal support to the Lebanese pro-Palestinian left. By then the Government was headed by the Phalange Party, which received instructions, weapons and troops from Israel. Notwithstanding the posting of a small contingent by the UN, Israel did not put a halt to its ground forays and bombings in southern Lebanon.

[30] The civil war was a catastrophe for the Lebanese, who witnessed the destruction of Beirut and the country's infrastructure. The tens of thousands of civilian dead included 20,000 Palestinians. In the period between 1975 and 1982, Lebanon's economic losses were almost total, but the oil boom of those years favored some business transactions that partially compensated for the deficit.

[31] In July 1981, the Israeli bombing of the PLO general headquarters in West Beirut caused the death of 300 civilians. Despite a conciliatory intervention by the US in June 1982, 60,000 members of the Israeli army invaded Lebanon within the framework of an operation by the Israeli government named 'Peace for Galilee'.

[32] In late August 1982, the PLO withdrew its troops from Beirut under the supervision of US, French and Italian soldiers. On 15 September, in retaliation for the assassination of Phalange President Bashir Gemayel - the sole candidate, elected a few days earlier - the city of Beirut was occupied by Israeli troops. The next day, the Lebanese militia, led by Israeli commanders, burst into the Palestinian refugee camps of Sabra and Shatila and murdered thousands of civilians.

[33] In June 1983, the new Lebanese president, Amin Gemayel (brother of the former president), signed an agreement with Israel which included establishing a 'security zone' (an area of 850 square kilometres, patrolled by Israel) in the south of Lebanon.

[34] In July 1984 the Lebanese currency, which had remained relatively stable since the beginning of the war in 1975, crashed, triggering unprecedented inflation. That year, the balance of payments showed a deficit of over $1.5 billion.

[35] The Israeli army officially pulled out of Lebanon in 1985 on the condition that the Christian militia would displace the Muslims from southern Lebanon. The Phalangists and the Christians, in general, as well as the Druze-Shi'a alliance and the PLO, divided into factions that supported or opposed Syrian leadership.

[36] In September 1988, at the end of Gemayel's presidential term, parliament was unable to reach an agreement on the selection of a new president. Despite continued popular demands that Selim al-Hoss become President, Gemayel appointed Maronite General Michel Aoun as Prime Minister. The country was governed by two rival administrations: Hoss from West Beirut and Aoun from the eastern

PROFILE

ENVIRONMENT
Lebanon has a fertile coastal plain, situated between the Mediterranean Sea and the Lebanon Mountains, where most of the population lives. The plain has a mild Mediterranean climate marked by winter rains. Between two parallel mountain ranges, the Lebanon Mountains (whose highest peak is Sauda, at 3,083 m) and the Anti-Lebanon Mountains, (with temperate forests on their slopes), lies the fertile Bekaa Valley. Despite the scarce rainfall, the soil is very fertile because of rich alluvial deposits. Along the coast, a dry bush - the *maquis* - grows, and wheat, cotton, olives, oranges and vineyards are cultivated. The cedar tree has become a national symbol. This wood was used to build the Phoenician fleet and temples. Nowadays there are only about 400 cedars left, ranging between 200 and 800 years old.

SOCIETY
Peoples: The Lebanese (80 per cent) are an Arab people. There is a significant Palestinian minority, mostly refugees. There are also Armenians (4 per cent), Syrians, Kurds, Europeans and others (2 per cent).
Religions: 55.3 per cent Muslim (34 per cent Shi'a and 21.3 per cent Sunni); 37.6 per cent Christian (25.1 per cent Catholic, 19 per cent Maronites and 4.6 per cent Greek-Catholics); 11.7 per cent Orthodox Christian (6 per cent Orthodox-Greeks, 5.2 per cent Apostolic-Armenians); 0.5 per cent Protestant; 7.1 per cent are Druze.
Languages: Arabic (official); French is widely spoken, Armenian and English are less common.
Main Political Parties: Rafik al-Hariri coalition (consisting of liberal and socialist parties); Resistance and Development bloc (consisting of *Hizbullah* -Party of God- and Syrian Social National Party); Aoun Alliance.
Main Social Organizations: Joined Lebanese Employees' and Workers' Syndicates Federation; Lebanon University National Union.

THE STATE
Official Name: al-Jumhouriyah al-Lubnaniyah.
Administrative Divisions: 6 governmental divisions.
Capital: Beirut (Bayrut) 1,792,000 people (2003).
Other Cities: Tripoli (Tarabulus) 206,500 people; Juniyah 77,400; Zahlé 74,300 (2000).
Government: Emile Lahoud, President since November 1998. Fouad Siniora, Prime Minister since June 2005. Unicameral Legislature: Assembly of Representatives, with 128 members elected for a four-year term by the religious communities.
National Holiday: 22 November, Independence Day (1943); 25 May, Resistance and Liberation Day (2000).
Armed Forces: 48,900 personnel.
Other: Hizbullah 3,000.

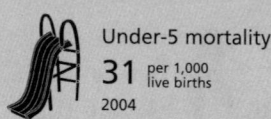
Under-5 mortality
31 per 1,000 live births
2004

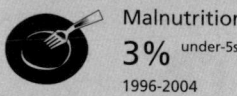
Malnutrition
3% under-5s
1996-2004

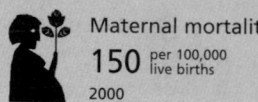
Maternal mortality
150 per 100,000 live births
2000

IN FOCUS

ENVIRONMENTAL CHALLENGES
Deforestation, desertification and soil erosion are pressing problems in various locations throughout the country. Coastal water is polluted by untreated sewage and occasional oil spills. In Beirut, air pollution from vehicle and industrial emissions is a problem.

WOMEN'S RIGHTS
Women have been able to vote and run for office since 1952. In 2005, women held five per cent of seats in Parliament and 7 per cent of ministerial or equivalent positions. In 2003, women comprised 30 per cent of the total labor force of 2 million.

Although 87 per cent of pregnant women receive prenatal care and 89 per cent of births are attended by qualified personnel*, in 2004 maternal mortality stood at 150 per 100,000 live births.*

CHILDREN
In general, child welfare continued to improve. Ninety-eight per cent of children aged 6-11 attended school, with no gender differentiation, and in 2004, 91 per cent of children between the ages of 3 and 5 were registered in preschool. That year, immunization against preventable diseases was 100 per cent. The mortality rate

had decreased considerably both for under-ones and under-fives, although it remained high in 2004 at 27 and 31 per 1,000 live births, respectively.*

INDIGENOUS PEOPLES/ ETHNIC MINORITIES
In Lebanon the main minorities are religious and/or political groups in addition to ethnic. There are Armenians (4 per cent of the population), Druze (6 per cent), Palestinians (11 per cent), Sunnis (20 per cent), Christian - Maronites (25 per cent), and Shi'a (32 per cent). The future of relations between the groups largely depends on Syria's influence on the country.

Arabic is the common language, and many Maronites also speak French. Maronites have enjoyed a privileged social and political status, especially after World War I, despite being fewer in number than the Shi'a, and a minority religious group as the others are all Muslim.

The Druze practise a branch of Islam that differs from that of the Shi'a and Sunnis in many respects, and have suffered social discrimination because of it in the past.

The '1943 National Pact' - never put down in writing - guarantees the representation of Maronites, Sunnis and Shi'as in the highest-

ranking positions in the country (the president is usually Maronite, the prime minister Sunni and the head of Parliament Shi'a). This agreement does not guarantee a position for the Druze, although the commander-in-chief of the armed forces has historically been Druze.

MIGRANTS/REFUGEES
Since the war against Iraq started in 2003, Lebanon (like Syria and Jordan) has received a constant inflow of Iraqi refugees for their temporary protection. Sudanese refugees are in the same situation. Both groups are settled mostly in cities. The situation in their home countries - especially in Iraq - made voluntary repatriation improbable, thus resettlement and integration policies seemed best. Lebanon also hosted, in 2004, more than 400,000 Palestinian refugees, in 12 camps throughout the country.

DEATH PENALTY
In 2006 the death penalty was still applicable, even in the case of common crimes.

* Latest data available in The State of the World's Children and Childinfo database, UNICEF, 2006.

⁴⁹ In the June 2005 election, the anti-Syrian alliance led by Saad al Hariri, son of the dead premier, won a majority in Parliament and Fouad Siniora, a Sunni allied with Hariri, became premier. Pro-Syrian Shi'a leader Nabih Berri was re-elected as speaker of Parliament.
⁵⁰ In July, Saad Hariri met with Syrian President Assad to seek an agreement that would normalize relations between both countries. In September, two pro-Syrian generals were accused of involvement in the murder of Rafiq Hariri. A call allegedly received by Lahoud, made by one of the main suspects minutes before the explosion, seriously implicated the President.
⁵¹ In July 2006, the kidnapping of two Israeli soldiers by Hizbullah was used as a pretext by Israel to launch air and sea attacks on targets in southern Lebanon. Tel Aviv declared its intention to dismantle the Lebanese resistance that was effectively ruling the southern part of the country and to wipe out its capacity to launch rockets against the northern cities of Israel.
⁵² While several analysts considered the attack as jointly planned by the US and Israel, Hizbullah's response - which even managed to hit an Israeli warship with its missiles - came as a surprise to Tel Aviv. Israeli troops entered Lebanon in August amidst increased international pressure for an immediate ceasefire.
⁵³ By mid-August, a very precarious ceasefire was achieved. Over one thousand Lebanese people - mostly civilians - and some 150 Israelis - mostly soldiers - had been killed. The damage caused to the Lebanese infrastructure and economy was beyond calculation. In Israel, losses were estimated at $5 billion, yet none of the initial goals of the bombing and invasion were fulfilled. Hundreds of thousands of displaced people started to return to their destroyed homes. Hizbullah celebrated what it regarded as a military victory.
⁵⁴ In October 2006, Hizbullah's leader Sheikh Hassan Nasrallah confirmed that indirect but serious talks with Israel about a prisoner exchange were under way.
⁵⁵ In November 2006, industry minister Pierre Gemayel was assassinated. He was a member of the country's most powerful Christian family and a prominent opponent of Syrian influence. He and other ministers had just approved UN plans to try suspects in the killing of former PM Rafiq Hariri and thereby prompted the resignation of pro-Syrian ministers and Hizbullah threats to bring down the Government. The assassination fostered fears of a descent into renewed civil war. ∎

portion of the city.
³⁷ In March 1989, Aoun declared what he called the 'liberation war', which aimed to end Syria's presence in Lebanon. In October, the Lebanese parliament met in Saudi Arabia to sign a national reconciliation agreement, granting greater power to the cabinet of ministers. It also determined that there would be an equal number of Christian and Muslim representatives in parliament, and stipulated the partial withdrawal of Syrian troops from Lebanon. General Aoun rejected the agreement because he considered it a 'Syrian ruse'.
³⁸ On 5 November 1989, René Moawad, a Maronite Christian inclined towards an opening to the Arab world, was unanimously elected president. However, Moawad was killed by a car-bomb 17 days before taking office. Two weeks later, Elias Hrawi, another Maronite, was elected president by a meeting of the Lebanese Parliament held in Syrian-controlled territory.
³⁹ Taking advantage of the new situation created by the Iraqi invasion of Kuwait in 1990, Syrian-backed forces launched an offensive in October against Aoun, who sought asylum in France upon his

defeat. In December, a government of national unity was formed, for the first time since the beginning of the civil war. It incorporated the Lebanese Forces (Christian militia), the Amal (Shi'a), the PSP (Druze), and pro-Syrian parties.
⁴⁰ The presidents of Lebanon and Syria signed a Brotherhood, Co-operation and Coordination Agreement in Damascus in May 1991. Syria recognized Lebanon as an independent state. Despite opposition by Israel, the Lebanese Parliament ratified the agreement.
⁴¹ In July that year, 6,000 Lebanese troops took over the territories occupied by the PLO in the south. Israel stated that it would not withdraw its forces from the security zone.
⁴² In mid-1992, violence and political turmoil brought about the fall of the pro-Syrian Government led by Omar Karame, who was succeeded by Rashid Al Sohl, a moderate Sunni. August parliamentary elections were boycotted by Christians. The new Parliament included new representatives from Amal and Hizbullah (or Party of God, a Shi'a group founded in 1982 backed by Iran, to fight Israeli occupation

in the south).
⁴³ In October, Sunni millionaire Rafiq Al Hariri was named Prime Minister.
⁴⁴ While the UN increased its intervention, Hizbullah and the Lebanese army resumed fighting against Israeli forces.
⁴⁵ In October 1998, the National Assembly elected General Emile Lahoud as president - who was backed by the army and Syria. In 2000, harassed by Hizbullah, Israeli troops withdrew from southern Lebanon.
⁴⁶ Hariri won a landslide victory in 2002 elections, thus forcing Lahoud - against his express will - to appoint him prime minister. In September 2003 he was replaced by Omar Karami.
⁴⁷ Hariri was killed by a car-bomb in February 2005 in Beirut. Karami's Cabinet fell after demonstrators protesting against the murder caused riots demanding the withdrawal of Syrian troops. In March there were massive demonstrations both for and against Syria.
⁴⁸ In April 2005, pro-Syrian moderate Najib Mikati became prime minister. That month, Syrian forces left the country - as required by the UN - after a 29-year presence.

Lesotho / Lesotho

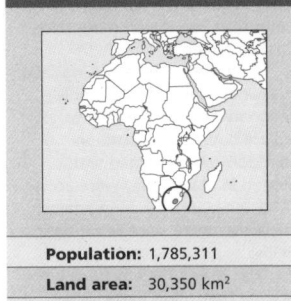

Population:	1,785,311
Land area:	30,350 km²
Capital:	Maseru
Currency:	Loti
Language:	Sotho and English

The San, first nomadic peoples of present-day Lesotho, populated southern Africa 2000 years ago. They came into contact with the Bantus, who migrated to these lands around the 4th century AD. By the 16th century, both groups dominated the Caledon river basin.

[2] The Zulu conquests, launched in 1818 by Shaka, affected many Bantu-speaking peoples, among them the North Sotho or Pedi who lived in what became Transvaal in northern South Africa. While some withdrew northwards, the head of the Bakwena tribe, Moshoeshoe, brought other Sothos and groups of dissident Zulus to the Drakensberg mountains. The lengthy war of resistance fought first against the Zulus, and then against the expansionist Boers, consolidated their bonds. These groups gave Moshoeshoe the title of 'Great Leader of the Mountain', and called themselves Basothos.

[3] In the mid-19th century, the Boers (Dutch colonizers) tried unsuccessfully to force the Basotho to work their land.

[4] Dutch colonization in South Africa seemed destined to fail until the discovery in 1867 of diamonds and, soon afterwards, of gold. The British began to arrive shortly thereafter and in 1868, British missionaries persuaded the Basotho traditional leader Moshoeshoe to turn his kingdom into a protectorate on the pretext of saving the people from the Boers' attempts to convert them into slaves. The territory was administered separately from South Africa, although both were controlled by Britain after the Boer War of 1899-1902.

[5] During World War II (1939-1945), 20,000 Basothos served in the British forces.

[6] Britain had promised the South African government that Basotholand (Lesotho), Bechuanaland (Botswana) and Swaziland would eventually become part of South Africa. However, when the South African Union broke all ties with London in 1961, consolidating apartheid, the British

preferred to grant the countries their independence. In 1956, a constitution was promulgated in Basotholand and in 1966 the country proclaimed independence as Lesotho.

[7] As an enclave within South Africa, Lesotho depended on the surrounding country as an outlet for its products. Its currency was the South African rand and South African companies controlled

the country's economy and communications.

[8] With imports 10 times higher than exports, the difference was offset by money that migrant workers sent home (45 per cent of the labor force worked in South Africa's gold mines).

[9] This economic situation enabled the opposition Congress Party to win the legislative elections of 1970. Prime Minister Leabua Jonathan

dissolved Parliament, and sent King Moshoeshoe into exile. He was allowed to return on the promise that he would refrain from political activity.

[10] After the 1976 student uprising in Soweto, South Africa, thousands of South Africans took refuge in Lesotho. When South Africa began its Bantustan policy, Lesotho refused to recognize the puppet state in Transkei. In early 1977,

PROFILE

ENVIRONMENT

This small country lies in the Drakensberg mountains, surrounded by South Africa. Landlocked and mountainous, its only fertile land is located in the west where corn, sorghum and wheat are grown. In the rest of the country cattle are raised. Erosion affects 58 per cent of the soil in the lowland areas. Two-thirds of all farmland belongs to migrant laborers; in their absence some is worked by people with little incentive to care for it. The Lesotho Highlands Water Project was developed to make greater use of water from the highlands - the country's main natural resource - through a hydroelectric project that would divert water to South Africa in return for electricity. The first phase was inaugurated in 2004 with the completion of the Mohale Dam. Except for small diamond deposits, there are no mineral resources.

SOCIETY

Peoples: Ethnically homogeneous, the country is inhabited by the Basotho (Sotho) people (85 per cent) and a Zulu minority (15 per cent). There are small communities of Asian and European origin. **Religions:**

Mainly Christian. Also traditional African beliefs.
Languages: Sesotho (Southern Sotho) and English (officials), Zulu, Xhosa.
Main Political Parties: Lesotho Congress for Democracy (LCD), Basotho National Party; National Independent Party.
Main Social Organizations: the Lesotho General Workers' Union (LGWU), founded in 1954, is the only central labor organization. Representative Students' Council.

THE STATE

Official Name: Kingdom of Lesotho. Administrative Divisions: ten Districts.
Capital: Maseru 170,000 people (2003).
Other Cities: Maputsoe 32,800 people; Mafeteng 29,400 (2000).
Government: King Letsie III, Head of State since February 1996. Pakalitha Mosisili, Prime Minister since May 1998. Parliamentary Monarchy. Parliament has two chambers: the 120-member National Assembly and the 33-member Senate.
National Holiday: 4 October, Independence Day (1966).
Armed Forces: 2,000 troops (2003).

Life expectancy
34 years
2005-2010

GNI per capita
$730
2004

Literacy
81% total adult rate
2000-2004

HIV prevalence rate
28.9% of population 15-49 years old
2003

IN FOCUS

ENVIRONMENTAL CHALLENGES

Erosion is the country's biggest environmental problem. The situation is aggravated by the Lesotho Highlands Water Project, which is diverting water to South Africa.

WOMEN'S RIGHTS

Women have been able to vote and stand for office since 1965. In 2002, 12 per cent of seats in the lower chamber and 36 per cent in the higher chamber were held by women, who also occupied 28 per cent of ministerial or equivalent positions.

In 2003 illiteracy among women over 15 was just over 9.5 per cent while for men of the same age group it was 26 per cent.

In 2004 HIV/AIDS represented the biggest health threat for women and also for the rest of the population: 170,000 women between 15 and 49 were living with the virus. In the capital the incidence of this disease among pregnant women aged 15 to 24 was 28 per cent. In 2004, 85 per cent of pregnant women received prenatal care and 60 per cent of births were attended by qualified personnel. According to the latest available data (2000) maternal mortality was 550 per 100,000 live births.

CHILDREN

HIV/AIDS obscures progress that has been made in various aspects of children's well-being. By the end of 2003, 29 per cent of 15 to 49-year-olds were living with the virus, as well as 22,000 under-15s. An estimated 100,000 children under 17 had been orphaned by AIDS. Under-one and under-five mortality diminished between 1990 and 2004. The former fell from 84 to 61 per thousand live births and the latter from 120 to 85 per thousand live births. Some 14 per cent of babies were born underweight and 46 per cent of under-fives measured below height norms*.

INDIGENOUS PEOPLES/ ETHNIC MINORITIES

The majority of the population is Sotho. The San, the original inhabitants of the area, were mainly killed or displaced by the British. Sotho cultural and social traditions remain.

MIGRANTS/REFUGEES

For geographical reasons the country has an extremely close relationship with South Africa. A very high percentage of the population have family links there and many have documents from both countries. Lesotho has provided hundreds of thousands of workers to South Africa's mining industry (according to official sources their number peaked in 1990 at 127,000). However, a new law restricting foreigners' access to South Africa has been devastating for Lesotho, leaving thousands of miners unemployed.

DEATH PENALTY

In force even for common offenses.

* Latest data available in The State of the World's Children *and* Childinfo *database, UNICEF, 2006.*

South Africa closed the Lesotho border in retaliation. The economic situation became dramatic and Lesotho appealed for international solidarity.

[11] After Zimbabwean independence in 1980, Lesotho joined the economic integration project promoted by the Front Line states, strengthening relations with Mozambique.

[12] South Africa retaliated by supporting groups opposed to the Jonathan Government, who sought help from the United Nations and the European Economic Community. The official Basotho National Party (BNP) also had to face the opposition of groups linked to the Basotho Congress Party (BCP), led by Ntsu Mokhele.

[13] Incidents triggered by South African military groups trying to prevent African National Congress (ANC) refugees from organizing in Maseru (see South Africa), led to an attack in December 1982 in which 45 people were killed, including 12 children. Three ANC leaders were killed; many of the other victims had no political affiliation.

[14] In 1982, the Government imposed emergency legislation. The army and the police were reinforced and a paramilitary group known as Koeko violently suppressed BCP supporters in the Drakensberg mountains. The BCP abandoned the nationalist stand it had endorsed in the 1970s and fell into deep crisis. Many leaders received aid from South Africa.

[15] In March 1983, an attempt to sabotage the country's main power plant sparked a border incident between troops from Lesotho and South Africa. There was growing South African pressure on Lesotho to sign a non-aggression treaty with Pretoria, similar to those which the apartheid regime had signed with Swaziland and Mozambique. Jonathan was against the treaty, but was forced to compromise due to Lesotho's economic dependence on South Africa.

[16] Towards the end of 1984, the South African Government retained weapons purchased in Europe by Lesotho, and delayed remittances sent home by Lesotho's 400,000 or so migrant workers in South Africa. It also delayed plans to build a dam on the Sengu River, on the border between the two countries. The pressure from South Africa was aimed at intimidating Lesotho's voters and strengthening opposition to the BNP, the conservative Basotho Democratic Party (BDA) and the (ANC-aligned) BCP.

[17] These measures did not satisfy Pretoria and on 20 January 1986 General Justin Lekhanya, head of Lesotho's paramilitary forces, overthrew the government of Leabua Jonathan and headed the military committee that replaced it.

[18] In 1988, workers living in South Africa sent home remittances totaling more than $350 million, equivalent to 500 per cent of the total value of Lesotho's exports.

[19] In March 1990, the military regime sent King Moshoeshoe into exile, accusing him of hindering the country's democratization program. His son Bereng Mohato Siisa replaced him as Letsie III. On 30 April 1991, another coup toppled Lekhanya's government; a Council was set up chaired by Colonel Elias P Ramaema.

[20] The South African Government blocked remittances from migrant workers in 1991. In May, a demonstration against foreign interference ended with 34 people dead and 425 arrests.

[21] In 1993, a new constitution appointed the King head of state, without granting him either legal or executive powers. In the July legislative elections, the BCP won all the seats. In August, the privatization of six state companies got under way with a loan from the IMF.

[22] The Government plan to integrate the armed wing of the BCP into the army led to the kidnapping and murder of the finance minister by disgruntled soldiers.

[23] In August, the King dissolved the Government and Parliament. Domestic opposition from Republicans together with international pressure forced Letsie III to abdicate in favor of his father. King Moshoeshoe was restored to the throne in January 1995.

[24] That year, part of the World Bank-financed project to increase the supply of water in the Vaal river valley in South Africa, using the water from the Maloti mountains, was completed.

[25] In January 1996, the King died in a car accident. The Assembly designated his son Letsie III to replace him. Four opposition politicians were charged with treason in March; they were accused of having planned a coup against the Government since September 1995.

[26] The May 1998 elections were won by the ruling party, now called the Lesotho Congress for Democracy (LCD). The opposition, which won one of the 80 disputed seats, complained that the elections were flawed. General Pakalitha Mosisili was elected Prime Minister. The Government asked South Africa for help in September when a sector of the army joined the protests. South Africa sent troops who brought the situation under control through the use of force. South Africa charged $1 million for the intervention.

[27] A body with government and opposition representatives was formed in December 1998 to organize new elections and revise the laws governing them. South African troops, along with a small contingent from Botswana, pulled out of the territory by May 1999.

[28] In April 2001, South African President Thabo Mbeki traveled to Lesotho to improve relations between the countries. The visit was not well received by the opposition due to the 1998 intervention by South African troops.

[29] In March 2002, after months of speculation, the King announced elections on 25 May. The LCD obtained 55 per cent of the vote, compared to the PNB's 22 per cent.

[30] Three years of major droughts affecting the country left thousands of people facing starvation. In February 2004 Mosisili declared a state of emergency and asked for international assistance.

[31] March saw the official opening of the multi-million dollar Lesotho Highlands Water Project, with its construction of enormous dams in the Maloti mountains and the diversion of the Senqu river towards South Africa.

[32] In April 2005 the first local elections since independence were boycotted by the opposition because of inadequate preparation time. In June, the European Commission agreed to give Lesotho a million euros in food aid to be distributed amongst the most needy.

[33] External Affairs Minister Monyane Moleleki was shot in January 2006 by an unidentified attacker whilst leaving his house. The attack may have been politically motivated as Moleleki was a possible successor to Mosisili in 2007. ∎

Liberia / Liberia

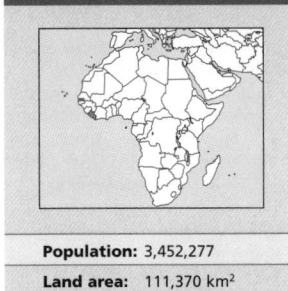

Population:	3,452,277
Land area:	111,370 km²
Capital:	Monrovia
Currency:	Liberian dollar
Language:	English

The territory of present-day Liberia - formerly known as the Grain Coast - was inhabited by 16 different ethnic groups. The Mande speaking peoples, including the Mandingo, lived in the east and northeast. After the arrival of the Portuguese, Mandingo traders and artisans played an important role as they spread throughout the territory, becoming the principal propagators of Islam.

2 Long before US President Abraham Lincoln freed the slaves in 1865, during the US civil war, emancipated blacks posed a social problem to US southern slaveholders. As a solution to the 'problem' some were 'repatriated'. On the assumption that blacks would feel at home in any part of Africa, plans were made to ship them to the British colony of Sierra Leone.

3 In 1821, the American Colonization Society purchased a portion of Sierra Leone and founded a city which was named Monrovia after James Monroe, president of the United States.

4 Only 20,000 US blacks returned to Africa. The local population distrusted these settlers whose language and religion were those of the colonizers. Supported by US Navy firepower, the newcomers settled on the coast and occupied the best lands. For a long time, they refused to mix with the 'junglemals', whom they considered 'savages'. Today only 20 per cent of the population speak English.

5 In 1841, the US Government approved a constitution for the African territory. It was written by Harvard academics, who called the country Liberia. Washington also appointed Liberia's first African governor: Joseph J Roberts. In July 1847, a Liberian Congress representing only the repatriates from the US, proclaimed independence. Roberts was appointed president of the country, which had a Harvard-made constitution and a flag which resembled that of the US.

6 The emblem on the Liberian coat of arms reads: 'Love

of liberty brought us here'. However, independence brought little freedom for the original population. For a long time, only landowners were able to vote. The 45,000 descendants of the former US slaves formed the core of the local ruling class, and they had close links to transnational capital. One of the principal exports, rubber, was controlled by Firestone and Goodrich under 99-year concessions granted in 1926. The same was true of oil, iron ore and diamonds. Resistance to this situation was suppressed on several occasions by US Marine interventions to 'defend democracy'.

7 The discovery of extensive mineral deposits, and the use of the Liberian flag by US ships, heralded a period of economic

growth beginning in 1960. This was instantly dubbed an 'economic miracle', but this so-called miracle only reached the American-Liberian sector of the population.

8 In 1979 a hike in the price of rice triggered demonstrations and unrest. A year later, Sergeant Samuel Doe overthrew the regime of William Tolbert, who was executed by a firing squad along with thirteen members of his government. Doe banned all political parties and suspended the Constitution.

9 In 1980, the beginning of a democratization process was announced, followed by the signing of the first agreement with the IMF.

10 Falling exports, increasing unemployment, wage cuts and spiraling foreign debt tipped the

country into an enormous crisis that fuelled popular discontent. Between 1980 and 1989, the Doe administration uncovered nine anti-government conspiracies.

11 Elections were held in 1985. With any viable political opposition banned, and accusations of fraud and imprisonment of opposition leaders abounding, Doe obtained 50.9 per cent of the vote. The Liberian People's Party (LPP) and United People's Party (UPP), which represented the major opposition forces, were not allowed to participate.

12 In 1987, most of the funding obtained by the Government came from Washington. The US maintained significant interests in Liberia: $450 million in capital goods, military bases, a regional radio station and the communications center for all US diplomatic services in Africa.

13 In December 1989, the National Patriotic Front of Liberia (NPFL) - unknown until then - launched an armed insurrection led by army officer Charles Taylor. In June 1990, a victory by the NPFL seemed imminent. But in the battle for Monrovia, the rebel front split and an Independent Patriotic Front of Liberia (INPFL) was formed, led by Prince Johnson.

14 In September 1990, President Samuel Doe was murdered by Johnson's troops. In the ensuing confusion, several interim presidents were announced simultaneously: Johnson, Taylor, Amos Sawyer and Raleigh Seekie, former head of Doe's presidential guard.

15 The Economic Community of West African States (ECOWAS) sent a peace force (ECOMOG), composed

Life expectancy
43 years
2005-2010

GNI per capita
$120
2004

Literacy
56% total adult rate
2000-2004

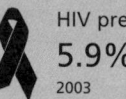
HIV prevalence rate
5.9% of population 15-49 years old
2003

IN FOCUS

ENVIRONMENTAL CHALLENGES
The tropical rain forest is subject to deforestation, soil erosion and particularly loss of biodiversity. There is frequent pollution of rivers from the dumping of iron ore tailings and of coastal waters from oil residue and raw sewage.

WOMEN'S RIGHTS
Women have been able to vote and stand for office since 1946. In 2005 women held 13 per cent of seats in the Lower Chamber and 17 per cent in the Higher Chamber.

Domestic violence was widespread throughout the country, but was not perceived as a problem by the Government, the courts or the media. At the beginning of 2006 more than a hundred women reported to UN authorities that Peacekeepers had used them for sex and as a result they had become pregnant, but the soldiers had not acknowledged paternity. Approximately 6,000 children had been born as a result, since 1990.

CHILDREN
One of the main challenges faced by Liberia after 14 years of civil war is the demobilization of children recruited (frequently for sexual purposes) by factions in the conflict. During the civil war an estimate of 50,000 children were killed and many others were wounded, orphaned, or directly abandoned. A considerable number of orphanages operate in Monrovia. Several hundred children fathered by UN peacekeepers were abandoned in the Economic Community of West African States' Military Observer Group's center for children. The UN is taking care of them under its food program, providing shelter, food and education. In 2004 under-one and under-fives mortality was one of the highest in the world, the former being 157 per thousand live births and the latter 235 per thousand live births. In the same year there were 8,000 children under 15 living with HIV/AIDS and 36,000 under-17s orphaned by the disease.

INDIGENOUS PEOPLES/ ETHNIC MINORITIES
Although the constitution bans ethnic discrimination, it also states that only blacks or descendants of blacks may be citizens or own land.

Formerly, the population was made up of more than 16 ethnic groups that spoke different dialects. None of these groups constituted a majority within the population.

Many members of the Muslim Mandingo minority encountered hostility when they sought to return, after the end of the civil war, mainly to villages in Lofa, Bong, and Nimba counties. Many of them were unable to reoccupy their homes, which had been taken over by members of the Lorma, Gio, and Mano minorities who held Mandingos responsible for atrocities committed during the war. Mandingos face arbitrary arrests, ethnic violence and many seek refuge in Guinea.

MIGRANTS/REFUGEES
In September 2005 there were more than 220,000 internally displaced people and almost 39,000 refugees. It is estimated that after peace agreements were signed in 2003 some 26,000 ex-combatants needed assistance for reintegration in society. At the same time, programs were established to help thousands of Liberian refugees in other West African countries return home. It was expected that all of the 223,000 internally displaced people would return to their places of origin. The 100,000 or so refugees from Algeria, Burkina Faso, Chad, Côte d'Ivoire, DR Congo, Iraq, Palestine, Rwanda, Somalia, Sudan and Togo were also expected to return to their countries of origin during 2006.

DEATH PENALTY
Liberia maintains and applies the death penalty to punish ordinary crimes.

* Latest data available in The State of the World's Children and Childinfo database, UNICEF, 2006.

Government agreed a ceasefire in seven months' time and general elections.

[18] In March 1994, the Council of State - a transitional body made up of representatives from the NPFL, ULIMO and the Sawyer administration - took office. Meanwhile, battles raged between rival armed groups and with ECOMOG. Negotiations continued in 1995, Charles Taylor became a member of the Council of State, and a new government was formed.

[19] Civil war broke out again in 1996 with violent combat, particularly in Monrovia. In September Ruth Perry became the new head of the Council of State with the backing of ECOMOG. In November peacekeeping forces began to disarm the rival factions. The war had claimed approximately 200,000 lives.

[20] In July 1997, Charles Taylor won the general elections with 75.3 per cent of the vote.

[21] In January 1999, Ghana and Nigeria accused Liberia of backing Sierra Leone's brutal Revolutionary United Front (RUF). Taylor accused Guinea of providing financial support to armed groups in the north of Liberia. In April, Guinean troops attacked Voinjama, in Lofa county.

[22] In December 2000 the UN concluded that Taylor was supplying weapons to RUF in exchange for stolen diamonds. It was estimated that Liberia exported more diamonds coming from Sierra Leone than from its own mines.

[23] In 2001 Amnesty International reported numerous cases of murder, rape and disappearance amongst human rights activists, journalists and politicians critical of the Government.

[24] That year, the UN imposed an arms embargo on Liberia, organized a boycott of diamonds exported from there and prohibited Taylor and members of his inner circle from traveling abroad. In September Taylor reopened the borders with Guinea and Sierra Leone, closed after accusations between the three neighboring governments that each was supporting rebel movements within the others' borders.

[25] In early 2002 conflict intensified between the Government and the rebel movement Liberians United for Reconciliation and Democracy (LURD), which had instigated the insurrection in Lofa.

[26] In March 2003, from outside Monrovia, LURD called for Taylor's resignation. In June a court in Sierra Leone accused Taylor of crimes against humanity and issued an international warrant for his arrest. Rebel groups held two-thirds of the territory but Taylor refused to resign. In July, Nigerian President Olusegun Obasanjo offered him safe exile.

[27] In August, Nigerian

peacekeeping troops entered the country, supported by US Marines. On 11 August, Taylor handed over power to Vice-President Moses Blah and left Liberia. The rebel groups signed a peace agreement in Ghana.

[28] In September the UN approved the creation of UNMIL, to be made up of 15,000 UN peacekeeping troops, the largest such force in the world. Its mission was to ensure the distribution of humanitarian aid and to facilitate the return home of 700,000 refugees living in neighboring countries and 450,000 internally displaced people, mostly from Monrovia.

[29] In October 2003 Gyude Bryant, a businessman seen as sufficiently neutral to lead a government of conciliation until elections were held, was appointed President.

[30] While UN peacekeepers maintained a relative peace in the country, in February 2004 the international community approved almost $500 million in aid for the reconstruction of Liberia after 24 years of civil war.

[31] In June 2005 the UN prohibited Liberian diamond exports, the main source of finance for the war. In September the Government made an agreement with the international community to impose strict controls on public finance and to fight corruption. A month later the country held its first elections since the end of the civil war. In the first round ex-footballer George Weah received the most votes with Ellen Johnson-Sirleaf coming second. However the second round in November saw Johnson-Sirleaf - called 'the iron lady' - winning by a large margin. She became Africa's first woman president. On assuming power she said: 'this is an opportunity to show the continent that women can lead and to help my nation recover from its brutal conflicts'.

[32] The Truth and Reconciliation Commission was set up in February 2006 by President Johnson-Sirleaf to investigate, in the President's words, 'flagrant violations of human rights and international law including massacres, rapes, murders, summary executions and economic crimes' perpetrated between 1979 and 2003 during the civil war.

[33] Ex-president Taylor was extradited from Nigeria in April 2006 to be tried on at least 17 charges of crimes against humanity during the armed conflict. The ex-president was captured just before Nigerian President Obasanjo traveled to Washington to meet US President George W Bush.

[34] In October 2006, a UN report criticized Liberia's high incidence of sexual violence. The country's Chief Justice rejected the report's call for a special court for trying rape cases to be established. ∎

of 10,000 troops from Nigeria, and supported a provisional government headed by Sawyer (of the LPP) in 1991.

[16] In September 1991, a new rebel movement was formed among Doe's followers: the United Liberation Movement of Liberia for Democracy (ULIMO) that launched attacks from

Sierra Leone against the NPFL, based in northeastern Liberia.

[17] In July 1993, a peace agreement was signed in Geneva under the auspices of ECOMOG and the UN, which had hindered military advances by means of an arms embargo. The two main armed groups and the provisional Sawyer

Libya / Libiyah

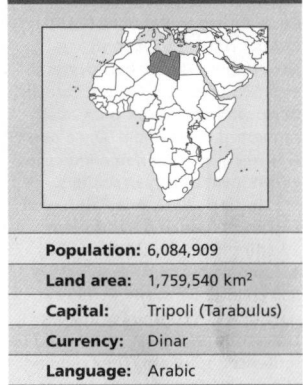

Population:	6,084,909
Land area:	1,759,540 km²
Capital:	Tripoli (Tarabulus)
Currency:	Dinar
Language:	Arabic

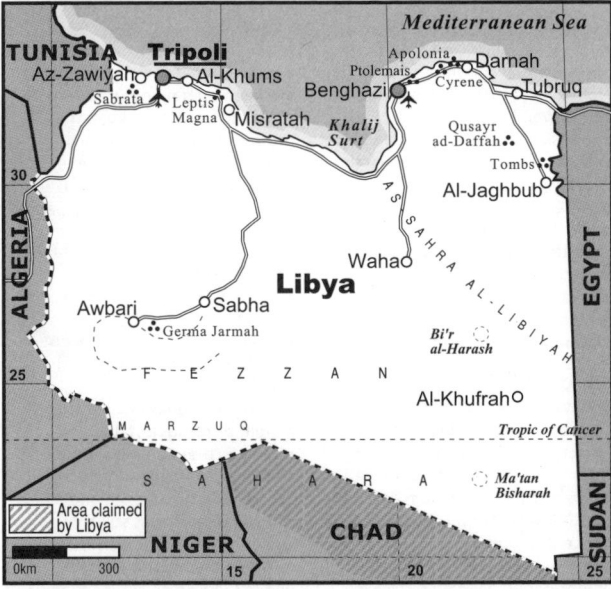

Libya has long been torn between the different political and economic centers of North Africa. The border with Egypt - where two Libyan dynasties ruled between the 10th and 8th centuries BC - permitted cultural contact, but did not lead to a unified state. The establishment of the Carthaginian and then the Roman empires on the western border further stressed this division. After the Arab conquest in the 7th century, Tunisia and Morocco on one side, and Egypt on the other, became the new centers of power, which left Libya's border situation unchanged.

2 The development of maritime trade and the ensuing piracy turned Tripoli (Tarabulus) into one of the major Mediterranean ports, leading to European and Turkish interventions. In 1551, Suleiman the Magnificent annexed the region to the Ottoman Empire. However, a weakened central authority gave increasing autonomy to the governors, precipitating independence movements. Piracy was used as a pretext to bomb Tripoli in 1804 (the first US foreign military intervention).

3 In 1837, Muhammad al-Sanussi founded a clandestine Muslim brotherhood (the Sanussi religious sect) which promoted resistance to Turkish domination, and was also active in Egypt. With the decline of the Ottoman Empire, Italy declared war on Turkey in 1911 and seized the Libyan coast, the last Turkish possession in North Africa. With the outbreak of World War I, Italy occupied the ports of Tripoli and Homs (Al-Khums) while the rest of the territory remained autonomous. At the end of the war, Italy faced the resistance led by Sidi Omar al-Mukhtar, which finally ended in 1931, when al-Mukhtar was captured and executed, and Libya was annexed by the Italian Empire.

4 From Egypt and Tunisia, the Sanussi brotherhood remained active and co-operated with the Allies in World War II. Muhammad Idris al-Sanussi, leader of the brotherhood, was recognized as Emir of Cyrenaica by the British. At the end of the war, the country was divided into a British zone (Tripolitania and Cyrenaica) and a French zone (Fezzan) governed from Chad. In 1949, a UN resolution restored legitimate union to the region and established the independent nation of Libya, with Idris al-Sanussi on the throne.

5 Idris based his power on religious authority and the support of powerful Turkish-Libyan families from the US and Britain (both holding military bases in the country) and transnational oil companies that had settled in the country, as the oil began to flow in great quantities in 1960.

6 In 1966, Muammar al-Qadhafi, the son of Bedouin nomads, founded the Union of Free Officers while studying in London. (He had joined the army as a young nationalist). He returned to Libya and on 1 September 1969 he led an insurrection in Sabha that swiftly overthrew the King.

7 Qadhafi's Revolutionary Council proclaimed itself Muslim, Nasserist and socialist; it eliminated all US and British military bases in Libya, and imposed severe limitations on the almost 60 transnational corporations operating in the country. The production of petroleum and its by-products was placed under state control, although the Government did not totally sever ties with the foreign corporations.

8 Qadhafi launched an ambitious development program, with special emphasis on agriculture. Each rural family was allotted 10 hectares of land, a tractor, a house, tools, and irrigation facilities. Over 1,500 artesian wells were drilled and two million hectares of desert began to be artificially irrigated.

9 Due to its rapid growth, Libya drew workers from other Arab countries and technicians from all over the world. In 1973, following publication of Qadhafi's Green Book - in which he expounds his ethical and political theories, rejecting capitalism and Marxism - he created a complex structure of popular participation through people's committees and a People's General Congress.

10 Qadhafi created a social security system in the cities providing free medical services and incentives to encourage large families. He gave industrial workers a 25 per cent share in company profits. After five years Libya was

PROFILE

ENVIRONMENT

Most of the country is covered by desert. The only fertile lands are located along the temperate Mediterranean coast, where most of the population live. There are no perennial rivers and rain is scarce. The country has significant oil deposits. Water is scarce and most of it is pumped from underground deposits. One of the largest hydraulic development projects in the world has partially solved this problem. Atmospheric pollution is caused by gases from oil refining. Desertification and erosion are growing.

SOCIETY

Peoples: The indigenous population was Berber. Today, Arabs account for 90 per cent of the population. There is a Berber minority that maintains its traditions. Key immigrant communities: Egyptian, Sudanese, Chadian, Italian, Greek, Pakistani, Turk, Korean and others.
Religions: Islam (official), Sunni. There is a small Christian minority.
Languages: Arabic (official); regional variations of Berber; languages of the immigrant communities; Italian and English.
Main Political Parties: The Socialist Party of Libya is in power. Other parties are banned. According to Qadhafi's 'Green Book', Libya is headed toward a direct democracy, in which there are no intermediaries. The new political organization is based on the Basic People's Congresses, directly elected, which elect the members of the 46 People's Congresses. These, in turn, select the members of the General People's Congress, the highest government body.
Main Social Organizations: Mass organizations of workers, peasants, students and women.

THE STATE

Official Name: Jamahiriya Al-Arabiya Al-Libiya Al-Shabiya Al-Ishtirakiya Al-Uzma (The Great Socialist People's Libyan Arab Jamahiriya).
Administrative Divisions: 3 provinces, 10 counties and 1,500 communes.
Capital: Tripoli (Tarabulus) 2,006,000 people (2003). In January 1987, Qadhafi appointed Hun, a village 650 Km to the southeast of Tripoli, as the administrative capital of the country.
Other Cities: Benghazi 1,041,000 people; Misrata 179,100; az-Zawiyah 175,100 (2000).
Government: Colonel Muammar al-Qadhafi, leader of the Revolution and commander-in-chief of the People's Armed Forces, has been Head of State since September 1969. Al-Baghdadi Ali al-Mahmudi has been Secretary of the People's Committee since 2006. Single-chamber legislature: the General People's Congress is the highest Government body, with 2,700 representatives of the People's Congresses.
National Holiday: 1 September, Revolution Day (1969).
Armed Forces: 76,000 (2003); 40,000 reservists, popular militia.

IN FOCUS

ENVIRONMENTAL CHALLENGES

Desertification and the scarcity of natural sources of drinking water are the most pressing problems to be dealt with. The Great Man-Made River Project, the first phase of which was inaugurated in August 1991 and involves more than 1,300 wells, will supply 6.5 million cubic meters of fresh water to the cities of Tripoli, Benghazi and Sirt. Colonel Qadhafi calls this project 'the eighth wonder of the world'.

WOMEN'S RIGHTS

Women have been able to vote and run for office since 1964. In 2003 women held 5 per cent of parliamentary seats and made up 25 per cent of the two-million-strong labor force. In 2004 maternal mortality was 77 per 100,000 births. In the same year 81 per cent of pregnant women received prenatal care and 94 per cent of births were attended by qualified personnel. Mistreatment of women is still a major problem. However, few complaints are filed, among other reasons because tradition exercises strong pressure to keep family affairs private. Some groups located far from urban areas continue to practice female genital mutilation on girls. Although the 1969 constitution ensures equal conditions for women, discrimination persists. For example, women need permission from their husband or another male relative to leave the country. Female emancipation is a generational phenomenon: women under 35 tend to have modern attitudes toward family and work, something not observed in older women. In 1979 a military academy for women was created and women play an active role in the army.

CHILDREN

Between 1990 and 2004 both infant and under-five mortality fell by 50 per cent. The former dropped from 35 to 18 per thousand live births and the latter from 41 to 20 per thousand live births. Some 15 per cent of under-fives measured moderately or severely below height norms. Military service is compulsory for adults over 18, and all teenagers over 14 receive pre-military training at school.

INDIGENOUS PEOPLES/ ETHNIC MINORITIES

The largest minority is the Berber, divided in three groups: Luata, Nefusa and Adassa. They live in the northeastern part of Libya, near the Mediterranean Sea; some live among the Jebel Nefusa plateaus and hills, as well as in the Fezzan Oasis in the southwest of Libya. The Government manipulates the clans, who are in need of funds and are keen on gaining government posts. It also tries to keep the groups separated from each other. There have been frequent allegations of discrimination, especially against the Tuareg and Berber.

MIGRANTS/REFUGEES

There are around 2.5 million foreign workers in Libya. Africans, especially, have been the object of resentment and violence.

The law does not grant asylum or refugee status. However, there are an estimated 30,000 Palestinians and 3,000 Somalis in Libya. At the end of 2005 UNHCR was still trying to get the Libyan Government to sign a memorandum of understanding on refugees and asylum seekers. This organization has registered and provides humanitarian assistance to some 12,000 refugees.

DEATH PENALTY

Libya retains the death penalty. Death sentences are still handed down, although there have been no recent reports of executions.

* Latest data available in The State of the World's Children and Childinfo database, UNICEF, 2006.

[10] no longer the poorest nation in North Africa and had the highest per capita income on the continent, $4,000 a year.

[11] In 1977, the country changed its name to the Socialist People's Libyan Arab Jamahiriya (meaning mass state in Arabic). But while Qadhafi achieved ample positive results internally, similar fruits were not achieved in the field of diplomacy. Attempts at integration with Syria, Egypt and Tunisia met with failure. Qadhafi criticized the rapprochement between Egypt and Israel, which led to friction with the Saudi monarchy, the Emirates and Morocco.

[12] From 1980, Libyan diplomacy and foreign relations focused on sub-Saharan Africa and Latin America. The Government supported the Polisario Front in Western Sahara and participated directly in the civil war in Chad, defending the Transitional Government of National Union, led by Goukouni Oueddei.

[13] The US linked Qadhafi to international terrorism and in August 1981 shot down two Libyan planes in the Gulf of Sidra. Qadhafi avoided any violent response thus gaining the political support of conservative Arab regimes previously hostile to Tripoli.

[14] In addition to imposing an economic embargo, the US bombed Tripoli and Benghazi in 1986, in an attempt to eliminate Qadhafi.

[15] In November 1991 US and British courts found the Libyan Government responsible for two terrorist attacks against commercial flights in 1988, one involving a Pan Am airliner over Lockerbie, Scotland, which left 270 dead including 189 Americans and the other a UTA plane over Nigeria, with a death toll of 170. In January 1992 Libya announced that it was willing to cooperate with the UN to determine where responsibility for these attacks lay.

[16] Despite his vow to cooperate, Qadhafi rejected a UN request for the extradition of Libyan agents linked to the attacks and unsuccessfully proposed that their trial take place in Tripoli. The UN gave an ultimatum demanding that Qadhafi explicitly renounce 'terrorism' by 15 April 1992. When this deadline expired the EC and G7 imposed economic sanctions. Qadhafi unsuccessfully appealed against these at the International Court of Justice. In 1994, the UN tightened the embargo.

[17] The isolation, however, did not impede the growth of the private sector or foreign investment, mainly in oil projects. One section of the pipeline designed to bring water to remote desert communities began to function in 1996.

[18] In 1998, the Movement of Non-Aligned Countries and the Organization of African Unity backed a Libyan request to try the two Lockerbie attack suspects in a neutral country. The UK and the US proposed that they be tried in The Hague by Scottish judges under Scottish law.

[19] In September 1999, more than 20 African and Arab leaders gathered in Tripoli to commemorate the 30th anniversary of the Libyan revolution. Fifteen years after having severed diplomatic relations, London sent its ambassador to Tripoli in December.

[20] Libya took advantage of this to launch a diplomatic offensive in the region, offering itself as mediator in the Sudan conflict and resuming relations with Chad. In March 2000, Washington sent a high level delegation to study the lifting of obstacles to investments and trips to Libya, banned since 1981.

[21] After the foiled coup in Central African Republic (CAR) in May 2001, the Qadhafi Government dispatched troops in order to protect President Patassé. In November, Patassé once again requested Libya's help. Once peace was restored, CAR demanded the withdrawal of Libyan soldiers.

[22] In April 2002, one Palestinian and six Bulgarian doctors were awaiting the verdict after being accused of deliberately infecting 400 children with HIV in 1999 as part of a CIA conspiracy against Libya. If found guilty, they could be sentenced to death. They claimed to have confessed under torture.

[23] In January 2003, Libya assumed the chair of the UN Commission on Human Rights, despite US opposition.

[24] After acknowledging its responsibility for the Lockerbie attacks in a letter to the UN Security Council in August 2002 the Libyan Government established a $2.7 billion indemnification fund for the victims' families. In September, the UN lifted sanctions against Libya.

[25] In December 2003, the Qadhafi Government announced that it would abandon its programs to develop weapons of mass destruction.

[26] In January 2004, Libya agreed to compensate the victims of a French aircraft shot down in the desert in 1989. In March, British Prime Minister Tony Blair went to Libya. It was the first visit by a British leader since 1943.

[27] Continuing with its new policy, in August Libya paid $35 million in compensation to the victims of a bomb attack on a night club in Berlin, Germany, in 1986.

[28] In January 2005, the greatest beneficiaries of the first gas and oil exploration license tender in four decades were American companies returning to the country after an absence of more than 20 years. In the second tender, in October, most of the contracts were given to Asian and European companies. In December the Supreme Court overruled the death penalty imposed on those accused of infecting Libyan children with HIV/ AIDS and a new trial began.

[29] In February 2006 cartoons published in a Danish journal satirizing the Prophet Muhammad provoked violent protests and police repression that left at least 10 dead.

[30] By May 2006 diplomatic relations with the US had greatly improved. External Affairs Minister, Abdel Rahman Shalgham, said that the normalization of relations between the two countries served not only bilateral interests but also international political stability. ∎

Liechtenstein / Liechtenstein

Population:	35,135
Land area:	160 km²
Capital:	Vaduz
Currency:	Swiss franc
Language:	German

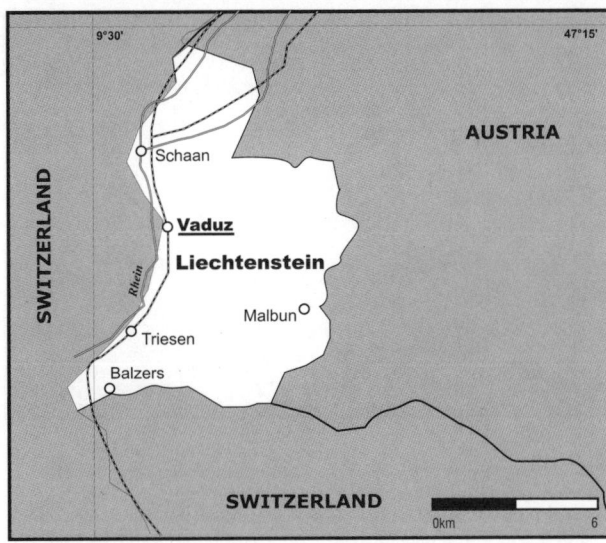

The territory of the present-day principality of Liechtenstein was inhabited by Neolithic times. The Rhaetians settled there in 800 BC, remaining until the Romans arrived in 15 BC. A Roman road traversed the country from north to south and was protected from German incursions by the fortifications of Schaan.

² In the 4th century AD, Saint Luzius brought Christianity to the province of Churrhaetia. A century later, the Germans invaded the country from the north and eliminated the Romans. Years later, the region passed into the hands of the Germanic Dukedom, forming part of the country of Lower Rhaetia, producing the two feudal domains of Vaduz and Schellenberg, ruled by the Counts of Werdenberg-Vaduz, the Barons of Brandis, the Counts of Sulz and the Counts of Hohenems.

³ Prince Johann Adam of Liechtenstein, founder of the current principality, bought the Schellenberg domain in 1699 and that of Vaduz in 1712. By combining hereditary rights from these Germanic domains he gained a seat and a vote in the Imperial Council of Princes.

⁴ The principality of Liechtenstein was established on 23 January 1719, when Emperor Charles VI of Germany converted the two counties into a 'principality of the empire' for his servant Anton Florian of Liechtenstein.

⁵ In 1806 Liechtenstein was made part of the Confederation of the Rhine - a league of 16 states belonging to the German Empire - by Napoleon. He guaranteed them independence and they recognized him as protector.

⁶ The principality entered the Germanic Confederation under the Vienna Congress in 1815 and remained there until this was dissolved in 1866, breaking the last juridical links with Germany. Liechtenstein has had no army since 1868. Between 1852 and 1918 (end of World War I) there was a tariff agreement with the Austro-Hungarian empire.

⁷ Beginning with the reign of Johann (1858-1929), Liechtenstein saw prosperity. Modern development of the country really began with the promulgation of the constitution of 1862 and was reinforced by the Democratic-Liberal Magna Carta of 1921, still in place today. Under Johann's leadership, Liechtenstein signed a tariff agreement with Switzerland in 1923 and in 1924 adopted the Swiss franc as legal currency.

⁸ Prince Franz Josef II began his reign in 1938 and became the first monarch to live permanently in Liechtenstein.

⁹ Liechtenstein - the only independent principality of the old Austro-Hungarian Empire - began its process of integration with the rest of Europe and the world after World War II.

¹⁰ In 1950 it became a member of the International Court of Justice at the Hague.

¹¹ The country entered the Council of Europe in 1975.

¹² Prince Franz Josef II died on 13 November 1989, after a reign of 51 years. He was succeeded by his oldest son, Hans-Adam II.

¹³ Liechtenstein joined the United Nations in 1990, the European Free Trade Area (EFTA) in 1991 and the European Economic Zone in 1995.

¹⁴ In 2000 a report by the German secret service (BND) stated that Liechtenstein was the leading money-laundering location in Europe, and among the leaders in the world. The Financial Action Task Force on Money Laundering (FATF) included the principality on a list of money-laundering havens.

¹⁵ New fiscal legislation enabled the principality to be removed from the FATF's black list in 2001. However, that year it received 107 requests from abroad for judicial co-operation on matters of economic or financial misconduct. For the first time, the principality's justice system enforced a legal sentence related to the drug trade.

¹⁶ Despite the new controls, Liechtenstein's neighbors and international organizations have pressed it to further modify its legislation, as they believe the State still maintains openings for money-laundering and tax evasion. But bankers - including the Prince himself, head of the LGT Bank, the biggest in Liechtenstein - said further banking reform would endanger most of the $70.3 billion held in the principality.

¹⁷ A March 2003 referendum granted new political powers to Prince Hans-Adam.

¹⁸ Prince Hans-Adam passed power on to his son, Prince Alois, in August 2004, although he remained as head of State.

¹⁹ In 2005, the Council of Europe criticized Liechtenstein's constitutional amendments, that increased the royal family's powers, granting them the right to veto new laws and to dismiss governors at the discretion of the monarch. Alois - who supported his father's position on the amendments - said Liechtenstein could leave the Council of Europe if the latter sought to monitor his country's democracy.

²⁰ In the March 2005 Parliamentary election, the Progressive Citizens Party (FBP) won with 48.7 per cent of the vote, followed by Patriotic Union (VU), with 38.2 per cent, and the environmentalist Free List (FL), with 13 per cent.

²¹ In April 2005, the final report by a commission created in 2001 by the Government to investigate the principality's banks' behavior regarding Nazi money during World War II concluded that the banks had acted according to the law.

²² In March 2006, the Organization for Economic Cooperation and Development kept Liechtenstein in a list of uncooperative countries regarding tax issues. The principality is one of the five countries from the industrialized North included in the list due to their policy of low taxes and bank secrecy, which hinders investigations into the origin of capital, which is often the fruit of illegal activities. ∎

PROFILE

ENVIRONMENT
This small principality lies between Switzerland and Austria, in the Rhine valley. Wheat, oats, rye, corn, grapes and fruit are produced. Some 38 per cent of the land is pasture. In recent years, the principality has been transformed into a highly industrialized country, producing textiles, pharmaceutical products, precision instruments and refrigerators among other items.

SOCIETY
Peoples: German 95 per cent; Italian and other 5 per cent.
Religions: Catholic 80 per cent; Protestant 6.9 per cent; other 5.6 per cent. **Languages:** German (official). **Main Political Parties:** The Progressive Citizens Party (FBP); Patriotic Union (VU); Free List, green. **Main Social Organizations:** Trades Union Association (artisans and traders), Agricultural Union.

THE STATE
Official Name: Fürstentum Liechtenstein. **Administrative Divisions:** 11 communes. **Capital:** Vaduz 5,000 people (2003).**Other Cities:** Schaan 5,143; Balzers 3,752; Triesen 3,586 (1995). **Government:** Liechtenstein is a constitutional monarchy. Prince Hans-Adam II, Head of State since 13 November 1989; to hand over power to his son, Prince Alois, in August 2004. Otmar Hasler, Head of Government since April 2001. Government functions are carried out by an FBP and VU coalition. Single-chamber legislature: Parliament with 25 members elected every four years. Diplomacy: A member of the European Council, Liechtenstein has a customs and monetary alliance with Switzerland, which is its representative abroad. **National Holiday:** 14 February.

Lithuania / Lietuva

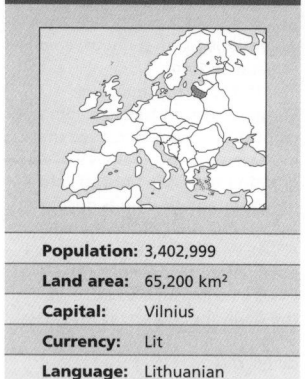

Population:	3,402,999
Land area:	65,200 km²
Capital:	Vilnius
Currency:	Lit
Language:	Lithuanian

Lithuanians have lived along the shores of the Baltic Sea since long before the Christian era. Protected by the forests, Lithuanian tribes fiercely resisted German efforts to subdue them in the 13th century, and united under the leadership of Mindaugas, who was crowned king by Pope Innocent IV in 1253.

[2] In the 14th century, Lithuania began its eastward and southern expansion, going into Belarusian lands. Gediminas built the Grand Duchy of Lithuania, which extended from the Baltic Sea to the Black Sea, with its capital at Vilnius. In 1386, Jagiello, Gediminas' grandson, married the Queen of Poland, thus uniting the two kingdoms.

[3] In 1480 with the coronation of Ivan III of Muscovy as the sovereign of all Russia, a new and greater threat emerged for historic Lithuania. Nevertheless, the Lithuanian-Polish union reached its peak in the 16th century, when it was unrivalled in Europe as a political system (see Poland), only to fall in the 17th century, in the course of a series of devastating wars with Sweden, Russia and Turkey as well as peasant rebellions within.

[4] In the 1772 and 1793 partitions of Poland among Russia, Prussia and Austria, Russia kept only Belarus. But the Polish state disappeared in 1795, and all of Lithuania was in Russian hands in

1815. That same year, the Congress of Vienna granted the Russian Emperor the additional titles of King of Poland and Grand Prince of Lithuania.

[5] The czarist regime treated Lithuania as though it were a part of Russia, calling it the Northwest Territory after 1832. Between 1864 and 1905, Russification extended to all aspects of life: books printed in Lithuanian had to use the Cyrillic alphabet, and Catholics were persecuted.

[6] During World War I, Germany occupied a major part of Lithuania. In 1915, a congress - authorized by the occupying Germans - elected the 20-member Council of Lithuania. The 214 delegates to the congress called for the creation of an independent Lithuanian state within its 'ethnic borders' and with Vilnius as its capital. On 16 February 1918 the Council declared Lithuania's independence and

terminated all political ties with other nations.

[7] In 1919, the Red Army entered Vilnius and formed a communist government, which was later forced to withdraw. The new head of the Polish State, Josef Pilsudski, tried to re-establish the former union, but failed. In the end, the League of Nations and the European powers agreed to the separation of Poland and Lithuania in 1923, but Lithuania refused to recognize this line of demarcation.

[8] In 1926, Lithuania and the Soviet Union signed a non-aggression treaty. Lithuania, Latvia and Estonia signed a treaty of good will and co-operation in Geneva in 1934.

[9] In September 1939, a secret German-Soviet non-aggression treaty brought Lithuania within the USSR's sphere of influence. In October, a mutual assistance treaty was signed in Moscow; according

to its terms, Lithuania was forced to accept the installation of Soviet garrisons and air bases on its soil. In 1940, the Soviet Army occupied Lithuania and a number of local political leaders were arrested and deported, while others fled toward Western Europe.

[10] In August 1940, during the term of Prime Minister Justas Paleckis, Lithuania was incorporated as a constituent republic of the USSR. After German occupation in 1941, the Baltic States and Belarus became the German province of Ostland.

[11] During the German occupation, 190,000 Jews were sent to concentration camps. Some 100,000 residents of Vilnius - a third of that city's population, most of them Jews - were killed. Vilnius was known as the 'Jerusalem of Lithuania' and had been considered to be one of the world's most important centers of Jewish culture.

[12] Vilnius was reconquered by the Red Army in 1944 and Lithuania was once again occupied by the Soviets. Almost 20,000 Lithuanians took refuge in Eastern Europe, while a new period of Sovietization began that included deportations to northern Russia and Siberia.

[13] With the democratization process initiated by Mikhail Gorbachev in the USSR, Lithuania began a period of intense political agitation. In June 1998, the Lithuanian Movement to Support Perestroika (restructuring) was founded; its Executive Committee adopted the name Sejm - the name of the Lithuanian Parliament at the time of independence - also known as Sajudis.

[14] In July, the Lithuanian Freedom

LAND USE

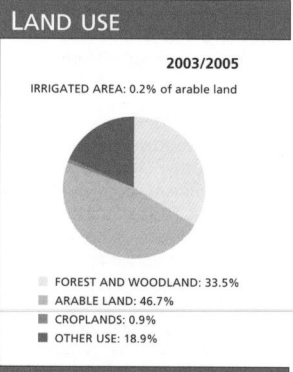

2003/2005

IRRIGATED AREA: 0.2% of arable land

- FOREST AND WOODLAND: 33.5%
- ARABLE LAND: 46.7%
- CROPLANDS: 0.9%
- OTHER USE: 18.9%

PROFILE

ENVIRONMENT
Situated on the eastern coast of the Baltic Sea, Lithuania is characterized by gently rolling hills and flat plains. The largest of the Baltic 'mini-states', it has more than 700 rivers and streams, abundant forests and around 3,000 lakes. The Nemunas river, which crosses the country from east to west, is an important shipping route. Some 49 per cent of the land is arable; the chief crops are grains, potatoes and vegetables. There are large numbers of livestock. Among the country's leading industries are the food and machine manufacturing industries, as well as the export of energy. Around 70 per cent of the country's energy is generated by a nuclear plant of the same type as the one in Chernobyl, Ukraine - the site of the 1986 nuclear accident. Since the 1980s, pollution has increased, especially bacterial pollution of rivers and lakes, which has been linked to the increase in infectious childhood diseases, particularly in the first years of life.

SOCIETY
Peoples: 83 per cent of the population is Lithuanian; the rest include Russians, 8.4 per cent; Poles, 7.0 per cent; Belarusians 1.5 per cent; Ukrainians 1.0 per cent and smaller populations of Germans, Jews, Latvians, Tatars, Armenians, Moldovans, Roma,

Uzbeks and Azeris.
Religion: Catholic majority (80 per cent); there are Protestant minorities. There are also Orthodox Russians; some evangelical minorities; plus Muslims, Jews and people with no religion.
Languages: Lithuanian (official); Russian, Polish.
Main Political Parties: Labour Party; Working for Lithuania Coalition; Homeland Union; Coalition of Rolandas Paksas for the Order and Justice.
Main Social Organizations: Lithuanian Green Movement, Human Rights Association.

THE STATE
Official name: Lietuvos Respublika.
Administrative Divisions: 44 Districts.
Capital: Vilnius 549,000 people (2003).
Other Cities: Kaunas 411,600 people; Klaipeda 202,400; Siauliai 146,300; Panevezys 133,600 (2000).
Government: Parliamentary republic. Valdas Adamkus, president since July 2004. Gediminas Kirkilas, Prime Minister since July 2006. Single-chamber parliament (Seimas) made up of 141 members.
National Holiday: 16 February, Independence Day (1918).
Armed Forces: 14,000 (2003).

Life expectancy	GNI per capita	Literacy	HIV prevalence rate
73 years 2005-2010	**$5,740** 2004	**100%** total adult rate 2000-2004	**0.1%** of population 15-49 years old 2003

League (LFL) emerged from underground activity. The LFL, which dates back to 1978, called for immediate withdrawal of Soviet troops from the country, and the independence of Lithuania, aiming in the long term at integration with the European Union (EU).

[15] In February 1989, the Sajudis demanded a free and neutral Lithuania, within a demilitarized area. That month the first secretary of the CP and the Sajudis attended the official commemoration of the country's independence, side by side. In December, Lithuania's Supreme Soviet did away with the article of the constitution which assigned the CP a leading role; the first decision of its kind within the USSR.

[16] In January 1990, Soviet President Mikhail Gorbachev announced in Vilnius that the details of a future relationship with the Union would be established through legislation. In March, the Lithuanian Parliament proclaimed the nation's independence, effective immediately. In September 1991, the new Council of State of the USSR accepted the independence of the three Baltic States, which were immediately recognized by several countries and by the UN.

[17] In August 1991, after the failed coup against Gorbachev in the USSR, the Lithuanian Parliament banned the Communist Party, the Democratic Workers' Party and the Lithuanian Communist Youth organization. The following month, President Vytautas Landsbergis issued a call, before the UN, for the withdrawal of Soviet troops from Lithuania.

[18] The new constitution was approved by a referendum on 25 October 1992. That year, GNP diminished by over a third and inflation reached almost 1,000 per cent.

[19] In February 1993, the leader of the former Communist Party, Algirdas Brazauskas, was elected President with 60 per cent of the vote. In 1993 and 1994, Brazauskas continued with the transition policy toward a market economy.

[20] In 1998, Valdas Adamkus was elected President, with a slight margin above his rival, former communist Arturas Paulauskas.

[21] Massive demonstrations against the sale of the state oil company to the US firm Williams International forced Prime Minister Rolandas Paksas to resign in October 1999. The President appointed former Vice-President Andrius Kubilius as Prime Minister, and went ahead with the sale.

[22] Paksas, leader of the right-wing Liberal Democratic Party, won the January 2003 presidential elections with 54.9 per cent of the vote, defeating Adamkus, the front-runner in most polls.

[23] In a referendum held in May 2003, Lithuania's incorporation into the EU was approved by 91 per cent of voters.

[24] In November 2003, President Paksas was accused in a military intelligence report of having ties with the Russian Mafia and secret services, of illegal arms sales and of financing international terrorism. The parliamentary commission appointed to investigate the accusations concluded that Paksas had violated the constitution and that he leaked secret information, thus posing a threat to national security.

[25] In February 2004 three Russian diplomats were accused of espionage and expelled from Lithuania. The foreign ministry said they were engaged in illegal activities, under the protection of their diplomatic positions, favoring privatizations and attempting to obtain secret information from members of parliament regarding President Paksas' impeachment. The following month, Lithuania joined the North Atlantic Treaty Organization (NATO).

[26] Parliament voted in April to impeach Paksas for violations to the Constitution, including leaking classified documents and granting Lithuanian citizenship to a Russian person in exchange for financial support. Paksas insisted he was innocent and that his mistakes did not deserve an impeachment. 'Impeachment is the political system's revenge against me, a vendetta due to my efforts in the fight against corruption in the country,' he said. In compliance with the Constitution, Paksas was replaced by his political rival, Arturas Paulauskas.

[27] Lithuania joined the EU in May 2004. In June, Valdas Adamkus was elected President in the second round with 52 per cent of the vote, followed by the 48 per cent of his rival Kazimeira Prunskiene. Adamkus said he would start immediate talks with the other candidates to form a Government.

[28] After the October general elections, Algirdas Brazauskas continued as Prime Minister. In November, Lithuania was the first member state to ratify the EU Constitution. The Labor Party (LP) stayed in the government coalition, although its leader, Viktor Uspaskich, resigned as economy minister in June 2005, due to accusations from right-wing parties that were ratified by a parliamentary Ethics Committee. He was accused of signing commercial agreements that favored private interests.

[29] One of the reactors at the Ignalina nuclear plant was shut down in December, in compliance with EU standards. The second was to be shut down in 2009.

[30] In May 2005, President Adamkus turned down Moscow's invitation to attend the commemoration of the end of World War II. In September, the crashlanding of a Russian military plane - carrying at least four missiles - over Lithuanian soil caused tension between Moscow and Vilnius. Finally, after investigations showed that the incident could be attributed to technical failure and human errors, the situation went back to normal and the pilot (who had remained under arrest in Lithuania) was freed and returned to Russia.

[31] Lithuania requested in January 2006 that the shutdown - scheduled for 2009 - of the Ignalina plant, one of the conditions for its entrance to the EU, be postponed until the country could find other power sources.

[32] A Lithuanian who had collaborated in the extermination of Jews during the country's occupation by the Nazis was found guilty but not sent to jail, due to his old age. Approximately 200,000 Jews were murdered in Lithuania during World War II. ∎

IN FOCUS

ENVIRONMENTAL CHALLENGES
Soil and groundwater have been significantly polluted by oil by-products and chemicals. Since 1980 an increase in pollution of the country's aquifers has been reported, especially of bacterial origin, which could threaten a spread of disease through rivers and lakes.

WOMEN'S RIGHTS
Lithuanian women have been able to vote and run for office since 1918. In 2004, women held 18 per cent of seats in Parliament, and 15 per cent of ministerial or equivalent positions. Women comprised 48 per cent of the two-million strong workforce in 2004; female unemployment reached 13 per cent, while male unemployment stood at 15 per cent.

Since the 1990s, Lithuania has been recognized as a country of origin, destination and transit of women forced into prostitution. The Government stated the need for international assistance at the Council of Europe.

CHILDREN
The infant mortality rate was 8 in every 1,000 live births in 2004, the same as for children under five.

The primary school enrolment rate stood at 91 per cent in 2004, with next to no gender gap. Most of the Lithuanian adult population is literate*. Online advertisements offering Lithuanian children for adoption, and explaining the legal procedures to be followed, are common.

INDIGENOUS PEOPLES/ ETHNIC MINORITIES
Poles and Russians make up national minorities, and the Roma/Gypsies are an ethnic minority. None of these groups has had serious problems with the Government or with Lithuanian society as a result of their ethnic or national origin. Since 1990, Lithuanian policies on minorities have been regarded as the most liberal in the area. A small group of organizations is working to promote Russian culture in Lithuania. Poles live mostly in Vilnius and stand out due to their language, culture and religion. Their most extreme demands include the reinstatement of property rights over land. Their most common demands are related to greater dissemination of Polish language and culture.

The greatest difficulty for the integration of the Roma into Lithuanian society is their nomadic character. Most of the adults lack identity documents and are illiterate. They generally speak Russian. Roma children are not encouraged to finish school and their linguistic differences cause them problems.

MIGRANTS/REFUGEES
In late 2002, about 30 Chechens were sent to Belarus, which the UNHCR considered a violation of the UN Convention on Refugees. Lithuanian law provides for asylum and refuge, according to the UN 1951 Convention regarding the status of refugees. It grants protection against the forced repatriation of those people who are persecuted in their countries of origin.

DEATH PENALTY
The death penalty was abolished for all crimes in 1998. The last execution took place in 1995.

*Latest data available in *The State of the World's Children* and *Childinfo* database, UNICEF, 2006.

Luxembourg / Luxembourg

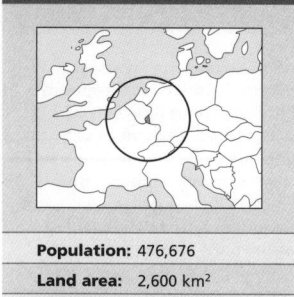

Population:	476,676
Land area:	2,600 km²
Capital:	Luxembourg
Currency:	Euro
Language:	Letzebuergish, French, German

Luxembourg, Belgium, the Netherlands, and part of northern France constitute the Low Countries, and until 1579 they shared a common history (see the Netherlands).

[2] In the war of the Low Countries against Spain, Luxembourg sided with the southern provinces, acknowledging the authority of Philip II. Luxembourg was conquered by France in 1684, but returned to Spain 13 years later, under the Treaty of Rijswijk. In 1713, it came under control of the Austrian Hapsburgs until the Napoleonic invasion of 1795, when it was annexed by the French Empire.

[3] In 1815, after the defeat of Napoleon, the Congress of Vienna handed over the Duchy to William of Orange, who incorporated it as his kingdom's 18th province. After the Belgian revolt in 1831, Luxembourg was divided.

[4] The largest section was given to Belgium, and the smallest to William as the Grand Duchy of Luxembourg, which he accepted in 1839. Thereafter the Duchy was administered independently until 1867. In 1866, the German Confederation was dissolved and the Treaty of London guaranteed the neutrality of the Grand Duchy.

[5] Germany occupied the country from 1914 to 1919 and again from 1940 until 1949. After World War II Luxembourg formed an alliance with Belgium and the Netherlands called Benelux.

[6] In 1949 Luxembourg abandoned its neutrality and became a founding member of NATO.

[7] In 1957 it became a founding member of the European Economic Community (EEC), which began to operate in 1958.

[8] In 1964 the Grand Duchess Charlotte abdicated in favor of her son, who became Grand Duke Jean.

[9] In the 1974 elections, the Christian Social Party (CSP) was replaced in power for the first time since the end of World War I by a coalition of the Socialist Workers' Party (SWP) and the Democratic Party (DP).

[10] In 1979 the CSP regained power, while the country experienced an economic recession.

[11] Jacques Santer became Prime Minister in 1984.

[12] In the June 1989 legislative elections the CSP, SWP and DP took 22 seats. Women - who won the right to vote in 1919 - began to be admitted to the armed forces in October.

[13] In 1990, the ambassador at NATO, Guy de Muyser, resigned amid accusations that he leaked classified information to the USSR. Border controls were abolished with Belgium, France, Germany and the Netherlands.

[14] Luxembourg signed in 1992 the Maastricht Treaty, which created the European Union (EU).

[15] In 1994, the CSP and the SWP were returned to power in the general elections, allowing Jacques Santer to continue as Prime Minister.

[16] In 1995, Santer became President of the European Commission and was replaced as Prime Minister by Jean-Claude Juncker.

[17] Unemployment rose to 3.7 per cent in late 1997, which was still the lowest rate among the 15 EU member states. In November, some 30,000 people from various European countries demonstrated in Luxembourg, calling for a Europe 'at the service of employment'.

[18] In March 1998, the Government announced it would increase funding for the plan to extend the rail network for the high-speed train (TGV) to Eastern Europe.

[19] Grand Duke Jean abdicated in September 2000. His son Henri, 45, took office in October.

[20] A report by a French parliamentary commission accused Luxembourg in January 2002 of standing in the way of the fight against money-laundering and financial corruption. Among the main obstacles, it mentioned the strict banking secrecy laws, and the administrative delays any time the authorities were asked to co-operate.

[21] The euro became the national currency in January 2002.

[22] On 1 March 2004 representatives of the workers, the Government and the EU met to define Luxembourg's stand on the Kyoto Treaty and emissions of greenhouse gases.

[23] In June, Jean-Claude Juncker's Social Christian Party (PSC) won a majority of votes in the general elections for the lower house. Juncker was re-elected Prime Minister and called for a new government coalition.

[24] The EU Constitution was approved by 57 per cent of voters in July 2005. Other countries - warned by the negative votes in France and the Netherlands - had agreed to take a one-year reflection period before their referendums, but the Luxembourg Parliament insisted on holding it on the scheduled date.

[25] In June 2005 Luxembourg chaired the EU's Council. In May 2006, Juncker, as chairman, said the negative votes against the Constitution and the 'reflection periods' the EU had granted its citizens - until mid 2007 - meant a stagnation that could continue during the following decade. ∎

PROFILE

ENVIRONMENT
Located on the southeastern side of the Ardennes, Luxembourg has two natural regions. The north is a sparsely populated region of valleys and woods, with a maximum altitude of 500 meters, where potatoes and grains are cultivated. The south (Gutland) is a low plain and the country's main demographic corridor, where most of the population, major industries (iron, steel and mining), and cities, including the capital, are located. This part of the country has problems with air and water pollution in urban areas.

SOCIETY
Peoples: Luxembourger 67.4 per cent; Portuguese 12.1 per cent; Italian 4.8 per cent; French 3.5 per cent; Belgian 2.8 per cent; German 2.3 per cent; other 7.1 per cent. **Religions:** No official religion. Catholic (94.9 per cent); Protestants (1.1%); Jewish (4%). **Languages:** Letzebuergish, French, German, Portuguese and Italian. **Main Political Parties:** Christian Social People's Party (center-right); Luxembourg Socialist Workers' Party (center-left); Democratic Party (center-left); The Greens (ecologist). **Main Social Organizations:** General Confederation of Luxembourg Workers; National Trade Union Council.

THE STATE
Official Names: Grand-Duché de Luxembourg; Grossherzogtum Luxemburg, Groussherzogtum Lëtzebuerg. **Administrative Divisions:** 3 districts and 12 cantons. **Capital:** Luxembourg-Ville 77,000 people (2003). **Other Cities:** Esch-sur-Alzette 25,500 people; Dudelange 17,000; Differdange 17,700 (2000). **Government:** Constitutional monarchy. Multi-party parliamentary system. Grand Duke Henri, Head of State since October 2000; Jean-Claude Juncker, Prime Minister and Head of Government since January 1995. Single-chamber legislature: Chamber of Deputies, with 60 members elected by direct popular vote, every 5 years. **National Holiday:** 23 June, National Day (1921). **Armed Forces:** 800. Other: 560 (Gendarmes).

Macedonia / Makedonija

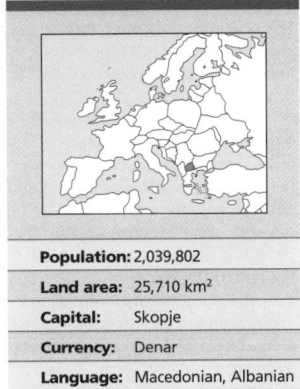

Population:	2,039,802
Land area:	25,710 km²
Capital:	Skopje
Currency:	Denar
Language:	Macedonian, Albanian

The area that was historically called Macedonia belongs to the present-day Republic of Macedonia and to the states of Serbia, Bulgaria and Greece. Archeological finds show evidence of human settlements between 7000 and 3500 BC. Semi-nomadic Indo-European peoples arrived then in the Balkan Peninsula. During the 1st millennium BC, the area was inhabited by Tracians, Illyrians, Dacians and Greeks.

2   Macedonia became the dominant power in Greece during the reign of Philip II (359-336 BC). Under Macedonian hegemony, the League of Corinth was created, linking all Greek city-states with the exception of Sparta.

3   Alexander III ('the Great'), Philip's son and a student of Aristotle, defeated the Persian Empire and led the Macedonian armies to northern Africa and the Arabic peninsula, crossing Mesopotamia and reaching as far east as India. Over 11 years, he built up the largest empire the world had ever seen up to that point. The aim of his empire was the urbanization of the Orient and the fusion of Greek culture with the cultures of the peoples he conquered, giving rise to what is known as Hellenism.

4   After Alexander's death in 323 BC, the succession struggle by his generals led to the division of the empire into three large kingdoms: Egypt, Macedonia and Asia. That period, which began with Alexander's death and lasted until the foundation of the Roman Empire, is known as the Hellenistic age.

5   By the 2nd century BC, the Romans began their expansion into the Balkans, where they arrived in search of metals, slaves and agricultural products.

6   In 168 BC, upon his defeat by the Romans, Perseus was forced to dissolve the kingdom of Macedonia which became a Roman province in 148.

7   The ethnic composition of the Macedonians was not significantly affected by the Goth, Hun and Avar invasions. However, when Slavs arrived in the Balkans, they established permanent settlements throughout Macedonia.

8   Between the 7th and the 14th centuries, Macedonia was successively subdued by the Bulgarian, Byzantine and Latin empires until they were almost completely dominated by the Serbs.

9   In 1389, after the Battle of Kosovo, Serbia recognized Turkish sovereignty, and in 1549 it joined the Ottoman Empire. The Ottomans seized the best lands for themselves and established a feudal system. Christian peasants either became vassals of Muslim lords, to whom they paid a tithe, or were driven onto the less fertile lands.

10   In 1864, the Ottoman Empire divided Macedonia into three provinces: Salonika, Monastir, including parts of Albania, and Kosovo, which extended into 'Old Serbia'. In 1878, Russia forced Turkey into accepting the creation of Bulgaria, which included most of Macedonia, but the other European powers returned this territory to the Ottomans. During the ensuing years, Bulgaria, Serbia and Greece all continued to lay claim to Macedonia.

11   Towards the end of the 19th century, a strong nationalist movement emerged in Macedonia. In 1893, the VMRO (Vatreshna Makedonska Revolutsionna Organizatsia) was created, with the slogan 'Macedonia for the Macedonians'.

12   In 1908, after the fall of the Ottoman Empire, the clamor for possession of Ottoman-Turkish territories in the region culminated in the two Balkan Wars of 1912 and 1913.

13   The Treaty of Bucharest (1913) put an end to the second Balkan War. Bulgaria lost Macedonia which was divided between Greece (Salonika and most of the Macedonian coastal area) and Serbia (central and northern parts of the territory). Albania became an autonomous principality.

14   The Balkans turned into the powder keg of Europe and finally set off World War I. The end of World War I saw the partition of 1913 confirmed and Slavic Macedonia was incorporated into the new Serbian, Croat and Slovenian kingdom.

15   In the inter-war period, Serbian domination deepened Yugoslav inter-ethnic conflicts. King Alexander, who assumed dictatorial powers in 1929, was assassinated in Marseilles in 1934 by Croatian nationalists. At the beginning of World War II, when Germany invaded Yugoslavia, these internal divisions meant that the invaders met little resistance.

16   The Yugoslav nationalist struggle intensified during the following years. Guerrillas led by the Yugoslav Communist League (YCL) seized power in May 1945, later proclaiming the Socialist Federal Republic of Yugoslavia, which included six republics (Slovenia, Croatia, Bosnia-Herzegovina, Montenegro, Serbia and Macedonia) and two autonomous regions belonging to Serbia (Kosovo and Voivodina).

17   That year, Tito (Josip Broz) was elected President. Commerce and banking were nationalized and agriculture was collectivized. He distanced himself from Moscow and launched the 'Yugoslav road to socialism'. On the foreign policy front, he was a prominent leader in the movement of non-aligned countries.

18   The Yugoslav system distinguished between 'constituent peoples' of the Federation (Serbs, Croats, Slovenes, Macedonians, etc.) and 'nationalities'. Since the latter had no State or their State of reference did not belong to the Federation, they were considered 'national minorities', regardless of how numerous they were in each region. Albanians were granted this minority status. They formed a clear majority in the province of Kosovo, although they represented a minority in the Republic of Serbia, of which Kosovo formed a part.

19   After Tito's death, conflict broke out among the republics that constituted the Federation. In 1989, the Federal Government withdrew all reference to minorities from the Constitution. In January 1990, a special Congress of the Yugoslavian Communist League did not accept the motion to grant greater autonomy to YCL branches in the republics. The Communist Leagues of Slovenia, Croatia and Macedonia decided to separate from the YCL and created the Communist League-Democratic Renewal Party.

20   The first republics to become independent were Slovenia and Croatia, at the expense of a conflict with Serbia. On 8 December 1991, a plebiscite was held in which

Life expectancy
74 years
2005-2010

GNI per capita
$2,420
2004

Literacy
96% total adult rate
2000-2004

HIV prevalence rate
<0.1% of population 15-49 years old
2003

Macedonians pronounced themselves in favor of independence. All Macedonian political parties, except for the Albanian ethnic minority, supported the decision.

[21] Greece refused to acknowledge the republic, claiming that the use of the name 'Macedonia' was a 'usurpation' of the name of a Greek province and of part of Greece's history and culture.

[22] On 12 January 1992, in a referendum, the Albanian minority of Macedonia voted for the creation of their own independent state. On 3 April, the territory of Macedonia was proclaimed the Independent Republic of Illirida (republic of Albanians resident in Yugoslavia). That year, the UN approved sending troops to control inter-ethnic conflicts.

[23] The new Yugoslav Federation withdrew its troops from the country. In July, the entire cabinet resigned, after failing to achieve international recognition. Social Democrat Branko Crvenkovski took over as Prime Minister in August and succeeded in obtaining recognition from Russia, Albania, Bulgaria and Turkey.

[24] In April 1993, the country was admitted as a member of the UN with the provisional name of The Former Yugoslav Republic (TFYR) of Macedonia.

[25] In 1996, a privatization plan led to the downfall of the ruling coalition. The October-November 1998 parliamentary elections were won by a new coalition, called the Internal Macedonian Revolutionary Organization-Democratic Party for Macedonian National Unity (VMRO-DMPNE), which obtained 28.1 per cent of the vote. The Social-Democratic League took 25.1 per cent, while the Democratic Alternative, with slightly more than 10 per cent, joined the governing coalition. The Albanian Democratic Prosperity Party came in third, with 19.3 per cent of the vote.

[26] To avoid trouble with the Albanian minority, while tension soared in neighboring Yugoslavia, the Government asked NATO to station troops on the border. When the bombing of Yugoslavia began in March 1999, Macedonia offered NATO the use of its troops and air space and opened the border to Albanian refugees coming from Kosovo. UNICEF reported that some 360,000 refugees had arrived in Macedonia during that month.

[27] Macedonia's frail economy was severely affected by the war. Yugoslavia was one of its largest markets and the main supplier of raw materials.

[28] A revolution broke out in March 2001 as people demanded more rights for the Albanian minority, resulting in a wave of refugees and the occupation of territory by the rebel National Liberation Army (NLA). In August, after the intervention of the international community, rebels agreed to hand over their weapons in exchange for recognition of the Albanian minority.

[29] In November, after delays and breaks in the ceasefire, Parliament passed constitutional reforms which granted Albanians broader rights. The operation 'Essential Harvest' was launched under NATO supervision in order to collect weapons from Albanian rebels.

[30] The international community decided in early 2002 to send more than 500 million euros, twice the initially foreseen amount, for economic reforms and reconstruction, in recognition of the stability achieved six months after the end of the war.

[31] In 2003, Amnesty International denounced the persistent abuse and ill-treatment of Albanians, especially by members of the 'Lions', a special all-Macedonian police unit set up by the Interior Ministry following the NLA uprising.

[32] In February 2004, Trajkovski died in a plane crash on his way to a conference in Mostar. Bosnian television blamed NATO forces for the accident. He was succeeded in May by Branko Crvenkovski, who was elected President with 63 per cent of the votes in a special election held in April - although his victory was questioned by the opposition. His ally, former Interior Minister Hari Kostov, was appointed Prime Minister in June.

[33] In July, some 20,000 people demonstrated in Skopje against proposals made by Parliament to redraw municipal borders with Albania and against giving more power in certain areas to Albanians living in Macedonia. Finally, in August, following approval by Parliament, the project on municipal borders was enacted.

[34] Prime Minister Kostov resigned in November 2004, due to the impossibility of implementing his reform plans. One month later, Defense Minister Vlado Bukovski was appointed as his successor and formed the new cabinet.

[35] In July 2005, Parliament passed a law that granted Albanians the right to fly their flag in those districts where they are a majority.

[36] In 2003, following Macedonia's signing of the US-Adriatic Charter together with Albania and Croatia, thus articulating their joint efforts to enter the North Atlantic Treaty Organization (NATO), a series of reforms were launched in order to enable accession by 2009. In May 2006, US Vice-President Dick Cheney expressed his support for the integration of these countries and praised the progress being made. He also praised the fact that these nations had been acting jointly with US-led coalition forces in Afghanistan and Iraq.

[37] The VMRO-DMPNE, the main opposition party, won the July 2006 election. The VMRO-DMPNE's candidate, Nikola Gruevski, received 40 per cent of the vote, while Bukovski's Social Democratic Union received 24 per cent. ∎

IN FOCUS

ENVIRONMENTAL CHALLENGES
Water and air pollution, as well as the generation of industrial waste, particularly in the metallurgical industry, have reached alarming levels.

WOMEN'S RIGHTS
Women have been able to vote and stand for office since 1946, in the former Yugoslavia, and since independence in 1991. In 2002, women held 19 per cent of parliamentary seats. They comprised 43 per cent of the workforce in 2003.

Although official documentation is sketchy and confusing, Macedonia is known to be a major transit route from countries of the former USSR and Eastern Europe for women and children trafficked into Western Europe and the Balkans. In addition, there is also growing evidence of internal trafficking and of the fact that Macedonia is becoming a final destination for human trafficking.

According to UNICEF, domestic violence is a growing problem, although official statistics don't reflect it.

CHILDREN
The progress of children and women is fragile and slow due to frequently changing governments, social tensions and ongoing economic problems*. A life skills curriculum adopted by many schools has had great success in improving the overall quality of education*.

Mortality rates for Roma children are almost double those for the general population. Nearly 25 per cent of Roma women give birth at home, without receiving any medical care*.

The movement of children and their families across the border since 2001 has proved that many lack identity documents and had not even been registered at birth. Most of these undocumented children were born outside the country's health system infrastructure and live in rural areas or belong to the Roma ethnic group. Preventing HIV/AIDS transmission among young people remains a major concern.

INDIGENOUS PEOPLES/ ETHNIC MINORITIES
The minority groups recognized in Macedonia - either due to ethnic or religious differences distinguishing them from the Macedonian majority - are the Albanians, Roma, Turks and Serbs. The Albanians and Macedonians have a long history of peaceful coexistence, although the two groups speak different languages and have different religions and traditions. When Macedonia became independent, the Albanians started to demand greater cultural and political rights.

The Roma are not concentrated in a particular region of the country. They do not enjoy the right to citizenship, they have no access to education in their own language (unlike other minority groups), they do not take part in politics and, as in many other areas in the region, they are among the poorest and most neglected groups.

The 40,000 Serbs in Macedonia are mainly living in the northern parts of the country, near the border with Serbia. They are demanding greater respect for their cultural rights, but their main demand is protection from the Albanians. During the NATO attack against Serbia, the Government of Macedonia was accused of serving the interests of the Albanian minority, and protests were held by Serbs to express solidarity with people in Serbia.

MIGRANTS/REFUGEES
There are about 9,000 internally displaced people who fled the 2001 conflict and have still not managed to return to their homes. During 2004, the Government received 1,000 refugees and just over 1,200 asylum applications, which had raised to 6,000 the number of people living as refugees by the end of the year. Some 4,000 Macedonians sought asylum abroad in 2004, most of them in Serbia and Montenegro (2,300) and the rest in different EU countries such as Austria, Belgium, Finland, France, Germany, Italy, Switzerland and Sweden.

DEATH PENALTY
Macedonian law does not provide for the death penalty.

* Latest data available in The State of the World's Children and Childinfo database, UNICEF, 2006.

Madagascar / Madagascar

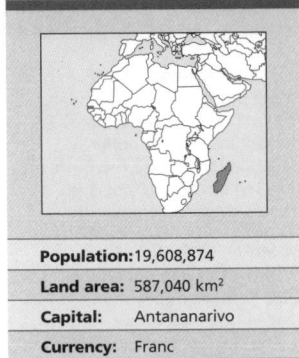

Twentieth century archeological research indicates Madagascar began to be inhabited around 700 AD. Even though this enormous island is geographically close to the area where African Bantu languages are spoken, Malagasy is derived from the Malay-Polynesian languages. The Malagasy people represent a unique mixture of Asian and African cultures. Prior to 1000 AD, important Afro-Arab influences expanded throughout Madagascar. Between the 1st and 5th centuries AD, Malay-Polynesian navigators (see Melanesians and Polynesians box), for reasons still unknown, repeatedly took upon themselves the challenge of crossing 8,000 kms of ocean to reach the African coast.

[2] During the 14th century, groups of Comoran traders established a series of ports in the northern region of the island, which were subsequently destroyed by the Portuguese. Madagascar is mentioned in the writings of Marco Polo; however, the first European to visit the island in 1500 was Diogo Dias, a Portuguese navigator. Portuguese explorers visited the valley of the River Matitana (in southeastern Madagascar) between 1507 and 1513, witnessing the arrival of an Afro-Arab group (Malindi Arabs). After one or two generations of that group's descendants, who married local Tompontany people, another group, the Antemoro, founded a theocratic state of Madagascar, the only state - in its time - to produce written texts. When the Portuguese found no gold, ivory or spices, they lost interest in the territory.

[3] In the 16th century the Sakalawas on the west coast and the Betsilios on the east coast established the first monarchies. In the 17th century the Merina kingdom or Imerina came into being on the eastern edge of the central plateau. A century later it was the Merinas, under their leader Nampoina, who initiated the process of unification which was completed later by Nampoina's son Radama I (1810-1828). Due to more frequent

contact with Arabs and Europeans, Radama organized a modern army and adopted the Latin alphabet for the Malagasy language. However, Nampoina's untimely death and the ensuing conflicts over succession paved the way for European occupation of the island by the end of the 19th century. Under the colonial system, huge swathes of forests were cleared to make way for sugarcane, cotton, and coffee plantations. The foreign colonists and companies seized the best lands and forced the peasants to work them in conditions of semi-slavery.

[4] Resistance to foreign domination, and the struggle for political rights and economic improvement, led to a major

uprising from 1947 to 1948, which was ruthlessly crushed by the French army with the loss of thousands of lives. The defeat of the insurrection enabled the colonial administration to control the transition to autonomy. Independence was finally proclaimed in 1960, with Philibert Tsiranana as president. In the first elections (1960), the Social Democratic Party (PSD) won, and Tsiranana became the first president of the republic, an office to which he was re-elected in 1965 and 1972.

[5] In 1972, after a series of serious disturbances, Tsiranana was forced to resign; he turned over full presidential powers to General Gabriel Ramanantsoa, who suspended the National Assembly

and the Senate. He also abolished the 1959 constitution. France withdrew its troops in 1973. After three years of instability, in 1975 Commander Didier Ratsiraka became president. He adopted socialist policies, and called a referendum that year, which overwhelmingly approved Ratsiraka's continuation as head of state for seven years, and a Charter from the Malagasy Socialist Revolution was adopted as the basis for a new constitution. On 30 December 1975 the country's name was changed to the Democratic Republic of Madagascar.

[6] In 1976, the 12-member Supreme Council of the Revolution was established. The Malagasy Revolutionary Vanguard, called Association for the Rebirth of Madagascar (AREMA, founded in 1975 in support of Ratsiraka) became the leading party within the National Revolutionary Front (union of peoples' parties). Colonel Joel Rakotomalala was appointed prime minister. Upon his death that same year, he was replaced by Justin Rakotoniaina. The legislative function was placed in the hands of a 144-member National Council. In 1977, Désiré Rakotoarijaona replaced Rakotoniaina as prime minister.

[7] In the 1989 general elections, President Ratsiraka was re-elected, with 67 per cent of the vote. His reform-oriented policies restored a multi-party system, and several opposition members were included in his cabinet. In 1991, the opposition united around the Committee of Living Forces (CFV), formed by 16 organizations. A series of street demonstrations and the

PROFILE

ENVIRONMENT

Madagascar is one of the world's largest islands, separated from the African continent by the Mozambique Channel. The island has an extensive central plateau of volcanic origin which overhangs the hot and humid coastal plains. These are covered with dense rainforest to the east and grasslands to the west. The eastern side of the island is very rainy, but the rest has a dry, tropical climate. The population is concentrated on the high central plateau. Rice and products for export (sugar, coffee, bananas, and vanilla) are cultivated along the coast. Stockbreeding is also an important activity throughout the island. The major mineral resources are graphite, chrome and phosphate.

SOCIETY

Peoples: The Malgaches, 98.9 per cent of the population, are made up of different ethnic groups of Malagasy-Afro-Indonesian origin. Immigrant communities: Indian and Pakistani, 0.2 per cent; French, 0.2 per cent; Chinese, 0.1 per cent, and others. **Religions:** Traditional beliefs 52 per cent; Christian 41 per cent (of which Roman Catholic 21 per cent, Protestant 19 per cent); Muslim 7 per cent. **Languages:** Malagasy and French (official). Hovba and other local dialects are also spoken.

Main Political Parties: I Love Madagascar (*Tiako I Madagasikara*); National Union (*Firaisankinam-Pirenena*); Pillar and Structure for the Salvation of Madagascar (*Andry sy Riana Enti-Manavotra an'i Madagasikara*); Economic Liberalism and Democratic Action for National Recovery (*Leader-Fanilo*). **Main Social Organizations:** Confederation of Malgache Workers (FMM); Confederation of Christian Trade Unions of Madagascar (SEKRIMA); Union of Independent Trade Unions of Madagascar (USAM) and Union of Workers' Trade Unions of Madagascar (FISEMA). National Council of Christian Churches.

THE STATE

Official Names: Repoblikan'i Madagasikara. République Démocratique de Madagascar. **Administrative Divisions:** 6 provinces, 10 districts, 1,252 sub-districts, and 11,333 towns. **Capital:** Antananarivo 1,678,000 people (2003). **Other Cities:** Toamasina 166,000 people; Fianarantsoa 131,600; Mahajanga (Majunga) 128,600 (2000). **Government:** Marc Ravalomanana, President and head of state since 2002. Bicameral Legislature: National Assembly with 160 members and Senate with 90 members. **National Holiday:** 26 June, Independence Day (1960). **Armed Forces:** 14,000 (2003). Other: 7,500 (Gendarmerie).

Life expectancy
56 years
2005-2010

GNI per capita
$290
2004

Literacy
71% total adult rate
2000-2004

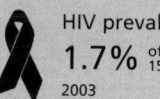

HIV prevalence rate
1.7% of population 15-49 years old
2003

IN FOCUS

ENVIRONMENTAL CHALLENGES
Deforestation is one of the most pressing environmental problems (the forests have been destroyed on 75 per cent of the land). Several aquifers have been polluted as a result of organic waste and limited sanitation. Madagascar is considered a great biological reserve; however, it is endangered due to the type of crops that have been cultivated for centuries. Destruction exceeds the jungle's capacity to regenerate.

WOMEN'S RIGHTS
Women have been able to vote and run for office since 1959. According to data published in 2003, women held 4 per cent of seats in Parliament and their representation in ministerial or equivalent positions stood at 6 per cent. In 2003, women made up 44 per cent of the country's

workforce of 8 million. The illiteracy rate among women stood at 35 per cent, compared to 24 per cent among men. Eighty per cent of pregnant women receive prenatal care and only 51 per cent of births are attended by qualified medical personnel*. The maternal mortality rate is 550 per 100,000 live births*.

CHILDREN
While major progress has been achieved in recent years in reducing under-five mortality rates, an average of 160 children continue to die each day in Madagascar from preventable causes, such as malaria, diarrhea and acute respiratory illnesses*. Half of all Madagascan children under five years old suffer from some form of malnourishment*.

On average, every day, five babies are born HIV-positive*. While infection prevalence rates are still low, UNAIDS estimated in

2004 that there were some 30,000 children orphaned by AIDS*. In spite of the efforts being made by different organizations in the country, stigma and discrimination against people living with HIV/AIDS make tackling the disease difficult*.

Over 20 per cent of children aged between 5 and 14 work, some of them in hazardous conditions, such as in mining and stone cutting*.

INDIGENOUS PEOPLES/ETHNIC MINORITIES
The Madagascan population is made up of 18 ethnic groups. This classification often reflects colonial stereotypes rather than the actual reality. The Merina, of Asian origin and light skinned, live in the highlands of the Pacific Area. The coastal peoples, known as *Cotiers*, are dark skinned people, of mixed African, Malayo-Indonesian, and Arab ancestry. Among them are

the Betsimisaraka, the Tsimihety in the north and the Antandroy in the south. The groups are demanding greater political participation at the center, while they remain alert to possible attacks by other communities.

MIGRANTS/REFUGEES
In spite of having signed the 1951 Refugee Convention, Madagascar has not yet ratified it. The Government has agreed with UNHCR on the admission of refugees to the country but has warned that no assistance can be provided. By the end of 2004, the country hosted 56 refugees.

DEATH PENALTY
The country has been de facto abolitionist since 1958.

* Latest data available in *The State of the World's Children* and *Childinfo* database, UNICEF, 2006.

occupation of the National Radio station led the Government to declare a state of emergency. The Committee called for Ratsiraka's resignation, appointing a transition government. Shortly after, following the detention of two ministers belonging to the transition cabinet, 400,000 people took to the streets and repeated the demand for Ratsiraka's resignation. The demonstration was repressed, with 31 people killed and hundreds injured. Guy Razanamasy took office as prime minister that year, and reopened dialogue with the opposition, appointing a government of national unity.
[8] In 1992, a multiparty forum was created to draw up a new constitution: the new constitution was approved by referendum that year, and presidential elections were set. Albert Zafy was elected president and took office in 1993. The economic and social situation of the country, one of the poorest in the world, was a disaster. Per capita income had grown hardly at all over 16 years, from $200 to $230 per year, while calorie consumption dropped from 108 per cent to 95 per cent. In 1994, the Government introduced a series of austerity measures recommended by the International Monetary Fund (IMF), which further inflamed social tensions. Massive demonstrations were held in opposition to these policies. In 1995, the governor of the Central Bank abandoned his post at the request of the IMF and the World Bank.
[9] In 1995, the Malagasy people approved increased powers for Zafy in a referendum. The debate over the structural adjustment policies

coincided with controversy over the use of the nation's natural resources, when the mining transnational RTZ Associates proposed opening a mine on the southern coast of the island to extract titanium dioxide. This sparked strong protests by environmentalists, since unique species of native flora and fauna would be destroyed. In 1996, the National Assembly approved a vote of no confidence in the Government, leading to the formation of a new cabinet. The motion was partly caused by comments made by IMF Managing Director Michel Camdessus, who said the lack of governmental cohesion meant compliance with agreements with the multilateral lender was not guaranteed. Accused of violating the constitution, Zafy resigned in 1996. In the subsequent presidential elections, Ratsiraka was re-elected.
[10] A new constitution, approved by referendum in 1998, gave broader powers to the President and granted economic autonomy to the country's six provinces. The CFV accused the Government of authoritarian tendencies. After that year's legislative elections, Tantely Andranarivo became the new prime minister, replacing Pascal Rakotomavo, both members of AREMA.
[11] In 2001, the police cracked down on demonstrators in the capital, Antananarivo. They were demanding freedom for Jean-Eugene Voninahitsy, deputy chair of parliament, who had been jailed for insulting Ratsiraka and for issuing bad cheques. After Voninahitsy's imprisonment, an opposition bloc was formed in Parliament, the

Unit for Defense of Democracy in Crisis. After 29 years, the Senate resumed operations, completing the framework agreed in the new Constitution, which included the presidency, the National Assembly and the High Constitutional Court (HCC). In the 2001 presidential elections, Ravalomanana claimed victory without a second round. But Ratsiraka did not accept defeat.
[12] In 2002, Ravalomanana and his followers organized a general strike in the capital and a series of protest demonstrations against the supposed manipulation of votes by Ratsiraka. Ravalomanana proclaimed himself president that year and appointed ministers, who took office in the capital. Due to a series of violent clashes, Ratsiraka declared martial law in Antananarivo. The country was at a standstill and, while Ratsiraka and his ministers left for Toamasina - second largest city and port in the country, which they named the new capital - an economic blockade was imposed on Antananarivo. Ravalomanana took over the Home Ministry, Ratsiraka's last stronghold in Antananarivo, and also proclaimed himself commander-in-chief of the armed forces.
[13] Later on, the HCC annulled the results of the elections and asked for a recount, which was accepted by both candidates. The HCC finally awarded Ravalomanana 51.46 per cent of the vote, and he took office as president. Ratsiraka - who obtained 35.9 per cent of the vote - ignored the verdict. The province of Toamasina declared its independence and Ratsiraka sought to isolate Antananarivo from the country's ports. After seven months

of political crisis, Ratsiraka fled to the Seychelles and then to France.
[14] In 2003, Ratsiraka - still in exile - was sentenced in absentia to ten years imprisonment with forced labor, on charges of misappropriation of public funds. The former Prime Minister Tantely Andrianarivo was sentenced to 12 years, accused of abuse of authority.
[15] Between February and March 2004, tropical cyclones Elita and Gafilo hit the country, forcing thousands of people on the coast to flee their homes. Strong winds caused major damage in northeastern Madagascar, while heavy rains resulted in flooding throughout its western region. According to preliminary studies by the Government, 700,000 people were affected and around 200,000 were left homeless.
[16] In March 2005, flooding due to heavy rains severely damaged the country's food production. Some 25 people were killed and 58,000 were affected by the floods and over 26,000 rice paddies were ruined by water. Fears mounted about a second year of food shortages in the provinces of Toamasina and Mahajanga. Madagascar was still recovering from rice shortages after being swept by cyclones in 2004 and February 2005.
[17] In March 2006, the Executive Board of the World Bank, the IMF and the African Development Bank (ADB) gave their approval to the cancellation of debts owed to them by 13 African countries, among them Madagascar. The writing-off was to come into effect on 1 July 2006, at the start of the Bank's new fiscal year. ∎

Malawi / Malawi

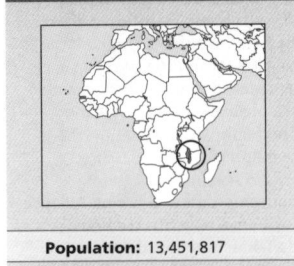

Population:	13,451,817
Land area:	118,480 km²
Capital:	Lilongwe
Currency:	Kwacha
Language:	Chewa, English

I n the 1st century BC, Bantu groups invaded the southeastern region of Africa, inhabited by Twa and Fulani groups. Between the 13th and 15th centuries, there were further migrations of Bantu people, who knew how to work with iron and used this knowledge to dominate the original inhabitants.

[2] In 1480, the Bantu groups formed several small states joined together in a federation which encompassed large parts of Zambia, Mozambique and all of present-day Malawi.

[3] In the 17th century, the first Portuguese explorers arrived from the area of the present-day Mozambique. From 1790 until 1860, the slave trade grew dramatically in the area.

[4] Around 1835, Zulu expansion (see South Africa) pushed the Ngoni-Ndwande to the shores of Lake Malawi, giving rise to 60 years of war.

[5] The country was explored by David Livingstone in 1859 and experienced a Portuguese attempt at colonization in 1890 which was checked by the British Government. Britain wanted to keep the territory which would eventually serve as a link in a continuous chain of colonies from South Africa to Egypt. In 1891, it became the protectorate of Nyasaland through Cecil Rhodes' British South African Company.

[6] In 1893, its name was

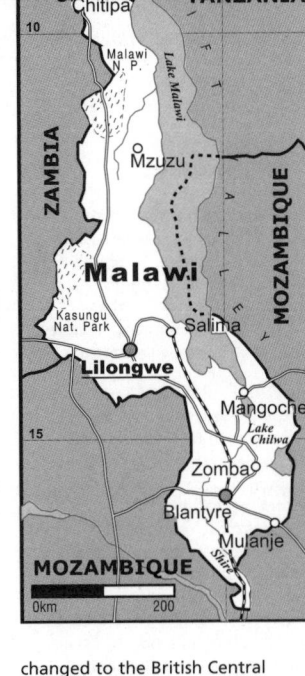

changed to the British Central African Protectorate. European settlers were offered land for coffee plantations at very low prices with large tax incentives. Africans worked the plantations in extremely difficult conditions.

[7] In 1907, the protectorate reverted to the name of Nyasaland.

[8] Nationalist leaders established the Nyasaland African Congress (NAC) in 1944. On 23 October 1953, concerned about their actions and those of white liberal activists, the British unified the territory with the Federation of Rhodesia.

[9] In 1958, Hastings Kamuzu Banda, 'the black messiah', returned from the US where he had graduated in medicine, to assume the leadership of the NAC.

[10] In 1959, the Malawi Congress Party (MCP), headed by Banda, was founded as a successor to the NAC. The party was pro-independence. Two years later, after the MCP scored a victory in the 1961 elections for a constituent assembly, Banda was appointed prime minister.

[11] In order to prevent internal divisions, Banda's authority in the Party was expanded. When the colony became independent on 6 July 1964, the MCP and the country remained under Banda's autocratic rule. He established close economic and diplomatic ties with the racist governments of South Africa and Rhodesia, and with the colonial administration in Mozambique.

[12] South Africa became the main market for Malawi's exports of tea and tobacco, while its investors built roads, railways and a new capital city. South African executives took charge of the airline, news services and development agencies and a large part of the State administration.

[13] In June 1978, in the first election in 17 years, all of the candidates had to belong to the MCP and pass an English test, which immediately excluded 90 per cent of the population.

[14] Zimbabwe's independence in 1980 changed things for Malawi. Banda lost his close economic relations with South Africa. Consequently, the Government

drew closer to the Front Line states, joining the SADCC association because of Malawi's dependence on the railway lines through Mozambique and Zimbabwe.

[15] This situation strengthened the Socialist League of Malawi (LESOMA) which favored breaking economic and political ties with South Africa and an end to Banda's dictatorship. In 1980 the party created a guerrilla force, while the Malawi Freedom Movement (MAFREMO), led by Orton Chirwa, gained strength.

[16] In 1983, Chirwa and Attati Mpakati, a LESOMA leader, were accused of conspiracy and sentenced to death. Shortly thereafter, Mpakati was assassinated by South African agents while visiting Harare. Chirwa and his wife were kidnapped in Zambia where they were living in exile, and imprisoned in Blantyre.

[17] Banda created a secret police force, called the Special Branch, with South African and Israeli advisers. The President also personally controlled the economy, owning 33 per cent of all businesses.

[18] In 1988, Amnesty International denounced the imprisonment of scholars and writers, among them Jack Mapanje, the country's foremost poet. The US cancelled $40 million of foreign debt in November 1989.

[19] The implementation of an IMF structural adjustment program brought down inflation and reduced the balance of payments deficit, while leading to increases in investment. However, the policies made conditions worse for the poor.

PROFILE

ENVIRONMENT
The terrain and the climate are quite varied. The major geological feature is the great Rift fault that runs through the country from north to south. Part of this large depression is filled by Lake Malawi, which takes up one fifth of the land area. The rest is made up of plateaus of varying altitudes. The most temperate region is the southern part, which is also the highest, containing most of the population and economic activities (basically farming). The lowlands are covered by grasslands, forests or rainforests, depending on the amount of rainfall they receive. In March 2002, the Shire River was discovered to be affected by a plague of water hyacinths, threatening to block the flow of the river or divert its course. The Shire, an outlet of Lake Malawi, pours into the Zambezi River and represents an important means of transport and source of food.

SOCIETY
Peoples: Maravi (including Nyanja, Chewa, Tonga, and Tumbuka) 58.3 per cent; Lomwe 18.4 per cent; Yao 13.2 per cent; Ngoni 6.7 per cent.
Religions: There is no official religion. Christian, 50 per cent (of which 20 per cent are Protestant and 18 per cent are Roman Catholic); Muslim, 20 per

cent. Many people follow traditional religions, but many of them also define themselves as Christian or Muslim.
Languages: Chewa and English (official languages); several Bantu languages - other than Chewa - are spoken by their respective ethnic groups.
Main Political Parties: United Democratic Front; Mgwirizano Coalition (which includes: Republican Party; Peoples Progressive Movement; Movement for Genuine Democratic Change); Malawi Congress Party; National Democratic Alliance.
Main Social Organizations: Trade Union Congress of Malawi. Union of Malawian Students.

THE STATE
Official Name: Republic of Malawi.
Administrative Divisions: 24 districts.
Capital: Lilongwe 587,000 people (2003).
Other Cities: Blantyre 518,800 people; Mzuzu 94,400 (2000).
Government: Presidential republic. Bingu wa Mutharika, President since May 2004. Legislature: single-chamber National Assembly, made up of 193 members.
National Holiday: 6 July, Independence Day (1964).
Armed Forces: 5,000 (2003). Other: 1,500 (elite police force).

LAND USE

2003/2005

IRRIGATED AREA: 2.2% of arable land

- FOREST AND WOODLAND: 36.2%
- ARABLE LAND: 26.0%
- CROPLANDS: 1.5%
- OTHER USE: 36.3%

Life expectancy
41 years
2005-2010

GNI per capita
$160
2004

Literacy
64% total adult rate
2000-2004

HIV prevalence rate
14.2% of population 15-49 years old
2003

IN FOCUS

ENVIRONMENTAL CHALLENGES
The degradation of the soil and deforestation are the main environmental problems. Drought intensified the lack of water in early 1998. Some water sources are polluted by industrial or agricultural waste as well as untreated sewage. Meanwhile, sedimentation in spawning areas is endangering marine species.

WOMEN'S RIGHTS
Women have been able to vote and stand for office since 1961. In 2004, 13.6 per cent of parliamentary seats were held by women, while their representation in ministerial or equivalent positions amounted to 14 per cent in 2003. That year, women made up 49 per cent of the country's labor force. Some 46 per cent of Malawian women are illiterate, while the rate for men stands at 25 per cent*.

A significant rise in maternal mortality has been registered in most health care centers, and most health facilities suffer from an acute shortage of staff and basic equipment*.

HIV/AIDS poses one of the most serious problems; it especially affects women of child-bearing age in rural areas. Only 34 per cent of women between the ages of 15 and 24 have an understanding of how to prevent infection*. In 2003, 18 per cent of pregnant women aged between 15 and 24 were living with HIV*. That same year, UNICEF estimated that some 460,000 women aged between 15 and 49 were living with HIV/AIDS in the country.

CHILDREN
The rural population has been hit hardest by Malawi's humanitarian crisis. According to UNICEF, more than 65 per cent of the population was living below the poverty line*. Malawi is among the 15 worst-ranked countries in terms of human development indicators.

The food crisis, a consequence of chronic poverty, unfavorable weather conditions and excessive reliance on one-crop farming has led to chronic malnutrition and endemic diseases (cholera, chronic infections, etc) that, together with the AIDS epidemic, paint an outlook of extreme vulnerability for the entire population, especially children.

Nearly 500,000 children under the age of 17 have been orphaned by HIV/AIDS*. Families affected by HIV/AIDS face increasing hardships to overcome the food crisis and, on the other hand, the number of households headed by children is on the rise. Since 1990, life expectancy has fallen almost 10 years, to the current 41 years*.

INDIGENOUS PEOPLES/ETHNIC MINORITIES
The main African ethnic groups make up 99.5 per cent of the population and are divided into the Chewa, Nyanja, Tumbuka, Yao, Lomwe, Sena, Tongo and Ngoni. The rest of the population, 0.5 per cent, includes Asian and European minorities.

MIGRANTS/REFUGEES
At the present time, Malawi hosts over 7,000 refugees and asylum-seekers, mostly from Rwanda. The new registration system, backed by UNHCR, provides more reliable information on the number of people that have entered the country and, at the same time, allows faster and more effective identification and aid. However, most refugees face the same food shortages that affect the country's population.

DEATH PENALTY
The death penalty is still applied.

* Latest data available in *The State of the World's Children* and *Childinfo* database, UNICEF, 2006.

---

[20] In 1990 and 1991, earthquakes and floods exacerbated food shortages among the rural populace, which made up 90 per cent of the total population.

[21] In February 1992, the Catholic Church wrote a pastoral letter criticizing the human rights situation and calling for greater political freedom. A popular uprising in Blantyre was harshly repressed.

[22] In April 1992, opposition leader Chafuka Chihana of the Alliance for Democracy was arrested when trying to return to the country. An international campaign prevented his execution.

[23] In May, a general strike called by textile workers was brutally put down, with 38 deaths and hundreds of injuries. In reprisal, the World Bank discontinued part of its financial aid.

[24] The ruling Congress Party of Malawi, the only party to participate in the June 1992 elections, obtained 114 seats in the National Assembly.

[25] In June 1993, the Public Affairs Committee (PAC) forced Banda to set the referendum for that month in order to choose between a one-party system and a multi-party system. Nearly two-thirds of voters chose a multi-party system. That month, Banda released Vera Chirwa, widow of the assassinated dissident, and Africa's oldest female political prisoner. Banda did not resign but promised presidential elections would be held in 1994.

[26] On 17 May 1994, four million Malawians elected a new president and 177 members of parliament in the first multi-party elections since the country's independence.

[27] Opposition member Bakili Muluzi won the elections, and his party, the United Democratic Front (UDF), won 84 of the 177 seats at stake. In September, having won only 55 seats, Banda decided to retire from political activity.

[28] Malawi suffered the consequences of an intense drought and famine in 1994. In the midst of an increasingly difficult social situation, the Government went ahead with its IMF-sponsored policy to cut public spending. In January 1995, ex-President Banda was arrested and charged with the murder of three former ministers.

[29] In 1996, Malawi entered negotiations to form a free trade area, along with 11 other African nations. The Government announced it would revoke the laws affecting foreign investors in rural areas.

[30] In 1997, the US started to train Malawian troops to create an African peacekeeping force.

[31] The drought, which affected vast regions of Africa, almost totally dried up Malawi's Shire River, one of the country's main watercourses.

[32] After flawed elections, Muluzi was re-elected in June 1999, while his party won 93 of the 192 legislative seats. In late February 2000, the President requested the resignation of his cabinet. Some of his ministers, among them Economic Minister Cassim Chilumpha, were not considered trustworthy by donor countries that provided development aid.

[33] In 2002, a state of national disaster was declared as a consequence of the great death toll caused by food shortages and crops lost due to drought and flooding. The Government was accused of selling the country's stocks of grain to Kenya and forcing the population to eat unripe grain. The country was facing a risk that the famine would stretch into the following year. Floods and the deterioration of roads and railways made food distribution more difficult. Seventy per cent of the population was going hungry and children and the elderly were those at greatest risk.

[34] In February 2003, Minister Thengo Magoya disclosed that he had lost three of his children to AIDS during the past 10 years and stated that it was high time Malawians accepted that the pandemic was decimating the population.

[35] In August 2003, the opposition party Genuine Alliance for Democracy was created by parliamentary dissidents.

[36] In early 2004, Vice-President Justin Malewezi resigned and joined the opposition at the start of the presidential election campaign.

[37] Bingu wa Mutharika, of the United Democratic Party, was declared winner of the May 2004 presidential elections. The MCP won 60 of the 193 seats in the parliamentary polls.

[38] That month, the Mgwirizano Coalition filed an appeal with the Supreme Court, alleging rigged results and other serious flaws, and demanding that the elections be re-run. EU observers stressed the need for transparency in vote counting and judged that elections had been free but not fair.

[39] At least four people were killed during protests after the elections.

[40] The second opposition group in Malawi, the Malawi Congress Party (MCP), filed another legal challenge, demanding the truth be revealed about the election's events so that the irregularities could never happen again.

[41] Education Minister Yusuf Mwawa was arrested in May 2005 for allegedly using public funds for his wedding ceremony. He was charged with abuse of public office, theft by a public servant, fraud and document forgery. Some senior officials, including former ministers, were arrested as part of an anti-corruption drive promoted by President Mutharika.

[42] In February 2006, the opposition accused Mutharika of using the Anti-Corruption Bureau (ACB) to persecute his opponents. Mutharika denied the accusations, saying his administration was not targeting the opposition and that anyone who was found to be corrupt would be dealt with, including those serving in his own Government.

[43] In April, following five years of drought and food crises, the Agriculture Ministry forecast a bumper maize harvest, which would allow the country to cushion the situation somewhat and bring relief to its decimated population.

[43] In May 2006, the Vice-President, Cassim Chilumpha, was arrested on a charge of treason, accused of having hired an assassin to murder President Mutharika. He had long been in dispute with Mutharika, who had tried to remove him from his position three months earlier. ■

Malaysia / Malaysia

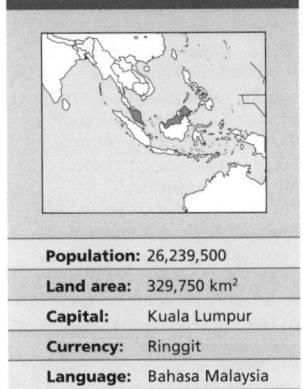

Population:	26,239,500
Land area:	329,750 km²
Capital:	Kuala Lumpur
Currency:	Ringgit
Language:	Bahasa Malaysia

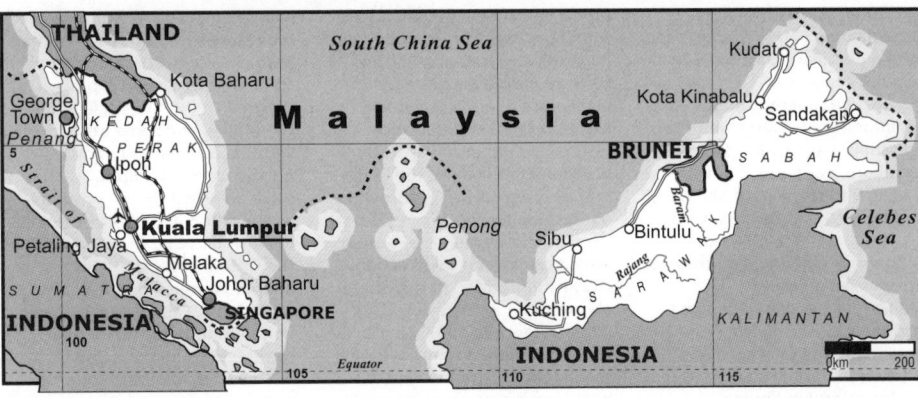

T he Malay Peninsula and the Borneo states of Sarawak and Sabah were first inhabited by the descendants of pre-Malay immigrants, the *orang asli*, who lived, as some still do, in the forests. The Malays probably came from southern China, via what is now Indonesia.

[2] Immigration from present-day China - made possible by the use of canoes during the second millennium BC - brought metal-working and agricultural techniques, particularly rice-farming, starting in the first millennium AD. The agricultural and fishing surplus contributed to the development of urban cultures, ports and trading with India. Indian influence was all-pervasive, bringing religion, political systems and the Sanskrit language.

[3] The Indianized kingdom of Funan was founded in the Mekong river area in the first century AD and Buddhist states eventually developed in the east, trading with China. In the 15th century, the port of Melaka (Malacca) was founded; its rulers, who were the first in the region to convert to Islam, stopped paying taxes to Siam (now Thailand). Trade with Islamic merchants brought prosperity to Melaka. The new faith spread across the rest of present-day Malaysia and Indonesia, replacing

Buddhism. At the beginning of the 16th century, Melaka attracted the Portuguese, who were competing with Arab merchants for the Indian Ocean trade routes.

[4] In 1511 the Portuguese viceroy of India, Alfonso de Albuquerque, seized the port by force. It was of vital strategic importance in the Portuguese struggle to maintain their monopoly on the spice-trade from the Moluccas Islands (which were traded for fabrics from India and silk and porcelain from China). After Melaka's displacement, the Malay model of a maritime, fishing and trading Muslim state emerged in the rival kingdoms of Johor, Aceh and Brunei. In the early 17th century, the Dutch, who were established in Batavia (present-day Jakarta) formed an alliance with Johor and drove the Portuguese out of Melaka and had no European competition during the following century.

[5] The British established back-up points for their trade with China in northern Borneo (Kalimantan), and in 1786 founded the port of George Town, on the island of Penang (Pulau-Pinang), off the western coast. The British model of free trade proved more successful than the Dutch trade monopoly and Penang attracted a cosmopolitan population of Malays, Sumatrans, Indians and Chinese. In 1819, the British founded Singapore,

but at the time they were more interested in safeguarding shipping than in the local spice trade, as their imports from China were being paid for with opium from India. Nevertheless, the Dutch and the British found it difficult to coexist in the region. A treaty drawn up in 1824 granted control of Indonesia to the Dutch, while Malaya was left in British hands.

[6] The colonies of Penang, Melaka and Singapore became the key points of the British colony. The British encouraged Chinese immigration to work in tin mines and ports along the Straits of Malacca in the early 19th century. They intermarried and became known as the Straits Chinese. The Malay peasants and fishing people continued their traditional activities. From 1870, the British began to sign protectorate agreements with the sultans and in 1895 they encouraged them to form a federation, with Kuala Lumpur as its capital. The sultanates of northern Borneo (Brunei, Sabah and Sarawak, the last ruled by James Brooke, and his heirs) also became British protectorates, administered from Singapore, but without any formal ties with the peninsula.

[7] Towards the end of the 19th century, the British introduced rubber with Hevea seeds smuggled from Brazil, so bringing about an end to the rubber

boom in the South American Amazon. They encouraged Tamil immigration from southern India, to get workers for the rubber plantations which faced growing demand from the incipient automobile industries.

[8] In the early 20th century, large numbers of Chinese arrived to work in the tin mines and urban service sector. The multi-ethnic society, with cultural, religious, and language differences, also had different educational systems for Malays, Indians and Chinese. On the economic front, the Malays worked in rural agriculture, the Chinese in the tin mines, and the Indians in the rubber estates. In the first decades of the 20th century, the Malays joined the Islamic reform movements of the Middle East, the Indians supported the struggles of Mahatma Gandhi, and the Chinese were ideologically influenced first by the nationalism of Sun Yat-Sen and then by the Communist Party. Although most sympathized with the Kuomintang, by 1927 there were Communist cells among the Chinese in the Straits Settlements.

[9] In 1942, during World War II, the country was occupied by the Japanese. The Japanese tried, like other Southeast Asian movements, to form alliances with the local nationalist movements to gain support against the European powers. The greatest resistance came from the Chinese, especially the Malayan Communist Party (CP), which organized guerrilla forces.

[10] At the end of the war it was clear that British domination could not continue without changes, but the difficulties of diverse ethnic interests, 'protected' sultanates and ports under direct colonial administration made it difficult to find a suitable political system. The British proposed a Malayan Union with equal citizenship for all. This threatened the position of the Malays, and Malay nationalists gathered around the symbolic figure of the sultans founding

LAND USE

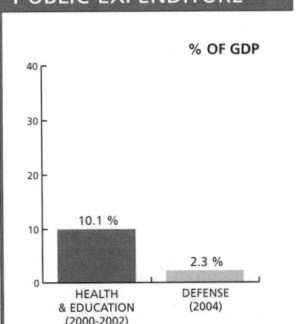

2003/2005

IRRIGATED AREA: 4.8% of arable land

- FOREST AND WOODLAND: 63.6%
- ARABLE LAND: 5.5%
- CROPLANDS: 17.6%
- OTHER USE: 13.3%

PUBLIC EXPENDITURE

% OF GDP

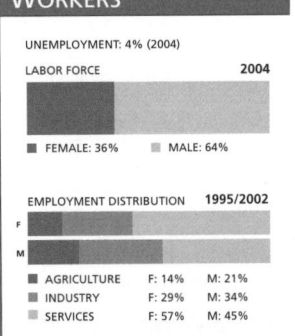

10.1 % — HEALTH & EDUCATION (2000-2002)

2.3 % — DEFENSE (2004)

WORKERS

UNEMPLOYMENT: 4% (2004)

LABOR FORCE 2004

- FEMALE: 36%
- MALE: 64%

EMPLOYMENT DISTRIBUTION 1995/2002

F
M

- AGRICULTURE   F: 14%   M: 21%
- INDUSTRY   F: 29%   M: 34%
- SERVICES   F: 57%   M: 45%

Life expectancy
74 years
2005-2010

GNI per capita
$4,520
2004

Literacy
89% total adult rate
2000-2004

HIV prevalence rate
0.4% of population 15-49 years old
2003

the United Malays' National Organization (UMNO), controlled by the dominant class but with popular support.

[11] In 1948, there was a communist-led insurrection that was suppressed by British forces. Its failure was partly due to the view of many poor Malays and Indians that the revolt was a Chinese effort and not the action of a unifying, anti-colonial progressive movement. The Marxist parties were outlawed and their leaders jailed. From 1948 until 1960, the CP waged guerrilla warfare in the northern Malayan Peninsula and in Borneo.

[12] In the early 1950s, journalist and activist Dato Onn made a new attempt at creating a pan-

ethnic party. He left UMNO to found the Malayan Independence Party. He was defeated in the 1952 municipal elections by an alliance of UMNO and the Malayan Chinese Association (MCA). The alliance was later broadened to include the Malayan Indian Congress (MIC), winning nationwide elections in 1955.

[13] Faced with the threat of an armed communist insurrection, the British decided to negotiate. This resulted in Malayan independence in 1957. Tunku Abdul Rahman, a prince who led the independence movement, became the first prime minister. A federation of 11 states was established with a parliamentary system and a monarch chosen

every five years from among the nine state sultans. A constitutional bargain was struck by the three communities: citizenship was granted to the non-Malays, but the Malays and some other groups were recognized as indigenous people, called Bumiputras. They were accorded special privileges in education and public-sector employment and Malay would be the official language. The country adopted a free market system, and foreign capital that had taken hold during the colonial times continued to be dominant.

[14] In 1963, the British colonial states of Singapore (south of Malaya), Sabah and Sarawak obtained independence and joined Malaya to form the Federation of Malaysia. There were major disagreements over ethnic policy and Singapore was expelled from the federation in 1965, becoming an independent republic.

[15] The conservative Alliance Party ruled with a large majority from 1957, but in the 1969 elections it lost many seats to the Islamic PAS Party, Gerakan and - mostly - to the Chinese-based Democratic Action Party (DAP). The parliamentary system was suspended and the country was ruled for two years by a National Operations Council.

[16] During the next two decades, the state companies that represented Bumiputras acquired shares in British property and mining companies, and became part of joint ventures with transnational corporations. The Bumiputra share of the economy rose to almost 20 per cent by 1989, while foreign ownership fell below 40 per cent. Because of these policies, Malaysians of Chinese descent claimed that they faced discrimination at work, in education and elsewhere.

[17] Dr Mahathir Mohamad took office as prime minister in 1981, and started to develop Malaysian industry. In the late 1980s he faced increasing challenges to his leadership from UMNO (the Front's main party). Some of his opponents left to form a new opposition party, Semangat 46. In 1990 they formed a loose opposition coalition with the Islamic PAS, the DAP and the small left-wing People's Party.

[18] Malaysian foreign policy moved from its pro-Western position in the 1960s through non-alignment in the 1970s to a pro-Third World stance in the 1980s. In 1990, Kuala Lumpur hosted the inaugural summit meeting of 15 Third World countries (aiming at fostering concrete South-South co-operation projects). It played a major role in the South Commission and actively

supported the Palestine Liberation Organization and South Africa's African National Congress.

[19] As hi-tech industries grew, the country experienced a shortage of skilled and semi-skilled labor. Wages increased significantly, accentuating the differences among social sectors.

[20] Tuanku Ja'afar ibni al-Marhum Tuanku Abdul Rahman became king in April 1994. In 1995, Prime Minister Mahathir's National Front coalition took 162 of the 192 seats in the Chamber of Representatives.

[21] Malaysia was hit by the 1997 economic crisis caused in part by the collapse of various regional currencies. In early September 1998, after months of debate on the situation, Mahathir removed Economy Minister Anwar Ibrahim from office, and accused him of 'sexual misconduct'. After his dismissal, Anwar led major protests against the Government and was arrested on 20 September. The opposition backed Anwar, seeing him as a liberal and a champion of foreign investment, and showed its support in protests. After he was imprisoned, the opposition leadership passed to his wife, Azizah Ismail.

[22] In April 1999, Anwar was sentenced to six years in prison, charged with 'sexual misconduct' and sodomy.

[23] The November 1999 legislative elections gave the absolute majority and more than two thirds of the seats to the ruling party, allowing Mahathir to stay in power until 2005. The Alternative Front, led by Azizah Ismail, received less than 20 per cent of the vote and complained that the Government had permitted just nine days of political campaigning. Meanwhile, the party of the Prime Minister, UMNO, suffered internal divisions, losing ground to its Islamic partner, PAS.

[24] In January 2000, once the trial against Anwar resumed, the Government detained four opposition political leaders and the former Economy Minister's defense lawyer, unleashing a wave of protests within and outside the country. Charges against the opposition included sedition and inciting racial violence. In March that year, the Government clamped down on press outlets that were pro-Anwar.

[25] In spite of protests and complaints from environmentalists, in February 2001 the Government went ahead with a mega-dam project in Bakun, Borneo, expected to become the largest in Southeast Asia. Apart from costing $5 billion, the dam would displace 10,000

PROFILE

ENVIRONMENT

The Federation of Malaysia is made up of peninsular Malaysia (131,588 sq km), and the states of Sarawak (124,450 sq km) and Sabah (73,711 sq km) in northern Borneo (Kalimantan), 640 km from the peninsula in the Indonesian archipelago. Thick tropical forests cover more than 70 per cent of the mainland area, and a mountain range stretches from north to south across the peninsula. Coastal plains border the hills on both sides. In Sabah and Sarawak, coastal plains ascend to the mountainous interior. There is heavy annual rainfall. Malaysia's economy is export-oriented. Tin and rubber, the traditional export products, have recently been replaced by petroleum and manufactured goods.

SOCIETY

Peoples: Bumiputra (Malays) and other indigenous peoples including Orang asli, Penan, Iban: 56 per cent; Chinese: 33 per cent; Indians: 11 per cent.
Religions: Islam, the official religion, is practiced by about half the population. Buddhist 17 per cent, Taoist 11 per cent, Hindu 7 per cent, Christian 7 per cent; animist.
Languages: Bahasa Malaysia (Malay) is the official language. Chinese languages, Tamil, English and many Orang Asli languages.
Main Political Parties: Barisan Nasional (ruling coaltion) includes United Malays National Organization (UMNO), Gerakan Rakyat Malaysia Party (PGRM), Malaysian Chinese Association (MCA), Malaysian Indian Congress (MIC), Sabah Progressive Party (SAPP), Sarawak United People's Party (SUPP). Main opposition parties include Democratic Action Party (DAP); Islamic Party of Malaysia (PAS).
Main Social Organizations: The leading labor organization is the Congress of Malaysia's Unions. Network of Indigenous Peoples of Malaysia.

THE STATE

Official Name: Persekutuan Tanah Malaysia
Administrative Division: 13 states, 3 federal territories and 130 districts.
Capital: Kuala Lumpur 1,352,000 people (2003).
Other Cities: Johor Baharu 691,000; Ipoh 552,800 people; Petaling Jaya 474,600; Melaka 126,100 (2000).
Government: Constitutional, parliamentary and federal monarchy. Tuanku Salehuddin Abdul Aziz Shah ibn al-Marhum Hisamuddin Alam Shah is the current king or *yang di-pertuan agong*, since April 1999. The sovereign is elected every five years from among the nine regents (sultans), and only the sultans can vote. Datuk Abdullah Ahmad Badawi, Prime Minister since 2003. Central bicameral parliament, 70-member Senate, 219-member Chamber of Deputies, with a constitution and a legislative assembly for every state.
National Holiday: 31 August, Independence Day (1957).
Armed Forces: 110,000 (2003).

Under-5 mortality
12 per 1,000 live births
2004

Poverty
<2% of population living on less than $1 per day
1997

Debt service
7.9% exports of goods and services
2003

Maternal mortality
41 per 100,000 live births
2000

IN FOCUS

ENVIRONMENTAL CHALLENGES

Indiscriminate logging and the use of highly toxic herbicides are causes for concern. There are fears that indigenous trees and crops may be irreparably harmed. There is significant air pollution from industrial and vehicular emissions and contamination of potentially drinkable water from the dumping of untreated sewage into rivers.

WOMEN'S RIGHTS

Women have been able to vote and stand for office since 1955. In 2004, women held 13 per cent of seats in Parliament, while their representation in ministerial or equivalent positions stood at 9 per cent. In 2003, women made up 38 per cent of the country's workforce. Malaysia is a country of origin, transit and destination for the trafficking of women and children into the sex trade, including women from Indonesia, the Philippines, Thailand, China, Taiwan, Singapore, Myanmar (Burma), Vietnam, Sri Lanka and Laos.

Some Chinese Malaysian women are involved in the sex trade in Hong Kong, Japan, Canada, the US, and Australia.

Rural women workers, mainly those who work in oil palm plantations, continue to be particularly poor and vulnerable. They are paid the lowest salaries and run the risk of sexual harassment and pesticide poisoning.

CHILDREN

In the last few years the economy has developed significantly and priority has been given to improvements in the fields of education and healthcare. The reductions in child and maternal mortality have been exceptional in recent decades and rates are now similar to those of many developed countries*. HIV/AIDS prevalence is increasing; the number of reported cases is doubling every three years*. Reports by different NGOs on violence against children and young people, including sexual abuse, are also increasing. It is estimated that there are over 640,000 economically active minors. Most of them are engaged in the urban informal sector, selling food, working in

night markets, minor industries and in rubber and oil palm plantations.

INDIGENOUS PEOPLES/ ETHNIC MINORITIES

The Dayaks are indigenous peoples making up over 40 per cent of the population of Sarawak state in northern Borneo. They are Christian, while most Malaysians are Muslim. Their main concerns are lack of political representation, the loss of their land, and social, urban and educational problems that hinder the community's development.

The Kadazans in Sabah, also in northern Borneo, are descended from diverse ancestral groups with different languages and a range of social customs that diverge from the prevailing Malaysian culture. They lack political representation.

People of Indian origin are geographically dispersed. Since most of them are Hindu, they are a religious as well as an ethnic minority. Chinese and Indians still face cultural restrictions, mainly due to the Government's aim of assimilating children in the Malay education system. They lack representation at Government

level and State policies do not sufficiently compensate for their disadvantage.

MIGRANTS/REFUGEES

Malaysia is not a signatory to the UN Convention on Refugees, but since 1998 it has allowed the UNHCR to give protection and assistance to refugees within the country. At the end of 2004, Malaysia hosted about 98,000 refugees and asylum-seekers, the overwhelming majority of whom were Muslims from the Philippines (nearly 60,000). That year, 24,900 applications for refugee status and 10,300 asylum requests were filed.

DEATH PENALTY

The death penalty is applied under the strictest interpretation of *Sharia* law. Drug possession and use, and on occasions alleged drug trafficking, are crimes punished by hanging.

* Latest data available in The State of the World's Children and Childinfo database, UNICEF, 2006.

people and flood large rainforest areas. A private company, jointly linked to the Government, began the preparatory logging.
26 In March 2001, the worst ethnic clashes between Indians and Malays since 1969 took place in Kuala Lumpur. The Defense Ministry minimized it, saying it was a misunderstanding among neighbors.
27 Tighter control over capital flows had enabled the Government to ease the effects of the financial crisis and avoid the special loans and rules imposed by international financial institutions (IMF and the World Bank). At UMNO's annual assembly in June, Mahathir lashed out at the West, alluding to the criticism he received for detaining Anwar and opposition members. The Premier accused Anwar Ibrahim's opposition party, Reformasi, of 'using Mafia-style strategies to reach its goals', and the Islamic PAS of being 'more interested in power than religious values'.
28 In April 2002 the Government gave illegal immigrants a certain period to 'give themselves up' and avoid punishments such as lashings, jail and fines, while Parliament passed amendments to the Immigration Act. After a conference in Bali, in which Asia-Pacific ministers agreed to take measures against human

trafficking, the authorities increased their actions against illegal residents (ten per cent of the labor force), most of them Indonesian and Filipinos.
29 Before a visit by Mahathir to Washington in May 2002, the police arrested 14 people suspected of being Islamic militants, including the wife of a man accused of aiding hijackers who carried out the 11 September 2001 attacks in the US. This police offensive was applauded by Washington, which thanked the Malaysian Government for the support it gave to the US-led 'war on terror'.
30 In June 2003, Mahathir expressed strong criticism of the US and Britain. In an obvious reference to the war on Iraq, he said that false allegations had been used to justify military action. He also claimed that the West had begun to invade and rule certain countries to exploit their wealth rather than because of security concerns - and that Malaysia could be a target. The speech, which followed categorical opposition to the war in Iraq voiced by the Non-Aligned Countries Conference in Kuala Lumpur in February 2003, angered US President George W Bush, who threatened Malaysia with economic sanctions.

31 Abdullah Ahmad Badawi became Prime Minister in October 2003, replacing Mahathir Mohamad after 22 years in power.
32 In March 2004, Abdullah Badawi's National Front (NF) coalition won the general elections and regained control of the state of Terengganu, located on the east coast, which had been ruled by the Islamic Party. The NF also won two-thirds of seats in Parliament.
33 In September, Anwar was set free when a court overturned his sodomy conviction.
34 In December, a tsunami devastated South Asia. The catastrophe killed over 5,000 people in Malaysia; most victims were near Phuket island, in the south of Thailand. Shielded by Sumatra, Malaysia's coastline was spared widespread devastation despite being close to the epicenter. Dozens of people were swept from beaches near the northern island of Penang. The states of Kedah and Perak also suffered severe damage.
35 A few weeks after the Government launched a big operation to deport migrant workers by the end of May 2005, Malaysia decided to soften immigration policies in order to allow formerly illegal workers to visit the country and seek work. The Government had intended to

make political capital out of its heavy-handed policy, since most migrants came from Indonesia, its rival country. However, it had to step back as a result of the workforce shortage, since undocumented foreigners made up ten per cent of the labor force.
36 At an international conference held in February 2006 in Kuala Lumpur, intended to promote dialogue between Western and Islamic thinkers, Malaysian Prime Minister Abdullah Badawi called on people to bridge the 'huge Islam-West divide'.
37 In May, the Government approved the deployment of 500 military and police personnel to take part in peacekeeping activities in East Timor, which had been undergoing a situation of violent unrest for several weeks, triggered by disgruntled former soldiers.
38 Also in May 2006, former prime minister Mahathir Mohamad accused his successor of surrendering sovereignty to Singapore over his handling of a bridge-building project. Malaysia had begun building its half of the bridge in January but abandoned it four months later after failing to reach agreement with Singapore. Mahathir said Prime Minister Badawi had shown Malaysia was a 'country with no guts'. ∎

Maldives / Dhivehi Raajje

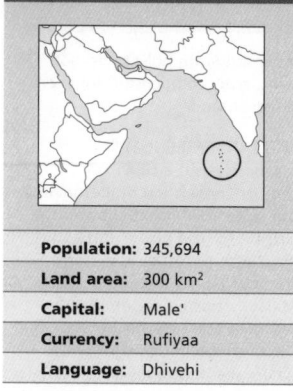

Population:	345,694
Land area:	300 km²
Capital:	Male'
Currency:	Rufiyaa
Language:	Dhivehi

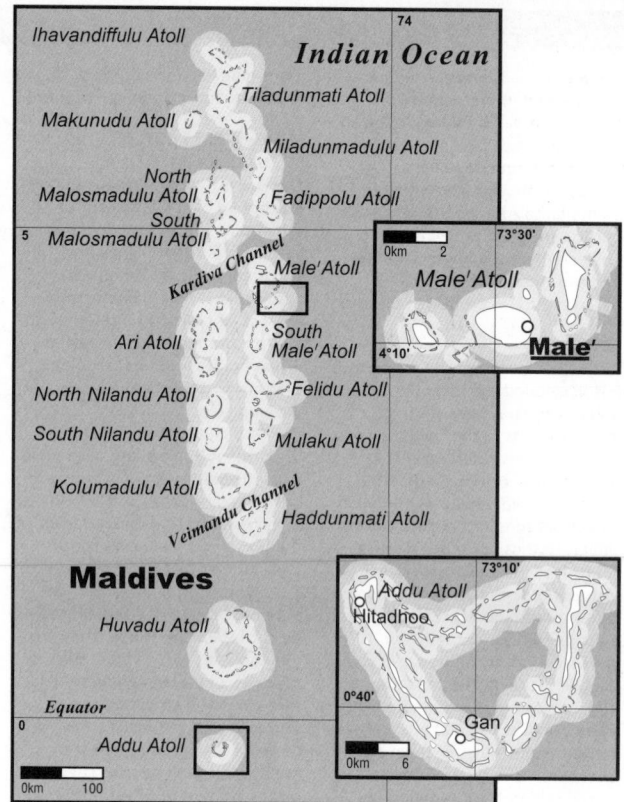

The Maldives archipelago was inhabited as early as the 5th century BC by Buddhist peoples who were probably from Sri Lanka and southern India. In their regions of origin, both groups spoke Indo-Aryan languages - Sinhalese and Dravidian, respectively - from which the Dhivehi language was later derived. Throughout the centuries, traders from Arab countries, Malaysia, Madagascar, Indonesia, and China visited the islands as a stopover destination.

2 Islam was adopted in 1153 AD. Ibn Battutah, a notable North African traveler who resided in the islands in the 1340s, remarked disapprovingly on the freedom enjoyed by women.

3 The Portuguese established themselves by force in Male' (today's capital) from 1558 until they were expelled by the local population in 1573.

4 In the 17th century, the islands were a sultanate under the protection of the Dutch rulers of Ceylon (present-day Sri Lanka). The British took possession of Ceylon in 1796, but it was not until 1887 that the incorporation of its territory to the British Crown was formalized.

5 The British had little economic interest in the natural resources of the islands (fishing and tropical fruits). However, the archipelago was of great importance to their maritime transport when they had control over the Suez Canal Company between 1875 and 1956. In order to maintain the transit route under British control at that time (from Gibraltar to Hong Kong), a naval base was set up on Gan Island, at the southern end of the Maldives.

6 Until 1932, absolute power over the local population rested with sultans, the only ones to benefit from the relationship with Britain. That year, a Constitution that granted some rights to Maldivians went into effect, although the country remained a sultanate. In 1953, a popular revolt overthrew the ruler and a republic was proclaimed. British troops intervened to 'restore order' and three months later the country reverted to a sultanate.

7 In 1957, Britain requested permission to enlarge the Gan naval base and install facilities for fighter planes to land there. The proposal triggered fierce opposition and Prime Minister Ibrahim Ali Didi was forced to resign. In 1959, rebellion broke out in the southern Maldives, which decided to break away under the name of the Republic of Suvadiva. One year later, the 20,000 Suvadivan republicans were restored to the sultanate with British help.

8 This intervention paved the way for the British to immediately sign a new agreement with the Sultan extending the protectorate, and maintaining and enlarging the military bases.

9 In 1965 the Maldives attained full independence from Britain, receiving immediate recognition from the UN. However, the Gan naval base was not dismantled until 1976, one year after the building of the US military installations on the neighboring island of Diego García.

10 The sultanate was abolished in a plebiscite held in 1968 and an Islamic republic was established. Amir Ibrahim Nasir, the Sultan's Prime Minister, became President. In March 1975, President Nasir accused Prime Minister Ahmed Zaki of leading a conspiracy; Zaki was immediately exiled to a desert island along with some of his supporters.

11 Once the British withdrew from the Gan naval base, Nasir proposed to rent the unused installations to transnational corporations, but his idea was rejected by the *Majilis* (legislative council). Nasir, who had been elected five years earlier, ended his second presidential term in 1978, without standing for re-election. Maumoon Gayoom, a former Minister of Transport during the Nasir administration who had studied in Sri Lanka and Egypt, was appointed President.

12 Until that time, the Maldives, with a population scattered across 200 islands, had practiced mainly traditional medicine and had only one hospital in Male'.

LAND USE

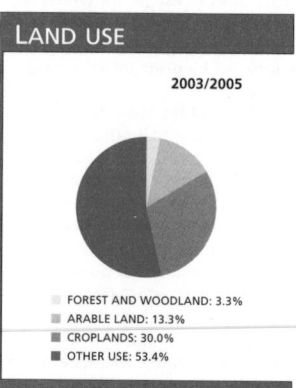

2003/2005

- FOREST AND WOODLAND: 3.3%
- ARABLE LAND: 13.3%
- CROPLANDS: 30.0%
- OTHER USE: 53.4%

PROFILE

ENVIRONMENT
There are nearly 1,200 coral islets in this archipelago, where the land is never more than 3.5 meters above sea level. Vegetation is sparse except for the plentiful coconut palms. A tropical monsoon climate prevails. There are no mineral or energy resources. Fishing is the main natural resource.

SOCIETY
Peoples: The population of the archipelago came from migrations of Dravidian, Indo-Aryan and Sinhalese peoples from India, later followed by Arab peoples.
Religions: Muslim (official and proclaimed universal).
Languages: Dhivehi (official), an Indo-Aryan language related to Sinhalese; English and Arabic are also spoken.
Main Political Parties: Political parties were allowed to register in June 2005; the first entrants are: Adhaalath (Justice) Party; Dhivehi Rayyithunge Party (Maldivian People's Party); Islamic Democratic Party; Maldivian Democratic Party.
Main Social Organizations: There are no social organizations.

THE STATE
Official Name: Dhivehi Jumhuriyya (Republic of the Maldives).
Administrative Divisions: 20 districts.
Capital: Male', 83,000 people (2003).
Other Cities: Hithadhoo, 10,800 people.
Government: Presidential republic. Maumoon Abdul Gayoom, President since 1978, re-elected in 1983, 1988, 1993, 1998 and 2003. Legislative Power: the Citizens' Council, with 50 members, eight of whom are elected by the President.
National Holiday: 26 July, Independence Day (1965).
Armed Forces: About 1,000. The force performs both army and police functions.

Life expectancy
69 years
2005-2010

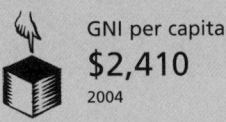

GNI per capita
$2,410
2004

Literacy
96% total adult rate
2000-2004

Half of the population had access to the Dhivehi-language schools *(makthabs)* which were focused on reading and reciting the Qur'an, while only English-language primary and secondary schools, attended by British descendants, were equipped to teach the standard curriculum.

[13] In 1979, Gayoom travelled to Europe, the Middle East and Cuba, where he attended the Sixth Summit Conference of the Non-Aligned Movement and became a member of the organization. In 1980, the Maldives signed a scientific and technological co-operation agreement with the Soviet Union. A similar agreement was signed with China in 1981, and also a trade agreement with India. That year, the Maldivian authorities refused to join ASEAN. Meanwhile, in 1982 the Maldives became a member of the British Commonwealth.

[14] During the first five years of his administration, Gayoom founded schools with educational programs following international standard curricula. During that period, the Government set up a fisheries corporation for the processing of canned and frozen tuna which became the main export product together with handmade clothing.

[15] In the early 1980s, tourism was promoted and the Maldives became a holiday destination for many Europeans. Within 20 years, the tourism industry grew by 1,600 per cent. In early 2000, 74 islets were set aside exclusively for tourist resort development which, on the negative side, led to increasing pollution of beaches.

[16] In August 1988, three months before he was elected president for the third time, Maumoon Gayoom - with the help of Indian troops - put down another coup attempt, allegedly promoted by Amir Nasir. Seventy-five people were arrested, most of them Sri Lankans.

[17] In 1990, in a cabinet reshuffle, Gayoom removed the Minister of Defense, Trade and Industry. The former minister was the president's brother-in-law and was considered the richest person in the Maldives.

[18] In 1991, the Maldives fell into an economic crisis caused by a chronic trade deficit. That year, the Government sold 25 per cent of its shares in the Bank of Maldives and adopted free market policies, although state control was maintained over exports of frozen and canned fish.

[19] President Gayoom began his fourth term of office in 1993 and remained as head of the

IN FOCUS

ENVIRONMENTAL CHALLENGES
Global warming and the resulting rise in the sea level are affecting the islands and coral reefs. Only ten per cent of the land is arable. A depletion of freshwater aquifers threatens drinking water supplies.

WOMEN'S RIGHTS
Women have been able to vote and stand for office since 1932. In 2005, women held 12 per cent of seats in Parliament.

In 2003, women made up 44 per cent of the islands' labor force. Primary and secondary enrolment rates were slightly higher among girls than boys*.

Eighty-one per cent of pregnant women receive prenatal care, while 70 per cent of births are attended by qualified health personnel*.

Traditionally, women have played a subordinate role in society, although lately they have become increasingly involved in public life.

CHILDREN
Due to the tsunami that hit most of the islands in December 2004 the country's statistics have not been updated.

Some 46 per cent of schools disappeared under the waves, while the rest suffered different degrees of damage*. People who were displaced to other islands in the atoll lack education or health infrastructure. The islands had 30 public libraries, which have now turned to rotting paper*.

The tsunami caused the death or disappearance of almost 110 people, mostly children, and amounted to a 20-year regression in the country's economic and social development. Only six days before the disaster the United Nations Development Programme had decided to take the Maldives off its list of least developed countries, after the country had paid off a major part of its debt.

INDIGENOUS PEOPLES/ ETHNIC MINORITIES
In early history - around 500 BC - the Buddhists replaced the ancient sun-worshipping religions, and in turn Buddhism was displaced in 1153 AD by Islam. The language of the Maldives is Dhivehi. It is related to Sinhala, a Sri Lankan language, but also contains Arabic and Tamil words. The chain of islands has stood out throughout history for being a cultural mosaic of Africans,

Arabs and sailors from Southeast Asia.

MIGRANTS/REFUGEES
The tsunami forced 12,000 people to leave the islands where they lived. Although relatively few lives were lost, all of the islands - except nine - were partially or totally covered by the waves. Approximately 8,000 people were displaced within their own island. It is estimated that a third of the population was directly affected by the disaster, which exacerbated poverty levels.

In 2005 the Canadian Government halted the deportation of approximately 4,000 people - including a considerable number of Maldivians - who had taken refuge in that country after the tsunami.

DEATH PENALTY
The death penalty has not been applied for any crime since 1952, the year in which the last execution took place.

* Latest data available in The State of the World's Children and Childinfo database UNICEF, 2006.

ministries of defense, national security and finance. That year, the Government evacuated five small islands that were under threat of being submerged by rising sea levels from the melting of the polar ice caps as a result of global warming.

[20] In 1997, one year prior to his fifth re-election, Gayoom opened the country's first university-level institute. At the same time, telephone connections were established in those islands still lacking in them.

[21] A new constitution was approved in 1998, establishing that the legislature would elect the president. Choosing from among five candidates, legislators decided in September to re-elect Gayoom; this was confirmed by referendum in November with 90 per cent of the vote. The Government had prohibited political parties from campaigning for their candidates, forcing all of them to run as individuals. The first Maldivian Republic set up the basis for a regulatory framework in which political pluralism was not reflected and government censorship was strengthened.

[22] Since the 1990s, sexual rights organizations have protested against the punishment - which

includes life imprisonment - for male homosexual acts. Likewise, Reporters Without Borders denounced the imprisonment of journalists who had published articles criticizing the Government. Amnesty International called upon the Maldivian Government to modify its laws to allow the free expression of ideas and respect the right to *habeas corpus*.

[23] In the October 2003 presidential referendum, Gayoom, the only candidate, won 90.3 per cent of the vote and started his sixth consecutive term of office.

[24] In June 2004 Gayoom promised constitutional amendments to shorten presidential mandates and allow for the formation of political parties.

[25] Mohammed Nasheed, leader of the Maldivian Democratic Party (MDP) and a harsh critic of Gayoom for more than 25 years, was arrested in August during a demonstration in Male'. Two weeks later he was charged with terrorism for saying in a public speech a month before that President Gayoom would be 'violently overthrown' unless he called for elections or resigned.

[26] Shortly after, 5,000

demonstrators demanded democracy for the Maldives and the release of all political prisoners. The Government declared a state of emergency which lasted two months.

[27] In December a tsunami which devastated South Asia left more than 80 dead and 20 people missing in the Maldives. Large areas of Male' were left under water. Most of the Maldives are only one meter above sea level.

[28] Amnesty International included Maldives in a list of countries which use hardware and software to censor and restrict access to the internet. The organization claims that many transnational companies are accomplices to these governments, since their operation in the countries is conditional upon their co-operation with this censorship.

[29] In June 2006 the International Press Freedom Mission urged the Government of Gayoom to put an end to arbitrary arrests, harassment and intimidation against the press and dissidents.

[30] In the same month, Parliament unanimously voted to back plans to introduce multi-party democracy for the first time in the country's history. ∎

Mali / Mali

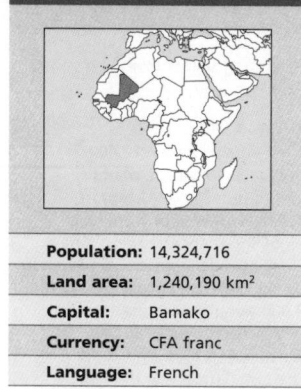

Population:	14,324,716
Land area:	1,240,190 km²
Capital:	Bamako
Currency:	CFA franc
Language:	French

The remains of rock paintings and carvings that date from before 5000 BC were found in the region of Malian Sahara (northern half of the territory). A human skeleton found near the city of Timbuktu in 1927 also dates from that era.

[2] Since the 3rd century, traders traveled in caravans across the Sahara desert from the Niger River, which was rich in gold deposits, to present-day Morocco and Algeria. They transported feathers, ivory and gold, as well as slaves to work in salt extraction.

[3] Between the 4th and 11th centuries, the Saharan trade routes across Mali were controlled by the Soninke kingdom of Ghana, between the Niger and Senegal rivers. Evidence of this black civilization are the terra-cotta statues found in the city of Djenne that were sculpted in the 8th century. In 1076, the Almoravids (a group of religious Muslim warriors and Berber dynasty) overthrew the Soninke empire.

[4] In the 12th century, the Almoravids were replaced by the kingdom of Mali (Malinke, black African group). From the middle and northern portions of the Niger River, the Malinke dominated the Saharan trade routes. In 1255, the empire covered what are now Senegal, Gambia, Guinea-Bissau, part of Guinea, half of Mauritania, southern Algeria and all of Mali.

[5] The decline of the Malinke empire began in the 15th century with the emergence of the Songhai empire that was settled in the area between Timbuktu and the present-day city of Gao and managed to spread its rule over all of Mali between the 15th and 16th centuries. Under Songhai influence, the cities of Djenne and Timbuktu flourished as centers of both trade and Islamic scholarship.

[6] In 1591, the Moroccan army of Ahmad al-Mansur took over the domain of the Songhai empire, imposing its rule for about two centuries. It was during this period that most of Mali's Berber groups such as the Tuareg - the most

numerous group, whose members are still nomadic - and the 'Moors' (Arab Berbers) came to the region.

[7] With the destruction of Songhai rule, a time of political chaos ensued. The Moroccans dispersed the manuscripts of Timbuktu's libraries and executed Songhai scholars. They also discontinued the trans-Sahara trade routes. New routes were established to supply gold and slaves to the European posts on the Atlantic coast.

[8] The ties between Moroccan authorities and invaders lapsed and in 1737 the Tuaregs managed to seize control of the Niger Bend. In 1833, the Muslim Fulani kingdom, which had spread to Cote d'Ivoire and Senegal since the 17th century, finally defeated the Moroccans. The Fulani were nomadic herders. Those who are still herders are widely

dispersed in West Africa; most have retained their original animistic beliefs.

[9] The 19th century in Mali was marked by French colonial penetration, from the west (Senegal), and at the same time, by the wars between Islamic groups (*jihads*) prompted by the establishment of a theocratic state by the Fulani dynasty of the Macina region.

[10] The French built their first fort in 1855. By combining military incursions and partial alliances with rival local groups, they managed to dominate the territory which would later be called French Sudan (present-day Mali, Burkina Faso, Benin and Senegal).

[11] The fact that France was growing weaker, combined with the democratic atmosphere that

prevailed after World War II, encouraged the emergence of anti-colonial organizations. In Africa, this was expressed by the formation of the African Democratic Assembly (RDA), at a conference in Bamako in 1945, led by Modibo Keita.

[12] In the following decade, under the impact of the defeat of French colonialism at Dien Bien Phu in Vietnam (1954) and the Algerian revolution (1954-1962), Paris embarked on a policy of gradual concessions that led to Mali's independence and to the proclamation of the Republic of Mali in August 1960.

[13] That same year, Modibo Keita became President and headed the Federation of Mali, which was joined by the presidents of Senegal (Senghor) and Côte d'Ivoire (Houphouet-Boigny). The three leaders, who had been educated in France, wanted a common program under the banner of 'African socialism'. However, their differences regarding their countries' individual relationship with France and with the USSR and China caused the project to fail.

[14] Keita decided to expropriate and nationalize all sectors of the economy. His administration neither managed to change the backward social and economic conditions nor build a strong political structure. In 1967, when the country was on the verge of political and financial collapse, Keita started negotiations with France. In November 1968, the Military Committee for National Liberation (CMLN), headed by Colonel Moussa Traoré, seized power in a coup.

[15] Traoré banned all political activity. Until that time, parliamentary democracy had never been fully exercised. In 1974, a new constitution was approved in

PROFILE

ENVIRONMENT

There are three distinct regions: the northern region which is part of the Sahara desert; the central region which consists of the Sahel grasslands, subject to desertification; and the southern region, with humid savannah vegetation, home to most of the population, and irrigated by the two major rivers: the Senegal and the Niger. There are significant gold deposits.

SOCIETY

Peoples: Among the many Malian ethnic groups, the largest are the Mande. Other significant ethnic groups are the Tuareg, Peul, Songhai, Moor, Senufo, Fulani, Dogon and Voltaic. In the north are found the lighter-skinned nomadic population, made up of Berber groups. **Religions:** 90 per cent Muslim, 9 per cent practice traditional African religions, and small Christian minority. **Languages:** French (official). Of the African languages, Bambara is the most widely spoken (80 per cent). Arabic and Tuareg are also spoken. **Main Political Parties:** Hope 2002 (which includes: Rally for Mali; National Congress for Democratic Initiative; Patriotic Movement for

Renewal; Rally for Labour Democracy); Alliance for Republic and Democracy (which includes: Alliance for Democracy in Mali-Pan-African Party for Liberty, Solidarity and Justice and Others).
Main Social Organizations: National Workers Union of Mali (UNTM); Alumni and Students Association of Mali (AEEM); National Union of Women of Mali (UNFM); National Union of Mali Youth (UNJM); Movements and Unified Fronts of the Azawads (MFUA), opposition Tuareg groups.

THE STATE

Official Name: République du Mali. **Administrative Divisions:** Eight regions and the district of Bamako. **Capital:** Bamako 1,264,000 people (2003). **Other Cities:** Ségou 132,400 people; Mopti 114,400; Sikasso 125,400; Gao 104,700 (2000). **Government:** Head of State: Amadou Toumani Touré, President since 2002. Head of Government: Ousmane Issoufi Maïga, Prime Minister since 2004. The Prime Minister appoints the members of the Cabinet. Legislature: single-chamber: National Assembly. **National Holiday:** 22 September, Independence Day (1960). **Armed Forces:** 7,000 (2003)

Life expectancy
49 years
2005-2010

GNI per capita
$330
2004

Literacy
19% total adult rate
2000-2004

HIV prevalence rate
1.9% of population 15-49 years old
2003

a referendum with 99.8 per cent of the vote, which provided for a six-year presidential term. Its Civil Code included a law that banned sodomy. The opposition was banned from taking part in the election and Keita's followers were imprisoned. Keita died in prison on 16 May 1977. In the largest mass demonstration ever seen in Bamako, the people followed Keita's body to the cemetery in open defiance of the military regime.

16 In 1979, Traoré's re-election triggered student demonstrations which were harshly repressed. As a result, three students were killed, around 100 arrested and 13 tortured. Immediately after, Traoré embarked upon an austerity program drawn up by the IMF and the international creditor banks, through the Democratic Union of Malian People (the military junta's puppet party).

17 In June 1985, President Traoré, running as the official party's only candidate, was re-elected with 99.94 per cent of the vote.

18 In 1988, the foreign debt amounted to 125 per cent of GDP, with debt servicing exceeding a quarter of export revenues. That same year, following IMF guidelines, the Government began the privatization of the banking system with French financial support. The authorities also announced a reduction in the number of public employees, as well as the decision to sell off state enterprises; at the same time, the cabinet was reshuffled. Students, teachers and public employees took to the streets to protest against these measures.

19 On 10 April 1991, a popular and military revolt against the Traoré regime carried Lt-Col Amadou Toumani Touré to power. He was the leader of the Council of Transition for the Salvation of the People (CTSP), which promised to transfer government to civilians early in 1992.

20 An uprising in June 1991 by the Tuaregs in the north and the Moors in the east exacerbated the social tensions. In July, an attempted coup by a portion of the armed forces led Toumani Touré to grant a 70 per cent wage raise to the armed forces and civil servants. In April 1992, the premier signed a peace agreement with the Azawad Unified Front and Movements, an umbrella organization linking four Tuareg opposition groups.

21 On 26 April 1992, Alpha Oumar Konaré, leader of the Alliance for Democracy in Mali (ADEMA), was elected president in the first multiparty elections since the country's independence. He continued the Government's economic policy.

22 An exodus of 120,000 Tuaregs was provoked by the persecution against them. They fled to Algeria, Mauritania, Niger and Burkina Faso

ENVIRONMENTAL CHALLENGES
The desertification process and soil erosion are aggravated by the lack of water and the deforestation by fire in order to open the way for crops. Poaching threatens many species. There is a growing lack of drinking water.

WOMEN'S RIGHTS
Women have been able to vote and stand for office since 1959. In 2003 women held 10 per cent of parliamentary seats and 19 per cent of ministerial or equivalent positions. In 2003, they made up 46 per cent of the total labor force.

The adult literacy rate among women stood at only 12 per cent, the third-lowest in the world after Burkina Faso (8 per cent) and Niger (9 per cent)*.

The overall fertility rate is seven children per woman*, a figure that has held broadly steady since 1960.

About 57 per cent of pregnant women receive prenatal care, while only 41 per cent of deliveries are assisted by qualified staff.*

Nearly 93 per cent of women between the ages of 15 and 49 have undergone some form of genital mutilation. No major differences (only 3 per cent more in rural areas) were found with respect to this practice between women from rural areas and those from urban areas or among those who belonged to different ethnic groups or religions.*

CHILDREN
More than half of Malians are under 18 years old (7,231,000)*. Of these, 2.54 million are under five years old*. In 2003, 15,000 were children under 14 were living with HIV/AIDS.* - but, in reality 75,000 children under 14 had lost both parents to AIDS*. Only 15 per cent of the population between 15 and 24 has some knowledge of how to prevent infection.* In 2004, 45 per cent of boys were enrolled in school, compared to 34 per cent of all girls*.

In 1998, the trafficking of thousands of children from Mali to northern Côte d'Ivoire was made public. The children were mostly boys and were trafficked to work in coffee, cotton and cocoa plantations or for domestic labor under abusive conditions.

INDIGENOUS PEOPLES/ETHNIC MINORITIES
There are 10 distinct ethnic groups in Mali. The Mande form the biggest group, accounting for 50 per cent of the population. The Tuareg and Senegalese - a group native to the region and a national minority, respectively - make up six and three per cent of the total population.

More than 600,000 Tuareg went through a period of rebellion in the mid-1990s. The government repression they suffered reduced the chance of further revolts in recent years, although the authorities remain unwilling or unable to implement development and educational projects to alleviate their social marginalization. The Tuareg had few opportunities to demand respect for their rights in the early 2000s.

MIGRANTS/REFUGEES
Mali hosted 12,300 refugees and asylum-seekers in late 2004. That year, about 6,200 people from Mauritania lived in the country in refugee-like circumstances. UNHCR also registered the presence of approximately 750 people from Côte d'Ivoire. Nearly 2,200 Malians sought asylum in Europe in 2004, mostly in France, Spain and Switzerland.

DEATH PENALTY
Mali is considered de facto abolitionist since the last execution dates back to 1980. However, the death penalty is still provided for by law.

* Latest data available in *The State of the World's Children* and *Childinfo* database, UNICEF, 2006

(Mali had confronted this latter country in a border war in 1985). In 2002, Amnesty International reported that thousands of Tuaregs had been executed and imprisoned without trial. In 1995, the Government pursued negotiations with Tuareg groups which led to the demobilization of 2,700 guerrillas in 1996 and to the gradual return of refugees.

23 In May 1997, Konaré was re-elected president with 95.9 per cent of the vote. The opposition complained that there was no guarantee of protection if they voted, and so they boycotted the 1998 general and local elections. That year, Moussa Traoré and his wife were charged with misappropriation of funds and abuse of power, and were sentenced to death, although the sentence was later commuted to life imprisonment.

24 A coup d'état was foiled in January 2000, but a few days later the President decided to include military officers in his cabinet. One month later, Mande Sidibe, an economist and former IMF official, was appointed prime minister.

25 The April 2002 elections were marred by allegations of fraud and gave the victory to Touré with 64.4 per cent of the vote. Touré was very popular for having overthrown Traoré and for having kept his promise to hand over power to civilians in 1992. However, the cabinet of the 'government of national unity' formed by Touré with 22 minority organizations, resigned en masse in October of that year. None of those involved gave a public explanation.

26 At the WTO ministerial conference held in Cancun in September 2003, Mali, together with three other African countries, filed a complaint against subsidies granted by the US Government to US cotton producers. The Malian representatives said it was a 'life or death' issue for the frail economy of the West African country.

27 At the end of 2003, the Government negotiated the release of 40 hostages (mainly German tourists) who had been kidnapped by an Islamic extremist group, the Salafist Group for Preaching and Combat, as part of an operation against the Algerian Government.

The Algerian organization to which the kidnappers belonged was included on Washington's list of terrorist groups.

28 In April 2004, after the Malian Army clashed with an Islamic group - supposedly allied with al-Qaeda - the Government resigned in full, at the request of President Touré. Ousmane Issoufi Maiga was named Prime Minister.

29 A scourge of locusts decimated the crops in November and a lack of rains affected the new crops. Together with the lack of response from the international community, this caused a major humanitarian crisis. Millions of Malians were threatened by the serious food insecurity and market turmoil. The locusts wiped out cereals and vegetables, and only rice (which grows under water) was spared by the insects' voracity. The lack of rains did the rest.

30 In March 2006 the World Bank, the IMF and the African Development Bank decided to write off Mali's debt, along with that of 12 other African nations. The measure would be effective from July 2006. ∎

Malta / Malta

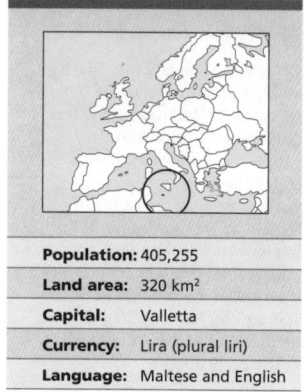

Population:	405,255
Land area:	320 km²
Capital:	Valletta
Currency:	Lira (plural liri)
Language:	Maltese and English

Malta's most valuable 'natural resource' is its geographic location, which has made it the historical focus of every conflict for domination of the Mediterranean. Being situated midway between Tunisia and Sicily, Malta is the key to the sea routes between east and west. In ancient times, Phoenicians, Greeks, Carthaginians, Romans and Saracens successively occupied the island.

[2] In 1090, the Normans conquered the island for the Kingdom of Sicily and 300 years later it fell to the Spanish Kingdom of Aragon. In the 16th century, the defense of the island was entrusted to the order of the Knights of St John of the Hospital (Knights Hospitalers). Kicked out of Palestine, they settled in Malta in 1574. They remained there for more than two centuries, known as the Knights of Malta, until they were driven out by the French in 1798. The Congress of Vienna in 1815 formally recognized the title of 'Sovereign Order of Malta', and consecrated English sovereignty over the island.

[3] From the early 20th century, the Maltese fought for their independence. After a popular uprising in 1921, London agreed to a certain degree of internal autonomy, which was revoked at the beginning of World War II.

[4] Malta was used as a base for the Allied counter-offensive against Italy. In 1947, London returned the island to a certain level of autonomy.

[5] Independence was formally declared on 8 September 1964 but Britain maintained a strong influence over politics on the island. In 1971, the Labour Party attained power and established broader diplomatic relations. NATO forces were expelled that year, and Malta later joined the Movement of Non-Aligned Countries.

[6] From the beginning of the Labour administration, measures to reduce the Church's power had been introduced. The bishops owned 80 per cent of all property on the island and controlled education. Conflict erupted in 1983 when the Government expropriated all Church property and made secular education obligatory in primary schools. In 1985, the Government and the Church signed an agreement providing for a gradual transition to secular education at the secondary level.

[7] The Nationalist Party came into power in May 1987. Edward Fenech Adami was named Prime Minister.

[8] Adami was re-elected in 1992, but he lost his position in the Government in 1996 when the Labour Party won the elections and Alfred Sant became Prime Minister. Sant pulled the country out of

NATO and asserted that Malta would use constitutional neutrality to promote stability and security in the Mediterranean region.

[9] The Government of Sant lasted for two years. In 1998, it collapsed following a vote of censure by opposition and new elections were called. The Nationalist Party's triumph reinstated Adami as Prime Minister. His first measure was to revive the request to join the European Union (EU), put on hold by the Sant administration.

[10] On 24 March 1999, Minister of Foreign Affairs Guido de Marco was promoted to President by Parliament. That same year, the EU formally re-accepted

Malta's application at its annual meeting in Brussels, and later authorized the start of final negotiations.

[11] Malta was the first country to abolish the death penalty in the new millennium when it passed the Law of the Armed Forces (Amendment) in March 2000.

[12] The debate surrounding accession to the EU was settled on 8 March 2003 in a referendum in which a narrow majority (53.6 per cent) voted in favor of membership. Malta formally joined the EU on 1 May 2004.

[13] In 2005 Malta received more than 1,500 asylum requests from Africans, which is considerable for a country of 400,000 inhabitants and with the third-highest population density in the world. This led the island to request the aid of other European countries. Meanwhile, anti-immigration sentiment led to the creation of a new far-right party, the Alleanza Nazionali Republikana, which organized the first demonstration against immigrants. European legislation states that asylum seekers must request asylum in the country where they arrive first. Most of the immigrants - Somalis, Liberians, Sudanese and others - had arrived on the island by mistake, trying to reach the mainland, and wished to leave Malta. ∎

PROFILE

ENVIRONMENT

Malta is an archipelago of small islands, none of which have permanent rivers or lakes. The inhabited islands are the largest ones: Malta, where the capital is located, 246 sq km, Gozo, 67 sq km and Commino. The archipelago is located in the central Mediterranean Sea, south of Sicily, east of Tunisia and north of Libya. The coast is high and rocky, with excellent natural harbors. The islands' environmental challenges are concentrated precisely in these areas. The problems are caused by the encroachment of urbanization, the development of tourism, the gradual abandonment of farmland, the increase in waste and the water pollution resulting from industrial activity.

SOCIETY

Peoples: The Maltese (95.7 per cent) come from numerous ethnic backgrounds, with strong Phoenician, Arab, Italian and British roots. There is a small British minority (2.1 per cent).
Religions: Catholic (official); there are minorities of Anglicans and Muslims.
Languages: Maltese and English, both official. Maltese is a Semitic language written in the Latin alphabet, with Italian elements.
Main Political Parties: The Nationalist Party; The Labour Party, Alternativa Demokratika/Alliance for Social Justice.

THE STATE

Official Name: Repubblika ta'Malta.
Capital: Valletta 83,000 people (2003).
Other Cities: Birkirkara 21,700 people; Qormi 18,100; Sliema 11,800 (2000).
Government: Parliamentary Republic. Fenech Adami, President since 2004. Lawrence Gonzi, Prime Minister since 2004. Unicameral Legislature: House of Representatives has a minimum of 65 members, elected for a five-year term. The number of seats can increase if no party wins an absolute majority.
National Holiday: 21 September, Independence Day (1964).
Armed Forces: 2,000 troops (2003).

LAND USE

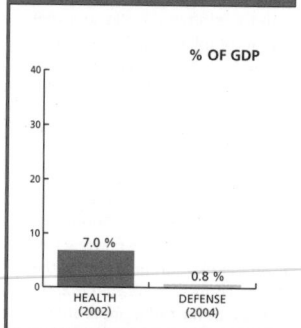

2003/2005

IRRIGATED AREA: 18.2% of arable land

- ◼ ARABLE LAND: 31.3%
- ◼ CROPLANDS: 3.1%
- ◼ OTHER USE: 65.6%

PUBLIC EXPENDITURE

% OF GDP

HEALTH (2002)	DEFENSE (2004)
7.0 %	0.8 %

Marshall Islands / Marshall Islands

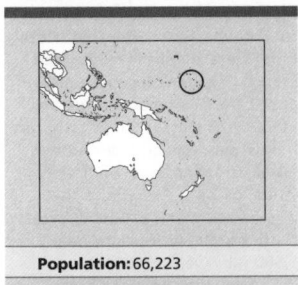

Population:	66,223
Land area:	180 km²
Capital:	Majuro
Currency:	US dollar
Language:	Marshallese and English

T he Kwajalein and Bikini atolls of the Marshall Islands were thrust into modern history in February 1944 when heavy bombing by combined US naval and air force units hit the islands. This was followed by a prolonged and bloody battle which ended with the defeat of the Japanese and the occupation of the islands. Countless lives were lost but the military command considered it was a fair price to pay for the islands given their strategic location.

2   In 1979, the US proposed making the Marshall Islands a self-governing territory in free association with the US. The Republic of the Marshall Islands (the official name since 1982) was granted jurisdiction over local and foreign affairs but the US declared the islands to be military territory for specific use. By doing so, they made a de facto situation official: between 1946 and 1958, nuclear tests had been performed on the Bikini and Kwajalein atolls which turned the Marshall Islands into the area with the highest level of radioactive contamination in the world.

3   In 1961, Kwajalein became the Pacific experimental missile target area, especially for intercontinental ballistic missiles launched from California, and early in the 1980s the US chose the atoll as a testing site for its new MX missiles. The local population was evacuated and entry was forbidden to civilians.

4   In the case of the Bikini atoll, 23 nuclear tests were performed there between 1946 and 1958, including the detonation of the first H-bomb. The inhabitants of the atoll insisted on returning to their homeland after having been transferred to the Rongelap atoll. In 1979, testing revealed that 130 of the total 600 inhabitants living on the Bikini atoll were contaminated with extremely dangerous levels of plutonium.

5   The inhabitants of Bikini, together with those living on Rongelap, sued the US Government for $450 million. The charges were filed together with reports compiled by US Government agencies demonstrating that local residents had been intentionally exposed to radioactivity in 1954 in order to study the effects of the bomb on humans. Reports revealed by the US Government in 1995 proved the dangers of exposure were known but this had never been passed on to the people of the Marshall Islands. The US has so far paid nearly $100 million in compensation for damage done by nuclear testing.

6   In October 1986, the US Congress signed the Compact of Free Association making the Marshall Islands responsible for its own internal political affairs. The US would be responsible for the defense of the new state for a period of 15 years, which would enable the US to set up an important air base on the island, in exchange for financial aid.

7   In the first elections of the new state, held in 1986, Amata Kabua was elected President. In 1988, the Marshall Islands were admitted to the South Pacific Regional Trade and Economic Co-operation Agreement.

8   In April 1990, the US announced that it would use the area as a site for destroying chemical weapons from Europe. Environmentalists denounced the idea, as well as plans to dispose of 25 million tons of toxic waste on one of the archipelago's atolls between 1989 and 1994. In 1991, the Marshall Islands were accepted as a member state of the UN.

9   Chancellor Tony de Brun founded the Ralik Ratak Democratic Party in June 1991, after distancing himself from President Kabua who, after being re-elected for a fourth consecutive term, died in December 1996. In January 1997, Parliament appointed Imata Kabua, cousin of the deceased, as the new President.

10   The hot issue in 1997 was the Government's plan for a feasibility study on the Marshall Islands acting as a dump for nuclear waste. In June, the Opposition won a suspension of the project to store nuclear waste. The Government announced the construction of a large hotel and casino complex, financed by South Korean capital.

11   In February 1998, Asian banks expressed satisfaction with economic reforms which, according to the Government, were linked to the end of US financial aid.

12   After the 2003 legislative elections, Kessai Note became President, taking office in January 2004. Note announced that his top priority would be fighting corruption, in response to reports identifying the Marshall Islands as one of the worst areas for money-laundering in the world.

13   In March 2006 the UN General Assembly passed a resolution creating a new Human Rights Council. Only four countries voted against: the Marshall Islands, the US, Israel and Palau. These four countries usually vote together, for example in favor of blockading Cuba or invading Iraq. ∎

Martinique / Martinique

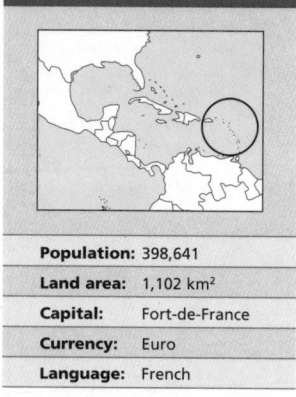

Population:	398,641
Land area:	1,102 km²
Capital:	Fort-de-France
Currency:	Euro
Language:	French

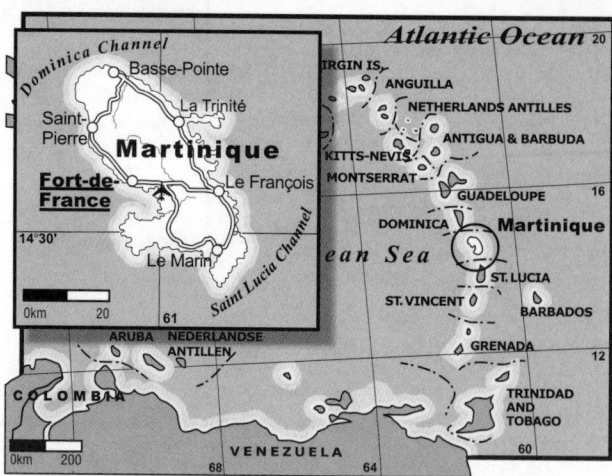

'**M**adinina' was the name given by the Carib Indians to the largest of the Lesser Antilles; the island came to be known as Martinique in the colonial period.

2   It was not until a century and a half after the French occupation in 1635 (interrupted by short periods of British rule) that economic activity began, due to the sparse population and an almost complete dearth of precious metals.

3   Late in the 17th century, sugar cultivation transformed the landscape, effectively ending the era in which fruits and vegetables were gathered rather than cultivated. The slave trade modified the system of production and African slaves replaced indigenous people on the plantations. Monoculture defined Martinique's role within the international labor structure, consolidating colonial ties with France.

4   A minority of 12,000 European land, sugar mill and business owners controlled 93,000 slave laborers, a situation that led to uprisings. The *Marronuage* - free zones within the colony created by collective rebellions of slaves - appeared. There are records of *quilombos* (see Quilombos in Brazil) on the island in 1811, 1822 and 1833, years when rebellions shook Martinique.

5   In the early 19th century, the crisis of capitalism put an end to the traditional plantation system, which was incapable of transferring capital to industry, leading to great social disturbances.

6   In 1937 the formation of Martinique's central labor union provided an organizational framework for social mobilization. Leftist Aimé Césaire was elected mayor of Fort-de-France in 1945 and, in 1946, Martinique's deputy in the National French Assembly. Césaire was co-founder, alongside Senegal's Léopold Sédar Senghor, of the *Négritude* movement that protested the imposition of French culture.

7   In 1948 the French Government created the Overseas Department. Martinique's middle-classes wanted the same rights as the French, but differences within the anti-colonialist movement made it impossible to reach a consensus.

8   While both emigration to France and French foreign aid had always provided support for the Martinique's economy, demands for independence resulted only in the granting of greater autonomy. The visits of Charles de Gaulle (1956, 1960, and 1964) did not smooth over the political unrest and by the late 1970s France decided to help Martinique become economically self-sufficient in preparation for independence. Despite liberation groups being responsible for several bombings in the 1980s, there were no major protests against colonialism, either in Paris or on the Caribbean islands.

9   In the March 1986 elections, left-wing parties took 21 out of 41 seats in the regional council. Aimé Césaire was re-elected president of the council. Rodolphe Désiré (Martinique Progressive Party) became the first left-wing Martinique representative in the French Senate.

10   The creation of a single European market led to a worsening of the economic situation in 1993. Elections resulted in a tie between the right and the pro-independence left coalition. In the plebiscite to ratify the Maastricht treaty, voter turnout was only 25 per cent in Martinique, indicating hostility or at least indifference towards deeper integration with the European Union.

11   In 1997, US pressure encouraging banana production in Central America continued, damaging Caribbean exports.

In early 1998, unemployment affected 40 per cent of the active population, an all-time record in the country.

12   French President Jacques Chirac visited the island in March 2000. The people of Martinique demanded greater trade with France and more autonomy in order to trade independently with its Caribbean neighbors.

13   Michel Cadot became commissioner, replacing Dominique Bellion in July 2000.

14   The new European currency, the euro, became legal tender in Martinique in January 2001, as it did in France.

15   In 2002, the French newspaper *Le Parisien* classified the operative conditions of the hotel industry in Martinique and Guadeloupe as a tourist cataclysm. The Accor Hotels Chief Gerard Pelission informed French President Jacques Chirac about the hostile and aggressive attitude of the staff in Martinican hotels. Workers also carried out work stoppages and went on strike, causing damage to the organization and productivity of those hotels. As a result, nearly 1,500 jobs were lost.

16   After the 'no' vote in the 2003 referendum proposing an Overseas Collectivity status for Martinique, the island continued under the administration of a General Council and a Regional Council.

17   The crisis in the banana industry worsened in early 2004, when growers announced that they could not afford to pay social contributions and taxes, and demanded a response from the Government or from France.

18   All the passengers on a plane that crashed in Venezuela in August 2005 were from Martinique. There were no survivors. The plane - and its crew - belonged to a Colombian airline and was chartered to fly from Panama to Martinique. Meanwhile, Colombian president Alvaro Uribe ordered an investigation into the crash to clarify if the company had met the necessary conditions to deliver the service for which it was hired. However, a year after, the investigation had still not published its conclusion. ∎

PROFILE

ENVIRONMENT

Martinique is a volcanic island, one of the Windward Islands of the Lesser Antilles. Mount Pelée, whose eruption in 1902 destroyed the city of St Pierre, is a dominant feature of the mountainous terrain. The fertile land is suitable for agriculture, especially sugarcane. The tropical, humid climate is tempered by sea winds.

SOCIETY

Peoples: Descendants of former African slaves (90 per cent). There is a minority made up of people of European origin (5 per cent). Indian, Lebanese and Chinese (5 per cent).
Religions: Catholic (95 per cent); Hinduism and traditional African religions (5 per cent).
Languages: French (official), and Creole.
Main Political Parties: Communist Party of Martinique, (PCM); Progressive Party of Martinique (PPM); Rally for the Republic/Union for French Democracy (RPR-UDF)
Main Social Organizations: General Confederation of Workers (CGT).

THE STATE

Official Name: Département d'Outre-Mer de la Martinique.
Capital: Fort-de-France 93,000 people (2003).
Other Cities: La Trinité 13,100 people; Le Marin 7,400 (2000).
Government: Head of State: President of France (Jacques Chirac). Head of government: Yves Dassonville, French-appointed Prfet (Commissioner) since February 2004. Claude Lise, President of the General Council since 1992, re-elected in 1998. The General Council has 45 members elected through universal suffrage for a six-year term. Martinique elects 2 members to the French Senate.

Mauritania / Muritaniyah

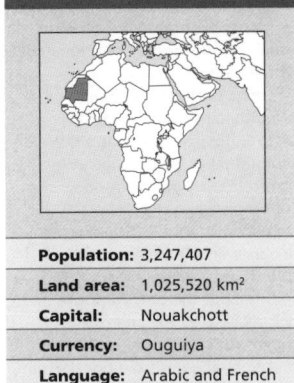

Population:	3,247,407
Land area:	1,025,520 km²
Capital:	Nouakchott
Currency:	Ouguiya
Language:	Arabic and French

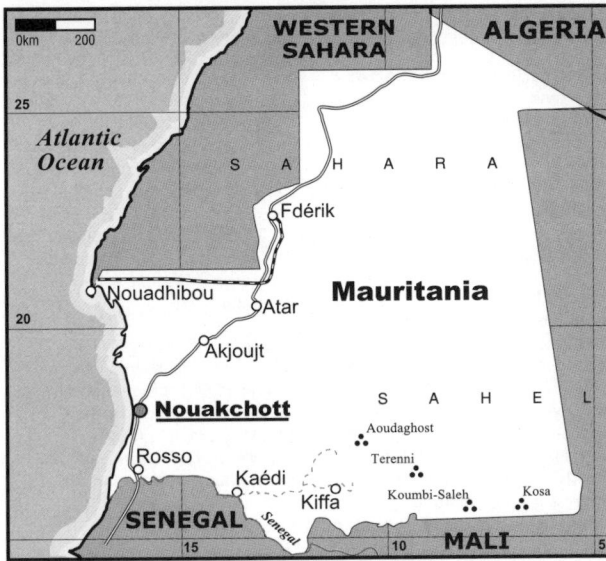

Numerous archeological remains have been discovered in northern Mauritania dating from the Lower Paleolithic and Neolithic periods. Mauritania was inhabited early on by sub-Saharan peoples and by the Sanhadja Berbers and was the cradle of the Berber Almoravid movement, which in the 11th century AD spread an austere form of Islam throughout the territory and neighboring areas. The caravan route that linked Mauritania with Morocco was followed by Arab peoples that formed several powerful confederations such as Trarza and Brakna, which dominated the Senegal River valley, Kunta in the east, and Rigaibat (Regeibat) in the north. Due to the Almoravid conquest in the first place and later to the Fulani migrations (see Cameroon), the population was integrated and unified.

[2] In the 14th century, the Beni Hilal, who had invaded North Africa three centuries before, reached Mauritania. For over 200 years, they plundered the region and fought with the Berbers throughout an area including present-day southern Algeria and Sahara, while southern Mauritania belonged to the Mali Empire. In 1644, all the Berber groups in the region joined to fight the Arabs, but the resulting conflict, the Cherr Baba War, ended 30 years later with the defeat of the Berbers. The Arabs became a warrior caste, known as the *Hassani*, monopolizing the use of weapons, while trade, education and other civilian activities were left in the hands of the local population. Beneath these two groups came the Haratan, African shepherds from the south, kept as semi-serfs. Though this rigid social stratification weakened with time, it is still intact among the Arab-Berbers, the Fulah and Soninke from the south.

[3] Towards the end of the 17th century various emirates arose, which failed to consolidate the country politically, due to internal rivalries and dynastic quarrels. Nevertheless, they provided a minimal degree of order within the region that led to a relative growth of trade caravans. This was due to the cultural unification being carried out by the Zuaias (group of Berber-Marabouts) who devised a simple system of Arabic writing, propagating it along with their religious teachings.

[4] In the 19th century, growing trade coincided with a French project to concentrate the commercial activities of the former French-Sudan (present-day Mali) in Senegal, which meant the elimination of trans-Saharan trade. Due to the frequent forays into Senegal, the French decided to invade Mauritania. The invasion began in 1858, under General Faidherbe, and the fighting continued until the 20th century. The resistance initially put up by the emirates of Trarza and Brakna was continued by Sheikh Ma al-Aini (see Western Sahara), his sons, and later his cousin Muhammad al-Mamun, the emir of Adrar. Pursued by the French almost 1,000 kilometers into the Sahara, Muhammad al-Mamun died in combat in 1934.

[5] After World War II, Mauritania became a French Overseas Province, with its own deputies in the French Parliament. In 1955, internal autonomy was granted and independence was declared on 28 November 1960. Since the country lacked its own infrastructure and administration, its organizational capacity was extremely limited. A French transnational corporation, MIFERMA, was more powerful than the Government, and its iron ore mines supplied 80 per cent of the country's exports and employed one in four wage-earning workers. The progressive wing of the Mauritanian People's Party (PPM), presided over by Moktar Ould Daddah, set in place the foundations for national independence. In 1965, Mauritania withdrew from the Common Afro-Mauritanian Organization (OCAM) through which the French maintained control over their former colonies. In 1966, SOMITEX, the state enterprise with a monopoly on imported consumer products, ended the monopoly held by French traders. Arabic culture began to be revived, and a customs system independent from Senegal was set up. The Arab-Mauritanian Bank was granted the monopoly on foreign trade and, for the first time, the country issued its own currency.

[6] In 1974, the iron mines were nationalized. The country started to replace French influence with closer links to Muslim countries, and finally became a member of the Arab League. Saudi Arabia, Kuwait, and Morocco supplied economic assistance. In 1975 Mauritania, fulfilling an old ambition to annex part of 'Spanish' Sahara (now Western Sahara), joined Morocco in

PROFILE

ENVIRONMENT
The Sahara Desert covers two-thirds of the country. The terrain consists of rocky, dry plateaus and vast expanses of dunes where an extremely dry climate prevails. In the south, the desert gradually gives way to the Sahel savannah, which has some rainfall and sparse vegetation. The southwestern region receives slightly more rain and is irrigated by a tributary of the Senegal River. This region holds most of the population and the main economic activities. Nomadic shepherds are scattered throughout the country. The most serious environmental problem is the process of desertification, which leads to erosion and lack of water.

SOCIETY
Peoples: Seventy per cent of all Mauritanians are descendants of the Moors, nomadic shepherds of north-west Africa, a mixture of Arab, Berber and African peoples. The other 30 per cent are from minority African groups in the south, the most important being the -fulani (known as 'Toucouleur' or 'Peul' by the French) and the Soninke. There are small groups of Wolof and Bambara.
Religions: Islam (official). Sunni Muslim 99.5 per cent, Catholics 0.2 per cent; other 0.3 per cent.
Languages: Arabic and French (official). The Moors speak Bassanya, an Arab dialect. In southern Mauritania, Peul-Fulani and Sarakole (of the Soninke) are also spoken.

Main Political Parties: Democratic and Social Republican Party; Rally for Democracy and Unity; Union for Democracy and Progress; Rally of Democratic Forces.-
Main Social Organizations: Mauritanian Workers' Union; General Organization of Workers of Mauritania.

THE STATE
Official Name: al-Jumhuriyah al-Islamiyah al-Muritaniyah.
Administrative Divisions: 12 regions and the district of Nouakchott.
Capital: Nouakchott 600,000 people (2003).
Other Cities: Nouadhibou 103,100 people; Kaédi 50,200; Kiffa 61,400; Rosso 46,700 (2000).
Government: Colonel Ely Ould Mohamed Vall, Chief of State, whose Military Council for Justice and Democracy deposed long-time President Maaouya Ould Sid Ahmed Taya in a coup on August 2005. Sidi Mohamed Ould Boubacar, Prime Minister since August 2005. Parliamentary republic with bicameral legislature: National Assembly made up of 81 members elected to a five-year term and Senate made up of 56 members elected to a six-year term and partially renewed every two years.
National Holiday: 28 November, Independence Day (1960).
Armed Forces: 16,000 (2003).

Life expectancy
54 years
2005-2010

GNI per capita
$530
2004

Literacy
51% total adult rate
2000-2004

HIV prevalence rate
0.6% of population 15-49 years old
2003

IN FOCUS

ENVIRONMENTAL CHALLENGES

The most pressing environmental problem is the process of desertification and the resultant erosion and growing shortage of water. Desertification is accelerated by overgrazing, deforestation and soil erosion aggravated by lengthy droughts. Mauritania has limited natural fresh water sources. The Senegal River, located along the border with the country of the same name, is the only perennial river.

WOMEN'S RIGHTS

Women have been able to vote and stand for office since 1961, but it was not until 1995 that they won their first seat in Parliament. In 2003 women held 4 per cent of parliamentary seats and 9 per cent of ministerial or equivalent positions. In the same year women made up 44 per cent of the total labor force.

The maternal mortality rate is among the world's 15 highest: 1,000 women die for every 100,000 live births*. Only 64 per cent of pregnant women receive medical care during pregnancy and 57 per cent of births are attended by skilled medical personnel*. The global fertility rate is 5.8 children per woman*.

NGOs say there are a large number of unreported cases of domestic violence. Generally, the cases of violence are resolved within the family unit.

Female genital mutilation is frequently practiced in spite of efforts by the Government and NGOs working in the region. It

is often performed after the first seven days of life and almost always before the age of six months. The percentage of women aged between 15 and 49 who have undergone genital mutilation varies depending on where they live: 77 per cent in rural zones and 65 per cent in urban areas.

CHILDREN

Infant mortality is 78 per thousand live births, while in under-fives it is 125 per thousand live births. In spite of a significant reduction over recent years Mauritania continues to have some of the highest child mortality rates in the world*.

There has been significant progress in child immunization, with between 64 and 83 per cent of children being immunized against the most common childhood diseases such as polio, measles and tetanus*.

In 2004 some 67 per cent of children had access to primary education, however attendance was somewhat lower, at 44 per cent*.

There are special programs for the protection and welfare of children, as well as programs to care for abandoned children, however, the lack of financial resources prevents these from having a major impact.

INDIGENOUS PEOPLES/ ETHNIC MINORITIES

Most 'Black Moors' were slaves in the past. At the present time, they are still discriminated against by the so-called 'White Moors' and other African peoples. They are dispersed throughout the country, although

their villages exist mainly in the southern regions.

The Kewri (farmers, shepherds and fisherfolk) are made up of three black ethnic groups: the Peul, the Soninke and the Wolof.-

During 1989 and 1990 the Government expelled thousands of Kewri in a dispute with Senegal.-

MIGRANTS/REFUGEES

In late 2004 Mauritania was hosting 29,500 refugees and asylum-seekers, most of them from Western Sahara. The refugees from Western Sahara fled to Mauritania during the 1970s to escape the Moroccan invasion of their country. Saharawi refugees in Mauritania are mostly self-sufficient and are not assisted by UNHCR or other aid agencies. At the same time, more than 4,500 Mauritanians sought asylum in other countries such as Belgium, Canada, the US and France. In March 2005 Mauritania allowed UNHCR operations in its territory. This organization, together with national and international NGOs, provided legal advice to asylum-seekers and basic humanitarian aid to refugees in the country.

DEATH PENALTY

The death penalty is applicable to a wide range of crimes.

* Latest data available in The State of the World's Children and Childinfo database, UNICEF, 2006.

an attempt to divide the Spanish possession. With logistical and military support from France, Mauritania and Morocco occupied Western Sahara. The government of Ould Daddah, independent Mauritania's first president (from 1961 to 1978) and the first Mauritanian to graduate from a university, paid a high price because after violent reprisals by the Polisario Front (see Western Sahara), Mauritania was virtually occupied by Moroccan troops. The economic crisis was intensified and popular discontent led to demonstrations and clashes with the police. The Mauritanian people, who felt a strong affinity with the Saharawis, condemned the intervention in the Western Sahara liberation war. The crisis exploded in 1978 and over the next six years, five coups were mounted. An attempt was made to Arabize the entire population, without recognizing other peoples

living in the southern part of the country.

[7] In 1984, Ould Haidalla (President since 1980) was ousted by a coup led by Maawiya Ould Sid'Ahmed Taya, an army colonel and chief of staff, who officially recognized the Democratic Saharawi Arab Republic. Steps were also taken to dismantle the 'clandestine economy' (only 50 per cent of companies kept legal accounting records). In 1985, Mauritania was granted a $12 million loan by the IMF, committing itself to the implementation of a stringent structural adjustment program. Due to the continuing desertification process, the country suffered from a persistent grain deficit, which in the 1980s reached 12,000 tons annually. The economic and social outlook deteriorated; the southern farming and grazing lands shrank as the desert expanded, and impoverished nomadic people were driven to the cities. In the fishing sector the

Government put the priority on preserving fish stocks and attempted to integrate the fishing industry into the rest of the economy. New licenses were no longer issued to foreign fishing fleets.

[8] In 1987, clashes broke out between farmers and cattle-herders in the border area along the Senegal River. In Nouakchott, angry crowds of Mauritanians attacked hundreds of unarmed Senegalese with sticks and stones in 1989. In Dakar, several Senegalese reported murders and mutilations of their compatriots in Mauritania. Radical groups in Senegal reacted by murdering Mauritanian shopkeepers, and looting their shops. Within the country, there have been instances of violence between black African Mauritanians from the south and Arabs and Berbers from the rest of the country: the army supported the Arabs and Berbers, killing hundreds of black African

Mauritanians. In 1989, the border incidents prompted Senegal to break off diplomatic relations with Nouakchott.

[9] In a 1991 referendum, Mauritanians voted in favour of a multiparty system and the establishment of a democratic government. The opposition, concentrated in the United Democratic Front (FDU), exerted pressure on the Government to introduce democracy in the country, but the only response was repression. In spite of the constitutional reform, social tensions caused by the crisis did not decrease and strikes, to which most of the population adhered, were held. In 1992, President Ahmed Taya was re-elected in the first multiparty elections, which were considered flawed by the opposition and international observers.

[10] That year, Mauritania re-established diplomatic ties with Senegal and Mali and resumed negotiations on border disputes and the situation of Mauritanian refugees in both neighboring countries. Foreign aid granted by China and France gave new impetus to the discredited Taya Government, accused of being responsible for the country's difficult social conditions. The application of an IMF-sponsored structural adjustment plan further aggravated the mounting tensions. Demonstrations were widespread and were harshly put down by the police. In 1994, the ruling party won the local elections, also considered fraudulent by the opposition. Sixty Islamic leaders were arrested, on charges of creating an atmosphere of fear.

[11] In 1995, thousands of demonstrators rocked Nouakchott, setting cars on fire and looting shops after a new value added tax caused a rise in prices. Mauritania's creditors accepted a re-negotiation of the public debt that was then partially cancelled. In 1996, Taya appointed former Fisheries Minister El Afia Ould Mohamed Khouna as Prime Minister.

[12] In 1996, an agreement with the EU gave Mauritania access to set quotas of credit for five years, in return for permission to fish in the country's territorial waters, considered the richest in the world. In 1997 Taya was re-elected with nearly 90 per cent of the vote. In 1998 human rights organizations demanded an end to slavery in Mauritania. Activists from these organizations were imprisoned for denouncing slavery practices.-

[13] The Paris Club agreed to reduce Mauritania's debt by $620 million, justifying its decision on the grounds that the country was making great efforts to reform its economy. The Government

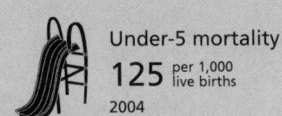

	Under-5 mortality		Poverty			Maternal mortality	
	125 per 1,000 live births 2004	**$**	**25.9%** 2000	of population living on less than $1 per day		**1,000** per 100,000 live births 2000

announced it would divide the energy and water company for separate privatization. Communications Minister Rachid Ould Salah reported that he would seek a majority partner for the former, while the water company would remain mostly state-owned. Mauritania also privatized the telecommunications sector following World Bank recommendations.

[14] In 2001, the parliamentary and municipal elections gave a majority in the National Assembly to the ruling PRSD. The opposition increased its representation in municipal government and in Parliament (from 1 to 10 seats). During the 1990s, despite a nominally multiparty system, minorities were in fact barred from participation. The new proportional representation system would make for greater electoral transparency. In 2002, the Government dissolved the opposition Action for Change party, which defended the rights of slaves' descendants and black Mauritanians, accusing it of posing a threat to national unity and of inciting intolerance.

[15] In 2003, Taya was forced to abandon the presidential palace after clashes with rebel soldiers in Nouakchott. The developing of very close ties with Israel since 1984 had apparently earned the Government the hostility of several Islamic countries. Finally, loyal government troops regained control over the capital. That same year, Taya was re-elected President in the general elections.

[16] In December, five years of ex-president Haidalla's prison sentence were suspended. He had been imprisoned on charges of organizing a coup d'état to overthrow Taya's government.

[17] In August 2004, members of the Mauritanian Navy were arrested on suspicion of attempting a coup d'état. Police Chief Sidi Ould Riha declared that the attempt was backed by Burkina Faso, with finance and materials.

[18] The armed forces mounted a successful coup d'état in early August 2005 and announced the creation of a military council that would lead the country for two years with the aim of preparing the way towards 'an open and transparent democracy'. The leaders of the coup appointed a 'Justice and Democracy Military Council' to govern, headed by Colonel Ely Ould Mohamed Vall, while keeping the existing Prime Minister and his Cabinet, who nevertheless resigned on 7 August. Many analysts claimed that Mauritania's entry into the select group of oil-producing nations during these years could be one of the hidden causes of the various coup attempts.

[19] Although slavery had been abolished in 1981 and the official position was that it did not exist in the country, the military junta leader, Ely Ould Mohamed Vall, acknowledged in May 2006 that it was still an ongoing problem and announced his support for ending this practice. 'I protest against any type of slavery and ask all Mauritanians to do the same', he declared (see box, 'Still enslaved').

[20] With the approach of elections in March 2007, political alliances and counter-alliances followed one after another as the different factions sought to ensure access to power.

[21] In June 2006 a referendum resulted in 97-per-cent approval for changes in the constitution limiting a serving president to two consecutive terms in office, imposing a maximum age limit of 75 and cutting each presidential term from six to five years. ∎

Still enslaved

THE MAURITANIAN upper class has practiced slavery for centuries. Members of the lower class, mostly poor black Africans living in rural areas, have traditionally been considered slaves. Although in recent years social attitudes have changed within the urban upper class, old class divisions endure in rural areas.

After three previous attempts, Mauritania's Government formally abolished slavery in 1980, following widespread protest against the public sale of a woman. Thus it became the last country in the world to abolish a practice that - in spite of the ban - continues to exist, mostly in rural areas. Neither organisms nor laws explicitly denounce slavery in Mauritania. A slave trader can act unpunished anywhere in the country, without being brought to justice.-

A report published in January 2005 by SOS-slaves states that, in spite of the absence of official data, approximately 30 per cent of the people are enslaved and exploited by the country's élites. Typically they work on plantations or as domestic servants, in a situation of complete dependency on their masters.

Anti-Slavery International (ASI), a UK-based non-governmental organization, has reported that there is no sanction against the employment of forced labor.

In fact, most former slaves remain in the same conditions as before abolition. Some of them were never freed (90,000, according to the Mauritanian League for Human Rights). In addition, ASI estimated that 300,000 freed slaves returned to their former masters, begging to be taken back as they had become economically and psychologically dependent through servitude.

OLD PRACTICE, NEW FACE
Slavery persists in new and less obvious ways, such as work being paid for in kind rather than in cash - a form of payment that curtails a person's independence and emphasizes the imbalance in their relationship with their employer. Another manifestation of slavery is the way that women and children (in particular) are traded like currency between families and ethnic groups.

The outlawing of slavery has not been backed up by the Government with information campaigns for enslaved people about their rights; no legal advicene was provided for enslaved people to help gain their freedom, nor legal protection for those slaves who escaped. Currently, a Mauritanian slave that manages to run away and reach the city 'can fill the appropriate official forms, but never receives attention and is ignored, without laws to protect' her or him, says SOS-slaves. Also, former slaves have no access to higher education or posts.

IN SEARCH OF POLICIES
In late 2005 the Government lifted a ban that prevented human rights organizations from operating in Mauritania. Thus groups that were previously considered illegal, such as SOS-slaves and ASI, could start working in the country.

The new regime of Colonel Ould Mohamed Vall allowed the independent press to operate with more freedom, which led to the rapid disappearance of national taboos such as talking about human rights abuses in general and slavery in particular. However, several global NGOs say that Colonel Vall has yet to announce a specific policy to address slavery. ∎

Mauritius / Mauritius

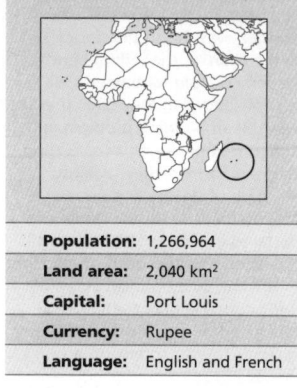

Population:	1,266,964
Land area:	2,040 km²
Capital:	Port Louis
Currency:	Rupee
Language:	English and French

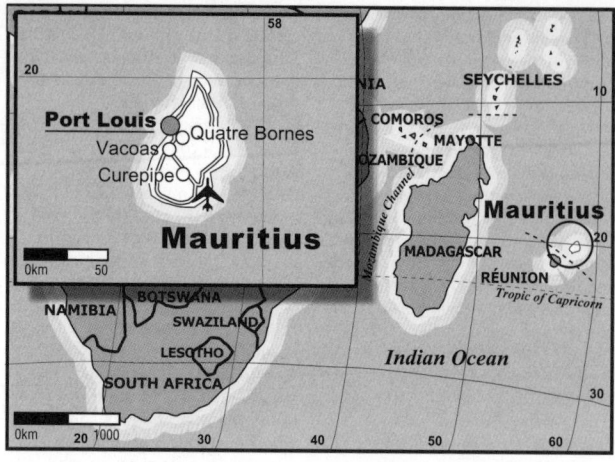

In the 10th century, Phoenician, Malay, Swahili, and Arab navigators visited the island of Mauritius, but did not settle there. In the 15th century, the Arabs named it Dina Robin - 'isle of silver'.

[2] In 1498, Vasco da Gama landed at the island during his trip around the Cape of Good Hope. In 1510, Portuguese seaman Pedro Mascarenhas used it as a stopover port, naming it Cirne.

[3] Around 1598 the Dutch claimed the island as their own, as an uninhabited island, and named it after their Head of State: Mauricio, Prince of Orange and Count of Nassau. From then until 1710, several attempts by the Dutch to colonize the island failed. In the meantime, pirates used it as a refuge.

[4] In 1721, Mauritius was colonized by the French Bourbons, who renamed it Ile de France. As rivalry surged between France and Britain over India, the island constituted an important strategic base.

[5] During the French Revolution (1789-1799), Mauritius gained a certain degree of autonomy, but in 1810 it fell into British hands. In 1814, after Napoleon's defeat, the Treaty of Paris recognized the island as a British colony. The British introduced sugarcane, which was to become the island's main economic resource right up

until the present day.

[6] In 1835 the emancipation of slaves, who constituted 70 per cent of the population, led to a serious labor shortage. Emancipation was opposed by the European landowners, who tried to alleviate the situation by bringing in, by the early 20th century, a total of more than 450,000 hired workers from India. In time, a majority of the population of Mauritius was comprised of people of Indian origin.

[7] By the early 20th century, local workers had begun to organize. In 1912, future Indian independence leader Mahatma Gandhi visited the island.

[8] In 1936, the labor movement (mainly made up of people of Indian descent), created the Labor Party, but a series of strikes were

brutally repressed, and the party leaders were killed, imprisoned or exiled. Seewoosagur Ramgoolam, a doctor, started his political career as the leader of the Advance group. In the late 1940s, with the support of the colonial administration, he became the head of the Labor Party of Mauritius (PLM).

[9] During World War II, the decline of British colonial power favored the independence movements. At the same time, US political and military influence increased. The inhabitants of Mauritius fought for and achieved representation within the British colonial government. In 1957, a new government structure was created, giving Mauritius its own Prime Minister. In 1959, the first elections with universal suffrage brought the PLM to power,

making Seewoo sagur Ramgoolam Prime Minister in 1961.

[10] From Mauritius, the British administration ran the islands of Rodrigues, Cargados-Carajos and the Chagos archipelago. In 1965, with the approval of Ramgoolam (knighted by the British Crown), Chagos and other islands became the British Indian Ocean Territories, and soon the United States established a major naval base on one of its islands, Diego García.

[11] Mauritius became independent in 1968, after a long decolonization process. Britain wanted merely to grant it limited autonomy, allowing it simply to run its domestic affairs.

[12] In 1971, the indigenous people of Diego García were secretly transferred to the outskirts of Port Louis (capital of Mauritius), a move that caused a scandal years later when it was discovered by the US Congress. The island was not returned to Mauritius and its inhabitants were not granted permission to return home.

[13] The political environment became tense in the 1970s when a new opposition group called the Militant Mauritius Movement (MMM) denounced the alliance between the Labor Party and the former French settlers. The legislative elections scheduled for 1972 were not held until 1976. Even though the MMM won the most votes, the Labor Party, through alliances with smaller groups, managed to stay in power. The workers' demonstrations

LAND USE

2003/2005

IRRIGATED AREA: 20.8% of arable land

- FOREST AND WOODLAND: 18.2%
- ARABLE LAND: 49.3%
- CROPLANDS: 3.0%
- OTHER USE: 29.5%

PROFILE

ENVIRONMENT

The archipelago is located in the Indian Ocean, 800 sq km east of Madagascar, is made up of the island of Mauritius (53 km wide and 72 km long); Rodrigues Island (104 sq km) with 23,000 inhabitants who farm and fish; Agalega (69 sq km) with 400 inhabitants, who produce copra; and Saint Brandon, 22 small islands inhabited by people who fish and collect guano. The islands are of volcanic origin, surrounded by coral reefs. The terrain climbs from a coastal lowland to a central plain surrounded by mountains. Heavy rainfall contributes to the fertility of the red tropical soil. Sugarcane is the main crop.

SOCIETY

Peoples: Indo-Mauritan 68.0 per cent; Creole (mix deriving from British and French with population from the east African coast) 27.0 per cent; Chinese 3.0 per cent.

Religions: 50.6 per cent Hindu; 27.2 per cent Christian (mostly Catholic, with an Anglican minority); 16.3 per cent Muslim; 0.3 per cent Buddhist; 2 per cent others.

Languages: English (official) less than 1 per cent, French (official) 3.4 per cent, Creole 80.5 per cent, Bhojpuri (a Hindi dialect) 12.1 per cent, French 3.4 per cent, other 3.7 per cent.

Main Political Parties: Alliance Sociale, which includes: Mauritian Labor Party; Mauritian Party of Xavier-

Luc Duval; The Greens; Republican Movement; Mauritian Militant Socialist Movement. Alliance MSM-MMM, which includes: Mauritian Militant Movement; Militant Socialist Movement; Mauritian Social Democrat Party.

Main Social Organizations: General Workers' Federation (GWF), affiliated with the Militant Mauritius Movement; Mauritian Women's Committee.

THE STATE

Official Name: State of Mauritius.

Administrative Divisions: 4 islands and 9 districts.

Capital: Port Louis 143,000 people (2003).

Other Cities: Curepipe 81,500 people; Quatre Bornes 78,900 (2000).

Government: Parliamentary republic. Anerood Jugnauth (MSM), President since 2003; Navinchandra Ramgoolam, Prime Minister since 2005. Legislature: single-chamber: National Assembly with 70 members (62 elected by popular vote, 8 appointed by the election commission to give representation to various ethnic minorities; members serve five-year terms).

National Holiday: 12 March, Independence Day (1968).

Armed Forces: non-existent. Other: 1,300 élite forces in charge of internal security.

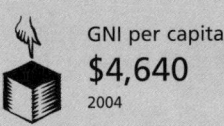
IN FOCUS

ENVIRONMENTAL CHALLENGES
Monoculture and the use of pesticides are degrading the soil and polluting water sources. That and the deterioration of the coral reefs are the main environmental problems.

WOMEN'S RIGHTS
Women have been able to vote and stand for office since 1956. In 2005 women held 17 per cent of parliamentary seats. In 2003 women made up 33 per cent of the workforce.

In 2004 the ratio of literate adult women (over 15) to literate adult men was 92 to 100*.

Some 98 per cent of births are attended by qualified personnel. Maternal mortality is 24 per 100,000 births*.

Domestic violence is a serious problem. There is a ministry in charge of guaranteeing women's rights, child development, and family welfare. The law punishes domestic violence and provides the judicial system with wide authority to fight against it.

CHILDREN
Significant strides in child welfare have been seen in the past 20 years.

Infant and under-five mortality continued to fall; the rates were 14 and 15 per thousand live births respectively in 2004 *.

There are pockets of poverty within the population, aggravated by domestic violence, sexual abuse and exploitation of women and children, as well as growing use of drugs and alcohol by young people. Commercial child sex exploitation is also considered a pressing problem.

INDIGENOUS PEOPLES/ ETHNIC MINORITIES
The majority of the population is Indo-Mauritian. The existing tension among the Hindu majority and Christian, Creole and Muslim minorities, persists. However, recent clashes have not been reported. Creoles and Muslims complain of a 'glass ceiling' in the public administration, beyond which they are not allowed to ascend.

MIGRANTS/REFUGEES
The law does not guarantee refugee status to people even if they meet the 1951 requirements or the current ones, established in 1967. Although the Government provides protection against the deportation of citizens to their countries of origin, it does not guarantee refuge or political asylum. The reason for this is that it is a small country with limited resources that wants to avoid becoming a shelter for refugees.

The Government collaborates with UNHCR in assistance to refugees by donating money to the organization.

Ever since around 4,000 people from the Chagos Islands were deported by Britain in the 1960s to make way for the US air base at Diego Garcia, they have lived in Mauritius. Following a legal ruling in the British courts in 2000 giving them right of return - overruled twice by the Blair Government under 'Queen's Orders-in-Council' but reasserted in 2006 - many islanders made a return visit to their homeland for the first time.

DEATH PENALTY
The death penalty was abolished in 1995. The last execution was carried out in 1987.

* Latest data available in The State of the World's Children and Childinfo database, UNICEF, 2006.

---

and the social unrest caused by unemployment increased, reaching their peak in 1979.

[14] In the 1982 elections the MMM allied with the Socialist Party of Mauritius (PSM) obtained 62 of the 66 seats and full control of the government. Anerood Jugnauth became Prime Minister. The MMM-PSM alliance promised to increase job opportunities and salaries, nationalize key sectors of the economy, reduce economic ties with South Africa, and demand that the US return Diego Garca.

[15] The rise in the price of oil, coupled with the drop in the international price of sugar, caused a deficit in the balance of payments equal to 12 per cent of the gross domestic product. The Government was forced to turn to the IMF, which approved five stand-by loans between 1979 and 1985.

[16] To receive these loans, the Government had to adopt austerity measures: to postpone part of the planned salary increase and job creation programs, to relinquish part of its control over public expenditure, to cut subsidies on basic foods and to devalue the Mauritian rupee.

[17] All these measures led to clashes between the MMM and the PSM, a situation which eventually resulted in a call for early elections, in August 1983. A new coalition was formed including labor, socialists and Duval's social-democrats. However, this apparent political stability was very weak and was followed by a series of governmental alliances, which failed to consolidate the coalition's position. Majority parties underwent serious divisions, giving rise to new political groups such as the Mauritius Socialist Movement (MSM), a splinter of the PSM, led by Anerood Jugnauth.

[18] In the legislative elections of December 1987, the coalition made up of the MSM, PSDM and Labor won a very narrow victory, giving it a majority in parliament. Anerood Jugnauth was re-elected Prime Minister, and the MMM became the main opposition force.

[19] That year, the newspapers reported scandals concerning the involvement of political leaders in drug-trafficking and money-laundering. At least one of several attempts to assassinate Jugnauth was blamed on the drug traffickers.

[20] In 1988, in agreement with the Organization of African Unity (OAU), Mauritius insisted on demanding the return of the island of Tromelin, administered by France, and of the Chagos archipelago, along with the demilitarization of the Indian Ocean, which is used for military maneuvers by the big powers. The demands were supported by environmental groups, due to the proliferation of nuclear weapons on the islands.

[21] In the September 1991 legislative elections, the ruling MSM succeeded in maintaining Jugnauth in the post of Prime Minister by reinforcing the traditional alliance with the MMM. In March 1992, Mauritius changed from a constitutional monarchy to a republic, and in June Cassam Uteem became the country's first President.

[22] In August 1993, Foreign Minister Paul Bérenger of the MMM withdrew from the cabinet, leaving Jugnauth without an absolute majority.

[23] In 1994, the economic performance of the island - whose foreign debt accounted for 25 per cent of GDP - was still regarded as satisfactory by the multilateral financial bodies.

[24] In January 1995, Jugnauth appointed representatives of the PMSD to his government. In the December legislative elections, an opposition coalition (MMM-Labor Party) led by Paul Bérenger and Nuvin Rangoolam took two-thirds of the seats. Rangoolam became Prime Minister.

[25] In June 1997, the MMM left the ruling coalition. Rangoolam took over the functions of foreign minister when Bérenger abandoned the post. A few days later, Parliament re-elected Cassam Uteem president.

[26] In September 2000 Jugnauth was re-elected as Prime Minister.

[27] In January 2001, making the most of preferential tariffs for the import of African textiles to the US - under the African Growth and Opportunities Act (AGOA) - China announced that it would build a textile factory in Mauritius.

[28] In February 2001, President Uteem refused to sign new anti-terrorism legislation; he resigned the same day. His office was transferred to vice-president Angidi Chettiar, who also refused to approve the law and likewise resigned. Arianga Pillay, the third president in one week, signed the new law. Amongst other changes, the law allowed for the detention of suspects for longer periods, and denied terrorism suspects the right to legal representation.

[29] In 2002, Karl Hoffman was elected president of the National Assembly. Information Technology and Telecommunications Minister Geehad Geelchand toured India, the UK, France, and Ireland promoting Mauritius as a new pole for cyber-development.

[30] On 11 September 2003, coinciding with the second anniversary of the terror attacks on the World Trade Center in New York and the Pentagon in Washington, President Anerood Jugnauth met US ambassador John Price in Port Louis, the capital. The fight against terrorism was the main topic of discussion.

[31] In early 2003, Anerood Jugnauth ceded his leadership of the MSM to his son Pravind. In September, Paul Bérenger became Prime Minister, the first non-Indian to hold the post. Jugnauth became President in October of the same year.

[32] In early January 2004, the local police force investigated deposits of $25 million in a Swiss bank, made in 1997. The money had apparently been destined to support the Labor Party, which

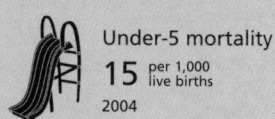

was in power at that time. Navim Ramgoolam, Prime Minister during that period, and two other leaders of the movement allegedly deposited the funds.

[33] Towards the end of the same month representatives of small island nations met in the Bahamas to prepare for the summit at the end of the year in Mauritius with more than 300 participants from the Caribbean, the Mediterranean, southern China, and the Pacific, Atlantic, and Indian oceans.

[34] In late February, workers demonstrated against a visit by World Trade Organization (WTO) Director-General Supachai Panitchpakdi, accusing the body of posing a threat to people's social rights.

[35] In October, Justice Ministers of francophone Africa, meeting in Mauritius, agreed to implement the universal instruments against terrorism and UN Conventions against Transnational Organized Crime and Corruption. The meeting was organized by the Government, the United Nations Office on Drugs and Crime (UNDOC) and the Intergovernmental Agency of the Francophonie. The Legislative Guide for the implementation of UNDOC's universal instruments against terrorism was also approved.

[36] Participants in the Small Island Developing States International Conference, held in Mauritius in January 2005, identified the adoption of renewable energy sources and cleaner technologies as a priority. They also emphasized that 'adapting to the adverse impact of climate change and rising sea-levels remains a major priority'.

[37] In April 2006 a group of 102 Chagosians living in Mauritius were finally allowed to visit their archipelago in the Indian Ocean, from where they had been evicted in the 1960s due to the construction of an American military base on Diego Garcia. The islanders were given three days to visit abandoned villages and cemeteries.

[38] In May 2006, concern over a chikungunya virus epidemic caused a substantial fall in the number of visiting tourists and as a consequence, a significant economic loss to the country. Chikungunya is a non-lethal disease that affects joints and is transmitted by the same mosquito as dengue.

[39] In June 2006, scientists announced that they had discovered part of the skeleton of a dodo, the large, flightless bird which became extinct over 300 years ago. ∎

Diego García Is.

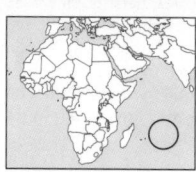

Population: 1,300
Land area: 52 km²
Language: English

The settlement of Diego García began in 1776 when the French Viscount de Souillac sent a ship there from Mauritius, trying to establish a French presence before the English could get there. French entrepreneurs obtained permission to exploit all of the island's riches: coconuts (for coconut oil), giant tortoises, fish and birds. In exchange. they established a leper colony on the island. With the defeat of Napoleon in 1815, the island passed into the hands of the British Crown together with Mauritius' other dependencies. During the 19th century, many workers came from India and Africa.

[2] Around 1900, there were approximately 500 inhabitants but the population increased radically over the next decades, with the arrival of Africans, Madagascans and Indians. Once they had settled, they developed a culture of their own. They spoke creole, a mixture of their local languages and they took part in the Tamul rituals (of Madagascan origin) even though they were mostly Roman Catholic. The indigenous Chagos community, the Ilois, lived in their traditional manner more or less unchanged until the 1960s.

[3] In 1965, Britain decided to remove Diego García from Mauritius jurisdiction, annexing it to the BIOT (British Indian Ocean Territories). Although this change was condemned by the UN, the British and Mauritius governments made a deal where a large sum changed hands. Two years later, Britain ceded the island to the US for 50 years in exchange for a discount on its purchase of nuclear arms. To make way for the construction of important US air and naval bases, the Ilois were deported to Mauritius in the early 1970s.

[4] The 4,000 or so uprooted Ilois were abandoned as soon as they arrived in Mauritius, finding themselves in a situation of total indigence. They had not been allowed to keep their possessions and even today they are deprived of nationality. Neither the British nor the Mauritius Government recognize them as citizens. The Ilois' attempts to return to their land became increasingly common, as they began to receive the support of international opinion. The situation reached the US Senate in 1975, but both the US and British governments have continued to ignore the problem until recently, each blaming the other for the situation and both spending intermittent sums of money, in order to temporarily alleviate the severity of the situation.

[5] The Ilois displaced from the Chagos archipelago have been filing lawsuits since the 1970s. In July 2000, they brought a case against the British crown. The 500 or so remaining Ilois wanted a judgement to allow them to return to their lands.

[6] In December 2001, they filed a suit in a federal court in Washington which involved US Defense Secretary Donald Rumsfeld and most of his predecessors at the Pentagon. The case was filed by three Ilois on behalf of the original inhabitants and their descendants. Charges included illegal deportations, racial discrimination, torture and genocide.

[7] Since 1960, the naval base has been used during the Cold War against the USSR, in the Gulf War in 1991 and in the Operation Desert Fox in 1998. In 2001 and 2002, the US used the island for refueling and also for B-1 and B-52 bombers during the war against Afghanistan.

[8] A British court decided in October 2003 that although the Ilois had received-'shameful' treatment in the past, their legal case was groundless.

[9] In March 2006, 40 years after being deported by the UK, a group of more than 100 Ilois from the Chagos archipelago returned for the first time, for a brief visit. They visited the abandoned plantations where they had lived and placed memorials. ∎

PROFILE

ENVIRONMENT
The Chagos Archipelago is located some 1,600 km southwest of India, in the middle of the Indian Ocean. The main islands are Diego García (8 km long by 6 km wide), Chagos, Peros, Banhaus and Solomon. The islands are made up of coral formations and are flat; they have a large number of coconut trees, which grow well here because of the tropical climate and year-round rain. The presence of the US naval base has seriously degraded the coral reefs, from the large-scale dredging and construction work.

SOCIETY
Peoples: At the present time, all the inhabitants are US or British military personnel.
Main Social Organizations: Lalit, local movement that demands the de-militarization of all the archipelago's islands and the devolution of Diego García to Mauritius.

THE STATE
Government: Rule is nominally exercised by a Commissioner from the Foreign Office in London.
Armed Forces: 25-strong Royal Navy detachment.

Mexico / México

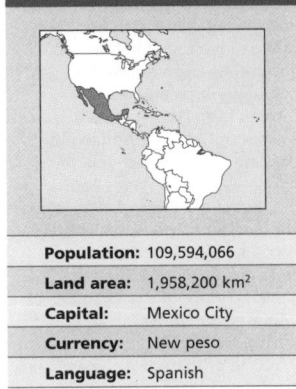

Population:	109,594,066
Land area:	1,958,200 km²
Capital:	Mexico City
Currency:	New peso
Language:	Spanish

The 20,000-year history of what we know today as Mexico includes over 2,000 years of urban life. Over this period, Meso-American peoples developed advanced civilizations such as the Olmec, Teotihuacan, Maya and Mexica. These cultures had complex political and social organizations, and advanced artistic, scientific, and technological skills. The Olmecs (1200-200 BC), Mexico's first established culture, developed in the coastal states of Veracruz and Tabasco. Despite the absence of a local supply of stone, they developed massive buildings (La Venta, San Lorenzo, Tres Zapotes) and also created an advanced calendar. By 600 AD the Mayan culture was in its heyday, most noted for its complex systems of mathematics and astrology, prolific city-building and complex architecture. By 1400 AD the Mayan state had splintered and almost disappeared, leaving an astonishing collection of ceremonial centers and ancient cities. In the valley of Oaxaca, around 900 BC, the Zapotecs, who were great city builders and artisans, rose; the Mixtec culture conquered the Zapotecs and developed around Mitla and Yagul. By the early 1400s the Mixtecs had become vassals of the mighty Aztec empire. These two cultures continue their existence today in the State of Oaxaca, which is inhabited by some 2 million of their descendants.

2   The Toltecs, a civilization of mighty warriors believed by some to have developed from the magnificent Teotihuacan culture, occupied the northern reaches of the Valley of Mexico from around 950-1300 AD. They built one of Mexico's most impressive cities (Tula), were master craftspeople, and strongly influenced later Mayan and Aztec cultures. The Aztec civilization dominated Mexico for nearly 200 years (1345-1521) and was flourishing when Spanish conquerors arrived in 1519. The Aztecs used an elaborate system of taxing and patronage to subjugate an enormous empire that stretched well into Central America. They borrowed heavily from their Olmec, Toltec and Mayan predecessors to develop a complex linguistic, religious, artistic, architectural and military heritage.

3   The empire came to a sudden end in 1521, when Spanish explorer Hernán Cortés took advantage of internal strife between the ruling Aztecs and other indigenous peoples who paid them tribute. After killing Moctezuma II, Emperor of the Aztecs,

Cortés laid siege to the capital, Tenochtitlan, with the help of a huge army led by the Tlazcaltecs. Once the city fell, Cortés initiated Christianization and Hispanization of the indigenous inhabitants.

4   In the 17th century, the major economic structures of the so-called 'New Spain' were laid down. The hacienda, a landed estate, emerged as the basic production unit, and mining became the basis of a colonial economy conceived to meet the gold and silver needs of the Spanish homeland. The American Indian population was exploited and decimated by hard labor and disease. By 1800, Mexico had become one of the world's richest countries, but with great poverty also.

5   After almost three centuries of colonial domination, the struggle for independence began in 1810, led by criollos (Mexicans of Spanish descent) such as Miguel Hidalgo and José Maria Morelos, two priests. The struggle became a broad-based national movement as Indians and Mestizos (people of mixed European and Indian descent) joined its ranks, but the rebels were soon crushed by the royal army. The liberal revolution in Spain radically changed the situation. Afraid to lose their privileges, the Spanish residents and the conservative clergy came to an agreement with the surviving revolutionaries. This pact became known as the Iguala Plan, trading independence for a guaranteed continuation of Spanish dominance. In 1821 General Iturbide proclaimed himself Emperor, but was rapidly replaced by General Antonio López de Santa Anna.

6   At the time, Mexico was the most extensive Spanish American country, covering 4.6 million square kilometers, including the

Central American provinces, but it was also stricken by economic, political and social problems. In 1824, a Constitution was approved establishing a federal republic, made up of 19 states, four regions and a federal district. In 1836, Santa Anna, elected President three years previously, passed a new Constitution which did away with all vestiges of federalism and the Mexican state of Texas, which had been settled by some 30,000 US citizens, called on the US for support and protection. Santa Anna led his army to victory against the Texans that year at the Alamo, but was defeated by US troops who took him prisoner, later releasing him for a large ransom.

7   In 1845, the US annexed Texas, which led to a break in diplomatic relations with Mexico. It also caused a frontier conflict, since the US claimed that the southern border of Texas was on the Rio Bravo or Rio Grande, not the Rio Nueces (further north) as was commonly accepted. In 1846, the US President James Polk tried to force through a frontier agreement and buy the state of California - both moves rejected by Mexico. Polk ordered the US army to occupy the disputed lands between the two rivers, leading to a two-year war which was won by the US. The victors annexed all Mexican territory north of the Rio Bravo or Rio Grande. More than half of the Mexican territory was controlled by the US at the end of the war.

8   Between 1821 and 1850, Mexico had 50 different governments. Political instability was not ended by the country's first elections. As in many other Latin American states, the Mexican bourgeoisie supported two political

LAND USE

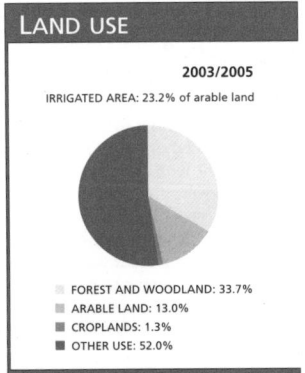

2003/2005

IRRIGATED AREA: 23.2% of arable land

- FOREST AND WOODLAND: 33.7%
- ARABLE LAND: 13.0%
- CROPLANDS: 1.3%
- OTHER USE: 52.0%

PUBLIC EXPENDITURE

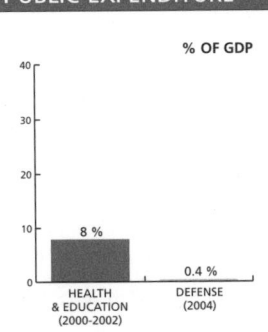

% OF GDP

8 % — HEALTH & EDUCATION (2000-2002)

0.4 % — DEFENSE (2004)

Life expectancy
76 years
2005-2010

GNI per capita
$6,790
2004

Literacy
90% total adult
rate
2000-2004

HIV prevalence rate
0.3% of population
15-49 years old
2003

parties: the liberals and the conservatives.

9   The Liberals' victory of 1857 strengthened the republic, instituting a free market economy, putting in place individual rights and guarantees, and expropriating the wealth of the clergy. But the conservatives, supported by the church, took up arms, and civil war - commonly known as the War of the Reform - broke out.

10   Mexico's first Native American president, Benito Juárez, was successful in re-establishing national unity in 1861. However, his decision to suspend payment on all foreign debts triggered armed intervention by France, Britain and Spain. Although Britain and Spain withdrew, France established a monarchy in an attempt to counterbalance US influence over the area. Maximilian of Austria was crowned Emperor of Mexico. The Mexican resistance soon brought the republican troops back together again, and Juarez was reinstated as president in 1867.

11   In 1871, Juan de Wata Rivera started the newspaper *The Socialist*. On 10 September, this newspaper published the general statutes of the 1st Marxist International for the first time in Latin America. General Porfirio Díaz, who fought with the Liberals against French intervention, seized power in 1876 and held onto it until 1911. During the 35 years of his dictatorship, the country opened its doors to foreign investment, the economy was modernized and social inequalities increased.

12   In 1910, Francisco Madero led the Mexican Revolution under the slogan 'effective suffrage, no re-election'. This was the first popular Latin American revolution of the century. In 1913, US ambassador Henry Lane Wilson participated in a conspiracy, resulting in the assassination of Madero. The people responded by taking up arms more vigorously than before. Peasants joined the revolt, led by Emiliano Zapata and Francisco (Pancho) Villa. The principles of this Revolution were set out in the Constitution of 1917, promulgated by Venustiano Carranza. It was the most socially-advanced constitution of its time and many of the principles are still in force today. However, conflict between the various revolutionary factions continued, resulting in the deaths of the major leaders.

13   In 1929, under President Plutarco Elias Calles, these factions joined to form the National Revolutionary Party. In 1934, General Lázaro Cárdenas took office. He embodied the continuation of the revolutionary process, and was one of the main driving forces behind its

accomplishments. The major reforms included land reform, the nationalization of oil (the founding of PEMEX), the expropriation of oil refineries, incentives for new industries, and a national education system. The National Revolutionary Party became the Institutional Revolutionary Party (PRI) and the deeply-rooted revolutionary socialist principles were gradually abandoned.

14   Conditions created by World War II accelerated the first phase of Mexico's industrialization which reached its peak during Miguel Aleman's term of office (1946-1952). The changes that took place during this period altered the former social balance. Mexico's population remained predominantly rural, with only 40 per cent in the cities, but the rapid development was not able to absorb the quickly growing population. Communal land ownership, which had stimulated solidarity and revolutionary feeling among peasants in the 19th century, was gradually replaced by a new type of individual land tenure causing the formation of large estates or *latifundios*.

15   During the following decade, Cardenas' successors, though not always loyal to his principles, stabilized the system by reinforcing those factors responsible for its success in a continent generally afflicted by underdevelopment and stagnation. Strong governmental influence, involving substantial public investment, kept the economy strong. A reasonable balance was maintained between heavy and light industry and tourism was encouraged.

16   Against the backdrop of the 1968 Mexico Olympics, the students' movement organized protests against the true social situation in the country. A student rally held on the Plaza de las Tres Culturas was dispersed by the army. The order to shoot to kill was given without warning, and hundreds of people died or were injured in the 'Tlatelolco Massacre'.

17   During the presidency of José Lopez Portillo (1976-1982), important oil deposits were discovered. This bound Mexico, as the US' main oil supplier, more closely to the US.

18   In 1982, Miguel de la Madrid assumed the presidency implementing an IMF economic adjustment plan. Cuts in subsidies and public spending, changes in the pattern of public investment, and a dual currency exchange rate caused public discontent and the PRI's first electoral defeat since its foundation. In the June 1983 elections, the PRI failed to impose its municipal candidates in the capital and two other major cities.

19   Under the pressure of foreign

debt the trends of 1983 continued: growing inflation; losses in real wages; reduction of public spending; falls in production, and rising unemployment.

20   The devastating earthquake of September 1985, in which more than 20,000 people were buried alive, further aggravated an already critical situation. The reduction in the oil quota by the US made it necessary to generate alternative sources of income. One of these was tourism, and another was the *maquiladoras* - foreign companies on the border with the US which are exempt from taxes and from paying their workers' social insurance, disposing of the products on the vast domestic market.

21   The internal situation became more difficult for the PRI and it was accused of rigging the 1986 municipal elections. The formation of an independent Labor Union Negotiation Board posed a threat to the PRI's labor wing, which had traditionally formed a bloc within the Labor Congress, the largest coordinating board of labor federations.

22   On 6 July 1988, several parties gained a significant vote in the national elections, something which had not happened since 1910. Carlos Salinas de Gortari, the PRI candidate, won the ballot with - according to official figures - 50 per cent of the vote (the lowest percentage in party history). The Left emerged for the first time as a real alternative to the PRI. Cuauhtemoc Cárdenas, the son of Lazaro Cardenas, led a coalition of groups operating as a single party, the FDN, representing the masses and opposing the Institutional Revolutionary Party. Cárdenas took second place with 31 per cent of the vote in an election marked by accusations of fraud and irregularities by the opposition. The abstention rate was 49.72 per cent.

23   In 1989, the FDN split, and Cárdenas founded the Democratic Revolution Party, made up of former members of the PRI, communists and members of smaller organizations. In the meantime, the People's Socialist Party (PPS), the Authentic Party of the Mexican Revolution (PARM) and the Party of the Cardenista Front for National Reconstruction (PFCRN) continued as autonomous organizations which voted with the PRI in Congress.

24   The PRI administration resolved to open up the country to foreign investment, and also announced a series of measures aimed at controlling inflation. Both these decisions were welcomed by the US Government. Mexico made overtures to the US to sign a free trade agreement, coinciding with Mexico's entry into GATT (now the

World Trade Organization - WTO) and the legal authorization for foreign investment in Mexican enterprises to go above the previously stipulated 49 per cent. In May 1990, President Salinas de Gortari privatized the banking system, which had been nationalized eight years earlier.

25   In the elections of 1991, the PRI proclaimed itself the winner with 61.4 per cent of the vote, amid accusations of fraud. It gained control of the Chamber of Deputies, and power to carry out constitutional reforms.

26   One of these was the agrarian reform approved in December 1991, which granted property rights to *campesinos* (peasants) who farm state lands known as *ejidos* (co-operative farms ceded by the Zapata Revolution in 1917). According to the PRI, the new system was designed to reduce the annual ten million tons of imported food. According to the opposition, the reform - which allowed campesinos to sell their lands - would bring about a transfer of small landholdings to larger investors.

27   On 17 December 1992, the Mexican, US and Canadian governments signed the North American Free Trade Association (NAFTA) Treaty.

28   During the Salinas administration, inflation was reduced from three figures to a rate of nine per cent in 1993. Between the end of 1988 and mid-1993, the State received some $21 billion from the privatization of state interests. Private foreign debt increased by $11 billion in 1993.

29   On 1 January 1994, the day the NAFTA agreement came into force, the Zapatista National Liberation Army (EZLN) - whose strength the Government had played down - occupied three towns in the southern state of Chiapas, declaring them a freed zone. Chiapas is one of the states with the largest population of Mayan Indians. It also has the highest level of Spanish-language illiteracy and the lowest incomes. But in addition it has large oil and gas reserves - 21 per cent of the nation's crude oil is extracted in Chiapas. In the poorest state, this band of indigenous people, called *Zapatistas*, rose up to claim the right to defend themselves against a government and a social order which offered them nothing but wanted their land.

30   The Government initially refused to recognize the scope of the uprising. When the number of dead rose above a thousand - the army had been deployed to the region - and the complaints of summary executions continued, the Government - at the insistence of Catholic Bishop Samuel Ruiz

Under-5 mortality	Poverty	Debt service	Maternal mortality
28 per 1,000 live births 2004	**4.4%** of population living on less than $1 per day 2002	**22.9%** exports of goods and services 2004	**83** per 100,000 live births 2000

and protests from national and international civil society - agreed to negotiate and unilaterally declared a ceasefire.

[31] In February, negotiations began in San Cristobal de las Casas with the rebels demanding reforms to electoral and agrarian statutes (approved in 1991) and to the Penal Code, besides other reforms to improve the quality of life of indigenous peoples.

[32] The PRI presidential candidate Luis Donaldo Colosio was murdered on 23 March in Tijuana. Three members of his bodyguard were involved. The PRI nominated Ernesto Zedillo as his replacement for the 21 August elections, and Zedillo was voted in by 49 per cent of the vote.

[33] In September, PRI secretary-general José Ruiz Massieu was assassinated, and in November his brother Mario resigned from his post as Attorney-General as he believed Party officials were blocking the criminal investigation. His brother's death increased suspicions that leading PRI figures, and perhaps the drug mafia, were also involved in the murders.

[34] The economy continued to grow until, on 20 December, the new government abandoned the policy of gradually depreciating the currency because of the acceleration of capital flight. By the end of the year, the peso had lost 42 per cent of its value. Its fall caused the stock market to collapse in a crisis that came to be known as 'The Tequila effect'. The crisis led to questioning of the free market and trade liberalization model prescribed by the IMF and threatened other economies in the region.

[35] After negotiations failed in Chiapas, in February, Zedillo launched a military offensive which was halted after protests from various international bodies. The investigations of the Colosio and Ruiz Massieu murders had still gotten nowhere.

[36] Raúl Salinas de Gortari, brother of the former president, was arrested in February accused of masterminding the Ruiz Massieu killing. At the same time, the Government called for the extradition of Mario Ruiz Massieu from the US, on charges of hampering the investigation into his brother's death. Raul Salinas' wife was arrested in Switzerland in November when using fake documents to transfer funds from her husband's account.

[37] In September 1996, seven months after the San Andrés accords on rights and indigenous culture had been signed by the federal and state governments and the EZLN, the peace negotiations stalled when the Zapatistas accused the Government of failing to live up to the agreements.

[38] Cuauhtemoc Cárdenas was elected mayor of Mexico City in July 1997, while in the same elections the Democratic Revolution Party (PRD) became the opposition party with most seats in the chamber of deputies, outdoing the National Action Party (PAN). These two parties, along with the Greens and the Workers Party, announced the creation of an opposition alliance against the PRI, from 1 September.

[39] On 22 December, in Acteal, Chiapas, 45 Tzotzil indigenous people of the Las Abejas (The Bees) pacifist group were massacred by paramilitaries with the compliance of the public security forces. This prompted the resignation of acting governor of the state, Julio Ruiz Ferro, and his secretary, Emilio Chuayffet.

[40] In August 1999, Cuauhtemoc Cárdenas's PDR and the right wing PAN led by Vicente Fox decided to form an alliance to confront PRI dominance, presenting a single candidate for the July 2000 elections.

[41] However, the coalition failed in December after the PAN rejected the proposal of a group of 'notables' on how to elect a common candidate for the coalition. The Right had opposed 25 observations made in a proposal by 14 independent figures who suggested primary elections and four surveys in order to designate a candidate for the eight opposition parties.

[42] The electoral reforms paved the way for the first truly clean elections in the history of the country. The PAN finally joined forces with the Green Party and in a landmark election, Fox was elected President in July, ending more than 70 years of PRI rule.

[43] When he took office in December 2000, Fox promised to fight corruption and sought, among other things, to eradicate economic inequity and provide education and health services to the population. His first tour of the Southern Cone region sought to open Mexican economy to foreign trade, especially with the Mercosur countries (Argentina, Brazil, Paraguay and Uruguay) and Chile.

[44] That month, Fox appointed former senator Luis Álvarez as commissioner for peace in Chiapas. By then, peace talks between the guerrillas and the Government were suspended. As he had promised in his campaign, Fox ordered the army to withdraw from indigenous communities. The President also lived up to the promise that the first bill he would send to the legislature would be the one based on the recognition of indigenous rights. EZLN leader Subcomandante Marcos agreed to resume negotiations with the

Government.

[45] Between 24 February and 11 March 2000, a caravan of 24 Zapatista delegates traveled from Chiapas to Mexico City to present its demands to Congress (legal guarantees, bilingual education, mass media that reflects indigenous cultures, autonomy in land use and customs, among others). After several discussions, legislators passed reforms on the Indigenous Act (Cocopa Law), but the EZLN rejected the new law because it did not take into account the indigenous peoples' demands, and announced it would not resume the peace talks.

[46] When the economic crisis

worsened in late 2001, the President's popularity fell drastically. It was felt he had not fulfilled his promises - among them to create 1.5 million jobs. Thousands of peasants marched throughout the country to demand solutions for their sector, one of the worst hit. In Mexico City alone, 30,000 peasants demonstrated at the ministries of Rural Development, Economy, Finance and Agriculture.

[47] In January 2002, the Supreme Court of Justice legalized abortion in cases of rape; when artificial insemination was applied without the woman's consent; and when the fetus had genetic or congenital defects.

PROFILE

ENVIRONMENT

The country occupies the southern portion of North America. It is mostly mountainous, with the Western Sierra Madre range on the Pacific side, the Eastern Sierra Madre on the Gulf of Mexico, the Southern Sierra Madre and the Sierra Neovolcánica Transversal along the central part of the country. The climate varies from dry desert wasteland conditions in the north to rainy tropical conditions in southeastern Mexico, with a mild climate in the central plateau, where the majority of the population lives. Due to its geological structure, Mexico has abundant hydrocarbon reserves, both on and off shore. The great climatic diversity leads to very varied vegetation. The rainforests in the southeastern zone and the temperate forests on the slopes of the Sierra Neovolcánica are strategic economic centers.

SOCIETY

Peoples: Mexicans are descended from Meso-American peoples and Spanish conquistadors. Indigenous peoples amount to 30 per cent of the national population. Out of the 56 indigenous groups, the main ones are: Tarahumara, Nahua, Huichol, Purepecha, Mixteco, Zapoteca, Lacandon, Otomi, Totonaca, and Maya.
Religions: Mainly Catholic.
Languages: Spanish (official). An estimated six million Mexicans speak indigenous languages.
Main Political Parties: Institutional Revolutionary Party (PRI), was in Government from its foundation in 1929 until 2000; National Action Party (PAN), conservative, founded in 1939; Party of the Democratic Revolution (PRD); Labor Party.
Main Social Organizations: The Labor Congress, a congregation of the large union organizations, was the PRI's support base. The National Peasants' Front was created in 1983 in a merger of the National Confederation of Farmers (CNC), the General Union of Workers and Farmers of Mexico (UGOCM), the National Confederation of Small Landowners (CNPP) and the Independent Farmers' Central Union (CCI). Authentic Labor Front (FAT); Mexican Network of Action against Free Trade (RMALC); National Coordinator of Mexican Students (CNEM); Zapatista National Liberation Front (FZLN) a civic-political group formed in 1995 by the Zapatista National Liberation Army (EZLN). Major indigenous groups: National Indigenous Congress, International Indigenous Press Agency, Plural National Indigenous Assembly for Autonomy.

THE STATE

Official Name: Estados Unidos Mexicanos.
Administrative Divisions: 31 states and a federal district.
Capital: Mexico City 18,660,000 people (2003).
Other Cities: Guadalajara 3,798,800 people; Monterrey 3,409,100; Puebla 2,495,100; Netzahualcóyotl 1,401,300 (2000).
Government: The system is both federal and presidential. Felipe Calderón, President since December 2006. The Congress of the Union (Legislature) has two chambers: the 500-member Chamber of Deputies, and the 128-member Senate.
National Holiday: 16 September, Independence Day (1810).
Armed Forces: 193,000 troops (2003), and 300,000 reserves. Other: 14,000 Rural Defense Militia.

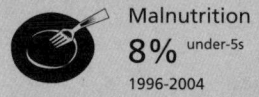
IN FOCUS

ENVIRONMENTAL CHALLENGES

The river system is not extensive and is unevenly distributed throughout the country and the pollution of some aquifers has aggravated the problem of access to drinking water. Air and land pollution are significant in industrial areas. Deforestation extends 6,000 sq km a year. The disposal of solid and liquid waste is inadequate. In the capital and in urban centers along the border with the US, land subsidence is a problem, due to the falling level of underground waters. The lack of drinking water and deforestation are considered a security issue by the State.

WOMEN'S RIGHTS

Women have been able to vote and stand for office since 1947. In 2003, 24 per cent of the Lower House seats were held by women, while they held 9 per cent of ministerial or equivalent positions. That year, women made up 34 per cent of the labor force of 44 million. Female and male unemployment stood both at 2.4 per cent. In 2004, 86 per cent of pregnant women received prenatal care and 95 per cent of the deliveries were assisted by qualified staff. The mortality rate stood at 65 per 100,000 births.

CHILDREN

Health, education and life quality of Mexican children depend on where they live. In the northern state of Nuevo Leon children have better access to sanitary, health and educational services than, for example, in southern Guerrero, where most of the population is indigenous. However, there

has been a great improvement since the 1990s. Starting in 2005, schools have implemented special courses that aim to reduce the influence of domestic violence on school desertion. This is linked to alcohol drinking and is added to the desertion caused by poverty, which forces many children to work in order to live. The infant mortality rate was, in 2004, 23 in every 1,000 live births; for children under 5 it stood at 28 per 1,000. Eight per cent of children were underweight at birth, and 18 per cent of children under 5 were severely or moderately stunted.

INDIGENOUS PEOPLES/ ETHNIC MINORITIES

With more than 50 ethnic groups and over 60 languages, Mexico is among the countries with the strongest indigenous presence in the Americas, with between 10 and 20 million indigenous people. The Aztecas, Choles, Mames, Mayas, Guajiros, Lacandons, Olmecas, Nahuas, Mixtecos, Apaches, Tlpanecos, Tarahumaras, Toltecas, Zapotecas and Zoques stand out on account of their number and cultural legacy.

There are unresolved agrarian conflicts, while indigenous groups face severe discrimination with respect to the administration of justice, as well as persistent threats of eviction and relocation which keep them in a permanent state of vulnerability. Excessive militarization in some indigenous communities and the presence of alleged paramilitary groups have been reported. Extreme poverty, lack of basic services, and difficulties in providing bilingual and intercultural education are key factors hindering the development of these communities.

Indigenous peoples, through

groups such as the Zapatistas (see History), are demanding the recognition of their fundamental rights and of the multicultural nature of the nation; the possibility of autonomous economic development, social and political representation; and the right to maintain and develop their cultural practices.

MIGRANTS/REFUGEES

In early 2005 there were 6,325 registered refugees and 492 asylum seekers in Central America and Mexico. The refugee situation in the region differs from others, where large numbers are on the move; here it refers to a continuous movement of individuals or families. Most of the 4,300 refugees and asylum seekers in Mexico live in urban areas. They come mostly from other countries in Latin America, Central Asia, Middle East and Africa. About 1,100 indigenous Guatemalans who settled long ago in rural areas in the south of Mexico became citizens in 2005.

It was estimated that 11 million Mexicans (born in Mexico) were living in the US in 2006. The Mexican community in the US, taking into account Mexicans born in the northern country, stood at more than 30 million people.

DEATH PENALTY

The death penalty is applicable in the case of exceptional crimes, such as those committed under military law or in wartime. The last execution was carried out in 1937.

* Latest data available in The State of the World's Children and Childinfo database, UNICEF, 2006.

Constitution, and would also enable federal authorities to intervene with haste in cases such as the hundreds of murders of women in Ciudad Juárez.

[53] Special prosecutor Ignacio Carrillo, who investigated the murders and other crimes during the 'dirty war' announced, in January 2005, that he would bring charges for genocide against 25 former government officials and military officers in relation to the 1968 massacre in Tlatelolco.

[54] Oscar Espinosa Villarreal, former mayor of Mexico city (1994-1997), was sentenced in June 2005 to seven-and-a-half years in prison for embezzling millions of dollars from public funds during his administration, becoming one of the highest-ranked Mexican officials to be convicted of a major offense.

[55] Hurricane Stan hit the south of the country in October 2005, killing 42 people.

[56] At least seven people died in May 2006 in violent clashes between peasants and the police due to a series of evictions and land expropriations in San Salvador Atenco, 24 km east of Mexico City. Subcomandante Marcos, of the EZLN, touring the country in what he called 'the other campaign' - an allusion to the July elections - declared his movement on 'red alert' due to the incidents and the detentions of many EZLN members during the clashes.

[57] During a tour of the US in late May, Fox said he did not support the emigration of undocumented Mexicans to the US, and requested a bilateral solution to the problem.

[58] In June, Washington began to place hundreds of cameras along the Mexican border with Texas to film and broadcast live images on the internet. Thus, anybody logged on to the web could become a 'virtual patrol' and inform the US authorities of any immigrants trying to cross the border. In October 2006, US President George W Bush signed a law enabling the construction of an 1,100 kilometer fence to prevent illegal Mexican immigration. Mexican President Fox compared the fence to the Berlin Wall.

[59] The conservative candidate Felipe Calderón won the presidential elections in July 2006, defeating the center-left candidate Andrés Manuel López Obrador by a very narrow margin (0,57 per cent of the vote). López Obrador filed a legal challenge to the election result, alleging widespread fraud and asking for a full manual recount. Supporters promised to swear him in as a 'parallel President' on 20 November, 11 days before Calderón is officially inaugurated. ∎

[48] In May 2003, Mexico's Electoral Tribunal (TEPJF) upheld a $90 million fine to the PRI for irregular financing activities during the presidential campaign of 2000. The TEPJF confirmed that the PRI had illegally received more than $45 million from the state-owned oil company PEMEX. The money found its way into the PRI's coffers through the union representing the company's employees.

[49] However, the PRI was the big winner of the parliamentary elections held in July of that year. The PAN's seats in the Chamber of Deputies went down from 207 to 155, while the PRI obtained 15 more seats than in the previous period. Likewise, President Fox's party lost six governors in an

election that was marked by a voter turnout of just 40 per cent, the lowest in Mexico's history.

[50] In August 2003, Amnesty International published a report on the ' Juárez femicide ': the abduction, rape, torture and murder of more than 370 women in Ciudad Juárez, in Chihuahua state which borders the US. The murdered women were triply vulnerable, being poor, young and female in a violent, patriarchal society. In spite of international pressure, the Mexican Government had not made progress in the investigation into the murders. On the contrary, in many cases it hampered and delayed investigations by performing inadequate or incomplete forensic

tests, falsifying evidence and allegedly torturing people into signing false confessions.

[51] In February 2004, the federal authorities arrested Miguel Nazar Haro, former Chief of the Federal Security Directorate (DFS), for his alleged participation in the 1975 forced disappearance of Jesús Piedra Ibarra, presumed member of a leftist guerrilla organization. This was the first arrest obtained by the special prosecutor that President Vicente Fox appointed in November 2001 to investigate and prosecute human rights violations committed under previous governments.

[52] In April, Fox proposed constitutional amendments that for the first time would guarantee human rights in the Mexican

Micronesia / Micronesia

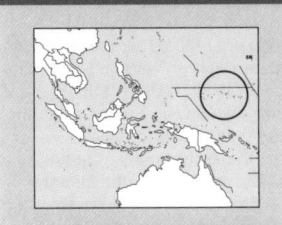

Population:	111,873
Land area:	700 km²
Capital:	Palikir
Currency:	US dollar
Language:	Kosraean, Yapese, Pohnpeian, Trukese and English

The name Micronesia is derived from the Greek meaning small islands. It refers to the Marshall Islands, the Marianas (including Guam) and the Caroline Islands.

2   Ferdinand Magellan, the first European to arrive in the Marianas, named them the Islands of Thieves, but they were later renamed after Queen Mariana of Austria, Spanish regent.

3   In 1885, the Germans tried to impose a protectorate on the islands. The Spanish appealed to the Vatican but lost Guam under the Treaty of Paris of 10 December 1898, which ended the war with the US. Finally, Spain chose to sell the archipelago to Germany for 25 million pesetas.

4   The Japanese occupied the islands in 1914, and kept the area demilitarized until 1935. Japan launched the attack on Pearl Harbor from Micronesia on 7 December 1941.

5   In 1947, an agreement with the United Nations allowed the US to keep the islands as a trust territory. A UN mandate forced the US to help develop a national awareness among the island's population to enable them to exercise their right to self-determination.

6   Under the Commonwealth system, the US was allowed to keep its military bases on the islands and take charge of their defense and foreign affairs. In 1975 a plebiscite was held and in 1978 the Northern Mariana Islands acquired 'Free Associate State' status.

7   In 1978, another plebiscite led to the creation of the Federated States of Micronesia. Four districts of the trust supported the motion, while Palau and the Marshall Islands remained autonomous states.

8   Nuclear experiments performed by the US on Bikini and Eniwetok atolls resulted in the permanent displacement of the local inhabitants. In 1977 water, fruit, and vegetables were still too radioactive for consumption.

9   In October 1982, the US signed a Free Association Agreement with the Marshall Islands and the Federated States of Micronesia, approved by both countries in 1983 and proclaimed in October 1986. The US would be responsible for defense (including the maintenance of military bases) in exchange for financial aid to the islands.

10   On 17 September 1990 Micronesia joined the United Nations.

11   After the March 1991 legislative elections, Bailey Olter was elected President in May.

12   In June 1993 Micronesia joined the IMF.

13   Leo A Falcam, who was appointed President in May 1999, issued an official statement reiterating Micronesia's opposition to the transportation of plutonium through the region by nations such as Britain, France and Japan, particularly through its own territorial waters.

14   Micronesia and 13 other nations from the Pacific Island Countries Trade Agreement (PICTA), met in August 2001 to discuss trade issues and demand that Australia and New Zealand/Aotearoa offer more financial support to the weaker economies.

15   After he was named President in May 2003, Joseph Urusemal stated that his country was threatened by the increase in frequency and intensity of Pacific storms due to global climate change.

16   That year, proposals for a new 20-year-agreement with the US and $1.8 billion in aid were made to Washington. Micronesia asked for fair compensation for the damage to the population from nuclear testing on its atolls, stating that the effects were unknown at the time the original agreement was signed.

17   In April 2004, Yap, an archipelago located to the west of the Caroline islands, was devastated by Typhoon Sudel. Virtually all of the island's infrastructure collapsed, and a state of emergency was declared.

18   In March 2006, during a UN conference on biodiversity protection, a program was launched to preserve wildlife in the Pacific islands, with a fund of $18 million allocated to protect sea life and Micronesian ecosystems. ■

PROFILE

ENVIRONMENT
Covers an area of 2,500 sq km, more than half of which is taken up by the island of Pohnpei (called Ponape until 1984). The terrain is mountainous, with a tropical climate and heavy rainfall. The Federation is made up of four states, including the Caroline islands, except Palau: Yap,119 sq km; Chuuk (called Truk before 1990), 127 sq km; Pohnpei, 345 sq km; and Kosrae, 100 sq km.

SOCIETY
Peoples: Trukese 41.1 per cent; Pohnpeian 25.9 per cent; Mortlockese 8.3 per cent; Kosraean 7.4 per cent; Yapese 6.0 per cent; Ulithian, or Woleaian, 4.0 per cent; Mokilese, 1.2 per cent and others.
Religions: Christianity is the predominant religion, with the Kosraeans, Pohnpeians, and Trukese being mostly Protestant and the Yapese mainly Roman Catholic.
Languages: local languages; English (official).

THE STATE
Official Name: Federated States of Micronesia.
Administrative Divisions: 4 states (Chuuk/Truk, Kosrae, Pohnpei, Yap).
Capital: Palikir, on the island of Pohnpei 7,000 people (2003).
Other Cities: Weno (Moen) 23,900 people; Kolonia 5,800 (2000).
Government: Head of State and Government Joseph Urusemal, President since May-2003; Legislature, single-chamber with 14 members (10 on 2-year terms, 4 on 4-year terms). There are no political parties.
National Holidays: 10 May, Proclamation of the Federated States of Micronesia (1979); 3 November, Independence (1986).
Armed Forces: External defense is guaranteed by the US.

Moldova / Moldova

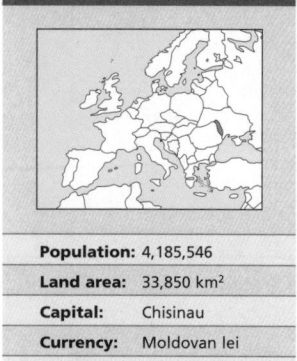

Population:	4,185,546
Land area:	33,850 km²
Capital:	Chisinau
Currency:	Moldovan lei
Language:	Moldovan

Moldovans are descended from the peoples of the southern part of Eastern Europe who had been subdued, and culturally influenced, by the Roman Empire. Byzantine chronicler John Skilitsa (976 AD) mentioned the Vlachs as their forebears. In the mid-14th century, Vlach peoples from the northeast formed their own state, independent of the Hungarian Kingdom, in the territory of South Bukovina. The first *gospodar* (governor) of the Moldovan Principality was Bogdan (1356-1374), although according to legend, it was Dragos who founded the principality.

2   During the second half of the 14th century, the Moldovans freed themselves from Hungarian domination, and from the Tatar khans. By the early 15th century, Moldova's borders were the Dnestr River (to the west), the Black Sea and the Danube (to the south) and the Carpathian mountains (to the west).

3   This small principality was subject to the influence and interest of larger states: Hungary, Poland, and the Ottoman Empire. The Christian Orthodox Church was the official church, and the language - known as Ecclesiastical Slav - was used for church liturgy, for official documents and education.

4   Moldova achieved its greatest political and economic success under the gospodars Alexander the Good (1400-1432) and Stephen the Great (1457-1504). During this period, Moldova went to war against Hungary, Poland and the Crimean Khanate, but its main threat came from the Turks, who in 1484 stripped it of key territories, gave it the Turkish name Akkerman, and created the *raya* - enclaves ruled over by the Turks.

5   In the early 16th century, Moldova lost its independence as a state, and recognized the power of the Turkish sultan (although still maintaining considerable autonomy within the Ottoman Empire). Turkish domination over Bukovina lasted until 1775, over Bessarabia until 1812, and over the rest of the Moldovan Principality until 1878. Turkey seized one Moldovan territory after another; by the mid-18th century, Moldova had lost half of its lands between the Prut and the Dnestr rivers.

6   Anti-Turkish sentiment grew with Moldova's territorial losses, the increase in tributes paid to the sultans, and invasions by Turkish and Tatar troops, who devastated Moldovan cities and towns. Gospodars Petra Rares (1527-38, 1541-46), Ioann Voda Liuti (1572-74) and Dmitri Kantemir (1710-11) all turned against Turkey. Moldova was forced to ally itself with the large powers that opposed Turkey (Hungary, Austria, Poland, and especially Russia). In 1711, Dmitri Kantemir and his Moldovan army joined forces with Russian czar, Peter the Great.

7   All the wars between Russia and Turkey in the 18th and 19th centuries were related to Moldova. Mistrusting the Moldovans, the Ottomans put Greek Phanariotes (from Phanar, a suburb of Istanbul), on the Moldovan throne, who ruled until 1821. There were bloody Russian -Turkish wars on Moldovan soil (1735-39; 1768-74; 1787-91), with many Moldovan volunteers fighting the Turks in Russian ranks.

8   Under the Treaty of Jassy (1792), the Russians obtained the left bank of the Dnestr, south of the Yagolik River. During the second partition of Poland between Russia, Prussia and Austria in 1793, Russia obtained the other part of the left bank of the Dnestr. After the Russo-Turkish War of 1806-1812 and the Peace of Bucharest, Russia seized the territory between the Prut and the Dnestr rivers (Bessarabia). The Muslim population was deported, thus putting an end to the Turkish invasions of Bessarabia.

9   During the 19th century, the population of Bessarabia grew from 250,000 to 2,500,000. By the end of the century, Moldovans made up half of the province's population. There were also a significant number of Ukrainians and Russians, as well as Bulgarians, Germans, Jews and Gagauz (Muslims). During the Russian-Turkish wars of 1828-29, 1877-78 and the Crimean War (1853-56), Bessarabia acted as a rearguard for the Russian army. Under the Treaty of Paris (1856), the part of Southern Bessarabia next to the Danube and the Black Sea was incorporated into the Moldovan Principality, which joined Walachia in 1859 to form the State of Romania. In 1878, the Treaty of Berlin returned this territory to Russia.

10   As of 1866 the Moldovan language was no longer taught. It was once again authorized after the Russian Revolution of 1905. On 2 December 1917, the People's Republic of Moldova was proclaimed. Romanian troops entered Bessarabia and ousted local Soviet authorities. Between December 1917 and January 1918, first the Soviets and then the Romanians gained control over Moldova. Toward the end of January, the independent Moldovan Republic was proclaimed.

11   In the 1920s and 1930s, the territory of modern Moldova was divided into two unequal parts. Bessarabia was part of the Romanian Kingdom, while the left bank of the Dnestr belonged to the USSR. On 12 October 1924, the Autonomous Soviet Socialist Republic (ASSR) of Moldavia was formed, in Ukraine. Its first capital was Balta, and after 1929, Tiraspol. Moldavia/ Moldova made certain gains in industrial and cultural development, while Bessarabia, as part of Romania, remained at a standstill.

12   On 28 June 1940 - with World War II already under way

Life expectancy
70 years
2005-2010

GNI per capita
$720
2004

Literacy
96% total adult rate
2000-2004

HIV prevalence rate
0.2% of population 15-49 years old
2003

IN FOCUS

ENVIRONMENTAL CHALLENGES
The intensive use of agrochemicals, including banned pesticides such as DDT, contaminates soils and underground water, of which 40 per cent has bacterial pollution. Also, 45 per cent of lakes and water bodies are polluted with chemicals. Extensive agriculture speeds erosion and degradation of fertile soils.

WOMEN'S RIGHTS
Women have been able to vote since 1978 (when Moldova was part of the USSR) and to stand for office since 1993, when a new Constitution was passed. In 2005, 22 per cent of seats in Parliament were held by women. In 2003, women made up 49 per cent of the workforce of 2 million*.

In 2004, 99 per cent of pregnant women received prenatal care and the same percentage of deliveries was assisted by qualified staff. Maternal mortality stood at 44 per 100,000 births.

Moldova is a source of women trafficked into forced prostitution in several countries, especially Italy, Turkey, Greece and Kosovo.

CHILDREN
One of the major effects on children of the poverty that affects the country is the large number of boys and girls who live in public institutions, without their family's care. It is estimated that at least one of the parents of 14,000 children is absent, having traveled abroad in search of better job opportunities. The low intake of iodine, especially during pregnancies, increases the incidence of brain damage, which affects one third of all children between 1 and 5. The incidence of anemia doubled in the 2001-2005 period.*

INDIGENOUS PEOPLES/ ETHNIC MINORITIES
The Gagauz and Slavs are identified as national minorities. In 1998, the former amounted to three per cent of the population, while the latter made up 27 per cent. The Slavs are a diverse group with Russian, Ukrainian and Bulgarian roots. Whether or not they should be identified as a single ethnic group is a question that is under debate. They are united by the Russian language, and have shown an ability to engage in unified political activity. In surveys from the 1990s, some 76 per cent of Russians had supported a unified USSR, including Moldova; as did 70 per cent of Ukrainians, 95 per cent of Gagauz and 89 per cent of Bulgarians. Slavs are geographically dispersed in Moldova, somewhat more concentrated in Dnestr, on the eastern border with Ukraine. Within the region, Russians and Ukrainians make up 53 per cent of the population, while Romanians make up 40 per cent. The Slav minority clashed with the Government over the leadership of Dnestr until 1998, after eight years of separatist fighting, when a definitive peace agreement was reached. Instead of fighting for independence, the Slavs focus their present demands on gaining more autonomy for the Dnestr region, within Moldova.

MIGRANTS/REFUGEES
Moldova, Belarus and Ukraine are in the process of creating and developing immigration policies in accordance to the standards of their new geopolitical reality, since they make up the eastern border of the expanded European Union. These three countries make up one of Europe's main migratory routes. A high number of Moldovans cross the borders in search of job opportunities, often never to return.

DEATH PENALTY
Abolished in 1995.

* Latest data available in *The State of the World's Children* and *Childinfo* database, UNICEF, 2006.

- the Soviet Government issued an ultimatum whereby Romania was forced to accept Soviet annexation of Bessarabia. On 2 August, the Moldavian Soviet Socialist Republic (Moldavian SSR) was founded as a part of the USSR, uniting the central part of Bessarabia and the ASSR of Moldavia. The northern and southern parts of Bessarabia, and the eastern region of the Moldavian ASSR, remained within Ukraine. In June 1941, Nazi troops invaded the USSR. Romania made an alliance with Hitler and recovered all of Bessarabia, as far as the Dnestr and Odessa. Three years later, the Red Army took on a weakened Germany, recovering Bessarabia and northern Bucovina.

[13] Soviet leader Leonid Brezhnev's political career began in Moldavia, where he was leader of the local Communist Party. He was later to become Secretary General of the Communist Party of the Soviet Union (CPSU) and President of the USSR, positions which he held simultaneously until 1983.

[14] After the liberalization process initiated by Soviet president, Mikhail Gorbachev, in 1985, political and ethnic problems began to emerge. In 1988, the Democratic Movement in Support of Perestroika (restructuring) began demanding the return to the use of the Latin alphabet instead of the Cyrillic alphabet for writing the Moldovan language. Nationalists called for an end to the political and economic privileges which Russian residents enjoyed, stating that if they could not do without them, they should return to their native land. On 10 November 1989, Parliament approved the Official Language Law, which established Moldavan as the country's official language for political, economic, social and cultural affairs, with Russian to be used only in the press and other mass media.

[15] On 27 August 1991, the country declared its independence from the USSR as Moldova. A month later, Dnestr (Transnistria) and Gagauz - opposing both Moldova's independence and the possibility of union with Romania - declared themselves independent republics.

[16] In December, Mircea Snegur won the country's first presidential elections. In March 1992, Moldova was admitted to the UN as a new member.

[17] The political spectrum split into those supporting unification with Romania and groups preferring independence. In parliamentary elections, pro-independence parties took an ample majority and in August the new Constitution declared Moldova an independent democratic state. Two months later an agreement was signed with Moscow for the withdrawal of Russian troops.

[18] In 1995, President Snegur accelerated the privatization of public companies and facilitated the inflow of foreign capital, but he failed to rally support and was defeated by Petru Lucinschi in the November 1996 elections. Economic reform continued.

[19] In 1999, Lucinschi began a new stage of privatizations, putting on sale the telecommunications monopoly and the electricity sector, which were both plagued by inefficiency.

[20] After the Communist Party won the elections, Vladimir Voronin became president in April 2001 and Vladimir Tarlev was appointed Prime Minister. One of the first measures implemented by the Government was to introduce Russian as the compulsory language, in a nation where 70 per cent of the population were Romanian speakers.

[21] In November 2003, before and even after President Voronin refused to sign an agreement with Russia to federalize the country and grant greater autonomy to Transnistria, nationalist opposition forces took to the streets in protest. The opposition accused Voronin of favoring rapprochement with Moscow (which has a 2,500-strong peace force and a large missile and artillery base in the region) and requested international peacekeeping forces. Meanwhile, Voronin's supporters accused Washington and the EU of using the area's fragility to expand their influence eastwards.

[22] In 2004 Moscow made the withdrawal of its troops from Transnistria conditional upon a negotiated solution to the conflict. In July, Transnistrian authorities closed several schools that used the country's official Latin alphabet, instead of the Cyrillic one considered official by those who advocate the region's independence. The central Government imposed economic sanctions on Transnistria.

[23] Shortly before the March 2005 elections, relations between Chisinau and Moscow deteriorated. Voronin expelled 20 Russian citizens, accusing them of spying, and did not allow the entrance of 100 Russian observers to the elections. Voronin's position - and that of his Communist Party, not an ally of the old Soviet ideology and leaning more toward the EU - was consolidated after winning the elections. Voronin was confirmed for a second period as President and Tarlev continued as Prime Minister.

[24] Chisinau asked Russia to withdraw its troops by the end of 2006. The declaration was issued after a special Moldovan Parliament session to discuss a regional plan proposed by Ukrainian President Viktor Yushchenko.

[25] In March 2006, in compliance with EU requirements on smuggling control, the Government started to demand that all products entering the country - including those entering Transnistria from Ukraine - should have the proper customs documentation. Transnistrian authorities considered the measure a disguised economic sanction. Meanwhile, Chisinau said that Moscow's decision to halt its Moldovan wine imports - citing sanitary reasons - was politically motivated. ■

Monaco / Monaco

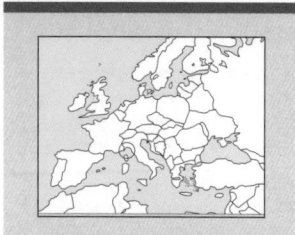

Population:	36,070
Land area:	2 km²
Capital:	Monaco
Currency:	Euro
Language:	French

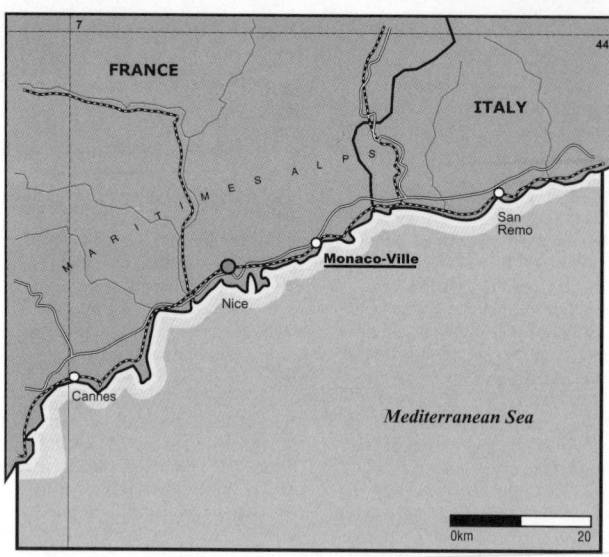

The territory of Monaco (located in a mountainous region on the Mediterranean coast) has been inhabited since the Stone Age. Monaco's Rock was a refuge for numerous peoples; the Ligures were the first settled inhabitants. Monaco's coast and harbor provided them with an outlet to the sea.

[2] The region was occupied by Phoenicians, Greeks and Carthaginians and by the end of the 2nd century BC by Romans. Monaco became a part of the Alpes-Maritime area. The Romans built La Turbie, the trophy of Augustus, who was celebrating his triumphant military campaigns. During that same period, Phoenician and Carthaginian sailors brought prosperity to the region. Monaco was annexed to Marseilles and converted to Christianity in the 1st century AD.

[3] With the fall of the Roman Empire (5th century), the region was invaded by different peoples. In the 7th century it was part of the Lombard kingdom and 100 years later it formed part of the kingdom of Arles. It was controlled by Muslims after the Saracen invasion of France. Starting in the 10th century, after the expulsion of the Saracens by the Count of Provence, the coast gradually became inhabited.

[4] In 1191, Monaco was ceded to Genoa as a colony. On 8 January 1297 the Grimaldis, a Genoese family of exiles, held on to the fortress and placed the founding stone of their present palace. Emperor Henry VI gave control of the land around the Rock of Monaco to their chief, Fulco del Castello. From then on, the Rock became an object of dispute between the two main parties of Genoa: the Ghibellines (followers of the Emperor) and the Guelfs (followers of the papacy) allied with the Grimaldis.

[5] In 1331 Charles I reconquered the Rock and acquired the wealth of the Spinola (allies of the Ghibellines), and gained control

over Menton and Roquebrune. He was considered by many to have been the real founder of the Principality, and first lord of Monaco. Charles died in 1357 and his son Rainier II fought against the Genoese until 1489. Then the King of France and the Duke of Savoy recognized the sovereignty of Monaco.

[6] In September 1641 Honorato II, Prince and Lord of Monaco since 1612, and Louis XIII of France signed the Treaty of Peroné, by which the kingdom of France guaranteed protection to the Prince of Monaco. The same year the Spanish were expelled from the Principality.

[7] During the French Revolution the Principality was annexed to France and proclaimed Protectorate of Sardinia from 1815 to 1860 under the Treaty of Vienna. Monaco's sovereignty was recognized by the Franco-Monegasque treaty of 1861.

[8] Prince Charles III of Monaco drew the international jet set when the first casino was opened in 1863 and the Monte Carlo center in 1866.

[9] Charles III ruled from 1856 to 1889; his son Albert I promulgated the first constitution in 1911.

[10] A treaty signed in 1918 contained provisions limiting French protection of Monaco, establishing that its policies as well as its military and economic interests would still be in line with those of France.

[11] Prince Rainier III succeeded his grandfather, Louis II, who died in 1949.

[12] A new constitution, proclaimed in 1962, abolished the death penalty, granted women the right to vote and appointed a Supreme Court in order to ensure basic freedoms.

[13] In May 1993 Monaco became an official member of the UN.

[14] Although not a European Union (EU) member, in 1999 Monaco adopted the euro as its official currency, and it participates in the EU market system through its customs union with France.

[15] Patrick Leclercq took over as Minister of State in January 2000, replacing Michel Lévêque. In October that year, France threatened to take parliamentary action against Monaco to pressure it to clamp down on money laundering. Paris accused the Principality of hiding

essential information by means of bank secrecy laws, but Monaco ignored the criticism.

[16] In October 2002 the Financial Action Task Force on Money Laundering (FATF) classified Monaco as 'somewhat co-operative' - the second category of three, based on the degree of commitment to the fight against money laundering - together with Barbados and Bermuda.

[17] In 2004, the British courts ruled that Stephen Troth (a member of the HSBC banking group) would face another two years in prison for siphoning more than $10 million from celebrity clients in Monaco including, among others, racing driver Michael Schumacher. A Monaco court had condemned him to four years in prison in 2002.

[18] Rainier died on 6 April 2005, at 81, due to lung, heart and kidney ailments. Upon his death, he was the longest-reigning monarch in Europe. His son Albert, 47, succeeded him.

[19] In April 2006, Prince Albert II carried out a 120-km journey on dog sleds from Barneo, a Russian Arctic base, to the North Pole. The trip, commemorating one of the four carried out 100 years before by his ancestor Albert I, was also aimed at underlining the Principality's official position against global warming. One of global warming's most serious effects is the increasing melt of polar ice. ■

PROFILE

ENVIRONMENT
The Principality of Monaco is located at the foot of the Alps and bordering the Mediterranean Sea. It shares borders with several French communities of the Alpes-Maritime département: Cap-d'Ail, La Turbie, Beausoleil and Roquebrune-Cap-Martin. Its territory covers an area of 195 hectares.

SOCIETY
Peoples: Around 15 per cent of Monacans are locals; French 47 per cent, Italian 15 per cent, others 12 per cent.
Religion: The State is Catholic. There are also Anglicans, Baha'i, Jews and Protestants.
Language: French is the official language, also Monegasque, English and Italian.
Main Political Parties: National and Democratic Union (UND); National Union for Monaco.

THE STATE
Official Name: Principality of Monaco **Administrative Divisions:** Four sections or *quartiers*, Monaco-Ville, the old town; La Condamine, the harbor area; Monte Carlo, residential area; Fontvielle, a newer zone reclaimed from the sea.
Capital: Monaco-Ville 34,000 people (2003).
Government: Parliamentary monarchy. Constitution effective since 17 December 1962. Prince Albert II, sovereign since November 2005, Head of State. Jean-Paul Proust, Minister of State and Head of Government since May 2005. Unicameral Legislature: National Council, with 18 members elected for a five-year term.
Armed Forces: Defense is the responsibility of France.

Mongolia / Mongol Uls

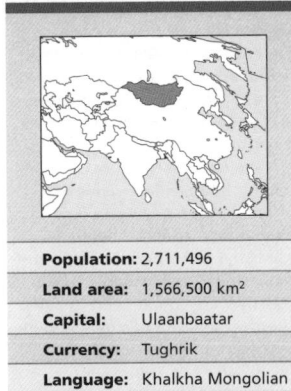

Population:	2,711,496
Land area:	1,566,500 km²
Capital:	Ulaanbaatar
Currency:	Tughrik
Language:	Khalkha Mongolian

The Mongols are one of the main ethnic groups of northern and eastern Asia, linked by cultural ties and a common language. Dialects vary from one part of the region to another, but most can be understood by a Mongolian.

2   Direct lineage from a male ancestor gives the family or clan its name, though there was an earlier tradition of female lineages. Intermarriage between members of the same clan was forbidden so there was great need for establishing alliances between clans, who formed tribal groups.

3   The Mongols were mostly nomadic, with the movement of livestock and campsites determined by pasturage needs throughout the year. Animals were owned individually, while grazing lands were collective property.

4   The most powerful clans tended to control the tribal groups' activities. The weakest families maintained their own authority and ownership of animals, but they were forced to pay tribute to the dominant clan. They moved, camped, grazed their livestock and went to war under that clan's orders.

5   Political and military organization was adapted to the needs of each clan or tribe. A person capable of handling a weapon could be a chief or a soldier, according to the needs of the moment. Capturing livestock,

women or prisoners from other tribes was a common means of acquiring wealth.

6   The Siung-nu, or Huns, were the earliest inhabitants of the Selenga valleys, joining Siberia to the heart of Asia, and they are thought to have settled in this region by 400 BC.

7   The Huns created a great empire in Mongolia when China was undergoing unification as an imperial state under the Ch'in and Han dynasties (221 BC-220 AD). The Hun Empire warred against China for centuries, until it disintegrated - perhaps due to internal conflicts - around the 4th century.

8   Some of the southern tribes surrendered to China and settled in Chinese territory, where they were eventually absorbed by the Chinese, while others migrated westward. In the 5th century, Attila's Huns conquered almost all of Europe, reaching Gaul and the Italian peninsula.

9   The Huns were subsequently displaced by the Turks who established themselves throughout the region. Social organization at the time did not consist only of nomadic tribal groups. The major Hun chieftains set up general headquarters, surrounded by cultivated lands where they bred

larger, stronger horses, capable of carrying a warrior in armor.

10   This led to a differentiation between aristocrats and traditional archers, who rode smaller horses. Agriculture also became more important to the economy.

11   The term 'Mongol' first appeared in records of different groups written during the T'ang Chinese dynasty. It then disappeared until the 11th century, when the Kidan became the rulers of Manchuria and northern China, controlling almost all of present-day Mongolia.

12   The Kidan established the Liao dynasty in China (907-1125) and ruled Mongolia, fostering division between the different groups.

13   The Kidan were succeeded by the Juchen, who were in turn succeeded by the Tatars, before the era of Genghis Khan (Temujin). Born in 1162, Temujin inherited several fiefdoms that had been seized from his family.

14   In 1206, because of his political and military prowess, Temujin was recognized as leader of all the Mongols, and given the title of Genghis Khan. His armies invaded northern China, reaching Beijing. By 1215, the Mongolian Empire extended as far as Tibet and Turkistan.

15   Upon Genghis Khan's death in 1227 disputes among his successors caused the Mongolian Empire to disintegrate, until the Chinese throne was left in the hands of the Ming dynasty in 1368. China invaded Mongolia and set fire to

Karakorum, the former imperial capital, though it was unable to bring the territory under control.

16   In the 15th and 16th centuries, controlling the areas beyond the Great Wall of China demanded military mobilization.

17   The Oyrat alliance between groups living in western Mongolia began gaining control of the territory. They conquered several oases in Sinkiang and the Tibet region, and added their own mercantile and administrative expertise to the Mongols' tribal organization.

18   The separation of the Oyrat from the Jaljas began during this period, with the latter forming the core of what was later to become Outer Mongolia. A tribal league was formed between the Khalkhas in the north and the Chahars in the south, while the leadership passed over to the Ordos, during the reign of Altan Khan (1543-83).

19   To keep their hold on power, the Mongolian princes thought it useful to be backed up by a religious ideology. They adopted the Tibetan Buddhist religion as Tibet posed no cultural threat, and the Tibetan script was easy to use.

20   Altan Khan proceeded to invite a Tibetan prelate, whom the Mongols called 'Dalai Lama' to lead the state religion. The merging of religious interests with those of the State was accomplished by claiming that an heir to the Khalkhas clan was the first 'reincarnation' of the Living Buddha of Urga.

21   In 1644, after consolidating their power in Manchuria, the Manchus seized the Chinese throne, with the help of Mongolian tribes from the far east. Before occupying Beijing, the Manchus took control of southern Mongolia, which was henceforth known as Inner Mongolia.

22   It took China almost a century to conquer Outer Mongolia. Meanwhile, Inner Mongolia became

LAND USE

2003/2005

IRRIGATED AREA: 7.0% of arable land

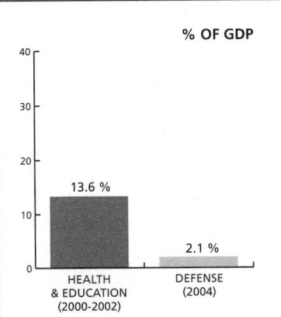

- FOREST AND WOODLAND: 6.5%
- ARABLE LAND: 0.8%
- CROPLANDS: 0.0%
- OTHER USE: 92.7%

PUBLIC EXPENDITURE

% OF GDP

13.6 % — HEALTH & EDUCATION (2000-2002)

2.1 % — DEFENSE (2004)

WORKERS

LABOR FORCE   2004

- FEMALE: 40%   ■ MALE: 60%

 Life expectancy
66 years
2005-2010

 GNI per capita
$600
2004

 Literacy
98% total adult rate
2000-2004

 HIV prevalence rate
<0.1% of population 15-49 years old
2003

a part of China, and the Khalkhas' desire to retain power in the south prevented the Oyrats from attaining reunification.

[23]   This was the final stage of the great wars among the Mongols; ending in their overall dispersal. Several groups of Khalkhas remained in the south; some Chahars settled in Sinkiang and the Oyrat dispersed in different directions, including czarist Russia.

[24]   In the Russian-Japanese War of 1904-05, both armies used Mongolian troops and staff. This served Japanese interests well as a resurgence of Mongolian nationalism could weaken both Russia and China. At the end of the war, Russia secretly recognized Inner Mongolia as belonging to Japan's sphere of influence.

[25]   With the outbreak of the Chinese Revolution in 1911, there was a pervading malaise in Mongolia. Until then, the region had been the object of disputes between Russia and Japan. However, the Mongolians' social and political discontent was directed against the Manchus and the local government.

[26]   Led by their Buddhist leader, Mongolia proclaimed independence from China and sought Russian support. However, because of its secret treaties with Japan and Britain, Russia could offer nothing more than mere 'autonomy'. After lengthy negotiations, this status was granted to Outer Mongolia.

[27]   This situation continued until the Russian Revolution in 1917. China sent in troops and made the Mongolians sign a request for aid from Beijing. But the region was invaded by retreating czarist troops, who expelled the Chinese and mistreated the Mongolians.

[28]   With the traditional leaders discredited because of their poor handling of the Chinese and White Russian interventions, some groups of Mongolian revolutionaries sought help from the Bolsheviks. Russian and Mongolian troops took the capital, Urga, in July 1921.

[29]   This was the beginning of the republic, although initially the Living Buddha acted as puppet king, only authorized to endorse the new regime's proposals. Upon his death in 1924, the People's Republic of Mongolia was proclaimed.

[30]   The Mongolian People's Revolutionary Party (MPRP), made up of conservatives and revolutionary nationalists, wavered between Beijing and Moscow until the defeat of the Chinese Revolution, at the hands of Chiang Kai-shek. At this time, Mongolia began to fall increasingly under the influence of the USSR, and Joseph Stalin.

[31]   The new republic proclaimed the right of women to vote.

[32]   Following the Soviet model, the MPRP Government tried to collectivize the economy in order to break the power of the feudal lords and the Buddhist priests. Between 1936 and 1938, the Mongolian regime purged the party and the army, executing many leaders.

[33]   In 1939, Japan invaded north eastern Mongolia, along the Siberian border. Mongolian troops resisted ferociously until Soviet help arrived.

[34]   The defeat of Japan was a severe blow to the Axis powers in Berlin and Tokyo. Mongolia and the USSR fought together in the Inner Mongolian and Manchurian campaign, two weeks before the end of World War II.

[35]   As part of the Yalta agreement, Chiang Kai-shek agreed to hold a plebiscite in Mongolia. Although the result favored independence, Mongolia failed to receive diplomatic recognition. In 1961, Mongolia was admitted to the UN.

[36]   In 1960, Government officials in Ulaanbaatar accused the Chinese Government of mistreating Mongolian citizens and of seeking territorial expansion, at Mongolia's expense.

[37]   Friction continued until 1986, when the Chinese deputy minister of the Council of Ministers visited Mongolia and re-established consular and commercial relations.

[38]   In March 1988, China and Mongolia signed a treaty aimed at defining the 4,655-kilometre border between the two countries. A year later, during Mongolian premier Tserenpylium Gombasuren's visit, the first in 40 years, relations between the two countries were returned to normal.

[39]   In 1989, within the framework of Soviet *perestroika* (restructuring), Moscow announced that three-quarters of its troops would be withdrawn in 1990. Shortly afterwards, both governments agreed to the complete withdrawal of all Soviet military personnel and equipment from Mongolian territory by the end of 1992.

[40]   Meanwhile, the MPRP leadership admitted that social and economic reforms were inadequate. The ruling party adopted democratic changes in internal elections.

[41]   In 1989 and 1990, several opposition groups emerged. One of the most active, the Democratic Union of Mongolia, was officially recognized in January 1990. In March, increasingly frequent public demonstrations against the Government triggered a new crisis within the MPRP.

[42]   The National Assembly approved a constitutional amendment withdrawing the reference to the MPRP as society's 'prime moving force' and approving new electoral legislation; however, no changes were made in relation to political party activity.

[43]   The legendary figure of Genghis Khan, whose name was forbidden for many years, was rehabilitated as an authentic expression of Mongolian pride and tradition, sentiments which until recently were condemned as being an expression of a narrow-minded 'nationalism'.

[44]   In spite of 65 years of Soviet aid, Mongolia's economy maintained vestiges of nomadism. In the early 1990s, urbanization was just beginning, and half of Ulaanbaatar's population lived in tents, with rudimentary electric and water supplies.

[45]   In the first months of 1991, there was a substantial reduction in Mongolia's foreign trade. There were acute shortages of food, medicine and fuel. The currency plummeted, and government income declined sharply, while expenditure steadily increased.

[46]   In May, Prime Minister Dashiun Byambasuren announced a new economic policy which included incentives to attract foreign investment, the establishment of a national stock exchange, the sale of two-thirds of the state's capital goods, deregulation of prices and changes in the banking system.

[47]   The chair of the Central Bank of Mongolia, Zhargalsaikhan, was arrested together with a group of new investors in December 1991, for an $82 million fraud as a result of which the country lost most of its reserves. At the same time, Deputy Prime Minister Cabaadorjiyn Ganbold was accused of secretly authorizing the transfer of 4,400 kilograms of gold to a branch of Goldman Sachs - a British merchant bank, as collateral for a $46 million loan, apparently earmarked for covering losses.

[48]   In 1992, Parliament approved a Government-proposed constitutional reform, adopting the official name 'Republic of Mongolia', and dropping the word 'People's'. The reform also established a pluralistic democratic system, replacing the socialist system.

[49]   In October 1992, after being defeated in the June elections which were won by a wide margin by the ruling MPRP, the opposition formed the Mongolian National Democratic Party (MNDP). The Social Democratic Party (SDP) preferred to remain independent.

[50]   In 1992, the withdrawal of Russian troops - begun in 1987 - was completed. Meanwhile, Otchirbat had a rapprochement with the MNDP and the SDP to prepare the

PROFILE

ENVIRONMENT

Comprises the northern area of Mongolia, also known as 'Outer Mongolia' (the southern part, 'Inner Mongolia', comes within Chinese territory under the name of Inner Mongolian Autonomous Region). At the center of the country lies the wide Gobi Desert, bordered on the north and south by steppes where there is extensive nomadic sheep, horse and camel raising. The Altai mountain region, in west Mongolia, is rich in mineral resources: copper, tin, phosphates, coal and oil.

SOCIETY

Peoples: Khalkha Mongol 78.8 per cent; Kazakh 5.9 per cent; Dörbed Mongol 2.7 per cent; Bayad 1.9 per cent; Buryat Mongol 1.7 per cent; Dariganga Mongol 1.4 per cent; other 7.6 per cent.
Religions: Buddhism.
Languages: Khalkha Mongolian.
Main Political Parties: Mongolian People's Revolutionary Party (MPRP); Motherland Democracy Coalition (consisting of the Democratic Party, the New Socialist Democratic Party and the Civic Will Republican Party); Republican Party.
Main Social Organizations: Central Council of Mongolian Unions; 'Blue Mongolia'; Mongolian Confederation of Free Unions.

THE STATE

Official Name: Bŭgd Nairamdach Mongol Ard Uls.
Administrative Divisions: 18 provinces and 1 municipality (Ulan Bator).
Capital: Ulaanbaatar (Ulan Bator) 812,000 people (2003).
Other Cities: Darhan 75,000 people; Erdenet 71,200; Choybalsan 37,700; Ölgiy 21,100 (2000).
Government: Parliamentary republic. Nambaryn Enkhbayar, President since June 1997, twice re-elected. Miyeegombo Enkhbold, Prime Minister since January 2006. Legislature: single-chamber Assembly with 76 members elected every 4 years.
National Holiday: 11 July, Independence Day (1921).
Armed Forces: 9,000 (2003).

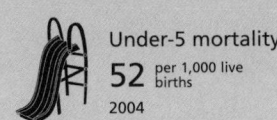
presidential elections in June 1993. Thanks to these former members of the opposition, the President was re-elected with almost 58 per cent of the vote and announced the 'Westernization' of the economy.

[51] Throughout 1994, disagreements between Otchirbat and the former communist majority in parliament were frequent. According to official estimates 26.5 per cent of the population lacked the minimum subsistence income.

[52] The June 1996 elections marked the end of communist dominion, with victory for the Democratic Union Coalition (DUC), a coalition formed by the SDP and the MNDP. The DUC took 50 of the 76 seats, while the former communist bloc shrank from 70 to 25 seats. In July, Parliament appointed Mendsayhany Enkhsaikhan as Prime Minister.

[53] After the election, the Government implemented reforms to switch quickly to a market economy. This process had a high social cost, increasing unemployment and poverty, and added to the damage caused by the lack of technical and economic assistance following the disappearance of the Soviet Union.

[54] In May 1997, Natsagiyn Bagabandi, of the MPRP, won the presidential election with 60.8 per cent of the vote. President Punsalmaagiyn Otchirbat, the DUC re-election candidate, took 29.8 per cent. The result was interpreted as a reaction against the 'shock therapy' applied in order to reach a liberalized economy.

[55] A law introduced in June 1998 required the use of surnames in legal documents. The new legislation, which responded to demands for modernization and greater integration with the rest of the world, caused confusion and concern amongst the people. For centuries, most of the population had been made up of nomadic herders, living in small groups, for whom surnames were irrelevant.

[56] The failure to tackle the economic crisis and strong criticism forced the Government to resign, and a new cabinet was formed, headed by Janlaviin Narantsatsralt. Seven months later, in July 1999, following a dispute related to the privatization of a copper mine co-owned by Mongolia and Russia, Narantsatsralt and his ten ministers all resigned. Parliament appointed a new member of the DUC, economist Rinchinnyamiin Amarjargal, as premier.

[57] Living conditions became considerably worse after the winter of 2000 - the coldest in 55 years. The Government declared more than half the country a disaster area. More than two million head of livestock were lost, the equivalent of $1.65 billion.

IN FOCUS

ENVIRONMENTAL CHALLENGES
Water is a scarce resource, especially in regions close to the Gobi Desert. Coal-burning power plants have severely polluted the air in the capital city. Deforestation and uncontrolled farming have led to soil erosion. Desertification is another symptom of environmental deterioration. In 2001 frequent forest fires and the *dzuds* - an environmental phenomenon in winter that combines extreme cold spells with blizzards and droughts - led to considerable loss of human life and damage to property.

WOMEN'S RIGHTS
Women have been able to vote and stand for office since 1924. In 2004, women held almost 7 per cent of seats in Parliament and 6 per cent of ministerial positions. They made up 46 per cent of the country's labor force of 2 million. In spite of the fact that 94 per cent of pregnant women received prenatal care and 97 per cent of births were attended by skilled health staff, maternal mortality amounted to 110 per 100,000 live births.

No effective measures have been taken to combat trafficking of women. Domestic violence is common but the culprits are not punished. Rape within marriage is not considered an offense in Mongolia.

CHILDREN
Despite the significant drop in infant and under-five mortality rates within the period 1990-2004, these remain high. The former was reduced from 78 to 41 per 1,000 live births and the latter fell from 108 to 52 per 1,000. Seven per cent of infants were born with low birth weight; 13 per cent of children under-five suffered from moderate or severe underweight and 25 per cent were moderately or severely stunted*.

During the harsh winters of 2002 and 2003 rural children and women were the hardest hit. Their incomes dropped and they had reduced access to food. These families were also psychologically, socially and emotionally affected, suffering in the post-*dzud* period from stress and immunological problems as well as malnutrition and fatigue.

In 2004, it was estimated that 30 per cent of children aged between 5 and 14 were working.

INDIGENOUS PEOPLES/ ETHNIC MINORITIES
Even though Mongolia is inhabited by several linguistic subgroups (Lhoton, Uriankhai, Zakhchin, Myangad, Oold and Torguud - all of them dialects derived from Mongolian) the Kazakh, who comprise six per cent of the country's population, are the only ethnic, religious or linguistic minority recognized by the Government. Most Mongols belong to the Khalka group, which comprises 79 per cent of the population.

MIGRANTS/REFUGEES
UNHCR closely monitors the situation of North Korean refugees in the country. Mongolia was taking the necessary steps at Government level to sign the 1951 UN Convention on Refugees, as well as the 1967 Protocol, during 2006. UNHCR fears that a significant increase in the number of North Koreans entering the country from China (a constant flow of small groups) might strengthen the position of those who oppose the signing of the 1951 treaty.

DEATH PENALTY
This is applicable even to ordinary crimes.

---

* Latest data available in *The State of the World's Children* and *Childinfo* database, UNICEF, 2006.

[58] In the July 2000 legislative elections the MPRP was elected with 72 of a total 76 seats.

[59] The winter of 2001 was even more severe than the previous one, with heavy blizzards and temperatures plunging below -50°C. More than 6 million head of livestock perished. China, the International Red Cross and the United Nations all appealed for support for the 75,000 families affected, and especially for the thousands of herders whose survival depends almost exclusively on their animals.

[60] In 2001 the IMF approved a $40 million low-interest loan over a three-year term. The loan was to be used in combating poverty, boosting the economy and investing in social plans in Mongolia.

[61] In November 2002, more than 50 demonstrators (members of the Democratic Party) were arrested while protesting against the land privatization act - in effect since May 2003 - which covered one per cent of Mongolian territory.

[62] In January 2004, after two years of negotiations, a new labor plan was implemented, allowing Mongolians to work in Taiwan. That month, Moscow wrote off $300 million of Mongolia's debt to Russia.

[63] In the June parliamentary elections, the opposition Democratic Coalition registered a considerable advance, obtaining 36 out of the 76 seats at stake. The remaining 40 seats were won by the ruling MPRP, which failed to repeat the results of 2000 elections, mainly due to its failure to fight poverty. In August, Tsakhiagiin Elbegdorj - of the Democratic Coalition - was appointed Prime Minister after long negotiations.

[64] In May 2005, Nambaryn Enkhbayar, candidate of the former Communist Party, won the presidential election. The June 2004 parliamentary elections, which had forced the setting up of a coalition government, had granted increased influence to the presidential role - normally ceremonial.

[65] In November 2005, George W Bush became the first US President to visit Mongolia. During his meeting with Enkhbayar, Bush thanked him for supporting the invasion of Iraq and for sending more than 100 troops.

[66] The MPRP dismantled the government coalition in January 2006 and continued to govern with the support of minority parties. Miyeegombo Enkhbold, of the MPRP, was appointed Prime Minister. Members of the opposition Democratic Party accused the MPRP of triggering the political crisis in order to prevent the investigation into corruption being carried out by the coalition from going ahead.

[67] In May, minority parties - the Healthy Society Movement, Mongolia's Green Party, the People's Party and the Civil Will Party - demonstrated outside the Government House against the new Election Committee, exclusively made up of MPRP and Democratic Party representatives. Protesters chanted slogans such as: 'Did the MPRP win the 2004 elections?' and 'The MPRP should end its anarchist point of view'.

[68] After an intense two-day debate in October 2006, Parliament voted down the demand of the Democratic Party for the resignation of the government of 'national unity'. The outcome was never in doubt but the debate was broadcast live on television on commercial as well as public channels and was keenly followed around the country. ∎

Montenegro / Crna Gora

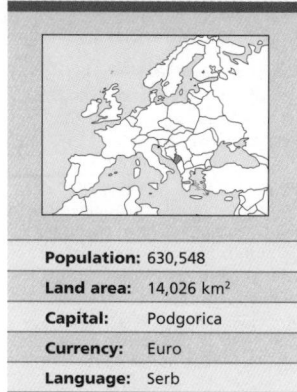

Population:	630,548
Land area:	14,026 km²
Capital:	Podgorica
Currency:	Euro
Language:	Serb

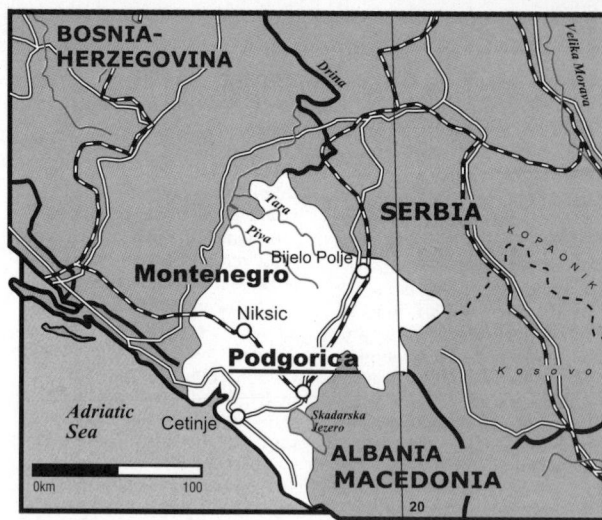

The name Montenegro ('black mountain') refers to the dark color of the forests that cover the Dinaric Alps and was given by Venetian sailors at the time of the Republic of Venice's hegemony over the eastern Adriatic coast, in the High Middle Ages. The country's name, *Crna Gora*, is a literal translation of these words into Serbian.

² Before Slavic peoples arrived in the Balkan peninsula in the 6th century, present-day Montenegro was inhabited by several tribes, including the Illyrians. The territory was conquered in the year 9 AD by the Romans, which annexed it to the province of Illyricum. As Roman power subsided, the region received successive invasions of semi-nomad peoples, mainly Goths by the end of the 5th century and Avars in the 6th. They were soon displaced by the Slavs, who by the mid-7th century were fully settled in the region.

³ In the 10th century, Slav tribes, mixed with Illyrians, Avars and Romans, formed the semi-independent dukedom of Duklja on the coasts of the Adriatic sea, in part of what is currently the Republic of Montenegro. Pope Gregory VII recognized its independence in 1077 and proclaimed its monarch, Mihailo, as *rex Docleae* (king of Duklja).

⁴ The princedom of Zeta, which corresponds more precisely to the current borders of Montenegro, appeared toward 1360 and was ruled by two dynasties: the Balsic House until 1421 and the Crnojevic House in 1421-1499. The Ottoman Empire had controlled the region since the 15th century but never fully conquered the princedom.

⁵ In 1516, prince Durad Crnojevic abdicated in favor of Archbishop Vavil, who turned Montenegro into a theocratic state ruled by the prince-bishop (*vladika*) of Cetinje. The post was filled for centuries by members of the Petrovic-Njegos family, from the Ridani tribe. In 1852, Danilo II Petrovic-Njegos - who had become vladika in 1851 - got married, left the ministry

and became *knjaz* (prince). Thus the territory became a secular princedom.

⁶ At the Congress of Berlin in 1878 the European powers recognized the independence of Serbia and Montenegro, which became kingdoms in 1882 and 1905 respectively (see history of Serbia).

⁷ In 1910, Prince Nikola I was crowned King of Montenegro. In 1912 he declared war on the Ottoman Empire. In the first of the Balkan wars (1912-1913), Montenegro formed an alliance with Bulgaria, Serbia and Greece, encouraged by Russia in order to expel the Turks from Macedonia. The Turks were defeated and had to withdraw to the north of Constantinople.

⁸ After the victory, Montenegro doubled its size upon receiving the former Ottoman territory known as Sanjak of Novi Pazar.

⁹ In World War I Montenegro, an Allied country, was occupied by troops from the Austro-Hungarian Empire. In 1918, the Assembly of Podgorica voted for union with the Kingdom of Serbia. However, pro-independence Montenegro forces rebelled against Serbia in December 1919. The uprising was controlled in 1924.

¹⁰ From 1919 onwards, Montenegro was part of the kingdom of Serbs, Croatians and Slovenians - renamed Yugoslavia (land of the Southern Slavs) in 1929 - until 1941.

¹¹ During World War II Montenegro was occupied first by Italy and then by Germany. The Balkans were liberated from Nazi occupation by the guerrillas of Croatian Josip Broz, known as Tito, and after a provisional government a Constituent Assembly proclaimed, in 1945, the Federal Socialist Republic of Yugoslavia, made up of six republics: Slovenia and Croatia in the northeast, Serbia

in the east, Bosnia-Herzegovina and Montenegro in the center and Macedonia in the south. The city of Podgorica replaced Cetinje as administrative capital and was named Titograd, as a tribute to Tito.

¹² In the following years the country underwent a strong urbanization and industrialization process. In the 1940s, seven-eighths of Montenegrins lived in rural areas. In the 1980s, the proportion had reversed and seven-eighths were living in cities.

¹³ Nationalism strengthened in 1969 when the Serbian Orthodox Church objected to the construction of a monument in tribute to Prince Petar Petrovic Njegos, one of the most prominent Montenegrin princes of the 19th century.

¹⁴ In a 1992 referendum, after the violent break-up of former

Yugoslavia (see history of Croatia, Macedonia, Slovenia, Serbia and Bosnia-Herzegovina), 95.96 per cent of Montenegrins voted for federation with Serbia. However, pro-independence movements and the Muslim and Catholic minorities boycotted the vote. That year the Yugoslav Federation was formed, made up of Serbia and Montenegro.

¹⁵ The break-up of the Yugoslav market and the economic embargo imposed by the UN on the Federation because of its role in the war crimes committed in Croatia and Bosnia caused the worst economic crisis in Montenegro since World War II. In January 1994 hyperinflation reached 3,000,000 per cent. The country replaced its currency with the German mark to control inflation and started to demand more autonomy in the economic field.

¹⁶ The Yugoslav Federation was replaced in 2003 by a new, more flexible union called Serbia and Montenegro. Each Republic was to have its own President, Defense and Foreign ministers and would handle its own economy. After three years, both would decide in a referendum whether they would declare independence.

¹⁷ Montenegrin Svetozvar Marovic, Vice-president of the Socialist Democratic Party, was elected by the Parliament of Serbia and Montenegro as the country's first President.

¹⁸ Once the period set by the Constitution was over, in a referendum in May 2006, 55.5 per cent of Montenegrins voted for independence, which was declared on 4 June. The result disappointed Serbia, which supported the union. ∎

PROFILE

ENVIRONMENT

The country has four geographical regions: the coast on the Adriatic sea, the rocky tablelands, a lowland - with the highest concentration of population and agricultural activities - and the northern high mountains (Dinaric Alps). The climate is Mediterranean at the coast and continental in the rest of the country. Tourism is an important source of income for Montenegro, which hosts roughly 900,000 visitors per year.

SOCIETY

Peoples: Montenegrin 43%; Serb 32%; Bosnian 8%; Albanian 5%; Croat 1%; Roma (Gypsy) 0.5%. **Religions:** Orthodox (74%); Muslim (18%). **Languages:** Serbian of the Ijekavian dialect (official), Montenegrin. **Main Political Parties:** Democratic Party of Socialists of Montenegro; Liberal Alliance; Socialist people's Party; Social Democratic Party.

THE STATE

Official Name: Republika Crna Gora (Republic of Montenegro). **Capital:** Podgorica 136,473 people (2003). **Other Cities:** Niksic 58,200; Bijelo Polje 32,000; Cetinje 15,137 (2003). **Government:** The country is a parliamentary republic. President: Filip Vujanovic, since May 2003. Prime Minister: Milo Djukanovic, since January 2003. Unicameral Parliament: Assembly of Montenegro, with 78 deputies (one deputy is elected per 6,000 voters). **National Holiday:** 13 July, recognition of Montenegro by the Congress of Berlin as the 27th independent State in the world (1878) and date of the first popular uprising against the occupying Axis powers in the country (1941).

Montserrat / Montserrat

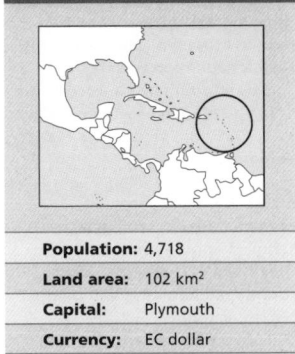

Population:	4,718
Land area:	102 km²
Capital:	Plymouth
Currency:	EC dollar
Language:	English

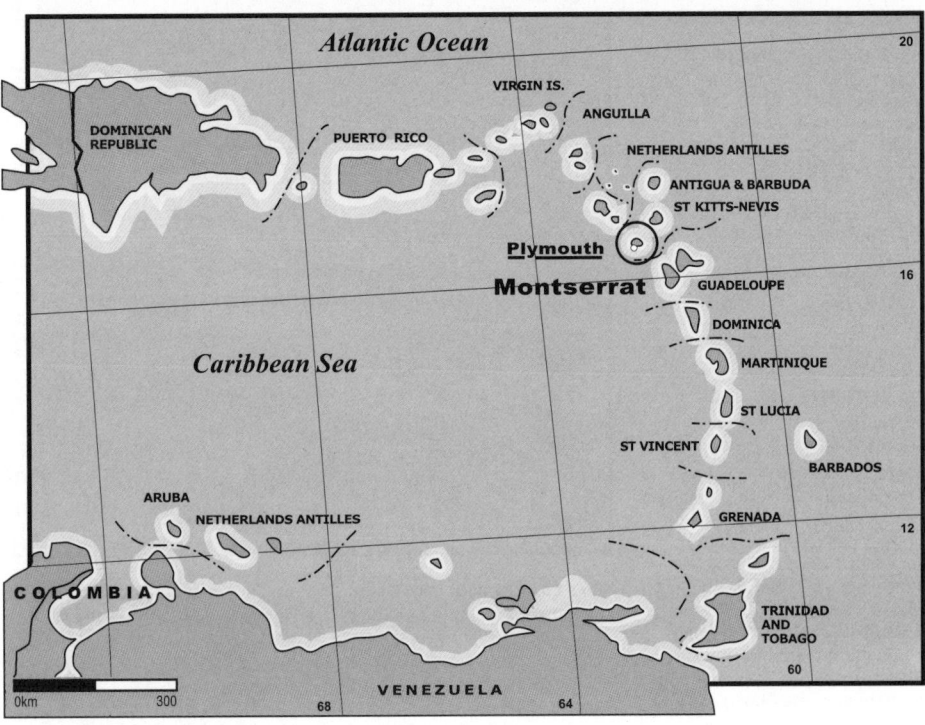

Montserrat, like the rest of the Lesser Antilles, was inhabited by the Caribs, who were wiped out by colonization. France and Britain fought at length for the island, which was also a haven for pirates, but Britain gained possession of it in 1632.

[2] Montserrat was colonized mainly by Irish people following their expulsion from nearby St Kitts. The island was covered by sugar and cotton plantations and large numbers of African slaves were transported to work these from 1651 onwards. In the mid-19th century the island had little more than 10,000 residents, of whom 9,000 were slaves.

[3] Montserrat has been a member of several Caribbean federations, the last being the West Indies Federation, dissolved in 1962. Since 1960 the island has been ruled by an Administrator (renamed Governor in 1971) appointed by the UK. The Democratic Progressive Party (DPP), founded by William Bramble and later chaired by his son, Austin Bramble, refused to alter this colonial arrangement when Montserrat joined the new grouping of West Indies Associate States in 1967.

[4] The DPP lost all its seats in parliament in November 1973 to the then recently created People's Liberation Movement (PLM). In 1979 the new chief minister, John Osborne, announced his plan for total independence for 1982 - though this was never put into practice. Those against independence argued that income from tourism and remittances of Montserrat nationals living abroad were not enough to cover the trade deficit, paid by the UK.

[5] In 1995 and 1996, the economy was seriously weakened by the constant expulsion of ashes by the Soufriere Hills volcano. The volcano erupted in mid-1997, killing more than 20 people, destroying the capital,-and making the greater part of Montserrat uninhabitable. Thousands of people were evacuated to areas in the north, to Britain, and to other Caribbean countries. Bertrand Osborne resigned as chief minister in August 1997, after residents complained at the way the evacuation of the local population was carried out.

[6] The new chief minister, David Brandt, requested a judicial investigation into the UK's decision to evacuate the population of Montserrat instead of aiding those who decided to stay. The results of that investigation held both officials in Montserrat and the UK responsible for contributing to the death of nine farmers, for not granting them land on which they could have taken refuge.

[7] Presenting the 2001 budget following another election, the new chief minister John Osborne listed some causes for the economic crisis which affected the island: lack of capacity to generate income, slow implementation of public sector projects, private sector hindrance of the Government, and a 6.3 per cent economic growth rate in fiscal year 2000, among others.

[8] In early 2002 the UK passed a bill granting British citizenship to its dependencies. This resulted in increased possibilities for emigration for the people of Montserrat and made the likelihood of winning independence more remote.

[9] In July 2003 Deborah Barnes Jones was appointed by the UK to take office as Governor in April 2004.

[10] The US revoked, in August 2004, the 'temporary protectorate' status it had granted the island in 1995.

[11] In July 2005 the new Gerald airport was opened, replacing Bramble, destroyed in 1997 by the Soufriere eruption.

[12] Initial results of the May 2006 elections showed that the Movement for Change and Prosperity had won, with a platform that included the protection of the population from natural disasters and education for all. It won 47 per cent of parliamentary seats. Lowell Lewis was sworn in as chief minister in June. ∎

PROFILE

ENVIRONMENT

Montserrat is one of the Leeward islands in the Lesser Antilles, located 400 km east of Puerto Rico, northwest of Guadeloupe. The terrain is volcanic in origin and quite mountainous, with altitudes of over 1,000 meters. The tropical, rainy climate is tempered by sea winds. The soil of the plains is relatively fertile and suitable for agriculture. Only a quarter the land is cultivated and population density in these areas is high. Half of the island is suitable for stockbreeding.

SOCIETY

Most of the population is of African descent, with a small minority of European descendants.
Religions: Mostly Christian; Anglicans, Catholics and Methodists predominate. **Languages:** English (official). Most people speak a local dialect.
Main Political Parties: New People's Liberation Movement (NPLM); Movement for Change and Prosperity (MCAP).
Main Social Organizations: The Montserrat Allied Workers' Union. There is also a Teachers' Union and Seamen and Waterfront Workers' Union.

THE STATE

Official Name: Montserrat.
Capital: Plymouth 2,000 people (1999).
Government: Deborah Barnes Jones, since April 2004. Lowell Lewis, chief minister since June 2006. Parliament: Legislative Council with 11 members, 9 of them elected every 5 years.

Morocco / Al Maghrib

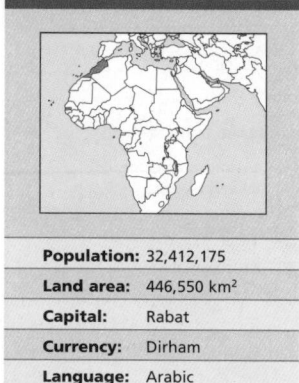

Population:	32,412,175
Land area:	446,550 km²
Capital:	Rabat
Currency:	Dirham
Language:	Arabic

Berber people lived in what is now Morocco long before the Phoenicians invaded the territory in 12th century BC. Although their origin is unknown, the Berber – a name given to them by the Arabs in the 7th century AD which means 'those that are not Arab' – are believed to be Euro-Asians. Throughout the centuries, three main tribes would form dynasties: the Sanhajah, the Masmouda and the Zenata.

2   The nomad and warrior Sanhajah formed the Almoravid dynasty and founded Marrakesh. The peaceful and farming Masmouda lived north and west of the Atlas mountain range and formed the Almohad dynasty. The horseriding and nomad Zenata controlled the area between Tafilalet and present Algeria and founded the Merinides dynasty.

3   Once the Mediterranean Phoenicians sighted North African lands in the 12th century BC, they gradually established trade posts throughout the eastern coast. Fish salting factories - topped by Roman ruins - and other traces of their occupation can be found in Liks, but also in Tangiers, Melilla, Chella, Rahat and Tamuda.

4   The Phoenicians were seafarers and traders, and scarcely conquerors. In the 8th century they founded their main city, Carthage, in present Tunisia, which grew to become a prosperous kingdom.

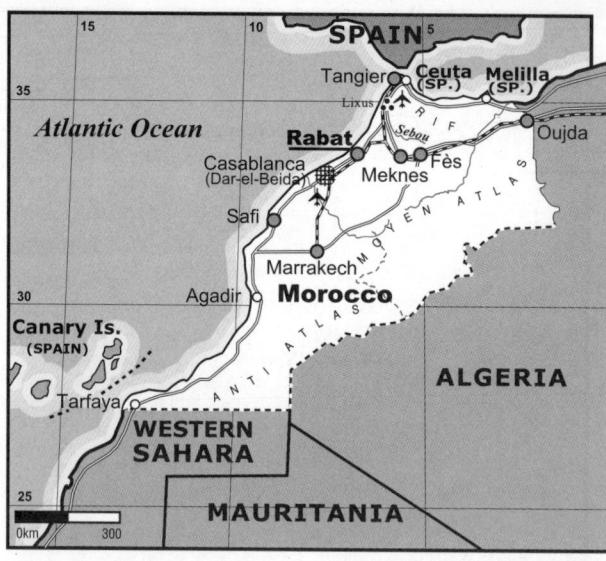

Some Carthaginians moved to the Moroccan coast, establishing prosperous communities, which took the name of the Garum, an anchovy paste that was to become a major export.

5   The Berber were greatly influenced by the Carthaginians. After the Second Punic war, when Carthage became an African province of Rome, thousands of Carthaginese fled from the Romans seeking refuge in friendly enclaves along the coast. Once Carthage had been taken, the Romans spread to the Berber kingdoms of Mauritania and Numidia (current Algeria).

6   Cities such as Volubilis, Sala Colonia or Tingis developed a mixed culture, of Mauritanian origin, partially Roman and even Christian. The Vandals and Goths went through the region, on the way to Carthage, without leaving traces.

7   In 683, the Arabs - led by Oqba Ben Nafi, head of the Umayed Dynasty from Damascus - brought Islam to the current Morrocan territory. Ben Nafi founded the city of Kairwan (current Tunis) and built the first mosque in the African continent. He called the land Maghreb al Aqsa. The Berber accepted the Qur'an and led, along with the Arabs, through the Almohades, Islamic expansion to the south. However, they held on to their language and customs.

8   In 703, the Berber backed the second great Umayed leader in the region, Musa Ibn Nouasser, in his Islamic expansion towards southern Spain and southern Morocco. Some Christian enclaves remained, but most Christians fled to the Iberian peninsula.

9   Idris Ben Abdallah, a descendant of prophet Muhammad, crossed Egypt, Tangiers and Volubilis, by then fully Islamic. The Berber kings proclaimed him King and pledged him their support.

10   When his father died, Idris II was crowned king at the age of 12. He founded Fès, which in 818 received 8,000 Arab families expelled by Spanish Christians from the Emirate of Cordoba. Seven years later, another 2,000 families came from Kairwan. The immigrants' sophistication and skills turned Fès into an intellectual and spiritual center of Islam.

11   After Idris II died, southern Morocco was dominated by the Almoravids, nomads without farming skills. For a century they imposed Islam on the black Saharan peoples. Ibn Tachafine founded Marrakech in 1062 and a large part of Spain joined the Almoravid Empire.

12   The 12th century is considered the golden age of Moroccan history, and coinciding with the emergence of the Almohad dynasty. Marrakech was spiritually led by Mohamed Ibn Toumart, founder of the Muwahiddin doctrine (utter union with God). Toumar, along with statesman Iacoub Al Mansour, led the country

LAND USE

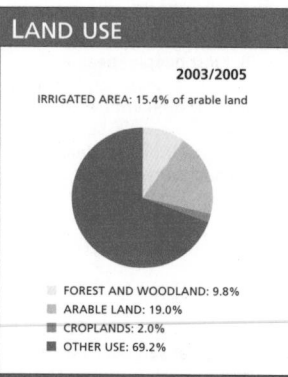

2003/2005

IRRIGATED AREA: 15.4% of arable land

- FOREST AND WOODLAND: 9.8%
- ARABLE LAND: 19.0%
- CROPLANDS: 2.0%
- OTHER USE: 69.2%

PROFILE

ENVIRONMENT
The country has an 800-km coastline. In the eastern part there are two mountain ranges (Atlas and Rif), covered with barren steppes and inhabited by nomadic Berbers. In the foothills lie irrigated lands where citrus fruit, vegetables and grain are cultivated. Stockbreeding is extensive on the western slopes of the Atlas Mountains (Grand Atlas and Anti-Atlas), which are rich in phosphate, zinc and lead deposits. Along the coastal plains, grapes and citrus fruit are grown. Fishing stocks are important, though mainly exploited by foreign fleets. Expansion of farms in marginal areas, over grazing of pastures by livestock, destruction of vegetation in the quest for firewood, and the conversion of forested areas into cultivated land are all factors that have led to soil erosion. Some efforts have been made to reverse this process, mostly through reforestation.

SOCIETY
Peoples: Arab 70 per cent; Berber 30 per cent.
Religions: Mainly Sunni Muslim (98.7 per cent). Christian (1.1 per cent) and Jewish (0.2 per cent).
Languages: Arab (official) and Berber variations. French and Spanish are also spoken.
Main Political Parties: Socialist Union of Popular Forces (progressive); Independence Party (*Istiqlal*), founded in 1943; Justice and Development Party (Islamic); National Rally of Independents (moderate).
Main Social Organizations: General Union of Moroccan Workers (UGTM); Moroccan Workers Union; Democratic Labor Confederation; Moroccan Employers Association; Organization of the Muslim Woman in Morocco; Islamic Educational Scientific and Cultural Organization (ISESCO).

THE STATE
Official Name: al-Mamlakah al-Maghribiyah.
Administrative Divisions: 37 provinces and two municipalities, Casablanca and Rabat.
Capital: Rabat 1,759,000 people (2003).
Other Cities: Casablanca (Dar-el-Beida) 3,292,100 people; Fès 900,900; Marrakech 736,500 (2000).
Government: Sayyidi Muhammad VI ibn al-Hasan, King since July 1999. Driss Jettou, Prime Minister since 9 October 2002. In September 1996, constitutional reform inaugurated a bicameral legislative regime. House of Deputies, *Majlis al-Nawwab*, with 325 members elected by direct vote for a 5-year term. The 270 members of the Senate, *Majlis al-Mustasharin*, an advisory body, are elected by in direct vote.
National Holiday: 2 March, Independence from France (1956).
Armed Forces: 196,000 (2003).

towards spiritual, intellectual and economic flourishing.

[13] Morocco was the cradle of the two North African empires that dominated the Iberian peninsula. It became one of the power centers of the region because of its geographic location: it was close to Spain, and at the northern end of the trans-Saharan trade routes. Although neither Fès nor Marrakech achieved the academic prestige of Cairo, their political influence was felt as far as Timbuktu and Valencia. Their close ties with Spain were culturally enriching during the Cordovan Caliphate, but they brought negative consequences to Morocco in the final stages of the 'Reconquest'. The war moved into Africa and the Spanish seized strongholds on the coast (Ceuta in 1415, Tangier in 1471, and Melilla in 1497). European naval dominance blocked Mediterranean and Atlantic routes to Morocco causing a decline in trade.

[14] Unlike Algeria and Tunisia, Morocco was not formally annexed to the Ottoman Empire, but it did benefit from the presence of Turkish corsairs who hampered Luso-Spanish expansion. This precarious balance allowed the sultans to remain autonomous until the 20th century. France's policy of economic penetration meant that France was supervising Moroccan finances, while arguing with Germany over who should have political sway over the area. The French finally won, securing agreements with Spain over the borders of the Spanish Sahara, and Sultan Muley Hafid ended his support of the Saharan rebels (see Western Sahara). In 1912, an agreement between France, Spain and Britain transformed Morocco into a French protectorate, giving Spain the Rif region, to the north (where Ceuta and Melilla are located), and the Ifni region to the south, near the Sahara. In exchange, Britain obtained French consent for its policies in Egypt and Sudan. The city of Tangier was declared an international free port and the sultan became a figurehead.

[15] The areas under Spanish control became sanctuaries for the nationalists unhappy with European domination. In 1921, it was on Spanish territory that the Berber revolt led by Emir Abdel Krim (Abd al-Karim al-Khattab) began. Backed by the Third International and the Pan-Islamic Movement, he proclaimed the Republic of the Confederated Tribes of the Rif, induced the inland peoples to rebel, and put Spain on the defensive. The French intervened causing the rebellion to spread throughout the entire

country. It took them until 1926 to force the Emir to surrender.

[16] In the south, Spanish rule was nominal, and French pressures to close down this 'sanctuary' for Algerian, Moroccan, Saharan and Mauritanian rebels failed (see Western Sahara).

[17] During World War II, there was sustained nationalist agitation. Demands for liberation were so pressing that Sultan Sid Muhammad Ben Youssef became the spokesperson for the cause. Growing tension led the French to depose the Sultan in 1953, but this only succeeded in making the nationalist movement more radical. The nationalists raised an army and fought until they achieved Muhammad's return to power as King Muhammad V. In 1956, the French were forced to acknowledge Morocco's independence.

[18] On 7 April 1956, Morocco

recovered Tangier, as well as the 'special zones' of Ceuta and Melilla, although the ports of these two cities still remain under Spanish control. Ifni was not returned to Morocco until 1969.

[19] The goal set by Muhammad V was 'to move forward slowly', gradually modernizing the country's economic and political structures. But his son, Hassan II, who succeeded him in 1961, had more conservative ideas. His family came from the lineage of Muhammad, the prophet, and his theocratic regime, based on a paternalistic system of favors and duties, prevented the development of authentic national commerce. The King also encouraged foreign investment, especially from France, to exploit the nation's natural resources.

[20] In 1965, Ben Barka, leader of the powerful National Union of Popular Forces (NUPF) was

assassinated on the orders of Hassan II. The NUPF worked for the social and economic welfare of workers and peasants.

[21] The death of Ben Barka in Paris was followed by a crackdown on popular organizations.

[22] In 1975, the conflicts underlying Moroccan society surfaced when King Hassan ordered the occupation of Spanish Sahara, as it was then called, unleashing a war that has brought about important political changes in North Africa.

[23] Funds for the military campaign, the fall in the price of phosphates on the international market, and the loss of financial aid from Saudi Arabia - in retaliation for Hassan's support of the Camp David agreements between Israel and Egypt - deepened the economic crisis.

[24] Severe drought in 1980 and 1981 drastically reduced food supplies, forcing the Government

IN FOCUS

ENVIRONMENTAL CHALLENGES
Soil erosion resulting from farming of marginal areas, destruction of vegetation and overgrazing have led to land degradation and desertification. Water supplies are contaminated by raw sewage, and pollution of coastal waters by oil spills.

WOMEN'S RIGHTS
Moroccan women have been able to vote and run for office since 1963, although it was not until 1993 that the first woman held a seat in Parliament. Thirty seats in the Chamber of Deputies are reserved for women under the new electoral code; thus, female representation grew from 1 per cent in 1995 to 10.8 per cent in 2003. That year, women held 65 per cent of positions at ministerial level.

The literacy rate among women over 15 amounted to only 38 per cent. In 2004, 68 per cent of pregnant women received prenatal care and 63 per cent of births were attended by skilled health staff. Maternal mortality stood at 220 per 100,000 live births*.

CHILDREN
Infant and under-five mortality rates were significantly reduced between 1990 and 2004. The former fell from 69 to 38 per 1,000 live births while the latter was reduced from 89 to 43 per 1,000. Eleven per cent of newborn babies were underweight and 24 per cent of children under-five were severely or moderately stunted. The practice of 'adoptive servitude',

in which urban families adopt rural girls to work as domestic servants (many are orphans; in other cases it is the girl's family who receives her salary), is socially accepted, and it is not regulated by the Government. A problem that has drawn recent attention is the situation of unaccompanied repatriated children. Following their deportation, mostly from Spain, they are subject to abuse on the streets. In December 2003, the Government signed a repatriation agreement with Spain, stating that Spain committed itself to helping to reunify children with their families and to provide education for them. In 2004, it was estimated that 11 per cent of children aged between 5 and 14 worked.

INDIGENOUS PEOPLES/ ETHNIC MINORITIES
Almost 60 per cent of the population regard themselves as being of Berber descent, including the royal family. Berber associations claim the Government does not defend their culture, since it refuses to register newborns with their traditional names, does not encourage public usage of their Tamazight language, limits the activities of members of these associations, and continues the Arabization of the names of cities, towns and geographic locations.

In September 2003, 317 primary and secondary schools started teaching the Berber language.

MIGRANTS/REFUGEES
Sixty-six per cent of the country's revenues come from remittances sent home by Moroccans living

abroad. In 2006, it was estimated that two million Moroccans lived in Western Europe, although the US and Canada were being increasingly chosen as destinations. From 1975, over 160,000 Saharawi have been living in refugee camps in the Algerian desert since the Moroccan invasion and occupation of Western Sahara. The refugees and Saharawis living in the occupied territory demand the right to self-determination but Morocco has consistently blocked a UN-administered referendum.

In late November 2003, Morocco set in motion an operation to expel thousands of undocumented sub-Saharans around Oujda, on the Algerian border. The border, closed since 1994, was crossed daily by smugglers of merchandise and people traffickers. In November 2005, 40 sub-Saharans with UNHCR documentation were hosted in the southern part of the country; another 45 were still waiting to be granted authorization to stay. Since 2000, a total of 265 people have been granted refugee status in Morocco.

DEATH PENALTY
The death penalty is applicable in the case of ordinary crimes. The last death sentence was handed down in August 2003, to four alleged terrorists.

* Latest data available in *The State of the World's Children* and *Childinfo* database, UNICEF, 2006.

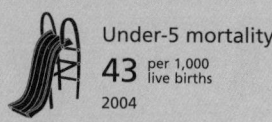
Under-5 mortality
43 per 1,000 live births
2004

Poverty
<2% of population living on less than $1 per day
1999

Debt service
14.0% exports of goods and services
2004

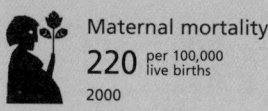
Maternal mortality
220 per 100,000 live births
2000

to increase food imports. This sent the country's foreign debt soaring to intolerable levels. The IMF assisted the monarchy with emergency loans, which carried a condition eliminating subsidies on food and housing, a measure that increased the hardships faced by the working classes. The Government failed to achieve its ambitious development aims, and the export of unemployed workers was limited by French immigration restrictions.

25   The situation worsened when several moderate opposition parties decided to break the tacit political truce. The Socialist Union of Popular Forces (USFP) staged anti-government demonstrations. In June 1981, Casablanca was the scene of bloody repression known as the 'Casablanca massacre', which led to an open conflict between the King and the leftist parties over the high cost of the Sahara war (over $1 million a day).

26   The stalemate on the battlefield in early 1983 made signs of dissent within the Moroccan armed forces visible. The tension became evident with the assassination of General Ahmed Dlimi, the supreme commander of the Royal Armed Forces.

27   In 1984, the Saharan Arab Democratic Republic (RASD), proclaimed by the Polisario Front's fighters in the former Spanish Saharan colony, was recognized as a full member of the Organization of African Unity (OAU). Morocco reacted by withdrawing from the pan-African organization.

28   In his capacity as religious leader, the Moroccan king became concerned about the rise of Islamic fundamentalism throughout the Arab world.

29   In 1987, the Moroccan monarch suggested to King Juan Carlos of Spain that both governments form a study group designed to consider the future of Ceuta and Melilla. The proposal was not well received in Spain, as there was an insistence on the 'historical nature' of Spain's presence in Ceuta and Melilla.

30   In May 1988, after 12 years of tension, Morocco and Algeria re-established diplomatic relations, through the mediation of Saudi Arabia and Tunisia. The cause of the disruption in their relations had been the war in Western Sahara, as Algeria had openly supported the Saharawi nationalists from the beginning. Better relations between Algeria and Morocco meant that a gas pipeline was built across the Strait of Gibraltar.

31   In Western Sahara, a UN peace plan announced in 1991 led to a ceasefire and plans for a referendum that would give RASD inhabitants the option for independence or integration with Morocco. The Moroccan Government had delayed the referendum, hoping to defeat the Polisario Front (see Western Sahara).

32   Torture and disappearances are common both in Western Sahara and Morocco. Nubier Amau, secretary general of the Democratic Confederation of Labor, was sentenced to two years' imprisonment, for slander against the regime. In February 1993, the Moroccan Human Rights Association announced the existence of 750 political prisoners.

33   The opposition won the first parliamentary elections following the 1992 reform, taking 99 of the 222 seats, while the ruling party took only 74.

34   Despite the constitutional reform, the King continued to dominate national politics, and in May 1994, he appointed one of his relations by marriage, Abd al-Latif Filali, as prime minister.

35   In early 1996, the Government announced it would submit proposals for constitutional reform to referendum. The changes, which basically aimed at the formation of a bicameral legislature, were approved in September. The King still had the right to dissolve the chambers. The privatization policies continued that year with the sale of several companies.

36   In September 1997 Morocco and the Polisario Front signed an agreement to re-launch a peace plan for Western Sahara, exchange prisoners, release political prisoners, allow refugees to return, and quarter troops. The long-postponed referendum on the status (independence or integration with Morocco) of the territory under dispute was announced.

37   In February 1998 King Hassan II appointed Abderrahmane El Youssoufi, leader of the Socialist Union of Popular Forces, as Prime Minister, and in March he appointed a whole new cabinet. The UN Secretary General proposed postponing the Western Sahara referendum - which had originally been scheduled for 1991 but was put off every time the date approached - until an indefinite date before 2002.

38   The death of Hassan II in July 1999 and the succession of his son Muhammad VI as king brought significant political changes. The first move of the new monarch was to free some 800 political prisoners. In a television address in August, he pledged to fight social inequalities, domestic violence, unemployment and rural emigration. The Polisario Front welcomed the King's first measures and his decision to go ahead with the Western Sahara self-determination referendum. In November Muhammad VI announced his decision to make some form of self-rule possible for the occupied zone.

39   That same month, the King dismissed Home Minister Driss Basri, who had served throughout King Hassan's reign of almost two decades. Muhammad VI announced the freeing of another 2,000 political prisoners as a goodwill gesture in January 2000, to celebrate the end of Ramadan. The Government's proposals to recognize some rights for women provoked demonstrations both for and against the measures during March 2000.

40   In May 2003 terrorist attacks in Casablanca left 45 dead. According to Government sources, the terrorists were members of the Sirat al-Mustaqim (part of he Salafiya-Jhadiya movement), but the alleged responsibility of al-Qaeda was not ruled out. The Moroccan parliament passed stringent anti-terrorism laws, which extend the definition of terrorism to cover any disturbance of the public order, as a consequence of the Casablanca incidents.

41   In February 2004 the Polisario Front unilaterally decided to release 100 Moroccan war prisoners as a 'humanitarian gesture' and in favor of peace. This new release raised to 1,743 the number of prisoners set free by Polisario since the year 2000.

42   On 24 February 2004 an earthquake measuring 6.5 on the Richter scale hit northeastern Morocco; its epicenter was located about 15 kms from Al Toxemias city. With an outcome of over 564 dead and 300 injured this was the worst earthquake in Morocco since the one that destroyed the city of Agadir (southwest) in 1960, leaving almost 12,000 dead.

43   In July, the first trade agreement between Morocco and the US came into effect. The deal eliminated more than 95 per cent of tariffs on consumer products and industrial goods. US farmers were expected to be among the biggest beneficiaries of the deal. That month, Morocco had hosted a major NATO military exercise involving naval and air forces, and had been granted US recognition for the country's support to the 'war on terror'.

44   In October 2004, Morocco agreed to the repatriation of 73 sub-Saharan immigrants living in Spain. King Juan Carlos of Spain had called Muhammad VI on three occasions to ask him for his 'help' with the massive assaults at the border fence between Morocco and the Spanish enclaves of Ceuta and Melilla. Meanwhile, the organization Médecins Sans Frontières reported that some 1,000 immigrants being transferred to southern Morocco were in urgent need of water, food and shelter. Morocco's Government did not authorize the EU technical mission to visit its side of the border with Ceuta and Melilla. At the same time, the Communication Minister and Government's Spokesman, Nabil Benabdellah, said that the country continued to contemplate 'a very broad autonomy (for Western Sahara) within the framework of Moroccan sovereignty and the respect to the territorial integrity of the Kingdom'.

45   Significant popular unrest in the occupied territory of Western Sahara from May 2005 onwards was termed by local activists their 'intifada' against Moroccan rule. The protests were ruthlessly suppressed - Saharawis are forbidden to fly their own flag or to refer to their own liberation movement, Polisario - and Saharawi human rights defenders were imprisoned and in some cases tortured. Fact-finding delegations from the Spanish Parliament were repeatedly refused access.

46   In August 2005 Polisario released all its remaining Moroccan prisoners of war. Some of the 404 prisoners had spent more than two decades detained in the Saharawi refugee camps in Algeria.

47   A truth commission - the first in the Arab world - set up to investigate human rights abuses during the rule of King Hassan II concluded in December 2005, after two years of investigation, that 592 people had been killed by the regime between 1956 and 1999.

48   In 2006, the Moroccan Government expected foreign investment to double. To hit the seven per cent economic growth-rate target set by the country, investments would have to exceed the $4.2 billion total expected for that year. Over the past decade, foreign investments had increased fivefold.

49   In April 2006, Muhammad VI ordered the release of 48 Saharawi activists arrested in 2005 for demanding the independence of Western Sahara. However, at the same time, he rejected UN-sponsored mediation in the conflict - especially the proposal by Secretary-General Kofi Annan involving a joint Government and a referendum on independence within five years.

50   In May 2006, 50 women were for the first time appointed as state preachers or *mourchidats* part of the Government's drive to promote a more tolerant version of Islam. The women will be able to give basic religious instruction in mosques. ■

Mozambique / Moçambique

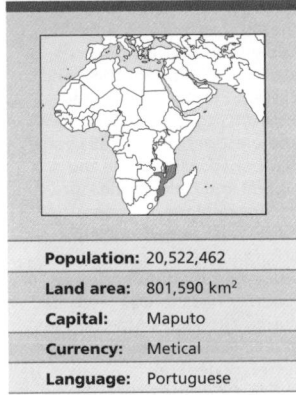

Population:	20,522,462
Land area:	801,590 km²
Capital:	Maputo
Currency:	Metical
Language:	Portuguese

A round the 3rd century AD, agricultural and cattle - herding communities moved into what is now Mozambique. Among these people were Bantu speakers from west-central Africa who introduced iron-making technologies and combined cultivation of some grains with knowledge of root and tree crops, providing them with sustenance and favoring their expansion along the Indian Ocean.

2   The city of Sofa (near present-day Beira) was founded by Shirazis towards the end of the 10th century. It became a point of contact between two of the most flourishing developed cultures in Africa: the commercial, Muslim cultures of the east coast and the metallurgical, animist culture of Zimbabwe. As with other civilizations on the continent, the Portuguese presence in Mozambique was fatal: they planned to seize control of the Eastern trade which had nourished the two civilizations for centuries. This led to the destruction of the ports and the stifling of Zimbabwean gold exports. The route to the gold mines was closed by the Changamiras of Zimbabwe. The Monomotapa or Mwene Mutapa - the title given from the 14th to the 17th centuries to a line of kings from an area of southeast Africa between what is now Zimbabwe and Mozambique, who were the leaders of the Karanga - declared allegiance to the Portuguese in 1629, following which they lost authority in the region.

3   When Zanzibar (now part of Tanzania) expelled the Portuguese from the territory under their control, the colonists turned to the slave trade as the only profitable option open to them. Attempts to connect Mozambique and Angola by land failed repeatedly and European control was confined to a coastal strip where their entire 'administration' was limited to granting *prazos* - concessions of huge areas of land - to Portuguese and Indian adventurers who plundered the land and enslaved the Africans. These *prazeiros* were given almost a free hand, but, in 1890, when the Portuguese had to prove their authority over the region, which the English challenged, threatening to occupy the territory, a long, hard struggle led to the forcible subjugation of the prazeiros. The conquest of the interior, however, was not completed until around 1920 when the ruler Mokombe, in the Tete region, was defeated.

4   Mozambique started to supply South African gold mines with migrant workers (up to one million every year) and its ports were open to South African and Rhodesian foreign trade. Portuguese colonialism controlled the country as an 'Overseas Province' and encouraged local group rivalries to prevent anti-European feelings from developing. Split into several movements, the nationalists staged strikes and demonstrations in their struggle for independence. In 1960, a spontaneous and peaceful demonstration in Mueda was fiercely repressed, leaving 500 people dead. This convinced many Mozambicans that peaceful negotiations with the colonial power were pointless.

5   In 1961, Eduardo Mondlane (a UN official) visited his home country and persuaded the pro-independence groups that they should unify. Finally, the Front for the Liberation of Mozambique (FRELIMO) was created in Tanzania in 1963. FRELIMO was made up of activists and organizations from all regions and ethnic groups in Mozambique. In 1964, FRELIMO, which had a high level of underground organizational and political activity, embarked on a guerrilla war to win 'complete and utter independence'. In 1965, FRELIMO controlled some areas of the country and by 1969, one-fifth of Mozambican territory was under their control. That year, Mondlane was assassinated by colonialist agents. Differences of opinion developed within FRELIMO about the desired form of independence; some wanting a mere 'Africanization' of the established system and others seeking to create a new popular democratic society. FRELIMO's Second Congress, held in the liberated areas, elected Samora Machel as president of the organization. Fighting intensified and spread to other areas.

6   The impossibility of winning the colonial wars in Africa led to the 1974 military uprising in Lisbon, ending the Salazar and Caetano regimes. A transitional government was established in Mozambique and in 1975, the People's Republic of Mozambique was founded. The first President of independent Mozambique and revolutionary leader, Samora Machel, announced that 'the struggle will continue' in solidarity with the freedom fighters in Zimbabwe and South Africa. The FRELIMO Government nationalized education, health care, foreign banks and several transnational corporations. Communal villages were promoted, bringing together the scattered rural population. Collective production methods were organized and health care, education and technical assistance were rationalized. In 1977, FRELIMO's Third Congress in Maputo adopted Marxist-Leninism as the Front's ideology.

7   Mozambique supported Zimbabwe's independence struggle, blockading imports and exports from Ian Smith's regime despite severe repercussions on the Mozambican economy. Zimbabwean freedom fighters were given permission to set up bases within Mozambican territory; the white minority regimes retaliated with air raids and invasions. Zimbabwe's independence in 1980 altered

LAND USE

2003/2005

IRRIGATED AREA: 2.6% of arable land

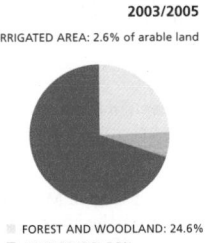

- FOREST AND WOODLAND: 24.6%
- ARABLE LAND: 5.5%
- CROPLANDS: 0.3%
- OTHER USE: 69.6%

PUBLIC EXPENDITURE

% OF GDP

40

30

20

10

0

4.1 % — HEALTH (2002)

1.2 % — DEFENSE (2004)

WORKERS

LABOR FORCE   **2004**

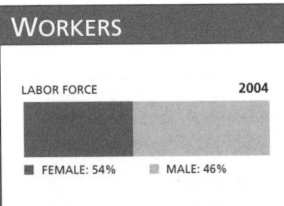

- FEMALE: 54%
- MALE: 46%

Life expectancy
42 years
2005-2010

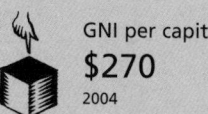
GNI per capita
$270
2004

Literacy
46% total adult rate
2000-2004

HIV prevalence rate
12.2% of population 15-49 years old
2003

the political outlook of the region, tightening the circle around apartheid and allowing Mozambique to revitalize its economy through greater integration with Zimbabwe, Malawi, Lesotho and Swaziland.
[8] In 1980, Machel launched a political campaign aimed at eliminating corruption, inefficiency and bureaucracy in state agencies and companies. A program for economic development was implemented, calling for investments in agriculture, transport and industry. All of these projects were affected by the increasing deterioration of relations with South Africa, which invaded Mozambique in 1981, attacking the Maputo suburb of Matola. They also backed the Movement of National Resistance (RENAMO), made up of former Salazar followers and mercenaries. South Africa attacked the anti-racist refugees living in Mozambique, while RENAMO aimed to sabotage economic objectives and intimidate the rural population. In 1982, the Government cracked down on the black market and launched a major

offensive against RENAMO in the Gorongoza region.
[9] In 1983, FRELIMO's Fourth Congress discussed major changes in the Government's economic program such as the reduction of large agricultural projects and the priority of granting concessions to minor investments. The delegates attending the Congress were mostly rural workers, while the number of women delegates had doubled since the 1977 Congress. The idea of creating small agricultural and industrial units was the result of a major reappraisal of all the large state-run farms. These were charged with excessive centralization, bureaucracy and economic inefficiency.
[10] In 1985, Mozambique suffered terrorist attacks by RENAMO and a severe drought which decimated cattle stock, causing a 70 per cent drop in production and reducing grain production by 25 per cent. Machel denounced South Africa's covert support for RENAMO, saying that it was a violation of the Nkomati agreements of March 1984, when the two countries had signed a non-

aggression treaty.
[11] Already tense economic and defense situations were compounded by the death of President Machel on 19 October 1986. His plane crashed while returning from a meeting of Presidents in Zambia. It has never been established whether the plane crash was an accident or an act of sabotage. Presidents Kenneth Kaunda (Zambia), Mobutu Sese Seko (Zaire, now DR Congo), José Eduardo Dos Santos (Angola) and Machel had debated joining forces to confront South Africa's aggression toward the independent countries of southern Africa and UNITA and RENAMO attacks in Angola and Mozambique. By the end of that year, FRELIMO's Central Committee had elected Joaquim Chissano (minister of Foreign Relations) as President and Commander-in-Chief of the Armed Forces.
[12] In 1987, the Mozambican Government reconsidered the economic strategy it had followed since independence. A more flexible foreign investment policy was adopted and local producers were encouraged to invest more. This was the first step toward the establishment of a mixed economy, a concept adopted by the FRELIMO Congress in 1989. The party dropped all references to its Marxist-Leninist orientation. Peace negotiations between RENAMO and the Government in Maputo began in 1990. These negotiations were made easier as the new Constitution admitted a multiparty system. The continuation of the single-party system had been one of the arguments used by the rebels to justify their terrorist activity.
[13] In 1991, the authorities of Manica province, one of the most fertile regions of the country, declared a state of emergency because of the drought which destroyed most of the crops. It was considered the worst drought for 40 years, causing enormous shortages for the 300,000 local inhabitants. That year, a peace protocol was signed by the Government of Mozambique and RENAMO in Rome. It foresaw the recognition of the rebel movement as a legal political party in a protocol that was considered the forerunner to a peace agreement. The refinancing of the $1.6 billion foreign debt was contingent upon the success of these accords. In addition to the promise to hold elections, this agreement introduced new laws regulating political parties and guaranteed the freedom of information, expression and association. Prime Minister Mario

da Graça Machungo explained that his country was suffering badly from the cessation of aid from the former USSR and Eastern European countries. RENAMO continued its actions and elections set for 1991 were postponed. Meanwhile the Liberal and Democratic Party of Mozambique, an opposition party, was created. A coup attempt by those opposed to peace negotiations ended in failure. Chissano was re-elected during FRELIMO's 6th Congress, and Feliciano Salamao was named secretary general.
[14] In 1992, the new political party regulations became one of the major hindrances to peace negotiations in Mozambique. Chissano offered RENAMO a special status, guaranteeing their members political rights but the rebels turned down the offer. The armed opposition also refused to accept the terms that established a minimum of a hundred registered members in each province, as well as in the capital, to qualify as a bona fide political party. That year, with Italy as mediator, Chissano and Alfonso Dhlakama (of RENAMO) signed a peace agreement in Rome, putting an end to 16 years of conflict which had caused over a million deaths and five million refugees. The terms of the agreement included the confinement of RENAMO and Government troops to pre-established areas, while weapons were to be turned over to UN soldiers charged with disarming both sides. The agreement also provided for the creation of an army of both Government and guerrilla forces. Also in 1992, differences between the two parties led to direct UN intervention in the elaboration of a new Peace Plan which included civilian observers and 7,500 peace-keeping troops
[15] In 1993, FRELIMO participated in joint military maneuvers with the US. This change of attitude towards the West favored Mozambique's request for foreign aid. The UN postponed elections until October 1994, hoping to restart the stalled peace process. After negotiations, RENAMO agreed to take part in the elections. Chissano was re-elected with over 53 per cent of the vote. In the parliamentary elections, FRELIMO won with 44.3 per cent, followed by RENAMO with 37.7 per cent. In 1995 the Paris Club promised to give Maputo $780 million for the country's reconstruction. The agricultural sector was devastated and fields were riddled with landmines.
[16] During 1996 the

PROFILE

ENVIRONMENT
The wide coastal plain, wider in the south, gradually rises to relatively low inland plateaus. The Tropic of Capricorn runs across the country and the climate is hot and dry. Two major rivers cross the country: the Zambezi in the center and the Limpopo in the south. Due to its geographic location, the country's ports are the natural ocean outlets for Malawi, Zimbabwe and part of South Africa. However, trade has been hampered by wars during the past two decades. Mineral resources are important though scarcely exploited.

SOCIETY
Peoples: The Mozambican population is made up of a variety of ethnic groups, mainly of Bantu origin. The main groups are: Makua 47.3 per cent; Tsonga 23.3 per cent; Malawi 12 per cent; Shona 11.3 per cent; Swahili 9.8 per cent; Yao 3.8 per cent; Makonde 0.6 per cent.
Religions: There is no official religion. The rural population practise traditional religions. Most of the urban population is Christian or Muslim. Islam prevails in the north.
Languages: Portuguese (official). Most of the population speaks Bantu languages, the main ones being Swahili and Macoa-Lomne.
Main Political Parties: Mozambique Liberation Front (FRELIMO), National Resistance of Mozambique (RENAMO); Party for Peace, Democracy and Development.
Main Social Organizations: Organization of Mozambican Women; Mozambican Youth Organization;

THE STATE
Official Name: República Popular de Moçambique.
Administrative Divisions: 10 provinces.
Capital: Maputo 1,221,000 (2003).
Other Cities: Matola 467,200 people; Beira 437,100; Nampula 333,700 (2000).
Government: Armando Gueguza, President since February 2005. Luisa Diogo, Prime Minister since February 2004. The Assembly of the Republic is the main political body.
National Holiday: 25 June, Independence Day (1975).
Armed Forces: 10,000 (2003).

Under-5 mortality
152 per 1,000 live births
2004

Poverty
37.8% of population living on less than $1 per day
1996

Debt service
4.5% exports of goods and services
2004

Maternal mortality
1000 per 100,000 live births
2000

Government's privatization program continued, with more than 900 of the 1,200 State companies being sold. Inflation fell to 5.8 per cent, the lowest figure since the World Bank and IMF began monitoring the national economy.

[17] In 1998 the establishment of trade relations with South Africa fostered economic growth, which at a rate of 11 per cent in 1999 was amongst the highest in the world. FRELIMO won that year's elections to the legislature and Chissano was re-elected President, but RENAMO claimed that the vote was fraudulent. In 2000 more than 40 people died during demonstrations staged by RENAMO to protest against the manipulation of electoral registers in the previous year's elections. Meanwhile, international observers declared that the elections had been fair and free.

[18] In 2000, several of the country's largest creditors approved a moratorium on Mozambique's debt repayment. In the same year floods devastated the country. Over a million people were displaced and more than 200 died in this catastrophe. In a gesture of solidarity, Germany proposed writing off Mozambique's debt of almost $1.5 billion. Journalist Carlos Cardoso, editor of the independent daily *Metical*, who was investigating political and corporate corruption, was murdered in 2000. Shortly after, six people - including two high level businesspeople - were formally accused of the murder.

[19] In 2002 FRELIMO elected Armando Guebuza, an independent, as their candidate for the 2004 presidential elections. Chissano declined to stand for a third term. In the same year two people accused of murdering journalist Carlos Cardoso in 2000, alleged during their trial that Nymphine Chissano, son of President Chissano, was linked to the crime.

[20] In February 2004 a Lutheran missionary, Doraci Edinger, was murdered in Nampula, northern Mozambique. She had previously been threatened after exposing, at the end of 2003, a network trafficking human organs in the region.

[21] In October a vaccination against malaria was presented by scientists and pharmaceutical companies after carrying out trials involving almost 1,600 children in Mozambique. According to a statement issued by the medical team these trials confirmed 'the safety of the vaccination for children aged one to four'.

IN FOCUS

ENVIRONMENTAL CHALLENGES
The war devastated the country's entire productive system, especially in the agricultural sector. The use of its mangrove forests for firewood has led to deforestation. Recurrent droughts in the hinterlands have prompted urban migration to coastal areas with adverse environmental consequences for cities there, which are now overcrowded. Another problem in recent years has been the illegal hunting of elephants for the black market trade in ivory.

WOMEN'S RIGHTS
Women have been able to vote and stand for office since 1975.

In 2004 women held 35 per cent of seats in Parliament and 13 per cent of ministerial positions.

In 2003 they made up 49 per cent of the country's ten million strong labor force. In 2004, 85 per cent of pregnant women received prenatal care but only 48 per cent of births were attended by qualified personnel. Maternal mortality was 1,000 per 100,000 live births. By the end of 2003 there were 670,000 women aged 15 to 49 living with HIV/AIDS.

Although there are no official statistics, according to the public health authorities and women's rights advocacy organizations, domestic violence against women - particularly rapes and beatings - is widespread. Many women believe that their husbands have the right to beat them and cultural pressures discourage them from taking legal action against abusive husbands.

CHILDREN
In spite of a significant fall in infant and under-five mortality rates during the period 1990-2005, they continued to be amongst the highest in the world. The former fell from 158 to 104 per thousand live births and the latter from 235 to 152 per thousand live births. Some 15 per cent of babies were born underweight and 41 per cent of under-fives measured moderately or severely below height norms. By the end of 2003 there were 99,000 children aged under 14 living with HIV/AIDS*.

INDIGENOUS PEOPLES/ ETHNIC MINORITIES
There are at least 10 ethnic groups in the country. Half of them make up more than 5% of the population, the rest are smaller minorities. Among the main ones are: the *Makua* (5.3 million people) – a *Bantu* people from the Great Lakes region; the *Tsonga*, who also are part of the *Tswa*, pastoralists; the *Shangaan*, who migrated from Central Africa some 500 years ago and the *Ronga*; the *Lomwe* in the northeast; the *Chwabo*, another Bantu group and the *Shona*, descending from the 'Great Zimbabwe' culture that flourished toward the 7th century. Among the minority groups are the Nyanja, a people formed with the migratory flow that went in the 14th-15th centuries from Congo to Malawi; the *Makonde* a very old group that historically never had a central authority; the *Swahili*, a linguistic community whose origin is not clear (some are considered descendants

from the Persian Empire, others from Arabs, Bantu or *Yemi*); the *Lemba*, who consider themselves to have a Jewish Semitic background, although this origin is not clear; and finally, the Zulu, descending from the Zulu patriarch of the *Nguni* people that lived in the central area of the Congo river toward the 16th century.

MIGRANTS/REFUGEES
Although reliable data on immigration into the country is not available, it is estimated that in 2005 approximately 6,000 refugees arrived from the Great Lakes region and there were around 4,000 asylum requests pending. In 2006 UNHCR was encouraging Mozambique to find a permanent solution to the situation of refugees for whom a return to their country of origin was not a viable option, principally those from DR Congo and Burundi, who made up 6,000 of the 7,100 refugees living in Maratane camp, in the north of the country. Since the establishment of this camp ethnic tensions have provoked severe disturbances, either among the inhabitants themselves or between refugees and local police.

DEATH PENALTY
Abolished in 1990. The last execution was carried out in 1986.

*Latest data available in *The State of the World's Children* and *Childinfo* database, UNICEF, 2006.*

[22] The FRELIMO candidate, Guebuza, won the December presidential elections with 64 per cent of the vote. However, RENAMO claimed that the elections had been fraudulent. According to the European Union Observation Mission (EOM) chief, Javier Poms, the election process was riddled with irregularities but not to a degree that would reverse the result. RENAMO's election office chief, Antonio Namburete, refused to concede the election. Chissano exhorted the opposition to occupy their parliamentary seats with dignity and contribute to the country's development.

[23] President Guebuza's new government took office in February 2005. In March, Prime Minister Luisa Diogo said in London that the Commission for Africa - convened by British Prime Minister Tony Blair to report to the G8 summit in Gleneagles, Scotland, in July 2005 - was not equipped for precise analysis of the continent and that the working group's paper was 'somewhat general because it is difficult to incorporate all aspects and particularities of each country in a report of this type'.

[24] In October, the construction of a bridge over the River Ruvuma to link Mozambique with Tanzania began. In December, more than five years after the crime, the accused murderer of journalist Carlos Cardoso, Anibal 'Anibalzinho' dos Santos, appeared in court after two previous postponements of the trial due to his mysterious escapes from prison. In January 2006 Anibalzinho was sentenced to 30 years in jail.

[25] In February 2006 the IMF raised questions about Mozambique's megaprojects such as MOZAL (an aluminum company), HCB (one of the biggest hydroelectric dams in Africa) and Sasol oil company. According to the IMF, these undertakings should be strictly controlled to avoid non-transparent procedures in returns distribution (a euphemism for corruption). It also concluded that as they were located in zones with special tax regimes they tended to provide very little benefit for the host country.

[26] In October 2006, donors again pledged massive support to Mozambique's state budget to the tune of $583 million for 2007 but emphasized that lack of progress in implementing the anti-corruption strategy remains a serious concern. ∎

Myanmar (Burma) / Myanma Naingngandaw

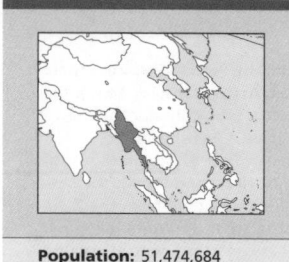

Population:	51,474,684
Land area:	676,580 km²
Capital:	Rangoon (Yangôn)
Currency:	Kyat
Language:	Burmese

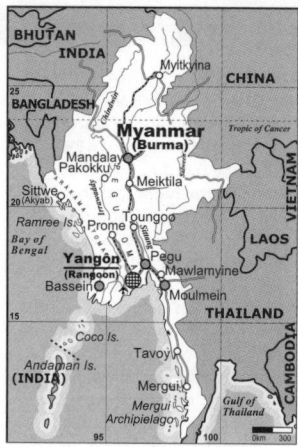

The first evidence of the existence of Burmese civilization dates from the 11th century when the Burmese established the state of Pagan.

2   The following two centuries represented the 'golden age' in Burmese thought and architecture. The Mongols attacked from the north, with aid from the Great Khan in Beijing. In 1283, the Mongol invasion ended the Pagan state and the Mongols remained in power until 1301.

3   Burma remained divided into small ethnic states until the 16th century when Toungoo local leaders reunified the territory. The second of these rulers, Bayinnaung, extended his domain to parts of present-day Laos and Thailand.

4   In 1740, a Toungoo ruler again achieved unification, with the help of the British. But when his successors continued the project of national reconstruction, they clashed with British interests in Assam, India, and a confrontation resulted with their former European allies. The Burmese fought three wars against the British throughout the 19th century. During the last war (1885-1886), Burma was annexed to the British viceroyalty of India and remained a part of it until 1937.

5   The 1930s began with a rising tide of nationalist movements; that of the Buddhist monk, U Ottama, inspired by Gandhi; Saya San's attempt to restore the monarchy; and uprisings organized by the University of Rangoon, bringing together Buddhists and Marxists.

6   Heavy taxes and the collapse of the world rice market in 1930 forced thousands of small farmers into debt and ruin. Discontent led to popular rebellions in 1938 and 1939.

7   When World War II broke out, a group of militant anti-colonialists, known as 'the 30 comrades', led by Aung San, formed the Burma Independence Army (BIA). They joined the Japanese against the British and invaded the capital in 1943. Minority groups of Karen, Kachin and Chin organized guerrilla groups, supported by the British, to combat both the BIA and the Japanese.

8   The Japanese granted Burma independence on 1 August 1943, but soon after friction developed between the Japanese and the socialist wing of 'the 30 comrades'. On 27 March 1945, the BIA declared war on Japan and was recognized by the British as the Patriotic Burmese Forces. On 30 May, they captured Rangoon, this time with the help of the British. Aung San was elected president in 1946. He organized a transition government, and in 1947, a constitution was drafted. On 19 July, a military commando assassinated Aung San and several aides in the palace, and U Nu stepped in as premier. On 4 January 1948, independence was proclaimed.

9   Over the next period the new government had to face several challenges: the rebellion of ethnic minorities; the presence of Chinese Kuomintang forces, and the armed insurrection of the People's Volunteer Organization, linked to the Communist Party.

10   In 1962, General Ne Win ousted the government in a coup, installing a military revolutionary council. He nationalized the banks, the rice industry (which accounted for 70 per cent of foreign earnings), and trade.

11   In 1972, a new constitution confirmed the ruling Burma Socialist Program Party (BSSP) as the only legal political organization.

12   The socioeconomic crisis unleashed protests and demands for democratization in 1987. A BSSP Congress appointed Sein Lwin as head of state, which triggered new protests that were repressed by the government. Lwin was forced to resign 17 days after he had taken office. His successor, Maung Maung, liberalized the regime.

13   After a coup in 1988 the military formed the State Law and Order Restoration Council (SLORC). They promised to hold free elections but in the meantime announced a state of emergency and suspended the constitution. They also changed the country's name to Union of Myanmar - Burma only refers to the majority ethnic group - and dropped the term 'Socialist'.

14   The National League for Democracy (NLD) won 80 per cent of the vote in the 1990 elections, while the ruling National Unity Party (ex-BSPP) retained only 10 of the 485 seats. The election outcome was disregarded by the Government, which banned opposition activities, imprisoned or banished its leaders, and harshly cracked down on street demonstrations.

15   In July 1989, the leader of the NLD, Aung San Suu Kyi, Nobel Peace Prize winner and the daughter of anti-colonial hero Aung San, was sentenced to house arrest.

16   The opposition was strengthened by agreements between students, Buddhist monks and some minorities. All opposition parties were dissolved or banned.

17   In April 1992, General Than Shwe took power, releasing 200 dissidents and permitting 31 universities and schools to reopen. In September, martial law was suspended.

18   In January 1993, the military invoked a National Convention to draw up a new Constitution.

19   An article of the new 1994 Constitution stipulated that presidential candidates could neither be married to foreigners nor bear children under foreign citizenship and they should have been residing in Myanmar for the last 20 consecutive years. The regulation was custom-made for Suu Kyi who was married to a British citizen and had lived abroad for several years.

20   In July 1995, Suu Kyi was released from house arrest and called on the SLORC to hold a dialogue. The SLORC refused, jailed dozens of dissidents and maintained the ban on political debate.

21   The fall in early 1996 of Manerplaw, headquarters of the rebel minority, was a major blow to the opposition because it was also a base for activities of other groups. In January, through a secret accord, the Government gained the surrender of Khun Sa, known as the 'opium king'. Later that year the regime also approved a law banning NLD political meetings and restricted Suu Kyi's freedom of movement.

22   Under growing international pressure, mainly from the US and

PROFILE

ENVIRONMENT

The country lies between the Tibetan plateau and the Malayan peninsula. Mountain ranges to the east, north and west surround a central valley where the Irrawaddy, Sittang and Salween rivers flow. Most of the population is concentrated in this area, where rice is grown. The climate is tropical with monsoon rains between May and October. Rainforest covers most of the country.

SOCIETY

Peoples: Burmese 69 per cent; Shan 8.5 per cent; Karen 6.2 per cent; Rakhine 4.5 per cent; Mon 2.4 per cent; Chin 2.2 per cent; Kachin 1.4 per cent.
Religions: Buddhist 89.1 per cent, Christian 4-9 per cent, Muslim 3.8 per cent; other 2.2 per cent.
Languages: Burmese (official) and the languages of the minority ethnic groups.
Main Political Parties: Political parties and their activities are generally illegal. The National League for Democracy (NLD) is the main opposition force. The military created the dictatorial National Unity Party (Taingyintha Silonenyinyutye) before the 1990 elections; the Government of the National Coalition of Burmese Unity is made up of parties that won the last elections but were not allowed to exercise power; it currently exists as a government in exile and as a leading opposition force. Karenni National Progressive Party; Arakan League for Democracy.
Main Social Organizations: All Burma-Students Democratic Front; All Burma-Young Monks' Union; Federation of Trade Unions - Burma (FTUB); The Federation of Trade Unions-Kawthoolei (FTUK).

THE STATE

Official Name: Pyidaungzu Myanma Naingngandaw.
Capital: Rangoon (Yangôn) 3,874,000 people (2003). The seat of government is in the process of being moved to the central town of Pyinmana.
Other Cities: Mandalay 1,037,300 people; Mawlamyine 360,400; Pegu 223,700; Akyab 161,200 (2000).
Government: General Than Shwe, President, Head of State (military junta - State Peace and Development Council) since April 1992. Soe Win, Prime Minister since October 2004. Legislature: suspended since 1988.
National Holiday: 4 January, Independence Day (1948).
Armed Forces: 378,000 (2003).

Life expectancy
62 years
2005-2010

Literacy
90% total adult rate
2000-2004

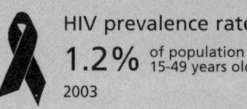

HIV prevalence rate
1.2% of population 15-49 years old
2003

---

the EU, the SLORC authorized the NLD to hold its first congress after seven years, although only half of its delegates were allowed to participate. By the end of the year, the military junta dissolved itself, appointing a State Peace and Development Council (SPDC) - though this was a cosmetic change that failed to disguise continuing military rule. In July, under the influence of neighboring countries India and China, Myanmar had been admitted as a member of the Association of Southeast Asian Nations (ASEAN).

[23] Intelligence chief Khin Nyunt visited Thailand in September 2001, in what was declared a sign of improvement in bilateral relations. Earlier that year, troops from Thailand and Myanmar had clashed along the common border, with Bangkok and Rangoon accusing each other of supporting the narcotics-producing Shan militias. In May the tensions had reduced due to the growing presence of special US troops on the Thai side of the border.

[24] In May 2002, after 19 months of house arrest, Suu Kyi was released and her right to participate in political activity was renewed, in what was seen as the beginning of a democratic transition process. The UN was an active mediator and proposed the immediate release of all prisoners of conscience, estimated at 1,500 to 2,000.

[25] In May 2003, members of the NLD who were travelling on party business in Upper Myanmar were attacked by pro-junta supporters. At least four people died and the exact whereabouts of over 100 detained people, including Suu Kyi, was at first unknown. Detentions continued throughout the year.

[26] After having been appointed Prime Minister in August 2003, Khin Nyunt promoted a reform of the constitution, of a more liberal nature, and the resumption of the National Convention, suspended since 1996. He also proposed to develop a 'road map' to democracy.

[27] From the 1988 uprising until 2003, military authorities have signed ceasefire agreements with more than 20 rebel ethnic armies, which include children fighters.-

[28] In January 2004, the Government and the Karen National Union (a guerrilla group that represents the Karen ethnic group) reached an agreement to put an end to hostilities. In April, Suu Kyi was still in 'protective custody'. After a 9-year suspension, the National Convention reconvened in May, in spite of the NLD's boycott.

[29] The US stated that Myanmar's political actions were hostile to its interests and that the Than Shwe Government threatened

IN FOCUS

ENVIRONMENTAL CHALLENGES
Deforestation has been responsible for the destruction of two-thirds of the country's tropical forest. Air, soil and water pollution is caused by industrial activity and inadequate solid waste disposal. The lack of sewage treatment, at both the industrial and domestic levels, poses a serious problem.

WOMEN'S RIGHTS
Women have been able to vote since 1935 and to stand for office since 1946 but there have been no elections since 1990 when the military refused to hand over power. There are no women holding senior political positions.

In 2003, women made up 44 per cent of the country's labor force of 27 million. In 2004, 76 per cent of pregnant women received prenatal care and 57 per cent of deliveries were assisted by qualified staff. Maternal mortality stood at 360 deaths for every 100,000 live births.

There are no independent organizations advocating for women's rights. Women have no representation at the governmental level, they are excluded from the army and there are no women in the courts, government ministries or councils. The most prominent woman in politics - in this case in the opposition - is Daw Aung San Suu Kyi, leader of the National League for Democracy, who was awarded the Nobel Peace Prize in 1991. In 2003, she was imprisoned and held incommunicado, like many

other pro-democracy activists who are subjected to this type of punishment by the government. The atrocities committed by the army against Karen and Shan women include tortures and gang-rapes, many times in front of their own husbands.

CHILDREN
Although Myanmar was not affected by the December 2004 tsunami in all of its territory, villages close to the coast were devastated, with the ensuing loss of lives, homes and education infrastructure. One year after the tsunami, thanks to international aid, schools were operational and children had the necessary materials to attend school. The experience left a very serious psychological impact on children.

Children, mainly those belonging to ethnic or religious minority groups, are used as porters by the army, and are forced to work from the age of 11. Rape of children by soldiers has been condemned.

Infant and under-five mortality rates have fallen between 1990 and 2004 but continue to be high. The former went from 91 to 76 per 1,000 live births and the latter from 130 to 106 in 1,000. Approximately 15 per cent of all children are born underweight, and 32 per cent of children under five are moderately or seriously stunted.

INDIGENOUS PEOPLES/ ETHNIC MINORITIES
The Kachin, Mon, Rohingya (Muslims) and Zomi are indigenous peoples that inhabit different regions of the country. There

are two ethnic-nationalist minorities, the Karen and the Shan, which are the biggest minority groups. With regard to religion, the Muslim minority is under close surveillance by the Government. Buddhism is the main religion.

The Karen and Shan are the minority groups that suffer the most attacks by the army; they are the victims of rape, torture, extortion and murder. Soldiers act with impunity and the Government neither takes steps to prevent abuses nor investigates complaints. Some minority groups, such as the Karen, have formed self-defense militias which occasionally aggravate the conflicts in which the civil population is caught up.

MIGRANTS/REFUGEES
The country's political situation makes it difficult for organizations like UNHCR to protect or even obtain reliable information on the situation of refugees in Myanmar. However, efforts are being made to achieve basic rights for the few refugees settled in Bangladesh and Thailand. In early 2006 hundreds of thousands of Burmese were internally displaced. That year, thousands of illegal Burmese immigrants were expelled from Thailand.

DEATH PENALTY
Myanmar maintains capital punishment even for ordinary crimes.

* Latest data available in *The State of the World's Children* and *Childinfo* database, UNICEF, 2006.

US national security, something Rangoon firmly denied. 'There are no weapons of mass destruction, terrorist organizations, missile development programs, expansionist ambitions, nor animosity toward the US,' stated Myanmar. 'The recent results in Afghanistan and Iraq are a classic example of how wrong interventions in the political history, culture and security of peoples can be when a foreign country tries to install democracy by external force,' it added. The military government said that, in spite of the NLD's absence, the Convention would legislate regarding the elections and that the new Constitution would allow Myanmar to turn into a stable and democratic nation.

[30] In August 2004 British companies Rolls-Royce and Lloyd's, along with 37 other transnationals, started a year-long campaign denouncing the 'dirty list' of 95 companies that financially support

the military regime in Myanmar, accused of systematic abuse of human rights. The campaign described the regime as one of the most brutal in the world. In October, Khin Nyunt left his post as Prime Minister due to health reasons, and was replaced by Soe Win. Nyunt was placed under house arrest.

[31] The tsunami that swept South Asia in December 2004 left more than 50 dead and 20 missing in Myanmar, although several aid organizations feared that the number of victims was much higher.

[32] In March 2005, the International Labour Organization (ILO), having estimated that there were some 800,000 forced labor victims in Myanmar, recommended the imposition of sanctions. The ILO decided to ask 'governments, employers and workers, as well as international agencies' to review their relations with Myanmar and adopt 'necessary' actions to

sanction the regime over its labor rights violations. A month earlier, Than Shwe had avoided a meeting with a high-level ILO mission visiting Rangoon.

[33] Rangoon lamented that the ILO was used by some countries to put pressure on the Myanmar regime and asserted that it was making every effort to put an end to forced labor. The army - the second-largest in Asia, with almost 400,000 troops - was the largest employer of forced labor.

[34] ASEAN rejected, in July 2005, Myanmar's candidacy to chair the group in 2006.

[35] In May 2006, the military regime, which in November 2005 had decided to move the seat of government to the central city of Pyinmana, extended Suu Kyi's house arrest, in spite of international pressure and the call of then UN Secretary General Kofi Annan for her freedom. ∎

Namibia / Namibia

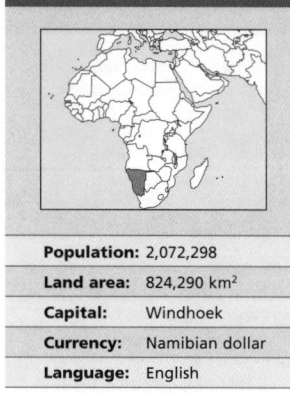

Population:	2,072,298
Land area:	824,290 km²
Capital:	Windhoek
Currency:	Namibian dollar
Language:	English

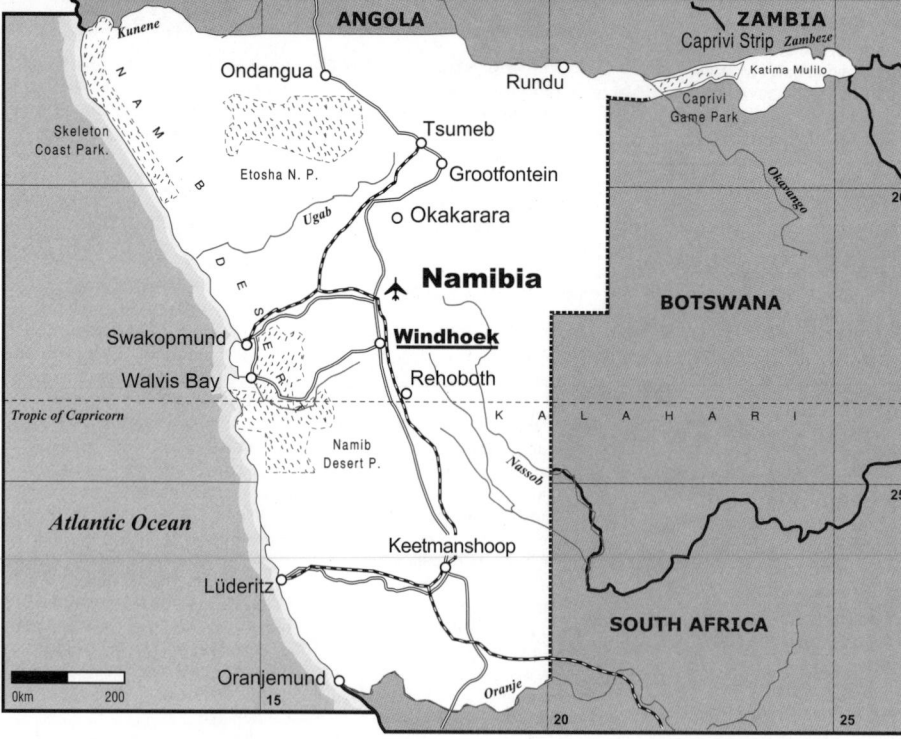

Southwestern Africa was occupied first by San (Bushmen) and Khoikhoi peoples, Ovambo in the north and the Bantu-speaking Herero. Later the Nama, known as the Red Nation, occupied lands in the south, maintaining close-knit ties between their clans and living from animal-herding. Closely linked to these peoples were the Damara, from Central Africa, who combined herding with hunting and copper-smelting. In the central and north-eastern regions, the Herero built clan alliances, often led by a supreme chief. The Ovambo, farmers who developed several kingdoms along the Kunene river, posed a constant threat to the unity of the Herero.

[2] Arid and scarcely populated, the Kalahari desert did not attract European colonization. In 1670, German missionaries, European traders and Norwegian whalers began to arrive on the coast and some ventured into the hinterland. Trade conflicts between the various clans became increasingly violent and facilitated conquest by the Oorlam-Nama, a group from the Cape that arrived in the early 19th century. Their military technology, which was modelled on that of the Afrikaners and included horses, rifles and an army of small commandos, allowed them to dominate the Nama and the Damara. Halfway through the century, a kingdom was set up by Chief Jonker Afrikaner, of Oorlam and Herero lineage, who had the full backing of the Damara and the Red Nation.

[3] In 1870, the British made a treaty with the Herero and flew their flag on the port of Walvis Bay. However, the German Second Reich of Bismarck and William I, occupied the area through dubious treaties, plunder and the strategy of dividing the African peoples. In 1884, Germany annexed the territory under the name of South West Africa. In 1885, Herero resistance forced the Germans back to Walvis Bay, until British support arrived.

[4] At the end of the 19th century, German colonists built railways from Swakopmund and Lüderitz. The country acquired strategic value for the Europeans with the discovery of large reserves of iron, lead, copper and diamonds, which were later augmented by metals of military interest: manganese, tungsten, vanadium, cadmium and great quantities of uranium.

[5] The great resistance led by the Herero, from 1904-1907, could barely be stopped by mass hangings and enforced detention in concentration camps, which led to a 90 per cent reduction in Herero numbers by 1910. The Nama, who had been slow to enter the conflict, were defeated in 1907. Only a third of the Nama population survived, confined to concentration camps. The rigid German control did not extend northwards significantly to trouble the Ovambo.

[6] During World War I, the British invaded the colony and, by the time the conflict ended, the region became a League of Nations trust territory. Administration of the territory, known as South West Africa, was assigned to the Union of South Africa.

[7] Over the years, the South African Boers and German colonists in Namibia overcame the initial resistance.

[8] In 1947, South Africa formally announced to the UN its intention to annex the territory. The UN, which had inherited responsibility for League of Nations trust territories, opposed the plan, arguing that 'the African inhabitants of South West Africa have still not achieved political independence'. Until 1961, the UN insisted on this point. Year after year it was systematically ignored by South Africa's racist regime.

[9] Between 1961 and 1968, the UN tried to establish the country's independence. Legal pressure was ineffective and the Namibian people, led by the South West Africa People's Organization (SWAPO), embarked on an armed struggle on 26 August 1966.

[10] In 1968, the UN finally proclaimed that South African occupation of the country now known as Namibia was illegal. A UN Council was set up as legal representative of the territory until sovereignty could be freely exercised by the people. However,

LAND USE

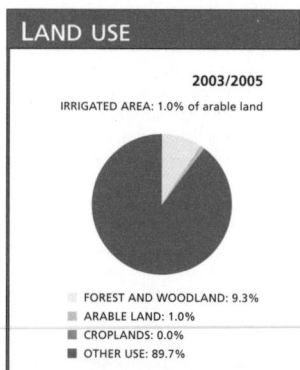

2003/2005

IRRIGATED AREA: 1.0% of arable land

- FOREST AND WOODLAND: 9.3%
- ARABLE LAND: 1.0%
- CROPLANDS: 0.0%
- OTHER USE: 89.7%

PUBLIC EXPENDITURE

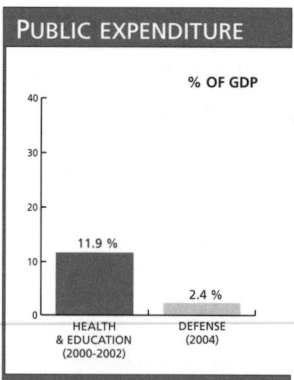

% OF GDP

11.9 % — HEALTH & EDUCATION (2000-2002)

2.4 % — DEFENSE (2004)

WORKERS

UNEMPLOYMENT: 31% (2004)

LABOR FORCE **2004**

- FEMALE: 44%
- MALE: 56%

EMPLOYMENT DISTRIBUTION **1995/2002**

- AGRICULTURE F: 29% M: 33%
- INDUSTRY F: 7% M: 17%
- SERVICES F: 63% M: 49%

Life expectancy
46 years
2005-2010

GNI per capita
$2,380
2004

Literacy
85% total adult rate
2000-2004

HIV prevalence rate
21.3% of population 15-49 years old
2003

attempts by most members of the UN General Assembly to follow this condemnation with economic sanctions systematically came up against the veto of the Western powers.

[11] Angolan independence, declared in 1975, affected Namibia's struggle for freedom, by providing SWAPO guerrillas with a friendly rearguard. The guerrilla war intensified and the Western powers started to put pressure on Pretoria to seek a 'moderate' solution.

[12] In December 1978, South Africa held elections in Namibia without UN observers and with no SWAPO participation, leading to a total lack of credibility in the results.

[13] The economy had grown in a sustained manner since World War II, reaching a peak of $1,000 per head in the 1970s ($20,000 for Europeans and $150 for black Namibians). Even while incomes in the white enclave soared, the salary of black workers was barely enough to live on. It took until halfway through this decade for workers to be trained to any significant degree in qualified tasks.

[14] The South African administration in Windhoek operated along customary colonial lines of dependence. The country exported corn, meat, fish, minerals and raw materials, while due to the lack of industries, all consumer products that were needed, such as wheat, rice and manufactured goods, had to be imported. Ninety per cent of goods in Namibia came from South Africa.

[15] The *status quo* preserved internal inequalities. Around 90 per cent of the population received 18.8 per cent of the GDP, while the rest, of European origin, received 81.2 per cent. Until independence, three-quarters of agricultural production was in the hands of white farmers. Although average per capita income was one of the highest in Africa - $1,410 - behind this figure were huge inequalities: while whites earned an average of $1,880, the rest of the population did not earn more than $108.

[16] Apart from boycotting the efforts to negotiate Namibia's independence, South Africa upped its military budget by 30 per cent for the conflict against the Namibian People's Liberation Army, SWAPO's armed wing. In the last months of 1982 and early 1983, the war intensified along the border with Angola and southwards, where SWAPO forces launched audacious attacks, even in the so-called 'iron triangle', located near the city of Grootfontein, where the main South African military units were concentrated.

[17] In February 1984, representatives of Angola and South Africa met in Zambia's capital, Lusaka, for peace negotiations. They agreed a planned withdrawal of South African troops from southern Angola, in return for a ceasefire. In May, delegates from SWAPO and other Namibian parties met a South African representative but negotiations failed. The South African withdrawal from Angola did not occur within the deadline and the situation was deadlocked again.

[18] In December 1988, after prolonged US-mediated negotiations, South Africa, Angola and Cuba reached an agreement whereby South African troops would leave Namibia and, at the same time, the Cubans would withdraw their 50,000 soldiers from Angola.

[19] In November 1989, the First Constituent Assembly was elected. Ten political parties stood in the UN-supervised elections. SWAPO won with 60 per cent of the poll, gaining control of the Assembly and appointing the 60-year-old leader Sam Nujoma as first President of Namibia.

[20] Independence was proclaimed on 21 March 1990. The presidential guard of honor was made up of troops from SWAPO and the SWA Territory Force which had protected the South African regime; both were to be integrated into the new National Army.

[21] The new government had to face up to the inequalities inherited from the South African apartheid system, above all in health and education.

[22] Namibia adopted English as its official language, replacing Afrikaans. The Government organized a Department of Informal Education, which together with UNICEF started to teach women to read and write, first in their own languages and then in English. A rehabilitation plan was developed for some 40,000 disabled people, mostly victims of 23 years of pro-independence guerrilla warfare.

[23] Prepared to play an important role in Southern Africa, Nujoma donated around $3 million in 1991 to the African National Congress (ANC), a sum which was handed over to ANC leader, Nelson Mandela, on a visit he made to Namibia on 31 January.

[24] In 1992 Namibia and South Africa agreed to return the port of Walvis Bay to Windhoek in 1994. One year later the National Council, the upper chamber of Parliament, came into operation. In 1994, the Government approved a land law, which aimed among other things to limit the concentration of wealth. It was estimated that one per cent of the population held 75 per cent of the nation's land.

[25] In December of that year, Nujoma was re-elected President with 70 per cent of popular support. However, in May 1995 his party, SWAPO, split - leading to the creation of the SWAPO for Justice group.

[26] In mid-1997, a SWAPO Congress proclaimed Nujoma President for a third term. That year, the Government refused to open an investigation into crimes committed during the period before independence, which created tensions with South Africa.

[27] In August 1999, some 200 rebels were detained and tortured after attempting to take over the small town of Katima Mulil. They 'confessed' to being supporters of independence for the Caprivi Strip, Namibia's panhandle territory, hundreds of kilometers long and barely four or five kilometers wide, that divides Botswana from Angola. The Caprivi Strip is inhabited by the Iozi ethnic group, who refused to be led by the country's Ovambo majority, which also dominates SWAPO.

[28] Also in 1999, the Constitution was modified so that Nujoma could run for a third Presidential term, which he won with 76 per cent of the vote.

[29] In January 2002, the Herero community demanded compensation of $2 billion for war crimes committed between 1904-1907 under German rule. Several companies - the best known of which is Deutsche Bank - were accused of forming a 'brutal alliance' with imperial Germany to exterminate more than 65,000 Herero.

[30] In March 2002, Nujoma

PROFILE

ENVIRONMENT

Mainly made up of plateaus in the desert region along the Tropic of Capricorn. The Namib desert, along the coast, contains rich diamond deposits and is only populated because of mining activities. To the east, the country shares the Kalahari desert with Botswana, an area populated by herders and hunters. The population is densest in the north, and in the central plateau, where rainfall is heaviest. There is a coastal fishing industry which, together with cattle raising, was the mainstay of the economy before mining began in the 1960s. The country has important reserves of copper, lead, zinc, cadmium and uranium.

SOCIETY

Peoples: The Namibian population is divided into 11 ethnic groups. The most numerous are the Ovambo 47.4 per cent; Kavango 8.8 per cent; Herero 7.1 per cent; and the Damara 7.1 per cent. There is a European minority (4.6 per cent).

Religions: There is no official religion. Many people practice traditional African religions although there are a large number of Christians (Lutheran 51.2 per cent, Catholic 19.8 per cent, Anglican 5 per cent).

Languages: English (official); Afrikaans and German are also widely spoken. There are six main indigenous languages: Oshivambo, Herero, Nama-Damara, Kwangali (Okavango region), Lozi (Caprivi region) and Tswana.

Main Political Parties: Southwest African People's Organization (SWAPO); Congress of Democrats; Democratic Turnhalle Alliance (DTA); United Democratic Front.

Main Social Organizations: National Union of Workers of Namibia.

THE STATE

Official Name: Republic of Namibia.

Administrative Divisions: 13 districts.

Capital: Windhoek 237,000 people (2003).

Other Cities: Rehoboth 33,800 people; Rundu 28,500; Swakopmund 28,300; Walvis Bay 24,500; Keetmanshoop 20,000 (2000).

Government: Hifikepunye Pohamba, President since March 2005. Nahas Angul, Prime Minister since 2005. The Constitution established a presidential regime and a multiparty system.

National Holiday: 21 March, Independence Day (1990).

Armed Forces: 9,000 (2003).

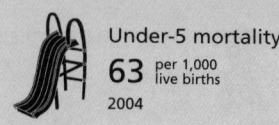
Under-5 mortality
63 per 1,000 live births
2004

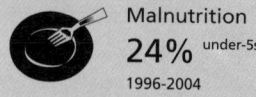
Malnutrition
24% under-5s
1996-2004

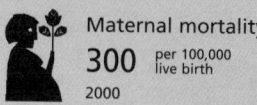
Maternal mortality
300 per 100,000 live birth
2000

ordered the arrest of all homosexuals and in a speech at the University of Namibia he declared that 'homosexuality or lesbianism are not allowed'.

[31] In January 2004, Germany apologized for the massacre of Herero people during its colonial occupation, but refused to pay any reparations. In August, Berlin apologized for the '1904 genocide' in the eastern region of Okakarara. During a ceremony on the centenary of the massacre, German Development minister Heidemarie Wieczorek-Zeul said: 'We Germans confess our historical, political, ethical and moral responsibility and our guilt in those times. I plead in the name of our Lord that you forgive our sins'.

[32] In May, Nujoma fired his Foreign Affairs Minister Hidipo Hamutenya during the SWAPO Congress held at Windhoek in which Nujomas successor for the November presidential elections was to be chosen. Hamutenya was then the favored candidate of the SWAPO party but Nujoma supported Hifikepunye Pohamba behind the scenes. After two days of congress, Pohamba won the second round of voting with 341 votes.

[33] That same month, a new bridge was opened across the Zambezi River (between Namibia and Zambia), which would provide a huge boost to regional trade. The bridge would also link the Democratic Republic of Congo to the Namibian port of Walvis Bay. The new bridge, financed by Germany, was part of a broader project that would turn Namibia into an important commercial axis.

[34] In November 2004, Pohamba won the fourth presidential elections since independence with 76 per cent of the vote. The opposing Democrats Congress, which received 9.9 per cent of the vote, claimed that the elections were faulty. SWAPO won 55 of the 72 Parliament seats.

[35] Some 20,000 people filled an open stadium in Windhoek in March 2005 to attend the swearing-in ceremony of the new President. Pohamba said land reform was too slow. Although more than once Nujoma had stated his admiration for the compulsory reform carried out in Zimbabwe (see Zimbabwe), in Namibia the process had remained within the limits of the supply and demand system.

[36] For years, black community farmers had pressed 4,000 commercial farmers, mostly white, to hasten the land reform in order to prevent what they called 'Zimbabwe-style farm invasions'. Only 35,000 Namibians had been relocated in fertile lands since 1990, because landowners refused to sell. In September 2005, the Pohamba Government started the expropriation of farms owned by whites.

[37] The fishing industry had grown strongly since 1998. However, the Fishing Ministry had to order a strong reduction of fishing quotas for the 2006-2007 season, which lowered to 50,000 tons the amount of hake fished. Most of the hake fished in Namibia was sent to Spain, from where it was distributed throughout Europe. Overberg Fishing, which fished near Namibia's Walvis Bay, stated that the drastic cut was 'a disaster' and 'catastrophic for the industry'. The chairman of the Fishing Association of Namibia, Denise van Bergen, said that, considering the state of fish stocks, the Ministry had little option, and that it was 'difficult to find a balance between saving the industry and protecting fish resources'.

[38] The Parliament passed in 2003 an Anti-Corruption Act that, among other initiatives, included the creation of an Anti-Corruption Commission. It was finally set up in February 2006. According to Pohamba, the commission was the final step to 'put an end to those who perpetrate frauds and robberies against the state'.

[39] The Legal Assistance Centre (LAC) of Namibia launched in March a campaign to decriminalize prostitution in the country, given the high proportion of HIV/AIDS cases (more than 20% of the population). The LAC argues that the legalization would help to carry out an effective campaign against HIV/AIDS aimed at prostitutes and their clients, and also to reduce violent and abusive acts linked with prostitution.

[40] In June, after a decade without any cases of poliomyelitis, 34 new cases appeared, seven of them fatal. The outbreak that affected mostly people over 20 was very unusual and worrisome, since the virus attacks adults more severely, causing them paralysis or, eventually, death. Given the potential magnitude of the outbreak, the Government requested the help of local and global organizations to launch a massive vaccination campaign.

[41] In September 2006, environmentalists expressed their opposition to the Government's revival of plans to build a controversial hydro power project on the Kunene River in northwest Namibia. Following an international outcry, the construction of the proposed Epupa Dam was halted eight years ago as the resultant flooding would have destroyed the livelihood of the semi-nomadic Ovahimba ethnic group.

[42] In October the Legal Assistance Center reported that the number of cases of rape and attempted rape have more than doubled since independence. ∎

IN FOCUS

ENVIRONMENTAL CHALLENGES
Long-term droughts have limited natural fresh water resources. Desertification and land degradation have spread to protected areas where wildlife poaching also poses a serious problem.

WOMEN'S RIGHTS
Women have been able to vote since 1989. In 2004, women held 27 per cent of Parliament seats and 19 per cent of ministerial posts. That year, women made up 42 per cent of the workforce, a percentage that has remained almost unchanged since 1980.

The illiteracy rate was 17 per cent for women and 13 per cent for men.*

Some 91 per cent of pregnant women received prenatal care, while 76 per cent of births were attended by skilled medical staff. However, the maternal mortality rate amounted to 300 per 100,000 live births. In late 2003 there were 110,000 women in the 15-49 age bracket living with HIV/AIDS.

CHILDREN
Inequity in access to education is worse among the most marginalized ethnic group, the San, and remains a serious problem among the rest of the population. Only 30 per cent of the children are enrolled in primary school and just 45 per cent of them make it to fifth grade. Malnutrition, especially in areas prone to flooding, and high HIV/AIDS prevalence, are the main health problems affecting children. When the flood waters subside there are other risks that threaten mostly women and children: malaria or water-borne diseases such as dysentery and cholera. In late 2003 there were approximately 57,000 orphans under 17 due to HIV/AIDS.

INDIGENOUS PEOPLES/ ETHNIC MINORITIES
The Namibian population is divided into 11 ethnic groups; the main ones are the Ovambo (47.4 per cent), Kavango (8.8 per cent), Damara (7.1 per cent) and Herero (7.1 per cent). The European minority represents 4.6 per cent of the population. The San were the earliest peoples to occupy Namibia. They are the most marginalized ethnic group in the country, showing high levels of alcoholism. They work on farms established on their own ancestral lands and earn very low wages. The Government of Botswana wants to remove them from central Kalahari in order to exploit mineral riches and tourist safaris. Out of a total of 55,000 San, 25,000 live in Namibia. They speak one of the varieties of click languages, which consists of sounds produced by different positions of the tongue. The situation of the San peoples was discussed at the World Conference Against Racism, held in South Africa in 2001, and a range of NGOs continue to denounce the discrimination suffered by them.

MIGRANTS/REFUGEES
In mid-2005, most of the 300,000 Angolan refugees in DR Congo, Zambia and Namibia had returned to their country, in a process that began after the 2002 peace agreements. In early 2006, Botswana attempted the repatriation of 1,200 Namibians from the Caprivi region which was most affected by floods. That year, permanent solutions were sought for refugees from the Great Lakes region. Thousands of people are displaced by the periodic flooding of the Zambezi river.

DEATH PENALTY
Namibia abolished the death penalty for all crimes in 1990, although Nujoma's Government has been consistently accused of human-rights violations such as extrajudicial executions and prison deaths.

* Latest data available in *The State of the World's Children* and *Childinfo* database, UNICEF, 2006.

Nauru / Nauru - Naoero

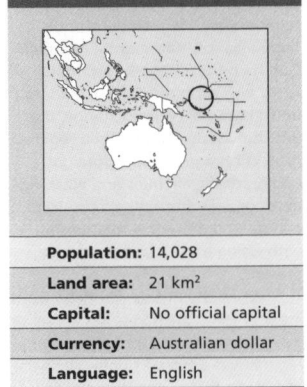

Population:	14,028
Land area:	21 km²
Capital:	No official capital
Currency:	Australian dollar
Language:	English

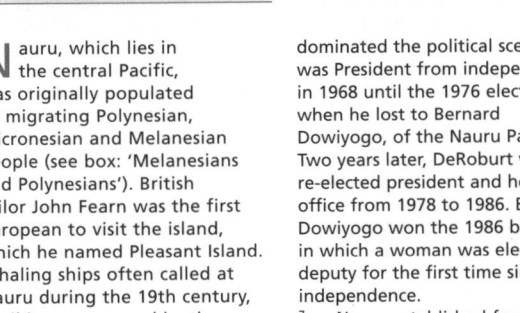

Nauru, which lies in the central Pacific, was originally populated by migrating Polynesian, Micronesian and Melanesian people (see box: 'Melanesians and Polynesians'). British sailor John Fearn was the first European to visit the island, which he named Pleasant Island. Whaling ships often called at Nauru during the 19th century, until it was annexed by the German Second Reich in 1888.

[2] In 1900, Australian Albert Ellis was sent a stone by a traveler, who thought it might make good marbles for children. Upon analyzing it, Ellis discovered the phosphate that would turn Nauru into a relatively wealthy country. In 1905, an Anglo-German company began mining on the island.

[3] At the end of World War I, Nauru became a trust territory administered by Australia. During World War II Nauru was under Japanese control and thousands of Nauruans were sent to forced labor camps on the island of Truk (in present-day Micronesia). At the end of the War, only 700 came back alive. Australia recovered the island, and mining was resumed.

[4] Traditionally fishing people, the Nauruans did not adjust well to mining and were soon replaced by immigrants, mostly Chinese. Immigrants were so numerous that in 1964, the Australian Government suggested that Nauruans should accept resettlement to a different island or elsewhere in Australia. Nauruans rejected the proposal, deciding to stay on their island and seek autonomy.

[5] The mining sector was nationalized in 1967, which resulted in a large increase in per capita income. Nauru declared independence on 31 January 1968, rejecting Australian attempts at domination. A year later it became a member of the Commonwealth.

[6] Hammer DeRoburt dominated the political scene and was President from independence in 1968 until the 1976 elections, when he lost to Bernard Dowiyogo, of the Nauru Party. Two years later, DeRoburt was re-elected president and held office from 1978 to 1986. Bernard Dowiyogo won the 1986 ballot in which a woman was elected deputy for the first time since independence.

[7] Nauru established formal diplomatic relations with Taiwan in 1980.

[8] Bernard Dowiyogo's second administration sued the government of Australia for its indiscriminate exploitation of the phosphate mines for 50 years. In 1991, Australia acknowledged the right of Nauru to be indemnified. Britain and New Zealand agreed to collaborate with Australia in the payment of $100 million to Nauru.

[9] In December 1997 then president Godfrey Clodumar re-established diplomatic relations with France, broken off two years previously in protest against nuclear tests in French Polynesia. At the same time, Nauru demanded a drastic cut in the production of greenhouse gases which lead to global warming and a rise in sea levels that would drown the island.

[10] Nauru citizens, who used to enjoy one of the highest per capita incomes in the world, suffered a sharp decrease in their standard of living due to bad investments of the Government. Dowiyogo lost Parliament's confidence. In April 1999 René Harris became President. On 14 September, Nauru joined the United Nations.

[11] In August 2001, 310 illegal immigrants - most of them Afghans rescued by a Norwegian cargo ship when the vessel carrying them sank - arrived at the island. Canberra offered to pay Nauru the expenses of those refugees.

[12] That year, the Financial Action Task Force (FATF) applied countermeasures against Nauru because it had not complied with its commitment to strengthen the laws against money-laundering; it was put on its list of non-cooperative countries and territories. It was taken off the list in 2005.

[13] Ludwig Scotty was elected President in May 2003, but shortly thereafter he received a vote of no confidence from Parliament. In August, Rene Harris was elected President once again.

[14] In December, UNHCR stated its concern over the situation of hundreds of refugees (mostly Afghans and Iraqis) detained in Nauru, some for more than two years.

[15] In June 2004, Harris received a vote of no confidence from Parliament and once again Scotty was elected President.

[16] In May 2005, Nauru restored its diplomatic ties with Taiwan, which had been severed in 2002. This angered China, which stated that the move was motivated by 'material gains'. In December, after the country could not pay its debt, the only airplane owned by Nauru was taken back by a US bank.

[17] A World Health Organization report stated in February 2006 that a diabetes epidemic affected Asia and would continue to do so for the following two decades. The report specified that Nauru had the highest incidence of that illness in the world. The easy riches derived from phosphate led Nauruans into a culture of increased inactivity and a diet of imported, processed food, and obesity remains a major problem.

[18] With the phosphate reserves almost exhausted, however, more and more Nauruans have had to resort to subsistence farming and fishing in order to survive. The country is increasingly dependent upon Australian aid and the money it receives from Australia for hosting the controversial refugee camp.

[18] By October 2006 there was just one Iraqi refugee left in the Australian refugee camp in Nauru - Mohammed Sagar, a Shi'a Muslim from Najaf, who was expressing concern about his mental state in his solitary condition. ∎

PROFILE

ENVIRONMENT

Nauru is a coral island, 6 km long by 4 km wide, located near Micronesia just south of the Equator. With sandy beaches and a thin belt of fertile land (300 meters wide), the island has a 60-meter high central plateau of guano (bird droppings), rich in phosphoric acid and nitrogen. These phosphate deposits have been intensively mined - leaving the interior reminiscent of a moonscape - and are now virtually exhausted.

SOCIETY

Peoples: Nauruans are descendants of Polynesian, Micronesian and Melanesian immigrants. There are also Australian, Aotearoan/ New Zealand, Chinese and European minorities, and workers from Tuvalu, Kiribati and other neighboring islands. Men outnumber women two to one.
Religions: Christian.
Languages: English (official) and Nauruan.
Main Political Parties: Nauru Party (NP); Democratic Party of Nauru.
Main Social Organizations: There are two social organizations: the Nauru Workers' Organization (NWO), founded in 1974 and affiliated to the NP; and the Phosphate Workers' Organization (PWO), founded in 1953.

THE STATE

Official Name: Republic of Nauru.
Administrative Divisions: 14 districts.
Capital: No official capital; administrative offices in Yaren.
Government: Ludwig Scotty, President since June 2004 (second time), elected for a three-year term by the Legislature. This is unicameral and has 18 members.
National Holiday: 31 January, Independence Day (1968).
Armed Forces: Australia is responsible for the island's defense.

Nepal / Nepal

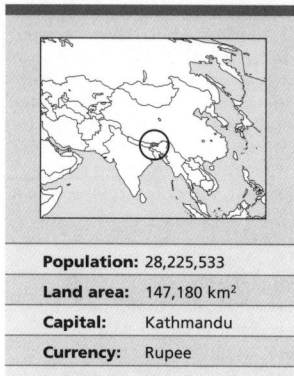

Population:	28,225,533
Land area:	147,180 km²
Capital:	Kathmandu
Currency:	Rupee
Language:	Nepali

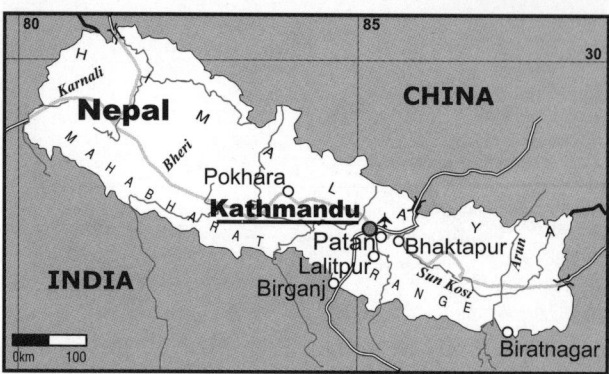

Nepal's rich prehistory lies mainly in the legendary traditions of the Newar, an indigenous people native to Nepal Valley (present day Kathmandu Valley). In Nepal Buddhism and Brahman Hinduism originated as related religions (although with different value systems), sharing holidays, events and myths. Nepal is situated between two of the world's most populated countries; India and China. References to Nepal Valley and Nepal's lower hill areas are found in the ancient Indian classics, suggesting that the Central Himalayan hills were closely connected culturally and politically to the Gangetic Plain at least 2,500 years ago. Lumbini, Gautama Buddha's birthplace in southern Nepal, and Nepal Valley also figure prominently in Buddhist accounts. There is substantial archeological evidence of Buddhist influence in Nepal, including a famous column inscribed by Ashoka (Emperor of India, 3rd century BC) at Lumbini and several shrines in the valley.

2   Although there are gaps, the sequence of the Nepal Valley's history can be traced along with the rise of the Licchavi dynasty in the 4th or 5th century AD. While the earlier Kirati dynasty had claimed the status of the Kshatriya caste of rulers and warriors, the Licchavis were probably the first ruling family in that area of Indian plains origin. This set a precedent for what became the normal pattern thereafter (Hindu kings claiming high-caste Indian origin ruling over a population much of which was neither Indo-Aryan nor Hindu).

3   The Licchavi dynastic chronicles, supplemented by numerous stone inscriptions, are detailed from AD 500 to 700. A powerful, united kingdom also emerged in Tibet in this period, and the Himalayan passes to the north of the valley were opened. Extensive cultural, trade and political relations developed, transforming the valley from a relatively remote backwater into the major intellectual and commercial center between South and Central Asia. Nepal's contacts

with China began in the mid-7th century with the exchange of several missions. But intermittent warfare between Tibet and China terminated this relationship; and while there were briefly renewed contacts in subsequent centuries, these were re-established on a continuing basis only in the late 18th century.

4   The middle period in Nepalese history roughly corresponds with the rule of the Malla dynasty (10th-18th century) in Nepal Valley and surrounding areas. Although most of the Licchavi kings were devout Hindus, they did not impose Brahmanic social codes or values on their non-Hindu subjects. The Mallas perceived their responsibilities differently, however. The Malla ruler Jaya Sthiti (reigned c.1382-1395) introduced the first legal and social code influenced by contemporary Hindu principles. His successor Yaksa Malla (reigned 1429-1482) divided his kingdom among his three sons, creating the independent principalities of Kathmandu, Patan and Bhaktapur (Bhagdaon) in the valley. Each state controlled the territory in the surrounding hill areas, with particular importance attached to the trade routes to Tibet (north) and to India (south) that were vital to their economies. In the western and eastern hill areas there were small independent principalities, which were sustained through a delicate balance of power based upon traditional inter-relationships and in common ancestral origins among the ruling families.

5   In the 16th century virtually all these principalities were ruled by dynasties claiming high-caste Indian origin whose members had left the hills in the wake of Muslim invasions of northern India. In the 18th century the principality of Gorkha (Gurkha) had a predominant role in the hills and even posed a challenge to the valleys. The Mallas, weakened by familial dissent and widespread social and economic discontent, were no match for the great Gorkha ruler Prithvi Narayan Shah, who conquered the valley in 1769 and moved his capital to Kathmandu, providing the foundation of the modern state of Nepal.

6   The Shah (or Saha) rulers faced persistent problems to centralize an area characterized by extreme diversity and ethnic and regional parochialism. They absorbed dominant regional and local élites into the central administration of Kathmandu, thus neutralizing potentially disintegrating political forces and involving them in national politics. This limited the center's authority in outlying areas because local administration was based upon a compromise division of responsibilities between the local élites and the central administration.

7   The British conquest of India in the 19th century posed a threat to Nepal that was left with no real alternative but to seek an accord with the British to preserve its independence. This was

accomplished by the Rana family regime after 1860. Under this de facto alliance, Kathmandu recruited Gurkha units for the British Indian Army and accepted British 'guidance' on foreign policy. In exchange, the British guaranteed the Rana regime protection against both foreign and domestic enemies and allowing autonomy in domestic affairs. Nepal maintained a friendly relationship with China and Tibet, for economic reasons.

8   When the British withdrew from India in 1947, the Ranas lost a vital external source of support and the regime was exposed to new dangers. Anti-Rana forces, composed mainly of Nepalese residents in India who had their political apprenticeship in the Indian nationalist movement, formed an alliance with the Nepalese royal family, led by the King. Nepal, India and the UK signed an agreement in Kathmandu in 1947, and Gurkha troops were used by India in the war against China (1961-62), Pakistan in 1965 and 1971, and by the UK against Argentina (1982).

9   In 1951, the Nepalese Congress overthrew the Rana regime with support from King Tribhuvan Bir Bikram Shah Deva, after which the country experimented with different forms of democracy. Political parties were legalized and a general election was held in 1959, based on the Constitution approved by King Mahendra Bir Bikran Shah

PROFILE

ENVIRONMENT

Nepal is a landlocked country in the Himalayas with three distinct geographical regions: the fertile, tropical plains of Terai, the central plateaus, covered with rainforest, and the Himalayan mountains, where the world's highest peaks are located. The climate varies according to altitude, from rainy and tropical, to cold in the high mountains. This diversity allows for the cultivation of rice, sugarcane, tobacco, jute and cereals. Most of the population lives from agriculture. Livestock - sheep and buffalo - is also important. Mineral and hydroelectric resources are as yet unexploited.

SOCIETY

Peoples: Nepali 53.2 per cent; Bihari (including Maithili and Bhojpuri) 18.4 per cent; Tharu 4.8 per cent; Tamang 4.7 per cent; Newar 3.4 per cent; Magar 2.2 per cent; Abadhi 1.7 per cent; other 11.6 per cent. **Religions:** Hinduism 86.2 per cent (official); Buddhism 7.8 per cent; Muslim 3.8 per cent; Christian 0.2 per cent; other 2 per cent. **Languages:** Nepali (official) is spoken by only half the population; many other languages are spoken, corresponding to the different cultural communities, Tibetan being the second most common language. **Main Political Parties:** The Nepalese Congress Party (NCP); the Communist Party of Nepal; National Democratic Party (NDM); Nepalese Goodwill Party; National People's Front.

THE STATE

Official Name: Nepál Adhirajya. **Administrative Divisions:** 5 regions, 14 zones and 75 districts. **Capital:** Kathmandu 741,000 people (2003). **Other Cities:** Biratnagar 203,300 people; Pokhara 157,700; Lalitpur 153,900; Birganj 118,300 (2000). **Government:** Parliamentary Monarchy. Gyanendra Bir Bikram Shah Dev, King (Maharajadhiraja) since June 2001. Girija Prasad Koirala, Prime Minister (4th time) since April 2006. Bicameral legislature: The House of Representatives with 205 members elected every five years and The House of the States with 60 members. **National Holidays:** 15 February, (Constitution) and 18 February, (Fatherland); 7 July, King's Birthday (1946). **Armed Forces:** 72,000 (2003).

Life expectancy
64 years
2005-2010

GNI per capita
$250
2004

Literacy
49% total adult rate
2000-2004

HIV prevalence rate
0.5% of population 15-49 years old
2003

Deva. Nepal became a member of the United Nations in 1955, and it has been an active member of the Non-Aligned Movement since the Bandung Conference. In 1960, the first elected Prime Minister, BP Koirala, was arrested. Parliament was dissolved, the Constitution was suspended and political parties were outlawed. In 1962 a non-party or Panchayat system was introduced.

[10] In 1979, student protest movements emerged in Kathmandu and other cities. King Birendra Bir Birkram Shah Deva responded by holding a plebiscite to choose between a multi-party system or a reformed Panchayat, the latter obtaining 55 per cent of the vote. The opposition considered the outcome a fraud. By July 1986, 75 countries endorsed a proposal by Birendra that Nepal should be declared a zone of peace. All of its neighbors except India and Bhutan endorsed the proposal. India is Nepal's leading trade partner. The two signed several trade and transit treaties between 1950 and 1989, when the two countries became involved in an undeclared trade war. Nepal clashed with India over its arms imports from China, India suspended all trade with Nepal and closed 19 of the 21 transit routes in 1989, seriously harming the Nepalese economy.

[11] The anti-Panchayat protests reached a peak in 1990. Birendra initiated talks with the opposition and agreed to political pluralism. The transition to a parliamentary democracy with a constitutional monarchy was announced by royal decree in April that year. In 1991 the first free elections were held in Nepal after 32 years of semi-monarchic rule. The Communist Party of Nepal (CPN) and the Nepalese Congress Party (NCP, monarchist and pro-Government) united for the elections, joining forces with other groups. The Communists won four of the five posts in the capital, but the national majority went to the NCP.

[12] That year, the new Prime Minister Girija Prasad Koirala (NCP), promised to launch a mixed economy, to earmark 70 per cent of national income to rural regions and to carry out agrarian reform. He also made primary education free. In 1993, to attract foreign investment, Koirala promoted the total convertibility of the Nepali rupee (national currency) with foreign currencies, and signed new trade agreements with India. Koirala resigned in 1994 due to lack of parliamentary support and infighting in the NCP, and Man Mohan Adhikari was named Prime Minister. In the November elections, the CPN won 88 seats in the Chamber of Representatives, surpassing the NCP's 83. In 1995, lacking majority political support,

Mohan Adhikari handed over control to NCP leader Sher Bahadur Deuba. Guerrilla groups describing themselves as Maoist emerged in 1996 to 'eradicate feudalism'.

[13] In 2001 Birendra, Queen Aishwarya and other members of the royal family were shot to death by their heir, Prince Dipendra Bir Bikram Shah Deva, 29, who shot himself in a suicide attempt, but survived in a state of coma. The Royal Council appointed Dipendra King of Nepal and Prince Gyanendra, Birendra's brother, as his Regent. Dipendra died and Gyanendra became King. That same year, Koirala resigned due to the growing violence between the insurgent Maoists and the security forces; Sher Bahadur Deuba took over as Prime Minister and declared a ceasefire.

[14] In the subsequent negotiations, the Maoists demanded the repeal of the 1990 Constitution, the creation of an interim government and the election of a Constituent Assembly to draft a new national charter that would end the monarchy and create a republic. The Government did not accept the demands and violence escalated. The King declared a state of emergency, categorizing the rebels as 'terrorists'. With constitutional guarantees suspended, the death toll among civilians and rebels alike rose. The conflict turned into a civil war.

[15] In May 2002 Deuba used the state of emergency as an excuse to dissolve Parliament, calling elections for November. In October,

Gyanendra dismissed Deuba, appointed Lokendra Bahadur Chand as Prime Minister and postponed the elections indefinitely.

[16] In February 2003 Maoist rebels and the Government agreed to a ceasefire. In August, after peace talks with the Government failed, rebels put an end to the seven-month truce, and called a general strike, which led to a major outbreak of violence, with clashes between students, activists and the police. In April 2004, opposition groups angry over the king's assumption of executive powers added their voices to the protests of the strikers. That month, Nepal joined the World Trade Organization.

[17] In August, the rebels staged a week-long blockade of Kathmandu, preventing supplies from reaching the city. In the same month, 12 Nepalese hostages in Iraq were murdered by their captors, sparking violent protests in Kathmandu. In September a two-day strike called by Maoist rebels paralyzed the country.

[18] In February 2005, Gyanendra sacked the Prime Minister and assumed direct power. This measure fulfilled one of the demands made by rebels, who had requested direct dialogue with the King, ruling out talks with representatives of the Executive. However, it was criticized by other countries, among them the US, France and India. In March, representatives of the main political parties agreed to call for a constitutional assembly

and overthrow Gyanendra. Thus, politicians joined the demands made by rebels who had been calling for the setting up of a republic and a national constitutional assembly.

[19] The state of emergency was lifted in April 2005. In September, rebels announced a truce that would last until January 2006. During that period, the rebels and main opposition parties agreed on a program intended to restore democracy.

[20] Weeks of mass political protest and strikes culminated in King Gyanendra agreeing to reinstate Parliament in April 2006. The veteran GP Koirala was appointed Prime Minister once again and announced his intention of bringing the Maoists into the political mainstream; the rebels called a three-month ceasefire. In May Parliament voted unanimously to curtail the King's political powers.

[21] In November 2006, an historic peace deal between the Government and the Maoist rebels promised to end the 10-year insurgency and mark 'a new era for Nepal'. The deal offered the Maoists a role in a temporary cabinet, sharing ministerial posts equally with each of the other main parties. In return the Maoist army was to be confined in camps and its weapons locked up under UN surveillance. An assembly due to be elected by June 2007 was to decide whether the monarchy will continue or Nepal will become a republic. ∎

IN FOCUS

ENVIRONMENTAL CHALLENGES
Wood-burning accounts for 90 per cent of energy consumption, resulting in deforestation and soil erosion. Urban areas are particularly affected by air and water pollution. The lack of sewerage systems in major cities has also contributed to the deterioration of the environment.

WOMEN'S RIGHTS
Nepalese women have been able to vote and run for office since 1951. In 2003, women held 6 per cent of seats in Parliament and 7 per cent of ministerial or equivalent posts. There are political restrictions for women of certain castes, but the Constitution states that at least 5 per cent of each party's candidates to the House of Representatives must be women. In 2003, women made up 40 per cent of the workforce, comprising 12 million people. In 2004, only 28 per cent of pregnant women received prenatal care and barely 15 per cent of births were attended by skilled health

staff. The maternal mortality rate amounted to 540 women per 100,000 births.
In 2004, 54 per cent of women had got married before their 15th birthday.

CHILDREN
The armed conflict has threatened the security of children in various ways: one of them is by preventing food and vitamin A distribution programs from reaching the most remote places. Without this aid, it is estimated that 12,000 children would die each year in addition to the 70,000 who are already killed by preventable diseases. It is estimated that 5,000 children die each year of measles every year. Besides, each day nearly 200 children under 5 years old die from diarrhea and acute respiratory infections. Some 31 per cent of children aged between 5 and 14 have to work to survive.

INDIGENOUS PEOPLES / ETHNIC MINORITIES
There are over 75 ethnic groups, speaking 50 different languages.

The Constitution grants each community the right to preserve and promote its language, writing and culture. The Government favours Hinduism and proselytizing is not allowed. Members of the lower castes are subject to widespread discrimination all over the country, except in Kathmandu, where differences are less marked.

MIGRANTS/REFUGEES
In early 2006, there were 105,000 people living in refugee camps, who had sought help upon the collapse of health and education services once the king assumed all state powers in February 2005. There were also an estimated 200,000 internally displaced; 20,000 settled Tibetans, an additional 3,000 in transit and over 600 urban refugees and asylum-seekers.

DEATH PENALTY
The death penalty for ordinary crimes was abolished in 1990 and for all types of crimes in 1997. The last execution took place in 1979.

Netherlands / Nederland

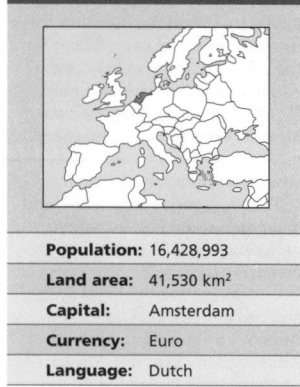

Population:	16,428,993
Land area:	41,530 km²
Capital:	Amsterdam
Currency:	Euro
Language:	Dutch

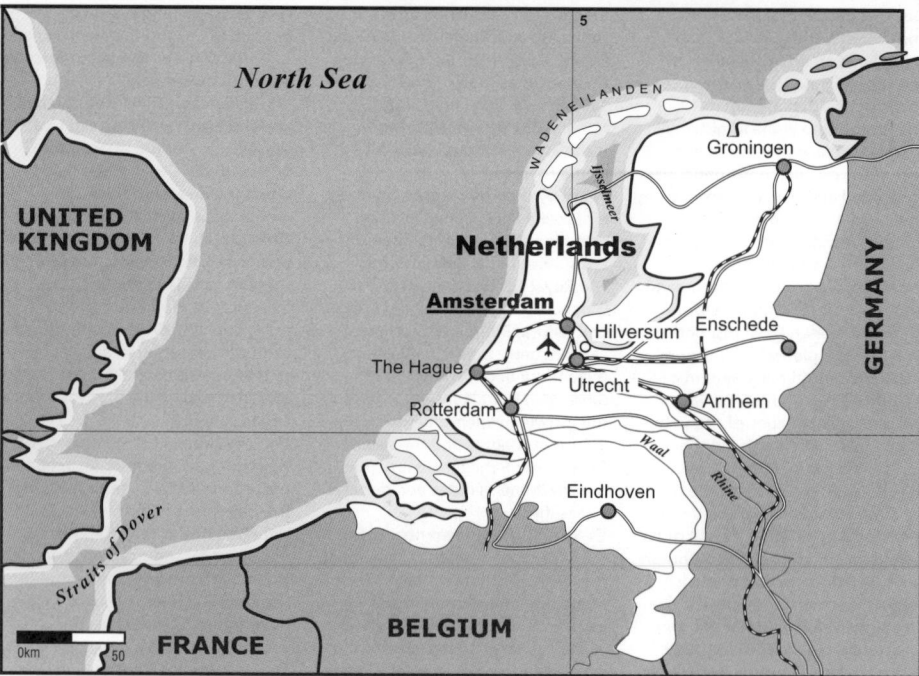

The borders of present-day Netherlands (Holland) were established in 1867. Until then, the region known as the Low Countries (made up of what are now Belgium, Luxembourg, Netherlands and the northern part of France) suffered several political divisions.

2 In the 4th millennium BC, the first pre-Celtic (Indo-European) tribes emigrated from present-day south-west Germany to the Low Countries. In the 2nd millennium BC, the high level of social and cultural development reached by these tribes found expression in their art which featured curves and arabesques. The Celts were conquered by Frisian and Batavian Germanic peoples in the 3rd century BC.

3 The Roman Empire never managed to occupy the land of the Frisians, who remained independent north of the Rhine until the 8th century. The Romans instead settled in the Rhine delta, where they founded the province of Gallia Belgica in 54 BC. There they worked the lands around their villas, while the Frisians continued to fish and raise cattle. In the 3rd century BC, a rise in sea levels affected production in the region.

4 In the 1st century AD, the Batavians rebelled against the Romans, but the latter only withdrew in the 4th century

with the invasions of the Saxons (Germanic), who came from eastern Germany, and the Franks (Germanic), who came from the north. The Low Countries, to the south of the Rhine, were split in two, with the southern half under Roman jurisdiction, using a Romance language, and the northern half under Germanic control.

5 The Franks subdued Roman Gallia (Gaul) in the 5th century and expanded the domain of the Western Empire, in alliance with the Catholic Church. In 695, the Frank Carolingian dynasties set up a bishopric in the city of Utrecht, encroaching on the Frisian region, which it control - led by the end of the following century. In 800, Charlemagne was crowned emperor.

6 In 911, the northern part of France was ceded to the Vikings from Scandinavia. The Normans, as they became known, founded the duchy of Normandy

and conquered other regions such as the Low Countries. As a result territorial sovereignty was divided between several fiefs. Through their expansion the Normans opened up routes for commercial and cultural exchange with the East.

7 Between the 10th and 13th centuries textile craft industries grew in several communities, accompanied by a growth in population. From the 11th century onwards, the Frisians developed a system to drain the sea water and reclaim land for pasture and agriculture. Cistercian and Premonstraten monks were active in the construction of dykes to reclaim land from the sea.

8 In 1302, Flanders became independent from France. The centers of production (such as Utrecht, The Hague, Amsterdam and Ghent) gradually expanded and perfected their economies, developing for example a

banking system, as well as winning political autonomy.

9 In the latter half of the 14th century, the dukes of Burgundy allied with the Flemish (from Flanders). Under this prosperous alliance, which lasted until 1477, the arts flourished with painters such as Jan Van Eyck (1395-1441), Rogier van der Weyden (1400-64) and Hugo van der Goes (1440-82) as well as Gothic Flemish sculpture and architecture. In this period, the Duchy of Burgundy's authority was extended throughout the Low Countries.

10 After the death of Charles the Bold in 1477, his daughter Mary of Burgundy married Maximilian of Austria and the Low Countries came under the control of the Habsburg dynasty, which expanded its hegemony throughout the Germanic Holy Roman Empire (962-1806).

11 Charles V (born in Ghent in 1500), son of the archduke of Austria, occupied the throne of Spain in 1517. He was succeeded in 1556 by his son Philip II who inherited the Habsburg lands in Austria, the Holy Roman Empire which included the Netherlands, and also Spain's overseas possessions. In 1566, political and religious struggle led by the Duke of Orange erupted in the Low Countries, where the Protestant Reformation took hold in the north (Netherlands). Two years later the Duke of Alva brutally crushed the revolt. In 1579, the seven northern provinces (Calvinist) proclaimed their independence in the Union of Utrecht and adopted the

LAND USE

2003/2005

IRRIGATED AREA: 59.9% of arable land

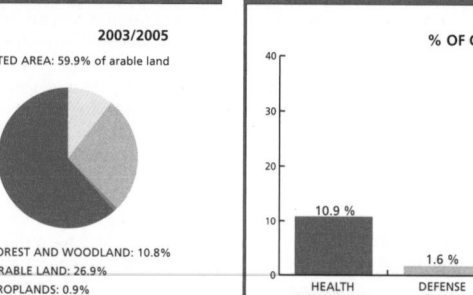

- FOREST AND WOODLAND: 10.8%
- ARABLE LAND: 26.9%
- CROPLANDS: 0.9%
- OTHER USE: 61.4%

PUBLIC EXPENDITURE

% OF GDP

10.9 % HEALTH & EDUCATION (2000-2002)

1.6 % DEFENSE (2004)

WORKERS

UNEMPLOYMENT: 4% (2004)

LABOR FORCE **2004**

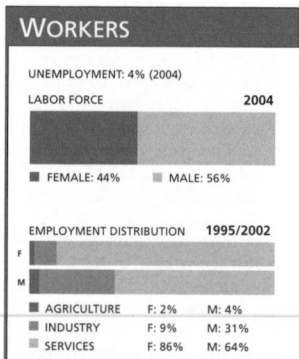

- FEMALE: 44%   ■ MALE: 56%

EMPLOYMENT DISTRIBUTION **1995/2002**

F

M

■ AGRICULTURE	F: 2%	M: 4%
■ INDUSTRY	F: 9%	M: 31%
■ SERVICES	F: 86%	M: 64%

name United Provinces of the Netherlands, while the southern provinces (Catholic) remained loyal to Spain.

[12] For most of the period 1568-1648 the United Provinces and Spain were at war. Their battles took place mostly along the sea routes to the Indies, where both colonial powers (like England and France) wanted to prevail.

[13] In 1602, the Netherlands - which had large shipyards along the North Sea coast and major investors - formed the Dutch East India Company, with agencies in Ceylon, India and Indonesia. In 1621, the United Provinces founded the Dutch West India Company. Its profits came largely from the slave trade, and smuggling from Spain's colonies.

[14] In 1648, the Dutch had three large settlements in America: one in the north for the fur trade, another in Brazil and another in Suriname, for the slave trade and smuggling with the Spanish. Of these settlements, in 1700 only the trading posts of Curaçao, Sint Eustatius and Sint Maarten remained, and the plantations in Dutch Guyana and Elmina as slave ports.

[15] During the Twelve Years' Truce with Spain (1609-21), controversies within the Union grew. The collaboration between the province of Holland and the House of Orange (German) gave way to a growing rivalry. In 1618, Maurice of Orange executed the leader of the main party in the Netherlands. At the end of the Truce, when the war with Spain was resumed, both rivals were forced to reunite until the signing of the Treaty of Munster in 1648, under which Spain recognized the independence of the United Provinces.

[16] In 1628 French philosopher Descartes - a critic of the dominant method of Scholasticism (based on comparing and contrasting the views of recognized authorities) moved to the Netherlands where he was able to publish *Discourse on Method* (1637).

[17] Dutch maritime power began to decline after the wars against Britain (1652-4 and 1665-7). The country used up part of its capital to buy bonds in foreign governments. Bankers in Amsterdam were among the most powerful in Europe.

[18] In the War of the Spanish Succession (1701-1714) the Dutch allied with the British against the French. In the latter part of the century wars continued between and among the Dutch, French, Spanish and British.

[19] After the French Revolution (1789), Holland became a French protectorate between 1794 and 1806, with a republican system of government. The United Provinces became the Batavian Republic. The new political and economic liberties were welcomed by various sectors of Dutch society, such as the anti-Orange Patriot movement which, under the impact of the first Industrial Revolution, rejected the monarchy.

[20] Between 1806 and 1814, under the Empire of Napoleon Bonaparte, the new Batavian Republic fell victim to political struggles. In 1814, Prince William I of Orange was called on to restore the monarchy and in 1815 he won sovereignty over all of the Low Countries. In 1830, a revolution led Belgium to secede from the Netherlands. In 1831, it annexed the eastern region of Luxembourg, while the Grand Duchy of Luxembourg became independent in 1867.

[21] Throughout the 19th century the Dutch expanded civil rights. The Constitution of 1848 became the foundation of the Netherlands' current democracy. Under its provisions arbitrary personal rule by the monarch was no longer possible and members of the first chamber of parliament, formerly appointed by the King, were thereafter elected by the provincial assemblies. Members of the assemblies and the second chamber of parliament were elected; only people who were tax-payers could vote.

[22] After decades of debate over the school system, Protestants and Catholics allied themselves against the liberals and, in 1888, the first private schools were opened. New political parties were founded, based on the religious ideas and ideologies of the time. To the Liberal, Protestant and Catholic parties were added the Protestant Conservative, the Socialist and the Communist parties. As none could obtain a majority, coalitions became commonplace.

[23] During World War I (1914-1918), the Netherlands declared its neutrality and political parties agreed to a truce in order to dedicate their energies to the domestic economy and foreign trade. The merchant navy had recovered and industry grew, in particular textiles, electronics, and chemicals.

[24] During the postwar period, the Netherlands was a member of the League of Nations, but it re-affirmed its neutrality, a symbol of which was the International Court of Justice at The Hague. During the Versailles negotiations, Belgium tried unsuccessfully to revive an old territorial claim against the Netherlands.

[25] During World War II (1939-45), Germany occupied the Netherlands (and attacked France from Dutch territory). Queen Wilhelmina formed a government-in-exile in London, and all political factions took part in anti-Nazi resistance.

[26] In 1945, an agreement was signed by the Government, companies and trade unions. It lasted 20 years and was aimed at controlling prices and salaries. The Netherlands underwent rapid industrialization especially in steel production, electronics and petrochemicals. In the following years, Dutch transnational companies were consolidated, such as Royal Dutch Shell (petrochemicals), Unilever (world's largest food and soap manufacturer), Philips (electronics), AKZO (chemicals), Heineken (beer), ABN AMRO (banking), and ING (banking and insurance).

[27] In the postwar years, with a Government composed of Labor Party-led (former socialists and Catholics) coalitions, the Netherlands joined NATO, the UN and the European Economic Community. Together with Belgium and Luxembourg it formed the Benelux economic alliance. It granted independence to Indonesia (1949), New Guinea (1963) and Suriname (1975).

[28] The 1960s witnessed struggles for the rights of women and sexual minorities, and public debates were held on the use of drugs. These pressure groups achieved significant changes in legislation and state medical insurance coverage. In the ensuing years abortion and prostitution were legalized, and use of marijuana and heroin decriminalized. To help addicts, information and medical assistance plans were implemented, including distribution of condoms, needles and the heroin substitute methadone.

[29] In the 1970s, the electorate voted in center-left governments, which reformed the tax system

PROFILE

ENVIRONMENT

The country is a large plain and 38 per cent of its territory is below sea level. Intensive agriculture and cattle-raising produce high quality milk products and crops (particularly flowers). Population density is amongst the highest in the world. Highly industrialized, the country is a major producer of natural gas, and is a dominant influence in petroleum activity. It has large refineries in the Netherlands Antilles and Rotterdam, the world center of the free, or 'spot', crude oil market.

SOCIETY

Peoples: Dutch 91 per cent; Turks 1.3 per cent; Moroccans 1 per cent; Germans 0.3 per cent; and others.
Religions: 31 per cent Catholic; Dutch Reformed Church 14 per cent; Calvinist 8 per cent; Muslim 3.9 per cent; other 4.1 per cent; no religion 39 per cent.
Languages: Dutch (official); Frisian and Saxon; Turkish, Arab, Kurdish and from other immigrant communities.
Main Political Parties: Christian-Democratic Appeal (CDA); Labor Party (PvdA), social-democratic, affiliated to the Socialist International; People's Party for Freedom and Democracy (VVD), conservative liberal; Socialist Party.
Main Social Organizations: Federation of Netherlands Trade Union Movements, which includes Socialist and Catholic trade unions; National Confederation of Christian Trade Unions.

THE STATE

Official Name: Koninkrijk der Nederlanden.
Administrative Divisions: 12 Provinces.
Capital: Greater Amsterdam 1,145,000 people (2003). Although the Government has its seat in The Hague, Amsterdam is still considered the capital. **Other Cities:** Rotterdam 1,125,500 people; The Hague (s'-Gravenhage) 443,700; Utrecht 232,900; Eindhoven 200,600 (2000).
Government: Constitutional and parliamentary monarchy. Queen Beatrix, Head of State since April 1980. Prime Minister Jan Peter Balkenende, since July 2002. Gerrit Zalm and Thom De Graaf, Deputy Prime Ministers since May 2003. The Legislature has two chambers: the First Chamber, with 75 members, and the Second Chamber, with 150 members.
National Holiday: 30 April, Queen's Day (1938).
Armed Forces: 53,000.
Dependencies: See Netherlands Antilles and Aruba.

Under-5 mortality
6 per 1,000
live births
2004

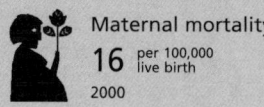

Maternal mortality
16 per 100,000
live birth
2000

and increased investment in social policies. However, the size of the defense budget and the installation of NATO atomic missiles in the country caused controversy.

[30] The Netherlands is one of the countries allocating the highest percentage of its GDP to aid for the Third World. It has also pursued policies defending human rights and took a stand against South Africa's apartheid system. Meanwhile, links with Israel have tended to estrange it from some Arab countries.

[31] In 1980, after an embezzlement scandal involving her husband, Queen Juliana abdicated in favor of her daughter Beatrix. In 1985, the Government authorized the installation of nuclear missiles, despite protests by the Dutch people, which continued until the end of the Cold War in 1991.

[32] In 1991, when it held the EU presidency, the Netherlands called on the heads of state and government meeting in Maastricht to condemn racism and to adopt legislation prohibiting xenophobic acts throughout Europe. However, in September 1993 the Netherlands enacted legislation restricting immigrants from outside the EU.

[33] In 1992, the Central Office of Statistics announced the 'green gross national product' (GGNP). This indicator evaluates losses of natural resources, in relation to the capacity for regeneration, and the effect on local communities.

[34] On 22 August 1994, Wim Kok, from the Labor Party,

became Prime Minister, in coalition with two other parties. Three months earlier, local election results had shown a growth in support for racist far-right parties. Kok would be re-elected in 1998.

[35] In April 2001 the first same-sex marriages took place in the Netherlands, and gay couples gained the right to adopt children.

[36] In January of the following year, the Netherlands became the first country to legalize euthanasia (a partial regulation had already been in place since 1993). The law permits ending a person's life under strict criteria: he/she must be in intolerable and constant pain and must have made repeated requests to be allowed to die; the doctor in charge has to obtain a second medical opinion. Finally, the death must occur in an appropriate medical way.

[37] The Government resigned on 16 April 2002 after admitting its responsibility in the Srebrenica massacre in Bosnia in 1995. The 7,600 page-long official report stated that Dutch security forces in the area - under UN command - failed to prevent the killing of 7,000 Bosnian Muslims by Bosnian Serbs. The victims were 'evacuated' from the UN security area by Dutch troops, which ultimately led to them being killed. The report also accused the UN of sending in the troops without a clear mandate or the weapons needed to defend the area. Dutch soldiers' use of force and weapons were limited to

self-defense. The Government's resignation took place weeks before the parliamentary elections set for 15 May.

[38] In early May 2002, far-right candidate Pim Fortuyn, who was riding high in the polls, was murdered in the city of Hilversum. Fortuyn had declared that Holland should close its borders to immigrants and that Islam was a reactionary religion. His electoral proposals included laying off 25 per cent of civil servants. The murderer, who received a prison sentence, stated that his victim was a menace to democracy.

[39] In the 22 January 2003 general elections, the Christian Democrats obtained a majority of the vote, while the VVD (People's Party for Freedom and Democracy) and the Democrats 66 Party (conservative) came in second and third place, respectively. The three parties agreed to form a coalition government headed by Jan Peter Balkenende, who sought to reduce fiscal deficit and unemployment by cutting State expenditure.

[40] In April 2004, more than 100 pictures that were first looted from Holland by the Nazis, who earmarked them for Hitler's private collection, and then in 1947 were stolen by the Red Army and taken to the former Soviet Union, were officially given back to the Netherlands by Ukrainian President Leonid Kuchma. The pictures had been stored in Kiev for more than 50 years.

[41] In October, more than 200,000 people from all over the

country protested in Amsterdam against an austerity plan that included public spending cuts and welfare reforms. It was the largest protest there for over two decades.

[42] The murder of film-maker Theo Van Gogh at the hands of a radical Islamist shook public opinion in November 2004. It took place just two months after his film 'Submission' - about the abuse of Muslim women - was shown on national TV.

[43] In a referendum held in June 2005, over 60 per cent of Dutch voters rejected the EU Constitution. The poll, with a voter turnout of over 60 per cent, took place only three days after French people voted against the continental charter.

[44] Following six months of wrangling and pressures by NATO, the UN and the US, Parliament agreed in February 2006 to send an additional 1,400 Dutch troops to bolster peacekeeping work in Afghanistan. Dutch public opinion was divided on the issue and many Dutch MPs demanded assurances that the country's troops would not have to work under command of US forces.

[45] Since temperatures started to be registered in the early 18th century, July 2006 was the hottest month in the history of the country, with an average 38 degrees Celsius. By mid-August, summertime in the Northern hemisphere, there had been more than 500 deaths due to the heat, which considerably surpassed previous records. ∎

IN FOCUS

ENVIRONMENTAL CHALLENGES
Underground water is highly polluted with nitrates, as a result of the widespread use of agrochemicals, and also with heavy metals. Major rivers carry all kinds of organic and industrial waste, especially from European countries. There are significant levels of air pollution, mostly from refineries and automobiles, leading to high levels of acid rain.

WOMEN'S RIGHTS
Women have been able to stand for office since 1917 and to vote since 1919.

In 2003, women held 37 per cent of seats in the Lower House and 26 per cent in the Upper House of Parliament. That year, of the 7 million people making up the labor force, 49 per cent were women.

Female unemployment stood at 3 per cent. The Netherlands has made more progress than any other country in preventing trafficking of women from Southern countries to the EU as well as in providing support for the trafficked women through social security programs and legal, medical and psychological aid, as well as temporary residence permits. At the end of 2003, some 3,800 women aged between 15 and 49 were living with HIV/AIDS.

CHILDREN
Educational, health and general welfare needs are covered for Dutch children, making poverty in this age group practically non-existent. In 2004, infant and under-5 mortality rates were among the lowest in the world: 5 and 6 per 1,000 live births, respectively. Traditionally mothers have been

responsible for early years child-care, but as in other countries, the Netherlands has seen a greater demand for nurseries. The Government is trying to address this issue.

On the other hand, children from ethnic minorities have more difficulties to become socially integrated. Those from Turkish, Moroccan or Surinamese families suffer more social segregation in schools, and have greater drop-out rates and learning difficulties.

INDIGENOUS PEOPLES/ ETHNIC MINORITIES
The main minority groups are Turks, Moroccans and Surinamese.

MIGRANTS/REFUGEES
In comparison to 2004, only 17 of the 50 countries that traditionally receive asylum applications registered an increase in the

number of requests in 2005; 4 registered no significant changes and 29 received fewer applications. The Netherlands was the country with the second-largest increase (26 per cent) in the number of asylum applications, following Greece (102 per cent). This meant a change in a trend that had been constantly falling since 2001. In 2005, the figure rose to 12,350. Most refugees arrived from Iraq (1,620), Somalia (1,315), Afghanistan (902), Iran (557), Burundi (419), China (356), Colombia (342), Sudan (339), Serbia and Montenegro (336) and Turkey (289).

DEATH PENALTY
The death penalty was abolished in 1982.

Netherlands Antilles / Nederlandse Antillen

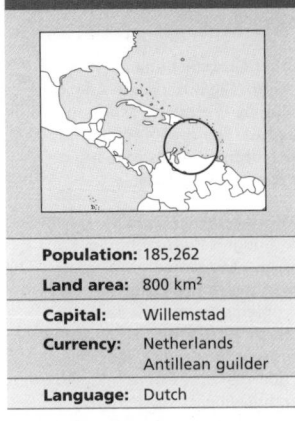

Population:	185,262
Land area:	800 km²
Capital:	Willemstad
Currency:	Netherlands Antillean guilder
Language:	Dutch

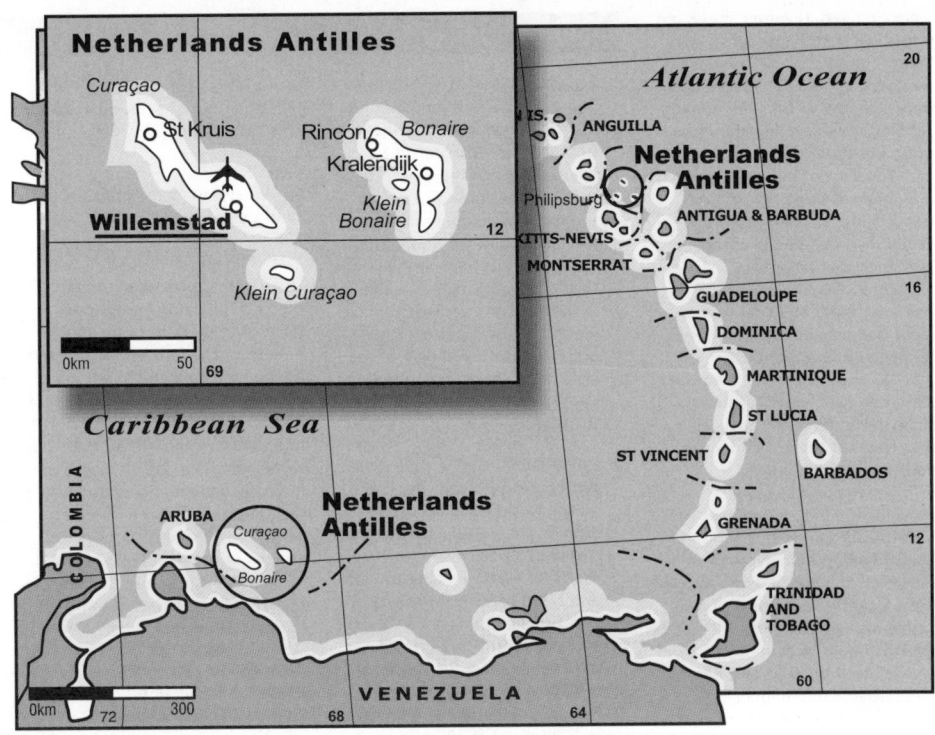

Both Caribs from the 'S' or 'Windward Islands' and Caiquetios (an Arawak nation and original inhabitants of what today are the islands of Aruba, Bonaire and Curaçao) were enslaved and taken to Hispaniola (present-day Haiti and Dominican Republic) by the end of the 19th century, soon after the arrival of Columbus in 1493 and then Alonso de Ojeda in 1499.

[2] Due to the lack of natural resources, only the Windward Islands (particularly Sint Maarten) acquired strategic importance as a port of entry into the Caribbean and also for salt extraction. Ports on the Iberian Peninsula were closed to Holland after the war with Spain and Portugal, so the Dutch were forced to seek alternative sources of salt in the Antilles.

[3] The traffic of Dutch ships was first repressed and then prohibited in 1606 by Spain. Holland created the West Indies Company to establish, manage and defend its colonies. The theft of Spanish ships and their cargoes became an important source of income.

[4] In 1634, the stockholders of the Company decided to invade Curaçao, which thus became a major international center of slave, salt and Brazilwood trading.

[5] In 1648, after three centuries of conflict, the Treaty of Westphalia granted Holland control over these islands. The native populations - particularly those of Curaçao and Bonaire - were replaced by African slaves to meet the needs of new agricultural operations, and thus became a minority on these islands. Slave rebellions were frequent in the 18th century and unleashed bloody

massacres at the hands of colonial troops.

[6] During the Napoleonic wars at the beginning of the 19th century, the islands passed into British hands for two brief periods. Although the slave trade had been prohibited in 1814, it was not abolished in the Dutch colonies until 1863. The lives of 'free' slaves did not change substantially. Many migrated to

the Dominican Republic, Panama, Venezuela and Cuba.

[7] In 1876, the Dutch Parliament proposed selling the islands to Venezuela, but negotiations fell through. With the rise of the oil industry at the beginning of the 20th century, refineries were installed because of the proximity to Lake Maracaibo.

[8] This new activity began in

the 1920s and radically changed the colony. It attracted thousands of immigrants from Venezuela, Suriname and the British West Indies and brought about the decline of agriculture. In the following decades, automation caused a significant reduction in the number of jobs.

[9] In 1937, political activity was launched when the first local parties

WORKERS

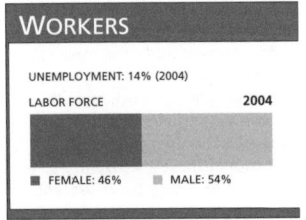

UNEMPLOYMENT: 14% (2004)

LABOR FORCE 2004

■ FEMALE: 46%   ■ MALE: 54%

PROFILE

ENVIRONMENT
The Netherlands Antilles are made up of two Caribbean island groups. The main group is composed of Bonaire (288 sq km) and Curaçao (444 sq km). Located near the coast of Venezuela, they are known (together with Aruba) as the 'Dutch Leeward Islands' or the 'ABC Islands'. The smaller group is made up of three small islands of volcanic origin: Sint Eustatius (21 sq km), Saba (13 sq km) and Sint Maarten (34 sq km - the southern part of the island of St Martin which is a dependency of the French island of Guadeloupe). These are known as the 'S Islands' or the 'Dutch Windward Islands', although in fact they are part of the Leeward group of the Lesser Antilles. In general the climate is tropical, moderated by ocean currents.

SOCIETY
Peoples: Most of the population descend from African slaves. There are also Carib Indians and descendants of Europeans and Asians.
Religions: Mainly Catholic; some Protestants, Jews, Seventh-Day Adventists.
Languages: Dutch (official). The most widely spoken language - in Curaçao and Bonaire - is Papiamento, a local dialect based on Spanish with elements of Dutch, Portuguese, English and some African languages. In St Eustatius, Saba and St Martin,

English is the main language. Spanish is also spoken.
Main Political Parties: Workers' Liberation Front/ Frente Obrero Liberashon (FOL); Antillean Restructuring Party (PAR); National People's Party (PNP); New Antilles Movement (MAN); Forsa Korsou.-
Main Social Organizations: National Confederation of Curaçao Trade Unions (AVVC); Central General Di Trahadonan di Corsow (CGTC); General Federation of Bonaire Workers (AFBW); Bonaire Labor Federation (FEDEBON).

THE STATE
Official Name: De Nederlandse Antillen
Capital: Willemstad 134,000 people (2003).
Other Cities: Kralendijk 7,900 people; Philipsburg 6,300 (2000).
Government: Frits Goedgedrag, Governor since July 2002, appointed by the Dutch Government. Emily S de Jongh-Elhage, Prime Minister since March 2006. Unicameral Legislature: the Staten (States), with 22 members elected every four years. The Netherlands Antilles are part of the Kingdom of the Netherlands, which handles its foreign relations. It has autonomy in internal affairs.
National Holiday: 30 May, Anti-colonial Movement (1969).
Armed Forces: The Dutch Crown is responsible for defense.

were founded. However, it was not until 1948 that the new post-war Dutch Constitution renamed what had been known as 'Curaçao and dependencies' as the 'Netherlands Antilles', thus bringing about the notion of statehood to the islands.

[10] The issue of independence has long been at the center of local political life. In 1954, a new law established the islands' autonomy over their internal affairs.

[11] On 30 May 1969, with unemployment at 20 per cent, a labor demonstration was broken up by the police, 300 Dutch marines and US marines from the American fleet, which happened to be anchored in the archipelago. As a result of these disturbances, Parliament was dissolved.

[12] The People's Electoral Movement (PEM), founded in 1971 in Aruba, proposed that since the islands had nothing in common (not even a name, because 'Antilles' is the generic designation of the whole region), each island should be free to choose its own constitution and set itself up as an autonomous republic. In a referendum in 1977, the majority of Arubans voted in favor of separation from the other islands.

[13] The Netherlands argued that the federation would ensure better economic prospects and political stability. The differences of opinion, also held among islanders themselves, delayed negotiations.

[14] The delay radicalized the electorate and in 1979 the New Antilles Movement (MAN) achieved a parliamentary majority in Curaçao (7 out of 12 seats). In coalition with the PEM and the UPB (Bonaire Patriotic Union) it formed the first left-of-center government in the islands. The MAN favored a federal formula with considerable autonomy for each of the islands but the PEM insisted on the total separation of Aruba.

[15] In 1980 it was agreed to set independence for 1990, on condition that each of the six islands submitted the issue to a referendum as soon as possible. Aruba chose to become an individual associated state, breaking away from the federation in January 1986 (see Aruba).

[16] The Netherlands Antilles became modern trading posts, totally dependent on transnational oil companies, which maintained a monopoly on all oil refining and processing. Exxon and Royal Dutch Shell (an Anglo-Dutch consortium), linked financially and commercially to branches of approximately 2,500 foreign firms registered on the islands, managed to have virtual decision-making authority over 85 per cent of the total imports, 99 per cent of the exports and 50 per cent of the islands' net income.

IN FOCUS

ENVIRONMENTAL CHALLENGES
There is little farming. There is a large Venezuelan oil refinery is located on Curaçao. The coastal areas of the islands have suffered the consequences of economic development. The soil has been polluted, and waste disposal is a serious problem due to the lack of proper facilities. Projects for the construction of landfill sites were developed but could not be implemented because of insufficient funding. Curaçao is the most polluted of the islands.

WOMEN'S RIGHTS
Although the islands depend on the Netherlands in several aspects, the population elects its own Parliament every four years. Women have been able to vote and stand for office since 1954.
Female illiteracy fell from 10.9 per cent in 1995 to 3.3 per cent in 2000 and has since remained at about that level. Female unemployment, which in 1990 stood at 18.1 per cent, dropped slightly to 16.2 per cent in 2000. Traditionally, women have made up slightly more than 40 per cent of the workforce. Some 99 per cent of births

take place in clinics or hospitals. Maternal mortality stood at 20 per 100,000 live births in 2000.

CHILDREN
School attendance is mandatory for children between 6 and 15 years old.
Child labor is prohibited by law for children under 14 with the two following exceptions: when they work in or for the benefit of the family in which they are being raised and in schools, provided these activities are of an educational nature and are not intended to generate a profit. The work should not be physically or mentally demanding or dangerous for the child. Children between 14 and 18 years old should not engage in night work or in work of a dangerous nature (including the risk of death, injuries and other dangers to health).

INDIGENOUS PEOPLES/ETHNIC MINORITIES
The Arawaks were the first inhabitants in the area. About 85 per cent of Curaçao's population is of African descent. The rest of the population is made up of a mixture of Dutch, Portuguese, North Americans and natives from

other Caribbean islands. The major religions include Anglican, Muslim, Protestant, Mormon and Baptist. The Jewish community is one of the oldest in the region, dating back to 1634.

MIGRANTS/REFUGEES
Although refugees should receive the same treatment as in the Netherlands, the problems are not the same. In March 2004, after being held in custody for nearly a year, five Cuban refugees were released and given permission to live on the island while authorities considered their request for political asylum.
It is estimated that 5 per cent of the island's population have emigrated in recent years due to economic difficulties. The main country of destination was the Netherlands.

DEATH PENALTY
As in the Netherlands, the death penalty for all crimes was abolished in 1982.

* Latest data available in *The State of the World's Children* and *Childinfo* database, UNICEF, 2006.

[17] At the end of 1984, the announcement that transnational oil companies were withdrawing from the islands spread panic among the people and local political leaders.

[18] The conflict was partially resolved in October 1985 when the Antilles Government purchased the Curaçao refinery and in turn rented it to Petróleos de Venezuela S.A. The deal did not include the Exxon plant on Aruba, which closed down.

[19] The separation of Aruba influenced the 1984 victory of a right-of-center coalition headed by María Liberia Peters of the National People's Party (PNP). This coalition was unable to sustain the minimum consensus necessary to stay in power. Therefore in January 1986, MAN leader, Domenico Martina, became Prime Minister once again. In 1988, María Liberia Peters returned to power.

[20] In 1990, the Government renewed its contracts with the Venezuelan oil company and introduced a series of austerity measures designed to cover the deficit generated by the Aruban withdrawal from the federation.

[21] In the general elections of March 1990, the PNP won seven seats while the MAN of Martina won only two. Peters was re-elected as Prime Minister.

[22] The election of the Island

Councils in 1991 was marked by the defeat of the Democratic Party, which had been in power for 40 years in Sint Maarten (Windward Islands group), as a result of internal divisions, financial irregularities committed by the administration and the fear of a downfall in the tourism boom.

[23] The Netherlands requested that each island made a separate proposal for constitutional reform in 1993. Accordingly, Curaçao received special status, Sint Maarten was made independent, while Bonaire, Saba and Sint Eustatius continued under Dutch control. The constitution also dictated that the Treaty of Strasburg, which predicted complete independence for 1999, would not be applied in the islands. A referendum carried out in 1994 supported the continuation of the Federation.

[24] In November 1999, Miguel Pourier was elected Prime Minister for the third time.

[25] The Netherlands announced a contribution equivalent to $6.24 million to subsidize the tourism industry, which suffered from the reduced air travel in the wake of the 11 September 2001 terrorist attacks in New York.

[26] Following the elections of January 2002, Antillean Restructuring party (PAR) leader

Etienne Ys became prime minster. He resigned in August 2003 and was replaced by Mirna Lousia-Godett but replaced her in office again during 2004.

[27] In late 2005 it was agreed with the Dutch Government that the federation will be dissolved by July 2007. Curaçao and Sint Maarten are set to become autonomous territories or 'associated states' of the Netherlands. Bonaire, Sint Eustatius and Saba, meanwhile, are to become 'kingdom islands', a new status that has yet to be fully defined, but will make them closely resemble Dutch municipalities (though the islands will not be obliged to adopt the euro).

[28] Following the parliamentary elections of January 2006, popular new PAR leader Emily de Jongh-Elhage became prime minister of the coalition government.

[29] In early 2006, due to the deteriorating relations between the Venezuela and US governments, and the constant remarks between their respective presidents Hugo Chavez and George W Bush, the media started to speculate that, given an eventual military attack from the US, the Netherlands Antilles could be used as an operations base. This could expose the country to a pre-emptive attack by Venezuela, according to these sources. ∎

New Zealand/Aotearoa

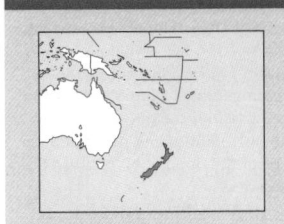

Population:	4,092,625
Land area:	270,530 km²
Capital:	Wellington
Currency:	NZ dollar
Language:	English and Maori

A otearoa, 'the land of the long white cloud', was settled around 1200 by Maoris, who may have arrived there from eastern Polynesia. Maori themselves speak of the arrival of *waka hourua* (voyaging canoes) from the legendary island of Hawiiki (see box 'Melanesians and Polynesians: surviving cultures'). Over the years a distinct culture developed, based on tribal organization and a strong affinity with the land. Maoris saw themselves as guardians of the land for future generations.

² In 1642 Abel Tasman, from Holland, reached the South Island, the larger of Aotearoa's two main islands. However, a misunderstanding with the indigenous population prevented him from going ashore. It was not until 1769 that James Cook from England surveyed the shores of the two larger islands. This opened the door for a growing colonization of the country that Tasman had named 'New Zealand'. The French also took an interest in this new land and set about purchasing land parcels most notably in an area called Akaroa, a small peninsula on the north-east coast of the South Island. In response, the British signed a 'Declaration of Independence' with 34 northern Maori tribes in 1835, thus declaring New Zealand an independent state under British rule. In response to increasing lawlessness, however, the Treaty of Waitangi was created and New Zealand was formerly annexed by the British Crown as a colony in 1840.

³ In the early 19th century, colonization increased with the arrival of British immigrants and missionaries. They brought with them new diseases, values and beliefs, which affected the traditional Maori way of life. Trading with settlers initially brought wealth to many Maori communities, but these gains were reversed with interest once the settlers began to alienate significant amounts of Maori land.

⁴ After 1840 the two larger islands were occupied under different legal arrangements: the South Island was incorporated by virtue of the right of 'discovery', and the North Island through the Treaty of Waitangi. According to the text of the treaty - which is different in its English and Maori versions - Maori chiefs accepted the presence of British settlers and the establishment of a government by the Crown to rule the settlers. In exchange, Maoris were assured absolute respect for their national sovereignty.

⁵ However, once the Treaty was signed, an extremely violent process of expropriation of Maori lands began. The so-called 'land wars' between the Maori and Europeans were essentially for sovereignty and guaranteed rights to the lands, forests, fisheries and other *taonga* (treasures).

⁶ Massive immigration, land confiscation and legal decisions led to the gradual annexation of Maori land, to the extent that out of the 27 million hectares they owned in 1840,

they now have little over a million left.

⁷ While the north of the country was involved in a series of wars, the South Island settlers went through a period of prosperity because of the discovery of gold. This discovery brought a massive flow of British, Chinese and Australian immigrants, which energized the region's economy.

⁸ After 1840 colonization increased. The Päkehä (non-Maori) usurped the right to fish in the area, thus depriving Maori people of one of their main activities. For the Maori, British dominance threatened their cultural extinction, due to the arbitrary imposition of European language, religion and costumes.

⁹ Maori opposition found new expression however, by the end of the 19th century. They organized petitions, delegations and submitted their claims before local courts and even before the British Crown itself, demanding compliance with the Treaty of Waitangi. These efforts were fruitless. The lands that had once belonged to the Maori were now used for farming, which had started to play a central role not only in the life of the settlers but also in the whole economy. New markets for dairy and meat products opened up in the 1880s, with the appearance of cold-storage systems, which made long-distance shipping possible. The rising price of these products constituted the basis of the country's economic development.

¹⁰ By the end of the century, the country's political scene was dominated by the Liberal government. They were the first in the world to grant the vote to women in 1893 and to establish measures to protect the rights of industrial workers, a group

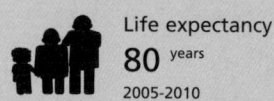

Life expectancy
80 years
2005-2010

GNI per capita
$19,990
2004

which was growing parallel to the development of cities and manufacturing industries.

[11] Until the 20th century there were no political movements to oppose the power of the Liberals who had become a coherent, organized political party after 20 years in power. The Labour Party (LP), with ample working-class and urban middle-class support, came to power for the first time in 1935. The Party also enjoyed the support of Maori even though there was still no legislative recognition of the Treaty of Waitangi.

[12] World War II marked the beginning of a new era for the country, for during that time Britain was unable to guarantee the security of its former colony. This led to closer ties between New Zealand/Aotearoa and the US. Through a series of political and military alliances, US presence was consolidated in the region. In the 1950s and 1960s New Zealand had to pay the price for this relationship, particularly when it found itself involved in the Vietnam War, a conflict that touched the political life of the country very deeply.

[13] Legislation in the 1950s forced many Maori from their land and the 1960s saw their increasing urbanization.

[14] In the 1970s, New Zealand/Aotearoa tried to diversify its production and to enter markets other than the UK and the US. Unemployment rose and inflation reached unprecedented levels due to the failure of the diversification scheme, the rise in the price of oil and financial loans. The year 1975 saw a drastic fall in the purchasing power earned from primary exports. This fall, combined with the foreign loans, increased foreign debt. That year the new National Party (NP) Government closed the doors on immigrants, mainly from the Pacific, whom it blamed for increasing unemployment.

[15] In 1975, growing Maori activism led to the formation of the Waitangi Tribunal to investigate Treaty claims. In 1986 the Labour Government gave it the power to hear claims dating back to the 1840 signing. The Tribunal had no binding powers over the Crown, and by 1992 less than 15 per cent of the recommendations had been implemented.

[16] During the 1980s, the Labour Government introduced a monetarist economic policy, which included the privatization of some public enterprises. These policies alienated many traditional Labour supporters and between1986 to 1999, Aotearoa was an international showcase for neo-liberalism. Public spending was slashed and labor laws restricted trade union activity.

[17] In 1987, New Zealand's nuclear-free status became law, prohibiting the entry of nuclear arms or vessels to the nation's ports, thus questioning the military presence of France and the United States, who used the South Pacific as a nuclear weapons testing ground. This decision suspended indefinitely the 1951 defense treaties between Australia, New Zealand and Australia (ANZUS), although no formal dissolution has ever been announced.

[18] The National Party won the October 1990 elections, but the change in administration did not affect the New Zealand economy. There was an increase in privatizations, protectionism was dismantled even further, and there were large cutbacks in health, education and social benefits. The result was falling inflation, but unemployment also rose.

[19] In 1994, New Zealand achieved its first budget surplus in 17 years, the currency became more robust, unemployment fell and inflation settled at two per cent.

[20] Between 1994 and 1995 the Government compensated the Tainui iwi in the North Island both with money and 15,400 hectares of land, for their claims over land colonized in the previous century. Britain's Queen Elizabeth apologized for the loss of life during the colonization of the islands and in 1996 the Crown compensated the major South Island Iwi, Ngai Tahu. Unemployment rates continued to fall, settling at six per cent in 1995. Young people, Maori and immigrants from other South Pacific Islands constituted the bulk of the unemployed and nearly ten per cent of the population received State subsidies.

[21] In 1996, New Zealand changed its electoral system for parliamentary elections from a relative majority system to a mixed-member proportional system based on the German model known as 'personalized proportional representation'. This system, which was fairer on political parties, encouraged electoral participation.

[22] In 1997, Jenny Shipley became New Zealand's first female Prime Minister, succeeding Jim Bolger after an internal National Party vote.

[23] Parliaments elected by proportional representation boosted the number of Maori and female representatives. The 1999 Parliament included 16 Maori deputies, two gay activists, as well as 35 women, including the first popularly elected female Prime Minister, Helen Clark.

[24] In April 2000 the Government announced that the titles of Sir and Dame, issued by the British Crown, would no longer be used but would be replaced by titles unique to New Zealand/Aotearoa.

[25] Also that year, to the alarm of consumers, environmentalist and Maori groups, which had organized a campaign to declare the country 'free from genetic engineering', the Royal Commission on Genetic Modification issued a report supporting the development and use of genetically modified organisms. To placate public concerns, the Government placed a moratorium on unauthorized GM field trials until October 2003.

[26] Queen Elizabeth II visited New Zealand in 2002, commemorating her Golden Jubilee. Prime Minister Helen Clark said at the time that it was 'inevitable that New Zealand shall become a republic and that this would reflect the reality that New Zealand is a 21st century state, completely sovereign and 20,000 kilometers away from the United Kingdom'.

[27] In July 2002, during the celebrations of the 40 years of Papua New Guinea's independence, the Prime Minister formally apologized for the treatment Papuans received from New Zealanders during the colonial period. That year, Clark was elected for a second term in office, albeit needing to form a coalition with the center-right party, United Future New Zealand, and the support of the Green Party. The election resulted in turmoil for the NP, which had the worst electoral showing in its 70-year history.

[28] In July 2004, Wellington

IN FOCUS

ENVIRONMENTAL CHALLENGES
The country faces problems such as deforestation and erosion but also the disappearance of autochthonous flora and fauna species.

However, the passing of the Resource Management Act in 1991 and the signing of the Kyoto Protocol in 1997 strengthened environmental protection to the point that any development or business proposal has to be previously approved and agreed on by all affected parties.

More recently there have been campaigns against genetic engineering in agriculture and unsustainable urban development in ecologically important areas.

WOMEN'S RIGHTS
New Zealand/Aotearoa was a pioneer in women's suffrage: women have voted since 1893 and been able to stand for election since 1919. In 2005, women held 32 per cent of seats in Parliament.

In 2004, of the 2 million people that made up the total labor force, 46 per cent were women. Female unemployment stood at 5 per cent.

The maternal mortality rate stood at 7 per 100,000 live births in the year 2000.

CHILDREN
In 2004, infant and under-five mortality rates stood at 5 and 6 deaths per 1,000 live births, respectively. Maori children amount to a quarter of the total number of minors in New Zealand.

One third of infant deaths are caused by Sudden Infant Death Syndrome; in 59 per cent of the cases the mother and/or father is a smoker. Among Maori children incidence rates were three times higher than among non-Maori children.

INDIGENOUS PEOPLES/ ETHNIC MINORITIES
Maori have inhabited the land they call Aotearoa for around 800 years. Tuberculosis, typhoid fever, chickenpox and other diseases unknown to the Maori, together with the introduction of firearms and the constant inter-tribal wars and wars against Europeans led to a significant decrease in their population (there were estimated to be 110,000 Maori in 1800).

In 2002 Maori or their descendants accounted for 16 per cent of the population (597,800 people) and their birth rate is higher than that of people of European origin, but they suffer disproportionately from poverty and unemployment.

Some of their current claims are: the creation of an independent Maori state, more participation in the government, protection of Maori land in particular, ownership of the foreshore, fisheries and other *taonga*, and increased promotion of their culture and their language.

MIGRANTS/REFUGEES
The conditions required by those who want to obtain asylum or refugee status in New Zealand/Aotearoa are not very flexible.

The number of asylum applications has been in steady decline between 2001 and 2005. In 2001, there were 13,970; 6,860 in 2002; 5,140 in 2003; 3,780 in 2004 and 3,560 in 2005. In 2005, most asylum applications were filed by refugees from Iran, Czech Republic, Bangladesh, Iraq, Nepal, China, India, Fiji, Syria and Somalia.

DEATH PENALTY
It was abolished in 1989; the last execution occurred in 1957.

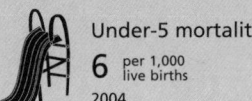

Under-5 mortality	Maternal mortality
6 per 1,000 live births 2004	**7** per 100,000 live births 2000

suspended high-level contacts with Israel after two Israeli agents working for the secret service Mossad were jailed for trying illegally to obtain New Zealand passports. Shortly after, several acts of desecration at Jewish cemeteries were condemned by the Government. The spies were deported in September 2004. Ties with Israel were reinstated in June 2005.
[29] A first round of talks on a potential free trade agreement between New Zealand and China was held in December 2004. In the same month, Parliament passed by a narrow margin the Civil Union bill which recognized unions between unmarried men and women, granting them the same rights as married couples in areas like child custody, tax and welfare. The bill also applied to same-sex couples.
[30] In elections held in September 2005, Prime Minister Clark secured a narrow victory over the National Party, but remained dependent on the support of minor parties.
[31] Also in May, the Government made the decision - to be ratified by Parliament - that the Waikato river (the longest in New Zealand/ Aotearoa) would be jointly managed with the Maori, who had been claiming their ancestral rights over the Waikato's natural resources for 30 years.
[32] In August 2006, the country declared an official week of mourning to mark the death of Maori queen Te Arikinui Dame Te Atairangikaahu. ∎

Niue

Population: 1,457
Land area: 260 km²
Capital: Alofi
Currency: NZ dollar
Language: English

Niue was settled by Samoans and Tongans. Captain James Cook named it 'Savage Island' when he visited it in 1774. The indigenous peoples' reputation for fierceness kept missionaries away (the island's first permanent mission dates from 1861), and also the slave traders who caused much suffering in other areas of the Pacific. Emigration to the phosphate mines on other islands in the area initiated an outflow that has continued ever since. In 1900 the island was declared a British protectorate and was annexed by New Zealand in 1901. It was administered along with the Cook Islands until 1904, when it became detached to form a separate possession. In 1974 it became an Autonomous Associated State of New Zealand. The United Nations recognized this as a legitimate decision and eliminated the 'Niue case' from the Decolonization Committee agenda. As well as economic support from New Zealand, Niue also received help with defense and international affairs.
[2] In April 1989, questions arose over the management of economic aid received from New Zealand/Aotearoa by Prime Minister Robert Rex's government. Despite this, the opposition lacked the votes necessary to approve a parliamentary motion to censure Rex's administration.
[3] In 1991, New Zealand announced a reduction in its financial aid to the island. Niue and Australia established diplomatic ties during 1992.
[4] The population of the island has been steadily declining due to emigration to New Zealand/Aotearoa; the current population estimated is 1,457 compared with a peak of 5,200 in 1966. ∎

ENVIRONMENT
Located in the South Pacific in southern Polynesia, 2,300 km northeast of New Zealand/Aotearoa, west of the Cook Islands and east of the Tonga archipelago. Of coral origin, the island is flat and the soil relatively fertile. Its rainy, tropical climate is tempered by sea winds.

SOCIETY
Peoples: The people of Niue are of Polynesian origin. **Religions:** Protestant. **Languages:** English (official), local Niue language (national). **Main Political Parties:** Niue People's Party (NPP); Niue People's Action Party (NPAP).

THE STATE
Official Name: Niue. **Capital:** Alofi 1,000 people (2003). **Government:** Autonomous associated state. Anton Ojala, Representative of New Zealand since February 2006. Young Vivian, Premier since April 2002 (second time). Single-chamber legislature: there is a 20-member Legislative Assembly which is headed by the Prime Minister. New Zealand/Aotearoa controls defense and foreign affairs. **National Holiday:** 6 February, Waitangi Day (1840). **Armed Forces:** New Zealand/ Aotearoa is responsible for the island's defense.

Cook Islands

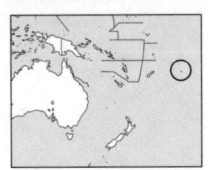

Population: 17,797
Land area: 236 km²
Capital: Avarua
Currency: NZ dollar
Language: English

The islands, which had already been explored and settled by Polynesians and Spaniards, received their name from the English navigator Captain James Cook, who drew up the first map of the archipelago in 1770.
[2] In 1821, Tahitian missionaries were sent to the islands by the London Missionary Society; a Protestant theocracy was established and all 'pagan' structures were destroyed, as were many of the traditional forms of social organization. The islands were declared a British Protectorate in 1888 and became part of New Zealand/ Aotearoa in 1901. The land rights of the indigenous Maori were recognized, and the sale of real estate to foreigners was prohibited. In 1965, the United Nations promoted and supervised a plebiscite, and the population voted against independence, and in favor of maintaining its ties to New Zealand.
[3] New Zealand's economic injections to the archipelago dropped in 1991 to 17 per cent of the local budget. Both governments decided that the auditors in the Cook Islands would take up supervision of the State finances in place of the New Zealand office.
[4] In the mid-1990s, Prime Minister Geoffrey Henry sharply reduced the State budget and launched a privatization plan. In late 1997, some ministries were closed down due to the lack of funds and the budget of the remaining entities was reduced by ten per cent.
[5] In March 2005, the islands were taken off an international list of countries with weak policies in the fight against money laundering.
[6] China granted in April 2006 $4 million in economic aid for infrastructural works. ∎

ENVIRONMENT
Area: 236.6 sq km. Archipelago located in the South Pacific 2,700 km northeast of New Zealand/Aotearoa, made up of 15 islands which extend over an ocean area of two million sq km. They are divided into two groups. The northern group is made up of six small coral atolls, which are low and arid, with a total area of 25.5 sq km. The southern group comprises eight larger and more fertile volcanic islands (211 sq km). The capital, Avarua, is located on Rarotonga, the largest of the islands. Every five years, for the past two decades, the islands have suffered terrible droughts.

SOCIETY
Peoples: The majority of the population is Maori. There are also some Europeans (2.4 per cent). **Religions:** Christian (Cook Islands Church). **Languages:** English (official), Cook Island Maori. Language and traditions similar to Maori in New Zealand. **Main Political Parties:** Democratic Party; The Cook Islands Party (CIP); Tumu Enua; Cook Islands First Party.

THE STATE
Official Name: The Cook Islands. **Capital:** Avarua 13,000 people (2003). **Government:** Autonomous associated state. Sir Frederick Goodwin, Representative of Queen Elizabeth II since February 2001. Jim Marurai, Prime Minister since December 2004. Single-chamber legislature; there is a Legislative Assembly, with 25 members elected by direct vote every five years.

Tokelau / Tokelau Islands

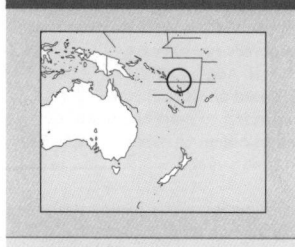

Population:	1,615
Land area:	10 km²
Capital:	Fakaofo
Currency:	NZ dollar
Language:	Tokelauan

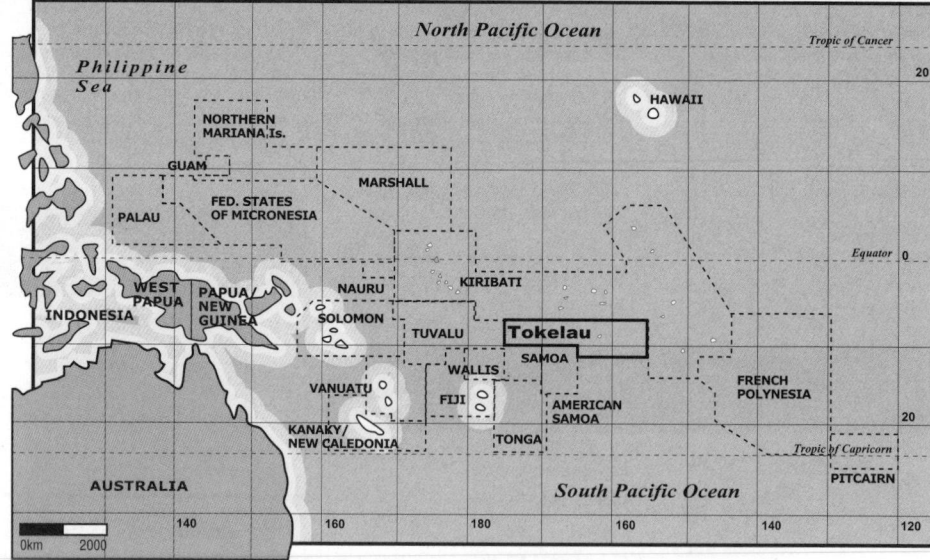

P olynesian peoples arrived in Tokelau around the 9th century, developing a culture in harmony with the land. They called themselves 'land guardians for the future'.

2   English explorer John Byron arrived in 1765. As there were no major resources to be exploited, the British only made the islands a protectorate in 1877. They were annexed in 1916, included as part of the colonial territory of the Gilbert and Ellice Islands (now Kiribati and Tuvalu). In 1925, Britain transferred administrative control of the islands to New Zealand/ Aotearoa. In 1946, the group was officially designated the Tokelau Islands, and in 1958 full sovereignty passed to New Zealand.

3   New Zealand adopted policies intended to maintain traditional customs, institutions and communal relations in Tokelau. Until 1990 there was only a single ship calling at the islands from Apia (Samoa) every two or three months. There are no adequate harbors for the development of tourism.

4   In the 1980s, farming underwent a crisis due to a number of adverse climatic conditions. Emigration to New Zealand/Aotearoa and Samoa rose considerably. The New Zealand/ Aotearoa Government tried to encourage Tokelau immigrants to return to their homeland, but without success.

5   In 1976 and 1981, UN envoys to Tokelau reported that the inhabitants wanted to maintain their existing relationship with New Zealand/Aotearoa. In December 1984, the UN Assembly decided that New Zealand/Aotearoa, as administrating power, should report on its management, but continue to administer the islands. However, before the UN Special Committee in June 1987, Tokelau expressed a wish to achieve greater political autonomy.

6   In February 1990, the country was devastated by Hurricane Ofa, which destroyed all the banana trees and 80 per cent of the coconut

plantations, as well as hospitals, schools, houses, and bridges. As a result, emigration increased.

7   In 1991 the first regular maritime transport service was established between the three atolls.

8   In May 1995, New Zealand Parliament approved extended powers for the local assembly. Satellite telephone arrived in 1997.

9   The New Zealand Official Development Aid (NZODA) set aside a budget of NZ$7.5 million for Tokelau in 2000.

10   The Minister of Pacific Islands implemented support programs in policy management and development in 2001, to help Tokelau initiate self-rule.

11   Between 2002 and 2003 the budget included additional funds for the construction of the future government buildings.

12   Australia has also maintained co-operation agreements with the islands. During 2002 and 2003, Australian funds were granted for education in Tokelau, as aid for students in the territory, or as scholarships within Australia.

13   With assistance provided by the Secretariat of the Pacific Community (SPC) and the Government of Samoa, Tokelau started to develop a community-based fisheries management plan which included the three atolls: Fakaofo, Nukunonu and Atafu.

14   The SPC and the Samoa Fisheries Department were in charge of training each community. The atolls still have traditional laws for the preservation of the sea coast and species for the next generations. The scientific methods of this plan would help communities to maximize the traditional institutions, knowledge and

fisheries regulatory measures.

15   Tokelau runs serious risk of disappearing under the ocean in the not-so-distant future due to global climate change. Its small size and flat terrain - its maximum height above water is 4.5 meters - make it particularly vulnerable

to the rising of the oceans.

16   In a referendum in February 2006, Tokelauans narrowly voted against becoming a self-governing territory rather than a colony of New Zealand. Another referendum on the same issue is due to take place in November 2007. ∎

PROFILE

ENVIRONMENT
A group of coral islands, comprising three atolls: Atafu 2.02 sq km and 577 people, Nakunono 5.46 sq km and 374 people, Fakaofo 2.63 sq km and 664 people. Geographically, Sivains atoll belongs to the group but is an administrative dependency of American Samoa. The group is located in Polynesia in the South Pacific, to the east of the Tuvalu islands. The islands are flat, with thin, not very fertile soil. Rainfall is erratic and there are frequent droughts. Fishing constitutes the traditional economic activity.

SOCIETY
Peoples: The population is of Polynesian origin.
Religions: Mostly Protestant, 70 per cent.
Languages: Tokelauan (official), English and local dialects.

THE STATE
Official Name: Tokelau.
Capital: Fakaofo 540 people (1999).
Other Cities: Fenua Fala, and small villages on every island.
Government: All executive and administrative functions are in theory vested in the Administrator of Tokelau, responsible to the New Zealand Minister of Foreign Affairs and Trade. In 1994 these powers were formally delegated to Tokelau's Fono (Parliament with 45 members, representing the three atolls), or to the Council of Faipule (cabinet) when the Fono is not in session. In 1996 the New Zealand Parliament amended the 1948 Act to confer limited legislative power on the Fono, giving Tokelau in practice, and in large measure, administrative and political autonomy. A referendum on whether Tokelau should become independent but in free association with New Zealand, supervised by the UN, was held in February 2006. Although a 60 per cent majority voted in favor of the proposal, a two-thirds majority was required for the referendum to succeed, so Tokelau remains a New Zealand territory. In June 2006, the Fono agreed to hold a similar referendum again in 2007 or 2008. Kolovei O'Brien, Head of Government since 2006. David Payton, Administrator since October 2006.
National Holiday: 6 February, Waitangi Day (1840).
Armed Forces: Defence is the responsibility of New Zealand.

Nicaragua / Nicaragua

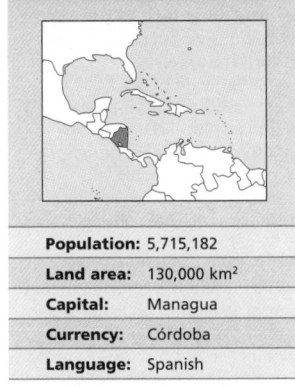

Population:	5,715,182
Land area:	130,000 km²
Capital:	Managua
Currency:	Córdoba
Language:	Spanish

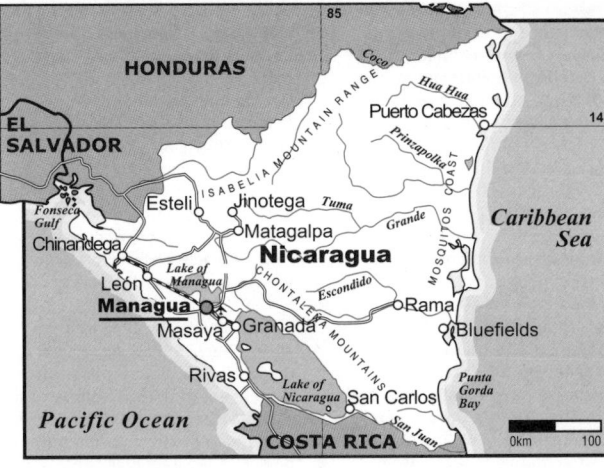

P resent-day Nicaragua was once a region of influence of the two great Central American cultures: the Chibcha (see Colombia) and the Maya. The Caribbean coast was inhabited by the Miskito, originally called 'Kiribi' - the oldest peoples in Nicaragua - and was visited by Christopher Columbus in 1502. By converting local leaders Nicoya and Nicarao to Christianity and crushing the resistance of Diriangen's armies, the conquistadors Gil Gonzales Dávila and Andrés Nino consolidated Spain's hold over the territory. In 1544 it was incorporated into the Captaincy-General of Guatemala.

² In 1821 Nicaragua became independent, together with the rest of Central America, joining the Mexican Empire. It withdrew in 1824 to form the Federation of the United Provinces of the Central America.

³ Nicaragua left the Federation in 1839, declaring itself an independent state. The country was divided between two groups: the coffee and sugar oligarchies, and the artisans and small landowners. The former would become the conservatives, and the latter, the liberals who favored free trade.

⁴ In 1856, 120 soldiers landed in Nicaragua under the command of William Walker, an American mercenary. With Washington's tacit support, he proclaimed himself President of Nicaragua. His purpose was to find new territories for slavery which was on the point of being abolished in the Union. Walker was defeated by the allied armies of Central America in 1857, and later executed. Nicaragua's ports were occupied by Germany in 1875 and Britain in 1895. Britain commandeered Nicaragua's Customs in order to collect on unpaid debts.

⁵ After 30 years of conservative rule the Liberal Party came to power in 1893, and José Santos Zelaya became President. The liberals refused to comply with demands made by the United States, as part of the 'dollar diplomacy' approach adopted by William M Taft's administration. In 1912, Taft sent in US Marines who killed the Liberal Party leader, Benjamín Zeledón, and remained in Nicaragua until 1925. In 1926 they returned to protect President Adolfo Díaz, who was about to be overthrown.

⁶ This second US occupation was resisted by Augusto C Sandino, who raised and commanded a popular army of about 3,000 troops. For more than six years they held out against 12,000 US marines who were supported by the airforce and ground troops of the local oligarchy. Sandino promised that he would lay down arms when the last marine left Nicaragua. He carried out his promise in 1933, but was assassinated by US-backed Anastasio Somoza García who, after seizing power, ruled despotically until he was killed by the patriot Rigoberto López Pérez in 1956.

⁷ During two decades of dictatorship, Somoza had achieved almost absolute control of the nation's economy. He was succeeded by his son, engineer Luis Somoza Debayle, who in turn handed the Government on to his son, Anastasio, a West Point graduate.

⁸ Anastasio Somoza outlawed all trade unions, massacred members of peasant movements and banned all opposition parties. In the 1960s the Sandinista National Liberation Front (FSLN) was founded. This organized and developed the guerrilla warfare which lasted for 17 years. When opposition leader Pedro Joaquín Chamorro, editor of the daily *La Prensa*, was assassinated on Somoza's orders in January 1978, a nationwide strike was called and there were massive protest demonstrations.

⁹ By March 1979, the Sandinista Front had united its three factions and formed the 'Patriotic Front '. In May 1979 the Front launched the 'final offensive', combining a general strike, a popular uprising, armed combat, and intense diplomatic activity abroad. On 17 July Somoza fled the country, bringing to an end a dynasty that had killed 50,000 people. The Junta for National Reconstruction, created a few weeks before in Costa Rica, was installed in Managua two days later.

¹⁰ The victorious revolutionaries nationalized Somoza's lands and industrial properties, which constituted 40 per cent of the economic resources of Nicaragua. They also replaced the defeated National Guard with the Sandinista Popular Army. The revolutionary government implemented a literacy campaign and began reconstruction of the devastated economy.

¹¹ In May 1980 two non-Sandinista members of the Junta, Violeta Barrios de Chamorro and Alfonso Robelo, resigned. The Government avoided a crisis by replacing them with Rafael Córdoba and Arturo Cruz, two 'moderate' anti-Somoza activists. The FSLN confirmed it would rule in a context of democratic participation, a non-aligned foreign policy and with respect for civil liberties.

¹² In 1981, US President Ronald Reagan announced his aim of destroying the Sandinistas. Between April and July 1982 deputy interior Minister Edén Pastora ('Commander Zero') deserted, and 2,500 former National Guards, supported by the US, invaded Nicaragua from Honduras. From then on Nicaragua was harassed without respite, forcing the authorities to extend the state of emergency, to institute compulsory military service and to ban pro-American political declarations.

¹³ In 1983, President Reagan admitted the existence of secret funds destined for covert CIA operations against Nicaragua. The funds were also used to aid counter-revolutionaries or *contras* operating from Honduran territory. Reagan referred to the contras as 'freedom fighters'.

¹⁴ Concerned at the serious threat of a war that might escalate throughout Central America, the governments of Colombia, Mexico, Panama and Venezuela sought a negotiated settlement to the conflict. As the 'Contadora Group', these countries' foreign ministers advanced peace plans which won great diplomatic support and prevented an invasion by US forces.

¹⁵ Contra attacks intensified steadily with open US backing. Elections were held in November 1984. Candidates were drawn from the FSLN, the Democratic Conservative Party, the Independent Liberal Party, the Popular Social Christian Party, the Communist Party, the Socialist Party and the Marxist-Leninist People's Action Movement. Over 80 per

LAND USE

2003/2005

IRRIGATED AREA: 2.8% of arable land

- FOREST AND WOODLAND: 42.7%
- ARABLE LAND: 15.9%
- CROPLANDS: 1.9%
- OTHER USE: 39.5%

PUBLIC EXPENDITURE

% OF GDP

- HEALTH & EDUCATION (2000-2002): 7 %
- DEFENSE (2004): 0.7 %

WORKERS

UNEMPLOYMENT: 8% (2004)

LABOR FORCE — 2004

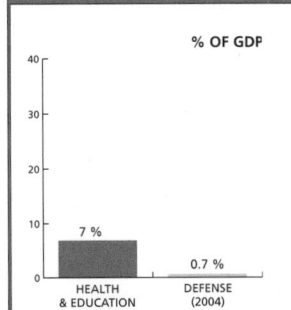

- FEMALE: 30%
- MALE: 70%

Life expectancy
71 years
2005-2010

GNI per capita
$830
2004

Literacy
77% total adult rate
2000-2004

HIV prevalence rate
0.2% of population 15-49 years old
2003

cent of Nicaragua's 1.5 million registered voters went to the polls and the FSLN obtained 67 per cent of the vote. In November Reagan was re-elected and in April 1985 he declared a trade embargo against Nicaragua and seized its assets.

[16] A new Constitution came into effect in January 1987. It provided for a presidential system, with a president elected by direct vote for a six-year term. Legislators would be elected on the basis of proportional representation.

[17] In February, with UN and OAS participation, the Central American presidents met for negotiations in Esquipulas, Guatemala. The Esquipulas II Accords stipulated an end to external support for armed opposition groups; the opening of internal dialogue in each of the countries, mediated by the Catholic Church; and an amnesty for those who lay down their arms, with guarantees of political representation.

[18] In Nicaragua a National Reconciliation Commission was formed. Contra leader Fernando Chamorro returned from exile; he was granted an amnesty after he renounced violence.

Press censorship was lifted and Violeta Chamorro's opposition daily *La Prensa* reappeared. On 7 October, a unilateral ceasefire went into effect in several parts of the country, although contra leaders announced that they would continue hostilities.

[19] Throughout 1988, US pressure and the effects of Hurricane Joan worsened the economic situation. Monetary reform and a cut in the Government budget in February did not halt spiraling inflation.

[20] In July 1988, the US ambassador to Managua was expelled on the accusation of encouraging anti-Sandinista activities. The US Government responded by expelling Nicaragua's representative in Washington.

[21] The Esquipulas II Accords seemed to be doomed, but when the five Central American presidents met at Costa del Sol, El Salvador, in February 1989, President Daniel Ortega embarked on fresh negotiations. The Sandinista proposal was to bring the elections forward to February 1990 and to accept proposed modifications to the 1988 electoral law. The condition was that the contras dismantle their

bases in Honduras within three months of an agreement. The US however insisted that the contras continue in Honduras, and President George Bush persuaded Congress to award them $40 million in 'humanitarian aid'.

[22] Daniel Ortega was the FSLN presidential candidate. The National Opposition Union (UNO), a 14-party coalition, nominated Pedro Joaquin Chamorro's widow, Violeta Barrios de Chamorro.

[23] All surveys showed the FSLN would win by a wide margin on 25 February 1990. Unexpectedly, the UNO won the elections with 55 per cent of the vote against the FSLN's 41 per cent. Ortega accepted defeat and pledged to hand over power to the new president, Violeta Chamorro.

[24] On 25 April, before she took office, the President and the FSLN signed a 'Transition Protocol'. This included respecting the standing Constitution and the social achievements of the revolution, and supporting disarmament of the contras. The new President announced that she would personally assume the Defense Ministry and maintain the Sandinista General Humberto Ortega as commander of the armed forces. This forced the UNO's Vice-President Virgilio Godoy and other members of the coalition to withdraw from the Government.

[25] In May 1990, public employees went on strike for wage increases of up to 200 per cent. The Government declared the strike illegal, and revoked the civil service law (under which civil servants could not be fired without just cause) as well as the agrarian reform law passed by the Sandinista Government. Workers responded by extending the strike over the whole country. After a week, the Government partially gave in to the workers' demands, and the strike ended.

[26] Since the mid-1990s, the Government has received several offers from international consortia interested in carrying out projects in some 270,000 hectares of tropical rainforests in northern Nicaragua, ranging from the creation of landfill sites for toxic waste, to the exploitation of the region's vast fishing, mineral and forestry resources.

[27] When the Government was accused of carrying out secret negotiations with a Taiwanese enterprise, the existence of large mineral deposits was inadvertently revealed. These included gold, silver, copper, tungsten and Central America's largest deposits of calcium carbonate, a raw material used in cement production.

[28] In 1991, President Chamorro agreed with the FSLN to recognize

agrarian reform and to set aside for workers at least 25 per cent of shares in state enterprises slated for privatization.

[29] Inflation fell from 7,000 per cent in 1990 to 3.8 per cent in 1992 due to an IMF and World Bank-sponsored adjustment program. Productive investments and spending in education and health were reduced. Unemployment rose to 60 per cent.

[30] Differences between the President and the UNO led them to split in 1993 after which Chamorro received support from the Sandinistas and the UNO's Center Group. The following month, the UNO expelled that group and changed its name to Political Opposition Alliance (APO).

[31] Bypassing the party's leadership, the FSLN parliamentary bloc presented its own bill against nepotism which banned presidential re-election and prohibited relatives of the standing presidents from running for president. This clause put an end to the political aspirations of Chamorro's son-in-law, minister Antonio Lacayo.

[32] The economic crisis was intensified by a drought which led to the loss of 80,000 hectares of crops and left 200,000 farmers without food. Malnutrition affected 300,000 children and some lost their sight through lack of vitamin A.

[33] In January 1994, the UNO, with less than half its founders and unable to obtain the support to set up a constituent assembly, ended a year of boycotting the National Assembly. Violence continued between the army, gangs of criminals and small guerrilla groups.

[34] In August the Assembly passed a new military law aimed at eliminating political involvement by the Sandinista Popular Army and increasing its dependence on civilian authority, although the power was actually left in the hands of a military council. General Humberto Ortega resigned.

[35] Debate on constitutional reform prevailed in 1995. In February, the Assembly proposed to change the army's name, ban compulsory military service and grant guarantees to private property. These measures were supported by President Chamorro but she did not agree with the shift of power from the executive to the legislative branch, regarding the right to raise taxes. The Assembly published the reforms unilaterally in February and began to implement them.

[36] In June, an agreement was reached on a general law for constitutional reforms which stated these had to be supported by a majority of 60 per cent in the

PROFILE

ENVIRONMENT

Nicaragua has both Pacific and Caribbean coastlines. It is crossed by two important mountain ranges: the Central American Andes, running from northwest to southeast, and a volcanic chain with several active volcanoes along the western coast. The Managua and Nicaragua lakes lie between the two ranges. On the eastern slopes, the climate is tropical with abundant rainfall, while it is drier on the western side where the population is concentrated. Cotton is the main cash crop in the mountain area, while bananas are grown along the Atlantic coast.

SOCIETY

Peoples: 69 per cent of Nicaraguans are mixed descendants of American Indians and Spanish colonizers; 17 per cent are of European descent; nine per cent are African descendants and five per cent belong to Indian minorities (Miskitos, Sunos and Ramas).
Religion: Catholics (85 per cent); Protestants (15 per cent)
Languages: Spanish (official and predominant). Miskito, Suno, English and Garifuna are spoken on the Atlantic coast.
Main Political Parties: The Constitutional Liberal Party, conservative (CLP); Sandinista National Liberation Front (FSLN); Conservative Party of Nicaragua (CPN).
Main Social Organizations: The Nicaraguan Labor Confederation (CTN); the Labor Unity and Action Confederation (CAUS); the Rural Workers' Association (ATC); the Workers' Front (FO); the Unified Labor Confederation (CUS); the National Employees' Union (UNE); the National Confederation of Professionals (CONAPRO); National Union of Nicaraguan Students.

THE STATE

Official Name: República de Nicaragua.
Administrative Divisions: nine regions, 15 departments, 143 municipalities.
Capital: Managua 1,098,000 people (2003).
Other Cities: León 153,200 people; Chinandega 120,400; Masaya 110,000; Granada 88,800; Matagalpa 73,400 (2000).
Government: Daniel Ortega Saavedra, President since January 2007. Legislature: National Assembly, with 93 members.
National Holidays: 15 September, Independence Day (1821); 19 July, Sandinista Revolution Day (1979).
Armed Forces: 14,000 troops (2003).

Under-5 mortality	Poverty	Debt service	Maternal mortality
38 per 1,000 live births 2004	**45.1%** of population living on less than $1 per day 2001	**5.8%** exports of goods and services 2004	**230** per 100,000 live births 2000

Assembly before being signed by the President, who concluded the agreement in July.

[37] Approval of the nepotism law was deferred. President Chamorro's son-in-law announced his plans to leave the Government in order to campaign for the November 1996 presidential elections.

[38] Conservative Arnoldo Alemán, ex-mayor of Managua, won 49 per cent of the vote, defeating Sandinista Daniel Ortega. The electoral law established that, having obtained more than 45 per cent of the vote, there was no need for a second round.

[39] On assuming the presidency, Alemán promised to create 500,000 new jobs and launched a plan to relieve the debt of the agricultural sector, estimated at $150 million.

[40] In April 1997, the Government and the Sandinista opposition accused each other of arming and training paramilitaries.

[41] In August, the Government announced it would not pass goods confiscated by the Sandinistas in 1979 on to Anastasio Somoza's heirs, in answer to a lawsuit filed by Lilian Somoza, the former dictator's daughter.

[42] According to estimates made in early 1998, the US had once again become Managua's main trading partner, with exports to there worth $375 million in 1997.

[43] After winning 53.3 per cent of the vote in the November 2001 elections, Liberal candidate Enrique Bolaños became President; Ortega received 45 per cent. Although he contested the number of seats obtained in the Assembly, which did not tally with those counted by the FSLN, OAS observers said the elections had been fair.

[44] A week after Bolaños took office in January 2002, former president Alemán was elected chairman of the National Assembly. Shortly after, the Assembly did not approve the courts' request to lift Alemán's parliamentary immunity to try him for fraud against the state ($1.3 million), embezzlement and unlawful association. Alemán had signed in 2000 a controversial agreement with the FSLN leadership to reform the Constitution, enabling him to share government positions with the Sandinistas, receive immunity and a life-long seat in the Assembly.

[45] In 2002, the courts ruled that the charges of sexual abuse and rape brought against former president Ortega in 1998 by his step-daughter Zoilamérica Narváez had lapsed, effectively absolving him. Ortega, an MP in 1998, was protected by National Assembly immunity. The Inter-American Commission on Human Rights accepted Narváez's complaint of lack of justice in Nicaragua.

[46] On 7 December 2003, former president Alemán was sentenced to 20 years' home arrest and made to pay a $17 million fine for crimes of corruption, including the use of $100 million of state funds for his electoral campaigns.

[47] On 17 December, Nicaragua, Guatemala, Honduras and El Salvador signed a free-trade agreement with the US to scale back tariffs and other trade barriers in agriculture, foodstuffs, investments, services and intellectual property sectors.

[48] Having qualified for the Highly Indebted Poor Countries (HIPC) initiative, after implementing anti-corruption and structural adjustment plans including privatizations, the World Bank 'pardoned' 80 per cent of the $6,500 million Nicaraguan external debt in January 2004. Critics of privatization stress it has not improved services - at least 50 per cent of people have no access to electricity or communications - and that government policies have not tackled serious social issues. Government spending on health fell from $50 per person in 1983 to $16 in 2000, malnutrition and infant mortality grew, and 50 per cent of the population still lived in poverty.

[49] Throughout 2005 tension grew with Costa Rica due to a dispute started in 1998 over the rights on the San Juan river, the border between the countries. A 19th century treaty granted Costa Rica the right to commercial navigation along the river. Managua stated that the treaty did not allow for Costa Rican police officers to patrol the river with weapons. San Jose argued that the area's high level of crime made it necessary to patrol the river with armed troops.

[50] In April 2006 a free trade agreement with the US, approved by the Nicaraguan Congress in 2005, came into force.

[51] In October 2006, Nicaragua announced an ambitious plan to build a new canal to link the Atlantic and Pacific Oceans that would rival the Panama Canal. Analysts were divided on whether the proposed $18 billion project was an impossible dream or a bold move that could deliver huge long-term benefits to one of the poorest countries in the region.

[52] In November 2006 Nicaragua's former leader and candidate of the Sandinista Front, Daniel Ortega, won the presidential election. The result was warmly welcomed by leftist leaders in the region such as Hugo Chávez of Venezuela and Fidel Castro of Cuba. It was less welcome in Washington, though Ortega claimed to be greatly changed since his revolutionary years, stressing his new commitment to the Catholic Church. Immediately after the election, with Ortega's support, President Bolaños signed into being a bill banning abortion on any grounds, even if the woman's life is in danger. ∎

IN FOCUS

ENVIRONMENTAL CHALLENGES
Approximately 40 per cent of the water and soil are polluted, and there is significant deforestation. Growing environmental deterioration is aggravating poverty. The country is prone to natural disasters such as earthquakes, volcanic eruptions, floods and droughts.

WOMEN'S RIGHTS
Women have been able to vote and stand for office since 1955. In 2003, 21 per cent of seats in Parliament were held by women, while female representation in ministries or equivalent positions reached 14 per cent.

Women made up 38 per cent of the two-million-strong work force.

In 2004, 86 per cent of pregnant women received prenatal health care, but only 67 per cent of deliveries were assisted by qualified staff*. The maternal mortality rate averaged 230 deaths per 100,000 deliveries, with the rate at its highest on the Atlantic coast.

CHILDREN
Children are affected by the country's extreme poverty and inequity. One in three has some degree of malnutrition, and one in nine suffers from chronic malnutrition. Although infant and under 5 mortality rates continue to be high, they have decreased between 1990 and 2004. The former went from 52 to 31 in every 1,000 live births and the latter from 68 to 38 in 1,000.

In 2002, only 29 per cent of children finished primary school, and it was estimated that 167,000 children and teenagers had to work. A new 'friendly schools' program implemented in 2003 has considerably improved the quality of education, with adequate facilities and equipments and a daily meal for all children. There is interest in addressing the issues of violence, working on conflict resolution through dialogue, and discrimination - boys share with girls educational tasks and also the cleaning and maintenance of the school. Children choose their own student government, voting through debates in which all can take part.

INDIGENOUS PEOPLES/ ETHNIC MINORITIES
There are two large ethnic minorities in Nicaragua: Afro-Nicaraguans (nine per cent of the population) and Miskitos (five per cent). Some Miskitos speak English and are Protestant, due to the British influence in the area between the 17th and 19th centuries, but most maintain their original language and culture.

There are other smaller communities: the Sumo, Rama and Garifuna (a mix of indigenous peoples and Afro-Nicaraguans). The Creoles are of African descent and live along the Caribbean coast. Most come from Jamaica and the Cayman Islands.

Although relations between Creoles and Miskitos have been tense throughout history, they share a common rivalry against the majority of the population, who are descendants of native peoples and Spanish colonizers, speak Spanish and are mostly Catholic.

To redress a history of exploitation and discrimination against the country's indigenous peoples, the Nicaraguan Constitution in 1987 provided for a degree of autonomy in the two regions on the eastern coast, granting indigenous communities there local powers and freedoms.

MIGRANTS/REFUGEES
Internal migration occurs from the countryside to Managua and toward peripheral areas of the country, caused by pressures from the expansion of agricultural cropland and the low agricultural returns in the dry areas.

In February 2006, the US Immigration Service extended for 12 additional months the coverage of its Temporary Protection Status for refugees that were already working in the country, allowing them to continue to do so. The measure was taken because of the serious situation caused by natural disasters in Central America. It benefited some 225,000 Salvadorans, 75,000 Hondurans and 4,000 Nicaraguans.

DEATH PENALTY
Abolished in 1979; the last execution took place in 1930.

* Latest data available in *The State of the World's Children* and *Childinfo* database, UNICEF, 2006.

Niger / Niger

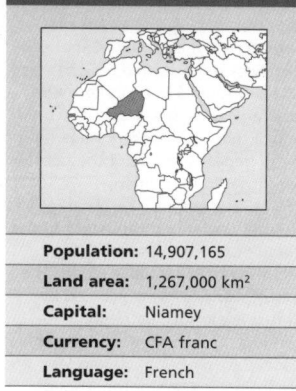

Population:	14,907,165
Land area:	1,267,000 km²
Capital:	Niamey
Currency:	CFA franc
Language:	French

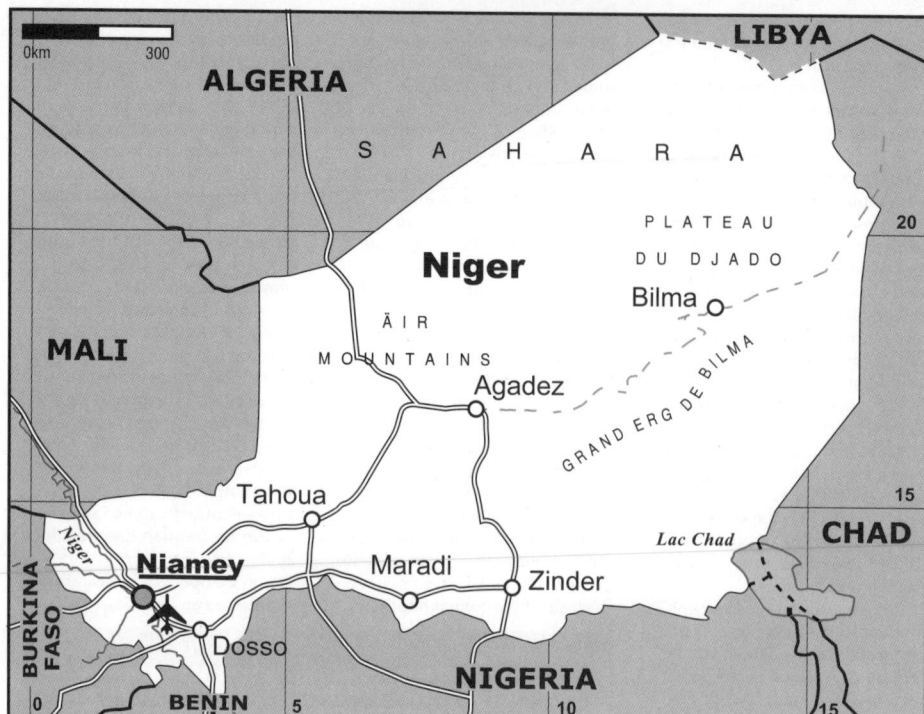

F ossil remains found in what is now Niger indicate that it has been inhabited since prehistoric times. The Nok Empire, reaching its peak in present-day Nigeria between the 15th century BC and the 5th century AD (see Nigeria), left its mark on this territory. From that time onwards, Niger was successively or simultaneously controlled by different kingdoms, empires and city-states in the region. During the 7th century AD, the western stretch of the country became part of the Songhai Empire created by the Berber, who were important propagators of Islam from the 11th century onwards.

2   Between the 14th and 19th centuries the eastern part of the territory belonged to the Kanem-Bornu State, which had been founded by the Kurani in the 8th century (see Chad). Meanwhile, during the 19th century, the Haussa states flourished in the south. The ancient *Hausa* territory covered an area of 140,000 km² in what today are Niger and Nigeria. These peoples have been there for the last 1,000 years, although it was only in the 12th century that they obtained complete control of the region. The country was divided in seven states: some farmed the land, others were traders, while others defended the empire's borders, usually besieged by

the kingdoms of Ghana and Songhai. The history of the Hausa is closely linked to that of Islam, when the Fulani took power away from them in the early 19th century. Afterwards colonization divided the country into a British and a French possession.

3   France dominated by force or agreements the various kingdoms that ruled the land. In 1922, Niger formally became a French colony. the traditional subsistence crops were changed for peanuts and cotton, meant to be exported. food shortages became prevalent among its inhabitants.

4   In the 1950s, within the context of decolonization in the region, Niger launched its independence movement, led by Hamani Diori. In 1960, the country's first constitution was

approved and Niger became an independent state.

5   When it broke its colonial ties, Niger was the poorest country in French West Africa, with 80 per cent of the population living in rural areas with persistent drought, soil erosion and population pressure which continue to threaten the country's agriculture and ecology.

6   In the new republic's first election in 1960, the Niger Progressive Party (NPP) candidate, Hamani Diori, was elected President. The new Government maintained close economic and political ties with France, to the point of allowing French troops to remain within its territory. In the first years of his presidency, Diori banned the opposition Sawaba (Freedom) party, forcing its leader, Djibo Bakari, into exile and creating a one-party state. Diori's Government was accused of corruption and of harshly repressing the growing political opposition.

7   In the early 1970s, in

response to the drought which hit the Sahel region, the army distributed food among the people there, thus becoming fully aware of their needs. On 13 April 1974, a Supreme Military Committee took power, suspending the constitution, and naming Lieutenant Colonel Seyni Kountche as head of state; Diori was detained. The first measures it took aimed to fix prices for agricultural products, increase salaries, stop nepotism, redirect investments, and develop education and sanitation services.

8   The new Government tried to establish a political base, particularly among young people, through the creation of 'samarias', a traditional form of social group, and signed bilateral agreements with France. The Military Council allowed Djibo Bakary, exiled leader of the Sawaba (Freedom) Party, to return to Niger on condition he did not engage in political activity. He was subsequently arrested and detained until 1984.

9   During the 1970s the country experienced an economic boom, based on an increase in the international price of uranium, of which Niger is the world's fourth largest producer. This mineral accounted for 90 per cent of the country's exports in 1980, when the so-called 'miracle' came to an abrupt end because of the decline in foreign demand as well as in the price of

LAND USE

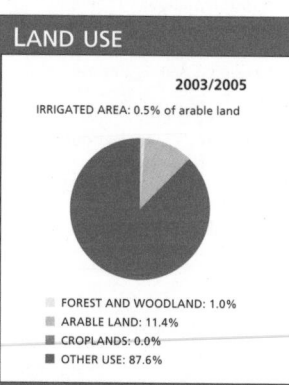

2003/2005

IRRIGATED AREA: 0.5% of arable land

- FOREST AND WOODLAND: 1.0%
- ARABLE LAND: 11.4%
- CROPLANDS: 0.0%
- OTHER USE: 87.6%

PUBLIC EXPENDITURE

% OF GDP

4.3 % HEALTH & EDUCATION (2000-2002)

0.9 % DEFENSE (2004)

WORKERS

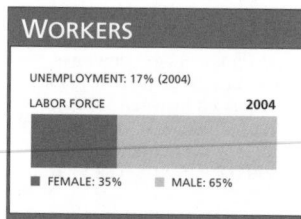

UNEMPLOYMENT: 17% (2004)

LABOR FORCE   2004

- FEMALE: 35%   ■ MALE: 65%

Life expectancy
45 years
2005-2010

GNI per capita
$210
2004

Literacy
14% total adult rate
2000-2004

HIV prevalence rate
1.2% of population 15-49 years
2003

this product. Economic strength and development programs helped divert attention from the absence of legal political activity.

[10] The country's foreign debt increased from $207 million to $1 billion between 1977 and 1983. In 1983 Kountche promoted an IMF structural adjustment program, but favorable uranium prices did not result and between 1984 and 1985 the perennial drought in the Sahel region worsened.

[11] In 1986 Kountche died and the Military Council appointed Ali Saibou as his successor. He appointed ten new ministers and declared an amnesty which provided for the return of political exiles; as a result, the leaders Djibo Bakari and Hamani Diori returned to political life. Diori died three years later.

[12] On 2 August 1988, the National Movement for a Developing Society (MNSD) was formed as the only government-authorized party, and the National Development Council drew up a new Constitution, which was approved in a plebiscite in 1989. In December 1989, Seibou was elected President in the first elections since 1960.

[13] The drought in the Sahel forced the Government to pay special attention to agricultural production and to people living in rural areas (80 per cent of the population). During the 1980-90 period it invested 32 per cent of the national budget in agriculture. A major environmental problem arose from the fact that most people used firewood for cooking. In a country suffering from deforestation and almost permanent drought, this almost exclusive use of wood for household energy had serious environmental consequences. Facing the problem head-on, the Government had solar and wind-powered generators installed, developed electric energy and sponsored the manufacture of low-cost energy-saving stoves. In order to stop the high consumption of wood, the Government also decided to increase sixfold the tariffs on the felling of trees.

[14] The new president, Ali Saibou, had a favorable economic outlook on his side, as there were 200,000 tons of surplus grain in 1989. He tried to link his project for a one-party system, capable of uniting Niger's different political tendencies, to his promise to initiate a genuine democratization process. Throughout 1990 there was

intense opposition from political actors, labor unions and students. Besides demanding salary increases and educational reform, support for a multi-party system was expressed through massive strikes and demonstrations which were harshly suppressed by the police.

[15] The Government launched a structural adjustment plan, imposed by the World Bank and the IMF, and announced a two-year freeze on public sector salaries. Workers and students reacted by calling a new series of strikes and holding more demonstrations. In late 1990, Seibou publicly announced his commitment to leading the country towards a multiparty democratic system and he created the National Conference to oversee the political transition.

[16] After four months, the National Conference decided to form a transitional government, headed by a new Prime Minister, Amadou Cheiffou. André Salifou was named president of the High Council of the Republic, the body holding legislative power during the transition period. The situation of Niger had never been so critical: the State was bankrupt, no resources were allocated to pay public sector salaries and student scholarships. In February 1992, Tuareg guerrillas rose up against the Government once more.

[17] In April's presidential election, Mahamane Ousmane was elected with 55.4 per cent. The efforts to reach an agreement to end the insurrection by Tuareg guerrillas in the north continued throughout 1993-1994. The fighting continued until an agreement was signed between the main guerrilla group, the Coordination of Armed Resistance, and the Government. The most important result was that the central Government granted autonomy to part of the country inhabited by some 750,000 Tuaregs.

[18] Amidst student protests calling for the payment of money owed in grants, the Government arrested 91 members of the opposition. In September 1994, Prime Minister Mahamadou Issoufou resigned after his party, the Nigerien Party for Democracy and Socialism, withdrew from the Government coalition, leaving it without an absolute majority in Parliament.

[19] In January 1995, an opposition coalition triumphed in the legislative elections and immediately replaced the Prime Minister, Amadou

Cissé, with Hama Amadou. The latter announced that his first move would be to introduce an economic austerity plan, reaching an agreement for settling payment of overdue civil service salaries.

[20] Tension continued to mount between the new Government and the President. In January 1996, a military coup toppled Ousmane, who was replaced by the National Salvation Council, headed by Colonel Ibrahim Baré Mainassara, who appointed Boukary Adji as Prime Minister. In July, Mainassara was elected President with 52 per cent of the vote. He dissolved the Independent National Electoral Commission, leading the main opposition parties to boycott the November legislative elections. In December, following the victory of Mainassara's supporters, Amadou Cissé was made Prime Minister.

[21] In October 1997, Ali Saibo, national coordinator for a coalition of eight opposition parties, the Front for the Restoration and Defence of Democracy, was detained for making hostile statements against Mainassara. In November 1997, Ibrahim Hassane Mayaki was appointed Prime Minister,

replacing Cissé.

[22] Political persecution continued in 1998 with the arrest of several members of the opposition. Social action was dominated by constant anti-government demonstrations. One year later, the Supreme Court annulled the March election results in some districts and called for a new round of ballots. On 4 April, after a tension-filled week, Mainassara was assassinated by members of his own presidential guard. Coup leader Daouda Malam Wanké was then named President and head of the National Reconciliation Council, who governed the country during the nine-month transition period. Prime Minister Hassane dissolved the National Assembly and political parties were temporarily suspended.

[23] The international community strongly pressured the country to return to democratic rule. In October 1999, Niger held the first round of general elections and, in the second round in November, retired military officer Tandja Mamadou, of the MNSD, won by a large majority over his rival, former prime minister and parliamentary leader Mahamadou Issoufou.

PROFILE

ENVIRONMENT
Most of Niger's territory is made up of a plateau with an average altitude of 350 m. The north is covered by the Sahara desert and the south by savannas. There are uranium, iron, coal and tin deposits, and possibly oil. Some 80 per cent of the population lives in rural areas. There are nomadic herders in the center of the country, and peanut, rice and cotton farming in the south. Around 85 per cent of all energy is provided by firewood.

SOCIETY
Peoples: Among the pastoralists of the central steppe, ethnic origins vary from the Berber Tuareg, to the Fulani, including the Tibu (Tubu). In the south there are West African ethnic groups; the Hausa, Djerma (Zarma), Songhai and Kamuri, among others.
Religions: Muslim; in the south there are traditional African religions and a Christian minority.
Languages: French (official) and several local languages.
Main Political Parties: National Movement for the Developing Society (MNSD); Nigerien Party for Democracy and Socialism; Rally for Democracy and Progress (RDP), a coalition made up of nine parties opposed to the MNSD; Democratic and Social Convention.
Main Social Organizations: Nigerien National Workers' Union (UNTN).

THE STATE
Official Name: République du Niger.
Administrative Divisions: 7 Départements.
Capital: Niamey 890,000 people (2003).
Other Cities: Zinder 185,100 people; Maradi 172,900; Tahoua 87,700 (2000).
Government: Tandja Mamadou, President since December 1999, re-elected in 2004. Hama Amadou, Prime Minister since January 2000. Parliament: National Assembly, with 83 members elected every five years.
National Holiday: 3 August, Independence (1960).
Armed Forces: 5,000 (2003).

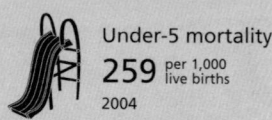

	Under-5 mortality		Poverty			Debt service			Maternal mortality	
	259 per 1,000 live births 2004		**60.6%** 1995	of population living on less than $1 per day		**7.5%** 2003	exports of goods and services		**1600** per 100,000 live births 2000	

IN FOCUS

ENVIRONMENTAL CHALLENGES

This region is affected by desertification from intensive grazing and deforestation. Strong winds produce considerable erosion. There is air and water pollution in densely populated urban areas. A range of animals, such as elephants, giraffes, hippopotamuses and lions, are in danger of extinction due to uncontrolled hunting and the destruction of their natural habitat.

WOMEN'S RIGHTS

Women have been able to vote and stand for office since 1948. In 2004, they held 12 per cent of seats in Parliament and 23 per cent of ministerial positions. Some 43 per cent of the five-million-strong labor force were women. Only 41 per cent of pregnant women received prenatal care and barely 14 per cent of births were attended by skilled health staff. The maternal mortality rate was 1,600 deaths per 100,000 live births. Among daily tasks performed by village women, the main one is to get water for their families and animals, often walking for many kilometers several times a day.

CHILDREN

Since 80 per cent of the population live in rural areas and just 59 per cent of them have access to safe drinking water it is almost inevitable that 80 per cent of under-five deaths are caused by waterborne diseases. In 2004, the country continued to have one of the world's highest infant and under-five mortality rates, in spite of the fact that they had declined since 1990. The first one had fallen from 191 to 152 per 1,000 live births and the latter from 320 to 250 per 1,000*. Some 13 per cent of newborns are underweight, 40 per cent of under-fives suffer from severe and moderate underweight and the same percentage are severely or moderately stunted.

There is evidence that the country is a destination for victims of trafficking, generally girls, boys and young women who end up in prostitution in the main urban centers. Some rural children are victims of domestic trafficking. They are sold by their families to work as servants in other homes.

INDIGENOUS PEOPLES/ ETHNIC MINORITIES

The Hausa, make up 56 per cent of the total population. Their economy has traditionally been based on herding, agriculture and trade. Most of the Hausa live in small rural communities, between 2,000 and 12,000 inhabitants each. Families are large and led by a male figure.

There are also minority ethnic groups. One of them, the Berber (14 per cent of the population) speak Tamazight - a language with more than 100 dialects - and have a writing system that dates back 2,500 years, although in practise they use the Latin or Arabic alphabet. There are also some 470,000 Dendi, who live off the land, farming being the most noble of activities for them; 80,000 Bororo, nomadic shepherds who speak a Fulani dialect; some 4,000 Buduma; 18,000 Daza, who raise camels and goats; 49,000 Gurma, a farming people whose men make the clothing for the whole family; 380,000 Kanuri, who build their walls around conical mud and straw huts; 148,000 Moros, descendants from Bedouin groups from Yemen; 50,000 Shuwa, mostly cattle breeders; 4,000 Soninke, who arrived around the 8th century fleeing from the Berber; 10,000 Teda, who continue to be nomads or semi-nomads, and 720,000 Tuareg, the only people that continue to use the Berber script.

MIGRANTS/REFUGEES

In early 2006, some 20,000 refugees from Rwanda, Congo and DR Congo were living as refugees in Benin, Burkina Faso, Niger and Togo, in addition to nearly 1,300 asylum-seekers. About 10,300 were due to return to their homes within the year.

DEATH PENALTY

Although this still applies, there have been no executions since 1976.

> * Latest data available in *The State of the World's Children* and *Childinfo* database, UNICEF, 2006.

24 Mamadou's Government faced several problems. One of them was the threat posed by the indiscriminate hunting of endangered species (giraffes, hippopotamuses and lions), a traditional activity in the northern deserts. In February 2001, Environment Minister Issoufou Assoumane warned that there had been a massacre of animals over the past ten years and that the number of hunting licenses sold would be limited. A year later the National Assembly also banned clitoridectomy, which was practised by some ethnic groups. According to a 1999 study, about 20 per cent of women in Niger had been subjected to this type of genital mutilation.

25 In spite of the agreements signed with the Tuaregs, the violence did not end. In July and August 2003 the Niger Delta was the site of armed confrontations among different gangs for control of the illegal trade in oil and derivatives. Between 50 and 100 people died as a result of fighting.

26 During 2003, Niger's government became involved in an international conflict when intelligence reports furnished by the US and the UK stated that Iraq had bought uranium from Niger to build atomic bombs. Mamadou demanded that the evidence proving these claims be produced. As this was not forthcoming, CIA Director George Tenet had to admit that the information on the sale of uranium was false. However, the British Government did not retract its accusations, even when a UN delegation of experts concluded that the information provided by the secret services was false.

27 The first-ever local elections were held in July 2004. On this occasion, parties backed Mamadou, who won a majority representation at government level. In December, Niger's Electoral Commission announced that Mamadou had been re-elected President in a second-round ballot. Mamadou won with 65 per cent of the votes, while the Socialist opposition candidate Mahamadou Issoufou polled 34.5 per cent. The vote was deemed by international observers to have been free and fair.

28 From April 2005 onwards, the Government systematically denied the existence of a food crisis and put pressure on journalists to toe the government line. However, independent media repeatedly published reports on it and private radio stations allowed people hit by the crisis to talk about it. By August, although admitting food shortages in certain areas, Mamadou continued to deny that the country was struck by famine and maintained that the idea was a fiction created by opposition parties and UN aid agencies. On the other hand, the World Food Program (WFP) denied that the scale of the problem had been exaggerated, highlighted the existence of pockets of severe malnutrition and started to distribute food in southern areas of the country. Other aid agencies stated that hunger was killing children on a daily basis.

29 In March 2006, the UN included Niger - the country with the world's worst poverty and human development indicators - in a fundraising appeal for African nations struck by food crises. Aid workers, who kept saying that the country was living in a 'culture of denial', pointed out that 1,000 children were admitted each week to feeding programs for the malnourished. Médecins Sans Frontières said that unless donors 'dug deep into their pockets' many in Niger could face a desperate future. A bad harvest year in Niger weakens the resources of much of the population for two or three years following.

30 As it turned out, months of good rainfall in mid-2006 boosted harvests and reduced the risk of further food shortages. Yet the plentiful rains brought other problems. In October 2006, over 46,000 people in Niger were displaced by flooding and needed food, blankets, mosquito nets and other international assistance, according to the UN. Niger was the worst affected of the four West African nations hit by the floods, the others being Burkina Faso, Mauritania and Guinea.

31 Also in October 2006, leaders of around 150,000 Arabs in Niger said they would fight in the courts the Government's attempts to expel them to Chad. The Government had ordered the Arabs, known as Mahamid, to leave the country, accusing them of wrongdoing, including theft and rape. ■

Nigeria / Nigeria

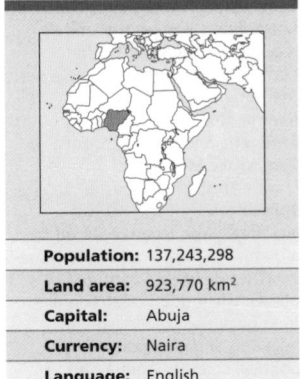

Population:	137,243,298
Land area:	923,770 km²
Capital:	Abuja
Currency:	Naira
Language:	English

A s heirs to the ancient Nok civilization, the Yoruba lived in walled cities with broad avenues. As early as the 9th century, they had a democratic system of urban administration, with a mayor and a municipal council elected by a citizens' assembly. As art, they produced beautiful ceramics and bronze sculptures. Between the 10th and 11th centuries, Ife, Oyo, Ilorin and Benin (not the present-day nation) were loosely confederated city-states extending their influence from the Niger River to present-day Togo.

2 The city of Ife has enjoyed a reputation as the main religious center of the nation since those times; the Oni of Ife was the High Priest of all Yorubans, whether Nigerian or not.

3 The city of Oyo, strengthened since the 16th century by the slave trade, maintained political power through the *Alafin* (ruler). The dependence on slavery caused its downfall when that institution was abolished.

4 In the northern part of the country, the Hausa states constituted a cultural center of very diverse character. In the southeast, the Igbo were active traders coming from the same group as the Yoruba; however, they did not develop urban civilizations.

5 In 1914, Britain unified all these territories under a single administration, interested in the exploitation of tin and agricultural and timber resources.

6 The British method of indirect colonial administration used the northern Muslim emirs as their agents. Consequently this area, populated by Hausa and Fulani, enjoyed greater political supremacy.

7 Independence in 1960 brought the Northern People's Congress to power, in an alliance with the National Council of Nigerian Citizens, an Igbo organization. However, the federal structure (four states) and the bicameral parliament, based on the British model, afforded the regional governors more effective power than that held by President Nnambi Azikiwe. Progressive parties were pushed aside in a succession of electoral frauds, while political leaders lost their national outlook, encouraging ethnic rivalries.

8 The army came to power when General Yacubu Gowon was appointed president in 1966. In 1967, the oil industry began to develop, just as France was inciting the separatist movement among the Igbo. The civil war in Ibarra that lasted from 1967 until 1970 was a bid for secession, but eventually failed.

9 With Nigeria as the world's 8th largest oil producer, Gowon expropriated 55 per cent of the transnational oil companies, allowing the consolidation of local entrepreneurs.

10 Real power was vested in the nationalist Supreme Military Council, with different presidents. The council closed US military and espionage installations. During Olusegun Obasanjo's presidency, Barclays Bank and British Petroleum assets were nationalized when these companies violated economic sanctions against apartheid South Africa.

11 In 1978, constitutional reform and a call for elections paved the way for a return to civilian government. The Federal Election Commission authorized only five parties, all representing the traditional financial and political elite. Parties with socialist or revolutionary perspectives were barred from the electoral process, under the pretext of avoiding political fragmentation. The National Party of Nigeria (NPN) won the election with 25 per cent of the vote; the Unity Party of Nigeria (UPN) obtained 20 per cent.

12 The new president Shehu Shagari launched a capitalist plan, based exclusively on petrodollars, to transform Nigeria into the development hub of sub-Saharan Africa. His promises included constructing a new capital, doubling elementary and high school enrollment and achieving self-sufficiency in food production via controversial 'green revolution' methods.

13 None of these proposals came to fruition. Economic indicators showed gloomy prospects, and there was an increase in contraband, large urban concentrations of immigrants and poor peasants, high unemployment and the reduction of workers' purchasing power. Another burden were the IMF conditions for refinancing foreign debt. Shagari was nevertheless re-elected in 1983 as the NPN candidate, amidst accusations of electoral fraud and military conspiracies.

14 On 1 January 1984, Muhamad Buhari staged a fourth coup, accusing the Government of corruption in the petroleum sector, which accounted for 95 per cent of export earnings. There were detentions at all levels, and civilian government officials were replaced by military personnel.

15 The aggravation of the crisis, an external debt of $15 billion, the repression and the expulsion of 600,000 illegal foreigners set the stage for another coup. On 26 August 1985, General Ibrahim Babangida was appointed president.

16 In December 1987, local elections were held with 15,000 independent candidates taking part. The appointed National Election Commission did not achieve an adequate level of organization. As a consequence of the violence, confusion and subsequent allegations of fraud, the elections were annulled.

17 On 7 December 1989, the military government announced that the presidential and legislative elections originally scheduled for the end of the month would be postponed until December 1990. Six months later, the ban on political activism had been lifted, in an attempt to monitor the transition from military to civilian government in 1992.

18 Babangida visited Britain in May 1990, where he obtained $100 million in aid for the Nigerian economy, which-is oriented to trade mainly with the US, the UK and France.

19 That same year, the creation of nine states to separate hostile ethnic groups ended in protests, repression, around 300 deaths and a curfew imposed by the Government.

20 Toward the end of 1991, the internal party elections held to select candidates for governor were

LAND USE

2003/2005

IRRIGATED AREA: 0.8% of arable land

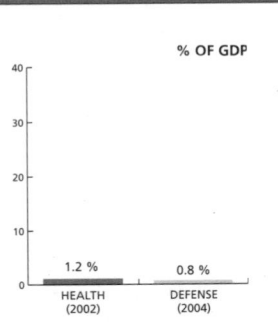

- FOREST AND WOODLAND: 12.2%
- ARABLE LAND: 33.5%
- CROPLANDS: 3.2%
- OTHER USE: 51.1%

PUBLIC EXPENDITURE

% OF GDP

1.2 % — HEALTH (2002)
0.8 % — DEFENSE (2004)

WORKERS

UNEMPLOYMENT: 17% (2004)

LABOR FORCE — **2004**

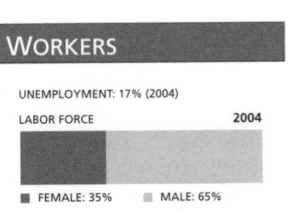

- FEMALE: 35%
- MALE: 65%

Life expectancy
44 years
2005-2010

GNI per capita
$430
2004

Literacy
67% total adult rate
2000-2004

HIV prevalence rate
5.4% of population 15-49 years old
2003

annulled due to alleged fraud. In November, a new census eliminated 20 million 'non-existent' voters from the electoral register.

[21] On 14 December, elections for governors were held; the Social Democratic Party (SDP, left-of-center) won in 16 states, and the National Republican Convention (NRC, right-of-center) won in 14. Opposition groups were granted a general amnesty, and 11 well-known dissidents were freed. A law prohibiting former government officials from running for office was revoked.

[22] Early in 1992, the imprisonment of 263 Muslim militants caused protests in the state of Katsina. During this period, there was also an escalation of inter-ethnic conflict between Hausa and Kataj in the state of Kaduna, and territorial conflict between Tiv and Jukin in the Taraba, leaving 5,000 dead.

[23] Legislative elections were held in July 1992. The SDP won 52 seats in the Senate and 314 in the Chamber of Representatives; the NRC won 37 and 275, respectively. The National Assembly was inaugurated in December, allowing the country to enter a transition period after 23 years of military regimes.

[24] In October 1992, the primary presidential elections with candidates from the SDP, the NRC and 23 other parties, were invalidated by President Babangida, claiming fraud.

[25] In November Babangida postponed until June the elections slated for January 1993. He also ratified the proscription of all the 1992 candidates.

[26] On 12 June 1993, the first presidential elections were held since 1983. The military government did not divulge election results until it had concluded an investigation of alleged fraud. The main contest was between the NRC and SDP, who had been authorized to take part in the election with alternative candidates.

[27] Babangida once again invalidated the election results on 23 June, accusing SDP and NRC candidates of 'buying votes'. Moshood Abiola, a Muslim millionaire who was the SDP candidate and who had apparently won the elections, asked for international condemnation of the regime in London.

[28] The US and Britain suspended their economic aid and military training and froze diplomatic relations. This external reaction provided Abiola with the impetus to launch a civil disobedience campaign. Massive protests broke out on the streets of Lagos, the former capital, where at least 25 people were killed by federal troops. Twenty-five opposition groups formed the 'Democracy Campaign'.

[29] Under pressure, the regime set new presidential elections for 14 August 1993 (with the express exclusion of Abiola and Othma Tofa of the NRC), and announced the transfer of power for 27 August.

[30] Clashes continued and on 26 August 1993 Babangida resigned, leaving the country in the hands of Ernest Shonekan, who promised to hold new elections.

[31] The following month, Abiola returned from London and labor unions called a general strike, demanding he be recognized as Nigerian president. Towards the end of 1993 the Minister of Defense, General Sani Abacha, overturned Shonekan, dissolved parliament and banned political activity.

[32] Abacha was a very influential member of the previous military regime and a key figure in the military coup that had ousted the Government in 1983. In one of his first statements, he announced he would abandon some of the liberal economic reforms adopted in the 1980s.

[33] Interest rates fell and a new foreign exchange control was established, at a time when any possibility of reaching an agreement with the IMF was increasingly remote. Popular support for Abiola mounted; his arrest in June 1994, triggered a 10-day-strike in the oil sector, the most important in the country.

[34] The execution of nine members of the Movement for the Survival of the Ogoni People in November 1994 resulted in the isolation of the military regime. Several countries, including the US, withdrew their ambassadors from Nigeria.

[35] In 1996 on the basis of a register of political parties drawn up by the National Electoral Commission, Abacha legalized five political groups: the United Nigerian Congress Party, the Committee for National Consensus, the National Central Party of Nigeria, the Democratic Party of Nigeria and the 'Grassroots' Democratic Movement.

[36] In April 1998, Abacha announced that the August elections were to be replaced by a plebiscite that would determine whether he was to continue in power. His sudden death on 8 June gave rise to widespread rejoicing and expectations of political change. Abiola also died shortly afterwards. General Abdusalam Abu-Bakar, appointed by the military junta as new President, promised to respect the democratic transition.

[37] The local elections held in February 1999 were won by the Popular Democratic Party (PDP) of former military ruler General Olusegun Obasanjo. General elections were set for March 1999 and were also won by Obasanjo. At first, the opposition attempted to appeal the results before the Electoral Court, but later it backed down. Upon taking office in May, Obasanjo called upon Nigerians to join him in a three-day fast to seek divine intervention that would ensure a positive presidential performance. During his first days in office, he deposed 30 military officers and confiscated millions of dollars which he said had been stolen from the public Treasury during previous administrations.

[38] In March 2000, Obasanjo visited Lagos, where violent ethnic fighting between the Ijaw and Ilaje peoples ended in hundreds of deaths. The President was able temporarily to end the fighting by establishing a peace committee with their leaders. In June of the following year, in the state of Nasarawa, neighboring the capital, ethnic violence was sparked off between the Azara and Tiv minorities by the death of an Azara community leader. Around 40,000 people fled the conflict.

[39] In August 2001, Nigeria's phone lines were so regularly intercepted that even international telephone calls by Cabinet members were interrupted without notice. At that time, the major mobile phone companies of the country, Johannesburg MTN and Wireless Econet promised a 'cable revolution'. However, due to frequent blackouts and excessive voltage, each company needed a generator for its transmission tower. In addition, each company had to pay license fees of $285 million to the Government. According to the 2001 Transparency International Index on perceptions of corruption

PROFILE

ENVIRONMENT

The country's extensive river system includes the Niger and its main tributary, the Benue. In the north, the *harmattan*, a dry wind from the Sahara, creates a drier region made up of plateaus and grasslands where cotton and peanuts are grown for export. The central plains are also covered by grasslands, and are sparsely populated. The southern lowlands, home to most of the country's population, receive more rainfall and have dense tropical forests. Cocoa and oil-palms are grown in this area. The massive delta of the Niger River divides the coast into two separate regions. In the east, oil production is concentrated around Port Harcourt, the homeland of the Igbo, who converted to Christianity and fought to establish an independent Biafra. To the west, the industrial area is concentrated around Lagos and Ibadan. Yoruba are the predominant western ethnic group, and some of them have converted to Islam.

SOCIETY

Peoples: Nigeria is the most populous country in Africa. The 250 or so ethnic groups fall into four main ones: the Hausa and Fulani in the north; the Yoruba in the southwest; and the Igbo in the southeast.

Religions: The north is predominantly Muslim, while Christians form the majority in the southeast. In the southwest are Muslims, Christians and followers of traditional African religions.

Languages: English (official). Each region has a main language depending on the predominant ethnic group, Hausa, Igbo or Yoruba.

Main Political Parties: People's Democratic Party; All People's Party; Alliance for Democracy; United Nigeria People's Party.

Main Social Organizations: National Labor Congress; National Association of Nigerian Students.

THE STATE

Official Name: Federal Republic of Nigeria.

Administrative Divisions: 30 States.

Capital: Abuja 452,000 people (2003).

Other Cities: Lagos 8,733,100 people; Ibadan 3,587,100; Kano 3,424,100; Ogobomosho 963,300 (2000).

Government: Olusegun Obasanjo, president since May 1999, re-elected in April 2003. Bicameral legislature: the House of Representatives, with 360 members, and the Senate, with 109 members.

National Holiday: 1 October, Independence Day (1960).

Armed Forces: 79,000 troops (2003).

in business, Nigeria was ranked the second most corrupt country after Bangladesh.

[40] In October, Obasanjo, together with President Thabo Mbeki of South Africa and Abdelaziz Bouteflika of Algeria formally launched the New Partnership for Africa's Development (NEPAD), which called on the rest of the world to become partners in the development of Africa. NEPAD commitments included instituting transparent and democratic governments in African states, respecting human rights and stopping wars in exchange for more foreign aid and the lifting of trade barriers to African exports.

[41] In 2001 and 2002, a radical version of the Islamic *Sharia* code, introduced in a dozen Muslim states, provoked great controversy and violent protests. Stoning, amputation and flogging were some of the punishments included in the law. In January 2002 a man was hanged in Katsina state in the first execution since Islamic law was introduced. In November, a confrontation between Christians and Muslims in Kaduna, during the Muslim Ramadan and on the eve of the celebration of the Miss World Contest in Nigeria, was caused by the publication of an article in *This Day* magazine, which suggested that the Prophet Muhammad would have been able to choose a wife among the contestants. The hundreds of dead and injured resulting from the strife forced the organizers to transfer the contest to London.

[42] Even before the riots, many countries had called for a boycott of the contest, due to the Katsina state Sharia court's decision to stone a woman, Amina Lawal, to death for committing adultery and having a child out of wedlock. Lawal was eventually freed in February 2004, when the Sharia court of appeal ruled that her conviction was invalid because she was already pregnant when the Islamic code was implemented in her home province.

[43] In April 2003 Obasanjo was re-elected president, in an election day marred by violence. The elections were deemed fraudulent by European Union observers, and the results were rejected by the opposition.

[44] The beginning of the March 2004 local election campaign was marked by political assassinations and a series of armed attacks. A state of emergency was declared in the Central Plateau state after violent religious confrontations. Previously in Yelwa, a town located in Kebbi state, in the northwest of Nigeria, 200 Muslims had been killed in attacks by

IN FOCUS

ENVIRONMENTAL CHALLENGES
Nigeria has lost between 70 and 80 per cent of its forests. Only 1.7 per cent of land lies within protected nature reserves. Deterioration of arable land, desertification and air and water pollution are a problem in urban areas. Some water sources have been seriously affected by oil spills.

WOMEN'S RIGHTS
Women have been able to vote and run for office in the south of the country since 1958, and in the north since 1978. In 2003, only 3 per cent of seats in Parliament and 10 per cent of ministerial positions were held by women. Of the 54 million people that made up the labor force, 34 per cent were women. Female enrolment rates stood at 60 per cent in primary education and 26 per cent at secondary education. In 2004, 58 per cent of pregnant women received prenatal care and just 35 per cent of births were attended by skilled health staff. That year, an estimated 19 per cent of Nigerian women between the ages of 15 and 49 had suffered genital mutilation.

CHILDREN
Notwithstanding the decline in infant and under-five mortality rates between 1990 and 2004, these remained high. The former fell from 120 to 101 per 1,000 live births and the latter from 230 to 197 per 1,000. Some 14 per cent of newborns were underweight, while 29 per cent of under-fives suffered from

moderate and severe underweight and 38 per cent were moderately or severely stunted*. At the end of 2005, Nigeria implemented Africa's largest-ever measles vaccination campaign, with the support of UNICEF, the WHO, the Red Cross and other members of the Measles International Partnership. A total of 30 million children, aged between 9 and 15, were vaccinated in Nigeria's 20 northern states. Phase two of the campaign, covering the southern part of the country, would be launched in 2006.

Nigeria is a country of origin and transit for child-trafficking for the sex-trade in western Africa, Asia and western Europe.

INDIGENOUS PEOPLES/ ETHNIC MINORITIES
The Ibo or Igbo have inhabited the region for thousands of years and are mainly concentrated in the southern states. During the 20th century and the beginning of the 21st century, the implementation of Islamic *Sharia* law has been opposed in the northern regions of the country.

The Ogoni are located on the Delta of the Niger River, in the southern part of the country (their ancestral lands). They have protested against oil companies such as Shell, demanding compensation for damages caused to their lands.

The Yoruba were excluded from political participation until the 1999 elections, when Obasanjo (of Yoruban origin) won the presidency, with the result that many restrictions against this group were lifted. However, until 2001 they were still suffered discrimination.

Some Yoruban organizations are banned, especially those considered most militant.

The Ijaw live mainly in the Niger Delta and have participated more in regional events during the past years. They pressure the Government to change its economic policy and practices that affect them, and have become involved in campaigns against the oil companies.

MIGRANTS/REFUGEES
In September 2005, following the signing of an agreement between the governments of Nigeria and Cameroon and UNHCR, 7,500 Nigerian refugees living in Cameroon returned to the country. The operation was to be completed in 2006, with the return of the remaining Nigerian refugees. It is estimated that as a result of the ethnic conflict in Tabara state, some 17,000 Nigerians had arrived in Cameroon by the end of 2002. In early 2006, Nigeria hosted some 8,000 refugees and other 1,100 asylum-seekers, mainly from Liberia, Sierra Leone, DR Congo, Chad and Sudan.

DEATH PENALTY
The death penalty is still applicable to all types of crimes and in regions where Sharia law is enforced it also applies in cases of adultery.

Latest data available in The State of the World's Children and Childinfo database, UNICEF, 2006.

Christian militia. Revenge attacks were launched by Muslims in Kano (located in northern Nigeria).

[45] In September 2004, Dokubo Asari, leader of the rebel group Niger Delta People's Volunteer Force (NDPVF), who had announced an 'all-out war' would be unleashed on the Nigerian state and foreign oil companies in the Niger Delta, said that he was willing to negotiate a halt to violence in the area. The rebel leader demanded 'control of resources and self-determination' for the Ijaw ethnic group, which was living in the most abject poverty. Asari stated that his mission was to protect the economic and political rights of Ijaw people.

[46] Oil economic profits in the Niger Delta region go directly to the Government or foreign companies and workers are

mostly foreigners. Oil spills in the area have ruined the livelihoods of a large number of locals. Programs launched by some oil companies aimed at improving the environment and living standards of local people have had almost no effect on poverty elimination.

[47] A total of 117 people, including the President's wife, died in October 2005 when an aircraft of the privately owned Nigerian Bellview Airlines crashed shortly after taking off from Lagos. Obasanjo immediately met with the Aviation Minister and promised stricter regulations would be implemented. This was the fourth air disaster in the country in 13 years.

[48] From January 2006, demanding more control over the region's oil wealth, militants in the Niger Delta began to attack

pipelines and other oil facilities and kidnap foreign oil workers. By April, helped by record oil prices, Nigeria had become the first African nation to pay off its debt to the Paris Club.

[49] In May, the Senate frustrated Obasanjo's intention to introduce changes to the Constitution which would have allowed him to be re-elected for a third term in the 2007 elections.

[50] In October 2006, Vice-President Atiku Abubakar appeared before a special court in Abuja on charges of corruption. He denied allegations that he diverted $125 million of public money into personal business interests. He said that his trial was part of a plot to prevent him running in the country's presidential elections in April 2007. Abubakar had helped block President Obasanjo in his quest for a third term. ∎

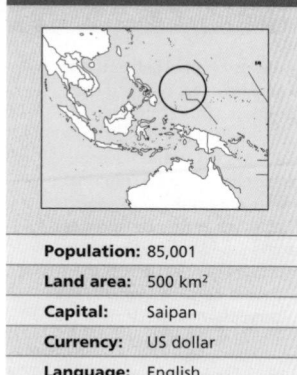

Population:	85,001
Land area:	500 km²
Capital:	Saipan
Currency:	US dollar
Language:	English

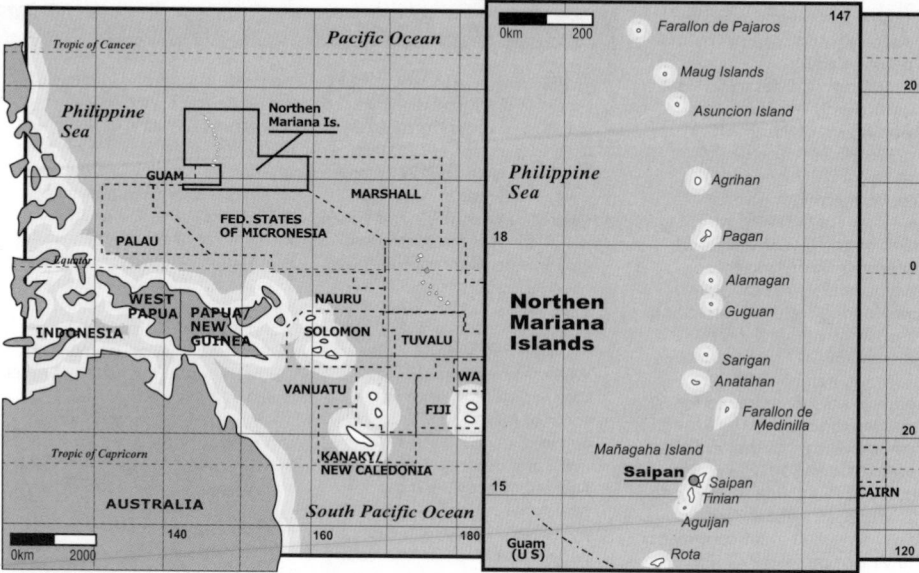

I n Saipan, the largest island of the Mariana archipelago, evidence has been found of human habitation from 1500 BC.

[2] During his first expedition around the world, Portuguese navigator Ferdinand Magellan sighted the islands in 1521 and claimed them for the Spanish Crown. They were held by the Spanish until ceded to Germany in 1899 as the Spanish empire declined.

[3] During World War I the islands came under the control of Japan, which fought with the Allies against Germany. The Japanese occupied the islands until World War II.

[4] In June 1944, after fierce fighting, the US took control of Saipan and Tinian because of their strategic location in the North Pacific, on the route between Hawaii and the Philippines. They finally came to form part of the Trust Territory of the Pacific Islands in 1947.

[5] The islands had this status until they opted to become a self-governing Free Associated State of the US in a referendum held in 1975.

[6] In the late 1970s, Washington developed a project to turn two-thirds of Tinian into a military air base and an alternative center for the storage of nuclear weapons. In 1984 the US Government started negotiations with local landowners.

[7] When news leaked out that cement deposits containing radioactive waste from Japanese nuclear plants had been dumped in this part of the Pacific, the alarm was sounded on similar US projects that would directly affect the Marianas Islands.

[8] In 1984 US President Reagan granted some civil and political rights to the islands' residents, such as equal employment opportunities in the federal government, the civil service and the US armed forces.

[9] The Northern Marianas were formally admitted to a Commonwealth arrangement in political union with the US in 1986. The inhabitants were granted US citizenship but not the right to vote in presidential elections. They have a representative in the US Congress with no voting rights.

[10] The country's main economic activities are fishing, agriculture (concentrated in smallholdings) and tourism which employs about 10 per cent of the workforce. Some of these activities have often been affected by the typhoons the islands suffer in the rainy season. In January 1988 Rota island was devastated by a typhoon, and the US declared a state of emergency. In January 1990, Typhoon Koryn hit the whole archipelago.

[11] In the 1989 local elections, Republicans retained the governorship. Larry Guerrero was elected governor after Pedro P Tenorio had decided to stand down.

[12] On 22 December 1990 the UN Security Council voted to dissolve the Trusteeship. Thus, the Northern Marianas became an independent state, associated to the US.

[13] In 1992, the US Supreme Court ratified the property ownership system, whereby only nationals could own land. In 1994, Froilan C Tenorio was elected governor.

[14] In 1995, there were some 22,600 foreign workers in the country, three times the number of Marianan workers. Unemployment amongst the latter stood at 15 per cent.

[15] Throughout 1999 local officials, backed by key figures from the Republican Party in the US Congress, argued intensely with President Bill Clinton's Democrat administration regarding the conditions of slavery and exploitation that immigrant workers were subjected to on the islands. Local Republican leaders claimed that the Marianas would not relinquish the exemptions on Federal customs, immigration and labor laws that it enjoyed. Both human rights groups and Washington officials stated that this exemption allowed for terrible working conditions and salaries.

[16] In March 2004, the Commonwealth Development Authority (CDA) dismissed an offer by Governor Babauta to invest $500,000 in the airline Palau Micronesia Air, considering it highly risky. That month, a modernization plan of Saipan airport set to cost several million dollars was launched.

[17] Between 1990 and 2005 more than 500,000 tourists - mainly Japanese - visited the country each year. Tourism employed 50 per cent of the country's workforce and contributed 25 per cent of the GDP.

[18] In 2006 Governor Benigno Fitial, Babauta's successor, was accused by the media of concentrating power in his own hands and behaving like a dictator. He had abolished the autonomy of several Government agencies, transferring their duties to the executive power. Fitial replied that drastic measures were needed after the excessive spending by the previous administration. ■

PROFILE

ENVIRONMENT

The Marianas archipelago, located in Micronesia, east of the Philippines and south of Japan, consists of 16 islands (excluding the island of Guam), of which only six are inhabited. The most important in size and population are: Saipan (122 sq km), Tinian (101 sq km) and Rota (85 sq km). Of volcanic origin, the islands are generally mountainous. The climate is tropical, with rain forest vegetation. In the northernmost islands, however, these conditions shade gradually into a more temperate climate and brush-like, herbaceous vegetation. Development plans endanger the local fauna. Toxic waste left by the US army has caused health problems in at least one of the country's villages.

SOCIETY

Peoples: The population is mostly indigenous, of Chamorro origin. There are also Japanese, Chinese, Korean, European, Filipino and Micronesian minorities. **Religions:** Catholic (majority); some traditional faiths persist. **Languages:** English (official); 55 per cent speak Chamorro; also Carolinian, Tanapag, Filipino and Japanese. **Political Parties:** Republican Party; Democratic Party; Reform Party.

THE STATE

Official Name: Commonwealth of the Northern Mariana Islands. **Capital:** Saipan 71,000 people (2003). **Other Islands:** Rota 2,100 people; Tinian 2,100 (2000). **Government:** By virtue of the US Commonwealth status, the President of the US is the Head of State. Benigno R. Fitial, Governor by direct election, since January 2006. There is a two-chamber legislature with nine senators and 18 representatives. **National Holiday:** Commonwealth Day, 8 January (1986). **Armed Forces:** The US is responsible for defense.

Norway / Norge

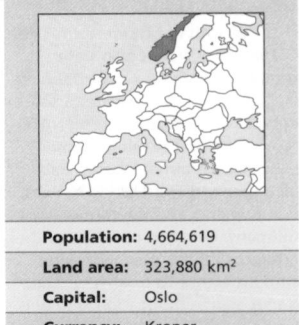

Population:	4,664,619
Land area:	323,880 km²
Capital:	Oslo
Currency:	Kroner
Language:	Norwegian

G ermanic groups are thought to have emigrated to Norway between the 9th and 7th millennia BC, when the glaciers receded from the northern European coasts. Cave drawings show that they were familiar with navigation and used skis for traveling over the snow. When the Scandinavians arrived around the 1st century AD, the Sami, shepherds that had lived on these lands for thousands of years, were expelled northwards.

[2] Historians believe that Norwegian nationality and conversion to Christianity began between 800 and 1030 AD. The Viking King Harald Harfagre, considered the founder of the nation, took control of a large part of the country after defeating his rivals in a naval battle at Hafrsfjord, near the city of Stavanger.

[3] Viking expeditions extended the Norwegian Empire to Greenland to the west and Ireland to the south. In 1002, Leif Erikson and his followers were the first Europeans to cross the Atlantic and reach North America, which they named Vinland.

[4] At the end of the Viking era, Norway was an independent kingdom in which four regional peasant assemblies (*lagting*) elected the monarch. Legitimate and illegitimate children of the king had equal rights to succession before the *lagtings*.

In the 10th and 12th centuries it was common for two kings to rule simultaneously without any conflict arising between them.

[5] King Magnus III Barfot (1093-1103) conquered the Scottish Orkney and Hebrides Islands. His three sons ruled together: they imposed a tithe, founded monasteries and built cathedrals.

At the beginning of the 12th century, a 100-year civil war broke out, as a result of the disputes between the monarchy and the church, lasting until the coronation of Haakon IV in 1217.

[6] The new King reorganized the public administration system, imposed a hereditary monarchy, and signed a treaty with Russia over the country's northern border. Greenland and Iceland agreed to a union with the King. With the Scottish islands and the Faeroes included the Norwegian Empire reached its maximum extent.

[7] The Black Death killed close to 50 per cent of Norway's population between 1349 and 1350. The upper classes were decimated; Danes and Swedes were hired to fill the positions left vacant in the higher levels of the Government and the church. However, the King lost control over his dominions and isolated regions organized autonomous administrations.

[8] The ascent of Queen

Margrethe of Denmark to the throne in 1387 enabled the union of the Scandinavian countries. In 1389, she was crowned Queen of Sweden and in 1397 her adopted nephew Erik was elected king of all Scandinavia in Kalmar, Sweden. With the Kalmar Union, Norway was gradually subordinated, ultimately becoming a province of Denmark.

[9] After 1523, Norway's administrative council sought greater independence from Denmark. However, the fact that power lay in the hands of the Catholic bishops made it difficult to gain Swedish support. At the end of the civil war, between 1533 and 1536, the council was abolished. In 1537, the Danish King made the Lutheran religion the country's official religion; the Norwegian Church has been a State church ever since.

[10] During this period, social conditions in Norway were better than in Denmark. Landlords in the

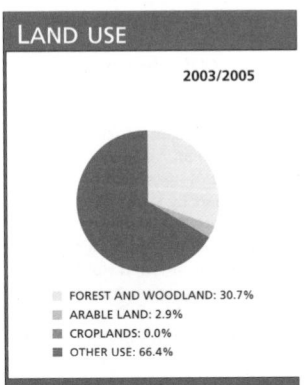

LAND USE

2003/2005

- FOREST AND WOODLAND: 30.7%
- ARABLE LAND: 2.9%
- CROPLANDS: 0.0%
- OTHER USE: 66.4%

PUBLIC EXPENDITURE

% OF GDP

15.6 %
HEALTH & EDUCATION (2000-2002)

1.9 %
DEFENSE (2004)

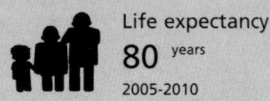

Life expectancy
80 years
2005-2010

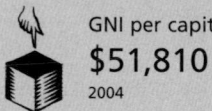

GNI per capita
$51,810
2004

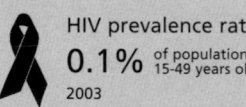

HIV prevalence rate
0.1% of population
15-49 years old
2003

countryside exploited the regional timber resources, and there was a large group of rural wage-earners. Most of Norway's population were peasants and fishing people, and cities no larger than 15,000 people.

[11] At the end of the Napoleonic wars, Denmark unilaterally surrendered control of Norway to Sweden. In 1814, Norway's constituent assembly proclaimed national independence. Sweden re-established its dominance by force, but in 1905 Norway regained its sovereignty without bloodshed.

[12] During the period of Swedish control, most of the laws enacted in 1814 remained in force. The Norwegian constitution is one of the oldest in the world, second only to that of the United States. It is based on the principles of national sovereignty, the separation of powers and the inviolability of human rights.

[13] Under a constitutional amendment in 1884, Norway adopted a parliamentary monarchy. The Danish Prince Carl was elected king of Norway, under the name Haakon VII, in 1905. Up until 1914, the country experienced rapid economic expansion, with the hydroelectric wealth of the region allowing large-scale industrial development.

[14] There was general concern that in 1906, 75 per cent of Norway's hydroelectric dams belonged to foreign investors. In 1909, Parliament passed laws for the protection of the country's natural resources.

[15] Universal suffrage, a term which applied to men only when it was passed in 1898, was extended to women by reforms approved in 1907 and 1913. One consequence of industrialization and universal suffrage was the growth of the Labor Party (LP).

[16] During World War I, Norway tried to remain neutral, but was obliged by the other powers to cut trade with Germany. Anti-German feeling was strong, particularly because of the various accidents caused by German submarines.

[17] Unlike other Western European social democracies, Norway's LP (with a left-wing majority) joined the Third Communist International in 1918. However, it could not agree with the centralization applied by the Soviet Communist Party, and cut its ties with the Comintern in 1923.

[18] Despite economic difficulties and serious labor conflicts (with unemployment reaching 20 per cent in 1938), Norway underwent vigorous industrial expansion in the inter-war years. The Government extended social legislation to include pensions, mandatory leave for workers and unemployment benefits.

[19] In 1940, at the beginning of World War II, Norway was invaded by Germany, after two months of fighting. King Haakon and the Government went into exile in London, co-ordinating the resistance from there. Vidkun Quisling became head of state and decreed martial law for resistance members. German troops left in May 1945 and Quisling was executed for treason, after King Haakon's return. Haakon died in 1957 and was succeeded by his son Olaf V.

[20] The LP governed continuously between 1935 and 1965 (except for the Nazi occupation), when it lost its parliamentary majority and Per Borten, the leader of the Center Party, was named Prime Minister. However, he resigned in 1971 when it was revealed that he had leaked confidential information during EC negotiations.

[21] After the War, Norway abandoned its policy of neutrality, joining NATO in 1949, the Nordic Council in 1952, and the European Free Trade Association (EFTA) in 1960. Large oil deposits were discovered in the North Sea in the late 1960s. Incorporation to the EEC was rejected in a referendum in 1972.

[22] Although its initiative to join the EEC was defeated, the LP remained in power throughout the 1970s and most of the 1980s, sometimes in alliance with the Socialist Left Party (SVP). When Prime Minister Oddvar Nordli resigned in 1981, the LP appointed Gro Harlem Brundtland, the first woman Prime Minister.

[23] From the beginning of the 1970s, Norway and the USSR disagreed over their rights to the Barents Sea. In 1977, Norway extended its territorial waters to 200 miles and designated a protected fishing zone in its territory of Svalbard, which had been dangerously overfished. A temporary agreement signed in 1978 defined a 'grey fishing area' to be jointly administered.

[24] In 1981 an agreement was signed with Iceland on mining and fishing rights. A similar dispute with Denmark, regarding Greenland, was taken to the International Court of Justice.

[25] The LP won the elections in May 1986, after four years of conservative governments, once more under Gro Harlem Brundtland, who appointed eight women to her 18-member cabinet.

[26] Ties with the US were strained in 1987 after Norwegian state company Kongsberg Vapenfabrikk (KV) exported 'heavy water' used in nuclear reactors, breaking NATO restrictions on the sale of those materials to former Warsaw Pact members and Third World nations.

[27] Romania and the former West Germany, who had re-exported the product to India, pledged, along with Israel, not to resell it without Norwegian authorization, and only to use it for peaceful purposes. The disagreement was overcome in 1988 when Norway banned all its 'heavy water' exports.

[28] That year, the US threat to sanction Norway over whaling was lifted when Norway agreed to limit whale hunting to scientific needs. But in 1990, Oslo announced it would resume traditional whaling.

[29] In 1988 fishing was seriously affected by a rise in the concentration of seaweed in the south and depredation by migrating seals in the north. However, after Sweden suspended fish imports in protest at Norway's hunting of seal pups, Norway banned the practice in 1989.

[30] Acid rain caused by industries in the Kola peninsula, to the east of Norway, and fires that broke out in Soviet nuclear submarines to the north - which the Soviets always denied - triggered protests. In 1989, Norway and the USSR agreed to share information on maritime accidents.

[31] The LP's vote fell from 41 per cent in 1985 to 37 per cent in 1989, due to the economic decline and the austerity measures taken by the Government. Incomes and sales fell, while unemployment reached six per cent, unheard of since the end of World War II.

[32] Gro Harlem Brundtland resigned in July 1989, after an agreement was reached between the Conservative, Center and Christian Democratic parties. In elections that year, the Labor and Conservative parties lost votes to the more radical parties. In November 1990 Brundtland was appointed Prime Minister once again.

[33] King Olaf V died in 1991, and was succeeded by his son Harald V.

[34] In January 1992, the Norwegian consulate in South Africa was upgraded to an embassy, after the ending of apartheid brought about in part by trade sanctions imposed by Norway and other countries. Minister of Foreign Affairs, Johan Juergen Holst, served as an intermediary in the Israeli-PLO negotiations.

[35] Despite serious disagreements among Norwegians, Oslo sought entry to the European Union

PROFILE

ENVIRONMENT

The Scandinavian mountain range runs north-south along the coast of the country. On the western side, glacier erosion has gouged out deep valleys that are way below actual sea level, resulting in the famous 'fjords', narrow, deep inlets walled in by steep cliffs. Maritime currents produce humid, mild winters and cool summers. The population is concentrated in the south, especially round Oslo. Nine-tenths of the territory is uninhabited.

SOCIETY

Peoples: Norwegians 96.3 per cent, Danish 0.4 per cent, British 0.3 per cent, Pakistani 0.2 per cent, Iranian 0.2 per cent, others (including 40,000 Sami, an indigenous people who live chiefly in the northern province of Finnmark) 1.9 per cent.
Religions: 88 per cent of the population belongs to the Church of Norway (Lutheran); there are Muslim, Evangelical and Catholic minorities.
Languages: Two forms of Norwegian are officially recognized; 80 per cent of school children learn the old form 'Riksmaal' (Bokmal, strongly influenced by Danish during the 434 year-long union of the two countries), and 20 per cent learn the neo-Norwegian 'Nynorsk' ('Landsmal', created out of the rural dialects). In the north, the Sami speak their own language.
Main Political Parties: Norwegian Labor Party; Conservative Party; Progress Party; Christian People's Party; Socialist Left Party.
Main Social Organizations: Norwegian Federation of Unions; Organization of Academics (AF); Organization of Trades (YS).

THE STATE

Official Name: Kongeriket Norge.
Administrative Divisions: 19 provinces (Fylker).
Capital: Oslo 795,000 people (2003).
Other Cities: Bergen 200,200 people; Trondheim 140,700; Stavanger 109,900 (2000).
Government: Parliamentary constitutional monarchy. Harald V, King since January 1991. Prime Minister: Jens Stoltenberg, since October 2006 (second time). Legislative power resides in the Storting, the 165-member unicameral Parliament.
National Holiday: 17 May, Constitution Day (1814).
Armed Forces: 27,000 (2003).

	Under-5 mortality			Maternal mortality	
	4	per 1,000 live births		**16**	per 100,000 live births
	2004			2000	

IN FOCUS

ENVIRONMENTAL CHALLENGES

The two main environmental issues are traditional whaling activity - opposed by environmentalists all over the world - and the oil industry, mostly in relation to joint projects with Russia (whose controls are slack) in the Barents Sea, which affect the fragile Arctic environmental balance. Fishing sectors also protest against these projects. In 1986, pollution in rivers and lakes in the south was blamed on sulphur dioxide emissions from the United Kingdom.

WOMEN'S RIGHTS

Women have been able to vote since 1907 with restrictions; which disappeared in 1913 when universal suffrage was implemented. In 2005, women held 38 per cent of seats in Parliament. In 2003, they made up 47 per cent of the total two million workforce. Female unemployment was less than four per cent. All births were attended by skilled health staff and maternal mortality was among the world's lowest: 16 per 100,000 live births.

CHILDREN

In 2004, five per cent of children had low birth weight*. Under-five mortality rates are among the world's lowest, at 4 per 1,000 live births. Education is compulsory and free up to the age of 16 and still free but optional up to 19.

INDIGENOUS PEOPLES/ ETHNIC MINORITIES

The Sami, a 40,000 member minority, live mostly in the northern province of Finnmark and belong to the ancient people of Lapland. Spread throughout Norway, Sweden, Finland and the Kola peninsula in Russia, and with a total population of 80,000, they speak a Uralic language (of the Finno-Ugric family). They arrived in these lands before the Scandinavians, Finnish or Russians, and although they used to practice shamanism, today they are mostly Protestants. The Icelandic sagas (first written down around the 13th century) include the first references to their existence. Although initially they were hunter-gatherers, for a long time reindeer farming has been their main activity. Only some 7,000 Sami currently herd reindeer, with a total of 500,000 animals. The arrival of hydroelectric, oil and fishing companies on their lands, as well as government efforts to integrate them into Norwegian society, has turned their reindeer-herding into more of a business than a way of life. They are represented in Norway by a Council or Parliament elected by the community.

MIGRANTS/REFUGEES

Between 2001 and 2005, Norway received 61,570 asylum requests. Most of them were filed in the first three years (48,220). In 2005, most refugees living in the country were from Iraq, Somalia, the Russian Federation, Serbia and Montenegro, Afghanistan, Iran, Eritrea, Turkey and Nepal.

DEATH PENALTY

The death penalty for common offenses was abolished in 1905 and for all types of crimes in 1979. The last execution took place in 1948.

* Latest data available in *The State of the World's Children* and *Childinfo* database, UNICEF, 2006.

(EU) in late 1992. Labor unions considered that joining the former European Community would threaten national sovereignty, while many entrepreneurs - particularly in the export sector - wanted full access to the EU market.

[36] Entry had to be approved by a referendum, and a date was set for November 1994. The campaign for this vote largely dominated political life for two years and revolved around oil exploitation, and regional and fishing policy.

[37] While the social democrats and conservatives supported joining the EU, 52.4 per cent of the electorate voted against integration, blocking Norway's membership.

[38] In 1994, the economy continued to expand and unemployment to fall. In 1995, unemployment reached 4.8 per cent and the trend was expected to continue.

[39] In October 1996, the Labor Party's Thorbjorn Jagland became head of government after Gro Harlem Brundtland resigned.

[40] Following the September 1997 elections, Jagland resigned. The ruling LP took 35 per cent of the vote and 65 seats out of the total 165 in play. The Prime Minister had announced he would stand down if the party could not equal the 36.9 per cent it had achieved in 1993.

[41] Christian Democrat Kjell Magne Bondevik was appointed Prime Minister. A coalition of three centrist parties took control but with only 43 seats in Parliament. Bondevik started by earmarking part of the oil income for investments in health and education.

[42] In 1997, the economy expanded for the fifth year running. However in early 1998, Norway began to suffer the consequences of falling international crude oil prices, following the crisis in Southeast Asia - the biggest oil-importing market in the world.

[43] The installation of natural gas plants became the center of a controversy that led Bondevik to resign in 2000. He maintained that the plants would release too much carbon dioxide into the atmosphere, while the opposition parties insisted on avoiding increased spending on fuel imports at all costs.

[44] Labor leader Jens Stoltenberg took office as the new Prime Minister in March 2000. Women were appointed to approximately half of Cabinet posts. Norway served as intermediary between the Government of Sri Lanka and Tamil separatists.

[45] In September 2000, the 'other Norwegian nation', the Sami, who for 20 years had tried to clarify their rights to the land, protested against a government plan to tap a gold mine in the northern region of Pasvik. The Sami Parliament – an advisory, non-executive body – demanded to be heard and said the use of the territory by the Government would endanger the community's survival, which was dependent on fishing and reindeer herds. For centuries the Sami had not been allowed to claim any land rights, because the land was restricted to speakers of Norwegian. The Sami language had been taught in schools since 1960.

[46] In early 2001, the Government decided to re-establish exports of whale meat and fat, thus lifting the ban imposed due to international pressure. Environmental organizations argued that endangered species would be affected. The killing of grey wolves, considered a threat to livestock, led to a clash with Sweden, which claimed the wolves were an endangered species. Meanwhile, the Government promoted seal hunting as a tourist attraction and proposed the killing of dolphins for scientific research.

[47] Benjamin Hermansen, a young Norwegian-Ghanaian anti-racist activist, was stabbed to death in January 2001 by six members of the neo-Nazi group Boot Boys, who were later arrested and tried.

[48] The LP did not gain a majority in the September 2001 general elections. A month later a three-party coalition was formed between Conservatives, the Christian People's Party and Liberals, who jointly supported the right-wing Progress Party and Kjell Magne Bondevik as the new Prime Minister.

[49] In 2002, environmentalists continued to try to stop whaling, warning consumers about the harm the hunting caused to several endangered Nordic species. Whale fat was exported mostly to Japan. The Government insisted that it applied controls on whaling.

[50] A strike by oil workers seriously affected Norway's oil production. The labor dispute to demand better pension rights and restrictions on temporary employment started in June 2004 at the state-owned company Statoil and later affected ExxonMobil and ConocoPhillips corporations. After five days of protest, Norway's oil output was cut by 375,000 barrels per day, that is 12 per cent of its daily average. Finally, upon the threat of halting all Norwegian oil production, the Government imposed compulsory arbitration.

[51] In June 2005, on the occasion of the 100th anniversary of the referendum that sealed Norway's independence from Sweden, the UN Secretary General, Kofi Annan, termed the event as an inspiration for all those who work for the world's peace.

[52] In the September elections, Bondevik was defeated by a center-left alliance led by Labor Party leader Jens Stoltenberg, who took office that month as Prime Minister. In his campaign, Stoltenberg, an economist, had pledged to spend more oil revenues on social welfare.

[53] In March 2006, the Norwegian Government decided to increase oil exploration in its Arctic waters, though it agreed to limit drilling in some areas until 2010 in order to protect the environment. Environmental groups criticized the move, which was considered insufficient. According to Samantha Smith, head of the World Wildlife Fund, the Norwegian Government should have limited exploration in a larger zone and for a longer period of time. 'A ground-breaking plan would have been permanent protection,' said Smith, adding that 2010 'may be a long time in politics but it's not long for nature.' ■

Oman / Uman

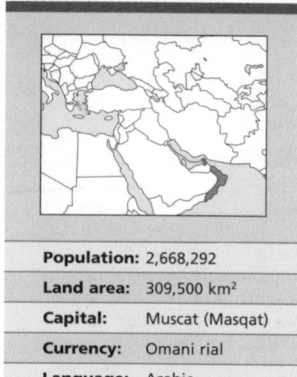

Population:	2,668,292
Land area:	309,500 km²
Capital:	Muscat (Masqat)
Currency:	Omani rial
Language:	Arabic

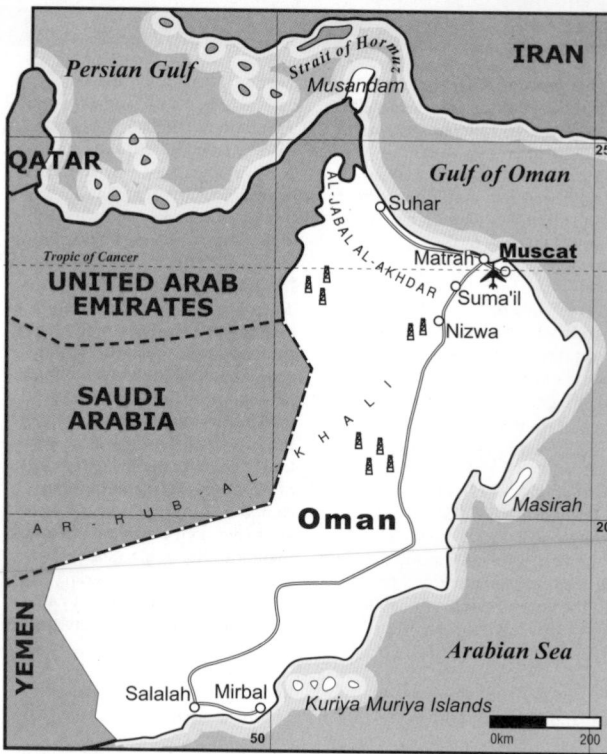

Sumerian clay tablets from the third century BC mention Oman as one of the outstanding markets in the economy of the Mesopotamian cities. Omani navigators became the lords of the Indian Ocean, connecting the Gulf to India, Indonesia and Indochina. In the 7th century they also played a major role in the peaceful propagation of Islam. Around 690 AD Abd Al-Malik decided to control the expansion of dissident sects. Consequently, some defeated leaders were forced to abandon the country. One of them, Prince Hamza, migrated to Africa where he founded Zanzibar, beginning a relationship between Oman and the African coast that would last until the 19th century.

[2] Towards 751, Oman took advantage of the dynastic strife in Damascus to elect an imam who gradually evolved from a spiritual leader to a temporal sovereign. Omani wealth gave the imam considerable power in the entire Gulf region, but also made Oman the object of successive invasions by the caliphs of Baghdad, the Persians, the Mongols and the local groups of central Arabia, which were all repelled. The Portuguese arrived in 1507, destroying the fleet and coastal fortifications, opening the way for the occupation of the principal cities and the control of the Strait of Hormuz. The Portuguese held control of the region for almost 150 years.

[3] In 1630, Imam Nasir ibn Murshid launched an inland struggle against the invaders. His son, Said, concluded this endeavor in 1650 with the expulsion of the Portuguese from Muscat (Musqat) and the recovery of Zanzibar and the African coast of Mombasa in 1698. Thus a powerful state was created which obtained the political unification of the African and Asian territories where a common culture and economy had developed.

[4] Sultan Said, the third in the Saiyid dynasty, expanded the African territories and moved the capital to Zanzibar in 1832. At the time of his death in 1856, the British presence was already being strongly felt on both continents. Said's sons argued and as a result, the African and Asian parts of the state were separated: the elder son, Thuwaini, kept the sultanate of Oman while his brother Majid took control of Zanzibar. The 1891 Canning Agreement virtually made Oman a British protectorate, weakening the Omani Kingdom.

[5] In 1913, inland peoples elected their own imam in opposition to the sultan's hereditary rule. Despite support from British troops, the sultan could not reconquer the rebel provinces. The struggle came to an end only in 1920, when a treaty was signed acknowledging the country's division in two: the Sultanate of Muscat and the Imamate of Oman. The imam persuaded the sultan to promise not to interfere in Imamate affairs, nor grant asylum to criminals and opposition members who fled Oman. In exchange Muscat was given control over customs and the right to set taxes on imports coming from the UK.

[6] Muscat was an extremely poor country, where arable lands accounted for less than one per cent of the total territory, and was artificially divided by colonialism. Between 1932 and 1970, it suffered the despotic rule of Sultan Said ibn Taimur, who fanatically opposed any foreign influence in the country, even in education and healthcare. This did not prevent him from granting control over the country's oil deposits to Royal Dutch Shell.

[7] Imam Ghaleb ibn Alim, elected in 1954, proclaimed independence and announced his intention to join the Arab League. In 1955, after putting down the nationalist movement, the British invaded the Imamate and reunited the country under the name of Sultanate of Muscat and Oman. Since then a liberation movement has been fighting against the monarchy, particularly in the southern province of Dhofar.

[8] Taimur was overthrown by his own son Qabus on 23 July 1970. Those who expected the young, Oxford-educated monarch to introduce modernizing changes soon realized that British domination was only being replaced by US domination. The US became a net importer of oil and began to develop an active interest in the area.

[9] Oil, produced commercially since 1967, provided more than half of the GDP. However more than half of Oman's labor force remained involved in agriculture, in the thin coastal strip that contains the country's only arable land.

[10] With US assistance, Qabus organized a mercenary army but when this force proved incapable of smashing the Popular Front for the Liberation of the Gulf, he signed an agreement with Iran's Shah Reza Pahlevi to secure Iranian intervention in the conflict.

[11] The guerrilla fighters were forced to retreat under the superior firepower of Iranian troops. Iranians also placed the Strait of Hormuz under their jurisdiction and stayed in the country establishing a virtual protectorate over Qabus's regime.

[12] At the Shah's downfall, Iranian soldiers were quickly replaced by Egyptian commandos and troops. The Sultan decided to give the US the Masirah island air base, and later the air bases at Ihamrit and Sib and the naval bases at Matrah and

---

PROFILE

ENVIRONMENT

With its 2,600 km coastline, Oman occupies a strategic position on the southeastern edge of the Arabian Peninsula and flanking the Gulf of Oman, where oil tankers leave the Persian Gulf. It is separated from the rest of the peninsula by the Rub al Khali desert that stretches into the center of the country. Local nomadic groups now live alongside petroleum and natural gas exploitation. Favored by ocean currents, the coastal regions enjoy a better climate. Monsoon summer rains fall in the north.

SOCIETY

Peoples: Omani Arab 73.5 per cent; Pakistani (mostly Baluchi) 18.7 per cent; other 5.5 per cent.
Religions: Muslim 86 per cent; Hindu 13 per cent; other one per cent.
Languages: Arabic, official and predominant, English, Baluchi and Urdu are also spoken.
Main Political Parties: There are no legal political parties.

THE STATE

Official Name: Saltanat 'Uman (Sultanate of Oman).
Administrative Divisions: 59 Districts.
Capital: Muscat (Masqat) 638,000 people (2003).
Other Cities: Nizwa 74,400; Suma'il 42,700; Salalah 163,600 (2000).
Government: Absolute Monarchy. Qabus bin Said, Sultan since July 1970 and Prime Minister since January 1972. Bicameral Legislature: the Consultative Assembly (Majlis al-Shura), with 82 elected members with only consultative tasks, and the Council of State (Majlis al-Dawla), with 40 appointed members.
National Holiday: 19 November, the Sultan's birthday.
Armed Forces: 43,500 (1996). Other: 3,900.

Life expectancy
75 years
2005-2010

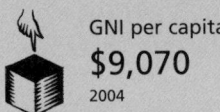
GNI per capita
$9,070
2004

Literacy
74% total adult rate
2000-2004

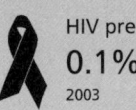
HIV prevalence rate
0.1% of population 15-49 years old
2003

IN FOCUS

ENVIRONMENTAL CHALLENGES
Rising soil salinity, coastal pollution from oil spills and very limited drinking water resources are the main environmental challenges the country is facing.

WOMEN'S RIGHTS
2003 saw the first universal elections to the *Majlis al-shura* (Consultative Assembly) for candidates over the age of 21, even though a restricted group of women have been able to run for election since 1994. In 2003, women held 7.8 per cent of parliament seats, and 10 per cent of ministerial or equivalent posts. That year, they comprised 20 per cent of the workforce. In 2004*, only 65 per cent of adult women were able to read and write, while 82 per cent of men were literate.

Maternal mortality stood at 87 in 100,000 live births*. The law bans discrimination by gender, ethnic origin, race, language, sect, place of residence or social status. But interpretations of Islamic law (Sharia) and some traditions affect women's rights to private property and state loans. Rape within marriage is not illegal, nor is female genital excision or mutilation, although this practice is declining.

CHILDREN
In 2004, a third of the population was under 18, including 300,000 children under 5*. Life expectancy has risen considerably since the 1970s, from 50 to 74.4 years*.

Teenagers over 15 are allowed to work, but not at night, nor weekends or holidays. It is still difficult to control child labor in small family businesses, particularly those in agriculture and fishing.

Government efforts have led to improvements in access to education, although the quality of preschool education remains poor.

INDIGENOUS PEOPLES/ ETHNIC MINORITIES
Citizens of African origin complained of job discrimination in both the public and private sectors. A Royal Decree ratified the International Convention on the Elimination of All Forms of Racial Discrimination in the year 2000.

Religious freedom is allegedly granted by law, but Hindu or Christian temples are only allowed to be built in locations determined by the Government. The Government prohibited non-Muslims from proselytizing, including publishing religious material, although religious material printed abroad could be brought into the country. Most Omanis are Ibadhi Muslims, followers of Abd Allah ibn Ibad. 25 per cent are Sunni Muslims, and there is a minor group of Shi'a Muslims, most of whom are of Iranian or Iraqi descent. Ibadhism is a form of Islam different in its hierarchical organization from the rest of the Muslim groups. The reason why Oman is in some aspects isolated from its neighbors is the practice of Ibadhism, according to Sunni and Shi'a Muslims.

MIGRANTS/REFUGEES
The extradition of political refugees is prohibited, and there have been no reports of the forced return of people to a country where they feared persecution.

Tight control over the entry of foreigners to the country has limited the entry of refugees and asylum-seekers.

Hundreds or thousands of illegal immigrants, mainly from Iran, Pakistan and Afghanistan are detained each year by the Royal Omani Police and held in special centers until their deportation can be arranged. The Government does not routinely grant protection to refugees or asylum-seekers.

DEATH PENALTY
The death penalty applies even for ordinary crimes.

* Latest data available in The State of the World's Children and Childinfo database, UNICEF, 2006.

Salalah. At that time, two-thirds of the national budget was committed to defense.

[13] In the 1980s, during the Iran-Iraq war, the US focused on Oman in its efforts to establish influence in the area.

[14] In 1989 oil prices became steady and in June, the major oil and gas company Petroleum Development Oman (PDO) discovered the most important natural gas deposits found within the last 20 years.

[15] That year Oman adopted a conciliatory policy towards Iran, establishing an economic cooperation agreement, on the condition that political stability be promoted in the country.

[16] Following the Iraqi invasion of Kuwait in March 1991, a member of the Gulf Cooperation Council (GCC), Oman suspended aid to Jordan and to the PLO.

[17] In 1991, the Government announced that the democratization process was under way; this included the creation of a parliament directly elected by the country's citizens.

[18] The Sultan launched a plan to diversify the country's economy, aiming at developing fishing, agriculture and tourism, among other sectors, faced with the prospect of the depletion of oil reserves before the year 2010.

[19] The fiscal deficits accumulated by the Government since 1981 led the World Bank to warn that the level of State expenditure was 'unsustainable'.

[20] Taking heed of the international financial organization's stand, in 1995 the Sultan announced a program of reforms which included a reduction of state spending, a series of privatizations and measures to attract foreign investment.

[21] In 1996, the Government announced a five-year plan to balance the budget by 2000. The project, aiming to free the economy from oil dependence, included privatizations and stimuli to increase foreign investment.

[22] That same year, the Sultan established a new succession mechanism. This meant that, if the royal family could not reach agreement on the appointment of a successor within three days of his death, the candidate chosen by the Sultan himself would be accepted.

[23] In June 1997 Sultan Qabus expanded women's political rights by royal decree, allowing women to stand for election. In the October 1997 elections, the Government selected two women to serve on the Consultative Council *(Majlis al-Shura)*. In December 1997, the Sultan appointed four women to the 41-member Council of State *(Majlis Al-Dawla)*.

[24] Oman became one of the first Arab countries to establish diplomatic and trade relations with Israel in January 1997. A month later, the rapprochement process came to a halt, when the Arab League questioned the Israeli decision to build new settlements in eastern Jerusalem.

[25] Oman and the United Arab Emirates signed an accord in May 1999 that defined part of the common border with the Abu Dhabi Emirate. Both parties agreed that, eventually, they would have to delineate more exactly the borders between Oman and the other emirates.

[26] In April 2001 the Government announced an amnesty for illegal workers. The amnesty allowed illegal residents not facing any criminal charge to leave Oman after paying a $125 fine, rather than the normal penalty of $25 a day for over-stayers.

[27] In October 2002, Muscat ratified the International Convention on the Elimination of Racial Discrimination. The Ministry of Social Development issued a decree for the formation of national social development committees. These committees hold the responsibilities of both promoting voluntary social organizations and through their activities, enhancing awareness of issues relating to childhood and disability, and also finding alternative funding for social programs.

[28] In November 2002 *Sultan Qabus* decreed an extension of voting rights to all citizens over the age of 21. Previously, those allowed to vote were selected from among local leaders, intellectuals and prominent entrepreneurs, with about a quarter of the state's 1.8 million people taking part in elections.

[29] In order to reduce unemployment, the Government tried to replace foreign workers with Omanis and encouraged job growth in the private sector. According to official figures, a quarter of the sultanate's population are foreigners.

[30] In October 2003 the first elections in which all citizens over the age of 21 were allowed to vote were held. Voters elected the 83 members of the *Majlis al-Shura* (Consultative Assembly). Two women were among those elected. In March that year, for the first time in Oman's history, a woman - Sheikha Aisha bint Khalfan bin Jameel al-Sayabiyah, aged 30 - was appointed as a minister, presiding over the National Authority for Industrial Craftsmanship.

[31] In May 2005, 31 people accused of 'conspiracy to overthrow the Government by the force of weapons' received prison sentences ranging from one to 20 years. A month later they were pardoned.

[32] In July, a human rights activist and former member of the Consultative Assembly, Taiba al-Mawali, was sentenced to six months in prison for criticizing the authorities in articles which she published on the internet, and another activist, Abdula Al Reyami, was detained for denouncing a lack of freedom in the country. In November, 1,300 Pakistanis who remained in jail for having entered the country illegally were repatriated.

[33] In January 2006, Oman signed a free trade agreement with the US. ■

Pakistan / Pakistan

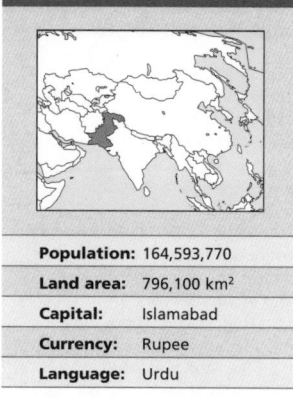

Population:	164,593,770
Land area:	796,100 km²
Capital:	Islamabad
Currency:	Rupee
Language:	Urdu

P akistan means the land of the pure, as it was religion (Islam) that bound together the people of different ethnic communities and languages. Poet-philosopher Mohammed Iqbal articulated the concept of Pakistan in 1931 when he proposed a separate state for the Muslims in the Indian subcontinent.

2   After the arrival in India of traders from Arabia and Persia, a permanent Muslim foothold was achieved with Muhammad ibn Qasim's conquest of Sind in 711 AD.

3   In 1296 Ala-ud-din Khalji proclaimed himself Sultan of Delhi and by 1311 the whole of India was under the Sultanate. In 1336 the Vijayanagara Empire, the kingdom of Hindu alliance, was founded with its capital at Hampi to counter the Muslim power. Over time uprisings divided the Empire and the Muslim Sultanates formed a new alliance. In 1565 the Sultanate coalition defeated the Vijayanagar army. As a result, the power in the region passed to Muslim rulers and later their kingdoms were annexed to the Mughal Empire (1529-1857).

4   The decline of Muslim power and the rise of the Hindu middle class took place during British colonialism. In October 1906, Muslim leaders met the British viceroy and demanded a reform of the electoral system with a separate system for Muslims. The All India Muslim League (ML) was founded in Dhaka (in what is now Bangladesh) to defend their political rights and interests. The British conceded the reform in the Government of India Act of 1909, confirming the ML's status as the representative of Indian Muslims.

5   In the 1930s there was a growing awareness of a common identity among Muslims as well as the need to preserve it within a separate territory. Under the leadership of Muhammad Ali Jinnah, the ML continued its campaign for the creation of Pakistan; a separate homeland in British India.-

6   The Hindu-Muslim relationship was affected by tensions and riots in different parts of India. This convinced the leadership of the Indian National Congress (representing mainly the nationalists) to accept Pakistan as a solution to the problems. On 3 June 1947, after the British withdrawal from India, a Partition Plan was announced and both the ML and the Congress accepted it. On 14 August of the same year, the new state of Pakistan was born comprising West Punjab, Sind, Baluchistan, North-West Frontier Province and East Bengal, surrounding northeastern and northwestern India.

7   Between 1948 and 1949, Pakistan annexed one third of the Indian province of Kashmir. The territory, mostly Muslim, had been annexed to India in 1947, in exchange for military support against Pakistani fighters.

8   Pakistan became a member of the Southeast Asian Treaty Organization (SEATO) in 1954 and the Central Treaty Organization (CENTO) in 1955, two strong military alliances led by the US. Pakistan later withdrew from these alliances, although bilateral relations with the US remained cordial.

9   Since gaining independence, Pakistan has suffered permanent political crises. The first constitution of March 1956 was abrogated by a coup on 7 October 1958, in which martial law was proclaimed. On 27 October 1958, General Ayub Khan introduced 'basic democracy', a system of local self-government and indirect presidential elections. Martial law came to an end in 1962 and a new constitution granted absolute power to the President and declared Pakistan an Islamic Republic. Ayub Khan was forced to resign on 25 March 1969, following a popular uprising. Martial law was again imposed and General Yahya Khan became President.

10   In the first general elections held between October and December 1970, the Awami League (AL) and the Pakistan People's Party (PPP) emerged victorious in East and West Pakistan, respectively. The AL won an absolute majority in the parliamentary elections, which gave it control of the federal government. When the opening of parliament was postponed in March 1971, the people of East Pakistan began the movement for an independent Bangladesh. The AL was banned and its leader Sheikh Mujibur Rahman was arrested. A civil war broke out and the AL formed a government in exile in India. The Indian army intervened and on 16 December 1971, Bangladesh was granted independence.

11   Zulfiqar Ali Bhutto, the leader of the PPP, formed a civilian government in 1972, following the resignation of General Yahya Khan. He encouraged strong public sector participation in the economy, followed a non-aligned foreign policy and approved radical land reforms. The PPP was victorious again in the general election of 1977, but the opposition parties accused the PPP of vote-rigging. At that point, General Zia-ul Haq overthrew the Bhutto Government and proclaimed martial law. Bhutto was arrested and sentenced to death on charges of conspiracy.

12   Pakistan strenuously opposed Soviet intervention in Afghanistan at the end of 1979.

13   Zia accelerated the process of Islamicization in all spheres of political and social life. Many political opponents were harassed and detained. A general election without political parties held in February 1985 partially legitimated Zia's Government. Zia was killed in a mysterious air crash in August 1988 and democracy was restored. Benazir Bhutto (PPP), daughter of the former President, took office following general election s in November 1988.

14   Benazir was the first woman to serve as head of state of a predominantly Islamic country, with two terms in office (1988-1990 and 1993-1996). Under Benazir, Pakistan became a signatory of the International Convention for the Elimination of Discrimination against Women, but some women felt their situation did not improve significantly. Some felt that she did not make sufficient efforts to push through laws to re-establish a quota of female seats in parliament. However her weak coalition government included conservative religious parties who opposed any such changes. Nonetheless, during her presidency women were appointed for the first time as high-court judges; a campaign against domestic violence began, and a women's bank was set up.

15   In mid-1990, there were about three million Afghan refugees living in Pakistan. During the Soviet occupation of Afghanistan, the US used Pakistani territory to supply arms to the *Mujahedin*, rebel groups fighting against the pro-Soviet regime in Kabul. This situation turned Pakistan into a

LAND USE

2003/2005

IRRIGATED AREA: 90.6% of arable land

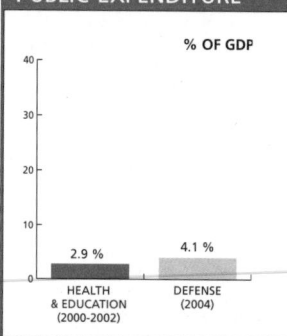

- FOREST AND WOODLAND: 2.5%
- ARABLE LAND: 25.2%
- CROPLANDS: 0.9%
- OTHER USE: 71.4%

PUBLIC EXPENDITURE

% OF GDP

2.9 %
HEALTH & EDUCATION (2000-2002)

4.1 %
DEFENSE (2004)

Life expectancy	GNI per capita	Literacy	HIV prevalence rate
65 years 2005-2010	**$600** 2004	**49%** total adult rate 2000-2004	**0.1%** of population 15-49 years old 2003

key ally for US regional policy, and resulted in the granting of significant economic assistance.

16 On 6 August 1990, President Ghulam Ishaq Khan dissolved Benazir's government, charging her administration with nepotism and corruption. The President suspended the National Assembly and named Ghulam Mustafa Jatoi, leader of the Combined Opposition Parties (COP) coalition, head of the interim government.

17 On 24 October 1990, Nawaz Sharif was elected Prime Minister, with the support of the Muslim League. The Pakistan People's Party (PPP) of Benazir Bhutto (who had also stood) denounced electoral fraud and launched an intense opposition campaign.

18 When the Gulf War broke out, following the Iraqi invasion of Kuwait, Pakistan sent troops to Saudi Arabia. Surveys showed that the population had strong pro-Iraqi tendencies, but the Government announced that the country's forces would only defend Islamic holy places, and would not take part in combat or go into Iraqi territory.

19 Shortly after taking office, Sharif approved a plan to encourage private investment that included the privatization of state-run companies. This plan was strongly resisted by the 300,000 public sector workers.

20 A process of re-Islamization of society, promoted by Sharif, included the introduction of *sharia* or Islamic law with consequent setbacks for women's social and legal status. The Government banned the media from making any reference to a woman's right to divorce.

21 In November 1991, the opposition accused Nawaz Sharif of embezzling public funds. Sharif was responsible for the bankruptcy of several cooperative credit institutions. Only the unconditional support of President Ishaq Khan prevented the fall of Sharif. The matter was referred to the judiciary and triggered a wave of demonstrations led by the PPP, to which the Government reacted with increased repression.

22 In February 1992, the ancient dispute over the border territory of Kashmir brought Pakistan and India to the brink of a new armed conflict. The Jammu and Kashmir Liberation Front, a Muslim group demanding the creation of an independent state, staged a protest march against the division of the territory between the two countries disputing it. The Pakistani Government ordered the army to shoot at the demonstrators, and several people were injured or killed.

23 When Pakistan developed a nuclear weapon construction project in 1992, the US suspended economic assistance and arms sales to Islamabad. Pakistan announced that China had guaranteed economic and technological support to continue the nuclear research program.

24 President Ishaq Khan accused Prime Minister Nawaz Sharif of poor administration, corruption and nepotism, forcing him to resign in April 1993. Sharif appealed to the Supreme Court and was reinstated in May. The dissolution of the Assembly was revoked and the call for elections cancelled. Both Khan and Sharif resigned in July.

25 Benazir Bhutto returned to power in October 1993. The PPP won 86 of the Assembly's 217 seats, against 72 obtained by Sharif.

26 Between 1994 and 1995, Bhutto attempted to democratize the country at a time of more political and ethnic violence than had been seen since the separation of Bangladesh in 1971. Karachi and the northern separatist areas were the center of disputes, in which over 3,500 died.

27 In November, Benazir was forced to resign after being accused of corruption. In the new parliamentary elections, the followers of former Prime Minister Nawaz Sharif won 136 seats of the 217 at stake. Miraj Khalid led the Government as Deputy Prime Minister until Sharif came to office in early 1997.

28 Tension with India intensified in May 1998, when that country carried out a series of nuclear tests and Pakistan responded with its own.

29 While the Indian Government kept claiming its authority over the Kashmir region, Pakistan insisted on calling for a referendum among Kashmiris to determine whether they were in favor of independence.

30 In October 1999, General Pervez Musharraf, who had been in charge of military operations in Kashmir and had been removed from his post by Sharif, led a coup and imprisoned the Prime Minister, accusing him of kidnapping, terrorism and attempted murder. Pakistan became the first nuclear power under military leadership.

31 On 10 December 1999, Sharif was granted a presidential pardon and exiled in Saudi Arabia in exchange for not returning to Pakistan for ten years and renouncing his personal fortune. He also promised not to participate in the country's politics for 21 years.

32 The first stage of local elections was held in late December, in 18 of the 106 districts. Analysts observed a change in control of the electoral scene, from political parties to local feudal families.

33 By March 2000, the second stage of local elections had been held in 20 districts, and the pressure had grown on the President to present a clear schedule to return to a democratic system. Meanwhile, the police detained more than 2,000 activists from the Alliance for the Restoration of Democracy and 22 of its leaders, who were planning a demonstration.

34 Musharraf became head of state and was appointed President of Pakistan in June 2001, announcing he would maintain the powers he had as head of the executive branch, and that he would remain as commander of the army. In October, Musharraf extended his mandate indefinitely as Supreme Commander of the Army.

35 After the terrorist attacks on 11 September 2001 against New York and Washington, the US announced it would lift economic and military sanctions it had imposed on India and Pakistan after the nuclear tests of 1998. The measure was motivated by Musharraf's support for the US military operation against Afghanistan's Taliban regime, allied to the Al-Qaeda terrorist network. In previous years, when the Taliban was fighting for power, it had received direct military, economic and ideological support from Pakistan.-

36 The US bombings over Afghanistan caused extensive protests from Islamic Pakistani groups. Musharraf made changes in the army shortly before the US military operations began in the neighboring country, releasing some pro-Islamic generals from strategic posts.

37 When US Secretary of State Colin Powell visited Islamabad in October 2001, Musharraf requested Washington's support in the Kashmir conflict, as well as aid for the economic consequences of the huge inflow of Afghan refugees crossing the border.

PROFILE

ENVIRONMENT

Pakistan is mountainous and semi-arid, with the exception of the Indus River basin in the east. This is virtually the only irrigated zone in the country, suitable for agriculture and vital to the local economy. The Indus rises in the Himalayas in the disputed province of Kashmir and flows into the Arabian Sea. The majority of the population live along its banks. The main agricultural products are wheat and cotton, grown under irrigation.

SOCIETY

Peoples: Most Pakistanis are of Indo-European descent, mixed with Persian, Greeks and Arabs (in the Indus Valley), and Turkish and Mongolian (in the mountainous areas). Much Indian immigration has occurred in recent years. There are five main ethnic groups: Punjabi, Sindhi, Pashtun, Mujahir and Baluch.
Religions: Islam is the official religion followed by more than 95 per cent of the population (most belong to the Sunni sect); two per cent are Christian; 1.6 per cent are Hindu; the remainder belong to other smaller sects.
Languages: Urdu (official, although it is spoken by only nine per cent of the population). Other languages are Punjabi, Sindhi, Pashto, Baluchi, English and more than 50 local languages.

Main Political Parties: Pakistan People's Party Parlamentarian; Pakistan Muslim League (Quaid-e-Azam); Muttahhida Majlis-e-Amal Pakistan (which includes: Islamic Assembly; Assembly of Islamic Clergy; Assembly of Pakistani Clergy and Movement for Islam).
Main Social Organizations: The National Pakistan Federation of Unions.

THE STATE

Official Name: Islam-i Jamhuriya-e Pakistan.
Administrative Divisions: Four provinces, one territory (Tribal Areas) and one capital territory (Islamabad).
Capital: Islamabad 698,000 people (2003).
Other Cities: Karachi 11,800,000 people; Lahore 5,470,000; Faisalabad 2,136,000 (2000).
Government: General Pervez Musharraf, self-appointed President since June 2001, confirmed by referendum in 2002. Shaukat Aziz, Prime Minister since August 2004.
National Holidays: 14 August, Independence Day (1947); 23 March, Proclamation of the Republic (1956).
Armed Forces: 619,000 (2003). Other: 275,000 (National Guard, Border Corps, Maritime Security, Mounted Police).

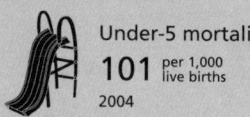

Under-5 mortality	Poverty	Debt service	Maternal mortality
101 per 1,000 live births 2004	**17%** of population living on less than $1 per day 2002	**21.2%** exports of goods and services 2004	**500** per 100,000 live births 2000

38 India blamed Pakistan for the suicide attack on the Indian Parliament in Delhi by Islamic activists on 13 December. Musharraf condemned the attack and denied any link between his Government and those responsible for it. On 26 December 2001, India deployed troops and planes on its border with Pakistan, increasing the tension in Kashmir. While Indian and Pakistani troops exchanged fire along the border, on 30 December Pakistani authorities detained Mohammed Saeed, the leader of one of the activist groups blamed for the attack on Parliament.

39 In April 2002, US journalist Daniel Pearl was murdered after being kidnapped in January by an Islamic group. Also in April, Musharraf called a referendum on his rule. With 97 per cent of the vote he was allowed to extend his presidency for another five years. All observers denounced the elections as fraudulent.

40 After tests of medium - range surface-to-surface missiles, capable of transporting nuclear warheads, were carried out by Pakistan, Musharraf declared that the country did not want war but was prepared to defend itself in the event of being attacked.

41 Opposition forces accused Musharraf of perpetuating his dictatorship, after he decided to grant himself new powers, among which was the right to dissolve a democratically elected parliament. In the October 2002 elections, the Government imposed restrictions on or proscribed important leaders such as Nawaz Sharif and Benazir Bhutto. The Pakistan People's Party (PPP), which supported the military, won by a slim majority. However, the most critical result was the growth of the Islamist parties, mainly in the areas bordering Afghanistan, which would render them a key element in any government coalition.

42 In November 2002, the National Assembly appointed Zafarullah Jamali (a close ally of Musharraf) as Prime Minister. In elections to the Senate in February 2003 (the final stage of what the President called transition to democracy) the ruling party won once again.

43 In June 2003, *Sharia* law was introduced in the North-West Frontier Province.

44 In November 2003, Pakistan declared a Kashmir ceasefire, which was swiftly matched by India. In December, Musharraf miraculously survived an attempt on his life when a bomb exploded seconds after his car passed.

45 In February 2004, the country's leading nuclear scientist Dr Abdul Qadeer Khan admitted to having

IN FOCUS

ENVIRONMENTAL CHALLENGES
Pakistan has significant water pollution from industrial and agrochemical waste. Soil degradation and desertification are also serious. Water sources that are fit to drink are limited, and most of the population lacks access to safe drinking water.

WOMEN'S RIGHTS
Women have been able to vote since 1947. The situation of gender inequality in the country has varied depending on political circumstances. Benazir Bhutto's governments made some advances that were reversed by military coups and fundamentalist surges in some provincial and national governments. In 2003, 20.5 per cent of Parliament seats were held by women.

Women's participation in the labor force - about 30 per cent - has remained constant since 1980.

In 2004*, the percentage of literate women was 56 per cent.

Forty-three per cent of pregnant women receive prenatal care and only 23 per cent of births are attended by trained health staff*. Maternal mortality rates are very high: 500 per 100,000 live births*.

CHILDREN
One third of Pakistan's population lives below the poverty line. In 2004, under-five mortality rate stood at 101 deaths per 1,000 live births* (the rate was 227 per 1,000 in 1960 and 130 per 1,000 in 1990) and infant mortality was at 80 per 1,000*. More than one third of children under-five are underweight.* Primary school enrolment rate was 76 per cent for boys and 57 per cent for girls, who are often victims of discrimination*. There is little information regarding HIV/AIDS incidence; in late 2003 UNICEF estimated there were 74,000 people under 50 living with HIV/AIDS*. At the end of 2001, around 2,200 children (under 14 years old) were living with HIV/AIDS and 25,000 were AIDS orphans.

INDIGENOUS PEOPLES/ ETHNIC MINORITIES
The Pashtun are the main ethnic minority. They account for 10 million people on the border with Afghanistan, where they make up the majority of the population. They speak Pashtu, a language of Indo-European origin. Women are considered to have almost no rights or value; they usually represent a form of payment for offenses.

The Hunzabut people, famous for their longevity, are divided into the Wakhi and the Burusho. They live on the banks of the river Hunza, number between 55,000 and 60,000 people and speak Burushaski, an Indo-European language, and are Muslims.

The Hazara, a group of between 110,000 and 220,000 Persian language speakers, live on the border with Afghanistan where they make up 9 per cent of the total population. They are warlike, self-sufficient people who trade almost exclusively with their Pashtun neighbors.

MIGRANTS/REFUGEES
The migratory flow from Afghanistan has been constant. In the late 1970s it was due to the Soviet occupation, in the mid-1990s it took place when the Taliban regime seized power and since January 2002, it has been caused by US attacks in the search for Osama bin Laden.

In 2002-2003, Pakistan received more than 2 million Afghani refugees. In September 2005, 391,000 Afghani refugees returned to their country. This takes to 2.6 million the total number of people who have returned to Afghanistan from Pakistan.

In 2004, 17,000 Pakistanis requested asylum in Canada and Germany. Another 5,000 requested asylum in Belgium, Cyprus, France, Greece, Poland, Ukraine, US, Sri Lanka, South Africa, UK, Italy, Nepal and China.

DEATH PENALTY
The death penalty is applicable even for ordinary crimes.

* Latest data available in *The State of the World's Children* and *Childinfo* database, UNICEF, 2006.

worked on secret projects for the development of nuclear weapons. He said that the technology had been transferred to Libya, North Korea and Iran.

46 In March-April 2004, after a 14-year break, test cricket was resumed between India and Pakistan - an important symbol of rapprochement between the two nations.

47 Prime Minister Chaudhry Shujaat Hussain resigned in March 2004 and was replaced by then Finance minister Shaukat Aziz.

48 In December, Musharraf announced he would continue as head of the army, in spite of having pledged to leave the post.

49 In the western province of Baluchistan, in 2004, nationalist forces had launched guerrilla attacks. They were seeking autonomy, a higher proportion of the earnings from their region's gas reserves and the withdrawal of Government troops from the region. By March 2005 the conflict had grown, with dozens of deaths among Pakistani military and paramilitary and militias. The attacks continued almost daily.

50 In July 2005 more than 300 alleged militant Islamists were detained in *madrassas* (religious schools) and offices of Muslim organizations, in a Government offensive against religious radicalism carried out days after the attacks against the public transportation system in London. Three of the four suicide bombers in London had recently visited Pakistan.

51 Approximately 74,000 people died when an earthquake hit the north of the country in October 2005, and more than 3 million Pakistanis were left without a home, with winter approaching. India and Pakistan agreed to open the Kashmir border to facilitate relief efforts.

52 In January 2006, 18 civilians died when US forces bombed a tribal area on the border with Afghanistan. The attack failed to reach its supposed target, al-Qaeda's second-in-command, Ayman al Zawahiri.

53 In February, Pakistan joined much of the rest of the Islamic world in protest over cartoons of Prophet Muhammad published by a Danish newspaper. The riots left several dead and injured. The Danish embassy in Islamabad was closed temporarily and the Pakistani ambassador in Copenhagen was called for consultations.

54 In March, on a visit to Islamabad, US President George W Bush praised Musharraf for his 'courageous decision' of joining the war against terrorism and reiterated the 'strategic partnership' of their respective countries in that fight. A few days before, a suicide attack in Karachi had killed a US diplomat and three Pakistanis.

55 In April, a suicide attack at a Sunni ceremony killed 57 people. In February, another attack during a Sh'ia ceremony in northwestern Hangu had left 31 dead.

56 In October 2006, a missile strike in the Bajaur area killed up to 80 alleged militant supporters of the Taliban and al-Qaeda. ■

Palau / Palau - Belau

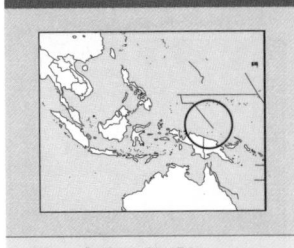

Population:	20,174
Land area:	460 km²
Capital:	Koror
Currency:	US dollar
Language:	English

F ive thousand years ago, sailors from Formosa and China peopled the islands of Micronesia, forming highly stratified societies where age, sex and military prowess defined rank and wealth.

[2] European colonization in the 19th century destroyed cultural diversity, but it did not totally eliminate the indigenous people of Palau.

[3] In 1914, the Japanese took the islands from Germany who had bought them from Spain in 1899. During World War II, Japan installed its main naval base in Palau and it soon became the scene of fierce combat.

[4] By the end of World War II, the indigenous population of Palau had been reduced from 45,000 to 6,000. Micronesia became a US trust territory. The US used the islands as nuclear testing grounds and reneged on promises of self-rule, instead fostering economic dependence. Palau's self-determination process was delayed until the late 1970s.

[5] In 1978, at the beginning of the transition to self-rule, the archipelago opted for separation from Micronesia. In January 1979, the new constitution banned all nuclear weapons installations, nuclear waste storage, and foreign land ownership. A further amendment established a 200-mile area of territorial waters under the UN-approved Law of the Sea, which the US tried to veto.

[6] Although the US pressed Parliament to modify the Constitution, to allow Palau to be a major military base, in July 1980 a referendum ratified the original text.

[7] In 1981, Washington announced it was willing to fund the construction of a 16-megawatt electric power plant, if Palau amended its Constitution. President Haruo Remliik held a plebiscite to approve the project jointly with a 'Compact of Free Association' with the US. The Compact failed to achieve enough

votes. In 1983, the President signed agreements for a loan totaling $37.5 million and in 1984, the power plant was built.

[8] In 1985 Remliik refused to call a new referendum when the US proposed to revoke the historic anti-nuclear Constitution. A few days later he was assassinated.

[9] Succumbing to pressure, the Palau Congress agreed to call a new referendum to amend the Constitution, and on 21 August 1987 the Compact was approved. However, the result was not valid because the minimum percentage of yes votes required by the Constitution was not reached.

[10] In 1987 and 1988 several attacks and threats on opposition groups took place.

[11] In August 1988 President Lazarus Salii was found dead from a gunshot wound. The official version was suicide.

[12] The status of Free Associated State was ratified by referendum in July 1993, with 68 per cent of the vote, becoming effective in October 1994.

[13] In November 1996, Kuniwo Nakamura was elected President in the first presidential poll since independence.

[14] Despite Chinese opposition, Palau established diplomatic relations with Taiwan in 1999. This coincided with an aggressive Taiwanese policy of winning recognition from small Pacific States in return for economic aid.

[15] In 2001, Tommy Remengesau was sworn in as President after winning the 2000 elections.

[16] That year, Remengesau authorized US naval bases in Palau.

[17] In March 2003 Palau joined the 'coalition of the willing' and

supported the US-led war on Iraq. Remengesau further offered the US use of Palau's facilities for its military operations.

[18] In November 2004, Remengesau was re-elected President with 66.5 per cent of the vote.

[19] Due to the agreement that enabled Palau's independence the US was to pay it $450 million over 15 years. In exchange, Washington would retain until 2044 the option of establishing a

military base on the islands. This enabled Palau's per capita GDP to stand in 2006 at an estimated $6,870, one of the highest among the small Pacific nations. But the imminent end of the transfer of US funds in 2009 created uncertainty on the archipelago's economic future.

[20] A report in October 2006 by the Philippine Overseas Labor Office indicated that at least 80 per cent of Filipino workers in Palau were undocumented. ■

PROFILE

ENVIRONMENT
Palau comprises almost 350 islands, with a total surface area of 494 sq km. A barrier reef to the west forms a large lagoon dotted with small islands. Coral formations and marine life in this lagoon are among the richest in the world with around 1,500 species of tropical fish and 700 types of coral and anemones. The maritime ecosystem is however being damaged by pollution.

SOCIETY
Peoples: Most are of Polynesian origin. Palauan (83.2 per cent), Filipinos (9.8 per cent), Micronesian (2.9 per cent), Chinese (1.2 per cent), 'white' (0.8 per cent), other (3.0 per cent).
Languages: English, Palauan, Sonsoralese, Tobi, Angaur.
Religion: No official religion. Catholics (40.7 per cent); Protestants (24.7 per cent); local religions, mainly modekngei (27.1 per cent), other (7.5 per cent).
Main Political Parties: Currently there are no political parties.

THE STATE
Official Name: Republic of Palau (Belu'u era Belau).
Administrative Divisions: 16 States. **Capital:** Koror 14,000 people (2003).
Other Cities: Meyungs 1,100 (2000).
Government: Tommy Remengesau, President since January 2001. The National Congress (Legislature) has two Chambers: the House of Delegates, with 16 members, and the Senate, with 9 members.
National Holidays: 9 July, Constitution Day (1979); 1 October, Independence Day (1994).
Armed Forces: The US is responsible for defense.

Palestine / Filestin

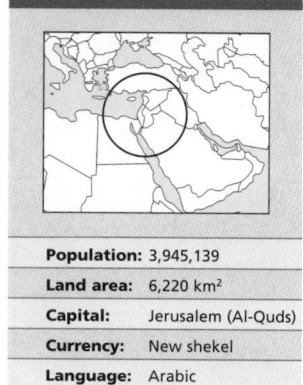

Population:	3,945,139
Land area:	6,220 km²
Capital:	Jerusalem (Al-Quds)
Currency:	New shekel
Language:	Arabic

Jewish State, UN Plan, 1947
Arab State, UN Plan, 1947
1 West Bank
2 Gaza Strip

Jewish State, UN Plan, 1947
Annexed to Israel in the 1948 war

Israeli borders before 1967
Occupied by Israel in 1967
Recovered by Palestine from 1994: Gaza Strip, Jericho, Nablus, Ramallah

B etween 3000 and 2500 BC, the Canaanites (from whom, together with the Philistines, the Palestinians are descended) settled in a land first known as Canaan and later as Palestine. The Jebusites, a Canaanite people, built a settlement they called Urusalim (Jerusalem) 'city of peace'.

2   Around 2000 BC another nomadic Semitic people, the Hebrews, led by Abraham, passed through Palestine. Seven centuries later, 12 Hebrew tribes returned from Egypt, following Moses. There was fierce fighting over possession of the land. The Bible records that 'The sons of Judah were unable to exterminate the Jebusites that dwell in Jerusalem' (Joshua 15, 63).

3   Four centuries later David managed to unify the Jewish kingdom. After the death of his son Solomon, the Hebrews split into two states, Israel and Judah. These later fell into the hands of the Assyrians, in 721 BC, and the Chaldeans, in 587 BC. In 587 Nebuchadnezzar destroyed Jerusalem and took the Jews into captivity in Babylon.

4   Palestine was conquered by Alexander the Great in 332 BC and became part of the Greek Empire. After his death, the territory returned to the Egyptian Empire of the Ptolomies. The country was subdued by the Seleucids from Syria before a rebellion, headed by Judas Maccabeus, restored the Jewish state in 67 BC.

WORKERS

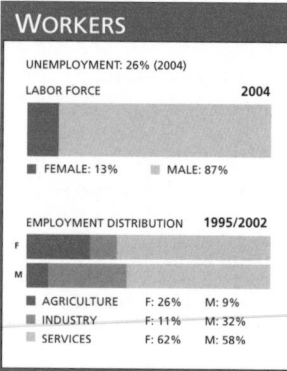

UNEMPLOYMENT: 26% (2004)

LABOR FORCE   2004

■ FEMALE: 13%   ■ MALE: 87%

EMPLOYMENT DISTRIBUTION   1995/2002

F
M

■ AGRICULTURE	F: 26%	M: 9%
■ INDUSTRY	F: 11%	M: 32%
■ SERVICES	F: 62%	M: 58%

5   In 63 BC, Palestine was annexed by the Roman Empire. The Maccabeans, Zealots and other Jewish tribes resisted the invaders but were fiercely subdued. The repression included the crucifixion of thousands of rebels, around 30 AD, during the time of Jesus of Nazareth.

6   In 70 AD the Roman Emperor Titus demolished Solomon's Temple. Years later, in 135 AD, the Jews were expelled from Jerusalem, and the Roman Emperor Hadrian founded a pagan city on the ruins.

7   From 330 AD Palestine came under the control of the Byzantine Empire. In 638 Omar Al-Khattaab entered Jerusalem, ending the era of Byzantine rule and marking the beginning of the Arab-Islamic era. According to Islamic tradition, the prophet Muhammad rose to heaven in Jerusalem, and so the city became a holy place for Muslims alongside Christians and Jews.

8   In 1516, Jerusalem was conquered by the Ottoman Empire, remaining under its control until the end of World War I.

9   From 1878 the Zionist movement began to establish the first Jewish settlements in Palestine. In 1895, the total population of Palestine numbered about 500,000 (453,000 Palestinian Arabs owned 95 per cent of the land and 47,000 Jews owned 5 per cent).

10   The Jewish National Fund, created by the Fifth Zionist

Congress, set about buying land and from 1904 to 1914 a second wave of immigrants arrived. In 1909 the first *kibbutz* (collective farm) was settled to the north of Jaffa.

11   When World War I broke out, the British promised independence to the Arab territories under Ottoman rule, including Palestine, in exchange for their support against Turkey, an ally of Germany.

12   In 1917, the British Foreign Secretary, in the so-called 'Balfour Declaration', agreed to give British support for the creation of a 'Jewish Homeland'. In 1919 the Palestinians held their first Conference at which they rejected the Balfour Declaration and demanded an independent Palestinian State, as the British had promised in exchange for their support during the war.

13   In 1920 the Conference of San Remo approved the British mandate over Palestine. Two years later the League of Nations Council also approved a mandate supporting the creation of a Jewish Homeland on the territory. For six months the Palestinians held strikes and demonstrations in protest at the confiscation of lands and illegal immigration, the objective of which was to increase the Jewish population and thus give force to their claim over the land.

14   The British Government issued a new 'White Paper', limiting

Jewish immigration and promising Palestine its independence after ten years. This was rejected by leading Zionists, who formed militias and launched a bloody campaign against the British and Palestinians. On 9 April 1948, the Irgun organization, led by Menahem Begin, raided the village of Deir Yassin, killing 254 civilians. Terrorized, thousands of Palestinians abandoned their lands.

15   At the end of World War II, the United Nations approved the partition of Palestine (Resolution 181). The Palestinians, who represented 70 per cent of the population and owned 92 per cent of the land, were restricted to 43 per cent of the territory. The rest was given to the Jews, who constituted 30 per cent of the population and owned eight per cent of the land. Jerusalem fell within the one per cent that remained under international control.

16   On 14 May 1948, the Jews proclaimed the State of Israel. The following day, the first Arab-Israeli war broke out. Palestine was split into three areas: the part occupied by Israel; the West Bank, which came under the control of Jordan; and Gaza, governed by Egypt. About 700,000 Palestinians fled their homes to neighboring countries and settled in refugee camps.

17   In 1964, the Palestine Liberation Organization (PLO) was created to defend the rights of the

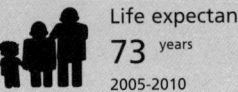  

Life expectancy
73 years
2005-2010

GNI per capita
$1,120
2003

Literacy
92% total adult rate
2000-2004

Palestinian people and reaffirm their identity both in the region and in the international arena. In 1969, Yasser Arafat was elected president of the PLO.

[18] The clandestine Palestinian organizations, like Al-Fatah, mistrusted this new organization that was being promoted by Arab governments and its emphasis on seeking a diplomatic solution. Convinced that they would only recover their lands by resorting to force, on 1 January 1965, they carried out their first armed attack in Israel.

[19] In 1967, the Six-Day War broke out: Israel took over Jerusalem, the Golan Heights in Syria, the Sinai desert in Egypt and the Palestinian territories in the West Bank and the Gaza Strip. The UN called on Israel to withdraw from the Arab territories that it had occupied by force, and declared the Palestinians' right to return to their lands and to self-determination.

[20] The defeat of the regular Arab armies strengthened the conviction that guerrilla warfare was the only way to reach their goals. In March 1968, during a battle in the village of Al-Karameh, Palestinians forced the Israelis to withdraw. The battle passed into folk history as the first victory of the Palestinian force. The armed groups joined the PLO and obtained the support of Arab governments.

[21] King Hussein of Jordan, who had been a representative and spokesperson of the PLO, regarded the increasing political and military strength of the Palestinians as a threat. In September 1970 this came to blows as Palestinian resistance elements in Jordan known as *fedayeen* (from the Arabic *fida'i*, 'one who is ready to sacrifice one's life for the cause') were attacked by King Hussein's largely Bedouin forces in response to a string of high profile terrorist hijackings of civilian aircraft by the Palestinian Front for the Liberation of Palestine led by George Habash, which drew international anger. A 10-day civil war led to an estimated 3,500 deaths and massive material destruction in Jordan. The PLO was expelled from Jordan and set up headquarters in Beirut.

[22] This new exile reduced the possibility of armed attacks on targets inside Israel, and new radical groups such as 'Black September', named after the fighting between Jordanian government forces and Palestinian *fedayeen*, directed their efforts towards Israeli institutions and businesses in Europe and other parts of the world. Palestinians, until then regarded by world opinion purely as refugees, quickly came to be identified by some as terrorists.

[23] PLO leaders promptly realized the need to change their tactics and, without abandoning armed struggle, launched a large-scale diplomatic offensive, starting to devote much of their energy to consolidating Palestinian unity and identity. The Algiers Conference of Non-Aligned Countries (1973) identified the Palestine problem, and not Arab-Israeli rivalry, as the key to the conflict in the Middle East for the first time.

[24] In 1974, an Arab League summit conference recognized the PLO as 'the only legitimate representative of the Palestinian people'. In October of the same year the PLO was granted observer status in the UN General Assembly, which recognized the right of the Palestinian people to self-determination and independence. On 10 November 1975 the UN General Assembly adopted, by a vote of 72 to 35 (with 32 abstentions), Resolution 3379, which stated that 'Zionism is a form of racism and racial discrimination'. The resolution was revoked on 16 December 1991, with a vote of 111 to 25 (with 13 abstentions).

[25] The PLO Charter agreed in 1968 called for a sustained revolutionary armed struggle against the 'Zionist occupation' to liberate all of Palestine including the internationally recognized pre-1967 borders of the State of Israel which the PLO rejected. 'Armed struggle is the only way to liberate Palestine. This it is the overall strategy, not merely a tactical

phase'. This necessarily implied the end of the present State of Israel. Without giving up this ultimate goal, the PLO has gradually come to accept the 'temporary solution' of setting up an independent Palestinian State 'in any part of the territory that might be liberated by force of arms, or from which Israel may withdraw'.

[26] In 1980, Israeli prime minister Menahem Begin, and Egyptian president Anwar Sadat signed a peace agreement, mediated by US president Jimmy Carter, at Camp David. Israel agreed to withdraw its forces from the Sinai peninsula, and return it to Egypt. Soon afterwards, Jewish settlements in the West Bank multiplied on appropriated Palestinian lands, increasing tension in the occupied territories. Successive UN votes against these measures, or for any action against Israel, were stripped of any practical value by the US using its veto in the Security Council.

[27] In July 1982, in an attempted 'final settlement' of the Palestine issue, Israeli forces invaded Lebanon. They sought to destroy the PLO's military structure, and to capture the greatest possible number of its leaders and combatants who had been staging attacks along Israel's northern border. A massacre took place at the refugee camps of Sabra and Shatila performed by the Southern Lebanese Army under orders from Israeli Defense Minister Ariel Sharon. International sympathy

with the plight of the Palestinian people surged. The headquarters of the organization were moved to Tunis and Yasser Arafat toured Europe receiving the honors due a head of state in various countries, in particular at the Vatican.

[28] The PLO leadership quietly participated in talks with Israeli leaders receptive to a negotiated settlement with the Palestinians. With the invasion of Lebanon, small but active peace groups emerged in Israel, demanding the initiation of a dialogue with the PLO. Palestinian radicals questioned these overtures, breaking with Yasser Arafat's policies. These factional divisions within the PLO at times led to violent confrontations.

[29] Following years of infighting, in 1987, the Palestinian National Congress, held in Algeria, managed to rebuild the PLO's unity.

[30] In December 1987, the funerals of several young Palestinians killed in a clash with Israeli military patrols led to further confrontations, general strikes and civil protests. The *Intifada* (uprising, lit. 'shaking off') began in the Gaza Strip, the West Bank and East Jerusalem. The Intifada marked a new stage in the Palestinian struggle: for the first time the population - young people, children, and the elderly - rose up against the occupying army. Many unarmed civilians resorted to throwing stones at Israeli troops, who responded with gunfire, a reaction that shocked the world.

PROFILE

ENVIRONMENT

Historically 'Palestine' is the 27,000 sq km territory west of the Jordan River which the League of Nations handed over to Britain's 'mandatory' power in 1918. This territory comprises: the area occupied by Israel before 1967, 20,073 sq km; Jerusalem and its surroundings, 70 sq km; the West Bank area, 5,879 sq km and the Gaza Strip, 378 sq km. Many nations now recognize Palestine to be the area under full and partial sovereignty of the Palestinian National Authority, with final status negotiations still in limbo. This is hotly disputed by most Palestinians who see the whole region as indisputably Palestine. It is a land of temperate Mediterranean climate, fertile on the coast and in the Jordan Valley. It is surrounded in the south and the northeast by the Sinai and Syrian deserts respectively.

SOCIETY

Peoples: The Palestinians are a group of mainly Arabic speakers who regard themselves as a distinct group of the Arabic-speaking peoples, with family origin in Palestine being the defining characteristic. As such, the designation is independent of nationality and religion. There are 700,000 Palestinians in Israel; 1,500,000 on the West Bank; 800,000 in the Gaza Strip, and the rest mostly living in Middle Eastern (Jordan 2,170,000; Lebanon 395,000; Syria 360,000; other Arab countries 517,000) and European countries. Thirty-three per

cent of the inhabitants of the occupied territories live in refugee camps. There are large Palestinian populations in the US, Chile, Brazil and other countries.
Religions: Muslims (mostly Sunni) 97 per cent; Christians (Eastern Orthodox) 3 per cent.
Languages: Palestinians speak a unique dialect of Levantine Arabic. Hebrew is a common second language, as is English.
Main Political Parties: Fatah (Palestinian National Liberation Movement). Palestine Liberation Organization (PLO). Hamas (Islamic Resistance Movement), currently in power; very influential in Gaza and the West Bank, is opposed to the Oslo process (1993) and autonomy agreements. Islamic Jihad.
Main Social Organizations: Palestinian Labor Federation; General Union of Palestine Women; Addamir (prisoners' association).

THE STATE

Official Name: As-Sulta Al-Watania Al-Filistiniya.
Capital: Jerusalem (Al-Quds), 668,000 people (1999). Almost universally seen by Palestinians as their capital; the Palestine National Authority (PNA) is in Jericho, 14,744 people (1997).
Government: Mahmoud Abbas, President since January 2005. Ismail Haniyeh, Prime Minister since March 2006. The Autonomous Council acts as parliament.
Armed Forces: No official data available.

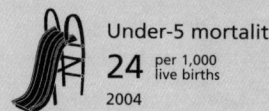

Under-5 mortality

24 per 1,000 live births

2004

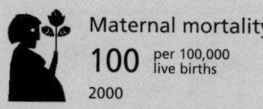

Maternal mortality

100 per 100,000 live births

2000

The Intifada lasted approximately five years and virtually destroyed the already fragile economies in the occupied territories.

31 On 14 November 1988, the Palestinian National Congress (parliament-in-exile), meeting in Algeria, proclaimed the Independent State of Palestine, in accordance with UN Resolution No. 181 of 1948, which divided Palestine into two states, one Jewish, the other Arab-Palestinian. This implied accepting the State of Israel. Ten days later, 54 nations officially recognized the new State of Palestine.

32 Arafat was elected president of Palestine and in that capacity gave an address to the UN General Assembly, in which he rejected terrorism, recognized the State of Israel, and demanded that international troops be sent into the occupied territories. As a result of his speech, US president Ronald Reagan decided to initiate talks with the PLO.

33 When the Gulf War broke out in 1991, it was clear that the Palestinian people were pro-Iraq. This position deprived the PLO of the financial backing of the rich Gulf emirates, who opposed the Iraqi regime.

34 In September 1991, Yasser Arafat was confirmed as the president of Palestine and of the PLO, and the Palestine National Council accepted the resignation of Abu Abbas, leader of the Palestine Liberation Front. Abbas was convicted in absentia by an Italian court and sentenced to life in prison for the 1985 hijacking of the Italian cruise ship Achille Lauro.

35 In 1991, the first Middle East Peace Conference was held in Madrid, sponsored by the US and the former USSR. Palestinians and Israelis agreed to mutual recognition.

36 In September that year, Israel and the PLO signed a Declaration of Principles at the White House, which established a five-year deadline for the withdrawal of Israeli forces from the occupied territories, and for negotiating the permanent settlement status of the Gaza Strip, the West Bank and East Jerusalem, and which would be followed by the establishment of an independent Palestinian State.

37 The Israeli Parliament ratified recognition of the PLO and the Declaration of Principles. The PLO Central Committee approved the text concerning autonomy.

38 The agreement was opposed by Hamas and Hizbullah on the Palestinian side, as well as by settlers in the occupied territories, and by far-right parties on the Israeli side. In a climate of hostility, the Israeli military withdrawal from Gaza and Jericho anticipated for December 13 was postponed.

39 In May 1994 Rabin and Arafat signed the 'Gaza and Jericho first' autonomy agreement, while Israeli withdrawal continued, enabling the return of several contingents of the Palestinian Liberation Army exiled in Egypt, Yemen, Libya, Jordan or Algeria.

40 After 27 years in exile Arafat arrived in Gaza in July as head of the Executive Council of the new Palestinian National Authority (PNA). Those regions under Palestinian control saw an influx of foreign and Palestinian investment, in addition to international aid, aimed at laying the foundations for the future State.

41 The struggle between the historic leader of the PLO and the Islamic fundamentalist opposition became increasingly violent. Arafat wanted Hamas to participate in the Palestinian general elections in January 1996, which would have further legitimized his leadership. However, the Islamic fundamentalists decided to boycott the elections. Arafat was elected president with 87 per cent of the vote and government candidates won 66 out of a total of 88 seats.

42 Right-wing Likud leader Binyamin Netanyahu's election as Israeli Prime Minister (see Israel) in May 1996 aggravated tension between the countries.

43 The difficult negotiations ended with the withdrawal of Israeli troops from the city of Hebron in 1997. In the same year, in accordance with the agreements between the two sides, Palestinian political prisoners were set free from Israeli jails. At the end of 1997, the peace talks stalled when Netanyahu ignored previous agreements and resumed construction of new illegal settlements. His action provoked violent confrontations and harsh international condemnation. Arafat declared that since the five-year deadline set by the commitments adopted had expired, he would proclaim an independent Palestinian State, with its capital in East Jerusalem.

44 In 2000, US president Bill Clinton invited Arafat and Prime Minister Ehud Barak (Labor Party) to a meeting at Camp David. The US and Israeli proposals for a final agreement did not meet basic Palestinian demands regarding the dismantling of illegal settlements on the West Bank, the return of refugees, and Palestinian border control. Jerusalem, a holy city for both Muslims and Jews, became the biggest stumbling block in the negotiations, since both sides wanted to install their capital city there.

45 Tension in the area increased when former Israeli Defense Minister Ariel Sharon visited Haram al-Sharif/Temple Mount in Al Quds/Jerusalem, considered a holy place by both Muslims and Jews. A new Intifada was declared, a string of suicide bombings in Israeli town centers led to numerous Israeli civilian deaths, and Israel resumed its bombing of Palestinian villages, leaving 400 civilians dead.

IN FOCUS

ENVIRONMENTAL CHALLENGES
The Israeli-Palestinian conflict has had a severely destructive impact on natural resources and fauna. Gaza suffers water shortages, and refugee camps are highly polluted by refuse and sewage water. Soil is affected by erosion, drains are inadequate and underground water is contaminated.

WOMEN'S RIGHTS
Polygamy is still allowed. At least 25 per cent of Palestinian women have suffered some form of abuse. There are no official statistics, since women beaten by their male relatives do not report these incidents, often because they do not perceive themselves as victims of violence, having internalised dominant male values that see such behavior as men exercising their 'right'. In general the authorities to whom they might report incidents hold the same values as the male relatives.

Women-headed households are increasing as a result of the death, imprisonment or unemployment of males. Female-led households are 1.3 times more likely to be poor than those headed by men, and close to 30 per cent of them fall below the poverty line.

Women face a disadvantage in terms of wages and social security benefits. According to the 2005 Social Watch Report, there are unequal barriers for women entrepreneurs in terms of property and inheritance rights, access to credit and penal liability. The average marrying age among Palestinian women is 18.

CHILDREN
Women, children and youth are the most vulnerable. Over half of the population are minors. In 2004, the infant mortality rate was 25 deaths for every 1,000 live births.

According to international human rights groups, the Israeli army has killed hundreds of children since 2002. Statistics show by late 2004, some 10,000 children had been caught in the crossfire. Many of these have been left permanently disabled.

In the meantime, an unspecified number of minors are being held in Israeli army prisons, with few activities and in many cases no family visits and no contact with the outside world.

The standard of the education received by Arab Palestinian children is, in almost all aspects, inferior to the education given to Israeli children. One of the consequences of this is lower achievement.

INDIGENOUS PEOPLES/ ETHNIC MINORITIES
Muslims, Jews, Christians, Bedouins, and Druze constitute the indigenous population. Most Druze and Bedouins live in Israel (see Israel). Among the Palestinian population there are mostly Muslims with a sizeable Christian minority. Christian Palestinians live mostly in Bethlehem and the surrounding area, some others have become Israeli citizens and dwell, among other places, in Jerusalem and Haifa. In both cities Arabs and Jews live side-by-side.

MIGRANTS/REFUGEES
In 2002, UNRWA (UN Relief and Works Agency for Palestine Refugees) estimated that there were slightly over four million Palestinians displaced from their homes between 1948 and 1968 and now living primarily in refugee camps in the Gaza Strip and West Bank. One third of these are in Jordan, Syria and Lebanon.

In 2004, over 360,000 Palestinians requested asylum and refugee status, including 240,000 in Saudi Arabia, 70,000 in Egypt, 23,000 in Iraq and 9,000 in Lebanon; the remainder sought refuge in various Western countries.

The number of Palestinians and their descendants who have emigrated to other countries in the world, including South America and Europe, is over seven million.

DEATH PENALTY
The Palestinian Authority applies the death penalty. Mistrials and flawed sentences have been denounced.

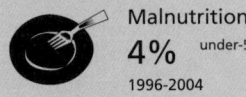
Malnutrition

4% under-5s

1996-2004

Water source

94% of population using improved drinking water sources

2002

Doctors

84 per 100,000 people

1990-2004

Primary school

86% net enrolment rate

2004

46   Sharon's victory in Israel's February 2001 elections was perceived as another blow to the weakened peace process. That month, the UN Secretary-General issued a document indicating that Israel's economic blockade of the West Bank and Gaza Strip had pushed Arafat's government to the verge of collapse due to lack of revenue.

47   The UN special envoy to the Middle East, Terje Road Larsen, warned that if other countries did not provide urgent financial support to the Palestinians (his report estimated that one billion dollars was needed for the rest of that year) violence would increase.

48   Fighting increased over the following months. Israel's offensive and the stalled negotiations increased resistance to the occupation. Sharon replied with targeted assassinations of suspected terrorists, and extended his offensive to attacking Palestinian communities and towns with helicopters and gunboats. Several hundred Palestinians died in the conflict, and military operations continued with the occupation of the territories that had been under relative Palestinian control.

49   After the 11 September 2001 attacks on New York and Washington, Sharon believed international public opinion and Western governments could turn in his favor, and he increased the offensive against the Palestinian uprising. The new US President, George W Bush, needing to attract allies to his anti-terrorist campaign against the Taliban regime in Afghanistan, preferred to keep his distance and avoid confrontations with the Arab world.

50   Numerous suicide attacks by radical Palestinian militants marked a new stage in the conflict. To strengthen security, Sharon limited the traffic of goods and people through the borders of the West Bank and Gaza Strip from the beginning of the uprising. The measure affected both Palestinian workers and businesses.

51   In December Sharon cut off all talks with Arafat. The new Israeli strategy no longer acknowledged the Palestinian leader as a valid counterpart for negotiations.

52   Restrictions on the movement of goods and people in Israel and the occupied territories after 18 months of uprising placed the Palestinian economy on the verge of collapse. The continued closure of border checkpoints caused irreparable harm. Unemployment tripled, affecting almost 30 per cent of the Palestinian labor force.

53   A summit of Arab countries was held in Beirut in March, which Arafat could not attend because Sharon kept him cornered in his Ramallah bunker for more than a month.

54   In spite of its initial chaos, the summit ended with the approval of a peace plan which included a historic decision. The signatory countries agreed to recognize the State of Israel, as long as it withdrew to its pre-1967 borders and allowed the return of three million Palestinian refugees, as well as the creation of a Palestinian state with a sector of Jerusalem as its capital. Israel called the proposal 'unacceptable'.

55   In April, Fatah, Hamas, Islamic Jihad, People's Front and Democratic Front for the Liberation of Palestine, agreed for the first time a common military strategy 'to confront any Israeli attack'. Most of the 82 suicide bombers in Israel and the Jewish settlements since the beginning of the Intifada were militants from these extremist organizations.

56   That month, Jenin refugee camp on the West Bank was the scene of bloody bombings by Israel, causing the death of hundreds of Palestinians. Terje Larsen, the UN envoy, called what happened in Jenin a 'morally repugnant humanitarian disaster'. Sharon declared him *persona non grata*. The Jenin camp was reduced to rubble. After the raids on Jenin and other areas under the relative control of the PNA, Israel detained some 5,000 Palestinians.

57   In June 2002 US President George W Bush called on Palestinians to reject Arafat's leadership and to choose a leader who was 'not committed to terrorism'. In December Arafat postponed elections, blaming Israel for the delay.

58   In March 2003, Mahmoud Abbas (a moderate politician also known as Abu Mazen) was elected prime minister of the Palestinian Authority. In April, Bush presented Sharon and Abbas with a new peace plan known as the 'Road Map', which was sponsored by the so-called Middle East Quartet (US, EU, UN and the Russian Federation). This plan proposed the creation of a Palestinian State and the resolution of all outstanding issues by 2005. Abbas resigned in July, accused by the radicals of making too many concessions to Israel.

59   The violence increased. Israel began building a 'separation wall' in the West Bank, allegedly as a means of keeping terrorists out. The Palestinians viewed the wall as an attempt to unilaterally demarcate the borders with a future Palestinian state, to Israel's advantage. The UN General Assembly demanded that Israel stop work on the wall, but the EU and US asked the International Court of Justice to refrain from giving an advisory opinion regarding its legality. The barrier deprived thousands of Palestinians access to basic services like water, medical care and education, as well as to their means of livelihood, such as farming and other jobs.

60   In March 2004, after a double suicide attack by Hamas in the port of Ashdod, Israel responded with a strategy of 'targeted assassinations' of leaders of radical Palestinian movements. Sheikh Ahmed Yassin, the 67-year-old spiritual leader of Hamas, was killed by an Israeli missile attack from a helicopter as he was leaving a mosque in Sabra (Gaza). Although his assassination was unanimously condemned by the international community, the US vetoed a UN Security Council motion of condemnation.

61   In April 2004, Sharon announced a 'Plan of unilateral disengagement from Palestinian areas' which included the evacuation of settlements in the Gaza Strip and the dismantling of six settlements in the West Bank. In exchange, Israel sought US support to keep 'settlement blocs' in the West Bank, home to most of the 230,000 Israeli settlers, and a declaration by President Bush denying Palestinian refugees the right to return.

62   In October, Israeli forces demolished the homes of hundreds of Palestinians, toppled infrastructure and killed over 70 people in the bloodiest attack in the Gaza Strip in years. The attack was launched after two Israeli children were killed in a Hamas rocket attack.

63   On 11 November 2004, Arafat died in Paris and there was a state funeral in Cairo, Egypt. He was buried in the Palestinian National Authority (PNA) compound in Ramallah, despite his wish to be buried in Jerusalem.

64   Rauhi Fattouh, the speaker of the Palestinian Legislative Council, took over as interim president of the PNA for 60 days, until elections were held, while Abbas was appointed chairman of the PLO executive committee.

65   In the elections held in early February 2005, Abbas, the Fatah candidate, was elected president of the PNA with 62 per cent of the votes, and immediately called on the radical groups Hamas and Islamic Jihad to suspend their attacks on Israel.

66   In February, Abbas succeeded in convincing Hamas and Islamic Jihad to agree to an unofficial temporary ceasefire. Against the backdrop of this fragile truce, Abbas and Sharon announced plans to meet in Egypt to initiate peace talks, but the meeting never took place.

67   In August, the Israeli army completed its withdrawal from Gaza, which included the evacuation - in many cases by force - of some 8,500 Jewish settlers, thus ending 38 years of occupation. The continuation of this process was cast into doubt when Sharon suffered a stroke and went into a coma in January 2006.

68   That same month, Hamas scored a surprise win in the Palestinian legislative elections, taking 76 of 132 seats in Parliament. Fatah refused to participate in the new government formed by Ismail Haniyeh, who took over as Prime Minister in February. Israeli acting Prime Minister Ehud Olmert announced that he would not negotiate with the new Government unless Hamas renounced violence and recognized the State of Israel. He also cut off the transfer of tax and customs revenues collected by Israel on behalf of the PNA.

69   The freezing of transfers of funds from Israel and the suspension of millions in aid from the US and EU brought the Palestinian Government to the verge of collapse. The Hamas administration resisted international pressure to recognize the State of Israel and asked for aid from Islamic countries in order to cover back-pay owed to public sector workers.

70   Throughout this period, Israel kept up its policy of 'targeted assassinations' of leaders and members of radical Palestinian organizations.

71   In May, clashes between police loyal to Fatah and a new security force created by Hamas sparked fears of a Palestinian civil war. Abbas, placed in a difficult situation by Hamas's succession to power, announced that he would attempt to resolve the crisis by calling a referendum for the population to decide upon the recognition of Israel and the viability of peaceful coexistence of two states - one Israeli, the other Palestinian - as a solution to the conflict.

72   In July, Israel launched an offensive on Gaza in order to rescue a captured Israeli soldier and reduce the number of rocket attacks by militants against Israeli towns. In the UN, the United States vetoed a resolution calling on Israel to halt its military operations in Gaza. Dozens of Palestinians, including civilians, died in the attacks.

73   At the end of October 2006 Israel renewed its onslaught on Gaza, targeting Hamas and Islamic Jihad militants. A tank attack killed 18 civilians in the town of Beit Hanoun, prompting Palestinian outrage and widespread international condemnation. Ehud Olmert said the strike on a civilian area was the result of a 'technical failure'. ∎

Panama / Panamá

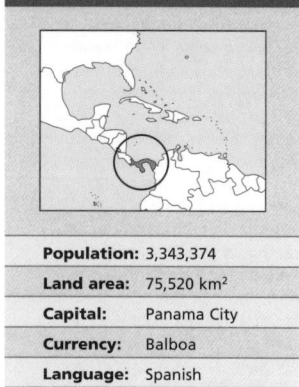

Population:	3,343,374
Land area:	75,520 km²
Capital:	Panama City
Currency:	Balboa
Language:	Spanish

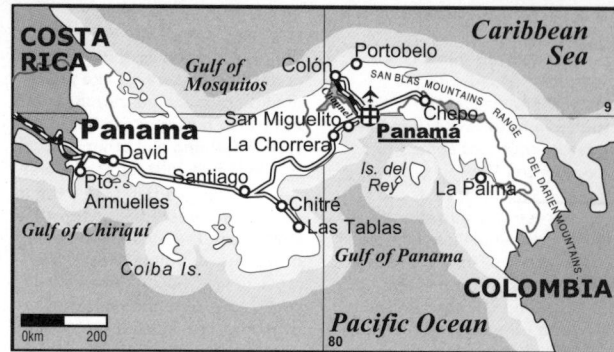

The Chibcha civilization, which developed in the isthmus of Panama, was one of America's great cultures. The Chibchans had a highly stratified society, developed elaborate architecture, crafted gold and had a wide scientific knowledge.

2   In 1508, Diego de Nicuesa was given the task of colonizing what was known as the Gold Coast (present day Panama and Costa Rica). The enterprise ended in complete failure. In 1513, Vasco Nuñez de Balboa was sent to look for what was assumed to be a 'South Sea', and on 25 September he found the Pacific.

3   The isthmus soon acquired great geopolitical significance due to the proximity of the Atlantic and Pacific oceans. Panama became an important commercial center for the Spanish monopoly. After sailing from Spain, ships arrived in Portobelo, in the Caribbean Sea, from where the cargo crossed the isthmus by mule to Panama City. Once in Panama, the merchandise was distributed throughout America from San Francisco to Santiago. The concentration of riches attracted English pirates and buccaneers; Francis Drake razed Portobelo in 1596 and Henry Morgan set fire to Panama City in 1671.

4   Panama was dependent upon the viceroyalty of Peru until 1717, when the Bourbons transferred it to the new viceroyalty of Granada, this was later to be a part of Greater Colombia when the country became independent from Spain in 1821.

5   In 1826, Panama was selected by Simón Bolívar as the site of the Congressional Assembly which was to seal the continent's unity. But the economic decadence of the end of the 18th century, coupled with the change in commercial routes meant that Panama did not maintain its strategic importance after breaking with Spain, and did not become an independent nation with the disintegration of Greater Colombia in 1830.

6   In 1831 Panama seceded from New Granada for a year, with the intention of forming a Colombian Confederation while maintaining autonomy. The state of Panama was created in 1855 within the Federation of New Granada (present-day Colombia).

7   The first direct reference to the US 'right' to intervene militarily in Panama, is the Mallarino-Bidlak Treaty of 1864, signed by the Governments of Washington and Bogotá. The document authorized the US to obtain a faster means of uniting the east coast with the west by building a railroad across the isthmus. It also allowed the US to offset the British presence in the area, especially in Nicaragua.

8   On 1 January 1880, a French company, the Universal Company of the Panama Canal led by Ferdinand de Lesseps, started to build the canal. In 1891, the company was accused of fraud in its dealings, causing its bankruptcy, though 33 km of the project had already been completed. In order to complete the project, three years later, the New Panama Canal Company was founded.

9   In 1902, the US bought out the French company, and in January 1903, the Hay-Herran Treaty was signed with a representative of the Colombian Government. The treaty spelled out the terms of the construction and administration of the canal, granting the US the right to rent a 9.5 km-wide strip across the isthmus in perpetuity.

10   The Colombian Senate rejected the treaty unanimously, considering it improper and an affront to Colombia's sovereignty. Only a revolution allowed the US to remain. The 'revolutionaries', supported by US marines, declared Panama's independence in November 1903, and the US recognized the new state within three days. While Theodore Roosevelt was President, the 'Big Stick' policy of sending troops into Central American states was common practice.

11   A new treaty, the Hay-Buneau Varilla Treaty, granted the US full authority over a 16 km-wide strip and the waters at either end of the canal in perpetuity. Buneau Varilla, a former shareholder of the canal company and a French citizen, signed as the official representative of Panama. He received payment for his services in Washington, and did not return to Panama. The canal, covering a distance of 82 km, was officially opened on 15 August 1914, and from then on was administrated and governed by the US.

12   The Canal Zone brought incalculable wealth to the US, not so much in toll fees but in time and distance saved by vessels traveling between California and the East coast. US military bases in Panama functioned as an effective means of control over Latin America. Against the backdrop of the Cold War, American military instructors lectured Latin American military officers on the National Security Doctrine, a politico-military system which ousted legally constituted governments and imposed military dictators. Also, the financial center created in the isthmus became an initial foothold for the expansion of US transnational corporations and money laundering.

13   In January 1964, 21 students who died in an attempt to raise Panama's flag in the Canal Zone were transformed into national martyrs. The demand for full sovereignty over the Zone was taken up by the Government of General Omar Torrijos. He rose to power in 1969, upon the dissolution of a three-member Military Junta which had overthrown President Arnulfo Arias in 1968. The diplomatic battle against the colonial enclave was waged in all the international forums and gained the support of the Latin American countries, the Movement of Non-Aligned Countries and the UN.

14   The struggle for sovereignty united Panamanians, stimulating nationalistic feelings repressed by decades of foreign cultural penetration, control of the economy and US military intervention. At the same time, the Torrijos Government initiated a process of transformation aimed at establishing a more equitable social order. The most important changes introduced included reforms in agriculture and education, and the nationalization of copper exploitation. A 'banana war' to obtain fairer prices was waged against transnational fruit companies such as the United Fruit Company, which was later named 'United Brands'.

15   The US finally agreed to open negotiations in favor of a new canal treaty, as the Panama issue was damaging its image in Latin America. The 1977 Torrijos-Carter Treaty abrogated the previous one and provided for the canal to become fully Panamanian from the year 2000. However, breaking the agreement, amendments introduced by the US Senate added provisions to the treaty giving the US the right to intervene 'in defense of the Canal' after 2000.

16   On 31 July 1981, Torrijos died in a suspicious airplane accident. Unconfirmed reports suggested that the plane's instruments were interfered with from the ground. President Aristides Royo, who succeeded Torrijos in 1978, lost the support of the National Guard and was forced to resign by his new commander-in-chief, Ruben Paredes. He started realigning the country's policies, adopting a pro-US stance. The role of the US in the Malvinas (Falklands) War and the launching of the Contadora Group - set up to try and mediate in the region's conflicts, with Panama as

LAND USE

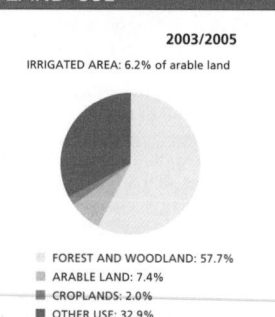

2003/2005

IRRIGATED AREA: 6.2% of arable land

- FOREST AND WOODLAND: 57.7%
- ARABLE LAND: 7.4%
- CROPLANDS: 2.0%
- OTHER USE: 32.9%

PUBLIC EXPENDITURE

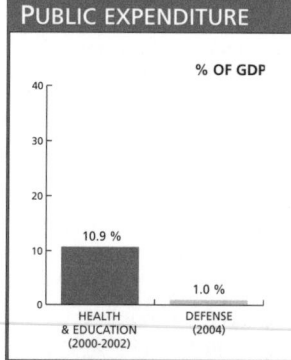

% OF GDP

10.9 %
HEALTH & EDUCATION (2000-2002)

1.0 %
DEFENSE (2004)

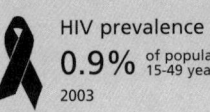
the first host - led to new frictions in relations between the two nations.

[17] In 1983, Paredes was replaced as commander-in-chief of the National Guard by General Manuel Noriega. The 1984 presidential and legislative elections were narrowly won by Nicolas Barletta, the candidate of the Revolutionary Democratic Party (RDP), which was founded by Torrijos and supported by the armed forces. The opposition, led by veteran politician Arnulfo Arias brought accusations of fraud.

[18] Barletta encountered growing opposition to his economic policies, and resigned toward the end of 1985. He was succeeded by Eric del Valle, but the driving force continued to be General Noriega, a former protégé of the US Government whom the US now aimed to overthrow for refusing to collaborate in plans to invade Nicaragua. A 'settling of accounts' began, in which Noriega was accused of links with drug-traffickers and other crimes. The opposition united behind the National Civil Crusade, made up of parties of the right and center with broad support of the business community.

[19] In 1987, Washington withdrew its economic and military aid. In 1988, it froze Panama's assets in the US and imposed economic sanctions, including the cessation of payments for Canal operations. In March all the Panamanian banks closed for several weeks, provoking a financial crisis. American military presence increased. Del Valle overthrew Noriega, but the National Assembly backed the commander-in-chief and removed the President, replacing him with the Minister of Education, Manuel Solís Palma.

[20] Elections were called for 5 May 1989. Amid interference from the White House, which discredited the electoral process and its results even before it had taken place, the results of the ballot were kept secret for several days and, because they favored the opposition candidate Guillermo Endara, from the Democratic Alliance of Civilian Opposition, the election was declared null and void.

[21] Solís declared that the US' objective was to set up a puppet government regardless of the election results and retain control of the Canal Zone, going back on the commitments it had assumed under the Torrijos-Carter Treaty.

[22] An anti-Noriega uprising by a group of young officers failed in October 1989. Economic sanctions were followed by a US military invasion of the country without warning or prior declaration of war. Endara was installed as President at Fort Clayton, the American base, at the start of the invasion.

[23] This was the largest US military operation since the Vietnam War (1964-1973), with the mobilization of 26,000 troops. Panamanian resistance proved to be stronger than the invaders had anticipated, which prolonged the military operation. Indiscriminate bombing damaged heavily-populated neighborhoods and killed many civilians - 560 according to official figures; between 4,000 and 10,000 according to the opposition. Losses were estimated at more than $2 billion and about 5,000 Panamanians were temporarily imprisoned. Noriega took refuge within the Vatican Embassy and was later extradited and transferred to the US.

[24] The Government of Guillermo Endara replaced the National Defense Force with a minor police agency, called the Public Force. In order to disarm the population, $150 was paid for each weapon that was turned in. US economic aid, which the new government had counted on, did not materialize and Endara himself began a hunger strike in order to obtain it. The Government had to accept the presence of US 'supervisors' in the ministries and the actions of the Southern Command troops outside the canal zone in order to fight drug trafficking and Colombian guerrilla warfare.

[25] Washington's interest in the region waned after the defeat of the Sandinistas in Nicaragua, and so the economic crisis in Panama at the time of the invasion was never overcome.

[26] The Organization of American States called the invasion 'deplorable' and called for a vote on the withdrawal of troops; there were 20 votes in favor, one against (the US) and six abstentions. Britain supported the invasion and France vetoed the UN Security Council denunciation. In Latin America, only El Salvador supported the US invasion.

[27] In March 1991, a Panamanian took over the administration of the canal for the first time.

[28] In April, Endara announced the end of his alliance with the Christian Democrat Party (PDC), dismissing five of their ministers. The Government's precarious stability was shaken by five coup attempts during its first two years in power.

[29] In the course of General Noriega's trial, in 1991 in Miami, it was disclosed that the former leader had close connections with the US Drug Enforcement Agency (DEA) and the CIA. That same year it was discovered that President Endara's legal office had connections with 14 companies that laundered drug money. The

DEA also disclosed that drug dealings had increased since the invasion. In June 1992, Noriega was sentenced to 40 years in prison.

[30] The Government suffered a serious setback when a referendum on constitutional reform on 15 November 1992 was defeated by 63.5 per cent of votes against 31.5 per cent. The people rejected, among other things, the formal abolition of the Defense Force (Public Force).

[31] Foreign Minister Julio Linares was forced to resign in August 1993 as he had been involved in the sale of weapons to Serbian forces in Bosnia through the Panamanian consulate in Barcelona.

[32] Economist Ernesto Pérez Balladares, former minister and admirer of Omar Torrijos, was elected President in 1994 with 34 per cent of the vote. These were the first general elections held after the US invasion.

[33] In June 1994, some indigenous groups rejected the construction of a road through their autonomous territory. They were supported by environmentalist groups and

the Catholic Church. The highway was to have crossed the 550,000 hectare rainforest called 'Darien's Barrier', declared a universal heritage by UNESCO. The 108-km highway would link Panama with Colombia. For years, this forest endured illegal logging of oak, cedar and mahogany. Likewise, the area was devoted to coca cultivation and arms smuggling.

[34] A plan to assassinate Pérez Balladares and several cabinet members was uncovered in January 1995. Ten National Police members were arrested on charges of conspiracy but the investigation was shelved for lack of evidence.

[35] The country continued to have an active role in arms and drug traffic as well as money laundering. The explosion of a package during a routine drug inspection killed three officials and injured 25. The explosives, grenades and ammunitions, were being sent to Ecuador, supposedly to guerrilla groups. Two arms deposits were found in the capital, belonging to a Colombian citizen.

[36] The reform of the labor code, aimed at attracting foreign

PROFILE

ENVIRONMENT

The country is bordered by the Caribbean in the north and the Pacific in the south. A high mountain range splits the country into two plains, a narrow one covered by rain-forests along the Atlantic slopes and a wider one with forests on the Pacific slopes. Canal navigation and the trading and financial activities connected to it constitute the main economic resource of the country. Tropical products are cultivated and copper is mined at the large Cerro Colorado mines.

SOCIETY

Peoples: 64 per cent of the inhabitants are descendants from native Americans and European colonist immigrants. 14 per cent are of African descent. The three main indigenous groups are the Cunas on the island of San Blas in the Caribbean, the Chocoes in the province of Darien and the Guaymies, in the provinces of Chiriqui, Veraguas and Bocas del Toro.

Religions: 80 per cent Catholic, 10 per cent Protestant (mainly Evangelist), 5 per cent Muslim, 1 per cent Baha'i, 0.3 per cent Jewish, 3.7 per cent others.

Languages: Spanish, official; several indigenous languages; most of the population speaks English.

Main Political Parties: Alliance New Motherland (which includes: Democratic Revolutionary Party and Popular Party); Solidarity Party; Alliance Vision of the Country (which includes: Arnulfista Party; Liberal Party and Molirena).

Main Social Organizations: Workers' Confederation of the Republic of Panama (CTRP); United Unions of Panama (SCS); Federation of Panamanian Students (FEP); Panamanian Workers' National Central (CNTP); Dobbo Yala Foundation (indigenous NGO).

THE STATE

Official Name: República de Panamá.
Administrative Divisions: 9 provinces and 4 indigenous territories.
Capital: Panama City 930,000 people (2003).
Other Cities: San Miguelito 331,692 people; David 79,100 (2000).
Government: Martín Torrijos, President since September 2004. Single-chamber parliament: Legislative Assembly, made up of 71 members, elected every 5 years by direct vote.
National Holiday: 3 November, Independence (1903).
Armed Forces: The National Guard was declared illegal in June 1991. US soldiers in the Canal Zone were withdrawn and military bases were returned to Panama. Other: 11,000 National Police.

| | Under-5 mortality | | Poverty | | Debt service | | Maternal mortality |
| | **24** per 1,000 live births 2004 | **$** | **6.5%** of population living on less than $1 per day 2002 | | **14.3%** exports of goods and services 2004 | | **160** per 100,000 live births 2000 |

investment, led to an atmosphere of social violence and strikes, since it would reduce labor security and the freedom to unionize and negotiate collectively. Confrontations of workers and students with the police left four dead and 86 injured in August. However, the law was passed.

[37] The announcement that US troops would remain in Panama was made after it became known that Balladares had received money from the Colombian Cali cartel to fund his 1994 election campaign. The President admitted the fact but denied having known of the origin of the funds. This scandal was followed by the forced closure of the Agro-Industrial and Commercial Bank due to its excessive debts and drug money-laundering activities.

[38] In September 1997, after 80 years' presence in Panama's territory, the Southern Command headquarters returned to the US. The Government guaranteed that the operation of the Panama Canal would not be affected by its return to national ownership.

[39] After Balladares was defeated in a referendum in which he proposed to change the Constitution to allow for his re-election, the main candidates for the May 1999 elections were Mireya Moscoso, the widow of Arnulfo Arias, and Martín Torrijos, son of General Omar Torrijos. Moscoso won on a manifesto of eradicating corruption and politicking in the administration of the Canal and allowing the participation of both workers and shipowners.

[40] On 14 December, a ceremony was held to celebrate the return of the Canal to Panamanian jurisdiction with the presence of Moscoso and former US President Jimmy Carter - who had signed the treaty agreeing to return the Canal with Omar Torrijos in 1977. Carter stated it was 'one of the most important historic occasions in the hemisphere' and that with the vigorous development of trade, Panama now had the chance to 'become the Singapore of the region'.

[41] Protests called in May 2001 by labor and student unions against the 66 per cent hike in public transport charges led to clashes with the police which left more than 100 injured and dozens arrested. The Government also raised taxes on essential services such as electricity, gasoline and telephones. Finally, after negotiations mediated by the Church, the measure was suspended.

[42] After the peace process with Colombian guerrillas ended, Panama increased its security forces along the border with Colombia, in case the conflict expanded to its territory. Panama also hosts hundreds of Colombian refugees who fled the guerrillas.

[43] In April 2002, the Truth Commission of Panama delivered its report. It had been formed by Moscoso to investigate what happened to those disappeared during the 1968-1989 military government, and clarified 110 of the 189 cases which had not been solved. The Commission carried out 35 excavations in old barracks and airports, finding 48 bodies. Some of the documents were supplied by the US Defense Department, which blamed the Defense Forces of the time for the crimes committed.

[44] Juan Jované, Director of the Social Security Fund, was removed from his post for his opposition to the Government's privatization plans (following the adjustment policies dictated by the IMF and the World Bank to counteract the serious crisis affecting the social security system). On 23 September 2003, workers replied with a general strike that paralyzed the country. More than 40 people were injured in clashes with the police.

[45] While it had military bases in Panama, the US took part in two world wars and fought several wars of its own (Korea, Vietnam, Persian Gulf, and the 1983 invasion of Grenada). The bases were used to train troops and for bombing practice - activities unrelated to the 'protection and defense of the Canal' - the pretext under which they had been created. The accumulation of unexploded missiles and ordnance during 96 years has resulted in dangerous environmental pollution. So, in spite of having recovered its territorial integrity after the withdrawal of foreign bases, Panama's economic activity is impeded.

[46] Martin Torrijos, from the RDP, won the May 2004 presidential elections. Voter turnout reached 80 per cent. Torrijos defeated Endara, who stressed he was 'happy because (Panamanian) democracy emerged... untouched.' from the elections.

[47] A general strike, started on 26 May 2005 by labor unions opposed to social security reform, forced President Torrijos to pledge a lowering of the minimum retirement age planned in the bill, as well as the contribution level of workers and employers. In June, Torrijos suspended the reforms so that Congress could analyze them once again, in order to 'normalize' the situation.

[48] In August, Panama reinstated diplomatic relations with Cuba, suspended by Havana in 2004 when President Moscoso amnestied four prisoners accused of plotting to assassinate Cuban President Fidel Castro.

[49] In April 2006, the Panama Canal Authority presented a long awaited project to widen the canal at a cost of $5.25 billion, including $2.3 billion in foreign loans. The Canal expansion project was overwhelmingly approved in a referendum in October 2006. Work on it is due to start in April 2007 and is expected to take until 2014 to be completed, at which point much bigger container ships than hitherto will be able to pass through. ▪

IN FOCUS

ENVIRONMENTAL CHALLENGES
Water pollution from agricultural run-off is threatening fishing. There is growing deforestation in areas of tropical rainforest. Land degradation and soil erosion are threatening to silt the Panama Canal. There is air pollution in urban areas and mining is damaging natural resources.

WOMEN'S RIGHTS
Women have been able to vote and stand for office since 1945. In 2004, women held 17 per cent of total seats in Parliament, and 14 per cent of ministerial or equivalent posts.

In 2003, women made up 37 per cent of the workforce of 1 million. Although the labor code prohibits gender-based discrimination, there are reports of women being sexually harassed in the workplace.

Illiteracy affects 7.6 per cent of the population (8.2 per cent of women and 7.1 per cent of men). In 1996-2004, 72 per cent of pregnant women received prenatal care, and 93 per cent of births were attended by trained medical staff*.

There were few convictions for domestic violence compared with the number of reported cases. In November 2003, 1,500 cases of domestic violence and more than 500 rapes had been registered.

In 2003 there were approximately 2,600 women aged 15 to 49 living with HIV/AIDS.

CHILDREN
Although the UN ranks Panama as medium human development status, it has the second widest income gap in the region. The eastern province of Darien, on the border with Colombia, is the poorest area.

In 2000, the Government estimated that 27,000 children between the ages of 12 and 14 were economically active. This figure rose to 83,244 among the 15-19 year-olds. According to the ILO, 66 per cent of child workers are in rural areas and 34 per cent in urban areas.

There is a major difference in quality between public and private education. Approximately 52 per cent of public primary schools are multigrade, with up to 20 students in different grades with one teacher in the same room.

INDIGENOUS PEOPLES/ ETHNIC MINORITIES
According to the *Social Watch 2005* report, the situation of the indigenous population is alarming; 95 per cent of them are poor and in some communities poverty reaches 93.4 per cent.

Indigenous people face discrimination as well as having reduced access to health services and low literacy levels.

The Guaymies (70,000) live mostly in the provinces of Bocas del Toro, Chiriqui and Veraguas, and today mainly work on the coffee plantations. The Cuna (45,000) are one of the most politically active indigenous groups in Latin America. Their main problems are poor healthcare and sanitation facilities. They preserve their culture but have freely mixed with Europeans and mestizos. The Chocoes, the smallest group with only a few thousand people, are the most seriously affected by the conflict in Colombia. Regular waves of refugees cross the border and frequently settle on their land.

The Panamanian Chinese community is fully integrated.

MIGRANTS/REFUGEES
Many Salvadorans and Nicaraguans have settled in the country since entering as refugees in the second half of the 1990s. In 2004 the country received some 2,000 refugees and asylum requests. Between 60,000 and 70,000 Colombians live in Panama in refugee-like circumstances; although not officially recognized as such, they have the same rights.

DEATH PENALTY
The death penalty has been abolished. It was applied for the last time in 1903.

* Latest data available in The State of the World's Children and Childinfo database, UNICEF, 2006.

Papua New Guinea / Papua New Guinea

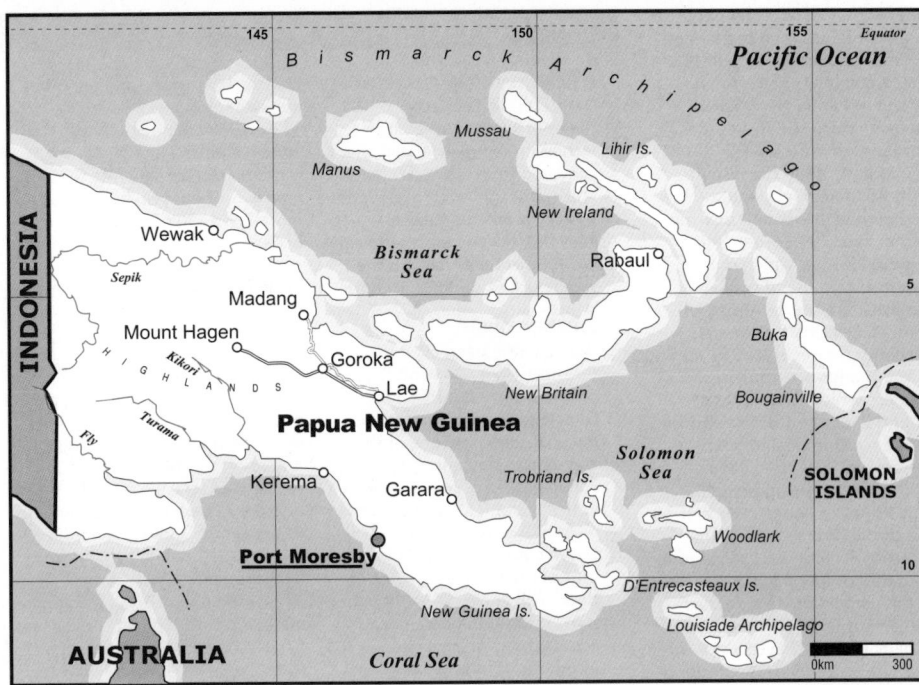

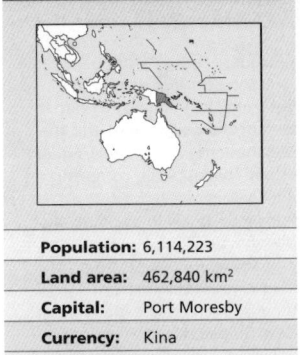

Population:	6,114,223
Land area:	462,840 km²
Capital:	Port Moresby
Currency:	Kina
Language:	English

The island of New Guinea, located on the eastern side of the Malay archipelago, was probably occupied 50,000 years ago by Melanesian peoples. With almost 800,000 sq km, it is the world's second largest island after Greenland. It has high mountains (the highest peak is over 5,000 meters above sea level) and deep valleys. The climate is tropical, with heavy rainfall, and climatic disasters, such as typhoons and tidal waves (*tsunamis*), are not uncommon. Traditionally, its population lived in groups scattered throughout the dense tropical jungle, cut off from the outside world. As a result, over 700 dialects are spoken on the islands of New Guinea.

2   The western part of the island (now West Papua) was explored by Indonesian and Asian navigators centuries before the arrival of Europeans. The first Europeans were the Portuguese and Spanish. In 1526, Portuguese sailor Jorge de Meneses named the island *Ilhas dos Papuas*. Twenty years later, Spaniard Iñigo Ortiz Retes renamed the island 'New Guinea', since he thought the islanders resembled the people of Guinea in Africa. Even though European sailors continued to visit and explore the area, they knew little about the inhabitants before the 19th century.

3   In the second half of the

19th century the island was disputed by Holland, Germany and the UK, as a result of which the territory was divided into areas called quadrants. England ceded its part to Australia in 1904 and it was renamed 'Territory of Papua'. After World War I, the League of Nations officially handed the German part over to Australia as a mandated territory. Australia regained control of the German and British areas after World War II and unified them under the name 'Territory of Papua New Guinea'. Holland continued to rule the western

part of the island, known as Irian Barat, which was annexed to Indonesia after a referendum in 1969 and became known as the province of Irian Jaya.

4   In 1971, the eastern territory was officially named Papua New Guinea (PNG), but continued under Australian rule until its independence in 1975. Michael Somare, leader of the Pangu party, who had won the first elections three years before, headed the separatist movement. Secessionist movements, operating both internally and from outside, had rocked the

country since independence. Even though Bougainville island had declared its independence some days before Papua New Guinea, Australia and PNG prevented it from gaining international recognition, leading to an armed conflict some years later.

5   In May 1988, Papua New Guinea together with Vanuatu and the Solomon Islands signed an agreement to defend and preserve traditional Melanesian cultures. Relations with Indonesia, which occupies the western portion of the island, are troubled. The Free Papua

LAND USE

2003/2005

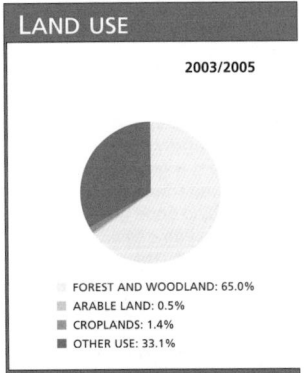

- ☐ FOREST AND WOODLAND: 65.0%
- ☐ ARABLE LAND: 0.5%
- ☐ CROPLANDS: 1.4%
- ■ OTHER USE: 33.1%

PROFILE

ENVIRONMENT
Located east of Indonesia, just south of the Equator, the country is made up of the eastern portion of the island of New Guinea (the western part is the Indonesian territory of West Papua) plus a series of smaller islands: New Britain, New Ireland and Manus, in the Bismarck archipelago; Bougainville, Buka and Mussau, which form the northern part of the Solomon Islands group; the Louisiade and D'Entrecasteaux archipelagos; and the islands of Trobriand/Kiriwina and Woodlark, southeast of New Guinea. The terrain is volcanic and mountainous, except for the narrow coastal plains. The climate is tropical and the vegetation is equatorial rainforest. The country suffers from deforestation, due to large-scale indiscriminate felling.

SOCIETY
Peoples: Papuans 85 per cent; Melanesians 15 per cent.
Religions: Many people follow local traditional religions but they also belong to Catholic (32.8 per cent) and Protestant (58.4 per cent) communities.
Languages: English (official). A local Pidgin, with

many English words and Melanesian grammar is widely spoken, as well as 700 other local languages.
Main Political Parties: National Alliance Party; People's Democratic Movement; People's Progress Party; Papua and Niugini Union Pati; People's Action Party.

THE STATE
Official Name: Independent State of Papua New Guinea.
Administrative Divisions: 20 Provinces.
Capital: Port Moresby 275,000 people (2003).
Other Cities: Lae 112,400 people; Madang 33,900; Wewak 27,400; Goroka 17,900 (2000).
Government: Queen Elizabeth II, Head of State, represented by Governor General Sir Paulias Matane since June 2004. Michael Somare, Prime Minister since August 2002. Single-chamber legislature: Parliament, with 109 members.
National Holiday: 16 September, Independence Day (1975).
Armed Forces: 3,700 troops (1996).

Life expectancy
57 years
2005-2010

GNI per capita
$560
2004

Literacy
57% total adult rate
2000-2004

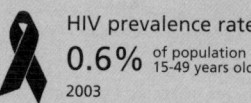
HIV prevalence rate
0.6% of population 15-49 years old
2003

independence movement operates in this province, and Indonesian military operations here in 1984 led 12,000 inhabitants to seek refuge in Papua New Guinea. Many of them still live on the border with Indonesia.

[6] In early 1989 the dispute with Bougainville island's secessionist movement became militarized. The Porgera mines in Bougainville, which exploit important gold and copper deposits, are under Australian control. Over half of export earnings are generated by mining in the islands of Papua New Guinea, which constitutes an essential element of the national economy. Oilfields are also important.

[7] In the following years PNG sent troops to Bougainville and the conflict worsened, but the secessionists retained control over Porgera's copper mine. It was in this context of violence that Parliament reinstated the death penalty in 1991, 34 years after it had been abolished. This decision was strongly rejected by different social movements. In October 1996, when peace envoy Theodore Miriung was murdered, the conflict with Bougainville deepened. In early 1997, Prime Minister Julius Chan resigned when it was revealed that he had contracted international mercenaries to quell the secessionist uprising.

[8] John Giheno replaced Chan on an interim basis, and a few weeks later Parliament appointed Bill Skate as Head of Government. In July 1997, in New Zealand/Aotearoa, the Government and secessionist rebels brought the Bougainville crisis to an end by signing a definitive peace agreement, including the demilitarization of the island and the deployment of UN peace-keeping troops. Australia promised funds for the rebuilding of Bougainville. Casualties of the nine-year conflict are estimated at 20,000.

[9] The political system of this young nation has been a concern from the very beginning. Formally a democratic and parliamentary monarchy, the country's party system was weak and fragmented. Switches in party loyalty were frequent, as was the creation of new coalitions and alliances, occasionally leading to governments being overturned through parliamentary votes of no confidence. From 1977 to 1988 Papua New Guinea had five different prime ministers.

[10] Skate's decision to ask Taiwan to grant a credit worth $2.5 billion prompted Parliament to request his resignation in 1999. He was succeeded by Mekere Morauta, who sought new loans

IN FOCUS

ENVIRONMENTAL CHALLENGES
Deforestation is affecting the rainforests, as a result of increasing commercial demand for wood from tropical climates. There are high levels of pollution caused by mining-related activities, but the most serious environmental problem is the recent lack of significant rainfall.

WOMEN'S RIGHTS
Women have been eligible for office since 1963 and have been able to vote since 1964. Female political participation is minimal, with only one of the 109 seats in parliament held by a woman. There are no women in the Supreme Court nor in the provincial governments.

Women made up 43 per cent of the total labor force of three million people. Although the Constitution guarantees equal protection regardless of gender or race, women commonly face discrimination in a range of areas. There are frequent incidents of gang rape and domestic violence (although the latter is a crime, in the majority of communities this type of aggression is regarded as a private issue).

It is estimated that 65 per cent of women prisoners have been convicted of assaulting or killing another woman, crimes that are common in those areas where polygamy is prevalent. Even in urban areas, Papuan women are often regarded as second-class citizens. Adultery is a crime and women tend to receive harsher sentences than men. In many rural communities women are treated as property,

being sold as wives, and are often denied their rights.

In 2004, the literacy rate among women stood at 51 per cent; the comparable rate for men was 63 per cent*. Maternal mortality is 300 per 100,000 live births*.

CHILDREN
Under-one mortality rate stands at 68 per 1,000 live births, while under-five rate is 24 deaths per 1,000 live births*. Preventable diseases such as measles, pneumonia and diarrhea are the major causes of child mortality*. According to UNICEF, there is widespread sexual abuse of minors. In 2004, 79 per cent of boys and 69 per cent of girls were enrolled in primary school, but the majority later drop out*.

By the end of 2003, it was estimated that 16,000 people aged between 0 and 49 were living with HIV/AIDS*. The Government does not assign sufficient resources to the rights and welfare of children. Following budget cuts, the Government stopped providing medical supplies to the population and free healthcare was suspended, hitting rural areas especially hard.

INDIGENOUS PEOPLES/ ETHNIC MINORITIES
The indigenous population is one of the most diverse in the world, comprising several hundred distinct communities, the majority of which number a few hundred people. The ethnic make-up of the population is very complex, with more than 700 groups, divided by language, traditions and customs. Many of these communities have been in conflict with their neighbors for

centuries.

The principal division is between the Papuans (80 per cent of the population) and the Melanesians (19 per cent). Small ethnic communities, comprising mainly Micronesians or Polynesians, live in the outlying islands.

There are over 715 languages in use, most of them spoken on the island of New Guinea, which is divided between Papua New Guinea and West Papua. Almost 650 languages have been identified and it was discovered that only between 350 and 450 are related. Most of these languages, spoken by hundreds or thousands of people, have extremely complex grammar. The non-Christian indigenous population practice different religions, as well as animism and ancestor worship.

MIGRANTS/REFUGEES
By the end of 2004, Papua New Guinea was home to approximately 8,000 refugees and asylum-seekers. Most of them came from neighboring West Papua; the rest were mainly Iraqis who had been unable to reach their final destination, Australia.

DEATH PENALTY
The death penalty was reinstated in 1991; since then, there have been no executions, although death sentences have been handed down.

*Latest data available in *The State of the World's Children* and *Childinfo* database,UNICEF, 2006.*

from the World Bank and the IMF, causing tension between Papua New Guinea and China.

[11] According to analysts, the November 2002 general elections were the most important since independence, for it was the first time a president had completed a full term in office, in a country marked by political instability and violence.

[12] The results of the chaotic elections were issued one month later, with the veteran Michael Somare obtaining the majority and becoming Prime Minister for the third time. One of his first resolutions as Prime Minister was to stop Morauta's privatization program, arguing that he needed more time to evaluate the conditions of the program and its advantages for the State. The main goals of the

new Government included the adoption of urgent measures against poverty, endemic unemployment and widespread crime.

[13] Given the Australian Government's rising concern over terrorism - 88 Australians died in October 2003 in an attack in Bali, Indonesia - Australia decided to intervene in neighboring states, such as the Solomon Islands and Papua New Guinea. An agreement between PNG and Australia signed in December 2003 stated that 230 police and 70 senior public officials from Australia would be deployed on the island in 2004. The first contingent started to arrive in December 2003.

[14] In May 2005, Australia withdrew its police after the Papuan Supreme Court ruled that

their deployment in the country was unconstitutional.

[15] The elections held in May and June 2005 were won by Joseph Kabui, of the Bougainville People's Congress, who became president of the autonomous Government with 54.7 per cent of votes.

[16] In 2005, China decided to finance a $650-million nickel and cobalt mining project.

[17] In July 2006, the Government declared a state of emergency and imposed a curfew in the energy-rich Southern Highlands province and deployed troops 'to restore law, order and good governance' according to Prime Minister Michael Somare. The province is vital to a proposed gas pipeline linking PNG and the Australian state of Queensland, which has been condemned by provincial governor Hami Yawari. ■

Paraguay / Paraguay

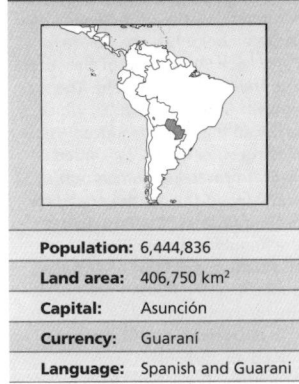

Population:	6,444,836
Land area:	406,750 km²
Capital:	Asunción
Currency:	Guaraní
Language:	Spanish and Guarani

B efore the 15th century, Guaraní Indians lived in the region between the Paraguay river and the Paraná river. The Guaranís spoke Tupí, the most widely used language in South America when the Spanish arrived. Women cultivated corn, cassava and sweet potatoes while men engaged in fishing, hunting and the defense of their villages.

2 In the late 15th century, persistent attacks mostly by Tupí-Payaguá groups who moved out of the barren region of Gran Chaco (a strip of forest along the borders between Paraguay, Argentina and Bolivia) were forcefully repelled by Guaranís, carrying the conflict into the margins of the Inca empire.

3 The Guaranís' peaceful reception of the Spanish explorers, who came through Brazil and the Río de la Plata (silver river) in the next century, was mainly due to their need to prevent new incursions by the Payaguás.

4 Although the region of Paraguay did not have precious metals, the Spanish colonizers made it their administrative center for South America because of its strategic location and lack of local opposition. In 1537, Juan de Salazar founded the fort of Nuestra Señora de Asunción, which was later to become the country's capital.

5 Spanish conquistadors and Guaranís lived together peacefully on the basis of a subsistence economy which was strengthened by the technical advances and work ethic introduced by the Jesuits in the early 17th century. The 32 missions where up to 100,000 Guaranís came to live became centers of religious conversion, agricultural production, manufacturing and trade, and within a few decades they comprised an autonomous administrative entity.

6 The Guaranís' military prowess allowed them to defend the missions from attacks launched by both Portuguese slave raiders and armies from Asunción. Between 1721 and 1735, the two colonial powers enforced a territorial partition of the mission settlements. In 1767, the Jesuits were finally expelled from South America and many Guaranís were sold into slavery.

7 The port of Buenos Aires became the epicenter of trade between Europe - whose industrial expansion fuelled a demand for leather, and later on wool, from the Río de la Plata region - and the southern cone of America. As a result, Buenos Aires was proclaimed capital of the Viceroyalty of Río de la Plata in 1776. When Buenos Aires proclaimed its independence from Spain in 1810, Paraguayans resisted the capital's authority, while at the same time taking advantage of the weakening of local Spanish authorities. On 14 May 1811, Paraguay proclaimed its independence under the leadership of Captains Caballero and Yegros.

8 Yegros established a junta with lawyer Gaspar Rodríguez de Francia, who imposed an isolationist policy toward both Buenos Aires and Brazil, in order to preserve territorial sovereignty and protect the Paraguayan economy from French and British products that were then flooding the continent. The Republic was proclaimed in 1813, and in 1816, the Congress appointed Francia dictator for life.

9 Francia ('El Supremo') secularized institutions, confiscated church property and made the State the nation's largest landowner. He also promoted production within the context of a strict autarchy. Upon Francia's death in 1840, Paraguayans enjoyed a healthy economy and an equitable distribution of wealth.

10 Carlos Antonio López became leader with popular backing, and after his participation in a civilian - military consulate between 1841 and 1844, Congress appointed him President of the Republic, and at the same time approved a presidential Constitution.

11 After the death of Francia and the fall of the Governor of Buenos Aires, Juan Manuel de Rosas - hostile to Britain and France - in 1852, the economic and military powers that were disputing control over the region (Britain, US, France and Brazil) developed new strategies for intervention in Paraguayan affairs.

12 Carlos Antonio López, threatened by the escalating attacks from Brazil, concentrated on developing the arms industry and communications in the country, with the support of the US. In 1858, the US navy sent a fleet and US representatives took part in several negotiations. Upon López's death, in 1862, Paraguay had the most developed arsenal and infrastructure (railroads, telegraph system and hospitals) in the southern cone of America.

13 Both the Brazilian Empire and the Government of Buenos Aires saw Paraguay not only as a territory to annex but also as a strong ally of the countries that opposed their attempts to dominate the region.

14 In 1864 Brazil helped the leader of Uruguay's Colorado Party to oust the Blanco Party opponent. In response, Paraguay's president Francisco Solano López (who succeeded his father Carlos Antonio López) went to war with Brazil which he saw as threatening the regional balance of power. Bartolomé Mitre, president of Argentina, then organized an alliance with Brazil and Colorado-controlled Uruguay (the Triple Alliance), and together they declared war on Paraguay on 1 May 1865.

15 The War of the Triple Alliance, which ended in 1870, was the bloodiest in South America's history. Paraguay's population fell from over half a million to around 220,000 and the country's state apparatus was destroyed. The end of the conflict was followed by six years of Brazilian occupation, during which time Paraguay became deeply indebted to British banks and allowed uncontrolled extraction of *quebracho* (used for tanning) and rubber by foreign companies.

16 Argentina and Brazil had annexed part of Paraguay and demanded reparations. In 1887, in the aftermath of defeat, the main Paraguayan parties were founded: the Colorado Party and the Authentic Liberal Radical Party (ALRP). The latter was founded by Faustino Sarmiento, President of Argentina during the War of the Triple Alliance, who settled in Paraguay in 1874 (until his death in 1888). Sarmiento promoted a new Paraguayan political class and, putting into practice the ideas laid out in his book *Civilization and Barbarism*, pushed for the prohibition of the Guaraní language and customs.

17 The Colorado Party ruled from 1887 until 1904 when the ALRP seized power, holding on to it during the next three decades. Since the War of the Triple Alliance,

LAND USE

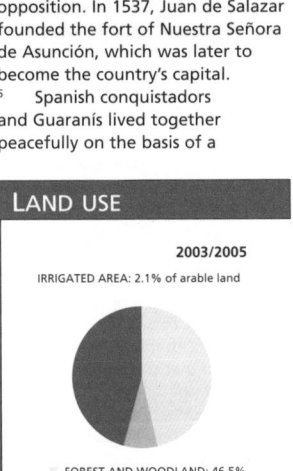

2003/2005

IRRIGATED AREA: 2.1% of arable land

- ◻ FOREST AND WOODLAND: 46.5%
- ◼ ARABLE LAND: 7.7%
- ◼ CROPLANDS: 0.2%
- ◼ OTHER USE: 45.6%

PUBLIC EXPENDITURE

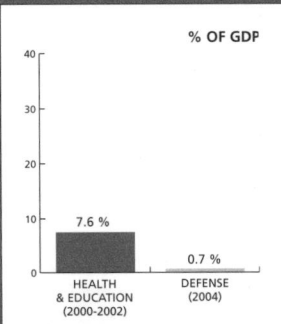

% OF GDP

HEALTH & EDUCATION (2000-2002): 7.6 %

DEFENSE (2004): 0.7 %

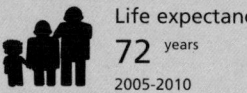

Life expectancy
72 years
2005-2010

GNI per capita
$1,140
2004

Literacy
92% total adult rate
2000-2004

HIV prevalence rate
0.5% of population 15-49 years old
2003

PROFILE

ENVIRONMENT

A landlocked country in the heart of the Río de la Plata basin, Paraguay is divided into two distinct regions by the Paraguay River. To the east lie fertile plains irrigated by tributaries of the Paraguay and Paraná rivers, and covered with rainforest. This is the main farming area, producing soybeans (the major export crop), wheat, corn and tobacco. The western region or Northern Chaco is dry savanna, with cotton and cattle.

SOCIETY

Peoples: Paraguayans are mostly mestizo (90 per cent), from the Spanish and indigenous peoples. The indigenous population (which comprises 80,000 people) belongs to the large Guaraní family, with linguistic and cultural variants. At present there are movements that defend the Guaraní ethnic identity. Immigration has produced German, Italian, Argentine and Brazilian minorities. Brazilians occupy a growing area on the border with their home country. A million Paraguayans live abroad, of whom approximately 200,000 emigrated for political reasons.
Religions: Mainly Catholic, official. Protestant groups.
Languages: Spanish and Guaraní (both official); most Paraguayans are bilingual.
Main Political Parties: Colorado Party; Authentic Radical Liberal Party; Beloved Fatherland Movement; National Union of Ethical Colorados.
Main Social Organizations: The Paraguayan Labor Confederation; the United Workers ' Central (CUT), uniting labor unions and the Rural Workers' Movement; the Paraguayan Women's Union; Guaraní Ñanduti Rogue; Paraguayan Students Union; Swindled Savers in Action.

THE STATE

Official Name: República del Paraguay.
Administrative Divisions: 17 departments and the capital city.
Capital: Asunción 1,639,000 people (2003).
Other Cities: Ciudad del Este 254,300 people; San Lorenzo 224,900; Lambaré 167,900; Fernando de la Mora 160,300 (2000).
Government: Head of state and government, Nicanor Duarte Frutos, President elected in April 2003 for a five-year term. Luis Castiglioni, Vice - President, elected in April 2003. The National Congress (Legislature) has two chambers: the Chamber of Deputies, with 80 members, and the Chamber of Senators, with 45 members.
National Holiday: 14 May, Independence Day (1811).
Armed Forces: 20,200 (1996). Others: 8,000 (Police).

foreign dependency and corruption were endemic to the Paraguayan State, as demonstrated by the border war with Bolivia (Chaco War, 1932-1935).

[18] The Chaco War cost Paraguay 50,000 lives and was orchestrated by rival oil multinationals - Standard Oil and Shell - disputing the territories of Gran Chaco that were supposedly rich in oil deposits. This war, as a result of which Paraguay won control over two-thirds of the Bolivian Chaco, came to an end through the diplomatic intervention of the US and Britain.

[19] Between 1936 and 1954, there were two coups (in 1936 and 1940, led by Generals Estigarribia and Morínigo, respectively), a civil war (in 1947, as a result of the ALRP rebellion) and seven popularly-elected presidencies which had no political continuity. In 1948, an alliance between a majority of the army and the Colorado Party was formed which brought General Alfredo Stroessner (then Commander-in-Chief of the armed forces) to the presidency in 1954 through a new coup.

[20] Stroessner (born in 1912) was backed by the US Government, within the framework of the National Security Doctrine, until the late 1970s. During his regime, which lasted until 1989, all freedoms were suppressed and state terrorism was applied (murder, disappearances, imprisonment without trial and torture). This was revealed in the 'Archives of Terror' found by the Paraguayan justice system in 1992.

[21] These archives provided clear evidence of thousands of human rights abuses both in isolated cases and as part of international plans of repression, such as Operation Condor (in the 1970s) in which Stroessner co-operated with the US and military regimes in Chile, Argentina, Brazil and Uruguay.

[22] Stroessner managed to orchestrate a puppet opposition and had himself re-elected seven times. Under his rule, Paraguay became home to World War II criminals, as well as the Nicaraguan dictator Anastasio Somoza, after the Sandinista revolution in 1979. Paraguay became a cocaine-trafficking center with the drug from Peru and Bolivia going on to Brazil, Argentina and the US.

[23] In the early 1980s, Stroessner ordered the creation of a tax-free area on the border with Brazil, the main center of which was Ciudad del Este. According to Brazilian media investigations, foreign exchange houses in Ciudad del Este have laundered about $100 million per year. Security forces assigned to this city and surrounding areas were then and still are inadequate to cover the hundreds of reports filed each year denouncing the organized groups trafficking in women, children, drugs, arms and stolen vehicles.

[24] The construction of the world's largest hydroelectric plant (Itaipú) near Ciudad del Este, opened by Stroessner in 1984, drew accusations of illegal expropriation of lands, as a result of investigative reports written mostly by Brazilian journalists, whose work was systematically obstructed by death threats and murders.

[25] Most of the companies and foreign settlers (nearly 400,000 in 2003) congregated around Ciudad del Este on a strip of land 1,200 km long and 65 km wide, where Portuguese and Brazilian currency are commonly used. At the end of 2001, the US Government labelled this region as an 'area of terrorist activity', where Hamas and Hizbulla groups purportedly operate.

[26] The democratization process in Latin America promoted by the US during the presidency of Jimmy Carter (1977 -1981), contributed to the emergence of opposition groups. In Paraguay it was supported by the Paraguayan Catholic Church as well as the Inter-Union Labor Movement, a Permanent Assembly of Landless Rural Workers (APCT), the Rural Women's Co-ordination Group and several indigenous people's organizations.

[27] On 3 February 1989, Stroessner was overthrown by a coup headed by his son's father-in-law, army commander General Andrés Rodríguez, who called elections in May. The elections were open to all political parties except the still-banned Communist Party.

[28] In the elections, General Rodríguez - who had been frequently linked to drug-trafficking, even by the US Drug Enforcement Administration (DEA) - was elected with 68 per cent of the vote, while the ALRP won 21 per cent. Despite the presence of foreign observers, voting was plagued by countless irregularities, attributed to the Colorado Party.

[29] In the elections for the National Constituent Assembly held in December 1991, the Colorado Party took 60 per cent of the vote and the ALRP, 29 per cent. The Constitution of June 1992, which replaced the one introduced by Stroessner in 1967, included clauses protecting human rights and banned the death penalty for ordinary crimes. The President of the Republic, the Supreme Court and Congress did not attend the public meeting held to promulgate the new constitution.

[30] In December 1992, Paraguay signed an agreement with Argentina, Brazil and Uruguay to create a common market (Mercosur) which came into operation in 1995. Nevertheless, the country's economic indicators (among the poorest in America) showed a steady decline.

[31] In the May 1993 elections, the ruling party's civilian candidate Juan Carlos Wasmosy was elected with 40 per cent of the vote. A year later, Parliament approved a law which banned the military from party politics. The Government and military high command took immediate action to have the law declared unconstitutional.

[32] In 1994, General Ramón Rozas Rodríguez, who had been appointed head of the anti-drugs campaign the previous year, was assassinated when he was supposed to present a report on illegal activities involving the military leadership, including General Lino Oviedo. In April 1996, the President retired eight high-ranking officers. General Oviedo resisted the order and holed up with a group of officers, while Wasmosy took refuge in the US embassy.

[33] In the Colorado Party internal elections in September 1997, Oviedo was elected presidential candidate with 36.8 per cent of the vote against 35 per cent for Luis María Argaña and 22.5 per cent for Wasmosy's candidate Carlos Facetti. In October, Oviedo was arrested on charges of sedition and in April 1998 he was sentenced to ten years in prison.

[34] The May 1998 elections, with Oviedo in prison, gave 46.8 per cent of the vote to the Colorado Party's Raúl Cubas and Luis María Argaña, against 38.2 per cent for the Democratic Alliance (ALRP and others). One of the first measures taken by the new president was to release Oviedo. However, in December, the Supreme Court ruled that this action was unconstitutional.

[35] In early March 1999, President Cubas found himself facing a possible vote of no confidence in Parliament. However, on 23 March, Vice President Argaña was assassinated by unknown troops. Following several days of street

Under-5 mortality
24 per 1,000 live births
2004

Poverty
16.4% of population living on less than $1 per day
2002

Debt service
13.5% exports of goods and services
2004

Maternal mortality
170 per 100,000 live births
2000

rioting, Cubas resigned and took refuge in Brazil, while Oviedo was granted safe passage to Argentina.

[36] The head of Congress, Luis González Macchi, from the Colorado Party, took over the presidency on 28 March 1999. A ruling by the Supreme Court allowed González Macchi to replace Cubas for the remainder of his term, due to end in 2003.

[37] In the late 1990s, the World Rain Forest Movement denounced the virtual extinction of the Ayoreo Indians, who originally occupied 2,800 hectares of the Gran Chaco. Since the 1970s, entrepreneurs belonging to the Paraguayan Mennonite community (about 25,000 people), as well as the companies Falabella and Veragilma - interested in the exploitation of *palosanto* wood (or holy wood) - have devastated and appropriated Ayoreo lands.

[38] In May 2000, the Government foiled a coup attempt by Oviedo, who was later arrested on the Brazilian border and placed under house arrest in Brasilia. Oviedo, who was accused of Argaña's assassination and of plotting the 1996 coup attempt, was placed near Stroessner's residence in Brasilia, where the former dictator had lived since 1989. In spite of the numerous charges filed against him, Stroessner received a regular pension as a former minister of the Paraguayan state until his death in 2006.

[39] In 2001, Amnesty International reported cases of torture of prisoners, including juveniles. In 2002, Human Rights Watch revealed that 112 soldiers had died in unexplained circumstances during the previous 13 years and demanded that the Paraguayan authorities establish a minimum age for conscription of 18 years.

[40] In July 2002, the Paraguayan Government decreed a state of emergency that lasted for a week, in order to control street protests. People demanded the resignation of President González Macchi for his alleged participation in the illegal investment of $16 million of state funds, which were wired from the Central Bank of Paraguay to his Citibank account in New York. Two people were killed and 300 were arrested as a result of the confrontations.

[41] In December that year, the Supreme Court of Justice passed a law preventing impeachment proceedings against President González Macchi over charges of alleged corruption. González Macchi, for his part, accused the Vice-President, Julio César Franco (of the ALRP), of having conspired with Oviedo to destabilize the country.

[42] In the April 2003 elections, the Colorado Party's candidates Nicanor Duarte Frutos and Luis Castiglioni

IN FOCUS

ENVIRONMENTAL CHALLENGES
Wide-scale deforestation, water pollution and the loss of wetlands are the country's most serious environmental problems. The inadequate waste treatment poses health risks for urban residents.

WOMEN'S RIGHTS
Women have been able to vote and stand for office since 1961. In 2003, they held 9.6 per cent of seats in Parliament and 31 per cent of ministerial or equivalent positions. That year, women made up 31 per cent of the country's labor force.

From 1990, the percentage ratio of women's earnings to men's was about 40 per cent in non-agricultural sectors. It is precisely in those sectors where women represent a majority that their salaries are considerably lower than men's. In 1990, 83.9 per cent of pregnant women received prenatal care; since then, considerable efforts started to be made to improve healthcare, thus increasing the rate of coverage to 94 per cent*. However, only 77 per cent of births are attended by trained health staff*. The maternal mortality rate is 170 deaths per 100,000 live births*.

CHILDREN
Over 460,000 children aged between 7 and 17 work; 36.7 per cent of the total do not attend school. Many children work on the family land. In urban areas, thousands of children under 12 work in the informal sector, selling newspapers, washing windscreens or collecting garbage; many of them are sexually abused or work as prostitutes.

Infant and under-five mortality rates have come down in the last 15 years and now they amount to 21 and 24 deaths per 1,000 live births respectively*.

Rural areas are worst affected by the country's economic problems. Half of the population here is poor and nearly one-third lives in extreme poverty. Only 15 per cent of peasants have access to piped drinking water, while in urban areas 60 per cent of people have access. This is a key factor for the prevention of infectious diseases, which are the main child killers. Rural illiteracy is double the urban rate. According to estimates, around 80 per cent of rural people are functionally illiterate.

INDIGENOUS PEOPLES/ ETHNIC MINORITIES
Ninety per cent of the population are mestizo, descending from the Spanish and indigenous peoples belonging to the Guaraní family. According to the last census carried out in 2001, the indigenous population comprises 80,000 people. There are 17 ethnic groups descended from five linguistic families: Tupí-Guraraní, Enxet-Maskoy, Mataguayo, Zamuco, and Guaicurú; the Tupí-Guaraní are the most numerous. Thirteen of these peoples live in the Chaco, where deforestation is destroying the indigenous habitat, while the other four live in the eastern region. While only 1.2 per cent identify themselves as indigenous, indigenous culture is deeply rooted in Paraguay where Guaraní is one of the two official languages and is spoken by most of the population, in both rural and urban areas.

MIGRANTS/REFUGEES
Over one million Paraguayans live abroad.

The Government cooperates with UNHCR in receiving refugees and asylum-seekers on a case-by-case basis. By the end of 2004, 41 people had entered the country as refugees and 6 asylum requests had been filed. The 41 refugees were assisted by UNHCR and Paraguayan Government offices for settlement purposes, while asylum requests were still under consideration.

DEATH PENALTY
The death penalty was abolished in 1992 and the last execution was in 1928.

* Latest data available in The State of the World's Children and Childinfo database, UNICEF, 2006.

won a majority with 37 per cent of the vote, followed by the ALRP with 24 per cent and Oviedo's National Union of Ethical Colorados with 13.5 per cent of votes. Duarte, who had been Minister of Culture between 1993 and 1997 and between 1999 and 2001, promised to fight corruption and promote a state investment policy in agriculture, the country's main productive sector.

[43] On leaving office in August 2003 González Macchi was subpoenaed and ordered to remain in the country to face public trial on a range of charges of violating the Constitution. In November that year, 20 accusations of constitutional offenses, conspiring to commit crimes and money-laundering were filed against six members of Paraguay's Supreme Court of Justice. By April 2004, three of the magistrates allegedly involved had handed in their resignations.

[44] In July 2004 more than 420 people died and hundreds were injured in a fire at Ikua Bolaños shopping mall, in Asuncion. After the police confirmed that there

were not enough emergency exits the owners were accused of voluntary manslaughter by the public prosecutor.

[45] In April 2005, a judge ordered the detention of former General Alejandro Fretes, former chief of staff of Paraguay's armed forces, who was accused of forced disappearances within the framework of the Condor Plan. Due to his age (84 years old) he was granted house arrest.

[46] By a law passed in May 2005, Paraguay granted Washington the right of free transit and also, allegedly, the possibility of US bases. In addition, it renounced its legal rights both to investigate crimes committed by foreign soldiers and to bring Washington before the International Criminal Court. Local political and human rights groups expressed their concern once the law became public, fearing the possibility that it might be opening the door for the setting up of a US base, such as those existing in Manta (Ecuador), Panama and Guantanamo (Cuba).

[47] In March 2006, some 40,000 people took to the streets to protest against Duarte's 'authoritarianism'. In spite of being rejected by the opposition and a large part of the public, Duarte was promoting a constitutional reform to allow for his re-election.

[48] That same month, thousands of Paraguayan demonstrators blocked the bridge linking Ciudad del Este with Brazil's Foz de Iguazu in protest over strict customs controls on the Brazilian side. The following month, within the framework of growing criticism of Mercosur's largest partners, the Paraguayan Government was visited in Asuncion by the Presidents of Uruguay and Bolivia - Tabaré Vázquez and Evo Morales - in order to agree on the joint construction of a regional gas pipeline. The 'gas summit' was construed by some analysts as a gesture of rejection of Argentinian-Brazilian 'paternalism' in terms of integration.

[49] In August 2006, former dictator Alfredo Stroessner died in exile. ∎

Peru / Perú

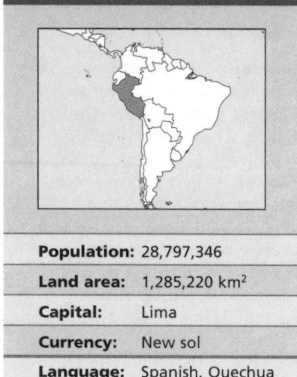

Population:	28,797,346
Land area:	1,285,220 km²
Capital:	Lima
Currency:	New sol
Language:	Spanish, Quechua and Aymara

E vidence of human life dating back over 15,000 years was found in caves near Ayacucho, in what is now Peru. The Chavin civilization, which reached its peak between 1400 and 200 BC, excelled at urban planning. The Paraca (700 to 100 BC) were skilled anatomists and embalmers. The Mochica built adobe temples in the Mocha valley, and it is thought their direct descendants were the Chimu (1000 to 1400 AD) who were great metalworkers. The Nazca culture (200 BC to 800 AD) developed agriculture with large-scale irrigation systems and built enormous calendars that are still discernible from the heights. The Tiahuanaco-Huari culture (600 BC to 1000 AD), based in what is currently Bolivia, expanded into the Peruvian highlands.

² The 12th century marked the zenith of the Inca empire, which politically united the various cultures and languages of the region, resettling many subjects in other parts of the empire and imposing Quechua as a common language. As with the rest of the Andean civilizations, the Incas' cultural legacy was destroyed by Spanish colonization, but its history was preserved down the generations by oral tradition and texts written after the conquest. The founders of the Inca dynasty, Manco Capac and Mama Oclo, settled in Cuzco, which later became the capital of the empire. In the 14th century, during the reign of the fourth Emperor (or Inca), Mayta Capac, they attacked neighboring populations.

³ Capac Yupanqui was the first to extend Inca influence beyond the Cuzco valley. With the eighth emperor, Viracocha Inca, the empire began a program of permanent conquest, establishing garrisons amongst the conquered peoples. In 1438, Pachacuti Inca Yupanqui, one of the sons of Viracocha Inca, usurped the throne from his brother Inca Urcon and the empire expanded beyond lake Titicaca, subjugating the Chanca, Quechua and Chimu peoples and taking over the kingdom of the Shiri – the Cara sovereigns - in Quito.

⁴ During the reign of Topa Inca Yupanqui (1471-1493) the Inca extended their power southwards, towards what is now central Chile. On the death of Topa Inca a war of succession broke out, which was won by Huayna Capac (1493-1525). He extended the northern frontier up to the river Ancasmayo (the current frontier between Ecuador and Colombia) before dying of a disease that was probably introduced by the Spanish. Tahuantisuyu, as the Inca Empire was known, governed around 13 million people.

⁵ The death of Huayna Capac caused another war of succession between Huascar, governor in Cuzco, and his younger brother Atahualpa, who ruled the northern part of the empire from Quito. In 1532, a group of 180 Spaniards led by Francisco Pizarro and Diego de Almagro disembarked in Tumbes. Recognizing Atahualpa as the legitimate ruler of the empire, they met with him in Cajamarca, then kidnapped him and demanded a large ransom in silver and gold. While imprisoned Atahualpa had Huascar killed, before he himself was strangled in 1533; the forces of the empire were thus paralyzed. The Spanish reached Cuzco, where they crowned Topa Hualpa, with the intention of reigning through an Inca emperor. However, Topa Hualpa was an ally of Huascar, and so the Europeans were committed to the faction they initially had not wanted to support.

⁶ Topa Hualpa died shortly afterwards and the Spanish reinforced their alliance with the pro-Huascar faction by putting his brother Manco Capac on the throne and dispersing the last of Atahualpa's army. In 1535, Pizarro prohibited Manco Capac from re-establishing control over the dominions along the coast and in the north, which were either still loyal to Atahualpa or lacked central control. Manco Capac then understood that the Spanish were a far greater threat than any of Atahualpa's followers and in 1536 he besieged Cuzco for a year. But his forces were finally disbanded by Diego de Almagro, returning from an expedition to Chile.

⁷ Manco Capac founded an independent Inca state in the Amazon regions which lasted until 1572, with the death by poisoning of Titu Cusi Yupanqui, the last Inca. The days of the Tahuantisuyu (the Inca Empire) were numbered from the moment Pizarro founded Lima on the coast in 1535, which operated as the center of Spanish power. The Spanish administration radically changed the property and land-use rules; the payment of tributes and forced labor broke up the bases of the old society and the old gods were officially replaced with Catholicism, although cults of minor deities did not disappear. Similarly, regions and cities of the old empire survived beyond the reach of the Spanish Crown for centuries. The most notable example of this was the fortress of Machu Picchu, 80 kilometers north-east of Cuzco, which was only re-discovered in 1911 by Hiram Bingham, a Yale University professor.

⁸ Due to the conflicts between the *conquistadores* (the Spanish conquerors), the Spanish Crown could not fully establish its authority for decades. The conquistadores, led by Gonzalo Pizarro - Francisco's brother - unhappy with the new laws passed by the king of Spain (which aimed to curtail feudalism, thereby threatening their wealth and power) rebelled in 1542, remaining in effect independent from the Crown until 1544, when Gonzalo Pizarro was defeated and executed.

⁹ It was only with the appointment of Viceroy Francisco de Toledo in 1569 that Spain consolidated its dominion in the region. The American institutions were adapted to Spanish authority and for a long time the chiefs of the various Andean nations administered the interests of their communities while collecting tributes and providing indigenous workforce for the mines. When Tupac Amaru (the son of Manco Capac) led the indigenous rural population in an uprising, the Crown had him captured and executed in 1571.

¹⁰ Once Toledo's administration was over, the Viceroyalty in Peru assumed the form it maintained until the 18th century, including all of South America except Venezuela and Brazil. The discovery of the silver mines in Potosí in 1545 was followed by those of Huancavelica in 1563. With the exception of the gold from New Granada (Colombia), mineral production was concentrated in Peru itself, or in Upper Peru (Bolivia). The Spanish Crown prioritized these areas, which became the most developed and richest parts of the continent.

¹¹ During the 16th and 17th centuries, Lima was the center of power and wealth for all of Spanish-controlled South America. Based on the labor of the indigenous workforce, the Court of Lima - where the King's justice was meted out - attracted the rich, religious orders, intellectuals and artists. It was in Lima that the tribunals of the Inquisition worked most avidly and cruelly. With the advent of the Bourbon dynasty in 1700, replacing the Hapsburgs as rulers of Spain, measures were taken to promote the development of the colonies and to achieve better government of the continent. The creation of the Viceroyalty of New Granada meant the Viceroyalty of Peru lost control over Quito as well as the territory constituting modern-day Colombia. The creation of the Viceroyalty of the River Plate in 1777 also removed control over Upper Peru and what are today Argentina, Paraguay and Uruguay.

¹² Reforms to the mercantile system, allowing the Pacific and

Life expectancy	GNI per capita	Literacy	HIV prevalence rate
71 years 2005-2010	**$2,360** 2004	**88%** total adult rate 2000-2004	**0.5%** of population 15-49 years old 2003

Atlantic ports to trade directly with Spain, weakened the condition of the Viceroyalty even further. In 1780, the *cacique* (indigenous leader) José Gabriel Condorcanqui had a *corregidor* (chief magistrate) arrested on charges of cruelty. He led a general uprising of indigenous people against the authority of the Viceroyalty in 1780 under the name of Tupac Amaru II, even gaining the support of some *criollos* (descendants of the Spanish). The rebellion, which spread to Bolivia and Argentina, lost support when it turned into a violent battle between the indigenous people and the whites. Tupac II was captured in 1781 and taken to Cuzco where, after being forced to witness the execution of his wife and children, he was quartered and beheaded. The revolution continued, however, until the Spanish Government approved a general pardon for the insurgents.

[13] The concentration of the Crown's military power in Lima, the conservative attitude of the local oligarchy, and the effective suppression of the indigenous uprisings meant that Peru remained loyal to Spain when the rest of its colonies in South America began the fight for independence between 1810 and 1821. General José de San Martín freed Chile in 1818 and used it as a base to attack Peru by sea with the aim of securing Buenos Aires' control over the mines of Upper Peru and the independence of the Argentine provinces. Toward the end of 1820, he occupied the port of Pisco and the Viceroy withdrew his troops into the interior of the country. San Martín entered Lima and declared independence on 28 July 1821.

[14] San Martín asked for the help of Venezuela's Simón Bolívar to attack the large Spanish contingents in the interior of Peru, but the latter would not agree to share the leadership. Bolívar (who had liberated the northern part of South America) took over in Peru to continue the fight. In the battles of Junín and Ayacucho, in 1824, the Spanish were defeated and Peru became politically independent. The first years of independence were spent in constant battles between the conservative oligarchy, yearning for the times of the viceroyalty, and the liberals. The wars with Colombia in 1827, and with Bolivia, took place against that backdrop. The unification of Peru with Bolivia, attempted by Bolivian President Andrés Santa Cruz in 1835, failed both socially and economically.

[15] Marshal Ramón Castilla, who ruled the country from 1845 to 1862, shaped the modern Peru after abolishing slavery and proclaiming the Constitution. In 1864, Spain attempted to establish enclaves on the Peruvian coast, but Peru, Chile,

Bolivia and Ecuador responded by declaring war on Spain, which was defeated in 1866.

[16] From 1845, with the silver mines exhausted, guano - bird feces used as fertilizer - became Peru's main export product. When the guano 'boom' was over, it was replaced by saltpeter from the southern deserts. This wealth was to bring about the Pacific War (1879-1883). Peru and Bolivia joined forces against Chile, which exploited the saltpeter, with the support of British companies. Peru and Bolivia lost the war and with it the provinces of Arica, Tarapac and Antofagasta.

[17] The 20th century marked the beginning of large-scale copper mining, particularly by the US-owned Cerro de Pasco Copper Corporation. Foreign capital was also involved in oil exploitation in the north, and sugarcane and cotton in the north and center. The anachronistic agrarian structures, however, continued unchanged. Within this context, the APRA (American Popular Revolutionary Alliance), a Marxist-inspired party committed to Latin Americanism, achieved widespread popular support. Víctor Haya de la Torre, its main leader, was in favor of merging class boundaries and debated with José Carlos Mariátegui, founder of the Peruvian Communist Party (PC). Triumphant in several elections, APRA never actually came to power due to successive military coups.

[18] In 1968, a military faction headed by General Juan Velasco Alvarado ousted President Fernando Belaúnde Terry and started a process of change by nationalizing oil production. This also included recovery of natural resources and fishing, co-operative based agrarian reform, worker participation in company ownership, the creation of socially-owned enterprises, the expropriation of the press - planning to hand the latter over to organized social sectors - and an independent non-aligned foreign policy. An ailing Velasco gradually lost control of the process and the trust of his allies. He was overthrown by his Prime Minister General Francisco Morales Bermúdez in 1975. Under pressure from the IMF and an oligarchy keen on regaining power, Morales called elections. Belaunde's Acción Popular (AP), which had boycotted the constituent elections, triumphed in the 1980 presidential elections and established IMF guidelines. That year armed violence reappeared with the *Sendero Luminoso* ('Shining Path') guerrillas and in 1984 with the Tupac Amaru Revolutionary Movement (MRTA).

[19] In the 1985 elections, APRA candidate Alan García came first with 46 per cent of the vote. With

ENVIRONMENT
The Andes divide the country into three regions. The desert coastal area, with large artificially irrigated plantations and some natural valleys has historically been the most modern and westernized. Half the population live in the *sierra* (highlands), between two ranges of the Andes. Numerous peasants here are still organized into *ayllus* (communities) with Incan roots.
There is subsistence farming of corn and potatoes, with traditional llama and alpaca rearing forced to move to higher slopes, due to expanding mining and sheep rearing. The eastern region, comprising the Amazon lowlands, with a tropical climate and rainforests, is sparsely populated. Peru is one of the largest producers of coca, a medicinal and energizing plant. Coca - traditionally consumed among Indians who chew its leaves - once chemically refined and developed into a different substance, becomes the basis of cocaine.

SOCIETY
Peoples: Nearly half of all Peruvians are of indigenous origin, mostly Quechua and Aymara living on the *sierra*. Along the coast, most of the population are mixed descendants of indians and Spaniards, and there are also small groups of descendants of African slaves. There are several indigenous groups in the East Amazon jungle. There are also Chinese and Japanese immigrant minorities.
Religions: Catholic (official), with syncretic expressions related to Indian beliefs.
Languages: Spanish, Quechua and Aymara (all official).
Main Political Parties: Peruvian Aprista Party; Union for Peru (includes Peruvian Nationalist Party).
Main Social Organizations: The Peruvian Workers General Central Union (CGTP), founded in 1928 and predominantly communist; the Peruvian Workers' Central Union (CTP), founded in 1944, linked to APRA; the independent National Workers' Confederation (CNT), founded in 1971; the Workers' Central of the Peruvian Revolution, founded in 1972 by Velasco backers. The National Agrarian Confederation (CNA), founded in 1972. The Peruvian Campesino Confederation (CCP), founded in 1974.

THE STATE
Official Name: República del Perú.
Administrative Divisions: 25 Departments, 155 Provinces and 1,586 Districts.
Capital: Lima 7,899, 000 people (2003).
Other Cities: Arequipa 720,400 people; Trujillo 590,200; Chiclayo 481,100; Cuzco 279,600 (2000).
Government: Alan García, elected president in June 2006. Unicameral legislature: The Congress, with 120 members.
National Holiday: 28 July, Independence (1821).
Armed Forces: 125,000 troops (65,000 conscripts), 188,000 reserves. Other: National Police: 60,000 members. Coast Guard: 600 members. Rondas Campesinas: the campesino self-defense forces, made up of 2,000 groups mobilized within emergency zones.

a foreign debt of $14 billion, García announced he would limit payments to ten per cent of the country's annual export income and would negotiate directly with the creditors, without IMF mediation. He ended his term disgraced by corruption charges and left the country in a full-blown crisis, with hyperinflation of 7,600 per cent annually and rampant guerrilla violence.

[20] The 1989 general elections were won by an unknown outsider, Alberto Fujimori, who was elected with 56.4 per cent of the vote. On assuming power, Fujimori implemented a severe anti-inflationary plan and, without parliamentary support, began to govern by decree. In 1992, Fujimori led a coup claiming that Parliament

was corrupt and inoperative and that the judicial system was obstructing national reconstruction. The imprisonment of Abimael Guzmán, founder and leader of Sendero Luminoso, dealt a major blow to the guerrilla group.

[21] In 1995, Peru and Ecuador waged an undeclared war along their common border at the Condor mountain range. Re-elected that year, Fujimori granted amnesty to members of the army and police who had been convicted for human rights violations in the fight against the guerrillas since 1980. In 1998, three years after the armed conflict, Peru and Ecuador agreed on a peace treaty based on new border lines proposed by Argentina, Brazil, Chile and the US.

	Under-5 mortality		Poverty			Debt service		Maternal mortality	

Under-5 mortality 29 per 1,000 live births 2004

Poverty 12.5% of population living on less than $1 per day 2002

Debt service 17.1% exports of goods and services 2004

Maternal mortality 410 per 100,000 live births 2000

22 In spite of being prohibited by the Constitution from running for a third term, Fujimori was an official candidate in the 2000 presidential elections. Alejandro Toledo, an economist and former shoeshine boy of indigenous descent, obtained 41 per cent of the vote against Fujimori's 48.7 per cent, and headed a large protest in Lima to demand a runoff vote. Toledo refused to take part in the second round unless their fairness was guaranteed. In the end, with no international observers and as the only candidate, Fujimori was declared president by the electoral court although 54 per cent of the population had voted against him. He was strongly criticized by the Organization of American States (OAS) and the US.

23 The OAS supported a schedule of institutional changes in Peru: freedom of the press, independence of the courts, modifications in the electoral system and civilian control of the army and intelligence services. Fujimori promised US Secretary of State Madeleine Albright he would implement the changes. In response to the scandal sparked by the release of a video showing National Intelligence director Vladimiro Montesinos - the true 'power behind the throne' - bribing a legislator, Fujimori declared that he would call new elections in 2001, in which he would not participate. In November 2000, during a trip to Japan, Fujimori announced his resignation, while at the same time in Lima, the Peruvian Congress voted to remove him from office. In 2001, Fujimori was charged by Congress with abandoning his post.

24 Having won the cleanest elections in many years, Toledo became the first freely elected President of indigenous origin. On taking office in 2001, he inherited a country that was heavily indebted, with serious fiscal problems, a severe recession and 54 per cent of the population living in extreme poverty. According to the estimates of Toledo's own economic team, poverty in Lima had increased from 35 to 45 per cent between 1997-2000. A Truth and Reconciliation Commission reported almost 70,000 deaths and disappearances that had occurred over the previous two decades. The authorities issued a second international arrest warrant (the first had been issued three months earlier) against Fujimori, on charges of corruption and human rights violations.

25 In 2002 there were violent protests against the privatization of two powerful Peruvian electricity companies, during which one person died and hundreds were repressed. After two weeks of conflict, the Minister of the Interior, Fernando Respigliosi,

resigned and Toledo suspended the privatizations.

26 In 2003 a court in Lima sentenced Vladimiro Montesinos to eight years in jail on the charge of embezzlement (in 2002 he had been incarcerated for abuse of power and for having illegally taken over the function of Intelligence Chief).

27 Prime Minister Beatriz Merino resigned in 2003 accused of involvement in corruption scandals, which she denied. Toledo requested Merino's resignation, and she was replaced by Carlos Ferrero, an experienced legislator from the Possible Peru party, founded by Toledo.

28 The National Commission for Development and Life Without Drugs reported in March 2005 that drug cartels take roughly $7 billion annually in Peru, an amount equivalent to 50 per cent of the

national budget. Some 60,000 hectares of land are used for growing coca leaves, of which 90 per cent go to the illegal drug trade. In the early months of the year, coca growers stepped up their protests, demanding that the eradication of their crops be carried out gradually and in cooperation with the farmers.

29 In November 2005, Fujimori was arrested after unexpectedly arriving in Chile - presumably with the intention of returning to Peru - and Lima asked for his extradition. The former president unsuccessfully attempted to register as a presidential candidate in the 2006 general elections, in which his daughter Keiko was elected to Congress with more votes than any other candidate.

30 In the runoff vote held in June, APRA candidate Alan García - considered by many to have been 'the worst president

in the history of Peru' - defeated nationalist former military officer Ollanta Humala, the candidate who had garnered most votes in the first round. Humala had led a failed coup against Fujimori in 2000, and his platform emphasized indigenous people's rights. Many voters are believed to have chosen García over Humala as the 'lesser of two evils'. As García was elected without a majority in Congress to back him, he had to form alliances and overcome the skepticism surrounding his abilities. Venezuelan President Hugo Chávez's vocal support for Humala during the election campaign sparked accusations of interference and diplomatic frictions between Peru and Venezuela.

31 In October 2006, former Sendero Luminoso leader Abimael Guzman was sentenced to life imprisonment for terrorism at his retrial. ∎

IN FOCUS

ENVIRONMENTAL CHALLENGES
Deforestation, particularly as a result of illegal industrial logging, is an increasingly serious problem. The country suffers from soil degradation. Some species of fish are endangered due to uncontrolled fishing. The coastline has been polluted by industrial and urban waste. Pollution from mining is a major problem, especially in the highlands. There are high levels of air pollution in Lima.

WOMEN'S RIGHTS
Women have been able to vote and run for office since 1955. In 2003, women occupied 18 per cent of seats in Parliament and 12 percent of ministerial or equivalent posts.

That same year, women made up 32 per cent of the total workforce of 11 million people. Most work for small businesses or in the informal sector and normally lack social security coverage. They earn roughly 40 per cent less than men for the same work.

Although boys and girls have equal access to primary education, there are marked gender differences in adult literacy rates: 82 per cent for women compared with 93 per cent for men.

While 84 per cent of pregnant women receive prenatal care, slightly over 40 per cent of births are not attended by qualified health personnel.* There are over 27,000 women living with HIV/AIDS.*

Violence against women is a chronic problem. Rape,

harassment by partners or others, and the sexual, physical and psychological abuse of Peruvian women are aggravated by the perpetrators' apparent impunity.

CHILDREN
The country is still characterized by huge socio-economic differences and exclusion, and child labor remains a serious problem. Almost two-thirds of people living in extreme poverty are minors; children are the most vulnerable. Under-five mortality is 29 deaths for every 1,000 live births, which reflects significant progress when compared with the rate of 80 deaths per 1,000 live births in 1990.* Roughly 25 per cent of children under five suffer stunted physical development.*

Child and adolescent labor has quadrupled since 1993; in 2003, there were over 1.8 million minors working in the country as a whole. Many of these children and adolescents are employed in mines, doing work that is considered high-risk even for adults.

INDIGENOUS PEOPLES/ ETHNIC MINORITIES
The majority of the indigenous population lives in the central and southern area of the Andes and comprises 38 per cent of the total population. There are nearly 80 languages including Aymara (22 per cent), Quechua (30 per cent), Arawak, Cahuapana, Harakmbet, Huitoto, Jibaro and Pano.

Traditionally their control over the land and the defense of their culture and traditions have been limited. Following the 1993 Constitution, a law abolishing the

inalienability of native lands was put into effect, as a result of which the sale of their lands is no longer prohibited.

Difficulties in access to the health care and education systems limit the possibility of better prospects for future generations of indigenous peoples.

MIGRANTS/REFUGEES
By the end of 2004, the country had received 1,000 requests for asylum from foreign refugees. The Peruvian Government estimates that over two million Peruvians live abroad, and that the vast majority are undocumented immigrants. This number has quadrupled over the last 20 years. The main destinations are the US, western European countries, Argentina, Chile and Japan.

In the 1980s and 1990s the conflict between Sendero Luminosa (Shining Path) and government forces devastated many rural areas. The conflict left 69,000 dead and 800,000 internally displaced people. Many returned to their original homes during the 1990s, but around 600,000 remained in their new homes.

DEATH PENALTY
The death penalty was abolished in 1979.

*Latest data available in *The State of the World's Children* and *Childinfo* database, UNICEF, 2006.

Philippines / Pilipinas

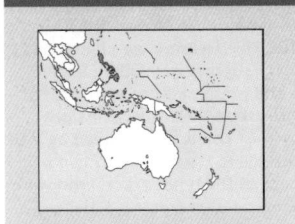

Population:	85,884,014
Land area:	300,000 km²
Capital:	Manila
Currency:	Peso
Language:	Filipino and English

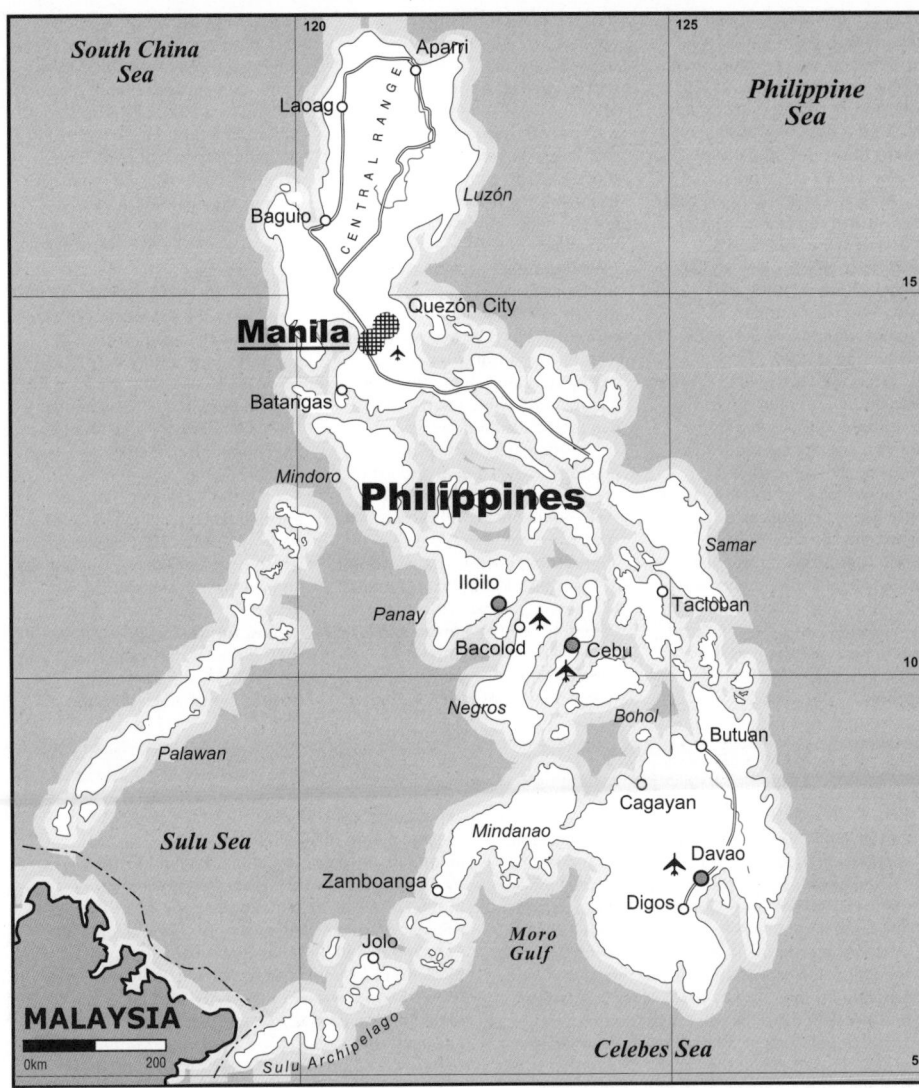

The Philippine archipelago was first inhabited in Paleolithic times, and the Neolithic culture on the islands began around 900 BC. Native peoples, such as the Aeta and the Igorot, probably subsisted without being assimilated by the later groups of migrants.

2   Between the 2nd and 15th centuries AD, migrants from Indonesia and Malaysia settled on the islands, gathering in clans. They were virtually uninfluenced by the classical Indian culture which had touched others in the region. Between the 11th and 13th centuries the coastal areas were raided by Muslim, Japanese and Chinese merchant ships, bringing traders and craftspeople to the islands. The southern islands adopted Islam and sultanates soon appeared.

3   The archipelago was 'discovered' by Ferdinand Magellan in 1521 but the explorer was killed on one of its beaches and Spanish possession of the islands, which were also coveted by the British and the Dutch, was not secured until 1564. The Igorot of the Cordillera region and the Islamic population of Mindanao were never fully incorporated by European colonization, and most of the rural population preserved their subsistence economy, never even paying tribute to the Europeans. Several uprisings by these communities and the Chinese were repressed by the Spaniards.

4   Spanish colonization in the Philippines followed similar patterns to that in the Americas. However, the Philippines had two distinguishing features; they were located on the oceanic trading routes, in a position that received merchandise from all over Southeast Asia on its way to Europe, and they were ruled by the Viceroyalty of Mexico.

5   Late in the 19th century, a local independence movement developed, led by the native bourgeoisie, who wanted the political power which was denied them. Other oppressed sectors soon

followed their lead. Anti-colonial revolution erupted in 1896 and independence was proclaimed on 12 July 1898. However, the US immediately started diplomatic negotiations to seize control of the archipelago, which resulted in the signing of the Treaty of Paris, on 10 December 1898, ending the Spanish-US war. In meetings, which barred Filipino delegates, Spain decided to cede the archipelago to the US in exchange for compensation.

6   Between 1899 and 1911, one million Filipinos died in the struggle against US occupying troops. During World War II, the archipelago was occupied by Japan but US troops returned after the end of the war. The archipelago was eventually granted formal independence in 1946 but the Philippines have remained under US economic domination ever since.

7   Nor did independence bring about any social changes. The

hacienda system - large estates farmed by sharecroppers - persisted in the country. More than half the population were peasants, and 20 per cent of the population owned 60 per cent of the land. Although the sharecropper was supposed to receive half of the harvest, most of the peasant's actual income went to pay off the debts incurred with the *cacique* (the landowner).

8   The Nationalist Party, a conservative party of landowners, remained in power until 1972, when Ferdinand Marcos, president since 1965, declared Martial Law. In 1986 a coalition of opposition forces rebelled against the continuous abuses of Ferdinand Marcos, paving the way for democratization. During his presidency repression grew both against armed movements (the Muslim independent groups of Mindanao and the New People's Army led by the Maoist Communist Party), and against political and trade union opposition. This

repression often had the military support of the US.

9   Due to its long presence in the country, the Catholic Church is deeply rooted in Filipino society. This is reflected in the fact that 75 per cent of Filipinos over the age of ten have learnt to read in institutions dependent on the Church. The Church played an active role in the denunciation of fraud when the 1976 referendum supported the imposition of martial law. Five years later, 45 political and trade union organizations united to boycott the fraudulent and unconstitutional elections that Marcos used to stay in power. In September 1981 thousands of people demonstrated in Manila, demanding an end to the dictatorship and the withdrawal of US military bases.

10   On 21 August 1983, opposition leader Benigno Aquino (People's Power party, social-democratic) was murdered at Manila airport as he stepped off the plane that had

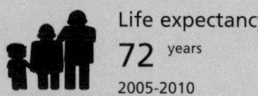
Life expectancy
72 years
2005-2010

GNI per capita
$1,170
2004

Literacy
93% total adult
rate
2000-2004

HIV prevalence rate
<0.1% of population
15-49 years old
2003

brought him back after a prolonged exile in the US. His murder was attributed to Marcos. More than 500,000 mourners followed his coffin to the cemetery. This event triggered a popular uprising which did not desist until the dictator was ousted.

[11] Amidst a scenario of increased violence and repression, a large section of the population put pressure on Marcos, demanding early elections in 1986, and supporting the candidacy of Corazón Aquino, widow of the assassinated leader.

[12] Elections were held in February 1986, but widespread fraud prevented 'Cory' Aquino from winning, and she subsequently called for civil disobedience. Marcos' Minister of Defense, Juan Ponce Enrile, attempted a coup against the dictator, but failed. A million supporters surrounded the rebels, led by Enrile, in the field where they had taken refuge. Marcos opted for exile and Corazón Aquino assumed the presidency with Enrile as her Minister of Defense.

[13] The new constitution was approved by a large majority in the February 1987 plebiscite. The charter granted autonomy to the Mindanao and Cordillera regions, thus paving the way for a truce with guerrilla groups operating in those areas. In early 1990 the New People's Army (NPA) representatives left the negotiating table, after several acts of provocation against mass organizations and attempts on the lives of civilian leaders. Agrarian reform, which should have been the cornerstone of the Government's plan for social transformation, was diluted after going through a legislature where many of the members were landowners.

[14] During 1991 increasing pressure from regional and ethnic groups, the urgent need for a more equitable distribution of land and wealth and, possibly, the approaching presidential elections of May 1992, led Corazon Aquino to create a Bureau of Northern Communities. The Bureau was concerned with the mountain and ethnic groups, particularly in Luzón. There was also a Bureau of

Southern Cultural Communities, excluding the Muslims. The staff of the Bureaus were recruited from within the communities in question.

[15] In June 1991, the eruption of Mount Pinatubo shook the country, claiming the lives of over 700 Filipinos, flattening entire villages, forcing the evacuation of over 300,000 people, and burying the evacuated Clark US air force base under the ashes.

[16] With its airbase unusable, and faced with the prospect of difficult negotiations as the contract approached its expiry date, the US opted to abandon the base of its own accord. On 26 November 1991, Clark airbase was formally abandoned. It had employed over 40,000 Filipinos.

[17] Of the 32 million Filipinos entitled to vote, 25 million took part in the May 1992 elections which were considered the calmest and cleanest in the country's history. The winner was Fidel Ramos, former Defense Minister in the Aquino administration.

[18] In 1994, the Ramos Government had to seek the

opposition's support to control evasion of the 10 per cent VAT tax. This measure won him the IMF's support - including a loan - and prompted five per cent growth in GNP. The campaign against crime, now headed by Vice President Estrada, led to two per cent of the police force, who were implicated in criminal activities, being discharged, while another five per cent were kept under investigation. The NPA communist guerrillas lost strength due to an amnesty for its members and to internecine conflicts regarding the amnesty.

[19] In 1995 Imelda Marcos was elected to the Chamber of Deputies, in spite of the many corruption charges against her. Swiss banks returned $475 million to the country which had been deposited by her husband Ferdinand Marcos during his dictatorship, but the Government was convinced that billions remained in other accounts.

[20] The elimination of restrictions on investments, the reduction in customs barriers and the presence of skilled and cheap labor, attracted investors which led to six-per-cent growth in GNP. Remittances totaling $2 billion from 4.2 million workers living abroad - mainly domestics - came into the country in 1995.

[21] In late 1995, there was an unprecedented food crisis, with a 70 per cent increase in rice prices. More than two-thirds of the population were estimated to be living below the poverty line. Farmers' organizations blamed the Government for their incoherent and corrupt agricultural policy, calling for agrarian reform, including the industrialization of rural activity, food self-sufficiency and protection of the environment.

[22] Despite protests from representatives of the Christian Filipinos, who make up the majority of the nation, the Government and the Muslim guerrillas signed a peace agreement on 30 September 1996. Nur Misuari, leader of the Moro Islamic Liberation Front (MILF), became governor of Mindanao, an autonomous region which covers around a quarter of national territory.

[23] In January 1998, thousands of children from various countries marched through the streets of Manila in protest against exploitative child labor. This sparked a worldwide campaign for better conditions for the world's 250 million child laborers.

[24] In May, Vice President Joseph Estrada was elected President with 37 per cent of the vote. In March 2000, Salamat Hashim, leader of the largest Islamic rebel

PROFILE

ENVIRONMENT

Of the 7,107 islands that make up the archipelago, spread over 1,600 kilometers from north to south, 11 account for 94 per cent of the total area and are home to most of the population. The archipelago is located approximately 100 kilometers southeast of the Asian continent; it is bordered on the east by the Philippine Sea, on the west by the South China Sea, and on the south by the Celebes Sea. Luzon and Mindanao are the most important regions. The archipelago is of volcanic origin, forming part of the 'Ring of Fire of the Pacific'. The terrain is mountainous with large coastal plains where sugarcane, hemp, copra and tobacco are grown. The local climate is humid and tropical. The mean annual temperature is around 26.5 C. Filipinos recognize three seasons: *Tag-init* or *Tag-araw* (summer; March to May), *Tag-ulan* (rainy season; June to November), and *Tag-lamig* (cold season; December to February). Abundant rains favor the growth of dense forests. It is the main producer of iron ore in Southeast Asia, and also has oil, chromium, copper, nickel, cobalt, silver and gold.

SOCIETY

Peoples: The vast majority of the population originates from the first migration waves from Malaysia and Indonesia. Some 200,000 Chinese traders settled there from the 11th century onwards. Islamic communities from Borneo entered the territory in the 15th century and resisted the evangelization of the Spanish conquistadors (who arrived in 1521), who had a significant cultural influence over the rest of the population. Some communities of Malaysian origin, in different stages of evolution, also resisted Christianization. After 1898, US colonization had a strong influence on Filipino society and culture.
Religions: Catholics 83 per cent; Muslims 5 per cent; Protestants 5 per cent; Independent Filipino Church 3 per cent; Animists, Buddhists and other 4 per cent.

Languages: Some 55 per cent of the population speaks Filipino (official), based on the Tagalog language, of Malaysian origin. English, spoken by 45 per cent, is also official, and is compulsory in the education system. But 90 per cent of the population speaks one of the following languages: Cebuano (6 million); Hiligaynon (3 million); Bicolano (2 million); Waray-Waray (1 million). Spanish and Chinese are minority languages.
Main Political Parties: Power-Christian and Muslim Democrats (Lakas-Christian and Muslim Democrats Koalisyon ng Katapatan at Karanasan sa Kinabukasan); Coalition of United Filipinos (Koalisyon ng Nagkakaisang Pilipino); Struggle for Democratic Filipinos (Laban ng Demokratikong Pilipino).
Main Social Organizations: Labor is divided between the left-wing Kihusan Mayo Uno (May Day Confederation) and the Trade Union Congress of the Philippines (TUCP), affiliated to AFL-CIO. The Philippines has over 700 voluntary organizations and church groups, which constitute the Green Forum.

THE STATE

Official Name: Republika g Pilipinas.
Administrative Division: 12 regions, 73 provinces.
Capital: Metro Manila 10,352,000 people (2003).
Other Cities: Cebu 1,172,800 people; Davao 1,145,600; Bacolod 739,600; Cagayan 407,800; Zamboanga 147,200 (2000).
Government: Presidential republic. Gloria Macapagal-Arroyo, President since January 2001. Bicameral Legislature: House of Representatives, with no more than 250 members; Senate, 24 members.
National Holiday: 4 July, (from the US, 1946); 12 July, Independence Day (from Spain,1898).
Armed Forces: 107,500 troops (1996). Others: National Police (Home Ministry): 40,500. Coast Guard: 2,000.

Under-5 mortality
34 per 1,000 live births
2004

Poverty
15.5% of population living on less than $1 per day
2002

Debt service
20.9% exports of goods and services
2004

Maternal mortality
200 per 100,000 live births
2000

IN FOCUS

ENVIRONMENTAL CHALLENGES
Indiscriminate deforestation, mostly due to lumber production, and soil erosion are the most important problems in non-urban areas. Manila has significant water and air pollution.

WOMEN'S RIGHTS
Women have been able to vote and stand for office since 1937. In 2004, they held 16 per cent of seats in Parliament, while they had 25 per cent of ministerial or equivalent posts the previous year. In 2003, women made up 38 per cent of the total workforce of 35 million people.

The maternal mortality rate is 200 deaths for every 100,000 live births.* While 88 per cent of pregnant women receive prenatal care, only 60 per cent of births are attended by skilled health personnel.*

At some point during their lives, 47.2 per cent of women have suffered at least one incident of violence at the hands of their spouses. Domestic violence is widespread partly because there are no laws to punish it (although there is currently a bill before Congress) and the majority of cases are not reported. Violence against women outside the home is also common.

Amnesty International has reported on repeated occasions that female prisoners face a high risk of rape, sexual assault and other forms of torture and mistreatment.

CHILDREN
Between 1990 and 2004, there was a 45 per cent reduction in the under-5 mortality rate, which fell from 62 to 34 deaths per 1,000 live births.*

The child population numbers over 35 million, almost half of the total population. According to the *Philippine Resource Network*, there are close to one million minors living on the streets, and over half suffer from malnutrition. The organization also reports that one in three children is abused, and that 60,000 are the victims of prostitution rings or organized crime groups. There are close to five million minors between the ages of 5 and 17 who work, and many are employed in the worst forms of child labor, such as prostitution, mining, domestic labor, firework manufacture, underwater fishing, drug trafficking and agriculture. At least three out of every five children are exposed to dangerous working environments, including physical and chemical hazards. Most children employed in agriculture work on planting, weeding, picking and fumigating with pesticides.

INDIGENOUS PEOPLES/ ETHNIC MINORITIES
Throughout the archipelago there are small concentrations of Negrito peoples. Aetas are the most discriminated among the indigenous groups. The terms Igorots and Cordilleras are used to refer collectively to a number of indigenous groups including the Bontoc, Kalinga, Ibaloy, Ifugao, Apayao/Isneg and Tinggians. Group members speak multiple languages and their customs differ from the Filipino majority. The Moro inhabit the Philippines' southern region, mainly the islands of the Sulu archipelago. Since the 11 September 2001 terrorist attacks, the Moro have remained under close surveillance and US soldiers have been deployed in the Philippines to assist in quashing them. Confrontations between Christian and Moro groups have decreased and several groups have chosen to disarm. The Moro have the country's lowest life expectancy and are the most disadvantaged group in terms of political and economic participation. They still demand the right to self-determination.

MIGRANTS/REFUGEES
At least 45,000 people remain internally displaced as a result of fighting between the armed forces and various insurgent groups. During 2004, the Philippines received 150 requests for asylum and refugee status, leading to a total of 2,000 people living in the country as refugees or asylum-seekers. There are 57,000 Filipino refugees living in Malaysia.

DEATH PENALTY
The death penalty was abolished in June 2006.

* Latest data available in *The State of the World's Children* and *Childinfo* database, UNICEF, 2006.

---

group in the Philippines, called for a referendum on the self-determination of Muslims in the South. Estrada declared that he would not give in to separatist demands.
25 Tens of thousands of Filipinos took to the streets of Manila to demand Estrada's resignation in October, while the opposition requested the President's impeachment, after a former crony denounced Estrada for receiving millions of dollars in kickbacks from an illegal gambling racket.
26 Parliament started impeachment proceedings against the President. The process revealed that Estrada had hundreds of millions of dollars in bank accounts under false names. In the midst of massive mobilizations, which led to Estrada's fall, Vice President Gloria Macapagal-Arroyo assumed the Presidency on 20 January 2001.
27 Corruption charges tainted the President when, in October 2001, her husband was accused of having accepted a bribe of more than $900,000 from a telecommunications firm, in exchange for lifting the presidential veto on a franchise agreement. Arroyo authorized a formal investigation into the case.
28 In April 2002, several bombs exploded in the city of General Santos, south of Mindanao, killing 14 people. The police blamed the MILF for the attacks.
29 In June the US Government pressed charges against five leaders of the Philippine Abu Sayyaf rebel group, purportedly linked to the al-Qaeda network and Osama bin Laden, for the kidnap and murder of two US citizens.
30 In October 2002, Abu Sayyaf - whose main objective was the creation of a Muslim state in the south of the Philippines - perpetrated a series of attacks against stores and a Christian temple which left 8 people dead and 170 injured. At least five people were arrested and taken to Manila.
31 On 23 January 2003, Rómulo Kintanar was murdered at a restaurant in Manila. Kintanar had been a Communist Party leader in the 1980s but no longer belonged to the institution. The Party claimed responsibility for the killing, attributing it to its armed wing, the New People's Army.
32 In a report published in January of that year, Amnesty International condemned the use of torture on political prisoners in Filipino prisons. Those most at risk of being tortured included alleged members of armed groups, their suspected sympathizers as well as ordinary criminals and members of poor or marginalized communities.
33 In March 2004, according to President Gloria Macapagal-Arroyo, four Abu Sayyaf members were arrested, while 36 kilos of high explosive trinitrotoluene (TNT) were confiscated, averting a terrorist bombing on the scale of the Madrid attacks perpetrated on 11 March. As Arroyo stated, one of the arrested men had claimed responsibility for the 27 February explosion aboard the SuperFerry 14 that killed over 100 people. The suspects, who had probably received military training from the terrorist network Jemaah Islamiah, linked to al-Qaeda, planned to launch attacks on trains and shops in Manila, which is home to ten million people.
34 Arroyo was elected to a second term in the May 2004 general elections, in which hundreds of thousands of emigrants were allowed to vote for the first time.
35 In July, the Government announced the early withdrawal of its small contingent of troops in Iraq, in response to the threats on the life of Ángelo Da Cruz, a Filipino truck driver taken hostage by Iraq insurgents. After the announcement, Da Cruz was released.
36 The MILF and the Government announced in April 2005 that during preliminary peace talks in Malaysia they had reached an agreement on the ancestral lands to which the rebels had been claiming the right for three decades - the key issue in the negotiations.
37 In June 2005 the Government reported that it had uncovered a plot to oust Arroyo by linking her to supposed irregularities in the previous year's elections. Opposition members were accused of hatching the plot, in which they would apparently use a tape supplied by US government sources. The US embassy in Manila denied any knowledge of the incident.
38 In November, fighting between the army and Abu Sayyaf rebels left 30 dead on the island of Jolo, in the southern Philippines.
39 In February 2006, Arroyo declared a state of emergency after a number of army officers were arrested and charged with plotting a coup against her. Four leftist lawmakers and another 12 opposition leaders were accused of complicity in the thwarted coup.
40 That same month, over 1,000 people were killed or disappeared when a mudslide buried the village of Guinsaugon on Leyte Island.
41 In June, Arroyo signed a law abolishing the death penalty. As a result, the sentences of some 1,200 inmates on death row were commuted to life imprisonment. ∎

Pitcairn / Pitcairn

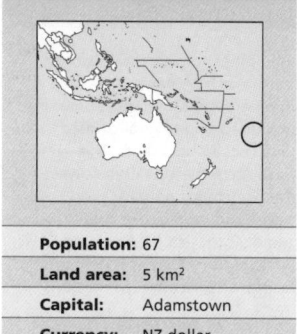

Population:	67
Land area:	5 km²
Capital:	Adamstown
Currency:	NZ dollar
Language:	English

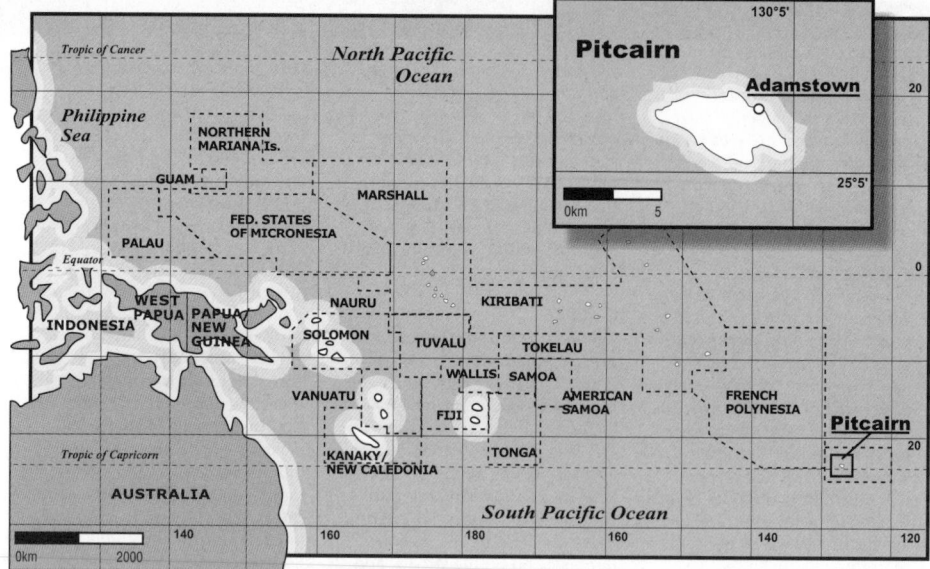

As on most of the islands in the region, Pitcairn's first settlers were Polynesians (see 'Melanesians and Polynesians' box). The first European to visit the island was the English voyager Robert Pitcairn who sailed along its coasts in 1767.

[2] In 1789, part of the crew of HMS Bounty mutinied on their return to Britain from six months in Tahiti. The captain and the rest of the crew were given a small boat and the other men returned to Tahiti. They stayed there a short time and then transferred to Pitcairn Island.

[3] The group, led by Fletcher Christian, was made up of eight crew members, six Tahitian men and 12 women. Ten years later, only one of the mutineers, John Adams, was still alive, with 11 women and 23 children. Adams christened the children and peopled Pitcairn, which later became a British colonial dependency.

[4] The population reached 200 in 1937, but decreased in recent times as young islanders emigrated to New Zealand/Aotearoa in search of work.

[5] Pitcairn was under the jurisdiction of the Governor of Fiji between 1952 and 1970, when it became a dependency of the British High Commissioner in New Zealand /Aotearoa.

[6] Despite the lack of communications, education in the islands was important and primary school was compulsory for all children between the ages of 5 and 15. A single teacher from New Zealand /Aotearoa is appointed for a two-year period and is also responsible for publishing the Pitcairn Miscellany, a four-page bulletin.

[7] There is no racial discrimination on the island and equal rights are ensured by the law. The Administrative Council is made up of 10 representatives of mixed descent, half of whom are elected.

[8] The islanders work almost exclusively at subsistence fishing and farming. The fertile valleys produce a wide variety of fruit and vegetables, including citrus, sugar cane, watermelons, bananas, potatoes, and beans. However, the island's main source of income is the export of postage stamps for sale to stamp collectors.

[9] In 1987, the British High Commissioner in Fiji, acting on behalf of Pitcairn, joined representatives from the US, France, New Zealand/Aotearoa and six Pacific island states in signing the South Pacific Regional Environment Protection Convention, designed to prevent the disposal of nuclear waste in the region.

[10] In early 1992, deposits of manganese, iron, copper, zinc, silver and gold were found. Exploitation of this mineral wealth could dramatically change the island's economy.

[11] The population of Pitcairn has been shrinking in recent years. In January 1998, there were just 30 people, only eight of whom were working. Ten had emigrated the year before.

[12] The survival of the island is dependent on the boats which ship in the necessary goods, as there is no airstrip. The lack of crews for the boats could accelerate depopulation. The UK aimed to avoid this happening by building an emergency air strip.

[13] The 44 inhabitants of Pitcairn discovered that the UK no longer had any interest in them in January 2000, when the Crown issued an edict to remove the last subsidies on electricity and the tariffs for unloading provisions, which had allowed the islanders to survive. The population then began to consider the possibility of becoming a French overseas colony.

[14] Richard Fell became the Governor in 2001.

[15] In 2002, 20 people from Pitcairn were involved in a case of sex abuse of girls aged 12 to 15. Island residents broke their silence and explained that the practice was a tradition, and that in Pitcairn children become sexually active at an early age.

[16] Finally, charges were brought against six of the accused. In 2004, four of them - including former mayor Steve Christian - were sentenced to 2 to 6 years in prison, while the rest received non-jail sentences. In May 2005, an appeals court rejected the six men's case, which was that British law was not applicable on the island, and confirmed the sentences. In May 2006 a Pitcairn court rejected a new appeal.

[17] In October 2006, the six men's appeal was rejected by the British Privy Council, the court of final appeal for UK overseas territories. The men were found guilty of child rape and indecent assault and will now serve their sentences, which range from six years in jail to community service. ■

Poland / Polska

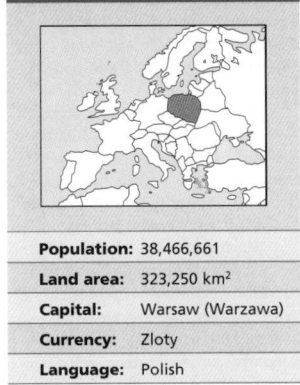

Population:	38,466,661
Land area:	323,250 km²
Capital:	Warsaw (Warzawa)
Currency:	Zloty
Language:	Polish

The name Poland comes from the Polanian people ('people of the plains') who lived in the heartland of what became Poland, ruled by the Piast dynasty.

[2] In the 10th century, the Polanians subdued the Kujavians, the Mazovians, the Ledzians, the Pomeranians, the Vistulans, and the Silesians. Mieszko I (960-992) Duke of the Piast, united neighboring peoples and thus founded the first Polish State.

[3] Poland was a hereditary monarchy until the 12th century, with an army of élite warriors and a large peasantry, who were mobilized as they were needed and paid taxes to support the system.

[4] Mieszko submitted in the face of the expanding German Empire, in exchange for recognition of his sovereignty. As compensation, he appealed for Papal protection and in 1,000 AD he founded the first Polish ecclesiastical city state.

[5] The Roman Catholic Church was a crucial element in the political structure of the Polish State until the 12th century, when the State began to fragment.

[6] During the feudal period, Poland was subdivided into several duchies, ruled by the Piasts, and some 20 overlords, who became increasingly autonomous as the power of the Church grew. This was also a period of great demographic growth.

[7] The arrival of German settlers changed the country's ethnic composition, for up to then the population had been of Slavic stock. From the 13th century, the population of the towns became increasingly German and Jewish, who brought in their own legal systems, their capital, their crafts and their agricultural skills.

[8] Under the reign of Casimir the Great (1333-70), Poland became a monarchy divided into estates, with the King acting as an arbiter between the nobility, the clergy, the bourgeoisie, and the peasants.

In 1399 the monarchy became elective.

[9] Through a royal marriage in 1386, Poland joined with Lithuania, although the differences between the two countries were upheld. In 1410, the Teutonic Order forces were defeated at Grunwald. This secured Poland's power, and at the same time left the Teutons weakened after the Peace of Torun in 1411.

[10] In 1466, after a new victory over the Teutons, Poland recovered Pomerania of Gdansk and Malbork, Elblag, and the Land of Chelm; it also gained the territory of Warmia. In recognition of their assistance during the war, Poland granted autonomy to Pomerania and some privileges to the towns. A period of economic prosperity and cultural renaissance began.

[11] During the 15th century the General Diet (parliament) of Poland and Lithuania was created. It had two houses: a lower house, comprising members of the nobility, and an upper house, or royal council, presided over by the King. The two states shared the King, the diet, and the management of foreign affairs, while administration, justice, finance and the army remained separate.

[12] The 16th or 'Golden' century is also known as the period of the Royal Republic, for the King had to consult the nobles before fixing taxes or declaring war. The rights of the bourgeoisie and the peasantry were curtailed in favor of the nobility and the clergy.

[13] In 1573, with the end of the Jagiellon dynasty, the Diet approved the free election of the King and guaranteed religious tolerance, at a time when Europe was being shaken by religious wars. King Stephen Bathory (1576-86) gave up his role as arbitrator and the nobility started to elect their own courts.

[14] During the 17th century, while Sweden fought Poland for control of the Baltic, and Russia entered into conflict with Lithuania, Turkish and Austrian ambitions in central Europe also put pressure on Poland.

[15] On the lower Dnepr, on the border with the Ukraine, free peasants and impoverished nobles became the first Cossacks, warriors who lived by pillaging. In 1648, they started a national revolt. The King made unsuccessful attempts to reach an agreement with the rebels, whose victories weakened

LAND USE

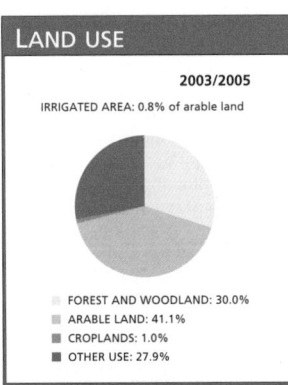

2003/2005

IRRIGATED AREA: 0.8% of arable land

- ▫ FOREST AND WOODLAND: 30.0%
- ▫ ARABLE LAND: 41.1%
- ▪ CROPLANDS: 1.0%
- ▪ OTHER USE: 27.9%

PUBLIC EXPENDITURE

% OF GDP

- HEALTH & EDUCATION (2000-2002): 10 %
- DEFENSE (2004): 1.9 %

Life expectancy
75 years
2005-2010

GNI per capita
$6,100
2004

HIV prevalence rate
0.1% of population
15-49 years old
2003

the republic.

16   The Cossacks formed occasional alliances with the Turks and the Russians. In 1654, Russian troops entered Polish territory. Sweden invaded the rest of the country a year later. King John Casimir fled to Silesia, and Austria aided Poland, while the peasants organized an armed resistance.

17   The Swedes and the Turks were expelled from the country and the Cossacks were defeated. Russia kept Smolensk and the Ukraine, the left bank of the Dnepr and the city of Kiev. The wars devastated the land, decimated the population and split the republic.

18   In 1772, Russia, Prussia, and Austria partitioned Poland. There was a second partition in 1793, after a new Russian invasion annulled the 1791 Constitution and put an end to attempts to reorganize the State.

19   A patriotic insurrection was crushed in 1794, and was followed by the third partition. The Polish State disappeared from the map, although the people retained a sense of national identity.

20   In the 19th century there were several attempts to free Poland. National conscience and Catholicism, both under persecution, became stronger. New political parties emerged (peasants', workers' and national) and resistance was expressed through art and culture.

21   The Russian Revolution in 1917 brought Poland the support of the Western powers. In 1918 a provisional government, led by Jozef Pilsudski, established an eight-hour working day and equal rights for men and women.

22   A sense of national identity had emerged in the Ukraine, Lithuania and Belarus. The creation of a federation failed as a result of the Soviet counter-offensive. The Peace of Riga, signed in 1921, granted the independence of the Baltic states and fixed Poland's eastern border at Zbrucz.

23   The 1921 Constitution adopted a parliamentary system.

24   Social and economic instability benefited the Communist Party, banned in 1923. Its main military leader, Jozef Pilsudski, staged a coup in 1926. The ensuing prosperity ended with the impact of the 1929 Wall Street crash.

25   The German-USSR non-aggression pact signed in August 1939 threatened Poland. Britain and France had a treaty with Poland which promised retaliation if any of them was attacked. When Germany invaded Poland on 1 September 1939 Britain and France declared war.

26   In the occupied territories millions of Poles died, especially Jews, some of whom were taken to German concentration camps. Many others starved or were executed.

27   The Polish government-in-exile led the resistance. A military contingent fought on the western front, while the Home Army carried out subversive actions. After the German invasion, the USSR accepted the creation of a Polish army under its jurisdiction.

28   The Soviet counter-offensive modified bilateral relations. The government-in-exile demanded an inquiry into the murders of Polish officers, and the USSR broke diplomatic relations and shifted to military occupation.

29   After Germany's defeat, the allies gathered at Yalta and agreed on a Provisional Polish Government of National Unity (made up by representatives from pro-Soviet and exiled groups) which was to call elections. The Polish Workers' Party dominated the Government.

30   In 1945 the provisional Government and the USSR signed an agreement establishing the Polish eastern border, along the Curso line. The allies fixed the eastern border along the Oder-Neisse line of Lusetia.

31   The Polish Workers' Party and the Socialist Party of Poland combined to form the Polish United Workers' Party (PUWP). The Polish Peasants' Party disintegrated, and elections were postponed.

32   The PUWP governed the country, modeling itself on the Soviet Communist Party (CPSU) in the USSR. Industry and commerce were nationalized, the State built great steel and metal works, and forcibly collectivized agriculture. Women were incorporated into the workforce.

33   The CPSU crisis, after the denunciation of Stalin's crimes in the 20th Congress of 1956, had repercussions on the PUWP. In November of that year Wladyslaw Gomulka was elected party first secretary and promised to take a 'Polish path towards socialism'. Gomulka freed Cardinal Stefan Wyszynski - head of the Catholic Church - stirring up popular expectations.

34   In 1970, West Germany recognized the Polish borders established after the War. East Germany had done so in 1950.

35   In 1970, strikes broke out due to an increase in prices. The Government gave orders to open fire on the workers and started another crisis within the PUWP. Gomulka was replaced by Edward Gierek, but the regime underwent new crises over corruption and internal fights within the party.

36   In 1976, new strikes broke out, which were repressed not through the use of firearms but by imposing long prison sentences. In 1979, the Polish Pope John Paul II visited his native land, and was welcomed by massive gatherings.

37   The strike at Gdansk's Lenin Dockyard in August 1980 was led by Lech Walesa and turned into a general strike. The Government was forced to negotiate and two months later recognized Solidarity, a workers' union with 10 million members. Rural Solidarity was created, to represent three million peasants.

38   The PUWP appointed Wojciech Jaruzelski, then Prime Minister, to the post of party first secretary. In December 1981, martial law was declared, Solidarity was banned and its leaders went underground.

39   Martial law was lifted in 1983, but the Constitution was modified to include a state of emergency. With the Catholic Church acting as mediator, government and Solidarity representatives went back to negotiations in 1989, while the USSR was embarking on *perestroika* (restructuring).

40   In the elections in June that year, the PUWP only obtained the number of representatives that had been agreed on with the opposition. Mazowiecki, a moderate member of Solidarity, was appointed the first president of a non-communist government in the East European bloc.

41   Poland re-established diplomatic relations with the Vatican and with Israel. The US and East Germany promised financial assistance. German reunification caused some alarm, but the negotiations ratified the postwar Polish borders.

42   In December 1989, the National Assembly approved reinstating the name the Republic of Poland. In January 1990, the PUWP was dissolved and the Social Democracy of the Republic of Poland and the Polish Social Democratic Union were created.

43   In January 1990, Poland started an economic adjustment program agreed with the IMF, requested entry to the Council of Europe and established relations with the EU. Poland's entry to NATO was made dependent on the results of its economic reforms and the upgrading of its military capability.

44   In May 1990, the first strike against the Government was held in Gdansk. Walesa accused Mazowiecki of having forgotten his days as a worker. Solidarity split into several political parties.

45   In the first direct presidential elections, held in December 1990, Lech Walesa won with 75 per cent of the vote. In August 1991, new Prime Minister Jan Krysztof Bielecki resigned, upsetting the precarious balance of political transition. The former Communist Party and a small peasant party were ready to accept his resignation, but Walesa backed the Prime Minister and insisted on giving him special powers, by threatening to dissolve the Diet.

46   In December 1991, Jan Olszewski was appointed Prime Minister. The cabinet was not

PROFILE

ENVIRONMENT

On the extensive northern plains, crossed by the Vistula (Wisla), Warta and Oder (Odra) rivers, there are coniferous woodlands, rye, potato and flax plantations. The fertile soil of central Poland's plains and highlands yield a considerable agricultural production of beet and cereals. The southern region, on the northern slopes of the Carpathian Mountains, is less fertile. Poland has large mineral resources: coal in Silesia; sulfur in Tarnobrzeskie; copper; zinc and lead. Major industries are steel, chemicals and shipbuilding.

SOCIETY

Peoples: Polish, 96 per cent; Ukrainian, 0.8 per cent; other (Belarusian, 0.8 per cent; German, 0.5 per cent; Swedish, 0.5 per cent). **Religions:** Catholic, 90.7 per cent; Orthodox, 1.4 per cent. Protestant and other (7.9 per cent). **Languages:** Polish. **Main Political Parties:** Law and Justice; Citizens Platform; Self-Defense of the Republic of Poland. **Main Social Organizations:** Union affiliation has decreased markedly. Two main groups with political affiliation: Poland Trade Union Alliance and Solidarity. Independent unions like the Central Union of Agricultural Groups.

THE STATE

Official Name: Polska Rzeczpospolita. **Administrative Divisions:** 49 provinces. **Capital:** Warsaw (Warzawa) 2,200,000 people (2003). **Other Cities:** Lódz 1,017,300 people; Kraków 784,800; Wroclaw 634,600; Poznan 580,200 (2000). **Government:** Lech Kaczynski, President since December 2005. Jaroslaw Kaczynsky, Prime Minister since July 2006. Bicameral Legislature: the *Diet*, with 460 members, and the Senate, with 100 members. **National Holiday:** 11 November, Independence Day (1918); 3 May, Constitution Day (1791). **Armed Forces:** 241,750 (1997). Other: 23,400 Border Guard, Police, Coast Guard.

Under-5 mortality
8 per 1,000 live births
2004

Poverty
<2% of population living on less than $1 per day
2002

Debt service
34.6% exports of goods and services
2004

Maternal mortality
13 per 100,000 live births
2000

ratified by the Diet until 23 December, and then only by a narrow margin of 17 votes.

[47] As of November 1991, Poland became the twenty-sixth member of the European Council, a Western European organization which also includes Turkey, the Czech Republic, and Hungary.

[48] In mid-June, at Walesa's request, Parliament deposed Olszewski and appointed as Prime Minister Hanna Suchocka, from the Democratic Union (DU), supported by a seven-party coalition.

[49] Suchocka applied strict monetary controls and promoted the August 1992 privatization act. Walesa, under pressure from the Church, revoked the right to abortion in February 1993.

[50] The September elections saw the return to power of those who had supported the communist regime: the Democratic Left Alliance (SLD), the Union of Labor (UP) and the Polish Peasants' Party (PSL) together won 73 of the 100 seats in the Senate. Walesa appointed PSL leader Waldemar Pawlak Prime Minister.

[51] In 1994, beset by conflicts with Parliament, Walesa slowed the pace of the reforms and economic liberalization policies to reduce their social impact.

[52] The former communists' return to power was concluded in November 1995 when Aleksander Kwasniewski won the second round of presidential elections, with 52 per cent of the vote.

[53] Prime Minister Josef Olesky, who replaced Pawlak, was forced to resign in January 1996, after the Interior Minister accused him of having been a collaborator with the Soviet KGB. He was replaced by Wlodzimierz Cimoszewicz.

[54] Right-wing factions of several parties formed a coalition headed by Marian Krzaklewski, called Solidarity Electoral Action (AWS).

[55] In the 1997 parliamentary elections, the AWS defeated the ruling Democratic Left Alliance (SLD). Jerzy Buzek was named Prime Minister.

[56] In February 1999 Parliament approved entry into NATO by 409 votes to seven. Preparatory reforms for EU accession left thousands of people unemployed. In September 1999 more than 30,000 farmers and workers held a protest march in Warsaw demanding that elections be brought forward.

[57] In October 2000 Kwasniewski, now leader of the SLD, comprising former communists, became the first president to be re-elected since the transition to democracy, with almost 54 per cent of the vote. Adrei Olechowski, his main opponent, won only 17.3 per cent. Walesa failed to get one per cent of the vote and retired from politics.

[58] The lowest vote in Poland's brief democratic history occurred a year later in the parliamentary elections with barely 41 per cent turnout. Corruption scandals under Solidarity and the poor state of the economy led to increased support for radical anti-Europeanists and ultra-Catholics.

[59] A coalition government formed by the SLD and the Polish Peasants' Party (PSL) - which took nine per cent of the vote - made Leszek Miller prime minister in October 2001. In December, the EU Summit included Poland on the list of ten countries that would join the block in January 2004.

[60] Skepticism about EU membership increased after it was revealed that citizens of new member states might have to wait up to seven years before being allowed to work in other EU countries due to fears of a massive influx of cheap labor. In rural parts there were fears of an 'invasion' of people from other EU states, particularly Germany, wanting to buy cheap land. In April 2002 Poland reached an agreement with Brussels to prevent foreigners from buying land in Poland for 12 years after it joined the EU.

[61] In March 2003 Miller expelled the PSL from the Government coalition, for not supporting his proposed tax reform.

[62] In a referendum held in June 2003 Polish citizens voted to join the EU.

[63] Poland, the only country of continental Europe to provide troops for the US-led invasion of Iraq, took command of one of Iraq's four reconstruction zones in early 2004.

[64] On 1 May 2004 Poland, along with nine other countries, became a full member of the EU, which thus expanded to 25 members. In the same month Prime Minister Miller resigned and was succeeded by ex Finance Minister Marek Belka.

[65] In January 2005 many world leaders gathered at the site of the Auschwitz Nazi extermination camp on the 60th anniversary of its liberation.

[66] With the promise of tax cuts, the protection of workers' rights and a 'moral renewal', Warsaw mayor, Lech Kaczynski, from the conservative Law and Justice Party, won the second round of presidential elections in October 2005 with 54.04 per cent of the vote, defeating Donald Tusk of the liberal Civic Platform Party. Having failed to reach agreement with the Liberals, the Conservatives formed a minority government in November with Kazimierz Marcinkiewicz as Prime Minister.

[67] In December Kwasniewski denied press accusations that the CIA kept secret prisons in Poland for suspected terrorists. That same month Kaczynski took office as President and soon after announced that he would continue the presence of Polish troops in Iraq until the end of 2006 but would reduce their number from 1,450 to 900 during the year.

[68] In March, 82 year old ex-communist leader Wojciech Jaruzelski was charged with imposing martial law against Solidarity in 1981.

[69] In May, the League of Polish Families (LPR), an anti-European, nationalist and ultra-Catholic movement, and the populist anti-liberal Samoobrona Party both joined the Government, thus giving it a parliamentary majority. ■

IN FOCUS

ENVIRONMENTAL CHALLENGES
The country has high levels of air pollution partly because of its location in the center of Europe. It absorbs polluted water and air 'in transit' from other countries. Sulfur dioxide emissions from coal-fired power plants, and the resulting acid rain, has damaged forests. The attempt to reduce pollution levels to EU - acceptable levels implies substantial costs for business and the Government.

WOMEN'S RIGHTS
Polish women have been able to vote since 1918. In 2005 they held 19 per cent of parliamentary seats and in 2003 some 6 per cent of ministerial or equivalent positions, which represented a significant reduction compared to 2000, when this figure was 17 per cent.

Women's participation in the workforce has been stable since 1980 at 46 per cent.

In 2004 the female enrolment rate was high in both primary and secondary education (98 and 93 per cent respectively). Ninety-nine per cent of girls finished 5th grade*.

All births are attended by qualified medical personnel*.

The law fails to provide either protection for women who suffer physical domestic violence or measures to avoid future abuse. Amnesty International highlights the lack of importance attached to domestic violence, pointing to inefficient investigative and judicial procedures and the absence of appropriate protection for victims against further acts of violence.

CHILDREN
Six per cent of newborn babies are underweight*. The under-one and under-five mortality rates are 7 per 1,000 live births and 8 per 1,000 live births respectively*

Education is practically universal (98 per cent of school-aged children are enrolled) and compulsory until 18.

Even though the Constitution guarantees the right to an education in accordance with the child's family values, religious education continues to be taught in public schools. A child is supposed to have an option between religious or ethical education but ethics classes proved unsustainable.

Commercial child trafficking and sexual exploitation is a growing problem.

INDIGENOUS PEOPLES/ ETHNIC MINORITIES
Poles are descendents of different ethnic groups (Slavonic, Polanie or 'people of the plain', Lithuanian, Finnish, Gothic and Celtic peoples). Seventy per cent of the population are Poles and the rest are Belarusian, Ukrainian, Ruten and Jewish. There are small Greek (114,000), Russian (60,000), Slovak (38,000) and Lithuanian (11,500) communities. The upheavals in Poland's political history created several linguistic communities. Polish Roma/Gypsies have several languages: Baltic Romany (30,000), Carpathian Romany, Sinte Romany and Vlach Romany (5,000).

MIGRANTS/REFUGEES
In late 2004 there were approximately 10,000 refugees and asylum-seekers in Poland. In that year there were 6,250 asylum requests mainly from Russians, Indians and Pakistanis.

During 2004 some 400 Poles requested asylum in Canada. Polish migration (mostly Jewish) to every part of the world was common throughout the 20th century.

DEATH PENALTY
The death penalty was abolished for all crimes in 1997. The last execution was in 1988.

* Latest data available in *The State of the World's Children* and *Childinfo* database, UNICEF, 2006.

Portugal / Portugal

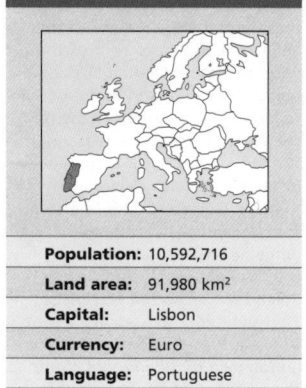

Population:	10,592,716
Land area:	91,980 km²
Capital:	Lisbon
Currency:	Euro
Language:	Portuguese

In ancient times Portugal was inhabited by Lusitanians, an Iberian people whose cultural influence extended over a vast area including the whole western shore of the Iberian Peninsula. The coastal areas were occupied by successive invaders from the Middle East.

² During the 2nd century BC the Romans settled in the territory, ruling over it until the fall of the Empire around the 5th century AD. Like the rest of Europe, Portugal was invaded by Northern European peoples (generically called Barbarians), who raided the Roman dominions. Among these peoples were the Visigoths. They had a developed culture and settled on the Iberian Peninsula dividing the territory into various kingdoms, and spreading the Christian faith. Their domination over the whole region lasted for nearly six centuries.

³ In the 8th century AD Arab peoples invaded the region, meeting resistance from the inhabitants. In spite of the Arab political and cultural dominance this was a period of religious tolerance and a flowering of art.

⁴ During the 11th century the Reconquest of the Lusitanian territory started, ending with the expulsion of the Arabs a hundred years later. With Muslim domination over, the territory was politically unified and

Portugal entered a period of great economic prosperity. This reached its height during the 15th and 16th centuries, with great maritime expeditions and conquests of vast territories in America, Africa, and the Far East.

⁵ Its maritime superiority enabled Portugal to develop active worldwide trade and achieve a privileged economic position within Europe. A long time elapsed before other nations like Britain and the Netherlands were in a position to threaten Portugal's naval supremacy.

⁶ Following a series of dynastic struggles, in 1581 the country was made subject to Philip II, King of Spain. The two kingdoms remained united until 1688 when Portugal succeeded in having its independence recognized in the Treaty of Lisbon. Unity with Spain brought the decline of Portugal's power. Most of the maritime

empire collapsed, besieged by the British and the Dutch, who started to control most of the trading routes and outposts.

⁷ By the time Portugal recovered its independence in 1640, it had been devastated by 30 years of war against Spain. The country was forced to look on while the new maritime powers seized most of its colonies in Africa and Asia. Brazil remained under Portuguese rule. The position of Britain as the leading maritime power became painfully obvious when Portugal was forced to sign the Treaty of Methuen, which established Portugal's political and economic dependence on the British. Pombal, an adviser of Jose I, carried out economic reforms. Like the Spanish Bourbons, Pombal had been influenced by the ideas of the French Enlightenment and he changed colonial management. The discovery and exploitation of

gold mines in Brazil enabled the country to enjoy a period of great economic prosperity. But in spite of Pombal, Portugal finally went into decline.

⁸ Dependence on Britain was further consolidated when Portugal was forced to seek support to end Napoleonic occupation, which lasted from 1807 to 1811. French domination led to the independence of Brazil; the Portuguese court had fled there in exile. Brazil had enjoyed a significant expansion in trade, in particular with Britain. At the end of the Napoleonic period in Europe, the rising Brazilian bourgeoisie was not ready to be displaced, so in 1821 Brazil declared independence. Meanwhile a civil war broke out in Portugal between those who wanted the restoration of absolutism and liberal groups preferring greater political participation.

⁹ While other countries embarked on rapid industrialization which would quickly put them in an economically powerful position, Portugal maintained its traditional agrarian structure. Thus, it reached the end of the 19th century economically stagnant, deprived of the richest and largest part of its colonial empire, and suffering from acute internal political crisis.

¹⁰ The monarchy was unable to ensure the stability needed to start economic recovery, and was definitively overthrown by liberal opposition forces in 1910. This started the Republican period. Once they had attained their objective, the alliance of Liberal and Republican groups started to fragment and internal differences prevented them from achieving a common governmental agenda. One of the few things they shared was active opposition to the Church, which had been a traditional ally of the *ancien régime* and had had important privileges and powers, including the control of education. The inefficiency of the Liberals, together with the ruthless persecution of representatives of the ancien régime, encouraged the formation of a vast opposition movement.

¹¹ Portugal sided with Britain in World War I. This only deepened the economic crisis and increased popular discontent. Political instability and economic stagnation were the most salient features of the period. In 1926 this led to a coup bringing a right-wing military group to power. They set up an authoritarian corporatist regime which they called the 'New State'. With a few changes, it was to rule the country for over 40 years. Political opposition was proscribed,

---

LAND USE

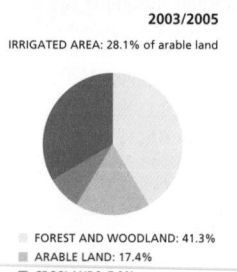

2003/2005

IRRIGATED AREA: 28.1% of arable land

- FOREST AND WOODLAND: 41.3%
- ARABLE LAND: 17.4%
- CROPLANDS: 7.9%
- OTHER USE: 33.4%

PUBLIC EXPENDITURE

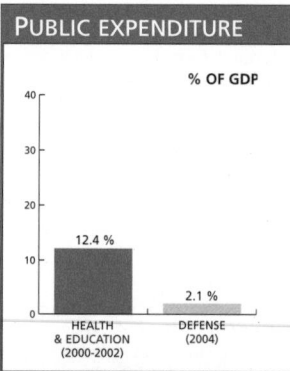

% OF GDP

12.4 % — HEALTH & EDUCATION (2000-2002)

2.1 % — DEFENSE (2004)

WORKERS

UNEMPLOYMENT: 7% (2004)

LABOR FORCE   **2004**

- FEMALE: 46%
- MALE: 54%

EMPLOYMENT DISTRIBUTION   **1995/2002**

F
M

- AGRICULTURE   F: 14%   M: 12%
- INDUSTRY   F: 23%   M: 44%
- SERVICES   F: 63%   M: 44%

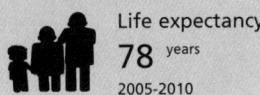

Life expectancy
78 years
2005-2010

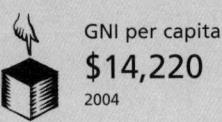

GNI per capita
$14,220
2004

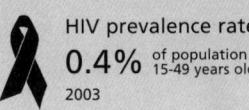

HIV prevalence rate
0.4% of population
15-49 years old
2003

with major figures imprisoned or exiled. Trade unions were dissolved and replaced by corporatist organizations similar to those in fascist Italy.

12 The most significant figure of this period and the true ruler behind the military was economist Antonio de Oliveira Salazar, who occupied various positions and dominated Portuguese political and economic life. The country remained neutral during the Spanish Civil War and World War II, which could both have jeopardized its barely stable economy.

13 The agricultural system remained unchanged throughout the whole period, causing much migration towards the major cities of Portugal and Europe. The 1950s saw the start of decolonization and the country was faced with the possibility of losing its last dominions in Africa. Salazar's regime sought to repress the rebellions in the colonies, which led to Portugal being isolated internationally.

14 The human and economic cost of these colonial wars accelerated the internal attrition of Salazar's government. Repressive measures had to be increased to halt the growing opposition. His death in 1970 and the deepening of the economic crisis showed that the end of the regime was close at hand.

15 In 1974, amidst the opposition of many social groups and political parties, a significant number of dissatisfied army officers gathered under the Armed Forces Movement (*Movimento das Forças Armadas* - MFA). In April, they staged a coup intending to end wars in Africa and start the democratization process.

16 The military government that emerged from this 'Carnation Revolution' had vast popular support. The new Government quickly recognized the independence of Angola, Mozambique, and Guinea-Bissau. Meanwhile it actively sought international recognition and attempted to improve the country's image abroad. It legalized left-wing political parties, decreed an amnesty for political prisoners and passed a series of land laws aimed at breaking up large rural estates and modernizing agricultural production.

17 After a year in office, dissent between the Socialist and Communist parties, which were the main supporters of the new regime, interrupted the process of democratization. In the 1976 general elections, the Socialist Party led by Mario Soares won the majority of votes, becoming Portugal's first democratic constitutional government in the 20th century. However, the

continuing economic crisis, Soares' harsh economic adjustment program and strong political and trade union opposition, wore the Socialist Government down very quickly.

18 The 1980s witnessed the continuation of the process of transition and political integration with Europe. The electorate approved a new constitution and eliminated all the special bodies created under military rule. Portugal joined NATO and the EEC in 1986. That year, the Socialist Party lost power again, this time to its one-time ally the center-left Social Democratic Party (PSD).

19 By the end of the 1980s Portugal was experiencing significant economic growth, but still far below the average for the rest of Europe. Changes accelerated after the electoral victory of the PSD, which used its ample parliamentary majority to liberalize the economy. The new economic policy received strong opposition, particularly from workers in the public sector, who saw the PSD's reforms as a threat to their jobs.

20 The trade union movement brought the country to a standstill on several occasions. It opposed the privatization of public companies, and the attempt to repeal labor and land reform legislation, enacted in 1974. In 1984, demanding respect for the achievements of the 'Carnation Revolution' an extreme left-wing group, called the Popular Forces of April 25 (FP-25), also started to take action against these measures.

21 In April 1987, the governments of Portugal and the People's Republic of China signed an agreement charging Portugal with the administration of Macau until 1999. Sovereignty was then transferred to China, under the 'one country, two systems' principle (see Macau).

22 In 1988, the PSD and the Socialist Party agreed to modify the constitution to allow the re-privatization of various companies nationalized during the 'Carnation Revolution' and to further reduce presidential powers. President Mario Soares opposed these reforms, which led to his distancing from the PS leadership and to a permanent clash with Prime Minister Anibal Cavaco Silva.

23 Portuguese politics became polarized between the ruling PSD and the PS. The latter was a more viable left-wing alternative after the collapse of real socialism. However, in the October 1991 parliamentary elections, the PSD won over 50 per cent of the vote. Cavaco Silvas' political victory was due to the social democratic slant with which he disguised

ENVIRONMENT

The country includes the Iberian continental territory and the islands of the Azores and Madeira archipelagos. The Tagus, the country's largest river, divides the continental region into two separate areas. The northern region is mountainous, with abundant rainfall and intensive agriculture: wheat, corn, vines and olives are grown. In the valley of the Douro, the major wine-growing region in the country, large vineyards extend in terraces along the valley slopes. The city of Oporto is the northern economic center. The South, Alentejo, with extensive low plateaus and a very dry climate, has large wheat and olive plantations and sheep farming. The cork tree woods, which made Portugal a great cork producer, are found here. Fishing and shipbuilding are major contributors to the country's economy. Mineral resources include pyrite, tungsten, coal and iron.

SOCIETY

Peoples: The Portuguese (99.5 per cent) came from the integration of various ethnic groups: Celts, Arabs, Berbers, Phoenicians, Carthaginians and others. Immigrants come from Africa (0.2 per cent) and the Americas (Brazilians 0.1 per cent, US Americans 0.1 per cent). There is substantial migration by Portuguese towards richer countries in the continent.

Religions: Catholic (94.5 per cent); Protestants (0.6 per cent); other Christians mostly Catholic Apostolic and Jehovah Witnesses (0.9 per cent); Jewish (0.1 per cent); Muslims (0.1 per cent).

Languages: Portuguese; there are two small areas where two dialects are spoken: 'Mirandes' (derived from Asturian- Leonese) and 'Barranquenho'.

Main Political Parties: Social Democrat Party; Socialist Party; Unitarian Democratic Coalition, Popular Party (PP).

Main Social Organizations: The General Confederation of Portuguese Workers (CGTP), a nationwide multi-union organization with 287 union members (represents 80 per cent of the organized workers); the General Union of Portuguese Workers (UGT-P), which combines 50 unions; National Agriculture Confederation (CNA).

THE STATE

Official Name: República Portuguesa.

Administrative Divisions: 18 districts, two autonomous regions (Azores and Madeira).

Capital: Lisbon 1,962,000 people (2003).

Other Cities: Oporto 1,206,800 people; Amadora 123,400; Vila Nova de Gaia 74,800 (2000). **Government:** Aníbal Cavaco Silva, President since March 2006. José Sócrates, Prime Minister since March 2005. Unicameral Legislature: Assembly of the Republic, with 230 members. National Holidays: 10 June, Portugal Day (1580); 5 October, Independence Day (1910); 25 April, Liberty Day (1975).

Armed Forces: 43,600 (2002). Other: 20,900 Republican National Guard; 20,000 Public Security Police; 8,900 Border Security Guard.

his orthodox liberal economic orientation.

24 In January 1992, Portugal took over the presidency of the European Community. The new President, Luis Mira de Amaral, Portuguese minister of industry and energy, announced he would promote industrial co-operation with Latin America, Africa and central Europe and the signing of the Maastricht Treaty between the members of the Community.

25 In August 1993, the Assembly restricted the right to seek asylum and enabled the expulsion of foreigners from the country. The legislation was based on the defense of the job market and was opposed by President Soares.

26 A plan financed by the EU was approved for the 1993-1997 period for the poorest members, including

Portugal, providing investment education, transport, industrial retrofitting and job creation.

27 The October 1995 general elections were won by the Socialist Party which gained an absolute majority at the Assembly. Antonio Guterres replaced Prime Minister Anibal Cavaco. After ten years of PSD dominance, oriented toward European integration and economic liberalism, the PS capitalized on domestic discontent with education and health and assured the financial market it would not interfere with the goals regarding monetary union and privatization.

28 The Socialist Jorge Sampaio took over the presidency of the country in March 1996. The Government brought in an economic plan in line with EU

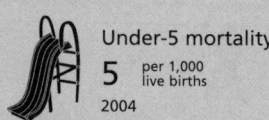

Under-5 mortality
5 per 1,000
live births
2004

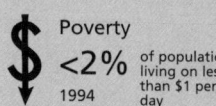

Poverty
<2% of population
living on less
than $1 per
day
1994

Maternal mortality
5 per 100,000
live births
2000

IN FOCUS

ENVIRONMENTAL CHALLENGES
Poor quality soils suffer the effects of erosion. The levels of air pollution are increasing in urban areas due to vehicle emissions, and in areas near cellulose and cement factories. Pollution has been reported in water sources and coastal areas.

WOMEN'S RIGHTS
Women have been able to vote and run for office since 1976, although a select group was allowed to vote from 1931.

In 2005, women held 21.3 per cent of seats in Parliament, while in ministerial or equivalent positions their representation stood at 17 per cent in 2003. That year, women made up 44 per cent of the country's total workforce of 5 million.

Women face discrimination in access to senior positions; the income gap between men and women doing similar jobs was 22.6 per cent in 2004. Cases of violence and oppression against women are registered, according to reports by different NGOs. In 2004, 6,459 cases of such aggression were reported, almost double the number in 2000.

Portugal is a transit and destination country, mostly for people from Eastern Europe. In 2003 there were several reports of women trafficked from Brazil and forced to work in prostitution. Prostitution is not illegal in Portugal, but human trafficking, procurement and the distribution of pornography involving minors are.

CHILDREN
In 2004, 2 million inhabitants were under the age of 18 and about 600,000 were under five. Life expectancy at birth rose from 67 years in 1970 to 78 years in 2004*.

The Government implemented programs for children's welfare and in defense of the rights of the child in 2003, mostly in public education and health services. Education is compulsory, free and universal up to the age of 15, and 99 per cent of school-age children were enrolled in 2004*. In addition, there is also free, public pre-school education for children from the age of 4. The number of children in pre-school grows annually. There are public nurseries for children 3 months to 3 years of age, which have been improving both in number and quality.

Portugal is a transit country for African children, particularly from Angola, some of whom are trafficked to other EU countries.

INDIGENOUS PEOPLES/ ETHNIC MINORITIES
Portugal's largest minority are the Roma or Gypsies with about 50,000 people; they are the group that most suffers discrimination, particularly at the hands of the police.

There have been reports of segregation of the Afro-Portuguese population. The Government has passed laws against racism and discrimination.

Five per cent of the total population are immigrants, who are also considered minorities.

MIGRANTS/REFUGEES
In 2002, some 2,500 people - former refugees from Guinea-Bissau's 1998 civil war - finally received their residence permits under an amendment of Portugal's laws on alien immigrants. These people were left in a legal void in 2000, when they lost their temporary protected status, but continued residing illegally in the country.

For some time, the Government has been carrying out tight immigration controls, aimed at dismantling the human trafficking networks operating in Western Europe. Most of the trafficking victims were men from Eastern Europe (plus the above mentioned cases of African children and Brazilian women).

During 2004, Portugal received some 400 refugees, while 360 Portuguese people requested asylum in other countries, mainly in Canada.

DEATH PENALTY
The death penalty was abolished for all kinds of crime in 1976.

* Latest data available in The State of the World's Children and Childinfo database, UNICEF, 2006.

including health.
[37] In late 2002 an investigative journalism report revealed a child-sex network, involving diplomats, politicians, sportsmen and journalists. The network had been covered up for two decades with the State's complicity, since the victims were children from Casa Pia (Pious House), Portugal's main state orphanage. Investigators stated that 128 boys and girls had been subjected to sexual abuse. Ten people had been arrested by the end of 2003, including Carlos Cruz, a famous Portuguese TV presenter; and Jorge Ritto, former Portuguese ambassador to South Africa.
[38] In August 2003 fire devastated almost 215,000 hectares of the country - an area the size of Luxembourg.
[39] Four bills to legalize abortion were rejected on 5 March 2004 in parliament with PSD and PP votes. According to the Health Ministry, 11,000 women had needed medical treatment after having illegal abortions, in 2002. The number of abortions is estimated at 30,000 a year.
[40] In mid-2004, Durão Barroso resigned as Prime Minister to become President of the European Commission. His succession led to a period of political instability, leading Sampaio to dissolve Parliament in November and call for early elections in February. The PS swept to victory with 45 per cent of the votes, while the Social Democrats came in second place with 28.8 per cent. In March, the Socialist José Sócrates took office as Prime Minister.
[41] In June 2005, the EU ordered Portugal to cut its budget deficit - which according to official estimates would reach 6.2 per cent that year - to fall in line with EU rules that set the maximum deficit limit at three per cent of GDP. The Government submitted a plan to cut spending aimed at having the deficit down to 2.8 per cent by 2008.
[42] Former center-right Prime Minister, Anibal Cavaco Silva, won the January 2006 elections with 50.54 per cent of the votes, defeating five leftist candidates. In March, upon taking office, the new President announced his intention to maintain a 'strategic co-operation' with the Socialist government during the new period of cohabitation.
[43] In June 2006, construction began near Serpa in the south of the country on what is set to be the world's largest solar power station. The $70 million plant will produce enough electricity for 8,000 homes when it comes on stream in 2007. ∎

demands, particularly regarding the budget deficit.
[29] A successful campaign against tax evasion meant increased spending on health, education and social policies. The privatization program was intensified, selling shares in telecommunications, electricity and roads. Unemployment fell to 6.7 per cent of the active population.
[30] In February 1998, Parliament approved a law legalizing abortion up to ten weeks of pregnancy. The Government called a referendum and voters rejected the law by 50.91 per cent to 49 per cent, with a turnout of 32 per cent.
[31] After 442 years' rule, Portugal handed Macao over to China on 20 December 1999. This act signified the end of the Portuguese empire, as well as the end of all European control in Asia. Portugal, which had been the first European power to control Asian territories, was also the last to withdraw.
[32] Sampaio visited Xanana Gusmao, East Timor's pro-independence leader, in February 2000 and promised to help the Timorese to restore their education system. This was the first visit of a Portuguese head of state since Portugal pulled out of Timor in 1974.
[33] In April 2001, Sampaio's presidential electoral victory confirmed his continuing popularity, consolidating the control of the Socialists led by Guterres.
[34] December saw the inauguration of the Alqueva hydroelectric project on the Guadiana river. It created the greatest artificial lake in Europe, and was condemned by several environmental groups as too big, destructive and unnecessary. Although the project would irrigate the southern wasteland of the country, it would also flood the habitat of its unusual wild life (including eagles, wild boar, falcons and Iberian lynxes) as well as submerging 160 Stone Age period rocks. Politicians endorsed the project arguing that it was essential for irrigating the wasteland areas of the country, but some environmentalists highlighted that only 48 per cent of the irrigated land would be useful for crops or pasture.
[35] The poor economic performance of Guterres' government led him to make repeated changes in the cabinet, which, together with charges of corruption within the Socialist Party, lost him popularity. In December 2001, after the drastic defeat in the local election, Guterres resigned and Parliament was dissolved.
[36] Elections were brought forward and, in March 2002, the social-democratic candidate Manuel Durão Barroso won. The new Prime Minister formed a center-right coalition government. At the time of his taking office he promised to cut corporate taxes and reduce public expenditure, as well as to privatize public services

Puerto Rico / Puerto Rico

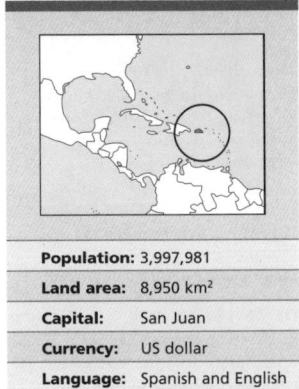

Population:	3,997,981
Land area:	8,950 km²
Capital:	San Juan
Currency:	US dollar
Language:	Spanish and English

I n 1508, 15 years after Christopher Columbus landed, the island of Borinquen (as Puerto Rico was called by the Arawaks/ Tainos who lived there) became a colony and has continued so up to the present day. Puerto Rico is the easternmost of the Greater Antilles and because of its strategic location at the entrance to the Caribbean, the island endured 400 years of Spanish rule. It also suffered repeated attacks by pirates and regular naval forces, whether British, Dutch or French, and finally remained under US administration after the Spanish-American War of 1898.

[2] As in neighboring islands, the local Taino people were exterminated by war, disease and overwork. African slaves were brought in to take their place in the fields where most food supplies for Spanish expeditions to the mainland were produced. Thus, Puerto Rican culture became a blend of its African and Spanish heritage.

[3] Spanish rule was continually challenged by external attacks and internal rebellions by both the Tainos and enslaved African workers. The latter rebelled successively in 1822, 1826, 1843, and 1848. The struggle for independence in the rest of Latin America had its counterpart in Puerto Rico's struggle for administrative reform (1812-1840),

but Spanish troops ruthlessly stifled the uprising.

[4] In 1868, five years before slavery was finally abolished, a group of patriots led by Ramón Emeterio Betances proclaimed Puerto Rico's independence in the town of Lares and took up arms to free the island. Despite their defeat, the Lares revolt signaled the birth of the Puerto Rican nation.

[5] The independence movement continued to gain strength in the following years. The Cubans were already up in arms in 1897, led by José Martí in a movement that reached Puerto Rico. US military intervention in the war against Spain, in 1898, hastened Spain's defeat but for Puerto Rico it only meant the imposition of a new ruler.

[6] US colonial administrations, first military and then civilian, imposed English as the official language and attempted to turn the island into a sugar plantation and military base. Puerto Ricans were made US citizens in 1917, though they were given no participation in the island's government. As a result, resistance to colonial rule grew. In 1922 the pro-independence Nationalist Party (PN) was founded. PN-led uprisings in 1930 and 1950 were harshly repressed.

[7] In 1947, intense internal and international pressure forced the US to allow Puerto Rico to elect its own governor. The 1948 elections gave the post to Luis Muñoz Marín, leader of the Popular Democratic Party (PPD), who favored turning the country into a

free associated state. Washington authorized the drafting of a new Constitution in 1959, which was approved by a plebiscite and later ratified by the US Congress.

[8] With the institution of Commonwealth status, US administrations were freed from the obligation of reporting on Puerto Rico's status to the UN Decolonization Committee. Moreover, in this way the UN tacitly endorsed the arrangement declaring the 'end' of colonial rule. In 1978 the situation changed when a UN General Assembly resolution defined Puerto Rico as a colony and demanded self-determination for its people.

[9] Muñoz Marín promoted industrialization on the island through massive US private investment enticed by government tax incentives. During the 1950s, Puerto Rico's system of agriculture was destroyed by an influx of US products resulting in over 50 per cent of the island's food being imported. The newly-formed labor reserve supplied cheap hands for the growing US corporate community, and Puerto Ricans soon began migrating en masse to the US, especially to New York, in search of work.

[10] With the great social upheavals of the 1960s, the struggle for independence flowered anew on the island. Despite the revival of the independence movement, a 1967 plebiscite confirmed the Commonwealth status. In 1976 Carlos Romero Barceló, from the New Progressive Party (PNP) that

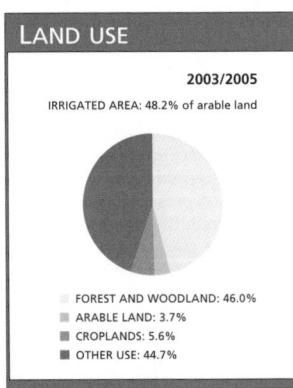
PROFILE

ENVIRONMENT
The smallest and easternmost island of the Greater Antilles. A central mountain range, covered with rainforests, runs across the island. In the highlands, subsistence crops are grown (corn, cassava/manioc); on the western slopes there are large coffee plantations and in the central region small tobacco farms. On the northern slopes citrus and pineapples are grown for export to the US. The main crop is sugarcane, which is cultivated on the best farmlands along the coastline. The islands of Vieques (43 sq km), Mona (40 sq km) and Culebra also belong to Puerto Rico.

SOCIETY
Peoples: Most Puerto Ricans are mestizos, descendants of Spanish colonizers, African slaves, Taino people and other small immigrant groups. In the 1940s, about three million Puerto Ricans emigrated to the US in search of better economic opportunities. This tendency began to decrease over the years until it has been almost reverted at the present time as a result of the favorable economic situation.
Religions: No official religion. Catholics (85.3 per cent), Protestants (4.7 per cent), others (10 per cent).
Languages: Spanish and English (both official). English is used in all matters related to political relations with the US. The forced introduction of

English in education, public administration and communications was given support in 1993, when it was made an official language alongside Spanish.
Main Political Parties: Popular Democratic Party; New Progressive Party; Puerto Rican Independence Party.
Main Social Organizations: The Labor Federation of Puerto Rico, which is affiliated to the US AFL-CIO. Federation of Pro-Independence University Students (FUPI); Committee for the Rescue and Development of Vieques.

THE STATE
Official Name: Estado Libre Asociado de Puerto Rico.
Capital: San Juan 2,332,000 people (2003).
Other Cities: Bayamon 205,400 people; Carolina 169,800; Ponce 156,500; Caguas 89,500 (2000).
Government: Aníbal Acevedo Vilá, governor since January 2005. Bicameral National Assembly: Chamber of Representatives with 51 members and the Senate with currently 29 members.
National Holiday: 25 July, Constitution (1952); 23 September, the Battle of Grito de Lares (start of the anti-colonial armed revolt in 1868); 4 July, US Independence (1776).
Armed Forces: 5,000 US Army troops (2003).

Life expectancy
77 years
2005-2010

GNI per capita
$10,950
2001

favored making the island the 51st state of the US, was elected governor and announced that if he were re-elected, he would call a referendum on annexation. Barceló was re-elected in 1980, but by such a slim margin that plans for a plebiscite were abandoned.

[11] Puerto Rico has one representative in the US Congress, but with no voting rights other than in committees. US citizenship only gave Puerto Ricans the right to participate in the 1980 presidential elections, although residents in the US are able to vote in all elections.

[12] Rafael Hernández Colón was again elected governor on 6 November 1984. He promised a 'four-year term of struggle against corruption and unemployment'. He renewed Puerto Rico's Commonwealth status thus rejecting his predecessor's intention to integrate into the Union. Hernández Colón was re-elected in 1988, with 48.7 per cent of the vote, against 45.8 per cent for those in favor of annexation by the US, and 5.3 per cent for those who favored independence.

[13] In April 1991, Governor Hernández Colón passed a law granting official status to the Spanish language. Soon after, the Puerto Rican people were granted the Prince of Asturias award by the Spanish crown, 'in recognition of the country's efforts to defend the Spanish language'.

[14] In the plebiscite carried out late in 1991, various strategies were proposed to promote development on the island. Hernández Colón succeeded in rallying moderate nationalists and supporters of independence, who campaigned together. They were in favor of self-determination, the end of subjection to US jurisdiction, the affirmation of Puerto Rican identity, regardless of any future referendum decisions, and the maintenance of US citizenship. These proposals were rejected by 55 per cent of voters who supported the PNP's position that a break with Washington had to be avoided.

[15] Pedro Roselló, a supporter of Puerto Rico's integration into the US, was elected governor in 1992. His plan to make English the only official language on the island - replacing Spanish - caused massive protest demonstrations. Finally in 1993 English was made an official language alongside Spanish.

[16] In a new referendum in November 1993 the proposal to maintain a 'Free Associated State' status won by a narrow margin with 48.4 per cent of the vote, against 46.2 per cent for the proposal to transform Puerto Rico into the 51st US state and 4.4

IN FOCUS

ENVIRONMENTAL CHALLENGES
The main environmental problems are soil erosion and occasional droughts which result in lack of water for human consumption.

WOMEN'S RIGHTS
Puerto Rico has its own constitution and local governmental autonomy. Women have been able to vote since 1928. In 2003 women made up 38 per cent of the country's two million labor force, with 70 per cent of them working in the service sector. That year, 10.9 per cent of women were unemployed as compared to 13.2 per cent of men.

In 2003 only 1.6 per cent of women aged 15 to 24 were illiterate. Boys and girls had virtually equal access to education.

There are over 21,000 recorded cases of domestic violence per year (almost 50 incidents per each 10,000 people). Victims of domestic violence are usually women (86 per cent) and particularly those between 20 and 29 years old (44 per cent).

HIV/AIDS is the most common cause of death in women aged 25 to 34.

CHILDREN
In 2000, 10.8 per cent of babies were born underweight and under-five mortality was nine per thousand live births.

A disturbing growth in child poverty is associated with ill-health, negative school-experiences and results, the use and abuse of drugs, teenage pregnancy and other social risk factors. There is concern about the number of child and adolescent murder victims (over 1,000 per year) particularly in the 15 to 19 age group.

INDIGENOUS PEOPLES/ ETHNIC MINORITIES
Traces of Taino physical characteristics can be found in Taino descendants who live in areas of Borinquen. Their written language took the form of petroglyphs which can still be found in some Puerto Rican caves. After 200 years' absence from official records, Tainos reappeared in a military census carried out in 1790. Approximately 2,000 natives are still living on the island of Mona, where they had been relocated by the Spanish after the conquest.

MIGRANTS/REFUGEES
Some 3.4 million Puerto Ricans are living in the US, an increase of 25 per cent over the last 15 years.

Approximately half had been born in Puerto Rico. Many Puerto Ricans are serving as US soldiers at military bases in Iraq and Afghanistan.

Often people from the Dominican Republic and Haiti trying to enter the US are intercepted in Puerto Rican territorial waters and deported.

DEATH PENALTY
Puerto Rico banned capital punishment two years after the execution of a man in 1927 and ratified this prohibition in the 1952 constitution, which confirmed Puerto Rico's self-governing commonwealth status. The question of capital punishment on the island is a matter of constant debate, due to its political relationship with the US. Virtually no local politician or public figure speaks out in favor of the death penalty. The nearly four million people that live in this US territory do not have representatives with voting rights in the US Congress, which passed a law reinstating the death penalty for drug barons in 1984 and broadened its reach to other types of crimes in 1992.

Latest data available in *The State of the World's Children* and *Childinfo* database, UNICEF, 2006.

per cent for independence. Five years later, a further referendum produced virtually the same result.

[17] Tens of thousands of people protested in February 2000 against the resumption of US military exercises on the Puerto Rican island of Vieques, which had been used for this purpose for 50 years. The exercises had been suspended after the accidental death of a civilian in April 1999.

[18] The controversy over US military training in Vieques continued until January 2003, the date set by US President George W Bush for the last military exercises by the US Navy on the island. Most of the 8,000 Vieques residents celebrated the Navy's departure, as did Calderón. However, the area remained under the control of the US Department of the Interior.

[19] Another source of tension between the US and Puerto Rican authorities was partially resolved in July 2003. A murder in Puerto Rico prompted US prosecutors to attempt to apply the death penalty, which had been abolished in 1929 in Puerto Rico, to the Puerto Ricans accused of the crime. The US initiative was strongly rejected by the islanders. Although

the alleged murderers were absolved, US judicial authorities continued their attempt to reinstate capital punishment on the island.

[20] On 2 November 2004 Puerto Ricans voted to elect their Governor, Resident Commissioner in Washington, 27 Senators, 51 Representatives and the mayors of the island's 78 municipalities. There were just over 2.5 million names on the electoral register. A technical draw between PNP's Pedro Roselló and PPD's Aníbal Acevedo provoked controversy and delays in the proclamation of the winner. Finally Acevedo was declared the winner by a margin of 3,566 votes.

[21] In the referendum of July 2005 Puerto Ricans voted in favor of replacing the Senate and the Chamber of Representatives with a single-chamber Parliament.

[22] In September pro-independence leader Filiberto Ojeda Ríos, wanted in the US for the theft of seven million dollars, was shot dead by US FBI agents in Hormigueros, south-east Puerto Rico. Thousands of people took to the streets of San Juan to express their indignation and the

Government announced an internal investigation into the operation.

[23] In April 2006, due to an acute fiscal deficit of $738 million and a dispute with Parliament, which denied approval for a loan to provide bridging finance, the Government announced the suspension of its operations for two months, which would cause the temporary closing of more than 1,500 schools and would leave more than 90,000 public employees without their salary. The measure was in force for two weeks until an agreement between the Government and the Opposition enabled activities to resume on 15 May.

[24] In October 2006, a jury sentenced a man to life in prison for murder instead of giving him the death sentence sought by US federal prosecutors. Puerto Rico abolished capital punishment in 1929 but US prosecutors claimed jurisdiction in the case because the killing took place at a federally run institution. The US decision to seek the death penalty caused a storm of popular criticism in Puerto Rico, as well as meeting opposition from the Governor and Catholic Church leaders. ■

Qatar / Qatar

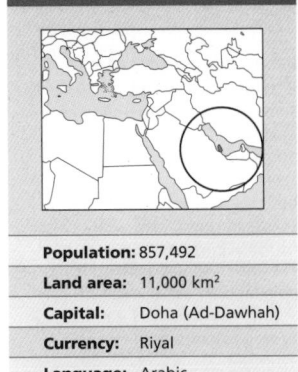

Population: 857,492
Land area: 11,000 km²
Capital: Doha (Ad-Dawhah)
Currency: Riyal
Language: Arabic

Like the neighboring island of Bahrain, ever since ancient times the Qatar peninsula has participated in the Persian Gulf trade between Mesopotamia and India. Islamicized in the 7th century (see Saudi Arabia), at the time of the Caliphate of Baghdad, Qatar had already obtained autonomy which was maintained until 1076 when it was conquered by the Emir of Bahrain. From the 16th century, after a brief period of Portuguese occupation, the country lived in great prosperity due to the development of pearl fishing which attracted immigrants. Settled on the coasts, under the leadership of the al-Thani family, these settlers succeeded in politically uniting the country in the 18th century, though it remained subject to Bahrain's sovereignty. The process of independence, begun in 1815 by Sheikh Muhammad and his son Jassim, culminated in 1868 with the mediation of the English; the al-Thanis agreed to end the war in exchange for guaranteed territorial integrity.

² The Turkish sultans, nominal sovereigns of the entire Arabian peninsula since the 16th century, feared the increasing British penetration in the Gulf. Consequently, they named the reigning Sheikh (Jassim al-Thani) governor of the 'province' of Qatar as a pretext for establishing a small military garrison in Dawhah (Doha). Neither Qatar nor Britain were concerned about this formal affirmation of sovereignty and the garrison remained until World War I without the slightest effect on British influence in the region.

³ In 1930, the price of pearls dropped when the Japanese flooded the market with a cheaper version of cultivated pearls. Consequently Sheikh Abdullah sold all the country's oil prospecting and exploitation rights, and granted a 75-year lease on its territorial waters, for £400,000. The Anglo-Iranian Oil Company discovered oil in 1939 but actual production only began after World War II, attracting other companies

that purchased parts of the original concession. Oil and tariff revenues increased the personal fortune of Sheikh Ahmad ibn Ali al-Thani by £15 million.

⁴ Shortly thereafter, Ahmad was ousted by his own family. They replaced him with his cousin Khalifa, giving him the task of 'removing any elements that are opposed to progress and modernization'.

⁵ Sheikh Khalifa created a Council of Ministers and an Advisory Council to share the responsibilities of his absolute power and promised social justice and stability. Redistributing the oil revenues, he exempted all inhabitants from taxation and provided free education and medical attention. His greatest achievement has perhaps been the subsidizing and promotion of productive activities at the beginning of his mandate.

⁶ To reduce Qatar's dependency on a single product the fishing industry was promoted, industrialization was accelerated and the country took advantage of its strategic position to provide commercial and financial services to the region. In addition, a large part of the country's financial surplus was invested abroad (in Europe and the US). In 1980, it was estimated that income from this 'exportation of capital' would eventually equal all oil revenues. In this way Qatar sought to ensure its future when the oil wells ran dry.

⁷ A state oil company - the Qatar

Petroleum Producing Authority (QPPA) - was set up in 1972. By February 1977, all foreign oil installations had been expropriated.

⁸ Qatar's economic expansion required the large-scale immigration of foreign technical experts and workers - the former were mainly European and American; the latter Iranian,

Pakistani, Indian and Palestinian. About 60 per cent of the economically active population in Qatar are foreigners. To avoid any profound transformation of the local culture, the Government has preferred and promoted immigration from Arab countries.

⁹ Since 1981, together with Bahrain, Kuwait, Oman, the United Arab Emirates and Saudi Arabia, Qatar has participated in the Gulf Cooperation Council (GCC), an organization designed to coordinate the area's policies on political, economic, social, cultural and defense issues.

¹⁰ Following OPEC policy, in 1982 the country cut crude oil production by 25 per cent and, as a consequence, exports decreased and industrial expansion stalled. Nevertheless, Qatar's iron and steel plant in the industrial center of Umm Said - producing 450,000 tons a year at the beginning of the decade - and a liquid gas plant made the Government decide to go ahead with a $6 billion natural gas project for use in its energy and desalinization programs.

¹¹ In April 1986, tensions flared between Bahrain and Qatar over the artificial island of Fasht ad-Dibal. This conflict was resolved through negotiations sponsored by the GCC. French troops were then recruited and under Qatar's uniform they participated in the defense of the Emirate. In November 1987, the Government renewed diplomatic relations with Egypt, which had been cut off when Egypt signed the Camp David accords.

PROFILE

ENVIRONMENT

The country consists of the Qatar Peninsula, on the eastern coast of the Arabian Peninsula in the Persian Gulf. The land is flat and the climate is hot and dry. Farming is possible only along the coastal strip. The country's main resource is its huge oil wealth on the western coast.

SOCIETY

Peoples: Qatari Arabs make up 20 per cent of the population. There are a further 25 per cent of Arabs as a result of Palestinian, Egyptian and Yemeni immigration. The remaining 55 per cent are immigrants, mostly from Pakistan, India and Iran. **Religions:** Muslim (official and predominant). The majority are Sunni, with mainly Shi'a Iranian immigrants. There are also Christian and Hindu minorities. **Languages:** Arabic (official and predominant). Urdu is spoken by Pakistani immigrants, and Farsi by Iranians. English is the business language. **Main Political Parties:** There are no organized political parties.

THE STATE

Official Name: Dawlat Qatar. **Capital:** Doha (Ad-Dawhah) 286,000 people (2003). **Other Cities:** Rayyan 183,000 people; Wakrah 22,900; Umm Said (Musay'id) 18,100 (2000). **Government:** Hamad ibn Khalifa al-Thani, Emir and Head of State since June 1995. Sheikh Abdullah ibn Khalifah al-Thani, Prime Minister since October 1996. Legislative Power: a Consultative Council with 45 member, 30 of them elected by universal suffrage for four-year terms and the rest appointed by the Emir. **National Holiday:** 3 September, Independence Day (1971). **Armed Forces:** 12,400 (2002).

Life expectancy
74 years
2005-2010

Literacy
89% total adult rate
2000-2004

[12] In March 1991, after the Iraqi invasion of Kuwait, the GCC suspended all economic aid to Jordan and the Palestine Liberation Organization (PLO). That month, the Foreign Ministers of Egypt, Syria, and the six Arab countries of the GCC signed in Saudi Arabia an agreement with the US that projected a common military strategy between the US and the Arab countries in the anti-Iraqi coalition, mechanisms to avoid arms proliferation, acceptance of a peace treaty by Israel and a new economic program for the development of the region.

[13] In September 1991 Qatar inaugurated North Field, an off-shore deposit of natural gas, and so became a major producer of this fuel, with an estimated reserve of ten billion cubic meters (five per cent of the world's total).

[14] By the end of 1991 Qatar and Bahrain became involved in another territorial dispute, this time over Hawar Island, and more especially underground rights to Dibval and Qitat, both potentially rich in oil.

[15] In June 1995, heir to the throne Hamad ibn Khalifa al-Thani overthrew his father to become Emir of Qatar. He promised to step up efforts to resolve the territorial disputes with Saudi Arabia and Bahrain.

[16] The al-Jazeera television network, founded at the personal initiative of the Emir in 1996, was an immediate success in Middle Eastern countries, with its pro-Arab and pro-Muslim stance and the broadcasting of information ignored by the major Western TV channels and the gagged state TV stations.

[17] In 1997, Qatar froze relations with Israel, applying the decision of Arab League of reactivating boycott against that country. However, it offered its capital Doha as the venue for the Economic Conference of the Middle East and North Africa to be held in November, as part of the peace process. In October, Sheikh Abdullah ibn Khalifa al-Thani was appointed Prime Minister, a post previously occupied by Emir Hamad.

[18] In November 1998 a plan for constitutional reform was unveiled, with the goal of creating a parliament elected by direct vote. At this time only men aged over 18 years old were eligible to vote. In the March 1999 local elections, women were allowed the vote, but none of the six female candidates was elected.

[19] The Emir was critical of US policy on Iraq and mediated in a regional dispute caused by United Arab Emirates' disapproval of Saudi

IN FOCUS

ENVIRONMENTAL CHALLENGES
Qatar suffered the polluting effects of the oil fires which occurred during the first Gulf War. Drinking water resources are scarce so that dependence on large scale desalinization units is increasing.

WOMEN'S RIGHTS
Women have been able to vote only since 2003 and none have been elected so far. In 2003 they held 8 per cent of ministerial or equivalent positions and made up just 18 per cent of the total workforce. There is an appearance of equal pay but in general women do not receive the same extra benefits such as travel and accommodation expenses. Even though the law does not impose restrictions on women traveling abroad alone it is deemed socially inappropriate that they do so without a male companion. No cases of rape and/or domestic violence were officially recorded. Qur'anic law punishes rapists with the death penalty, unless the rape is within marriage when it is not seen as rape.-Some employers have ill-treated their female domestic employees, most of whom come from Southern Asia or

the Philippines. Foreign countries' embassies have been refuges for female employees, victims of their bosses' abuse. In several cases, charges were not laid for fear of losing a job. The legal system allows men to commit 'honor murders' if they feel offended by the immodest behavior of their partners.

CHILDREN
In recent years Gulf states have made major progress with regard to the situation of women and children. Economic prosperity during the 1970's and the early 1980s was used to establish socio-economic infrastructures, which have been the basis for the development, distribution and availability of basic services such as education, health services, drinking water, sanitation and electricity-generating power plants. Education is compulsory and free for every child from 6 to 18 years*.

Youngsters between 15 and 18 are allowed to work under family supervision, but there is no data available on the exact number of children who work.

Forced labor is forbidden, but the plight of some children forced to work as camel-jockeys or of some girls in domestic service is well known.-

INDIGENOUS PEOPLES/ ETHNIC MINORITIES
Non-citizens are discriminated against in terms of their jobs, education, housing and health services. While Qataris get free basic services, non-citizens have to pay for electricity, water, health services and education. They are also not allowed to own land. The major groups are Indian, Pakistani and Iranian, as well as Arabs from other countries.

MIGRANTS/REFUGEES
Qatar has not signed the 1951 UN Convention or the 1967 Protocol on refugees. The Government does not provide help or protection to asylum-seekers from neighboring countries and does not co-operate with UNHCR. In exceptional circumstances it has granted temporary asylum as in the cases of an Algerian political activist and members of the former Iraqi regime.-

DEATH PENALTY
Death penalty applies according to the interpretation of the Sharia (Muslim law) even for ordinary crimes.

Arabia and other GCC countries improving their relations with Iran while the UAE had ongoing territorial disputes with that country.

[20] The rebels responsible for a 1996 coup attempt against the Emir were sentenced to life imprisonment in February 2000. Amongst them was the Emir's cousin, Sheikh Hamad ibn Jassem al-Thani.

[21] In March 2001, Qatar solved its border conflicts with Bahrain and Saudi Arabia. The International Court of Justice at The Hague found in favor of Bahrain's ownership of Hawar islands, while it acknowledged the rights of Qatar over the city of Zubarah. Saudi Arabia and Qatar signed a demarcation agreement for their 60 km long common sea and land border.

[22] In May 2001 the Appeal Court issued death sentences to Hamad ibn Jassem al-Thani and 18 others accused of taking part in the attempted coup of 1995. At that time, Qatar had not applied the death penalty for a decade.

[23] During the 2002 US bombing on Afghanistan, while the US established its largest operation center in the Gulf in Qatar, Al-Jazeera was the only TV channel

authorized by the (Afghan) Taliban regime to broadcast in the areas under its control and to spread propaganda messages and threats by terrorist network al-Qaeda leader, Osama bin Laden.

[24] Even though the summit of the Islamic Organization Conference held in Doha on 5 March 2003 condemned the imminent strike on Iraq, the Emir allowed the use of the US military base in Qatar as a planning center for the attack. Meanwhile, Al-Jazeera reported the invasion and broadcasted tapes by Iraq's leader Saddam Hussein - whose whereabouts were unknown - calling for resistance.

[25] The new constitution was approved on 29 April 2003, by 96 per cent of the votes. The reforms, which were less than expected, did not create a parliament, but a new *Majlis ash-Shura* (Consultative Assembly). Women obtained the right to vote and to hold public posts, and information and religious freedoms were obtained, but political parties remained banned.

[26] In February 2004 Chechen ex-president and separatist leader, Zelimkhan Yanderbiyev, died in an explosion in Doha, where he had lived for the previous three

years. Qatar accused Russian secret services of the murder (Russia had requested Yanderbiyev's extradition) but Moscow denied the accusation.

[27] In March 2004 a car bomb suicide attack in Doha left more than 12 wounded and a British citizen dead. The Ministry of the Interior identified the attacker as Abdullah Ahmad Ali, an Egyptian, but did not identify the organization responsible for the attack.

[28] In June 2005 the new Constitution came into force.

[29] In November Doha and Washington announced a joint project to build the world's biggest liquid gas refinery in the Emirate, that would involve an investment of $14 billion. Most of the gas would be exported to the US.

[30] In April 2006 it was announced that the first elections for the legislature in the history of the Emirate would be held in early 2007, to elect members of the Consultative Assembly. While making this announcement, Foreign Affairs Minister Sheikh Hamad Bin Jassem Bin Jabr Al-Thani took the opportunity to criticize Washington for trying to 'impose' democracy on the Middle East. ∎

Réunion / Réunion

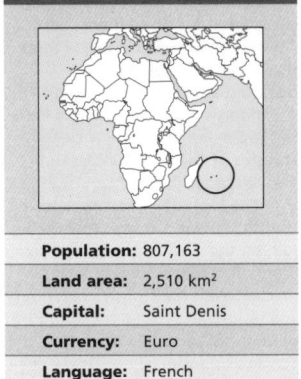

Population:	807,163
Land area:	2,510 km²
Capital:	Saint Denis
Currency:	Euro
Language:	French

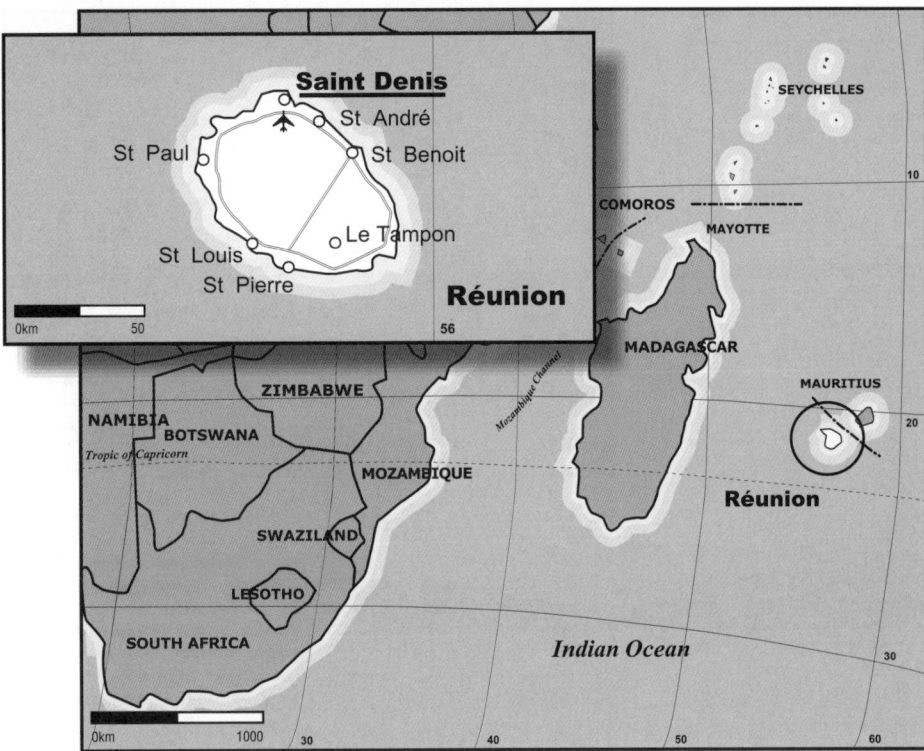

The island of Réunion was uninhabited until the beginning of the 17th century when Arab explorers arrived, calling it Diva Margabin. Its subsequent name changes reflect the colonial power struggle: the Portuguese renamed it Ilha Santa Apolonia in 1513; French settlers called it Bourbon in 1600. After the French Revolution it was given its current name of Réunion.

2   The French colonial regime replaced subsistence farming with the export commodities of sugarcane and coffee, cultivated by slave labor. The island's fish stocks were depleted by the over-exploitation of several species. From the 19th century the island was an important French military base in the Indian Ocean, together with other dependencies such as Djibouti and Comoros Islands.

3   In March 1946 Réunion was designated a French Overseas Department. This was backed by the Left, both in Réunion and in France. The local middle class and the French colonists supported the integration of Réunion into French territory.

4   Later, conservative political changes in France led to a realignment of political forces on the island. While sectors linked to colonial interests defended the island's Department status, the Réunion Communist Party (PCR) changed its stance, and in 1959 campaigned for partial autonomy. For ten years, the PCR was the sole supporter of gradual independence.

5   In 1978, the UN supported full independence for Réunion. The dispute between the island and France became an international issue. That year, in a meeting with other anti-imperialist and anti-colonialist organizations in the Indian Ocean area, a Permanent Liaison Committee was formed as part of their common struggle against foreign domination. In June 1979, during elections for the European Parliament, Paul Vergés - PCR leader - raised the issue of

dual colonialism, claiming that Réunion was not only dominated by the French, but forced to serve the interests of the entire European Community.

6   A demographic explosion on the island, whose population grew by 20 per cent in a decade, caused a production crisis and a rise in unemployment. Increasing social unrest led to frequent rioting. In 1991, eight people died in clashes with the police in protest over the closure of the underground television station Tele Free-DOM, run by Camille Sudre. Sudre, a French national settled in Réunion, won the regional elections in 1992. Backed by the PCR, he occupied one of the two main executive posts on the island.

7   However, France's supreme court annulled the elections for alleged irregularities and banned Sudre from standing again. The 1993 regional elections were won by the Free-DOM party, led by Marguerite Sudre, Camille's wife.

8   In 1996, unemployment soared to an all-time high of 40 per cent, primarily affecting young people, who make up the largest sector of the population.

9   In 2001 Gonthier Friederici took office as prefect. In 2002 the island was devastated by cyclone Dina, with major damage to agriculture and housing.

10   Although sugar exports continue to be the island's main economic resource, the service sector has grown considerably in recent years. Nevertheless, the

economy is still largely dependent on France.

11   In the early 21st century, the gap between the rich and poor was a constant source of social tensions. While the white and Indian communities enjoyed economic conditions close to European standards, those of African descent faced poverty and unemployment levels typical of the poorest countries in Africa.

In 2006, more than 160,000 people were infected with the disease chikungunya, a viral fever that attacks the joints: its name comes from a Swahili word meaning 'that which bends up', in reference to the stooped posture of those affected. The disease, from East Africa, is spread by the same mosquito as dengue. While it is not considered fatal, there is no known cure for it. ∎

PROFILE

ENVIRONMENT
Located in the Indian Ocean, 700 km east of Madagascar, Réunion is a volcanic, mountainous island. It has a tropical climate, heavy rainfall and numerous rivers. These conditions favor the growth of sugarcane, the main economic activity. Tourism also contributes to the economy.

SOCIETY
Peoples: Mostly of African descent (63.5 per cent); Indian 28.3 per cent; Europeans 2.2 per cent; Chinese 2.2 per cent.
Religion: Mainly Catholic (94 per cent). Muslim, Hindu and Buddhist minorities.
Languages: French (official) and Creole.
Main Political Parties: Free-DOM Movement; Communist Party of Réunion; Union for French Democracy; Socialist Party.

THE STATE
Official Name: Département d'Outre-Mer de la Réunion.
Administrative Divisions: 5 Arrondisements.
Capital: Saint Denis 178,000 people (2003).
Other Cities: St Paul 91,600 people; St Pierre 72,000; Le Tampon 63,000; St André 44,800 (2000).
Government: Laurent Cayrel, Prefect appointed by the French Government in July 2005 (representing French President Jacques Chirac). There are two local councils: the 47-member General Council, and the 45-member Regional Council. The island has five representatives and three senators in the French parliament.
National Holiday: 20 December, Abolition of slavery (1848).
Armed Forces: 4,000 French troops (1995).

12

Romania / România

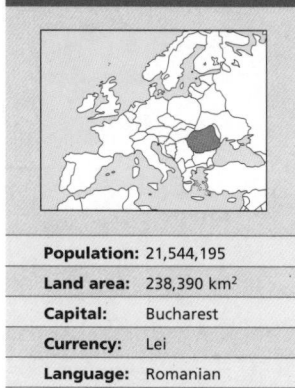

Population:	21,544,195
Land area:	238,390 km²
Capital:	Bucharest
Currency:	Lei
Language:	Romanian

The origins of Romanians date back to the Dacians or Getae (sold as slaves in Athens in the 4th century BC), after whom the Roman province of Dacia was named. The province was located in the Carpathian mountains and in Transylvania, in the northeastern territory of present-day Romania.

[2] In the first century BC, the Dacians established a powerful kingdom, and in alliance with other peoples they fiercely resisted the Romans, until Rome finally triumphed in 106 AD, annihilating them or expelling them to the north.

[3] The province was first a consulate and then subdivided into Upper Dacia and Lower Dacia. Emperor Marcus Aurelius withdrew from the region by the year 270, but the influence of Roman culture and language still remains.

[4] Between the 3rd and the 12th centuries, the region underwent successive invasions by Germanic tribes, Slavs, Avars and others. Bulgarian rule, which lasted over 200 years, established a certain social organization and introduced Greek Orthodox Christianity. In the late 9th century, the Magyars expelled the Bulgarians.

[5] Hungary conquered Transylvania during the 11th century, but the Tatar-Mongol invasion of 1241 wiped out all trace of the first inhabitants of this region. The Vlachs from Transylvania reappeared in the 13th century to the south of the Carpathian Mountains, in Walachia and Moldova.

[6] The state of Walachia was created in 1290, when a prince from Fagaras and his followers crossed the mountains and established themselves in Cimpulung, moving on later to Curtea de Arges. The emigration southwards is attributed to the arrival of Germanic peoples and to the consolidation of Hungarian feudal power in Transylvania.

[7] The new principality fought for its independence from Hungary until the rule of Mircea the Elder (1386-1418), when the Ottoman Empire became a greater threat.

[8] After defeating the Serbs in Kosovo in 1389, the Ottoman Empire began closing in on Walachia, with intensifying pressure after the fall of Bulgaria in 1393. Walachia became a vassal state of Sultan Mehmed I in 1417, though Prince Mircea maintained his claim to the throne and the Christian religion remained intact.

[9] King Mircea's death in 1418 was followed by a rapid succession of princes, until the Turks appointed a Romanian prince of their choice to the throne. After the Battle of Mohacs in 1526, the Turks ruled Walachia through an imperial governor.

[10] In 1594, the Turkish inhabitants of Walachia were massacred by Prince Michael in alliance with Moldova, and with Transylvania's support he went on to invade Turkish territory. Faced with the collapse of his counter-offensive, the Sultan had no choice but to recognize the sovereignty of Walachia, which subsequently became linked to Hungary.

[11] In 1600, Michael conquered Moldova, proclaiming himself regent. Rudolf II recognized the claim, although he later tried to take over Transylvania and Moldova.

[12] In the 17th and 18th centuries, Walachia and Moldova fell under Turkish rule again and were administered by the Greeks for certain periods, until Russia occupied the region in 1769. Austria forced Russia to return the principalities to the Sultan in 1774.

[13] In 1806, Russia invaded the region, but under the Treaty of Bucharest in 1812, it only retained the southeastern part of Moldova-Bessarabia.

[14] Another war broke out between Russia and Turkey in 1828. The following year, the Treaty of Adrianopolis maintained the principalities as tributaries of the Sultan, but under Russian occupation. Russian troops remained in the region, and the princes began to be appointed for life.

[15] The local nobility drew up a constitution known as the 'Règlement organique', which was passed in Walachia in 1831 and in Moldova in 1832. After the Sultan's approval in 1834 Russia withdrew.

[16] During the European revolutions of 1848, nationalist sentiment in Moldova and Walachia was stimulated by peasant rebellions. These reached a climax in May with the protests at Blaj, which were put down by Turkish and Russian troops, restoring the 'Règlement organique'.

[17] Russian troops extended their occupation for three more years. During the Crimean War the three Romanian principalities were occupied alternately by Russian and Austrian troops. The Treaty of Paris in 1856 maintained the ancient statutes of the principalities until it was revised in 1857.

[18] The local delegates proposed that the provinces be autonomous, joining together under the name 'Romania'; that a foreign king be elected, with the right to hereditary succession; and that the country be neutral. In August 1858, despite the Sultan's opposition, the Treaty of Paris created a commission to carry out the unification process.

[19] In 1859, the principalities elected a single prince, Alexandru Ion Cuza, who was recognized by the major powers and by the Sultan in 1861. The Constitution of 1863 established a bicameral legislative body, granting property holders greater electoral power.

[20] When the war between Russia and Turkey resumed in 1877, Russia rejected the alliance with Romania and threatened to invade its territory. Romania authorized the transit of Russian troops through its territory in April and declared war on Turkey in May. Romania contributed to the Russian victory, but it was not admitted to the subsequent peace talks.

[21] The 1878 Treaty of Berlin respected Romania's independence, but failed to return Bessarabia to Romania, instead giving it Dobrudja, a small region on the Danube delta. The Government reinforced its loyalty to the Romanian crown.

[22] When the Balkan War broke out in 1912, tension from territorial disputes in past wars persisted between Romania and its neighbors. After the first few battles, Bucharest demanded a ratification of its borders in Dobrudja. The St Petersburg Conference of 1913 gave Romania Silistra, much to Bulgaria's displeasure.

[23] Romania took advantage of the second Balkan War to shore up its position. The Treaty of Bucharest gave Romania the southern part of Dobrudja. At the beginning of World War I, Romania wavered between taking Bessarabia or Transylvania, finally opting for the latter.

[24] In 1916, Romania allied itself with Britain, France, Russia and Italy, declaring war on Austria and Hungary. After the occupation of Bucharest, King Ferdinand and the Romanian Government and army took refuge in Moldova, under the Czar's protection. The defeat of the Central Powers in 1918 made it possible for Romania to incorporate Transylvania, Bessarabia, Bucovina

LAND USE

2003/2005

IRRIGATED AREA: 31.2% of arable land

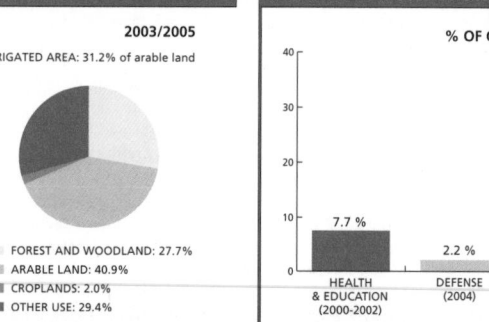

- FOREST AND WOODLAND: 27.7%
- ARABLE LAND: 40.9%
- CROPLANDS: 2.0%
- OTHER USE: 29.4%

PUBLIC EXPENDITURE

% OF GDP

7.7 %
HEALTH & EDUCATION (2000-2002)

2.2 %
DEFENSE (2004)

Life expectancy
72 years
2005-2010

GNI per capita
$2,960
2004

Literacy
97% total adult rate
2000-2004

HIV prevalence rate
<0.1% of population 15-49 years old
2003

and Banat. That year, the King made voting obligatory for men over the age of 21, and introduced the secret ballot.

25   Social upheaval, and the landowners' fear of having their lands expropriated, led General Averescu (the hero of two wars) to take harsh measures. The general strike of 1920 was put down and the Communist Party was declared illegal in 1924.

26   In the 1928 election, the National Peasant Party (NPP) obtained 349 out of 387 seats. The Government abolished martial law and press censorship, also decentralizing public administration, a measure supported by the ethnic minorities. It also authorized the sale of land, and foreign investment in the country.

27   The council of regency which had been formed upon Ferdinand's death, was dissolved when King Carol assumed the throne in 1930, a succession agreed to by the major political parties. The King took advantage of the emergence of the Iron Guard, a fascist Moldovan group, to weaken the traditional parties.

28   In 1938, after a fraudulent plebiscite, Carol passed a new corporative constitution. Seeking closer ties with Germany, he met Hitler in November and upon his return he had 13 officials of the Iron Guard assassinated, along with its leader. Carol founded his own party, the National Renaissance Front (NRF).

29   Carol affirmed that he was being forced by Hitler and obtained French and British assurances of the country's territorial integrity. By the time of Poland's invasion in 1939, Romania had not only renounced the mutual defense treaty which it had signed with Warsaw, but had also detained Polish authorities as they fled across Romanian territory.

30   Between June and September 1940, Romania was forced to turn Bessarabia and Bucovina over to the USSR, Transylvania to Hungary, and Dobrudja to Bulgaria. Because of his disastrous foreign policy, Carol had to abdicate in September, leaving his son Michael on the throne and turning over the Government of the country to General Antonescu. Romania was occupied by German troops, and was proclaimed a 'national legionary state'.

31   Romanian troops cooperated with the abortive German offensive against the USSR, but when the Red Army mounted the counter-offensive, Bucharest abruptly changed sides. In March 1944, the Peasant, Liberal, Social Democratic and Communist Parties created the National Bloc and in August, King Michael ousted Antonescu and declared war on Germany.

32   In September, with most of its territory occupied since late August by Soviet troops, Romania signed an armistice with the Allies. At the Potsdam Conference, the Allies decided to resume relations with Romania, provided its government was 'recognized and democratic'. The USSR granted it immediate recognition while the US and Britain adopted a 'wait-and-see' attitude. The Government arrested, prosecuted and sentenced leaders and members of the Social Democratic and National Peasant parties. In 1947, King Michael was forced to abdicate.

33   In 1948, the Communists and some Social Democrats formed the Romanian Workers' Party (RWP) which joined the Ploughers' Front and the Hungarian People's Union to form the People's Democratic Front (PDF). In the March election that year, the PDF won 405 of the 414 seats of the National Assembly. In April the People's Republic of Romania was proclaimed, and a socialist constitution was adopted. In June, the Government adopted centralized economic planning.

34   Between 1948 and 1949, Bucharest signed friendship and cooperation treaties with the European socialist bloc, and joined the Council for Mutual Economic Assistance. In 1955, Romania joined the Warsaw Pact, but in 1963 it began to drift away from the Soviet fold.

35   In 1951 the First Five Year Plan was started, aimed at socialist industrialization of steel, coal, and oil. In 1952, the new regime started to consolidate under President Groza and Prime Minister Gheorghiu-Dej, head of state from 1961.

36   In 1962 the land collectivization policy was ended, and trade links with the US, France, and Germany were established.

37   In 1965, when Gheorghiu-Dej died, Nicolae Ceausescu was elected First Secretary of the RWP, which changed its name to the Romanian Communist Party (RCP) in June. The Constitution was reformed, and the National Assembly changed the name of the country to the Socialist Republic of Romania. In 1967, Ceausescu was elected President of the State Council.

38   Romania established diplomatic relations with West Germany and, unlike other members of the Warsaw Pact, it did not break off relations with Israel in 1967, nor did it participate in the Soviet invasion of Czechoslovakia in 1968.

39   In the 1970s and early 1980s, Ceausescu was re-elected several times as president of the country.

40   From 1987, difficult living and labor conditions triggered marches and strikes. These were put down by the security forces. In 1988 and 1989, several government scandals broke out; various cabinet ministers and government authorities were subsequently tried and dismissed.

41   With the beginning of *perestroika* in the USSR and the crisis of the European socialist bloc, Ceaucescu became progressively more discredited in world opinion. Towards the end of 1989, confrontations between civilians and the army in Timisoara left many dead or injured; the international press spoke of hundreds of deaths, and the news had strong repercussions within Romania. The Government declared a state of emergency, but a faction within the regime carried out a coup with massive popular support. Accused of 'genocide, corruption and destruction of the economy', Ceausescu and his wife were secretly executed by army soldiers. The National Salvation Front (NSF) assumed control of the government.

42   NSF leaders were called into question and resistance towards the new government increased, leading to violent confrontations in the streets. In the May 1990 elections, the NSF obtained 85 per cent of the vote, but international observers confirmed fraud allegations.

43   In September 1991 Prime Minister Peter Roman and his entire cabinet were forced to resign, under pressure from thousands of miners who marched towards Bucharest to protest against the Government's privatization policy. Demonstrators were heavily repressed during the three-day demonstration and it is estimated that there were at least three deaths and over a hundred people injured.

44   In 1991, 77 per cent of the electorate approved the new Constitution, which turned Romania into a multi-party presidential democracy. In Transylvania, however, it received scant support.

45   In the early 1990s, Western countries and international financial organizations continued

PROFILE

ENVIRONMENT

The country is crossed from north to center by the Carpathian Mountains, the westernmost peaks of which are known as the Transylvanian Alps. The Transylvanian plateau is contained within the arc formed by the Carpathian Mountains. The Moldavian plains extend to the east, while the Walachian plains stretch to the south, crossed by the Danube, which flows into a large delta on the Black Sea. The mountain forests supply raw material for a well-developed timber industry. With its abundant mineral resources (oil, natural gas, coal, iron ore and bauxite), Romania has begun extensive industrial development. Its economy still depends to a great extent on the export of raw materials and agricultural products. It is one of Europe's largest oil producers.

SOCIETY

Peoples: Romanian 89.5 per cent, Hungarian 6.6 per cent, German 0.3 per cent, Ukrainian 0.3 per cent, Turkish, Greek and Croatian. These data do not include the Roma people, who represent between five and ten per cent of the population, but who are usually considered as Romanian in the census.
Religions: Mainly Romanian Orthodox (86.8 per cent). There are Catholic (five per cent) and Protestant (3.5 per cent) minorities.

Languages: Romanian (official language, spoken by the majority); ethnic minorities often speak their own languages, particularly Hungarian and Romany.
Main Political Parties: National Union PSD+PUR (which includes: Social Democratic Party and Humanist Party of Romania); Justice and Truth Alliance (which includes: National Liberal Party and Democratic Party); Greater Romania Party.
Main Social Organizations: General Union of Trade Unions.

THE STATE

Official Name: România.
Administrative Divisions: 41 Districts and the Municipality of Bucharest.
Capital: Bucharest 1,853,000 people (2003).
Other Cities: Timisoara 338,900 people; Constanza 339,300; Iasi 353,600 (2000).
Government: Traian Basescu, President and Head of State since December 2004. Calin Popescu-Tariceanu, Prime Minister and Head of Government since December 2004. Legislature, bicameral: Senate, 137 members; Deputies, 314 members, 18 representing ethnic minorities.
National Holiday: 1 December, Unification Day (1918).
Armed Forces: 228,400 (1996). Other: 43,000 Border Guard, Gendarmes, Construction Troops.

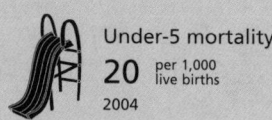
Under-5 mortality
20 per 1,000 live births
2004

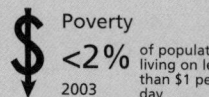
Poverty
<2% of population living on less than $1 per day
2003

Debt service
17.2% exports of goods and services
2004

Maternal mortality
49 per 100,000 live births
2000

to voice their discontent with the Government for its alleged sluggishness in implementing economic reforms. Thirty per cent of the land still belonged to the State in 1994. After two years of deliberation, Bucharest passed a law to privatize state enterprises in June 1995.

[46] After years of fruitless negotiations, Bucharest and Budapest signed a treaty in September 1996 regarding the 1.6 million Hungarians living in Romania. Hungary had to agree to accept Romania's commitment to 'guarantee the rights of the minority', and give up its demand of 'autonomy' for Transylvanian Hungarians. Although ultranationalist tendencies diminished with the defeat of Iliescu in the November elections (Hungary installed a consulate in Cluj, the capital of Transylvania) problems arose again when the Romanian senate voted against a university education project for the minorities.

[47] The new government of Emil Constantinescu, the President elected in November 1996, announced it would attack corruption and organized crime. Prime Minister Victor Ciorbea implemented an economic and structural adjustment program based on IMF prescriptions: balanced state finance, increased privatizations and decentralization of the administration.

[48] After disagreements between members of the government coalition and faced with growing popular discontent, Ciorbea resigned in March 1998. He was replaced by Radu Vasile, who proposed accelerating the privatization of state enterprises, wiping out corruption and reducing bureaucracy.

[49] The new government coalition failed to achieve economic or political stability. Minister of Finance Daniel Daianu resigned.

[50] In December 1999, a mining strike and internal disagreements in the coalition caused the Cabinet to collapse. Vasile was forced to resign on 14 December. His place was temporarily filled by Alexandru Athanasiu, and later by Mugur Isarescu.

[51] In January 2000, tens of thousands of liters of cyanide spilled into the Somes River, a tributary of the Tisza and the Danube, killing flora and fauna along hundreds of kilometers of the rivers. The accident occurred at the Aurul gold mine, located near Baia Maresobre. Drinking water supplies in Hungary, Ukraine, Yugoslavia and Bulgaria were affected by the contamination.

[52] Promising to speed up reforms that would help the country enter the EU, Iliescu easily won the

IN FOCUS

ENVIRONMENTAL CHALLENGES
Soil erosion and degradation, water and air pollution by solid or gaseous industrial wastes and the contamination of the Danube delta wetlands are the most significant environmental challenges.

Copsa Mica, in the center of Romania, is considered to be one of the areas with the highest levels of industrial pollution in Europe.

WOMEN'S RIGHTS
Women have been able to vote and run for office since 1946. Female suffrage, albeit with some restrictions, had already been introduced in 1929. In 2004, 11.2 per cent of Lower House seats and 9.5 per cent of Upper House seats were held by women. Romania is a country of origin and transit for girls and women who are sexually exploited in Turkey, Italy, Greece and other Balkan countries. This phenomenon, already in existence during the 1990s, has been steadily increasing since the year 2000. The maternal mortality rate stood at 49 per 100,000 live births in 2004.

CHILDREN
Although the HIV/AIDS incidence is low, the pandemic is concentrated mostly on girls and boys infected between 1986 and 1991. An increase in the number of cases was expected from 2003/2004, as minors start having sex at an early age and have little knowledge on how the disease is transmitted. The number of children living in orphanages fell in 2005, while there was a rise in the number of Roma children enrolled in primary school. In late 2005 Government policies addressing domestic violence, abuse and trafficking of children started to show results. The number of children abandoned by their parents remained high, as well as the number of deaths due to unsafe abortion practices.

INDIGENOUS PEOPLES/ ETHNIC MINORITIES
Ethnic minorities are recognized by law. There are 18 political organizations that represent different ethnic groups. Hungarians (1,434,377 according to the 2002 census) have parliamentary representation, while Roma/Gypsies do not have as much due to internal troubles and the low level of voting. The Roma are territorially dispersed and have relatively little organization. Originally they were brought to the country as slaves, and they have to face discrimination even today. The National Committee against Discrimination fined two companies that did not allow Roma to enter, and some schools were regarded as discriminatory.

MIGRANTS/REFUGEES
Since July 2005, 439 Uzbeks at risk in Kyrgyzstan were granted temporary asylum in Romania. Of them, 11 were deported to Kyrgyzstan in September 2005. In May 2006, another 200 were relocated to other countries and the remaining 228 were deported. Romania's location, on the EU's eastern border, puts it on the route of migration flows. Organizations such as UNHCR claim that this situation makes it necessary to adapt refugee laws to European standards.

DEATH PENALTY
The last execution was carried out in 1989, when the death penalty was abolished for all crimes.

*Latest data available in The State of the World's Children and Childinfo database, UNICEF, 2006.

December presidential elections, beating the ultra-nationalistic and xenophobic Corneliu Vadim Tudor, of the Greater Romania Party. Adrian Nastase, also a Social Democrat, was appointed Prime Minister.

[53] In January 2001, nearly 250,000 state buildings and properties, which had been nationalized during the Communist regime, were included in a property restoration list approved by Parliament.

[54] A debate on the limits on respect for cultural traditions and human rights was set up in October 2003 by the forcible marriage of the under-14-year-old daughter of a Roma Gypsy King, which was annulled by the Government four days afterwards, as a result of the pressures exerted by the EU.

[55] In 2003, while Transparency International ranked Romania among the three most corrupt countries in Europe, where citizens complained of having to bribe doctors and nurses at hospitals so that their relatives could be operated on, three ministers resigned. The Health Minister was expelled from the university where he taught for plagiarizing medical textbooks; the Minister for European Integration was accused of improper usage of EU funds by her family and a third minister resigned because several of his

aides had accepted bribes.

[56] That year, several railway and mine workers' strikes protested against the laying off of 20,000 employees.

[57] Romania, along with Bulgaria, Slovakia, Slovenia, Lithuania, Latvia and Estonia joined the North Atlantic Treaty Organization (NATO) in March 2004. Romania's strategic location and its navy and air bases on the Black Sea, as well as its support for the US in the war in Iraq, made the country an attractive candidate for NATO membership.

[58] In October, Iliescu admitted that Romania had been an accomplice of the Nazis in the holocaust. Hundreds of thousands of Romanian Jews, Roma and other peoples were murdered in Germany during World War II.

[59] In the second electoral round in December 2004, Traian Basescu, the opposition leader and former mayor of Bucharest, was elected President with 51.2 per cent of the vote. Adrian Nastase was second, with 48.8 per cent. Basescu would carry out numerous reforms leading to the country's goal of joining the EU in 2007.

[60] In April 2005 Romania signed the treaty to join the EU. Opinion polls showed that most of the population believed the move

would bring more benefits than disadvantages. Among the latter are a sharp increase in the tariffs of certain services, the elimination of state subsidies for non profitable sectors and higher unemployment.

[61] Prime Minister Calin Popescu-Tariceanu announced his resignation in June 2005, but then changed his decision and sought a vote of confidence, which he obtained later that month.

[62] In late 2005, Romanian Foreign minister Mihai Razvan Ungureanu and US Secretary of State Condoleezza Rice signed in Bucharest an agreement to install US military bases in Romania.

[63] In May 2006, the European Commission conditioned Romanian entry into the EU on 1 January 2007 on more reforms, especially in the judicial system and the fight against corruption. The conditions imposed on the new EU entrants, Romania and Bulgaria, were stricter than for all previous new members.

[64] Sixty years after its expropriation by the Communist administration, one of the country's most popular tourist sites, the Bran Castle - alleged home of fabled Count Dracula - was returned by the Government to its former owner in June 2006 - on condition that the museum remained open to the public at least until 2009. ■

Russia / Rossiya

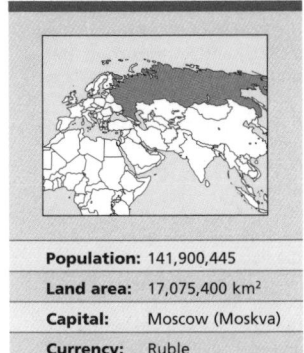

Population:	141,900,445
Land area:	17,075,400 km²
Capital:	Moscow (Moskva)
Currency:	Ruble
Language:	Russian

The southern part of present-day Russia was inhabited from ancient times by Sarmatians and Scythians. The Slavs (Indo-Europeans) invaded the northern regions during the first centuries of the Roman Empire. In the 3rd century, the Goths conquered the region situated between the Baltic and Black seas but were later expelled by the Huns in the 4th century. Russia was then invaded by different nomadic peoples from the north who later brought down the Roman Empire.

² In the early 9th century, the Varangians (Vikings from Sweden) established small kingdoms in the region of Lake Ladoga, constituting a commercial and warrior aristocracy that subjugated the local population. The territories under Varangian control were unified into the kingdom of Novgorod under the leadership of Norman chief Rurik.

³ In 882, Oleg 'the Wise' united Novgorod (in the north) and Kiev (in the south) and created the first Russian State, known as the 'Ancient Rus' with Kiev as its capital. In the 9th century, the route between the Baltic and Black seas, known as 'the route of the Varangians and the Greeks', became important for European trade.

⁴ In 944 after a failed incursion into Constantinople, Igor (913-945) signed a commercial treaty with Byzanz and opened the principality to Christian influence.

⁵ Vladimir (980-1015) consolidated the judicial and territorial organization of the Russian State. In 988, he was converted to orthodox Christianity and made it the official religion.

⁶ Svyatopolk, Vladimir's successor, killed three of his brothers in order to consolidate his own power, but Yaroslav, the fourth brother, ousted Svyatopolk and assumed power in Kiev. After his death in 1054, Russia was divided into a series of principalities that were dominated by conflicts and rivalries.

⁷ The 12th and 13th centuries were characterized by a political, economic and cultural decline. Except for Novgorod and Pakov, all the principalities were devastated by Teutons and Mongols. In 1242, Prince Alexander of Novgorod defeated the Teutons in the famous 'battle on ice' on Lake Chudskoye, near the Neva River, winning the title Prince of Nevsky as a result.

⁸ In 1245, Mongol (Tatar) rule over Russia was consolidated. In 1380, Muscovite Prince Dmitry defeated the Mongols in the battle of Kulikov, near the Don river, marking the beginning of the liberation process.

⁹ In 1439, the Russians abandoned the Greek Orthodox Church and established their own church (Russian Orthodox).

¹⁰ In 1480, Ajmat, the last of the Golden Horde's khans, retreated from a confrontation with the troops of Prince Ivan III, who completed the process of unification of Russian lands under Muscovite authority.

¹¹ In 1547, Ivan IV came to the throne. He was known as Ivan 'the Terrible' for his tyrannical behavior. He established serfdom (whereby peasants were bound to the land by debt) murdered or deported members of boyar clans (nobility) and attempted to create a universal empire. On his death, the kingdom fell into anarchy as a result of the struggles over succession.

¹² In 1598, after the death of Ivan's son, Fyodor, the ruling dynasty came to an end. The boyar Boris Godunov was elected as Czar and a period marked by famine and turmoil began, which became known as the 'Time of Troubles' (1605-1613).

¹³ In 1612, a patriotic uprising expelled the Polish troops who had occupied Moscow. The *Zemsky Sobor* (parliament) elected boyar Michael Romanov as the new Czar, thus beginning the Romanov dynasty.

¹⁴ Under the Romanovs, Russia became an absolute monarchy, administrated by an efficient bureaucracy and an oligarchy (made up of nobles, merchants and bishops) which was integrated into the Government structure. Under Czar Alexis, a schism took place in the Russian Orthodox Church when Patriarch Nikon tried to adapt it to the Greek orthodoxy in 1653. Members of the traditional clergy known as the 'Old Believers' refused to accept the changes and their spokesperson, Archpriest Habacuc, was burnt at the stake.

¹⁵ During the 17th century the economy grew rapidly, as a result of territorial expansion, and the exploitation of Siberia's natural resources. A market also developed for Russia's forest products and semi-manufactured goods, primarily in Britain and Holland.

¹⁶ Under Peter I, who reigned from 1689 to 1725, Russia entered the Modern Age. The Czar attempted to 'Westernize' Russia in his despotic manner. He established an espionage network in his administration, essential for maintaining his strict autocracy. He put down the boyars in Moscow, and had his own son, Alexei, tortured and executed for joining them. In 1703, Peter founded St Petersburg, where he established the imperial capital.

¹⁷ In foreign policy, Russia gained access to the sea after the Great Northern War against Sweden (1700-1721). After winning the war against Persia, Peter extended Russia's southern borders as far as the Caspian Sea. The territorial, economic and commercial expansion which characterized this period made Russia one of the major European powers but also created a mosaic of ethnic and cultural groups which could not easily be assimilated into a single unit.

¹⁸ In 1762, Catherine II 'the Great' came to the throne. She disseminated the principles of the Enlightenment but also signed the Charter of the Nobility which legalized the privileges of the nobles. The Empire's conquests continued, with the annexation of parts of Ukraine, Poland, Lithuania and Crimea.

¹⁹ While the Russian State and nobility grew richer - court expenses amounted to 50 per cent of the state budget - impoverishment of the peasants and serfs increased.

²⁰ The French Revolution and the fight against absolutism influenced the Russian intelligentsia, which began demanding freedoms and social equality. Emperor Pavel I (1796-1801) reacted with extreme severity and imposed cultural censorship, internal exile and even banned trips to foreign countries.

²¹ Alexander I began his reign with the implementation of liberal reforms. In 1812, Napoleon's troops invaded Russia. The 'War for the Motherland' in which peasant fighters were also involved, ended with the triumph of the Russian army commanded by Marshall Kutuzov. This victory transformed Russia into the continent's major power.

²² As a reaction to absolutism, secret societies were formed that fought for the liberation of the peasantry, the distribution of land and the enacting of a constitution. Nikolai I, 'the policeman of Europe', started his government by brutally repressing the 'Decembrist revolt' that became a symbol for young revolutionaries. Since the Czar strongly believed in the divine right of the monarchy, he was intent on perpetuating the class privileges of the aristocracy and preventing the advance of liberalism.

²³ His foreign policy was mainly focused on the suppression of revolutionary movements that emerged in Poland, Germany and Hungary in 1848 as well as on the division of Turkey which prompted

the Crimean War against Britain and France (1853-1856). The war was badly managed on both sides and revealed the weakness of the Russian administration, army and economy.

24 Czar Alexander II began a series of reforms (1856-1874): he abolished serfdom which affected 40 million peasants, reformed the judicial system, reduced censorship and accepted autonomy for the University. However, the peasantry continued to suffer from lack of land and poverty, which fomented revolutionary ideas. Several secret societies emerged such as the 'Land and Freedom' or 'People's Will' groups. The latter assassinated Czar Alexander II in 1881 and its main leaders were hanged.

25 Alexander III maintained autocratic rule in Russia, supported by the church and the political police (ochrana), which kept most institutions, from schools to the judicial system, under strict control. His policy was also oriented towards a radical russification of border areas.

26 Socialist ideas reached Russia by way of Plekhanov who was in touch with Marx and Engels and founded the first social democratic group. In 1898, the Social Democratic Workers' Party of Russia (SDWPR) was banned but was later reorganized abroad around exiles Plekhanov, Vera Zasulich, Pavel Axelrod and Vladimir Ilich Ulyanov (known as 'Lenin').

27 In the second Congress held at Brussels and London in 1903, the Party was split into two factions: the 'Mensheviks' who shared the idea of an evolutionary socialism and the 'Bolsheviks', led by Lenin, who proclaimed insurrection and rule by the proletariat.

28 The war against Japan (1904-1905) led to the First Russian Revolution. On 9 January 1905, the army opened fire on thousands of demonstrators in St Petersburg. The uprising became generalized. There were strikes and rebellions such as that by sailors from the battleship Potemkin in Odessa and the revolt in the Kronstadt garrison.

29 Nicholas II promised to enact a new Constitution and convene a national Parliament (Duma). The Duma's political and social demands were rejected by the Czar who dissolved it and put down the revolution.

30 In 1907, the second Duma was elected, with the participation of the SDWPR. The main political issue continued to be land. Prime Minister Stolipin promoted agrarian reform, in order to create a class of land-owning peasants (kulaks). However, he was unsuccessful and was assassinated. The second Duma was also dissolved. An electoral reform guaranteed that the Duma

committees which followed also had a conservative majority.

31 Russia's entry into World War I precipitated a crisis within the regime. The losses brought about by war, and the lack of food, simply deepened popular discontent and ended in the February 1917 revolution. Soldiers joined the revolution and the first Workers' and Soldiers' Council (Soviet) was created in St Petersburg.

32 Nicholas II abdicated and the Duma committee established a new Provisional Government, while local soviets multiplied. Lenin returned to Russia in a sealed railway car with the help of the German Government. Once in Russia he launched his April Thesis ('All power to the soviets'), proclaiming the constitution of a socialist republic of soviets, the nationalization of the banking system and the abolition of private ownership. Soon after, the Bolsheviks' influence within the soviets began to grow, and a power split emerged between the councils and Alexander Kerensky's Provisional Government.

33 The government decision to continue the war together with the difficulties suffered by the population made the Provisional Government lose credibility. The Bolsheviks exhorted the people to 'turn the world war into a civil war'. On 25 October, Lenin led the uprising which brought down the Government, and the first socialist republic was established. In early 1918, the Bolsheviks dissolved the Constituent Assembly, in which the revolutionary socialists held a majority.

34 The Soviet Government (Council of People's Commissars) approved a peace 'without annexation or indemnities', the abolition of private ownership of land (150 million hectares were expropriated without indemnities), which was turned over to the peasantry, and the nationalization of the banking system. Other measures were approved, including the control of factories by their workers, the creation of a militia and of revolutionary tribunals, the abolition of class-based privileges and inheritance rights, the separation of the Church and the State and equal rights for men and women.

35 In July 1919, the Russian Soviet Federal Socialist Republic (RSFSR) was created with the adoption of a Constitution based on the system of soviets and the dictatorship of the proletariat. That same month, Czar Nicholas II and his family were executed.

36 Immediately afterward, anti-Bolshevik groups ('White Russians') led by former Czarist generals attempted to restore the previous

regime with the support of Germany, France and Britain, which prompted the outbreak of civil war (1918-1920). The Red Army under the command of Leon Trotsky, the People's Commissar for Defense, defeated the foreign intervention force and brought the war to an end.

37 During this period, the Soviet Government imposed the policy of 'war communism' which resulted in the nationalization of the means of production and the centralization of economic planning. The failure of this policy, which brought about a fall in industrial and agricultural production and threatened the country with economic collapse, prompted the 10th Party Congress of the Russian Communist Party to approve the New Economic Policy (NEP). This consisted of a return to the laws of a market system ('state capitalism'); freedom was granted to determine salaries, constitute small private companies and develop domestic trade and foreign

investment. The State retained control over foreign trade, heavy industry and infrastructure (state property). The Communist Party dictatorship was strengthened with the prohibition of all opposition within the Party in 1921 and the creation of the Tcheka (Soviet secret police).

38 In April 1922, Joseph Stalin became secretary general of the Party. After Lenin's death in 1924, Trotsky and Stalin vied for power; Stalin effectively became the ruler. He abandoned the NEP and re-established the system of centralized planning (five-year plans) and forcibly imposed the collectivization of agriculture by means of the system of kolkhoses and sovkhoses (collective farms).

39 In December 1922, the Union of Soviet Socialist Republics (USSR) was founded, comprising Russia, Ukraine, Belarus and the Transcaucasia Federation (Azerbaijan, Armenia and Georgia). Stalin ruled as absolute dictator

PROFILE

ENVIRONMENT

The largest country in the world, Russia is divided into five vast regions: the European region, the Ural area, Siberia, Caucasia and the Central Asian region. The European region is the richest, lying between Russia's western border and the Ural Mountains (the conventional boundary between Europe and Asia); it is a vast plain crossed by the Volga, Don and Dnepr rivers. The Urals, which extend from north to south, have important mineral and oil deposits in their outlying areas. The third region, Siberia, lies between the Urals and the Pacific coast. It is rich in natural resources, but sparsely populated because of its harsh climate. Caucasia is an enormous steppe which extends northward from the mountains of the same name, between the Black and Caspian Seas. Finally, the Central Asian region is a large depression of land made up of deserts, steppes and mountains. Grain, potatoes and sugar beet are grown on the plains; cotton and fruit in Central Asia; tea, grapes and citrus fruit in the subtropical Caucasian and Black Sea regions. The country's vast mineral resources include oil, coal, iron, copper, zinc, lead, bauxite, manganese and tin, found in the Urals, Caucasia and Central Siberia.

SOCIETY

Peoples: Russians, 81.5 per cent; Tatars, 3.8 per cent; Ukrainians, 2.9 per cent; over 100 other nationalities, including Chuvash, Bashkirs, Belarusians, Moldovans, and Chechens. (1996). **Religions:** Christian Orthodox is predominant. There are also Muslim, Protestant and Jewish minorities. **Languages:** Russian (official); there are almost as many languages as nationalities. **Main Political Parties:** United Russia; Communist Party of the Russian Federation; Liberal Democratic Party of Russia; Motherland-National Patriotic Union. **Main Social Organizations:** Federation of Independent Labor Unions of Russia (FNPR), with more than 40 million workers; All-Russian Confederation of Labor; Russian Confederation of Labor.

THE STATE

Official Name: Rossiyskaya Federatsiya. **Administrative Divisions:** The federation is made up of 26 autonomous republics. **Capital:** Moscow (Moskva) 10,469,000 people (2003). **Other Cities:** St Petersburg 4,656,900 people; Nizhnij Novgorod 1,351,800; Novosibirsk 1,397,800 (2000). **Government:** Parliamentary republic. Vladimir Putin, President since 1999, re-elected in 2004; Mikhail Fradkov, Prime Minister since March 2004. Bicameral Legislature: the Federal Assembly is formed by the State Duma, with 450 members, and the Federation Council, with 178 members. **National Holiday:** 12 June, Russia Day (1990). **Armed Forces:** 1,270,000. Other: 220,000.

Under-5 mortality
21 per 1,000 live births
2004

Poverty
<2% of population living on less than $1 per day
2002

Debt service
9.8% exports of goods and services
2004

Maternal mortality
67 per 100,000 live births
2000

The Chechens: always resisting

LOCATED IN southwestern Russia, the Chechen Republic of Ichkeria (Chechnya) is an area of 15,000 sq km on the eastern flank of the northern Caucasus. It borders North Ossetia and Ingushetia to the west, Dagestan to the east, and Georgia to the south. The northern and western regions are lowlands. Agriculture is mainly concentrated in the Terek and Sunzha river valleys. Chechnya and the region are rich in oil and gas.

CHECHEN PEOPLE
The Chechens were one among the several ethnic groups that made up the Alan state from the 8th century until its destruction by the Mongols in the 13th century. About 200 years later they descended to the plains, where they fought and traded with Russia and Georgia. Sunni Islam has been the main religion since the 18th century.

Chechens call themselves Nokhchi and their language, Chechen, belongs to the Nakh branch of the northeast Caucasian language family. Chechen was written in Arabic script until the 1920s when it was replaced with Russian Cyrillic script with the rise of the Soviet Union; the Latin alphabet was also used for written Chechen. With Soviet dominance, the Cyrillic alphabet replaced the Latin alphabet in 1938. After the declaration of the Chechen Republic in 1991 the Latin alphabet became standard again.

HISTORY OF RESISTANCE
From the 16th to the 18th century, Caucasia was fought over by Russian Czars, the Ottoman Empire and Persia. It was then that the resistance movement against invaders came into existence under the leadership of Sheikh Mansur (the legendary national hero). Mansur was captured by the Russians in 1791 and died some years later.

In spite of having trade links with Russia, Chechnya resisted czarist attempts at domination. From 1840, Imam Shamil led a rebellion of Caucasians that held off the imperial troops sent by Nicholas I and Alexander II for over a decade. But eventually, in 1859, the Russian Empire annexed Chechnya. Shamil was captured and later died in exile.

SOVIET PERIOD
After the 1917 Russian Revolution, the Chechens fought locally against both the Cossacks - who tended to be anti-Communist - and the Communists. Once Soviet authority was established, the Chechens joined with other Caucasian peoples to form the Republic of the Mountain Peoples in 1920. The Soviets abolished the Republic in 1924, after separating the Chechen Autonomous Oblast from it in 1922. Many Chechens were forced onto collective farms in the 1930s and suffered religious persecution. In 1934 the Soviets united the Chechens and Ingush in the

Chechen-Ingush Autonomous Oblast, which became an autonomous republic in 1936.

During World War II, Josef Stalin accused the Chechens and Ingush of collaboration with the Nazis. They were deported to Central Asia in 1944 and the Republic was abolished. It was not restored until 1957 when its former inhabitants were allowed to return from exile.

FIRST CHECHEN-RUSSIAN WAR
The Caucasian region's rich oil deposits lie at the heart of the modern conflict. Russia wants to keep control over the production and transport of oil and gas from the region in the face of Chechen secessionists and Western corporations.

In October 1991, General Dzhokhar Dudayev, who had expelled the Communist Government in Grozny, won a resounding victory in the elections, and Chechnya declared itself independent. However, Moscow refused to recognize it; the Ingush set up their own republic, and Dudayev did not manage to obtain international support.

In December 1994, Boris Yeltsin launched an invasion of Chechnya. Grozny was almost completely destroyed as a result of bombings and the Russian Federation army finally occupied it in February 1995, with thousands dead. Dudayev was forced into hiding, but his rebel forces continued to fight.

In May 1996, Yeltsin and Chechen President Zelimkhan Yandarbiyev agreed to a ceasefire. By this time, some 40,000 people including civilians had been killed and 300,000 Chechens had fled. Later, Chechnya's autonomy was recognized by the Russian Federation, but it was not allowed to secede. While some Chechens wanted to support the settlement, the rebels continued the fight for full independence. In August the Chechens launched a major offensive, defeating the Russian forces to retake Grozny.

Aslan Maskhadov won Chechnya's elections of January 1997. In May, a peace treaty was signed along the lines of the 1996 agreement. Maskhadov lost control to more radical rebels and anarchy followed.

The ending of the first war allowed the Russians to repair and reopen the Chechen section of the oil pipeline from Baku to Novorossiysk. This pipeline is vital to Russia, for not only does the Russian 'Transneft' monopoly earn transit fees of up to $300 million annually from it, but its operation will determine the routes of future oil and gas pipelines. The pipeline was reopened in October 1997.

SECOND WAR
In August 1999, Vladimir Putin became Russia's Prime Minister. In that period, Islamic

secessionist guerrillas entered Chechnya from Dagestan, occupying some villages and declaring the region an Islamic area. Russian forces took back the villages, but Chechen fighters joined the struggle for an Islamic state. A series of bomb attacks in Moscow was blamed on Chechen Islamic terrorists by Putin and used as the pretext for a new assault on Grozny. Rebel leader Shamil Basayev denied responsibility for the attacks.

Putin reinforced Russia's pressure to keep Chechnya in the Federation. Bitter fighting between the Russian army and Chechen rebels continued until early February 2000. By then, Grozny was in ruins, but under Russian control. Around half the surviving rebels fled to the mountains of southern Chechnya, where they continued to fight.

The capital was transferred to Gudermes. Captured Chechens were sent to the detention centers in Chernokosovo, in northern Chechnya, and Mozdok, in Ingushetia. According to various human rights organizations, they were tortured.

The conflict continued through 2001 amid allegations of human-rights abuses by Russian troops. After 11 September attacks on the US, Putin re-framed his war in Chechnya as an 'anti-terrorist operation' against Islamic extremists linked to al-Qaeda. In this way, he gained the backing of some Western countries that had previously ignored or opposed the war.

In October 2002, a group of 50 Chechen guerrillas took 800 civilians hostage at the Duvrovka Theater in Moscow, threatening to kill them if the Russian army did not withdraw from Chechnya. (See Russia's history).

After this crisis, Putin called a referendum for May 2003. The majority of Chechens voted in favor of granting Chechnya the status of a republic within the Russian Federation. However, the rebels rejected the decision and vowed to continue the struggle for full independence.

The first parliamentary elections since the start of Russian military control over Chechnya were held on 27 November 2005, amidst widespread criticism. Several human rights organizations deemed the electoral proccess 'a farce', considering that the current levels of violence made it impossible to hold free and fair elections.

In June 2006, rebel leader Abdul Khalim Saydullayev - who had been appointed president of the Chechen Republic by separatists - was killed during an armed confrontation between rebels and police forces in Argun, 30 kilometers to the east of Grozny. ■

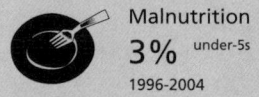

Malnutrition

3% under-5s

1996-2004

Water source

96% of population using improved drinking water sources

2002

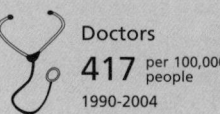

Doctors

417 per 100,000 people

1990-2004

and eliminated all opposition. Trotsky was expelled from the USSR and took refuge in Mexico where he was murdered in 1940. It is estimated that during the purges of 1935-1938 alone, nearly ten million people died. The victims of the different waves of purges totaled at least 20 million.

40   In 1939, a secret agreement with Germany (the Molotov-Ribbentrop Pact), allowed the USSR to occupy part of Poland, as well as Romania, Estonia, Latvia and Lithuania. In 1941, Hitler launched a large-scale attack against Moscow, sending in thousands of troops, as well as German air power. At a cost of between 25 and 30 million lives, the Red Army was able to repel the German troops and finally took Berlin in May 1945.

41   In 1945, at the Yalta conference, the Western powers and the USSR 'carved up' their respective areas of influence in Europe. In those countries occupied by the Red Army (Bulgaria, Hungary, Romania, Czechoslovakia, Poland and East Germany), the Communists took power and proclaimed first 'people's republics', then socialist republics, following the model of the Communist Party of the Soviet Union (CPSU).

42   In 1956, at the 20th Party Congress of the CPSU, Nikita Krushchev began a de-Stalinization process (denouncing the Stalinist personality cult and dogmatism), which came to an abrupt end when Leonid Brezhnev ousted Krushchev in October 1964.

43   The strategy of the Cold War devised by the US in the post-war period fuelled the arms race. The Warsaw Pact between the USSR and its Eastern European allies was created in 1955. The East-West confrontation eventually included nuclear weapons and the control of outer space, where the US and the USSR actively pursued their own space programs in the 1960s and 1970s, with neither one actually taking a lead.

44   In December 1979, the USSR embarked on the war in Afghanistan. It was the first time since World War II that the Soviet army had taken part in a conflict outside Eastern Europe. The military action turned into a disaster for the invaders: the invasion was condemned by 104 votes in the UN General Assembly; 55 countries boycotted the Moscow Olympic Games; 13,300 Soviet soldiers died in Afghanistan and the USSR did not manage to win the war.

45   In 1985, Mikhail Gorbachev became Secretary-General of the CPSU. He initiated a period of drastic changes based on the restructuring of the economy (*perestroika*) and increased transparency in cultural and

political affairs (*glasnost*). His reform program included opening up the country to a free-market economy and to foreign capital, internal democratization of the Party and constitutional reform to allow for a multi-party system. In the area of foreign policy, he withdrew Soviet troops from Afghanistan, improved relations with China, signed arms control accords with the US and in 1990 co-operated with Washington in driving Iraq - its former ally - out of Kuwait during the first Gulf War.

46   In April 1986, there was a serious accident at the Chernobyl nuclear power plant in Ukraine. Some 135,000 people were evacuated and by 1993 7,000 people had died (see Ukraine).

47   In June 1991, Boris Yeltsin was elected President. After a failed coup, in August, the CPSU was dissolved after 70 years in power. The changes within the USSR unleashed similar processes in Eastern Europe.

48   Trouble broke out in the Chechen-Ingush region (Northern Caucasia). Toward the end of October, parliamentary and presidential elections were held there. General Dzhojar Dudaev, leader of the Chechen nationalist movement, seized power. In early November he proclaimed the independence of the Chechen Republic. An economic embargo was promptly announced by Moscow. Meanwhile, Boris Yeltsin took office as the Russian head of state and Ruslan Khasbulatov (of Chechen nationality) became chairman of the Russian Parliament.

49   In 1991, Deputy Prime Minister Yegor Gaidar continued the rapid liberalization of the economy ('shock therapy') with the support of the US and international financial institutions. The effects were devastating for ordinary people: between 1990 and 1999 the number of people living on less than $2 a day more than tripled.

50   On 8 December, Yeltsin and the rulers of Belarus and Ukraine laid the USSR to rest, proclaiming the Commonwealth of Independent States (CIS) to take its place. Russia assumed the formal representation of the former USSR in foreign affairs. Latvia, Estonia and Lithuania withdrew and were recognized as separate countries by the UN.

51   President Yeltsin declared that the US was no longer a 'strategic rival', and continued the reform of the economy that Gorbachev had begun, including the liberalization of prices, and the privatization of industry, agriculture and trade.

52   In March 1993, the Parliament tried unsuccessfully to limit Yeltsin's powers. Following further disagreements, Yeltsin dissolved Parliament on 21 September.

Communists and nationalists staged violent demonstrations in Moscow. With US blessing, Yeltsin put down the revolt: tanks surrounded the Russian parliament building which was bombed and taken by force, causing the deaths of 138 people. Opposition leaders including Vice-President Rutskoi, and the chair person of Parliament, Ruslan Khasbulatov, were arrested.

53   A few days later, Yeltsin called for new elections and organized a referendum to increase his own powers. The December elections marked the defeat of those sectors faithful to Yeltsin, but 60 per cent of the voters approved the constitutional reform which granted him greater powers.

54   The tension between Moscow and the pro-independence groups of Chechnya, a mostly Muslim republic, became more serious and in December 1994, Yeltsin ordered military intervention. Despite protests both within Russia and abroad, the President maintained the military attacks on Grozny, the Chechen capital, which was almost totally destroyed in 1995.

55   In December 1995 the Communist Party led by Zyuganov won the legislative elections. The communist victory led Yeltsin to fear a defeat in the 1996 presidential elections. He tried to modify his policies, stopping the privatizations and nominating Yevgeny Primakov - a diplomat from the Soviet era who was an ally of Gorbachev - as foreign minister. In the electoral campaign all the opposition candidates from Gorbachev to the Communists, criticized the unlimited financial speculation, corruption and 'clannishness' of Yeltsin and his allies. In the second round of the July elections, Yeltsin took 53.8 per cent of the vote through an unexpected alliance with Alexandr Lebed, an opposition candidate who had received 11 million votes.

56   Lebed was appointed as Security Council Secretary and immediately started negotiations to end the war in Chechnya which had caused the death of 80,000 people. When it was announced in September that Yeltsin would undergo surgery, Lebed and Prime Minister Chernomyrdin became the two opposite candidates to fill the eventual power vacuum.

57   Yeltsin returned in March 1997, following a long absence. He reformed his cabinet and launched a far-reaching plan to cut back State spending and privatizations. The living conditions of the population continued to worsen. The chaotic change to a market economy damaged the production mechanisms, dismantled the social protection systems, and fed the rise of the mafias. In that year 73

per cent of the banking sector was under mafia control, and one of their most lucrative lines was the trafficking of nuclear material.

58   In September 1998, Eugeni Primakov was appointed Minister of Foreign Affairs. Primakov, an economist, refinanced debts with international organizations, without promising concrete changes. Fiscal controls were reinstated and the Government began to intervene in the economy. In the international field, Primakov introduced a policy less dependent on Washington and stood up to the US and Britain after their attacks against Iraq in December 1998.

59   Primakov's popularity, which outshone Yeltsin's, convinced the President that a change in the administration was needed. In May 1999, during the bombing of Yugoslavia, he decided to replace Primakov with Sergei Stepashin. In August, Stepashin was replaced by Vladimir Putin, a former member of the State Security Committee (KGB) during the Soviet period.

60   Putin vowed to 'recover' Chechnya for the Russian Federation and used every possible means to conquer the territory, but even so, after eight months of fighting, he was barely able to control the capital Grozny and the central area, while the guerrillas remained strong in the mountains.

61   On 31 December 1999, Yeltsin resigned unexpectedly. Putin won the elections of March 2000 which were marked by irregularities denounced by the *Moscow Times*.

62   In August 2000, the Russian nuclear submarine *Kursk* sank in the Barents Sea after two explosions, killing all 118 people on board.

63   Between 1992-2000, the Russian population had fallen by 2.8 million. In 2001 it fell by a further 700,000. Vladimir Zhirinovsky, a far right leader who had in the past proposed stimulating population growth through polygamy - 'allowing men to have up to four women each' - called for a halt to the 'reduction in population' and for abortion to be made illegal.

64   According to official statistics, almost one in every three Russians was living in poverty, the average pension had fallen below basic survival levels and more than 60 per cent of pensioners were at risk. An Amnesty International Report published in 2002 revealed that 800 million Russians were living in poverty and that the income gap also had a geographical dimension: the income of a Muscovite was 17 times higher than that of a resident of Ingushetia. The same report stated that economic and business activity was still being developed with little regard for the law, transparency and honesty.

65 In December 2001, the Duma passed the first non-deficit budget since the demise of the Soviet Union. However, the fall in gas and oil prices after the September 2001 terrorist attacks in the US forced the Ministry of Finance to review its estimates, since the price of oil had fallen since then. Russia had become the second oil exporter in the world, after Saudi Arabia.

66 In January 2002, the TV station TV-6, the last national independent station, was forced to shut down at the request of a minor stockholder. The Supreme Court of Moscow refused an appeal. The station's largest stockholder was Boris Berezovsky, one of Russia's new oligarchs and a critic of Putin. Journalists and opposition politicians claimed it was another attempt by the President to gain control of the country's independent media.

67 With parliamentary approval in July, and for the first time since the 1917 Revolution, the sale of farmlands was allowed in the country, as long as the purchasers were Russian.

68 In October 2002, a group of Chechen rebels occupied a theater in Moscow, holding 800 hostages, and demanding a change in Moscow's policy and tactics on Chechnya. Three days later, elite Russian troops stormed the theater after pumping into the building a poisonous gas which killed 50 rebels and 120 hostages and severely affected many of the survivors. Amid an outcry as to why so many perished, Russians and the international community wanted the Government to reveal what gas had been used. However, the Government issued no information.

69 According to analysts, the financial interests of transnational oil corporations, such as Chevron, Exxon Mobil and Unocal, were also driving the Russian-Chechen conflict. Since the collapse of the USSR, these companies have fought their way to the Caspian Sea. The fight for control of Russian oil also led the US to support the construction of a 1,750 km pipeline that would transport one million barrels of oil per day from Baku to Turkey, without going through Iran.

70 In March 2003, Russia opposed the intervention in Iraq led by the US and supported by the UK and other allies. According to Russian Foreign Minister Igor Ivanov, intervention in Iraq without UN approval would weaken the anti-terrorist coalition formed after the 11 September 2001 attacks in the US.

71 During the first six months of the year, 12,700 Russian citizens requested asylum at foreign embassies.

72 In March 2004, Putin was re-elected President with over 70 per cent of the vote. He defined economic growth and a reduction of at least 12 per cent in poverty levels as his Government's goals. In the area of foreign policy, he insisted on the need to withdraw coalition troops from Iraq.

73 Between January and March 2004, 43 people went missing in Chechnya. Some of the disappearances, like that of human rights activist Aslan Davletukaev, involved military armed forces and vehicles. His body was found on 16 January on a highway.

74 On 9 May 2004, the President of the separatist Chechen republic (see box on Chechnya), Akhmad Kadyrov, was assassinated in Grozny. Prime Minister Sergey Abramov took office as interim President.

75 That month, the EU agreed to support Russia's World Trade Organization (WTO) application. Russia - sole economic power left out of the WTO - still had to negotiate its incorporation with the US and China. Putin said that Moscow would hasten the process to ratify the Kyoto Protocol, which aims to cut greenhouse gas emissions.

76 On 1 September, an armed group kidnapped more than 500 children, parents and teachers at a school in Beslan, southern Russia. The kidnappers demanded the release of Chechen militias incarcerated in the neighboring town of Ingushetia and the withdrawal of Russian troops from Chechnya. Putin was on vacation but returned to Moscow to face the crisis.

77 After violent clashes between Russian military forces and the kidnappers and two days of national mourning, official figures showed more than 300 dead and at least 200 missing.

78 In January-February 2005, hundreds of thousands of people demonstrated throughout Russia against Putin's wage policies, in the largest protests in five years in the country. Putin's popularity started to decline, as did his support from the army and the police.

79 The Memorial Human Rights Center reported in April that more than 3,000 Chechens had died and 1,543 had been kidnapped - of whom 892 remained missing - in Chechnya since 2000.

80 In May 2006, Putin accused the US of blocking Russia's incorporation to the WTO. Moscow said that Washington demanded harsher conditions than for other countries. The Kremlin warned that US oil companies that wished to invest in Russia could lose business deals if the US block remained in place.

81 That month, Moscow Mayor Yuri Luzhkov and a city court - citing the need to prevent possible violence between religious and gay rights groups and to keep traffic flowing - banned what would have been the first Gay Pride march in Russia, marking the date when homosexuality was decriminalized in 1993. Gay groups tried to carry out the march in spite of the ban, and were confronted by hard-line religious groups. More than 70 people were arrested.

82 Campaigning journalist Anna Politovskaya was found shot dead in October 2006. She was known as a fierce critic of the Kremlin's actions in Chechnya and prosecutors said they believed her death could be linked to her investigative reporting. Her murder caused widespread concern about the declining space for independent journalism in Russia. ∎

IN FOCUS

ENVIRONMENTAL CHALLENGES
The country's main rivers are being polluted by the indiscriminate use of toxic agrochemicals and the lack of treatment of waste and sewage waters. Another risk factor is the huge stock of old and unused pesticides that are a serious threat to the environment. Chelyabinsk, a city in the Ural Mountains, has high levels of radioactivity, due to leaks in its plutonium plant. Lake Baikal is contaminated by the dumping of industrial waste.

WOMEN'S RIGHTS
Women have been able to vote and stand for office since 1918. In 2003, women held 10 per cent of seats in the Lower House and 3.4 per cent in the Upper House of Parliament. That year, women made up 49 per cent of the labor force of 78 million. Female unemployment stood at 9 per cent. In early 2006, in spite of the continued improvement of Russian economy, poverty was widespread, especially outside Moscow and St Petersburg, affecting mostly women and children. HIV/AIDS continued to be a pressing issue. In late 2003 it was estimated that 290,000 women between 15 and 49 were living with HIV/AIDS.

CHILDREN
The mortality rate for children under 1 and 5 in 2005 stood at 17 and 21 per 1,000 live births respectively. Approximately 6 per cent of newborns are underweight, while 13 per cent of under-fives are moderately or seriously stunted.

There are reports of children being kidnapped or purchased from their parents for sexual abuse, child pornography or organ trafficking. The government has verified foreign adoptions of Russian children who are later forced to work in the sex industry.

INDIGENOUS PEOPLES/ ETHNIC MINORITIES
Russia is home to more than 100 ethnic groups. Less than eight per cent of the country's population is made up of the peoples living in the Volga River basin and the Urals area - Bashkirs, Kalmyks, Komis, Maris, Mordovans, Tatars, Udmurts and Chuvash. Of these almost half are Tatars, the second largest nationality in Russia. The traditional religion of the Tatars and Bashkirs is Islam, the Kalmyks practice Buddhism, and the most common religion among the rest of the population is Orthodox Christianity.

The peoples living in the North Caucasus - Abazians, Adygeis, Balkars, Ingushetians, Kabardins, Karachayevs, Ossetians, Cherkessians, Chechens, and the ethnic groups from Dagestan (Avars, Aguls, Dargins, Kumyks, Laks, Lezgins, Nogays, Rutuls, Tabasarans and Tsakhurs) - make up less than three per cent of the Russian population.

The peoples living in Siberia and the far North - Altais, Buryats, Tuvas, Khakass, Shors, Yakuts - and the 30 or so groups in the far North constitute 0.6 per cent of the entire population.

MIGRANTS/REFUGEES
The number of displaced people fell by approximately 50 per cent between 1996 and 2003 (when there were 492,000 people living in that condition). According to official statistics, in 2003 the Government granted refugee status to some 14,000 people (compared to 290,000 in 1996) of which 11,500 were from South Ossetia, a republic that declared its independence from Georgia although it has not been recognized by the international community. In early 2006 there were 1.5 million Afghans living in Russia. The actual number of refugees from the different former Soviet republics living in Russia is unknown.

DEATH PENALTY
Russia is considered abolitionist in practice by Amnesty International. The last known execution was carried out in 1999.

Rwanda / Rwanda

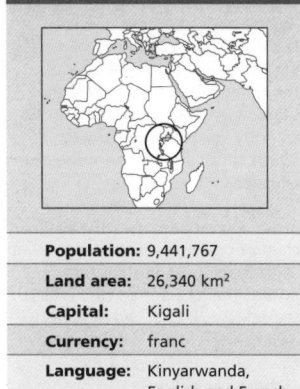

Population:	9,441,767
Land area:	26,340 km²
Capital:	Kigali
Currency:	franc
Language:	Kinyarwanda, English and French

U ntil the 14th century, the only human beings to occupy the territory of what is now Rwanda were the Twa (or Batwa) Pygmies. The Twa lived as hunter-gatherers and potters and were subjugated by Bantu ethnic groups from Central Africa and the Watutsi (or 'Tutsi') from Ethiopia, who migrated to Rwanda, Burundi, Tanzania, Zaire and Uganda in several waves over the course of two centuries.

[2] Before the German occupation of Rwanda, Burundi and Tanzania in 1897, mixed marriages between the Bantu and Watutsi led to integration between these groups.

[3] Academic studies of the Kinyarwanda language - the only language spoken by 80 per cent of people in present-day Rwanda - in which the word *Tutsi* originally meant 'cattle-raiser' and the word *Hutu* 'a subordinate', as well as studies of the coexistence of present-day myths and cults among all Rwandans (with the exception of the Twas) conclude that the pre-colonial society of their ancestors had effective mechanisms of social cohesion and mobility.

[4] This oral culture - rich in poetry, rhetoric, songs and dances - lasted in its original form until the mid-1920s. In 1916, Belgium occupied the region encompassing Rwanda and Burundi and seven years later it took official control of these territories by mandate of the League of Nations.

[5] Until the 1920s, Europeans used Rwandan territory - which had small gold deposits - as a transit zone for precious stones, metals and ivory that were extracted in neighboring regions. For this purpose, they used slave labor and classified people based on their height and skin color, according to an ethnocentric scale, ignoring the idiosyncratic native system.

[6] Under this system, Rwanda was ruled and administered through participation in the main productive sectors: agriculture and livestock farming. The shortage of pasture lands (by then Rwanda already had a high population density) meant they were assigned temporarily to individuals who were in permanent competition on the basis of their productive output. So, the 'Tutsi' status, which meant that the person had a reputation for being a productive worker, was granted or withdrawn by village courts (*gacacas*) that operated on a weekly basis.

[7] The Tutsis won the right to create armies in order to expand their territories. Kigeri Rwabugiri was the king to achieve the greatest power between 1860 and 1895.

[8] In the mid-1920s, persistent threats to the Tutsis who had been invested with a degree of authority in trading posts by the Belgians (they were Bantus, chosen because of their greater height) led the settlers to establish a tutelary monarchy and put an end to the local system of administration. Belgium also sent missionaries to educate the Rwandan people (who were first redistributed geographically and banned from entering the jungle) through religious instruction and servile labor in coffee plantations.

[9] Together with the clergy, this new Tutsi elite (15 per cent of the population) was eager to inculcate

PROFILE

ENVIRONMENT

Known as the country of a thousand hills on account of its geographical location between two mountain ranges, Rwanda lies at the heart of the African continent. The terrain is mountainous and well-irrigated by numerous rivers and lakes supporting varied wildlife. Rwanda's forests are home to one of the few remaining populations of mountain gorillas. Most people live in the highlands where the economic mainstay is subsistence agriculture.

SOCIETY

Peoples: 84 per cent are Hutu (Bantu-speaking people), 15 per cent Tutsi (Hamitic) and 1 per cent Twa (descendants of the Pygmies). There is a minority of European origin, most of whom are Belgian.
Religions: 69 per cent profess traditional African religions, 20 per cent are Catholic, 10 per cent Protestant and 1 per cent Muslim.
Languages: Kinyarwanda, English and French (all official).
Main Political Parties: Rwandan Patriotic Front (FPR); National Revolutionary Movement for Development (MRND); Republican Democratic Movement (MDR); Liberal Party (PL).

THE STATE

Official Names: Repubulika y'u Rwanda; République Rwandaise; Republic of Rwanda.
Capital: Kigali 656,000 people (2003).
Other Cities: Butare 32,400 people; Ruhengeri 33,500; Gisenyi 24,800 (2000).
Government: Paul Kagame, President since 2000, re-elected in 2003. Bernard Makuza, Prime Minister since 2000. Bicameral Parliament: Senate (26 members) and Chamber of Deputies (80 members).
National Holiday: 1 July, Independence Day (1962).
Armed Forces: 33,000 (1996).

LAND USE

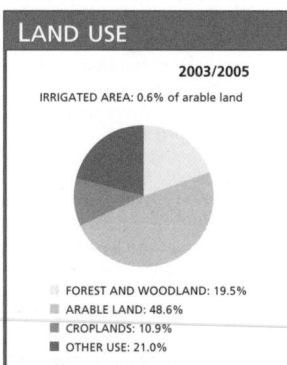

2003/2005

IRRIGATED AREA: 0.6% of arable land

- FOREST AND WOODLAND: 19.5%
- ARABLE LAND: 48.6%
- CROPLANDS: 10.9%
- OTHER USE: 21.0%

PUBLIC EXPENDITURE

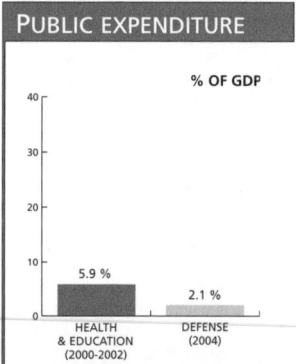

% OF GDP

5.9 %
HEALTH & EDUCATION (2000-2002)

2.1 %
DEFENSE (2004)

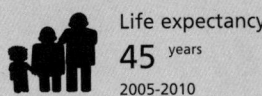
Life expectancy
45 years
2005-2010

GNI per capita
$210
2004

Literacy
64% total adult rate
2000-2004

HIV prevalence rate
5.1% of population 15-49 years old
2003

a distorted version of the history of Rwanda among the inhabitants of the Belgian protectorate of Rwanda-Urundi. In this way, the first written history of Rwanda laid the foundations for the segregation of the Hutus and Twas.

[10] From 1930, the Hutus and Twas were identified at birth and deprived of all rights except for access to primary schools and Catholic churches. Likewise, until 1961, Hutus were only authorized to study and pursue an ecclesiastical career in which they came to represent 30 per cent of members.

[11] Also from that year, the ratio of Hutus to Tutsis among priests and nuns was inverted (80 per cent Hutus to 20 per cent Tutsis). This change within the Rwandan Catholic Church - which had the largest congregation on the African continent until the late 20th century - paralleled political events that led to the replacement of the Tutsi monarchy by a racist Hutu dictatorship.

[12] During the 1950s, in line with a plan for democratization and the subsequent withdrawal of the European authorities, the UN urged Belgium to increase the participation of Hutus in public life.

[13] By 1959, the Belgians had replaced with Hutus half of Tutsi high-level officials, who had enjoyed uninterrupted participation in the trafficking of diamonds to the metropolis. After gaining access to the diamond market, the Hutus were able to obtain arms.

[14] In the same year, the Hutus founded the Party of the Hutu Emancipation Movement (Parmehutu) and the monarchic Tutsis created the National Rwandese Union (UNAR), amidst violent confrontations.

[15] In early 1960, two years before the proclamation of the Republic of Rwanda and its separation from Burundi, the UN supervised elections in which the Hutu-dominated party won a large majority.

[16] Between 1961 and 1973, the Parmehutu killed 20,000 Tutsis and forced 300,000 to flee to Burundi, Uganda, Tanzania and what was then Zaire. The remaining Tutsis had their lands confiscated and were excluded from all state institutions, as well as being demonized for allegedly collaborating in ten invasions carried out by exiled Tutsi groups belonging to the UNAR during that period.

[17] In 1973, a coup was plotted by factions of the Parmehutu that disagreed with President Gregoire Kayibanda's foreign policy and unequal distribution of privileges. On 5 July that year, Kayibanda was ousted by his Minister of Defense, Colonel Juvenal Habyarimana.

[18] Some days later, Habyarimana dissolved the Parmehutu and ordered the execution of 50 members of Kayibanda's Government. In 1975, the new President of Rwanda officially created a one-party state with himself at the head of the National Revolutionary Movement for Development (MRND).

[19] From that date, all Rwandans were compulsorily enrolled in the MRND. In this way, and through massive propaganda campaigns, Habyarimana strictly controlled and manipulated the circulation of people and property, while the Tutsis were subjected to a covert segregation campaign that induced 300,000 of them to go into exile.

[20] In 1978, the one-party Constitution was ratified in a referendum and Habyarimana was elected President. Five years after taking power, Habyarimana had an army of 7,000 troops, 1,300 elite squads, and a personal guard of 1,000 men.

[21] By the late 1980s, the sudden fall in coffee prices, which represented 75 per cent of Rwanda's exports, brought about a socio-economic collapse as well as a wave of denunciations by journalists, claiming the existence of nepotism and abuse of power, which Habyarimana did not manage to silence.

[22] In 1988, Habyarimana was re-elected and women won 15 per cent of seats in the Chamber of Deputies.

[23] In October 1990, 7,000 Rwandan Patriotic Front (FPR) soldiers trained in Uganda (their country of asylum) invaded Rwanda.

[24] President Habyarimana repelled the Tutsi offensive thanks to the intervention of troops from Belgium (traditionally linked to the Tutsis), France (a beneficiary of the Hutu hegemony) and Zaire, which had a Hutu President at that time, and also through the diplomatic intervention of Belgian Prime Minister, Wilfried Martens. Although Belgium and Zaire withdrew their troops the

Rwanda - 1994: The West's genocidal negligence

ON 7 APRIL 1994, Rwandan official radio called for the annihilation of all Tutsis - 'the nation's enemies', as they had been called since Rwandan independence - and of all those who were unwilling to kill them. The Army and the militias, together with Hutu and Twa civilians murdered more than 10,000 people a day, until 15 July 1994, when Paul Kagame's Tutsi troops took the capital city (Kigali). Abandoned to their fate by the Catholic and Anglican churches, as well as by the UN, between 800,000 and one million Rwandans died in that period, 90 per cent of them Tutsis, under horrendous circumstances. Only those who took refuge in the 250

Rwandan mosques managed to survive. Three months before the genocide was launched, Canadian general Romeo Dallaire, leading the Belgian, French and US blue helmet forces present in Rwanda on a UN peace mission, personally informed Kofi Annan, the UN Secretary General, that lists of names were circulating in Rwanda in order to implement the murder of 1,000 Tutsis every 20 minutes. Dallaire's report and his request for reinforcements obtained no response from the UN. On the contrary, the organization ordered the withdrawal of all its troops in Rwanda, just hours after the assassination of President Juvenal Habyarimana - his

plane was shot down on 6 April 1994, in an incident that was never fully explained and that triggered the massacres. The International Criminal Tribunal for Rwanda - created in 1995 by the UN, after the organization apologized for not having intervened during the genocide - and the Rwandan courts, which in 1996 had only 81 judges, had only ruled on 21 cases by 2004. This suggests that the identification and trial of those who orchestrated the genocide could take several decades. The Red Cross estimated that half of all Rwandan homes in 2004 were headed by children under 15, and that a quarter of those homes were headed by widows. ∎

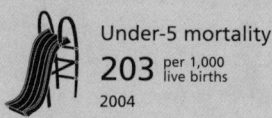
Under-5 mortality
203 per 1,000 live births
2004

Poverty
51.7% of population living on less than $1 per day
2000

Debt service
11.2% exports of goods and services
2004

Maternal mortality
1,400 per 100,000 live births
2000

following month, the French troops remained in Rwandan territory.

[25] In June 1991, the President signed a new Constitution which provided for a multi-party system, established the separation of state powers and limited presidents to two terms in office.

[26] In August 1992, in view of the systematic violation of the Constitution, and repeated armed confrontations and insistent attempts by Tutsi refugees to return to their country, the UN (represented by Belgium, France and the US) negotiated a ceasefire and called for peace talks to be held in early 1993 in Arusha (Tanzania).

[27] On 6 April 1994, Habyarimana and the President of Burundi (also a Hutu) died when the plane carrying them was attacked. A few hours later, the UN troops - who had suffered ten casualties - abandoned Rwanda together with 600 French citizens resident in the capital.

[28] The death of President Habyarimana triggered a genocide against the Tutsis, as well as moderate Hutus and Twas, which cost the lives of between 800,000 and a million people, 90 per cent of whom were Tutsis, within a period of 100 days.

[29] In early July, the FPR army seized control of Kigali and the UN sent in troops again to guarantee the establishment of a government of national unity. That same month, Major General Kagame became both Vice-President and Minister of Defense while Pasteur Bizimungu (Hutu) was appointed President.

[30] From July 1994, skirmishes between the Rwandan army and Hutu militias caused hundreds of deaths per year. Amnesty International (AI) denounced the participation of children in the fighting.

[31] In November 1995, the UN described the 1994 killings in Rwanda as 'genocide' and in 1996 an International Criminal Tribunal for Rwanda (ICTR) was established and started proceedings against military officers, clergy, political leaders and directors of the state-run radio.

[32] In 1998, Kagame ordered the occupation of half of DR Congo's territory. Laurent Kabila, who had succeeded Mobutu, did not manage to neutralize the Democratic Liberation Forces (FDLR) of Rwandan Hutu refugees as Kagame expected.

[33] In April 2000, President Bizimungu resigned while Prime Minister Pierre Rwigema (a moderate Hutu) was replaced by Bernard Makuza (another moderate Hutu).

IN FOCUS

ENVIRONMENTAL CHALLENGES
Excessive grazing in lowlands has caused erosion and a considerable loss of natural vegetation. Ninety per cent of the energy consumed by Rwandans comes from natural firewood resources, leading to deforestation. Illegal hunting threatens wildlife.

WOMEN'S RIGHTS
Women have been able to vote and stand for office since 1961. In 2003, they held 49 per cent of seats in the lower house and 35 per cent of seats in the upper house of Parliament. Women made up 50 per cent of the country's labor force of five million. In 2004, 92 per cent of pregnant women received prenatal care, but only 13 per cent of births were attended by trained medical staff*. The maternal mortality rate stood at 1,400 per 100,000 births. Women continue to face social discrimination. Traditionally, they have played a fundamental role in subsistence activities in rural areas and since the 1994 genocide have had to assume greater responsibilities as heads of household.

CHILDREN
In the period 1990-2004, there was an increase in under-1 and under-5 mortality rates. The former rose from 103 to 118 deaths per 1,000 live births and the latter from 173 to 203 per 1,000. Nine per cent of newborns were underweight and 41 per cent of children under five were moderately or severely undersized. Apart from leaving a huge number of children with physical and psychological scars, the 1994 genocide decimated the fragile economy, severely impoverishing the population, particularly women, and eroding the country's human resource base. The spread of HIV/AIDS has had a devastating impact on life expectancy. By the end of 2003, 22,000 children under 14 were living with HIV. In 2004, 31 per cent of children aged between 5 and 14 were working.

INDIGENOUS PEOPLES/ ETHNIC MINORITIES
The Hutu are the largest ethnic group, while the number of Tutsis can not be accurately estimated since they have intermarried for generations. The 1994 mass killings and migrations affected the ethnic composition of the population. The Twa (or Batwa), descendants of the Pygmies, number approximately 23,000. The Twa demand access to land, housing and education, as well as the eradication of discrimination against them. There have been bloody conflicts between Hutus and Tutsis since 1962, resulting in hundreds of thousands of deaths. In 1994, a genocide of Tutsis was carried out under the direction of the dominant Hutus.

MIGRANT/REFUGEES
In early 2006, an estimated 50,000 Rwandans continued to live as refugees in 19 African countries. Some 10,000 of them fled Rwanda towards Burundi in mid-2005. The country hosted approximately 44,000 refugees. The three main groups were either from DR Congo and Burundi or Rwandans that had returned from foreign countries. The 36,500 Congolese refugees were living in Kiziba camp, in the southwest, and Gihembe camp, in the north of the country. Ninety four per cent of them came from South Kivu. About 10,000 of them were expected to choose voluntary repatriation in 2006. Early in that year, the intention was to settle some 1,500 refugees from Burudi in Kigeme camp, in southern Rwanda. Most of the 3,500 urban refugees living in the outskirts of Kigali also come from Burundi and Congo DR, although there is a small number who come from other 10 African countries. Nearly 20,000 Rwandans would be repatriated at the end of 2005, and a similar number in 2006.

DEATH PENALTY
The death penalty is still applied, in some cases even for ordinary crimes.

*Latest data available in *The State of the World's Children* and *Childinfo* database, UNICEF, 2006.

[34] That same month, Paul Kagame was elected President by the National Transitory Assembly.

[35] In 2000, the UN, which had apologized to Rwanda and Belgium in 1999 for its lack of action during the genocide, blamed Kagame for the attack that killed Habyarimana in 1994.

[36] In 2001, Kabila was assassinated while Rwandan troops controlled most of the territory of DR Congo. Under a peace treaty signed in December 2002, Kagame agreed to withdraw Rwandan forces and DR Congo promised to disarm the FDLR.

[37] In August 2003, Kagame was elected President in a plebiscite with 95.5 per cent of the vote. In October of that year, the FPR won an absolute majority in the first multi-party parliamentary elections (marred by irregularities according to observers).

[38] Rwandan media remained under state control and Kagame banned several political parties.

[39] On 24 April 2004, Kagame violated the peace treaty signed in 2002 when he sent Rwandan troops to attack DR Congo.

[40] In June, former president Pasteur Bizimungu was sentenced to 15 years in jail for embezzlement, inciting violence and associating with criminals.

[41] In November, AI urged the Government to do its utmost to foster freedom of the press and not to intervene in judicial decisions. AI congratulated those judges and magistrates who were struggling to withstand political pressures and the temptation of corruption to uphold the rule of law.

[42] In March 2006, the Executive Board of the World Bank, in collaboration with the IMF and African Development Bank (ADB) gave the 'green light' to the cancellation of debts owed to them by 13 African countries, including Rwanda. This was to come into effect on 1 July 2006, at the start of the Bank's new fiscal year.

[43] On 3 June 2006, Joseph Serugendo, former board member of the Radio Television Libre during the 1994 genocide, was sentenced to six years in prison for his participation and technical assistance in broadcasting messages to incite genocide. Serugendo had been arrested in Gabon in September 2005 and sent to Rwanda to be tried by the ICTR. After being informed of the court's judgment he pointed out that: 'this may encourage others to acknowledge their involvement in the genocide and contribute to national reconciliation'.

[44] In October 2006, the ruling RPF party announced that it would press for a law abolishing the death penalty. Outlawing capital punishment would clear the way for the Tanzania-based UN International Criminal Tribunal for Rwanda and Western countries to extradite genocide suspects to Rwanda. Party spokesperson Servilien Sebasoni denied suggestions that the decision was taken less out of principle than to facilitate the extradition of suspects to Rwanda. ∎

St Helena / Saint Helena

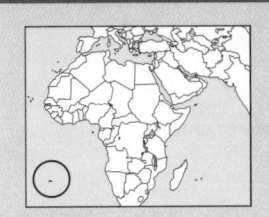

Population:	5,019
Land area:	122 km²
Capital:	Jamestown
Currency:	Pound sterling
Language:	English

St Helena was uninhabited when Portuguese navigators arrived in 1502. In 1659 it became a British colony when an outpost of the British East India Company was established on the island. Of scant economic interest, the island acquired notoriety as the location of Napoleon's second exile, from 1815 until his death in 1821.

² The Malvinas/Falklands War put it back on the map, a century and a half later. A British representative stated: 'It was only with the help of Ascension Island and the labor force provided by St Helena, that we could recover the Falklands'. This may justify the expensive maintenance of this British enclave through the Overseas Development Administration.

³ In December 1984, the UN General Assembly urged Britain to bolster the fishing industry, handcrafts and reforestation on the island and to foster awareness of the right to independence. Washington and London both

voted against the resolution. The UN also questioned the existence of the military base on Ascension Island since there should be no bases in non-autonomous territories.

⁴ In January 1989 a new Constitution was instituted, conferring greater powers on the members of the Legislative Council and enabling civil servants to stand for election with the approval of the Governor. The new Constitution also lowered the voting age to 18.

⁵ The island's only export is fish, but there has been a decline in the total catch in recent years.

⁶ St Helena is of scientific interest because of its rare flora and fauna. The island has some 40 plant species unknown in the rest of the world.

⁷ In 1997 unemployment rose to 18 per cent. In 1999 the UK announced it would grant citizenship to the residents of dependent territories.

⁸ The British Government had decades earlier acquired a ship, the RMS St Helena, to link the islands to the rest of the world. This vessel has been the only means of reaching St Helena. Plans have been made to build an airport, although some islanders fear it could be bad for the environment.

⁹ In May 2002 the island celebrated the 500th anniversary of its discovery with a series of maritime and cultural events, including the opening of a museum in Jamestown, the capital city. On display in the museum are graphic panels telling the island's history.

¹⁰ In late 2003 the British Crown

appointed a new Governor. Michael Clancy became Governor and Commander-in-Chief in October 2004, replacing David Hollamby.

¹¹ In April 2004 the UK Government rejected as non-viable four public-private tenders for

development of air access. Most islanders would prefer air access after the St Helena ship is retired in 2010. In October, Clancy insisted that the future airport would mean a turning-point in the history of the island and would foster its development. ■

Ascension Island

Ascension Island is of volcanic origin, with a surface area of 88 sq km. Its importance derives from its strategic location in the South Atlantic, 1,200 km northwest of St Helena. It is a communication relay center between South Africa and Europe, and the United States maintains a missile tracking station - Wideawake Airfield - there under an agreement with the UK. The island's naval installation and airbase were vital to Britain during the Malvinas/Falklands war (April-June 1982) and afterwards as a base for the ships and planes that supplied the British troops occupying the islands claimed by Argentina. The island is used by the BBC, the Royal Air Force (RAF), USAAF, Cable and Wireless and other networks as a relay station. There is an attempt to save the sea bird population, which is at risk of becoming extinct.

There is no indigenous population and the majority of residents are employees of the St Helena Government. In 1988, out of the 1,099 inhabitants, 765 were from St Helena, 222 were British, 102 were American and 10 were of other nationalities. These figures do not include British military personnel. The main religion is Protestant and the official language is English. In June 2006 there were 40 UK troops. The island received regular flights from the RAF and a military cargo ship every six weeks. ■

Tristan da Cunha

The most important of a group of South Atlantic islands 2,400 km west of Cape Town, South Africa, and under the administration of St Helena. The islands total 201 sq km (Tristan da Cunha 98 sq km; Inaccessible Island 10 sq km, 32 km west of the main island; Nightingale Islands 25 sq km, 32 km south of Tristan da Cunha, and Diego Alvarez or Gough Island 91 sq km, 350 km south of the main island).

The approximately 300 inhabitants (in 2000) are concentrated on Tristan da Cunha, the majority employed by the Government and in a lobster-processing factory. There were volcanic eruptions in 1961 and the island was evacuated, though the population returned in 1963. On Diego Alvarez there is a small weather station run by the South African Government. The main religion is Protestant and the official language is English. The administrator, Mike Hentley, represents the Government of St Helena. An advisory council with executive and legislative duties, comprises eight elected and three appointed members. Of the eleven government departments in Tristan Da Cunha, three are headed by women.

A ten-year contract to operate Tristan da Cunha's lucrative lobster-fishing concession, which was awarded in 1996 to a South African firm, went into effect on 1 January 1997. The residents of Tristan da Cunha - like those of St Helena - demanded British citizenship during the annual visit by St Helena's Governor.

In June 2006, a storm hit the islands' coasts, causing severe damage in port infrastructure and blocking roads for days. According to the World Conservation Union (IUCN) catalogue, issued in November 2003, some native species of fauna in Tristan da Cunha and Ascension Island are disappearing because of the loss of their natural habitat through encroachment by domestic animals. ■

St Kitts-Nevis / Saint Kitts-Nevis

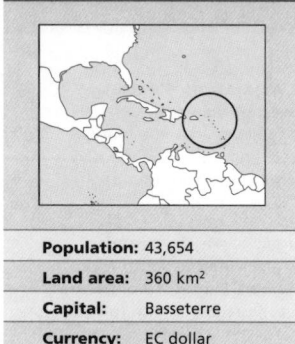

Population:	43,654
Land area:	360 km²
Capital:	Basseterre
Currency:	EC dollar
Language:	English

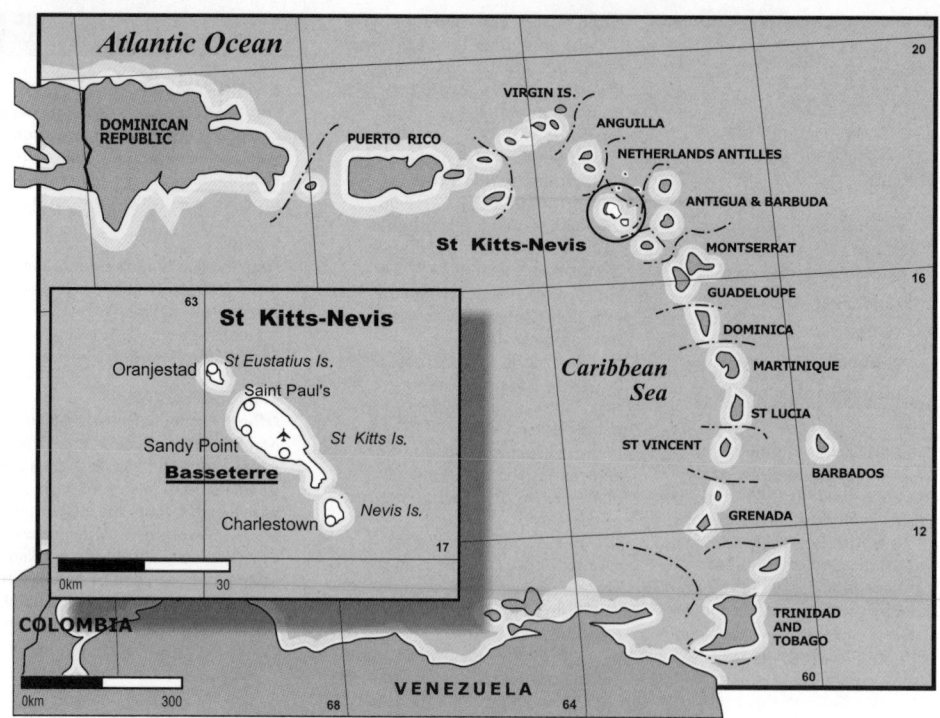

The island of Liamuiga, or 'fertile land' in the language of the Carib Indians who originally lived there, was renamed St Christopher by Columbus on his second voyage to America, in 1493. It was colonized by the Europeans in 1623, when the adventurer Thomas Walker established the first English settlement in the Caribbean. The neighboring island of Nevis was colonized five years later. After the rapid extermination of the Caribs, the English started to grow plantation crops, especially sugarcane, for which they used slaves from Africa.

[2] In the 20th century, the decolonization process following World War II gave the islands total internal autonomy, while foreign relations and defense were left to the colonial capitals. These islands joined the Associated States of the West Indies. In 1980, Anguilla formally separated from St Kitts and Nevis (see Anguilla); after this the islands were governed by a Prime Minister and a Parliament, both elected by universal suffrage.

[3] The Labor Party had been in office since 1967, but suffered a major defeat in the 1980 election at the hands of an opposition coalition of the People's Action Movement (PAM) and the Nevis Reformation Party (NRP). Kennedy Alphonse Simmonds became Prime Minister. The opposition victory meant that independence,

planned for June 1980, had to be postponed, as the NRP opposed a post-independence federation with St Kitts. The 1976 plebiscite showed that 99.4 per cent of the population of Nevis favored separation. In the 1984 elections, Kennedy Simmonds and his government increased their parliamentary representation. Simmonds was re-elected in March 1989, in line with the US interests in the region.

[4] In 1992, the Concerned Citizens' Movement won the election in Nevis, ousting the Nevis Reformation Party (NRP) led by Daniel Simeon. Together with the People's Action Movement, the NRP made up the main coalition. Weston Paris, Governor-General Sir Clement Athelston's representative on Nevis, died in unclear circumstances.

[5] The November 1993 election was inconclusive. Further elections were held in July 1995, and were won by the St Kitts-Nevis Labor Party led by Denzil Douglas. In 1998 the Concerned Citizens' Movement, led by Vance Amory, Prime Minister of Nevis, failed to win the necessary two-thirds of the vote to achieve independence.

[6] Hurricane George, in late 1998, caused serious damage to 80 per cent of homes. Reconstruction on the islands took many months.

[7] In late 2000, UNESCO declared the Fort in Brimstone Hill National Park a World Heritage Site.

[8] Prime Minister Douglas was awarded the 2001 Gandhi-King-Ikeda Peace Prize, by the Martin

Luther King International Chapel at Morehouse College, for his work for unity and peace.

[9] In March 2004, Douglas visited the Dominican Republic to take part in a regional meeting on AIDS organized by the Caribbean Community (CARICOM). As part of an emergency plan to address the pandemic in the Caribbean, the US would contribute $15 million over five years. Douglas said 'stigma is a challenge that blocks the

progress of AIDS programs in the Caribbean', and that 'discrimination and exclusion leave many without access to treatment'.

[10] The Miracle Mission, a program co-ordinated with the Government of Cuba and started in August 2005, had assisted in March 2006 more than 8,000 patients with eyesight problems. The overwhelming majority of operations performed under this program had been successful. ■

LAND USE

2003/2005

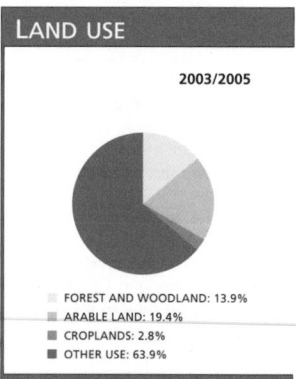

- FOREST AND WOODLAND: 13.9%
- ARABLE LAND: 19.4%
- CROPLANDS: 2.8%
- OTHER USE: 63.9%

PROFILE

ENVIRONMENT
The territory is divided between St Kitts, 168.4 sq km, and Nevis, 93.2 sq km. The two islands are in the Windward Islands of the Lesser Antilles. They are of volcanic origin, hilly, with a rainy tropical climate, tempered by sea winds which make the land fit for plantation crops, especially sugarcane.

SOCIETY
Peoples: the majority of the population are people descended from African slaves and European colonizers (mulattos). There are British, Indian and Pakistani minorities. **Religions:** Mainly Protestant (76.4 per cent, of which Anglican 36.2 per cent, Methodist 32.3 per cent); Roman Catholic 10.7 per cent. **Languages:** English (official). **Main Political Parties:** Saint Kitts and Nevis Labor Party (SKNLP), People's Action Movement (PAM); Concerned Citizens' Movement (CCM).

THE STATE
Official Name: Federation of St Christopher (St Kitts) and Nevis. **Administrative divisions:** 14 parishes. **Capital:** Basseterre (St Kitts) 13,000 people (2003). **Other Cities:** Charlestown (Nevis) 1,300 people; Saint Paul's 1,200 (2000). **Government:** Queen Elizabeth II has been Head of State since February 1952. Cuthbert Montroville-Sebastian, Governor-General appointed by the British Crown in January 1996. Denzil Douglas, Prime Minister since July 1995, re-elected in 2000. There is a National Assembly with 15 members. **National Holiday:** 19 September, Independence (1983). **Armed Forces:** Royal St Kitts and Nevis Police Force; Defense Force, and the Coast Guard.

St Lucia / Saint Lucia

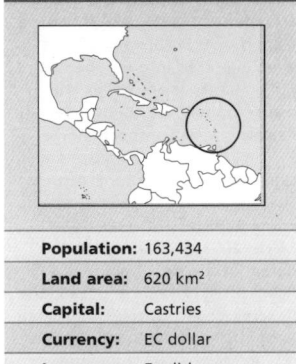

Population:	163,434
Land area:	620 km²
Capital:	Castries
Currency:	EC dollar
Language:	English

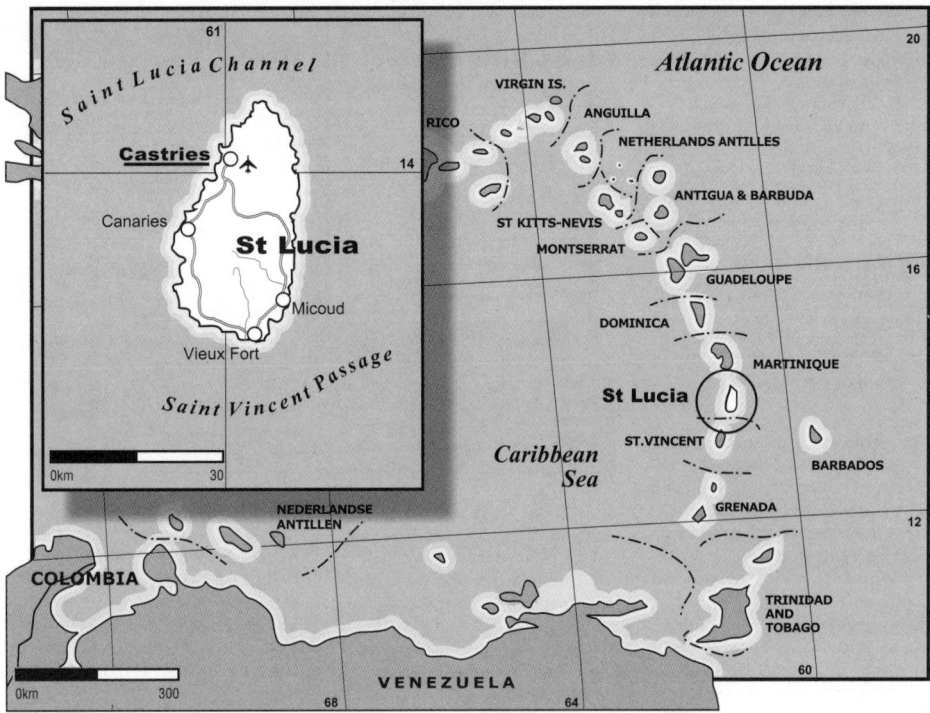

B efore Christopher Columbus named it Santa Lucia in 1502, the East Caribbean island had already been conquered by the Caribs, who had expelled the first inhabitants, Arawak indians from South America.

2 Neither the Spaniards nor the British defeated local resistance. In 1660 the French settled on the island, starting a dispute with Britain which lasted 150 years. Over this period, the flag of St Lucia changed 14 times.

3 In 1814, the Treaty of Paris transferred the island from France to Britain, which ruled until independence in 1978. France left the legacy of *patois*, a pidgin language of mixed African and French.

4 Under British rule, St Lucia became one big sugarcane plantation populated by African slave laborers. Agriculture is still the main economic resource but sugar gave way to banana cultivation. There are also cocoa and coconut crops.

5 The island was part of the Colony of the Windward Islands, and between 1959 and 1962 St Lucia formed part of the West Indies Federation. In 1967 the island became more autonomous and adopted a new constitution as one of the Federated States of the Antilles, negotiating its independence separately.

6 In 1979, in the first elections held as an independent nation, the St Lucia Labor Party (SLP) beat Prime Minister John G M Compton and his United Workers Party (UWP).

7 The new Prime Minister, Allen Louisy, promised to help workers and peasants and to encourage small business as a means of curbing

unemployment. George Odlum, Deputy Prime Minister and also leader of the SLP's 'new left' wing, promoted the country's entry to the Non-Aligned Movement and established diplomatic relations with Cuba and North Korea.

8 After repeated political crises, the UWP won the 1982 and 1987 elections. Compton returned to power with a conservative platform: a market economy and adjustment measures recommended by the IMF. The increase in exports and tourism revenues was not enough to leave economic crisis behind, which continued through the 1990s.

9 The years 1994 and 1995 were marked by protests from workers on plantations growing bananas - the island's main export - and also by dock employees, demanding higher wages. In 1996, Vaughn Allen Lewis (UWP) was elected Prime Minister.

10 In 1997, the SLP won the elections. Upon taking office as Prime Minister, Kenny Anthony formed a commission to investigate corruption during the UWP administration.

11 The summit of Caribbean nations in 1998 decided to remove custom tariffs between member countries to compensate for the reduction of US support. During the summit Anthony stated his 'deep discomfort' over the US policy of not including Caribbean textile industries in the North American Free Trade Agreement (NAFTA).

12 St Lucia, supported by other countries, withdrew from the talks at the WTO in Geneva (1999), after the WTO refused to discuss

US sanctions against the European Union (EU) regarding the special treatment given by the EU to Caribbean banana exports, as against those from Latin America.

13 In 2000, churchgoers in Castries were attacked with machetes by a group who murdered a nun, injured 13 others and then set fire to the building. The attackers said that 'God had told them to carry out the attack, because of the corruption within the Catholic Church'.

14 In 2004, two British subjects visiting the island were sentenced to a six-year prison term for trying to smuggle 2.5 kilos of cocaine from St Lucia on a flight to London.

15 In July 2005, the Government announced that it would promote an alliance with Venezuela to jointly fight drug trafficking. Prime Minister Kenny Anthony and Foreign Affairs Minister Petrus Compton would chair talks with Caracas. ■

WORKERS

LABOR FORCE	2004

■ FEMALE: 41%   ■ MALE: 59%

PROFILE

ENVIRONMENT
One of the volcanic Windward Islands of the Lesser Antilles, south of Martinique and north of St Vincent. The climate is tropical with heavy rainfall, tempered by ocean currents. The soil is fertile; bananas, cocoa, sugarcane and coconuts are grown.

SOCIETY
Peoples: Most inhabitants descend from African slaves and their integration with European colonists. There is also a minority of Europeans. **Religions:** Roman Catholic 79 per cent; Protestant 15 per cent, of which Seventh-Day Adventist 6.5 per cent, Pentecostal 3 per cent; other 5.5 per cent. **Languages:** English (official) and a local patois derived from French and African elements. **Main Political Parties:** Saint Lucia Labour Party; United Workers Party.

THE STATE
Official Name: St Lucia.
Capital: Castries 14,000 people (2003).
Other Cities: Vieux Fort 4,600 people; Micoud 3,700 (2000).
Government: Queen Elizabeth II has been Head of State since February 1952. Calliopa Pearlette Louisy, Governor-General appointed by the British Crown in September 1997. Kenny Anthony, Prime Minister since March 1997. Bicameral Parliament: Senate with 11 members and House of Assembly with 17 members.
National Holiday: 13 December, Independence Day (1978), and discovery by Christopher Columbus.

St Vincent and The Grenadines / Saint Vincent

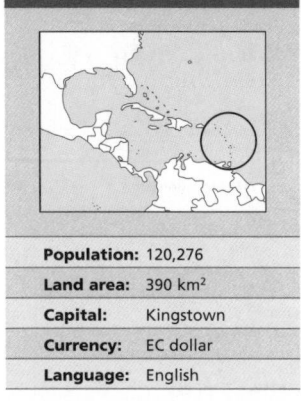

Population:	120,276
Land area:	390 km²
Capital:	Kingstown
Currency:	EC dollar
Language:	English

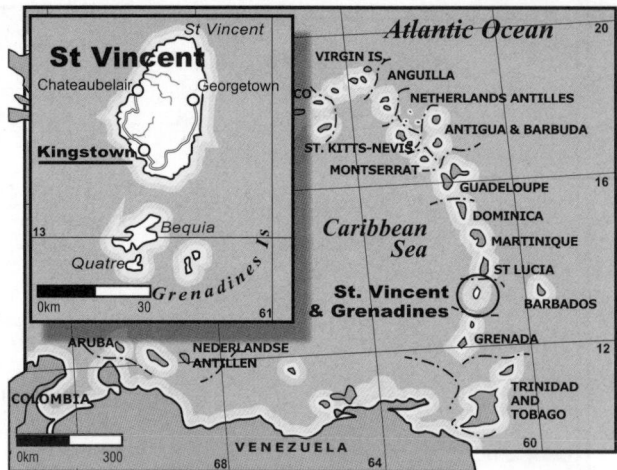

The island's first inhabitants were Arawaks; they were displaced by the Caribs, who lived on the island when Columbus arrived in 1498. In 1783, St Vincent became a British colony. However, the local people resisted European conquest. Former slaves who had rebelled on the neighboring islands and had taken refuge on St Vincent joined the Caribs to oppose the invaders. In 1796, they were defeated and exterminated or deported.

2   The island developed a plantation economy using slave labor, the chief crops being sugarcane, cotton, coffee and cocoa, and in 1833 became part of the Windward Islands colony. In 1960, together with the Grenadine Islands, it was granted a new constitution with substantial internal autonomy.

3   St Vincent became a self-governing state in association with the UK in 1969. The post of head minister - similar to that of Prime Minister, but with more limited powers - was held then by Milton Cato, together with the pro-US St Vincent Labor Party (SVLP). Defense and foreign relations continued to be controlled by Britain. Independence was declared in October 1979.

4   The elections in December 1979 reinforced the predominance of the SVLP, while Ebenezer Joshua's neo-colonial People's Political Party (PPP) received only 2.4 per cent of the vote.

5   The new government faced an armed rebellion of Rastafarians led by Lennox 'Bumba' Charles on Union Island. This rebellion was quickly put down by troops from Barbados.

6   In 1980, the Government faced a serious socio-economic crisis, enabling the popular movements to gain ground. In May 1981, the National Committee in Defense of Democracy was formed, supported by several opposition parties, the labor unions and other organizations. Several days later, the Government attempted to impose repressive legislation designed to maintain 'public order', triggering mass protests.

7   Cato's government supported the US invasion in Grenada and sent a police detachment to join the occupation forces (see Grenada). Cato called early elections, but the economic crisis resulted in the New Democratic Party (NDP) winning the election. James Mitchell became Prime Minister.

8   In the May 1989 election, James Mitchell (NDP) was re-elected, going on to sign an agreement with the Prime Ministers of Dominica, St Lucia and Grenada to create a new state of the four islands in 1990 (see Dominica).

9   Given the Grenadinians' secessionist feelings, which had already erupted in violence in 1980, Mitchell created a Ministry of Grenadine Affairs and appointed Herbert Young, a Grenadinian, as Minister of Foreign Affairs.

10   Pressure from Washington led to the eradication of marijuana crops by an army battalion in 1998. Cannabis growers complained that they would lose their livelihoods through the destruction of their plantations. In 1999, more than 25,000 banana growers protested because of a case brought to the World Trade Organization by the US, which stated that the arrangement under which bananas from former European colonies were imported into Europe was unfair. Bananas are St Vincent's main crop and the country depends heavily on the European market.

11   In 2000 the unemployment rate reached 30 per cent. Arnhim Eustace was appointed the new Prime Minister in October. The March 2001 elections led to the defeat of the New Democratic Party (NDP) after 17 years in power, and its replacement by the Unity Labor Party (ULP), led by Ralph Gonsalves.

12   In March 2002 the island declared as its main national hero Caribbean leader Joseph Chatoyer, 200 years after his death. Chatoyer led the nationalist movement against English colonization and fought until his death.

13   In March 2004 Deputy Prime Minister Louis Straker met with Cuban officials to sign technical cooperation agreements - training of qualified personnel in Cuba to promote educational development in St Vincent - and to intensify already existing cooperation in the agricultural and construction sectors.

14   The Cabinet was re-organized in May 2005. Mike Browne became Foreign Affairs Minister, replacing Louis Straker, who was appointed Deputy Prime Minister. In December, the SVLP won the general elections and Gonsalves was re-elected for a second term as Prime Minister. ■

PROFILE

ENVIRONMENT
Comprises the island of St Vincent (345 sq km) and the northern part of the 32 Grenadines islands (43 sq km) including Bequia, Canouan, Mustique, Matreau, Quatre, Savan and Union. They are part of the Windward Islands of the Lesser Antilles. Of volcanic origin, the islands have fertile rolling hills. The climate, tropical with heavy rainfall and tempered by ocean currents, is fit for plantation crops. St Vincent is a leading arrowroot producer, a plant with starch-rich rhizomes, used in the manufacture of a type of paper used in electronics. Bananas are the main export. The population is mainly concentrated on the island of St Vincent. There are severe pollution problems in the coastal waters.

SOCIETY
Peoples: Descendants of African slaves 82 per cent, mixed 14 per cent; there are also European, Asian, and indigenous minorities.
Religions: Anglican and other Protestant; Catholic.
Languages: English (official); also a local dialect.
Main Political Parties: The New Democratic Party (NDP); the Unity Labor Party, social-democratic (ULP); the People's Independent Movement.

THE STATE
Official Name: St Vincent and The Grenadines.
Capital: Kingstown 29,000 people (2003).
Other Cities: Georgetown 1,600 people; Byera 1,300 (2000).
Government: Head of State, Queen Elizabeth II. Frederick Ballantyne, Governor General appointed by Britain, since 2002. Ralph Gonsalves, Prime Minister, elected in 2001, re-elected in 2005. There are 21 members in the House of Assembly.
National Holiday: 27 October, Independence Day (1979).

LAND USE

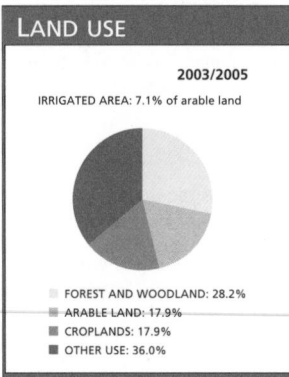

2003/2005

IRRIGATED AREA: 7.1% of arable land

- FOREST AND WOODLAND: 28.2%
- ARABLE LAND: 17.9%
- CROPLANDS: 17.9%
- OTHER USE: 36.0%

PUBLIC EXPENDITURE

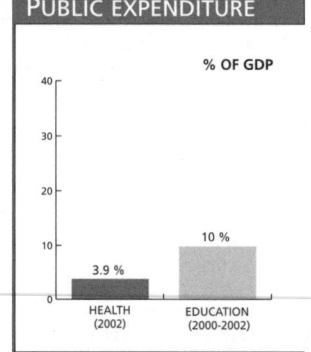

% OF GDP

3.9 % HEALTH (2002)

10 % EDUCATION (2000-2002)

Samoa / Samoa

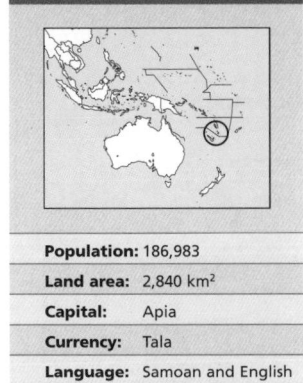

Population:	186,983
Land area:	2,840 km²
Capital:	Apia
Currency:	Tala
Language:	Samoan and English

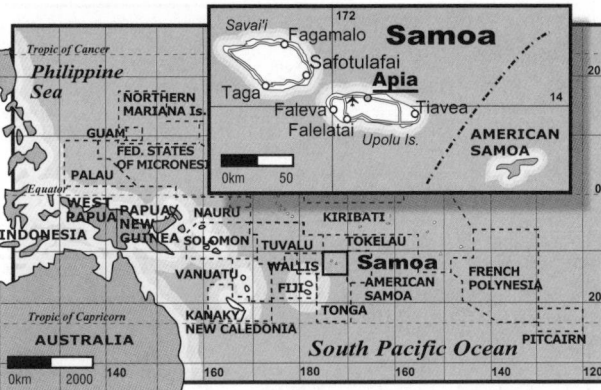

The archipelago of Samoa has been inhabited since at least 1000 BC. The first Samoans of the Polynesian ethnic group (See box on Melanesians and Polynesians) developed a complex social structure organized around the family and its heads, the Matai. Four of these local groups still hold a privileged position: the Malietoa, the Tamasese, the Mataafa and the Tuimalealiifano. Their heads self-designate themselves as being 'descended from kings' (*tama aiga*).

2   The Dutch were the first Europeans to visit the islands in 1722, but colonization did not begin until the end of the 19th century.

3   For decades the US, Britain and Germany were in dispute over Samoa. In 1855, Germany finally occupied the islands. German merchants bought copra with Bolivian and Chilean currency valued at ten times less than its real worth. In 1889 a new treaty recognized US rights over the part of Samoa located east of meridian 171; 'rights' which the US still retains. The western half remained under German rule.

4   In 1914, upon the outbreak of World War I, New Zealand occupied the German part of the island, which was later granted to New Zealand/Aotearoa by the League of Nations as a trust territory.

5   In 1920, an influenza

epidemic killed 25 per cent of the population. The 'Mau' movement that began to spread throughout the archipelago, in resistance to the foreign governments, carried out a nine-year-long civil disobedience campaign, which eventually became a vigorous pro-independence movement.

6   In 1961 after intense protests and pressure from the UN, a plebiscite was held for Samoans to vote on independence. This was achieved the following year, with a Constitution based largely on the traditional social structure and the executive power in the hands of two rulers, Tupua Tamasase Meoble and Malietoa Tanumafili.

7   After being elected Prime Minister in 1970, Tupua Tamasese Lealofi launched a battle against the Matai, in favor of foreign corporations. The family leaders opposed the establishment of those companies on the islands.

8   The 1976 elections were won by the opposition. Tupuola Tais became Prime Minister. In 1979 he retained office by only one vote in parliament.

9   In February 1982, Va'al Kolone, leader of the Human Rights Protection Party (HRPP), became Prime Minister. In September he was removed from government, amid accusations of corruption and abuse of power.

10   In April 1988, Tofilau Eti Alesana came to power. His political party won an absolute majority in the legislature in the 1985 elections when it obtained 31 of the 47 seats.

11   The 1991 constitutional reform extended the parliamentary term from three to five years, and

increased the number of seats from 47 to 49. Fiame Naomi became Minister of Education that year, the first woman to be appointed to the cabinet.

12   Prime Minister Tofilau Eti Alesana kept his post in the 1996 election and continued to liberalize the economy. Health reasons led him to resign in November. He was replaced by Tuilaepa Sailele Malielegaoi.

13   In July 1997, a constitutional amendment changed the name of the country to Samoa, eliminating the adjective 'Western'.

14   In April 2000, Leafa Vitale and Toi Aukuso, respectively the Minister for Women's Affairs and a former Minister of

Communications, were sentenced to death by hanging for murdering a Cabinet colleague. According to the charges, they had killed the Minister of Public Works, Levaula Kamu, during a political rally held the previous year.

15   After a close-run electoral race, Tuilaepa Sailele Malielegaoi was re-elected in March 2001 for a second term.

16   In 2001, the Organization for Economic Cooperation and Development (OECD) demanded that Samoa eliminate bank secrecy and introduce financial controls. The OECD categorizes Samoa as a 'black box' country because it hides illegal transactions that are evading taxes in their home countries.

17   The Prime Minister of New Zealand, Helen Clark, apologized officially in June 2002 for the 'brutal treatment' of Samoans by her country in colonial times. That treatment 'was little known in New Zealand, but very well known by Samoa', and was due to the incompetence and ineptitude of the New Zealand administration of that time, she said.

18   The HRPP - the ruling party for the last 24 years - won the April 2006 legislative elections with 30 of the 49 Parliament seats, gaining eight more than in 2001. Tuilaepa was confirmed as Prime Minister for a new term. ∎

PUBLIC EXPENDITURE

% OF GDP

HEALTH (2002)	EDUCATION (2000-2002)
4.7 %	4.8 %

WORKERS

LABOR FORCE   2004

■ FEMALE: 32%   ■ MALE: 68%

American Samoa / American Samoa

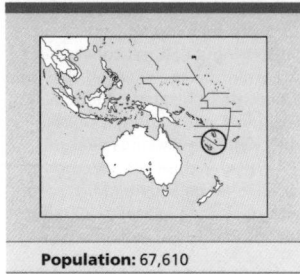

Population:	67,610
Land area:	200 km²
Capital:	Pago Pago
Currency:	US dollar
Language:	Samoan and English

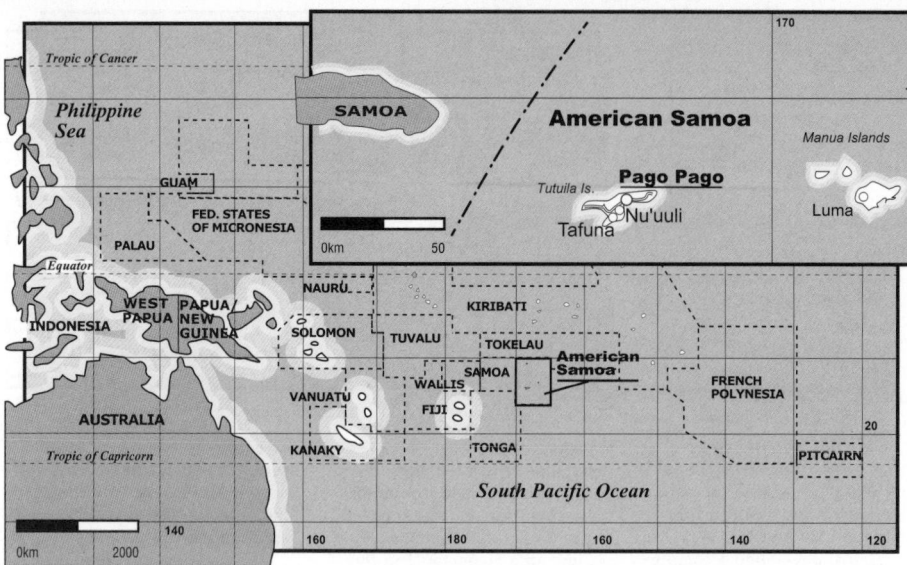

Inhabited since the 7th century BC by Melanesians (see box: 'Melanesians and Polynesians'), the island was reached by Europeans in the 18th century. The colonial powers Germany, Britain and the US were in dispute over its possession 150 years later. A treaty in 1899 settled the conflict, granting the United States the seven islands east of meridian 171. Traditional social structures were maintained but agriculture was not stimulated so the population became totally dependent on the external colonial economy. This situation resulted in an increasing number of emigrants; over half of the Samoan population currently lives in Hawaii and other parts of the United States.

² On 5 December 1984, the UN General Assembly considered Eastern Samoa's right to self-determination and independence. A unanimous vote reiterated that factors such as territory, geographic location, population and meager resources should not hinder independence. The US, in its role as administrative power, was urged to hasten the decolonization process and to implement an educational program to assure Samoans' full awareness of their rights. The islanders, however, seemed to be content with their existing status, which allowed them to emigrate to the US without restrictions. There were no organized pro-independence groups.

³ In 1984, Governor Coleman (elected in the first elections for governor held in 1977) submitted proposals for a new Constitution in American Samoa for ratification by the US Congress. The proposals were withdrawn in May of the same year, because it was feared that they would be harmful to the interests of US citizens. In November, AP Lutali was elected Governor and Faleomavaega Eni Hunkin became Vice-Governor.

⁴ In October 1988, the delegate to the US House of Representatives, Fofo Sunia, was sentenced to a five to 15 week term in prison for fraud. Hence, Eni Hunkin replaced Sunia. In November, Coleman was re-elected for his third term and Galeani Poumele replaced Hunkin as Vice-Governor.

⁵ Despite reforms to the 1967 Constitution during the 1980s, proposed changes were not ratified by the US Congress. According to the Constitution currently in effect, in addition to a governor, who is elected for a four-year term, there is also a legislature, or *Fono* with an 18-member Senate elected every four years by the *matai*, or clan heads. There is also a 20-member House of Representatives elected by direct popular vote for two-year terms. Women do not have the right to vote. Samoans are considered US 'nationals', but not 'citizens' of that country, that is without the right to vote. They do send a delegate to Congress - also without the right to vote.

⁶ After his re-election as Governor in 1992, Lutali took measures to cut public spending, especially by reducing the number of government employees. The projected social security reform in the US and its dependencies led to a debate in the second half of 1996 about the consequences for the inhabitants of American Samoa.

⁷ In November 1996, Tauese Pita Sunia was elected Governor with 51 per cent of the vote, to replace AP Lutali. Togiola Tulafano became Vice-Governor. Both took office in January 1997.

⁸ In November, the Government in Pago Pago imposed a curfew between 9:00pm and 6:00am, after expressing concern over the apparent rise in crime rates. In February 1998, the Governor was accused of diverting funds for his own benefit. In April, Republican Eni Faleomavaega was re-elected senator to the US with 86 per cent of the vote.

⁹ The Government had to impose austerity measures in order to reduce the deficit, including a shortened workweek, a rise in taxes and cost reductions.

¹⁰ In November 2000, Governor Tauese P Sunia, from the Democrat Party, was elected with 50.7 per cent of the vote.

¹¹ In March 2003, Sunia died before ending his mandate and in April his deputy, Togiola Tulafona, took up the post.

¹² In May 2006 a Hawaiian Airlines flight from Honolulu toward Pago Pago had to return after a Samoan attacked a crew member. The man was detained by the FBI in Honolulu until he was taken before a federal court.

¹³ That month, Governor Togiola appointed the delegates for the Political Status Study Committee, made up by 11 members: four from Parliament, four named by the Governor, one by the Higher Education Council, one by deputy Faleomavaega and one by Chief Justice Michael Kruse. The committee's report was to be ready that year. ■

PROFILE

ENVIRONMENT

The island occupies 197 sq km of the eastern part of the Samoan archipelago, located in Polynesia, slightly to the east of the International Date Line, northwest of the Fiji islands. The most important islands are Tutuila (where the capital is located), Tau, Olosega, Ofu, Annuu, Rose and Swains. Of volcanic origin, the islands are mountainous with fertile soil on the plains. The climate is rainy and tropical, tempered by sea winds. There is dense, woody vegetation and major streams of shallow waters. The main export product is fish, especially tuna. Bananas and crafts are also exported.

SOCIETY

Peoples: Samoans are mostly Polynesians (89 per cent); Tongans 4 per cent; other 5 per cent. **Religions:** Christian (Protestant 50 per cent, Catholic 20 per cent); others 30 per cent. **Languages:** Samoan, predominant, and English are the official languages. **Main Political Parties:** Democratic Party; Republican Party

THE STATE

Official Name: Territory of American Samoa. **Capital:** Pago Pago (on Tutuila) 52,000 people (2003). **Other Cities:** Tafuna 7,100 people; Nu'uuli 5,300; Fagatogo 3,800 (2000). **Government:** Head of State, President George W Bush. Togiola Tulafono became acting Governor in April 2003, following the death of Tauese P Sunia - who had been re-elected by popular vote in 2000 - until the 2004 elections. There is a bicameral legislature: the House of Representatives with 21 members and the Senate composed of 18 members. Samoans are considered US 'nationals', but do not have the right to vote in US presidential elections while living on the islands. **National Holiday:** 17 April, Territorial Flag Day (1900). **Armed Forces:** Defense is the responsibility of the US.

San Marino / San Marino

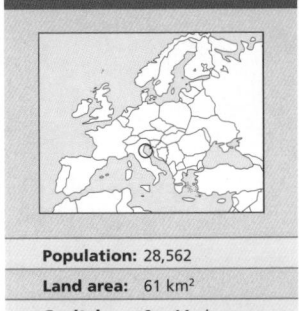

Population:	28,562
Land area:	61 km²
Capital:	San Marino
Currency:	Euro
Language:	Italian

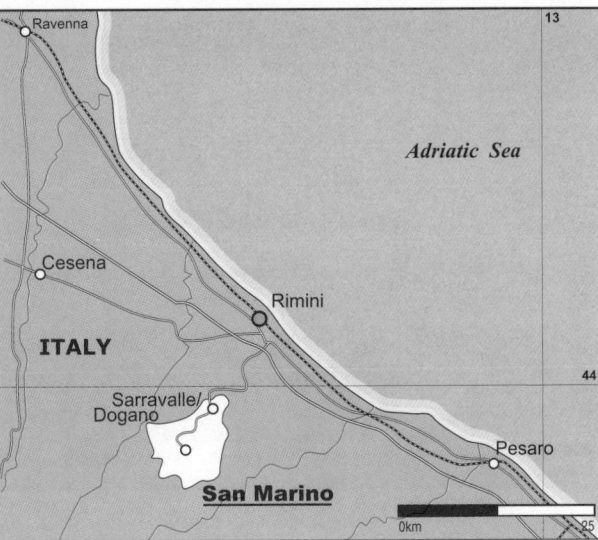

According to archaeological evidence, the territory of San Marino, located between the provinces of Romana and Marca, was inhabited in prehistoric times. The city was founded in 301 AD by a Dalmatian stonemason from the island of Arbe, called Marinus, who after converting to Christianity and escaping religious persecution by emperor Diocletian, took refuge with other Christians on Mount Titano, in the Apennines. Marinus built a community there that, in time, acquired the features of a small state. In memory of the stone cutter, the land was christened 'Land of San Marino', later to be called 'Community of San Marino', until it finally received its current name.

[2] The original system of government was made up of an assembly, the *Arengo*, consisting of the heads of each family. In 1243, two posts of Captain-Regent (*Capitani Reggenti*) were created as joint heads of state. After the fall of the Roman Empire (5th century AD) the lack of a central power favored self-rule for several Italian cities from the 12th century, as well as the development of trade, manufacturing and crafts. San Marino followed this path, with the difference being that it is the only city to remain independent today.

[3] In the 15th century, San Marino's territory grew with support from the Duke of Urbino and through an alliance against Sigismondo Pandolfo Malatesta, lord of Rimini. When the alliance won, Pope Pius II awarded San Marino the castles of Fiorentino, Montegiardino and Serravalle.

[4] The Constitution of San Marino dates from the year 1600, and the Sammarinese regard themselves as the oldest republic in the world.

[5] In 1739 Cardinal Giulio Alberoni invaded the territory, as part of his campaign to recover Italian possessions, but civil disobedience acts and secret messages sent to the Pope appealing for justice achieved their goal of papal recognition of San Marino's autonomy and the restoration of its independent status.

[6] San Marino remained on the sidelines of Italian unification (1830-1870) and stayed independent, signing a Friendship Treaty with Italy in 1862.

[7] The state did not take part in World War I (1914-1918) but the conflict affected the country's economy. At the end of the war, unemployment - already high - and inflation rose considerably.

[8] In 1923, two years after Benito Mussolini came to power, the General Council was dissolved, giving way to the fascist Sovereign and Supreme Council. However, San Marino survived the expansion of Italian fascism and remained neutral during World War II (1939-1945). German troops bombed the republic in July 1944, but Nazi occupation ended after a large demonstration, putting an end to the Supreme Council and paving the way for new elections. During the War, San Marino received more than 100,000 refugees.

[9] After the War, the Communists came to power. In 1945 the Communist Party and the San Marino Socialist Party (PSS) formed a coalition which held power for 12 years.

[10] In 1957 a centrist alliance dominated by the San Marino Christian Democratic Party (PDCS) won control of the government until 1973.

[11] A PDCS-PSS coalition governed for the next five years until in November 1977 the Socialists accused the Christian Democrats of not solving the country's economic problems. They then formed a coalition with the Communists. Early elections were called in May 1978, and the leftist coalition ruled San Marino setting up a highly advanced social welfare system.

[12] San Marino reinforced its links with the West and joined the European Union (EU) and the European Council in 1988. In 1992 it became a member state of the United Nations, and joined the IMF.

[13] In 1990 the Communist Party renamed itself the Progressive Democratic Party, and remained in the coalition with the Christian Democrats, who in 1992 formed a new coalition with the Socialists. This coalition won the 1993 and 1998 elections.

[14] In 2000, San Marino had an enviable standard of living and healthcare system, considered one of the world's best. The Government sought to maintain these through an economic development program to support traditional craftspeople and agriculture.

[15] In October 2004, Giuseppe Arzilli and Roberto Raschi were elected Captains-Regent, while Fiorenzo Stolfi was appointed Secretary of State and Minister of Foreign Affairs.

[16] The Party of Socialists and Democrats significantly increased its representation in Parliament in the June 2006 elections, taking a total of 20 seats, just one less than the ruling PDCS. One of the main challenges faced by the new government coalition was to increase transparency and efficiency within the Parliament itself and in its dealings with the cabinet and the Captains-Regent. ∎

PROFILE

ENVIRONMENT

Located in Italy, San Marino is an independent enclave, south of Rimini on the Adriatic coast. The hilly terrain is dominated by the Apennine peak of Mount Titano (738 m). The climate is Mediterranean. The little industrial activity (mostly in construction) is concentrated in the capital, San Marino, a population center which spreads across Titano's western slope. The main source of income is tourism, including crafts and stamps. Also significant are the remittances sent by the emigrant population, settled mostly in Italy and other neighboring countries.

SOCIETY

Peoples: Sammarinese; Italians.
Languages: Italian and a local dialect.
Religions: Catholic 95 per cent.
Main Political Parties: San Marino Christian Democratic Party (PDCS); Party of Socialists and Democrats (PSD); Popular Alliance (AP); United Left (SU).
Main Social Organizations: Unity Trade Union Central, Democratic General Confederation of Workers, General Confederation of Labor.

THE STATE

Official Name: Serenisima Repubblica di San Marino.
Capital: San Marino, 5,000 people (2003).
Administrative divisions: 9 municipalities (castelli); Acquaviva, Borgo Maggiore, Chiesanuova, Domagnano, Faetano, Fiorentino, Monte Giardino, San Marino, Serravalle. **Other Cities:** Serravalle/Dogano 4,726 inhabitants; Borgo Maggiore 2,366 inhabitants (1996).
Government: Presidential republic. The executive branch consists of the Council of State, with 10 members, chaired by two Captains-Regent Antonio Carattoni and Roberto Giorgetti (since October 2006); Augusto Casali is Secretary of State and Minister of Foreign Affairs (2003). Unicameral legislature: the Great General Council, with 60 members elected for a five-year term by proportional representation.
National Holiday: 3 September, Foundation of the Republic (301 A.D.)
Armed Forces: Volunteers. Military service is voluntary, but all citizens aged between 16 and 65 may be drafted if the State requires them for national defense. Police force.

São Tomé and Príncipe / São Tomé e Príncipe

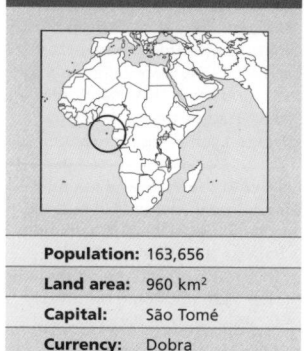

Population:	163,656
Land area:	960 km²
Capital:	São Tomé
Currency:	Dobra
Language:	Portuguese

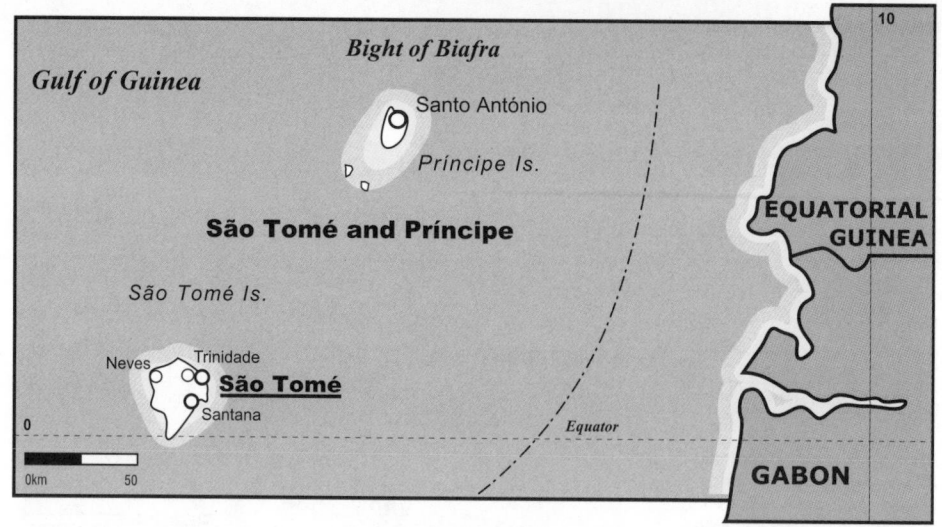

These islands were probably uninhabited when first visited by European navigators in the 1470s. Thereafter, the Portuguese began to settle convicts and exiled Jews there and established sugar plantations, using slave labor from the African mainland. Strategically located 300 kilometers off the African coast, the islands' natural ports were used by the Portuguese as supply stops for ships in the 15th century. Dutch, French, Spanish, British and Portuguese slave traders bought enslaved African laborers to be sold in the American colonies. Some of the slaves remained on the islands, and later became the leading African producers of sugarcane.

[2] Rebellions broke out and a slave named Amador led a revolt that succeeded in taking over two-thirds of the island of São Tomé, where he proclaimed himself ruler.

[3] Soon defeated, the rebels hid in *quilombos* (guerrilla shelters in the forest) after burning their crops.

[4] Agriculture virtually disappeared for three centuries. In the 17th century the islands were briefly held by the Dutch before reverting to Portuguese control. The islands were used to hold slaves in transit, until they recovered their prosperity in the late 19th century with the cultivation of cocoa and coffee. Even after abolition was declared in 1869, slavery continued in a disguised manner ('free' workers signed contracts for nine years at fixed salaries), leading to revolts and an international boycott against the 'cocoa slavery' of the Portuguese colony in the early 20th century.

[5] This neo-slavery system continued until the mid-1900s. A Society for Immigration of São Tomé organized the modern slave trade, 'hiring' plantation workers in other Portuguese colonies: Angola, Cape Verde, Guinea and Mozambique. This flow 're-Africanized' the country as the *filhos da terra* (sons of the earth), the result of several centuries of intermixing between the native people and the Portuguese, mixed with the African immigrants. During the colonial regimes of Salazar and Caetano, repression was particularly harsh. In February 1953, over 1,000 people were killed in Batepá in less than a week.

[6] This massacre demonstrated the need for the rebels to join forces, and in 1969 the Movement for the Liberation of São Tomé and Príncipe (MLSTP) was founded, with two main objectives: independence and land reform.

[7] Foreign companies owned 90 per cent of São Tomé land and, despite the fertile soil, most food was imported due to the island's monoculture policy. Rural workers were one of the major pillars of the MLSTP, and held a 24-hour strike in August 1963, which paralyzed all the plantations.

[8] As the island's terrain did not favor guerrilla warfare, the MLSTP launched an intensive underground political campaign that resulted in its recognition by the OAU (Organization of African Unity) and the Non-Aligned Nations. Together with the MPLA of Angola, the PAIGC of Guinea and Cape Verde, and FRELIMO of Mozambique, the MLSTP joined the Conference of National Organizations of the Portuguese Colonies. It was the only legitimate group in existence when, after the 1974 revolution,

LAND USE

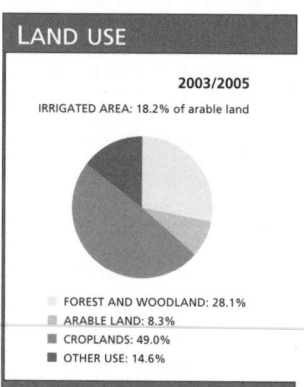

2003/2005

IRRIGATED AREA: 18.2% of arable land

- FOREST AND WOODLAND: 28.1%
- ARABLE LAND: 8.3%
- CROPLANDS: 49.0%
- OTHER USE: 14.6%

PROFILE

ENVIRONMENT
The country comprises the islands of São Tomé (857 sq km) and Príncipe (114 sq km), and the smaller islands of Rólas, Cabras, Bombom and Bone de Joquei in the Bay of Biafra of the Gulf of Guinea, facing the coast of Gabon. The islands are mountainous, of volcanic origin, with dense rainforests, a tropical climate and heavy rainfall. Cocoa, copra and coffee are the main export crops.

SOCIETY
Peoples: Most are Africans of Bantu origin traditionally classified in five groups formed as a result of different migratory waves: the *Filhos da terra* (sons of the earth), descendants of the first enslaved workers brought to the islands and intermingled with the Portuguese; the Angolares, thought to descend from Angolans who came to the islands in the 16th Century; the Fôrros, descendants of freed slaves when slavery was abolished; the Serviçais, migrant workers from Mozambique, Angola and Cape Verde; and Tongans. Since independence, these categories have begun to disappear.
Religions: Roman Catholic, about 80.8 per cent; remainder mostly Protestant, predominantly Seventh-Day Adventist and an indigenous Evangelical Church.

Languages: Portuguese (official); Fôrro; Crioulo, a dialect with Portuguese and African elements, is widely spoken.
Main Political Parties: MDFM-MPCD Coalition (which includes: Force for Change Democratic Movement-Liberal Party; Democratic Convergence Party-Reflection Group); Movement for the Liberation of São Tomé and Príncipe-Social Democratic Party; Independent Democratic Action.
Main Social Organizations: Women's, Youth and Pioneers Organizations linked to the Movement for the Liberation of São Tomé and Príncipe.

THE STATE
Official Name: República Democrática de São Tomé e Príncipe.
Administrative divisions: 7 Districts.
Capital: São Tomé 54,000 people (2003).
Other Cities: Trinidade 14,200 people; Santana 7,700; Neves 7,400 (2000).
Government: Fradique de Menezes, President since September 2001, re-elected in 2003. Tome Vera Cruz, Prime Minister since April 2006. Unicameral Legislature: National Assembly, with 55 members.
National Holiday: 12 July, Independence Day (1975).

Life expectancy
64 years
2005-2010

GNI per capita
$390
2004

Portugal began to free its colonies.

[9] The MLSTP joined a transition government in 1974 and in the following year declared independence. Its accomplishments were impressive: banks and farms were nationalized, medicine was socialized, a national currency was created, a major administrative reform was launched to reorganize public administration, and numerous centers of popular culture, based on the culture-building educational methods of Brazilian Paulo Freire, were created as part of a literacy campaign.

[10] Opposing these reforms was a right wing faction led by Health Minister Carlos da Graça, who fled to Gabon to plot a mercenary invasion of the islands in early 1978. The MLSTP's first congress in August, aimed at strengthening the rank and file, eventually led to the formation of a People's Militia and the creation of mass organizations to 'defend the revolution'.

[11] In March 1986, two opposition groups based outside the country - the São Tomé e Príncipe Independent Democratic Union (UDISTP) and the more radical São Tomé e Príncipe National Resistance Front (FRNSTP), founded by Carlos da Graça - announced the formation of an alliance called the Democratic Opposition Coalition. Its aim was to put pressure on the Government to hold free elections. A month later, a fishing vessel with 76 members of the FRNSTP on board arrived in Walvis Bay, the South African enclave in Namibia. They asked the Pretoria Government to supply military aid needed to destabilize the São Tomé Government. These events led to Carlos da Graça's resignation as president of the FRNSTP. In May, he announced his willingness to cooperate with the Government, on condition that Cuban and Angolan troops stationed in the country be withdrawn.

[12] In 1985, in the midst of the worst drought in the country's history, the Government sought to open up the economy: new legislation was designed to promote foreign investment. Gradually the State relinquished economic control, previously heavily dependent upon such key products as cocoa, coffee and bananas. The Government sought ways to attract foreign capital to the agricultural, fishing and tourism sectors.

[13] In March 1990, the People's National Assembly approved amendments to the Constitution,

IN FOCUS

ENVIRONMENTAL CHALLENGES
Deforestation and soil erosion are significant, as are dwindling natural resources. São Tomé and Príncipe has environmental protection legislation.

WOMEN'S RIGHTS
Women have been able to vote and stand for election since 1975. In 2003, women held 9 per cent of seats in Parliament and 14 per cent of ministerial positions. That year, 91 per cent of pregnant women received prenatal care and 76 per cent of births were attended by skilled health staff*. Domestic violence is a problem, including rape within marriage. However, the scope of the problem is unknown, since disputes tend to remain within the home.

The Constitution grants the same political, economic and social rights to men and women.

CHILDREN
By mid-2005, 80 per cent of the population was at risk due to an outbreak of cholera. Nearly half of the first reported cases were among children. The cholera outbreak was the result of improper hygiene and sanitation practices, which resulted in contaminated food and water sources. Child labor is common throughout the country, especially on the plantations, where conditions are very harsh. In 2004, it was estimated that 14 per cent of children aged between 5 and 14 were working. In 2004, infant and under-five mortality rates remained the same as in 1990. The former amounted to 75 per 1,000 live births and the latter to 118 per 1,000.

One-fifth of newborn babies were underweight while 29 per cent of under-fives were moderately or severely stunted.

INDIGENOUS PEOPLES/ ETHNIC MINORITIES
Since the 15th century, the population has been based on intermarriage between Portuguese convicts, Jews that arrived after 1496, when King Emmanuel I of Portugal ordered their expulsion, and slaves from the African coast that were used as labor.

Notwithstanding the fact that their traditions were banned, *congo* music - a dance to incite people to revolt - and *lundum* - metaphorical songs - created a certain cultural unity among those being colonized.

MIGRANTS/REFUGEES
According to data published in December 2000 by the Embassy of São Tomé and Príncipe in Lisbon, there were 14,251 São Tomé citizens registered in the Consulate, mainly working in the construction, health or education sectors. Five per cent are professionals, and many are doctors and nurses who practice in Portugal. The Embassy estimates there are some 20,000 citizens living in Portugal illegally.

DEATH PENALTY
It was abolished in 1990.

* Latest data available in *The State of the World's Children* and *Childinfo* database, UNICEF, 2006.-

later to be submitted to a referendum. These changes made possible a move to a multi-party system. Independent candidates were admitted in the legislative elections, and the tenure of the President was limited to no longer than two five-year terms.

[14] The first parliamentary elections after independence were held in January 1991. The opposition Democratic Convergence of Leonel d'Alva was voted into power. In March, the former prime minister Miguel Trovoada returned from exile and was unopposed in the presidential elections.

[15] The social and economic situation of the country worsened as the result of an IMF and World Bank imposed austerity plan. Public sector salaries were frozen, a third of the 5,000 civil servants were dismissed and the local currency was devalued by 80 per cent.

[16] The November 1998 elections marked the MLSTP's return to the Government. Upon taking office in January 1999, Prime Minister Guilherme Posser da Costa stated that his center-left administration would have to implement an 'austerity package' while seeking to reactivate the economy through oil exploration and agricultural development.

[17] In April 2002 Gabriel Costa was sworn in as Prime Minister after winning the March

elections. The Government was shared by the MLSTP and a coalition called MDFM/MPCD.

[18] In April 2003 Fradique Menezes, who was first voted in 2001, was re-elected as President. In July, Maria das Neves was elected Prime Minister.

[19] In July 2003, while President Menezes was on visit to Nigeria, military troops took control of the archipelago, arrested the main government authorities and set up a 'national salvation junta'.

[20] Nine days later, the coup's leaders accepted international mediation and returned power to the President. An amnesty was agreed for the coup leaders, as well as the establishment of a new government and new elections.

[21] In September 2003, Menezes launched a complete revamping of the armed forces and publicly criticized the Army for not having protected democracy more effectively during the brief but intense *coup d'état*.

[22] In July, Menezes agreed with his Nigerian counterpart, Olusegun Obasanjo, to develop the exploration and exploitation of São Tomé's crude oil reserves. During a meeting held in Rivers State (the heart of Nigeria's hydrocarbon industry), both Presidents signed an agreement that would promote oil-industry activity in the area, hoping it

would also serve as a model of cooperation for other African countries.

[23] In September 2004, after a series of corruption scandals, President Menezes replaced Prime Minister Das Neves with Damiao Vaz d'Almeida, of the opposition MLSTP party, who came from the small Príncipe island - which usually feels displaced by the government of the archipelago.

[24] In February 2005, São Tome, jointly with Nigeria, US oil corporations Chevron Texaco and Exxon Mobil and Nigerian Dangote Energy Resources, signed an oil and natural gas exploration and production-sharing agreement. In June, amidst allegations of corruption directly involving President De Menezes himself, Vaz d'Almeida handed in his resignation and the MLSTP withdrew from the Government.

[25] In the April 2006 parliamentary elections, the ruling MDFM/MPCD won the majority of seats - 23 out of 55; the MLSTP obtained 19 and the Independent Democratic Alliance 12.

[26] At the end of May, oil exploration firms announced the discovery of oil and natural gas in the joint development zone and the drilling of the first eight wells. Whether the reserves are present in commercially exploitable quantities still remains to be seen. ∎

Saudi Arabia / Al Arabiyah as Saudiyah

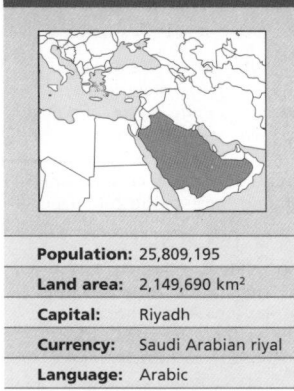

Population:	25,809,195
Land area:	2,149,690 km²
Capital:	Riyadh
Currency:	Saudi Arabian riyal
Language:	Arabic

Arabia was drawn into the orbit of western Asiatic civilization toward the end of the 3rd millennium BC. Caravan trade between South Arabia and the Fertile Crescent began about the middle of the 2nd millennium BC. The domestication of the camel around the 12th century BC made desert travel easier and gave rise to a flourishing society in South Arabia, centered around the state of Saab (Sheila). In eastern Arabia the island of Dolman (Bahrain) had become a thriving entrepot between Mesopotamia, South Arabia, and India as early as the 24th century BC. With the discovery by the Mediterranean peoples of the monsoon winds in the Indian Ocean, Roman and Byzantine seaborne trade between the northern Red Sea ports and South Arabia flourished, extending to India and beyond. In the 5th and 6th centuries AD, successive invasions by Christian Ethiopians and counter-invasions by the Sasanian kings disrupted the states of South Arabia.

² In the 6th century Quraysh - the noble and holy house of the confederation of the Hejaz controlling the sacred enclave of Mecca - contrived a series of agreements with the northern and southern peoples. Under this aegis, caravans moved freely from the southern Yemen coast to Mecca and thence northward to

Byzantium or eastward to Iraq. As a result of the Quraysh's dominant position, members of the house of 'Abd Manaf concluded pacts with Byzantium, Persia, and rulers of Yemen and Ethiopia, promoting commerce outside Arabia. Quraysh had some sanctity as lords of the Meccan temple (the *Ka'bah*) and were themselves known as the Protected Neighbors of Allah. The people on pilgrimage to Mecca were called the Guests of Allah.

³ The Ka'bah, through the additions of other cults, developed into a pantheon, the cult of other gods perhaps being linked with political agreements between Quraysh - worshippers of Allah - and the other clans.

⁴ Muhammad, the prophet of Islam, was born in 570 of the Hashimite branch of the noble house of 'Abd Manaf. Though orphaned at an early age, he never lacked protection by his clan. Marriage to a wealthy widow improved his position as a merchant, but he began to make his mark in Mecca by preaching the oneness of Allah. Rejected by the Quraysh lords, Muhammad sought

affiliation with other groups; he was unsuccessful until he managed to negotiate a pact with the chiefs of Medina. He obtained their protection and became theocratic head and arbiter of the Medina confederation (*ummah*). Those Quraysh who joined him there were known as *muhajirun* (refugees or emigrants), while his Medina allies were called *ansar* (supporters). The Muslim era dates from the *hijrah* (hegira) - Muhammad's move to Medina in AD 622.

⁵ Muhammad's supporters attacked a Quraysh caravan in AD 624, thus breaking the vital security system established by the 'Abd Manaf house, and hostilities broke out against his Mecca kin. In Medina he faced both the necessity to enforce his role as arbiter and to raise supplies for his moves against Quraysh. He overcame internal opposition and, externally, his rising power was demonstrated after Quraysh's failure to overrun Medina, when he declared it his own sacred enclave. Muhammad foiled Quraysh offensives and marched back to Mecca. After taking Mecca in AD 630 he became lord of the two sacred sites. However, even though he broke the power of some Quraysh lords, he then sought reconciliation with his Quraysh kin.

⁶ After Muhammad's entry into Mecca the groups linked with Quraysh came to accept Islam; this meant little more than giving up their local deities and worshipping only Allah. They had to pay the tax, but this was not new since the chiefs had already been taxed to protect the Meccan enclave. From then on Islam was destined for a world role.

⁷ Under Muhammad's successors, the expansionist urge of different

groups, temporarily united around the nucleus of the two sacred sites, coincided with the weakness of Byzantium and Sasanian Persia. Those who converted to Islam launched a career of conquest that promised to satisfy the mandate of their new faith as well as the desire for booty and lands. With families and flocks, they left the peninsula. Such large population movements affected all Arabia; in Hadhramaut they possibly caused neglect of irrigation works, resulting in erosion of fertile lands. In Oman, too, when Arabs evicted the Persian ruling class, its complex irrigation system suffered.

⁸ As the conquests far beyond Arabia poured loot into the Holy Cities (Mecca and Medina), they became wealthy centers of a sophisticated Arabian culture. Medina became a center for Qur'anic (Koranic) study, the evolution of Islamic law, and historical record. Under the caliphs - Muhammad's successors - Islam began to assume its characteristic shape. Paradoxically, outside the cities it made little difference to Arabian life for centuries. After the Prophet's death, the second caliph Omar led the Arab conquest. Within ten years the Arabs occupied Syria, Palestine, Egypt and Persia. With Muawiya, the caliphate became hereditary in the family of the Ummaias and the Arabs became a privileged caste which ruled over the conquered nations.

⁹ In the 8th century, the borders of the Arabian Empire reached from North Africa and Spain to the west, to Pakistan and Afghanistan in the east. Upon moving the capital to Damascus, Syria became the cultural, political and economic center of the Empire. Greco-Roman, Persian and Indian components blended into a mix in which science played an important role. Contrary to Muhammad's expectations, the Arabian peninsula was to remain on the sidelines within the enormous empire, except in religious matters. Mecca, although failing to match Baghdad or Damascus in socio-economic and cultural importance, continued to be the center of Islam and the destination for large pilgrimages from all over the world.

¹⁰ This situation remained unchanged for centuries. The Empire split up; the capital moved to Baghdad and the power of the caliphs passed to the viziers, while culturally Arabic civilization attained the highest standards in all fields of knowledge and artistic creation. Arabic became the language of scholars from Portugal to India. In the peninsula, nomadic groups continued to herd their flocks, the settled population

LAND USE

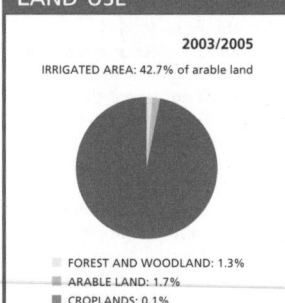

2003/2005

IRRIGATED AREA: 42.7% of arable land

- FOREST AND WOODLAND: 1.3%
- ARABLE LAND: 1.7%
- CROPLANDS: 0.1%
- OTHER USE: 96.9%

PUBLIC EXPENDITURE

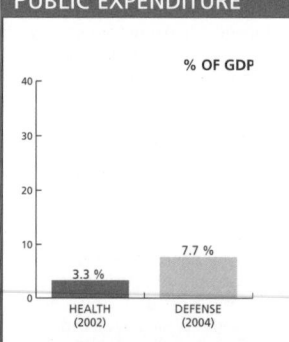

% OF GDP

HEALTH (2002): 3.3 %
DEFENSE (2004): 7.7 %

Life expectancy
73 years
2005-2010

GNI per capita
$10,140
2004

Literacy
79% total adult rate
2000-2004

PROFILE

ENVIRONMENT

The country occupies up to 80 per cent of the Arabian peninsula. There are two major desert regions: the An Nefud in the north, and the Rub al-Khali in the south. Between the two lies the Nejd massif, of volcanic origin, and the plain of El Hasa, the country's only fertile region, where wheat and dates are cultivated. The country's cultivated land amounts to less than 0.3 per cent; 90 per cent of agricultural products consumed are imported. Oil extraction, concentrated along the shores of the Persian Gulf shores, is the source of its enormous wealth.

SOCIETY

Peoples: Saudis are mainly of Arab origin. In recent years there has been large immigration of Iranians, Pakistanis and Palestinians who have settled in the new eastern industrial areas, bringing the number of foreigners to an estimated five million (1992).
Religions: Islam, Sunni orthodox Wahhabism (official) 95 per cent; Shi'a 3 per cent; Christians 1 per cent.
Languages: Arabic, with dialect variations; languages of other communities, particularly of foreign workers, include Farsi and Urdu.
Main Political Parties: Not permitted, but there exist opposition groups, most of them in the diaspora.
Main Social Organizations: Not permitted.

THE STATE

Official Name: al-Mamlakah al-'Arabiyah as-Saudiyah.
Administrative divisions: 13 regions (Al-Baha, Al-Jouf, Asir, Eastern, Hail, Jizan, Madinah, Makkah, Najran, Western Border, Qasim, Riyadh and Tabouk)
Capital: Riyadh (Ar-Riyad) (royal capital) 5,126,000 people (2003).
Other Cities: Jeddah (administrative center) 2,604,500 people; Mecca (Makkah - religious center) 1,229,200 (2000).
Government: Absolute monarchy. King Abdullah bin Abdul Aziz al-Saud, Head of State and of the Government since August 2005. The Consultative Assembly (*Majlis al-Shura*) has advisory role.
National Holiday: 23 September, National Unification (1932).
Armed Forces: 105,000 (plus 57,000 active National Guard). Other: 10,500; Coast Guard, 4,500; Special Security Force, 500.

kept up their commerce, and rivalries between the two were frequently settled through war. As in Muhammad's time, demographic growth was channeled towards conquest, with the emigration of whole communities, such as the Bani Hilals in the 11th century. Trading caravans carrying supplies to Mecca became much more frequent, while the ports became more active as a result of trade with Africa. The peninsula was governed from Egypt, first by Saladin and later by the Mamelukes. The Turks ruled from the 16th century to the 20th century, without introducing any major changes in the socio-economic pattern of the Arab nation.

[11] Under Turkish rule the provinces of Hidjaz and Asir on the Red Sea had some autonomy due to the religious prestige of the *shereefs* of Mecca, descendants of Muhammad. The interior, with Riyadh as the main urban center, became the Emirate of Najd at the end of the 18th century, through the efforts of the Saud family supported by the Wahabite sect (known as the Islamic Puritans). In the 19th century, with Turkish assistance, the Rated clan forced Abd al-Rahman ibn Saud out of power; the ousted leader sought exile in Kuwait. In 1902, his son Abd al-Aziz, again backed by the Wahabites, organized a religious-military sect, the Ikhwan, in which he enlisted nearly 50,000 Bedouins to reconquer Najd. Twelve years later the Saudis defeated the Rachidis and added the Al-Hasa region on the Persian Gulf which had been controlled by the Turks, against whom the Saudi forces fought during World War I. At the end of the conflict, Britain - the major power in the area - faced a difficult situation. In exchange for Abd al-Aziz's continued anti-Turk campaign, Britain had promised to guarantee the integrity of his state. But for the same reason it had also promised to make Hussein ibn Ali (the shereef of Mecca) king of a nation that would encompass Palestine, Jordan, Iraq and the Arabian peninsula.

[12] The Emir of Najd thought that Britain would not keep its word to Hussein. Such a powerful kingdom, ruled by the Prophet's family with the capital in the holy city, would alter the regional balance of power. But in 1924 Hussein proclaimed himself caliph (see Jordan). Abd al-Aziz invaded his territory immediately, despite British opposition, and in January 1926 was declared King of Hidjaz and Sultan of Nadj in the great mosque of Mecca. Six years later the 'Kingdom of Hidjaz, of Nedj and its dependencies' was formally unified under the name of Saudi Arabia.

[13] In 1930, the monarch gave US companies permission to drill for oil. When he died in 1953, his son Saud squandered the Kingdom's Aramco oil company revenues on his playboy lifestyle. In 1964 the country was on the verge of bankruptcy when Saud was ousted by his brother Faisal, an able diplomat who had also proved a valiant soldier in the wars. Monogamous, deeply religious and very austere, Faisal gave new life to the country's economy and began to invest petrodollars in ambitious development programs, though maintaining the traditional feudal structure headed by the autocratic ruler. Under him, the Emirs ruled the provinces, with the support of chiefs and their desert armies. Other sectors of the population had no say in government.

[14] Faisal rejected the Soviet Union and any other system linked with atheism, including Nasser's nationalism in Egypt, as well as Iraqi or Syrian Ba'athism. His strategic alliance with the US was seen as 'natural', but was undermined by US support for Israel after the 1967 war and rivalries with neighboring Iran, which under the Shah Pahlevi also played watchdog for Washington's interests in the Gulf.

[15] During the 1973 Arab-Israeli war, Faisal supported an oil embargo on the countries backing Israel, including the US. The sudden oil shortage allowed the Organization of Petroleum Exporting Countries (OPEC) to hike oil prices rapidly, heralding a new era in international relations. In 1975, Faisal was murdered by an apparently insane nephew. His brother Khaled was named as his successor. However, due to the latter's ill health, his brother Crown Prince Fahd ibn Abd al-Aziz became the ruler.

[16] Oil revenue, which amounted to $500 million a year when Faisal was crowned in 1964, had grown to almost $30 billion when he died. New cities, universities, hospitals, freeways and mosques sprouted up everywhere. Yet there was surplus money available. Instead of planning oil production to meet the country's needs, which would have avoided the fall in prices and the weakening of OPEC during the 1980s, fortunes accumulated in Western banks. Thus, Saudi Arabia tied its fortune to the industrialized capitalist world. In addition it created a surplus of money in circulation, which the banks lent to various Third World countries for some questionable projects and ventures. When the countries defaulted on their huge debts, this created the 1984-85 foreign debt crisis, with a rise in interest rates.

[17] Muslim fundamentalist groups denounced the Saudi dynasty for allegedly betraying Islam, leading to violent confrontations in 1979. The progressive rise to power of new members of the ruling family, trained in European and US universities and military academies, rather than in the traditional desert-tent Qur'anic schools, was often seen as contradicting the theological basis which legitimized Saudi monarchy.

[18] After the overthrow of the Shah of Iran in 1979, the Saudi Government drew closer to the US. After king Khaled's death, in 1982, his brother Fahd - the architect of Saudi Arabia's modernization - became king.

[19] The previous year Fahd had drafted a peace plan for the Middle East which had been approved by several Arab countries, the PLO and the US, though it collapsed after Israeli opposition. This plan proposed the creation of a Palestinian State with Jerusalem as its capital, the withdrawal of Israel from the occupied Arab territories and the dismantling of Jewish settlements established since 1967.

[20] Saudi Arabian links with Washington continued to strengthen through large weapon supplies and the construction of two naval bases in Jubail and Jiddah. Also, investments and bank deposits of the Kingdom were closely linked with the performance of the US economy.

[21] The 1985-89 five-year plan promised a 'wide income redistribution'. Yet that intention coincided with the first signs of economic trouble in the Kingdom. In 1984, due to another drop in the price of crude, the official budget closed with a deficit for the first time. The Minister of Industry, Ghazi Al Gosaibi, was forced to resign after writing a poem which made reference to corruption.

[22] During the Iran-Iraq war (1980-1990), Saudi Arabia backed Iraq financially, afraid that the Iranian Islamic revolution might spread through the Gulf. Fahd changed his title to Guardian of the Holy Sites, but every year some pilgrims to Mecca protested about the alliance between Riyadh and Washington and by what they considered a commercialization of holy places, surrounded today by shopping centers, highways and other symbols of transnational

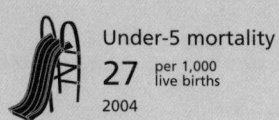

Under-5 mortality
27 per 1,000
live births
2004

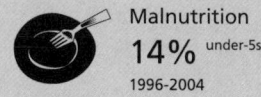

Malnutrition
14% under-5s
1996-2004

culture. In 1987, Saudi police fired on a march by Iranian women and disabled war veterans in Mecca, killing hundreds of pilgrims. In retaliation, the embassies of Saudi Arabia and Kuwait in Tehran were attacked and burned, and relations between both countries grew very tense, remaining so until 2000, when an economic co-operation treaty was signed.

[23] In March 1992, Fahd issued several decrees called 'The Basic System of Government', aimed at decentralizing political power. It established an Advisory Council, with the right to review all matters of national policy, and the *mutawein*, religious police whose mandate was to ensure the observance of Islamic customs.

[24] The hostility of a segment of the population toward the US grew stronger, reaching a critical point with an attack on the US Al Khobar base that killed 19 soldiers in June 1996.

[25] Due to his failing health, in 1996 Fahd transferred power to his brother Abdullah bin Abd al-Aziz al Saud - who had been confirmed as heir four years earlier.

[26] In September 2002, with the prospect of a second Gulf War, Prince Saud Al-Faisal, the Saudi Foreign Minister, said that Saudi Arabia would only allow the use of its territory for military action against Iraq if that action was supported by a UN Security Council resolution. In October the borders between Saudi Arabia and Iraq were officially opened for the first

time since the invasion of Kuwait in 1990.

[27] In May 2003, the explosion of several car bombs killed over 30 people in a residential area of Riyadh. In November, another attack killed 17 people. Terrorist violence continued throughout 2004 and the first attack on a government building took place in April, when a car bomb was detonated next to the agency in charge of the General Security in the Kingdom. The following month there was a clash between attackers identified as terrorists and Saudi security forces in the port city of Yanbu.

[28] A series of terrorist attacks in Riyadh, carried out by suspected al-Qaeda members in May and June 2004, caused turmoil in the country and resulted in an oil price rise, which reached record levels. The country's capital witnessed violent attacks against the Al-Khobar Petroleum Center, the headquarters of the Organization of Arab Petroleum Countries, and also against the luxurious Oasis Resort, which was considered as a heavily fortified site.

[29] Calls for political reform mounted, backed by the West who saw this as the way to shore up their allies in Saudi Arabia, faced with growing unrest and terrorist attacks. In late 2003, the authorities announced that elections (municipal) would be held for the first time in the Kingdom's history. According to the official statement, the measure aimed to 'increase

citizen participation in local political management through the empowering of municipal councils'. Rising demands for reform resulted in the announcement in March 2004 that women would be able to vote and run for office, although the announcement was made through the Saudi embassy in London, and the statement said it was hoped that both men and women 'would have the chance to vote'.

[30] In the same month, King Fahd issued a limited amnesty offer for those suspect terrorists who turned themselves in before the end of July. This offer was announced after al-Qaeda leader in Riyadh, Abdul Aziz al-Muqrin, was shot dead in a clash with security forces.

[31] In August, the Government announced that Saudi Arabia would hold municipal elections in November. These elections would be the first step towards a democratization of the country since the establishment of an absolute monarchy 70 years ago. There were no stipulations regarding the minimum age to vote or women's participation. Foreign correspondents pointed out that the political reform was accelerated after pressure from the US.

[32] At least seven people died - four of them Saudi guards - in an attack on the US consulate in the city of Jeddah, located in the west of the country. In January 2005, two car bomb explosions following an attack on the Ministry of the Interior were attributed to al-Qaeda.

[33] In February 2005, the municipal elections were held without the participation of women. In August, King Fahd died after a decade of incapacitating ill-health. Crown Prince Abdullah, who had been acting as Regent, was sworn in as new King and Prime Minister and he named Defense Minister Prince Sultan as Crown Prince and heir to the throne. Among the greatest challenges to be faced were the threats of Islamic militants and a reduction in the high expenditure of the royal family in a kingdom with a rising unemployment rate.

[34] In November, following 12 years of talks, Saudi Arabia - the world's largest oil exporter - became the 149th member of the World Trade Organization. The Commerce and Industry Minister, Hashim Yamani, stated that this was a 'high point in the program of economic and structural reform undertaken by Saudi Arabia'.

[35] In January 2006, 360 pilgrims were crushed to death during a stone-throwing ritual in Mecca. In a separate incident, more than 70 pilgrims died when the building they occupied collapsed.

[36] The Government announced in February that it had foiled a planned suicide bomb attack - attributed to al-Qaeda - on a major oil-processing plant at Abqaiq, in the east of the country, when the National Guard opened fire against two vehicles that were carrying explosives. The explosion of the vehicles killed two terrorists and injured two guards. ■

IN FOCUS

ENVIRONMENTAL CHALLENGES
Oil production has increased water pollution levels, which grew during the first Gulf War, when 640 kilometers of coast and wetlands were affected by a 4.5 million-barrel oil spill, killing thousands of fish and birds. Water resources are being depleted by a vast agricultural irrigation system.

WOMEN'S RIGHTS
In late 2003, the authorities announced that municipal elections would be held for the first time. In March 2004 it was announced that women would be able to vote and run for office, but this has not yet been implemented. In 2005, there were no women in Government. Saudi Arabia signed and ratified in 2000 the Convention on the Elimination of All Forms of Discrimination Against Women, but there are no reports on its implementation. No measures were taken against

discrimination in legislation, education or daily life (women are not represented in the *Majlis al Shura* (National Assembly), cannot travel abroad without written permission from a male relative, cannot drive since 1990, and are arrested if they break the dress code). Women make up 16 per cent of the labor force. Between 1995 and 2000, 9.7 per cent of women between 15 and 24 years of age could not read or write. In 2004, 90 per cent of women had access to prenatal care and 91 per cent of the births were assisted by skilled health staff. The maternal mortality rate was 23 per 100,000 births.

CHILDREN
Infant and under-five mortality have decreased in the period 1990-2004. The former decreased from 35 to 21 per thousand live births and the latter from 44 to 27 per thousand. The proportion of infants with low birth weight was 11 per cent while 14 per cent of under-fives suffered from severe and

moderate low weight. Moderate or severe stunting affected 20 per cent of the children. The primary school attendance rate reached 55 per cent for boys and 54 per cent for girls.

INDIGENOUS PEOPLES/ ETHNIC MINORITIES
The Shi'a, a minority religious group, which made up 15 per cent of the population in 2000, continued to be subject to political, social and economic discrimination. Until 2000 their political organization was restricted, as well as their freedom of speech, and they did not have access to the same public positions as the Sunni majority.
  Shi'a advocacy groups make different types of demands from the Saudi Government. The Reform Movement acts from abroad and seeks Shi'a recognition as an Islamic sect, freedom of religion, the provision for Shi'a education in their region, freedom of speech, an end to government harassment

and the same powers for Shi'a and Sunni courts in issues such as marriage, divorce, and inheritance.

MIGRANTS/REFUGEES
Between 2002 and 2003, 4,800 Iraqis (refugees since the first Gulf War) were repatriated. In early 2006, 440 Iraqi refugees remained in Rafha camp, located on the Arabia-Iraq border. The country also hosted about 240,000 Palestinian refugees. There is an indeterminate number of Bedouin refugees, considered stateless because of their nomadic life, who move between Saudi Arabia and Kuwait.

DEATH PENALTY
The death penalty is applied, even for ordinary offences.

* Latest data available in *The State of the World's Children* and *Childinfo* database, UNICEF, 2006.

Senegal / Sénégal - Sounougal

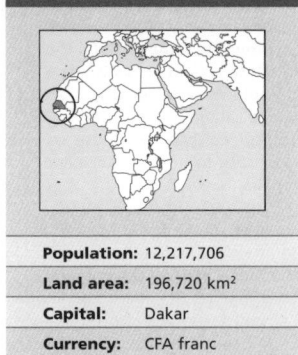

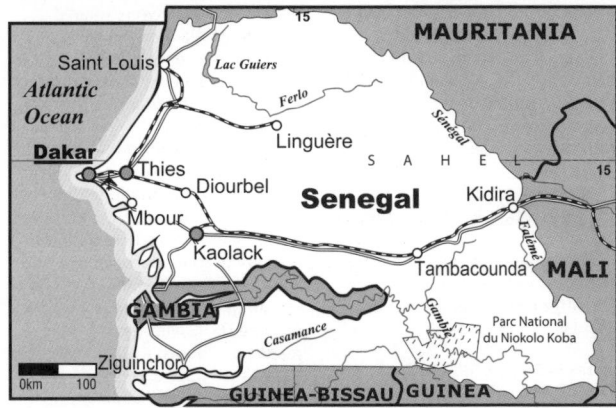

Population:	12,217,706
Land area:	196,720 km²
Capital:	Dakar
Currency:	CFA franc
Language:	French

The banks of the Senegal River were inhabited early on by peoples who had converted to Islam through contact with neighboring Arab countries; parts of those countries make up the region known as the Sahel. The Wolof (who constitute more than a third of the population), Fulani (see Cameroon), Pulaar and other peoples, all lived within the area of modern Senegal. When the French occupied it during the 17th century, Senegal was incorporated into a triangular world trade pattern whereby European manufactured goods were exchanged in Africa for slaves. The slaves were sold in the Caribbean for rum and sugar to go back to Europe.

[2] After slavery was abolished by the French Revolution of 1848, the Senegalese became 'second class citizens' of the French Empire, with one political representative in Paris. Senegal exported thousands of tons of peanuts per year and supplied the French army with soldiers. During the second half of the 19th century, there were frequent rebellions among the Muslim leaders, and it was only in 1892 that the French managed to fully 'pacify' the country.

[3] The Pan-African movement inspired Senegalese Léopold Sédar Senghor and Martinique's Aimé Césaire, to create in 1933 the concept of *négritude* against the imposition of French culture, which proclaimed itself as 'universal'. As 'an effective liberation tool', *négritude*, in Senghor's view, had to escape 'picturesquism' and achieve a specific political identity.

[4] Senghor had been a member of the French resistance during World War II, and in 1945 was elected representative to the French National Assembly. In 1948 he led the foundation of the Senegalese Democratic Bloc (later on Senegalese Progressive Union), which demanded greater autonomy for the colonies, although not independence.

[5] Finally, on 4 April 1960, Senegal declared its independence, and on 5 September adopted a republican system. Senghor was elected president and through successive re-elections he stayed in power for two decades. He applied an ideology of African socialism based on the collectivist essence of traditional African agrarian society. Senghor believed that 'socialism' already existed on the continent and therefore did not need to be imposed. In the global economy, this 'collectivism' actually served to provide cheap labor for the export-oriented production of peanuts and cotton, while 82 per cent of the nation's industry was French-controlled.

[6] In an effort to keep up with political change in France, Senghor sought membership of the Socialist International, and reformed the constitution establishing a three-party system: a 'liberal democratic' party (the Senegalese Democratic Party), a 'Marxist-Leninist' party (the African Party for Independence) and his own, renamed the Senegalese Socialist Party (PSS).

[7] In the early 1980s the Government gave in to pressure from the US, France, the World Bank and the IMF to accept a structural adjustment plan. The elimination of agricultural subsidies triggered a rise in production costs and prices of basic consumer goods. The country has faced a series of serious droughts, seen as part of the desertification process caused by climatic conditions and the French-imposed substitution of export crops for traditional food crops.

[8] In 1981 Senghor was replaced by then prime minister Abdou Diouf. The new President brought flexibility to the political system. Fourteen new parties were created, preventing the regime's main rival, Abdoulaye Wade's Senegalese Democratic Party (PDS), from uniting all the Government opposition forces. Meanwhile, in the southern province of Casamance a new separatist movement emerged in 1982: the Casamance Movement of Democratic Forces (MFDC), led by abbot Austin Diamacoune Senghor. The people in Casamance, mainly Dioles, were proud of their independence and resistance to the Islamic hierarchical societies of the north. In 1983, after successive confrontations with the Senegalese police forces, Diamacoune and other separatist leaders were detained.

[9] In 1988 the PSS won another electoral victory, obtaining 73 per cent of the vote, while the PDS - which obtained a mere 26 per cent - questioned the legitimacy of the election. Several PDS leaders were detained, and Wade was forced to go into exile in Paris, returning to Senegal in 1989. That year, a border conflict broke out with Mauritania triggered by violence between peasants and farmers, causing hundreds of casualties and forcing some 70,000 refugees to enter Senegal. The conflict stopped integration with Gambia. Both countries had planned on creating the Senegambia Federation. Diouf criticized the neighboring country for harboring the Casamance guerrillas and for signing a defense treaty with Nigeria (see Gambia).

[10] In 1991, Diouf was elected President of ECOWAS (Economic Community of the Western African States), which includes 17 countries in the region. That year, the US Government wrote off $42 million of Senegal's foreign debt after Diouf supported the allies in the Gulf War and contributed to the 'peace troops' stationed in Liberia. The IMF approved a new $5.7 million loan. The office of Prime Minister was restored and members of opposition parties were appointed to two cabinet posts. After negotiations with opposition parties, a consensus was reached to reform the electoral laws.

[11] In 1993, amid allegations of widespread fraud, Diouf won the presidential elections in the first round ballot, with 58.4 per cent of the vote. In the legislative elections, his PSS party maintained control of the National Assembly, with 84 of the 120 seats. The main opposition party - the PDS - won 27 seats.

PROFILE

ENVIRONMENT
Located on the west coast of Africa, embracing Gambia, its northern border formed by the Senegal River, the country's population is concentrated in the less arid western part, close to Dakar. The Senegal Valley is still underpopulated as a result of slave trade which was very intense in this region.

SOCIETY
Peoples: Wolof 42 per cent; Serer 14.9 per cent; Diola 10 per cent; Peul (Fulani) 9.3 per cent; Malinke (Mandingo) 3.6 per cent. Approximately 3 per cent are immigrants from non-African countries, mainly France, Lebanon and Syria.
Religions: 94 per cent Muslim, 5 per cent Christian, traditional African religions and others 1 per cent.
Languages: French (official). The most widely spoken indigenous languages are Wolof, Peul and Ful.

Main Political Parties: Senegalese Democratic Party; Socialist Party of Senegal; Alliance of Progress Forces; Union for Democratic Renewal.
Main Social Organizations: National Federation of Senegalese Workers (CNTS); Union of Free Senegalese Workers (UTLS).

THE STATE
Official Name: République du Sénégal.
Administrative divisions: 10 Districts.
Capital: Dakar 2,167,000 people (2003).
Other Cities: Thies 255,200 people; Kaolack 221,400; Ziguinchor 200,700; Saint-Louis 144,100 (2000).
Government: Parliamentary Republic. Abdoulaye Wade, President and Chief of State since April 2000. Macky Sall, Prime Minister and Head of Government since April 2004. Unicameral National Assembly with 120 members elected by direct popular vote to serve five-year terms.
National Holiday: 4 April, Independence Day (1960).
Armed Forces: 14,000 (2003).

Life expectancy
57 years
2005-2010

GNI per capita
$630
2004

Literacy
39% total adult rate
2000-2004

HIV prevalence rate
0.8% of population 15-49 years old
2003

[12] Senegal's economic and financial situation worsened during 1993, partly because international prices for Senegalese export products fell considerably.
[13] The 100 per cent devaluation of the CFA franc agreed by France and the IMF accentuated social tensions in early 1994. The opposition Co-ordination of Democratic Forces organized an anti-government demonstration which ended in confrontations with the police. Six police died and dozens of people were injured. Among people arrested were opposition leaders Wade and Landing Savané, who were declared innocent and freed after five months in prison.
[14] Wade joined Diouf's cabinet in 1995, after negotiating with the Government. International finance institutions supported the President when he announced legislation to encourage foreign investment and accelerate privatization.
[15] In spite of support from France and the US, Government troops made no progress against the Casamance guerrillas. Observers said the MFDC was popular among Casamance youth and that the region's geography made it impossible for Dakar to win.
[16] After 40 years of one-party rule, Wade won the 2000 presidential elections with a campaign denouncing the corruption and inefficiency of the PSS government and calling for a change. Upon taking office, Wade appointed opposition leader Mustafa Niasse as Prime Minister. The Government's main priority was to put an end to the separatist conflict by peaceful means, but violence increased in Casamance with successive attacks by MFDC factions.
[17] A referendum passed a new Constitution in 2001. This limited the president's term of office from seven to five years, reducing the presidency to a maximum of two consecutive periods. It also enabled the President to dissolve the National Assembly (lower chamber) without the support of a majority of its members. Shortly after, Wade dissolved the Assembly, preparing for parliamentary elections that year: the 'Sopi' - or 'Change' - coalition, formed by parties loyal to Wade and led by the PDS, won by a landslide. The new Senegalese Prime Minister - the first woman to hold that office in the country - Mame Madior Boye, announced the composition of her new government. Ten of the posts went to PDS members.
[18] In 2001, Wade announced he would hand over Hissene Habré, former president of Chad facing war crime charges, if a third country would be willing to give him a fair trial. Human rights groups held Habré responsible for some 40,000 executions and the torture of 200,000 people during his rule between 1982-1990. Jean-Marie Francois Biagui resigned as Secretary-General of the MFDC, claiming lack of loyalty from its members.
[19] In 2002 the Joola, a ferry built in 1990 by German shipbuilder Neue Germersheimer Schiffswerft and meant to solve the isolation problem in Casamance, sank off Gambia's coast, as a result of its severely deteriorated condition, according to some experts. The death toll stood at 1,863. After the shipwreck, Wade dismissed the whole cabinet. A 1996 decree had authorized the Joola to use 'its own resources' for maintenance and repairs.
[20] In 2003, Sidy Badji, MFDC founder, died a few days before the start of peace talks between the rebels and the Government. That year, MFDC leader Jean-Marie Francois Biagui announced the end of the separatist war before hundreds of rebel delegates in Ziguinchor.
[21] The armed wing of the MFDC did not take part in the talks and assured that peace would only be possible when the strongest factions allowed it to happen.
[22] MFDC's veteran leader Abbe Augustin Diamacoune was replaced by Jean-Marie Francois Biagui in September 2004. In October, the Government guaranteed funding for the electrification of the region of Louga, located in the north-west of Senegal, as a part of a new project to improve rural electrification in the country.
[23] The Government and MFDC rebels reached a final peace agreement in December after 20 years of civil war in Casamance - west Africa's longest-running internal conflict - which had left over 3,500 dead and tens of thousands of displaced people. For the last two years the region had shown relative calm.
[24] Opposition leader Abdourahim Agne was arrested in May 2005 and charged with inciting to rebellion after a peaceful demonstration against President Wade. Agne could be sentenced to five years in prison if convicted.
[25] In June, an MFDC faction which had failed to acknowledge the peace agreement signed in December 2004, launched new attacks against the Government in Casamance.
[26] Former president of Chad Hissene Habre - exiled in Senegal since his ousting in 1990 - whose regime was accused of 40,000 executions and the torture of 200,000 people - was arrested in November. However, a Senegalese court declared itself-not competent to rule on the extradition of the former dictator to Belgium, where he was to be tried.
[27] In March 2006, the rebel group that continued to campaign for the independence of Casamance clashed along the border with troops from Guinea-Bissau.
[28] In May, the UN Committee against Torture gave Senegal 90 days to put Habre on trial or send him to Belgium. ∎

IN FOCUS

ENVIRONMENTAL CHALLENGES
Deforestation and desertification are the main problems. A hydroelectric dam project in a valley north of the Senegal river poses a serious threat for environmental balance. There are few regulations governing the use of most natural resources: there is poaching, excessive fishing and shepherding.

WOMEN'S RIGHTS
Women have been able to vote and run for office since 1945.

In 2003, women held 19 per cent of seats in Parliament and 21 per cent of ministerial positions. They made up 43 per cent of the country's total workforce of five million.

Although the Constitution includes gender equity, discrimination is rampant. Women have the right to choose their husbands, but in practice almost half are in polygamous marriages.

During 2005, a program designed to eliminate female genital mutilation succeeded in making more than 1,600 villages - 30 per cent of the total - abandon this practice. In 2004, 79 per cent of women received prenatal care and 58 per cent of births were attended by skilled health staff. Maternal mortality amounted to 560 deaths per 100,000 births*.

CHILDREN
The gender gap in education is significant: girls tend to drop out of primary school more frequently than boys and only 52 per cent of boys and 45 per cent of girls attend school.

The Ministry of Family, Social Development and Solidarity is responsible for the promotion of child welfare. The Government built more schools and programs were chosen to encourage the enrollment and attendance of girls in school. In 2004, one of the causes of school drop-out was the fact that 33 per cent of children aged between 5 and 14 were working. Infant and under-five mortality rates stood at 78 and 137 deaths per 1,000 live births respectively. At the same time, 18 per cent of newborns were underweight and 25 per cent of children under-five suffered from moderate or severely stunted growth.

INDIGENOUS PEOPLES/ ETHNIC MINORITIES
The Wolof are the largest ethnic minority (42 per cent) and 75 per cent of the population speak their language. The Lebu, who basically live off fishing, live in the Cape Verde and Saint Louis peninsula. The Serer constitute almost 15 per cent of the total population and are Muslims, except for a group living along the Petite-Côte.

The Mandingo live in Eastern Senegal; the Sininkes in the border zones with Mali and Mauritania.

The Diola comprise 10 per cent of the population and live mostly in Casamance. They continue to suffer repression and exclusion from political processes. The Government makes few efforts to improve their conditions or meet their demands for better infrastructure, education and economic opportunities.

MIGRANTS/REFUGEES
In early 2006, nearly 8,000 Senegalese continued to live as refugees in Gambia and Guinea-Bissau, out of a total of 15,000 which had fled the country during the conflict in Casamance province, which lasted until late 2002. Furthermore, an estimated 5,000 people were internally displaced in Senegal.

In early 2006, some 20,000 Mauritanians remained located in northern Senegal; not in traditional refugee camps but in about 30 small informal villages along a 600 kilometer stretch of the Senegal River valley. No assistance was provided for these people, whose status was uncertain.

DEATH PENALTY
Senegal is a de facto abolitionist country since, although the death penalty is still in force, the last execution took place in 1967.

*Latest data available in *The State of the World's Children* and *Childinfo* database, UNICEF, 2006.

Serbia / Srbija

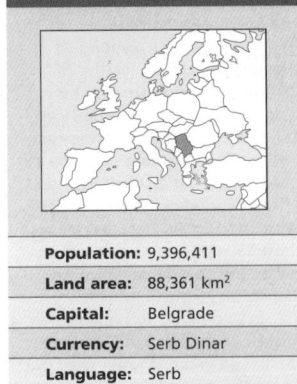

Population:	9,396,411
Land area:	88,361 km²
Capital:	Belgrade
Currency:	Serb Dinar
Language:	Serb

In the 4th century BC, the Balkan peninsula and the Adriatic coast were inhabited by Illyrian, Thracian and Pannonian tribes, and were also the site of Greek colonies. In the 2nd century BC Rome defeated an alliance of Illyrian peoples and colonized the new province of Illyria. Important Roman cities were built, such as Emona (currently Ljubljana), Mursa (Osijek) and Singidunum (Belgrade). When the Roman Empire split into Eastern and Western regions, the border between both ran through the Balkans. Christianity was established at the end of Roman domination.

[2] In the 5th and 6th centuries AD, these territories were invaded by several nomad tribes (Visigoths, Huns, Ostrogoths, Avars, Bulgars, and Slavs) which imposed their own religious beliefs. Christianity gradually took hold again between the 9th and 11th centuries. Several feudal states were formed from the 7th to the 13th centuries. Serbs, who had settled in the 7th century and accepted Christianity in the 9th, were unable to resist external pressure. Bosnia was conquered by Hungary, and the rest - as far as the state of Ducla - by Byzantium; Macedonia was divided up between Byzantium and Bulgaria (see Bosnia-Herzegovina and Croatia).

[3] In the mid-11th century, under the reign of Stephen Nemanja (1168-1196), Serbia freed itself from Byzantine domination. The Serbian rulers of the Nemanja dynasty fought the non-Christian religions which had spread in the Balkans, in an attempt to get a royal title from the Pope. They received it in 1217, but the hoped-for propagation of the Catholic faith did not follow. In 1219, the Serbian Orthodox Church was founded, and mass began to be celebrated in Serbian. Under the reign of Stefan Dusan (1331-1355) the medieval Serbian State reached its apogee, occupying Albania and Macedonia.

[4] The Ottoman Empire began its conquest of the Balkans in the mid-14th century, after the Battle of Kosovo in 1389. A 489-year period

of Ottoman domination began. In the 14th and 15th centuries, the first of a series of migrations began from Serbia and Bosnia to neighboring Slav regions, as far as Russia. In 1395, all of Macedonia came under the Ottoman Empire. Bosnia which had been a part of the Hungarian kingdom since the 12th century was conquered in 1463 and the Slav population became Muslim. Herzegovina fell in 1465. Previously, Venice had annexed the coast territories from Neretva to Zetina. The city-state of Dubrovnik came under Hungarian control, and then in 1526 became a part of the Ottoman Empire.

[5] Between the 16th and 18th centuries, all the territories of what would become Yugoslavia had been divided up. Serbia, Bosnia, Herzegovina, Montenegro and Macedonia belonged to the Ottoman Empire; Croatia, Slovenia, Slavonia, part of Dalmatia and Voivodina belonged to the Hapsburgs; and Istria and Dalmatia belonged to the Venetian Republic. After the 1690 revolution was put down in Old Serbia, some 70,000 people took refuge in the Hapsburg Empire. The Ottoman Empire transferred Albanian Muslims to the abandoned territories of Kosovo and Metohija.

[6] After the Russian-Turkish war of 1768-74, Russia obtained the right to sponsor the Orthodox population of the Ottoman Empire, through the Treaty of Kuchuk-Kainardzhi. Austria seized the Balkans in 1797, as a result of the Napoleonic Wars. The Balkan *pashalik* (the north of Serbia), which belonged to the Ottoman Empire after the first Serbian uprising (1804-13), the Russian-Turkish war (1806-12) and the second Serbian uprising (1815), was granted internal autonomy. The political and military

leaders Gueorgui Cherny (Karadjordje) and Milos Obrenovic founded Serbia's ruling dynasties. In 1829, Serbia became an independent princedom within the Ottoman Empire, with Milos Obrenovic as its Prince.

[7] At the 1878 Congress of Berlin, the European Powers recognized the full independence of Serbia and Montenegro, which became kingdoms in 1882 and 1905 respectively. In the Balkan wars of 1912 and 1913, Serbia, Montenegro, Greece, Romania and Bulgaria formed first an alliance against the Ottoman Empire, and then warred against each other. The result was Macedonia's partition between Serbia, Greece and Bulgaria, while Serbia and Montenegro expanded their territories.

[8] Serbian resistance to the Austro-Hungarian Empire led to the assassination of the Austrian Archduke Franz Ferdinand in 1914, in Sarajevo, the event that marked the beginning of World War I. After the war, which saw the end of Austria-Hungary's empire, a kingdom of Serbs, Croats and Slovenes was founded in 1918, including Serbia, Montenegro and the territories of Slovenia, Croatia, Slavonia, Bosnia and Herzegovina.

[9] In 1929, the kingdom began to be called Yugoslavia: the land of the southern Slavs. The Government remained in the hands of Serbians, and under the reign of Alexandr Karagueorgevich, it became a dictatorship. The regime's nationalist policies gave rise to a strong anti-Serbian movement among Croats and other ethnic minorities, which led to the King's assassination in Marseilles in 1934.

[10] At the beginning of World War II, Yugoslavia was neutral. In 1941, when German and Italian forces

attacked, the country was internally divided and was easily subdued within a few days. The King and the government fled to London, and the German Command carried out a policy of extermination against the Serbian and Muslim population.

[11] Two rival groups launched the resistance movement: the nationalists loyal to the King - called *chetniks* - led by Draza Mihajlovic, and the partisans under the leadership of Josip Broz, a Croat better known by his *nom de guerre*, Tito. This group was made up of anti-Nazi communists in favor of a united Yugoslavia, and anti-Nazi forces from all the republics with the exception of Serbia. Later on it was to become the Yugoslav League of Communists (YLC).

[12] After bloody fighting against occupation troops and the Croatian *ustashi* (fascist) movement allied to the Germans, Tito's guerrillas emerged victorious. Following the liberation of the country in May 1945, a Provisional Government was formed, led by Tito and supported by the Soviet Union and the UK.

[13] During the War, two million Yugoslavians were killed and 3.5 million were left homeless. The country was in ruins.

[14] On 29 November 1945, a Constituent Assembly abolished the monarchy, proclaiming Yugoslavia a federation made up of six republics: Slovenia and Croatia in the northwest, Serbia in the east, Bosnia-Herzegovina and Montenegro at the center, and Macedonia in the south. It also had two autonomous provinces: Voyvodina and Kosovo, to the northeast and southwest of Serbia, respectively.

[15] The YLC joined the Cominform (Communist and Workers' Parties' Information Bureau) in 1947, but withdrew in 1948 over disagreements with the Soviet CP leadership. The USSR imposed an economic embargo on Yugoslavia, which led to the strengthening of the country's ties with the West and the Third World.

[16] After a phase of economic centralization which included the forced collectivization of agriculture in 1950, Tito introduced the concept of self-management. Its goals were to ensure workers' direct participation in all decision-making processes concerning their living and working conditions, and to protect social democracy against the technocratic distortions and abuses of 'statism'.

[17] Tito was one of the founders of the Non-Aligned Movement. The Yugoslav leader defined non-alignment as the process whereby countries which were not linked to political or military blocs could take part in international issues, without being satellites of the major powers.

[18] The Yugoslavian regime was

charged with 'revisionism' and isolated from the international Communist movement because of its neutral foreign policy and heterodox model of social and economic organization. Relations with the USSR slowly returned to normal after Stalin's death in 1953.

[19] That year, a Land Reform Law was passed authorizing private farming, and 80 per cent of the land returned to private hands. The combination of private economic activity and the self-management system led to an average annual GNP growth rate of 8.1 per cent between 1953 and 1965. In 1968, industrial production was 12 times bigger than in 1950 and its impact upon the national economy was three times as important as agricultural production.

[20] The country's growth rate declined slightly toward the end of the 1960s. Even so, until the end of the 1970s, the annual growth rate exceeded five per cent.

[21] President Tito was aware of inter-ethnic tensions in the country, as well as the sharp contrasts between the socio-economic situation of the industrialized north and that of the underdeveloped south. In 1970, he announced that after he stepped down, the country's leadership should be exercised by a body made up of representatives of the federated republics and the autonomous provinces.

[22] In 1971 and 1972 ethnic conflicts worsened, especially between Serbs and Croats. Croatia presented a formal complaint against the confederated system. After 1974, there was an increase in separatist activity by Kosovo's Albanian majority.

[23] After Tito's death in April 1980, executive power was vested in a collective presidential body, made up of a representative of each republic plus the autonomous province and the president of the YLC. This body would have a rotating annual presidency.

[24] In March and April 1981, there were riots in the autonomous province of Kosovo (bordering on Albania). These recurred in 1988 and 1990. In Kosovo, 90 per cent of the population (1.9 million) was of Albanian origin. It was the poorest region of Yugoslavia, with unemployment reaching 50 per cent in 1990 and a per capita GNP of $730 while Serbia's was $2,200.

[25] The federal Government accused nationalist forces and separatist extremists, instigated from abroad, of seeking the secession of Kosovo. Many Serbs and Montenegrins left the area. Repression of the uprisings in Kosovo left a number of people dead and injured and led to mutual diplomatic recrimination between Belgrade and Tirana. Kosovo's governor, Jusuf Zejnullahu, resigned in March 1990. There was also tension in other republics due to the growth of militant Muslim and Catholic groups.

[26] Ethnic conflicts and inflation - which reached 90 per cent in 1986 and four figures in 1989 - were considered by the YLC to be rooted in deeper contradictions. During these years, several lawsuits dealing with government corruption exposed the fact that the system was crumbling. The Communist parties of Slovenia and Croatia announced their withdrawal from the YLC. In its January 1990 congress, the YLC renounced its political monopoly, granted by the Constitution, and called on Parliament to draft a new Constitution, abolishing the one-party system and the leading role assigned to the League in all spheres of public life.

[27] In April 1990, in the first multiparty elections to be held in Yugoslavia since World War II, nationalist groups demanding either secession or a confederated regime won in all of the republics except Serbia and Montenegro.

[28] In 1989, Yugoslavia's last Socialist prime minister, Ante Markovic - a Croat and a neo-liberal - launched a series of structural reforms in order to put an end to the crisis and eliminated custom tariffs on 90 per cent of imports, which grew 11.9 per cent, while exports grew 5.6 per cent.

[29] As of 1990, the situation deteriorated steadily. Industrial production fell by 18-20 per cent in Serbia - while the gross foreign debt of the Federation as a whole reached $16.295 million. Of this, more than $5.5 million arose from Serbia and Montenegro.

[30] With mounting social pressures because of the economic situation and the disintegrating State, two fundamentally opposed concepts were introduced. Milan Kucan, the reformist communist president of Slovenia, argued for decentralization to relieve the wealthier regions of the obligation to subsidize the more backward areas, while Slobodan Milosevic, president of Serbia and leader of the Socialist Party of Serbia (SPS, former Serbian Communist League), proposed greater centralization and solidarity within the Federation.

[31] In December 1990, the Croatian Parliament adopted a new Constitution which established the right to withdraw from the Federation. At the same time, a referendum in Slovenia endorsed independence. In the following months, disagreement over the reform of the federal system and the appointment of the President caused an unsolved crisis in the Yugoslavian collective presidency.

[32] On 8 September 1991, Croatia and Slovenia declared their independence. The Serbian population within Croatia declared its intention of separating from Croatia. The federal army, whose officials answered primarily to Serbia, intervened in Slovenia and Croatia, stating that separation was a threat to Yugoslavia's integrity (see Croatia and Bosnia-Herzegovina).

[33] War broke out, causing thousands of civilian deaths in Croatia and the destruction by federal troops of entire cities - like Osijek, Vukovar and Karlovac - as well as the occupation of nearly a quarter of Croatia's territory. Areas affected included the territories of West and East Slavonia, as well as the area known as Krajina, which in late 1991 proclaimed itself the 'Republic of Serbian Krajina'.

[34] In December, both the president of the governing council, Stjepan Mesic, and Prime Minister Markovic - both Croats - resigned. They were the last representatives of a united government. By the end of 1991, there were 550,000 refugees in the country, 300,000 of them from occupied areas of Croatia.

[35] On 15 January 1992, the European Community (EC) recognized Croatia and Slovenia as sovereign states. On 27 April, the Parliament of Serbian and Montenegrin deputies announced the foundation of the new Federal Republic of Yugoslavia, a federation between Serbia and Montenegro with a parliamentary system of government.

[36] In the meantime, fighting continued in the regions characterized by inter-ethnic strife. From April 1992, the heart of the fighting was the republic of Bosnia-Herzegovina, which was recognized by the EC on 7 April. In order to divest itself of responsibility for the aggression in Croatia and Bosnia-Herzegovina, in early May Belgrade announced that it was no longer in control of troops from what had formerly been the federal army, now fighting in the independent republics. Far from having the desired effect, Serbia's position led the EC to declare a trade embargo against Yugoslavia on 28 May.

[37] In 1993 a series of purges began within Serbian government institutions. The intransigence of the Serbian Government was evident in Kosovo, where any attempt at independence was met with repression.

[38] As Milosevic and the ruling SPS party grew more authoritarian, international pressure increased and the social and economic crisis deepened. Milosevic's main aim was to achieve the suspension of the sanctions imposed by the UN. On 24 September 1994, the international organization decided to partially lift the measures for 100 days, allowing international flights, and cultural and sports exchanges. This diplomatic coup by Milosevic - who was previously accused of being the main instigator of the war in the Balkans - contributed to his increasing popularity in the Federation.

[39] Milosevic's popularity soared

again following the signing of a Bosnian peace agreement in the US city of Dayton, Ohio, in November 1995, though his major triumph came on 14 December 1995 when the US suspended the sanctions on Yugoslavia as a result of the signing of the Paris agreement. During the embargo, per capita income had fallen by 50 per cent and more than one-and-a-half million people were unemployed.

[40] The year 1997 was marked by confrontations between the federal army and the Albanian population of Kosovo. Milosevic called a referendum, in all of Serbia, on the need for foreign mediation on this conflict. Foreign mediation was rejected by 75 per cent of the voters, although the Albanian-speaking population of Kosovo boycotted the referendum.

[41] Confrontations in Kosovo in June 1998 led Britain and other NATO countries to issue a call for intervention in the region. After a brief truce in the second half of the year, which the Kosovo Liberation Army (KLA) guerrilla forces used to establish a policy of persecuting the Serb minority, the Yugoslav army re-entered the territory.

[42] In March, 1999 the Alliance launched a series of air raids against Yugoslav targets in Serbia, Montenegro and Kosovo. The UK and the US, the main military forces, also had air, naval and logistic support from Germany, Italy, France and Turkey amongst others. Bombardments, practiced at a height beyond the reach of anti-aircraft fire, caused hundreds of deaths amongst Kosovars and Serbs. The air attack unleashed a campaign of revenge attacks against Kosovars, who had to flee in tens of thousands to neighboring countries and Montenegro. NATO air attacks with 'smart' bombs destroyed hundreds of civilian buildings, including a State TV station, schools, a hospital and the Chinese embassy in Belgrade.

[43] The bombardment ended on 10 June and Yugoslav forces accepted the stationing in Kosovo of a UN force, KFOR, made up of some 35,000 troops. According to the agreement, Yugoslav sovereignty over the province was recognized, but in practice it was a protectorate under NATO military control and UN political coverage.

[44] Ethnic cleansing of Serbs and Roma led to dozens of deaths after KFOR arrived. The UN force was unable to totally disarm the KLA, and integrated it into the police. Several thousand members of these communities had to move from their homes, while a large number of Kosovar refugees returned to the province.

[45] Zeljko Raznjatovic - better known as Arkan - one of the Serb paramilitary leaders and president of the Serbian Unity Party, was assassinated in January 2000 in a hotel in Belgrade. This prompted a cycle of political killings that also took the life of the Defense Minister, Montenegrin Pavle Bulatovic, in February.

[46] In June, the Yugoslav legislature passed constitutional changes by which the President would be elected by popular vote and no longer by the Assembly. After the elections, the opposition claimed that their candidate, Vojislav Kostunica, had won by a majority and that a second-round election was not needed. But after several days of silence, the electoral commission gave Kostunica 48.22 per cent of the vote, to Milosevic's 40.23 per cent, making a run-off necessary. The US and the EU denounced the election as fraudulent, and demanded that Milosevic acknowledge defeat.

[47] A popular uprising, organized by the opposition and joined by the police and army forced Milosevic to flee. After intense negotiations with the former president's supporters, Kostunica was finally sworn in as President. A new Parliament, elected in early November, appointed a Milosevic supporter, Zoran Zizic, as Prime Minister.

[48] In January 2001, after the Socialist Party lost the parliamentary elections by a broad margin to the Democratic Opposition of Serbia (DOS) coalition, Milosevic was placed under house arrest. In April he was transferred to a prison in Belgrade. The following month, US President George W Bush said economic aid to Yugoslavia was conditional upon Milosevic being handed over to the International Criminal Tribunal for the Former Yugoslavia (ICTFY) at The Hague. On 28 June the former president was extradited, prompting Premier Zizic to resign in protest. In his first appearance before the ICTFY, Milosevic refused to recognize the legitimacy of the court and said the trial's goal was 'to produce false justifications for the war crimes that NATO committed in Yugoslavia'. The trial would continue until Milosevic died in his prison cell at The Hague without having been sentenced, in March 2006.

[49] In March 2002, Montenegro and Serbia signed an accord to suppress the name 'Yugoslavia', provisionally creating a looser federation called 'Serbia and Montenegro'. By the end of the year these semi-independent states were to vote on new, separate constitutions with their own President. After three years, each one was to hold a referendum to decide whether or not to seek independence.

[50] Parliament approved the constitutional charter for the new union of Serbia and Montenegro in February 2003. The two republics would keep using separate currencies. The issue of Kosovo, an international protectorate which was still legally a part of Serbia, remained unresolved.

[51] In February 2006, talks began with the United Nations to define the future status of Kosovo. In March, the nomination of Albanian rebel leader Agim Ceku as Prime Minister of the province caused a bitter reaction in Belgrade, which issued an arrest warrant against the new premier, accusing him of having committed crimes against Serbians both in Kosovo and Croatia.

[52] In May, once the three-year period established by the 2003 Constitution was over, Montenegro held a referendum on the future of the country. Those who favored secession won by a slim margin, and Montenegro announced its independence. In response, Serbia announced its own independence as a sovereign State.

[53] In November 2006, the UN announced that it was postponing its key report on the future status of Kosovo until after Serbia's election in January 2007. ■

IN FOCUS

ENVIRONMENTAL CHALLENGES
Many environmental problems derive from the period of armed conflict, especially the NATO bombings. One of the worst problems is contamination from the depleted uranium in NATO missiles. Pollution in coastal waters, especially in tourist areas like Kotor Bay, and air pollution in Belgrade and other industrial cities, reach alarming rates. Industries discharge their waste into the Sava river, a tributary of the Danube.

WOMEN'S RIGHTS
In 2003, women held eight per cent of Parliament seats. Patriarchal notions regarding gender roles - especially in rural areas - are discriminatory. In minority communities women have no right to property. In 2003, women made up 43 per cent of the labor force of four million people. Almost all deliveries are assisted by qualified personnel. In 2004 maternal mortality stood at only 7 per 100,000 deliveries.

CHILDREN
Although statistics suggest that children are well provided for in terms of health and education, there are major disparities. The system guarantees nine years of free and compulsory education. However, economic hardship afflicts Roma children, who do not attend pre-school education. The most significant problem for Roma children is that often they do not learn Serbian, due to their lack of schooling. Some of those that do not pass the standard exam taken by all Serbian children have been sent to schools for children with emotional problems. In 2004, the infant and under-five mortality rates stood at 13 and 15 per 1,000 live births respectively.

INDIGENOUS PEOPLES/ ETHNIC MINORITIES
Although the Protection of Rights and Freedoms of National Minorities Act was passed in February 2003, discrimination against the Roma continues. Apart from being more affected by unemployment than the rest of the society, Roma are frequently attacked by racist groups who remain unpunished by the authorities. Most of the Roma that fled from Kosovo after July 1999 continued to suffer major problems, exacerbated by their difficulties to register as citizens.

MIGRANTS/REFUGEES
In 2004 the country continued to be the epicenter of displacement in the Balkans region. It sheltered more than 534,000 people displaced due to the various wars suffered by the region in recent years. In Kosovo there were more than 248,000 refugees of Serbian ethnic origin, 189,400 Croatians, 99,700 Bosnians and 1,500 Macedonians of Albanian ethnic origin. Serbia also had 252,000 internally displaced people and 70,100 of its citizens - most of them Kosovars - were seeking asylum or refuge abroad. Serbia and Montenegro sheltered more than 220,000 internally displaced Kosovars, of whom 70 per cent were of Serbian origin, 13 per cent were Roma, and 6 per cent Montenegrin. Another 22,200 Kosovars were displaced within Kosovo itself. Kosovo also sheltered 5,000 people of Albanian origin, displaced from the Presevo Valley in southern Serbia, mostly during the conflict in the valley, that ended in 2001.

DEATH PENALTY
In January 2003 it was abolished for all crimes.

Statistical information used in this box applies to the former federal republic of Serbia and Montenegro, which broke up in May 2006.

Seychelles / Seychelles

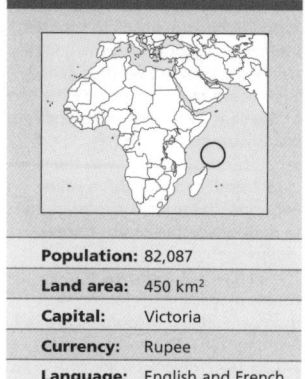

Population:	82,087
Land area:	450 km²
Capital:	Victoria
Currency:	Rupee
Language:	English and French

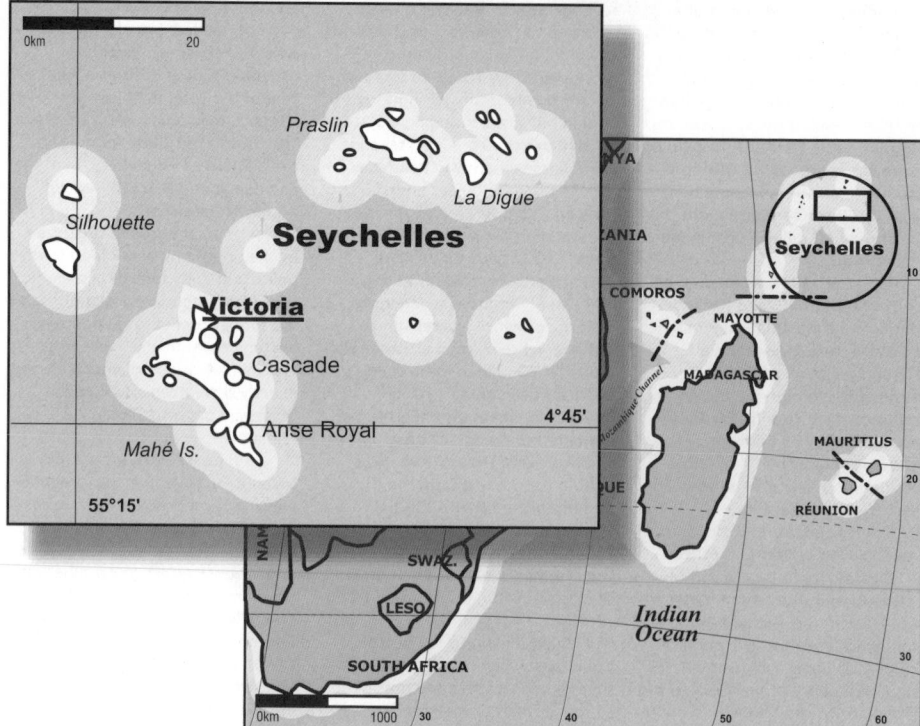

Prior to exploration by Portuguese sailors in the 16th century, the archipelago had already been visited by Persian Gulf merchants. French and British colonists fought violently over this Indian Ocean colony during the 18th century. After expelling the French in 1794, the British paid little attention to it, and it was governed from the island of Mauritius until 1903. The Seychelles archipelago gained strategic significance within the context of European colonial expansion during the second half of the 19th century.

2   The Seychelles People's United Party (SPUP) founded in 1964 gave the local population - most of them descendants of enslaved Africans and Indian workers - a new sense of nationalism. Meanwhile, the colonial interests organized themselves into the Seychelles' Taxpayers Association. It was later renamed the Seychelles Democratic Party and, led by James Mancham, it opposed independence.

3   In the legislative elections of April 1974, the SPUP won 47.6 per cent of the vote. However, the peculiar colonial 'democratic' system awarded the Party only two of the 15 seats and Mancham remained Prime Minister. In spite of this, it was too late to stem the tide of nationalism and in 1976 Mancham agreed to the British Foreign Office's suggestion that he become the first President of the Republic of Seychelles. Shortly before independence, he had agreed to 'return' the strategic British Indian Ocean Territory (BIOT) islands to the UK. The British in turn had promised to pass them on to the US, which planned to set up the important Diego Garcia naval base there.

4   Aware that the people would not accept this deal, Mancham postponed the elections until 1979, arguing that they were not necessary as all of the parties were in favor of independence.

5   Mancham's foreign policy was directed at cementing a strong alliance with South Africa, the source of most of its tourists, while domestic policy destroyed tea and coconut plantations to make room for new five-star hotels, owned by foreign companies. Entire islands were sold off to foreigners like Harry Oppenheimer, the South African gold magnate, and the actor Peter Sellers.

6   In 1977, while Mancham was maneuvering to postpone the elections yet again, the SPUP took over the country 'with the collaboration of the local police force' while Mancham was out of the country. Accused of 'leading a wasteful life while his people worked hard', Mancham was replaced by SPUP leader Albert René.

7   René renewed his support for the Non-Aligned Movement, which had recognized the SPUP as a legitimate liberation movement prior to independence. The new Government turned to socialism, promising to reorganize tourism, give priority to self-sufficiency in agriculture and fishing, increase education, and reduce the high unemployment rates which were affecting nearly half the working population.

8   In mid-1978, in response to the changing political context, the SPUP became the People's Progressive Front of Seychelles (FPPS). In June 1979, the FPPS won the national elections with 98 per cent of the vote. After the victory, President René demanded that the US base on Diego Garcia be closed and the island returned to Mauritius.

9   In August 1978, a land reform law was passed calling for the expropriation of all uncultivated land. René also nationalized the water and electricity services, the construction industry and transportation, which led to a rapid economic recovery in the Seychelles. In spite of not having any mineral or oil wealth, by 1979-1980 they had the largest per capita income of any of the islands in the region.

10   This economic growth was

PROFILE

ENVIRONMENT
An archipelago of 92 islands: Mahé, Praslin, and La Digue are the largest islands, made of granite; the rest are of coral. The climate is tropical with plentiful vegetation and heavy rainfall. Only the largest islands are inhabited, but some economic use is made of the other islands.

SOCIETY
Peoples: Most Seychellois are descended from Africans and Europeans. There are minorities of European, Chinese and Indian origin.
Religions: Roman Catholic 88.6 per cent; other Christian (mostly Anglican) 7.7 per cent; Hindu 0.7 per cent; other, 3 per cent (1996).
Languages: English and French (official); most of the population speak Creole, a local dialect with European and African influences.
Main Political Parties: People's Progressive Front of Seychelles (FPPS); Seychelles National Party (SNP), continuation of the United Opposition (OU, centrist), in turn formed by the National Alliance Party; Rally of the People of Seychelles for Democracy; Democratic Party (PD).
Main Social Organizations: The National Workers' Union.

THE STATE
Official Name: Repiblik Sesel.
Capital: Victoria 25,000 people (2003).
Other Cities: Anse Royal 3,700 people; Cascade 2,500 (2000).
Government: James Michel, President since April 2004. Single-chamber legislature: National Assembly, with 34 members.
National Holiday: 28 June, Independence Day (1976).
Armed Forces: 300 (1996). Other: 1,300 National Guard.

GNI per capita
$8,190
2004

Literacy
92% total adult rate
2000-2004

encouraged and supported by tourism and effective administration. At this time the Seychelles received an average of 80,000 tourists a year. During Mancham's time in office most of them were South Africans, but after René became president more Europeans came.

[11] The opponents of Albert René, mainly the South African apartheid regime, did not give up their efforts to overthrow the socialist government of the Seychelles. In November 1981, a group of 45 mercenaries led by former colonel Mike Hoare tried to invade the island and oust the Government. The coup plotted by former President Mancham failed and the mercenaries had to hijack an Indian airliner to escape to South Africa.

[12] After the unsuccessful invasion, the Seychelles Government declared a state of emergency and imposed a curfew.

[13] The failed invasion and the economic recession in Europe caused tourism to decline by approximately ten per cent. In August 1982, the situation deteriorated further when military personnel plotted an unsuccessful rebellion, strengthening rumors of another possible mercenary conspiracy. European conservative groups organized a smear campaign to support these rumors. In June 1984, Albert René was re-elected with 93 per cent of the vote.

[14] In September 1986, there was another coup attempt. Albert René, who was in Zimbabwe at a Non-Aligned Countries summit, returned immediately and put down the rebellion. Most of those responsible were arrested, including the main leader, Minister of Defense Colonel Ogilvy Berlouis.

[15] The Seychelles Government promoted the creation of a peace zone in the Indian Ocean and demanded that all warships wishing to call at its ports be 'nuclear-free'. As a result, the US and British naval fleets stopped calling.

[16] In the presidential elections of June 1989, René was the only candidate and was re-elected for a third term, with 96 per cent of the vote.

[17] The opposition forced President René to accept a multi-party system. The first experiment was carried out in July 1992, when a commission was elected to draw up the new constitution. The People's Progressive Front of Seychelles (FPPS), the ruling party, took 13 of the 23 seats. Former president Mancham's Democratic Party (PD) won eight seats.

[18] The Government continued with its authoritarian ways. The

IN FOCUS

ENVIRONMENTAL CHALLENGES
Global warming is one of the most pressing problems Seychelles faces today. Several plant and animal species, such as birds, giant turtles, coconut palms and others have already been affected by the rising temperatures and are increasingly endangered. Much of the coral in coastal waters has died. The supply of water is precarious.

WOMEN'S RIGHTS
Women have been able to vote and stand for office since 1948. In 2003, women held 29 per cent of seats in Parliament and 13 per cent of ministerial positions.

Domestic violence against women continues to be a problem. In the few cases that reached court, the perpetrator was released or given only a light sentence. In general, acts of domestic violence are not considered criminal offences. Concern about such violence

has prompted various NGOs to sponsor awareness campaigns for women and girls. Although prostitution is illegal, its incidence is growing. Seychelles' society is largely matriarchal, and does not discriminate against single mothers.

CHILDREN
In 2004, the infant and under-five mortality rates amounted to 12 and 14 deaths per 1,000 live births respectively. In February 2006, an outbreak of *chikungunya* - a tropical mosquito-borne disease - began to spread rapidly across the Indian Ocean. In Reunion alone, there were 160,000 people affected. The disease, which in spite of not being fatal is incurable, causes very high fever and swelling of the body and can even cripple its victims.

INDIGENOUS PEOPLES/ ETHNIC MINORITIES
There is no indigenous culture in the Seychelles, but some customs of African origin have survived.

A large number of Seychellois are Catholic; however, there is widespread belief in old spirits. Witchcraft is outlawed but there are still many traditional witch doctors. The majority of the population is of mixed African and European descent. There are minority groups of European, Chinese and Indian origin.

MIGRANTS/REFUGEES
A growing number of immigrants have been admitted since 2003 - particularly Chinese, Indians, Filipinos, Thai and Madagascans - who work in the fishing sector. Frequent reports allege abuse in terms of working conditions, as these workers are paid lower wages and forced to work longer hours than citizens, without receiving corresponding compensation.

DEATH PENALTY
It was abolished in 1993. No executions have been carried out since independence.

PD soon withdrew from the commission and the first draft constitution was rejected. The project was suspended until the PD returned to the commission in June 1993, and the text was written with support from both parties; 73.6 per cent of the voters approved the new constitution.

[19] The constitution officially established a multi-party system, a 33-member National Assembly and a five year presidential term. René won the July 1993 elections and his party had an overall majority in the National Assembly.

[20] Tourism continued to be the main source of income. The annual number of visitors was higher than the national population in 1993 and 1994. Oil derivatives and tinned tuna made up more than 80 per cent of the islands' exports during this period.

[21] Although according to the United Nations Development Program (UNDP) Seychelles ranked in 1998 as the African country with the highest human development index, it was only 56th in the world.

[22] With a view to proving the country's political stability to investors, René called elections two years early. In September 2001, he was re-elected to his third consecutive term since the establishment of the multi-party system. Wavel Ramkalawan, the opposition candidate of the Seychelles National Party, challenged the results saying voters had been intimidated or

had been bribed to support the FPPS.

[23] In January 2002, the Seychelles National Party's bid to annul the previous year's presidential election on the grounds of alleged irregularities was denied by the Constitutional Court.

[24] In July 2003, the Government implemented an economic reform program aimed at reducing the country's budget deficit. Three overseas embassies were closed, the country withdrew from the Southern African Development Community (SADC) and a new tax was placed on imports as well as on local goods and services. The police arrested four members of the Seychelles National Party, including Jean-François Ferrari, of the independent newspaper *Regar*, for collecting signatures for a petition against the new tax.

[25] In September 2003, the body of Thérese Blanc-Payet, Ferrari's sister-in-law, was found on a beach. In response, the European Parliament's Committee for Development and Co-operation requested that President René provide a report on the political and human-rights situation in Seychelles.

[26] After nearly three decades in power, President René stepped down in April 2004 and was replaced by his Vice-President James Michel. René continued as head of his party. The new President announced that he would open new spaces

for dialogue - particularly on economic issues.

[27] Seychelles' fishing industry and artisanal fishing sector suffered extensive damage when a tsunami devastated South Asia in December 2004. A great number of fishing boats were damaged or destroyed. The FAO launched assistance programs for the repair and replacement of fishing vessels and landing facilities and for the restoration of sustainable livelihoods in the sector.

[28] In August 2005 Seychelles announced the impossibility of rejoining the SADC due to financial problems that prevented the country from fulfilling its commitments within the organization. Notwithstanding this, the official position changed in December and preparations to rejoin the SADC began. According to President Michel: 'The Ministry of Foreign Affairs has a key role to play in defending and promoting Seychelles' interests in an increasingly globalized world. In this context I am pleased to announce that Seychelles will be rejoining SADC in 2006'.

[29] In May 2006, there was increasing concern over the spread of chikungunya - a very aggressive mutant virus, transmitted by a mosquito found in Africa (see In Focus) - across the Indian Ocean region. Some 275,000 people had already been infected.

[30] In July 2006, President Michel won the presidential elections with 54 per cent of the vote. ∎

Sierra Leone / Sierra Leone

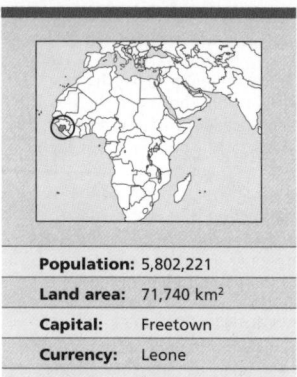

Population:	5,802,221
Land area:	71,740 km²
Capital:	Freetown
Currency:	Leone
Language:	English

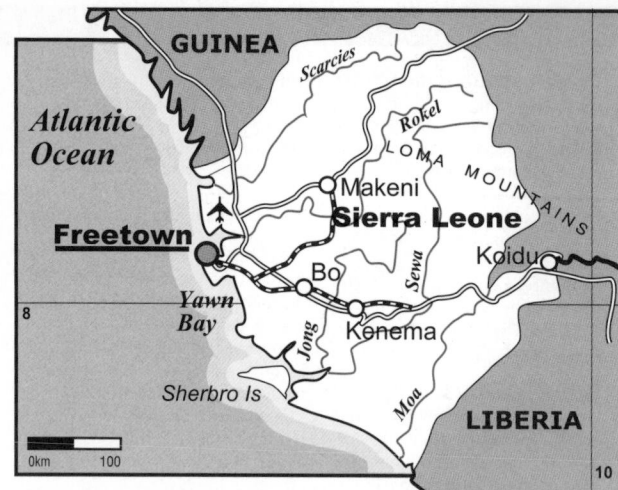

The Portuguese reached Sierra Leone ('lion mountains') in 1462. In the early 16th century it was a regular stop for European traders of cloth and metal goods, ivory, timber, and some slaves. Around that time, Mande-speaking people migrated from present-day Liberia, and they eventually established the states of Bullom, Loko, Boure, and Sherbro. In the 17th century British traders arrived. A century later, Fulani and Mande-speaking traders from the Fouta Djallon region of what is now Guinea converted numerous Temne to Islam, which became firmly established in the north and spread through the rest of the country.

[2] In the late 18th century, Britain decided to 'return' runaway and freed slaves from the Caribbean to Africa, and selected the recently acquired territory of Sierra Leone for the purpose. Abolitionist leader Granville Sharp purchased an area of 250 sq km for £60 from the local rulers, where he established an agricultural society based on democratic principles, but which quickly became a British colonial company in 1791. Over the next 50 years, the population of Freetown swelled with the arrival of 70,000 slaves, in addition to migration from the interior of the country.

[3] The country became a crown colony in 1808, and in 1821, it was merged with Gambia and Gold Coast (present-day Ghana) to create the British West African Territories.

[4] The creoles sought to emulate European culture and considered themselves superior to the 'savages' of the interior, acting as brokers for British colonialism. In 1896 Sierra Leone became a British protectorate.

[5] In 1898 resistance leader Bai Buré rallied most of the inland population, already angered by a British tax on their dwellings. However, British military superiority overwhelmed the insurgents after nearly a year of fighting.

[6] In 1960, the British wanted to withdraw from Sierra Leone, and an agreement was negotiated with the traditional leaders to protect their interests. In 1961, Sir Milton Margai became the first Prime Minister of an independent Sierra Leone.

[7] Although they lost their political power, the creoles, together with British and Syrian-Lebanese merchants, retained control over the economy.

[8] When Margai died in 1964, his brother Albert headed a government tainted by widespread corruption. The diamond trade and crime became major sources of income.

[9] In 1967 the All Peoples' Congress (APC) party led by Siaka Stevens won the elections. The conservative creoles, traditional leaders, and British neo-colonialists united to prevent change. Stevens was overthrown by a military coup and forced into exile in Guinea (Conakry).

[10] In April 1968, a group of low-ranking officers took power through the so-called Sergeants' Revolt. They brought back Stevens who in 1971 broke all ties with Britain and declared Sierra Leone a republic, becoming its first President.

[11] Stevens nationalized the lumber industry and the State seized majority control of the diamond trade. To protect the price of iron ore and bauxite, Sierra Leone joined the associations of iron and bauxite producing countries.

[12] In 1978 a plebiscite approved the establishment of a single-party political system. The APC incorporated prominent members of the SLPP within its ranks, granting them posts within the Government.

[13] The economic and political crisis peaked in 1979, as a result of a fall in exports, growing inflation and declining standards of living, together with authoritarian measures and government corruption, which undermined Stevens' popularity.

[14] In September 1981 the Sierra Leone Labour Congress (SLLC) declared a general strike, demanding immediate changes in economic policy. The strike spread throughout the country and nearly toppled the Government, which had to make several concessions.

[15] In urban areas, the scarcity of food and services became chronic. Smuggling grew as the purchasing power of salaried workers fell by 60 per cent. A dynamic black market emerged.

[16] Powerful Lebanese traders controlled the black market and more than 70 per cent of the country's exports. Gold and diamond smuggling was estimated at nearly $150 million per year while official exports amounted to $14 million in 1984.

[17] In November 1985, Siaka Stevens handed over the presidency to Joseph Momoh, a member of his cabinet; the change did not affect the country's critical economic situation.

[18] The Government declared a state of economic emergency in 1987. This gave the State the sole right to market gold and diamonds; a 15 per cent surcharge on imports, and a cut in the salaries of public employees.

[19] In March 1991, rebel forces operating from Liberia occupied two border towns. Guerrilla groups from Burkina Faso, Liberia, and Sierra Leone joined them, occupying one third of the country.

[20] A referendum held in August approved a new constitution establishing a multi-party system, while the economic crisis and corruption continued.

[21] In 1992, the Government launched an IMF-imposed structural adjustment program. James Funa, a former World Bank executive, was named Finance Minister. He implemented monetary control, incentives for natural resource exploration by foreign companies, widespread privatization, and an overhaul of the state apparatus affected by widespread corruption.

[22] That year, Captain Valentine Strasser seized power through

PROFILE

ENVIRONMENT
The country is divided into three regions. The coastal strip, nearly 100 kilometers long, is a swampy plain that includes the island of Sherbro.

SOCIETY
Peoples: The Temne and Mende account for nearly one-third of the population. The Lokko, Sherbro, Limba, Sussu, Fulah, Kono and Krio are other important groups. The Krio - whose name comes from the English word 'creole' - are descendants of African slaves freed in the 19th century who settled in Freetown. There are also Arab, European, Chinese and Indian minorities.
Religions: Most of the people practise traditional African religions; nearly one-third are Muslims, concentrated in the north; the Catholic minority is located in the capital.
Languages: English (official). The most widely spoken native languages are Temne, Mende and Krio. The latter serves as the language of commerce in the capital.
Main Political Parties: Sierra Leone People's Party; All People's Congress; Peace and Liberation Party.
Main Social Organizations: Sierra Leone Labour Congress.

THE STATE
Official Name: Republic of Sierra Leone.
Capital: Freetown 921,000 people (2003).
Other Cities: Koidu 109,900 people; Bo 79,700; Kenema 69,900 (2000).
Government: Ahmad Tejan Kabbah, President since March 1996, re-elected in 2002. Unicameral Legislature: House of Representatives, with 124 members.
National Holiday: 19 April, Republic Day (1971).
Armed Forces: 13,000 troops (2003).

Life expectancy
42 years
2005-2010

GNI per capita
$210
2004

Literacy
30% total adult rate
2000-2004

a coup. He suspended the constitution, created the National Provisional Governing Council and confirmed Funa in his post.

23   The United Liberation Movement for Democracy in Liberia used the eastern part of Sierra Leone as a base from which to carry out attacks against Charles Taylor's forces (see Liberia). In the meantime, the Revolutionary United Front of Sierra Leone (RUF) was operating in the southeast.

24   Guerrilla activity led to an abrupt decline in legal mining.

25   The RUF extended its armed struggle to the rest of the country in 1995. Government forces won back the Sierra Rutile titanium mine, but seemed unable to defeat the guerrillas.

26   In January 1996, following a bloodless coup, Strasser was replaced by Brigadier-General Julius Maada Bio. The presidential elections held in February as planned were won by Ahmad Tejan Kabbah of the SLPP in the second round with nearly 60 per cent of the vote.

27   Rebel troops led by Major-General Johnny Paul Koroma ousted President Kabbah in May 1997. The Organization of African Unity, meeting in Namibia, criticized the coup and began negotiations to force the leaders step down.

28   In September 1997, former president Kabbah asked the UN for help to re-establish his government. In March 1998, the UN ECOMOG troops - paradoxically mostly Nigerians who at that time were ruled by dictator Sani Abacha - took the main cities and regions of Sierra Leone, forcing out Koroma and his military junta.

29   The Kabbah administration achieved stability and in March 1998 it ordered a halt to all gold and diamond mining - which had been largely controlled by foreign companies for the last 60 years - except by Sierra Leonean concerns. In the final months of the year another rebel offensive took control of part of Freetown, the capital. The arrival of Nigerian troops brought the conflict to an impasse. The two sides signed a ceasefire in January 1999.

30   The new rebel leader, Foday Sankoh, signed a peace treaty with the Government in July. Under the treaty, Sankoh was appointed director of the Strategic Minerals Commission.

31   Rebels resumed fighting in May 2000 and Foday Sankoh was incarcerated. In August, he was succeeded by Issa Sesay. Sankoh was informed of this by Kabbah, who together with presidents Olusegun Obasanjo of Nigeria and Alfa Oumar Konare of Mali held

IN FOCUS

ENVIRONMENTAL CHALLENGES

The central rainforest, crossed by many rivers, has been cut down to create land for agriculture. In the eastern plateau there are diamond reserves. There is widespread deforestation and soil exhaustion due to the rapid growth of the population. Civil war resulted in the depletion of natural resources.

WOMEN'S RIGHTS

In 2003, women held 15 per cent of seats in Parliament and 13 per cent of ministerial positions. That year, they made up 37 per cent of the country's total labor force of 2 million.

In 2004, 68 per cent of pregnant women received pre-natal care, and 42 per cent of births were attended by skilled health staff*. Maternal mortality stood at 1,800 deaths per 100,000 births.

CHILDREN

The 11-year civil war in Sierra Leone destroyed much of the country's educational infrastructure. In early 2006, it was estimated that 375,000 children (60 per cent of the total) living in remote areas, mostly girls, had no access to education. A plan for the construction of community schools managed to establish 400 facilities that benefit some 19,000 students. The number of community schools established throughout the country is

expected to reach 1,300 by 2007. In spite of a slight decrease in infant and under-five mortality rates within the period 1990-2004, they remained very high. The former was reduced from 175 to 165 per 1,000 live births and the latter from 302 to 283 per 1,000 live births. Nearly a quarter of newborns are underweight; 9 per cent of under-fives suffer from moderate or severe low weight and 34 per cent are moderately or severely stunted.

INDIGENOUS PEOPLES/ ETHNIC MINORITIES

The Bande, Mende and Temne are the dominant ethnic groups, but there are many more minority groups. In Freetown some 5,000 residents speak Bassa (no relation to the Basa in Nigeria, Ghana, Cameroon and Benin) and practise a traditional religion. The Fulani, herders and nomadic traders of Caucasian origin, comprise five per cent of the population and are spread throughout the land, but mostly concentrated in the north. The Gola live on the border with Liberia in the provinces of Kenema and Pujehun. They number about 9,000, speak Gola and practice Islam (75 per cent) and traditional religions. The Kissi make up three per cent of the population, living mainly in the north-west. Eight per cent are Muslim. The Krio Fulah, descendants of freed slaves live in the western peninsula, and on the Banana, York and Bonthe islands. They were the most politically

powerful group during colonial times, despite being only three per cent of the total population. The Limba, 9 per cent, shared power with the Krio in the late 1960s. Mandingo people, around 2.5 per cent of the population, live in the Kabala region. Their Maninka language is spoken quite widely. They became part of a national state only when they joined the Mali Empire in the 13th century. The Sussu live in the northern province, and the Vai live on the Atlantic coast, numbering 18,000.

MIGRANTS/REFUGEES

In early 2006, 270,000 refugees were expected to return to Sierra Leone from different countries in the region. On the other hand, a solution was being sought to the situation of 30,000 Liberian refugees who remained in Sierra Leone, in spite of the fact that another 20,000 (including 2,500 urban refugees) had returned to their country under UNHCR programs. The possibilities for those that still remained in the country were either repatriation or final integration in Sierra Leone.

DEATH PENALTY

The death penalty is still applicable.

*Latest data available in *The State of the World's Children* and *Childinfo* database, UNICEF, 2006.

meetings with the imprisoned leader.

32   In January 2001, the Government postponed elections scheduled for February and March. For the first time, UN troops began a peaceful deployment in rebel territory in March, and in May the disarmament of the 45,000 rebel soldiers began. It was completed by January 2002.

33   General elections were held in May 2002. Kabbah won with 70 per cent of the vote. His SLPP also won 83 of the 112 seats in Parliament.

34   In July, UN British troops started to leave Sierra Leone, but two months later the UN Security Council decided to extend the military mission's stay in the country on request from President Kabbah, who was concerned about instability caused by the civil war in neighboring Liberia.

35   Foday Sankoh died in July 2003, while awaiting trial for war crimes.

36   In March a UN-backed war

crimes tribunal began hearing cases to try top leaders from both sides of the conflict. Defense lawyers of former Liberian president Charles Taylor - investigated for his alleged involvement in war crimes - stated that the court's jurisdiction should not extend beyond Sierra Leone's borders.

37   In May 2004, the first local elections in more than three decades were held amidst confusion over the voting system, which resulted in a large number of void ballots. In July, the UN-backed tribunal began hearing cases against RUF members, whose numerous acts of terror against civilians during the civil war included hacking off legs and arms.

38   In September 2004, Fanny Ann Eddy, Sierra Leone's most outstanding activist for the rights of sexual minorities, was murdered by a former employee of the Sierra Leone Lesbian and Gay Association (founded in 2002).

39   UN troops withdrew from Sierra Leone in December 2005 and handed over control to local forces, marking the end of a five-year mission.

40   Former Liberian president Taylor - captured in Nigeria, where he had been exiled since 2003 - was placed under UN custody in March 2006. Taylor, charged with crimes against humanity over his alleged role in the civil war in Sierra Leone, was immediately transferred to Freetown upon request by UN Secretary-General Kofi Annan. In June 2006 Taylor was transferred to The Hague to stand trial there for war crimes and crimes against humanity. He is still the responsibility of the Special Court for Sierra Leone but will be tried at the premises of the International Criminal Court because of concern that his supporters could destabilize Sierra Leone if the prosecution went ahead in Freetown. ■

Singapore / Singapore - Singapura

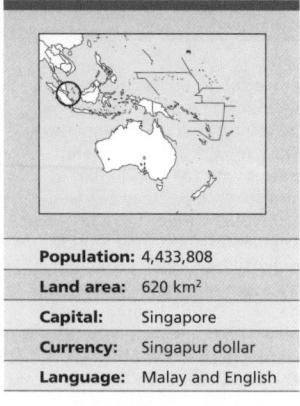

Population:	4,433,808
Land area:	620 km²
Capital:	Singapore
Currency:	Singapur dollar
Language:	Malay and English

One of the earliest references to Singapore as Temasek, or Sea Town, was found in the Javanese 'Nagarakretagama' of 1365. The name was also mentioned in a Vietnamese source at around the same time. By the end of the 14th century, the Sanskrit name, Singapura (Lion City), became commonly used. By that time Singapore was caught in the struggles between Siam (now Thailand) and the Java-based Majapahit Empire for control of the Malay Peninsula. According to the Malay Annals, Singapore was defeated in one Majapahit attack, but Iskandar Shah, or Parameswara, a prince of Palembang, later killed the local chieftain and installed himself as the island's new ruler. Shortly after, he was driven out and fled north to Muar in the Malay Peninsula, where he founded the Malacca Sultanate.

2   From 1819 Singapore became an extremely important base for the British, when Sir Thomas Stamford Raffles established the local headquarters of the British East India Company there.

3   In 1824, the island of Singapore and the adjacent islets were purchased as a single lot by Raffles from the Sultan of Johore (see Malaysia). The Company appointed Prince Hussein as the new ruler of Singapore. In gratitude, he granted the

Company royal authorization to improve the port. Chinese immigrants soon constituted the majority of the local population.

4   Singapore was part of the British colony called 'the Straits Settlements', together with the ports of Penang and Melaka (Malacca). In 1946, Penang and Melaka joined the Malayan Union and Singapore became a crown colony.

5   Japan occupied Singapore in February 1942 but was defeated in 1945 by British forces and a strong internal resistance group, organized by a revolutionary movement and led by the Communist Party of Malaya (CPM). The name Malaya included both Singapore and the Malay peninsula; the separation of the latter was always questioned by the Left.

6   Once the war was over, the Malay sultanates and the former Straits Settlements attempted to form a union or federation, with a view to attaining independence for the territory. However, the conflicting interests of the Chinese and Malay communities, and the conservative and progressive forces, made progress

difficult. In January 1946, the Singapore Labor Union declared a general strike and in 1948 the Communist Party led an anti-

colonial uprising which failed to gain the support of the Malays and the poorer sectors of the Indian population. Marxist parties were outlawed and had to take refuge in the forests, where they resorted to guerrilla warfare.

7   As the first step towards the self-government of the city-state, municipal elections were held in 1949. Only English-speaking people were allowed to vote until 1954, when the People's Action Party (PAP) was founded. Anti-imperialism advocated by the PAP brought citizens of British and Chinese backgrounds together for the first time. In 1959, the Chinese were allowed to vote; full internal autonomy was granted, and the PAP obtained an overwhelming victory. Lee Kuan Yew, founder of the party, became Prime Minister, campaigning on a platform of social reforms and independence. He planned a federation with Malaya, which had been independent since 1957.

8   The PAP split into a socialist

PROFILE

ENVIRONMENT

Singapore consists of one large island and 54 smaller adjacent islets. The country is connected to Malaysia by a causeway across the Johore Strait. The terrain, covered by swampy lowlands, is not conducive to farming and the population traditionally works in business and trade. The climate is tropical with heavy rainfall. In its strategic geographical location, central to the trade routes between Africa, Asia and Europe, the island has become a flourishing commercial center. Economically, the main resources of the country have been its port, the British naval base and, more recently, industrial activity; textiles, electronic goods and oil refining.

SOCIETY

Peoples: 76 per cent of Singaporeans are of Chinese origin. Malaysians account for 15 per cent and 6 per cent are from India and Sri Lanka.
Religions: Buddhist 28 per cent; Christian 19 per cent; Islam 13 per cent; Taoist 13 per cent; Hindu 5 per cent. There are Sikh and Jewish minorities.
Languages: Malay, English, Chinese (Mandarin) and Tamil are the official languages. Malay is considered the national language but English is spoken in the public administration and serves as a unifying element for the communities. Various Chinese dialects are spoken, as are Punjabi, Hindi, Bengali, Telegu and Malayalam in the Indian communities.
Main Political Parties: Peoples Action Party; Workers' Party of Singapore; Singapore Democratic Alliance (which includes: National Solidarity Party; Pertubuhan Kebangsaan Melayu Singapura - Singapore Malay National Organisation; Singapore Justice Party; Singapore People's Party).
Main Social Organizations: The major national trade union is the Congress of Singapore Labor Unions.

THE STATE

Official Name: Hsin-chia-p'o Kung-ho-kuo (Mandarin); Republik Singapura (Bahasa Malaysia); Singapore Kudiyarasu (Tamil); Republic of Singapore.
Capital: Singapore 4,253,000 people (2003).
Government: Parliamentary republic. Sellapan Ramanathan, President since September 1999, re-elected - unopposed - in 2005. Unicameral Legislature, with 84 members.
National Holiday: 9 August, Independence Day (1965).
Armed Forces: 55,500, inc. 34,800 conscripts. Other: 11,600 Police and Maritime Police (estimate). 100,000 Civil Defense Force.

LAND USE

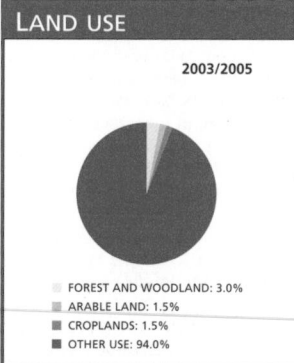

2003/2005

- FOREST AND WOODLAND: 3.0%
- ARABLE LAND: 1.5%
- CROPLANDS: 1.5%
- OTHER USE: 94.0%

PUBLIC EXPENDITURE

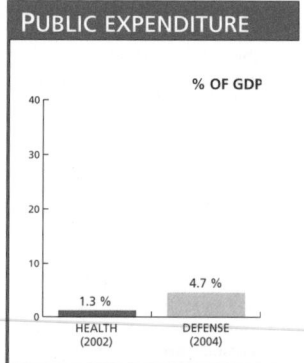

% OF GDP

HEALTH (2002): 1.3 %
DEFENSE (2004): 4.7 %

Life expectancy	GNI per capita	Literacy	HIV prevalence rate
79 years 2005-2010	**$24,760** 2004	**93%** total adult rate 2000-2004	**0.2%** of population 15-49 years old 2003

faction, led by Lim Chin Siong, and the 'moderates' of Lee Kuan Yew, who encouraged the promotion of private enterprise and foreign investment.

9   In 1961, the left wing of the PAP founded *Barisan Sosialis* (the Socialist Front), which opposed the project of uniting Singapore and Malaya under British control. In September 1963, the Federation of Malaysia, consisting of Singapore, the Malay peninsula, Sarawak and Sabah (both in the north and northeast of Borneo) was set up, after those opposed to the Federation had been conveniently purged.

10   As a Federation member, Singapore depended heavily on the peninsula; even its water supply came from there. Integration at that time was not feasible because of profound disagreements between Singapore's Chinese community and the Malay community of the rest of the Federation. On 9 August 1965, after several ethnic conflicts, Prime Minister Lee decided to withdraw from the Federation.

11   In 1965, there were serious internal conflicts caused by the mistreatment of the Malay population and other ethnic minorities. There were also intense struggles with the left-wing opposition, whom the Government classified as 'communist subversives'. The island became an independent republic, and a Commonwealth member. A mutual assistance and defense treaty was signed with Malaysia, and on 15 October of the same year Singapore became a member of the United Nations.

12   The first years of independence witnessed substantial economic growth. The island was turned into an 'enclave' for the export of products manufactured by transnationals, and into an international financial center which controlled the regional economy.

13   In 1974 the oil crisis upset Singapore's export scheme. Singapore was the fourth largest port in the world, with the second highest per capita income in Asia, after Japan. The ensuing economic deterioration brought about public demonstrations by students and workers. These protests were fiercely repressed, to the extent that the Socialist International expelled the PAP from its ranks in 1975.

14   In order to placate criticism from the opposition, the regime approved a reform which allowed the incorporation of two representatives from the Workers' Party and the Democratic Party of Singapore, respectively.

IN FOCUS

ENVIRONMENTAL CHALLENGES
Industrialization has caused serious air and water pollution. There is little land available, making the final disposal of solid waste difficult. There are periods of smog, resulting from frequent forest fires in Indonesia.

WOMEN'S RIGHTS
Women have been able to vote and stand for election since 1947.
Female representation in politics is scanty. In 2003 only 12 per cent of Parliament seats were held by women, and no women held ministerial or equivalent positions. That year women made up 36 per cent of the workforce of 2 million.
In 2003, most of the country's sex workers were foreign, especially from Malaysia, Thailand, Philippines, China, Indonesia, Vietnam, India and Sri Lanka. Most had come to the country for that purpose, but the authorities investigated several cases of women forced into the sex trade that year.

CHILDREN
Singapore has the lowest infant and under-five mortality rates in the world - lower than those in any industrialized country. Between 1970 and 2004 the infant mortality rate fell from 22 to 3 deaths per 1,000 live births and the under-five mortality rate from 27 to 3 per 1,000.-In 2003, 8 per cent of newborns and 14 per cent of under-fives were underweight.
The Government has shown it is strongly committed to children's rights and welfare, especially by investing in health and education. Six years of primary public education are compulsory by law.
In 2002, there were more than 70 cases of prostitution among minors, most of whom were foreigners aged under 18. Sexual intercourse between adults and females under 16 is banned, but concessions are made for those girls 'aware' of being involved in the sex trade at 16 or 17.

INDIGENOUS PEOPLES/ ETHNIC MINORITIES
More than 75 per cent of the population is Chinese, while the rest are Malaysians, Indonesians, Pakistanis and Indians. A small number of Europeans are concentrated in the capital and nearby urban centers.

MIGRANTS/REFUGEES
From its creation Singapore's population was determined by immigration, which diminished throughout the 20th century. The 1947 census showed that 56 per cent of the population was born in the country, while in the 1980 census the proportion had grown to 78 per cent.
Most of the immigrant population is from Malaysia, Thailand, Philippines, Sri Lanka and India, employed in the labor sectors, while immigrants with a university education come mostly from Japan, Eastern Europe, North America and Australia.

DEATH PENALTY
It is applicable even for ordinary offences.

* Latest data available in *The State of the World's Children* and *Childinfo* database, UNICEF, 2006.

15   Despite the sustained economic growth from 1987, the Government expelled thousands of Thai and Filipino workers, blaming them for taking over the jobs of natural citizens.

16   The decision to ban foreign publications judged detrimental by the Government, the imprisonment of opposition members and a Security Law allowing imprisonment without trial for two years, renewable indefinitely, raised countless reports of human rights violations.

17   In the 1988 elections the opposition vote grew, but due to the electoral system its parliamentary representation diminished. In 1991, the PAP again won an overwhelming majority of seats. In November, Goh Chok Tong replaced Lee Kuan Yew as Prime Minister, but the latter nevertheless retained considerable political weight. At his initiative, Singapore offered Washington the possibility of installing bases in the country when the Filipino Congress decided to close US military bases in the Philippines.

18   In 1994 and 1995 relations were strained between Singapore and the US, Philippines and the Netherlands when a Dutch engineer was hanged on heroin trafficking charges, a US man was sentenced to a beating with a rattan cane for vandalism, and a Filipina maid was executed for murdering a colleague. In the

case of the domestic worker, the crime was proved and diplomatic relations were re-established. The US citizen's sentence was reduced in terms of the number of strokes and months in prison.

19   In June 1996 the 'Speak Mandarin' campaign was challenged by ethnic minorities, concerned that the language was becoming a condition of employment. The Government's concern shifted to the declining standard of English spoken by the population, and in 1999 launched a campaign against *singlish*, the local linguistic version of English.

20   In the August 1999 presidential elections, only Sellapan Ramanthan Nathan, the PAP candidate, was declared eligible to run. Nathan took office in September.

21   The free trade accord signed by Singapore and Japan in January 2002 was seen as a 'milestone' by Singaporean authorities because it marked Japan's commitment to the region. Singaporean investors won access to the Japanese service market. Under the terms of the agreement, companies based in Singapore could freely transfer investment in and out of Japan.

22   Lee Hsien Loong, son of former prime minister Lee Kuan Yew, became Prime Minister in August 2004.

23   Malaysia and Singapore

announced in January 2005 an agreement over the narrow strip of sea that separates them. Both countries agreed to safeguard navigation security in what they called a 'shared body of water'.

24   President SR Nathan was re-elected in August 2005 for a second six-year term. He was the sole qualified candidate. The electoral commission had refused to award the 'Candidate Certificate' to three other applicants, arguing that they lacked the 'experience and ability to manage financial matters and carry out the obligations and responsibilities of the president'. This was the third presidential election of the city-State since a constitutional amendment stated in 1992 that the President should be chosen by the people.

25   Mas Selamat Kastari, an alleged leader of the Singapore branch of Jemaah Islamiah - a terrorist network that seeks to establish Islamic fundamentalist theocracies in Southeast Asia, especially in Singapore, Indonesia, Brunei, Malaysia, Thailand and Philippines - arrived in Singapore in February 2006, after being arrested in Indonesia.

26   Lee Hsien's ruling People's Action Party won the first 'real' general elections - that is, those with more than one certified candidate - in May 2006. The Prime Minister's party won 82 of the 84 parliamentary seats. ∎

Slovakia / Slovensko

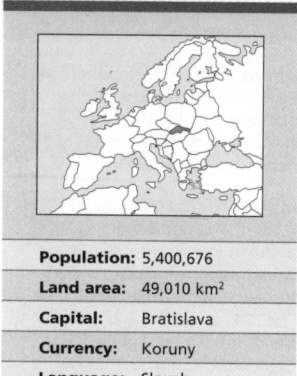

Population:	5,400,676
Land area:	49,010 km²
Capital:	Bratislava
Currency:	Koruny
Language:	Slovak

The Slovak territory was populated between 500 and 100 BC by the Cotini, Celts from western Europe. Later, between 100 BC and 400 AD, the Quadi, a Germanic people, formed satellite states of the Roman Empire north of the Danube, remaining part of Bohemia with the Marcomani, until they were pushed out by the Huns led by Attila around 400 AD.

[2] Slovaks from the eastern region of the Vistula who were closely related to the Czechs, began to settle the territory between the 6th and 7th centuries. They soon had to defend themselves from the Avars, nomads from lower Panonia, until the Frankish trader Samo united the Slavs and was chosen as their King. In 805, the Christian King, Charlemagne formed an alliance with the Czech leaders of Bohemia and Moravia to defeat the Avars. In return for their help, Charlemagne distributed dukedoms amongst the Czechs, who took control over the regions of Moravia, Bohemia and Slovakia. The first monarch of the kingdom of Moravia, Mojmir I, ruled from 830 to 846, and Christianity was adopted under his rule.

[3] His nephew Rotislav I succeeded him, ruling from 846 to 870. He expanded the kingdom to include all of Bohemia and founded Great Moravia, unifying the Slav territories in the region for the first time. Rotislav consolidated relations with the Frankish Empire, which answered to Rome, and maintained contact with the Byzantine Empire. In 863, the Byzantine Emperor sent the monks Constantine (Cyril) and Methodius to the region. The monks translated the gospels and designed the first Slav alphabet (see Czech Republic).

[4] During the reign of Svatopluc (870-894) the frontiers of Moravia were extended to include the western part of present-day Hungary and southern Poland. Svatopluc terminated relations with the Byzantine Empire, and Methodius' disciples were forced to abandon the kingdom, taking refuge in the Balkans (see Czech Republic).

[5] Great Moravia came to an end in 906, when it was destroyed by the German king Arnulf in alliance with the Magyars, a nomadic tribe from the Upper Volga who controlled most of the territories of modern-day Hungary. The western part of the old kingdom remained in the possession of the Czech dukes of Bohemia, while the area between the Carpathians and the Danube - the Slovakia of today - was occupied by the Magyars. Despite repeated attempts to conquer it by the dukes of Bohemia, for ten centuries it continued to be Hungarian.

[6] The Slovaks continued to have cultural links with the Czechs. During the 15th century, the University of Prague exercised considerable influence. The Hussites of Bohemia (see Czech Republic) repeatedly invaded Hungary, and introduced the custom of conducting the liturgy in the national language instead of Latin in Slovakia.

[7] The Hussite incursions prepared the ground for the advent of Protestantism, based on the teaching of the Kralice Bible translated by the Bohemian Brethren. In the early 16th century most of the Slovaks had adopted Calvinism. However, when Hungary was invaded by the Ottoman Empire in 1525, Slovakia was governed by the House of Austria, which Germanized Slovak culture to a large extent and strengthened the Counter-Reformation throughout the region.

[8] In 1620, the Magyars recovered Slovakia for the Kingdom of Hungary. The conquest of central Hungary by the Ottoman Empire increased Magyar influence in Slovakia as Hungarian nobles fleeing Turkish power settled in Slovak villages and initiated Hungarian customs.

[9] In the 17th century Turkish dominion in Hungary was replaced by the House of Hapsburg. At the end of the 18th century the Emperor Joseph II changed the Kingdom of Hungary by Germanizing the bureaucracy and limiting the power of the Hungarian authorities. This helped the Slovaks recover their Slav origins and their cultural links with the Czechs.

[10] The nationalist fever across Europe following the Napoleonic Wars also reached the Austrian territories and caused conflict between Slovaks and Hungarians. In 1834, Magyar replaced Latin as the official language. In 1848, the Slovaks, allied with Czechs and German Republicans, rose up against the Magyars, who had in turn rebelled against the Austrians. Within this period, the Slovaks took over control of their secondary education and founded their first scientific society, the Matica Slovaka.

[11] In the late 19th century, the Hungarian authorities banned the Slovak language from public life, replacing it with Magyar. Slovak leaders, especially editors of journals, were persecuted and many were imprisoned. In 1907, the Appony law made Slovak-speaking primary schools adopt Magyar.

[12] The Slovaks supported the Allies in World War I. Tens of thousands of Slovak soldiers forced to serve in the Hungarian army joined the Allies against the Austro-Hungarian Empire. In 1915, the Czech Alliance and the Slovak League (in the US) reached an agreement in Cleveland proclaiming the liberation of the Czech and Slovak nations and their federate union, with complete independence for Slovakia, which would have its own parliament and administration with Slovak as its official language.

[13] Once the War had ended with the Allied's victory in 1918, the

PROFILE

ENVIRONMENT

The terrain is mountainous, dominated by the Carpathian mountain range. Mountainous areas are covered by forests which supply an important timber industry. Agriculture is concentrated in the fertile plains of the Danube and Uh rivers. There are abundant mineral resources, including copper, zinc, lead and mercury, as well as oil and natural gas deposits. Industrial pollution and sulfur emissions are the main environmental problems.

SOCIETY

Peoples: Slovak 85.7 per cent; Hungarian 10.6 per cent; Roma 1.6 per cent; Czech 1.1 per cent; Ruthenia, 0.3 per cent; Ukrainian 0.3 per cent; German 0.1 per cent; other 0.3 per cent (1994).
Religions: Roman Catholic 60.3 per cent; non-religious and atheist 9.7 per cent; Protestant 7.9 per cent, of which Slovak Evangelical 6.2 per cent and Reformed Christian 1.6 per cent; Greek Catholic 3.4 per cent; Eastern Orthodox 0.7 per cent; other 18 per cent.
Languages: Slovak is the official language; Hungarian is also spoken.
Main Political Parties: Smer ('Direction') - leftist; Slovak Democratic and Christian Union (centre right); Slovak National Party (extreme right); Hungarian Coalition Party; Movement for a Democratic Slovakia; Christian Democratic Movement.
Main Social Organizations: Confederation of Trade Unions of the Slovak Republic (KOZ-SR), Slovak Union of Nature and Landscape Protectors (environmentalist).

THE STATE

Official Name: Slovenska Republika.
Administrative divisions: 3 regions divided into 38 municipalities and the capital zone.
Capital: Bratislava 425,000 people (2003).
Other Cities: Kosice 244,400 people; Presov 95,300; Zilina 87,600; Nitra 87,400; Banská Bystrica 84,400 (2000).
Government: Parliamentary republic, according to the constitution effective since January 1993. Ivan Gasparovic, President since 2004. Robert Fico, prime minister since 2006.
Unicameral legislature: National Council of the Slovak Republic, with 150 members elected for a four-year term by proportional representation. **National Holiday:** 1 September, Constitution Day (1992). **Armed Forces:** 20,000 (2003). Other: Border Guards: 600; Internal Security Forces: 250; Civil Defense Troops: 3,100.

Life expectancy
75 years
2005-2010

GNI per capita
$6,480
2004

Literacy
100% total adult rate
2000-2004

HIV prevalence rate
<0.1% of population 15-49 years old
2003

nationalist efforts of the Slovak doctor Thomas G Masaryk, the scientist Milan Stefanik (a Slovak living abroad), and the Czech Eduard Benes, working with the opposition forces in Czech and Slovak lands, led to the creation of the Republic of Czecho-Slovakia on 28 October of that year.

[14] In November 1918, Masaryk was elected President of the new Republic, a position he held until 1935. During his term in office - and also that of his successor, Eduard Benes - the Slovaks felt they were relegated within a State controlled by the Czechs.

[15] The occupation of Czechoslovakia by Nazi troops in 1939 put the history of the Republic on hold. With the occupation of the Sudetenland in 1938, Benes was forced to resign and go into exile in London. Czechoslovakia was dismembered: Bohemia became a German province and Carpathian Russia was taken by the Hungarians. In March 1939, Slovak independence was proclaimed, with Hitler's puppet President Joseph Tiso in power.

[16] After the Allied victory in 1945 and with Soviet forces in the territory, Benes returned to the presidency. Unity was guaranteed by membership of the Soviet bloc (see under Czech Republic) until, in 1991, with the fall of the Soviet regime and the system of alliances, the Czech and Slovak peoples divided.

[17] In February 1993 Michal Kovac was elected President of the new Republic of Slovakia. Vladimir Meciar, leader of the Movement for a Democratic Slovakia (MED) and the architect of Czechoslovakian separation, was named Prime Minister.

[18] Meciar's administration was marked by controversy. He renationalized the newspaper *Smena*, created a compulsory television slot for the broadcasting of Government news and propaganda, and opposed the teaching of Hungarian in schools.

[19] Jozef Moravcik was appointed Prime Minister in March 1993. The new Government attempted to keep Meciar out of power and accept the general principles of European democracy. However, the coalition's inner differences weakened this project and, in the October 1994 elections, Meciar once again became Prime Minister (with 35 per cent of the vote) and cancelled the privatization policy begun by his predecessor.

[20] In late 1997 five opposition parties founded the Slovak Democratic Coalition (SDK). President Kovak, whose relations with Meciar were already strained, joined in criticizing the Prime Minister, urging him in January 1998 to improve relations with

IN FOCUS

ENVIRONMENTAL CHALLENGES
Pollution from industry (chemicals, machinery and the paper industry) and sulfur emissions are the main environmental problem, affecting the health of half the population. Acid rain is damaging the forests.

WOMEN'S RIGHTS
Women have been able to vote since 1920. In 2003, women held 19 per cent of parliamentary seats, while there were no women in ministerial or similar posts.

Women represented 48 per cent of the labor force of three million. In 2003, the life expectancy of women stood at 78 years, compared to 70 for men. The literacy rate stood at 100 per cent for both men and women in the 1990-2004 period, according to World Bank data.

Prenatal health care reached 98 per cent of women and 99 per cent of births were assisted by qualified staff.

CHILDREN
In 2006, infant and under-five mortality stood at 6 and 9 per 1,000 live births, respectively*. Seven per cent of newborns were underweight. Primary

school enrolment rates stood at 85 per cent for boys and 86 per cent for girls.

INDIGENOUS PEOPLE/ ETHNIC MINORITIES
Almost 11 per cent of the population are Hungarian, the largest minority in the country. Although minority rights are mentioned in the Constitution, the law allows for unequal treatment of foreigners. A law regarding the official language, for instance, was transformed in practice into a legal tool for discrimination, as it banned the use of Hungarian in official documents, such as school certificates, as well as in any oral communication between civil servants and the public. This implied, for example, that a police officer had to address a civilian in Slovak (although both of them might be Hungarian). The same applied to doctors, teachers or any other official. In 1996, after lengthy negotiations between Hungary and Slovakia an agreement was reached to modify the situation, authorizing the use of minorities' languages in any situation, public or private, oral or written and also by the media.

There are also Roma/Gypsy (1.6 per cent), Czech (1.1 per cent), Ruthenian (0.3-per cent), Ukrainian (0.3 per cent) and German (0.1 per cent) minorities.

The relationship between the Government and the Roma is especially problematic. In February 2004, in response to unrest in the east of the country the police raided Roma settlements, leading to complaints of maltreatment, racial violence, solitary confinement of those arrested. Organizations such as Amnesty International have demanded that the Slovak Government carry out independent investigations and adhere to international Human Rights resolutions.

MIGRANTS/REFUGEES
In 2005, 773 Slovakians took refuge in Belgium and 711 in the Czech Republic. Slovakia received more than 1,000 refugees from the Russian Federation, 564 from India, 310 from Moldova, 270 from Bangladesh, 244 from Georgia, 243 from China, 194 from Pakistan, 109 from Afghanistan, 99 from Vietnam and 55 from Palestine.

DEATH PENALTY
The death penalty was abolished for all crimes in 1990.

*Latest data available in *The State of the World's Children* and *Childinfo* database, UNICEF, 2006

the US and the European Union. After his defeat in Parliamentary elections, Meciar resigned and the two main opposition parties, the Slovak Democratic Opposition and the Democratic Left, formed a new government in October, with Mikulás Dzurinda as Prime Minister.

[21] The planned installation of a nuclear reactor in Mochovce, near the Austrian frontier in June 1998, led to protests from Vienna.

[22] In May 1999, Rudolf Schuster won the first direct presidential elections, which had been established through constitutional reform in January.

[23] Protesting against discrimination, some 1,000 Slovakian Roma people sought asylum in Finland in July 1999. President Schuster recognized the demand as legitimate. The Government adopted immediate measures to improve this minority's situation and avoid a large outflow.

[24] The Czech and Slovak Governments reached an agreement in November putting an end to financial differences arising from the separation. Prague promised to deliver 4.5 tons of gold to Bratislava and acknowledged having a $1.5 billion debt with Slovakia.

[25] In February 2001 Parliament approved important reforms to the constitution, preparatory to joining the EU and NATO. The new constitution decentralized power, granted more authority to the office that monitors and audits the Government, and strengthened the independence of the judiciary.

[26] In January 2002, eight new regional parliaments were established, continuing the reforms in readiness for entry to the EU.

[27] Mikulás Dzurinda's center-right Slovak Democratic Coalition (SDK) won the September 2002 elections in the second round. During the November Prague summit, Slovakia was formally invited to join NATO.

[28] In December 2002 in Copenhagen, the country was invited to join the EU in 2004.

[29] In May 2003, following a referendum with a vote just above the required 50 per cent, entry to the European Union was approved.

[30] In April 2004 Slovakia gained full membership of NATO. In presidential elections the same month, Ivan Gasparovic was elected with 59.9 per cent of the vote.

[31] In 1 May 2004 Slovakia joined the EU along with nine other

countries, taking to 25 the number of members of the bloc.

[32] In November, Slovakia joined the European Exchange Rate Mechanism, which pegged its currency, the Slovakian crown, to the euro. This first step - also taken by Poland, Czech Republic and Hungary - was an essential preliminary if it was to enter the euro zone in future.

[33] In April 2006, the overflowing of the Danube river caused serious floods in the country's central region and around Bratislava. More than 8,600 hectares of land were submerged and the cost of the damage was estimated at more than $4 million.

[34] In June 2006, the leftist Smer Party won the election and its leader Robert Fico became Prime Minister, aiming to form a coalition that would push through a leftist program after years of right-wing reforms.

[35] In October 2006, the alliance of socialists in the EU took an unprecedented decision to expel Fico's Smer party for having formed a governing coalition with the far-right Slovak National Party led by Jan Slota. The vehemently anti-Hungarian Slota had been blamed for an upsurge in attacks on the Hungarian minority in Slovakia. ∎

Slovenia / Slovenija

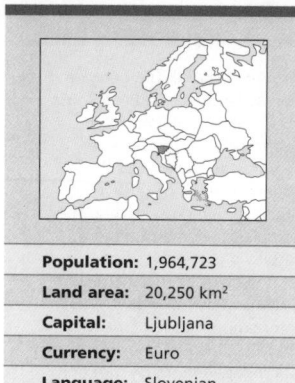

Population:	1,964,723
Land area:	20,250 km²
Capital:	Ljubljana
Currency:	Euro
Language:	Slovenian

One of the southern Slav groups, the Slovenes occupied what is now Slovenia and the land to the north of this region in the 6th century AD. Subdued by the Bavarians around the year 743, they were later incorporated into the Frankish Empire of the Carolingians. With the division of the Empire in the 9th century, the Slovenes were reduced to serfdom and the region north of the Drava River was completely dominated by the Germans.

2   The Slovene people preserved their cultural identity because of the educational efforts of their intellectuals who were mostly Catholic monks and priests. The House of Austria gradually established itself in the region, from the latter part of the 13th century onwards.

3   Between the 15th and 16th centuries, the Slovenes participated in several peasant revolts - some, like the 1573 revolt, in conjunction with the Croats - leading the Hapsburgs to improve the system of land tenure.

4   After 1809, a large part of the territory fell within the Napoleonic Empire's Illyrian provinces. After Napoleon's defeat in 1814, Hapsburg (House of Austria) rule was restored within the region. With the 1848 Revolution, the Slovenes called for the creation of a united Slovene province within the Austrian Empire. The first glimmer of hope for a union of southern Slavs (Slavs, Serbs and Croats) emerged in the 1870s.

5   In the 1890s, the Slovene People's Party (Catholic), and the Progressive (Liberal) and Socialist parties were formed. Members of the Catholic clergy also promoted a large-scale organization of peasants and artisans into co-operatives.

6   In 1917 the Austrian Parliament, representing Slovenes and other southern Slav peoples, defended the unification of these territories into a single autonomous political entity, within the Hapsburg realm.

7   At the end of World War I, amid widespread enthusiasm over the fall of the Austro-Hungarian Empire, the Slovenes supported the creation of a kingdom of Serbs, Croats and Slovenes, known since 1929 as Yugoslavia (land of the southern Slavs). Nevertheless at the Paris Peace Conference the victorious powers handed Gorica to Italy despite the presence of a large Slovene population.

8   The St Germain Treaty, signed between the victorious powers and Austria, gave Yugoslavia only a small part of southern Carintia. Two plebiscites were to define the future of the remainder. However, when the southern region opted to join Austria in 1920, the second plebiscite was not held, and both regions remained part of Austria.

9   Serbian hegemony within the Yugoslav kingdom gave rise to some resentment among Slovenes, although less than among the Croats, and that led to an anti-Serbian movement there. In World War II, Slovenia was partitioned between Italy (the southwest), Germany (the northeast) and Hungary (a small area north of the Mura River). The most prominent group within the Slovene resistance movement was the Liberation Front, led by the Communists.

10   The communist guerrillas fought on two fronts at the same time: against the foreign invaders and against anti-communist military units, organized by the occupation with the participation of the local population. After the defeat of the Axis powers (Germany, Italy and Japan), the major part of old Slovenia was returned to Yugoslavia.

11   Upon the foundation of the Federated People's Republic of Yugoslavia, in 1945, Slovenia became one of the Federation's six republics, with its own governing and legislative bodies. Legislative power was made up of a republican council, elected by all citizens, and the council of producers, elected from among Slovenian industrial workers and officials.

12   Although such entities did not add up to an autonomous government, Slovenia managed to maintain a high degree of cultural and economic independence through this self-management brand of socialism (led by the Yugoslav League of Communists). In 1974, changes in the Yugoslavian federal constitution made Slovenia a Socialist Republic.

13   Slovenia became one of the most industrialized of the Federation's republics, especially in the area of steel production and the production of heavy equipment. Yugoslavia's first nuclear power plant was completed in 1981 in Krsko, with the assistance of a private US firm.

14   In the late 1980s, influenced by the changes in Eastern Europe, Slovenia evolved toward a multi-party political system. In January 1989, the Slovene League of Social Democrats was founded, the country's first legal opposition party, and in October Slovenia's National Assembly approved a constitutional amendment permitting Slovenia to secede from Yugoslavia.

15   The Slovene League of Communists left the Yugoslav League in January 1990, becoming the Democratic Renewal Party. In April, in the first multiparty elections to be held in Yugoslavia since World War II, the victory went to Demos, a coalition of sectors united in their aim of achieving separation from the federation.

16   Slovenia and Croatia declared their independence on 25 June 1991. In the hours that followed, federal troops attacked Slovenian territory and occupied frontier posts. After fierce fighting and the bombing of Ljubljana airport, Belgrade announced that it controlled the federation's borders. However the 21,000-strong Slovenian territorial

PROFILE

ENVIRONMENT

Bordered in the west by Italy, in the north by Austria, in the northeast by Hungary and in the south and southeast by Croatia, Slovenia is characterized by mountains, forests and deep, fertile valleys. The Sava River flows from the Julian Alps (highest peak: Mt Triglav, 2,864 meters), in the northwest of the country, to the southeast, crossing the coal-mining region. The Karavanke mountain range is located along the northern border. The region lying between the Mura, Drava, Savinja and Sava rivers is known for its vineyards and wine production. To the west and southwest of Ljubljana, all along the Soca river (known as 'Isonzo' on the Italian side), the climate is less continental, and more Mediterranean. The capital has an average annual temperature of 9°C, with an average of -1°C in the winter and 19°C in summer. The country's main mineral resources are coal and mercury, which contribute to the country's high level of industrialization.

SOCIETY

Peoples: Slovenian 87.8 per cent; Serbs 2.4 per cent; Croats 2.8 per cent; Bosnian 1.4 per cent, Roma 1. 7 per cent. The Hungarian (0.4 per cent) and Italian (0.1 per cent) minorities are officially recognized, 10,000 Roma people live in the country. Significant German presence. **Religions:** Christian-Catholics are a majority (83.6 per cent), including followers of traditional Catholic church of Slovenia; Christians from the Eastern Orthodox Church (16.4 per cent). **Languages:** Slovenian (official), Serbo-Croatian, Hungarian, Italian, German, Czech, Romany. **Main Political Parties:** Liberal Democracy of Slovenia; Slovenian National Party. **Main Social Organizations:** Two large trade unions, and an independent one, successor of the communist trade union (Association of Free Trade Unions).

THE STATE

Official Name: Republika Slovenija. **Administrative divisions:** 62 Districts. **Capital:** Ljubljana 256,000 people (2003). **Other Cities:** Maribor 97,800 people; Celje 38,300; Kranj 35,500; Velenje 26,400 (2000). **Government:** Janez Drnovsek, President since December 2002. Janez Jansa Prime Minister since November 2004. Unicameral Legislature: the Assembly of Slovenia is formed by 90 members, 40 elected by direct vote and 50 by proportional representation. The State Council is a consultative body with limited legislative powers (40 members). **National Holiday:** 25 June, Independence Day (1991). **Armed Forces:** 9,000 (2002). Other: 4,500 Police (2002).

 **Life expectancy**
77 years
2005-2010

 **GNI per capita**
$14,770
2004

 **Literacy**
100% total adult rate
2000-2004

 **HIV prevalence rate**
<0.1% of population 15-49 years old
2003

IN FOCUS

ENVIRONMENTAL CHALLENGES

Mineral and chemical plants have caused severe pollution of the Sava River and the coast, as well as deforestation due to acid rain. Untreated domestic sewage contributes to this water pollution.

WOMEN'S RIGHTS

Women have been able to vote and stand for office since 1945. In 2004, they held 12 per cent of seats in the lower house and 8 per cent of seats in the upper house of Parliament. In 2003, they also held 6 per cent of ministerial positions. That year, women made up 43 per cent of the country's labor force of one million. Female unemployment reached 6.3 per cent.

Although domestic violence against women is not always reported, social awareness of violence and abuse within marriage has increased as a result of work by civil society organizations. Trafficking in

women - to and from the country - for sexual exploitation remains, however, a serious problem. In 2003, some 98 per cent of pregnant women received pre-natal care and 100 per cent of births were attended by skilled health staff. The maternal mortality rate stood at 17 per 100,000 births.

CHILDREN

Infant and under-five mortality rates have registered a sharp decrease between 1990 and 2004. The former was reduced from 8 to 4 deaths per 1,000 live births and the latter from 10 to 4 per 1,000. Six per cent of newborns were underweight. Over the past few years, the Government has committed itself to the protection of children's rights and welfare. There is free, universal and compulsory education from the age of 6 to 15. According to the Ministry of Education, in 2003, all school-aged children were enrolled and attended school.

Children are able to work from

the age of 16, although during the harvest or for other kinds of farm work, some younger children may work during the year. Urban employers are more regulated and in the cities the age limit is more respected.

There is some segregation of Roma children and in general they attend separate schools; frequently these are for children with learning difficulties.

INDIGENOUS PEOPLES/ ETHNIC MINORITIES

The Constitution provides for special rights and protection of Italian and Hungarian ethnic minorities, and they have representation in Parliament. Unlike these minorities, the Roma/Gypsies have no special rights. Although Roma representatives have participated for several years in talks to improve their situation, no substantial progress has been made. It is estimated that 40 per cent of the Roma population came from Serbia, Croatia, Bosnia and Albania.

Discriminatory attitudes towards the Roma have been denounced by human rights organizations.

MIGRANTS /REFUGEES

In 2004, over 1,000 people took refuge in the country: 379 from Serbia and Montenegro, 193 from Albania, 187 from Turkey, 106 from Bosnia and Herzegovina, 62 from Macedonia, 32 from Georgia, 31 from Moldova, 26 from Iraq, 18 from Algeria and 17 from Bangladesh. In 2005, the country hosted 520 refugees from Serbia and Montenegro, 230 from Turkey, 222 from Bosnia and Herzegovina, 159 from Bangladesh, 143 from Albania, 66 from Macedonia, 61 from Moldova, 34 from India, 28 from Pakistan and 14 from Iraq.

DEATH PENALTY

This was abolished for all crimes in 1989.

---

army caused considerable Yugoslav losses over the next ten days.

[17] On 7 July 1991 a cease-fire went into effect, brokered by the European Community (EC) on the Yugoslav island of Brioni. The agreement reaffirmed the sovereignty of Yugoslav peoples, the federal army agreed to withdraw from Slovenia and Ljubljana promised a three-month freeze of the independence process. In October 1991, the Slovenian Parliament finally approved the end of its official commitment to Yugoslavia. Slovenia implanted its own currency, the Tolar, its national institutions and applied various measures to establish its independence.

[18] While it was part of Yugoslavia, the Slovenian population amounted to a mere eight per cent of the total population, but accounted for 25 per cent of the country's industrial production.

[19] In January 1992, the EC recognized Slovenia and Croatia as independent states, although civil war continued in Croatia. The homogeneity of the Slovenian population made its secession the least painful in the Yugoslavian dissolution process. International recognition was among the clearest as the country controlled its borders, maintained its own armed forces and issued its own national currency. Following withdrawal of the Yugoslav troops, the Government headed by Milan Kucan reinitiated the task of economic reconstruction without

interference from the former Yugoslav Government.

[20] In April 1992, the centrist leader Janez Drnovsek was appointed Prime Minister. In the December elections, Drnovsek's Liberal Democrat Party won and he formed a coalition Government with the Christian Democrats.

[21] With a view to getting closer to the EU, Slovenia began to define itself as a 'European, not Balkan' State. Its shortfall in tax revenue dropped to two per cent of GDP, inflation was below five per cent and the balance of payments had a deficit of only $70 million.

[22] Slovenia requested association with the EU. This was opposed by Italy which demanded compensation for the nationalization of property belonging to 150,000 Italians between 1945 and 1972. The Catholic Church also demanded the return of property that had been nationalized by the communist regime. In June 1996 Slovenia signed an agreement of association with the EU, with the condition that Slovenia would allow foreigners to purchase property in the country.

[23] In January 1997, Parliament ratified Janez Drnovsek as Prime Minister. President Milan Kucan was re-elected for a further five-year period in November.

[24] During the bombings of Kosovo and Serbia in 1999, Slovenia allowed NATO the use of its airspace. According to a Slovenian research group, RIS, approximately ten per cent of the working population - some 80,000 people - worked via

the Internet that year.

[25] In April 2000, Drnovsek lost the confidence of Parliament. He was replaced by Andrej Bajuk, leading a conservative coalition. However, the October parliamentary elections returned Drnovsek to power.

[26] In June 2001, Presidents Putin, of Russia, and Bush, of the US, met for the first time in Slovenia, in what used to be the residence of President Tito, the former Yugoslavian leader.

[27] At a summit meeting held in Copenhagen in November 2002, Slovenia was formally invited to join the EU.

[28] On 1 December 2002 Drnovsek won the second round of the presidential elections with 56.4 per cent of the votes. The Minister of Finance, Anton Rop replaced him as Prime Minister.

[29] On 23 March 2003 the Slovenians voted in two plebiscites in favor of joining the EU and NATO. Over 92 per cent of the voters supported joining the EU - the most favorable result among countries joining in 2004 - and over 60 per cent voted for joining NATO. Slovenia was the only country to submit the NATO decision to a referendum. It also refused to help the US during the attack on Iraq; 80 per cent of Slovenians were against the war.

[30] In February 2004, a constitutional court ordered restitution of the right of residence and other civil rights to 18,000 Croats, Bosnians and Serbians, who had effectively been erased from the records following

independence in 1992.

[31] On 29 March 2004, Slovenia and six other countries from the former Communist bloc joined NATO, the largest increase in membership in the history of the now 26-member-strong organization.

[32] In April 2004, in a referendum promoted by the Right, a strong majority voted in favor of revoking Parliament's decision to restore civil rights to ethnic minorities, embarrassing the Government as it prepared for EU accession.

[33] On 1 May 2004 the EU welcomed ten new members, among them Slovenia, the only country of the former Yugoslavia that had been invited to join the bloc.

[34] Surprisingly, the October 2004 elections gave the victory to the SDS. The center-right party won 30 per cent of the votes, finishing six per cent ahead of Rop's Liberal Democratic Party. The new Prime Minister Janez Jansa, who would have to seek support from other right-wing minority parties to achieve a majority in Parliament, stated that policies from the previous government that were doing well would remain unchanged.

[35] The Slovene Parliament ratified the European Constitution by an overwhelming majority in 2005.

[36] In May 2006, the EU gave Slovenia the green light to join the eurozone in January 2007, after a trial period during which its currency, the tolar, remained pegged to the euro. ∎

Solomon Islands / Solomon Islands

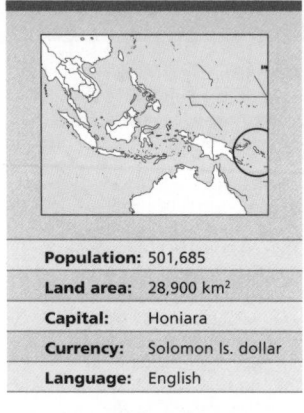

Population:	501,685
Land area:	28,900 km²
Capital:	Honiara
Currency:	Solomon Is. dollar
Language:	English

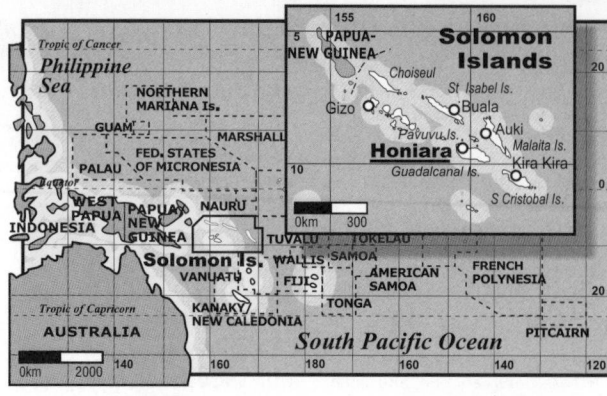

The earliest known occupation of the Solomon Islands is circa 28,000 BC by Australoid hunter-gatherers. The islanders developed a complex social organization based on the *wantok* or extended family system.

[2] The Spaniard Alvaro de Mendaña arrived in 1567, searching for *El Dorado* - the land of gold. In the 18th and 19th centuries, the islands were used as a source of slave labor for the sugar plantations of Fiji and Australia.

[3] Solomon Islands was declared a British Protectorate after World War I. It was occupied by the Japanese in 1942. After World War II, the archipelago was recovered and divided in two. The eastern part, some 14,000 sq km, was placed under Australian administration, and later annexed to Papua New Guinea.

[4] The struggle for independence led by the People's Progressive Party (PPP), headed by Solomon Mamaloni, and the Solomon Islands United Party (SIUPA) led by Peter Kenilorea, ended in 1976 when the islands were granted internal autonomy. Independence was finally declared on 7 July 1978, with Kenilorea as Prime Minister.

[5] Mamaloni was elected Prime Minister in 1981 and established five ministries dedicated to regional affairs.

[6] After the 1984 election, Kenilorea led a coalition government. He patched up relations with the US, but resigned in 1986 amid allegations of accepting French aid for repair work on his home village, damaged by a cyclone. He named Ezequiel Alebua (SIUPA) as his successor. After taking office, Alebua followed Kenilorea's policies.

[7] In 1989, Mamaloni, whose party was now called the People's Alliance Party (PAP), was re-elected Prime Minister.

[8] Francis Billy Hilly succeeded him in 1994 but was forced to step down in October after a vote of no confidence. The new Prime Minister, Mamaloni, ordered the felling of all trees on Pavuvu island, resettling its inhabitants on other islands.

[9] In 1997, Prime Minister Bartholomew Ulufa'alu (of the Solomon Islands Liberal Party) nationalized the logging industry and ordered an investigation into the use of development funds by the previous administration.

[10] In 1999, the Isatabu Freedom Movement (IFM) in Guadalcanal, the largest island of the archipelago, demanded compensation from the Government for the use of their lands for extensive plantations. Upon the Government's refusal, the IFM evicted farmers and forced businesses to close - their targets were islanders from Malaita, who controlled power structures in Honiara. The Malaitans retaliated by forming a private army, the Malaita Eagle Force (MEF), which seized arms and gunboats from the police force.

[11] In June 2000 Ulufa'alu was kidnapped in the capital, Honiara, by members of the Malaita Eagles militia, working in co-operation with paramilitary police, and he was forced to resign in exchange for his release. An emergency vote saw Mannasseh Sogavare of the Solomon Islands National Unity, Reconciliation and Progressive Party take over in July. In October a peace agreement was reached between the IFM and MEF militias, to be enforced by Australia and New Zealand/Aotearoa.

[12] In December 2001, Allan Kemakeza of the PAP became Prime Minister. Australia warned it would not grant any economic aid until the islands achieved significant political stability (compliance with the law, restoring order), and budgetary and economic reforms.

[13] In July 2003 an armed intervention by the Australian-led Regional Assistance Mission to Solomon Islands (RAMSI) put an end to four years of confrontation. Days before the operation, Harold Keke, leader of the Isatabu guerrillas (by now styling themselves the Guadalcanal Liberation Army), declared a ceasefire and freed some hostages.

[14] In October 2003, Keke gave himself up to RAMSI in Weathercoast, Guadalcanal. He was accused of various crimes, including the murders of a cabinet minister and a group of Anglican missionaries.

[15] In November 2003, after declaring the area of Weathercoast safe, RAMSI forces sharply reduced the number of military troops deployed in the country and hundreds of people who had been displaced by violence returned to their homes.

[16] In March 2005, Keke and two of his associates were sentenced to life in prison for the murder of Father Augustine Geve, who was a cabinet minister, in 2002.

[17] The appointment in April 2006 of former finance minister Snyder Rini as Prime Minister sparked riots by people claiming his appointment had been fixed and would benefit both local and overseas Chinese interests. The capital, Honiara, was left in ruins by the rioters. After eight days in office, Rini resigned and Parliament elected Manasseh Sogavare to replace him. ∎

LAND USE

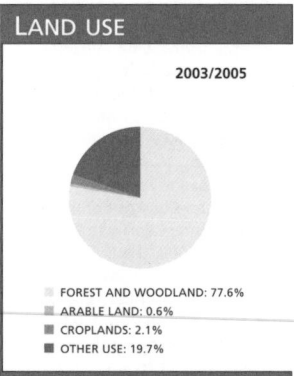

2003/2005

- FOREST AND WOODLAND: 77.6%
- ARABLE LAND: 0.6%
- CROPLANDS: 2.1%
- OTHER USE: 19.7%

PROFILE

ENVIRONMENT

Solomon Islands comprises most of the island group of the same name, except for those in the northwest which belong to Papua New Guinea, the archipelago of Ontong Java (Lord Howe Atoll), the Rennell Islands and the Santa Cruz Islands. The Solomon Islands are part of Melanesia, east of New Guinea. The major islands, of volcanic origin, are: Guadalcanal (with the capital Honiara), Malaita, Florida, New Georgia, Choiseul, Ysabel and Makira. The land is mountainous and there are several active volcanoes. Fishing and subsistence agriculture are the traditional economic activities. Deforestation is severe. Heavy rains cause soil erosion, particularly in exposed areas. Coral reefs are being heavily damaged.

SOCIETY

Peoples: Most of the population is of Melanesian origin (93 per cent). There are also Polynesians (four per cent), Micronesians (1.5 per cent) and Chinese and European minorities. **Religions:** Anglicans (45 per cent), Catholic (18 per cent), Methodist and Presbyterian (12 per cent) Baptist (nine per cent), Seventh Day Adventists (seven per cent). Other Protestant five per cent, local traditional beliefs four per cent. **Languages:** English (official, only spoken by 1-2 per cent of the population), pidgin (local language derived from English) and over 120 local languages and dialects. **Main Political Parties:** National Party; Rural Advancement Party; Peoples Alliance Party. **Main Social Organizations:** The Solomon Islands Council of Trade Unions (SICTU), formed in 1986, made up of six trade unions.

THE STATE

Official Name: Solomon Islands. **Administrative divisions:** 8 provinces and the capital. **Capital:** Honiara 56,000 people (2003). **Other Cities:** Gizo 7,000 people; Auki 5,000; Kira Kira 3,800; Buala 3,000 (2000). **Government:** Queen Elizabeth II, Head of State; Nathaniel Waena Governor-General since July 2004, appointed by the British Government. Manasseh Sogavare, Prime Minister since May 2006. The National Parliament, has 50 members elected for a four-year term. **National Holiday:** 7 July, Independence Day (1978).

Somalia / Soomaaliya

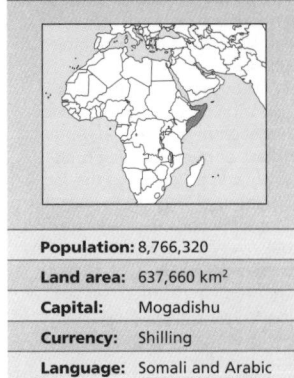

Population:	8,766,320
Land area:	637,660 km²
Capital:	Mogadishu
Currency:	Shilling
Language:	Somali and Arabic

A round 2000 BC, Egypt began trading with the Somali territory called Puntland. Centuries later, the Romans called it 'the land of aroma' because of the incense produced there.

² This commercial tradition took on new dimensions from the 8th century when Arab refugees founded settlements on the coast.

³ In the 13th century, after converting to Islam and led by Yemeni immigrants, Somalis founded a state which they called Ifat. It paid tribute to Ethiopia, and its principal center was in Zeila. Ifat quickly consolidated its independence and annexed new territories, becoming the Sultanate of Adal.

⁴ Links with Arab markets and the southern coast of Zandj (East Africa) contributed to a vigorous commercial activity. At the same time, the sultans tried to enlarge their dominion, at the expense of the Ethiopian Empire. As from 1439, the religious implications of conflict prompted the Abyssinians to seek help from European Christians.

⁵ However, it was not until 1541 that the Portuguese Government, in an attempt to monopolize Indian Ocean trade, sent its fleet. Backed up by the Ethiopian army, the Portuguese razed the city of Zeila, going on to destroy Mogadishu (Muqdisho), Berbera and Brava.

⁶ The Portuguese destroyed but did not occupy the area, though

the presence of their armada hindered economic reconstruction. Adal was divided into a series of minor sultanates, the northern ones controlled by the Ottoman Empire, while the ones in the south accepted the sovereignty of the Sultan of Zanzibar after the expulsion of the Portuguese in 1698.

⁷ The Suez Canal gave new strategic value to the 'Horn of Africa' (now Somalia, Somaliland, Puntland, Djibouti, Eritrea and Ethiopia). In 1862, the French bought the port of Obock,

leading to the creation of present-day Djibouti. In 1869 the Italians settled in Aseb and later extended their control over Eritrea. The British took Zeila and Berbera in 1885. In 1906, in compensation for their defeat in Ethiopia, the Italians obtained Somalia's southern coast.

⁸ In 1889, Italy created a protectorate over central Somalia and southern territories given up by the Sultan of Zanzibar.

⁹ The British colony was the main center of resistance to occupation. Sheikh Muhammad

bin Abdullah Hassan organized an Islamic revolutionary movement, which defeated the British troops on four occasions between 1900 and 1904. The British finally gained control of the territory in 1920 using airplanes for the first time in Africa.

¹⁰ In 1925, the lands to the east of the River Jubba passed from Kenya to become an Italian protectorate.

¹¹ With the union of a Somali-speaking part of Ethiopia, Italian East Africa was created in 1936.

¹² In 1940, the Italians occupied the British part of the country, and in 1941, the British conquered the Italian part.

¹³ In 1950, Italian Somalia became a UN territory under Italian control. In 1956 it obtained internal autonomy, under the name of Somalia.

¹⁴ The British and Italian regions became independent and were merged as the United Republic of Somalia. Aden Abdullah Osman Daar was the first president.

¹⁵ Between 1963 and 1964 the country broke off relations with Britain and disputed its frontiers with Kenya and Ethiopia.

¹⁶ Abdi Rashid Ali Shermarke was elected president in 1967. Two years later he was assassinated and Muhammad Siad Barre took on the presidency. In 1970 Somalia adopted socialism, nationalizing part of the economy. In 1974, the country entered the Arab League.

¹⁷ In July 1976, Somalia invaded Ogaden (Ethiopia), supposedly in support of the Front for the Liberation of Western Somalia. The Ethiopian army, with the support of Cuban troops repelled the invasion. Somalia broke off relations with Cuba.

LAND USE

2003/2005

IRRIGATED AREA: 18.7% of arable land

- FOREST AND WOODLAND: 11.4%
- ARABLE LAND: 1.7%
- CROPLANDS: 0.0%
- OTHER USE: 86.9%

PROFILE

ENVIRONMENT

Somalia is a semi-desert country with a large nomadic population. The north is mountainous, descending gradually from the Galia-Somali plateau to the coastal strip bathed by the Gulf of Aden. The south is almost entirely desert with the exception of a fertile area crossed by the rivers Juba and Shebeli. Important crops include banana, soy and maize.

SOCIETY

Peoples: Somalis are Hamitic and their largest ethnic groups (Hawiye, Darod, Issaq, Dir and Digil-Mirifle) have a cultural and linguistic unity that is uncommon in Africa. Arabs and Bantu-speaking Africans (descendents of slaves). Italian minority.
Religions: Islam (official), mostly orthodox Sunni. There are a few Christians in Mogadishu.
Languages: Somali (a language without an alphabet until 1973, when it was adapted to the Latin alphabet) and Arabic, both official. English, Italian and Swahili are spoken.
Main Political Parties: None. **Main Social Organizations:** Armed groups and also civil associations such as CEPADO (Somali Organization

for the Defense of Nature and Against Deforestation).

THE STATE

Official Name: Jamhuriaydda Soomaaliya.
Administrative divisions: Somalia is divided into 18 regions or provinces. The northern part of the country has declared itself independent as the Republic of Somaliland.
Capital: Mogadishu 1,175,000 people (2003).
Other Cities: Hargeysa 231,00 people; Bervera 213,400; Kismayo 201,600.
Government: Abdullahi Yusuf Ahmed, Transitional Federal President since October 2004. Ali Mohamed Ghedi, Prime Minister since December 2004. The Transitional Federal Institutions (TFIs), a transitional governing entity with a five-year mandate, was established in October 2004. The TFI relocated to Somalia in June 2004, but its members remain divided between Mogadishu and Jowhar and the Government continues to struggle to establish effective governance in the country. Unicameral National Assembly with 275 members.
National Holiday: 1 July, Independence Day (1960).
Armed Forces: Regular armed forces have not existed since rebel forces ousted the Government in 1991.

[18] The war and the 1978-79 drought brought the country to the brink of collapse.

[19] In October 1980 Barre declared a state of emergency and reinstated the Supreme Revolutionary Council, which had ceased functioning in 1976.

[20] In 1985, problems with Ethiopia increased because of the dispute over the Ogaden plains and the flow of refugees into Somali camps.

[21] Said Barre was re-elected in December 1986 by 99 per cent of the vote. In 1998 peace was signed with Ethiopia.

[22] In January 1991, the opposition formed the United Somali Congress (USC) and ousted the President, replacing him with Ali Mahdi Mohammed, of the Hawiye clan, who represented business interests. He fled the capital in November following confrontations between USC factions. The capital remained in the hands of General Mohamed Farah Aideed, leader of the military wing of the USC and of another subdivision of the same clan. It was the start of faction warfare, which was to leave thousands of people dead or in exile.

[23] In March 1992, nine *ugas* or kings of the Hiraan region met for the first time in over a hundred years to discuss how to bring peace to their country. The territory of the former British colony declared independence in May as the Republic of Somaliland. Without international recognition, it suffered from the same inter-clan violence.

[24] That year, under the UN's Operation Restore Hope, 28,000 US troops were deployed on the pretext of re-establishing food supply and encouraging the factions to disarm. In 1993 UN soldiers from Pakistan, India and other countries took over and US troops withdrew in 1994. UN forces left one year later. The intervention was seen as disastrous by many, especially in the US, and continues to affect US foreign policy.

[25] In 1995, Somaliland continued to function; elsewhere the factions regrouped around the Somali Salvation Alliance (SSA), led by Ali Mahdi Mohammed, and the SNA, led by Farah Aideed, both claiming power. Farah Aideed died in August 1996 and was succeeded by his son, Hussein. Farah's former right-hand, Osman Hassan Ali ('Ato'), emerged as a new force, though associated with Ali Mahdi.

[26] In January 1997 in Ethiopia, political leaders affiliated with the SSA, with the support of the Organization of African Unity, established a National Salvation Council. Both Hussein Aideed and Mohammed Ibrahim Egal, re-elected President of Somaliland in March, refused to recognize the decision.

[27] In June 1998 a conference of 300 northeastern leaders in the Garowe district elected Colonel Abdullah Yussuf Ahmed as President and Mohammed Abdi Hashi as Vice-President. The so-called Puntland administration would include the Garowe, Bari and Galkayo areas, with Garowe as a possible capital.

[28] In August 2000, a peace conference in Djibouti, involving several factions, elected the 245 members of Parliament, and chose Abdiqasim Salat Hassan as the new President. In October, in Mogadishu the new Prime Minister, Ali Khalif Galaydh, announced his Cabinet of 25 ministers, all men and representing the various Somali clans, which he called a 'reconciliation government'.

[29] In May 2001 a referendum voted by a large majority for the separation of Somaliland. The country still lacked recognition from the international community.

[30] In May 2002 Mohammed Egal died in a hospital in South Africa, leading to fears that his death could prompt the re-emergence of old rivalries. Dahir Riyale Kahin was designated President of Somaliland.

[31] In October 2002, the 21 factions at war and the Government agreed to a ceasefire, while further negotiations took place.

[32] By a close margin, Dahir Riyale was elected President of Somaliland in April 2003. In July the

Somaliland

Population: 3,500,000
Land area: 137,600 km²
Capital: Hargeysa
Currency: Somali shilling
Language: Somali

Somaliland declared its independence from Somalia in 1993, claiming authority within the old British colonial borders and differentiating itself from the Italian colonial territories of Somalia (see history).

[2] Since then, the autonomous government has pursued its demand for international recognition. However, the various UN bodies refer to this territory as 'Somalia's Northwest Zone', in an attempt to preserve the illusion of a future reunification of the former country, devastated by more than 12 years of civil war, with neither strong government institutions, nor representative or stable social organizations.

[3] Within this reality, the Government of Somaliland has made enormous efforts to achieve a multiparty democratic organization and lasting peace on the domestic front, as well as building diplomatic and commercial relations with the rest of the world.

[4] The death of President Mohamed Ibrahim Egal, in May 2002, threatened to halt the independence process. Egal's successor, Dahir Riyale Kahin, took office and continued the democratization policy and peace process.

[5] The party of President Kahin won the second round of the April 2003 elections by just 80 votes in a ballot of 488,543. The peaceful acceptance of the result by the main opposition party was evidence of a significant advance towards cementing the country's democracy.

[6] In spite of the relative freedom of expression, in June 2003, General Jama Mohamed Ghalib, a former police chief who advocated the return of Somaliland to a federal Somalia, was detained and deported. Several of his supporters were arrested after a shoot-out with Somaliland security forces and remained detained without charge or trial at the end of the year.

[7] The murder of several international aid workers in October 2003 dented the image of a safe and peaceful country that the Government wanted to portray to the rest of the world. The authorities accused foreign agents of infiltrating the country and trying to upset the efforts made during many years in order to gain international recognition.

[8] The disputes with Puntland (another northern semi-autonomous territory) have prevented Somaliland from defining its borders. Military incursions led by Colonel Abdullahi Yussuf in the Sool and Sanaag areas, are aimed at taking over an area where the population belongs to the Darod clan, while in Somaliland most of the population are Isak members.

[9] In January 2004, Puntland forces advanced to within eight kilometers of the Sool regional capital, Las Anod, and kidnapped the brother of the Minister for Rural Affairs, Fou'ad Adan Ade. The Government's weak response to the attacks drew criticism from the opposition and the press.

[10] In May 2004, President Kahin set parliamentary elections for March 2005. According to the authorities, this will be the last step in the democratization process and will help Somaliland obtain recognition as an independent state.

[11] Somaliland not only faces difficulties arising from its pursuit of recognition and State-building, but also the problems of being located in one of the world's poorest areas.

[12] It is hard to engage people's interest in the complex political issues beyond those of group and clan loyalties, among a population of 3.5 million people, of whom 90 per cent led nomadic lives until a short time ago, and who are still suffering the consequences of a long civil war.

[13] The gross national income per capita is $120 (estimated for Somalia by the World Bank). The infant mortality rate is 133 per 1,000 live births. Barely 34 per cent of births are attended by skilled personnel and 95 per cent of girls suffer genital mutilation.

[14] Most people are Muslim. Women have been traditionally relegated and semi-secluded. However, their key role in the rebuilding of post-war civil society has recently begun to be recognized.

[15] The literacy rate is 24 per cent. These rates fall abruptly in rural areas where people's contact with schools and modern lifestyles is sporadic. However, at least there is now relative peace, which allows the country to concentrate on achieving the economic development necessary for dealing with the serious crisis faced by its population. ■

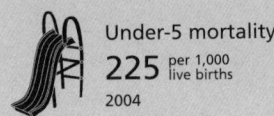
Under-5 mortality
225 per 1,000
live births
2004

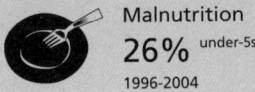
Malnutrition
26% under-5s
1996-2004

Maternal mortality
1,100 per 100,000
live births
2000

IN FOCUS

ENVIRONMENTAL CHALLENGES
The effects of the most recent droughts have been exacerbated by overgrazing. The sharp increase in livestock numbers has triggered desertification. Fishing using explosives has damaged coral reefs and aquatic vegetation. The destruction of the habitat of several fish species could threaten future catches. There are 74 endangered species in Somalia, including mammals, plants and birds. There is serious water pollution.

WOMEN'S RIGHTS
Women have been able to vote since 1956. In 2004, women held eight per cent of parliamentary seats. That year, women comprised 43 per cent of the labor force of four million people. Pre-natal medical care reaches 32 per cent of pregnant women and qualified staff assists 25 per cent of births*. The maternal mortality rate is 1,100 per 100,000 live births*.

CHILDREN
In 2004, the infant and under-five mortality rates were unchanged since 1990, at 133 and 225 per 1,000 live births respectively. Some 17 per cent of under-fives are moderately or seriously emaciated and 23 per cent are moderately or seriously stunted*. Primary school attendance averaged just 10 per cent for girls and 12 per cent for boys between 1996 and 2004*. Approximately 32 per cent of children aged between 5 and 14 worked.

INDIGENOUS PEOPLE/ ETHNIC MINORITIES
Somalis make up 64 per cent of the population. This ethnic group is divided into various subclans: Dir, Darood, Issaq, Hawiye, Rahanweyn, Digil. The latter comprise 16 per cent of the population and speak three different languages, which until 1921 were grouped under the name of Somali: Maay, Tunni y Jiddu.

The Afars number 60,000. They have occupied the territory for the past 2,800 years. They are mainly nomadic, and speak Afaraf, an Afro-Asiatic language.

In the Gedo region there are some 3,000 Boranas. They are united by the concept of *nagya borana* (Borana peace), and try to maintain internal harmony despite external conflicts. Their social organization, known as *gada*, is divided into age groups. They are camel herders and farmers.

A small number of Tikuu-speaking people, the Bajun, a fishing community, are mostly located in Kenya.

Gosha people, one per cent of the population, live in the south, in the Jamaame district, Jubba region and in the urban centers of Kismaayo and Mogadishu. In Somali the name means forest people. They are descendents of the Bantu-speaking slaves imported to Somalia from Tanzania and Mozambique.

MIGRANTS/REFUGEES
From January to August 2005, 7,000 Somali refugees were repatriated by the UNHCR, taking to 486,000 the total of those that were voluntarily repatriated without the support of an organization. In 2006 some 1.25 million of people who had returned and 400,000 internally displaced were being assisted in 34 different areas, 250,000 of them in Mogadishu.

DEATH PENALTY
This is still applied.

*Latest data available in *The State of the World's Children and Childinfo* database,UNICEF, 2006.

Government of Somalia continued with the peace negotiations in Kenya.

[33] In January 2004, the talks produced an agreement between the military chiefs and the politicians to set up a new Parliament, seen as a major step towards lasting peace.

[34] Two factions of the SNF (Somali National Front) started fighting over control of Bula Hawo in June 2004. The town is of key importance to the commercial routes that link Mogadishu with Kenya and Ethiopia. The Kenyan Government reacted by arresting suspects in the area and police forces were deployed along the border to prevent militias from crossing over.

[35] Throughout 2004, peace talks went on in Nairobi, Kenya, in an effort to create a new government. In October, warlord Abdullahi Yusuf - supported by Ethiopia - was elected President. Yusuf immediately requested aid from the international community to disarm the militias.

[36] The new President confirmed Prime Minister Ali Mohammed Ghedi in December, in spite of accusations against him. That month, the tsunami that hit Southeast Asia reached Somali coasts, killing hundreds of people on the island of Hafn and displacing thousands more.

[37] In May 2005, while Ghedi was delivering a speech in Mogadishu, a bomb went off killing at least 10 people and injuring another 10. In June, the Government started returning from its exile in Kenya, although it was not decided where it would settle in Somalia.

[38] A new assassination attempt on Ghedi took place in November, when shots were fired against the official motorcade in Mogadishu. The Prime Minister escaped unharmed, but six people died.

[39] More than a year after its creation in Kenya, Parliament finally met in Somali territory - in the central city of Baidoa - in February 2006. Although 205 of the 275 MPs attended, many warlords - allies of speaker Sharif Hassan Sheikh Adan, unhappy with President Yusuf - were absent. The meeting was possible due to talks both leaders had held a month earlier in Yemen.

[40] The Islamic Courts militia took control of Mogadishu in June 2006. This movement - the strongest and most popular in the country - was made up of 11 autonomous courts of the capital, ruthlessly fighting for the imposition of Islamic law so as to eradicate pornography, drugs and everyday crime on the streets of the city. That month, the UN managed to implement a precarious ceasefire. Ethiopia declared that it would not allow the Islamic Courts to take over Baidoa - seat of the weak transitional government - and sent troops to stay near that city, while arms were arriving - supposedly from Eritrea, Egypt, Iran and Libya - to support the Islamic Courts.

[41] Somali Islamic leaders threatened to wage a holy war against Ethiopia if that country refused to withdraw its troops. ■

Puntland

Population: 2,500,000
Land area: 300,000 km²
Capital: Garowe
Currency: Somali shilling
Language: Somali

Puntland, in northeastern Somalia, declared itself a federal regional State in 1998. Its internal organization is precarious, even more so than Somalia's. Its poor economic situation, reflecting that in Somalia, as well as the violence - both internal and arising from border disputes with Somaliland - pose major obstacles to resuming normal daily life and to arousing the population's interest in participation in their new autonomy.

[2] Puntland took part in peace talks for Somalia and supported a federal constitution. In May 2003, a peace agreement was signed between Puntland President Abdullahi Yusuf Ahmed and an armed opposition group, the Puntland Salvation Council, headed by General Mahamoud Musse Hersi ('Ade'). Opposition political leaders were integrated into the government and all captured soldiers were released.

[3] The situation showed a slight improvement in late 2003. Independent expert Ghanim Alnajjar submitted a report to the UN on his visit to Kenya and Somalia in August-September that year. He met the Minister of Commerce, UN members in the country and local NGOs; he also visited the police headquarters, the port, the main prison and camps for internally displaced people. He reported that the country was relatively peaceful at this time, which allowed authorities to concentrate on urgent internal issues.

[4] In early 2004, President Abdullahi Yusuf Ahmed travelled to Italy, Malaysia and Libya. In Italy, he discussed Somalia's peace situation with government authorities. In Malaysia, he signed bilateral trade agreements and in Libya, he met President Muammar al-Qadhafi, who supported the fight against the Siad Barre regime (see history of Somalia). ■

South Africa / South Africa - Suid-Afrika

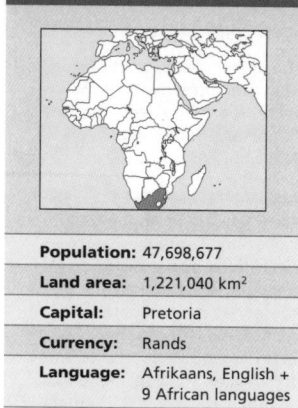

Population:	47,698,677
Land area:	1,221,040 km²
Capital:	Pretoria
Currency:	Rands
Language:	Afrikaans, English + 9 African languages

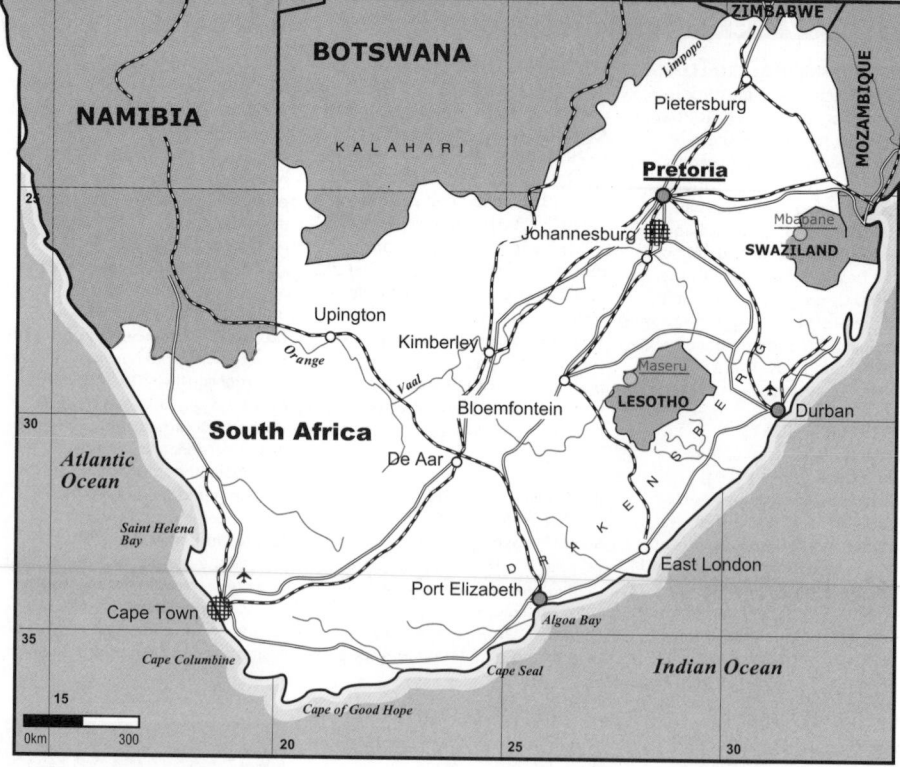

F or most of the past 100,000 years, the territory of present-day South Africa has been occupied by the *San*, small mobile groups of hunting and gathering people who expressed their beliefs, rituals and activities in richly abundant rock art. They were gradually displaced by the *Khoikhoi*, agro-pastoralists whose presence goes back 2,000 years. About the year 500 AD, immigrant *Bantu*-speakers began to work the soil, mainly in the river valleys of south-eastern Africa where summer rainfall predominates. Techniques first learnt further north came to be applied to the growing of edible crops such as millet and squash.

2   Domestication of cattle created new possibilities as societal and political systems arose. The Bantu-speaking chiefs expanded their power through control of their women as producers, and the youth as workers and soldiers. Wealth in cattle made patronage possible, and was also used as *lobola*, bride-price, by the bridegroom's people to the father or guardian of the bride. Metallurgical skills gave the chiefs an additional valuable trading commodity and greater military potential.

3   The first Dutch settlers arrived at Cape Town in 1652, more than 150 years after the Portuguese sailor Vasco da Gama rounded the Cape of Good Hope. Jan van Riebeeck was the first of the Dutch to challenge the *Khoikhoi*. He landed in Cape Town and established a colony supplying ships on their way to Indonesia. In 1688, nearly 600 farmers had settled there, dividing their energies between farming and the war against the Khoikhoi. Being such a small minority, the first Dutch colonists were fiercely united and aggressive, two characteristics which pervaded *boer* (farmer) society in southern Africa.

4   The Dutch who worked for the Dutch East India Company (VOC) were not allowed to trade with the local people and had to deliver all their production to the Company's ships. They gradually came into conflict with their overseas bosses, who would not loosen the grip of their monopoly. The Boers won the dispute, and towards the end of the 17th century the so-called free colonists or burghers were in the majority. The population of European origin split between those linked to foreign trade and those who moved inland in search of new lands.

5   In 1806, with the Dutch colonial empire on the wane, the British settled in Cape Town. They moved ahead on agreements to trade goods, incorporated the African leaders as intermediaries and ended slavery. This rapidly led to conflict with the slavery maintained by the intransigent Boers, who had begun calling themselves Afrikaners, to distinguish themselves from the colonists. In 1834, nearly 14,000 Afrikaners emigrated to the interior, starting the Great Trek which would take them to what became Transvaal, the Orange Free State and Natal, endeavoring to exploit slave labor without foreign interference. They established the state of Transvaal in 1852 and Orange Free State in 1854.

6   The British recognized the independence of the two regions, as the settling of new lands by Europeans increased Cape Town's security. Furthermore, the Boers necessarily had to trade through ports operated by the British. In their expansion northward, the Afrikaners confronted Xhosas and Zulus. The latter, led by military genius Shaka, blocked the colonists' advance over a period of 50 years. Shaka became the head of a great empire which collapsed shortly before the Great Trek, due to internal strife over the royal succession.

7   Peace between the Boers and the British Crown ended in 1867 when rich gold and diamond fields were discovered in Transvaal. Certain that the region held great economic and strategic value, Britain proposed a federation between the Cape Colony and the two Free States. The Boers refused, leading to war in 1899. Britain was supported by most of its colonies, while the Boers had German backing. After three years of war, with nearly 50,000 Afrikaners dead and double that number confined to concentration camps, they surrendered, accepting British domination, while maintaining

LAND USE

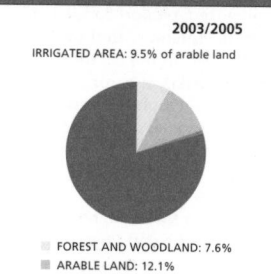

2003/2005

IRRIGATED AREA: 9.5% of arable land

▫ FOREST AND WOODLAND: 7.6%
▪ ARABLE LAND: 12.1%
▪ CROPLANDS: 0.8%
■ OTHER USE: 79.5%

PUBLIC EXPENDITURE

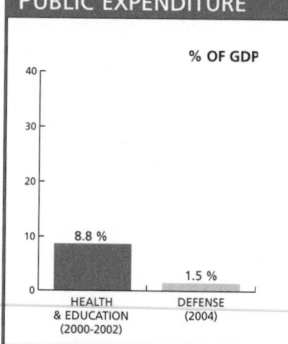

% OF GDP

8.8 %   1.5 %

HEALTH   DEFENSE
& EDUCATION   (2004)
(2000-2002)

WORKERS

UNEMPLOYMENT: 28% (2004)

LABOR FORCE   2004

■ FEMALE: 38%   ■ MALE: 62%

Life expectancy
44 years
2005-2010

GNI per capita
$3,630
2004

Literacy
82% total adult rate
2000-2004

HIV prevalence rate
21.5% of population 15-49 years old
2003

a certain independence for their regions. The British victory signaled the end of the hegemony of landed Boer farmers in the Orange Free State and Transvaal and the beginning of mining's importance in the economy.

8   For some Boers, the African peoples were 'savages' who had to be tamed and made into slaves. White supremacy and racial segregation were established to justify the subjugation of the black population and to guarantee a supply of cheap farm labor from tenants. Boer farms could not compete with British farming as practised in the Cape and Natal and therefore they needed very cheap labor.

9   The British focus on trade and liberalism made them see slavery as a restraint on the creation of consumer markets. This did not prevent them from erecting rigid barriers to exclude black South Africans from economic and social advancement. The labor legislation from 1809 imposed severe controls on worker mobility. The 1843 Master and Servant Act made it a criminal act to break a work contract.

10   Around 1850, the British also contracted black workers in the territories of present-day Mozambique, Lesotho, Botswana, as well as Indians and Chinese. These 'imported' workers were not allowed to bring their families; pay was poor and they had to return to their own countries if they lost their jobs.

11   In 1894, work taxes were introduced, payable in cash, unless a person could certify having worked outside their home districts for a given time. This was decreed to force Africans to work for a salary far lower than that paid to those of European origin. Another law levied an annual tax on peasants, also payable in cash, which they could only get by selling their produce to the Europeans. Salaries were kept low, destroying traditional African ways of life.

12   When the gold and diamond mines began to be exploited, the European capitalists had to employ qualified white workers. Most of them were former Boer farmers who had lost everything in the war. Others came from Europe attracted by 'gold fever'. Both of these groups were used to the workings of the industrial capitalist system, and made demands for better pay and conditions. The mining companies promised benefits to these white workers as long as they fell in with the exploitation of the black

workforce.

13   In 1896, the so-called color bar was in place in the mining sector and in the urban centers where the British were in the majority. In 1910, the Constitution of the Union of South Africa - a federation of Cape Province, Natal, the Orange Free State and Transvaal - deprived most black people of the right to vote or to own land. In 1930, nine-tenths of the arable land was in the hands of Europeans or their descendents.

14   From 1910, segregationist legislation increased. The Native Labor Act pushed urban workers into a system of submission similar to that operating on rural estates. The 1913 Native's Land Act, earmarked seven per cent of national territory for the blacks - the so-called Bantustans - as reserves which became home to 75 per cent of the population. The remaining 93 per cent of the land was reserved for whites who only made up ten per cent of the population. In the overcrowded black reservations subsistence agriculture was the main activity. The rest, under white control, was exploited under intensive farming with the reserves providing a permanent source of cheap labor. The 1923 Native Urban Act tightly regulated blacks' lives in cities which were considered white strongholds. The movements of black people became subject to absolute control.

15   From the time South Africa started on the road to independence in 1934 until 1984, political participation was limited to less than 17 per cent of the population. Constitutional Reform in 1984 extended the vote to the Asians - mainly Indians - and the 'Colored' or mixed race groups. Black South Africans - almost two-thirds of the population - remained without the right to vote.

16   At the outset of World War I, the white economy was based on mining and intensive agriculture. The post-war recession obliged the large mining companies to hire blacks, leading to racial confrontations within the workforce. The Rand strike in 1922 was harshly put down by the Government. Most of the strikers were poor whites, descendants of both English and Boers. Frustrated by their defeat in the war and the loss of their lands, with no easy way of entering the nascent industrial structure, Afrikaners were attracted by the ultra-nationalist propaganda of the far right.

17   The Nationalists, triumphant in the 1924 elections along with

their English-speaking allies, broke with the traditional liberal economic policy and imposed protectionism. State capitalism promoted by the Nationalists - with steel works, railways and electricity - made rapid national growth possible, something many saw as an 'economic miracle'.

18   Toward the end of the 1920s the falling gold price on the international market led to a crisis between the Nationalists and the Labor party. The Nationalists joined forces with previously despised foreign capital and maintained

the racial segregation system which guaranteed cheap workers. The ensuing industrial take-off brought with it an increased number of black employees, leading to racial conflict. A secret society, the *Afrikaner Broeder Bond* - Brotherhood - became the bastion of white right-wing politics.

19   The recession following World War II led to a repeat of events: poor whites, threatened by unemployment, rebelled. Racism flourished under the slogan 'Gevaar KKK' (Beware blacks, Indians and communism - Kafir, Koelie, Komunism). In

PROFILE

ENVIRONMENT

Located on the southern tip of the African continent, with coastlines on the Indian and Atlantic Oceans, South Africa has several geographic zones. A narrow strip of lowland lies along the east coast, with a hot and humid climate and large sugarcane plantations. In the Cape region there are vineyards and fynbos vegetation. The vast semi-arid and arid Karoo, with cattle and sheep ranching, makes up over 40 per cent of the total territory, extending inland. The Highveld extends to the north and is the richest arable area. It surrounds the Witwatersrand, a mining area in Gauteng (formerly Transvaal) where large cities and industries are found. The country's economic base lies in the exploitation of mineral resources: South Africa is the world's largest producer of gold and diamonds; the second largest producer of manganese; and the eighth largest producer of coal.

SOCIETY

Peoples: Over 76 per cent of the population is of African origin, of which Zulu 22 per cent, Xhosa 18 per cent, Pedi 9 per cent, Sotho 7 per cent, Tswana 7 per cent, Tsonga 3.5 per cent, Swazi 3 per cent, Ndebele 2 per cent, and Venda 2 per cent. There are also descendants of whites, slaves and Khoisan, called 'Colored'. European descendants account for 13 per cent of the total. Asian groups, predominantly Hindu, make up less than 3 per cent.
Religions: Christianity predominant (68 per cent) including African independent churches. African beliefs (28 per cent); Islam (2 per cent).
Languages: Eleven official languages: Afrikaans, English, isi Ndebele, Sepedi, Sesotho, siSwati, Xitsonga, Setswana, Tshiven da, isi Xhosa, isi Zulu.
Main Political Parties: African National Congress (ANC), Democratic Alliance (DA); African Christian Democratic Party; Freedom Front (FF) representing the Afrikaner minority.
Main Social Organizations: South African Students' Congress (SASCO), the Congress of South African Trade Unions (COSATU), the National Congress of Trade Unions. Azanian Peoples Organization, advocating black identity.

THE STATE

Official Name: Republic of South Africa.
Administrative divisions: 9 provinces.
Capital: Pretoria (administrative) 1,209,000 people; Cape Town (legislative) 2,967,000 (2003); Bloemfontein (judicial).
Other Cities: Johannesburg 4,927,200 people; Durban 2,314,100; Port Elizabeth 1,029,400 (2000).
Government: Thabo Mbeki, President since June 1999, re-elected in April of 2004. Bicameral Legislature: the National Assembly, with 400 members, and the National Council of Provinces, with 90 members.
National Holiday: 27 April, Freedom Day (1994).
Armed Forces: 56,000 (2003).

Under-5 mortality
67 per 1,000 live births
2004

Poverty
10.7% of population living on less than $1 per day
2002

Debt service
6.4% exports of goods and services
2004

Maternal mortality
230 per 100,000 live births
2000

1948, the Nationalists formed a government by themselves, imposing even harsher restrictions on the black population.

[20] The first national political organization of South African blacks had appeared in 1912. The African National Congress (ANC) was created by a group of former students from schools run by missionaries and by people who had studied or gained degrees in North American and European universities. They believed the Afrikaners could be persuaded of the unfairness of the racial segregation laws and that the Anglophile liberals would allow blacks to participate in politics.

[21] In the 1940s, the failure of this first strategy led the ANC to adopt a strategy of non-violent resistance to the race laws. In 1955, the anti-racist front was broadened with the Freedom Charter, proclaimed at a multiracial gathering in Kliptown. The Charter included a radical denunciation of *apartheid* (separateness) and called for its abolition, along with wealth redistribution.

[22] In 1958, sectors of the ANC that disagreed with its multiracial policy created the Pan African Congress (PAC), which in 1960 held a demonstration in the city of Sharpeville to protest against the pass laws that restricted the movement of black workers in areas reserved for whites. The march was brutally repressed, leaving 70 dead.

[23] Following this incident, the PAC, ANC and the Communist Party were all outlawed. The African National Congress (ANC) formed an armed group, the *Umkhonto we Sizwe* (Spear of the Nation), while the PAC set up another, *Poqo* (Only Us). In 1963, the main leaders of the ANC were arrested; Nelson Mandela was sentenced to life imprisonment and Oliver Tambo, in exile, took over leadership of the movement. The Government's repressive violence and the lack of supply bases in neighboring countries - dominated by regimes allied to the Afrikaners - prevented the guerillas from making enough progress to attract greater numbers of recruits.

[24] The racist system was largely upheld by the interests of international capitalism in the region, attracted by cheap labor. Foreign investments, especially from the US, increased five times in value between 1958 and 1967. The Afrikaners' protectionist policies created the infrastructure necessary for large industries to be set up, with the aim of developing an industrial center capable of supplying all southern Africa.

[25] During the 1960s the number of black rural workers migrating to the cities increased, driven by the poverty of the 'Homelands' or Bantustans, some of which had only poor quality soils, and the lack of social services. This affected the expectations of other urban sectors, such as the Colored people, who saw their hopes of integration into the white economy threatened.

[26] In 1976, the black community in the suburbs of Johannesburg erupted. The youth rebellion in Soweto - the South West Township - made the whites realize the crisis had reached the cities where they had previously felt safe. In 1970, 75 per cent of workers in agriculture, mining or the services were black, and the participation of non-whites in specialized jobs had tripled over the last 20 years though blacks earned five to ten times less than whites for the same work. The governing minority proposed making some reforms to apartheid, in an attempt to prevent further conflicts among migrant workers in the cities.

[27] The Pretoria regime declared four Bantustans - Transkei, Ciskei, Venda and Bophuthatswana - to be 'independent states' hoping to halt the internal migration of the unemployed. Eight million people were thus deprived of their South African nationality and converted into foreigners by decree. However very few countries recognized these newly 'independent states'.

[28] The independence of Angola and Mozambique in 1975 and that of Zimbabwe in 1980 radically affected the situation in southern Africa. The ANC found the support bases it desperately needed in these countries, and in the other Frontline states of Botswana, Tanzania and Zambia. South Africa, with an economy three times bigger than that of those independent countries combined, initiated a destabilization campaign which included economic pressure, sabotage, support for rebel movements and invasion. All this was in order to force them to deny support to the anti-apartheid movement and block attempts by the newly independent countries to escape South African domination.

[29] One of the main arenas of the conflict in southern Africa was Namibia, a former German colony which South Africa occupied during World War I and later annexed. In 1966, the UN ruled that South Africa had to grant Namibia its independence - a demand which the Organization of African Unity (OAU) and the Frontline countries continued to make, despite the delaying tactics of South Africa and the Western powers. It took until March 1990 for Namibia to become independent.

[30] To enable South Africa to impose its economic and military strength on southern Africa, the support it received from the US was crucial. Close on 400 US companies had interests in the country, and US capital and technology were vital for developing its industrial and military might.

[31] On the domestic front, PW Botha, Prime Minister from 1978 to 1989, began reforming the apartheid system. Between 1982 and 1984, he brought in constitutional reform granting the vote to Indians and colored people and creating two more chambers in parliament for these groups. Blacks were still excluded, with their participation limited to a local level. Many non-whites boycotted the reform, abstaining from voting.

[32] The gradual liberalization of apartheid was widely opposed. Repression against blacks did not diminish and was further complicated by inter-ethnic confrontations. In July 1985, the Government declared a state of emergency in 36 districts. By the end of 1986, more than 750 had died and several thousands of the Government's opponents were in prison.

[33] Public opinion in the US and Europe forced Western governments and an increasing number of companies and banks to limit their activities in South Africa. The US Congress lifted the veto imposed by President Reagan on economic sanctions, forcing a change in his policy of 'constructive engagement'. Within South Africa, political opposition led to the creation of the United Democratic Front (UDF) which brought together more than 600 organizations working together within the law.

[34] From early 1988 the Botha Government came down more heavily on the opposition, outlawing all the constituent groups and imprisoning religious leaders opposed to apartheid, including the black Archbishop, Desmond Tutu, a Nobel Peace Prize winner.

[35] In August 1989, cornered by an internal crisis in his party - that had governed for 41 years - Botha resigned. He was replaced by Frederik de Klerk, who declared himself in favor of changing South Africa's racist image. In September, parliamentary elections were held under the State of Emergency which had been operating since 1986.

[36] The Mass Democratic Movement (MDM), an anti-apartheid coalition of legal organizations, called a general strike. Despite police raids and threats, three million black South Africans stopped work in the largest protest ever held there. A few days later, the first mass legal demonstration against apartheid since 1959 took place. The growing momentum was accompanied by repression and killings. But by now increasing numbers of the white minority were joining the protests.

[37] The opposition agreed to establish the principle of 'one person, one vote' for any negotiations with the Government.

[38] In February 1990, De Klerk legalized the ANC and other opposition groups. Nelson Mandela was released on 11 February after 27 years in prison. A period of negotiations began. Mandela resumed his role as leader of the black majority, a post not without its difficulties. Some were related to confrontations - which had caused 5,000 deaths since 1986 - between the ANC and members of the Zulu Inkatha movement. Inkatha had government backing providing arms, funds and training.

[39] In May, Mandela announced an agreement between the ANC and the Government to end the violence and bring political life back to normal. He called on the international community to maintain economic sanctions and other forms of pressure on the South African Government. He renounced the policy of creating Bantustans - ten had been set up - and abolished racial segregation in hospitals and everywhere else. In December ANC president Oliver Tambo returned to the country after 30 years of exile.

[40] In April 1991, the European Community (EC) considered lifting the economic blockade and set 30 June as the deadline for starting democratization. On that day, the Government abolished the Population Registration Act and the Land Acts, which had prevented blacks from owning land. De Klerk also promised to begin negotiations for a new Constitution. The US went ahead and lifted the blockade. The EC planned to follow suit, but Denmark and Spain, which had received a visit from Mandela, vetoed the move.

[41] In 1993 the Inkatha Freedom Party (IFP), the Afrikaaner National Front (NFA)

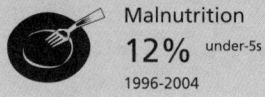

Malnutrition
12% under-5s
1996-2004

Water source
87% of population using improved drinking water sources
2002

Doctors
69 per 100,000 people
1990-2004

Primary school
89% net enrolment rate
2004

and the Conservative Party abandoned negotiations on the Constitution and attempted to boycott the electoral process. Bophuthatswana President Lucas Mangope - in the midst of a strike by public employees - declared he would join the boycott and received military support from the ultra-right Afrikaner Resistance Movement (AWB). The resistance of black civilians and local forces obliged them to withdraw. Mangope was deposed and the South African army took control.

42 Meanwhile Inkatha boycotted ANC activities and clashed bitterly with Mandela's supporters. Its leader, Mangosuthu Buthelezi tried unsuccessfully to control another Bantustan, also in Natal; however, he did get the constitution to recognize his nephew Goodwill Zwelethini as King of the Zulus, and he finally agreed to participate in the election.

43 In October, the UN lifted sanctions against the South African regime. The United States immediately withdrew its financial restrictions.

44 The provisional Constitution created a 400-seat National Assembly and a Senate with 90 members. The President would be elected by the Assembly for a five-year term. The country was newly divided into nine provinces, each with a governor and legislature, absorbing the ten abolished Bantustans.

45 The first multiracial elections in South Africa were held in April 1994. The ANC won 63 per cent of the vote.

46 The Government of National Unity (GNU) included members of the NP and IFP. The minister of finance and the governor of the South African Reserve Bank from the previous government were retained in their posts.

47 Despite the removal of the barriers of apartheid, economic and cultural obstacles remained. Black workers earned nine times less than whites and unemployment was 33 per cent and 3 per cent respectively. The overall infant mortality was 50 per 1,000 live births (1998) but the rate for blacks was far higher.

48 Among the measures to be applied at the beginning of his term, Mandela proposed free health care for children aged under six and for pregnant women, a basic diet for schoolchildren and the provision of electricity to 350,000 homes. New legal guidelines for education were established. In October it was announced that 3.5 million people would be given access to water services

over the next 18 months. The first GNU budget gave 47 per cent to social services, education took 26 per cent, investment in housing doubled and military spending was reduced.

49 Ambitious land reform was implemented by Land Affairs Minister Derek Hanekom, a farmer. A labor relations act was approved, guaranteeing the right to strike and set up discussion fora in the work place. There were far fewer strikes than in previous years.

50 In January 1995, the ANC withdrew immunity guaranteed before the elections to two former cabinet ministers and 3,500 police officers who were to be investigated by the Truth and Reconciliation Committee (TRC). The trial of a former police colonel for 121 murders, kidnappings and frauds, provided new evidence of police incitement of political violence during the former regime. Prominent Inkatha leaders were implicated in payments made to the security police. A report by the Goldstone Commission was presented to De Klerk in 1994, reiterating these charges. In June, the under-secretary of the IFP was arrested for murders committed in 1987.

51 The local elections of November 1995 favored the ANC throughout the country. In May 1996, the NP left the Government to join the opposition, for the first time since 1948.

52 The National Assembly approved a new Constitution which attempted to consolidate the transition to democracy. During the production of the new text, demonstrations attended by thousands of workers and businesspeople led to the elimination of a clause in the final text which gave bosses the right to close their factories.

53 In 1995 the Truth and Reconciliation Commission was set up, under the presidency of former Archbishop Desmond Tutu, collecting evidence of human rights violations committed between 1960 and 1993. During the investigations, several police officers admitted the use of torture in the 1980s and the hiring of mercenaries. Those responsible were offered an amnesty provided they clarified their part in the events.

54 The Government announced a strategy aimed at creating 800 thousand jobs by the year 2000. Throughout 1996, GDP grew three per cent. By November, some two million hectares of land had been redistributed under the Government agrarian reform program.

IN FOCUS

ENVIRONMENTAL CHALLENGES
The use of polluted water resources causes health problems. Excessive farming causes deforestation, soil erosion and desertification.

WOMEN'S RIGHTS
White women have been able to vote since 1930, mixed race and Asian women since 1984 and black African women since 1994. In 2004, women held 33 per cent of seats in Parliament and 41 per cent of ministerial or similar positions. In 2003, women accounted for 38 per cent of the labor force of 19 million people. Although in 2004 prenatal care covered 94 per cent of pregnant women and 84 per cent of births were assisted by qualified staff, maternal mortality stood at 230 per 100,000 live births. This rate is notoriously higher for the black population than for the white. That year, the incidence of HIV/AIDS among pregnant women stood at 28 per cent. More than three-quarters of people living with HIV/AIDS aged between 15 and 24 were women. In spite of this, the quality of life for poorer women has improved markedly since *apartheid* was abolished.

CHILDREN'S RIGHTS
In 2004, 40 per cent of deaths of children under five were due to HIV/AIDS. The disease had caused an increase in mortality rates for that group (67 per 1,000 live births) and for infants (54 per 1,000). The pandemic had left a million orphans in the country. Only 50 per cent of children are registered at birth, which limits their chances of being assisted by social services. Although the Government has made efforts to address child prostitution, it remains a widespread problem. More than

one million children did not attend school, either because the nearest one was too far away or due to malnourishment or lack of resources.

INDIGENOUS PEOPLES/ ETHNIC MINORITIES
The word ethnic is representative of a conflictive time in the history of the country; the concept of tribe possesses derogatory connotations related to *apartheid*. In addition to the Zulu, who represent 20 per cent of the total population, other large ethnic groups are the Hottentot, Ndebele, Sotho, Swazi, Tsonga, Tswana, Venda and Xhosa.

During the last elections in April 2004, Zulus and Xhosas were murdered in political confrontations. President Mbeki is Xhosa and the Minister of the Interior, Buthelezi, is Zulu.

MIGRANTS/REFUGEES
In early 2006 South Africa sheltered 12,000 refugees and 23,000 asylum seekers from DR Congo, 9,000 refugees and 10,000 asylum seekers from Somalia, 5,700 refugees and 6,000 asylum seekers from Angola and 10,300 refugees and more than 59,000 asylum seekers from other countries in the region. Refugees - mostly economic immigrants who seek asylum to legalize their status in the country - are concentrated in the larger cities, mostly Johannesburg, Pretoria, Durban and Cape Town.

DEATH PENALTY
It was abolished for common crimes in 1995 and all types of crimes in 1997.

* Latest data available in The State of the World's Children and Childinfo database, UNICEF, 2006.

55 In October 1997, Mandela visited Libya to mediate in the conflict between Tripoli, Washington and London over the 1992 embargo against Libya resulting from the Lockerbie airplane bombing (see Libya). Mandela supported Libya's stance in calling for a trial in a neutral country, although he made it clear he did not back the unconditional lifting of the embargo.

56 During his June 1998 farewell message at the Organization of African Unity, Mandela demanded 'the right and the duty to intervene whenever

behind sovereign borders people are being massacred to protect tyranny'. These statements contradicted the founding principle of the OAU of non-intervention in the internal affairs of member countries.

57 That year, during hearings of the Truth and Reconciliation Commission, a plan was revealed, created by apartheid scientists to undermine Mandela's health when he was imprisoned. The plan also included the development of fertility-inhibiting chemical agents and also diseases which would attack the black population.

External debt
$597 per capita
2004

Imports **(millions)**
$57,888 goods and services
2004

Exports **(millions)**
$56,734 goods and services
2004

Received Aid
$14 per capita
2003

[58] Attacks on white farmers broke out again as the black population began to express its anger over the slow changes. During the ANC congress, Mandela and his deputy, Thabo Mbeki, warned that the era of formal reconciliation would end at the same time as Mandela's term of office, and that a second ANC government would take tougher measures.

[59] Although the June 1999 elections gave the ANC solid control of Parliament, it did not achieve the two-thirds majority needed to unilaterally amend the constitution. The opposition leadership in the National Assembly fell into the hands of the mainly white Democratic Party (DP). Thabo Mbeki became the new President and appointed Jacob Zuma, also ANC, as deputy President.

[60] Although the Constitution had banned the main forms of discrimination since 1994, a new law approved by Parliament in January 2000 implemented, for the first time, non-discrimination in relations between individuals. The law also prohibited all discrimination based on age, sexual orientation, culture, pregnancy, marital status, matters of conscience and language.

[61] In spite of the fact that since 1994 South Africa had announced restriction policies with regard to the export and import of weapons, and was one of the world leaders in the implementation of the Convention on the Ban, Use, Storage, Production and Transfer of anti-Personnel Mines and their Destruction in 1997, at the end of 2000 the organization Human Rights Watch accused Pretoria of selling weapons to countries with human rights abuses and where the flow of arms could imply an increase of such abuse. In May 2001 an official panel began to investigate accusations of corruption in a weapons deal which had involved Pretoria with British, German, French, Swedish and South African firms. On concluding the process, the government was acquitted of any responsibility.

[62] Reacting to an official who said the Indian community was unable to carry out certain responsibilities, in early 2001 Mandela accused members of the black majority of using their political power to frighten ethnic minorities and urged the ANC to change the situation.

[63] In September 2001, the World Conference Against Racism took place in Durban, where, faced with the demand by African countries for economic reparation by the former colonizing countries, most of the European states - which at first had been in favor - considered the demand 'unreasonable'.

[64] Over two thousand minors were reported as having been raped during 2001. To face this problem the government set up special police units, installed legal protection for the victims of rape and established special courts to deal with sexual offenses.

[65] In April 2002, South African justice absolved Wouter Basson, known by the South African media as 'Doctor Death'. He was notorious for having developed the Costa Project, seeking to create 'intelligent' bacteria that would only kill black people. He had accumulated sufficient cholera and anthrax reserves to cause an epidemic. The weapons he invented included sugar containing salmonella, cigarettes containing anthrax, chocolates containing botulism and whisky containing weed-killer. During the hearings, Basson claimed he was innocent and maintained that he had only obeyed orders, although he did not show any remorse.

[66] Between five and seven million South Africans will die of AIDS by 2010, according to a report issued by the South African Council of Medical Research, and life expectancy by then may be only 36 years. Mbeki disputed the research and tried to prevent its publication. South Africa has one of the highest rates of HIV infection in the world, with: 5.3 million people affected. In sub-Saharan Africa as a whole, there are more than 12 million AIDS orphans.

[67] In March 2001 the leaders of the Democratic Alliance asked Mbeki to declare a state of emergency so HIV carriers could receive generic medicines. The South African law that allows the import of such drugs in case of emergency had never been implemented, due to a lawsuit filed by major multinational pharmaceutical firms, which make the drugs. However, international pressure led the 39 companies to withdraw their lawsuit - a decision marking an important precedent for poor countries needing to import cheap drugs to fight the pandemic.

[68] South Africa's economy continued to struggle, except for some exports - wine, arms, vehicles - which benefited from a weak national currency. Political turmoil in neighboring Zimbabwe reduced foreign investment in the region. Drought in 2002 exacerbated food shortages in southern Africa, and poverty was compounded by HIV/AIDS.

[69] In July 2002, the Constitutional Court ruled - in a case brought by the Treatment Action Campaign (TAC) started in 1998 - that the Government is now legally obliged to implement a comprehensive nationwide program for the prevention of mother-to-child transmission of HIV, to include provision of the anti-retroviral drug nevirapine.

[70] July 2002 saw the inauguration, in Durban, of the African Union (AU). The AU replaced the Organization of African Unity, seen as 'the dictators' club', whereas the focus of the AU was placed on people's progress and good governance. The 53-member AU was loosely modeled on the EU. It has the right to intervene in the affairs of its member states, in cases of genocide and war crimes and would have a peace-keeping force and a court of justice. Mbeki was appointed as the AU's first president, while South Africa became a key piece in the organization's Peace and Security Council.

[71] In April 2004, after a decade of democracy, the ANC won the elections for the third consecutive time. Mbeki took up office for the second time.

[72] That same year, the Provincial Council (NPC) found itself - for the first time in its history - under the dominion of an ANC majority, leaving all the provinces under the effective control of the governing party.

[73] During the first decade of democracy, the Government built 1.6 million houses for underprivileged people, 70 per cent of which had electricity.

[74] In May 2004, former Haitian President Jean Bertrand Aristide was given asylum by the South African Government.-

[75] Also in May, Mbeki promised that his government would grant electricity and running water to every one of the South African households that had been expecting these services for five to eight years.

[76] Mbeki was being strongly criticized for not doing enough to combat AIDS. The Government, acknowledging criticism, promised to lay emphasis on the improvement of home-based care and to provide increasingly cheaper medicines.

[77] In January 2005 former President Nelson Mandela announced that his older and only surviving son had died from AIDS, and insisted that the only way to combat the disease was to talk about it openly. In March, for the first time, the Truth and Reconciliation Commission unearthed the remains of one of the hundreds of people that had gone missing during the apartheid regime.

[78] Achabir Shaik, financial advisor to Vice President Jacob Zuma, was sentenced to prison in June 2005 for corruption. Shaik had used the name of the Vice President - who denied any link to the case - to benefit from deals with a French firm. Almost unanimously the press demanded Zuma's resignation. He resigned after Mbeki requested it.

[79] In May 2006, a court acquitted Zuma from rape charges presented against him by a 31-year old woman. The verdict did not prevent the leader from losing much of his popularity, which had made him a potential future president. Zuma accused the media of having sentenced him before the judge gave his verdict. 'An accused person is innocent until proven otherwise; that is one of the golden rules of our Constitution, but the press has broken that rule,' he stated.

[80] Meanwhile, Archbishop Tutu, in total disagreement with the court's verdict of not guilty, stated in London that the trial had been 'one of the worst instances in South Africa's democratic life'.

[81] On 9 August, some 20,000 women demonstrated in Pretoria next to the government building demanding effective measures against domestic violence. South Africa's rate of domestic violence was one of the highest in the world.

[82] Mike Heywood, one of the most renowned researchers in HIV/AIDS and promotor of multiple campaigns to stop the havoc it wreaks in Africa, criticized Pretoria for not taking effective measures to reduce the spread of HIV. He also had harsh words for global political leaders for not questioning the inaction of the Mbeki administration. According to a government survey, an eighth of people with HIV/AIDS in the world were born in South Africa.

[83] In September 2006, Archbishop Desmond Tutu warned that South Africa was in danger of losing its moral compass. He said it had failed to sustain the idealism that ended apartheid and warned of growing ethnic divisions. Delivering the Steve Biko memorial lecture in Cape Town, he referred to the high murder rate and said that the African reverence for life had been lost. He said he opposed the idea of ex-Vice President Jacob Zuma becoming President due to his 'moral failings'. ■

Spain / España

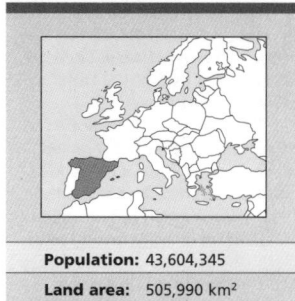

Population:	43,604,345
Land area:	505,990 km²
Capital:	Madrid
Currency:	Euro
Language:	Spanish

The territory of what is present-day Spain has been inhabited by numerous cultures. Originally, the peninsula was settled by peoples from northern Africa and western Europe. From 1100 BC, the peninsula attracted seafaring civilizations, such as Phoenicians, Greeks and Carthaginians, who founded settlements and trading posts, especially along the eastern and southern coasts. The seafarers found a diversity of peoples, collectively called Iberians (name probably derived from the Greek name for the river Ebro), who lacked a common culture or language.

² Between the 9th and 7th centuries BC, the Celts settled in the center and west of the peninsula. Later on, during the 6th and 5th centuries BC, the Iberian culture developed in the southern part of the territory and the fusion of these two cultures produced what is known as the Celtiberian civilization. They were then colonized by the Carthaginians, who in the 3rd century BC took over most of the peninsula. The Carthaginians were expelled by the Romans, who by the 1st century BC had, for the first time, established a unified authority governing the whole of Iberia. The Iberian elites adopted the Roman culture and became citizens of Rome, especially in the south, where the imperial presence was stronger.

³ The fall of the Roman Empire around 350 AD coincided with the spread of Christianity and, more importantly, with the invasion of Europe by groups from the north. The Iberian peninsula was occupied by the Visigoths who ruled the area for 300 years.

⁴ In 700 AD the peninsula was invaded by Arabs who defeated Rodrigo, the last Visigoth king, marking the start of the era of Muslim domination. The descendants of the Visigoths lived in the north of the territory and set up kingdoms like Castile, Catalonia, Navarre, Aragon, Leon and Portugal. Over the centuries these kingdoms gradually unified, a process that culminated with their joint military stand against the Arabs.

⁵ The Arabs called the lands in the south of the Iberian peninsula al-Andalus, a region that reached its peak during the 10th century. In contrast with the rest of impoverished rural Europe, its cities - and Córdoba in particular - prospered through active trade with the East. Religious tolerance enabled Muslims, Jews, and Christians to live side by side, and science, medicine, and philosophy developed. Copies and translations of the Greek classics were made, paving the way for the 15th century European Renaissance.

⁶ In 1492, a triple process of national unification took place in Spain, through the marriage between Isabel of Castile and Fernando of Aragon, the expulsion of the Moors, and the conquest and subsequent colonization of the new American territories. The uniting of the country's political power and the creation of the Kingdom of Spain were carried out at the expense of the Jews (and members of other cultures) who were expelled from Spain after having lived there for many centuries. Both the Inquisition and centralized power were institutionalized under the new system, while the new American colonies supplied precious metals, sustaining three centuries of economic bonanza. The Crown imposed Christianity on the indigenous population in America. Many native peoples died as a result of the exploitation they were subjected to through forced labor, and also from European diseases for which they had no immunity.

⁷ The economic prosperity provided by the colonies was reflected in a period of great cultural development in Spain. Literature in particular developed extensively during the 16th and 17th centuries, which were dubbed the Spanish Golden Age. Portugal was annexed to the Spanish Kingdom for the period 1580-1688.

⁸ In the 18th century, the Bourbons came to the Spanish throne. They reorganized the domestic and colonial administrations, ruling in accordance with the principles of the Enlightenment, as the liberal ideas of the French Revolution spread throughout Europe and America. Combined with Napoleonic expansion, this view contributed to the disintegration of the Spanish colonies after the wars of independence.

⁹ At the end of the Napoleonic era, there was great conflict between the liberal sectors seeking political and economic modernization, and the absolutists who wished to preserve the traditional order. The disputes between groups weakened the power of the Empire, making way for independence movements in Spanish America.

¹⁰ By the end of the 19th century Spain had renounced its last American territories and had come to terms with the loss of its privileges.

¹¹ At the beginning of the 20th century, Spain was plunged into a deep political, social, and economic crisis; this was exacerbated by World War I (1914-1918). In an atmosphere of extreme polarization, Primo de Rivera's dictatorship, following a coup in 1923, attempted to halt any further demands from workers or regional groups seeking autonomy. The dictatorship, closely resembling the Italian fascist model, retained power until 1931. Its end came about as a result of existing contradictions within the Church, the armed forces and industry, who were all fascist supporters, rather than because of the continuous opposition from political and labor organizations. The end of the dictatorship marked the end of the monarchy, and the dawn of a new republican era.

¹² The 'Second Republic' was born into a series of complex political and economic difficulties. In 1936, after two moderate governments, the People's Front, of socialists, republicans, communists and anarchists, won a narrow electoral victory, causing friction with their political opponents.

¹³ Immediately after the elections, the army, Church and powerful sectors of Spanish economy started working to overthrow the Government, which was weakened by internal differences. In 1936 a sector of the army, led by General Francisco Franco, rose up against the Republic and a three-year long civil war ensued. The republican Government waited in vain for help from the European democracies, but these opted for a policy of non-intervention. Only the Soviet Union provided material support, and many volunteers from America and Europe joined the ranks of the republican army.

¹⁴ Franco's forces won the civil war in March 1939, aided by the republicans' internal divisions, the military superiority of his troops, and German and Italian support.

¹⁵ When the Civil War ended, Franco became head of the new state. He set up an authoritarian regime along fascist lines, with a corporatist state, a personality cult, and extreme nationalism. Franco ruled over a deeply divided society and an economy that had been devastated by the Civil War.

¹⁶ From the beginning of the Cold War in the early 1950s, the US tried to secure Spanish support and Spain became a member of the United Nations in 1955, confirming a change in Franco's foreign policy towards improving the country's international image.

¹⁷ In the 1960s, Franco opened Parliament to other groups and movements. During those years key figures from Opus Dei, an ultra-conservative Catholic movement, occupied important posts in the Government and influenced economic policies.

¹⁸ Spain ended its economic isolation and liberalized its economy by abolishing some state control. The urban middle classes enjoyed improvements in their standard of living, which led to a political relaxation. However, the peasants were still extremely poor and many migrated to the major Spanish and European cities.

¹⁹ Franco died in 1975, and power was handed to his successor, the heir to the Spanish throne, Juan Carlos I of Bourbon. The

Life expectancy
80 years
2005-2010

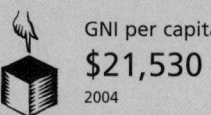

GNI per capita
$21,530
2004

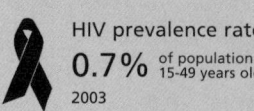

HIV prevalence rate
0.7% of population
15-49 years old
2003

new monarch immediately started negotiations with the political opposition to re-establish the democratic system overthrown in 1939.

20 Between 1976 and 1981 Adolfo Suarez, the leader of the Center Democratic Union (UCD), was Prime Minister. In December 1978 during his term in office, a plebiscite was held which turned Spain into a parliamentary monarchy, re-establishing political freedom, and guaranteeing the right of autonomy to some Spanish regions. Several politicians, intellectuals and artists were able to return to the country after up to 40 years in exile.

21 In February 1981, a group of Civil Guard officers took the *Cortes* (parliament) by force. The firm reaction of all the democratic political groups and in particular of King Juan Carlos, who had the support of the army, guaranteed the failure of the coup and the consolidation of the democratization process.

22 The Spanish Socialist Workers' Party (PSOE) won the October 1982 elections, with a solid majority in the Cortes. Felipe González became President of the Government, the equivalent of Prime Minister in other countries. He was subsequently re-elected in 1988 and 1993.

23 Despite some setbacks, the PSOE's parliamentary majority enabled it to push through an ambitious adjustment and

growth plan which deeply transformed the Spanish economy and gave large social sectors access to unprecedented levels of consumption. However, this modernization resulted in high unemployment and social tensions, which led to a split between the Government and the UGT, the trade union that had supported the PSOE.

24 The Spanish Government was active in international affairs, joining the European Economic Community and NATO in 1986. While it was in the opposition, PSOE had opposed Spain's becoming a member of NATO, but once in power the party defended the decision, which was confirmed through a plebiscite.

25 Spain spent $10 billion in 1992 to celebrate the 500th anniversary of the conquest of America, while it intensified its political shift toward Europe.

26 ETA, the Basque separatist group which has often resorted to violence to achieve its political goals, suffered serious setbacks in 1993. Co-operation between French and Spanish security forces led to the arrest of some of its leaders, and the discovery in Bayonne, France, of the organization's main arsenal.

27 After a series of corruption scandals, in 1995 the socialist Government lost a key sector of its parliamentary backing. Although he defended his own and his Government's record in office,

Ceuta

Population: 75,000
Land area: 19 km^2
Currency: Euro
Language: Arabic and Spanish

This is an enclave on the Mediterranean coast of Morocco facing Gibraltar (Spain). The climate is Mediterranean with hot summers and moderate winters. There is sparse rainfall in winter.

2 Occupied by troops of Portuguese King John in 1415, Ceuta was transferred to Spain in 1688 and retained by the Spanish after Moroccan independence in 1956. Despite many claims and negotiations before the UN Decolonization Committee, backed by the Organization of African Unity (OAU), the port is still in Spanish hands, and most of the territory is used for military purposes.

3 Estimates indicate that in the period 1995-1999 more than 3,000 people - seeking better work conditions - died while trying to cross the Strait of Gibraltar to go to Europe. Between 2001 and 2003, Amnesty International accused the Ceuta authorities of the repeated abuse of underaged immigrants.

4 The continued flow of undocumented immigrants from Morocco entering Spain through Ceuta caused a diplomatic crisis between Madrid and Rabat in 2001. In December 2003, the Moroccan authorities accepted the reforms of the Spanish Law on Foreign Persons, which called for the immediate return of unaccompanied minors found in Spanish territory.

5 In September 2004, new confrontations with the police occurred in the border between Ceuta and Morocco. The frontier police reported fires and stoning of people and vehicles. Al-Jaima, a human rights organization in Tanger (a city in the north of Morocco) denounced systematic violations of immigrants' human rights on both sides of the border, resulting from the combined repression of the Spanish police and the Moroccan military.

6 In 2005, after an increase in the trafficking of people from Morocco to Spain, both countries agreed on the deployment of a greater number of troops along the common border. ∎

SOCIETY
Peoples: 80 per cent of the population is Spanish, born in Ceuta.
Religions: Catholic (majority), Muslim and others. There are also Muslims and Jews. **Languages:** Arabic and Spanish.

THE STATE
Government: Civil authority is exercised by the Spanish Home Affairs Ministry. Military authority lies in the hands of a General Commander. The territory has one representative in the Spanish Parliament.

Melilla

Population: 69,184
Land area: 12 km^2
Currency: Euro
Language: Arabic and Spanish

Melilla is a small peninsula on the Mediterranean coast of Morocco with two adjacent island groups. The climate is similar to that of Ceuta. Melilla is an ancient walled town built upon a hill with a modern European-style city on the plain. It is an important port which, like Ceuta, hopes to be a tourist attraction.

2 Founded by Phoenicians and successively held by Romans, Goths and Arabs, Melilla was occupied by Spain in 1495. It was repeatedly besieged by the Riffs, a Berber group that opposed French and Spanish domination, most recently in 1921. It has been claimed by Morocco and as Ceuta, is used for military purposes.

3 On the northern Moroccan coast the Spanish also hold Peñón (rock) de Vélez de la Gomera, which had 60 inhabitants in 1982; Peñón de Alhucemas with 61 inhabitants in 1982, (with the islets of Mar and Tierra); and the Chafarinas Archipelago (Islands of Congreso, Isabel II and Rey), 1 sq km in area and with a population of 191 inhabitants in 1982.

4 The Spanish Government approved a statute of autonomy for Melilla in 1995, replacing the city council with an assembly, similar to those of other autonomous regions of Spain.

5 In 2000, Melilla received 16.95 per cent of the regional fund that Catalunya, the Balearic Islands, as well as Basque and Navarre provided to support other regions. Ceuta and Extremadura also benefited from this fund.

6 In October 2005, more than 2,000 African immigrants crossed the fences separating Melilla from Morocco. Spanish and Moroccan troops positioned in the border opened fire at the immigrants, leaving at least ten dead and dozens injured.

7 Some 500 immigrants that hadn't been able to cross the border were abandoned by Moroccan authorities in the Sahara desert, without food or water. ∎

SOCIETY
Peoples, languages, religions and other features of the population are similar to those of Ceuta.

THE STATE
Government: A representative of the Spanish Government is responsible for administrating the territory's civilian affairs. There is a military command in charge of military affairs. Like Ceuta, Melilla has a representative in the *Cortes*, the Spanish parliament.

González called elections a year early.

[28] The conservative People's Party (PP) won the Parliamentary elections in March 1996 with 38.9 per cent of the vote. On 5 April, the PP leader, José María Aznar, took office as Prime Minister.

[29] In August 1997, an ETA cell kidnapped Miguel Ángel Blanco, a PP activist and councilor in the Basque Country. ETA warned that he would be executed within 48 hours if the Spanish Government did not order the transfer of ETA prisoners to Basque jails. The deadline passed and Blanco was murdered, sparking huge countrywide protest demonstrations. The Government and opposition, with the exception of Herri Batasuna (HB), a political group close to ETA, reached an agreement not to condone violence.

[30] ETA declared a ceasefire in September 1998. All the Basque nationalist groups, including the Basque Nationalist Party and former members of Herri Batasuna, signed the Estella Pact, an alliance of Basque nationalist parties that, under the slogan 'Self-rule for Basques', won local elections in 1999. ETA resumed armed actions in September - exploding several bombs and killing three people - which ended the non-violence pact.

[31] The fall in unemployment achieved by the Aznar Government's economic policy and Spain's entry into the European currency system, helped the PP to win an overall majority in March 2000.

[32] The Foreigners' Act, passed in August 2000, was strongly criticized by political parties and non-governmental organizations, and caused wide protest. The law sought to regulate the rights and obligations of foreign residents, including the right to freedom of movement, education, employment and social security, trade union membership and so on. As a result, by January 2001, between 30,000-100,000 foreigners were left in disadvantaged legal and labor conditions.

[33] Tensions with Morocco came to the fore as a result of Madrid's reaction to the 11 September 2001 attacks in New York and Washington. Spain tightened controls on its enclaves of Ceuta and Melilla (on Morocco's coast), and suggested that Rabat 'do more' to limit the flow of illegal immigrants to Spain. Likewise, signs in 2002 of an agreement between Madrid and London regarding the future status of Gibraltar, intensified the discontent of Morocco, which claims sovereignty over Ceuta and Melilla.

[34] In January 2002 Spain took over the presidency of the EU. Anti-terrorism became a priority of Aznar's European agenda.

[35] On 16 March 2003, in line with his policy of increasing closeness to Washington, Aznar attended a summit with US president George W Bush and British Prime Minister Tony Blair in the Azores. There, the three leaders gave the UN Security Council 24 hours to adopt a resolution demanding the immediate disarming of Iraq - which they accused of having weapons of mass destruction - as an alternative to an invasion led by the US. The Security Council rejected the ultimatum and the invasion that followed was endorsed by Aznar's Government, which provided logistical support and later 1,300 soldiers to the occupation forces, although surveys had shown that 90 per cent of Spaniards were against the war.

[36] On 11 March 2004, three days before the general elections, bombs on passenger trains exploded in and near the Madrid station of Atocha. According to official figures, 190 people died, of whom a quarter were foreigners. The next day, over 11 million Spaniards took to the streets to demonstrate their rejection of terrorism. Although the al-Qaeda network claimed responsibility for the attacks in retaliation for Spain's support for the US, the Government and some media insisted that ETA was responsible. On the eve of the elections, thousands of demonstrators gathered at PP premises around the country to demand information on the investigations.

[37] The turnout for the elections on 14 March was far higher than anticipated before the bombings and resulted in unexpected victory for José Luis Rodríguez Zapatero, the PSOE candidate. However, he did not win an absolute majority. He promised to realign Spanish foreign policy towards Europe, emphasize social issues in his economic policies, and withdraw Spanish troops from Iraq unless the UN took over command of the occupation. The withdrawal from Iraq was one of the first measures implemented by the new President of the Government on taking office in April 2004.

[38] In February 2005, the explosion of a car bomb in Madrid, which injured over 40 people, was regarded by Prime Minister Zapatero as a proof of ETA's operative capability. He also stated that 'they will never achieve any of their objectives with violence. And this feeling is overwhelmingly the majority view, not just among Spaniards, but also among all Basque citizens'. That

PROFILE

ENVIRONMENT

Spain comprises 82 per cent of the Iberian Peninsula and includes the Balearic and Canary Islands. The center of the country is a plateau which rises to the Pyrenees in the north, forming a natural border with France. The Betica mountain ranges extend to the south. Inland the Central Sierras separate the plateaus of New and Old Castile. In the Ebro River basin, to the north, lie the plains of Cataluña and Valencia, with Murcia in the south-east. The Guadalquivir River basin, to the south, forms the plains of Andalusia. The climate is moderate and humid in the north and north-west, where there are many woodlands. In the interior, the climate is dry in the south and east. Forty per cent of the land is arable. Approximately five per cent of the total land area is under environmental protection. Natural resources include coal, some oil and natural gas, uranium and mercury. Industry is concentrated in Cataluña and the Basque Provinces. Per capita emissions of air and water pollutants exceed Western European averages. Since 1970, the use of nitrogen fertilizers has doubled. Nitrate concentrations in the Guadalquivir River are 25 per cent above 1975 levels. The percentage of the population serviced by sewage systems rose from 14 per cent in 1975, to 48 per cent in the late 1980s. This fact, together with industrial wastes from oil refining plants and natural gas production, has increased the level of pollution in the Mediterranean. The Government has launched a reforestation plan to increase production and stop erosion. With more trees, there has been an exponential increase in forest fires. There is considerable mono-cropping, particularly of eucalyptus, that does not take biodiversity into account.

SOCIETY

Peoples: Castilians, Asturians, Andalusians, Valencians, Catalans, Aragonese, Extremadurans, Basques and Galicians, of mixed descent from the Iberian people of the Mediterranean, the Celts of Central Europe and the Arabs of North Africa. There is a Roma minority and immigrant communities, especially Latin Americans, North Africans and Asians.
Religions: The vast majority of Spaniards are Catholic (95.2 per cent). Muslims 1.2 per cent.
Languages: Spanish or Castilian (official national); there are also official regional languages, such as Basque, Catalan, Valencian and Galician. Non-official languages: Aragonese, Asturian, Castuo, Canarian, Caló, Romany.
Main Political Parties: Spanish Socialist Workers' Party (PSOE); Peoples' Party (PP); United Left. Political platforms of nationalist movements represented in Parliament: Convergence and Union, Basque Nationalist Party, Galician Nationalist Bloc, Canarian Coalition, Andalusian Party, Republican Left of Cataluña, Basque Solidarity, Aragonese Junta.
Main Social Organizations: The largest labor federations are the General Union of Workers (UGT) and Workers' Confederation (CCOO), along with the anarchist National Labor Confederation (CNT) and General Labor Confederation (CGT). COAG farmers' union, environmental and women's groups.

THE STATE

Official Name: Reino de España (Kingdom of Spain).
Administrative divisions: Spain is divided into 17 autonomous regions: the Basque Country, Cataluña, Galicia, Andalusia, the Principality of Asturias, Cantabria, La Rioja, Murcia, Valencia, Aragon, Castile, La Mancha, the Canaries, Navarre, Extremadura, the Balearic Islands, Madrid and Castilla-León. Each region has its own local authorities, including an executive branch and a unicameral legislature.
Capital: Madrid 5,103,000 people (2003).
Other Cities: Barcelona 3,855,300 people; Valencia 1,406,600; Seville 1,130,600; Zaragoza 605,900 (2000).
Government: Hereditary monarchy, with King Juan Carlos I of Bourbon as Head of State. Prime Minister José Luis Rodríguez Zapatero since April 2004. Bicameral parliament (*Cortes*): the Senate, with 255 members, and the Chamber of Deputies with 350 members elected by proportional representation.
National Holiday: 12 October, Hispanic Day.
Armed Forces: 151,000 (2003). Other: Civil Guard 66,000 (3,000 conscripts).

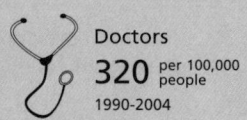

Doctors
320 per 100,000 people
1990-2004

Primary school
100% net enrolment rate
2004

month, a law was passed to grant a legal amnesty for up to 800,000 undocumented immigrants.

[39] Also in February, Spaniards approved the EU Constitution in a referendum, with 77 per cent voting 'yes'.

[40] In April 2005, the National Court in Madrid sentenced former Argentine naval officer Adolfo Scilingo to 640 years in prison for murders, illegal detentions and torture committed during the Argentine military dictatorship.

Scilingo was sentenced to 21 years in prison for each of the 30 murders he was found guilty of, plus five years for illegal detention and another five years for torture. However, under the Spanish penal code only a maximum of 40 years in prison was to be served, which in practical terms amounted to no more than 25, with time off for good behavior and work.

[41] Zapatero announced in May 2005 that he had held talks with ETA aimed at the disarmament of

Gibraltar

Population:	27,982
Land area:	6 km²
Capital:	Gibraltar
Currency:	Gibraltar pound
Language:	English and Spanish

A British colony located on a rocky peninsula on the southern coast of Spain, Gibraltar is only 32 kilometers from Morocco and overlooks the strait which joins the Mediterranean Sea with the Atlantic Ocean. Given its strategic location, the UK maintains a highly fortified naval base there.

[2] Human settlement in the area can be traced back to the Phoenicians around 950 BC. Semi-permanent settlements were later established by the Carthaginians and Romans. After the collapse of the Roman Empire, it was briefly controlled by the Vandals, an East Germanic tribe, and would later form part of the Visigothic Kingdom of Hispania that collapsed due to the Muslim conquest in 711 AD. At that time, Gibraltar was named as one of the Pillars of Hercules, set up by the mythical hero to mark the edge of the known world.

[3] The Rock has been a symbol of British naval strength since the 18th century. It was captured by England in 1704 and formally ceded by Spain in 1714 following the Treaty of Utrecht. Since 1964 Spain has laid claim to Gibraltar on a number of occasions despite the opposition of the majority of the population, which have voted to retain links with the UK in several referendums.

[4] Gibraltarians (people born in the colony before 1925 and their descendants) make up two-thirds of the population.

[5] In a referendum held in 1967, Gibraltarians voted overwhelmingly to remain a British dependency. In 1968, a resolution adopted by the UN determined that the UK should relinquish sovereignty over the Rock. However, the 1969 Constitution reaffirmed the link with the UK. Spain reacted by closing the border completely until 1985.

[6] A series of talks have been held since 1972 in search of a solution to the dispute.-Since 1996, Chief Minister Peter Caruana has pursued a twin track approach, working towards the establishment of a new Constitution in talks with the British Governments while simultaneously seeking dialogue with both Britain and Spain to improve relations and develop areas of co-operation.

[7] In July 2002, British Foreign Secretary Jack Straw declared that the UK was willing to share sovereignty with Spain, but in a new referendum held in November, only 187 of the 17,900 voters accepted the proposal. This referendum brought joint sovereignty to an end.

[8] As of May 2006, the UK maintained in Gibraltar a military staff of 558 people. By then, formal constitutional negotiations between the UK Government and a Gibraltar all-party group had been concluded, in order to modernize Gibraltar's constitutional relationship with Britain so that it ceases to be colonial in nature while retaining British sovereignty of the Rock.-The new Constitution is due to be submitted to referendum in the near future. ∎

IN FOCUS

ENVIRONMENTAL CHALLENGES
The Mediterranean is polluted by sewage and emissions from overseas oil and gas production plants, and there has been a proliferation of illegal toxic and dangerous waste dumps near population centers, mainly in rural areas of the peninsula. Environmental pollution, deforestation and desertification affect various areas of the country.

WOMEN'S RIGHTS
Spanish women participate in political life; they have been able to vote and stand for office since 1931. In 2004 the Vice-President was Maria Teresa Fernández de la Vega, and women held half of the 16 ministerial posts. That year, female representation in Parliament stood at 36 per cent in the Lower Chamber and 23 per cent in the Upper Chamber. In 2003, women made up 38 per cent of the country's total labor force of 18 million. Female unemployment reached 16 per cent. By the end of that year, 27,000 women aged between 15 and 49 were living with HIV/AIDS. Violence against women continues to be a problem. The Government has implemented protective measures, including shelters and civil guards, and a telephone hotline to support threatened women.

CHILDREN
Children's rights and welfare are promoted through the public education and health systems. Education is compulsory up to the age of 16 and free up to 18. However, many Roma children do not go to school. The infant and under-five mortality rates are among the lowest in the world, at 3 and 5 deaths per 1,000 live births respectively.

INDIGENOUS PEOPLE/ETHNIC MINORITIES
Following the banishment of the two largest minorities - the Jews, exiled by the Catholic Monarchs in 1492, and the Moors, exiled by Philip II in 1609 - the Spanish population has been homogeneous in terms of religion. Ceuta and Melilla are home to communities from Morocco and other African countries.

Roma/Gypsy citizens continue to face discrimination and exclusion, especially in access to jobs, education and housing. It is estimated that almost 46 per cent of the Roma population are unemployed. The Basque country, Galicia and Valencia have laws that require the promotion of their respective languages in schools and other institutions. In Cataluña, a percentage of radio and TV broadcasts must be in Catalan.

MIGRANTS/REFUGEES
By the end of 2004, there were 8,400 refugees living in the country. Between 1995 and 2004, over 65,000 people filed asylum requests. Such status was granted to 3,266 of them. Most requests were filed by citizens from Nigeria, Algeria, Colombia, DR Congo, Côte d'Ivoire, Liberia, Cameroon, Guinea, Russian Federation, Iraq and Cuba. It was also estimated that some 56,000 illegal immigrants (mainly Moroccans, Algerians and people from Sub-Saharan countries) arrived in Spain through the enclaves of Ceuta and Melilla.

DEATH PENALTY
The death penalty was abolished for all crimes in 1995. The last execution took place in 1975.

the organization and also admitted to having strong differences with the US Government over the situation in Iraq. The Prime Minister stated that 'ETA's only destiny is to dissolve itself and lay down its arms'. With regards to the withdrawal of Spanish troops from Iraq - which he implemented immediately upon taking office - he said: 'today, our soldiers are where Spaniards want them to be'.

[42] In June 2005, Parliament defied the Catholic Church by passing a law that legalized same-sex marriage and also granted adoption rights to homosexual couples. According to

a report by the Spanish Sociologic Research Center, 68 per cent of the population believed that homosexuals should enjoy the same rights as heterosexuals and 66 per cent approved the new legislation. The Church expressed that such regulation 'fails to address the truth of marriage and therefore fails to do it justice'.

[43] In March 2006, ETA declared a permanent ceasefire. In a pre-recorded statement, three members of the organization announced that: 'ending the conflict, here and now, is possible. This is the desire and the will of ETA'. ∎

Sri Lanka / Sri Lanka

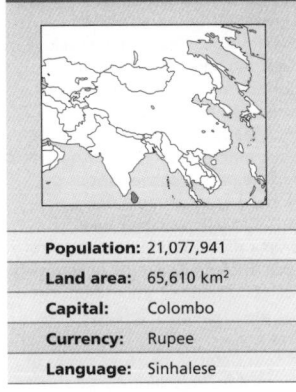

Population:	21,077,941
Land area:	65,610 km²
Capital:	Colombo
Currency:	Rupee
Language:	Sinhalese

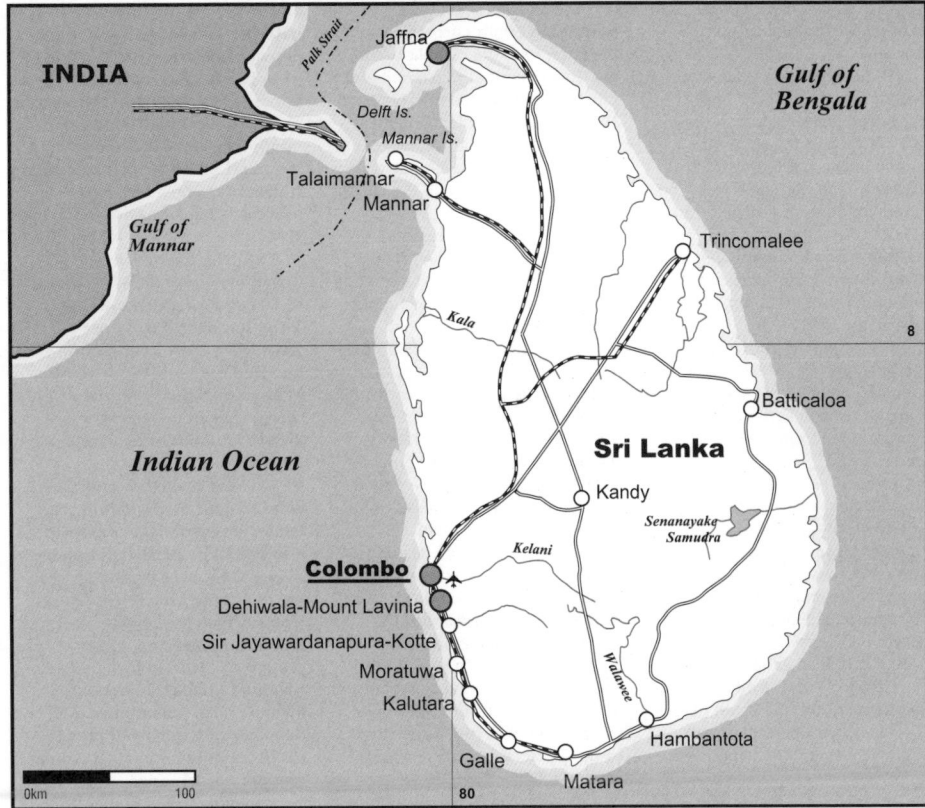

The island of Ceylon was populated by the indigenous Vedda in ancient times. It was then successively invaded by the Sinhalese, Indo-Europeans and Tamils, who laid the foundations of an advanced civilization. When the Portuguese arrived in 1505, the island was divided into seven autonomous local societies.

2 The Dutch expelled the Portuguese from their coastal trading posts 150 years later, but it was the British - already in possession of neighboring India - who finally made the island a colony in 1796. Even then, it took them until 1815 to subdue all the local governments, who fought hard to remain autonomous. The British then introduced new export crops such as coffee and tea-products that gave Ceylon a worldwide reputation because of their excellent quality.

3 In the 20th century, a strong nationalist movement developed. In 1948, Ceylon became independent and joined the British Commonwealth, with nationalist leader Don Stephen Senanayake of the United National Party (UNP) as its first prime minister.

4 In subsequent years, under the leadership of Sir John Kotelawala of the UNP and, from 1956, of Solomon Bandaranaike, the country pursued a vigorous anti-colonial foreign policy. In August 1954, Bandaranaike met in Colombo with India's Nehru, Muhammad Ali of Pakistan, U Nu of Burma and Indonesia's Sastroamidjojo. The meeting was of great political importance as it led to the 1955 summit conference of Afro-Asian countries in Bandung, heralding the Movement of Non-Aligned Countries.

5 One of the first acts of the Bandranaike Government was to pass a law making Sinhalese the only official language, which alienated the Tamil minority. In the late 1950s, the Tamil minority staged a series of secessionist uprisings and in September 1959 the Prime Minister was assassinated. His widow Sirimavo Bandaranaike led the Sri Lanka Freedom Party to electoral victory at the beginning of 1960, although she had no previous political experience.

6 Sirimavo Bandaranaike became the first woman in the world to head a government. In coalition with the Communist and Trotskyist parties, in 1962 she nationalized various US oil and other companies. In 1965, she was defeated by a right-wing coalition but regained power in 1970, in a landslide election victory.

7 She was faced with a 'Guevarist' (after Che Guevara) guerrilla uprising which she crushed ruthlessly. Consistent with her anti-imperialist stand in 1972 she declared Sri Lanka a republic, cutting all ties with the British Commonwealth and launching a land reform program which nationalized British-owned tea plantations but did not substantially change the standard of living of the rural population.

8 Conflict between the Sinhalese majority and the Tamil minority, descended from the Dravidians of South India, has persisted throughout all of the island's history. The Sinhalese account for 74 per cent of the country's population, while the Tamils comprise 22 per cent and are divided into two groups: the Sri Lankan Tamils and the Indian Tamils. The Tamils reached the island some 2,000 years ago. They settled principally in the northern and eastern provinces. The Indian Tamils are more recent immigrants. Both groups, possessing common characteristics, sought regional autonomy or even the formation of a separate Tamil nation. The Tamil United Liberation Front (TULF) founded on 4 May 1972, joined together three Tamil parties: the Federal Party, the Tamil Congress and the pro-Indian Ceylon Workers' Congress.

9 Bandaranaike organized the Non-Aligned Conference in Colombo in 1976 and was appointed president of the Movement. However, the difficult economic situation, accusations of nepotism, and censorship of the press and emergency measures in force since 1971, weakened her government and enabled the opposition to win the July 1977 election. In spite of his socialist leanings, the new prime minister, Junius Jayewardene of the United National Party (UNP) opened the door to transnational capital.

10 After a presidential commission had found her guilty of 'abuse of authority' during her coalition government, between 1970 and July 1977, in October 1978 Bandaranaike was expelled from Parliament and deprived of her political rights for seven years.

11 One month later a constitutional reform made Jayewardene Sri Lanka's first president. In November 1980, a series of International Monetary Fund (IMF)-approved economic

LAND USE

2003/2005

IRRIGATED AREA: 38.8% of arable land

- FOREST AND WOODLAND: 29.9%
- ARABLE LAND: 14.2%
- CROPLANDS: 15.5%
- OTHER USE: 40.4%

PUBLIC EXPENDITURE

% OF GDP

	HEALTH (2002)	DEFENSE (2004)
	1.8 %	2.8 %

Life expectancy
75 years
2005-2010

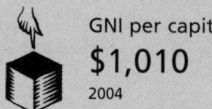
GNI per capita
$1,010
2004

Literacy
90% total adult rate
2000-2004

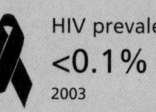
HIV prevalence rate
<0.1% of population 15-49 years old
2003

measures began to be applied, with disastrous consequences for the country.

[12] Sri Lanka's first presidential election, in October 1982, gave Jayewardene a clear victory, with 52.5 per cent of the vote. His campaign was backed by the state apparatus and benefited from the division and internal strife of the Freedom Party. In some parts of the country, the political upheaval meant that elections had to be held under state of emergency conditions.

[13] The project to turn Sri Lanka into an export center like Hong Kong or Taiwan led to the creation of a free zone in Latunyabe where considerable foreign investment was recorded.

[14] In early 1982, in spite of earlier denials and the Government's outspoken commitment to non-alignment, the US Navy was granted permission to use Sri Lanka's refuelling facilities in Trincomalee, a vital spot linking eastern and western sea routes through the Suez Canal.

[15] Early in 1983, the ethnic conflict worsened. In July, the death of 13 soldiers in a Liberation Tigers of Tamil Eelam (LTTE) ambush led to a wave of Sinhalese violence against the Tamils - according to the latter with the complicity of the Government - in Colombo and Jaffna. It is estimated that between 400 and 3,000 people died and over 100,000 were left homeless. More than 40,000 Tamils fleeing the conflict sought refuge in the Indian state of Tamil Nadu.

[16] The ethnic war worsened during 1985, discouraging foreign investors and affecting tourism, one of the main sources of income for this 'fiscal paradise'.

[17] In late July 1987, presidents Rajiv Gandhi (of India) and Junius Jayewardene signed an accord in Colombo granting a certain autonomy to the Tamil minority of the northern and eastern provinces of Sri Lanka, providing for the merger of the two provinces under a single government, and giving Tamil the status of a national language.

[18] In November 1988, the official party's hold on power was ratified by 50.4 per cent of the votes. Jayewardene, at the age of 82, ceded his post to Ranasinghe Premadasa, who was prime minister at the time. Political violence was so pervasive that only 53 per cent of the electorate actually voted.

Elections were boycotted by both the Tamil guerrillas and the Popular Liberation Front (made up of Sinhalese who were violently opposed to any kind of concessions to ethnic minorities). Political opposition to the Government increased, partly fuelled by a strong student movement which was eventually harshly repressed in early 1989.

[19] In 1990, the Indian Government withdrew the last of its 60,000-strong peacekeeping force, stationed there since 1987. More than 1,000 troops had died on the island. Amnesty International reported that in 1990, the Sri Lankan Government had killed thousands of civilians in the region.

[20] In May 1991, the Tamil Tigers were accused of murdering the Indian President Rajiv Gandhi in a suicide mission. He had become their enemy after the Indian peace forces attacked the rebels. The Tigers denied any involvement.

[21] In the November 1994 presidential elections, the Popular Alliance candidate Chandrika Bandaranaike Kumaratunga, daughter of two former prime ministers, became the first woman president of Sri Lanka.

[22] In August 1995 Kumaratunga submitted to Parliament a plan for State reform supported by Tamils. Among other measures, the plan stipulated the transformation of Sri Lanka into a federation of eight regions. Approval of the project was postponed indefinitely due to intensified military clashes.

[23] The city of Jaffna, on the peninsula of the same name, was at the center of the fighting from October 1995 onwards.

[24] The Government offered to suspend the military offensive and discuss proposals to increase the autonomy of regional councils administered by Tamils and Muslims. However, it demanded the surrender of the LTTE, something the guerrillas considered unacceptable.

[25] Despite efforts at mediation, the civil war worsened in March 1998, with dozens killed on both sides. In December, government troops carried out a major offensive and forced the LTTE to retreat to inhospitable areas and abandon several northern cities.

[26] Clashes between government troops and Tamil rebels intensified at the end of 1999. The LTTE took the northern city of Oddusudan in November. In December, just four days after surviving an assassination attempt, Kumaratunga was re-elected for a new term, beating her main rival, Ranil Wickremesinghe, from the UNP.

[27] While Kumaratunga and Wickremesinghe met in March

2000 to discuss a possible political agreement to put an end to the ethnic conflict, Tamil rebels captured a key government position on the Jaffna peninsula after a 12-hour battle.

[28] The LTTE openly rejected the government peace plan, but agreed to take part in conversations if Norway acted as mediator.

[29] By May the rebels had practically all Jaffna under control. Lack of commitment by government forces was one of the reasons cited for their repeated defeats in the peninsula.

[30] In October, to avoid a vote of censorship, President Kumaratunga dissolved Parliament and called for elections to be held on 5 December-in which the Popular Alliance lost their majority.

[31] For the first time since the beginning of the civil war, in November 2001 the LTTE publicly renounced its demand for independence of the Tamil provinces. Velupillai Prabhakaran, leader of the rebels, said that autonomy allowing the Tamils to make political and economic decisions would be enough.

[32] The alliance led by Wickremesinghe won the December 2001 parliamentary elections with promises to try to negotiate an end to the conflict with the separatists and to reactivate the economy.

[33] In January 2002 the Government eased the embargo on food and medicine for the northern areas, after a seven-year blockade. The measure was a condition the LTTE had imposed for the peace talks. In February, under Norwegian mediation, the Government and the rebels signed a permanent ceasefire agreement.

[34] After 12 years, early in 2002 the highway uniting the Jaffna peninsula with the rest of Sri Lanka was reopened and flights to the peninsula started up again.

[35] In September Prime Minister Wickremesinghe lifted the ban that had prevented the LTTE from taking part in negotiations on an equal footing since 1998. In December in Oslo (Norway) the Government and the LTTE agreed on the establishment of a federal system within a united Sri Lanka.

[36] Tamils suspended peace talks in May 2003, complaining of the Government's lack of interest. However they later stated their willingness to resume negotiations and submitted a proposal for a transitional autonomous administration. This idea was not welcomed by the parliamentary opposition party, Kumaratunga's Popular Alliance.

[37] On 4 November, President Kumaratunga dismissed the Ministers of Defense, Interior and Information, ordered the

PROFILE

ENVIRONMENT
An island in the Indian Ocean, southeast of India, separated from the continent by the Palk Channel. The land is flat, except for a central mountainous region. These mountains divide the island into two distinct regions and also block the monsoon winds which are responsible for the tropical climate. The southwest area receives abundant rainfall, and the remainder of the island is drier. Large tea plantations cover the southern mountain slopes, and other major crops are rice, for domestic consumption, and rubber, coconuts and cocoa, for export. Deforestation and soil erosion are important environmental problems. Air pollution is also on the increase, due to industrial gas emissions.

SOCIETY
Peoples: 74 per cent of Sri Lankans are Sinhalese; Tamils are the largest minority group, 18 per cent, and there are Arab (7.7 per cent) and Vedda (1 per cent) minorities.
Religions: 70 per cent Buddhist; 15.5 per cent Hindu; 7.6 per cent Muslim; 7.5 per cent Christian; and 1 per cent other religions.
Languages: Sinhalese (official). Also Tamil and English.
Main Political Parties: United Peoples' Freedom Alliance (UPFA), United National Party (center), United Socialist Party.
Main Social Organizations: Workers' Congress, Trade Union Federation, Trade Union Council, Labor Federation of Workers. Pressure groups exist, such as the Buddhist clergy, the insurgent Liberation Tigers of Tamil Eelam/LTTE ('Tamil Tigers') and Sinhalese groups such as the Nationalist Anti-Terrorism Movement.

THE STATE
Official Name: Sri Lanka Prajathanthrika Samajavadi Janarajaya.
Administrative divisions: 9 Provinces, 24 Districts.
Capital: Colombo 648,000 people (2003).
Other Cities: Dehiwala-Mount Lavinia 218,500 people; Moratuwa 204,500; Maha Nuwara 150,700; Kotte 127,000 (2000).
Government: Mahinda Rajapakse, President since November 2005. Ratnasiri Wickremanayake, Prime Minister since November 2005. Unicameral Legislature: National Assembly, with 225 members.
National Holiday: 4 February, Independence Day (1948).
Armed Forces: 118,000-123,000 (including reservists) (2001). Para-military forces: 88,600.

Under-5 mortality
14 per 1,000
live births
2004

Poverty
5.6% of population
living on less
than $1 per
day
2002

Debt service
8.5% exports
of goods and
services
2004

Maternal mortality
92 per 100,000
live births
2000

IN FOCUS

ENVIRONMENTAL CHALLENGES

Deforestation and soil erosion are considerable. Air pollution is on the increase due to industrial gas emissions. The living conditions of the indigenous population are affected by poaching and increasing urbanization. Mining activities have degraded the coasts and water resources are contaminated by untreated industrial solid and liquid wastes. The levels of air pollution are high in the capital.

WOMEN'S RIGHTS

Women have been able to vote and stand for election since 1932. In 1960 Sirimavo Bandaranaike became the first woman in the world to head an elected Government. In 2004, women held five per cent of seats in Parliament. They made up 36 per cent of the 9 million-strong workforce. In early 2006, Sri Lankan society was continuing to recover from the trauma of the tsunami which just a year earlier had devastated several countries in Southeast Asia. Women played a key role in re-organization and getting back to 'normal life', mainly supporting the recovery of children who were directly or indirectly affected by the disaster. A group of volunteer teachers called *akkis* (elder sisters), provided psychological support for the most affected, trying - among other things - to rebuild their relationship with the sea, so closely entwined with life in a country like Sri Lanka.

CHILDREN

Just like armed conflicts, natural disasters of the magnitude of the 2004 tsunami especially affect children. Either because they had been among the direct victims of the disaster, losing their parents, siblings and friends or because they had been left without schools or social care centers, in early 2006 Sri Lankan children were still trying to get their lives back to normal with the assistance of international organizations and Sri Lankan authorities and volunteers. One of the priorities was to prevent the spread of diseases among the thousands of refugees and displaced people that were living in camps or in precarious conditions. Thus, 91,000 anti-malaria bed nets and vitamin A capsules for 380,000 children were distributed, as well as safe water on a daily basis, in nearly 220 camps and settlements. In just over a year, 110 new schools had been built and another 104 had been repaired, while psychological support had been provided for some 32,000 children throughout the country.

INDIGENOUS PEOPLES/ETHNIC MINORITIES

The Tamils originating from Sri Lanka are a majority in the northeast of the country and account for 12 per cent of the population. The group has remained in its region of origin, with scant migratory movements, but has been very much affected by the migratory flow of Sinhalese towards their traditional lands of residence over the past few years. This group is distinct from the Tamils of Indian descent, accounting for 6 per cent of the total population, who reside in central regions. They are considered to be two distinct groups although they share a common language and religion (both are Hindus and communicate in Tamil, while the Sinhalese are generally Buddhists and have their own language).

The problems of the Tamils of Indian origin are different from those originating in Sri Lanka (formerly Ceylon). Since independence, the former have been denied rights of citizenship. In general, they are perceived by the majority of the Sinhalese as foreigners. Efforts have been made between India and Sri Lanka to give these people the rights of citizenship in one or the other country since the Bandaranaike-Gandhi agreement in 1974. In 1980, 500,000 Tamils of Indian origin were repatriated to India. Most of those remaining in the country were recognized as having Sinhalese citizenship, but in 2000 there were still 200,000 stateless Tamils. Furthermore, this community has suffered economic problems based on discrimination against the group, expressed as restrictions on access to certain jobs, in addition to the historical political discrimination.

In an effort to protect their culture and ensure equal rights, the Tamils started exerting pressure to achieve autonomy. Political parties, (such as TULF) used conventional methods such as participation in Government coalitions to achieve their objectives; however, more militant Tamils sought the creation of a totally independent state in the northeast.

MIGRANTS / REFUGEES

By the time of the tsunami, at the end of 2004, there were over 350,000 displaced people and 130,000 people who had sought refuge abroad as a result of the internal conflict. Of the latter, some 70,000 were settled in refugee camps in the state of Tamil Nadu, in southern India. The tsunami caused the displacement of another 570,000 people who had to stay with friends or relatives or in public buildings allocated for such purpose. By mid-2005, about 55,000 people continued to live in temporary shelters and were still waiting for solutions. By early 2006 there were still over 400,000 people - either refugees, returnees or displaced - living in a precarious situation.

DEATH PENALTY

In practice the country is considered as abolitionist, as no executions have been reported since 1976. In June 2003, the Government promised to continue automatically commuting all death sentences.

---

deployment of troops, suspended Parliament and declared a state of emergency, considering that Wickremesinghe had endangered the 'country's territorial security, stability and integrity' through his concessions to the rebels. Parliament was restored two weeks later, but negotiations with the Tamils were suspended.

[38] The third legislative elections in four years in Sri Lanka took place in April 2004. Kumaratunga's UPFA won 105 of the 225 seats, but did not achieve an absolute majority. Mahinda Rajapakse, a Buddhist lawyer, became Prime Minister.

[39] In an unexpected turn of events, the Popular Alliance recognized the Tamil Tigers, tacitly endorsing them as the sole representatives of the Tamil minority. Peace talks were due to start up again in June.

[40] In December 2004, a massive undersea earthquake in the Indian Ocean resulted in a tsunami that devastated the southern and eastern coasts of Sri Lanka, killing at least 35,000 people and leaving nearly one million homeless. At least 400,000 people lost their jobs. The Government launched a $3.5 billion reconstruction campaign but disputes over distribution of aid with Tamil rebels in areas under their control delayed the undertaking. A deal was reached in June, although this resulted in a political tsunami. The government coalition was broken when the Marxist People's Liberation Front (PLF) and the National Unity Alliance pulled out, calling the deal a 'handover of sovereignty' to the LTTE, calling for massive protests and threatening to start legal procedures. The Jathika Hela Urumaya, represented in Parliament by eight Buddhist monks, announced its support for an eventual impeachment motion against Kumaratunga, while Islamic parties, which demanded increased participation in the distribution of billions of dollars in aid, called a strike.

[41] In August 2005, a state of emergency was declared after Foreign Minister Lakshman Kadirgamar was assassinated at his home. Kumaratunga said that the murder was due to political reasons but stopped short of accusing the Tamil Tigers.

[42] After an unusually calm electoral campaign, with few violent incidents and a low death toll compared with previous elections, Mahinda Rajapakse became President, beating his main rival, Ranil Wickremesinghe, by a narrow margin in November 2005. The campaign had been focused on three key issues: the economy, the peace process and reconstruction efforts after the tsunami. The Tamil population, following LTTE's guidelines, boycotted 'an election that was of no interest to Tamil people'.

[43] In April 2006, the explosion of two bombs in the northeastern port city of Trincomalee left 16 dead and dozens injured, marking the beginning of a new escalation of violence. That month, a suicide bomber attack killed at least eight people in Colombo.

[44] In May, a fresh attack on a naval convoy near Jaffna killed 18 sailors. It was followed by an attack on Tamil civilians, which left 16 dead and was blamed on the navy. According to observers, this represented the final breakdown of the 2002 truce.

[45] In a military offensive in October, at least 129 Sri Lankan soldiers were killed in one day of fighting and more than 300 soldiers injured - the worst single day of casualties for the military since the 2002 ceasefire. In a separate Tamil Tiger suicide attack more than 90 sailors were reported dead.

[46] In October 2006, peace talks in Geneva ended in deadlock. Despite a cordial atmosphere, both sides refused to budge from their starting position, though they did both promise to abide by the faltering ceasefire agreement. ■

Sudan / As-Sudan

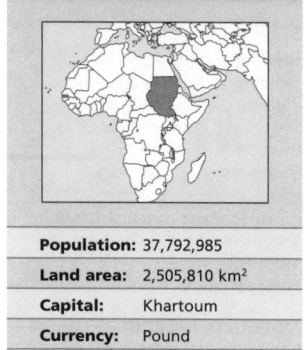

Population:	37,792,985
Land area:	2,505,810 km²
Capital:	Khartoum
Currency:	Pound
Language:	Arabic

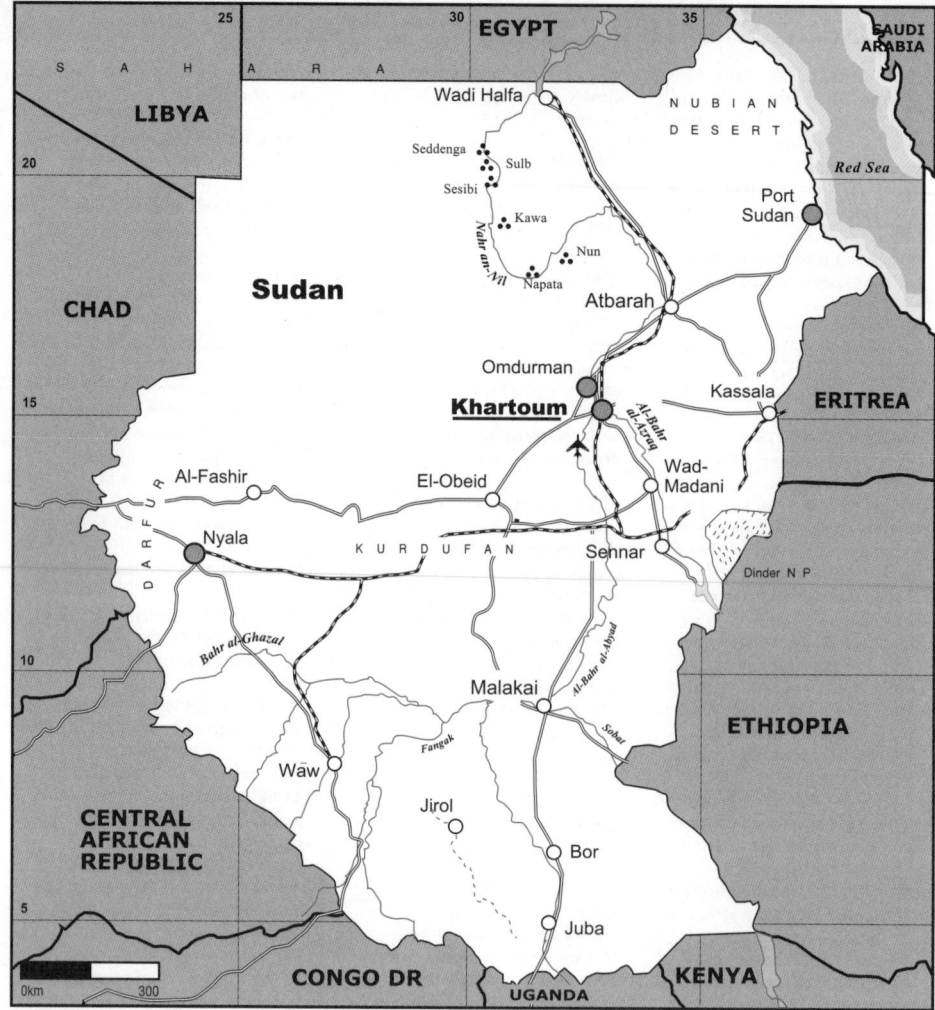

The first people settled in what is now Sudan during the Paleolithic period (30,000-7,000 BC). Their descendants began domesticating animals between 10,000 and 3,000 BC. There was trade in gold, slaves, ivory and granite from this region along the Nile to the Mediterranean.

2   At the end of the 4th millennium BC, North Sudan was colonized by the First Dynasty of the Old Egyptian Empire, causing the black people of the Nile to assimilate elements of their imperial culture.

3   Between 2181 and 1938 BC, immigrants from what is now Libya began farming in this region.

4   A new Nubian (Kushite) culture, which emerged in 2150 BC, arose from the amalgamation of these three ethnic groups, after the decline of the Old Egyptian Empire.

5   In 1580 BC the Egyptians again took over the region, remaining there for 500 years. Despite the presence of imperial Egypt, Nubian traditions were maintained; their art - with its distinctive African slant - flourished.

6   Nubians occupied prominent positions in the Pharaonic bureaucracy that divided Nubia/ Kush into two districts: Wawat in the north, and Kush in the south, where gold and emeralds could be found, and where a new syncretic Nubio-Egyptian culture forged its own alphabet.

LAND USE

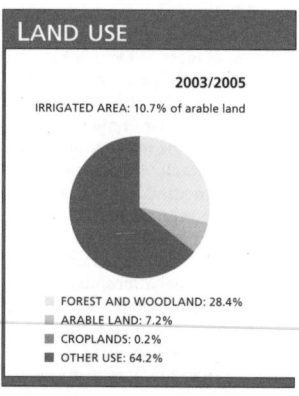

2003/2005

IRRIGATED AREA: 10.7% of arable land

- FOREST AND WOODLAND: 28.4%
- ARABLE LAND: 7.2%
- CROPLANDS: 0.2%
- OTHER USE: 64.2%

7   In the 11th century BC, the viceroys of Kush took advantage of the decline of the New Egyptian Empire to obtain their virtual independence. In 748 BC, their descendants conquered Egypt, remaining until they were expelled in 591 BC.

8   The Persians controlled Egypt and the Kush region from 500 BC.

9   From 300 AD, the inhabitants began to convert to Christianity. Missionaries Nobatia, Dongola and Alodia brought the religion to the black kingdoms of the mid-Nile river region before withdrawing in 675 AD in the face of the Islamic invasion that established the Fatimid dynasty in the 7th century. Despite Muslim influence, the Nubians mainly remained Christian. The Fatimids were conquered by the Ottoman Turks around 1300.

10   Towards the 15th century, recurrent looting by nomadic Arab groups (Bedouins), together with confrontations between the latter and the Ottoman Empire and the Mamelukes (Egyptian oligarchy, 1250-1517), led to the devastation of Nubia.

11   Between the 13th and 15th centuries, when the Christian kingdoms collapsed, there was a massive immigration of Muslim Arab groups, who in turn became the majority of the population in the northern Sudan territory.

12   From that period until 1820, Sudanese territory fell into two main regions: that of the Muslims, where the Sufi brotherhood were in charge of Islamization, and that of the Fujis (non-Muslims from Ethiopia), whose Islamized aristocracy had governed the central region since the beginning of the 17th century.

13   In 1820, the Egyptian Viceroy under the Ottoman Empire, Muhammad Ali, sent in troops searching for gold and slaves. By 1876, his successors controlled the entire Sudanese territory and had established a centralized bureaucracy in Khartoum. They also implemented a taxation system that constituted a virtual confiscation of gold and agricultural produce, and established commercial routes.

14   The appointment of British General Charles Gordon as

Governor of Sudan in 1877 by the Egyptian Viceroy, was as much due to the latter's financial commitments with Britain as to corruption amongst occupying Egyptian authorities.

15   Gordon set out to enforce compliance with an 1877 Convention, to end the lucrative slave trade, with a view to establishing a capitalist economy in Sudan.

16   The loss of this source of income, the arbitrary repression by British troops and the general discontent among the Sudanese brought about by the imposition of taxes and of foreign religious practices (Egyptian Orthodox Islam and British-style Christianity), paved the way for Sufi spiritual leader Muhammad Ahmad, who proclaimed himself *Mahdi* (savior) of his people in 1881.

17   Ahmad's popular forces rose up, took the city of Khartoum, expelled Gordon from Sudanese territory, and established the first nationalist theocracy.

18   In 1898, the Egyptian authorities under British military

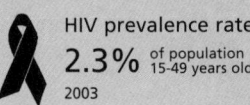
intervention, together with the British Crown, made the territory an Anglo-Egyptian Condominium. Egyptian troops occupied it again, under a policy of *closed districts,* which prevented any contact between north and south.

¹⁹ Early in this period, the British introduced extensive cotton cultivation (still Sudan's main crop), and expanded its communications. At the same time, they allowed freedom of worship, in order to eliminate religion as a source of unrest. They also opened elementary and polytechnic schools and, in 1902, inaugurated the Gordon Memorial College (later the University of Khartoum), where an élite began acquiring a British curricular education. Many were appointed to key posts and belonged to the Graduates' General Congress, which evolved into a fledgling political organization.

²⁰ In 1936, Britain demanded that Egypt sign an agreement prohibiting entry to Sudan by any Egyptian military suspected of fomenting unrest with Egyptian nationalists or Sudanese groups. The Congress wanted to participate in these negotiations.

²¹ When it was not recognized by the British, Congress divided into two groups: one, with the majority, was moderate, well-disposed toward Britain; the other, led by Ismael al-Azhari, was radical, tending toward Egypt.

²² Towards 1943, Azhari and his followers won the majority vote in the congress and constituted the first Sudanese political party, Ashiqa (Brothers - National Unionist Party - as of 1951). Shortly after, the moderate group organized under the Ummah (Nation) Party led by Arman al-Mahdi, son of the *mahdi* deposed by the British in 1898.

²³ In 1951, Egypt revoked the Anglo-Egyptian Condominium Treaty and proclaimed its sovereignty over Sudanese territory. It was hostile to Britain's threat to grant some independence to the Christian and animist south, while not to the Arab and Muslim north. Despite Egypt's objections, in 1953 Sudan received a degree of autonomy.

²⁴ In 1955 parliamentary elections were held and the National Unionist Party, backed by nationalist Egyptian president Nasser, won by a wide margin over the Ummah. This was followed in 1956 with a declaration of independence by Azhari and his parliamentary majority. The provisional constitution consolidated the northerners' position and reneged on promises of a federation.

²⁵ The southern Christians and animists, whose hopes of representation in the Assembly

were negated by the Constitution, initiated a civil war which continued until 1972.

²⁶ In 1958, General Ibrahim Abbud seized power in a coup. After freeing the price of cotton and dissolving political parties, he installed a Supreme Council that ensured compliance with orthodox Islamic laws throughout Sudanese territory, where he imposed Arabic. In 1962, he evicted the Christian missionaries from southern Sudan schools.

²⁷ In October 1964, Abbud was forced to resign and a transitional government took over.

²⁸ The 1965 elections brought Muhammed Mahjud, Ummah Party leader, to the presidency. In his four years in office, he did not improve Sudan's economic situation. At the same time, the different factions of the parliament were irreconcilable and the southerners launched new offensives, following the failure to fulfill promises of political participation.

²⁹ In 1969, General Gaafar al-Nimeiry seized power through a coup.

³⁰ The escalating civil strife in the south was draining the country's resources. This led Nimeiry to negotiate with the southern insurgent groups. Talks culminated in the 1972 peace agreement signed in Addis Ababa, Ethiopia, which brought an end to 17 years of civil war and granted administrative autonomy to the Christian and animist peoples of the south.

³¹ The 1972 peace and the subsequent rise in oil prices attracted investment from various Arab countries, which was directed to agricultural and infrastructure development. In 1977, Nimeiry was re-elected, but his government's incompetence and corruption had sunk Sudan in debt. This reached $8 billion that year and sealed its bankruptcy in 1978, after the suspension of all IMF credit.

³² In 1983, when experts from the US company Chevron discovered oil deposits in the south, Nimeiry revoked the Addis Ababa Agreement and, under the influence of the (Sunni) Muslim Brotherhood of the National Islamic Front (NIF), imposed *Sharia* (Islamic law), which establishes the death penalty for homosexuality, among other human rights violations.

³³ These measures sparked renewed fighting in the south. An offensive by John Garang's Sudanese People's Liberation Army (SPLA), the most powerful of the 12 organizations in the region, forced the withdrawal of all foreign companies prospecting for oil.

³⁴ The NIF and the northern opposition parties - on the one hand - and the international

PROFILE

ENVIRONMENT
The largest African country, Sudan has three distinct geographic regions: the Sahara and Nubian deserts in the north, the flatlands of the central region and the rainforests of the south. Most of the population lives along the Nile (Nahr an-Nil), where cotton is grown. Port Sudan (Bur Sudan), on the Red Sea, handles all the country's foreign trade. Desertification has affected nearly 60 per cent of the territory. Industrial waste has contaminated coastal areas and some rivers.

SOCIETY
Peoples: There are over 600 ethnic groups. Arabs, who live in the center and north of the country, together with Nubians, account for nearly half of the population. Among the other groups, the most important are the Nilote, Nilo-Hamitic and some Bantu-speaking peoples.
Religions: Islam is the predominant religion among Arabs and Nubians with a majority of Sunni Muslims. In the south, traditional African religions are practiced and there are Christian communities in both north and south.
Languages: Arabic (official and spoken by most of the population); the different ethnic groups speak over 400 different languages.
Main Political Parties: National Congress Party; Working People's Force Alliance.
Main Social Organizations: Political associations were allowed in 1998. Democratic Unionist Party; National Democratic Alliance; Sudan People's Liberation Movement; Ummah Party.

THE STATE
Official Name: Jumhuriyat as-Sudan.
Administrative divisions: 9 states, 66 Provinces and 281 Local Government Areas.
Capital: Khartoum (Al-Khartum) executive and ministerial, 4,286,000 people (2003); Omdurman (Umm-Durman), legislative, 1,599,300 people (2000).
Other Cities: Port Sudan (Bur Sudan) 384,100; Kassala 295,100 (2000).
Government: General Omar Hassan Ahmad al-Bashir, President since June 1989, after overthrowing the civilian government; re-elected in 1996 and 2000. Bicameral Legislature. The National Assembly (*Majlis Watani*) with 450 appointed members who represent the government, former rebels, and other opposition political parties. The Council of States (*Majlis Welayat*) has 50 members who are indirectly elected by state legislatures.
National Holiday: 1 January, Independence Day (1956).
Armed Forces: 105,000 personnel (2003). Other: 30,000 to 50,000 (People's Defense Force).

financial organizations - on the other - intensified their criticism of Nimeiry's application of the Sharia, with its restraints on political freedom and effects on the financial systems.

³⁵ In April 1985, while Nimeiry was in the US, his minister of defense and army chief of staff, Abdul al-Dahab seized power and called for elections the following year.

³⁶ The Party of the People (Ummah) won the April 1986 elections, and its leader Sadiq al-Mahdi was elected prime minister.

³⁷ The SPLA then demanded Mahdi's resignation and the formation of a provisional government. Its 12,000 guerrillas were besieging government garrisons in the southern provinces. They took control of the region, frequently blocking aid to the most desperate people affected by

the violence who lacked food and medicines.

³⁸ In June 1989, in the midst of the war between the Sudanese People's Liberation Movement (SPLM) (the armed wing of the SPLA) and the government army, General Omar al-Bashir ousted the regime then in power, dissolved political parties and created a military junta with the participation of the National Islamic Front (NIF), renamed National Congress Party (NCP).

³⁹ In 1995, when the civil war had taken over one million lives and forced three million people to flee to neighboring countries, the African Rights humanitarian organization accused Khartoum of the genocide of the Nubians.

⁴⁰ In January 1998, after proving that Sudan had sheltered the al-Qaeda terrorist network leader, Osama bin Laden, at the beginning

Under-5 mortality
91 per 1,000 live births
2004

Malnutrition
17% under-5s
1996-2004

Debt service
6.0% exports of goods and services
2004

Maternal mortality
590 per 100,000 live births
2000

of the 1990s, the US announced an economic embargo on Sudan. After the bombings of US embassies in Tanzania and Kenya, it accused Khartoum of supporting international terrorism, and a few months later bombed a supposed terrorist target (in fact a chemical plant) near the capital.

[41] In 1999, al-Bashir declared a state of siege and renewed his Cabinet.

[42] That same year China - importer of 55 per cent of all Sudanese exports in 2004 - in addition to a Malaysian and a Canadian company, agreed to finance an oil pipeline to the Red Sea. This was expected to yield a net annual revenue to Sudan of $500 million, as of 2003.

[43] In February 2001, al-Bashir took office once again, having obtained 86.5 per cent of the vote in the December 2000 elections, which were boycotted by most of the opposition parties.

[44] In December 2001, after a six-month campaign by human rights organizations, the Khartoum authorities reported they had released over 14,500 slaves.

[45] One month later, the SPLA signed an alliance with its southern rival, the Sudanese People's Defense Force, forming a common front against the Government.

[46] In October, the start of peace negotiations in Kenya between the Sudanese Government and the SPLA marked the end of 19 years of civil war that had taken the lives of around two million people. On that occasion, US Secretary of State Colin Powell, whose officials had declared access to African oil to be a 'matter of national interest', threatened to triple the US contribution to the SPLA to $300 million and to maintain the embargo on Sudan, if peace were not reached by March 2003. Largely as a result of war, in 2003 92 per cent of Sudanese were living below the poverty line.

[47] For his part, SPLA head Colonel John Garang wanted the vice-presidency of Sudan in place of Osman Ali Taha. He also wanted to reclaim the southern provinces of Nuba, Abyei and Blue Nile, which had fallen under northern jurisdiction in 1972, but this remained unresolved.

[48] Between April and December 2003, the Sudanese Government and the SPLA made a pact to combine their troops in a 39,000-strong army; to share oil profits as of January 2004; to draw up a new constitution during 2004; to award administrative autonomy to the south as of that same year and to call for a referendum in 2010 on southern independence. Hassan al-Turabi, leader of the NIF who had been imprisoned several years earlier, was released in October 2003, at the same time as the proscription on his party was lifted.

[49] Whilst peace between north and south was being brokered, government troops launched an offensive in January 2004 in Darfur in western Sudan, an area under both northern and southern jurisdictions. They attacked the Sudanese Liberation Movement/Army (SLM/A, formerly Darfur Liberation Movement).

[50] The SLM/A had been founded the year before, in response to systematic attacks on the Fur region by groups of Arab pastoralists belonging to the Janjawid people who had been driven out of the Sahel (their region of origin) by desertification, and who wanted to evict the Muslim black ethnic groups (Masaalit, Fur and Zaghawa) from their well-irrigated lands.

[51] In March 2004, al-Bashir once again ordered the arrest of Turabi and his political and military supporters. In April, the UN Commission on Human Rights abstained from applying sanctions against the Sudanese Government. For their part, humanitarian organizations denounced the Government for obstructing the distribution of food and medicine.

[52] In March 2005 it was estimated that 180,000 people had been killed in the Darfur conflict over the previous 18 months and that two million had left their villages seeking refuge in cities, while another 200,000 had fled to Chad. A UN commission of inquiry on Darfur concluded that the Sudanese Government was not guilty of genocide - a crime that would have obliged the international community to intervene - although it had committed 'serious violations of human rights and international humanitarian law' that could be prosecuted as crimes against humanity.

[53] On 5 May 2006, in Abuja, Nigeria, the SLM/A and the Sudanese Government signed an African Union-brokered peace accord. However, an SLM/A splinter group and the Justice and Equality Movement rejected the agreement, demanding greater influence in post-war Darfur. A 31 May deadline was set for all parties to join the accord and for the disarmament process to begin.

[54] The World Food Programme (WFP) reported in mid-2006 that a shortage of resources continued to increase malnutrition and worsen living conditions for millions of people in Sudan. The situation was described as the 'the world's worst humanitarian crisis.' The WFP initially had to cut daily food rations for refugees in half due to lack of funding from donor governments; its appeal led to some new funds being committed but food rations were still only 85 per cent of the recommended minimum. ∎

IN FOCUS

ENVIRONMENTAL CHALLENGES
Around 60 per cent of the land is affected by desertification. There are periodic droughts and the soil is extremely eroded. Industrial waste has contaminated coasts and rivers, thus endangering fresh water reserves. Indigenous people are affected by indiscriminate hunting.

WOMEN'S RIGHTS
Women have been able to vote and run for office since 1964. In 2005, 15 per cent of the seats in the lower house of Parliament and four per cent in the upper house were held by women.

Women make up 30 per cent of the country's total workforce of 13 million people. The situation of girls and female adolescents in the southern regions (which faced decades of armed conflict until peace was achieved in 2005) is much more complex than in the rest of the country. Extreme poverty and daily violence make it extremely difficult for young women to lead normal lives. Educational opportunities are cut off by teenage marriage. One in every five adolescent girls in this region is already a mother, and barely seven per cent of girls attend school. The practice of teenage marriage is sometimes used as a means to free parents from the economic burden of raising a daughter (young brides are usually purchased with cattle or money); at other times, it is meant to provide girls with a man's protection against the constant threat of harassment and rape. Meanwhile, one in every nine women in southern Sudan dies during pregnancy or childbirth.

CHILDREN
Civil war in southern Sudan - which ended in 2005 - and, more recently, the conflict that broke out in 2003 in western Darfur have led to a drastic deterioration in living conditions for Sudanese children. Violence and the forced recruitment of minors in the armed conflict continued to be a serious problem in early 2006. The infant and under-five mortality rates in 2004 were 63 and 91 deaths per 1,000 live births, respectively*. The average life expectancy at birth was a mere 57 years. In addition, 31 per cent of children were underweight at birth, and 17 per cent of under-fives suffered moderate to severe weight deficiency. There were also 21,000 minors living with HIV/AIDS.

INDIGENOUS PEOPLES/ ETHNIC MINORITIES
There are more than 600 ethnic groups in the country, with over 400 languages. There is wide-scale ethnic cleansing by the government-backed militias - Janjawid, Murahelin and the People's Defense Forces - against the inhabitants of Darfur. These are the Masaalit, Zaghawa, Tama, Tanjur and Dajo people, who constitute the Fur ethnic group. Human Rights Watch has accused the Government of 'crimes against humanity'. Killings, bombings, mass rape of women and girls and blocking of humanitarian aid have continued in Darfur ever since 2004.

MIGRANTS/REFUGEES
The conflict in Darfur left almost two million internally displaced people living in camps, totally dependent on international aid and with little or no prospect of returning home due to the insecurity and destruction in their homeland areas, according to the UNHCR. Meanwhile, another 200,000 became refugees in neighbouring Chad.

In early 2006, the situation of the roughly 113,000 Eritrean refugees in Sudan appeared to be worsening, as the possibilities of voluntary repatriation remained slim. In fact, their numbers were expected to grow by a further 3,000 by the end of the year. During 2006 there were plans for the repatriation of 200 of the 14,800 Ethiopian refugees and 500 of the 7,700 Ugandan refugees living in the country. By the end of the year the number of refugees and asylum-seekers from different countries in the region was expected to fall from 35,000 to 14,000.

DEATH PENALTY
The death penalty is still enforced, even as a punishment for common crimes.

Latest data available in The State of the World's Children and Childinfo database, UNICEF, 2006.

Suriname / Suriname

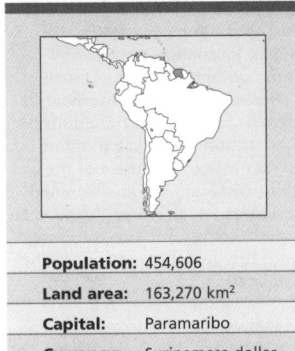

Population:	454,606
Land area:	163,270 km²
Capital:	Paramaribo
Currency:	Surinamese dollar
Language:	Dutch

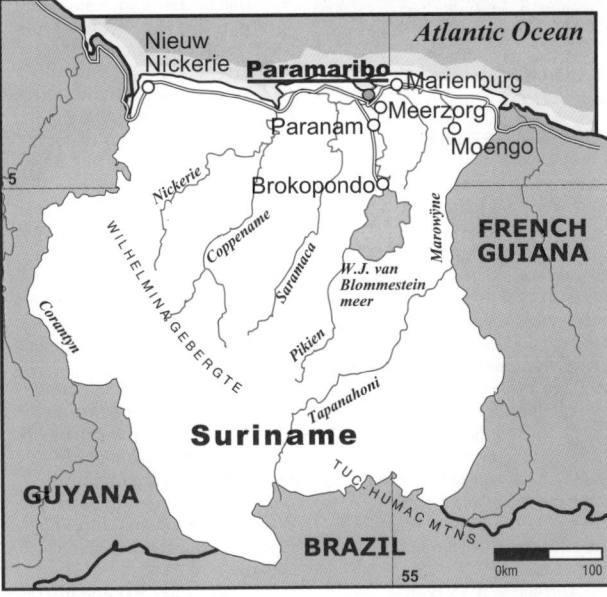

B efore the arrival of the Europeans, the region was inhabited by the Caribs. They were warriors, living in small communities where they existed by hunting, fishing and small-scale farming.

2 Dutch traders arrived in the region in the 17th century, but the first colonies were established by the British, who bought slaves to work in their plantations. In the 19th century Suriname came under definitive Dutch control.

3 In 1863, when slavery was abolished in Dutch territories it was replaced by another source of cheap labor, Asian Indian and Javanese immigrants. This gave rise to a complex ethnic structure in Guyana, with a majority of Indians strongly attached to their cultural heritage. Then there were smaller groups: the Creoles or Afro-Americans, the Javanese, the bush people whose ancestors had rebelled against slavery and fled the plantations, indigenous people and a small European minority.

4 These ethnic, cultural and linguistic differences hindered the development of a national identity. The Creoles formed the NPK (National Party Combination), a coalition of four center-left parties, and they led the fight for independence after World War II. Jaggernauth Lachmon's Vatan Hitakarie represented the Indian population of shop owners and

business people, and they sought to postpone independence.

5 In October 1973, the independence faction won the legislative elections. Hanck Arron, leader of the NPS (National Surinamese Party) became the first Prime Minister of the local Government, which had enjoyed a certain degree of autonomy since 1954.

6 Independence was finally proclaimed in 1975. Many middle-income Surinamese took advantage of their status as Dutch citizens to emigrate to the ex-colonial power. Nearly a third of the population left, causing a serious shortage of technical, professional and administrative personnel. The country lost nearly all its much-needed qualified labor,

with the sole exception of the workers employed by SURALCO and Billiton, two transnationals which monopolized local bauxite mining and indeed, the country's economic life. Economic activity decreased and agriculture declined to dangerously low levels.

7 In February 1980, the Prime Minister was overthrown by a coup (the sergeants' revolution). The National Military Council (NMC) summoned opposition leaders to form a government and several leftwing leaders took up cabinet posts.

8 Another coup in February 1981 led to Lt-Colonel Desir Delano (Desi) Bouterse taking power. The Government established relations with Cuba, in the face of domestic opposition

and external opposition from the US and from the Netherlands.

9 Labor unions, merchants and professional groups began to express their discontent in 1982. In December of that year, 15 journalists, intellectuals and trade union leaders were executed without trial in Fort Zeeland for allegedly conspiring against the State. The December murders are considered to be the most traumatic event in the country's history.

10 In January 1983, Bouterse formed a new government, which appointed Errol Halibux, a nationalist and member of the Farmers and Labor Union, as Prime Minister. After the US invasion of Grenada, the Suriname Government did an about-face in its relations with Cuba, asking Havana to recall its ambassador and suspending all agreements for co-operation between the countries.

11 In an effort to reduce Suriname's isolation, the Government joined the Caribbean Community (CARICOM) as an observer and re-established relations with Cuba, Grenada, Nicaragua, Brazil and Venezuela.

12 In 1986 violence broke out again. On 29 November, a special military unit attacked the village of Moiwana, burning down the home of the leader of the armed opposition, Ronnie Brunswijk and killing 35 people, mostly women and children. Years later, in 1990 the investigation was reopened and Inspector Herman Gooding, responsible for ordering the new investigation, was murdered and his body was left outside Colonel Bouterse's offices. In April 1987 the National Assembly approved a Constitution, providing for a

LAND USE

2003/2005

IRRIGATED AREA: 75.0% of arable land

- FOREST AND WOODLAND: 94.7%
- ARABLE LAND: 0.4%
- CROPLANDS: 0.1%
- OTHER USE: 4.8%

PROFILE

ENVIRONMENT
The coastal plain, low and subject to floods, is suitable for agriculture. Rice, sugar and other crops are grown. Land has been reclaimed from the sea by means of drainage and dykes. Inland, the terrain is hilly with dense tropical vegetation, rich in bauxite deposits. Year-round heavy rainfall feeds an important system of rivers, some of which are used to generate hydro-electric power for the aluminum industry.

SOCIETY
Peoples: Suriname Creole 30 per cent; Indo-Pakistani 33 per cent; Javanese 16 per cent; 'Bush Negro' 10 per cent; Amerindian 3 per cent; other 3 per cent.
Religions: Christian 44 per cent (21.6 per cent Catholic, 18 per cent Protestant); Muslim 18.6 per cent; Hindu 26 per cent; other 15.8 per cent.
Languages: Dutch (official), English (for business), Hindi, Javanese, and a form of Creole are spoken. This, called either Taki-Taki or Sranang-Tongo, is based on African languages mixed with Dutch, Spanish and English.

Main Political Parties: New Front for Democracy and Development (which includes: National Party of Suriname; Progressive Reform Party; Pertjajah Luhur; Surinamese Labour Party); National Democratic Party.
Main Social Organizations: Suriname Trade Union Federation; Central Organization of Civil Servants.

THE STATE
Official Name: Republiek Suriname.
Administrative Divisions: 9 Districts.
Capital: Paramaribo 253,000 people (2003).
Other Cities: Nieuw Nickerie 13,100 people; Meerzorg 6,400; Marienburg 4,300 (2000).
Government: Parliamentary Republic. Ronald Venetiaan, President since August 2000, re-elected in 2005. Ram Sardjoe, Vice President since August 2005. Unicameral Parliament: 51-member National Assembly.
National Holiday: 25 November, Independence Day (1975).
Armed Forces: 2,000 (2003).

	Life expectancy		GNI per capita		Literacy		HIV prevalence rate

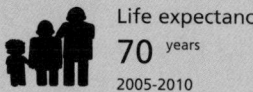
Life expectancy
70 years
2005-2010

GNI per capita
$2,230
2004

Literacy
88% total adult rate
2000-2004

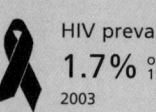
HIV prevalence rate
1.7% of population 15-49 years old
2003

IN FOCUS

ENVIRONMENTAL CHALLENGES
Progressive deforestation is occurring because of extensive logging for timber exportation. Some rivers are polluted due to small-scale mining activities.

WOMEN'S RIGHTS
Women have been able to vote and stand for election since 1948. In 2005, they held 25 per cent of Parliament seats and made up 34 per cent of the labor force. In 2003, female unemployment stood at 20 per cent.

Although 91 per cent of pregnant women receive prenatal care and 85 per cent of deliveries are assisted by qualified staff, the maternal mortality rate stands at 150 per 100,000 births.

No mechanisms exist to control domestic violence. The police are not inclined to intervene in domestic cases and the victims do not always file a complaint.

CHILDREN
Between 1990 and 2004, the infant mortality rate fell from 35 to 30 per 1,000 live births and the under-five mortality rate from 48 to 39 per 1,000 live births. In 2004, 13 per cent of the newborn were underweight and 10 per cent of under-fives were moderately or severely stunted. In 2004, primary school attendance rates stood at 91 per cent for girls and 88 per cent for boys, although these figures were higher in cities than in rural areas. Economic pressure forces children to leave school in order to work.

INDIGENOUS PEOPLES / ETHNIC MINORITIES
Close to 13 per cent of the Surinamese are traditional peoples or their direct descendants.-Between 3 and 5 per cent are indigenous peoples. There are 9 distinguishable indigenous groups, the largest of which are the Arawak, the Caribs, the Trio and the Wayanas. They are divided into 35 communities all over the country. The Caribs and the Arawak are located in coastal zones and in the savannah belt between the coast and the inland tropical forest. The Trio and the Wayanas live in the southern tropical forest zone.

Furthermore, six groups of Maroones people live in the country, including the Saramaka, the N'djuca, the Matawi, the Kwinti, the Aluku and the Paramaka. These make up almost 15 per cent of the country's total population. Since the 18th century they have maintained a culture based on a mixture of African and Amerindian traditions.

All these ethnic groups claim their ancestral territories. Suriname is the only country in the western hemisphere that does not recognize the basic rights of its indigenous population.

DEATH PENALTY
The last execution was carried out in 1982.

[18] The Netherlands issued an international warrant in April 1997 for former dictator Desi Bouterse, under suspicion of links with drug-trafficking. In response, President Wijdenbosch appointed the former dictator State Councillor, granting him diplomatic immunity.

[19] Toward the end of that year, a failed coup ended in the arrest of 17 low-ranking officers. The coup attempt was related to the working conditions of the troops, which had deteriorated, with low salaries and outdated equipment.

[20] Social unrest and an unprecedented economic crisis intensified during the first months of 1999. In June, after the biggest general strike in Suriname's history and months of massive protests, Parliament deposed the Wijdenbosch Cabinet, blaming it for the country's economic collapse.

[21] In May 2000, Venetiaan's NF won the legislative elections and in August he was elected President by the National Assembly by 37 votes out of 51.

[22] Decades of tension between Suriname and Guyana due to a dispute over territorial waters reached a peak in June 2000 when a Surinamese vessel forced the withdrawal of the Canadian company CGX Energy, to which Guyana had granted oil exploration rights. In July, after the leaders of the two countries failed to reach an agreement in several days of negotiations in Jamaica, the Canadian company called off the project.

[23] In November 2000, the High Court of Amsterdam ruled that coup leader Desi Bouterse would be prosecuted - again in absentia - for leading a cocaine-smuggling ring during his time in office and for the 1982 assassinations. Suriname had begun investigating the executions and asked the Netherlands for cooperation and assistance. Because there was no extradition treaty between the two countries, Suriname was not obliged to send Bouterse to Amsterdam for the trial.

[24] During a handing over of command ceremony held in mid-2001, the former National Army commander, Glenn Sedney, offered his apologies to the Surinamese community for the wounds and divisions caused in the past by the military.

[25] Financial misfortune added to low market prices led to State banana companies closing down in April 2002, triggering protests and demands by the workers.

[26] In May 2002, President Venetiaan stated the need to monitor respect for freedom of expression and to recognize the fact that during the 1980s and 1990s, journalists, newspaper directors and radio broadcasting companies had been intimidated. Venetiaan signed the Chapultepec Declaration regarding freedom of expression. Members of the Journalist Association welcomed the measure, stressing the need to reform some laws in accordance with the guidelines set out in the Declaration.

[27] In January 2004, in a maneuver seeking to strengthen the economy, the Surinamese dollar was established as the valid currency, replacing the Dutch guilder.

[28] No candidate reached the two-thirds of parliamentary votes needed to win the presidency in the first round of the July 2005 elections. The same thing happened in the second round. Finally, the United People's Assembly, a regional organ formed by 891 members of Parliament and district representatives, re-elected Venetiaan as President with 560 of the 879 votes.

[29] In May 2006, strong rains caused severe flooding. More than 30,000 square km were covered by water and 175 towns were practically 'erased from the map' covered by up to two meters in mud. Some 25,000 people lost all their possessions. The Government described the situation as a 'disaster' and requested the immediate aid of international agencies.

[30] The yearbook *The indigenous World 2006*, published by the International Work Group for Indigenous Affairs, reported that a Mining Bill under consideration by the National Assembly was racially discriminatory. If approved, the bill would force several northern indigenous communities to leave their homes in order to pave the way for new mines. The International Work Group for Indigenous Affairs says the residents of nearby areas would be too exposed to mercury used in the mines, which could lead to birth defects and poisoning in adults.

[31] In October 2006, the killings of homeless men while they slept on the streets of the capital, Paramaribo, shocked the country. Two had been shot in February and two others had been doused with gasoline and set on fire in May. The latest victim was murdered with an axe, leaving police wondering if a serial killer was responsible. The number of homeless people in the capital has increased in recent years, most of them suffering from mental illness. ∎

return to institutional government. The draft constitution was supported by the three main political parties and the army.

[13] The United Front for Democracy and Development won the 1988 elections. In July 1989, President Ramsewak Shankar granted an amnesty to the guerrilla movement, allowing them to keep their weapons as long as they remained in the forests. Bouterse and the NDP (National Democratic Party) opposed this agreement, arguing that it legalized an autonomous military force.

[14] The New Front (NF) won the May 1991 elections held to choose the National Assembly. One of its major proposals was the re-establishment of relations with the Dutch Government.

[15] On 16 September, Ronald Venetiaan of the NF was elected President and in October launched a cut in spending on defense and the armed forces. A peace process, including UN-sponsored guerrilla disarmament, was started under the supervision of Brazil and Guyana.

[16] In 1993 the country suffered the consequences of a drop in the price of bauxite and between 1980 and 1990 the economy shrank at an average annual rate of 2.6 per cent. The new civil government adopted a harsh structural adjustment program, leading to considerable discontent among the population.

[17] Poverty and unemployment in the rural agricultural communities formed the background for the occupation of the Afobakka dam, 100 kilometers south of Paramaribo, in March 1994. The rebels, who called for the resignation of the Government, were expelled by government troops after a four-day occupation.

Swaziland / Swaziland

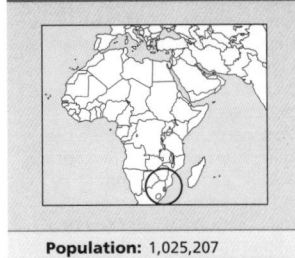

Population:	1,025,207
Land area:	17,360 km²
Capital:	Mbabane
Currency:	Emalangeni
Language:	English

Swaziland was born as a nation in the 19th century, when groups of very different origins joined together. The danger posed by Zulu expansion led Sobhuza, head of the Dlamini local groups, to bring together several peoples - including Zulu deserters and Bushmen or San who remained in the region. They became a powerful force in the northeastern part of the present-day South African province of Natal. At Sobhuza's death, shortly after the Zulu defeat by the Boers (1839), his son, Mswati, was left with the task of keeping the nation together, in the face of constant threats from the Afrikaners. The nation was named after this king, who led the nation in 30 years of resistance and had to ally with the British shortly before his death so as to avoid defeat.

² In 1867, Swaziland became a British protectorate, like Basutoland (Lesotho) and Bechuanaland (Botswana). When Britain defeated the Boers and imposed its dominion over the whole of South Africa, these countries remained under separate colonial administrations. Native authorities were formally recognized in accordance with the British policy of making use of 'local' intermediaries.

³ In 1961, when the Union of South Africa broke off relations

with Britain and toughened racial segregation policies, London accelerated the decolonization process in the region. Swaziland was granted internal autonomy in 1967 and formal independence the following year. Sobhuza II was recognized as head of State. In April 1973, he dissolved the bicameral Parliament, suspended the Constitution inherited from the British, banned all political parties and proclaimed himself absolute monarch. Sobhuza also announced a state of emergency that is still in place.

⁴ In 1978 the Swazi Liberation Movement (SWALIMO) was founded, led by Ambrose Swane. The strengthening of the opposition resulted in a rapid growth of the armed forces, from 1,000 in 1975 to more than 5,000 in 1979.

⁵ Mounting opposition was encouraged among other things by the consolidation of a socialist regime in Mozambique, which led Swaziland to develop closer military relations with South

Africa and Israel. Sobhuza II was one of only three African rulers who never severed diplomatic relations with Tel Aviv.

⁶ A new Constitution was introduced in 1978 without approval from or consultation with the electorate. Following

tradition, the actual approval process consisted of consultation with the heads of the 40 clans a fortnight before its enforcement. It banned opposition parties and established a Parliament.

⁷ After 1980, the economic situation in Swaziland was affected by world recession. There was an increase in the prices of imported goods and a drop in prices for corn, sugar and wood exports. Mineral exports fell from 40 per cent of total exports to only ten per cent due to the depletion of iron ore reserves. Various new coal deposits were discovered in 1980 but remained hardly exploited due to the lack of resources available for this.

⁸ In August 1982, when Sobhuza II died, his successor, Prince Makhosetive, was only 15 years old. The King's death sparked a power struggle within the royal family and Prince Mabandla Dlamini was deposed from his position as Prime Minister. He was succeeded by Bhekimpi Dlamini, a pro-South African conservative, who began his rule by persecuting South African refugees.

⁹ In August 1983, Ntombi, one of Sobhuza's widows, overthrew Queen Dzellue and took power, strengthening the conservative

PROFILE

ENVIRONMENT

The country is divided into three distinct geographical regions known as the high, middle and low veld (plain), all approximately the same size. The western region is mountainous with a central plateau and flatlands to the east. The main crops are sugarcane, citrus fruits and rice (irrigated), cotton, maize corn (the basic foodstuff), sorghum and tobacco.

SOCIETY

Peoples: 84.3 per cent of the population are Swazis. Zulus account for 9.9 per cent; Tonga and Shangaan another 3 per cent; there are Indian (0.8 per cent), Pakistani (0.8 per cent) and Portuguese (0.2 per cent) minorities.

Religions: 77 per cent of the population are Christian, the rest follow traditional African religions.

Languages: Swazi and English (official); ethnic minorities speak their own languages.

Main Political Parties: The new constitution came into effect in February 2006 and the 1973 decree by King Sobhuza II that banned all political parties and activities was suspended. Political parties are still illegal but political activity is allowed.

Main Social Organizations: Swaziland Federation of Trade Unions (SFTU); Swaziland Youth Congress.

THE STATE

Official Name: Umbuso wakaNgwane. **Administrative divisions:** 210 Tribal Areas, including 40 traditional communities. **Capital:** Mbabane 70,000 people (2003).

Other Cities: Manzini 22,500 people; Big Bend 14,300; Lobamba 14,000 (2000).

Government: King Mswati III, crowned on 25 April 1986. Absalom Themba Dlamini, Prime Minister since November 2003. Parliament has two chambers: The House of Assembly has 65 members and the Senate has 30 non-partisan members.

National Holiday: 6 September, Independence Day (1968).

Armed Forces: 14,000 (2003).

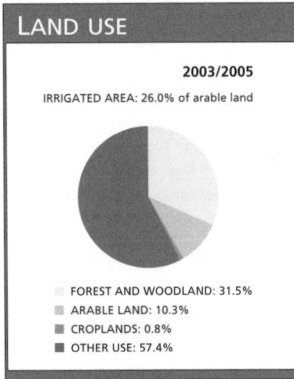

LAND USE

2003/2005

IRRIGATED AREA: 26.0% of arable land

- FOREST AND WOODLAND: 31.5%
- ARABLE LAND: 10.3%
- CROPLANDS: 0.8%
- OTHER USE: 57.4%

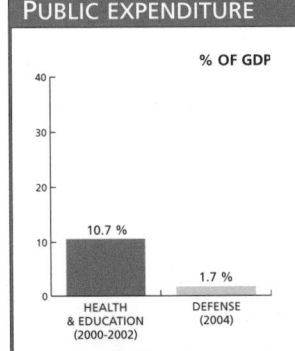

PUBLIC EXPENDITURE

% OF GDP

10.7 % — HEALTH & EDUCATION (2000-2002)

1.7 % — DEFENSE (2004)

faction. Two months later, people voted for a new Parliament through a complicated indirect system which elects administrative representatives.

10 The repression of anti-apartheid militants increased in 1984, with their detention and return to the Pretoria Government.

11 During the second half of 1984, the Government closed down the university. Dlamini's authoritarianism stimulated the revival of SWALIMO, led by Clement Dumisa Dlamini, a well-known nationalist leader, former secretary-general of the Progressive Party, who was exiled in England.

12 In April 1986, Prince Makhosetive was crowned, taking the name King Mswati III. He maintained the conservative policy of his predecessor and replaced Prime Minister Bhekimpi with Sotsha Dlamini.

13 In September 1987, King Mswati dissolved Parliament, announcing elections for November, a year ahead of schedule. The 40 members of Parliament and ten senators were elected by an electoral college. However, the King disputed the election of senators, demanding that the process be repeated.

14 From the late 1980s the country's economic situation improved noticeably. The economy grew and foreign investment continued.

15 There was growing civil unrest throughout 1990. In 1992, the Popular United Democratic Movement (PUDEMO) became the official opposition force after being joined by the Swaziland United Front of Matsapa Shongue, and the Swaziland National Front of Elmond Shongue, thus forcing the creation of Royal Discussion Committees (which should propose the political reforms).

16 In 1992-3, a drought destroyed the corn crop and generated further unemployment.

17 In September 1993 the country's first multi-party elections resulted in Prime Minister Obed Dlamini failing to win a seat in Parliament. In November Jameson Mbilini Dlamini became Prime Minister.

18 During 1994, protests against Mswati III continued. In February 1995, after an arson fire in the seat of Parliament for which the Swaziland Youth Congress claimed responsibility, 40,000 people took part in a demonstration in support of a two-day general strike.

19 In mid-1996, the King called for an end to the upheaval, promising that he would reconsider the situation of political parties.

20 The situation did not change substantially in 1997. In March the King let down the opposition by not sending delegates to the agreed negotiations. In July, he created a 30-member committee, charged with drafting a Constitution, asking for all related initiatives to be referred to this body. These delaying tactics intensified the protests. Security forces were ordered to fire on demonstrators. The clashes caused large numbers of wounded people and the detention of several leaders.

21 In October 1997, fresh strikes broke out in strategic areas, like the sugar sector. The unions also called for the Constitutional Revision Committee to be dissolved.

22 In October 2000, the Government evicted 40 families from their land, which was then given to the king's brother. This caused demonstrations that paralyzed the country for two days, as protestors demanded their civic and labor rights.

23 In mid-2001 a royal decree empowered Mswati to ban any publication not in line with 'Swazi morality and ideals'. Two opposition publications were immediately banned.

24 In March 2002, Swaziland made an urgent appeal for food aid for 200,000 people at risk of dying due to a famine. The crisis arose as the countries of southern Africa were suffering variously from drought, flooding, poor governance and devastated economies.

25 In October 2003, new elections were held and the King appointed Absalom Themba Dlamini as Prime Minister. Since most parties were banned from participating in the vote, opposition leaders called for a boycott of the election.

26 At the beginning of 2004, Mswati asked for $15 million to build a palace for each one of his 11 wives. At the same time, the Prime Minister announced that the country was facing a humanitarian crisis because of scant rainfall over the previous three years.

27 In June 2004, the European Commission approved two million euros in aid for Lesotho and Swaziland, to help 100,000 victims of droughts which had affected both countries over the last two years. A state of emergency had been declared in Swaziland in February. Funds were allocated to the distribution of food, aiming at covering basic needs.

28 As of the end of 2005, Swaziland was the world's largest recipient of economic aid. Its unemployment rate neared 40 per cent and 70 per cent of the population was living on less than $1 a day.

29 A wave of detentions of pro-democracy activists and members of banned political parties for their alleged possession of bombs and involvement in attacks, which started in December 2005 and lasted until January 2006, led numerous activists to flee the country for fear of being arrested. Amnesty International accused the Government of carrying out arbitrary detentions and torturing prisoners to force them to plead guilty.

30 In April 2006, Mswati said that his country was not ready for political parties and criticized foreign governments for meddling in Swaziland's internal affairs. Although the King had lifted a royal decree banning political activity in the kingdom, he stated that the nation's economy had to improve before parties could be allowed.

31 According to global research on freedom of the press carried out by Freedom House (FH), in 2006 Swaziland ranked among states that are 'not free'. The country has 32 communication media laws that curtail freedom of the press, such as a ban on publishing information relating to the royal family or whatever the government deems confidential. According to FH, protection for journalists and communication workers was virtually non-existent. ■

IN FOCUS

ENVIRONMENTAL CHALLENGES
In the low-veld areas, water-borne infections result in high mortality rates. Wildlife was exterminated by European hunters in the first half of this century and the remaining animals are being killed by poachers. Swaziland is suffering a serious problem of soil erosion due to overgrazing and deforestation and also suffers from a shortage of drinking water.

WOMEN'S RIGHTS
Women have been able to vote and stand for election since 1968. In 2003, women held 3 per cent of seats in Parliament and 13 per cent of ministerial positions. That year, they made up 36 per cent of the country's total labor force. Although 90 per cent of pregnant women received prenatal care and 74 per cent of births were attended by skilled health staff in 2004, the maternal mortality rate amounted to 230 per 100,000 births. At the end of 2003, it was estimated that 110,000 women aged between 15 and 49 were living with HIV/AIDS. Swaziland has the highest HIV prevalence in the world. Nearly 40 per cent of people aged between 15 and 49 are HIV-positive.

CHILDREN
Infant and under-five mortality rates worsened between 1990 and 2004: the former increased from 78 to 108 per 1,000 live births and the latter from 110 to 156 per 1,000*. Due to the huge HIV/AIDS prevalence, in 2004 Swazi life expectancy stood at just 31 years. Due to the seriousness of the situation, care centers for children orphaned by the epidemic have been opened, supporting some 30,000 boys and girls. The hospitals are generally overcrowded. Staff and medication are also scarce, particularly in rural areas.

INDIGENOUS PEOPLES/ ETHNIC MINORITIES
There is both social and government discrimination against non-ethnic Swazis (both white and of mixed origin). Although there are no official statistics on this, it is estimated that approximately 2 per cent of the population are non-ethnic Swazis. They have difficulty in obtaining official documents, such as a passport.

MIGRANTS/REFUGEES
The law grants the status of refugee or asylum-seeker to individuals who are protected under the 1951 UN Convention and the 1967 Protocol. In early 2006, there were over 700 refugees and some 300 asylum-seekers in a refugee camp located in the east of the country.

DEATH PENALTY
It continues to be applied to all types of crimes.

* Latest data available in The State of the World's Children and Childinfo database, UNICEF, 2006.

Sweden / Sverige

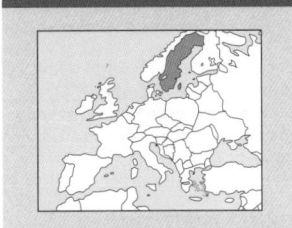

Population:	9,095,374
Land area:	449,960 km²
Capital:	Stockholm
Currency:	Kronor
Language:	Swedish

According to archeological research, the first inhabited area in Sweden is thought to have been the southern part of the country, with occupation dating back to 10,000 years BC. Between 8,000 and 6,000 BC, the region was inhabited by peoples who made a living by hunting and fishing, using simple stone tools. The Bronze Age (1,800-500 BC) brought with it cultural development, reflected in particular in the richness of the tombs of that period.

2   In 500 AD, in Lake Malaren valley, the Sveas created the first important center of political power. From the 6th century BC until the year 800 the population went through a migration period, later becoming settled, with agriculture becoming the basis of economic and social activities.

3   Between the 9th and 11th centuries, the Swedish Vikings reached the Baltic shores on trade expeditions as well as pirate raids and also went as far as what is now Russia, reaching the Black and Caspian Seas. There they established relations with the Byzantine and Arab empires.

4   During the same period, Christian missions from the Carolingian empire (led by the missionary, Ansgar) converted most of Sweden. However, the gods of the ancient local mythologies survived into the 12th century. Sweden had its first archbishop in 1164.

5   Between 1160 and 1250, the fiefdoms of Sverker and Erik alternated in power as each fought to gain control of the Swedish kingdom. The feudal chieftains remained relatively autonomous until the second half of the 13th century, when the King enforced nationwide laws and annexed Finland.

6   The Black Death brought the country's growth to a standstill in 1350, and it did not revive until the second half of the 15th century. In that period the foundries in the central region became important. During the 15th and 16th centuries the German Hanseatic League dominated Swedish commerce and

encouraged the founding of cities.

7   In 1397, the royal power of Norway, Sweden and Denmark was handed over to Danish Queen Margaret, who proclaimed the Union of Kalmar. The ensuing conflicts between the central Danish power and the rebellious Swedish nobility, townspeople and peasants ended in 1523, with the accession of Gustav Vasa to the throne of Sweden.

8   Under the reign of Vasa, the monarchy ceased to be elected by the nobility and became hereditary. A German administrative model was adopted and the foundations were laid for a nation state. The possessions of the Church went to the state, in the wake of the Protestant Reformation. From then on, Sweden aspired to becoming the main power in the Baltic region.

9   In 1630, after intervening successfully in the Thirty Years' War,

Sweden waged two more wars to conquer the Danish regions of Skåne, Halland, Blekinge, and the Baltic island of Gotland, as well as the Norwegian islands of Bohuslan, Jamtland, and Harjedalen.

10   After its defeat in the Great Northern War (1700-21), the Swedish Empire lost most of the provinces to the south and east of the Gulf of Finland. It was reduced to the territories roughly corresponding to modern Sweden and Finland, with Finland being ceded to Russia during the Napoleonic Wars.

11   In 1718, after the death of Charles XII, a parliament (*Riksdag*) made up of nobles did away with the absolute monarchy, assuming power itself. However, the new king Gustav II staged a coup in 1772, and finally re-established full monarchic powers in 1789.

12   Inspired by the success of

the Dutch and British East India Companies, the Swedish East India company was founded in 1731 to trade in east Asia. It was Sweden's largest company in the 18th century, before its demise in 1813.

13   In compensation for the losses incurred during the Napoleonic Wars, Norway was ceded to Sweden. After a short war, it was forcibly annexed by Sweden in 1814. After a series of conflicts, the union dissolved peacefully in 1905.

14   In the second half of the 19th century, Sweden continued to be a poor country, with 90 per cent of the population engaged in agriculture. At this point, a great emigration movement began: one million out of a total of five million Swedes left, mainly for North America.

15   During this period the liberal majority in parliament, supported by King Oscar I, established universal education (1842), the free enterprise system and the liberalization of foreign trade (1846). Legislation was also passed establishing sexual equality in inheritance law (1845), the rights of unmarried women (1858), and religious freedom (1860).

16   Several social movements emerged, such as the temperance league, women's rights advocates, and especially the workers' movement, which grew with industrialization and influenced the Government through the creation of the Social Democratic Party (SAP) in 1889.

17   From 1890 onwards, with the support of foreign capital, industrialization accelerated in Sweden. The country had one of the most thriving economies in that part of Europe. Finished products using Swedish technical innovations quickly became the country's main exports.

18   Alfred Nobel had a major influence in this process. A renowned scientist and inventor, up to 300 patents were registered under his name, including the patent for dynamite in 1867. Most of the fortune from his inventions and companies was set aside, after his death in 1896, for the Foundation that carries his name and which every year awards the Nobel prizes in areas such as physics, medicine, literature and peace.

19   In the early 20th century, the SAP became a major political force. In 1917 its leaders won public posts. The party consolidated its power in the 1930s, although this process was not without conflicts. In 1931, in the northern mining city of Ådalen, firepower was used to repress the Workers Day demonstration and five workers died, in what came to be known as the Ådalen Tragedy.

20   The SAP's rise to power was accompanied by a policy of seeking consensus for a major social reform program (including state pension funds, free education and public

Life expectancy
81 years
2005-2010

GNI per capita
$35,840
2004

HIV prevalence rate
0.1% of population
15-49 years old
2003

healthcare) and the establishment of a Welfare State, characterized by strong state intervention in the economy and a social security system that aimed to give 'cradle-to-grave' protection. A crucial step towards that goal took place in 1938 in Saltsjöbaden, when employers and workers agreed to resolve their differences peacefully and through institutional channels. This deal symbolized the birth of the 'Swedish Model'.

[21] The model was progressively implemented from the 1930s until 1976, when the SAP lost its first elections since 1936. Prime ministers Per Albin Hansson and Tage Erlander played a major role in that historical process. Erlander led three consecutive governments, from 1946 until 1969.

[22] Trade routes were disrupted during World War II, resulting in serious food shortages. This provoked a protectionist agricultural policy, for strategic security reasons, that is maintained even today. Swedish neutrality was upheld during the War, but at the price of allowing some one million Nazi soldiers to cross the territory to invade and occupy Norway.

[23] Sweden favored a thaw in East-West relations during the Cold War, and worked actively for international disarmament. One of the cornerstones of its foreign policy is its support for the UN. After Dag Hammarskjöld became UN Secretary-General in 1953, the UN played an important role as mediator in several international crises, such as the Suez Canal crisis in 1956. Hammarskjöld died in a plane crash in 1961, in Africa.

[24] In the 1970s there was a slow-down of economic growth, due partly to the increasing cost of oil imports to satisfy half the country's energy needs. Employers and political parties representing the interests of the bourgeoisie used the economic crisis to express their discontent with the Swedish Model.

IN FOCUS

ENVIRONMENTAL CHALLENGES
Coastal areas on the North and Baltic seas are highly polluted. Extremely high levels of acid rain damage the soil and pollute sources of drinking water.

WOMEN'S RIGHTS
All women have been able to vote and stand for election since 1921; female suffrage was first granted in 1919, albeit with restrictions. In 2003, women held 45 per cent of seats in Parliament and 52 per cent of ministerial positions.

As of 2003, women made up 48 per cent of the country's total labor force of five million. Female unemployment stood at five per cent. Every year, between 200 and 500 women arrive in Sweden from the Baltic countries, Central Europe and Russia, to work as domestic employees or in the sex industry. In 2004, the annual fertility rate was 1.7 children per woman.

CHILDREN
Successive governments have been committed to welfare policies and to protecting children's rights - though

how this is affected by the policies of the new center-right government remains to be seen. Child pornography, above all on the internet, had become quite a serious problem by the end of 2005. It was estimated that between 20,000 and 30,000 Swedish internet users had access to some 1,100 webpages that published pornographic images of children. Although the possession of these images is an offense punishable by imprisonment, their viewing is not. Ecpat, an organization fighting child pornography, asserted that most of these internet users were not pedophiles but rather encountered these images when searching for common pornography.

INDIGENOUS PEOPLES/ ETHNIC MINORITIES
The Sami descend from nomadic peoples that for thousands of years migrated throughout the north of Scandinavia. The largest community is in Norway, but they are also a minority in Sweden. Their ethnic origin and language (as with Basque people in Spain and France) is uncertain and their roots have

become lost in time. Until a few years ago, reindeer were the basis of their economy. However, the abandonment of their nomadic lifestyle has substantially altered their customs.

MIGRANTS/REFUGEES
Within the period 2001-2005, 128,580 refugees arrived in the country: 23,520 in 2001, 33,020 in 2002, 31,350 in 2003, 23,160 in 2004 and 17,530 in 2005. In 2004, most of them came from Serbia and Montenegro (4,022), Iraq (1,456), the Russian Federation (1,288), Azerbaijan (1,041), Somalia (905), Afghanistan (903), Bosnia and Herzegovina (785), Iran (660) and Bulgaria (567). In 2005, the figures were: 2,944 from Serbia and Montenegro, 2,330 from Iraq, 1,057 from the Russian Federation, 751 from Bulgaria, 582 from Iran, 451 from Libya, 435 from Afghanistan, 431 from Azerbaijan and 427 from Burundi.

DEATH PENALTY
Capital punishment was abolished for ordinary offenses in 1921 and for all offenses in 1972.

[25] In 1976 the SAP lost power to a coalition of three moderate, liberal and conservative parties. Six years later, the Social Democrats returned to power and were re-elected in 1988. In 1986, charismatic leader Olof Palme was shot dead in a murder case that has still not been solved. Ingvar Carlsson succeeded him as prime minister.

[26] In the years that followed, Parliament investigated an alleged case of bribery - involving the Palme administration and Bofors, an arms manufacturer - in connection with the sale of weapons to the Middle East and India. The country's laws

banned such sales, as well as trade with areas where there are military tensions or with countries at war. The investigation found that SAP members had carried out illegal espionage operations, embezzled funds and spread false information regarding Palme's assassination.

[27] In the September 1991 elections, the Social Democrats lost their parliamentary majority. Led by Prime Minister Carl Bildt, a conservative administration sought to dismantle the Welfare State and abandoned the full employment policy that had been implemented for decades by the SAP. The crisis, reflected in

high unemployment rates, was the background to a wave of xenophobia and assaults on foreigners, unprecedented in Sweden.

[28] On 26 August 1993, the King opened a new parliament for the Sami population of Lapland who numbered 17,000, out of a total Sami population of 60,000, resident in Norway (40,000), Finland, Russia and Sweden. A major dispute with the Government arose over the abolition of the Sami's exclusive hunting rights over their lands.

[29] The SAP were re-elected in 1994 after a contentious electoral campaign. In a referendum that year, 52 per cent of voters approved Sweden's entry into the EU. A significant part of the political establishment and a number of entrepreneurs supported entry, while left-wing groups and environmentalists were against it. Sweden joined the EU on 1 January 1995. A year later, former Finance Minister Göran Persson replaced Carlsson as Prime Minister.

[30] In 1996 it was revealed that some 60,000 people, mostly women, had been sterilized between 1935 and 1976, as part of a government plan to prevent 'inferior' humans from reproducing. The concept of the 'purity' of the Swedish race had been defined in 1922 by the Uppsala Institute of Racial Biology. The Government authorized a parliamentary investigation into the

PROFILE

ENVIRONMENT
Sweden lies on the eastern side of the Scandinavian Peninsula. In the wooded northern part of the country are iron mines and paper mills. The central region has fertile plateaus and plains. The main industrial area is located in the south of the country where there is also agricultural production of wheat, potatoes and sugar-beet as well as cattle. The southern region is also the most densely populated.

SOCIETY
Peoples: Swedes 89.4 per cent; Finn and Sami minorities. There are refugees from Iran and former Yugoslavia. **Religions:** Lutheran (official) 89 per cent; Catholics 1.8 per cent; Pentecostal Church 1.1 per cent. Other 10.6 per cent. **Languages:** Swedish (official); Finnish and Lapp. **Main Political Parties:** Social Democrats; Moderates; Liberal Peoples Party; Christian

Democrats. **Main Social Organizations:** Confederation of Swedish Trade Unions, Confederation of Professional Associations, Central Wage Earners' Organization.

THE STATE
Official Name: Konungariket Sverige. **Administrative divisions:** 24 provinces. **Capital:** Stockholm 1,697,000 people (2003). **Other Cities:** Göteborg 744,300 people; Malmö 242,700; Uppsala 125,400 (2000). **Government:** Hereditary constitutional monarchy. Sovereign: Carl XVI Gustaf, since 15 September 1973. Fredrik Reinfeldt, Prime Minister since October 2006. Unicameral Legislature: the *Riksdag*, with 349 members. **National Holiday:** 6 June, Swedish Flag Day. **Armed Forces:** 16,000 annual conscripts and a regular force of 20,000 officers (2003). Other: Coast Guard: 600 (1993).

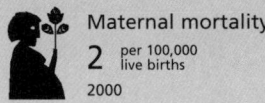

Under-5 mortality	Maternal mortality
4 per 1,000 live births 2004	**2** per 100,000 live births 2000

sterilization issue.

[31]   Persson was re-elected in 1998 with 36 per cent of the vote, the lowest obtained by the SAP in almost 60 years. A bridge between Malmö and the Danish capital Copenhagen was built in 2000, linking the two countries by a short car journey.

[32]   Persson was elected yet again in the 2002 elections. A new referendum was called to decide if Sweden should join the European Monetary Union (EMU) and adopt the euro as its currency, replacing the kronor. Foreign Affairs Minister Anna Lindh was stabbed to death in September while shopping. Her death was initially linked to the controversial referendum, to be held a few days later. However, the murderer, Mijailo Mijailovic, the son of Serbian immigrants, was found to be mentally unstable.

[33]   Mijailovic was sentenced to life in prison in March 2004. Although psychiatric tests had concluded that he was sane when he carried out the attack on Lindh, new medical tests determined that the murderer was suffering from a mental illness. An Appeals Court overturned the life prison sentence and he was transferred to a psychiatric hospital.

[34]   The Swedes voted against joining the EMU, which was a major setback for the Persson administration and had repercussions throughout Europe and especially in the UK, which also opposed the proposed single European currency.

[35]   In 2004, the issue of terrorism was placed on the country's political agenda. In January, the Israeli ambassador in Stockholm, Zvi Mazel, destroyed a work of art, which he considered to be antisemitic, in one of the city's museums. The incident sparked a diplomatic row between the two countries.

[36]   In February 2005, the Minister of Sustainable Development, Mona Sahlin, declared that the Government expected the Swedish economy to break its dependency on oil by 2020, without building new nuclear plants and aiming at renewable energy sources. Sweden has been investing in renewable energy projects, which in 2003 generated 26 per cent of all energy consumed in the country. In the other EU countries, renewable energy only accounted for an average of six per cent.

[37]   In May 2005, the UN found Sweden guilty of violating the International Convention against Torture for refusing to grant political asylum to Egyptian citizen Ahmed Agiza, who had requested it in 2001. Egypt had sentenced Agiza, in his absence, to 25 years in jail for being a former member of the Egyptian Islamic Jihad. According to the UN there was relevant information suggesting that Agiza could be a victim of torture in his country.

[38]   A report on gender equality issued in May by the World Economic Forum placed Sweden at the top of the list of countries with the smallest gender gap.

[39]   The September 2006 general election saw a center-right alliance, led by the Moderate Party's Fredrik Reinfeldt, end 12 years of Social Democrat rule. Reinfeldt proposed reforms to Sweden's welfare state, including cutting taxes for the lowest earners and reducing unemployment benefits. ■

Ombudsman: arbiter between government and citizens

THE WORD OMBUDSMAN is of Swedish origin, coming from umbodhsmadhr, and has several related meanings: 'representative', 'trustworthy commissioner', 'agent who looks after the interests of a group or business' and 'one who speaks on behalf of another'. In a classical sense, it was originated in Sweden in the 18th century, though a similar kind of post already existed in Turkey at about the same time. King Charles XII is credited with its creation and it has been speculated that he was influenced by the Turkish model after spending several years in Turkey. The Ombudsman institution soon spread throughout Scandinavia.

The Swedish constitution of 1809 established the Ombudsman's office for the purpose of respecting people's dignity. Through the Ombudsman's office it sought to exert additional control over the fulfillment of laws, to supervise how these laws were really being applied by the administration and to create a new and agile way, without formalisms, by means of which individuals could claim against the abuses and violations committed by state authorities and officials. In the early 19th century, Swedish citizens faced an administration that neither acknowledged nor corrected its faults, with a very slow and difficult-to-access judicial system and a Parliament overloaded with functions and responsibilities, and difficulty in dealing with individual cases and requirements.

All these factors contributed to the progressive introduction of the Ombudsman institution, which prompted the resolution of citizens' claims in a flexible and expeditious way.

SWEDEN AND BEYOND
In recent decades, there have been three instances that suggest this institution could usefully be more widely adapted internationally. On the one hand, the adoption of this institution by two countries that were removing long dictatorial regimes, such as Portugal in 1975 and Spain in 1981. On the other hand, following the fall of the Berlin Wall in 1989, a significant number of Eastern European countries set up similar offices.

Finally, in Latin America, a region disrupted by the systematic violation of human rights and with a slow process of democratic transition in countries such as Guatemala, Colombia, Argentina, Peru, Honduras, Mexico and El Salvador, the creation of Ombudsman Offices meant there was a new impetus to the consolidation of this institution.

FUNCTIONS
The Ombudsman in countries like Sweden, Spain, Germany, Guatemala, Peru, Honduras, and the US is appointed by Parliament, although he/she acts with total independence, and his/her mission is to defend the rights of citizens and to supervise acts of the authorities. The Ombudsman intervenes upon another person's request and should do it without having any personal interest in the matter he/she was called upon to deal with.

The Ombudsman should be independent and should arbitrate between government and citizens. In spite of having a wide jurisdiction, he/she can only perform an advisory role. The Ombudsman can suggest the government make changes but cannot order them. In all countries that have adopted this institution, it lacks executive authority.

Traditionally, the Ombudsman has flourished in parliamentary systems where on account of its functions and fields of competence it finds a natural constitutional harmony. In those presidential systems that have established this institution, a parliamentary association is also maintained since, in most of cases, the person to hold such office is appointed by Parliament.

An Ombudsman receives a large number of complaints per year, apart from those he/she investigates on his/her own initiative. Most of these complaints are rejected without any investigation being carried out. In a significant number of cases, citizens are unable to indicate a specific claim. In other cases, the complaint falls outside the jurisdiction of the Ombudsman, who can only advise the citizen about where to make the complaint. At the same time, he/she serves as an advocate for poorer people in the area of administrative law.

In view of the neutrality of his/her function, the Ombudsman shall maintain strict confidentiality about matters that are brought to his/her attention, unless given permission to do otherwise. The only exceptions that are left at the sole discretion of the Ombudsman are situations that present an imminent threat of serious harm. The Ombudsman should take all reasonable measures to prevent anybody, including the administration, from having access to confidential records and files. ■

Sources: *Encyclopedia Britannica, U.N., University of Guanajuato* (http://www.ugto.com)

Switzerland / Schweiz - Suisse - Svizzera

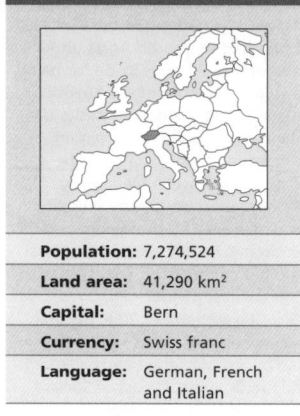

Population:	7,274,524
Land area:	41,290 km²
Capital:	Bern
Currency:	Swiss franc
Language:	German, French and Italian

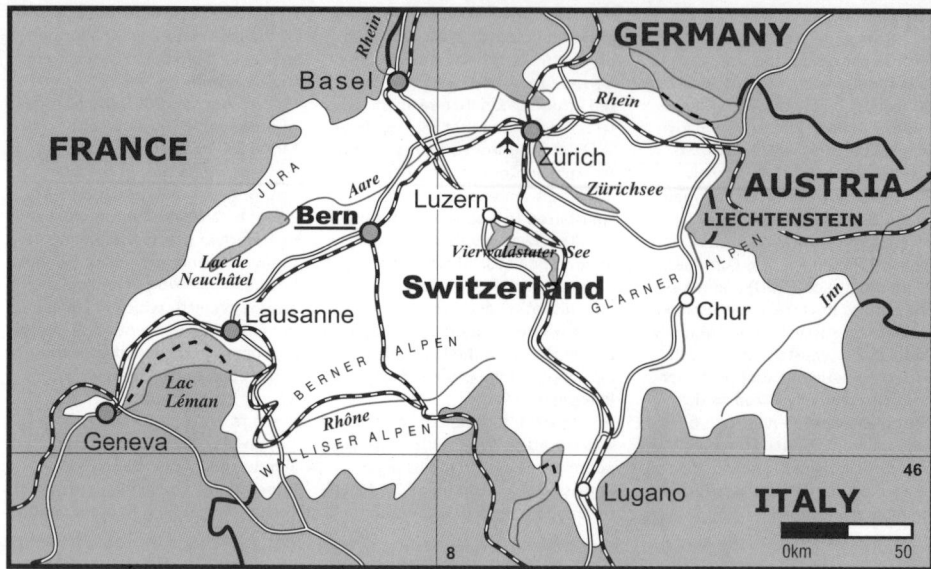

C eltic peoples, the most significant being the Helvetians, occupied the territory of what is now Switzerland before Roman colonization. The Alpine valleys north of the Italian peninsula were conquered by Julius Caesar in 58 BC because of their strategic importance for Rome as an access to parts of the Empire.

2 The Germanic peoples north of the Rhine invaded from the year 260 onwards. Between the 5th and 6th centuries the Germans permanently occupied the region east of the Aar river, together with Burgundian and Frankish groups. By 639 they had founded the kingdoms that would later become France.

3 The Christian survivors from Roman times had completely disappeared when St Columba and St Gall arrived in the 6th century. These missionaries created the dioceses of Chur, Sion, Basel, Constance and Lausanne. Monasteries were built, in Saint-Gall, Zurich, Disentis and Romainmotier.

4 Until the partition of Verdun in 843 these territories belonged to Charlemagne's empire. Thereafter, the region west of the Aar was allotted to Lothair, while the east remained in the hands of Louis the German. French and German influence formed a peculiar blend with the Latin tradition of the Roman Catholic Church.

5 Around 1033, for dynastic and political reasons, Helvetia became a part of the Holy Roman Empire, remaining so during the Middle Ages. In the 11th century the region was divided after the re-establishment of imperial authority and its disputes with the Papacy. Dukes, counts, and bishops exerted virtually autonomous local power.

6 Walled cities served as administrative and commercial centers, and protected powerful families seeking to expand their possessions through wars against other lords and kingdoms. In the 13th century, Rudolf IV of Habsburg conquered most of the territories of Kyburg and became the most powerful lord in the region.

7 In the cities independence developed in opposition to the nobility. However, it was stronger among the peasant communities in the most inaccessible valleys who practiced economic cooperation to survive the harsh conditions, rejecting forced labor and payment of tithes in cash or kind.

8 In 1231 the *canton* (area) of Uri fell under the authority of the Holy Roman Empire, and in 1240 Schwyz

and Nidwald were subjected to Emperor Frederick II, although retaining the right to choose their own magistrates. The Habsburg overlords questioned this freedom and uncertainty remained until Rudolf of Habsburg was crowned King of Germany in 1273. He exercised his imperial rights in Uri and inherited rights over Schwyz and Unterwald until his death in 1291. These regions thereafter constituted the Perpetual League.

9 This was an agreement for dispute arbitration, putting law above armed strength. The honorary magistrates had to be residents of those cantons.

10 The league of the Uri, Schwyz, and Unterwald cantons was joined by the city of Zurich, constituting the first historic antecedent of the Swiss Confederation. This confederation was consolidated with the victory of Margarten in 1315, defeating an army of knights sent to impose imperial law in the region by the Habsburgs.

11 The Confederation was supported by new alliances. In 1302 the League signed a pact with the city of Luzern, previously dependent on Vienna. In 1315 Zurich reaffirmed its union and in 1353 it was joined by Bern, followed by the Glarus and Zug cantons, thus forming the core of an independent state within the Germanic Empire.

12 During the second half of the 14th century, the rural oligarchy was defeated and their lands and laws given over to city councils. This democratic rural movement gave birth to the 'Landesgemeinde', a sovereign assembly of canton inhabitants, and a similar movement was led by the city guilds. From then on, the Confederation launched into territorial conquest. During the 15th century the union grew to 13 cantons, it made alliances with other states, and the institution of government known as the *Diet* was formed where each canton was represented by two seats and one vote.

13 In 1516, after the defeat of the Helvetians, the King of France forced a peace treaty with the cantons. In 1521 an alliance gave France the right to recruit Swiss soldiers. Only Zurich refused to sign this alliance, maintaining military and economic links with the Old Confederation until its end in 1798.

14 The Reformation came to Switzerland with Huldrych Zwingli, a priest who preached against the mercenary service and the corruption and power of the clergy. Popular support for Zwingli strengthened the urban bourgeoisie. The Reformation became more radical in rural areas where harsh repression re-established the domination of cities over peasants.

15 Zwingli's attempt to alter the federal alliance to benefit the reformed cities was frustrated by the military victory of the Catholic rural areas. The second national peace of Kappel, signed in 1531 granted the Catholic minority advantages over the Protestant majority.

16 The areas where both religions coexisted were subject to constant

LAND USE

2003/2005

IRRIGATED AREA: 5.8% of arable land

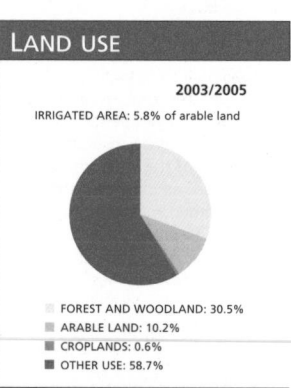

- ☐ FOREST AND WOODLAND: 30.5%
- ☐ ARABLE LAND: 10.2%
- ☐ CROPLANDS: 0.6%
- ■ OTHER USE: 58.7%

PUBLIC EXPENDITURE

% OF GDP

12.3 % HEALTH & EDUCATION (2000-2002)

1.0 % DEFENSE (2004)

WORKERS

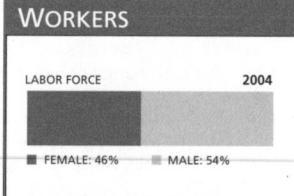

LABOR FORCE **2004**

- ■ FEMALE: 46%
- ☐ MALE: 54%

Life expectancy
81 years
2005-2010

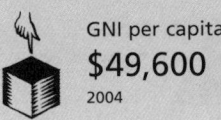

GNI per capita
$49,600
2004

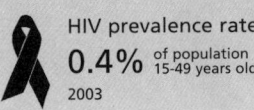

HIV prevalence rate
0.4% of population
15-49 years old
2003

tension, but cooperation was required to preserve the union of the federation. In Catholic regions agriculture prevailed, while in Protestant areas trade and industry flourished, aided by French, Italian and Dutch refugees.

17 The ownership of real estate, trade and industry, together with the recruiting of mercenary troops, gave great wealth and power to a small group of families, while the small peasants had no rights, and were obliged to work mediocre lands or as farm laborers.

18 Popular consultation disappeared in the 17th century. The power of the cities caused uprisings, such as the great peasant revolt of 1653, which were harshly repressed. Three years later when a further war ensued the prerogatives of the Catholic cantons were re-established.

19 During the European conflicts of the 17th and 18th centuries Switzerland remained neutral because of its religious division and its mercenary armies. Neutrality became a condition for the Confederation's existence. The policy of armed neutrality, which still holds, was first formulated by the Diet in 1674.

20 In 1712 the Protestant victory in the second battle of Villmergen ended religious struggles, ensuring the hegemony of cities which were undergoing industrial expansion. Switzerland became the most industrialized country in Europe. Industry was based on labor at home, completely transforming work in the countryside.

21 Throughout the 18th century, a series of popular revolts against the urban oligarchy called for the reform of the Swiss Constitution. In March 1798, the Old Confederation fell under pressure from Napoleon's army. The Helvetic Republic was proclaimed 'whole and indivisible' with sovereignty for the people. Between the unitary Republic and the 1848 Federal Constitution, Switzerland was shaken by coups, popular revolts, and civil wars. The new federal pact marked a final victory for liberalism in the country. Two legislative bodies were established guaranteeing the rights of the small Catholic cantons.

22 A state monopoly was created for custom duties and coin minting, while weights and measures were standardized, so satisfying the industrial and commercial bourgeoisie's economic requirements. The 1848 Constitution thus removed the obstacles to capitalist expansion.

23 Nepotism and the concentration of capital benefited only the few and fuelled growing opposition to the institutional system. The 1874 Constitution partially addressed these issues,

and introduced the mechanism of referendum as an element of direct democracy.

24 Expansion of the home labor system delayed workers' organization as the country industrialized. The Swiss Workers' Federation, created in 1873, had only 3,000 members, and the Swiss Workers' Union, which replaced it in 1880, only exceeded this figure ten years later. The first achievement of the workers' movement was factory legislation, passed by parliament in 1877. The working day was limited to 11 hours with improved working conditions, until then men, women and children worked 14 hours without even basic hygiene and safety conditions.

25 The creation of the Socialist Party in 1888 prompted liberals in 1894 and conservatives in 1894 and 1912 to organize all over the country. For several decades, the Socialist Party's main demand was the incorporation of proportional representation.

26 In 1910, 15 per cent of the workers in Switzerland were foreign. Many were anarchists and socialists who had suffered persecution in their own countries and they encouraged radical positions in the workers' movement.

27 World War I brought great internal tensions to Switzerland, especially between the French and German-speaking regions. Under the leadership of Ulrich Wile, the Swiss army cooperated with Germany. Tension only decreased after the French victory, when Switzerland formally approached the allies and became a member of the League of Nations. The 1918 general strike, although lifted three days later under pressure from the armed forces, led the bourgeoisie to form an anti-Socialist bloc. That year proportional representation was introduced.

28 The elections in 1919 marked the end of the liberal hegemony, in place since 1848. The Socialists obtained 20 per cent of the vote, leading liberals to ally with the peasants who had 14 per cent, while the conservatives became the second power in the Federal Council.

29 The 48-hour week was included in factory legislation, while in 1925 an article on old-age pensions was added to the Constitution. Assistance to the unemployed improved and collective work contracts became more common.

30 During World War II, the European powers recognized Swiss armed neutrality, and the country stayed outside the conflict.

31 During the Cold War, Switzerland sided with the West

but did not join the UN, in order to maintain its neutrality.

32 The Swiss economy expanded greatly during the postwar period. The chemical, food, and machinery exporting industries became large transnational corporations. In 1973 Switzerland was placed fourth in direct foreign capital investments, after the US, France, and Britain.

33 Due to its political neutrality, Switzerland did not join the European Economic Community in 1957. However, it has been a member of EFTA (European Free Trade Association) since 1960.

34 In 1959, the socialists joined the Federal Council with two representatives. Since then the Executive has remained practically unchanged with 80 per cent of the electorate represented in Government.

35 The 1980s saw new groups arise, including the feminist movement and campaigners against nuclear power, which in 1981

incorporated equal rights for men and women in the Constitution - as well as violent demonstrations by youth organizations against the consumer society.

36 Increased poverty and rising numbers of immigrants encouraged the growth of the extreme right. The small Swiss Democratic Party and the Party of Drivers, xenophobic and opposed to social policies, gained support in the early 1990s.

37 In May 1992, a plebiscite approved Switzerland's integration into the IMF and World Bank. In June 1993, Parliament approved the idea of incorporating Swiss troops into the UN peacekeeping forces. This represented a change in the traditional policy of Swiss neutrality. However, most voted against this proposal in a 1994 referendum.

38 In July 1997, Swiss banks - the targets of international lawsuits filed by individuals - released a list of names of account holders with

PROFILE

ENVIRONMENT
A small landlocked state in continental Europe, Switzerland is a mountainous country made up of three natural regions. To the northwest, on the French border, are the Jura mountains, an agricultural and industrial area. Industry is concentrated in the Mitteland, a sub-Alpine depression between the Jura and the Alps, with numerous lakes of glacial origin. It is also an agricultural and cattle-raising region. The Alps cover more than half of the territory and extend in a west-east direction with peaks of over 4,000 meters. The main activities of this region are dairy farming and tourism.

SOCIETY
Peoples: Two-thirds of the population are of German origin, while 18 per cent and 13 per cent are French and Italian, respectively. 17.1 per cent of the country's citizens or permanent residents are Italian, from former Yugoslavia, Portuguese, German, Turkish or other nationalities.
Religions: 47.1 per cent Catholic, 40 per cent Protestant, 2.2 per cent Muslim, 1 per cent Orthodox Christian.
Languages: German, French and Italian. A very small minority (0.4 per cent) in certain parts of the Grisons canton to the east speak Rhaeto-Romanic (or Romansch), of Latin origin.
Main Political Parties: Swiss People's Party; Social Democratic Party of Switzerland; Free Democratic Party of Switzerland; Christian Democratic People's Party of Switzerland.
Main Social Organizations: The Swiss Federation of Trade Unions; the Confederation of Christian Trade Unions; the Federation of Swiss Employees' Societies; environmentalist and anti-economic globalization organizations.

THE STATE
Official Name: Confoederatio Helvetica (Romansch); Schweizerische Eidgenossenschaft (German); Confédération Suisse (French); Confederazione Svizzera (Italian).
Administrative divisions: 20 Cantons, 6 Sub-Cantons.
Capital: Bern (administrative) 320,000 people; Lausanne (judicial) 114,600 (2003).
Other Cities: Zürich 958,100 people; Geneva 172,900; Basel 163,600 (2000).
Government: Parliamentary republic with strong direct democracy. Moritz Leuenberger, President since January 2006. Micheline Calmy-Rey, Vice President since January 2006. The Federal Assembly (Legislature) has two chambers: the National Council, with 200 members, and the Council of States, with 46 members. The Federal Council (executive collegiate), made of seven members appointed by the United Federal Assembly for a four-year term. The presidency rotates each year among the coalition in power's councillors.
National Holiday: 1 August, Foundation of the Swiss Confederation (1291).
Armed Forces: 3,400 regular troops (1995) 28,000 annual conscripts (15-week courses). Other: 480,000 Civil Defense.

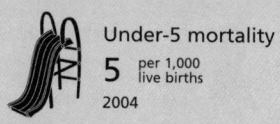
Under-5 mortality
5 per 1,000 live births
2004

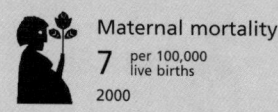
Maternal mortality
7 per 100,000 live births
2000

ENVIRONMENTAL CHALLENGES
There is considerable water pollution due to fertilizer and pesticide use. There is air pollution caused by vehicle emissions, acid rain and loss of biodiversity.

WOMEN'S RIGHTS
Women have been able to vote and stand for office since 1971. In 2003, they held 25 per cent of the seats in the lower house of Parliament and 24 per cent in the upper house, as well as 14 per cent of ministerial posts. That year, women made up 41 per cent of the four million labor force.

Abortion has been legalized; women are able to end pregnancy during the first 12 weeks. The authorities consider that between 1,500 and 3,000 women are victims of human trafficking. They generally come from Latin America and Eastern Europe.

In early 2006 Amnesty International launched a campaign against domestic violence in Switzerland. This was a growing practice that caused the death of at least 40 women per year and against which all efforts had been in vain.

CHILDREN
The public education system and basic health care are provided by the Government. Education is free. Almost all children attend primary school.

Some cases of child abuse have been reported, but it is not a generalized problem. In 2002, laws increasing sentences for this type of crime came into force. An anti-pedophile operation was also carried out: 600 cases have already gone to court, 63 men have been imprisoned and another 163 men have had to pay fines. Some 400 cases are still pending. Cycos is a program that receives complaints by individuals who come across child pornography on the internet; approximately 500 complaints a month are recorded. The police can remove certain material from a site.

INDIGENOUS PEOPLES/ ETHNIC MINORITIES
The canton of Jura is located in the northeast of the country. It was created on separation from a larger canton - Bern - in 1979 in an attempt to attenuate the language and religious differences between the German Protestants who remained in what is now Bern and the Roman Catholic Jurassians, of mainly French origin. There is still some tension between the two groups, as some

Germans remained in the south of Jura. Furthermore, not all the Francophone and Catholic inhabitants were included in the new canton, causing conflict between the cantons of Jura and Bern.

MIGRANTS/REFUGEES
In 2004 the number of immigrants in Switzerland grew 1.6 per cent and reached almost 1.5 million people. The growth was due to the arrival that year of 24,000 new immigrants from EU countries, mostly Germany and Portugal. The number of immigrants that enter the country each year has been growing steadily since 1980. Of the 1.5 million foreigners that lived permanently in Switzerland in December 2005, slightly more than 284,000 were in Zurich, 183,000 in Vaud, 142,000 in Geneva, 116,000 in Bern and 114,000 in Aargau. The rest were distributed between 23 other cities.

DEATH PENALTY
The death penalty was abolished for ordinary crimes in 1942, and the last execution was in 1944. Abolition for any type of crime finally came into force in 1992.

(made up of nine members from Switzerland, UK, US and Israel) which had been established in 1996 by the Swiss Parliament to investigate the country's relationship with the Axis powers, reported that the Swiss authorities had held secret talks with Nazi Germany that helped prolong World War II, that it had refused to give refuge to thousands of Jews - although aware of the existence of concentration camps - and contributed to the expansion of the Nazi economy through the establishment of commercial and financial agreements with Germany.

[49] In 2003 an initiative to reform national legislation on asylum was rejected by a small majority. This initiative would have made the Swiss asylum system one of the most restrictive in the industrialized world.

[50] Amnesty International criticized police treatment of demonstrators near Geneva and Lausanne in June 2003. This led AI to request guarantees for the welfare of demonstrators at the World Economic Forum in Switzerland in January 2004.

[51] Ironically, that same year the 60th UN Human Rights session was held in Geneva.

[52] In the June 2005 referendum, 54 per cent voted in favor of the Schengen Treaty for closer cooperation with the EU in security and asylum issues. Integration into the Schengen area meant that Switzerland accepted eliminating systematic checks of identity documents at its borders. In exchange, the country would have access to an electronic database throughout Europe about people sought and missing, immigrants and illegal property. The right-wing People's Party and the isolationist Campaign for an Independent and Neutral Switzerland had led the opposition to the agreements and gathered enough signatures to force the issue to be put to a referendum. They argued that integration to a passport-free area would cause an inflow of foreign criminals and would compromise Swiss sovereignty.

[53] In March 2006 Amnesty International stated its concern that a large number of intellectuals and members of the Swiss Government justified openly or indirectly the use of torture in the context of the 'war on terror'. Only 147 of the 246 members of Parliament signed a statement against any form of torture.

[54] The Secretary of State for the Economy issued in June statistics showing that unemployment in the country stood at 3.3 per cent, the best rate since 2002. ∎

funds untouched since World War II. Most of these belonged to Jews who had been exterminated by the Nazis. The World Jewish Congress, the main plaintiff, said the presentation was only a symbolic gesture compared with the profit the banks had made by holding the money for 50 years.

[39] A scandal that involved the embezzlement of millions of dollars by a former intelligence officer for organizing a clandestine army led the Government to suspend the military intelligence chief on charges of masterminding the operation. Defense Minister Adolf Ogi was in charge of the investigation.

[40] Under Ogi's leadership, the Democratic Union of the Center (also known as the Swiss People's Party) won 44 of the 200 parliamentary seats in the October 1999 elections. The Social Democratic Party also won 51 seats, the Freethinking Democratic Party won 43 seats, while the Christian Democrat People's Party won 35 seats. The four parties formed the Confederation, which has an annual rotating presidency, held by Ogi beginning in January 2000.

[41] A 1998 government report established that antisemitism had re-emerged in Switzerland due to

the controversy over its relations with Nazi Germany and also over the question of what Swiss banks had done with the accounts of the Holocaust victims. In January 2000, a study revealed that 16 per cent of the Swiss population had antisemitic views, an increase during the previous decade.

[42] On taking office as President in January 2001, Social Democrat Moritz Leuenberger faced serious criticisms from the press and politicians, mainly because of the Government's crack-down on anti-capitalist protesters who surrounded the World Economic Forum meeting of world economic and business leaders in Davos, a winter resort where the conference had been held annually since 1971.

[43] In March 2001, a referendum overwhelmingly rejected the 'Yes to Europe' proposal that had been presented by the Socialist Party and youth groups, with 77 per cent voting against. The governing coalition opposed the plan, saying that negotiations to become an EU member should not begin before the 2003-2007 legislative period.

[44] A new referendum in June approved a measure allowing Swiss soldiers to carry weapons during peace missions abroad. The

electorate also decided that the armed forces could cooperate in military training exercises under NATO. That month, Swiss troops were serving in Kosovo and were themselves protected by Austrian troops because they were not allowed to carry weapons. The Government wanted its army to be able to work on equal footing with the other NATO forces.

[45] In 2001 the Swiss economy was shaken by restructuring measures, causing thousands of layoffs, with unemployment reaching two per cent. In October, Swissair, the country's flagship airline, filed for bankruptcy protection after an expansion plan failed. The Government, banks and several private companies launched a multi-million-dollar rescue package to create a new national airline, arising from Swissair regional subsidiary Crossair. Kaspar Villiger, of the Freethinking Democratic Party, took office as President.

[46] In 2002, 72 per cent of the population voted in favor of legalizing abortion.

[47] That same year, 55 per cent voted to join the UN and the country became the 190th member of the organization.

[48] The Bergier Commission

Syria / Suriyah

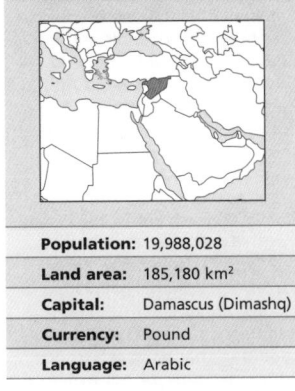

Population:	19,988,028
Land area:	185,180 km²
Capital:	Damascus (Dimashq)
Currency:	Pound
Language:	Arabic

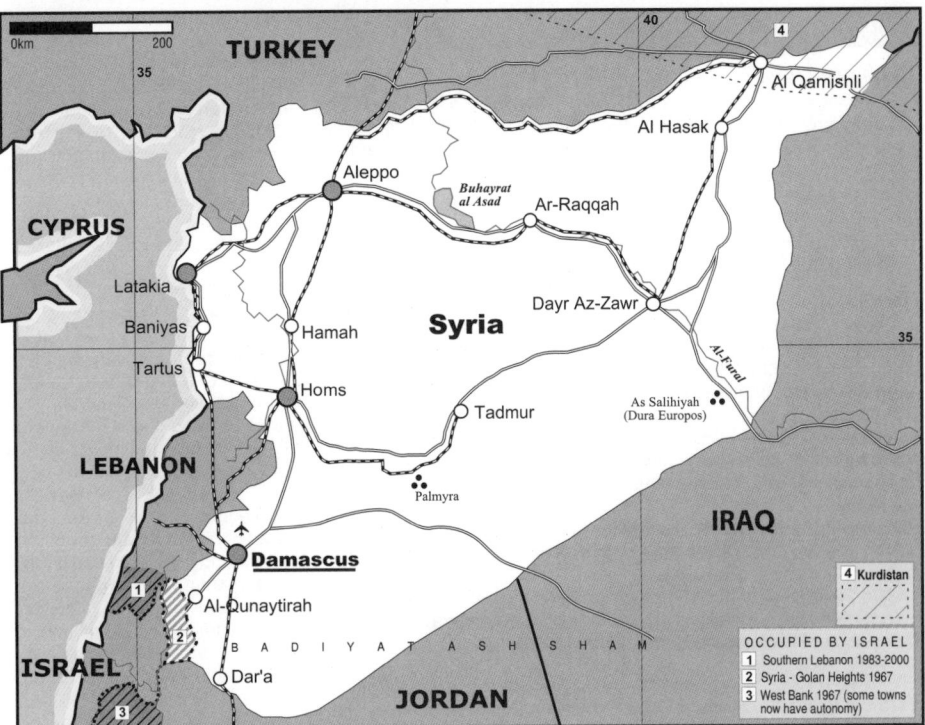

S yria was once the name for the entire region between the peninsulas of Anatolia (Turkey) and Sinai, including the fertile crescent. Ancient civilizations coveted the territory: the Egyptians wanted it as a port, while the Persians considered the region a bridge to their plans for a universal empire.

[2] Between the 12th and 7th centuries BC, the Canaanite civilization - known as Phoenicians by the Greeks - developed on the central coastal stretch of the territory; a society of sailors and traders without expansionist aims. Phoenician cities were always independent, although some exercised temporary hegemony over others, and they developed the world's first commercial economy.

[3] The Canaanites invented the alphabet - the first lineal system of writing dates back to 1600 BC, was found in 1928 in Ugarit, on the Mediterranean coast of present Syria, and has 30 signs. They also constructed ocean-going ships, practised large-scale ceramic and textile manufacturing, expanded and systematized geography, sailing around the coast of Africa. Their propagation throughout the Mediterranean helped form what would later be called 'Western civilization', of which the Greeks were the main exponents.

[4] After the death of Alexander the Great in 323 BC, his vast empire was divided and Syria became the center of a Seleucidan state (named for Seleucus Nicator, one of Alexander's generals) that initially stretched as far as India. The eastern part was later lost to the Parthians. In the Roman era the province of Syria was a border zone constantly shaken by fierce local wars.

[5] The Arabization of the territory was carried out by the Ummaia Caliphs, who, between the years 660 and 750, turned Damascus (Dimashq) into the capital of the empire (see Saudi Arabia), and fostered a strong national spirit. When the Abbas defeated the Ummaias, the capital was transferred to Baghdad, where the new caliphs enjoyed greater support. Although still economically and culturally important, the loss of political power proved significant in the 11th century. When Europeans invaded during the Crusades, the caliphs of Baghdad reacted with indifference. Local emirs were left to their own resources, and disagreements among them allowed a small Christian force to conquer the area, leading to 200 years of occupation.

[6] In the 13th century, the Egyptians initiated the process of driving out the Europeans. One result was that Syria became a virtual Egyptian province and center stage for a confrontation with Mongol invaders. In the 16th century, the country became a part of the Ottoman Empire.

[7] The Crusaders left behind a significant Christian community, especially with the Maronites, serving as sufficient cause for European interference from the 17th century onwards. The Egyptian Khedive Muhammad (Mehemet) Ali conquered Syria in 1831, and heavy taxes and compulsory military service provoked revolt among both Christian and Muslim communities. The European powers used the repression of Christians as an excuse for intervention. Ali's offensive was suppressed and the 'protection of Syrian Christians' was entrusted to the French. A withdrawal of Egyptian troops took place in 1840, along with the restoration of Ottoman domination and the establishment of Christian missions and schools subsidized by Europeans.

[8] In 1858, Maronite Christians gathered in the mountainous region between Damascus and Jerusalem rebelled against the ruling class and eliminated the traditional system of land ownership. Their Muslim neighbors, particularly the Druze, moved to repress the movement before it spread further, triggering a conflict that culminated in the deaths of a large number of Christians in June 1860.

[9] A month later, French troops disembarked in Beirut and forced the Turkish Government to create a separate province called 'Little Lebanon'. This was to be governed by a Christian appointed by the Sultan but with the approval of the European powers, with its own police force, and traditional privileges were abolished in the territory. The social conflict thus became a confrontation between confessional groups with the Christians in 'Little Lebanon' placed in a position of superiority over the local Muslim population.

[10] When an Arab rebellion broke out during World War I (see Saudi Arabia, Jordan and Iraq), Emir Faisal was proclaimed king

LAND USE

2003/2005

IRRIGATED AREA: 24.6% of arable land

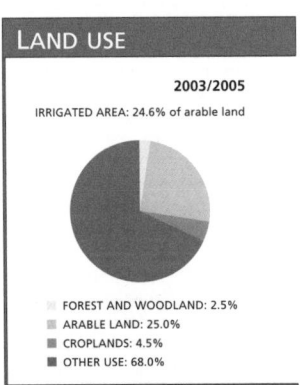

- FOREST AND WOODLAND: 2.5%
- ARABLE LAND: 25.0%
- CROPLANDS: 4.5%
- OTHER USE: 68.0%

PUBLIC EXPENDITURE

% OF GDP

	HEALTH (2002)	DEFENSE (2004)
	2.3 %	7.0 %

WORKERS

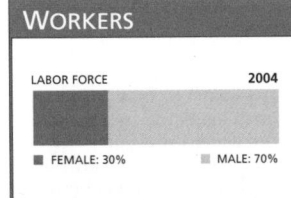

LABOR FORCE   2004

- FEMALE: 30%
- MALE: 70%

PROFILE

ENVIRONMENT

To the west, near the sea, lies the Lebanon mountain range. To the south there are semi-desert plateaus and to the north low plateaus along the basin of the Euphrates River. Farming - grains, grapes and fruit - is concentrated in the western lowlands that receive adequate rainfall. In the south, the volcanic plateaus of the Djebel Druze are extremely fertile farmlands, as are some of the oases surrounding the desert, the main one being that of Damascus. Cotton and wool are exported. Oil is, however, the country's chief industry.

SOCIETY

Peoples: Syrians are mostly Arabs, with minority ethnic groups in the north: Kurds, Turks and Armenians. At the end of 2002, more than 400,000 Palestinian refugees living in the country were registered with UN agencies. The Jewish population was authorized to emigrate in 1992.
Religions: Mainly Muslims, mostly Sunni, followed by Alamites, Shi'a and Ismailites. Also minor communities of eastern Christian religions.
Languages: Arabic (official). Minority groups speak their own languages.
Main Political Parties: National Progressive Front, which includes: Arab Socialist Ba'ath Party; Arab Socialist Movement; Arab Socialist Union.
Main Social Organizations: The General Federation of Labor Unions unites ten workers' federations; Committee for the Defense of Democratic Freedoms and Human Rights in Syria.

THE STATE

Official Name: Al-Jumhouriya al Arabiya as-Suriyah.
Administrative divisions: 14 Districts.
Capital: Damascus (Dimashq) 2,228,000 people (2003).
Other Cities: Aleppo (Halab) 2,319,800 people; Homs 698,800; Latakia 391,300; Hamah 350,900 (2000).
Government: Bashar al-Assad, President since July 2000. Muhammad Naji al-Otari, Prime Minister since September 2003. Unicameral Legislature: People's Assembly, with 250 members.
National Holiday: 17 April, Independence Day (1946); 16 November, Revolution Day (1978); 25 May, Resistance and Liberation Day (2000).
Armed Forces: 297,000 troops (2003). Other: 8,000 Gendarmes.

of Syria. At the time French and British intentions were unknown, but the Sykes-Picot agreement divided the fertile crescent giving Syria (with Lebanon) to France, and Palestine (including Jordan) and Iraq to Britain.

[11] In 1920, France occupied Syria forcing Faisal to retreat. Two months later, Syria was divided into five states: Greater Lebanon (adding other regions to the province of 'Little Lebanon'), Damascus, Aleppo, Djabal Druza and Alawis (Latakia). The latter four were reunified in 1924.

[12] Throughout 1932, Syria experienced a period of relative stability. A Syrian President and Parliament were elected that year, but France made it clear that autonomy was unacceptable. This attitude engendered political agitation and confrontation, which only ended with a 1936 agreement with the French. France recognized certain Syrian demands, chiefly reunification with Lebanon. However, the French Government never ratified the agreement, and this led to new waves of violence, which culminated in the 1939

resignation of the Syrian President and a French order to suspend the 1930 Constitution that governed both Syria and Lebanon.

[13] In 1941, French and British troops occupied the region to flush out Nazi collaborators. In 1943 Chikri al-Quwatli was elected President of Syria and Bechara al-Kuri President of Lebanon. Bechara al-Kuri proposed elimination of the mandate provisions from the constitution; however, he and his cabinet members were imprisoned by the ever-present French troops. Violent demonstrations followed in both Lebanon and Syria, and the British pressed for withdrawal of the French. In March 1946, the UN finally ordered the European forces to withdraw and the end of the French mandate was declared.

[14] In 1948 Syrian troops fought to prevent the partition of Palestine, and in 1956 joined Egypt in the battle against Israeli, French and British aggression. This aggression was the answer to Egyptian President Gamal Abdel Nasser's decision to nationalize

the Suez Canal.

[15] In 1958, Syria joined Egypt in founding the United Arab Republic, but Nasser's ambitious integration project collapsed in 1961. Ten years later the scheme was reactivated with greater flexibility, and the Federation of Arab Republics was created including Libya.

[16] In 1963, after a revolution, the Ba'ath Arab Socialist Party, founded in 1947 by Christian leader Michel Aflaq, came to power. Its main tenet was that the Arab countries were merely 'regions' of a larger Arab Nation. In November 1970, General Hafez al-Assad became President. He launched a modernization campaign, including a series of social and economic changes. The subsequent party congress named Assad party leader and proposed 'accelerating the stages towards socialist transformation of different sectors'. This guideline was adopted and became part of the new constitution which was approved in 1973.

[17] Syria took an active part in the Arab-Israeli wars of 1967 and 1973, during which Israeli troops occupied the Golan Heights. Syria also resisted US efforts to impose a 'settlement' in the Middle East, together with Algeria, Iraq, Libya, Yemen and the Palestine Liberation Organization (PLO). They also opposed the Camp David agreement (see Egypt). Syrian troops formed a major part of the Arab Deterrent Force that intervened in Lebanon in 1976 to prevent partition of the country.

[18] In 1978, the Syrian and Iraqi branches of the Ba'ath Party drew closer, but negotiations for creation of a single state disintegrated. In late 1979, the Syrian branch censured the Muslim Brotherhood (an Islamic movement), labeling their members 'Zionist agents'.

[19] In 1982 the army launched an offensive; thousands of Brotherhood members were killed and the Syrian Government blamed Iraq for arming the rebels. In April, the border between the two countries was closed.

[20] The virtual alliance formed in 1980 between Saudi Arabia, Iraq and Jordan and tensions between those three countries and Syria were exacerbated by the outbreak of the Iran-Iraq War. Assad charged Iraq with being the aggressor and diverting attention from what he called the major regional issue - the Palestinian question. Toward the end of the year, Syrian accusations of Jordanian support for the Brotherhood brought the two countries to the verge of war.-

[21] In 1981 the 'missile crisis'

broke out in Syria, when the Christian Phalangist Movement sought to extend their area of authority to include the area around the Lebanese city of Zahde. An Arab Deterrent Force, led by Syria, attempted to prevent the advance. Syria installed Soviet missiles, triggering an Israeli reaction. The crisis was finally averted, but in 1983 Israel invaded Lebanon, and destroyed the Syrian missile bases. Syrian forces (aproximately 30,000 troops) remained in Lebanon and only agreed to retreat on condition that all Israeli troops were previously withdrawn.

[22] The fall in oil prices further aggravated the economic problems caused by the war, which forced the Government to set up strict austerity measures in 1984.

[23] In 1985, President al-Assad won a new seven-year term with 99.8 per cent of the vote (a similar percentage to those of the 1971 and 1978 elections). In spite of this, in 1987 a political crisis broke out which forced Prime Minister Abdul Rauf al-Kassem to resign amid charges of corruption.

[24] In May 1990, Syria finally re-established diplomatic relations with Egypt. Some observers considered this a result of a reduction in USSR military support to Damascus.

[25] When Iraq invaded Kuwait, Syria immediately sided with the anti-Iraqi alliance and sent troops to Saudi Arabia. Diplomatic relations with the US improved noticeably. During the crisis Syria increased its influence over Lebanon and strengthened the allied government in that country; they were also successful in disarming most of the autonomous militias.

[26] In May 1991 Syria and Lebanon signed a cooperation agreement whereby Syria recognized Lebanon as an independent and separate state, for the first time since both countries gained independence from France.

[27] On 2 December 1991, al-Assad was re-elected for the fourth time, by 99.98 per cent of the vote, in elections in which he was the sole candidate.

[28] Syria stayed away from the first stages of the regional peace process, which facilitated the establishment of a limited autonomy for Palestine and the signing of agreements between Israel and Jordan in July 1994.

[29] In June 1995, official negotiations with Israel failed to return the Golan Heights to Syria. In October, a Hizbullah ambush of Israeli troops in southern Lebanon complicated negotiations again.

Under-5 mortality
16 per 1,000 live births
2004

Malnutrition
7% under-5s
1996-2004

Debt service
3.5% exports of goods and services
2004

Maternal mortality
160 per 100,000 live births
2000

30 As part of a policy to stimulate the Syrian private sector economy, key state sectors, including electricity generation, cement production and pharmaceuticals were opened up to private capital.

31 In November 1997, Damascus unexpectedly strengthened relations with Baghdad at the threat of fresh US military intervention in Iraq - a strategy designed to work against the rapidly consolidating Turkish-Israeli alliance. Iran joined the Syrian-Iraqi negotiations on security issues in April 1998.

32 Al-Assad was re-elected for his seventh consecutive five-year presidential period in 1999. In March 2000, all 37 cabinet members resigned and Muhammad Mustafa Mero, a veteran leader of the Ba'ath party, was appointed the new Premier.

33 The sudden death of al-Assad on 10 June plunged the country into mourning for the only leader most Syrians had ever known. He was succeeded by his son, Bashar al-Assad, who took office in July.

34 Among the first measures of the new President was the April 2001 authorization to establish a private banking system and shortly after, a private radio station was granted broadcast rights for music only, no political content was allowed.

35 In May, Pope John Paul II visited Syria, and at the reception ceremony al-Assad strongly criticized Israel, comparing the suffering of the Arabs to the persecution of Christ. In response, John Paul II called upon all parties in the conflict to seek a new understanding, including respect among Christians, Muslims and Jews.

36 With the unanimous support of Asian and African countries, Syria obtained a seat at the UN Security Council in October 2001, in spite of Israel's opposition.

37 Damascus engaged intensely in international relations in 2001. After heavy pressure from the Lebanese Government, Syrian troops withdrew from Beirut and were deployed in other parts of Lebanon. In August, Syria's Premier Miro visited Iraq in the first high-level visit to that country since relations had cooled as a result of Syrian support for Iran during the 1980-1988 Iran-Iraq war.

38 The November release of dozens of political prisoners belonging to the Muslim Brotherhood, after more than two decades behind bars, was applauded by Amnesty International as a 'satisfactory step towards respect for human rights in Syria'. Nearly all the prisoners had been kept

IN FOCUS

ENVIRONMENTAL CHALLENGES
The dumping of toxic substances from oil refineries pollutes Syria's water and threatens the scarce drinking water resources. Overgrazing, desertification and soil erosion pose further environmental problems which affect large regions of the country.

WOMEN'S RIGHTS
Women have been able to vote and stand for election, subject to conditions and restrictions, since 1953. In 2003, they held 10 per cent of seats in Parliament and 6 per cent of ministerial posts. Women make up 28 per cent of the labor force of 6 million people. Domestic violence occurs but there are no figures; very few cases are reported and victims are reluctant to seek assistance outside the family. There are a few private shelters for battered women. Rape is a felony; however, there are no laws against spousal rape. The punishment for adultery by a woman is twice as hard as for a man committing the same crime. Polygamy is legal. In 2004, 71 per cent of pregnant women received prenatal care and 77 per cent of pregnancies were assisted by qualified staff.

CHILDREN
The infant and under-five mortality rates fell sharply in the 1990-2004 period. The former dropped from 35 to 15 per 1,000 live births and the latter from 44 to 16 per 1,000. Six per cent of children had low birth weight and 18 per cent of under-fives had moderate or serious stunting. School net enrolment rates are high; 96 per cent for girls and 100 per cent for boys*. However, geographic disparities persist. The Government launched plans to reduce the number of girls who drop out of school - and to reincorporate those who had already dropped out. The law provides for severe penalties for those found guilty of abuses against children.

INDIGENOUS PEOPLES/ ETHNIC MINORITIES
Alawis gave themselves this name which means 'those who adhere to the teachings of Ali', the son-in-law of the Prophet Muhammad. They were formerly called the *Nusayris*, a name that accentuates their differences from traditional Islam practices. This name is still used by those who are unsympathetic to them.

Kurds are the largest minority group. They are prevented from fully enjoying their own culture and language. Their freedom of expression is restricted and they allegedly suffer violations of their basic human rights.

MIGRANTS/REFUGEES
Syria has not signed the 1951 Refugees Convention, nor the 1961 Protocol, and does not officially recognize refugee status as defined by the UNHCR. However, it was estimated that in 2005 hundreds of thousands of Iraqi refugees were in the country being offered temporary protection after the situation in Iraq deteriorated in 2004. The international community showed no interest in these refugees, many of whom had to resort to prostitution and child labor to subsist. There were, in addition, 420,000 Palestinian refugees and a high number of asylum seekers from Sudan, Somalia, Afghanistan and several North African countries.

DEATH PENALTY
Syria maintains the death penalty as legal punishment, even for ordinary crimes.

* Latest data available in *The State of the World's Children* and *Childinfo* database, UNICEF, 2006.

incommunicado in degrading conditions and subjected to torture and mistreatment.

39 In April 2002, the Syrian radar station in Lebanon was bombed by Israeli aircraft in response to an attack by Hizbullah guerrillas. The offensive raised fear of a military escalation which finally did not take place.

40 In May, US senior official John Bolton included Syria in a list of states that made up the so-called 'axis of evil', accusing Damascus of trying to obtain weapons of mass destruction. In April 2003, when the Iraq invasion was already in progress, Washington threatened Syria with economic and diplomatic sanctions, alleging that it was helping fugitive Iraqis. The Syrian Government denied US allegations.

41 In January 2004, al-Assad became the first Syrian leader to visit Turkey in a trip that marked the end of frosty relations with Ankara.

42 On 8 March, the Committee for the Defense of Democratic Freedoms and Human Rights in Syria organized a rare protest in Damascus to demand democracy and freedom for political prisoners. Two members of the

organization, Ahmad Jazen and Hassan Wattfa, were arrested and spent two months in prison.

43 In April 2004, there was an explosion in a disused UN building in Damascus; in the subsequent shooting, one civilian, one police officer and two of the four activists involved were killed. The Government blamed the attack on Islamic fundamentalists.

44 In May, the US imposed economic sanctions on Syria over what it called Syria's support for terrorism and failure to stop militants entering Iraq.

45 After former Lebanese premier Rafik Hariri was assassinated in Beirut in February 2005, there was growing pressure from the US, France, the UN and Lebanese opposition for Syrian troops and intelligence agents to withdraw from Lebanon immediately. Al-Assad, in a summit with Lebanese President Emile Lahoud, agreed on a partial schedule for all Syrian troops to leave Lebanon before that country's general elections in May.

46 In early February 2006 Syrian demonstrators set fire to the building hosting the embassies of Denmark and Norway in Damascus, during a protest

against cartoons satirizing the Prophet Muhammad which had appeared in a Danish daily. The embassies of Chile and Sweden, located in the same building, suffered minor damage. A week later, Denmark closed its embassy in the country and accused Syrian authorities of not having guaranteed a minimum level of security for Danish employees.

47 In June, a clash between Syrian security forces and ten Islamic militants, members of a *takfiri* group (an extremist Sunni ideology) near the Defense ministry in Damascus left four militants and one police officer dead.

48 Analysts said that horror at the situation in Iraq had accomplished what President Assad's regime had previously been unable to do, by silencing public demands for democratic reforms in Syria. Iraq's descent into sectarian warfare had, they said, reinforced the case of traditionalists and bureaucrats that the stability offered by the Assad regime was invaluable.

49 In October 2006, Syria announced that Washington's sanctions against the country had prompted Damascus to begin switching its foreign currency surplus from US dollars to euros. ■

Tajikistan / Tojikiston

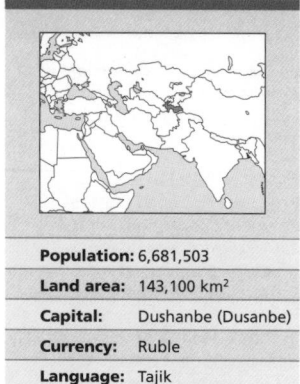

Population:	6,681,503
Land area:	143,100 km²
Capital:	Dushanbe (Dusanbe)
Currency:	Ruble
Language:	Tajik

A round 500 BC, Central Asia and southern Siberia's first centers of settled civilization arose in what is now Tajikistan. The Bactrian State was located in the upper tributaries of the Amu Darya. In the Zeravshan river basin and the Kashkadarya river valley lay the nucleus of another State, Sogdiana. The inhabitants of these early civilizations built villages, with adobe and stone houses all along the rivers that they used for irrigating their crops. These included wheat, barley and millet, and a variety of fruits. Navigation was well developed and the cities that lay along the route of the caravans uniting Persia, China and India became important trading centers.

[2]   In the 6th century BC these lands were annexed by the Persian Achaemenid Empire. In the 4th century BC, Alexander the Great conquered Bactria and Sogdiana. With the fall of his empire in the 3rd century, the Greco-Bactrian State and the Kingdom of Kushan

emerged, and subsequently fell to the onslaught of the advancing Yuechzhi and Tojar steppe tribes. In the 4th and 5th centuries AD, Sogdiana was invaded by the Eftalites, and in the 6th and 7th centuries, by Central Asian Turkish peoples. In the 7th century, Tajikistan came under the control of the Arab Caliphate. After its demise, the region was incorporated into the Tahirid and Samanid kingdoms. In the 9th and 10th centuries, the Tajik people emerged as an identifiable ethnic group.

[3]   From the 10th to the 13th centuries AD, Tajikistan formed part of the Gaznevid and Qarakhanid empires, as well as the realm of the Shah of Khwarezm. In the early 13th century, Tajikistan was conquered by Genghis Khan's Mongol Tatars. In 1238, Tarabi, a Tajik artisan, led a popular revolt. From the 14th to the 17th centuries, the Tajiks were under the control of the Timurids and the Uzbek Shaybanid dynasty. From the 17th to the 19th centuries, the land was

divided into small fiefdoms whose chieftains alternately revolted against the khans of Bukhara.

[4]   In the 1860s and 1870s, the Russian Empire conquered Central Asia, and annexed the northern part of Tajikistan. The Tajik population of Kuliab, Guissar, Karateguin and Darvaz became a province (Eastern Bukhara) of the Bukhara Khanate. Oppression by the Russian bureaucracy and the local feudal lords triggered a wave of peasant revolts toward the end of the 19th century and beginning of the 20th century, the most important of which was the 1885 uprising led by Vose.

[5]   In 1916, during World War I, the population of Central Asia and Kazakhstan revolted over the mobilization of their people for rearguard duty with the Russian army. After the triumph of the Bolshevik Revolution in October 1917, Soviet power was established in Northern Tajikistan. In April 1918, this territory became a part of the Soviet Republic of Turkistan. Nevertheless, a large number of

Tajiks remained under the power of the Emirate of Bukhara, which existed until 1921. In early 1921, the Red Army took Dushanbe, but in February it was forced to withdraw from Eastern Bukhara.

[6]   Alim Khan's resistance was broken in 1922 and Soviet power was proclaimed throughout Tajikistan. On 16 November 1929 Tajikistan became a federal republic of the Soviet Union.

[7]   After World War II, the Soviet regime carried out a series of large construction projects, including a water system linked to the neighboring republic of Uzbekistan, meant to develop the region's cotton crops. In the 1970s and 1980s, the effects of mismanagement and economic stagnation were felt in Tajikistan, one of the poorest regions in the USSR. It suffered a high rate of unemployment, especially among the young, who made up most of the population.

[8]   From 1985, the changes promoted by President Mikhail Gorbachev gave way to the expression of long-suppressed ethnic and religious friction in Tajikistan. In February 1990, there were violent incidents in the capital, with more than 30 people killed. The Government decreed a state of emergency, which remained in effect during that year's elections to the Supreme Soviet (Parliament) in which the Communist Party won 90 per cent of the seats at stake.

[9]   In August 1990, Parliament passed a vote of no confidence against President Majkamov, accusing him of supporting those responsible for the Moscow coup, and forcing him to resign. He was succeeded by Kadridin Aslonov as interim president of the republic.

[10]   In September 1991, Parliament passed the declaration of independence and the new Constitution, decreed a state of emergency and banned the Islamic Renaissance Party, which pledged a State respectful of political and religious freedoms, but based on Islam, and advocated the enforcement of the Sharia, the religious, moral and legal Islamic code.

[11]   Tajikistan's religious revival was stronger than in the rest of the former Soviet Union republics. In the early 1980s, there were 12 mosques; ten years later there were 128 mosques, 2,800 places of prayer, an Islamic institute and five centers offering religious instruction. From 1991, the Government instituted the celebration of several important Muslim holy days.

[12]   In November 1991, in the first presidential elections Rajmon Nabiev was confirmed in his post by 58 per cent of the vote.

[13]   On December 21 Tajikistan

PROFILE

ENVIRONMENT

The Tian Shan, Guissaro-Alai and Pamir mountains occupy more than 90 per cent of Tajikistan's territory. This republic is located in the southeastern part of Central Asia between the Syr Darya River and the Fergana Valley in the north, the Pamir and Paropaniz mountain ranges in the south, the Karakul lake and the headwaters of the Murgab in the east, and the Guissar and Vakhsh valleys in the southwest. The Turkistan, Alai and Zeravshan mountains cross Tajikistan from north to south, joining the Pamir plateau. The highly cultivated valleys (lying at an altitude of 1,000-2,000 meters) have a warm, humid climate. Lower mountains and valleys, located in northern and southeastern Tajikistan, have an arid climate. It is bounded by Kyrgyzstan in the north, Uzbekistan in the west, China in the southeast and Afghanistan in the southwest. Abundant mineral resources include iron, lead, zinc, antimony and mercury, as well as important uranium deposits.

SOCIETY

Peoples: Tajik 64.9 per cent; Uzbek 25 per cent; Russian 3.5 per cent (diminishing due to emigration); Tatar 1.4 per cent; Kyrgyz 1.3 per cent; Ukrainian 0.7

per cent; German 0.3 per cent; other 2.0 per cent.
Religions: there is no state religion. Most are Muslim (Sunni 85 per cent; Shi'a 5 per cent); Russian Orthodox (1.5 per cent); Jews (0.8 per cent).
Languages: Tajik (official), Uzbek, Russian.
Main Political Parties: People's Democratic Party of Tajikistan; Communist Party; Islamic Revival Party.

THE STATE

Official Name: Jumhuri Tochikiston.
Administrative Divisions: 3 Regions, 57 Districts.
Capital: Dushanbe (Dusanbe) 523,000 people (1999).
Other Cities: Khujand (formerly Leninabad) 205,200 people; Kulob 97,100; Qurgonteppa 76,200; Uroteppa 60,700 (2000).
Government: Imomali Rakhmonov, President since November 1992, re-elected in 1999. Akil Akilov, Prime Minister since December 1999. Bicameral Legislature: Assembly of Representatives, with 63 members, and the National Assembly, with 33 members.
National Holiday: 9 September, Independence (1991).
Armed Forces: 8,000 (2003).

| Life expectancy **64** years 2005-2010 | GNI per capita **$280** 2004 | Literacy **99%** total adult rate 2000-2004 | HIV prevalence rate **<0.1%** of population 15-49 years old 2003 |

entered the Commonwealth of Independent States (CIS). At the same time, the autonomous region of Gorno-Badakhshan requested the status of an autonomous province. Like neighboring Afghanistan, the majority of this region's inhabitants are Shi'a Muslims.

[14] In March 1992 large anti-government protests erupted in Dushanbe. These spread outside the capital in April and took on civil war proportions. In this context, a protest demonstration against Nabiev led to his resignation on September 8. But pro-communist forces in the north launched a major offensive, leading to the fall of the capital and the formation, in December, of a new government controlled by the Popular Front (a paramilitary group). Some 300,000 people fled to other CIS republics and Afghanistan. The persecution and massacre of members of the opposition only diminished in February 1993.

[15] Moscow recognized the new regime, led by Imomali Rakhmonov. In March bombing started from Afghanistan and there were incursions of opposition detachments across the frontier controlled by Russian troops. Rakhmonov proposed a draft for a new Constitution, approved by referendum in November. The President was re-elected at the same time, but the Fundamental Islamic opposition accused the government of fraud.

[16] The civil war officially ended on 27 June 1997, with the signing of a peace agreement between the Government and the opposition - grouped together in the United Tajik Opposition (UTO). The agreement guaranteed the opposition 30 per cent of posts in the cabinet, presence in the legal authorities and an amnesty for all those accused of war crimes. The conflict left 20,000 dead and 600,000 internally displaced while another 300,000 fled to Afghanistan, Russia and other CIS countries.

[17] In February 1998 the Government announced that it would speed up privatization programs.

[18] Akbar Turayonzoda - second in command in the UTO - returned from exile in Iran to become the first deputy minister. He proposed allowing the IRP to join political life with full rights, including participation in elections. Parliament passed a law banning the creation of parties based on religious movements, and while the president vetoed the law, he made it clear he would not allow an Islamic government in the country.

[19] The elections slated for November 1999 went ahead despite strong protests from Usmon, the only opposition candidate. Other candidates could not raise the

ENVIRONMENTAL CHALLENGES
Since 1960, the area of irrigated land has increased by 50 per cent, but as in other countries where cotton is the only crop, soil salinity has also increased. Pesticides are overused for farming and industrial pollution affects the air and water. Sewage and drinking water supply systems are inadequate.

WOMEN'S RIGHTS
Women have been able to vote and stand for election since 1924. In 2005, 18 per cent of seats in the Lower Chamber and 24 per cent in the Higher Chamber were held by women. Female workers made up 45 per cent of the three-million-strong labor force. Violence against women is common. The kidnapping of young women who are forced to have sex with, and even marry the kidnappers, is frequent. Trafficking of women for sexual exploitation is growing. In 2004, 71 per cent of pregnant women received prenatal care and the same percentage of births were attended by qualified personnel. Maternal mortality was 100 per 100,000 live births.

CHILDREN
Infant and under-five mortality decreased slightly between 1990 and 2004. The former fell from 99 to 91 and the latter from 128 to 118 per 1,000 live births. Funding for child social support systems continued to be inadequate. State schools were run down and parents preferred to send their children to private schools or hire private teachers.

Health care is no longer free, due to legislation changes. The quality and quantity of medical services is extremely limited.

Some 15 per cent of babies are underweight at birth and 36 per cent of under-fives measure moderately or seriously below height norms*. Widespread poverty has contributed to an increase in child trafficking.

INDIGENOUS PEOPLES/ ETHNIC MINORITIES
The largest ethnic minorities are Uzbeks and Russians. Uzbeks and Tajiks have been considered separate nationalities only since 1929, and this differentiation was consolidated in 1991, with the fall of the Soviet Union. Uzbeks are concentrated to the north of the capital, in the Ferghana valley. There are other Uzbek communities in the province of Khatlon, a rural region to the southwest which is one of the poorest in the country. They suffer severe social discrimination and some official restrictions. Their language and culture are considered marginal and are not adequately represented in the political system. Uzbeks do not usually have access to high political posts.

After the peace treaty that ended the civil war was signed in 1997, killings of Uzbeks continued in the Pani district. While the Government has tried to improve relations between Tajiks and Uzbeks, these murders have not been taken to court.

Russians are not oppressed by the Government and are not politically organized. Although they maintain a group identity, this has not translated into political action.

MIGRANTS/REFUGEES
In mid-2005 attempts were being made to resolve the situation of some 12,000 Tajik refugees living in Turkmenistan. At the same time a successful resettlement program enabled more than 1,300 Afghan refugees living in Tajikistan to be accommodated in Canada.

DEATH PENALTY
It is applicable even for ordinary offenses. In July 2003 Parliament passed a law banning capital punishment for women and reducing the number of applicable cases for men. In April 2004, the President declared a moratorium on the death penalty.

*Latest data available in *The State of the World's Children* and *Childinfo* database, UNICEF, 2006.

number of signatures required for them to run for office, and the opposition blamed this on government pressure. Rakhmonov took 96 per cent of the vote.

[20] In September 2001, Tajikistan's culture minister, Abdurakhim Rakhimov, was assassinated in Dushanbe. Habib Sanginov, deputy Minister of the Interior, had also been killed in April.

[21] That month, following the terrorist attacks in New York and Washington, Tajikistan offered its support for the anti-terrorism coalition led by the US. In addition, Rakhmonov closed the border with Afghanistan to prevent terrorist infiltration. In February 2002, Tajikistan became the last former soviet republic to join NATO.

[22] In a referendum held in June, 90 per cent of voters approved 50 amendments to the Constitution. Of these, the most controversial allowed Rakhomonov to stand for re-election.

[23] In July 2003, the deputy president of the Islamic Revival Party (IRP), Shamsiddin Shamsiddinov, was sentenced to 16 years in prison for organizing crime rings and other serious crimes. His party claimed it was a political arrest, rather than religious. Concerned about fundamentalism, Rakhmonov jailed more than 200 members of the radical Hizb ut-Tahrir in 2003, and confiscated tons of texts which promoted the creation of an Islamic caliphate in Central Asia.

[24] In October 2004, Russia formally established a military base in the Tajik mountains, replacing the one that existed in the Soviet era, and took over the air traffic control center. Russian President Vladimir Putin visited Tajikistan to meet Rakhmonov and mark the re-launching of Russian military presence in the country. Two months later the leader of the opposition Democratic Party, Mahmadruzi Iskandarov, was arrested in Moscow. The Tajik public prosecutor's office accused him of being involved in terrorism, dealing in illegal arms and corruption. However, Iskandarov's supporters claimed that this detention was a political ploy.

[25] In the February 2005 parliamentary elections - which according to observers lacked an acceptable level of transparency - the governing party won a landslide victory. Iskandarov was freed in Moscow in April, but was later arrested again in Tajikistan.

[26] In June, Russia withdrew all its troops from the border with Tajikistan, leaving frontier security exclusively under the control of Tajik forces. In October, Iskandarov was found guilty of leading a terrorist organization and was sentenced to 23 years in jail. The opposition again claimed that there was a political motive behind the court's decision.

[27] In May 2006 a heavily armed group attacked and killed three border guards when crossing from Tajikistan to Kyrgyzstan. The Batken region, where the borders of these two countries and Uzbekistan meet, has been very unstable since the fall of the Soviet Union and several Islamic fundamentalist groups are based there.

[28] In the November election President Rakhmonov defeated his four obscure opponents by a wide margin following an election campaign that utterly failed to enthuse voters, especially the young. ∎

Tanzania / Tanzania

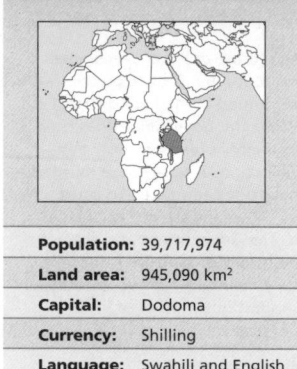

Population:	39,717,974
Land area:	945,090 km²
Capital:	Dodoma
Currency:	Shilling
Language:	Swahili and English

The oldest human fossils were found in the Olduvai gorge, in the north of Tanzania. These remains date from millions of years ago, yet little is known about life on most of modern Tanzania's mainland before the seventh century.

2   A mercantile civilization, heavily influenced by Arab culture, flourished in this region between 695 and 1550 but was eventually destroyed by Portuguese invaders. Some 150 years later, under the leadership of the Sultan of Oman, the Arabs drove out the Portuguese, but the rich cultural and commercial life of past times did not return. As the slave trade grew, Kilwa and Zanzibar became the leading trade centers.

3   Between 1698 and 1830, Zanzibar and the coast were under the rule of the Oman Sultanate, with the Sultan living in Zanzibar. His sons, under British pressure, later split the inherited domain, dividing the old Sultanate in two.

4   In the late 19th century a German adventurer set up a company which immediately received imperial endorsement, leasing out the mainland coastal strip to the Sultan of Zanzibar. The British had also made a similar deal, so the Congress of Berlin, where the European powers distributed Africa among

themselves, agreed upon the 'cession of German rights' in favor of Britain.

5   The areas of influence were delimited in 1886. Tanganyika, Rwanda and Burundi were recognized as German possessions while Zanzibar formally became a British protectorate in 1890. German troops and British warships joined efforts to stifle a Muslim rebellion on the coast of Tanganyika in 1905.

6   After Germany's defeat in World War I, the League of Nations placed Tanganyika under British mandate. Resistance came from traditional chiefs and also from those opposed both to the British and indigenous authorities. The Tanganyika African Association (TAA) was established in 1929 as a forum for trade unionists and co-operative

farmers who opposed British rule. In 1951 the British rulers began to implement a program to remove African homesteaders to make way for post-war British settlers, a policy that swelled an incipient nationalist movement.

7   Nationalist feelings were later channeled into TANU (Tanganyika African National Union), a party founded in 1954 by Julius Nyerere, a primary school teacher known by the people as *Mwalimu*, the teacher.

8   After seven years of organizing and fighting against racial discrimination and the appropriation of lands by European settlers, independence was achieved in 1961, and Nyerere became President, elected by an overwhelming majority.

9   Meanwhile, in Zanzibar, two nationalist organizations, which had been active since the 1930s, merged to form the Afro-Shirazi Party in February 1957. In December 1963, the British transferred power to the Arab minority, and a month later this

government was overthrown by the Afro-Shirazi. In April, Tanganyika and Zanzibar formed the United Republic of Tanzania.

10   Under Nyerere's leadership, Tanzania based its foreign policy on non-alignment, standing for African unity, and providing unconditional support to liberation movements, particularly FRELIMO in neighboring Mozambique.

11   In February 1967, TANU proclaimed socialism as its objective in the Arusha Declaration, which laid down the principle of self-sufficiency and gave top priority to the development of agriculture, on the basis of communal landownership, a traditional system known in Swahili as *ujamaa*, which means community.

12   Ten years later TANU and the Afro-Shirazi Party merged into the Chama Cha Mapinduzi (CCM), which officially incorporated the aim of building socialism on the basis of self-sufficiency.

13   In October 1978, Tanzania was invaded by Ugandan troops in an attempt by dictator Idi Amin to distract attention from his internal problems and divert Tanzanian energies from supporting the liberation struggles in Southern Africa. The aggression was repelled in a few weeks, and Tanzanian troops co-operated closely with the Ugandan National Liberation Front to overthrow Amin.

LAND USE

2003/2005

IRRIGATED AREA: 3.6% of arable land

- FOREST AND WOODLAND: 39.9%
- ARABLE LAND: 4.5%
- CROPLANDS: 1.2%
- OTHER USE: 54.4%

PUBLIC EXPENDITURE

% OF GDP

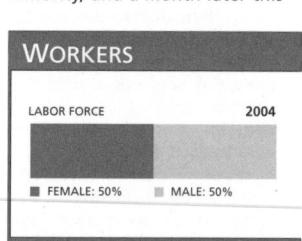

	2.7 %	3.0 %
	HEALTH (2002)	DEFENSE (2004)

WORKERS

LABOR FORCE   2004

- FEMALE: 50%
- MALE: 50%

Life expectancy
47 years
2005-2010

GNI per capita
$320
2004

Literacy
69% total adult rate
2000-2004

HIV prevalence rate
8.8% of population 15-49 years old
2003

14 The cost of mobilizing the army and maintaining troops in Uganda weighed heavily on the State budget, which was facing serious difficulties by the end of the 1970s. The falling price of Tanzania's main exports - coffee, spices, cotton, pyrethrum and cashew nuts - plus the growing cost of imported products, caused serious financial imbalances.

15 Ujamaa villages were conceived as the nucleus of the Tanzanian economy and were designed to be self-sufficient. Despite Nyerere's enormous efforts, various factors hindered the progress of ujamaa. The communal enterprises continued to depend upon imported foodstuffs and, as Government aid was cut, many of them collapsed.

16 In 1983 Edward Sokoine was appointed Prime Minister and immediately launched a campaign against corruption while adopting a more flexible policy towards foreign investment. Sokoine also arranged for Tanzania and Kenya to resume relations that had broken down in 1977 when the East African Economic Community of Tanzania, Kenya and Uganda was dissolved.

17 A national debate was organized to discuss constitutional reform to reorganize the executive power, grant greater political participation to women, strengthen democracy along CCM lines and prohibit more than two successive presidential terms.

18 On 5 November 1985, after 24 years as head of state, President Julius Nyerere passed power on to Ali Hassan Mwinyi, elected with 92.2 per cent of the vote.

19 In 1986, an economic recovery plan went into effect, following IMF and World Bank guidelines. The measures emphasized the reduction of tariff barriers on imports and included incentives for private capital. Agricultural production improved and some industrial enterprises increased their profits.

20 Meanwhile, the ujamaa village model was in crisis due to diminishing yields and to people's increasing resistance to resettlement, sometimes compulsory, of entire villages. Economic recovery now depended on the loans promised by international institutions in exchange for the introduction of structural reforms.

21 The development of private capital and incentives to create such capital caused new problems. According to UNICEF studies, half the children in the country at that time were malnourished. Tanzania remained among the 30 poorest countries in the world, although it managed to escape the famine that hit other Central African nations.

22 Agricultural tasks have traditionally been carried out chiefly by women. While nearly half the workforce is made up of women, they do 85 per cent of all agricultural work. In the outlying areas around the major cities, the female population is increasingly opting for work in the underground economy.

23 At the beginning of 1990, former president Julius Nyerere abandoned his opposition to a multiparty democracy, arguing that the absence of an opposition party contributed to the fact that the CCM had abandoned its program and its commitments. In February 1991 a commission was formed to oversee the country's transition period.

24 In its 1991 report, Amnesty International disclosed the existence of at least 40 political prisoners on the island of Zanzibar. Mwinyi's government denied that any political arrests had been made, and invited the human rights organization to prove its claims.

25 In December, after a 23-year exile in England, opposition leader Oscar Kambona announced his plan to return to Tanzania and lead the fight for a multiparty system. He announced the founding of the Democratic Alliance of Tanzania party. However, the national elections held in April 1993 confirmed yet again the predominance of the ruling CCM, which obtained 89 per cent of the vote.

26 The Government promised the IMF it would implement a strict program of economic adjustment that included the elimination of 20,000 public sector jobs and a reduction of the budget deficit. It thus reduced education spending from 30 per cent in 1960 to five per cent of total public expenditure and in February1994, authorized a 68 per cent raise in electricity prices and a 233 per cent increase in several local taxes.

27 In March, the World Bank praised Tanzania as its second-best African student after Ghana. The country's harsh social conditions were aggravated by an influx of Rwandan refugees, fleeing from the genocide that killed over 500,000.

28 1995 was dominated by the multiparty legislative and presidential elections held in October, where the CCM triumphed again due to Nyerere's support. Benjamin Mkapa became the new President and appointed Frederick Sumaye Prime Minister.

29 In December 1996, the Government decided to expel the majority of the 540,000 Rwandan refugees from the country despite the risk to them from the same conflicts they had fled three years before. Tanzania still had 230,000 Hutu refugees from Burundi, and some 50,000 from the Democratic Republic of the Congo (formerly Zaire).

30 A bomb at the US embassy in Dar es Salaam killed 11 and injured 80 people in August 1998. It was linked with the global terror network al-Qaeda.

31 The death of Julius Nyerere in October 1999 brought tributes from political leaders all over the world. A draft plan for semi-autonomous local government on the mainland presented in December - to run alongside national government and Zanzibar's own semi-autonomous regime - led to conflict between the President and defenders of the project. The idea of a federation of autonomous regions had been criticized by Nyerere but his death paved the way for renewed discussion. Such a system would allow the mainland population equal rights to the islanders, who already had semi-autonomous government.

32 Mkapa was re-elected with 72 per cent of the vote in the October 2000 election. In Zanzibar the CCM candidate Amani Karume was victorious, but flagrant irregularities in the monitoring on the island meant the opposition did not recognize the result. After violent protests where at least 30 people died, and a re-run of elections in two districts, Karume was sworn in as President of Zanzibar in November.

33 In March 2001, the CCM signed an agreement with the main opposition party in Zanzibar, the United Civic Front (UCF), aiming to end the political violence in the semi-autonomous islands of

PROFILE

ENVIRONMENT
The country is made up of the former territory of Tanganyika plus the islands of Zanzibar and Pemba. The offshore islands are made of coral. The coastal belt, where a large part of the population live, is a flat lowland along the Indian Ocean with a tropical climate and heavy rainfall. To the west lies the central plateau, dry and riddled with tsetse flies. The north is a mountainous region with slopes suited for agriculture. Around Lake Victoria, a heavily populated area, there are irrigated farmlands. Large plantations of sisal and sugarcane stretch along the coastal lowlands. Mount Kilimanjaro, the highest peak in Africa at 6,000 m, is located in the northern highlands.

SOCIETY
Peoples: Tanzanians are mostly of Bantu origin, subdivided into around 120 ethnic groups. On the mainland there are also Nilo-hamitic groups in the west; there is a Shirazi minority of Persian origin in Zanzibar. Both regions have groups of Arab, Indian, Pakistani and European immigrants.
Religions: Muslim 35 per cent; traditional religions 35 per cent; Christian 30 per cent. Zanzibar is 99 per cent Muslim, Shi'a and Sunni.
Languages: Swahili (official - the *lingua franca* of Central and Eastern Africa of Bantu origin). English (official), Arabic and more than 100 local languages.
Main Political Parties: Revolutionary State Party; Civic United Front; Party for Democracy and Progress; Tanzania Labour Party.
Main Social Organizations: Organization of Tanzanian Trade Unions (OTTU); Union of Women of Tanzania (UWT); Tanzanian Association of NGOs (TANGO), National Union of Students of Tanzania (MUWATA).

THE STATE
Official Name: Jamhuri ya Muungano wa Tanzania.
Administrative Divisions: 25 Divisions.
Capital: Dodoma 324,347 people (2002). Official capital since 1974, functions as seat of legislature.
Other Cities: Dar es Salaam (former capital) 2,372,200 people; Mwanza 291,100; Zanzibar 247,500; Tanga 202,900 (2000).
Government: Jakaya Kikwete, President since December 2005. Edward Lowassa, Prime Minister since December 2005. Amani Karume, the second Vice-President since November 2000, reelected in 2005, is also President of Zanzibar and Pemba. Unicameral Legislature: 274-member National Assembly.
National Holiday: 26 April, Union Day (1964).
Armed Forces: 27,000 (2003). Other: 1,400 Rural Police, 85,000 Militia.

Under-5 mortality
126 per 1,000 live births
2004

Poverty
57.8% of population living on less than $1 per day
2000

Debt service
5.3% exports of goods and services
2004

Maternal mortality
1,500 per 100,000 live births
2000

IN FOCUS

ENVIRONMENTAL CHALLENGES

The need to increase exports has led to intensification of farming even in semi-arid areas, causing increasing soil erosion. Forestry continues apace despite the resultant severe desertification. The destruction of coral reefs threatens sea habitats. Animals are threatened by hunting and illegal trade, especially by the ivory trade.

In 2004, wildlife authorities investigated the death of over 10,000 flamingos at the Lake Manyara National Park (to the east of the Rift valley). Reports stated that deaths were caused by toxins in algae. Among the more than 300 bird species living at the lake, there are over three million flamingos.

WOMEN'S RIGHTS

Women have been able to vote and stand for election since 1959, and in 2005, they held 30 per cent of parliamentary seats. In 2004, female workers made up 49 per cent of the country's 19-million-strong labor force. Prenatal care reached 94 per cent of pregnant women but only 46 per cent of births received qualified assistance. Although it had dropped over the previous decade, maternal mortality was still high at 580 per 100,000 births. In late 2003 an estimated 840,000 women between the ages of 15 and 49 were living with HIV/AIDS. Domestic violence continues to be a serious problem, especially because there are no laws against it. Some 18 per cent of women between the ages of 15 and 49 are subjected to genital mutilation.

CHILDREN

Although under-five and infant mortality rates decreased between 1990 and 2004, they remained high: the former dropped from 102 to 78 per 1,000 live births and the latter from 161 to 126 per 1,000 live births*. In late 2003, according to estimates, 140,000 children under 14 years of age lived with HIV/AIDS and 980,000 children under 17 were orphans

as a result of this disease. Life expectancy at birth was just 46 years, mainly due to the pandemic.

INDIGENOUS PEOPLES/ETHNIC MINORITIES

The people of Zanzibar and Pemba islands, which became part of Tanzania in 1964, are of Arab, or of African or of mixed (Shirazi) descent. Before independence, Arabs dominated trade and political life, in spite of comprising less than 20 per cent of the total population in Zanzibar, and only two per cent of the whole country's population. Many Tanzanians are Muslim, but not necessarily Arab. Most residents of mainland Tanzania are Christian and speak Swahili. Africans and Arabs on the islands do not always get along, but not necessarily for ethnic reasons - conflicts arise between those who wish to separate from Tanzania and those who do not. Most Arabs would prefer an independent state, in contrast with the African community.

MIGRANTS/REFUGEES

In mid-2005, Tanzania hosted 566,000 refugees in 10 camps and three settlements. According to official sources there were 200,000 Burundian and Congolese refugees living in small villages along the border. Thanks to successful elections in their country, the return of Burundians rose from 2,500 in July 2005 to 4,500 in August of the same year. At the same time, 10,000 repatriates arrived in DR Congo from Tanzania, according to the UNHCR. Due to its relative stability compared to other countries in the region, Tanzania has traditionally been the country with the highest number of refugees. In recent years this has been a cause of instability, particularly in the northeast.

DEATH PENALTY

It is still applicable, even for ordinary offenses.

*Latest data available in *The State of the World's Children* and *Childinfo* database, UNICEF 2006.

Bulyanhulu was taken over by the Canadian corporation Barrick Gold and, after investing $280 million, production forecasts were made of 11,300 kilograms of gold per year.

[35] In November 2001 Mkapa founded an East African regional parliament and court with the presidents of Uganda and Kenya in Arusha. The agreement between Tanzania, Uganda and Kenya was seen as a step toward the creation of a common market across three countries. In December, the government was strongly criticized for spending millions on the purchase of a military air traffic control system from the United Kingdom.

[36] In early 2002, the governor of the Central Bank, Daoud Ballali, said there had been great progress in macroeconomic stability over the previous seven years. According to Ballali, inflation had fallen from 30 per cent in 1995 to 4.7 per cent in 2002 and by this date the economy was growing by more than five per cent per year (above the African average). In the financial sector there were an increasing number of banks - there had been only one before 1993, and 20 by early 2002 - complemented by a further 12 non-banking financial institutions.

[37] In January 2004 Zanzibar celebrated the 40th anniversary of its independence, while the UCF accused the Government of economic mismanagement. The average income of Zanzibaris was $0.60 per day.

[38] In March, Tanzania signed a customs union treaty with Kenya and Uganda, forming a bloc of 90 million people with a GDP of $25 billion.

[39] In April 2004, Tanzania celebrated the 40th anniversary of the union between Zanzibar and the mainland. President Mkapa issued a call to keep the union at all costs. Shortly before this, Zanzibar's autonomous government had announced the island would have its own flag by the end of the year, alleging that the Constitution allowed it to 'have its own identity'.

[40] In May, an official website called Parliamentary Online Information System (POLIS) was launched in Dar es Salaam (the city with the largest number of internet cafés in the country). Its goal is to explain the Parliament's activities and make politics more understandable and accessible to the public. The government declared it expected POLIS to be one of the conuntry's most visited sites. POLIS was set up by the UN Development Programme (UNDP).

[41] Between March and April 2005 there were serious disturbances in Zanzibar before the voters' registration for the October election. The UCF denounced them as a political maneuver organized by the ruling CCM to avoid being defeated in the polls.

[42] President Karume was re-elected in the October election in Zanzibar with 53 per cent of the vote. On polling day, UCF members had several clashes with the police.

[43] Foreign Affairs Minister Jakaya Kikwete, candidate of the of CCM, obtained 80 per cent of the vote in the October presidential election. Kikwete replaced Mkapa, who stepped down after two terms in power. The UCF won on the island of Pemba. According to Kikwete, this proved that the election had been well run, contradicting the denunciations made by Augustine Mrema, leader of the Labor Party, who obtained only one per cent of the vote. In his inaugural speech, the new President said that 'those who expect radical changes in policy and direction are mistaken and lost'.

[44] In March 2006, a Zanzibar court started hearing a case challenging the legality of the 1964 treaty that allowed the union of mainland Zanzibar to the archipelago. The case was filed by inhabitants of the archipelago who claimed that the treaty favored only one side and was prejudicial to Zanzibar.

[45] In June, Chinese prime minister Wen Jiabao arrived in Tanzania to sign cultural and trade cooperation agreements between the two countries. A Chinese economic aid programme was implemented focusing on Tanzania's communication and transport sectors.

[46] In September 2006, health officials on the island of Zanzibar predicted a drastic drop in the number of malaria cases on the island after a successful mosquito control campaign. The island had met the target of 90-per-cent coverage of the area with residual spraying during the 54-day program.

[47] Data collected by the Tanzania Media Women Association showed a strong correlation between HIV/AIDS, early school exit, teenage marriage and pregnancy. Tanzanian law allows girls aged as young as 15 to get married with parental consent, and between 20 per cent and 40 per cent do so before reaching adulthood, according to the United Nations Population Fund. ∎

Zanzibar. The parties agreed to form a joint committee to re-establish peace and encourage the 2,000 or so refugees in Kenya to return home. In April Dar es Salaam witnessed the first joint demonstration by opposition parties in decades. The opposition pressed for

constitutional reform. The Labor Party demanded a new constitution and independent electoral commissions.

[34] An enormous new gold mine - the Bulyanhulu mine - opened in Tanzania in July 2001, which made the country the third biggest African producer.

Thailand / Prathet Thai

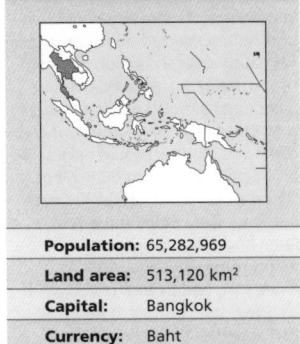

Population:	65,282,969
Land area:	513,120 km²
Capital:	Bangkok
Currency:	Baht
Language:	Thai or Siamese

Archeological evidence indicates almost continuous human occupation of Thailand for the last 20,000 years. Thai-speaking peoples migrated southward from China around the 10th century. By the 13th century, the Thais had emerged as a dominant force in the region, slowly absorbing the weakened empires of the Mons and Khmers. By 1238, the first Thai kingdom, Sukhothai, was established. King Ramkamhaeng the Great, noted as an administrator, legislator and statesman, is credited with the invention of the Thai script. The Sukhothai period saw the Thais for the first time developing a distinctive civilization with their own administrative institutions, art and architecture. Sukhothai Buddha images, characterized by refined facial features, lineal fluidity, and harmony of form, are considered to be the most beautiful and original examples of Thai artistic expression.

[2]   In 1350 the more powerful state of Ayutthaya exerted its influence over Sukhothai; formerly a vassal state, it usurped administrative power, leaving Sukhothai a deserted kingdom. The Ayutthaya kings became very powerful, moving east to take Lopburi - a former Khmer stronghold - and then in 1431 to Angkor, the capital of the Khmer Empire. The Ayutthaya kings adopted Khmer court customs, language and culture. Unlike the paternalistic rulers of Sukhothai, Ayutthaya's kings were absolute monarchs and assumed the title *devaraja* or God King. Ayutthaya became one of the greatest and wealthiest cities in Asia. From the early 16th century the Portuguese established trade, supplied mercenaries and taught the Thais cannon foundry and musketry.

[3]   The Burmese were the most powerful rival of the Ayutthaya kingdom and in 1569 they defeated Ayutthayan forces, occupied their capital, and ruled the kingdom for 15 years. Ayutthayan rule ended in 1767 when, after a siege of 14 months, a new Burmese invasion sacked the city, burning and looting and melting down the gold from Buddha images and taking 90,000 captives. After

expelling the Burmese occupying forces, Thai general Phya Taksin moved his capital to the west bank of the Chao Phraya river, known as Thonburi, and was proclaimed king. During his reign Phya Taksin liberated Chiang Mai and the rest of northern Thailand from the Burmese.

[4]   Rama I (1782-1809), founder of the Chakkri dynasty which still reigns in Thailand, moved his capital across the Chao Phraya river to Bangkok (Krung Thep), which at the time was still a small village. By the mid-19th century, Bangkok had become a city of some 400,000 people, swollen by the huge numbers of Chinese who had poured into Siam (as Thailand was then called) during those years. Rama II re-established relations with the West, allowing the Portuguese to construct the first Western embassy in Bangkok. His successor, Rama III, extended the Thai Empire south along the Malay Peninsula, north into Laos and southeast into Cambodia, while he continued to reopen the country to foreigners, successfully promoting trade with China.

[5]   The name Thailand means 'land of the free', and throughout the

country's 800-year history, the Thai people have never been colonized. While Rama IV (1851–68) opened the country to Western influence and initiated reforms and modern development, control of Siam was disputed by the French and the British. In 1896, the two powers agreed to leave the state formally independent. In World War I, Siam fought alongside the Allies and later joined the League of Nations.

[6]   On 24 June 1932, a coup curtailed the power of the monarchy, creating a parliament elected by universal suffrage. The country's name was changed to Thailand with the advent of a democratic government in1939, but the democratic experience was short-lived and in 1941, during World War II, the Bangkok Government allowed Japan to use its territory, becoming a virtual satellite of that country. During the conflict, Siamese troops occupied part of Malaysia, but were forced to abandon it in 1946, after the allied victory.

[7]   In June that year, King Ananda Mahidol was assassinated in mysterious circumstances. US manoeuvring succeeded in putting

his brother, Rama IX, into power. Born in the US, the new king had never hidden his pro-US leanings. Since then, Thailand has remained under Washington's patronage.

[8]   US interests in Thailand were essentially strategic as its geographical location and plans for a projected channel through the Kra Isthmus made the country critical to US military policy in the area. In 1954, the Southeast Asian Treaty Organization (SEATO), a military pact designed to counterbalance the growing power of revolutionary forces in the region, established its headquarters in Bangkok.

[9]   In 1961, large numbers of US troops entered the country in reaction to an insurrection in Laos. The US military maintained its presence in Thailand for 14 years, resulting in strong ties being forged between the Thai and US armed forces. In exchange for their participation in the anti-communist struggle, the Thai military enjoyed greater political influence and impunity from their corrupt activities, which included control of the drug trade from the famous 'golden triangle' in the north.

[10]   Between 1950 and 1975, US support of military regimes in Thailand cost it over $2 billion. On 14 October 1973, a popular uprising by students brought down the Government. As a result, in 1975 the first civilian government in 20 years was formed. Elections brought Prince Seni Pramoj to power. He demanded the immediate withdrawal of US troops, the dismantling of military bases and an improvement of relations with neighboring revolutionary governments.

[11]   The Thai military disagreed with the new government, and in October 1976 Pramoj was overthrown in a bloody coup planned by right-wing navy officers. Thousands of students and intellectuals joined a guerrilla struggle led by the Communist Party in the rural areas, and Thai relations with neighbors became extremely tense. Finally, in October 1977, a second coup brought the 'civilized right wing' of the military to power. They took a liberal line in policy-making as they were eager to attract new transnational investments. In 1979, Thailand granted asylum to Cambodian refugees. Refugee camps along the border became the rearguard of Cambodian strongman Pol Pot's Khmer Rouge guerrillas; a large part of international humanitarian aid was unwittingly channeled to them. Thailand thus served US interests in the region, as a base for new attacks against communist Vietnam which supported the anti-Khmer forces.

[12]   On 1 April 1981, another military coup shook Bangkok. This time it was led by General Sant Chitpatima with the support of

Life expectancy
72 years
2005-2010

GNI per capita
$2,490
2004

Literacy
93% total adult rate
2000-2004

HIV prevalence rate
1.5% of population 15-49 years old
2003

young middle-ranking officers demanding institutional democracy and social change. The King and Prime Minister General Prem had been asked to lead the coup and had apparently accepted. However, they finally opposed the uprising, and it was put down after three days of great tension.

[13] The young officers' revolt, rooted in the military establishment, was basically a reaction to government measures bringing inforced retirement for certain senior generals. Nevertheless, the short-lived coup gained strong support from trade unions and student groups - and this support in itself was extraordinary since the coup was instigated by the military.

[14] At the same time, the left-wing opposition was severely weakened by an internal split in the Communist Party into the pro-Vietnamese and pro-Chinese factions.

[15] On 20 April 1983, elections were held and General Prem Tinsulanonda was appointed Prime Minister for a second successive term.

[16] Towards the end of 1984, the currency was sharply devalued, sparking discontent from General Arthit Kamlang-Ek and other hardline generals, who threatened to withdraw military support for Prem. Some timely political maneuvering by the Prime Minister, offering

incentives to the officers for their support, allowed Prem to isolate Arthit.

[17] After a further coup attempt in September 1985, Prem relieved Arthit of his command and designated General Chaovalit Yongchaiyut as the head of the armed forces.

[18] Political instability continued in Bangkok, with several cabinet changes and requests for early legislative elections on two separate occasions. In May 1986, the Democrat Party obtained enough of a majority to form a coalition, presided over once again by General Prem as Prime Minister.

[19] While General Chaovalit's stature continued to grow, enhanced by his forceful attacks on corruption, the Government called for early elections in April 1988, to avoid Prem being censured. The main criticisms of his administration were his questionable management of public funds and overall incompetence, particularly in the handling of the border war with Laos several months earlier, which had escalated from a dispute over the control of three villages.

[20] In the first general election since 1976, amid massive vote-buying campaigns (a practice considered normal by almost all candidates), the Thai Nation Party won the election. King Bhumibol Adulyahed asked

General Chatichai Choonhavan to form a new government, which he did by arranging a six-party coalition.

[21] Thailand's new leading force and its successful entrepreneur, Chatichai, introduced important policy changes, breaking with the traditional focus on internal affairs and security. His main idea was to convert what had once been the Indochinese battlefield into a huge regional market. Tempted by the possibility of Cambodia and Laos opening up their markets, he invested considerably in those countries.

[22] This formula further fuelled the Thai economy, and in 1989 the growth rate exceeded 10 per cent. Thailand then challenged Western Europe and its subsidy policies for agricultural export products. There were also disagreements with the United States over Bangkok's refusal to accept North American trade criteria on intellectual property, especially with regard to computer programs.

[23] In March 1991, the military carried out another coup led by General Sunthorn Kongsompong, who presented King Bhumibol Adulyahed with a draft for a new constitution. The latter approved the draft, and justified the military coup on the grounds of 'growing corruption' within the civilian government. The King also agreed with the military on the need for calling new elections. As an indirect result of the military coup in Thailand, peace negotiations in neighboring Cambodia came to a standstill, with the Cambodian Government denouncing Thailand for its renewed support of the Cambodian armed opposition.

[24] Throughout 1991, Thailand remained under the command of the National Peace Keeping Council (NPKC), a body of the military commanded by Sunthorn. In December, the King approved the new constitution which stipulated that elections to replace the NPKC Government would be held within 120 days. However, the military junta reserved the right to directly appoint 270 senators out of a total of 360, giving it total control over the new government.

[25] In early April, General Suchinda Kraprayoon, commander-in-chief of the army at the time, became Prime Minister, backed by a small majority made up of five pro-military parties. In his 49-member cabinet, Suchinda included 11 former ministers who had been accused of embezzlement during Chatichai's government.

[26] At the end of May, what began as an anti-government demonstration ended in a massacre, with hundreds of people killed and injured. Army troops fired into a crowd gathered at the Democracy Monument and political leaders

such as Chamlong Srimuang were imprisoned. The protests continued until an unexpected television appeal for national reconciliation was made by the Thai King. Suchinda, in the meantime, announced his support for a constitutional amendment whereby the Prime Minister would have to be an elected member of Parliament, a clause which would disqualify even Suchinda from holding the post. Srimuang was subsequently freed, and an amnesty announced for those who had been arrested. With a curfew still in effect in Bangkok, Parliament initiated discussion of the constitutional amendment and the King appointed General Prem Tinsulanonda to supervise the process.

[27] On 24 May, Suchinda resigned and his Vice President, Mitchai Ruchuphan, became interim president. Constitutional amendments approved in June reduced military participation in the Government and King Bhumibol named Anand Panyarachun Prime Minister. Panyarachun, who enjoyed a great deal of prestige in Thailand, had been Prime Minister after the1991 military coup.

[28] Parliament accepted the King's nomination of Panyarachun, who appointed a number of technocrats to cabinet positions and requested the resignations of the 12 military officers responsible for the May massacre.

[29] In June 1992, the King dissolved Parliament and called early elections to be held in September, which were won by the Democratic Party (DP) with 79 seats in the House of Representatives. On 23 September, a parliamentary majority was formed (177 seats), comprising the DP, the Palang Dharma (PD) and the New Aspiration Party (NAP), which appointed anti-military leader Chuan Leekpai as Prime Minister. Shortly after, the pro-military Social Action Party joined this coalition.

[30] Although in 1994 the economy of Thailand was among the fastest growing in the world, it was in political turmoil. The first parliamentary session of 1995 brought an exchange of accusations and the tense political atmosphere led to the dissolution of Parliament. On 2 July early elections took place. The Thai Nation Party obtained 25 per cent of the seats and the Democrats 22 per cent. Banham Silapa-archa, leader of the winning party, formed a government supported by several small parties.

[31] In November 1996, Chavalit Yonchaiyudh (NAP) was elected Prime Minister, with 125 of the 393 seats in the Chamber of Representatives. Chuan Leekpai's DP, took 123 seats. A coalition with the Chart Pattana (52), Social Action (20), Prachakorn Thai (18) plus two minor parties, allowed Chavalit to secure a majority weeks

PROFILE

ENVIRONMENT

Thailand is located in central Indochina. From the mountain ranges in the northern and western zones, the Ping and Nan Rivers flow down to the central valley and then through extensive deltas, into the Gulf of Thailand. The plains are fertile with large commercial rice plantations. The southern region occupies part of the Malay Peninsula. Severe deforestation of the area has resulted in decreased production of rubber and timber, and has been responsible for migration of part of the native population.

SOCIETY

Peoples: The Thai group constitutes the majority of the population. The most important minority groups are the Chinese, 12 per cent and, in the south, the Malay, 13 per cent. Other groups are Khmer, Karen, Indians and Vietnamese.

Religions: Most people (94 per cent) practise Buddhism. Muslims, concentrated in the south, make up about 4 per cent of the population. There is a Christian minority.

Languages: Thai or Siamese (official). Minority groups speak their own languages.

Main Political Parties: Phak Thai Rak Thai (TRT-Thais Love Thais Party); Democratic Party; Thai Nation Party; National Development Party.

Main Social Organizations: Forum of the Poor; Thai Network for Community Rights and Biodiversity (BIOTHAI).

THE STATE

Official Name: Muang Thai, or Prathet Thai (Kingdom of Thailand).
Administrative Division: 5 regions and 73 provinces. **Capital:** Bangkok (Krung Thep) 6,355,144 (2000). **Other Cities:** Ratchasima 207,500 people; Chiang Mai 170,300; Khon Kaen 143,200; Nakhon Pathorn 122,500 Thanyaburi 115,500 (2000). **Government:** King Bhumibol Adulyadej, Head of State, since June 1946. Surayud Chulanont Prime Minister since October 2006. **National Holiday:** 5 December, the King's birthday (1927). **Armed Forces:** 307.000 troops (2003). Other: 141,700 (National Security Volunteer Force, Air Police, Frontier Police Patrols, Province Police and Hunter Soldiers).

Under-5 mortality
21 per 1,000 live births
2004

Poverty
<2% of population living on less than $1 per day
2002

Debt service
10.6% exports of goods and services
2004

Maternal mortality
44 per 100,000 live birth
2000

later.

32   Thailand was hit by a financial crisis in 1997, exacerbated by scandals and attempts to salvage the banks - leading to the resignation of the economy minister and the closure of businesses - and by Chavalit's proposal for constitutional reform in order to avoid a vote of no confidence lodged by the opposition.

33   In August 1997, the constitutional reform project was completed. It aimed to limit corruption, put citizens' rights into law, and eliminate army influence from the Senate. Alarmed by the perceived loss of power, the governing parties initially rejected this project, but they were forced to accept it given the agreement between the army and King Bhumibol. The new Constitution was passed in a General Assembly on 11 October.

34   When the new prime minister Chuan Leekpai took office in November 1997, the Thai currency, the baht, had devaluated 37 per cent. The crisis lasted throughout the following year, which closed with an eight per cent drop in Gross National Product. Unemployment affected 1.8 million people (six per cent of the population), while the Government proposed making 200,000 of its 1.5 million employees redundant. The $17.2-billion loan granted by the IMF involved a series of counterpart demands. As well as redundancies in the state sector, the Government put 59 state companies up for sale, including telecommunications, electricity, petrochemicals and finances.

35   The March 2000 elections for the direct appointment of senators were plagued with irregularities. The electoral authority disqualified 78 of the 200 legislators elected, accusing them of corruption and fraud. Two of the disqualified senators were the wives of the Interior and Justice ministers. It was the first time senators had been elected by popular vote.

36   The Thais Love Thais Party (TRT), headed by media magnate Thaksin Shinawatra, won the January 2001 parliamentary elections but did not obtain an absolute majority. The TRT took 248 of the 500 seats in contention. Although the elections took place under a new Constitution, designed to reduce electoral fraud, accusations of irregularities forced new balloting in 62 districts. In February, Parliament elected Shinawatra Prime Minister.

37   In early 2001, Thai and Burmese troops clashed along the shared border - the Mae Sai-Tachilek crossing - and Bangkok and Rangoon accused each other of supporting drug-producing militias. Shinawatra had made combating drug-trafficking a government priority. In May 2001 the tensions

IN FOCUS

ENVIRONMENTAL CHALLENGES
There is increasing air pollution, caused by exhaust emissions from vehicles in cities, industrial pollution and lack of disposal mechanisms for waste matter, combined with an unrestrained industrial development.

Waterways are polluted due to the dumping of organic and industrial waste; tree felling continues to aggravate soil erosion. Several animal species are endangered by massive illegal hunting.

WOMEN'S RIGHTS
Women have had the vote and been eligible for office since 1932. In 2005 they held 11 per cent of parliamentary seats and made up 47 per cent of the workforce, which totaled 37 million people. In 2004, 92 per cent of pregnant women had access to prenatal care and 99 per cent of births received specialized assistance. Maternity-related mortality was 24 per 100,000 births*.

CHILDREN
The December 2004 tsunami in Southeast Asia had a devastating effect on thousands of Thai children. Many of them lost one or both parents when the waters swept through the west coast.

In the small north-eastern village of Nok Kok alone, 10 children were orphaned. According to estimates, the total number of children who became orphans after the disaster is 1,200. The tsunami also had an impact on thousands of people who worked in the tourist or fishing industries. Relief programs aim to help these children cope with tragedy and recover a sense of normalcy as soon as possible.

INDIGENOUS PEOPLES/ ETHNIC MINORITIES
The largest minorities are the Chinese (12 per cent) and South Malaysians. The former arrived in Thailand in the late 19th and early 20th centuries and currently control 85 to 90 per cent of the country's businesses. Additionally, since the early 1990s, they have acquired political power and currently occupy 86 of a total of 347 parliamentary seats.

The South Malaysians (Muslims) are a religious minority in a country where Buddhism is the official religion. Successive military régimes have attempted to foster generalized nationalism in the country, among other things teaching only in Thai and encouraging Buddhist practices. Malaysians were thus left out, as 80 per cent do not speak Thai. Furthermore, the Malaysians

have practically no political representation, they are perceptibly poorer, and their access to health services is less secure than for the rest of the population.

MIGRANTS/REFUGEES
In early 2006, approximately 140,000 refugees from Myanmar were living in 9 camps along the border. Of them, roughly 20,000 were unregistered refugees. Some of them had been living there for 20 years, but the conditions in Myanmar made a voluntary repatriation impossible. Although Thailand is not a signatory to the 1951 Convention, its policy towards refugees and asylum seekers has always been friendly. In 2006, a successful plan was being implemented to resettle 15,500 Hmong refugees from Laos who were living in the province of Saraburi, in the center of the country. A similar plan was being developed to resettle 5,000 urban refugees from Myanmar.

DEATH PENALTY
Applicable for all kinds of crime.

* Latest data available in *The State of the World's Children* and *Childinfo* database, UNICEF, 2006

escalated as the presence of special US troops increased on the Thai side of the border. Their stated mission was to train the Thai army in anti-drug-trade techniques. That same month, near the border with Burma/ Myanmar, more than 20,000 US and Thai soldiers carried out their annual joint maneuvers. In June, Shinawatra traveled to Rangoon to discuss the border problems, resulting in the reopening of Mae Sai-Tachilek.

38   The jewelry and precious stones industry suffered a major retraction in early 2002. US consumption, which represented 50 per cent of the global market for cut gems, had collapsed in the wake of the September 2001 terrorist attacks in New York and Washington. For Thailand, one of the leading world exporters, the consequences were severe. By March 2002, the recession in that market had left some 200,000 Thais out of work.

39   In May the army sent 500 members of the deployment force on a 'search and destroy' mission against 5,000 Muslim militants in the south. Two 1,000-strong battalions were sent as backup. In October, after a demonstration in the south, 78 Muslim militants died, mostly by suffocation, while in the custody

of the police.

40   In December a tsunami devastated wide areas of South Asia. The cataclysm killed approximately 5,300 Thais. The southwest coast was severely affected, including tourist areas such as Phuket and the Phi Phi islands, and the coast villages of Phang Nga and Krabi.

41   In the legislative elections held in February, the Thais Love Thais (TRT) party of Prime Minister Thaksin Shinawatra won 376 of the 500 seats at stake. The opposition Democrat Party of Thailand was second with 97 seats. Shinawatra started his second term as Prime Minister.

42   In July 2005, the Government increased the resources allocated to fight violence in the south of the country, which had caused 800 deaths since 2004. Clashes between Muslim minorities and security forces became more frequent.

43   Following weeks of mounting protests demanding Shinawatra's resignation, the Prime Minister declared in March 2006 that the dispute was to be solved democratically and called an election for April.

44   The elections were bycotted by opposition parties, which considered them pointless, and were annulled

by the Constitutional Court. While the opposition denounced Shinawatra for his intolerance and for having sold that year part of the national assets to favor his family and foreign investors, high military officials accused him of using the army to repress Thai Muslims. In early September, in the midst of growing demonstrations against him, Shinawatra denounced the existence of a military plot to murder him and stage a coup d'état.

45   On the 19th, while the Prime Minister was in New York, where he addressed the United Nations General Assemby, troops under the command of General Sondhi Boonyaratglin – belonging to the Muslim minority – took control of Bangkok.

46   King Bhumibol gave his support to the coup and appointed Boonyaratglin head of the Government junta – that included another five generals. While Shinawatra remained in exile in London, the junta broke up Parliament and announced the appointment of retired general Surayud Chulanont as Prime Minister. A new Constitution and a civilian government was promised within a year. ∎

Timor-Leste / Timor Leste

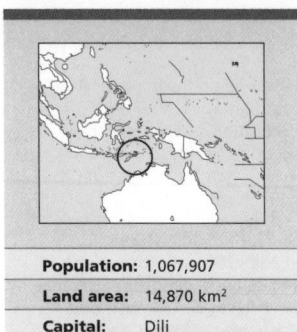

Population:	1,067,907
Land area:	14,870 km²
Capital:	Dili
Currency:	Indonesian rupiah
Language:	Tetum

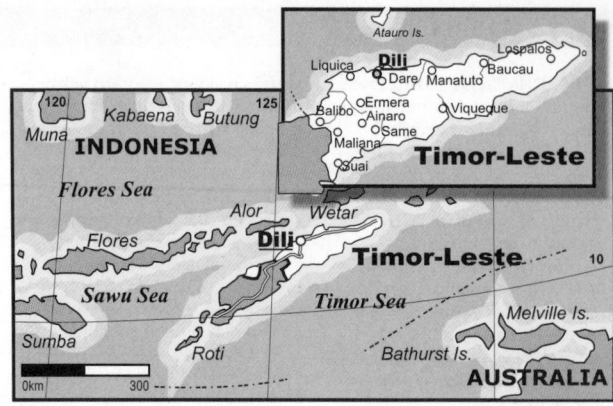

Before the arrival of Vasco da Gama, the Chinese and Arabs knew Timor as an 'inexhaustible' source of precious woods which were exchanged for axes, pottery, lead and other goods of use to the local inhabitants.

[2] Timor's traditional society consisted of five categories: the *Liurari* (kings and chiefs), the *Dato* (nobles and warriors), the *Ema-reino* (freemen) the *Ata* (slaves) and the *Lutum* (nomadic shepherds).

[3] The local population opposed colonialism with armed insurrections in 1719, 1895 and 1959, all of which were put down. In 1859, Portugal and the Netherlands agreed to divide the territory between them. The Portuguese kept the eastern part, under an accord ratified in1904. Passive resistance by the Maubere enabled their culture to survive five centuries of colonialism. Since the devastation of the forests of precious woods, the cultivation of coffee became Timor's economic mainstay.

[4] In the mid-1970s, an independence movement was organized and the struggle for national liberation began, bringing together nationalist political forces and several social organizations.

[5] In April 1974, when the clandestine struggle against colonial rule had already grown and gained broad support, the 'Carnation Revolution' took place in Lisbon. With the fall of the fascist colonial regime in the metropolis the political scene in Timor changed and the patriotic movement was legalized. In September, the Revolutionary Front for the Independence of Timor-Leste (FRETILIN) was created.

[6] The new Portuguese Government promised independence but the colonial administration favored the creation of the Democratic Union of Timor (UDT), which supported the colonial status quo and 'federation' with Portugal. At the same time, the Indonesian consulate in Dili encouraged a group of Timorese to organize the Timor Popular Democratic Association (APODETI) which supported integration with Indonesia.

[7] In August 1975, the UDT attempted a coup causing FRETILIN to issue a call for general armed insurrection, and the Portuguese administration withdrew from the country. FRETILIN achieved territorial control and declared independence on 28 November 1975, proclaiming the Democratic Republic of Timor-Leste. Portugal's withholding of official recognition had important diplomatic and political implications, which continue until the present time.

[8] On 7 December, Indonesia invaded Timor. A few hours earlier, US President Gerald Ford had visited Jakarta, capital of Indonesia, where he had probably learned of, and endorsed, Indonesian President General Suharto's expansionist plans. FRETILIN was forced to withdraw from the capital, Dili, and from the major ports. On 2 June 1976, a so-called 'People's Assembly', made up of UDT and APODETI members, approved Timor's annexation as a province of Indonesia. The annexation was not recognized by the United Nations Decolonization Committee, which still regarded Portugal as the island's ruler.

[9] In December 1978, Nicolás dos Reis Lobato, President of the Republic and FRETILIN leader, died in combat. Despite this blow, the liberation movement continued the struggle.

[10] In an attempt to set the Maubere against each other, the Indonesian army recruited young Timorese. But the young Timorese answered calls from the Front and mutinied to join FRETILIN. In 1983 a ceasefire was signed between FRETILIN commander in chief Xanana Gusmão and head of the expeditionary corps Colonel Purwanto. Suharto did not recognize the agreement and guerrilla warfare continued.

[11] In 1988, rapprochement between FRETILIN and the TDU culminated in the creation of a joint Nationalist Convergence and Gusmão was confirmed as commander in chief of the liberation army. This body was instrumental in Portugal, actively raising the Timorese question with the European Parliament and the European Commission (EC). Europe rejected Indonesian occupation and backed Maubere self-rule along with a negotiated solution to the conflict.

[12] In October 1989 the UN Human Rights Sub-commission approved a motion condemning Indonesian occupation and repression in Timor-Leste. When Pope John Paul II visited Dili that month a group of youths unrolled a FRETILIN banner, a few meters away from the platform where he was leading Mass. Indonesian security forces repressed the students shouting anti-occupation slogans.

[13] Timorese families were forced to hang a list of household members' names on their doors so that occupation forces could check who was present at any time of day or night. Thousands of Maubere women were sterilized against their will. The Indonesian authorities applied policies aimed at reducing the Maubere people to a minority. Mass graves of Maubere proved the occupation forces had carried out wide-scale executions.

[14] On November 12 1991, a large funeral procession accompanying the remains of a young student who had been killed turned into a bloodbath when the army machine-gunned the crowd, killing at least 50 and leaving countless people injured. When the Portuguese Government heard the news, it called on the European Community to cut trade with Indonesia - which had a preferential trade agreement - and call a UN Security Council meeting. Portugal produced veiled criticism of the body for not confronting Indonesia and acting as it had when Iraq invaded Kuwait in August 1990.

[15] In late 1991, claims came to light in Portugal that Jakarta and Canberra had signed a contract with 12 companies for the extraction of a billion barrels of crude oil in the

PROFILE

ENVIRONMENT
Located between Australia and Indonesia, Timor-Leste comprises the eastern portion of Timor Island, the dependency of Oecusse, located on the northwestern part of the island, the island of Atauro to the North, and the islet of Yaco to the East. Of volcanic origin, the island is mountainous and covered with dense rainforest. The climate is tropical with heavy rainfall, which accounts for the extensive river system. The southern region is flat and suitable for farming. Agriculture is the basis of its export-oriented economy, and copra, coffee, rice, cotton, tobacco and sandal are its main crops.

SOCIETY
Peoples: The Maubere people are descended from Melanesian and Malayan populations. In 1975 there was a Chinese minority of 20,000, who had arrived during the 20th century, as well as 4,000 Portuguese. Amnesty International estimates that 210,000 people have died as a result of the Indonesian occupation. There are 6,000 Maubere refugees in Australia and 1,500 in Portugal. **Religions:** Most of the population follows traditional practises. 30 per cent are Catholic. **Languages:** Tetum is the national language. There are some 40 dialects. Indonesian occupation had banned the use of these languages in education, and virtually all the teaching was done in Bahasa Indonesia, the Indonesian language. This situation was reversed after official Independence in 2002. A minority also speaks Portuguese. **Main Political Parties:** Revolutionary Front for the Independence of Timor-Leste (FRETILIN); Democratic Party (PD); Social-Democratic Party (PSD); Timorese Social-Democratic Association (ASDT).

THE STATE
Official Name: Republic of Timor-Leste. **Capital:** Dili 65,000 people (1999). **Other Cities:** Dare 17,500 people; Baucau 14,500 (2000). **Government:** Xanana Gusmão, President elected in April 2002, took office on 20 May that year, when the country became formally independent. That same day the Transitional Authority imposed by UN in 1999 ceased functioning. José Ramos Horta, Prime Minister since June 2006. Unicameral National Parliament. **National Holiday:** 28 November, Independence Day (1975).

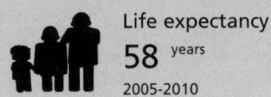
Life expectancy
58 years
2005-2010

GNI per capita
$550
2004

IN FOCUS

ENVIRONMENTAL CHALLENGES

The felling and clearing of trees have led to deforestation and land erosion. As a result of the independence struggles, sandalwood forests were destroyed, lagoons polluted and domestic animals hunted. More sandalwood forests, the only ones still existing in the South Pacific, were devastated during the 20 years of Indonesian occupation than during the 450 years of Portuguese colonization.

WOMEN'S RIGHTS

In 2003 women held 26 per cent of Parliament seats and made up 45 per cent of the workforce. Since independence in 2002, its main priorities were to rebuild

infrastructure destroyed by the war and to train staff for education and health sectors, to replace the many Indonesians who had worked in these fields before leaving the country. Female literacy has become a priority, especially in rural areas. In 2004 female primary school enrolment reached 74 per cent. The health system is patchy. Some 61 per cent of pregnant women receive prenatal care and only 18 per cent of all births are attended.

CHILDREN

Government immunization campaigns coordinated by UNICEF, with the aid of Médecins Sans Frontières International and the World Health Organization (WHO), have reduced child mortality rates.

In the 1990-2004 period, the infant mortality rate went from 130 to 64 per 1,000 live births, while the under-5 mortality rate went from 172 to 80 per 1,000. In 2004, 12 per cent of all children had low birth weight, 15 per cent of under 5s were moderately or seriously under weight; 12 per cent were moderately undernourished, and 49 per cent had moderately and seriously stunted growth. Some 4 per cent of children between 5 and 14 worked.

MIGRANTS/REFUGEES

It is estimated that 250,000 Timorese left the country and a similar number were internally displaced during the violent conflict in 1999, when anti-independence militias intimidated and attacked the population. In the

years following independence, the militias had significant influence within West Timor refugee camps. Complicated repatriation schemes that were only completed in late 2004 resulted in more than 225,000 voluntary returns since 1999. Some 28,000 people decided to remain in West Timor and confirm their Indonesian nationality. In late 2004, 4,500 children separated from their parents during the 1999 conflict were returned home. However, some remained in Indonesia to continue their studies, generally with parental consent.

DEATH PENALTY

It was abolished in 1999.

Timor Sea. The list included British, Dutch, US, Australian and Japanese firms, and the transnationals Phillips Petroleum, Marathon and Enterprise Oil Company. Opposition leaders accused the companies' home countries of collaborating with Jakarta in playing down the genocide and silencing the international press for the benefit of their own economic interests.

[16] In December 1996, exiled activist José Ramos Horta and Catholic bishop Carlos Filipe Ximenes Belo received the Nobel Peace Prize in Oslo. Indonesian authorities planned to boycott the ceremony, but the problem of Timor-Leste hit the headlines all around the world.

[17] Suharto's resignation in Jakarta in June 1998 and his replacement by Bacharuddin Jusuf Habibie acted as a catalyst for instability on the island. Habibie announced a plan to give Timor-Leste greater autonomy and to this end, a date was set for a referendum.

[18] The referendum was finally held on 30 August 1999, amidst a campaign of violence by Indonesian paramilitaries, backed by some of the army. 78.5 per cent of the voters were in favor of independence, and this outcome spurred a fresh wave of paramilitary violence which caused thousands of deaths.

[19] Following discussion with Jakarta, the UN decided to send a peace mission headed by Australian troops. This force (known as Interfet) arrived in Dili two weeks after the referendum, ousting paramilitaries from the capital and surrounding areas, and setting up posts along the frontier. Sovereignty was finally handed over on 25 October.

[20] The UN made Sergio Vieira de Melo head of the Timor Transition Authority. Vieira created a 15-member Consultative Council, which included Gusmão. In February 2000,

Interfet ceded control of large areas of the territory to the Transition Authority, which remained under Vieira's responsibility.

[21] In August 2001, FRETILIN won the parliamentary elections with 57 per cent of the vote, taking 55 of the 88 seats in the Assembly. This new body would draw up the Constitution and become the first Parliament when Timor-Leste proclaimed independence on 20 May 2002.

[22] In January 2002, Timor-Leste inaugurated the Truth and Reconciliation Commission (TRC) to investigate crimes committed by the Indonesian forces during the 25 years of occupation. In Vieira's words, the TRC would give an 'official hearing' to the population's complaints, recognizing past suffering and providing an opportunity for genuine and lasting reconciliation between victim and persecutors. However, no amnesty was offered to those admitting crimes like murder or rape. These cases would be transferred to the regular courts so that criminals could be tried. Between 100,000 and 200,000 Timorese died in the first years of the occupation, many through starvation or disease.

[23] Also in January 2002, Jakarta officially inaugurated a Human Rights court to try military officers and others involved in atrocities committed in Timor-Leste following the 1999 pro-independence vote there. Indonesia had been under international pressure to try at least 18 military and paramilitary leaders accused of encouraging pro-Indonesian forces in Timor-Leste to commit murder.

[24] In February 2002, the assembly approved a draft Constitution for the country to adopt on formal independence. The project divided the authorities into executive,

legislative and judicial bodies and stipulated that the military must be politically neutral.

[25] That month, Indonesia and the UN administration in Timor-Leste signed two agreements in order to 'smooth' relations between the two countries after Timor-Leste's independence. The two parties reached agreement in a meeting in Bali, Indonesia.

[26] In April, Gusmão won the presidential elections with 82.7 per cent of the vote. He would take office as the country's first President on 20 May. Shortly after this, Gusmão met with Indonesian President Megawati Sukarnoputri in Jakarta, and offered her a personal invitation to the independence celebrations.

[27] In early May, a UN Development Program report revealed that after independence Timor-Leste would be the poorest country in Asia and it would rank among the 20 poorest nations of the world. At that time, most Timorese depended on subsistence fishing and agriculture. The country had no industry and practically no exportable products. Per capita GDP stood at $478 in early May and half the population earned less than 55 cents per day.

[28] In August 2002, Abilio Soares was sentenced to three years in prison. He had been Governor of the former Indonesian province during violence after the 1999 independence referendum. He was found guilty on two counts of gross rights violations.

[29] On 5 August 2003, General Adam Damiri, former Timor-Leste military commander, was sentenced to three years in prison by the Human Rights Court in Jakarta. Damiri was one of the 18 high-ranking Indonesian officers accused by this court for their alleged role in atrocities committed in Timor-Leste during the 1999 referendum. He also

became the highest-ranking military officer to be tried and condemned by such a court.

[30] Offshore production of gas in Bayu-Undan, a territory in the Timor Sea some 500 km from Darwin and 250 km south of Timor-Leste, started in February 2004. This activity will generate an income of $100 million a year.

[31] In May, the UN reduced its peace mission force from 3,000 to 700.

[32] After two years, an Indonesian Human Rights court concluded the trial of 18 people accused of atrocities during the transition to independence. Rebel leader Eurico Guterres was the only one whose charges were not lifted. He was freed until the appeals trial resumed.

[33] In January 2005 Timor-Leste and Indonesia agreed to set up a truth and friendship commission to investigate the 1,500 murders and human rights violations during the 1999 referendum.

[34] A strike by 590 soldiers from the west of the country, who demanded better working conditions, was launched in April 2006. Their dismissal on 27 April caused violent protests in Dili, harshly repressed by the military. At least five people died and buildings and cars were set on fire. Fearing the clashes might become increasingly violent, some 100,000 people fled the capital to take refuge in the mountains.

[35] In July 2006, Nobel Peace Prize winner Jose Ramos-Horta became Prime Minister. Former premier Mara Alkatiri had resigned in June, having been widely blamed for sparking factional violence that caused at least 21 deaths and necessitated the advent of an Australian-led peacekeeping force. Ramos Horta was seen as a potentially unifying figure and was to hold office until elections in mid-2007. ∎

Togo / Togo

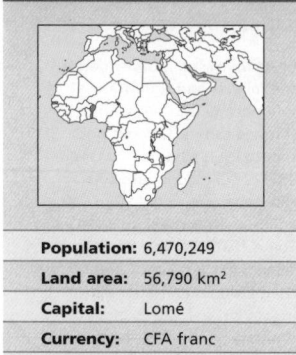

Population:	6,470,249
Land area:	56,790 km²
Capital:	Lomé
Currency:	CFA franc
Language:	French

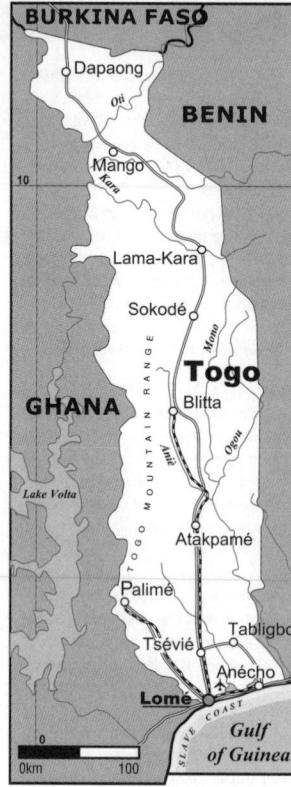

Between the 4th and 10th centuries, Togo was peopled by the Ewe (of the same origin as the Igbo and Yoruba of Nigeria and the Ashanti of Ghana). The Ewe were a relatively poor and peaceful people with simple social structures like the Dagomba kingdoms in the north.

2   The slave trade gave the region the name of Slave Coast, and millions of people were shipped to slavery between the 16th and 19th centuries.

3   Togo became a German colony in 1884 under 'treaties' signed with local chieftains. Occupied by Anglo-French troops during World War I, the territory was divided between the two powers with the endorsement of the League of Nations. The western part was annexed to Ghana in 1956, while the eastern part remained under French rule and became an Overseas Territory.

4   In 1958, Sylvanus Olympio, (a member of the Togolese Unity party) won the elections. The 1960 proclamation of independence seemed to overlook a contract signed in 1957 with the Benin Mines Company, in which the French consortium had seized control of Togo's phosphate reserves, its main natural resource. When Olympio confronted this situation in 1963, proposing a series of basic reforms - doubtlessly influenced by Nkrumah's radical pro-independence program in Ghana - he was assassinated in a military revolt with the participation of army officer Etienne Gnassingbé Eyadéma.

5   Olympio's successor, the neo-colonialist Nicholas Grunitzky, from the opposition Togolese Progress Party, was overthrown by a further coup in 1967, led by (now) General Gnassingbé Eyadéma, who became the new head of state.

6   In 1969, the Togo People's Group (RPT) brought in a one-party system with Eyadéma as President. His government adopted nationalist measures which were later modified. In 1972 a law gave the State 35 per cent of shares of the mining company, which was raised to 51 per cent in 1975 and in 1976 the production and export of phosphate was nationalized.

7   In December 1979, a new presidential Constitution was approved and Eyadéma was re-elected for a seven-year term. His government was confident of improving the economy, basing its optimism on tourism, oil and a rise in world phosphate prices. However in 1981 phosphate prices dropped by 50 per cent and a worldwide economic recession reduced the number of European tourists. A serious balance of payments deficit increased the foreign debt to $1 billion.

8   In June 1984 a refinancing agreement with stringent conditions was reached with the Paris Club and the IMF. This brought about salary freezes, a large reduction of government investments and more taxes, including a so-called 'solidarity tax' which took five per cent of the population's incomes.

9   In January 1985 the Lomé III agreements were signed in Togo's capital, regulating cooperation between ACP (Africa, Caribbean and Pacific) countries and the EEC.

10   In January 1986 Eyadéma was re-elected with 99.95 per cent of the vote. In 1988, Togo signed new agreements with the IMF. In 1990, the Government began a program of privatization of state-run companies, to tackle the trade deficit and a foreign debt of $1.27 billion.

11   In 1991, more than 10,000 peasants in the northern district of Keran Oti lost their lands to create an 80 sq km game reserve for hunting.

12   In April 1991, more than a thousand people took over one of Lomé's most important neighborhoods, demanding Eyadéma's resignation. A few days later, opposition parties were legalized and Eyadéma announced an amnesty for political prisoners.

13   In August 1991, a National Conference named Kokou Koffigoh - a human rights leader - as provisional Prime Minister. A legislative assembly was established which ousted Eyadéma as head of the armed forces and blocked his candidacy in the upcoming 1992 elections.

14   The dissolution of the RPT by the legislative assembly prompted a military coup in November 1991. The armed forces took over the Government, dissolved the legislative assembly and kidnapped Koffigoh for several hours.

15   Negotiations were launched in the midst of political, economic and social turmoil (the latter caused by tribal conflicts). A timetable was established for a return to democracy and all citizens were re-registered to vote.

16   The opposition split from Koffigoh in January 1993 and nominated a government-in-exile

LAND USE

2003/2005

IRRIGATED AREA: 0.3% of arable land

- FOREST AND WOODLAND: 7.1%
- ARABLE LAND: 46.1%
- CROPLANDS: 2.2%
- OTHER USE: 44.6%

PROFILE

ENVIRONMENT

The country is a long, narrow strip of land with distinct geographical regions. In the south, a low coastline with lakes, typical of the Gulf of Guinea; a densely populated plain where manioc, corn, banana and palm oil are produced. In the north, subsistence crops are gradually giving way to coffee and cocoa plantations. The Togo Mountains run through the country from northeast to southwest.

SOCIETY

Peoples: The main ethnic groups are the Ewe (43.1 per cent), Kabye (26.7 per cent), Gurma (16.1 per cent), Kebu (3.8 per cent) and Ana (Yoruba, 3.2 per cent). The descendants of formerly enslaved Africans who returned to Togo from Brazil are called Brazilians. They form a caste with great economic and political influence. The small European minority (0.3 per cent) is concentrated in the capital.
Religions: The majority follow traditional African religions (50 per cent). There are Christian (35 per cent) and Muslim (15 per cent) minorities.
Languages: French (official). The main local languages are Ewe, Kabye, Twi and Hausa.

Main Political Parties: Rally for the Togolese People; Union of Forces for Change supported by Alliance of Democrats for Integral Development; Action Committee for Renewal; Democratic Convention of African Peoples; Socialist Pact for Renewal.
Main Social Organizations: National Confederation of Togolese Workers (CNT).

THE STATE

Official Name: République Togolaise.
Administrative Divisions: 5 Regions and 21 Prefectures.
Capital: Lomé 790,000 people (1999).
Other Cities: Sokodé 115,100 people; Palimé 47,100; Atakpamé 40,300 (2000).
Government: Faure Gnassingbé, President since February 2005. Yawovi Agboyibo, Prime Minister since September 2006. Unicameral Legislature: 81-member National Assembly.
National Holiday: 27 April, Independence Day (1960).
Armed Forces: 9,000 (2003). Other: 750 Gendarmes.

Life expectancy
56 years
2005-2010

GNI per capita
$310
2004

Literacy
53% total adult rate
2000-2004

HIV prevalence rate
4.1% of population 15-49 years old
2003

ENVIRONMENTAL CHALLENGES

There are signs of serious deforestation attributable to agriculture and the use of wood for fuel. Water pollution poses health risks and a threat to the fishing industry. Air pollution is a growing problem in cities.

WOMEN'S RIGHTS

Women have been able to vote since 1945. In 2003, seven per cent of seats in Parliament and 20 per cent of ministerial and equivalent positions were held by women. That year, female workers made up 40 per cent of the two-million-strong labor force. In 2004, for every 100 literate men there were just 56 literate women. While the gender gap in education has been reduced, it still remains a significant obstacle. In 2004, 85 per cent of pregnant women received prenatal care, but only 61 per cent of births were attended by skilled health personnel. The maternal mortality rate was 570 deaths for every 100,000 live births*. In late 2003, there were an estimated 54,000 women between the ages of 15 and 49 living with HIV/AIDS.

CHILDREN

There was only a slight reduction in infant and under-five mortality rates between 1990 and 2004, as the former fell from 88 to 78 deaths per 1,000 live births and the latter from 152 to 140. In 2004, 18 per cent of infants were born with low birth weight. Among children under five, 25 per cent were moderately or severely underweight, 12 per cent suffered from moderate or severe emaciation and 22 per cent from moderate or severe stunting. Although vaccination campaigns had improved immunization coverage rates, these were still well short of reaching all the country's children. In late 2003, it was estimated that some 9,300 children under 15 were living with HIV/AIDS.

INDIGENOUS PEOPLES/ ETHNIC MINORITIES

The Akposso people inhabit southern Togo and number about 102,000 people who practise a traditional religion based on a very rich mythology.

Gurma people make up three per cent of the population, living mainly in northern Togo. They are Muslim; mostly pastoralists leading a semi-nomadic life.

The Mamprusi have a population of about 8,000 people. They combine Islam with traditional religion and belong to an ethnic group that is mostly found in Ghana.

The Dagomba have a population of about 30,000 and inhabit the western central area. They settled in the 16th century and after a long dispute with Gonja people, managed to occupy what is now their territory. They practise Islam and traditional religions.

There are 11,400 Hausas (the largest group in Central Africa with more than 40 million Hausa-speakers). The largest number of Hausa live in Nigeria and Niger.

The Mossi, in the northeast, have a population of some 23,000; most Mossi people live in Burkina Faso.

MIGRANTS/REFUGEES

In mid-2005, there were 23,000 Togolese refugees in Benin and another 15,500 in Ghana, while around 10,000 people were internally displaced as a consequence of the election-related conflict in April of that year. Roughly 9,000 of the refugees living in Benin were staying in camps in Come and Agame - in the southwestern region of the country - while the remainder were hosted by friends and relatives. In early 2006, the UNHCR was providing protection and assistance to 20,000 refugees and 1,300 asylum-seekers from Rwanda, Congo and DR Congo, distributed among Benin, Burkina Faso, Niger and Togo. By 2006, the number of Togolese refugees in Benin had reached 27,500, with the number in Ghana still around 15,500, according to the UNHCR.

DEATH PENALTY

Capital punishment has been abolished in practice.

*Latest data available in *The State of the World's Children* and *Childinfo* database, UNICEF, 2006.

in Benin. In this same month the presidential guard killed some hundred demonstrators in Lomé, which led to thousands of people fleeing to Ghana and Benin.

[17] In an atmosphere of civil war, Eyadéma won the elections with 96.5 per cent of the vote, in a poll denounced as fraudulent by the opposition. The protests in the streets intensified, and in January 1994, Eyadéma survived unharmed an assassination attempt which left 67 people dead.

[18] The opposition triumphed in the February legislative elections, but Eyadéma barred Prime Minister Edem Kodjo from forming a government without ruling party members.

[19] The President, in power since 1967, was pressed - even by the military - to liberalize the political system. The June 1997 presidential elections, considered fraudulent by the opposition, saw Eyadéma re-elected with 52 per cent of the vote. Over the following months, the capital was rocked by constant demonstrations disputing the election results.

[20] In January 2000, the West African Economic and Monetary Union (WAEMU) member countries - eight West African states that use the CFA franc as national currency (Benin, Burkina Faso, Côte d'Ivoire, Guinea-Bissau, Mali, Niger, Senegal and Togo) - created a customs union within the bloc.

[21] The meeting of Francophone nations of West Africa in March 2000 was a diplomatic success for the Eyadéma Government, as Benin, Niger, Burkina Faso and Côte d'Ivoire all attended. In September, Agbeyome Messan Kodjo was made Prime Minister.

[22] In February 2001, an international commission denounced grave and systematic human rights violations in Togo.

[23] In August 2001, Eyadéma announced he would respect the Constitution and end his term in 2003. Prime Minister Kodjo supported the possibility of constitutional change allowing Eyadéma to stand again.

[24] After planned legislative elections failed to be held in October, in March 2002 the Joint Commission of Inquiry - created from ruling party and opposition members - met in Lomé to 'revive political dialogue.' The opposition had refused to participate in the National Electoral Commission until attorney Yawovi Agboyibo - leader of the Togolese opposition imprisoned since 3 August 2001 - was freed. In March 2002, Agboyibo was finally liberated.

[25] In April 2002, WAEMU member countries signed a trade agreement with the US.

The countries of the union had previously had closer trade relations with the European Union (EU).

[26] The ruling party won the parliamentary elections held in October 2002. In December, Parliament reformed the Constitution, removing a clause which barred Eyadéma from seeking a third term as president.

[27] In February 2003, the EU offered support to help organize the June election provided the Government showed a clear commitment to ensuring free and fair elections with the participation of all political groups.

[28] Again amid accusations of fraud, Eyadéma was re-elected in the June 2003 election with 57 per cent of the vote.

[29] In September 2003, Togo sent around 150 soldiers to Liberia to reinforce the western African peacekeeping forces there.

[30] After almost 40 years in power, Eyadéma died in February 2005 at the age of 69. Faure Gnassingbe, the late president's son and minister of Communications, took over the country's leadership with the support of the army - in order to prevent a power vacuum, according to official spokespeople - and the borders and airports were closed. This move was interpreted by analysts as a new coup d'état.

[31] Opposition leader Yawovi Agboyibo, head of the Action Committee for Renewal, called for a demonstration. Two days of protests followed. For its part, the African Union threatened to impose sanctions on Togo. In the face of this international pressure, the new leader announced presidential elections.

[32] Gnassingbe won the election held in April 2005. The opposition denounced electoral fraud and took to the streets to demonstrate. Violent clashes left hundreds dead or imprisoned, while many opposition supporters fled the country.

[33] In June, Gnassingbe designated Edem Kodjo as Prime Minister, thereby returning the opposition leader to the post he had held in the 1990s. Kpatcha Gnassingbe, the President's brother, was named Minister of Defense.

[34] In March 2006, opposition leader Harry Olympio was accused by the Government of having orchestrated a firebomb attack on police headquarters a month earlier. Olympio denied any involvement in the attack, and said he had received anonymous calls warning him that the security forces were planning to kill him. ∎

Tonga / Tonga

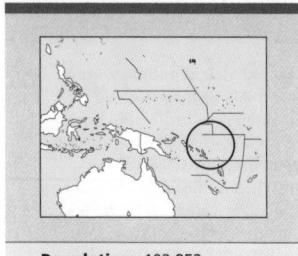

Population:	102,852
Land area:	750 km²
Capital:	Nuku'alofa
Currency:	Pa'anga
Language:	Tongan and English

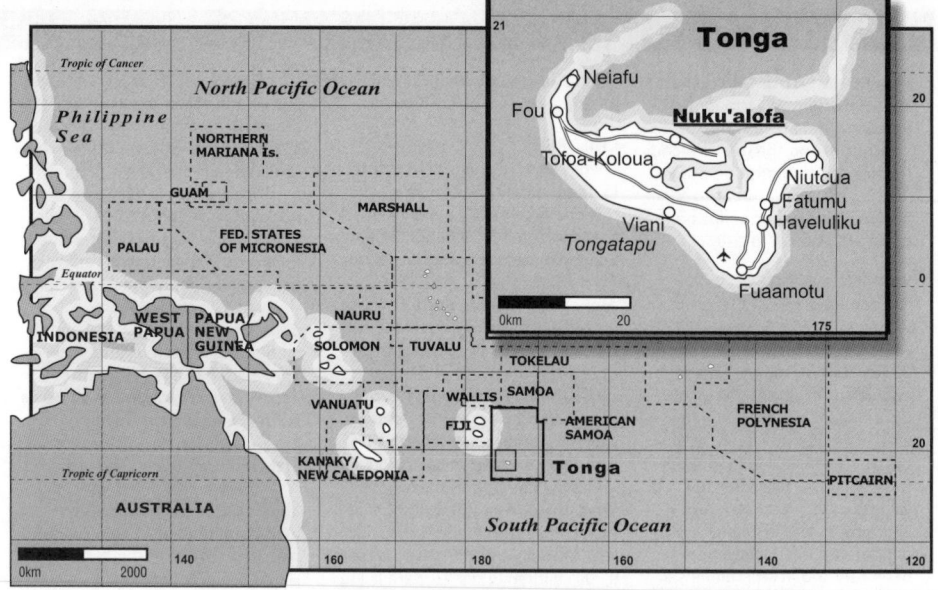

Tonga was inhabited over 1,000 years ago by immigrants from Samoa who created a complex society, with a monarch in charge. The first ruler was Ahoeitu, in the latter half of the 10th century. Towards the 15th century, religious and social roles were separated. That is why the Dutch found two rulers when they came to the islands in 1616. In 1773 British explorer James Cook named the archipelago the 'Friendly Isles'.

[2] In the mid-19th century, after a civil war, King Taufa'ahau Tupou (who after converting to Christianity became George I) secured political union, which had not existed since the late 18th century. Backed by European missionaries, he seized Vavau and Tongatopu and introduced a parliamentary system and a land reform which granted each adult male in the country 3.3 hectares of arable land.

[3] In 1889, Britain and Germany signed a treaty which gave Tonga over to Britain. In 1890 year the archipelago was made a 'protectorate' of the British crown, though the monarchy was kept with limited powers.

[4] Queen Salote, great-grand-daughter of George I, was crowned in 1918 and in 1960 she gave women the right to vote in legislative elections.

[5] The British transformed the country's agriculture and fishing,

orienting it towards copra and banana crops for export.

[6] Present King Taufa'ahau Tupou IV was crowned in 1967, and in 1970 Tonga obtained independence. A social security system which included free education and medical services for all was implemented.

[7] Opposition groups founded the Pro-Democracy Movement (PDM) in 1992. The following year, it won six of the nine seats in the Legislative Assembly filled through general elections.

[8] In 1994, the PDM became the Tonga Democratic Party, led by 'Akilisi Pohiva.

[9] Demonstrations in favor of political liberalization in 1997 led to the arrest of several political leaders and journalists.

[10] Prince Ulukalala Lavaka Ata was appointed Prime Minister in January 2000, by his father.

[11] A Constitutional amendment in October 2003 granted more powers to the King and increased state control over the media.

[12] In March 2005 there were legislative elections. Tonga's Legislative Assembly is made up of nine members elected by the people, nine chosen by the country's 33 hereditary nobles and 12 ministers appointed by the King. Seven of the nine representatives elected by the people were supporters of the pro-democracy movement.

[13] A public sector strike to demand higher wages paralyzed the country during July and August 2005, and sparked violent clashes between strikers and police in the capital. After an agreement with the Government was reached in early September, protests continued throughout the month to demand democratic reforms.

[14] After a decade of negotiations, Tonga was accepted as a member of the World Trade Organization (WTO) in December 2005. 'We will continue to work together to facilitate the effective involvement of small economies such as Tonga in the work of our Organization', said WTO Director-General Pascal Lamy.

[15] In March 2006, Fred Sevele became the first 'commoner' ever designated Prime Minister

of Tonga. He was appointed following the sudden resignation of Prince Lavaka Ata 'Ulukalala, who was forced to step down by popular protests demanding a lesser role for the royal family in government. Sevele said his designation proved that rules were changing in the country.

[16] In September 2006, King Taufa'ahau Tupou the IV died at the age of 88 and was succeeded by his son, George Tupou V. ■

LAND USE

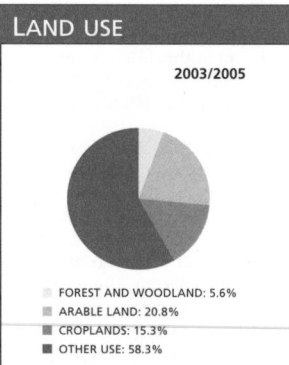

2003/2005

- FOREST AND WOODLAND: 5.6%
- ARABLE LAND: 20.8%
- CROPLANDS: 15.3%
- OTHER USE: 58.3%

PROFILE

ENVIRONMENT

The archipelago was known as the 'Friendly Isles' and is located in western Polynesia, east of the Fiji Islands, slightly north of the Tropic of Capricorn. It comprises approximately 169 islands, only 36 of which are permanently inhabited. It includes three main groups of islands: Tongatapu, the southernmost group where more than half of the population live; Vavau, to the north, and Haapai in between. The volcanic islands are mountainous while the coral ones are flat. The climate is mild and rainy with very hot summers. The fertile soil is suitable for growing banana, copra and coconut trees.

SOCIETY

Peoples: Polynesian, European (around 300). It is estimated that 20 per cent of Tongans now live abroad. **Religions:** Free Wesleyan, 43.6 per cent; Roman Catholic, 16.0 per cent; Mormon, 12.1 per cent; Free Church of Tonga, 11.0 per cent; Church of Tonga, 7.3 per cent. **Languages:** Tongan and English are official. **Main Political Parties:** Human Rights and Democracy Movement (HRDM, formerly People's Party).

THE STATE

Official Name: Pule'anga Fakatu'i o' Tonga. Kingdom of Tonga. **Administrative Divisions:** 23 Districts. **Capital:** Nuku'alofa 37,000 people (1999). **Other Cities:** Neiafu 4,000 people; Haveluliku 3,200; Vaini 2,800; Tofoa-Koloua 2,400 (2000). **Government:** Monarchy, limited by the power of the nobles (the five ministries are lifetime terms). George Tupou V, King since 2006; Dr. Feleti Sevele, Prime Minister since February 2006. Unicameral Legislative Assembly that includes five ministers, governors of Hapai and Vavau, seven nobles (elected by their 33 peers) and seven deputies elected by male taxpayers aged over 21. **National Holiday:** 4 June, Independence Day (1970). **Armed Forces:** Tonga Defense Services, comprising the Royal Marines, the Royal Guards and the Maritime Force.

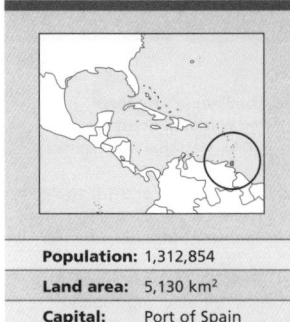

Population:	1,312,854
Land area:	5,130 km²
Capital:	Port of Spain
Currency:	Trinidad dollar
Language:	English

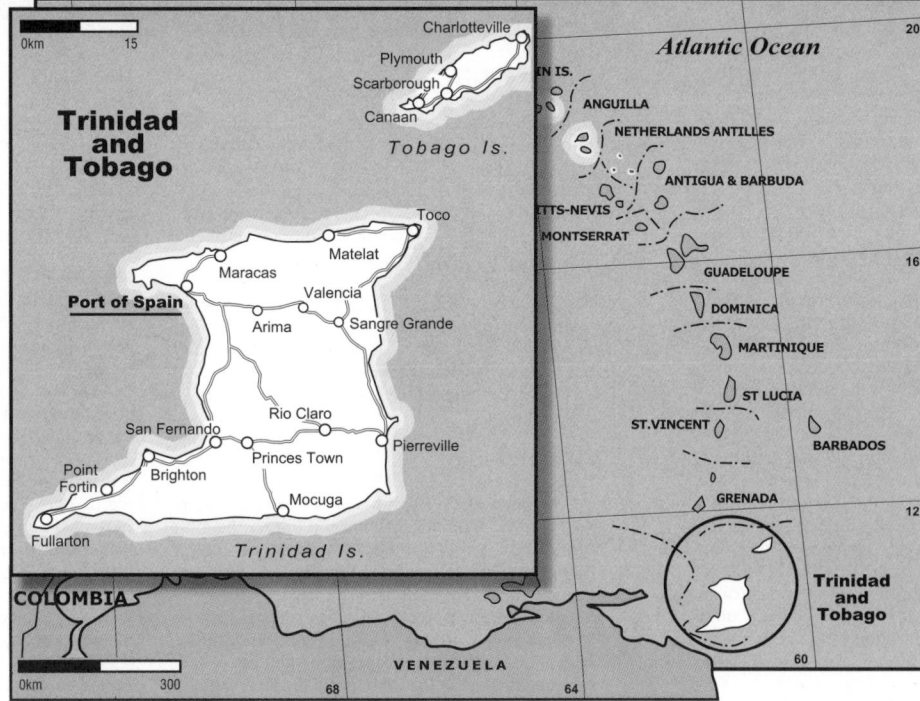

A lthough the islands of Trinidad and Tobago form a single nation, their histories are different. Trinidad, 12 kilometers from the mouth of the Orinoco River, was claimed by Columbus for Spain in 1498. Tobago had been inhabited by Carib indians, but when the Dutch arrived in 1632, they found it uninhabited. Shortly thereafter the Spanish took the island to prevent the Dutch from using it as a base for exploring the Orinoco, where there was thought to be gold.

[2]   Like the region's other colonies, the islands underwent numerous Dutch, French and British invasions. Instability and minimal demographic growth were the rule: in 1783 the population consisted of 126 Europeans, 605 Africans (of whom 310 were slaves) and 2,032 Amerindians. Trinidad became a British colony in 1802, as did Tobago in 1814.

[3]   Once slavery was abolished in 1834, Africans were replaced by workers from India and China on the sugar-cane plantations, which were central to the economy. As a result most of the African population became urban workers, while the majority of rural workers were Indian. Some black workers still living in the countryside developed a mutual aid system called *gayap*, similar to those of other Latin American communities with Amerindian, African or mixed-blood roots.

[4]   1924 saw the first moves toward autonomy, and the colonial administration allowed limited suffrage for certain minor positions. Trade unions were organized at this time, and they raised the issue of independence.

[5]   The sugar-cane based economy began to decline in the early 20th century and sugar was gradually replaced by oil, which had become the main economic activity by 1940.

[6]   In 1950, internal autonomy having been won, the People's National Movement (PNM) won the elections and Dr Eric Williams was appointed Prime Minister.

[7]   After a brief period as a part of the West Indian Federation (1958-1962), Trinidad and Tobago became independent in 1962.

[8]   Dramatic increases in oil prices during the 1970s changed the economy and society radically. Worker mobilization also grew.

[9]   In 1975 worker protests gave way to major strikes which brought together workers from the oil and sugar-cane sectors, overcoming ethnic rivalry. The movement was defeated when the Government brought out the army to distribute gasoline.

[10]   In August 1976 a new constitution proclaimed Trinidad and Tobago a republic. At the time oil-industry nationalization was initiated, and prices were brought into line with OPEC's, while investment from transnationals was encouraged.

[11]   In May 1981 Dr Eric Williams died, after 31 years as Prime Minister. George Chambers took over from him.

[12]   From 1982 onwards dependence on oil brought about instability and other serious problems. Factors causing the crisis were the international recession and fall in oil prices, as well as falling demand, competition from refineries on the south and east coasts of the US and declining production.

[13]   In October the Chambers Government opposed the US invasion of Grenada, and did not contribute to the expeditionary force from six Caribbean nations. The rival National Alliance for Reconstruction (NAR) won the December 1986 elections, taking 33 of 36 parliamentary seats.

[14]   Arthur Napoleon Robinson's government proposed a five-year 'readjustment' plan as of December 1990, designed to hasten integration with CARICOM (Caribbean Community and Common Market) countries.

[15]   An austerity plan agreed upon with the IMF, from whom the Government received 110 million dollars in 1988 and a standby loan of 128 million dollars in 1989, led to major strikes in the oil sector, as well as a general strike in March 1989.

[16]   In July 1990 the country's first attempted coup d'état took place when some 100 Muslims occupied parliament and demanded the Prime Minister's resignation. On 1 August the rebel group surrendered and the Government granted them amnesty.

[17]   In December 1991 the PNM won the elections with 46 per cent of the vote. Patrick Manning became Prime Minister. The United National Congress (UNC) took 26 per cent of the vote while Robinson's NAR took 25 per cent.

[18]   The Government's economic adjustment and privatization plans sparked massive protest demonstrations in January 1993. Manning called out the army to keep matters under control.

[19]   Believing that the economic and political situation was favorable to him, Manning called early elections in November 1995. He was, however, mistaken, for the PNM took 17 parliamentary seats, as did Basdeo Panday's opposition UNC.

[20]   After making an alliance with

LAND USE

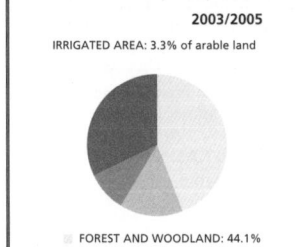

2003/2005

IRRIGATED AREA: 3.3% of arable land

- FOREST AND WOODLAND: 44.1%
- ARABLE LAND: 14.6%
- CROPLANDS: 9.2%
- OTHER USE: 32.1%

PUBLIC EXPENDITURE

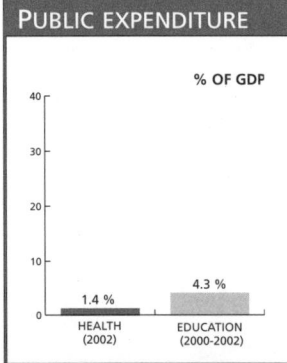

% OF GDP

HEALTH (2002) 1.4 %

EDUCATION (2000-2002) 4.3 %

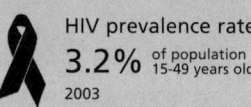

Life expectancy
70 years
2005-2010

GNI per capita
$8,730
2004

Literacy
98% total adult rate
2000-2004

HIV prevalence rate
3.2% of population 15-49 years old
2003

the NAR, Panday became Trinidad and Tobago's first prime minister of Indian immigrant stock.

[21] In early 1997 two PNM members of parliament joined the government coalition as independents. This desertion left the PNM only 15 seats to the coalition's 21.

[22] In February 1997 former prime minister Robinson was elected President.

[23] In May that same year British legal authorities prevented the hanging of nine people charged with murders committed in 1996. The Privy Council was still the highest legal authority.

[24] The December 2001 general elections ended with a tie between the UNC and the PNM (with 18 parliamentary seats each). Panday and Manning agreed that Robinson should decide who would be Prime Minister. The President chose Manning, but Panday (claiming that the ruling party should continue in power in the case of a draw) requested new elections in six months' time.

[25] Manning called on parliament to appoint a spokesperson to break the deadlock that had been paralyzing the political system for almost four months. However, UNC and PNM members of parliament had still not managed to do so by April 2002, and Panday requested new elections due to the lack of parliamentary support. Manning meanwhile claimed that simply opening parliament would be

PROFILE

ENVIRONMENT
The country is an archipelago located near the Orinoco River delta off the Venezuelan coast, at the southern end of the Lesser Antilles in the Caribbean. Trinidad, the largest island (4,828 sq km), is crossed from east to west by a mountain range that is an extension of the Andes. One-third of the island is covered with sugar and cocoa plantations. Petroleum and asphalt are also produced. Tobago, 300 sq km with a small central volcanic mountain range, is flanked by Little Tobago (1 sq km), the islet of Goat and the Bucco Reef. In the archipelago the prevailing climate is tropical with rains from June to December, but tempered by the sea and east trade winds. Rivers are scarce, but dense forest vegetation covers the mountains.

SOCIETY
Peoples: There is a large minority of African origin (43 per cent), and a slight majority descended from Asian Indians (40 per cent) brought during the 19th century as contract workers. Mestizo (14 per cent), European (one per cent) and Chinese (one per cent) groups make up a small minority.
Religions: There is no official religion. Catholic 29.4 per cent; Protestant 29.7 per cent; Hindu 23.7 per cent; Muslim 5.9 per cent; other 11.3 per cent.
Languages: English (official). Hindi, Urdu, French

and Spanish.
Main Political Parties: The People's National Movement (PNM); United National Congress (UNC); Citizen Alliance (CA); National Alliance for Reconstruction (NAR).
Main Social Organizations: Trinidad and Tobago Labor Congress (TTLC) is the only trade union center, with 80,000 members; Jamaat-al-Muslimeen.

THE STATE
Official Name: Republic of Trinidad and Tobago.
Administrative Divisions: seven Counties, four Cities with own government, one semi-autonomous island, Tobago.
Capital: Port of Spain 55.000 people (2003). Scarborough is the main town on Tobago.
Other Cities: San Fernando 29,600 people; Arima 26,400; Point Fortin 17,500 (2000).
Government: Maxwell Richards, President since March 2003. Patrick Manning, Prime Minister since December 2001, re-elected in October 2002. The Parliament of the Republic of Trinidad and Tobago (Legislature) has two chambers: the House of Representatives, with 36 members, and the Senate, with 31 members.
National Holiday: 31 August, Independence Day (1962).
Armed Forces: 3,000 (2003). Other: 4,800 Police.

sufficient to meet constitutional requirements.

[26] Manning won the October 2002 elections, the third in under two years. After a hotly contested race, the PNM won 20 parliamentary seats, while Panday took the remaining 16.

[27] Maxwell Richards was sworn in as President in March 2003, after having been appointed by the electoral college.

[28] In February 2004, the Government proposed a motion to declare Port of Spain bilingual before the Free Trade Area of the Americas (FTAA) was launched - Spanish is the official language in 18 of the 34 American countries and the city was nominated to become the organism's main seat.

[29] In September 2005, Manning requested the aid of London's Metropolitan Police and the US

Federal Bureau of Investigation (FBI) in order to fight crime, which had reached unacceptable levels. At least 10,000 people took part in the March of Death in October to protest about the growing number of violent crimes.

[30] Former Prime Minister Panday was sentenced to two years in jail in April 2006 for having concealed his foreign bank accounts while he was in office. ∎

IN FOCUS

ENVIRONMENTAL CHALLENGES
Waters are polluted by agricultural chemicals, industrial waste and raw sewage. Deforestation and soil erosion are also present. Geographic proximity to a major route for maritime traffic leading from the Caribbean and the Gulf of Mexico out into the Atlantic has caused petroleum pollution on the coast.

WOMEN'S RIGHTS
Women have had the vote and been eligible for office since 1946. In 2003, women held 19 per cent of seats in Parliament and 18 per cent of ministerial level positions. Female participation in the workforce of one million people stood at 38 per cent. In 2004, prenatal care was available to 92 per cent of pregnant women, and 99 per cent of deliveries were attended by qualified personnel. The

maternal mortality rate stood at 160 per 100,000 live births. In late 2003 it was estimated that 14,000 women between 15 and 49 were living with HIV/AIDS.

CHILDREN
The infant mortality rate dropped from 28 per 1,000 live births in 1990 to 18 in 2004 and the child mortality rate from 33 per 1,000 live births in 1990 to 20 in 2004. In 2004, 23 per cent of newborn babies and 7 per cent of under-fives were underweight. The number of children living with HIV/AIDS went from 300 in 2001 to 700 in 2004. As of 2004, two per cent of minors aged between 5 and 14 worked.

INDIGENOUS PEOPLES/ ETHNIC MINORITIES
The Yaio, Nepuyo, Chaima, Warao, Kalipuna, Carinepogoto, Garani and Arawak tribes, generically known as Caribs, inhabited the

islands some 6,000 years before the arrival of the first Europeans. There were some 40,000 when the Spanish settled in 1592. In 1699, the Indian Revolt of Arena (led by Chief Hyarima) was the first major insurrection in favor of the islands' independence. In 1783 they were forced off their land, to make way for sugar plantations worked by African slaves.

Many Carib names are still current, as names of rivers (Caroni and Oropouche), of mountains (Tamana and Aripo), and of locations (Arima, Paria, Arouca, Caura, Tunapuna, Tacarigua, Couva, Mucurapo, Chaguanas, Carapichaima, Guaico, Mayaro, Guayaguayare).

Currently some 12,000 descendants of the original inhabitants live in northeast Trinidad. St Rose Carib Community brings together people who try to preserve ancestral customs and lifestyles.

MIGRANTS/REFUGEES
In April 2004, seven Liberian citizens sought asylum in Port of Spain, claiming that they ran risk of death in Liberia. These applicants were arrested and confined in isolation in Golden Grove jail, in the south of Tobago island. Amnesty International received reports that the prisoners were mistreated and held in inhuman conditions. Although Trinidad and Tobago is a signatory state of the 1951 Convention on the Statute of Refugees and its 1967 Protocol, the country was apparently not complying with the international agreement.

DEATH PENALTY
Still applicable.

* Latest data available in *The State of the World's Children* and *Childinfo* database, UNICEF, 2006.

Tunisia / Tunisie

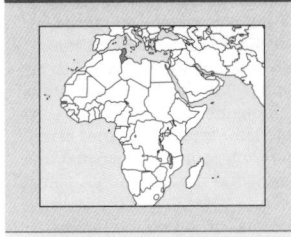

Population:	10,318,640
Land area:	163,610 km²
Capital:	Tunis
Currency:	Dinar
Language:	Arabic

From the 12th century BC the Phoenicians had ports in North Africa. Carthage was founded in the 8th century BC and by the 6th century the Carthaginian kingdom encompassed most of present-day Tunisia. Carthage became part of Rome's African province in 146 BC after the Punic Wars. Roman rule lasted until the Muslim Arab invasions in the mid-7th century AD.

2   In Tunisia the Arabs met the strongest resistance to their advance, but this region eventually became one of the best-cultivated and developed of their cultural centers; the city of Kairuan is associated with some of the most outstanding names in Islamic architecture, medicine and historiography. During the dissolution of the Almohad Empire, Tunisia attained independence under the Berber dynasty of the Hafsids who, between the 13th and 16th centuries, extended their power over the Algerian coast.

3   European maritime trade attracted Turkish corsairs. The most famous, Khayr ad-Din (known as Red Beard), set up his headquarters in Tunisia, placing the Tunisian-Algerian coast under the authority of the Ottoman sultans. The inland regions, however, remained in the hands of the Berbers, allied to Constantinople (now Istanbul).

The need to work with them allowed the Bey (designated governor) to act with a large degree of autonomy and to become in practice a hereditary ruler. The Murad family ruled between 1612 and 1702, and from 1705 until after independence in 1957, this role was filled by the Husseinite family.

4   After the French occupation of Algeria (1830), European economic penetration became increasingly evident as did indebtedness. In 1869 the burden of the foreign debt forced the Bey to allow an Anglo-French-Italian commission to supervise the country's finances.

5   In 1882, 30,000 French soldiers invaded the country, under an agreement whereby Britain, which had just occupied Egypt, 'transferred its rights' to Tunisia, to compensate France for its loss of control over the Suez Canal. In 1883 the country formally became a French protectorate.

6   The Tunisians launched a campaign in 1925 for a Constitution, which would bring autonomy to the country.

7   Habib Bourguiba, a lawyer, founded the pro-independence party Neo-Destur in 1934. He was imprisoned by the French for 11 years, and again at the end of World War II.

8   In 1942, during World War II, German troops arrived to fight against the Allies in Algeria; in 1943 the last troops withdrew from the country.

9   After the War, the Neo-Destur party grew and a series of demonstrations and anti-colonial uprisings led to armed struggle between 1952 and 1955. In 1955, Bourguiba was released and France granted home rule under the Bey's regime. The Bey was deposed in 1957 by a constituent assembly. A republic was proclaimed with Bourguiba as President. He started an energetic campaign against the French presence at the Bizerte naval base, finally dislodging them in 1964. The Neo-Destur Party became the Destur Socialist Party (PSD) and until 1981 was the only legal political organization.

10   Between 1963 and 1969, there was a program of collectivization of small farms and trading companies, and nationalized foreign enterprises. Tunisia opposed recognition of Israel by the Arab League in 1968.

11   In 1969, the collectivization process was aborted and the Tunisian economy was opened up to foreign investment. A 1972 law effectively turned the whole country into a duty-free zone for export industries. Habib Bourguiba, the 'Supreme Warrior', was appointed president for life.

12   Towards the end of the 1970s, the economy suffered the effects of declining phosphate exports, and protectionist measures applied in the European Economic Community against textile imports. In January 1978,

PROFILE

ENVIRONMENT

Tunisia is the northernmost African state. The eastern coastal plains are heavily populated and intensively cultivated with olives, citrus fruit and vineyards. The interior is dominated by the mountainous Tell and Aures regions populated by nomadic shepherds. The Sahara desert, in the south, has phosphate and iron deposits, while dates are cultivated in the oases.

SOCIETY

Peoples: 93 per cent of Tunisians are Arab, 5 per cent are Berber and 2 per cent are European.
Religions: 99.4 per cent Sunni Muslim. There are also Jewish and Catholic groups.
Languages: Arabic (official). French, Tamazight (Berber).
Main Political Parties: Democratic Constitutional Rally; Party of People's Unity; Renewal Movement Ettajdid.
Main Social Organizations: General Union of Tunisian Workers (UGTT).

THE STATE

Official Name: Al-Jumhuriyah at-Tunisiyah.
Administrative Divisions: 25 Government Regions.
Capital: Tunis 699,700 people (2003).
Other Cities: Safaqis (Sfax) 262,000 people; Aryanah 203,500; Sousse 149,200 (2000).
Government: Zine al-Abidine Ben Ali, who became President after a bloodless coup in November 1987; he was elected as President in April 1989, re-elected in 1994, 1999 and 2004. Mohamed Ghannouchi, Prime Minister since November 1999. Unicameral Legislature: Chamber of Deputies, with 189 members.
National Holiday: 20 March, Independence Day (1956).
Armed Forces: 35,000 (2003). Other: 13,000, National Police; 10,000, National Guard.

LAND USE

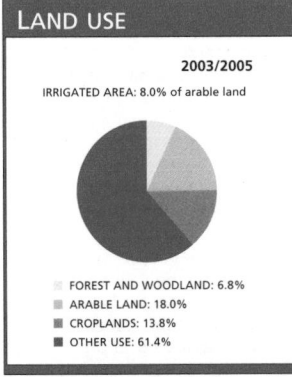

2003/2005

IRRIGATED AREA: 8.0% of arable land

- FOREST AND WOODLAND: 6.8%
- ARABLE LAND: 18.0%
- CROPLANDS: 13.8%
- OTHER USE: 61.4%

PUBLIC EXPENDITURE

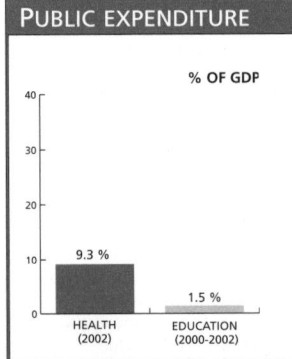

% OF GDP

HEALTH (2002) 9.3 %
EDUCATION (2000-2002) 1.5 %

Life expectancy
74 years
2005-2010

GNI per capita
$2,650
2004

Literacy
74% total adult rate
2000-2004

HIV prevalence rate
<0.1% of population 15-49 years old
2003

IN FOCUS

ENVIRONMENTAL CHALLENGES
Inadequate toxic waste dumps are causing serious environmental damage. There is water pollution, deforestation and overgrazing, which causes soil erosion and desertification.

WOMEN'S RIGHTS
Women have been able to vote since 1959. In 2005 they held 23 per cent of seats in the lower chamber and 13 per cent of seats in the upper chamber of Parliament. In 2003, female workers made up 33 per cent of the total labor force of four million. In 2004, prenatal healthcare covered 92 per cent of pregnant women, and 90 per cent of births were attended by qualified personnel. Maternal mortality stood at 69 per 100,000 deliveries.

CHILDREN
Between 1990 and 2004 the infant and under-five mortality rates were reduced by half. The former fell from 41 to 21 and the latter from 52 to 25 per 1,000 live births*. The priority is to reduce repetition and drop-out rates in primary education, and to guarantee quality education throughout the country.

INDIGENOUS PEOPLES/ ETHNIC MINORITIES
The Bedouins, located in the city of-Gafsa (Qafsa) in the center of the country and the southern region, number around 2,100,000. They have mostly adopted a semi-nomadic form of life and in the winter raise cattle and carry out traditional trading with caravans across the desert. In the summer they crop-farm on the edges of the desert. They are Muslim, mostly Sunni, although some have adopted Sufism.

The Berbers, who speak Tamazight, amount to some 4,500 inhabitants, mostly living on the island of Jerba, in the Gabes Gulf, to the southwest of Tunis, where they have lived for thousands of years. Europeans and Jews make up two per cent of the population.-

MIGRANTS/REFUGEES
The country grants refugee status and considers requests for asylum in line with international recommendations. Extradition of political refugees is against the law. In the few cases in which it was requested, extradition was refused and asylum-seekers were not obliged to return to countries where they had suffered persecution. In early 2006, the UNHCR assisted 100 refugees from Algeria, Burundi and Iraq, while it attempted to find a solution to their situation.

DEATH PENALTY
In practice, the country has abandonned the death penalty. The last execution took place in 1991.

* Latest data available in *The State of the World's Children* and *Childinfo* database, UNICEF, 2006.

prisoners. The PSD, still playing a dominant role, was renamed the Democratic Constitutional Rally (RCD).
[18] The elections of April 1989 were considered by observers to be the freest since independence. In the leglislative election, the ruling RCD won 80 per cent of the vote and all the seats; the Hezb Ennahda Islamic movement, although illegal, obtained 15 per cent through independent candidates. In the presidential election, Ben Ali was elected with 99 per cent of the vote.
[19] In 1991, religious parties were banned. A restrictive law of association was adopted in March 1992 and in July members of Hezb Ennahda were sentenced to life imprisonment.
[20] In November 1993, Ben Ali passed another law limiting 'basic freedoms'. In this political climate, he was re-elected with 99 per cent of the vote in the March 1994 general elections.
[21] The plan for economic liberalization and hard-line political policy was continued. One of the main opposition political leaders, Mohamed Moada, was sentenced to 11 years in prison in October 1995 for having published an account of the curtailment of freedom in Tunisia and for maintaining secret relations with Libya.
[22] At the first multi-party elections, held in October 1999, Ben Ali took 99.4 per cent of the vote. One of the first measures passed by the new administration was the release of 600 political prisoners, mainly those from the al-Nahda Movement and the Workers' Communist Party.
[23] The death of Habib Bourguiba in April 2000 brought together political leaders from Europe and the Arab countries, including Presidents Jacques Chirac of France, Abdelaziz Bouteflika of Algeria and Yasser Arafat of Palestine.
[24] In March 2001,-Amnesty International (AI) called on Tunisia to stop the escalation of harassment and pressure on human rights activists, which started when the Tunisian League of Human Rights was suspended in November 2000.-
[25] Although the Constitution limited presidential rule to three terms in office, in September 2001 the DCR Central Committee again elected Ben Ali as presidential candidate for 2004.-
[26] At the beginning of 2003, President Ben Ali made several appeals for a peaceful solution to the Iraq problem and supported the UN Security Council resolutions on this issue.
[27] In 2004, General Habib Ammar was designated president of the organizing committee for the World Summit on the Information Society (WSIS) to be held in Tunis in 2005. Ammar - a commander in the National Guard and later Interior Minister after Ben Ali's coup - was accused by the World Organization Against Torture (OMCT) and the Swiss Association Against Impunity (TRIAL) before the Geneva General Prosecutor of torturing opposition activists.
[28] In 2004, Ben Ali and his party won with 95 per cent of the vote in the presidential and legislative elections, deemed fraudulent by opposition parties. An independent human rights group was not authorized to monitor the elections, but there were observers from the Cairo-based Arab League in several voting districts.
[29] According to the opposition, the Government orchestrated the elections to cover up a police state which assaulted dissidents, kept hundreds of political activists in jail and restricted the work of the media. The Government rejected the allegations. Nejib Chebbi, the top opposition leader, who had boycotted the elections, said 'these are not the results of a democratic country, but of a totalitarian regime'.
[30] Unexpectedly, French Prime Minister Jean-Pierre Raffarin called for democratic reform during his official visit to Tunisia in January 2005, the first by a French premier in two decades. His statements were received with surprise in Tunisian circles.
[31] In a letter sent to the United Nations in October 2005 for the International Day of Solidarity with the Palestinian People, Ben Ali asked Israel to respond to international peace efforts, in order to turn the Middle East into a peaceful, secure and stable region.
[32] The WSIS was held in Tunis in November 2005. More than 18,000 Government, civil society and international organization representatives met to discuss general access to the internet and other information and telecommunication technologies. AI had said that the Government's continual repression of human rights activists and its intolerance of dissent threatened to make a mockery of the summit.
[33] An AI member was arrested in May 2006 while attending an annual meeting of his organization. He was expelled from the country for breaking the law and inciting public disorder, according to the authorities. ■

the General Union of Tunisian Workers-(UGTT) - the oldest labor union in Africa - called a general strike-and street fighting left dozens dead. Union leaders, including the President, Habib Achour, were arrested.
[13] Tunisia, like other Arab states, broke off diplomatic relations with Egypt following the Camp David accords-in 1979 which were seen as a betrayal of the Palestinians.
[14] Appointed Prime Minister in 1980, Mohammed Mzali initiated a program of-liberalization. Political parties were allowed to reorganize and labor unions and the UGTT were revived. General elections were held in November 1981 in which the ruling National Front won 94 per cent of the vote and all of the seats. Irregularities were reported.
[15] The country received-Palestinian militants expelled-from Beirut in 1982, and since then has hosted the official headquarters of the PLO (Palestine Liberation Organization).
[16] In January 1984, the Government decided to end some food subsidies. The price of bread rose 115 per cent and violent demonstrations left more than 100 dead. Bourguiba cancelled the price increases. In 1985, there was further labor unrest and the UGTT was placed under government control. There were also violent confrontations with an emerging Islamic fundamentalist movement that ended with some death sentences.
[17] Colonel (later General) Zine al-Abidine Ben Ali began his rise to power in 1986.-He was appointed Prime Minister in 1987 and in November that year he removed President Bourguiba, who was declared mentally and physically unfit to govern. Ben Ali took over as President. The period of national reconciliation meant greater press freedom and the liberation of hundreds of political

Turkey / Türkiye

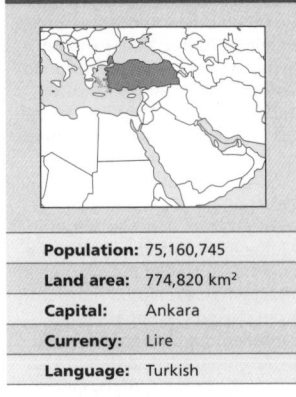

Population:	75,160,745
Land area:	774,820 km²
Capital:	Ankara
Currency:	Lire
Language:	Turkish

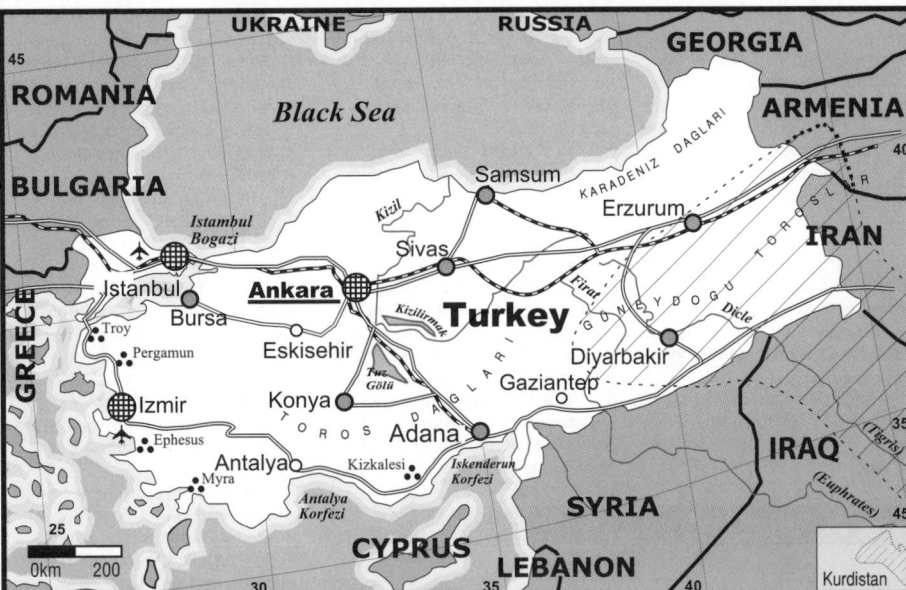

The territory that in 1923 became the Turkish Republic has been inhabited by many different peoples since before the 10th millennium BC.

2   Animal figures carved in caves near Anatolia (the name given to the Asian territory of modern Turkey, as well as Asia Minor) are the oldest traces of human habitation here. Representations of humans from the 7th millennium BC have been found in Hacilar.

3   Wall paintings in the prehistoric city of Çatalhöyük show the customs, dress and deities of an agricultural indigenous people from the mid-7th and 6th centuries BC, who raided the Mediterranean Sea area.

4   Between 5500 and 4500 BC the fortification, sculpture and painting on pottery from Mersin reached a peak of inventiveness. In the latter half of that period, people's lives were revolutionized by the development of metallurgy in copper, bronze, iron, silver and gold.

5   From the 4th millennium BC until 2300 BC Indo-European invaders - mostly Achaeans from Thessalia, in the north of modern Greece - settled on the southeast coast and in the center of the territory of modern Turkey. The kingdoms of Troy (now Hihsarlik), Alisar Huyuk (in central Anatolia), Beycesultan and Cilicia (in the southeast) were prominent.

6   Anatolian architecture and artforms - which included a repertoire of cuneiform symbols arising from their diplomatic and trade relations with Assyria and Mesopotamia - inspired the later foundation of the Mycenaean culture (15th century BC) by the Achaeans in modern Greece.

7   In the 17th century BC, the Hatti culture of Alisar Huyuk was extinguished by the settling of the Indo-European Hittite Empire, which shared hegemony with the Arameans (nomadic semitic groups) as it expanded into Syria. The region of Cilicia remained under the control of the Armenian kingdom of Urartu, until 250 BC, when it was conquered by Parthians (Iranian). The latter were displaced by the Romans in 224.

8   The Hittite empire collapsed in the 12th century BC, as a result of several invasions from Balkan Phrygians and Thracians, among others, who had occupied Greece. In the 8th and 7th centuries BC their expeditions from Greece set forth throughout the Mediterranean, and they founded Byzantium - now Istanbul - among many other cities.

9   In the 7th century BC, the Thracians founded the kingdom of Lydia (now southwest Turkey). Its prosperous gold trade led to the introduction of currency in Greece.

10   Between 546 BC and 334 BC, Anatolia remained under control of the Persian Achaemenian empire, which forced its mostly Greek inhabitants to fight against Athens during the Peloponnesian War.

11   Alexander the Great invaded Asia Minor in 334 BC, but his empire there was short-lived, since some regions were fiercely defended by local princes and others by the Syrians. In the two following centuries, Celtic peoples arrived in the area, while the Romans also started to take territories. The Romans annexed Anatolia as a province of the Roman Empire in the 1st century BC.

12   Asia Minor remained under Christianity's political, religious and cultural dominion from the reconstruction of Byzantium when it became Constantinople, the capital of the Eastern Roman Empire (334/340-1453), until the partial occupation of Asia Minor by Turkish-Mongolian Oguz peoples in the 11th century.

13   The (orthodox) Persian Sunni Muslim house of the Seleucids had Islamized the Oguz during the 10th century. However, the Oguz - like the old inhabitants of Asia Minor - refused to accept the bureaucratic regulations the Persians tried to impose along with their religious practices.

14   In 1299, the Mongolian Turkish leader Osman founded an independent state in Anatolia which, under the name of the Ottoman Empire, would later extend to Asia, Europe - through the Balkan peninsula - and Africa.

15   In 1375, after repelling an invasion by the Mameluke army (in power in Egypt and Syria between 1250 and 1517), the Ottomans defeated the Byzantine Empire (weakened by the Crusades) by taking Constantinople in 1453. It was renamed Istanbul by Mehmed II, the Conqueror, as the capital of his empire.

16   Sultan Selim I, when appointed Caliph (king and religious chief of Islam), extended the Ottoman Empire to Syria, Egypt and part of Mesopotamia, between 1512 and 1520.

17   The reign of Selim's successor, Sultan Suleyman the Magnificent (1520-1566) marked the pinnacle of the Ottoman Empire. During this period, his court developed a culture that blended the Byzantine and Seleucid traditions, as shown in the architecture of the city of Sinan, (Mimar). The Empire became a French ally and thus dominated the Mediterranean

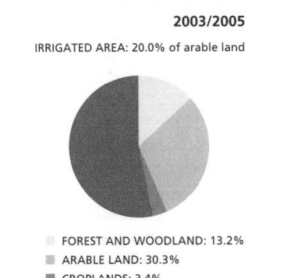

LAND USE

2003/2005

IRRIGATED AREA: 20.0% of arable land

- FOREST AND WOODLAND: 13.2%
- ARABLE LAND: 30.3%
- CROPLANDS: 3.4%
- OTHER USE: 53.1%

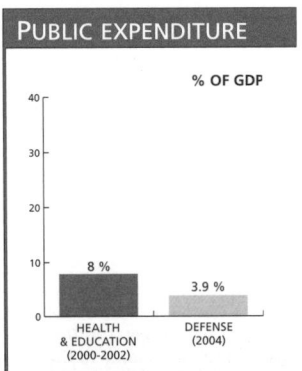

PUBLIC EXPENDITURE

% OF GDP

HEALTH & EDUCATION (2000-2002): 8 %

DEFENSE (2004): 3.9 %

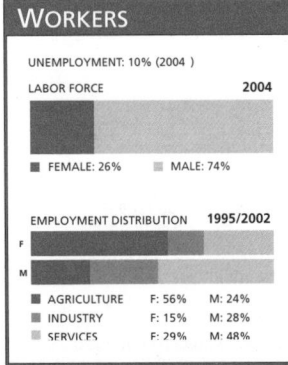

WORKERS

UNEMPLOYMENT: 10% (2004)

LABOR FORCE **2004**

- FEMALE: 26%
- MALE: 74%

EMPLOYMENT DISTRIBUTION **1995/2002**

- AGRICULTURE   F: 56%   M: 24%
- INDUSTRY   F: 15%   M: 28%
- SERVICES   F: 29%   M: 48%

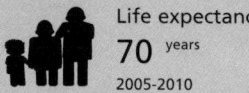

Life expectancy
70 years
2005-2010

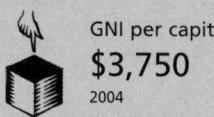

GNI per capita
$3,750
2004

Literacy
88% total adult rate
2000-2004

region, occupying Morocco, Algeria, Tunisia, Libya, Hungary, and even laying siege to Vienna.
[18] After Suleyman's death, the Ottoman Empire declined, in part because of the prosperity brought to Western Europe by American gold and silver and the expansion of British and Dutch trade.
[19] In 1571, the victory at the sea battle of Lepanto signalled the triumph of the Christians through the Holy League - led by John of Austria, with Spanish and Venetian and papal support - and effectively ended the Ottoman navy's control of the Mediterranean. The Ottomans tried to take Vienna once again in 1683. Under the treaty of Karlowitz (1699) Lepanto was returned to them but they

had to give Hungary over to Austria, as well as part of Ukraine and Podolia to Poland. Also, they lost their territories north of the Black Sea in several defeats against Russia.
[20] In the 19th century, the remains of the Ottoman Empire went through permanent crisis: its Christian regions sought independence, while many regional *Pashas* (administrative authorities) rebelled. The only thing that prevented the Empire from falling apart was the rivalry between Russia and Britain over the control of Christian insurgents.
[21] Greek independence (1831-32) was followed by Egypt's (1839-41). The treaty of San Stefano ended the Russian-Turkish wars (1877-78),

with great territorial losses for the Ottomans. Also at the Congress of Berlin (1878), European power forced the Turks to withdraw from Serbia, Montenegro, Romania, Bulgaria, Bosnia, Herzegovina and Cyprus, which remained under British control.
[22] In the Tanzimat ('reorganization') period from 1839-1876, the Ottomans borrowed large sums from the British for revitalizing the economy. The debts undermined the reforms and suppression followed. These plans were driven by the Tanzimat, which supported a Parliamentary monarchic Constitution, accepted by the Sultan in 1876 but never promulgated.
[23] In 1908, a reform movement of university and military academy students - the Young Turks - rebelled, demanding the implementation of the 1876 Constitution. In 1909, Sultan Abdul Hamid was-deposed by Muhammad V, but the sultans' power was now limited. The Young Turks' reforms included the secularization of Muslim schools and courts and the introduction of women's rights during World War I (1914-1918). The modern state apparatus of the Tanzimat was democratized, management of the economy, industry and agriculture were developed.
[24] In 1911-12, Italy occupied Libya, Turkey's last African possession, and Rhodes, among other islands in the Aegean sea. After the Balkan Wars in 1912-13, Turkey was left with only Eastern Thrace in Europe.
[25] In World War I (1914-18), the Ottoman Empire sided with Germany and the Austro-Hungarian Empire. Minority ethnic groups that remained in Turkey were brutally repressed by the Young Turk government, which killed some one million Armenians in 1915 in what was to be the first genocide of the 20th century. Armenians and Greeks had controlled much industry and business before the War.
[26] In 1919, a Turkish nationalist movement organized by military leader Mustafa Kemal (later known as Ataturk: father of all Turks) formed a revolutionary government in Ankara, leading the Turkish War of Independence (1918-1923).
[27] The Treaty of Sèvres (1920) took away all the Ottoman Asian dependencies, Eastern Thrace (except Istanbul and its suburbs), Gallipoli, its Aegean islands and Smyrna, while the Bosphorus strait and the Dardanelles were declared international waters. The Treaty also granted autonomy to the Turkish region of Armenia and the

area inhabited by Kurds, which was then called Kurdistan (spread between Turkey, Iraq, Iran and Syria), where oil had been found. The area holds 100 per cent of Turkish oil and 74 per cent of Iraqi oil. The Treaty was not recognized by Ataturk's movement.
[28] After repulsing a Greek incursion into Asia Minor in 1922, Ataturk deposed Sultan Muhammad VI and negotiated a new treaty, signed in Lausanne in 1923, which exempted his country from paying war reparations, cancelled the privileges enjoyed by foreign traders, and set Turkey's current borders, ignoring Kurdistan's autonomy.
[29] In 1923, Ataturk proclaimed the Republic of Turkey and was elected President by the National Assembly, where Ataturk's Republican People's Party (CHP) held a majority until 1950.
[30] Ataturk banned the two opposition parties that were formed before his death. The Kurdish separatist movement, that rebelled in 1925, 1930 and 1937, was the only organization that challenged his regime.
[31] Implementing his party's program (identical to the 1937 Constitution) between 1924 and 1937, Ataturk imposed a new national identity that sought to undermine imperial traditions and build on the Young Turks' reforms. Measures included the use of Turkish as the official language and prohibition of other languages, use of the Latin alphabet and Gregorian calendar.
[32] Turkey, which had regained the Bosphorus and Dardanelles in the 1936 Montreux Conference, declared its neutrality when World War II (1939-45) began.
[33] The centralized economy of Ataturk's autocratic regime exacerbated the low productivity that had contributed to the fall of the Ottoman Empire. After his death in 1938, his successor Ismet Inonu (1884-1973) saw the risk of losing territory near the Black Sea to the USSR after World War II, and accepted US military bases and loans from 1947. The Turkish President adopted liberalization measures in exchange.
[34] In 1947, Inonu reinstated the Democratic Party (DP - split from the CHP), the National Party (NP - pro-Ataturk) and some newspapers that operated under censorship. Like Kemal, he also repressed separatist, religious, socialist and communist activity. Among those jailed was poet Nazim Hikmet (1902-63) whose work - known in Turkey after his death - synthesized Turkish oral poetry.
[35] In the 1950 elections, the DP candidate Celal Bayar was chosen

PROFILE

ENVIRONMENT

The country is made up of a European part, Eastern Thrace, and an Asiatic part, the peninsula of Anatolia and Turkish Armenia, separated by the Dardanelles, the Sea of Marmara and the Bosphorus. Eastern Thrace, located in the southeast of the Balkan Peninsula, makes up less than one-thirtieth of the country's total land area, including an arid steppe plateau, the Istranca mountains to the east, and a group of hills suitable for farming. Anatolia is a mountainous area with many lakes and wetlands. The Ponticas range in the north and the Taurus range in the south form the natural boundaries of the Anatolian plateau, which extends eastward to form the Armenian plateaus. The east is occupied by the Armenian massif, around the lake region of Van, where there is much volcanic activity and occasional earthquakes. Parallel to the Taurus there are a number of ranges known as Antitaurus, which run along the borders of Georgia, together with the Armenian mountains. The country is mainly agricultural. The lack of natural resources, and absence of capital and appropriate infrastructure, have been major obstacles to industrialization.

SOCIETY

Peoples: Most are descendants of ethnic groups from Central Asia that began to settle in Anatolia in the 11th century. The largest minority is Kurdish (20 per cent), followed by Arabs (1.5 per cent), Jews, Greeks Georgians and Armenians (0.3 per cent). Their cultural autonomy is limited.
Religions: Mainly Islamic (80 per cent Sunni, 20 per cent Shi'a, of which 14 per cent are Alevi, non-orthodox Shi'a); Christians (0.2 per cent).
Languages: Turkish (official) and some 30 languages of ethnic minorities, such as Kurmanji spoken by the Kurdish minority.
Main Political Parties: Justice and Development Party (AKP); Republican People's Party (CHP); True Path Party; Nationalist Action Party.
Main Social Organizations: Confederation of Public Sector Unions; Confederation of Revolutionary Workers Unions; Moral Rights Workers Union; Turkish Confederation of Employers' Unions; Women for Women's Rights.

THE STATE

Official Name: Türkiye Cumhuriyeti.
Administrative Divisions: 74 Provinces.
Capital: Ankara 3,582,000 people (2003).
Other Cities: Istanbul 9,500,000 people; Izmir (Smyrna) 2,272,500; Bursa 1,164,400; Adana 1,137,100; Gaziantep 778,200 (2000).
Government: Head of State: Ahmet Necdet Sezer, President since May 2000. Head of Government: Recep Tayyip Erdogan, Prime Minister since March 2003. The Grand National Assembly of Turkey has 550 seats.
National Holiday: 29 October, Republic Day (1923).
Armed Forces: 515,000 troops (2003). Other: 70,000 Gendarmes-National Guard; 50,000 Reserves.

Under-5 mortality **32** per 1,000 live births 2004	Poverty **3.4%** of population living on less than $1 per day 2003	Debt service **35.9%** exports of goods and services 2004	Maternal mortality **70** per 100,000 live births 2000

president with 54 per cent of the vote.

[36] The Bayar administration (1950-60) increased Turkey's economic and political dependence on the US. Turkey joined NATO in 1952. Annual average inflation rose to 15 per cent in that period, when Turkish emigration to Europe began.

[37] The 1959 British-Turkish-Greek agreement on Cyprus included British withdrawal but did not determine the issue of sovereignty over the island in a way that satisfied both Turks and Greeks. That, added to a series of corruption accusations against the DP cabinet and growing unemployment, caused widespread protests that led to a military coup on 28 April 1960.

[38] After the coup, the military imposed martial law and formed the National Unity Committee, where Muslims and non-religious wrote a secular Constitution which was approved the following year by referendum.

[39] The October 1961 elections were marked by confusion and fear. They were won by the successors to the banned DP (Justice Party - JP, Party of the New Turkey and Peasants' Party), which had been heavily rejected 18 months earlier. General Gursel was elected president, with a coalition government led by Inonu.

[40] Civil liberties continued to improve slightly until the Government resigned in 1965. In 1964, legislation was brought in to prevent the consolidation of the JP in Parliament. Between 1962 and 1964, the Turkish Government amnestied the DP's political prisoners and legalized trade unions.

[41] The JP won the 1965 elections with 51 per cent of the vote. The new president, Suleiman Demirel (JP - re-elected in 1969) was able to reconcile opposing interests and attracted foreign investment, but the army forced him to resign in March 1971 after an escalation of violence between the leftist Federation of Turkish Revolutionary Youth and the Turkish People's Liberation Army (founded in 1969), on one side, and radical groups from the JP that in 1970 had formed the DP, on the other.

[42] In April 1971, a new government coalition formed by the army decreed martial law until the 1973 elections, when the army replaced military courts by 'special security courts' that were still active in 2004.

[43] No party obtained a majority in the 1973 elections. Bulent Ecevit, of the Social Democratic People's Party, founded in 1972 by Ismet Inonu, was the

IN FOCUS

ENVIRONMENTAL CHALLENGES

Istanbul is polluted with high levels of sulfur dioxide and the Marmara Sea is contaminated with mercury. Uncontrolled logging also contributes to environmental degradation.

WOMEN'S RIGHTS

Women have been able to vote and run for office since 1930. In 2003, they held 4.4 per cent of seats in Parliament and 4 per cent of ministerial positions. Female workers made up 39 per cent of the country's total labor force of 34 million people. Some 56 per cent of them worked in the agricultural sector, 29 per cent in the service sector, and 15 per cent in the industrial sector. Prenatal care is received by 81 per cent of pregnant women, and 83 per cent of deliveries are attended by skilled health workers.*

Spousal abuse is considered an extremely private matter, involving societal notions of family honor. Few women go to the police or other institutions seeking help. Close to 60 per cent of Turkish women are victims of violence, and 95 per cent of them suffer this violence within their own homes.

'Honor killings' - the murder of women by male family members when they are suspected or proven to be unfaithful or of having sex before marriage - continue in rural areas. Every year there are at least 30 murders of this kind,

and they are more frequent among Kurdish families in the southeast.

Spousal rape is not considered a crime. Rapists can evade punishment if they agree to marry their victims. Punishment for sexual assault is more severe if the victim is married rather than single or not a virgin.

CHILDREN

Children in the country's rural areas lag behind their urban counterparts in almost every social and economic indicator, from infant and under-five mortality rates to school enrolment.* Thousands of girls receive no schooling whatsoever. In rural areas, the shortage of schools and classrooms means that many teachers have more than 100 students in a single class. The infant mortality rate is 28 deaths per 1,000 live births and the under-five mortality rate is 32 deaths per 1,000 live births.* Some 16 per cent of newborns are underweight and 12 per cent are undersized.* Between 77 and 88 per cent of children under one year old are immunized against the most common childhood diseases such as poliomyelitis, measles and tetanus.*

While 88 per cent of girls and 89 per cent of boys attend primary school, the enrolment rate for girls is significantly lower than that for boys in rural areas.

INDIGENOUS PEOPLES/ ETHNIC MINORITIES

Kurds comprise 20 per cent of the country's population. Most of them live in the southeast, although

thousands have given up public expressions of their culture, have been assimilated into Turkish society, and live in Istanbul. Kurds living in Turkey are mostly Sunni Muslims, although a Shi'a minority also exists. Southeastern Kurds depend on agriculture for their livelihoods and continue to be semi-nomadic.

Kurds in Turkey face a great deal of social, cultural, economic and political discrimination. For a long time the Government has marginalized the southeastern region of the country by allocating it meager budgets. Until 1991, speaking Kurdish in public was outlawed. In 2003 the Turkish Government, pressured by EU demands, allowed a few Kurdish language courses and broadcasts.

MIGRANTS/REFUGEES

In late 2004 there were close to 10,000 refugees and asylum seekers in Turkey, mostly from Iran, Iraq, Afghanistan and Somalia. At the same time, just over 33,000 Turks sought asylum in different parts of the world, primarily in Western European countries.

DEATH PENALTY

Although it is still in force, there have been no executions since 1984.

*Latest data available in *The State of the World's Children* and *Childinfo* database, UNICEF, 2006.

candidate with the most votes. The pro-Ataturk Inonu and his CHP followers had proclaimed themselves social-democrats, calling upon their traditional following among Turkish peasants. Inonu sought to modify his public image after his close association with army. The JP and the Islamic National Salvation Party (NSP) came in second and third.

[44] Turkish military intervention in Cyprus caused that island to divide in July 1974, provoking cabinet splits. The cabinet was replaced with a crisis committee which alternated in government with Ecevit and Demirel, while clashes with guerrillas and right-wing extremists intensified in the interior.

[45] On 12 September 1980 General Kenan Evren dissolved Parliament and applied martial law with a brutality that resulted in thousands of accusations of human rights violations by several

Western European organizations.

[46] In 1982, the official return to democracy was heralded with a new constitution that included a one-chamber Parliamentary system. However, 200 candidates were banned from the 1983 elections.

[47] The 1983 elections were won by Turgut Ozal (a former World Bank official), the candidate for the new center-right coalition, the Motherland Party.

[48] In 1984, Kurdish separatists founded the Kurdistan Workers Party (PKK), launching an armed struggle in southeast Turkey. That year, the Government recognized the Northern Republic of Cyprus.

[49] In 1987, an election year, inflation reached 87 per cent. However, Ozal was helped by the fact that Turkey's application to join the European Economic Community (EEC) was accepted that same year. He obtained 36 per cent of the vote in the elections,

followed by Demirel with 29 per cent and the True Path Party (TPP).

[50] The EEC made Turkey's admission conditional on its ratification of human rights treaties, the normalization of its relations with Greece (which implied negotiations on the status of Cyprus and the Aegean sea oil), and the reduction of unemployment.

[51] In 1988, Istanbul was freed from eight years of martial law, after the Government pledged itself to enforce human rights at the European Council and the UN.

[52] In October 1989, Ozal was re-elected President, in spite of having been defeated in the March municipal elections due to accusations of corruption.

[53] In August 1990, when Iraq was blockaded after the invasion of Kuwait, Turkey interrupted the flow of Iraqi oil to the Mediterranean by blocking the oil pipeline through its territory. It

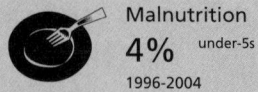 **Malnutrition**
4% under-5s
1996-2004

 Water source
93% of population using improved drinking water sources
2002

 **Doctors**
124 per 100,000 people
1990-2004

Primary school
90% net enrolment rate
2004

also allowed the use of its military airports and US bases for the bombing of Iraq.

[54] In October 1991, 20,000 Turkish soldiers entered northern Iraq in order to attack PKK bases. Kurdish representatives accused the Turkish Government of bombing the civilian population.

[55] The October 1991 parliamentary elections were won by Demirel's True Path Party (TPP) with 27 per cent of the vote. Demirel sought an alliance with Erdal Inonu's Social Democratic Populist Party (SDPP), which came in third, to become Prime Minister. The Motherland Party (ANAP), which obtained 24 per cent of the vote, formed the opposition, although it would support any measure leading to Turkey's admission into the EEC.

[56] In mid-March 1992, the banned PKK announced the formation of a war government and a national assembly in the territory they claimed as the core of Kurdistan. In April, Turkey and Syria announced an agreement to fight against the PKK. Syria closed its PKK training camps and carried out stricter controls along its borders.

[57] In 1992, the Council of Europe urged the Government to reduce repression against the Kurdish community. The Turkish authorities subsequently granted an amnesty to 5,000 political prisoners and authorized the use of the Kurdish language in public places. In November, the EEC set 1996 as the date for Turkish admission to the Customs Union, a first step toward eventual membership.

[58] Upon the death of President Turgut Ozal in April 1993, Demirel was chosen as his successor. Tansu Çiller, minister of economic affairs, assumed the leadership of the DYP and was named Prime Minister. Çiller, the first woman to head a government in Turkey, announced she would cut back state spending. In July, some 700,000 civil servants carried out strikes and demonstrations for several days in Ankara, Istanbul and Izmir.

[59] In 1995 the Government used 35,000 soldiers to launch its largest offensive against the PKK, with support from Iranian Kurds from the Kurdistan Democratic Party in order to dismantle PKK bases and attack Kurdish towns. By 2004, some 4,000 Kurdish towns had been destroyed.

[60] The TPP and ANAP formed a government coalition, led by Mesul Yilmaz from the ANAP, who took office in March 1996. In April Turkey signed a military agreement with Israel. On 24 April, a few days before a Muslim festival, Ankara shut off the water supply to Damascus (Syria),

alleging technical difficulties at one of its dams.

[61] In June 1996, the alliance was dissolved and the TPP chose to rule with the fundamentalists (PP). Necmettin Erbakan became the first Muslim head of government in Turkey since 1923. The PP had obtained 158 of the 550 seats, pledging to create Islamic organizations to balance the influence of NATO and the European Union (EU) in Turkey's domestic affairs.

[62] In January 1998, the Constitutional Court, which had accused Erbakan the previous year of leading the country to the brink of civil war and conspiring against the secular regime (while the army claimed the PP leader was connected to underground Islamic organizations), banned him from politics for five years and dissolved his party. The PP then became the Islamic Virtue Party (VP).

[63] In July 1999, PKK leader Abdullah Ocalan was sentenced to death after his extradition five months earlier from the Greek embassy in Kenya, where he had requested asylum. That year, Greece lifted its veto on Turkey's admission to the EU and announced it was willing to negotiate the unification of Cyprus.

[64] Two earthquakes in the northwestern city of Ismit killed some 20,000 people that year.

[65] The Party of the Democratic Left (DSP) led by Ecevit won 22 per cent of the vote, and Ahmet Necder Sezer became President.

[66] Sezer, an independent, began a seven-year term in May 2000. Although he defended the military's secularism, Sezer had been in favor of eliminating clauses that curbed civil liberties from the 1982 Constitution in order to facilitate Turkey's integration into the EU. The following month, Ecevit was appointed Prime Minister.

[67] In July 2001 a new pro-Islamic party, called Saadet - 'happiness' - was founded by members of the VP, which had been banned in June by the Constitutional Court.

[68] That year, the European Human Rights Constitutional Court declared Turkey was guilty of human rights violations against the Greek population of Cyprus during its occupation of that island's northern sector. In March 2002, Turkey allowed a gas pipeline through its territory to supply Greece.

[69] In the November 2002 elections, the Islamic vote, through the Party for Justice and Development (AKP), won 365 of the 550 Parliament seats with 34.3 per cent of the vote, followed by

the CHP with 19.4 per cent.

[70] Although the AKP leader, Erdogan, was not allowed to take part in the 2002 elections, his party campaigned for the elimination of secular principles from the Constitution.

[71] Abdullah Gul (AKP), who became Prime Minister in November 2002, handed over the post to Erdogan in March 2003, after legislative reforms enabled this move.

[72] Erdogan's popularity was based on his self-made image: he started life in a poor family, went to university and became mayor of Istanbul. Erdogan said that as Prime Minister he would direct his administration in such a way as to allow Turkey to become a full member of the EU by 2012.

[73] The year before, inflation had reached 45 per cent, while 50 per cent of the government's spending was used to pay the interest on foreign debts. A new IMF loan was contracted, on the condition that public spending would be reduced and taxes would be increased.

[74] When Erdogan took office in March 2003, Parliament prevented the US from using its military bases in Turkey to attack Iraq, although it allowed US planes to fly over Turkish territory.

[75] In November 2003 a car bomb next to a synagogue in Istanbul killed 25 people and injured more than 200. Two days later, two coordinated attacks against the British consulate and a British bank killed another 25 people.

[76] In February 2004 the European Council presented Ankara with a project for the effective enforcement of Kurdish rights based on an Amnesty International report that accused the Turkish authorities of more than 30,000 deaths, thousands of 'disappearances', torture, sexual violence against women and so on.

[77] In April 2004, the Turkish and Greek inhabitants of Cyprus voted in a referendum to decide on unification between south (Greek) and north (Turkish), and on integration into the EU. The Greek population voted against the first issue - in effect preventing unification - and in favor of the second one, while the Turkish residents voted for both. The EU promised to study the implementation of an economic development plan for the Turkish population of Cyprus, which can only trade with Turkey and is almost totally isolated.

[78] In the last two decades the Turkish drug cartels have been in control of most of the heroin used by some 30 million Europeans. They have turned Turkey into the main drug-dealing hub between

Asia and Europe.

[79] In June 2004 Istanbul hosted a NATO summit under heavy security. Four people died and several were injured in attacks during the summit, which provoked protest from pacifists and environmentalists. The US did not persuade Turkey to intervene in Iraq. Ankara only offered bilateral help to its neighbor country. Germany's refusal and especially French president Jacques Chirac's, was decisive in this matter (see history of Iraq).

[80] On 1 January 2005, Turkey adopted a new currency, the new Turkish lira or YTL, in an attempt to promote greater economic stability. The new lira was equivalent to one million old Turkish lira (TRL).

[81] In March 2005, two days before International Women's Day, police used clubs and tear gas to disperse several hundred demonstrators - mostly women - who were demanding political reforms to guarantee women's rights in Turkey.

[82] The day before, the EU had agreed to initiate formal talks with Istanbul to pave the way for Turkey's accession. The agenda included the demand for negotiations between Turkey and Cyprus' Greek government, which was not recognized by Istanbul, and the adoption of democratic reforms, particularly with regard to the rights of women. The EU immediately expressed its concern over the violence used to quash the demonstration demanding respect for women's rights.

[83] In May 2006, ten people were killed in armed clashes between Turkish soldiers and PKK separatist rebels in the southeastern province of Sirnak. The victims were five soldiers, three village guards and two rebels.

[84] In June 2006, a methane gas explosion in a Turkish coal mine left 17 workers dead and another eight injured. In response to the various versions of the cause of the accident, Energy Minister Hilmi Guler declared that the explosion had not been the result of negligence.

[85] In November 2006, the European Commission warned Turkey to open its ports to Cypriot ships by mid-December or face a possible freeze of its EU membership negotiations. The Commission also expressed grave concern over allegations of torture and human rights infringements. It insisted on the immediate repeal or reform of article 301 of the penal code, under which writers had been prosecuted for 'insulting Turkishness'. ∎

Turkmenistan / Türkmenistan

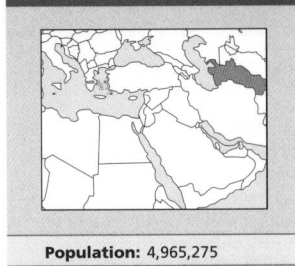

Population:	4,965,275
Land area:	488,100 km²
Capital:	Ashkhabad
Currency:	Manat
Language:	Turkmen

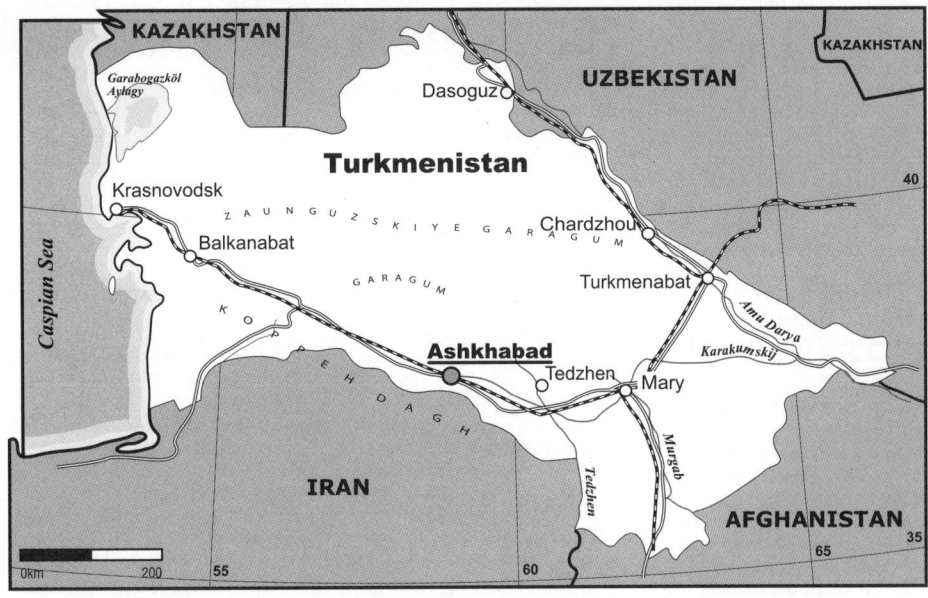

F rom 1000 BC onwards, the area now known as Turkmenistan formed part of different states: first the Persian Empire (controlled by the Achaemenid dynasty) and later the empire of Alexander the Great. In the third century AD it was conquered by the Sassanids, an Iranian dynasty.

2 Between the 5th and 8th centuries, there were successive invasions by the Eftalites, the Turks and the Arabs. From the 6th to the 8th century, the whole Caspian Sea area was controlled by the Arab Caliphate. When that declined in the 9th and 10th centuries, the territory became a part of the Tahirid and Sassanid states. In the mid-11th century, the Seljuk Empire was formed in Turkmenistan. During this period, the Turkmens emerged as an ethnic group through the fusion of Oguz Turks and local groups, and towards the end of the 12th century they were conquered by the Khwarazm Shah dynasty.

3 In the early 13th century, Genghis Khan invaded Turkmenistan and his heirs divided it. The northern regions were taken by the Mongol Tatars. In the 14th and 15th centuries, the country fell under the Timurids (of Tamerlane), who were succeeded by the Uzbek khans of the Shaybani dynasty. From the 16th to the 18th centuries, Turkmenistan

was divided among the khanates of Khiva and Bukhara, and the Persian Safavid state.

4 In the 1880s, the territory was conquered by the Russians and became part of the Trans-Caspian region and the province of Turkistan, but the lands inhabited by the Turkmens passed to Khiva and Bukhara, which were Russian protectorates.

5 Resistance to Russian domination in Turkmenia lasted

until the Battle of Geok-Tepe in 1881, in which the rebels were defeated. The Turkmens were active participants in the 1916 uprising against the Czar. In the city of Tedzhen, several Russian residents and government officials were executed by the local population.

6 The fall of the Czar in 1917 left the Trans-Caspian region under the control of the Russian Provisional Government.

In December 1917, after the Bolshevik take-over in St Petersburg, Worker Council (Soviet) power was proclaimed.

7 In July 1918, Britain re-established the Provisional Government in Trans-Caspia. After two years of civil war, Soviet power was reinstated in 1920 and on 14 February 1924, the Soviet Socialist Republic of Turkmenistan was founded, as a member of the Soviet Union (USSR).

LAND USE

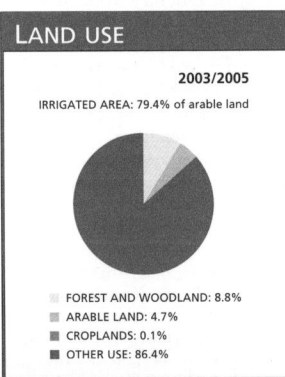

2003/2005

IRRIGATED AREA: 79.4% of arable land

- FOREST AND WOODLAND: 8.8%
- ARABLE LAND: 4.7%
- CROPLANDS: 0.1%
- OTHER USE: 86.4%

PROFILE

ENVIRONMENT

Turkmenistan is in an arid zone, with a dry continental climate, located in the southeastern part of Central Asia between the Caspian Sea to the west, the Amu Darya River to the east, the Ustiurt mountains to the north and the Kopet-Dag and Paropamiz mountain ranges to the south. The land is flat and most of the territory (80 per cent) lies within the Kara-Kum desert. Topographically, 90 per cent of Turkmenistan is sandy plain. On the eastern shore of the Caspian Sea lie the Major and Minor Balkan ranges, of relatively low altitude. The Amu Darya river crosses Turkmenistan from east to west. The Kara-Kum Canal diverts the waters of the Amu Darya to the irrigation systems of the Murgab and Tedzhen oases, as well as those of the Mary and Ashkhabad areas. Turkmenistan is bounded by Kazakhstan to the north-west, Uzbekistan to the east and Afghanistan and Iran to the south. There are rich mineral deposits, including natural gas and oil.

SOCIETY

Peoples: Turkmenis 77 per cent; Russians 6.7 per cent; Uzbeks 9.2 per cent; Kazakhs 2 per cent; Tatars 0.8 per cent; other 6.6 per cent (1996).
Religions: Sunni Muslim (87 per cent); Russian Orthodox 6.4 per cent. Under the 1996 religion law, only these two were allowed to register. Unregistered religious activity is a criminal offense.
Languages: Turkmen (official), Russian.
Main Political Parties: Democratic Party of Turkmenistan. Formal opposition parties are

outlawed; unofficial, small opposition movements exist underground or in foreign countries.
Main Social Organizations: Independent labor unions in process of formation.

THE STATE

Official Name: Türkmenistan Jumhuriyäti.
Administrative Divisions: 3 provinces and 1 Dependent Region (Ashkhabad).
Capital: Ashkhabad (Aschabad) 574.000 people (2003).-
Other Cities: Chardzhou (Cardzou) 166,000 people, Tedzhen.
Government: Saparmurad Niyazov Turkmenbashi, President since October 1990. The Constitution adopted in 1992 gave the President the powers of Head of State and Head of Government. The epithet 'Turkmenbashi' - 'Leader of the Turkmen people' - was conferred on the President in 1993. In 1999 Parliament decided that 'Turkmenbashi will be president for life'. In November 2005 the People's Council voted down Niyazov's suggestion to hold presidential elections in 2009. Bicameral Legislature: a unicameral People's Council (supreme legislative body of up to 2,500 delegates, some of whom are elected by popular vote and some of whom are appointed; meets at least yearly) and a unicameral Parliament (with 50 members elected by direct election every five years, membership is scheduled to be increased to 65 members).
National Holiday: 27 October, Independence Day (1991).
Armed Forces: 18,000.

Life expectancy
63 years
2005-2010

GNI per capita
$730
2001

Literacy
99% total adult rate
2000-2004

HIV prevalence rate
<0.1% of population 15-49 years old
2003

8　Up until that time, Turkmenistan had never had national political unity. Clan membership was the only form of social organization, and most of the population were nomadic. The Soviet regime imposed secularization, industrialization and collective agriculture. During the 1920s and 1930s there were several armed uprisings against Moscow's measures.

9　After World War II, the economy grew, accompanied by increases in oil and gas production and cotton farming. But during Leonid Brezhnev's administration from 1963-83, political problems worsened and the economy entered a period of stagnation (particularly in those republics where the Soviets had adopted a mono-cropping policy, as in the case of Turkmenistan with the cultivation of cotton).

10　From 1985, the changes promoted in the USSR by President Mikhail Gorbachev led to an Islamic renaissance, which found expression in the construction of a number of mosques.

11　After the coup attempt in the USSR in August 1991, the Communist Party of Turkmenistan lost its legitimacy to govern. President Saparmurad Niyazov called a plebiscite that led to the declaration of independence in October and the adoption of a presidential system. In November, Turkmenistan joined the Commonwealth of Independent States (CIS), and the Communist Party changed its name to the Democratic Party.

12　Niyazov gave priority to relations with Turkey - the country which, according to him, had the closest cultural links with Turkmenistan. He also moved closer to Iran, among other things through the construction of the Askhabad-Tehran railway.

13　In December 1997, the first oil pipeline between Turkmenistan and Iran was inaugurated. It was aimed at exporting Turkmen oil and gas to Mediterranean and Persian Gulf countries.

14　In 1999 Parliament voted in favor of making Niyazov president for life, but he announced that he would step down in 2010, when he turned 70.

15　In October that year, Niyazov said that the country would not privatize oil and gas industries for at least the next 10 to15 years, adding that he expected the industry to continue as the economic mainstay of the country. In February 2001, shortly after gas supplies to Russia had been stopped due to disagreements over pricing, Turkmenistan agreed to supply the Russian company Itera with ten billion cubic meters

IN FOCUS

ENVIRONMENTAL CHALLENGES

There is contamination of soil and groundwater from pesticides and agricultural chemicals, tree loss due to poor irrigation, and pollution in the Caspian Sea. The re-routing of the Am Darya river for irrigation means the Aral Sea does not receive water. There is desertification. Only 55 per cent of the rural population has access to drinking water*.

WOMEN'S RIGHTS

Women have been able to vote since 1927. In 2004, 16 per cent of parliamentary seats were held by women. In 2003, they made up 46 per cent of the country's total labor force of two million.

Only three per cent of women aged between 15 and 24 have comprehensive information about HIV/AIDS; less than half of the youth population is aware of how to avoid becoming infected with HIV*.

CHILDREN

Acute respiratory infections and diarrhea are the leading causes of infant mortality*. Some 12 per cent of under-fives are moderately or severely underweight and 22 per cent suffer from moderately or severely stunted growth*. Under-five mortality stands at 103 per 1,000 live births, and infant mortality at 80 per 1,000 live births*.

INDIGENOUS PEOPLES/ ETHNIC MINORITIES

The Russian minority amounts to almost seven per cent of the population. Although the Constitution guarantees equal rights for all inhabitants, in practice Russians are discriminated against and do not feel represented, for example in the civil service or the army. The lack of social and political organizations prevents an organized struggle for their rights. Use of the Russian language is severely restricted and it is barely taught in schools. Its use is almost exclusively limited to the business community. Furthermore,

the Russian language media has gradually been closed down.

MIGRANTS/REFUGEES

By the end of 2004, Turkmenistan hosted 13,250 refugees, mostly coming from Tajikistan (12,150). In August 2005, the President of Turkmenistan passed a decree granting citizenship and residence permits to more than 12,000 Tajik refugees. The more than 1,000 that failed to be reached by the decree remained under UNHCR assistance. A decree issued in 2002 authorized the internal deportation of people who are suspected of causing unrest, engaging in 'immoral' conduct or not working to help the country's economy.

DEATH PENALTY

Capital punishment was abolished for all crimes in 1999.

*Latest data available in *The State of the World's Children* and *Childinfo*dat abase,UNICEF, 2006.

at the price originally agreed.

16　In May 2001 the exploitation of natural resources in the Caspian Sea led to conflict with Azerbaijan, which was working in oilfields claimed by Turkmenistan. Negotiations between experts from both nations could not resolve the dispute and in June Turkmenistan withdrew its ambassador from Baku.

17　After the September 2001 terrorist attacks on New York and Washington and the US military response in Afghanistan, Turkmenistan stood by its neutrality and refused to allow its airspace to be used by allied troops to carry out attacks. However, Niyazov did allow international humanitarian aid through to Afghanistan. The Imam-Nazar frontier crossing became a central point on transport routes.

18　A cabinet reshuffle started in 2000 gathered pace in 2001. By the end of that year few ministers had been in their post for more than a year. One of the most significant changes was the concentration of power in the hands of the President of the National Security Committee, who was made Deputy Prime Minister, responsible for defense, law and order, and foreign affairs, as well as special legal advisor to the President.

19　In late 2001 and early 2002 the Government hoped that the relative stability in Afghanistan could provide Turkmenistan with

a pipeline route to the coast of Pakistan. Despite cordial relations with Moscow, Turkmenistan's continuing lack of economic development was partly due to Russia's refusal to allow gas exports to more competitive markets.

20　In November 2002, the President survived a machine-gun attack on his vehicle. Exiled opposition leaders were accused of planning the operation. Former Minister of Foreign Affairs, Boris Shikhmuradov, was detained together with 40 other activists, accused of conspiracy and condemned to life imprisonment.

21　Niyazov's visit to Moscow in April 2003 led to the signing of an agreement whereby Russia agreed to purchase 60,000 million cubic meters of natural gas annually from Turkmenistan. Diplomatic relations between the two countries were soured by Niyazov's decision to cancel dual nationality, allowed since a common agreement in 1993.

22　Early in 2004, a series of bizarre laws were enacted, including a ban on unkempt beards and hair, and a tax for foreigners wishing to marry Turkmen women.

23　In April in Geneva the UN denounced the human-rights situation in Turkmenistan. Allegations were made about political assassinations designed to keep President Niyazov in power.

24　According to reports issued in 2006 by Freedom House and the

Committee to Protect Journalists, Turkmenistan is among the world's worst performers regarding press freedom. The role of the press is restricted to serve as a tribune for the Government, and the access of citizens to other media, such as the internet, is controlled.

25　In late May 2006, Niyazov decreed the construction of a seismological station near the border with Iran in order to monitor possible nuclear tests to be conducted by its neighbor country. This took place within the framework of the Preparatory Commission for the Comprehensive Nuclear-Test-Ban Treaty Organization (CTBTO) - a UN initiative to promote the non-proliferation of weapons of mass destruction and nuclear tests.

26　In October 2006, Niyazov announced all citizens of the energy-rich country would be provided with natural gas and power free of charge right through to 2030. The President said the decision would 'help ensure a carefree life for our people. Turkmenistan is the second-biggest natural-gas producer in the former Soviet Union after Russia. The country's proven commercial reserves amount to 2.8 trillion cubic meters. The President, first ordered in 1993 that all residents of the ex-Soviet republic receive gas, electricity and water free of charge for a decade. In 2003, the promise was extended through 2020. ■

Turks and Caicos / Turks and Caicos Islands

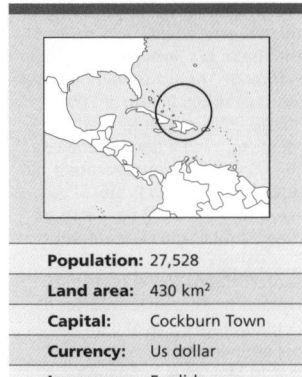

Population:	27,528
Land area:	430 km²
Capital:	Cockburn Town
Currency:	Us dollar
Language:	English

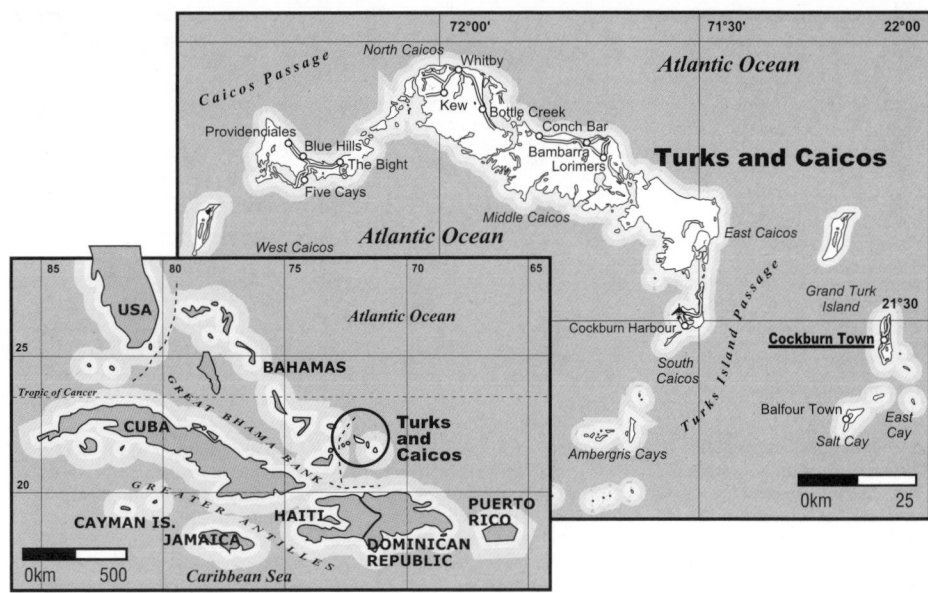

Turks and Caicos are two island chains separated by a deep water channel, lying approximately 150 km north of Haiti at the southernmost tip of the Bahamas chain. Like many of the smaller Caribbean islands, these were first inhabited by Arawak peoples. Some researchers claim it was East Caicos or Grand Turk that Columbus first set foot on when he reached the New World in 1492. The first Europeans to settle on the islands were salt-rakers from Bermuda in 1678.

2   During the following century, Turks and Caicos faced several invasions by both French and Spanish forces. The islands were a refuge during this period for both pirates and their merchant-vessel victims, Spanish galleons carrying American wealth to Europe. By 1787, colonial settlers had established cotton plantations and imported African slaves, and British domination was consolidated. Both archipelagos remained British colonies.

3   Turks and Caicos were administered from the Bahamas until the Separation Act of 1848. After 1874, the islands were annexed to Jamaica, remaining a dependency until independence in 1962, when they again became a separate colony. During World War II, the United States built an airstrip on South Caicos and in 1951 the islands' authorities signed an agreement permitting the US to establish a missile base and a Navy base on Grand Turk island.

4   After the Bahamas' independence in 1972, Turks and Caicos received their own governor. Further autonomy achieved through the 1976 constitution provided for a Governor, a Legislative Council, a Supreme Court and a Court of Appeals.

5   The pro-independence People's Democratic Movement (PDM) won the 1976 elections, defeating the pro-US Progressive National Party (PNP).

6   In 1980, an overwhelming electoral victory of the PNP was attributed to the PDM's failure to resolve the economic crisis and the local population's fear that the state of the economy could worsen with independence.

7   The new head minister, Norman Saunders, convinced Britain to participate in a tourism project with the French company Club Méditerranée. He also aimed at the development of light industry and offshore banking and finally reached an agreement with BCM Ltd for the construction of an oil refinery.

8   Saunders and Stafford Missik, his Development Minister, were arrested in Miami in March 1985. They were attempting to create an international drug network, using the islands as a bridge between the US and South America.

9   The British Government suspended the ministerial system until 1988, when a general election was held. The PDM's Oswald Skippings won the election, marking the end of direct British administration of the islands.

10   Martin Bourke was appointed Governor in 1993, and Derek H Taylor became Prime Minister in 1995.

11   The economy of the islands is mainly supported by tourism, fishing and international financial services.

12   Haitian refugee ships reaching the islands in 1998 and 1999 were systematically rejected by the archipelago's maritime authorities. One of these ships was sunk in June 1998 and six people died as a result of a confrontation with the coastguard.

13   The 1999 Legislative Council elections resulted in a majority for the PDM. Mervyn Jones was appointed Governor in January 2000.

14   A new financial scandal arose in April that year when a lawyer disappeared leaving no trace of the money from a Christian foundation that he had managed. There are 7,000 offshore financial companies registered on the islands, which deal with business considered shady by other countries. The UK, the US and several European countries have complained that this financial system enables the laundering of millions of dollars produced daily by drug dealing and other criminal activities.

15   In February 2004, UN Secretary-General Kofi Annan called for an end to the colonization still suffered by 16 territories, including Turks and Caicos. According to UN resolution 1541, colonies have three options: free association, integration with another State or independence. Canada has expressed interest in integration with Turks and Caicos.

16   On 11 July 2005, Richard Tauwhare took office as Governor.

17   According to an assessment made by the organization *Action Atlas* in 2006, pollution - both organic and chemical - has seriously damaged the coral reefs that surround the islands and threatens their survival. ■

PROFILE

ENVIRONMENT

An archipelago of more than 30 islands comprising the southeastern part of the Bahamas. Only eight are inhabited: Grand Turk, Salt Cay, South Caicos, Central Caicos, North Caicos, Providenciales, Pine Cay and Parrot Cay. The climate is tropical with heavy rainfall, moderated by ocean currents. The islands are on the hurricane path and were particularly devastated in 1928, 1945 and 1960. Fishing is the main economic activity.

SOCIETY

Peoples: The majority is of African descent. There is a minority of European descent and a significant number of Haitian immigrants.
Religions: Baptists 41.2 per cent; Methodists 18.9 per cent; Anglicans 18.3 per cent; Seventh Day Adventists 1.7 per cent; other, 19.9 per cent. **Languages:** English. **Main Political Parties:** People's Democratic Movement; Progressive National Party.

THE STATE

Official Name: Turks and Caicos Islands. **Capital:** Cockburn Town on Grand Turk Island 4,900 people (1999). **Other Cities:** The islands with the largest populations are: Providenciales 7,900 people; Grand Turk 5,000; North Caicos 1,900; South Caicos 1,700 (2000). **Government:** Queen Elizabeth II has been Head of State since February 1952. Richard Tauwhare, Governor appointed by Britain, since July 2005. Michael Misick, Chief Minister since August 2003. Unicameral Legislative Council: 19 members of which 13 are popularly elected; members serve four-year terms. **National Holiday:** 30 August, Constitution Day (1976). **Armed Forces:** Defense is the responsibility of the United Kingdom.

Tuvalu / Tuvalu

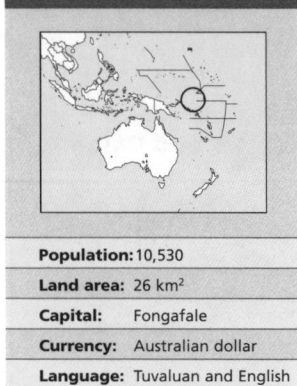

Population:	10,530
Land area:	26 km²
Capital:	Fongafale
Currency:	Australian dollar
Language:	Tuvaluan and English

Around 30,000 BC, peoples from Southeast Asia started their expansion toward the Pacific islands. By the 9th century AD, they had spread throughout practically all of Polynesia. The inhabitants of the archipelago then known as 'Funafuti' came there from the islands of Samoa and Tonga.

[2] In the 16th century the Europeans reported seeing the islands but did not settle on them due to the lack of exploitable resources. Years later, a European named the atolls the Ellice Islands.

[3] Between 1850 and 1875, thousands of islanders were captured by slave traders and sent to the phosphate (guano) works in Peru and the saltpeter mines in Chile. Within a few years, the population had shrunk from 20,000 to 3,000 inhabitants. As a source of slave labor, the Polynesian islands offered direct access to markets on the Pacific coast.

[4] A new wave of invasions began in 1865, with the arrival of British and North American missionaries. By 1892 the islands had become a British protectorate. In 1915, with their neighbors, the Gilbert Islands (see Kiribati), the British formed the Colony of the Gilbert and Ellice Islands. This arbitrary merger was decided on administrative grounds. In a 1974 referendum, 90 per cent of the Ellice islanders voted in favor of separate administrations. The split became official in October 1975, and the first elections in independent Tuvalu were held in August 1977. Toaripi Lauti was named Prime Minister.

[5] The archipelago became independent on 1 October 1978, adopting the name Tuvalu, which in the local language means 'united eight', symbolizing the eight inhabited islands which make up the country. The islands became autonomous from London, but they fell under the economic influence of Australia, which had already started to hold considerable sway over the economy.

[6] Under a friendship treaty of 1979 Washington relinquished its claim over the islands of Nurakita, Nukulaelae, Funafuti and Nukufetau.

[7] Tuvalu is the smallest of the Less Developed Countries, in both population and land area. One of its main sources of foreign exchange is the sale of stamps, which are valuable collectors' items, and the granting of fishing licenses to foreign fleets which earns some $100,000 per year.

[8] Faced with a lack of natural resources, a sparse population and an almost total lack of internal sources of savings or investment, the country depends heavily on foreign aid, mostly from Australia, to finance its regular and its development budget. In 1989, the State was able to meet only about ten per cent of its expenses. Most of this money was sent home by the quarter of the islands' population who lived on neighboring islands working in the phosphate mines.

[9] A 1989 report by the UN included Tuvalu in a list of countries most likely to disappear under the sea in the 21st century as a result of global warming.

[10] In August 1991 the Government announced it would seek compensation from the UK for damages caused when that country authorized the US to build landing strips on the islands during World War II. The army dug trenches that left 40 per cent of the land area in Funafuti unfit to live on.

[11] During the Independence Day celebrations, in October 1995, a new flag was hoisted, replacing the British Union Jack. In 1996, the new Prime Minister, Bikenibeu Paeniu, once again made the British flag official.

[12] A fire swept through a residential village for students in Vaitupu island, causing the death of 18 girls and their supervisor in March 2000.

[13] In September, Tuvalu was formally admitted to the UN as its 189th member, one week after having joined the British Commonwealth.

[14] Lagitupu Tuilimu replaced Ionatana Ionatana as Prime Minister after the latter died in December 2000.

[15] The Government hired out its internet domain name - identified by the letters 'tv' - to a US company for $50 million and 15 per cent of its shares.

[16] Some Tuvaluans are buying land in Fiji, Australia and New Zealand/Aotearoa, anticipating the moment when - as a result of the rise in sea levels due to global warming - the water will no longer be fit to drink. In March 2001 Tuvalu, jointly with Kiribati and Maldives, filed a lawsuit against the US Government because it did not sign the Kyoto Protocol that seeks to reduce the emissions of gases causing global warming.

[17] Former finance minister Saufato Sopoanga was elected Premier in August 2002, after defeating Koloa Telake in the general elections.

[18] In May 2004 Sopoanga announced that he would hold a referendum on whether to turn Tuvalu into a republic, thus replacing the Queen of the UK as head of state. In October, Maatia Toafa took office as Prime Minister after winning the parliamentary elections held that month.

[19] Filoimea Telito was appointed as Governor General in April 2005, after Faimalaga Luka's death.

[20] Reports issued in 2006 confirm that Tuvalu is likely to be one of the first islands to disappear beneath the waves of the Pacific should sea levels keep rising due to global warming. ■

Uganda / Uganda

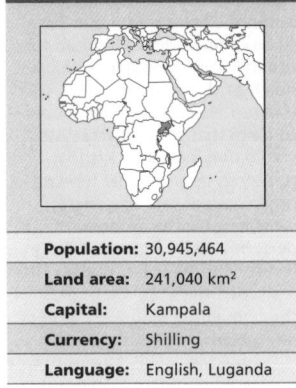

Population:	30,945,464
Land area:	241,040 km²
Capital:	Kampala
Currency:	Shilling
Language:	English, Luganda

In present-day Uganda the ruined mud walls of the Kingdom of Bigo show evidence of urban civilizations dating from the 10th century.

2   The ruins of sizeable hilltop fortresses still remain. These fortifications mark the lines of penetration from the North of the Bacwezi, a nation of Nilotic herders, who subdued the Bantu peoples of the area around the 13th century. Their fortresses - in some cases up to 300 meters in diameter - were built to protect their cattle - their main source of wealth and status. Gradually, the conquerors mixed with the local people, adopted a Bantu language, and came to be called Bahima but continued their nomadic lifestyle.

3   Between the 17th and 18th centuries, the kingdoms of Bunyoro, Buganda, Busoga and Ankole were founded. A dispute for supremacy arose between Bunyoro, supported by the Swahili traders, and Buganda, linked to the 'Shirazis' of Zanzibar. At the beginning of the 19th century Bunyoro lost some of its allies, who formed the independent state of Toro, leading to the undisputed hegemony of Buganda.

4   In the mid-19th century Buganda was governed by *Kabakas* or traditional leaders, who were in theory absolute

rulers but in practice were limited by the *Lukiko*, a council representing the higher castes. Buganda had a standing army which guaranteed its regional autonomy. It had a relatively equitable society, in which caste privileges were more honorary and political than economic, and a sound agricultural economy that allowed it to survive the decline of the slave trade.

5   HM Stanley, the British adventurer and journalist, arrived in 1875. He denounced the spread of Islam in the region and reported an alleged request by Kabaka Mutesa I asking Europe to send missionaries to halt Egyptian-Sudanese religious infiltration.

6   These missionaries soon arrived: English Protestants in 1877 and French Catholics in 1879. They quickly converted part of the Bugandese hierarchy, splitting the power élite into three factions. Two of these reflected the rivalry between missionaries - in local dialect the 'Franza' and 'Ingleza' parties - while the third (moderate and

Islamic) assumed the defense of national interests. The main impact was the consolidation of the European presence.

7   The missionaries succeeded in deposing the Muslim Kabaka Mwanga in 1888, and shortly afterwards the Imperial British East Africa Company (IBEA) arrived, a typical colonial trading company, forerunner for the British Government.

8   The 1886 Anglo-German agreements had ceded the states in the lakes area to the British, who established them as a Protectorate in 1893.

9   The other organized local groups were forced to adopt political systems similar to Buganda's, as the British thought that the Lukiko resembled their own parliamentary system. With the intention of developing a ruling élite to serve as intermediaries for the colonial power, the British undertook 'land reform', privatizing communally-owned land, which left the rural population landless and benefited the Lukiko-based bureaucracy.

10   In 1894, Buganda became a British protectorate. An agreement signed in 1900 transformed it into a constitutional monarchy, controlled by the Protestants.

11   In 1902 the western province

of the country became part of Kenya.

12   The cultivation of cotton as a cash crop started in 1904.

13   During the first half of the 20th century, until the end of World War II, the country was governed indirectly, through the local power structures. London allowed the creation of labor union-type organizations, which brought together the more active militants. Their modern nationalist tendencies subsequently led to the development of anti-colonialist feelings.

14   The severe disruption in food production caused by the confiscation of lands was aggravated by the introduction of export cash crops in the post-War period. Many export crops were new to the region, and their prominence in agricultural production resulted in a steady decline in people's living standards until the 1960s, when the decolonization movement brought about Ugandan independence. Kabaka Mutesa II of Buganda was the first President and Dr Milton Obote the Prime Minister.

15   In 1965, Obote reformed the Constitution, assuming greater powers, and eliminated the federal system imposed by the British. He also adopted policies that favored the poorest sectors, arousing fierce opposition from the Asian population - a minority of some 40,000, holding British passports - who controlled almost all commercial activity in the country.

16   Obote supported regional economic integration with Tanzania and Kenya and the East African Community was established (1967-1977).

17   In January 1971, Obote was overthrown in a coup led by deputy commander of the army, Idi Amin Dada. Obote took refuge in Tanzania. In the economic crisis the Government faced opposition from the Asian minority (which he expelled en masse in 1972) and transnational corporations.

18   Amin's attitudes and measures were controversial: he maintained trade relations with the US and Britain, but also cultivated good relations with the socialist world. Similarly, while he supported several African liberation movements, he opposed Angola's bid for membership of the Organization of African Unity (OAU) and adopted a permanently hostile attitude toward Nyerere's government in Tanzania. He also expropriated land and property from members of the Jewish

LAND USE

2003/2005

IRRIGATED AREA: 0.1% of arable land

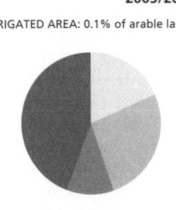

- ☐ FOREST AND WOODLAND: 18.4%
- ☐ ARABLE LAND: 26.4%
- ☐ CROPLANDS: 10.9%
- ■ OTHER USE: 44.3%

PUBLIC EXPENDITURE

% OF GDP

	2.1 %		2.5 %	
	HEALTH (2002)		EDUCATION (2000-2002)	

WORKERS

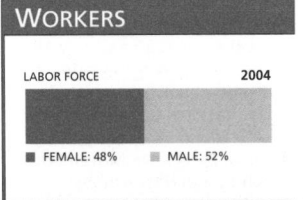

LABOR FORCE   **2004**

■ FEMALE: 48%   ■ MALE: 52%

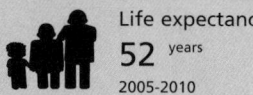
Life expectancy
52 years
2005-2010

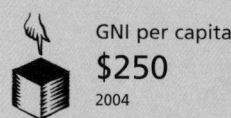
GNI per capita
$250
2004

Literacy
69% total adult rate
2000-2004

HIV prevalence rate
4.1% of population 15-49 years old
2003

PROFILE

ENVIRONMENT

The land is made up of a number of plateaus, gently rolling towards the Nile River in the northwest. There are volcanic ranges and numerous rivers, the largest of which is the Nile. Nearly 18 per cent of the territory is covered by rivers, lakes and swamps. The climate is tropical, tempered by altitude. Timber is taken from the rainforest that covers 6.2 per cent of the land. In addition to subsistence farming of rice and corn/maize, coffee, cotton, tea and tobacco are cultivated as cash crops. Lake Victoria is one of the largest fish reservoirs in the world.

SOCIETY

Peoples: Most Ugandans descend from a mix of various African ethnic groups, mainly the Baganda, Bunyoro and Batoro, and some San/Bushmen, and Sudanese. There are also minorities of Indian and European origin.
Religions: More than half the people are Christian (62 per cent), 19 per cent practise traditional religions and 15 per cent are Muslims. Others: 1 per cent.
Languages: English, the official language, is spoken by a minority. Luganda is the most widely spoken language.
Main Political Parties: National Resistance Movement (NRM), led by Yoweri Museveni; Forum for Democratic Change; Democratic Party.
Main Social Organizations: National Organization of Trade Unions (NOTU); women's groups and other civil society organizations.

THE STATE

Official Name: Republic of Uganda.
Capital: Kampala 1,246,000 people (2003).
Other Cities: Jinja 85,200 people; Mbale 70,600; Masaka 64,900 (2000).
Government: Yoweri Museveni, President since January 1986, elected as President in 1996 and re-elected in 2001 and 2006. Apolo Nsibambi (NRM), Prime Minister since April 1999. Unicameral Legislature: National Parliament, with 303 members.
National Holiday: 9 October, Independence Day (1962).
Armed Forces: 50,000 (2003).

community and made overtures toward the Arab countries.

[19] In the mid-1970s Amin declared himself President-for-Life. In 1978, he provoked a war with Tanzania by annexing land in the north. In April 1979, he was forced to flee Kampala, after a joint offensive launched by Tanzanian troops and opposition activists united in the Ugandan National Liberation Front (FNLU), an umbrella movement uniting efforts aimed at ending Idi Amin's reign of terror.

[20] The new Government's main body, a National Advisory Council led by Yusuf Lule, a politically inexperienced university professor with conservative tendencies, lasted 68 days. Lule was replaced by the FNLU leader, Godfrey Binaisa. However, Binaisa was unable to reconcile the conflicting tendencies within the movement. He was even less capable of confronting the growing prestige of Milton Obote, whose Uganda People's Congress party (UPC) continued to enjoy wide popular support.

[21] The President brought forward the elections scheduled for 1981, and tried to ban Obote's candidacy. This fuelled a crisis that exploded in May 1980 when the army replaced Binaisa with a Military Commission entrusted with maintaining the

electoral schedule and enforcing the democratic principles of the movement that had overthrown Amin. The Commission, under the orders of General David Oyite Ojok, supervised the elections in December 1980, and as predicted the UPC won an overwhelming majority, with Obote as President.

[22] Obote inherited a bankrupt country. The copper mines had not been worked for several years and corruption was rife. Despite the reaffirmation of his support in the elections, in 1981 the defeated parties initiated a destabilization campaign that turned into a guerrilla movement.

[23] The Government authorized the return of Asian businesses, regulated the participation of foreign capital and embarked upon a reorganization of the economy, fighting corruption and speculation. Despite the intensification of political violence in 1985, it achieved the withdrawal of Tanzanian troops who had been in Uganda since the fall of Amin.

[24] Between 1981 and July 1985, major military offensives were launched against the strongholds of the National Resistance Army (NRA) - the military wing of the National Resistance Movement (NRM) founded by former president Yusuf Lule and led by

Yoweri Museveni - and other opposition groups.

[25] In 1981 the Government took steps to prevent cattle smuggling by the nomadic peoples across the Kenyan border, which was causing starvation among thousands in Karamoja province.

[26] The anticipated victory of the UPC in the general elections (scheduled for December 1985) was thwarted by General Bazilio Olara Okello's coup in July 1985. The new president (from the Acholi ethnic group) accused Obote (of Lango origin) of unilateral tribal domination and called elections to form a broad-based government within six months.

[27] After the coup, the National Resistance Army intensified its actions, occupying Kampala in January 1986. On 30 January, NRA leader Yoweri Museveni assumed the presidency and in March announced the fall of the northern town of Gulu, the last bastion of forces loyal to Okello.

[28] Museveni was faced with the reconstruction of a country virtually destroyed, with almost a million dead, two million refugees, 600,000 injured and incalculable property damage.

[29] Uganda's foreign debt rose to $1.2 billion in 1987. In an attempt to establish economic independence and avoid the International Monetary Fund (IMF) Museveni resorted to exchange arrangements with other African states. Some Western countries disapproved of Uganda's relations with Cuba and Libya. The US pressed Tanzania and Rwanda into ending the exchange operations that they had with Uganda.

[30] In February 1992, local human rights organizations pushed for multiparty democracy in the country. The Government's response was that it wanted to build a democracy based on traditional ethnic structures, and that political parties were therefore unnecessary.

[31] Faced with pressure from the opposition and some international agencies, in February 1993 the Government announced the election of a Constituent Assembly for 1995, charged with studying the draft of a new Constitution. The text drafted by the Government was criticized by the opposition (Democratic Party and former President Obote's party, the UPC) for maintaining the partial ban on political parties for a seven more years.

[32] In the March 1994 elections Museveni's supporters won around half the seats, and the direct appointment of some of the posts gave Museveni a broad majority in the new assembly. Continuing with his policy of

restoring local authorities, he authorized the creation in June of an independent kingdom for the Bunyoro, a people in the north of the country.

[33] In 1995 Museveni continued to claim that a multiparty system would only exacerbate 'tribal divisions'. International funding organizations said they were satisfied with the economic performance of Uganda. Foreign investment grew, but budget cuts worsened the situation of most of the population who were already living below the poverty line.

[34] On 9 May 1996, Museveni was re-elected president by more than 75 per cent of the electorate, with a 72.6 per cent turnout, defeating Paul Semogerere and Muhammad Mayanja.

[35] Museveni's economic reforms meant Uganda topped the World Bank's list of aid to 20 debtor countries in 1997. It was estimated that $24 million would be needed to tackle hunger in the country. The Franco-Australian LaSource company paid Uganda for the right to exploit its cobalt mines, and the nation received loans from the European Union and North Korea for the construction of a hydroelectric plant.

[36] In mid-1998, the Ugandan army entered neighboring Democratic Republic of Congo (DRC) and joined the rebels fighting President Laurent Kabila. In October 1999, Ugandan Defense Minister Stephen Kavuma stated that the troops would remain in DRC until peace was restored.

[37] In November 1999, in the Tanzanian city of Arusha, the presidents of Kenya, Tanzania, and Uganda signed a treaty that established the East African Community (EAC) in 2001.

[38] Museveni hosted the African Development Forum 2000, held in Addis Ababa. At the Forum Museveni made reference to the alarming rates of HIV/AIDS in sub-Saharan Africa. The prevalence of HIV in Uganda was one of the highest in the world, but the country managed to considerably reduce the spread of the epidemic through information and prevention campaigns.

[39] A referendum was held in June 2000 to decide on the establishment of a multiparty system. In spite of a boycott by the Democratic Party (DP) and other political groups, 80 per cent of voters (50 per cent of the registered electorate) supported the 'democracy without parties' formula defended by Museveni in the campaign, which meant that Uganda maintained its system of government, unique in the continent. This did not prevent,

Under-5 mortality

138 per 1,000 live births
2004

Malnutrition

23% under-5s
1996-2004

Debt service

6.9% exports of goods and services
2004

Maternal mortality

880 per 100,000 live births
2000

IN FOCUS

ENVIRONMENTAL CHALLENGES
Environmental problems include deforestation and soil erosion caused by intensive cattle grazing. Lake Victoria is being throttled by water-hyacinths. Poaching is widespread. Swamplands are being indiscriminately drained for agricultural use.

WOMEN'S RIGHTS
Women have been able to vote since 1962. Their political representation has doubled in recent years. Women's share of seats in Parliament rose from 12 per cent in 1990 to 24 per cent in 2003, when 23 per cent of ministerial level positions were also occupied by women.

Between 1980 and 2003, women consistently accounted for 47 per cent of the total labor force.

While the female illiteracy rate has steadily decreased, it remains much higher than the rate for men. For every 100 literate men there are only 75 literate women.*

Although 92 per cent of pregnant women receive prenatal care, only 39 per cent of deliveries are attended by skilled health personnel.* The maternal mortality rate is 880 deaths for every 100,000 live births.*

CHILDREN
The under-five mortality rate is 138 deaths per 1,000 live births and the infant mortality rate 80 per 1,000 live births*. Some 12 per cent of all newborn babies are underweight, 23 per cent of under-fives are moderately or seriously underweight and 39 per cent are moderately or seriously stunted*. Of the 530,000 people living with HIV/AIDS at the end of 2003, 84,000 were children aged between 0 and 14. Close to one half of the two million orphans in Uganda were the children of people who died of AIDS. The number of orphans is predicted to reach 3.5 million by 2010.* Some 20,000 minors are infected with HIV/AIDS every year through mother-to-child transmission.*

According to official data, only 64 per cent of schoolchildren reach the 5th grade of primary education.

Approximately 34 per cent of children between the ages of 5 and 14 work.

Since 1986, the Lord's Resistance Army (LRA) has kidnapped more than 25,000 boys and girls. Roughly 80 per cent of those displaced by the armed conflict are children and women.*

INDIGENOUS PEOPLES/ ETHNIC MINORITIES
There are at least 43 different ethnic groups in Uganda. Among the smaller groups are the Batwa, native to the region. The estimated 2,100 survivors of this Banda-speaking group in Uganda have been evicted from the lands in the Bwindi and Mgahingà Parks, to free new areas for cultivation. At present they are almost all living as beggars in the cities. Some 18,000 Nubians live in the Bombo area, 60 km from Kampala. They claim descent from the Kingdom of Kush and the Egyptian Kushites, and speak Kenuzi-Dongola as well as Arabic. They arrived in the region at the end of the 19th century, when they received land in Uganda and Kenya in return for helping the British suppress revolts in Sudan and Egypt. Among the larger groups are the Baganda people whose original lands were to the north-east and north of Lake Victoria. Their highly structured society made them an ideal instrument for British domination. The Baganda and Bunyoro have elected kings, in line with government policy to reintroduce traditional systems of local rule.

MIGRANTS/REFUGEES
Approximately 1.4 million Ugandans have been forced to flee their homes and resettle in other areas of the country due to the armed conflict between the Government and the LRA. In 2004, Uganda took in 252,000 refugees and asylum seekers, most of them from three neighboring countries also engulfed in armed conflict. Over 230,000 of them were from Sudan (198,000), Rwanda (20,000) and DR Congo (13,000). The remainder came from other countries in the region like Burundi, Eritrea, Ethiopia and Somalia. In the meantime, 29,000 Ugandans sought refuge or asylum in other countries such as Sudan, DR Congo, Kenya, South Africa and the United Kingdom.

DEATH PENALTY
The death penalty is still applied.

*Latest data available in *The State of the World's Children* and *Childinfo* database, UNICEF, 2006.

---

in practice, the NRM (now simply called the 'Movement') from acting as a state party. The system gave absolute power to its creator, Museveni.
40   In the March 2001 presidential elections, Museveni was re-elected with 69.3 per cent of the vote, followed by 27.8 per cent for Kizza Besigye, a former colonel of the NRM. International observers estimated that there had been up to 15 per cent electoral fraud, and confirmed that the elections had taken place in a climate of intimidation.
41   In July 2001, Museveni held a meeting with the political leader of the Democratic Republic of Congo, Laurent Kabila's son Joseph, in Dar es Salaam, accelerating the withdrawal of the Ugandan army from that country and the stationing of UN soldiers along the front lines.
42   A report by the World Trade Organization (WTO) indicated that the economic reforms carried out by the Museveni administration, including liberalization of the trade regime, had attracted foreign investment and contributed to the country's growth. GDP had grown, until 2001, by about six per cent per year; the fiscal deficit and inflation were reduced, improving the economic outlook. The agricultural sector accounted for 42 per cent of GDP and provided 80 per cent of jobs. The European Union was Uganda's main trading partner. Regional integration favored an increase in trade between Uganda and other sub-Sahara African countries.
43   In March 2002, Uganda signed an agreement with Sudan to fight the Lord's Resistance Army (LRA), a group led by the 'prophet' Joseph Kony, whose aim was to govern the country according to the Ten Commandments. The rebel movement, located along the border between the two countries, used systematic kidnapping of thousands of Ugandan children as part of its tactics.
44   In October 2002, the escalation of the conflict with the LRA led the army to evacuate over 400,000 citizens from the combat area. In December, after five years of negotiations, a peace treaty was signed with rebel movement Uganda National Rescue Front (UNRF II), made up of Amin's former soldiers.
45   In May 2003, the last Ugandan troops withdrew from DR Congo and tens of thousands of citizens sought asylum in Uganda. In August the former dictator, Idi Amin Dada died in hospital in Jeddah, Saudi Arabia.
46   In February 2004 at least 200 people were killed by LRA rebels in a refugee camp in the north of the country. The President apologized for mistakes in coordination that resulted in the army's failure to prevent the massacre.
47   LRA commander Charles Okullu Bongomin surrendered to government forces in June 2005 following a raid on his hideout in Lalak Bamboo Forest, in the district of Kitgum. Museveni called on rebel leader Joseph Kony to turn himself in to government forces, pledging that he would be granted the same fair treatment and immunity as other former LRA commanders.
48   In a referendum held in July 2005, Ugandans voted for a return to multi-party politics.
49   After the Constitution was amended in 2005 to raise the maximum allowed number of consecutive presidential terms from two to three, Museveni was re-elected in the February 2006 elections, despite serious accusations of fraud before, during and after the vote. This was the first multiparty election since Museveni took over power in 1986.
50   In late March 2006, the World Bank executive council, in coordination with the International Monetary Fund (IMF) and the African Development Bank (ADB), gave the 'green light' for the cancellation of the debt of 13 African countries, including Uganda. The cancellation was scheduled to come into effect on 1 July, at the beginning of the World Bank's fiscal year.
51 In September 2006, the LRA demanded an urgent review of its ceasefire agreement with the Ugandan Government, claiming that the army had opened fire on rebel fighters on their way to an assembly point in southern Sudan. LRA fighters have been moving towards the border in anticipation of a forthcoming peace deal with the Government.
52 In October 2006, the Minister for Water and Environment, Maria Mutagamba, said that Uganda would soon start following its own environmental policies rather than those of the World Bank and the African Development Bank. ▪

Ukraine / Ukrayina

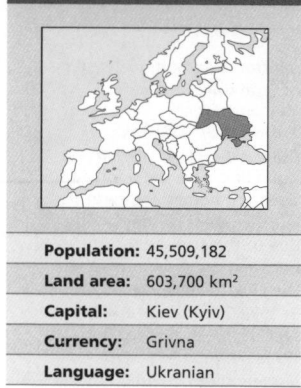

Population:	45,509,182
Land area:	603,700 km²
Capital:	Kiev (Kyiv)
Currency:	Grivna
Language:	Ukranian

Between the 9th and 12th centuries AD, most of present-day Ukraine belonged to the Kievan Rus, which grouped together several alliances of Eastern Slavic peoples. Its nucleus was the Russian alliance, with its capital at Kiev. The ancient Russian people gave rise to the three main eastern Slav nations: Russia, Ukraine and Belarus. In the 12th century, the Kievan Rus separated into the principalities of Kiev, Chernigov, Galich and Vladimir-Volynski, all in what is now Ukrainian territory. In the 14th century, the Grand Principality of Lithuania annexed the territories of Chernigov and Novgorod-Seversky, Podolia, Kiev and a large part of Volin. The Khanate of Crimea emerged in the southern part of Ukraine and Crimea, and expanded into Galicia and Podolia. After the 11th century, Hungary began seizing the Transcarpathian territories.

² Ukrainians emerged as an identifiable people in the 15th century. Their name was derived from *krai*, meaning border, which in 1213 was the name given to the territories along the Polish border. In the 16th century, the use of the name was extended to the entire Ukrainian region. Historically, there were close ties between Ukrainians and Russians, who had fought together against the Polish and Lithuanian feudal kingdoms and against the Tatars in Crimea. The Ukrainian territories (Volin, eastern Podolia, Kiev and part of the left bank of the Dnepr) were incorporated into the Rzecz Pospolita (the union of Poland and Lithuania), which imposed Roman Catholicism.

³ During the first half of the 17th century, the struggle for independence from Poland and Lithuania intensified. The war of the Ukrainian people (1648-1654) under Bohdan Khmelnytsky ended with the unification of Ukraine and Russia. In March 1654, Ukrainian autonomy within the Russian Empire was ratified.

⁴ In 1783, the Khanate of Crimea - home of the Tatars - was annexed by Russia. After the partition of Poland between Russia, Prussia and Austria

(1793-95), the right bank of the Dnepr became a part of Russia and Ukraine's autonomy was abolished at the end of the 18th century. In 1796, the left bank of Ukraine became the Province of Malo-Rossiya (Little Russia).

⁵ After the end of czarist rule in 1917, a dual system emerged in Ukraine, with power being divided between the Provisional Government of Saint Petersburg and the Ukrainian Central Rada (council) in Kiev. In December, after the Bolshevik Revolution, the First Congress of Ukrainian Soviets formed a government. The Ukrainian Central Rada supported the Austro-German troops which invaded the country in the spring of 1918. Until 1920, Ukraine was the scene of major fighting between the Soviets and their internal and external enemies. In December 1922, Ukraine attended the first All-Union Congress of the Soviets, held in Moscow, where the founding of the Union of Soviet Socialist Republics (USSR) was ratified.

⁶ In the period between the world wars, the Soviet Government carried out rapid industrialization and

collectivization of agriculture.

⁷ The secret clauses of the 1939 Soviet-German non-aggression pact incorporated western Ukraine into the USSR. In 1940, the Ukraine was enlarged through the addition of Bessarabia and Northern Bukovina. Germany invaded Ukraine in 1941,when a strong guerrilla resistance began. By the end of World War II, all areas inhabited by ethnic Ukrainians became part of the USSR. Ukraine participated in the founding of the UN as a charter member.

⁸ In 1954, Crimea - which had formerly belonged to the Russian Federation - was turned over to Ukraine by the Soviet centralized authority. The leader of the Soviet Communist Party at the time was Nikita Krushchev, formerly first secretary of the Ukrainian Communist Party (KPU).

⁹ On 26 April 1986, the nuclear plant at Chernobyl - 130 kilometers north of Kiev - was the scene of the worst nuclear accident in history when one of its reactors exploded, affecting an area inhabited by 600,000 people. By 1993, 135,000

were evacuated and 7,000 had died of radiation-related diseases. The reactor was sealed in cement. Chernobyl's radioactive fallout contaminated Ukraine, Russia, Belarus, Poland and regions of Sweden and Finland. In the following years, researchers recorded an increase in cancers and other radioactivity-related diseases.

¹⁰ In 1985 within the framework of the reforms in the USSR, Communist leaders and Ukrainian nationalists founded the Ukrainian People's Movement for *Perestroika* (restructuring) (RUKH), which demanded greater political and economic autonomy. In the March 1990 legislative elections, RUKH candidates received massive support from the population. On 16 July 1990, the Ukrainian Supreme *Soviet* (Parliament) proclaimed the sovereignty of the republic. On 24 August 1991, the Ukrainian Parliament approved the republic's independence, and convened a plebiscite to ratify or reject the decision.

¹¹ In December 1991, 90 per cent of Ukrainians ratified their independence and elected Leonid Kravchuk, formerly first secretary of the KPU, as President.

¹² On 8 December1991, the presidents of Ukraine, the Russian Federation and Belarus pronounced the end of the USSR, founding the Commonwealth of Independent States (CIS).

¹³ In 1992,Ukraine deregulated prices, created a new currency, the karbovanets, made bids for arms factories and encouraged foreign investment.

¹⁴ On 5 May, the Crimea peninsula declared independence, but it was vetoed by the Ukrainian Parliament. Crimea yielded and withdrew the declaration. Russia reacted to the

PROFILE

ENVIRONMENT

Ukraine is bordered by Poland, Slovakia, Hungary, Romania and Moldova in the west and southwest; by Belarus in the north and Russia in the east and northeast. The Black Sea and the Sea of Azov (Acovsko More) are located in the south. It is mostly made up of flat plains and plateaus, with the Carpathian Mountains (maximum altitude 2,061m) along the country's southwestern borders, and the Crimean Mountains (maximum altitude 1,545m) in the south. The climate is moderate and mostly continental. There is black soil; both wooded and grassy steppes in the south. Much of the north is made up of mixed forest areas (such areas occupy 14 per cent of the republic's total land surface).

SOCIETY

Peoples: Ukrainians, 72.7 per cent; Russians, 22.1 per cent; Belarusians, 0.9 per cent; Moldovans, 0.6 per cent, Poles, 0.4 per cent. **Religions:** Mainly Christian Orthodox. Catholics, Protestants, Jews **Languages:** Ukrainian (official), Russian. **Main Political Parties:** Party of Regions; Bloc Yuliya Tymoshenko (which includes: All-Ukrainian United Fatherland; Ukrainian Social Democratic Party); Bloc Our Ukraine (which

includes: People's Union Our Ukraine; Party of Industrialists and Entrepreneurs of Ukraine; People's Movement of Ukraine; Christian Democratic Union; Ukrainian Republican Party Assembly; Congress of Ukrainian Nationalists). **Main Social Organizations:** Confederation of Free Trade Unions of Ukraine (KVPU, with 18 trade unions); League of Ukrainian Women; Ukrainian National Committee of Youth Organizations (UNKMO).

THE STATE

Official Name: Ukrayina. **Administrative Divisions:** 25 Regions, the Republic of Crimea has special status as well as great internal autonomy. **Capital:** Kiev (Kyiv) 2,660,401 people (2005). **Other Cities:** Kharkov (Char'cov) 1,692,700 people; Donetsk 1,764,000; Dnipropetrovsk 1,483,300; Odessa 1,121,500 (2000). **Government:** Viktor A. Yushchenko, President since January 2005. Viktor Yanukovych, Prime Minister since August 2006. Legislature, single-chamber: Supreme Council, with 450 members. **National Holiday:** 24 August, Independence (1991). **Armed Forces:** 273,000 (2003). Other: 72,000 National Guard and Border Guard.

Life expectancy
66 years
2005-2010

GNI per capita
$1,270
2004

Literacy
99% total adult rate
2000-2004

HIV prevalence rate
1.4% of population 15-49 years old
2003

situation in June, annulling the 1954 decree by which it had ceded Crimea to Ukraine, demanding that it be returned. Kiev refused, but granted Crimea economic autonomy.

[15] Prime Minister Vitold Fokin resigned in September 1993 over the failure of his economic policies. He was replaced by Leonid Kuchma, the former president of the Union of Industrialists and Entrepreneurs.

[16] The liberal policies of the new government and its privatization scheme soon came up against the dual obstacles of the Supreme Council - dominated by former communists - and worker resistance.

[17] In June, in a direct challenge to Kravchuk's moderate foreign policy, the Supreme Council announced the appropriation of the entire ex-USSR nuclear arsenal in Ukraine. With the disintegration of the Soviet Union, Ukraine became the world's third most important nuclear power.

[18] Finding himself politically vulnerable, in September 1993 Kravchuk agreed to cede Ukraine's Black Sea fleet to Russia, in compensation for debts incurred through oil and gas purchases from Moscow. In addition, he accepted help from Russia in dismantling the 46 intercontinental SS24 missiles which Ukraine had wanted to keep as a last bastion against any possible future expansionist schemes on the part of Russia. However, opposition in Kiev led to the invalidation of the settlement.

[19] In the meantime, the economy went out of control with inflation reaching 100 per cent per month and Kuchma resigned.

[20] The first presidential elections of the post-Soviet era took place in June and July 1994. Former Prime Minister Leonid Kuchma defeated Kravchuk with 52 per cent of the vote, after which he declared his intention to strengthen links with Russia and enter fully into the Commonwealth of Independent States (CIS).

[21] The KPU obtained 113 seats (24.7 per cent) in the March 1998 Parliamentary elections, effectively becoming a left and center-left parliamentary majority. President Kuchma, an independent, was re-elected in the second round of the December 1999 presidential elections. International observers reported that the elections were far from free and fair.

[22] In late December, after striking a deal with the opposition, Kuchma appointed Viktor Yushchenko, then head of the National Bank of Ukraine, as Prime Minister. In February 2000, Yushchenko announced Ukraine would restructure its foreign debt with tougher fiscal policies and a massive privatization plan.

[23] After the disappearance of a journalist critical of the regime, opposition politicians alleged that there was a tape in which the President, prior to the journalist's

IN FOCUS

ENVIRONMENTAL CHALLENGES
Air and water pollution are significant; deforestation is becoming a serious problem. Approximately 2.8 million people live in areas contaminated with radiation from the 1986 accident in the Chernobyl atomic plant, in the northeast.

WOMEN'S RIGHTS
Women have been able to vote and stand for election since 1919. The first female members of Parliament were elected in 2002, when women won 24 of the 450 seats in the Lower Chamber. In 2003 they held five per cent of ministerial or equivalent positions. Women comprised 49 per cent of the total labor force in 2003. Prenatal and maternity care are universal*. The maternal mortality rate is 35 per 100,000 live births*.

Domestic violence affects women of all social classes in every region of the country. Approximately 20 per cent of women suffer mistreatment from their partner*. A study carried out at the beginning of the 2000s showed that most people involved in law enforcement in the country considered domestic violence a private affair and not an indictable crime. There is only one State-financed shelter for female victims of violence

disappearance, discussed the 'solution' to the problematic journalist with security officials. In February 2001 massive demonstrations requested Kuchma's resignation.

[24] In April 2001 Parliament removed Yushchenko, the best ally of the market reforms demanded by the West, causing a serious political crisis. This revealed the tensions among three sectors: pro-Russian forces trying to return Kiev to the Kremlin's sphere, pro-Western forces who wanted to continue with the reforms and movement towards the EU and NATO, and organized crime, whose goal was to maintain the instability that benefited it. That month, Ukraine signed a technology and military cooperation agreement with Russia.

[25] Ukraine - although still wanting to join the EU and NATO - agreed in September to form a Common Economic Space with Russia, Belarus and Kazakhstan, despite domestic opposition and fears that the agreement would hinder its pro-European integration policy.

[26] In November 2004 the Ukrainian Central Election Commission officially declared Viktor Yanukovych winner of the controversial presidential elections. According to the Commission, the

in the country. Trafficking in women, especially for the sex industry, is a growing problem. The International Organization for Migration estimates that around 500,000 women are trafficked out of the country every year, generally to other European countries. Ukraine is also a transit country for trafficked women.

CHILDREN
The infant mortality rate is 14 per thousand live births and the under-five mortality rate 18 per thousand live births*.

More than 100,000 children - 14 per cent of them under seven years old - live on the streets. Most have left their homes fleeing from domestic violence or seeking a better financial situation. According to a survey made by the State Family and Youth Institute, 43 per cent of minors say they have suffered some type of violence. Most of the abuses are related to child prostitution.

Ukraine is on the brink of an HIV/AIDS epidemic. Most of those infected are under 30 and there is concern over the increase in mother-child transmission. High-risk behavior among young people, such as the injection of drugs, also contributes to HIV/AIDS propagation*. Some 360,000 people aged below 50 were living with HIV/AIDS in 2003*.

pro-Russian candidate Yanukovych had won with 49.6 per cent of the vote. Most international observers, including those appointed by the US and the Organization for Security and Cooperation in Europe, announced that there were irregularities in the election process.

[27] Shortly before the results were announced, Yanukovych claimed that he did not need a fraudulent victory that could spark violence, adding that he would wait until the legitimacy of the result was legally proved. Opposition leader Yuschenko declared that he was ready to stand in a second round if an 'honest election commission' could be guaranteed. European Commission President José Manuel Durão Barroso warned that there would be consequences for relations between the EU and Ukraine unless an objective and balanced investigation of the election process and results was carried out.

[28] On 3 December the Supreme Court invalidated the second round results. After five days of deliberation, the Supreme Court's 21 members recommended a re-run of the second round. The new poll was held on Sunday 26 December, with the same two candidates participating.

[29] On 20 January 2005 the

INDIGENOUS PEOPLES/ ETHNIC MINORITIES
Russians comprise slightly more than 22 per cent of the population. They live mostly in the eastern regions of Ukraine. They have a strong group identity and are politically mobilized, constantly demonstrating for union with Russia. They have formed political parties, but have not achieved their aims. Since separation from the USSR, the Government has established Ukrainian as the official language and closed all Russian-speaking education centers.

MIGRANTS/REFUGEES
Ever since a ministerial resolution in 1996, Ukraine has a temporary protection regulation for all refugees from war. Between 1,200 and 1,400 people request asylum in Ukraine every year. In 2004 there were 1,364 new applications and a total of more than 2,300 pending requests. Many see the country as a door to the EU. In early 2005 some 2,500 people were living in the country as refugees, almost half of them (45 per cent) in Kiev.

DEATH PENALTY
Abolished in 2000.

*Latest data available in *The State of the World's Children* and *Childinfo* database, UNICEF, 2006.

Ukrainian Supreme Court confirmed Yuschenko's victory in the second round.

[30] In September Yuschenko dismissed Prime Minister Yulia Timoshenko and appointed Yurii Yekhanurov to succeed her. Timoshenko declared that by removing her from office, Yuschenko had destroyed their political alliance and the country's future.

[31] In January 2006 Russia cut gas supplies to Ukraine because of Kiev's refusal to pay the 460 per cent price rise decided by Moscow. Ukraine claimed that Russia had cut the supply in retaliation for Kiev's attempts to become more independent from Moscow and to develop closer links with Europe. Gazprom, the Russian state gas company, was charging subsidized prices to former USSR countries.

[32] Yanukovych's Party of Regions won the March 2006 parliamentary elections, getting 186 out of the 450 seats at stake. The 'Timoshenko Block' came second with 129 seats and Yuschenko's party, New Ukraine, won just 81 seats.

[32] In August 2006, President Yuschenko made his former rival Yanukovych Prime Minister, heading a pro-Russian coalition. ∎

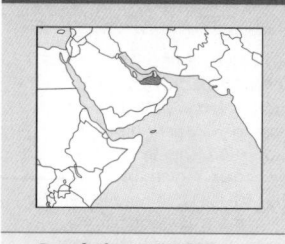

Population: 4,775,260
Land area: 83,600 km²
Capital: Abu Dhabi
Currency: Dirham
Language: Arabic

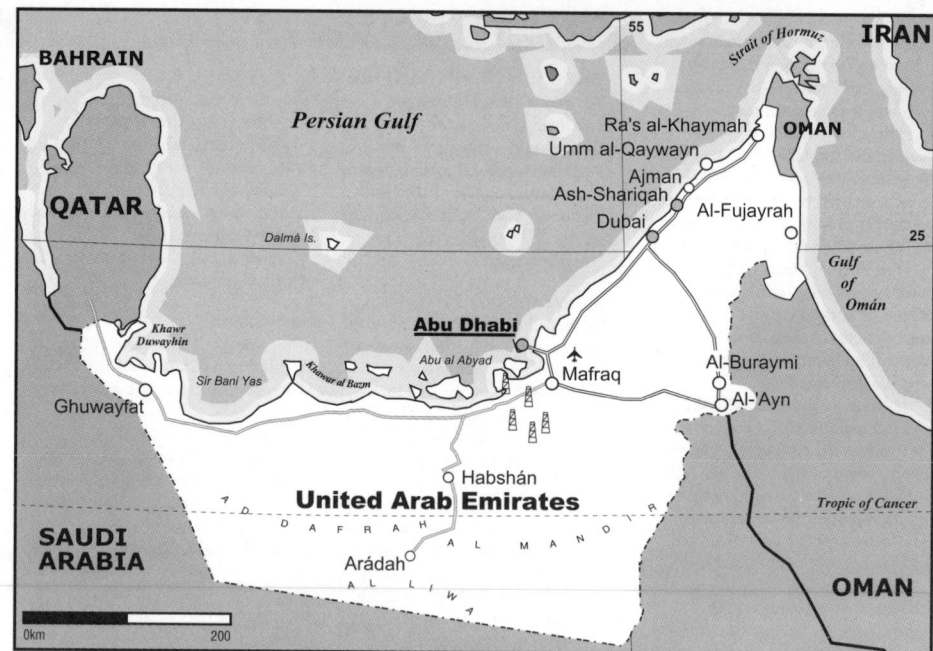

In the southeastern corner of the Arab peninsula, the Rub al-Khali desert occupies part of the territory of present-day Saudi Arabia and almost all that of the United Arab Emirates.

[2] By the 6th century the oases supplied water to a small stable population that spoke different Arabic dialects. Some were farmers, others were merchants or craftspeople from small villages and others (generically known as 'Bedouins') were nomads who raised camels, sheep and goats. Along with these activities, the coastal peoples also fished in Gulf waters.

[3] The Bedouins - a mobile, tribally organized and armed-group - together with the merchants, dominated the farmers and craftspeople.

[4] Among herders and farmers, religion had become another type of social control. Local gods were identified with the heavenly bodies and could manifest themselves taking the form of rocks, trees or animals. Some families, by interpreting the language of the gods, managed to exercise control over others.

[5] Up until the 7th century, the Byzantine and Sassanid empires waged a long war in the peninsula, although this did not directly involve the territory of the present-day Emirates. Such activity, as well as the opening of trade routes, attracted merchants, dealers and craftspeople who brought knowledge of the foreign world and its cultures.

[6] Islam was adopted during the prophet Muhammad's lifetime. Tribal chiefs secured their power without significantly changing the lifestyle of the few inhabitants.

[7] Upon the Prophet's death, various groups disputed his spiritual inheritance. The Ibadis (who claimed to be his direct descendants), created the Uman (Oman) imanate in the mid-8th century. At the end of the 9th century it was suppressed by the Abassids, caliphs who claimed a universal authority and whose capital was in Baghdad.

[8] After the 11th century, the Sunni form of Islam gradually spread from being the religion of the ruling groups, to reach the population at large. The Ibadis continued to exert strong religious authority until the 15th century.

[9] The Gulf ports were used for trading in textiles, glass, porcelain and spices from China, which were transported to the Red Sea through a chain of oases.

[10] During the 17th and 18th centuries, the Ibadis reinstated their imanate under a Yaribi dynasty, remaining on the borders of the Ottoman Empire. To the north, Bahrain was under Iranian domination.

[11] While the Ottomans were occupied with constant wars in Europe, Africa and Asia, the southeastern region was devoted to trade. Ruling families linked directly with merchants appeared and piracy developed, benefiting from the natural advantages of the coasts (known since that period as the 'Pirate Coast').

[12] When European fleets increased their use of the maritime route round the Cape, British influence grew. British ships used the Gulf ports as harbors on the way to India, and helped to combat piracy.

[13] At the beginning of the 19th century, Britain gained complete control over the region through

PROFILE

ENVIRONMENT
Located in the southeastern part of the Arabian peninsula, stretching from the Qatar peninsula toward the strait of Hormuz, the land is mostly desert with few oases and *wadis* (dry and rocky river beds). The coastal areas are very hilly lowlands with coral islands offshore and sand dunes. This is where most of the operating oilfields are, and therefore, where the highest levels of coastal pollution occur. The country is one of the world's main oil producers.

SOCIETY
Peoples: UAE nationals make up just 19 per cent of the population. Most of the population are expatriate workers: around 45 per cent come from South Asia, 17 per cent from Iran and 13 per cent from other Arab countries (mainly Egypt). Some 80 per cent live in Dubai, attracted by the oil wealth.
Religions: Muslim 94.9 per cent (Sunni 80 per cent, Shi'a 14.9 per cent); Christian 3.8 per cent; others 1.3 per cent.
Languages: Arabic (official); Persian, Hindi and Urdu are also spoken. English is spoken among immigrants and for business reasons.

Main Political Parties: No political parties or trade unions are allowed.

THE STATE
Official Name: Daulat al-Imarat al-'Arabiya al-Muttahida.
Administrative Divisions: seven emirates: Abu Dhabi; Dubai; Sharjah; Ajman; Umm al-Qaiwain; Ras al-Khaimah and Fujairah.
Capital: Abu Dhabi (Abu Zaby) 475,000 people (2003).
Other Cities: Dubai 1,401,100 people; Ash-Shariqah 402,000; Al-'Ayn 283,800; Ajman 143,700 (2000).
Government: Federation of monarchies, with the president of the federation as Head of State: Khalifa bin Zayed Al Nahayan, since November 2004. Vice President and Prime Minister: Sheikh Mohammed bin Rashid Al Maktoum, since January 2006. Unicameral Legislature: Federal National Council with 40 members appointed by the emirs.
National Holiday: 2 December, Independence Day and Proclamation of the Union (1971).
Armed Forces: 51,000 (2003).

 Life expectancy
79 years
2005-2010

 GNI per capita
$23,770
2004

Literacy
77% total adult rate
2000-2004

agreements reached with the local chiefs and small governors of the ports. The Trucial States, (the Pirate Coast's new name) included Abu Dhabi, Dubai and Sharjah. Relations with Britain continued in the same way until the first decades of the 20th century.

[14] Around 1914, the Saudi state re-emerged in Central Arabia. Russia, France and Germany also sought to intensify their presence in the area. This led the British to formalize relations with the Trucial States of Bahrain, Oman and Kuwait, which let the Government in London handle their external affairs.

[15] World War I did not alter these relationships. Britain was the real power behind Abd al-Aziz's government in the new kingdom of Saudi Arabia, controlling the southern and southeastern coasts of the peninsula. With the development of air routes, the Gulf's airfields and those of Egypt, Palestine and Iraq took on an important role.

[16] After World War II, relations among Arab countries changed. The League of Arab States was formed in 1945 by those countries that had some form of independence.

[17] At the beginning of the 1960s, the Middle East's oil deposits were known to be among the largest in the world. The United States joined Britain in keeping control over the Gulf States, whose revenues depended almost entirely on oil.

[18] The growing influence of the Pan-Arabist Egyptian President Gamal Abdel Nasser led Britain to allow greater local participation in the governments of several states of the Protectorate. In 1968, it withdrew its military forces from the region. That same year, the OPEAC - a branch of OPEC (Organization of Oil Exporting Countries) - was created, formed exclusively by oil-exporting Arab states.

[19] In 1971, Abu Dhabi began a large-scale exploitation of its oil wells. The clear establishment of borders between the territories became indispensable. Under British influence, the United Arab Emirates were created that year without the participation of Qatar or Bahrain.

[20] The new state had to face a conflict with Iran which, claiming historical rights, occupied the islands of Abu Mussa, Tunb al-Cubra and Tunb al-Sughra on the Strait of Hormuz. During the first decade, oil production rose (mainly in: Abu Dhabi, Dubai and Sharjah). National participation in the control of oil exploitation also grew.

[21] When-OPEC decided in 1973 to raise the price of oil by 70 per cent and reduce supply by five per cent, a new era in the oil-rich states' relations with the world began. The results of this policy were explosive. The UAE's annual growth rate in the 1970s was over ten per cent due to oil revenues.

[22] There was a rapid growth of cities with state-of-the-art highways, oil pipelines and banks, and many immigrants were attracted by the region's possibilities. Little was left of the ancient pursuits of fishing or diving for pearls on the coast.

[23] The 1980s began with the Iran-Iraq war. Although the United Arab Emirates maintained an apparently neutral stand, they gave economic support to Iraq to avoid a possible 'Iranization' of the region. Once the conflict had ended, the UAE had become the Middle East's third largest oil producer, after Saudi Arabia and Libya.

[24] As from 1981 the Government tried to develop other industrial fields to reduce dependence on oil production.

[25] The country was a member of the Non-Aligned Movement and supported Palestinian claims. In late 1986, diplomatic relations were established with the Soviet Union and the People's Republic of Benin. In 1987, relations with Egypt - broken off after the Camp David agreements with Israel - were re-established.

[26] During the Gulf War (1991), the Emirates supported the fight against Iraq, which had occupied Kuwait.

[27] In March 1991, the Gulf Co-operation Council signed an agreement with the US which included a common military strategy and mechanisms to prevent arms proliferation in the zone.

[28] In 1992, with Syrian mediation, Iran modified its claims over the islands of the Strait of Hormuz. The conflict was placed under international arbitration, following pressure by the Emirates.

[29] The influence of Islamic fundamentalism increased in the UAE between 1993 and 1996. Sheikh Zayed, President of the Union, was keen to extend 'integral' Islam. In February 1994, he decided to extend *sharia* (Islamic law) to criminal cases which had previously been dealt with by civil courts.

[30] In 1997, in view of the US threat of an armed intervention in Iraq, Sheik Zayed stated that the Iraqi people deserved a 'new chance' and that measures of this type would be 'unacceptable'.

IN FOCUS

ENVIRONMENTAL CHALLENGES
Desalinization plants compensate for the lack of fresh water. There is desertification and oil spills cause coastal pollution.

WOMEN'S RIGHTS
Neither women nor men have the vote; the main governmental posts are granted by the President.-In 2003, 6 per cent of these posts were held by women. That year, women comprised only 14 per cent of the country's labor force. Some 86 per cent of female laborers work in the service sector and 14 per cent in industry.

The net enrolment rate stood at 82 per cent for primary education, while 19 per cent of adult women were illiterate.

Approximately 97 per cent of pregnant women receive prenatal care and specialized personnel assist 99 per cent of births*.

CHILDREN
The infant mortality rate is 7 per 1,000 live births, and the under-five mortality rate 9 per 1,000.* Some 15 per cent of newborns are underweight*.

Primary school enrolment stands at 83 per cent*.

Children continue to be trafficked for work as camel jockeys, although using children for races has been expressly prohibited since 1993.

INDIGENOUS PEOPLES/ ETHNIC MINORITIES
Only 19 per cent of the population is originally from the country. The ready availability of jobs generated by the oil industry in previous decades has meant that 45 per cent of the citizens are from South Asia and 13 per cent from other Arab countries. Islam is practised by 95 per cent of the population while the remaining inhabitants are Christians or Hindus.

MIGRANTS/REFUGEES
United Arab Emirates does not grant asylum. Occasionally, foreigners have been detained in the Emirates while awaiting asylum in a third country.-

There is a large number of illegal immigrants. Since they lack documents, they cannot defend their rights. Although the Government refuses to recognize them, they are necessary for the labor-hungry economy.

DEATH PENALTY
It is widely enforced, as is flogging.

*Latest data available in *The State of the World's Children* and *Childinfo* database, UNICEF, 2006.

[31] In the field of domestic policy, economic liberalization continued, leading among other things to the establishment of a free trade zone in 1998 in the city of Ra's al-Khaymah; the first in the Middle East. In April 2000, the UAE embassy in Iraq was reopened.

[32] The fall in oil prices was the main reason that GDP growth in the Emirates dropped in 2001. The country achieved some diversification of its economic activity in sectors not related to oil - such as aluminum, tourism, telecommunications and aviation - which still contribute two thirds of the GDP and 30 per cent of exports.

[33] In March 2004 the US Embassy in Abu Dhabi and the US General Consulate in Dubai suspended their operations citing a 'specific threat'. Washington warned of a strong risk of attacks against US interests in the Middle East and North Africa after the death of Hamas leader Ahmed Yassin.

[34] On 2 November, President Sheikh Zaid ibn Sultan Al Nahayan died. Vice-President and Prime Minister Sheikh Maktum ibn Rashid Al Maktum took office as interim president, while Sheikh Khalifa ibn Zaid Al Nahayan succeeded his father, Zaid ibn Sultan al-Nahyan, as emir of Abu Dhabi.

[35] A cabinet reshuffle in late November 2004 included a woman for the first time in the Government. Shaikha Lubna al-Qasimi was appointed Economy Minister.

[36] On 5 January 2006, Sheikh Muhammad ibn Rashid Al Maktum became Prime Minister.

[37] In May, the British Broadcasting Corporation (BBC) launched a report on internet censorship in the United Arab Emirates. The BBC claims it is practically impossible in the Emirates to access websites with information about and/ or criticism of Persian Gulf governments. United Arab Emirates is one of the fastest-developing countries in the world, but that 'development is more economic than political', stated the report. ■

United Kingdom / United Kingdom

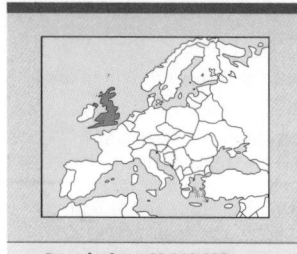

Population:	60,018,293
Land area:	242,910 km²
Capital:	London
Currency:	Pound sterling
Language:	English

1 The first known inhabitants of what is now Britain were Paleolithic hunters, following herds of wild animals. After the final ice age, agriculturalists began to settle on the island. Over thousands of years, these people and the many others who migrated from the continent evolved increasingly complex social systems.

2 In 44 AD the Romans invaded southern Britain. In 90 AD they created the province of Britannia, and founded London between 70 and 100 AD. In the early 5th century they abandoned the island, leaving it largely defenseless against the raids of Anglos, Saxons and Jutes. These Germanic peoples pushed the Celts westwards, taking over the southern part of the island and establishing Anglo-Saxon kingdoms.

3 During the 5th century, the inhabitants of Ireland and Wales adopted Christianity. In the 7th century, the British church came under the power of Rome.

4 During the 7th and 9th centuries Danish invaders overran the eastern part of England. In the 11th century, the Normans, led by William the Conqueror, invaded England and secured the throne. Successive Anglo-Norman kings maintained their power by establishing various forms of vassalage over the feudal lords, though under John (1199-1216), these barons, in alliance with the clergy, were able to restrict the power of the monarchy through the Magna Carta, signed in 1215.

5 The Magna Carta laid the foundations of the British parliamentary system. It also marked the beginning of a continuous power struggle between the monarchy and the nobility. The growing power of the land-owning class and later the bourgeoisie eventually led to the consolidation of a parliamentary monarchy. The Welsh came under English control in 1382.

6 Frequent dynastic conflicts, disputes over territories in France belonging to the English Crown, commercial rivalry between England and France in Flanders, and French aid to Scotland in its wars with England paved the way for the Hundred Years' War (1337-1453), which culminated in the loss of the English possessions on the continent.

7 The negative effects of the War increased the unpopularity of the monarchy which faced at the same time an anti-Papal movement led by the followers of Wycliffe (a precursor of Luther) and a peasant rebellion. The peasants, led by Wat Tyler, rose up against the payment of tribute and the power of the feudal lords. In 1381, Tyler and his followers managed to enter London and negotiate directly with the King, Richard II. The Peasants' Revolt was unsuccessful however and Tyler was later executed.

8 The period following the Hundred Years' War was dominated by a long struggle for control of the throne between the royal Houses of Lancaster and of York. This led to the War of the Roses which ended with the coming to power of the Welsh House of Tudor in 1485. The Tudor period is considered the beginning of the modern British state. One of the Tudor kings, Henry VIII (1509-47), broke away from the Church of Rome, founding the Anglican Church. The desire to extend English authority and the religious Reformation to Ireland led to the subjugation of Ulster by Henry's daughter Elizabeth I (1558-1603).

Tudor involvement in Ireland laid the foundations for centuries of religious and political conflict in the country.

9 Under the reign of Elizabeth I, poetry and the theatre flourished with playwrights such as Ben Jonson, Marlowe and Shakespeare. Industry and trade developed, and the country embarked upon its 'colonial adventure', the beginning of its future empire. After defeating the Spanish fleet - the 'Invincible Armada' - in 1588, the British Navy 'ruled the waves', with no other fleet capable of opposing it.

10 British merchant ships involved in the slave trade, or laden with colonists or belonging to pirates and privateers sailed the oceans freely. Markets multiplied, demand grew rapidly, and producers were forced to seek new techniques in order to accelerate production. It was a prelude to the industrial revolution which was to take place in the country at the beginning of the 18th century.

11 In 1603, the crowning of James I (James VI of Scotland) put an end to the independent Scottish monarchy. The religious intolerance of James' son Charles I, led to a Scottish uprising and increasing discontent in England. This culminated in the English Civil War, which broke out in 1642. The deteriorating political situation led to the Puritans forming an army supported by Parliament; led by Oliver Cromwell, they

defeated the royal forces in 1646 and again in 1648.

12 In 1649, Parliament executed the King and proclaimed Cromwell 'Lord Protector', establishing a republic known as the Commonwealth. Radical egalitarian ideas came to the fore within the parliamentary movement, notably among the Levellers, who advocated political democracy and the abolition of the English class system. But their ideas were suppressed and their leaders imprisoned by Cromwell. After Cromwell's death, in 1658, the monarchy was restored with Charles II.

13 The priorities of the new regime were the colonization of North America and trade with America, the Far East and the Mediterranean. The slave trade - the kidnapping, trafficking and selling of slaves from Africa to buyers in America and other places - which had started in the 16th century, became one of the main sources of income for the empire.

14 The absolutism of Charles II's successor, James II, and his espousal of Catholicism were opposed by the Protestant Parliament which deposed James through the 'Glorious Revolution'. Parliament invited the Dutch prince William of Orange to assume the English throne. William III was forced to sign the Declaration of Rights (1689), limiting royal powers and guaranteeing the supremacy of Parliament.

15 In this period John Locke summarized revolutionary ideals, proposing that people have basic natural rights: to property, life, liberty and personal security. Government, created by society to protect these rights, must fulfill its mission; if it fails to do so, the people have the right to resist its authority.

16 In 1707, the parliaments of Scotland and England were joined together, creating the United Kingdom of Great Britain. Britain intervened in the war of succession in Spain, obtaining Minorca, Gibraltar and Nova Scotia through the Treaty of Utrecht (1713). In 1765, increased taxes imposed by the Stamp Act triggered the rebellion and secession of the American colonies, which declared their independence in 1776.

17 During this period, the two large political parties were formed: the Conservatives (Tories), representing the interests of the large landowners, and the Liberals (Whigs), representing the merchant class. The ideas forming the basis of economic liberalism were developed at this time by Adam Smith. The liberal doctrine provided the political ideology for British imperialism, which used the concept of 'free trade' as a justification for forcing open the ports and markets of the

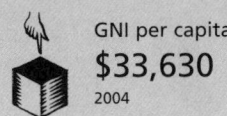
Third World, often with the use of naval force. Perhaps the most notorious example of this was the Opium Wars fought against China in the mid-19th century.

18 After the crushing of a nationalist rebellion in Ireland in 1798, the United Kingdom of Great Britain and Ireland was created in 1801 with the dissolution of the Irish Parliament (see Ireland).

19 The 18th century gave rise to the agricultural 'revolution', which introduced important innovations in farming techniques, as well as major changes in land tenure. The large landowners enclosed their properties, eliminating communal lands which had hitherto been used by small farmers, and introducing a more capitalist agricultural economy.

20 At the same time, the industrial revolution began, with the textile manufacturers being the first to confront the problem of meeting a growing demand for cloth overseas. The introduction of machinery changed the way in which work was done, and the medieval shop was replaced by the factory. On the heels of the textile industry came mining and metallurgy. The mechanization process was consolidated with the invention of the steam engine, the use of coal as a fuel and the substitution of first iron, then steel, for wood in construction.

21 This period was characterized by population growth (up from 10,900,000 in 1801 to 21,000,000 in 1850), increasing demand and expanding trade, improvements in the transport system, capital accumulation, the creation of a vast colonial empire, scientific advances and the golden age of the bourgeoisie. Britain became the world's premier manufacturing nation. Its colonial policy helped to prevent competition against its factories; for example it established regulations which destroyed the Indian textile industry.

22 The United Kingdom obtained new territories from its wars with France, particularly its triumph over Napoleon at Waterloo (1815).

23 One result of the industrial revolution was increasing discontent among the rapidly growing working class, due to low salaries, unhealthy working conditions, unsatisfactory housing, malnutrition, job insecurity and the long and tiring working days to which men, women and children were subjected. In many cases, popular uprisings were characterized by violence, and were met with equally violent repression.

24 In the early stages of the industrial revolution, spontaneous movements arose, like the 'Luddites' - textile workers who destroyed machinery to prevent it from destroying their cottage industry. Trade unions began to appear later.

25 In 1819 a demonstration in Manchester was ruthlessly put down (the Peterloo massacre), and repressive legislation followed, limiting the right of association and freedom of the press. Nevertheless, resistance movements continued their activity. One of the main movements of this period was the nationalist Irish Association led by Daniel O'Connell.

26 The most important of the mass movements was the Chartist Movement, made up primarily of workers. It took its name from the People's Charter, published in 1838 at a mass assembly in Glasgow, Scotland. This movement brought a number of issues to the fore, both political - universal suffrage, use of the secret ballot, reform of voting registers - and social - better salaries and better working conditions. After its demonstrations and strikes, 'Chartism' faded away. However, it had a far-reaching influence and its grievances were subsequently taken up by some members of Parliament.

27 Robert Owen (1771-1858), considered to be the founder of socialism and the English co-operative movement, argued that the predominance of individual interests led to the impoverishment of the masses. From 1830 on, he devoted himself to the establishment of co-operatives and the organization of labor into trade unions.

28 During the long reign of Queen Victoria (1837-1901), the traditional nobility strengthened its alliance with the industrial and mercantile bourgeoisie, and the first socialist movements emerged. Trade unions were legalized in 1871 and shortly afterwards some labor legislation was approved.

29 Beginning in 1873, the rising population numbers led to a food shortage, making imports necessary. At the same time industry began to feel the competition from the US and Germany. Britain increased its imperial activities in Africa, Asia and Oceania, not only for economic reasons, but also because of the political ambition to build a great empire. The Boer or South African War (1899-1902), fought to secure control over southern Africa, was the most expensive regional conflict of the 19th century.

30 The first quarter of the 20th century saw the birth of the women's liberation movement. The militancy of the suffragettes led to some women obtaining the right to vote in 1917. The most famous example of their militancy was the suicide of Emily Davison, who threw herself in front of the King's horse during a race in 1913.

31 In Ireland, the majority Catholic population were stripped of their lands, restricted in their civil rights because of their religion, and deprived of their political autonomy.

PROFILE

ENVIRONMENT

The country comprises Britain (England, Scotland and Wales), Northern Ireland and several smaller islands. The Pennines, a low mountain range, run down the northern center of England. The Grampian mountains are located in Scotland and the Cambrian mountains in Wales. The largest plains are in the southeast. The climate is temperate. Farming is highly mechanized and is now a subsidiary activity. The service industry, especially insurance, finance and tourism, are big income earners. The huge coal and iron ore deposits which made the Industrial Revolution possible have nearly run out, and the gas and oil reserves in the North Sea that turned the UK into an exporter of these products from the 1970s are also fast diminishing.

SOCIETY

Peoples: English, Scots, Welsh and Irish. In the 2001 Census 92.1 per cent defined themselves as white; 1.8 per cent Indian; 1.3 per cent Pakistani; 1.2 per cent mixed race; 1.0 per cent Black Caribbean; 0.8 per cent Black African; 0.5 per cent Bangladeshi; 0.4 per cent Chinese. There has been a large influx of people from eastern Europe, particularly from Poland, since the expansion of the European Union. **Religions:** Christian 71.6 per cent; No religion/not stated 23.2 per cent; Muslim 2.7 per cent; Hindu 1.0 per cent; Sikh 0.6 per cent; Jewish 0.5 per cent; Buddhist 0.3 per cent. **Languages:** English (official), Welsh, Gaelic and the languages of various immigrant groups. **Main Political Parties:** Labour Party, social democrat, led by Tony Blair, in government. Conservative Party, opposition. Liberal Democrat Party, center-left. In Northern Ireland, Democratic Unionist Party (DUP), ultra conservative; Sinn Fin, Irish republican; Social Democratic and Labor Party (SDLP), liberal-left; Ulster Unionist Party, conservative. **Main Social Organizations:** The Trade Union Congress (TUC) has 12 million members. National Alliance of Women's Organizations, which unites organizations defending women's rights.

THE STATE

Official Name: United Kingdom of Great Britain and Northern Ireland. **Administrative Divisions:** 39 Counties and 7 Metropolitan Districts. **Capital:** London 7,429,000 people (2001). **Other Cities:** Birmingham 992,000 people; Leeds 720,000; Glasgow 578,000; Sheffield 516,000; Edinburgh 454,000; Liverpool 445,000; Manchester 437,000 (2001). **Government:** Constitutional parliamentary monarchy. Elizabeth II, Queen and Head of State since February 1952. Tony Blair, Prime Minister since May 1997, re-elected in 2001 and 2005. Bicameral Legislature: the House of Commons, with 659 members elected for a five-year term, and the unelected House of Lords, with 703 members, 586 life peers and 91 hereditary members and 26 bishops (2004). **Armed Forces:** 206,380 (2004). **Dependencies:** Anguilla, Bermuda, Gibraltar (contested by Spain), the Falkland Islands/Malvinas (contested by Argentina), British Virgin Islands, Northern Ireland, Montserrat, Cayman Islands, Guernsey, Jersey, Isle of Man, Turks and Caicos, St Helena, British Territories in the Indian Ocean and Pitcairn (Oceania).

Millions emigrated, and political unrest periodically resulted in violent uprisings. Not until 1867 were the privileges of the Anglican Church eliminated; at the same time, measures were taken to improve the situation of the peasants. The 1916 Easter Rising in Dublin was ruthlessly put down by the British, but the Crown forces were unable to win the ensuing guerrilla war which began in 1918, and Britain finally granted Ireland independence in 1921. Six counties in the north-east, with Protestant majorities, remained under British control with a devolved administration in Belfast.

32 Economic and political rivalry between the European powers led to the outbreak of World War I (1914-18). The Central Powers of Austro-Hungary and Germany, joined subsequently by Turkey and Bulgaria, fought against the Allied powers of France, Britain, Russia, Serbia and Belgium, with Italy, Japan, Portugal, Romania, the United States and Greece joining during the course of the War.

33 Despite its victory, Britain emerged from the War in a weakened condition. It had invested $40 billion in military expenditure, mobilized 7,500,000 troops, suffered a loss of 1,200,000 soldiers and acquired an enormous foreign debt. The deep economic depression in the post-War years led to renewed unrest among workers, which reached its height in the General Strike of 1926. The Conservative Government declared the strike illegal, but did not take any measures to revive British industry. In the elections of 1929, the Labour Party came to power for the first time.

34 In the aftermath of World War

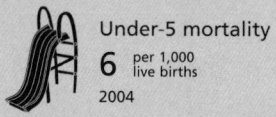

Under-5 mortality
6 per 1,000
live births
2004

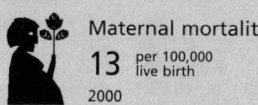

Maternal mortality
13 per 100,000
live birth
2000

IN FOCUS

ENVIRONMENTAL CHALLENGES
This highly industrialized country
has serious environmental
contamination problems,
principally air pollution. The
Government plans to implement
policies that comply with the
Kyoto Protocol, which obliges the
country to reduce its greenhouse
gas emissions by 12.5 per cent,
compared to emissions in
1990. The volume of domestic
garbage is another important
environmental problem,
addressed by recycling plans that
aim to recycle 33 per cent of all
garbage by 2015.

WOMEN'S RIGHTS
Women have been able to vote
and run for office since 1918.
In 2005, women held 18.5 per
cent of seats in Parliament and
29 per cent of ministerial and
similar positions. In 2003, women
comprised 44 per cent of the
country's total labor force: 88 per
cent worked in services, 11 per
cent in industry, and 1 per cent in
agriculture.
Isolated cases of female
genital mutilation have been
registered.
The law allows for equal
opportunities; however, on
average women earn 83 per
cent of men's hourly salary for
the same job. Domestic violence
amounts to 25 per cent of all
violent crimes in the United
Kingdom. The country is also
used as a destination by gangs
trafficking in women, mainly
girls and adolescents, to work

in prostitution. It is estimated that
each year some 1,500 women and
girls are trafficked into the UK for
this purpose.-

CHILDREN
The infant mortality rate is 5 deaths
per 1,000 live births; among under-
fives, it is 6 per 1,000 live births*.
Eight per cent of newborns are
underweight*. Young people under
16 are not allowed to work for more
than a few hours a week except as
part of educational programs.
Pedophilia is an increasing
concern and police personnel
have been trained to combat its
encouragement via the internet.
A growing number of African
children are kidnapped and sent
to other countries via the UK,
mostly from Nigeria, Sierra Leone,
Congo, Malawi, Angola, Ethiopia,
South Africa, Somalia, Kenya and
Uganda. Apart from being forced
into prostitution, they are used as
household servants or slaves for
gangs linked to drug trafficking, and
are also forced to work in factories,
restaurants and/or as beggars.-

**INDIGENOUS PEOPLES/
ETHNIC MINORITIES**
Afro-Caribbean migration to the
UK increased after World War II.
The largest communities are located
in the south of England, mainly in
London. In spite of their proficiency
in the English language, the Afro-
Caribbeans have had to endure
much discrimination and violence.
Asian immigrants started
arriving in great numbers in the
1950s. They are mostly from

South Asia (India, Pakistan and
Bangladesh) and are concentrated
in the large English cities. They
suffer great social discrimination
at the hands of the majority
groups and encounter obstacles
which prevent their access to high
government positions.

MIGRANTS/REFUGEES
In 2004, the UK received 299,000
asylum and refuge requests.
Approximately 49,200 were
granted and the rest sere still
pending. Most of the granted
requests were from Iraq (5,065),
Iran (4,550), Somalia (4,300),
Pakistan (3,385) and Zimbabwe
(3,185).-
The applicants must fill
in forms at the borders, at the
time of entry into the UK. These
must be completed in English;
otherwise, they are refused.
Amnesty International reported
that the UK Immigration Act
has led to the detention of tens
of thousands of asylum seekers.
In many cases, the detention
was lengthy, inappropriate,
disproportionate and illegal.

DEATH PENALTY
It was abolished in 1965 but
technically retained for treason,
piracy and crimes committed
under military jurisdiction. The UK
became completely abolitionist
in 1998.

* Latest data available in *The State
of the World's Children* and *Childinfo*
database, UNICEF, 2006.

I, Britain supported the creation
of the League of Nations. In 1931,
the British Community of Nations
(Commonwealth) was established
under the Statute of Westminster.
This formally recognized the
independence of Canada, Australia,
New Zealand and South Africa.
[35] On 1 September, 1939,
Germany invaded Poland and two
days later Britain declared war on
Germany, marking the beginning
of its participation in World War II
(1939-45). In May 1940 a coalition
cabinet was formed, with Winston
Churchill as Prime Minister. From
1939 to 1941, Britain and France
were ranged against Germany which
was joined by Italy in 1940. Hungary,
Romania, Bulgaria and Yugoslavia
participated in the war as 'lesser'
allies of the Nazis.
[36] In 1941, the Soviet Union, Japan
and the US entered the conflict. On 8
May 1945, Germany surrendered. The
UK, US and the USSR emerged as the
major victors from the war. However,
the British Empire was eclipsed by

the rising power of the US, which
became the undisputed economic,
technological and military leader.
[37] In May 1945, the Labour
Government of Clement Attlee, who
won the elections with the slogan 'We
won the war, now we will win the
peace', nationalized the coal mines,
the Bank of England and the iron and
steel industries. It also established the
National Health Service, which offered
free health care to all.
[38] Pakistan was formed and
India became independent in 1947,
although both remained members of
the British Commonwealth. During
the following decade, most of
Britain's overseas colonies obtained
their independence. Britain was a
founder member of NATO in 1949.
[39] The Franco-British military
intervention in the Suez Canal Zone
in 1956, which failed due to a lack
of US support, was met by strong
criticism from both inside and outside
Britain (see Egypt). The following
year, the UK detonated its first
hydrogen bomb in the Pacific Ocean.

[40] The general election of 1964
was won by the Labour Party under
the leadership of Harold Wilson.
His government faced serious
problems, such as the declaration of
independence by Southern Rhodesia
(today Zimbabwe), and the severing
of diplomatic relations with nine
other African countries.
[41] In 1967, having been denied
entry to the Common Market, and
faced with economic problems and
rapidly increasing unemployment,
Wilson withdrew British troops from
South Yemen, evacuated all bases
east of Suez except for Hong Kong,
discontinued arms purchases from
the US and implemented a savage
austerity budget.
[42] In Northern Ireland in 1969, the
latent conflict erupted. A number
of people were killed and wounded
in riots between Catholics and
Protestants. The Catholics demanded
equal political rights, and better
access to housing, schools and social
security. The Protestant-controlled
Northern Irish Government

responded by sending in their
armed police reserves against the
Catholic demonstrators. The British
Government sent in their troops
to separate the two sides and took
control of police and reserve forces
away from the Belfast Government.
[43] In August 1971, Prime Minister
of Northern Ireland Brian Faulkner
opened internment camps and
authorized the detention of
suspects without trial. Protests
against these measures resulted
in more than 25 deaths. On 30
January, 1972, 'Bloody Sunday',
British soldiers opened fire on
a peaceful protest march in
Derry (Londonderry), killing 13
Catholics and injuring hundreds
more. The Irish Republican Army
(IRA) responded with numerous
assassinations.
[44] In January 1973 a majority of
those taking part in a referendum
voted in favor of joining the
European Economic Community
(EEC). In March 1973, the people
of Northern Ireland voted in a
referendum to remain within
the UK rather than join a united
Ireland. There was a high abstention
rate of 41.4 per cent.
[45] In the 1970s, social conflict
in Britain intensified, and Edward
Heath's Conservative Government
(1970-74) was faced with strikes
in key public sectors (dockers, coal
miners and railway workers), which
led to Labour Party victories in two
elections in 1974.
[46] In 1979, voters in Scotland and
Wales turned down autonomy for
their regions in referenda organized
by James Callaghan's Labour
Government.
[47] In May of that year, after the
'winter of discontent' characterized
by strikes, the Conservative Party
won the election, with Margaret
Thatcher as its leader. The new
Prime Minister brought in a severe
monetarist policy to bring down
inflation. She began to reverse the
nationalization process carried out
under Labour, and returned to a
free market policy.
[48] In April 1982, Thatcher sent a
Royal Navy force, including aircraft
carriers and nuclear submarines,
to the Malvinas islands (Falklands)
which had been occupied by
troops from the military junta in
Argentina. After 45 days of fighting
the British recovered the islands for
the Crown (see Argentina).
[49] In October 1983, the British
Government decided to withdraw
its troops from Belize. The following
year, in agreement with a treaty
dating back to the First Opium War,
Britain ceded sovereignty over Hong
Kong to the People's Republic of
China, with effect from June 1997.
[50] During the Thatcher
Administration, the trade union
movement suffered serious setbacks,
hampered by increasingly restrictive

Doctors
166 per 100,000 people
1990-2004

Primary school
100% net enrolment rate
2004

laws and the loss of affiliates in the industrial sector, itself in decline. The 1984/85 miners' strike culminated in defeat for the union after a year of violent internal strife and confrontation with the police.

[51] In 1987, Thatcher was elected to her third consecutive term in office. She continued her policies as before: radical economic liberalization, privatization of state corporations, tax reform and opposition to union demands. In foreign affairs, Britain opposed greater European Community integration and continued to align itself closely with the US.

[52] In February 1990, the UK and Argentina renewed diplomatic relations and their representatives met in Madrid to negotiate the issue of the return of the Malvinas/Falklands.

[53] In November 1990 Thatcher was replaced as head of the Government and Tory leader by her former minister, John Major. On taking office, Major declared himself to be in favor of capitalism with a human face, thus setting himself apart from the 'Iron Lady's' (Thatcher) harsher version of capitalism.

[54] In European affairs, the Prime Minister distanced himself from his predecessor. In 1991, London gave its backing to European agreements on monetary union. However, the fidelity which British diplomacy showed towards the US remained unaltered, as proven by British participation alongside the US in the Gulf War against Iraq.

[55] The Government managed to bring down inflation and interest rates, but economic activity remained stagnant. In 1991, industrial production declined and numerous small businesses failed.

[56] Despite this growing unpopularity, Major's Conservative Party unexpectedly won the general election in 1992 - the Party's fourth consecutive victory - plunging progressive forces into despair.

[57] From 1993 the Conservatives began to suffer a series of electoral defeats in local by-elections against a backdrop of economic recession and three million unemployed. On 15 December, London signed a joint declaration with Dublin on Northern Ireland, paving the way for peace talks (see Ireland). A series of scandals in 1994, such as the illegal funding of a dam in Malaysia, further sullied the Tories' image. Meanwhile, Parliament lowered the age of consent for homosexuals from 21 to 18, ignoring calls to match the legal age of 16 for heterosexual sex.

[58] Successive Labour local election victories in 1996 heralded the general election victory in May 1997, which saw Tony Blair become Prime Minister. The crushing defeat of the Tories, who only obtained 30 per cent of the vote, against Labour's 43.1 per cent (the biggest landslide victory of the century), forced leadership changes within the party.

[59] Referenda in Wales and Scotland in 1997 supported greater devolution to the regions. In early 1998, talks on Northern Ireland brought a new peace plan. A referendum in Northern Ireland in May gave 70 per cent support for the plan. In London, that same month, people gave their approval for direct election to a new post, the Mayor of London.

[60] Under the 'Good Friday' agreements, Northern Ireland would have its own legislative assembly, directly elected by the population, as in Wales and Scotland. A referendum in Ireland put an end to territorial claims over the North. The nomination of Northern Ireland MPs David Trimble (Ulster Unionist) and John Hume (Social Democratic and Labour Party, sympathetic to the nationalists) for the Nobel Peace Prize in October 1998 helped create the feeling of a common goal between Protestant and Catholic communities.

[61] The Spanish courts called for Scotland Yard to arrest General Augusto Pinochet - the former Chilean dictator - as he recovered from a back operation in hospital in London. After a year and a half of legal wrangling, Pinochet was freed. Home Secretary Jack Straw said Pinochet's poor health meant he would be unable to stand the strain of extradition to Spain and the ensuing trial.

[62] During 1999 the Northern Irish unionist and nationalist parties made commitments on joint government, which culminated with the installation of a power-sharing executive in December that year. Sovereignty, which had been devolved to the province in December, returned to London.

[63] In June 2001 Blair's Labour Government won a second massive victory at the general election, prompting further soul-searching and another change of leader in the Conservative Party.

[64] In October 2001, in order to stop the spread of an outbreak of foot and mouth disease, 3,915,000 animals were slaughtered. This epidemic followed the outbreak of 'mad cow disease' (Bovine Spongiform Encephalopathy - BSE) which had peaked in 1996 and forced the health authorities to kill 4.5 million cows. Exports of British beef dropped dramatically and have not regained the levels attained before 1996, when exports to the EU were banned.

[65] Following the September 2001 attacks on Washington and New York, Britain offered support for the US-launched war on terror. Blair spearheaded the US offensive on Afghanistan, under the application of Article 5 of the NATO members' mutual defence clause, which states that member states must support another member who comes under attack. Similarly, in March 2002, Blair declared jointly with US vice-president Dick Cheney that Iraq was a threat to world stability, making it possible for Britain to support Washington in attacks on Iraq.

[66] In March 2002 Elizabeth, the Queen Mother of Elizabeth II, died at 101 years old. She had earned her popularity when she stayed in London during the Nazi bombings in World War II.

[67] In July 2002, Foreign Secretary Jack Straw announced that the UK was to supply parts to the US for its F-16 jet fighters destined for Israel. This outraged activists who saw the move as another nail in the coffin of Labour's supposed 'ethical' foreign policy.

[68] In January 2003 Blair stated that an attack on British soil was inevitable and assured he had evidence that would link Saddam Hussein and the al Qaeda network. Notwithstanding France and Germany's reluctance to accompany Washington in the war against Iraq, and the fact that the surveys indicated most British people were opposed, Blair decided to go out on a limb with Bush, prompting severe criticism even within his own party. The anti-war demonstration in London on 16 February 2003 was the biggest such protest in British history, involving at least one million people.

[69] The US, UK and coalition forces invaded Iraq in March 2003 in spite of the strong international opposition to the war. Both nations were the only permanent members of the UN Security Council in favor of the invasion.

[70] The controversy surrounding the arguments wielded by Blair to lend support to the invasion of Iraq was reawakened in July 2003 with the suicide of David Kelly, a scientist and adviser to the Ministry of Defense. Kelly had been pinpointed as the main source of a press investigation accusing the British Government of distorting intelligence reports to exaggerate the threat posed by Baghdad. The death of the scientist, added to the impossibility of finding the weapons of mass destruction in Iraq that London and Washington accused Saddam Hussein of possessing, caused Blair's popularity to plunge, even though an official investigation concluded that there had been no information manipulation on the part of the government.

[71] Confronted with opinion polls indicating that a great number of British people were opposed to the adoption of a common constitution for the European Union, Blair announced that a referendum would be held regarding this matter in 2005.

[72] In July 2004, Lord Butler of Brockwell said, in what was known as the 'Butler report', that London based its participation in the Iraq war on a 'limited' intelligence with 'real quality problems'. In his report, Butler did not directly accuse Blair and estimated that the Government did not act with bad faith nor was there a deliberate attempt to distort information from the secret service.

[73] Following the report's publication, Blair appeared in the House of Commons, where he accepted 'responsibility' for the mistakes. According to the Prime Minister, the Butler report clarified that 'nobody lied' in his Government regarding the information of intelligence services. However, Conservative opposition leader Michael Howard asked Blair to consider whether he had any credibility left as head of the British Government.

[74] The Labour party won a third mandate in the May 2005 elections with 36 per cent of the vote, the lowest ever for a British government. According to analysts, this was due to Blair's sinking popularity and to the unpopularity of the Iraq war.

[75] On 7 July, 56 people died and 700 were injured when suicide bombers set off four bombs in London's transportation system. Three of the four suspects were British citizens of Pakistani origin. Two weeks later there was a failed attempt to detonate another four bombs on London's buses and subway trains.

[76] In November, Blair experienced his first defeat in Parliament since his election in 1997 when a new security bill - which would have legalized the incarceration without charge of suspects for 90 days - was voted down. Blair announced that he would not stand in the next election (due in or before 2010).

[77] In late 2005 homosexual couples were granted the same legal rights as heterosexual couples. The new law stopped short of legalizing their marriage but enabled them to register a 'civil partnership' with the same legal status as a traditional marriage.

[78] In April 2006, the New Economics Foundation and the Open University published a report stating that in a typical calendar year at current levels of natural resource use Britain would be in 'ecological debt' to the rest of the world by 16 April - compared with ecological debt days of 14 May in 1981 and 9 July in 1961.

[79] In October 2006, Blair announced that he would be definitely standing down as Prime Minister and Labour leader some time before October 2007. ∎

United States / United States

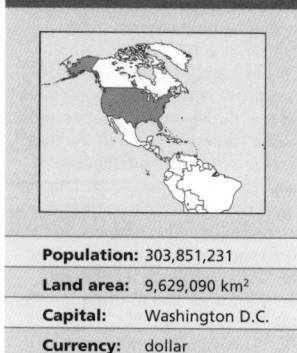

Population:	303,851,231
Land area:	9,629,090 km²
Capital:	Washington D.C.
Currency:	dollar
Language:	English

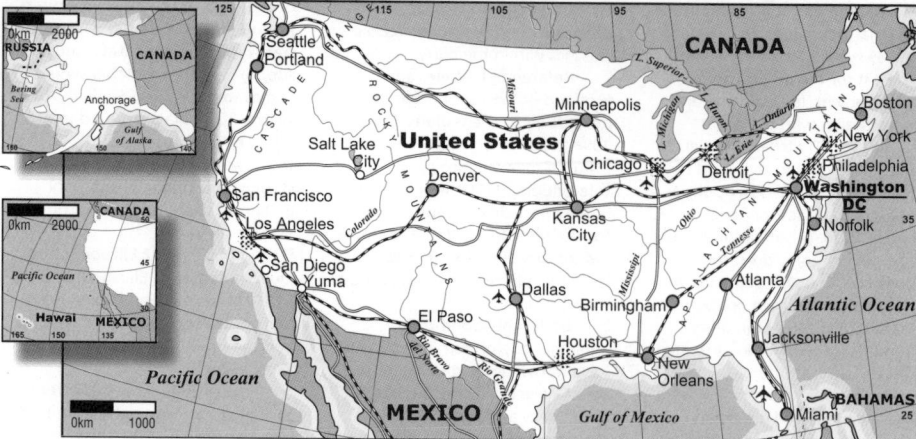

The continental territory occupied by the United States was inhabited 30,000 years before the arrival of the Europeans by peoples who probably came from Asia across the Bering Strait. The various Native American groups spread across the landmass, adapting their lifestyle to local conditions from the desert areas in the southwest through the plains to the eastern woodlands.

2   In the deserts and on the plains, they were primarily hunters and gatherers, living in small tribes with a simple social structure. Where the lands were more fertile, agriculture developed and relatively large towns were established (Cahokia, close to present-day St Louis, had 40,000 people in the year 1000 AD).

3   The religious beliefs of these peoples were rooted in a cosmic conception in which the Earth belongs to the Universe, and is considered a living being, with both material and spiritual powers. The shamans, calling upon these forces, could foretell the future, lead their people or heal the sick.

4   The first Europeans to come to America - Scandinavian - did not settle permanently in the region. After the voyage of Christopher Columbus in 1492, Spaniards established the colonies of St Augustine in Florida and Santa Fe in New Mexico; they also explored Texas and California. Then came the British, French and Dutch, all bent on territorial conquest.

5   In 1540, Hernando De Soto wrote in his journals of having found among the Cherokee an advanced agricultural society, linked to the peoples of the Ohio, the Mississippi and even the Aztecs.

6   There were an estimated 1,500,000 Native Americans in the 15th century. Two centuries later, the large plantations of the South began buying slaves, and by 1760 there was a total of 90,000 Africans - twice the number of whites in that part of the country. The total number of British settlers on the Atlantic coast at the time was 300,000, far outnumbering the French in the Mississippi Valley.

7   Most British immigrants left their country fleeing from poverty, religious persecution and political instability. Yet the birth of the colonies was marked by war against the native peoples and other European colonists. By 1733, there were 13 English colonies, whose chief economic activities were agriculture, fishing and trade.

8   In 1763, with the European wars behind them, France ceded its colonies east of the Mississippi to Britain, while its possessions west of that river went to Spain.

9   War with Britain broke out in 1775. The Declaration of Independence, which marked the birth of the United States, was signed on 4 July 1776. The war continued, but in the end the United States won (aided by its ally, France). England recognized US sovereignty in 1783.

10   In 1787, the Philadelphia Constitutional Convention drew up the first federal constitution, which came into effect in 1788. George Washington, commander of the Continental Army, was elected President in April 1789. In 1791, ten amendments dealing with individual freedoms and 'states' rights' were added to the original constitution.

11   The West was conquered simply by staking claims on the land, without previous ownership. The presidents in power at the time justified this as being the US 'Manifest Destiny' to become a great nation.

12   In 1803, the Louisiana purchase (from the French) doubled the size of the Union. Between 1810 and 1819, the US went to war against Spain in order to annex Florida. In 1836, the Texans rebelled against Mexico and set up a republic, subsequently joining the Union in 1845. After a new war, the US took over half of the Mexican territory. California became a state in 1850, and Oregon in 1853.

13   Westward expansion meant another tragedy for the original inhabitants of the region, who were decimated by successive waves of gold fever and multiple treaties that were ignored by new settlements in native lands. In 1838, 14,000 Cherokee people were forced off their lands by the army, 4,000 perishing in a march to their new territory.

14   In 1850, of the six million inhabitants of European origin in the South, only 345,525 were slave owners. However, most whites were pro-slavery, remembering the slave rebellions which had taken place in South Carolina (1822) and Virginia (1800 and 1831).

15   The Civil War (or War of Secession), from 1861 to 1865 revolved around the question of slavery, but it was in fact a struggle between the two economic systems prevailing in the country. While the industrial North sought to free an important source of labor and protect the domestic market, the slave-owning and agricultural South's interest was to maintain its cheap labor force and continue to enjoy free access to foreign markets.

16   With the election of Abraham Lincoln in 1860, the Southern states seceded. Committed to preserving the Union, and with a superior industrial base and superior weapons, the North finally triumphed over the South in the Civil War, although a million people were killed on both sides. Slavery was abolished, but racial discrimination and ill-feeling between the two regions persisted, leading to the assassination of Lincoln in 1865.

17   After the War, the Native Americans of the Great Plains, especially the Lakota (Sioux), fought for their lands. The sovereignty treaties of 1851 and 1868 were ignored after the discovery of gold in the area. In spite of a major Lakota victory at Little Big Horn in 1876, the occupation of their territory was completed by 1890, when the native peoples were finally defeated.

18   In the 1880s, the remaining Native Americans were confined to reservations on arid, barren lands. Years later, when uranium, coal, oil, natural gas and other minerals were discovered on some of the reservations, the issue of 'rights' to the land was once again brought up by mining companies.

19   Between 1870 and 1920, the population of the United States grew from 38 to 106 million, and the number of states increased from 37 to 48. It was a period of rapid capitalist expansion, triggered by the growth of the railroads. An

LAND USE

2003/2005

IRRIGATED AREA: 12.8% of arable land

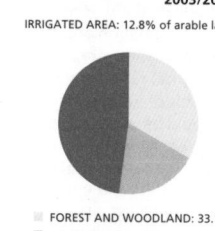

- FOREST AND WOODLAND: 33.1%
- ARABLE LAND: 18.9%
- CROPLANDS: 0.2%
- OTHER USE: 47.8%

PUBLIC EXPENDITURE

% OF GDP

12.3 % — HEALTH & EDUCATION (2000-2002)

4.0 % — DEFENSE (2004)

Life expectancy
78 years
2005-2010

GNI per capita
$41,440
2004

HIV prevalence rate
0.6% of population 15-49 years old
2003

agrarian country was transformed into an industrial society.

20 A two-party system of government had been established, with the Republican and Democratic parties alternately in power. Despite their different traditions, both parties have historically maintained a large degree of consensus on major national and international issues, leading to a highly coherent foreign policy. The long struggle for women's suffrage began in 1889, being achieved in 1920.

21 Having reached domestic stability, the US ventured onto the international scene. The US justified its interventions with the Monroe Doctrine and the slogan 'America for the Americans'. France was forced to withdraw its troops that protected Emperor Maximilian in Mexico, and Britain had to drop a territorial dispute with Venezuela. In 1890, the first Pan-American conference was held, paving the way for the inter-American system later set up.

22 The Spanish-American War in 1898, fought in Cuba and the Philippines, marked the beginning of a US imperial era. The occupation of Panama, and subsequent construction of the Panama Canal and a series of military bases turned Central America - an area of 'vital security interest' to the US - into a kind of protectorate.

23 During World War I, the US broke its neutrality in 1917 when it declared war on Germany, Austria and Turkey. In 1918, President Woodrow Wilson was one of the architects of the Treaty of Versailles, which established the framework for a new European peace. He also sought to guarantee peace by establishing the League of Nations. However, in 1920 US entry into the League of Nations was blocked by Congress.

24 The 1921-1933 period comprised the so-called lawless years because gangsterism practically governed many major cities and the Ku Klux Klan (KKK) terrorized the black population in the southern states. A controversial law that banned the production and consumption of alcoholic beverages - Prohibition - gave way to smuggling, clandestine production and gangsterism.

25 In 1924, J Edgar Hoover was appointed Director of the Bureau of Investigations, a special police force created by Attorney General Charles Bonaparte in 1908 during the Theodore Roosevelt administration. Hoover tried to obtain more power for his agents - who were not allowed to carry firearms - in their fight against crime.

26 The Great Crash on Wall Street (the US Stock Exchange) in 1929 led many banks to fail, affected industry and trade severely and

produced a global economic crisis. Unemployment figures rose to 11 million. Under Franklin Roosevelt's presidency (1933-45), the Government managed to bring the financial crisis under control through the New Deal policy, which involved massive investment in public works.

27 In 1935 Hoover's aspiration gave way to a Congress resolution creating the Federal Bureau of Investigations (FBI), with national jurisdiction and allowing its agents to carry firearms.

28 In 1935, while Europe prepared for another war, Congress passed a law proclaiming US neutrality. Roosevelt amended the law in order to allow arms shipments to France and Britain. In 1941, the Japanese attack on the US military base at Pearl Harbor in Hawaii precipitated US entry into World War II.

29 The War acted as a dynamo for the US economy. With 15 million soldiers at the Front, employment grew from 46.5 million to 53 million with the military industry. To meet the demand for workers, six million people migrated from

the countryside to the cities and many women worked outside the home. In spite of a labor 'truce', there were 15,000 strikes during the War, which led Congress to pass a law limiting the right to strike.

30 At that time, the FBI persuaded President Roosevelt to investigate a Soviet conspiracy against the US which included massive espionage.

31 With Germany defeated in 1945, President Harry Truman (who had assumed the presidency upon Roosevelt's death in 1944) wanted to put an end to the war with Japan. Thus, on 6 and 9 August, he gave the order to wipe out the cities of Hiroshima and Nagasaki with history's first atomic bombs. That same year, at Yalta and Potsdam, Britain, the United States and the Soviet Union divided up the areas which would come under their respective spheres of influence.

32 Truman presided over the opening of the United Nations in 1946, and was re-elected President in 1948. As the number one Western power, and in the context of the Cold War, the US

assumed a global confrontation with the USSR. The Inter-American Treaty of Reciprocal Assistance was signed, and the North Atlantic Treaty Organization (NATO) was created. The US took it upon itself to safeguard the global capitalist system, with the support of international institutions such as the World Bank and the IMF, as well as armed intervention around the world.

33 Addressing the country's new needs, in 1947 Truman created the Central Intelligence Agency (CIA), whose mandate was to provide the President with reliable information on all domestic or foreign activities related to national security. The Marshall Plan was implemented in 1948 to reactivate the post-war European economies, at a cost of $13 billion over four years.

34 Hoover took advantage of the domestic and global situation to increase his influence, and in 1950 Senator Joseph McCarthy began to persecute Communist sympathizers in Washington. The FBI opened files on politicians, intellectuals, journalists and ordinary citizens suspected of un-American activities

PROFILE

ENVIRONMENT

There are four geo-economic regions. The East includes New England, the Appalachian Mountains and part of the Great Lakes and the Atlantic coast, a sedimentary plain which stretches from the mouth of the Hudson River to the peninsula of Florida. To the west are the Appalachian mountains, where mineral deposits (iron ore and coal) abound. This is the most densely populated and industrialized area, where the country's largest steel plants are located. High-technology agriculture provides food for the large cities. The Midwest stretches from the western shores of Lake Erie to the Rocky Mountains, also including the middle Mississippi. Formed by the grasslands of the central plain, the Midwest is the country's largest agricultural area; horticulture and milk production predominate in the north, while wheat, corn and other cereals are cultivated in the south, side by side with cattle ranches where cows and pigs are raised. Major industrial centers are located near the Great Lakes, near the area's agricultural production and large iron ore and coal deposits. The South is a subtropical flatland area, comprising the south of the Mississippi plain, the peninsula of Florida, Texas and Oklahoma. Large plantations (cotton, sugarcane, rice) predominate here, while there is extensive cattle-raising in Texas. The region is also rich in mineral deposits (oil, coal, aluminum, etc.). The West is a mountainous, mineral-rich area (oil, copper, lead, zinc). There is considerable horticultural production along the fertile valleys of the Sacramento and San Joaquin rivers in California. Large industrial centers are located along the Pacific coast. In addition, the US has two states outside its original contiguous area: Alaska, on the continent's northwest where Mt McKinley is located (Mt 'Denali', in the indigenous Atabasco language), the highest peak in North America, and Hawaii, an archipelago in the Pacific Ocean.

SOCIETY

Peoples: There are 1.9 million Native Americans, half of whom live in 300 reservations. The white population were originally immigrant Europeans including British, German, Irish, Russian and Italian, now mixed with immigrants from all parts of the world. The largest minorities are of African origin, 11 per cent of the total population; Hispanic ten per cent; and Asian eight per cent.
Religions: Protestant (58 per cent); Catholic (26 per cent); Jewish (two per cent); Muslim (two per cent); other (two per cent); non-religious (ten per cent). **Languages:** English; Spanish; Native American languages and those of each immigrant group.
Main Political Parties: Republican Party; Democratic Party; Independent Reform Party; Libertarian Party; Constitution Party; Green Party.
Main Social Organizations: The American Federation of Labor - Congress of Industrial Organizations (AFL-CIO) is the country's largest workers' organization, with 13,500,000 members. Many of the country's rural laborers, especially those of Mexican origin, are organized in the United Farm Workers (UFW) labor union, founded by César Chávez.

THE STATE

Official Name: United States of America.
Administrative Divisions: Federal State, 50 States and one Federal District, Columbia.
Capital: Washington, DC 4,098,000 (2003).
Other Cities: New York 8,008,278; Los Angeles 3,694,820; Chicago 2,896,016; Houston 1,953,631; Philadelphia 1,517,550 (2000).
Government: Presidential government, federal system. George W Bush, President since January 2001. Dick Cheney, Vice-president since January 2001. There is a bicameral Congress: the House of Representatives, with 435 members, and the Senate, with 100 members.
National Holiday: 4 July, Independence (1776).
Armed Forces: 1,434,000 (2003). Other: 68,000 Civil Air Control.

Under-5 mortality
8 per 1,000 live births
2004

Maternal mortality
17 per 100,000 live births
2000

IN FOCUS

ENVIRONMENTAL CHALLENGES

Air pollution causes acid rain. The US is the world's largest emitter of contaminating gases, such as carbon dioxide, produced by burning fossil fuels. Pesticides and fertilizers have polluted the water. Water resources are limited, especially in the west. There is desertification.

WOMEN'S RIGHTS

Women have been able to stand for election since 1788, and to vote since 1920. In 2004, women held 14 per cent of seats in Congress. That year they comprised 47 per cent of the labor force.

The percentage of men and women was similar at all education levels, except in higher education, where the enrolment rate among women amounted to 83 per cent, compared to 63 per cent among men. Prenatal health care covers 99 per cent of women, and 99 per cent of births are attended by qualified personnel*.

CHILDREN

There are no significant malnutrition problems among children but levels of obesity cause increasing concern. Under-five mortality stands at 8 per 1,000 live births*. Primary school enrolment stands at 94 per cent*.

Since 1985, some 9,300 children have been infected with HIV/AIDS, and approximately 5,000 have died. More than 90 per cent became HIV-positive during pregnancy or at birth.

INDIGENOUS PEOPLES/ ETHNIC MINORITIES

The country comprises multiple territories that were originally under indigenous control. After the bloody repression of indigenous resistance in the 19th century, most of the ancestral ethnic groups have descendants among the present population. Many reside in reservations negotiated with the Government or directly established by it, but others are relatively integrated into US society. Cherokees are a special case, being a wealthy nation with a high quality of life.

The main ethnic groups are: Cherokee (16 per cent of total Native American population), Navajo (12), Chippewa (5), Lakota (5), Choctaw (4), Pueblo, Apache, Iroquois, Lumbee, Creek (3), and Blackfoot (2).

The states with indigenous reservations are: Washington, Idaho, Montana, North Dakota, South Dakota, Minnesota, Wisconsin, Michigan, New York, Oregon, Wyoming, Nebraska, California, Nevada, Utah, Colorado, Kansas, Arizona, New Mexico, Oklahoma, North Carolina, Mississippi and Florida.

The US has not been able to eradicate discrimination towards African-Americans, Hispanics, indigenous groups, Asian or Arabs, among other ethnic minorities. African Americans have three times less chance of finding a job than white workers with similar skills.

MIGRANTS/REFUGEES

The US Patriot Act 2001 authorizes the detention and deportation of non-citizens suspected of terrorist activity, extending the refusal of asylum to the spouse and children of the detained. Persons without documentation who are detained in Caribbean waters are taken to the Guantanamo (Cuba) naval base, according to new government regulations which particularly affect Haitians.

In 2004, the country hosted approximately one million immigrants with legal residence. That year 420,854 new asylum and refuge requests were filed, in addition to the more than 263,700 asylum requests that were still pending. Most asylum seekers came from Bosnia-Herzegovina, Somalia, Vietnam, Iran, Liberia, China, Iraq, Afghanistan, Cuba and Colombia. By late 2004, 49,638 refuge requests had been granted, while 33,100 had been rejected.

DEATH PENALTY

Of the 50 states in the Union, the death penalty remains effective in 38. The country continues to violate international standards by executing people who were under 18 at the time they committed their crime.

*Latest data available in *The State of the World's Children* and *Childinfo* database, UNICEF, 2006.

(mostly Communist). So-called McCarthyism, through the House Committee on Un-American Activities (HUAC) and backed by the McCarran-Walter Act, carried out a major witch-hunt which saw a red and potential Soviet spy in any dissident. The FBI files were filled with thousands of pages of information on citizens from all walks of life.

[35] In 1952, General Dwight Eisenhower, commander-in-chief of American forces in Europe during the War and head of NATO, was elected President. The Korean War (1950-53), the partition of Germany, popular uprisings in Poland and Hungary, and the delicate balance of nuclear weapons served to maintain tension with the USSR. In 1954, the Senate put an end to McCarthyism, censoring McCarthy, but the FBI continued to maintain its files on citizens.

[36] In 1956, the US offered South Vietnam's government military support. In 1960, a summit meeting between Eisenhower and Soviet leader Khruschev was cancelled when an American U-2 spy plane flying over Soviet territory was brought down.

[37] The election of Democratic presidential candidate John F Kennedy in 1960 brought hope for relief from domestic and foreign tensions. However, influenced by CIA reports, Kennedy supported the Bay of Pigs invasion of Cuba in 1961, and initiated the economic blockade of that country. He also supported the National Aeronautics and Space Administration's (NASA) drive to overtake the USSR (which had launched the first satellite and put the first man into orbit) in the space race and to put a man on the Moon before the decade was over.

[38] Kennedy was assassinated in Dallas, Texas, in 1963. Although Lee Harvey Oswald was jailed for the crime and later murdered, other suspects for the assassination were the Mafia, Cuban agents (pro and anti-Castro), the CIA and the Army.

[39] With regard to Latin America, President Kennedy had launched the Alliance for Progress in Uruguay in 1961, in an attempt to prevent a new Cuba. However, the funds earmarked for this project were insufficient to effect real change. Faced with growing guerrilla activity in the region, the US, through the CIA (and its global presence) supported the regional armies.

[40] Lyndon Johnson was elected President in 1964. The war in Vietnam escalated, causing a wave of protests throughout the US. Racial segregation led to increasing confrontation, underlined by the assassinations of black civil rights leaders Martin Luther King and Malcolm X.

[41] In 1968, the American Indian Movement (AIM) was founded by two Chippewa leaders. In 1969, AIM occupied the abandoned prison on Alcatraz Island in San Francisco to call attention to their demands and denounce the mistreatment of their people. The hippie movement and student protests reflected a deep cultural renewal.

[42] In 1968 Richard Nixon (Eisenhower's Vice-president and fomer HUAC chairman) was elected President. In 1969, two astronauts from NASA's Apollo program set foot on the Moon, leading the US to win the space race.

[43] In 1972, Nixon visited Moscow and Beijing. Also that year, after 48 years as head of the FBI, Hoover died, leaving behind him millions of files of citizens under investigation, including his own and those of the eight presidents he had worked for.

[44] In 1973, Nixon was re-elected and in 1974 he signed the final withdrawal of US troops from Vietnam - the effective defeat by the North Vietnamese forces was seen as a major blow to the American psyche. Nixon was forced to resign the same year, following the discovery that Republicans had spied on Democrat election campaign headquarters located in the Watergate Hotel.

[45] Latin American dictatorships - initially supported by Washington - had their heyday in the latter half of the 1970s. However, the Democratic administration of Jimmy Carter signified the beginning of the end for them. His pressure on the issue of human rights undermined the dictatorships.

[46] On the international front, Carter organized the meeting between Egyptian President Mohammed Anwar al Sadat and Israeli Prime Minister Menachim Begin, in Camp David, signed the Salt II agreement to restrict the number of nuclear weapons held by the US and the USSR, and established full diplomatic relations with China. However, economic inflation and the lengthy US hostage crisis in Iran caused his electoral defeat in 1980.

[47] During the Republican administration of Ronald Reagan (1980-88) the military-industrial complex energized the whole economy and partly made up for lagging behind Japan and Western Europe on other fronts. In 1983 the US invaded Grenada in response to its rapprochement with Cuba.

[48] The Irangate scandal erupted in 1986, with the illegal use of weapons to support counter-revolutionary forces in Nicaragua. In spite of his peace through strength doctrine, Reagan agreed with USSR leader Mikhail Gorbachev to limit the number of mid-range missiles.

[49] George Bush won the 1988 elections. In 1989 he invaded Panama, toppled the government and arrested its leader, the former CIA informer Manuel Noriega. In February 1991, the US led the

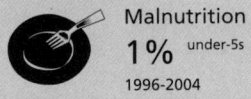 **Malnutrition**
1% under-5s
1996-2004

 Water source
100% of population using improved drinking water sources
2002

Doctors
549 per 100,000 people
1990-2004

 **Primary school**
94% net enrolment rate
2004

multinational force which expelled Iraq after its invasion of Kuwait. The Gulf War consolidated US military supremacy.

50   Bill Clinton, Democrat Governor of Arkansas, was elected President in November 1992, with a majority in both chambers of Congress. The North American Free Trade Agreement (NAFTA) with Mexico and Canada went into operation in January 1994. The economy recovered and unemployment fell.

51   For the first time in 40 years, the Democratic Party lost the election to both chambers of Congress in 1994. Clinton led a military intervention in Bosnia-Herzegovina and imposed the Dayton (Ohio) agreements in November 1995. That year, a bomb in Oklahoma killed 160 people, in the worst terrorist attack in the country up to that point. In October 1996, Clinton supported talks between Israelis and Palestinians in Washington. In November, he was re-elected with 49.2 per cent of the vote.

52   In 1998 Clinton had to contest the charge of perjury brought against him by independent counsel Kenneth Starr. In February 1999 the Senate voted against impeaching Clinton. Between March and June, the US led the NATO forces that bombed Yugoslavia in response to Serb persecution of Albanian citizens in Kosovo.

53   The November Presidential elections marked a critical point for the electoral system. One of the tightest elections ever, scarred with irregularities, pitched Republican candidate George W Bush, governor of Texas and son of the former president, against Democratic Vice-President Al Gore. The uncertainty went on for weeks after the state of Florida, which had had 6 million votes, ordered a limited recount.

54   On December 2000 the US Supreme Court ruled in favor of Bush. The difference between electoral votes and the popular vote underlined the need to revise the country's electoral system.

55   On 11 September 2001 the US suffered the most serious terrorist attack ever made on its territory. Four commercial airplanes were kidnapped by suicide bombers, of which two crashed into the World Trade Center twin towers in New York, which collapsed, and another into the Pentagon building in Washington. More than 3,000 died and direct economic destruction amounted to an estimated $1 billion. The US population was traumatized.

56   The Bush Government reacted by declaring a global war on terror. Washington increased the military budget by 20 per cent, and then a further 15 per cent. The US military

budget was thus larger than the sum total of the military budgets of all of its 18 allies in NATO.

57   The Bush offensive on terror started in October 2001 in Afghanistan, where the US deposed the Taliban regime although it could not find its leader, Mullah Muhammad Omar, nor al-Qaeda leader Osama Bin Laden.

58   In January 2002, the US vowed to confront an 'axis of evil' made up by North Korea, Iran and Iraq. Iraq was chosen as the first target. Arguments for the attack were based on two claims: Iraqi President Saddam Hussein's 'will' to develop weapons of mass destruction and his possible link with international terrorists. In early September 2002, during the UN's 57th General Assembly in New York, Bush asked the international organism to act or allow the US to do it alone.

59   Although in October Hussein agreed to allow inspections of 'sensitive' locations, the UK and US rejected this offer. That month, Washington revealed that North Korea had admitted that it was developing a nuclear weapons program and demanded that it was dismantled.

60   The US, the UK and a coalition of forces led by Washington and London invaded Iraq in March 2003. Other permanent members of the Security Council did not support the war, and Germany was openly opposed.

61   The Allied forces rapidly advanced on Baghdad. On 1 May, on board the USS Abraham Lincoln, Bush announced the end of the first stage of the conflict and the start of Iraq's democratic reconstruction.

62   In April 2004 a scandal broke after the publication of photographs and videos showing torture, ill treatment and sexual abuse of Iraqi prisoners by the US army in the Abu Ghraib prison, on the outskirts of Baghdad.

63   George W Bush was re-elected for another four-year term in November 2004. He won with 51 per cent of the popular vote, receiving 3.5 million votes more than his Democratic rival, John Kerry, in contrast to to the 2000 poll, when he was elected without getting more votes than the Democratic candidate.

64   Hundreds died in September 2005 when Hurricane Katrina - the most destructive storm to hit the US in decades - devastated the Gulf of Mexico states. A large part of New Orleans was covered by water and a political crisis soon ensued. The Government was harshly criticized for its slow reaction and lack of foresight (the administration's top officials - including Bush - were on holiday).

65   In March 2006, Bush achieved an extension to the controversial

Patriot Act which had been in effect since 2001, tightening restrictions on individual liberties in the name of national security.

66   Millions of demonstrators took to the streets nationwide in April and May 2006 to protest against an immigration reform law targeting illegal immigrants. Protesters carried banners saying 'we are a country of immigrants' and described the demonstrations as 'our generation's struggle for civil rights'.

67   Democratic Senator Robert F Kennedy Jr published an article in the June issue of *Rolling Stone* magazine reporting an alleged fraud committed by the Republican Party in the 2004 elections, which prevented Kerry from getting to the White House. According to Kennedy, more than 350,000 people in Ohio were not able to vote because their names were missing from the registers, although they had registered as voters. The article also questioned

the operation of the voting machines. The state of Ohio was crucial to Bush's victory.

68   In October 2006, Bush signed a law allowing construction of more than 1,100 kilometers of fence to prevent Mexicans illegally entering the US. Mexican President Vicente Fox compared the fence to the Berlin Wall.

69   In the November 2006 mid-term elections, the Democratic Party swept to victory, regaining control of both the Senate and the House of Representatives and dealing a bloody nose to the Bush Administration. The increasing unpopularity of the Iraq war, in which more than 2,800 US soldiers had by then died, was widely blamed for the political sea change. President Bush immediately replaced Defense Secretary Donald Rumsfeld with former CIA director Robert Gates, who was seen as a less gung-ho figure. ∎

US DEPENDENCIES

JOHNSTON:
Coral atoll of 2.6 sq km, made up of the Johnston island and the Sand, East and North islets, located approximately 1,150 km west-southwest of Honolulu (Hawaii). Once mined for guano by the US, the US Navy took over in 1934 and the US Air Force in 1948. The site was used for nuclear tests in the 1950s and 1960s and later as a disposal site for chemical weapons. The military facility closed in 2004, leaving just 200 people from the US Air Force, US Fish and Wildlife Service and some civilian contractors.

MIDWAY:
A round atoll which comprises two islands: Sand and Eastern, with five sq km of total area and an almost exclusively military population. In 1867, they were annexed by the US and are currently administered by its Navy. The islands are used for military purposes and operate as a refueling base for trans-Pacific flights. In 1942 during World War II, an important sea battle took place here.

WAKE:
Together with neighboring Wilkes and Peale, they make up an atoll with an area of 6.5 sq km and an almost exclusively military population. Located between Midway and Guam, the island was seized by the United States during the 1898 war with Spain. It has a large airport which used to be a stopover for trans-Pacific flights, but nowadays it is not used for business purposes. Since 1972 it has been administered by the Air Force which now uses it as a missile testing station.

HOWLAND, JARVIS & BAKER:
Located in central Polynesia, in the Equatorial Pacific they were occupied by the US in the middle of the 19th century. The islands have been uninhabited since the end of World War II as major phosphate deposits had been depleted. There is a lighthouse in Howland. The islands are administered by the US Fish and Wildlife Service.

PALMIRA & KINGMAN:
The northernmost islands of the Line archipelago. Palmira is an atoll surrounded by more than 50 coral islets covered with exuberant tropical vegetation. It was annexed in 1898 during the war with Spain and is now a private property, dependent on the US Department of the Interior. Kingman is a reef, located north of Palmira, annexed by the US in 1922. The total area of both islands is seven sq km. Although uninhabited, the US-Japanese project to turn them into deposits of radioactive waste raised protests throughout the Pacific region in the 1980s.

Uruguay / Uruguay

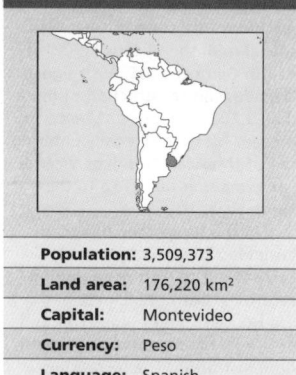

Population:	3,509,373
Land area:	176,220 km²
Capital:	Montevideo
Currency:	Peso
Language:	Spanish

The territory of what is now part of Uruguay was discovered by Spanish explorer Juan Díaz de Solís in 1516. At that time, the country was inhabited by the Charrúa, the Chaná and the Guaraní. The Charrúa were nomadic hunters, while the Chaná developed a rudimentary form of agriculture and the Guaraní practiced agriculture and mastered ceramics and navigation of rivers.

[2] Solís sailed into the Río de la Plata which he called the *Mar Dulce* (Freshwater Sea). Ambushed by Indians, Solís and all the other members of the expedition were killed.

[3] Spanish colonization was slow since settlers considered the region as a 'land of no profit' because it did not have the mineral wealth of Mexico and Peru. In 1611, the Governor of Asunción, Hernando Arias de Saavedra (Hernandarias) introduced cattle and horses, which - thanks to the good pasture and climate - thrived and transformed the region into what was known as the *Vaquería del Mar* (Cattle Ranch of the Sea). The Jesuits, a Christian religious order which had begun to establish missions in Paraguay in the 17th century, started to expand towards the eastern side of the Uruguay River in 1632. From 1667, seven missions were founded to the east of the river.

[4] The cattle attracted the *faeneros* - leather workers - from Brazil and Buenos Aires. The spread of cattle gave rise to the skilled horsemen/cowboys known as *gauchos*, who were usually of mixed European and indigenous ancestry.

[5] The widespread cattle-ranching led to the extinction of some indigenous mammals, a reduction of plant diversity and soil degradation. The indigenous people were displaced and moved into the Jesuit missions further north. In 1831, the Charrúa were exterminated in Salsipuedes by the first independent Government of Uruguay. The survivors of this ethnic group were taken as slaves and a few of them, who became known as 'the last Charrúas' were sent to France and

exhibited there as fairground-type curiosities.

[6] In 1680, the Portuguese founded Colonia do Sacramento on the shore opposite Buenos Aires. This settlement caused constant friction between Portugal and Spain. The border disputes gave rise to the city of Montevideo, the only natural port in the Río de la Plata, which was founded in 1724 by the Governor of Buenos Aires, Bruno Mauricio de Zabala.

[7] The May Revolution which broke out in Buenos Aires in 1810 was rejected by the Montevideans, who supported the trade monopoly with Spain. In contrast, small and medium-scale producers from the countryside and the landless people rose up in arms. The rebels were led by the creole captain of the Spanish army, José Artigas. His republican and federal ideas as well as his close relationship with Indians and slaves and his land distribution program turned him into the leader of the insurrection. Artigas also headed the Federal League which comprised what is now Uruguay and the Argentine provinces of Córdoba, Corrientes, Entre Ríos, Misiones and Santa Fe.

[8] Uruguay was invaded by the Portuguese in 1816, with the tacit approval of Buenos Aires and Montevideo, whose governments were alarmed by the events known as the 'Artigan chaos'. Once defeated and betrayed, Artigas sought refuge in Paraguay in 1820, where he died 30 years later.

[9] When Brazil became independent from Portugal in 1823, Uruguay became the Brazilian 'Cisplatine Province'.

[10] In 1825 the 'orientales' renewed their independence campaign

against the Brazilian Empire, which ended on 25 August with the declaration of independence and the decision to return to the United Provinces of the Río de la Plata.

[11] Since the British diplomatic service was interested in preventing both Río de la Plata territories from belonging to Argentina in order to protect Britain's commercial interests, it promoted the creation of a small state within the region. In 1828, through the mediation of Lord Ponsonby, the war was brought to an end and the independence of the country was approved.

[12] The first Constitution of Uruguay was adopted in 1830. The new Constitution established a republican government but most of the population (who were illiterate, rural workers) were denied the right to vote.

[13] The 19th century was characterized by civil wars between the two traditional political parties: the Blanco (White) or National Party, linked to the country and to land owners, and the Colorado (Red but not communist), linked to European capital and liberal ideas. The successive uprisings that followed were favored by a weak state and rural workers who were turned into professional soldiers.

[14] In 1865, Colorado dictator Venancio Flores signed an agreement with Brazil and Argentina and created the Triple Alliance, waging war on Paraguay with European support - and forcing the country to open its borders to foreign trade (see Paraguay).

[15] In 1876, Colonel Lorenzo Latorre ushered in a period of militarism. In that period (1876-1879), state power was strengthened thus preventing armed uprisings

by monopolizing the use of force. Also, the countryside was enclosed with fences, resulting in the land being completely appropriated by the private sector, and leaving no room for the gaucho, who was transformed into a hired hand.

[16] In 1903 the Colorado José Batlle y Ordoñez assumed the presidency and was determined to modernize the state. The last rural uprising, led by the *caudillo* (leader) of the Blancos Aparicio Saravia, took place one year later. Saravia died during the confrontation and Batlle laid the foundations for the modern Uruguayan state.

[17] The State became the main employer. A large liberal middle-class developed, educated in State schools.

[18] The Church and the State were separated and divorce legalized. A collegiate system of government was introduced in 1917 and women's suffrage was enacted in 1932. This open legislation earned the country the title 'the Switzerland of South America'.

[19] Uruguayan exports grew during both World Wars. Meat and its products were supplied first to the Allies, who were fighting against Nazism and Fascism in Europe, and later to US troops fighting in Korea.

[20] Trade balance surpluses secured the country large foreign currency reserves. The welfare policy of subsidies encouraged the emergence of relatively strong import-substitution industries. Meanwhile, a prosperous building industry helped maintain high employment. However, the cattle and sheep sector, which generated most exports, did not expand but remained at the production levels of 1908.

[21] Land owners invested their profits abroad, engaging in financial speculation and superfluous consumption. In the 1950s, the industrial sector stagnated - a situation that proved impossible to reverse.

[22] The first Blanco Government, in 1959, accepted IMF economic guidelines which accelerated the recession. Social conflict was intensified in 1968 when the Colorado Government of Jorge Pacheco Areco, curbed the spending power of wage-earners and tried to eliminate the trade unions' bargaining power. A broad movement led by the National Workers' Convention and the student organizations opposed such policies. At the same time the Tupamaro (MLN) guerrilla movement was active throughout the country. In 1971 the Frente Amplio (Broad Front), a left-wing coalition, was founded. It promoted a progressive government program, and nominated a retired general, Liber Seregni, in the general elections of that year.

Life expectancy
76 years
2005-2010

GNI per capita
$3,900
2004

Literacy
98% total adult rate
2000-2004

HIV prevalence rate
0.3% of population 15-49 years old
2003

23 Juan María Bordaberry became President in 1972. Parliament declared a state of emergency which allowed homes to be searched without warrants, the suspension of habeas corpus and the referral of civilians to military courts. In 1972 the Tupamaros were defeated. A campaign including the systematic use of torture rapidly dismantled the clandestine organization. In June 1973, President Bordaberry and the armed forces staged a coup. Parliament was dissolved, and a civilian-military government was formed. Left-wing parties and unions were banned; torture and arbitrary detentions of people opposed to the regime were commonplace.

24 During the ensuing dictatorship the concentration of wealth in transnational corporations increased. Salaries lost 50 per cent of their purchasing power, whilst foreign debt reached $5 billion.

25 In 1980, the Government submitted an authoritarian constitution to a referendum. The defeat suffered by the military marked the beginning of the end of the dictatorship.

26 Social organizations sprang up again in 1983: the struggle against the military regime became open and took to the streets. One of the most important protest mobilizations was called by the newly-formed Inter-union Plenary of Workers (PIT) which celebrated May Day for the first time since 1973, under the banner of Freedom, Work, Salary and Amnesty.

27 That same month, the military started negotiations with the three political parties recognized as legal, excluding the Broad Front, whose president Líber Seregni had been in prison since the coup, and the Blanco Party leader Wilson Ferreira Aldunate, who was to be arrested on his return to the country after an 11-year exile.

28 In the November 1984 elections some leaders and political parties remained proscribed. The elections were won by the conservative leader of the Colorado Party, Julio María Sanguinetti, who became President. The new government restored freedoms and political rights and, in response to popular demand, Parliament approved an amnesty law whereby all political prisoners were released.

29 A parliamentary commission was created to investigate the fate of those Uruguayans who had disappeared both in the country and outside. The civilian courts summonsed the officers allegedly responsible for human rights abuses. Then, Parliament approved the Expiry Law of the Punitive Powers of the State, which exempted all military and police personnel responsible for human rights crimes from punishment. A national referendum ratified the law with 56 per cent of the vote.

30 In 1989, a structural adjustment policy was implemented, marked by the deregulation of the market. The Government signed a secret agreement with the World Bank, in exchange for rescheduling Uruguay's debt. The Government committed itself to reduce expenditure on social security; to privatize bankrupt banks absorbed by the State, and to reform public companies making them profitable and attractive for privatization.

31 In 1989, the Blanco/National Party won the elections and Luis Alberto Lacalle took office as President. The Broad Front won in Montevideo with its candidate Tabaré Vázquez. The Left assumed responsibility for municipal administration for the first time in the history of the country.

32 Lacalle carried out his neoliberal policy: taxes were increased and the privatization of state-run companies was encouraged. In March 1991, Argentina, Brazil, Paraguay and Uruguay approved the Common Market of the South (Mercosur) agreement.

33 A committee convened by the labor union movement, constituted by members of several parties, managed to submit the state enterprise privatization law, previously approved by Parliament, to a plebiscite. In December 1992, the population voted in a referendum to repeal the law.

34 In the 1994 elections Sanguinetti was elected President once again, although by a small majority. In the first round of the 1999 elections, the leftist Progressive Encounter-Broad Front received the most votes. To prevent it from winning the second round, the National Party allied itself with the Colorado Party and voted for the Colorado candidate, Jorge Batlle, who was elected President with 52 per cent of the vote in November.

35 In March 2000, President Batlle declared that building peace and trust among Uruguayans was essential. In August, the Peace Commission was created by presidential decree, and charged with taking all possible steps to investigate the fate of Uruguayans who 'disappeared' - within the country and abroad - during the military dictatorship. The work of the Commission, which was questioned by human rights organizations, ended with a final report approved in April 2003. This report confirmed the 'action of state agents' in the illegal repression.

36 The country's poor economic performance in 2002 caused a considerable downturn in most people's quality of life. The Argentine economic crisis and the shrinking of the Brazilian market posed serious problems for Uruguay, raising the possibility of a default within the short term.

37 In January, thousands of Uruguayans marched toward Punta del Este, the main beach resort, to protest against the Government's economic policy. The Government banned the demonstrators from entering all the Atlantic resorts and ordered special security measures around the Punta del Este peninsula.

38 In March 2002, the IMF approved a $443 million loan to help Uruguay out of the recession and to protect it from the Argentine economic crisis. Uruguay agreed to achieve fiscal balance in 2004. The IMF urged Uruguay to privatize its state monopolies, such as electricity, oil, telecommunications and railroads.

39 That month, Sara Méndez, a Uruguayan woman who had been kidnapped in Buenos Aires in 1976 and illegally transported to Uruguay, managed to find her son, Simón Riquelo, who had been kidnapped with her shortly after his birth and had been disappeared. The boy had been given to an Argentinian chief of police who had brought him up as his own child. This ended Méndez's 25-year search for her son - something she had done with the help of human rights organizations but without the cooperation of Uruguayan justice. A huge crowd gathered in Montevideo to celebrate the reunion of mother and son.

40 In April, Uruguay sponsored a UN resolution against Cuba regarding the human rights situation on the island. The motion caused protests from Havana against President Batlle, whom Fidel Castro called a 'wretched Judas'. As a result, both countries broke off diplomatic relations.

41 That same month, for the first time in 20 years, employers and workers joined in a major demonstration against the Government's economic policy. The national strike was called by the so-called Agreement for Growth. The demonstrators protested against the bankruptcy filed by

PROFILE

ENVIRONMENT

Uruguay has a gently rolling terrain, crossed by characteristic low hills - an extension of Brazil's southern plateau - belonging to the ancient Guayanic-Brazilian massif. Its average altitude is 300 meters above sea level. This, together with its location and latitude, determines its temperate, subtropical, semi-humid weather, with rainfall throughout the year. The vegetation is made up almost entirely of natural grasslands, suitable for cattle and sheep raising. The territory is well irrigated by many rivers, and has over 1,100 km of navigable waterways, in particular on the rivers Negro and Uruguay, and on the Plata estuary. The coast is made up of many sandy beaches that attract a large number of tourists.

SOCIETY

Peoples: Most Uruguayans are descendants of Spanish, Italian and other European immigrants. Recent historical and genetic research show that a significant part of the population also has American Indian ancestry. Afro-American descendants make up about eight per cent of today's population.
Religions: Catholics 66 per cent; Protestants two per cent; Jews two per cent. Afro-Brazilian cults are also practiced.
Languages: Spanish.
Main Political Parties: Broad Front (which includes: People's Participation Movement; Uruguay Assembly; Socialist Party; Communist Party; Artiguista Angle); National/White Party (Blanco); Red Party (Colorado).
Main Social Organizations: PIT-CNT (Inter-Union Workers' Bureau - National Workers' Convention). FUCVAM (Uruguayan Federation of Housing Construction by Mutual Help). Federation of University Students (FEUU).

THE STATE

Official Name: República Oriental del Uruguay.
Administrative Divisions: 19 Departments.
Capital: Montevideo 1,237,000 people (1999).
Other Cities: Salto 86,600 people; Paysandú 76,400; Las Piedras 70,700; Rivera 69,400; Maldonado 40,600 (2000).
Government: Presidential system. Tabaré Vázquez, President since March 2005. Bicameral Legislature: the Chamber of Senators, with 31 members, and the Chamber of Deputies, with 99 members. The Vice-President chairs the Senate.
National Holiday: 25 August, Independence Day (1825); 18 July, Constitution Day (1830).
Armed Forces: 24,000 (2003). Other: 700 Metropolitan Guard, 500 Republican Guard.

IN FOCUS

ENVIRONMENTAL CHALLENGES
The projected installation of two pulp mills on the banks of the Uruguay River brought conflict with Argentinian environmentalists who argued that environmental damage would be huge. Rivers and streams are contaminated by agro-chemicals. Furthermore, the meat industry contributes to river pollution by discharging animal waste - generally in bad condition - into their waters. An increasing loss of ecosystems on the plains and in the eastern wetlands has been observed, due to monocultures of forestry and rice, respectively.

WOMEN'S RIGHTS
Women make up 52 per cent of the country's population. They have been able to vote and stand for office since 1932. In 2006, female representation in Parliament amounted to 10.76 per cent. On the other hand, 3 out of the 13 ministries were headed by women (Social Development, Defense and Health). They make up 43 per cent of the country's labor force; 84 per cent work in services, 14 per cent in industry and 2 per cent in agriculture. On average, women's salaries are 30 per cent lower than those earned by men performing the same task. About 68 per cent of university students are women, although they are under-represented in professional positions.

One of every two women heads of household in the economically active age group and with children under the age of five is poor*. Although HIV prevalence among women (20 per cent of cases) has risen in the country, it is much higher among men (80 per cent). Uruguay has the highest suicide rate in Latin America, and consistently has one of the seven highest suicide rates in the world. Some 500 Uruguayans kill themselves every year.

CHILDREN
The infant mortality rate is 15 deaths per 1,000 live births, while the under-5 mortality rate is 17 deaths per 1,000 live births*. Over 57 per cent of children under six years old are affected by poverty, which rose considerably after the economic crisis that started in 2002*.

While not yet a major problem, child sexual exploitation has increased in recent years, particularly in the areas bordering Brazil and Argentina as well as in tourist centers such as Punta del Este, Maldonado and Montevideo.

Approximately 50,000 children aged between 5 and 17 work, mostly in the informal sector (which accounts for 40 per cent of the total employment in the country). HIV/AIDS is increasingly affecting young people, particularly those aged between 15 and 24.

Although the secondary school enrollment rate stands at 74 per cent, 48 per cent of young people aged between 15 and 19 drop out before completing it*.

INDIGENOUS PEOPLES/ ETHNIC MINORITIES
The Afro-Uruguayan minority, estimated at 8 per cent of the population, faces societal discrimination. A study carried out by the NGO Mundo Afro found that the illiteracy rate among Afro-Uruguayan women was twice the national average. Some 70 per cent of economically active Afro-Uruguayan women work in domestic service and their salaries are up to 20 per cent lower than those earned by non-Afro women performing the same job. Afro-Uruguayans are virtually unrepresented in the bureaucratic and academic sectors.

MIGRANTS/REFUGEES
Emigration is competing against population growth rate and as a result some experts predict that the country may start experiencing negative population growth in a few years' time. Most emigrants are between 20 and 29 years old. The emigrants have an average or higher education in comparison to those Uruguayans of the same age group that remain in the country. The main destinations for emigrants are: the United States (33.3 per cent), Spain (32.6 per cent), Argentina (8.5 per cent) and Italy (4.7 per cent).

DEATH PENALTY
This was abolished for all crimes in 1907.

*Latest data available in *The State of the World's Children* and *Childinfo* database, UNICEF, 2006.

Batlle's Colorado Party, with former Interior Minister Guillermo Stirling as candidate, only obtained 10.36 per cent of votes. For the first time, a left-wing candidate managed to win the presidency.

[48] Thousands of people gathered in the streets to celebrate Vazquez's inauguration on 1 March 2005. On the domestic front, the new President's first step was to launch a two-year Emergency Plan - one of the campaign's main banners - to address the food, health and education needs of extremely poor people. On the international front, the first measures included the signing of an agreement with Venezuela to exchange oil for food products exported at low prices, and the restoration of diplomatic relations with Cuba.

[49] On 4 November 2005, social organizations carried out a march against capitalism and globalization, coinciding with the Summit of the Americas held in Mar del Plata, Argentina. The march resulted in smashed cars and windows and spray-painted messages. After these events, judge Fernández Lecchini charged four demonstrators with the crime of sedition, based on a Penal Code article imported from Italian legislation in the 1930s. After broad criticism of the sentence by public opinion, the four demonstrators were freed one month later.

[50] During the first months of 2006, the possibility of signing a Free Trade Agreement (FTA) with the US further weakened internal cohesion within the Broad Front. Friction had already arisen as a result of the economic policy carried out by Economy Minister, Danilo Astori, which was criticized for being a continuation of the previous government's policy.

[51] The construction of two pulp mills on the banks of the Uruguay River - a natural border between Uruguay and Argentina - provoked criticism and protests from inhabitants of the Argentinian city of Gualeguaychú and from the Argentinian Government. During January and February, the months of greatest tourist inflow, residents of Gualeguaychú blocked access to the bridges over the river, thus preventing vehicles from entering Uruguay and causing the country significant economic losses.

[52] Following allegations exchanged between both governments, Argentina appealed to the International Court of Justice at The Hague to halt construction of the mills while further environmental impact studies were carried out. In July, the Court ruled out the immediate halting. This decision meant that Uruguay could continue building the plants while the judges went on considering the overall case for or against the mills. ∎

35 per cent of businesses since 1998 and the closing of 15 ranches every day, overcome by debt. They also pointed out that over 75,000 workers had been laid off.

[42] The value of the dollar rocketed on 20 June, increasing by 40 per cent, after the Minister of Economy Alberto Bensión unexpectedly announced the decision to change the exchange rate policy. The free flotation of the US dollar was praised by the US and the IMF, but also sparked several protest demonstrations.

[43] On 30 July, the Central Bank of Uruguay declared a bank holiday - the first one in several decades - which was extended until 5 August to prevent the outflow of deposits that was precipitating the collapse of the banking system. A month earlier, a large private bank had been ordered to suspend activity, for illiquidity and an alleged diversion of funds.

[44] In October 2002, former Minister of Foreign Affairs, Juan Carlos Blanco, was prosecuted for his part in the disappearance of schoolteacher Elena Quinteros, kidnapped from the Venezuelan Embassy 26 years previously. Blanco was the only civilian or military officer during the dictatorship to be prosecuted for the human right violations committed during that period. The former Minister was released eight months later when Carlos Ramela, Batlle's representative on the Peace Commission, appeared in court to state that, according to investigations, the schoolteacher had been murdered without the knowledge of the Armed Forces or the government of the time.

[45] In August 2003, in a historic decision, the Supreme Court ruled to proceed with the prosecution of former dictator Bordaberry for his alleged responsibility in the June 1973 coup and in the murder of eight communist workers in April 1972.

[46] In December 2003, citizens voted in a referendum to confirm or repeal a law that abolished the fuel import monopoly of the state-run company ANCAP and opened it up to investment by private companies. The 'yes' option to repeal the law was supported by 63 per cent of voters. For the first time, an electoral option supported by both the Blanco and Colorado Parties was defeated at the polls.

[47] In the October 2004 presidential and parliamentary elections, Tabaré Vázquez, candidate of the Progressive Encounter-Broad Front-New Majority (EP-FA-NM), was elected President in the first round with 50.45 per cent of the vote. In its worst defeat ever, President

Uzbekistan / Özbekiston

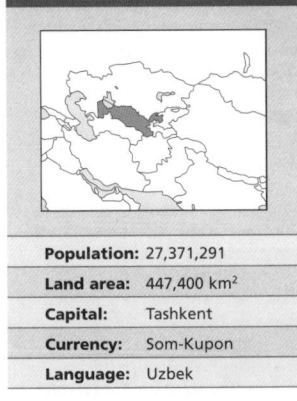

Population:	27,371,291
Land area:	447,400 km²
Capital:	Tashkent
Currency:	Som-Kupon
Language:	Uzbek

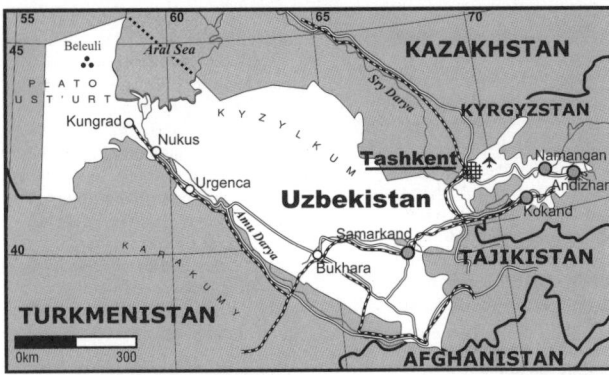

In the 10th century BC agricultural centers such as Khwarezm (on the lower Amur-darya), Ma Wara-an-Nahr (between mid Amur-darya and Sir-Darya) and the Fergana Valley became the first Indo-European-speaking states. Between the 6th century BC and the 7th century AD they successively formed part of the Persian Achaemenid empire, Alexander the Great's empire, the Greco-Bactrian kingdom and the white Eftalite-Hunnish Kushan kingdom.

2   Nomadic Turks annexed most of Central Asia to their Turkish Khanate from the 6th to the 8th century. Turkish speakers arrived in Uzbek territory, and intermarried with local inhabitants. In the mid 8th century they were conquered by the Arabs, who spread Islam, especially in the cities.

3   From the 9th century to the early 13th century, when the caliph's power declined and that of the local Samanids, Karajanids and the shas of Khwarezm increased, the Muslims reached a significant level of development in agriculture and crafts. The cities of Bukhara, Samarkand and Urgenca were prosperous trading centers for caravans following the Great Silk Route, from China to Byzantium.

4   Between 1219 and 1221 Khwarezm was devastated by the Mongols and handed over to Genghis Khan's oldest son. His second son, Chagatai, took control of Ma Wara an-Nahr and Fergana. Inhabitants were now called Chagatais. Turkish and Mongol tribes took refuge in the steppes. In the second half of the 14th century, Timur, the head of one of these tribes, occupied Ma Wara an-Nahr and made Samarkand his imperial capital.

5   The union of the Uzbek nomadic peoples took place in the 15th century in Central Kazakhstan. The new arrivals gave their name to all the country's inhabitants. Once the State of Shaybani was dissolved, the khanates emerged (feudal theocracies made up of Uzbeks, Turkmen, Tajiks, Kyrgyz and Karakalpaks). In 1512 the Khanate of Khiva emerged, whose military elite were from the Kungrats, an Uzbek people. In 1806 their leader, Mohammed Amin, founded-the dynasty that was to govern Khiva until 1920.

6   In the mid-16th century the Khanate of Bukhara emerged, headed by the Uzbek Manguite military elite. In 1753 the Manguite leader Muhammad Rajim founded a dynasty that ruled until 1920. The Bujara Khanate reached its apogee during the reign of the Khan Nasrula (1826-1860). At the beginning of the 19th century the Emirs of Fergana, of the Ming Dynasty, founded the Khanate of Kokand.

7   These states, which had no fixed borders, were unable to command the complete loyalty of their regional leaders. The Emirs of Khiva and Bukhara exercised nominal sovereignty over the Turkish groups of the Kara Kum Desert (slave-traders in Iran). Although they were at their highest level of organization, they were unable to resist the advancing European expansion in the heart of Central Asia, where there was a clash of British and Russian interests over cotton.

8   The 1860 Russian offensive was hampered by the states' geographical isolation, such as Khiva, located in the middle of the desert. In 1867 the Czar created the Province of Turkistan, with its center in Tashkent, belonging to Kokand, which he annexed in 1875. In the late 19th century the province included Samarkand, Syr-Darya and Fergana. In August 1873 the Khan of Khiva, and then in September the Emir of Bukhara, accepted Russian protectorate status. The harsh living conditions imposed by Moscow triggered several uprisings, such as those in Andizhan in 1898 and Central Asia in 1916.

9   After the fall of the Czar in 1917, power passed first to the Committee of the Provisional Government and the *soviets* and then, after the success of the October revolution in Petrograd, to the Soviet of Tashkent. In 1918, the

PROFILE

ENVIRONMENT

Uzbekistan is bordered to the north and northwest by Kazakhstan, to the southwest by Turkmenistan, to the southeast by Tajikistan, to the northeast by Kyrgyzstan and to the south by Afghanistan. There are two main rivers and more than 600 streams, some of which are diverted for irrigation, while others are used for hydroelectric projects. In the northwest and center of the country there are plains (the Ustyurt Plateau, the Amu Darya Valley and the Kyzylkum desert) and there are mountains in the southeast (the Tien-Shan and Gissar and Alay ranges). The climate is hot and dry on the plains and more humid in the mountains. There are large deposits of natural gas, oil and coal. Among the most pressing environmental problems are the salinization of the soil as a result of monoculture, desertification, and contamination of drinking water and air pollution.

SOCIETY

Peoples: Uzbeks 75.8 per cent; Russians 5 per cent; Tajiks 4.8 per cent; Kazakhs 4.1 per cent, Kyrgyz 0.9 per cent; Ukrainians 0.6 per cent; Turks 0.6 per cent; other 7.2 per cent (1995).
Religions: Muslim (Sunni), 88 per cent; Orthodox, one per cent; other (mostly non-religious), 11 per cent.
Languages: Uzbek (official), Russian, Tajik.
Main Political Parties: Uzbekistan Liberal Democratic Party; Uzbekistan People's Democratic Party; Self-Sacrifice National Democratic Party; Uzbekistan National Revival Democratic Party.
Main Social Organizations: There are no independent trade unions. State enterprise workers are members of the Trade Union Federation of Uzbekistan. Human Rights Society of Uzbekistan (HRSU).

THE STATE

Official Name: Özbekiston Jumhuriyati.
Administrative Division: 12 regions, one capital city and one autonomous republic (Qoraqalpoghiston).
Capital: Tashkent 2,155,000 people (2003).-
Other Cities: Samarkand 362,000 people; Namangan 333,000; Andizhan 313,000; Bukhara 235,000 (1995).
Government: Islam Karimov, President since March 1990, elected in 1991 (shortly after Uzbek independence) and re-elected in 2000. Shavkat Mirziyayev, Prime Minister since December 2003. Bicameral Legislature: Supreme Assembly or Senate, with 100 members; Legislative Chamber, with 120 members.
National Holiday: 1 September, Independence Day (1991).
Armed Forces: 52,000 (2003). Other: (National Guard) 700.

LAND USE

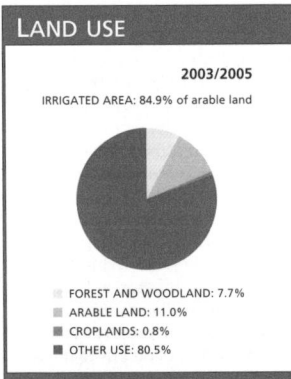

2003/2005

IRRIGATED AREA: 84.9% of arable land

- ☐ FOREST AND WOODLAND: 7.7%
- ☐ ARABLE LAND: 11.0%
- ■ CROPLANDS: 0.8%
- ■ OTHER USE: 80.5%

PUBLIC EXPENDITURE

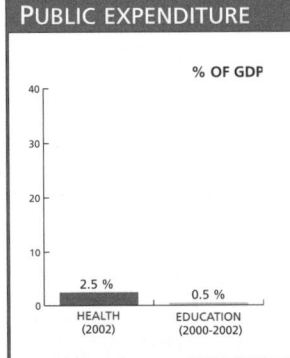

% OF GDP

2.5 %   HEALTH (2002)

0.5 %   EDUCATION (2000-2002)

Life expectancy
67 years
2005-2010

GNI per capita
$450
2004

Literacy
99% total adult rate
2000-2004

HIV prevalence rate
0.1% of population 15-49 years old
2003

IN FOCUS

ENVIRONMENTAL CHALLENGES
The Aral Sea is shrinking after years of intensive irrigation for agriculture, with resultant concentration of chemical pesticides (including DDT) and salt. Desertification and soil salinization are increasing. Industrial waste pollutes water sources, endangering human health. Monoculture, mainly cotton, has degraded the soil.

WOMEN'S RIGHTS
Women have been able to vote since 1938. In 2005, women held 16 per cent of seats in Parliament.

Between 1980 and 2003, female participation in the workforce remained steady at around 47 per cent.

Net female primary school enrollment stands at 80 per cent*.

Prenatal health care coverage is 97 per cent and 96 per cent of births are attended by skilled health personnel*.

More than half of women of child-bearing age are anemic*.

CHILDREN
The infant mortality rate is 57 per 1,000 live births, while the under-five mortality rate amounts to 69 per 1,000 live births*. About 25 per cent of children under three years of age suffer from stunted growth. Over 98 per cent of children are immunized against childhood diseases such as measles, polio and tetanus*.

Some 15 per cent of children aged between 5 and 14 work. Child labor rates in rural areas are double those in urban areas.

INDIGENOUS PEOPLES/ ETHNIC MINORITIES

The dominant ethnic group are the Uzbeks who, since the fall of the communist regime, have been attempting to recover their cultural identity. This ethnic group extends to Afghanistan and Turkey, and was a part of Genghis Khan's Mongol forces, playing a key role in his conquests. Uzbeks make up 80 per cent of the population and are one of the groups who originally settled in the territory. During the era of Soviet domination, Moscow used their traditional enemies, the Tajiks, to keep them under control. Russians, who settled in the country during the Soviet period and still enjoy great political influence, account for five per cent of the population. The remaining 15 per cent is made up by Tajiks, Kasajos and Tatars.

MIGRANTS/REFUGEES
Uzbekistan is neither signatory to the 1951 Refugee Convention and the 1967 Protocol, nor has any legislation on refugees.

At the end of 2004, Uzbekistan hosted 44,455 refugees and 477 asylum-seekers receiving assistance from UNHCR. Most of them came from Afghanistan and Tajikistan.

On the other hand, just over 500 Uzbeks requested asylum in Sweden and the US during that year.

DEATH PENALTY
The death penalty may be applied even for common crimes.

*Latest data available in *The State of the World's Children* and *Childinfo* database, UNICEF, 2006.

up the Aral Sea, the world's fourth-largest inland body of water.

[12] Sharaf Rashidov governed from 1956 to 1983, a period of great stability for the republic.

[13] When Leonid Brezhnev took over the CP leadership in the USSR in 1983, he appointed new people to the Uzbekistan Government. The new local first secretary revealed that official figures for earlier cotton crops had been false.

[14] The resulting scandal led to arrests, proceedings against 4,000 public employees and expulsions from the ruling party, but no structural changes.

[15] From 1985, the reforms introduced by Mikhail Gorbachev, together with the deteriorating economic situation and the diminishing centralized authority of the USSR CP, led to ethnic and religious clashes in Uzbekistan, stemming from the majority Sunni Muslims' resistance to the USSR CP's anticlericalism.

[16] The Soviet invasion of Afghanistan (1979-1989), in which Sunni Muslims fought, had increased hostility towards Moscow and the Russian minority resident in Uzbekistan. The most serious consequences were the conflict in Fergana in June 1989 and clashes in Namangan in December 1990.

[17] Islam Karimov, who became First Secretary of the Uzbekistan CP in 1989, was appointed president of the SSRU in 1990. In August 1991 the Uzbek Soviet approved the Independence Law and in December, in Almaty (Kazakhstan), the Uzbek delegation signed the creation of the Commonwealth of Independent States (CIS). That same month, with most opposition parties proscribed, Karimov was elected president.

[18] Karimov's style proved authoritarian, stamping out dissidence, and he adopted the Southeast Asian model as the basis for Uzbekistan's new path to development, moving towards a market-based economy. A privatization plan was approved in January 1994 and the price of basic foodstuffs and electricity increased by up to 300 per cent.

[19] Attempts were made to limit Russian and Iranian influence in the region, in particular after 1995. Karimov supported the US embargo against Iran and called for the creation of a 'common Turkistan' to counter Russia's imperialist pretensions. In late 1997, Uzbekistan voted with the US and Israel against ending the economic embargo against Cuba in the UN General Assembly.

[20] In November 1998 Karimov supported Russia in the war against Chechnya. The Islamic Movement of Uzbekistan (IMU), led by a member of the Taliban, began

an offensive with the objective of creating an Islamic state in the north-east of the country. The Government accused the IMU of setting off several car-bombs in Uzbekistan during 1999.

[21] On 9 January 2000, Karimov was re-elected with 92 per cent of the vote. The Organization for Cooperation and Security in Europe (OCSE) had refused to send observers, criticizing the electoral system, while the US described the elections as 'neither free nor fair'.

[22] Uzbekistan's entry to the Shanghai Five Group - China, Russia, Kazakhstan, Kyrgyzstan and Tajikistan - in June 2001 signaled the creation of the Shanghai Cooperation Organization (SCO), which agreed to fight ethnic and religious fundamentalism.

[23] Karimov was a key ally when the US attacked Afghanistan in 2001, allowing US forces to use Uzbek airspace and military bases for deployment. After substantial international pressure was brought to bear, Karimov opened the border with Afghanistan to international aid from the UN and other humanitarian organizations.

[24] Five years after it was founded, in March 2002, the Human Rights Society of Uzbekistan (HRSU) was officially recognized as the country's first fully independent human rights organization.

[25] In March 2004, the Uzbek Government blamed Islamic extremists for the death of some 20 people in a bomb and gun attack.

[26] In November, the Turkmen and Uzbek presidents signed a declaration of friendship in the city of Bukhara (situated in the south of Uzbekistan, near the border with Turkmenistan), thus putting an end to years of mistrust between the two countries. Karimov and Saparmyrat Niyazov said that all bilateral issues would be solved. Furthermore, the two leaders agreed to share water resources and ease the existing travel restrictions between the two countries.

[27] A popular uprising in May 2005 - the largest since the country's independence in 1991 - led to harsh government repression, leaving hundreds dead in the eastern city of Andijan. Thousands of armed demonstrators plunged the city into chaos, releasing prisoners and clashing with security forces.

[28] In May 2006, Amnesty International reported that the human rights situation in Uzbekistan was dire. According to the organization, the country's leaders were turning a blind eye to massive violations of human rights. ∎

Red Army thwarted the attempt to establish an autonomous Muslim government in Kokand and crushed the rebellion by the 'Turkistan Union for the Struggle Against Bolsheviks'. The army occupied Khiva in April 1920 and Bukhara in September. Land reform started in 1921. Military operations continued until mid-1922, when the impact of the reforms led to a loss of support for the rebels.

[10] In 1924 Moscow reorganized Central Asian frontiers along ethnic lines, creating the Soviet Socialist Republic of Uzbekistan (SSRU). In May 1925 Uzbekistan became part of the Union of Soviet Socialist Republics (USSR). Tajikistan formed part of Uzbekistan (as an autonomous republic) until 1929,

when it became part of the USSR. During Josef Stalin's regime several Uzbeks were sentenced to death, including the prime minister, Fayzullah Khodzhayev, and the first secretary of the Communist Party (CP), Akmal Ikramov (both were later rehabilitated in 1953). During the 1930s the capital was transferred from Samarkand to Tashkent.

[11] Reforms designed to develop the region's agricultural potential (since 1956 based on cotton monoculture) through the construction of huge irrigation canals and dams turned the country into the USSR's main cotton supplier and the third-largest producer in the world. However, in less than 30 years this intensive irrigation dried

Vanuatu / Vanuatu

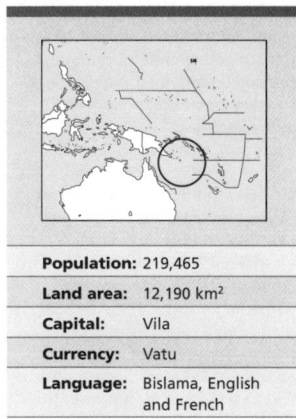

Population:	219,465
Land area:	12,190 km²
Capital:	Vila
Currency:	Vatu
Language:	Bislama, English and French

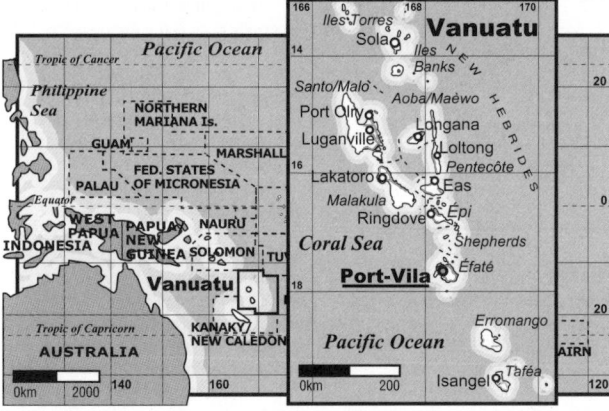

The first colonization of Polynesia and Melanesia is still unclear to anthropologists and historians (see box 'Melanesians and Polynesians'). Sailing westwards, Polynesians reached Vanuatu around 1400 BC. These navigators crossed and populated the entire Pacific Ocean from Antarctica to Hawaii and as far as Easter Island on the eastern edge of the Pacific Ocean. Their culture was highly developed; they domesticated animals and developed some subsistence crops; they manufactured ceramics and textiles, organized their societies into a caste system and, in some cases, possessed a historical knowledge which had been orally transmitted down the generations for centuries.

[2] On 29 April 1605 the Portuguese-Spanish navigator, Pedro Fernández de Quiros, was the first European to sight mountains which he believed to be part of the Great Southern Continent for which he was searching; he named the place 'Tierra del Espíritu Santo' (Land of the Holy Spirit).

[3] A century and a half later, Frenchman Louis Antoine de Bougainville sailed around the region and demonstrated that it was not part of Australia but rather a series of islands. In 1774, British captain James Cook drew the first map of the archipelago, calling it the New Hebrides.

[4] Shortly thereafter traders arrived and felled the forests and the islands became the source of a semi-enslaved labor force. The workers were either taken by force or purchased from local leaders in exchange for tobacco, mirrors and firearms.

[5] During almost all of the 19th century, the archipelago was on the dividing line between the French (in New Caledonia) and British (in the Solomon Islands) zones of influence, and the two nations finally decided to share the islands. In 1887, a Joint Naval Commission was established, and in 1906, the Condominium was formalized, which envisaged joint

provision of some basic services: post, radio, customs, public works, but left each power free to develop other services. Consequently, there were two police forces, two monetary systems, two health services and two school systems ruled by two representatives on the islands.

[6] The local inhabitants, Melanesians, were relegated to being 'stateless' in their own country. They were not considered citizens until a legislative assembly was established in the territory in 1974. Until then only British or French people were entitled to citizenship and land ownership.

[7] In the 1970s most of the neighboring archipelagos achieved independence. This triggered the formation of the New Hebrides National Party in 1971 (now called the Vanuaaku Pati (VP) - Party of Our Land). The party organized grassroots groups based on the Protestant parochial structure on all the islands. When it won two-thirds of the vote in 1979, the British agreed to grant the islands' independence.

[8] Independence was declared on 30 July 1980. Measures were immediately taken to return the land held by foreigners to the Melanesians; the school system was unified and a national army was created.

[9] In 1996 Ombudsperson Marie-Noelle Ferrieux-Patterson accused the Government of diverting aid for cyclone victims, issuing false passports and misappropriating money from pension funds.

[10] In July 2002 former Prime Minister Barak Sope was sentenced to three years in prison for issuing false government guarantee certificates but was pardoned three months later due to his poor health.

[11] Alfred Maseng Nalo was elected President in April 2004 but was removed in May by the Supreme Court, because of a rule banning people with criminal records from being elected. (He had been convicted of criminal

complicity and embezzlement). In May, Prime Minister Edward Natapei lost his majority in Parliament. Acting president Roger Abiut dissolved Parliament and called elections for July.

[12] In August, Kalkot Mataskelekele was elected president by 58 members of the Electoral College, defeating another 15 candidates. Serge Vohor was appointed Prime Minister. Mataskelekele called

on the Government, the Church, traditional leaders and all citizens to unite and promote the country's development. Three months later Parliament passed a motion of no-confidence in Vohor's administration due to his controversial efforts to establish diplomatic relations with Taiwan. Ham Lini was appointed to succeed him.

[13] With the Monte Manaro volcano threatening to erupt - it had begun to spew ash and smoke in December 2005 - thousands of people had to be evacuated from Ambae island.

[14] In May 2006, Parliament met in an extraordinary session to study a set of bills submitted by the Prime Minister. The proposed measures were related to prison administration, the creation of a new agricultural bank and amendments to the judicial system, among other issues. In June a new five-year development strategy between the Government and NZAID (New Zealand's international development agency) was unveiled, focusing on poverty reduction, particularly in rural areas. ∎

PROFILE

ENVIRONMENT

Vanuatu is a Melanesian archipelago of volcanic origin, comprising more than 70 islands and islets, many of them uninhabited. It stretches for 800 km in a north-south direction in the South Pacific about 1,200 km east of Australia. Major islands are: Espiritu Santo, Malekula, Epi, Pentecost, Aoba, Maewa, Paama, Ambrym, Efate, Erromango, Tanna and Aneityum. Active volcanoes are found in Tanna, Ambrym and Lopevi and the area is subject to earthquakes. The land is mountainous and covered with dense tropical forests. The climate is tropical with heavy rainfall. The subsoil of Efate is rich in manganese and the soil is suitable for farming. Fishing is a traditional economic activity. The land tenure system has contributed to general soil depletion due to deforestation and erosion. Rising sea levels will affect both inhabited and uninhabited coastal areas. There is also a risk that rising tides will seep into ground water, threatening water supplies.

SOCIETY

Peoples: Most people are Melanesian (98 per cent), with one per cent European (British and French) and smaller groups from Vietnam, China and other Pacific islands. **Religions:** Presbyterian (36.7 per cent), Anglican (15 per cent), Catholic (15 per cent), indigenous religions (7.6 per cent), other (15.7 per cent). **Languages:** Bislama, English and French are official. More than 100 Melanesian languages are also spoken. **Main Political Parties:** VP-VNUP (which includes: Vanuaaku Pati (VP)/Party of Our Land; Parti national uni / Vanuatu National United Party); Union of Moderate Parties; Vanuatu Republican Party. **Main Social Organizations:** Vanuatu Trade Union Congress (VTUC); Vanuatu Association of Non Governmental Organizations (VANGO).

THE STATE

Official Name: Ripablik blong Vanuatu. République de Vanuatu, Republic of Vanuatu. **Administrative Divisions:** 6 provinces: Malampa, Penama, Sanma, Shefa, Tafea and Torba. **Capital:** Vila, on Efate Island, 34,000 people (2003). **Other Cities:** Luganville (Santo) 8,400 people; Port Olry 1,200; Isangel 1,200 (2000). **Government:** Kalkot Mataskelekele, President since August 2004. Ham Lini, Prime Minister since December 2004. Unicameral Legislature with 52 members elected for a four-year term. **National Holiday:** 30 July, Independence Day (1980).

Vatican City / Città del Vaticano

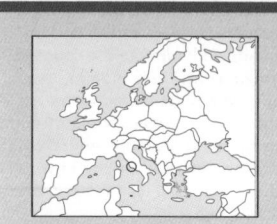

Population:	932
Land area:	0.44 km²
Capital:	Vatican City
Currency:	Euro
Language:	Italian and Latin

In ancient times, the territory now known as the Vatican City, to the west of the Tiber river, was known as the Ager Vaticanus (Vatican fields). Some sources say the name Vatican comes from a former Etruscan town on the site called Vaticum, others say it derives from the Latin *vates* (seer), and that in the past there was a hill called Vatican inhabited by soothsayers who predicted the fate of passers-by.

2   Due to the persecution of Christians and the destruction of Church documents in Rome by Emperor Diocletian in 303, today there are few traces of the first Christians' presence in the area.

3   Emperor Constantine the Great (307-337) made peace with the Church, allowing Christianity to abandon its clandestine status and hold a privileged juridical standing.

4   In the 4th century construction began, at the foot of Vatican Hill, of what would later be known as St Peter's Basilica. According to archaeological evidence, the first pope was buried there. Medieval pontiffs purchased the territory and built a bridge, the Pons Aelius, to link the lands with Rome.

5   The landscape and building architecture were developed by each pope. The pontiffs became rulers of the city of Rome and the surrounding areas.

6   In the year 756 this domain was officially granted to Pope Stephen II by Pippin the Short, King of the Franks, in appreciation for having appointed him king. Papal possessions increased through donations, acquisitions and conquest and in this way the future Papal States, legally established by Charlemagne in the ninth century, covered almost all of the central zone of Italy.

7   In 847, Pope Leon IV erected a large wall, named Leonine, to defend the Vatican against the Saracens. This wall turned the St Peter's area into walled grounds protecting the Basilica and its treasures, smaller churches, monasteries, homes of the clergy, and the orchards of its inhabitants.

It turned the city into a *sui generis* district, different from the rest of Rome.

8   Between 1309 and 1377, the popes lived mostly in Avignon, due to constant conflicts in Rome. Pressed by Philip IV of France, Pope Clement V moved the pontiff's capital to Avignon, then belonging to the pope's vassals and which in 1348 became property of the pontiffs. The seven popes of that period were French, as well as 111 of the 134 cardinals.

9   After Gregory XI re-established the pontifical capital in Rome, Clement VII led the cardinals who in 1378 declared invalid the election of Urban VI, and was elected anti-pope, occupying the empty throne in Avignon. Europe was divided in its support for both pontiffs: while France backed Clement, England supported Urban, a dispute that would continue throughout the 100 Years War (1337-1453) and would lead to the period known as the Great Schism, in which several anti-popes were named, and which would finally end in 1417.

10   Most of the annexations to the Vatican territory were kept under pontifical rule until 1797, when Napoleon Bonaparte took the territory, creating the Roman Republic.

11   In 1801 Pope Pius VII recovered some of his power and in 1815 after the fall of Napoleon, the Congress of Vienna returned almost all of the papacy's possessions.

12   The first Vatican Council was held in 1869 and decreed the dogma of papal infallibility. A year later, the Papal States were finally dissolved when Victor Emmanuel II annexed them to the unified kingdom of Italy, including Rome. The papacy's jurisdiction was limited to the Vatican, in which each of the successive pontiffs remained, in protest, as self-imposed prisoners. This continued until 1929 when the Lateran Treaty, signed between the Holy See and the Kingdom of Italy - ruled then by Benito Mussolini - acknowledged the sovereignty and international status of the Vatican City State. This state is different from the Holy See, since the latter is the executive organ of the Catholic Church, and the former is the physical territory over which that government rules. Thus, the pontiff's political authority was consolidated.

13   Throughout the centuries, and especially during the Renaissance Papal patronage turned the Vatican into one of the world's most important cultural hubs, with architectural works such as St Peter's, the Sistine Chapel, decorated by Michelangelo, Botticelli and other artists, and the *Stanza della Segnatura* frescoes by Raphael.

14   During World War II, Pope Pius XII took the Vatican's definition of 'neutrality' to the limit, provoking criticism to this day of the relationship with Hitler's Germany and in particular the Vatican's knowledge of and response to the Holocaust.

15   In 1982, banker Roberto Calvi was found dead in London. This led to the collapse of his huge, privately-owned bank, Banco Ambrosiano, revealing a 'black hole' in its balance sheet of some $1.3 billion. A large part of the missing money was later found in accounts owned by the Vatican Bank. The death of Calvi - 'God's banker' - exposed links with masons, mafia and Vatican fraud.

16   The Vatican's resources came from the 1,750 million lira that the Lateran Treaty determined as compensation for the territories lost in 1870, and donations from all over the world, especially from the US and Germany. These funds are administered by the Institute for Religious Works - better known as the Vatican Bank.

17   During the last decade of the 20th century, scandals arising from allegations of pedophilia against Catholic priests rocked the Vatican. The victims received a total of $119.6 million in damages, the largest compensation awarded in the history of sexual abuse cases. In a memorandum issued in April 2002, after a meeting with US cardinals, the Pope undermined the prospects of a 'zero-tolerance' policy with pedophile priests.

18   In July, the Vatican launched a global campaign against gay marriages in an attempt to reverse progress towards their legalization in Europe and the Americas.

19   In April 2005, John Paul II - who had been suffering Parkinson's disease and arthritis for a decade - died from a cardio-circulatory collapse. Thousands of people gathered in St Peter's Square upon hearing the news. Several countries declared days of mourning.

20   Taking the name Benedict XVI, the 78-year-old German cardinal, Joseph Ratzinger, was appointed Pope. For many years, he had been head of the Congregation for the Doctrine of the Faith (formerly the Inquisition).

21   In April, a cardinal announced that the Vatican was soon about to authorize the use of condoms by those Catholic people suffering from sexually transmitted diseases - particularly HIV/AIDS. Until then, the Vatican had preached abstinence as the best prevention.

22   In September 2006, the Pope caused widespread offence among Muslims worldwide when he quoted a 14th-century Christian emperor's opinion that the prophet Muhammad had brought the world only 'evil and inhuman things'. The Pope apologized but in the ensuing furore an Italian nun was killed in Somalia and churches were attacked in Palestine. ▪

PROFILE

ENVIRONMENT

Located within the city of Rome, next to the Tiber river, it includes St Peter's basilica and square, Vatican palaces and gardens, the church and palace of St John in Lateran, the papal 'villa' of Castelgandolfo, and 13 buildings outside this area which are considered extraterritorial.

SOCIETY

Peoples: Vatican citizens are members of the Papal and Catholic Church administration who live there because of their work. Most of the permanent officials are Italians, a large number are Swiss and the rest come from different countries. **Religions:** Catholic. **Languages:** Italian (the state) and Latin (official).

THE STATE

Official Name: Stato della Città del Vaticano. **Administrative Divisions:** Two parallel administrations: Holy See (supreme organ of the Catholic Church); Vatican City (physical headquarters of the church). **Capital:** Vatican City, 932 residents (est 2003). **Government:** monarchy elected for life. Pope Benedictus XVI (original name Joseph Alois Ratzinger), sovereign elected by the College of Cardinals (in secret meeting) on 19 April 2005. The equivalent of the head of government is the Secretary of State, Cardinal Angelo Sodano since 1991, who chairs a commission of five cardinals. The Pope is also the Bishop of Rome and supreme head of the Catholic Church. Sodano was to be replaced in September 2006 by cardinal Tarcisio Bertone. The Church administration is advised by the College of Cardinals and the Bishops' Synods, who meet when the pontiff wishes. The Church's administrative bodies are nine Holy Congregations, three Secretariats and several commissions, prefectures or courts, which are jointly known as the Roman Curia.

Venezuela / Venezuela

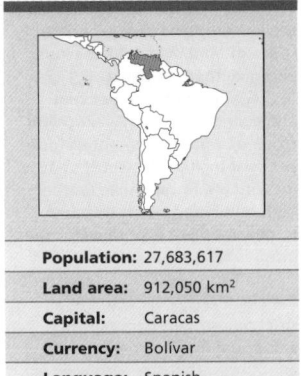

Population:	27,683,617
Land area:	912,050 km²
Capital:	Caracas
Currency:	Bolívar
Language:	Spanish

Cumanagotos, Tamaques, Maquiritares, Arecunas, and other Carib groups inhabited the northern tip of South America when the Spanish arrived in 1498. Local buildings, constructed on stilts, reminded the Spaniards of Venice, so they named the country Venezuela (Little Venice).

2 During the colonial period, Venezuela was organized as a Captaincy General of the Viceroyalty of New Granada. Its agricultural economy, based on cocoa and slave labor, forged a society dominated by a local aristocracy of *mantuanos*, with a majority of *pardos* (African slaves and their descendants).

3 The first phase of the revolution for freedom started in April 1810 with Francisco de Miranda. He sought independence from Spain and the creation of a vast American Confederacy to be known as Colombia, which would crown an Inca emperor. Miranda failed, was captured by the Spaniards in 1811 and later died in prison.

4 The second revolutionary stage was initiated by Simón Bolívar who, backed by the *mantuano* oligarchy, installed a government in Caracas. His plans for independence did not include changes in social structure, and he was not supported by the *pardos*, most of whom hated their white creole owners. Their liberation movement was led by the

Spanish loyalist General José Tomás Boves, who defeated Bolívar in 1814. Boves abolished slavery and redistributed the land among the people. It was the end of the First Republic.

5 Bolívar went into exile and on returning to Venezuela, championed popular demands and won mass support. Accompanied by other military leaders like Antonio José de Sucre, Santiago Mariño, José A Páez and Juan B Arismendi, he carried out successful military campaigns in the northern half of the continent.

6 In 1819, the Angostura Congress created the new republic of 'Gran Colombia', uniting Colombia, Ecuador, Panama and Venezuela. In 1830, shortly before Bolívar's death, General José Antonio Páez declared Venezuela's secession from Gran Colombia.

7 For decades, Venezuelan politics revolved around the *caudillo* or leader, Páez. His political successor, Antonio Guzmán Blanco, was determined to modernize the country. He succeeded to a certain extent, introducing new technology, new means of communication, and reforming the legal code.

8 Juan Vicente Gómez took power in 1908, ruling for 17 years

as a dictator. He gave free access to the foreign oil companies, which operated primarily in the Lake Maracaibo oil fields. In 1935 General Eleazar López Contreras took office. He was succeeded in 1941 by General Isaías Medina Angarita, who laid the foundations for greater political activity by legalizing the Democratic Action (AD) party. He supported the Allies during World War II.

9 In 1945 a civilian-military movement took power, led by Rómulo Betancourt, the leader of AD, and General Marcos Pérez Jiménez. The country's first free elections were held in 1947. Writer Rómulo Gallegos (AD) was elected President but overthrown in 1948 by yet another military coup, installing the harsh dictatorship of Marcos Pérez Jiménez.

10 In 1958, Pérez Jiménez was overthrown and the country then entered a stable 'democratic' period under a coalition government formed by the Punto Fijo Pact (1958) which enabled the AD and COPEI (the Christian Democrats) to alternate in power for decades. This stability was largely achieved because of massive oil revenues, improved relations with the US, and expanded political rights.

11 Unfortunately, the ensuing economic growth brought little change to the lives of the poor majority. Popular discontent resulted in guerrilla warfare led by the Communist Party, the Movement of the Revolutionary Left (which split from AD), and other groups.

12 In 1960, Venezuela sponsored the formation of the Organization of Petroleum Exporting Countries (OPEC). Sixteen years later, during the presidency of social democrat Carlos Andrés Pérez, Venezuelan oil was nationalized. Pérez also supported the creation of the Latin American Economic System (SELA)

and argued in favor of a New International Economic Order.

13 Although Venezuela was the world's third largest oil exporter and received its highest prices during the governments of Pérez and Christian Democrat Luis Herrera Campins, the Government was unable to manage the enormous amounts of money coming into the country. Huge state-run companies were created for the manufacture of iron, aluminum, cement, and for hydroelectric power, while most private companies were being subsidized.

14 In 1982, a sharp decrease in oil revenues, foreign debt, and a flight of private capital forced the Government to take control of exchange rates and foreign trade. Inflation, unemployment, and housing shortages started to rise, while extreme poverty rose too.

15 The 1983 presidential elections were won by the AD's Jaime Lusinchi, who gained 56 per cent of the vote, defeating COPEI's Rafael Caldera.

16 In response to the economic problems, Lusinchi's policy comprised an austerity plan which had meager results, and a flawed 'social pact' between employers and unions and, towards the end of his term, increasing state control over the economy.

17 The AD's candidate in the 1988 elections, Carlos Andrés Pérez, won the presidency backed by the Confederation of Venezuelan Workers (CTV).

18 Social tension increased and 25 days later the poorest people took to the streets in a wave of riots and looting. Police repression left more than 1,000 dead (246 according to the Government), and 2,000 wounded or jailed.

19 IMF-backed economic restructuring was initiated, which lost the Government popular support. In December 1989, these policies were blamed for abstention rates of nearly 70 percent and broad gains by the Christian Democrats and left-wing parties, when Venezuelans elected their 20 state governors and 369 mayors for the first time.

20 Although the Government recognizes indigenous peoples' land rights, there has not been adequate protection of the indigenous population. They suffer persecution from landowners, farmers and government officials, while Brazilian miners continue to invade their land prospecting for gold within Venezuelan territory.

21 Early in 1992, the popularity of the Pérez Government was at its lowest, with the AD party distancing itself from imposing further economic measures. Congress exercised its controlling function by investigating crimes,

LAND USE

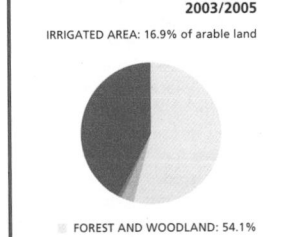

2003/2005

IRRIGATED AREA: 16.9% of arable land

- FOREST AND WOODLAND: 54.1%
- ARABLE LAND: 2.9%
- CROPLANDS: 0.9%
- OTHER USE: 42.1%

PUBLIC EXPENDITURE

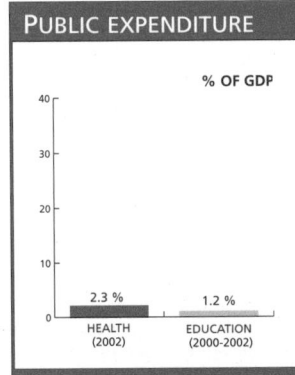

% OF GDP

HEALTH (2002)	EDUCATION (2000-2002)
2.3 %	1.2 %

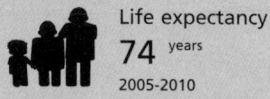

Life expectancy
74 years
2005-2010

GNI per capita
$4,030
2004

Literacy
93% total adult rate
2000-2004

HIV prevalence rate
0.7% of population 15-49 years old
2003

PROFILE

ENVIRONMENT

The country comprises three main regions. In the north and west are the Andes and other mountain chains, and there are more high mountains to the south. The central Orinoco Plains are a livestock farming area. In the southeast, highlands of ancient rock and sandstone extend to the borders with Brazil and Guyana, forming Venezuelan Guyana. This is a sparsely inhabited area with thick forests, savannas, rivers, and some peculiar features: the tepuyes or plateau mountains, and the rare Sarisarinama depths. Most of the population lives in the hilly north. The main oil basins are: Lake Maracaibo (Zulia), Orinoco River Basin (Delta Amacuro, Monagas, Gurico), Falcn Basin (Falcn), Apur-Barinas Basin (Apure and Barinas states) and Cariac Basin (Sucre). In the Gulf of Paria and the center of the State of Anzotegui there are natural gas reserves. The country produces iron ore, manganese, bauxite, tungsten and chrome, gold and diamonds.

SOCIETY

Peoples: Venezuelans are descended from the integration of indigenous peoples, Afro-Caribbean and European settlers. Today, indigenous peoples and Afro-Caribbeans each account for less than seven per cent of the population. In recent times, Venezuela has received more immigrants than any other South American country. **Religions:** Mainly Catholic, 92.7 per cent. **Languages:** Spanish, official and predominant; 31 local languages are spoken. **Main Political Parties:** Movement for the Fifth Republic; For Social Democracy; Fatherland for Everybody; People's Electoral Movement. **Main Social Organizations:** For 40 years The Confederation of Venezuelan Workers (CTV) was the main trade union grouping but tended to collaborate with management. After CTV leaders supported the 2002 military coup and led a 63-day strike aimed at toppling Chávez, the National Workers' Union was founded in April 2003. It has grown fast and now is the principal labor federation, with an estimated 1.2 million workers affiliated, compared with just 0.2 million in the CTV.

THE STATE

Official Name: República Bolivariana de Venezuela. **Administrative Divisions:** 21 states with partial autonomy (including the Federal District), 2 Federal territories. **Capital:** Caracas 3,226,000 people (2003). **Other Cities:** Maracaibo 1,847,000 people; Valencia 1,719,500; Barquisimeto 1,027,700; Guyana City 748,200 (2000). **Government:** Presidential System. Hugo Chávez, President since February 1999, re-elected in 2002. Unicameral Legislature: National Assembly, with 167 members. **National Holiday:** 5 July, Independence Day (1811). **Armed Forces:** 82,000 (2003). Other: Co-operation Army 23,000.

but due to the corruption and inefficiency of the judiciary these investigations very rarely led to prosecutions.

[22] On 4 February 1992, a military coup against the President was led by Francisco Arias. While the attempt failed, it showed that administrative corruption and the economic crisis were the causes of instability. Arias and parachute commander Hugo Chávez, both belonging to the Bolívar-200 Military Movement, were arrested.

[23] The Government suspended all constitutional rights. An agreement was reached with the teachers' unions, putting an end to a fortnight of strikes and police repression of students and teachers.

[24] Another coup attempt by Chávez took place in November 1992. The air force played a key role in controlling the rebels.

[25] President Pérez was suspended from office on 21 May 1993, charged with misappropriation of

funds. Shortly after he was tried for corruption and sentenced to house arrest. After serving time, he returned to politics in 1996.

[26] Ramón Velázquez became deputy President until former President Rafael Caldera won the 1994 general elections. The abstention rate reached 40 per cent.

[27] In the 1990s, Caracas became one of Latin America's most violent cities. A feeling of insecurity prevailed as the number of murders rose. In 1995 it was estimated that ten per cent of Caracas residents carried a weapon.

[28] The economic crisis worsened in February 1993, with the fall of the Banco Latino, the country's second commercial bank. In August 1995, 18 out of the 41 private banks had been investigated and 70 per cent of the deposits were being managed by the Government.

[29] President Caldera suspended constitutional guarantees regarding

real estate, private property and business. He also restricted trips abroad, meetings and the immunity against arbitrary arrests. Despite Congress' vote to reinstate these rights, the President restricted them once more to prevent speculation and capital flight.

[30] On 6 December 1998, Hugo Chávez was elected constitutional president with 56.5 per cent of the vote. In his inaugural speech, he announced he would replace what he described as a 'dead'constitution through a 'peaceful revolution' to fight poverty and restructure foreign debt, standing at more than $23 billion. In order to do this there would be a National Constituent Assembly (NCA) which would draw up a new constitution in six months.

[31] In early 1999, Chávez publicized his 'peaceful revolution' through his diplomatic corps and through a television program which received and directly 'resolved' the concerns of the public. On July 25, in elections for the NCA, 75 per cent of the population voted for the Chávez' Movement for the Fifth Republic, giving it the absolute majority in the body.

[32] In September, the country hosted the second meeting between Colombia's National Liberation Army (ELN) and the Colombian governmental authorities. Chávez, who played an important role in the Colombian conflict, turned down a US request to use his air bases to 'combat drug trafficking' there. It was widely believed the US aimed to invade Colombia and destroy insurgent groups.

[33] Although there was a 54-per-cent abstention rate and opposition from the Catholic Church, the business sector and followers of the old political system, 70 per cent of those who turned out to vote approved the new constitution. Thus the 'Bolivarian Republic of Venezuela' was born.

[34] In June 2000 Chávez was re-elected by an overwhelming majority for a term ending in 2006. In August, he met the leaders of the main oil exporting countries. The re-elected President faced harsh criticism from the US for meeting Iraqi president Saddam Hussein in Baghdad. Chávez was the first democratically elected head of state to visit Iraq since the 1991 Gulf War.

[35] Chávez announced measures in early 2002 aimed at stimulating the economy and achieving social justice. These included reforms of the Lands Law and fossil fuels legislation. The latter doubled the royalties paid by foreign investors from 16 per cent to 30 per cent, and reserved for the state at least 51 per cent of the shares of joint

ventures. The Lands Law allowed the Government to expropriate tracts of land over 5,000 hectares and give them to peasants.

[36] GDP grew by 2.8 per cent and economic activity - excluding crude oil - jumped by about four per cent in 2001. However Chávez was unable to capitalize on the high oil prices to shift revenues to development programs for the most vulnerable people.

[37] In February 2002, to confront the crisis, Chávez enacted a strict adjustment package that was applauded by the IMF. He announced that the currency exchange rate would be floated, after parity with the dollar had been tied at a ten-per-cent annual depreciation of the bolívar, which represented a devaluation of more than 25 per cent.

[38] As a result of his ties with Cuba, Iraq and Libya, and his hesitation to condemn the September 2001 attacks on the US, Chávez ended up in direct confrontation with the superpower.

[39] The President moved to take control of the state oil company Petróleos de Venezuela (PDVSA) - nominally owned by the Government, but actually in thrall to the foreign operators. He removed several of its managers and appointed a new board of directors. In April 2002 a general strike was called by employers and trade unions, which staged massive demonstrations.

[40] The armed forces joined civilian protests. Army commander Efraín Vázquez announced he was no longer loyal to the President and Chávez was overthrown. He was kidnapped and sent to Orchila island, where - it was later revealed - they planned to assassinate him.

[41] The coup put Pedro Carmona, head of the country's main business association, into the presidency. He dissolved Parliament and the Supreme Court, annulled the Constitution, called presidential elections to take place in a year and convened legislative elections for December, under a decree of 'reorganizing the public powers'. These announcements came with the figures on the previous day's violence: 15 dead and 350 injured, according to the fire department. On 14 April, at midnight, troops loyal to Chávez's legitimate government entered the Miraflores presidential palace and waited for the operation that would rescue Chávez and return him to power. Carmona was forced to step down and was arrested.

[42] The US did not condemn the coup against Chávez until he returned to power. The British *Guardian* newspaper reported that US intelligence had been studying the possibility of removing Chávez

Under-5 mortality	Poverty	Debt service	Maternal mortality
19 per 1,000 live births 2004	**8.3%** of population living on less than $1 per day 2000	**16.0%** exports of goods and services 2004	**96** per 100,000 live births 2000

from office for over a year; also that on the day of the coup, the US navy on the Venezuelan coast had tuned in to communications in Caracas. Washington's stance on Venezuela fed fears of 'coup contagion' throughout the region.

⁴³ In July former US President James Carter travelled to Venezuela and met Chávez to mediate in negotiations between those involved in the conflict, but the opposition did not show up. In October, a group of 14 high-ranking military officers-called for 'civil and military disobedience' and demanded either the President's resignation or a referendum on his removal.

⁴⁴ In late October the Secretary-General of the Organization of American States(OAS), former Colombian president César Gaviria, arrived in Caracas to try and mediate in the military-political crisis. Negotiations got nowhere. The business association, the CTV, and the opposition coalition Democratic Coordinator (CD), called for a general strike to press for a referendum to decide whether or not Chávez should remain in office. The Government rejected the referendum arguing that it was unconstitutional to call it before the presidential mid-term.

⁴⁵ Although the 62-day strike left the country on the brink of bankruptcy, Chávez managed to stay in power. Former US President Carter had proposed to shorten the presidential term via a constitutional amendment.

⁴⁶ In May 2003 the Government and the CD signed a 19-point agreement, including a referendum to revoke the Presidency, the creation of the National Electoral Council (CNE), the disarmament of civil society and the setting up of a Truth Commission.

⁴⁷ The opposition gathered 3.4 million signatures demanding a referendum, of which the CNE validated just 1.8 million. The opposition protested in the streets against what it deemed a fraud, and met with stern Government repression. In June 2004, the CNE accepted 2.4 million signatures, thus enabling the referendum.

⁴⁸ Since the beginning of the political crisis, the Government had withstood strong pressure from the Venezuelan mass media (dominated by the Cisneros group), which had played a decisive role in the 2003 coup.

⁴⁹ In May 2004, 88 Colombian paramilitaries were detained in Venezuela and charged with plotting to overthrow Chávez. Former President Carlos Andrés Pérez's home was raided. From Miami, Pérez stated that Chávez had to be overthrown by force.

⁵⁰ Finally, the recall referendum held in August 2004 gave Chávez the victory with 58.2 per cent of the vote, while the CD obtained 41.7 per cent. Although Henry Ramos Allup, CD's general coordinator, denounced electoral fraud and called for a manual recount of votes, César Gaviria, OAS's observer, said that until fraud evidence emerged, the OAS was not going to question the results. Meanwhile, Jimmy Carter, who was acting as observer, stated that the President's victory was fair and highlighted the massive and clear margin in favor of the Chávez administration. According to this result, the President would remain in power until 2007. That month, Gaviria issued a report by the OAS's electoral observation mission confirming that 'most Venezuelans decided not to revoke the mandate of President Hugo Chávez Frías'.

⁵¹ In the November municipal elections, the ruling party won the mayoralty of Caracas and retained control of the adjacent municipalities of Libertador and Sucre, while the opposition kept those of Chacao, Baruta and El Hatillo.

⁵² In January 2005, the President launched a new land reform campaign aimed at handing over uncultivated lands or idle farms and large landed estates to small farmers. A controversial 2002 land law was still not being implemented.

⁵³ Venezuela offered to supply the nations of the region facing energy shortages with cheap fuel. The announcement was criticized by the opposition, which accused Chávez of using oil to increase its diplomatic influence.

⁵⁴ The opposition pulled out of the December parliamentary elections at the last minute, citing mistrust of the electoral council and lack of transparency. However, international observers pointed out that the security and transparency measures were 'in line with the most advanced international practice'. Meanwhile, the turnout amounted to just 25 per cent out of the 14 million registered voters. The ruling party and its allies won all 167 parliamentary seats at stake.

⁵⁵ In March 2006, the army began to implement a plan aimed at training civilians to resist the eventuality of an invasion of Venezuela through guerrilla-type tactics. It was estimated that once the plan was completed about two million people would have been trained.

⁵⁶ In April 2006, Peru recalled its ambassador to Venezuela, accusing Chávez of interfering in Peru's internal affairs. In the middle of the election campaign, the Venezuelan President had said that candidate Alan García - who was later elected as President of Peru - was a thief. In June, the delivery of 100,000 Russian assault rifles - out of a total of 170,000 ordered from Moscow - heightened tensions between the US and Venezuela. Washington stated that Venezuela's arms purchase was a threat and part of an attempt by Chávez to throw his weight around in the region. Caracas replied that the acquisition was part of efforts to modernize its military equipment.

⁵⁷ In July 2006, Venezuela joined the South American trade bloc Mercosur, which had hitherto comprised Brazil, Argentina, Uruguay and Paraguay. ■

IN FOCUS

ENVIRONMENTAL CHALLENGES
Deforestation and soil degradation are key environmental problems. A lack of water treatment in the main urban and industrial centers has increased the pollution of lakes Maracaibo and Valencia, and the Caribbean Sea. Rainforests are seriously threatened by mining activity.

WOMEN'S RIGHTS
Women have been able to vote and stand for election since 1947. In 2005, 17 per cent of seats in Parliament were held by women. They made up 36 per cent of the country's total labor force of 11 million. In 2004, 94 per cent of pregnant women received prenatal care and the same proportion of births were attended by skilled health staff. The maternal mortality rate was 96 per 100,000 births*. Domestic violence remained a widespread problem in Venezuelan society.

CHILDREN
In early 2005, at least 23 states in Venezuela were hit by severe floods caused by heavy rains during the dry season, directly affecting about 175,000 people. Over 120,000 of them were children. In addition to the loss of human lives, education infrastructure was severely hit (180 centers were considerably damaged). Many schools that remained in good condition were used as shelters for more than 20,000 people that had been left homeless. The spread of diseases such as pneumonia, diarrhea and skin infections posed an additional threat to children living in the affected states.

A decline was registered in infant and under-five mortality rates between 1990 and 2004. The former fell from 24 to 16 per 1,000 live births and the latter from 27 to 19 per 1,000.

INDIGENOUS PEOPLES/ ETHNIC MINORITIES
There are at least two million Afro-Venezuelans, who suffer mostly social discrimination. Most Afro-Venezuelans live in the Barlovento region, on the Caribbean coast, where they were brought between the 16th and 19th centuries to work as slaves. There are approximately 27 different indigenous groups, of which only four communities have more than 10,000 residents. Indigenous peoples, since the conquest of America, have lived in poverty: most lack access to health care and education, and are undernourished. During the 1990s there were considerable achievements in protecting indigenous rights, both cultural and political. In spite of this, in the political realm there are several communities without access to the system. For many years indigenous groups have called for a system of proportional representation that allows ethnic minority participation. As they are excluded from the Venezuelan Congress, they have created the Congress of Indigenous Peoples of Venezuela to protect and claim their rights.

MIGRANTS/REFUGEES
In 2005, according to official figures, some 270,000 Colombians were living in Venezuela. In 2004-2005, during the Government's implementation of the program called 'Mission Identity', over 400,000 undocumented foreigners were registered, many of them from Colombia. In spite of this, only about 400 people have been granted refugee status and asylum-seekers registered by the National Refugee Commission barely amount to 4,000. In 2005, the implementation of a program called 'Missions' helped thousands of Colombian refugees who were living in the border region.

DEATH PENALTY
Capital punishment was abolished in 1863.

* Latest data available in *The State of the World's Children* and *Childinfo* database, UNICEF, 2006.

Vietnam / Viêt Nam

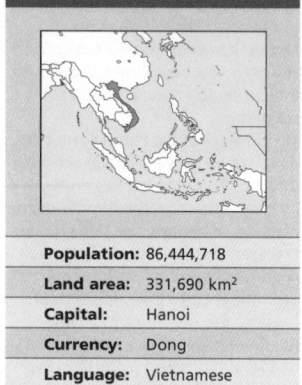

Population:	86,444,718
Land area:	331,690 km²
Capital:	Hanoi
Currency:	Dong
Language:	Vietnamese

The Vietnamese nation was born out of centuries of war: in the 9th century against the Chinese Han dynasty, ending nearly a thousand years of subjugation; against the Chams in the 11th and 12th centuries; driving back Genghis Khan and his grandson Kublai Khan in the 13th century; the Chinese Ming and Ching dynasties in the 15th and 17th centuries; and the Khmers in the 18th century.

2   When France began its conquest of Indochina in 1860, it encountered sporadic but disorganized and poorly armed resistance, delaying victory by 30 years. By around 1900 the French had consolidated their position in the peninsula and Vietnam was divided into Tonkin in the north, Annam in the center and Cochinchina in the south. To counter Chinese influence, the use of *Quoc Ngu* - Vietnamese written in Latin script - was encouraged. This provided nationalists with a powerful tool to popularize knowledge and modernize the culture, which reduced printing costs and made reading easier, compared with publishing in Chinese script.

3   In 1929 several Marxist-Leninist parties emerged, which were unified the following year by Nguyen Ai Quoc (Ho Chi Minh) in the Indochina Communist

Party. Later this party split into three national sections, corresponding to Laos, Cambodia and Vietnam. The last was called the Workers' Party until 1976, when it took the name of Communist Party of Vietnam (CPV).

4   During World War II the communists organized the resistance to the Japanese occupation, collaborating with the Allies. Meanwhile, in 1941, Ho Chi Minh founded the Viet Minh or League for Independence.

5   After the defeat of Japan, the Viet Minh had a powerful army and enjoyed widespread popular support. August 1945 saw the start of a general insurrection. Two weeks later the revolutionaries, who controlled Hanoi, proclaimed independence for the country and the founding of the Republic. Emperor Bao Dai abdicated and offered his services as adviser to the new regime.

6   The March 1946 agreement with the French (who controlled Cochinchina) recognized the Vietnamese Government and granted the country free state status in the French Union. However France proclaimed Cochinchina an independent republic in June 1946 and united it with the North, proclaiming the Associated State of Vietnam and

appointing-former emperor Bao Dai head of state. The Viet Minh waged a nine-year-long guerrilla war and were finally victorious in 1954 with the defeat of French forces at Dien Bien Phu.

7   The 1954 Geneva Agreement stipulated French withdrawal and general elections in Vietnam for 1956. The Viet Minh was to redeploy north of Parallel 17. In the South, the US set up the Ngo Dinh Diem regime in Saigon, and the promised elections did not take place. In 1960, democrats, socialists, nationalists and Marxists united in the National Liberation Front (NLF) (Vietcong), headed by Nguyen Huu Tho, a lawyer. The 'second resistance' was launched, against successive military governments in Saigon and the US, who first sent Saigon arms and advisors, and later troops (580,000 in 1969). A larger tonnage of bombs was dropped than during the whole of World War II, and chemical and bacteriological weapons were experimented with. Ho Chi Minh died in 1969.

8   Over the course of the 15-year-long war, the US spent $150 billion, destroyed 70 per cent of northern villages and left ten million hectares of land barren. Nevertheless, Saigon was taken by the Vietcong in April 1975 and renamed Ho Chi Minh City. On 2 July 1976 the territory was reunited under the name Socialist Republic of Vietnam.

9   Vietnam did not enjoy long-lasting peace. In January 1979 it was forced to go to war with Pol Pot's Cambodian Government which claimed part of its territory (see History of Cambodia). In the same year, after Vietnamese protégé Heng Samrin replaced Pol Pot, China (a staunch ally of the defeated Pol Pot) invaded the North, supposedly to protect the Chinese population from a 'Vietnamization' campaign. New frontier skirmishes in 1980 demonstrated Vietnamese military power, and frontiers remained unaltered.

10   From 1981, President Ronald Reagan blocked a UN assistance program and prevented a US charity from making shipments to Vietnam.

11   After 1985 Hanoi freed several political prisoners and grew closer to ASEAN and the US, becoming more receptive to US requests to search for unidentified war dead and for the children of US soldiers.

12   Le Duan, secretary-general of the CPV since 1969 and a close associate of Ho Chi Minh, died in 1986. Congress appointed former Vietcong strategist Nguyen Van Linh his successor.

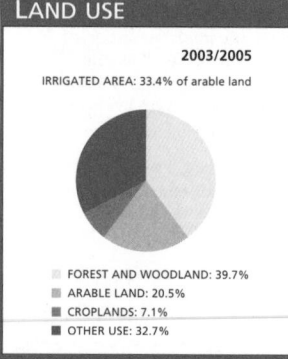

LAND USE

2003/2005

IRRIGATED AREA: 33.4% of arable land

- FOREST AND WOODLAND: 39.7%
- ARABLE LAND: 20.5%
- CROPLANDS: 7.1%
- OTHER USE: 32.7%

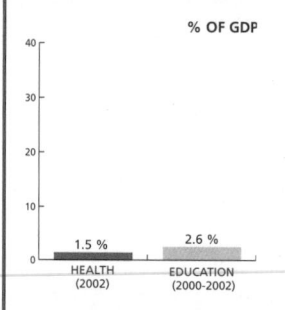

PUBLIC EXPENDITURE

% OF GDP

1.5 % HEALTH (2002)

2.6 % EDUCATION (2000-2002)

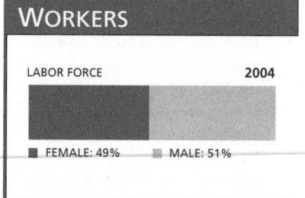

WORKERS

LABOR FORCE   2004

- FEMALE: 49%
- MALE: 51%

| | Life expectancy
72 years
2005-2010 |  | GNI per capita
$540
2004 | | Literacy
90% total adult
rate
2000-2004 | 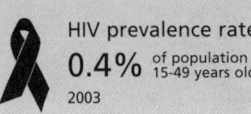 | HIV prevalence rate
0.4% of population
15-49 years old
2003 |

[13] Political change in Eastern Europe in the late 1980s led to internal pressure to open up the country politically. As of 1989 dissenting voices within Congress were calling for change towards a multi-party-parliamentary system.

[14] In August 1989 Nguyen Van Linh called for a rejection of bourgeois liberalization, political pluralism and non-socialist opposition parties. Even so, non-communist candidates were allowed to run in the 1989 National Assembly elections. In 1990 the Vietnamese Veterans' Association was formed with government authorization. The dominant role of the CPV was reaffirmed, with the arrest of many defenders of political pluralism and the dismissal of 18,000 public employees charged with corruption.

[15] In 1989 the British started deporting the thousands of Vietnamese refugees living in shelters in Hong Kong, until international pressure brought the measure to a halt.

[16] The 1991 CPV Congress reiterated its commitment to socialism, considering it 'the only option'. Secretary-general Van Linh was replaced by Du Muoi who continued the renewal (Doi Moi) initiated in 1986.

[17] After the disintegration of the USSR the CPV declared that the one-party system would be maintained, but decided to go ahead with some political and economic changes, including conditional acceptance of private enterprise and foreign investment.

[18] In October 1991 a UN initiative ended the deadlock with Cambodia, with an agreement signed in Paris. This also represented the first step toward the normalization of relations with China, which had been cool following the 1963 break between Beijing and Moscow, and were broken off completely with the overthrow of the Khmer Rouge in Cambodia (who were supported by China).

[19] In November 1991 the British resumed the deportation of 64,000 Vietnamese in Hong Kong camps.

[20] The new April 1992 constitution allowed independent candidates to run for election. Even so, in July, 90 per cent were CPV members. The new president named by the Assembly in September was General Le Duc Anh, an ally of prime minister Vo Van Kiet. In November a delegation of US senators arrived to discuss the issue of more than 2,000 missing POWs.

[21] Privatizations and the liberalization of foreign investment led to 8.3 per cent growth in GDP in 1992.

[22] The boom in rice production mostly benefited the South, and emphasized the inequality between the two regions. With over five million inhabitants (almost twice as many as Hanoi), Ho Chi Minh city was the economic capital and seemed to adapt to the new model better than the North. The Government granted long-term loans and lease contracts, tax breaks and the right to inherit up to three hectares of land.

[23] In the mid 1990s a large part of the country was still devastated by napalm or the defoliant 'Agent Orange'. Telecommunications systems, highways and electric power facilities were in the process of recovery, with loans from the World Bank and the Asian Development Bank. Undetonated landmines were a serious problem, particularly in the central region.

[24] In 1995 diplomatic relations were established with the US. President Bill Clinton expressed concern about the 2,000 US citizens still unaccounted for in Southeast Asia.

[25] The 1996 rapprochement between Hanoi and Washington led many transnationals to express their interest in the Vietnamese market once the liberalization of the economy was consolidated.

[26] In September 1997 Tran Duc Luong was elected president, and Phan Van Khai prime minister. In December three CPV leaders were replaced, former president Le Duc Anh, former prime minister Vo Van Kiet and veteran leader Du Muoi, whose position as secretary-general was taken over by Le Kha Phieu.

[27] In April 1998 drought destroyed 7,000 of 260,000 hectares of coffee plantations, from which produce had been exported to Europe and the US since 1980.

[28] The 1998 regional crisis caused a relatively modest drop in growth, from 8.8 per cent in 1997 to 6.1 per cent, but in order to maintain export competitiveness two currency devaluations were deemed necessary. Foreign direct investment fell 70 per cent. In September the country became a full member of Asia-Pacific Economic Cooperation (APEC).

[29] A typhoon and exceptionally heavy rains in 1998, followed by floods in 1999, affected the country's economy. Environmental organizations also pointed out that matters had been aggravated by the indiscriminate felling of forests. Prime Minister Khai pointed out that the crisis had been influenced by low consumer demand, growing production stocks and the inefficiency of State enterprises.

[30] A land-frontier agreement with China was signed in December 1998, after an eight-year negotiation process. The settling of sea-borders was left for later talks.

[31] Corruption was recognized as a national problem in 1998 and identified as a by-product of economic liberalization.-In January three Tamexco employees with close CPV ties were executed for corruption, including former director Pham Huy Phuoc. In 1999 the CPV demanded the resignation of deputy prime minister Ngo Xuan Loc (who returned to the cabinet five months later) and punishment of former Central Bank governor Cao Sy Kiem for misappropriation of funds.

[32] The CPV politburo acknowledged in 2000 that relatives of its own members were involved in corruption and that this was commonplace throughout party ranks. It became compulsory for leaders to declare their wealth and an autonomous monitoring body

PROFILE

ENVIRONMENT

A long narrow country covering the eastern portion of Indochina along the Gulf of Tonkin and the South China Sea. The monsoon-influenced climate is hot and rainy. Rainforests predominate and there is a well-supplied river system. The northern region is comparatively higher. There are two river deltas: the Song Koi in the north, and the Mekong in the south. Most of the people are farmers, and rice is the main crop. The north is rich in anthracite, lignite, coal, iron ore, manganese, bauxite and titanium. Textile manufacture, food products and mining are the major economic activities. The felling of trees for domestic use (firewood) and construction have contributed to deforestation. However, the most significant losses - particularly in the northern part of the country - are a result of the Vietnam War, specifically from the use of such chemical defoliants as 'Agent Orange'. Vietnamese Government policy - started after 1975 and reaching its peak in the mid-1980s - that moved millions of former North Vietnamese to what was thought to be the relatively underpopulated central highlands took a heavy toll on the environment, including widespread deforestation.

SOCIETY

Peoples: Vietnamese make up most of the population. The remainder is made up of mountain people, consisting of several ethnic groups – Tho, Hoa, Tai, Khmer, Muong, Nung – and descendants of Chinese.

Religions: Mainly Buddhist; traditional religions. There are some two million Catholics and three million followers of the Hoa-Hao and Cao-Dai sects.

Languages: Vietnamese (official) and languages of the ethnic minorities.

Main Political Parties: Vietnamese Fatherland Front, which includes: Communist Party of Vietnam; mass organizations and affiliated; non-party candidates.

Main Social Organizations: The Federation of Unions of Vietnam (Tong Cong Doan Vietnam), founded in 1946, is the only union confederation and a WFTU member; the Vietnamese Women's Union, founded in 1930.

THE STATE

Official Name: Công hòa xâ hôi chu' nghi'a Viêt Nam.

Administrative Divisions: 39 Provinces, including the urban areas of Hanoi, Haiphong and Ho Chi Minh City.

Capital: Hanoi 3,977,000 people (2003).-

Other Cities: Ho Chi Minh City (formerly Saigon) 5,566,900 people; Haiphong 1,763,300; Da-Nang 762,800 (2000).

Government: Nguyen Minh Triet, President and head of state since June 2006. Nguyen Tan Dung, Prime Minister and head of government since June 2006. Nong Duc Manh, secretary-general of the Communist Party since April 2001. Unicameral Legislature: National Assembly with 498 members. Executive power is exercised by a council of ministers.

National Holiday: 2 September, Independence Day (1945).

Armed Forces: 484,000 troops (2003). Other: 5,000,000 (Urban Defense Units, Rural Defense Units).

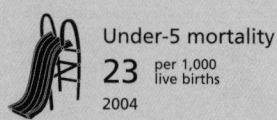
Under-5 mortality
23 per 1,000 live births
2004

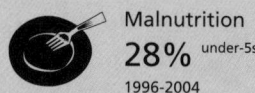
Malnutrition
28% under-5s
1996-2004

Debt service
6.0% exports of goods and services
2002

Maternal mortality
130 per 100,000 live birth
2000

IN FOCUS

ENVIRONMENTAL CHALLENGES

Slash-and-burn methods in use in agriculture have caused deforestation and soil degradation. Pollution and overfishing threaten marine life. Underground waters are contaminated, limiting availability of drinking water. Growing industrialization and migration to cities damage the environment, principally in Hanoi and Ho Chi Minh City. The most important damage, especially in the north, was originally caused by the war, including the use of chemical defoliants (Agent Orange).-

WOMEN'S RIGHTS

Women have been able to vote since 1946. In 2003, women held 27 seats in Parliament and 12 per cent of ministerial positions.

Female workers made up 49 per cent of the country's total labor force of 43 million.

In 2004, the female primary school attendance rate was 96 per cent, one point below the male rate*. That year, 86 per cent of pregnant women received prenatal care and 85 per cent of births were attended by skilled health staff. The maternal mortality rate was 170 per 10,000 births. In 2004, it was estimated that some 65,000 women aged between 15 and 49 were living with HIV/AIDS.

CHILDREN

Infant and under-five mortality rates decreased by more than half between 1990 and 2004. The former fell from 38 to 17 per 1,000 live births and the latter from 53 to 23 per 1,000. In 2004, 9 per cent of newborns had low birth weight, while 28 per cent of under-fives suffered from severe or moderate low weight and 32 per cent were severely or moderately stunted*.-

Mountain regions in Vietnam render access to rural schools difficult. On account of this, shelters have been built on school grounds so that children can stay there during the week and avoid walking every day the long distances (and rough terrain) which often lead to school dropout. Rural school enrollment rates are much lower than urban rates.

INDIGENOUS PEOPLES/ ETHNIC MINORITIES

The Chinese minority numbers around 500,000 and is relatively well integrated into society although many have been through difficult periods (including significant migration within the country or abroad). After the 1975 unification of the country they had to adapt to the communist regime which hindered their commercial activity, although the situation improved for those in the north, especially peasant farmers near the Chinese border.

The minority which has challenged the Government most has been the original *Montagnard* inhabitants from the highlands in the center of the country. Historically the relationship between the indigenous population and the Government has been violent, in particular since the mid-1960s when the United Front for the Oppressed Minorities' Struggle was founded.

MIGRANTS/REFUGEES

In early 2006, UNHCR continued to develop programs aimed at the repatriation of thousands of *Montagnards* to Vietnam's central region. Furthermore, negotiations were still under way between UNHCR and those countries that hosted Vietnamese refugees, in an attempt to regularize their situation. A solution was also being sought for some 2,300 Cambodians, who - together with other thousands of stateless persons - were still living as refugees in Vietnam.

DEATH PENALTY

Capital punishment is still in force.

* Latest data available in *The State of the World's Children* and *Childinfo* database, UNICEF, 2006.

many Montagnards were sick and exhausted after having been hiding in the trees, with a shortage of food, water and shelter. The official report pointed out that Montagnards survived by eating leaves and wild mushrooms. Those who were found in the most critical conditions were sent to Phnom Penh to receive medical treatment.

[44] Vietnam and Argentina signed a series of trade agreements in November 2004. Argentinian President Néstor Kirchner considered dialogue with his counterpart Tran Duc Luong as 'very enriching' and underscored Vietnam's 'increased participation in the international arena'.

[45] In April 2005, tens of thousands of people took part in celebrations in Ho Chi Minh City to mark the 30th anniversary of the end of the war. Prime Minister Phan Van Khai said the victory of 30 April 1975 had been 'forever written' in the country's history, but he added that Vietnam was still facing many challenges and should move on from self-congratulation for the past and look to the future. More than 7,500 prisoners, including some political detainees, had been previously released as part of the celebration.

[46] Prime Minister Phan Van Khai's official visit to the US in June - the first one since the end of the war - was regarded by Vietnamese people as an extremely positive event, which gave a significant boost to the country's social and economic status.

[47] In June 2006, Parliament opened an extensive investigation into corruption cases - involving bribery, nepotism and illegal gambling - in several ministries. The main allegations were targeted at Transport Minister Dao Dinh Binh. In April, Dao had resigned and accepted responsibility for the embezzlement of millions of dollars from infrastructure projects mainly funded by Japan.

[48] Also in June, 72-year-old Prime Minister Phan Van Khai, President Tran Duc Luong (aged 69) and National Assembly President Nguyen Van An (aged 68) resigned. This was the most important renewal within the communist regime in decades.

[49] Nguyen Minh Triet became President and Nguyen Tan Dung became Prime Minister. Nong Duc Manh had taken over as new secretary-general of the Communist Party in April.

[50] In November 2006, Vietnam was given the go-ahead to become the 150th member of the World Trade Organization. ∎

was set up to gather complaints. This body reported judicial lack of independence, lack of information about legislation in force and the need for an independent parliament, with a true voice in politics.

[33] In April 2001 reformer Nong Duc Manh (former National Assembly president) was appointed CPV secretary-general.

[34] During his official visit in November 2000, President Clinton promised more US assistance in clearing the territory of undetonated explosives. In June 2001 Tran Duc Luong signed an agreement with the US in Hanoi during a donors' meeting.

[35] In April 2002, 59 people faced charges of bank fraud for $100 million in Ho Chi Minh City, involving hundreds of individuals and organizations. The trial was regarded by Transparency International as part of the Government's effort to change the situation in a country ranked among the most corrupt on the planet.

[36] In May 2002 the Russians gave up their naval base in Cam Ranh Bay, once the biggest Soviet facility outside the Warsaw Pact.

[37] President Tran Duc Luong was re-elected by the Assembly for a second term in July 2002. This meant a second term for Prime Minister Phan Van Khai, too.

[38] Six organized crime bosses from Ho Chi Minh City were sentenced to death in June 2003, including *capo* Nam Can, and government officials were sentenced to long prison terms.

[39] In November, for the first time since the war, a US vessel entered Vietnamese waters near Ho Chi Minh City.

[40] In January 2004, the resurgence and spread in Vietnam (and Thailand) of a pandemic known as avian flu, a subtype of the influenza virus, which registered the first case of human infection in HOng Kong in 1997, put the government's helath mechanisms on alert.

[41] In March, weeks after Hanoi had declared the eradication of 'bird flu', the Vietnamese Ministry of Agriculture reported the detection of the virus in chickens on a farm located in southern Vietnam. That same month, the virus death toll hit 16. Both WHO and FAO had warned Hanoi about not reopening poultry farms and not declaring itself free of disease too soon.

[42] In June, the UNHCR - UN agency for refugees - found more than 40 Vietnamese *Montagnards* (mountain people and a Christian minority group) hiding in the jungle in Cambodia. Human rights organizations claimed that a large number of them were still living secretly in the province of Ratanakiri, in north-east Cambodia. Phnom Penh declared them illegal immigrants and refused to grant them humanitarian aid. Under pressure from the king of Cambodia, Norodom Sihanouk, diplomats, human rights groups and the UNHCR, the latter was allowed to reopen its office in the city of Banlung (capital of Ratanakiri) to aid Montagnards.

[43] According to the UN,

Virgin Is (US) / Virgin Islands

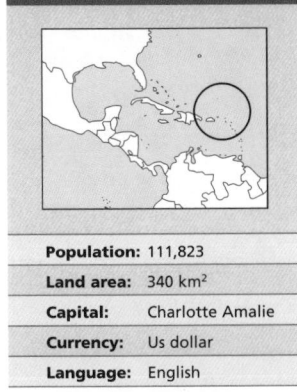

Population:	111,823
Land area:	340 km²
Capital:	Charlotte Amalie
Currency:	Us dollar
Language:	English

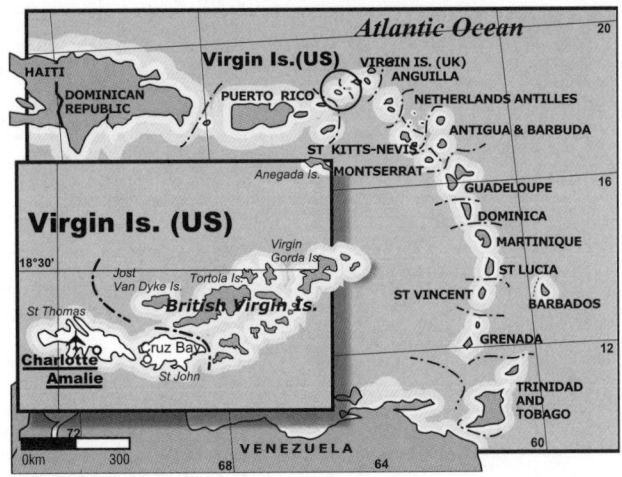

As is the case in other Caribbean islands, the Virgin Islands were originally inhabited by Carib and Arawak indians. The only gold on the islands was seen in the ornaments worn by the inhabitants and was of little economic interest. From the time of Columbus' arrival in 1493 onwards, the local population was persecuted and massacred, and was destroyed by the second half of the 16th century.

[2] The western part of the archipelago was claimed by other countries after the departure of the Spaniards. The Netherlands gained control in the 18th century and organized crop cultivation; first of sugar cane and then of cotton. They imported African slave workers as in many other parts of Latin America and the Caribbean. At the height of Dutch colonization, there were 40,000 African slaves on the islands.

[3] During Abraham Lincoln's second presidential term, the United States failed to acquire the islands. But in 1917, after World War I when the US wanted to consolidate its presence in the area, $25 million was paid to the Dutch for the islands and the 26,000 former African slaves who inhabited them.

[4] From that time on, the US initiated a series of changes in the territorial administration, ranging from the continuation of the legislation established

by the Dutch, to the new law approved in 1969. This instituted the election of a Governor and Vice-Governor, the election of a non-voting Congressional delegate and the right of its inhabitants to vote in American elections. In a 1981 referendum, a proposed Constitution was voted against by 50 per cent of the registered voters. In December 1984, the UN General Assembly reiterated the right of the islands to self-determination, and urged the local population to exercise their right to independence.

[5] The local economy depended on the US. It largely revolved around an oil refinery owned by US-Ameranda Hess Corporation on the island of St Croix. The refinery is the largest of its kind in the world and produces over 700,000 barrels a day. The company's influence on local politics was very important, as it had 1,300 permanent and 1,700 temporary employees and supplied all the

islands' energy needs. In spite of this, tourism was still the island's main source of income.

[6] A referendum scheduled for November 1989 was to decide the future status of the islands. However, it was not held because in September 1989 the islands were severely damaged by Hurricane Hugo. According to official estimates, 80 per cent of the buildings were destroyed. The disaster caused an outbreak of looting and unrest. Armed bands wandered the streets; some members of the police and National Guard also took part in the pillaging. The US sent more than 1,000 troops to the area, but the disturbances had virtually ceased by the time they arrived. In the words of local people, there was nothing left to steal.

[7] An aluminum oxide plant, which had been closed in 1985, was purchased in 1989 by an international group trading in raw materials. The plant has a

capacity for an annual production of 700,000 metric tons. Shipments to the US and Europe began in 1990.

[8] Alexander A Farrelly, who had governed the islands since 1987, quit his post in 1995 and was replaced by Roy L Schneider. In the following elections, in November 1998 Charles Turnbull won with 58.9 per cent of the vote, while Schneider took the remaining 41.1 per cent.-

[9] Fourteen counts of embezzlement, fraud and falsifying documents against former Governor Schneider and three top officials of his administration were dismissed by Judge Ive Swan in March 2000.

[10] The US Virgin Islands are among the 16 territories of the world that still remain a colony. In May 2004, the UN launched a campaign of reaffirmation of the principle of equality of rights and self-determination of peoples. However, neither the US Virgin Islands, nor the British ones, have expressed the desire to become independent, particularly fearing a breakdown of the lucrative tourist industry and a rise of taxes, since any product or raw materials from the islands are admitted tax-free into the US.

[11] Each year the islands receive two million tourists, almost all of them from the US, who are involved in a large number of automobile accidents because they are not used to driving on the left side of the road.

[12] The Government aimed to improve tax discipline and support private construction projects, as well as improving tourism infrastructure. The main challenges for 2006 were the reduction of crime and protection of the environment. ∎

LAND USE

2003/2005

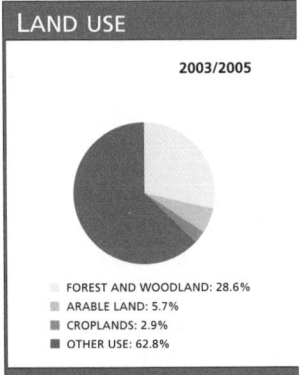

- FOREST AND WOODLAND: 28.6%
- ARABLE LAND: 5.7%
- CROPLANDS: 2.9%
- OTHER USE: 62.8%

PROFILE

ENVIRONMENT
The western part of the Virgin Islands group, east of Puerto Rico, includes three main islands - St Thomas, St John and St Croix - and approximately 50 uninhabited small islands. The mountainous terrain is of volcanic origin. The climate is tropical, but irregular rainfall makes farming difficult. A small amount of fruit and vegetables are grown in St Croix and St Thomas. Tourism is an important economic activity. There is a large oil refinery in St Croix, supplying the US market.

SOCIETY
Peoples: Most of the population is English-speaking and of African origin with a small Spanish-speaking Puerto Rican minority. 35 to 40 per cent of the inhabitants are from other Caribbean islands; 10 per cent from the US.
Religions: Protestant and Catholic.
Languages: English, official; also Spanish and Creole.
Main Political Parties: There are local representatives of the US Republican and

Democratic parties and an Independent Citizens Movement (ICM).

THE STATE
Official Name: Virgin Islands of the United States.
Capital: Charlotte Amalie 12,000 people (1999).
Other Cities: Anna's Retreat 13,500 people; Charlotte Amalie West 8,000 (2000).
Government: Head of State, President of the US, George W Bush, since January 2001. Charles Turnbull, Governor since January 1999. The unicameral legislature has 15 members elected by popular vote (seven from St Thomas, seven from St Croix and one from St John).
National Holiday: 27 March, Transfer Day (from The Netherlands to the US, 1917).
Armed Forces: The US is responsible for the defense of the islands. The naval bases have been under local jurisdiction since 1967, but the US retains the right to occupy them at any time, as well as to recruit local residents for US armed forces.

Virgin Is (British) / British Virgin Islands

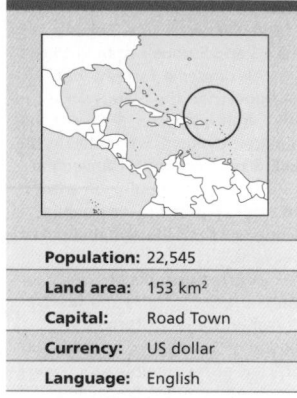

Population:	22,545
Land area:	153 km²
Capital:	Road Town
Currency:	US dollar
Language:	English

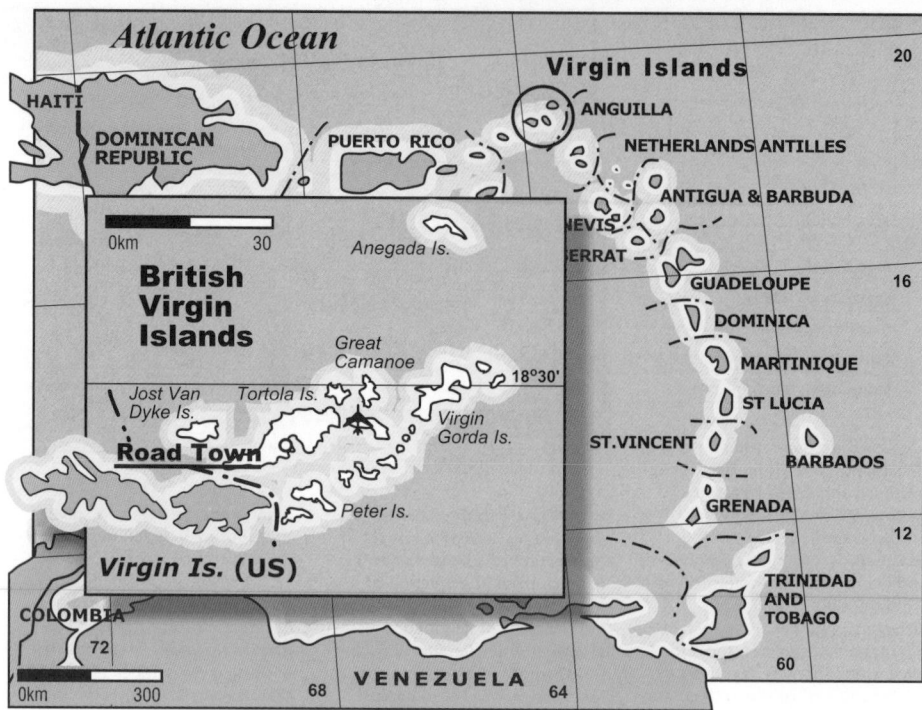

The Virgin Islands were named by Christopher Columbus in 1493 in honour of St Ursula. At that time the islands were inhabited by Caribs and Arawaks, but by the mid-16th century they had all been exterminated by the Europeans.

2   In the 18th century, the British gained control of the easternmost islands of the archipelago and, using African slave labor, they began cultivating sugarcane, indigo and cotton. By the mid-18th century, the islands' slave population had reached 7,000, outnumbering the European colonists six to one. Slavery was eventually abolished there in the 1830s.

3   In 1872 the islands joined the British colony of the Leeward Islands, which was administered under a federal system. This federation was dissolved in July 1956, but the Governor of the Leeward Islands continued to be responsible for the administration of the Virgin Islands until 1960. In that year direct control over the islands passed into the hands of an administrator designated by the British Crown.-

4   The colony was governed under several constitutions during the 20th century. The ministerial form of government began in 1967 with the first open elections, after which Hamilton Lavity Stoutt of the Virgin Islands Party (VIP) became the first Chief Minister. In 1977 the Constitution was modified, granting greater autonomy and carrying out changes in the electoral system. Responsibility for defense, internal security and foreign affairs remained in the hands of the British-appointed governor.

5   In the 1980s tourism accounted for 45 per cent of the national income, while fishing was the traditional economic activity. Gravel and sand were important exports, and a 'tax-haven' banking facility was established.

6   In August 1986, the British Government dissolved the

Legislative Council, calling a new election several days after the opposition presented a vote of no confidence in the Chief Minister Cyril Romney. In September the VIP won five of the nine seats and the United Party (UP) won two. Lavity Stoutt, the VIP leader, was again named Chief Minister and remained so until his death in 1995.

7   The British Crown appointed Frank Savage Governor in 1998, and in the May 1999 elections the VIP took eight seats with 38 per cent of the vote. The National Democratic Party took five with 36.9 per cent of the vote, while the remainder of the vote went to the CCM (four per cent). Independent candidates took 13.4 per cent of the vote, but none had enough support to win a seat on the Legislative Council.

8   The Government started road works and other infrastructural projects on the outer islands. In December 2001 Chief Minister Ralph T O'Neal presented the 2002 budget and underscored the importance of these infrastructural works for developing tourism in Virgin Gorda, Anegada and Jost Van Dyke.

9   In May 2004, a new agency was launched in order to fight money laundering in the islands. Chief Minister Orlando Smith of the National Democratic Party, who had been elected into office the previous year, stated that the Financial Investigation Agency strengthened its

administration with the strong purpose of upholding honesty and transparency in business.

10   The economy of the islands, one of the most successful in the

Caribbean, depends mostly on tourism, which stands for 45 per cent of the GDP. Some 350,000 tourists - most of them from the US - visit the islands each year. ■

PROFILE

ENVIRONMENT

The eastern portion of the Virgin Islands comprises 36 islands and reefs, 16 of which are inhabited. The larger ones are Tortola, Anegada, Virgin Gorda and Jost Van Dyke, which are home to most of the population. The islands' rolling terrain is of volcanic origin. Agricultural production is limited to fruit and vegetables as a result of erratic rainfall. The major economic activities are fishing and tourism. The coral reefs are threatened by tourist activities.

SOCIETY

Peoples: Most are descendants of African slaves. There is also a small British minority. 86 per cent of the population are concentrated on Tortola Island, nine per cent live on Virgin Gorda, three per cent on Anegada and two per cent on Jost Van Dyke.
Religions: Mainly Protestant (45 per cent Methodist, 21 per cent Anglican, seven per cent Church of God, five per cent Seventh-Day Adventists, four per cent Baptist, two per cent Jehovah's Witnesses, two per cent other Protestant); six per cent Roman Catholic; six per cent other religions (1981).
Languages: English (official).
Main Political Parties: National Democratic Party; Virgin Islands Party.

THE STATE

Official Name: British Virgin Islands.
Capital: Road Town 9,400 people (2004).
Other Cities: East End-Long Look 4,900 people (2000).
Government: David Pearey, British-appointed Governor since 2006. Orlando Smith is the Chief Minister, since June 2003. The Legislative Council has 15 members.
National Holiday: 1 July, Territory Day; 7 March, Lavity Stoutt Day.
Armed Forces: Defence is the responsibility of the United Kingdom.

Western Sahara / Sahara Occidental

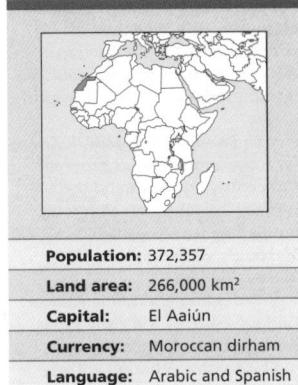

Population:	372,357
Land area:	266,000 km²
Capital:	El Aaiún
Currency:	Moroccan dirham
Language:	Arabic and Spanish

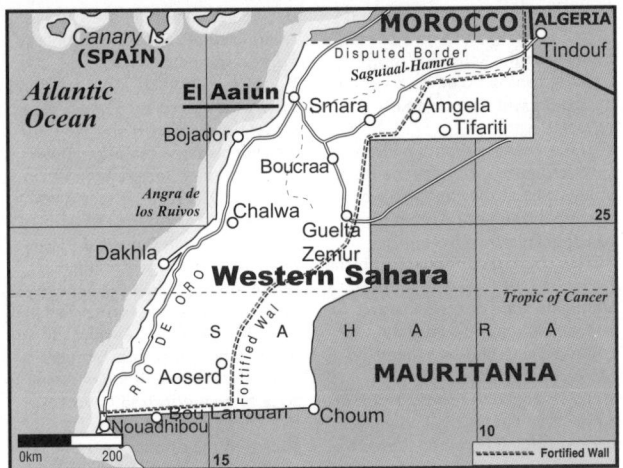

From the 5th century, the far west of the Sahara has been populated by Moors, Tuaregs and Tubus, resulting from migratory flows prompted by desertification, which has affected the region since the Neolithic Era. Their presence is documented by the Tassili stone carvings and other relics. In the 9th century, waves of migrants arrived from Yemen and intermarried with the local population. Four hundred years on, the first confederation of Saharawi peoples appeared.

2   The arrival of the Spanish on this coast had a strategic motive: the defense of the eastern coast of the Canary Islands. Colonization was mostly limited to Villa Cisneros (present-day Dakhla) until 1886, when, as a result of the Berlin Conference, Madrid resolved not to let an 'empty space' fall to another power. However, after an agreement with France was reached in 1904, establishing the borders of Spanish Sahara, the situation returned to what it had been. Colonialism divided the territory into four countries, where the nomadic ethnic groups continued to live completely independently, ignoring the frontiers imposed.

3   In 1895, Sheikh Ma al-Aini founded the Smara citadel and fought the Franco-Spanish presence with the support of the Sultan of Morocco until 1910, when the latter gave in to European pressure and suspended assistance to the rebels,

who expanded their actions into Morocco and even threatened Marrakech. The French counter-attack consisted of an invasion of 'Spanish' territory and the conquest of Smara in 1913, though resistance continued until 1920.

4   The French pressed Spain to increase its control over the territory, and in 1932 El Aaiún was founded. In 1933, the victory of the forces under Mohammedal Mamun, cousin to Ma al-Aini and Emir of Adrad, forced a change in colonialist tactics. France occupied the rebel base at Tindouf oasis and advanced into Algeria, Mauritania and Morocco, while Spanish troops took Smara, overcoming the insurgents in 1934.

5   When the French deposed Sultan Mohammed V, the National Liberation Army (ALN) was created in Morocco, and its Southern Division operated in close cooperation with the Saharawi people. Following Morocco's independence in 1956 and the dissolution of the ALN, the Saharawis were left to face Spain and the French air force on their own, leading to a withdrawal of resistance in 1958.

6   At that time, phosphate exploitation began at Boucraa, where 10 billion tons of what was considered to be the best quality phosphate in the world had been discovered. Transnational capital, with the consent of Spain's fascist Franco regime, invested more than $160 million, transforming the country, particularly its population distribution. In 1959, El Aaiún had 6,000 residents, but had reached 28,000 by 1974, while the nomadic population dropped from 90 per cent to 16 per cent of the country's total in that same period.

7   With the decline of the traditional nomadic way of life, ties and relationships began to weaken, though the colonial administration maintained latent divisions through political recognition of the *shiuj*

(clan chiefs) and notables of the different groups and also specified in the national identity document which people or faction the holder belonged to. However, a new national identity slowly developed, transcending traditional divisions.

8   The Saharawis founded the Al Muslim movement in 1967, and the Sahara Liberation Front a year later. In 1973, the revolutionary leadership decided to pursue armed struggle, creating the Polisario Front (the Popular Front for the Liberation of Saguia al Hamra and Rio de Oro), which was led by Elwali Mustafa Sayed until he died in combat. The

war and the resolutions of the UN, which favored the independence movement, prompted the Franco Government to prepare to withdraw from the colony, recognizing the right to self-determination and organizing a census prior to a referendum. The census showed that there were 73,497 Saharawis living in the territory.

9   In 1974, the World Bank classified Western Sahara as potentially the richest territory in the Maghreb region for its extensive fisheries and phosphate reserves.

10   With one eye on the fishing and phosphate reserves and the other on the advantage of unifying his troubled nation around an external cause, King Hassan II of Morocco claimed sovereignty over Western Sahara. The International Court of Justice at The Hague overruled its claims and ordered decolonization. King Hassan responded by organizing what was dubbed the 'Green March', a propaganda move that mobilized 350,000 Moroccans to head south, crossing the border, to press for a reversal of The Hague ruling. Within days these were replaced by Moroccan soldiers. As General Franco lay on his deathbed, Spain signed a secret agreement which handed over the territory to Morocco and Mauritania.

11   Tens of thousands of ordinary Saharawis fled into the desert from

LAND USE

2003/2005

IRRIGATED AREA: 3.6% of arable land

- FOREST AND WOODLAND: 26.5%
- ARABLE LAND: 7.5%
- CROPLANDS: 0.9%
- OTHER USE: 65.1%

PROFILE

ENVIRONMENT

The country is almost completely desert and is divided into two regions: Saguia el Hamra in the north and Río de Oro in the south. It has one of the world's most abundant fishing reserves, but the principal source of wealth is mining, especially for phosphate deposits. Oil companies are interested in prospecting for deposits both on and offshore.

SOCIETY

Peoples: The Polisario Front estimates the dispersed Saharawi population at one million. These are traditionally nomadic groups that differ from the Tuaregs and Berbers in their social and cultural organization.
Religion: Islam.
Languages: Arabic and Spanish (official). Many Saharawis also speak Hassania, one of the purest dialects of Arabic.
Main Political Parties: The People's Liberation Front of Saguia al-Hamra and Rio de Oro (Polisario Front) founded on 10 May 1973 by Elwali Mustafa Sayed.
Main Social Organizations: The Saguia al-Hamra and Rio de Oro General Workers' Union (UGTSARIO).

THE STATE

Official Name: Saharawi Arab Democratic Republic.
Capital: El Aaiún 187,000 people (2003).
Other Cities: Dakhla 40,200 people; Smara 36,100; Boucraa 27,800 (2000).
Government: Muhammad Abdelaziz, President of the Republic since 1982, is also Secretary-General of the Polisario Front. Abdelkader Taleb Oumar, Prime Minister since October 2003. The National Assembly (parliament) has 101 members, elected by local and regional conferences and acts as a check on the executive branch. Polisario is committed to multi-party democracy when it gains independence.
National Holiday: 27 February, Proclamation of the Republic (1976).

the invading Moroccan forces, setting up their own makeshift refugee camps. Many, like the 25,000 who gathered at Guelta Zemmour, were determined to stay on Western Saharan territory but were repeatedly bombed by Moroccan planes using napalm. In the face of this onslaught the refugees had to walk hundreds of kilometers across the desert to the Algerian town of Tindouf. There the Algerian Government ceded effective control over a swathe of its own territory to Polisario, which built and administered its own refugee settlements.

12   On 27 February 1976 the Saharawi proclaimed the Saharawi Arab Democratic Republic (SADR). The new African republic was born in Bir Lahlu, a desert post in Saguia El Hamra, a few kilometers from the Mauritanian border. Just hours earlier, in El Aaiún, the last representative of the colonial administration had officially announced the end of the Spanish presence.

13   Several countries recognized the new nation, but it triggered a war against Morocco and Mauritania. In 1979, Mauritania,

on the verge of collapse, decided to halt fighting and sign a peace treaty with the Polisario Front. Hassan's troops, however, stepped up attacks with French and US support.

14   The Polisario Front's military victories led to a diplomatic success in July 1980 at the conference of the Organization of African Unity (OAU) in Freetown, where 26 African countries announced official recognition of the SADR as the legitimate state of the Saharawi people. Four months later the UN issued a resolution requesting Morocco's withdrawal. In 1981 Morocco began the construction of a defensive sand wall - known as a *berm* - dividing occupied Western Sahara from the liberated zones. Eventually the berm ran the length of the country and was over 1,500 kilometers long (longer than the Great Wall of China); it was guarded by 120,000 soldiers and two million mines.

15   The SADR was accepted in November 1984 as a full member of the OAU. Morocco withdrew from the organization, as it had announced it would. On 14 November1985 the UN

Decolonization Committee recognized the Saharawi people's right to self-determination.

16   In August 1988, Moroccans and Saharawis agreed on a peace plan presented by the UN and the OAU, which included a ceasefire and a referendum on self-determination.

17   In July 1990, representatives from Morocco and the Polisario Front debated in Geneva a procedural code for holding a referendum to allow the Saharawis to decide their own future. The greatest challenge was how to define who could vote. Morocco wanted its people in the occupied zone to be authorized to vote.

18   On 29 April 1991, the UN approved the establishment of MINURSO (United Nations Mission for the Referendum in Western Sahara) and established 6 September of the same year as the date for the ceasefire to come into force and 26 January 1992 for the referendum.

19   MINURSO was entrusted with drawing up the electoral roster based on the 1974 census. This implied that an undetermined number of Saharawis could not vote in the referendum, but nor could

the Moroccans who immigrated after 1976. The Saharawis of voting age who were living in refugee camps in Algeria would be transported back to their towns of origin.

20   By January 1992, MINURSO was far from completing its program for identifying voters and the Saharawi repatriation plan could not be finalized. Meanwhile, 60,000 Moroccan soldiers remained in Western Sahara.

21   The refugee settlements near Tindouf were, by the early 1990s, well established and efficiently run by Polisario, though entirely dependent on international food aid due to the barrenness of this part of the desert.

22   In March 1997 UN Secretary-General Kofi Annan appointed former US Secretary of State James Baker as his personal envoy to Western Sahara, entrusting him with relaunching the peace process. Morocco and the Polisario Front agreed on 16 September 1997 to reactivate the peace plan for Western Sahara, exchange prisoners and release political prisoners. The referendum date was set for 7 December 1998.

Western Sahara:
Fishy business in the European Union

IN MAY 2006 the European Parliament voted in favor of a Fisheries Partnership Agreement with Morocco that will allow European ships to fish off the coast of illegally occupied Western Sahara, despite claims that this violates international law. Amendments by Green and left-wing groups achieved close to 200 votes, but were voted down, despite support from campaigners and the governments of Sweden, Finland and Ireland.

The EU-Morocco Fisheries Partnership Agreement is similar to a host of deals being signed down the West African coast, allowing European fishing in African waters to make up for the over-fishing of European waters in recent decades. But it is different in one crucial way - it fails to define Morocco's southern border. Instead it allows Morocco to decide where to apply the Agreement, knowing full well that Morocco will apply it to Saharawi waters. This gives tacit approval to Morocco's occupation of Western Sahara, which has continued despite international legal rulings and UN pressure since its invasion in 1975 (see history). The Saharawi government-in-exile, Polisario,

has denounced the decision as 'a flagrant violation of international law'. It continued: 'Spain and France have drawn the rest of the Member States of the European Union into their reckless adventure that consists in drawing up and signing an agreement that is nothing but an act of international banditry and plundering, which will undoubtedly represent a dark page in the history of the European Union.'

This is a profitable arrangement for the EU. Western Sahara's continental platform of is one of the richest fishing grounds in the world. It comprises an area of more than 150,000 sq km with a great diversity of species. There are 200 varieties of fish, 60 types of mollusks and several species of cephalopods and crustaceans. This marine wealth allowed Morocco to develop an export trade without assigning large amounts of capital or investment - which could have had beneficial spin-offs for Western Sahara's economy - to the occupied territory. El Aaiún, Western Sahara's capital, accounts for 40 per cent of Morocco's total fish catch. By comparison, Saharawis will see almost no benefit from

the Agreement. It is the corporations that control fishing in Western Sahara, mostly Moroccan or Spanish, that will gain most. Even the employment that filters down to ordinary workers will go mostly to Moroccan settlers, and not to Saharawis. The new fisheries deal follows in the wake of years of disputes over oil exploration in Saharawi territory, both offshore and on land. Following immense pressure from Saharawi groups and solidarity activists worldwide, companies that had been prospecting in the area by illegal arrangement with Morocco - such as the US corporation Kerr-McGee - withdrew. Polisario has now issued its own licences for oil exploration to British and South African companies. However, in June 2006 Dallas-based corporation Kosmos Energy, previously a minority partner to Kerr-McGee, gained its own exploration permit from the Moroccan regime despite the clear ruling of the UN that Western Sahara is a non-self governing territory and that any exploitation of its resources without the permission of its indigenous people is illegal. ∎

23 Under the UN plan, Morocco and the Front agreed to accept that if Morocco won, the UN would disarm the Saharawi combatants, while if the Polisario Front won, it would supervise the withdrawal of Moroccan troops and administration from Western Sahara.

24 The death of King Hassan II of Morocco in July 1999 and accession of his son, Muhammad VI, brought some promised political changes. The Polisario Front welcomed the new King's first measures and his decision to carry on with the referendum on Western Saharan self-determination. In November, Muhammad VI decided he would allow the occupied zone to have autonomy within Morocco.

25 The UN Security Council again suspended the referendum scheduled for July 2000. Morocco expressed its intention of negotiating with the Polisario Front to grant it a certain degree of autonomy, but closed the door to the referendum.

26 Having previously seemed to be firmly in favor of a referendum on self-determination, Baker and the UN started to put pressure on the Saharawis to accept some form of autonomy within Morocco. In 2001, following the failure of several attempts at negotiation in London and Berlin, the Polisario Front rejected the Framework Agreement for the Sahara Statute, known as the Baker Plan, which granted a certain degree of autonomy to the zone, but under Moroccan sovereignty. The referendum had already been postponed 12 times.

27 In October 2001, King Muhammad VI made his first trip to the Western Sahara region. This trip coincided with an agreement between Morocco and transnational oil corporations - the French company Total Fina Elf and the US company Kerr-McGee - to explore along the Sahara coast, where there are thought to be substantial reserves of oil and natural gas. A UN legal judgement made it clear that Morocco had no right to make such agreements in relation to a territory still not decolonized. Polisario accused Morocco of signing these deals in a backdoor attempt to legitimize their invasion.

28 In February 2002 Algeria rejected the Baker Plan and proposed that the UN should administer Western Sahara. Kofi Annan proposed four options to solve the conflict: continue with the Arrangement Plan which included the self-determination referendum; continue with the Framework Agreement, with slight changes; start negotiations for the participation of the territory; and the withdrawal of MINURSO.

IN FOCUS

ENVIRONMENTAL CHALLENGES
Heat and drought are problems inherent to the zone, causing a scarcity of arable land. The sirocco blows during the fall and winter; this dry Sahara wind blows towards the coast of western Africa between November and March, causing frequent fogs, severely restricting visibility. Morocco exploits fishing (with the support of agreements signed with Spain and France, and now the EU), phosphates and oil, recently discovered.

WOMEN'S RIGHTS
The experience of Saharawi women is deeply intertwined with their fight against Moroccan occupation. Women in the occupied territories are subjected to torture and rape, which are constantly reported by their organizations. Women's role in society has been transformed by life in refugee exile. Most men are required to serve in the Polisario resistance army, leaving women to take primary responsibility for running the refugee 'towns' outside Tindouf, in western Algeria. Women traditionally received no education in Saharawi society or Spanish colonial schools but the younger generation are among the most literate, educated girls and women in the Muslim world. Many visit or study in foreign countries. There are, as yet, very few women in positions of political power but this is changing as the more educated generation gains more influence.

INDIGENOUS PEOPLES/ ETHNIC MINORITIES
The Saharawis are one of the nomadic tribes descending from African slaves, Arab Bedouins and Berbers from Sanhanja. They are mainly Sunni Muslims, and speak Hassania, a pure dialect of the Arabic language. Ethnic status is an explosive political issue in Western Sahara. In occupied Western Sahara the Saharawi face both discrimination and severe repression from the Moroccan authorities. Moroccans have for three decades been offered financial inducements by their Government to settle in Western Sahara so as to entrench colonial control - and to confound the workings of a referendum on self-determination which was originally only supposed to include Saharawis resident at the time of the 1975 invasion. Recently Saharawi leaders have reluctantly accepted a new UN plan which would allow Moroccan settlers to vote - but Morocco continues to block the referendum regardless.

MIGRANTS/REFUGEES
In early 2006 there were 165,000 Saharawi refugees living in camps near Tindouf, in western Algeria. The camps are self-ruled and are not under control of an international agency or the host country, though they do depend on food supplies from the UN. Algeria effectively cedes autonomy over the area to the Saharawi Arab Democratic Republic, a fellow founding member of the African Union. Organization of the camps is efficient and both education and health services have been prioritized. Some refugees have not seen their families in 30 years, though the UN has recently arranged for some exchange visits between the camps and the Occupied Territory.

DEATH PENALTY
Saharawis may face the death penalty under Moroccan rule in the occupied territories but not in the areas governed by the Saharawi Arab Democratic Republic.

29 In July 2002 the African Union (AU, successor to the OAU) was formally launched, with the SADR (ruled in exile by Polisario) as a founder member. Morocco refused to join the new continental body in protest.

30 In January 2003, Baker submitted a new plan under which Rabat would gain control of Western Sahara for an interim period. At the end of this a referendum for self-determination would be held, in 2007 or 2008, in which the Saharawis included in the Spanish census of 1974 and the settlers who have arrived since would vote for independence or integration. Despite the compromise involved in accepting a majority of Moroccan settlers as voters in the proposed referendum, the Polisario Front accepted the proposal and freed 243 Moroccan prisoners. Morocco refused to consider it.

31 In June 2004 Baker stepped down as the UN Secretary General's Personal Envoy for Western Sahara. The decision was widely interpreted as a blow for the Saharawis, given that Baker represented the best chance of bringing US pressure to bear on Morocco to agree to a referendum.

32 In September 2004 South Africa formally recognized the SADR. Given South Africa's prominence on the continent and within the AU, the recognition was seen as a huge boost for the cause of Saharawi independence and South African foreign minister Clarice Dlamini Zuma followed up by visiting the Saharawi refugee camps in Algeria in April 2005. In June 2005 Kenya also recognized the SADR while in December 2005 Sudan became the first state formally to support Moroccan sovereignty over Western Sahara.

33 In May 2005, on the 30th anniversary of Morocco's annexation of Western Sahara, there were street protests in all the main cities of the Occupied Territory in what was claimed by Saharawis as a new 'intifada' against Moroccan rule. Many were arrested and even those released within a few days complained of having been 'savagely tortured'. On 17 June security forces swooped all over Morocco and the Occupied Territory in an attempt to stop reports about the repression leaking out. Fact-finding delegations from the Spanish Parliament were repeatedly refused access. Between 3 August and 29 September 37 of those imprisoned maintained a hunger strike in protest at their treatment, 14 of whom had to be hospitalized.

34 In July 2005 Dutch diplomat Peter van Walsum was appointed the UN Secretary General's new Personal Envoy for Western Sahara. He launched into a series of fact-finding visits to the region. In August Polisario released all its remaining 404 prisoners of war, many of whom had been held in the refugee camps for almost two decades.

35 In April 2006 the UN further extended MINURSO's mission but Annan's report suggested that the UN was giving up its efforts to resolve the Western Saharan issue. According to the Secretary General 'nobody was going to force Morocco to give up its claim of sovereignty over Western Sahara'. He proposed that the UN took 'a step back' and that it was up to the parties involved to find a solution, which might involve 'a compromise between international legality and political reality'.

36 In May 2006 the European Parliament voted to endorse a controversial deal with Morocco that involved fishing in Western Saharan waters (see box opposite). ∎

West Papua / Papua

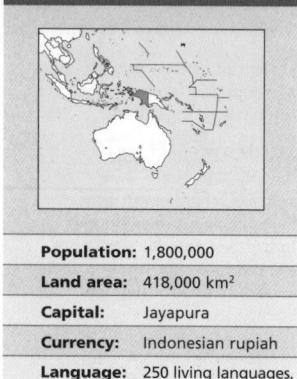

Population:	1,800,000
Land area:	418,000 km²
Capital:	Jayapura
Currency:	Indonesian rupiah
Language:	250 living languages, Bahasa Indonesian

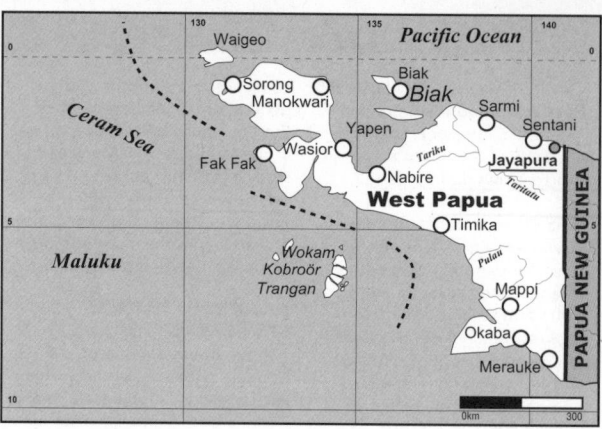

Humans first settled New Guinea at least 50,000 years ago. It was connected to Australia by a land bridge until about 10,000 years ago. There is evidence of agriculture from possibly as early as 9,000 years ago, which makes the island one of the earliest independent centers for the domestication of several varieties of sweet potato and fruit-bearing trees.

2 The British were the first Europeans who attempted to settle the western part of the island (which the indigenous people now call-West Papua). In 1793 a short-lived British colony was set up near present-day Manokwari but it was evacuated within two years. A unilateral proclamation was made by the Dutch on the 24th of August 1828 at the newly founded Fort du Bus on West Papua's south coast: the natives of the western half of New Guinea were to be subjects of the King of the Netherlands from this point forward.

3 The Dutch were not initially concerned with colonizing West Papua, but opened Fort du Bus to protect their lucrative trade in the surrounding spice islands (the Dutch East Indies) from other European powers. Fort du Bus was abandoned after only ten years. Until the turn of the twentieth century, the Dutch governed West Papua indirectly through the Sultan of Tidore, and did not establish a continuous settlement in West Papua until 1897, nor did they engage in any substantial development within the country until the 1950s.

4 On 27 November 1949 the Dutch ceded sovereignty of the Dutch East Indies to the Indonesian Republic, but excluded Dutch New Guinea (West Papua). Throughout the 1950s, the Dutch argued that West Papua was geographically and ethnically different from Indonesia and the Papuans should - over time - be granted self-determination. By contrast, the Indonesians argued that Dutch New Guinea had already been transferred to them in 1949, and had achieved independence then. In 1962, 1,500 Indonesians 'invaded'.

5 Under the auspices of the UN, the US urged Indonesia and the Netherlands to the negotiating table. Retired US diplomat Ellsworth Bunker drew up a plan to transfer the administrative authority for West Papua from the Netherlands to a neutral administrator, and thence to Indonesia. No West Papuans were involved in these negotiations. This 'New York Agreement' was signed by the Indonesians and the Dutch at UN headquarters on 15 August 1962. It fell well short of guaranteeing a referendum on independence, instead requiring Indonesia to make vague arrangements for West Papuans to 'exercise freedom of choice'.

6 In 1968, a UN team arrived to 'assist, advise and participate' in the 'Act of Free Choice', which took place the following year. Only 1,025 West Papuans were chosen (many handpicked and intimidated by Indonesian troops) to vote in open meetings. UN Secretary General U Thant later reported that without dissent the West Papuans had pronounced themselves in favor of remaining with Indonesia.

7 It is estimated that at least 100,000 people have been killed by the Indonesian armed forces since then. Indonesia's harsh regime is responsible for murders, torture, forced migrations and the 'export' of Muslim Indonesians to the island. Most of the profit obtained from the world's biggest gold mine and the third biggest copper deposit ends up in Indonesia. Only a fifth of this revenue has been returned to West Papua.

8 In February 2000, 400 delegates including representatives of the armed wing of West Papua's longest-standing separatist movement (the OPM) met in Sentani and openly discussed a strategy to take West Papua towards independence. This meeting rejected the 1969 Act of Free Choice as fraudulent and illegal.

9 Throughout these developments, Jakarta has consistently opposed independence. In 2001, pro-independence leader Theys Eluay, then President of the Papuan Council, was murdered by Indonesian troops after holding negotiations with Indonesian authorities for the independence of West Papua. Two years later, the assassins were found guilty by an Indonesian military court but received light sentences. In 2003, seven military officers were expelled from the army and received prison sentences ranging from two to three and a half years for the death of Eluay.

10 In June 2006, in a demonstration against mining companies in Abepura, five police officers and a demonstrator were killed. Indonesian police immediately raided student dormitories and vehicles; 57 people, mostly students, were arrested and - according to Papuan sources - tortured. Sixteen students were put on trial, facing charges of murder and subversion.

11 In late July, the chair of the OPM Revolutionary Council, Nikolaus Ipo Hau, announced in an interview with Australian Radio that the OPM had agreed to give up armed struggle and 'give peace and talks with the army a chance'.

12 The Asian Development Bank (ADB), located in Manila, approved in August a 350-million dollar loan to finance, jointly with British Petroleum, a program to develop liquefied gas. The project included the export of gas to China, North Korea and the western coast of North America. According to the ADB, the environment of Papua, one of the planet's last virgin territories, 'would not be affected in the least' and the funds collected would be used for social development and to curb poverty. However, analysts and environmental activists believe the consequences could be catastrophic for the environment. ∎

PROFILE

ENVIRONMENT
Located in the Pacific Ocean directly north of Australia, West Papua, together with the eastern half of its island landmass, Papua New Guinea, comprises the second largest island in the world and contains rainforests second only in size to those of the Amazon. West Papua has an abundance of natural resource wealth such as oil, gold, copper and wood. While it comprises 21 per cent of the total landmass of Indonesia, it is home to only one per cent of its population.

SOCIETY
Peoples: The indigenous population is of Melanesian (South Pacific Islands) descent. Total population: 2.1 million (2.5 million)*. Indigenous: 62 per cent (60 per cent). Migrants and transmigrants born in other parts of Indonesia (called Javanese): 17 per cent (40 per cent). **Religions:** Reflecting an exposure to Dutch missionaries, Christianity is threaded through the belief system of many indigenous West Papuans. Indonesian migrants and transmigrants are predominantly Muslim. **Languages:** Bahasa Indonesia (official). There are 253 tribal languages. West Papua and its neighbour, Papua New Guinea, contain 15 per cent of all the world's known languages.

THE STATE
Official Name: Papua, formerly Irian Jaya (West Papua). **Capital:** Jayapura. **Government:** Effectively controlled by Indonesia. The Papua Council and its Presidium (Executive), which once acted as a figurehead for the independence movement, no longer has the backing of all the organizations advocating independence in West Papua. The local legislature, with a native Papuan upper house, has limited real power: it cannot propose legislation and has limited veto rights. Effective law-making power is retained in Jakarta where Indonesia's parliament retains control over revenue collection and distribution, the military and the police. **National Holiday:** West Papuans celebrate their Independence Day on 1 December. **Armed Forces:** The military presence has more than doubled in the last five years to 12,000.

(*) NOTE: The information from Indonesia often differs from that from West Papuans. Where there are discrepancies, the Indonesian version is given first, then the West Papuan version in brackets.

Yemen / Al Yaman

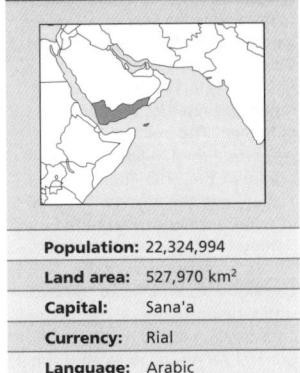

Population:	22,324,994
Land area:	527,970 km²
Capital:	Sana'a
Currency:	Rial
Language:	Arabic

G eographically located on the Indian, African and Mediterranean trade routes, Yemen was famous in antiquity because of its trade in incense, perfume and myrrh. Its main cities (Ma'in, Marib, Timna, and Najzan), stretched along the caravan routes that brought aromatic fragrances from Dhufar (presently part of Oman) and Punt (Somalia). These extensive trade routes continued along the coast of the Red Sea as far as the Mediterranean markets, and from Taima on towards Mesopotamia.

2   These cities were united in kingdoms, first Mina and later the better known Saba, cited in the Bible. In the year 20 BC the Romans, who had already conquered Egypt, unsuccessfully attempted to extend their dominion to Saba.

3   At the beginning of the 2nd century, the Greek geographer and mathematician Ptolemy started to refer to this zone as 'Happy Arabia' because of its vegetation (distinguishing it from the rest of the peninsula) and its wealth, resulting from trade.

4   In the 3rd century, the Kingdom of Saba fell under the control of the Himyarite dynasty (Kingdom of Himyar). A century later, Christian missionaries started arriving in Yemen.

5   The last Himyarite king was Jewish. He launched a violent persecution of the Christian community, leading to the intervention and occupation in 522 by the Ethiopian King of Aksum, who was a Christian. The Kings of Aksum spoke Greek and had been converted to Christianity in 333, establishing the traditional Christian foundations of the future Abyssinia (Ethiopia).

6   In 572, the Persians invaded Arabia, driving out the Ethiopians and turning Yemen into a Persian *satrapy* (province).

7   When Islamic domination reached the country in the 7th century, the country had suffered almost three centuries of conflict and invasions, resulting in the loss of its splendor. It was governed from Damascus by the Omeyas and later from Baghdad by the Abasids.

8   By the end of the 8th century the borders of the Arabian Empire reached from North Africa and Spain in the west, to Pakistan and Afghanistan in the east. Damascus in Syria became the capital of the empire, where the foundations of a new culture were laid. Greco-Roman, Persian and Indian components blended to form the new dominant culture, with the Arabs reaching high levels of scholarship and philosophy. The Arabs formed the social élite, the ruling class, though little changed in the lives of the Yemenis and other subject peoples.

9   In the 15th century the Portuguese arrived in Arabia, blocking trade routes in the Red Sea and controlling the spice route where Yemen held a strategic position. In 1516 they conquered Aden and established themselves there until 1538. With Aden in its power, Portugal controlled the entry to the Red Sea.

10   Then, in the 16th century, the Ottoman expansion started. The Turks occupied a few coastal spots on the Red Sea, leaving inland areas and the southern coast independent, governed by an *imam* (religious leader). In 1618, the British arrived and established the East India Company in the port of al-Mukha (*Mocha*: origin of the name for a type of coffee).

11   In the 19th century the British expanded their presence. As a consequence of Muhammad Ali's conquest of the country, the British occupied the entire extreme south-western region (see Egypt) and took Aden (in 1839), the best harbor in the region, to monitor Turkish activities. Meanwhile the Turks consolidated their inland dominion, which was finally achieved in 1872. Concessions were made to allow the imam to retain his position and also make the post hereditary rather than elective. This division set in train the process by which Yemen and the Yemeni people were split into two countries.

12   Towards 1870, with the opening of the Suez Canal and the consolidation of Turkish domination over the northern part of Yemen, the Aden settlement acquired new importance in British global strategy; it was a key port on the Red Sea and ultimately gave them access to the new canal.

13   In the early 20th century, Turkey and Britain made a border between their territories, which came to be called North and South Yemen, respectively. Friendship and protectorate treaties were gradually signed with local leaders, but it was a slow process, only completed in 1934 when the British gained control of the southern territory, as far as the border with Oman.

14   In 1911, Imam Yahya Hamid ad-Din led a nationalist rebellion; and two years later the Turks recognized his authority over the territory in exchange for the formal acceptance of Turkish sovereignty.

15   During World War I the Imam allied himself with the Ottoman Empire and remained loyal to it

LAND USE

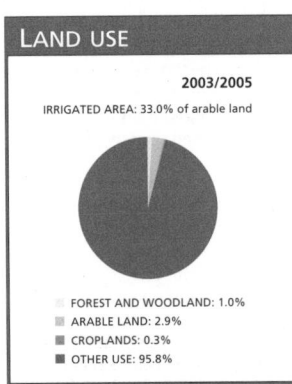

2003/2005

IRRIGATED AREA: 33.0% of arable land

- FOREST AND WOODLAND: 1.0%
- ARABLE LAND: 2.9%
- CROPLANDS: 0.3%
- OTHER USE: 95.8%

PUBLIC EXPENDITURE

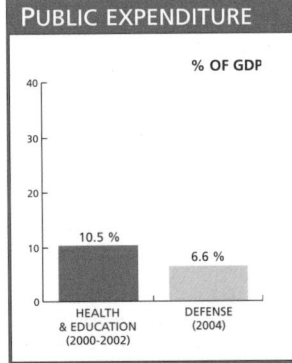

% OF GDP

10.5 % — HEALTH & EDUCATION (2000-2002)

6.6 % — DEFENSE (2004)

WORKERS

LABOR FORCE   2004

- FEMALE: 28%   ■ MALE: 72%

Life expectancy
63 years
2005-2010

GNI per capita
$550
2004

Literacy
49% total adult rate
2000-2004

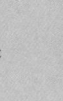

HIV prevalence rate
0.1% of population 15-49 years old
2003

until the end of the war. When the Empire broke up, Yemen recovered its independence (November 1918) and Hamid ad-Din was proclaimed King.

[16] In 1925, Britain recognized Yemen's independence and in the 1934 peace treaty, its sovereignty was guaranteed. Taking advantage of treaties with peoples in the surrounding areas, the British first turned Aden into a protectorate and in 1937, into a British colony.

[17] The birth of the nationalist Free Yemeni Movement in the mid-1940s was provoked by the autocratic rule of the imams. In 1945, North Yemen became a founding member of the Arab League and two years later it became a member of the United Nations (UN).

[18] There was an aborted uprising in 1948 in which Imam Yahya was killed, followed in 1955 by a coup against Imam Ahmad. In 1958,

six governors of South Yemen established the Federation of South Arabia, which by 1965 with British support, comprised the region's 17 states.

[19] That same year Imam Ahmad ash-Shams who ruled North Yemen, joined the United Arab Republic formed by Egypt and Syria, remaining in it until 1961. He was succeeded by his son Muhammad al-Badr, who was deposed by the Nasserist military in 1962, when the Yemen Arab Republic (YAR) was proclaimed under the leadership of Abdullah al-Sallal.

[20] The former Imam received both Saudi and British support, initiating a long civil war against the republican government that was backed by Egypt. A coup within the Republicans placed the moderate al-Iryani in power. Meanwhile, in South Yemen, the National Liberation Front, established in 1963, took the port

of Aden in 1967 and proclaimed independence, launching a socialist revolution.

[21] South Yemen became the People's Republic of Yemen, closed all the British bases in 1969, took control of the banking system, foreign trade and the shipbuilding industry and initiated an agrarian reform. Its foreign policy closely linked it to the USSR.

[22] In October 1972, despite ideological and political differences between North and South Yemen, al-Iryani signed a treaty with the revolutionary government of what was now called the People's Democratic Republic of Yemen (PDRY) for a future merger of the two states.

[23] This ran counter to Saudi strategy and in June 1974 Colonel Ibrahim al-Hamadi forced al-Iryani to resign and installed himself as ruler in Sana'a. Although initially accepted by Saudi's King Faisal, the young officer made a powerful enemy when he challenged the landlords of the north in an effort to centralize power. After surviving three assassination attempts, he was killed on 11 October 1977 together with his brother.

[24] A junta led by Lt Colonel Ahmed al-Gashmi, with Prime Minister Aziz Abdel Ghani and Major Abdul al-Abdel Aalim, took power vowing to continue their predecessor's policies. Al-Gashmi was killed in a bomb attack in June 1978.

[25] In October 1978 the National Liberation Front (NLF) established the Yemen Socialist Party at a congress with considerable support from the population. In December the first general elections were held since independence, to appoint the 111 members of the People's Revolutionary Council.

[26] Saudi Arabia's constant hostility increased when it claimed parts of PDRY, precisely those areas where oil-fields had been discovered. Tension heightened with the increasing US military presence in Saudi Arabia.

[27] Major Ali Abdullah Saleh was appointed President of YAR in 1978 but was unable to prevent internal dissent leading to armed conflict in January 1979. The National Democratic Front, which included the nation's progressive sectors, was on the brink of taking power, so with Saudi provocation the conflict was diverted into a war with the South (PDRY). Syria, Iraq and Jordan intervened and their mediation led to a cease-fire and negotiations for the unification of the two Yemeni states, which had been suspended since 1972.

[28] In January 1986 civil war broke out in PDRY. The confrontation was brief but left 10,000 dead. Muhammad al-Hasani was ousted

and replaced by former Prime Minister Haydar Bakr al-Attas, who became elected president in October 1986.

[29] Finally, on 22 May 1990, the republics united as the Republic of Yemen. The political capital was established in Sana'a (former capital of the Arab Republic of Yemen) and the economic capital in Aden (former capital of the Democratic Republic of Yemen).

[30] In a joint session of the Legislative Assemblies of the two states, held in Aden, a Presidential Council was elected, made up of General Ali Abdullah Saleh (former president of YAR/North Yemen), Abdel Karim Abdullah al-Arashi, Salem Saleh Mohammed and Abdul Aziz Abdel Ghani. The Council elected Ali Abdullah Saleh as president of the united Republic.

[31] In May 1991, the Constitution was ratified in a national referendum: an overwhelming majority voted for freedom of expression and political pluralism. Islamic fundamentalist groups opposed to unification called for a boycott, finding the absence of *shari'a* (Islamic law) unacceptable, and also the introduction of voting rights for women.

[32] A few months after its installation, the provisional Government of the Republic of Yemen protested over the presence of foreign armies, massing in Saudi Arabia to prevent the invasion of Kuwait. In retaliation, the Saudi Arabian Government expelled 850,000 Yemeni migrant workers, whose return worsened the state of the national economy.

[33] In March 1993, Ali Abdullah Saleh's General People's Congress (GPC) won the parliamentary elections.

[34] In order to weaken Yemen, seen as a 'bad example' by the region's monarchies, Saudi Arabia supported the fight for secession led by vice-president Ali al-Beidh. In May 1994, secessionists proclaimed a southern Yemen democratic republic, but they were defeated by forces loyal to the Government. In July, the council of ministers adopted a plan of general amnesty in order to protect political pluralism. In September, Socialist Party members were forced to leave the Government, while the Islamist Yemeni Congregation for Reform (Islah) obtained six new places in the cabinet. The Constitution was modified to make shari'a law the basis of all Yemeni legislation.

[35] In February 1995, 11 parties formed a new alliance, the Democratic Opposition Coalition. The Government signed a draft agreement with Saudi Arabia in which both states expressed their will to set permanent common borders and promote bilateral relations.

PROFILE

ENVIRONMENT

The Republic of Yemen is formed by the union of the People's Democratic Republic of Yemen (South) and the Yemen Arab Republic (North). The north has the most fertile lands of the Arabian Peninsula. For that reason the country, together with the Hadhramaut Valley, used to be called 'Happy Arabia'. Beyond a semi-desert coastal strip along the Red Sea, lies a more humid mountainous region where the agricultural lands are found (sorghum is grown for internal consumption and cotton for export). The traditional coffee crop has been replaced by *qat*, a narcotic herb. The climate is tropical with high temperatures especially in Tihmah - where rainfall is heavy - and in the eastern region. The country has no mineral resources. The southern territory is dry, mountainous and lacks permanent rivers. Two-thirds of the land area is either desert or semi-desert. Agriculture is concentrated in the valleys and oases (1.2 per cent of the country). Fishing is an important commercial activity. Yemen's territory includes the island of Socotra, which is important strategically because of its location at the entrance to the Gulf of Aden. This island, which became part of South Yemen in 1967, has 17,000 inhabitants spread over 3,626 sq km.

SOCIETY

Peoples: Nearly all Arab. A small Persian minority lives along the coast.
Religions: Islam, official (Shi'a, 53 per cent and Sunni, 47 per cent).
Languages: Arabic, official.
Main Political Parties: General People's Congress (GPC); Yemeni Congregation for Reform (Islah); Yemeni Socialist Party.
Main Social Organizations: General Federation of Yemen Workers' Trade Unions, Association of Yemeni Students.

THE STATE

Official Name: al-Jumhuriya al-Yamaniya.
Administrative Divisions: 16 Provinces.
Political Capital: Sana'a 1,747,627 people (2004).
Economic Center: Aden.
Other Cities: Aden 562,000 people; Al-Hudaydah (Hodeida) 246,000; Ta'izz 290,107 (1995).
Government: Ali Abdullah Saleh, President since May 1990, re-elected in 1994 and 1999. Abdul Kader Bajammal, Prime Minister since March 2001. Bicameral Legislature (created by constitutional amendment in 2001): Shura Council, with 111 members appointed by the president; House of Representatives, with 301 members elected by popular vote.
National Holiday: 22 May, unification (1990).
Armed Forces: 67,000 (2003).

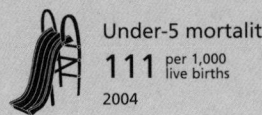

Under-5 mortality	Poverty	Debt service	Maternal mortality
111 per 1,000 live births 2004	**15.7%** of population living on less than $1 per day 1998	**3.5%** exports of goods and services 2004	**570** per 100,000 live births 2000

36   The landing of Eritrean forces on the Hanish Islands in the Red Sea in December led to a war. In March 1996, Yemen and Eritrea accepted international arbitration to resolve the conflict.

37   In 1997, the GPC won the parliamentary elections and on 15 May the new Prime Minister, Faraj Said ibn Ghanem took office.

38   That year the Government implemented a structural adjustment plan to revive an economy suffering the effects of civil war and the reduction in foreign aid. The privatizations earned the Government funding from the World Bank and the International Monetary Fund.

39   Abdul Karim al-Iryani took over as Prime Minister in May 1998. A Saudi attack on a Yemeni detachment on the island of Duwaima, in the Red Sea, in July, further worsened relations between the two neighbors. In October the international court in The Hague ruled in favor of Yemeni ownership of the Hanish islands, also claimed by Eritrea.

40   The September 1999 elections gave a clear victory to Ali Abdullah Saleh, who was re-elected with 96.3 per cent of the vote.

41   The abduction of 28 tourists - mostly British - in December 1998 raised awareness about the common practice of this method by some clans to get government attention and force it to give in to local demands. In the 1990s some 200 foreigners were kidnapped. In February 2000, a new law was introduced making the abduction of foreigners a capital offence.

42   In October 2000, 17 US soldiers were killed in Aden in a suicide attack on the US warship USS Cole. The radical Islamist Osama bin Laden, a Saudi of Yemeni descent, was blamed for masterminding the attack. That same month, a bomb exploded at the British embassy. Four Yemenis were imprisoned after declaring they had carried out the attack in solidarity with the Palestinian cause.

43   Violence marked the municipal elections of February 2001, which were accompanied by a referendum on extending the presidential term to seven years. Clashes among factions and with the police, as a result of the vote recount, left 30 people dead. The recount resulted in approval for the constitutional reform and Saleh's term in office was extended.

44   In May, Wahiba Fare took over the human rights portfolio and became the first woman in Yemen history to head a ministry.

45   In November, Saleh traveled to Washington to reassure his US counterpart that Yemen would participate in the coalition of nations that Washington was

IN FOCUS

ENVIRONMENTAL CHALLENGES
Fresh water sources are very limited, the use of groundwater beyond its replacement capacity has led to a drop in its level. Extensive grazing has caused soil erosion and very acute desertification.

WOMEN'S RIGHTS
Women have been able to vote and stand for election since 1967. In 2003, only one per cent of seats in Parliament and three per cent of ministerial positions were held by women. That year, they made up 29 per cent of the country's total labor force of six million. In spite of progress made, inequality in education between men and women remained a serious problem. In 2004, female primary school attendance barely amounted to 41 per cent, while the male rate was 68 per cent. Only 41 per cent of pregnant women received prenatal care and 27 per cent of births were attended by skilled health staff. The maternal mortality rate was 570 per 100,000 births. In spite of being against the law, female genital mutilation was still practised and 22 per cent

of Yemeni women had undergone it*.

CHILDREN
Notwithstanding the decline in child mortality rates between 1990 and 2004, they remained high. The infant mortality rate fell from 98 to 82 per 1,000 live births and the under-five mortality rate was reduced from 142 to 111 per 1,000*. In 2004, 32 per cent of newborns had low birth weight; 15 per cent of under-fives suffered from moderate or severe low weight; 12 per cent showed moderate or severe emaciation and 53 per cent were moderately or severely stunted. In 2005, a major vaccination campaign aimed at five million children was launched in an attempt to stop a polio outbreak in West Africa (just across the Gulf of Aden).

INDIGENOUS PEOPLES/ ETHNIC MINORITIES
Yemen is a Muslim country with both *Sunni* and *Shi'a*. Most people are Sunni from the *Shafi'i* school (an orthodox branch of Muslim jurisprudence; its name comes from *ash-Shafi'i*, an Islamic jurist who lived between 767 and 820), while the Zaydi minority, living in the northern provinces around *Sa'da*, is *Shi'a*.

There is also a small group of Ismailites (a *Shi'a* sect), representing around 1 per cent of the population. Differences between the various religious factions are based on different interpretations of the *sharia* or Islamic law. There are small non-Muslim minorities: Christians (around 4 per cent), Hindus and Jews, who remained when their communities left the country after independence.

MIGRANTS/REFUGEES
In early 2006, Yemen continued to receive refugees and asylum-seekers from the Horn of Africa region, mostly Somalis. Over 7,500 new arrivals were registered in the first half of 2005 alone. In early 2006, Yemen hosted 70,000 Somalis, 2,000 Ethiopians and nearly 1,900 refugees and asylum-seekers from other countries within the region.

DEATH PENALTY
It is applicable to all types of crimes.

* Latest data available in The State of the World's Children and Childinfo database, UNICEF, 2006.

organizing to fight terrorism. This alliance resulted in the detention of many Yemenis and foreigners in the following years.

46   Amnesty International reported that Yemen's war on terrorism had led to a failure to protect human rights. The organization deplored intimidation of journalists and widespread use of torture, in addition to the continuation of the death penalty, flogging and mutilation.

47   In March 2002, the Government expelled more than 100 Islamic students, including French and British citizens, as part of its campaign against terrorism. The US agreed to send military advisers to train Yemeni forces.

48   At least 5,000 people staged a protest in April, burning Israeli and US flags and demanding the closure of the US embassy, which had recently been the target of an explosion and a grenade attack.

49   In October 2002, the French oil tanker Limburg exploded off the coast of Yemen. To avoid any US intervention and ensure continuation of financial support, the Government promoted a campaign to capture members of the al-Qaeda network, led by bin Laden.

50   In November, a missile launched on Yemen by an

unpiloted aircraft killed six people, among them Salim Sinah al-Harethi, considered by Washington to be al-Qaeda's leader in the country. In May 2004, Yemeni security forces arrested another two suspected al-Qaeda leaders.

51   In May 2003 a court in Yemen sentenced alleged al-Qaeda member Abed Abdulrazzak to death for killing three US Christian missionaries in 2002.

52   Between June and August 2004, government troops battled members of the 'Faithful Youth' Zaidi sect - followers of the dissident Shi'a cleric Hussein al-Houthi - in the north of the country. Zaidis - members of a moderate Shi'a Muslim sect - are a majority in the northern region although they represent a minority in what is a mainly Sunni country. According to different estimates, confrontations left between 80 and 600 dead.

53   In August 2004, 15 men were sentenced on terror charges, including bombing the Limburg tanker in 2002. One of the defendants was sentenced to death for killing a police officer while the rest received jail terms of up to ten years. Most of the defense lawyers boycotted the proceedings, which they deemed unfair.

54   Authorities announced in

September that Hussein al-Houthi had been killed by government forces. More than 200 people were killed between March and April 2005 in a resurgence of fighting in the north. President Saleh accused the Al-Haq and the Public Powers Union (PPU) parties of forming military wings to overthrow the Republican regime, in what he called a 'foreign conspiracy' that included Iran. In May, Saleh stated that the new leader of the rebellion in the north had agreed to renounce the campaign in return for a pardon. Notwithstanding this, minor clashes continued.

55   In July, at least 36 people were killed across the country in clashes between police and demonstrators protesting against a rise in fuel prices.

56   In May 2006, Saudi-born Muhammad Hamdi al-Ahdal, suspected of being al-Qaeda's number two in Yemen, was sentenced to three years and one month in prison, charged with 'being part of an armed gang to attack foreign interests' and 'financing criminal acts'.

57   In September 2006, President Saleh was elected to another seven-year term with 77 per cent of the vote in an election international monitors concluded was fair. ■

Zambia / Zambia

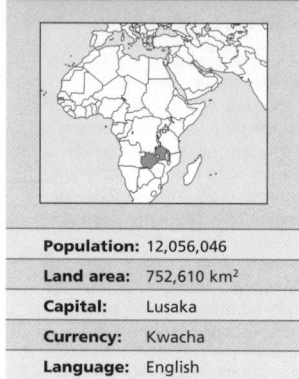

Population:	12,056,046
Land area:	752,610 km²
Capital:	Lusaka
Currency:	Kwacha
Language:	English

Evidence of *Homo sapiens rhodesiensis* going back 100,000 years has been found near Broken Hill, Kabwe. Like many African countries whose boundaries were demarcated in the colonial period, modern-day Zambia includes the descendants of many different peoples, including San/Bushmen. Migrants from the north - possibly Bantu-speaking groups - settled in the region with their agriculture, use of metal, pottery and domesticated animals in the 1st millennium AD. Over time, the settled Bantu-speakers displaced some of the hunting and gathering peoples. Around the year 1,000, a copper currency was in use. Burial objects, including gold ornaments, signal the existence of social hierarchies towards the 14th century.

[2] Waves of Bantu-speaking immigrants - the Luba and Lunda peoples - arrived from today's DR Congo and Angola in the 18th century. By the 19th century, before colonization, they - like the Bemba who settled in the northeast and the Lozi (Barotse) along the Zambezi river - formed powerful centralized state structures. In the 19th century, Portuguese attempts to establish authority over the area were resisted, but many communities were disrupted by Arab and Mozambican slave traders.

[3] In1851, David Livingstone, a British missionary and explorer, reached the Victoria Falls. He opened the way for merchants and adventurers. Cecil Rhodes and his British-South Africa Company (BSA) were looking to expand northwards from South Africa, and established a mining and trade monopoly in the area in 1889. A year later, Rhodes signed a treaty with the Lozi ruler Lewanika and the resulting protectorate was soon transformed through colonial domination into Northern Rhodesia.

[4] The BSA protectorate won support from the British because it prevented the Portuguese from linking their colonies of Angola and Mozambique. The BSA's prime concern in Northern Rhodesia was to exploit the rich copper deposits in the north. By 1909 a railway line was built linking the copper belt with the capital, Lusaka, and the southern African network. In 1924, Britain assumed direct control of the region. South African and US mining companies increased their investment in copper. The miserable working conditions of black miners led to the formation of labor unions, which gave rise to the first independence movements, such as the Northern Rhodesia African National Congress (NRANC).

[5] In 1952, primary school teacher Kenneth Kaunda became NRANC Secretary-General. In 1953, the British created a federation of Northern Rhodesia (now Zambia), Southern Rhodesia (now Zimbabwe) and Nyasaland (now Malawi). Zambia's nationalists felt that their country's copper wealth was being used to underpin the settler-controlled federation, and they campaigned for it to be dismantled. They fought for independence and against racial discrimination. Kaunda broke away from the NRANC to form the Zambia African Congress (ZAC).

[6] The ZAC refused to cooperate with the British on a gradual transfer of power. It was outlawed and Kaunda was arrested in 1959. The ZAC became the United National Independence Party (UNIP), which Kaunda chaired when he was released in 1960. As the Party gained widespread support, violence broke out in 1961.

[7] One year later, the British announced a constitutional review. The nationalist platform was thus strengthened and succeeded in dissolving the Federation in 1964. The UNIP won the elections and proclaimed independence in October 1964.

[8] Independent Zambia had one of the largest mining sectors in Africa at this time and was becoming one of the most urbanized countries. It nationalized its copper reserves, was a founding member of the Organization of Copper Exporting Countries (OCEC), and hosted the 3rd Summit Meeting of the Non-Aligned Countries Movement in 1970. It actively supported liberation movements in neighboring countries.

[9] As a land-locked country, trapped in a railway network controlled by the white minority governments of South Africa and Rhodesia, Zambia (together with Tanzania) accepted Chinese aid to construct a major new line. In 1974, with the inauguration of the Tan-Zam Railway, Zambia gained access to Indian ocean ports. In the same year, the Portuguese colonial-fascist regime was overthrown and Zambia's neighbors Angola and Mozambique embarked on their paths to independence.

[10] By the late 1970s Zambia's economic fortunes had taken a downturn. The OPEC oil crisis hit the country hard and copper prices

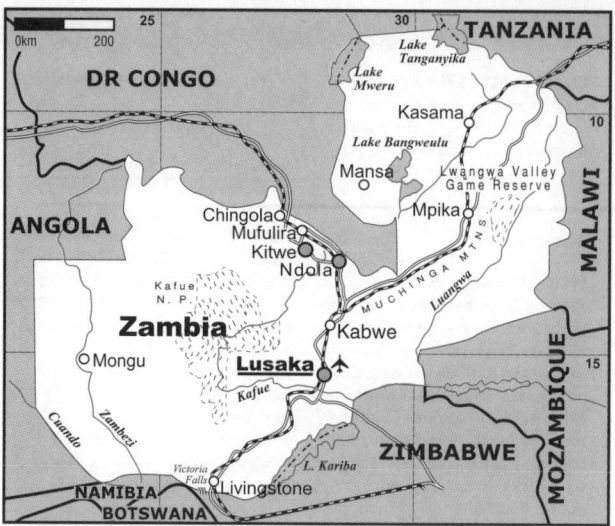

LAND USE

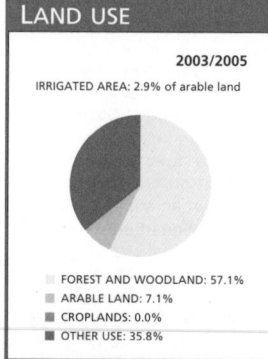

2003/2005

IRRIGATED AREA: 2.9% of arable land

- FOREST AND WOODLAND: 57.1%
- ARABLE LAND: 7.1%
- CROPLANDS: 0.0%
- OTHER USE: 35.8%

WORKERS

LABOR FORCE	2004

- FEMALE: 42%
- MALE: 58%

PROFILE

ENVIRONMENT
A high plateau extends from Malawi in the east to the swamp region along the border with Angola in the west. The Zambezi River flows from north to south and provides hydroelectric power at the Kariba Dam. The climate is tropical, tempered by altitude. Mining is the main economic activity, as there are large copper deposits. Shanty towns account for 45 per cent of housing in Lusaka, with attendant problems like the lack of drinking water and adequate health care, factors which contributed to the 1990 and 1991 cholera epidemics.

SOCIETY
Peoples: 98 per cent of Zambians are descendants of Bantu-speaking migrants, in some 70 ethnic groups. There are about 15,000 Europeans and a small Asian population.
Religions: Traditional African religions are practiced. There are Christian and Muslim and Hindu minorities.
Languages: English (official). The 70 or so local languages include Bemba, Kaonda, Lozi, Lunda, Luvale, Nyanja and Tonga.
Main Political Parties: Movement for Multiparty Democracy (MMD); United Party for National Development (UPND); Forum for Democracy and Development (FDD); United National Independence Party (UNIP).
Main Social Organizations: The Trade Union Congress of Zambia comprises 16 unions. The newly-created Workers' Trade Union has two million members.

THE STATE
Official Name: Republic of Zambia.
Capital: Lusaka 1,394,000 people (2003).
Other Cities: Ndola 346,500 people; Kitwe 762,700; Mufulira 130,400 (2000).
Government: Levy Mwanawasa, President since January 2002. Unicameral Legislature: National Assembly, with 159 members.
National Holiday: 24 October, Independence Day (1964).
Armed Forces: 18,000 (2003).

Life expectancy	GNI per capita	Literacy	HIV prevalence rate
39 years 2005-2010	**$400** 2004	**68%** total adult rate 2000-2004	**16.5%** of population 15-49 years old 2003

IN FOCUS

ENVIRONMENTAL CHALLENGES
Wildlife is threatened by poaching, as well as by the lack of resources with which to maintain protected areas. Mining has damaged the environment. Soil erosion and loss of fertility are associated with the overuse of fertilizers. Air pollution results in acid rain in mining areas.

WOMEN'S RIGHTS
Women have been able to vote since 1962. In 2003, they held 12 per cent of seats in Parliament and 25 per cent of ministerial positions. They made up 43 per cent of the country's total labor force of 4 million. In 2005, the first lady, Maureen Mwanawasa, founded an NGO aimed at improving social care, health and education for children, with special emphasis on increasing the participation of girls in the education system. In 2004, in spite of the fact that 93 per cent of pregnant women received prenatal care, only 43 per cent of births were attended by skilled health staff.

The maternal mortality rate was 750 per 100,000 births. At the end of 2003, it was estimated that 470,000 women aged between 15 and 49 were living with HIV/AIDS.

CHILDREN
Infant and under-five mortality rates remained almost unchanged (with a slight increase in both cases) over the 1990-2004 period. The former rose from 101 to 102 per 1,000 live births and the latter from 180 to 182 per 1,000. In 2004, 12 per cent of newborn babies were underweight and 49 per cent of children under five were severely or moderately stunted. It was estimated that some 85,000 children under 15 were living with HIV/AIDS and 710,000 children under 18 had been orphaned by the disease.

INDIGENOUS PEOPLES/ ETHNIC MINORITIES
Zambia, like other African nations, is ethnically very diverse. The Bemba and the Nyanja are the main peoples.

The Kavango people, living mainly near the borders with

Namibia and Angola, number around 10,000 in Zambia. They are cultivators and herders who practise traditional religions.

The Lozi/Barotse (nearly 600,000) are located in the western and southern provinces and practise the traditional religion. They have a complex hereditary system (known as Hawaiian) which combines the matrilineal and patrilineal systems.

The Luba people (254,000) live in northwestern Zambia. They are the descendants of a powerful empire which dates back to the 15th century. Their close neighbors are the Luchazi (68,000) who combine Christianity with traditional religions. The Tabwa peoples are in the north and practise Islam, Christianity and traditional religions. They comprise some 17,000 descendants of several ethnic groups who migrated centuries ago to Central Africa.

The Chokwe peoples are also located in the northwestern province and have a population of 56,000. They are descendants of the Mbundi and Mbuti pygmies, who emigrated from the upper reaches of the Kasai river towards

1600. They practise traditional religion.

MIGRANTS/REFUGEES
In early 2006, Zambia hosted some 138,000 refugees (mostly from Angola and the Great Lakes region) who were distributed among five refugee camps and several urban settlements (mainly in Lusaka). It was estimated that nearly 71,000 refugees were from DR Congo, 48,800 from Angola, 5,750 from Rwanda and just over 13,000 from other countries within the region. All this in spite of the fact that by October 2005, the UNHCR had managed to repatriate more than 20,000 Angolans (a third of them had voluntarily returned to their country).

DEATH PENALTY
Capital punishment is still in force.

* Latest data available in The State of the World's Children and Childinfo database, UNICEF, 2006.

---

fell. Zimbabwean independence in 1980 was welcomed by Zambia. However, white-ruled South Africa (SA) punished Kaunda's support for his neighbor by boycotting Zambia. SA also sent in troops because Zambia sheltered members from the liberation movements of Namibia (SWAPO) and South Africa (ANC).

[11] In October 1980, after a failed coup attempt, which was supported by South Africa, the Government declared a state of emergency and imposed a curfew. Government officials, business people and foreigners were arrested. After a long trial, seven were hanged.

[12] Kaunda surrounded himself with former comrades-at-arms. His critics questioned the need for a state of emergency and the one-party system. Allegations of government corruption grew. The tension diminished after the 1983 elections, in which President Kaunda won 93 per cent of the vote.

[13] Kaunda was re-elected in 1988. But in the 1991 elections, which were brought forward two years because of economic crisis and followed a constitutional reform establishing a multiparty system, Kaunda was defeated by Frederick Chiluba, of the Movement for Multiparty Democracy (MMD). Chiluba, a former union leader, won 81 per cent of the vote and 125 of the 150 seats in Parliament. Kaunda resigned as UNIP party leader in 1992.

[14] Chiluba declared Christianity the official religion and banned the creation of an Islamic fundamentalist party in spite of there being more than two million Muslim Zambians (both Shi'as and Sunnis) in the country.

[15] Accusations of corruption and the agricultural crisis led the President to request his Lands Minister to step down and the other ministers to declare their incomes. Shortly afterwards, he removed the director of the Bank of Zambia for suddenly devaluing the currency, the kwacha, by more than 20 percent. Chiluba blamed the crisis on foreign debt payments, which consumed 40 per cent of the country's GDP.

[16] In March 1996 the Paris Club cancelled 67 per cent of Zambia's debt. The IMF/World Bank structural adjustment program in 1997 increased rural poverty levels and at least 150,000 workers lost their jobs as a result of the privatization of state enterprises.

[17] In November 1997, dozens of people - including Kaunda, disqualified from possible re-election by a constitutional amendment - were arrested for allegedly taking part in a failed coup. A state of emergency was imposed and later extended to March 1998. Kaunda was released from house arrest in June and the charges against him were withdrawn.

[18] In 1999, a high court sentenced

59 soldiers to death after they were found guilty of treason for the failed coup attempt in 1997.

[19] In May 2001, political divisions - resulting from the debate on whether Chiluba should run for a third term as President - prompted some 80 MMD members to create the Forum for Democracy and Development (FDD).

[20] In July 2001, Paul Tembo, previously Chiluba's campaign manager and a member of the FDD, was murdered shortly before he was to testify in a high-level corruption case.

[21] Also in July, Zambia issued an urgent appeal for food aid to feed some two million people in the wake of poor harvests caused by flooding and drought in different parts of the country.

[22] In January 2002, MMD candidate Levy Mwanawasa was declared President amid controversy arising from the close presidential and parliamentary elections of December 2001. Ten opposition parties protested over alleged fraud. Mwanawasa won 28.7 per cent of the vote, while Anderson Mazoka, of the United Party of National Development (UPND), won 26.7 per cent.

[23] Parliament, encouraged by President Mwanawasa, voted to lift former president Chiluba's immunity from prosecution in July 2002.

[24] In February 2003, former president Chiluba was arrested and

charged on 59 counts, including abuse of office and the theft of $30 million. In December, he went on trial on corruption charges. That same month, the Supreme Court confirmed the death sentences of 40 out of the 59 soldiers involved in the 1997 failed coup.

[25] In September 2004, many charges of corruption against Chiluba were dropped, but within hours he was re-arrested on six new charges.

[26] The Government admitted in January 2005 that a racket perpetrated by public officers and contractors had cost the state over $940 million.

[27] In April, the World Bank approved a $3.8 billion debt relief package which implied the cancellation of more than 50 per cent of Zambia's external debt.

[28] In November, Mwanawasa declared a national disaster in the country as a result of drought and appealed for international aid to help nearly one million Zambians who faced food shortages.

[29] In June 2006, the Minister of Tourism and Environment warned that the country was losing 800 hectares of forest every year due to excessive logging.

[30] In October 2006, President Mwanawasa once again scored a narrow victory in a disputed election. Analysts said he had benefited from a huge rise in the price of copper - Zambia's main export. ■

Zimbabwe / Zimbabwe

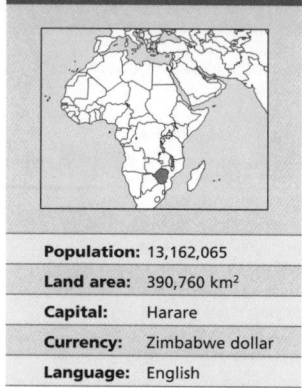

Population:	13,162,065
Land area:	390,760 km²
Capital:	Harare
Currency:	Zimbabwe dollar
Language:	English

In the region known today as Zimbabwe there are many signs of ancient African civilizations, including mines, irrigation and terracing. Among the major archaeological sites are Mapungubwe, and Great Zimbabwe, with its monumental walled enclosure. The ancestors of today's Shona people, Bantu-speaking iron workers who had settled in the region before the 5th century, built these walls in later centuries.

2   They discovered gold, copper and tin deposits, and developed sophisticated techniques for working these metals. They traded with Arab-influenced centers on the coast, such as Sofala in present-day Mozambique, and this trade facilitated in an expansion of their culture. The Mutapas, or kings, extended their influence over most of the region.

3   The height of Great Zimbabwe was 14th-16th century. This civilization established trading connections as far as Asia. By the time the Portuguese conquered the coastal settlements in the 16th century, Great Zimbabwe was in decline and the center of gravity of a more developed Zimbabwean culture moved northwards under the Rozvi kings. Khami, in the south, was another major center after the fall of Great Zimbabwe.

4   Shona society was deeply disrupted in the1830s by the invasion of the Ndebele from Zululand (in what is now South Africa), escaping the military might of Zulu King Shaka.

5   The Zulu-speaking Ndebele established a kingdom in the southeast by conquering and incorporating the local, mainly Shona people. In the first half of the 19th century, the territory was divided between the Shona people in the northeast, and the Ndebele. When white settlers arrived in the late 19th century they negotiated with Lobengula, the Ndebele king. He granted exclusive rights for the exploitation of the country's mineral resources and land to imperial entrepreneur Cecil Rhodes' British-South Africa Company (BSA) in exchange for money.

6   Britain authorized the BSA to take control over the territory and open it for settlers, and a 'pioneer column' moved into a fortified camp called Salisbury.

7   In 1893, the Ndebele rebelled. Rhodes' 'police' attacked Lobengula's capital Bulawayo and destroyed it.

8   In 1896-7 a major uprising by Shona people was brutally repressed and the country (Southern Rhodesia) came under the secure control of the BSA.

9   In 1923 it became a settler-run state which emulated South Africa in many aspects of racial segregation. Its intensive agriculture and gold mines made it the second richest country in colonial Africa.

10   In 1953 Southern Rhodesia formed a Central African Federation with its neighbors Northern Rhodesia (now Zambia) and Nyasaland (now Malawi). Through the Federation the British hoped to create a counterweight to Afrikaaner-dominated apartheid in South Africa.

11   When the decolonization process began in Africa in the late 1950s, Europeans (five per cent of the population) owned around half the land. Zambia and Malawi gained independence in 1964. In Southern Rhodesia, the African National Congress (ANC) also intensified the struggle for self-determination. The settler government of Ian Smith refused to countenance black rule or British attempts to compromise and announced a unilateral declaration of independence in 1965. This provoked the turn to armed struggle by the African liberation movements ZAPU (Zimbabwe African Peoples Union, led by Joshua Nkomo) and ZANU (Zimbabwe African National Union, led by Robert Mugabe from 1975).

12   Smith's rebel Rhodesian Front regime faced a UN embargo, though the blockade was systematically violated with the co-operation of the white government in South Africa. Following Mozambique's independence in 1975, the armed struggle was stepped up and Zimbabwean guerrillas penetrated from camps outside the country. Smith retaliated against Zambia and Mozambique. These countries together with Angola, Botswana and Tanzania, had suffered significant political and economic destabilization for having formed a common front against the region's white regimes.

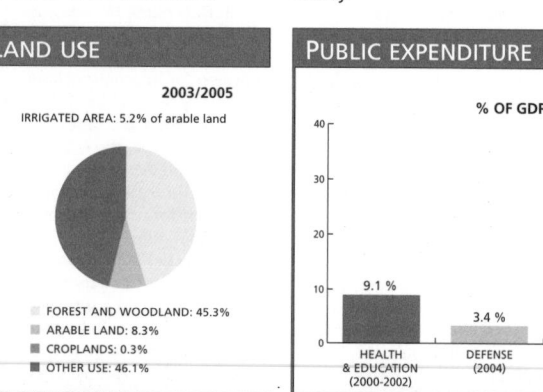

LAND USE

2003/2005

IRRIGATED AREA: 5.2% of arable land

- FOREST AND WOODLAND: 45.3%
- ARABLE LAND: 8.3%
- CROPLANDS: 0.3%
- OTHER USE: 46.1%

PUBLIC EXPENDITURE

% OF GDP

9.1 %
HEALTH & EDUCATION (2000-2002)

3.4 %
DEFENSE (2004)

WORKERS

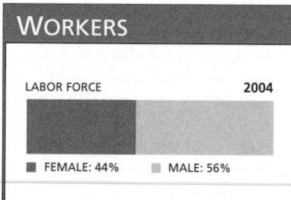

LABOR FORCE   2004

- FEMALE: 44%   MALE: 56%

Life expectancy
37 years
2005-2010

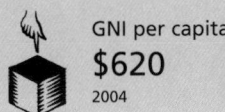
GNI per capita
$620
2004

Literacy
90% total adult rate
2000-2004

HIV prevalence rate
24.6% of population 15-49 years old
2003

13   Under increasing international pressure, in1978 Smith and some African leaders signed an internal settlement. However, this was unacceptable to the liberation movements whose success forced the Government to negotiate. London supervised the February 1980 free elections, won by Robert Mugabe. In line with the April 1980 Lancaster House agreements, Britain transferred power to ZANU. The white Zimbabweans retained some privileges, including their land and protected seats in Parliament.

14   Mugabe began by abolishing racist legislation and reconstructing the economy, affected by seven years of war, which had reduced cattle to one third, devastated roads, and exacerbated diseases such as malaria. Mugabe advocated reconciliation and his cabinet included ZAPU and white leaders. He set an ambitious Development Plan and in the first few years there was rapid growth, especially in farm production. However, Mugabe was hampered by South Africa's blockade on Zimbabwe's agricultural exports and political dissent between ZANU and Nkomo's ZAPU, whose mainly Ndebele former guerrillas were unhappy with ZANU and Shona dominance. A serious drought in 1983 was an additional setback.

15   African farmers' hopes for a true agrarian reform clashed with the limitations imposed by the Lancaster House agreement, which ensured that white land had to be purchased (not expropriated) by the Government.

16   In late June 1985, Mugabe's ZANU obtained a comfortable victory in parliamentary elections throughout the country, except in the Ndebele area of Matabeleland. Most whites voted for the Rhodesian Front.

17   In 1986, 4,500 farmers (most of them white) still owned 50 per cent of the country's most productive land, employing black workers who lived on the farms. Over four million Africans were squeezed onto communally owned tribal trust lands, which were poor, situated on dry regions and lacking infrastructure and communications.

18   The (white) Commercial Farmers Union generated 90 per cent of all agricultural production, paid a third of Zimbabwe's wages and exported 40 per cent of the country's goods. The organization blocked many Government land reform and redistribution initiatives.

19   Two constitutional reforms were enacted in September 1987, through which the 30 parliamentary seats protected for whites were abolished and executive authority was transferred to the president, elected by Parliament for a six-year period.

20   In December1987, Mugabe and Nkomo reached a unification agreement (ratified in April 1988) creating the Patriotic Front of the Zimbabwe African National Union (ZANU-PF).

21   In the March 1990 elections, ZANU-PF won 116 of the 119 parliamentary seats. Mugabe interpreted this result as the people's support for a one-party system. However, there was only a 54 per cent turnout and the newly formed opposition Zimbabwe Unity Movement (ZUM) obtained 15 per cent of the vote.

22   In 1990, Parliament approved a land reform law authorizing the Government to expropriate land held by whites at a price fixed by the State, and to redistribute it among black Zimbabweans. Zanu-PF moved away from Marxist-Leninist doctrine, maintaining a social democratic and mixed economy in which whites had an important role.

23   In 1992, the Mozambican conflict came to an end which removed the threat of destabilization for Zimbabwe. The April 1996 elections (with an abstention rate of 68 per cent) gave Mugabe the victory with 93 per cent of the votes.

24   IMF-imposed structural adjustment programs cutting public spending, and economic stagnation, made it increasingly difficult for the Government to provide its supporters with basic services. Although about 30 per cent of white-owned land had been redistributed, pressure mounted from the war veterans who felt they had been neglected since Independence. In June 1996 they forcibly took over six farms east of Harare. In July the Government announced a new plan to settle 100,000 families on five million hectares to be purchased from white farmers. The IMF and EU, although recognizing the need for land reform, rejected the plan, saying it was too ambitious, and agreed only to support a two-year pilot project. Mugabe changed his mind yet again in November, ordering the expropriation of 841 farms.

25   In October 1999 a special three per cent tax was imposed on salaries. The Government stated it would be used to treat people with HIV/AIDS (25 per cent of the country's adult population). The measure was seen by many as a scheme to finance the 11,000 troops that Mugabe sent to back the Laurent Kabila regime in the DR Congo. Corruption by the ruling elite was widely reported in the media.

26   Mugabe suffered a shock defeat by a new opposition party - the Movement for Democratic Change (MDC) - in a constitutional referendum held in February 2000. This had been designed to enhance presidential powers and allow for the confiscation of white farms without compensation.

27   In March 2000, 420 white-owned farms were occupied by black war veterans. The Commercial Farmers Union accused Mugabe of orchestrating the occupations to hide the failure of the land distribution projects. The Government owned nearly two million acres of uncultivated farmland. Mugabe responded that the country could not pay for the division of land into smaller plots or to provide minimal infrastructure, such as waterworks and roads.

28   In May, as a platform for the June parliamentary elections, Mugabe launched a manifesto stating that he would not give in to international pressure and that he would not reverse or prevent the land invasions.

Anxious to prevent foreign scrutiny, his government decided to slash the number of election observers. Despite allegedly rigging the elections, ZANU won only five more seats than the trade-union backed and largely urban opposition Movement for Democratic Change (MDC). In the elections, Mugabe encountered real opposition for the first time since 1980, and determined to suppress it.

29   In 2001, the Mugabe regime was accused of violating human rights, including the right to information. The British media and several NGOs reported that many judges and journalists had been forced to leave Zimbabwe and that several opposition leaders had been assassinated.

30   Renewed international pressure led Mugabe to slow the illegal land occupations in September 2001 in exchange for major financial aid from Britain to compensate the white farmers whose land had been expropriated. The reluctance of companies to invest in the country had fueled unemployment, and inflation reached 70 per cent by the end of the year.

31   The Government, which had censored the state-owned

PROFILE

ENVIRONMENT
The country consists mainly of a high rolling plateau. Most of the urban population live in the High Veld, an area of fertile land, with moderate rainfall and mineral wealth. The climate is tropical, tempered by altitude.

SOCIETY
Peoples: The majority of Zimbabweans (70 per cent) are of Bantu origin from the Shona (founders of the first nation in the region) and Ndebele (20 per cent) a Zulu-speaking people that arrived in the 19th century. Others include Venda, Shangaan, Tsonga and San (Bushmen).
Religions: African traditional beliefs (55 per cent); Christian (45 per cent).
Languages: English (official); Shona, Zulu and other languages.
Main Political Parties: Zimbabwe African National Union-Patriotic Front (ZANU-PF); Movement for Democratic Change (MDC).
Main Social Organizations: Organization of Rural Associations for Progress; Council of Trade Unions of Zimbabwe, National Students Union of Zimbabwe.

THE STATE
Official Name: Republic of Zimbabwe.
Administrative Divisions: Eight Provinces.
Capital: Harare 1,600,000 people (2006).
Other Cities: Bulawayo 794,600 people; Chitungwiza (outside Harare) 390,600; Mutare 168,100 (2000).
Government: Robert Mugabe, President since December 1987, re-elected in 1990, 1996 and 2002. Bicameral Parliament: the Senate, with 66 members (50 of them directly elected, six appointed by the president and 10 by traditional chiefs), and the House of Assembly, with 150 members.
National Holiday: 18 April, Independence Day (1980).
Armed Forces: 29,000 (2003).

Under-5 mortality
129 per 1,000 live births
2004

Poverty
56.1% of population living on less than $1 per day
2001

Maternal mortality
1,100 per 100,000 live births
2000

IN FOCUS

ENVIRONMENTAL CHALLENGES
There is deforestation, soil erosion and degradation, and air and water pollution. Poaching is decimating the herds of black rhinoceros (among the largest in the world). Small-scale mining produces heavy metal pollution and toxic waste.

WOMEN'S RIGHTS
Women have been able to vote since 1957 and run for office since 1978. In 2005 they held 16 per cent of Lower House seats and 32 per cent of Upper House seats in Parliament. In 2003, female workers comprised 44 per cent of the total labor force of 6 million.

A high incidence of HIV/AIDS affects Zimbabwean women. In late 2003, it was estimated that 930,000 women between 15 and 49 were living with HIV/AIDS. Although 93 per cent of pregnant women received prenatal healthcare and 73 per cent of births were attended by qualified personnel in 2004, maternal mortality stood at 1,100 deaths per 100,000 births.

CHILDREN
Infant and under-five mortality rates increased significantly in the 1990-2004 period. The former went from 53 to 79 per 1,000 live births and the latter from 80 to 129 per 1,000*. In 2004, 11 per cent of newborns were underweight, 13 per cent of children under five were moderately or seriously underweight and 27 per cent were moderately or seriously stunted. In late 2003 it was estimated that 120,000 children under 15 were living with HIV/AIDS and 980,000 children under 18 had been orphaned by the disease.

INDIGENOUS PEOPLES/ ETHNIC MINORITIES
Only about 10 per cent of the population are not Shona or Ndebele people. These other ethnic groups include Shangaan, Venda and Tsonga people. There are also some Tswa (San/Bushmen). Shona remain the largest population group (approximately 70 per cent), to which most politicians in power, including President Robert Mugabe, belong. Ndebeles from Matabeleland constitute approximately 20 per cent.

MIGRANTS/REFUGEES
In 2005 the country sheltered 10,800 refugees, of which 1,500 were in the Tongogara camp (in the southeast, near the border with Mozambique) and the rest were urban refugees. That year, the Government launched an 'operation to restore order' and destroyed settlements and huts belonging to urban refugees. As a result, approximately 1,000 were resettled in the Tongogara camp - which was overpopulated by the new arrivals - and 8,000 more were displaced. In early 2006, the Government sought to achieve the voluntary repatriation of the Rwandan refugees that remained in the country.

DEATH PENALTY
Capital punishment is still applied.

* Latest data available in The State of the World's Children *and* Childinfo *database, UNICEF, 2006.*

broadcast media, introduced a bill in January 2002 for an-Information Access and Privacy Protection Act. This sought to silence the independent press and foreign journalists before the upcoming presidential elections. National and international pressure led the Government to withdraw the bill - but it promptly returned an amended version to Parliament.

[32] Mugabe won the March 2002 presidential elections with 56.2 per cent of the vote, beating Morgan Tsvangirai of the opposition MDC, which won 41.9 per cent of votes. The opposition, independent election observers and the international community again disputed the result. Weeks after the elections, the Commonwealth expelled Zimbabwe for one year, following recommendations by electoral observers that the voting had been obstructed by government-inspired violence. The EU and the US announced that they would impose severe sanctions against Zimbabwe in 2002. Denmark closed its embassy in Harare. Like the EU, Switzerland froze the assets of some senior government figures and refused them entry.

[33] In August 2002, 3,000 white farmers were told to quit their lands and in September Parliament passed a series of laws to speed up the expropriation process, which was to end in a month's time. Some 2,500 farmers decided not to obey the order, defying the Government, and violence ensued. Inflation soared in response to the economic crisis. Morgan Tsvangirai was charged with treason, allegedly for plotting to overthrow the President.

[34] In March 2003, observers reported unprecedented levels of repression in the widespread confrontations over land expropriation. The trade unions and the MDC attempted to stage mass strikes and actions to pressure Mugabe to retire early. The protests met with severe government repression.

[35] Mugabe stepped up his rhetoric against Northern countries. The *Daily News*, the only independent daily, was closed down in October 2003. Many journalists have been imprisoned and beaten. In December 2003, the Commonwealth extended Zimbabwe's expulsion indefinitely.

[36] In March 2004, the EU renewed sanctions and extended its list of officials banned from entry to its states. In June, the World Association of Newspapers (WAN) condemned Zimbabwe's press laws.

[37] That same month, several Zimbabwean civil rights groups expressed discontent over a decision by the African Union (AU), which during the summit held in Ethiopia suspended publication of a report that denounced human rights abuses in the country, such as torture and arbitrary arrest of opposition members of Parliament and human rights lawyers. African leaders stated that the suspension was aimed at giving the country time to respond to the allegations.

[38] In October, Tsvangirai was declared innocent of treason in an alleged attempt to assassinate Mugabe, although he faced other charges of treason. Joyce Mujuru was sworn in as vice-president in December.

[39] In January 2005, Mugabe removed several top members of the ZANU-PF and jailed Philip Chiyangwa, one of the country's wealthiest men, accused of spying. That month, systematic attacks and tortures against MDC followers - previous to the March elections - were reported.

[40] Also in January the US described Zimbabwe as one of the six 'outposts of tyranny' in the world, along with Belarus, North Korea, Cuba, Iran and Myanmar.

[41] In March, ZANU-PF won legislative elections with two-thirds of the vote. Tsvangirai reported 'massive frauds' once again.

[42] In May-June, tens of thousands of shanties and illegal street vending posts were demolished in an 'urban clean up' operation that offered no alternative solutions to their owners. The UN estimated 700,000 people were left homeless.

[43] In August, pending treason charges against Tsvangirai were lifted.

[44] In November, ZANU-PF obtained an overwhelming majority in elections called to make up the new Senate, reinstated after six years of single-chamber Parliament.

[45] After a visit in December, Jan Egeland, UN Under Secretary-General for Humanitarian Affairs, said that Zimbabwe was in 'meltdown'. In April 2006, after seven years of economic recession, inflation had reached a yearly 1,043 per cent - the highest in the world - unemployment topped 70 per cent, and the World Food Program said that the survival of more than four million of the country's thirteen million inhabitants depended on food aid.

[46] The Famine Early Warning System (FEWSNET), an organization located in the US, warned in February 2006 that domestic access to basic food products had fallen considerably in the region and that a large number of people were in need of food aid. According to FEWSNET, food insecurity in Zimbabwe would get worse in a matter of months due to the 1,200 ton shortage in cereal production in 2005.

[47] The IMF issued in March its latest report regarding Zimbabwe's non-performing loans with the agency, as well as its possible sanctions on the country. The IMF's Executive Board decided not to restore Zimbabwe's voting right in the institution, nor the possibility to use the Fund's general resources.

[48] Although the Mugabe Administration blamed the crisis on the multilateral financial institutions, international analysts said the country's crisis was due simply to the government's 'constant economic failure'.

[49] In October 2006, women's organizations were outraged by an opposition parliamentarian who urged the national assembly not to pass a bill aimed at stamping out domestic violence, because women were inferior to men.

[50] In November 2006, a report by Human Rights Watch claimed that violent repression of civil-society organizations had intensified in Zimbabwe over the previous three years. The report documented alleged systematic abuses against rights activists, including excessive use of force by police during protests, arbitrary arrests and detention during 2006. ∎

Bibliography

AMARASINGAM, SP.- *The Industrialized Nations of the West and the Third World*. Colombo, Tribune, 1982.

AMIN, S., et al.- *Nuevo Orden Internacional*. México, Nueva Política, 1977.

AMNESTY INTERNATIONAL.- *Guía de la Carta Africana de los Derechos Humanos y de los Pueblos*. Madrid, 1991.

ATLASECO *du Monde 1997*. Paris, Les Editions EOC, 1996.

BARDINI, R.- *El Frente Polisario y la lucha del pueblo saharaui*. Tegucigalpa, CIPAAL, 1979.

BARRACLOUGH, G.- *Introducción a la Historia Contemporánea*. Barcelona, Anagrama, 1975.

BASSOLS BATALLA, A.- *Geografía, subdesarrollo y regionalización*. México, Nuestro Tiempo, 1976.

BASTIDE, R.- *Las Américas negras*. Madrid, Alianza, 1969.

BECKFORD, G.- *Persistent Poverty, Underdevelopment in Plantations of the Third World*. 2a.ed. London, 1983.

BEDJAQUI, M.- *Towards a new International Economic Order*. Paris, UNESCO, 1979.

BELFRACE, C & ARONSON, J.- *Something to Guard*. New York, Columbia Univ. Press, 1978.

BELLER, WS., ed.- *Environmental and Economic Growth in the Smaller Caribbean Island*. Washington, Department of State, 1979.

BENOT, I.- *Ideologías de las independencias africanas*. Barcelona, 1973.

BENZ, W.- *El siglo XX: Problemas mundiales entre los dos bloques de poder*. Madrid, Siglo XXI, 1984.

BIANCO , L.- *Asia contemporánea*. México, Siglo XXI, 1980.

BOUSTANI, R & FARGUES, P.- *Atlas du Monde Arabe. Géopolitique et Societé*. Paris, Bordas, 1990.

BRANDT, W.et al.- *Das Ueberleben sichern*. Kiepenheur & Witsch, Koln, 1980.

BRITTAIN, V., & SIMMONS, M.- *Third World Review*. London, Guardian, 1987.

BROWN, L.- *Food or Fuel: New Competition for the Worlds Cropland*. Washington, Worldwatch Institute, 1980.

BUETTNER, T., et al.- *Afrika: Geschichte von den Anfaengen bis zur Gegenwart*. Koln, Pahl-Rugenstein, 1979.

BURCHETT, WG.- *Otra vez Corea*. México, ERA, 1968.

BURGER, J.- *The Gaia Atlas of first peoples. A future for the indigenous world*. London, Gaia, 1990.

CALCAGNO, AE & JAKOBOWICZ, JM.- *El monólogo Norte-Sur y la explotación de los países subdesarrollados*. México, Siglo XXI, 1981.

CALDWELL, M.- *The Wealth of Some Nations*. London, Zed, 1977.

CASTRO, F.- *La crisis económica y social del mundo*. Havana, OPCE, 1983.

CAVALLA, A.- *Geopolítica y Seguridad Nacional en América*. México, UNAM, 1979.

CHALIAND, G.- *Revolution in the Third World*. Middlesex, Penguin, 1979.

CHEE, Y.- *How big powers dominate the Third World*. Penang, Third World Network, 1987.

CHOMSKY, N., STEELE, J. & GITTINGS, J. *Super Powers in Collision*. Penguin, 1984.

CHRISTENSEN, Ch.- *The Right to Food: how to guarantee*. New York, 1978.

CIPOLLA, C.- *La explosión demográfica* (entrevista). Barcelona, Salvat, 1973.

CLAIRMONTE, FF & CAVANAGH, H.- *Transnational Corporations and Services: The final frontier*. Geneva, UNCTAD, 1984.

CLARK, C.- *Crecimiento demográfico y utilización del suelo*. Madrid, 1968.

COLCHESTER, M & LOHMANN, L.- *The tropical forestry action plan; what progress?* 1990.

COMANDANTE *de los pobres: testimonios sobre Omar Torrijos*. Madrid, Centro de Estudios Torrijistas, 1984.

COMISION INTERNACIONAL SOBRE PROBLEMAS DE LA COMUNICACION.- *Un solo mundo, voces múltiples*. México, Fondo de Cultura, 1980.

CONNELL-SMITH, G.- *Los Estados Unidos y la América Latina*. México, 1977.

CORDOVA, A., et alii.- *El Imperialismo*. México, Nuestro Tiempo, 1979.

CORM, G.- *Le Proche-Orient éclaté*. Paris, La Découverte, 1983. Le COURIER des Pays de l'est. Paris, 1986.

DAWISHA, K.- *Eastern Europe, Gorbachev and reform: the great challenge*. 2a.ed. New York, Cambridge Univ., 1990.

DEBRAY, R.-*La crítica de la armas*. Madrid, Siglo XXI, 1975.

DECRAENE, P.- *El Panafricanismo*. Bs.As., EUDEBA, 1962.

DEGENHARDT, H, comp.- *Political dissention international guide to dissent, extraparliamentary, guerrillas and illegal political movements*. London, Longman, 1983.

DE LA COURT, T; PICK, D & NORDQUIST, D.-*The nuclear fix. A guide to nuclear activities in the Third World*. Amsterdam, Wise, 1982.

DENVERS, A.- *Points choc: l'environnement dans tous ses états*. Editions 1, 1990.

DER FISCHER WELT ALMANACH 1993. Fischer Taschenbuch Verlag. Bonn, 1992.

DIFRIERI, H., et al.- *Geografía Universal*. Bs.As., ANESA, 1971.

DORE, F.- *Los regímenes políticos en Asia*. México, Siglo XXI, 1976.

DRECHSLER, H.-*África del Sudoeste bajo la dominación colonial alemana*. Berlin, Verlag, 1986.

DREIFUSS, R.-*A Internacional capitalista*. Rio de Janeiro, Espaço e Tempo, 1986.

ECKHOLM, E.- *The dispossessed of the earth: Land reform and Sustainable Development*. Washington, Worldwatch Institute, 1979.

L'ECONOMIE *de la drogue*. Le Monde; Dossiers et Documents; Paris, feb.1990.

ELLIOT, F.- *A Dictionary of Politics*. Middlesex, Penguin, 1974.

EMMANUEL, A., AMIN, S, et al.- *Imperialismo y comercio Internacional (El intercambio desigual)*. Madrid, Siglo XXI, 1977.

ENZENSBERG, H.-*Zur Kritik der politischen Okologie*. Berlin, Kursbuch, 1973.

L'ETAT *des religions dans le Monde*. Paris, La Découvert, 1987.

L'ETAT *du Tiers Monde*. Paris, La Découverte, 1989. The EUROPA Yearbook. London, Europa Pub., 1984.

FABER, G.- *The Third World and the EEC*. Netherlands, 1982.

FALK, R & WAHL, P.- *Befreiungsbewegungen in Afrika*. Pahl-Rugenstein, Koln, 1980.

FANON, F.- *Os condenados da Terra*. Rio de Janeiro, Civilizaçao Brasileira, 1979.

FERNANDEZ, W.- *El gran culpable.La responsabilidad de los Estados Unidos en el proceso militar uruguayo*. Montevideo, Atenea, 1986.

FIELDHOUSE, DK.- *The Colonial Empires. A comparative survey from the eighteenth century*. 2a.ed. London, MacMillan1982.

FIGUEROA ALCACES, E, ed.- *Antología de geografía histórica moderna y contemporánea*. México, UNAM, 1974.

FISHLOW, A, et al.- *Rich and Poor Nations in the World Economy*. New York, McGraw Hill, 1978.

FOREST *Resources crisis in the Third World*, Malaysia, Sep.1986. Proceeding. Malaysia, 1987.

FRENTE POLISARIO.-*VII Anniversaire du Declenchement de la Lutte de Liberation nationale*. Dep. Informations, 1980.

FRETLIN *Conquers the Right to Dialogue*. London, 1978.

GALTUNG, J. & O'BRIEN, P., et al.- *Self-reliance. A Strategy for development*. London, IDS, 1980.

GAZOL SANTAGE, A.- *Los países pobres*. México, FCE, 1987.

GENOUD, R.- *Sobre las revoluciones parciales del Tercer Mundo*. Barcelona, Anagrama, 1974.

GEORGE, S.- *How the Other Half Dies*. Middlesex, Penguin, 1980.

GREINER, B.- *Amerikanische Aussenpolitik von Truman bis heute*. Koln, Pahl-Rugenstein, 1980.

GROUSSET, R.- *Historia de Asia*. Buenos Aires, EUDEBA, 1962.

GRUNEBAM, GE, ed.- *El Islam. Desde la caída de Constantinopla hasta nuestros días*. México, Siglo XXI, 1975.

GUIM, JB.- *Compendio de geografía Universal*. México, Bouret, 1875.

HALLIDAY, F.- *Arabia without Sultans*. Middlesex, Penguin, 1974.

HALPERIN DONGHI, T.- *Historia contemporánea de América Latina*. Madrid, Alianza, 1983.

HAMELINK, C.- *The Corporate Village*. Rome, IDOC, 1977.

HAYES, MD.- *Dimensiones de seguridad de los intereses de Estados Unidos en América Latina*. México, CIDE, 1981.

HERRERA, L & VAYRYNEN, R, ed.- *Pace, development and New International Economic Order*. Tampere, IPRA, 1979.

HUIZER, G.- *El potencial revolucionario del campesino en América Latina*. México, Siglo XXI, 1976.

HUMANN, K. & BRODERSEN, I.- *Welt Aktuell'86*. Hamburg, Roro, 1985.

IBGE.- *Tabulaçoes Avançadas do Censo Demográfico*. Rio de Janeiro, 1982.

IDB.- *External Debt and Economic Development in Latin America*. Washington, 1984.

INSTITUTE OF RACE RELATIONS.- *Patterns of Racism*. London, 1982.

INTERNATIONAL COMMISSION OF JURISTS AND CONSUMERS' ASSOCIATION OF PENANG.- *Rural development and Human Rights in South East Asia*. Penang, 1982.

JALEE, P.- *El Tercer Mundo en la economía Mundial*. México, Siglo XXI, 1980.

JAULIN, R.- *La des-civilización. Política y práctica del etnocidio*. México, Nueva Imagen, 1979.

KENT, G.- *Food Trade: The Poor Feed the Rich en the Ecologist Ecosystem*. United Kingdom, 1982.

KETTANI, M.- *Ali, the Muslim Minorities*. Leicester, Islamic Foundation, 1979.

KHADAFI, M.- *O livro verde*. Tripoli, EPOEPD.

KHOR, K.P.- *Recession and the Malaysian Economy*. Penang, Masyarakat, 1983.

KIDROM, M & SEGAL, R.- *The State of the World Atlas*. London, Pan Books, 1981.

KI-ZERBO, J.- *Historia del Africa Negra*. Madrid, Alianza, 1980.

KLARE, M, et al.- *Supplying Repression*. Washington, IPS, 1981.

KUPER, L.- *Genocide: Its political use in the Twentieth Century*. Middlesex, Penguin, 1981.

LAINO, D.- *Paraguai: Fronteiras e penetraçao brasileira*. Sao Paulo, Global, 1979.

LEWIS, D.- *Reform and revolution in Grenada: 1950-1981*. Havana, Casa de las Américas, 1984.

LEWYCKY, D.& WHITE, S.- *An African abstract*. Council for International Cooperation. Manitoba, 1979.

LICHTHEIM, G.- *Imperialism*. Middlesex, Penguin, 1974.

LINHARES, MY.- *A luta contra a metrópole (Asia e Africa)*. Sao Paulo, Brasilense, 1981.

LIPSCHUTZ, A.- *El problema racial en la conquista de América*. México, Siglo XXI, 1963.

MAGDOFF, H.- *La empresa multinacional en una perspectiva histórica*. Barcelona, 1980.

MAX-NEEF, M.- *La economía descalza*. Lima, 1985.

MAYOBRE MACHADO, J.- *Información dependencia y desarrollo;la prensa y el nuevo orden económico internacional*. Caracas, Monte Avila, 1978.

MEILE, P.- *Historia de la India*. Buenos Aires, EUDEBA, 1962.

MOITA, L.- *Os Congressos da Frelimo do PAIGC e do MPLA*, CIDAC. Lisbon, 1979.

MOORE LAPPE, F & COLLINS, J.- *El hambre en el mundo. Diez mitos*. México, Copider/Fonapas, 1980.

MYERS, N.- *The Gaia Atlas of future worlds: challenge and opportunity in age of change*. London, Gaia, 1990.

MYLLYMAKI, E.& DILLINGER, B.- *Dependency and Latin American Development*. Finnish Peace Research Asociation. Tampere, 1977.

NKRUMAH, K.- *Africa debe unirse*. Buenos Aires, EUDEBA, 1965

NEARING, S & FREEMAN, J.- *La diplomacia del dólar*. México, SELFA, 1926.

NYERERE, JK.- *Freedom and Socialism*. London, Oxford Univ.Press, 1968.

NEWLAND, K.- *The Sisterhood of Man*. New York, WW Norton, 1979.

ORTIZ MENA, A.- *América Latina en desarrollo*. Washington, IDB, 1980.

ORTIZ QUESADA, F.- *Salud en la pobreza*. México, CEESTEM-Nueva Imagen, 1982.

OSBORNE, M.- *Region of Revolt. Focus of Southeast Asia*. Middlesex, Penguin, 1970.

OSMANCZYK, EJ.- *Enciclopedia Mundial de relaciones internacionales y Naciones Unidas*. Madrid, Fondo de Cultura, 1976.

PAQUE, R., ed.- *Afrika antwortet Europa*. Berlin, Ullstein, 1976.

PEASE GARCIA, H., ed.- *América Latina 80: Democracia y movimiento popular*. Lima, Desco, 1981.

PERROT, D.& PREISWEK, R.- *Etnocentrismo e Historia*. México, Nueva Imagen, 1979.

Le PETROLE, *les matières de base et le développment*. Algeria, Sonatrach, 1974.

PIACENTINI, P.- *O Mundo do petróleo*. Lisbon, Tricontinental, 1984.

PIERRE-CHARLES, G.- *El caribe contemporáneo*. México, Siglo XXI, 1981.

PUTZGER, FW.- *Historicher Weltatlas*. Berlin, Velhagen, 1961. *Quality of Life, from a common people's point of view*, by PapyRossa Verlags GmbH & Co. KG, K+ln & World Data Research Center, Ernst Fidel FŸrntratt-Kloep, 1995, Hackús, Sweden.

RAGHAVAN, Ch.- *Recolonization. GATT, the Uruguay round and the Third World*. Penang, Third World Network, 1990

RAHMAN, MA.- *Grass-Roots Participation and self-reliance: experiences in South and South East*. Asia, ILO, 1984.

RAMA, C.- *Historia de América Latina*. Barcelona, Bruguera, 1985.

REFORMS *in foreigns economic relations of Eastern Europe and the Soviet Union*. Proceedings. New York, ECE, 1991. (Economic Studies, 2)

RETURN *to the good earth*. Third World Network Dossier, Penang, 1983.

RIBEIRO, D.- *Las Américas y la*

civilización. Buenos Aires, Centro Editor, 1973.

RIBEIRO, S.- Sobre a unidade no pensamento de Amilcar Cabral. Lisbon, Tricontinental, 1983.

RODNEY, W.- How Europe Underdeveloped Africa. Washington, Howard Univ., 1974.

ROGER, M.- Timor - Hier la colonisation portugaise, aujourd'hui la résistance. Paris, L'Harmattan, 1977.

SAID, E.- Orientalism. Middlesex, Penguin, 1985.

SCHLESINGER, R.- La Internacional Comunista y el Problema Colonial. México, Pasado y Presente, 1974.

SCHMIEDER, O.- Geografía de América Latina. México, FCE, 1975.

SCHUON, F.- Understanding Islam. London, Mandala Books, 1979.

SEAGER, J & OLSON, A.- Women in the world. An International Atlas. London, Pan Books, 1986.

SEGAL, G.- The world affairs companion. New York, Touchstone, 1991.

SHINNIE, M.- Ancient African Kingdoms. New York, Mentor, 1980.

SHIVA, V.- The violence of the green revolution, ecological degradation and political conflict in Punjab. India, 1989.

SIVARD, RL.- World Military and Social Expeditures. Virginia, World Priorities, 1979.

SOCIAL WATCH - Instituto del Tercer Mundo, Montevideo, 2006.

STANLEY, D.- Eastern Europe on a shoestring. Australia, Lonely Planet, 1989.

STATE OF THE WORLD World Watch Institute, New York, Norton & Company.

STAVRIANOS, LS.- Global Rift. The Third World Comes of Age. New York, William Morow, 1981.

SUAREZ, L.- Los Países No Alineados. México, FCE, 1975.

SUB-SHARAN AFRICA: From crisis to sustainable growth, a long-term perspective study. Washington, World Bank, 1989.

SUTER, K.- West Irian, East Timor and Indonesia. London, 1979.

SWEEZY, P.- Teoria do Desenvolvimento Capitalista. Sao Paulo, Abril, 1983.

TERRE DES FEMMES. Panorama de la situation des femmes dans le monde. Paris, La Découverte, 1982.

THE INDIGENOUS world 2002-2003; International Work Group for Indigenous Affairs, Copenhagen, 2003.

THIRD WORLD FOUNDATION.- Third World Affairs 1987. London, 1987.

THOMAS, EJ., ed.- Les travailleurs immigrés en Europe: quel status? Paris, UNESCO, 1981.

TIMERMAN, J.- Israel: La Guerra más larga. Madrid, Mochnik, 1983.

TITO, JB.- La misión histórica del Movimiento de No Alineación. Beograd, CAS, 1979.

TOMLINSON, A. & WHANNEL, G., ed.- Five-ring circus: Money, power and politics at the Olympic Games. London, Pluto Press, 1984.

TORRIELLO GARRIDO, G.- Tras la Cortina de Banano. Havana, Ciencias Sociales, 1979.

TORRIJOS, O.- La quinta frontera. Costa Rica, Ed.Univ.Centroamericana, 1981.

TORRIJOS, O.- Figura, tiempo, faena. Panamá, Lotería Nacional, 1981.

TOWARDS Socialist Planning. Tanzania, UCHEMI, 1980.

TRIBUNAL PERMANENTE DOS POVOS. Sessao sobre Timor-Leste. Lisbon, 1981.

UL HAQ, M.- La cortina de la pobreza. México, FCE, 1978.

UN.- Resolutions on the Palestine question. Beirut

UN.- Terminology Bulletin. New York, 1979.

UN.- UNCTAD VII. Actas. Geneva 1987.

UNESCO.- Geografía de América Latina. Barcelona, 1975.

UNESCO.- Recherches en matière du relations raciales. Paris, 1965.

UNDP.- Human Development Reports 1980- 2006.

UNICEF.- The State of the World's Children Report 2006. New York 2006.

US ARMS CONTROL AND DISARMAMENNTAGENCY.- World Military Expenditure and Arms Transfer 1972-1982. Washington, 1982.

VARELA BARRAZA, H.- África: Crisis del poder político. México, CEESTEM, 1981.

VERDIEU, E & BWATSHIA, K.- Les Eglises face au Nouvel Ordre Economique National e International. Quebec, CECI, 1980.

VERDIEU, E & BWATSHIA, K.- Cooperation Technique des Pays en Development;est-ce Possible? Quebec, CECI, 1980.

VERDIEU, E & BWATSHIA, K.- Liberation et Autonomie Collectives. Québec, CECI, 1980.

VIVO ESCOTO, JA.- Geografía Humana y Económica. México, Patria, 1975.

WEISSMAN, S. et al.- The Trojan Horse. A radical Look at Foreign Aid. San Francisco, Ramparts, 1974.

WETTSTEIN, G.- Subdesarrollo y geografía. Mérida, Universidad de los Andes, 1978.

WHANNEL, G.- Blowing the whistle: The politics of sport. London, Pluto, 1983.

WIENER, D.- Shalom, Israels Friedensbewegung. Hamburg, Rowohlt, 1984.

WILLIAMS, N.- Chronology of the modern World 1763-1965. Middlesex, Penguin, 1975.

WOLFE, A.et al.- La cuestión de la Democracia. México, UILA, 1980.

WORLD BANK.- World Development Indicators 2006. Washington, 2006.

WORLD Directory of Minorities; Minority Rights Group International, London, 1997.

WORLD FACTS and maps. Rand McNally. USA, 1994.

WORLD MIGRATION 2006: Managing Migration, Challenges and Responses for People on the Move; International Organization for Migration, Geneva 2003.

WORLD Military Expenditure and Arms Transfer 1972-1982. Washington,1984.

WORLD RESOURCES INSTITUTE. The Environmental Almanac 1993. Houghton Mifflin Company, Boston & New York, 1992.

WORSLEY, P.- El tercer Mundo: una nueva fuerza en los asuntos internacionales. México, Siglo XXI, 1978.

ZERAOUI, Z.- El mundo arabe: imperialismo y nacionalismo. México, CEESTEM-Nueva Imágen, 1981.

ZERAOUI, Z.- Irán-Irak: guerra política y sociedad. México, Nueva Imágen, 1982.

ZIEGLER, J.- Main Basse sur l'Afrique. Paris, Du Seuil, 1978.

PERIODICALS

Africa News, Durham, NC, EUA

Afrique Mass-Media, Budapest, Hungary

Afrique Nouvelle, Dakar, Senegal

Afrique-Asie, Paris, France

ALAI, Montreal, Canada

ALDHU, Quito, Ecuador

Altercom, México,

AMPO Japan-Asia Quarterly Review,Tokyo,

Análisis, Santiago de Chile,

APSI, Santiago de Chile, Chile

AQUI, La Paz, Bolivia

Barricada Internacional, Managua, Nicaragua

Bohemia, La Habana, Cuba

Boletín de Namibia, UNITED NATIONS, New York

Bulletin of Concerned African Scholars, Charlemont, MA, EUA

Caribbean Monthly, Río Piedras, Puerto Rico

CEAL, Brussels, Belgium

Central America Update, Toronto, Canada

CERES, Rome, Italy

CILA, México

Comercio Exterior, México,

Contextos, México,

Counterspy, Washington, DC, EUA

CovertAction, Washington, EUA

CRIE, México,

Cuadernos del Tercer Mundo, Brasil,

Descolonización, UNITED NATIONS, New York

Diálogo Social, Panama,

Documentos FIPAD, Ginebra, Switzerland

Economie et Politique, Paris, France

El Caribe Contemporáneo, México,

Eritrea in Struggle, New York, EUA

Facts & Reports, Amsterdam, Netherlands

Far Eastern Economic Review, Asia 1990, Yearbook. Japan

Fortune International, Los Angeles, CA, EUA

Freedomways, New York, EUA

Gombay, Belize

IFDA, Nyon, Switzerland

Informe R - CEDOIN , Bolivia.

Indian and Foreign Review, New Delhi, India

Internews, Berkeley, EUA

Isis, Rome, Italy

Ko-Eyú, Caracas, Venezuela

L'Economiste du Tiers Monde, Paris, France

Lateinamerika Nachrichten, Berlin

Latin America Weekly Report, London, United Kingdom

Le Monde Diplomatique, Paris, France

Monthly Review, New York, US

Mujer-Fempress, Santiago, Chile.

Multinational Monitor, Washington, US

NACLA Report on the Americas, New York, US

New Outlook, Dar-es-Salaam, Tanzania

New Internationalist, Oxford, UK

Newsfront International, Oakland, US

Noticias Aliadas, Lima, Peru

Novembro, Luanda, Angola

O Correio da UNESCO, Paris-Rio de Janeiro,

Onze Wereld, Den Haag, Netherlands

Palestine, Beirut, Lebanon

Pensamiento Propio, Managua, Nicaragua

Philippine Liberation Courier, Oakland, US

Politica Internazionale, Rome, Italy

Política Internacional, Belgrade, Yugoslavia

Resister (Bulletin of the Committee on South African War Resistance), London, UK

Revista del Centro de Estudios del Tercer Mundo, México.

Revista del Sur, ITeM, Montevideo, Uruguay.

Sharing, Geneva, Switzerland

SIAL, Italy

Soberanía, Managua, Nicaragua

Southern Africa, New York, US

Statesman's Yearbook, 1988-1989.

Tempo, Maputo, Mozambique

Tercer Mundo Económico, ITeM, Montevideo, Uruguay

The Black Scholar, Sausalito, CA, US

The Ecologist, United Kingdom.

The CTC Reporter, UNITED NATIONS, New York.

Third World Quarterly, London, United Kingdom.

Third World Resurgence No.165/166, Penang, Malaysia.

Tiempos Nuevos, Moscow

Tigris, Madrid, Spain

Tribune, Sri Lanka

Tricontinental, Havana, Cuba

Two Thirds, A journal of underdeveloment studies, Toronto, Canada

WISE (World Information Service on Energy), Netherlands

YEKATIT Quarterly, Addis Ababa, Ethiopia

NEWS AGENCIES

Interpress - IPS, INA, WAFA, SALPRESS, ANN, Prensa Latina, Angop, AIM, Shihata.

ALAI (Latin American Information Agency).

ONLINE RESOURCES

NGOs, FOUNDATIONS, RESEARCH INSTITUTES, ENCYCLOPEDIAS

Institute of Latin American Studies: http://lanic.utexas.edu/

Rulers: http://www.rulers.org

Choike: http://www.choike.org/nuevo/

Enciclopedia Britannica: www.eb.com

The American Society of International Law: http//:www.asil.org

Migrants and Refugees: http://www.refugees.org/

Landmine monitor: http://www.icbl.org/lm/

World Organization Against Torture: http://www.omct.org/

The International Work Group for Indigenous Affairs: http://www.iwgia.org/sw619.asp

Minorities at Risk: http://www.cidcm.umd.edu/inscr/mar/home.htm

Women's suffrage: www.idea.int

Amnesty international: www.amnesty.org

http://web.amnesty.org/pages/deathpenalty-index-eng

World of Information: www.worldinformation.com.

INTERNATIONAL ORGANIZATIONS

Inter-american Development Bank: http://www.iadb.org/

United Nations: http://www.un.org/

United Nations' Human Development Report 2003: http://www.undp.org/hdr2003/

http://devdata.worldbank.org/genderstats/home.asp http://www.un.org/womenwatch/daw/cedaw/ www.unifem.undp.org/ www.unicef.org

www.childinfo.org

http://www.ecpat.net

http://www.unhcr.ch/cgi-bin/texis/vtx/home

International Organization for Migration: http://www.iom.int/

Joint United Nations Programme on HIV/AIDS: http://www.unaids.org/en/default.asp

Food and Agriculture organization of the United Nations: http://www.fao.org

International Labour Organization: http://www.ilo.org

Council of Europe: http://www.coe.int

NEWS RESOURCES

Le Monde Diplomatique online: http:// www.ina.fr/CP/MondeDiplo/

Folha de Sao Paulo: http://www.uol.com.br/ fsp/

OneWorld News Service: http://www.oneworld.org/news/index.html

OMRI Daily Digest: http://www.omri.cz/Publications/Digests/DigestIndex.html

Il Manifesto in Rete: http://www.mir.it/ mani/index.html

The Washington Report's Resources Page: http://www.washington-report.org

New Internationalist magazine: http:// www.newint.org

BBC online: http://news.bbc.co.uk/

Indymedia: www.indymedia.org

El Mundo online: www.el-mundo.es/

Eurasiannews: http://eurasiannews.com/

Himalayannews: http://www.himalayannews.com/

Commondreams: www.commondreams.org

African News Agency: http://afrol.com/

AllAfrica Global Media: http://allafrica.com

Index

The world in figures

	AREA (SQUARE KM) [1]	TOTAL POPULATION (THOUSANDS) [3]	DEMOGRAPHIC GROWTH 1985-2000 (%) [3]	DEMOGRAPHIC GROWTH 2000-2015 (%) [3]	TOTAL POPULATION ESTIMATE FOR YEAR 2015 (THOUSANDS) [3]	POPULATION DENSITY (INHABITANTS PER SQ KM) [4]	URBAN POPULATION (% OF TOTAL) [3]	URBAN POPULATION, ANNUAL GROWTH (%) [4]	2015 URBAN POPULATION AS % OF TOTAL POPULATION [3]
	2006	2004	2004	2004	2004	2007	2004	2005-2010	2004
AFGHANISTAN	652,090	32,254	3.6	3.7	41,401	49	25	5.9	31
ALBANIA	28,750	3,163	0.2	0.6	3,325	110	46	1.9	51
ALGERIA	2,381,740	33,861	2.1	1.5	38,085	14	61	2.4	65
ANDORRA	453	67	2.3	0.2	68	149			
ANGOLA	1,246,700	16,867	2.7	2.8	20,947	14	39	4.9	45
ANGUILLA	96	13	2.8	1.5	14	131			
ANTIGUA	440	84	1.2	1.2	92	190			
ARGENTINA	2,780,400	39,531	1.3	1.0	42,676	14	91	1.2	92
ARMENIA	29,800	2,999	-0.5	-0.2	2,970	101	64	-0.3	64
ARUBA	190	101	2.5	0.9	106	533			
AUSTRALIA	7,741,220	20,576	1.3	1.0	22,250	3	93	1.3	95
AUSTRIA	83,860	8,218	0.4	0.2	8,288	98	66	0.3	67
AZERBAIJAN	86,600	8,536	1.3	0.7	9,083	99	50	0.9	51
BAHAMAS	13,880	332	1.7	1.3	365	24	90	1.5	92
BAHRAIN	710	751	3.2	1.6	852	1058	90	1.9	91
BANGLADESH	144,000	147,059	2.2	1.8	168,158	1021	26	3.5	30
BARBADOS	430	271	0.3	0.2	276	630	54	1.4	59
BELARUS	207,600	9,645	0.0	-0.6	9,218	46	72	-0.1	75
BELGIUM	33,100	10,453	0.3	0.2	10,540	316	97	0.2	98
BELIZE	22,960	280	2.6	1.9	321	12	49	2.4	52
BENIN	112,620	8,971	3.3	3.0	11,217	80	48	4.6	53
BERMUDA	50	65	0.6	0.3	65	1291			
BHUTAN	47,000	2,260	1.9	2.2	2,684	48	10	5.7	13
BOLIVIA	1,098,580	9,525	2.2	1.8	10,854	9	65	2.5	69
BOSNIA-HERZEGOVINA	51,130	3,920	-0.5	0.1	3,893	77	46	1.3	51
BOTSWANA	581,730	1,753	2.3	-0.2	1,690	3	53	0.5	58
BRAZIL	8,547,400	191,341	1.6	1.2	209,401	22	85	1.8	88
BRUNEI	5,770	390	2.7	2.0	453	68	79	2.8	83
BULGARIA	110,910	7,616	-0.8	-0.7	7,156	69	71	-0.3	74
BURKINA FASO	274,000	14,042	2.8	3.0	17,678	51	19	5.3	23
BURUNDI	27,830	8,141	1.9	3.3	10,617	293	11	7.3	15
CAMBODIA	181,040	14,638	3.0	1.9	17,066	81	21	5.0	26
CAMEROON	475,440	16,874	2.6	1.7	19,040	35	54	3.0	60
CANADA	9,970,610	32,852	1.1	0.9	35,051	3	82	1.2	84
CAPE VERDE	4,030	530	2.3	2.2	628	132	59	3.6	65
CAYMAN IS.	260	47	4.3	1.6	51	179			
CENTRAL AFRICAN REP.	622,980	4,151	2.3	1.4	4,647	7	45	2.8	50
CEUTA	19	75				3947			
CHAD	1,284,000	10,303	3.0	3.0	12,832	8	27	4.6	31
CHILE	756,630	16,635	1.6	1.0	17,926	22	88	1.3	90
CHINA	9,598,050	1,331,356	1.2	0.6	1,392,980	139	42	2.8	50
CHRISTMAS IS.	135	0.4				3			
COCOS	14	0.6				43			
COLOMBIA	1,138,910	46,952	1.9	1.4	52,086	41	78	2.0	81
COMOROS	2,230	841	2.9	2.5	1,019	377	38	4.4	43
CONGO D.R.	2,344,860	61,174	3.2	3.1	78,016	22	34	5.2	40
CONGO R.	342,000	4,238	2.9	3.0	5,441	12	55	3.9	59
COOK IS.	236	18	0.2	-0.6	17	75			
COSTA RICA	51,100	4,468	2.5	1.6	4,983	87	63	2.4	67
CÔTE D'IVOIRE	322,460	18,770	3.1	1.7	21,553	58	47	2.8	51
CROATIA	56,540	4,555	0.1	-0.1	4,454	81	61	0.7	65
CUBA	110,860	11,317	0.7	0.2	11,437	102	76	0.5	78
CYPRUS	9,250	854	1.3	1.1	927	92	70	1.3	72
CZECH REPUBLIC	78,870	10,198	0.0	-0.1	10,066	129	75	0.0	76
DENMARK	43,090	5,461	0.3	0.3	5,560	127	86	0.4	87
DIEGO GARCÍA IS.	52	1.3				25			
DJIBOUTI	23,200	820	3.8	1.8	930	35	85	2.0	88
DOMINICA	750	80	0.4	0.7	87	107			
DOMINICAN REPUBLIC	48,730	9,148	1.7	1.4	10,124	188	61	2.1	65
EAST TIMOR	14,870	1,068	0.6	4.8	1,486	72	8	7.3	10
ECUADOR	283,560	13,611	2.0	1.4	15,144	48	64	2.2	68
EGYPT	1,001,450	76,853	2.0	1.8	88,175	77	43	2.3	45
EL SALVADOR	21,040	7,116	1.8	1.6	8,017	338	61	2.3	64
EQUATORIAL GUINEA	28,050	527	2.4	2.2	627	19	52	4.0	58
ERITREA	117,600	4,708	1.9	3.3	5,840	40	22	5.5	27
ESTONIA	45,100	1,321	-0.7	-0.4	1,292	29	70	-0.1	71
ETHIOPIA	1,104,300	81,176	3.1	2.3	97,155	74	17	4.3	20
FAEROE IS.	1,400	48	0.0	0.6	50	34			
FIJI	18,270	861	0.9	0.7	903	47	55	2.0	60
FINLAND	338,150	5,274	0.4	0.2	5,359	16	61	0.3	62
FRANCE	551,500	60,940	0.5	0.3	62,339	110	77	0.6	79
FRENCH GUIANA	90,000	196	4.2	2.3	232	2	76	2.5	78
FRENCH POLYNESIA	4,000	264	2.0	1.4	291	66	52	1.4	53
GABON	267,670	1,429	3.0	1.5	1,605	5	86	2.2	89
GAMBIA	11,300	1,594	3.5	2.4	1,889	141	26	2.8	28
GEORGIA	69,700	4,396	-0.8	-0.8	4,183	63	51	-0.9	52
GERMANY	357,030	82,729	0.4	0.0	82,513	232	89	0.2	90
GHANA	238,540	22,995	2.6	1.9	26,562	96	47	3.0	51
GIBRALTAR	6	28	0.3	0.1	28	4664			

DEMOGRAPHY

	AREA (SQUARE KM) [1]	TOTAL POPULATION (THOUSANDS) [3]	DEMOGRAPHIC GROWTH 1985-2000 (%) [3]	DEMOGRAPHIC GROWTH 2000-2015 (%) [3]	TOTAL POPULATION ESTIMATE FOR YEAR 2015 (THOUSANDS) [3]	POPULATION DENSITY (INHABITANTS PER SQ KM) [4]	URBAN POPULATION (% OF TOTAL) [3]	URBAN POPULATION, ANNUAL GROWTH (%) [4]	2015 URBAN POPULATION AS % OF TOTAL POPULATION [3]
	2006	2004	2004	2004	2004	2007	2004	2005-2010	2004
GREECE	131,960	11,160	0.7	0.2	11,233	85	62	0.7	65
GREENLAND	341,700	57	0.4	0.3	58	0.2			
GRENADA	340	105	0.6	1.0	119	310			
GUADELOUPE	1,705	455	1.3	0.6	472	267	100	0.6	100
GUAM	550	175	1.8	1.5	194	318	94	1.6	95
GUATEMALA	108,890	13,230	2.3	2.3	15,869	122	48	3.4	52
GUINEA	245,860	9,808	3.0	2.3	11,890	40	38	4.3	44
GUINEA-BISSAU	36,120	1,682	2.8	3.0	2,133	47	37	5.2	44
GUYANA	214,970	752	-0.1	0.0	742	3	39	1.3	44
HAITI	27,750	8,773	1.7	1.4	9,751	316	40	3.1	45
HONDURAS	112,090	7,521	2.9	2.1	8,780	67	47	3.1	51
HUNGARY	93,030	10,045	-0.2	-0.3	9,802	108	67	0.3	70
ICELAND	103,000	300	1.0	0.8	319	3	93	1.0	94
INDIA	3,287,260	1,135,614	1.9	1.4	1,260,366	345	29	2.5	32
INDONESIA	1,904,570	228,121	1.5	1.1	246,813	120	50	3.3	58
IRAN	1,648,200	71,220	2.1	1.2	79,917	43	69	2.3	74
IRAQ	438,320	30,291	2.9	2.5	36,473	69	67	2.3	67
IRELAND	70,270	4,267	0.5	1.4	4,674	61	61	1.8	64
ISRAEL	21,060	6,967	2.6	1.7	7,838	331	92	1.8	92
ITALY	301,340	58,173	0.1	0.0	57,818	193	68	0.2	69
JAMAICA	10,990	2,672	0.8	0.4	2,748	243	52	0.7	54
JAPAN	377,800	128,325	0.3	0.1	127,993	340	66	0.3	68
JORDAN	89,210	5,966	4.1	2.2	6,956	67	80	2.3	81
KANAKY-NEW CALEDONIA	18,580	245	2.2	1.7	277	13	62	2.1	65
KAZAKHSTAN	2,724,900	14,802	-0.3	-0.1	14,877	5	56	0.2	58
KENYA	580,370	36,012	3.0	2.4	44,194	62	44	5.1	52
KIRIBATI	730	103	2.4	1.9	118	141			
KOREA, NORTH	120,540	22,670	1.1	0.4	23,299	188	62	1.0	66
KOREA, SOUTH	99,260	48,142	0.9	0.3	49,092	485	81	0.6	83
KUWAIT	17,820	2,839	1.7	2.8	3,381	159	97	2.6	97
KYRGYZSTAN	199,900	5,386	1.4	1.1	5,852	27	34	1.3	35
LAOS	236,800	6,193	2.5	2.2	7,306	26	23	4.6	27
LATVIA	64,600	2,284	-0.6	-0.5	2,191	35	66	-0.6	66
LEBANON	10,400	3,653	1.3	1.0	3,965	351	88	1.3	90
LESOTHO	30,350	1,785	1.3	-0.2	1,744	59	19	0.9	21
LIBERIA	111,370	3,452	2.3	2.4	4,381	31	49	4.2	54
LIBYA	1,759,540	6,085	2.3	1.9	7,018	3	87	2.2	89
LIECHTENSTEIN	160	35	1.3	0.9	37	220			
LITHUANIA	65,200	3,403	-0.1	-0.4	3,288	52	67	-0.4	68
LUXEMBOURG	2,600	477	1.1	1.2	523	183	93	1.4	94
MACEDONIA, TFYR	25,710	2,040	0.6	0.1	2,055	79	60	0.4	62
MADAGASCAR	587,040	19,609	2.9	2.6	23,813	33	28	3.8	31
MALAWI	118,480	13,452	3.1	2.2	15,998	114	18	4.8	22
MALAYSIA	329,750	26,240	2.6	1.7	29,558	80	66	2.6	71
MALDIVES	300	346	3.1	2.4	416	1152	31	4.1	35
MALI	1,240,190	14,325	2.6	2.9	18,093	12	35	5.0	41
MALTA	320	405	0.9	0.4	419	1266	93	0.6	94
MALVINAS-FALKLANDS	11,410	3	3.1	0.5	3	0.3			
MARSHALL IS.	200	66	2.1	3.1	83	331			
MARTINIQUE	1,102	399	0.8	0.3	404	362	97	0.4	97
MAURITANIA	1,025,520	3,247	2.5	2.7	3,988	3	67	4.4	74
MAURITIUS	2,040	1,267	1.0	0.8	1,344	621	44	1.5	47
MAYOTTE IS.	400	163				408			
MELILLA	12	69				5750			
MEXICO	1,958,200	109,594	1.8	1.2	119,146	56	77	1.5	79
MICRONESIA	700	112	1.5	0.5	116	160	31	2.1	35
MOLDOVA	33,850	4,186	0.1	-0.3	4,114	124	47	0.4	50
MONACO	2	36	1.1	1.1	40	18035			
MONGOLIA	1,566,500	2,711	1.8	1.2	2,988	2	57	1.6	60
MONTSERRAT	102	5	-7.1	1.6	5	46			
MOROCCO	446,550	32,412	1.8	1.4	36,152	73	60	2.5	65
MOZAMBIQUE	801,590	20,522	2.0	1.8	23,513	26	40	4.6	48
MYANMAR-BURMA	676,580	51,475	1.7	0.9	54,970	76	32	3.1	38
NAMIBIA	824,290	2,072	3.5	1.1	2,248	3	35	2.7	40
NAURU	21	14	2.7	1.5	15	668			
NEPAL	147,180	28,226	2.4	2.0	32,747	192	17	4.6	20
NETHERLANDS	41,530	16,429	0.6	0.4	16,812	396	68	1.1	71
NETHERLANDS ANTILLES	800	185	-0.2	0.6	193	232	71	1.0	73
NEW ZEALAND-AOTEAROA	270,530	4,093	1.1	0.8	4,302	15	86	0.8	87
NICARAGUA	130,000	5,715	2.3	1.9	6,637	44	59	2.8	63
NIGER	1,267,000	14,907	3.2	3.3	19,283	12	24	5.9	30
NIGERIA	923,770	137,243	2.7	2.1	160,931	149	50	3.6	56
NIUE	260	1	-2.7	-0.1	2	6			
NORFOLK IS.	36		1.9			53			
NORTHERN MARIANAS	500	85	6.8	2.2	98	170			
NORWAY	323,880	4,665	0.5	0.5	4,841	14	82	1.3	86
OMAN	309,500	2,668	3.1	1.7	3,173	9	80	2.8	83
PAKISTAN	796,100	164,594	2.7	2.0	193,419	207	36	3.3	40
PALAU	460	20	2.2	0.6	21	44			

DEMOGRAPHY

	AREA (SQUARE KM) [1]	TOTAL POPULATION (THOUSANDS) [3]	DEMOGRAPHIC GROWTH 1985-2000 (%) [3]	DEMOGRAPHIC GROWTH 2000-2015 (%) [3]	TOTAL POPULATION ESTIMATE FOR YEAR 2015 (THOUSANDS) [3]	POPULATION DENSITY (INHABITANTS PER SQ KM) [4]	URBAN POPULATION (% OF TOTAL) [3]	URBAN POPULATION, ANNUAL GROWTH (%) [4]	2015 URBAN POPULATION AS % OF TOTAL POPULATION [3]
	2006	2004	2004	2004	2004	2007	2004	2005-2010	2004
PALESTINE	6,220	3,945	3.8	3.1	4,996	634	73	3.7	76
PANAMA	75,520	3,343	2.0	1.6	3,774	44	58	2.3	62
PAPUA NEW GUINEA	462,840	6,114	2.5	1.9	7,013	13	13	2.5	14
PARAGUAY	406,750	6,445	2.8	2.2	7,613	16	60	3.3	64
PERU	1,285,220	28,797	1.9	1.4	32,172	22	75	1.9	78
PHILIPPINES	300,000	85,884	2.2	1.6	96,840	286	64	2.8	69
PITCAIRN	5	0	0.6	0.0	0	13			
POLAND	323,250	38,467	0.3	-0.1	38,110	119	62	0.2	64
PORTUGAL	91,980	10,593	0.1	0.4	10,827	115	57	1.3	61
PUERTO RICO	8,950	3,998	0.8	0.5	4,157	447	98	0.8	99
QATAR	11,000	857	3.5	3.1	972	78	93	2.1	94
RÉUNION	2,510	807	1.8	1.3	886	322	93	1.7	95
ROMANIA	238,390	21,544	-0.2	-0.4	20,871	90	55	-0.1	56
RUSSIA	17,075,400	141,900	0.1	-0.5	136,696	8	73	-0.4	74
RWANDA	26,340	9,442	1.9	2.3	11,262	358	25	9.9	41
SAHARA, WESTERN	266,000	372	3.3	3.8	526	1	94	4.9	95
SAMOA	2,840	187	0.8	0.5	190	66	23	1.2	25
SAMOA, AMERICAN	200	68	2.5	2.0	78	338			
SAN MARINO	61	29	1.1	0.7	30	468			
SÃO TOMÉ AND PRÍNCIPE	960	164	2.0	2.1	192	170	38	2.7	40
SAUDI ARABIA	2,149,690	25,809	3.4	2.4	30,828	12	89	2.8	91
SENEGAL	196,720	12,218	2.7	2.3	14,538	62	52	3.7	58
SERBIA AND MONTENEGRO*	102,000	10,493	0.5	-0.1	10,416	103	53	0.5	56
SEYCHELLES	450	82	0.8	0.9	88	182			
SIERRA LEONE	71,740	5,802	1.5	2.8	6,897	81	42	3.9	48
SINGAPORE	620	4,434	2.6	1.2	4,815	7151	100	1.2	100
SLOVAKIA	49,010	5,401	0.3	0.0	5,385	110	59	0.5	61
SLOVENIA	20,250	1,965	0.3	-0.1	1,942	97	51	0.2	53
SOLOMON IS.	28,900	502	2.9	2.4	596	17	18	4.3	21
SOMALIA	637,660	8,766	0.5	3.0	10,970	14	37	4.8	43
SOUTH AFRICA	1,221,040	47,699	2.1	0.3	47,902	39	59	1.0	63
SPAIN	505,990	43,604	0.4	0.6	44,372	86	77	0.6	78
SRI LANKA	65,610	21,078	1.3	0.8	22,293	321	21	1.2	23
ST HELENA	122	5	-0.8	0.7	5	41			
ST KITTS-NEVIS	360	44	-0.2	1.1	47	121			
ST LUCIA	620	163	1.3	0.8	174	264	32	2.4	37
ST PIERRE AND MIQUELON	242	6	0.0	0.7	6	24			
ST VINCENT	390	120	0.7	0.5	124	308	62	2.0	69
SUDAN	2,505,810	37,793	2.3	1.9	44,035	15	43	4.3	49
SURINAME	163,270	455	0.8	0.6	472	3	78	1.2	82
SWAZILAND	17,360	1,025	2.4	-0.2	992	59	24	0.6	27
SWEDEN	449,960	9,095	0.4	0.3	9,315	20	84	0.4	84
SWITZERLAND	41,290	7,275	0.6	0.2	7,334	176	68	0.2	69
SYRIA	185,180	19,988	2.9	2.3	23,802	108	51	2.7	52
TAIWAN	36,960	22370				605			
TAJIKISTAN	143,100	6,682	2.0	1.4	7,605	47	24	1.1	24
TANZANIA	945,090	39,718	3.0	1.8	45,598	42	40	4.3	47
THAILAND	513,120	65,283	1.3	0.8	69,064	127	33	1.9	37
TOGO	56,790	6,470	3.1	2.5	7,847	114	38	4.4	43
TOKELAU	12	1	-0.8	0.6	2	117			
TONGA	750	103	0.5	0.2	104	137	35	1.2	38
TRINIDAD AND TOBAGO	5,130	1,313	0.6	0.3	1,338	256	77	0.8	80
TUNISIA	163,610	10,319	1.8	1.0	11,140	63	65	1.6	68
TURKEY	774,820	75,161	1.8	1.3	82,640	97	68	2.0	72
TURKMENISTAN	488,100	4,965	2.2	1.3	5,498	10	46	2.1	50
TURKS AND CAICOS	430	28	4.8	2.9	30	64			
TUVALU	26	11	1.1	0.4	11	405			
UGANDA	241,040	30,945	3.3	3.6	41,918	128	13	4.9	14
UKRAINE	603,700	45,509	-0.2	-1.1	41,849	75	68	-0.9	69
UNITED ARAB EMIRATES	83,600	4,775	5.6	3.6	5,588	57	86	2.5	87
UNITED KINGDOM	242,910	60,018	0.3	0.3	61,417	247	89	0.4	90
UNITED STATES	9,629,090	303,851	1.0	0.9	325,723	32	81	1.3	84
URUGUAY	176,220	3,509	0.7	0.6	3,676	20	93	0.8	94
UZBEKISTAN	447,400	27,371	2.1	1.4	30,651	61	36	1.4	37
VANUATU	12,190	219	2.5	1.8	252	18	25	3.8	29
VATICAN	0	1	0.3	0.0	1	1780			
VENEZUELA	912,050	27,684	2.3	1.7	31,330	30	89	1.9	90
VIETNAM	331,690	86,445	1.9	1.3	95,029	261	28	3.2	32
VIRGIN IS. (AM.)	340	112	0.4	0.0	111	329	95	0.2	96
VIRGIN IS. (BR.)	153	23	2.9	1.2	24	147			
WALLIS AND FUTUNA IS.	274	16	0.5	1.2	18	58			
YEMEN	527,970	22,325	3.9	3.1	28,480	42	27	4.8	31
ZAMBIA	752,610	12,056	2.7	1.7	13,841	16	37	2.7	41
ZIMBABWE	390,760	13,162	2.3	0.6	13,804	34	37	2.0	41

* All data available are previous to the political separation of Serbia and Montenegro.

1. World Development Indicators 2006, World Bank.
3. World Population ProspectsThe 2004 Revision. United Nations.
4. Calculated from World Population ProspectsThe 2004 Revision. United Nations.

EDUCATION

	ADULT LITERACY RATE (%) [5]	ADULT MALE LITERACY RATE (%) [5]	ADULT FEMALE LITERACY RATE (%) [5]	PRIMARY SCHOOL NET ENROLMENT RATE (%) [1]	MALE PRIMARY SCHOOL NET ENROLMENT RAT (%) [1]	FEMALE PRIMARY SCHOOL NET ENROLMENT RATE (%) [1]	SECONDARY SCHOOL NET ENROLMENT RATE (%) [1]	MALE SECONDARY SCHOOL NET ENROLMENT RATE (%) [1]	FEMALE SECONDARY SCHOOL NET ENROLMENT RATE	TERTIARY GROSS ENROLMENT RATE (%) [1]	PRIMARY PUPIL TO TEACHER RATIO [1]
	2000-2004	2000-2004	2000-2004	2004	2004	2004	2004	2004	2004	2004	2004
AFGHANISTAN										1	65
ALBANIA	99	99	98	96 2003	96 2003	95 2003	74 2003	73 2003	75 2003	16 2003	22 2003
ALGERIA	70	79	60	97	98	96	66	68	65	20	27
ANDORRA				89	90	87	71	72	71	9	13
ANGOLA	67	82	54							1 2003	
ANGUILLA											
ANTIGUA											19 2000
ARGENTINA	97	97	97	100 1999	100 1999	100 1999	81 2002	83 2002	78 2002	61 2002	17 2002
ARMENIA	99	100	99	97	95	99	89	90	88	26	22
ARUBA				98	98	97	74	75	74	29	19
AUSTRALIA				95 2003	95 2003	95 2003	85 2003	86 2003	85 2003	74 2003	
AUSTRIA							89 2003	89 2003	90 2003	49 2003	13 2003
AZERBAIJAN	99	99	98	84	85	83	77 2003	76 2003	77 2003	15	14
BAHAMAS				84	83	85	74	78	70		20
BAHRAIN	88	92	83	97	96	97	90	93	87	34	16 2002
BANGLADESH	41	50	31				48 2003	51 2003	46 2003	7 2003	54 2003
BARBADOS	100	100	100	100	100	99	95	98	93	38 2001	16
BELARUS	100	100	99	95	97	94	87	88	87	61	15
BELGIUM				100 2003	100 2003	100 2003	97 2003	97 2003	96 2003	61 2003	12 2003
BELIZE	77	77	77	99	99	100	71	73	70	3	23
BENIN	34	46	23	83	93	72	17 2001	11 2001	23 2001	3 2001	52
BERMUDA							86 2002			62 2002	9 2002
BHUTAN											40 2002
BOLIVIA	87	93	80	95	95	96	74	73	74	41	24
BOSNIA-HERZEGOVINA	95	98	91								
BOTSWANA	79	76	82	82 2003	81 2003	84 2003	60 2003	63 2003	56 2003	6	26 2003
BRAZIL	88	88	89	97 2002	98 2000	91 2000	75 2002	78 2002	71 2002	20 2002	24 2002
BRUNEI	93	95	90							13	13
BULGARIA	98	99	98	94 2003	95 2003	94 2003	88 2003	87 2003	89 2003	41 2003	17 2003
BURKINA FASO	13	19	8	41	46	35	10	8	11	2 2003	49
BURUNDI	59	67	52	57	60	54	8	7	9	2	51
CAMBODIA	74	85	64	98	100	96	25 2003	19 2003	30 2003	3	55
CAMEROON	68	77	60							5	53
CANADA				100 2001	99 2001	100 2001	94 1999	94 1999	94 1999	57 2002	17 2002
CAPE VERDE	76	85	68	92	92	91	55	58	52	6	27
CAYMAN IS.				87	89	85	91	96	87	19 2001	13
CENTRAL AFRICAN REP.	49	65	33							2 2000	
CEUTA											
CHAD	26	41	13	57 2003	68 2003	46 2003	11 2003	5 2003	16 2003	1 2001	69
CHILE	96	96	96	86 2003	86 2003	85 2003	78 2003	78 2003	77 2003	43 2003	34 2003
CHINA	91	95	87							15 2003	21 2003
CHRISTMAS IS.											
COCOS											
COLOMBIA	94	94	95	83	83	84	55	58	52	27	28
COMOROS	56	63	49	55 2000	60 2000	51 2000				2	35
CONGO D.R.	65	80	52							1 1999	26 1999
CONGO R.	83	89	77							4 2003	83
COOK IS.											
COSTA RICA	96	96	96	92	91	92	50	52	49	19 2003	22
CÔTE D'IVOIRE	48	60	38	56 2003	62 2003	50 2003	20 2002	15 2002	26 2002	7 1999	42 2003
CROATIA	98	99	97	87 2003	88 2003	87 2003	85 2003	86 2003	84 2003	39 2003	18 2003
CUBA	100	100	100	96	98	95	87	87	86	54	10
CYPRUS	97	99	95	96 2003	96 2003	96 2003	93 2003	94 2003	91 2003	32 2003	19 2003
CZECH REPUBLIC				87 2003	87 2003	87 2003	90 2003	92 2003	89 2003	37 2003	17 2003
DENMARK				100 2003	100 2003	100 2003	95 2003	96 2003	93 2003	67 2003	10 2001
DIEGO GARCÍA IS.											
DJIBOUTI				33	36	29	19	15	22	2	34 2002
DOMINICA				88	87	88	90	92	89		19
DOMINICAN REPUBLIC	88	88	87	86	85	87	49	54	45	33	21
EAST TIMOR							20 2001			10 2002	51 2002
ECUADOR	91	92	90	100	99	100	52	53	52		23
EGYPT	56	67	44	94 2003	96 2003	93 2003	79 2002	77 2002	81 2002	29 2003	22 2003
EL SALVADOR	80	82	77	91	91	91	48 2003	49 2003	47 2003	18	
EQUATORIAL GUINEA	84	92	76	59	61	58	24 2001			3 2000	30
ERITREA				48	52	44	19	16	23	1	47
ESTONIA	100	100	100	95 2003	95 2003	94 2003	88 2003	90 2003	86 2003	65 2003	14 2002
ETHIOPIA	42	49	34	46	49	44	25	19	31	3	65
FAEROE IS.											
FIJI	93	94	91	96	97	96	78 2002	81 2002	75 2002	15	28
FINLAND				100 2003	100 2003	100 2003	94 2003	95 2003	94 2003	87 2003	16 2003
FRANCE				100 2003	100 2003	100 2003	95 2003	96 2003	94 2003	55 2003	19 2003
FRENCH GUIANA											
FRENCH POLYNESIA											
GABON				77 2001	77 2001	77 2001				7 1999	36
GAMBIA				73 2001	76 2001	70 2001	33 2003	27 2003	39 2003	1	38 2001
GEORGIA				93	93	93	69	69	70	42	22
GERMANY										50 2003	14 2003
GHANA	54	63	46	58	62 2003	62 2003	36	33	39	3	32
GIBRALTAR											

EDUCATION

	ADULT LITERACY RATE (%) [5]	ADULT MALE LITERACY RATE (%) [5]	ADULT FEMALE LITERACY RATE (%) [5]	PRIMARY SCHOOL NET ENROLMENT RATE (%) [1]	MALE PRIMARY SCHOOL NET ENROLMENT RAT (%) [1]	FEMALE PRIMARY SCHOOL NET ENROLMENT RATE (%) [1]	SECONDARY SCHOOL NET ENROLMENT RATE (%) [1]	MALE SECONDARY SCHOOL NET ENROLMENT RATE (%) [1]	FEMALE SECONDARY SCHOOL NET ENROLMENT RATE	TERTIARY GROSS ENROLMENT RATE (%) [1]	PRIMARY PUPIL TO TEACHER RATIO [1]
	2000-2004	2000-2004	2000-2004	2004	2004	2004	2004	2004	2004	2004	2004
GREECE	91	94	88	98 2003	98 2003	98 2003	85 2003	87 2003	83 2003	72 2003	12 2003
GREENLAND											
GRENADA											19 2003
GUADELOUPE											
GUAM											
GUATEMALA	69	75	63	93	95	91	34	32	35	10 2002	31
GUINEA				64	69	58	21	14	28	2	45
GUINEA-BISSAU				45 2001	53 2001	38 2001	9 2001	6 2001	11 2001	0 2001	44 2001
GUYANA				100	100	99				9	27 2003
HAITI	52	54	50								
HONDURAS	80	80	80	91	90	92				16	34
HUNGARY	99	99	99	89 2003	90 2003	89 2003	92 2003	92 2003	92 2003	52 2003	10 2003
ICELAND				99 2003	100 2003	98 2003	86 2003	88 2003	85 2003	62 2003	11 2003
INDIA	61	73	48	87 2003	90 2003	84 2003				12 2003	41 2003
INDONESIA	88	92	83	96 2003	97 2003	95 2003	55 2003	55 2003	55 2003	16 2003	20 2003
IRAN	77	84	70	89	89	88	78	76	80	23	20
IRAQ				88	94	81	38	31	44	15	21
IRELAND				96 2003	96 2003	96 2003	85 2003	88 2003	82 2003	55 2003	19 2003
ISRAEL	97	98	96	99 2003	99 2003	99 2003	89 2003	88 2003	89 2003	57 2003	15 2003
ITALY				99 2003	100 2003	99 2003	91 2003	92 2003	91 2003	59 2003	11 2003
JAMAICA	88	84	91	89 2003	88 2003	89 2003	75 2003	77 2003	74 2003	19 2003	30 2003
JAPAN				100 2003	100 2003	100 2003	100 2003	100 2000	99 2000	52 2003	20 2003
JORDAN	90	95	85	93 2003	92 2003	94 2003	82 2003	83 2003	81 2003	35 2003	20 2003
KANAKY-NEW CALEDONIA											
KAZAKHSTAN	100	100	99	98	99	98	92	92	93	48	18
KENYA	74	78	70	76	76	77					40
KIRIBATI				92 2000	92 2000	92 2000					27 2003
KOREA, NORTH											
KOREA, SOUTH				100	100	100	88	88	88	89	30
KUWAIT	83	85	81	86	85	87	78 2002	80 2002	76 2002	22	13
KYRGYZSTAN	99	99	98	90	90	90				40	24
LAOS	69	77	61	84	87	82	30 2005	33 2005	27 2005	3	31
LATVIA	100	100	100	87 2003	87 2003	86 2003	87 2003	88 2003	87 2003	71 2003	14 2003
LEBANON				93	94	93				48	14
LESOTHO	81	74	90	86	83	89	23	28	18	3 2003	44
LIBERIA	56	72	39	66 2000	74 2000	58 2000	17 2000	12 2000	22 2000	16 2000	38 2000
LIBYA	82	92	71							56 2003	
LIECHTENSTEIN											9 2003
LITHUANIA	100	100	100	92 2003	92 2003	92 2003	94 2003	94 2003	94 2003	69 2003	16 2003
LUXEMBOURG				90 2003	90 2003	91 2003	80 2003	83 2003	77 2003	12 2003	12 2003
MACEDONIA, TFYR	96	98	94	92 2003	92 2003	92 2003	81 2002	80 2002	82 2002	27 2003	20 2003
MADAGASCAR	71	76	65	89	89	89	11 1999	11 1999	11 1999	3	52
MALAWI	64	75	54	95	93	98	25	23	27	0	63 1999
MALAYSIA	89	92	85	93 2002	93 2002	93 2002	70 2002	74 2002	66 2002	29 2002	19 2002
MALDIVES	96	96	96	90 2002	89 2002	90 2002	51 2002	55 2002	48 2002	0 2003	18 2003
MALI	19	27	12	47	50	43				2	52
MALTA	88	86	89	94 2003	95 2003	94 2003	86 2003	87 2003	85 2003	30 2003	18 2003
MALVINAS-FALKLANDS											
MARSHALL IS.				84 2002	85 2002	84 2002	65 2002	66 2002	64 2002	17 2002	17 2002
MARTINIQUE											
MAURITANIA	51	60	43	74	75	74	14	13	16	4	45
MAURITIUS	84	88	81	95	94	96	75	78	72	17	22
MAYOTTE IS.											
MELILLA											
MEXICO	90	92	89	100 2003	100 2003	100 2003	62 2003	63 2003	62 2003	23 2003	27 2003
MICRONESIA										14 1999	
MOLDOVA	96	97	95	78	78	77	69	70	67	32	19
MONACO											16 2001
MONGOLIA	98	98	98	84	84	85	82	88	77	39	35
MONTSERRAT											
MOROCCO	51	63	38	87	90 2003	84 2003	35 2003	32 2003	38 2003	11	28
MOZAMBIQUE	46	62	31	71	75	67	4	4	5	1	65
MYANMAR-BURMA	90	94	86	85 2003	85 2003	86 2003	34 2003	33 2003	35 2003	11 2002	33 2003
NAMIBIA	85	87	83	74 2003	71 2003	77 2003	38 2003	43 2003	32 2003	6 2003	28 2003
NAURU											
NEPAL	49	63	35	66 2000	72 2000	59 2000				6	36 2003
NETHERLANDS				99 2003	100 2003	99 2003	89 2003	89 2003	89 2003	58 2003	
NETHERLANDS ANTILLES							77 2003	81 2003	73 2003	24 2002	20 2003
NEW ZEALAND-AOTEAROA				100 2003	100 2003	100 2003	92 2003	94 2003	91 2003	72 2003	18 2003
NICARAGUA	77	77	77	88	89	87	41	43	38	18 2003	35
NIGER	14	20	9	39	46	32	7	5	8	1	44
NIGERIA	67	74	59	88	95	81	28 2003	25 2003	31 2003	10	36
NIUE											
NORFOLK IS.											
NORTHERN MARIANAS											
NORWAY				100 2003	100 2003	99 2003	95 2003	96 2003	95 2003	80 2003	10 2003
OMAN	74	82	65	78	77	79	75	75	74	13	19
PAKISTAN	49	62	35	66	76	56				3	47
PALAU				97 2001	98 2000	95 2000				41 2002	16 2000

EDUCATION

	ADULT LITERACY RATE (%) [5]	ADULT MALE LITERACY RATE (%) [5]	ADULT FEMALE LITERACY RATE (%) [5]	PRIMARY SCHOOL NET ENROLMENT RATE (%) [1]	MALE PRIMARY SCHOOL NET ENROLMENT RAT (%) [1]	FEMALE PRIMARY SCHOOL NET ENROLMENT RATE (%) [1]	SECONDARY SCHOOL NET ENROLMENT RATE (%) [1]	MALE SECONDARY SCHOOL NET ENROLMENT RATE (%) [1]	FEMALE SECONDARY SCHOOL NET ENROLMENT RATE	TERTIARY GROSS ENROLMENT RATE (%) [1]	PRIMARY PUPIL TO TEACHER RATIO [1]
	2000-2004	2000-2004	2000-2004	2004	2004	2004	2004	2004	2004	2004	2004
PALESTINE	92	96	87	86	86	86	89	92	87	38	27
PANAMA	92	93	91	100	100	100	64	67	61	46	24
PAPUA NEW GUINEA	57	63	51								36 2003
PARAGUAY	92	93	90	89 2002	89 2002	90 2002	51 2002	53 2002	50 2002	26 2002	27 2002
PERU	88	93	82	100 2002	100 2002	100 2002	69 2002	68 2002	70 2002	32 2001	25 2002
PHILIPPINES	93	93	93	94 2003	93 2003	95 2003	59 2003	65 2003	54 2003	29 2003	35 2003
PITCAIRN											
POLAND				98 2003	98 2003	98 2003	92 2003	93 2003	91 2003	60 2003	13 2003
PORTUGAL							82 2003	87 2003	78 2003	56 2003	11 2003
PUERTO RICO											
QATAR	89			90	90	90	87	86	88	18	9
RÉUNION											
ROMANIA	97	98	96	90 2003	90 2003	90 2003	81 2003	82 2003	80 2003	36 2003	18 2003
RUSSIA	99	100	99							65 2003	17 2003
RWANDA	64	70	59	73	72	75				3	62
SAHARA, WESTERN											
SAMOA	99	99	98	94 2002	94 2002	94 2002	66 2002	70 2002	62 2002	8 2001	27 2002
SAMOA, AMERICAN											
SAN MARINO											5 2000
SÃO TOMÉ AND PRÍNCIPE				85 1999	85 1999	85 1999	29 2002	26 2002	31 2002	1 2002	33 2002
SAUDI ARABIA	79	87	69	53	54	53	52	51	54	28	12
SENEGAL	39	51	29	66	68	65	15	13	18	5	43
SERBIA AND MONTENEGRO*	96	99	94	96 2001	96 2001	96 2001				36 2001	20 2001
SEYCHELLES	92	91	92	100 2002	100 2002	99 2002	98 2003	97 2003	100 2003		14 2003
SIERRA LEONE	30	40	21							2 2002	37 2001
SINGAPORE	93	97	89								
SLOVAKIA	100	100	100	85 2003	85 2003	85 2003	88 2003	88 2003	88 2003	34 2003	18 2003
SLOVENIA	100	100	100	96 2003	97 2003	96 2003	95 2003	96 2003	95 2003	70 2003	13 2003
SOLOMON IS.				80	80	79	27 2002	24 2002	29 2002		19 1999
SOMALIA											
SOUTH AFRICA	82	84	81	89 2003	88 2003	89 2003	62 2000	65 2000	58 2000	15 2003	34 2003
SPAIN				100 2003	100 2003	99 2003	95 2003	97 2003	93 2003	64 2003	14 2003
SRI LANKA	90	92	89	99 2003	99 2003	98 2003					23 2003
ST HELENA											
ST KITTS-NEVIS											17
ST LUCIA	90	90	91	98	99	96	63	62	63	14	23
ST PIERRE AND MIQUELON											
ST VINCENT				94	95	93	62	63	62		17
SUDAN	59	69	50	43 2000	47 2000	39 2000				6 2000	29 2003
SURINAME	88	92	84	92 2003	90 2003	96 2003	63 2003	74 2003	53 2003	12 2002	20 2003
SWAZILAND	79	80	78	77 2003	76 2003	77 2003	29 2003	32 2003	26 2003	4 2003	31 2003
SWEDEN				100 2003	100 2003	100 2003	98 2003	99 2003	98 2003	82 2003	11 2003
SWITZERLAND				94 2003	94 2003	94 2003	83 2003	79 2003	86 2003	45 2003	13 2000
SYRIA	83	91	74	98	100	96	58	56	60		18
TAIWAN											
TAJIKISTAN	99	100	99	98	100	96	79	73	86	16	22
TANZANIA	69	78	62	86	87	85				1	58
THAILAND	93	95	91	87	88	86				41	21 2003
TOGO	53	68	38	79	85	72	22 2000	14 2000	30 2000	4 2001	44
TOKELAU											
TONGA	99	99	99	98	100	96	71 2001	75 2001	68 2001	3 2001	20
TRINIDAD AND TOBAGO	98	99	98	92	92	92	72	74	70	12	18
TUNISIA	74	83	65	97 2003	97 2003	97 2003	64 2003	67 2003	61 2003	26 2003	22 2003
TURKEY	88	96	81	90 2003	92 2003	87 2003				28 2003	
TURKMENISTAN	99	99	98								
TURKS AND CAICOS											
TUVALU											
UGANDA	69	79	59	98	98	99	15	14	16	3	50
UKRAINE	99	100	99	86	86	86	84	84	83	66	19
UNITED ARAB EMIRATES	77	76	81	71	72	70	62	64	61	23 2003	15
UNITED KINGDOM				100 2003	100 2003	100 2003	96 2003	97 2003	94 2003	63 2003	17 2003
UNITED STATES				94 2003	93 2003	95 2003	89 2003	89 2003	88 2003	83 2003	15 2003
URUGUAY	98	97	98	90 2002	90 2002	91 2002	73 2002	77 2002	70 2002	38 2002	21 2002
UZBEKISTAN	99	100	99							15 2003	
VANUATU	74			94	95	93	39	36	42	5	20
VATICAN											
VENEZUELA	93	93	93	92	92	92	61	65	57	39 2003	20
VIETNAM	90	94	87	93 2002	97 2001	92 2001	62 2001			10 2003	23
VIRGIN IS. (AM.)											
VÍRGIN IS. (BR.)											
WALLIS AND FUTUNA IS.											
YEMEN	49	69	29	75	87	63	34 2000	21 2000	46 2000	9	30 1999
ZAMBIA	68	76	60	80	80	80	24	21	27	2 2000	49
ZIMBABWE	90	94	86	82 2003	81 2003	82 2003	34 2003	33 2003	35 2003	4 2003	39 2003

* All data available are previous to the political separation of Serbia and Montenegro.

1. World Development Indicators 2006, World Bank.
5. The State of the World's Children 2006, UNICEF.

	LIFE EXPECTANCY AT BIRTH (YEARS) [3]	LIFE EXPECTANCY AT BIRTH, MALE (YEARS) [3]	LIFE EXPECTANCY AT BIRTH, FEMALE (YEARS) [3]	TOTAL FERTILITY RATE (CHILDREN PER WOMAN) [5]	CRUDE BIRTH RATE (PER 1,000 PEOPLE) [3]	CRUDE DEATH RATE (PER 1,000 PEOPLE) [3]	WOMEN 15-49 YEARS IN UNION USING CONTRACEPTIVES (%) [5]	MATERNAL MORTALITY (PER 100,000 LIVE BIRTHS) [5]	BIRTHS ATTENDED BY TRAINED HEALTH PERSONNEL (%) [5]	INFANT MORTALITY RATE (PER 1,000 LIVE BIRTHS) [5]
	2005-2010	2005-2010	2005-2010	2005-2010	2005-2010	2005-2010	1996-2004	2000	1996-2004	2004
AFGHANISTAN	48	47	48	7.1	48	18	10	1900	14	165
ALBANIA	74	72	77	2.2	17	7	75	55	98	17
ALGERIA	72	71	74	2.4	21	5	57	140	96	35
ANDORRA										6
ANGOLA	42	40	43	6.4	47	21	6	1700	45	154
ANGUILLA										
ANTIGUA							53		100	11
ARGENTINA	75	72	79	2.3	18	8	74	82	99	16
ARMENIA	72	68	75	1.4	12	10	61	55	97	29
ARUBA										
AUSTRALIA	81	78	83	1.8	12	7	76	8	100	5
AUSTRIA	80	77	82	1.4	9	10	51	4	100	5
AZERBAIJAN	67	64	71	1.9	16	8	55	94	84	75
BAHAMAS	72	69	75	2.2	18	7	62	60	99	10
BAHRAIN	75	74	77	2.3	16	3	62	28	98	9
BANGLADESH	65	64	66	3.0	25	7	59	380	13	56
BARBADOS	76	73	79	1.5	11	8	55	95	98	10
BELARUS	69	63	75	1.2	10	15	50	35	100	9
BELGIUM	80	76	83	1.7	10	10	78	10	100	4
BELIZE	72	70	74	2.8	25	5	56	140	83	32
BENIN	56	55	57	5.4	40	12	19	850	66	90
BERMUDA										
BHUTAN	65	64	66	3.8	29	8	31	420	37	67
BOLIVIA	66	63	68	3.5	27	8	58	420	67	54
BOSNIA-HERZEGOVINA	75	72	78	1.3	9	10	48	31	100	13
BOTSWANA	34	35	33	2.9	25	28	48	100	94	84
BRAZIL	72	68	76	2.2	19	7	77	260	96	32
BRUNEI	77	75	80	2.3	22	3		37	99	8
BULGARIA	73	70	76	1.2	8	15	42	32	99	12
BURKINA FASO	49	48	50	6.3	46	16	14	1000	38	97
BURUNDI	46	44	47	6.8	47	18	16	1000	25	114
CAMBODIA	58	55	61	3.7	30	10	24	450	32	97
CAMEROON	46	46	47	4.1	33	17	26	730	62	87
CANADA	81	78	83	1.5	10	7	75	6	98	5
CAPE VERDE	72	68	74	3.4	29	5	53	150	89	27
CAYMAN IS.										
CENTRAL AFRICAN REP.	40	39	40	4.6	36	22	28	1100	44	115
CEUTA										
CHAD	44	43	45	6.7	49	19	8	1100	16	117
CHILE	79	75	82	1.9	15	5	56	31	100	8
CHINA	73	71	75	1.7	13	7	87	56	96	26
CHRISTMAS IS.										
COCOS										
COLOMBIA	73	70	76	2.5	20	5	77	130	86	18
COMOROS	65	63	67	4.3	33	6	26	480	62	52
CONGO D.R.	45	44	46	6.7	50	19	31	990	61	129
CONGO R.	54	52	55	6.3	44	12		510		81
COOK IS.							44		98	18
COSTA RICA	79	76	81	2.1	18	4	80	43	98	11
CÔTE D'IVOIRE	46	46	47	4.5	35	17	15	690	68	117
CROATIA	76	72	79	1.3	9	12		8	100	6
CUBA	79	77	80	1.6	11	7	73	33	100	6
CYPRUS	79	77	82	1.6	12	7		47	100	5
CZECH REPUBLIC	76	73	79	1.2	9	11	72	9	100	4
DENMARK	78	76	80	1.8	11	11	78	5	100	4
DIEGO GARCÍA IS.										
DJIBOUTI	54	53	55	4.5	33	12		730	61	101
DOMINICA							50		100	13
DOMINICAN REPUBLIC	69	65	72	2.6	23	6	70	150	99	27
EAST TIMOR	58	57	59	7.2	51	12	10	660	18	64
ECUADOR	75	72	78	2.6	21	5	66	130	69	23
EGYPT	71	69	73	3.0	25	6	60	84	69	26
EL SALVADOR	72	69	75	2.7	23	6	67	150	92	24
EQUATORIAL GUINEA	42	41	42	5.9	43	21		880	65	122
ERITREA	56	54	58	5.0	38	10	8	630	28	52
ESTONIA	73	67	78	1.4	10	14	70	63	100	6
ETHIOPIA	49	48	49	5.4	39	15	8	850	6	110
FAEROE IS.										
FIJI	69	66	71	2.7	22	6	44	75	99	16
FINLAND	79	76	82	1.7	10	10	77	6	100	3
FRANCE	80	77	83	1.9	12	10	75	17	99	4
FRENCH GUIANA	76	73	79	3.0	22	4				
FRENCH POLYNESIA	74	72	77	2.3	18	5				
GABON	53	53	54	3.5	29	13	33	420	86	60
GAMBIA	58	56	59	4.2	32	11	18	540	55	89
GEORGIA	71	67	75	1.4	11	12	41	32	96	41
GERMANY	79	76	82	1.3	8	11	75	8	100	4
GHANA	58	58	59	3.8	30	10	25	540	47	68
GIBRALTAR										

	UNDER 5 MORTALITY RATE (PER 1,000 LIVE BIRTHS) [5]	INFANTS WITH LOW BIRTH WEIGHT (2500 GM) (%) [5]	MALNUTRITION (% OF CHILDREN UNDER 5 YEARS OLD) [5]	UNDERNOURISHED PEOPLE AS % OF TOTAL POPULATION) [9]	WOMEN WHO BREASTFEED FOR SIX MONTHS (%) [5]	DAILY CALORIE CONSUMPTION (PER CAPITA) [8]	DOCTORS PER 100,000 PEOPLE [6]	NURSES (PER 100,000 PEOPLE) [7]	ACCESS TO IMPROVED WATER SOURCES (% OF TOTAL POPULATION) [5]	ACCESS TO SANITATION SERVICES (% OF POPULATION) [5]
	2004	1998-2004	1996-2004	2000-2002	1996-2004	2003	1990-2004	2004	2002	2002
AFGHANISTAN	257		39					22 2001	13	8
ALBANIA	19	3	14		6	2874	139	362 2003	97	89
ALGERIA	40	7	10	5	13	3055	85	199 2002	87	92
ANDORRA	7							311 2003	100	100
ANGOLA	260	12	31	40	11	2089	8	115 1997	50	30
ANGUILLA										
ANTIGUA	12	8	10			2313	17	328 1999	91	95
ARGENTINA	18	8	5	<2,5		2959	301	80 1998		
ARMENIA	32	7	3		30	2357	353	435 2003	92	84
ARUBA										
AUSTRALIA	6	7				3135	249	910 2001	100	100
AUSTRIA	5	7				3732	324	938 2003	100	100
AZERBAIJAN	90	11	7		7	2727	354	711 2003	77	55
BAHAMAS	13	7				2709	106	447 1998	97	100
BAHRAIN	11	8	9		34		160	404		
BANGLADESH	77	36	48	30	36	2193	23	14	75	48
BARBADOS	12	10	6			3123	121	370 1999	100	99
BELARUS	11	5				2885	450	1163 2003	100	
BELGIUM	5	8				3634	418	583 2003		
BELIZE	39	6	6		24	2876	105	126 2000	91	47
BENIN	152	16	23	15	38	2574	6	72	68	32
BERMUDA						2235				
BHUTAN	80	15	19				5	14	62	70
BOLIVIA	69	7	8	21	54	2219	73	319 2001	85	45
BOSNIA-HERZEGOVINA	15	4	4		6	2668	134	413 2003	98	93
BOTSWANA	116	10	13	32	34	2196	29	265	95	41
BRAZIL	34	10	6	9		3146	206	384 2000	89	75
BRUNEI	9	10				2845	101	267 2000		
BULGARIA	15	10				2885	338	375 2003	100	100
BURKINA FASO	192	19	38	19	19	2516	4	37	51	12
BURUNDI	190	16	45	68	62	1647	5	19	79	36
CAMBODIA	141	11	45	33	12	2074	16	61 2000	34	16
CAMEROON	149	11	18	25	21	2286	7	160	63	48
CANADA	6	6				3605	209	995 2003	100	100
CAPE VERDE	36	13	14		57	3216	17	87	80	42
CAYMAN IS.										
CENTRAL AFRICAN REP.	193	14	24	43	17	1932	4	26	75	27
CEUTA										
CHAD	200	10	28	34	2	2147	3	24	34	8
CHILE	8	5	1	4	63	2872	109	63 2003	95	92
CHINA	31	4	8	11	51	2940	164	105 2001	77	44
CHRISTMAS IS.										
COCOS										
COLOMBIA	21	9	7	13	26	2567	135	55 2002	92	86
COMOROS	70	25	25		21	1760	7	61	94	23
CONGO D.R.	205	12	31	71	24	1606	7	53	46	29
CONGO R.	108		14	37	4	2183	25	84	46	9
COOK IS.	21	3			19			272 2001	95	100
COSTA RICA	13	7	5	4	35	2813	173	92 2000	97	92
CÔTE D'IVOIRE	194	17	17	14	5	2644	9	46	84	40
CROATIA	7	6	1		23	2795	237	505 2003		
CUBA	7	6	4	3	41	3286	591	744 2002	91	98
CYPRUS	5					3246	298	376 2002	100	100
CZECH REPUBLIC	4	7	1			3308	343	971 2003		
DENMARK	5	5				3472	366	1036 2002	100	
DIEGO GARCÍA IS.										
DJIBOUTI	126		18			2239	13	26	80	50
DOMINICA	14	10	5			2785	49	417 1997	97	83
DOMINICAN REPUBLIC	32	11	5	25	10	2281	188	184 2000	93	57
EAST TIMOR	80	12	46		31	2819		179	52	33
ECUADOR	26	16	12	4	35	2641	148	157 2000	86	72
EGYPT	36	12	9	3	30	3356	212	198	98	68
EL SALVADOR	28	7	10	11	24	2556	124	80 2002	82	63
EQUATORIAL GUINEA	204	13	19		24		25	43	44	53
ERITREA	82	21	40	73	52	1520	3	55	57	9
ESTONIA	8	4				3222	316	850 2000		
ETHIOPIA	166	15	47	46	55	1858	3	20 2003	22	6
FAEROE IS.										
FIJI	20	10	8		47	2974	34	196 1999		98
FINLAND	4	4				3143	311	1433 2002	100	100
FRANCE	5	7				3623	329	724		
FRENCH GUIANA										
FRENCH POLYNESIA						2912				
GABON	91	14	12	6	6	2671	29	479	87	36
GAMBIA	122	17	17	27	26	2288	4	113 2003	82	53
GEORGIA	45	7	3		18	2646	391	347 2003	76	83
GERMANY	5	7				3484	362	972 2003	100	
GHANA	112	16	22	13	53	2680	9	74	79	58
GIBRALTAR										

	LIFE EXPECTANCY AT BIRTH (YEARS) [3]	LIFE EXPECTANCY AT BIRTH, MALE (YEARS) [3]	LIFE EXPECTANCY AT BIRTH, FEMALE (YEARS) [3]	TOTAL FERTILITY RATE (CHILDREN PER WOMAN) [5]	CRUDE BIRTH RATE (PER 1,000 PEOPLE) [3]	CRUDE DEATH RATE (PER 1,000 PEOPLE) [3]	WOMEN 15-49 YEARS IN UNION USING CONTRACEPTIVES (%) [5]	MATERNAL MORTALITY (PER 100,000 LIVE BIRTHS) [5]	BIRTHS ATTENDED BY TRAINED HEALTH PERSONNEL (%) [5]	INFANT MORTALITY RATE (PER 1,000 LIVE BIRTHS) [5]
	2005-2010	2005-2010	2005-2010	2005-2010	2005-2010	2005-2010	1996-2004	2000	1996-2004	2004
GREECE	79	76	81	1.3	9	11		9		4
GREENLAND										
GRENADA							54		100	18
GUADELOUPE	79	76	82	2.0	14	6				
GUAM	76	73	78	2.7	19	5				
GUATEMALA	68	65	72	4.2	33	6	43	240	41	33
GUINEA	54	54	54	5.5	40	13	7	740	56	101
GUINEA-BISSAU	45	44	47	7.1	49	19	8	1100	35	126
GUYANA	65	62	68	2.1	19	8	37	170	86	48
HAITI	53	53	54	3.6	29	13	27	680	24	74
HONDURAS	69	67	71	3.3	27	6	62	110	56	31
HUNGARY	74	70	78	1.3	9	13	77	16	100	7
ICELAND	81	80	83	1.9	14	6		0		2
INDIA	65	63	67	2.8	23	8	47	540	43	62
INDONESIA	69	67	70	2.2	19	7	57	230	72	30
IRAN	72	70	73	2.0	20	5	74	76	90	32
IRAQ	61	60	63	4.2	32	8	44	250	72	102
IRELAND	78	76	81	1.9	15	7		5	100	5
ISRAEL	81	78	83	2.7	19	6	68	17	99	5
ITALY	81	77	84	1.4	9	11	60	5		4
JAMAICA	71	69	73	2.3	19	8	66	87	97	17
JAPAN	83	79	86	1.4	9	9	59	10	100	3
JORDAN	72	71	74	3.1	25	4	56	41	100	23
KANAKY-NEW CALEDONIA	76	74	79	2.3	17	5				
KAZAKHSTAN	64	59	70	1.9	16	11	66	210	99	63
KENYA	50	51	49	5.0	40	14	39	1000	42	79
KIRIBATI							21		85	49
KOREA, NORTH	64	62	67	1.9	14	11	62	67	97	42
KOREA, SOUTH	78	74	82	1.2	9	6	81	20	100	5
KUWAIT	78	76	80	2.3	19	2	50	5	98	10
KYRGYZSTAN	68	64	72	2.5	21	7	60	110	98	58
LAOS	56	55	58	4.3	33	11	32	650	19	65
LATVIA	73	67	78	1.3	9	14	48	42	100	10
LEBANON	73	71	75	2.2	18	7	63	150	89	27
LESOTHO	34	34	34	3.3	27	26	30	550	60	61
LIBERIA	43	42	43	6.8	49	20	10	760	51	157
LIBYA	75	73	77	2.7	23	4	45	97	94	18
LIECHTENSTEIN										4
LITHUANIA	73	68	79	1.3	9	12	47	13	100	8
LUXEMBOURG	79	76	82	1.7	12	8		28	100	5
MACEDONIA, TFYR	74	72	77	1.4	11	9		23	99	13
MADAGASCAR	56	55	57	4.9	37	11	27	550	51	76
MALAWI	41	42	41	5.7	42	20	31	1800	61	110
MALAYSIA	74	72	76	2.6	20	5	55	41	97	10
MALDIVES	69	69	68	3.8	30	6	39	110	70	35
MALI	49	49	50	6.6	48	16	8	1200	41	121
MALTA	79	77	81	1.5	10	8		0	98	5
MALVINAS-FALKLANDS										
MARSHALL IS.							34		95	52
MARTINIQUE	79	76	82	1.9	12	8				
MAURITANIA	54	53	56	5.5	40	13	8	1000	57	78
MAURITIUS	73	70	76	1.9	15	7	76	24	98	14
MAYOTTE IS.										
MELILLA										
MEXICO	76	74	79	2.1	19	4	73	83	95	23
MICRONESIA	68	68	69	4.2	29	6	45		88	19
MOLDOVA	70	66	73	1.2	11	11	62	36	99	23
MONACO										4
MONGOLIA	66	64	68	2.2	21	7	69	110	97	41
MONTSERRAT										
MOROCCO	71	69	73	2.6	22	6	63	220	63	38
MOZAMBIQUE	42	42	42	5.1	38	20	17	1000	48	104
MYANMAR-BURMA	62	59	65	2.1	18	9	34	360	57	76
NAMIBIA	46	47	45	3.5	26	16	44	300	76	47
NAURU										25
NEPAL	64	63	64	3.3	28	8	38	740	15	59
NETHERLANDS	79	76	82	1.7	11	9	79	16	100	5
NETHERLANDS ANTILLES	77	74	80	2.0	13	7				
NEW ZEALAND-AOTEAROA	80	78	82	2.0	13	7	75	7	100	5
NICARAGUA	71	69	73	2.9	27	5	69	230	67	31
NIGER	45	45	45	7.5	52	20	14	1600	16	152
NIGERIA	44	44	44	5.3	40	18	13	800	35	101
NIUE									100	
NORFOLK IS.										
NORTHERN MARIANAS										
NORWAY	80	78	83	1.8	11	9	74	16	100	4
OMAN	75	74	77	3.2	24	3	32	87	95	10
PAKISTAN	65	65	65	3.7	29	8	28	500	23	80
PALAU							17		100	22

	UNDER 5 MORTALITY RATE (PER 1,000 LIVE BIRTHS) [5]	% OF INFANTS WITH LOW BIRTH WEIGHT (<2500 GM) [5]	% OF MALNUTRITION (UNDER 5 YEARS OLD) [5]	UNDERNOURISHED PEOPLE AS % OF TOTAL POPULATION [9]	% OF WOMEN WHO BREASTFEED FOR SIX MONTHS [5]	DAILY CALORIE CONSUMPTION PER CAPITA [8]	DOCTORS PER 100,000 PEOPLE [6]	NURSES PER 100,000 PEOPLE [7]	% OF POPULATION WITH ACCESS TO IMPROVED WATER SOURCES [5]	% OF POPULATION WITH ACCESS TO SANITATION SERVICES [5]
	2004	1998-2004	1996-2004	2000-2002	1996-2004	2003	1990-2004	2004	2002	2002
GREECE	5	8				3666	440	386 2000		
GREENLAND										
GRENADA	21	9			39	2990	50	370 1997	95	97
GUADELOUPE										
GUAM										
GUATEMALA	45	12	23	24	51	2227	90	405 1999	95	61
GUINEA	155	16	21	26	23	2447	9	52	51	13
GUINEA-BISSAU	203	22	25		37	2051	17	67	59	34
GUYANA	64	12	14	9	11	2764	48	229 2000	83	70
HAITI	117	21	17	47	24	2109	25	11 1998	71	34
HONDURAS	41	14	17	22	35	2373	83	129 2000	90	68
HUNGARY	8	9	2			3552	316	885 2003	99	95
ICELAND	3	4				3275	347	1363 2003	100	
INDIA	85	30	47	21	37	2473	51	80	86	30
INDONESIA	38	9	28	6	40	2891	16	57 2003	78	52
IRAN	38	7	11	4	44	3096	105	119	93	84
IRAQ	125	15	16		12			125	81	80
IRELAND	6	6				3717	237	1520		
ISRAEL	6	8				3554	391	626 2003	100	
ITALY	5	6				3675	606	544 2003		
JAMAICA	20	10	4	10		2690	85	165 2003	93	80
JAPAN	4	8				2768	201	779 2002	100	100
JORDAN	27	10	4	7	27	2680	205	324	91	93
KANAKY-NEW CALEDONIA						2782				
KAZAKHSTAN	73	8	4		36	2858	330	601 2003	86	72
KENYA	120	10	20	33	13	2155	13	118 2002	62	48
KIRIBATI	65	5	13		80	2846		236 1998	64	39
KOREA, NORTH	55	7	23	36	65	2178		385 2003	100	59
KOREA, SOUTH	6	4		<2,5		3035	181	175 2003	92	
KUWAIT	12	7	10	5	12	3061	153	391 2001		
KYRGYZSTAN	68	7	11		24	3173	268	614 2003	76	60
LAOS	83	14	40	22	23	2338	59	103 1996	43	24
LATVIA	12	5				3014	291	527 2003		
LEBANON	31	6	3	3	27	3164	325	118 2001	100	98
LESOTHO	82	14	18	12	15	2626	5	62 2003	76	37
LIBERIA	235		26	46	35	1930		18	62	26
LIBYA	20	7	5	<2,5		3337	129	360 1997	72	97
LIECHTENSTEIN	5									
LITHUANIA	8	4				3372	403	762 2003		
LUXEMBOURG	6	8					255	916 2003	100	
MACEDONIA, TFYR	14	6	6		37	2852		519 2001		
MADAGASCAR	123	17	42	37	67	2056	9	20	45	33
MALAWI	175	16	22	33	44	2125	1	59	67	46
MALAYSIA	12	9	11	<2,5	29	2867	70	135 2000	95	
MALDIVES	46	22	30		10	2558	78	270	84	58
MALI	219	23	33	29	25	2237	4	45	48	45
MALTA	6	6				3521	293	583 2003	100	
MALVINAS-FALKLANDS										
MARSHALL IS.	59	12			63			298 2000	85	82
MARTINIQUE										
MAURITANIA	125		32	10	20	2786	14	56	56	42
MAURITIUS	15	14	15	6	21	2970	85	360	100	99
MAYOTTE IS.										
MELILLA										
MEXICO	28	8	8	5	38	3171	171	90 2000	91	77
MICRONESIA	23	18			60			383 2000	94	28
MOLDOVA	28	5	3			2729	269	606 2003	92	68
MONACO	5							1419 1995		
MONGOLIA	52	7	13	28	51	2250	267	313 2002	62	59
MONTSERRAT										
MOROCCO	43	11	9	7	31	3098	48	72	80	61
MOZAMBIQUE	152	15	24	47	30	2082	2	21	42	27
MYANMAR-BURMA	106	15	32	6	15	2912	30	20	80	73
NAMIBIA	63	14	24	22	19	2290	30	306	80	30
NAURU	30							545 1995		
NEPAL	76	21	48	17	68	2483	5	22	84	27
NETHERLANDS	6					3495	329	1373 2003	100	100
NETHERLANDS ANTILLES						2592				
NEW ZEALAND-AOTEAROA	6	6				3199	223	816 2001		
NICARAGUA	38	12	10	27	31	2291	164	107 2003	81	66
NIGER	259	13	40	34	1	2170	3	19	46	12
NIGERIA	197	14	29	9	17	2714	27	103 2003	60	38
NIUE		0						550 1996	100	100
NORFOLK IS.										
NORTHERN MARIANAS										
NORWAY	4	5				3511	356	1484 2003	100	
OMAN	13	8	24				126	350	79	89
PAKISTAN	101	19	38	20	16	2316	66	31	90	54
PALAU	27	9			59			144 1998	84	83

	LIFE EXPECTANCY AT BIRTH (YEARS) [3]	LIFE EXPECTANCY AT BIRTH, MALE (YEARS) [3]	LIFE EXPECTANCY AT BIRTH, FEMALE (YEARS) [3]	TOTAL FERTILITY RATE (CHILDREN PER WOMAN) [5]	CRUDE BIRTH RATE (PER 1,000 PEOPLE) [3]	CRUDE DEATH RATE (PER 1,000 PEOPLE) [3]	WOMEN 15-49 YEARS IN UNION USING CONTRACEPTIVES (%) [5]	MATERNAL MORTALITY (PER 100,000 LIVE BIRTHS) [5]	BIRTHS ATTENDED BY TRAINED HEALTH PERSONNEL (%) [5]	INFANT MORTALITY RATE (PER 1,000 LIVE BIRTHS) [5]
	2005-2010	2005-2010	2005-2010	2005-2010	2005-2010	2005-2010	1996-2004	2000	1996-2004	2004
PALESTINE	73	72	75	5.0	35	4	51	100	97	22
PANAMA	76	73	78	2.6	21	5	58	160	93	19
PAPUA NEW GUINEA	57	57	58	3.6	28	10	26	300	41	68
PARAGUAY	72	70	74	3.5	28	5	57	170	77	21
PERU	71	69	74	2.7	22	6	69	410	59	24
PHILIPPINES	72	69	74	2.8	23	5	49	200	60	26
PITCAIRN										
POLAND	75	71	79	1.2	10	10	49	13	100	7
PORTUGAL	78	75	81	1.5	10	11	66	5	100	4
PUERTO RICO	77	73	81	1.9	14	8				
QATAR	74	72	77	2.8	17	3	43	140	99	18
RÉUNION	76	72	80	2.4	19	6				
ROMANIA	72	69	76	1.3	10	13	64	49	99	17
RUSSIA	65	59	72	1.4	11	16		67	99	17
RWANDA	45	43	46	5.2	41	18	13	1400	31	118
SAHARA, WESTERN	66	64	68		27	7				
SAMOA	71	69	75	3.9	25	5	30	130	100	25
SAMOA, AMERICAN										
SAN MARINO										3
SÃO TOMÉ AND PRÍNCIPE	64	63	65	3.6	32	8	29		76	75
SAUDI ARABIA	73	71	75	3.6	26	4	32	23	91	21
SENEGAL	57	56	58	4.5	35	11	11	690	58	78
SERBIA AND MONTENEGRO*	74	72	76	1.6	11	11	58	11	93	13
SEYCHELLES										12
SIERRA LEONE	42	41	43	6.5	46	22	4	2000	42	165
SINGAPORE	79	78	81	1.3	8	6	74	30	100	3
SLOVAKIA	75	71	79	1.2	9	10	74	3	99	6
SLOVENIA	77	74	81	1.2	9	10	74	17	100	4
SOLOMON IS.	63	63	64	3.8	30	7	11	130	85	34
SOMALIA	49	48	50	6.0	43	16	1	1100	25	133
SOUTH AFRICA	44	44	44	2.6	22	21	56	230	84	54
SPAIN	80	77	84	1.3	11	9	81	4		3
SRI LANKA	75	73	78	1.9	15	6	70	92	96	12
ST HELENA										
ST KITTS-NEVIS							41		99	18
ST LUCIA	73	72	75	2.2	19	7	47		100	13
ST PIERRE AND MIQUELON										
ST VINCENT	72	69	75	2.2	20	7	58		100	18
SUDAN	57	56	58	4.0	31	11	7	590	87	63
SURINAME	70	67	73	2.4	20	7	42	110	85	30
SWAZILAND	30	31	29	3.5	28	31	48	370	74	108
SWEDEN	81	79	83	1.7	11	10	78	2	100	3
SWITZERLAND	81	78	84	1.4	9	9	82	7		5
SYRIA	74	72	76	3.1	27	3	48	160	77	15
TAIWAN										
TAJIKISTAN	64	62	67	3.3	28	7	34	100	71	91
TANZANIA	47	46	47	4.4	35	16	26	1500	46	78
THAILAND	72	69	75	1.9	15	7	79	44	99	18
TOGO	56	54	57	4.8	37	11	26	570	61	78
TOKELAU										
TONGA	73	72	74	3.2	22	6	33		95	20
TRINIDAD AND TOBAGO	70	68	73	1.6	14	8	38	160	96	18
TUNISIA	74	72	76	1.9	16	5	66	120	90	21
TURKEY	70	67	72	2.3	20	7	71	70	83	28
TURKMENISTAN	63	59	68	2.5	22	8	62	31	97	80
TURKS AND CAICOS										
TUVALU							32		100	36
UGANDA	52	51	53	7.1	51	13	23	880	39	80
UKRAINE	66	61	73	1.1	9	17	89	35	100	14
UNITED ARAB EMIRATES	79	77	82	2.4	15	1	28	54	99	7
UNITED KINGDOM	79	77	81	1.7	11	10	82	13	99	5
UNITED STATES	78	75	81	2.0	14	8	76	17	99	7
URUGUAY	76	73	80	2.2	16	9	84	27	100	15
UZBEKISTAN	67	64	70	2.5	23	7	68	24	96	57
VANUATU	70	68	72	3.7	29	5	28	130	88	32
VATICAN										
VENEZUELA	74	71	77	2.5	21	5	77	96	94	16
VIETNAM	72	70	74	2.1	19	6	79	130	85	17
VIRGIN IS. (AM.)	79	75	83	2.1	13	7				
VÍRGIN IS. (BR.)										
WALLIS AND FUTUNA IS.										
YEMEN	63	61	64	5.7	39	7	23	570	27	82
ZAMBIA	39	40	39	5.2	40	21	34	750	43	102
ZIMBABWE	37	38	36	3.2	29	23	54	1100	73	79

HEALTH

	UNDER 5 MORTALITY RATE (PER 1,000 LIVE BIRTHS) [5]	% OF INFANTS WITH LOW BIRTH WEIGHT (<2500 GM) [5]	% OF MALNUTRITION (UNDER 5 YEARS OLD) [5]	UNDERNOURISHED PEOPLE AS % OF TOTAL POPULATION [9]	% OF WOMEN WHO BREASTFEED FOR SIX MONTHS [5]	DAILY CALORIE CONSUMPTION PER CAPITA [8]	DOCTORS PER 100,000 PEOPLE [6]	NURSES PER 100,000 PEOPLE [7]	% OF POPULATION WITH ACCESS TO IMPROVED WATER SOURCES [5]	% OF POPULATION WITH ACCESS TO SANITATION SERVICES [5]
	2004	1998-2004	1996-2004	2000-2002	1996-2004	2003	1990-2004	2004	2002	2002
PALESTINE	24	9	4		29	2242	84		94	76
PANAMA	24	10	7	26	25	2287	168	154 2000	91	72
PAPUA NEW GUINEA	93	11	35		59		5	53 2000	39	45
PARAGUAY	24	9	5	14	22	2524	117	169 2002	83	78
PERU	29	11	7	13	67	2579	117	67 1999	81	62
PHILIPPINES	34	20	28	22	34	2480	116	169 2000	85	73
PITCAIRN										
POLAND	8	6				3366	220	490 2003		
PORTUGAL	5	8				3747	324	436 2003		
PUERTO RICO										
QATAR	21	10	6		12		221	494 2001	100	100
RÉUNION										
ROMANIA	20	9	6			3582	189	389 2003	57	51
RUSSIA	21	6	3			3118	417	805 2003	96	87
RWANDA	203	9	27	37	84	2071	2	42	73	41
SAHARA, WESTERN										
SAMOA	30	4				2921	70	202 1999	88	100
SAMOA, AMERICAN										
SAN MARINO	4							9548 1990		
SÃO TOMÉ AND PRÍNCIPE	118	20	13		56	2468	47	155	79	24
SAUDI ARABIA	27	11	14	3	31	2840	140	297		
SENEGAL	137	18	23	24	24	2374	8	25	72	52
SERBIA AND MONTENEGRO*	15	4	2		11	2703		464 2002	93	87
SEYCHELLES	14		6			2484	132	793	87	
SIERRA LEONE	283	23	27	50	4	1943	7	31	57	39
SINGAPORE	3	8	14				140	424 2001		
SLOVAKIA	9	7				2779	325	677 2003	100	100
SLOVENIA	4	6				2954	219	721 2002		
SOLOMON IS.	56	13	21		65	2260	13	80 1999	70	31
SOMALIA	225		26		9			19 1997	29	25
SOUTH AFRICA	67	15	12		7	2962	69	408	87	67
SPAIN	5	6				3421	320	768 2003		
SRI LANKA	14	22	29	22	84	2416	43	120	78	91
ST HELENA										
ST KITTS-NEVIS	21	9			56	2713	118	502 1997	99	96
ST LUCIA	14	8	14			2975	518	228 1999	98	89
ST PIERRE AND MIQUELON										
ST VINCENT	22	10				2626	88	238 1997		
SUDAN	91	31	17	27	16	2260	16	51	69	34
SURINAME	39	13	13	11	9	2697	45	162 2000	92	93
SWAZILAND	156	9	10	19	24	2343	18	424	52	52
SWEDEN	4	4				3208	305	1024 2002	100	100
SWITZERLAND	5	6				3545	352	1075 2000	100	100
SYRIA	16	6	7	4	81	3057	140	194 2001	79	77
TAIWAN										
TAJIKISTAN	118	15			50	1907	218	458 2003	58	53
TANZANIA	126	13	22	44	41	1959	2	37 2002	73	46
THAILAND	21	9	19	20	4	2425	30	282 2000	85	99
TOGO	140	18	25	26	18	2358	6	37	51	34
TOKELAU										
TONGA	25	0			62		34	316 2001	100	97
TRINIDAD AND TOBAGO	20	23	7	12	2	2788	79	287 1997	91	100
TUNISIA	25	7	4	<2,5	47	3247	70	258	82	80
TURKEY	32	16	4	3	21	3328	124	170 2003	93	83
TURKMENISTAN	103	6	12		13	2840	317	904 2002	71	62
TURKS AND CAICOS										
TUVALU	51	5						264 2002	93	88
UGANDA	138	12	23	19	63	2360	5	57	56	41
UKRAINE	18	5	1		22	3054	297	762 2003	98	99
UNITED ARAB EMIRATES	8	15	14	<2,5	34	3238	202	418 2001		100
UNITED KINGDOM	6	8				3450	166	1212 1997		
UNITED STATES	8	8	1			3754	549	937 2000	100	100
URUGUAY	17	8	5	4		2883	365	85 2002	98	94
UZBEKISTAN	69	7	8		19	2312	289	982 2003	89	57
VANUATU	40	6	20		50	2604	11	235 1997	60	50
VATICAN										
VENEZUELA	19	9	4	17	7	2272	194		83	68
VIETNAM	23	9	28	19	15	2617	53	56 2001	73	41
VIRGIN IS. (AM.)										
VIRGIN IS. (BR.)										
WALLIS AND FUTUNA IS.										
YEMEN	111	32	46	36	12	2020	22	64	69	30
ZAMBIA	182	12	23	49	40	1975	7	156	55	45
ZIMBABWE	129	11	13	44	33	2004	6	72	83	57

* All data available are previous to the political separation of Serbia and Montenegro.

3. World Population ProspectsThe 2004 Revision. United Nations. / 5. The State of the World's Children 2006, UNICEF. / 6. Human Development Report 2005 UNDP. / 7. WHOSIS WHO Statistical Information System, Web Site WHO 2006. 8. FAOSTAT Statistical Database FAO Web site 2006 / 9. The State of Food Insecurity in the World 2005, FAO.

	NEWSPAPERS PER 1,000 PEOPLE [1]	RADIOS (PER 1,000 PEOPLE)	TV SETS (PER 1,000 PEOPLE)	TELEPHONES MAINLINES (PER 1,000 PEOPLE) [1]	COMPUTERS (PER 1,000 PEOPLE) [1]
	2000	1996-2001	1998-2003	2004	2004
AFGHANISTAN		114 1997	14 2001	1.7	
ALBANIA		260 1997	318 2002	90.0	11.7 2002
ALGERIA	27.21998	244 1997	114 2001	70.7	9.0
ANDORRA			462 2000		
ANGOLA	11.31998	78 2000	52 2001	6.2	3.2
ANGUILLA					
ANTIGUA		523 1997	452 2001	474.5	
ARGENTINA	40.52000	697 1997	326 2001	226.7	96.4
ARMENIA		264 1997	229 2001	192.5	66.1
ARUBA			226 2001		
AUSTRALIA	161.02000	1996 2001	722 2003	540.6	682.2
AUSTRIA	309.02000	763 1997	637 2002	460.	418.4
AZERBAIJAN	9.91999	22 1997	334 2003	118.4	17.9
BAHAMAS		744 1997	248 2001	439.0	
BAHRAIN		78 2000	428 2002	267.6	169.0
BANGLADESH		49 1997	59 2002	5.9	11.9
BARBADOS		749 2000	328 2001	504.8	126.4
BELARUS		199 2001	362 2001	328.8	
BELGIUM	153.02000	793 1997	541 2002	456.	348.0
BELIZE		578 1997	182 2001	119.5	132.0 2002
BENIN	5.41998	445 2001	12 2001	8.9	3.7
BERMUDA		1311 1997	1084 2001	870.9	528.8 2002
BHUTAN		50 1997	27 2001	33.0	12.3
BOLIVIA	98.81998	671 1997	121 2000	69.4	35.5
BOSNIA-HERZEGOVINA		243 1999	116 2001	239.4	44.6 2002
BOTSWANA	24.71998	150 1997	44 2001	77.1	45.2
BRAZIL	45.92000	433 1997	369 2003	230.4	105.2
BRUNEI		297 1997	629 2001	251.7	84.8
BULGARIA	172.92000	543 1997	453 2000	356.9	59.4
BURKINA FASO	1.31998	433 2001	12 2003	6.3	2.2
BURUNDI	2.51998	220 2000	35 2003	3.4	4.7
CAMBODIA		113 1997	8 2001	2.7	2.8
CAMEROON	6.31998	161 1997	75 2001	6.9	10.0
CANADA	167.92000	1047 1997	691 2000	634.5	700.2
CAPE VERDE		181 2001	101 2001	148.3	96.9
CAYMAN IS.			223 2001		
CENTRAL AFRICAN REP.	1.71998	80 1997	6 2002	2.5	2.8
CEUTA					
CHAD	0.21998	233 1997	2 2002	1.4	1.6
CHILE		759 2000	523 2002	205.8	132.6
CHINA	59.32000	339 1997	350 2002	241.1	40.9
CHRISTMAS IS.					
COCOS		548 2001	319 2003		
COLOMBIA	26.41999			195.2	66.7
COMOROS		174 1997	4 2001	23.0	8.5
CONGO DR	2.81998	385 1997	2 2001	0.2	
CONGO R.	6.31998	109 1997	13 2001	3.6	4.4
COOK IS.					
COSTA RICA	70.02000	816 1999	231 2000	315.8	238.4
CÔTE D'IVOIRE	15.61998	185 2001	61 2001	12.6	14.7
CROATIA	133.82000	330 1997	293 2000	424.9	189.5
CUBA	53.62000	185 2001	251 2001	68.3	26.7
CYPRUS	69.51997	526 2001	386 2001	506.5	301.5
CZECH REPUBLIC		803 1997	538 2001	337.7	239.8
DENMARK	283.22000	1400 2001	859 2001	643.0	655.6
DIEGO GARCÍA IS.					
DJIBOUTI		83 2000	78 2001	14.3	27.0
DOMINICA		639 1997	225 2001	293.4	125.9
DOMINICAN REPUBLIC	27.52000	181 1997	97 1998	106.8	0.52002
EAST TIMOR					
ECUADOR	98.22000	422 2001	252 2003	123.6	55.5
EGYPT	31.31999	339 1997	229 2002	130.3	31.7
EL SALVADOR	28.51998	481 1999	233 2001	131.3	43.9
EQUATORIAL GUINEA	4.61998	425 1997	116 1998	20.0	14.2
ERITREA		464 2001	53 2003	9.3	3.5
ESTONIA	191.62000	1136 2001	507 2003	329.2	920.7
ETHIOPIA	0.41998	189 2000	6 2001	6.3	3.2
FAEROE IS.			1022 2000	419.0	
FIJI		681 1999	117 2003	122.4	52.3
FINLAND	445.02000	1624 2000	679 2003	452.9	481.1
FRANCE	142.12000	950 1997	632 2001	560.9	487.1
FRENCH GUIANA					
FRENCH POLYNESIA		571 1997	224 2003	215.3	308.7
GABON	29.01998	488 1999	308 2002	28.4	29.4
GAMBIA	1.71998	394 1999	15 2001	27.4	15.6
GEORGIA	4.92000	568 1997	357 2002	151.2	42.5
GERMANY	291.02000	570 2001	675 2003	661.1	561.1
GHANA	13.91998	695 1999	53 2001	14.5	5.2
GIBRALTAR					
GREECE		466 1997	519 2001	466.5	89.2
GREENLAND		482 1997	410 1998	447.5	107.5 1995
GRENADA		573 1997	370 2001	309.3	151.3
GUADELOUPE					
GUAM		1511 1997	725 2001	506.5	
GUATEMALA		79 1997	145 2001	92.1	18.8
GUINEA		52 2001	47 2001	2.9	4.8
GUINEA-BISSAU	4.81998	178 2001	36 2001	7.1	
GUYANA	74.82000	560 2000	98 2001	136.9	36.0
HAITI		18 2001	60 2003	16.7	
HONDURAS		411 1997	119 2002	52.7	15.6
HUNGARY	162.32000	690 1997	475 2001	353.9	146.0
ICELAND	322.42000	1081 1999	509 2000	652.1	472.4
INDIA	60.01998	120 1997	83 2001	40.7	12.1
INDONESIA	22.91999	159 1997	153 2001	45.9	13.9
IRAN		281 1998	173 2002	219.5	109.6
IRAQ		222 1997	83 2000	36.9	7.5 2002
IRELAND	147.72000	695 1997	694 2002	496.3	494.3
ISRAEL		526 1997	330 2001	441.3	741.0
ITALY	109.02000	878 1997	494 2000	450.9	315.3
JAMAICA		795 1998	374 2001	189.1	62.8
JAPAN	566.02000	956 1997	785 2002	460.1	541.6
JORDAN	74.21998	372 1997	177 2002	113.5	55.1
KANAKY-NEW CALEDONIA		560 1999	502 2003	231.7	
KAZAKHSTAN		411 1997	338 2001	166.7	
KENYA	8.31999	221 2001	26 2001	8.9	13.2
KIRIBATI		388 2001	44 2003	47.2	10.2
KOREA, NORTH		154 1997	160 2003	44.0	
KOREA, SOUTH		1034 1997	458 2003	541.9	544.9
KUWAIT		570 1999	418 2001	202.1	183.0
KYRGYZSTAN		110 1997	49 2001	78.6	17.1
LAOS		148 1997	52 2001	9.7	3.8
LATVIA	137.82000	700 1999	859 2003	272.8	216.6
LEBANON	63.32000	182 2001	357 2001	178.0	113.0
LESOTHO	9.01998	61 1997	35 2002	20.7	
LIBERIA	14.21998	274 1997	25 2000	2.2	
LIBYA	14.11998	273 1997	137 2000	133.2	23.6 2002
LIECHTENSTEIN			512 2002	587.7	
LITHUANIA	30.92000	524 1999	487 2002	238.7	155.1
LUXEMBOURG	275.72000	392 2000	598 2003	800.2	653.0

	NEWSPAPERS PER 1,000 PEOPLE [1]	RADIOS PER 1,000 PEOPLE [10]	TV SETS PER 1,000 PEOPLE [10]	TELEPHONES (LANDLINES) PER 1,000 PEOPLE [1]	COMPUTERS PER 1,000 PEOPLE [1]
	2000	1996-2001	1998-2003	2005	2004
MACEDONIA, TFYR	53.52000	205 1997	282 2000	308.2	68.9
MADAGASCAR	4.51998	216 1997	25 2002	3.4	5.0
MALAWI	2.41998	499 1998	4 2001	7.4	1.6
MALAYSIA	95.32000	420 1997	210 2002	178.6	196.8
MALDIVES		108 2000	131 2002	98.1	112.1
MALI	1.11998	180 2001	33 2002	5.7	3.2
MALTA		666 1997	566 2002	522.0	314.0
MALVINAS-FALKLANDS					
MARSHALL IS.				75.6	81.7
MARTINIQUE					
MAURITANIA		148 1997	44 2003	13.2	14.1
MAURITIUS	116.42000	379 2000	299 2001	286.7	278.7
MAYOTTE IS.					
MELILLA					
MEXICO	93.52000	330 1997	282 2001	174.1	108.0
MICRONESIA		71 1996	25 2003	109.4	
MOLDOVA	153.41998	758 1999	296 2001	204.7	26.6
MONACO			761 2001		
MONGOLIA	17.62000	50 2001	81 2003	55.7	124.1
MONTSERRAT					
MOROCCO	29.12000	243 1997	167 2001	43.9	20.8
MOZAMBIQUE	2.51998	44 1997	14 2002	4.1	5.8
MYANMAR-BURMA	8.71998	66 2001	7 2003	8.5	6.5
NAMIBIA	17.21998	134 1997	269 2002	63.7	109.5
NAURU					
NEPAL		39 1997	8 2001	15.1	4.4
NETHERLANDS	279.52000	980 1997	648 2002	482.8	682.4
NETHERLANDS ANTILLES		1036 1997	334 2000	460.8	
NEW ZEALAND-AOTEAROA	202.22000	991 1997	574 2003	443.4	473.8
NICARAGUA		270 1998	123 2001	39.9	37.2
NIGER	0.21998	122 2000	10 2001	1.8	0.7
NIGERIA	25.41998	200 1997	103 2001	8.0	6.7
NIUE					
NORFOLK IS.					
NORTHERN MARIANAS					
NORWAY	569.02000	3324 2001	884 2001	668.7	572.8
OMAN		621 1997	553 2002	94.9	46.6
PAKISTAN	39.32000	105 1997	150 2001	29.6	4.9 2003
PALAU					
PALESTINE			148 2002	101.9	48.2
PANAMA		300 1997	191 2002	118.4	40.9
PAPUA NEW GUINEA		86 1997	23 2003	12.1	63.6
PARAGUAY		188 1997	218 2000	50.4	59.2
PERU	22.71998	269 1997	172 2002	74.4	97.6
PHILIPPINES	66.11997	161 1997	182 2002	42.1	45.1
PITCAIRN					
POLAND	101.62000	523 1997	229 2003	321.8	192.8
PORTUGAL	102.42000	299 1997	413 2001	403.5	133.5
PUERTO RICO		761 1997	339 2001	285.5	
QATAR		488 1997	426 2003	245.7	171.2
RÉUNION					
ROMANIA		358 2000	697 2002	202.4	113.0
RUSSIA		418 1997	538 2000	255.8	132.2
RWANDA	0.11998	85 1997	0 1998	2.6	
SAHARA, WESTERN					
SAMOA		1063 1997	148 2003	72.9	6.6 2002
SAMOA, AMERICAN			274 2001		
SAN MARINO			863 2002	738.9	857.1
SÃO TOMÉ AND PRÍNCIPE		318 1998	93 2002	46.6	
SAUDI ARABIA		326 1997	265 2002	154.3	353.9
SENEGAL		126 2001	78 2001	20.6	21.3
SERBIA AND MONTENEGRO*		297 1997	282 2000	329.6	47.7
SEYCHELLES		543 1997	202 2001	253.4	179.3
SIERRA LEONE		259 1998	13 2001	4.9	
SINGAPORE	272.92000	672 1997	303 2002	439.6	763.2 2003
SLOVAKIA	130.82000	965 1998	409 2001	232.3	296.0
SLOVENIA	168.42000	405 1997	366 2002	407.0	352.5
SOLOMON IS.		147 1997	10 2003	13.7	42.9
SOMALIA		60 1997	14 2001	25.1	6.3
SOUTH AFRICA	25.42000	336 1997	177 2002	105.2	82.2
SPAIN	98.22000	330 1997	564 2001	415.8	256.7
SRI LANKA	28.82000	215 2000	117 2001	51.0	27.3
ST HELENA					
ST KITTS-NEVIS		687 1997	239 2001	532.1	234.1
ST LUCIA		742 1997	296 2001	321.2	158.9
ST PIERRE AND MIQUELON					
ST VINCENT		690 1997	234 2000	160.6	135.1
SUDAN		461 2001	386 2002	29.0	17.1
SURINAME	67.01998	726 1997	261 2001	182.1	45.7 2001
SWAZILAND		162 1997	34 2002	41.8	32.1
SWEDEN	409.52000	2811 2001	965 2001	708.1	763.0
SWITZERLAND	371.72000	1002 1997	552 2001	710.5	826.2
SYRIA		276 1997	182 2002	143.1	32.3
TAIWAN					
TAJIKISTAN		141 1997	357 2001	38.6	
TANZANIA		406 2001	45 2001	4.0	7.4
THAILAND	196.91998	235 1997	300 2001	106.7	58.3
TOGO	2.22000	263 2000	123 2002	10.4	28.6
TOKELAU					
TONGA		653 1999	70 2003	110.7	49.0
TRINIDAD AND TOBAGO		534 1997	345 2001	246.9	105.3
TUNISIA	18.92000	158 1999	207 2001	121.2	47.5
TURKEY		470 2001	423 2002	266.6	51.6
TURKMENISTAN	6.82000	279 1997	182 2001	80.1	
TURKS AND CAICOS					
TUVALU					
UGANDA	2.72000	122 1997	18 2002	2.6	4.3
UKRAINE	174.82000	889 1997	456 2000	255.9	28.0
UNITED ARAB EMIRATES		309 1997	252 2001	274.9	115.7
UNITED KINGDOM	326.52000	1445 1997	950 2001	562.9	599.5
UNITED STATES	196.32000	2109 1997	938 2001	606.0	749.2
URUGUAY		603 1997	530 2000	290.7	125.0
UZBEKISTAN		456 1997	280 2001	66.5	
VANUATU		346 1997	13 2003	32.6	14.5
VATICAN					
VENEZUELA		292 2000	186 2001	128.1	82.1
VIETNAM	5.81999	109 1997	197 2002	70.3	12.7
VIRGIN IS. (AM.)		996 1997	663 2000	626.5	
VIRGIN IS. (BR.)					
WALLIS AND FUTUNA IS.					
YEMEN		65 1997	308 2002	39.3	14.8
ZAMBIA	21.92000	179 2001	51 2001	7.6	9.8
ZIMBABWE		362 1999	56 2001	24.5	77.3

* All data available are previous to the political separation of Serbia and Montenegro.

1. World Development Indicators 2006, World Bank.
10. World Development Indicators 2005, World Bank.

ECONOMY

	POPULATION LIVING UNDER ONE DOLLAR PER DAY (%) [1]	GNI PER CAPITA ($, ATLAS METHOD) [1]	GDP PER CAPITA ($, ATLAS METHOD) [1]	GDP, ANNUAL GROWTH (%) [1]	AVERAGE ANNUAL INFLATION (%) [1]	CONSUMER PRICE INDEX (ALL ITEMS 195=100) [1]	TOTAL EXTERNAL DEBT (MILLION $) [1]	PER CAPITA EXTERNAL DEBT ($) [2]	EXTERNAL DEBT (% OF EXPORTS OF GOODS AND SERVICES) [1]
	1994-2004	2004	2004	2004	2004	2004	2004	2004	2004
AFGHANISTAN				7.5	13.8				
ALBANIA	<2 2002	2,120	4,978	5.9	5.8	2.3	1,549	490	2.6 2003
ALGERIA	<2 1995	2,270	6,603	5.2	10.2	3.6	21,987	649	
ANDORRA									
ANGOLA		930	2,180	11.1	42.2	37.3	9,521	564	14.8 2004
ANGUILLA									
ANTIGUA		9,480	12,586	4.1	9.4				
ARGENTINA	7.0 2003	3,580	13,298	9.0	9.2	4.4	169,247	4,281	28.5 2004
ARMENIA	<2 2003	1,060	4,101	7.0	0.6	8.1	1,224	408	8.0 2004
ARUBA				6.0 1994	8.5 1994	2.5			
AUSTRALIA		27,070	30,331	3.0	3.5	2.3			
AUSTRIA		32,280	32,276	2.2	1.9	2.1			
AZERBAIJAN	<2 2002	940	4,153	10.2	6.4	6.7	1,986	233	5.2 2004
BAHAMAS		15,100 2002	17,072 2002	0.7 2002	1.3 2002	0.5			
BAHRAIN		14,370	20,758	5.4	7.7	1.6 2003			
BANGLADESH	36.0 2000	440	1,870	6.3	4.2	3.2	20,344	138	5.2 2004
BARBADOS		8,670 1999		2.6 1999	1.8 1999	1.4	702	2,593	5.2 2004
BELARUS	<2 2002	2,140	6,970	11.0	21.8	18.1	3,717	385	2.1 2004
BELGIUM		31,280	31,096	2.9	2.3	2.1			
BELIZE		3,940	6,747	4.2	2.6	3.1	959	3,419	62.5 2004
BENIN	30.9 2003	450	1,091	2.7	1.4	0.9	1,916	214	7.6 2003
BERMUDA		35,590 1997		3.1 1997	2.1 1997				
BHUTAN		760		4.9	4.8	4.6	593	263	
BOLIVIA	23.2 2002	960	2,720	3.6	8.5	4.4	6,096	640	18.6 2004
BOSNIA-HERZEGOVINA		2,040	7,032	6.2	2.9		3,202	817	3.7 2004
BOTSWANA		4,360	9,945	4.9	4.8	6.9	524	299	1.2 2003
BRAZIL	7.5 2003	3,000	8,195	4.9	8.2	6.6	222,026	1,160	46.8 2004
BRUNEI									
BULGARIA	<2 2003	2,750	8,078	5.6	4.2	6.3	15,661	2,056	17.1 2004
BURKINA FASO	27.2 2003	350	1,169	3.9	0.9	-0.4	1,967	140	11.9 2001
BURUNDI	54.6 1998	90	677	5.5	6.4	12.6	1,385	170	66.0 2003
CAMBODIA	34.1 1997	350	2,423	7.7	5.3	3.9	3,377	231	0.8 2004
CAMEROON	17.1 2001	810	2,174	4.3	0.4	2.8 2002	9,496	563	2004
CANADA		28,310	31,263	2.9	3.0	1.8			
CAPE VERDE		1,720	5,727	5.5	2.5	-1.9	517	975	5.3 2003
CAYMAN IS.				5.3 1994	3.1 1994				
CENTRAL AFRICAN REP.		310	1,095	1.3	-1.9	-2.1	1,078	260	
CEUTA									
CHAD		250	2,090	29.8	13.4	-5.4	1,701	165	
CHILE	<2 2000	5,220	10,874	6.1	6.6	1.1	44,058	2,649	24.2 2004
CHINA	16.6 2001	1,500	5,896	10.1	6.9	4.0	248,934	187	3.5 2004
CHRISTMAS IS.									
COCOS									
COLOMBIA	7.0 2003	2,020	7,256	4.1	7.1	5.9	37,732	804	33.0 2004
COMOROS		560	1,943	1.9	2.5		306	364	2004
CONGO DR		110	705	6.3	5.9	4.1	11,841	2,794	
CONGO R.		760	978	3.6	6.9	2.4	5,829	95	4.0 2004
COOK IS.									
COSTA RICA	2.2 2001	4,470	9,481	4.2	11.6	12.3	5,700	1,276	7.3 2004
CÔTE D'IVOIRE	14.8 2002	760	1,551	1.6	0.8	1.4	11,739	625	6.9 2004
CROATIA	<2 2001	6,820	12,191	3.8	3.3	2.1	31,548	6,926	27.2 2004
CUBA				1.1 2002	2.6 2000				
CYPRUS		16,510	22,805	3.7	2.3	2.3			
CZECH REPUBLIC	<2 1996	9,130	19,408	4.4	3.0	2.8	45,561	4,468	10.5 2004
DENMARK		40,750	31,914	2.4	1.6	1.2			
DIEGO GARCÍA IS.									
DJIBOUTI		950	1,993	3.0	3.0		429	522	
DOMINICA		3,670	5,643	2.0	2.9	2.3	226	2,808	9.7 2002
DOMINICAN REPUBLIC	2.5 2003	2,100	7,449	2.0	51.2	51.5	6,965	761	6.4 2004
EAST TIMOR		550		1.8	-0.8				
ECUADOR	15.8 1998	2,210	3,963	6.9	4.1	2.7	16,868	1,239	36.0 2004
EGYPT	3.1 2000	1,250	4,211	4.2	11.5	11.3	30,291	394	7.6 2004
EL SALVADOR	19.0 2002	2,320	5,041	1.5	4.3	4.5	7,250	1,019	8.8 2004
EQUATORIAL GUINEA		710 2001	19,304 2001	10.0	-8.3	36.4 1994	291	552	
ERITREA		190	977	1.8	20.3		681	145	3.1 2000
ESTONIA	<2 2003	7,080	14,555	7.8	3.1	3.0	10,008	7,579	15.7 2004
ETHIOPIA	23.0 2000	110	756	13.1	9.5	3.3	6,574	81	5.3 2004
FAEROE IS.									
FIJI		2,720	6,066	4.1	2.9	2.8	202	234	
FINLAND		32,880	29,951	3.7	0.8	0.2			
FRANCE		30,370	29,300	2.3	1.6	2.1			
FRENCH GUIANA									
FRENCH POLYNESIA		16,070 2000	24,538 2000	4.0 2000	1.0 2000				
GABON		4,080	6,624	1.4	7.0	0.4	4,150	2,904	10.9 2003
GAMBIA	59.3 1998	280	1,991	8.3	15.1	14.2	674	423	
GEORGIA	6.5 2003	1,060	2,844	6.2	9.6	5.7	2,082	474	11.2 2004
GERMANY		30,690	28,303	1.6	0.4	1.7			

ECONOMY

	TOTAL NET OFFICIAL DEVELOPMENT ASSISTANCE RECEIVED ($ MILLION) [6]	TOTAL NET OFFICIAL DEVELOPMENT ASSISTANCE RECEIVED ($ PER CAPITA) [6]	TOTAL NET OFFICIAL DEVELOPMENT ASSISTANCE RECEIVED (% OF GDP) [6]	TOTAL NET OFFICIAL DEVELOPMENT ASSISTANCE DISBURSED (% OF GNI) [6]	TOTAL NET OFFICIAL DEVELOPMENT ASSISTANCE DISBURSED ($ MILLION) [6]	ENERGY USE CONSUMPTION (OIL EQUIVALENT) PER CAPITA (KG) [1]	ENERGY IMPORTS (% OF CONSUMPTION) [1]	PUBLIC HEALTH EXPENDITURE (% OF GDP) [6]	PUBLIC EDUCATION EXPENDITURE (% OF GDP) [6]	DEFENSE EXPENDITURE (% OF GDP) [1]
	2003	2003	2003	2003	2003	2003	2003	2002	2000-2002	2004
AFGHANISTAN										
ALBANIA	342	108	5.6			673.6	56.9	2.4		1.2
ALGERIA	232	7	0.3			1,035.5	-394.8	3.2		3.3
ANDORRA										
ANGOLA	499	37	3.8			605.8	-456.6	2.1	2.8	9.1
ANGUILLA										
ANTIGUA	5	64	0.7					3.3	3.8	
ARGENTINA	109	3	0.1			1,574.8	-40.9	4.5	4	1.0
ARMENIA	247	81	8.8			659.8	65.5	1.3	3.2	2.9
ARUBA										
AUSTRALIA				0.3	1.2	5,668.2	-125.1	6.5	4.9	1.8
AUSTRIA				0.2	505.0	4,086.0	69.8	5.4	5.7	0.7
AZERBAIJAN	297	36	4.2			1,492.6	-61.3	0.8	3.2	1.8
BAHAMAS	4	12	0.1					3.4		
BAHRAIN	38	53				10,252.7	-116.0	3.2		4.3
BANGLADESH	1,393	10	2.7			158.7	19.1	0.8	2.4	1.2
BARBADOS	20	73	0.8					4.7	7.6	
BELARUS	32	3	0.2			2,612.6	86.4	4.7	6	1.2
BELGIUM				0.6	1.9	5,701.3	77.4	6.5	6.3	1.4
BELIZE	12	47	1.2					2.5	5.2	1.4 1997
BENIN	294	44	8.5			291.7	31.4	2.1	3.3	
BERMUDA										
BHUTAN	77	88	11.1					4.1	5.2	
BOLIVIA	930	104	11.8			503.8	-73.6	4.2	6.3	1.6
BOSNIA-HERZEGOVINA	539	130	7.7			1,136.5	30.2	4.6		2.4
BOTSWANA	30	18	0.4					3.7	2.2	3.6
BRAZIL	296	2	0.1			1,065.3	11.4	3.6	4.2	1.4
BRUNEI	1	1				7,495.4	-691.6	2.7	9.1	
BULGARIA	414	53	2.1			2,493.9	48.4	4	3.5	2.4
BURKINA FASO	451	37	10.8					2		1.4
BURUNDI	224	31	37.6					0.6	3.9	5.8
CAMBODIA	508	38	12.0					2.1	1.8	2.2
CAMEROON	884	55	7.1			428.9	-79.7	1.2	3.8	1.5
CANADA				0.2	2.0	8,240.3	-47.8	6.7	5.2	1.2
CAPE VERDE	144	306	18.0					3.8	7.9	0.7
CAYMAN IS.										739.5 1996
CENTRAL AFRICAN REP.	50	13	4.2					1.6		1.1 2003
CEUTA										
CHAD	247	29	9.5					2.7		1.1
CHILE	76	5	0.1			1,646.8	68.3	2.6	4.2	3.9
CHINA	1,325	1	0.1			1,093.9	2.0	2		1.9
CHRISTMAS IS.										
COCOS										
COLOMBIA	802	18	1.0			641.5	-162.1	6.7	5.2	4.3
COMOROS	25	41	7.6					1.7	3.9	20.4 2003
CONGO DR	5,381	101	94.9			292.9	-4.2	1.2		1.0 2000
CONGO R.	70	19	2.0			272.7	-1,078.2	1.5	3.2	1.4 2003
COOK IS.										
COSTA RICA	28	7	0.2			880.0	55.8	6.1	5.1	
CÔTE D'IVOIRE	252	15	1.8			373.6	-1.7	1.4	4.6	1.6 2003
CROATIA	121	27	0.4			1,976.5	57.3	5.9	4.5	1.7
CUBA	70	6				999.7	40.6	6.5	18.7	
CYPRUS	19	24	0.2			3,279.0	98.4	2.9	6.3	1.5
CZECH REPUBLIC	263	26	0.3			4,324.3	25.2	6.4	4.4	1.8
DENMARK				0.8	1.7	3,852.7	-37.3	7.3	8.5	1.5
DIEGO GARCÍA IS.										
DJIBOUTI	78	110	12.5					3.3		4.3 2002
DOMINICA	11	154	4.2					4.6		
DOMINICAN REPUBLIC	69	8	0.4			922.5	80.6	2.2	2.3	
EAST TIMOR	151	186	44.2					6.2		
ECUADOR	176	14	0.6			708.4	-159.4	1.7	1	1.9
EGYPT	894	13	1.1			734.6	-16.5	1.8		2.8
EL SALVADOR	192	29	1.3			675.4	46.7	3.6	2.9	0.7
EQUATORIAL GUINEA	21	43	0.7					1.3	0.6	2.1 1995
ERITREA	307	70	40.9					3.2	4.1	19.4 2003
ESTONIA	85	63	0.9			3,631.3	25.5	3.9	5.7	1.8
ETHIOPIA	1,504	22	22.6			298.9	7.8	2.6	4.6	4.3
FAEROE IS.										
FIJI	51	61	2.5					2.7	5.6	1.2
FINLAND				0.4	558.0	7,203.9	57.5	5.5	6.4	1.2
FRANCE				0.4	7.3	4,519.4	49.9	7.4	5.6	2.5
FRENCH GUIANA										
FRENCH POLYNESIA										
GABON	-11	-8	-0.2			1,256.2	-637.0	1.8	3.9	0.3 1998
GAMBIA	60	42	15.1					3.3	2.8	0.4
GEORGIA	220	43	5.5			597.4	49.5	1	2.2	1.4
GERMANY				0.3	6.8	4,205.4	61.2	8.6	4.6	1.4
GHANA	907	44	11.9			400.4	29.5	2.3		0.8

	POPULATION LIVING UNDER ONE DOLLAR PER DAY (%) [1]	GNI PER CAPITA ($, ATLAS METHOD) [1]	GDP PER CAPITA ($, ATLAS METHOD) [1]	GDP, ANNUAL GROWTH (%) [1]	AVERAGE ANNUAL INFLATION (%) [1]	CONSUMER PRICE INDEX (ALL ITEMS 1995=100) [1]	TOTAL EXTERNAL DEBT (MILLION $) [1]	PER CAPITA EXTERNAL DEBT ($) [2]	EXTERNAL DEBT (% OF EXPORTS OF GOODS AND SERVICES) [1]
	1994-2004	2004	2004	2004	2004	2004	2004	2004	2004
GHANA	44.8 1999	380	2,240	5.8	14.1	12.6	7,035	306	6.6 2004
GIBRALTAR									
GREECE		16,730	22,205	4.2	3.4	2.9			
GREENLAND									
GRENADA		3,750	8,021	-2.8	2.9	1.1 2002	433	4,108	15.9 2002
GUADELOUPE									
GUAM									
GUATEMALA	13.5 2002	2,190	4,313	2.7	8.2	7.4	5,532	418	7.4 2004
GUINEA		410	2,180	2.6	16.4		3,538	361	19.9 2004
GUINEA-BISSAU		160	722	4.3	2.3	0.9	765	455	16.1 2003
GUYANA	<2 1998	1,020	4,439	1.5	6.7	4.7	1,331	1,769	5.8 2004
HAITI	53.9 2001	400 2003	1,844 2003	0.4 2003	27.0 2003	22.8	1,225	140	4.0 2003
HONDURAS	20.7 1999	1,040	2,876	4.6	7.7	8.1	6,332	842	7.8 2004
HUNGARY	<2 2002	8,370	16,814	4.6	4.6	6.8	63,159	6,288	25.2 2004
ICELAND		37,920	33,051	5.2	2.4	2.8			
INDIA	34.7 2000	620	3,139	6.9	5.3	3.8	122,723	108	18.9 2003
INDONESIA	7.5 2002	1,140	3,609	5.1	7.1	6.2	140,649	617	22.1 2004
IRAN	<2 1998	2,320	7,525	5.6	16.6	14.8	13,622	191	10.7 2000
IRAQ				46.5	13.2 2003				
IRELAND		34,310	38,827	4.9	3.5	2.2			
ISRAEL		17,360	24,382	4.4	-0.2	-0.4			
ITALY		26,280	28,180	1.2	2.6	2.2			
JAMAICA	<2 2000	3,300	4,163	0.9	12.6	13.6	6,399	2,394	14.8 2004
JAPAN		37,050	29,251	2.7	-2.1	0.0			
JORDAN	<2 2003	2,190	4,688	7.7	5.2	3.4	8,175	1,370	8.2 2004
KANAKY-NEW CALEDONIA		14,020 2000	22,140 2000	2.1 2000	-0.6 2000				
KAZAKHSTAN	<2 2003	2,250	7,441	9.4	9.8	6.9	32,310	2,183	38.0 2004
KENYA	22.8 1997	480	1,140	4.3	6.9	11.6	6,826	190	8.6 2004
KIRIBATI		970		1.8	-8.1				
KOREA, NORTH									
KOREA, SOUTH	<2 1998	14,000	20,499	4.6	2.7	3.6			
KUWAIT		22,470	19,384	7.2	23.1	1.2			
KYRGYZSTAN	<2 2003	400	1,935	7.1	4.8	4.1	2,100	390	14.2 2004
LAOS	27 2002	390	1,953	6.3	10.3	10.5	2,056	332	9.0 2001
LATVIA	<2 2003	5,580	11,653	8.3	7.2	6.2	12,661	5,544	21.1 2004
LEBANON		6,010	5,837	6.3	2.9	6.8 1994	22,177	6,070	
LESOTHO	36.4 1995	730	2,619	2.3	1.7	6.7 2003	764	428	4.5 2004
LIBERIA		120		2.4	2.1		2,706	784	
LIBYA		4,400		4.5	19.9	-2.2			
LIECHTENSTEIN									
LITHUANIA	<2 2003	5,740	13,107	6.7	3.3	1.2	9,475	2,784	14.3 2004
LUXEMBOURG		56,380	69,961	4.5	2.5	2.2			
MACEDONIA, TFYR	<2 2003	2,420	6,610	2.9	1.5	-0.4	2,044	1,002	10.5 2004
MADAGASCAR	61.0 2001	290	857	5.2	14.3	13.8	3,462	177	6.0 2003
MALAWI	41.7 1998	160	646	6.7	11.6	11.4	3,418	254	7.6 2002
MALAYSIA	<2 1997	4,520	10,276	7.1	6.2	1.5	52,145	1,987	7.9 2003
MALDIVES		2,410		10.8	-1.6	6.4	345	997	4.6 2004
MALI	72.3 1994	330	998	2.2	-0.5	-3.1	3,316	232	5.8 2003
MALTA		12,050	18,879	0.4	2.0	2.8			
MALVINAS-FALKLANDS									
MARSHALL IS.		2,320		1.5	1.6				
MARTINIQUE									
MAURITANIA	25.9 2000	530	1,941	6.9	7.9	10.4	2,297	707	
MAURITIUS		4,640	12,027	4.2	6.0	4.7	2,294	1,811	7.4 2004
MAYOTTE IS.									
MELILLA									
MEXICO	4.4 2002	6,790	9,803	4.4	6.1	4.7	138,689	1,265	22.9 2004
MICRONESIA		2,300		-3.8	1.4				
MOLDOVA	22.0 2001	720	1,729	7.3	8.0	12.5	1,868	446	12.1 2004
MONACO									
MONGOLIA	27.0 1998	600	2,056	10.7	18.1	8.2	1,517	559	2.9 2004
MONTSERRAT									
MOROCCO	<2 1999	1,570	4,309	4.2	1.5	1.0	17,672	545	14.0 2004
MOZAMBIQUE	37.8 1996	270	1,237	7.2	12.6	12.7	4,651	227	4.5 2004
MYANMAR-BURMA				9.7 2001	22.6 2001	4.5	7,239	141	3.8 2004
NAMIBIA		2,380	7,418	6.0	2.9	4.1			
NAURU									5.5 2004
NEPAL	24.1 2004	250	1,490	3.5	4.5	2.8	3,354	119	
NETHERLANDS		32,130	31,789	1.4	1.2	1.3			
NETHERLANDS ANTILLES						1.4			
NEW ZEALAND-AOTEAROA		19,990	23,413	4.4	3.8	2.3			
NICARAGUA	45.1 2001	830	3,634	5.1	10.2	8.4	5,145	900	5.8 2004
NIGER	60.6 1995	210	779	0.9	1.6	0.3	1,950	131	7.5 2003
NIGERIA	70.8 2003	430	1,154	6.0	19.9	15.0	35,890	262	8.2 2004
NIUE									
NORFOLK IS.									
NORTHERN MARIANAS									

ECONOMY

ECONOMY	TOTAL NET OFFICIAL DEVELOPMENT ASSISTANCE RECEIVED ($ MILLION) [6]	TOTAL NET OFFICIAL DEVELOPMENT ASSISTANCE RECEIVED ($ PER CAPITA) [6]	TOTAL NET OFFICIAL DEVELOPMENT ASSISTANCE RECEIVED (% OF GDP) [6]	TOTAL NET OFFICIAL DEVELOPMENT ASSISTANCE DISBURSED (% OF GNI) [6]	TOTAL NET OFFICIAL DEVELOPMENT ASSISTANCE DISBURSED ($ MILLION) [6]	ENERGY USE CONSUMPTION (OIL EQUIVALENT) PER CAPITA (KG) [1]	ENERGY IMPORTS (% OF CONSUMPTION) [1]	PUBLIC HEALTH EXPENDITURE (% OF GDP) [6]	PUBLIC EDUCATION EXPENDITURE (% OF GDP) [6]	DEFENSE EXPENDITURE (% OF GDP) [1]
	2003	2003	2003	2003	2003	2003	2003	2002	2000-2002	2004
GIBRALTAR										
GREECE				0.2	362.0	2,708.9	66.8	5	4	4.1 2003
GREENLAND										
GRENADA	12	112	2.7					4	5.1	
GUADELOUPE										
GUAM										
GUATEMALA	247	20	1.0			607.8	25.0	2.3		0.4
GUINEA	238	30	6.5					0.9	1.8	2.9 2002
GUINEA-BISSAU	145	98	60.8					3		3.1 2001
GUYANA	87	113	11.7					4.3	8.4	0.8 1996
HAITI	200	24	6.8			269.9	25.2	3		
HONDURAS	389	56	5.6			521.8	53.9	3.2		0.7
HUNGARY	248	25	0.3			2,600.4	60.5	5.5	5.5	1.7
ICELAND						11,694.1	27.4	8.3	6	0.0
INDIA	942	1	0.2			519.9	18.1	1.3	4.1	2.3
INDONESIA	1,744	8	0.8			752.5	-54.7	1.2	1.2	1.4
IRAN	133	2	0.1			2,055.1	-94.5	2.9	4.9	3.4
IRAQ						943.1	-165.8			
IRELAND				0.4	504.0	3,777.2	87.4	5.5	5.5	0.6
ISRAEL	440	66	0.4			3,085.8	96.4	6	7.5	9.3
ITALY				0.2	2.4	3,140.3	84.7	6.4	4.7	1.9
JAMAICA	3	1	(.)			1,543.1	88.5	3.4	6.1	
JAPAN				0.2	8.9	4,053.4	83.6	6.5	3.6	1.0
JORDAN	1,234	233	12.5			1,026.8	94.8	4.3		7.6
KANAKY-NEW CALEDONIA										
KAZAKHSTAN	268	18	0.9			3,342.2	-111.8	1.9	3	1.0
KENYA	484	15	3.4			494.0	16.6	2.2	7	1.6
KIRIBATI										
KOREA, NORTH						895.5	5.9			
KOREA, SOUTH	-458	-10	-0.1			4,290.6	82.0	2.6	4.2	2.5
KUWAIT	4	2	(.)			9,565.9	-426.6	2.9		7.5
KYRGYZSTAN	198	39	10.4			528.1	48.7	2.2	3.1	2.9
LAOS	299	53	14.1					1.5	2.8	1.7
LATVIA	114	49	1.0			1,881.4	54.8	3.3	5.8	
LEBANON	228	51	1.2			1,699.6	95.8	3.5	2.7	3.8
LESOTHO	79	44	6.9					5.3	10.4	2.6
LIBERIA										7.5 2002
LIBYA	10	2				3,191.3	-331.4	1.6		1.9
LIECHTENSTEIN										
LITHUANIA	372	108	2.0			2,585.2	41.6	4.3	5.9	1.7
LUXEMBOURG				0.8	194.0	9,472.2	98.6	5.3		0.9
MACEDONIA, TFYR	234	114	5.0						3.5	2.5
MADAGASCAR	540	32	9.9					1.2	2.9	7.2 2001
MALAWI	498	45	29.1					4	6	0.8 2001
MALAYSIA	109	4	0.1			2,318.4	-48.0	2	8.1	2.3
MALDIVES	18	61	2.5					5.1		
MALI	528	45	12.2					2.3		1.9
MALTA	10	25	0.2			2,235.6		7		0.8
MALVINAS-FALKLANDS										
MARSHALL IS.										
MARTINIQUE										
MAURITANIA	243	90	22.2					2.9		1.2
MAURITIUS	-15	-12	-0.3					2.2	4.7	0.2
MAYOTTE IS.										
MELILLA										
MEXICO	103	1	(.)			1,563.5	-51.6	2.7	5.3	0.4
MICRONESIA										
MOLDOVA	117	28	5.9			772.1	98.1	4.1	4.9	0.4
MONACO										
MONGOLIA	247	100	19.4					4.6	9	2.1 2002
MONTSERRAT										
MOROCCO	523	17	1.2			378.0	94.2	1.5	6.5	4.5
MOZAMBIQUE	1,033	55	23.9			430.3	2.5	4.1		1.2
MYANMAR-BURMA	126	3				276.4	-34.2	0.4		1.9 2001
NAMIBIA	146	73	3.4			635.5	75.6	4.7	7.2	2.4
NAURU										
NEPAL	467	19	8.0			335.9	10.9	1.4	3.4	1.7
NETHERLANDS				0.8	4.0	4,981.7	27.7	5.8	5.1	1.6
NETHERLANDS ANTILLES						9,209.8				
NEW ZEALAND-AOTEAROA				0.2	165.0	4,333.0	24.2	6.6	6.7	1.0
NICARAGUA	833	152	20.4			588.3	41.8	3.9	3.1	0.7
NIGER	453	39	16.6					2	2.3	0.9 2002
NIGERIA	318	2	0.5			776.6	-119.4	1.2		0.8
NIUE										
NORFOLK IS.										
NORTHERN MARIANAS										
NORWAY				0.9	2.0	5,100.4	-898.9	8	7.6	1.9
OMAN	45	17				4,975.3	-378.9	2.8	4.6	10.4

ECONOMY

	POPULATION LIVING UNDER ONE DOLLAR PER DAY (%) [1]	GNI PER CAPITA ($, ATLAS METHOD) [1]	GDP PER CAPITA ($, ATLAS METHOD) [1]	GDP, ANNUAL GROWTH (%) [1]	AVERAGE ANNUAL INFLATION (%) [1]	CONSUMER PRICE INDEX (ALL ITEMS 1995=100) [1]	TOTAL EXTERNAL DEBT (MILLION $) [1]	PER CAPITA EXTERNAL DEBT ($) [2]	EXTERNAL DEBT (% OF EXPORTS OF GOODS AND SERVICES) [1]
	1994-2004	2004	2004	2004	2004	2004	2004	2004	2004
NORWAY		51,810	38,454	2.9	4.9	0.5			
OMAN		9,070	15,259	3.1	9.1	0.4	3,872	1,451	6.9 2004
PAKISTAN	17.0 2002	600	2,225	6.4	7.8	7.4	35,687	217	21.2 2004
PALAU		6,870		2.0	0.5				
PALESTINE		1,120 2003		-1.7 2003	7.0 2003				
PANAMA	6.5 2002	4,210	7,278	6.2	0.5	0.4	9,469	2,832	14.3 2004
PAPUA NEW GUINEA		560	2,543	2.5	0.7	2.1	2,149	351	12.7 2001
PARAGUAY	16.4 2002	1,140	4,813	4.0	9.2	4.3	3,433	533	13.5 2004
PERU	12.5 2002	2,360	5,678	4.8	5.7	3.7	31,296	1,087	17.1 2004
PHILIPPINES	15.5 2000	1,170	4,614	6.1	6.1	6.0	60,550	705	20.9 2004
PITCAIRN									
POLAND	<2 2002	6,100	12,974	5.4	2.9	3.6	99,190	2,579	34.6 2004
PORTUGAL	<2 1994	14,220	19,629	1.0	2.5	2.4			
PUERTO RICO		10,950 2001	24,915 2001	5.6 2001	5.3 2001				
QATAR					6.8				
RÉUNION									
ROMANIA	<2 2003	2,960	8,480	8.3	15.8	11.9	30,034	1,394	17.2 2004
RUSSIA	<2 2002	3,400	9,902	7.1	18.1	10.9	197,335	1,391	9.8 2004
RWANDA	51.7 2000	210	1,263	4.0	12.6	11.9	1,656	175	11.2 2004
SAHARA, WESTERN									
SAMOA		1,840	5,613	3.1	7.2	16.3	562	3,006	
SAMOA, AMERICAN									
SAN MARINO				2.3 2002	-1.7 2002				
SÃO TOMÉ AND PRÍNCIPE		390		4.5	8.5		362	2,213	25.6 2002
SAUDI ARABIA		10,140	13,825	5.2	11.0	0.3			
SENEGAL	22.3 1995	630	1,713	6.2	1.9	0.5	3,938	322	2004
SERBIA AND MONTENEGRO*		2,680		8.2	8.9		15,882	1,514	
SEYCHELLES		8,190	16,652	-2.0	4.0	3.8	615	7,490	8.1 2004
SIERRA LEONE		210	561	7.4	15.9	14.2	1,723	297	10.9 2004
SINGAPORE		24,760	28,077	8.4	3.5	1.7			
SLOVAKIA	<2 1996	6,480	14,623	5.5	4.6	7.5	22,068	4,086	13.8 2003
SLOVENIA	<2 1998	14,770	20,939	4.6	3.0	3.6			
SOLOMON IS.		560	1,814	5.5	7.1	7.1	176	351	
SOMALIA							2,849	325	
SOUTH AFRICA	10.7 2000	3,630	11,192	3.7	5.9	1.4	28,500	597	6.4 2004
SPAIN		21,530	25,047	3.1	4.1	3.0			
SRI LANKA	5.6 2002	1,010	4,390	5.4	9.4	7.6	10,887	516	8.5 2004
ST HELENA									
ST KITTS-NEVIS		6,980 2003	12,376 2003	2.1 2003	1.7 2003	2.3	316	7,248	24.5 2002
ST LUCIA		4,180	6,324	3.5	6.3	4.7	413	2,529	7.9 2002
ST PIERRE AND MIQUELON									7.3 2002
ST VINCENT		3,400	6,398	6.0	0.5	2.9	257	2,136	
SUDAN		530	1,949	6.0	10.6	8.5	19,332	512	6.0 2004
SURINAME		2,230		4.6	9.1	23.0 2003			
SWAZILAND		1,660	5,638	2.1	5.3	7.3 2003	470	459	1.7 2004
SWEDEN		35,840	29,541	3.6	0.8	0.4			
SWITZERLAND		49,600	33,040	2.1	0.5	0.8			
SYRIA		1,230	3,610	2.0	10.5	1.0 2002	21,521	1,077	3.5 2004
TAIWAN									
TAJIKISTAN	7.4 2003	280	1,202	10.6	16.9		896	134	6.8 2004
TANZANIA	57.8 2000	320	674	6.3	4.0	0.0	7,799	196	5.3 2004
THAILAND	<2 2002	2,490	8,090	6.2	3.3	2.8	51,307	786	10.6 2004
TOGO		310	1,536	3.0	3.4	0.4	1,812	280	2.0 2003
TOKELAU									
TONGA		1,860	7,870	4.3	11.3	11.0	81	788	2.5 2002
TRINIDAD AND TOBAGO		8,730	12,182	6.2	12.5	3.7	2,926	2,228	3.8 2003
TUNISIA	<2 2000	2,650	7,768	5.8	3.0	3.6	18,700	1,812	13.7 2004
TURKEY	3.4 2003	3,750	7,753	8.9	9.9	8.6	161,595	2,150	35.9 2004
TURKMENISTAN		730 2001	4,315 2001	20.4 2001	13.7 2001				
TURKS AND CAICOS									
TUVALU									
UGANDA		250	1,478	5.7	6.0	3.3	4,822	156	6.9 2004
UKRAINE	<2 2003	1,270	6,394	12.1	15.1	9.0	21,652	476	10.7 2004
UNITED ARAB EMIRATES		23,770	24,056	8.5	8.5				
UNITED KINGDOM		33,630	30,821	3.1	2.2	3.0			
UNITED STATES		41,440	39,676	4.2	2.6	2.7			
URUGUAY	<2 2003	3,900	9,421	11.9	7.4	9.2	12,376	3,526	34.9 2004
UZBEKISTAN		450	1,869	7.7	15.0		5,007	183	
VANUATU		1,390	3,051	3.0	1.9	1.4	118	539	1.4 2003
VATICAN									
VENEZUELA	8.3 2000	4,030	6,043	17.9	31.2	21.8	35,570	1,285	16.0 2004
VIETNAM		540	2,745	7.7	7.9	7.8	17,825	206	6.0 2002
VIRGIN IS. (AM.)									
VIRGIN IS. (BR.)									
WALLIS AND FUTUNA IS.									
YEMEN	15.7 1998	550	879	2.7	14.4	10.8 2003	5,488	246	3.5 2004

* All data available are previous to the political separation of Serbia and Montenegro.

ECONOMY

	TOTAL NET OFFICIAL DEVELOPMENT ASSISTANCE RECEIVED ($ MILLION)[6]	TOTAL NET OFFICIAL DEVELOPMENT ASSISTANCE RECEIVED ($ PER CAPITA)[6]	TOTAL NET OFFICIAL DEVELOPMENT ASSISTANCE RECEIVED (% OF GDP)[6]	TOTAL NET OFFICIAL DEVELOPMENT ASSISTANCE DISBURSED (% OF GNI)[6]	TOTAL NET OFFICIAL DEVELOPMENT ASSISTANCE DISBURSED ($ MILLION)[6]	ENERGY USE CONSUMPTION (OIL EQUIVALENT) PER CAPITA (KG)[1]	ENERGY IMPORTS (% OF CONSUMPTION)[1]	PUBLIC HEALTH EXPENDITURE (% OF GDP)[6]	PUBLIC EDUCATION EXPENDITURE (% OF GDP)[6]	DEFENSE EXPENDITURE (% OF GDP)[1]
	2003	2003	2003	2003	2003	2003	2003	2002	2000-2002	2004
PAKISTAN	1,068	7	1.3			466.9	19.9	1.1	1.8	4.1
PALAU										
PALESTINE	972	289	28.1							
PANAMA	31	10	0.2			835.8	73.6	6.4	4.5	1.0 1999
PAPUA NEW GUINEA	221	40	6.9					3.8	2.3	0.6 2003
PARAGUAY	51	9	0.8			678.7	-66.0	3.2	4.4	0.7
PERU	500	18	0.8			441.9	21.3	2.2	3	1.2
PHILIPPINES	737	9	0.9			525.5	46.6	1.1	3.1	0.9
PITCAIRN										
POLAND	1,192	31	0.6			2,452.2	14.6	4.4	5.6	1.9
PORTUGAL				0.2	320.0	2,468.9	83.2	6.6	5.8	2.1
PUERTO RICO										
QATAR	2	3				20,725.9	-335.5	2.4		
RÉUNION										
ROMANIA	601	27	1.1			1,794.2	25.8	4.2	3.5	2.2
RUSSIA	1,255	9	0.3			4,424.1	-73.0	3.5	3.8	3.9
RWANDA	332	40	20.3					3.1	2.8	2.1
SAHARA, WESTERN										
SAMOA	33	186	12.3					4.7	4.8	
SAMOA, AMERICAN										
SAN MARINO										
SÃO TOMÉ AND PRÍNCIPE	38	240	63.3					9.7		
SAUDI ARABIA	22	1	(.)			5,606.8	-308.1	3.3		7.7
SENEGAL	450	45	6.9			287.2	45.4	2.3	3.6	1.4
SERBIA AND MONTENEGRO*						1,991.4	29.3			3.4
SEYCHELLES	9	110	1.3					3.9	5.2	1.9
SIERRA LEONE	297	56	37.5					1.7	3.7	1.6
SINGAPORE	7	2	(.)			5,358.6	99.4	1.3		4.7
SLOVAKIA	160	30	0.5			3,442.8	65.4	5.3	4.4	1.7
SLOVENIA	66	34	0.2			3,518.1	53.2	6.2	6.1	1.6
SOLOMON IS.	60	132	23.8					4.5	3.4	
SOMALIA										
SOUTH AFRICA	625	14	0.4			2,587.2	-30.3	3.5	5.3	1.5
SPAIN				0.2	2.0	3,240.3	75.8	5.4	4.5	1.0
SRI LANKA	672	35	3.7			421.2	47.1	1.8		2.8
ST HELENA										
ST KITTS-NEVIS		0						3.4	7.6	
ST LUCIA	15	92	2.1					3.4	7.7	
ST PIERRE AND MIQUELON										
ST VINCENT	6	58	1.7					3.9	10	
SUDAN	621	19	3.5			476.7	-62.3	1		2.2 2003
SURINAME	11	25	0.9					3.6		
SWAZILAND	27	25	1.5					3.6	7.1	1.7 2001
SWEDEN				0.8	2.4	5,753.9	38.6	7.8	7.7	1.7
SWITZERLAND				0.4	1.3	3,689.2	55.7	6.5	5.8	1.0
SYRIA	160	9	0.7			986.4	-90.1	2.3		7.0 2003
TAIWAN										
TAJIKISTAN	144	23	9.3			501.1	54.5	0.9	2.8	2.2
TANZANIA	1,669	47	16.2			464.6	6.6	2.7		3.0
THAILAND	-966	-16	-0.7			1,405.7	45.6	3.1	5.2	1.2
TOGO	45	9	2.5			445.2	27.9	1.1	2.6	1.5
TOKELAU										
TONGA	28	269	16.9					5.1	4.9	
TRINIDAD AND TOBAGO	-2	-2				8,553.3	-159.9	1.4	4.3	
TUNISIA	306	31	1.2			837.4	21.7	2.9	6.4	1.5
TURKEY	166	2	0.1			1,116.6	70.1	4.3	3.7	3.9
TURKMENISTAN	27	6	0.4			3,662.0	-240.4	3		2.9 1999
TURKS AND CAICOS										
TUVALU										
UGANDA	959	38	15.2					2.1		2.5
UKRAINE	323	7	0.7			2,772.4	43.0	3.3	5.4	2.6
UNITED ARAB EMIRATES	5	1				9,707.0	-305.8	2.3	1.6	2.8 2003
UNITED KINGDOM				0.3	6.3	3,893.0	-6.1	6.4	5.3	2.6
UNITED STATES				0.2	16.3	7,842.9	28.5	6.6	5.7	4.0
URUGUAY	17	5	0.1			737.5	53.9	2.9	2.6	1.4
UZBEKISTAN	194	8	2.0			2,023.2	-6.7	2.5		0.5 2003
VANUATU	32	154	11.4					2.8	11	
VATICAN										
VENEZUELA	82	3	0.1			2,112.1	-231.2	2.3		1.2
VIETNAM	1,769	22	4.5			544.3	-23.2	1.5		2.6 1994
VIRGIN IS. (AM.)										
VÍRGIN IS. (BR.)										
WALLIS AND FUTUNA IS.										
YEMEN	243	13	2.2			289.2	-284.3	1	9.5	6.6
ZAMBIA	560	54	12.9			592.3	5.0	3.1	2	0.6 2000
ZIMBABWE	186	14				751.6	11.8	4.4	4.7	3.4

1. World Development Indicators 2006, World Bank. / 2. Calculated from World Development Indicators 2006, World Bank. / 6. Human Development Report 2005 UNDP.

	LABOR FORCE (% OF TOTAL POPULATION) [11]	UNEMPLOYMENT (% OF LABOR FORCE) [1]		WOMEN IN LABOR FORCE (%) [1]	FEMALE EMPLOYMENT IN AGRICULTURE (% OF FEMALE LABOR FORCE) [6]	FEMALE EMPLOYMENT IN INDUSTRY (% OF FEMALE LABOR FORCE) [6]	FEMALE EMPLOYMENT IN SERVICES (% OF FEMALE LABOR FORCE) [6]	MALE EMPLOYMENT IN AGRICULTURE (% OF MALE LABOR FORCE) [6]	MALE EMPLOYMENT IN INDUSTRY (% OF MALE LABOR FORCE) [6]	MALE EMPLOYMENT IN SERVICES (% OF MALE LABOR FORCE) [6]
	2004	2004		2004	1995-2002	1995-2002	1995-2002	1995-2002	1995-2002	1995-2002
AFGHANISTAN										
ALBANIA	42.7	15	2003	42						
ALGERIA	38.2	27	2001	30						
ANDORRA										
ANGOLA	40.5			46						
ANGUILLA										
ANTIGUA										
ARGENTINA	45.4	16	2003	42		12	87	1	30	69
ARMENIA	42.6	36	1997	49						
ARUBA		7	1997							
AUSTRALIA	49.5	5		45	3	10	87	6	30	64
AUSTRIA	48.0	5		44	6	14	80	5	43	52
AZERBAIJAN	47.1			47	43	7	50	37	14	49
BAHAMAS	46.4	10		50	1	5	93	6	24	69
BAHRAIN	44.2			19						
BANGLADESH	42.4	3	2000	37	77	9	12	53	11	30
BARBADOS	57.0	11	2003	47	4	10	63	5	29	49
BELARUS	49.5			49						
BELGIUM	42.7	7		43	1	10	82	3	36	58
BELIZE	39.5	10	2002	34	6	12	81	37	19	44
BENIN	35.5			39						
BERMUDA										
BHUTAN				35						
BOLIVIA	42.3	6	2002	43	3	14	82	6	39	55
BOSNIA-HERZEGOVINA	52.1			48						
BOTSWANA	35.3	19	2001	42	17	14	67	22	26	51
BRAZIL	47.0	10	2003	42	16	10	74	24	27	49
BRUNEI	41.1			34						
BULGARIA	41.3	14	2003	46						
BURKINA FASO	40.1			47						
BURUNDI	44.9			52						
CAMBODIA	45.2	2	2001	51						
CAMEROON	36.6	8	2001	40						
CANADA	52.9	7		46	2	11	87	4	33	64
CAPE VERDE	30.2			34						
CAYMAN IS.		4	1997							
CENTRAL AFRICAN REP.	43.4			46						
CEUTA										
CHAD	34.6			47						
CHILE	38.7	7	2003	35	5	13	83	18	29	53
CHINA	57.7	4	2002	45						
CHRISTMAS IS.										
COCOS										
COLOMBIA		14	2003	44	7	17	76	33	19	48
COMOROS	29.2			40						
CONGO D.R.	36.4			41						
CONGO R.	34.6			40						
COOK IS.										
COSTA RICA	42.3	7	2003	34	4	15	80	22	27	51
CÔTE D'IVOIRE	35.5			29						
CROATIA	43.0	14	2003	45	15	21	63	16	37	47
CUBA	47.2	3	2002	37						
CYPRUS	47.5	4	2003	45	4	13	83	5	31	58
CZECH REPUBLIC	50.8	8		45	3	28	68	6	50	44
DENMARK	52.1	5		47	2	14	85	5	36	59
DIEGO GARCÍA IS.										
DJIBOUTI	37.5			39						
DOMINICA		23	1997		14	10	72	31	24	40
DOMINICAN REPUBLIC	41.0	16	2001	35	2	17	81	21	26	53
EAST TIMOR	32.9			37						
ECUADOR	45.3	11	2003	42	4	16	79	10	30	60
EGYPT	29.0	11	2003	22	39	7	54	27	25	48
EL SALVADOR	38.1	7	2003	40	4	22	74	34	25	42
EQUATORIAL GUINEA	36.6			37						
ERITREA	36.5			41						
ESTONIA	50.2	10	2003	49	4	23	73	10	42	48
ETHIOPIA	38.1	8	1999	45						
FAEROE IS.										
FIJI	44.1	5	1995	38						
FINLAND	50.1	9		48	4	14	82	7	40	53
FRANCE	44.2	10		46	1	13	86	2	34	64
FRENCH GUIANA										
FRENCH POLYNESIA	41.3			38						
GABON	40.8			43						
GAMBIA	40.1			42						
GEORGIA	51.9	12	2003	44	53	6	41	53	12	35
GERMANY	49.3	10		45	2	18	80	3	44	52
GHANA	41.6	8	2000	48						
GIBRALTAR										

LABOR FORCE

	LABOR FORCE (% OF TOTAL POPULATION) [11]	UNEMPLOYMENT (% OF LABOR FORCE) [1]	WOMEN IN LABOR FORCE (%) [1]	FEMALE EMPLOYMENT IN AGRICULTURE (% OF FEMALE LABOR FORCE) [6]	FEMALE EMPLOYMENT IN INDUSTRY (% OF FEMALE LABOR FORCE) [6]	FEMALE EMPLOYMENT IN SERVICES (% OF FEMALE LABOR FORCE) [6]	MALE EMPLOYMENT IN AGRICULTURE (% OF MALE LABOR FORCE) [6]	MALE EMPLOYMENT IN INDUSTRY (% OF MALE LABOR FORCE) [6]	MALE EMPLOYMENT IN SERVICES (% OF MALE LABOR FORCE) [6]
	2004	2004	2004	1995-2002	1995-2002	1995-2002	1995-2002	1995-2002	1995-2002
GREECE	45.5	10	41	18	12	70	15	30	56
GREENLAND									
GRENADA		15 1998		10	12	77	17	32	46
GUADELOUPE									
GUAM	44.0		39						
GUATEMALA	29.9	3 2003	31	18	23	56	50	18	27
GUINEA	44.1		46						
GUINEA-BISSAU	36.8		41						
GUYANA	43.4	9 2001	37						
HAITI	40.7	7 1999	42	37	6	57	63	15	23
HONDURAS	39.9	5 2003	37	9	25	67	50	21	30
HUNGARY	41.9	6	45	4	26	71	9	42	49
ICELAND	57.9	3	47	3	10	85	12	33	54
INDIA	37.6	4 2000	28						
INDONESIA	46.1	10	38	43	16	41	43	19	38
IRAN	36.8	12 2003	33						
IRAQ		28 2003	19						
IRELAND	47.0	4	42	2	14	83	11	39	50
ISRAEL	38.4	11 2003	47	1	12	86	3	34	62
ITALY	41.2	8	40	5	20	75	6	39	55
JAMAICA	43.6	11	44	10	9	81	30	26	45
JAPAN	52.2	5	41	5	21	73	5	37	57
JORDAN	30.3	13 2000	24						
KANAKY-NEW CALEDONIA	39.0	19 1996	37						
KAZAKHSTAN	53.7	9 2003	49						
KENYA	42.0		44	16	10	75	20	23	57
KIRIBATI									
KOREA, NORTH	46.8		39						
KOREA, SOUTH	50.1	4	41	12	19	70	9	34	57
KUWAIT	46.7		25						
KYRGYZSTAN	41.2	10 2003	44	53	8	38	52	14	34
LAOS	36.9		41						
LATVIA	47.9	11 2003	49	12	16	72	18	35	47
LEBANON	37.6	9 1997	30						
LESOTHO	35.7	39 1997	45						
LIBERIA	34.3		40						
LIBYA	37.0		26						
LIECHTENSTEIN									
LITHUANIA	47.8	12 2003	49	12	21	67	20	34	45
LUXEMBOURG	41.8	5	42						
MACEDONIA, TFYR	42.1	37 2003	39						
MADAGASCAR	42.5	5 2002	48						
MALAWI	43.2	1 1998	50						
MALAYSIA	40.9	4	36	14	29	57	21	34	45
MALDIVES	32.3	2 2000	38	5	24	39	18	16	55
MALI	37.1		47						
MALTA	41.1	8 2003	33	1	21	78	3	36	61
MALVINAS-FALKLANDS									
MARSHALL IS.		31 1999							
MARTINIQUE									
MAURITANIA	36.0		40						
MAURITIUS	44.4	10 2003	35	13	43	45	15	39	46
MAYOTTE IS.									
MELILLA									
MEXICO	38.7	3	35	6	22	72	24	28	48
MICRONESIA									
MOLDOVA	51.1	8 2003	48	50	10	40	52	18	31
MONACO									
MONGOLIA	43.1	14 2003	40						
MONTSERRAT									
MOROCCO	33.7	11	25	6	40	54	6	32	63
MOZAMBIQUE	44.5		54						
MYANMAR-BURMA	52.3		45						
NAMIBIA	30.9	31 2001	44	29	7	63	33	17	49
NAURU									
NEPAL	36.3	1 1999	40						
NETHERLANDS	52.1	4 2003	44	2	9	86	4	31	64
NETHERLANDS ANTILLES	44.2	14 2000	46						
NEW ZEALAND-AOTEAROA	51.7	4	46	6	12	82	12	32	56
NICARAGUA	34.4	8 2003	30						
NIGER	38.5		42						
NIGERIA	34.0	17 1995	35	2	11	87	4	30	67
NIUE									
NORFOLK IS.									
NORTHERN MARIANAS									
NORWAY	53.6	4	47	2	9	88	6	33	58
OMAN	35.3		16						
PAKISTAN	33.1	8 2002	26	73	9	18	44	20	36
PALAU									

LABOR FORCE

	LABOR FORCE (% OF TOTAL POPULATION) [11]	UNEMPLOYMENT (% OF LABOR FORCE) [1]	WOMEN IN LABOR FORCE (%) [1]	FEMALE EMPLOYMENT IN AGRICULTURE (% OF FEMALE LABOR FORCE) [6]	FEMALE EMPLOYMENT IN INDUSTRY (% OF FEMALE LABOR FORCE) [6]	FEMALE EMPLOYMENT IN SERVICES (% OF FEMALE LABOR FORCE) [6]	MALE EMPLOYMENT IN AGRICULTURE (% OF MALE LABOR FORCE) [6]	MALE EMPLOYMENT IN INDUSTRY (% OF MALE LABOR FORCE) [6]	MALE EMPLOYMENT IN SERVICES (% OF MALE LABOR FORCE) [6]
	2004	2004	2004	1995-2002	1995-2002	1995-2002	1995-2002	1995-2002	1995-2002
PALESTINE	18.7	26 2003	13	26	11	62	9	32	58
PANAMA	42.7	14 2003	38	6	10	85	29	20	51
PAPUA NEW GUINEA	40.9	3 2000	48						
PARAGUAY	43.1	8 2001	43	20	10	69	39	21	40
PERU	45.1	10 2003	42	6	10	84	11	24	65
PHILIPPINES	41.8	10 2001	39	25	12	63	45	18	37
PITCAIRN									
POLAND	45.1	19	46	19	18	63	19	40	40
PORTUGAL	52.0	7	46	14	23	63	12	44	44
PUERTO RICO	36.2	12 2003	41						
QATAR	52.0	4 2001	14						
RÉUNION									
ROMANIA	48.2	7 2003	46	45	22	33	40	30	30
RUSSIA	51.5	9 2002	49	8	23	69	15	36	49
RWANDA	43.2	1 1996	51						
SAHARA, WESTERN									
SAMOA	34.6		32						
SAMOA, AMERICAN									
SAN MARINO		3 2003							
SÃO TOMÉ AND PRÍNCIPE	29.1		29						
SAUDI ARABIA	29.7	5 2002	15						
SENEGAL	36.6		42						
SERBIA AND MONTENEGRO*	37.4	15 2003	42						
SEYCHELLES									
SIERRA LEONE	39.3		38						
SINGAPORE	48.7	5 2003	40		18	81		31	69
SLOVAKIA	49.2	18	45	4	26	71	8	48	44
SLOVENIA	52.1	7 2003	46	10	29	61	10	46	43
SOLOMON IS.	37.6		39						
SOMALIA	38.9		39						
SOUTH AFRICA	40.0	28 2003	38	9	14	75	12	33	50
SPAIN	46.7	11	41	5	15	81	8	42	51
SRI LANKA	39.3	9 2003	30	49	22	27	38	23	37
ST HELENA									
ST KITTS-NEVIS									
ST LUCIA	46.9	25 2003	41	16	14	71	27	24	49
ST PIERRE AND MIQUELON									
ST VINCENT	46.1		40						
SUDAN	27.2		25						
SURINAME	33.1	14 1999	35	2	1	97	8	22	64
SWAZILAND	32.4	25 1997	33						
SWEDEN	51.2	7	47	1	11	88	3	36	61
SWITZERLAND	57.2	4	46	3	13	84	5	36	59
SYRIA	36.4	12 2002	30						
TAIWAN									
TAJIKISTAN	31.6		44						
TANZANIA	47.6	5 2001	50						
THAILAND	54.0	2	46	48	17	35	50	20	30
TOGO	36.4		37						
TOKELAU									
TONGA	38.3		38						
TRINIDAD AND TOBAGO	47.2	10 2002	39	3	13	84	11	36	53
TUNISIA	36.3	14 2003	27						
TURKEY	35.3	10	26	56	15	29	24	28	48
TURKMENISTAN	43.1		47						
TURKS AND CAICOS									
TUVALU									
UGANDA	37.1	3 2003	48						
UKRAINE	49.3	9	49	17	22	55	22	39	33
UNITED ARAB EMIRATES	54.0	2 2000	13		14	86	9	36	55
UNITED KINGDOM	50.6	5	46	1	11	88	2	36	62
UNITED STATES	50.6	6	46	1	12	87	3	32	65
URUGUAY	49.2	17 2003	44	2	14	85	6	32	62
UZBEKISTAN	40.6		45						
VANUATU	47.2		47						
VATICAN									
VENEZUELA	44.8	17 2003	40	2	12	86	15	28	57
VIETNAM	49.9	2	49						
VIRGIN IS. (AM.)	47.3		46						
VÍRGIN IS. (BR.)									
WALLIS AND FUTUNA IS.									
YEMEN	25.5	12 1999	28	88	3	9	43	14	43
ZAMBIA	40.3	12 1998	42						
ZIMBABWE	43.2	8 2002	44						

* All data available are previous to the political separation of Serbia and Montenegro.

1. World Development Indicators 2006, World Bank. / 6. Human Development Report 2005 UNDP.
11. Calculated from World Development Indicators 2006, World Bank and World Population Prospects 2004.

	FOREST AND WOODLAND (% OF LAND AREA)	ARABLE LAND (% OF LAND AREA)[1]	CROPLANDS (% OF LAND AREA)[1]	OTHER USES OF THE LAND (% OF LAND AREA)[1]	IRRIGATED AREA (% OF ARABLE LAND)[1]	FERTILIZER USE (KGS PER HA)[1]
	2004	2005	2005		2005	2005
Afghanistan	1.3	12.1	0.2	86.4	33.8	26
Albania	29.0	21.1	4.4	45.5	50.5	612
Algeria	1.0	3.2	0.3	95.5	6.9	130
Andorra	34.0	2.1		63.9		
Angola	47.4	2.6	0.2	49.8	2.2	0
Anguilla						
Antigua	20.5	18.2	4.5	56.8		
Argentina	12.1	10.2	0.4	77.3	5.4	265
Armenia	10.0	17.7	2.1	70.2	51.1	228
Aruba		10.5		89.5		
Australia	21.3	6.2	0.0	72.5	5.3	472
Austria	46.8	16.9	0.9	35.4	0.3	1,498
Azerbaijan	11.3	21.6	2.7	64.4	72.3	99
Bahamas	51.4	0.8	0.4	47.4	8.3	1,000
Bahrain		2.8	5.6	91.6	66.7	500
Bangladesh	6.7	61.3	3.4	28.6	56.1	1,780
Barbados	4.7	37.2	2.3	55.8	29.4	507
Belarus	38.0	26.8	0.6	34.6	2.3	1,334
Belgium	20.3	26.6	0.6	52.5	4.5	3,310
Belize	72.5	3.1	1.4	23.0	2.9	671
Benin	21.3	24.0	2.4	52.3	0.4	188
Bermuda	20.0	20.0		60.0		1,000
Bhutan	68.0	2.3	0.4	29.3	31.3	0
Bolivia	54.2	2.8	0.2	42.8	4.1	45
Bosnia-Herzegovina	42.7	19.6	1.9	35.8	0.3	327
Botswana	21.1	0.7	0.0	78.2	0.3	122
Brazil	56.5	7.0	0.9	35.6	4.4	1,303
Brunei	52.8	2.3	0.9	44.0	5.9	0
Bulgaria	32.8	30.0	1.9	35.3	16.6	495
Burkina Faso	24.8	17.7	0.2	57.3	0.5	4
Burundi	5.9	38.6	14.2	41.3	1.5	26
Cambodia	59.2	21.0	0.6	19.2	7.1	0
Cameroon	45.6	12.8	2.6	39.0	0.4	59
Canada	34.1	5.0	0.7	60.2	1.5	572
Cape Verde	20.8	11.4	0.7	67.1	6.1	48
Cayman Is.	46.2	3.8		50.0		
Central African Rep.	36.5	3.1	0.2	60.2	0.1	3
Ceuta						
Chad	9.5	2.9	0.0	87.6	0.8	49
Chile	21.5	2.6	0.4	75.5	82.4	2,296
China	21.2	15.3	1.3	62.2	35.3	2,777
Christmas Is.						
Cocos						
Colombia	58.5	2.2	1.5	37.8	23.4	3,016
Comoros	2.2	35.9	23.3	38.6		38
Congo D.R.	58.9	3.0	0.5	37.6	0.1	16
Congo R.	65.8	1.4	0.2	32.6	0.4	5
Cook Is.						
Costa Rica	46.8	4.4	5.9	42.9	20.6	6,736
Côte d'Ivoire	32.7	10.4	11.3	45.6	1.1	330
Croatia	38.2	26.1	2.2	33.5	0.7	1,177
Cuba	24.7	27.9	6.6	40.8	23.0	398
Cyprus	18.8	10.8	4.3	66.1	28.6	1,541
Czech Republic	34.3	39.6	3.1	23.0	0.7	1,202
Denmark	11.8	53.4	0.2	34.6	19.7	1,303
Diego García Is.						
Djibouti	0.3	0.0		99.7		0
Dominica	61.3	6.7	21.3	10.7		1,086
Dominican Republic	28.4	22.7	10.3	38.6	17.2	819
East Timor	53.7	8.2	4.6	33.5		

	FOREST AND WOODLAND AS % OF LAND AREA	ARABLE LAND AS % OF LAND AREA[1]	CROPLANDS AS % OF LAND AREA[1]	OTHER USES OF THE LAND AS % OF LAND AREA[1]	IRRIGATED AREA AS % OF ARABLE LAND[1]	FERTILIZER USE (KGS PER HA)[1]
	2004	2005	2005		2005	2005
Ecuador	39.2	5.9	4.9	50.0	29.0	1,417
Egypt	0.1	2.9	0.5	96.5	99.9	4,342
El Salvador	14.4	31.9	12.1	41.6	4.9	838
Equatorial Guinea	58.2	4.6	3.6	33.6		0
Eritrea	15.4	5.6	0.0	79.0	3.7	65
Estonia	53.9	12.9	0.4	32.8	0.7	441
Ethiopia	13.0	11.1	0.7	75.2	2.5	151
Faeroe Is.		2.1		97.9		
Fiji	54.7	10.9	4.7	29.7	1.1	615
Finland	73.9	7.3	0.0	18.8	2.9	1,332
France	28.3	33.5	2.0	36.2	13.3	2,151
French Guiana						
French Polynesia	28.7	0.8	6.0	64.5	4.0	4,347
Gabon	84.5	1.3	0.7	13.5	1.4	9
Gambia	47.1	31.5	0.5	20.9	0.6	25
Georgia	39.7	11.5	3.8	45.0	44.0	355
Germany	31.7	33.9	0.6	33.8	4.0	2,200
Ghana	24.2	18.4	9.7	47.7	0.5	74
Gibraltar						
Greece	29.1	20.9	8.8	41.2	37.9	1,491
Greenland						
Grenada	11.8	5.9	29.4	52.9		
Guadeloupe						
Guam	47.3	3.6	18.2	30.9		
Guatemala	36.3	13.3	5.6	44.8	6.3	1,307
Guinea	27.4	4.5	2.6	65.5	5.4	31
Guinea-Bissau	73.7	10.7	8.9	6.7	4.5	80
Guyana	76.7	2.4	0.2	20.7	29.4	373
Haiti	3.8	28.3	11.6	56.3	8.4	179
Honduras	41.5	9.5	3.2	45.8	5.6	470
Hungary	21.5	50.1	2.1	26.3	4.8	1,087
Iceland	0.5	0.1		99.4		25,554
India	22.8	54.0	3.1	20.1	32.9	1,008
Indonesia	48.8	11.6	7.4	32.2	13.1	1,460
Iran	6.8	9.9	1.3	82.0	41.9	860
Iraq	1.9	13.1	0.6	84.4	58.6	1,111
Ireland	9.7	17.2	0.0	73.1		5,236
Israel	7.9	15.8	4.0	72.3	45.3	2,384
Italy	33.9	27.1	9.3	29.7	25.7	1,729
Jamaica	31.3	16.1	10.2	42.4	8.8	1,287
Japan	68.2	12.1	0.9	18.8	54.7	2,906
Jordan	0.9	3.3	1.2	94.6	18.8	1,136
Kanaky-New Caledonia	39.2	0.3	0.2	60.3	100.0	1,800
Kazakhstan	1.2	8.4	0.1	90.3	15.7	29
Kenya	6.2	8.2	1.0	84.6	2.0	310
Kiribati	2.7	2.7	47.9	46.7		
Korea, North	51.4	22.4	1.7	24.5	50.3	986
Korea, South	63.5	16.7	2.0	17.8	47.6	4,149
Kuwait	0.3	0.8	0.2	98.7	72.2	700
Kyrgyzstan	4.5	6.8	0.3	88.4	78.5	211
Laos	69.9	4.1	0.4	25.6	16.9	
Latvia	47.4	29.3	0.5	22.8	1.1	273
Lebanon	13.3	16.6	14.0	56.1	33.2	2,319
Lesotho	0.3	10.9	0.1	88.7	0.9	342
Liberia	32.7	4.0	2.3	61.0	0.5	0
Libya	0.1	1.0	0.2	98.7	21.9	341
Liechtenstein	43.8	25.0		31.2		
Lithuania	33.5	46.7	0.9	18.9	0.2	662
Luxembourg						
Macedonia, TFYR	35.6	22.3	1.8	40.3	9.0	394

	FOREST AND WOODLAND (% OF LAND AREA)[1]	ARABLE LAND (% OF LAND AREA)[1]	CROPLANDS (% OF LAND AREA)[1]	OTHER USES OF THE LAND (% OF LAND AREA)[1]	IRRIGATED AREA (% OF ARABLE LAND)[1]	FERTILIZER USE (KGS PER HA)[1]		FOREST AND WOODLAND AS % OF LAND AREA[1]	ARABLE LAND AS % OF LAND AREA[1]	CROPLANDS AS % OF LAND AREA[1]	OTHER USES OF THE LAND AS % OF LAND AREA[1]	IRRIGATED AREA AS % OF ARABLE LAND[1]	FERTILIZER USE (KGS PER HA)[1]
	2004	2005	2005		2005	2005		2004	2005	2005		2005	2005
MADAGASCAR	22.1	5.1	1.0	71.8	30.6	31	SAN MARINO		16.7		83.3		
MALAWI	36.2	26.0	1.5	36.3	2.2	839	SÃO TOMÉ AND PRÍNCIPE	28.1	8.3	49.0	14.6	18.2	
MALAYSIA	63.6	5.5	17.6	13.3	4.8	6,833	SAUDI ARABIA	1.3	1.7	0.1	96.9	42.7	1,059
MALDIVES	3.3	13.3	30.0	53.4			SENEGAL	45.0	12.8	0.2	42.0	4.8	136
MALI	10.3	3.8	0.0	85.9	5.0	90	SERBIA AND MONTENEGRO*	26.4	33.2	3.2	37.2	0.9	906
MALTA		31.3	3.1	65.6	18.2	778	SEYCHELLES	87.0	2.2	13.0	-2.2		170
MALVINAS-FALKLANDS							SIERRA LEONE	38.5	8.0	1.0	52.5	4.7	6
MARSHALL IS.		11.1	44.4	44.5			SINGAPORE	3.0	1.5	1.5	94.0		24,180
MARTINIQUE							SLOVAKIA						
MAURITANIA	0.3	0.5	0.0	99.2	9.8	59	SLOVENIA	62.8	8.6	1.4	27.2	1.5	4,160
MAURITIUS	18.2	49.3	3.0	29.5	20.8	2,500	SOLOMON IS.	77.6	0.6	2.1	19.7		
MAYOTTE IS.							SOMALIA	11.4	1.7	0.0	86.9	18.7	5
MELILLA							SOUTH AFRICA	7.6	12.1	0.8	79.5	9.5	654
MEXICO	33.7	13.0	1.3	52.0	23.2	690	SPAIN	35.9	27.5	10.0	26.6	20.2	1,572
MICRONESIA	90.0	5.7	45.7	-41.4			SRI LANKA	29.9	14.2	15.5	40.4	38.8	3,103
MOLDOVA	10.0	56.1	9.1	24.8	14.0	55	ST HELENA						
MONACO							ST KITTS-NEVIS	13.9	19.4	2.8	63.9		2,429
MONGOLIA	6.5	0.8	0.0	92.7	7.0	37	ST LUCIA	27.9	6.6	23.0	42.5	16.7	3,358
MONTSERRAT							ST PIERRE AND MIQUELON						
MOROCCO	9.8	19.0	2.0	69.2	15.4	475	ST VINCENT	28.2	17.9	17.9	36.0	7.1	3,047
MOZAMBIQUE	24.6	5.5	0.3	69.6	2.6	59	SUDAN	28.4	7.2	0.2	64.2	10.7	43
MYANMAR-BURMA	49.0	15.3	1.4	34.3	17.0	134	SURINAME	94.7	0.4	0.1	4.8	75.0	983
NAMIBIA	9.3	1.0	0.0	89.7	1.0	4	SWAZILAND	31.5	10.3	0.8	57.4	26.0	393
NAURU							SWEDEN	67.1	6.5	0.0	26.4	4.3	1,000
NEPAL	25.4	16.5	0.9	57.2	47.0	377	SWITZERLAND	30.5	10.2	0.6	58.7	5.8	2,275
NETHERLANDS	10.8	26.9	0.9	61.4	59.9	3,668	SYRIA	2.5	25.0	4.5	68.0	24.6	703
NETHERLANDS ANTILLES	1.3	10.0		88.7			TAIWAN						
NEW ZEALAND-AOTEAROA	31.0	5.6	7.0	56.4	8.5	5,686	TAJIKISTAN	2.9	6.6	0.9	89.6	68.3	300
NICARAGUA	42.7	15.9	1.9	39.5	2.8	280	TANZANIA	39.9	4.5	1.2	54.4	3.6	18
NIGER	1.0	11.4	0.0	87.6	0.5	3	THAILAND	28.4	27.7	7.0	36.9	28.2	1,072
NIGERIA	12.2	33.5	3.2	51.1	0.8	55	TOGO	7.1	46.1	2.2	44.6	0.3	68
NIUE							TOKELAU						
NORFOLK IS.							TONGA	5.6	20.8	15.3	58.3		0
NORTHERN MARIANAS							TRINIDAD AND TOBAGO	44.1	14.6	9.2	32.1	3.3	434
NORWAY	30.7	2.9		66.4		2,113	TUNISIA	6.8	18.0	13.8	61.4	8.0	368
OMAN	0.0	0.1	0.1	99.8	90.0	3,219	TURKEY	13.2	30.3	3.4	53.1	20.0	727
PAKISTAN	2.5	25.2	0.9	71.4	90.6	1,371	TURKMENISTAN	8.8	4.7	0.1	86.4	79.4	529
PALAU							TURKS AND CAICOS						
PALESTINE							TUVALU						
PANAMA	57.7	7.4	2.0	32.9	6.2	524	UGANDA	18.4	26.4	10.9	44.3	0.1	18
PAPUA NEW GUINEA	65.0	0.5	1.4	33.1		536	UKRAINE	16.5	56.1	1.6	25.8	6.6	181
PARAGUAY	46.5	7.7	0.2	45.6	2.1	507	UNITED ARAB EMIRATES	3.7	0.8	2.3	93.2	29.9	4,667
PERU	53.7	2.9	0.5	42.9	27.8	741	UNITED KINGDOM	11.8	23.4	0.2	64.6	3.0	3,113
PHILIPPINES	24.0	19.1	16.8	40.1	14.5	1,269	UNITED STATES	33.1	18.9	0.2	47.8	12.8	1,111
PITCAIRN							URUGUAY	8.6	7.8	0.2	83.4	14.9	941
POLAND	30.0	41.1	1.0	27.9	0.8	1,162	UZBEKISTAN	7.7	11.0	0.8	80.5	84.9	1,602
PORTUGAL	41.3	17.4	7.9	33.4	28.1	1,262	VANUATU	36.1	1.6	7.0	55.3		
PUERTO RICO	46.0	3.7	5.6	44.7	48.2		VATICAN						
QATAR		1.6	0.3	98.1	61.9	0	VENEZUELA	54.1	2.9	0.9	42.1	16.9	1,155
RÉUNION							VIETNAM	39.7	20.5	7.1	32.7	33.4	2,993
ROMANIA	27.7	40.9	2.0	29.4	31.2	347	VIRGIN IS. (AM.)	28.6	5.7	2.9	62.8		3,000
RUSSIA	49.4	7.5	0.1	43.0	3.7	119	VIRGIN IS. (BR.)						
RWANDA	19.5	48.6	10.9	21.0	0.6	137	WALLIS AND FUTUNA IS.						
SAHARA, WESTERN							YEMEN	1.0	2.9	0.3	95.8	33.0	75
SAMOA	60.4	21.2	24.4	-6.0		583	ZAMBIA	57.1	7.1	0.0	35.8	2.9	124
SAMOA, AMERICAN	90.0	10.0	15.0	-15.0			ZIMBABWE	45.3	8.3	0.3	46.1	5.2	342

* All data available are previous to the political separation of Serbia and Montenegro.

1. World Development Indicators 2006, World Bank.

	IMPORTS OF GOODS AND SERVICES (CURRENT MILLION $) [1]	EXPORTS OF GOODS AND SERVICES (CURRENT MILLION $) [1]	CEREAL IMPORTS (METRIC TONS) [8]	FOOD PRODUCTION PER CAPITA INDEX (1999-2001=100) [1]	FOOD IMPORTS (% OF MERCHANDISE IMPORTS) [1]	IMPORTS OF CONVENTIONAL WEAPONS ($ MILLION, 1990 PRICES) [6]	EXPORTS OF CONVENTIONAL WEAPONS ($ MILLION, 1990 PRICES) [6]
	2004	2004	2004	2005	2005	2004	2004
AFGHANISTAN							
ALBANIA	2,586 2003	1,167 2003	501,439	105.1	19.3	6	
ALGERIA			7,013,842	116.8	21.9	282	
ANDORRA					18.2		
ANGOLA	10,635	13,798	873,237	112.9		5	0
ANGUILLA							
ANTIGUA	510 2002	437 2002	27	107.8	21.7 2000		
ARGENTINA	28,152	39,702	26,032	102.0	2.3	129	0
ARMENIA	1,514	985	458,860	140.6	22.5	68	
ARUBA	3,782	3,955	17,475		23.1		
AUSTRALIA	131,417	112,514	95,129	91.9	4.7	334	52
AUSTRIA	155,304	161,062	642,150	102.2	6.3	46	1
AZERBAIJAN	6,312	4,235	1,236,467	121.1	11.7	0	
BAHAMAS	3,033	2,714	19,812	104.8	17.5 2001	0	
BAHRAIN	7,069	9,179	97,535	99.0	8.1	10	0
BANGLADESH	13,089	9,234	3,324,530	104.6	18.9	26	
BARBADOS	1,820	1,517	70,222	100.8	20.1		
BELARUS	17,019	15,666	911,336	116.0	10.4	0	50
BELGIUM	284,718	297,953	5,850,093	101.0	8.2	12	0
BELIZE	626	506	20,174	116.6	16.6 2003	0	
BENIN	1,073 2003	713 2003	526,034	137.4	23.9 2002	0	
BERMUDA			1,738		20.5 1997		
BHUTAN			35,358	94.5	17.9 1999	0	
BOLIVIA	2,319	2,546	345,293	110.3	11.6	1	
BOSNIA-HERZEGOVINA	7,111	2,914	683,700	98.0		0	0
BOTSWANA	2,780 2003	3,689 2003	93,222	104.3	13.9 2001	10	
BRAZIL	80,069	109,059	6,317,078	124.3	5	38	100
BRUNEI			42,003	121.3	17.6 2003	0	
BULGARIA	16,465	13,975	420,395	107.7	5.4	12	0
BURKINA FASO	650 2001	260 2001	217,901	115.2	12	0	
BURUNDI	175 2003	43 2003	102,443	104.4	9	0	
CAMBODIA	3,663	3,243	107,069	105.4	7.9	0	0
CAMEROON	1,608 1995		583,567	104.7	18.3	0	
CANADA	336,733	377,646	2,576,421	101.6	5.8	340	543
CAPE VERDE	547 2003	277 2003	37,190	91.8	30.9 2003	0	
CAYMAN IS.			481				
CENTRAL AFRICAN REP.	244 1994		29,700	108.2	23.4 2003	0	
CEUTA							
CHAD	411 1994		65,837	112.2	24.3 1995	0	
CHILE	29,542	37,981	1,453,901	112.8	7.4	43	0
CHINA	606,543	655,827	16,109,703	117.8	3.8	2	125
CHRISTMAS IS.							
COCOS							
COLOMBIA	19,929	19,496	3,531,098	109.7	10.6	17	
COMOROS	103 1995		34,681	104.6	21.9 2000		
CONGO D.R.			361,434	97.5		0	
CONGO R.	995 2003	1,546 2003	206,280	108.8	20.8 1995	0	
COOK IS.			1,175				
COSTA RICA	9,140	8,610	872,096	99.4	8.6	0	
CÔTE D'IVOIRE	6,181	7,650	1,160,467	101.2	21.7 2003	14	
CROATIA	20,180	17,828	260,305	96.7	8.4	8	0
CUBA			1,314,138	109.6	18.4 2001	0	
CYPRUS	7,853	7,376	493,413	105.0	12	0	0
CZECH REPUBLIC	76,966	76,569	135,146	104.6	5	18	0
DENMARK	98,925	111,355	967,722	101.4	11.8	194	6
DIEGO GARCÍA IS.							
DJIBOUTI	292 1995		87,115	109.6		0	
DOMINICA	156 2002	123 2002	6,939	98.2	21.8		
DOMINICAN REPUBLIC	9,049	9,283	1,182,231	102.6	12.1 2001	21	
EAST TIMOR			8,893	112.9			
ECUADOR	9,306	8,734	951,248	107.2	9	22	
EGYPT	26,915	26,516	6,815,135	110.9	22.2	398	0
EL SALVADOR	7,029	4,301	773,173	104.8	17.7	0	0
EQUATORIAL GUINEA	477 1996		19,645	93.4		0	
ERITREA	500 2000	98 2000	574,576	86.3		382	0
ESTONIA	9,674	8,794	130,234	102.1	8.8	5	0
ETHIOPIA	3,778	1,684	893,652	112.1	21.5 2003	162	
FAEROE IS.			9,428		18.1		
FIJI	1,043 1999		120,981	95.5	17.7	0	
FINLAND	60,636	71,099	225,845	103.6	5.6	57	17
FRANCE	526,635	531,488	1,428,564	101.6	8.2	89	2
FRENCH GUIANA							
FRENCH POLYNESIA			40,676	111.7	20.7		
GABON	1,882 2003	3,351 2003	117,965	101.7	24.2	0	
GAMBIA	282 1997		101,512	69.0	37.6 2003	0	
GEORGIA	2,491	1,631	1,186,611	100.8	20.9	0	20
GERMANY	912,587	1,051,303	3,637,812	102.9	6.8		
GHANA	5,356	3,487	767,641	121.0	20.8	27	
GIBRALTAR							

	IMPORTS OF GOODS AND SERVICES (CURRENT MILLION $) [1]	EXPORTS OF GOODS AND SERVICES (CURRENT MILLION $) [1]	CEREAL IMPORTS (METRIC TONS) [8]	FOOD PRODUCTION PER CAPITA INDEX (1999-2001=100) [1]	FOOD IMPORTS (% OF MERCHANDISE IMPORTS) [1]	IMPORTS OF CONVENTIONAL WEAPONS ($ MILLION, 1990 PRICES) [6]	EXPORTS OF CONVENTIONAL WEAPONS ($ MILLION, 1990 PRICES) [6]
	2004	2004	2004	2005	2005	2004	2004
GREECE	61,380	48,824	1,807,429	95.3	11.4	1	0
GREENLAND					17.2 2002		
GRENADA	270 2002	175 2002	22,802	98.7	17.9 2003		
GUADELOUPE							
GUAM			2,675	107.0			
GUATEMALA	8,483	4,608	1,085,137	104.4	12.4	0	
GUINEA	964	811	210,490	113.8	23.1 2002	0	
GUINEA-BISSAU	102 2003	71 2003	60,016	109.7	43.6 1995	0	
GUYANA	782	748	34,739	105.2	14.1	0	
HAITI	1,375 2003	469 2003	661,699	100.6			
HONDURAS	4,430	3,066	524,673	111.0	15.9 2003	0	
HUNGARY	69,425	66,351	215,926	111.9	3.9	15	0
ICELAND	5,250	4,517	68,065	104.3	9.9		0
INDIA	93,918 2003	82,735 2003	14,959	104.7	4.4	2	22
INDONESIA	79,116	89,789	6,463,986	117.4	10.5	85	50
IRAN	17,503 2000	29,727 2000	3,985,349	115.4	10.5 2003	283	1
IRAQ							
IRELAND	124,724	152,172	788,756	98.4	7.9	25	
ISRAEL	52,040	51,445	3,798,471	108.2	6.1	724	283
ITALY	423,241	435,871	9,916,430	98.1	9.3	317	261
JAMAICA	5,272	3,899	635,244	99.4	15.2 2002	0	
JAPAN	542,380	636,611	25,942,764	97.7	11.6	195	0
JORDAN	9,407	5,983	2,142,869	118.2	17.2	132	72
KANAKY-NEW CALEDONIA			37,552	103.0	13.6		
KAZAKHSTAN	18,800	22,602	10,268	103.1	7.2	27	5
KENYA	5,115	4,202	882,984	104.3	10.4	0	
KIRIBATI	45 1994		6,160	107.3	36.8 1999		
KOREA, NORTH			1,805,036	109.7			
KOREA, SOUTH	269,782	299,174	12,102,671	92.1	5	737	50
KUWAIT	18,510	33,543	664,021	125.9	15.5 2001	0	0
KYRGYZSTAN	1,135	942	109,574	97.9	13.5	5	0
LAOS		708	37,343	116.8		0	
LATVIA	8,180	6,001	80,057	117.4	10.7	14	0
LEBANON			911,373	100.8	18.5 2003	0	0
LESOTHO	1,398	771	21,141	106.0		1	
LIBERIA			179,895	97.3			
LIBYA	10,532	17,862	2,458,326	104.3	16.8	74	0
LIECHTENSTEIN							
LITHUANIA	13,321	11,751	132,733	112.2	8.1	31	0
LUXEMBOURG	37,859	46,853	74,909		10.7	0	
MACEDONIA, TFYR	3,247	2,080	189,110	108.5	13.8	0	29
MADAGASCAR	1,654 2003	1,126 2003	254,680	107.6	13.5	0	
MALAWI	795 2002	472 2002	43,713	95.6	12.7	0	0
MALAYSIA	96,820 2003	118,577 2003	6,926,676	120.0	5.6	277	0
MALDIVES	725	688	40,316	115.0	18.3	0	
MALI	1,471 2003	1,152 2003	109,262	109.6	16.2 2001	0	
MALTA	4,428	4,022	115,306	106.8	11.4	0	10
MALVINAS-FALKLANDS							
MARSHALL IS.							
MARTINIQUE							
MAURITANIA	471 1998		291,101	108.8	26.2 1996	0	
MAURITIUS	3,603	3,460	309,236	105.9	17.5	0	
MAYOTTE IS.							
MELILLA							
MEXICO	216,589	202,003	12,977,083	107.8	6.3	265	
MICRONESIA			11,883	100.1			
MOLDOVA	2,122	1,331	126,394	115.7	11.7	0	0
MONACO							
MONGOLIA	1,405	1,211	196,752	93.6	14.4 2003		
MONTSERRAT							
MOROCCO	19,860	16,632	4,076,672	132.1	10.9	0	
MOZAMBIQUE	2,381	1,759	838,216	104.0	10.6 2002	0	
MYANMAR-BURMA	2,458	3,181	110,587	115.4		65	
NAMIBIA	2,495	2,310	80,913	114.0	14.9 2003	53	
NAURU			542				
NEPAL	2,186	1,224	53,856	110.5	17.3 2003	32	
NETHERLANDS	341,622	388,899	7,851,751	95.1	10.4	183	211
NETHERLANDS ANTILLES	2,803	2,655	45,463		12.4 1995		
NEW ZEALAND-AOTEAROA	28,791	28,305	467,877	116.4	7.7	42	1
NICARAGUA	2,851	1,653	199,644	123.1	16.6	0	0
NIGER	681 2003	415 2003	117,394	118.4	33.5 2003	0	
NIGERIA	16,064	26,993	4,023,406	106.2	15.5 2003	10	0
NIUE			34				
NORFOLK IS.			167				
NORTHERN MARIANAS							
NORWAY	73,557	109,104	333,147	99.5	7.2	1	51
OMAN	10,613	14,175	528,730	92.1	13.6	123	0
PAKISTAN	22,057	16,079	112,369	110.6	10.5	344	10
PALAU							

	IMPORTS OF GOODS AND SERVICES (CURRENT MILLION $) [1]	EXPORTS OF GOODS AND SERVICES (CURRENT MILLION $) [1]	CEREAL IMPORTS (METRIC TONS) [8]	FOOD PRODUCTION PER CAPITA INDEX (1999-2001=100) [1]	FOOD IMPORTS (% OF MERCHANDISE IMPORTS) [1]	IMPORTS OF CONVENTIONAL WEAPONS ($ MILLION, 1990 PRICES) [6]	EXPORTS OF CONVENTIONAL WEAPONS ($ MILLION, 1990 PRICES) [6]
	2004	2004	2004	2005	2005	2004	2004
PALESTINE			638,170			0	
PANAMA	9,172	8,859	481,323	103.7	14.4 2003	0	
PAPUA NEW GUINEA	1,594 2001	2,098 2001	400,163	107.7	16.5 2003	0	
PARAGUAY	3,540	3,397	8,831	115.0	8.7	4	
PERU	12,581	14,530	2,716,808	110.2	12.7	14	5
PHILIPPINES	50,492	42,829	3,298,750	115.5	6.3	59	
PITCAIRN							
POLAND	99,935	95,333	881,310	106.7	5.7	256	86
PORTUGAL	65,411	51,899	3,064,343	98.9	12.2	59	0
PUERTO RICO				98.2			
QATAR			132,837	143.9	7.5	0	0
RÉUNION							
ROMANIA	34,029	27,099	1,426,107	123.2	6.3	276	0
RUSSIA	130,144	203,741	3,022,217	111.4	16.6	0	6
RWANDA	493	201	42,875	113.2	11.7 2003	0	
SAHARA, WESTERN							
SAMOA	140 1999		13,181	103.3	23.4		
SAMOA, AMERICAN			29,921				
SAN MARINO							
SÃO TOMÉ AND PRÍNCIPE	41 2002	19 2002	10,094	109.2			
SAUDI ARABIA	66,746	131,849	6,720,450	118.6	16.2 2003	838	0
SENEGAL	2,657 2003	1,826 2003	1,220,215	81.6	28.3	0	
SERBIA AND MONTENEGRO*			145,454	114.2	11.4 2002		
SEYCHELLES	629	625	20,722	91.6	28.3	0	
SIERRA LEONE	342	215	73,306	113.5	22.5 2002	0	
SINGAPORE	206,796	238,522	661,121	70.2	2.9	456	70
SLOVAKIA	25,649 2003	25,241 2003	70,020		5	0	0
SLOVENIA	19,927	19,519	476,687	108.5	5.9	14	
SOLOMON IS.	198 1999		7,470	143.2		0	
SOMALIA							
SOUTH AFRICA	57,888	56,734	2,658,851	105.9	5	8	35
SPAIN	307,365	269,030	9,072,743	105.9	9.5	261	75
SRI LANKA	9,108	7,284	1,358,150	95.6	12.3	6	
ST HELENA			277				
ST KITTS-NEVIS	257 2002	155 2002	1,544	100.0	16.1 2003		
ST LUCIA	402 2002	328 2002	6,111	91.8	22.9		
ST PIERRE AND MIQUELON			107				
ST VINCENT	217 2002	177 2002	39,796	104.0	22		
SUDAN	4,651	3,822	1,536,379	107.8	16.4 2003	270	
SURINAME	1,011	924	44,492	100.6	17.6 2001	0	
SWAZILAND	2,448	2,438	4,086	105.9	18.2 2002	0	
SWEDEN	134,855	163,934	203,173	99.4	7.6	13	260
SWITZERLAND	146,291	181,568	550,218	99.6	5.8	125	154
SYRIA	7,915	8,175	1,881,750	121.7	16.7	0	0
TAIWAN							
TAJIKISTAN	1,445	1,220	118,313	145.8	10.2 2000	0	
TANZANIA	3,196	2,179	929,909	105.6	15	0	
THAILAND	107,512	114,019	1,149,967	104.7	4.9 2003	105	5
TOGO	959 2003	693 2003	133,268	104.2	18.4	0	
TOKELAU							
TONGA	105 2002	41 2002	7,314	102.2	32.5 2000	0	
TRINIDAD AND TOBAGO	4,283 2003	5,890 2003	331,575	117.5	9.5 2003	0	
TUNISIA	14,099	13,308	1,992,400	101.6	8.6	0	
TURKEY	102,199	91,048	2,551,669	103.9	3.3	418	18
TURKMENISTAN	1,680 1997		394	131.0	11.7 2000	20	
TURKS AND CAICOS							
TUVALU			1,464				
UGANDA	2,154	1,153	382,016	109.2	16.8	19	
UKRAINE	34,846	39,719	767,073	115.4	6.4 2002	29	452
UNITED ARAB EMIRATES			2,235,147	63.7	10.7 2001	1	3
UNITED KINGDOM	604,562	533,167	2,868,873	98.0	9	171	985
UNITED STATES	1,769,031	1,151,448	4,310,661	107.5	4.4	533	5
URUGUAY	3,673	4,008	149,791	115.5	8.6	0	
UZBEKISTAN			54,289	105.2		0	170
VANUATU	146 2003	122 2003	20,611	97.3			
VATICAN							
VENEZUELA	22,042	39,846	1,678,004	98.3	14.8	12	1
VIETNAM	21,458 2002	19,654 2002	1,282,840	124.4	6.3 2003	247	
VIRGIN IS. (AM.)							
VIRGIN IS. (BR.)			1,512				
WALLIS AND FUTUNA IS.							
YEMEN	4,918	5,045	2,202,315	110.5	28.2	309	
ZAMBIA	1,318 2000	871 2000	64,696	108.0	6.5	0	0
ZIMBABWE	2,515 1994		612,180	86.4	18.7	0	

* All data available are previous to the political separation of Serbia and Montenegro.

1. World Development Indicators 2006, World Bank.
6. Human Development Report 2005 UNDP.
8. FAOSTAT Statistical Database FAO Web site 2006.

	FEMALE PROFESSIONAL AND TECHNICAL WORKERS (% OF TOTAL)[6]	FEMALE LEGISLATORS, SENIOR OFFICIALS AND MANAGERS (% OF TOTAL)[6]	RATIO OF ESTIMATED FEMALE TO MALE EARNED INCOME[6]	WOMEN IN GOVERNMENT AT MINISTERIAL LEVEL (% OF TOTAL)[6]	SEATS IN PARLIAMENT HELD BY WOMEN (% OF TOTAL)[6]
	1992-2003	1992-2003	1991-2003	2005	2005
AFGHANISTAN					
ALBANIA			0.56	5.3	6.4
ALGERIA			0.31	10.5	6.2
ANDORRA					
ANGOLA			0.62	5.7	15
ANGUILLA					
ANTIGUA				15.4	10.5
ARGENTINA	55	25	0.37	8.3	33.7
ARMENIA			0.70	0	5.3
ARUBA					
AUSTRALIA	55	36	0.72	20	24.7
AUSTRIA	49	27	0.35	35.3	33.9
AZERBAIJAN			0.58	15	10.5
BAHAMAS	51	40	0.64	26.7	20
BAHRAIN	19	10	0.31	8.7	0
BANGLADESH	25	8	0.54	8.3	2
BARBADOS	71	45	0.61	29.4	13.3
BELARUS			0.65	10	29.4
BELGIUM	48	31	0.54	21.4	34.7
BELIZE	52	31	0.24	6.3	6.7
BENIN			0.69	19	7.2
BERMUDA					
BHUTAN				0	8.7
BOLIVIA	40	36	0.45	6.7	19.2
BOSNIA-HERZEGOVINA			0.46	11.1	16.7
BOTSWANA	53	31	0.61	26.7	11.1
BRAZIL	62		0.43	11.4	8.6
BRUNEI				9.1	
BULGARIA	34	30	0.67	23.8	26.3
BURKINA FASO			0.73	14.8	11.7
BURUNDI			0.72	10.7	18.4
CAMBODIA	33	14	0.76	7.1	9.8
CAMEROON			0.45	11.1	8.9
CANADA	54	35	0.64	23.1	21.1
CAPE VERDE			0.48	18.8	11.1
CAYMAN IS.					
CENTRAL AFRICAN REP.			0.61	10	
CEUTA					
CHAD			0.59	11.5	6.5
CHILE	52	24	0.39	16.7	12.5
CHINA			0.66	6.3	20.2
CHRISTMAS IS.					
COCOS					
COLOMBIA	50	38	0.51	35.7	12
COMOROS			0.55		3
CONGO D.R.			0.56	12.5	12
CONGO R.				14.7	8.5
COOK IS.					
COSTA RICA	40	29	0.37	25	35.1
CÔTE D'IVOIRE			0.37	17.1	8.5
CROATIA	52	26	0.56	33.3	21.7
CUBA				16.2	36
CYPRUS	47	18	0.47	0	16.1
CZECH REPUBLIC	52	26	0.64	11.1	17
DENMARK	51	26	0.73	33.3	36.9
DIEGO GARCÍA IS.					
DJIBOUTI				5.3	10.8
DOMINICA				0	19.4
DOMINICAN REPUBLIC	49	31	0.36	14.3	17.3
EAST TIMOR				22.2	25.3
ECUADOR	40	26	0.30	14.3	16
EGYPT	31	9	0.26	5.9	2.9
EL SALVADOR	44	32	0.44	35.3	10.7
EQUATORIAL GUINEA			0.40	4.5	18
ERITREA			0.51	17.6	22
ESTONIA	69	35	0.64	15.4	18.8
ETHIOPIA			0.52	5.9	7.7
FAEROE IS.					
FIJI	9	51	0.37	9.1	8.5
FINLAND	53	28	0.72	47.1	37.5
FRANCE			0.59	17.6	12.2
FRENCH GUIANA					
FRENCH POLYNESIA					
GABON			0.59	11.8	9.2
GAMBIA			0.59	20	13.2
GEORGIA	63	28	0.42	22.2	9.4
GERMANY	50	36	0.54	46.2	32.8
GHANA			0.75	11.8	10.9
GIBRALTAR					
GREECE	48	26	0.45	5.6	14
GREENLAND					
GRENADA				40	26.7
GUADELOUPE					
GUAM					
GUATEMALA			0.33	25	8.2
GUINEA			0.68	15.4	19.3
GUINEA-BISSAU			0.49	37.5	14
GUYANA			0.39	22.2	30.8
HAITI			0.56	25	3.6
HONDURAS	36	22	0.37	14.3	5.5
HUNGARY	61	34	0.62	11.8	9.1
ICELAND	55	29	0.69	27.3	30.2
INDIA			0.38	3.4	8.3
INDONESIA			0.52	10.8	11.3
IRAN	33	13	0.28	6.7	4.1
IRAQ					
IRELAND	50	29	0.41	21.4	13.3
ISRAEL	54	29	0.55	16.7	15
ITALY	45	21	0.46	8.3	11.5
JAMAICA			0.66	17.6	11.7
JAPAN	46	10	0.46	12.5	7.1
JORDAN			0.31	10.7	5.5
KANAKY-NEW CALEDONIA					
KAZAKHSTAN			0.64	17.6	10.4
KENYA			0.93	10.3	7.1
KIRIBATI					
KOREA, NORTH					
KOREA, SOUTH	39	6	0.48	5.6	13
KUWAIT			0.35	0	0
KYRGYZSTAN			0.65	12.5	
LAOS			0.65	0	22.9
LATVIA	64	40	0.62	23.5	21
LEBANON			0.31	6.9	2.3
LESOTHO			0.39	27.8	11.7
LIBERIA					
LIBYA					
LIECHTENSTEIN					
LITHUANIA	70	39	0.68	15.4	22
LUXEMBOURG			0.39	14.3	23.3
MACEDONIA, TFYR	51	27	0.56	16.7	19.2
MADAGASCAR			0.59	5.9	6.9
MALAWI			0.68	14.3	14
MALAYSIA	40	23	0.47	9.1	9.1
MALDIVES	40	15		11.8	12

	FEMALE PROFESSIONAL AND TECHNICAL WORKERS (% OF TOTAL)[6]	FEMALE LEGISLATORS, SENIOR OFFICIALS AND MANAGERS (% OF TOTAL)[6]	RATIO OF ESTIMATED FEMALE TO MALE EARNED INCOME[6]	WOMEN IN GOVERNMENT AT MINISTERIAL LEVEL (% OF TOTAL)[6]	SXEATS IN PARLIAMENT HELD BY WOMEN (% OF TOTAL)[6]		FEMALE PROFESSIONAL AND TECHNICAL WORKERS (% OF TOTAL)[6]	FEMALE LEGISLATORS, SENIOR OFFICIALS AND MANAGERS (% OF TOTAL)[6]	RATIO OF ESTIMATED FEMALE TO MALE EARNED INCOME[6]	WOMEN IN GOVERNMENT AT MINISTERIAL LEVEL (% OF TOTAL)[6]	SXEATS IN PARLIAMENT HELD BY WOMEN (% OF TOTAL)[6]
	1992-2003	1992-2003	1991-2003	2005	2005		1992-2003	1992-2003	1991-2003	2005	2005
MALI			0.60	18.5	10.2	SAUDI ARABIA	6	31	0.21	0	0
MALTA	39	18	0.39	15.4	9.2	SENEGAL			0.55	20.6	19.2
MALVINAS-FALKLANDS						SERBIA AND MONTENEGRO*					
MARSHALL IS.						SEYCHELLES				12.5	29.4
MARTINIQUE						SIERRA LEONE			0.42	13	14.5
MAURITANIA			0.56	9.1	3.7	SINGAPORE	45	26	0.51	0	16
MAURITIUS			0.37	8	5.7	SLOVAKIA	61	35	0.65	0	16.7
MAYOTTE IS.						SLOVENIA	56	33	0.62	6.3	12.2
MELILLA						SOLOMON IS.			0.66	0	0
MEXICO	40	25	0.38	9.4	22.6	SOMALIA					
MICRONESIA						SOUTH AFRICA			0.45	41.4	32.8
MOLDOVA	66	40	0.65	11.1	15.8	SPAIN	47	30	0.44	50	36
MONACO						SRI LANKA	46	21	0.51	10.3	4.9
MONGOLIA	66	30	0.66	5.9	6.8	ST HELENA					
MONTSERRAT						ST KITTS-NEVIS				0	0
MOROCCO			0.40	5.9	10.8	ST LUCIA				8.3	11.1
MOZAMBIQUE			0.68	13	34.8	ST PIERRE AND MIQUELON					
MYANMAR-BURMA						ST VINCENT				20	22.7
NAMIBIA	55	30	0.51	19	25	SUDAN			0.32	2.6	9.7
NAURU						SURINAME	51	28		11.8	19.6
NEPAL			0.51	7.4	5.9	SWAZILAND	61	24	0.39	13.3	10.8
NETHERLANDS	48	26	0.53	36	36.7	SWEDEN	51	30	0.69	52.4	45.3
NETHERLANDS ANTILLES						SWITZERLAND	45	28	0.90	14.3	25
NEW ZEALAND-AOTEAROA	52	36	0.68	23.1	28.3	SYRIA			0.29	6.3	12
NICARAGUA			0.45	14.3	20.7	TAIWAN					
NIGER			0.57	23.1	12.4	TAJIKISTAN			0.62	3.1	
NIGERIA			0.41	10	4.7	TANZANIA	32	49	0.71	15.4	21.4
NIUE						THAILAND	52	26	0.61	7.7	10.6
NORFOLK IS.						TOGO			0.47	20	6.2
NORTHERN MARIANAS						TOKELAU					
NORWAY	50	30	0.75	44.4	38.2	TONGA					0
OMAN			0.19	10	2.4	TRINIDAD AND TOBAGO	54	38	0.46	18.2	19.4
PAKISTAN	26	2	0.34	5.6	21.3	TUNISIA			0.37	7.1	22.8
PALAU						TURKEY	30	6	0.46	4.3	4.4
PALESTINE	34	12				TURKMENISTAN			0.63	9.5	
PANAMA	50	40	0.51	14.3	16.7	TURKS AND CAICOS					
PAPUA NEW GUINEA			0.57		0.9	TUVALU					
PARAGUAY	54	23	0.33	30.8	10	UGANDA			0.67	23.4	23.9
PERU	47	23	0.27	11.8	18.3	UKRAINE	63	39	0.53	5.6	5.3
PHILIPPINES	62	58	0.59	25	15.3	UNITED ARAB EMIRATES	25	8		5.6	0
PITCAIRN						UNITED KINGDOM	45	33	0.62	28.6	18.1
POLAND	61	34	0.62	5.9	20.2	UNITED STATES	55	46	0.62	14.3	15
PORTUGAL	52	32	0.54	16.7	19.5	URUGUAY	53	35	0.53	0	12.1
PUERTO RICO						UZBEKISTAN			0.66	3.6	17.5
QATAR				7.7		VANUATU				8.3	3.8
RÉUNION						VATICAN					
ROMANIA	57	31	0.58	12.5	11.1	VENEZUELA	61	27	0.42	13.6	9.7
RUSSIA	64	39	0.64	0	9.8	VIETNAM			0.68	11.5	27.3
RWANDA			0.62	35.7	48.8	VIRGIN IS. (AM.)					
SAHARA, WESTERN						VÍRGIN IS. (BR.)					
SAMOA				7.7	6.1	WALLIS AND FUTUNA IS.					
SAMOA, AMERICAN						YEMEN	15	4	0.31	2.9	0.3
SAN MARINO						ZAMBIA			0.56	25	12.7
SÃO TOMÉ AND PRÍNCIPE				14.3	9.1	ZIMBABWE			0.58	14.7	10

* All data available are previous to the political separation of Serbia and Montenegro.

6. Human Development Report 2005 UNDP.